ORGANIC CHEMISTRY

ORGANIC CHEMISTRY

Stephen J. Weininger
Worcester Polytechnic Institute

Frank R. Stermitz
Colorado State University

ACADEMIC PRESS, INC.
(Harcourt Brace Jovanovich Publishers)
Orlando San Diego San Francisco New York
London Toronto Montreal Sydney Tokyo São Paulo

Cover art by Debbie Morse

Academic Press, Inc.
Orlando, FL 32887

United Kingdom Edition Published by Academic Press, Inc.
(London) Ltd. 24/28 Oval Road, London NW1 7DX

ISBN: 0-12-742360-5

Library of Congress Catalog Card Number: 82-71940

Printed in the United States of America

Preface

This textbook is intended to provide a modern, comprehensive, and comprehensible introduction to organic chemistry for instructors and students in the 1980s. In its preparation, we have considered the needs of both the instructor and the student and have tried to organize the text in a manner satisfactory to both.

Instructors need a textbook organized in such a way that all the essential subject matter may be covered within one academic year of 28 weeks. We have achieved this by dividing the material into an approximate one-chapter-per-week format. In addition to covering the basics of organic chemistry, modern advances in the field are integrated into appropriate chapters rather than set apart as "special topics." We have also incorporated current organic synthesis theory, spectroscopy, and industrial chemistry into almost every chapter. The great majority of illustrative reactions are taken directly from the literature, and yields of products are usually given.

Instructors also require a text that is useful and interesting to the large numbers of nonmajors, without slighting the needs of the chemistry major. To this end we have made a particular effort to show how an understanding of organic mechanisms illuminates cellular processes at the molecular level. We introduce spectroscopy early in the text (Chapter 8) and use it consistently throughout the remainder of the book. Spectral methods are of central importance to the chemistry major and are becoming more and more important to the nonchemist as well. They provide one of the best tools for teaching organic structure and structural relationships.

Students need a textbook organized so that the more difficult material can be readily assimilated. To achieve this goal we have incorporated several pedagogical devices that provide students with tools to simplify learning the material. Worked-out examples, straightforward in content, appear in every chapter. They are immediately followed by in-chapter problems dealing with similar information. These problems emphasize the previous section's major concepts so students can quickly assess their understanding of them. Students should be able to solve these problems before proceeding to the next section. Additional problems at the end of each chapter mix straightforward exercises with more involved and demanding ones relating to material covered throughout the chapter. Their arrangement reflects the order in which the topics are introduced. Brief answers to many of the in-chapter and end-of-chapter problems are included at the end of the text for immediate feedback. Boxed inserts highlighting historical, biographical, or current topics of interest have been added to each chapter to better illustrate organic chemistry as a living science and to enhance readability by providing some contrasting material.

A detailed *Student Solutions Manual,* by Thomas J. Cogdell, accompanies the text. This study guide supplies complete answers to all the problems in the book and provides insight into the mechanics of problem-solving in organic chemistry. It also lists new reactions and reaction mechanisms as they are encountered in each chapter.

We believe we have achieved our goal of providing instructors and students with a textbook that each will find current, comprehensive, and pedagogically useful, while at the same time interesting and readable. We hope it will also instill in both a bit of the excitement generated by the continuing importance of organic chemistry in today's society.

Stephen J. Weininger
Frank R. Stermitz

Acknowledgments

In the course of writing this book we benefited from the suggestions, criticisms, and insights of many thoughtful reviewers. Most took time from already busy schedules to review our manuscript, and we are indebted to all of them. Particularly comprehensive reviews were carried out by R. J. Artz, T. J. Cogdell, R. S. Givens, S. G. Levine, R. A. Olofson, T. A. Spencer, L. G. Wade, Jr., and P. P. Wickham.

We also received much valuable comment from T. A. Bryson, F. A. Carey, J. Coke, H. Ensley, P. K. Freeman, U. Hollstein, J. E. Leonard, E. McNelis, H. Meislich, S. F. Nelsen, R. J. Panicci, M. Rathke, A. Santoro, P. J. Stang, and B. Vittimberga.

The staff of Academic Press provided material and moral support at all stages of this venture, and never so much as when it was sorely needed. We thank Randi Kashan for her creativity and commitment and André Spencer for her skill and persistence.

Finally, we gratefully acknowledge the boundless understanding and unstinting support of our families.

Contents

6 The Interconversion of Alkyl Halides, Alcohols, and Ethers: Substitution Reactions 175

7 The Structure of Alkenes: Their Synthesis by Way of Elimination Reactions 226

8 Physical Methods of Chemical Structure Identification: Basic Mass and Infrared Spectroscopy 267

18 Aldol Condensation Reactions: Enols and Enolates 680

19 Alkylation and Acylation of Enolates 713

20 Sugars, Nucleosides, and Nucleic Acids 757

21 Amines, Phosphines, and Sulfides: Basicity, Nucleophilicity, and Valence Shell Expansion 798

22 Amino Acids, Peptides, and Proteins: Natural and Synthetic Condensation Polymers 847

ORGANIC CHEMISTRY

Introduction

The Nature of Organic Chemistry

There are many impulses which drive modern scientists to study nature. Two very important ones are the desire to understand how nature works, and the desire to acquire control over it. They are related, but distinct, motivations. Both these themes are prominent in the history of organic chemistry. Organic chemists have added greatly to our understanding of the entire domain of chemistry, and have provided crucial insights into biology and medicine as well. In addition, they have laid the foundation for several of the major industries which sustain the global economy.

Organic chemistry is the science that explores the structures, reactions, and syntheses of carbon compounds. Its scope is enormous; it is fundamental to our understanding of life and life processes, and it also deals with many of the most important man-made artifacts of everyday life, such as gasoline, synthetic fibers, plastics and resins, and drugs and medicines.

In the chemical industry, the ultimate sources of carbon for these many products are natural gas, petroleum, and coal. Nature, on the other hand, uses carbon dioxide as its primary carbon source. When plants absorb CO_2 from the atmosphere, they convert it to a multitude of organic compounds which are consumed by all plant-eating animals, including man. If we understand the structures and reactions of these organic compounds, we can better explain all the transformations that take place in our bodies, such as the production and use of fats, carbohydrates, and proteins. Reproduction, life processes, disease, aging and dying can be looked upon *in part* as just complex transformations of various organic compounds, taking place under highly organized conditions.

The quality of human life has been greatly influenced by man-made organic compounds. The control of many infectious diseases, first through sulfa drugs and then through naturally occurring antibiotics, is one of the major triumphs of organic medicinal research. Other advances include medicines that relieve pain without addiction, control diabetes, and lessen arthritic damage. Cancer chemotherapy has had some successes, particularly in treating childhood leukemia, where both synthetic drugs and those isolated from plants have proved their value.

The value and excitement of organic chemistry can be measured by the number of vital problems that have barely been attacked. What organic molecules might be designed to capture energy from the sun and store it for later use? What practical gasoline additive or alternative fuel source could be developed which would completely eliminate vehicle-caused pollution? Can a medicine be found which will dissolve arterial cholesterol deposits? What were the first organic compounds on Earth and what processes led from that beginning to the appearance of life? If life exists elsewhere in the universe, will it be similar to what we know here, or different? Can simple life forms be synthesized in the laboratory?

The point of this discussion is that organic chemistry has already enriched modern life materially and intellectually, and has the capacity to continue doing so. Yet, it has also degraded our lives to some extent—a paradox not unique to organic chemistry, but one that it shares with the rest of our sophisticated scientific–technological enterprises. Thus, the same chemical industry that provides anticancer drugs can also release cancer-causing chemicals into the environment. The development of plastics and synthetic fibers has allowed us to improve upon nature and also to conserve valuable resources, but nondegradable synthetic materials from our refuse now litter the mountain wilderness and the open sea. Organic pesticides improve food production by eliminating harmful insects, but in the process they have poisoned birds, fish, and animals.

How then can we maximize the beneficial effects of organic chemistry while minimizing the harmful ones? A technical discussion of the subject, such as ours, cannot answer that question. It can demonstrate the power of organic chemistry but is not able to indicate whether or how that power should be used. To come to grips with those questions we have to take account of the social, political, and economic forces in the societies in which organic chemistry operates. We cannot, for instance, even begin to decide whether the risks of DDT use outweigh its benefits, without first asking, *to whom?* Science tells us only about *physical* possibilities and constraints. An understanding of them is necessary, but not sufficient, for the humane development of society.

Many of you are probably interested primarily in the applications of organic chemistry. As in all science, applications are usually based on a good understanding of fundamental principles. It is the intent of this text to provide those principles in the field of organic chemistry. As the text develops, the relationship of the material to biological or industrial processes may not always be obvious. Therefore, many examples of direct relevance have been included. In this way, the importance of organic chemistry to biological and industrial problems can be judged on its own merits.

The Historical Distinction Between Organic and Inorganic Chemistry

Over the centuries people have become familiar with various substances created by the natural processes of living organisms. Several natural substances, such as alcohol and acetic acid, had been obtained long ago in a reasonably pure state. During the eighteenth and nineteenth centuries, when scientists began to organize chemical knowledge systematically, they decided to classify compounds from living organisms as **organic** compounds.

The early scientists noted several qualitative differences between organic compounds and inorganic compounds, such as salt and iron oxide. All organic compounds seemed to contain carbon and hydrogen, and very few other elements except oxygen and nitrogen. Inorganic compounds, however, appeared to have been formed from a wide variety of elements. Most inorganic compounds were solid crystalline substances that were quite stable when heated, while many organic compounds were difficult to crystallize and could be easily altered or destroyed by heating.

The most striking difference between organic and inorganic chemistry concerned the number of compounds that could be constructed from a certain set of elements. The situation in early inorganic chemistry is typified by the nitrogen/hydrogen compounds ammonia (NH_3) and hydrazine ($H_2N—NH_2$). These are the *only* compounds containing just nitrogen and hydrogen that can be isolated. By contrast, there are enormous numbers of compounds composed of just carbon and hydrogen, called **hydrocarbons.** Even in the early days of organic chemistry many different hydrocarbons were known, and most of them seemed to have different ratios of carbon to hydrogen.

These differences led chemists of the eighteenth and early nineteenth centuries to believe that there was a fundamental separation between organic and inorganic substances. Many inorganic compounds had already been synthesized from their constituent elements or from other compounds. It was believed that synthesis of organic compounds could only take place in living organisms that possessed a **vital force** necessary to bring about these syntheses.

Wöhler Demonstrates the Unity of Organic and Inorganic Chemistry

One of the earliest experiments leading to the abandonment of the vital force theory was conducted by the German chemist Friedrich Wöhler. He was studying the salts of cyanic acid, HOCN, and in 1828 he reported on his attempts to make ammonium cyanate, $NH_4^+OCN^-$. Wöhler expected that the following reaction would occur:

$$\underset{\text{Lead cyanate}}{Pb(OCN)_2} + 2NH_3 + H_2O \rightarrow PbO\downarrow + \underset{\text{Ammonium cyanate}}{2NH_4OCN}$$

He allowed lead cyanate to react with aqueous ammonia, filtered off the lead oxide which had formed, and isolated the other product by boiling off the water. He obtained a white crystalline product which did not have the properties expected for ammonium cyanate. Wöhler showed that his isolated compound was urea, which had previously been isolated only from human urine. In the presence of water and heat, the expected ammonium cyanate had been converted to urea:

$$\underset{\text{Ammonium cyanate}}{NH_4^+OCN^-} \xrightarrow[\text{heat}]{H_2O} \underset{\text{Urea}}{H_2N—\overset{\overset{\displaystyle O}{\|}}{C}—NH_2}$$

Wöhler had created an *organic* compound in the laboratory using *inorganic* starting materials and without the intervention of any living organism. Scientists of the time were deeply impressed by Wöhler's results, which seriously undermined the vital force hypothesis. Evidence continued to accumulate that contradicted the predictions of this theory, and by 1860 few prominent scientists believed in vitalism.

Wöhler himself had doubts about whether or not his results constituted a refutation of vitalism. The cyanate used in his study had been produced by oxidation of cyanide that had, in turn, been obtained from a reaction in which the hooves and horns of animals constituted some of the ingredients. Wöhler remarked that a defender of vitalism could maintain that the vital force had in fact been carried over into the cyanide from the horns and hooves, and was actively responsible for the formation of urea.

This objection was circumvented in 1845 when Hermann Kolbe synthesized acetic acid, CH_3CO_2H, the substance responsible for the taste of vinegar. Kolbe's material was made in several steps from carbon and water, and so was of indisputably inorganic origins.

The laboratory synthesis of organic molecules found in nature (molecules much more complex than urea) is a flourishing branch of organic chemistry called **natural product** synthesis. The final aim of workers in this field is often a *total* synthesis of the compound in question. The desired compound is constructed in stages from simpler compounds, and these may be related by a series of known reactions back to the constituent elements. Interest in total synthesis may be, in part, a psychological legacy of the vitalistic controversy.

Why Organic Chemistry is a Separate Discipline

When vitalism died, the term "organic chemistry" came to mean simply "chemistry of carbon compounds," since it was apparent that the great majority of compounds from living systems contained carbon as the major element. It was no longer necessary to believe that carbon-containing compounds were created by forces different from those responsible for the synthesis of molecules lacking carbon atoms. What, then, was the justification for singling out this element and elevating its study into an entire branch of chemistry?

In fact, carbon is unique among the elements. Carbon atoms form fairly strong, unreactive bonds with each other, producing long chains or complex rings with durable carbon atom skeletons. Vast numbers of organic compounds have been discovered or synthesized to date. They outnumber all the known inorganic compounds. The mere number of compounds that fall under the organic heading has served to encourage specialization in the study of these substances.

Organic chemistry is also primarily the study of **covalent compounds** as opposed to ionic substances. Covalent compounds are those held together by the sharing of electrons in a bond, rather than by the attraction of positive and negative ions as in inorganic salts. The majority of inorganic reactions are carried out in water solution, since water dissolves ionic substances. The majority of covalent organic compounds are, however, not water soluble and their reactions are usually carried out in inert organic solvents. These differences provide further reasons for separating inorganic and organic chemistry.

Carbon is almost always tetravalent in its compounds—that is, each carbon atom forms four electron-pair bonds to other atoms. This tetravalency explains the great variety of spatial arrangements of atoms found in organic molecules. The three-dimensional character of these molecules directly affects their chemical behavior, particularly in systems of biological interest. Many organic chemists are devoted solely to studying the architecture of organic molecules as it relates to their chemical reactivity.

Modern Biochemistry and Organic Chemistry

Modern organic chemists are still exploring that area which intrigued the first organic chemists—the chemistry of living systems. Numerous investigations throughout the nineteenth and twentieth centuries established the fact that most physiologically active compounds are organic. The enormous chemical complexity of living systems frustrated early efforts to understand the molecular basis of biological function. But as we began to discover what happened to molecules when they were digested, for instance, it seemed more and more likely that chemical reactions which take place in cells are similar to ones well known in the laboratory. The same *principles* are involved in test tube organic chemistry and in the reactions of molecules in a biological system. This understanding stimulated the study of **biochemistry** because it meant that the chemical insight gained by *pure* organic chemists could serve as a guide to those interested in physiological systems.

While many chemists seek to apply organic chemistry to the study of living systems, this is by no means true of all chemists. The following sections describe some of the different types of investigations typical of organic chemistry. The chemist may be a specialist in one or more of these areas, or he or she may use all these different techniques in solving a particular problem.

Organic Synthesis

Building larger units from smaller ones, more complex structures from simpler ones, is a natural human activity. The building blocks can be words, stones, girders, or atoms and molecules. Organic chemistry is a particularly enticing field in this respect, because it permits a huge number and variety of molecular structures. Thus, many organic chemists work at the technology of **organic synthesis**—devising new reactions useful in building complex molecules. A related activity is the purposeful construction of new molecules which have interesting chemical and/or biological properties.

For instance, medicinal research has inspired some novel applications of organic synthesis. The first penicillin antibiotics were excellent antibacterial drugs, but some had narrow activity ranges and were ineffective against certain bacteria. As the use of penicillin continued, bacteria evolved mutant strains and became resistant to the drug. Synthetic chemists then modified the structure of penicillin by removing one part of the molecule and replacing it with a slightly different arrangement of atoms. This "fooled" the mutant strains of bacteria, thus allowing the modified penicillin to attack them. Such synthetic chemistry, where the compound in question is produced naturally and then modified, is called *semisynthesis.* A total synthesis of penicillin has been accomplished, but too many steps are involved to make the total synthesis economically useful. On the other hand, total synthesis of some steroidal and prostaglandin hormones is a useful and economic way of obtaining new drugs.

Many drugs have more than one type of physiological action, and synthetic methods used to modify drug structures are often used to separate the desired properties from harmful ones. For example, much effort has been directed in medicinal synthesis towards separating the analgesic (pain-killing) effects of morphine-type drugs from their addicting properties. The modern trend in synthesis is thus to prepare compounds of better and better specificity.

Our discussions of industrial and laboratory syntheses will be particularly important to chemistry majors, many of whom will eventually be employed in the field of synthesis. The students of biological science, however, may be more interested in another aspect: how organic molecules are synthesized in nature by plants and animals (**biosynthesis**).

One reason for studying the biosynthesis of a particular compound is that it often leads to practical results. The pain-relieving action of aspirin was a mystery for decades until a series of hormones called prostaglandins was discovered. One of these hormones is important in determining the human response to pain and it was found that aspirin blocks one of the chemical steps in the body's synthesis of that hormone. Knowing the molecular basis for aspirin effectiveness will allow new and more effective analgesics to be developed.

Reaction Mechanisms

The previous section stressed the fact that chemical reactions are carried out in order to synthesize useful compounds. In addition, when we conduct a reaction, we want to get the maximum yield of the desired product(s) as rapidly as possible with the minimum yield of by-products. To achieve this goal we have to choose a reaction temperature; decide which solvent, if any, to use; decide which catalyst is to be added, if any; and so on. Making rational choices in all these areas is much easier if we understand the **reaction mechanism**. This is a detailed description of all the structural and energetic changes that occur as reactants are converted to products. So, one reason for studying reaction mechanisms is to give us increased control over chemical reactions. Another reason is simply to satisfy our curiosity about how the world works.

Let's take a familiar example and explain what a reaction mechanism is. Equation 1 shows what happens when carbon dioxide dissolves in water. It is partially converted to carbonic acid, which gives soda water its characteristic flavor. The reaction mechanism can be described in terms of changes in bonding.

$$\text{O{=}C{=}O} + \text{H—OH} \rightleftarrows \text{O{=}C(OH)}_2 \qquad (1)$$

Carbon dioxide — Carbonic acid

To convert CO_2 to carbonic acid new C—O and O—H bonds must be formed and one O—H bond must be broken. One possibility is that all these bonding changes take place simultaneously. This is not the only reasonable possibility, however. We know that H^+ and OH^- ions are always present in aqueous solution. It might happen that CO_2 reacts first with OH^- to produce bicarbonate ion, HCO_3^-, as a reaction **intermediate**. The bicarbonate ion could then react with H^+ to produce carbonic acid.

How do we decide between these two possible reaction courses and perhaps other plausible reaction mechanisms? One of the most effective approaches is by studying reaction rates. That is, we find out how many molecules of CO_2 are converted to carbonic acid per second, at a specified temperature and pressure, at pH 7. We might then increase the concentration of OH^- while maintaining the same temperature. If the rate increases, then we have support for the suggestion that OH^- and CO_2 are reacting together. Or, we might experiment with changing the temperature or labeling a reactant with an isotope to see how these changes affect the reaction rate. From experiments of this type we can usually piece together a fairly convincing reaction mechanism.

Knowing the mechanism for this reaction enables us to predict which other starting materials, chemically similar to CO_2 and H_2O, would react by way of a similar reaction mechanism. Thus, learning mechanisms of reactions enables us to think about general types of reactions rather than having to memorize an endless string of specific cases. There are literally millions of different reactions, but only a limited number of reaction mechanisms.

Many organic chemists make mechanistic studies their prime concern. They belong to the discipline of **physical organic chemistry**, which came into being about 50 years ago when chemists realized the need to classify the tremendous number of known laboratory reactions.

Determining the Structures of Organic Molecules: Analytical Organic Chemistry

The analytical chemist has always sought to determine the number and arrangement of atoms in a particular molecule. In the early history of organic chemistry, structural information about an unknown compound was inferred from the kind of chemical reactions it underwent. This was an involved and time-consuming process. For example, morphine, an addicting drug noted for its analgesic properties, was first obtained pure from the opium poppy in 1803. However, due to the complexity of this molecule (morphine has a molecular formula of $C_{17}H_{19}NO_3$), a correct structure was not proposed until 1925.

The time lag between discovery and structure determination of new compounds has been decreasing steadily. The antitumor drug vinblastine (molecular formula $C_{46}H_{58}N_4O_9$) was first isolated from the periwinkle plant in 1958 and its complete structure was established by 1965. Currently, new natural products of equal complexity can usually be isolated and have their structures determined in a period of weeks. Part of this increased efficiency is due to the basic fund of knowledge about organic compounds which has accumulated over the years. The major advance, however, has been in the development of spectroscopic methods.

Spectroscopy is the study of how molecules absorb radiant energy. A sample of a compound is treated with electromagnetic radiation, such as visible or infrared light. The energy of the light is varied while, at the same time, the amount of light absorbed by the compound is measured. These operations are carried out automatically by instruments called **spectrometers.** A recorder on the spectrometer provides us with a **spectrum** of the compound, and from this spectrum it is possible to tell a good deal about the structure of the molecule. Spectroscopy is applied to less complex molecules as well. For example, an infrared spectrum of ethanol (ethyl alcohol, CH_3CH_2OH) shows a band in a region known to be representative of an O—H bond. It is a simple matter of a five-minute, nondestructive experiment to tell whether or not a molecule contains an OH group. More sophisticated spectroscopic techniques have shown the presence of ethanol in interstellar gas clouds.

A more direct method for determining the structure of an unknown molecule is through **x-ray diffraction studies**. A map of electron densities, much like a geographical contour map, is obtained by computer analysis of the pattern of scattered x rays emerging from a single irradiated crystal of the unknown. From this map a picture of the molecule can be constructed with exact placement of each atom in relation to all the others. Small molecules not easily obtainable as crystals can be studied in the gas phase, with similar results, by means of electron diffraction. X-ray diffraction pictures of simple molecules can now be completed in less than a week, whereas larger molecules, such as complex biological compounds, may take months or even years.

Spectroscopic methods can be made quantitative, so that they can tell us not only the structure of a compound but how much of it is present. When these methods are combined with efficient techniques for separating compounds, the result is a set of powerful analytical tools.

Separation techniques have become as sophisticated, and as heavily automated, as spectroscopic ones. Complex mixtures can be resolved into their pure components, including those present in minute quantities. The principle method for the separation of mixtures is **chromatography** (Greek *chroma,* meaning "color"). Chromatography was originally developed for the isolation of plant pigments like chlorophyll, but can now be applied to most types of organic compounds, colored or not. There are numerous types of chromatography, but in all of them the initial mixture is dispersed in a moving liquid or gas. This mobile phase flows by and contacts a stationary phase (liquid or solid). The individual compounds in the mixture exchange rapidly but unequally between the two phases because of differences in charge, size, or polarity. For this reason, they travel through the system at different speeds and become separated.

These analytical tools are used by chemists in all branches of organic chemistry, whether they are engaged in research, production, or clinical work. They are even found in institutions very different from academic or industrial laboratories, such as hospitals and forensic laboratories.

> The chemist does not always or only stand and wait, but chemistry can hold its head as high as any branch of philosophical inquiry has ever done. It rejoices, first of all, in its own perfection, and even when its works serve others, its primary and proper end is not servitude, but its own prosperous independence. . . . But the great virtue of the chemical artist is that he produces not just isolated

speculation, but a useful and beneficial praxis for society, appreciating that no art is worth its salt which does not produce practical consequences from its theory.

A. Libavius (1595)

"Are you ever afraid of what those chemicals might come up with next?" Greer said.

"They're never evil things," Miss Hawkline said. She was going to say malicious, but she changed her mind.

"What are those chemicals supposed to do when they're finished?" Greer said.

"We don't know what they're supposed to do," Miss Hawkline said. "Our father told us when The Chemicals were completed that the answer to the ultimate problem facing mankind would be solved."

Richard Brautigan, *The Hawkline Monster,* Simon & Schuster, Inc., New York, N.Y., 1974.

Summary

The three topics just outlined briefly (synthesis, reaction mechanisms, and structure determination) form the backbone of organic chemistry. In each chapter, when new types of organic compounds are discussed, we will discuss their detailed structure (the exact arrangement of atoms and types of bonds), how they can be prepared in the laboratory, and the manner in which they react to form new materials.

Before getting into details of organic reactions, we will give an overview of the kinds of molecules you will encounter in organic chemistry (Chap. 1). This will be followed by a chapter on bonding, most of which will be a review of what you learned in your general chemistry course. We will then be in a position to begin our study of organic chemistry in depth.

Supplementary Reading

L. N. Ferguson, Organic chemicals: Angels or goblins, *Journal of Chemical Education* (*J. Chem. Ed.*), **55,** 591 (1978).

J. Lipscombe and B. Williams, *Are Science and Technology Neutral?,* Butterworths, Boston, Mass., 1979.

A Survey of Organic Structures: Carbon Skeletons and Functional Groups

1

This chapter is a review of some fundamental chemical concepts. These include empirical and molecular formulas, structural formulas based on electron pair bonds, and acidity and basicity. We assume that you have studied these subjects before in a course in general chemistry. If you haven't, you will probably find the treatment in this chapter too concise for complete understanding. We recommend that you consult a general chemistry textbook; several are listed among the Supplementary Readings at the end of this chapter.

A good deal of attention is paid here to writing and visualizing molecular structures. These skills are essential for understanding the material in the next three chapters. They are also indispensable for a secure understanding of organic chemistry.

1:1 Empirical and Molecular Formulas

Organic molecules vary tremendously in their composition. They range from the very simple (methane—CH_4) to the complex (strychnine—$C_{21}H_{22}N_2O_2$) to the fantastic (vitamin B_{12}—$C_{63}H_{88}CoN_{14}O_{14}P$). Whether simple or complex, compounds of unknown structure are routinely encountered in all branches of organic chemistry. A natural product chemist may extract a new substance from a plant leaf, or a synthetic chemist may obtain a previously unknown material as a reaction product. Since the appearance and smell of a compound usually give little information about its structure, it is necessary to perform a number of physical and chemical operations on it.

We begin by establishing the **empirical formula** of the compound, which tells us what elements it contains and in what proportions. At this early stage, we also determine the **molecular formula,** which tells us how many atoms of each type are present in each molecule. For example, the compound ethane, a major component of natural gas, has the empirical formula CH_3. The formula tells us that ethane is a hydrocarbon—a compound containing only carbon and hydrogen—with a carbon to hydrogen ratio of one to three. Ethane's molecular formula is C_2H_6, which indicates that each ethane molecule contains two carbon atoms and six hydrogen atoms.

When determining the empirical formula, it is first necessary to analyze the unknown compound for the presence of various elements. Carbon, hydrogen, nitrogen, and oxygen are those elements most commonly found in organic compounds. A quantitative chemical analysis is used to establish the relative percentages of C, H, and/or N in the unknown. The presence or absence of rarer elements, such as the halogens (F, Cl, Br, I), sulfur, or phosphorus, is established qualitatively. If none of these is present, then the percentage of oxygen in the compound is obtained by subtracting from 100 the percentages of C, H, and N found. A direct analysis for oxygen is rarely done because in the standard analytical procedure, a weighed sample of the compound is oxidized catalytically with excess O_2. Carbon in the compound is converted to CO_2, hydrogen to H_2O, and nitrogen to N_2, and the number of moles of each is determined. This procedure is called a *combustion analysis* and is performed with automatic techniques in specialized laboratories.

A typical set of analytical data might be obtained as follows: C, 45.80%; H, 10.57%; N, 13.24%. The combined total is 69.61%, leaving 30.39% of the weight of the material to be accounted for. We can assume this to be due to the presence of oxygen if the absence of other elements has been proven. These percentages are weight values equal to the weight of each element in 100 g of sample. By dividing by the appropriate atomic weight we obtain the relative *molar* proportion of each element. Thus, 100.0 g of this unknown compound contains 3.813 mol of carbon atoms, 10.49 mol of hydrogen atoms, 0.9450 mol of nitrogen, and 1.899 mol of oxygen atoms.

$$\mathrm{C} = \frac{45.80\ \mathrm{g}}{12.01\ \mathrm{g/mol}} = 3.813\ \mathrm{mol} \qquad \mathrm{N} = \frac{13.24\ \mathrm{g}}{14.01\ \mathrm{g/mol}} = 0.9450\ \mathrm{mol}$$

$$\mathrm{H} = \frac{10.57\ \mathrm{g}}{1.008\ \mathrm{g/mol}} = 10.49\ \mathrm{mol} \qquad \mathrm{O} = \frac{30.39\ \mathrm{g}}{16.00\ \mathrm{g/mol}} = 1.899\ \mathrm{mol}$$

In order to obtain the empirical formula for the unknown, we compute the molar abundance of each element relative to that of the least abundant element. Nitrogen is the least abundant element in this case, so we divide the molar proportion of each element by 0.9450 mol. Thus, the relative molar abundance of carbon equals 3.813 mol C/0.9450 mol N. The value of this proportion, 4.03, equals the ratio of moles of carbon to nitrogen in the sample, and thus the ratio of carbon to nitrogen atoms in the molecule. Similarly, the hydrogen to nitrogen ratio is 10.49 mol H/0.9450 mol N = 11.1, and the oxygen to nitrogen ratio is 1.899 mol O/0.9450 mol N = 2.01. Thus, the empirical formula given by analysis is $C_{4.03}H_{11.1}N_{1.00}O_{2.01}$. Allowing for experimental error in the original weight percent determination, this formula can be assumed to correspond to $C_4H_{11}NO_2$.

Since the molecular formula could be any whole multiple of $C_4H_{11}NO_2$ (e.g., $C_8H_{22}N_2O_4$), an independent method for establishing the molecular weight of the unknown is necessary. The most common and best way is to use the **mass spectrometer**. This instrument removes an electron from the molecule to form a molecular ion (M^+). The instrument then measures the mass (or mass/charge ratio, commonly designated m/z) of the molecular ion. The charge is usually +1. Equation 1.1 illustrates the ionization process for H_2O. The mass, m, of the ion is 18; its charge is +1.

$$\mathrm{H{-}\ddot{O}{-}H} \xrightarrow[\text{spectrometer}]{\text{mass}} \mathrm{H{-}\dot{O}^{+}{-}H} + e^- \tag{1.1}$$

Molecular ion, M^+
$m/z = 18$

For molecular weight measurements either of two types of mass spectrum can be used. A *low-resolution* mass spectrum gives the molecular weight to the nearest whole number. The unknown discussed above (whose empirical formula we determined) would give a M^+ at $m/z = 105$ if the molecular formula were $C_4H_{11}NO_2$, and a M^+ at $m/z = 210$ if the formula were $C_8H_{22}N_2O_4$. (In Section 8:1 we will show how a more accurate molecular formula can be obtained using *high-resolution* mass spectrometry.)

1:2 Deriving Structural Formulas from Molecular Formulas

Obtaining a molecular formula for a compound is only the first step in determining its structure. The molecular formula must be translated into a **structural formula**, which shows the order in which the individual atoms are connected to one another. To do this we must know the usual valences for each atom in the molecule.

Structural formulas for organic compounds are written in several different ways. One type of structural formula, the **electron dot**, or **Lewis** structure, is seldom used since it is very cumbersome for large molecules. Therefore, shorthand methods of designating structure are necessary. For example, line bond structures show the **bonding electron pairs** as lines and the unshared valence electrons which make up the octet as dots. (See the line bond formula for water in eq. 1.1.) Unshared electron pairs, called **nonbonding electron pairs**, are valence electrons not used in covalent bonds. Nonbonding electron pairs found in oxygen, nitrogen, and sulfur compounds are commonly included in the structural formula because they sometimes take part in reactions. Nonbonding electron pairs also occur on other atoms, such as the halogens (F, Cl, Br, and I), but they are usually unreactive and are therefore not included in the formulas.

Let's see what these normal valences are for the main group elements, those found in groups 1A to 7A of the Periodic Table (see endpapers). The major factor controlling valency in these elements is their tendency to attain the same electron configuration as the nearest noble gas (group 8A elements). Since all noble gases except helium have eight electrons in their valence shells, most atoms react in such a way as to attain an octet of valence electrons (the *octet rule*). A few light elements, such as hydrogen and lithium, achieve a stable configuration with two valence shell electrons, which corresponds to the electronic configuration of the noble gas helium.

We can apply these concepts in calculating the valences of those elements commonly found in organic compounds. Thus, atomic carbon, which has four valence electrons, requires four additional electrons to complete its octet. It can get them by sharing four electrons provided by other atoms, so we say that the valence of carbon is four. Similar reasoning suggests that N needs three electrons, S and O each need two electrons, and each of the halogens needs one electron to obtain a complete octet. Since the hydrogen atom has one electron, it needs only one additional electron to fill its valence shell. Using these normal valences we can predict the structural formulas corresponding to the molecular formulas C_2H_5Cl and CH_5N.

```
  H H  ..            H  ..
H:C:C:Cl:          H:C:N:H
  H H  ..            H H
```

Chloroethane (C_2H_5Cl) — Methylamine (CH_5N)

Table 1–1. Structural Formulas of Some Single-Bonded Organic Molecules

	Methane	Ethane	Methylamine	Fluoromethane *(methyl fluoride)*★
Electron dot (Lewis) structure	H H:C:H H	H H H:C:C:H H H	H H:C:N̈:H H H	H H:C:F̈: H
Line bond (modified Lewis) structure	H | H—C—H | H	H H | | H—C—C—H | | H H	H | H—C—N̈—H | | H H	H | H—C—F | H
Condensed structural formula	CH_4	CH_3CH_3	CH_3NH_2	CH_3F

★ Many organic compounds have more than one name. One name is usually of historical origin; these so-called common names are given here and throughout the text in italics and parentheses. The other name is a systematic one, based upon a naming convention discussed in Section 3:2.

Table 1–1 shows structural formulas for some simple organic compounds containing only single bonds. Drawing line bond or electron dot structural formulas can become somewhat tedious with even moderate size organic compounds, but this will not be necessary after the fundamentals of molecular structure have been learned. In much of this book we will write **condensed formulas** in which the arrangement of atoms is implied but bonds are not explicitly shown. For example, a condensed formula for ethane would be CH_3CH_3. When you start out, be very careful when using condensed formulas—use line bond formulas until you feel comfortable with them. In the condensed formula CH_3CH_3, for example, the second carbon looks as if it were bonded to three hydrogens on one side and three hydrogens on the other. Actually, the two carbons are bonded together, as shown by the line bond structure in Table 1–1. When translating line bond structures into condensed formulas, we usually leave out other single bonds such as C—C and O—H. Exceptions are made when you wish to draw attention to a specific bond and when a structure is *branched* (branched chains are discussed in sec. 1:5).

Example

Write electron dot and line bond structures for ammonia (NH_3). Write electron dot, line bond, and condensed formulas for a compound having the molecular formula CH_4O.

Answers

Nitrogen has five valence electrons: ·Ṅ:. It needs three more electrons to reach the stable valence configuration of eight and can get these from three hydrogen atoms: 3H·. Therefore the electron dot formula is

H
H:N̈:
H

The line bond structure replaces each *bonding* electron pair with a line. The nonbonding electron pair is indicated with dots.

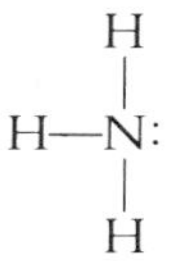

```
    H
    |
H—N:
    |
    H
```

For the compound CH_4O, we make use of the normal valence electron configuration of four for carbon, one for hydrogen, and six for oxygen.

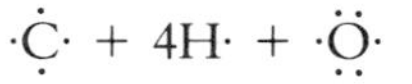

$$\cdot\dot{\underset{\cdot}{C}}\cdot + 4H\cdot + \cdot\ddot{\underset{\cdot\cdot}{O}}\cdot$$

These are joined together so that carbon and oxygen will each have eight valence electrons in the molecule

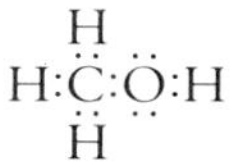

```
  H
H:C:Ö:H
  H  ··
```

To draw the line bond structure, replace all bonding electron pairs with a line.

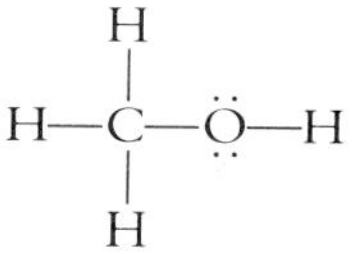

```
    H
    |   ··
H—C—O—H
    |   ··
    H
```

A rational (and acceptable) condensed formula derived directly from the line bond formula would be H_3COH. However, it is a more common (but confusing) convention that the condensed formula be written with the hydrogens *after* the carbon, which gives us CH_3OH.

Problems

1–1 Write electron dot and line bond structures for the following compounds:

a. chloroform ($CHCl_3$)
b. carbon tetrachloride (CCl_4)
c. dichloromethane (CH_2Cl_2)
d. borane (BH_3)
e. hydroxylamine (NH_2OH)

1–2 Write electron dot, line bond, and condensed formulas for compounds having the following molecular formulas:

a. C_2H_5Br
b. C_3H_8
c. C_2H_5F
d. $C_2H_4Br_2$
e. C_2H_7N
f. $C_2H_4O_2$
g. $C_2H_3Cl_3$

Many organic compounds contain **multiple bonds**, in which a pair of atoms shares more than two electrons. It is common practice to show multiple bonds explicitly, even in condensed formulas. When condensed formulas are used, the multiple bond is usually indicated by lines. Table 1–2 provides examples of simple organic compounds containing double or triple bonds.

Reasonable structures for molecules with multiple bonds are readily arrived at by employing the common valences for each of the constituent

Table 1–2. Structural Formulas of Some Organic Molecules Containing Multiple Bonds

	Ethene (*ethylene*)★	Ethyne (*acetylene*)★	Methanal (*formaldehyde*)★	Ethanenitrile (*acetonitrile*)★
Electron dot (Lewis) structure	H. .H :C::C: H˙ ˙H	H:C:::C:H	H. :C::Ö: H˙	H H:C̤:C:::N: H
Line bond (modified Lewis) structure	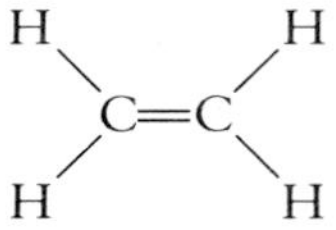 H H \\ / C=C / \\ H H	H—C≡C—H	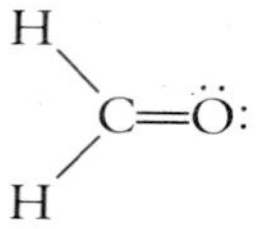 H \\ C=Ö: / H	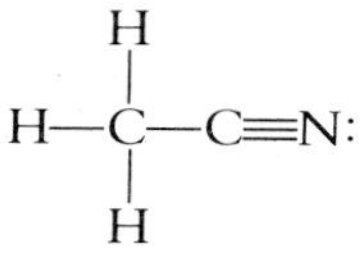 H | H—C—C≡N: | H
Condensed structural formula	CH_2=CH_2	HC≡CH	CH_2=O or CH_2O	CH_3C≡N or CH_3CN

★ See footnote to Table 1–1.

atoms. Remember also that atoms such as nitrogen, oxygen, and the halogens will have enough nonbonding electron pairs to satisfy the octet rule. Given that the covalency of carbon is four, of oxygen is two, and of hydrogen is one (Table 1–3), and knowing the octet rule requirements, you can see that the structure of formaldehyde, CH_2O, can only be the one shown below. The structures can be verified by calculating the total number of

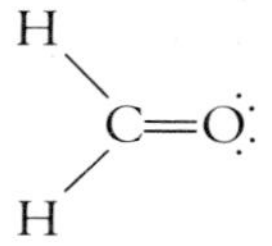

```
H
 \
  C=Ö:     Formaldehyde
 /
H
```

Table 1–3. Covalent Bonding Patterns Most Commonly Found in Organic Compounds

Element	Normal Valence (number of bonds most commonly formed)	Covalent Bonding Patterns★
C	4	—C— (with bonds above and below), —C= (with bond above), =C=, —C≡
N	3	—N̈— (with bond below), —N̈=, :N≡
O	2	—Ö—, .Ö=
H	1	H—
X (halogens)	1	:Ẍ—

★ The formation of covalent bonds to achieve stable noble gas configurations. Each bond represents a pair of electrons.

valence electrons available and comparing that number with the number shown in the structure. In the case of formaldehyde, 12 valence electrons are available—four from carbon, six from oxygen, and two from the two hydrogens. This number is in complete accord with a structure containing four bonding and two nonbonding electron pairs.

Example

Write electron dot and line bond structures for C_2H_3Cl, which contains a multiple bond.

Answer

The result may be arrived at by using the rules of normal valence (Table 1–3). Since hydrogen and chlorine both have normal valences of one, they can together satisfy only four of the eight valences required by the two carbon atoms. The carbon-carbon double bond then becomes a mathematical necessity.

The correct line bond structure is shown below. The reason this molecule assumes this geometry is discussed in Section 2:1.

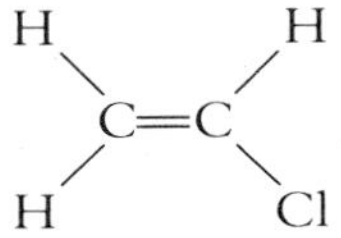

```
H       H
 \     /
  C=C
 /     \
H       Cl
```

Problem

1–3 The following molecules all contain multiple bonds. Write their electron dot and line bond structures.

a. C_2H_2 c. H_2CNH e. CH_3CHO g. CH_3COCH_3
b. HCN d. CH_3NO f. CH_2CO h. NH_2CONH_2

1:3 Organic Ions and Radicals

Much of organic chemistry deals with uncharged molecules, yet there are many important organic charged molecules or *ions.* Within these ions, charges appear to be preferentially concentrated on certain atoms. The charge on an atom, called its **formal charge**, is associated with an unusual number of valence electrons and/or an unusual valence. Some neutral molecules also contain atoms with formal charges. In these neutral molecules, a "+" charge on one atom is balanced by a "−" charge on another atom in the same molecule.

To calculate the formal charge on an atom in a neutral molecule or ion, compare the number of electrons around the atom in the molecule with the number it would have if it were a neutral, isolated atom. The number of valence electrons for the neutral isolated atom can be taken from the Periodic Table of elements. Shown here are the valence electrons of some representative atoms.

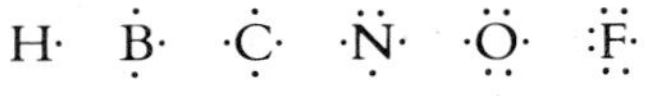

H· Ḃ· ·Ċ· ·N̈· ·Ö· :F̈·

In a molecule, the atom is considered to "own" one-half of the electrons it shares with other atoms and to "own" *all* of its nonbonding electrons. Consider the familiar cases of water and hydronium ion:

$$\begin{array}{c} \text{H} \\ \text{H:}\underset{\cdot\cdot}{\ddot{\text{O}}}\text{:} \end{array} \qquad\qquad \left[\begin{array}{c} \text{H} \\ \text{H:}\underset{\cdot\cdot}{\ddot{\text{O}}}\text{:H} \end{array}\right]^{+}$$

Water　　　Hydronium ion

In both molecules, the hydrogens each share two electrons and so each "owns" one. Since each hydrogen atom would have one electron if it were an isolated atom, these bonded hydrogens bear no charge. In water, the oxygen "owns" two nonbonded pairs (four electrons) and shares two pairs for a total of six "owned" electrons. This is the same number as that found in the neutral isolated atom of oxygen, and the oxygen in the water molecule is therefore uncharged. Now look at the hydronium ion. Here oxygen shares three pairs (it "owns" one-half of six, or three) and "owns" one nonbonded pair for a total of five electrons "owned." Five is one less than the required six of the neutral atom, and hence the *oxygen* in the hydronium ion has a charge of +1. Being able to identify the formally charged atom in a molecule or ion is important because their charge is often what gives the molecule or ion its reactivity.

Let's look next at the organic molecule methane. The carbon atom has no nonbonding electrons and shares four electron pairs with the hydrogens.

$$\begin{array}{c} \text{H} \\ \text{H:}\underset{\cdot\cdot}{\ddot{\text{C}}}\text{:H} \\ \text{H} \end{array}$$

Methane

It therefore "owns" four electrons, the same number that is in the valence shell of an isolated carbon atom. We conclude that the carbon atom in methane is uncharged. Since each hydrogen atom in the molecule "owns" one electron (by sharing two), the hydrogens are uncharged as well.

Removing a hydrogen atom (H·) from methane results in the formation of the **methyl radical** (eq. 1.2). A *radical* is a species with one or more

(1.2)
$$\begin{array}{c} \text{H} \\ \text{H:}\underset{\cdot\cdot}{\ddot{\text{C}}}\text{:H} \\ \text{H} \end{array} \rightarrow \begin{array}{c} \text{H} \\ \text{H:}\underset{\cdot\cdot}{\ddot{\text{C}}}\cdot \\ \text{H} \end{array} + \text{H}\cdot$$

Methyl radical

unpaired electrons. Carbon has one nonbonded electron to itself and shares three pairs with the hydrogens. Since it "owns" a total of four electrons the carbon atom in the methyl radical should be uncharged, which is what we find. Loss of a hydride ion (H:^-) from methane yields the **methyl cation** (eq. 1.3). The carbon in the methyl cation shares three electron pairs with the

(1.3)
$$\begin{array}{c} \text{H} \\ \text{H:}\underset{\cdot\cdot}{\ddot{\text{C}}}\text{:H} \\ \text{H} \end{array} \rightarrow \begin{array}{c} \text{H} \\ \text{H:}\underset{\cdot\cdot}{\ddot{\text{C}}}^{+} \\ \text{H} \end{array} + \text{H:}^{-}$$

Methyl cation

hydrogens and has no nonbonding electrons. It "owns" only three electrons, one less than a neutral carbon atom. The carbon atom therefore carries a formal charge of +1. The cation as a whole has a charge of +1, which equals the sum of the formal charges on all the atoms.

Note that Equations 1.2 and 1.3 are each charge balanced. That is, the charges add up to the same value on both sides of the arrow. There has been no *net* creation of charge in the course of either reaction. (Neither reaction is a very likely process *as written*. In the laboratory, methyl radicals and cations are formed by other reactions.)

Positive and negative charges attract one another, and the closer together they are, the lower their energy. Thus it takes work to separate charges, and as a result uncharged molecules are more stable than charged molecules of the same molecular formula. Consider the molecular formula CHN. By using normal valences we would choose the structural formula HC≡N (hydrogen cyanide) for this molecule. An alternative structure that satisfies the octet rule and fits this molecular formula is **1**. (Verify the formal charges

$$\mathrm{H{-}\overset{+}{N}{\equiv}\overset{-}{C}{:}}$$

1

for yourself by using the procedure just discussed.) It turns out that **1** and hydrogen cyanide are two distinct compounds that are in chemical equilibrium. That equilibrium heavily favors HCN, substantiating our assertion that the structure without formal charges is the one of lower energy. Compounds such as HCN and **1** with the same molecular formula but different bonding sequences for their atoms are discussed in Section 1:5.

Note that in structure **1** the valences of carbon and nitrogen are "abnormal," although both have filled octets. This example illustrates why we have stressed the normal valences of atoms; they are the ones that lead to structures with the least number of formal charges.

Problems

1–4 The methyl anion CH_3^- contains a negatively charged carbon. Draw a line bond structure for the methyl anion.

1–5 What would be the charge (if any) on the underlined species in each of the following equations:

a. $:NH_3 + H^+ \rightarrow \underline{NH_4}$

b. $:NH_3 + H^- \rightarrow \underline{:\ddot{N}H_2} + H_2$

c. $CH_3Cl \rightarrow \underline{:CH_2} + HCl$

d. $CH_3\overset{\overset{:O:}{\parallel}}{C}{-}\ddot{\underset{\cdot\cdot}{O}}{-}H + H_2O \rightarrow \underline{CH_3\overset{\overset{:O:}{\parallel}}{C}{-}\ddot{\underset{\cdot\cdot}{O}}:} + \underline{H_3O}$

e. $\underline{CH_3\ddot{\underset{\cdot\cdot}{O}}H} + OH^- \rightarrow \underline{CH_3\ddot{\underset{\cdot\cdot}{O}}:} + H_2O$

1–6 For each of the following pairs of structures calculate the formal charges (if any) on the carbon, nitrogen, and oxygen atoms. Then decide which structure, if either, is lower in energy.

a. H₂C=O (H—C(—H)=O) or H—C=O—H

b. H—O—N=O or H—N=O with :O: bonded to N

c. H—O—C≡N: or H—N=C=O

d. H—O—C=O—H or H—C—O—H with :O: double-bonded to C

1:4 Using Molecular Models

In 1962 Francis Crick and James Watson received the Nobel prize for presenting the first three-dimensional picture of the double helix structure of deoxyribonucleic acid (DNA). The success of their work was dependent upon building *molecular models*. Many organic molecules have a three-dimensional structure that is difficult to visualize from the printed page. There are, however, various types of three-dimensional molecular models commercially available as visual aids. Figure 1–1 shows photographs of models of the molecule propane (C_3H_8) built with three basic styles of model kits: space-filling, ball and stick, and framework models. The space-filling models give an idea of the relative size of each atom in a molecule and are useful in studying structures where various atoms or parts of a molecule may collide with one another. Ball and stick and framework models do not give such an accurate picture of relative sizes of atoms, but they allow easier observation of geometrical relationships among bonded atoms.

The value of three-dimensional models becomes obvious when you examine the structures in Figure 1–1. Line bond and condensed formulas **2** and **3** give the false impression that the three carbon atoms of propane lie in a straight line, whereas observation of the models in Figure 1–1 tells us that the molecule is in fact angular. When building a model is inconvenient, we can always resort to drawing the molecule in a way that more accurately conveys the three-dimensional structure. Two commonly used conventions, struc-

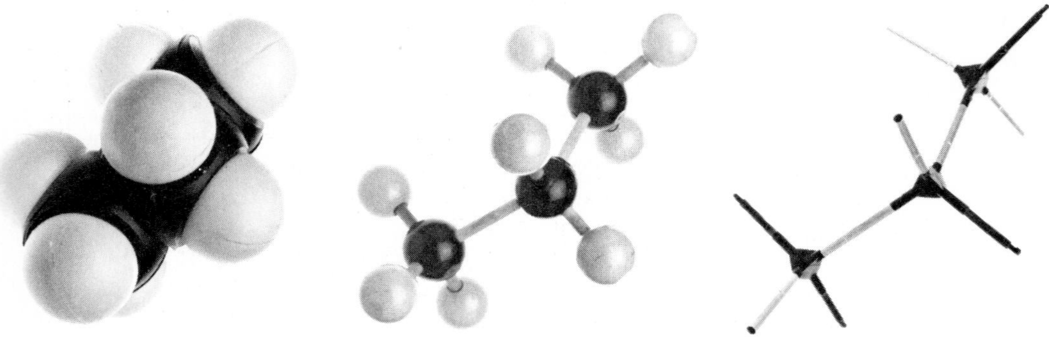

Figure 1–1. *Molecular models of propane.*

Box 1–1. Molecular Models and the Discovery of the Double Helix

In his highly personal account of the discovery of the structure of DNA★, James D. Watson commented on the importance of Linus Pauling's proposed α-helix polypeptide structure:

"I soon was taught that Pauling's accomplishment was a product of common sense, not the result of complicated mathematical reasoning. Equations occasionally crept into his argument, but in most cases words would have sufficed. The key to Linus' success was his reliance on the simple laws of structural chemistry . . . the essential trick was to ask which atoms like to sit next to each other. In place of pencil and paper, the main working tools were a set of molecular models superficially resembling the toys of preschool children."

"We could thus see no reason why we should not solve DNA in the same way. All we had to do was to construct a set of molecular models and begin to play. . . ."

★ James D. Watson, *The Double Helix*. Copyright © 1968 by James D. Watson. Reprinted with the permission of Atheneum Publishers.

tures **4** and **5**, show the three-dimensional structure around the central carbon in propane. In **4**, a standard width line (or bond) in a drawing means that

```
    H   H   H                                projecting          projecting
    |   |   |                                below the           above the
H — C — C — C — H      CH3CH2CH3               page                page        H     H
    |   |   |                                      H           H                \  /
    H   H   H                                       `.  ◄    ▲`                  C
                                                        C                       / \
                                                       / \                   CH3   CH3
                                                    CH3   CH3
        2                  3                            4                          5
```

the atoms which the line connects lie in the plane of the paper. Dashed lines are drawn to an atom that is behind the plane of the paper, and a wedge-shaped bond is drawn to an atom that is in front of the plane of the paper. We will use this convention because of its clarity. The second convention, **5**, which has a heavy line replacing the wedge, is also found frequently in chemical literature.

Problem

1–7 a. With a molecular model set, build a model of butane, $CH_3CH_2CH_2CH_3$.

b. Arrange your model so that all the carbon atoms lie in one plane, in a zigzag pattern. Make a drawing of this model using the wedge–dash convention shown in structure **4**.

1:5 Isomerism

For a small number of simple organic compounds such as methane (CH_4) there is only one reasonable structural formula that can be written for a particular molecular formula. As the molecular formula becomes more complex, however, it becomes possible to write more than one structure that fits the molecular formula. For instance, there are two different carbon frameworks that satisfy the formula C_4H_{10}. The four carbons may be joined in a straight chain (**6**), or in T-shaped branched chains (**7**). Both of these structures are in fact found in nature; we can isolate two distinct compounds, each

$CH_3CH_2CH_2CH_3$

or

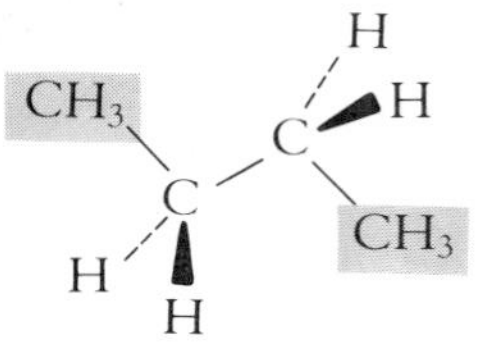

CH_3—C(H_2)—C(H_2)—CH_3 (perspective)

6

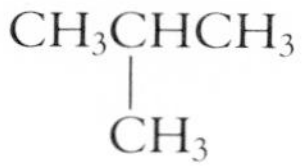

CH_3CHCH_3 with CH_3 below the central carbon

or

CH_3, CH_3, CH_3, H on C (perspective)

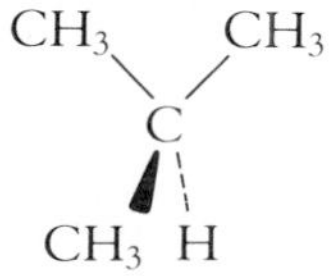

7

having its own chemical and physical properties yet sharing the molecular formula C_4H_{10}. Compounds such as these, which possess identical molecular formulas but can be distinguished from one another by chemical or physical means, are called **isomers** (from the Greek word *isomeres*, meaning "of the same parts").

The term *isomerism* covers the many differences that exist between molecules having the same molecular formula but different structures. How can we describe the difference between structures **6** and **7**? One approach is to refer to the sequence in which the atoms are connected to one another. In **6**, the two carbon atoms of the CH_3 groups are each connected to *one* other carbon, while the remaining carbon atoms (from CH_2) are each connected to *two* other carbons. Linear hydrocarbons such as **6**, in which no carbon atom is bonded to more than two others, are classified as *normal* or *straight-chain* hydrocarbons. In contrast, **7** has three carbon atoms that are each connected to one carbon, and one carbon atom connected to *three* other carbons. Compound **7** is a *branched-chain* hydrocarbon. In terms of their patterns of bond connections, molecules **6** and **7** are constituted differently and are therefore labelled **constitutional isomers**.* Figure 1–2 shows two pairs of constitutional isomers.

A word of caution. It is possible to convince yourself that isomers exist for a particular molecular formula when in fact they do not. The problem arises because of the way in which we *write* chemical structures. For instance, **8** and **9** represent line bond structures for chloroethane, C_2H_5Cl. They may appear to be isomeric structures, but they are not. The fact that there is only

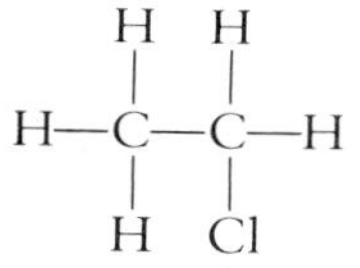

H—C(H)(H)—C(H)(Cl)—H **8**   H—C(H)(H)—C(H)(H)—Cl **9** CH_3CH_2Cl **10**   Cl, H, H on C—C with H, H, Cl (perspective) **11**

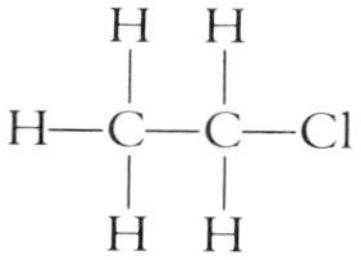

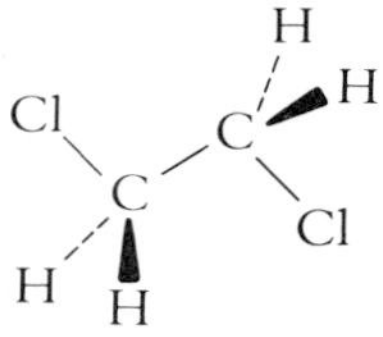

one C_2H_5Cl structure is made clear if we use the condensed (**10**) or perspective (**11**) method of representation. It may also appear that constitutional isomers can be produced by rotating the entire structure on the page, but this is not so; thus, structures **10** and **12** depict the same compound. Rotating one

* Constitutional isomers are also widely referred to as **structural isomers**. Because it is not always clear what amounts to a structural difference between two compounds, the term "constitutional isomerism" is preferred.

CH_3OCH_3	CH_3CH_2OH	$H_2C=CH(CH_3)$	cyclopropane
Dimethyl ether	Ethanol	Propene	Cyclopropane
C_2H_6O	C_2H_6O	C_3H_6	C_3H_6

Figure 1–2. *Two pairs of constitutional isomers.*

part of the structure with respect to the rest of the structure does not produce a new constitutional isomer either, since the order in which the atoms are connected is not affected. For example, structures **11** and **13** both depict the same molecule, chloroethane. Furthermore, structures **8** and **9** are just planar

$ClCH_2CH_3$
12

13

projections of structures like **11** and **13**, and like them do not represent constitutional isomers. (At this point we assume that rotation about single bonds is free and unhindered. More will be said about single bonds in Chapter 3.)

Because extended line bond structures (**8** and **9**) can be misleading, as well as clumsy, we try to avoid using them, and suggest the same to you. We will instead rely mainly on condensed formulas, in which a few bonds may be shown explicitly to draw attention to their chemical reactivity. When it is necessary to emphasize the spatial relations among atoms, then perspective structures such as **11** and **13** will be used.

As the number of atoms in a molecule increases, the number of possible constitutional isomers increases sharply. There are seven isomers of formula $C_4H_{10}O$ and 17 isomers of formula $C_6H_{13}Cl$. Distinguishing among isomers is one of the major problems in organic chemistry, and this is why specialized diagnostic techniques such as spectroscopy (Chaps. 8 and 9) are so important.

Example

Draw structures for all the constitutional isomers having the molecular formula C_3H_7Br.

Answer

There are only two isomers that differ in the position of attachment of the functional group to the carbon chain. Note the identity of many structures written below. Use molecular models if you are not convinced of these identities. The condensed formulas would be $CH_3CH_2CH_2Br$ and $CH_3CH(Br)CH_3$.

```
                      H  H  H                          H  H  H
                      |  |  |                          |  |  |
                   H—C—C—C—Br                       H—C—C—C—H
                      |  |  |                          |  |  |
                      H  H  H                          H  Br H
                        |||                               |||
                      H  H  Br                         H  Br H
                      |  |  |                          |  |  |
                   H—C—C—C—H                        H—C—C—C—H
                      |  |  |                          |  |  |
                      H  H  H                          H  H  H
                        |||                               |||
                      H  H  H                                H
                      |  |  |                                |
                   H—C—C—C—H                          H H—C—H
                      |  |  |                          |     |
                      H  H  Br                      H—C————C—Br
      H                 |||                            |     |
      |                                                H     H
 H H—C—Br             H  H  H                            |||
 |    |               |  |  |                            etc.
H—C————C—H  ≡  Br—C—C—C—H
 |    |               |  |  |
 H    H               H  H  H
                        |||
                        etc.
```

Problem

1–8 Draw the structures of all the constitutional isomers having each of the following molecular formulas.

a. C_3H_8O (3 isomers)
b. C_4H_9Br (4 isomers)
c. C_5H_{12} (3 isomers)
d. C_4H_8 (5 isomers)
e. $C_4H_8Br_2$ (9 isomers)

1:6 Functional Groups and Carbon Skeletons

It is sometimes useful to conceptually divide molecules into two parts—the **carbon skeleton** and the **functional group(s)**. A functional group consists of an atom, or a small group of atoms, which undergoes the same chemical reactions *regardless of the structure of the remainder of the molecule*. After all the functional groups in the molecule are accounted for, the remaining carbon atoms with their attached hydrogens define the carbon skeleton.

The division between molecular skeleton and functional groups is not always hard and fast. Nevertheless, it is a very effective aid for a student of organic chemistry, because it focuses attention on the part of the molecule that is most likely to be chemically reactive. It will also help you to decide whether or not two different compounds will show similar chemical behavior.

Table 1–4. Different Representations of Some of the Major Classes of Hydrocarbons

	Aliphatic Hydrocarbons		Aromatic Hydrocarbons
	Acyclic	Alicyclic	
General name	alkane	cycloalkane	arene
Specific name	butane	cyclohexane	benzene
Condensed structural formula	$CH_3CH_2CH_2CH_3$	CH_2 H_2C CH_2 H_2C CH_2 CH_2	H C HC CH HC CH C H
Line–angle formula*			

* In line–angle formulas the vertex of each angle represents a carbon atom, with a sufficient number of attached hydrogen atoms to bring its total valence to four. The termini of the linear formulas also represent carbon atoms, each attached to three hydrogen atoms (CH_3 groups).

Let's first consider the carbon skeletons. They can be divided into two broad classes: **aliphatic** and **aromatic** (Table 1–4). (The word "aliphatic" is derived from the Greek *aleiphar,* meaning "oily," and refers to the slippery feel of many aliphatic compounds.) The aliphatic class may be further subdivided into **open chain** (or **acyclic**) compounds and **cyclic** (or **alicyclic**) compounds. The fundamental hydrocarbons of the acyclic class are the **alkanes**, which are built of *chains* of carbon atoms. The fundamental alicyclic compounds are the **cycloalkanes** such as cyclohexane, which contain *rings* of carbon atoms.

Aromatic hydrocarbons, called arenes, are always cyclic. Their fundamental unit is the *benzene ring,* which is a six-membered ring containing three double bonds. Arenes consist of one or more benzene rings.

Replacement of one or more hydrogen atoms in any of these hydrocarbons by a functional group results in formation of a *substituted* alkane, cycloalkane, or arene. Thus, substituting the bromine functional group for one of the hydrogen atoms in butane gives us bromobutane, an *alkyl bromide.* A similar operation on a cycloalkane or arene produces a *cycloalkyl bromide* or *aryl bromide,* respectively. Table 1–5 contains several examples of substituted hydrocarbons. When we want to designate *any* aliphatic bromide, regardless of the carbon skeleton, and focus only on the functional group, we use the generic symbol RBr. For aromatic bromides the designation is ArBr. The fact that we use two different generic symbols for these two types of bromide emphasizes the fact that their chemistry is often different.

Table 1–5. Some Substituted Aliphatic and Aromatic Hydrocarbons

$CH_3CH_2CH_2CH_2Br$ 1-bromobutane (an alkyl bromide)	bromocyclohexane (a cycloalkyl bromide)	bromobenzene (an aryl bromide)
$CH_3CH_2CH_2CH_2OH$ 1-butanol (an open-chain alcohol)	cyclohexanol (an alicyclic alcohol)	phenol (an arenol)
$CH_3CH_2CH_2CH_2\overset{\overset{O}{\parallel}}{C}OH$ pentanoic acid (an open-chain carboxylic acid)	cyclohexanecarboxylic acid (a cycloalkylcarboxylic acid)	benzoic acid (an arylcarboxylic acid)

When the functional group consists of one or two atoms, such as Br or OH, it is easily recognized. However, functional groups may consist of many atoms, some of which can be carbon. How then do we identify them? In Table 1–6 we have collected some representative organic functional groups and indicated some of the properties that make them important. All functional groups exhibit one or both of the following features:

1. The functional group may contain one or more **heteroatoms** (atoms other than C and H).
2. The functional group may contain one or more double or triple bonds.

If you look at the functional groups in Table 1–6 you can verify that each group meets one or both of the criteria listed above. The table also shows

Table 1–6. Structures, Names, and Examples of Some Important Functional Groups, Arranged by Properties

General structure	How Designated in Compound Name: Prefix	Suffix	Compound class	Specific example
Acidic Functional Groups				
—C(=O)—OH	carboxy-	-oic acid	carboxylic acid	CH_3—C(=O)OH ethanoic (*acetic*) acid
—S(=O)(=O)—OH	sulfonato-	-sulfonic acid	sulfonic acid	CH_3—S(=O)(=O)OH methanesulfonic acid
Basic Functional Groups				
—NH_2	amino-	-amine	primary amine	CH_3—NH_2 methylamine
—NHR	alkyl(aryl) amino-	-amine	secondary amine	$(CH_3)_2NH$ dimethylamine
Reactive Neutral Functional Groups				
—OH	hydroxy-	-ol	alcohol	CH_3CH_2—OH ethanol (*ethyl alcohol*)
>C=C<	alkenyl-	-ene	alkene	CH_2=CH_2 ethene (*ethylene*)
C—C(=O)—C	oxo-	-one	ketone	CH_3—C(=O)—CH_3 propanone (*acetone*)

that there is a relationship between the prefixes and suffixes in the *name* of a compound and the *functional groups* that are present in it. A good reason for learning proper organic nomenclature is that it requires you to analyze molecular structures in terms of functional groups.

It is generally true that compounds that contain the same functional group exhibit similar chemical and physical properties, provided that their carbon skeletons are not radically different. As an example, consider the constitutional isomers of formula C_3H_8O, shown in structures **14**, **15**, and

16. Compound **14** boils at 97°, compound **15** at 82°, and compound **16** at 7°.

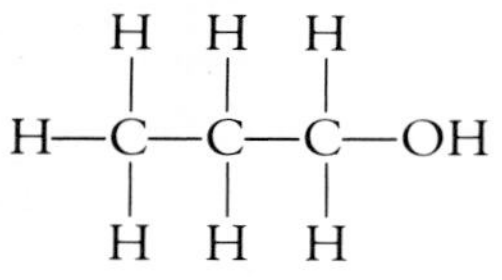

```
   H  H  H              H  H  H              H        H  H
   |  |  |              |  |  |              |        |  |
H—C—C—C—OH          H—C—C—C—H          H—C—O—C—C—H
   |  |  |              |  |  |              |        |  |
   H  H  H              H  OHH               H        H  H
```

($CH_3CH_2CH_2OH$) — 1-Propanol **14**

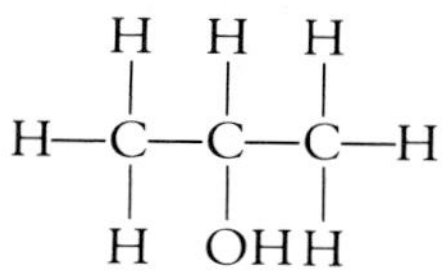

(CH_3CHCH_3 with OH on the middle carbon) — 2-Propanol **15**

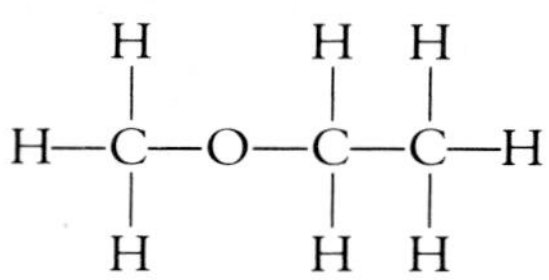

($CH_3OCH_2CH_3$) — Methyl ethyl ether **16**

Compounds **14** and **15** react with metallic sodium to release hydrogen while compound **16** is inert toward sodium. The formulas show that compounds **14** and **15** contain the hydroxy (OH) group, and thus they both belong to a class of compounds called alcohols. Compound **16** contains an alkoxy (OR) group and is a member of the ether class. Thus, both alcohols undergo a common reaction which is a property of the OH functional group. Because the OH group is not present in an ether, it does not react this way. The alcohols have boiling points close to each other, while the boiling point of the ether is much lower.

Problem

1–9 Circle all the functional groups in each of the following molecules. Draw line–angle formulas for each. Use Table 1–6 to identify the classes to which each compound belongs.

a. $CH_3CH_2CH_2I$

b. $CH_3CHCH_2CH_2CH_3$ (with NH_2 on the second carbon)

c. $ClCH_2CH_2OH$

d. CH_3CHCH_2COH (with Br on the second carbon and O double-bonded to the carbonyl carbon)

e. $CH_3OCH_2CH_2NHCH_3$

f. Cyclopentane ring bearing F and H on one carbon; CCH_3 (with =O) and H on the adjacent carbon; H and CH_2CH_3 on another carbon.

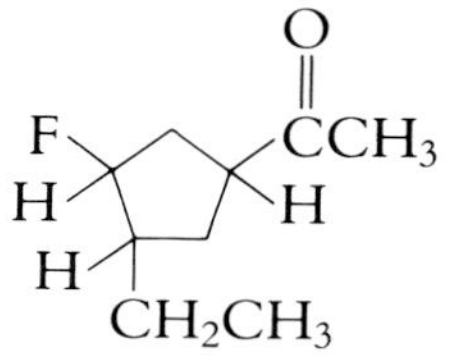

1:7 The Chemical and Physical Effects of Functional Groups: Molecular Acidity and Basicity

The nature of the functional groups present has a decisive effect on the physical and chemical properties of most compounds. In this section we will look at some examples which illustrate these effects. The concepts discussed here will reinforce our earlier statements about the importance of learning functional group properties and will be returned to many times in this text.

Butane (Table 1–4) is a low-boiling hydrocarbon (bp −0.5°) that has virtually no solubility in water. Introduction of a bromine atom, as in 1-bromobutane (Table 1–5), brings about a substantial increase in the boiling point (bp 101°). Like butane, however, 1-bromobutane is not very soluble in water. By contrast, the hydroxyl group in 1-butanol (Table 1-5) both increases the boiling point (bp 117°) *and* substantially increases water solubility. The increase in water solubility comes about because the hydroxyl group, like water itself, is both a donor and an acceptor of *hydrogen bonds.* Recall that hydrogen bonds are electrostatic bonds between partially positive hydrogen atoms and electronegative atoms, such as N and O, having lone pairs. Thus, both the O and H atoms in a hydroxyl group can take part in hydrogen bonding. These examples indicate the range of physical effects which depend on the character of the functional group present. We consider next an important chemical property which is dependent on functional group character.

Acidity and basicity are molecular properties that are highly significant, both chemically and biologically. Organic molecules that are acids or bases usually contain acidic or basic functional groups. We'll demonstrate this fact first by turning our attention to some common organic acids. In this discussion we will treat only those acids that act by donating protons (Brønsted-Lowry acids).

The most important organic acids are the **carboxylic acids,** compounds which contain the carboxyl group

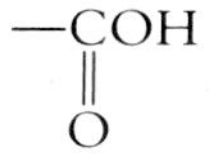

$$-\underset{\underset{O}{\|}}{C}OH$$

When these compounds dissolve in water, they partially dissociate by donating the carboxyl proton to a water molecule (eq. 1.4). Loss of a proton from a carboxylic acid produces a *carboxylate ion,* which is the *conjugate base* of the acid.

$$\text{(1.4)} \qquad CH_3\overset{\overset{O}{\|}}{C}\ddot{O}H + H_2\ddot{O}: \overset{K_a}{\rightleftharpoons} CH_3\overset{\overset{O}{\|}}{C}\ddot{O}:^- + H_3\ddot{O}^+$$

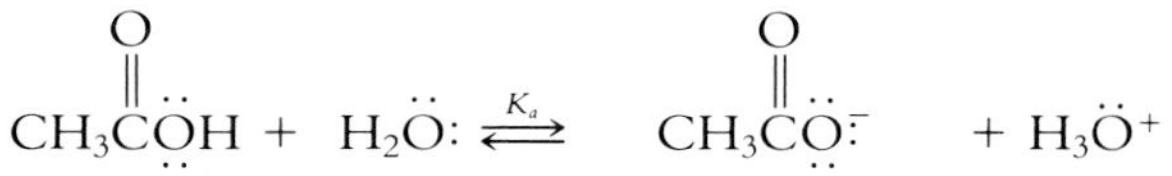

Acetic acid — Acetate ion (conjugate base)

The distinguishing structural feature of the carboxyl group is the hydroxyl (OH) group attached to a carbon doubly bonded to oxygen. This structural arrangement occurs in other types of acids, including inorganic ones in which heteroatoms replace carbon. Many of these acids are probably familiar to you, though you may not have thought about them in precisely this way. Some representative ones are shown here.

$CH_3-\overset{\overset{O}{\|}}{C}-OH$	$:\overset{\overset{O}{\|}}{N}-OH$	$HO-\underset{\underset{O}{\|}}{\overset{\overset{O}{\|}}{S}}-OH$	$HO-\underset{\underset{O}{\|}}{\overset{\overset{O}{\|}}{Cr}}-OH$
Acetic acid	Nitrous acid	Sulfuric acid	Chromic acid

The structural grouping

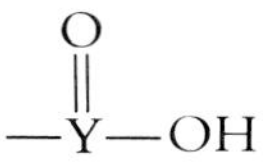

$$-\overset{\overset{O}{\|}}{Y}-OH$$

is not a requirement for acidity, as clearly demonstrated by HCl, HBr, and

HCN. However, when we do encounter a molecule with this structural grouping we can reasonably expect it to be an acid.

Like most carboxylic acids, acetic acid is not completely ionized in water. In fact, a majority of the molecules are undissociated at equilibrium. For this reason acetic acid is classified as a weak acid. Because the terms weak and strong are very qualitative, we need a quantitative measure of acidity in order to assess the effect of molecular structure on acidity. We can obtain this by measuring the **dissociation constant**, K_a, of each acid.

Equation 1.5 defines K_a for acetic acid. (Remember that the concentration of water is not changed measurably by the ionization process so it does not appear in the K_a expression.) Since K_a's are often numerically inconvenient, it is common practice to report acidities as pK_a's. The pK_a of acetic acid is defined and its numerical value given in Equation 1.6. It is instructive to

$$(1.5) \qquad K_a = \frac{[CH_3\overset{\overset{\displaystyle O}{\|}}{C}O^-][H_3O^+]}{[CH_3COOH]} = 1.76 \times 10^{-5} \qquad (25°)$$

$$(1.6) \qquad pK_a = -\log K_a = \log \frac{[CH_3COOH]}{[CH_3\underset{\underset{\displaystyle O}{\|}}{C}O^-]\,[H_3O^+]} = 4.75$$

compare the K_a and pK_a values for acetic acid (as given in eqs. 1.5 and 1.6) with the corresponding quantities for a strong acid such as HCl. HCl is almost completely ionized in water so its K_a is very large, about 10^7, and its pK_a is approximately −7. Thus strong acids are characterized by small pK_a's and weak acids by large ones (Table 1–7).

Our contention that functional groups are the principal determinants of chemical behavior is reinforced by comparing pK_a's for different carboxylic acids. Acetic acid and benzoic acid (Table 1–5) have quite different carbon skeletons, yet their pK_a's differ by only 0.5 units. So we see that it is the structure of the carboxyl group (the functional group) that is the major influence on acidity in these compounds.

Table 1–7. Structures and Acidities of Some Common Weak and Strong Acids

Name	Structure	K_a	pK_a
hydrobromic acid	HBr	10^9	−9
hydrochloric acid	HCl	10^7	−7
sulfuric acid	H_2SO_4	2×10^5	−5.2
nitric acid	HNO_3	2×10^1	−1.4
hydrofluoric acid	HF	6×10^{-4}	3.2
benzoic acid (Table 1–5)	C_6H_5COOH	6×10^{-5}	4.2
acetic acid	CH_3COOH	2×10^{-5}	4.8
hydrocyanic acid	HCN	8×10^{-10}	9.1
phenol (Table 1–5)	C_6H_5OH	1×10^{-10}	10.0

Organic bases represent another example of functional group reactivity. Most organic bases react by simply accepting a proton from an acid (Brønsted-Lowry bases). Equation 1.7 shows how a common organic base, an amine, accepts a proton from water. An equilibrium constant, K_b, analogous to the definition for K_a, can be defined for this reaction. Here again, it is convenient to use the quantity pK_b to characterize the strength of the base.

$$(1.7) \qquad \underset{\text{Methylamine}}{CH_3\ddot{N}H_2} + H_2O \overset{K_b}{\rightleftarrows} CH_3\overset{+}{N}H_3 + HO^{:-} \qquad K_b = 4.5 \times 10^{-4} \quad pK_b = 3.34$$

From Equation 1.7 you can see that it is the nonbonded electron pair on the nitrogen of methylamine that is responsible for the functional group property of basicity. Theoretically, any functional group with a nonbonded pair can accept a proton and act as a base. Organic amines are the most useful of such bases.

Summary

1 Organic chemistry is organized around covalent compounds of carbon bonded to hydrogen (hydrocarbons). Compounds containing heteroatoms, particularly oxygen and nitrogen, are also of prime importance. An empirical formula for an unknown organic compound is established from an elemental analysis. The percentages of carbon, hydrogen, and heteroatoms (except oxygen) in the compound are determined. The percentage of oxygen is calculated by the difference between the percentages of C, H, and N and 100. The molecular formula can then be established from the empirical formula and the molecular weight, which is usually determined by mass spectrometry (sec. 1:1).

2 Structural formulas show the bonding pattern among the atoms in organic compounds. They can be written once the molecular formula is known, using electron dots or lines to represent the bonding electrons. Structural formulas are deduced by relying on the normal valences of the constituent atoms (sec. 1:2). Charged species (cations or anions) and radicals can also be depicted by these methods and their formal charges calculated (sec. 1:3). A given molecular formula may fit several constitutional isomers which differ in bond connectivity order (sec. 1:5).

3 Most organic compounds consist of one or more functional groups attached to a carbon skeleton (sec. 1:6). Functional groups contain heteroatoms and/or multiple bonds. These groups usually make the main contribution to the reactivity of the molecule. An example is provided by compounds that show acidity and basicity (sec. 1:7). Compounds with the general structure

$$—\overset{\overset{\displaystyle O}{\|}}{Y}—OH$$

are acids. The most important class of organic acids is the carboxylic acids

$$R—\overset{\overset{\displaystyle O}{\|}}{C}—OH$$

These acids lose a proton to water and are converted to their conjugate bases. If the acid is relatively strong, the conjugate base will be relatively weak, and vice versa. Acid strengths are measured by means of dissociation constants, K_a's or pK_a's. The larger the K_a, the greater the extent of dissociation, and the stronger the acid. Organic bases such as amines accept a proton from water and generate OH^- ions.

Problems

1–10 Calculate empirical formulas for compounds having the following experimental elemental analyses.
a. C, 70.94%; H, 6.40%; N, 6.90% b. C, 65.92%; H, 7.75%

1–11 An unknown compound gives the following elemental composition upon analysis: C, 76.31%; H, 10.18%; N, 13.52%. Allowing for an error of ±0.3% in each determination, list three or four empirical formulas that fit the analytical data.

1–12 Calculate the mass of the molecular ion (molecular weight) for compounds having the following molecular formulas:
a. C_7H_{14} b. $C_6H_{10}O$ c. C_5H_6O

1–13 Draw pictures of the following molecules showing their three-dimensional structure. Use the wedge–dash convention. Build models of the compounds to aid you in visualizing their structures.
a. CH_3CH_2OH (ethanol)
b. CH_3NH_2 (methylamine)
c. $CH_3C(=O)CH_3$ (acetone)
d. $(CH_3)_2C(CH_3)_2$ (neopentane)
e. [pentagon] (cyclopentane)

1–14 Draw uncharged nonidentical structures (constitutional isomers) that could correspond to the following molecular formulas:
a. $C_2H_4F_2$ (two structures)
b. C_4H_{10} (two structures)
c. $C_2H_6O_2$ (four structures)
d. C_3H_4 (three structures)
e. C_4H_6 (six structures)
f. C_3H_9N (four structures)

1–15 Circle and name the functional groups on the following organic molecules:

a. $HCCl_3$ (H–C bonded to three Cl) (chloroform: a solvent and anesthetic)

b. CF_2Cl_2 (F–C(F)(Cl)–Cl) (Freon-12: an aerosol propellant)

c. $C_6H_5CH_2CH(CH_3)NHCH_3$ (methamphetamine: the drug "Speed")

d. 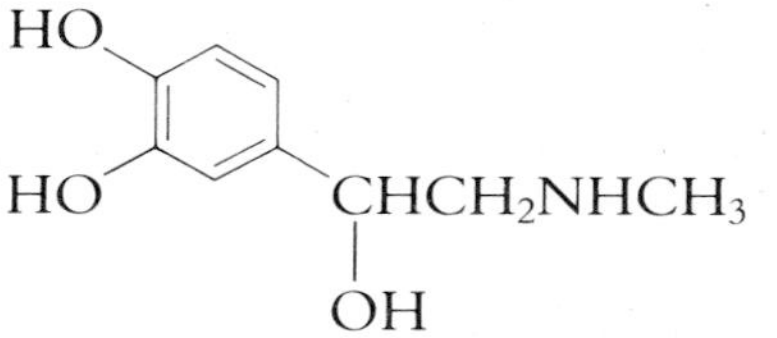 (adrenaline: an adrenal hormone)

e. 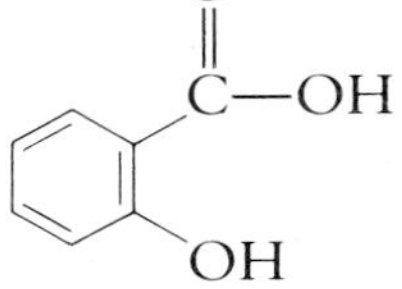 (salicylic acid: precursor of aspirin; also present as derivatives in many plants such as wintergreen)

f. 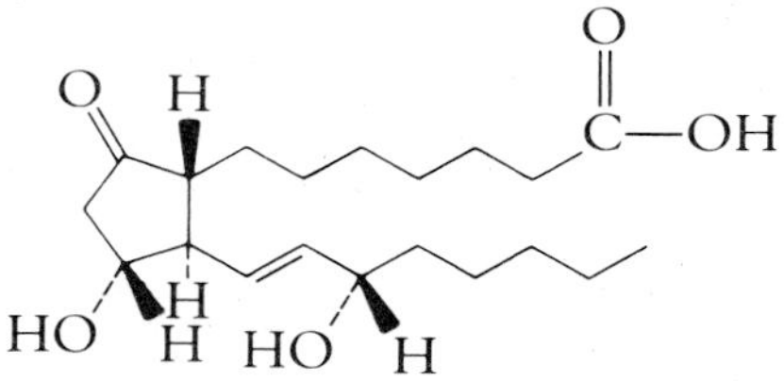 (prostaglandin E_1 [PGE_1]: a hormone that lowers blood pressure and causes uterine contractions)

1–16 Draw structures for the following organic ions:
a. $C_3H_7^-$ (two structures)
b. CH_5O^+ (one structure)
c. $C_3H_5^+$ (four structures)
d. $C_2H_8N^+$ (two structures)

1–17 Write the formulas of the conjugate bases of each of the acids listed in Table 1–7 (sec. 1:7).

1–18 The salt sodium acetate, $CH_3COO^-Na^+$, dissociates completely into sodium and acetate ions when it dissolves in water. When sodium acetate is added to an aqueous solution of HF (Table 1–7), a reaction takes place between the acid and the salt. What are the products? Explain why the reaction occurs.

1–19 Write a structural formula for formic acid, CH_2O_2, showing all the bonding and nonbonding electron pairs. In a 1 *M* solution of formic acid in water (at 25°), about 1.3% of the molecules are dissociated. Use this information to calculate approximate values of K_a and pK_a for formic acid.

1–20 Explain why 1-bromobutane has a higher boiling point than butane (sec. 1:7). Based on your explanation indicate what you would expect the boiling point of 1-chlorobutane to be.

Supplementary Reading

O. T. Benfey, *From Vital Force to Structural Formulas,* Houghton Mifflin, Boston, Mass., 1964.

W. H. Trahanovsky, *Functional Groups in Organic Compounds,* Prentice-Hall, Englewood Cliffs, N.J., 1971.

O. T. Benfey, *The Names and Structures of Organic Compounds,* John Wiley & Sons, New York, N.Y., 1966.

B. H. Mahan, *University Chemistry,* Addison-Wesley, Reading, Mass., 1975.

T. L. Brown and H. E. LeMay, Jr., *Chemistry: The Central Science,* Prentice-Hall, Englewood Cliffs, N.J., 1981.

J. C. Bailar, Jr., *et al., Chemistry,* Academic Press, New York, N.Y., 1978.

G. E. Maciel, D. D. Traficante, and D. Lavallee, *Chemistry,* D. C. Heath, Lexington, Mass., 1978.

2 Molecular Bonding: Bond Angles, Bond Energies, and Bond Lengths

Our aim throughout this book is to enable you to understand and to predict the course of organic reactions. To do this you need a knowledge of the structure of organic molecules. In Chapter 1 we reviewed the method for calculating the number and placement of covalent electron-pair bonds in molecules. We need now to explain the forces that make those bonds possible, and which determine three dimensional molecular shapes. The simple electron-pair bonding model can be used to rationalize the shapes of molecules. After doing that we'll delve more deeply into the nature of chemical bonding, which will carry us beyond the Lewis electron pair theory.

2:1 How Valence Shell Electron Pair Repulsions Determine Molecular Shapes

An atom in a molecule bonds to the other atoms by sharing electron pairs with them. Since all electrons are negatively charged they repel one another. Thus, the various electron pairs occupying the valence shell of an atom will tend to be as mutually far apart as possible, thereby minimizing repulsion. Suppose that we have four valence electron pairs around a particular atom, such as the carbon atom in methane, CH_4. What geometry is predicted to give the minimum repulsion among them? That geometry in which the four pairs point toward the four corners of a regular tetrahedron. Experimental determination of methane's geometry reveals that it is indeed tetrahedral (Fig. 2–1). The four hydrogen atoms are equivalent and occupy the corners of the tetrahedron, with carbon at the center. The H—C—H **bond angles** are all equal to 109.5°. (The bond angles are defined by straight lines connecting the hydrogen atoms to the carbon atom; Fig. 2–1.)

The fact that methane is tetrahedral supports the idea that molecular

Figure 2–1. *The tetrahedral structure of methane. The hydrogen atoms lie at the vertices of the tetrahedron and the carbon atom is at its center. All H—C—H bond angles equal 109.5°.*

shape is determined by *valence shell electron pair repulsion* (VSEPR). However, since one experimental result is not sufficient to establish a hypothesis, we need to look at more examples before accepting VSEPR theory. With a three-electron-pair atom we would expect minimum repulsion when the electron pairs point toward the corners of an equilateral triangle which has the atom at its center. Once again, prediction and experiment agree. Measurements on boron trifluoride, BF_3 (**1**), show that it has trigonal planar geometry, with F—B—F bond angles of 120°.

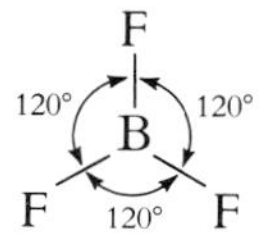

1 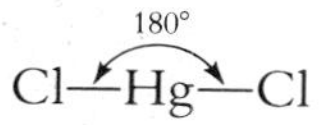 **2**

As a final test we consider a molecule in which there are two valence shell electron pairs around the central atom. Theory predicts that the central atom and the electron pairs will be colinear. The covalent molecule mercuric chloride, $HgCl_2$ (**2**), is found to have a bond angle of 180°, once again demonstrating complete agreement between theory and experiment.

Problem

2–1 What would the bond angles be in methane if the four C—H bonds all lay in the same plane? Explain why this planar methane would be higher in energy than the observed tetrahedral methane.

The examples just given suggest some general guidelines:

1. The bond angles around a particular atom are determined by the number of electron pairs in the valence shell of that atom.
2. Two electron pairs result in a 180° bond angle; three electron pairs bring about 120° bond angles; and four electron pairs require bond angles of 109.5°. These bond angle values are *ideal* values.

BF_3 and CH_4 each have the ideal bond angle values for the number of valence electron pairs. This is so because in BF_3 all B—F bonds are equivalent and in CH_4 all C—H bonds are equivalent.

A molecule in which the electron-pair bonds are *not* all the same is chloromethane, CH_3Cl. The C—Cl bond is not equivalent to the three C—H bonds, so the bond angles for H—C—Cl and H—C—H bonds differ (**3**). These bond angle differences depend on several factors—differences in bond lengths, in atomic size, and in ability to attract electrons. However, the major point is that chloromethane resembles a tetrahedron sufficiently for us to classify it as such.

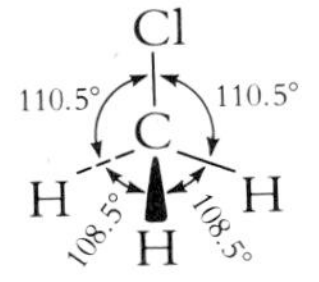

3  **4**

What happens in the case of molecules with atoms having nonbonding pairs (sec. 1:2)? An example is ammonia, $:NH_3$, which has three bonding and one nonbonding valence pairs. The actual H—N—H bond angle is 107° (**4**), which is less than the ideal 109.5°, a tetrahedral angle. This indicates that a nonbonding electron pair repels the electron pair of an H—N bond more strongly than does another H—N bonding pair. Lone pair repulsion forces the three bonding pairs closer together, thereby decreasing the H—N—H bond angle.

The electron pair repulsion theory also works for molecules with multiple bonds. The two electron pairs that make up a double bond may be treated like a single pair. Therefore, we would predict that in formaldehyde, $H_2C{=}O$, the geometry would be close to trigonal. The actual structure (**5**) confirms the prediction.

Formaldehyde
5

The Lewis electron pair theory is very useful in explaining much of what is known about molecular bonding and geometry. However, more sophisticated treatment of electron distribution in atoms and molecules is necessary to appreciate the full range of organic chemical phenomena. Such a treatment is outlined in the following sections.

Problems

2–2 The H—O—H angle in water is found to be 104.5°. What would be the ideal bond angle for a molecule with this electron distribution? Explain the difference between the ideal and experimental bond angle values.

2–3 Give values for all the bond angles in each of the following molecules and ions. If you cannot propose a precise value, then give approximate answers such as "less than 120°" or "greater than 109.5°."

a. CCl_4 (tetrachloromethane)
b. NH_4^+
c. CH_2Cl_2 (dichloromethane)
d. $HBCl_2$
e. CH_3OH (methanol)
f. BF_4^-
g. $ZnCl_2$
h. C_2H_4 (ethylene)
i. C_2H_2 (acetylene)
j. CH_3^+ (eq. 1.3)
k. CH_3^-: (Prob. 1–4)

2:2 How Electrons are Distributed in Atoms: Electron Levels and Orbitals

Atoms are held together by the attraction between their negatively charged electrons and positively charged nuclei. All electrons in the atom do not have the same energy. Some are closer to the nucleus and are lower in energy than those farther away. The idea of electrons arranged in levels, or shells, that are at different distances from the nucleus gives a physical explanation for the modern Periodic Table.

We know that in an atom the electron energies are **quantized** (have certain specific values), which means that electrons are found at specific distances from the nucleus. Because of the charge interaction between the positive nucleus and the negative electron, the closer an electron is to the nucleus, the lower the electron's energy. Electrons of lowest energy occupy the first level, denoted by the quantum number n which equals 1. The electrons of next highest energy occupy the second level, for which $n = 2$, and so on.

Within each energy level we find **atomic orbitals** that define the *specific* energy state of an electron and the *specific* region an electron occupies. Thus, each electron is in a specified energy state, called an **orbital.** It is very useful to create a physical picture of an orbital because orbitals are crucial to an understanding of bonding. To do that we need to examine the nature of the electron itself.

It is common to think of electrons as tiny, negatively charged particles. Many properties of electrons fit this model, such as the fact that they can be "fired" from electron "guns" like the ones that generate the picture on your TV screen. This so-called *particle model* of the electron allows us to think in terms of the probability of locating an electron as it moves in its orbit. Modern physical theory has shown that some of the electron's properties are best explained, however, by treating the electron as a *wave.* We will use the wave model often in our discussion of atomic structure and bonding. This does not imply that the wave model is "better" than the particle model; each is effective in explaining some electron properties but not others. The wave–particle duality is not confined to electrons; it applies to protons and neutrons as well, and even, as we'll soon see, to radiant energy.

Like a water wave, an electron wave is spread out in space and is sometimes referred to as a *charge cloud.* As an example, consider the first electron level which contains only one orbital, the *s* orbital. The 1*s* orbital is spherical in shape, with the nucleus at its center (Fig. 2–2a). The 1*s* electron is spread out within this sphere, but not uniformly. Electron waves have regions of greater and lesser amplitude, leading to variations of electron density within the orbital. Where the electron wave has its highest amplitude, the electron

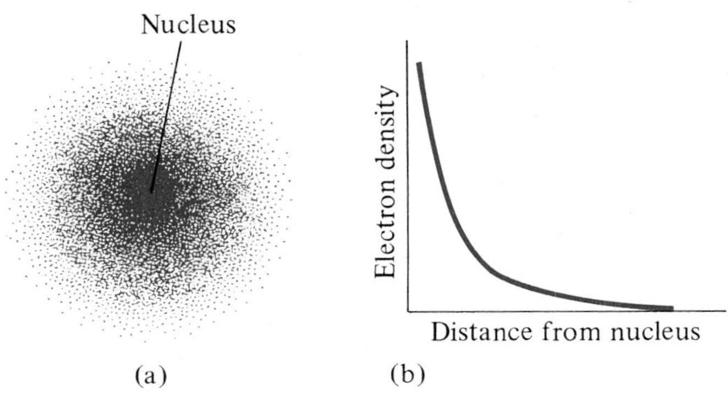

Figure 2–2. *(a) The electron density distribution in a 1*s *orbital. (b) A plot of electron density vs. distance from the nucleus for a 1*s *electron. The electron density is highest at the nucleus and decreases uniformly in all directions.*

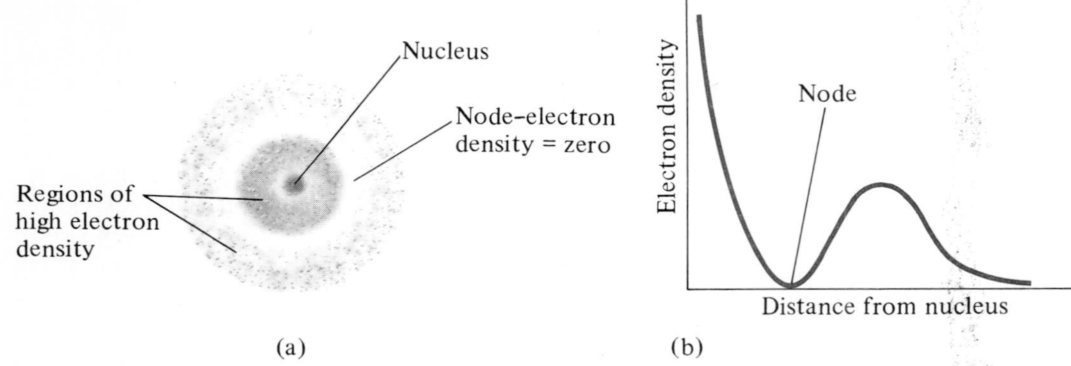

Figure 2–3. *(a) The electron density distribution in a 2*s *orbital. (b) A plot of electron density vs. distance from the nucleus for a 2*s *electron. This orbital contains a spherical node, a region where the electron density goes to zero.*

density (concentration of negative charge) is greatest. In a 1*s* orbital the electron density is greatest right next to the nucleus and drops off steadily as the distance from the nucleus increases (Fig. 2–2b). If we were treating the electron as a particle, we would say that the location of highest probability for a 1*s* electron is next to the nucleus.

Note that when orbitals are depicted with line drawings, such as a circle to represent a 1*s* orbital, the line is usually chosen to include 90 or 95% of the electron density. The orbital has no actual border; the electron density just keeps diminishing with increased distance from the atomic center.

The second electron level contains a 2*s* orbital which is also spherical, but differs from a 1*s* orbital in one important respect. Within it is a spherical surface where the electron density is zero (Fig. 2–3). Such a surface is called a **node** (the electron wave has an amplitude of zero). In addition to the 2*s* orbital the second electron level also contains 2*p* orbitals. They lack the spherical symmetry of the *s* orbitals and are much more directional in shape. The 2*p* orbitals each consist of two lobes separated by a nodal plane that

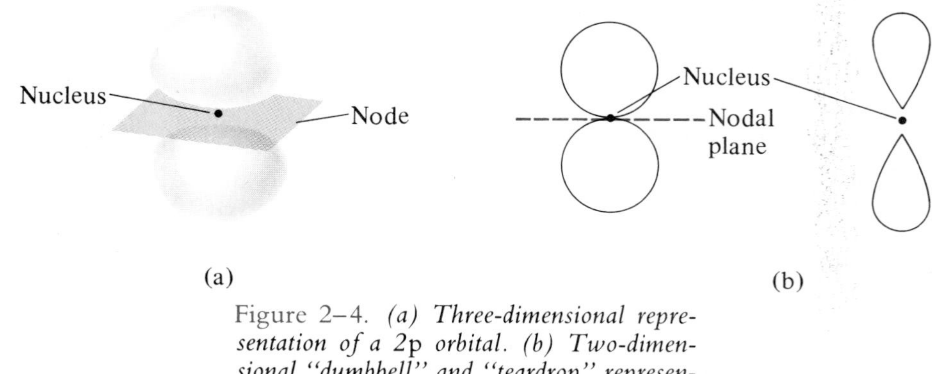

Figure 2–4. *(a) Three-dimensional representation of a 2*p *orbital. (b) Two-dimensional "dumbbell" and "teardrop" representations of a 2*p *orbital.*

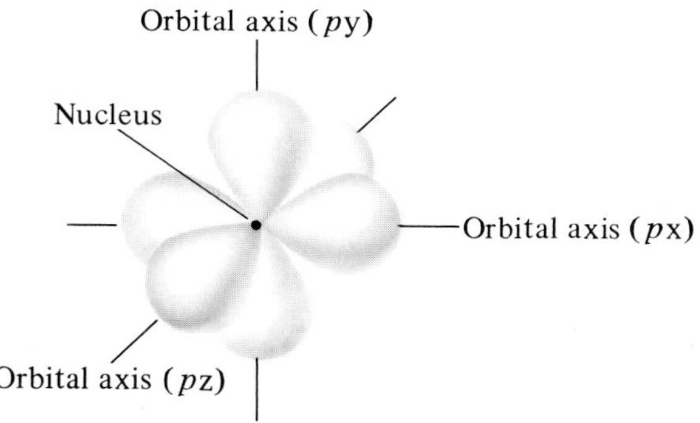

Figure 2–5. *Arrangement of the three* 2p *orbitals. The axes of the three orbitals are mutually perpendicular.*

passes through the nucleus (Fig. 2–4). There are three 2*p* orbitals, oriented so that their axes are mutually perpendicular (Fig. 2–5). The 2*s* and 2*p* orbitals differ in energy, the *p* orbitals having the higher energy. This difference is much smaller than that between 1*s* and 2*s* orbitals.

Each orbital can accommodate a maximum of two electrons. Knowing this restriction and the number of orbitals in each level allows us to explain the rules governing the use of Lewis electron pair structures. The lowest energy state of an atom, its **ground state**, is that in which all its electrons are in the orbitals of lowest possible energy. Figure 2–6 shows the electronic configuration corresponding to the ground states of the elements hydrogen through neon. Here the electrons are represented by arrows (see p. 40).

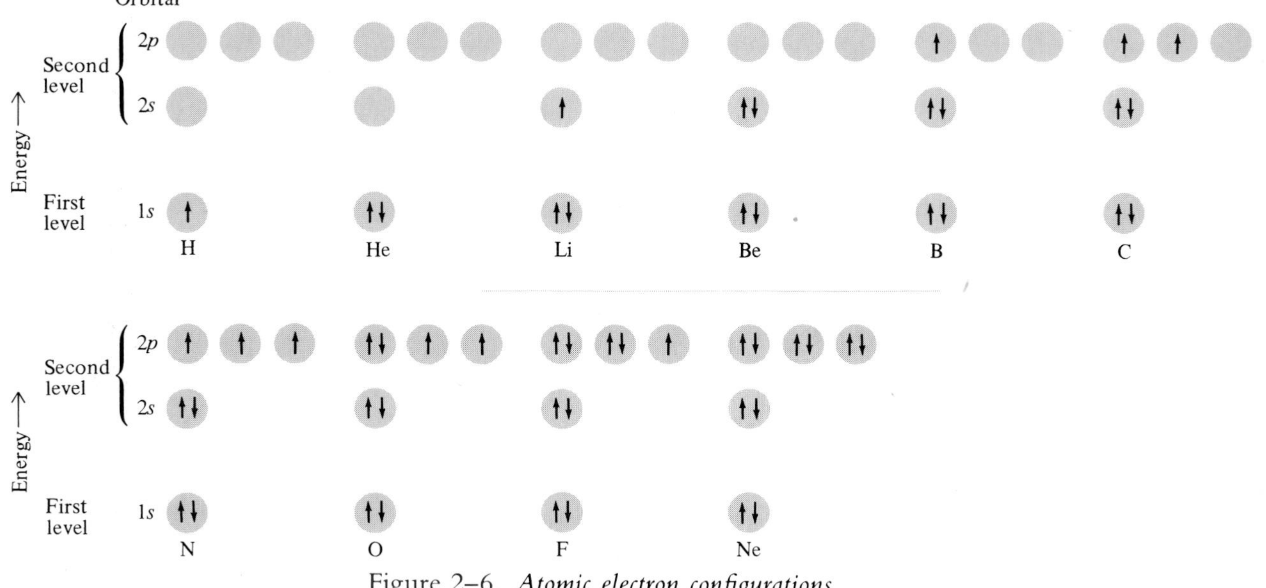

Figure 2–6. *Atomic electron configurations of the first 10 elements.*

Hydrogen tends to acquire an additional electron by sharing, thus filling its 1*s* orbital and therefore its first level. Helium is unreactive because its 1*s* orbital is full. The elements lithium through fluorine have partially filled second levels which govern their chemistry. Lithium can lose its 2*s* electron and attain the stable helium configuration. Carbon and fluorine, on the other hand, can fill their 2*s* and 2*p* orbitals by sharing or acquiring additional electrons. It is the electron capacity of the 2*s* and 2*p* orbitals which accounts for the octet rule (see sec. 1:2).

Each atom has only one ground state (Fig. 2–6) but a very large number of possible **excited states**. An excited state corresponds to any electron configuration other than the ground state one. For instance, one possible excited state for carbon has the electron configuration $1s^22s2p^3$.

Problems

2–4 The **ionization energy** of an atom is the *minimum energy* needed to remove an electron from that atom and convert it to an ion (eq. 1).

$$A \xrightarrow{\text{energy}} A^+ + e^- \tag{1}$$

a. The ionization energy of He is 567 kcal/mol, while that of Li is 124 kcal/mol. Explain why the second number is so much smaller than the first. Show the electronic configurations of the helium and lithium ions that are produced.

b. It is also possible to remove an electron from Li at very high energy, 1260 kcal/mol. What process does this correspond to? Show the electron configuration of the lithium ion produced by this process. What is its relation to the lithium ion produced by the low energy ionization?

2–5 Write out the electron configurations of the elements sodium through argon.

The rule stating that each orbital can accommodate a maximum of two electrons is subject to an important restriction—the two electrons must be of opposite **spin** (*Pauli exclusion principle*). Spin is a property assigned to electrons on the basis of their magnetic properties, and is derived from the particle model. According to this model, the spinning electron acquires a magnetic moment when placed in an external magnetic field. The orientation of this magnetic moment in the external field is quantized—it can only be aligned with or against the external field (Fig. 2–7). We symbolize these two different spin states by means of arrows which point up ($\uparrow$) or down ($\downarrow$), as shown in Figure 2–6.

The tendency for electrons to become spin paired ($\uparrow\downarrow$) arises naturally because it allows the maximum number of electrons to enter the lower energy orbitals. When spin-paired electrons occupy the same orbital, their magnetic moments cancel, lowering their energy compared to the (forbidden) situation in which electrons with the same spin ($\uparrow\uparrow$) would occupy a single orbital. The only circumstance that favors unpairing of spins is that in which two or more electrons have two or more **degenerate orbitals** available to them. Degenerate orbitals have identical energies. The three 2*p* orbitals are a familiar example of orbital degeneracy. The lowest energy electron

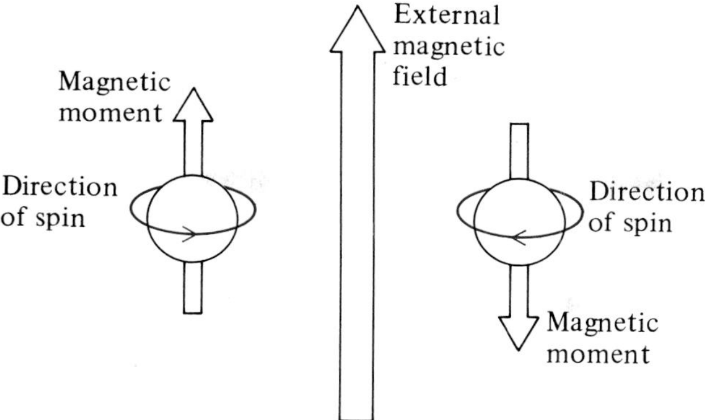

Figure 2–7. *Allowed orientations of the electron magnetic moment in an external magnetic field.*

configuration in that case has the degenerate orbitals singly occupied by electrons of the same spin (*Hund's rule;* see C and N, Fig. 2–6).

We must draw attention to one more property of electron orbitals that is vital to understanding molecular bonding. The mathematical equations that describe the various orbitals are called *wave functions,* because they have the same form as equations that describe hydrodynamic waves such as water waves. For orbitals with nodes, the wave function has a different sign ("+" or "−") on opposite sides of the nodal surface. In Figure 2–8, for instance, the upper lobe of the 2*p* orbital has been given a positive sign, in which case the wave function would be given a negative sign in the lower lobe. The "+" and "−" signs do *not* refer to charge; they are used only to indicate the different lobes.

The change of sign on opposite sides of a nodal plane can be directly related to the behavior of ordinary water waves. Like water waves, a typical electron wave has crests and troughs, or *phases,* which can also be labelled "+" and "−" (Fig. 2–9). When electron waves cross a node they change sign. The wave function signs have relatively little significance in the isolated

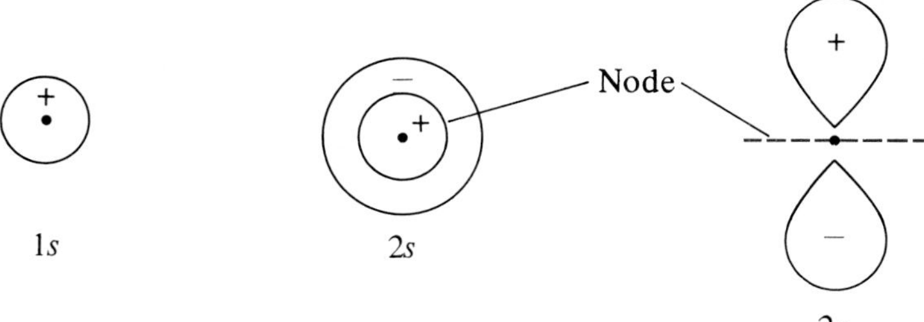

Figure 2–8. *Nodes and boundaries of first and second level orbitals showing the change in wave function sign when a node is crossed. The "+" and "−" signs do not signify charge. They indicate different sides of the nodal surface, also indicated by color.*

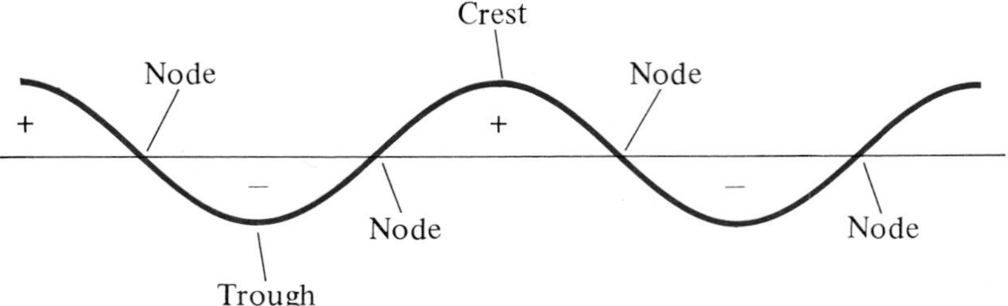

Figure 2–9. *Phases of a standing wave. Note the change in sign of amplitude on opposite sides of a node.*

atoms, but great significance when atoms combine to form molecules, as we will see in Section 2:3.

In the third electron level we find, in addition to 3*s* and 3*p* orbitals, five 3*d* orbitals. The *d* orbitals have important roles to play in the chemistry of phosphorus, sulfur, and the transition elements (chaps. 21 and 26).

2:3 How Molecular Orbitals Are Formed from Atomic Orbitals

When atoms join to form molecules, their atomic orbitals overlap one another in space. *This process is the key to bond formation.* The overlapping atomic orbitals combine to form **molecular orbitals.** A molecular orbital is a region of high electron density, corresponding to a specific electron energy, that encompasses two or more nuclei. Figure 2–10 illustrates the formation of a molecular **σ orbital** (sigma orbital) that accompanies the reaction of two H atoms to yield molecular H_2. Sigma orbitals don't have a node in the bonding region between the atoms.

Molecular orbitals, like atomic orbitals, can accommodate a maximum of two electrons, and their spins must be paired ($\uparrow\downarrow$). The two electrons that were in the 1*s* orbitals of the H atoms are now in the σ orbital of the H_2 molecule. Electrons have less energy (greater stability) in the molecular orbital than they did in the atomic orbitals, and hence, bond formation is favorable. One reason for this increased stability is that electron density can accumulate in the region between the two positive nuclei, where it is subjected to favorable attraction by both of them. Also, the larger size of the molecular orbital compared to the atomic orbital allows each electron to be more spread out upon molecule formation. This spreading out, or **delocalization**, of electron density has an extremely important stabilizing effect, which we will refer to again later.

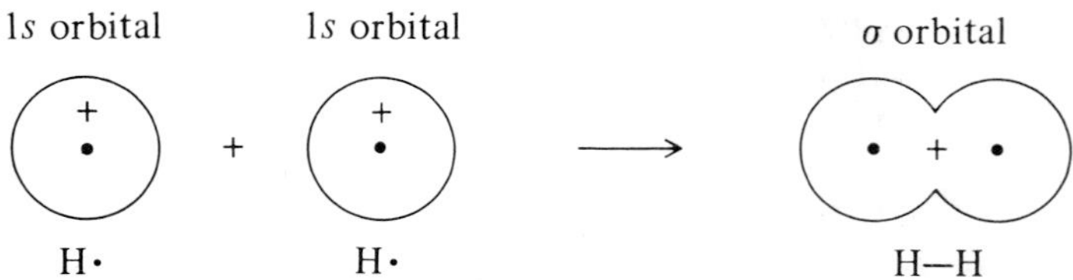

Figure 2–10. *Formation of the H_2 σ orbital by overlap of two 1s orbitals. The molecular orbital encompasses both hydrogen nuclei.*

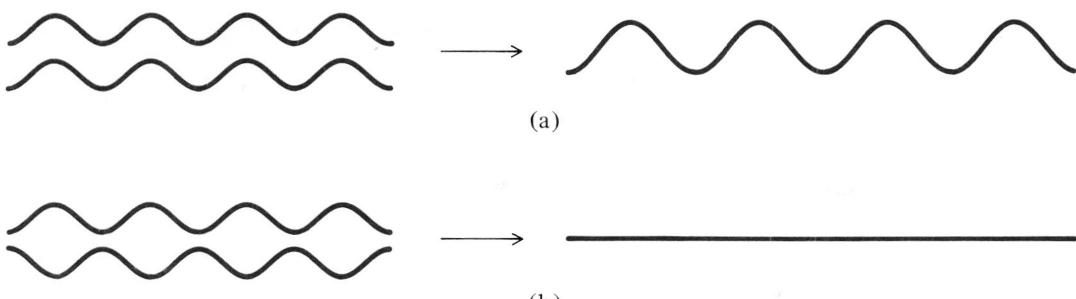

Figure 2–11. *(a) Reinforcing effect of two waves in phase. (b) Cancelling effect of two waves out of phase.*

Formation of a bond between the two atoms is subject to an important restriction. *The two atomic orbitals must have the same sign in the internuclear region where they overlap.* To understand why, we again compare the wave nature of electrons in atomic orbitals to water waves. If two water (electron) waves are in phase, that is, if their crests and troughs occur in the same place at the same time, then they reinforce each other and a new wave of greater magnitude is formed (Fig. 2–11a). This corresponds to two atomic orbitals in phase combining to form a bond between the two atoms. But if a crest of one wave interacts with a trough of the second wave, then there is interference between the two and they will cancel each other out if they are of identical amplitude (Fig. 2–11b). Similarly, bonding cannot occur between two out-of-phase atomic orbitals.

Sigma bonds can also result from the overlap of an *s* orbital with one lobe of a *p* orbital, as in the formation of HF (Fig. 2–12). Two *p* orbitals can also yield a σ orbital by allowing the end-to-end overlap of one lobe of each with the same sign, as in the formation of F_2 (Fig. 2–13). Two *p* orbitals on neighboring atoms may also combine so that *both* lobes on one atom can overlap with the two lobes on the other. This side-by-side overlap of two *p*

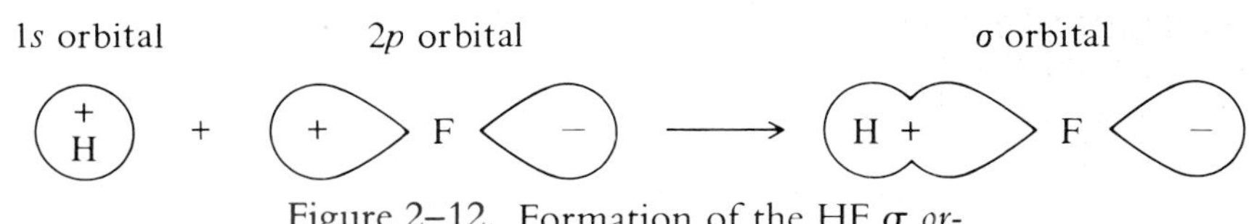

Figure 2–12. Formation of the HF σ *orbital by overlap of an H 1s orbital with one lobe of a F 2*p *orbital.*

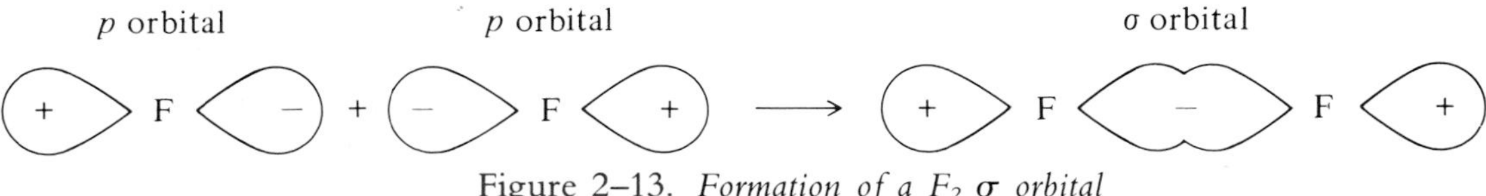

Figure 2–13. *Formation of a F_2 σ orbital by overlap of one lobe of each of two F 2*p *orbitals.*

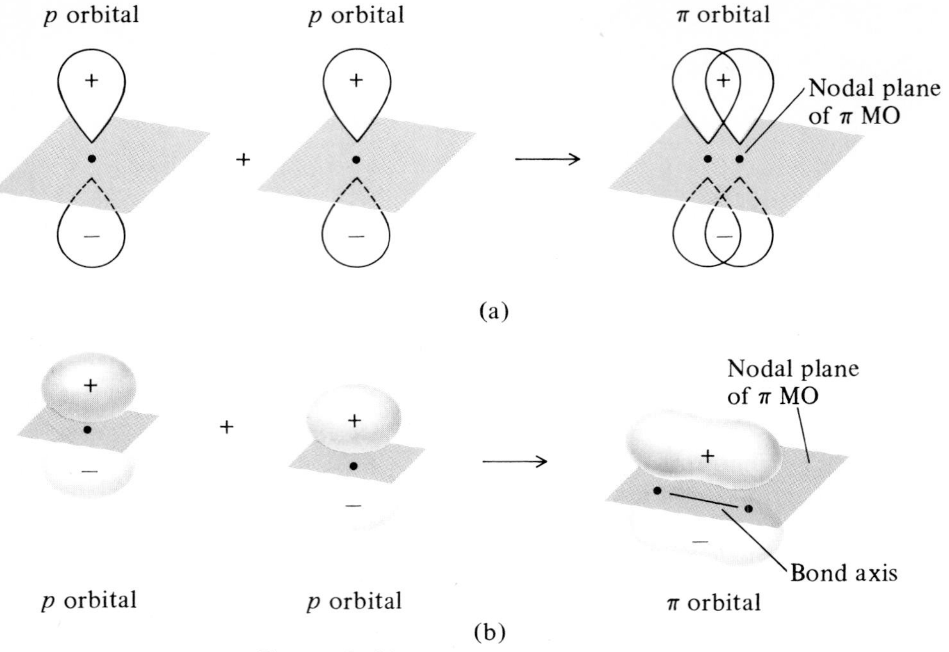

Figure 2–14. *(a) Cross-sectional and (b) spatial representations of π orbital formation by side-by-side overlap of two p orbitals. The molecular orbital (MO) has two lobes separated by a nodal plane.*

orbitals (Fig. 2–14) creates a new type of molecular orbital called a **π orbital** (pi orbital). The π orbital has a planar node that incorporates the bond axis. We classify σ and π orbitals as **bonding molecular orbitals** (MO).

In each of the above examples (Figs. 2–10 and 2–12 through 2–14) two atomic orbitals have been combined to form one molecular orbital. This is not the entire picture presented by MO theory. A basic postulate of the theory states: *When atomic orbitals combine to form molecular orbitals, the number of molecular orbitals formed equals the number of atomic orbitals used.* This means, for example, that there is a second molecular orbital in addition to the bonding σ orbital that is formed when two hydrogen atoms come together to produce H_2. This second orbital is of higher energy and does not contribute to bonding. It results from two out-of-phase electron waves canceling each other (Fig. 2–11b), and is therefore called an **antibonding molecular orbital** (MO★). In the case of H_2 a σ★ orbital (sigma star orbital) is formed, which contains a node in the bonding region. Figure 2–15 shows how both orbitals are formed in the hydrogen molecule. Note that both electrons go into the bonding orbital in the ground state and the antibonding orbital is vacant. However, if the H_2 molecule was to absorb energy, such as light from an ultraviolet lamp, then one electron could be promoted from a bonding orbital to the antibonding orbital. The resultant molecule, which is in an excited state, has extra energy and can decompose or react. Plant photosynthesis, for example, is initiated by a photoexcited molecule of chlorophyll, and many chemical reactions can be performed in the laboratory by

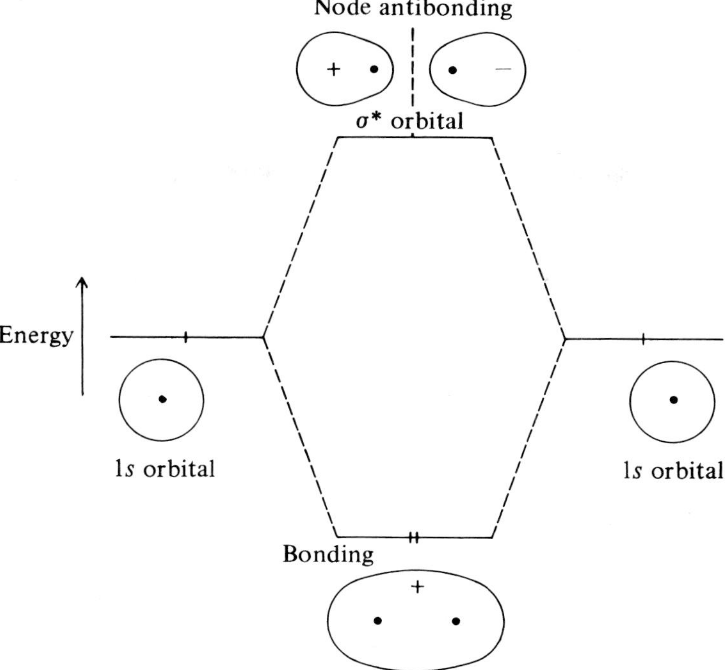

Figure 2–15. *Formation of bonding (σ) and antibonding (σ★) molecular orbitals by overlap of hydrogen 1*s *orbitals. The relative energy levels of the atomic and molecular orbitals are shown.*

irradiating molecules with ultraviolet light. **Photochemistry** is the study of reactions that occur when molecules are excited by the absorption of ultraviolet or visible light energy.

Example

Draw and label the atomic orbital from each atom that is used to form the molecular σ orbital of the diatomic molecule Li_2.

Answer

From Figure 2–6 you can see that a lithium atom has a 2*s* electron available for sharing in a covalent bond; hence the Li—Li molecular orbital will be similar to that of H—H in Figure 2–10.

2*s* + 2*s* ⟶ σ orbital

Li· + Li· ⟶ Li—Li

Problem

2–6 For each of the molecules below, draw and label the atomic orbital on each atom that is used to form each bonding molecular σ orbital.

a. Na_2 b. Cl_2 c. HBr d. ICl e. H_2S f. H_2O

Box 2–1. A Faithful Couple: The Electron Pair

Since the time of Michael Faraday (1791–1867), the Royal Institution in London has invited well-known scientists to give *discourses,* which are presentations of scientific ideas to non-specialists. In 1977 a discourse entitled "A Faithful Couple: The Electron Pair" was given by the theoretical chemist L. Salem of the University of Paris.* Salem used everyday terms to describe electron-pair bonding and the wave nature of electrons. Perhaps an excerpt will aid in your understanding of what we have been saying in the past few sections.

"Let us now consider the encounter of two atoms—and of their waves—to form a molecule. Let us, for instance, bring two ball-shaped waves together. If these were waves in the sea, there would just be one big splash. Here, however, two new waves are formed which extend over both atoms. One wave is a large crest which covers the entire molecule. Electron couples are particularly happy to enter such a crest, where they find additional stability. In a sense the favorable nature of this wave is due to its larger extension. At the same time a second wave is formed, in which, as if by magic (but nothing is magic in science), one crest has become a trough. Such a wave is extremely unfavorable for accommodating electron couples. Electron couples will do their utmost to avoid having to enter such a wave, whose effect is disastrous.

We might say that the first type of wave is an "auspicious" wave, or simply a "good" wave. The second type of wave we could call "ill-fated" or simply "bad." The creation of these two waves is due to an "interference" effect. It is somewhat similar to the result of throwing two stones in a pond: the ripples may add, or they may destroy each other."

* *J. Chem. Ed.*, **55**, 344 (June 1978).

2:4 Explaining Molecular Geometry Using MO Theory: The Need for Hybrid Orbitals

VSEPR theory is a good tool for predicting molecular shape, but it does not provide us with any insights into molecular reactivity. In this section we will first show that molecular orbital theory can predict geometries as well as VSEPR theory can. We will then show that molecular orbital theory gives us a more detailed look at electron distributions in molecules, which enables us to understand and predict chemical reactivity. We begin with a discussion of *localized molecular orbitals*—MO's confined to only two atoms. In Chapter 13 we introduce *delocalized molecular orbitals,* which extend over more than two atoms.

In order to explain molecular geometry, we focus our attention on the *p* orbitals, since they have definite spatial orientations (Fig. 2–5). Let's assume that methane (CH_4) bonding involves the overlap of four hydrogen 1*s* orbitals with the three 2*p* orbitals and one 2*s* orbital of carbon. Unfortunately, such an approach cannot explain the empirical reality of methane's tetrahedral geometry (Fig. 2–1). This assumption would predict H—C—H bond angles of 90°, which is the angle between the *p* orbital axes (Fig. 2–5). It would also predict that the C(2*s*)—H bond should be different from the C(2*p*)—H bonds. However, *all physical and chemical evidence suggests that the four C—H bonds of methane are equivalent.*

This reality forces us to reject the assumption that the electron configuration of an *isolated* carbon atom is the one that is optimal for bonding. For instance, when a σ bond is formed between an *s* orbital and a *p* orbital, only one lobe of the *p* orbital is used (Fig. 2–12). From the bonding standpoint, half the *p* orbital is wasted. What we require then is a bonding state of carbon that makes maximum use of the valence orbitals and produces four equivalent C—H bonds.

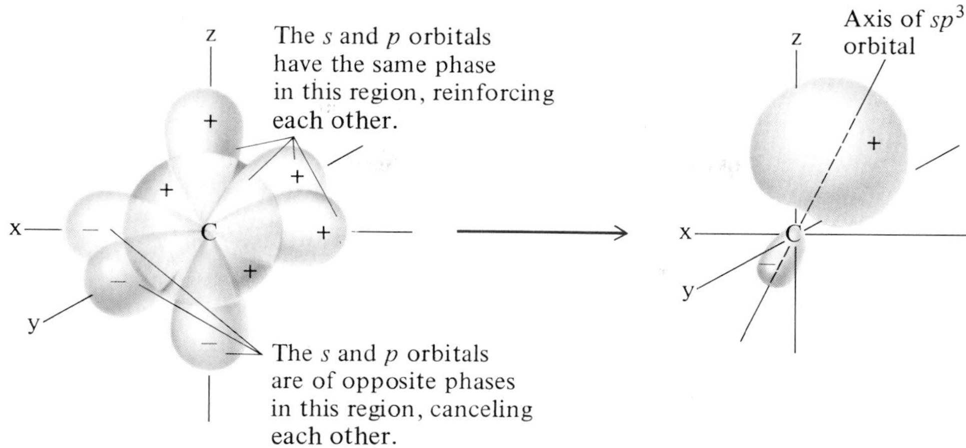

Figure 2–16. *Formation of a hybrid* sp^3 *orbital by mixing of the spatially overlapping portions of the 2*s *and 2*p *orbitals.*

This goal is achieved by the formation of **hybrid orbitals**. Hybrid orbitals are atomic orbitals that can be derived by mixing together the pure *s*, *p*, and *d* orbitals of the isolated atom. The mixing process is governed by the wave character of the electrons. Atomic orbitals can only be mixed if they overlap in space. Furthermore, there will be a reinforcement of the electron waves when they are in phase but a diminution when they are out of phase (Fig. 2–11). Let's apply these ideas to the 2*s* and 2*p* orbitals of carbon.

We see in Figure 2–16 that the in-phase lobes of the p_x, p_y, and p_z orbitals can mix successfully with the part of the *s* orbital that is bounded by them. They produce a new positive lobe strongly concentrated along a line at 45° to each axis. The out-of-phase *p* orbital lobes cancel the part of the *s* orbital in their vicinity, yielding a small negative lobe opposite the large one. These two lobes together constitute a hybrid ***sp*^{3} orbital**. Since there are four unhybridized atomic orbitals, they will combine to produce four *sp*^{3} orbitals, whose axes are oriented at 109.5° to one another (Fig. 2–17). The four hybrid orbitals are equal in energy, lying between the 2*s* and 2*p* orbitals (Fig. 2–18). Each orbital has 25% *s* character and 75% *p* character, and lies three-quarters of the way between *s* and *p* orbitals on a relative energy scale.

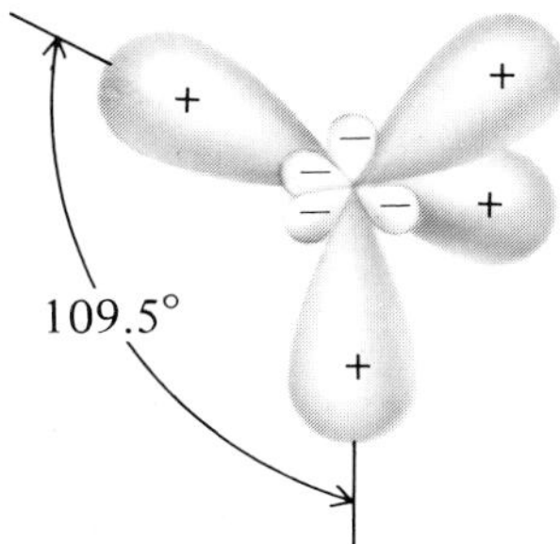

Figure 2–17. *Four* sp^3 *orbitals, whose axes make angles of 109.5° with one another.*

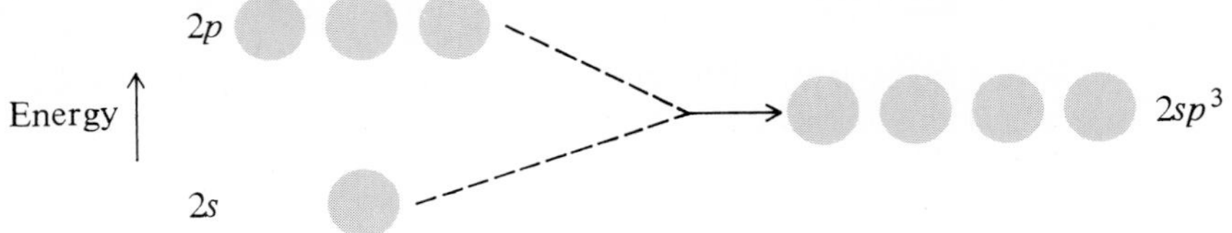

Figure 2–18. *Energy level diagram for* s, p, *and* sp³ *orbitals. The hybrid orbitals lie three quarters of the way along the relative energy scale from* s *to* p.

Overlap of the *larger* lobe of each sp^3 orbital with a hydrogen 1*s* orbital yields the four equivalent C(sp^3)—H(1*s*) localized MO's of methane (Fig. 2–19). The hybrid orbitals allow a large buildup of electron density in the bonding region. The C—H bonds formed this way are stronger than they would have been if pure carbon 2*p* orbitals had been used.

The hybridization scheme just presented is not a physical process. It is a description of a mathematical approach that we can use in calculating the minimum energy molecular orbitals for methane. The molecular geometry arrived at is the correct one and is identical to that predicted by VSEPR theory. It follows that we can assign sp^3 hybridization to any atom that has a tetrahedral arrangement of valence electron pairs. Thus, chloromethane, ammonia, and water (see structures **3**, **4** and Prob. 2–2, respectively, in sec. 2:1) each contain a central atom that is sp^3 hybridized. In the case of ammonia and water, some of the sp^3 orbitals are occupied by nonbonding electrons. Since none of the three molecules just mentioned are as symmetrical as methane, their central atoms are not exactly sp^3 hybridized. However, their hybridization states are close enough to sp^3 to allow us to label them as such without serious error.

At this point it may seem that the molecular orbital approach has not provided much more information than VSEPR theory. However, we'll show in the following section that the localized MO model provides us with insights not available from the VSEPR model.

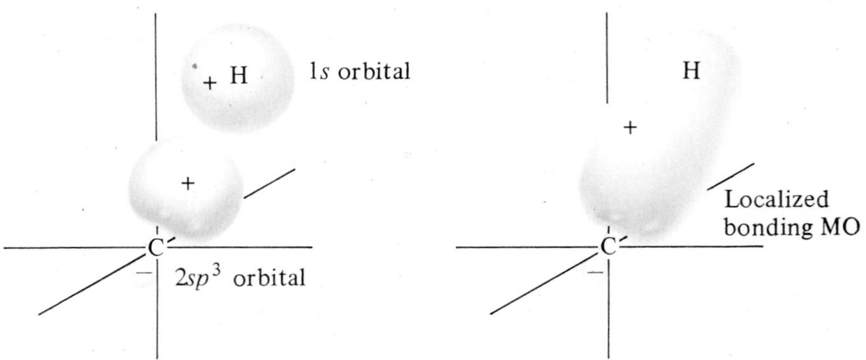

Figure 2–19. *Formation of a localized bonding MO between carbon and hydrogen. The hydrogen* 1s *orbital overlaps with the larger lobe of the carbon* 2sp³ *orbital.*

Example

What is the hybridization state of the carbon atom in CF_4? What type of orbitals are contributed from the carbon and fluorine atoms to form the C—F bond?

Answer

Tetravalent carbon forms four equivalent hybridized orbitals that are available for single bonding. The carbon atom in CF_4 is sp^3 hybridized. From Figure 2–6 you can see that F has electrons in *p* orbitals available for bonding (also shown in Figs. 2–12 and 2–13). The C—F bond is therefore the result of overlap between an sp^3 orbital from the carbon and a *p* orbital from the fluorine. The C—F bonds in CF_4 are described as being sp^3-*p* bonds.

Problem

2–7 Indicate the hybridization state of the central atom in the following species. Label each bond according to the type of orbitals contributed by each atom. Show all the valence electrons around the atom(s), and illustrate the geometry.

a. CCl_4
b. CH_2Cl_2
c. BF_4^-
d. $CH_3NH_3^+$ (both C and N)
e. $(CH_3)_2O$ (both C and O)
f. $(CH_3)_2S$ (both C and S)

2:5 The Molecular Orbital Description of Double and Triple Bonds: Division into σ and π Electrons

Although electrons repel one another, the formation of double and triple bonds requires that more than one electron pair occupy the same bonding region between two nuclei. VSEPR theory does not tell us how this can be accomplished, but MO theory does. Consider a typical σ orbital (Figs. 2–12 and 2–13). Note that electron density is concentrated along a line connecting the two atoms. Compare that to a typical π orbital (Fig. 2–14). There the line connecting the atoms is in the nodal plane, where the π-electron density is zero. Thus, one way of having a double bond in which the two electron pairs stay out of each other's way is to have it consist of a σ bond and a π bond.

Let's consider ethylene, $CH_2{=}CH_2$, as an example. In order to generate a π bond there must be a *p* orbital available on each carbon atom (Fig. 2–14). The remaining two *p* orbitals and the *s* orbital at each carbon mix to produce three hybrid sp^2 orbitals (Fig. 2–20) which are used to form σ bonds. The sp^2

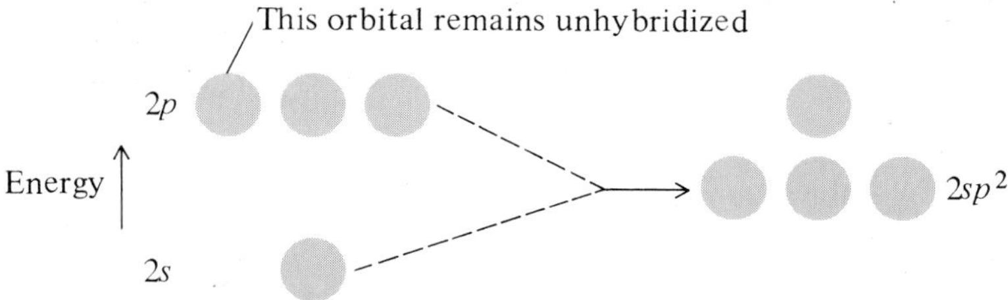

Figure 2–20. *Energy level diagram for* s, p, *and* sp² *orbitals. The hybrid orbitals lie two-thirds of the way along the relative energy scale from* s *to* p.

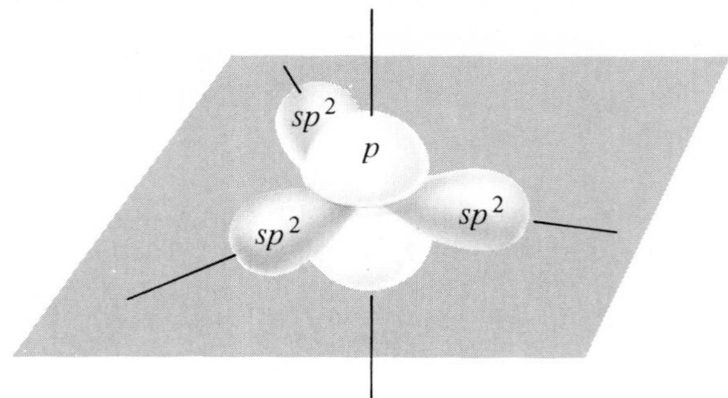

Figure 2–21. *Atomic orbitals at a* sp^2*-hybridized atom. The axes of the* sp^2 *orbitals make 120° angles with one another and lie in the nodal plane of the* p *orbital.*

orbitals are arranged in a trigonal planar manner, with the *p* orbital set at right angles to their plane (Fig. 2–21). Ethylene therefore contains C—H and C—C σ bonds utilizing carbon sp^2 orbitals, as well as a C—C π bond constructed from carbon *p* orbitals (Fig. 2–22).

A similar analysis leads to the conclusion that the triple bond in acetylene (HC≡CH) consists of one σ bond and two π bonds. Each carbon atom requires two *p* orbitals for π bond formation. The remaining valence orbitals mix to yield two hybrid *sp* orbitals (Fig. 2–23). The *sp* orbitals are linearly oriented and 180° apart, and are used in the C—H and C—C σ bonds (Fig. 2–24).

It is not necessary for a molecule to contain multiple bonds for its atoms to be sp^2 or *sp* hybridized. Any atom whose valence electrons are in a trigonal arrangement, such as the boron in BF_3 (structure **1**, sec. 2:1), will adopt sp^2 hybridization. Analogously, a linear arrangement of valence electrons, as in $HgCl_2$ (structure **2**, sec. 2:1), usually indicates *sp* hybridization. The hybridization of any atom may change in the course of a reaction if the geometry of its valence electron changes (eqs. 2.1 to 2.3); however, a change of hybridization is not inevitable in a chemical reaction (eq. 2.4).

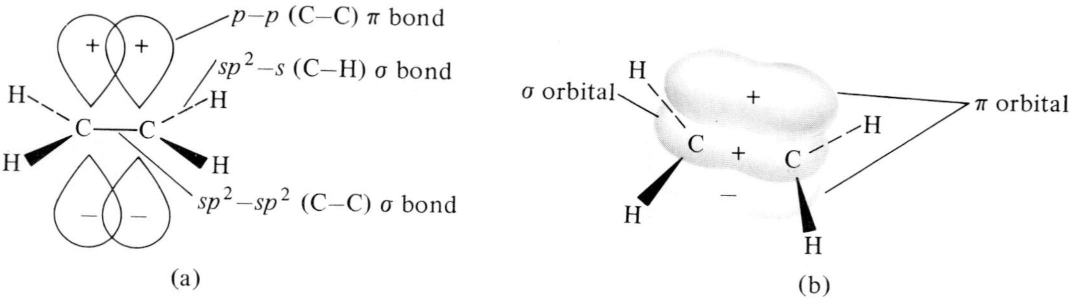

Figure 2–22. *(a) Two-dimensional and (b) three-dimensional representations of the* σ *and* π *bonds in ethylene.*

These orbitals remain unhybridized

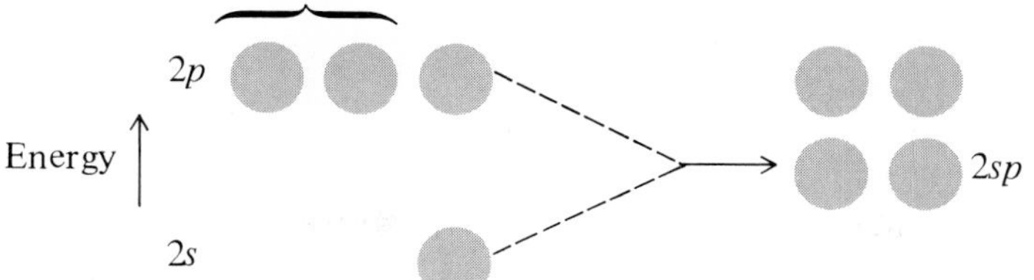

Figure 2–23. *Energy level diagram for* s, p, *and* sp *orbitals.*

H—C≡C—H

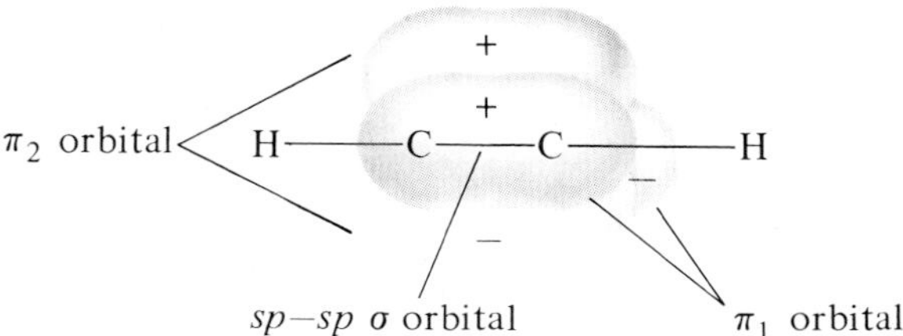

Figure 2–24. *σ- and π-bonding in acetylene. The triple bond consists of six electrons. Two are in an* sp–sp *σ orbital and two are in one* p-p *π orbital. The remaining two electrons are in a second* p–p *π orbital that is perpendicular to the first.*

(2.1) Cl—Hg—Cl + 2Cl$^-$ ⟶ $HgCl_4^{2-}$

Linear *sp* ⟶ Tetrahedral sp^3

(2.2) BF_3 + F$^-$ ⟶ BF_4^-

Trigonal sp^2 ⟶ Tetrahedral sp^3

(2.3) CH_3Br ⟶ CH_3^+ + Br$^-$

Tetrahedral sp^3 ⟶ Trigonal sp^2

$$\text{(2.4)}\quad H_3\ddot{N} + H^+ \longrightarrow NH_4^+$$

Tetrahedral sp^3 → Tetrahedral sp^3

Problem

2–8 Indicate the hybridization state of each multivalent atom in the following compounds. Show which atomic orbitals are involved in each of the bonds.

a. BCl_3
b. $CH_2{=}\ddot{N}H$
c. CO_2
d. N_2
e. $(CH_3)_3B$
f. HONO
g. $COCl_2$
h. $(CH_3)_2Hg$

2:6 Bond Strengths and Bond Lengths

Bond angles are only one of several parameters which describe the structures of molecules. Different types of bonds also result in different **bond lengths.** A bond length is the distance between the nuclei of bonded atoms in a molecule. Bond length is related to another very important molecular parameter, bond strength. Chemistry is, after all, primarily a science of changes; reactions occur which involve bonds breaking and bonds forming. This section will first explore the strengths of bonds and then look briefly at the effect of hybridization on bond lengths.

Bond strengths are usually defined in terms of the amount of energy needed to break a given bond in a molecule. More energy is required to break a stronger bond than a weaker one, and it is often useful to compare differences in bond strengths by comparing **bond dissociation energies** or **average bond energies** of various molecules.

Equation 2.5 shows the **homolysis** (from the Greek *homos,* meaning the same, plus *lysis,* meaning loosening or breaking) of the A—B bond. In a

$$\text{(2.5)}\qquad A{-}B + \text{energy} \xrightarrow[\text{phase}]{\text{gas}} A\cdot + B\cdot$$

homolytic reaction, the bonding electrons are divided equally between the two fragments A and B. *The energy needed to break one mole of A—B bonds homolytically is the bond dissociation energy (BDE) of the A—B bond.* This quantity is measured for reaction in the gas phase at 25°C. Table 2–1 lists bond dissociation energies for some diatomic molecules with single bonds.

The average bond energy in more complex molecules (Table 2–2) is simply an average of the bond dissociation energy for a particular type of bond, such as

$$-\overset{|}{\underset{|}{C}}-H \qquad \text{or} \qquad -O-N\!\!<$$

in a number of molecules. The actual bond dissociation energies that are averaged may vary over a wide range, as is illustrated for various C—H bonds in Equations 2.6 through 2.8. Therefore, the values listed in Table 2–2 are not exact for any particular molecule.

Table 2–1. Bond Dissociation Energies of Some Diatomic Molecules (kcal/mol)

Molecule	Bond Dissociation Energy
H_2	104
Li_2	26
F_2	39
Cl_2	58
Br_2	46
I_2	36
HF	136
HCl	103
HBr	87.5
HI	71
FCl	61
ICl	50

$$CH_3{-}H \rightarrow CH_3\cdot + H\cdot \qquad \text{BDE} = 104 \text{ kcal/mol} \tag{2.6}$$

$$CH_3CH_2{-}H \rightarrow CH_3CH_2\cdot + H\cdot \qquad \text{BDE} = 98 \text{ kcal/mol} \tag{2.7}$$

(drawn as $CH_3C(H)_2{-}H$)

$$(CH_3)_3C{-}H \rightarrow (CH_3)_3C\cdot + H\cdot \qquad \text{BDE} = 91 \text{ kcal/mol} \tag{2.8}$$

Table 2–2. Average Bond Energies of Some Single Bonds (kcal/mol)*

Bond	Average Bond Energy	Bond	Average Bond Energy	Bond	Average Bond Energy
C—H	97	N—N	50–70	C—N	79
N—H	101	O—O	45	C—Cl	81
O—H	104	Si—Si	70	C—Br	68
S—H	88	C—F	106	C—I	51
Si—H	92	C—O	86	N—O	55
C—C	84	C—Si	80		

* Assuming sp^3 hydridization at C and N.

We can use these bond strength data to make qualitative predictions about the course of various reactions. For instance, when methane is heated to a sufficiently high temperature, it begins to decompose by rupture of a C—H bond (eq. 2.9). When ethane, CH_3CH_3, is heated to a high temperature, two different modes of cleavage can occur, involving either C—C or C—H rupture. Table 2–2 shows that C—C bonds are considerably weaker than C—H bonds, and so we would expect Equation 2.10 to be the major reaction.

$$CH_3{-}H \xrightarrow{\text{heat}} CH_3\cdot + H\cdot \tag{2.9}$$

$$CH_3{-}CH_3 \xrightarrow{\text{heat}} 2CH_3\cdot \tag{2.10}$$

Problem

2–9 Use the average bond energies listed in Table 2–2 to calculate the total bonding energy of dimethyl ether, CH_3OCH_3, and ethanol, CH_3CH_2OH. According to your calculations, which isomer has the lower energy?

Table 2–3. Average Equilibrium Bond Lengths

Single Bonds	Length (Å)	Multiple Bonds	Length (Å)
—C—H (sp^3–s)	1.09	C═C (aromatic)	1.39
═C—H (sp^2–s)	1.08	C═C	1.34
≡C—H (sp–s)	1.06	C═N	1.30
—C—C— (sp^3–sp^3)	1.54	C═O	1.23
—C—C═ (sp^3–sp^2)	1.53	C≡C	1.20
═C—C═ (sp^2–sp^2)	1.49	C≡N	1.16
≡C—C≡ (sp–sp)	1.37		
C—N	1.47		
C—O	1.43		
C—F	1.38		
C—Cl	1.76		

Table 2–4. Average Bond Energies of Some Double Bonds (kcal/mol)

C═C	146
N═N	100
C═N	142
C═O	176

The strength of a bond is directly related to its length. This correlation will become clear when we compare trends in bond strengths with trends in bond lengths, which are usually measured in angstroms, Å (1Å = 10^{-8} cm). Molecular bond lengths cannot be measured precisely because the atoms in the molecule continually move with respect to one another. Nonetheless, using x-ray or electron diffraction (see p. 8) it is possible to measure *equilibrium bond lengths* for each bond in a molecule.

Table 2–3 lists a number of *average* equilibrium bond lengths for several different types of bonds. By comparing the bond lengths in Table 2–3 with the bond strengths in Table 2–2 we see that shorter bonds are generally stronger bonds. C—F is shorter and stronger than C—O, which is shorter and stronger than C—C. (C—N is an apparent exception.) Double bonds are shorter than the corresponding single bonds, and they are also stronger (Table 2–4). Table 2–3 also shows that bond lengths between a given pair of atoms vary with the hybridization states of the bonded atoms. Note that C—H and C—C bonds get shorter (and therefore stronger) as the percentage of *s* character in the hybrid orbital increases.

2:7 Electronegativity and Bond Polarization

Covalent bonds are formed when atoms share electrons. It is not necessary, however, for the bonding electrons to be shared equally between the atoms. Many covalent bonds are **polarized**. A polarized bond is one in which the bonding electrons spend more time near one of the atoms rather than the other.

In order to account for bond polarization, individual atoms are assigned an **electronegativity** value. The electronegativity of an atom measures its tendency to polarize electrons in a covalent bond. Electronegativities are defined values rather than experimentally measured ones, although they can be calculated from experimental data. The higher the electronegativity of an atom, the greater the tendency of shared electron density to accumulate near that atom. Table 2–5 presents a set of electronegativity values for some common elements important in organic chemistry.

The *absolute* electronegativities shown in Table 2–5 are not very significant. More important is the electronegativity *difference* between two covalently bound atoms. The larger the electronegativity difference between the atoms, the more strongly polarized the bond. Since the more electronegative atom has the greater share of the electron pair, it will have a partial negative charge.

Table 2–5. Electronegativities of Some Main Group Elements

1A	2A	3A	4A	5A	6A	7A
H 2.1						
Li 1.0	Be 1.5	B 2.0	C 2.5	N 3.1	O 3.5	F 4.1
Na 1.0	Mg 1.2	Al 1.5	Si 1.7	P 2.1	S 2.4	Cl 2.8
K 0.9						Br 2.7
Rb 0.8						I 2.2

Figure 2–25 shows how bond polarization can be depicted for the water molecule. Since oxygen is more electronegative than hydrogen it carries a fractional negative charge. In one representation, the symbols $\delta-$ and $\delta+$ stand for fractional negative and positive charges, respectively (δ is the lower case Greek delta). Another method uses an arrow to represent a polarized bond. The head of the arrow is connected to the negative atom and the tail to the positive atom, indicating the direction of electron polarization.

The following scale of decreasing electronegativities can serve as a guide to bond polarities: $F > O > N > Cl > Br > C \sim S > I > H$. A useful general rule is that these atoms form covalent bonds with one another. However, when atoms from the extreme left columns (1A and 2A) and right columns (6A and 7A) of the Periodic Table (Table 2–5) combine, complete electron transfer takes place. In these cases positive and negative ions, rather than covalent molecules, are formed.

Problem

2–10 Use the $\delta+$ and $\delta-$ notation of Figure 2–25 to show how the bonds are polarized in the following molecules.

a. NH_3 c. CH_3—O—H
b. CH_3—Cl d. CH_3—S—H

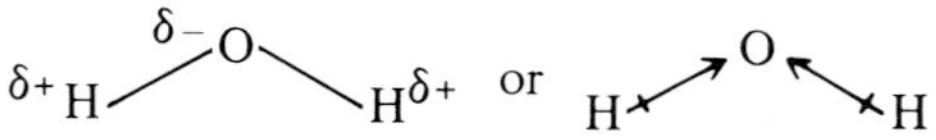

$\delta+$H—O($\delta-$)—H$\delta+$ or H→O←H

Figure 2–25. *Two representations of bond polarization in the water molecule.*

2:8 A Practical Example of Relative Electronegativities: Organic Acid Strengths

The concept of electronegativity is a very useful one for explaining many qualitative trends in organic chemistry. In this section we introduce an example in which electronegativity is used this way, and we also call attention to its limitations.

Alcohols (ROH; Table 1–6, p. 27) are very weak acids in water. Their K_a's are comparable to, or somewhat smaller than, that of water itself (sec. 1:7). Equation 2.11 depicts the ionization of ethanol in water. The equation shows that the proton given up by ethanol comes from the OH group. Why doesn't ethanol donate a proton from the CH_3 or CH_2 groups instead (eq. 2.12)?

(2.11) $CH_3CH_2\ddot{\underset{..}{O}}H + H_2\ddot{O}: \rightleftarrows CH_3CH_2\ddot{\underset{..}{O}}:^- + H_3\ddot{O}^+$ $\qquad K_a = 1 \times 10^{-16}$, $pK_a = 16$

Ethanol $\qquad$ Ethoxide ion

(2.12) $CH_3CH_2\ddot{\underset{..}{O}}H + H_2\ddot{O}: \rightleftarrows {}^{-}\!:CH_2CH_2\ddot{\underset{..}{O}}H + H_3\ddot{O}^+$ $\qquad K_a \sim 10^{-55}$, $pK_a \sim 55$

As you can see, the equilibrium constant for loss of a proton from the CH_3 group is about 10^{-55}! It would take a tremendously powerful base to remove such a proton.

Why is there such a great difference in acidity between the OH and CH_3 groups? In equilibria such as those in Equations 2.11 and 2.12, the important governing factor will usually be one that helps stabilize the anion product. Since the reactants are the same in both cases, only effects on the products need be considered. The $CH_3CH_2O^-$ anion is much more stable than the $^-CH_2CH_2OH$ anion because *in $CH_3CH_2O^-$ the negative charge is on oxygen, which is a much more electronegative atom than carbon*. The more electronegative an atom is, the more readily it bears a negative charge. Therefore, equilibrium 2.11 goes farther to the right than 2.12.

When we measure the K_a for ethanethiol (eq. 2.13), however, we obtain a result that doesn't follow this rule. Although sulfur is less electronegative than oxygen (Table 2–5), ethanethiol ionizes to a greater extent than ethanol. This apparent contradiction is largely due to the difference in size between sulfur and oxygen. Because of its greater size, sulfur can spread out, or delocalize (sec. 2:3), the negative charge of the anion better than oxygen can.

(2.13) $CH_3CH_2\ddot{\underset{..}{S}}H + H_2O \rightleftarrows CH_3CH_2\ddot{\underset{..}{S}}:^- + H_3O^+$ $\qquad K_a = 1 \times 10^{-11}$, $pK_a = 11$

Ethanethiol $\qquad$ Ethanethiol-ate ion

As our comparison of the relative acidities of the CH_3 and OH groups proved, electronegativity differences can be used to explain chemical trends. But, as we saw when the SH group was brought in, we also must take into account other parameters that may be equally important, such as the size of the atoms or groups being compared.

Problems

2–11 Based on the examples just given, which would you expect to be the stronger acid, HF or HCl? Explain your answer.

2–12 Write the equilibrium equation for CH_3NH_2 (methylamine) acting as an *acid* in water. Which atom of methylamine, carbon or nitrogen, will be the proton donor?

Summary

1 Covalent bonds between atoms consist of valence shell electron pairs. The shape of the molecule is governed by the fact that these electron pairs repel one another, and therefore try to stay as far from one another as possible. For an atom with two electron pairs in its valence shell, this repulsion leads to a bond angle of 180°. Three valence shell electron pairs will arrange themselves at 120° angles, and four electron pairs at 109.5° angles. These are ideal values and hold only when all electron pairs are equivalent. Electron pairs become nonequivalent when they are nonbonding or are connected to different atoms. Then the bond angles differ from the ideal ones. Double and triple bonds are treated as a single electron pair in VSEPR theory (sec. 2:1).

2 Electrons in atoms are arranged in energy levels. The first level is closest to the nucleus, and the succeeding ones are increasingly farther from it. The electrons in different levels have different energies, with the energy increasing as the electron is placed further from the nucleus. Within each level are orbitals that correspond to allowed energy states of the electrons. Physically, an orbital is a region of space within which the electron wave, or charge cloud, is spread out.

Atomic orbitals have different shapes and sizes, but each accommodates a maximum of two electrons, which must have paired spins. Atomic *s* orbitals are spherical, while *p* orbitals have two lobes separated by a nodal plane. Each time an orbital node is crossed, the sign of the wave function changes (sec. 2:2).

3 The formation of a covalent bond may be pictured as the overlap of atomic orbitals on different atoms. Two molecular orbitals are produced, each of which can hold two spin-paired electrons. The molecular orbital is a bonding orbital if the atomic orbitals have the same sign in the region of overlap. If not, it is an antibonding orbital. Electrons in bonding molecular orbitals have less energy than they did in atomic orbitals. Overlap of *s* and *p* orbitals or end-to-end overlap of *p* orbitals results in the formation of molecular σ orbitals. Side-by-side overlap of *p* orbitals leads to π-orbital formation. When two atomic orbitals overlap, they yield one bonding (σ or π) and one antibonding ($\sigma^\star$ or $\pi^\star$) orbital. The ground states of most molecules have the bonding orbitals filled and the antibonding orbitals empty (sec. 2:3).

4 By postulating hybrid atomic orbitals we are able to calculate molecular geometries that agree with experimental results. Hybrid orbitals can be conceived of as arising from the mixing of atomic orbitals on the same atom. The hybrid orbitals of greatest importance in organic chemistry are the *sp* (two orbitals arranged 180° apart), sp^2 (three orbitals in a plane at 120°), and

sp^3 (four orbitals disposed tetrahedrally). Hybrid orbitals form stronger bonds than unhybridized orbitals (sec. 2:4).

Double or triple bonds consist of two types of molecular orbitals. A σ orbital is formed from hybrid *sp* or sp^2 orbitals, and one or two π bonds are formed from unhybridized *p* orbitals (sec. 2:5).

5 Bond dissociation energies are the experimental energies needed to homolytically dissociate specific bonds in specific molecules. Average bond energies are calculated using bond energies for a specific type of bond in several different molecules. Double bonds are stronger and shorter than corresponding single bonds. There is a general relationship between bond length and bond strength—shorter bonds are stronger bonds. Even single bonds between two given elements will vary in length and strength, depending on the hybridization states of the atoms (sec. 2:6).

6 The electronegativity of an atom is a measure of its tendency to polarize electrons it is sharing with another atom. Electronegativity increases from left to right across a row of the Periodic Table, and from bottom to top in each column of the table. When two elements of different electronegativity form a covalent bond, that bond is polarized. The more electronegative atom acquires a partial negative charge and the less electronegative one a partial positive charge. When the electronegativity difference between two atoms is fairly large, the bond is ionic rather than covalent (sec. 2:7).

The more electronegative an atom is, the more stable it is when carrying a negative charge. The effect shows up when we compare the acidity of Y—H groups. For the comparison to be valid, however, the Y atoms must be of comparable size (sec. 2:8).

Problems

2–13 For each of the following pairs, indicate which will have the greater H—Y—H bond angle. Explain your reasoning.

a. CH_4 or $:CH_3^-$
b. NH_3 or NH_4^+
c. $H_2C(OH)_2$ or $H_2C{=}O$
d. CH_3^+ or $:CH_2$
e. CH_3Cl or CH_2Cl_2

2–14 For each of the following atoms designate the number of electrons in the valence level. Then draw the structure of the stable, covalent compound formed between that element and hydrogen and specify the probable hybridization state of the atom in the compound.

a. C
b. B
c. Be
d. S
e. P
f. Al
g. Si

2–15 What is the most likely hybridization state of the designated atom in the following ions and molecules?

a. $\underline{Si}Cl_4$
b. $\underline{N}H_4^+$
c. $H_3\underline{O}^+$
d. $\underline{B}Br_3$
e. $\underline{Al}H_3$
f. $(CH_3)_3\underline{P}$
g. $:\underline{C}H_3^-$
h. $:\underline{C}H_2$
i. $\underline{C}H_3^+$
j. $\underline{C}H_3\underline{C}H_2Cl$

2–16 Complete each of the following equations by showing the structure of the unknown product indicated by a letter. Show all nonbonding electrons and charges in the products. Assume that all atoms of the reactants not explicitly accounted for are present in the products.

a. $CH_4 + H\cdot \rightarrow H_2 + \mathbf{A}$

b. $CH_3OH + H^+ \rightarrow \mathbf{B}$

c. $CH_3OOCH_3 \rightarrow 2\mathbf{C}$

d. $NH_3 + CH_3\overset{\overset{\displaystyle O}{\|}}{C}OH \rightarrow NH_4^+ + \mathbf{D}$

e. $CH_3O^- + CH_3SH \rightarrow CH_3OH + \mathbf{E}$

2–17 When different substituents are attached to a given atom, the resultant bond angles may deviate somewhat from the ideal values given by hybridization theory, due to interactions among the substituents.

a. All the carbon atoms in propane, $CH_3CH_2CH_3$, may be considered to be sp^3 hybridized. The

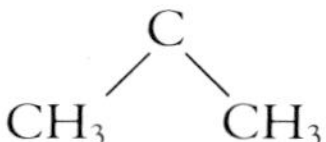

angle, however, equals 112° rather than 109.5°. Explain.

b. The H—O—H angle in water is found to be 104.5°, while the C—O—C angle in dimethyl ether, CH_3OCH_3, equals 111°. Rationalize the differences between each of these and the theoretical tetrahedral angle (109.5°).

2–18 How many isomers of CH_2Cl_2 would be possible if

a. the four bonds from the central carbon atom to the H and Cl atoms extended to the corners of a square;

b. the four bonds from the central carbon atom to the H and Cl atoms extended to the four corners of a rectangle?

2–19 For each of the following compounds indicate the type of molecular orbital or atomic orbital occupied by each electron pair. Indicate the hybridization state of each carbon and heteroatom.

a. NH_3

b. CH_3OH

c. $(CH_3)_2Cd$

d. HCN

e. $CH_3CH{=}CH_2$

f. $CH_3\overset{\overset{\displaystyle O}{\|}}{C}Cl$

2–20 In each of the following compounds designate the bond or bonds that would most easily undergo homolytic cleavage if the compound were heated in the gas phase.

a. CH_3CH_3

b. CH_3F

c. CH_3OH

d. CH_3OOCH_3

e. $(CH_3)_2N{-}N(CH_3)_2$

f. $CH_3{-}N{=}N{-}CH_3$

g. $CH_3\overset{\overset{\displaystyle O}{\|}}{C}CH_3$

h. $HC{\equiv}CH$

i. $SiH_3{-}SiH_3$

2–21 Calculate the net change in bond energies for each of the following reactions. (Make sure you balance the equation first.)

a. $CH_4 + Cl_2 \rightarrow CCl_4 + HCl$
b. $CH_4 + F_2 \rightarrow CH_2F_2 + HF$
c. $SiH_3{-}SiH_3 + H_2 \rightarrow SiH_4$
d. $I_2 + Cl_2 \rightarrow ICl$

2–22 Calculate the difference in total bond energy between each of the following pairs of isomers, using Tables 2–2 and 2–4. In each case indicate which member of the pair is the more stable, remembering that it *requires* energy to break a bond.

a. $CH_3CH{=}CHCH_2CH_3$ and cyclopentane (ring: CH_2 / H_2C — CH_2 / $H_2C{-}CH_2$)

b. $CH_2{=}CH{-}OH$ and $CH_3CH{=}O$

c. $CH_2{=}N{-}NH_2$ and $CH_3{-}N{=}N{-}H$

2–23 In each of the following gaseous molecules the atoms are joined by electron-pair bonds. By using the symbols δ+ and δ−, indicate how the bond is polarized in each case, if at all.

a. Li_2
b. LiH
c. HBr
d. FCl
e. Cl_2
f. H_2O
g. H_2S
h. PCl_3
i. CF_4

2–24 Of the sets of pairs of atoms, which pair will form the most highly polarized bond? Indicate by δ+ and δ− the positive and negative ends of the bond.

a. Li—F or C—F
b. C—H or C—O
c. C—H or Cl—H
d. C—F or C—Cl
e. Br—F or I—F
f. B—C or B—O
g. B—C or Al—C
h. B—C or Be—C
i. Li—C or K—C

2–25 Which of the following compounds would you expect to be ionic in structure, and which covalent (in the solid state)? For those that you believe to be ionic, draw the structure of the ions.

a. CH_4
b. CO_2
c. CF_4
d. $LiCH_3$
e. LiF
f. HF
g. H_2O
h. NH_3
i. NaCl
j. NaBr
k. $MgCl_2$
l. $MgBr_2$
m. CCl_4

2–26 Use your knowledge of electronegativities to explain why methylamine, CH_3NH_2, is a stronger base than methanol, CH_3OH (sec. 1:7).

2–27 In Section 1:6 we showed that the isomers of formula C_3H_8O could be differentiated on the basis of their reaction with sodium metal. 1-Propanol and 2-propanol react with sodium, while ethyl methyl ether does not. Indicate what reaction takes place between the metal and the two alcohols, and what structural feature of the alcohols permits this reaction to occur.

$CH_3CH_2CH_2OH$	CH_3CHCH_3 (with OH on central C)	$CH_3CH_2OCH_3$
1-Propanol	2-Propanol	Ethyl methyl ether

Supplementary Reading

L. Salem, A faithful couple: the electron pair, *J. Chem. Ed.,* **55,** 344 (1978). A description of bonding based upon orbital wave function signs (sec. 2:2). Its intent is "to present some of the simplest facts of Theoretical Chemistry in a language which is as close as possible to our everyday language."

G. C. Pimentel and R. D. Spratley, *Chemical Bonding Clarified through Quantum Mechanics,* Holden-Day, Inc., San Francisco, Calif., 1969.

W. F. Luder, The electron repulsion theory of the chemical bond, *J. Chem. Ed.,* **44,** 206 (1967). An interesting approach to chemical bonding, substantially different from the one given in this text.

M. Orchin and H. H. Jaffe, *The Importance of Antibonding Orbitals,* Chaps. 1 and 2, Houghton-Mifflin Co., Boston, Mass., 1967.

3

Alkanes and Cycloalkanes: The Three-Dimensional Structure of Hydrocarbons and Their Derivatives

In Chapter 2 you learned that the nature of the bonding in organic molecules determines their shapes. In this chapter we will examine the three dimensional structures of simple but important compounds of carbon and hydrogen, and show you how to name them.

Organic molecules are not static objects, but are continually undergoing internal and external movement. This chapter discusses how the changing shapes of molecules, together with their internal motions, produces dynamic effects. Using some fundamental thermodynamic relationships, we will show how these molecular properties can be determined quantitatively.

3:1 Natural Sources of Hydrocarbons

The most important sources of commercial organic compounds are coal, petroleum, and natural gas. The many pure compounds that can be separated from these sources are either used directly or as raw materials that can be converted to other substances by synthetic procedures. The compounds obtained from coal are mostly aromatic, that is, benzene derivatives or similar substances (sec. 1:6). (The chemistry and uses of these are taken up in later chapters.) Petroleum and natural gas are the ultimate sources of most aliphatic compounds (sec. 1:6).

Coal, petroleum, and natural gas are nonrenewable natural resources whose major uses are as fuels. Because petroleum and natural gas are so important as a source of organic compounds, many people feel they are too valuable a commodity to be burned as a fuel. Petroleum (Latin for *rock oil*) was formed any time from 50 to 500 million years ago. Its main precursor was apparently marine plant life, although it included some animal life as well. The animal and vegetable materials were trapped in sedimentary rock where they were degraded by anaerobic microbes and subjected to moderate temperatures and pressures. Under these conditions, most of the original nitrogen and oxygen content was lost, while the carbon and hydrogen were retained. Thus, petroleum consists mainly of hydrocarbons—compounds containing only carbon and hydrogen. This chapter is concerned with those hydrocarbons called **alkanes**, also known as *paraffins* (from the Latin *parum affinis,* meaning having little affinity) because of their relative unreactivity.

The compounds that comprise the alkane family are **homologs** of one another. Homologs are compounds that differ by the addition or subtraction of a CH_2 group. The molecular formulas of the alkanes all fit the general

Box 3–1. Ode to Petroleum

"That viscous, tarry liquid nature laid beneath the ground
A hundred million years before man came upon the scene
Can be transformed to marvelous new products we have found—
Like nylon, orlon, polyesters, polyethylene,
Synthetic rubber, plastics, films, adhesives, drugs, and dyes,
And other things, some that we now can only dream about.
Who knows what wondrous products man might some day synthesize.
From oil! Except, alas, that our supplies are running out.
The time is near when earth's prodigious flow of oil may stop,
(We've taken so much from the ground, with no way to return it);

Meanwhile we strive to find and draw out every precious drop,
. . . And then, incredibly, we take the bulk of it and burn it."

D. Kolb and K. E. Kolb, *J. Chem. Ed.*, **56,** 465 (1979). Used by permission.

formula C_nH_{2n+2} ($n = 1, 2, 3, \ldots$). Alkanes are classified as **saturated** hydrocarbons because each molecule contains the maximum number of hydrogen atoms possible for that number and arrangement of carbon atoms. Saturated compounds do not contain double or triple bonds.

Alkanes with four or fewer carbon atoms are gases at ordinary temperatures, and gas pockets containing them are often found in connection with petroleum deposits. Petroleum is a complex mixture of hydrocarbons that must be at least partially separated and purified before it can be used as a fuel or source of chemicals. Liquid petroleum is subjected to **fractional distillation**—it is heated to boiling and the vapors that form are condensed and collected. This is done continuously as the temperature is steadily raised. The distillation procedure divides the petroleum into several fractions on the basis of their boiling point (bp) ranges. Each fraction consists of a mixture of compounds of similar volatility and molecular weight; therefore, each fraction is partially purified and contains molecules with similar numbers of carbon atoms.

In a representative petroleum sample from an Oklahoma field, 50% of the hydrocarbons had boiling points of 230° C or less. Table 3–1 shows the composition of the volatile fractions from the Oklahoma petroleum. The highest boiling point material, not listed in Table 3–1, is the source of light and heavy lubricating oils and paraffin wax. The most valuable liquids in petroleum are those found in the gasoline and naphtha fractions. In order to

Table 3–1. Volatile Constituents of Oklahoma Crude Petroleum

Name of Fraction	Natural Gas	Gasoline	Naphtha
Boiling point range (°C)	<40	40–180	180–230
Number of carbon atoms in constituents	1–5	6–10	11 and 12
Percentage of total petroleum sample	4	33	13

increase their yield, the higher boiling fractions are subjected to **catalytic cracking**, which means they are brought into contact with an acidic catalyst at elevated temperatures. The large, high molecular weight molecules are broken down, or cracked, into smaller ones. The catalyst is not consumed in the reaction but serves to speed it up considerably, so that contact times of the order of seconds are sufficient. This allows large quantities of material to be processed per day. Equation 3.1 shows a representative cracking product distribution.

(3.1) C_{21} to C_{50} (approx) compounds $\xrightarrow[500°]{Al_2O_3—SiO_2 \text{ cat.}}$ C_1 to C_4 compounds
(heavy distillate, bp 350–550°) (petroleum gas, 19%)

+ C_5 to C_{12} compounds
(gasoline and naphtha, bp 30–220°, 36.5%)

+ C_{13} to C_{20} compounds
(kerosene and gas oil, bp 220–350°, 15%)

+ residue
(asphalt, 29.5%)

Since petroleum and natural gas are nonrenewable resources, alternate energy sources must become available so that petroleum and natural gas can still be used as chemical sources for many years. With our future in mind, chemists are looking for ways to obtain so-called petrochemicals from renewable resources such as fast growing and easily cultivated plants. Another possible chemical source may be microbes that feed on garbage and industrial waste and are able to convert these to some of the same compounds found in natural gas and petroleum.

3:2 Naming Alkanes: The Semisystematic Method

There are two nomenclature systems now used for naming alkanes. The **semisystematic** method is the older system, and is generally used only for low molecular weight compounds. Semisystematic names are being steadily replaced by **IUPAC** names. The term IUPAC is an acronym for the International Union of Pure and Applied Chemistry, which invented the modern system of nomenclature. The IUPAC method is gaining ground because it allows us to give unambiguous names to all types of compounds, including very complex ones. By contrast, the semisystematic method is much less versatile. Nonetheless, we begin with a description of the semisystematic method because it is still in use, and parts of it are incorporated into the IUPAC method.

There is only one structure for those alkanes having one, two, or three carbon atoms, the names of which are given here. There are *two* isomeric

CH_4	CH_3CH_3	$CH_3CH_2CH_3$
Methane	Ethane	Propane

four-carbon alkanes, however (sec. 1:5), and each must be given a distinctive name. It was decided that the "normal" structure for an alkane is the one in which all carbon atoms are connected in a single continuous chain. The semisystematic name for the four-carbon alkane with this structure is *n*-butane, while its branched chain isomer is named isobutane.

$$CH_3CH_2CH_2CH_3 \qquad \begin{array}{c} CH_3\underset{|}{C}HCH_3 \\ CH_3 \end{array} \text{ or } (CH_3)_2CHCH_3$$

n-Butane Isobutane

Among the five-carbon alkanes, called pentanes, there is an *n*-pentane and an isopentane which are homologs of *n*-butane and isobutane. In addition, there is a third pentane isomer called neopentane.

$$CH_3CH_2CH_2CH_2CH_3 \qquad \begin{array}{c} CH_3\underset{|}{C}HCH_2CH_3 \\ CH_3 \end{array} \qquad \begin{array}{c} CH_3 \\ | \\ CH_3CCH_3 \\ | \\ CH_3 \end{array}$$

n-Pentane Isopentane Neopentane

All the names so far presented end with the suffix *-ane,* which indicates that the compounds are alkanes. For the C_1 through C_4 alkanes, the roots of their names, such as meth- and prop-, have a purely historical origin. The name methane derives from a Greek term meaning "wine wood" because methane is related to methanol, which was first obtained by distillation of wood ("wood alcohol"). The name ethane is derived from the Greek verb meaning "to burn." From C_5 onward, the root words indicate the number of carbons in the molecule. For instance, pent- comes from the Greek word for "five," hex- from the Greek word for "six," and so on.

Problem

3–1 Draw the structures of *n*-hexane and all of its isomers.

Before we leave the semisystematic nomenclature method, we must describe an important part of it which is still in use. If one hydrogen atom of an alkane is removed, the remaining structure is known as an alkyl group (sec. 1:6). Alkyl groups are named by dropping the suffix *-ane* from the name of the parent alkane and replacing it with *-yl*. The C_1 through C_4 alkyl groups are found so often in organic molecules that they are often represented by standard abbreviations. Table 3–2 presents the structures, names and abbreviations of some of the most common alkyl groups.

The prefixes *sec-* and *tert-*, or *s-* and *t-*, are abbreviations for secondary and tertiary. These terms are part of a nomenclature system which specifies the bonding pattern of the carbon atom *carrying the free bond in an alkyl group*. A **primary** carbon is one that is bonded to *one* other carbon atom. Thus ethyl and propyl groups are primary alkyl groups because the carbon atom carrying the free bond is a 1° carbon atom (see Table 3–2). **Secondary** carbons are those which are bonded to *two* other carbon atoms. You can see that the isopropyl and *s*-butyl groups are secondary alkyl groups. Tertiary carbons

Table 3–2. Some Common Alkyl Groups

Structure	Name	Abbreviation
CH_3—	methyl	Me
CH_3CH_2—	ethyl	Et
$CH_3CH_2CH_2$—	propyl	Pr
CH_3 | CH_3CH—	isopropyl	i-Pr
$CH_3CH_2CH_2CH_2$—	butyl	Bu
CH_3 | CH_3CH_2CH—	*sec*-butyl or *s*-butyl	*s*-Bu
CH_3 | CH_3CHCH_2—	isobutyl	i-Bu
CH_3 | CH_3C— | CH_3	*tert*-butyl or *t*-butyl	*t*-Bu

are bonded to *three* other carbon atoms, as illustrated by the *t*-butyl group. Quaternary carbons are connected to *four* other carbon atoms, which means that quaternary alkyl groups cannot exist. The symbols 1°, 2°, 3°, and 4° are used for primary, secondary, tertiary, and quaternary.

2° Carbon

$CH_3CH_2CH_2CH_3$

1° Carbon

CH_3
|
CH_3CHCH_3

3° Carbon

CH_3
|
CH_3CCH_3
|
CH_3

4° Carbon

Problems

3–2 Classify each of these alkyl groups as primary, secondary, or tertiary (remember that the entire group is defined by the carbon atom carrying the free bond).

a. $(CH_3)_2CHCH_2$—

b. $CH_3CH_2C(CH_3)_2$
|

c. $CH_3CHCH(CH_3)_2$
|

d. $(CH_3CH_2)_2CH$—

e. $CH_3CH_2CHCH_2$—
|
CH_3

3–3 Label each carbon atom in each molecule as primary (1°), secondary (2°), tertiary (3°), or quaternary (4°).

a. $CH_3(CH_2)_4CH_3$

b. $(CH_3)_2CHCH_2CH_2CH_3$

c. 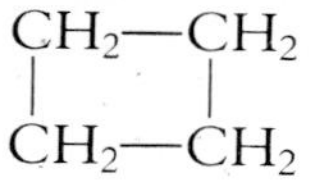
```
CH2—CH2
 |    |
CH2—CH2
```

d. 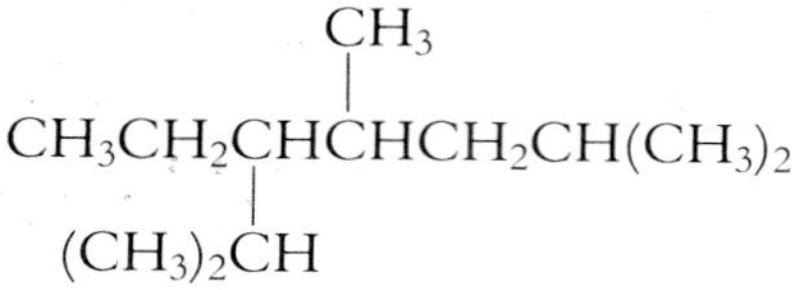
```
                CH3
                 |
CH3CH2CHCHCH2CH(CH3)2
         |
     (CH3)2CH
```

e. 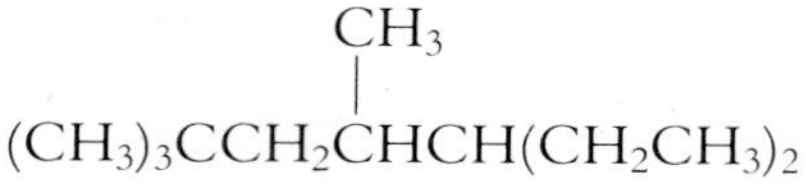
```
                 CH3
                  |
(CH3)3CCH2CHCH(CH2CH3)2
```

f. 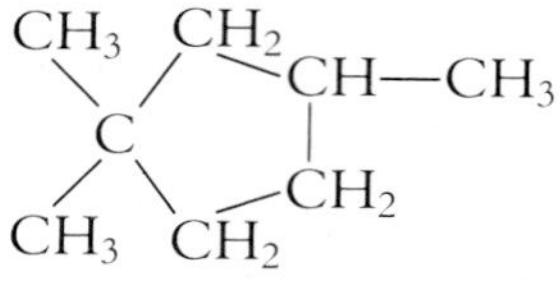
```
CH3  CH2
  \  /   \CH—CH3
   C       |
  /  \    /CH2
CH3  CH2
```

3–4 Draw line–angle formulas (Table 1–4, p. 25) for the compounds shown in Problem 3–3.

3:3 Naming Alkanes: The International (IUPAC) System

The semisystematic naming method becomes cumbersome with complex structures and cannot be relied upon to give a unique name for every compound. A completely **systematic** naming method was therefore devised and agreed upon by an international group of chemists. Scientific journals and indexing systems try to use the IUPAC naming method exclusively.* This is not always achieved in practice and hence it is necessary to learn semisystematic names as well as the IUPAC nomenclature.

In the IUPAC system, the root name for the alkane is based upon the *longest continuous carbon chain*, rather than the total number of carbon atoms in the molecule. Branches on the longest chain are numerically designated according to their position on the chain and then a group name is given to the branch. Numbering can begin from either end of the chain, but the numbering is done so as to give the smallest number to the branch substituent. Here are two simple examples which show how the IUPAC method works.

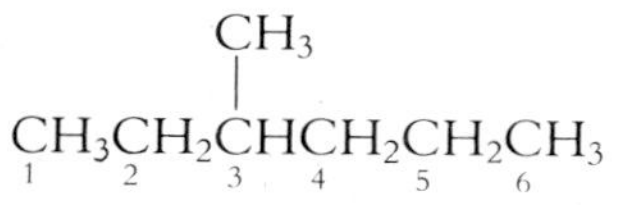

```
        CH3
         |
CH3CH2CHCH2CH2CH3
1   2   3  4   5   6
```

3-Methylhexane
(*not* 4-methylhexane)

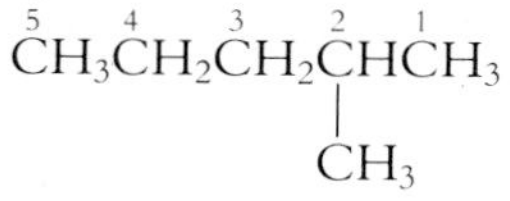

```
5   4   3   2  1
CH3CH2CH2CHCH3
             |
            CH3
```

2-Methylpentane
(*not* 4-methylpentane)

In order to apply the IUPAC system to more complex molecules, the following general procedure can be used:

1. Identify the longest continuous carbon chain. The root of the name is based on the number of carbon atoms in this chain; for example, if it contains six carbons, the compound is a hexane. The names of several alkane chains are given in Table 3–3. The longest continuous chain may

---

* Chemical Abstracts, the most important indexing system, frequently uses its own nomenclature system.

Table 3–3. Names and Physical Properties of Some Straight-Chain Alkanes

Number of Carbon Atoms (MW)	IUPAC Name	Boiling Point (°C)	Melting Point (°C)
C_1 (16)	methane	−162	−182
C_2 (30)	ethane	− 89	−182
C_3 (44)	propane	− 42	−188
C_4 (58)	butane	− 1	−138
C_5 (72)	pentane	36	−130
C_6 (86)	hexane	69	− 95
C_7 (100)	heptane	98	− 91
C_8 (114)	octane	126	− 57
C_9 (128)	nonane	151	− 57
C_{10} (142)	decane	174	− 30
C_{11} (156)	undecane	195	− 25
C_{12} (170)	dodecane	216	− 3
C_{13} (184)	tridecane	243	− 5
C_{14} (198)	tetradecane	253	+ 6
C_{16} (266)	hexadecane	287	+ 18
C_{20} (282)	eicosane	343	+ 37
C_{30} (423)	triacontane	~425	+ 65

not always be written in a straight line across the page. (Note that the prefix *n*- is *not* used in the IUPAC system.)

2. Any smaller chains are treated as substituents on the principal chain. These smaller chains are named as alkyl groups, such as the ones listed in Table 3–2.

$$\begin{array}{l} \overset{1}{C}H_3\overset{2}{C}H_2\overset{3}{C}HCH_3 \\ \qquad\quad | \\ \qquad\ {}^4CH_2 \\ \qquad\quad | \\ \qquad\ {}^5CH_3 \end{array}$$

3-Methylpentane

$$\begin{array}{l} \overset{1}{C}H_3\overset{2}{C}H_2\overset{3}{C}H_2\overset{4}{C}H\overset{5}{C}H_2\overset{6}{C}H_2\overset{7}{C}H_3 \\ \qquad\qquad\qquad | \\ \qquad\qquad\quad CH_2 \\ \qquad\qquad\qquad | \\ \qquad\qquad\quad CH_3 \end{array}$$

4-Ethylheptane

3. The carbons of the principal chain are numbered consecutively, starting from either end. The starting point is chosen so that the first substituent encountered will be on the lowest numbered carbon. We use steps 1–3 to name the following compound:

$$\begin{array}{l} \text{Methyl} \longrightarrow CH_3\overset{5}{C}H\overset{4}{C}H_2\overset{3}{C}H\overset{2}{C}H_2\overset{1}{C}H_3 \\ \text{group} \qquad\qquad | \qquad\qquad | \\ \qquad \underset{8}{C}H_3\underset{7}{C}H_2\underset{6}{C}H_2 \quad CH_2CH_3 \longleftarrow \text{Ethyl group} \end{array}$$

3-Ethyl-5-methyloctane

4. If there is more than one branch of a given type, this must be indicated. The presence of two ethyl groups is indicated by the substituent name *di*ethyl, while *tri*propyl indicates that there are three propyl substituents, and so on. If there are two substituents on the same carbon atom, the number must be used twice, even if the substituents are identical.

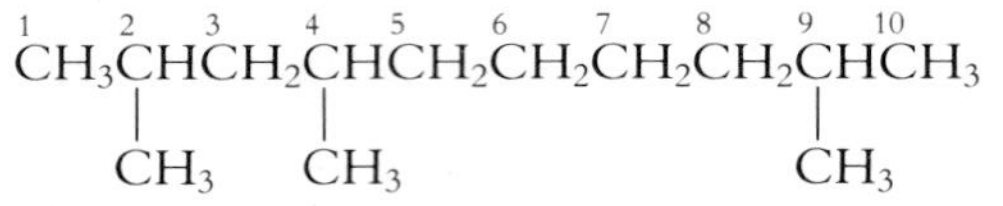

$$\begin{array}{cccccccccc} \scriptstyle 1 & \scriptstyle 2 & \scriptstyle 3 & \scriptstyle 4 & \scriptstyle 5 & \scriptstyle 6 & \scriptstyle 7 & \scriptstyle 8 & \scriptstyle 9 & \scriptstyle 10 \\ CH_3 & CH & CH_2 & CH & CH_2 & CH_2 & CH_2 & CH_2 & CH & CH_3 \\ & | & & | & & & & & | & \\ & CH_3 & & CH_3 & & & & & CH_3 & \end{array}$$

2,4,9-Trimethyldecane

$$\begin{array}{c} CH_3 \\ | \\ CH_3CCH_3 \\ | \\ CH_3 \end{array}$$

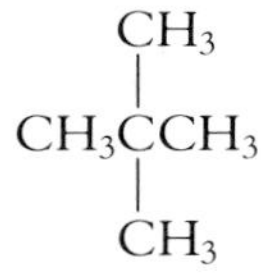

2,2-Dimethylpropane

5. In the compound name the order of substituent names is alphabetically determined (butyl before ethyl, methyl before propyl, etc.). The prefixes di, tri, and so forth are ignored when determining alphabetical order.

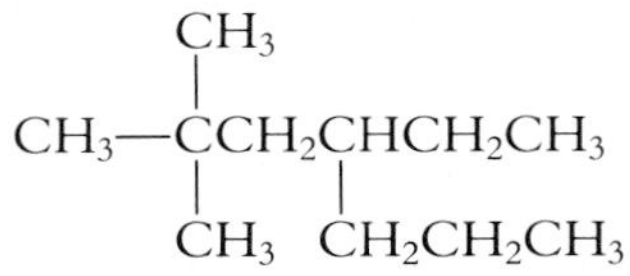

$$\begin{array}{llll} & CH_3 & & \\ & | & & \\ CH_3— & C & CH_2 & CHCH_2CH_3 \\ & | & & | \\ & CH_3 & & CH_2CH_2CH_3 \end{array}$$

4-Ethyl-2,2-dimethylheptane (alphabetical)

6. If numbering from either end of the principal chain provides the same number for the first substituent, choose the sequence that will provide the lowest number for the second substituent, and so forth (see 2,4,9-trimethyldecane, step 4).

Finally, we must touch upon the IUPAC system for naming alkyl substituents. Substituents with four or fewer carbon atoms are almost always denoted by the alkyl group names given in Table 3–2. Larger substituents are named by a system patterned on main chain nomenclature. The carbon atom of the substituent that is bonded to the main chain in numbered 1.

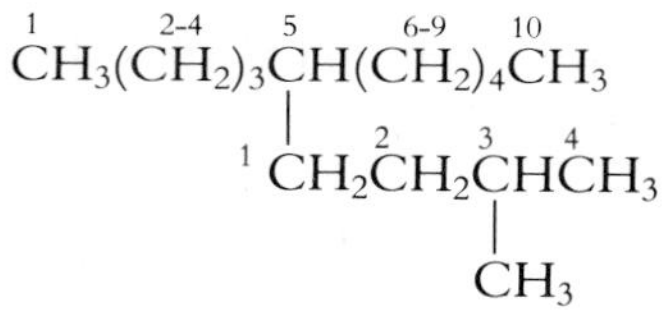

$$\begin{array}{lllll} \scriptstyle 1 \quad 2\text{-}4 & \scriptstyle 5 & \scriptstyle 6\text{-}9 & \scriptstyle 10 & \\ CH_3(CH_2)_3 & CH & (CH_2)_4 & CH_3 & \\ & | & \scriptstyle 2 \quad 3 & \scriptstyle 4 & \\ & {}^{1}\,CH_2 & CH_2CH & CH_3 & \\ & & | & & \\ & & CH_3 & & \end{array}$$

5-(3-Methylbutyl)decane

When you are first learning to number and name organic compounds it is often helpful to rewrite condensed structures so as to make explicit all the chain branches. This is particularly useful in calculating the length of the principal chain. Rewriting the structure $(CH_3)_3CCH_2CH_2CH(CH_3)_2$ as shown below, for example, makes clear that one methyl group in each set of parentheses is to be counted as part of the principal chain.

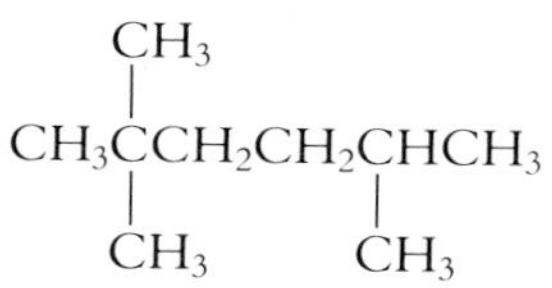

$$\begin{array}{llll} & CH_3 & & \\ & | & & \\ CH_3 & C & CH_2CH_2 & CHCH_3 \\ & | & & | \\ & CH_3 & & CH_3 \end{array}$$

2,2,5-Trimethylhexane

Example

(a) Name the following compound.
(b) Draw the structure of 2,3-dimethylbutane.

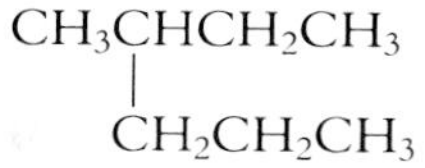

$$\begin{array}{l} CH_3\underset{|}{C}HCH_2CH_3 \\ \quad\; CH_2CH_2CH_3 \end{array}$$

Answer

(a) Number the carbon atoms in such a way as to find the longest chain:

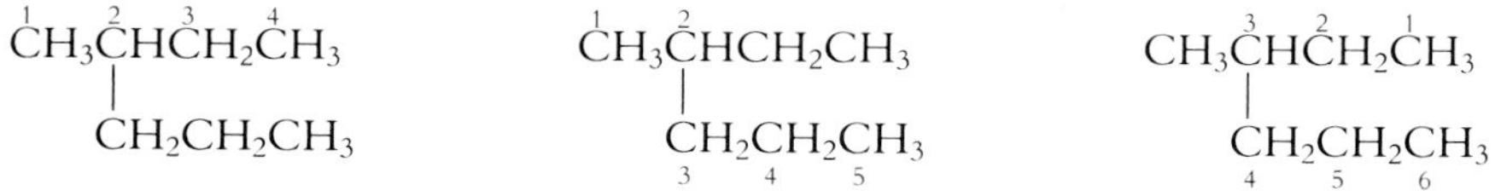

$$\begin{array}{l} \overset{1}{C}H_3\overset{2}{C}H\overset{3}{C}H_2\overset{4}{C}H_3 \\ \quad\; | \\ \quad\; CH_2CH_2CH_3 \end{array} \qquad \begin{array}{l} \overset{1}{C}H_3\overset{2}{C}HCH_2CH_3 \\ \quad\; | \\ \quad\; \underset{3}{C}H_2\underset{4}{C}H_2\underset{5}{C}H_3 \end{array} \qquad \begin{array}{l} CH_3\overset{3}{C}H\overset{2}{C}H_2\overset{1}{C}H_3 \\ \quad\; | \\ \quad\; \underset{4}{C}H_2\underset{5}{C}H_2\underset{6}{C}H_3 \end{array}$$

The last numbering sequence provides the longest chain and shows that the compound is a hexane. There is a methyl group on carbon 3, so the correct name is 3-methylhexane. (Note that if the main chain were numbered in the reverse order, the carbon atom bearing the methyl group would be numbered four, which is incorrect.)

(b) The best method for drawing the structure of a compound whose name is given is to work backward in the name. Since 2,3-dimethylbutane is a butane, write four carbons in a row:

$$C{-}C{-}C{-}C$$

Now put methyl groups on carbons 2 and 3:

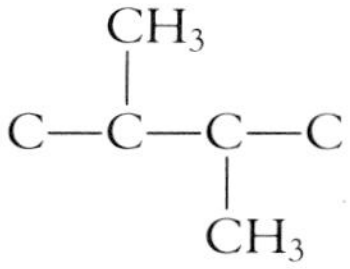

$$\begin{array}{c} \quad CH_3 \\ \quad | \\ C{-}C{-}C{-}C \\ \qquad\quad | \\ \qquad\quad CH_3 \end{array}$$

(It doesn't matter if you put both methyls on the same side or on opposite sides—the two structures are equivalent.)

Now fill in the proper number of hydrogen atoms:

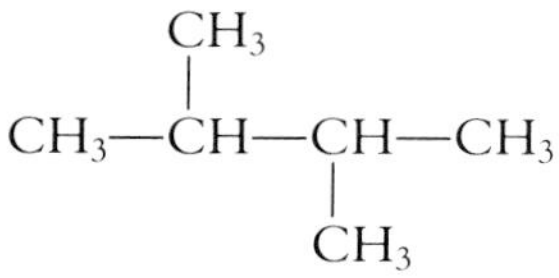

$$\begin{array}{c} \quad CH_3 \\ \quad | \\ CH_3{-}CH{-}CH{-}CH_3 \\ \qquad\qquad | \\ \qquad\qquad CH_3 \end{array}$$

2,3-Dimethylbutane

Problem

3–5 Name the following compounds according to the IUPAC system:

a. $CH_3CH_2CH(CH_3)_2$

b. $\begin{array}{l} \quad\; CH_3 \\ \quad\; | \\ CH_3CHCH_2CH(CH_3)_2 \end{array}$

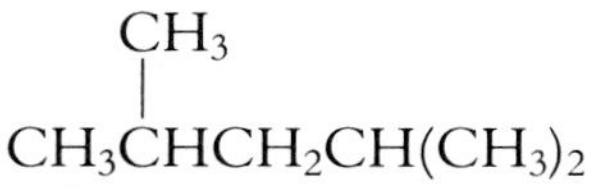

c. $\begin{array}{l} CH_3CH_2CHCH_2CH_2CH(CH_3)_2 \\ \qquad\qquad | \\ \qquad\qquad CH_2CH_3 \end{array}$

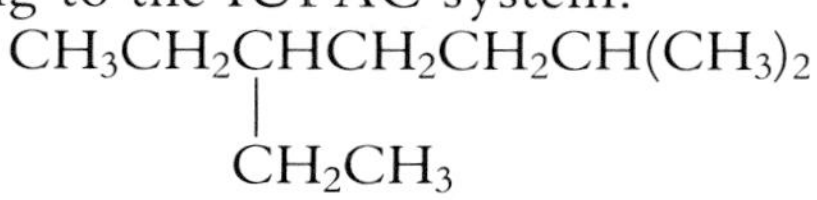

d. $\begin{array}{l} CH_3CH(CH_2)_3CH_3 \\ \qquad | \\ \qquad (CH_2)_4CH_3 \end{array}$

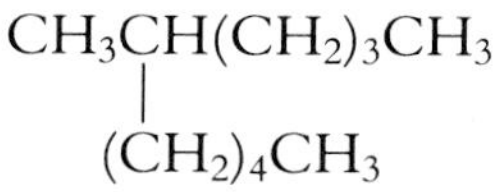

3:4 Properties of Alkanes

Table 3–3 records the names and physical properties of some straight-chain alkanes. The boiling points and melting points generally rise with increasing molecular weight. The irregularity in melting point difference reflects a difference between compounds with odd and even numbers of carbon atoms. Table 3–4 shows a comparison of some branched alkanes with their straight-chain isomers. Highly symmetrical compounds often melt at abnormally high temperatures because the molecules pack well in the crystal lattice (see the C_5 compound, 2,2-dimethylpropane, and the C_8 compound, 2,2,3,3-tetramethylbutane). There are several reasons why boiling point differences among alkane isomers are less dramatic than melting point differences. One very important reason is that the average distance between molecules is greater in the liquid than in the solid state, so molecular shapes have relatively little influence on intermolecular attraction in the liquid.

Liquid alkanes are colorless and flow easily. They are less dense than water and do not dissolve in it. Alkanes find their main use in the laboratory as solvents, since they are inexpensive and easily purified. Pentane and hexane are particularly popular because of their low boiling points, which make them easy to remove by evaporation. This property also makes them highly flammable, so they must be handled with care. Alkanes are chemically unreactive to most common reagents, such as acids and bases, but they do undergo characteristic reactions with certain reactive intermediates such as radicals (Chap. 5). Alkanes are usually biologically, as well as chemically, inert, but certain strains of bacteria can feed on them. These bacteria represent a potential means for turning petroleum into edible protein.

A mixture of alkanes, especially rich in straight-chain compounds, is found in the surface layer of most plant leaves. The compounds $C_{29}H_{60}$ and $C_{31}H_{64}$ are particularly abundant. (These waxy solids seem to help protect the leaf.) Lower plants, such as algae, also contain numerous alkanes, with a wide range of molecular weights. These findings support the theory of the origin of petroleum (sec. 3:1). They have also led to investigations of the possibility of specific cultivation of certain plants as sources of hydrocarbons.

Table 3–4. Comparison of Physical Properties of Straight-Chain and Branched Alkanes

Number of Carbon Atoms (MW)	IUPAC Name	Boiling Point (°C)	Melting Point (°C)
C_5 (72)	pentane	36	−130
C_5 (72)	2-methylbutane	28	−160
C_5 (72)	2,2-dimethylpropane	10	− 20
C_8 (114)	octane	126	− 57
C_8 (114)	2,2,3,3-tetramethylbutane	106	+101

3:5 Alkane Conformations: The Preferred Shapes of Alkane Molecules

In Chapter 2 we explored the connection between molecular bonding and molecular shapes. By using the electron pair repulsion or hybridization theories the geometry around individual atoms can be predicted and explained. For example, we used both theories to explain and predict the shape of methane, CH_4, but what happens when two methyl groups bond to give ethane, $CH_3—CH_3$? We find that there is an infinite number of possible shapes for the molecule because of the free rotation about the C—C bond. We want to know if any of these molecular shapes are preferred, and, if so, why.

You will find it very helpful to pursue this topic with the aid of molecular models, such as those shown in Figure 3–1. All of these models, which depict ethane, allow free rotation about the carbon-carbon bonds. It is by

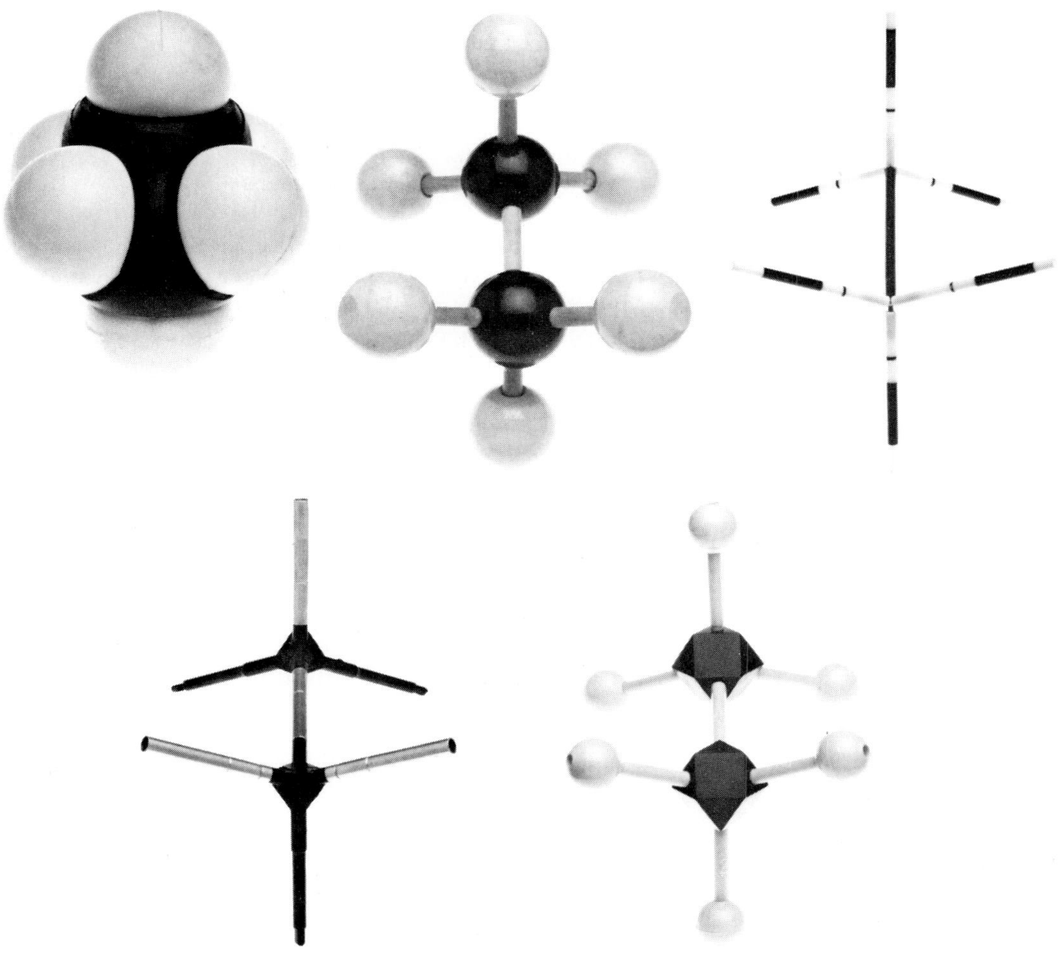

Figure 3–1. *Molecular models of ethane.*

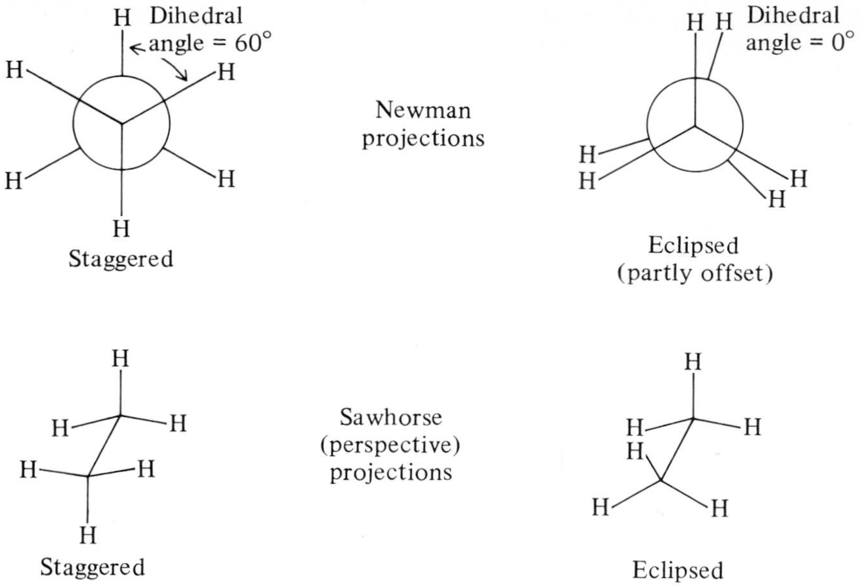

Figure 3–2. *Newman and sawhorse projections of the eclipsed and staggered conformers of ethane.*

means of this rotation that the ethane molecule changes shape. If rotation in the actual molecule is truly free, as it is in the models, then all molecular shapes would be equal in energy. To test this hypothesis, detailed calculations of the thermodynamic properties of ethane were carried out. The results indicated that rotation about the carbon-carbon bond is *not* completely free because some rotational arrangements, called **conformations,** are more stable than others. Conformational isomers, or *conformers,* are structures that can be interchanged by rotation about one or more single bonds. The most stable conformation of ethane is the **staggered conformer,** depicted in Figure 3–2. This figure also illustrates the least stable conformation, the **eclipsed conformer.** The conformers can be defined in terms of their *dihedral angles* which are the angles between C—H bonds on neighboring carbon atoms, measured when the molecule is viewed right down the C—C bond. In the staggered conformer the dihedral angles are 60°, 180°, and 300°, while in the eclipsed conformer they are 0°, 120°, and 240°.

In Figure 3–2 two different conventions are used to represent the three-dimensional structure of ethane. In the **Newman projection** we are looking directly down the carbon-carbon bond (an end view). The point at the center of the circle stands for the front carbon atom, while the circle itself stands for the rear carbon atom. In the Newman projection of the eclipsed conformation (Fig. 3–2), the central bond has been rotated slightly so that all H's are visible. In reality, an eclipsed conformation in a Newman projection has bonds on neighboring carbons arranged *exactly* one behind the other. In the **sawhorse** (or **perspective**) **projection** the line of sight intersects the carbon-carbon bond at a 45° angle. The carbon atom closest to us is at the lower left and the farther one is at the upper right. You will better appreciate these

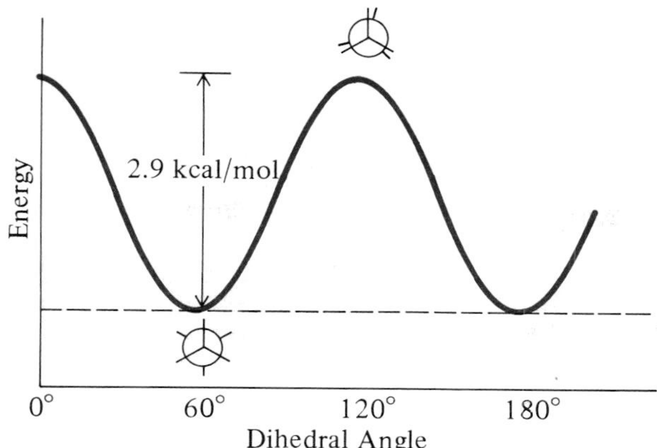

Figure 3–3. *This graph shows how the internal energy of ethane changes as the dihedral angle between adjacent C—H bonds changes. The energy is a minimum for dihedral angles of 60°, 180°, and 300° (staggered conformers). The energy of an eclipsed conformer, relative to a staggered one, equals the energy of the rotation barrier. If there were no barrier, the curve would be flat (dashed line).*

projections when you look at molecular models of eclipsed and staggered ethane from several different viewing angles.

An ethane molecule can go from a staggered to an eclipsed conformation by means of a C—C bond rotation. In doing so, its energy increases by 2.9 kcal/mol. We can therefore say that there is an *energy barrier* to rotation about the C—C bond in ethane. The height of this barrier is 2.9 kcal/mol. If there were free rotation in ethane there would be no rotational barrier.

Figure 3–3 shows how the internal energy of ethane varies as one methyl group is rotated while the other is kept fixed. At a dihedral angle of 0° the molecule is in the eclipsed conformation; at 60° it is in the staggered conformation. The dashed line shows what the situation would be if there were no energy barrier. In this case every conformation of ethane would have the same internal energy.

Where does the energy barrier come from? Finding the answer to this question is a source of continuing investigation and controversy. We know that atoms not directly bonded together can still attract and repel one another. These forces are called *nonbonded interactions.* The data for ethane and many other compounds suggest that in hydrocarbons the repulsive interactions are dominant. For example, the stable staggered conformer places adjacent C—H bonds as far apart as possible, supporting the theory that repulsion effects are predominant. When highly polarizable atoms such as the halogens are present, attractive forces become relatively more important and the relative energies of different conformers may be determined by attractive, rather than repulsive, forces.

Problems

3–6 What would the energy/dihedral angle curve look like if an attractive force between adjacent C—H bonds outweighed the repulsive ones?

Stronger repulsion than that between two hydrogens

Rotation around this bond

(a) (b)

Figure 3–4. *(a) Rotation of one methyl group of propane. (b) Newman projection of ethane showing eclipsing of methyl and hydrogen groups.*

3–7 How many *equivalent* staggered conformations are there for ethane? How many eclipsed ones? Is either of these conformers favored on statistical grounds?

Now consider propane (build a model), the next homolog of ethane. If one of the methyl groups of propane is rotated while the rest of the molecule is held fixed (Fig. 3–4a), we obtain an energy diagram that is similar to the one for ethane (Fig. 3–3). In the case of propane, however, the energy difference between the staggered and eclipsed conformation is affected by the eclipsing of a bulky methyl group and a hydrogen (Fig. 3–4b). This

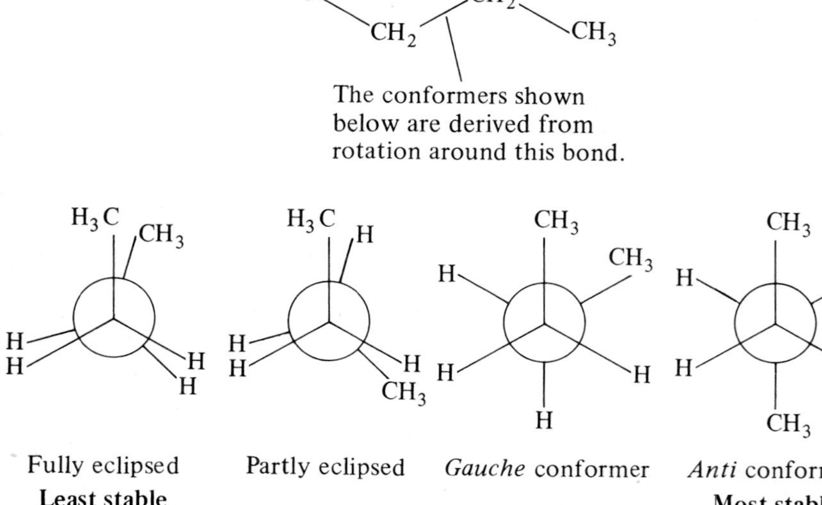

Figure 3–5. *Staggered and eclipsed conformers of butane.*

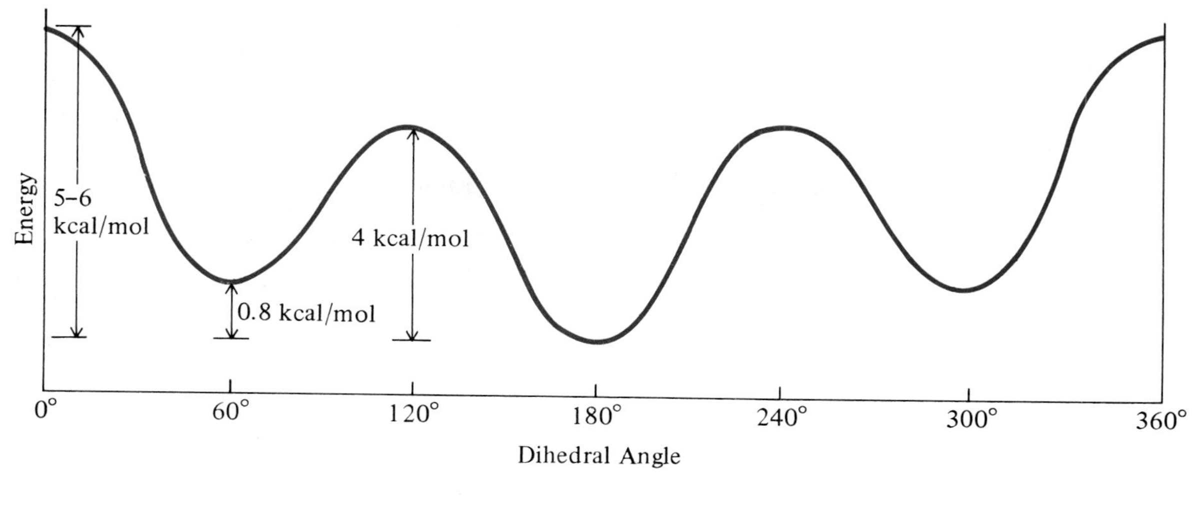

Figure 3–6. *Energy relationships between the various staggered and eclipsed conformations of butane.*

results in more repulsion (higher energy) than does a hydrogen-hydrogen eclipsing, and thus the rotational barrier in propane is 3.3 kcal/mol.

The energy diagram describing the rotation about the central carbon-carbon bond in butane is more complex than the diagrams for ethane or propane. There are *two* different staggered conformers and two eclipsed conformers (Fig. 3–5). The so-called *anti* conformer, in which the bulky methyl groups are as far apart as possible, is lowest in energy. Next comes the *gauche* (pronounced gōsh) conformer, also in a staggered conformation, in which the dihedral angle between methyl groups is only 60° (Fig. 3–6). This conformer is 0.8 kcal/mol higher in energy than the *anti* conformer. The conformation of highest energy is that in which the two bulky methyl groups eclipse one another (Figs. 3–5 and 3–6).

In all of the molecules we have considered the energy barrier is not large enough to prevent rotation at room temperature (25°C). At this temperature enough energy is available from intermolecular collisions to overcome energy barriers below about 15 kcal/mol. At room temperature, therefore, all the possible conformers are in rapid equilibrium. However, only the stable conformations are present in appreciable concentrations. Stable conformations are those that correspond to minima in the potential energy curve, such as the *gauche* and *anti* conformers of butane (Fig. 3–6).

Long-chain alkanes are quite flexible. Their most stable conformation is a zigzag one in which there is an *anti* conformation about each carbon-carbon

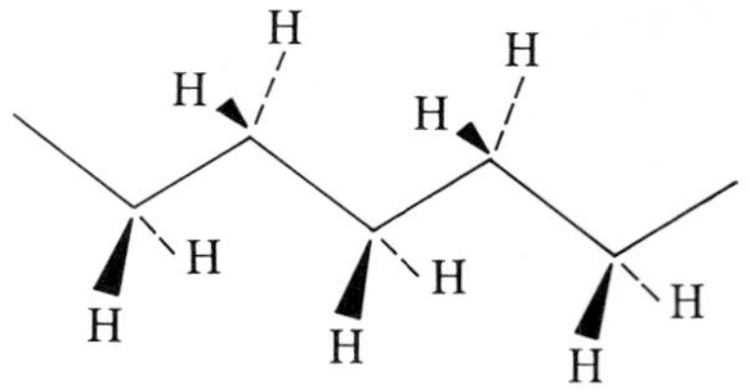

Figure 3–7. *Zigzag conformation in a straight-chain alkane.*

bond (Fig. 3–7). They are not confined to this conformation in solution, but may bend back on themselves.

Problems

3–8 Draw an energy diagram for internal rotation in $ClCH_2CH_2Cl$ (1,2-dichloroethane). Show the structures of the eclipsed and staggered conformers.

Box 3–2. Gasolines and Octane Numbers

Gasoline is a mixture of alkane hydrocarbons from the 40–180° boiling point fraction of petroleum. Some hydrocarbons are burned more efficiently than others in automobile engines and the performance of the gasoline can thus depend upon which alkanes are present in the greatest amounts. For example, branched-chain alkanes are superior to straight-chain compounds; too much of the latter will cause an engine to "knock" or "ping."

In order to measure relative efficiencies of gasoline it was decided to make up some standard gasolines based on pure alkanes. One of the poorest is *n*-heptane, while one of the best is 2,2,4-trimethylpentane, also known as isooctane. If a gasoline burns as well as 2,2,4-

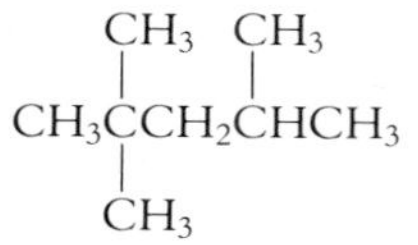

$$CH_3C(CH_3)_2CH_2CH(CH_3)CH_3$$

2,2,4-Trimethylpentane
(*isooctane*)

trimethylpentane in an engine, that gasoline is said to have an *octane number* of 100. A mixture of 80% 2,2,4-trimethylpentane and 20% *n*-heptane would have an octane number of 80; so would any gasoline blend with the same combustion properties.

The combustion process depends heavily on the presence of free radicals—species with unpaired electrons (sec. 1:3)—because they react so readily with oxygen. As you will soon see, branched-chain alkanes give rise to free radicals much more readily than do straight-chain alkanes. Thus, the higher the percentage of branched-chain alkanes in gasoline, the higher its octane number and the better its combustion characteristics. Unfortunately, the alkanes obtained directly from petroleum are mostly straight-chain compounds that must be converted to branched-chain hydrocarbons by chemical means. Therefore, increasing the octane number of gasoline means also increasing its price.

Another way to increase the desirable burning properties of a gasoline is to include small amounts of certain compounds called additives. Among the first additives discovered was tetraethyllead. This was used for

$$(CH_3CH_2)_4Pb$$

Tetraethyllead

many years to increase the octane number of gasolines until it became apparent that PbO_2 from the combustion was being deposited in ever greater amounts in the environment. Other additives are being developed to replace tetraethyllead in unleaded gasolines. One of the newest is methyl *t*-butyl ether, whose structure is similar to that of isooctane.

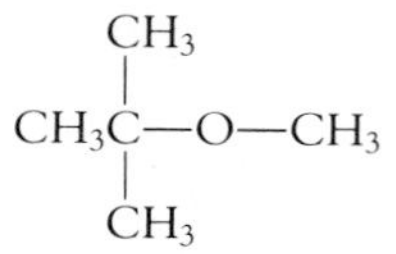

$$CH_3C(CH_3)_2—O—CH_3$$

Methyl *t*-butyl ether

What assumptions have you made in constructing your energy diagram (see comments on the origin of the energy barrier, p. 75)?

3–9 Draw the structures of the stable conformers that arise from rotation about the bond between carbons 2 and 3 of 2-methylbutane. Indicate which is (are) the most stable.

3–10 In 1-chloropropane, CH_3—$\overset{2}{CH_2}$—$\overset{1}{CH_2Cl}$, there are three staggered conformers that can be obtained by rotation about the C_1—C_2 bond.

a. Draw the structures (Newman projections) of the staggered conformers and label them *gauche* or *anti*. Which ones are equivalent?

b. The free energy difference between the *gauche* and *anti* conformers is 0.5 kcal/mol, with the *gauche* having the lower free energy. Draw an approximate energy diagram for internal rotation in 1-chloropropane.

3:6 Energy and Entropy Differences Between Conformers: Their Influence on Conformational Equilibrium

The 0.8 kcal/mol which separates the *gauche* and *anti* conformers of butane is an **enthalpy** difference, ΔH, which is the difference in heat content between these two conformers. This indicates that the two conformers differ with respect to bond energies (sec. 2:6) or nonbonded interactions (sec. 3:5). That enthalpy difference influences the equilibrium between them, as does the **entropy** difference, ΔS, between the two structures. Entropy differences reflect the fact that one structure or state is more *disordered* than another. Structures or states that are disordered are more probable than ones that are ordered. The more disordered state has the higher entropy.

Let's examine several processes involving butane to see how order and entropy are related. Solid butane is highly ordered because the molecules are arranged in fixed relationships to one another in the crystal lattice. Furthermore, individual butane molecules have little rotational freedom in the solid. When butane melts, the molecules move relatively easily in three dimensions (translational motion), and they also rotate among their various conformations (Fig. 3–8). Thus the liquid state of butane is more disordered, and therefore more probable, than the solid state. For this reason the entropy of butane *increases* when it melts (positive ΔS), which provides a strong driving force for the melting process.

Looking now at the liquid state of butane, we see that the molecules will be equilibrating between the *gauche* and *anti* conformers. Here again entropy plays an important role. There are *two* equivalent *gauche* conformers

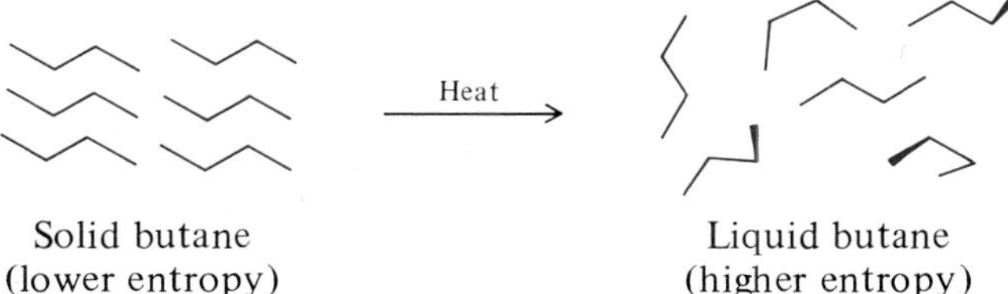

Figure 3–8. *Translational and conformational changes that take place when butane melts. In solid butane the molecules are locked in the* anti *conformation and are arranged in an ordered three-dimensional array. After melting they are free to move with respect to one another and are also conformationally mobile.*

of butane, but only *one anti* conformer. Thus the *gauche* → *anti* process goes from a less ordered to a more ordered state, or from a state of higher entropy to a state of lower entropy. Equation 3.2 summarizes the enthalpy and entropy contributions to the *gauche/anti* conversion. Since *anti*-butane is

(3.2) *gauche*-butane → *anti*-butane

ΔH is negative (favors *anti*)
ΔS is negative (favors *gauche*)

lower in energy while *gauche*-butane is more probable, how do we know which will predominate at equilibrium? The answer lies in the difference in **free energy,** ΔG, between the two. The free energy difference is a thermodynamic function that incorporates *both* the enthalpy and entropy values (eq. 3.3, where T is absolute temperature). A negative free energy value means

$$\Delta G = \Delta H - T\Delta S \tag{3.3}$$

that the products have less free energy than the reactants and that formation of products is favored. Thus a negative ΔG value for the reaction of Equation 3.2 means that formation of *anti*-butane is favored and that the *anti*-butane conformer predominates at equilibrium. A positive ΔG value means that the *gauche* conformer predominates at equilibrium.

Equation 3.3 tells us that the free energy difference changes with temperature because the relative importance of the ΔH and ΔS factors varies with changing temperature. At low temperature the $T\Delta S$ term is small, so the ΔH term is the predominant influence on ΔG. Under these conditions, then, ΔH is the principal factor determining the position of equilibrium. At low temperatures, therefore, we expect the *anti* conformer to be the major one, which is precisely what is found. At high temperatures (large T) the $T\Delta S$ factor outweighs the ΔH factor, making the *gauche* the favored conformer. This prediction is also confirmed by experiment. The relative proportion of each conformer varies smoothly as the temperature changes. At room temperature the *anti* to *gauche* ratio is about 2:1.

The energy, entropy, and temperature relationships which we derived for butane apply to chemical equilibria in general. That is why we have examined them in some detail. This discussion will prove useful when we discuss the dynamics of chemical reactions (Chap. 5) and periodically throughout the remainder of the text.

The rest of this chapter is devoted to the cycloalkanes. Their chemical and physical properties are very similar to those of the alkanes, as is their conformational behavior. You can apply to them the principles you have already learned.

Problems

3–11 Refer back to the stable conformers of 2-methylbutane (Prob. 3–9). Indicate which will predominate at low temperature and which at high temperature. Explain your answer.

3–12 Each of the following processes is characterized by a positive ΔS. Explain why, using the entropy/disorder correlation.

a. $CH_3O{-}OCH_3 \rightarrow 2CH_3O\cdot$
b. $NaCl(s) + nH_2O \rightarrow Na^+(hydr) + Cl^-(hydr)$
c. *anti*-1-chloropropane → *gauche*-1-chloropropane (Prob. 3–10)

3:7 Drawing and Naming Cycloalkanes

If one hydrogen atom were clipped from each end of a straight-chain alkane and the ends then joined, the resulting compound would be a **cycloalkane**. These compounds form a homologous series having the empirical formula $(CH_2)_n$ where $n = 3, 4, 5, \ldots$. Note that this empirical formula requires *two* fewer hydrogens than the formula for the alkanes. In polycyclic compounds (compounds that contain more than one ring) the number of hydrogen atoms is two fewer *per ring* than the number of hydrogen atoms in an alkane with the same number of carbon atoms.

Cycloalkanes are frequently known as alicyclic compounds — ring compounds that belong chemically to the aliphatic group (Table 1–4, p. 25)★. The chief function of this nomenclature is to distinguish cycloalkanes from aromatic compounds such as benzene, C_6H_6. The latter are chemically quite distinct but are also cyclic (sec. 1:6).

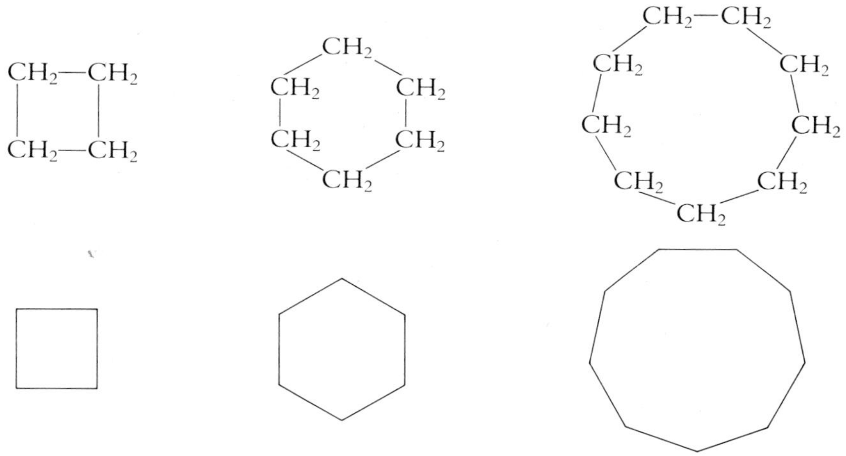

Cyclobutane, C_4H_8 Cyclohexane, C_6H_{12} Cyclononane, C_9H_{18}

Naming the cycloalkanes is a refreshingly simple matter—merely tack the prefix *cyclo* onto the name of the alkane having the same number of carbon atoms. Drawing the cycloalkanes is also quite simple. The molecule is represented by a regular plane polygon, with the same number of vertices as there are carbon atoms in the ring. It is understood that each vertex represents a CH_2 group unless substituents are specified. These representations are a type of shorthand and it is important to remember that they are not meant to imply that the molecules are planar. In fact, you will see shortly

★ Do not confuse "alicyclic" with "acyclic," which means *noncyclic*.

that most cycloalkanes form nonplanar rings.

Methylcyclopentane   1,3-Diethylcyclobutane

3:8 Angle Strain in Cycloalkanes: Why Most Cycloalkanes are Not Flat

Compounds with the cyclohexane structure are widespread in nature. For example, all steroids and many antibiotics and other medicinals contain cyclohexane rings. We will consider the structure and properties of cyclohexane in some detail because what you learn about cyclohexane can be applied to more complex compounds.

At one time it was thought that certain ring compounds could not be synthesized because of unfavorably strained bond angles. Once these rings had in fact been synthesized and also found in naturally occurring substances, attempts were made to formulate a structural theory that would account for their properties.

In 1885, Adolph von Baeyer proposed a theory based on two assumptions:

1. that the ideal bond angles at carbon in saturated compounds were tetrahedral (109.5°); and
2. that the monocycloalkanes (single-ring structures) were regular plane polygons.

If Baeyer's hypotheses were true and cyclopentane was planar, the C—C—C bond angles would actually be 108°. Since the ideal value for those bond angles is 109.5°, there would not be a great deal of discrepancy between the actual situation and the ideal one. Planar cyclohexane, on the other hand, would have C—C—C bond angles of 120°, a significant deviation from the ideal 109.5° (Fig. 3–9). As the ring size increased, so would the gap between the real and the ideal values of the bond angle. Baeyer proposed that bonds forming nonideal angles (angles other than 109.5°) would be

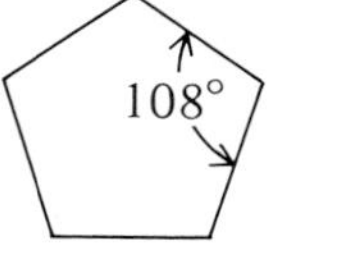

Planar cyclopentane

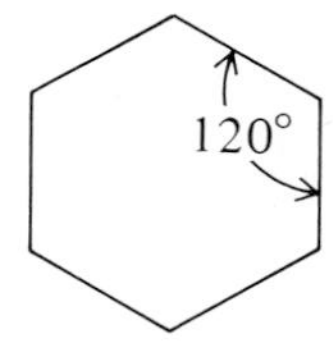

Planar cyclohexane

Figure 3–9. *C—C—C bond angles of planar cyclopentane and planar cyclohexane.*

"strained" and that this strain would result in a higher energy (lower stability) for the entire molecule. Furthermore, Baeyer proposed that the difference between the ideal and actual values would be a measure of the strain in that molecule.

In order to test Baeyer's theory it was necessary to conduct an experiment to detect the strain energy in a molecule. One way would be to determine the total amount of energy in a molecule because a strained molecule would have more total energy than an unstrained one. The amount of energy present in compounds such as cycloalkanes can be determined by burning them in oxygen to produce CO_2 and H_2O (eq. 3.4, where n is the number of CH_2 groups in the cycloalkane). This reaction always gives off heat (it is **exothermic**) and the amount of heat liberated is called the **heat of combustion.** When the molar heat of combustion is divided by n, then the heat of combustion per CH_2 group is obtained and the energy content of two cycloalkanes can be compared.

$$(CH_2)_n + \frac{3n}{2} O_2 \rightarrow nCO_2 + nH_2O + \text{heat} \tag{3.4}$$

For example, the molar heat of combustion for cyclopropane is 500 kcal/mol. Since cyclopropane contains three CH_2 groups, the heat of combustion per CH_2 is 500/3 or 167 kcal/mol. The molar heat of combustion of cyclopentane is 794 kcal/mol and the heat of combustion per CH_2 would be 794/5 or 159 kcal/mol. Cyclopropane gives off more heat per CH_2 than cyclopentane because it contains more energy, mainly because of the strain in the three-membered ring, which requires a C—C—C bond angle of 60°. In other words, the gap between ideal and actual values for the bond angle for cyclopropane is 49.5° (109.5° - 60°), while for cyclopentane it is only 1.5° (109.5° − 108°). Figure 3–10 shows how the heat of combustion per CH_2 varies with ring size.

The strain energy in *any* ring can be quantitatively measured by comparing its heat of combustion per CH_2 group to that of an unstrained molecule (a standard). The standard that is used is a long-chain n-alkane. The heat of

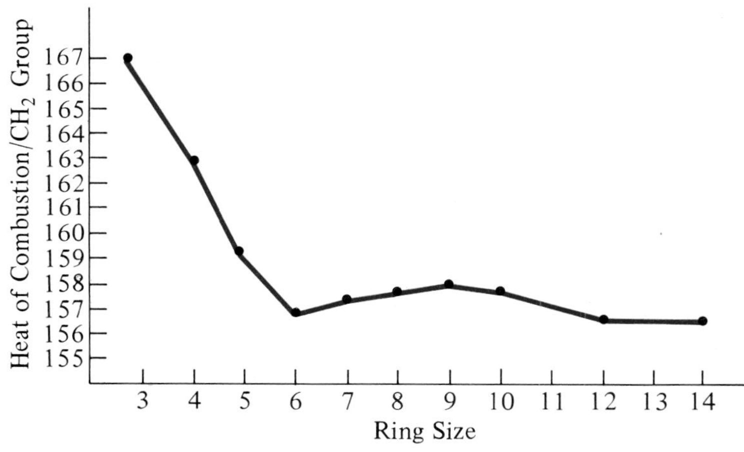

Figure 3–10. *Variation of heat of combustion per CH_2 group with ring size.*

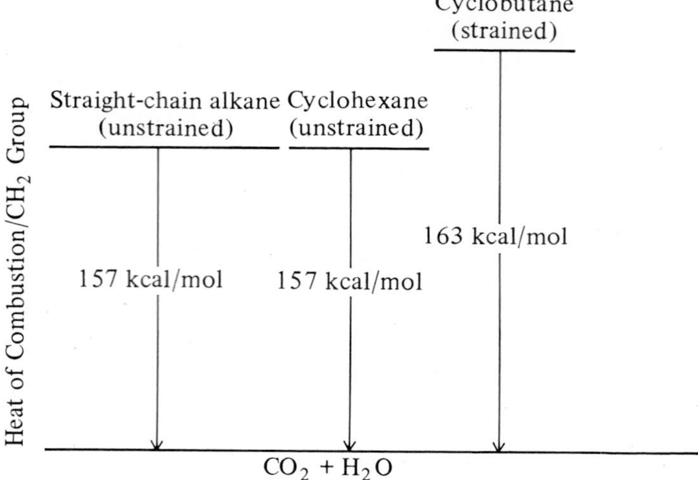

Figure 3–11. *Heat of combustion per CH_2 group of strained and unstrained cycloalkanes.*

combustion per CH_2 group for long-chain alkanes is approximately 157 kcal/mol. Figure 3–11 shows how the comparison is then made between a strained and unstrained cycloalkane. Since cyclopentane has a heat of combustion per CH_2 of 159 kcal/mol, it is slightly strained when compared with the standard (157 kcal/mol). The heat of combustion of cyclohexane is 942 kcal/mol or 157 kcal/mol per CH_2 group (942/6). Thus, by this experimental test, there is *no* strain energy in cyclohexane at all since it has the same heat of combustion per CH_2 group as an unstrained straight-chain alkane. We would have to conclude that there is a flaw in Baeyer's original theory.

In 1890 the German chemist H. Sachse proposed an alternative model for cyclohexane. Sachse noted that it was possible to construct a cyclohexane ring in which all the C—C—C bond angles equaled 109.5°, *as long as the ring could be puckered rather than flat.* He identified two distinctly different conformations that such a puckered ring could adopt—the **chair conformer** and the **boat conformer** (Fig. 3–12). (We *strongly* recommend that you construct models of these conformers, become familiar with their geometry, and be able to draw them. Study the models from several different viewing angles.)

Direct proof of the existence of nonplanar cyclohexanes came from electron and x-ray diffraction studies in 1943. For their work in confirming and elaborating the consequences of Sachse's theory, the Norwegian chemist O. Hassel and the British chemist D. H. R. Barton were jointly awarded the Nobel Prize for Chemistry in 1969. This field of investigation is now known as **conformational analysis**.

Problem

3–13 Suppose you have a jet engine which runs on a hydrocarbon fuel. You can use any alkane or cycloalkane that you wish. Which would you choose to get maximum power? Justify your answer.

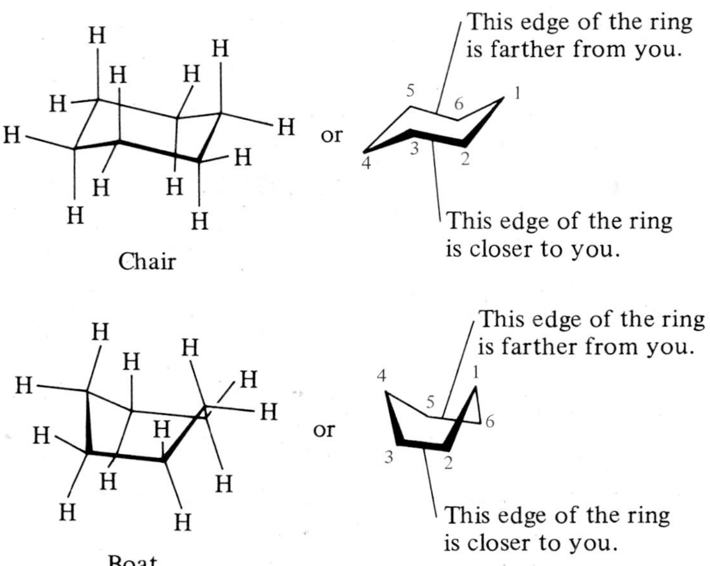

Figure 3–12. *Chair and boat conformations of cyclohexane.*

The remainder of this chapter will illustrate the application of conformational analysis to a variety of ring sizes. The greatest emphasis will be on cyclohexane, reflecting the importance of the six-membered ring with respect to all other ring sizes. It is extremely abundant in nature because of its great stability.

3:9 The Conformations of Cyclohexane and How They Interconvert

Figure 3–13 is a Newman projection (an end view) of cyclohexane in the

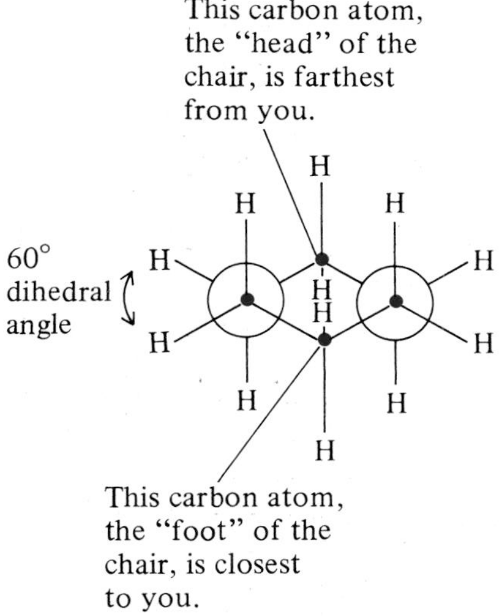

Figure 3–13. *Newman projection of chair conformation of cyclohexane.*

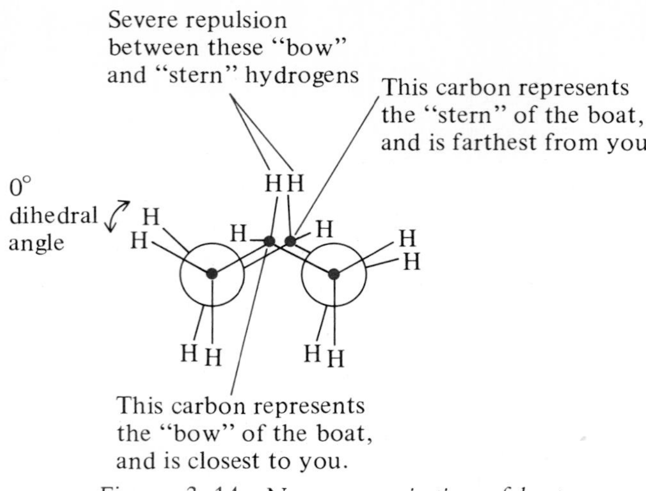

Figure 3–14. *Newman projection of boat conformation of cyclohexane (seen from the "bow").*

chair conformation. Note that all the C—H and C—C bonds are staggered, very much like the *gauche* conformer of butane (Fig. 3–5). Since all the CH_2 groups are equivalent, the same projection is obtained if the ring is rotated around its central axis. A Newman projection of cyclohexane in the boat conformation (Fig. 3–14, seen "bow on") indicates that there is extensive eclipsing of hydrogen atoms, and resembles the fully eclipsed form of butane. A side view of the boat form (Fig. 3–15) reveals that there is a particularly strong repulsion between the inside hydrogens on the "bow" and "stern" atoms.

With molecular models you can see that the boat form is quite flexible, in contrast to the chair, which is rigid. Some of the repulsion energy in the boat may be lost if the "bow" twists to one side while the "stern" moves in the opposite direction (Fig. 3–16). The resulting **twist boat** conformation is of

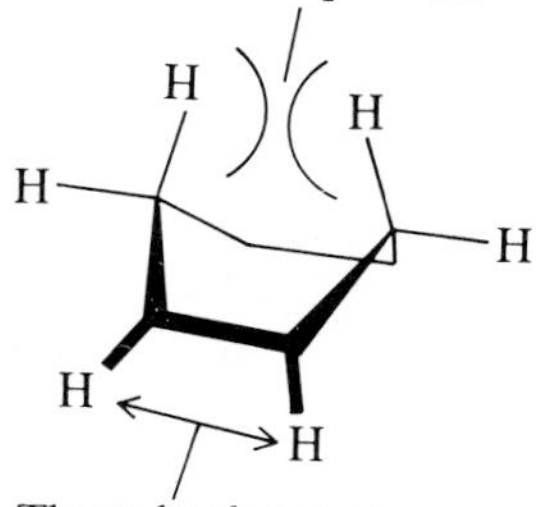

Figure 3–15. *Boat conformation (side view) showing severe repulsion between "bow" and "stern" hydrogen atoms.*

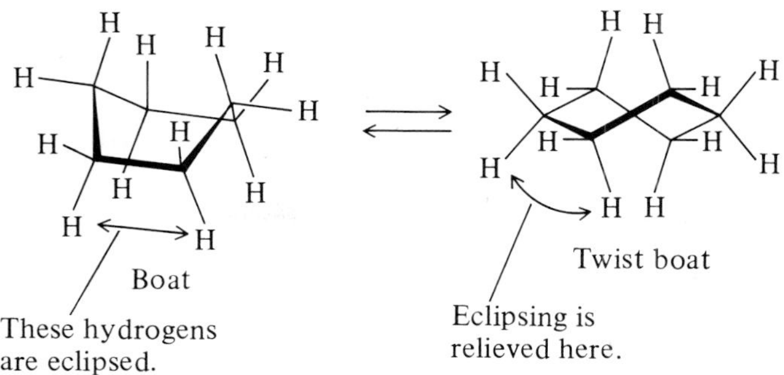

Figure 3–16. *Boat ⇆ twist boat conformational equilibrium of cyclohexane.*

somewhat lower energy than the pure boat but still has about 4.4 kcal/mol more free energy than the chair at room temperature (25° C).

The chair conformation for cyclohexane and most of its derivatives is the predominant conformer (at room temperature) because it has the lowest free energy of the three conformations and thus is the most stable. For this reason we will concentrate almost exclusively on chair conformers. It will be important for you to be able to draw them in a recognizable form. Although there are inexpensive templates available for this purpose, it is desirable (and sometimes necessary) to be able to draw them freehand. Figure 3–17 illustrates a step-by-step procedure for doing so.

Problems

3–14 a. Draw a cyclohexane chair conformer with the opposite orientation of that shown in Figure 3–17.

b. Draw a projection structure in which two cyclohexane rings share a common edge.

3–15 Equation 1 shows the enthalpy and free energy difference between the twist boat and chair conformers of cyclohexane at 25°. Which conformer has the greater entropy (sec. 3:6)? Build models of both conformers, and use them to explain the entropy difference. (Make sure you pay attention to the mechanical differences between the two.)

(1)

$$\Delta H = -5.5 \text{ kcal/mol}$$
$$\Delta G = -4.4 \text{ kcal/mol}$$

Cycloalkanes have much less internal rotational freedom than do acyclic alkanes. Nonetheless, restricted rotation about C—C single bonds is still possible, allowing interconversion of the different ring conformers, as the following section shows.

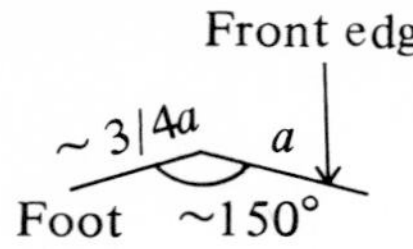

Draw roof-shaped figure with fairly large angle, one side (length *a*) longer than the other

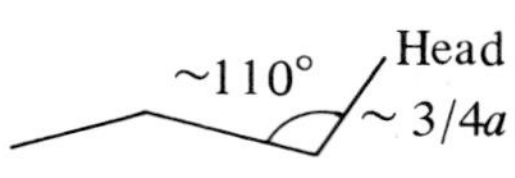

Draw bond from front edge to "head," making angle slightly larger than right angle

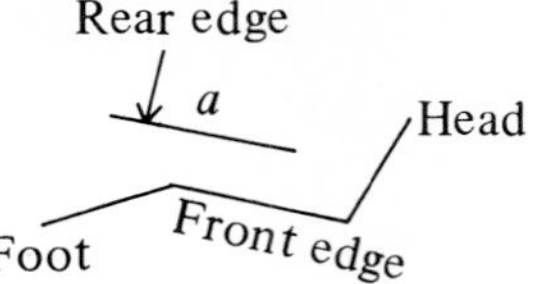

Draw line parallel to front edge, above it and displaced to the left

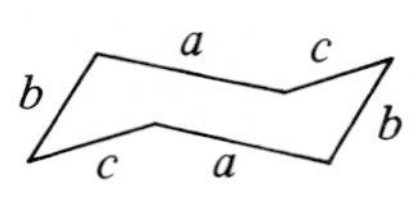

Connect remaining line ends. Sides marked with same letters should be parallel

Figure 3–17. *A step-by-step procedure for drawing a cyclohexane chair conformer.*

3:10 How the Axial and Equatorial Hydrogens of Cyclohexane Interconvert: The Preferred Conformations of Monosubstituted Cyclohexanes

If a model of chair cyclohexane is held in a horizontal orientation and viewed from the side, as in Figure 3–18, two different types of hydrogen atoms may be identified:

1. Six hydrogen atoms are connected to the ring by vertical bonds; these **axial** hydrogen atoms lie alternately above and below the plane of the ring★ (see Fig. 3–18).
2. Six hydrogen atoms form a belt around the periphery of the ring. They are connected to the carbon atoms by bonds that lie close to the ring plane. These are **equatorial** hydrogen atoms (Fig. 3–18). (Note that the C(1)—H equatorial bond is parallel to the C(2)—C(3) bond, the C(2)—H bond is parallel to the C(3)—C(4) bond, and so on.)

The models of cyclohexane show that *two* different kinds of hydrogens, axial and equatorial, are present in the molecule. Cyclohexane is, however, continually undergoing an equilibrium process that rapidly interconverts axial and equatorial hydrogen atoms. This process is illustrated in Figure 3–19. Chair 1 cyclohexane is converted very rapidly at room temperature to twist boat cyclohexane (sec. 3:9). When the twist boat reverts back to the chair conformation, it can yield the identical chair from which the process started (Chair 1), *or* it can yield an equivalent chair (Chair 2). All the axial hydrogens in Chair 2 were equatorial in Chair 1, and all the equatorial hydrogens in Chair 2 were axial in Chair 1. Since the two chairs are interconverting so rapidly, the axial and equatorial hydrogens cannot be distinguished chemically.

★ Although the chair is puckered, it is not far from planar. Orienting the ring in a horizontal position, as in Figure 3–18, makes the vertical position of the axial bonds obvious.

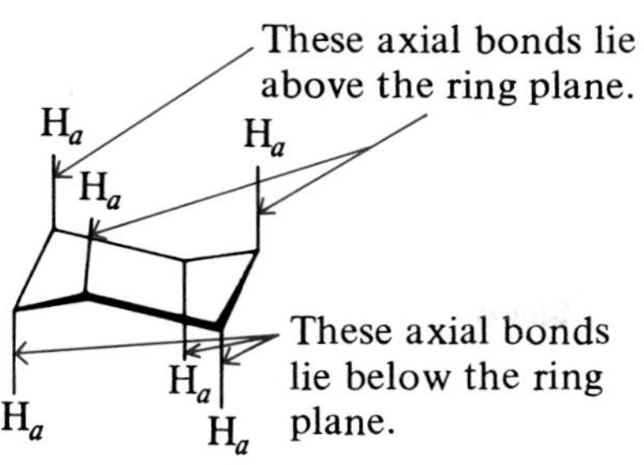

Axial hydrogens
All axial bonds lie perpendicular to the ring plane.

(a)

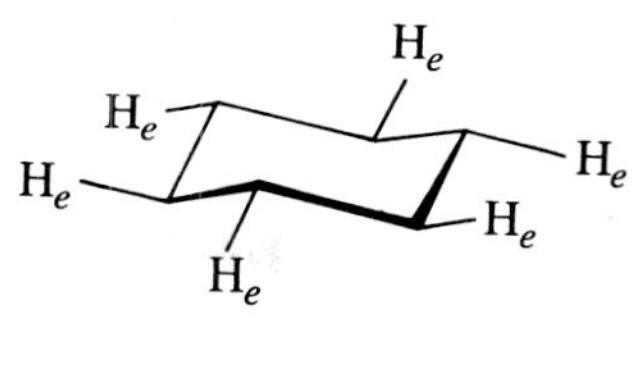

Equatorial hydrogens
All equatorial bonds lie approximately in the ring plane.

(b)

Figure 3–18. *The axial (a) and equatorial (b) orientations of the hydrogen atoms of chair cyclohexane.*

The rapid interconversion of chair conformers is more easily understood when we examine substituted cyclohexanes. Consider methylcyclohexane. Since the methyl group can be axial or equatorial, we would expect to isolate two isomeric methylcyclohexanes. In fact, only one methylcyclohexane is found at room temperature and it is a mixture of the axial and equatorial conformers. The two conformers are interconverted (eq. 3.5) when the ring undergoes its normal chair–chair "flip" (Fig. 3–19). This interconversion is so fast at 25° that neither conformer can be isolated, so we are always dealing with a mixture of the two. The mixture contains mostly the equatorial conformer because it has a lower free energy than the axial conformer.

(3.5)

H, CH_3 (fast / at 25°) H, CH_3

More stable (~95%)

Less stable (~5%)

3:10 Interconversion of Axial and Equatorial Hydrogens

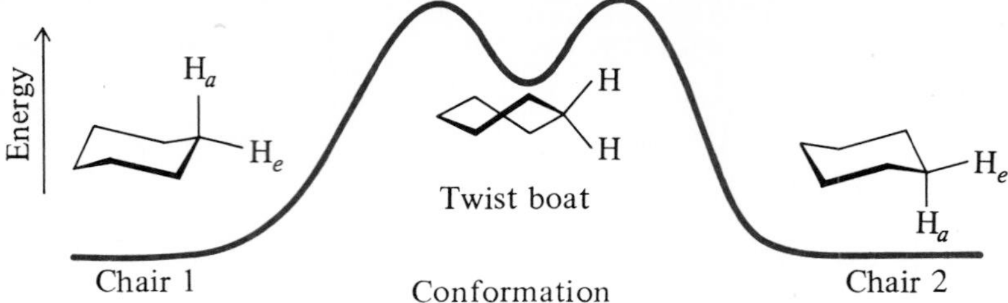

Figure 3–19. *Energy diagram of the equilibrium interconversion of cyclohexane's equatorial and axial hydrogens.*

Figure 3–20 shows a Newman projection of methylcyclohexane in the axial conformation as one looks down the C_1—C_2 bond. This projection shows that the axial methyl group is separated by a dihedral angle of only 60° from the methylene (CH_2) group at the C_3 carbon. The interaction between the methyl and the methylene is very much like the *gauche* interaction in butane (Fig. 3–5). Figure 3–21 shows that when the methyl group is equatorial then the system resembles the more stable *anti* conformer of butane. Equatorial methylcyclohexane has 1.7 kcal/mol less free energy than the axial conformer, and a sample of methylcyclohexane at room temperature consists of about 95% equatorial conformer.

The origin of the repulsive interaction with axial substituents can be better appreciated with the aid of a model (Fig. 3–22). You can see that when the methyl group rotates, it approaches closely the axial hydrogens at C-3 and C-5. These are known as **1,3-diaxial interactions**. They will obviously become more severe as the substituents on the ring become larger. In *t*-butylcyclohexane, which has the very large *t*-butyl substituent, essentially all the molecules are in the equatorial conformation at room temperature.

Problem

3–16 a. With the aid of models draw an equation illustrating the equilibrium between the two chair forms of *t*-butylcyclohexane, with the *t*-butyl group in the axial and the equatorial positions. For the axial conformer show the destabilizing interaction between the *t*-butyl group and the appropriate axial ring hydrogens.

b. Do the same for isopropylcyclohexane.

c. The destabilizing effect of axial isopropyl is much less severe than that of axial *t*-butyl. Why?

Figure 3–20. *Newman projection of a cyclohexane ring with an axial methyl group.*

Figure 3–21. *Newman projection of a cyclohexane ring with an equatorial methyl group.*

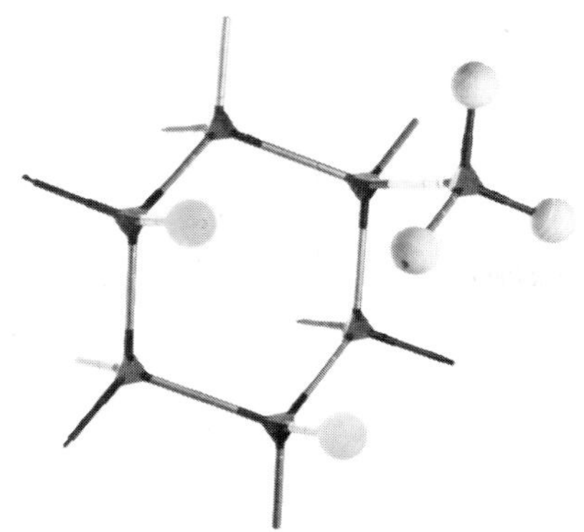

Figure 3–22. *Molecular model of methylcyclohexane.*

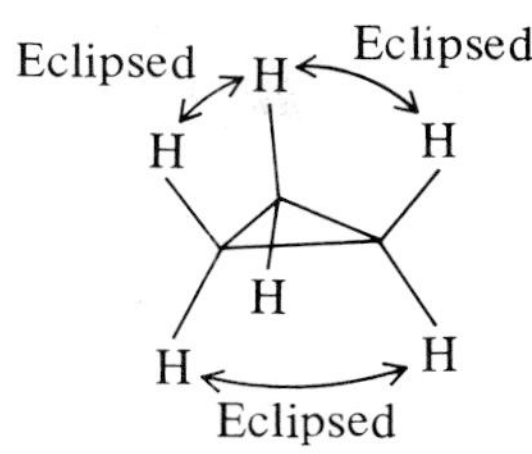

Figure 3–23. *Hydrogen eclipsing in cyclopropane.*

Rings larger than cyclohexane have several low energy conformations, and thus do not adopt one that is predominant. Since the rings are also very flexible they equilibrate rapidly among the favored conformations. There is relatively little strain energy in C_7 and larger cycloalkanes, so their chemistry, for the most part, is similar to that of cyclohexane. By contrast, some small rings experience considerable strain (Fig. 3–10). Because this strain has important chemical consequences, the small rings will be examined individually in the next section.

3:11 The Shapes of Cyclopropane, Cyclobutane, and Cyclopentane: Strained Hydrocarbons

In cyclopropane the ring carbons lie in a plane so that all the C—H bonds are eclipsed (Fig. 3–23). This eclipsing, as well as the required bending of the C—C—C bond angle from the preferred 109° to 60°, makes cyclopropane highly strained. There are no hybridization states of carbon that have 60° angles between orbitals. The overlap of orbitals on adjacent carbons must therefore be incomplete because the orbitals are not directed along the bond axis (Fig. 3–24). The weakness of the resulting C—C bonds, along with the repulsion between eclipsed C—H bonds, shows up in the large strain energy of cyclopropane. This amounts to 10 kcal/mol per CH_2 group, or 30 kcal/mol total, using cyclohexane as the unstrained standard (Fig. 3–10).

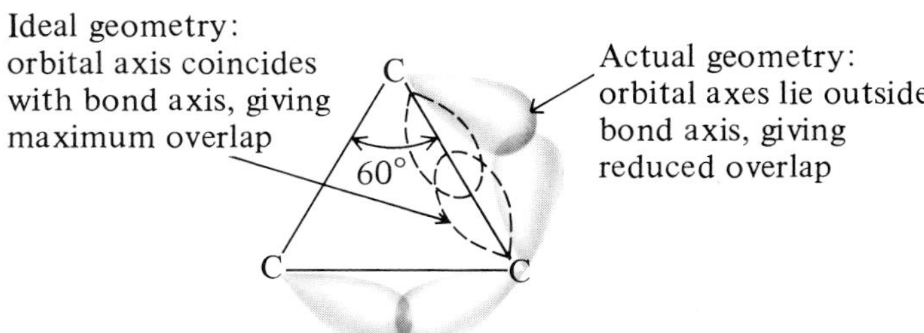

Figure 3–24. *Reduced orbital overlap in cyclopropane, resulting in weak C—C bonds.*

The chemical consequence of this strain is that three-membered rings are rather easily ruptured, under conditions where rings of larger sizes are stable. You will see this effect when we discuss the oxiranes, three-membered rings that consist of two CH_2 groups and an oxygen atom (sec. 6:13).

Bond angle distortion is not quite as severe in cyclobutane as it is in cyclopropane, but, again, all C—H bonds would be eclipsed if cyclobutane were planar. Cyclobutane therefore exists as a mixture of rapidly interconverting, equivalent folded conformations (eq. 3.6).

(3.6)

As Baeyer correctly predicted (sec. 3:8), planar cyclopentane experiences very little angle strain. However, he did not know that the planar conformation is destabilized by 10 sets of eclipsed C—H bonds. Cyclopentane therefore exists preferentially as a mixture of nonplanar conformers that rapidly interconvert (Fig. 3–25).

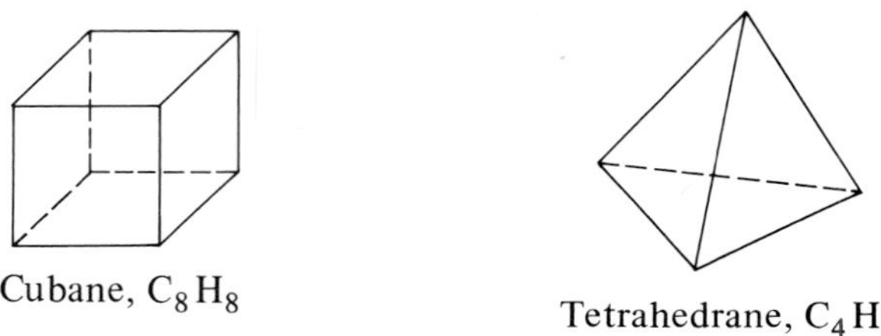

Cubane, C_8H_8

Tetrahedrane, C_4H_4

The synthesis of highly strained compounds has led to some interesting information about bond lengths and bond angles. For example, in cubane all the C—C bond lengths are close to those of "normal" C—C bonds, that is, about 1.54 Å, while the bond angles are near 90°. Lengthening bonds would decrease the strain, but apparently this is a very high energy process. Instead, the molecule adapts to 90° bond angles, which are considerably smaller than the expected 109.5° tetrahedral bond angles. An even more strained molecule, tetrahedrane, has up to now never been synthesized in spite of many attempts.

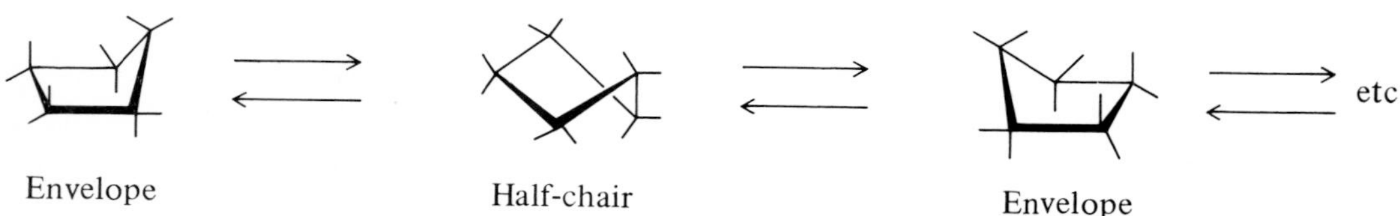

Figure 3–25. *Equilibrium among the stable conformers of cyclopentane. There are five equivalent envelope conformations and five equivalent half-chair conformations.*

3–17 a. Make a model of cyclobutane and, from your model, draw planar cyclobutane showing the eclipsing C—H bonds.
b. Use the heat of combustion data from Figure 3–10 to estimate the total strain energy of cyclobutane.

3:12 *cis–trans* Stereoisomerism in Disubstituted Cycloalkanes

The presence of one substituent on cyclic molecules does not present any special structural problems; however, the presence of *two* substituents on a cycloalkane does introduce a structural complexity we have not previously encountered. In this section we discuss disubstituted derivatives of small rings, and in section 3:13, the disubstituted cyclohexanes.

Two isomeric 1,2-disubstituted cyclopropane derivatives can exist, such as those shown in Figure 3–26 for 1,2-dimethylcyclopropane. The structure showing two methyl groups on opposite sides of the ring is called the ***trans isomer*** (from the Latin *trans,* meaning "across"). The other structure has the two methyl groups on the same side of the ring and is called the ***cis isomer*** (from the Latin *cis,* meaning "on the same side"). *Cis* and *trans* isomers such as these are *not* constitutional isomers of one another (sec. 1:5) because they have the same bond connectivities. The two compounds differ, however, in the way their bonds are directed in space. Such isomers are called **stereoisomers**. *Cis–trans* isomers are also called **configurational isomers** because they differ in spatial configuration.

There is a third dimethylcyclopropane, 1,1-dimethylcyclopropane, that is not a stereoisomer of the other two. In this structure the methyls are on

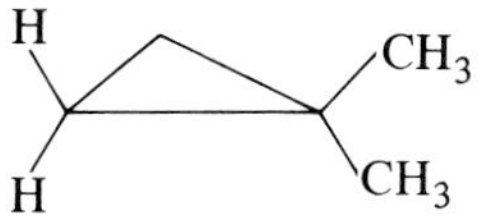

H, H, CH_3, CH_3

1, 1-Dimethylcyclopropane

opposite sides of the plane of the ring but are attached to the same carbon. This is not considered a *trans* compound since there is no corresponding *cis* isomer. 1,1-Dimethylcyclopropane is a constitutional isomer of *cis*-1,2-dimethylcyclopropane and of *trans*-1,2-dimethylcyclopropane.

When determining the number of stereoisomers for a given structure, treat the compound *as if it were* planar and not undergoing conformational equilibrium. Disubstituted cyclobutanes, for example, have two pairs of *cis–*

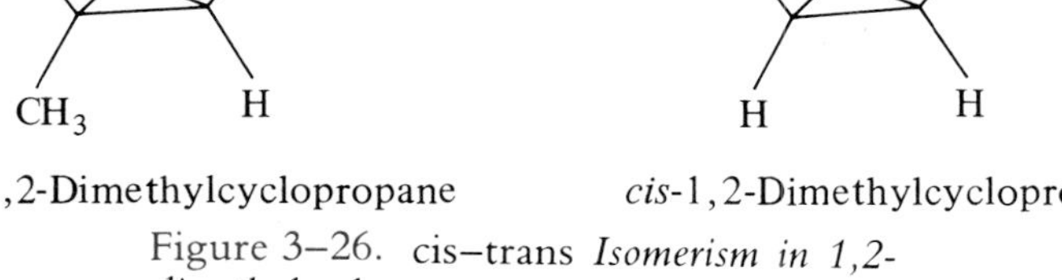

H, CH_3, CH_3, H

trans-1,2-Dimethylcyclopropane

CH_3, CH_3, H, H

cis-1,2-Dimethylcyclopropane

Figure 3–26. cis–trans *Isomerism in 1,2-dimethylcyclopropane.*

trans isomers; one pair for the 1,2 derivative and one pair for the 1,3 compound.

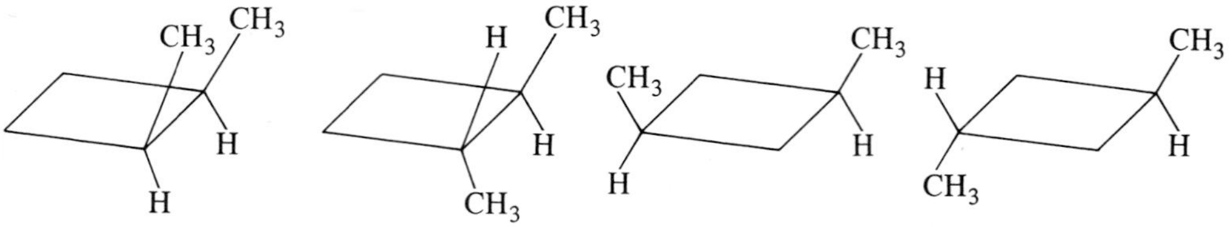

cis-1,2-Dimethylcyclobutane

trans-1,2-Dimethylcyclobutane

cis-1,3-Dimethylcyclobutane

trans-1,3-Dimethylcyclobutane

Problems

3–18 The heat of combustion of *cis*-1,2-dimethylcyclobutane is greater than that of the *trans* isomer. Offer an explanation for this observation.

3–19 Draw three-dimensional representations for all the *cis–trans* isomers for dimethylcyclopentane. Which pairs are stereoisomers?

3:13 *cis–trans* Isomers of Disubstituted Cyclohexanes

The number of *cis* and *trans* isomers for disubstituted cyclohexanes can also be determined by ignoring the conformational equilibrium and treating the cyclohexane ring *as if it were* planar. In addition to the 1,2 disubstituted cyclohexanes (Fig. 3–27), there are *cis* and *trans* isomers of 1,3 and 1,4 disubstituted cyclohexanes. Figure 3–28 shows several *cis* and *trans* isomers of dimethylcyclohexane in which the wedge–dash convention (sec. 1:4) is used. Remember that a wedge-shaped bond is above the plane of the paper and a dashed line bond goes to a substituent below the plane of the paper. Ordinary lines represent bonds which lie in the plane of the paper.

When conformational analysis is applied to *cis* and *trans* isomers of cyclohexane systems, we obtain some interesting results. *Each* stereoisomer can exist in two conformations. Equations 3.7 and 3.8 depict this conformational interconversion for the *trans* and *cis* isomers of 1,2-dimethylcyclohexane, respectively.

(3.7) *trans-ee* ⇌ *trans-aa*

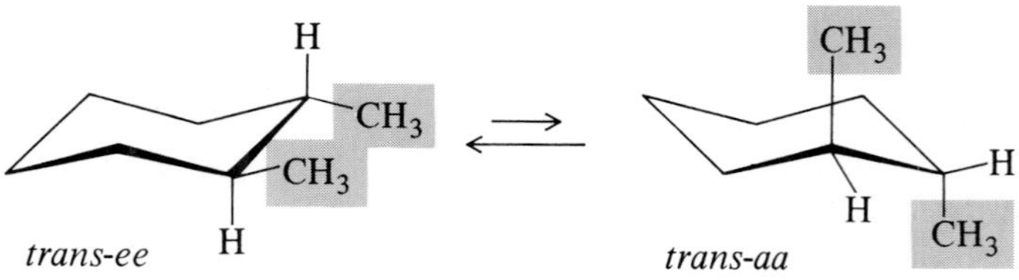

(3.8) *cis-ea* ⇌ *cis-ae*

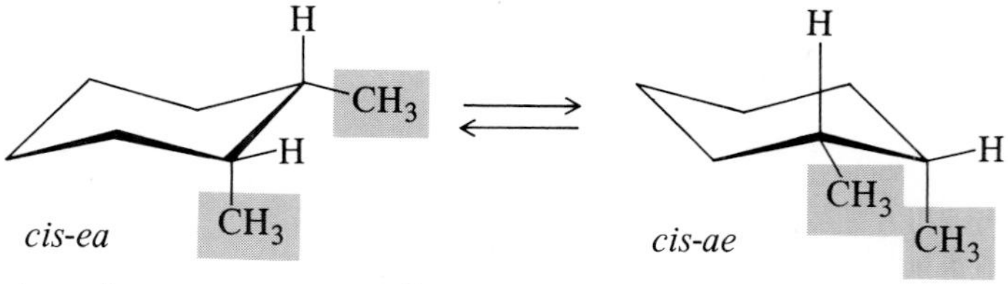

In the *trans* isomer, one conformer (the *ee*) is much lower in energy than the other, since both of its substituents are *equatorial*. The other conformer, with its two *axial* substituents, is relatively high in energy (eq. 3.7). Axial

cis-1,2-Dimethylcyclohexane *trans*-1,2-Dimethylcyclohexane

Figure 3–27. cis–trans *Isomerism in 1,2-dimethylcyclohexane.*

groups have destabilizing 1,3 interactions, as we pointed out in Section 3:10. Because of this energy difference, the *trans* isomer consists almost entirely of the *ee* conformer. Furthermore, because the *cis* isomer consists of equivalent *ea* conformers (eq. 3.8), the *trans* isomer is more stable than the *cis* isomer.

The situation with the 1,3-dimethylcyclohexanes is just the reverse of that described above. The *cis*-1,3-isomer is the more stable of the two since it can exist as the *ee* conformer.

With only a conformational drawing, it is not always easy to visualize whether the substituents are on the same or opposite sides of the ring. Specifically, in Equation 3.7 it is difficult to realize that the *ee* isomer has *trans* methyl groups. One useful approach is to "flatten" the chair mentally until you get the planar representation shown in Figure 3–27. This is done by pushing up the "foot" of the chair and pushing down its "head" (Fig. 3–29;

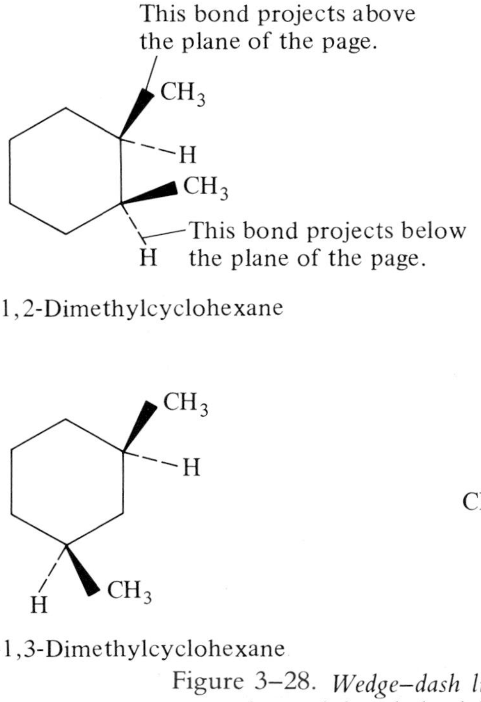

cis-1,2-Dimethylcyclohexane *trans*-1,2-Dimethylcyclohexane

cis-1,3-Dimethylcyclohexane *trans*-1,4-Dimethylcyclohexane

Figure 3–28. *Wedge–dash line representations of several dimethylcyclohexanes.*

Figure 3–29. *A "flattened" chair conformer of cis-1,2-dimethylcyclohexane, showing that the methyl groups lie on opposite sides of the ring.*

see arrows). You will see that one methyl group lies above the plane and one below it.

Problems

3–20 Draw *trans*-1,3-dimethylcyclohexane and *cis*-1,4-dimethylcyclohexane stereoisomers using the wedge–dashed line structures of Figure 3–28.

3–21 Draw the structures for all the chair conformers of 1,3- and 1,4-dimethylcyclohexane. For each constitutional isomer, indicate which stereoisomer is lowest in energy.

3:14 Configuration Versus Conformation: Two Different Types of Stereoisomerism

We defined stereoisomers as isomers that differ only in the *spatial* arrangement of their atoms (sec. 3:12). Axial and equatorial methylcyclohexane fit this definition. So do *cis*- and *trans*-1,2-dimethylcyclohexane. Nonetheless, there is an important difference between these two pairs of stereoisomers. Axial and equatorial methylcyclohexane interconvert rapidly at room temperature (25° C) and cannot be individually isolated at this temperature. Stereoisomers like these are called conformational isomers. (Other conformational isomers, or conformers, were discussed in sec. 3:5.) In contrast, *cis*-

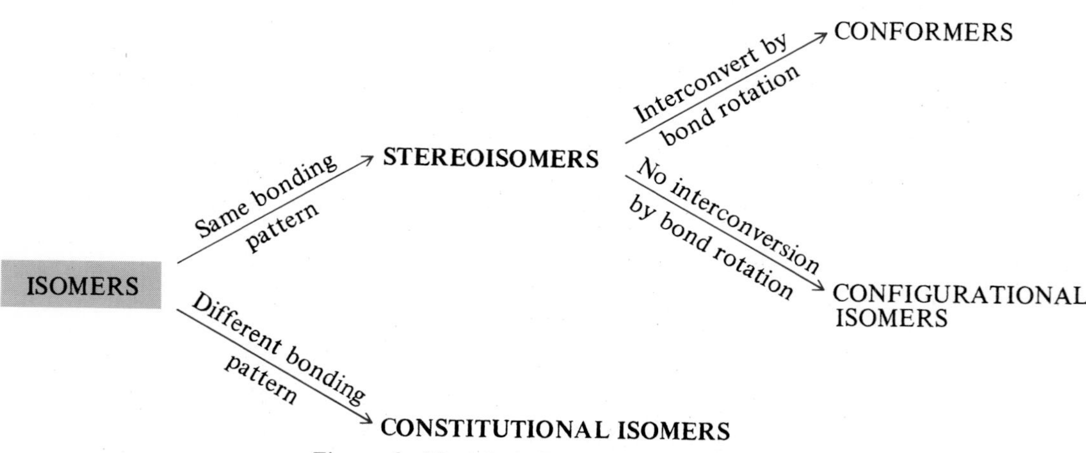

Figure 3–30. *The relationship between conformational and configurational isomers.*

and *trans*-1,2-dimethylcyclohexane do not interconvert at room temperature. Such molecules are called **configurational isomers**. To change one configurational isomer to another (e.g., *cis* to *trans*) usually requires the breaking of one or more chemical bonds. Since bond breaking processes need a large amount of energy, they are exceedingly slow at room temperature.

In our future discussions of stereoisomers we will refer almost exclusively to isomers that can be isolated at room temperature—configurational isomers. The relationship between conformers and configurational isomers is summarized in Figure 3–30.

3:15 Rings That Share Carbon Atoms: Their Occurrence in Steroids, Physiologically Active Polycyclic Compounds

When two rings *share* two or more carbon atoms they are said to be **fused**. A large number of naturally occurring molecules—in particular, many that are important physiologically—fall into this class. One polycyclic ring system found frequently in natural products has the common name **decalin**. The decalins consist of two cyclohexane rings sharing a common edge (Fig. 3–31). The prefixes *cis* and *trans* refer to the relative positions of the hydrogen atoms attached to the shared (*bridgehead*) carbon atoms (Fig. 3–32). Each configurational isomer adopts a conformation in which both of its rings are in a chair form. Chair–chair interconversion is possible in *cis*-decalin but not in the *trans* isomer, which is conformationally "locked."

The decalin ring system is incorporated in the class of biologically potent compounds known as **steroids**. A large number of our bodily hormones that regulate growth, sexual characteristics, and many other functions are steroids. It requires relatively small amounts of these remarkable substances to cause a biological response—in some cases as little as one *billionth* the weight of the organism receiving the steroid.

Figure 3–33 contains two representations of androstane, a steroid consisting of a polycyclic ring system common to all steroids. It incorporates a decalin ring fused to a cyclohexane ring which is itself fused to a cyclopen-

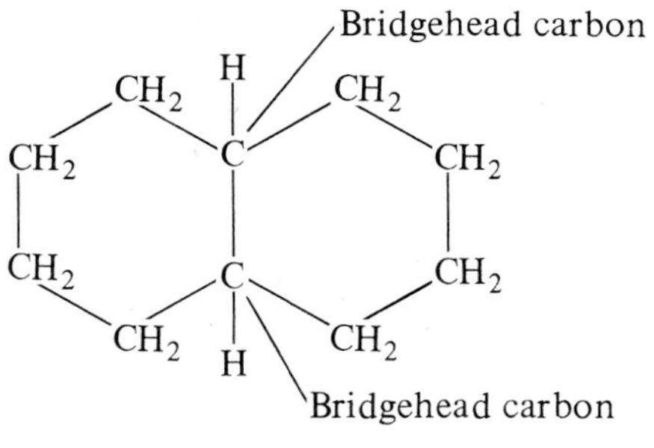

Bridgehead carbon

Bridgehead carbon

Bicyclo[4.4.0]decane
(decalin)

Figure 3–31. *The structure of the bicyclic (double-ringed) alkane bicyclo[4.4.0]decane, commonly called* decalin. *The two carbon atoms which are shared by the two rings are called bridghead carbons.*

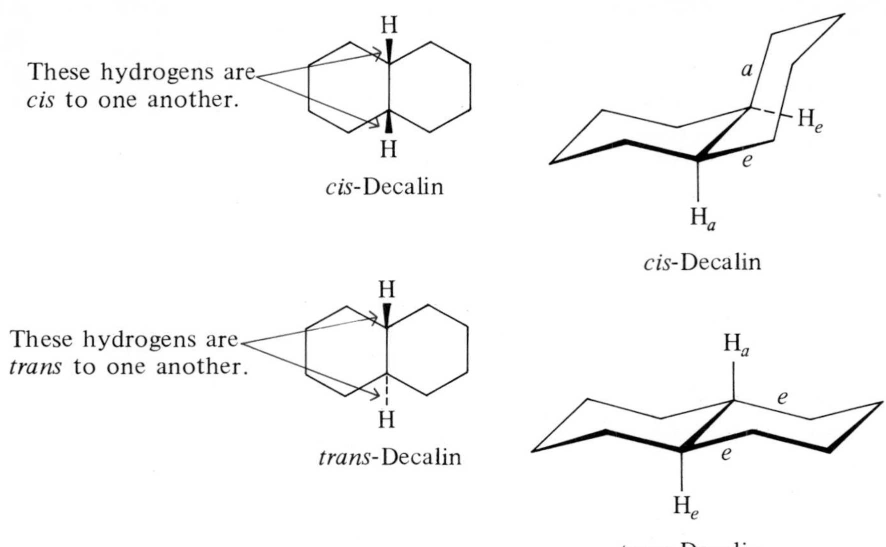

Figure 3–32. *Planar and conformational projections of* cis- *and* trans-*decalin.*

tane ring. The ring fusions in androstane are all *trans,* as they are in most natural steroids. One of the most widespread natural steroids, especially in animal tissue, is cholesterol (Fig. 3–34), which serves as the precursor of many steroid hormones. When a steroid such as cholesterol is presented in the planar representation, its substituents will be projecting either above or below the plane of the paper. Groups that project above, such as the OH at C-3, are designated β and groups that project below the skeletal plane, such as the cholesterol hydrogens at C-9 and C-14, are designated α. Conformational analysis of six-membered rings such as steroids has been invaluable in helping us understand the relationship between structure and physiological activity. For example, Figure 3–35 gives the conformational representation of androsterone, a steroid of the *androgen* group. Androgens are male sex

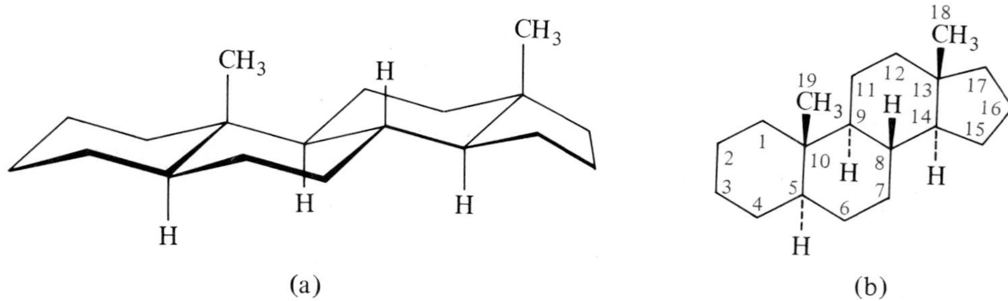

Figure 3–33. *Conformational (a) and planar (b) projections of the steroid androstane. The numbering system is one specifically designed for steroids. By convention, the androstane structure is always drawn with the five-membered ring at the upper right.*

Cholesterol

Figure 3–34. *Planar structure of the animal steroid cholesterol.*

hormones—compounds which promote primary and secondary sex characteristics in humans and animals. The androgenic activity is reduced considerably in epiandrosterone (partial structure, Fig. 3–35), which differs from androsterone only in the configuration of the hydroxyl group at C-3. By changing the ring juncture from *trans,* as in androsterone, to *cis,* as in 5β-androsterone (partial structure, Fig. 3–35), androgenic activity is lost completely. You can see that these configurational changes alter the spatial relationship between the hydroxyl group and other parts of the molecule.

Androsterone
(high physiological activity)

Epiandrosterone (partial structure)
(reduced physiological activity)

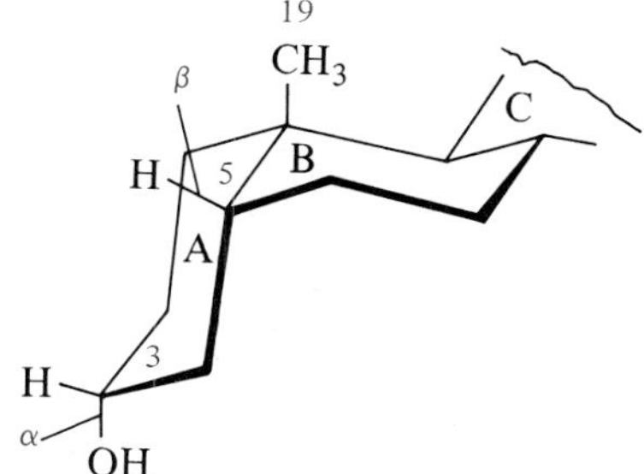

5β-Androsterone (partial structure)
(no physiological activity)

Figure 3–35. *Conformational projection of androsterone and partial structures of the related steroids epiandrosterone and 5β-androsterone. Note how the spatial characteristics of ring A affect the physiological activity.*

Summary

1 Alkanes are hydrocarbons—compounds of only carbon and hydrogen—and are obtained in large quantities from natural gas and by the fractional distillation of petroleum (sec. 3:1). Alkanes can be open-chain or cyclic molecules. The open-chain compounds have the general molecular formula, C_nH_{2n+2}.

2 The semisystematic nomenclature method is generally used only for C_1 through C_5 alkanes and their derivatives. The root of the name tells us the total number of carbons in the molecule. If the prefix iso- is used, the alkane chain ends in a $(CH_3)_2CH—$ group. Alkyl substituents (Table 3–2) are named by dropping the -ane suffix from the name of the corresponding alkane and adding -yl (sec. 3:2).

The IUPAC system of nomenclature is based upon the longest continuous chain of carbon atoms; the number of atoms *in this principal chain* determines the root of the alkane name. Carbon atoms not in the principal chain are named as substituents on the chain. If there is more than one way to choose the principal chain, preference is given to the most highly substituted chain. The carbon atoms of the principal chain are numbered so as to provide the lowest possible number for the first substituent. Substituent names are listed alphabetically (sec. 3:3).

Alkanes are unreactive, nonpolar compounds that are often used as solvents and are consumed in enormous quantities as fuels (sec. 3:4).

3 Rotation about carbon–carbon single bonds in alkanes is not unhindered. There is a barrier to rotation that arises from the greater internal energy of the eclipsed conformer compared to the staggered conformer. For ethane, the staggered conformer is lower in energy by 2.9 kcal/mol. Hydrogen–hydrogen repulsions are weaker than hydrogen–methyl repulsions, which in turn are weaker than methyl–methyl repulsions. At ordinary temperatures, alkanes exist as mixtures of staggered conformers; because of the low energy barriers, interconversion of conformers is very rapid (sec. 3:5).

The proportion of each conformer present at equilibrium depends on the free energy difference (ΔG) between them. The free energy difference is dependent upon the enthalpy difference (ΔH) and the entropy difference (ΔS). Entropy is a measure of the randomness or statistical probability of a conformation. Thus, the *gauche* conformers in butane or 1,2-dichloroethane have about twice the entropy of the *anti* conformers of the same compounds, since there are two equivalent *gauche* forms and only one *anti* form. The effect of entropy on the equilibrium increases or decreases as the temperature is raised or lowered (sec. 3:6).

4 If we assume that cycloalkanes are planar, then we would anticipate that rings both larger and smaller than cyclopentane would exhibit angle strain. This expectation is contradicted by the results of heat of combustion studies on the larger rings, which are all puckered. The puckering not only avoids angle strain, but it also relieves eclipsing interactions between hydrogen atoms on adjacent carbons (sec. 3:8).

5 For cyclohexane, there are three important puckered conformations: the chair, the boat, and the twist boat. The chair has by far the lowest energy of the three, and cyclohexane exists almost exclusively in this conformation at room temperature (sec. 3:9).

There are two types of C—H bonds in chair cyclohexane: axial and equatorial. The equatorial hydrogens lie approximately in the plane of the ring, while the axial hydrogens project at right angles to that plane. At room temperature a given cyclohexane molecule rapidly alternates between the two possible chair conformations. In the course of this equilibration a particular hydrogen atom is shifted quickly from an axial to an equatorial orientation and back again. A substituent on a cyclohexane ring is also rapidly alternating between axial and equatorial orientations. In general, the conformation in which the substituent is equatorial is the preferred one. When the substituent is axial, it experiences repulsion by the axial hydrogens at C-3 and C-5 (sec. 3:10).

6 Cyclopropane has a planar carbon skeleton, which results in eclipsing of all its C—H bonds. This eclipsing, coupled with bond angle distortion, results in a great deal of strain in cyclopropane. The strain facilitates rupture of the three-membered ring. In order to minimize eclipsing effects, cyclobutane and cyclopentane adopt slightly puckered conformations, so they are less strained than cyclopropane (sec. 3:11).

7 When cycloalkanes carry two substituents on different carbon atoms, *cis–trans* isomerism becomes possible. *Cis–trans* isomers are configurational isomers (sec. 3:12).

In the cyclohexane system, *cis–trans* isomerism is complicated by the fact that each isomer can exist in two chair conformations. These conformers are in equilibrium with one another, but the isomers do not interconvert at room temperature. The most stable isomer will be the one that has the conformer with the largest number of bulky substituents in equatorial positions (sec. 3:13 and 3:14).

8 Polycyclic alkanes are those that contain two or more rings. When two rings share a side they are said to be fused. Fused ring systems, such as decalin, are found in a number of physiologically active molecules, such as steroids. The ring junctures in these molecules exhibit *cis–trans* isomerism (sec. 3:15).

Problems

3–22 Write the structures for the following:

a. 2-methylpropane
b. 2,2-dimethylpentane
c. 6-ethyl-4-isobutyl-2-methyloctane
d. 6-isopropyl-4-propylnonane
e. 5-butyl-6-pentyldodecane
f. 3,3-di-*t*-butyl-2,2,4,4-tetramethylpentane
g. *cis*-1,2-dimethylcyclopentane
h. The most stable conformer of *trans*-1,3-dimethylcyclohexane
i. The most stable conformer of *cis*-1-*t*-butyl-2-methylcyclohexane

3–23 Provide IUPAC names for the following structures:

a. $CH_3(CH_2)_3CHCH_3$ with CH_2CH_3 attached to the CH

b. $\underset{\substack{|\\CH_3CHCH_3}}{\overset{\substack{CH_3\\|}}{CH_3CHCH(CH_2)_2CH_3}}$

c. $(CH_3)_2CH(CH_2)_4CH(CH_2CH_3)CH(CH_3)CH_2CH_3$

(structure shown with CH_3 above and CH_2CH_3 below: $(CH_3)_2CH(CH_2)_4\underset{\substack{|\\CH_2CH_3}}{CH}\overset{\substack{CH_3\\|}}{CH}CH_2CH_3$)

d. cyclohexane ring bearing $CH_2CH(CH_3)_2$ and H on the same carbon

3–24 Draw the structures of all the C_7H_{16} isomers and name each using the IUPAC method.

3–25 a. Draw the structures of all the C_6H_{14} isomers that contain just three methyl groups. (Note that butane has two methyl groups.)
b. Draw the structure of the isomer that has the maximum number of methine (CH) groups.

3–26 Draw the structure of a $C_{10}H_{22}$ isomer that has three branches in its chain, and give its IUPAC name.

3–27 The sex attractant of the tsetse fly, *Glossina morsitans,* is 15,19,23-trimethylheptatriacontane ($C_{40}H_{82}$). Give its condensed structural formula.

3–28 Draw an energy diagram for rotation around the $C_1—C_2$ bond in 1-chloro-2-methylpropane. Show the relative heights of the peaks and valleys (you need not assign numerical values). Draw the structures of the eclipsed and staggered conformers, and show their positions on the energy diagram.

3–29 Draw an energy diagram for the internal rotation of 2-bromo-2-chloro-1,1,1-trifluoroethane (fluothane, a widely used anesthetic).

3–30 In compounds of the type XCH_2CH_2Y the energy difference between the *gauche* and *anti* conformers depends on the nature of X and Y. When X and Y are both methyl the *anti* conformer is the more stable of the two. However, when X and Y are halogens then in some cases the *gauche* conformer is lower in energy than the *anti*. What factor(s) account(s) for the difference in behavior of these systems?

3–31 When linear alkanes such as pentane melt, they undergo a substantial increase in entropy. This entropy increase is not nearly as great when a highly branched isomer, such as neopentane, melts. Explain the difference in entropy of melting. (Take note of the degree of change in conformational mobility in each case.)

3–32 Linear alkanes can be thought of as chains of CH_2 groups with an extra hydrogen atom at each end. These extra hydrogens contribute to the total heat of combustion. How would you determine the heat of combustion per CH_2 group in linear alkanes?

3–33 The heats of combustion of three gaseous cycloalkanes are 11.9, 11.7, and 11.3 kcal/g. One of these compounds is cyclobutane, one is cyclopentane, and the third is cyclopropane. Which is which? Calculate the heat of combustion *per mole* for each of these compounds and then the heat of combustion per mole *per methylene group*. If the "normal" heat of combustion per methylene group is 157.4 kcal/mole, calculate the total strain energy in cyclopropane, cyclobutane, and cyclopentane.

3–34 The equatorial conformers of monosubstituted cyclohexanes are always lower in energy than the axial conformers. What effect does lowering the temperature have on the axial ⇄ equatorial equilibrium for these compounds? Could this effect be put to use in an attempt to isolate a pure conformer? Explain your answer.

3–35 Indicate whether the *cis* or *trans* isomer of 1,4-dimethylcyclohexane will have the greater heat of combustion and why.

3–36 Draw the most stable chair conformer of each of the following compounds. Use Figure 3–12 as a guide.

a. ethylcyclohexane
b. 1,1-dichlorocyclohexane
c. *cis*-1,2-diethylcyclohexane
d. *trans*-1-chloro-3-methylcyclohexane
e. The isomer of 1,2,3,4,5,6-hexahydroxycyclohexane in which *all but one* of the hydroxyl groups is in the most stable configuration. (This is *myo*-inositol, a compound found in all cells.)

3–37 Which isomers of dibromocyclohexane can have a chair conformer in which *neither* of the bromine atoms is in the axial configuration?

3–38 1-Isopropyl-2-methylcyclohexane can exist in *cis* and *trans* isomeric forms.

a. Draw the conformers corresponding to each isomer.
b. Which is the more stable conformer of each pair?
c. Which is the more stable isomer?
d. Carry out the same analysis on the 1,3 compound.

3–39 The free energy difference between twist and chair cyclohexanes is usually so large that the twist form is not detectable. In *trans*-1,3-di-*t*-butylcyclohexane, however, the *predominant* conformer is the twist. *cis*-1,3-Di-*t*-butylcyclohexane, on the other hand, behaves like the great majority of cyclohexanes and overwhelmingly favors the chair conformation. Explain these results.

3–40 The structural formula of the 2-alkyl-1-cyclohexanols is shown here. When R is methyl, ethyl, or isopropyl, then the *trans* isomer is more stable than the

R

OH

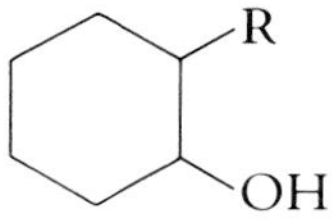

cis. When R is *t*-butyl, however, then the *cis* isomer is more stable than the *trans*.

a. Draw the major chair conformer of each *cis* or *trans* isomer for all four alkyl groups.
b. Explain why the *cis*-*t*-butyl compound is more stable than the *trans*, when the reverse is true with all the other alkyl groups.

3–41 Draw structures of all the possible cycloalkanes of formula C_5H_{10} and name each one.

3–42 A compound of formula C_6H_{12} is known to be a substituted cyclobutane. Show all the isomers, constitutional and geometric, that fit the data.

3–43 Compound **A,** $C_6H_{11}Cl$, is treated chemically so that the Cl atom is replaced by an H atom. No other changes occur. The product of this transformation is methylcyclopentane. Show all the possible structures for **A** that fit the data.

3–44 Show the structure that corresponds to each name.
a. cyclopentylcyclopentane
b. 1,1,3,3-tetrafluorocyclobutane
c. 1,1,2,2-tetracyclopropylethane
d. *trans*-1,4-dichlorocyclooctane
e. *cis*-1,4-diisopropylcyclohexane

3–45 Strained cycloalkanes may have their rings ruptured by reaction with hydrogen (eq. 2).

$$CH_3\text{–}\underset{H\quad H}{\triangle}\text{–}CH_3 + H_2 \xrightarrow{\text{catalyst}} CH_3(CH_2)_3CH_3 + \text{heat} \qquad (2)$$

The reaction is exothermic. If *cis*- and *trans*-1,2-dimethylcyclopropane are each subjected to this cleavage, which reaction will be more exothermic?

3–46 Draw the structure of the androstane derivative with a doubly-bonded oxygen (C=O) at C-3, a double bond between C-4 and C-5, and a β-hydroxyl group at C-17. (This is the structure of the male sex hormone testosterone.)

3–47 a. *trans*-Decalin is more stable than *cis*-decalin. Why? (It may help to focus on one ring in each isomer and treat it as a 1,2-disubstituted cyclohexane.)
b. *cis*-Decalin is conformationally mobile. Both rings may undergo conversion to their equivalent chair conformations. *trans*-Decalin, on the other hand, is rigid and cannot undergo this interconversion. Explain the behavioral difference of these two stereoisomers (models may be very helpful).

Supplementary Reading

J. E. Banks, *Naming Organic Compounds,* W. B. Saunders Co., Philadelphia, Pa., 1976. C. R. Johnson, *Organic Nomenclature: A Programmed Study Guide,* Worth Publishers, Inc., New York, N.Y., 1976. Both these paperback texts are programmed guides which cover all organic nomenclature.

E. L. Eliel, *Elements of Stereochemistry,* John Wiley & Sons, New York, N.Y., 1969.

C. A. Russell, "The Origins of Conformational Analysis," *van't Hoff-Le Bel Centennial,* O. B. Ramsay, ed., American Chemical Society, Washington, D.C., 1975.

4 Stereoisomerism and Chirality: Right- and Left-Handed Chemistry

Monosodium glutamate, the food seasoning, can be prepared from the natural amino acid glutamic acid. Monosodium glutamate can also be prepared from totally synthetic glutamic acid, but the synthetic compound has only one-half the potency of the natural substance. The heart stimulant epinephrine (adrenaline), which is excreted by the adrenal gland, is twice as potent as epinephrine prepared in the laboratory. These discrepancies occur because the natural substances are pure compounds, while the synthesized materials are 50:50 mixtures of two isomers. These are neither constitutional (structural) isomers, nor *cis-trans* isomers; rather, they are examples of another important type of stereoisomerism called **mirror image isomerism**.

$Na^+OOC—CH(NH_2)CH_2CH_2COOH$

Monosodium glutamate

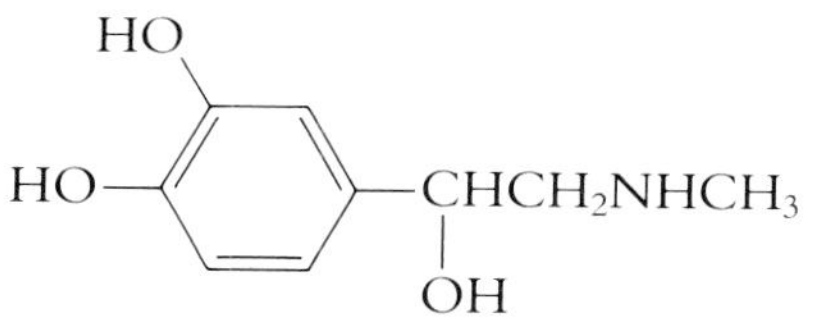

HO, $HO—C_6H_3—CH(OH)CH_2NHCH_3$

Epinephrine (adrenaline)

4:1 Enantiomers and Diastereomers: Different Types of Stereoisomers

Chapter 1 introduced constitutional isomerism, which is isomerism based on differences in bond connectivity. In Chapter 3 a second kind of isomerism was introduced; stereoisomers (sec. 3:12), such as **1** and **2**, are isomers that have the same number, type, and sequence of bonds, but have the bonded atoms arranged differently in space. These two dibromocyclobutanes are configurational isomers in which the relative spatial arrangements of certain groups in the molecule differ. The two bromine atoms are on the same side of the molecular plane in **1** but on opposite sides of this plane in **2**. If each

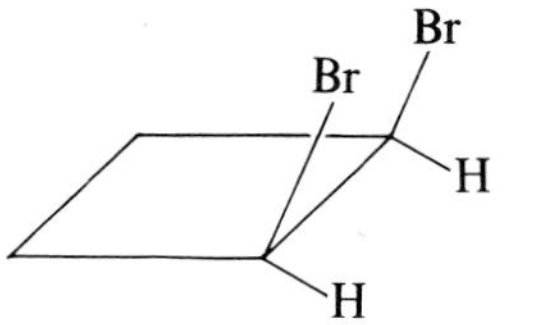

1

cis-1,2-Dibromocyclobutane

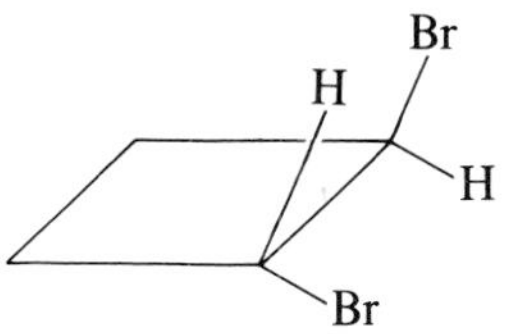

2

trans-1,2-Dibromocyclobutane

member of your class was told to construct models of *cis*-1,2-dibromocyclobutane, the models would all correspond to structure **1**. If, however, they constructed models of *trans*-1,2-dibromocyclobutane, about half of the models would correspond to structure **2**. The remainder would have structure **3**, *which is not the same as structure* **2**.

H Br Br H

3

There is an interesting relationship between **2** and **3**: they are mirror images of one another, but they are *not* identical. The easiest way to verify this is to build models of the two structures and attempt to superimpose them—that is, to place one model on top of the other so that equivalent atoms coincide. You will see that **2** and **3** are not superimposable, and therefore are not identical. Nonidentical (nonsuperimposable) mirror image isomers are called **enantiomers** (from the Greek *enantios,* meaning "the opposite of one another").

If models are not available, you can use drawings to check for enantiomerism. For instance, we draw structure **2** and next to it its mirror image (Fig. 4–1a). Then the mirror image is rotated 180° to give structure **3**. A visual comparison of **2** and **3** shows that they are not identical; since they are mirror images, however, they must be enantiomers. When the same procedure is applied to structure **1** (Fig. 4–1b), it is found that **1** and its mirror

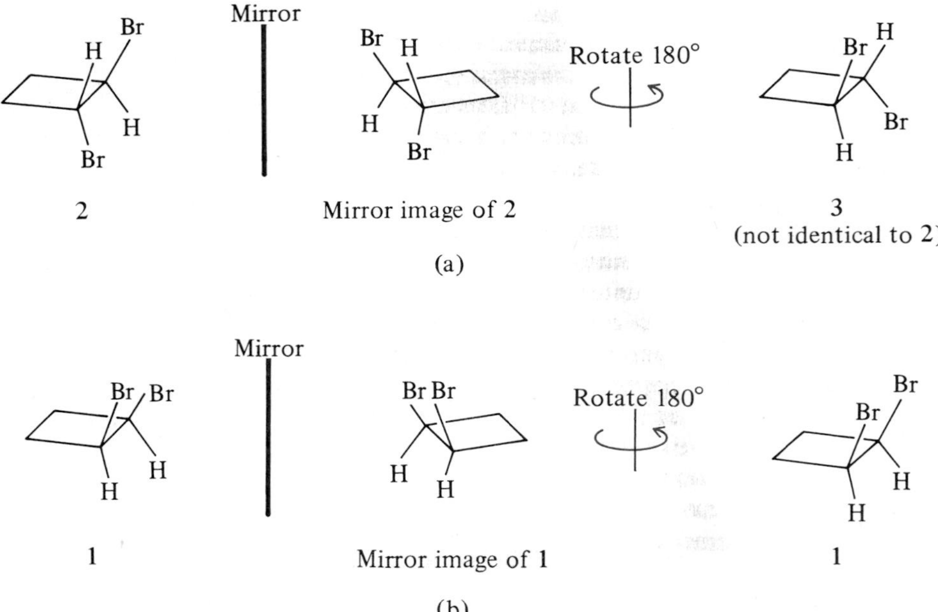

Figure 4–1. *(a) Nonidentical mirror images (enantiomers) of* trans-*1,2-dibromocyclobutane. (b) Identical mirror images of* cis-*1,2-dibromocyclobutane.*

image *are* identical (superimposable). We can therefore conclude that enantiomerism is not possible with *cis*-1,2-dibromocyclobutane.

Molecules that can exist as enantiomers, for example, **2** and **3**, are called **chiral** (pronounced *kyrul*) molecules, while molecules that cannot, such as **1**, are called **achiral**. The word chiral comes from the Greek *cheir,* meaning "hand." Some chemists speak of "handedness" when referring to chiral molecules because the relationship between enantiomers is the same as that between your left hand and your right hand: they are nonidentical mirror images. Thus, *trans*-1,2-dibromocyclobutane is a chiral molecule, while the *cis* isomer is achiral.

Many analogies to molecular chirality and achirality can be found in the macroscopic world. Consider shoes and socks. A sock is analogous to an achiral molecule. It is identical with its mirror image and cannot exist in two forms, that is, there are no right and left socks, and any sock fits either foot. A shoe is analogous to a chiral molecule. It is different from its mirror image and can exist in two forms analogous to enantiomers: right and left shoes. The mirror image of a right shoe is identical to a left shoe, not a right shoe. Right and left gloves and right and left hands can be considered chiral. Right and left shoes or gloves are *specific* for right or left hands or feet. You will soon see that chiral molecules display very specific behavior when undergoing reaction with other chiral species. This specificity has important consequences for chemistry and biology.

We have shown that **2** and **3** are enantiomers of one another, but what is the relationship between each of them and **1**? The *trans* isomers **2** and **3** are *not* enantiomers of the *cis* isomer **1**, yet they *are* each stereoisomers of **1**. Thus, **1** and **2** (and **1** and **3**) *are* stereoisomers but are *not* mirror images. Isomeric pairs that fall into this category are called **diastereomers** (from the Greek *dia,* meaning "apart"). Figure 4–2 presents a summary of the different kinds of isomerism (refer back to Fig. 3–30, p. 96).

Using molecular models is the most clear-cut way of identifying various types of isomers and is strongly advised as an aid in visualizing relationships.

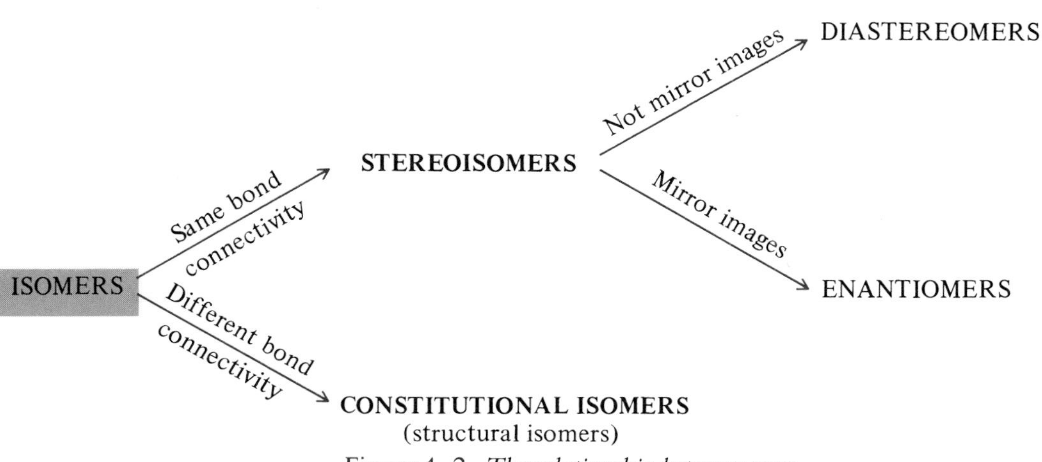

Figure 4–2. *The relationship between enantiomers and diastereomers, the two classes of stereoisomers.*

You must also learn to discern the possibility of stereoisomerism from examination of written structural formulas. There are certain structural elements that automatically confer chirality or achirality on molecules. We need to be able to look at a molecule and quickly tell whether or not it is chiral. Although our initial examples were cyclic molecules, the analysis will be applied to all molecules, cyclic and acyclic. The following sections show how chirality in molecules can be readily identified.

Problems

4–1 Make drawings (and/or models) of *cis-* and *trans*-1,3-dibromocyclobutane. Are either of these molecules chiral (not identical with their mirror images)? What is the isomeric relationship between *cis-* and *trans*-1,3-dibromocyclobutane? What is the isomeric relationship between *cis*-1,2- and *cis*-1,3-dibromocyclobutane?

4–2 Draw all the cyclic structures that have the molecular formula C_4H_7Cl. Describe the isomeric relationship (constitutional, enantiomeric, diastereomeric) of each compound to all of the others (ignore conformers).

4–3 Which of the following common objects are chiral and which are achiral? (Ignore letters, numbers, and decorations, and consider only the structural features of the object.)

a. a fork
b. a cup
c. a telephone
d. a shirt with buttons
e. an automobile body
f. an automobile, including the interior
g. a screw
h. a pair of scissors

4:2 Symmetry Planes and Chiral Centers: Symmetry Features That Determine Chirality and Achirality

The surest test for chirality is to build a model of the molecule and of its mirror image to see if one is superimposable on the other. Just take one model and turn or rotate it in all possible ways to see if it will exactly fit the other in all respects. (This procedure may be difficult with large molecules.) It is sometimes possible to draw the mirror image of a structure, and then to manipulate the two structures to see if they are identical (see Fig. 4–1). However, this approach is usually possible only with relatively simple molecules.

An easier test for chirality involves looking for certain *symmetry features* in a molecule. If these are present, then the molecule is *not* chiral. The symmetry feature that is most likely to be present, indicating achirality, is a **plane of symmetry**. A symmetry plane is one that cuts the molecule into two halves that are mirror images of each other. Figure 4–3a shows that there is a symmetry plane in the achiral *cis*-1,2-dibromocyclobutane. This figure also shows that the *trans* isomer, which is chiral, has *no* symmetry plane. These examples illustrate the following general rule:

If a molecule possesses at least one plane of symmetry, it is achiral (identical with its mirror image).

This rule can be applied even to compounds that assume many different conformations, such as cyclohexanes. *If at least one of the conformations has a*

cis-1,2-Dibromocyclobutane

trans-1,2-Dibromocyclobutane

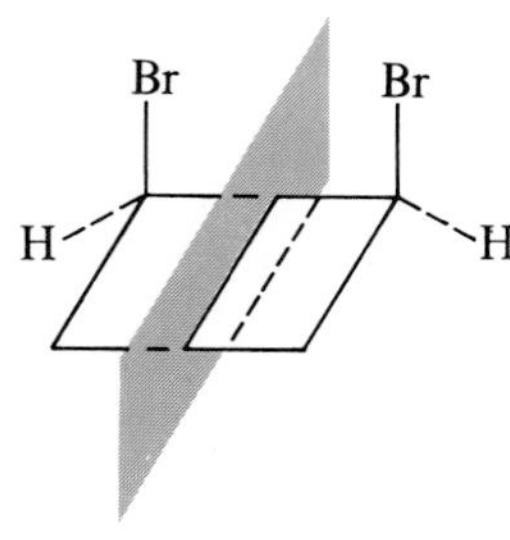

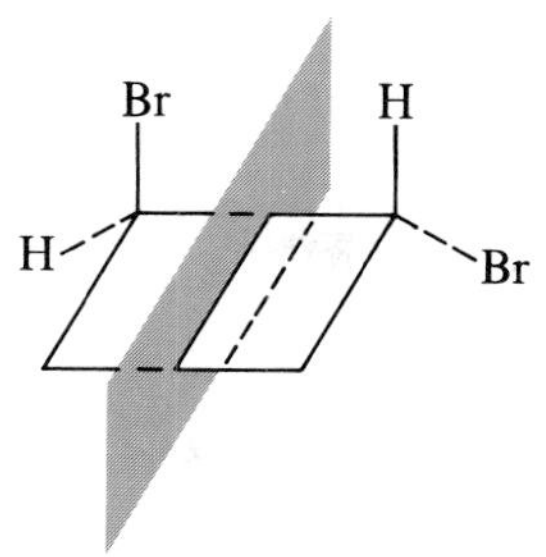

(a)

Plane of symmetry
Achiral

Not a plane of symmetry
Chiral

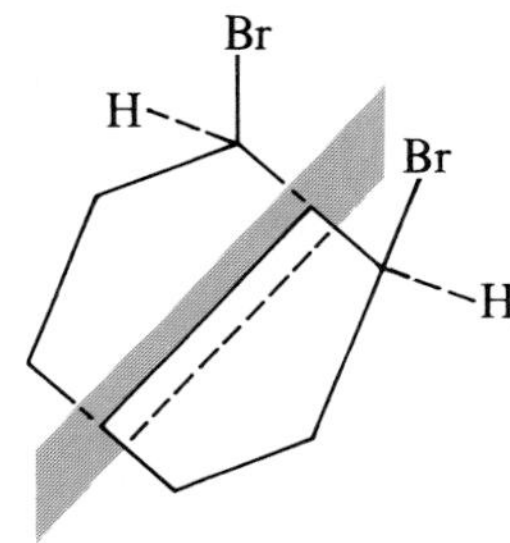

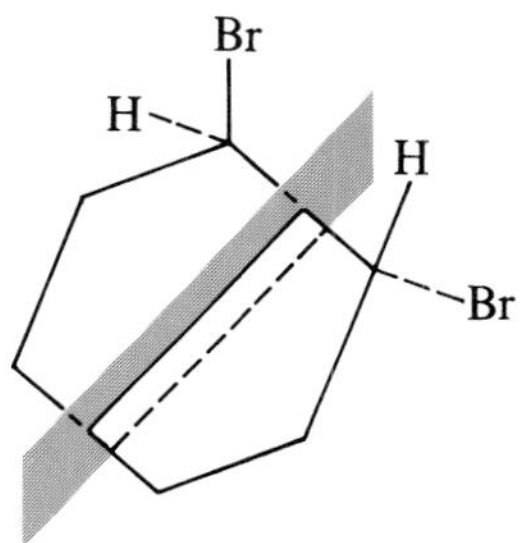

cis-1,2-Dibromocyclohexane

trans-1,2-Dibromocyclohexane

(b)

Figure 4–3. *The relationship between chirality and symmetry planes in (a) 1,2-dibromocyclobutane and (b) 1,2-dibromocyclohexane.*

symmetry plane, then the molecule is achiral. The conformation does not have to be an energetically favorable one. For instance, the high energy planar conformation of *cis*-1,2-dibromocyclohexane has a symmetry plane (Fig. 4–3b), so this compound is achiral. By contrast, the *trans* isomer has no conformer, including the planar conformation, that contains a symmetry plane and it therefore exists in two enantiomeric forms.

The presence of a symmetry plane confirms achirality in open chain, as well as cyclic, molecules. Figure 4–4a shows the plane of symmetry that passes through one conformer of 1,1-dichloroethane. Note that this plane can cut through atoms as well as bonds; the hydrogen atoms are shown as spheres in order to emphasize this fact. The presence of the symmetry plane correlates with the fact that 1,1-dichloroethane exists in only one form. The closely related molecule 1-bromo-1-chloroethane lacks a symmetry plane, however, and is found in two enantiomeric forms (Fig. 4–4b).

The presence of a symmetry plane proves that a molecule is achiral, but the absence of such a plane is *not* proof of chirality. In these cases we must consider other elements of symmetry to test for chirality. One such element is a **center of symmetry**. Structure **6** shows a molecule that lacks a plane of

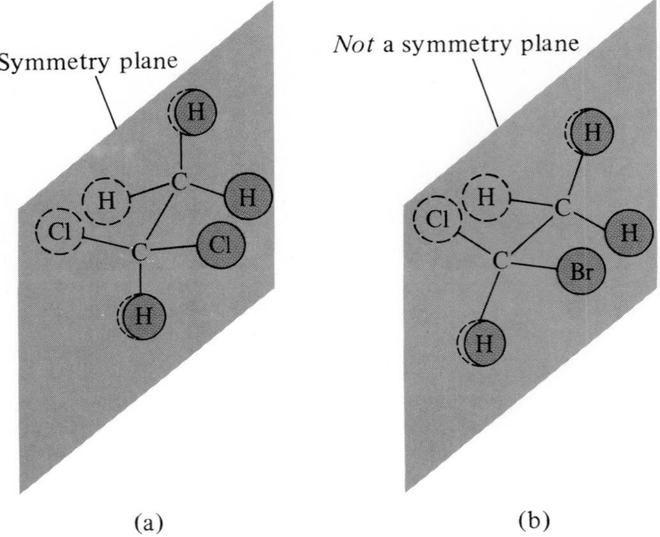

Figure 4–4. *(a) Symmetry plane in 1,1-dichloroethane, an achiral molecule. (b) Absence of a symmetry plane in 1-bromo-1-chloroethane, a chiral molecule.*

symmetry but is nevertheless achiral since it possesses a center of symmetry. In a molecule possessing a center of symmetry, all straight lines passing through this center will encounter identical atoms at identical distances from it.

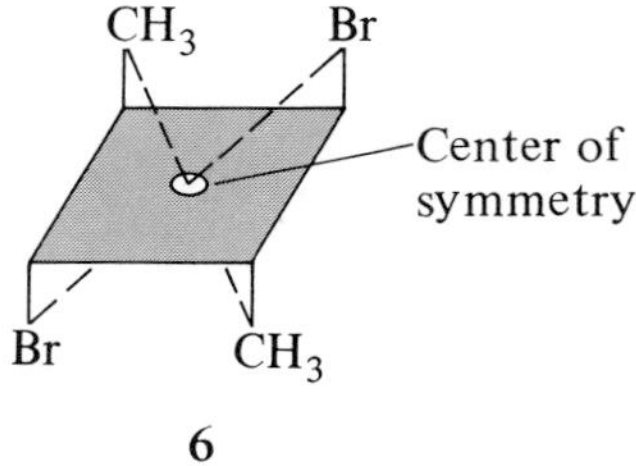

6

There is one other symmetry element used to test for chirality, but it is found in very few molecules and will not be treated in this book.

Problems

4–4 Find the plane(s) of symmetry (if any) in the following molecules. Models may be helpful.

a. *trans*-1,3-dibromocyclobutane
b. *cis*-1,3-dibromocyclobutane
c. *trans*-1,3-dichlorocyclopentane
d. *cis*-1,3,-dichlorocyclopentane

4–5 What symmetry elements (plane of symmetry or center of symmetry), if any, are present in the following structures?

a. the conformer of butane with eclipsed methyl groups
b. the *anti* conformer of butane
c. a *gauche* conformer of butane

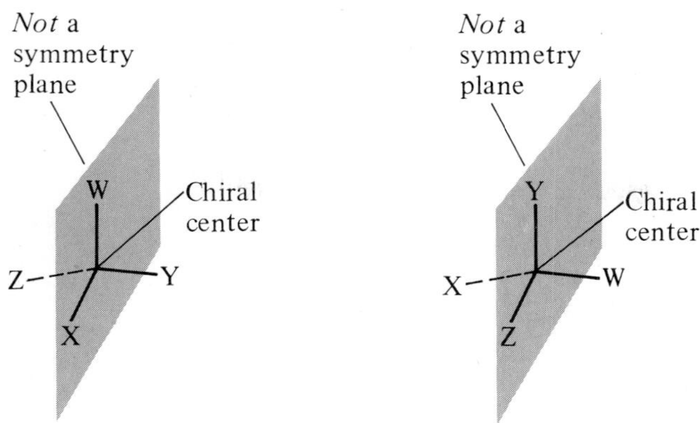

Figure 4–5. *Four different groups (W, X, Y, Z) attached to the same center by bonds which do not all lie in a plane. A plane of symmetry cannot pass through the center, which is therefore chiral.*

Confirming the presence or absence of planes or centers of symmetry constitutes a direct test for chirality. In many cases, however, it is difficult to envision these elements when confronting a structural formula that does not indicate the three-dimensional placement of atoms. In such cases it is always possible to find out whether the molecule contains any **chiral centers** (also called **asymmetric centers**). An *atom* is considered a chiral, or asymmetric, center if all the groups attached to it are different, and if the attached groups and the central atom do not all lie in one plane. With this geometry, no symmetry plane through the center can be found (Fig. 4–5). (The terms chiral atom or asymmetric atom are sometimes loosely applied to a chiral center.) It follows from the above that *any tetrahedral carbon attached to four different groups is a chiral center.* A compound that contains one chiral center is itself chiral. (We will consider molecules with more than one chiral center in the next section.)

To illustrate this concept, we will compare two closely related compounds, 2-propanol and 2-butanol (*sec*-butyl alcohol).

H | CH_3CCH_3 | OH	H | $CH_3CCH_2CH_3$ | OH
2-Propanol	2-Butanol

2-Propanol does not have a chiral carbon since no carbon atom is bonded to four *different* groups. As a confirmation of this molecule's achirality, Figure 4–6 shows that 2-propanol has a plane of symmetry. By contrast, in 2-butanol the second carbon has four different groups attached (CH_3, CH_3CH_2, OH, and H). Figure 4–7 shows that 2-butanol has no symmetry plane since the CH_3 on one side is not equivalent to the CH_3CH_2 on the other side of the plane. (Even isotopic differences are enough to render two groups such as CH_3 and CD_3 nonequivalent.)

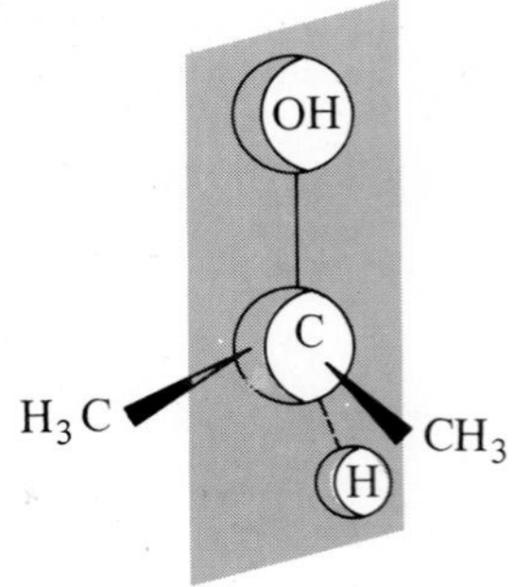

Figure 4–6. *Achiral 2-propanol, with a symmetry plane cutting the H, C, and OH groups.*

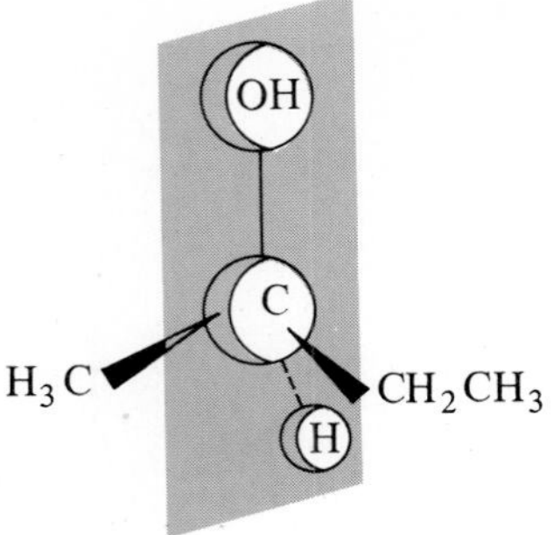

Figure 4–7. *Chiral 2-butanol, with a plane (not a symmetry plane) cutting the H, C, and OH groups. C-2 is a chiral center.*

Chiral and achiral atoms are easily identified, even in condensed structural formulas. For example, from the definition of a chiral center it is clear that the carbon atoms in methyl, CH_3, and methylene, CH_2, groups *cannot* be chiral, nor can an sp^2 carbon. Chiral atoms need not be carbon atoms. Structures **7** and **8** contain chiral silicon and nitrogen atoms, respectively.

$$\begin{array}{c} CH_3 \\ | \\ CH_3CH_2SiH \\ | \\ Cl \end{array} \qquad\qquad \begin{array}{c} CH_3 \\ |^{+} \\ CH_3CH_2CH_2NCH_2CH_3 \\ | \\ CH(CH_3)_2 \end{array}$$

7   **8**

The major ideas developed in this section may be summarized as follows:

1. A chiral molecule is one that is not identical to its mirror image.
2. A molecule that lacks both a plane and a center of symmetry is not identical to its mirror image and is therefore chiral.
3. A molecule that contains one chiral center is not identical to its mirror image and is therefore chiral.

Example

Is the following molecule chiral? If so, identify the chiral centers.

$$\begin{array}{l} \quad\;\; H_3C \;\; H \;\; H \qquad Br \\ \qquad\;\; | \quad\;\; | \quad\; | \qquad / \\ H_3C{-}C{-}C{-}C{=}C \\ \qquad\;\; | \quad\;\; | \qquad\quad \backslash \\ \qquad\;\, H \;\; OH \qquad\;\; Cl \end{array}$$

Answer

The molecule is chiral because it contains one carbon with four different substituents:

The substituents attached to sp^2-hybridized carbons lie in the same plane as the carbon, so these carbons cannot be chiral centers.

H_3C H H Br

H_3C—C—C—C=C

H OH Cl

This carbon has two identical CH_3 groups and is not chiral.

This carbon is chiral.

Problem

4–6 Which of the following compounds are achiral molecules and which are chiral? Identify the chiral centers in the chiral molecules.

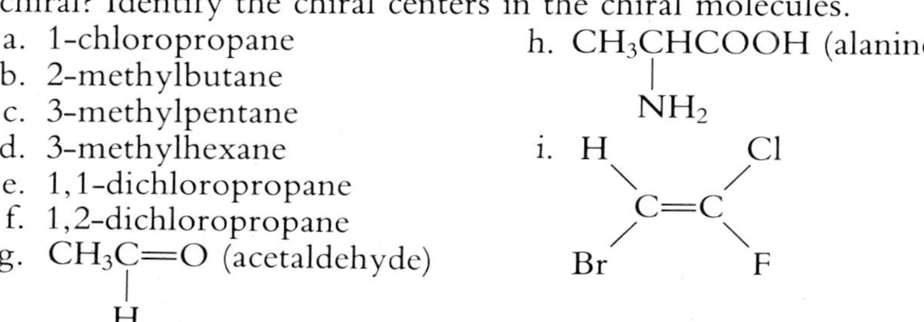

a. 1-chloropropane
b. 2-methylbutane
c. 3-methylpentane
d. 3-methylhexane
e. 1,1-dichloropropane
f. 1,2-dichloropropane
g. CH_3C=O (acetaldehyde) with H on C
h. $CH_3CHCOOH$ (alanine) with NH_2 on CH
i. H, Cl, C=C, Br, F

4:3 Compounds That Contain More Than One Chiral Center: *meso*-Isomers

In the last section we showed that the presence of one chiral center in a molecule results in the existence of one pair of enantiomers. Earlier we discussed the dibromocyclobutanes, which contain more than one chiral center, but we did not establish any relationship between the number of chiral centers and the number of stereoisomers. We examine that relationship here by analyzing a compound with two chiral centers. Since each center can have two stereoisomeric arrangements, the total number of stereoisomers should be $2 \times 2 = 4$. This example can be generalized by saying that *a molecule possessing* n *chiral centers has a* maximum *of* 2^n *stereoisomeric forms*. We'll look at a molecule that illustrates this rule and then see why certain structures yield fewer than the predicted maximum number of stereoisomers.

1-Bromo-2-chlorocyclobutane is a molecule with two chiral centers. Let's draw one possible *cis* isomer (**9a**, Fig. 4–8) and its mirror image (**9b**). Rotation of **9b** allows us to verify that it is not identical with **9a**. Thus, these two structures make up an enantiomeric pair. Similar analysis of the *trans* compound (**10a** and **b**, Fig. 4–8b) shows that it also consists of a pair of enantiomers. Thus, there are four stereoisomers of 1-bromo-2-chlorocyclobutane in accord with the general rule just given.

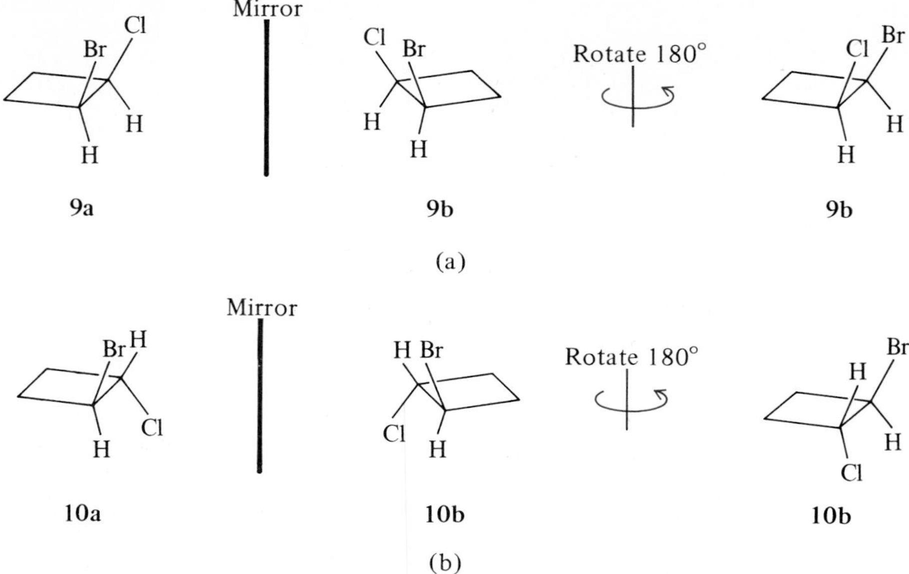

Figure 4–8. *Nonidentical mirror images (enantiomers) of both (a)* cis- *and (b)* trans-*1-bromo-2-chlorocyclobutane.*

It is very instructive to compare stereoisomerism in this compound with that found in 1,2-dibromocyclobutane (Fig. 4–1). The dibromo compound also contains two chiral centers, yet gives rise to only three stereoisomers. This result arises from the fact that both chiral centers are bonded to the *same* set of four groups (H, CH_2, CHBr, Br), which permits the *cis* isomer to have a plane of symmetry (Fig. 4–3). Because the *cis* compound is achiral, the number of stereoisomers is one less than the theoretical maximum. *cis*-1,2-Dibromocyclobutane is an example of a *meso* compound. Meso *compounds have more than one chiral center, yet the entire molecule has a plane of symmetry and is therefore achiral.*

A brief review of chirality and achirality is useful at this point. If a compound lacks a chiral center, it is *usually* superimposable on its mirror image and is therefore achiral. A compound that contains chiral centers is usually not superimposable on its mirror image and is therefore chiral. *Meso* compounds are exceptions to this last generalization. The easiest way to detect a *meso* compound is to look first for two chiral centers that are identically substituted. If you find them, then see if the molecule has a plane or center of symmetry.

Meso isomerism is found in open-chain compounds as well as in cyclic ones. 2,3-Dichlorobutane is an acyclic compound that exists in two diastereomeric forms (Fig. 4–9). One diastereomer (**11**) has an eclipsed conforma-

$$\begin{array}{c} CH_3\underset{|}{C}H\underset{|}{C}HCH_3 \\ \;\;Cl\;\;\;Cl \end{array}$$

2,3-Dichlorobutane

tion with a symmetry plane, and its stable *anti* conformer has a center of

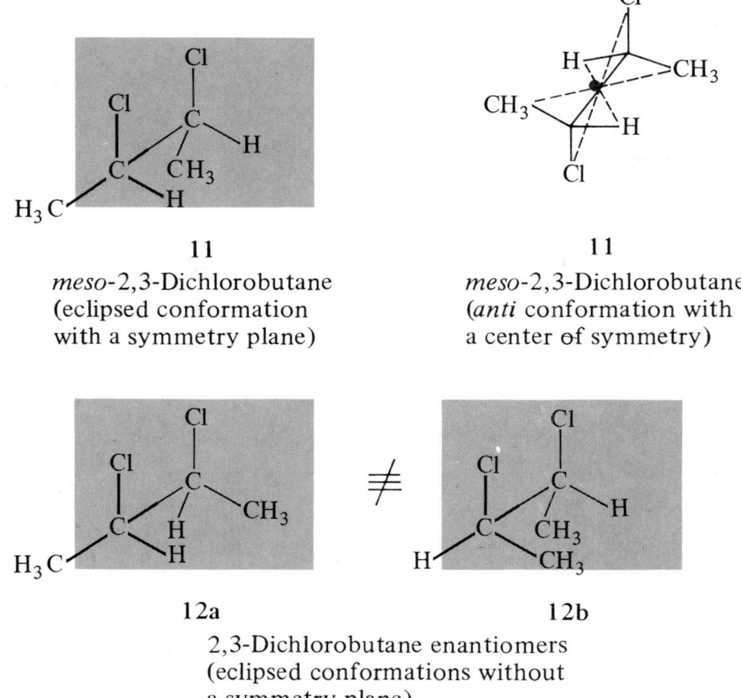

11
meso-2,3-Dichlorobutane
(eclipsed conformation
with a symmetry plane)

11
meso-2,3-Dichlorobutane
(*anti* conformation with
a center of symmetry)

12a $\neq$ **12b**

2,3-Dichlorobutane enantiomers
(eclipsed conformations without
a symmetry plane)

Figure 4–9. Meso *and chiral stereoisomers of 2,3-dichlorobutane.*

symmetry. The other diastereomer (either **12a** or **12b**) has no symmetrical conformations and is therefore chiral.

Sections 4:2 and 4:3 described the conditions that confer molecular chirality, but they did not indicate how chirality can be detected experimentally. If we are dealing with a chiral compound, how do we know whether the sample contains one or both enantiomers? How are the individual enantiomers to be characterized, named, and described? These questions are answered in the following sections.

Example

What is the maximum number of stereoisomers possible for the molecule shown below? Draw structures for the stereoisomers and tell if any of them represent a *meso* molecule. What are the relationships among the different stereoisomers?

$$\underset{\text{F}}{\overset{}{CH_3\underset{|}{C}H}}\text{—}\underset{\text{OH}}{\underset{|}{C}HCH_3}$$

Answer

There are two chiral centers in the molecule, so the maximum number of stereoisomers will be 2^2 or four stereoisomers. The next step is to draw the stereoisomers in three-dimensional form (or construct models) and look for a plane of symmetry in each structure. One way to do this would be with the sawhorse projections given in Figure 4–9. Here, one Cl would be replaced by a F and the second Cl would be replaced by an OH to give the molecule above.

Another way to draw structures is to use the wedge–dash convention. First draw one structure and then its mirror image:

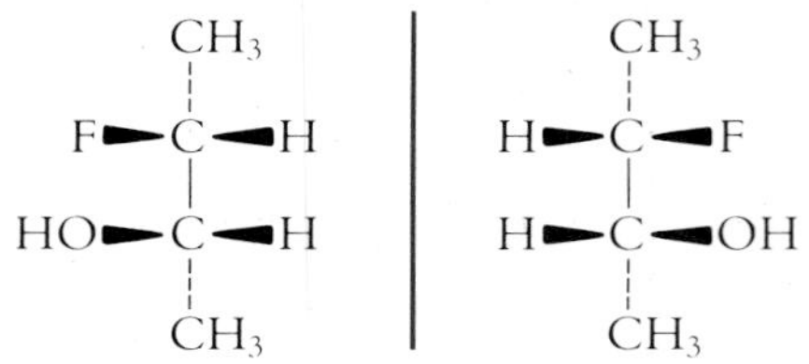

Mirror

Now change the configuration at *one center only* by interchanging two groups (H and F) and redraw:

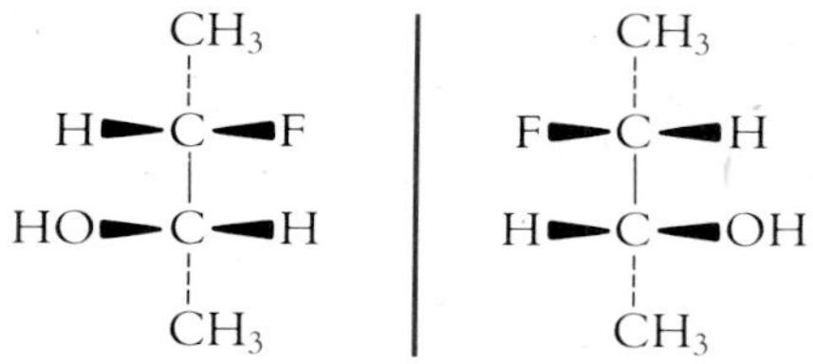

Mirror

These are the four stereoisomers. None has a plane of symmetry, so none can be called a *meso* molecule. The four represent two pairs of enantiomers. Either of the bottom pair is a diastereomer of either of the top pair.

Problems

4–7 For each of the following compounds, indicate which ones will have *meso* isomers. Draw the structure of the *meso* isomer in each case. For those compounds which will not have *meso* isomers, indicate why this type of isomerism is not possible.

a. methylcyclopentane
b. 1,3-dimethylcyclopentane
c. 2,3-dimethylpentane
d. 3,4-dimethylhexane
e. 2,3-dichlorohexane
f. 1,4-dimethylcyclohexane

4–8 The stable *anti* conformer of *meso*-2,3-dichlorobutane (**11**, Fig. 4–9) has a center of symmetry and is achiral. Are the *gauche* conformers of this molecule chiral or achiral? What is their relationship to one another?

4:4 Optical Activity: Experimental Evidence for Chirality

Because pairs of diastereomers (such as structures **1** and **2** in sec. 4:1, or **11** and **12**, Fig. 4–9) have sufficiently different structures, we would expect their physical properties to differ. This is precisely what we observe experimentally. For instance, *meso*-2,3-dichlorobutane (**11**) has a boiling point of 116°. Its chiral diastereomer (**12**) has a boiling point of 120°. The two diastereomers also differ in melting point, density, and a number of other physical properties. It is usually possible to separate diastereomers by crystallization, distillation, or other standard separation techniques.

Enantiomers, by contrast, have identical physical properties. Both enantiomers (**12a** and **12b**) of the chiral 2,3-dichlorobutane have boiling points of 120°. Furthermore, they have identical melting points, densities, and so forth. Enantiomers *cannot* be separated or distinguished by the usual separation techniques.

In order to distinguish between enantiomers they must be placed in a *chiral environment*. To pursue our earlier analogy, shoes (the chiral environment) may be used to distinguish between feet (the enantiomeric compounds) because a right shoe fits a right foot much better than it does a left one. However, a sock (an achiral environment) cannot be used to distinguish a left foot from a right one, since it fits both feet equally well. The first chiral environment deliberately used for distinguishing between enantiomers was polarized light, a mixture of left- and right-handed helical light beams.*

A ray of light consists of a number of individual waves traveling along the same line of propagation. The waves are planar, and the wave planes are oriented at all possible angles to one another (Fig. 4–10). Several materials are known to subject light waves to **plane polarization**, a process whereby all those waves not vibrating in one particular plane are filtered out. It was discovered early in the nineteenth century that certain naturally occurring substances could *rotate* plane-polarized light. The light remained polarized but its plane of polarization was now tilted at an angle to what it was before passing through the rotating medium (Fig. 4–11). Any substance possessing this property is called **optically active**. Some optically active materials are levorotatory—they rotate the plane of polarization to the left (counterclockwise); other optically active materials are dextrorotatory—they rotate

* Since helices are chiral (think of ordinary wood screws or bolts), polarized light has chiral components.

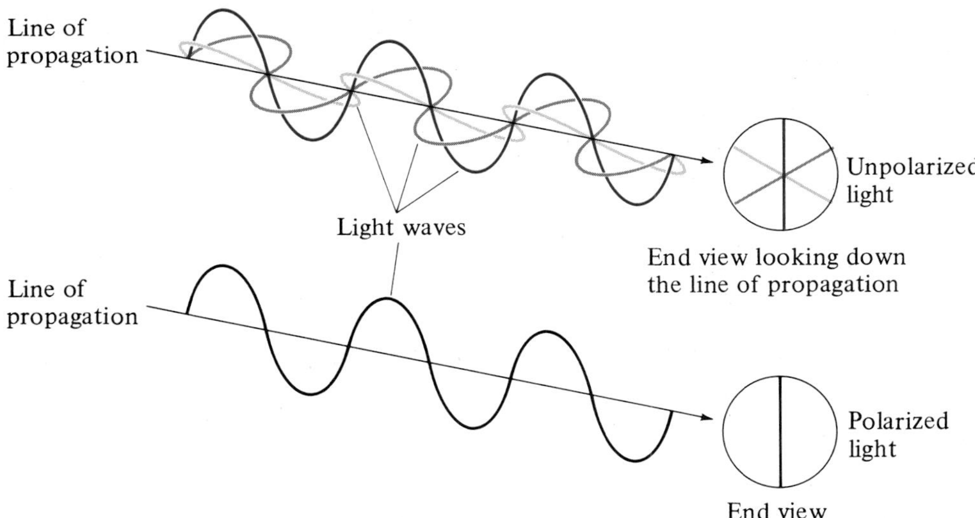

Figure 4–10. *Planes of vibration of light waves in unpolarized and plane-polarized light.*

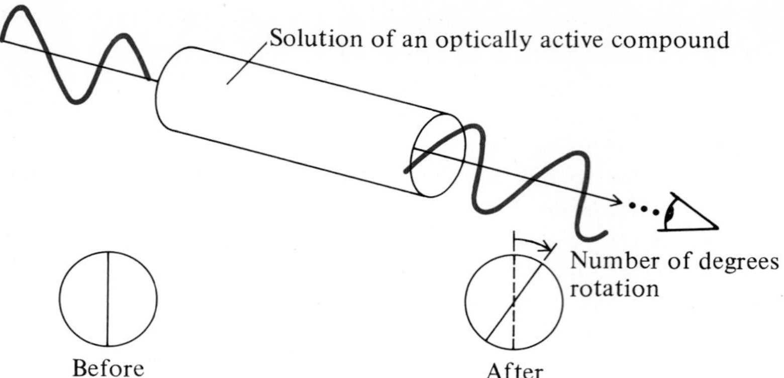

Figure 4–11. *Rotation of the plane of polarization by an optically active substance.*

the plane of polarization to the right (clockwise). *Chiral molecules are optically active, while achiral molecules are not.*

It will be easier to understand the significance of optical rotation if we consider a concrete example, such as the two enantiomers of lactic acid (**13** and **14**, Fig. 4–12). Lactic acid (from the Latin *lac,* meaning "milk") is found in biological substances such as muscle tissue and milk. As the data in Figure 4–12 show, the two enantiomers cannot be differentiated by means of their physical properties. Suppose, however, that we have a sample of the lactic acid enantiomer shown in structure **14**. We dissolve this compound in an achiral solvent such as water, ethanol, or chloroform, and place the solution in a tube in an instrument called a **polarimeter**, which measures the rotation of polarized light.

A typical polarimeter has a light source, usually a sodium lamp, emitting light centered around a wavelength (yellow) of 589 nm (nanometers; 1 nm = 10^{-7} cm). The lamp beam is first passed through a prism-quartz filter system that allows only light polarized in one plane to pass (Fig. 4–10b). This light is then focused on the tube containing the chiral substance. This substance absorbs and refracts some of the polarized light, but these effects are unequal for the two helical components of that beam. The net result is a rotation of the plane of polarization as the polarized light

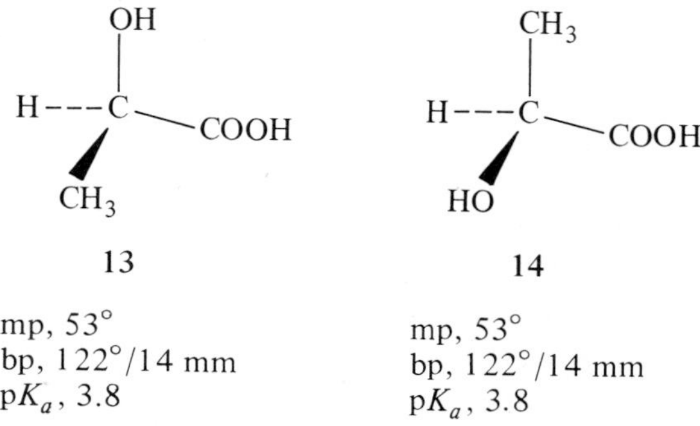

13	14
mp, 53°	mp, 53°
bp, 122°/14 mm	bp, 122°/14 mm
pK_a, 3.8	pK_a, 3.8

Figure 4–12. *The two enantiomers of lactic acid and some of their properties.*

encounters the chiral molecules. Note that structure **14** *rotates the plane of the transmitted light a certain number of degrees to the right.*

The extent of rotation (measured in degrees) is dependent upon the number of molecules encountered by the light beam. Thus, it is proportional to the concentration of the solution in the cell and the length of the cell through which the light beam passes. Lactic acid of configuration **13** will rotate the plane of polarized light to the left while **14**, its enantiomer, will rotate the plane exactly the same number of degrees to the right (**13** and **14** are used in the same concentration in the same cell). If an exactly 50:50 mixture of the two enantiomers were in the cell, the two opposite rotations would cancel and there would be no net rotation of the plane of polarization.

A *pure* optical isomer (a sample of one enantiomer) is usually characterized by its **specific rotation**, which is the rotation per unit weight per unit pathlength. This number is dependent on the type of solvent, the temperature of the solution, and the wavelength of light used, but once these have been fixed, the specific rotation does not vary. Specific rotation, symbolized by $[\alpha]_\lambda^T$, is defined by Equation 4.1. Rotatory power may also be calculated

(4.1) $$[\alpha]_\lambda^T = \frac{\alpha}{l \cdot c}$$ where T = temperature (°C)

λ = wavelength of light
l = pathlength of light (dm)
c = concentration (g ml^{-1})
α = actual rotation under given conditions

on a molar rather than a weight basis. Equation 4.2 defines the molecular rotation $[M]$; MW stands for molecular weight.

(4.2) $$[M] = \frac{[\alpha] \cdot \text{MW}}{100}$$

A pair of enantiomers may now be distinguished by means of their specific rotations. The absolute value of the specific rotation for each enantiomer will be identical, but the *signs* will be opposed; one will be dextrorotatory (+), while the other will be levorotatory (−).

The structures of the two lactic acid enantiomers are reproduced again below. They are accompanied by values of the specific rotation, the only physical parameter which serves to differentiate the two isomers. (The 589 nm wavelength is that for the sodium D spectral line. The symbol $[\alpha]_D$ is therefore often used.) It is not usually possible to predict the sign of rotation

OH / H---C(CH₃)—COOH

13
$[\alpha]_{589}^{25} = -3.8°$

CH₃ / H---C(HO)—COOH

14
$[\alpha]_{589}^{25} = +3.8°$

for any particular enantiomer. For instance, we could not tell from the structure which enantiomer of lactic acid, **13** or **14**, would be the dextrorotatory one. For some molecules, the sign of rotation is known to change with changing wavelength of light, p*H*, solvent, or temperature.

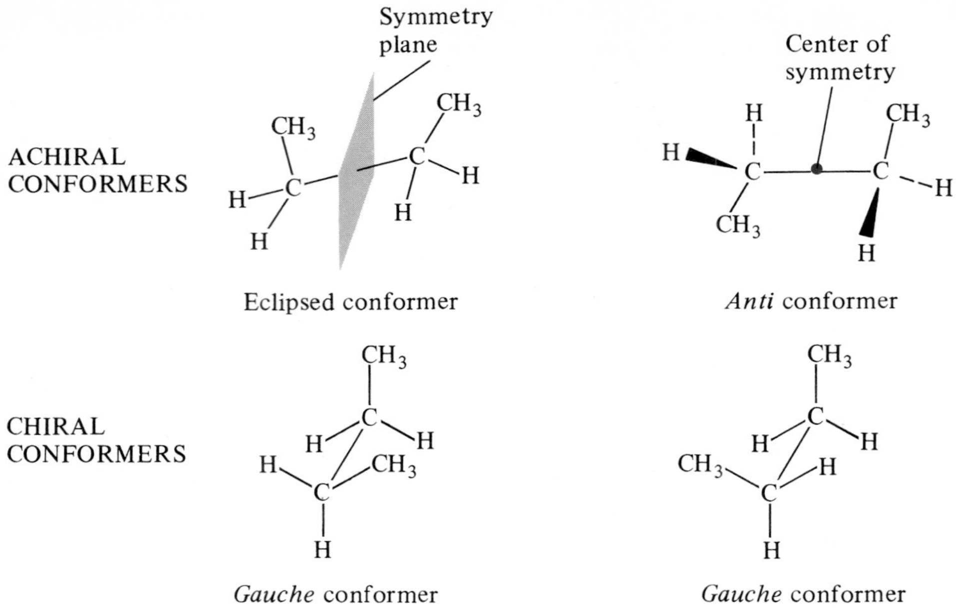

Symmetry plane

Center of symmetry

ACHIRAL CONFORMERS

Eclipsed conformer

Anti conformer

CHIRAL CONFORMERS

Gauche conformer

Gauche conformer

Figure 4–13. *Chiral and achiral conformers of butane. The eclipsed and* anti *conformers are symmetric and therefore achiral. The* gauche *conformers, which are chiral, are enantiomers.*

The major principle of this section—that chiral mirror image structures have equal but opposite rotations—gives us further insight into achiral molecules, and their lack of optical activity. Consider the hydrocarbon butane (Prob. 4–5). Its eclipsed and *anti* conformers possess a plane and a center of symmetry, respectively (Fig. 4–13), and so both are achiral. But what about its other conformers, such as the *gauche*? They are in fact chiral, *but are mirror images of each other.* Thus their optical rotations cancel each other, resulting in a zero optical rotation for butane. This phenomenon is a general one. Achiral molecules may have many chiral conformations, but they always occur as pairs of enantiomers whose optical rotations cancel each other (see Prob. 4–8).

Problems

4–9 In Section 4:2 we pointed out that only the high energy planar conformation of *cis*-1,2-dibromocyclohexane has a plane of symmetry. Draw the structures of the energetically favored chair conformers of this compound, and use them to explain why it is not optically active.

4–10 Most naturally occurring compounds, such as cocaine, are optically active. If we dissolve 4.0 g of cocaine in 100 ml of chloroform at 20°, measure the optical rotation in a 1-dm cell with a polarimeter (sodium D line light source), and find the rotation to be 0.64° to the left, what is the $[\alpha]_D^{20}$ for cocaine?

Box 4–1. Pasteur and the Chiral Basis of Life

In 1815, the French scientist Biot discovered the phenomenon of optical activity. He found it in crystals, in extracts of resin from pine trees, and in solutions of certain compounds. Biot strove in vain for more than 30 years to establish a clear link between chemical structure and optical activity. He was convinced that the study of optical rotation was one of the surest ways to discover the molecular constitution of pure substances. This conviction was not shared generally by other chemists of that day. Finally, in 1848, the connection between optical activity and molecular structure was made by the great French chemist Louis Pasteur. Pasteur graduated in chemistry with the equivalent of a C average, but ended up being one of the very few scientists who made fundamental discoveries in several different sciences. He is best known for his research which led to the pasteurization of milk.

Pasteur was studying the crystal structures of salts of an organic compound, tartaric acid, when he discovered that what looked like a collection of similar crystals was actually a mixture of two different kinds of crystals. They were irregular polyhedral (many sided) structures that were nonidentical mirror images of each other. Pasteur used a microscope and a pair of tweezers to painstakingly separate the two kinds of crystals. He then dissolved each group separately in water and passed plane-polarized light through each solution. The solutions were optically active, and they rotated light in opposite directions, but by an equal amount for the same concentration of crystals. Because optical activity occurred when the crystals were dissolved in solution, Pasteur reasoned that it was not the crystals, but the molecules that make up the crystals, that caused the optical activity. He further suggested that the molecules of each kind of tartaric acid should be nonidentical mirror images, just as the two kinds of crystals were.

Pasteur performed this experiment in front of Biot and, as Pasteur related in an 1860 lecture, he never forgot the visible emotion with which the old scientist greeted this vindication of his own early theories. It is worthwhile to quote from this lecture, since it shows that Pasteur fully grasped the fundamentals of enantiomer reactivity in both laboratory and biological chemistry.

"All artificial or natural chemical compounds, whether mineral or organic, must be divided into two great classes: non-asymmetric [achiral] compounds and asymmetric [chiral] compounds. . . . If the two tartaric acids are placed in contact with non-asymmetric compounds such as potash, ammonia, alcohol, they behave identically. If, on the contrary, they are submitted to the action of products themselves asymmetric (quinine, strychnine, sugar), all is changed in an instant. The solubility is no longer the same. If combination takes place, the crystalline form, the specific weight, the more or less easy destruction by heating, all differ as much as in the case of two of the most distantly related isomers. Here, then, the molecular asymmetry of a substance obtrudes itself as a powerful modifier of chemical affinities. Towards the two tartaric acids, quinine [a base] does not behave like potash, simply because it is asymmetric and potash is not. . . . We now reach a final experiment which does not yield in interest to any of those which precede, in respect to the manifest proof it will give us of the influence of asymmetry intervening in the phenomena of life. We have just seen asymmetry intervening to modify chemical affinities: but the reactions involved were purely artificial and we know all the prudence that must be observed in the application of the results of the laboratory to the phenomena of life. On this account I kept to myself almost all the views expressed in this Lecture till the moment that I recognized in the most certain manner that molecular asymmetry exhibited itself as a modifier of reactions of physiological character." [Pasteur then describes an experiment in which he treats "right" and "left" tartaric acids with a fermentation yeast. The result was that the yeast fermented "right" tartaric acid, while "left" tartaric acid remained unchanged. He continues as follows.] "Here, then, the molecular asymmetry proper to organic substances intervenes in a phenomenon of a physiological kind, and it intervenes in the role of a modifier of chemical affinity. . . . Thus we find introduced into physiological principles and investigations the idea of the influence of the molecular asymmetry of natural organic products, of this great character which establishes perhaps the only well marked line of demarcation that can at present be drawn between the chemistry of dead matter and the chemistry of living matter."

4–11 The compound with structure **I** has specific rotation $[\alpha]_D^{25} = -13.51°$. Draw the structure of the compound with $[\alpha]_D^{25} = +13.51°$.

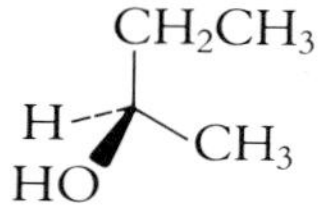

CH_2CH_3

H CH_3
HO

I

4:5 Finding and Designating the Absolute Configurations of Chiral Centers

Suppose we are given a pair of enantiomers, such as those of 2-bromobutane, **A** and **B** (Fig. 4–14). We know that one of them is dextrorotatory and the other is levorotatory. However, we have no way of knowing which enantiomer, **A** or **B**, is the dextrorotatory one. What we need is a means of determining and designating the **absolute configuration** at a chiral center. By absolute configuration we mean *the actual arrangements of the atoms in space without reference to any other center.* The problem was solved in 1951 by the Dutch chemist J. M. Bijvoet. By using x-ray crystallography Bijvoet and his co-workers showed that (+)-tartaric acid has the structure shown below.

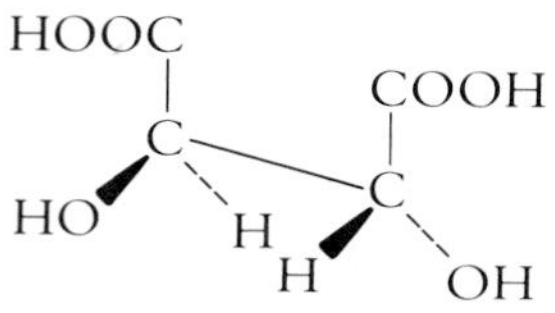

HOOC COOH
C C
HO H H OH

(+)-Tartaric acid

Since the configurations of many chiral molecules were known relative to that of (+)-tartaric acid, Bijvoet's experiment provided a standard with which we can compare all other molecules and thus determine their absolute configurations. For example, we now know that enantiomer **A** is the levorotatory one and enantiomer **B** is the dextrorotatory one.

Once chemists were able to determine the absolute configurations at chiral centers, it was necessary to *designate* those configurations in an unambiguous way. The system of designation has to be independent of any physical property so that it can be used for chiral compounds whose physical properties are not known.

A system for designating absolute configuration was published in 1956 by R. S. Cahn, C. K. Ingold, and V. Prelog. The Cahn-Ingold-Prelog

CH_2CH_3 CH_3CH_2

H CH_3 H_3C H
Br Br

A **B**

Figure 4–14. *Two enantiomers of 2-bromobutane. Before 1951 it was not known which of these was dextrorotatory* (**B**) *and which was levorotatory* (**A**).

Bromochlorofluoromethane 2-Chlorobutane

Figure 4–15. *Assigning priorities in bromochlorofluoromethane and in 2-chlorobutane.*

method permits us to assign an absolute configuration to each chiral center in a molecule, and to then distinguish between isomers on this basis. In order to assign an absolute configuration to a chiral center, we must assign a *priority number* to each group attached to that chiral center.

Priority is assigned on the basis of the atomic number of the *atom directly connected to the asymmetric center;* the higher the atomic number, the higher the priority. If ① denotes the substituent of highest priority, and ④ the lowest priority substituent, then bromochlorofluoromethane has the priorities shown in Figure 4–15 because in atomic number Br > Cl > F > H. In 2-chlorobutane two of the atoms directly attached to the asymmetric center are identical (the carbons of the methyl and ethyl groups). In such cases, priority is assigned by referring to the *next* closest atoms. The carbon of the methyl group is attached to three hydrogens, while the carbon of the ethyl group is attached to two hydrogens and one carbon. Therefore, ethyl takes priority over methyl since carbon has a higher priority than hydrogen. If two isotopes of the same element are compared, the heavier isotope takes precedence.★ Table 4–1 shows the relative priorities of some other common functional groups.

★ Remember that isotopes are different forms of the same element that differ only in their nuclear masses. Thus ${}^{1}_{1}H$ and ${}^{2}_{1}H$ are the two stable isotopes of hydrogen. The nucleus of ${}^{1}_{1}H$ consists of only a proton; the nucleus of ${}^{2}_{1}H$ has a proton and a neutron. The heavier isotope, deuterium (D), is given priority over protium (H) in the assignment of absolute configuration.

Table 4–1. Relative Priorities of Some Common Functional Groups Attached to a Chiral Center

$I— > Br— > Cl— > F— > R_3C—$

$RO— > R_2N— > R_3C—$

$(CH_3)_3C— > (CH_3)_2CH— > CH_3CH_2— > CD_3— > CH_3—$

$HC{\equiv}C— > (CH_3)_3C— > CH_2{=}CH— > (CH_3)_2CH— > CH_3CH_2—$

$RC(=O)— >$ (phenyl ring)— $> HC{\equiv}C— > CH_2{=}CH— > (CH_3)_2CH— > CH_3CH_2—$

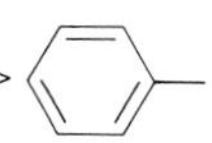

(C_6H_5)

In general, *priority is given to the group with the atom of higher atomic number found at the first point of difference.* Thus, the substituent —CH_2CH_2Cl has a higher priority than

$$\text{—}CH_2\underset{\displaystyle CH_3}{\underset{|}{C}}HCH_2Cl$$

Both substituents begin with —CH_2 groups. In —CH_2CH_2Cl, the next carbon is bonded to a Cl atom, whereas in the other substituent, the next carbon is bonded to two carbon atoms and a hydrogen atom.

When a substituent contains a double bond, the atoms at each end of the bond are considered to be connected to two atoms of the type found at the other end. For example, $CH_2{=}CH$— is equivalent to

$$\underset{\displaystyle C}{\underset{|}{CH_2}}\text{—}\underset{\displaystyle C}{\underset{|}{CH}}\text{—}$$

for the purposes of assigning priority in the Cahn-Ingold-Prelog system. Analogously, $>C{=}O$ is equivalent to

$$\text{—}\overset{|}{\underset{\displaystyle O}{\underset{|}{C}}}\text{—}O$$

Once priority has been assigned, the projection of the molecule is manipulated so that the group of lowest priority is projecting away from you and behind the plane of the paper, and the other three groups are sitting in front of the plane of the paper (models can be a great help here). *If the order of descending priority, ① to ② to ③, travels clockwise around the substituents, then the absolute configuration of that asymmetric center is* R (from the Latin *rectus,* meaning "right"). *If the priority descends in a counterclockwise sense, the absolute configuration at that center is* S (from the Latin *sinister,* meaning "left"). (See Fig. 4–16.)

The following summary of the operations used to assign absolute configuration according to the *R,S* system will be of value to you.

1. Assign a priority based upon atomic number to each atom attached to the chiral (asymmetric) center (① > ② > ③ > ④). If two atoms have the same atomic number, priority is assigned by comparing groups attached to those atoms until a difference is reached.
2. Draw a three-dimensional figure or hold a model so that the group with lowest priority (④) is placed behind the plane of the paper or projects away from you.
3. Draw or imagine a circular arrow from the substituent of highest priority (①) to that of the next highest priority (②).
4. If the circular arrow points clockwise, the asymmetric atom has the *R* configuration. If the arrow points counterclockwise, the configuration is *S*.

Now, let's consider a specific example. Figure 4–17 shows how the absolute configuration is assigned to (−)-lactic acid from its printed projec-

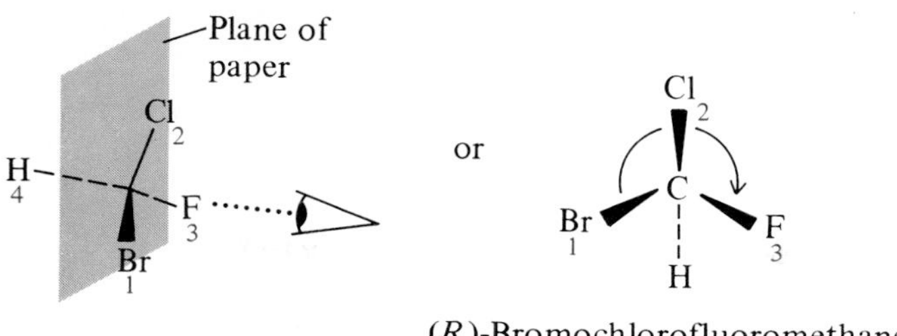

(*R*)-Bromochlorofluoromethane

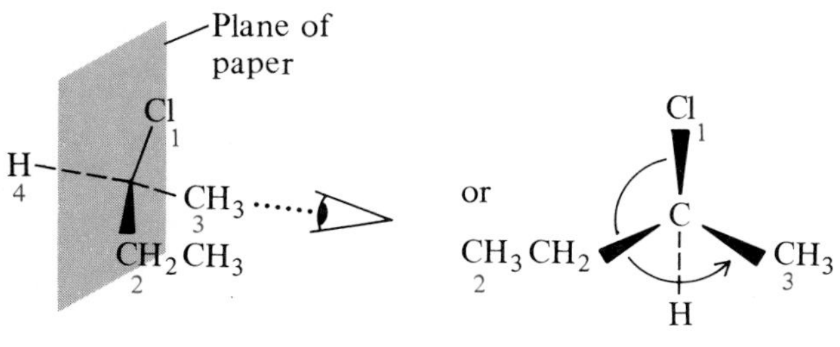

(*S*)-2-Chlorobutane

Figure 4–16. *Proper view for assigning* R *or* S *configuration to a chiral center.*

tion formula, **15**. Notice that the atom of lowest priority is the one projecting behind the reference plane (dashed bond). Priorities are first assigned to the groups, and the line of descending priority is then drawn. Since this line proceeds clockwise, we assign (−)-lactic acid the configuration *R*.

Projection formulas must sometimes be manipulated so that a convenient orientation is obtained. One method involves interchanging the substituents. Each time a pair of groups attached to an asymmetric atom is interchanged, the absolute configuration at that center is reversed. That is because interchanging a pair of substituents is equivalent to reflection of the molecule in a mirror. In order not to change configuration in the new orientation, *two* such interchanges must be carried out (Fig. 4–18).

A convenient memory device is to consider the bond pointing toward the atom of lowest priority as the steering column of an automobile and the circle described by the arrow as the steering wheel. The wheel is grasped and turned in the direction from the group of highest priority (①) to that of the

15 (*R*)-(−)-Lactic acid

Figure 4–17. *Assigning absolute configuration to a lactic acid enantiomer.*

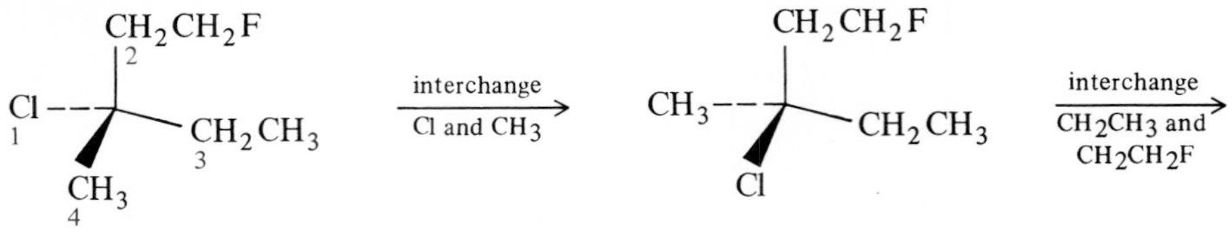

CH_2CH_2F (2), Cl (1), CH_2CH_3 (3), CH_3 (4) —interchange Cl and CH_3→ CH_2CH_2F, CH_3, CH_2CH_3, Cl —interchange CH_2CH_3 and CH_2CH_2F→

15
(lowest priority group in front)

(enantiomer of original compound)

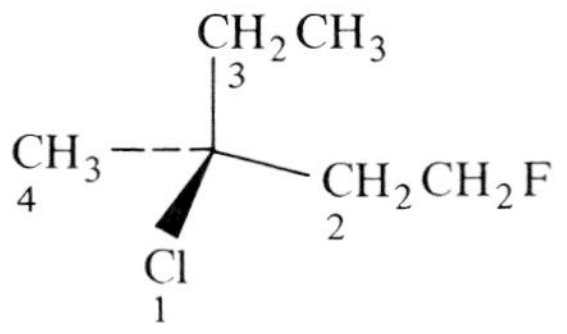

CH_2CH_3 (3), CH_3 (4), CH_2CH_2F (2), Cl (1)

15
(lowest priority group in back,
and thus easy to assign as *S*)

Figure 4–18. *Manipulating a projection formula in order to assign absolute configuration.*

next priority (②). If the automobile turns right, the configuration is *R*. If it turns left, the configuration is *S*.

It is important to remember that *R* and *S* assignments are based on a *convention*. Chiral centers having opposite designations may still have corresponding spatial arrangements for their substituents. For instance, the compounds (*R*)-2-chlorobutane and (*S*)-2-chloro-1-butanol, although opposite in designation, have corresponding absolute configurations because all three of their common substituents (Cl, CH_2CH_3, and H) coincide when the two molecules are superimposed.

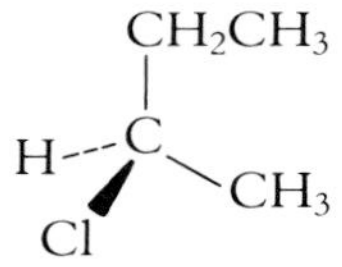

CH_2CH_3, H, C, CH_3, Cl — (*R*)-2-Chlorobutane

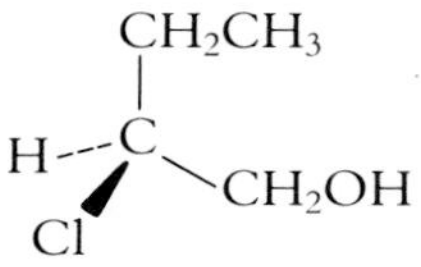

CH_2CH_3, H, C, CH_2OH, Cl — (*S*)-2-Chloro-1-butanol

Determining *R* and *S* configurations is a simple matter with molecular models, and we strongly recommend their use.

Example

Assign the configuration (*R* or *S*) at the chiral carbon of

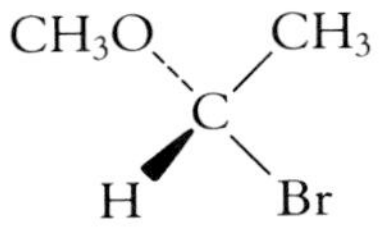

CH_3O, CH_3, C, H, Br

Answer

Step 1: Assign priorities to the groups on the chiral carbon. The atomic number order is Br > O > C > H, so the priorities are

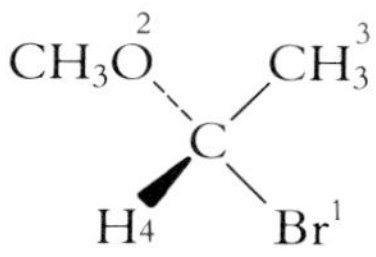

Step 2: If you are working the problem with a molecular model, arrange the molecule so that the group of lowest priority is facing away from you. Turn the molecule like a car steering wheel from group 1 to group 2. You should have turned it counterclockwise (so that your "car" would have made a left-hand turn), and found the configuration to be *S*.

Step 3: If you are not working with a model, you will have to redraw the molecule. Changing two substituents will change the configuration. If this is done *twice,* your new drawing will have the same configuration as the original:

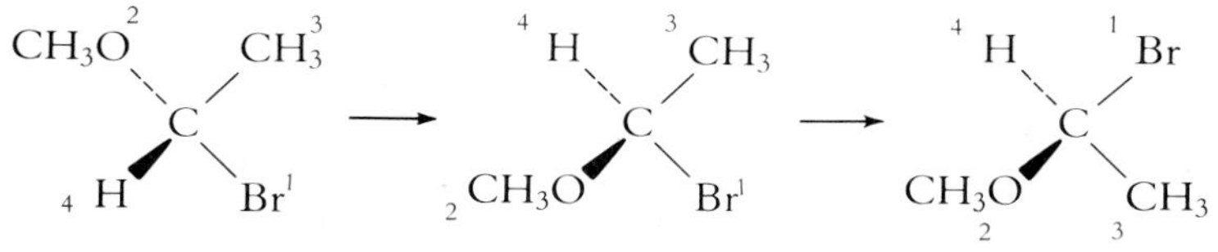

Now draw a circular arrow from priority group 1 to 2 to 3:

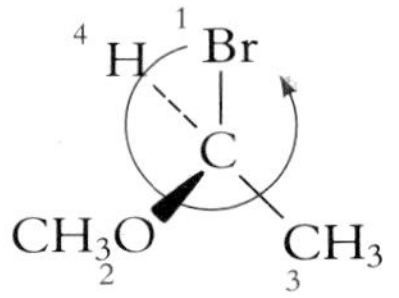

The arrow points counterclockwise; therefore the configuration is *S*.

Problems

4–12 Assign the configuration at each chiral center of each of the following compounds:

4–13 Draw an unambiguous stereochemical structure for each of the following compounds:

a. (*R*)-3-methylhexane
b. (*R*)-2-chloropentane
c. (*S*)-CH_3CHDOH
d. (*R*,*S*)-1,2-dibromocyclobutane

4–14 Assign configurations to the chiral centers of the isomers shown in Figure 4–8.

4:6 Racemic Mixtures

In Section 4:5 we showed that the major physical property that distinguishes enantiomers is their optical activity. One enantiomer is dextrorotatory and the other is levorotatory, while the absolute value of their specific rotation is identical. It follows from this relationship that a 50 : 50 mixture of two enantiomers will display an optical rotation of exactly zero, because the rotatory effect of each enantiomer is canceled by the other. Such a mixture is called **racemic** (from the Latin word *racemes,* meaning "a bunch of grapes"). The first racemic substance characterized as such was a mixture of (+)- and (−)-tartaric acids, obtained as a byproduct of grape juice fermentation (hence the name). This mixture was initially thought to be a single compound and was given the name racemic acid. Its true composition was revealed when Pasteur separated a salt of racemic acid into its components, the enantiomerically pure tartrate salts (sec. 4:4). The process of separating pure enantiomers from a racemic mixture is known as **resolution**.

It is important to distinguish between a mixture that is optically inactive because it is racemic (equal amounts of two enantiomers), and a compound that is optically inactive because it is achiral. We can use the various kinds of tartaric acids to illustrate this point. We now know that (+)-tartaric acid has the configuration given in structure **16**, while its enantiomer (−)-tartaric acid is **17**. (*R* and *S* assignments are shown for the chiral carbons.) Racemic acid is the equimolar mixture of the two. There is yet another tartaric acid stereoisomer. *meso*-Tartaric acid, **18**, which has both a plane and a center of symmetry, is optically inactive and *cannot* be resolved since it has no enantiomer. It is therefore a single substance, not a mixture. *meso*-Tartaric acid is a diastereomer of **16** and **17**. It has different physical properties from each of them and from the racemic mixture.

16
(+)-(*R,R*)-Tartaric acid

17
(−)-(*S,S*)-Tartaric acid

18
(*R,S*)-Tartaric acid
(*meso*)

Sometimes a crystalline racemic mixture (known as a racemate) has different physical properties than the pure enantiomers. Racemic acid melts at 206°, while (+)- and (−)-tartaric acids both have melting points of 168–170°. *meso*-Tartaric acid has a melting point of 159–160°. Melting points are partly a reflection of how the molecules pack together in a crystal. This packing is different for crystals composed of one enantiomer as opposed to crystals composed of equal numbers of each enantiomer.

It is useful to have a nomenclature that will distinguish between enantiomeric compounds and racemic mixtures. The most common way to refer to racemic mixtures is by the symbol (±). "Racemic acid" is now known as (±)-tartaric acid.* If there is a single chiral center, the *R,S* terminology can

* In early chemical literature, the symbol *d* was used for (+) and *l* for (−), while racemic mixtures were denoted by *dl*. These symbols for enantiomers are no longer in use, but racemic mixtures are still often called *dl*-mixtures.

be used. Thus, a racemic mixture of lactic acids (Fig. 4–12) can be designated (*R*,*S*)-lactic acid.

4:7 How Racemic Mixtures Are Resolved

In nature, chiral compounds often occur as pure enantiomers. For instance, by using different natural sources, Pasteur obtained pure (+)-tartaric acid, as well as "racemic acid," (±)-tartaric acid. On the other hand, most chiral compounds synthesized in the laboratory are available only as racemic mixtures. For this reason it is necessary to have general methods for resolving racemic mixtures into their component enantiomers. Pasteur was able to resolve a racemic tartrate salt simply by crystallizing it from water (sec. 4:4). Unfortunately, such spontaneous resolutions are rare. There is, however, a widely applicable method for carrying out resolutions that is based on some fundamental acid–base chemistry (sec. 1:7) and makes use of chiral compounds available from nature.

The key to this method is to take a racemic mixture of enantiomeric acids and convert it to a separable mixture of diastereomeric salts. This is done by reacting the racemic acid with a chiral base.

As an example, consider the resolution of (*R*,*S*)-lactic acid. Suppose we react this mixture with a chiral base, designated (*R*)-base. Two different salts are formed, as shown in Equation 4.3. *The two salts are diastereomers, not enantiomers,* and hence should be separable by crystallization.⋆ (The enantiomer of [(*R*)-baseH$^+$ (*R*)-lactate$^-$], for example, would be [(*S*)-baseH$^+$ (*S*)-lactate$^-$].)

$$(4.3)\quad \underbrace{(R)\text{-lactic acid} + (S)\text{-lactic acid}}_{\text{Racemic mixture}} \xrightarrow{(R)\text{-base}} [(R)\text{-baseH}^+(R)\text{-lactate}^-] + [(R)\text{-baseH}^+(S)\text{-lactate}^-]$$

After separation of the two salts, each can be treated with a strong aqueous mineral acid to regenerate the pure components (eqs. 4.4 and 4.5). In these reactions, the freed lactic acids can be separated from the solution, while the protonated (*R*)-base and Cl$^-$ remain in solution.

$$(4.4)\quad [(R)\text{-baseH}^+(R)\text{-lactate}^-] \xrightarrow[H_2O]{HCl} (R)\text{-lactic acid} + (R)\text{-baseH}^+ + Cl^-$$

$$(4.5)\quad [(R)\text{-baseH}^+(S)\text{-lactate}^-] \xrightarrow[H_2O]{HCl} (S)\text{-lactic acid} + (R)\text{-baseH}^+ + Cl^-$$

Pasteur's biological experiment (see Box 4–1, sec. 4:4) in which he treated (±)-tartaric acid with a fermentation yeast, is actually another method of resolution that is based upon this same principle. The yeast, as is true of nearly any biological substance, is composed of chiral compounds

⋆ Although diastereomers always have different physical properties, the difference may not be large enough to be useful in separating them. When resolving a racemic acid, for instance, we may have to try several different chiral bases until we find one that produces salts with large solubility differences.

Box 4–2. Chiral Molecules in Living Systems

Life is inseparably linked to chiral molecules. Most molecules of fundamental biological importance are chiral and usually only one enantiomer occurs naturally. For instance, the protein components of cells are made up of amino acids, all of which have the same enantiomeric configuration in higher animals. Primitive bacteria, on the other hand, often contain amino acid derived compounds of the opposite enantiomeric configuration.

In the course of chemical evolution, how did this discrimination in favor of one enantiomer arise? There is still no definitive answer to this question even though Pasteur recognized the problem in 1860:

"If the mysterious influence to which the asymmetry of natural products is due should change its sense or direction, the constitutive elements of all living beings would assume the opposite asymmetry. Perhaps a new world would present itself to our view. Who could foresee the organization of living things if cellulose, right as it is, became left; if the albumen of the blood, now left, became right? These are mysteries which furnish much work for the future, and demand the most serious consideration from science."

Given the time scale of chemical evolution (beginning some four to five billion years ago), it is not easy to perform decisive experiments that establish any theory as most probable. If it is true that chiral synthesis occurs only in a chiral environment then biological chirality may reflect an inherent dissymmetry in the universe. On the other hand, it is known that crystallization of racemic mixtures sometimes produces crystals that contain an enantiomeric excess of one isomer. This process is called *spontaneous resolution*. If this had happened in a primordial soup, and was then amplified by further reaction of the chiral precipitate, the present predominance of one isomer would be accidental and reflect no inherent universal asymmetry.

and therefore reacts at different rates with (+)-tartaric acid and (−)-tartaric acid. Indeed, many biological reactions are highly specific and, as Pasteur pointed out, the yeast only ferments (+)-tartaric acid. This destroys that enantiomer, but it does leave pure (−)-tartaric acid unreacted. The mixture has been resolved even though we are only able to obtain one enantiomer through this procedure.

Knowing the effects of chirality on biological reactions may be the most important part of this chapter for many of you, so further examples are given in the next section.

Problems

4–15 One salt of Equation 4.3 is [(*R*)-baseH^+(*S*)-lactate$^-$]. What is the enantiomer of this salt?

4–16 A sample of (*R,S*)-1-phenylethylamine, **II**, is prepared in the laboratory. (C_6H_5 denotes a benzene ring, which is achiral.) A sample of (*R*)-lactic acid is also available. Show how the amine can be resolved, and draw structures of the resolved enantiomers, showing their absolute configurations.

$$\begin{array}{c} C_6H_5\text{—}\underset{\displaystyle NH_2}{\underset{|}{C}}HCH_3 \end{array}$$

II

4:8 Chiral Discrimination in Biochemical Reactions

In the beginning of this chapter, we mentioned that monosodium glutamate and epinephrine derived from natural sources had twice the potency of the corresponding laboratory synthetics. This is because the natural substances are pure enantiomers, while the synthetic ones are 50 : 50 mixtures of two enantiomers. It is usually although not invariably true that only one of two enantiomers exhibits a physiological effect. Sometimes each enantiomer has an effect, but one may be greatly reduced or even completely changed. These are examples of **chiral discrimination**.

Some interesting chiral discriminations are those shown by our senses. For instance, (*S*)-leucine, a naturally occurring amino acid, is without taste. Its unnatural (*R*)-isomer has a sweet taste. Both (*R*)- and (*S*)-carvone are isolated from natural sources, the *R*-enantiomer from oil of caraway and the *S*-enantiomer from oil of spearmint. They possess distinctly different odors.

$(CH_3)_2CHCH_2CH(NH_2)COOH$

Leucine

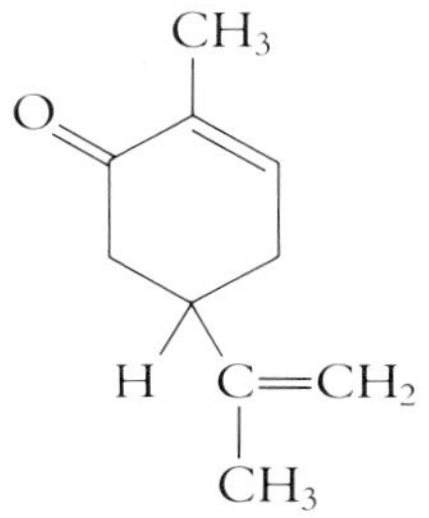

Carvone

Chiral discrimination in internal bodily processes is well-known and could be illustrated with a bookful of examples. One example from the area of clinical chemistry is the compound **dopa**. About 10 years ago dopa (an acronym for **d**ihydr**o**xy**p**henyl**a**lanine) was found to be very effective in combating Parkinson's disease, which afflicts the elderly with tremors and muscle rigidity. Large doses of dopa were required, and toxic side effects soon became apparent. The initial clinical trials were conducted with racemic dopa. Further investigation showed that the *S*-enantiomer was more effective, and less toxic, than its *R*-isomer. An established treatment for this disease now is the administration of pure *S*-dopa (usually called L-dopa, or levodopa).

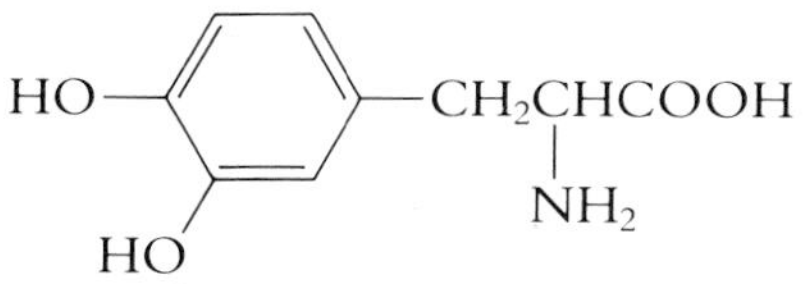

$(HO)_2C_6H_3CH_2CH(NH_2)COOH$

Dopa

Biological systems also characteristically convert achiral molecules to chiral molecules in the course of a reaction. An example is given in Equation 4.6. When reactions (in this case, hydrogenation) are carried out in the laboratory with common reagents, the product is racemic. By contrast, the corresponding biochemical reaction usually produces one enantiomer only. The chiral discrimination ability of enzymes will be discussed further in Sections 10:12 and 11:12.

$$\text{(4.6)}\quad CH_3CH_2\underset{\underset{O}{\|}}{C}CH_3 \xrightarrow{2H} \begin{cases} \xrightarrow{\text{laboratory}} CH_3CH_2\underset{\underset{OH}{|}}{C}HCH_3 \text{ (racemic mixture)} \\ \xrightarrow{\text{enzyme}} \text{H, OH, } CH_3CH_2\text{, } CH_3 \text{ on C (single enantiomer)} \end{cases}$$

Although common laboratory syntheses starting from achiral reactants always lead to achiral products, chemists have discovered ways of avoiding this. One method is to use a chiral catalyst. This method resembles enzymatic processes, and leads to chiral products.

Finally, we call your attention to one more situation involving chiral molecules which illustrates the differences between chemical and biochemical systems. A chiral center already present in a reactant can remain chiral in the product. While this is sometimes true in laboratory reactions, it is almost invariably true in biological ones. If the chiral atom has the same configuration after the reaction as it had before, then that particular chiral center **retained** its configuration during the reaction. If, on the other hand, the configuration of the product was the opposite of that in the starting material, then the chiral center is said to have undergone **inversion of configuration**. This type of information is important to an understanding of the mechanism of the reaction. An example of such a study is given in Problem 4–35 at the end of this chapter.

We have now completed our introduction to molecular structures. Among organic chemists, the study of molecular structure is almost always linked to the study of molecular reactivity, because a knowledge of structure is indispensable to a clear understanding of reactivity. In Chapter 5 we will begin to study organic reactions using our foundation of molecular structure.

Summary

1 Stereoisomers that are nonidentical mirror images are called enantiomers. Stereoisomers that are not enantiomers are called diastereomers.

Enantiomers are chiral molecules. To test a molecule for enantiomerism, make a model or drawing of its mirror image and see if it is nonsuperimposable on the original (sec. 4:1).

2 Molecules that contain a plane of symmetry or a center of symmetry are identical with their mirror images. They are therefore achiral. On the other hand, the presence of one chiral center in the molecule results in chirality. A chiral center is one bonded to four different groups not lying in a plane. Certain molecules can contain more than one chiral center and still be achiral. This can occur only when the chiral atoms have identical substituents and a plane or center of symmetry results. Compounds that have such symmetry elements, even though they possess chiral centers, are *meso* isomers (secs. 4:2 and 4:3).

3 Enantiomers are physically alike in all respects except when they interact with chiral substances or chiral media, such as plane-polarized light. If one enantiomer is dextrorotatory (+), the other will be levorotatory (−) to the same extent. This optical activity is evidence for the presence of chiral molecules. It is a function of the number of chiral molecules present, and of their inherent specific rotation. There is no simple theory correlating structure and sign of rotation (sec. 4:4).

4 The configuration at a chiral center can be designated independently of optical activity. The groups attached to the chiral center are assigned priorities based on the atomic number of the atom attached to the center. Then the molecule is oriented so that the group of lowest priority (lowest atomic number) points away from the observer. The three remaining groups are connected by a line of descending priority, which moves in either a clockwise (*R*) or counterclockwise (*S*) direction.

In order to assign configuration from a projection formula, it is necessary that the group of lowest priority be shown projecting behind the plane of the paper. If the structural formula is not shown in this orientation, it can be obtained by exchanging the positions of two pairs of substituents (sec. 4:5).

5 A racemic mixture (±) is one that contains equal amounts of two enantiomers. Racemic mixtures show no optical activity, but they can be resolved into optically active components. Resolution is often accomplished by treating the racemic mixture with a pure chiral reagent, yielding two diastereomeric products. The diastereomers are physically separated and then converted to pure enantiomers. Racemic acids can be resolved with pure chiral bases, and vice versa (secs. 4:6 and 4:7).

6 Two enantiomers sometimes display different smells and tastes. They usually show very different reactivity in biochemical systems, one enantiomer being far more effective than the other. When achiral molecules are converted to chiral molecules in the laboratory, the product is usually racemic. When this process takes place in biochemical systems, however, the product is almost always optically active (sec. 4:8).

Problems

4–17 Write the structure of each compound and indicate each chiral center. There may be one or more, or none, in any one of the compounds.

a. 2-methylhexane
b. 2-chloro-2-methylhexane
c. 1-chloro-2-methylhexane
d. 2,4-dibromohexane
e. *s*-butylcyclobutane
f. 3-iodo-4-methylheptane
g. *trans*-1,3-diethylcyclobutane
h. *trans*-1,3-diethylcyclopentane

4–18 A student was studying the reaction of butane with chlorine (sec. 5:6). She isolated all the dichlorobutanes ($C_4H_8Cl_2$). Some of them were chiral and some of them were not. Indicate which compounds belong to each category. For those that are chiral, write the structures of both enantiomers.

4–19 One of the isomers of pentane, C_5H_{12}, is allowed to react with Br_2. In this reaction a Br atom can replace any one of the H atoms on pentane. A number of products are obtained, all of molecular formula $C_5H_{11}Br$. It is found that *two* of the products are chiral. Using this information, deduce the structure of the starting pentane isomer.

4–20 Which of the following molecular pairs are identical? Which are enantiomers? Which are diastereomers?

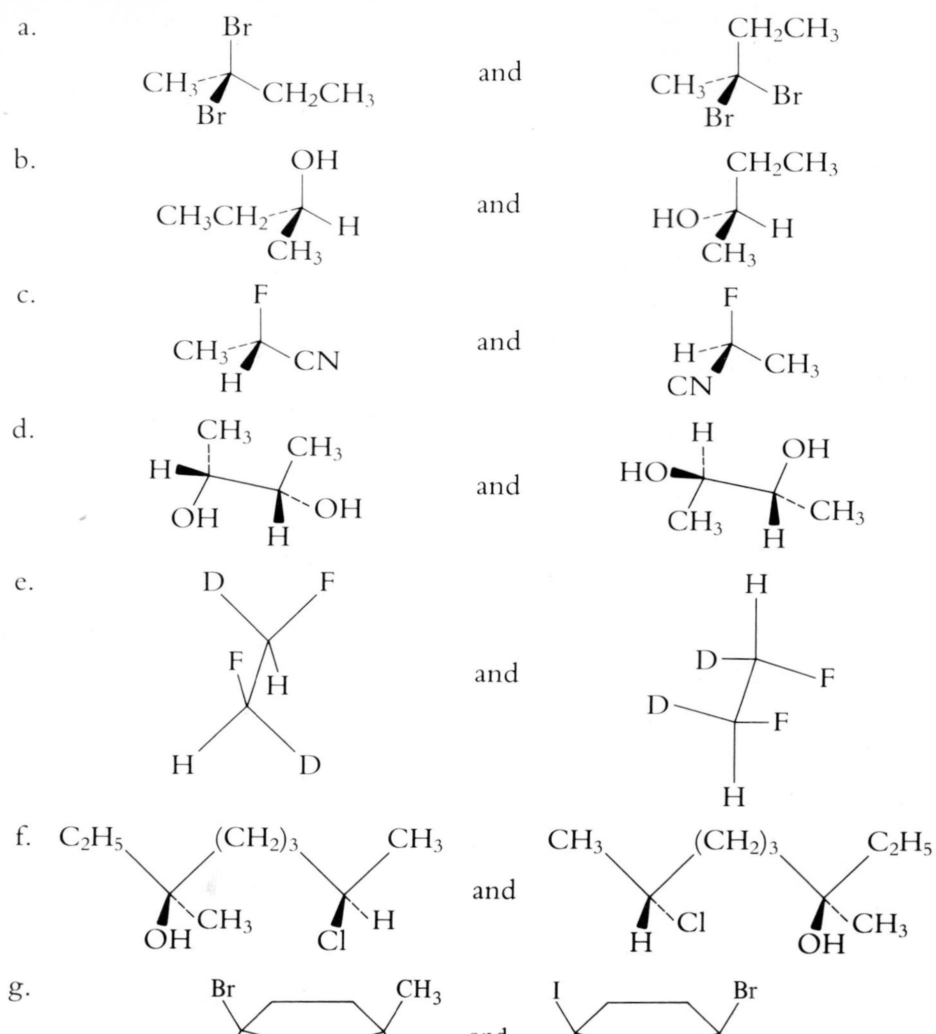

4–21 a. Draw the two chair conformers of *cis*-1,2-dimethylcyclohexane (you may need to review some material in Chap. 3). What is the isomeric relationship between these two conformers? Would it be possible (at least in theory) to isolate enantiomeric forms of this molecule?

b. Draw the stable diequatorial conformer of *trans*-1,2-dimethylcyclohexane. Now draw its mirror image. Are they identical? Are they interconvertible by means of bond rotation? Would it be possible (at least in theory) to isolate enantiomeric forms of this molecule?

c. Now draw planar projection formulas for the two 1,2-dimethylcyclohexanes. According to these formulas, which of the compounds should be achiral and which are chiral? Do the predictions you arrived at using the planar projections agree with those arrived at in (a) and (b)?

4–22 An unknown liquid is found to have the following elemental composition: C, 68.1%; H, 13.7%; O, 18.2%. Its mass spectrum shows that it has a molecular weight of 88. Other evidence shows that the molecule contains an OH group. The compound is found to be optically active by polarimetry. Suggest one or more structures that fit all the data.

4–23 a. If 5.68 g of sucrose, $C_{12}H_{22}O_{11}$, is dissolved in 20 ml of water, the rotation of the solution (measured at 589 nm) in a 10-cm tube at 20° is +18.9°. What is the $[\alpha]_D^{20}$ for sucrose?

b. A solution of sucrose whose concentration was unknown gave a rotation of +5.37° under the above conditions. What was its molar concentration?

4–24 What is the meaning of each of the following terms and symbols?

a. +
b. −
c. levorotatory
d. dextrorotatory
e. *R*
f. *S*
g. *meso*
h. $[\alpha]_D^{25}$
i. enantiomers
j. diastereomers
k. absolute configuration
l. resolution

4–25 Designate the absolute configuration (*R* or *S*) at each chiral center in each of the following molecules:

a. CH_3 (up), Br (dashed), H (wedge), OCH_3 (right)

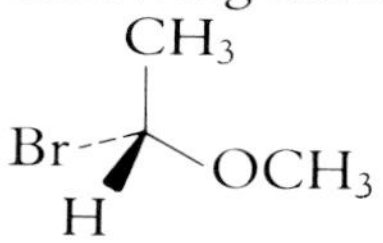

b. Br (up), H (dashed), CH_3 (wedge), OCH_3 (right)

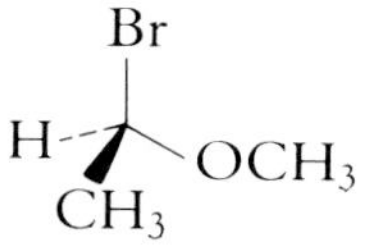

c. CH_2CH_3 (up), $(CH_3)_3C$ (dashed), CH_3 (wedge), CH_2F (right) on N^+

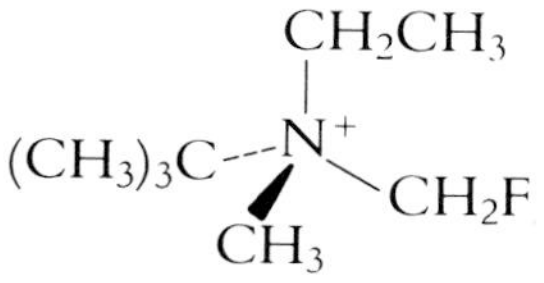

d. OCH_3 (up), $(CH_3)_2CH$ (dashed), $OCH_2CH_2CH_3$ (wedge), CH_2CH_3 (right) on P^+

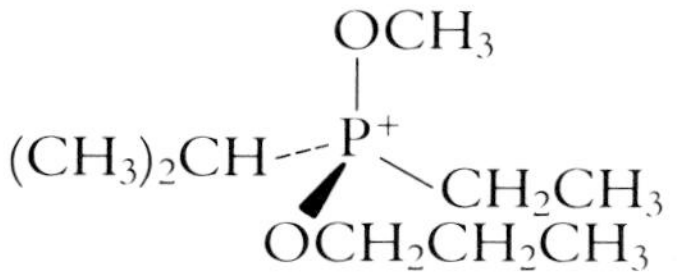

e. COOH (up), HO (dashed), HC=O (wedge), CH_2OH (right)

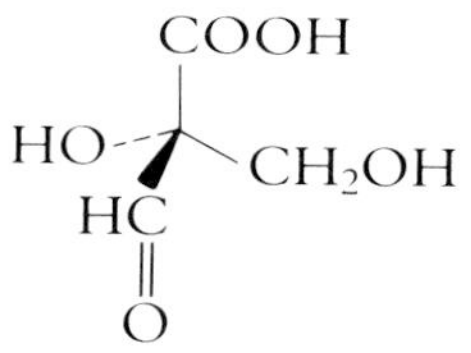

f. cyclopentenyl (with CH_3 on ring) bearing H (wedge) and CH_3 (dashed)

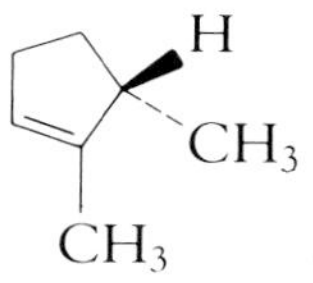

g. cyclohexane bearing H, CH_3; Cl, Br; H, CH_3

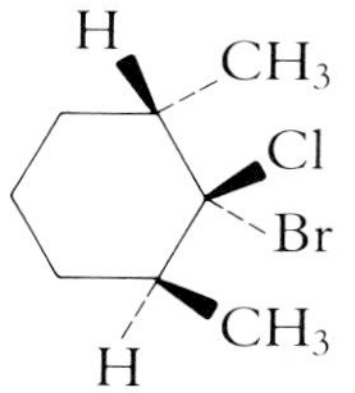

4–26 Assign the correct designation, *R* or *S*, to each compound.

a. Cl (up), CH_3 (dashed), H (wedge), $CH(CH_3)_2$ (right)

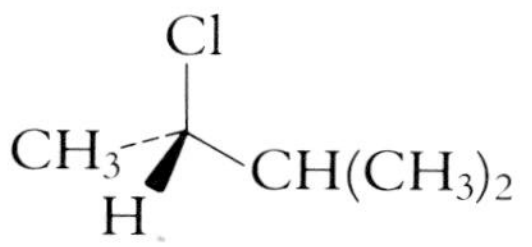

b. CH_2CH_3 (up), $BrCH_2$ (dashed), I (wedge), CH_3 (right)

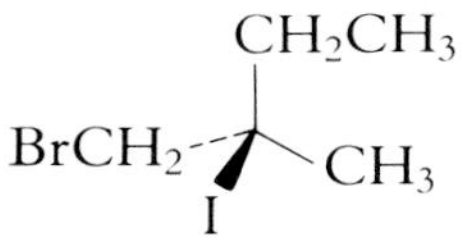

c. CH_2Cl (up), cyclopentyl (dashed), $CH_3(CH_2)_3$ (wedge), OH (right)

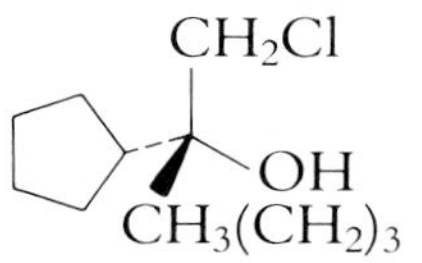

d. F (up), Br (dashed), CN (wedge), I (right)

4–27 Write structures for each compound. Show the stereochemistry unambiguously.

a. (*R*)-3-methylhexane
b. (*S*)-1,4-dichloropentane
c. (*R*)-2-fluorobutane
d. (1*S*),(3*R*)-1-bromo-3-ethylcyclopentane

4–28 Which of the following compounds has a *meso* isomer? For those that do, show the plane of symmetry.

a. 1,2-dichlorohexane
b. 1,6-dichlorohexane
c. 1,1-dichlorohexane
d. 2,5-dichlorohexane
e. 1,6-dichloro-2,5-dimethylhexane
f. 15,19,23-trimethylheptatriacontane (see Problem 3–27, p. 102)

4–29 Consider the trihydroxydicarboxylic acid with the constitutional formula **III**

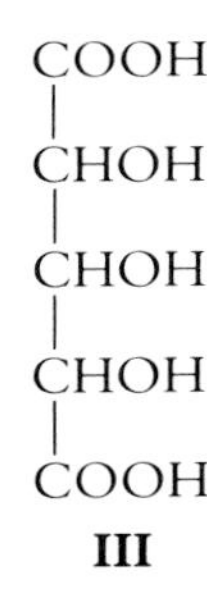

a. How many *meso* isomers of **III** are possible? How many optically active stereoisomers may be found? Draw the structures of all these compounds.
b. If one of the hydroxyls next to a carboxyl group were changed to $—OCH_3$, how many stereoisomers would be possible?

4–30 Usnic acid, a compound isolated from lichens, has $[\alpha]_D = +510°$. If 0.6 g of usnic acid is dissolved in 50 ml of chloroform and then placed in a 2-dm polarimeter tube, what would be the measured rotation?

4–31 It was necessary to determine the weight of a sample of cholesterol but all the lab balances were broken. The polarimeter was working, however, so the sample was dissolved in 100 ml of $CHCl_3$. The rotation of this sample in a 1-dm tube was $-2.65°$. The literature showed that for cholesterol $[\alpha]_D = -39°$. What was the weight of the sample?

4–32 A sample of 2-iodobutane found on the shelf has a rotation of $[\alpha]_D = +8.2°$. The literature shows that pure (*R*)-2-iodobutane has $[\alpha]_D = -15.9°$.

a. Does the shelf sample of 2-iodobutane contain a pure enantiomer? If so, which one? If not, which enantiomer is present in larger amounts?
b. Calculate the percentages of the (*R*) and (*S*) enantiomers in the sample.

4–33 A student finds a reagent bottle with a partially decomposed label. All that can be read says "dibromocyclopentane, $[\alpha]_D = +20°$." Show the structures of all the isomers that *could* be present in the reagent bottle.

4–34 The microbe *Pseudomonas putida* is capable of oxidizing racemic 3-methylcyclohexene to 3-methylcyclohexane-1,2-diol (**IV**), as shown in Equation 1. Analysis shows that the only isomers of **IV** that are produced are those in which the two hydroxyl groups are *cis* to one another.

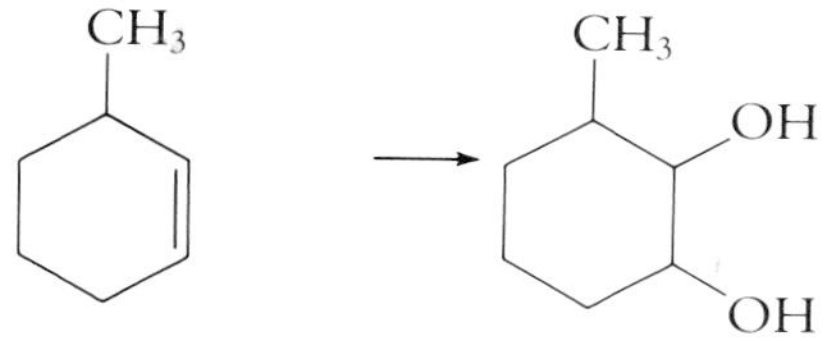

CH_3 CH_3 OH OH (1)

3-Methylcyclohexene **IV**

a. Draw the structures of the four stereoisomers which correspond to compound **IV** with *cis*-hydroxyl groups. Which of these are enantiomeric and which are diastereomeric?

b. Two diastereomeric alcohols were separated from the reaction mixture. Each of these diastereomers was optically active, suggesting that each consisted of only *one* enantiomer. Chemical correlation established that the *cis,cis*-diastereomer contained the (1*S*),(2*R*)-dihydroxy-(3*R*)-methylcyclohexane enantiomer. The *cis,trans* isomer was shown to consist of the (1*S*),(2*R*)-dihydroxy-(3*S*)-methylcyclohexane enantiomer. Write the structures of these two compounds. Show the most stable chair conformer of each.

c. If the microbe had been given only one enantiomer of 3-methylcyclohexene, how many of the products mentioned in (b) would you expect to have been produced?

4–35 Microbial oxidation of oleic acid, $CH_3(CH_2)_7CH{=}CH(CH_2)_7CO_2H$, converts it to chiral 9,10-epoxystearic acid (**V**). Equation 2 shows that when compound **V** then reacts with water, again under the influence of the microbe, it is converted to the chiral 9,10-dihydroxystearic acid (**VI**).

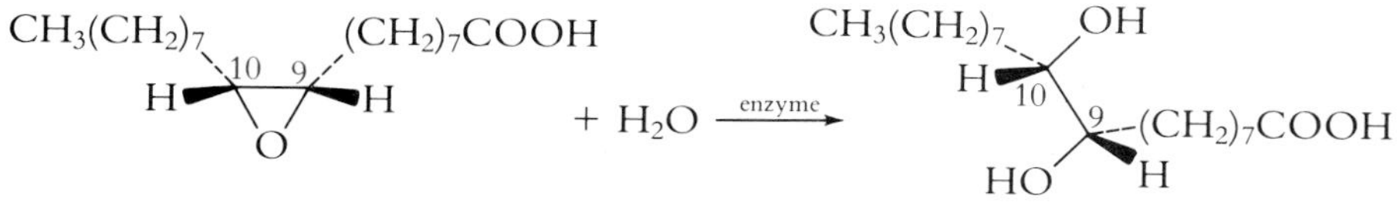

$CH_3(CH_2)_7$ $(CH_2)_7COOH$ 10 9 H H O + H_2O enzyme $CH_3(CH_2)_7$ OH H 10 9 $(CH_2)_7COOH$ HO H (2)

V **VI**

a. Assign the configuration at C-9 and C-10 in both reactant (**V**) and product (**VI**). Did each of them undergo retention or inversion of configuration?

b. In a nonenzymatic version of the reaction shown above, the retention/inversion pattern was reversed for half the molecules. What was the stereochemical outcome?

4–36 Compound **A**, $C_8H_{18}O$, is shown by spectroscopic means to contain the functional grouping

$$\underset{\displaystyle OH}{R{-}\underset{|}{C}H{-}R'}$$

(R and R′ are alkyl groups). A sample of **A** is obtained and found to be optically active. **A** can be oxidized to a ketone, **B**, which is also optically active (eq. 3).

$$\underset{\text{Optically active}}{\mathrm{R{-}\underset{\displaystyle OH}{\underset{|}{C}H}{-}R'}} \xrightarrow{\text{oxidation}} \underset{\text{Optically active}}{\mathrm{R{-}\underset{\displaystyle O}{\underset{\|}{C}}{-}R'}} \tag{3}$$

Give at least one structure for **A** that is compatible with the data. In addition, give the structure of an isomeric compound which is ruled out by the data. In each case explain why your structure does or does not fit the data.

Supplementary Reading

G. B. Kauffman, The resolution of racemic acid, *J. Chem. Ed.,* **52,** 777 (1975).

S. F. Mason, The foundations of classical stereochemistry, *Topics in Stereochemistry,* **9,** 1 (1976).

J. J. McCullough, Diastereomers, geometrical isomers, and rotation about bonds, *J. Chem. Ed.,* **59,** 37 (1982).

O. Cori, Complementary rules to define *R* or *S* configuration, *J. Chem. Ed.,* **49,** 461 (1972).

W. L. Alworth, *Stereochemistry and its Application in Biochemistry,* Wiley-Interscience, New York, N.Y., 1972.

M. Orchin *et al, The Vocabulary of Organic Chemistry,* Wiley-Interscience, New York, N.Y., 1980, Chapter 5.

W. E. Elias, The natural origin of optically active compounds, *J. Chem. Ed.,* **49,** 448 (1972); R. E. Pincock and K. R. Wilson, Spontaneous generation of optical activity, *J. Chem. Ed.,* **50,** 455 (1973).

A. Ault, Selectivity and specificity in organic reactions, *J. Chem. Ed.,* **54,** 614 (1977).

5 Reactive Intermediates: Formation of Haloalkanes by Way of Free Radicals

We study chemical reactions for several reasons and from several viewpoints. The reaction products are often commercially valuable or useful in the laboratory for further research or synthesis. The reaction itself is also of interest to us. You may remember from the Introduction that the reaction mechanism is a description of all energetic and structural changes the reactants experience as they are converted to products. Why should we want to investigate a reaction in such great detail? For one thing, knowing the reaction mechanism allows us to take efficient measures to raise the yield of a product, or to reduce the reaction time. Furthermore, by studying the mechanisms of a number of different reactions we can develop an appreciation for the physical factors which govern all reactions. Recognition of these mechanisms suggests an additional direct benefit to the student of organic chemistry, who can learn to predict what will happen in a given chemical situation, rather than having to memorize a huge list of specific cases. In this chapter we'll study the reactions of molecular halogens with alkanes and cycloalkanes for all of these reasons, and then apply what we have learned to such problems as food spoilage, radiation damage, and aging.

5:1 The Role of Intermediates in Reaction Mechanisms

Equation 5.1 shows the commercially important products which result from the chlorination of methane. The relative proportions of the products are

(5.1) $CH_4 \xrightarrow[\text{heat or light}]{Cl_2}$ CH_3Cl + CH_2Cl_2 + $CHCl_3$ + CCl_4 + HCl

- CH_3Cl: Chloromethane (*methyl chloride*)
- CH_2Cl_2: Dichloromethane (*methylene chloride*)
- $CHCl_3$: Trichloromethane (*chloroform*)
- CCl_4: Tetrachloromethane (*carbon tetrachloride*)

dependent on reactant concentrations and conditions. All of them serve as starting materials for the preparation of other compounds. The three polychloromethanes have also been used as solvents, although the toxicity of carbon tetrachloride has led to its being withdrawn for that purpose. Carbon tetrachloride still serves as a starting material for the manufacture of the refrigerant dichlorodifluoromethane (eq. 5.2), which was also used as an aerosol propellant. This compound is no longer used as a propellant, however, because of its potential threat to the atmosphere ozone layer, and is being replaced in part by dichloromethane.

$$(5.2) \qquad CCl_4 + 2HF \xrightarrow{SbCl_5} \underset{\substack{\text{Dichlorodifluoromethane} \\ (\textit{Freon 12})}}{CCl_2F_2} + 2HCl$$

Equations 5.1 and 5.2 tell us *what* happens when certain reactants are mixed. To find out *how* it happens, we have to turn to the reaction mechanisms. One of the most basic mechanistic questions is whether or not **reactive intermediates** have been formed during the reaction. *A reactive intermediate is a species that is generated while the reaction is taking place, but does not last long enough to be isolated.* Thus, the intermediates are consumed by further reaction shortly after they are formed. When reactive intermediates are involved in a reaction, then the overall process must be a multistep one. It takes one or more steps to convert the reactants to intermediates, and then one or more additional steps to convert the intermediates to products.

How do we know when reactive intermediates are involved in a reaction? We can often make an educated guess based on our knowledge of the chemical properties of the reactants. For example, in order to initiate the reaction between chlorine and methane, it is necessary to heat the mixture or to irradiate it with light (eq. 5.1). It has been found previously that heating or irradiating Cl_2 results in its partial dissociation into chlorine atoms, :C̤̈l·. We might provisionally suppose, therefore, that chlorine atoms are intermediates in the chlorination of methane. If we were to test this hypothesis experimentally, we would find that the data strongly support it.

As you'll see, halogen atoms are always intermediates in the halogenation of alkanes. The unpaired electron is responsible for their reactivity. Halogen atoms in turn generate carbon species having unpaired electrons. These species are called free radicals (sec. 1:3). Radicals are intermediates in many different organic and biological reactions, so we'll be able to apply what we learn here to other areas.

```
     H
     |
  H—C·
     |
     H
```

Methyl radical

A very useful way of classifying organic reactions is according to the type of intermediate, if any, involved in the reaction. Although many reactions go by way of radical intermediates, they are not the only intermediates important in organic chemistry. The reaction in equation 5.2, for example, involves a carbocation intermediate, in which there is a positive charge on carbon and all electrons are paired (sec. 1:3). In Chapter 6 we discuss carbocations in some detail, and introduce yet another important class of intermediate, the **carbanions**. They also have all electrons paired but contain a

```
     H                 H
     |                 |
  H—C⁺              H—C:⁻
     |                 |
     H                 H
```

Methyl cation (a carbocation) — Methyl anion (a carbanion)

negatively charged carbon atom. Because intermediates help us to understand and to classify large numbers of reactions, we'll devote considerable attention to them in the chapters to come.

5:2 The Importance of Free-Radical Chemistry

Despite the chemical instability of most free radicals, they are intermediates in many reactions of industrial, laboratory, or biological importance. Take, for example, the industrial conversion of alkanes to haloalkanes (eq. 5.3). Using petroleum hydrocarbons as starting materials, the worldwide chemical industry manufactures billions of pounds of haloalkanes every year. Similar reactions are carried out in the laboratory to synthesize halogenated hydrocarbons.

$$\text{R—H} + X_2(X = \text{F, Cl, Br}) \rightarrow \underset{\substack{\text{A haloalkane} \\ (\textit{alkyl halide})}}{\text{R—}X} + \text{H}X \tag{5.3}$$

It is also known that radicals are present in a variety of cells. It is therefore likely that they are intermediates in normal cellular chemistry. In addition, a number of abnormal cellular reactions appear to proceed by way of radicals. Many of these abnormal reactions are related to pathological changes in the cell. For instance, there is circumstantial (but not definitive) evidence linking free radicals to aging and carcinogenesis (conversion of a normal cell to a cancer cell).

Although most radicals are uncharged, both radical cations and radical anions are known. Radical ions, and their role in mass spectrometry, are discussed in Chapter 8.

5:3 Different Ways of Generating Radicals

Since reactive intermediates cannot be purchased in a bottle, they must be generated in the course of a reaction. Radicals are often produced by homolytic cleavage (sec. 2:6) of a covalent bond in a molecule (eq. 5.4). Homolysis is easily brought about if the bond is relatively weak (average bond energy $\leq$

$$Y{:}Z \rightarrow Y\cdot + Z\cdot \tag{5.4}$$

40–50 kcal/mol; see Table 2–2, p. 53). This is shown in Equation 5.5 for homolysis of an alkyl peroxide, in Equation 5.6 for homolysis of molecular chlorine, and in Equation 5.7 for decomposition of the gasoline octane booster, tetraethyllead.

$$\underset{\text{Alkyl peroxide}}{\text{R—}\ddot{\underset{..}{\text{O}}}{:}\ddot{\underset{..}{\text{O}}}\text{—R}} \xrightarrow{\text{heat}} \underset{\text{Alkoxy radicals}}{\text{R}\ddot{\underset{..}{\text{O}}}\cdot + \cdot\ddot{\underset{..}{\text{O}}}\text{R}} \tag{5.5}$$

$$\underset{\text{Molecular chlorine}}{:\ddot{\underset{..}{\text{Cl}}}{:}\ddot{\underset{..}{\text{Cl}}}{:}} \xrightarrow[\text{light}]{\text{heat or}} \underset{\text{Chlorine atoms}}{:\ddot{\underset{..}{\text{Cl}}}\cdot + \cdot\ddot{\underset{..}{\text{Cl}}}{:}} \tag{5.6}$$

$$\underset{\text{Tetraethyllead}}{\text{CH}_3\text{CH}_2\text{—}\overset{\overset{\displaystyle \text{CH}_2\text{CH}_3}{|}}{\underset{\underset{\displaystyle \text{CH}_3\text{CH}_2}{|}}{\text{Pb}}}\text{—CH}_2\text{CH}_3} \xrightarrow{\text{heat}} \underset{\text{Ethyl radical}}{4\text{CH}_3\text{CH}_2\cdot} + \text{Pb} \tag{5.7}$$

Another way to generate radicals is to break bonds so that a particularly stable molecule is formed, such as N_2 in Equation 5.8 and CO_2 in Equation 5.9.

$$\text{(5.8)}\quad \underset{\text{Azomethane}}{CH_3{-}N{=}N{-}CH_3} \xrightarrow[\text{light}]{\text{heat or}} \underset{\substack{\text{Methyl}\\\text{radicals}}}{2CH_3\cdot} + N{\equiv}N$$

$$\text{(5.9)}\quad \underset{\substack{\text{Acetyl}\\\text{peroxide}}}{CH_3C({=}O){-}O{-}O{-}C({=}O)CH_3} \xrightarrow[\text{light}]{\text{heat or}} \underset{\text{Acetoxy radicals}}{CH_3C({=}O)O\cdot + \cdot OC({=}O)CH_3} \longrightarrow \underset{\substack{\text{Methyl}\\\text{radicals}}}{2CH_3\cdot} + 2CO_2$$

Radicals may also be formed by the *reduction* of carbocations (eq. 5.10) or by the *oxidation* of carbanions (eq. 5.11). These **redox** processes sometimes

$$\text{(5.10)}\quad R_3C^+ + e^- \rightarrow R_3C\cdot$$

$$\text{(5.11)}\quad R_3C\!:^- \rightarrow R_3C\cdot + e^-$$

occur in the course of ionic reactions, resulting in the production of radicals not anticipated by the chemist. On the other hand, redox reactions are sometimes deliberately carried out, using **electrochemical** methods. In an electrochemical cell, reactants can pick up electrons at the cathode (eq. 5.10) or release them at the anode (eq. 5.11) to form radicals.

5:4 The Chlorination of Methane: The Free-Radical Chain Mechanism

Free-radical reactions have many features in common, regardless of the complexity of the system involved. Some of these basic features will be more readily apparent if we first consider a reaction involving structurally simple reactants. Such a reaction is the chlorination of methane (eq. 5.1). For commercial purposes the reaction is carried out so as to produce highly chlorinated compounds, such as CCl_4, as the major products. However, by adjusting the ratio of CH_4 to Cl_2 we can shift the product distribution to favor formation of CH_3Cl.

Equation 5.12 shows the reaction of chlorine with an excess of methane to give mainly chloromethane and hydrogen chloride. If the reactant gases

$$\text{(5.12)}\quad CH_4 + Cl_2 \xrightarrow[\text{light}]{\text{heat or}} \underset{\text{Chloromethane}}{CH_3Cl} + HCl + \text{small amounts of polychloromethanes}$$

are mixed at room temperature and kept in the dark, no reaction takes place. It is necessary to heat the reaction mixture or irradiate it with certain wavelengths of light in order to initiate reaction. Once underway, the reaction proceeds rapidly and evolves considerable amounts of heat.

Similar reactions take place between methane and fluorine (eq. 5.13) and methane and bromine (eq. 5.14). However, the three halogens exhibit some striking differences in behavior. Fluorination proceeds even more rapidly

(5.13) $$F_2 + CH_4 \xrightarrow[\text{fast}]{\text{very}} \underset{\text{Fluoromethane}}{CH_3F} + HF$$

(5.14) $$Br_2 + CH_4 \xrightarrow{\text{slow}} \underset{\text{Bromomethane}}{CH_3Br} + HBr$$

than chlorination, with explosive violence. (Special methods are needed to synthesize fluoroalkanes.) Bromination, on the other hand, is much less rapid and vigorous than chlorination. Iodine (I_2) does not react with methane to a measurable extent.

One more finding of mechanistic significance is the *inhibitory,* or slowing down effect of molecular oxygen on halogenation. If O_2 is added to the CH_4/Cl_2 mixture, the production of chloromethane is slowed. If enough oxygen is added, the chlorination can be suppressed completely.

We can now formulate a likely reaction mechanism based on these empirical data. First note that the reaction requires heat or light in order to start. This means that one or both reactants need to be "energized" or "activated" before they can react.

A closer look at the light-initiated chlorination can tell us a good deal about this activation process. Visible light, in common with all forms of radiation, consists of discrete units of energy called **quanta**, or **photons**. A molecule will absorb a photon and move to a higher energy (more reactive) state. For example, a molecule of chlorine can absorb a photon, be promoted to a higher energy state, and then dissociate into chlorine atoms (Fig. 5–1).

The wavelengths of light that can initiate chlorination of methane have energies of about 60 kcal/mol. Table 2–1 (sec. 2:6, p. 53) shows us that light of this energy can cause dissociation of the Cl—Cl bond but not the C—H bond. Since we also know that Cl_2 is more easily decomposed by heat than CH_4, we can assume that the first step in the chlorination reaction involves the dissociation of Cl_2 to chlorine atoms (eq. 5.15).

(5.15) $$:\ddot{\underset{..}{Cl}}:\ddot{\underset{..}{Cl}}: \xrightarrow[\text{photon}]{\text{heat or}} 2:\ddot{\underset{..}{Cl}}\cdot$$

What happens to the chlorine atoms? We know that they will tend to undergo any reaction that will provide them with an octet of valence electrons. It is also known that HCl is a reaction product. A reasonable sugges-

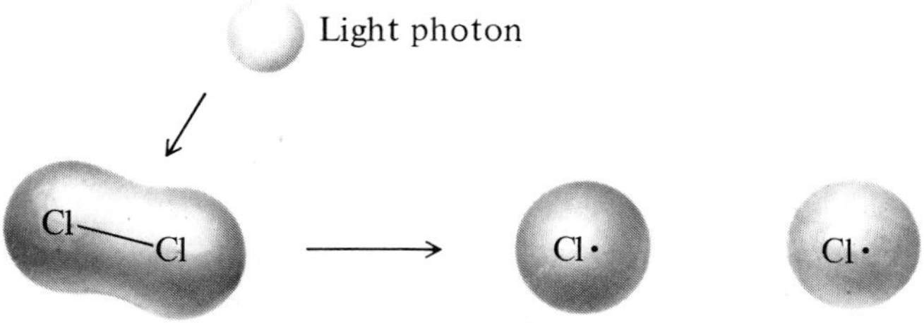

Figure 5–1. *Molecular chlorine absorbs a photon of light and dissociates into chlorine atoms.*

tion is that Cl· *abstracts* a hydrogen atom from methane, giving HCl and a methyl radical (eq. 5.16). In this way, the valence shell of chlorine is filled, at the price of creating a carbon with an unfilled shell.

(5.16) $$:\ddot{\underset{..}{Cl}}\cdot + H{:}CH_3 \rightarrow :\ddot{\underset{..}{Cl}}{:}H + \cdot CH_3$$

Chlorine atom — Methyl radical

The reaction in Equation 5.16 may also be viewed as one in which a chlorine radical becomes bonded to hydrogen by *displacing* a methyl radical. Both abstraction and displacement are accurate terms used to describe reactions such as the one shown in Equation 5.16. The attacking group (chlorine atom) and the displaced group (methyl radical) share a common electronic feature—an unpaired electron. In the next chapter we'll see that the concept of displacement is a general one that extends beyond radical reactions.

The methyl radical will tend, like the chlorine atom, to participate in reactions that give it a complete octet of electrons. One such reaction would be the displacement of a chlorine atom from a chlorine molecule, Cl_2, which produces chloromethane (eq. 5.17).

(5.17) $$CH_3\cdot + :\ddot{\underset{..}{Cl}}{:}\ddot{\underset{..}{Cl}}: \rightarrow CH_3{:}\ddot{\underset{..}{Cl}}: + \cdot\ddot{\underset{..}{Cl}}:$$

Methyl radical — Chlorine atom

As an alternative to this reaction you may propose the reaction in Equation 5.18:

(5.18) $$CH_3\cdot + :\ddot{\underset{..}{Cl}}\cdot \rightarrow CH_3{:}\ddot{\underset{..}{Cl}}:$$

Although the combination of a methyl radical and a chlorine atom is a very fast reaction, it is rather improbable. Because radicals are highly reactive, they tend to be present in very low concentrations. Thus, the reaction in Equation 5.18 is not important compared to Equation 5.17 because of the low probability of a methyl radical encountering a chlorine atom. Similarly, the reaction of two chlorine atoms to form Cl_2 is relatively improbable compared to the reaction of a chlorine atom with methane.

Our tentative mechanism comprises three steps, shown in Equations 5.15 to 5.17. Note that in the last step (eq. 5.17) a chlorine atom is formed as a product. This chlorine atom can serve as reactant in a repeat of the second step (eq. 5.16). Thus, the process, once started, can keep itself going. A process of this type is known as a **chain reaction**. For methane chlorination, the chain reaction includes an *initiation step* (eq. 5.15) and two *chain-propagating steps* (eqs. 5.16 and 5.17). By adding together the two chain-propagating steps, you get the equation for the overall reaction (eq. 5.12).

$Cl_2 \rightarrow 2Cl\cdot$	Initiation
$Cl\cdot + CH_4 \rightarrow HCl + CH_3\cdot$	Chain propagation
$CH_3\cdot + Cl_2 \rightarrow CH_3Cl + Cl\cdot$	
$Cl_2 + CH_4 \rightarrow CH_3Cl + HCl$	Overall reaction (eq. 5.12)

Reaction chains can be broken by the combination of two radicals, since such a step reduces the net number of radicals. One *chain-terminating step* is shown in Equation 5.18. Two others are given in Equations 5.19 and 5.20.

$$\left.\begin{array}{ll} (5.19) & CH_3\cdot + CH_3\cdot \rightarrow CH_3CH_3 \\ (5.20) & Cl\cdot + Cl\cdot \rightarrow Cl_2 \end{array}\right\} \text{Chain termination}$$

As we just noted, the combination of two radicals is a relatively rare occurrence. For that reason many chain-propagating reactions will take place for each chain-terminating one. This feature of the chain reaction mechanism nicely explains an important experimental observation, namely, that the absorption of a single photon results in the formation of 10^2 to 10^3 molecules of product. The high probability of chain-propagating reactions, compared to chain-terminating ones, means that each chlorine atom formed by photon absorption gives rise to many chain-propagating product-forming steps.

The radical chain mechanism can readily explain the inhibiting effect of oxygen. Molecular oxygen is one of a small number of molecules that are ground state **diradicals** (species having two unpaired electrons). Since it is difficult to represent O_2 adequately by Lewis structures, the abbreviated formula ·O—O· will be used. Due to its radical character, oxygen is very reactive toward other free radicals. It combines with a methyl radical to yield a methylperoxyl radical (eq. 5.21). The methylperoxyl radical is much more stable (and less reactive) than the methyl radical. The net effect of the reaction in Equation 5.21 is to remove methyl radicals from circulation and thus interfere with one of the chain-propagating steps (eq. 5.17).

$$(5.21)\qquad \underset{\substack{\text{Methyl}\\ \text{radical}\\ \text{(reactive)}}}{CH_3\cdot} + \cdot O{-}O\cdot \rightarrow \underset{\substack{\text{Methylperoxyl}\\ \text{radical}\\ \text{(unreactive)}}}{CH_3{-}O{-}O\cdot}$$

Problems

5–1 A mixture of bromine and methane does not react when kept in the dark at moderate temperatures. When the mixture is irradiated with visible light, however, bromomethane and hydrogen bromide are formed (eq. 5.14). It has been shown that a number of product molecules are formed for each quantum of light absorbed. Write a mechanism for bromination of methane that is in accord with all the experimental facts.

5–2 When methane is not in large excess, the chlorination reaction produces substantial amounts of di-, tri-, and tetrachloromethane. Show the mechanisms by which each of these is produced successively from chloromethane.

The radical chain mechanism for methane halogenation accounts for all the data so far presented, and others not mentioned. It seems highly probable that it is the correct mechanism, but it is not possible to say this with *complete certainty*. In general, it is usually possible to show a given mechanism to be conclusively wrong, but it is never possible to prove a mechanism to be conclusively correct. This is because the possibility always remains that there are other, unthought of mechanisms which fit all the data equally well.

Nonetheless, many mechanisms are known with sufficient certainty to be useful guides to general reaction patterns and to have predictive value.

In order to gain more insight into halogenation reactions and their mechanisms, it is important to study their enthalpies and reaction rates. These are the subjects of the next several sections.

5:5 The Thermodynamics of Methane Halogenation

Reaction enthalpies, $\Delta H°$ (sec. 3:6), can readily be calculated from bond dissociation energies (sec. 2:6). This section shows how it is done, and applies the method to the case of methane halogenation. All the data needed for these calculations are contained in Table 5–1.

Energy is required to break chemical bonds, and energy is released when bonds are made. Therefore, the enthalpy of reaction should equal the sum of the bond energies for all bonds broken, minus the sum of the bond energies for all bonds made. This postulate is exactly correct for gas-phase reactions, since no solvent interactions have to be accounted for.

In Equation 5.22 we show how to use bond energy values to calculate $\Delta H°$ for the chlorination of methane. The fact that $\Delta H°$ is negative means

$$\underset{104}{CH_3{-}H} + \underset{58}{Cl{-}Cl} \rightarrow \underset{83.5}{CH_3{-}Cl} + \underset{103}{H{-}Cl} \tag{5.22}$$

$$\Delta H° = (104 + 58 - 83.5 - 103) = -24.5 \text{ kcal/mol}$$

that the products have less enthalpy (energy) than do the reactants, so heat should be given off as the reaction proceeds. That is what we see in an

Table 5–1. Bond Dissociation Energies*

H—H	104	CH_3—OH	91.5
F—F	38	CH_3—F	108
Cl—Cl	58	CH_3—Cl	83.5
Br—Br	46	CH_3—Br	70
I—I	36	CH_3—I	56
H—F	136	CH_3CH_2—H	98
H—Cl	103	$(CH_3)_2CH$—H	94.5
H—Br	87.5	$(CH_3)_3C$—H	91
H—I	71	$HOCH_2$—H	93.5
H—NH_2	110	CH_2=CH—H	103
H—$NHCH_3$	103	CH_3—CH_3	88
H—OH	119	$(CH_3)_2CH$—CH_3	84
H—OCH_3	104	$(CH_3)_3C$—CH_3	80
H—CH_3	104	CH_3CH_2—CH_2CH_3	82
HO—OH	51	CH_2=CH—CH_3	92

* The data are in kilocalories per mole and refer to the dissociation of 1 mol of gaseous molecules, in the manner indicated, at 25° C and 1 atm. Take care when comparing these data with the results of thermochemical experiments, since many of the molecules listed here are not gases at 25° C and 1 atm and energy has to be supplied to vaporize them. The values given refer in all cases to *homolytic* bond cleavage. The bond shown for each molecule is the one being broken.

experiment. Note also that there are equal numbers of reactants and product molecules, so $\Delta S°$ is small (sec. 3:6). When $\Delta S°$ is small, the free energy change, $\Delta G°$, is nearly equal to $\Delta H°$ (eq. 5.23). Hence we can judge the favorability of the reaction on the basis of enthalpy change ($\Delta H°$) alone; in the chlorination of methane we therefore expect a high conversion of reactants to products.

$$\Delta G° = \Delta H° - T\Delta S° \qquad (5.23)$$

Similar calculations can be carried out for the fluorination and bromination of methane (eqs. 5.24 and 5.25). All three halogenations are *exothermic* (heat producing), but to widely varying degrees. Fluorination is extremely exothermic, while bromination of methane is only moderately so.

$$\underset{104}{CH_3—H} + \underset{38}{F—F} \rightarrow \underset{108}{CH_3—F} + \underset{136}{H—F} \qquad \Delta H° = -102 \text{ kcal/mol} \qquad (5.24)$$

$$\underset{104}{CH_3—H} + \underset{46}{Br—Br} \rightarrow \underset{70}{CH_3—Br} + \underset{87.5}{H—Br} \qquad \Delta H° = -7.5 \text{ kcal/mol} \qquad (5.25)$$

Example

Calculate $\Delta H°$ for the following reaction:

$$(CH_3)_3C—H + Cl\cdot \rightarrow (CH_3)_3C\cdot + H—Cl$$

Answer

This is done by comparing the bond dissociation energies of the bond being broken in the reactant with the bond being formed in the product. The values are given in Table 5–1.

$(CH_3)_3C—H$ — Bond being broken: 91 kcal/mol

$H—Cl$ — Bond being formed: 103 kcal/mol

$$\Delta H° = -103 + 91 = -12 \text{ kcal/mol}$$

Problem

5–3 a. Using the bond dissociation energies given in Table 5–1, calculate $\Delta H°$ for the iodination of methane.

b. Explain why no detectable amounts of iodomethane are formed when iodine/methane mixtures are heated or irradiated.

We noted earlier that of all the halogens, fluorine reacts most rapidly with methane, followed by chlorine and then bromine. This is also the order of decreasing exothermicity of the halogenations. For this series of related reactions there seems to be a correlation between the *rate* of a reaction and its *exothermicity*. In other words, the reaction that produces the most stable products (relative to reactants) proceeds at the fastest rate. While correlations of this kind have been found for many sets of related reactions, the relationship does not hold for all reactions. Sections 5:12 and 6:7 will explore the factors behind this correlation and indicate when it can be expected to hold and when it is likely to fail.

Finally, it is worthwhile to examine briefly the thermodynamics of the individual steps of the radical chain mechanism. Equations 5.26 and 5.27 give values of $\Delta H°$ for the two chain-propagating steps in chlorination.

$$(5.26)\quad \underset{104}{CH_3\text{—}H} + Cl\cdot \rightarrow CH_3\cdot + \underset{103}{H\text{—}Cl} \qquad \Delta H° = 1 \text{ kcal/mol}$$

$$(5.27)\quad CH_3\cdot + \underset{58}{Cl\text{—}Cl} \rightarrow \underset{83.5}{CH_3\text{—}Cl} + Cl\cdot \qquad \Delta H° = -25.5 \text{ kcal/mol}$$

Although the first step is slightly **endothermic** (heat is absorbed as the reaction proceeds), the large exothermicity of the second step allows the overall process to be exothermic. A similar phenomenon is observed in bromination of methane (eqs. 5.28 and 5.29). In this case the first step is substantially endothermic, yet the overall process is again exothermic. For all the halogenation reactions, the first propagation step, hydrogen abstraction, is thermodynamically more difficult than the second.

$$(5.28)\quad \underset{104}{CH_3\text{—}H} + Br\cdot \rightarrow CH_3\cdot + \underset{87.5}{H\text{—}Br} \qquad \Delta H° = 16.5 \text{ kcal/mol}$$

$$(5.29)\quad CH_3\cdot + \underset{46}{Br\text{—}Br} \rightarrow \underset{70}{CH_3\text{—}Br} + Br\cdot \qquad \Delta H° = -24 \text{ kcal/mol}$$

You may be wondering why we consider the enthalpy change in only the chain-propagating steps when determining the overall endothermicity or exothermicity of a free-radical halogenation (see Prob. 5–4). The reason we do not have to consider the chain-initiating or chain-terminating steps is that the energy expended in the former is effectively recovered in the latter.

Problems

5–4 Show how the enthalpy change for the overall chlorination reaction (eq. 5.22) can be derived from the enthalpies of the chain propagating steps (eqs. 5.26 and 5.27). Do the same for the bromination reaction.

5–5 Calculate the enthalpies for the individual chain-propagating steps in the iodination of methane. Can you think of a free radical route to iodomethane that does *not* involve a reaction between methane and iodine? (You may want to review sec. 5:3 for the various methods of generating radicals.)

Since all the methane hydrogens are equivalent, the chemistry of this compound is relatively simple. It gives only one monochlorination product, one monobromination product, and so forth. More complex alkanes can, by contrast, give several products upon halogenation. We are now capable of accurately predicting the relative yields of the various products. Such knowledge is very useful to anyone planning a synthesis, whether in the laboratory or in a chemical plant. The studies that enable us to predict the product yields also allow us to define those reaction conditions which will give the highest yield of a desired product. This will be the subject of the following sections.

5:6 Selectivity in the Halogenation of Higher Alkanes

Chlorination of ethane can give only one monochlorinated product, CH_3CH_2Cl. Chlorination of propane, however, leads to two isomeric products, 1-chloropropane and 2-chloropropane (eq. 5.30), because propane has two different kinds of hydrogen atoms.★

(5.30) $$CH_3CH_2CH_3 + Cl_2 \xrightarrow[\text{gas phase}]{\text{light, 25° C}} \underset{\substack{\text{1-Chloropropane} \\ (40\%)}}{CH_3CH_2CH_2Cl} + \underset{\substack{\text{2-Chloropropane} \\ (60\%)}}{CH_3\underset{\substack{| \\ Cl}}{C}HCH_3} + HCl$$

It contains six primary (1°) hydrogen atoms and two secondary (2°) hydrogen atoms. (The terms primary, secondary, and tertiary were defined in sec. 3:2 and are illustrated here; R = alkyl or aryl group.)

$R{-}CH_2{-}H$ (C bearing R, H, H, H)	$R{-}CH{-}R$ (C bearing R, R, H, H)	$R{-}C{-}H$ (C bearing R, R, R, H)
Primary (1°) hydrogen atoms	Secondary (2°) hydrogen atoms	Tertiary (3°) hydrogen atom

Replacement of a 1° hydrogen produces 1-chloropropane, while replacement of a 2° hydrogen yields 2-chloropropane. If 1° and 2° hydrogens were replaced with equal ease, there would be three times as much 1-chloro as 2-chloro isomer, since that is the ratio of 1° : 2° hydrogens. Surprisingly, the actual product mixture contains more 2-chloropropane than 1-chloropropane (eq. 5.30). The product distribution is determined in the first step of the reaction when Cl· abstracts a hydrogen from propane (eq. 5.31). In this **competition reaction** the 1° and 2° hydrogens compete for the chlorine atom.

(5.31) $CH_3CH_2CH_3$ (2° hydrogens on the central carbon; 1° hydrogens on the end carbons) $\xrightarrow{Cl\cdot}$

$$\underset{\text{1-Propyl radical}}{CH_3CH_2CH_2\cdot} \xrightarrow{Cl_2} \underset{\text{1-Chloropropane}}{CH_3CH_2CH_2Cl}$$

$$\underset{\text{2-Propyl radical}}{CH_3\dot{C}HCH_3} \xrightarrow{Cl_2} \underset{\text{2-Chloropropane}}{CH_3\underset{\substack{| \\ Cl}}{C}HCH_3}$$

We must conclude that Cl· is *selectively* abstracting 2° hydrogens in preference to 1° hydrogens. Under the reaction conditions Cl· will abstract a 2° hydrogen 4.6 times more often than it will abstract a 1° hydrogen. (The 2°/1°

★ Equation 5.30 depicts a very common situation in organic chemistry—an unbalanced chemical equation. We have not balanced the equation because its main purpose is to indicate the course of the reaction. Furthermore, balancing is not really practical because the relative *proportions* of the products usually vary with the reaction conditions.

selectivity factor of 4.6 is corrected for the different numbers of each type of hydrogen and is applicable to any unstrained alkane.)

Chlorine atoms abstract tertiary hydrogens even more selectively, as Equation 5.32 shows.

$$\text{(5.32)}\quad CH_3\text{—}\underset{\displaystyle}{\overset{\displaystyle CH_3}{\overset{|}{C}H}}\text{—}\overset{\displaystyle CH_3}{\overset{|}{C}H}\text{—}CH_3 + Cl_2 \xrightarrow[\text{gas phase}]{\text{light, 25° C}}$$

2,3-Dimethylbutane

$$CH_3\text{—}\overset{\displaystyle CH_3}{\overset{|}{C}H}\text{—}\overset{\displaystyle CH_3}{\overset{|}{C}H}\text{—}CH_2Cl \quad + \quad CH_3\text{—}\overset{\displaystyle CH_3}{\overset{|}{C}H}\text{—}\underset{\displaystyle Cl}{\underset{|}{\overset{\displaystyle CH_3}{\overset{|}{C}}}}\text{—}CH_3$$

1-Chloro-2,3-dimethylbutane (50%) (from 12 1° hydrogens)

2-Chloro-2,3-dimethylbutane (50%) (from two 3° hydrogens)

If no selectivity were operating, the 1- and 2-chloro isomers would have been formed in the ratio 6:1. The data show that Cl· will abstract a 3° hydrogen six times as often as a 1° hydrogen.

The experimentally observed selectivities are a result of reaction rate differences. The reaction of Cl· with 2° hydrogens, for instance, is faster than its reaction with 1° hydrogens. Since we want to know the reason for these rate differences, we can ask if they are connected with energy differences. The data in Table 5–1 show that 2° and 1° C—H bonds do indeed differ in energy, and that the reactions of Cl· with these bonds differ in their $\Delta H°$ values (eqs. 5.33 and 5.34).

$$\text{(5.33)}\quad CH_3\underset{\displaystyle \textcircled{H}}{\underset{|}{C}}HCH + Cl\cdot \rightarrow CH_3\underset{\cdot}{C}HCH_3 + \textcircled{H}\text{—}Cl \qquad \Delta H° = -8.5 \text{ kcal/mol}$$

94.5 103

$$\text{(5.34)}\quad CH_3CH_2CH_2\text{—}\textcircled{H} + Cl\cdot \rightarrow CH_3CH_2CH_2\cdot + \textcircled{H}\text{—}Cl \quad \Delta H° = -5 \text{ kcal/mol}$$

98 103

Taken together, the thermodynamic calculations and experimental selectivities indicate that the more exothermic reaction (2° H abstraction) goes faster than the less exothermic one (1° H abstraction). A similar correlation was uncovered when we looked at the reaction of different halogens with methane (sec. 5:5).

The preceding examples established the fact that *the degree of radical selectivity depends partly on the difference in bond strengths of the C—H bonds being broken.* As the following example (eq. 5.35) shows, selectivity also depends on the nature of the attacking radical. A comparison of Equations 5.35 and 5.30 makes evident the fact that Br· is much more selective (or discriminatory) than Cl· (also see Table 5–2).

Table 5–2. Relative Selectivity of Some Halogen Atoms Toward C—H Bonds (Gas Phase)*

Atoms	Temperature (°C)	Type of C—H Bond 1°	2°	3°
F·	25	1	1.2	1.4
Cl·	25	1	4.6	6.0
	150	1	3.3	3.9
Br·	98	1	250	6300
	146	1	140	1800

* These are *relative* selectivities. These data do *not* imply that F·, Cl·, and Br· all react at the same rate with 1° C—H bonds.

$$\text{(5.35)}\quad CH_3CH_2CH_3 + Br_2 \xrightarrow[\text{gas phase}]{100^\circ} \underset{\substack{\text{1-Bromopropane}\\(1\%)}}{CH_3CH_2CH_2Br} + \underset{\substack{\text{2-Bromopropane}\\(99\%)}}{CH_3\underset{Br}{\underset{|}{C}}HCH_3} + HBr$$

The difference between Cl· and Br·, which involves the reaction rate/reaction energy correlation discussed above, will be explained in Section 5:12.

Problems

5–6 a. Write the structures of the monochlorination products of pentane.
b. Chlorine atoms show the same 2°/1° selectivity toward pentane that they do toward propane. Given that fact, predict the major product of monochlorination of pentane, and explain your reasoning.

5–7 You want to convert propane to 2-propanol, $(CH_3)_2CHOH$, in high yield. The overall route (eq. 1) involves halogenation of propane, isolation of the 2-halopropane, and conversion to the desired product by replacing the halogen by hydroxyl (a nonradical process; see Chap. 6). Which halogen would you use for the first step? Would you operate at 25° or at 150°? Justify your answers.

$$CH_3\text{—}CH_2\text{—}CH_3 + X_2 \longrightarrow CH_3\text{—}\underset{X}{\underset{|}{CH}}\text{—}CH_3 \xrightarrow{H_2O} \underset{\text{2-Propanol}}{CH_3\text{—}\underset{OH}{\underset{|}{CH}}\text{—}CH_3} \quad (1)$$

Free-radical reactions bring about replacement of hydrogen atoms without altering the carbon skeleton, except in rare circumstances. Thus, chlorination of isobutane yields only products with branched carbon chains and no straight-chain isomers (eq. 5.36). In Chapter 6 we'll show that certain kinds

$$\text{(5.36)}\quad CH_3\underset{CH_3}{\underset{|}{C}}HCH_3 \xrightarrow{Cl_2} \underset{\substack{\text{2-Chloro-2-}\\\text{methylpropane}}}{CH_3\overset{Cl}{\overset{|}{\underset{CH_3}{\underset{|}{C}}}}CH_3} + \underset{\substack{\text{1-Chloro-2-}\\\text{methylpropane}}}{CH_3\underset{CH_3}{\underset{|}{C}}HCH_2Cl} \text{ (}\textit{no}\text{ 1- or 2-chlorobutanes)}$$

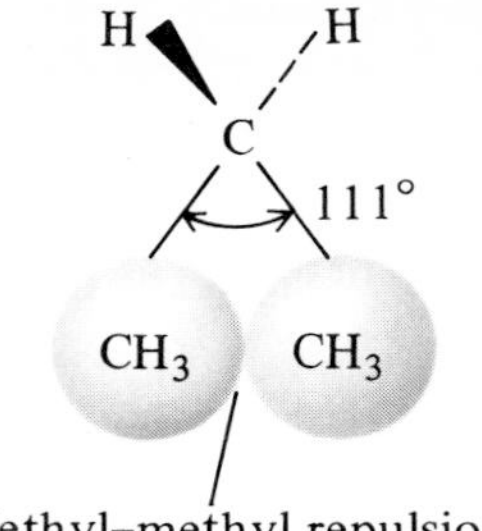

Figure 5–2. *Steric strain arising from the repulsion between methyl groups in propane. The steric strain causes the C—C—C angle to increase to 111°.*

of nonradical substitution reactions can result in carbon skeleton rearrangement.

5:7 How Steric Strain Accounts for the Rates of Formation of 1°, 2°, and 3° Radicals

You've just seen that 1° and 2° radicals are formed at different rates because different amounts of energy are required to break 1° and 2° C—H bonds. In order to account for this difference we'll have to look again at the starting alkane and observe the structural and energetic changes that occur as it is converted to the various possible radicals.

One factor that affects the energy of any molecule is the degree of **steric strain** it experiences. Steric strain arises when atoms not bonded together come close enough to repel one another. It is thus a measure of intramolecular crowding, and raises the energy of a molecule. Figure 5–2 shows the repulsion between the methyl groups of propane. The resulting steric strain causes the central bond angle to expand from an ideal 109.5° to 111°. A chemical reaction which *relieves* some of the steric strain in propane will be made energetically easier. Removal of a hydrogen atom from C-2 is such a reaction, because it allows the bond angle to expand to ~120° (sp^2 hybridization). This lessens the repulsion between the methyl groups (Fig. 5–3) by moving them farther apart.

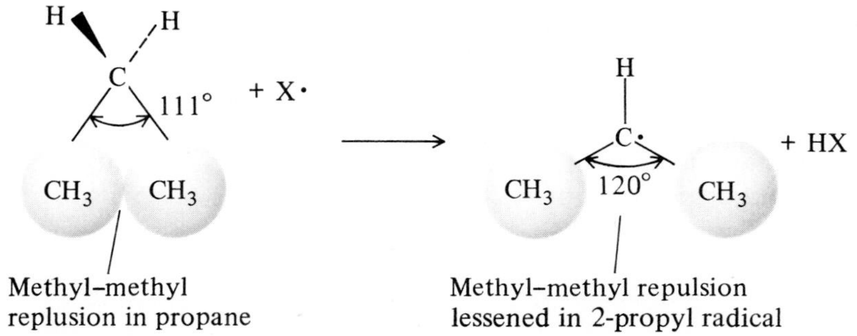

H H C 111° + X· → H C· 120° + HX

CH_3 CH_3 → CH_3 CH_3

Methyl–methyl replusion in propane

Methyl–methyl repulsion lessened in 2-propyl radical

Figure 5–3. *Lessening of the methyl–methyl repulsion when propane is converted to 2-propyl radical. The C—C—C angle expands to 120°, moving the methyl groups apart.*

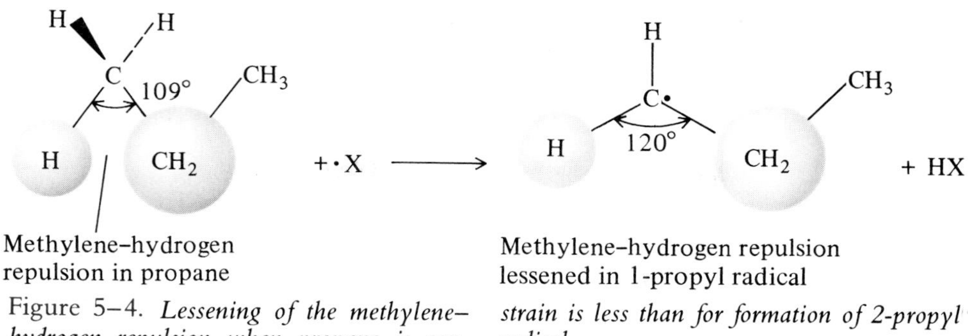

Figure 5–4. *Lessening of the methylene–hydrogen repulsion when propane is converted to 1-propyl radical. The relief of steric strain is less than for formation of 2-propyl radical.*

Does the removal of a hydrogen atom from C-1 of propane lead to some relief of steric strain? It does, but the effect is not as great as at C-2 (Fig. 5–4). This is because there is less crowding around the methyl carbons than around the methylene carbons, because each of the methyl carbons are bonded to three small hydrogen atoms. A comparison of C—H bond breaking at C-1 and C-2 can be made apparent by means of an energy diagram (Fig. 5–5). It shows that less energy is required to remove a hydrogen atom from C-2 than from C-1 because C-2 hydrogen removal leads to greater relief of intramolecular crowding (steric strain).

Steric strain is not the only factor that affects relative rates of hydrogen atom abstraction. Electronic factors also play a role and become particularly

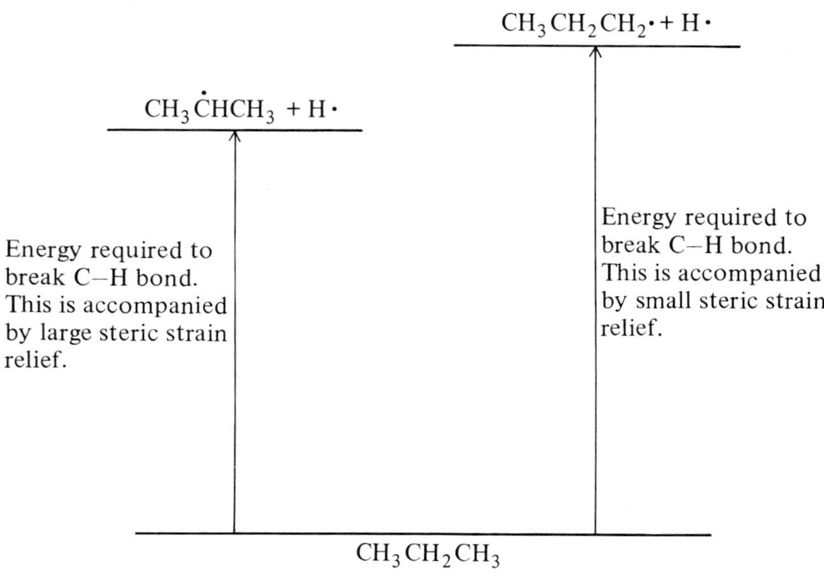

Figure 5–5. *A comparison of the energy required to break C—H bonds at C-1 and C-2 of propane. Less energy is required to break the bond at C-2 because it is aided more by relief of steric strain.*

important when the reactant contains nitrogen, oxygen, or halogen atoms. In their absence the relative rates of formation of different radicals, shown here, is correctly predicted from a consideration of steric strain relief.

$CH_3\cdot$	$RCH_2\cdot$	$R_2CH\cdot$	$R_3C\cdot$
Methyl radical	Primary radical	Secondary radical	Tertiary radical

$\longrightarrow$

Increasing Rate of Formation

The analysis we've just applied to the rate of formation of alkyl radicals can also apply to the relative *stability* of alkyl radicals. Consider a series of isomeric radicals, differing in degree of substitution. We expect the most stable radical to be the one with the greatest number of (bulky) alkyl groups attached to the trivalent carbon *because the trivalent carbon has the largest bond angles*. Thus the most highly substituted radical should be the most stable, as well as the most rapidly formed.

In Chapter 6 we'll see that steric strain can help explain the relative rates of formation of carbocations. It is a basic determinant of molecular behavior with widespread ramifications.

Example

What two different radicals are formed when a hydrogen atom is abstracted from butane? Are they primary, secondary, or tertiary radicals? Which is the more stable radical and why?

Answer

Because there are two different kinds of hydrogen in butane, it can give rise to two different radicals

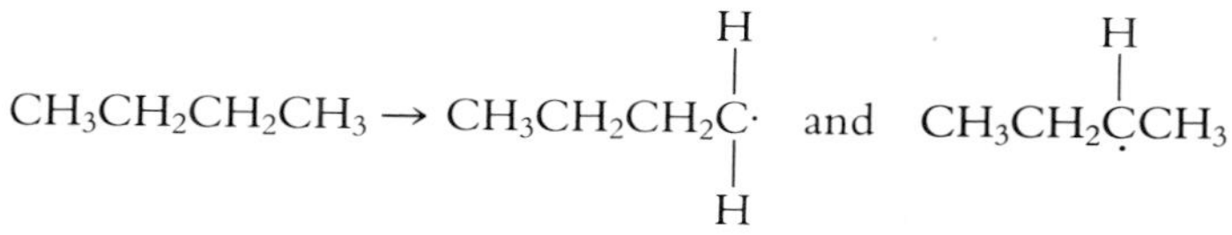

$$\underset{\text{Butane}}{CH_3CH_2CH_2CH_3} \rightarrow CH_3CH_2CH_2\overset{H}{\underset{H}{C}}\cdot \quad \text{and} \quad CH_3CH_2\overset{H}{\dot{C}}CH_3$$

One product is a primary radical (the carbon that bears the radical is attached to *one* other carbon). The other is a secondary radical (the carbon that bears the radical is attached to *two* other carbons).

The secondary radical is more stable than the primary one. The main reason for this is that removal of a hydrogen atom from C-2 of butane leads to a greater relief of steric crowding than does removal of an H atom from C-1.

Problems

5–8 Classify each of the following radicals as primary (1°), secondary (2°), or tertiary (3°).

a. $(CH_3)_2CHCH_2\cdot$ b. $(CH_3)_3C\cdot$ c.  [cyclohexyl radical] d. 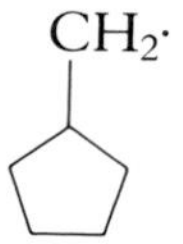 [cyclopentyl–$CH_2\cdot$]

e. 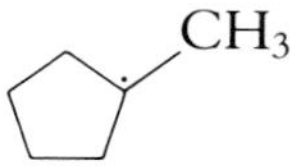 (cyclopentyl radical structure with CH_3) f. 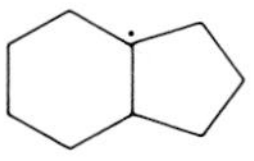 (bicyclic radical structure) g. $CH_3CHD\cdot$ h. $(CH_3)_3C\dot{C}HCH_3$

5–9 Arrange the radicals of Problem 5–8 in order of *decreasing* rate of formation.

5–10 Draw the structures of the two isomeric radicals formed by hydrogen abstraction from isobutane. Show their approximate bond angles. Use these structures to help explain why a 3° hydrogen is abstracted faster than a 1° hydrogen.

5:8 Nomenclature of Alkyl Halides

In the IUPAC system, halogen substituents on an alkane are treated exactly like alkyl substituents. They are denoted by prefixes attached to the name of the longest carbon chain, the prefixes being arranged in alphabetical order. As before, the locator numbers are chosen so that the lowest possible number will be assigned to the first substituent. When there are two ways of doing this, the substituent with alphabetical priority is given the smaller locator number (see 3-bromo-4-ethylhexane).

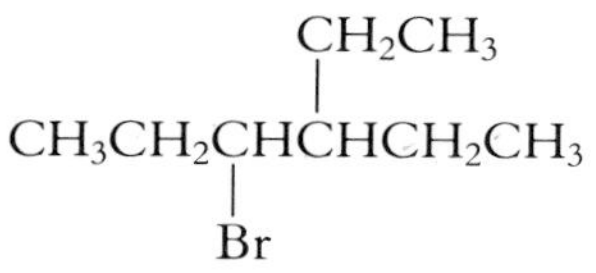

$CH_3CH_2CH(Br)CH(CH_2CH_3)CH_2CH_3$

3-Bromo-4-ethylhexane
(not 3-ethyl-4-bromohexane)

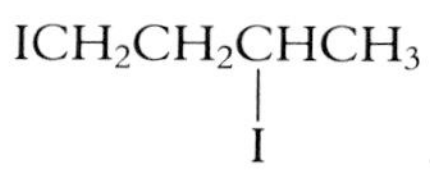

$ICH_2CH_2CH(I)CH_3$

1,3-Diiodobutane

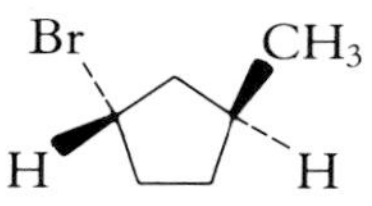

trans-1-Bromo-3-methylcyclopentane

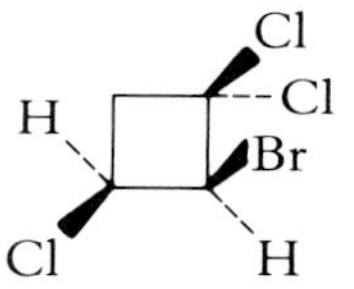

cis-2-Bromo-1,1,3-trichlorocyclobutane

There is also an older, so-called radicofunctional nomenclature in which organic halides are named in a way analogous to the naming of inorganic halides. The first word gives the name of the alkyl substituent and the second word the name of the halogen. This system is mostly confined to compounds of low molecular weight. Some examples are given here.

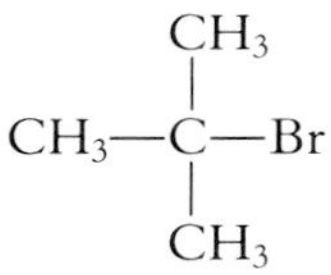

$CH_3—C(CH_3)_2—Br$

t-Butyl bromide

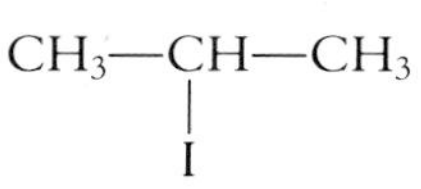

$CH_3—CH(I)—CH_3$

Isopropyl iodide

$CH_3—CH(CH_3)—CH_2F$

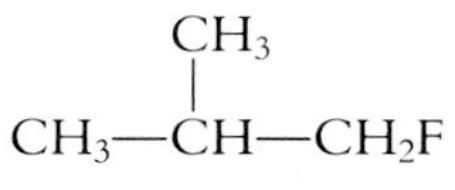

Isobutyl fluoride

Problems

5–11 Provide IUPAC names for each of the three compounds whose structures and radicofunctional names were just given.

Box 5–1. Halogenated Pesticides

In the 1940s it was discovered that many pests such as insects, nematode ground worms, and weed plants could be controlled by a variety of chemicals, particularly halogenated compounds. One of the most effective was DDT, which was particularly potent against mosquitoes and helped lower the malaria incidence throughout the world. Aldrin was one of the few compounds effective in the control of locust plagues, which periodically devastated regions of the Middle East. Ground worms in agricultural land could often be controlled with 1,2-dibromo-3-chloropropane (DBCP).

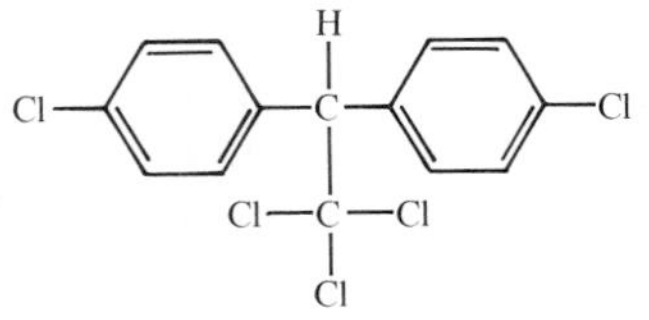

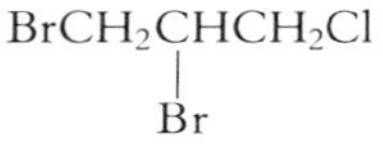

DDT (dichlorodiphenyltrichloroethane)

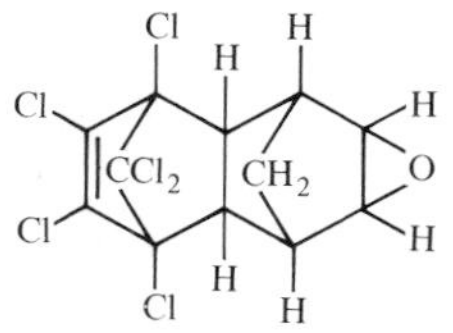

Aldrin

$$BrCH_2CH(Br)CH_2Cl$$

DBCP

Use of many such pesticides was later discontinued when undesirable side effects occurred. DBCP, for example, contributes to low sperm counts in male agricultural workers, while their wives experienced greater than average miscarriage rates. Although DDT was not a particularly poisonous compound to mammals, it had devastating effects on some birds' egg shell durability. Brown pelicans and ospreys were particularly hard hit.

The pesticide controversy shows dramatically that the "goodness" or "badness" of chemicals depends not only on their chemical behavior, but on the social and economic climate in which they are used. Under conditions in which diseases such as malaria and bubonic plague are not immediate threats, the harmful effects of the pesticides that help control them stand out. However, when diseases such as these are imminent dangers and few other affordable remedies are available, these harmful effects are much less cause for concern.

It seems that not all living creatures are adversely affected by pesticides. One species of grasshopper sequesters toxic defensive chemicals in specific glands. One such chemical is 2,5-dichlorophenol, which is so different from normal metabolites that it is believed to come from pesticide-sprayed leaves which were ingested by the grasshopper.

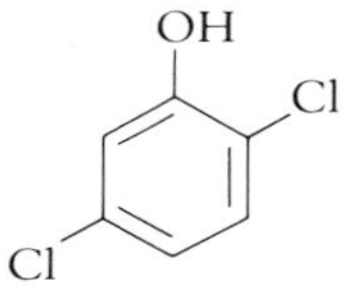

2,5-Dichlorophenol

5–12 Draw the structures of all the monobromination products of the following compounds and give their IUPAC names.
a. chlorocyclobutane b. 3-fluoropentane c. 2-methylpentane

5:9 Rate Equations, Rate Constants, and Energy Barriers: Some Fundamental Information About Chemical Reactions

Thus far in this chapter we have shown how alkyl halides are formed from alkanes, how these reactions actually proceed, and how and why mixtures of products are formed. In the course of these discussions we have referred several times, in a qualitative way, to rates of reaction. We will now see that it is possible to get a *quantitative* measure of reaction rates from which important mechanistic information can be extracted. The measurement of rates of reaction and the significance of the data derived from them is the subject of the next few sections.

By the middle of the 19th century, chemists had recognized that the rate, or velocity, of a reaction usually depends on some power of the reactant concentrations. Equation 5.37 is the general rate equation for a reaction in which the reactants are A and B, with concentrations (mol/L) of [A] and [B].

$$\text{rate} = k[\text{A}]^x[\text{B}]^y \tag{5.37}$$

The quantity k is called the **rate constant**. This parameter depends on the reaction temperature, pressure, and solvent, but when these are fixed, its value is always the same for a particular reaction, regardless of concentrations. For this reason we characterize a reaction by its rate constant, rather than by its rate. This is because the rate changes constantly during the reaction, since the reactant concentrations are constantly changing.

The exponents x and y are found experimentally. This is done by changing the concentrations of the individual reactants and observing the effects of these changes on the rate. For example, doubling the concentration of hydroxide ion in the reaction below (eq. 5.38) results in a doubling of the initial rate, showing that the rate varies as the first power of OH^-.

$$OH^- + CH_3Br \xrightarrow[CH_3OH]{} CH_3OH + Br^- \tag{5.38}$$

An analogous rate doubling is found when the concentration of bromomethane is doubled. These findings allow us to write the experimental rate equation for this reaction (eq. 5.39). In this case the exponents x and y each

$$\text{rate} = k[OH^-][CH_3Br] \tag{5.39}$$

equal one, and we say that the reaction rate is **first order** with respect to OH^- and first order with respect to CH_3Br. Overall, the reaction is *second order,* since the total reaction order equals the sum of x and y. A study in which the rate constant and kinetic orders of a reaction are measured is called a **kinetic study** (from the Greek *kinetikos,* meaning "of motion").

It turns out that for the reaction in Equation 5.38 the stoichiometric coefficients of the reactants are identical to their kinetic orders in the rate equation. This relationship is fortuitous. Kinetic orders *cannot* be deduced from the reaction stoichiometry. For example, the reaction in Equation 5.40

$$\underset{\text{Acetone}}{CH_3\overset{\overset{O}{\|}}{C}CH_3} + \underset{\text{Formaldehyde}}{CH_2{=}O} \xrightarrow[H_2O]{OH^-} \underset{\text{4-Hydroxy-2-butanone}}{CH_3\overset{\overset{O}{\|}}{C}CH_2CH_2OH} \tag{5.40}$$

follows the rate equation

$$\text{rate} = k[CH_3\overset{\overset{O}{\|}}{C}CH_3][OH^-] \tag{5.41}$$

Notice that the rate is *zero order* with respect to the reactant formaldehyde. Furthermore, the rate is first order with respect to hydroxide ion even though this species is a catalyst and does not take part in the reaction stoichiometrically.

Example

Ethanol is converted to diethyl ether in the presence of an acid:

$$\underset{\text{Ethanol}}{2CH_3CH_2{-}OH} \xrightarrow{H^+} \underset{\text{Diethyl ether}}{CH_3CH_2{-}O{-}CH_2CH_3} + H_2O$$

When the concentration of ethanol is doubled, the rate increases by a factor of four. Doubling the acid concentration leads to a doubling of the rate. Write the experimental rate equation for this reaction.

Answer

Since the increase in reaction rate is *the square of* the increase in ethanol concentration, the rate must have a second-order dependence on ethanol concentration.

$$\text{rate} = k[CH_3CH_2OH]^2$$

The rate is also first order in hydrogen ion, because its increase is proportional to the increase in hydrogen ion concentration.

$$\text{rate} = k[CH_3CH_2OH]^2[H^+]$$

Problems

5–13 When the concentration of $C_2H_5O^-$ is doubled in Equation 2, the rate doubles. When the concentration of acetophenone is doubled, the rate increases by a factor of four. Write the experimental rate equation for the reaction.

$$\underset{\text{Acetophenone}}{C_6H_5\overset{\overset{\displaystyle O}{\|}}{C}CH_3} + \underset{\text{Acetophenone}}{C_6H_5\overset{\overset{\displaystyle O}{\|}}{C}CH_3} \xrightarrow[C_2H_5OH]{C_2H_5O^-} C_6H_5\overset{\overset{\displaystyle O}{\|}}{C}CH{=}\overset{\overset{\displaystyle CH_3}{|}}{C}C_6H_5 + H_2O \quad (2)$$

5–14 When the concentration of hydrogen ion or *t*-butyl alcohol in Equation 3 is halved, the rate drops by a factor of two. When the chloride ion concentration is halved, the rate remains the same. Write the empirical rate equation for this reaction.

$$\underset{t\text{-Butyl alcohol}}{(CH_3)_3COH} + H^+ + Cl^- \xrightarrow[H_2O]{} (CH_3)_3CCl + H_2O \quad (3)$$

You'll see in Chapter 6 that rate equations are important sources of information about reaction mechanisms. The rate constant itself can also provide mechanistic information, particularly when we study its variation with temperature.

5:10 The Temperature Dependence of the Rate Constant Reveals an Energy Barrier to Reaction

In 1889 the Swedish chemist Svante Arrhenius (Nobel Prize, 1903) showed that the rate constant increased exponentially with temperature for many nonenzymatic reactions (eq. 5.42).

$$\ln k \propto T(°K) \quad (5.42)$$

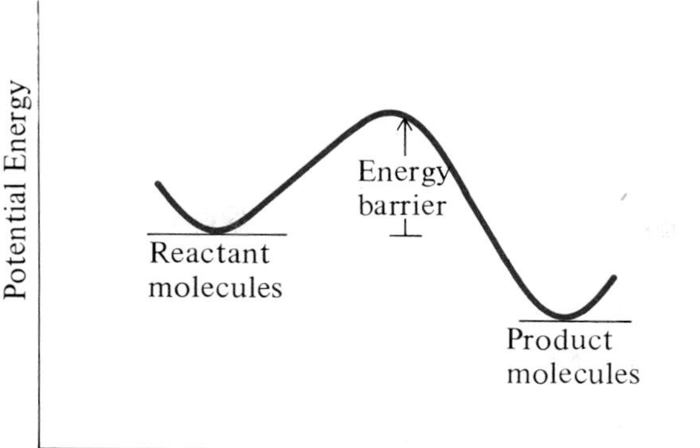

Figure 5–6. *Potential energy changes occurring during a single-step reaction. The potential energy must increase before reaching its final value; this is the energy barrier of the reaction.*

He interpreted this temperature dependence to mean that there is an **energy barrier** separating reactants and products. To understand the significance of this conclusion, let's examine the potential energy changes that accompany a single reaction step (Fig. 5–6). The figure shows that the potential energy of the reactants must increase before they can be converted to products. This is true even when the products have lower energies than the reactants. The minimum energy increase necessary before a reaction can take place corresponds to the energy barrier. The larger the energy barrier, the slower the reaction.

The horizontal coordinate of Figure 5–6 is called the *reaction coordinate*. It is a function of all the bond lengths and bond angles of the reacting molecules, and it measures the progress of the reaction.

In most reactions there will be substantial changes in bond angles, bond lengths, and electron distributions as molecules change from reactants to products. Figure 5–7 illustrates some of the changes that take place in the

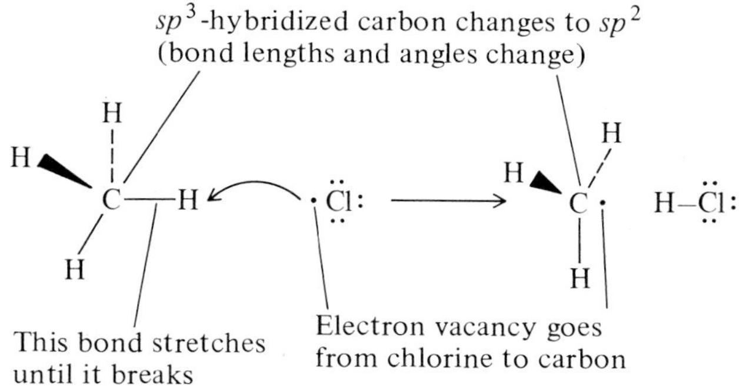

Figure 5–7. *Some of the molecular changes occurring in the reaction of Cl· with CH_4, which account for the energy barrier.*

reaction between Cl· and CH_4. The energy barrier arises because extra potential energy is required by the reacting molecules to initiate these changes.

How do reactant molecules acquire the extra potential energy needed for reaction? They do it by colliding with one another, in the course of which some of their kinetic energy (energy of motion) is converted into potential energy. At any given moment a fraction of the colliding molecules will acquire enough additional potential energy to surmount the energy barrier. If we reflect on this we can understand why raising the temperature increases the rate constant. Raising the temperature increases the average velocity and kinetic energy of the reactants, allowing more of them to acquire the extra potential energy needed for reaction in a unit time.

5:11 Transition States—The Key to Understanding Reaction Rates

Although rate constants can usually be determined experimentally, chemists would like to be able to calculate them directly from a knowledge of molecular properties. In the 1930s the American chemist Henry Eyring and his coworkers used the concept of the energy barrier to develop the *theory of absolute reaction rates* for this purpose. Unfortunately, we can not yet apply the theory accurately to any except the simplest reactions. Nonetheless, it has given us great qualitative insight into how molecules react.

Absolute reaction rate theory rests on the following essential ideas:

1. There will generally be an energy barrier in each step of a multistep reaction.
2. The critical part of each reaction step involves reaching the **transition state**, or **activated complex**. The transition state refers to the configuration of the reactants when they are at the top of the energy barrier (Fig. 5–8).
3. The transition state can be treated as if it is a normal chemical species, in equilibrium with the starting material, for which there is a standard equilibrium constant ($K^‡$, eq. 5.43).

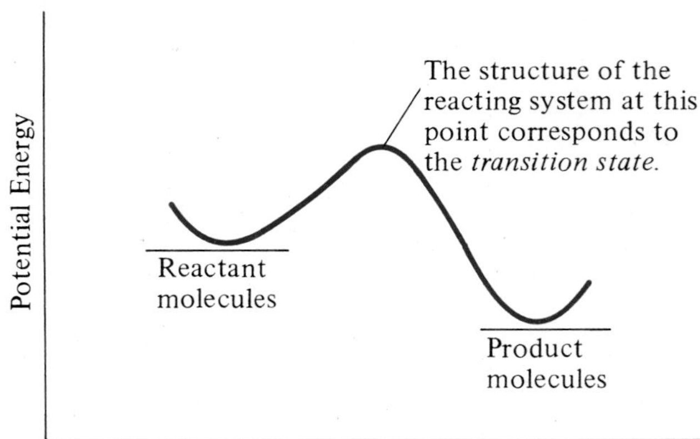

Figure 5–8. *An energy diagram for a single reaction step, showing the transition state at the potential energy maximum.*

$$(5.43) \qquad \text{reactants} \overset{K^\ddagger}{\rightleftarrows} \text{transition state} \rightarrow \text{products}$$

Using these ideas Eyring derived an expression for the rate constant in terms of the transition state properties (eq. 5.44).

$$(5.44) \qquad \ln k = \ln \frac{\mathbf{k}T}{h} - \frac{\Delta G^\ddagger}{RT}$$

where $\mathbf{k}$ = Boltzmann's constant (R/N)
h = Planck's constant
R = gas constant
T = absolute temperature (°K)

$\Delta G^\ddagger$ represents the **free energy of activation** (kcal/mol), the difference in free energy between the transition state and the reactants. $\Delta G^\ddagger$ is always positive; the larger its magnitude, the smaller the equilibrium concentration of transition states. Therefore, when $\Delta G^\ddagger$ is large, the rate constant (eq. 5.44) is small.

We know that free energy differences are a combination of enthalpy and entropy differences (sec. 3:6). Using the relationship that ties these three thermodynamic functions together (eq. 3.3, p. 80) allows us to rewrite the rate constant expression (eq. 5.45). $\Delta H^\ddagger$ is the **enthalpy of activation**, the

$$(5.45) \qquad \ln k = \ln \frac{\mathbf{k}T}{h} - \frac{\Delta H^\ddagger}{RT} + \frac{\Delta S^\ddagger}{R}$$

enthalpy difference between transition state and reactants. Analogously, $\Delta S^\ddagger$ represents the **entropy of activation**, the entropy difference between transition state and reactants. The magnitude of $\Delta H^\ddagger$, which is always positive, is directly related to the size of the energy barrier. A large $\Delta H^\ddagger$ signifies a large energy barrier, which slows down the reaction. The sign and magnitude of $\Delta S^\ddagger$ give information about the *probability* of formation of the transition state, which we will say more about shortly.

The theory of absolute reaction rates highlights the importance of the transition state in determining the rate constant. To understand reaction kinetics, we could try to isolate the transition states of a number of representative reactions, or at least study them spectroscopically. Alas, neither alternative is possible. Transition states have very short lifetimes, on the order of

Box 5–2. Sheepherding, Enthalpy, and Entropy

Chemistry is the science of getting reactant molecules over an enthalpy and entropy barrier so they can become products on the other side. This science can be likened to guiding a herd of sheep over a mountain to greener pastures.

How rapidly one can herd sheep over a mountain depends on the height of the mountain pass, whether it is broad or narrow, and on the number of sheep being herded. We can think of the height of the pass as the enthalpy term and its width as the entropy term. Other things being equal, you can get sheep over a lower mountain pass faster than you can over a higher mountain pass. Unfortunately, in sheepherding as in chemistry, other things are rarely equal.

For example, can you get 1000 sheep (or molecules) over a low, but very narrow pass faster than you can over a higher, but very broad pass? If the low pass is so narrow that sheep can only go through single file (or molecules can react only in a restricted conformation), then the answer is no, because the overall rate of such a process might be too slow. Although the enthalpy term is favorable, the entropy term is unfavorable. It might be better to herd the sheep (or molecules) over a somewhat higher pass through which many can cross at the same time.

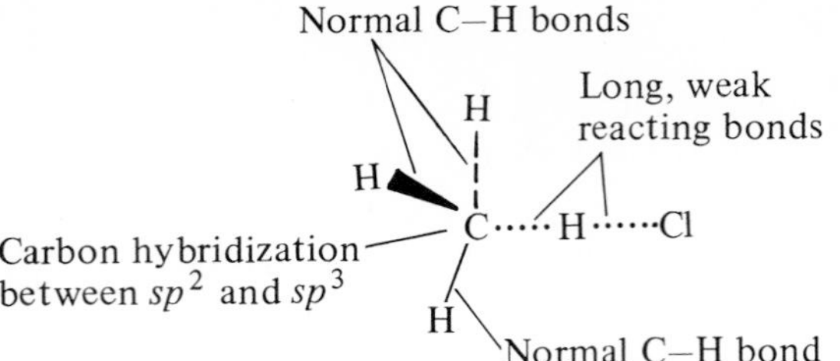

Figure 5–9. *Transition state for the reaction of methane with a chlorine atom. The bonds being broken and made, C—H and H—Cl, are longer and weaker than normal.*

10^{-13} s, and their concentration is very small as well. However, we can do a good job of inferring the properties of transition states by collecting large quantities of kinetic and stereochemical reaction data and supplementing them with theoretical calculations.

In most respects transition states are like ordinary molecules, with standard bond lengths and bond angles. The exceptions are the *reacting bonds,* the ones being made or broken during the reaction. These reacting bonds are longer and weaker than nonreacting bonds. In other words, the reacting bonds are only partly formed or partly broken in the transition state.

The transition state for the reaction of CH_4 with Cl· (eq. 5.16) provides a good illustration. The H—Cl bond is long and only partly formed, and the reacting C—H bond is long and only partly broken (Fig. 5–9). The nonreacting C—H bonds, by contrast, have normal bond lengths. Their bond angles are between 109.5° and 120°, since in the transition state the carbon hybridization is between sp^3 and sp^2. Figure 5–10 shows the position of the transition state on a reaction energy diagram.

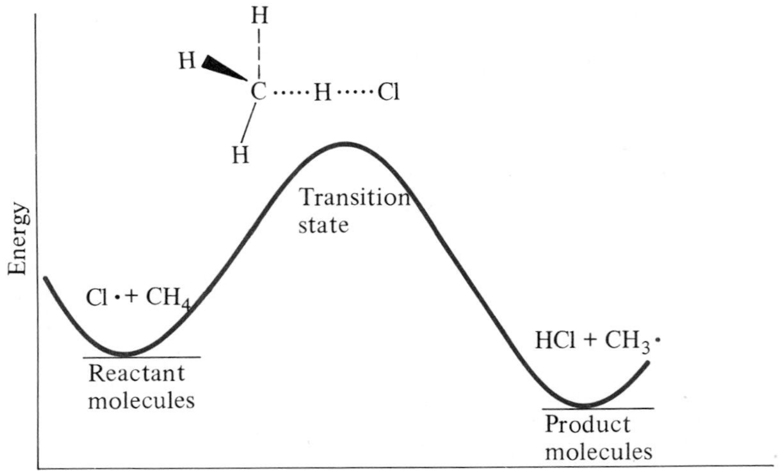

Figure 5–10. *An energy diagram for the reaction of Cl· with CH_4, showing the position of the transition state.*

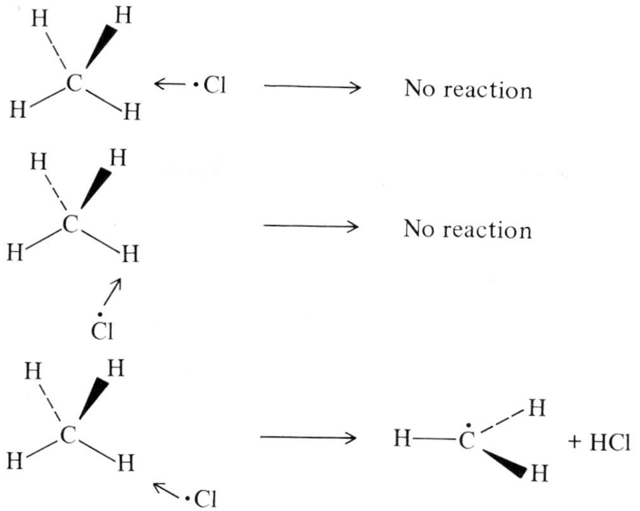

$CH_4 \leftarrow \cdot Cl$ ⟶ No reaction

CH_4 (Cl· approaching from below) ⟶ No reaction

CH_4 (·Cl approaching along the C—H bond) ⟶ $H—\dot{C}H_2$ + HCl

Figure 5–11. *Favorable and unfavorable collision orientations in the reaction of Cl· and CH_4. Only a fraction of the collisions lead to product formation.*

$\Delta H^\ddagger$ for formation of this transition state is about 3.2 kcal/mol. The small value reflects the fact that the energetic cost of breaking the C—H bond is being compensated for by the energetic gain from making the H—Cl bond. The fairly negative $\Delta S^\ddagger$ for the reaction, −16 cal/mol °K, also lends support to the proposed transition state (Fig. 5–9) because it shows that the reactants have lost some of their freedom of motion going to the transition state. In order for the transition state to be reached, chlorine and methane must collide so that the C···H···Cl bond angle is approximately 180°. In actuality only a fraction of the chlorine–methane collisions occur with the correct geometry (Fig. 5–11), so the probability of reaction is relatively low. This low probability is indicated by the negative value of $\Delta S^\ddagger$.

From the preceding discussion you should have a qualitative idea of why chemical reactions do not take place instantly, that is, as soon as the reactant molecules collide. It is because the rate constant depends on the *product* of two probabilities. One is the probability that the reactants will, upon collision, have enough energy to get over the energy barrier. The other is the probability that the colliding reactants are properly oriented for reaction.

The overall enthalpy of reaction (ΔH, sec. 5:5) and the enthalpy of activation ($\Delta H^\ddagger$) are often most informative when we compare a series of related reactions. The next section shows how that can be done graphically, and what conclusions it can lead to.

5:12 Comparative Energy Diagrams for Radical Reactions: Reactivity and Selectivity

One of the most convenient formats for presenting the results of thermodynamic and kinetic reaction studies is an energy diagram, such as Figure 5–6. We'll make use of such diagrams to compare two similar reactions, hydrogen abstraction from CH_4 by F· and by Br·.

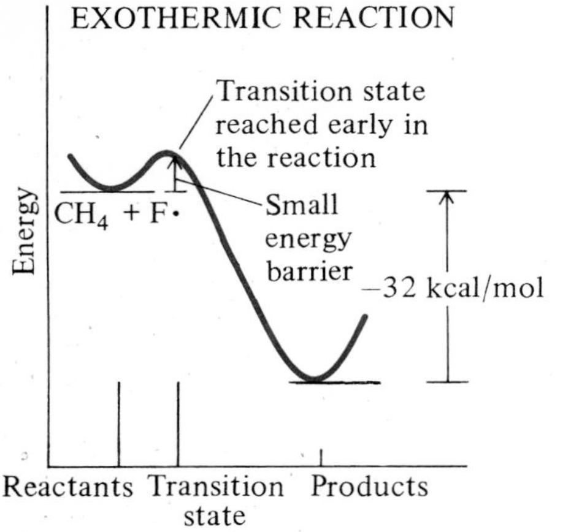

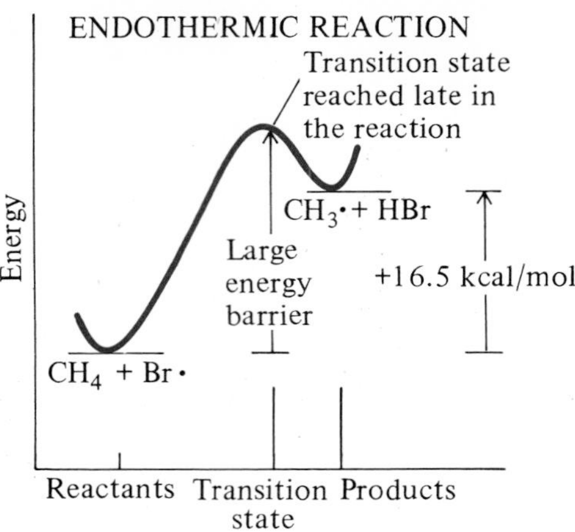

Figure 5–12. *Comparative energy diagrams for the reaction of F· and Br· with CH_4. The energy barrier is small for the exothermic fluorine reaction and large for the endothermic bromine reaction.*

The enthalpies for each reaction (ΔH) are calculated from the appropriate bond dissociation energies (Table 5–1). They equal −32 kcal/mol for the fluorine reaction and +16.5 kcal/mol for the bromine reaction (eq. 5.28). The enthalpies of activation ($\Delta H^\ddagger$) are obtained from kinetic studies and are used to approximate the energy barriers. The barrier is very small for the fluorine reaction ($\Delta H^\ddagger < 1$ kcal/mol), and fairly substantial for the bromine reaction ($\Delta H^\ddagger = 18$ kcal/mol). These kinetic and thermodynamic relationships between the two reactions are shown in Figure 5–12. The figure shows that when we compare two similar reactions, we find that the more exothermic one (smaller ΔH value) has the greater rate (smaller $\Delta H^\ddagger$). This indicates a correlation between ΔH and the size of the energy barrier to reaction. The ΔH and $\Delta H^\ddagger$ values for the Cl·/CH_4 reaction lie between those for F·/CH_4 and Br·/CH_4, in accord with the correlation. Thus, we have found a relationship between the overall energy change for a set of similar reaction steps and their relative rates.

Another significant aspect of the halogenation reactions is also revealed by Figure 5–12. Note that for the exothermic reaction (fluorine atom attack) the transition state is formed early. It therefore resembles the reactants because it is not very far from them along the reaction coordinate. Thus, for

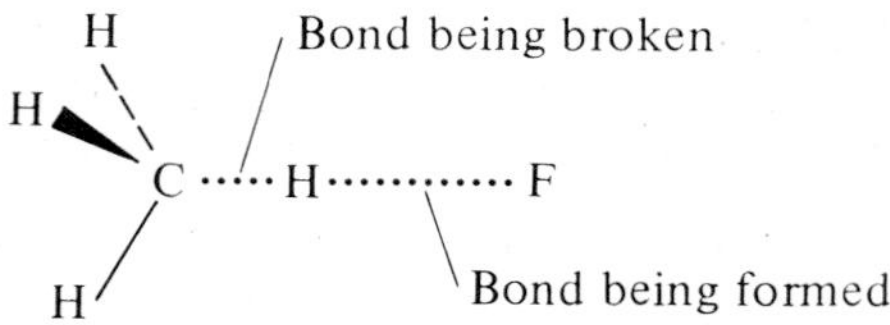

Figure 5–13. *The transition state in the reaction of fluorine with methane. The C—H bond is only slightly broken and the H—F bond is hardly formed.*

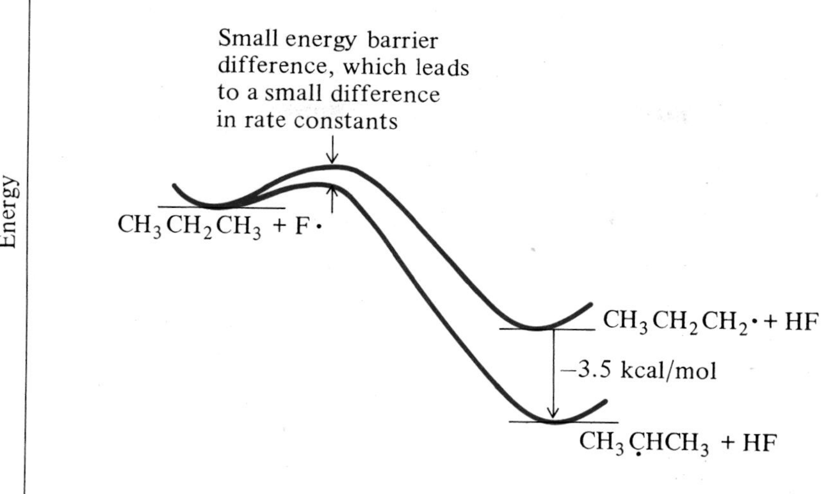

Figure 5–14. *Energy diagrams for competitive abstraction of 1° and 2° hydrogens of propane by fluorine atoms. There is only a small difference in energy barriers for the two reactions, so their rate constants are very close. Thus fluorine is not very selective in hydrogen abstraction.*

the F·/CH_4 transition state, the C—H bond will hardly be broken and the H—F bond will just be starting to form (Fig. 5–13). Since the C—H bond is hardly broken, the transition state energy does not depend much on what type of C—H bond is being broken (e.g., 1°, 2°, or 3°), so we would expect that F· is not a very discriminatory reactant, which fits the experimental results exactly (sec. 5:6).

The lack of selectivity by F· is made evident if we plot two energy diagrams, one for attack on a 1° and one on a 2° hydrogen atom, on the same coordinates (Fig. 5–14). Here we see that the *difference* in barrier heights is small when the transition states come early, so the rates of the two reactions are not very different. This is what is meant by low selectivity.

Bromine attack on methane goes by way of a late transition state (Fig. 5–12) resembling the products. The C—H bond is almost completely broken and the H—Br bond nearly fully formed (Fig. 5–15). Thus the difference in

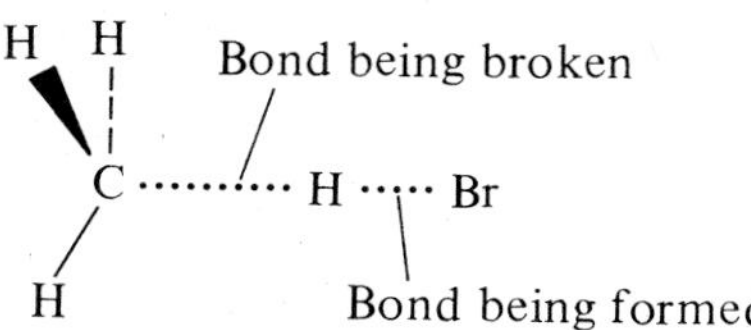

Figure 5–15. *Transition state for the reaction of bromine with methane. The C—H bond is largely broken and the H—Br bond is largely formed.*

5:12 Comparative Energy Diagrams for Radical Reactions

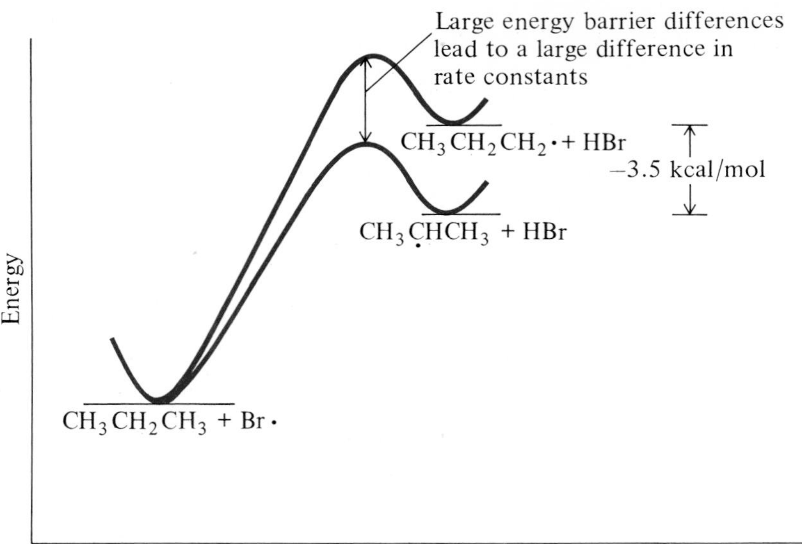

Figure 5–16. *Energy diagrams for competitive abstraction of 1° and 2° hydrogens of propane by bromine atoms. The large difference in energy barrier heights leads to large differences in rate constants, which results in bromine being highly selective.*

barrier heights is substantial for different types of C—H bonds (Fig. 5–16). In general, then, endothermic reactions such as Br· attack on C—H bonds will show substantial selectivity (Table 5–2, sec. 5:6).

The two major conclusions that have been developed in this section are

1. for a series of similar reactions, the most exothermic will have the largest rate constant, and the least exothermic the smallest rate constant; and
2. for a series of similar reactions, the most exothermic will show the least selectivity, the least exothermic the most selectivity.

These generalizations have proved quite accurate because we have taken a narrow view of the term "similar reactions." The examples discussed all involved halogen atom attack on C—H bonds, and thus have similar transition state structures. If the attacking radical were substantially different from a halogen atom, the correlations might be less firm. Similarly, a change in the atom to which the hydrogen is bonded might also result in a weakening of the correlations. Finally, it should be noted that all the examples had similar $\Delta S^{\ddagger}$ values. This means that the differences in rate constants among the various reactions are due mainly to differences in $\Delta H^{\ddagger}$. Similarity of $\Delta S^{\ddagger}$ is usually easier to insure in the gas phase than it is in solution, because solvent effects often show up as entropy differences.

If we keep in mind the limitations just discussed, then correlations between exothermicity and rate and exothermicity and selectivity can be very useful explanatory and predictive tools. We will use them as such in subsequent chapters.

5 Reactive Intermediates

5–15 Draw a structure for the transition state of the second propagating step in methane chlorination, $CH_3\cdot$ reacting with Cl_2.

5–16 a. Draw energy diagrams for the reactions of $CH_3\cdot$ with Cl_2 and with Br_2. Make sure your diagrams reflect the relative enthalpies of these two processes. Based on these enthalpies, estimate the relative sizes of the energy barriers for your diagrams.

b. In which reaction would you expect the $CH_3\cdots X$ bond to be more formed at the transition state? Explain your answer.

5–17 a. The H—OH bond strength is 119 kcal/mol. Use this value to calculate ΔH for the reaction in Equation 4.

$$CH_4 + \cdot OH \rightarrow CH_3\cdot + H_2O \qquad (4)$$

b. Using your answer to (a), estimate how the selectivity of ·OH will compare to the selectivity of the halogen atoms (Table 5–2). Explain your answer.

5:13 The Free-Radical Chemistry of Oxygen: Some Biological and Nonbiological Examples

This chapter has emphasized the importance of free-radical chemistry in the preparation of industrial chemicals. The basis of this chemistry is the rapid reaction between halogen atoms and hydrocarbons. Other radicals can also attack hydrocarbons and molecules containing alkyl groups. Among the most important are oxygen-centered radicals such as alkoxyl radicals, RO·, and hydroxyl radicals, HO· (Prob. 5–17). Since these species are often derived from molecular oxygen, we will take a brief look at the free radical chemistry of O_2 in the presence of alkanes and aklyl groups.

One place where both alkyl groups and O_2 come together is the living cell. These cells are surrounded by membranes which hold the cell contents together and allow selective chemical exchange between the cell and its surroundings. Cell membranes contain a high proportion of compounds rich in alkyl and cycloalkyl groups, called **lipids**. (Lipids are cell constituents which are not soluble in water, such as fats, waxes, and oils.) These alkyl and cycloalkyl groups are susceptible to radical attack.

Most cells require oxygen for energy production. One dangerous side reaction involving oxygen is its reduction to hydrogen peroxide, which can be further reduced to give hydroxyl radicals (eq. 5.46). The hydroxyl radi-

$$(5.46) \qquad O_2 + 2H^+ + 2e^- \rightarrow \underset{\text{Hydrogen peroxide}}{H\ddot{\underset{..}{O}}—\ddot{\underset{..}{O}}H} \xrightarrow{e^-} \underset{\text{Hydroxyl ion}}{H\ddot{\underset{..}{O}}:^-} + \underset{\text{Hydroxyl radical}}{\cdot\ddot{\underset{..}{O}}H}$$

cals attack the C—H bonds in the membrane lipids, giving rise to alkyl radicals (eq. 5.47). The further reactions of these alkyl radicals ultimately yield products that are detrimental to cell membrane functioning.

$$(5.47) \qquad R—H + \cdot\ddot{\underset{..}{O}}H \rightarrow R\cdot + H_2\ddot{O}: \qquad \Delta H < -15 \text{ kcal/mol}$$

Deterioration of its membranes is one characteristic sign of cell aging, and there seems to be a strong link between cell aging and the occurrence of free-radical processes in the cell. The seriousness of the problem is underlined by the fact that living systems have evolved mechanisms to protect themselves from radical attack. These include *antioxidants* such as vitamins C and E, which are good hydrogen atom donors (Chap. 24). Antioxidants can intercept alkyl and hydroxyl radicals and reduce them to harmless products. Analogously, synthetic antioxidants retard the effects of alkyl group/oxygen reactions responsible for such undesirable nonbiological oxidations as the drying of oil-base paints, the aging of rubber and various plastics, and the deterioration of food.

While the free-radical chemistry of oxygen can be detrimental to cells and various man-made products, it also has its beneficial side. The combustion of hydrocarbon fuels is a prime example. Combustion takes place by way of very rapid and very complex free radical chains. Oxygen-centered radicals are vitally important in the combustion mechanism for alkanes. Because of the large quantities of heat per gram alkanes yield when burned (eq. 5.48), they are in heavy demand as fuels.

$$CH_4 + 2O_2 \rightarrow CO_2 + 2H_2O \qquad \Delta H = -212 \text{ kcal/mol} = -13.2 \text{ kcal/g} \tag{5.48}$$

Problem

5–18 Hydroxyl radicals ($HO\cdot$) are so energetic that they react nearly at the same rate at which they collide with other molecules (10^9 reactions per second). This may explain why they are so detrimental biologically.

a. Consider the reaction of $HO\cdot$ with *ethyl alcohol* (ethanol, eq. 5). What are the possible radical products that can be formed?

$$HO\cdot + \underset{\text{Ethanol}}{CH_3CH_2{-}OH} \rightarrow H_2O + ? \tag{5}$$

b. Calculate ΔH for each reaction and predict the major radical product (use Table 5–1).

Summary

1 Many reaction mechanisms involve highly reactive, transient species called intermediates. In the commercially important halogenation of alkanes, the intermediates are halogen atoms and alkyl radicals. Their reactivity is due to their unpaired electrons. Radicals are usually generated by the homolysis of single bonds, which can be brought about by heat or light. (secs. 5:1, 5:2, and 5:3).

2 Alkanes can react with molecular fluorine, chlorine, or bromine to yield halogenated alkanes and hydrogen halides. The halogenations of practical importance, chlorination and bromination, require initiation by light or heat.

The generally accepted mechanism for halogenation involves dissociation of a small percentage of halogen molecules into atoms. This is followed by abstraction of hydrogen atoms from the alkanes by the halogen atoms. The chain is maintained by the attack of the resulting alkyl radicals on molecular halogen. Oxygen inhibits the reaction, apparently by capturing alkyl radicals (sec. 5:4).

3 The enthalpies of halogenation of methane can be calculated by subtracting the sum of the bond energies of bonds being formed from the sum of the bond energies of bonds being broken. Fluorination turns out to be highly exothermic, while iodination is endothermic. This trend correlates with the observed reaction rates, fluorination being fastest and iodination slowest. The first chain-carrying step, hydrogen abstraction by a halogen atom, is always less exothermic than the second step, halogen abstraction by a methyl radical (sec. 5:5).

Halogenation of higher alkanes shows that 3° (tertiary) hydrogens are abstracted preferentially, followed by 2° (secondary) hydrogens and then 1° (primary) hydrogens. This result shows that the more exothermic reaction will proceed more rapidly than the less exothermic one. The tendency of different halogen atoms to discriminate among 1°, 2°, and 3° hydrogens lies in the order Br > Cl > F (sec. 5:6). The different rates of formation among alkyl radicals is connected to the differing amounts of strain energy released. The relief of steric strain is greatest for the generation of 3° radicals, which are formed faster than 2° or 1° radicals. Relief of steric strain is the least for 1° radical generation, which has the slowest rate (sec. 5:7).

4 In the IUPAC system halogens are named as substituents on the principal carbon chain. Along with alkyl substituents, their names are arranged alphabetically as prefixes to the name of the principal chain. In radicofunctional nomenclature the name of the alkyl group is followed by the name of the halide (sec. 5:8).

5 The rate of a chemical reaction generally depends on the concentration of its reactants, each raised to some power. These powers, which are the kinetic orders for each reactant, cannot be deduced from the reaction stoichiometry, but must be found by experiment. Rates change during the course of a reaction, but rate constants do not (sec. 5:9).

6 The rate constants of most nonenzymatic reactions increase exponentially with absolute temperature. Arrhenius interpreted this phenomenon to mean that in every reaction there is an energy barrier, because the potential energy of the reactants must increase before they can be converted to products. Reactant molecules overcome the barrier by converting kinetic energy to potential energy (sec. 5:10).

The structure of the reacting system at the peak of the energy barrier is called the transition state. $\Delta H^\ddagger$ is the molar enthalpy difference between the transition state and the reactants from which it is derived. A small value of $\Delta H^\ddagger$ indicates a small energy barrier and, usually, a large value of k. $\Delta S^\ddagger$ is the molar entropy difference between the transition state and the reactants, and measures the relative probability of transition state formation. Transition states resemble normal molecules in most respects, except that their reacting bonds are longer and weaker than normal (sec. 5:11).

Comparisons among similar reactions reveal that the most exothermic (most negative ΔH) usually has the smallest energy barrier (smallest $\Delta H^\ddagger$). Furthermore, the most exothermic reaction reaches the transition state earliest, while the most endothermic reaction has the latest transition state. Because of their different transition state positions, endothermic processes tend to show much greater selectivity than exothermic ones (sec. 5:12).

7 Oxygen-centered radicals are derived from O_2. These radicals readily attack hydrocarbons and alkyl groups, and are a vital part of the combustion mechanism for alkanes. In cells, the free-radical chemistry of oxygen can lead to deterioration of the cell membrane (sec. 5:13).

Problems

5–19 Using condensed formulas, show all the hydrogens and the odd electron on the trivalent carbon of the following radicals:

a. ethyl
b. isopropyl
c. neopentyl
d. *t*-butyl
e. 2-pentyl
f. 3-methyl-2-hexyl

5–20 Give the IUPAC name for each of the following compounds:

CH_3
a. $CH_3CHCH_2CHCH_3$
Br

CH_3
b. $CH_3CHCH_2CH_2CHCl_2$

CH_3
c. $CH_3CHCH_2CHCH_2F$
Br

d. Cl
CH_3 CH_3

e.
Br Br F

5–21 Draw the structure of each of the following compounds, showing the stereochemistry.

a. 4-bromo-1,1-diethylcycloheptane
b. an (*R*)-isomer of formula $C_5H_{11}F$
c. *cis*-1-ethyl-3-iodocyclohexane
d. *trans*-1,1-dichloro-3,5-diiodocyclohexane
e. (*R*)-1,1,5-tribromo-3-chloropentane
f. (*S*)-3-chloro-1,1-diisopropylcyclohexane

5–22 The Pb—C bond energy in $(CH_3)_4Pb$ (tetramethyllead) is rather low, amounting to only 49 kcal/mol. A small amount of $(CH_3)_4Pb$ is added to a mixture of CH_4 and Cl_2. The usual chlorination reaction takes place, except that a lower temperature is needed to initiate the reaction than is needed in the absence of $(CH_3)_4Pb$. Show how tetramethyllead can initiate the chloriation reaction, and explain why lower temperatures suffice.

5–23 An alternative to the accepted mechanism for the chlorination of methane is given in Equations 6 and 7. As far as can be determined, this mechanism does *not* operate to any significant extent under normal conditions.

$$Cl\cdot + CH_4 \rightarrow CH_3Cl + H\cdot \qquad (6)$$

$$H\cdot + Cl_2 \rightarrow HCl + Cl\cdot \qquad (7)$$

Using the data in Table 5–1, calculate the enthalpies of each step and compare them with the enthalpies in Equations 5.26 and 5.27. Then suggest a reason why this mechanism cannot compete with the accepted one.

5–24 Write and name all the products of monochlorination of the following compounds. Which products will be isolated as racemic mixtures? Are there any diastereomeric pairs?

a. 2-methylbutane
b. cyclobutane
c. 3-methylpentane
d. 2,2-dimethylpentane
e. 2-chlorobutane

5–25 Show all the monobromination products formed when each of the following alkanes is treated with Br_2 and light.

a. heptane
b. 2,2,3,3-tetramethylbutane
c. 1,1-dimethylcyclohexane
d. 3-methylhexane
e. 2,5-dimethylhexane

5–26 Using the data given, deduce the possible structures (there may be one or more) for each of the following hydrocarbons:

a. $C_4H_{10} \xrightarrow[\text{light}]{Cl_2}$ two C_4H_9Cl isomers

b. $C_5H_{10} \xrightarrow[\text{light}]{Cl_2}$ one C_5H_9Cl isomer

c. $C_6H_{14} \xrightarrow[\text{light}]{Cl_2}$ two $C_6H_{13}Cl$ isomers

d. $C_7H_{16} \xrightarrow[\text{light}]{Cl_2}$ three $C_7H_{15}Cl$ isomers

e. $C_6H_{12} \xrightarrow[\text{light}]{Cl_2}$ three $C_6H_{11}Cl$ isomers

5–27 Indicate which hydrogen atom in each compound will be most rapidly replaced during radical chlorination.

a. 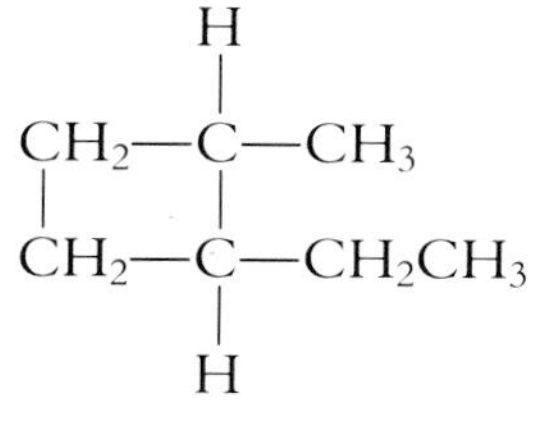
```
        H
        |
CH2—C—CH3
|       |
CH2—C—CH2CH3
        |
        H
```

b. $(CH_3)_3C—CH_2CH(CH_3)_2$

c. 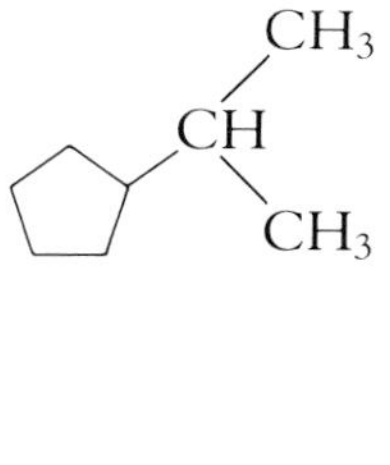 cyclopentyl–CH(CH_3)(CH_3)

d. 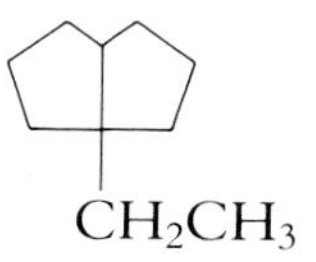 cyclopentane with CH_2CH_3 substituent

e. $CH_3—O—CH_2CH_3$

f. 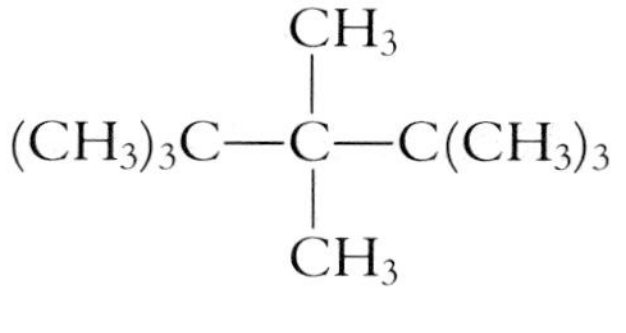 $(CH_3)_3C—C(CH_3)_2—C(CH_3)_3$

g. 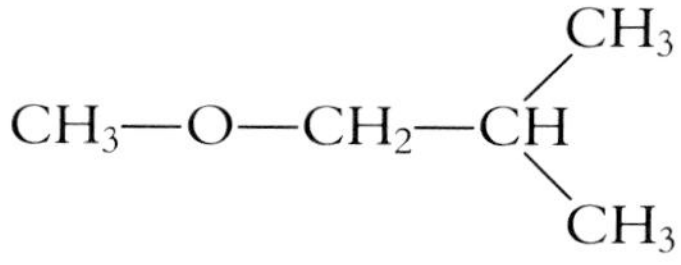 $CH_3—O—CH_2—CH(CH_3)_2$

5–28 a. Write the structures of all the possible monobromination products of methylcyclopentane. Include all *cis–trans* isomers.

b. It is found that there is more *trans*-1-bromo-2-methylcyclopentane in the product mixture than there is *cis* isomer. Propose a mechanism for the formation of each of these compounds, and use it to explain the preferential formation of the *trans* isomer.

5–29 Many free-radical reactions are difficult to start at convenient temperatures. In those cases it is common practice to add a small quantity of an **initiator** to the reaction mixture. An initiator is a substance that decomposes readily into free radicals, which will in turn start the chain reaction (also see Prob. 5–22). Organic peroxides, ROOR, are often used as initiators because of their weak O—O bonds. Show how di-*t*-butyl peroxide, $(CH_3)_3COOC(CH_3)_3$, could initiate the bromination of cyclohexane.

5–30 Chlorination or bromination of a chiral alkane leads to racemic products when a hydrogen atom at the chiral center is removed (eq. 8). Explain how the structure of the alkyl radical intermediate is responsible for the stereochemical result (see, for instance, Fig. 5–7).

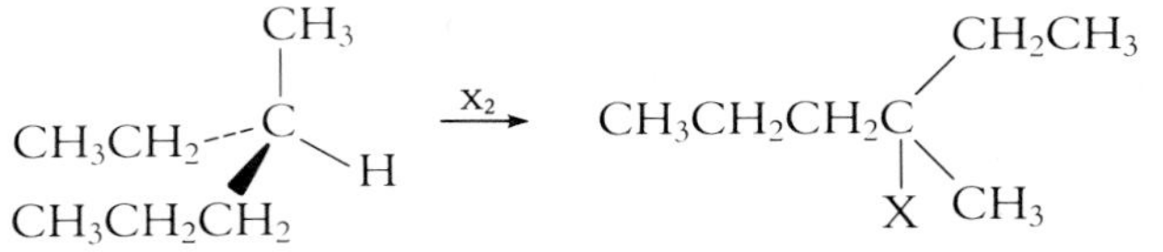

CH_3 / CH_3CH_2, $CH_3CH_2CH_2$, H on C $\xrightarrow{X_2}$ $CH_3CH_2CH_2C(X)(CH_3)CH_2CH_3$ (8)

(*R*)-3-Methylhexane → (*R,S*)-3-Halo-3-methylhexane (racemic)

5–31 Using the C—H bond energy data in Table 5–1, predict the major product formed when 1 mol of methane and 1 mol of ethane are allowed to react with 0.01 mol of Cl_2 and light at 25°. Explain your answer.

5–32 Each of the following compounds is heated until its bonds begin to break homolytically. In each case show which radical(s) are most likely to be formed by homolysis. Explain your answer.

a. CH_3CH_3

b. $(CH_3)_3CCH_2CH_3$

c. 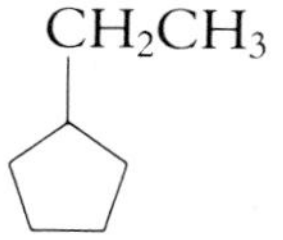 CH_2CH_3 on cyclopentane (ethylcyclopentane)

d. 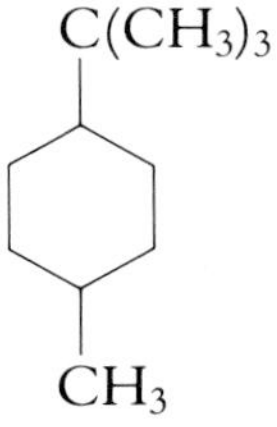 $C(CH_3)_3$ and CH_3 on cyclohexane (1,4-positions)

e. $(CH_3)_3CCH_2I$

5–33 Monochlorination of ethane-*d* (C_2H_5D) leads to a mixture consisting of 93% C_2H_4DCl and 7% C_2H_5Cl.

a. What conclusions can you draw about the relative reactivity of C—H and C—D bonds toward chlorine atoms?

b. How would you actually go about finding out the relative amounts of the two chloroethanes in the product?

5–34 A number of halogenating agents other than the molecular halogens are used to chlorinate and brominate alkanes. This is because they may be easier to handle, or they give less objectionable by-products. Several of them are shown below (eqs. 9–11). Write a plausible mechanism for the reaction of each one with an alkane.

a. $$R\text{—H} + \underset{\text{Bromotrichloromethane}}{CCl_3Br} \xrightarrow{\text{light}} R\text{—Br} + CHCl_3 \qquad (9)$$

b. $$R\text{—H} + \underset{\text{Sulfuryl chloride}}{Cl\text{—}\overset{\overset{O}{\|}}{\underset{\underset{O}{\|}}{S}}\text{—}Cl} \xrightarrow{\text{heat}} R\text{—Cl} + SO_2 + HCl \qquad (10)$$

c. $$R\text{—H} + \underset{t\text{-Butyl hypochlorite}}{(CH_3)_3CO\text{—}Cl} \xrightarrow{\text{heat}} R\text{—Cl} + \underset{t\text{-Butyl alcohol}}{(CH_3)_3CO\text{—}H} \qquad (11)$$

5–35 Chlorination and bromination of alkanes usually lead to the breaking of C—H bonds, but almost never to C—C cleavage. An exception is the low temperature photobromination of cyclopropane, which leads to 1,3-dibromopropane by way of a free-radical chain process (eq. 12). Explain why cyclopropane behaves differently from other alkanes and cycloalkanes. (Trying to build a model of cyclopropane may help you answer this question). On the same energy diagram compare the formation of 3-bromo-1-propyl radical (**I**) from cyclopropane and the formation of 1-propyl radical from propane and Br·.

$$Br\cdot + \triangle \xrightarrow[\text{low temperature}]{\text{light}} \underset{\mathbf{I}}{BrCH_2CH_2CH_2\cdot} \xrightarrow{Br_2} Br(CH_2)_3Br + Br\cdot \qquad (12)$$

5–36 Show how many products can be formed when each of the following compounds is monochlorinated:

a. (*R,S*)-2-chlorobutane
b. 3-chloropentane
c. chlorocyclopentane
d. *meso*-3,4-dichlorohexane

5–37 The following alkanes were each treated with Br_2 at 100°. Write the structures of the *major* (>10% yield) monobromination product(s) in each case.

a. methylpropane
b. hexane
c. octane
d. methylbutane
e. 1,2-dimethylcyclobutane
f. 3-ethyl-4-methylhexane

5–38 Oxygen, O_2, contains two unpaired electrons. The O—O bond length is 1.21 Å, while that of an O—O bond in peroxides is 1.48 Å. What electron dot formulas can be written for O_2? Can any of them explain the given physical properties?

5–39 When organic materials burn they create flames. The propagation of these flames depends on the presence of H· and ·OH radicals. Organic fibers, such as those used in clothing and curtains, can be made flame resistant by introducing Br and Cl groups into the fiber. These halogenated materials give off HBr and HCl when heated. Show how these compounds could act to squelch the flame, using the bond energies in Table 5–1 to support your analysis.

Supplementary Reading

W. A. Pryor, *Introduction to Free Radical Chemistry,* Chaps. 1, 6, and 7, Prentice–Hall, Inc., Englewood Cliffs, N.J., 1966.

R. A. Jackson, *Mechanism: An Introduction to the Study of Organic Reactions,* Oxford Univ. Press, London, 1972.

W. A. Pryor, (ed.), *Free Radicals in Biology,* Academic Press, New York, N.Y., 1976.

L. Packer and J. Walton, Antioxidants vs. aging, *Chemtech,* 276 (1977).

J. Bland, Biochemical consequences of lipid peroxidation, *J. Chem. Ed.,* **55,** 151 (1978).

6 The Interconversion of Alkyl Halides, Alcohols, and Ethers: Nucleophilic Substitution Reactions

In Chapter 5 you saw how to prepare alkyl halides from alkanes, which are readily available from petroleum. In this chapter we show how alkyl halides can be converted to alcohols and ethers, which cannot be obtained directly from alkanes. These three classes of compounds are treated together because their functional groups contain only single bonds, making their chemistry similar in many ways. In particular, they all undergo a characteristic substitution reaction in which any one of these functional groups can be converted to the others.

In many instances substitution reactions are accompanied by a competing reaction, elimination. Substitution and elimination are closely related not only with respect to reactants, but also with regard to mechanism. For clarity we will devote this chapter to substitution reactions, and elimination reactions will be treated in Chapter 7. Try to remember as you study the equations in this chapter that they do not necessarily show *all* the products that could be formed from the given reactants.

6:1 Definitions and Conventions

Equations 6.1 and 6.2 are two examples of functional group interchange. The first equation shows an alkyl halide being converted to an alkyl cyanide, or **nitrile**; the second shows an alcohol being converted to an alkyl halide. The organic starting materials are called **substrates**.

(6.1) $CH_3{-}\ddot{B}r{:} + {:}CN^- \longrightarrow CH_3{-}CN + {:}\ddot{B}r{:}^-$

Substrate   Nucleophile

(6.2) $CH_3CH(\ddot{O}H)CH_3 + {:}\ddot{I}{:}^- \xrightarrow{H^+} CH_3CH(\ddot{I}{:})CH_3 + H\ddot{O}H$

Substrate   Nucleophile

The ions CN^- and I^- are **nucleophiles** (from the Greek, meaning "nucleus-loving"). Nucleophiles possess one or more nonbonding electron pairs that can be used to establish an additional bond to an atom in the substrate. Nucleophiles may be neutral molecules as well as ions, as long as they have available electron pairs. If the substrates are alkyl compounds, then the nucleophiles are said to undergo **alkylation**—attachment of an alkyl group to a reactant.

An arrow is sometimes used in the equation to indicate the behavior of the reacting electron pairs. In Equation 6.3 the arrow from O to C shows that a nucleophile electron pair is being used to make a new bond to the substrate. The other arrow shows that a bond in the substrate is being broken.

$$\text{(6.3)} \qquad \text{H}\ddot{\underset{\cdot\cdot}{\text{O}}}\!:^- + \text{CH}_3\text{—}\ddot{\underset{\cdot\cdot}{\text{Br}}}: \rightarrow \text{HO—CH}_3 + :\ddot{\underset{\cdot\cdot}{\text{Br}}}:^-$$

Nucleophile Substrate Leaving group

Both electrons in this bond are being carried off by the **leaving group**. Reactions of the type shown in Equations 6.1 to 6.3 are known as **nucleophilic substitutions**, or **nucleophilic displacements** (these two terms are used interchangeably). All of them are characterized by a bond-breaking step and a bond-forming step. One of the important mechanistic questions is whether or not these steps take place simultaneously, as implied in Equation 6.3.

Alkanes and cycloalkanes are examples of molecules that lack functional groups. As you learned in Chapter 5, they can be functionalized by allowing them to react with molecular halogens under *free-radical conditions*. This process is not completely satisfactory, however, because it is not very specific. That is, the reaction of a long-chain alkane with chlorine or bromine yields a mixture containing a large number of different haloalkanes. By contrast, *nucleophilic substitution* reactions of substituted alkanes and cycloalkanes are quite specific. As Equation 6.2 shows, the only product obtained from 2-propanol is 2-iodopropane (no 1-iodopropane is produced). In order to learn more about such substitutions we begin with reactions in which alkyl halides serve as substrates for the synthesis of other alkyl halides.

Problem

6–1 Indicate which of the following ions and molecules are nucleophilic and which are not. Justify your answers.

a. Cl^-
b. Br^-
c. SH^-
d. Na^+
e. Mg^{2+}
f. NH_3
g. H_2O
h. CH_4
i. BH_3

6:2 Converting One Alkyl Halide to Another: How Changes in Alkyl Group Structure Affect the Reaction Rate

Equations 6.4 and 6.5 depict the reactions of haloalkanes with halide salts to produce new haloalkanes. Equation 6.4 shows a chloroalkane reacting with a fluoride salt to produce a fluoroalkane. Equation 6.5 shows the reaction between an iodide salt and a bromoalkane to yield an iodoalkane. Ethylene glycol and acetone are the solvents for these reactions.

$$\text{(6.4)} \qquad CH_3(CH_2)_5Cl + KF \xrightarrow[\text{ethylene glycol, } 200^\circ]{} CH_3(CH_2)_5F + KCl$$

1-Fluorohexane (54%)

$$(6.5) \qquad CH_3Br + NaI \xrightarrow[\text{acetone, 56°}]{} \underset{\substack{\text{Iodomethane} \\ (100\%)}}{CH_3I} + NaBr \downarrow$$

Why are substitutions of this type important? First of all, these reactions are synthetically useful. Fluoroalkanes and iodoalkanes are usually *not* made by direct halogenation of alkanes. Fluorine is too reactive and iodine too unreactive to use in this way (sec. 5:5). Instead, fluoroalkanes and iodoalkanes are synthesized by way of nucleophilic substitutions, such as the ones shown in Equations 6.4 and 6.5.★

Beyond their synthetic utility, the reactions of inorganic halides with haloalkanes yield much information about the mechanisms of *all* nucleophilic displacements. This information is obtained by changing the structure of either of the reactants to see if there is a change in the reaction rate constant. For instance, if we substitute NaF for KF in Equation 6.4, we find that the rate constant remains the same. This result indicates that only the nucleophilic halide ion (F^- in this case) takes part in the reaction and that the cation does not. Our choice of a particular salt for any desired nucleophilic substitution is usually dictated only by the questions of availability and solubility. Sodium iodide is highly soluble in acetone, for example, and this combination is often used for the conversion of alkyl chlorides and bromides to iodides. It has an additional advantage in that the product salts (NaCl and NaBr) are not soluble, so they precipitate and the reaction is driven to completion (eq. 6.5).

While the identity of the metallic cation does not usually affect the substitution reaction, the structure of the alkyl group definitely does. We can measure this effect by allowing a series of alkyl halides, all having the same leaving group, to react with a single nucleophile. Table 6–1 contains rate constants and activation enthalpies for the reaction of iodide ion with a set of alkyl bromides in acetone. Some definite trends can be identified in the data. For instance, as the carbon atom holding the leaving group becomes more heavily substituted, the reaction slows down. Therefore, CH_3Br is the

Table 6–1. Rate Constants and Activation Energies for the Reaction

$$R{-}\ddot{\underset{..}{Br}}{:} + {:}\ddot{\underset{..}{I}}{:}^- \xrightarrow[\text{acetone}]{} R{-}\ddot{\underset{..}{I}}{:} + {:}\ddot{\underset{..}{Br}}{:}^-$$

RBr	$k(\text{mol/L})^{-1}\text{sec}^{-1}$	$\Delta H^{\ddagger}$(kcal/mol)
CH_3Br	2.5×10^{-1}	15.7
CH_3CH_2Br	1.7×10^{-3}	18.2
$CH_3CH_2CH_2Br$	1.4×10^{-3}	18.1
$(CH_3)_2CHCH_2Br$	6.0×10^{-5}	19.6
$(CH_3)_2CHBr$	1.3×10^{-5}	19.9

★ Fluoroalkanes are valued for their low reactivity and have many commercial uses. Iodoalkanes are used more in the laboratory where they are valued for their high reactivity.

most reactive substrate, followed by the 1° haloalkanes CH_3CH_2Br and $CH_3CH_2CH_2Br$. The least reactive substrate is the 2° halide $(CH_3)_2CHBr$. Furthermore, the trend in decreasing rates is paralleled by a trend in increasing activation enthalpies. As the carbon atom undergoing displacement becomes more highly substituted, the barrier to reaction increases (Fig. 6–1).

Very highly substituted alkyl halides yield complex results when subjected to nucleophilic substitution in acetone. Consider the 3° substrate 2-bromo-2-methylpropane [*t*-butylbromide, $(CH_3)_3CBr$)]. Its reaction with iodide ion is very slow. When chloride ion instead of iodide is used as the nucleophile, two reactions take place simultaneously—the expected substitution, and the competing **elimination** of HBr from the substrate to yield an alkene (eq. 6.6). We will comment on elimination reactions again in

(6.6)

$$(CH_3)_3C\text{—}\ddot{\underset{..}{Br}}\text{:} + \text{:}\ddot{\underset{..}{Cl}}\text{:}^- \xrightarrow[\text{acetone}]{\text{very slow}} (CH_3)_3C\text{—}\ddot{\underset{..}{Cl}}\text{:} + \text{:}\ddot{\underset{..}{Br}}\text{:}^-$$

Substitution products

$$(CH_3)_3C\text{—}\ddot{\underset{..}{Br}}\text{:} + \text{:}\ddot{\underset{..}{Cl}}\text{:}^- \xrightarrow[\text{acetone}]{\text{very slow}} CH_2{=}C(CH_3)_2 + H\text{—}\ddot{\underset{..}{Cl}}\text{:} + \text{:}\ddot{\underset{..}{Br}}\text{:}^-$$

Elimination products

Section 6:12, and discuss them in detail in Chapter 7. For now, it is enough to note that elimination often competes with substitution, especially with 3° substrates.

The rate constants and activation enthalpies in Table 6–1 show us that substitution at the reaction center strongly affects reactivity. This finding is

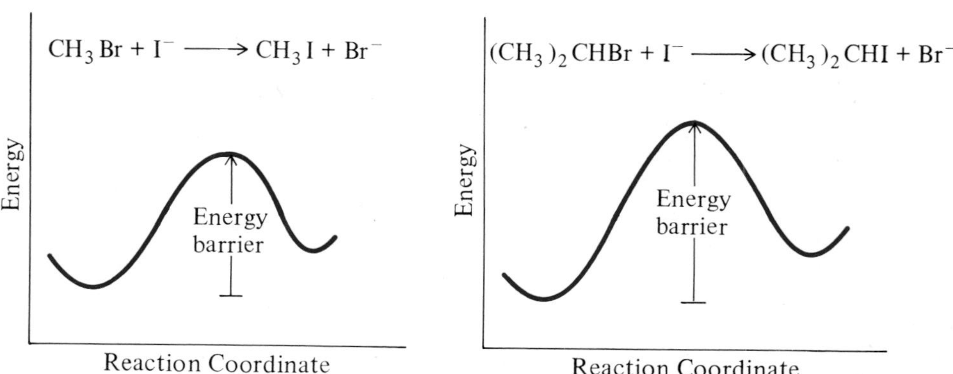

Figure 6–1. *Energy diagrams for the reactions of CH_3Br and $(CH_3)_2CHBr$ with I^- in acetone, showing the lower energy barrier to CH_3Br reaction.*

an important clue to the reaction mechanism. Additional mechanistic information can be obtained from further analysis of the kinetic data.

Problem

6–2 Arrange the following substrates in order of decreasing reactivity. Show the substitution products when each is treated with NaI in acetone.

a. 2-chlorobutane
b. 2-chloro-2-methylbutane
c. chlorocyclopentane
d. chloromethane
e. 1-chlorobutane

6:3 Kinetic and Stereochemical Studies of the Reactions Between Alkyl Halides and Halide Ions

The enthalpies of activation quoted in Table 6–1 were obtained from kinetic studies of the various reactions. From these studies we also obtain the rate equation for each reaction (eq. 6.7; also see sec. 5:9).

$$\text{rate} = k[A]^x[B]^y \qquad (6.7)$$

For the reactants shown in Table 6–1, both coefficients, x and y, were found to equal one, so the reaction is first order in each reactant and second order overall. With these reactions there is a correlation between the rate constant and the structure of the alkyl group. *For second-order nucleophilic substitutions in which the leaving group is constant, the usual order of alkyl halide reactivity is as follows:*

$$\underset{\text{Methyl halide}}{CH_3X} > \underset{1^\circ\text{ Halide}}{RCH_2X} > \underset{2^\circ\text{ Halide}}{R_2CHX} \gg \underset{3^\circ\text{ Halide}}{R_3CX}$$

Thus far our examples of second-order substitutions have been confined to the reactions between iodide ion and bromoalkanes. However, the second-order rate equation (eq. 6.8) applies to all reactions of halide ions (F^-, Cl^-, Br^-, I^-) with haloalkanes (RCl, RBr, RI; RF is too unreactive) in acetone and similar solvents (secs. 6:6 and 6:10). They all support the generalization that the more highly substituted the carbon undergoing substitution, the slower the reaction.

$$\text{rate} = k[\text{RBr}][I^-] \qquad (6.8)$$

The kinetic results for these substitutions could be explained by a **concerted** mechanism—one in which *bond-breaking and bond-making take place simultaneously* (eq. 6.3). Both the nucleophile and the substrate have to take part in this one step, so the mechanism is classified as *bimolecular*. The fact that the concentrations of both nucleophile and substrate appear in the empirical rate equation (eq. 6.8) lends support to the single step, bimolecular mechanism. Further evidence in support of the mechanism comes from stereochemical studies of the substitution reaction. These studies use either cycloalkyl halides (Chap. 3) or chiral alkyl halides (Chap. 4) as substrates, and tell us a good deal about the geometry of the transition state.

Whether we use a cycloalkyl halide (eq. 6.9) or a chiral alkyl halide (eq. 6.10), second order nucleophilic substitutions give the same stereochemical results—**inversion of configuration**.

(6.9)

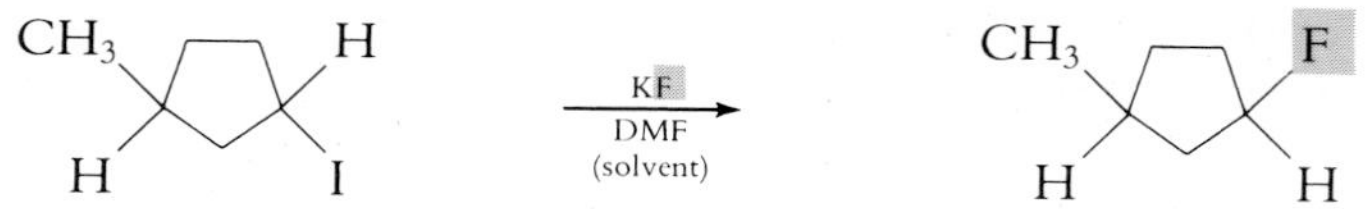

trans-1-Iodo-3-methylcyclopentane → (KF, DMF (solvent)) cis-1-Fluoro-3-methylcyclopentane (inverted configuration)

(6.10)

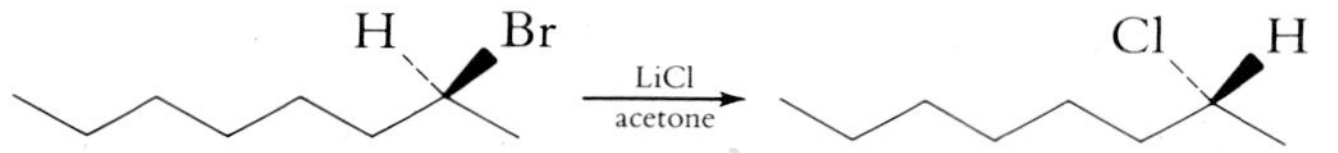

(S)-2-Bromooctane → (LiCl, acetone) (R)-2-Chlorooctane (inverted configuration)

In Equation 6.9 the relative configuration changes from *trans* in the substrate to *cis* in the product. (The structure of the solvent DMF is shown in Table 6–7, sec. 6:10.) In Equation 6.10 the spatial orientation of the substituents around the chiral center is reversed as one goes from substrate to product. As you can see from the projection formulas, the relative configuration of the product is inverted compared to the reactant. The absolute configuration also changes during this reaction from *S* to *R* because inversion has occurred while the priority order of the substituents remained the same.

The reactions shown in Equations 6.9 and 6.10 are **stereospecific**. A stereospecific reaction is one in which each stereoisomer of the starting materials always yields a specific stereoisomer of the product. For instance, if the starting material in Equation 6.10 had been (*R*)-2-bromooctane, then the product would have been (*S*)-2-chlorooctane.

Example

What are the structure and stereochemistry of the product in the following reaction?

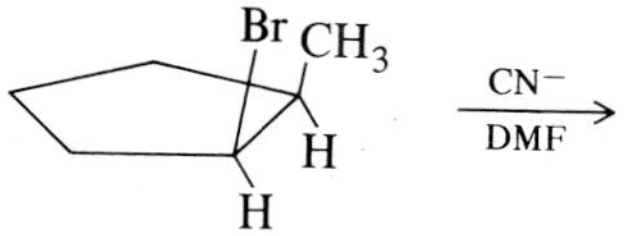

cis-1-Bromo-2-methylcyclopentane → (CN^-, DMF)

Answer

The substrate, *cis*-1-bromo-2-methylcyclopentane, contains a good leaving group, Br^-, attached to a 2° carbon atom. In addition, the reaction mixture contains a good nucleophile, CN^-. These conditions favor a bimolecular nucleophilic substitution reaction. This reaction characteristically involves inversion of configuration, so we expect to obtain the *trans* product. The nucleophile will directly displace the leaving group as follows:

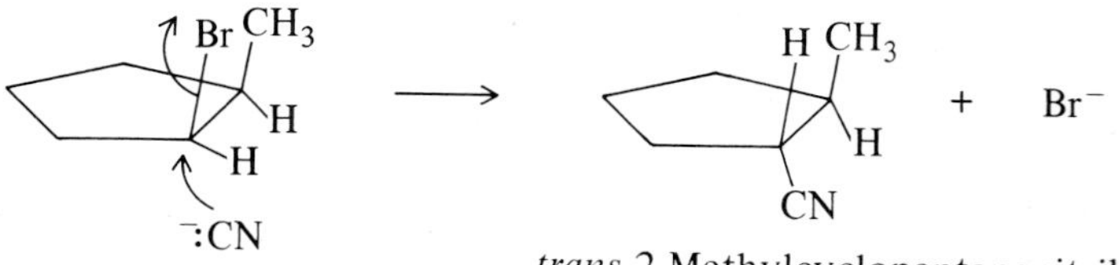

$^-$:CN + cis-1-bromo-2-methylcyclopentane → *trans*-2-Methylcyclopentanenitrile + Br^-

Table 6–2. Nucleophilic Displacements by Some Inorganic Anions

CH_3—Br: + :CN^- → CH_3—CN + :Br:$^-$ Cyanide ion → Acetonitrile (*methyl cyanide*)		(Chap. 17)
CH_3CH_2—I: + :N_3^- → CH_3CH_2—N_3 + :I:$^-$ Azide ion → Azidoethane (*ethyl azide*)		(Chap. 21)
Cyclopentyl—Cl: + :SH$^-$ → Cyclopentyl—SH + :Cl:$^-$ Hydrosulfide ion → Cyclopentanethiol (*cyclopentyl mercaptan*)		(Chap. 21)
$CH_3(CH_2)_2$—Br: + :OH$^-$ → $CH_3(CH_2)_2$—OH + :Br:$^-$ Hydroxide ion → 1-Propanol (*n-propyl alcohol*)		(Chap. 6)

Problem

6–3 Show the structure and stereochemistry of the product when each of the following substrates is treated separately with NaI in acetone and with KF in DMF.

a. *cis*-1-bromo-2-methylcyclohexane
b. *trans*-1-chloro-4-ethylcyclohexane
c. (*R*)-2-bromobutane
d. (*S*)-1-chloro-2-methylbutane

The fact that some nucleophilic substitution reactions proceed stereospecifically is not just an isolated observation. Rather, it is correlated with some of the other characteristic features of these reactions. In particular, *all nucleophilic displacement reactions that display inversion of configuration also follow second-order kinetics*. We can see, then, that there is an intimate connection among kinetic order, stereochemical change, and structural effect on reaction rate. By taking account of the relationship among *all* of these factors, it becomes possible to construct a very convincing mechanism for these reactions, including a model of their transition states, which will be done in the next section.

Finally, there are a number of inorganic ions that behave like halide ions toward organic substrates. When salts of these inorganic ions are allowed to react with alkyl halides, substituted alkanes are produced. A few representative reactions are given in Table 6–2 (note chapters in which these reactions are discussed in more detail.)

6:4 The Transition State for Second-Order Nucleophilic Substitution Reactions

We've seen that the rate law for second order nucleophilic substitutions is compatible with a single step, bimolecular mechanism. The enthalpies and entropies of activation are also in accord with this mechanism. For the reactions listed in Table 6–1 the enthalpies of activation are moderate, lying between 12 and 20 kcal/mol. These values reflect the fact that the energetic cost of breaking the C—Br bonds is partly compensated for by the energy released during formation of the C—I bond. The entropies of activation are on the order of −10 cal/mol °K, consistent with a transition state more ordered, and therefore less probable, than the reactants (sec. 5:11). A transition state formed from two independent reactant molecules would fit this description.

We can imagine the reaction taking place in the following way. Iodide ions collide with the alkyl bromide in a random fashion. When an iodide approaches from behind the carbon bearing the bromine, the iodide orbitals can begin to overlap with the small lobe of the carbon sp^3 orbital and initiate bond formation (eq. 6.11). As the I—C bond is being formed, the C—Br bond is being broken. That the nucleophile must approach the substrate on the side *opposite* the leaving group is shown by the result of stereochemical studies. Inversion of configuration at the reacting center is always observed in such reactions.

(6.11) $$I^{-} + \text{C—Br} \longrightarrow \text{I—C} + :Br^{-}$$

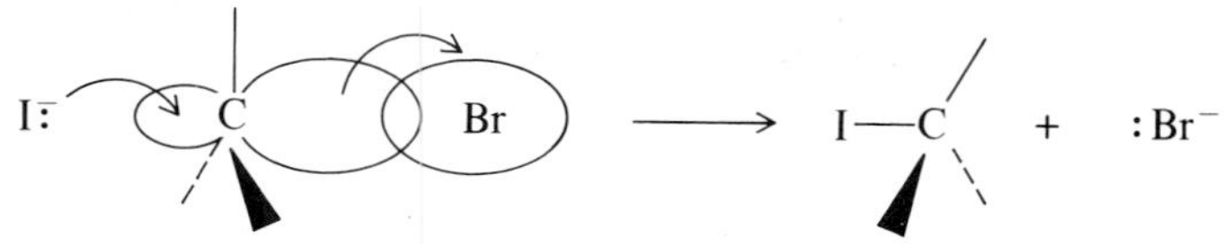

Figure 6–2 shows how backside attack accounts for the inversion of configuration observed in the reaction in Equation 6.9. We will assume that inversion is characteristic of all bimolecular nucleophilic substitutions.

We can now put together the kinetic, thermodynamic, and stereochemical data to create a general picture of the transition state in second order nucleophilic substitutions. The reaction of I^- with CH_3Br (Fig. 6–3) will be taken as a representative example. The incoming nucleophile (I^-) and the leaving group (Br^-) are both partially bonded to carbon, sharing the negative charge. Although the carbon atom in the transition state exceeds its normal valency, the total number of its valence shell electrons is equal to a normal octet.

Problem

6–4 Draw an energy diagram for the reaction of chloride ion with 1-bromoethane. Show the location of the reactants, products, and transition state. Draw a likely structure for the transition state.

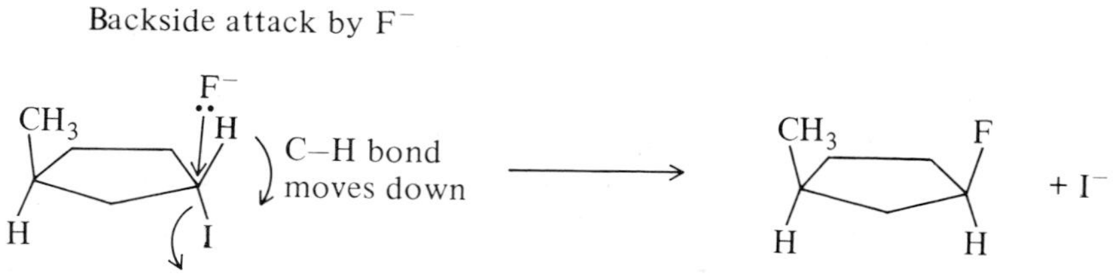

Figure 6–2. *Conversion of* trans-*1-iodo-3-methylcyclopentane to* cis-*1-fluoro-3-methylcyclopentane. The inversion of configuration is accounted for by backside attack of F^- on the C—I bond.*

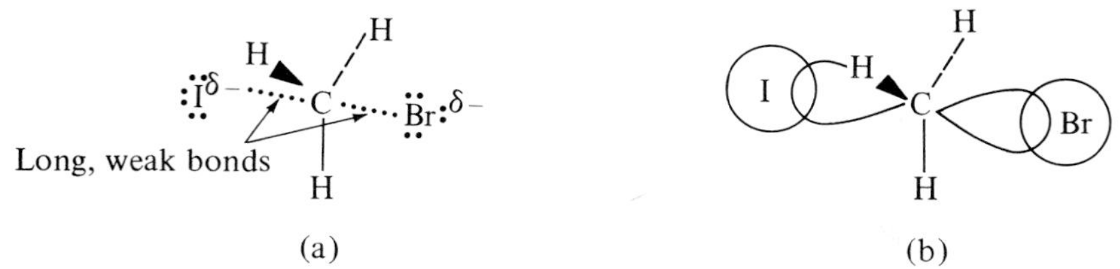

Figure 6–3. *(a) Transition state for the reaction of I^- with CH_3Br. (b) Carbon and halogen bonding orbitals for the transition state.*

6:5 The S_N2 Reaction: A Summary of Its Features

The reactions between halide ions and alkyl halides in acetone are all bimolecular; they have been classified as S_N2—**substitution, nucleophilic, bimolecular.**★ The transition state and reaction coordinate diagram for a typical S_N2 reaction are pictured in Figure 6–4, which we will use to explain some of its experimental aspects.

★ S_N2 reactions are examples of concerted processes in which bond breaking and bond making are taking place simultaneously.

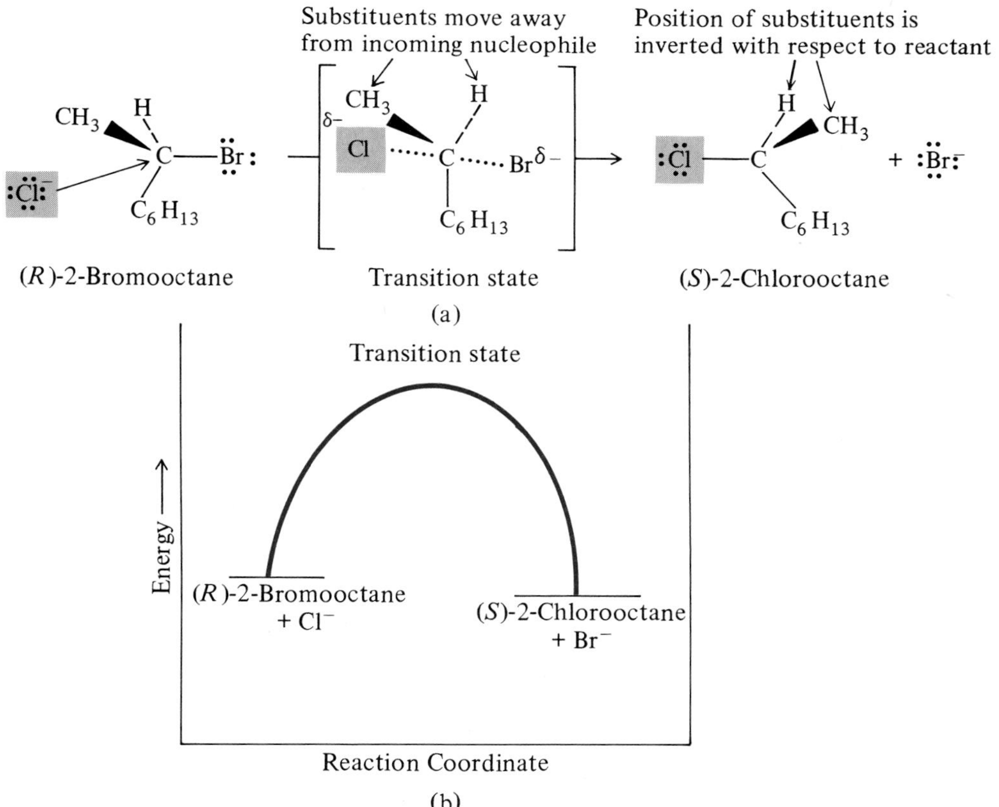

Figure 6–4. *(a) Inversion of configuration in the reaction of Cl^- with 2-bromooctane. The transition state is formed by backside nucleophilic attack. (b) Reaction coordinate diagram. This concerted reaction involves no intermediate.*

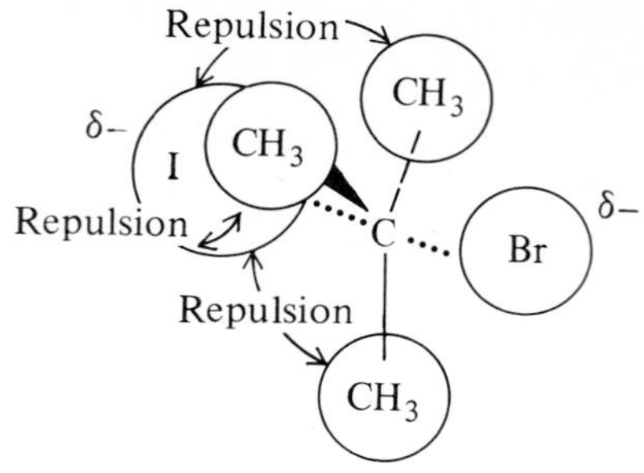

Transition state

Figure 6–5. *Steric hindrance to the reaction of I^- with* t-butyl bromide. *The transition state is very crowded, resulting in repulsion between the halogen and methyl groups. This repulsion raises* $\Delta H^{\ddagger}$ *markedly and thereby slows the reaction significantly.*

An essential feature of the transition state is the approach of the nucleophile to the substrate on the side opposite the leaving group. Note particularly what happens to the substituents—H, CH_3, C_6H_{13}—attached to the carbon atom undergoing substitution. They "flip" from one side of the carbon to the other, like the ribs of an umbrella turning inside out in a high wind.

The reactivity order of alkyl halides—$CH_3X > RCH_2X > R_2CHX \gg R_3CX$ (sec. 6:2)—can be explained on the basis of the S_N2 transition state. The transition state is one in which carbon is bonded (at least partially) to five groups, making it more crowded than the substrate. This means that bulky groups, such as CH_3, will destabilize the transition state even more than they destabilize the substrate. Reactions that go by way of crowded transition states are therefore very slow. When a reaction is slowed by crowding in the transition state, the effect is called **steric hindrance** (Fig. 6–5).

Problem

6–5 Neopentyl bromide (1-bromo-2,2-dimethylpropane) is a 1° bromide. Despite that, it reacts very slowly with iodide ion in acetone (eq. 1).

$$CH_3\overset{\displaystyle CH_3}{\underset{\displaystyle CH_3}{\overset{|}{\underset{|}{C}}}}CH_2Br + I^- \xrightarrow[\text{acetone}]{\text{very slow}} CH_3\overset{\displaystyle CH_3}{\underset{\displaystyle CH_3}{\overset{|}{\underset{|}{C}}}}CH_2I + Br^- \quad (1)$$

The kinetics of this reaction are bimolecular, however, and in most ways it behaves like a typical S_N2 reaction. How would you explain the low rate? (It may be helpful to construct a model.) Note that the related molecule, $(CH_3)_2CHCH_2Br$, also reacts more slowly than other 1° halides (Table 6–1).

The characteristics of S_N2 reactions can be summed up as follows: they are concerted (single step), bimolecular nucleophilic substitutions that proceed with inversion of configuration. In the next section we shall discuss a group of nucleophilic displacements that exhibit different kinetic and stereochemical behavior from that of S_N2 reactions.

6:6 How Substitution Mechanisms Are Affected by the Solvent: The S_N1 Mechanism

Up to now we have treated the solvent as nothing but a medium that allows reactants to interact. We find, in fact, that the solvent can substantially influence the reaction mechanism and in many cases serve as a reactant as well. This is particularly true when the solvent is *hydroxylic,* which means that it contains a hydroxyl group (OH). It is common in nucleophilic substitutions to use a mixture of a hydroxylic and a nonhydroxylic solvent, such as water and acetone. This procedure allows us to achieve maximum solubility for both the inorganic salt and the organic substrate.

Equation 6.12 shows the reaction between sodium azide and bromomethane in a solvent containing water and acetone (aqueous acetone). Note

(6.12)
$$Na^+N_3^- + CH_3{-}Br \xrightarrow[\text{acetone}]{H_2\ddot{O}:} \underset{\substack{\text{Azidomethane}\\(\textit{methyl azide})}}{CH_3{-}N_3} + Na^+Br^-$$

that the only substitution product is azidomethane, even though water contains nonbonding electrons and is therefore a potential nucleophile. Azide ion is a much more powerful nucleophile than water (sec. 6:10), which accounts for the fact that no CH_3OH, the product of nucleophilic attack by water, is isolated. The reaction in Equation 6.12 follows second-order kinetics (eq. 6.13) and shows all the characteristics of an S_N2 reaction.

(6.13)
$$\text{rate} = k[CH_3Br][N_3^-]$$

If we now use *t*-butyl bromide as the substrate instead of methyl bromide, the results bring some surprises. In aqueous acetone, CH_3Br and $(CH_3)_3CBr$ react at comparable rates, while in pure acetone CH_3Br is far more reactive (sec. 6:2). Furthermore, the major product from the *t*-butyl bromide reaction is *t*-butyl alcohol, $(CH_3)_3COH$ (eq. 6.14). Thus, for this substrate, water turns out to be the major nucleophile, rather than azide ion.

(6.14)
$$(CH_3)_3C{-}Br + NaN_3 \xrightarrow[\text{acetone}]{H_2\ddot{O}:} \underbrace{(CH_3)_3C{-}N_3 + NaBr}_{(5\%)} + \underbrace{(CH_3)_3C{-}OH + HBr}_{(95\%)}$$

Equally surprising is the experimental rate equation we derive for the reaction in Equation 6.14. Since water is present in large excess, it is very difficult to determine the kinetic order with respect to water. We can, however, determine the kinetic order for formation of *t*-butyl azide. We find that the reaction is first order in *t*-butyl bromide (substrate) and *zero order* in the nucleophile, azide ion (eq. 6.15; see sec. 5:9).

(6.15)
$$\text{rate} = k[(CH_3)_3CBr]$$

The experimental data strongly suggest that 3° substrates such as *t*-butyl bromide do not react by way of the S_N2 mechanism in aqueous solvents. In order to discover just what mechanism is operating, let's look at some stereochemical results of a reaction between a chiral 3° halide and water. Since the water is both solvent and reactant, the reaction is called a **solvoly-**

sis (eq. 6.16). The most striking feature of this solvolysis is that the resulting

(6.16) (R)-Chloride (100%) $\xrightarrow{H_2\ddot{O}:}$ (R)-Alcohol (40%) + (S)-Alcohol (60%) + H^+ + $:\ddot{Cl}:^-$

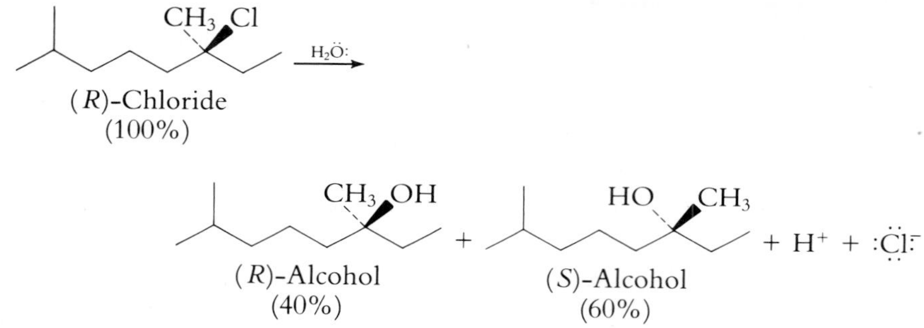

alcohol mixture is nearly racemic. The reaction is only slightly stereospecific. You should contrast this result with the stereochemical course of S_N2 reactions (eqs. 6.9 and 6.10), which are completely stereospecific.

S_N2 reactions are stereospecific because they are concerted—bond-breaking and bond-making are taking place simultaneously, thus fixing the relationship of the nucleophile and leaving group. We can account for the first-order kinetics and the low stereospecificity of 3° halide solvolysis by proposing that bond-making and bond-breaking are *not* simultaneous, that these two processes take place in *separate* steps. We would then have a multistep mechanism in which *a reactive intermediate is formed and then undergoes further reaction* (sec. 5:1).

The intermediate which best fits the data is a **carbocation**, or **carbonium ion**. Carbocations are organic ions that contain a trivalent, positively charged carbon atom with only six electrons in its valence shell (sec. 1:3). Carbocations are formed in the first step of the mechanism, in which the bond to the leaving group is broken (eq. 6.17).

(6.17) $(CH_3)_3C-\ddot{Br}: \xrightarrow{\text{bond breaking}} (CH_3)_3C^+ + :\ddot{Br}:^-$

t-Butyl cation

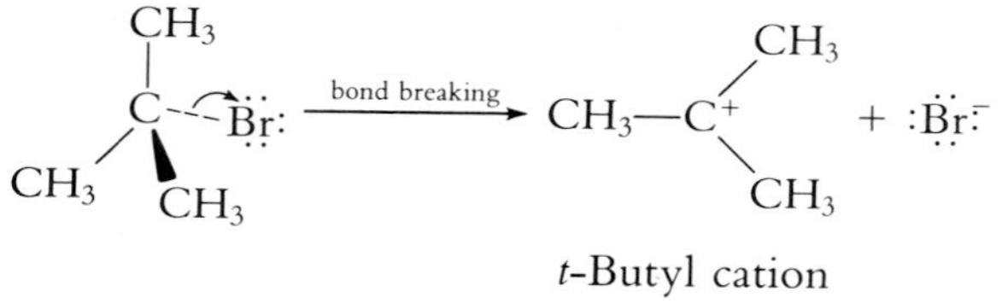

The carbon skeleton of the carbocation is planar, with an sp^2-hybridized central carbon atom carrying an empty *p* orbital (Fig. 6–6). Because of their

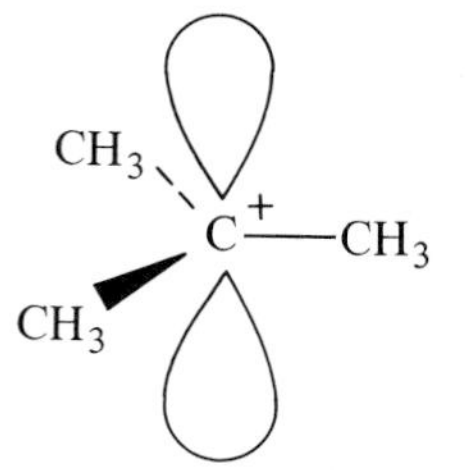

Figure 6–6. *The* t*-butyl cation. The trivalent carbon is* sp^2 *hybridized and the* p *orbital is empty.*

unfilled valence shells, carbocations are highly reactive, indiscriminately reacting with whatever nucleophiles are in their immediate vicinity. The *t*-butyl cation (eq. 6.17) is surrounded mainly by water molecules that stabi- [illegible] [illegible]al bond-making reaction is [illegible] to formation of *t*-butyl alcohol [illegible]. 6.14). A minority of the carbocations

[illegible]) $CH_3{-}\overset{+}{C}(CH_3)_2 + :\ddot{O}H_2 \xrightarrow{\text{bond making}} (CH_3)_3C{-}\overset{+}{\ddot{O}}H_2 \underset{\text{fast}}{\overset{\text{very}}{\rightleftharpoons}} (CH_3)_3C{-}\ddot{O}{-}H + H^+$

survive long enough to react with the much scarcer azide ion (eq. 6.19).

(6.19) $CH_3{-}\overset{+}{C}(CH_3)_2 + \bar{:}\ddot{N}{=}\overset{+}{N}{=}\ddot{N}\bar{:} \rightarrow (CH_3)_3C{-}:N{=}\overset{+}{N}{=}\ddot{N}\bar{:}$

The two-step mechanism nicely accounts for the experimental observations. Since the first step involves the formation of both an anion and a cation (eq. 6.17), it requires a solvent capable of solvating both ions. Water is much more effective at this than acetone (sec. 6:10), so the mechanism is observed in aqueous acetone but not in pure acetone. The low stereospecifity can be explained in terms of the geometry of the carbocation (Fig. 6–6), which has several symmetry planes and is therefore achiral (sec. 4:2). The nucleophile H_2O can attack either lobe of the *p* orbital equally well, giving both enantiomers of the product (Fig. 6–7). We will explain shortly why the product is not *completely* racemic.

Perhaps the most unusual aspect of the 3° halide reaction is that it follows first-order kinetics (eq. 6.15), in which the rate is determined only by substrate concentration. To see why the nucleophiles do not affect the reaction rate, we have to focus on the properties of the carbocations. These intermediates are extremely reactive and are therefore consumed almost as fast as they are formed. Thus, *the critical part of the mechanism is the first step,* in which the carbocations are generated by bond-breaking. The final products cannot be produced any faster than the rate at which carbocations are produced, no matter how much nucleophile is present. So the rate of the overall reaction is controlled by the rate of the first step, which is the **rate-determining step**. Because the nucleophiles are not involved in the rate-determining step, we don't find the concentrations of the nucleophiles in the rate equation (eq. 6.15). The rate-determining step is **unimolecular**—only one molecule takes part. For this reason reactions which follow the carbocation mechanism are labelled S_N1—**substitution**, **nucleophilic**, **unimolecular**.

In the S_N1 mechanism, *k* is smaller and $\Delta H^\ddagger$ is larger for the first step than for the second. Figure 6–8 shows an energy diagram for an S_N1 reaction. You can see that the major energy hurdle is the formation of the

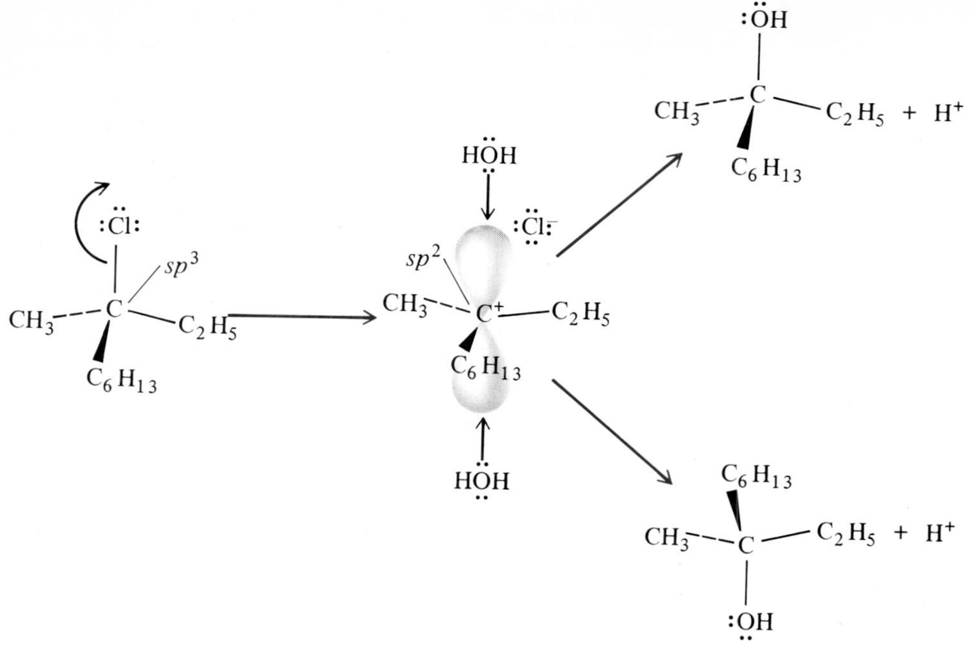

Figure 6–7. *Reaction of a 3° halide with H_2O to yield both enantiomers of the product.*

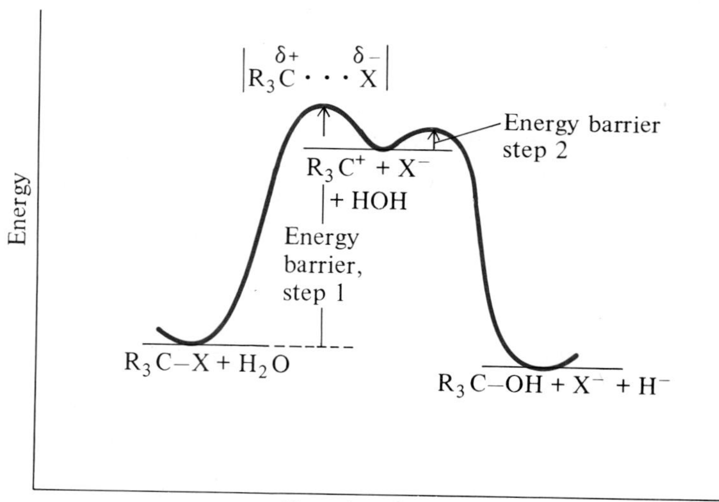

Figure 6–8. *Energy diagram for a two-step reaction with a carbocation intermediate. The first step is rate determining because it involves a much larger energy barrier than the second step.*

Table 6–3. Comparisons of S_N1 and S_N2 Reactions

Characteristic	S_N2	S_N1
Mechanism	concerted displacement	stepwise (carbocation)
Kinetics	second order (rate = $k[RX][Y^-]$)	first order (rate = $k[RX]$)
Solvent	nonpolar or polar	polar and hydroxylic
Stereochemistry	stereospecific (inversion)	nonstereospecific (racemization)
1° RX	fast	slow
2° RX	medium	medium
3° RX	very slow	fast

carbocation. Since the mechanism consists of two steps, there will be two transition states. Figure 6–8 shows the transition state for rate-determining carbocation formation. The C—X bond is stretched, and positive and negative charges are building up on C and X, respectively.

S_N1 reactions are not restricted to water. They can take place in any solvent that solvates both carbocations and leaving groups. Solvents containing hydroxyl groups, such as alcohols (eq. 6.20) and acids (RCOOH), are usually effective S_N1 solvents.

(6.20) 1-chloro-1-methylcyclopentane $\xrightarrow[S_N1\ \text{reaction}]{CH_3\ddot{O}H}$ 1-methoxy-1-methylcyclopentane + H^+ + Cl^-

Table 6–3 compares the major characteristics of S_N1 and S_N2 reactions. There are two important factors listed there with which we have not yet dealt—why 3° substrates favor the S_N1 mechanism, and what the preference of 2° substrates is. The next section addresses these questions.

Example

What products would be formed from the following reaction?

1-Bromo-1-methylcyclohexane $\xrightarrow{C_2H_5OH}$

Answer

These are typical S_N1 conditions and hence, after the carbocation is formed, it can react with any nucleophiles present:

$$\text{1-bromo-1-methylcyclohexane} \longrightarrow \text{(3° carbocation)} + Br^- \xrightarrow{C_2H_5\ddot{O}H} \text{(oxonium ion, } \overset{+}{O}(H)C_2H_5\text{)} \longrightarrow$$

$$\text{(1-methylcyclohexyl)}\ddot{O}C_2H_5 + H^+$$

1-Ethoxy-1-methylcyclohexane

The substrate, 1-bromo-1-methylcyclohexane, has a good leaving group, Br^-, attached to a 3° carbon atom. Since the solvent, C_2H_5OH, is hydroxylic, conditions favor an S_N1 reaction. The substrate dissociates to a 3° carbocation, which then reacts with any nucleophiles present. The solvent itself is a nucleophile, since it has nonbonding electron pairs on the —$\ddot{O}$H group.

Problems

6–6 3-Chloro-3-methylpentane is dissolved in a mixture of ethanol (C_2H_5OH) and water containing sodium azide (eq. 2). The chloro compound reacts to give three substitution products. Draw the structures of these products and a mechanism for their formation.

$$CH_3CH_2\overset{CH_3}{\underset{Cl}{C}}CH_2CH_3 \xrightarrow[H_2O,\ C_2H_5OH]{Na^+N_3^-} \text{products} \qquad (2)$$

6–7 By increasing the concentration of NaN_3 in the reaction in Equation 6.14 we can increase the proportion of *t*-butyl azide relative to that of *t*-butyl alcohol. However, the rate of formation of total product (azide plus alcohol) remains constant as the concentration of NaN_3 is increased. Explain both of these observations in terms of the S_N1 mechanism.

6–8 a. When 2-bromo-2-methylbutane is warmed in acetic acid (CH_3COOH), it is converted to 2-methyl-2-butyl acetate (eq. 3). Suggest a probable mechanism for this reaction.

$$CH_3CH_2\overset{CH_3}{\underset{Br}{C}}CH_3 \xrightarrow{CH_3\overset{:O:}{\overset{\|}{C}}\ddot{O}H} CH_3CH_2\overset{CH_3}{\underset{:\ddot{O}\underset{\underset{:O:}{\|}}{C}CH_3}{C}}CH_3 + HBr \qquad (3)$$

2-Methyl-2-butyl acetate

b. Adding a small quantity of methanol (CH_3OH) to the acetic acid results in less 2-methyl-2-butyl acetate being formed. In its place a new product appears—2-methoxy-2-methylbutane (**I**). The addition of methanol does not affect the rate of disappearance of the alkyl bromide. Show how your mechanism from (a) can account for the formation of **I** and for the kinetic results.

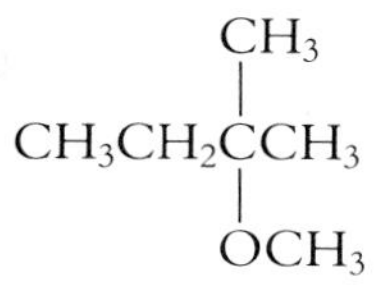

$$\underset{\textbf{I}}{CH_3CH_2\overset{CH_3}{\overset{|}{C}}(OCH_3)CH_3}$$

6–9 Write the substitution products from each of the following reactions. Show the structures of the intermediate carbocations.

a. 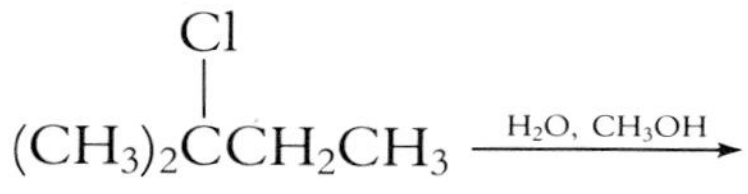 $(CH_3)_2C(Cl)CH_2CH_3 \xrightarrow{H_2O,\ CH_3OH}$

b. 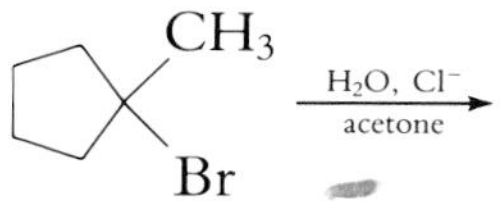 1-bromo-1-methylcyclopentane (CH_3, Br) $\xrightarrow[\text{acetone}]{H_2O,\ Cl^-}$

c. 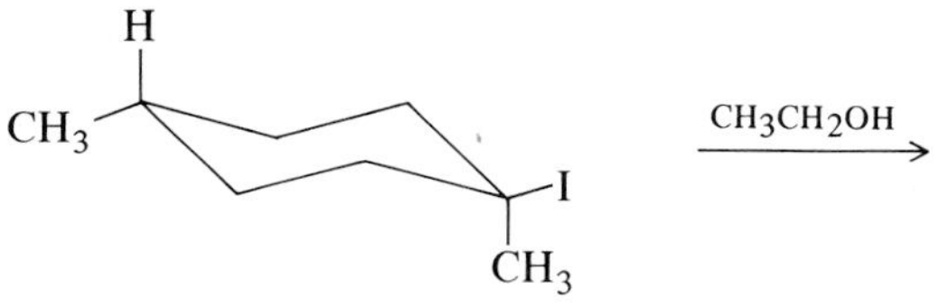 cyclohexane ring with CH_3 and H on one carbon, I and CH_3 on another $\xrightarrow{CH_3CH_2OH}$

d. 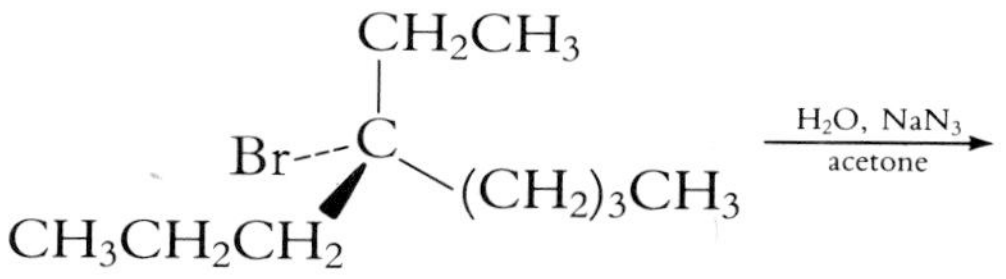 Br---C(CH_2CH_3)($CH_2)_3CH_3$)($CH_3CH_2CH_2$) $\xrightarrow[\text{acetone}]{H_2O,\ NaN_3}$

e. 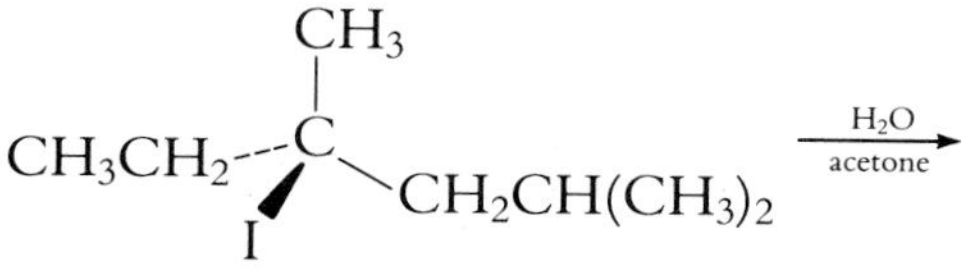 CH_3CH_2---C(CH_3)(I)$CH_2CH(CH_3)_2$ $\xrightarrow[\text{acetone}]{H_2O}$

6:7 How Carbocation Stability Affects the S_N1 Reaction

As we mentioned in the last section, carbocations are very reactive and therefore usually have very short lifetimes in solution. The energy required to form carbocations in the gas phase can be measured in a mass spectrometer.

Figure 6–9 shows the relative gas phase energies of a series of carbocations, R_3C^+, where R is either H or CH_3. You can see that ion stability increases as the degree of substitution increases. This is the same trend that was found for alkyl radicals (sec. 5:7). The effect of replacing H by CH_3 is much more dramatic for carbocations than it was for radicals. There is a difference of 70 kcal/mol between CH_3^+ and $(CH_3)_3C^+$.

Why does replacement of H's by CH_3 groups cause such an increase in stability? It is because the electrons in an alkyl group, such as CH_3, are more polarizable than those around a single hydrogen atom. (Polarizability refers to the tendency of electrons to be attracted to or repelled by neighboring

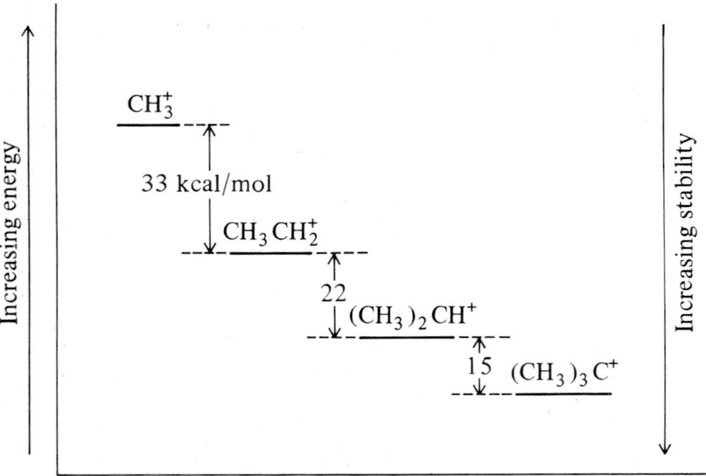

Figure 6–9. *Gas phase energy differences among carbocations.*

positive or negative charges.) Their polarizability allows the C—C and C—H bonding electrons of the alkyl groups to be pulled closer to the positively charged trivalent carbon. As this happens some of the positive charge of the trivalent carbon is *delocalized* onto the alkyl groups (Fig. 6–10). Carbocations in which charge is delocalized are lower in energy and therefore more stable than carbocations in which charge is concentrated on one atom. Charge delocalization by means of σ-bond polarization is known as the **inductive effect** (Fig. 6–10). Since alkyl groups exert a stronger inductive effect than do hydrogens, alkyl substitution increases the stability of carbocations (Fig. 6–11).

There is an additional contributory factor that favors formation of 3° carbocations over 2° ones, and formation of 2° carbocations over 1° ones. That factor is relief of steric strain (sec. 5:7). Loss of a leaving group from a 3° carbon atom results in a substantial decrease in crowding. That decrease is progressively less when the leaving group is lost from a 2° or 1° carbon atom. The steric strain effect and the inductive effect both favor 3° carbocation formation most, and 1° carbocation formation least. Of the two, the inductive effect is usually the more important.

In Section 6:6 you saw that 3° carbocations are formed much more

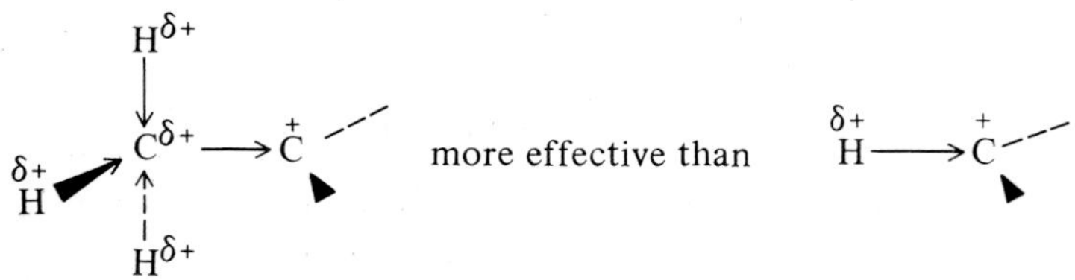

Figure 6–10. *A positively charged carbon polarizing the C—C and C—H bonds of its substituent groups (the inductive effect).*

Methyl cation	1° Carbocation	2° Carbocation	3° Carbocation
$H{-}\overset{H}{\underset{+}{C}}{-}H$	$\overset{\delta+}{CH_3}\rightarrow\overset{H}{\underset{+}{C}}{-}H$	$\overset{\delta+}{CH_3}\rightarrow\overset{H}{\underset{+}{C}}\leftarrow\overset{\delta+}{CH_3}$	$\overset{\delta+}{CH_3}\rightarrow\overset{\overset{\delta+}{CH_3}\downarrow}{\underset{+}{C}}\leftarrow\overset{\delta+}{CH_3}$

INCREASING SUBSTITUTION →
INCREASING INDUCTIVE STABILIZATION

Figure 6–11. *Inductive stabilization of carbocations by alkyl substituents.*

rapidly than 1° carbocations. In this section we have shown that 3° carbocations are much more stable than 1° carbocations. As you may have guessed, there is a correlation between the rate at which a carbocation is formed and its stability. We found such a rate/stability correlation once before, when investigating free-radical reactivity (sec. 5:12). What free radicals and carbocations have in common is that they are relatively high energy intermediates, formed in *endothermic* reactions. Because of this circumstance, more stable carbocations (or radicals) are formed faster than less stable carbocations (or radicals). By examining the appropriate reaction energy diagrams we can explain this trend.

Figure 6–12 shows energy diagrams for the formation of 1° and 3° carbocations from standard substrates. The 1° cation is the less stable of the two. Note that the transition state leading to the 1° cation also lies above the

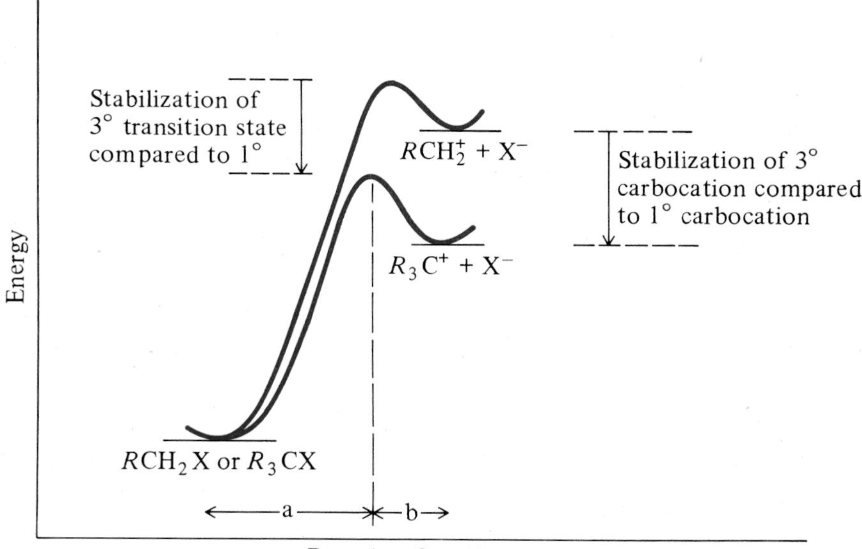

Figure 6–12. *Energy diagram for ionization of 1° and 3° alkyl halides. The transition states occur far along the reaction coordinate and so resemble the carbocations in structure. The energy difference between transition states is close to the energy difference between carbocations.*

transition state leading to the 3° cation. The transition state relationship explains the rate relationship. Since the 1° transition state is less stable than the 3° transition state, it requires more energy to reach the 1° transition state. Thus, 1° carbocations form more slowly than 3° carbocations.

Why is the 1° transition state less stable than the 3° one? Here is where the endothermic character of the reaction comes into play. For endothermic processes, the transition state is reached late in the reaction—on the reaction coordinate of Figure 6–12, distance a ≫ distance b. Since the transition state is so close to the product, we conclude that it strongly resembles the product in structure. Thus, a high energy product like a 1° carbocation can only be reached via a high energy transition state. The more stable 3° carbocation, by contrast, will require a less energetic transition state. For endothermic reactions in general, structural changes that affect the energy of the product will also strongly affect the stability of the transition state. The American chemist George S. Hammond applied these ideas to various organic reactions and set forth the general principle which states: *in highly endothermic reactions the transition state will be like the products* (the *Hammond postulate*). The more endothermic the reaction, the more valid is this postulate.

Finally, we need to say a word about secondary alkyl substrates. In hydroxylic solvents like H_2O and ROH, secondary alkyl halides can dissociate to carbocations. However, as Figure 6–11 shows, 2° carbocations have substantially more energy than 3° carbocations. As a result, a secondary substrate tends to dissociate to an **ion pair** (**1**) in which the carbocation is stabilized by staying close to the anionic leaving group (eq. 6.21). The ion

(6.21)

CH_3, H, Br, C_6H_{13} — (*R*)-2-Bromooctane → {CH_3, H, C^+ Br^-, C_6H_{13}} Ion pair, 1 —EtOH, inversion→ CH_3, EtO, C, H, C_6H_{13}, 2 — *S* (major product)

Ion pair, 1 ↓ Free carbocation (CH_3, C^+, H, C_6H_{13}) —EtOH, racemization→ *S* (major product) and CH_3, H, C, OEt, C_6H_{13}, 3 — *R* (minor product)

pair can then react further in two ways. It can be attacked directly by a nucleophile at the backside, giving only the product of inverted configuration. If a nucleophile doesn't attack, then the ion pair will dissociate to a free carbocation. The carbocation reacts with the nucleophile in the usual way, giving enantiomeric products of both inverted (**2**) and retained (**3**) configuration (remember that Et stands for CH_3CH_2).

Ion pairing is a general phenomenon that occurs even with 3° carbocations. That is why tertiary alkyl halides do not give completely racemic products. Equation 6.16 shows a chiral 3° alkyl chloride hydrolyzing to an alcohol product, 20% of which is formed by inversion. This 20% probably arises by nucleophilic attack on an ion pair.

6:8 Carbocation Rearrangements

A common property of carbocations is their tendency to undergo **rearrangement**. Equation 6.22 gives an example. A 1° carbocation rearranges to a 2° carbocation by transfer of a hydrogen, with its bonding electrons, from

$$\text{(6.22)}\quad \underset{\text{1-Propyl cation}}{CH_3{-}\overset{H}{\overset{|}{C}}H{-}\overset{+}{C}H_2} \xrightarrow[\text{phase}]{\text{gas}} \underset{\text{2-Propyl cation}}{CH_3{-}\overset{+}{C}H{-}CH_3} \qquad \Delta H = -21.4\ \text{kcal/mol}$$

C-2 to C-1. Carbocation rearrangements are rapid, even compared to their rate of reaction with nucleophiles, when the arrangement leads to a more stable carbocation (as in eq. 6.22). Rearrangements also take place between carbocations of equal stability (2° → 2°, 3° → 3°). Rearrangement to produce a less stable carbocation is not observed when nucleophiles are present.

Rearrangements can also involve the shift of an alkyl group. For example, in Equation 6.23, a methyl group with its bonding electrons is being transferred:

$$\text{(6.23)}\quad \underset{2^\circ\ \text{Cation}}{CH_3{-}\underset{CH_3}{\overset{CH_3}{C}}{-}\overset{+}{C}HCH_3} \longrightarrow \underset{3^\circ\ \text{Cation}}{CH_3{-}\overset{CH_3}{\underset{+}{C}}{-}\underset{CH_3}{C}HCH_3}$$

When we find rearranged products from a nucleophilic substitution, we take it as evidence that carbocations were reaction intermediates such as in

$$\text{(6.24)}\quad CH_3\underset{CH_3}{\overset{CH_3}{C}}{-}\underset{I}{C}HCH_3 \rightarrow CH_3{-}\underset{CH_3}{\overset{CH_3}{C}}{-}\overset{+}{C}HCH_3 + I^- \rightarrow$$

$$CH_3{-}\overset{CH_3}{\underset{+}{C}}{-}\underset{CH_3}{C}HCH_3 \xrightarrow{H_2\ddot{O}:} \underset{\text{Rearranged product}}{CH_3\overset{CH_3}{\underset{OH}{C}}{-}\underset{CH_3}{C}HCH_3} + H^+$$

eq. 6.24. This is because carbocations are much more likely to rearrange than are free radicals or carbanions. If we can show that rearrangement would have involved an increase in carbocation stability, then we have good reason to believe that carbocations took part in the reaction. S_N2 reactions, by contrast, do not lead to products with rearranged carbon skeletons.

Box 6–1. Biological Effects of Alkylating Agents

Compounds that react with nucleophiles by S_N1 or S_N2 reactions are called alkylating agents because they transfer an alkyl group to the nucleophile.

$$Nu\colon^- + CH_3\text{—}I \rightarrow Nu\text{—}CH_3 + I^-$$

Alkylating agent

Biological organisms contain many nucleophilic sites on important molecules such as amino acids, proteins, and nucleic acids (DNA and RNA). Because of this, alkylating agents have powerful effects in living organisms. Methyl sulfate, for example, rapidly alkylates the nitrogen atom of one of the heterocyclic bases in DNA, thereby preventing DNA replication and ultimately cell growth and reproduction.

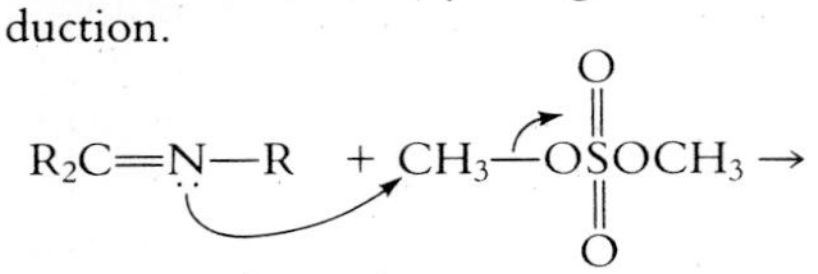

$$R_2C{=}\ddot{N}\text{—}R + CH_3\text{—}OSOCH_3 \rightarrow$$

(with O above and below the S of $OSOCH_3$)

Portion of DNA molecule — Methyl sulfate

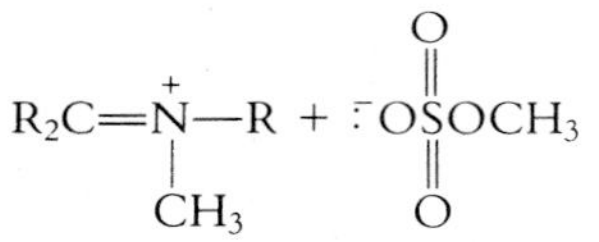

$$R_2C{=}\overset{+}{N}(CH_3)\text{—}R + {}^-\colon OSOCH_3$$

(with O above and below the S of $OSOCH_3$)

Alkylated DNA

This toxic effect of alkylating agents is useful in the treatment of some cancers. Rapidly growing cancer cells can be attacked by alkylating agents during certain stages of the mitotic cycle. Compounds like methyl iodide and methyl sulfate are too indiscriminate to be used as medicines, but other more complex alkylating agents have proven useful against cancer.

One of the first useful groups of chemotherapeutic alkylating agents were the *nitrogen mustards* such as mechlorethamine. (The noxious mustard gases of World War I were similar compounds containing sulfur rather than nitrogen.) More valuable (and less toxic) compounds such as cyclophosphamide still incorporate the nitrogen mustard moiety in their structures.

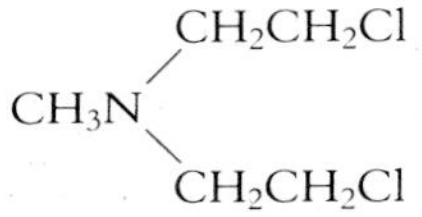

$$CH_3N(CH_2CH_2Cl)_2$$

Mechlorethamine

(Ring: —NH—P(=O)—O— in a six-membered ring; P bears $N(CH_2CH_2Cl)_2$)

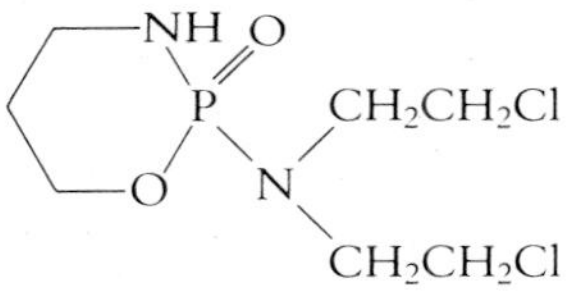

Cyclophosphamide

Problems

6–10 Propose a mechanism for the reaction shown in Equation 4.

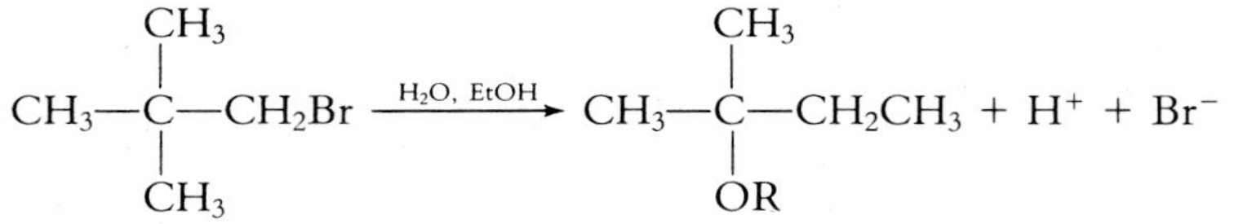

$$CH_3\text{—}C(CH_3)_2\text{—}CH_2Br \xrightarrow{H_2O,\ EtOH} CH_3\text{—}C(CH_3)(OR)\text{—}CH_2CH_3 + H^+ + Br^- \quad (4)$$

(R = H or Et)

Why didn't this reaction follow an S_N2 pathway (refer back to Prob. 6–5)?

6–11 4,4-Dimethylcyclohexyl tosylate is heated in acetic acid (tosylate, OTs^-, is a good leaving group, like Br^- or Cl^-; sec. 6:11). The major substitution product is 4,4-dimethylcyclohexyl acetate, but there is also a lesser amount of 3,3-dimethylcyclohexyl acetate (eq. 5). Suggest a mechanism that accounts for the formation of both products.

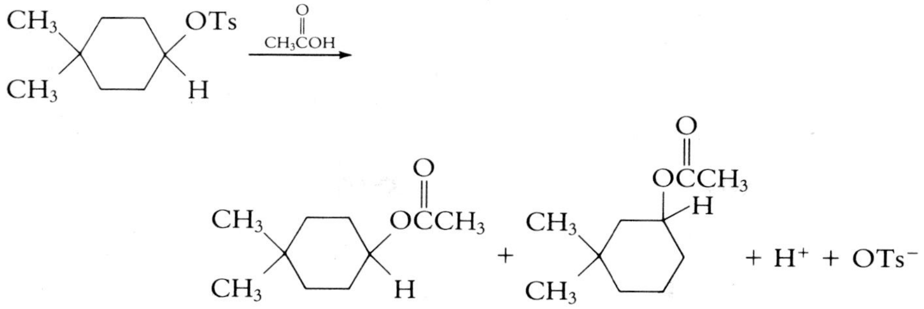

$$\text{(CH}_3)_2\text{C}_6\text{H}_9\text{OTs} \xrightarrow{\text{CH}_3\text{COOH}} \text{(CH}_3)_2\text{C}_6\text{H}_9\text{OCCH}_3 + \text{(CH}_3)_2\text{C}_6\text{H}_9\text{OCCH}_3 + \text{H}^+ + \text{OTs}^- \quad (5)$$

Most of the nucleophilic displacements that we have shown thus far have involved alkyl halides as both substrates and products. In the following sections the focus is on alcohols and ethers as substrates, products, and solvents in nucleophilic substitution. We've already noted that the choice of solvent can have a decisive influence on the mechanism of nucleophilic substitution. We'll carry this one step further to show how a solvent can be chosen to maximize the yield of the desired product in a substitution reaction.

6:9 Alcohols and Ethers: Their Names, Physical Properties, and Chemistry

There are both IUPAC and radicofunctional nomenclature systems for alcohols and for ethers. With alcohols, all but the simplest are named by the IUPAC method. In this system the final "e" of the hydrocarbon name is replaced by "ol," which denotes the presence of the functional group —OH. If there are several nonequivalent carbon atoms in the molecule, then a number is written immediately before the name to indicate the position of the hydroxyl group. The name of the alcohol is based on the longest carbon chain *that contains the hydroxyl group,* numbered so that the carbon atom bearing this group has the lowest possible number. Some examples of IUPAC names are given here.

CH_3OH	CH_3CH_2OH	$CH_3CH_2CH_2OH$	$CH_3CH(OH)CH_3$	$CH_3CH_2CH(OH)CH_3$
Methanol	Ethanol	1-Propanol	2-Propanol	2-Butanol

$CH_3CH(OH)CH(CH_2CH_2CH_3)CH_2CH_3$
3-Ethyl-2-hexanol

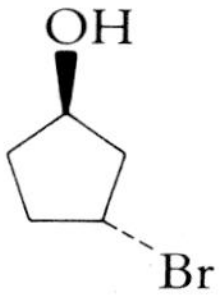

trans-3-Bromocyclopentanol

When more than one hydroxyl group is present in the compound, the hydrocarbon name is retained, the positions of the hydroxyls are numbered, and a suffix denoting the number of hydroxyls is appended.

$$\underset{\displaystyle OH}{CH_3\overset{|}{C}HCH_2OH}$$
1,2-Propanediol

$$\underset{\displaystyle OH}{HOCH_2\overset{|}{C}HCH_2OH}$$
1,2,3-Propanetriol

The radicofunctional system is similar to that employed for alkyl halides. The name of the alkyl radical is followed by "alcohol." This system is used only for low molecular weight compounds (usually C_1 to C_4).

$$\underset{\displaystyle OH}{CH_3\overset{|}{C}HCH_2CH_3}$$
s-Butyl alcohol

$$\underset{\displaystyle CH_3}{CH_3\overset{|}{C}HCH_2OH}$$
Isobutyl alcohol

$$CH_3-\underset{\displaystyle CH_3}{\overset{\displaystyle CH_3}{C}}-OH$$
t-Butyl alcohol

Table 6–4 gives the structures of some alcohols, their IUPAC and radicofunctional names, and their physical properties.

Ethers are oxygen-containing molecules whose oxygen atom is bonded to two alkyl or aryl groups. Ethers are symmetrical (ROR) if both groups are identical, or unsymmetrical (ROR′) if the groups are different. In the IUPAC system, one alkyl group and the oxygen atom are taken together as an *alkoxy* substituent on the other alkyl group. In unsymmetrical ethers, the alkoxy group contains the smaller alkyl group. Some examples are given here.

Table 6–4. Names and Physical Properties of Some Alcohols

IUPAC Name	Radicofunctional or Trivial Name	Structure	Boiling Point (°C)
methanol	methyl	CH_3OH	65
ethanol	ethyl	CH_3CH_2OH	78
1-propanol	*n*-propyl	$CH_3CH_2CH_2OH$	97
2-propanol	isopropyl	$(CH_3)_2CHOH$	82
1-butanol	*n*-butyl	$CH_3(CH_2)_2CH_2OH$	118
2-butanol	*s*-butyl	$\underset{\displaystyle OH}{CH_3CH_2\overset{|}{C}HCH_3}$	100
2-methyl-1-propanol	isobutyl	$(CH_3)_2CHCH_2OH$	108
2-methyl-2-propanol	*t*-butyl	$(CH_3)_3COH$	83
1-pentanol	*n*-amyl	$CH_3(CH_2)_3CH_2OH$	138
2-pentanol		$\underset{\displaystyle OH}{CH_3(CH_2)_2\overset{|}{C}HCH_3}$	119
3-pentanol		$(CH_3CH_2)_2CHOH$	116
1-octanol	*n*-octyl	$CH_3(CH_2)_7OH$	195
1-dodecanol	*n*-dodecyl	$CH_3(CH_2)_{11}OH$	257
1,2-ethanediol	ethylene glycol	$HOCH_2CH_2OH$	198

$CH_3CH_2OCH_2CH_3$ Ethoxyethane (*diethyl ether*)

$CH_3CH_2CH_2OCH_2CH_3$ 1-Ethoxypropane (*ethyl propyl ether*)

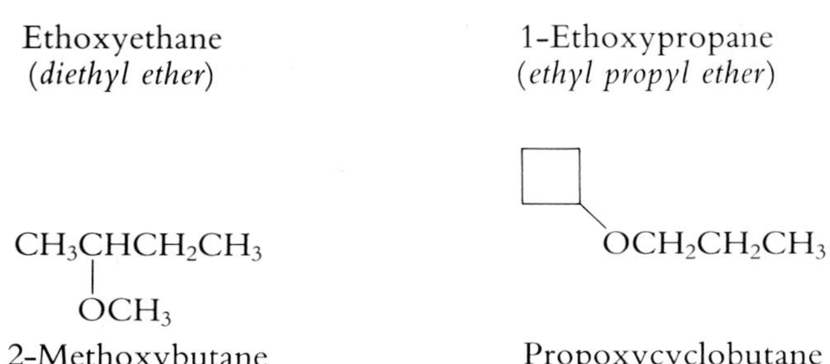

2-Methoxybutane

Propoxycyclobutane

For the simplest ethers, we use the radicofunctional naming system (IUPAC names for these ethers are used in certain compendium indexes). The alkyl group on either side of the oxygen is named and "ether" is added. Thus, $CH_3OCH_2CH_3$ is ethyl methyl ether and $CH_3CH_2OCH_2CH_3$ is diethyl ether. Diethyl ether is so widely used that it is often known simply as ether and is abbreviated Et_2O. Some important cyclic ethers have their own IUPAC names, but their trivial names are still in common use. These cyclic ethers and their physical properties are shown in Table 6–5, along with those for several simple straight-chain ethers.

Table 6–5. Names and Physical Properties of Some Ethers

IUPAC Name	Radicofunctional or Trivial Name	Structure	Boiling Point (°C)
methoxymethane	dimethyl ether	CH_3OCH_3	−24
ethoxyethane	diethyl ether	$CH_3CH_2OCH_2CH_3$	35
1-ethoxypropane	ethyl propyl ether	$CH_3CH_2OCH_2CH_2CH_3$	62
1-butoxybutane	dibutyl ether	$CH_3(CH_2)_3O(CH_2)_3CH_3$	142
oxirane	ethylene oxide	CH_2—CH_2 \ O /	11
oxetane	trimethylene oxide	O	50
oxolane	tetrahydrofuran (THF)	O	65
oxane	tetrahydropyran (THP)	O	88
1,4-dioxane	—	O O	101

6–12 Give IUPAC names for the alcohols (a)–(c) and structural formulas for alcohols (d)–(f).

a. $CH_3CH_2CHCH_2CHCH_3$ (with OH on C-3 and CH_3 on C-5)

b. $BrCH_2CHCH_2CH_2CH_3$ (with OH on C-2)

c. $CH_3CHCH_2CH_2OH$ (with OH on C-2)

d. 2,2-dimethyl-1-butanol

e. 3-fluoro-3-methyl-2-butanol

f. 2-methyl-1,4-butanediol

6–13 Give IUPAC and "ether" names to structures (a)–(c) and structural formulas for ethers (d)–(g).

a. $CH_3OCH_2CH_2CH_3$

b. $CH_3CHOCH_2CH_3$ (with CH_3 on C-2)

c. $(CH_3)_2CHO$–(cyclobutyl)

d. *cis*-1-bromo-2-methoxycyclohexane

e. 4-ethoxy-1,2-butanediol

f. cyclopentyl methyl ether

g. *cis*-2,5-dimethyltetrahydrofuran

Suppose we now compare the physical properties of some of the isomeric alcohols and ethers listed in Tables 6–4 and 6–5. Diethyl ether has a boiling point of 35°, while its isomer 1-butanol has a boiling point of 118°. The consistently higher boiling point of alcohols, compared to their isomeric ethers, is due to hydrogen bonding. Since hydroxyl

$$\overset{\delta-}{(\text{—}\ddot{\underset{\cdot\cdot}{O}}} \overset{\delta+}{\text{—H})}$$

groups are both hydrogen bond donors and acceptors, alcohol molecules are held together by intermolecular hydrogen bonds (Fig. 6–13). Ethers, on the other hand, lack such donor groups and therefore lack hydrogen bonds, and so have lower boiling points. Functional groups that have hydrogens capable of hydrogen bonding are called *protic* groups. Alcohols are protic, while ethers are *aprotic*—they are not capable of hydrogen bonding.

The high electronegativity of oxygen also makes alcohols weak acids (pK_a's = 16 to 19). While they are too weak to affect litmus paper, they are strong enough to react with electropositive metals (Li, Na, K, Mg, Al) to release hydrogen and produce metal **alkoxides** (eqs. 6.25 and 6.26). Ethers do not react with metals, due to their lack of acidic hydrogens.

(6.25) $$2CH_3CH_2\ddot{\underset{\cdot\cdot}{O}}\text{—H} + 2Na \rightarrow 2CH_3CH_2\ddot{\underset{\cdot\cdot}{O}}{:}^-\,Na^+ + H_2$$

Sodium ethoxide

(6.26) $$2(CH_3)_3C\ddot{\underset{\cdot\cdot}{O}}\text{—H} + 2K \rightarrow 2(CH_3)_3C\ddot{\underset{\cdot\cdot}{O}}{:}^-\,K^+ + H_2$$

Potassium *t*-butoxide

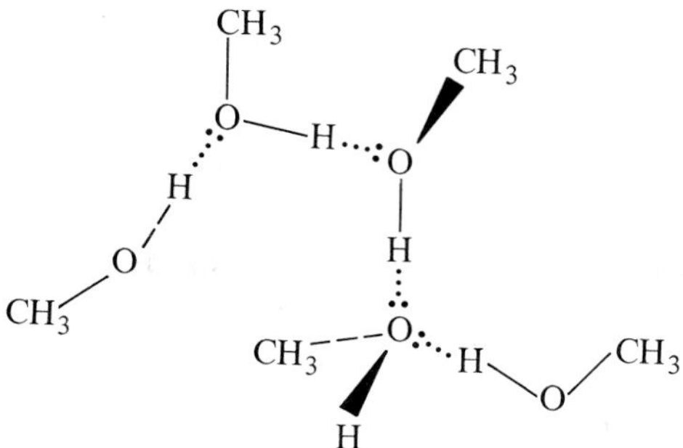

Figure 6–13. *A cluster of hydrogen-bonded methanol molecules.*

Problems

6–14 Dimethyl sulfide, CH_3SCH_3, and ethanethiol, CH_3CH_2SH, have nearly identical boiling points (38° and 37°, respectively). Using these data, compare oxygen and sulfur with respect to their hydrogen bonding abilities.

6–15 When sodium metal is dropped into a mixture of 2-propanol and water, a reaction takes place and gas is evolved. The metal is completely consumed. Show the species that are present at the end of the reaction.

6:10 Protic and Aprotic Solvents: How They Affect Reaction Rates

Alcohols and ethers are commonly used as solvents for organic reactions. You have already seen that the choice of a solvent can play a decisive role in nucleophilic substitutions—for instance, the S_N1 mechanism only makes its appearance in certain solvents (sec. 6:6). This section deals with the relationship between the properties of a solvent and its molecular structure.

Solvents are commonly classified according to their *polarity*. Polar solvents are those that possess one or more polar bonds, and have a net dipole moment (see Chap. 8). Polar bonds arise when atoms of substantially different electronegativity are joined together (sec. 2:7). Water is a highly polar solvent, and low molecular weight alcohols are also fairly polar, due to their C—O and H—O bonds. Ethers are much less polar than alcohols, as they contain only C—O bonds. Hydrocarbons are essentially *nonpolar*.

$$\text{H—O} > \text{C—O} > \text{C—H} > \text{C—C}$$

Decreasing bond polarity

One important quantitative indicator of solvent polarity is the **dielectric constant**, ϵ. The dielectric constant measures the relative ability of a substance to interfere with the attractive force between unlike charges, such as those on positive and negative ions. Highly polar solvents, which have high dielectric constants, are able to dissolve salts because they facilitate the separation of the positive and negative ions. Table 6–6 shows the structure and dielectric constants for several common solvents. Note the correlation between the size of the dielectric constant and the presence or absence of polar bonds.

In order to understand how a solvent affects nucleophilic substitution reactions, we have to look beyond its dielectric constant, which is a macroscopic property. It's necessary to consider specific interactions between sol-

Table 6–6. Dielectric Constants of Some Common Solvents

Solvent	Structure	Dielectric Constant (ϵ), 25° C
water	HOH	78.5
methanol	CH_3OH	33
ethanol	CH_3CH_2OH	24
acetone	$(CH_3)_2C{=}O$	21
tetrahydrofuran	 (five-membered ring with O)	7
diethyl ether	$(CH_3CH_2)_2O$	4
benzene	 (benzene ring)	2
pentane	$CH_3(CH_2)_3CH_3$	2

vent molecules and the nucleophile. Nucleophilic reactivity is enhanced in anions because the nucleophile has to transfer a negative charge to the leaving group. Thus, anions are stronger nucleophiles than closely related neutral molecules (compare eqs. 6.27 and 6.28). These anionic nucleophiles, particularly small ones such as F^-, Cl^-, CN^-, and OH^-, are among the strongest hydrogen bond acceptors (Fig. 6–14). These hydrogen-bonding interactions

(6.27) $$CH_3{-}I + CH_3\ddot{\underset{..}{O}}{:}^- \rightarrow CH_3\ddot{\underset{..}{O}}CH_3 + I^- \qquad k^{rel} = 10^6\star$$

(6.28) $$CH_3{-}I + CH_3\ddot{\underset{..}{O}}H \rightarrow CH_3\underset{\underset{H}{|}}{\ddot{O}^+}CH_3 + I^- \qquad k^{rel} = 1$$

★ k^{rel} is a *relative* rate constant, used for comparing a series of rate constants, when the absolute values do not matter.

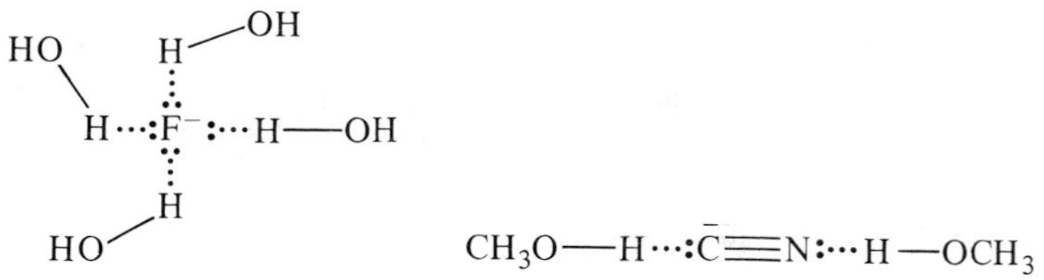

(a) HO—H···:F⁻:···H—OH, with two further H—OH hydrogen bonds to F⁻ (above and below)

(b) $CH_3O{-}H{\cdots}{:}\bar{C}{\equiv}N{:}{\cdots}H{-}OCH_3$

Figure 6–14. *Anions strongly solvated by protic solvents. (a) Fluoride ion in water; (b) cyanide ion in methanol.*

Table 6–7. Structures and Dielectric Constants of Some Useful Dipolar Aprotic Solvents

Solvent	Structure	Dielectric Constant
dimethyl sulfoxide (DMSO)	$(CH_3)_2S{=}O$	49
N,N-dimethylformamide (DMF)	$(CH_3)_2N\overset{\overset{\displaystyle O}{\|}}{C}H$	37
acetonitrile	$CH_3C{\equiv}N$	36
acetone	$(CH_3)_2C{=}O$	21

increase the solubility of the anion, but at the same time decrease its reactivity. For example, when a fluoride ion in water acts as a nucleophile it must "shake loose" one or more water molecules attached to its lone pairs (Fig. 6–14a). This process requires energy and, in turn, slows down the nucleophilic reactions of fluoride ion.

To avoid these hydrogen bonding effects, chemists developed a series of *dipolar aprotic solvents.* These solvents contain polar multiple bonds, have relatively high dielectric constants, and are therefore able to dissolve salts

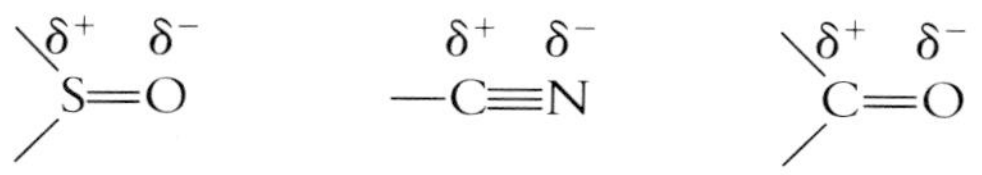

$$\overset{\delta^+}{S}{=}\overset{\delta^-}{O} \qquad -\overset{\delta^+}{C}{\equiv}\overset{\delta^-}{N} \qquad \overset{\delta^+}{C}{=}\overset{\delta^-}{O}$$

Some polar multiple bonds

(Table 6–7). However, they lack OH or NH groups and thus cannot hydrogen-bond to anions. For this reason anions are highly reactive in dipolar aprotic solvents, making them good solvents for S_N2 reactions (eqs. 6.9 and 6.10). They are particularly effective at solvating cations, as we show for a sodium ion in DMSO:

$$(CH_3)_2\overset{\delta+}{S}{=}\overset{\delta-}{O}{:}\cdots Na^+$$

Problems

6–16 The reaction shown in Equation 6 is carried out in both methanol and DMF (dimethylformamide, Table 6–7). The rate constant is 10^6 larger in DMF than in methanol. Explain this result.

$$CH_3I + K^+F^- \xrightarrow[\text{solvent}]{} CH_3F + K^+I^- \tag{6}$$

6–17 Show the substitution products of each of the following reactions:

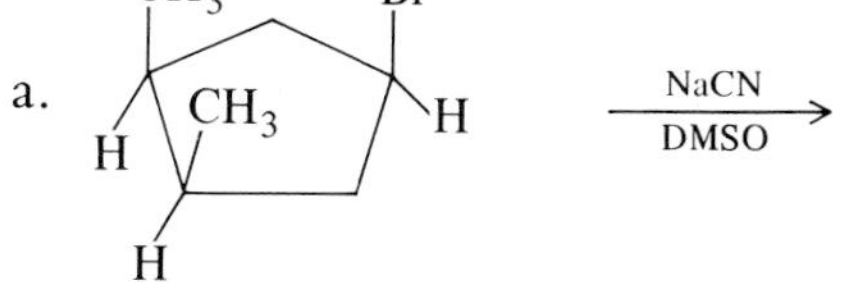

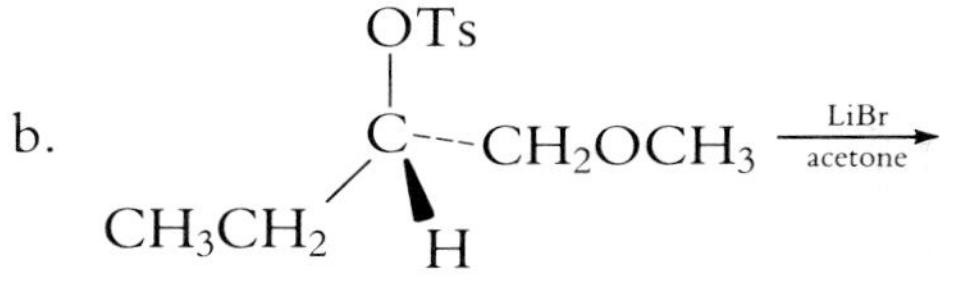

a. [cyclopentane with CH_3, CH_3, Br, H, H, H substituents] $\xrightarrow[\text{DMSO}]{\text{NaCN}}$

b. CH_3CH_2–C(OTs)(H)–CH_2OCH_3 $\xrightarrow[\text{acetone}]{\text{LiBr}}$

c. 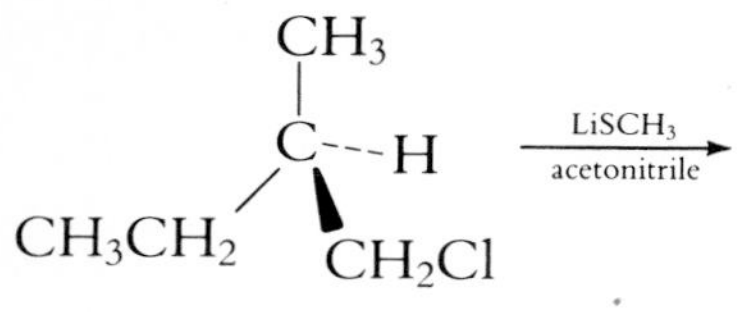 [structure: CH_3 / C---H / CH_3CH_2 / CH_2Cl] $\xrightarrow[\text{acetonitrile}]{\text{LiSCH}_3}$

d. 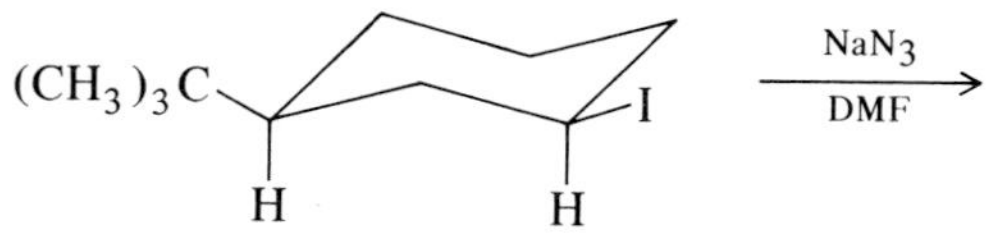 $(CH_3)_3C$ [cyclohexane ring with H, H, I] $\xrightarrow[\text{DMF}]{\text{NaN}_3}$

6–18 Draw the structures of all the substitution products from each of the following reactions, paying attention to stereochemistry. Indicate which mechanism operates in each reaction.

a. (R)-2-chloropentane $\xrightarrow[\text{DMSO}]{\text{OH}^-}$

b. (R)-2-iodobutane $\xrightarrow[\text{acetone}]{\text{H}_2\text{O}}$

c. *trans*-1-bromo-3-ethylcyclobutane $\xrightarrow[\text{DMF}]{\text{F}^-}$

d. 2-bromo-4-chloro-2-methylpentane $\xrightarrow[\text{acetone}]{\text{I}^-}$

Box 6–2. Crown Ethers and Ionophores: Synthetic and Natural Cation Complexes

The goal of dissolving salts without using protic solvents has also been achieved with special reagents called **crown ethers**, which completely enclose the cation. These large ring ethers are prepared so that only specific cations fit into the central cavity (Fig. 6–15). Thus, 12-crown-4 binds Li^+ but not K^+, while dicyclohexyl-18-crown-6 binds K^+ but not Li^+ or Na^+. In the name, the first number denotes the total number of atoms in the ring and the last number denotes the number of oxygen atoms. When salts are dissolved by crown ethers, the uncomplexed anions are highly reactive (Fig. 6–15).

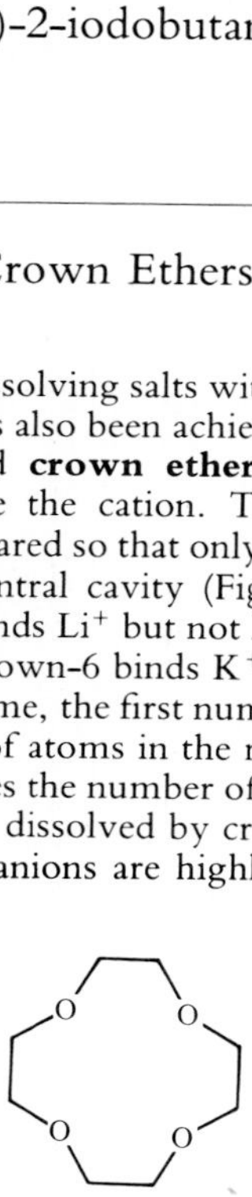

12-Crown-4

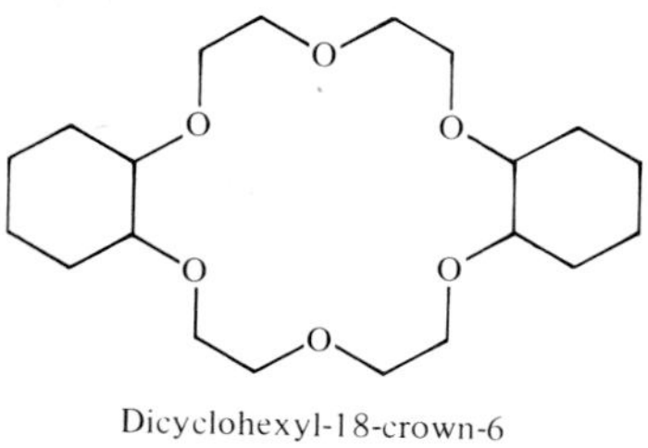

Dicyclohexyl-18-crown-6

Although this is a very cleverly designed aid to synthetic reactions, it turns out that Nature did it first. Only recently have we discovered that a certain group of antibiotics called **ionophores** can enclose cations in much the same way. The ionophore lasalocid is used

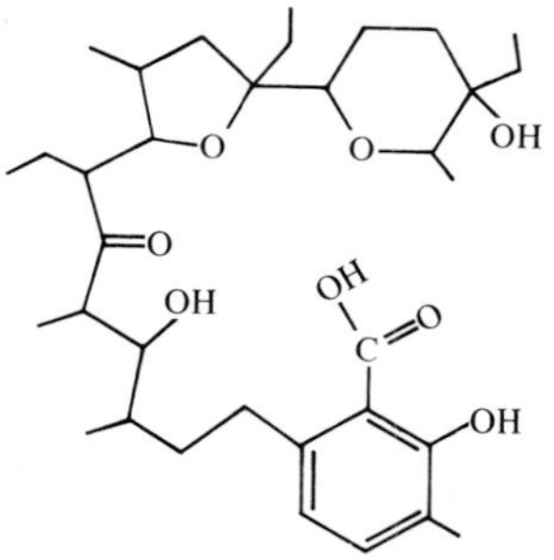

Lasalocid
(an ionophore antibiotic)

against coccocidial infections in chickens and may be of value in treating some human cardiovascular diseases. Another natural ionophore is so constituted that it only holds Ca^{2+}. Its function is to remove these ions from solution on one side of a cell membrane and transport them into the cell.

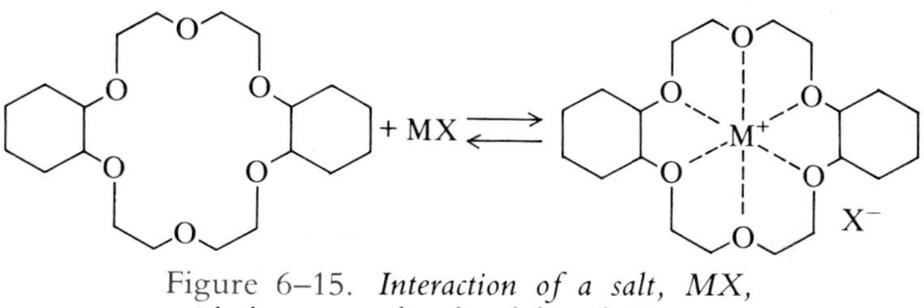

$$\text{dicyclohexyl-18-crown-6} + MX \rightleftharpoons [\text{crown}\cdot M^+]\ X^-$$

Figure 6–15. *Interaction of a salt, MX, with the crown ether dicyclohexyl-18-crown-6. The crown ether solvates the cation M^+ completely, but leaves the anion X^- free and reactive.*

6:11 Leaving Group Effects: Conversion of Alcohols and Ethers to Alkyl Halides

The halide ions (I^-, Br^-, and Cl^-) are good leaving groups in nucleophilic displacements, but they are all not equally effective. Equation 6.29 shows that Br^- is a better leaving group than Cl^-. As you can imagine, these differences are synthetically useful.

$$\text{(6.29)}\qquad \underset{}{Cl(CH_2)_3Br} + \underset{(1\text{ eq})}{K^+CN^-} \xrightarrow[C_2H_5OH,\ \text{heat}]{} \underset{(85\text{ to }90\%)}{Cl(CH_2)_3CN} + K^+Br^-$$

The relative effectiveness of a series of similar leaving groups is inversely related to their basicity, as seen from the halide ions. Basicity is the most

$$I^- > Br^- > Cl^- \gg F^-$$

Effectiveness as a leaving group

reliable general guide to leaving group ability. Figure 6–16 shows some

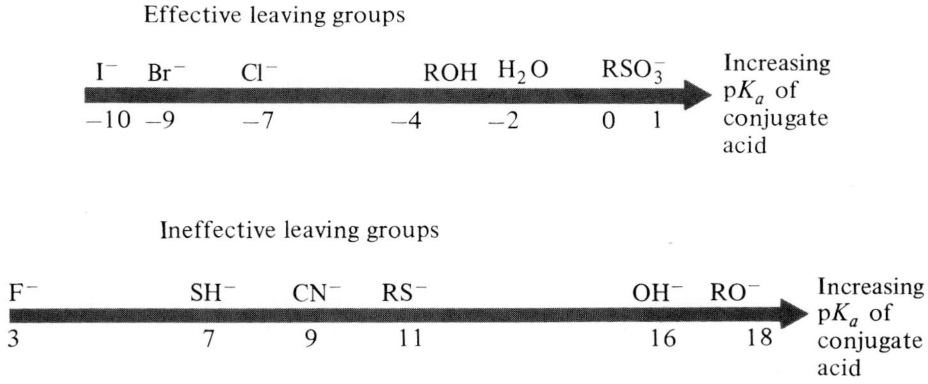

Effective leaving groups

I^-	Br^-	Cl^-	ROH	H_2O	RSO_3^-	
−10	−9	−7	−4	−2	0	1

Increasing pK_a of conjugate acid

Ineffective leaving groups

F^-	SH^-	CN^-	RS^-	OH^-	RO^-
3	7	9	11	16	18

Increasing pK_a of conjugate acid

Figure 6–16. *Effective and ineffective leaving groups in nucleophilic substitution reactions of alkyl groups. Leaving group ability is inversely related to basicity, as measured by the pK_a of the conjugate acid of the leaving group.*

widely used leaving groups in nucleophilic substitution, as well as some common molecules and ions that are ineffective as leaving groups, such as those that are basic in water. The conjugate acids of the effective leaving groups all have pK_a's less than two (sec. 1:7), which makes them strong acids and the leaving groups themselves weak bases.

Strong base — Weak base — Nonbase

$\longrightarrow$

Decreasing basicity/Increasing leaving group ability

The inverse correlation between leaving group ability and basicity arises as follows. Strong bases are good electron pair *donors,* while effective leaving groups must be good electron pair *acceptors.* That is because the leaving group departs with the bonding electron pair it shared with the carbon atom. Good electron pair acceptors are those that can spread out (delocalize) the electrons over a large volume. That is the case for large anions such as I^-, and for polyatomic ions such as the sulfonate ions, RSO_3^-. In the sulfonate ions the charge is shared among the three oxygen atoms. Finally, note that water is a much better leaving group than the negatively charged hydroxide ion.

We can conclude from this analysis that OH^- cannot be displaced from alcohols and that OR^- cannot be displaced from ethers. These conclusions are supported by experiment. It *is* possible to perform nucleophilic displacements on alcohols if the reactions are carried out *under acidic conditions.* Equations 6.30 and 6.31 show that concentrated solutions of hydrogen halides convert alcohols and ethers to alkyl halides.

(6.30) $$(CH_3)_3COH + HCl\ (conc) \xrightarrow[25^\circ]{} (CH_3)_3CCl + H_2O$$
(85%)

(6.31) $$(CH_3CH_2CH_2)_2O + HBr\ (conc) \xrightarrow[125^\circ]{} 2CH_3CH_2CH_2Br + H_2O$$
Dipropyl ether

When alcohols are substrates, their OH groups are rapidly and reversibly protonated by the acid to yield the intermediate **alkyloxonium ions** (eq. 6.32).

(6.32) $$CH_3{-}\ddot{O}H + H_3O{:}^+ \overset{\text{fast}}{\rightleftharpoons} CH_3{-}\overset{+}{\ddot{O}}H_2 + H_2\ddot{O}{:}$$
Methyloxonium ion

Oxonium ions are good intermediates in nucleophilic substitutions because they release a good leaving group, H_2O. Primary alkyloxonium ions undergo S_N2 substitution (eq. 6.33). Secondary and tertiary alkyloxonium ions

(6.33) $$CH_3(CH_2)_2CH_2{-}\ddot{O}H \underset{100^\circ}{\overset{\text{HBr (conc)}}{\rightleftharpoons}}$$

$$CH_3(CH_2)_2CH_2{-}\ddot{O}H_2^+ \ (:\ddot{Br}:^-) \xrightarrow{\text{slow}} CH_3(CH_2)_2CH_2{-}\ddot{Br}{:} + H_2\ddot{O}{:}$$
(90%)

undergo substitution by way of carbocations (eqs. 6.34 and 6.35). For this reason, 3° alcohols are unusually reactive toward hydrogen halides (eq. 6.30).

(6.34) [1-methylcyclopentanol (CH_3, $\ddot{O}H$)] $\xrightleftharpoons{\text{HCl (conc)}}$

[1-methylcyclopentyl-$\ddot{O}H_2^+$] $+ Cl^-$ $\xrightarrow{\text{slow}}$ $H_2\ddot{O}: +$ [1-methylcyclopentyl cation, $^+$—CH_3] + $:\ddot{Cl}:^-$ $\longrightarrow$ [1-chloro-1-methylcyclopentane (CH_3, $:\ddot{Cl}:$)] $+ H_2\ddot{O}:$

(6.35) $CH_3CHC(CH_3)(\ddot{O}H)CH_3$ (i.e., $CH_3CH(OH)C(CH_3)_2CH_3$ with HO: on C-2 and two CH_3 on C-3) $\xrightarrow{\text{HBr (conc)}}$ $CH_3CH(\overset{+}{H_2O}{:})C(CH_3)_2CH_3 + Br^-$ $\longrightarrow$ $CH_3\overset{+}{C}HC(CH_3)_2CH_3\ Br^- + H_2\ddot{O}:$

$\downarrow$ CH_3 shift

$CH_3CH(CH_3)\overset{+}{C}(CH_3)CH_3$ ($:\ddot{Br}:^-$) $\longrightarrow$ $CH_3CH(CH_3)C(:\ddot{Br}:)(CH_3)CH_3$

Only bromide formed

Protonation of ethers results in the formation of intermediate dialkyloxonium ions, which undergo nucleophilic substitution just like the monoalkyloxonium ions (eq. 6.36).

(6.36) $CH_3CH_2—\ddot{O}—CH_2CH_3 \xrightarrow{\text{HI (conc)}} CH_3CH_2—\overset{+}{\underset{|}{O}}(H)—CH_2CH_3 \longrightarrow$ (with $:\ddot{I}:^-$)

$$CH_3CH_2—\ddot{I}: + CH_3CH_2—\ddot{O}H \xrightarrow{\text{HI}} 2CH_3CH_2I + H_2O$$

Another way to convert the hydroxyl group of alcohols into a good leaving group is to transform it to a new functional group, as in Equation 6.37. This new functional group will be displaced by a nucleophile if the

(6.37)
$$R—O—H \longrightarrow R—O—\overset{|}{\underset{|}{Z}}—$$

following requirements are met:

1. the Z—O bond should be stronger than the R—O bond;
2. the Z group should be able to strongly attract the electron pair from the R—O bond.

Z groups containing sulfur and phosphorus meet these criteria. The procedures we are about to outline are useful mainly with 1° and 2° alcohols, since they involve S_N2-type processes.

Primary and secondary alcohols react with *p*-toluenesulfonyl chloride TsCl) to yield **toluenesulfonate esters** (*tosylates,* eq. 6.38).

(6.38) $CH_3(CH_2)_3—O—H + Cl—SO_2—C_6H_4—CH_3 \xrightarrow[H_2O]{OH^-}$

BuOH　　*p*-Toluenesulfonyl chloride

$CH_3(CH_2)_3—O—SO_2—C_6H_4—CH_3 + Cl^- + H_2O$

Butyl tosylate (BuOTs)

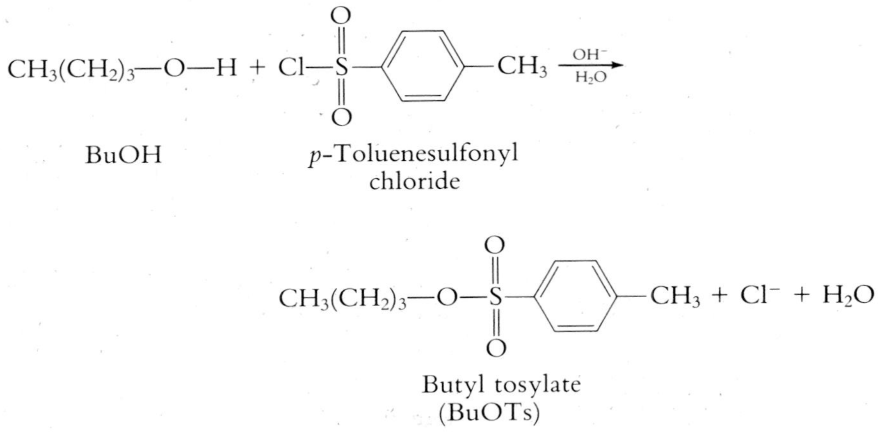

Although the molecule appears complex, its crucial portion is the sulfonyl group

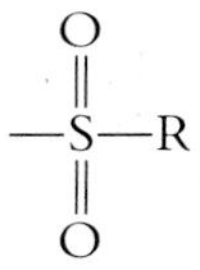

$$—\overset{O}{\underset{O}{\overset{\|}{\underset{\|}{S}}}}—R$$

This group binds strongly to the alcohol oxygen, thus meeting the first criterion mentioned above. When a 1° or 2° tosylate is treated with a nucleophile in an aprotic solvent, the tosylate ion is displaced with complete inversion of configuration by an S_N2 reaction (eq. 6.39).

(6.39) $:\ddot{Cl}:^-$ + (*S*)-*sec*-BuOTs [C bearing CH_2CH_3, H, CH_3, $\ddot{O}Ts$] $\xrightarrow{DMSO}$ (*R*)-*sec*-BuCl [$:\ddot{Cl}$—C bearing CH_2CH_3, H, CH_3] $+ :\ddot{O}Ts^-$

(*S*)-*sec*-BuOTs　　(*R*)-*sec*-BuCl

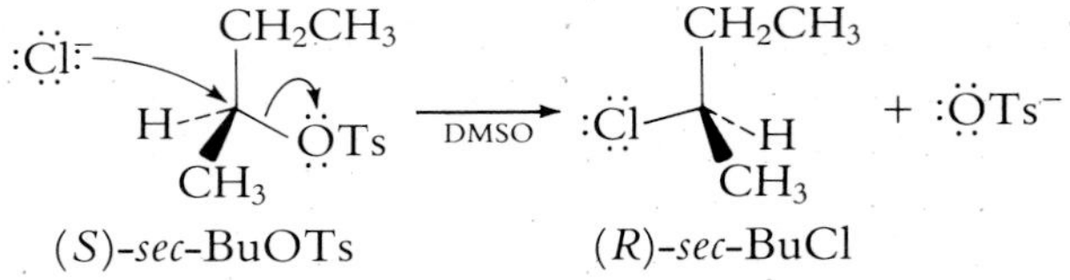

Thionyl chloride, $SOCl_2$, is another reagent used to convert primary and secondary alcohols to chloroalkanes. Thionyl chloride transforms hydroxyl groups to chlorosulfite groups, which are allowed to decompose to chlorides and sulfur dioxide (eq. 6.40). If the alcohol group is bonded to a chiral center, thionyl chloride in DMF (as in eq. 6.40) converts it to chloride *with retention of configuration.* The chlorosulfite ester gives an ion pair that goes to product by way of front-side attack.

$$(6.40)\quad (CH_3CH_2)_2CH{-}OH + Cl{-}\overset{\overset{O}{\|}}{S}{-}Cl \xrightarrow[\text{DMF, }100^\circ]{}$$

Thionyl chloride

$$HCl + \left[(CH_3CH_2)_2CH{-}O{-}\overset{\overset{O}{\|}}{S}{-}Cl\right] \longrightarrow HCl + SO_2 + (CH_3CH_2)_2CH{-}Cl$$

Chlorosulfite ester (not isolated) — 3-Chloropentane (73%)

In addition to these commonly used sulfur-containing reagents (TsCl and $SOCl_2$), there are several phosphorus-containing reagents that can convert alcohols to alkyl halides. Among the most widely used is phosphorus tribromide, PBr_3 (eq. 6.41).

$$(6.41)\quad 3CH_3CH_2\underset{\underset{OH}{|}}{C}HCH_3 + PBr_3 \xrightarrow{0-25^\circ} 3CH_3CH_2\underset{\underset{Br}{|}}{C}HCH_3 + H_3PO_3$$

(80%)

With optically active 2° alcohols, the reactions are carried out at low temperatures for long periods. This procedure gives almost complete inversion.

We've treated alkyl halides, alcohols, and ethers together for several reasons. One reason is the fact that each of these classes of compounds can be converted to the others by means of nucleophilic substitution reactions. The conversion of alcohols and ethers to alkyl halides, summarized in Table 6–8, was the subject of this section. The next section completes the loop by

Table 6–8. Methods for Converting Alcohols and Ethers to Alkyl Halides

Equation	Reagents
Alcohols to Alkyl Halides	
(1) $ROH + HX \rightarrow RX + H_2O$	HI, HBr, HCl
(2) $ROH + SOCl_2 \rightarrow RCl + SO_2 + HCl$	$SOCl_2$ in DMF
(3) $3ROH + PX_3 \rightarrow 3RX + P(OH)_3$	PBr_3, PI_3
(4) $ROH + TsCl \xrightarrow{OH^-} ROTs \xrightarrow{X^-} RX + OTs^-$	TsCl and OH^-, followed by a salt containing a halide ion
Ethers to Alkyl Halides	
$ROR' + 2HX \rightarrow RX + R'X + H_2O$	HBr (conc), HI

showing how alkyl halides are used as starting materials in the synthesis of alcohols and ethers.

Problems

6–19 Write the structures of all the organic products expected from each of the following reactions:

a. cyclohexanol $\xrightarrow[\text{heat}]{\text{HBr (conc)}}$

b. 1,6-hexanediol $\xrightarrow[\text{heat}]{\text{KI, } H_3PO_4}$

c. 2-methyl-2-pentanol $\xrightarrow[25°]{\text{HCl (conc)}}$

d. *cis*-2-methylcyclopentanol $\xrightarrow[\text{heat}]{\text{HCl (conc)}}$

e. propyl tosylate $\xrightarrow[\text{DMF}]{\text{KF}}$

f. cyclohexanol $\xrightarrow[\text{DMSO}]{\text{KCN}}$

g. 4-methyl-2-pentanol $\xrightarrow[0°]{PBr_3}$

6–20 Show how each alkyl halide could be formed from an appropriate alcohol. Comment on any difficulties which you might anticipate.

a. 1-chlorohexane
b. 2-bromohexane
c. fluorocyclobutane
d. (*R*)-1-chloro-2-methylbutane
e. 2-iodo-3,3-dimethylpentane

6–21 (*R*)-2,2-Dimethyl-3-hexanol is treated with concentrated hydrochloric acid at 25°. A small quantity (6%) of (*S*)-3-chloro-2,2-dimethylhexane, **A**, is formed, along with large quantities of 2-chloro-2,3-dimethylhexane, **B**, and 3-chloro-2,3-dimethylhexane, **C**. Products **B** and **C** are formed in equal amounts and both are racemic. Write a mechanism that accounts for the formation of all the products.

6:12 Synthesis of Alcohols and Ethers from Alkyl Halides

It is possible to reverse the processes shown in the last section and synthesize alcohols and ethers from alkyl halides and tosylates. The hydroxide ion converts 1° substrates to alcohols in good yields (eqs. 6.3 and 6.42).

$$(6.42)\quad CH_3(CH_2)_4CH_2\text{—}\ddot{\underset{..}{Br}}\text{:} + \text{:}\ddot{\underset{..}{O}}H^- \xrightarrow[H_2O,\ \text{dioxane}]{} CH_3(CH_2)_4CH_2\text{—}\ddot{\underset{..}{O}}H + \text{:}\ddot{\underset{..}{Br}}\text{:}^-$$

When 2° halides and tosylates are used, there is a strong tendency for elimination rather than substitution to occur. The two reactions compete to yield products such as those shown in Equation 6.43. (This reaction is discussed in detail in Chapter 7).

$$\text{(6.43)}\quad CH_3CH_2\underset{\displaystyle Br}{\underset{|}{C}}HCH_3 + OH^- \longrightarrow$$

$$CH_3CH_2\underset{\displaystyle OH}{\underset{|}{C}}HCH_3 + \underbrace{CH_3CH_2CH{=}CH_2 + CH_3CH{=}CHCH_3}$$

Substitution product — Elimination products

Tertiary alkyl halides give elimination products almost exclusively when treated with OH^- or OR^-. Tertiary alcohols can be synthesized from 3° halides by allowing them to solvolyze in water (eq. 6.16).

Ethers are synthesized from 1° and 2° halides and tosylates when alkoxides are used as nucleophiles. Sodium or potassium is added to an excess of alcohol to yield a solution of the alkoxide (eqs. 6.27 and 6.28). The alkyl halide (or tosylate) is then added to this solution. Two examples of ether synthesis, commonly called the *Williamson synthesis,* are given in equations 6.44 and 6.45.

$$\text{(6.44)}\quad (CH_3)_2CHCH_2\ddot{\underset{..}{O}}H \xrightarrow{Na} (CH_3)_2CHCH_2\ddot{\underset{..}{O}}{:}^- Na^+$$

Sodium isobutoxide

$$\xrightarrow{CH_3OTs} (CH_3)_2CHCH_2\ddot{\underset{..}{O}}CH_3 + Na^+OTs^-$$

Isobutyl methyl ether (90%)

$$\text{(6.45)}\quad C_6H_{11}\text{—}OH \xrightarrow{Na} C_6H_{11}\text{—}O^-Na^+ \xrightarrow{CH_3CH_2Br} C_6H_{11}\text{—}OCH_2CH_3$$

Ethoxycyclohexane (cyclohexyl ethyl ether)

Since these are S_N2 reactions, the alkyl halide or tosylate should be as little substituted as possible. This factor should be considered when choosing reagents for the synthesis of unsymmetrical ethers. For instance, in Equation 6.44 the yield would be lower if the reactants were sodium methoxide and isobutyl tosylate.

Example

Synthesize $CH_3CH_2OCH_2CH_2CH_3$.

Answer

Use the Williamson synthesis—reaction of an alkoxide with an alkyl halide—to synthesize the ethers. Either of two reaction schemes are equally valid here:

$$CH_3CH_2\ddot{O}:^- + CH_3CH_2CH_2—X \longrightarrow CH_3CH_2OCH_2CH_2CH_3 + X:^-$$
(X = Br, Cl)

or

$$CH_3CH_2CH_2\ddot{O}:^- + CH_3CH_2—X \longrightarrow CH_3CH_2CH_2OCH_2CH_3 + X:^-$$

Problems

6–22 Two students are assigned the synthesis of *t*-butyl ethyl ether. One uses the process shown in Equation 7, while the other tries the approach outlined in Equation 8.

$$CH_3CH_2OH \xrightarrow{Na} CH_3CH_2O^-Na^+ \xrightarrow{(CH_3)_3CBr} (CH_3)_3COCH_2CH_3 \quad (7)$$

$$(CH_3)_3COH \xrightarrow{Na} (CH_3)_3CO^-Na^+ \xrightarrow{CH_3CH_2Br} (CH_3)_3COCH_2CH_3 \quad (8)$$

Only one student obtains the desired product. Which one is it? Explain why only one of these syntheses works.

6–23 Diethyl ether is made commercially by heating ethanol to 140° in the presence of an acid catalyst (eq. 9).

$$2CH_3CH_2\ddot{O}H \xrightarrow[140°]{H^+} CH_3CH_2\ddot{O}CH_2CH_3 + H_2\ddot{O}: \quad (9)$$

a. Write a possible mechanism for this reaction, paying particular attention to the role of the acid catalyst.
b. Would it be sensible to synthesize ethyl propyl ether in a similar fashion, using a mixture of ethanol and 1-propanol? Justify your answer.

6–24 Devise syntheses of the following ethers from appropriate haloalkanes and alcohols. Show all steps.
a. *n*-pentyl isopropyl ether
b. *trans*-1-methoxy-2-methylcyclohexane
c. 1,4-diethoxybutane

6:13 Formation of Epoxides by Intramolecular Nucleophilic Displacement: Some Reactions of Epoxides with Nucleophiles

Epoxides, which are three-membered cyclic ethers, can be synthesized by way of an *intramolecular* nucleophilic displacement. Because of the ring strain inherent in their small size (sec. 3:8), epoxides are much more reactive than any other ethers, which accounts for their being implicated in carcinogenesis. Commercially, epoxides are widely used in adhesives.

When 2-haloalcohols are treated with bases they yield epoxides (eq. 6.46).

$$(6.46)\quad Cl—CH_2CH_2—OH + OH^- \underset{H_2O}{\rightleftharpoons} Cl—CH_2CH_2—O^- + H_2O \longrightarrow \text{oxirane} + Cl^-$$

2-Chloroethanol — Oxirane (ethylene oxide)

The reaction shown in Equation 6.46 takes place much more rapidly than do nucleophilic substitutions of simple alkyl chlorides. This reaction is an example of **neighboring group participation**, in which a nucleophilic functional group in a molecule assists the departure of a leaving group. Neighboring group participation is most likely to occur when the cyclic transition state involves a three-, five-, or six-membered ring.

Ethylene oxide is the simplest member of a class of compounds known as **oxiranes**, or **epoxides**. They are also called alkene oxides because they can be synthesized by direct oxidation of alkenes (Chap. 10).

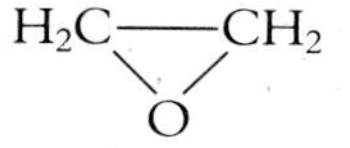

H_2C—CH_2 (bridged by O)
Oxirane
(*ethylene oxide*)

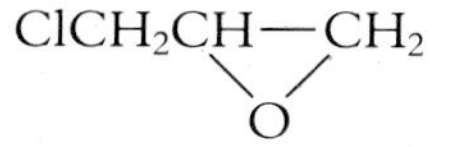

$ClCH_2CH$—CH_2 (bridged by O)
Chloromethyloxirane
(*epichlorohydrin*)

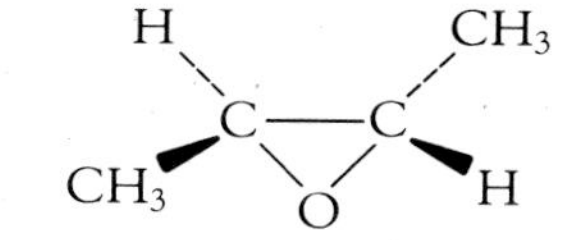

H, CH_3 / C—C (bridged by O) / CH_3, H
trans-2,3-Dimethyloxirane

Epoxides are unusually reactive. While other ethers will not react with nucleophiles unless they are first protonated, unprotonated epoxides react readily with nucleophiles (eq. 6.47). The mechanism is given in Equation 6.48; note that the nucleophile attacks the less hindered carbon.

(6.47) Methyloxirane + CH_3CH_2OH $\xrightarrow[80°]{NaOH\ (0.3\%)}$ $CH_3CH(OH)CH_2OCH_2CH_3$

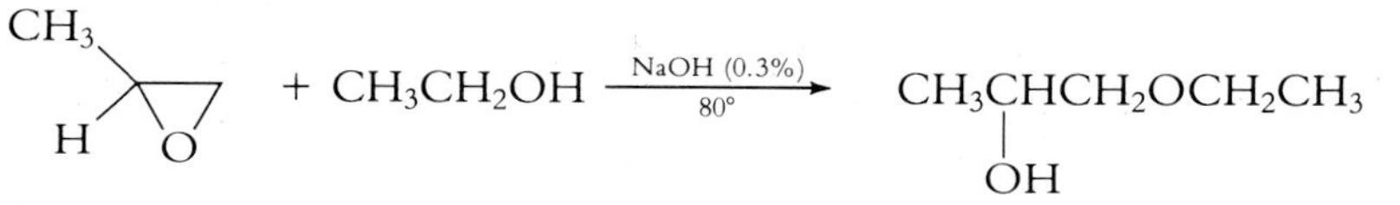

Methyloxirane — 1-Ethoxy-2-propanol (81%)

(6.48) $C_2H_5\ddot{O}H \overset{:\ddot{O}H^-}{\rightleftharpoons} C_2H_5\ddot{O}:^- + H_2\ddot{O}:$ $\xrightarrow{CH_2—CHCH_3\ (\text{O bridge})}$ $C_2H_5\ddot{O}CH_2CH(:\ddot{O}:^-)CH_3$

$\xrightarrow{H_2\ddot{O}:}$ $C_2H_5\ddot{O}CH_2CH(:\ddot{O}H)CH_3 + :\ddot{O}H^-$

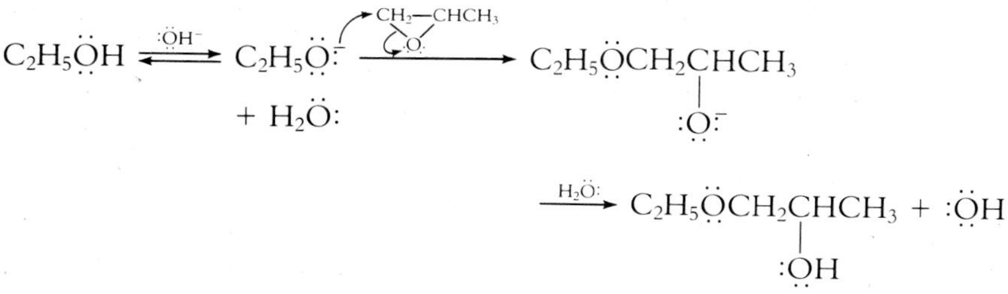

As is true of all S_N2 reactions, nucleophilic attack on epoxides involves backside attack (eq. 6.49).

(6.49) Cyclohexene oxide + H_2O $\xrightarrow{OH^-}$ *trans*-1,2-cyclohexanediol

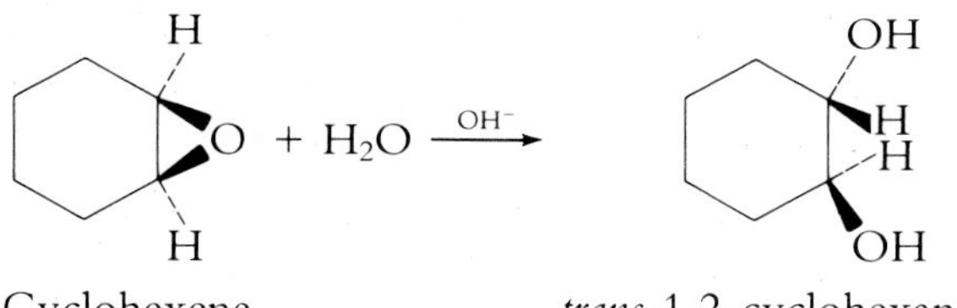

The unusual reactivity of epoxides is due largely to their substantial strain energy (see cyclopropane discussion, sec. 3:11). This property also makes them biologically hazardous (sec. 6:15).

Problem

6–25 A chemist attempts to prepare 1,2-propanediol, **II**, by treating 1,2-dichloropropane with two equivalents of OH^-. However, the major product isolated is methyloxirane (**III**; eq. 10). Write a mechanism for this reaction, and explain why it took the course that it did.

$$\underset{\displaystyle Cl}{CH_3\underset{|}{C}HCH_2Cl} + 2OH^- \xrightarrow[H_2O\text{-dioxane}]{} \underset{\displaystyle OH}{CH_3\underset{|}{C}HCH_2OH} + \text{(methyloxirane: } CH_3, H \text{ on ring C; O)} \quad (10)$$

II Minor product — III Major product

6:14 Organometallic Compounds: Sources of Carbon Nucleophiles

The previous sections have emphasized the usefulness of alkylation reactions in syntheses. Alkylation of oxygen nucleophiles yields alcohols and ethers, alkylation of halogen nucleophiles yields alkyl halides, and so on. If we could generate carbon nucleophiles, such as carbanions, then we might expect to be able to alkylate them and produce alkanes (eq. 6.50).

(6.50) $$\underset{\text{A carbanion}}{R_3C{:}^-} + R'{-}X \rightarrow R_3C{-}R' + X^- \qquad R' = \text{alkyl group}$$

Unfortunately, simple carbanions in which R is an H atom or an alkyl group, cannot be prepared as free ions in solution. They are extremely basic and react with common solvents. However, many **organometallic compounds**, R_nM (M = metal), which have carbon–metal bonds, behave like carbanions. This is because the metal atom is less electronegative than carbon (Table 2–5). The carbon–metal bond is polarized $M^{\delta+}{-}C^{\delta-}$, leaving a substantial negative charge on carbon. Before we consider the role of organometallic compounds in alkane synthesis, let's review some of their basic chemistry.

Organometallic compounds are readily formed when alkyl halides (RCl, RBr, and RI) react with metals of low electronegativity, in particular, Mg and Li (eqs. 6.51 and 6.52). The magnesium derivatives are known as **Grignard reagents**, named after their discoverer, the French chemist V. Grignard, who shared a Nobel Prize in 1912 for their development.

(6.51) $$CH_3CH_2Br + Mg \xrightarrow[\text{ether, 35°}]{} \underset{\text{Ethylmagnesium bromide (97\%)}}{CH_3CH_2MgBr}$$

(6.52) $$(CH_3)_3CCl + Mg \xrightarrow[\text{ether, 35°}]{} \underset{t\text{-Butylmagnesium chloride (80\%)}}{(CH_3)_3CMgCl}$$

Grignard reagents are unstable except in solution. Coordination of the magnesium to the nonbonding electrons on ether greatly stabilizes the compounds (Fig. 6–17).

Organolithium compounds are prepared similarly to Grignard reagents

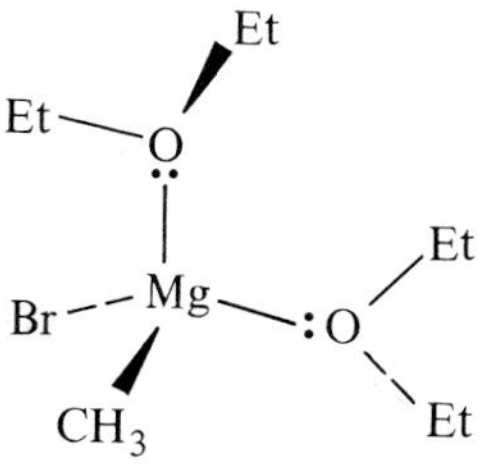

Et—O(:)(Et)—Mg(Br)(CH$_3$)—:O(Et)(Et)

Figure 6–17. *Methylmagnesium bromide in ether solution. Note how the ether oxygens solvate the magnesium atom, thereby stabilizing the compound.*

(eq. 6.53). They are even more reactive than Grignard reagents because Li is more electropositive than Mg (Table 2–5).

(6.53) $$CH_3CH_2CH_2CH_2Br + 2Li \xrightarrow[\text{ether, } -10°]{N_2,} CH_3CH_2CH_2CH_2Li + LiBr$$

Butyllithium (80–90%)

Organolithium and Grignard reagents are excellent nucleophiles toward numerous classes of compounds. For example, the reaction between a Grignard reagent and oxirane provides an efficient way of obtaining alcohols from alkyl halides (eq. 6.54).

(6.54) $$RBr \xrightarrow{Mg} RMgBr$$

$$\overset{\delta^-}{R}\overset{\delta^+}{Mg}Br + \underset{\diagdown O \diagup}{CH_2{-}CH_2} \rightarrow RCH_2CH_2OMgBr \xrightarrow{H_3O^+} RCH_2CH_2OH$$

Unfortunately, organolithium and Grignard reagents do not react with alkyl halides in a synthetically useful way. There is much elimination, and radical processes take place because of electron transfer (sec. 5:3). One way to avoid these problems is to use organocopper reagents with alkyl halides. These reagents are synthesized from alkyllithiums and copper(I) salts (eq. 6.55) to which the alkyl halide is then added. Reactions such as the one shown in Equation 6.55 are not simple nucleophilic substitutions, and their mechanisms are under investigation.

(6.55) $$2BuLi + CuI \xrightarrow[THF]{} LiI + Bu_2CuLi$$

Dibutylcopper lithium

$$\xrightarrow[25°]{CH_3(CH_2)_4{-}I} CH_3(CH_2)_4{-}(CH_2)_3CH_3$$

Nonane (98%)

6:15 Alkylations in Biology

The amino acids, proteins, and nucleic acids that are constituents of all cells contain many nucleophilic groups. Alkylation of these groups occurs regularly in normal cell chemistry. However, it can also lead to pathological changes in the cell.

The biosynthesis of the amino acid methionine, for example, is a normal biological alkylation reaction. Methionine is essential for human growth and development, but our bodies are unable to produce it. We can, however, take in methionine with our food because many lower species present in foodstuffs, such as microbes, synthesize methionine for us. The last step in microbial synthesis of methionine is an alkylation—the methylation of the sulfur atom of homocysteine by betaine (eq. 6.56).

(6.56)

$$\text{Methyltransferase enzyme}-B\!: + H\ddot{S}(CH_2)_2\underset{^{+}NH_3}{CHCOO^-} \rightarrow$$

Homocysteine

$$\text{Methyltransferase enzyme}-\overset{+}{B}H + {}^{-}\!:\!\ddot{S}(CH_2)_2\underset{^{+}NH_3}{CHCOO^-}$$

$$^{-}OOCCH_2\overset{+}{N}(CH_3)_2-CH_3 + {}^{-}\!:\!\ddot{S}(CH_2)_2\underset{^{+}NH_3}{CHCOO^-} \xrightarrow{S_N2}$$

Betaine

$$^{-}OOCCH_2\ddot{N}(CH_3)_2 + CH_3\ddot{S}(CH_2)_2\underset{^{+}NH_3}{CHCOO^-}$$

Methionine

This reaction takes place at a useful rate only in the presence of an **enzyme**. Enzymes are large, complex molecules (**macromolecules**) that act as catalysts for biochemical reactions. In methionine biosynthesis the catalyst performs two important functions: it holds the reactants together in a favorable arrangement, and it removes the SH proton, making the sulfur more nucleophilic. (A more extended introduction to enzymes is found in Chapter 10.)

The methylation of homocysteine (eq. 6.56) is shown as a typical S_N2 reaction. How do we know that biosynthetic reactions follow the same mechanisms that laboratory reactions do? Might not the action of the enzyme result in biosynthetic reactions following different paths? The accumulating evidence indicates that this is not so, and that laboratory reactions are good models for biological ones. As an example, consider the reaction of haloacid anions with hydroxide ion (eq. 6.57). This reaction is catalyzed by

(6.57)

$$Cl-CH_2COO^- + OH^- \xrightarrow[\text{enzyme}]{\text{bacterial}} HO-CH_2COO^- + Cl^-$$

Chloroacetate ion → Hydroxyacetate ion

an enzyme found in certain bacteria. When an optically active haloacid is used, the product is formed stereospecifically with inversion of configuration (eq. 6.58). The result strongly suggests an S_N2 mechanism.

(6.58) $OH^- + $ (S)-H_3C–CH(Cl)–COO^- $\xrightarrow[\text{enzyme}]{\text{bacterial}}$ (R)-H_3C–CH(OH)–COO^- $+ Cl^-$

While alkylation is a perfectly normal biosynthetic process, not all alkylation reactions are biologically desirable. In particular, there are often harmful effects when the substrate comes from outside the organism and the reaction is therefore not under enzyme control. For instance, many simple alkylating agents, including haloalkanes such as CH_3I, are known to be **mutagenic**—they produce mutations in cells. A mutation is a structural change which the cell passes on genetically to its offspring. Some of these alkylating agents are also *carcinogenic*. They induce a genetic change which causes cells to proliferate wildly, without the normal restraints.

Alkylating agents attack nucleophilic oxygen, sulfur, and nitrogen atoms (eq. 6.59). Since the biomolecules that contain these atoms play critical roles in cell chemistry, their alkylation can lead to serious malfunction.

(6.59) $\ddot{N} + CH_3{-}X \rightarrow N^+{-}CH_3 \;\; X^-$

For example, the polymeric DNA molecules that transmit the cell's genetic information are composed of thousands of organic bases. These bases have numerous nitrogen atoms that are susceptible to alkylation. Alkylation of these bases can lead to fragmentation of the long-chain DNA molecule which would interfere with its ability to pass on the correct genetic information (See Box 6–1).

Epoxides are potent alkylating agents because of their high reactivity toward nucleophiles (eq. 6.60), and they therefore constitute a health hazard.

(6.60) $\ddot{N}{-}H$ + epoxide (O) → $H{-}N^+$–C–C–O^- → $\ddot{N}$–C–C–OH

For instance, synthetic epoxides used in powerful adhesives are made from epoxy resins that can bind to skin. Epoxides are also synthesized in various human tissues as part of their normal metabolic processes (sec. 14:11), but these tissues also contain enzymes that render the epoxides harmless by catalyzing their hydrolysis to diols (eq. 6.49). There is evidence, however, that cigarette smoke inhibits the action of these hydrolytic enzymes, thus permitting an accumulation of the toxic epoxides.

Chapter 6 has shown that an alkane carrying a good leaving group will undergo substitution in the presence of a nucleophile. As we noted earlier, such a substituted alkane can also undergo elimination. This is the subject of the next chapter.

Summary

1 Nucleophiles contain nonbonding electron pairs that can be used to form additional covalent bonds; examples include Cl^-, Br^-, I^-, CN^-, and NH_3. Formation of a new bond requires that a leaving group be displaced from the substrate. Reactions of alkyl halides with halide ions in acetone follow second-order kinetics, and are subject to steric hindrance from the alkyl group (Table 6–1). These second-order nucleophilic displacements also show inversion of configuration when chiral or cyclic substrates are used. These conditions define the S_N2 reaction (secs. 6:1 to 6:3).

R—X +	:Nu	→ R—Nu +	:X	rate = k[RX][Nu]
Substrate	Nucleophile	Product	Leaving group	

2 In the transition state for an S_N2 reaction, the entering and leaving groups are on opposite sides of the carbon undergoing substitution and are connected to it by weak bonds (sec. 6:4). The transition state fully accounts for the stereochemical and steric characteristics of the reaction (sec. 6:5).

3 In hydroxylic solvents such as water, 3° halides dissociate slowly to carbocations. These ions then react rapidly with any nucleophile present, particularly the solvent, which is present in highest concentration. The reaction rate depends only on the substrate concentration in these S_N1 reactions, because formation of the carbocation is rate determining. Optically active substrates give partly racemic products, in contrast to S_N2 reactions, which are completely stereospecific (sec. 6:6 and Table 6–3).

$$\text{R—X} \xrightarrow{\text{slow}} \text{R}^+ + :\text{X}^- \xrightarrow[\text{fast}]{:\text{Nu}} \text{R—Nu} + :\text{X}^- \qquad \text{rate} = k[\text{RX}]$$

Carbocation stabilities lie in the order 3° > 2° > 1°. The stability order depends on how effectively the positive charge can be dispersed from the trivalent carbon onto the neighboring atoms in the ion. Alkyl groups have polarizable electrons that are attracted by the positively charged carbon atom, thus dispersing charge onto the alkyl groups. That is why cation stability increases as the degree of substitution increases. Relief of steric strain also favors 3° carbocations most and 1° carbocations least.

The more stable the carbocation, the faster it forms, because ion formation is endothermic and the transition states resemble the ions. Carbocations may rearrange by shifting hydride or alkyl groups. Rearrangements are fast if they result in increased ion stability (secs. 6:7 and 6:8).

4 Alcohols are named by an IUPAC method based on the longest chain containing the OH group. The OH group is given the lowest possible number. The radicofunctional "alcohol" system is used only with low molecular weight compounds. Ethers are named as alkoxyalkanes in the IUPAC system. In the radicofunctional system the name of the two alkyl groups precedes "ether."

Alcohols boil at a higher temperature than the isomeric ethers, due to hydrogen bonding. Alcohols also react with highly electropositive metals, while ethers do not (sec. 6:9).

5 Polar solvents consist of molecules containing polar bonds. They have large dielectric constants (Table 6–6) and are good solvents for ions. Polar solvents containing OH groups solvate anions by hydrogen bonding, thus decreasing their nucleophilicity. These solvents promote S_N1 reactions. Dipolar aprotic solvents (Table 6–7) dissolve salts by solvating their cations but they cannot hydrogen bond to the anion. The anions are left free, thus promoting S_N2 reactions (sec. 6:10).

6 Leaving group ability is inversely related to basicity (Fig. 6–16). Very basic groups, such as OH^- and OR^-, are ineffective as leaving groups. They can be made into effective leaving groups by protonation. Heating alcohols and ethers with solutions of hydrogen halides results in their being converted to alkyl halides. Alcohols can be converted to alkyl halides by PBr_3 and by $SOCl_2$, while TsCl converts the OH group to OTs, which can be easily displaced. Rearrangements can usually be avoided by use of these reagents (sec. 6:11).

Primary and secondary halides and tosylates react with hydroxide ions (S_N2) to yield 1° and 2° alcohols. To produce 3° alcohols, 3° alkyl halides are allowed to solvolyze in water (S_N1). Alkoxide ions react with 1° and 2° halides and tosylates to give ethers. When unsymmetrical ethers are being synthesized, the more highly substituted alkoxide is allowed to react with the less highly substituted halide in an S_N2 reaction (sec. 6:12).

Intramolecular S_N2 reactions in 2-haloalcohols lead to epoxides. Because of ring strain, epoxides are very reactive toward nucleophiles (sec. 6:13).

7 Organometallic reagents serve as the equivalent of carbon nucleophiles. Their bonds are polarized in the $M^{\delta+}$—$C^{\delta-}$ sense. Organolithium and Grignard reagents are made by allowing alkyl halides to react with Li and Mg metals, respectively. These reagents react with oxirane to yield 1° alcohols. Alkylcopper compounds react with alkyl halides to give alkanes, R—R′ (sec. 6:14).

8 Enzyme-controlled alkylations are found in many normal biosynthetic processes. They obey the same mechanisms as do the nucleophilic substitutions studied in the laboratory. Nonenzymatic alkylation reactions, involving nonbiological alkylating agents, can cause pathological changes in the cell (sec. 6:15).

Problems

6–26 If (+)-2-iodooctane is allowed to stand in a solution of sodium iodide in acetone, it is eventually converted to racemic (±)-2-iodooctane. Explain this result. Why isn't the product (−)-2-iodooctane?

6–27 A sample of 2,2-dimethylpropanol-1-*d* (**IV**; *d* represents deuterium) was obtained optically pure. C-1 in this compound is chiral because it carries two different isotopes of hydrogen. The alcohol was converted to a tosylate, **V**, by treatment with tosyl chloride and base, and the tosylate was then allowed to react with LiCl to give a chloride, **VI**. Write the structures of the tosylate and chloride, showing their absolute configurations. Explain your reasoning.

$$\underset{\textbf{\textit{S}-IV}}{\mathrm{H{-}C(D)(OH){-}C(CH_3)_3}} \xrightarrow[OH^-]{TsCl} \underset{\textbf{V}}{(CH_3)_3C{-}CHDOTS} \xrightarrow[DMF]{LiCl} \underset{\textbf{VI}}{(CH_3)_3C{-}CHDCl}$$

6–28 The hydrolysis of 1-chloropropane is a very slow reaction (eq. 11). It can be speeded up by the addition of a small quantity of sodium iodide. The iodide is a true catalyst; it is not consumed in the course of the reaction. Explain in detail how sodium iodide catalyzes this process.

$$CH_3CH_2CH_2Cl + H_2O \rightarrow CH_3CH_2CH_2OH + H^+ + Cl^- \qquad (11)$$

6–29 A student wishes to make some $Br(CH_2)_6CN$ by the action of CN^- on 1,6-dibromohexane. What problems would arise in the course of this synthesis?

6–30 Draw the structure of the carbocation that would be formed by loss of a halide ion from each of the following compounds. Classify the carbocation as 1°, 2°, or 3°. If the ion can undergo energetically favorable rearrangement(s), show the structure of the rearranged ion(s) as well.

a. 2-chlorobutane
b. 3-bromo-3-methylpentane
c. 1-chloro-2-methylpentane
d. 1-iodopentane
e. 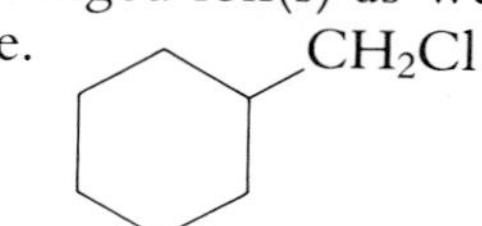

f. 3-bromo-4,4-dimethylhexane
g. 3-ethyl-2-iodopentane
h. 1-bromo-2-ethyl-2-methylhexane

6–31 *t*-Butyl chloride is warmed in a solution of potassium iodide, KI, in a mixture of methanol, ethanol, and water. Product analysis shows that four products are formed. Give their structures.

6–32 When 2-methyl-3-ethyl-2-bromopentane is warmed with propanoic acid,

$$CH_3CH_2C(=O)OH$$

two isomeric products of formula $C_{11}H_{22}O_2$ are formed. Give the ir structures and explain their formation.

6–33 Give the IUPAC name for each of the following compounds:

a. $CH_3C(CH_3)(CH_2F){-}CH(F){-}CH_2CH(CH_2CH_3)CH_3$

b. (cyclopropyl)–CH(OH)–(cyclopropyl)

c. (cyclobutyl)–CH(CH_3)–CH_2Br

d. $CH_3O{-}C(Cl)(H){-}CH_2{-}CH(OCH_2CH_3)CH_3$

e. 1,2-disubstituted cyclohexane with OCH_3 (wedge) and OCH_3 (dash)

6–34 a. Show the structures of all the cyclic alcohols of formula C_4H_8O.
b. Show the structures of all the cyclic compounds of formula C_4H_8O that are *not* alcohols.

6–35 Show the structures of all the tertiary alcohols of formula $C_7H_{16}O$.

6–36 Draw the structures of all the alcohols of formula $C_5H_{12}O$ and name them by the IUPAC system. Classify them as primary, secondary, or tertiary alcohols.

6–37 Anhydrous potassium hydroxide dissolves easily in dry ethanol. What species will be present in the solution? Write an expression for the equilibrium constant that relates to the equilibrium among these species.

6–38 What factors are likely to affect the position of equilibrium for the system shown in Equation 12? In particular, explain how the solvent could alter the equilibrium in the manner found.

$$I^- + CH_3Br \rightleftarrows CH_3I + Br^- \qquad (12)$$
$$K(H_2O) = 15.4$$
$$K(\text{acetone}) = 0.62$$

6–39 a. Ethanol is a moderately good solvent for salts, since it can solvate both anions and cations. Show how ethanol can solvate the ions from lithium bromide.
b. Dimethyl ether, an isomer of ethanol, is less effective than ethanol as a solvent for salts. Is it a poorer solvent for cations or anions? Explain your answer.

6–40 The reaction in Equation 13 was studied in water and in acetone. In water the rate constants decreased in the order $k^{I^-} > k^{Br^-} > k^{Cl^-}$. In acetone, however, the reverse order was observed: $k^{Cl^-} > k^{Br^-} > k^{I^-}$. Explain why the relative nucleophilicities reversed when the solvent was changed.

$$X^- + CH_3OTs \xrightarrow[\text{solvent}]{} CH_3X + OTs^- \qquad (X = Cl, Br, I) \qquad (13)$$

6–41 CH_2Cl_2 is used on a very large scale for paint removal and as an aerosol propellant. U.S. production amounted to about 550 million pounds in 1979. It is often obtained from the mixture resulting from the free-radical chlorination of methane (sec. 5:1). When the economics are favorable, dichloromethane may also be synthesized, in two steps, from methanol. Suggest a route for this synthesis.

6–42 Both *t*-butyl chloride and 4-chloro-2,2,4,6,6-pentamethylheptane are 3° chlorides. However, the chloroheptane solvolyzes 500 times faster than *t*-butyl chloride in aqueous ethanol. Draw a projection structure of the chloroheptane and use it to explain the rate difference. A model may be helpful.

6–43 Show all the products that are formed when 1 mol of PBr_3 is treated at room temperature with 3 mol of each of the following alcohols:
a. methanol
b. cyclobutanol
c. 2-pentanol
d. 1,4-butanediol
e. 2-methyl-1-butanol
f. *cis*-3-ethylcyclopentylmethanol

6–44 When (*R*)-2-methyl-1-butanol is heated in 48% aqueous HBr, two isomeric products, **A** and **B**, of formula $C_5H_{11}Br$, are formed in 80% and 20% yield, respectively. **A** is found to be optically active, while **B** is optically inactive. Write out this chemistry and explain the result.

6–45 When cyclopentylmethanol is boiled with concentrated HCl, three isomeric products of formula $C_6H_{11}Cl$ are formed. Give the structures of these products and explain the mechanism of their formation.

6–46 Describe methods for effecting the following conversions. Specify all reagents and conditions.

a. 1-butanol → 1-bromobutane (two ways)
b. cyclohexanol → cyclohexyl chloride
c. *trans*-3-methylcyclopentanol → *trans*-3-methylcyclopentyl cyanide
d. (*S*)-3-heptanol → (*S*)-3-chloroheptane
e. 1-pentanol → dipentyl ether
f. isopropyl bromide → isopropyl methyl ether
g. isobutyl alcohol → cyclohexyl isobutyl ether

6–47 Show how you would make the following ethers by removing water from two molecules of appropriate alcohols, using an acid catalyst (see Prob. 6–23).

a. dimethyl ether
b. diisopropyl ether
c. dibutyl ether

6–48 Show the best Williamson synthesis route to each of the following ethers:

a. cyclopentyl ethyl ether
b. *s*-butyl propyl ether
c. [cyclohexane bearing CH_3 and CH_2OCH_3 on the same carbon]
d. butyl cyclopropyl ether

6–49 a. Draw the structures of all the alkoxide ions of formula $C_4H_9O^-$ and name each. Which of these can exist in enantiomeric forms?

b. Show the product of the reaction of each anion of (a) with methyl iodide. Name each product.

6–50 Clearly show the stereochemistry of the ether derived from each pair of the following reactants:

a. potassium (*R*)-2-pentoxide and (*R*)-2-bromopentane
b. sodium *cis*-3-methylcyclohexoxide and (*S*)-2-bromopentane
c. *trans*-3-ethylcyclobutyl bromide and lithium (*R*)-2-butoxide

6–51 Give the structure of the substitution products from each of the following reactions. Specify their stereochemistry.

a. (*R*)-2-chloropentane $\xrightarrow[\text{DMSO}]{\text{KCN}}$

b. (*S*)-3-methyl-3-heptanol $\xrightarrow[\text{H}_2\text{O, heat}]{\text{HI}}$

c. (*S*)-3-heptanol $\xrightarrow[\text{H}_2\text{O, heat}]{\text{HI}}$

d. *trans*-2-methylcyclobutanol $\xrightarrow[0^\circ]{\text{PBr}_3}$

e. (+)-2-bromobutane $\xrightarrow[\text{acetonitrile}]{\text{LiCl}}$

f. (+)-*cis*-3-ethylcyclohexanol $\xrightarrow[\text{DMF}]{\text{SOCl}_2}$

g. (R)-$CH_3\underset{|}{C}HCH_2OH$ $\xrightarrow[\text{DMSO}]{\text{KCN}}$
Cl

6–52 In the presence of acid catalysts straight-chain primary alcohols give unrearranged bromides and iodides. If there is a branch at the carbon atom *next to* the primary carbon atom (the β carbon), as in isobutyl alcohol, then rearrangement becomes much more probable. Neopentyl alcohol (2,2-dimethylpropanol) is notorious for the ease with which it rearranges upon attempted conversion to neopentyl halides. If the hydrocarbon branch is further away, then the tendency toward rearrangement disappears. How would you explain these findings?

6–53 A sample of 2-octanol has $[\alpha]_D = -9.9°$. When a portion of this alcohol is allowed to react with sodium and the salt is then treated with ethyl iodide, ethyl 2-octyl ether is isolated; it has $[\alpha]_D = -17.5°$. A second portion of the same 2-octanol is treated with tosyl chloride in base, leading to the isolation of 2-octyl tosylate, $[\alpha]_D = -6.8°$. When this tosylate is solvolyzed in dry ethanol, ethyl 2-octyl ether is produced; it has $[\alpha]_D = +17.5°$. What conclusions can *rigorously* be drawn (without unfounded assumptions) from these experiments?

6–54 A mixed ether such as *t*-butyl ethyl ether can be prepared in high yield by heating a mixture of ethanol and *t*-butyl alcohol (2-methyl-2-propanol) in the presence of acid. Write a mechanism for this synthesis, and explain why it gives the mixed product almost exclusively (i.e., virtually no diethyl ether or di-*t*-butyl ether is formed; see Prob. 6–23).

6–55 Diethylene glycol, **VIIa**, triethylene glycol, **VIIb**, and their dimethyl ethers are important solvents. Their synthesis uses oxirane (ethylene oxide) and ethylene glycol as starting materials. Show how **VIIa** and **VIIb** might be synthesized.

$HOCH_2CH_2OCH_2CH_2OH$ **VIIa**

$HOCH_2CH_2OCH_2CH_2OCH_2CH_2OH$ **VIIb**

6–56 a. 3-Bromo-1,2-epoxybutane reacts with sodium methoxide to give a new oxirane, 1-methoxy-2,3-epoxybutane (eq. 14). Suggest a mechanism for this reaction.

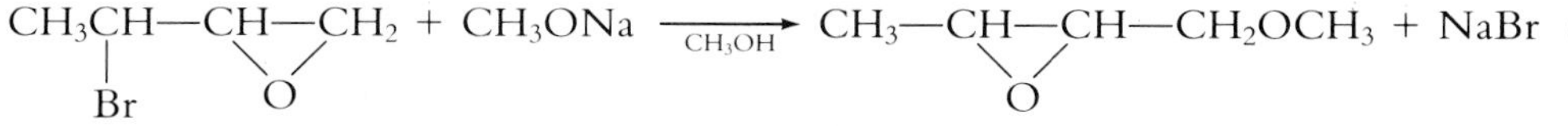

(14)

b. Predict the stereochemistry of the product when the reactant is (2R, 3R)-3-bromo-1,2-epoxybutane, **VIII**.

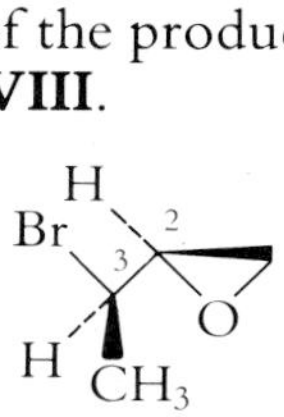

VIII

6–57 When 4-chloro-1-butanol is treated with NaOH it is converted to **A**, C_4H_8O. Chemical tests show that **A** does not contain either an OH or a C=C group. Treating 5-chloro-1-pentanol in a similar way gives rise to a similar product **B**, $C_5H_{10}O$. Like **A**, **B** lacks both OH and C=C groups. Give structures for **A** and **B** and show how they are formed.

Higher homologs of the chloroalcohols, such as 6-chloro-1-hexanol, 7-chloro-1-heptanol, and so on do *not* give products analogous to **A** and **B** when treated with NaOH. What factor(s) make formation of **A** and **B** so favorable?

6–58 1,4-Butanediol is an important intermediate for the manufacture of various polymers (Chap. 22). The diol can be obtained by the acid-catalyzed reaction between a common ether and water. Which ether is involved? Show how it reacts with water under acid catalysis.

6–59 When (2*R*, 3*R*)-*trans*-2,3-dimethyloxirane, **IX**, is treated with concentrated HBr it yields 3-bromo-2-butanol. When this alcohol is allowed to react with PBr_3 it gives an optically active 2,3-dibromobutane. Write out all the chemistry, showing the absolute configuration of each asymmetric carbon atom.

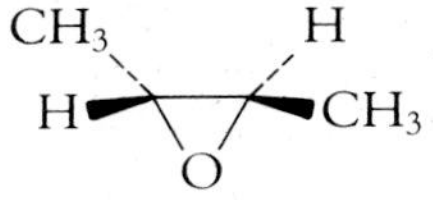

CH_3 H
H CH_3
O

IX

6–60 1,4-Butanediol may be dehydrated to yield the cyclic ether tetrahydrofuran. Analogously, 4-methyl-1,4-hexanediol can be dehydrated to 2-ethyl-2-methyltetrahydrofuran (eq. 15). A series of experiments is carried out with optically active (*S*)-4-methyl-1,4-hexanediol.

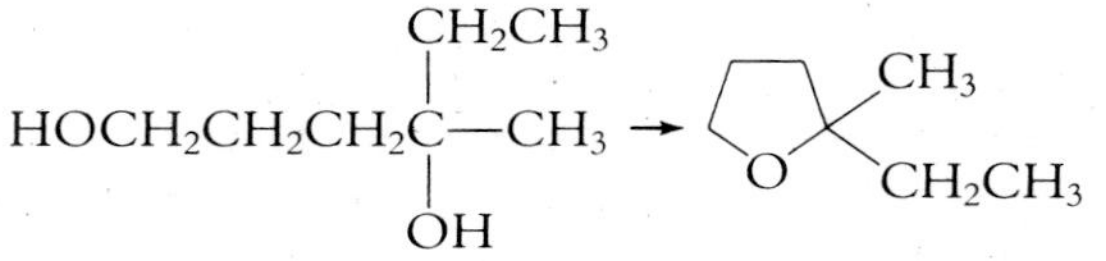

$$HOCH_2CH_2CH_2C(CH_2CH_3)(OH)—CH_3 \rightarrow \text{2-ethyl-2-methyltetrahydrofuran } (CH_3, CH_2CH_3) \quad (15)$$

a. When the (*S*)-diol is heated with acid, the 2-ethyl-2-methyltetrahydrofuran is completely racemic. Write a mechanism that accounts for this finding.
b. The (*S*)-diol is treated with one equivalent of tosyl chloride to form a monotosylate. Allowing the monotosylate to react with OH^- gives (*S*)-2-ethyl-2-methyltetrahydrofuran. Write a mechanism for this reaction.
c. Explain why the ring closure takes two different courses under the different conditions.

6–61 Using the relative electronegativities of C, N, and O, explain why the reactions in Equations 16 and 17 go rapidly to completion.

$$CH_3—MgBr + CH_3O—H \rightarrow CH_3—H + CH_3O—MgBr \quad (16)$$

$$CH_3—MgBr + CH_3NH—H \rightarrow CH_3—H + CH_3NH—MgBr \quad (17)$$

6–62 Several years ago it was shown that an ether is not needed as a solvent to make Grignard reagents; instead, a tertiary amine such as triethylamine,

$(CH_3CH_2)_3N$, can be used. Explain why tertiary amines are good solvents for Grignard reagents.

6–63 2-Chloroethyldimethylamine, **X**, is a member of a class of highly reactive alkylating agents called *nitrogen mustards* (see Box 6–1). Kinetic studies show that **X** forms, in the slow step, a very reactive intermediate which then reacts rapidly with a nucleophile (eq. 18). Compounds like **X** are used to combat cancer, since the rapidly growing cells are sensitive to chemical attack.

$$\underset{\mathbf{X}}{(CH_3)_2\ddot{N}CH_2CH_2Cl} \xrightarrow[\text{slow}]{} \text{intermediate} \xrightarrow[\text{fast}]{OH^-} (CH_3)_2\ddot{N}CH_2CH_2OH \qquad (18)$$

a. Suggest a structure for the intermediate and explain its high reactivity toward nucleophiles. (Review the discussion of neighboring group effects, sec. 6:13.)

b. The ammonium salt **XI**, closely related to **X**, is not unusually reactive toward nucleophiles. Show how this result fits your answer in (a).

$$CH_3-\overset{\overset{CH_3}{|}}{\underset{\underset{CH_3}{|}}{N^{+}}}-CH_2CH_2Cl \qquad Cl^-$$

XI

6–64 Methionine (eq. 6.56) can be alkylated on sulfur by the important biological molecule adenosine triphosphate, ATP (eq. 19). The product (*S*)-adenosylmethionine, **XII**, can be used to methylate other molecules. For instance, **XII** converts 2-aminoethanol to the nerve transmitter choline (eq. 20). Write a reasonable mechanism for choline formation.

$$\underset{\text{Methionine}}{^-OOC\underset{\underset{NH_3^+}{|}}{C}H(CH_2)_2-\overset{\overset{CH_3}{|}}{\ddot{S}}:} + \underset{\text{ATP}}{RCH_2-OP_3O_9H_4} \xrightarrow{\text{enzyme}}$$

$$\underset{\mathbf{XII}}{^-OOC\underset{\underset{NH_3^+}{|}}{C}H(CH_2)_2-\overset{\overset{CH_3}{|}}{\ddot{S}^+}-CH_2R} + OP_3O_9H_4^- \qquad (19)$$

$$H_2\ddot{N}CH_2CH_2OH + 3\mathbf{XII} \xrightarrow[\text{(several steps)}]{\text{enzyme}} \underset{\text{Choline}}{CH_3-\overset{\overset{CH_3}{|}}{\underset{\underset{CH_3}{|}}{N^{+}}}-CH_2CH_2OH} \qquad (20)$$

Supplementary Reading

W. H. Saunders, Jr., *Ionic Aliphatic Reactions*, Chap. 2, Prentice-Hall, Inc., Englewood Cliffs, N.J., 1965.

L. M. Stock. The origin of the inductive effect, *J. Chem. Ed.* **49,** 400 (1972).

D. Whittaker, *Stereochemistry and Mechanism*, Oxford Univ. Press, London, 1973.

L. N. Ferguson, Cancer: How can chemists help? *J. Chem. Ed.* **52,** 495 (1975).

O. T. Benfey, *Introduction to Organic Reaction Mechanisms,* McGraw-Hill, New York, N.Y., 1970.

7 The Structure of Alkenes: Their Synthesis by Way of Elimination Reactions

In Chapter 6 alkenes were introduced briefly as the elimination by-products of nucleophilic substitution reactions. Alkenes are more than mere by-products, however—they are a very important class of organic compounds in their own right. They are versatile intermediate substances used for laboratory syntheses, and many of them play crucial roles in various biological and industrial processes. The usefulness of alkenes stems from their high reactivity, which is due to their double bonds. In particular, the π bond is relatively weak (compare Tables 2–2 and 2–4, pp. 53 and 55, respectively) and therefore susceptible to attack by many different reagents.

You've already seen that elimination and nucleophilic substitution reactions take place simultaneously with many substrates. In this chapter we will show how elimination can be promoted and substitution suppressed, so that the maximum yield of alkenes is obtained. We begin by explaining some crucial differences between carbon–carbon single and double bonds, and then we describe how the latter are formed by way of elimination reactions.

7:1 Carbon–Carbon π-Bonding: How It Creates Geometrical Isomerism in Alkenes

The carbon–carbon double bond contains two different types of electron-pair bonds (sec. 2:5). One is the familiar σ bond in which two hybrid atomic orbitals overlap one another head-to-head. The second type is a π bond, which uses two unhybridized *p* orbitals arranged for side-by-side overlap (Fig. 7–1). The doubly-bonded carbon atoms are sp^2 hybridized, so that they

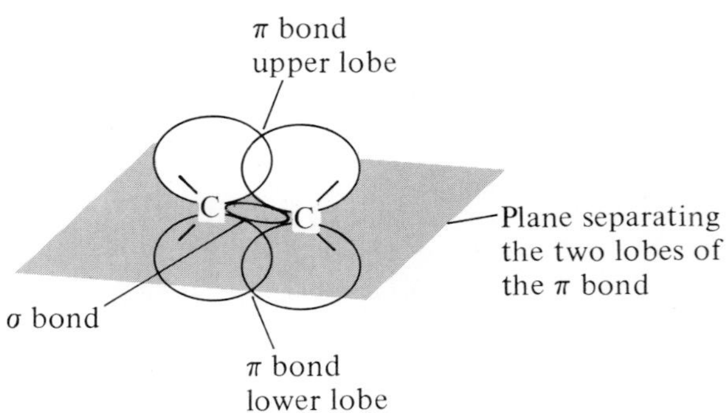

Figure 7–1. *The σ and π bonds, which together form the carbon–carbon double bond.*

and the two atoms attached to each carbon all lie in one plane. This plane divides the π orbital into two lobes.

There are important physical differences between the σ and π bonds. The π electrons are, on the average, farther away from the nucleus than the σ electrons and therefore less tightly held. As a result, *π bonds are usually weaker and more reactive than σ bonds* (Chaps. 10 and 11). Another difference is that the bond energy is affected by the geometry of the groups attached to the double bond. When these groups all lie in the same plane (Fig. 7–1), the *p* orbitals experience maximum overlap and the π bond is strongest. If one *p* orbital is rotated with respect to the other, however, then the overlap decreases and the π bond weakens (Fig. 7–2). The result is that rotation about carbon–carbon double bonds is very difficult, requiring an activation energy of about 60 kcal/mol. At room temperature (25° C), therefore, double bonds are rotationally rigid.

The rigidity allows a type of isomerism not found with open-chain alkanes and their derivatives. Consider as examples the dichloro derivatives

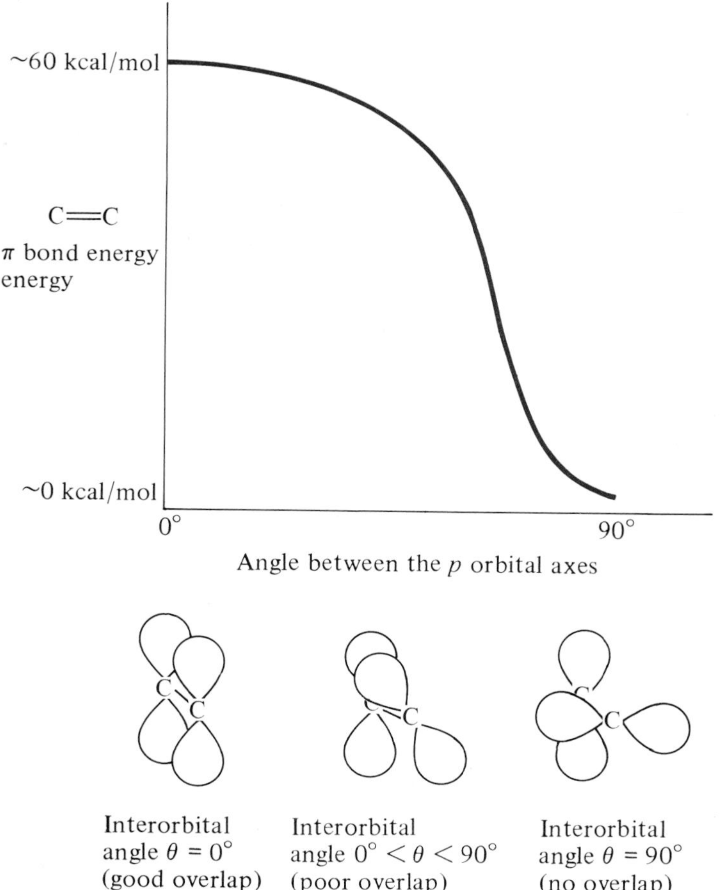

Figure 7–2. *Changes in the C═C π bond energy as one* p *orbital is twisted with respect to the other.*

of ethylene, $CH_2{=}CH_2$. There are three isomers—one 1,1-dichloroethylene and *two* 1,2-dichloroethylenes. Because of the high barrier to rotation, the

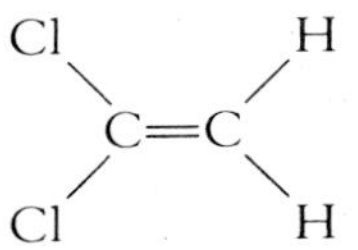

$$\underset{\text{1,1-Dichloroethylene (bp 32°)}}{(Cl)_2C{=}CH_2} \qquad \underset{\text{1,2-Dichloroethylene (bp 60°)}}{\text{cis-}ClHC{=}CHCl} \qquad \underset{\text{1,2-Dichloroethylene (bp 48°)}}{\text{trans-}ClHC{=}CHCl}$$

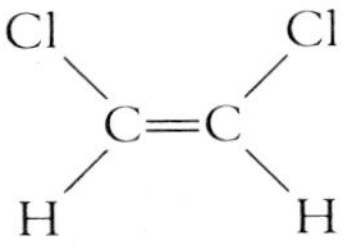

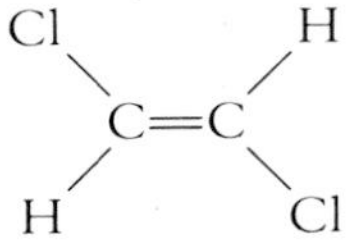

1,2-dichloroethylenes do not interconvert at room temperature. Since they differ only in the spatial arrangement of their atoms, they are stereoisomers (sec. 4:1) of the *cis–trans* type. More specifically, they are configurational isomers that exhibit *cis–trans* isomerism (sec. 3:12). Since *cis* and *trans* isomers are diastereomers, they have different physical properties and can be separated by conventional physical means. Note that the dichloroethylenes do not exhibit optical isomerism (enantiomerism). They are achiral because the sp^2 hybridization of the carbon atoms results in their being planar (sec. 4:2).

7:2 Naming Alkenes and Alkene Configurational Isomers

The simplest alkene is ethylene. The IUPAC name for this compound is ethene, but the older name is so entrenched that we will use it throughout this book. We will use the standard IUPAC system, however, for all other alkenes named here.

The name of each alkene is based on the longest carbon chain *that contains the double bond*. The root (eth-, prop-, and so forth) indicates the number of carbon atoms in this chain, while the suffix-*ene* indicates the presence of a double bond. If the double bond can occupy more than one nonequivalent position in the molecule, its location is given by a numerical prefix. The chain is numbered so that the double bond occupies the lowest numbered position—this rule takes precedence over assigning low numbers to alkyl, alkoxyl, and halogen substituents. Only one locator number is used for the double bond, that for the doubly-bonded carbon at the lower numbered position. Thus, in 1-butene the double bond lies between carbons 1 and 2, while in 2-hexene, it connects carbon 2 and 3. The following examples illustrate how the system works:

$CH_2{=}CH_2$ — Ethene

$\overset{1}{C}H_2{=}\overset{2}{C}H\overset{3}{C}H_2\overset{4}{C}H_3$ — 1-Butene

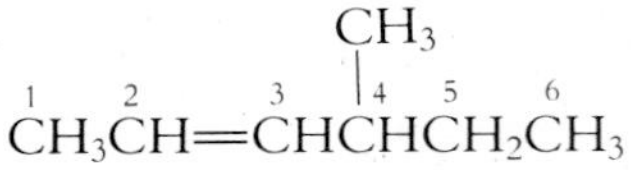

$\overset{1}{C}H_3\overset{2}{C}H{=}\overset{3}{C}H\overset{4}{C}H(CH_3)\overset{5}{C}H_2\overset{6}{C}H_3$ — 4-Methyl-2-hexene

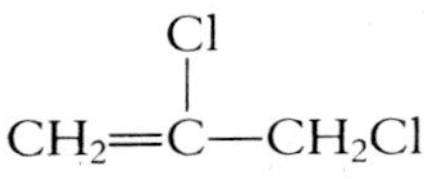

$CH_2{=}C(Cl){-}CH_2Cl$ — 2,3-Dichloropropene

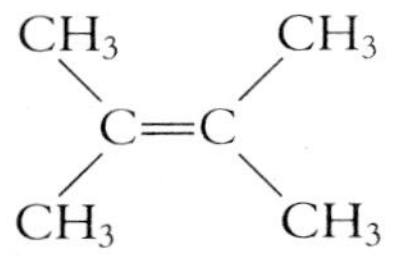

$(CH_3)_2C{=}C(CH_3)_2$ — 2,3-Dimethyl-2-butene

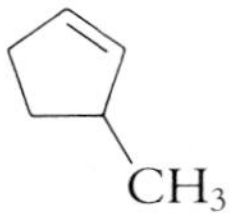

3-Methylcyclopentene

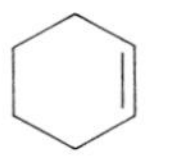

Cyclohexene

If a molecule contains both a double bond and a hydroxyl group, it is named as an alkenol (the name ends in "ol") and the hydroxyl-bearing carbon atom is given the lowest number.

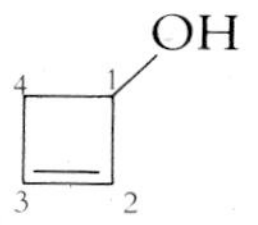

2-Cyclobuten-1-ol

The group $CH_2{=}CH{-}$ is a common feature in organic compounds. It has been given the trivial name *vinyl.* For this reason hydrogen atoms attached to double bonds are sometimes referred to as *vinyl hydrogens.* The vinyl terminology is now used primarily in simple structures, such as

$CH_2{=}CHBr$	$CH_2{=}CHOCH_2CH_3$
Bromoethylene (*vinyl bromide*)	Ethoxyethylene (*ethyl vinyl ether*)

Problem

7–1 Draw structures corresponding to the following names:

a. 3,3-dimethyl-1-butene
b. 2-ethyl-1-butene
c. 1,3-dichlorocyclopentene
d. *trans*-4,5-dibromocyclohexene
e. 2,3-dimethyl-4-penten-2-ol

Before the IUPAC system was adopted the collective name for alkenes was **olefins**, because some common chemicals synthesized from ethylene were rather oily. The terms olefin and alkene are now used interchangeably.

Compounds containing double or triple bonds are classified as *unsaturated. Saturated* compounds have no multiple bonds and therefore contain the maximum number of hydrogen atoms compatible with their carbon skeleton. Unsaturated alkenes add hydrogen to their double bonds and become converted into saturated alkanes (sec. 7:3). Note that both alkenes and cycloalkanes fit the same general formula, C_nH_{2n}. However, only alkenes are unsaturated and readily add hydrogen.

In order for there to be two configurational (*cis–trans*) isomers of a single alkene, *it is necessary that each of the doubly-bonded carbon atoms be attached to two different groups.* The designations *cis* and *trans* refer to a plane drawn lengthwise through the double bond and perpendicular to the nodal plane (Fig. 7–3).

For 1,2-disubstituted double bonds, the *cis* isomer is simply the one in which both substituents are on the same side of the double bond. The *trans* isomer has the substituents on opposite sides of the double bond.

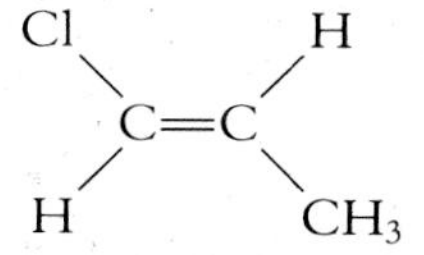

trans-1-Chloro-1-propene

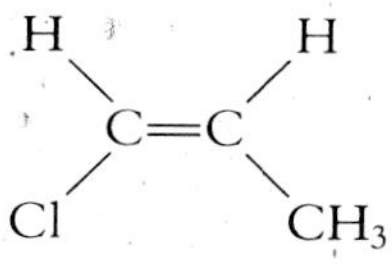

cis-1-Chloro-1-propene

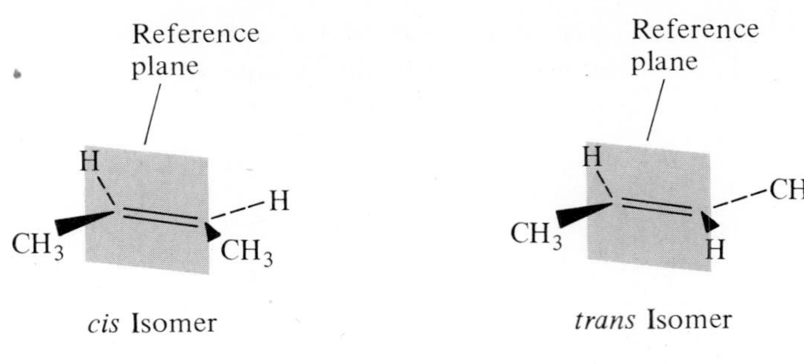

Principal carbon chain lies on one side of reference plane

Principal carbon chain crosses the reference plane

Figure 7–3. Cis *and* trans *isomers of 2-butene, showing the reference plane. The stereochemical designation depends on whether or not the principal chain crosses the reference plane.*

Problem

7–2 Apply the *cis–trans* nomenclature to the two 1,2-dichloroethylenes discussed in Section 7:1.

The *cis–trans* nomenclature system can only be applied to a small number of alkenes. For example, it does not provide a way to distinguish between the two isomers of 2-bromo-1-chloro-1-fluoroethylene (ClFC=CHBr). An unambiguous naming system for alkene isomers was therefore devised. In the modern system, priorities are assigned to each of the two groups attached to each of the doubly-bonded carbons. We use the same priorities (based on the atomic number of the substituents) as those developed for the *R,S* absolute configuration designations (sec. 4:5, Table 4–1, p. 123).

When the higher priority groups *at each carbon* are on the *same* side of the double bond, the isomer is designated *Z* (from the German *zusammen,* meaning "together"). If the groups of higher priority at each carbon are on opposite sides of the double bond, the molecule is designated as an *E* isomer (from the German *entgegen,* meaning "opposite"; see Fig. 7–4).

(*Z*)-2-Bromo-1-chloro-1-fluoroethane

(The two groups of higher priority, Br and Cl, are on the same side.)

(*E*)-2-Bromo-1-chloro-1-fluoroethane

(The two groups of higher priority, Br and Cl, are on opposite sides.)

Figure 7–4. (Z) *and* (E) *isomer designations of 2-bromo-1-chloro-1-fluoroethene.*

Table 7–1. Physical Properties of Some Alkenes

	Melting Point (°C)	Boiling Point (°C)
ethylene	−169	−104
propene	−185	−48
1-butene	−185	−6.3
(*Z*)-2-butene	−139	3.7
(*E*)-2-butene	−106	0.9
(*Z*)-2-pentene	−178	38
(*E*)-2-pentene	−135	36
(*Z*)-3-hexene	−135	67
(*E*)-3-hexene	−113	67
1-octene	−102	121
1-dodecene	−35	213

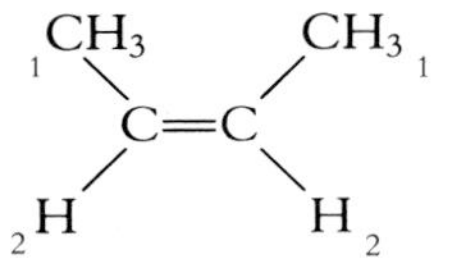

(*Z*)-2-Butene

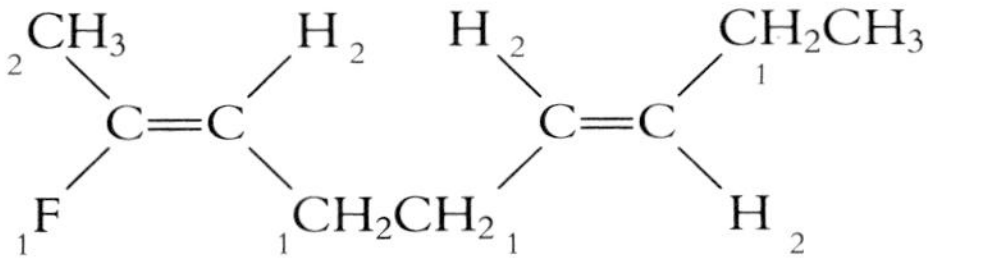

(2*Z*,6*E*)-2-Fluoro-2,6-nonadiene

Problem

7–3 Write the structures of all the open-chain compounds having the molecular formula C_4H_7Cl. Pay attention to *cis–trans* as well as constitutional isomerism. Name each isomer (you should find 11 if you do not distinguish between optical isomers). For *cis–trans* isomers use the *Z–E* nomenclature system.

In Section 7:1 we showed that *cis* and *trans* isomers are diastereomers and therefore have different physical properties. This means that they can usually be separated from one another by conventional techniques such as distillation and crystallization. Melting points may differ significantly for a pair of *cis–trans* isomers because their different shapes will strongly influence how they pack into a crystal lattice (Table 7–1). Their boiling points are much closer to one another because the isomers have identical molecular compositions.

7:3 Determining the Relative Stabilities of Alkene Isomers: Heats of Hydrogenation

When alkenes are prepared in the laboratory we usually end up with a mixture of product isomers, instead of a single isomer. This situation has led chemists to devise different reaction conditions which maximize the yield of the desired isomer. To do this we need to know a good deal about the alkenes, particularly the relative stabilities of isomeric compounds. In this section we show how those relative stabilities are measured.

Chapter 3 demonstrated how stability differences between alkane isomers are determined by comparing their heats of combustion. The same method could be applied to alkenes; however, heats of combustion are large, while energy differences between alkene isomers are small, and it is difficult to accurately measure small differences when comparing large numbers. As an alternative, we take advantage of the exothermic reaction between hydrogen and alkenes that gives alkanes. The heat released during this reaction is called the **heat of hydrogenation**. Heats of hydrogenation are much smaller and can be measured more precisely than heats of combustion. If two alkenes yield the same alkane after hydrogenation, then the difference in their heats of hydrogenation equals their difference in stability. The alkene isomer with the larger heat of hydrogenation is the less stable of the pair.

Equations 7.1 to 7.3 show the hydrogenation of several butene isomers to butane and give the heats of hydrogenation. The data are summarized in Figure 7–5. The hydrogenation reaction is exothermic, yet it is very slow in the absence of a catalyst.

$$(CH_3CH_2)(H)C{=}C(H)(H) + H_2 \xrightarrow{Pt} CH_3CH_2CH_2CH_3 \quad \Delta H = -30.3 \text{ kcal/mol} \tag{7.1}$$

1-Butene

$$(CH_3)(H)C{=}C(CH_3)(H) + H_2 \xrightarrow{Pt} CH_3CH_2CH_2CH_3 \quad \Delta H = -28.6 \text{ kcal/mol} \tag{7.2}$$

(*Z*)-2-Butene

$$(CH_3)(H)C{=}C(H)(CH_3) + H_2 \xrightarrow{Pt} CH_3CH_2CH_2CH_3 \quad \Delta H = -27.6 \text{ kcal/mol} \tag{7.3}$$

(*E*)-2-Butene

The most effective catalysts for these reduction reactions are the transition metals Pt, Ni, and Pd. These metals are insoluble in organic liquids and so the reaction mixture is heterogeneous. For heterogeneous reductions the substrate may be present as a gas, as a pure liquid, or in solution. Heterogeneous reduction is the most common procedure both commercially and in the laboratory. However, in recent years homogeneous reduction has become increasingly important. The catalysts for homogeneous reduction are organometallic complexes (sec. 26:8), and both they and the substrates are soluble in a number of organic solvents.

By comparing Equation 7.1 with Equations 7.2 and 7.3, you can see that 1-butene is less stable than either of the 2-butenes. Similar results are ob-

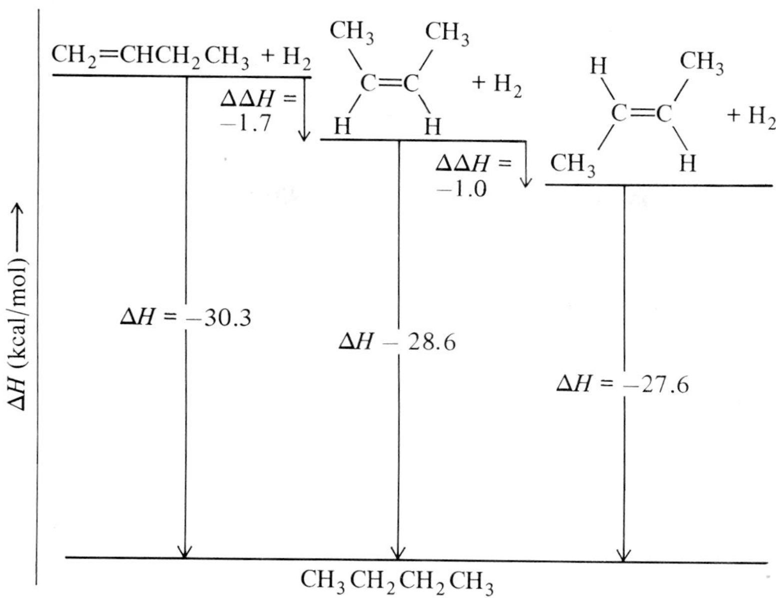

Figure 7–5. *The heats of hydrogenation, ΔH, and the isomer energy differences, ΔΔH, of the three linear butenes.*

tained with isomeric linear pentenes, hexenes, and so on. These results have been generalized in a rule that states that *internal alkenes* (those with a double bond in the interior of the chain) *are more stable than terminal alkenes* (those with a double bond at the end of the chain). The reason(s) for this difference in stability has been a controversial subject. One explanation rests on the idea that an sp^2 carbon is more electronegative than an sp^3 carbon. This seems reasonable since π electrons are a greater distance from the carbon nucleus than are σ electrons. The sp^2 carbon thus wants to draw electrons away from the substituents to better balance the nuclear positive charge. Methyl groups are electron-donating (Fig. 6–10, p. 192) and hence substitution of methyl groups on the double bond will stabilize it. Hydrogen atoms are not electron donors and do not help stabilize a double bond. Because the double bond of 1-butene is substituted by three hydrogens and one ethyl group, it will not be as stable as the double bond of 2-butene. Here the double bond is substituted with two hydrogens and two methyl groups. Thus, the more highly substituted a double bond is, the more stable it is.

A comparison of Equations 7.2 and 7.3 suggests that *cis* isomers are less stable than *trans* isomers. The relative stability of configurational isomers of alkenes is determined by a balance among repulsive and attractive forces between nonbonded atoms. These are the same forces that control the relative stabilities of conformational isomers of alkanes (sec. 3:5). As was the case with alkanes, repulsive forces seem to be dominant with alkenes. Thus the repulsion between the neighboring methyl groups destabilizes (*Z*)-2-butene relative to the *E* isomer (Fig. 7–6). This explanation applies to other open-chain alkenes as well.

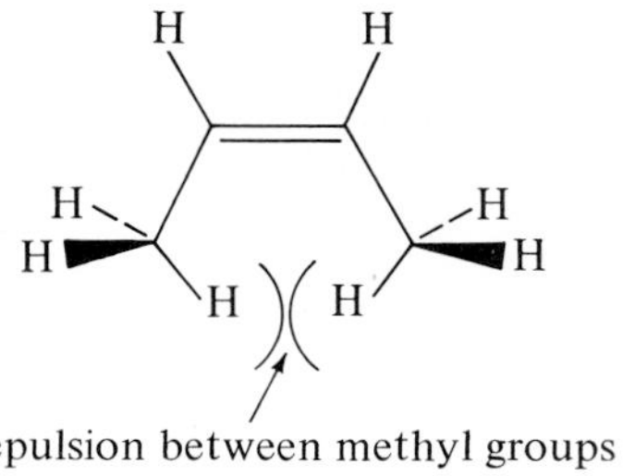

Figure 7–6. *Destabilization of (Z)-2-butene by repulsive interactions between the methyl groups.*

Problem

7–4 a. 2-Methylpropene (*isobutylene*) is an isomer of the 1- and 2-butenes. However, the stability of 2-methylpropene cannot be directly compared with those of the butenes by simply taking the difference in their heats of hydrogenation. Why can't we make this direct comparison?

b. Equations 1 and 2 give data for the hydrogenation of 2-methylpropene and the isomerization of butane, respectively. Use these data to calculate the stability difference between 2-methylpropene and 1-butene and draw a complete energy diagram for the system.

$$(CH_3)_2C{=}CH_2 + H_2 \xrightarrow{Pt} (CH_3)_2CHCH_3 \quad \Delta H = -28.1 \text{ kcal/mol} \qquad (1)$$

$$CH_3CH_2CH_2CH_3 \longrightarrow (CH_3)_2CHCH_3 \quad \Delta H = -2.0 \text{ kcal/mol} \qquad (2)$$

The "rules" given here concerning alkene stability are derived mainly from observation of a large number of cases. As with most generalizations in organic chemistry, the rules are not infallible. Particular structural features of a special molecule may impose conditions not accounted for in the generalization. As an example, consider the relative stability of *cis–trans* isomers in cyclic systems.

Although *trans* isomers are usually more stable in open-chain systems, the reverse is true when the double bond has to be incorporated into a relatively small ring (3 to 7 C atoms). In a ring this size, there are not enough carbon atoms to reach from one end of the bond to the *trans* position at the other end, without twisting the double bond. Such twisting destabilizes the double bond (Fig. 7–2). For cyclopropene through cycloheptene, only the *cis* isomers have been isolated at room temperature. *trans*-Cyclooctene is the smallest *trans*-cycloalkene stable at 25°.

Problem

7–5 The heat of hydrogenation for (Z)-2,2,5,5-tetramethyl-3-hexene is −36.2 kcal/mol, while that for (E)-2,2,5,5-tetramethyl-3-hexene is −26.9 kcal/mol. Draw the structure for each alkene and explain why there is such a large difference in the heats of hydrogenation.

7:4 Introduction to Alkene Preparation by Elimination

The most common method of synthesizing alkenes is to remove the elements H—Y from adjacent carbon atoms in a substituted alkane (eq. 7.4).

(7.4)

$$\text{H}-\overset{|}{\underset{|}{\text{C}}}-\overset{|}{\underset{|}{\text{C}}}-\text{Y} \rightarrow \text{C}=\text{C} + \text{H}^+ + :\text{Y}^-$$

In this *elimination* process, two bonds, C—H and C—Y, must be broken. The relative timing of these two bond cleavages is what distinguishes different mechanisms from one another. For instance, the Y group can depart with its bonding electrons while the C—H bond remains undisturbed. A carbocation that can lose a proton and be converted to an alkene is an intermediate (eq. 7.5).

(7.5)

$$\text{H}-\overset{|}{\underset{|}{\text{C}}}-\overset{|}{\underset{|}{\text{C}}}-\text{Y} \xrightarrow{\text{slow}} :\text{Y}^- + \text{H}-\overset{|}{\underset{|}{\text{C}}}-\text{C}^+ \xrightarrow{\text{fast}} \text{H}^+ + \text{C}=\text{C}$$

Carbocation

This is an observed process, and many of the carbocations produced in S_N1 reactions give alkenes as well as substitution products (Chap. 6).

If the C—H and C—Y bonds break simultaneously, as in Equation 7.6, then the starting material is transformed into the product in one step without any intermediate being formed. This process requires the participation of a base (B:) to break the C—H bond.

(7.6)

$$\text{B:}\ \text{H}-\overset{|}{\underset{|}{\text{C}}}-\overset{|}{\underset{|}{\text{C}}}-\text{Y} \xrightarrow{\text{concerted}} \text{BH}^+ + \text{C}=\text{C} + :\text{Y}^-$$

The third possibility is that the C—H bond breaks first, again with the participation of a base (eq. 7.7). A carbanion intermediate is formed which

(7.7)

$$\text{B:}\ \text{H}-\overset{|}{\underset{|}{\text{C}}}-\overset{|}{\underset{|}{\text{C}}}-\text{Y} \rightarrow \text{BH}^+ + -\overset{..}{\underset{|}{\text{C}}}^{-}-\overset{|}{\underset{|}{\text{C}}}-\text{Y} \rightarrow \text{C}=\text{C} + \text{Y}^-$$

Carbanion

then expels the leaving group. Either step in Equation 7.7 may be the rate-determining one, depending on substrate structure and reaction conditions.

Most simple derivatives of alkanes undergo elimination by way of either the first or second mechanism (eqs. 7.5 and 7.6). In Section 7:5 we devote our attention to the first, called the *E*1 mechanism.

7:5 Elimination by Way of Carbocations: The *E*1 Mechanism

In hydroxylic solvents, 3° halides undergo substitution by way of carbocation intermediates (S_N1 mechanism; see sec. 6:6). The substitution product is often accompanied by alkenes that are elimination products, as when *t*-butyl bromide is solvolyzed in ethanol (eq. 7.8).

$$CH_3-\underset{CH_3}{\overset{CH_3}{\underset{|}{\overset{|}{C}}}}-Br \xrightarrow[25^\circ]{C_2H_5OH} \tag{7.8}$$

$$(CH_3)_2C{=}CH_2 \quad + \quad CH_3-\underset{CH_3}{\overset{CH_3}{\underset{|}{\overset{|}{C}}}}-OC_2H_5 \quad + HBr$$

2-Methylpropene (19%) (elimination product) 2-Ethoxy-2-methylpropane (81%) (substitution product)

We know the mechanism of ether formation; we want to find out whether the alkene is formed by way of a related mechanism or an entirely different one.

When the substrate is changed from $(CH_3)_3CBr$ to $(CH_3)_3CI$, the rate constant increases; when it is changed to $(CH_3)_3CCl$, the rate constant decreases. However, *regardless of the leaving group, the proportion of alkene remains the same,* about 20%. These results show that the double bond is formed *after* the leaving group has departed. Thus the mechanisms for both substitution and elimination share the same first step—loss of the leaving group (eq. 7.9).

$$CH_3-\underset{CH_3}{\overset{CH_3}{\underset{|}{\overset{|}{C}}}}-Br \xrightarrow{slow} CH_3-\overset{+}{C}(CH_3)_2 + Br^- \tag{7.9}$$

The *t*-butyl cation formed in this step can then react with ethanol in two ways. Ethanol can act as a base, removing a proton from the carbocation and

$$CH_3-\overset{+}{C}(CH_3)-CH_2-H \;\; :\ddot{O}(H)-C_2H_5 \xrightarrow{fast} CH_3-\overset{CH_3}{\overset{|}{C}}{=}CH_2 + C_2H_5\overset{+}{O}H_2 \tag{7.10}$$

Acting as a base

converting it to an alkene (eq. 7.10). Competitively, ethanol can react as a nucleophile and convert the cation to an ether (eq. 7.11).

$$(7.11)\qquad CH_3-\overset{+}{C}(CH_3)_2 \longleftarrow :\ddot{O}(H)-C_2H_5 \xrightarrow{\text{fast}} CH_3-C(CH_3)_2-\overset{+}{\ddot{O}}(H)-C_2H_5$$

Reacting as a nucleophile

This mechanism is further confirmed by the observation that addition of the base $NaOC_2H_5$ in low concentration causes no increase in rate. Since $C_2H_5O^-$ is a much stronger base than C_2H_5OH, we would expect it to bring about a rate increase *if* the proton was being removed before or during the rate-determining step. However, in our mechanism the proton is removed after the rate-determining step (eq. 7.10). Thus the nature of the base has no kinetic effect, and the reaction rate depends only on the concentration of the substrate (eq. 7.12).

$$(7.12)\qquad \text{rate} = k[(CH_3)_3CBr]$$

(When the base is present in high concentrations, it does take part in the rate-determining step, and a different mechanism is operative; see Section 7:7.)

The elimination reaction just discussed is a unimolecular one, since the rate-determining step (the slow step) involves only the substrate (eqs. 7.9 and 7.12). It is therefore labelled *E*1, which stands for **elimination, unimolecular**. Figure 7–7 shows an energy diagram for the reaction. Remem-

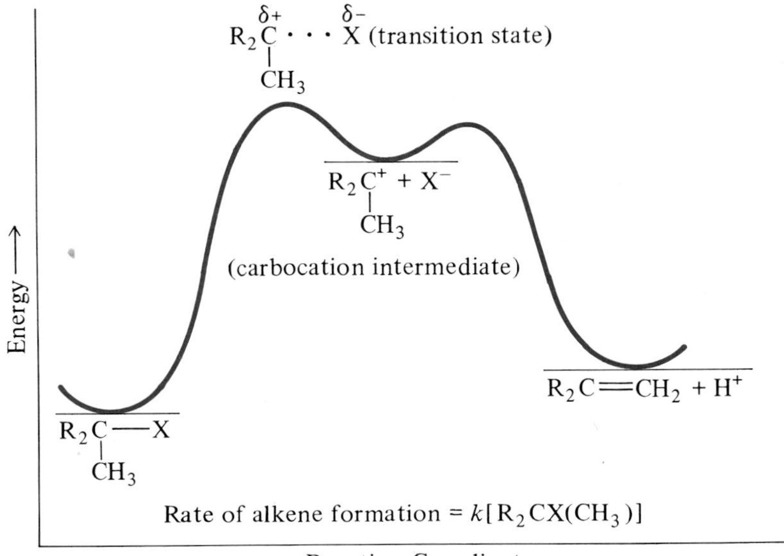

Figure 7–7. *Energy diagram for an* E1 *reaction.*

ber that *E*1 and S_N1 reactions usually occur simultaneously, since they both depend on rate-determining formation of a carbocation (sec. 6:6). In addition, the same substrates that rapidly undergo S_N1 substitution, such as 3° halides and tosylates, also undergo *E*1 elimination. The tendency toward *E*1 elimination decreases as we go to secondary derivatives and is even less for primary substrates. For elimination to occur, there must be a hydrogen adjacent to the leaving group.

The reaction in Equation 7.8 is not a very practical procedure for preparing 2-methylpropene because of the high yield of substitution product. This is due to the presence of a high concentration of ethanol, which is reasonably nucleophilic. Furthermore, the relatively low temperature favors substitution at the expense of elimination. The next section shows how both of these problems can be avoided.

7:6 Dehydration of Alcohols to Alkenes: A Practical Application of the *E*1 Reaction

To undergo an *E*1 reaction a substrate must be capable of forming a stable carbocation and of releasing a good leaving group. Alcohols will not undergo elimination under neutral or basic conditions because OH^- is not a good leaving group. Under acidic conditions, however, alcohols are protonated on oxygen to give oxonium ions (sec. 6:11) which can release H_2O, a good leaving group (eq. 7.13). Thus, the reaction of 2° and 3° alcohols with

$$\text{(7.13)}\quad \text{H}-\overset{|}{\underset{|}{\text{C}}}-\overset{|}{\underset{|}{\text{C}}}-\text{OH} \xrightleftharpoons{H^+} \text{H}-\overset{|}{\underset{|}{\text{C}}}-\overset{|}{\underset{|}{\text{C}}}-\overset{+}{\text{O}}\text{H}_2 \xrightarrow{\text{slow}} \text{H}-\overset{|}{\underset{|}{\text{C}}}-\overset{+}{\text{C}}\!\!\!< \;+\; H_2O \longrightarrow \;>\!\text{C}=\text{C}\!< \;+\; H^+$$

Oxonium ion

an excess of a strong acid is a standard procedure for synthesizing alkenes (eqs. 7.14 to 7.16). Sulfuric and phosphoric acids are usually used because their conjugate bases, HSO_4^- and $H_2PO_4^-$, are poor nucleophiles and give very little substitution product. These eliminations are examples of *dehydration* reactions. The elevated temperatures speed up the reaction and promote

$$\text{(7.14)}\quad \text{cyclohexanol} \xrightarrow[165\text{–}170°]{85\%\ H_3PO_4} \text{cyclohexene} + H_2O$$

$$\text{(7.15)}\quad CH_3(CH_2)_3CH_2\overset{CH_3}{\underset{CH_3}{\overset{|}{\underset{|}{C}}}}OH \xrightarrow[120°]{50\%\ H_2SO_4} \underset{(\sim 60\%)}{CH_3(CH_2)_3CH=C(CH_3)_2} + \underset{(\sim 40\%)}{CH_3(CH_2)_4\overset{CH_3}{\overset{|}{C}}=CH_2}$$

$$(7.16)\quad \mathrm{CH_3{-}\underset{\substack{|\\CH_3}}{\overset{\substack{CH_3\\|}}{C}}{-}\overset{\substack{OH\\|}}{C}HCH_3} \xrightarrow[80^\circ]{85\%\ H_3PO_4}$$

$$\underset{(0.4\%)}{\mathrm{CH_3{-}\underset{\substack{|\\CH_3}}{\overset{\substack{CH_3\\|}}{C}}CH{=}CH_2}} + \underset{(20\%)}{\mathrm{CH_2{=}\overset{\substack{CH_3\\|}}{C}{-}\underset{\substack{|\\CH_3}}{\overset{\substack{H\\|}}{C}}{-}CH_3}} + \underset{(80\%)}{\mathrm{(CH_3)_2C{=}C(CH_3)_2}}$$

elimination. They also allow the alkene to distill out as formed, preventing side reactions.

Look again at Equations 7.15 and 7.16. In both reactions, more than one alkene is formed from the intermediate carbocation. *When alkene mixtures are formed by way of a carbocation* (E1 *mechanism*), *the most stable alkene usually predominates.* Equations 7.15 and 7.16 show that the more highly substituted, and therefore more stable, alkene is usually the major isomer, despite the statistical factor favoring the less stable isomer (there are more primary than secondary or tertiary hydrogen atoms).

The major products in Equation 7.16 were formed by rearrangement (sec. 6:8), which shows conclusively that carbocations were intermediates in the reaction. Loss of water from the protonated alcohol gives rise to a 2° carbocation, which rearranges faster than it loses a proton (eq. 7.17). The 3° (more stable) carbocation formed by rearrangement is the source of the major products.

$$(7.17)\quad \mathrm{(CH_3)_3C\underset{\substack{|\\:\ddot{O}H}}{C}HCH_3} \overset{H^+}{\rightleftharpoons} \mathrm{(CH_3)_3C\underset{\substack{|\\\dot{O}H_2^+}}{C}HCH_3} \longrightarrow \mathrm{CH_3\underset{\substack{|\\CH_3}}{\overset{\substack{CH_3\\|}}{C}}{-}\overset{+}{C}H{-}CH_3 + H_2\ddot{O}:}$$

$$\mathrm{CH_3\underset{\substack{|\\CH_3}}{\overset{\substack{CH_3\\|}}{C}}{-}\overset{+}{C}H{-}CH_3} \xrightarrow{\text{H}^+\text{ loss}} \mathrm{CH_3\underset{\substack{|\\CH_3}}{\overset{\substack{CH_3\\|}}{C}}{-}CH{=}CH_2}$$

$$\mathrm{CH_3\underset{\substack{|\\CH_3}}{\overset{\substack{CH_3\\|}}{C}}{-}\overset{+}{C}H{-}CH_3} \xrightarrow{\text{rearrangement}} \mathrm{CH_3{-}\underset{\substack{|\\CH_3}}{\overset{+}{C}}{-}\overset{\substack{CH_3\\|}}{C}H{-}CH_3} \xrightarrow{\text{H}^+\text{ loss}} \mathrm{CH_2{=}\underset{\substack{|\\CH_3}}{C}{-}CH(CH_3)_2} + \mathrm{(CH_3)_2C{=}C(CH_3)_2}$$

Example

What products will be formed from the following reaction:

$$CH_3\underset{\underset{Cl}{|}}{\overset{\overset{CH_3}{|}}{C}}CH_2CH_2CH_3 + EtOH,\ H_2O \rightarrow ?$$

Answer

First we must evaluate what the most likely mechanism would be, since it is virtually impossible to memorize products from a reaction such as this. Here we are dealing with a tertiary halide and a very polar, protic solvent mixture. It is most likely that a carbocation intermediate will be formed:

$$CH_3\underset{+}{\overset{\overset{CH_3}{|}}{C}}CH_2CH_2CH_3 + Cl^-$$

The carbocation can react with any nucleophiles in solution and it can also undergo *E*1 elimination:

$$CH_3\underset{+}{\overset{\overset{CH_3}{|}}{C}}CH_2CH_2CH_3 \xrightarrow{H_2\ddot{O}:} CH_3\underset{\underset{\overset{+}{O}(H)H}{|}}{\overset{\overset{CH_3}{|}}{C}}CH_2CH_2CH_3 \rightarrow CH_3\underset{\underset{OH}{|}}{\overset{\overset{CH_3}{|}}{C}}CH_2CH_2CH_3$$

$$CH_3\underset{+}{\overset{\overset{CH_3}{|}}{C}}CH_2CH_2CH_3 \xrightarrow{Et\ddot{O}H} CH_3\underset{\underset{\overset{+}{O}(H)Et}{|}}{\overset{\overset{CH_3}{|}}{C}}CH_2CH_2CH_3 \rightarrow CH_3\underset{\underset{OEt}{|}}{\overset{\overset{CH_3}{|}}{C}}CH_2CH_2CH_3$$

$$CH_3\underset{+}{\overset{\overset{CH_3}{|}}{C}}CH_2CH_2CH_3 \rightarrow \underset{\text{Major (more stable)}}{CH_3\overset{\overset{CH_3}{|}}{C}{=}CHCH_2CH_3} + \underset{\text{Minor (less stable)}}{CH_2{=}\overset{\overset{CH_3}{|}}{C}CH_2CH_2CH_3}$$

Problems

7–6 Write mechanisms for the dehydration reactions shown in Equations 7.14 and 7.15.

7–7 Show all the products of each of the following reactions, including geometrical isomers:

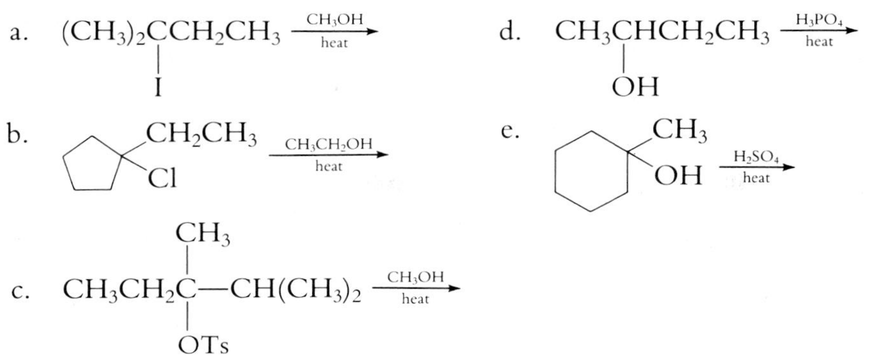

a. $(CH_3)_2\underset{\displaystyle I}{C}CH_2CH_3 \xrightarrow[\text{heat}]{CH_3OH}$

b. 1-chloro-1-ethylcyclopentane $\xrightarrow[\text{heat}]{CH_3CH_2OH}$ (ring with CH_2CH_3 and Cl)

c. $CH_3CH_2\overset{\displaystyle CH_3}{\underset{\displaystyle OTs}{C}}—CH(CH_3)_2 \xrightarrow[\text{heat}]{CH_3OH}$

d. $CH_3\underset{\displaystyle OH}{C}HCH_2CH_3 \xrightarrow[\text{heat}]{H_3PO_4}$

e. 1-methylcyclohexanol (ring with CH_3 and OH) $\xrightarrow[\text{heat}]{H_2SO_4}$

7–8 Go back to Problem 7–7 and consider only the alkene products. In each case indicate which product will be found in highest yield and which in lowest yield.

7–9 2-Cyclobutyl-2-propanol, **I**, gives several products when heated at 100° in concentrated HBr. Two of these are alkenes, 2-propenylcyclobutane, **II**, and 1,2-dimethylcyclopentene, **III**.

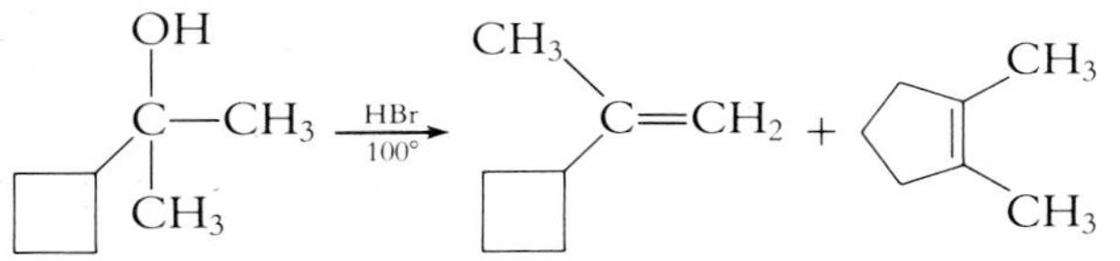

$$\underset{\textbf{I}}{\text{cyclobutyl}—\overset{\displaystyle OH}{\underset{\displaystyle CH_3}{C}}—CH_3} \xrightarrow[100^\circ]{HBr} \underset{\textbf{II}}{\text{cyclobutyl}—\overset{CH_3}{C}{=}CH_2} + \underset{\textbf{III}}{\text{1,2-}(CH_3)_2\text{-cyclopentene}}$$

a. Write a mechanism accounting for the formation of both **II** and **III**.
b. The reaction mixture contains 20 times more **III** than **II**. How would you explain this finding?

The nineteenth-century Russian chemist A. Zaitsev (or Saytzeff) was the first to notice that when a mixture of alkenes is obtained by elimination, the most highly substituted alkenes are formed in greatest amounts. Reactions that give such product distributions are called, accordingly, *Zaitsev (Saytzeff) eliminations*. A modern restatement of the Zaitsev rule would be that *in an elimination reaction, the thermodynamically more stable alkenes are formed in larger amounts than the less stable ones*. This means that internal alkenes will predominate over terminal alkenes, and *trans* isomers over *cis* isomers.

*E*1 reactions generally obey the Zaitsev rule and give the most stable alkene as the major product. In the following sections we will study a type of elimination in which the base is involved in the rate-determining step and so does not follow the *E*1 mechanism. In these base-promoted eliminations we can get either the most stable alkene (Zaitsev product) or the least stable alkene (Hofmann product; see sec. 7:8). Which one predominates will usually depend on the structure of the reactants and on the reaction conditions. We tend to have more control over the product distribution in base-promoted eliminations than we do in *E*1 reactions. This factor gives base-promoted eliminations some preparative advantages.

7:7 Formation of Alkenes by Concerted Elimination: The *E*2 Mechanism

Secondary alkyl halides and tosylates, unlike 3° alkyl substrates, do not produce free carbocations at a useful rate. Nonetheless, they do give alkenes when treated with bases (eqs. 7.18 to 7.20; eqs. 7.18 and 7.19 are examples of *dehydrohalogenation* reactions).

(7.18) $CH_3CH(Br)CH_3 + Na^+OH^- \xrightarrow{C_2H_5OH,\ 55°}$

$CH_3CH{=}CH_2$ (71%) $+ CH_3CH(OH)CH_3$ (29%) $+ Na^+Br^- + H_2O$

(7.19) $CH_3CH_2CH(Br)CH_3 + C_2H_5O^-Na^+ \xrightarrow{C_2H_5OH,\ 25°}$

$CH_3CH_2CH{=}CH_2$ (16%) $+ CH_3CH{=}CHCH_3$ (*cis* and *trans*, 66%) $+$

$CH_3CH_2CH(OC_2H_5)CH_3$ (18%) $+ Na^+Br^- + C_2H_5OH$

(7.20) $CH_3CH_2CH(OTs)CH_3 + C_2H_5O^-K^+ \xrightarrow{C_2H_5OH,\ 55°}$

$CH_3CH_2CH{=}CH_2$ (35%) $+ CH_3CH{=}CHCH_3$ (*cis* (22%), *trans* (43%)) $+$

$CH_3CH_2CH(OC_2H_5)CH_3$ (small quantity) $+ K^+OTs^- + C_2H_5OH$

Kinetic studies of these elimination reactions show that they are second order—first order in substrate *and* first order in base (eq. 7.21). The kinetic data indicate that the base is present during the rate-determining step, and

(7.21) $$\text{rate} = k[\text{R—X}][\text{base}]$$

therefore that the C—H bond is being broken during this step (eq. 7.22). These kinetic studies also show that the reaction rate depends on which leaving group is present. This means that the bond to the leaving group is also being broken in the rate-determining step.

(7.22) $H\ddot{O}{:}^- \;\; H{-}CH_2{-}CH(:\ddot{Br}:)CH_3 \rightarrow H\ddot{O}H + CH_2{=}CHCH_3 + :\ddot{Br}:^-$

These reactions are classified as *E*2—**elimination, bimolecular**. The reaction pathway and the energy diagram for an *E*2 reaction are given in Figure 7–8. When a strong base is present, even 3° halides and tosylates react mainly by way of the *E*2 mechanism. Thus, in a concentrated solution of sodium ethoxide in ethanol, *t*-butyl bromide is converted mainly to 2-methylpropene (eq. 7.23; compare with eq. 7.8).

$$(7.23)\quad (CH_3)_3CBr + NaOC_2H_5 \xrightarrow[C_2H_5OH]{} \underset{(93\%)}{(CH_3)_2C{=}CH_2} + \underset{(7\%)}{(CH_3)_3COC_2H_5} + NaBr + C_2H_5OH$$

The bases used to promote elimination contain nonbonding electron pairs and are therefore also nucleophiles. In fact, except for 3° systems, bimolecular nucleophilic substitution (S_N2 reaction) almost always competes with *E*2 reactions. Equations 7.18 to 7.20 demonstrate this. For a given base, the highest yields of elimination products will be obtained with 3° substrates. At the other end of the scale, 1° substrates usually give mostly substitution product and little elimination product.

In order to suppress substitution and promote elimination, large sterically hindered bases are used. These bases are very often the anions of tertiary alcohols, such as the *t*-butoxide ion $(CH_3)_3CO^-$ (eq. 7.24). These anions can be generated by adding potassium metal to an excess of the alcohol. The alkoxide salts may also be dissolved in dipolar, aprotic solvents

$$(7.24)\quad CH_3CH_2\underset{\substack{|\\ Br}}{C}HCH_3 + (CH_3)_3CO^-K^+ \xrightarrow[(CH_3)_3COH]{} \underset{(53\%)}{CH_3CH_2CH{=}CH_2} + \underset{\substack{cis\ (18\%)\\ trans\ (29\%)}}{CH_3CH{=}CHCH_3} + (CH_3)_3COH + K^+Br^-$$

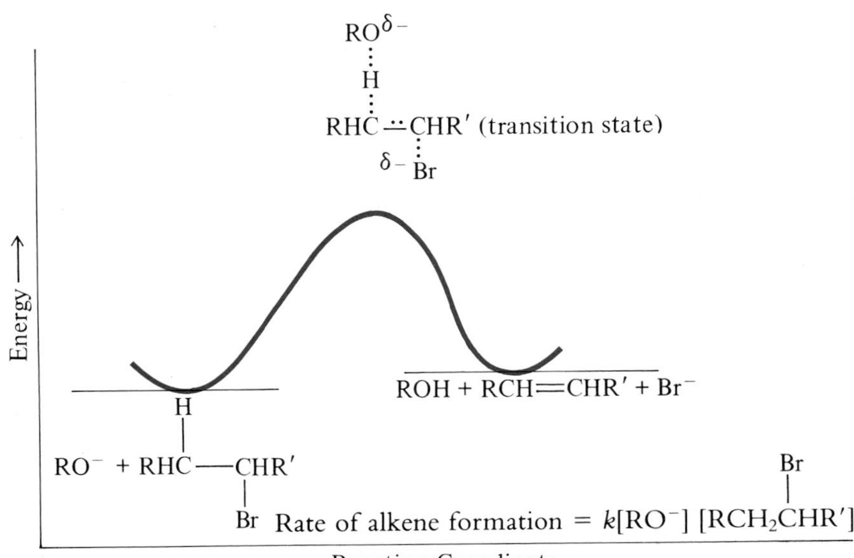

Figure 7–8. *Energy diagram for an* E2 *reaction.*

Table 7–2. A Comparison of $E1$ and $E2$ Reactions

Characteristic	$E1$	$E2$
Mechanism	two step; carbocation intermediate	concerted elimination
Kinetics	first order (rate = k[RX])	second order (rate = k[RX][base])
Rearrangement	sometimes	never
Substrates	halides, tosylates, alcohols + acid	halides, tosylates, tetra-alkylammonium hydroxides (sec. 7:8)
Favorable structural features	3° RX, polar solvent	high base conc., bulky base
Competing substitution process	S_N1 reaction	S_N2 reaction

such as dimethyl sulfoxide (DMSO, $(CH_3)_2S{=}O$; Table 6–7, p. 203). Product distributions depend strongly on the base (compare eqs. 7.19 and 7.24). As the base gets bulkier, less highly substituted alkene products are increasingly favored. In addition, bulky bases discourage nucleophilic substitution. The large t-BuO$^-$ group is thus a poor nucleophile but a powerful base (eq. 7.25).

$$(7.25)\qquad CH_3(CH_2)_5CH_2CH_2Br \xrightarrow[(CH_3)_3COH,\ 80^\circ]{(CH_3)_3CO^-K^+} \underset{\text{1-Octene (85\%)}}{CH_3(CH_2)_5CH{=}CH_2} + \underset{t\text{-Butyl octyl ether (12\%)}}{CH_3(CH_2)_7OC(CH_3)_3}$$

The information in this and the previous section can be summarized as follows:

With alcohol substrates, maximum yields of alkene are obtained with concentrated acids at elevated temperatures. With halide and tosylate substrates, maximum yields of alkene are obtained with hindered alkoxide bases at elevated temperatures.

One of the most important advantages $E2$ reactions have over $E1$ processes is that *they do not involve carbocations, so they do not produce rearranged products*. Thus, we have more control in obtaining the desired product. For instance, the $E2$ reaction shown in Equation 7.26 does not produce any 1-

(7.26) [cyclohexane with CH_2Br and H] $\xrightarrow[(CH_3)_3COH]{(CH_3)_3CO^-K^+}$ [cyclohexane with $=CH_2$] + [cyclohexane with $CH_2OC(CH_3)_3$ and H]

Methylenecyclohexane

methylcyclohexene, although this isomer is more stable than the methylenecyclohexane actually formed. Table 7–2 summarizes the major characteristics of $E1$ and $E2$ reactions.

Example

What alkenes can be formed from the following $E2$ elimination reactions?

a. $CH_3CH_2CH_2CH_2CH_2Cl + EtO^-, EtOH \rightarrow ?$

b. $CH_3CH_2CH_2CH(Cl)CH_3 + EtO^-, EtOH \rightarrow ?$

Answers

$E2$ elimination reactions are also known as 1,2-eliminations because they involve loss of a leaving group (halogen, tosylate) from one carbon and a proton from the neighboring carbon.

a. $CH_3CH_2CH_2\overset{2}{C}H_2\overset{1}{C}H_2Cl + EtO^-, EtOH \rightarrow CH_3CH_2CH_2CH{=}CH_2 + EtOH + Cl^-$

b. In this case, elimination can take place in two directions and in one of the directions, both *cis* and *trans* isomers can be formed:

$CH_3CH_2\overset{2}{C}H_2\overset{1}{C}H(Cl)\overset{2}{C}H_3 + EtO^-, EtOH \rightarrow$

$(CH_3CH_2)(H)C{=}C(CH_3)(H)$ + $(CH_3CH_2)(H)C{=}C(H)(CH_3)$ +

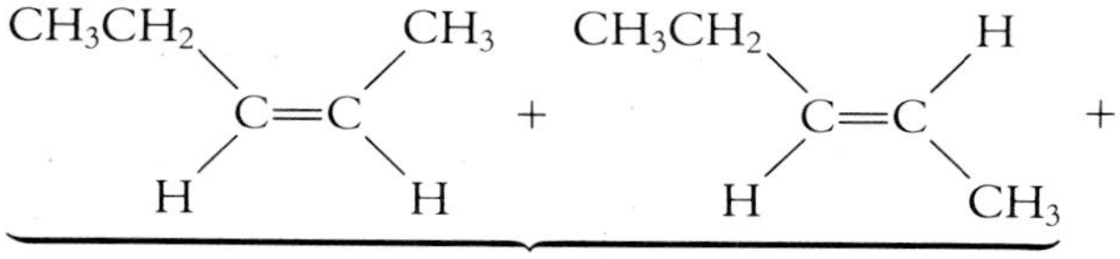

Major products

$CH_3CH_2CH_2CH{=}CH_2$

Minor product

Problems

7–10 Show the structures of all products (from both elimination and substitution) in each of the following reactions, taking note of the possibilities for configurational and geometrical isomerism.

a. 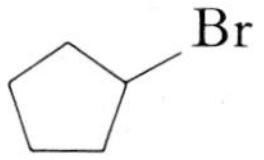 Bromocyclopentane (cyclopentyl–Br) $+ CH_3O^-Na^+ \xrightarrow{CH_3OH}$

b. $(CH_3CH_2)_2CHCH_2I + K^+OH^- \xrightarrow{DMSO}$

c. $CH_3CH_2CH(OTs)(CH_2)_4CH_3 + (CH_3)_2CHO^-K^+ \xrightarrow{(CH_3)_2CHOH}$

d. $(CH_3)_3CCH(Cl)CH_2CH_2CH_3 + (CH_3)_2C(O^-K^+)CH_2CH_3 \xrightarrow{DMSO}$

e. [cyclohexane ring bearing CH_3 and Br on the same carbon] $+ K^+OH^- \xrightarrow{CH_3CH_2OH}$

7–11 A chemist has some 3-methyl-2-pentanol from which he wants to obtain substantial amounts of 3-methyl-2-pentene and 3-methyl-1-pentene, both fairly pure. The sample is divided in half and each half is treated in such a manner as to get the highest yield of one of the alkenes. Show how each half of the sample is treated in order to get the two alkenes.

Box 7–1. Custom Synthesis In the "Good Old Days"

There was a problem [at Du Pont]. H. (the director of development) was impossible to see. He was surrounded by three concentric [rings] of secretaries who protected him from salesmen.

I got through immediately by telling the first that I was Dr. H.'s natural son. In his sanctum, he listened carefully to my speech, in which I informed him that Du Pont was in serious trouble—that they had too many chemists. I proposed to help them by using my 'small, efficient company' to make the chemicals they needed at rock-bottom prices. He was interested and called in others . . . I told them too that Du Pont was in serious trouble . . . they nodded their heads in agreement without listening to the rest of my pitch.

Dr. H. wanted to know if I was familiar with the synthesis of allene. . . . I told him that the best synthesis, to my knowledge, consisted of the dehydrochlorination of 1,2,3-trichloropropane, followed by treatment with a suspension of zinc in methyl alcohol to remove the remaining chlorine. He . . . suggested that we make it a joint project; we would dehydrochlorinate and they would dechlorinate. I told him that I would prefer it the other way around. He reminded me that I had told him that my visit was to make Du Pont's life easier; we agreed to run *both* reactions. My promise was complicated by the fact that I had never run either.

When one treats 1,2,3-trichloropropane with alkali and a little water, the reaction is violent; there is a tendency to deposit the reaction product, the raw materials, and the apparatus on the ceiling and [on] the attending chemist. I solved this by setting up duplicate 12-liter flasks, each equipped with double reflux condensers, and surrounding each with a half dozen large tubs. In practice, when the reaction 'took off,' I would flee through the door or window and battle the eruption with water from a garden hose. The contents flying from the flasks were deflected by the ceiling and collected under water in the tubs. I used towels to wring out the contents [that] separated, shipping the lower layer to Du Pont. They complained of solids suspended in the liquid, but accepted the product and ordered more. I increased the number of flasks to four, doubled the number of washtubs, and completed the order.

They ordered a 55-gallon drum. At best, with myself as chemist and supervisor, I could make a gallon a day, arriving home with skin and lungs saturated with 2,3-dichloropropene. I needed help. An advertisement in the local newspaper resulted in an interview with a former producer of illicit spirits named Preacher, who had just done penance at the local penitentiary.

He listened carefully and approved of my method of production, which he said might be improved with copper coils. Immediately, he began to enlarge our production room by removing a wall, putting in an extra table, and increasing the number of washtubs and reaction setups. It was amazing to see Preacher in action (I gave him encouragement through the window) . . . he would walk up the aisles from setup to setup, putting in first the caustic, then the water, then fastening on the rubber stoppers and condensers, then using the hose. At this stage, the room was a swirling mass of steam and 2,3-dichloropropene. We made a vast amount of material and shipped the complete order to Du Pont—on schedule.

As part of our contract with them, we had agreed to supply details of the production as well as innovations we might have discovered. I wrote them a complete description, giving credit to my indefatigable coworker. Two weeks later, I received a telegram with only two words: "Ship Preacher!"

Max G. Gergel, *Excuse Me Sir, Would You Like to Buy a Kilo of Isopropyl Bromide?* Pierce Chemical Co., 1979. Reprinted with permission.

7:8 Hofmann Eliminations: A Route to Less Stable Alkenes

When one starting material can give two or more alkenes, $E1$ conditions lead to predominant formation of the most stable isomer (eqs. 7.15 and 7.16). $E2$ eliminations, however, do not always produce such clear results. Small bases, such as OH^-, CH_3O^-, and $C_2H_5O^-$, favor the more stable products (eqs. 7.19 and 7.20), but as the base gets bulkier, the percentage of less stable isomer in the product increases (eq. 7.24). In addition, the leaving group can strongly affect the product distribution. The nineteenth-century German chemist A. W. Hofmann found that when the leaving group was an amine, elimination gave very high yields of the least stable product. Ammonium hydroxides, which have an amine leaving group, are often used as the substrate in reactions of this type. This section deals specifically with eliminations using *tetraalkylammonium hydroxides,* which are organic salts. These salts decompose to an alkene, an amine, and water when heated in a concentrated aqueous solution. This process, shown in Equation 7.27, is called a *Hofmann elimination.*

(7.27) $CH_3CH_2CH(H)\text{—}CH_2\text{—}\overset{+}{N}(CH_3)_3 \;\; H\ddot{O}{:}^- \xrightarrow[130°]{\text{conc aq soln}}$

Butyltrimethylammonium hydroxide

$CH_3CH_2CH{=}CH_2 + {:}N(CH_3)_3 + H_2\ddot{O}{:}$

1-Butene   Trimethylamine

The salts themselves are made by a two-step process: an alkyl halide and an amine undergo an S_N2 reaction to give an ammonium halide, and the halide ion is then replaced by a hydroxide ion (eq. 7.28).

(7.28) $CH_3(CH_2)_2CH_2\text{—}\ddot{I}{:} + {:}N(CH_3)_3 \rightarrow CH_3(CH_2)_2CH_2\text{—}\overset{+}{N}(CH_3)_3\ {:}\ddot{I}{:}^-$

Trimethylamine

$\xrightarrow{Ag_2O,\ H_2O} CH_3(CH_2)_2CH_2\text{—}\overset{+}{N}(CH_3)_3\ {:}\ddot{O}H^-$

Hofmann also found that when two or more alkenes were formed from an ammonium hydroxide, the *least* substituted one predominated (eq. 7.29). This was usually the least stable alkene. Therefore *an elimination reaction that yields the least stable alkene as the major product* follows the *Hofmann rule.* A comparison of Equations 7.19 and 7.29 shows a change from Zaitsev to Hofmann elimination as the leaving group is changed from Br^- to $(CH_3)_3N$. Thus, Zaitsev elimination is favored by good leaving groups such as Br^-, while Hofmann elimination is favored by poor leaving groups such as

(7.29) $CH_3CH_2CH(\overset{+}{N}(CH_3)_3)CH_3\ \ {:}\ddot{O}H^- \xrightarrow{\text{conc aq soln}} CH_2CH_2CH{=}CH_2$ (95%)

$+\ CH_3CH{=}CHCH_3 + {:}N(CH_3)_3 + H_2\ddot{O}{:}$

cis and *trans* (5%)

$(CH_3)_3N$. (Trimethylamine is a weak base in water whose conjugate acid, $(CH_3)_3NH^+$, has a pK_a of 9.8; see sec. 6:11.)

How does the leaving group character affect the product ratio? If the leaving group is poor, then the main driving force for the elimination comes from breaking the C—H bond. It becomes necessary to accumulate negative charge on carbon in order to help "push out" the poor leaving group. The mechanism begins to resemble the one shown in Equation 7.7 and the carbon losing the proton begins to resemble a carbanion. The isomer distribution then depends on how easily protons are removed from different carbon atoms. Proton removal depends on steric factors and on charge factors. Protons at the end of the chain are more accessible than ones in the middle because there is less hindrance to approach of the base. Charge effects arise because negative charge accumulates on the carbon that is losing the proton (Fig. 7–9). Negative charge accumulation requires strong solvation, which is easiest at the least substituted carbon atom. It follows that methyl carbanions are the most stable and 3° carbanions the least stable. This order is opposite to that of carbocation stability (sec. 6:7).

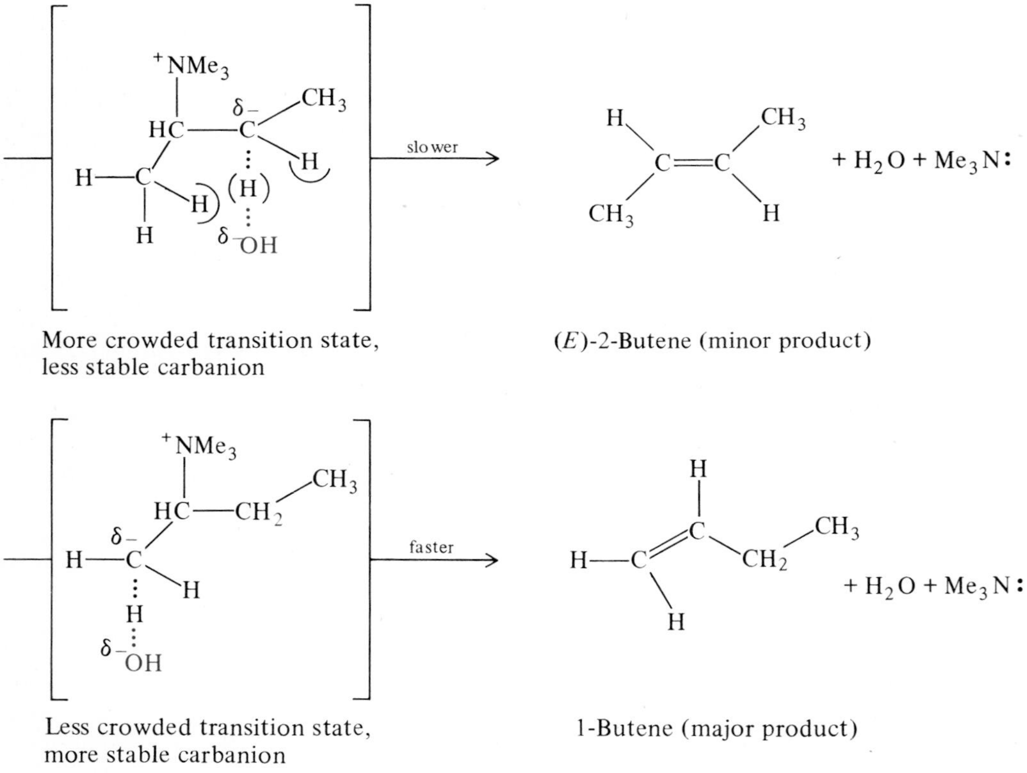

Figure 7–9. *The effect of steric hindrance and carbanion stability on the relative rate of alkene isomer formation in the Hofmann elimination. Protons are removed preferentially from the less hindered positions, leading to preferential formation of the least substituted alkene.*

Table 7–3. A Comparison of Zaitsev and Hofmann Eliminations

	Zaitsev	Hofmann
Favored product	most substituted alkene	least substituted alkene
Leaving group	good (I^-, Br^-, OTs^-)	poor (F^-, $(CH_3)_3N$)
Base size	small	large
Major factor determining product distribution	relative stability of alkene products	relative ease of proton removal

$$\ddot{C}H_3^- > R\ddot{C}H_2^- > R_2\ddot{C}H^- > R_3\ddot{C}^-$$

Order of carbanion stability
reflecting relative ease of solvation

To summarize, removal of a proton from the least substituted carbon is favored by *both* steric accessibility and carbanion stability in the Hofmann elimination. The result is that the less highly substituted alkene is generated faster than the more highly substituted one (Fig. 7–9).

Table 7–3 lists the factors that favor either Zaitsev or Hofmann product distributions in *E*2 reactions. It reminds us that we can influence the product distribution by our choice of base and leaving group. Because we have this input and because no rearrangements occur, *E*2 reactions are often preferred over *E*1 for alkene synthesis.

Problem

7–12 a. Trimethylamine, $(CH_3)_3N\!:$, is commonly used to synthesize the tetraalkylammonium salts needed for the Hofmann elimination (eq. 7.28). The related compound triethylamine, $(CH_3CH_2)_3N\!:$, is never used for this purpose. What problem would occur if triethylamine were used? (Consider the elimination products that could be formed in the final step.)

b. Ethyldimethylpropylammonium hydroxide (**IV**) is subjected to Hofmann elimination. Show the structures of all elimination products, and indicate which ones will be the major products. Explain your choices.

$$CH_3CH_2CH_2\overset{\displaystyle CH_3}{\overset{|}{\underset{\underset{\displaystyle CH_3}{|}}{N^+}}}CH_2CH_3 \quad OH^-$$

IV

7:9 The Stereochemistry of *E*2 Reactions

You have already seen how stereochemical studies increase our understanding of nucleophilic substitutions (secs. 6:3 and 6:6). Likewise, we can better appreciate the mechanism of elimination reactions by studying their stereochemistry. This section deals specifically with the *E*2 reaction. The basic approach involves carrying out an elimination reaction on each of two diastereomers and then comparing the products formed.

Consider 2,3-dibromobutane as an example. When treated with hydroxide ion it gives 2-bromo-2-butene as the predominant product (eq. 7.30). Because 2,3-dibromobutane contains two chiral atoms it exists in two diastereomeric forms (sec. 4:3) that undergo stereospecific elimination. The *meso* isomer (*R,S* compound) gives only the *E* product (eq. 7.31), while the racemic isomer (a mixture of *R,R* and *S,S* enantiomers) produces only the *Z* product (eq. 7.32).

$$CH_3CH(Br)-CH(Br)CH_3 \xrightarrow[H_2O,\ EtOH]{OH^-} CH_3CH{=}C(CH_3)Br \qquad (7.30)$$

(7.31) 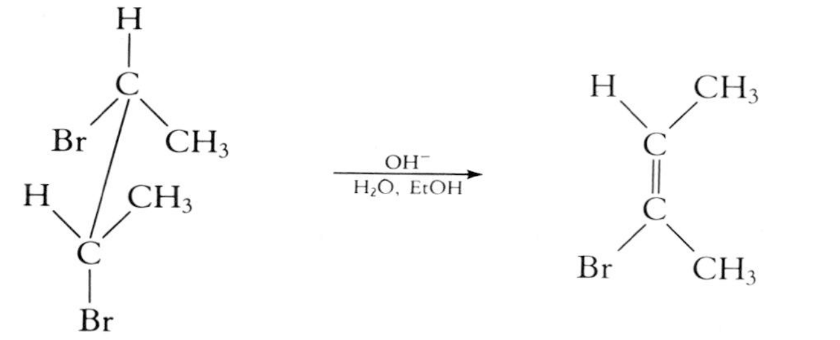 (*R,S*)-2,3-Dibromobutane $\xrightarrow[H_2O,\ EtOH]{OH^-}$ (*E*)-2-Bromo-2-butene

(7.32) 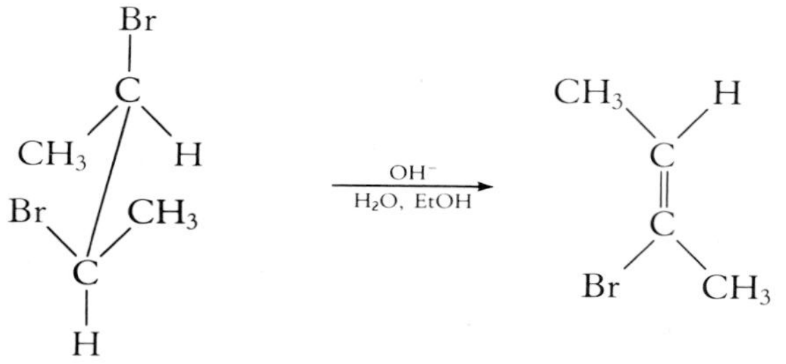 (*R,R*)-2,3-Dibromobutane $\xrightarrow[H_2O,\ EtOH]{OH^-}$ (*Z*)-2-Bromo-2-butene

The stereochemical results are consistent with a transition state in which the H atom being removed and the departing Br atom are on opposite sides of the molecule. Because of the relationship of the proton and the leaving group, this mechanism is labelled an *anti* elimination. Figure 7–10 shows how the *anti* elimination mechanism accounts for the stereospecific conversion of *meso*-2,3-dibromobutane to (*E*)-2-bromo-2-butene.

Problems

7–13 a. Show that the *anti* elimination mechanism accounts for the formation of (*Z*)-2-bromo-2-butene from both (*R,R*)- and (*S,S*)-2,3-dibromobutane.

b. Regardless of which 2,3-dibromobutane diastereomer is reacting, the base preferentially removes a proton from C-3, giving 2-bromo-2-butene. A competing base attack at C-1, to give 3-bromo-1-butene, is much slower. Explain this result.

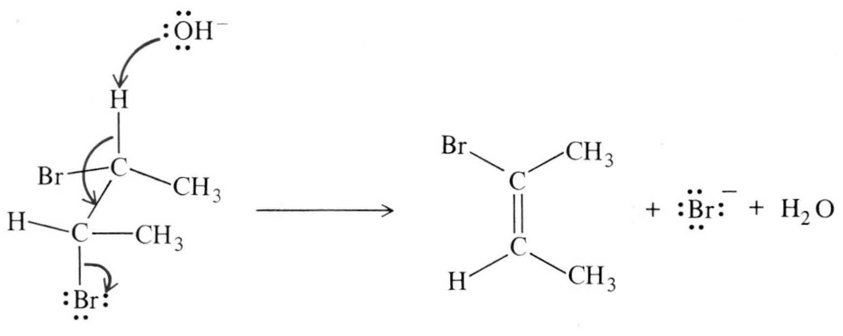

meso-2, 3-Dibromobutane (*E*)-2-Bromo-2-butene

Figure 7–10. Anti *elimination from* meso-*2,3-dibromobutane, resulting in stereospecific formation of* (E)*-2-bromo-2-butene.*

7–14 Base treatment of (2*R*,3*S*)-2-bromobutane-3-*d*, **V** (eq. 3) gave, in addition to the deuterated 1-butene, only (*Z*)-2-butene-2-*d* and (*E*)-2-butene. Use three-dimensional structures (perspective drawings or Newman projections) to explain this result.

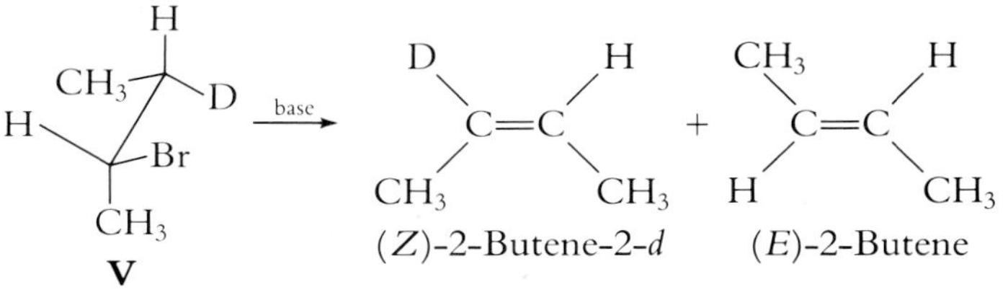

V $\xrightarrow{\text{base}}$ (*Z*)-2-Butene-2-*d* + (*E*)-2-Butene (3)

The use of deuterium allows us to differentiate between the two hydrogens at C-3 of 2-bromobutane and thereby study the elimination stereochemistry.

You can better appreciate the significance of these results by noting that (*E*)-2-bromo-2-butene is less stable than the *Z* isomer. Despite this stability difference, the *meso* starting material gives only the *E* alkene (eq. 7.31). We must conclude, therefore, that the *anti* elimination mechanism is highly favorable. Why? First, because it minimizes interference between the attacking base and the leaving group by placing them as far from one another as possible (180°; see Fig. 7–10). Second, by requiring that the C—H and C—X bond orbitals lie in the same plane, it aligns them correctly for conversion into π bond orbitals (Fig. 7–11). Most, although not all, *E*2 reactions of open-chain compounds follow this *anti* elimination mechanism.

Anti elimination mechanisms are also very common for cyclohexyl derivatives. Adjacent C—H and C—X bonds on a cyclohexane ring are perfectly aligned for *anti* elimination, *provided that they are both axial.* We know that monosubstituted cyclohexanes prefer the conformation that has an equatorial C—X bond (sec. 3:10). Nonetheless, the ring undergoes rapid inversion so that the equatorial and axial conformers are in equilibrium.

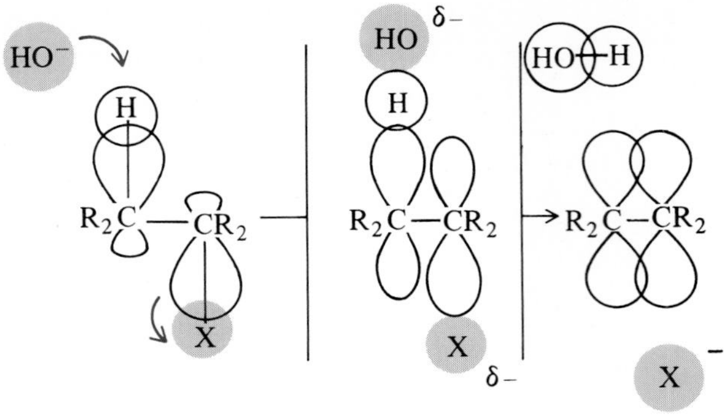

Figure 7–11. *Pi* (π) *bond formation is facilitated by parallel orientation of the C—H and C—X bonds in the* E2 *transition state.*

Enough axial conformer is therefore present to allow rapid *anti* elimination (eq. 7.33).

(7.33)

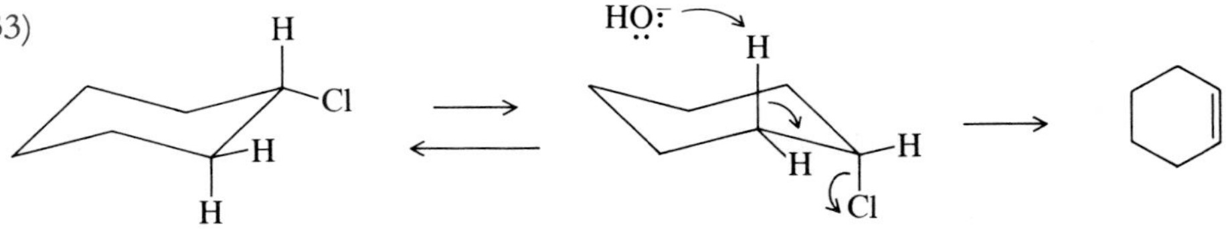

Equatorial conformer　　　　Axial conformer

Problems

7–15 3-Methyl-2-pentyl tosylate gives mostly 3-methyl-2-pentene when treated with $C_2H_5O^-Na^+$ in C_2H_5OH. The diastereomers of the tosylate each react stereospecifically to give different alkene diastereomers. Thus, the (2*S*,3*R*)-tosylate, **VI**, yields only the *Z* alkene, while the (2*S*,3*S*)-tosylate, **VII**, is converted only to the *E* alkene.

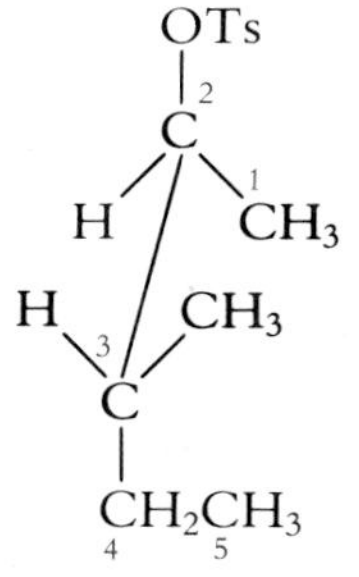

VI

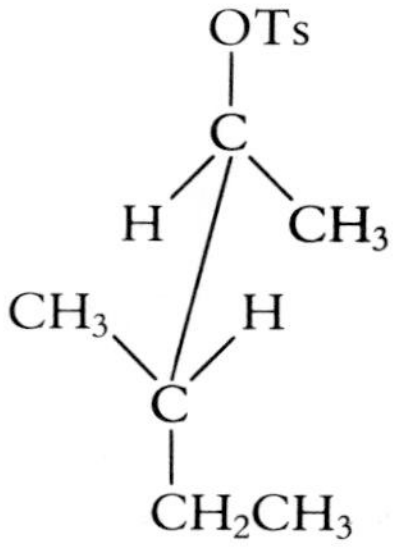

VII

Write mechanisms for each of these reactions, showing all the products. Are they *anti* eliminations?

7–16 The *cis* and *trans* isomers of 1,2-dibromocyclohexane are treated separately with NaOH in aqueous ethanol, leading predominantly to elimination products. The *cis* isomer gives only 1-bromocyclohexene (eq. 4), while the *trans* isomer is converted to 3-bromocyclohexene, which underwent further elimination to 1,3-cyclohexadiene (eq. 5). Explain why the two bromocyclohexanes follow different reaction paths (you may want to review cyclohexane conformational changes, sec. 3:13).

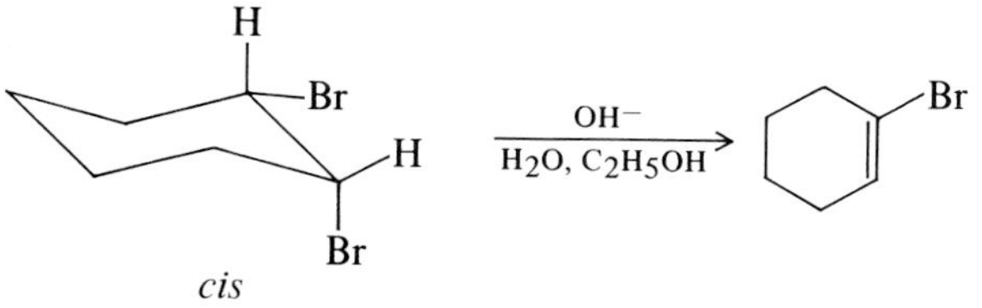

$$\text{cis-1,2-dibromocyclohexane} \xrightarrow[H_2O,\ C_2H_5OH]{OH^-} \text{1-bromocyclohexene} \quad (4)$$

cis

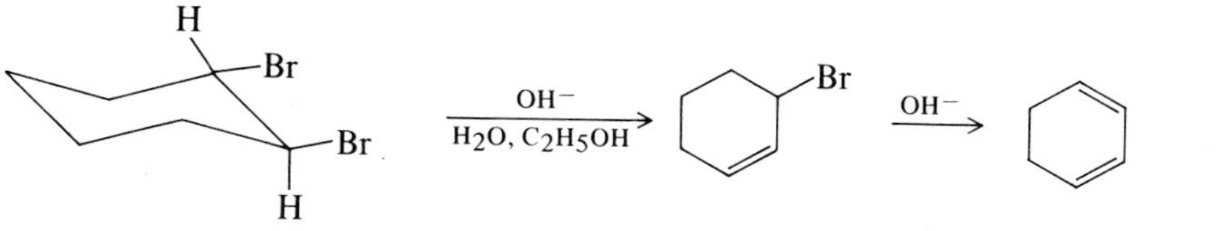

$$\text{trans-1,2-dibromocyclohexane} \xrightarrow[H_2O,\ C_2H_5OH]{OH^-} \text{3-bromocyclohexene} \xrightarrow{OH^-} \text{1,3-cyclohexadiene} \quad (5)$$

trans

1,3-Cyclohexadiene

Not every compound that can undergo elimination can meet all the stereochemical requirements for *anti* elimination. Of these requirements, the most important is that the C—H bond and the C—X bond be in the same plane. This arrangement is sometimes best achieved by having an angle of 0° (instead of 180°) between C—H and C—X. This geometry results in *syn eliminations*. The transition state for a *syn* elimination places the H atom being removed and the leaving group on the same side of the molecule.

Syn elimination is most often found in rings other than cyclohexyl, and when poor leaving groups are present. A good example is Hofmann elimination in cyclobutane (eq. 7.34), which involves mainly a *syn* mechanism.

(7.34)

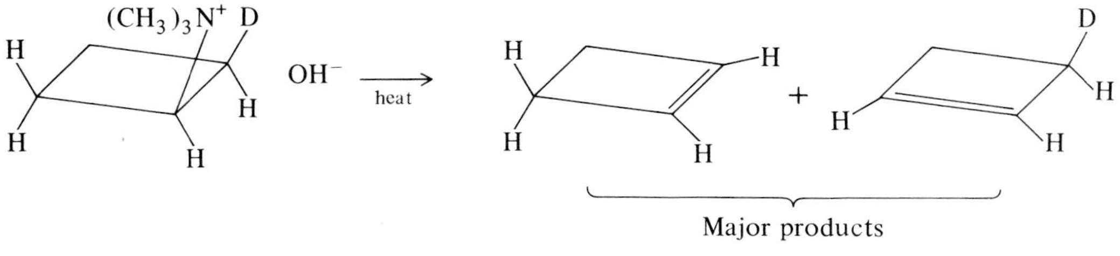

$$\text{cyclobutyl-}N^+(CH_3)_3\ (D\ cis) \xrightarrow[\text{heat}]{OH^-} \text{cyclobutene (H)} + \text{cyclobutene (D)}$$

Major products

The mechanism was investigated by using a substrate which contained a deuterium atom *cis* to the $-\overset{+}{N}Me_3$ group. Since it was found that a substantial portion of the deuterium was lost during the reaction, it was concluded that the elimination had followed a *syn* pathway.

7:10 An Overview of Substitution and Elimination

As we noted at the beginning of Chapter 6, substitution and elimination usually take place concurrently. We have been discussing them separately for reasons of clarity, but we will now consider them together.

One of the most important factors affecting the competition between substitution and elimination is the structure of the substrate. *For any given set*

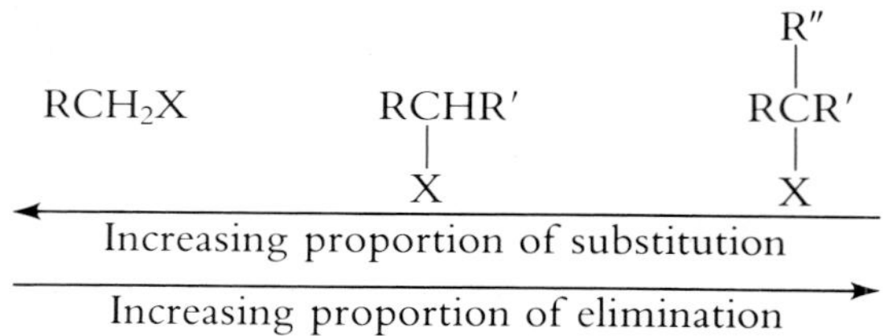

of conditions, substitution is greatest with 1° halides and tosylates, while elimination is greatest with 3° halides and tosylates. There are several reasons for this; here are two very important ones:

1. Since alkyl substituents stabilize double bonds, elimination to form a double bond will be favored at a highly substituted carbon atom. At the same time, those substituents will interfere with nucleophilic substitution (Fig. 7–12).
2. When a carbon atom changes hybridization from sp^3 (alkane) to sp^2 (alkene), the substituents attached to it move farther apart (Fig. 7–13). This relieves steric strain, the effect being greatest for highly crowded carbon atoms.

There are additional factors which also markedly influence the competition between substitution and elimination. Other things being equal, the presence of strong bases, such as alkoxides, promotes elimination. By contrast, very weak basic nucleophiles, such as I^-, Br^-, and N_3^-, favor substitution rather than elimination.

Finally, it is important to remember that *E*1 and S_N1 reactions usually accompany one another, as do *E*2 and S_N2 reactions. Since *E*1 and S_N1 products both come from carbocation intermediates, they are most likely to be found under conditions that favor these intermediates: a 2° or 3° substrate, a good leaving group, and a polar solvent. In line with the carbocation

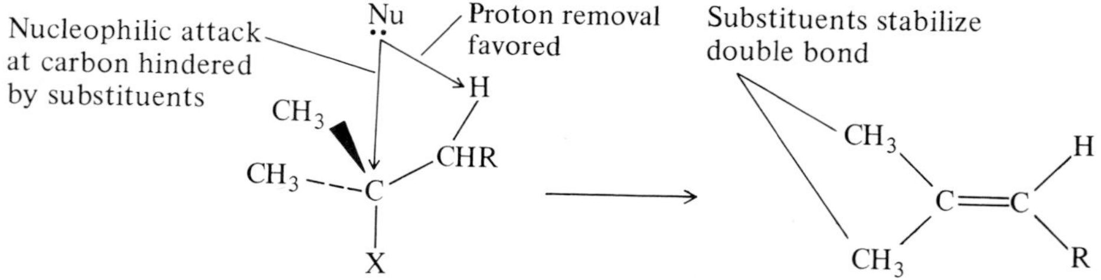

Figure 7–12. *Highly substituted substrates tend to undergo elimination because the substituents hinder nucleophilic attack and stabilize the double bond as it is formed.*

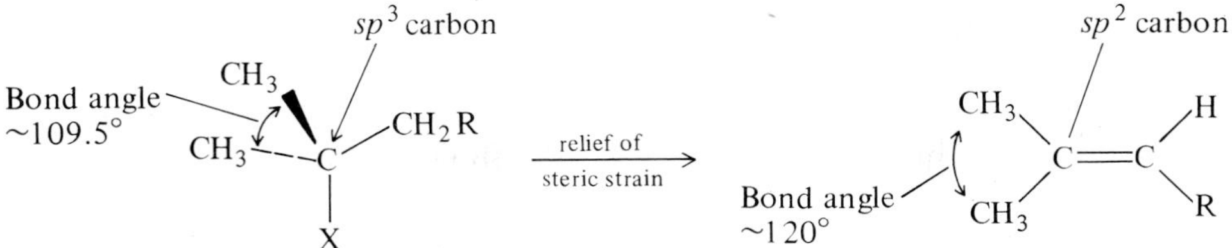

Figure 7–13. *Steric congestion is relieved when a heavily substituted carbon changes hybridization from* sp^3 *to* sp^2. *This is one reason why 3° alkyl halides tend to form alkenes readily.*

mechanism, we can anticipate rearranged substitution products or alkenes from S_N1 or *E*1 processes, respectively.

In similar fashion, the S_N2 and *E*2 reactions are tightly linked. The presence of a strong nucleophile and/or base is necessary for both. Neither reaction gives rearranged products, and both are highly stereospecific. S_N2 reactions always result in inversion, while *E*2 reactions usually involve *anti* stereochemistry.

Now that we've studied the fundamental characteristics of the elimination and substitution reactions, we'll see how they can be used in multistep syntheses.

7:11 Introduction to Multistep Syntheses

In Chapters 5, 6, and 7 we demonstrated how new functional groups can be introduced into molecules. These reactions allow us to take a readily obtainable starting material and convert it into some desired product. When the necessary starting material is not easily or economically available, however, multistep pathways are needed to synthesize the product from those reagents which are at hand. In this section you will see how the reactions which you have learned so far can be combined into multistep synthetic sequences.

Consider as a first example the synthesis of cylcopentene from cyclopentane. Since we have not encountered any reactions that directly convert alkanes to alkenes, this process will be a multistep one (eq. 7.35). In order to plan an efficient process involving the smallest number of steps, the individual reactions are arrived at by working *backward* from the product. Applying

(7.35) several steps

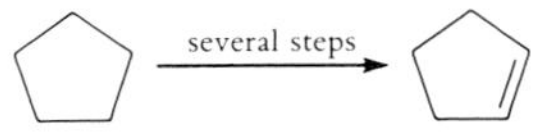

this approach to Equation 7.35 leads to the conclusion that the last step in the sequence will be an elimination reaction (eq. 7.36). It follows, then, that a

(7.36) X, H; elimination

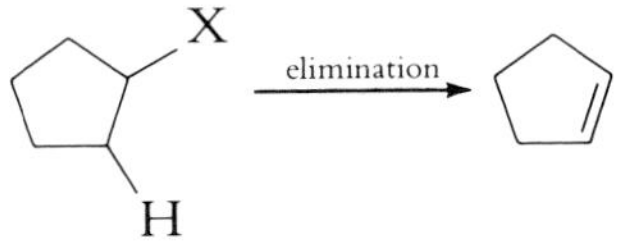

cyclopentyl alcohol or halide will have to be an intermediate, since these are the types of compounds that easily undergo elimination.

Looking again at the starting material reminds us that there are reactions which convert alkanes into alkyl halides (eq. 7.37). Finally, combining

(7.37) cyclopentane $\xrightarrow{X_2,\ h\nu}$ cyclopentyl–X + HX

Equations 7.36 and 7.37 provides us with a complete synthetic sequence (eq. 7.38; we've substituted a specific halogen for the general one designated X).

(7.38) cyclopentane $\xrightarrow{Br_2,\ h\nu}$ cyclopentyl–Br $\xrightarrow[(CH_3)_3COH]{(CH_3)_3CO^-K^+}$ cyclopentene

Using a similar analysis, we can find a pathway from cyclopentane to cyclopentyl cyanide (eq. 7.39). The introduction of the cyano group usually

(7.39) cyclopentane $\xrightarrow{\text{several steps}}$ cyclopentyl–CN

involves a substitution reaction in which CN^- is the nucleophile (eq. 7.40).

(7.40) cyclopentyl–X $\xrightarrow{CN^-}$ cyclopentyl–CN + X^-

We combine Equations 7.37 and 7.40 to get the overall synthetic sequence (eq. 7.41).

(7.41) cyclopentane $\xrightarrow{Cl_2,\ h\nu}$ cyclopentyl–Cl $\xrightarrow[EtOH]{NaCN}$ cyclopentyl–CN

The aim of any synthesis is to provide a high overall yield of *pure* product. Our concern for yield and purity requires that we take account of the likelihood of isomer and by-product formation in each step. Isomers are particularly troublesome because they are often difficult to separate from the desired product, due to their similarity in physical properties.

For example, in the synthesis of 1-pentene from 1-pentanol (eq. 7.42), we must take isomer formation into account. It would appear that the simplest procedure would be to dehydrate the alcohol by heating with acid.

(7.42) $$CH_3(CH_2)_4OH \rightarrow CH_3CH_2CH_2CH{=}CH_2$$

Such reactions usually involve carbocations, however, and a carbocation rearrangement will result in a mixture of 1- and 2-pentenes (eq. 7.43). To

(7.43) $$CH_3(CH_2)_2CH_2CH_2OH \xrightarrow{H^+,\ heat}$$

$$CH_3(CH_2)_2CH_2CH_2^+ \rightarrow CH_3(CH_2)_2\overset{+}{C}HCH_3 \rightarrow CH_3CH_2CH{=}CHCH_3$$

2-Pentene

(both cations ↓)

$$CH_3(CH_2)_2CH{=}CH_2$$

1-Pentene

avoid rearrangement, it is necessary for the elimination to take place under $E2$ conditions. This can be accomplished by first converting the hydroxyl group to a good leaving group, and then treating the intermediate with a strong base (eq. 7.44).

$$(7.44)\quad CH_3(CH_2)_4OH \xrightarrow[OH^-]{TsCl} CH_3(CH_2)_4OTs \xrightarrow[(CH_3)_3COH]{(CH_3)_3CO^-K^+} CH_3(CH_2)_2CH{=}CH_2$$

Example

Prepare the following compounds, starting with 2-methyl-3-pentanol.

$$(CH_3)_2CHCH{=}CHCH_3,\quad (CH_3)_2CH\underset{\substack{|\\ CN}}{C}HCH_2CH_3$$

Answer

We know that alkenes can be made from alcohols by acid-catalyzed dehydration. In this case, however, that route leads to the Zaitsev product, which we do not want.

$$(CH_3)_2CH\underset{\substack{|\\ OH}}{C}HCH_2CH_3 \xrightarrow[heat]{H_2SO_4} \underset{\text{Major product}}{(CH_3)_2C{=}CHCH_2CH_3}$$

We turn then to base-promoted elimination, which requires a good leaving group, such as a halide ion. Use of a strong base favors the Hofmann product.

$$(CH_3)_2CH\underset{\substack{|\\ Br}}{C}HCH_2CH_3 \xrightarrow[DMSO]{(CH_3)_3CO^-K^+} \underset{\substack{\text{Major product}\\ (\textit{cis–trans}\text{ mixture})}}{(CH_3)_2CHCH{=}CHCH_3} + Br^-$$

Putting in a weaker leaving group increases the yield of Hofmann product, but requires extra steps.

$$(CH_3)_2CH\underset{\substack{|\\ Br}}{C}HCH_2CH_3 \xrightarrow{(CH_3)_3N:} (CH_3)_2CH\underset{\substack{|\\ (CH_3)_3N^+\ Br^-}}{C}HCH_2CH_3 \xrightarrow[H_2O]{Ag_2O}$$

$$(CH_3)_2CH\underset{\substack{|\\ (CH_3)_3N^+\ OH^-}}{C}HCH_2CH_3 \xrightarrow{heat} \underset{\textit{cis–trans}\text{ mixture}}{(CH_3)_2CHCH{=}CHCH_3} + (CH_3)_3N$$

Cyano groups are introduced into aliphatic molecules by way of S_N2 reactions with halides. The bromide shown above will serve nicely:

$$(CH_3)_2CH\underset{\substack{|\\ Br}}{C}HCH_2CH_3 \xrightarrow[C_2H_5OH]{K^+CN^-} (CH_3)_2CH\underset{\substack{|\\ CN}}{C}HCH_2CH_3 + K^+Br^-$$

The bromide itself is obtained from the alcohol and PBr_3:

$$(CH_3)_2CH\underset{\displaystyle OH}{\underset{|}{C}H}CH_2CH_3 \xrightarrow{PBr_3} (CH_3)_2CH\underset{\displaystyle Br}{\underset{|}{C}H}CH_2CH_3$$

Problem

7–17 Propose efficient synthetic sequences for bringing about the following conversions:

a. 2-methylpropane → 2-methylpropene
b. cyclohexane → iodocyclohexane
c. tetrahydrofuran → 1,3-butadiene
d. 2-methyl-1-pentanol → 2-methyl-1-pentene
e. 2-methyl-1-pentanol → 2-methyl-2-pentene

7:12 A Biological Elimination Reaction

All living cells require energy. One of the chief sources of cellular energy is the controlled oxidation of various organic compounds. The six-carbon sugar glucose is one of the most common of these cellular fuels. Its oxidation makes available substantial quantities of free energy (eq. 7.45). A biological

(7.45) $$\underset{\text{Glucose}}{C_6H_{12}O_6} + 6O_2 \xrightarrow{25^\circ} 6CO_2 + 6H_2O \quad \Delta G^\circ = -686 \text{ kcal/mol}$$

cell can capture more than 50% of this energy, provided that the oxidation is carried out in many steps. Let's examine one of those steps, which involves an elimination reaction, and compare it to similar laboratory reactions.

Glucose is first divided into two three-carbon lactic acid molecules by a series of 10 reactions (eq. 7.46). The lactic acid is then oxidized to CO_2 and

(7.46) $$\underset{\text{Glucose}}{C_6H_{12}O_6} \xrightarrow[\text{steps}]{10} \underset{\text{Lactic acid}}{2CH_3\underset{\displaystyle OH}{\underset{|}{C}H}\overset{\displaystyle O}{\overset{\|}{C}}OH}$$

H_2O. During this second stage most of the energy is recovered. The reaction we will focus on is the dehydration of 2-phosphoglycerate (eq. 7.47)—one of the 10 reactions in the path from glucose to lactic acid. This elimination of water is catalyzed by the enzyme *enolase*. The enzyme requires a divalent

(7.47) $$\underset{\text{2-Phosphoglycerate}}{HO\overset{3}{C}H_2{-}\underset{\displaystyle OPO_3H^-}{\underset{|}{\overset{2}{C}H}}{-}\overset{\displaystyle O}{\overset{\|}{C}}O^-} \xrightarrow[Mn^{2+}]{\text{enolase}} \underset{\text{Phosphoenolpyruvate}}{\overset{3}{C}H_2{=}\overset{2}{C}(CO_2^-)(OPO_3H^-)} + H_2O$$

metal cation such as Mg^{2+} or Mn^{2+}. The cation appears to function as a

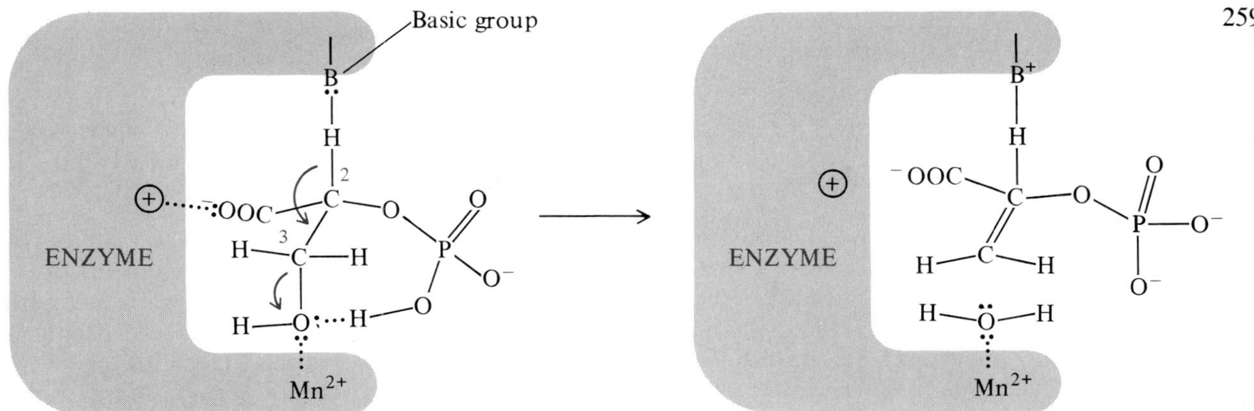

Figure 7–14. *Enzyme-catalyzed dehydration of 2-phosphoglycerate, an* E2 *reaction proceeding with* anti *stereochemistry.*

Lewis acid, coordinating with a lone pair on the hydroxyl group and helping it to leave (Fig. 7–14). A proton on the phosphate group is also involved in converting the hydroxyl group into a good leaving group. In addition, the enzyme contains a basic group which removes the proton from C-2 of the substrate.

Comparing this biological elimination with its laboratory counterparts reveals many similarities. The biological reaction proceeds with *anti* stereochemistry (Fig. 7-14), which is the predominant stereochemistry found in *E*2 eliminations (sec. 7:9). Biological eliminations are dependent on the presence of an adequate leaving group, as are nonbiological ones, and the enzyme plays an important role in creating this leaving group. The common requirement for a base to remove a proton is also apparent in both biological and nonbiological processes. Overall, the conversion of 2-phosphoglycerate to phosphoenolpyruvate nicely illustrates how our understanding of biochemical reactions is aided by our investigations of related reactions conducted in the laboratory.

Summary

1 Carbon–carbon double bonds consist of one σ bond and one π bond. The π bond is most stable when the axes of its two p orbitals are parallel. As a result, there is a powerful barrier to rotation about double bonds, which makes geometrical isomerism possible. The geometrical isomers are diastereomers (sec. 7:1).

The nomenclature systems for geometrical isomers (*cis–trans* and *Z–E*) refer to an imaginary plane that bisects the π bond and passes through both sp^2 carbon atoms (Fig. 7–3). In the older system a *cis* isomer is one in which the principal carbon chain lies on one side of this plane. A *trans* isomer is one in which the principal chain crosses the reference plane. The current nomenclature system requires that a priority, based on atomic number, be assigned to each of the two groups attached to each end of the double bond. The *Z* isomer has the two higher priority groups on the same side of the reference plane; if the two higher priority groups are on opposite sides, the isomer is called *E* (sec. 7:2).

2 Hydrogenation of double bonds is exothermic, and heats of hydrogenation can be used to measure stability differences among alkenes. For open-chain alkenes, *trans* isomers are found to be more stable than *cis* isomers, but the reverse is true with small and medium ring cycloalkenes (sec. 7:3).

3 When halides and sulfonates undergo solvolysis, alkenes are usually produced by way of carbocations (*E*1 reactions). The yield of alkene is often low because of the competing S_N1 reaction (sec. 7:5). Heating alcohols with strong acid results in high yields of alkenes, produced by way of carbocations. All these *E*1 reactions can result in rearrangement (sec. 7:6).

When carbocations are not easily formed or when a good base is present in high concentrations, then the base takes part in the rate-determining step (*E*2 reactions). *E*2 and S_N2 processes compete. The use of bulky bases favors elimination over substitution. Rearrangements are absent in *E*2 reactions (sec. 7:7; Table 7–2).

*E*2 reactions involving good leaving groups (Br^-, OTs^-) usually give more internal alkenes than terminal alkenes. These reactions, whose product distributions reflect the relative stability of the alkenes, are termed Zaitsev eliminations. When poor leaving groups such as trimethylamine are involved, then terminal alkenes are heavily favored. The classic example of this is the Hofmann elimination, in which tetraalkylammonium hydroxides decompose to alkenes and trialkylamines. The product distribution in this reaction reflects the relative ease of proton removal from different carbon atoms (sec. 7:8; Table 7–3).

4 Open-chain and cyclohexyl derivatives usually undergo *E*2 elimination by way of a transition state with *anti* stereochemistry. This arrangement allows the C—H and C—X bonds to be coplanar and the base and leaving group to be as far from one another as possible. If the *anti* transition state is too high in energy, coplanarity of these two bonds is achieved with *syn* stereochemistry (sec. 7:9).

5 In order to synthesize a desired product from available starting materials, multistep reaction pathways must often be used. In order to plan these it is often most efficient to work backward from the desired product, one reaction at a time (sec. 7:11).

Problems

7–18 Name the following alkenes by means of the IUPAC system:

a. $(CH_3)_3CCH_2CH_2CH{=}CH_2$

b. $Br(CH_2)_3$ and H on one carbon, CH_3 and H on the other of C=C:

$$\begin{matrix} Br(CH_2)_3 & & CH_3 \\ & C{=}C & \\ H & & H \end{matrix}$$

c.

$$\begin{matrix} CH_3CH(Cl)CH_2 & & CH_3 \\ & C{=}C & \\ Cl & & (CH_2)_4CH_3 \end{matrix}$$

d. 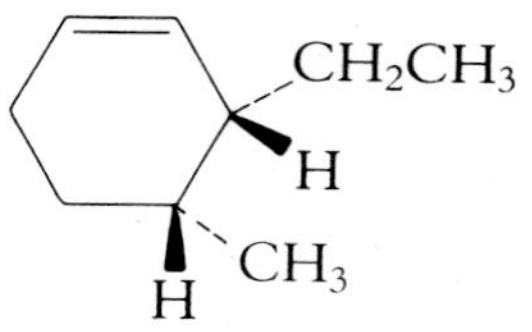 (cyclohexene ring with CH_2CH_3 / H and CH_3 / H substituents)

e. 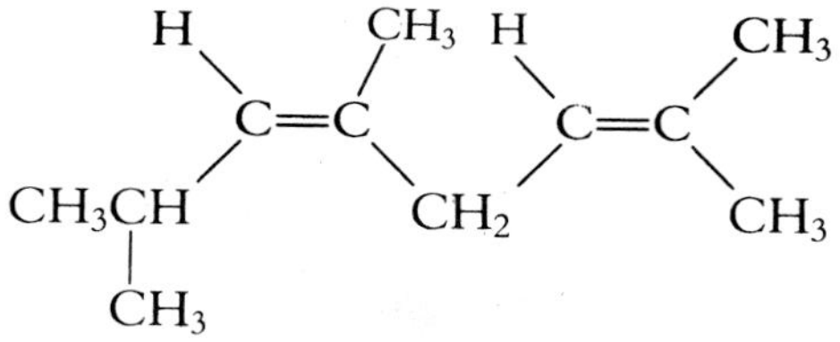 H, $CH_3CH(CH_3)$ — C=C — CH_3, CH_2 — H, C=C — CH_3, CH_3

7–19 Write the structures of all the alkene isomers (including optical and *cis–trans* isomers) of the following compounds and name them.

a. 1-hexene
b. 2-hexene
c. 2-methyl-2-pentene
d. chloropropene
e. 4-methyl-2-hexene

7–20 Draw the structures of the following compounds:

a. 3-ethyl-3-hexene
b. 1,1,4,4-tetrachloro-1-butene
c. 3,3-dimethylcyclopentene
d. 1,2-diisopropylcyclobutene
e. 6-hepten-1-ol
f. perfluoropropene ("perfluoro" means all H's are replaced by F's)

7–21 Each of the following names is *wrong*. If the structure corresponding to the name is chemically permissible, then give it the correct IUPAC name. If not, indicate why the structure is flawed.

a. 1-ethylpropene
b. 2-cyclopentene
c. 2,2-dimethyl-2-butene
d. *trans*-1,1-dichloro-1-butene
e. pentachloroethylene

7–22 Write structures and give names for all the alkenes having the following molecular formulas:

a. $C_2H_2D_2$
b. C_3H_6
c. C_4H_8
d. C_6H_{12} (straight chain)
e. C_6H_{12} (branched chain)

7–23 Give the IUPAC name, including the *E*- or *Z*- prefix where needed, for each of the following structures:

a. 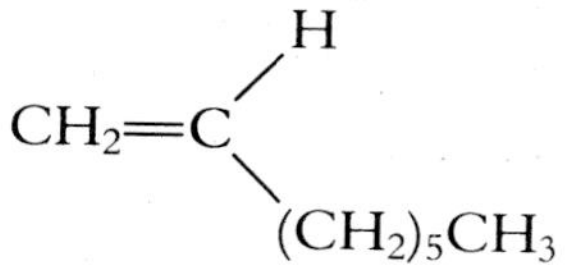 $CH_2{=}C(H)(CH_2)_5CH_3$

b. $CH_3CH_2CH_2$, H — C=C — Cl, Br

c. $(CH_3)_2CH$, $CH_3{-}C(CH_3)H$ — C=C — Br, CH_3

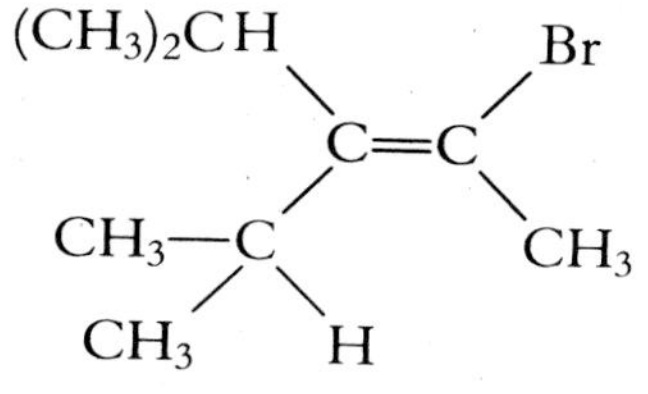

d. 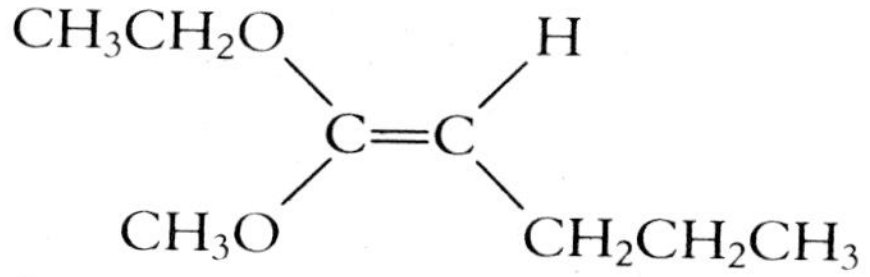 CH_3CH_2O, CH_3O — C=C — H, $CH_2CH_2CH_3$

e. 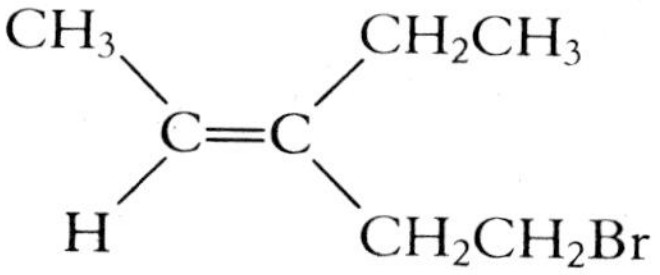 CH_3, H — C=C — CH_2CH_3, CH_2CH_2Br

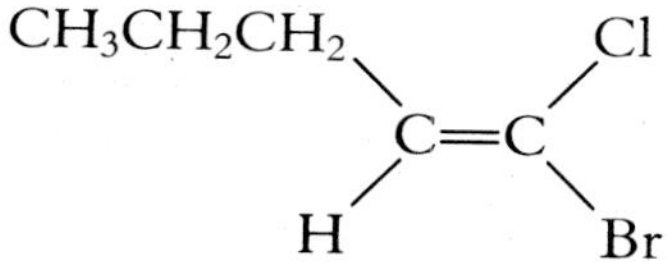

7–24 Listed here are some heats of hydrogenation for a series of *cis* and *trans* cycloalkenes. What conclusions can be drawn about the relative stabilities of the isomers? Why does relative stability vary with ring size?

	cis	*trans*
cyclooctene	−22.98 kcal/mol	−32.24 kcal/mol
cyclononene	−23.62 kcal/mol	−26.94 kcal/mol
cyclodecene	−20.67 kcal/mol	−24.01 kcal/mol

7–25 The heats of isomerization shown in Equations 6 and 7 have been determined from the heats of hydrogenation of each of the four alkenes.

$$\text{methylenecyclobutane} \rightarrow \text{1-methylcyclobutene} \quad \Delta H = -0.9 \text{ kcal/mol} \tag{6}$$

$$\text{methylenecyclopentane} \rightarrow \text{1-methylcyclopentene} \quad \Delta H = -3.9 \text{ kcal/mol} \tag{7}$$

a. Explain why both enthalpies are negative and why the enthalpy for Equation 7 is much more negative than that for Equation 6.
b. Would you expect ΔH for Equation 8 to be positive or negative? Explain your answer.

$$\text{cyclopropane}{=}CH_2 \rightarrow \text{cyclopropene}{-}CH_3 \tag{8}$$

7–26 For each of the following sets of isomers indicate which isomer is the most stable, and which the least. (In some cases two or more isomers may not differ appreciably in energy.)

a. $CH_3(CH_2)_2C(CH_3){=}CH_2 \rightleftharpoons CH_3CH_2CH{=}C(CH_3)_2$

b. cis-$CH_3CH{=}CHCH_2CH(CH_3)_2$ (H and CH_3 on one carbon, H and $CH_2CH(CH_3)_2$ on the other, CH_3 and $CH_2CH(CH_3)_2$ trans) $\rightleftharpoons$ CH_3 and $CH_2CH(CH_3)_2$ cis

c. 1-methylene-3-methylcyclopentane $\rightleftharpoons$ 1,4-dimethylcyclopentene (CH_3 on the double-bond carbon) $\rightleftharpoons$ 1,3-dimethylcyclopentene (CH_3 on the double-bond carbon)

d. 1-methylene-2-methylcyclohexane (CH_2, CH_3) $\rightleftharpoons$ 1,2-dimethylcyclohexene (CH_3, CH_3) $\rightleftharpoons$ 1,6-dimethylcyclohexene (CH_3, CH_3)

7–27 The equilibrium between 2,4,4-trimethyl-1-pentene and 2,4,4-trimethyl-2-pentene favors the 1-pentene, even though it is the less highly substituted isomer. How would you rationalize this result? (These alkenes are intermediates in the manufacture of 2,2,4-trimethylpentane, commonly called "iso-octane." Isooctane is one of the standards used to assign octane numbers to fuels for internal combustion engines; see Box 3–2, p. 78.)

7–28 The heats of hydrogenation given in Equations 9 and 10 have been measured experimentally. Use them to calculate ΔH for the isomerization shown in Equation 11. Is this isomerization favorable or unfavorable? Why?

$$(CH_3)_2CH{-}\overset{\overset{\displaystyle CH_3}{|}}{C}{=}CH_2 + H_2 \rightarrow (CH_3)_2CHCH(CH_3)_2 \quad \Delta H = -28.00 \text{ kcal/mol} \qquad (9)$$

$$(CH_3)_2C{=}C(CH_3)_2 + H_2 \rightarrow (CH_3)_2CHCH(CH_3)_2 \quad \Delta H = -26.63 \text{ kcal/mol} \qquad (10)$$

$$(CH_3)_2CH{-}\overset{\overset{\displaystyle CH_3}{|}}{C}{=}CH_2 \rightarrow (CH_3)_2C{=}C(CH_3)_2 \qquad (11)$$

7–29 Acetic acid, CH_3COOH, is often used as a solvent for organic reactions (sec. 6:6). It is reasonably effective at stabilizing ions, but it is not very nucleophilic. What products will be formed when 2-pentyl tosylate is heated in acetic acid? Show the mechanism of their formation.

7–30 When 2,2-dimethylpropanol is heated with sulfuric acid it gives a mixture of two isomeric alkenes (C_5H_{10}).

a. Draw the structures of these alkenes and write a mechanism for their formation.

b. Draw an energy diagram for this reaction, showing the relative energies of the intermediates and the products.

7–31 Give the structures and names of all the alkenes that you would expect to be produced by the acid-catalyzed dehydration of the following alcohols:

a. 1-propanol
b. 2-propanol-1-*d*
c. 1-pentanol
d. 2-pentanol
e. 2-methylcyclohexanol

7–32 Show how each of the following alkenes can be synthesized by dehydration of an alcohol. If you expect more than one alkene to be formed, show all of their structures.

a. 2-methylpropene
b. 3-ethyl-2-pentene
c. *trans*-3-heptene
d. 1-methylcyclopentene
e. 4-methylcyclohexene
f. 3-butylcyclohexene

7–33 a. Heating 1,2-dimethylcyclopentanol with aqueous acid yields three isomeric alkenes, **A**, **B**, and **C**. Upon reduction with H_2 and Pt all three alkenes give the same dimethylcyclopentane. Show the structures of **A**, **B**, and **C** and the mechanism of their formation.

b. When 2,2-dimethylcyclopentanol is heated under the same conditions it yields four alkenes: **D**, **E**, **F**, and **G**. Reduction of **D**, **E**, and **F** all give the same dimethylcyclopentane, **H**. Reduction of **G** gives **I**, which is an isomer of **H**. Propose structures for **D**, **E**, **F**, and **G**, and a mechanism for their formation.

7–34 The commercially important compound 1,1-dichloroethylene (*vinylidene* *chloride*) is manufactured by removing HCl from 1,1,2-trichloroethane with NaOH or $Ca(OH)_2$ (eq. 12).

$$ClCH_2CHCl_2 \xrightarrow[100°]{NaOH\ or\ Ca(OH)_2} CH_2{=}CCl_2 \qquad (12)$$

Propose a mechanism for the reaction in Equation 12. Use it to explain why virtually *no* 1,2-dichloroethylene is formed.

7–35 2-Chloropentane is allowed to react with sodium isopropoxide $[(CH_3)_2CHO^-Na^+]$ in isopropyl alcohol. Write all the products that could be obtained from this reaction, and name all the alkenes.

7–36 Show how to carry out the following multistep conversions in good yield.
a. ethane → ethylene
b. propane → $CH_3CH(D)CH_2D$ (propane-1,2-d_2)
c. 3-methoxy-1-propanol → 3-methoxypropene
d. $Br(CH_2)_4Br \rightarrow CH_2{=}CH{-}CH_2CH_2CN$

7–37 Give a possible structure for the starting chloroalkane in each case.
a. $C_4H_9Cl \xrightarrow[C_2H_5OH]{KOH}$ only one alkene
b. $C_5H_{11}Cl \xrightarrow[C_2H_5OH]{KOH}$ only one alkene
c. $C_5H_{11}Cl \xrightarrow[C_2H_5OH]{KOH}$ two alkenes

7–38 Show how the following conversions could be carried out:
a. cyclopentanol → cyclopentene
b. 3-pentanol → 2-pentene
c. bromocyclobutane → cyclobutene
d. 1-iodo-2-methylbutane → 2-methyl-1-butene
e. 1-hexanol → 1-hexene
f. 3-methyl-3-pentanol → 3-methyl-2-pentene (mostly)
g. 3-methyl-3-pentanol → 2-ethyl-1-butene (mostly)

7–39 Alkane **A** (C_6H_{14}) is subjected to free radical chlorination. It gives only two monochloro products, **B** and **C** ($C_6H_{13}Cl$). **B** is treated with $(CH_3)_3CO^-K^+$ in DMSO, and a single alkene is formed (**D**, C_6H_{12}). **C** is treated with base in the same way, and it yields two isomeric alkenes, **D** and **E**. Heating pure **D** in the presence of a catalyst results in its being converted mostly to **E**. Propose structures for compounds **A** to **E** and show the chemistry of their formation.

7–40 The Hofmann elimination was used extensively in the past for determining the structure of amines, particularly naturally occurring compounds. The strategy was to decompose larger molecules to smaller ones, which could then be identified by comparison with known samples. Many of these naturally occurring amines, called alkaloids, are cyclic. Equation 13 shows how the cyclic amine pyrrolidine is broken down by way of a Hofmann elimination.

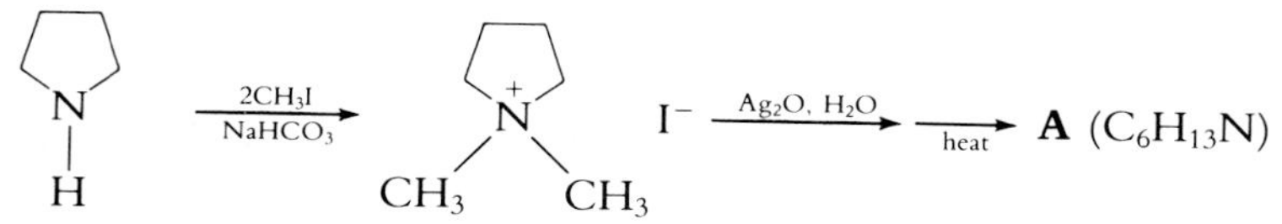

Pyrrolidine $\xrightarrow[\text{NaHCO}_3]{2\text{CH}_3\text{I}}$ N,N-dimethylpyrrolidinium I^- $\xrightarrow{Ag_2O,\ H_2O}$ $\xrightarrow{\text{heat}}$ **A** $(C_6H_{13}N)$ (13)

a. Propose a structure for compound **A**.
b. The product **A** can be further degraded, as shown in Equation 14. Propose a structure for **B**.

$$\mathbf{A} \xrightarrow{CH_3I} \xrightarrow{Ag_2O,\ H_2O} \xrightarrow{\text{heat}} \mathbf{B} \quad (14)$$

c. Compare the type of product obtained in Equations 13 and 14. Is there a general difference between the type of product obtained from cyclic and noncyclic amines? Explain.
d. 3-Methylpyrrolidine is subjected to Hofmann degradation by a series of reactions similar to those in Equation 13. Show all the possible products and indicate which is the major one.

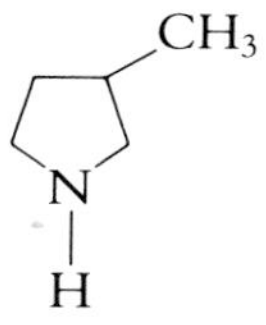

3-Methylpyrrolidine

7–41 Neopentyltrimethylammonium hydroxide, **VIII**, is subjected to the customary Hofmann conditions. The major products are an amine and a volatile alcohol. Give their structures and suggest a mechanism for their formation. Compounds similar to these are often by-products in many Hofmann eliminations.

$$(CH_3)_3CCH_2\overset{+}{N}(CH_3)_3 \quad OH^-$$
VIII

7–42 When compound **IX** is treated with $(CH_3)_3CO^-K^+$ in DMSO it gives rise to two bromoalkenes, **A** and **B** (both C_5H_9Br; eq. 15). **A** and **B** each contain only *one* vinylic hydrogen.

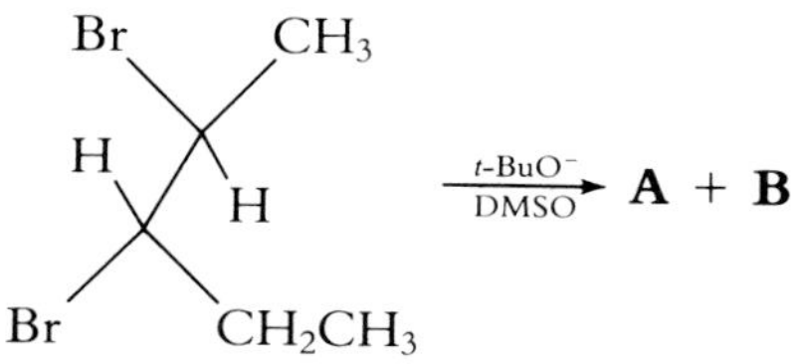

IX (Br, CH_3 / H, H / Br, CH_2CH_3) $\xrightarrow[\text{DMSO}]{t\text{-BuO}^-}$ **A** + **B** (15)

a. Write structures for **A** and **B**, showing their stereochemistry.
b. Explain why only **A** and **B** are formed in this reaction and no other bromoalkenes.

7–43 When *cis*-2-methylcyclohexyl tosylate is treated with sodium methoxide it gives a mixture of 1-methyl- and 3-methylcyclohexene as the principal products. The more stable 1-methylcyclohexene is the major isomer (93%).

Starting with *trans*-2-methylcyclohexyltosylate gives reversed results—the major alkene isomer is 3-methylcyclohexene (92%). Explain why the two tosylate isomers yield such very different alkene mixtures (reference to the stable conformations of the tosylates will be of help in solving this problem).

7–44 Cyclohexane derivatives that contain the bulky *t*-butyl substituent remain almost completely in the conformation with an equatorial *t*-butyl group. These compounds are useful for studying the stereochemistry of elimination reactions. *cis*-4-*t*-Butylcyclohexyl tosylate undergoes $E2$ reaction very easily when treated with $C_2H_5O^-Na^+$ in C_2H_5OH. Its *trans* diastereomer, by contrast, will not undergo $E2$ reaction; it reacts instead by way of an $E1/S_N1$ path. Show the reaction products in both cases, and explain why the two isomers behave differently.

7–45 Devise efficient pathways for synthesizing the products from the indicated starting materials.

a. cyclobutane → iodocyclobutane

b. butane ⟶ $CH_3CH_2CH(SH)CH_3$

2-Butanethiol

c. pentane → (*E*)- and (*Z*)-2-pentene

d. 1,6-hexanediol → 1,5-hexadiene

Supplementary Reading

E. A. Walters, Models for the double bond, *J. Chem. Ed.*, **43,** 134 (1966).

R. J. Baumgarten, Substitution products in the Hofmann elimination, *J. Chem. Ed.*, **45,** 122 (1968).

W. H. Saunders, Jr., *Ionic Aliphatic Reactions*, Chaps. 5 & 6, Prentice-Hall, Englewood Cliffs, N.J., 1965.

W. L. Mock, The reactivity of noncoplanar double bonds, *Tetrahedron Letters*, 475 (1972).

8 Physical Methods of Chemical Structure Identification: Basic Mass and Infrared Spectroscopy

The advent of various kinds of spectrometers has revolutionized structural chemistry. Spectrometric techniques allow us to determine the structure of unknown compounds without performing chemical reactions. In addition, unstable intermediates such as carbocations or radicals, which cannot be isolated, are sometimes sufficiently stable in solution so that they can be detected spectroscopically.

This chapter introduces two important types of spectroscopy: *mass* spectroscopy (*ms*), which gives information about a molecule's molecular formula (sec. 1:1) and structural sub-units, and *infrared* spectroscopy (*ir*), which provides clues as to which functional groups a molecule contains. In Chapter 9 we will introduce nuclear magnetic resonance spectroscopy (nmr), which allows us to determine the number, type, and arrangement of carbon and hydrogen atoms within a molecule.

In Chapter 13, a method for probing the electronic structure of compounds with multiple bonds is introduced: ultraviolet-visible spectroscopy (uv-vis). Each of these spectroscopic methods gives its own specific type of information about structures and when several or all are applied to a given compound, they provide a powerful investigative tool.

8:1 Determination of Molecular Weight and Molecular Formulas by Mass Spectroscopy: The Importance of Isotopes

A mass spectrometer★ contains a vacuum chamber into which a minute amount (10^{-6} g or less) of a compound is volatilized. There the molecules are bombarded with high energy electrons that cause the ejection of an electron from the neutral molecules (sec. 1:1). This creates a positively charged ion and, since the electron that is lost has a negligible weight, a **molecular ion** (M^+) is created whose mass is equal to the molecular weight (MW) of the compound.† (The molecular ion is sometimes called the parent ion.)

A low-resolution mass spectrometer measures ion masses accurate to one mass unit. A high-resolution spectrometer measures ion masses accurate

★ Before you begin studying this section, it might be helpful to review the essentials of mass spectrometry given in Section 1:1.

† Occasionally, a molecular ion is *not* observed for a particular compound. This is because the molecular ion formed breaks apart before it can be detected. Special techniques are available to obtain the molecular weight of these sensitive compounds.

Table 8–1. Exact Isotopic Masses of Some Common Elements

Isotope	Mass (atomic mass units)
^{1}H	1.0078
^{12}C	12.0000
^{14}N	14.0031
^{16}O	15.9949
^{19}F	18.9984
^{31}P	30.9738
^{32}S	31.9721
^{35}Cl	34.9689
^{79}Br	78.9183
^{127}I	126.9045

to thousandths of a mass unit. From these measurements molecular formulas and molecular weights can be determined. For example, the compounds $C_4H_{11}NO_2$ and C_4H_8FNO each have nearest whole number molecular weights of 105. In a low-resolution mass spectrometer each compound would show a molecular ion at $m/z = 105$. (Remember from Section 1:1 that the spectrometer measures the mass/charge ratio, m/z, for ions. Most ions have a charge of +1, so m/z usually is the same as the mass number.) In a high-resolution mass spectrometer, molecular ions can be measured more accurately. The $C_4H_{11}NO_2$ compound would give a molecular ion at $m/z =$ 105.079, while the molecular ion for C_4H_8FNO would come at $m/z =$ 105.059. These molecular weight values can be calculated from the exact isotopic masses given in Table 8–1. Because of the accuracy of the high-resolution mass spectrometer and the fact that it requires only micrograms of material, it is now commonly used to determine molecular formulas of unknown substances.

Problem

8–1 A compound of formula $C_3H_7NO_3$ has a molecular weight of 105. Would high-resolution mass spectrometry distinguish it from the compounds of formula $C_4H_{11}NO_2$ and C_4H_8FNO given above?

Isotopes play a critical role in mass spectrometry, which is based on differences in mass. Virtually every compound, no matter how chemically pure, is a mixture of isotopic species. Consider methane, CH_4. This hydrocarbon consists mainly (>98%) of $^{12}CH_4$, MW = 16. However, about 1% of any methane sample is comprised of $^{13}CH_4$, MW = 17. Another 0.2% has the composition $^{12}CH_3D$, MW = 17, and there are much smaller quantities of molecules containing two minor isotopes, such as $^{13}CH_3D$, $^{12}CH_2D_2$, and so on.

Table 8–2. Isotopic Composition of the Elements Most Common in Organic Compounds

Element	Stable Isotopes	Percentage of Element
Hydrogen	^{1}H (protium)	99.985
	^{2}H or D (deuterium)	0.015
Carbon	^{12}C	98.89
	^{13}C	1.11
Nitrogen	^{14}N	99.63
	^{15}N	0.37
Oxygen	^{16}O	99.76
	^{17}O	0.04
	^{18}O	0.20
Fluorine	^{19}F	100
Phosphorus	^{31}P	100
Sulfur	^{32}S	95.0
	^{33}S	0.76
	^{34}S	4.22
Chlorine	^{35}Cl	75.53
	^{37}Cl	24.47
Bromine	^{79}Br	50.54
	^{81}Br	49.46
Iodine	^{127}I	100

Compounds are isotopic mixtures because most of the elements themselves are isotopic mixtures. Table 8–2 lists the isotopic composition of the elements most prominent in organic compounds. These isotopic proportions are fairly uniform regardless of the origin of the sample, allowing us to make certain deductions about molecular composition from low resolution mass spectra.

For example, the mass spectrum of methane contains, in addition to the molecular ion peak (M) at $m/z = 16$, an *isotope peak* at $m/z = 17$ ($M + 1$). This peak is due mainly to $^{13}CH_4$, since ^{13}C is the most abundant minor isotope in methane. The ratio $(M + 1)/M$ is 1.1%, because carbon contains 1.1% ^{13}C (Table 8–2). In the mass spectrum of ethane, the isotope peak ($m/z = 31$) would be 2.2% as intense as the molecular ion peak ($m/z = 30$) because the molecule contains two carbon atoms, each with a 1.1% probability of being ^{13}C. The $(M + 1)/M$ ratio could therefore help us to distinguish the mass spectrum of ethane from that of formaldehyde (CH_2O), for which the ratio $m/z = 31$ to $m/z = 30$ would be about 1.1%.

Some elements give characteristic isotope peaks because they have two or more abundant isotopes. These are the most valuable for purposes of deducing molecular formulas. For instance, natural chlorine consists of two isotopes, ^{35}Cl and ^{37}Cl, in a 3 to 1 ratio. Figure 8–1 shows a graphical representation of the mass spectrum of chloroethane (C_2H_5Cl). Because of the two isotopes, there are two molecular ion peaks, one at $m/z = 64$ (for $C_2H_5{}^{35}Cl$) and one at $m/z = 66$ (for $C_2H_5{}^{37}Cl$). The intensity of these two

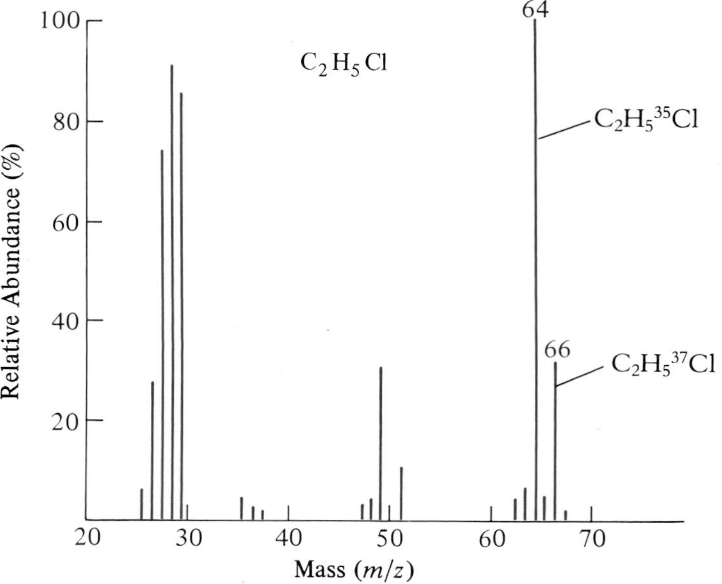

Figure 8–1. *The mass spectrum of chloroethane, showing the molecular ion peaks containing* ^{35}Cl *and* ^{37}Cl. *(This is a line drawing taken from an actual recorded mass spectrum.)*

peaks is in the ratio 3 to 1. If a mass spectrum exhibits peaks two mass units apart in the ratio 3 to 1, then there is quite possibly an atom of chlorine in the molecule.

The chemical significance of isotopes extends far beyond molecular formula measurements. Compounds containing a single isotope of a particular element are now available commercially, such as D_2O (99.9% D) and $^{18}O_2$ (99% ^{18}O). Because of the availability of minor isotopes in states of high purity, they are extensively used as "markers" in chemical reactions. A reactant is synthesized with one specific position isotopically substituted. It is subjected to reaction and the distribution of the isotope in the products is then determined. In addition, kinetic studies can show to what extent the presence of the isotope altered the rate constant. Information of this kind gives us unique insights into reaction mechanisms. It is available because mass spectrometry and nmr spectrometry allow us to distinguish among molecules that differ only in isotopic composition.

8:2 Fragment Ions: Mass Spectra of Alkanes

The bombardment of neutral molecules to form molecular ions takes place in the *ion source* chamber of the mass spectrometer (Fig. 8–2). The bombarding electrons may have energies up to 70 electron volts (1 eV = 23 kcal/mol), and some of this energy is transferred to the molecular ion. As a result, the molecular ions break apart and produce lower mass **fragment ions**. The molecular and fragment ions are drawn out of the ion source and into a magnetic field where they are separated according to mass. By changing the

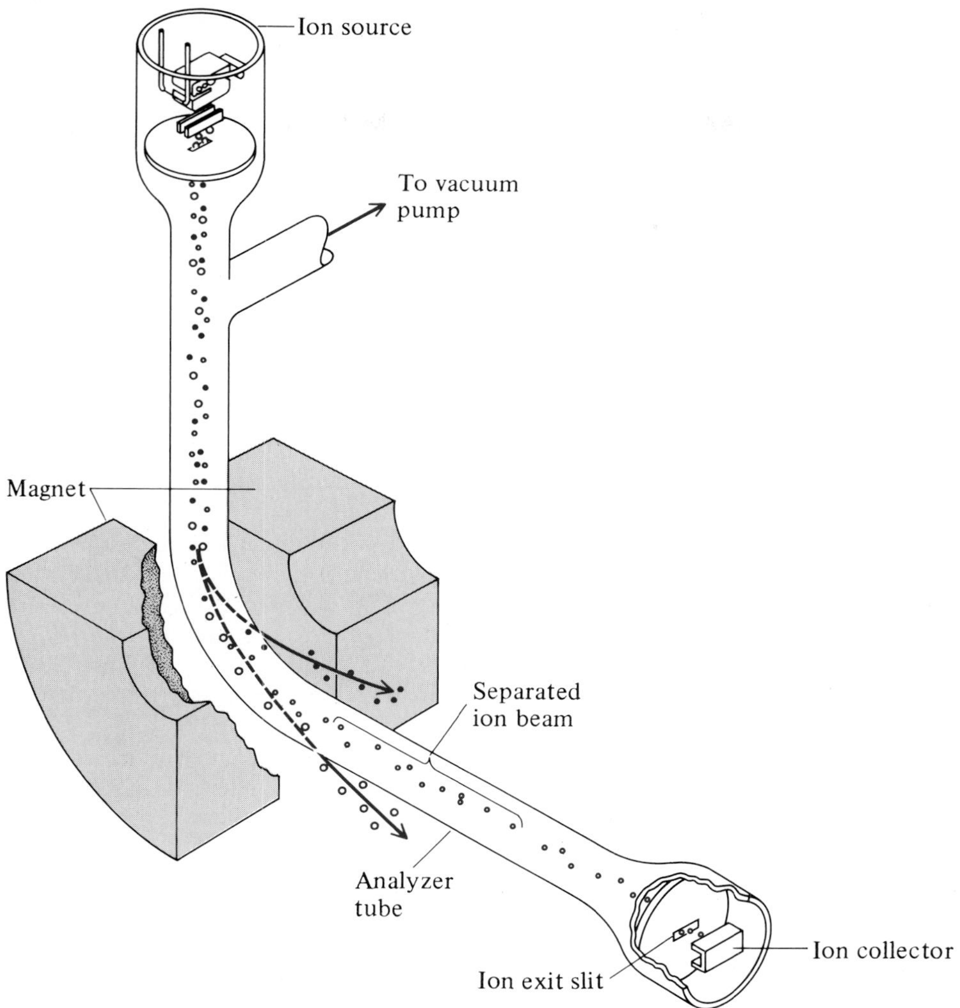

Figure 8–2. *Schematic diagram of a single-focusing mass spectrometer. A stream of ions of different masses flows out of the ion source. When this mixed beam enters the magnetic field, most of the ions are forced into trajectories which lead them into the wall of the analyzer tube. Only a beam of one particular mass is focused onto the ion exit slit and collected for a particular magnetic field strength. (From F. W. McLafferty,* Interpretation of Mass Spectra. *Used by permission.)*

magnetic field strength or the ion velocity, ions of different mass can be made to pass successively through a narrow collector slit. They fall onto a collector and cause an electrical response proportional to the number of ions being collected.

By measuring the intensity of all the ions of different mass, the spectrometer produces a **mass spectrum** of the compound. The molecular ion peak gives us information about the molecular weight and molecular formula of a compound. If we want to find out something about the *structure* of that compound, we have to look at the fragment ion peaks. Their masses and intensities constitute the *fragmentation pattern,* which is nearly unique for each compound. In common with other types of spectra, mass spectra have two general uses:

1. to *confirm* the suspected structure of a compound by showing that the spectrum of that compound is identical to the spectrum of an authentic sample; and
2. to *deduce* the structure of a new compound by interpretation of its spectrum.

In order to interpret the mass spectrum (or fragmentation pattern), you must first understand the fragmentation process. In general, fragmentation of the molecular ion will proceed so as to give the most stable fragment ions. The molecular ion is a **radical cation**, since it arises from loss of a single electron from a neutral electron-paired molecule. (Some mass spectrometers also give radical anions, but negative ion mass spectrometry is not commonly used.) The removal of one electron from a molecule will leave a positive charge (cation) and one unpaired electron (radical). A review of the chemistry of radicals (Chap. 5) and carbocations (Chap. 6) will help you understand the chemistry of molecular ions.

Figure 8–3 shows the mass spectrum of butane, C_4H_{10}, presented in two different formats. The major peaks are due to the carbocations $C_3H_7^+$ (m/z = 43) and $C_2H_5^+$ (m/z = 29) and to ions derived from them. Figure 8–4 shows a fragmentation scheme that explains the butane mass spectrum. Fragmentation takes place mainly by way of C—C bond cleavage since C—C bonds are weaker than C—H bonds. The mass spectrum of butane is characteristic of the mass spectra of straight-chain alkanes. A branched-chain alkane will

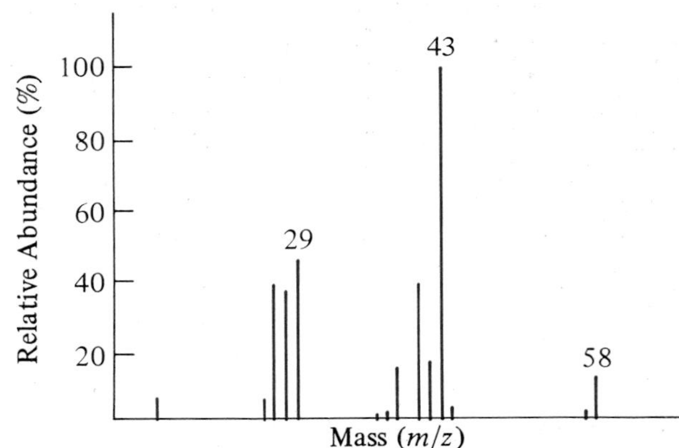

Mass	15	26	27	28	29	37	38	39	40	41	42	43	44	50	51	57	58*
Relative Abundance	5.3	6.2	37	32	44	1.0	1.9	12.5	1.6	28	12	100	3.3	1.3	1.1	2.4	12

*Molecular ion

Figure 8–3. *Low-resolution mass spectrum of butane in graphic and tabular form.*

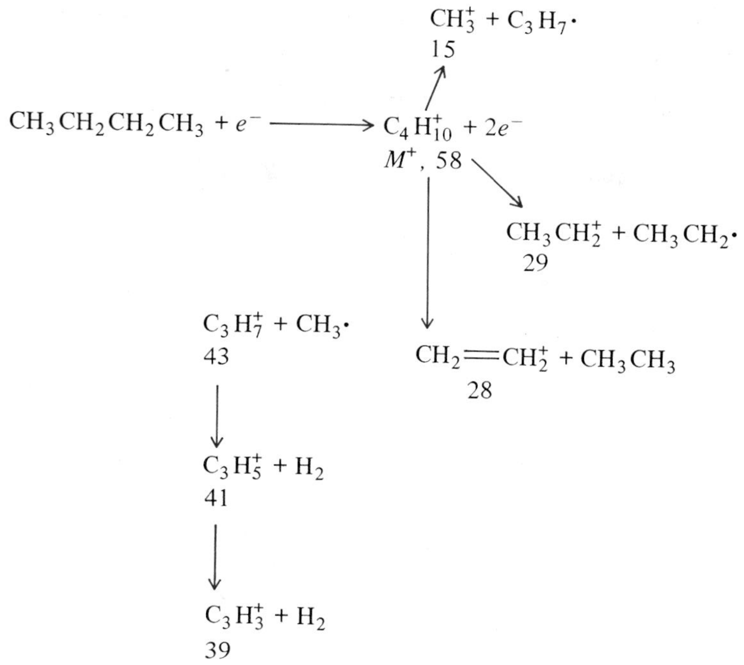

Figure 8–4. *Decomposition of butane after electron bombardment. (The* m/z *values of the ions are shown.)*

tend to cleave preferentially near the branch point since more highly substituted (more stable) radicals and carbocations will be produced (eq. 8.1). The uncharged fragments are not seen in the spectrum.

(8.1)

$$\left[\begin{array}{c} CH_3 \\ | \\ CH_3CH_2CHCH_2CH_3 \end{array}\right]^{+\cdot} \quad m/z = 86$$

$$\longrightarrow CH_3\overset{+}{C}HCH_2CH_3 + CH_3CH_2\cdot \qquad m/z = 57$$

$$\longrightarrow CH_3CH_2^+ + CH_3\underset{\cdot}{C}HCH_2CH_3 \qquad m/z = 29$$

$$\longrightarrow CH_3CH_2\overset{+}{C}HCH_2CH_3 + CH_3\cdot \qquad m/z = 71$$

Problem

8–2 Assign structures to the fragment ions at $m/z = 49$ and $m/z = 51$ in the mass spectrum of chloroethane (Fig. 8–1).

8:3 Mass Spectra of Alcohols and Ethers

Alkane molecular ions fragment by fairly random cleavage of all their C—C bonds. Molecular ions that contain heteroatoms, on the other hand, fragment by more specific pathways. These specific fragmentations are consistent with the idea that the electron is lost from a lone pair around the heteroatom. For example, the ionization of ethanol can be represented by Equation 8.2.

$$CH_3CH_2\ddot{O}H \xrightarrow[\text{source}]{\text{ion}} CH_3CH_2\overset{\bullet +}{O}H + e^- \tag{8.2}$$

Radical cation

The presence of an unpaired electron in the radical cation leads to a homolytic cleavage (sec. 2:6) of an adjoining C—C bond [eq. 8.3; the single-headed arrow or "fishhook" (↷) shows the movement of one electron].

$$\underset{\beta}{CH_3}\text{—}\underset{\alpha}{CH_2}\text{—}\overset{\bullet +}{O}H \rightarrow CH_3\cdot + CH_2\text{=}\overset{+}{O}H \tag{8.3}$$

$m/z = 46$ $\qquad$ $m/z = 31$ (a stable ion; C and O have electron octets)

Note that the breaking of the C—C σ bond is compensated for by the formation of a new C—O π bond.

The α,β-cleavage (*alpha–beta* cleavage) of a C—C bond is very characteristic of alcohols. This fragmentation produces ions that can provide information about the position of the hydroxyl group. The decomposition of the molecular ions of 1-butanol (eq. 8.4) and 2-butanol (eq. 8.5) show how fragmentation patterns are indicative of molecular structure, and enable us to distinguish between these two isomers.

$$CH_3CH_2\overset{\beta}{C}H_2\text{—}\overset{\alpha}{C}H_2\overset{\bullet +}{O}H \xrightarrow{\alpha,\beta\text{-cleavage}} CH_3CH_2CH_2\cdot + CH_2\text{=}\overset{+}{O}H \tag{8.4}$$

$m/z = 74$ $\qquad$ $m/z = 31$

$$CH_3\overset{\beta}{C}H_2\text{—}\overset{\alpha}{C}H(H\overset{+}{O}\colon)\text{—}\overset{\beta'}{C}H_3 \; (m/z = 74) \xrightarrow{\alpha,\beta\text{-cleavage}} CH_3CH_2\cdot + H\overset{+}{O}\text{=}CHCH_3 \; (m/z = 45) \tag{8.5}$$

$$\xrightarrow{\alpha,\beta'\text{-cleavage}} CH_3\cdot + CH_3CH_2CH\text{=}\overset{+}{O}H \; (m/z = 59)$$

Ethers, like alcohols, readily undergo α,β-cleavage of a C—C bond. Equations 8.6 and 8.7 show how the isomeric molecular ions from diethyl ether and methyl propyl ether will give different fragment ions.

$$CH_3CH_2\overset{\bullet +}{O}CH_2CH_3 \rightarrow CH_3CH_2\overset{+}{O}\text{=}CH_2 + CH_3\cdot \tag{8.6}$$

$m/z = 74$ $\qquad$ $m/z = 59$

$$CH_3\overset{\bullet +}{O}CH_2CH_2CH_3 \rightarrow CH_3\overset{+}{O}\text{=}CH_2 + CH_3CH_2\cdot \tag{8.7}$$

$m/z = 74$ $\qquad$ $m/z = 45$

One great advantage of mass spectral analysis is that it can be conducted at trace levels. For this reason, mass spectrometers have been used in the search for life on other planets (see Box 8–2), for the study of trace metabolites in human medicine, and in the analysis of the chemical components of flavors, aromas, and air pollutants.

Example

How would you distinguish between 1-propanol and 2-propanol by mass spectrometry?

Answer

First, draw the molecular ion structure. Then, use fishhook arrows to predict fragmentation patterns. The structures of the stable ions result from α,β-cleavage, characteristic of alcohols.

$$\overset{\beta}{CH_3CH_2}—\overset{\alpha}{CH_2}—\ddot{O}H \rightarrow CH_3CH_2—CH_2—\overset{+\cdot}{\ddot{O}}H \rightarrow CH_3CH_2\cdot + CH_2{=}\overset{+}{\ddot{O}}H$$

$m/z = 31$

1-Propanol — Molecular ion — Fragments

$$\overset{\beta}{CH_3}—\overset{\alpha}{CH}(CH_3)—\ddot{O}H \rightarrow CH_3—CH(CH_3)—\overset{+\cdot}{\ddot{O}}H \rightarrow CH_3\cdot + CH_3CH{=}\overset{+}{\ddot{O}}H$$

$m/z = 45$

2-Propanol — Molecular ion — Fragments

Thus, 1-propanol would have a base peak (highest peak in the spectrum) at $m/z = 31$, while 2-propanol would yield a base peak at $m/z = 45$.

Problems

8–3 What would be the principal fragment ions in the mass spectrum of *t*-butyl alcohol (2-methyl-2-propanol)? Could you use the spectrum to distinguish it from 1- and 2-butanol?

8–4 a. What prominent fragment ions would result from the α,β-cleavage of a C—C bond in the molecular ions of 1-, 2-, and 3-pentanol? Could mass spectrometry be used to differentiate among these isomeric alcohols?

b. An isomer of these three pentanols shows prominent mass spectral peaks at $m/z = 59$ and $m/z = 73$. Propose a structure for this isomer.

8–5 Show the structures of the principal fragment ions expected from the decomposition of the isomeric molecular ions of 1-methoxybutane and 1-ethoxypropane.

8:4 Spectroscopy and Electromagnetic Radiation

In the mass spectrometer, the compound to be studied is bombarded with high energy electrons. In most other types of spectrometric analysis, the compound is bombarded with *photons,* minute bundles of electromagnetic radiation (sec. 5:4). Visible light is a form of electromagnetic radiation, as are microwaves and radio waves. These different forms of radiation have different energies. *Photons of different energies bring about characteristic changes within molecules*. If a beam of radiation is directed at a collection of molecules, the radiation may be absorbed *if* the energy of the photon exactly matches the energy of some excitation process that can occur in the molecule.

For example, infrared light is absorbed when its energy exactly matches the energy needed to cause bonds to stretch or bend. This gives rise to ir spectroscopy (to which we will devote the following sections). Ultraviolet-visible light is absorbed when the energy associated with that light exactly matches the energy needed to promote an electron from one orbital to another one of higher energy. This gives rise to uv-vis spectroscopy (Chap. 13).

Table 8–3 outlines the different types of electromagnetic radiation available for spectroscopy, along with the energy associated with the various wavelengths of radiation. The energy of a photon at any wavelength of radiation can be calculated by applying Planck's equation (8.8).

$$(8.8)\quad E = h\nu = \frac{hc}{\lambda}$$

Where E = energy of a photon
h = Planck's constant = 6.62×10^{-27} erg sec/photon
c = velocity of light = 2.99×10^{10} cm/sec
λ = wavelength of radiation in cm
ν = frequency of radiation in hertz (cycles/sec)

Equation 8.8 shows that the amount of energy (E) in a photon of light (also called a *quantum* of light) is related to the frequency (or wavelength) of the radiation: the higher the frequency (shorter the wavelength) of the radiation, the greater the energy in a photon.

We can demonstrate the use of Planck's equation by calculating the wavelength of radiation needed to bring about a particular molecular change. In this example it is the splitting of molecular bromine into bromine atoms which, as we know, is the initiation step for free-radical bromination of alkanes (Chap. 5). The bond strength of Br_2 is 46 kcal/mol, which is then the minimum energy the light must have to cleave the molecule. Since Planck's equation has the units of ergs/photon, we must first convert 46 kcal/mol to these units (eq. 8.9). Then we can substitute the value of the

$$(8.9)\quad E = 46\text{ kcal/mol} \times \frac{1}{6.023 \times 10^{23}\text{ photons/mol}} \times 4.18 \times 10^{10}\text{ ergs/kcal} = 3.2 \times 10^{-12}\text{ ergs/photon}$$

Table 8–3. Energies and Wavelengths of Various Types of Electromagnetic Radiation

Type of Radiation	γ Rays	X Rays	Ultraviolet	Visible	Infrared	Radio
Approximate wavelength range (nm)	10^{-5}–10^{-4}	10^{-3}–1	100–350	350–1000	10^{3}–10^{6}	10^{6}–10^{8}
Energy (kcal/mol)	$\sim 10^{8}$	$\sim 10^{5}$	286–81.6	81.6–28.6	28.6–0.0286	$\sim 10^{-3}$

energy per photon into Equation 8.8 and calculate the wavelength (eq. 8.10).

$$\lambda = \frac{hc}{E} = \frac{6.62 \times 10^{-27} \text{ erg sec/photon} \times 2.99 \times 10^{10} \text{ cm/sec}^{-1}}{3.2 \times 10^{-12} \text{ ergs/photon}^{-1}} \quad (8.10)$$

$$= 6.2 \times 10^{-5} \text{ cm} = 620 \text{ nm}$$

The calculated wavelength, 620 nm, lies at the red end of the visible spectrum, showing that bromine can be split by ordinary light. Equation 8.11 is a compact formula for converting energy in kilocalories/mole to wavelength in nanometers.

$$\lambda(\text{nm}) = \frac{28{,}600}{E(\text{kcal/mol})} \quad (8.11)$$

Problem

8–6 Use the data of Tables 8–3 and 2–2 (p. 53) to answer the following questions:

a. Is there sufficient energy in visible light to break the C—C bond of ethane?
b. Is there sufficient energy in visible light to break the O—O bond of a peroxide (ROOR)?
c. What is the longest wavelength of light that could be used to dissociate Cl_2 into Cl· atoms, as in the photochlorination of alkanes (sec. 5:4)?
d. Is there sufficient energy in infrared light to break any bonds in organic molecules?
e. Why are X rays and some ultraviolet rays damaging to living systems?

8:5 Infrared Spectroscopy: The Instrument and What It Measures

If we shine infrared light on a compound (as a gas, as a pure liquid, or dissolved in an inert solvent), the light might pass through or it might be absorbed. Whether or not infrared light is absorbed depends on the exact frequency of the light and on the types of bonds or functional groups present in the molecule. An infrared *spectrophotometer* measures the amount of light absorbed or transmitted at various frequencies and displays this on a recorder. An O—H bond, for example, absorbs light in a given frequency range. If the recorder display (spectrum) shows an absorption in this frequency range, then the molecule contains an O—H bond. If no absorption is noted, then there is no O—H bond.

Why do molecules absorb infrared light? The forces holding atoms together in a molecule are not rigid like rods, but elastic like springs (Fig. 8–5). These atoms are always in motion, vibrating with respect to one another. The carbon–carbon single bond distance (i.e., internuclear distance) is given as 1.54 Å in Table 2–3 (p. 54). If we were to take an instantaneous look at a collection of CH_3—CH_3 molecules, not all the C—C distances would equal 1.54 Å. Some bonds would be longer than this and some shorter at any given instant. The bond lengths measured experimentally are only average values, and it is important to realize that all chemical bonds are continually contracting and expanding.

These vibrating bonds can absorb energy and thereby increase their *amplitude* of vibration. That is, the bonds stretch and contract (or bend) to a

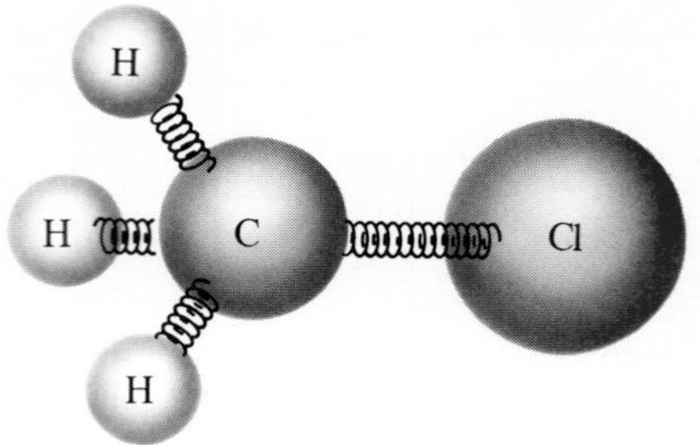

Figure 8–5. *A ball-and-spring model of chloromethane. In this model the bonds are elastic, as they are in the real molecule.*

greater degree after absorbing the energy than they did before. Figure 8–6 indicates this process for the >C=O stretching motion in formaldehyde. Changes in molecular vibrational energy are quantized—only by absorption of an infrared quantum of the correct frequency can a change in vibrational state take place. It turns out that the *different types of bonds in a molecule absorb radiation from different parts of the ir spectrum,* which is why ir spectroscopy is useful. It is therefore possible to speak of frequencies that are characteristic of a functional group. Tables 8–4 and 8–5 show some of the important infrared absorption regions for common functional groups.

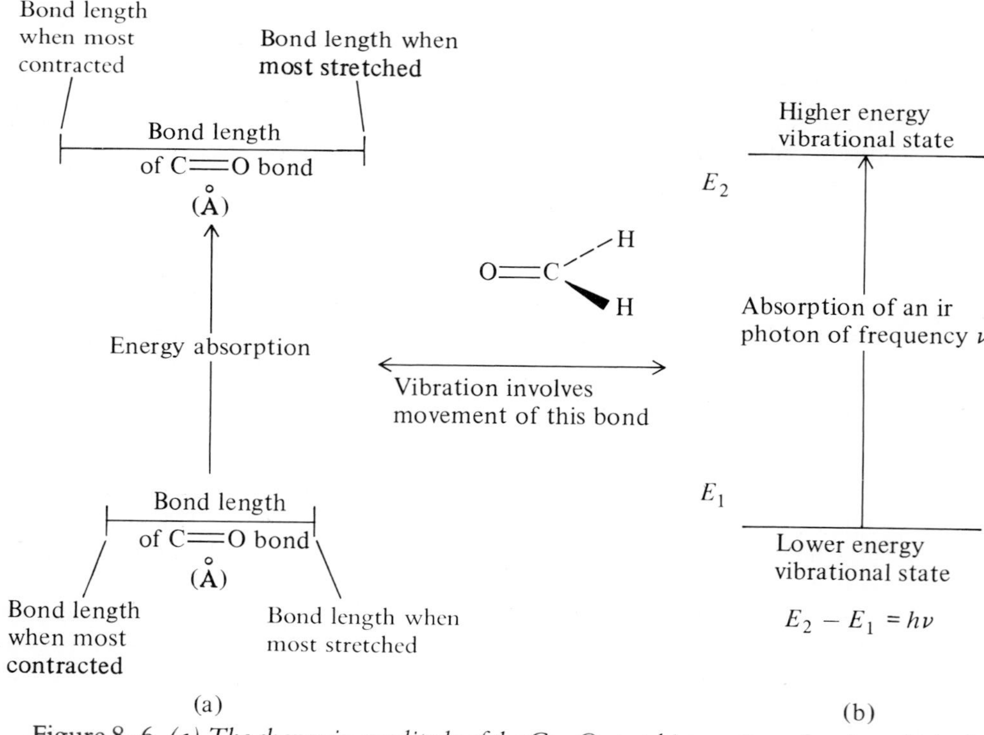

Figure 8–6. *(a) The change in amplitude of the C=O stretching motion when formaldehyde absorbs infrared radiation. (b) The quantized nature of the absorption process.*

Table 8–4. Stretching Frequencies of Some H—Y Bonds

H–Y Bond	Frequency Range (cm^{-1})	Average Bond Energy (kcal/mol)
H—O—	3650–3300	104
H—N(—)—	3500–3300	101
H—C≡(*sp* hybridized)	3300	125
H—C(—)=(*sp*^{2} hybridized)	3100–3000	104
H—C(—)(—)—(*sp*^{3} hybridized)	3000–2850	97

Molecular vibration is necessary for the absorption of infrared radiation, but it is not always sufficient. In order for ir absorption to take place, the vibration must cause a change in the *polarity* of the molecule. As we've seen, polar molecules have polar bonds, bonds that connect atoms of different electronegativity (sec. 2:7). A polar bond has separate centers of positive and negative charge, as in the carbonyl group,

$$>\overset{\delta+}{C}=\overset{\delta-}{O}$$

If the molecule *as a whole* has separate centers of positive and negative charge, then the molecule possesses a **dipole moment**, which is a measure

Table 8–5. Stretching Frequencies of Some Multiple Bonds

Multiple Bond	Frequency (cm^{-1})
—C≡N	2260–2215
—C≡C—	2250–2100
>C=O	1900–1675
>C=N—	1690–1630
>C=C<	1680–1600

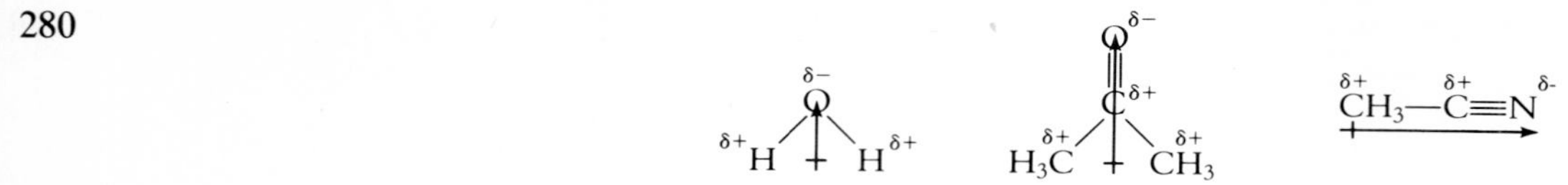

Figure 8–7. *Molecules with polar bonds that possess dipole moments. The arrow shows the location of the dipole. The head coincides with the center of negative charge, and the tail coincides with the center of positive charge.*

of its polarity (Fig. 8–7). Most molecules that have dipole moments are good ir absorbers because their vibrations bring about changes in the magnitude of the dipole moment. The most intense ir absorptions are those arising from vibrations involving the most polar bonds, such as C=O and O—H.

Some symmetrical molecules that have polar bonds still have average dipole moments of zero because the individual bond dipoles cancel one another. As a result, their centers of positive and negative charge coincide (Fig. 8–8). These molecules are usually weak ir absorbers. Nonetheless, they do show some ir absorption peaks because some of their vibrations change the dipole moment from zero to a finite value.

Box 8–1. IR Spectrometry

The infrared spectrophotometer (Fig. 8–9) contains a *source* that generates a beam of light containing all the frequencies of infrared radiation whose absorption is to be measured. The beam of light is then passed through the *sample* and on to the *monochromator*. The monochromator selects one frequency at a time and passes it on to the *detector,* which puts out an electronic signal proportional in strength to the amount of light falling on it. The detector actually receives two light beams—one that has passed through the sample and one that passes through only the solvent (if the sample is present as a solution). In this way the spectrophotometer can directly measure the fraction of light absorbed (or transmitted) by the sample alone at any wavelength. A recorder continuously plots the amount of light transmitted or absorbed for each frequency and the resulting plot is called the *infrared spectrum* of the compound.

A very common way of obtaining the ir spectrum of a solid or liquid is as a solution in a nonpolar solvent such as $CHCl_3$ or CCl_4. The solvents have few ir absorptions of their own, and exert minimal effect on the absorptions of the solute. Liquids that are not soluble in the standard ir solvents can be spread into thin films between NaCl plates and their ir spectra can then be measured. This "neat film" technique is also used with soluble samples to obtain the part of the spectrum that is obscured by the solvent absorptions.

Solids which are not soluble in the standard ir solvents can be dispersed in thin discs of solid KBr. The sample and the KBr are ground together, and the mixture is then converted to a thin, translucent disc by being subjected to high pressure in a die. The KBr does not absorb ir radiation, so when it is placed in the spectrometer, only the sample spectrum is recorded. KBr has the disadvantage that, being a salt, it is highly polar and can alter the position and intensity of the sample peaks. It is nonetheless often possible to compare solution and KBr spectra since the effects of different media on many functional group absorptions are known.

$^{\delta-}\text{O}{=}\overset{\delta+}{\text{C}}{=}\text{O}^{\delta-}$

Center of positive and negative charge

CCl_4: $\text{Cl}^{\delta-}$ (four), $\text{C}^{\delta+}$

Center of positive and negative charge

$^{\delta-}\text{N}{\equiv}\overset{\delta+}{\text{C}}{-}\overset{\delta+}{\text{C}}{\equiv}\text{N}^{\delta-}$

Center of positive and negative charge

Figure 8–8. *Molecules with polar bonds that lack a dipole moment. In these molecules, the centers of positive and negative charge coincide.*

8:6 Functional Group Detection by IR Spectroscopy

The power of infrared spectroscopy becomes more apparent if some actual infrared spectra are studied. Figure 8–10 shows the ir spectra of 2-propanol, diethyl ether, and acetone. On the bottom of each spectrum is a frequency scale representing the incident light. The unit used, cm^{-1} (called wave numbers), is obtained by dividing the light wave frequency, ν, in hertz (Hz) by c, the velocity of light. Most organic compounds absorb in the region 4000 to 600 cm^{-1}.★

★ At the top of each spectrum is a wavelength scale in microns, μ ($1\mu = 10^{-4}$ cm). The frequency scale is the more modern and more widely used of the two, but the wavelength scale is still used sometimes so we show both. The relationship between μ and cm^{-1} is a simple one: frequency (cm^{-1}) = 10,000/wavelength (μ). The frequency in cm^{-1} corresponds to the reciprocal of the number of wavelengths in one centimeter.

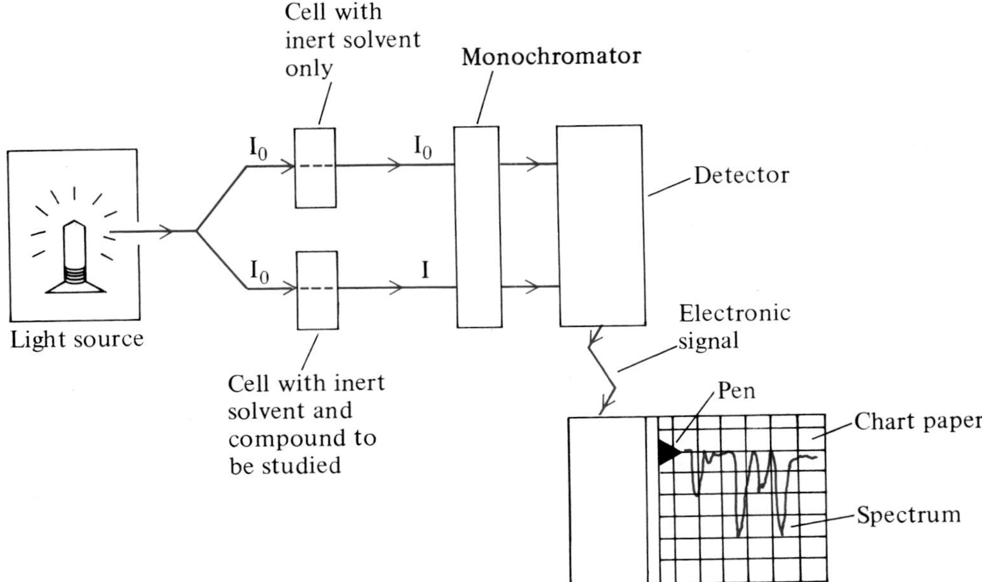

Figure 8–9. *Schematic diagram of an infrared spectrophotometer.*

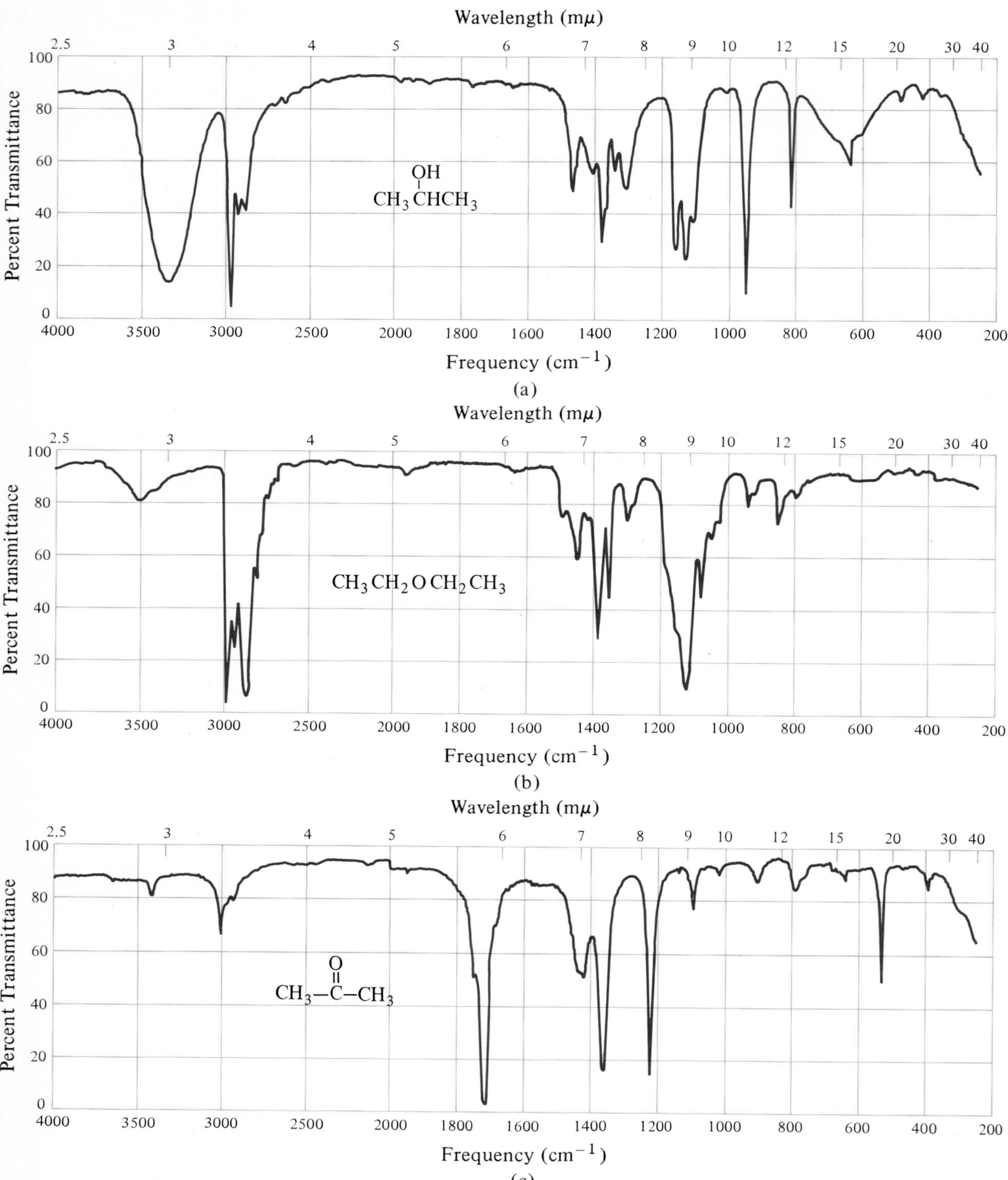

Figure 8–10. *Infrared spectra of (a) 2-propanol, (b) diethyl ether, and (c) acetone.*

The left hand vertical scale is "percent transmittance." At 100% transmittance—the top of each spectrum—all of the ir light is passing through the sample and none is being absorbed. At 0% transmittance, all of the light is being absorbed. The spectrum is usually recorded from left to right, starting with the highest frequency radiation.

First examine the spectrum of 2-propanol. The frequency of light striking the sample decreases from 4000 to about 3600 cm^{-1} without any absorption taking place. There are no bonds in 2-propanol that have vibrational energies corresponding to these frequencies. As the irradiation frequency decreases further, a strong absorption takes place that reaches its maximum at about 3400 cm^{-1}. This absorbance occurs because the O—H bond of the hydroxyl group stretches and contracts with a frequency of 3400 cm^{-1}. (The absorption is broad because hydrogen-bonding causes a range of O—H bond vibrations, which in turn cause a range of frequency values.) As the frequency further decreases, absorptions occur that are due to stretching and contracting of C—H bonds (3000 cm^{-1} region) and of the C—O bond (1200 to 1000 cm^{-1} region). Many of the other absorptions are caused by other bond movements, such as bending of bonds, but these are generally not as useful as stretching absorptions for functional group identification.

Now examine the ir spectrum of diethyl ether. Diethyl ether has C—H bonds and C—O bonds and hence has absorptions in some of the same regions as 2-propanol. However, it does not have an O—H bond, so there is no strong ir absorption at about 3400 cm^{-1}. (Note that there is, however, a *small* broad absorption at about 3500 cm^{-1} in the ir spectrum of diethyl ether. This is the characteristic position of the O—H stretch in H_2O. The sample used for the spectrum in Figure 8–8 undoubtedly contained a trace of water.)

The ir spectrum of acetone shows C—H bond stretching and bending vibrations, and a strong band at 1715 cm^{-1}. This is within the region for C═O (carbonyl) stretching vibrations. We will study many compounds (acids, esters, ketones, aldehydes, and amides) that contain the C═O group and hence give rise to a characteristic peak in the C═O bond stretching region of the ir spectrum.

The two most important uses of ir spectroscopy are

1. to aid in the structure determination of an unknown compound by showing which functional groups are present or absent, and
2. to identify an unknown by showing that its ir spectrum is identical to that of a known compound.

We have already discussed examples of the first use. To understand the second use, note that although the spectra in Figure 8–10 contain many absorptions in the same general regions they are nonetheless not identical. The region from 800 to 1400 cm^{-1} is often referred to as the *fingerprint region* because it is rich in distinctive peaks. Many of these cannot be assigned to standard functional groups; rather they are characteristic of the particular structure of each compound. Peaks in this region are therefore valuable for establishing the identity or nonidentity of two compounds that share the same functional groups. When analytical chemists want to establish the identity of an unknown substance, for example, cocaine in drug analysis or DDT in pesticide analysis, they take an ir spectrum of the unknown and compare it with the ir of a standard known sample. If the two spectra are superimposable, the analyst has established the most important criterion of identity.

Example

How would we use infrared spectroscopy to distinguish between cyclohexanol and cyclohexanone?

OH — Cyclohexanol   O — Cyclohexanone

Answer

Cyclohexanol has an OH functional group and hence its infrared spectrum would show an absorption in the 3400 cm^{-1} region (similar to that of the OH peak in the spectrum of 2-propanol, Fig. 8–8a).

Cyclohexanone has a C═O functional group and its infrared spectrum would have a peak in the 1715 cm^{-1} region (similar to the C═O peak in the spectrum for acetone, Fig. 8–8c).

Problem

8–7 A chemistry student is running a reaction overnight but forgets to attach a cooling water hose tightly. The next morning the student arrives to find a flooded laboratory and all the bench drawers filled with water. Labels have been washed off three identical vials which contain compounds **I**, **II**, and **III**.

$CH_3CH{=}CHCH_2OH$	$CH_3\overset{O}{\overset{\|\|}{C}}CH_2CH_3$	H_2C—CH_2, H_2C CH_2, O (ring)
I	**II**	**III**
(2-Buten-1-ol)	(2-Butanone)	(Tetrahydrofuran, THF)

To identify the compounds, ir spectra were taken and are given below. Which spectrum belongs to which compound? Which peaks, by their presence or *absence*, lead you to make the assignments?

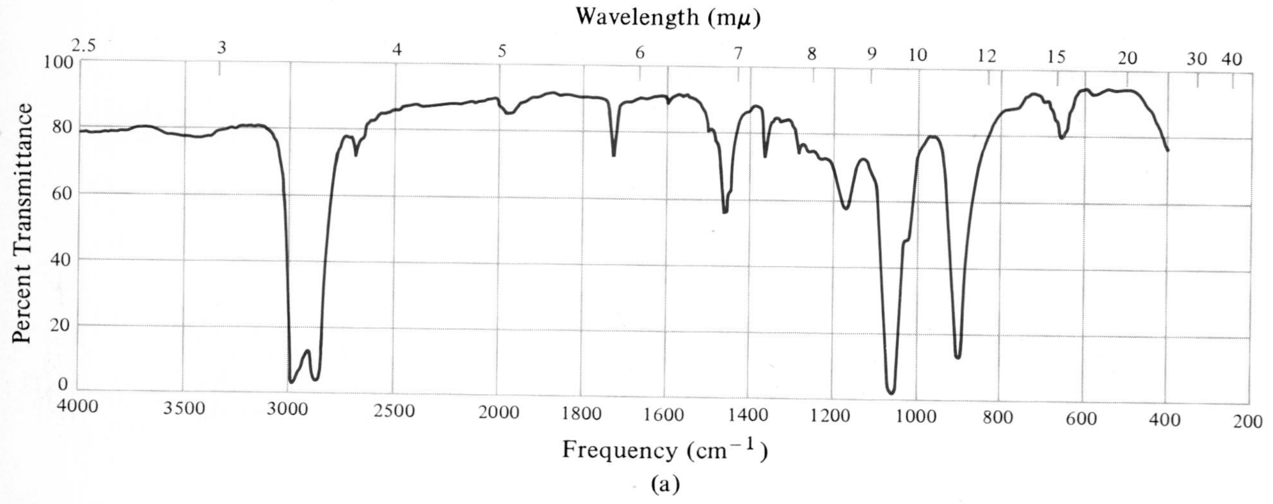

(a)

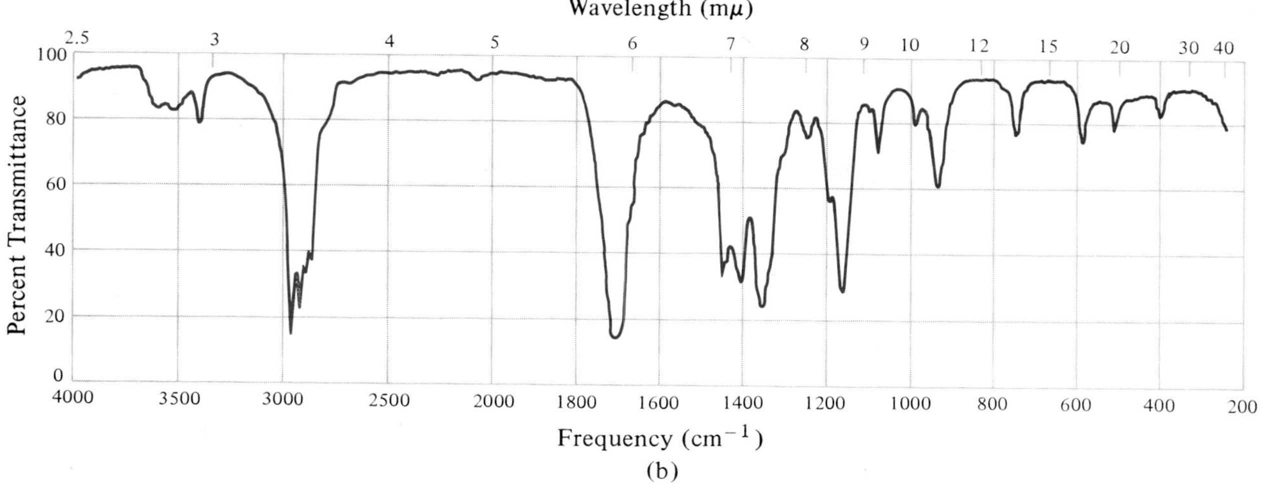

(b)

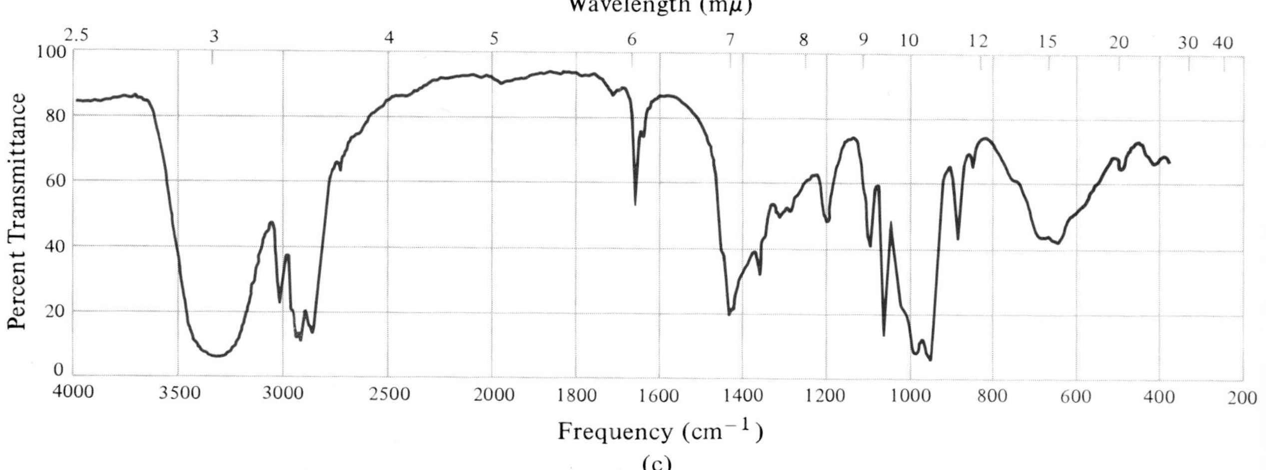

(c)

8:7 Infrared Absorptions Due to Vibrations Involving Hydrogen

Let's look in greater detail at the cause of ir absorptions. We have already mentioned that atoms in a molecule are bound together in an elastic manner. You might imagine them as a set of hard, spherical atoms joined by springs (Fig. 8–5). This model fits molecular behavior quite well and is an excellent framework for understanding molecular vibrations.

When two particles are bound together by a spring, their frequency of vibration depends on the strength of the spring and on the masses of the particles. If one particle is much lighter than the other, then the vibrational motion is not equally divided between them. The lighter particle moves a great deal, the heavier one very little. This situation describes the vibrations of hydrogen atoms bonded to C, N, O, and so on. Virtually all the movement in these vibrations is due to the hydrogen atom. As a result, the frequency differences among C—H, N—H, O—H, and similar vibrations are due almost solely to differences in bond strengths, because the bond corresponds to the spring.

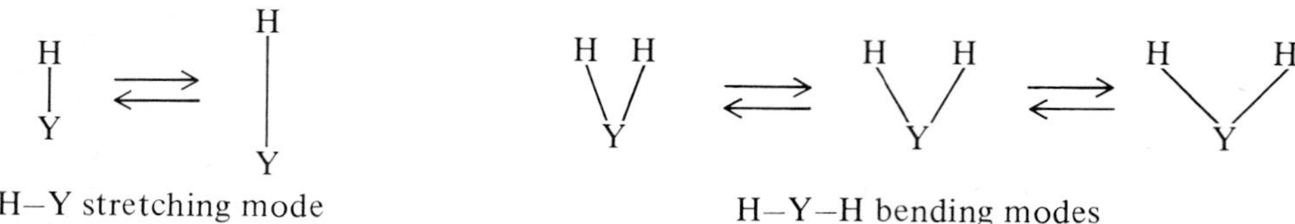

Figure 8–11. *Stretching and bending motions for bonds to hydrogen.*

There are two different ways in which a hydrogen atom can move relative to another atom. The bond may be *stretched* so that H and the other atom move further apart, or it may be *bent* so that the bond length remains the same but the bond angles change (Fig. 8–11). It takes more energy to stretch the bond than to bend it.

The stretching vibrations of hydrogen atoms lie at the high frequency end of a normal ir spectrum. Table 8–4 gives the ranges of a number of hydrogen stretching frequencies. The values in Table 8–4 reveal a rough correlation between bond energies and force constants.

Since many of the bending vibrations of H—Y bonds occur mixed together in the fingerprint region, they are not easy to assign. However, some of them give rise to characteristic peaks worth noting. A CH_3 group attached to a saturated carbon gives a C—H bending peak at about 1375 cm^{-1}. A CH_2 group attached to a saturated carbon gives rise to an analogous peak at about 1460 cm^{-1}. Both these frequencies can shift if the CH_3 or CH_2 group is bonded to N, O, or halogen.

8:8 Infrared Absorption by Carbon–Heteroatom Bonds: Band Positions and Intensities

Stretching vibrations involving hydrogen give rise to absorptions in the 2500 to 4000 cm^{-1} region, depending on the atom to which the hydrogen is bonded. At lower frequencies, we find bands due to the stretching of single and multiple bonds between heavier elements. Triple bonds characteristically give absorptions between 2000 and 2300 cm^{-1}, while the spectral region of 1600 to 2000 cm^{-1} is dominated by absorptions due to double bonds (Table 8–5). Figure 8–12 (see p. 288) shows some infrared spectra of compounds that have single or double bonds from carbon to heteroatoms.

Bonds from carbon or hydrogen to heteroatoms are often distinguished by a high *intensity* of absorption. Examine the spectra of 2-propanol (Fig. 8–10a) and 2-ethylcyclohexanone (Fig. 8–12). The area of the O—H absorption in 2-propanol is much larger than that of the C—H absorption even though there are seven C—H bonds and only one O—H bond.

Analogously, the C=O absorption in 2-ethylcyclohexanone is more intense than the C—H absorption, although the C—H groups numerically outweigh the C=O group by a large factor. The explanation for these observations lies with the high polarity of the O—H and C=O bonds (sec. 8:5). As these bonds stretch and contract, their polarities change substantially; this increases their probability of absorbing an infrared photon. By contrast, C—H and C—C bonds have low polarities because of the similar or identical electronegativities of the bonded atoms. Thus, C—H and C—C bonds do not change their polarity very much when they vibrate, and are therefore weak infrared absorbers.

Example

Shown below is the infrared spectrum of a compound whose molecular formula is C_3H_5ClO. What can you learn about the structure of the compound from its infrared spectrum?

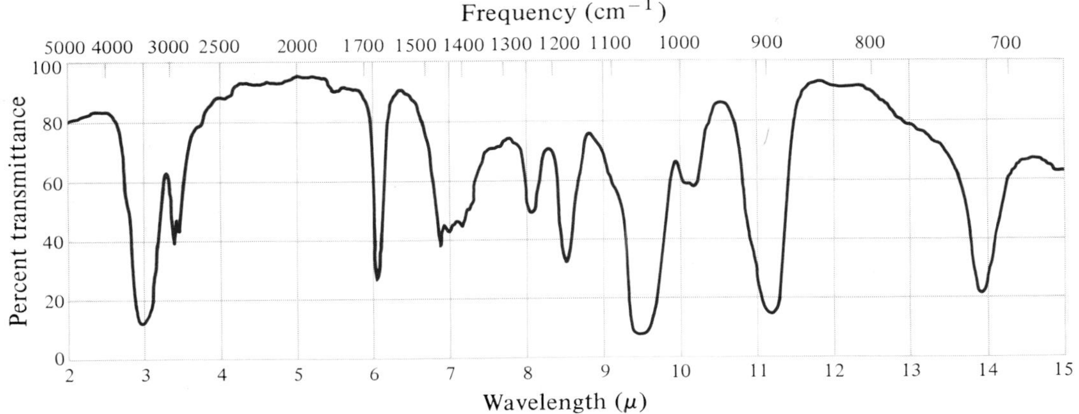

Answer

There is a strong, broad peak in the 3600 to 3400 cm^{-1} region; therefore the compound must be an alcohol. There is a peak at about 1650 cm^{-1}, which is in the double bond region. It is a little low for a carbonyl, and we have already used the one oxygen of the compound in the alcohol functional group. The peak at 1650 cm^{-1} must therefore be due to a C=C group. If it were a carbonyl, it would also probably be the most intense peak in the spectrum.

The spectrum is actually that of 2-chloro-2-propen-1-ol (2-chloroallyl alcohol).

$$\begin{array}{c} H_2C{=}\underset{\displaystyle Cl}{\underset{|}{C}}{-}CH_2OH \end{array}$$

2-Chloro-2-propen-1-ol

Problems

8–8 a. Table 8–5 shows that C≡C bonds absorb at a higher frequency than do C=C bonds. Does this observation fit the bond strength/force constant correlation discussed in Section 8:7? Explain.

b. Based on your answer to (a), would you expect C=O or C=N bonds to have the higher characteristic frequency? (Table 2–4, p. 55, may be helpful.) Which bond should show the higher absorption intensity?

8–9 The ir spectrum of a compound of molecular formula C_3H_5N contains a sharp peak at 2225 cm^{-1} and no absorption beyond 3000 cm^{-1}. What is its likely structural formula? (Table 8–5 may be helpful.)

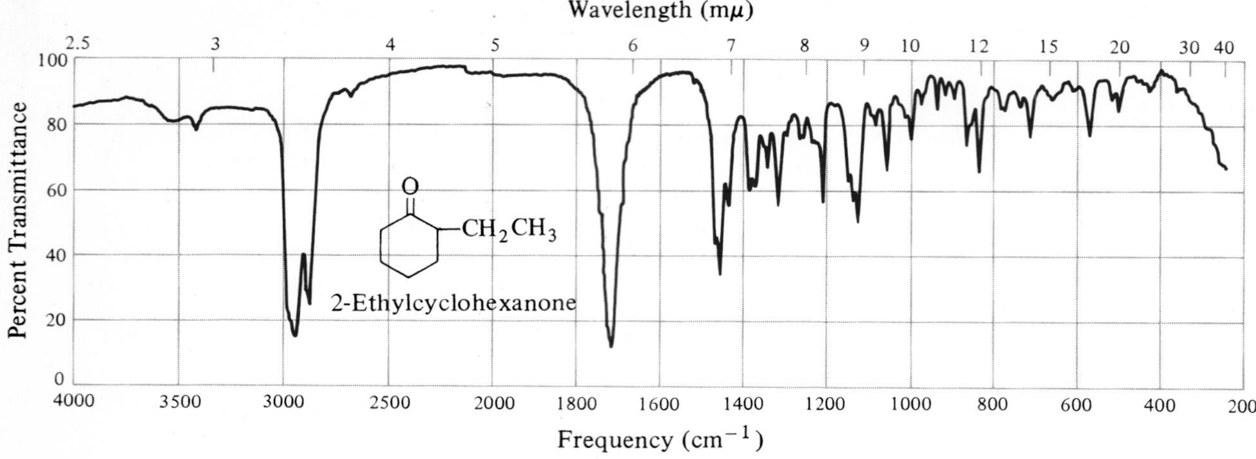

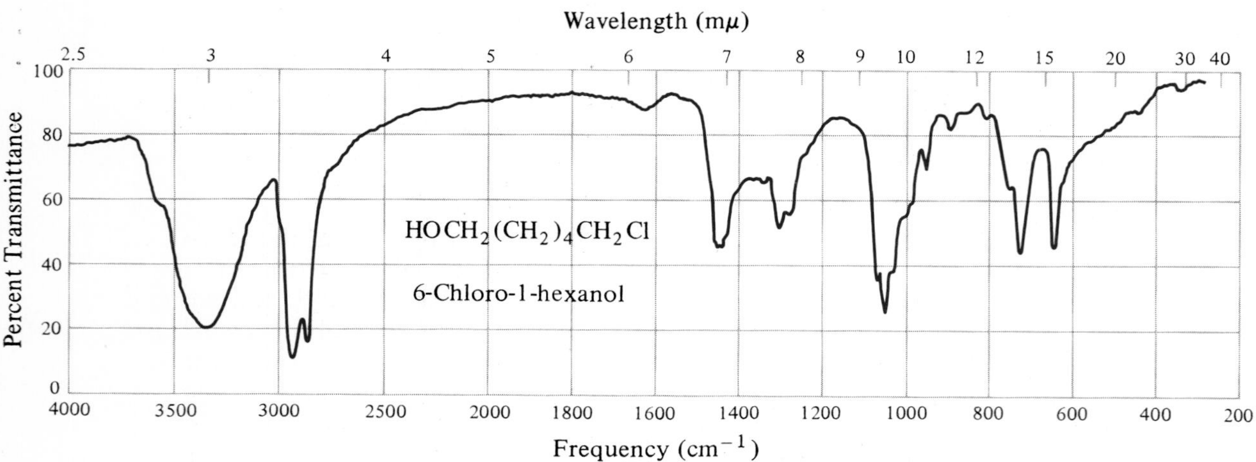

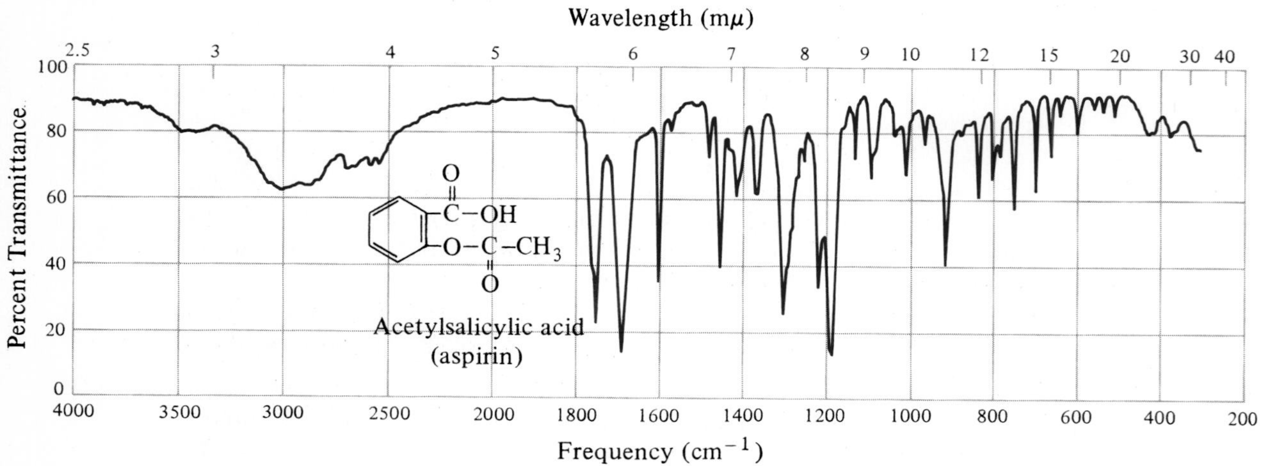

Figure 8–12. *Infrared spectra of some compounds containing singly- and doubly-bonded heteroatoms.*

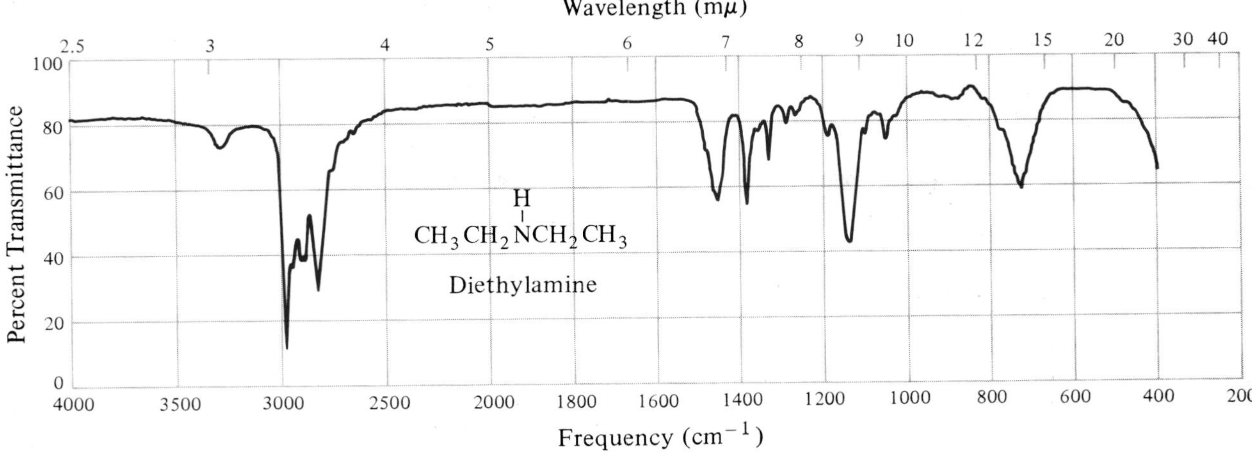

(Fig. 8–12 cont.)

8–10 A compound of molecular formula C_4H_8O gives an ir spectrum containing the following important peaks: 3090, 2980, 2960, 1650, 1450, 1400, 1270, 1110, 990, 930, and 880 cm^{-1}.

a. Is this compound identical to any of the compounds, the spectra and structures of which were given in Problem 8–7.

b. If not, propose a structure for this compound. Give an explicit account of your reasoning.

8:9 IR Spectroscopy and Structure Determination of Alkanes, Alkenes, Ethers, and Alcohols

In subsequent chapters, we will discuss the spectroscopy of each new type of compound or functional group as they are introduced. Since the chemistry of alkanes, alkenes, alcohols, and ethers has already been examined in previous chapters, their ir spectral properties are summarized here.

Alkanes

Alkanes contain only C—H and C—C bonds, and since these are common to all organic molecules, little structural information can be obtained from an ir spectrum of an alkane or cycloalkane. However, the *absence* of peaks characteristic of functional groups such as C═O, O—H, and C═C is the most informative aspect of alkane ir spectra.

Alkenes

The ir stretching frequencies of hydrogens bonded to sp^2-hybridized carbon atoms, such as those in alkenes, lie between 3000 and 3100 cm^{-1}. The carbon–carbon double bond stretching frequency is in the 1600 to 1680 cm^{-1} region. This absorption is sometimes weak due to the lack of polarity difference between the two carbon atoms. Infrared spectroscopy can sometimes be used to distinguish between *cis–trans* isomers and between terminal and internal double bonds. The *trans*-1,2-disubstituted alkenes show ir bands in the 965 to 990 cm^{-1} region, while the *cis* isomers usually (but not always) give bands in the 675 to 730 cm^{-1} area. The $=CH_2$ grouping, characteristic of terminal alkenes, gives rise to absorptions in the 890 to 910 cm^{-1} region. Sample ir spectra, those of 1-hexene and a mixture of 2-hexenes, are given in Figure 8–13.

Ethers

Ethers are distinguished from alkanes by the presence of strong bands in the region 1050 to 1150 cm^{-1}. These are attributable to C—O stretching vibrations (Fig. 8–10b). The remainder of the spectrum is hydrocarbon-like. Unknown ethers are often identified as such by default. If the compound is known to contain oxygen, but has no bands in the O—H or C=O region, it

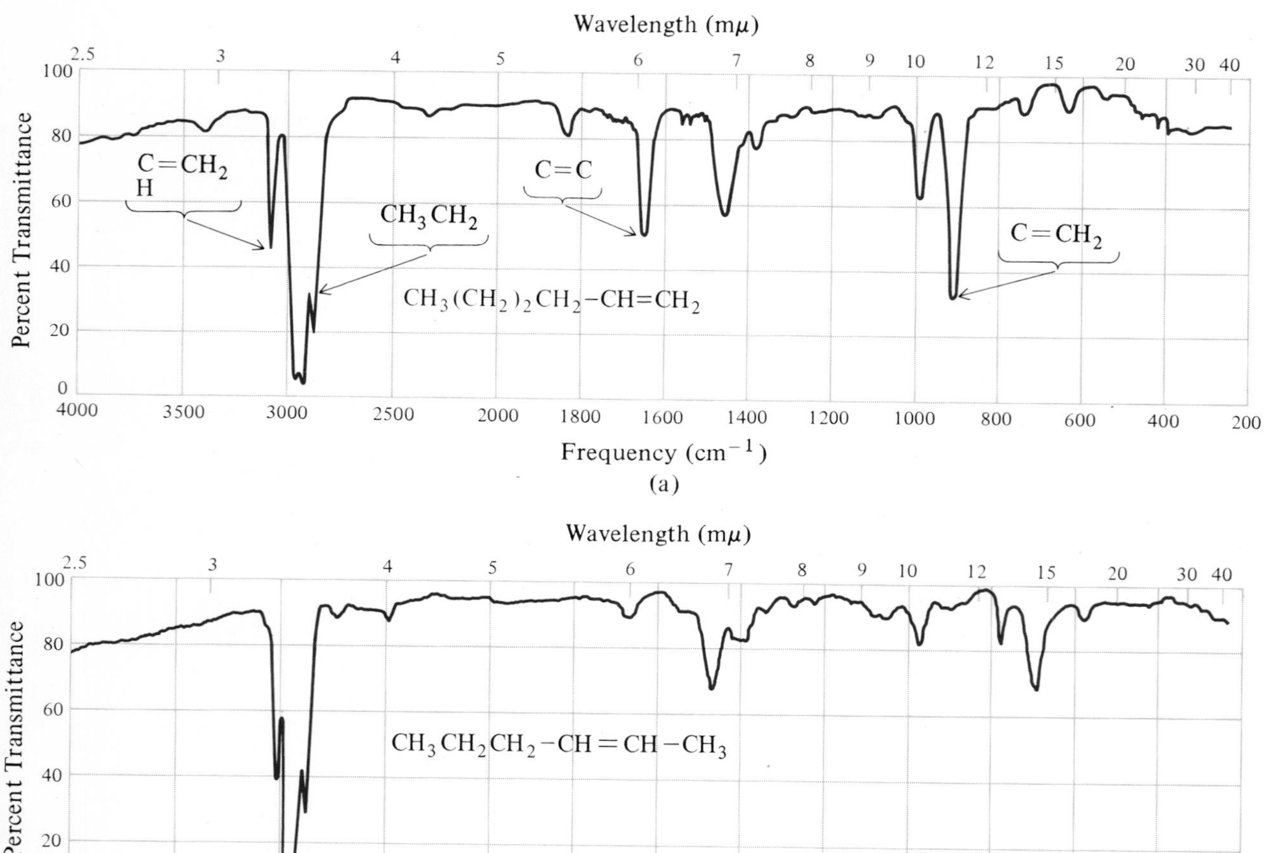

Figure 8–13. *Infrared spectra of (a) 1-hexene and (b) mixed 2-hexenes.*

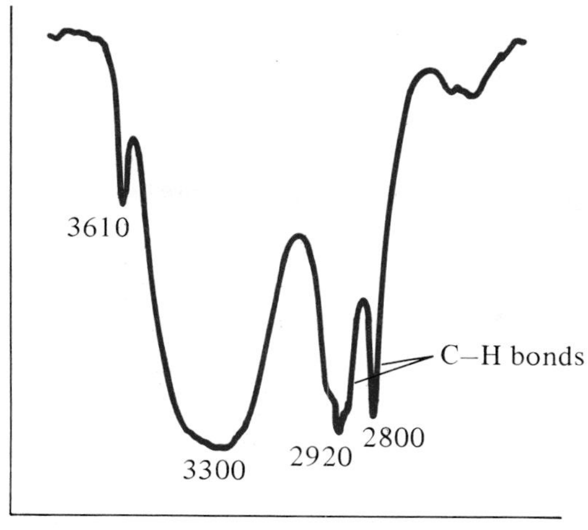

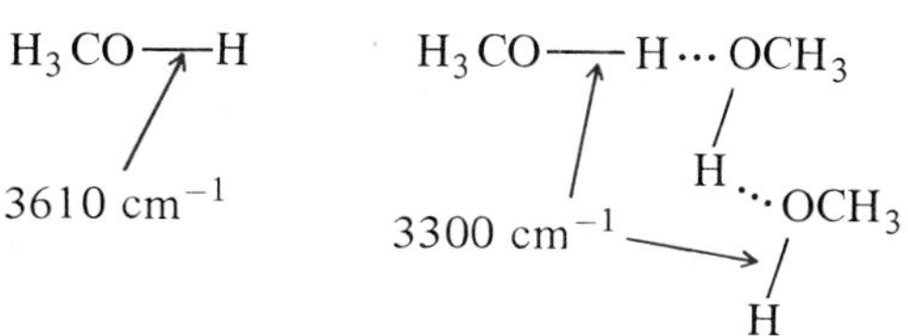

Figure 8–14. *The O—H stretching region of a methanol ir spectrum.*

is usually assumed to be an ether. More crudely, but quite effectively, ethers can be distinguished from hydrocarbons by their more pungent odors, which are unlike the "gasoline-like" odors of hydrocarbons. The solubility of ethers in concentrated sulfuric acid also sets them apart from alkanes.

Alcohols

The O—H stretching region (3650 to 3400 cm^{-1}) is an important identifying feature in ir spectra of alcohols. The tendency of alcohol molecules to associate by hydrogen bonding is evident from their infrared spectra. A partial spectrum of a solution of methanol (Fig. 8–14) shows two bands in the OH region. The sharp band centered at 3610 cm^{-1} is from the unassociated alcohol molecules, while the broad band centered at 3300 cm^{-1} is due to hydrogen-bonded dimers and higher clusters. The relative intensity of these two bands is a function of concentration. The one due to "free" alcohol molecules (3610 cm^{-1}) gets relatively more intense as the solution is diluted, and vice versa. As with ether spectra (Fig. 8–10b), alcohol spectra contain C—O bands in the 1000 to 1200 cm^{-1} region. The remainder of the spectrum is hydrocarbon-like.

Problem

8–11 Why is the ir band for the unassociated OH group at higher frequency ($\simeq$3600 cm^{-1}) than that for the hydrogen-bonded OH group ($\simeq$3300 cm^{-1})?

Box 8–2. Organic Compounds in Space

United States' interplanetary missions, such as the Mars *Viking* lander and the *Voyager* Jupiter and Saturn fly-by explorations, rely on instrumentation for the detection and analysis of organic compounds. A knowledge of what organic compounds, if any, exist on these planets would provide information on planetary evolution and perhaps tell us if life exists or could exist in these environments.

The Mars *Viking* lander contained a miniaturized mass spectrometer that performed analyses on soil samples. Water and CO_2 were detected, but no organic compounds. The Jupiter and Saturn missions had to use different spectrometric techniques because in a fly-by mission, samples cannot be collected at the surface. These two spaceships carried infrared *emission* spectrometers that could detect infrared energy given off when a vibrationally excited molecule returns to its ground state (the principle is similar to that of infrared *absorption* spectrometry). Methane, ethane, and acetylene ($HC{\equiv}CH$) were detected on Jupiter. In addition to these molecules, ethylene, HCN, methyl acetylene ($CH_3C{\equiv}CH$), and propane were detected on Saturn and Titan, a Saturn satellite.

8:10 Determining the Number of Multiple Bonds and/or Rings in Unknown Compounds

A common problem in structure determination is that more than one structural formula may fit the spectroscopic data. We try to alleviate this problem by narrowing the range of possible structures that have to be considered. One way to do this is to use chemical reactions that are specific for particular functional groups. Another source of useful structural information is the molecular formula, as we'll now see.

Since an acyclic alkane has the formula C_nH_{2n+2} and the presence of a ring requires two less hydrogens, an alkane with the formula C_nH_{2n} must have one ring. The formula C_nH_{2n-2} would indicate the presence of two rings, and so forth. *Two hydrogens are removed from the original acyclic alkane formula for each ring present.* Hydrocarbons with one double bond (alkenes) also have two hydrogens fewer than the acyclic saturated alkane. Thus, C_nH_{2n} as a molecular formula could denote either a cycloalkane *or* an acyclic compound containing one double bond. It is often said that such a formula has one *degree of unsaturation* (sec. 7:2). A compound with the molecular formula C_nH_{2n-2} could have two rings, two double bonds, one ring and one double bond, or one triple bond. Some of the C_4H_6 isomers (structures **1** to **5**) illustrate these possibilities.

1	**2**	**3**	**4**	**5**
(bicyclobutane)	▷=CH_2	$CH_2{=}CH{-}CH{=}CH_2$	$CH_3{-}C{\equiv}C{-}CH_3$	(cyclobutene)

If an unknown compound contains only carbon, hydrogen, oxygen and sulfur, the same formula relationships apply (that is, the number of oxygens or sulfurs can be ignored). As an example, we can calculate the degree of unsaturation for C_4H_8O, the formula in Problem 8–10:

$$
\begin{array}{rr}
\text{Saturated acyclic } (C_nH_{2n+2}) \text{ for } C_4: & C_4H_{10}O \\
\text{Subtract the formula of the unknown:} & -\ C_4H_8O \\
\hline
\text{Remainder:} & H_2
\end{array}
$$

Since the unknown lacks two hydrogens, there is one degree of unsaturation. This could be a double bond or a ring.

When an unknown contains nitrogen or halogens, the situation is modified because nitrogen (if neutral) has three bonded substituents and halogens have one. For compounds with one halogen, the acyclic saturated formula is C_nH_{2n+1}. For compounds with one nitrogen, the acyclic *saturated* formula is $C_nH_{2n+3}N$, and for compounds with two nitrogens the formula is $C_nH_{2n+4}N_2$. As a result, a neutral molecule with an odd number of nitrogens has an odd molecular weight, while a compound with an even number of nitrogens has an even molecular weight. As an example, consider an unknown C_4H_7N:

$$
\begin{array}{rr}
\text{Saturated acyclic } (C_nH_{2n+3}) \text{ for } C_4: & C_4H_{11}N \\
\text{Subtract the formula of the unknown:} & -\ C_4H_7N \\
\hline
\text{Remainder:} & H_4
\end{array}
$$

The unknown has two degrees of unsaturation and has either two rings, two double bonds, one double bond and one ring, or one triple bond. The ir spectrum, along with other spectroscopic data from techniques we will discuss in Chapter 9, will usually allow you to make a choice among these alternatives.

Example

Calculate the degrees of unsaturation (number of rings and/or double bonds) in the following compounds:

a. $C_8H_{10}O$ b. $C_7H_{11}Br$

Answer

a. A saturated acyclic compound with eight carbons would have the formula

$$C_nH_{2n+2} = C_8H_{2x8+2} = C_8H_{18}$$

We now compare that formula with the unknown:

$$
\begin{array}{r}
C_8H_{18}O \\
-\ C_8H_{10}O \\
\hline
H_8
\end{array}
$$

Thus, there are eight less hydrogens in the unknown than in a saturated acyclic compound of eight carbons. The number of rings and/or double bonds is therefore 8/2 or 4 because each ring or double bond in the molecule would eliminate two hydrogens from the saturated acyclic formula.

b. A saturated acyclic compound with one halogen would have the formula

$$C_nH_{2n+1} = C_7H_{2x7+1} = C_7H_{15}$$

Compare this with the unknown:

$$
\begin{array}{r}
C_7H_{15}Br \\
-\ C_7H_{11}Br \\
\hline
H_4
\end{array}
$$

The number of rings and/or double bonds will be 4/2 or 2.

Problems

8–12 Propose four more structures, in addition to **1** to **5** (p. 292), that fit the molecular formula C_4H_6.

8–13 Write the structures of all compounds that fit the molecular formula C_2H_3N. Indicate which specific bands you would look for in the ir spectrum of C_2H_3N to decisively identify its structure.

8–14 Given below is the infrared spectrum of a compound with the molecular formula C_3H_6O.

a. Calculate the degree of unsaturation of the unknown.
b. Note the major peaks in the spectrum. By comparing them with the sample spectra already given, and with the infrared correlation chart, attempt to assign these peaks to functional groups or structural units.
c. Are there functional groups compatible with the molecular formula that you can *eliminate*?
d. Suggest one or more structures that are in accord with the results of (a) through (c).

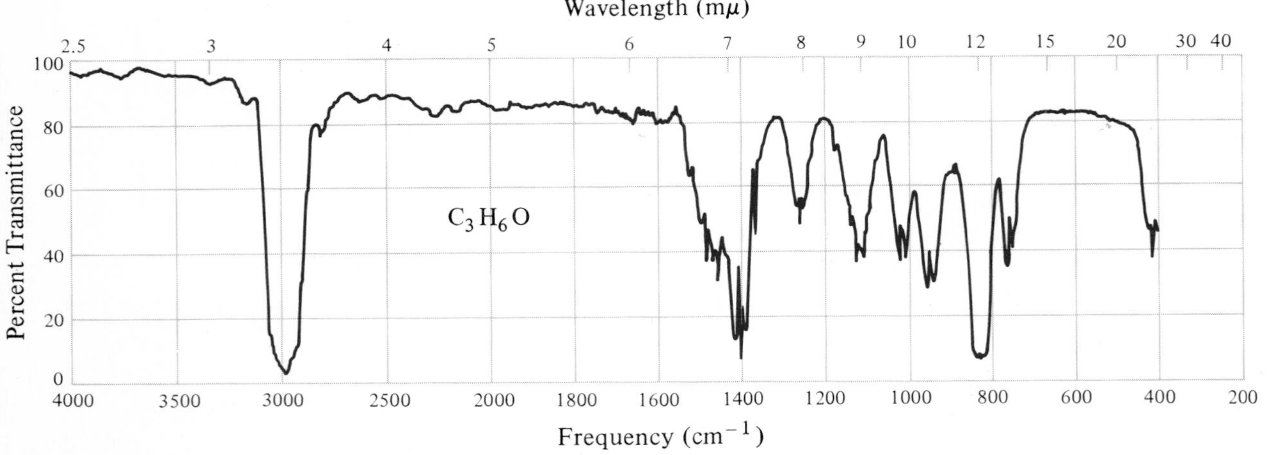

Summary

1 Mass spectrometry utilizes a beam of electrons to form cation radicals by loss of an electron from neutral organic molecules in the gas phase. High-resolution mass spectroscopy allows the measurement of ion masses to thousandths of a mass unit. Such measurements of the molecular ion mass can provide exact molecular formulas. The relative intensities of isotope peaks can also be used to obtain data about the molecular formula of an ion (sec. 8:1).

Molecular ions decompose to give stable fragment ions. Alkane molecular ions break apart by way of random C—C bond cleavages (sec. 8:2). Alcohols and ethers, in contrast, show a preference for cleaving the C—C bond(s) closest to the oxygen atom (sec. 8:3). These fragmentation patterns provide structural information about the original molecule.

2 Absorption spectroscopy is an instrumental technique that yields structural information about a molecule. Electromagnetic radiation that passes through a molecule will be absorbed if the energy of the radiation is equal to the energy of some excitation process that can occur in the molecule. The relationship between the energy of the excitation and the light being absorbed is given by $E = h\nu$, or $E = hc/\lambda$ (sec. 8:4).

3 Functional groups of molecules absorb infrared light selectively. When the energy of the incident light exactly equals the energy needed to excite (stretch or bend) a bond, an absorption will occur (sec. 8:5). Some of the most useful ir absorptions are those characteristic of the O—H (3650 to 3300 cm^{-1}), C—H (3300 to 2850 cm^{-1}), C≡C (2250 to 2100 cm^{-1}), and C=O (1900 to 1675 cm^{-1}) stretching vibrations (Tables 8–4 and 8–5; secs. 8:6 to 8:8).

Alcohols are readily detected by the presence of OH stretching bands. A very dilute solution of an alcohol in an inert solvent will show only a sharp peak in the 3600 cm^{-1} region. In more concentrated solutions, hydrogen-bonding occurs, giving rise to an additional broad O—H band at about 3300 cm^{-1}. Both alcohols and ethers contain C—O stretching bands in the 1000 to 1200 cm^{-1} region (sec. 8:9).

4 If the molecular formula of an unknown compound is determined, we can calculate its degree of unsaturation, that is, the number of rings and/or multiple bonds it contains. An examination of the ir spectrum will then tell what functional groups the unknown has or what functional groups are absent (sec. 8:10).

Problems

8–15 In addition to the main oxygen isotope, ^{16}O, there is an isotope two mass units heavier, ^{18}O, that is 0.2% as abundant as ^{16}O. Similarly, there is a minor sulfur isotope, ^{34}S, that is 4.4% as abundant as the main sulfur isotope, ^{32}S. Using this information, how could you distinguish between $C_2H_4O_2$ and C_2H_4S with a low-resolution mass spectrometer?

8–16 a. Suggest structures for the major ions (relative intensity > 20%) in the mass spectrum of chloroethane (Fig. 8–1).

b. One useful way to interpret mass spectra is to try and identify the neutral species lost when one ion decomposes to another. Suggest structures for each of the neutrals formed when the molecular ion of chloroethane decomposes to give each of the major fragment ions.

8–17 Diethyl ether and tetrahydrofuran (structure **III**, Prob. 8–7) have almost identical molecular formulas. Yet the mass spectrum of diethyl ether shows a weaker molecular ion peak and more intense fragment ion peaks than does the mass spectrum of tetrahydrofuran. Explain this observation.

8–18 Calculate the exact masses for the following compounds:

a. $CH_3C(CH_3)(OH)CH{=}CH_2$ (MW = 86)

```
     CH3
     |
CH3CCH=CH2
     |
     OH
```

b. $CH_3CH(CH_3)CH(CH_3)CH_3$ (MW = 86)

```
       CH3
       |
CH3CHCHCH3
     |
     CH3
```

8–19 Predict the mass spectral fragmentation pattern for the following compounds and give "fishhook" mechanisms (sec. 8:3) for decomposition of the molecular ions.

a. 3-methyl-3-hexanol b. 2-ethoxy-3-methylbutane

8–20 It was mentioned in Section 8:1 that some molecular ions are so short-lived that they are not detected by the mass spectrometer. Of the three isomers of formula C_5H_{12}, two show molecular ions in their mass spectra while the third does not. The mass spectrum of the third isomer has its highest mass peaks at m/z = 57 and 58. Give the structure of this isomer and explain the absence of a molecular ion in its mass spectrum.

8–21 Write structures for all the isomers that correspond to each of the molecular formulas listed below. Could the different isomers in each set be distinguished by means of infrared spectroscopy? Which ir bands in particular would aid in assigning structures to each isomer?

a. CH_3ClO b. C_2H_6O c. $C_2H_6O_2$ d. C_4H_{10}

8–22 a. The infrared spectrum of carbon monoxide consists of a single band centered at 2143 cm^{-1}. On the basis of the data supplied in Tables 8–4 and 8–5, what sort of bond does there seem to be between the carbon and oxygen atoms? Draw a structure for carbon monoxide, showing all the valence electrons, in accord with your answer.

b. Compound **IV** shows a strong band at 1848 cm^{-1} in its ir spectrum, characteristic of the carbonyl group. The ion **V** does not show such a peak. Ion **V** does, however, show a very strong band at 2299 cm^{-1}. Draw a probable structure for **V**, showing all the valence electrons.

$$\underset{\textbf{IV}}{CH_3\text{—}\overset{\overset{\displaystyle O}{\|}}{C}\text{—}F} + BF_3 \rightarrow \underset{\textbf{V}}{CH_3CO^+} + BF_4^-$$

8–23 Propose structures for the compounds with the following molecular formulas and ir spectra. (Remember to first calculate the degree of unsaturation.)

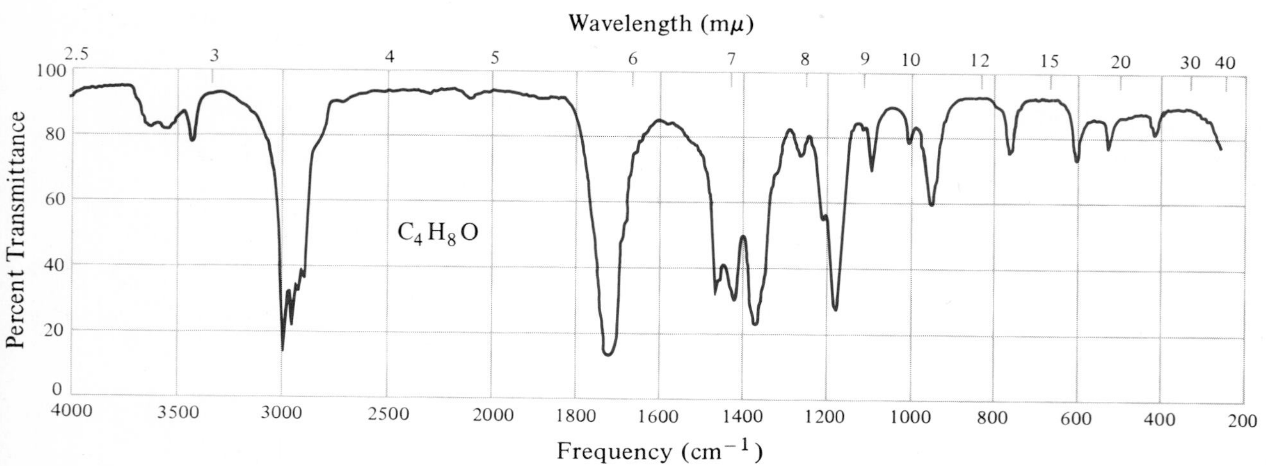

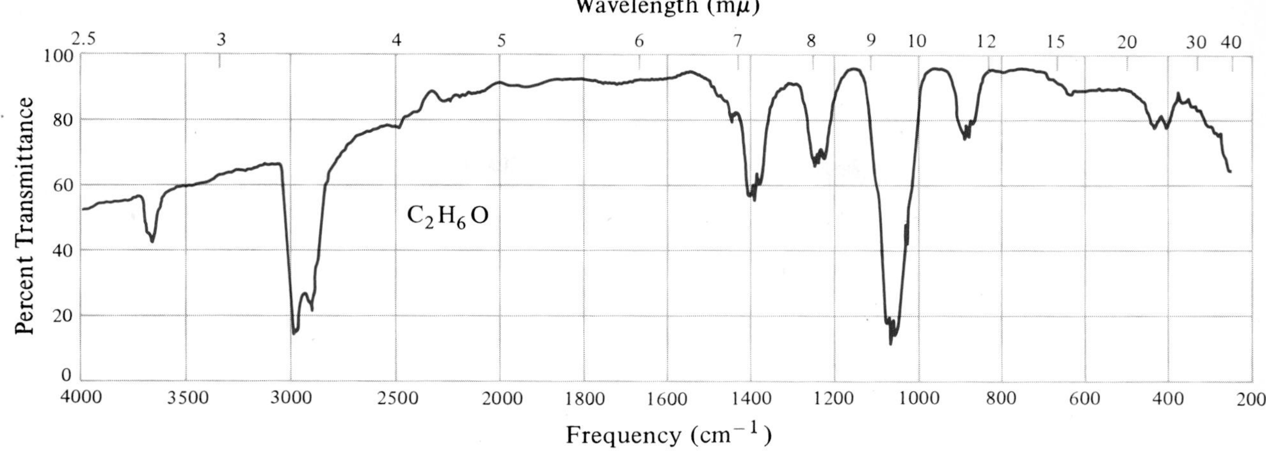

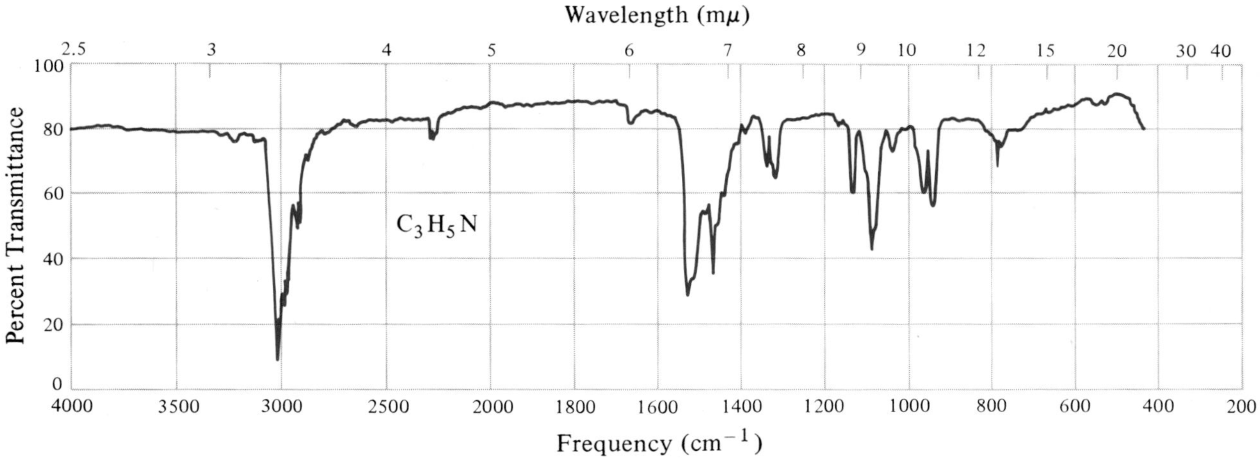

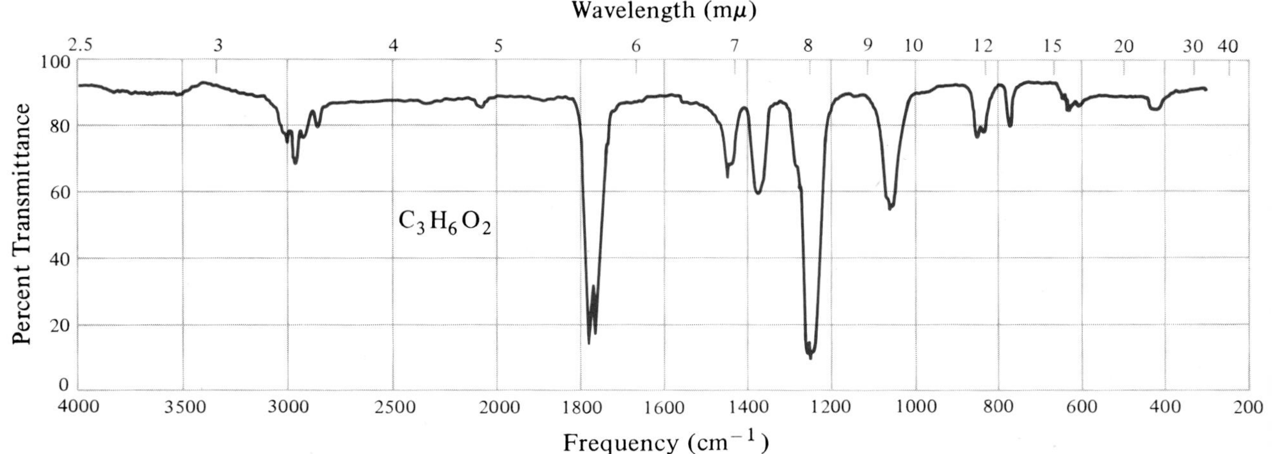

8–24 Explain each of the following observations:

a. For a given concentration, the relative size of the hydrogen-bonding band is much larger in the ir spectrum of *n*-butyl alcohol than it is in the spectrum of *t*-butyl alcohol.

b. The ir spectrum of a dilute solution of *trans*-cyclopentane-1,2-diol shows only one peak in the —OH stretching region. A spectrum of *cis*-cyclopentane-1,2-diol at the same concentration exhibits two peaks in this region. The relative intensities of these two peaks do not change with concentration.

Supplementary Reading

F. W. McLafferty, *Interpretation of Mass Spectra,* 3rd ed., University Science Books, Mill Valley, Calif., 1980.

M. A. Campbell and O. Runquist, Fragmentation mechanisms in mass spectrometry, *J. Chem. Ed.,* **49,** 104 (1972).

K. Nakanishi and P. H. Solomon, *Infrared Absorption Spectroscopy,* 2nd ed., Holden-Day, San Francisco, Calif., 1977.

R. M. Silverstein and G. C. Bassler, *Spectrometric Identification of Organic Compounds,* 4th ed., John Wiley & Sons, New York, N.Y., 1981.

D. L. Pavia, G. M. Lampman, and G. S. Kriz, Jr., *Introduction to Spectroscopy,* W. B. Saunders, Philadelphia, Pa., 1979.

J. R. Dyer, *Organic Spectral Problems,* Prentice-Hall, Englewood Cliffs, N.J., 1972.

L. J. Bellamy, *The Infra-red Spectra of Complex Molecules,* Vol. 1, 3rd ed.; Vol. 2, 2nd ed., Methuen, Inc., New York, N.Y., 1975, 1981.

9

Proton and Carbon Nuclear Magnetic Resonance Spectroscopy

Chapter 8 showed the usefulness of mass and infrared spectroscopy in chemical structure determination. With infrared we can establish the presence or absence of various functional groups. With mass spectrometry we can establish a molecular formula and, by analyzing the fragmentation pattern, learn something about the carbon skeleton. Much more knowledge about the carbon skeleton and about the numbers and kinds of hydrogen atoms in the unknown comes from a technique called *nuclear magnetic resonance (nmr)* spectroscopy. With ir, mass, and nmr spectra combined, we can often establish the exact structure of an unknown compound.

Nuclear magnetic resonance can also provide more subtle details of molecular structure, such as bond angles and electron densities. Because nmr is easily used at low temperatures, molecules that are unstable or very reactive at room temperature can be studied. Using nmr, we can examine conformational equilibria, including those of complex, biologically important molecules. Biological activity is often highly dependent on the manner in which one conformation of a compound may fit into a certain site on an enzyme. Nuclear magnetic resonance studies in solution can show which conformation of a biologically active compound is most favorable and thus give information on the enzyme shape.

In the first few sections of this chapter we will discuss the basic theoretical background of nmr spectroscopy. The latter sections will be devoted to the more detailed aspects of nmr.

9:1 Introduction

The name nuclear magnetic resonance is derived from the fact that this type of spectroscopy involves energy changes of atomic *nuclei,* and that a *magnetic field* is necessary to observe these energy changes. The word resonance is analogous to the word absorption in ir spectroscopy. It denotes the absorption of energy connected with a change in nuclear energy states.

The two elements most commonly present in organic compounds are carbon and hydrogen. For this reason, the two nuclei most often studied in organic nmr experiments are the hydrogen nucleus, ^{1}H, and the nucleus of the minor isotope of carbon, ^{13}C. (The word *proton* is used interchangeably with *hydrogen nucleus,* so ^{1}H nmr is often called proton magnetic resonance.) In order to be magnetically active and capable of generating a magnetic resonance signal, a nucleus must have an *odd* number of either protons or neutrons or both. Both ^{1}H and ^{13}C satisfy this condition. Other nuclei that give rise to nmr spectra useful to organic chemists include ^{2}H, ^{15}N, ^{17}O, ^{19}F, and ^{31}P. Note that the widely distributed isotopes ^{12}C and ^{16}O contain *even* numbers of both neutrons and protons, and therefore cannot produce nmr spectra.

Neutrons and protons behave as if they are spinning, just as electrons do (sec. 2:2). Two neutrons in the same nucleus will tend to pair their spins so that they cancel one another out; the same effect occurs with protons. An odd proton or neutron, however, can confer spin on the entire nucleus and since a nucleus is charged, it generates a magnetic field as it spins. When the nucleus is placed in an external magnetic field, such as that generated by the magnet of an nmr spectrometer, the nuclear magnetic field interacts with the external magnetic field. Under these conditions the nucleus behaves like a small bar magnet, the north–south axis of which can be aligned in different orientations with respect to the external field (Fig. 9–1). In common with ordinary bar magnets, nuclear magnets have different potential energies in these different orientations. Some of these nuclei will *change* their orientations if radiant energy is available from a suitable source. The energy in a single photon of the radiation must exactly match the energy difference between two different orientations of the nucleus in a magnetic field. When this occurs we have *resonance* between the nuclei and the radiation—the nuclei absorb some of the energy and are raised to a higher energy state. By detecting this energy absorption we create a spectroscopic tool.

Nuclear magnetic resonance spectroscopy is of practical value only if two different ^{1}H (or ^{13}C) nuclei in a molecule absorb energy at two different spectrometer settings, so that one nucleus can be distinguished from the other. There are two factors that control energy absorption—the frequency of the incoming radiation, and the strength of the magnetic field. Usually nmr spectrometers operate at a fixed frequency and with a variable magnetic field. As the magnetic field strength changes, the energy difference between

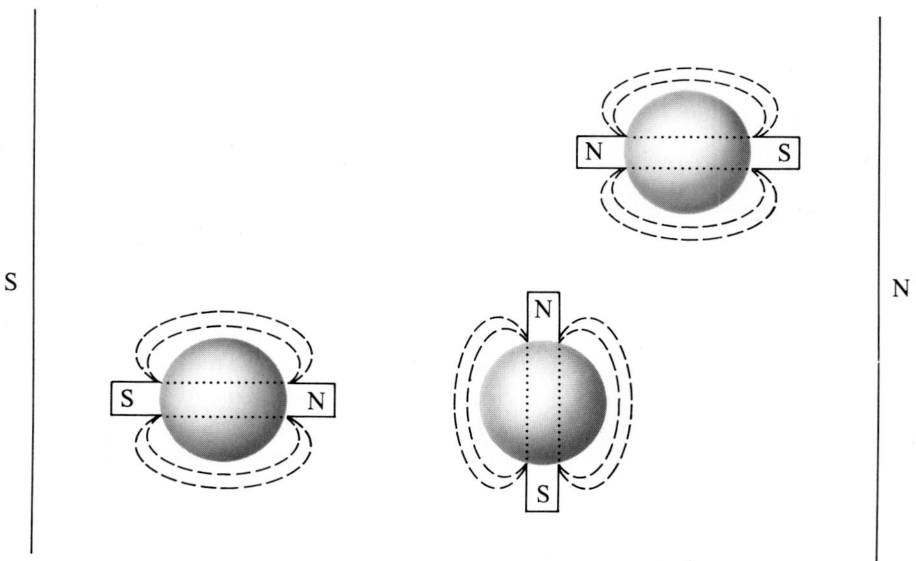

Figure 9–1. *Magnetic nuclei behave like tiny bar magnets when placed in an external magnetic field. Different nuclear orientations have different energies.*

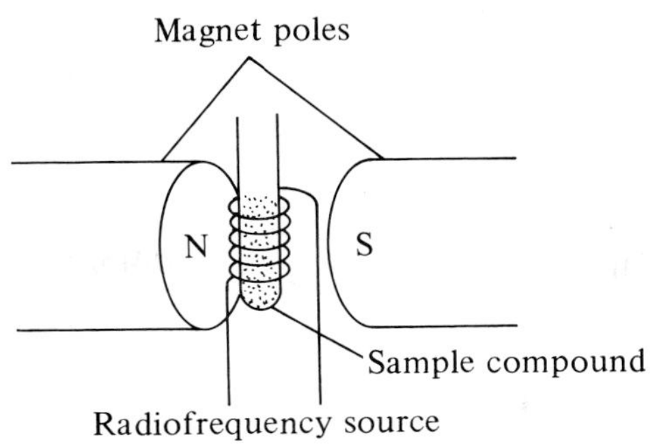

(a)

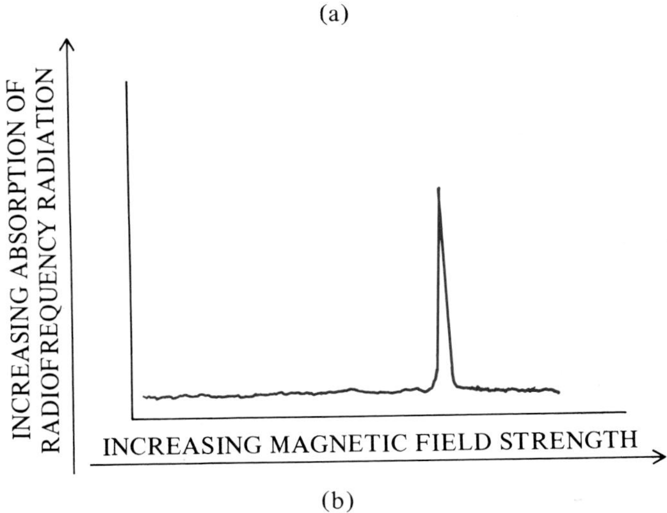

(b)

Figure 9–2. *(a) A portion of an nmr spectrometer, showing the sample tube in the radiofrequency coil between the magnet poles. (b) A simplified general nmr spectrum.*

nuclear orientations (Fig. 9–1) also changes. When that difference exactly matches the radiation energy, resonance is achieved and energy is absorbed. For instance, when a sample of CH_3CH_2Br is subjected to these conditions in an ^1H nmr spectrometer, we observe energy absorption at *two* different magnetic field settings corresponding to CH_3 and CH_2 protons. We obtain a spectrum of peaks that can be directly related to the structure of the molecule. In order to better appreciate the process, let's look briefly at the spectrometer and its output.

Figure 9–2a shows a highly simplified diagram of a nuclear magnetic spectrometer. The sample compound to be studied is placed in a glass tube between the poles of a very strong magnet. A coil around the tube surrounds the sample with radiation in the wavelength range 10^6 to 10 nm (radiofrequency range), where most nuclei absorb. Figure 9–2b shows the general type of spectrum obtained as a compound is subjected to a continually increasing magnetic field strength until an absorption peak is observed.

9:2 Explanation of a ^{1}H NMR Spectrum

In order to become familiar with some of the terminology used in nmr spectroscopy, it is helpful to examine an actual spectrum. Figure 9–3 shows the ^{1}H nmr spectrum of a common ketone, *pinacolone* (more properly called 3,3-dimethyl-2-butanone or *t*-butyl methyl ketone). Study this spectrum carefully as it is described in the following paragraphs.

$$\begin{array}{ccccc} & O & CH_3 & \\ & \| & | & \\ CH_3— & C— & C— & CH_3 \\ & & | & \\ & & CH_3 & \end{array}$$

Pinacolone

First, we'll look at the spectrum itself and then we'll discuss the measurement units. The magnetic field strength increases from left to right, with the left side called "downfield" and the right side, "upfield." The base line is flat on the downfield (left) side because there is no change in energy being absorbed. As the magnetic field increases, an absorption occurs at 2.10 ppm (parts per million, the unit of measurement) and another (about three times as intense) is observed at 1.15 ppm. These two absorptions are due to the sample, pinacolone. There is also a small peak at 0.00 ppm which comes from a *standard,* tetramethylsilane (TMS). The standard is added to the sample just before the spectrum is recorded.

$$\begin{array}{ccc} & CH_3 & \\ & | & \\ CH_3— & Si & —CH_3 \\ & | & \\ & CH_3 & \end{array}$$

Tetramethylsilane (TMS)

The standard is used to help specify the absorption positions of each type of proton in the sample. We could try and directly measure the absolute magnetic field strength for each of these absorptions, but this is difficult to do accurately. Instead, we add a standard and *arbitrarily* assign its absorption position the value 0.00. We can then measure the sample's absorptions *relative* to the TMS absorption, with a high degree of accuracy. For instance, pinacolone has a peak at 1.15 ppm. This means that the magnetic field that causes the pinacolone absorption is 1.15 ppm, or 0.000115% weaker than the magnetic field that caused the TMS protons to absorb. [The Greek letter δ (delta) is sometimes used to indicate ppm with respect to TMS.] TMS is used as the standard because its protons produce a sharp absorption peak at higher magnetic field strength than those of most other organic compounds. Hence, TMS absorptions will not interfere with the sample absorptions.

Figure 9–3 also shows an absorption scale graded in hertz (Hz) just above the spectrum. Energy absorption depends on the presence of both radiofrequency radiation and a magnetic field. When specifying the position of a sample absorption relative to TMS, we can convert a magnetic field strength

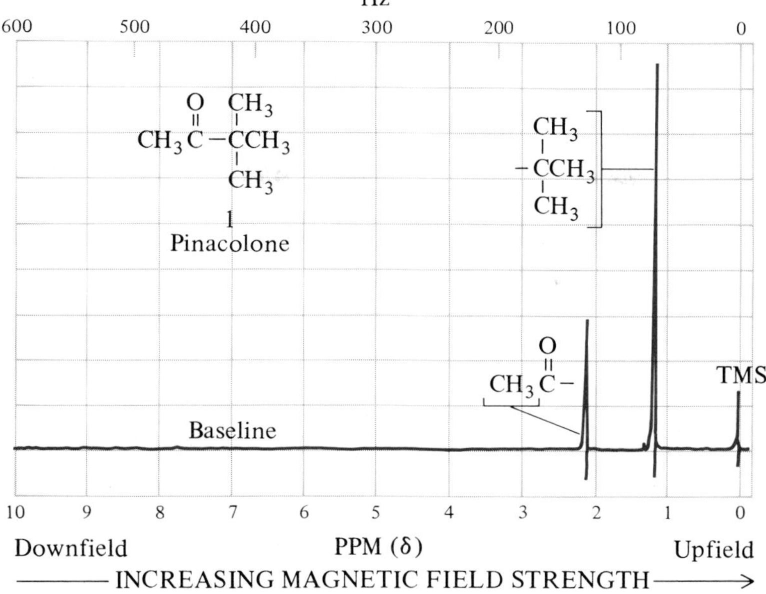

Figure 9–3. *[1]H nmr spectrum of pinacolone.*

difference into a frequency difference. In Section 9:3 we will introduce a quantitative expression that connects frequency and magnetic field strength.

We are now ready to look more closely at the pinacolone absorptions. One is 1.15 ppm downfield from TMS on the magnetic field scale and the second is at 2.10 ppm. The relative position of an absorption in an nmr spectrum has a special name: the **chemical shift**. Using nmr terminology, we would say that pinacolone has two absorptions, one at chemical shift 1.15 ppm (or δ 1.15) and a second at chemical shift 2.10 ppm (or δ 2.10). *The fact that pinacolone has two different absorption peaks means that it contains two different kinds of hydrogens*. Furthermore, the presence of only two peaks indicates that all three methyl protons of the CH_3CO group are equivalent, and that all nine protons of the $(CH_3)_3C$ group are also equivalent.

The *intensity* of each absorption is directly related to the *numbers* of each kind of proton. The peak at 1.15 ppm is three times as intense as the peak at 2.10 ppm; therefore there are three times as many protons giving rise to the 1.15 peak as give rise to the 2.10 peak. The intensity of an nmr absorption is determined by the total area under the peak. The relative intensities (areas) are usually measured electronically by the nmr spectrometer. In the spectrum of Figure 9–3, the peaks are unusually sharp and the relative intensities are very close to the relative heights of the absorptions.

Problem

9–1 The ^1H nmr spectrum of an unknown compound is given on the next page.

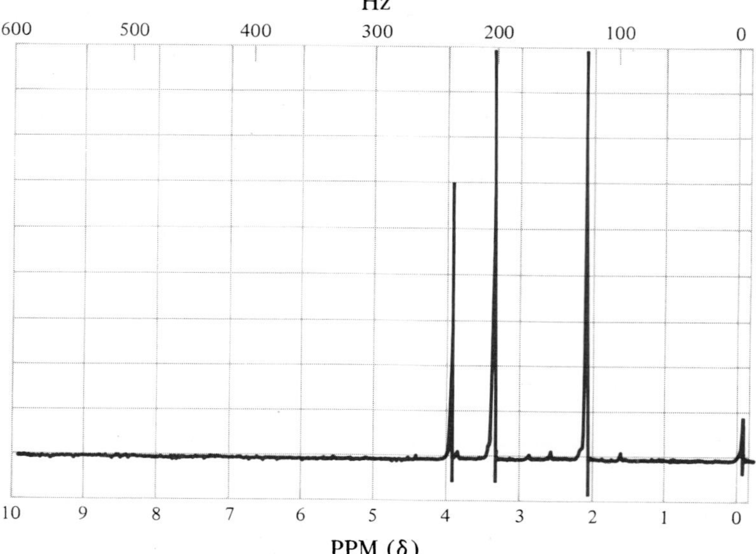

a. How many different kinds of hydrogens are present? What is the relative ratio (based on peak heights) among the different hydrogens?
b. Give the chemical shift for each absorption, in Hz and in ppm (δ) from TMS.
c. Which of the following structures could produce the given ^{1}H nmr spectrum?

$$CH_3\overset{\overset{\displaystyle O}{\|}}{C}CH_2CH_2OH \qquad CH_3-\underset{\underset{\displaystyle CH_3}{|}}{\overset{\overset{\displaystyle OH}{|}}{C}}-\overset{\overset{\displaystyle O}{\|}}{C}CH_3 \qquad CH_3OCH_2\overset{\overset{\displaystyle O}{\|}}{C}CH_3$$

9:3 A More Detailed Look at NMR Theory

In Section 9:1 we discussed briefly the behavior of spinning nuclei in order to introduce the phenomenon of nuclear magnetic resonance. That discussion did not contain enough detail to allow some important practical results to be fully understood. For example, in Figure 9–3, why do the two kinds of methyl groups have protons that absorb energy at different magnetic field values? The answer will become clear if we look a little more closely at nmr theory.

In Section 9:1 we compared magnetic nuclei to bar magnets. This analogy is useful but it breaks down in one important way. A real bar magnet can assume any orientation when placed in an external magnetic field (Fig. 9–4a), but for a nucleus there are only a few allowed orientations. For example, both the ^{1}H and ^{13}C nuclei have only two allowed orientations when put in an external magnetic field—they either align *with* the magnetic field or *against* it (Fig. 9–4b). Those nuclei whose magnetic axes are aligned with the external field have slightly lower energy than the ones whose axes are aligned against the field. For instance, when we place a group of protons in a magnetic field they become sorted into two groups which differ in spin orientation and energy. A slight majority of them are found in the lower

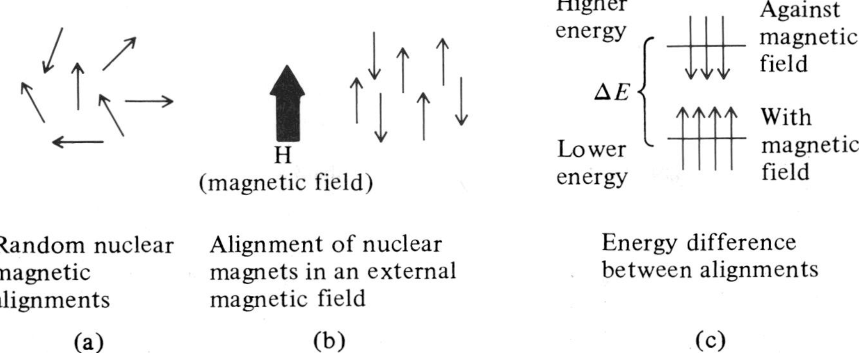

Figure 9–4. *(a) Random alignment of nuclear magnets. (b) Two allowed orientations of nuclear magnets, with and against an applied magnetic field. (c) Energy difference between nuclear magnets aligned with and against an applied magnetic field.*

energy group (Fig. 9–4c). The energy difference between the two groups, ΔE, is determined by the magnetic field strength. We can increase or decrease ΔE by making the field stronger or weaker.

When the energy gap between the high and low energy nuclei exactly matches the energy in the radiofrequency photons, resonance between nuclei and photons takes place. Some of the low energy nuclei absorb photons, "flip their spins," and enter the higher energy state (Fig. 9–5). Equation 9.1 gives the resonance relationship between the photon frequency and the magnetic field strength: ν is the frequency (Hz), H^{nucleus} is the field strength in gauss (G) felt by the nucleus, and γ (Greek gamma) is the *gyromagnetic ratio*, which is different for each isotope.

$$\nu = \frac{\gamma H^{\text{nucleus}}}{2\pi} \tag{9.1}$$

For ^{1}H nmr spectrometers a radiofrequency (ν) of 60 MHz (60×10^6 Hz) is common. When we put this value into Equation 9.1, along with the value for γ_{proton}, we calculate a magnetic field strength H of about 14,000 G. Suppose that our spectrometer is set up so that the TMS protons resonate at a field strength of exactly 14,000.000 G. If we add to the TMS a sample that gives a peak at 1.00 ppm, then the magnetic field strength for the sample

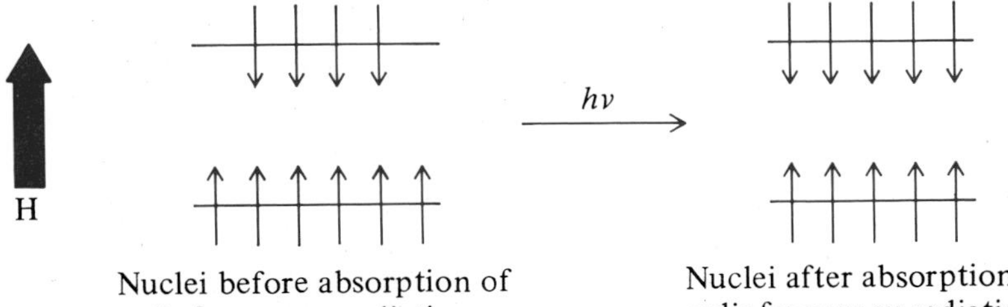

Figure 9–5. *Absorption of radiofrequency photons causes nuclei in a lower energy state to "flip their spins" and enter a higher energy state.*

resonance equals 13,999.986 G! The spectrometer, then, must be able to make very small changes in the magnetic field accurately and reproducibly.

If the only magnetic field source were the spectrometer itself, then all the nuclei in a sample would experience exactly the same magnetic field at any moment. Equation 9.1 tells us that in this case all nuclei of the same isotope, say ^{1}H, would absorb at exactly the same time, which would not be very useful. It turns out, however, that the molecular *electrons,* which are charged and in motion, also generate magnetic fields as a result of their circular motions. This electron motion gives rise to a weak induced magnetic field *that opposes the applied field from the spectrometer* (Fig. 9–6). Thus, a proton in a molecule experiences a weaker magnetic field than would a bare proton under the same conditions. We say that the circulating molecular electrons **shield** the proton from the applied magnetic field. Furthermore, different protons in a single molecule are shielded to different extents, because their electronic environments are different.

In order to appreciate how shielding affects the nmr process, remember that all protons must experience the *same net* magnetic field (H^{nucleus}) in order to absorb energy (eq. 9.1). If a proton is shielded, then the applied field must be made stronger than H^{nucleus} in order to overcome the shielding. Suppose we have a spectrometer sample with several structurally different protons in it, and we start to change the magnetic field strength from lower to higher. The weakly shielded protons will absorb energy first, since they resonate at low field. As we increase the field strength, more and more highly shielded protons absorb until finally we see an absorption peak due to TMS. It is the difference in shielding of different protons which produces their different chemical shifts.

With this theory in hand, let's see if we can analyze the spectrum in Figure 9–3 (pinacolone). The peak at lowest field is due to the methyl protons next to the carbonyl group (CH_3CO). The carbonyl group, $^{\delta+}C{=}O^{\delta-}$, tends to draw electrons away from neighboring groups because of the electronegative oxygen atom (the inductive effect; sec. 6:7). By drawing electrons away from the neighboring methyl group the carbonyl group reduces its shielding—the carbonyl group *deshields* the methyl protons. The deshielding effect of the carbonyl group on the *t*-butyl protons, $(CH_3)_3C$, is

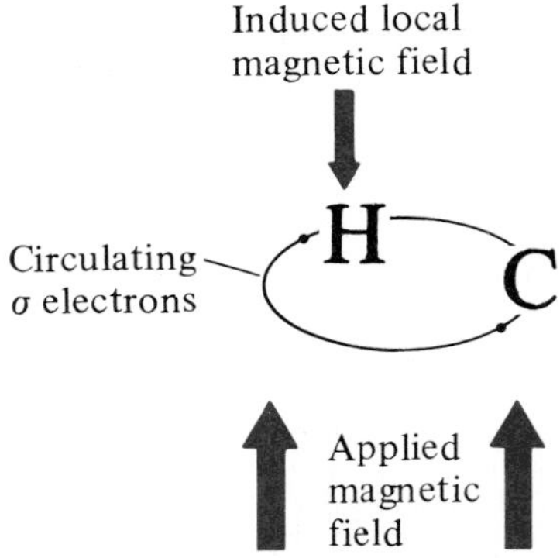

Figure 9–6. *Under the influence of an applied magnetic field, C—H σ electrons create a local magnetic field which opposes the applied field near the proton.*

less because they are further from the carbonyl group than is the methyl group. Thus, the *t*-butyl protons absorb at higher field than do the methyl protons. Finally, the high field absorption of the TMS protons is understandable because silicon is less electronegative than carbon and therefore donates electrons to the methyl groups, increasing their shielding.

Here is a brief summary of the major conclusions developed in Sections 9:2 and 9:3:

1. When an organic compound containing hydrogen is placed in a magnetic field, all the hydrogen nuclei align in either one of two orientations, which differ in energy.
2. If radiation of the correct energy is supplied then some of the hydrogen nuclei will absorb the energy and change their orientation. This is the resonance phenomenon.
3. Hydrogens in different chemical environments require different energies to undergo resonance absorption because they are shielded differently from the magnetic field by neighboring electrons.
4. The energy needed to cause resonance in one particular group of hydrogens is measured *relative* to the energy required to bring about resonance absorption in a standard compound, TMS.

9:4 A Systematic Look at Chemical Shifts

It is difficult to correlate precisely the chemical shift position with simple physical parameters. There are, however, a number of generalizations that can be made, most of which relate ultimately to the electron density near the nucleus being observed. This is most easily seen by looking at the effects produced by a neighboring electronegative atom or charged group. Table 9–1 gives some typical absorption ranges for various CH_3 and CH_2 groups. Note that the protons of a methyl group bonded to another carbon absorb in the 0.8 to 1.9 ppm region, while a CH_3 bonded to nitrogen is at 2.1 to 3.1 ppm, and a CH_3 bonded to oxygen is at 3.2 to 4.1 ppm. These positions can be explained by the fact that nitrogen is more electronegative than carbon, and oxygen is more electronegative than nitrogen. The electronegative atom withdraws electrons from the methyl group, thus reducing the magnetic shield around the ^1H nuclei of the methyl. A positive charge will have the same effect: $CH_3\overset{+}{N}R_3$ methyls absorb in the 3.5 to 4.0 ppm region.

As another example of electronegativity effects, consider the methyl halides, CH_3X. All three hydrogen atoms are identical and, therefore, each methyl halide will give a ^1H nmr spectrum consisting of only one peak. The chemical shifts are, in ppm units, CH_3F, 4.26; CH_3Cl, 3.05; CH_3Br, 2.69; CH_3I, 2.10. These values parallel the order of decreasing electronegativity: $F > Cl > Br > I$. All halogens cause a downfield shift relative to the CH_4 absorption. The electronegative atom draws electron density away from carbon. The carbon atom, in turn, attracts electrons in the carbon–hydrogen bonds so that these electrons are less available to shield the hydrogen nuclei from the magnetic field.

These effects of electronegative atoms are approximately additive. Thus, the presence of two halogens, as in H_2CX_2, would deshield the protons

Table 9–1. Chemical Shift Ranges for Some Representative CH_3 and CH_2 Protons

Functional Group	Usual Chemical Shift Range (ppm)
$CH_3—\overset{|}{\underset{|}{C}}—$	0.8 to 1.9
$CH_3—\overset{O}{\overset{\|}{C}}—$	1.9 to 2.2
$CH_3—N<$	2.1 to 3.1
$CH_3—O—$	3.2 to 4.1
$—CH_2—\overset{|}{\underset{|}{C}}—$	1.0 to 2.0
$—CH_2—\overset{O}{\overset{\|}{C}}—$	2.0 to 2.4
$—CH_2—N<$	2.3 to 3.6
$—CH_2—O—$	3.3 to 4.5

almost twice as much as one halogen, and the position of the CH_2 absorption would be that much further downfield. Note from Table 9–1 that $—CH_2—$ absorptions are slightly lower than CH_3 values. It has also been observed that chemical shifts for CH groups will be slightly downfield from those for CH_2 groups. A more complete table of chemical shifts appears in the end-papers at the back of the book.

One of the largest effects on chemical shift is that due to hybridization. Protons bonded to sp^3 carbons generally absorb at much higher fields than do protons bonded to sp^2 carbons. Thus, alkane methyls show ^1H absorptions near 1 ppm (high field), while protons on a C=C double bond absorb in the 5 to 6 ppm region (low field). One reason for this is that an sp^2 carbon is more electronegative than an sp^3 carbon. Other properties of π bonds may also contribute to the low field chemical shift, as we will see in later chapters.

Problem

9–2 A chemist isolates a substance of molecular formula $C_3H_6O_2$ and has good reason to believe that it is one of three compounds (**I**, **II**, or **III**).

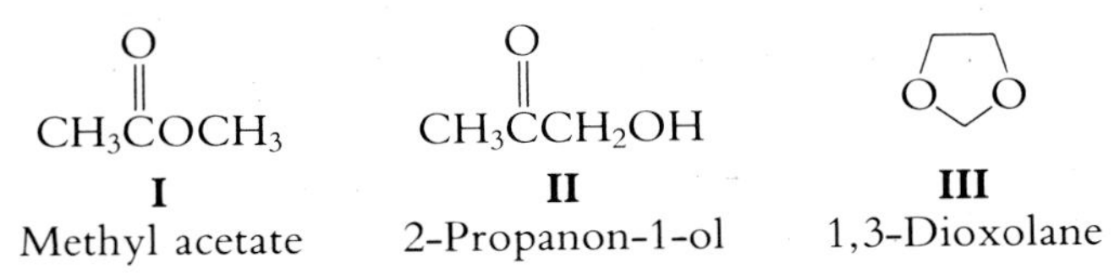

The nmr spectrum is shown here. Using the data of Table 9–1 and your estimates of peak areas, decide which structure is correct. Explain your reasoning.

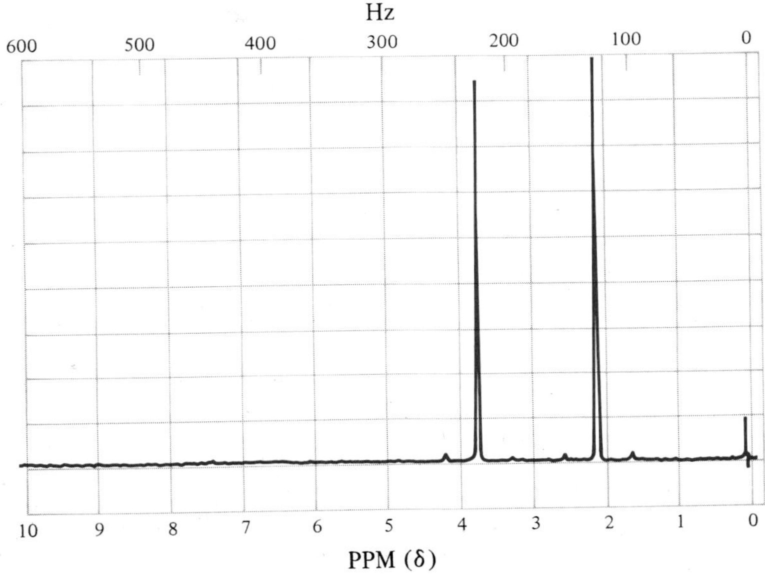

9:5 Spin–Spin Splitting

We used the compound pinacolone (Fig. 9–3) to introduce ^{1}H nmr because its spectrum is very simple. One reason for its simplicity is that in pinacolone, none of the methyl groups is bonded to a carbon that is itself attached to a proton.

If two structurally different kinds of protons are on neighboring carbons, there is an interaction between the nuclear spins of the neighboring protons that causes increased complexity in the spectrum.

Consider the spectrum of bromoethane, CH_3CH_2Br. Based only on information given thus far, we would expect two absorptions with areas in the ratio of 3:2. The two-proton absorption would exhibit a downfield chemical shift due to the electronegativity effect of the Br. Figure 9–7 shows the actual ^{1}H nmr spectrum of bromoethane. The spectrum shows seven peaks (not including the TMS peak at 0.0 ppm) rather than two, but the seven peaks appear in *two* groups. A group of three closely spaced peaks is centered at 1.65 ppm. This group is called a **triplet**. The second group consists of four closely spaced peaks centered at 3.37 ppm, and it is called a **quartet**. When we compare the *total* area of the triplet group with the total

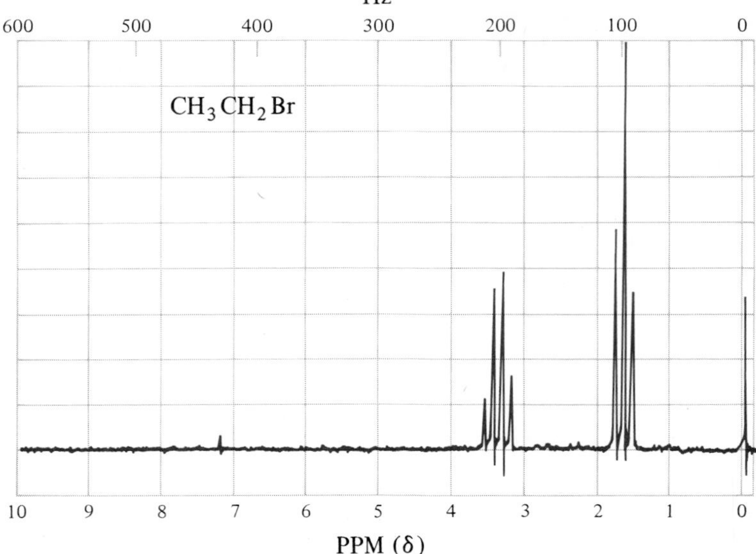

Figure 9–7. *¹H nmr spectrum of bromoethane.*

area of the quartet group, we find that the triplet has an area 1.5 times the area of the quartet. Therefore there is a 3:2 ratio in areas for the 1.65 and 3.37 ppm absorptions. The chemical shift positions and the relative areas both indicate that the triplet absorption is due to the CH_3 protons of CH_3CH_2Br, while the quartet of peaks is due to the CH_2 protons. The methyl group absorption is said to be *split* into a triplet and the CH_2 absorption is *split* into a quartet. When one set of protons gives rise to a group of absorption peaks, that group is called a **multiplet**. Triplets and quartets are multiplets. The TMS peak, on the other hand, is a **singlet**.

Splitting is a result of interactions between the proton spins on neighboring carbons; the effect is called **spin–spin splitting**. If neighboring protons cause spin–spin splitting they are said to be **coupled**. The distance between the chemical shifts of the individual peaks of the split multiplets is called the **coupling constant**, or ***J*-value**. Figure 9–8 shows a magnification of the methylene quartet from the bromoethane spectrum (Fig. 9–7). Note that all four peaks are equally spaced at about 7 Hz (the grid lines of the spectrum are 6 Hz apart). This is a typical *J*-value for neighboring C—H protons in alkane derivatives.

The methyl peak is split because the resonance of the methyl protons is sensitive to the spin arrangements of the protons on neighboring carbon atoms. These neighbors create their own magnetic field that contributes to the overall field experienced by the methyl protons. The particular field a given proton generates depends on whether its magnetic moment is aligned with or against the main magnetic field arising from the instrument. A system consisting of a single proton has only two possible spin arrangements—either up or down.

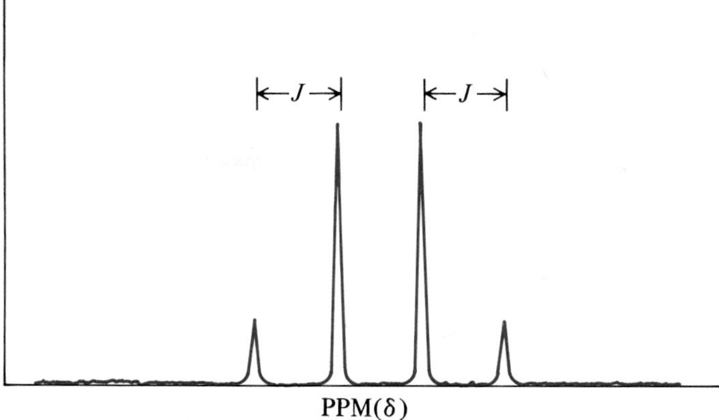

Figure 9–8. *An expansion of the 3 to 4 ppm region of the bromoethane ^1H nmr spectrum.*

A two-proton system has three possible spin states—both up, both down, or one up and one down (Fig. 9–9). Since there are two equivalent ways of arriving at a one up, one down arrangement, this configuration is twice as probable as the other two. What, then, will a spectrometer "see" as it scans a large collection of ethyl bromide molecules in the region where the methyl protons absorb energy? In 25% of the molecules the methyl group will be next to a methylene (CH_2) in which both proton spins are up; 50% of the molecules will contain methyl groups whose neighboring protons have one spin up and one down; and 25% of the molecules will contain methylene groups with both proton spins down (Fig. 9–10). In each of the three cases the methyl absorption comes at a slightly different field value, and that is why the methyl absorption of bromoethane actually appears as a triplet, the components of which have areas in the ratio of 1 : 2 : 1 (Fig. 9–7).

Similar reasoning tells us that the methylene proton peak will be affected by the spin arrangements of the methyl protons. The three methyl proton spins may be combined in four independent ways (Fig. 9–11). As a result, the *methylene* proton peak appears as a quartet, the area of its component peaks standing in the ratio of 1 : 3 : 3 : 1. *The absorption of an observed hydrogen will be split into (*n + 1*) peaks if there are* n *equivalent protons on the neighboring carbons.* This is the (n + 1) rule.

These concepts of nmr analysis may be confusing at first, so let us restate the analysis of the spectrum of Figure 9–7 in another way. An absorption

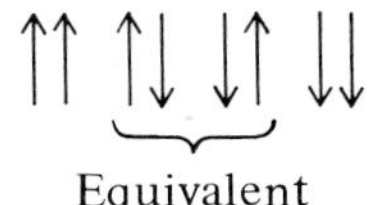

Figure 9–9. *Possible nuclear spin arrangements of a methylene (CH_2) group.*

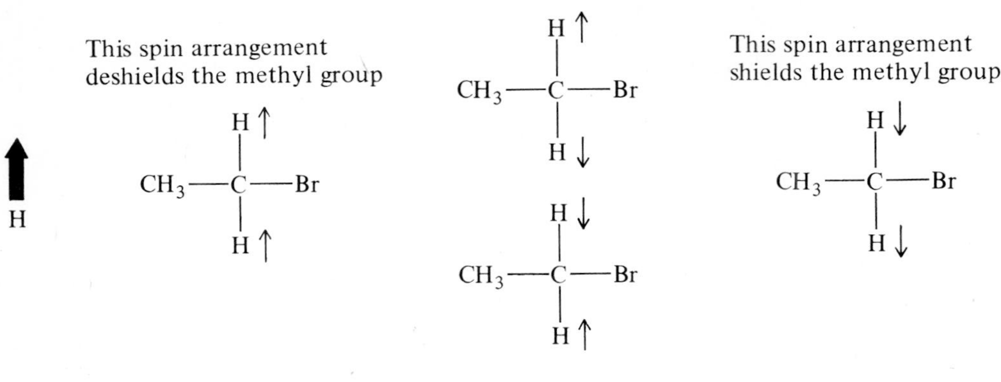

Figure 9–10. *The effect of methylene (CH_2) proton spins on neighboring methyl (CH_3) proton absorptions in bromoethane.*

appears in the region of 1.65 ppm. This is a high-field region and typical of CH_3 groups bonded to other sp^3 carbons. The absorption occurs as a triplet with component peak intensities in the ratio 1:2:1. This tells us that the *neighboring* carbon atom is bonded to two equivalent protons. Another absorption appears at 3.37 ppm. This chemical shift means that the protons on that carbon are less shielded than those producing the CH_3 absorption. They are less shielded because the electronegative (electron-withdrawing) Br group removes electrons from bonds near the CH_2 protons. The absorption at 3.37 ppm is split into four peaks (a quartet) in the ratio 1:3:3:1 because the *neighboring* carbon atom is bonded to three equivalent protons (Fig. 9–12).

Chemically *equivalent* protons on two neighboring carbons do not cause spin–spin splitting among themselves. Thus, molecules such as ethane (CH_3—CH_3), 1,2-dibromoethane ($BrCH_2CH_2Br$), and cyclohexane (—CH_2CH_2—) show only a single-proton absorption. Furthermore, pro-

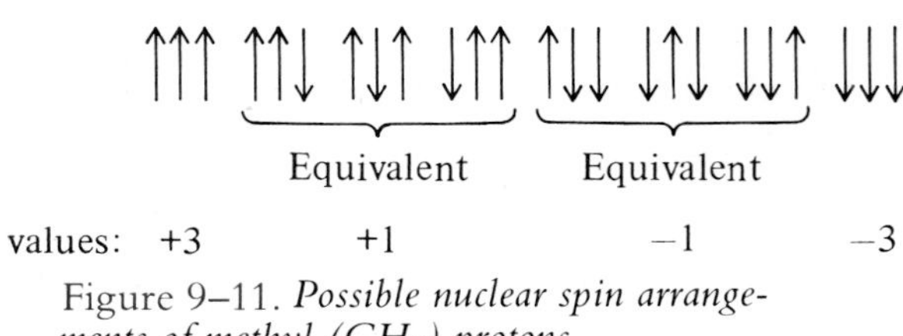

Figure 9–11. *Possible nuclear spin arrangements of methyl (CH_3) protons.*

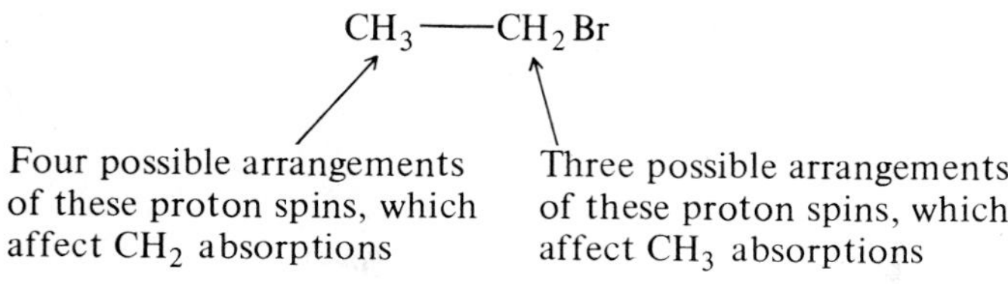

Figure 9–12. *The effect of proton spins on neighboring proton absorptions in bromoethane.*

tons do not have to be on the same or neighboring carbons to be equivalent, as shown by the spectrum of 1,3-dichloropropane, $ClCH_2CH_2CH_2Cl$ (Fig. 9–13). The four hydrogen atoms at the ends of the molecule are chemically equivalent and act as a single unit, even though they are separated by a carbon atom. The areas of the resulting triplet and quintet are in the ratio 2 : 1.

You will know that you understand these concepts and are ready to proceed if you can explain why the nmr spectrum of 2-bromopropane

$$\underset{\displaystyle\overset{|}{Br}}{CH_3CHCH_3}$$

consists of a doublet (two peaks) centered at 1.8 ppm and a septet (seven peaks) centered at 3.2 ppm, with relative areas 6 : 1.

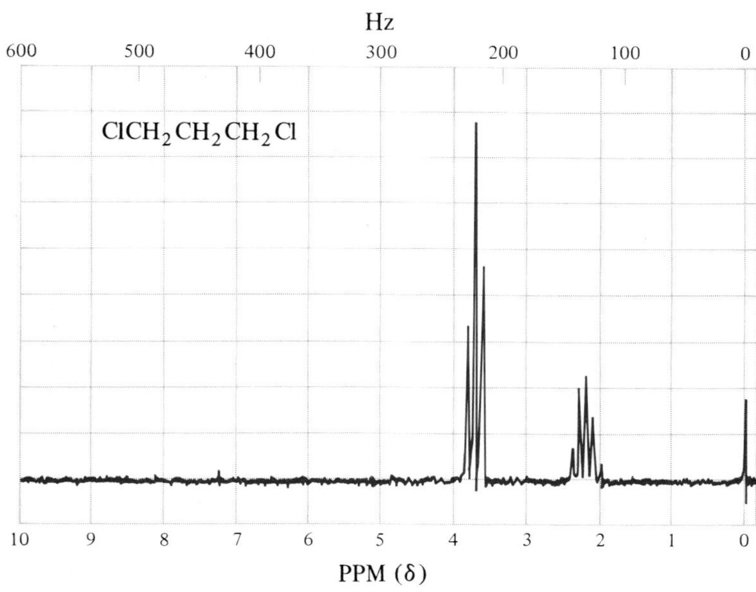

Figure 9–13. *[1]H nmr spectrum of 1,3-dichloropropane.*

Example

Draw the ^1H nmr spectrum you would expect for chloromethyl ethyl ether.

$$CH_3CH_2OCH_2Cl$$

Answer

The protons on the CH_3 part of the CH_3CH_2 group would absorb near 1 ppm since they are on a carbon bonded to another carbon. There are *two* protons on the neighboring carbon, so the CH_3 absorption would appear as a triplet [$(n + 1)$ rule where $n = 2$].

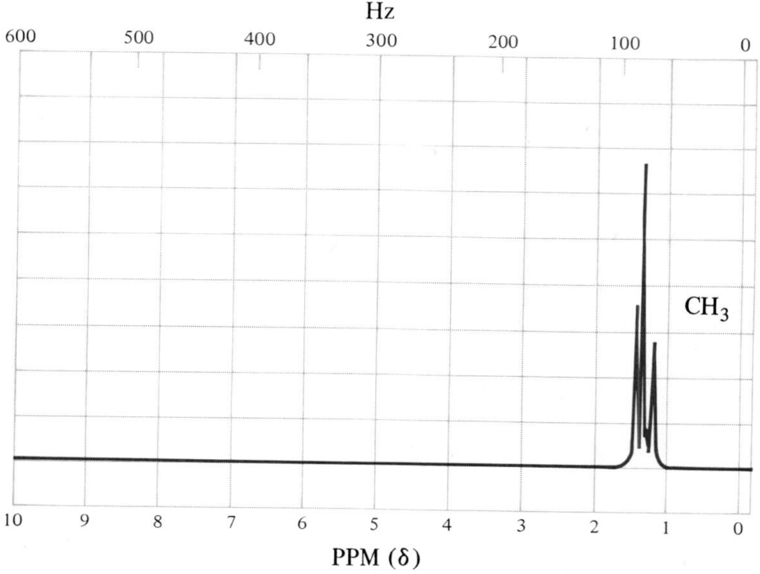

The CH_2 protons next to the CH_3 are on a carbon bonded to oxygen, and they would appear downfield near 4 ppm. The absorption would be a quartet because there are three protons on the neighboring carbon [$(n + 1)$ rule where $n = 2$].

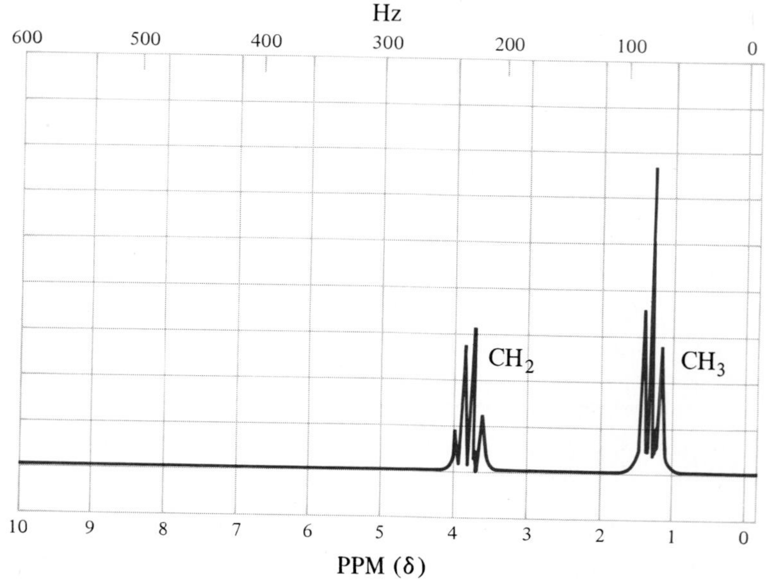

There is a CH_2 group between an oxygen and a chlorine. These electronegative atoms would cause the proton absorption to be far downfield at about 5.5 ppm. Since there are no hydrogens on neighboring atoms, the absorption would be a singlet.

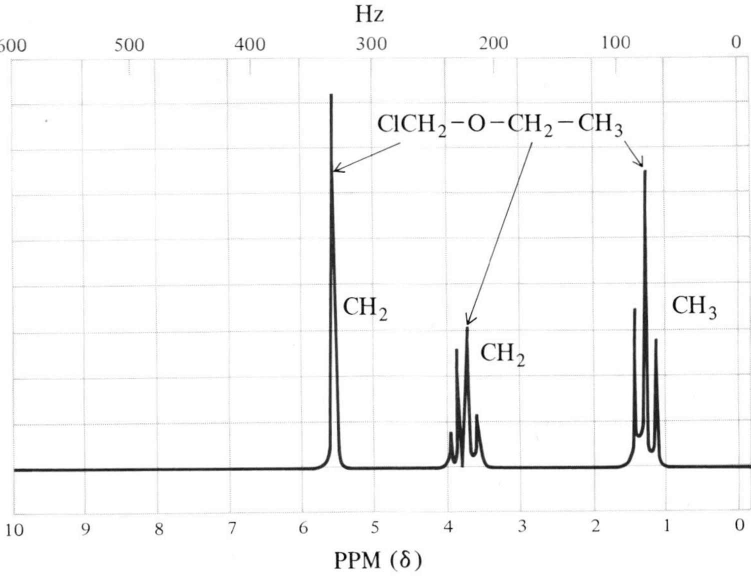

Problems

9–3 Indicate how many different types of hydrogens there are in each of the following compounds, and what the splitting pattern will be in their proton nmr spectra.

a. $ClCH_2CH_2I$

b. $CH_3CCl_2CH_2CH_3$

c. CH_2—CH_2 (ring through O)

d. $BrCH_2\overset{\overset{\large O}{\|}}{C}CH_2CH_2Cl$

9–4 a. There are two isomers corresponding to the formula $C_2H_4Br_2$. Write their structures.

b. The nmr spectrum of one of the structures consists of one singlet. Which isomer is it?

c. Draw a schematic nmr spectrum of the other isomer.

In order to determine the relative numbers of protons present in an unknown compound, it is important to have an accurate measure of the relative areas under the peaks. Experimentally, this is done with an electronic integrator, which is built into the nmr spectrometer. Figure 9–14 shows a ^1H nmr spectrum of cyclopentanol with an *integrator trace*. The integrator trace is the heavy line that begins just above the base line at about 5 ppm. Where there are no peaks, the line remains horizontal; as the integrator crosses a peak, the trace rises vertically by an amount proportional to the

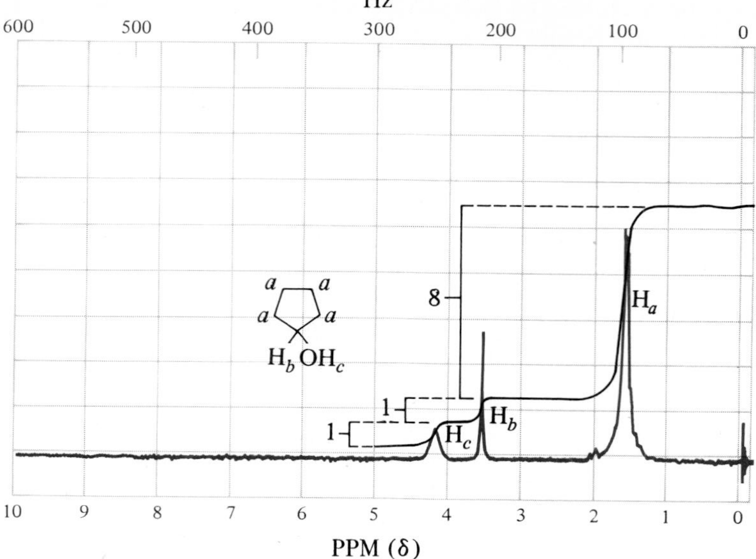

Figure 9–14. *Electronically integrated* [1]*H nmr spectrum of cyclopentanol with an integrator trace.*

area of the peak (indicated by brackets). By measuring the rises and taking their ratios, the chemist can obtain the ratios of the peak areas, which tell the relative numbers of each type of hydrogen. In Figure 9–14 the broad absorbance at 4.21 ppm and the sharp absorbance at 3.58 ppm cause identical vertical rises in the integrator trace, while the peak at 1.61 ppm causes a vertical rise that is eight times that of the others.

Box 9–1. Nuclear Magnetic Resonance in Medical Diagnosis

One goal of medical researchers is to develop a noninvasive, risk-free method of seeing what is happening inside the human body. The use of X rays and CAT (computer assisted tomography) scanners has contributed to the attainment of this goal, but they subject the patient to hazardous high energy ionizing radiation. A recent development of great promise has been the use of ^{1}H nmr to produce an image of an organ or even the whole body. This technique uses a magnet large enough to enclose the body, or a portion such as the brain, and two magnetic fields. The image obtained looks somewhat like an X ray but is actually a map of proton densities. These densities vary from tissue to tissue and often distinguish healthy tissue from unhealthy tissue. For example, some of the first experiments showed that normal cell images were different from cancer cell images.

Brain images showing the presence of tumors or other abnormalities have been particularly well-studied. At one laboratory an apparently normal visitor asked to see how the instrumentation worked. A ^{1}H nmr image of his brain was obtained, and an abnormality was seen behind one of his eyes. It turned out to be a benign tumor that had unknowingly been the cause of occasional severe headaches.

The energy of the magnetic fields which cause spin excitations is sufficiently low so that there should be no damage to the organs being studied. Some researchers have spent over 12 hours in such magnetic fields with no apparent ill-effects.

9:6 Comparison of ¹H NMR and IR Spectroscopy

The following list summarizes the kind of information that can be obtained from a proton nmr spectrum:

1. The number of peaks (or groups of peaks) indicates the number of different types of hydrogens in the molecule.
2. The position (chemical shift) of a peak in ppm (δ) units indicates the type of carbon atom to which the protons are attached.
3. The area under the peak is a measure of the relative number of each type of hydrogen.
4. The splitting pattern indicates the number of hydrogens on neighboring carbons.

Because there are usually more CH bonds than any other type of bond in organic molecules, proton nmr spectroscopy is one of the most valuable sources of structural information. For example, suppose we want to know whether an unknown C_3H_8O compound is $CH_3CH_2CH_2OH$ (1-propanol), $CH_3CHOHCH_3$ (2-propanol), or $CH_3OCH_2CH_3$ (ethyl methyl ether). The first two can be distinguished from the last by infrared spectroscopy, since only they would show OH infrared absorption. An ir spectrum would not easily differentiate between the two alcohols, however. On the other hand, a single ^1H nmr spectrum would immediately distinguish among all three molecules.

There are certain cases where ir, alone, is more valuable than nmr. For example, both

$$CH_3\overset{\overset{\displaystyle O}{\|}}{C}CH_3$$

(acetone) and CH_3CN (acetonitrile) show only one singlet in the ^1H spectrum, and their chemical shifts are very similar. Therefore ^1H nmr would not absolutely distinguish between the two. Infrared spectroscopy, on the other hand, would immediately settle the question, since the C=O stretch for acetone is at 1715 cm^{-1} and the C≡N stretch is at 2250 cm^{-1}.

9:7 First-Order Spectra and Beyond: Complications in NMR Interpretation

The spectra so far presented have consisted of either singlets or nicely symmetrical multiplets because they provide clear illustrations of the principles of nmr. Unfortunately, not all organic compounds give rise to simple spectra like these. In this section we discuss some of the molecular features that produce nmr complications. With some study and practice, many of the more complex spectra also prove amenable to interpretation.

Consider first the case in which three different groups of protons interact, as in 2-propanol (Fig. 9–15). Note that this spectrum was obtained with an unusually pure, anhydrous sample of 2-propanol, the significance of which will shortly become apparent. Concentrate on the proton at C-2 coupled to the six equivalent methyl protons and to the hydroxyl proton. Its absorption peak, centered at 4 ppm and shown in expanded form, contains

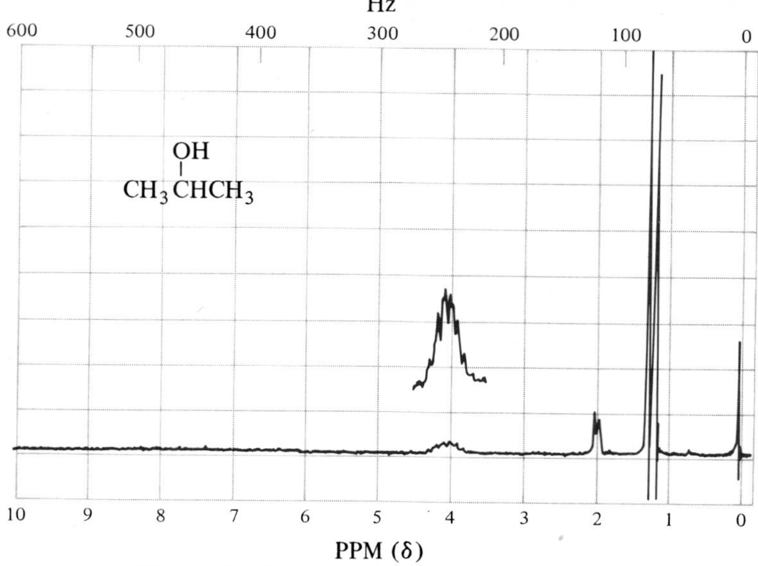

Figure 9–15. *^1H nmr spectrum of purified 2-propanol.*

14 lines (not all are easily visible). This splitting pattern arises as follows. The six methyl protons split the C-2 proton absorption into seven components. The hydroxyl proton further splits each component in two, yielding a total of $2 \times 7 = 14$ lines (Fig. 9–16).

Based on our analysis of the 2-propanol spectrum we can propose an

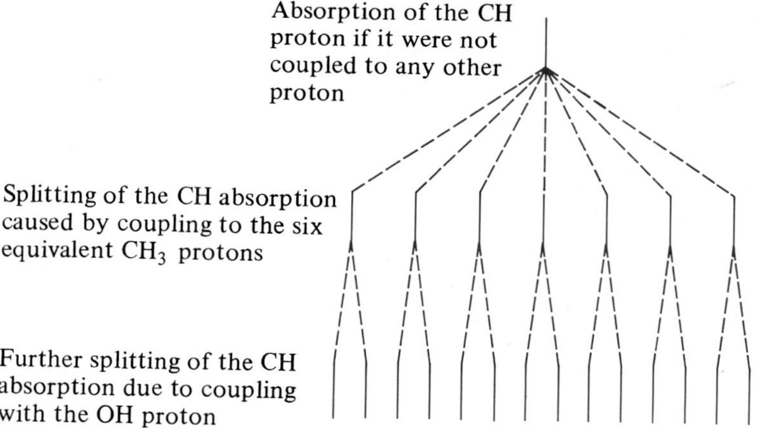

Figure 9–16. *Origin of the splitting pattern of the CH absorption in 2-propanol arising from coupling to the CH_3 and OH protons.*

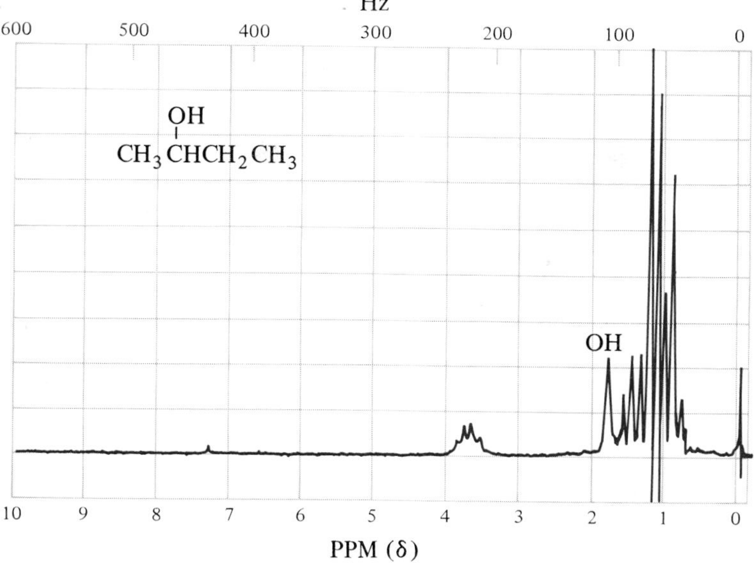

Figure 9–17. *[1]H nmr spectrum of 2-butanol.*

extension of the $(n + 1)$ rule (sec. 9:5). If a proton has n neighboring protons of one kind and m neighboring protons of another, its absorption will be split into a maximum of $(n + 1)(m + 1)$ components. Spectra which can be completely interpreted by applying the $(n + 1)(m + 1)$ rule are called *first-order spectra*. In the following examples we'll find cases where both more and fewer absorption peaks are found than predicted by first-order analysis.

In the nmr spectrum of 2-butanol (Fig. 9–17) the **CH**—OH absorption occurs at 3.75 ppm. First-order analysis suggests that this absorption should be extensively split, since the proton has three different types of neighboring groups—CH_3, CH_2, and OH. In fact, it is split into only six components (a sextet). An important clue to this puzzle is the OH absorption at 1.8 ppm, which appears as a *singlet*. We must therefore conclude that the OH and CH protons are not coupled to one another. The lack of coupling between OH protons and neighboring CH is what is *usually* found in alcohol nmr spectra, for reasons set out in Section 9:12. Coupling with alcohol protons is observed only when the sample is highly purified and anhydrous, as in Figure 9–15, or when it is dissolved in certain solvents.

If the CH proton is coupled to other carbon-bound hydrogens, then the observed sextet implies that the five neighboring hydrogens are equivalent. However, we know that the CH_3 and CH_2 hydrogens are *not* chemically equivalent, and we will shortly see that the two CH_2 hydrogens are not even equivalent to one another (sec. 10:12). What is happening is that the *coupling constant* between the CH proton and its five neighbors is nearly identical in each case, so the five *behave as if they are equivalent*. Thus, the fact that there is no CH—OH coupling and that the CH—CH coupling constants are nearly identical serves to simplify the nmr spectrum of 2-butanol.

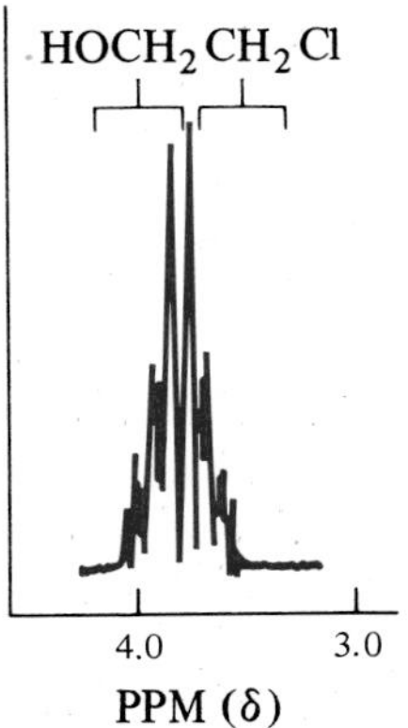

Figure 9–18. *[1]H nmr spectrum of the —CH_2CH_2— region of 2-chloroethanol.*

Finally, we consider a compound that gives a more complex spectrum than expected, 2-chloroethanol ($ClCH_2CH_2OH$). Figure 9–18 shows the CH region of its nmr spectrum. First-order analysis would lead us to expect two triplets of equal intensity here. We find instead an aggregation of many peaks that bears little resemblance to a triplet. This departure from first-order behavior is due to the very similar chemical shifts of the CH_2Cl and CH_2OH groups. When chemical shift differences are small between coupled protons they interact in multiple ways and generate multiple absorption peaks. Spectra like that in Figure 9–18 usually cannot be deciphered by inspection, but can be readily cracked with the aid of a computer.

There is a standard system used to designate coupled protons that give rise to characteristic nmr patterns. If the protons have similar chemical shifts they are designated *A*, *B*, and *C*. Thus, the 2-chloroethanol spectrum would be classified as an A_2B_2 spectrum. When the coupled protons have well-spaced chemical shifts they are labeled *A*, *M*, and *X*. The bromoethane spectrum is an example of an A_3X_2 spectrum. We will use this method of designation in future nmr discussions.

Problems

9–5 Assign the peaks upfield of 1.8 ppm to various protons in 2-butanol (Fig. 9–17). Start by predicting what patterns and relative chemical shifts you would expect from the protons on carbons 1, 3, and 4. Note that adjacent multiplets may experience overlap of some of their component peaks.

9–6 a. A compound, C_3H_6O, gives the following ^1H nmr spectrum: 1.19 ppm (triplet); 2.44 ppm (eight peaks); 9.80 (triplet), with relative areas 3 : 2 : 1. What is its structure? (Use chemical shift data from the endpapers to assign the 9.80 ppm peak.)

b. The ^1H spectrum of yet another C_3H_6O isomer is given at the top of the next page. Draw its structure. (The numbers indicate *relative* numbers of hydrogens.)

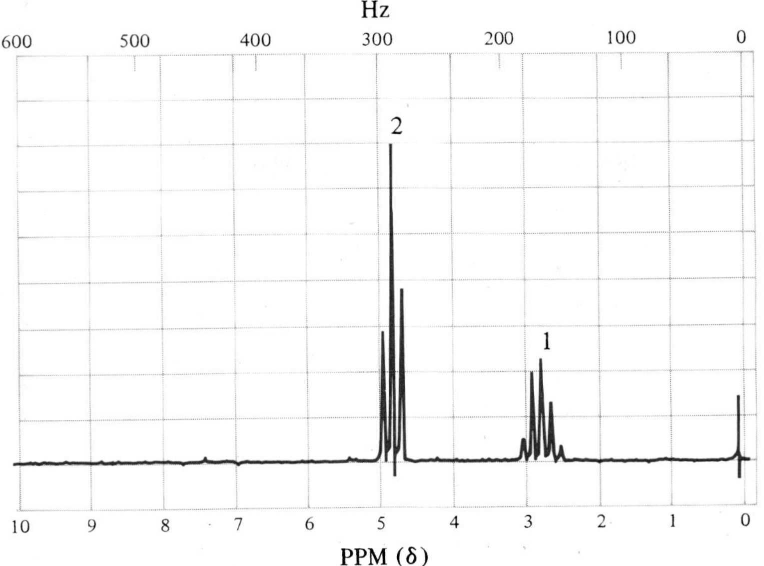

9–7 The spectrum below is that of 1,3-dibromobutane. Assign each multiplet to one type of proton in the molecule, and explain its splitting pattern.

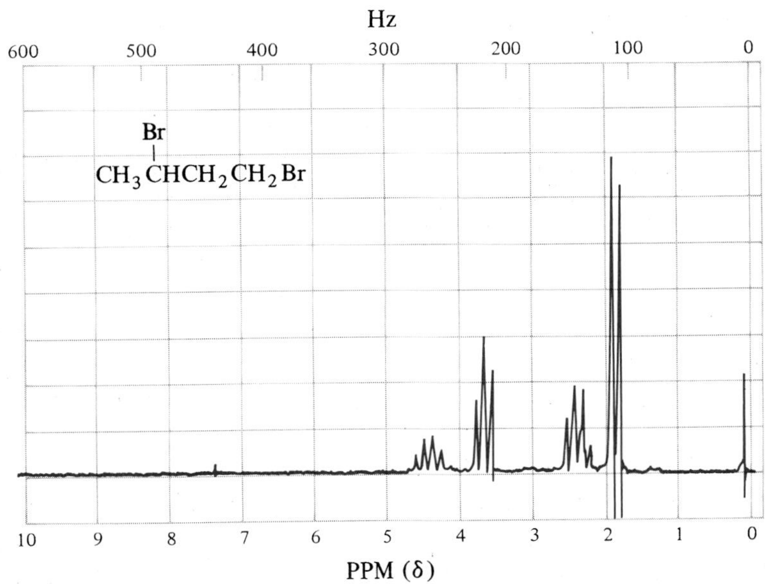

9:8 Spin Decoupling

Spin–spin coupling is a very useful phenomenon because of the information it provides about neighboring protons. Its effects are not always beneficial, however. As the 2-chloroethanol spectrum (Fig. 9–18) shows, when the chemical shifts of coupled groups are close together, the spectrum is no longer first-order, and the result is sometimes a featureless "blob" of absorption. In other cases, the molecule may be so complex that independent multiplets (not coupled together) may coincidentally have similar chemical shifts. Multiplets may occur so close together that individual splitting patterns cannot be extracted from the many overlapping peaks.

One of the instrumental techniques used to simplify such complex spectra is **spin decoupling**. To see how this works in practice, consider again the spectrum of bromoethane (Fig. 9–7). Suppose that the sample is irradiated with a *second* radiofrequency signal while the spectrum is being taken. This signal can be tuned exactly to the resonance frequency of the methyl hydrogens at 1.65 ppm. The resulting spectrum will have *no* peak at all at 1.65 ppm and the methylene peak at 3.37 ppm will be a singlet. Similarly, if we irradiate at 3.37 ppm, no resonance will appear there and the 1.65 resonance will be a singlet. The second radiofrequency removes couplings between neighboring hydrogens by exciting one nucleus so that it no longer affects its neighbor.*

As another example, consider the spectrum expected for 1,2-dibromopropane. The absorptions due to the C**H**Br and C**H**$_2$Br protons would be

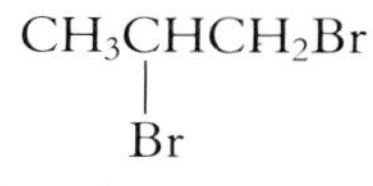

1,2-Dibromopropane

overlapping and it would be difficult to pick out the 12 lines expected for the C**H**Br proton. (Remember the $(n + 1)(m + 1)$ rule. Here, the 12 lines are the result of couplings to both the CH_3 and the CH_2Br protons: $(3 + 1)(2 + 1) = 12$.) If we were to irradiate at the CH_3 proton absorption position with a second radiofrequency, the C**H**Br proton would then appear only as a triplet. All couplings to the CH_3 protons would have been removed. This type of simplification can be of enormous help in unraveling the coupling pattern of a complex spectrum. It is of particular importance in ^{13}C nmr spectroscopy, the topic to be considered next.

9:9 ^{13}C Nuclear Magnetic Resonance

Although the major carbon isotope, ^{12}C, is not magnetically active, the active minor isotope, ^{13}C, is sufficiently abundant in all carbon compounds so that its resonance can be observed. Because the natural abundance of ^{13}C is only about 1.1%, signals arising from the ^{13}C nuclei are weak and need to be enhanced. This is done with a small computer, often built into the instrument, which allows spectra to be taken repeatedly over a short period of time. The signals are processed by a mathematical operation called a **Fourier transformation**. Taking *Fourier transform* ^{13}C spectra is now almost as routine as obtaining the instrumentally simpler ^{1}H spectra. The resonances of ^{13}C nuclei vary over a considerably larger range of field strength than do proton resonances. Most proton absorptions fall between 0 and 10 ppm, while ^{13}C absorptions vary from about 0 to 220 ppm. A few ^{13}C absorptions even occur upfield from TMS.

* Essentially, the second radiofrequency signal causes the protons being irradiated to undergo very rapid transitions between the two energy states. As a result, the *neighboring* hydrogens "see" only a single, average spin distribution, and their absorption signal is therefore unsplit.

Spin–spin coupling occurs between a ^{13}C nucleus and the protons bonded to it. Thus the ^{13}C absorption peak for a $—CH_3$ would appear as a quartet, that for a $—CH_2—$ would be a triplet, and so on, in accordance with the $(n + 1)$ rule. Figure 9–19a shows a coupled ^{13}C nmr spectrum of dihydropyran. There are triplets at 20, 24, and 67 ppm downfield from the ^{13}C resonance of TMS, and doublets at 101 and 147 ppm. The latter two resonances are in the region of alkene carbons, with the far downfield peak at 147 ppm assignable to the alkene carbon next to oxygen. The ring CH_2 carbon next to the double bond gives rise to the lowest field triplet (67 ppm), while the other two ring CH_2 carbon triplets are at high field positions.

If the proton nuclei are irradiated with a wide-band radiofrequency wave, we obtain the spin-decoupled spectrum for dihydropyran (Fig. 9–19b). This irradiation removes all the proton spin interactions with the carbon nuclei, and a single peak is obtained for each different carbon atom. Note that in ^{13}C nmr the peak heights are *not* always directly related to the number of carbon atoms. Carbons with attached hydrogens tend to give higher peaks than those without, but there are additional factors which also contribute to this discrepancy in peak heights. The ^{13}C chemical shifts roughly parallel those of protons: sp^3 carbons absorb at high field and sp^2 carbons at low field. Electronegativity effects in ^{13}C nmr are also similar to those in ^1H nmr (see table of ^{13}C chemical shift values in the endpapers).

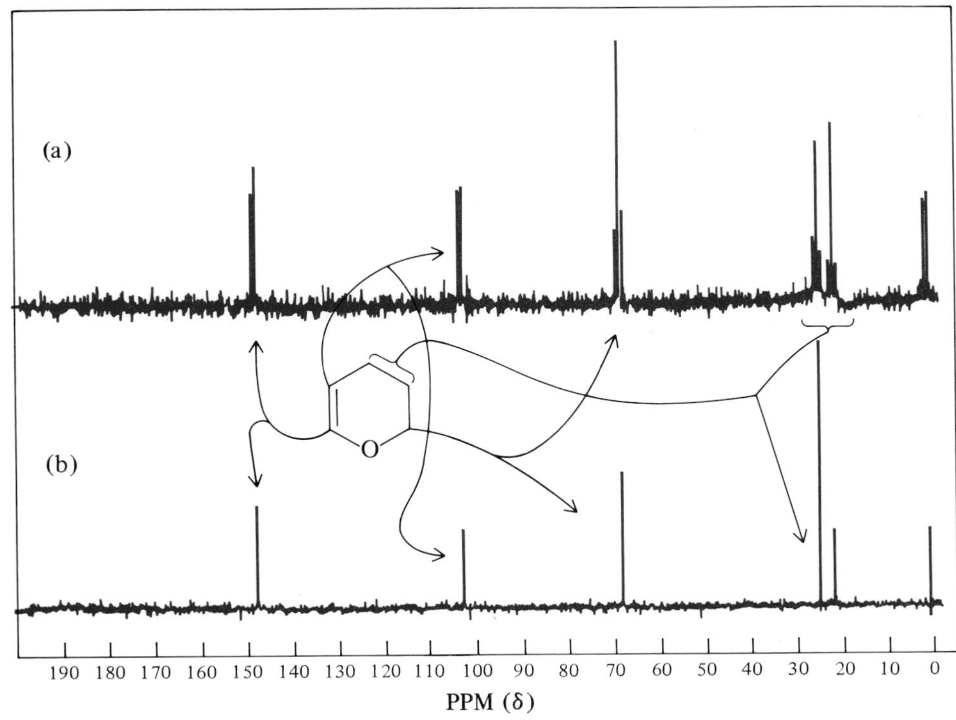

Figure 9–19. *(a) Proton-coupled ^{13}C nmr spectrum of dihydropyran. (b) Spin-decoupled ^{13}C nmr spectrum of dihydropyran.*

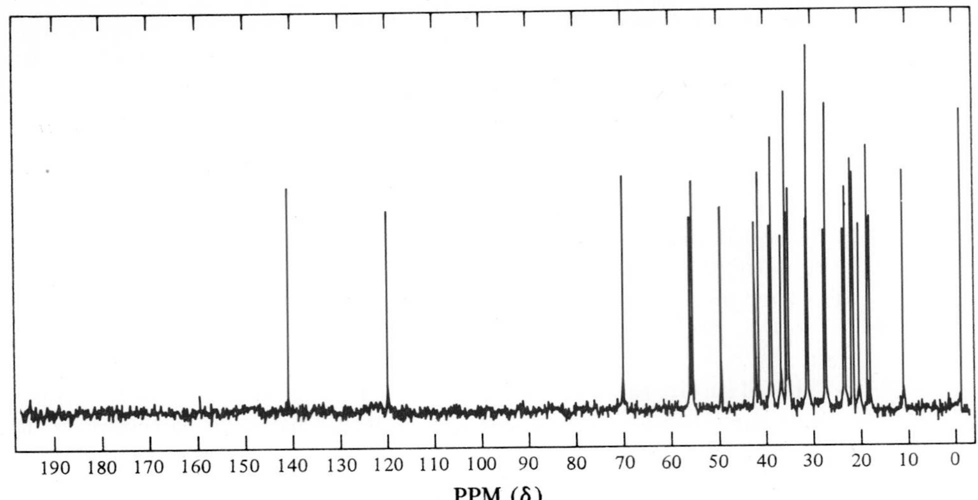

Figure 9–20. *Decoupled* ^{13}C *nmr spectrum of cholesterol.*

Figure 9–20 shows the decoupled ^{13}C nmr spectrum of cholesterol. Even in very complex molecules such as this, absorbances for nearly every carbon atom can be seen—there are 26 peaks and 27 carbons in cholesterol.★ The ^{1}H nmr spectrum of cholesterol is much less informative, as you can see from Figure 9–21.

★ One peak must consist of two overlapping ^{13}C resonances.

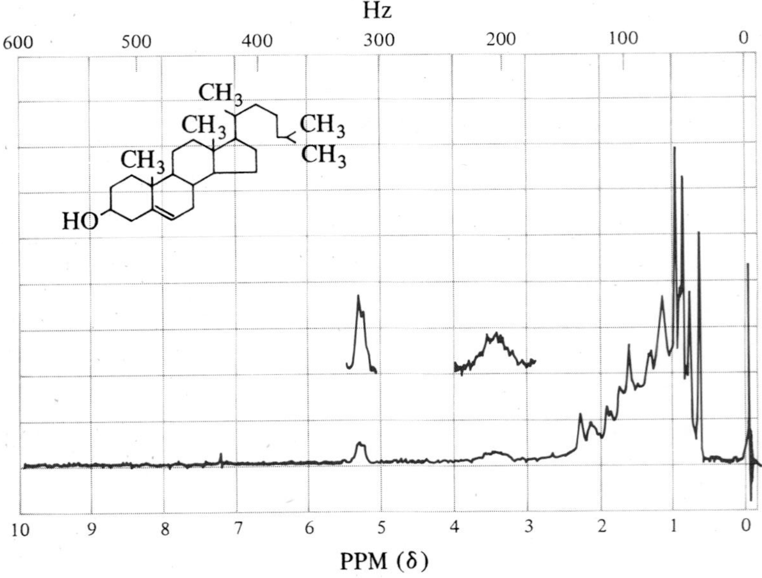

Figure 9–21. ^{1}H *nmr spectrum of cholesterol.*

Although this discussion has been brief, ^{13}C nmr is an instrumental tool as important as ^{1}H nmr. Throughout the text, further examples of the use of ^{13}C nmr spectra will be given.

Example

The compound below gives a *decoupled* ^{13}C nmr spectrum with single peaks as indicated. Assign the peaks to the individual carbons and tell what their multiplicity would be in the *coupled* spectrum.

$$\begin{array}{ccc} BrCH_2 & & H \\ & C{=}C & \\ H & & \underset{\displaystyle \overset{\|}{O}}{C}OCH_3 \end{array}$$

28 ppm
52 ppm
125 ppm
142 ppm
166 ppm

Answer

All sp^2 carbons absorb above 100 ppm and the highest absorption is usually due to the carbon of the carbonyl (C=O) group. Therefore the 166 ppm peak would be due to the carbonyl carbon and the 125 and 142 peaks would be due to the carbons of the C=C group. The carbon of the CH_3 group next to oxygen would be at 52, while the remaining CH_2 carbon must be that next to the bromine.

The carbonyl carbon peak would be a singlet in the coupled spectrum since it bears no hydrogens. Each carbon of the C=C group bears one hydrogen, so the carbon peaks at 125 and 142 would be doublets. The CH_3 carbon at 52 would be a quartet (three attached hydrogens), and the CH_2 peak at 28 ppm would be a triplet (two attached hydrogens).

Problem

9–8 Given here are the ^{1}H and ^{13}C spectra for an unknown compound whose combustion analysis gives the following results: C, 35.06%; H, 6.62%; Br, 58.32%. What is the compound? Assign the peaks.

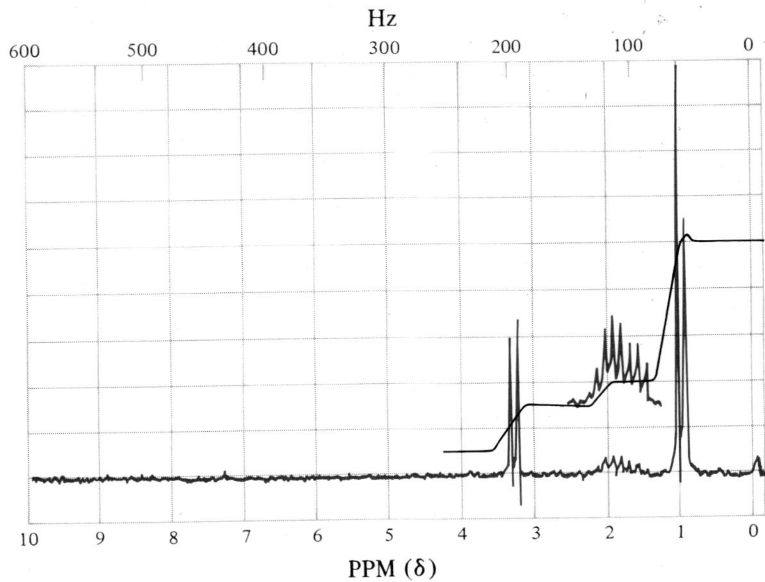

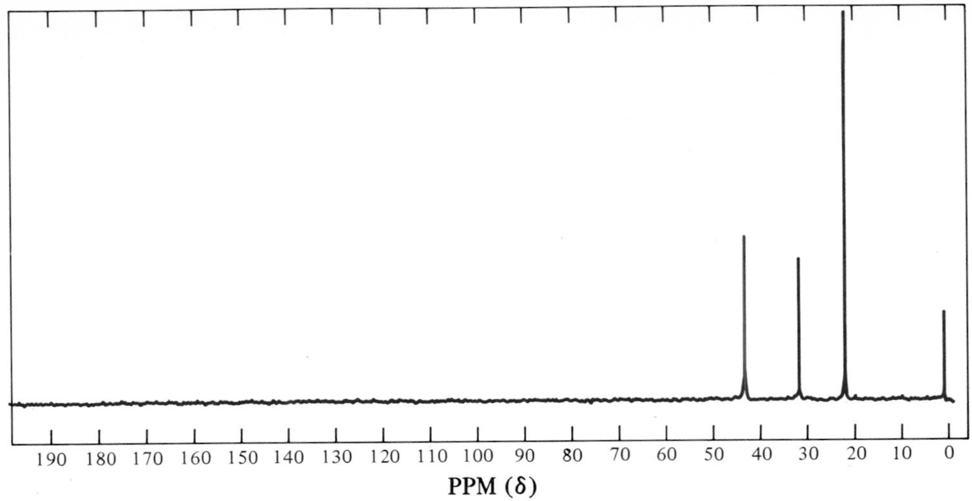

9:10 Practical Details of NMR Spectroscopy

Nuclear magnetic resonance spectra are usually obtained from solutions of the sample compound. Interactions between the solvent and the sample can affect the chemical shifts, so the choice of solvent is important. The solvent should not, of course, produce a peak coinciding with peaks due to the sample. One of the best solvents for proton nmr spectroscopy is carbon tetrachloride, CCl_4, which contains no protons. One drawback, however, is that many compounds are not sufficiently soluble in CCl_4 to allow their spectra to be obtained. Organic solvents that contain hydrogen can be used if the hydrogens are replaced by deuterium, D. Deuterium nuclei undergo transitions at different field strengths than protons do. The effect of substituting deuterium for hydrogen is to blank out the peaks that normally arise from those protons. Two common nmr solvents are deuterochloroform, $CDCl_3$, and dimethyl sulfoxide-d_6, CD_3SOCD_3 (DMSO-d_6). If the sample is insoluble in organic solvents, then heavy water, D_2O, can be used. Ordinary water is an excellent solvent for ^{13}C nmr, while $CDCl_3$★ is used for water-insoluble compounds.

After the sample is dissolved in a solvent, about 1% of TMS is added directly to the solution. The solution is transferred to a narrow glass tube that is then placed in a cavity directly between the poles of a magnet (Fig. 9–2a). The tube is spun rapidly while the spectrum is being recorded, in order that the sample may "feel" as uniform a magnetic field as possible.

In ^1H nmr the radiofrequency radiation is usually at 60 MHz with the magnetic field being about 14,000 G. The ^1H spectra in this book were taken under these conditions. If higher frequencies (and higher magnetic fields) are employed, then the resolution of the instruments is increased greatly. Many instruments operate at 100 MHz (and about 23,000 G), and particularly valuable (and more expensive) instrumentation is available at even higher magnetic fields.

★ There is a ^{13}C—^2D coupling which produces a triplet centered at 77 ppm when $CDCl_3$ is used as a solvent.

60 MHz spectra that are complex and difficult to interpret may attain first-order simplicity when taken at higher field. Figure 9–22 illustrates the better resolution and simplicity achieved in this manner for 1-butanol. The overlapping and complex CH_2 signals present at 60 MHz become simple and first-order at 360 MHz.

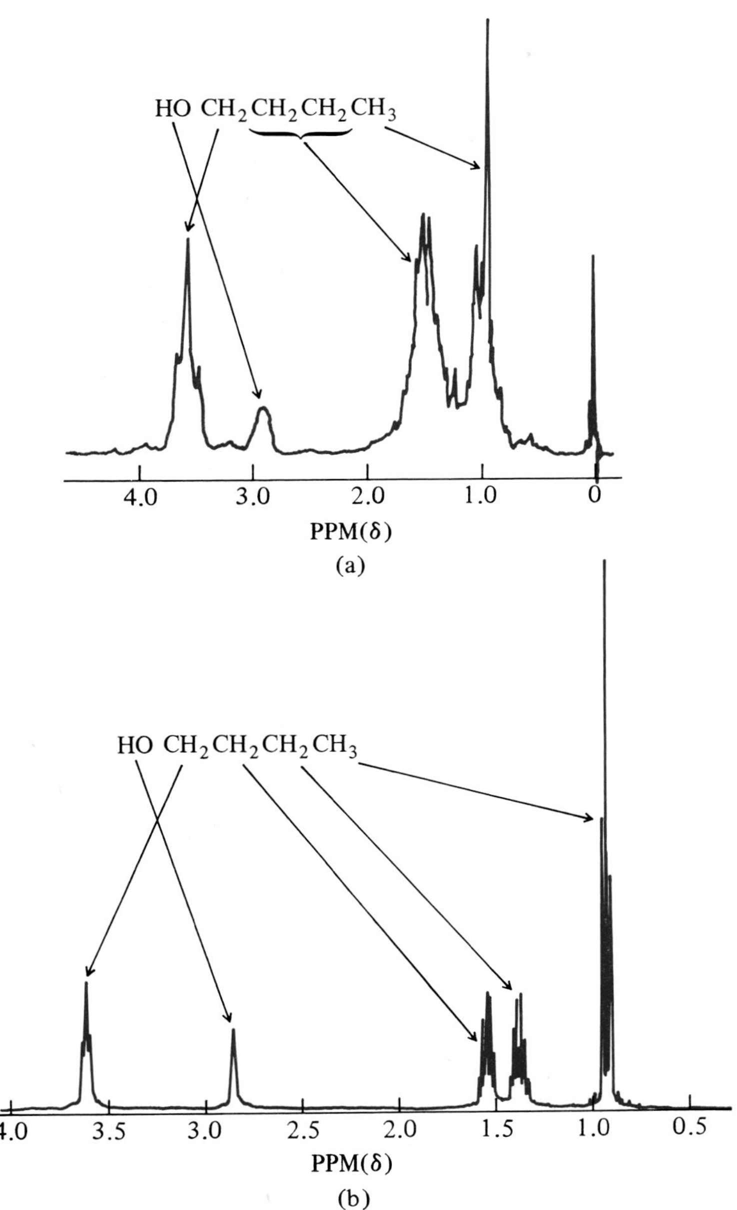

Figure 9–22. *(a) ¹H nmr spectrum of 1-butanol taken at 60 MHz. (b) ¹H nmr spectrum of 1-butanol taken at 360 MHz.*

9:11 NMR Spectroscopy of Alkanes, Alkenes, Ethers, and Alcohols

Alkanes

For alkanes, ^{1}H nmr spectroscopy is not very informative, since proton absorbances are grouped closely in the 0.8 to 1.2 ppm region. Both ^{1}H and ^{13}C nmr have, however, been applied to conformational analysis of substituted cyclohexanes. For this purpose, ^{13}C nmr is particularly useful. Figure 9–23 shows how the ^{13}C nmr spectrum of *cis*-1,2-dimethylcyclohexane changes with temperature. At $-20°$ (and all temperatures above that), the spectrum of Figure 9–22a is obtained. The equilibrium between the two conformers is rapid (eq. 9.2), so that the methyl carbons become equivalent

(9.2) 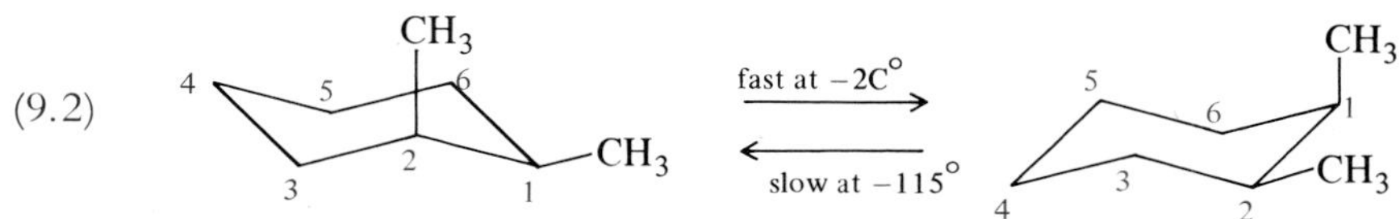

on the nmr time scale, as do carbons 1 and 2, 3 and 6, and 4 and 5. Thus each of these carbon pairs give rise to a single ^{13}C nmr peak. The chemical shift of the methyl carbon peak is an average of the values for axial and equatorial methyl groups. At $-115°$ (Fig. 9–23b), the spectrum shows six peaks for the six different ring carbons. Equatorial and axial methyl groups no longer interchange and a sharp signal is seen for each. Rate constants and activation parameters can be calculated from these temperature-dependent nmr spectra. For example, the $\Delta G^{\ddagger}$ value of 9.3 kcal/mol for the inversion barrier in *cis*-1,2-dimethylcyclohexane was determined in this manner.

Alkenes

Nuclear magnetic resonance spectroscopy is very useful when applied to structural analysis of compounds containing double bonds, such as alkenes. Alkene ^{1}H absorptions are found between 4.5 and 7.5 ppm, but a large majority are confined to the 4.5 to 6.5 ppm region where few other absorp-

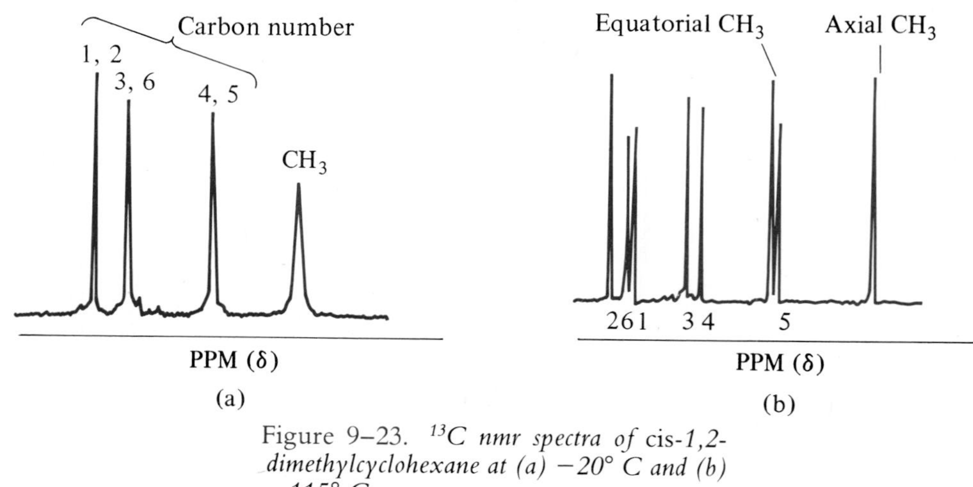

Figure 9–23. *^{13}C nmr spectra of* cis-*1,2-dimethylcyclohexane at (a) $-20°$ C and (b) $-115°$ C.*

tions occur. Protons on benzene rings generally absorb from 7 to 8 ppm and hence are usually distinguishable from alkene protons. In ^{13}C nmr spectra, alkene carbon absorptions are usually found between 110 and 150 ppm.

Let's look at the ^1H nmr spectrum in Figure 9–24 for an illustration of some general principles. Notice that this is an *ABC*-type spectrum (sec. 9:7) of three interacting protons. The spectrum of vinyl pivalate shows two interesting absorptions at 4.8 and 7.2 ppm. These two multiplets of four peaks each are due to the vinyl group ($CH{=}CH_2$). To interpret these absorptions, we will have to refer again to the coupling constant, *J*. As you may recall from Section 9:5, this value defines the distance between splittings in coupled multiplets. Note that the H_a absorption is a doublet with $J = 14$ Hz, while the H_b absorption is a doublet with $J = 6$ Hz. Both are doublets because they are coupled to H_c, but since H_a is *trans* to H_c and H_b is *cis* to H_c, the coupling constants (*J*-values) are different. Note that the *J* for H_a coupled to H_c is called J_{ac}, while the *J* for H_b coupled to H_c is called J_{bc}. For two hydrogens *trans* to each other across a double bond, the coupling constant *J* is typically in the 12 to 18 Hz range. For two hydrogens *cis* to each other, the coupling constant is in the 6 to 11 Hz range.

Also note that all four peaks at 4.8 ppm are actually not singlets, but are split further into barely visible doublets. This is because H_a and H_b are not equivalent and hence can be coupled to one another, even though they are both on the same carbon atom. Up to this point all examples of coupling have been between protons on adjacent carbons, but any two nonequivalent protons can, in theory, give rise to splittings. Two protons on the same carbon are called *geminal protons*. The coupling constant between geminal vinyl protons is very weak (0 to 3 Hz) which is why the splitting is barely observable.

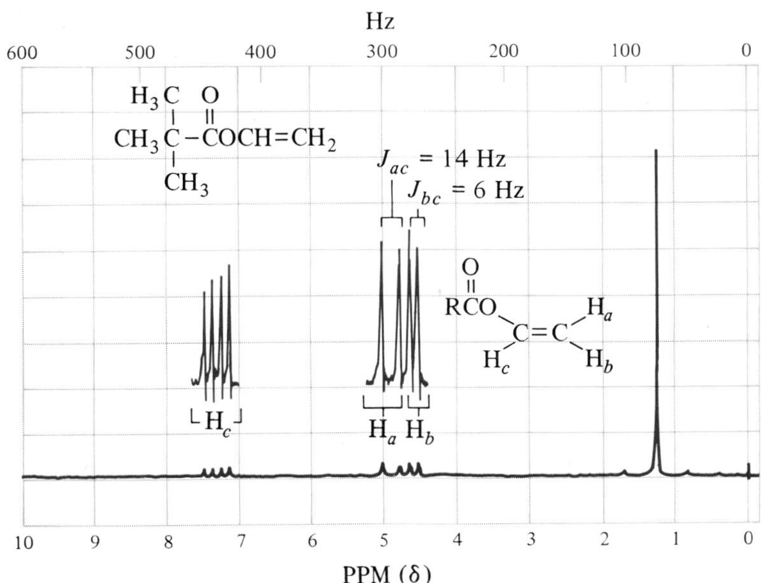

Figure 9–24. *^1H nmr spectrum of vinyl pivalate.*

Now consider the four peaks at 7.2 ppm due to H_c. As we just mentioned, H_c is coupled differently to H_a and H_b. If J_{ac} = 14 Hz and J_{bc} = 6 Hz, then J_{ca} must be 14 Hz and J_{cb} must be 6 Hz. This reciprocal relationship between two coupled protons is evidence of their interaction. The best way to analyze the peaks for H_c is with a diagram such as that given in Figure 9–25. H_c would give rise to a singlet if there were no couplings and to a doublet (J = 14 Hz) if it were only coupled to H_a. Since it is also coupled to H_b (J = 6 Hz), the result is four peaks of equal intensity, as shown in Figure 9–25.

Alcohols

One of the most significant features of the ^{1}H nmr spectra of alcohols is the fact that the OH proton is usually not coupled to neighboring CH protons (sec. 9:7). The reason for this is that most alcohol samples contain traces of acid, which catalyze the rapid exchange of hydroxyl protons (eq. 9.3). A CH

$$\mathbf{ROH} + \mathrm{ROH} \underset{\text{fast}}{\overset{H^+}{\rightleftarrows}} \mathbf{R}\mathrm{OH} + \mathrm{RO}\mathbf{H} \tag{9.3}$$

proton next to the OH group does not "feel" the hydroxyl proton long enough to tell whether its magnetic field is slightly shielding or deshielding. It simply experiences an average, which results in no coupling. This effect is shown in Figure 9–26a, which contains the spectrum of ethanol that has not been highly purified. When a pure sample is used, CH_2—OH coupling is observed (Fig. 9–26b). Note the large difference in OH chemical shift be-

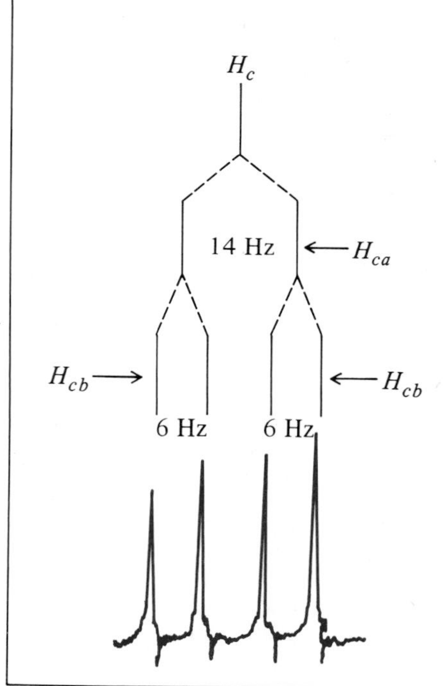

Figure 9–25. *The H_c proton of vinyl pivalate (see Fig. 9–24) gives rise to a quartet as a result of its coupling to protons H_a and H_b.*

tween the two ethanol spectra. The chemical shift of this absorption depends strongly on the solvent used and the concentration of the sample in it. Note how the inductive effect of the OH shifts the CH_2 absorption downfield.

For any unknown sample, we can confirm the presence of an OH absorption in the ^1H nmr spectrum by adding a drop of D_2O to the sample tube

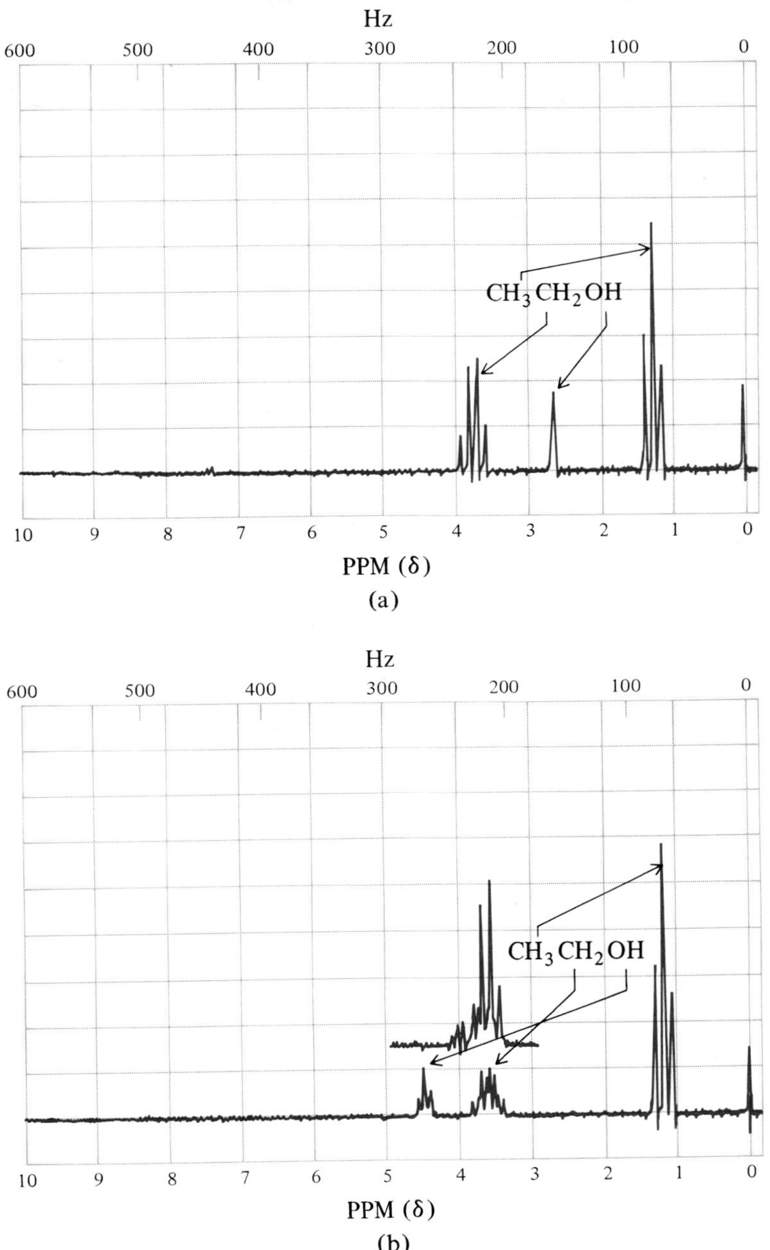

Figure 9–26. *^1H nmr spectra of ethanol. (a) Ordinary sample in $CDCl_3$; (b) highly purified sample in CCl_4.*

after the spectrum has been taken. The sample is then shaken and another spectrum is obtained, but the original OH absorption is no longer visible. The OH will have exchanged with D_2O and been converted to OD; hence, no OH absorption appears.

Ethers

Both ^1H and ^{13}C nmr can be applied to the structural identification of ethers. In ^1H nmr, protons on the carbon next to oxygen (C**H**—O) characteristically appear in the 3.2 to 4.5 ppm region, since the electron-withdrawing character of the oxygen causes a large deshielding effect (just as in alcohols). The ^{13}C absorptions for sp^3 carbons next to oxygen are in the 40 to 60 ppm region.

PROBLEM

9–9 When alcohols are dissolved in DMSO-d_6 (sec. 9:10) they hydrogen-bond strongly to the solvent. The hydrogen bonding slows down proton–proton exchange (eq. 9.3), allowing CH—OH coupling to be observed. Indicate how DMSO-d_6 can be used to distinguish among 1°, 2°, and 3° alcohols.

9:12 Electron Spin Resonance Spectroscopy

Electron spin resonance (esr) is another type of spectroscopy, similar to nmr, used to detect the presence of unpaired electrons in molecules. It can, therefore, provide structural information about *radicals*.

Like ^{13}C and ^1H nuclei, electrons assume two different energy states when placed in a magnetic field. By providing radiofrequency energy we can cause some electrons to jump from the lower to the higher state. As in nmr, the magnetic fields of neighboring protons will slightly alter the energy

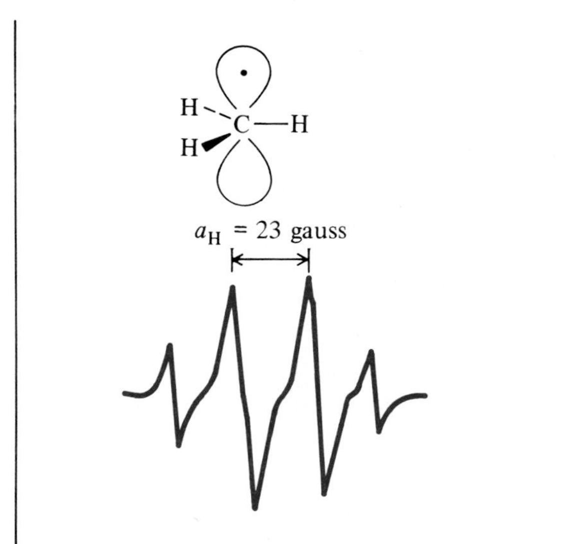

Figure 9–27. *Esr spectrum of methyl radical, $CH_3\cdot$.*

of this transition. If there is an unpaired electron on a carbon that bears n equivalent hydrogen atoms, the esr spectrum will consist of $(n + 1)$ peaks. Figure 9–27 shows the esr spectrum of a methyl radical, $CH_3\cdot$. The spectrum is displayed in a somewhat different manner than nmr spectra,★ but the four peaks show that the unpaired electron is interacting with three equivalent protons. The distance between the peaks, a_H (electron–proton coupling constant), is a characteristic of the chemical environment of the carbon bearing the unpaired electron. Thus, a_H is the equivalent in esr of J in nmr. A real advantage of esr spectroscopy is its sensitivity, which is far greater than that of nmr. Although the concentration of radicals in a reaction is usually miniscule, they can often be detected by esr spectroscopy.

Summary

1 Magnetically active nuclei such as ^1H and ^{13}C assume different energy spin states when placed in a magnetic field. By absorbing energy of the correct frequency, some of these nuclei go from a lower to a higher energy state (secs. 9:1 and 9:3). The number of absorption peaks in a nuclear magnetic resonance (nmr) spectrum is related to the number of different kinds of hydrogens (^1H nmr) and carbons (^{13}C nmr) in the molecule (secs. 9:2 and 9:9).

2 The induced circulation of molecular electrons around a nucleus generates a local magnetic field at that nucleus opposed to the applied field. The induced field shields the nucleus in question, so that an increase in applied field strength is needed to cause it to resonate. The resonance positions of protons in different chemical environments are different, because their shielding is different. There is often a correlation between the chemical shifts of various protons and the electronegativity of nearby atoms (Table 9–1), since inductive effects alter shielding. Factors such as hybridization also have an effect on the chemical shift (secs. 9:3 and 9:4).

3 Protons on neighboring nuclei are magnetically coupled to one another. The different spin orientations of one set of equivalent nuclei result in slightly different chemical shifts for the neighboring nuclei. In general, n equivalent nuclei will split the resonance of neighboring protons into $(n + 1)$ components. If a proton has two sets of neighbors, n of one kind and m of another, its absorption peak should contain $(n + 1)(m + 1)$ components. First-order spectra are those which satisfy this rule. There may be fewer than the predicted number of peaks when one of the neighboring groups is OH, or when chemically different protons happen to display identical coupling constants (secs. 9:5 and 9:8).

If chemical shift differences are about as small as spin–spin splittings, the spectrum becomes complicated. It can no longer be analyzed by use of the $(n + 1)$ rule. Such spectra can be simplified by electronically decoupling protons on adjacent atoms (secs. 9:7 and 9:8).

4 The use of ^{13}C nmr allows the analyst to determine the number and kinds of carbon atoms present in an unknown molecule. Although there are some

★ This representation (Fig. 9–27) is a derivative spectrum.

exceptions, many carbon absorptions in ^{13}C nmr appear in the same relative positions as would the corresponding hydrogen absorptions in ^{1}H nmr.

Protons on a given carbon are coupled to the ^{13}C nucleus. One can observe coupled ^{13}C nmr spectra or, by decoupling, produce ^{13}C spectra having a single peak for each different carbon atom (sec. 9:9).

5 Both ^{1}H and ^{13}C nmr spectroscopy can be used in the structural analysis of alkanes, alkenes, alcohols, and ethers. These spectra can also provide information about conformational equilibria and thermodynamic parameters (sec. 9:11).

6 Unpaired electrons behave like protons when placed in a magnetic field, sorting themselves into two groups with different energies. They undergo transitions between the two energy levels, and the transition energies are affected by nearby magnetic nuclei. This is the basis of electron spin resonance (esr) spectroscopy, which provides structural information about radicals (sec. 9:12).

Problems

9–10 In Problem 8–21 (p. 296), infrared spectra were predicted for all the isomers of compounds (a–d). Now draw the expected ^{1}H and decoupled ^{13}C nmr spectra for the same compounds. Show approximate chemical shifts and ^{1}H nmr multiplicities.

a. CH_3ClO b. C_2H_6O c. $C_2H_6O_2$ d. C_4H_{10}

9–11 Propose structures for the compounds whose molecular formulas and ^{1}H nmr spectra are given here.

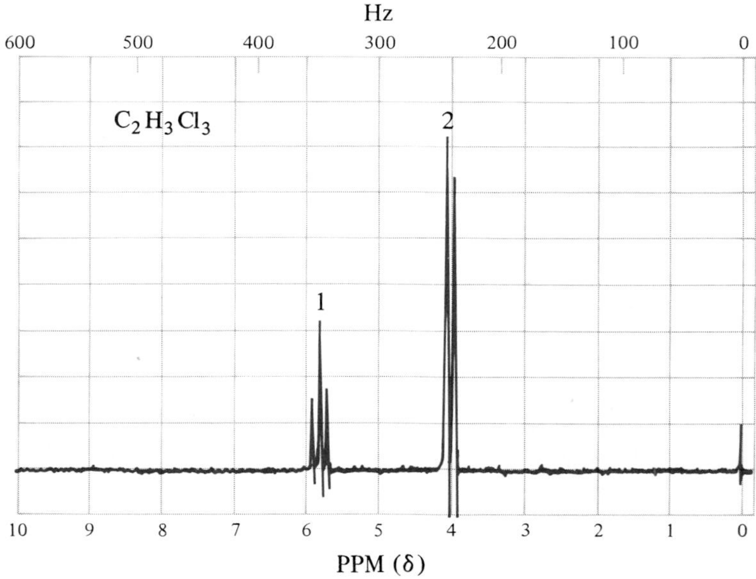

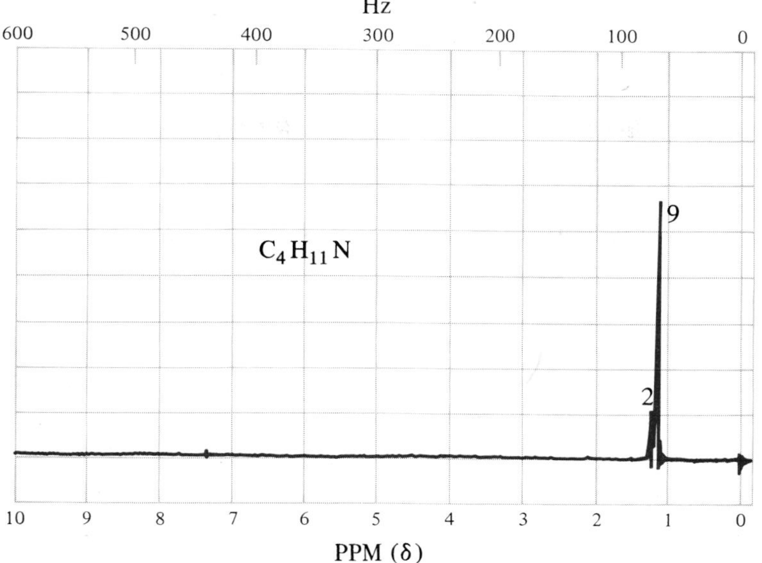

Hz
600
500
400
300
200
100
0
$C_4H_{11}N$
9
2
10
9
8
7
6
5
4
3
2
1
0
PPM (δ)

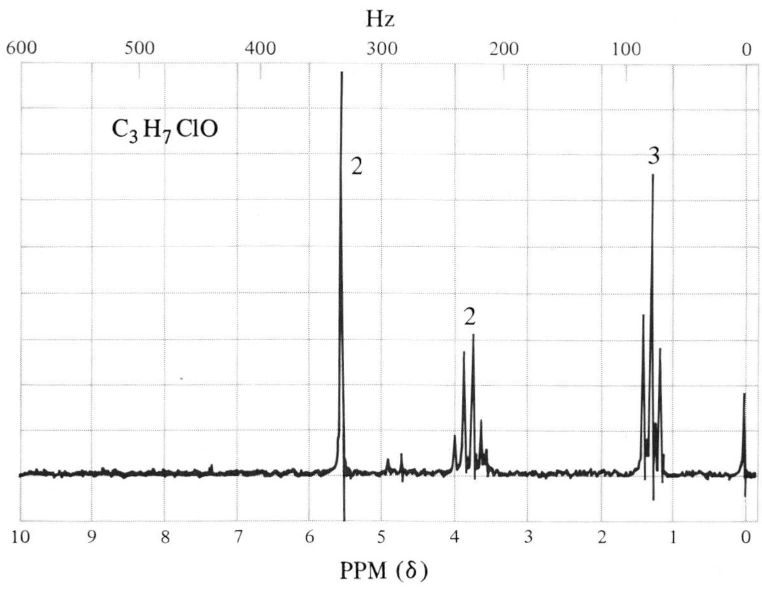

Hz
600
500
400
300
200
100
0
C_3H_7ClO
2
2
3
10
9
8
7
6
5
4
3
2
1
0
PPM (δ)

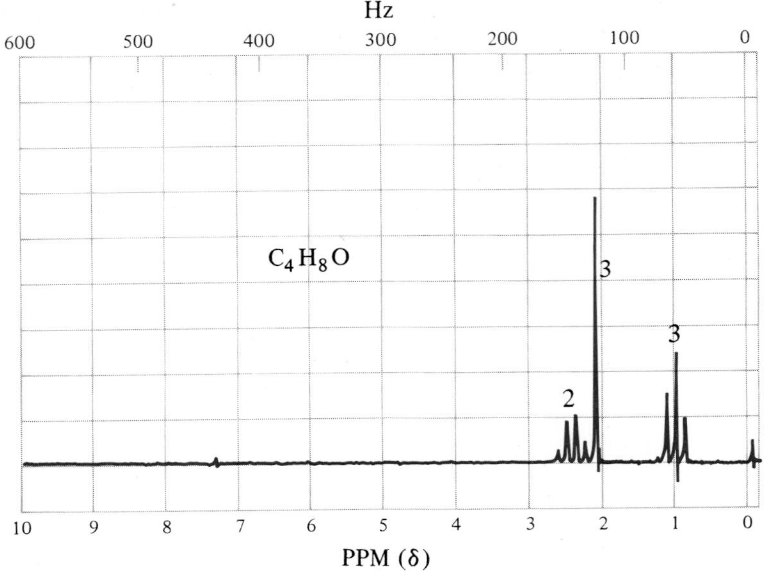

9–12 Shown here are the pairs of ir and ¹H nmr spectra for three different compounds. For each of the compounds, deduce structures that fit the molecular formulas and that are in harmony with the spectral data.

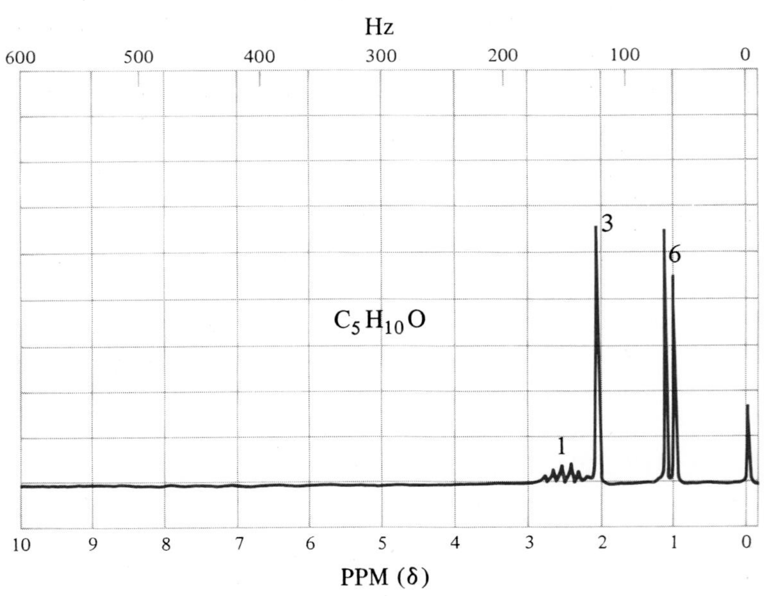

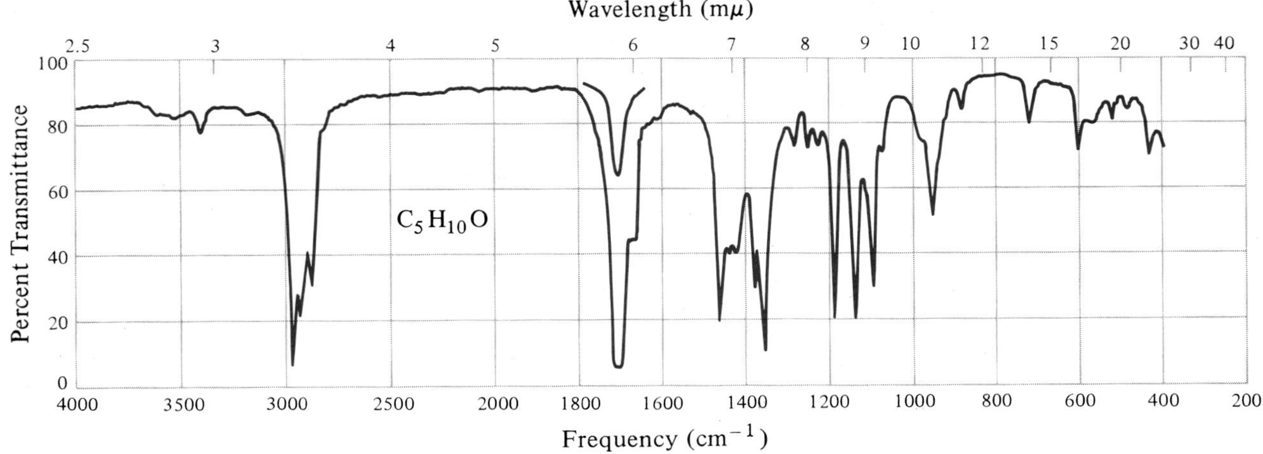

Wavelength (mμ)
Percent Transmittance
$C_5H_{10}O$
Frequency (cm^{-1})

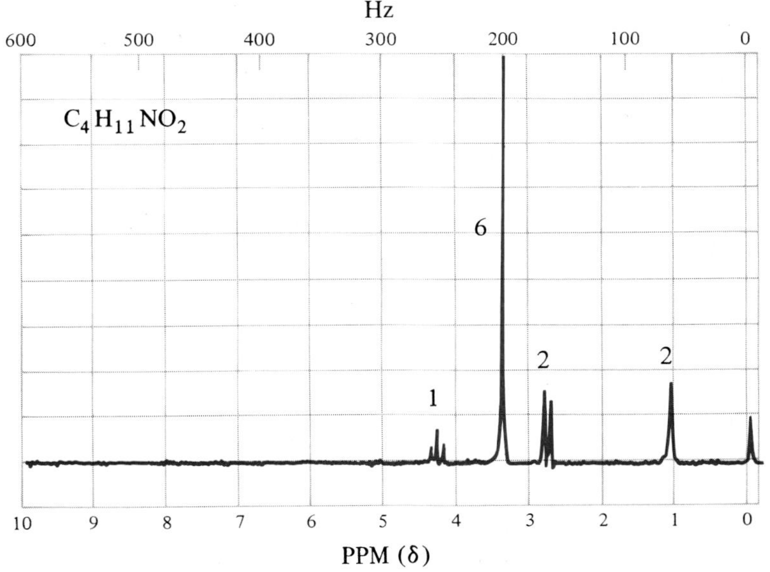

Hz
$C_4H_{11}NO_2$
6
1
2
2
PPM (δ)

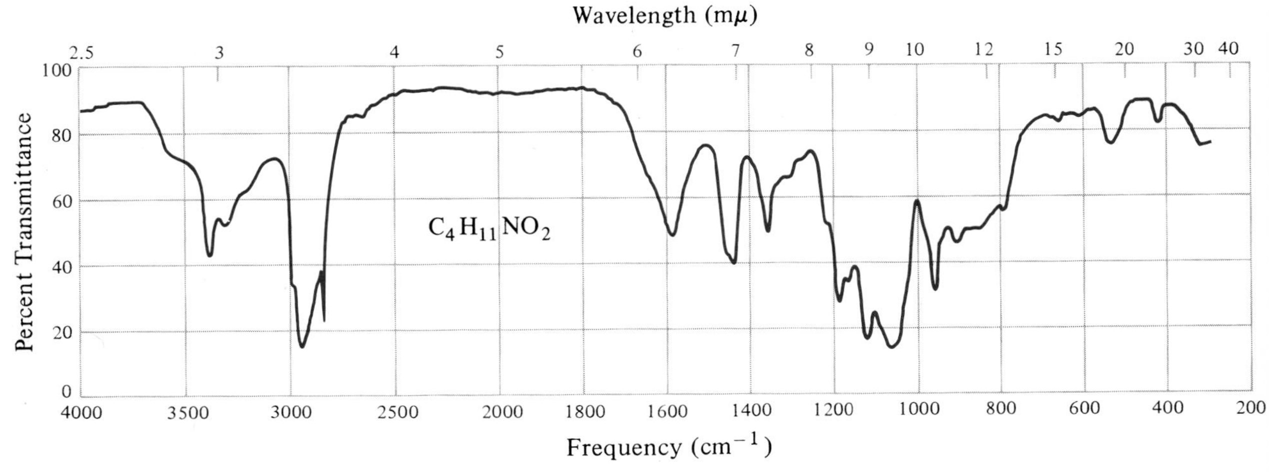

Wavelength (mμ)
Percent Transmittance
$C_4H_{11}NO_2$
Frequency (cm^{-1})

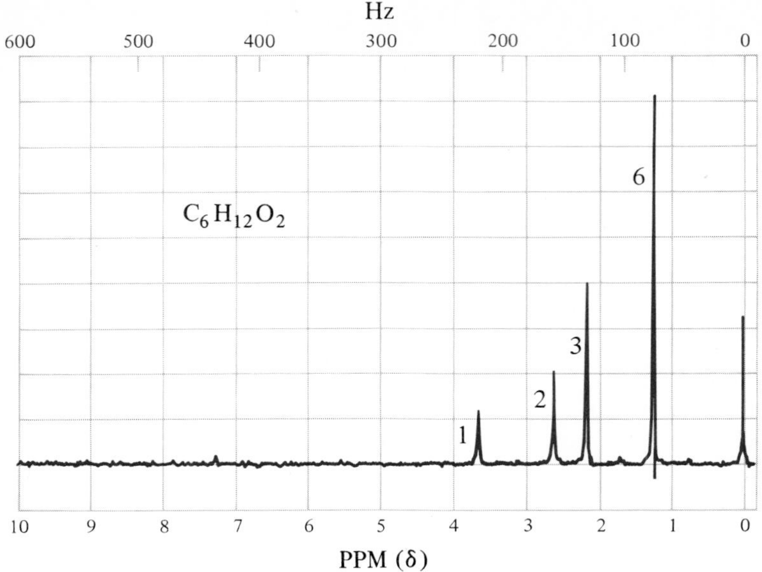

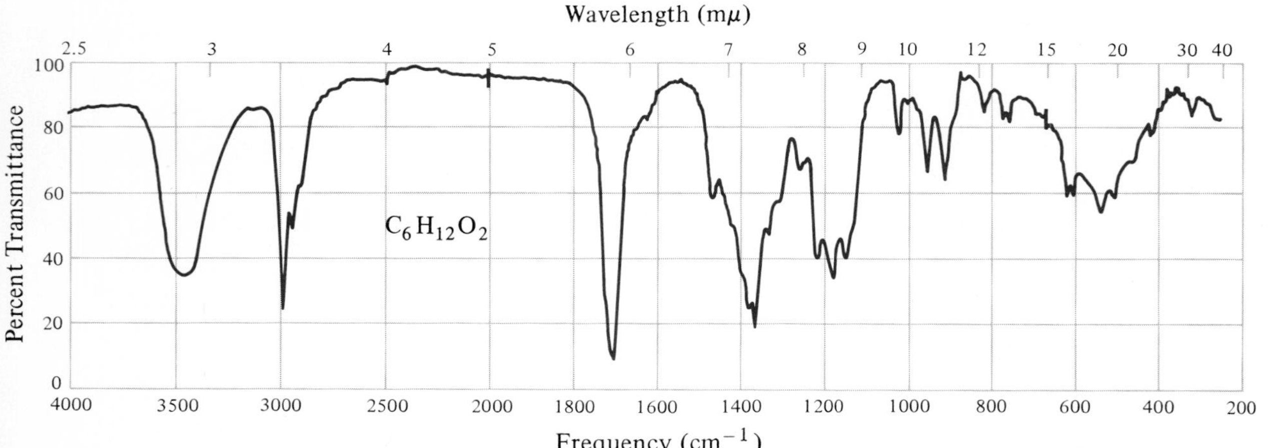

9–13 In the ^1H nmr spectrum of chlorocyclohexane, $C_6H_{11}Cl$, the hydrogen atom at C-1 gives rise to an easily distinguishable peak. When a sample of this compound is frozen and dissolved in a solvent at $-120°$, its spectrum shows only one absorption for the C-1 proton. As the solution stands, a second absorption appears at the expense of the first, due also to the C-1 proton. The two peaks are separated by 0.7 ppm. After the system comes to equilibrium the two peaks have an intensity ratio of about 7:1. Explain all these observations.

9–14 Shown here are the ^{1}H and decoupled ^{13}C nmr spectra of an unknown compound containing carbon, hydrogen, and oxygen. Suggest a structure for it.

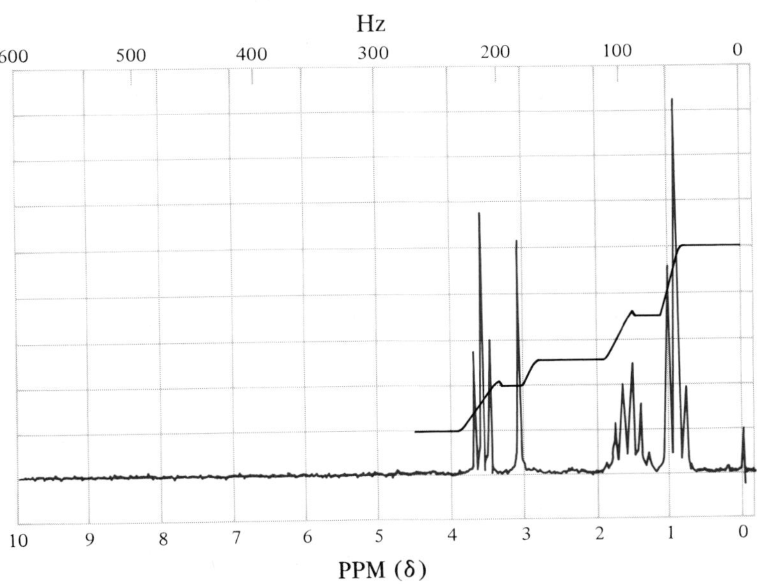

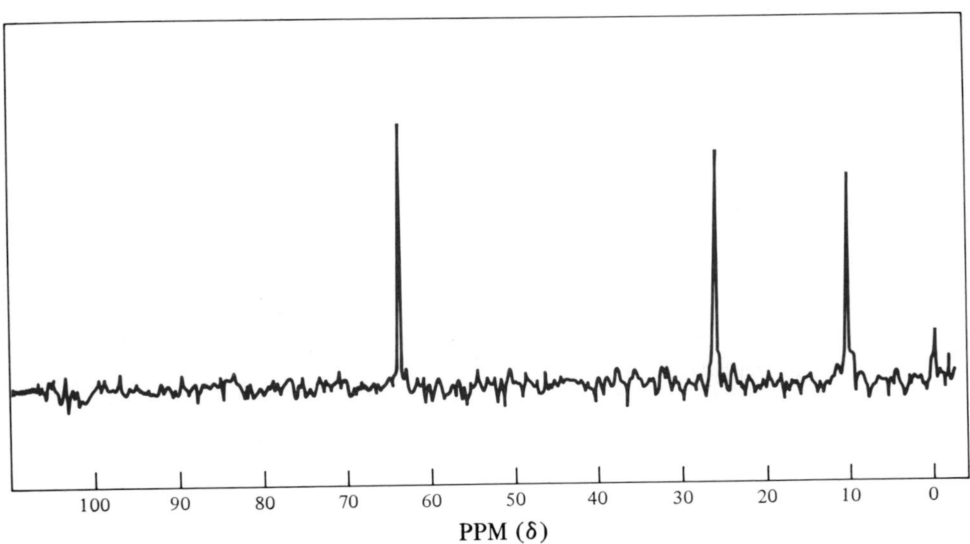

9–15 A very pure unknown compound ($C_3H_5Cl_3O$) has the ^1H nmr spectrum given here. When the spectrum is retaken after D_2O is added to the sample tube, the doublet at 3.08 ppm disappears and the quintet at 4.25 ppm becomes a quartet. Suggest a structure for the unknown, and explain the observations.

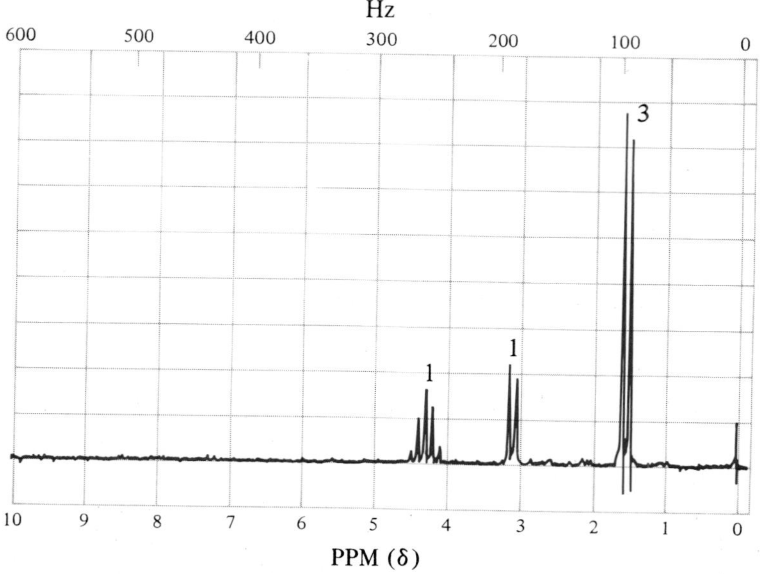

9–16 An unknown compound, C_3H_6O, has the ^1H nmr spectrum given in spectrum (a). When the unknown is heated with 48% HBr and then distilled, the resultant compound, $C_3H_6Br_2$, gives the ^1H nmr spectrum shown in (b). What are the structures of the two compounds? Write a mechanism for the reaction.

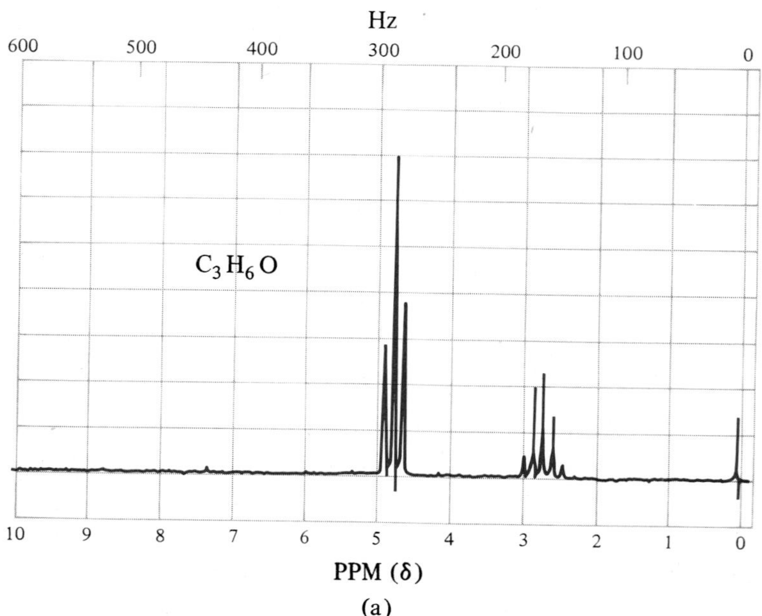

(a)

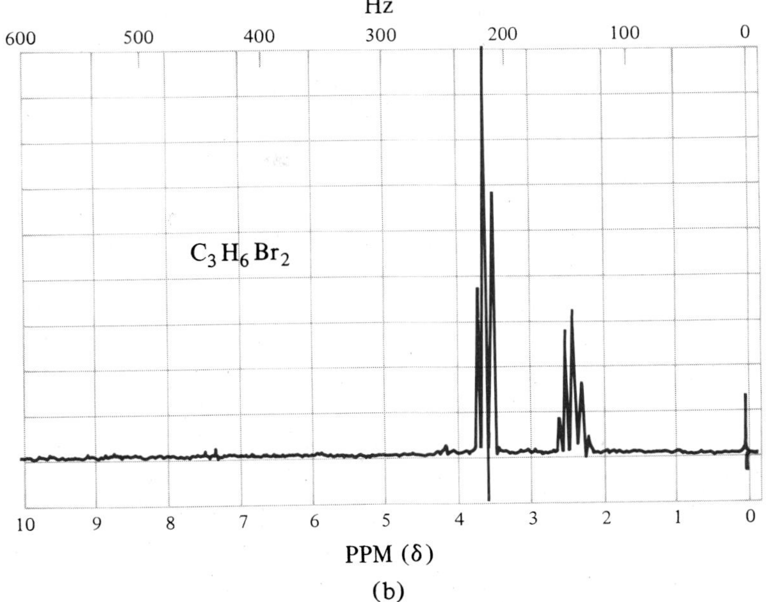

(b)

9–17 An unknown compound has prominent mass spectral peaks at $m/z = 43$, 73, and 102. Its ^{1}H nmr spectrum contains the following peaks: 0.95 (triplet), 1.60 (sextet), and 3.40 (triplet) ppm, relative areas 3:2:2. Suggest a structure for the unknown and explain the spectral data.

9–18 Shown here is the decoupled ^{13}C nmr spectrum of a compound containing alcohol, ether, alkene, and alkane carbons. Beside each peak is given the multiplicity that would be seen in a coupled spectrum. Assign each resonance to the appropriate carbon.

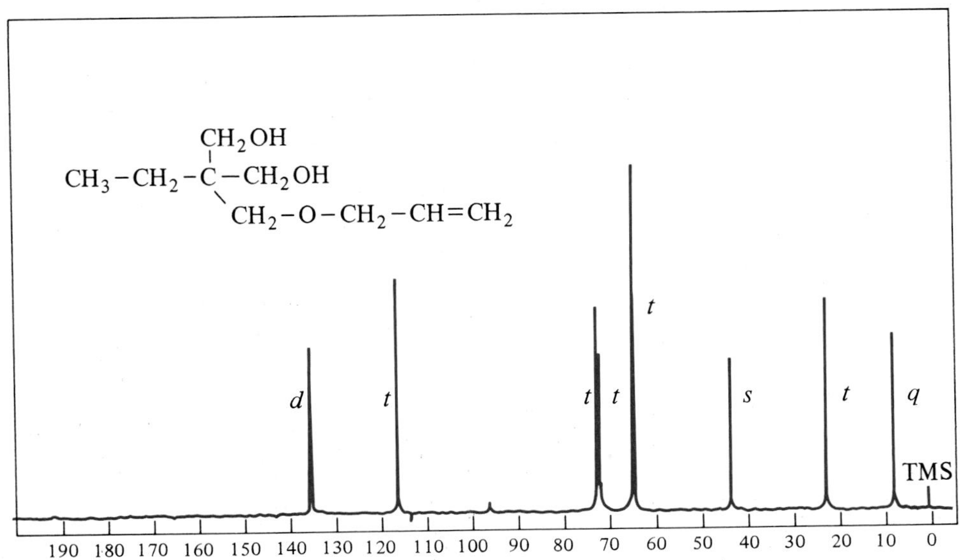

9–19 Identify the two compounds, C_4H_8O and $C_5H_{10}O$, whose ir and ^{1}H nmr spectra are given here.

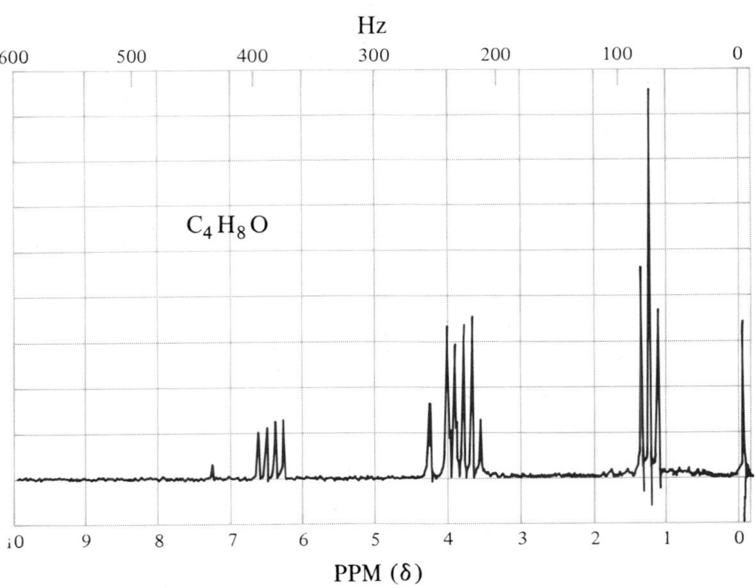

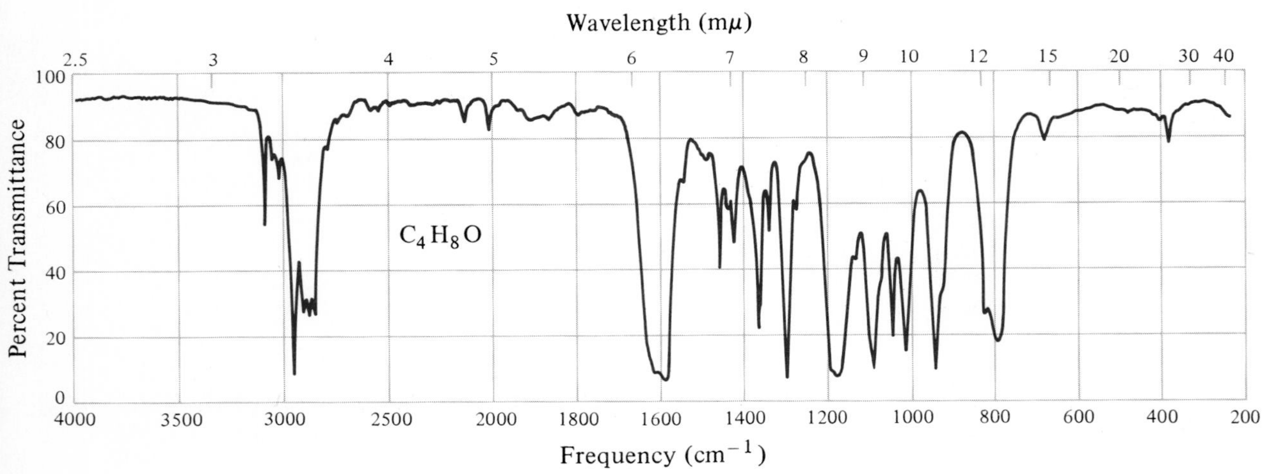

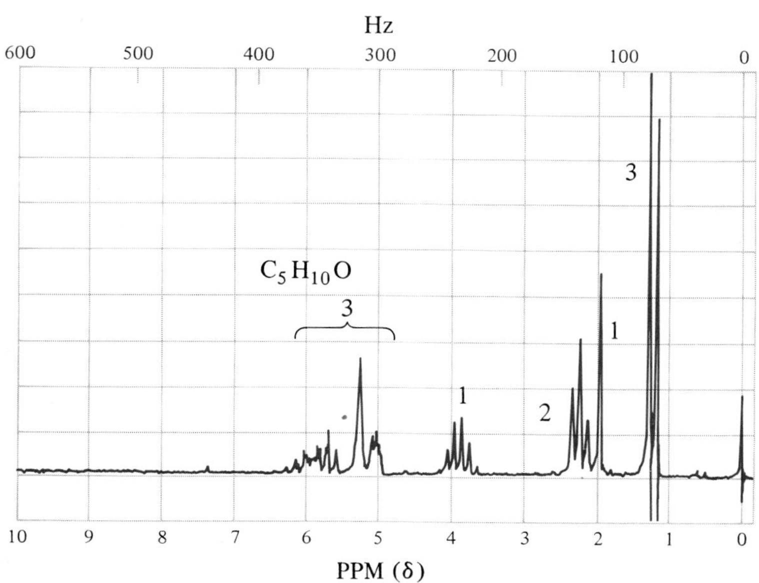
Hz
600
500
400
300
200
100
0
$C_5H_{10}O$
3
1
2
1
3
10
9
8
7
6
5
4
3
2
1
0
PPM (δ)

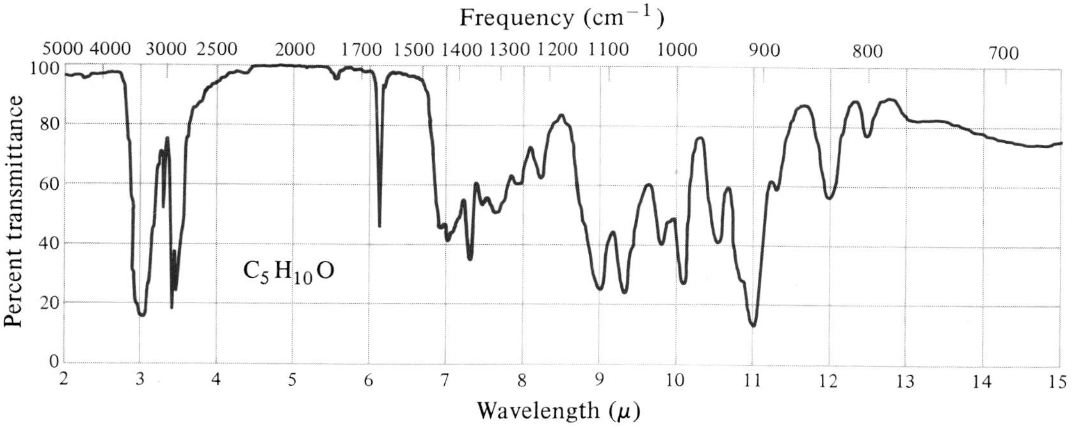
Frequency (cm^{-1})
5000 4000 3000 2500 2000 1700 1500 1400 1300 1200 1100 1000 900 800 700
Percent transmittance
100
80
60
40
20
0
$C_5H_{10}O$
2
3
4
5
6
7
8
9
10
11
12
13
14
15
Wavelength (μ)

9–20 A compound C_5H_8 has the decoupled ^{13}C nmr spectrum shown here. Suggest a structure and assign the ^{13}C absorptions

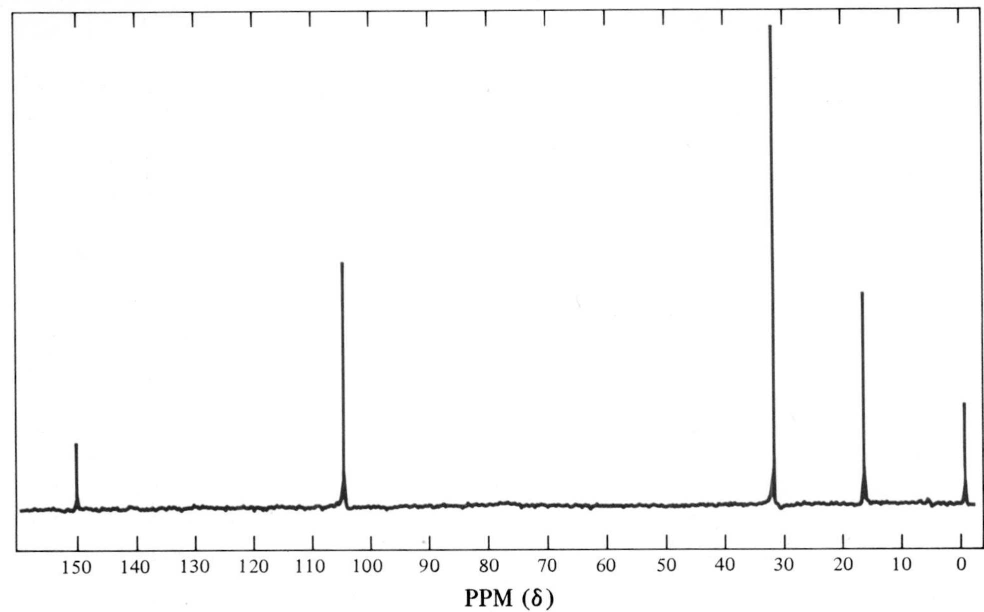

9–21 A hydrocarbon (**A**) is isolated from a plant. It is known to have the molecular formula C_6H_{10}. **A** is optically active, and its ^1H nmr spectrum contains a 2H multiplet at 5.7 ppm (and no other peaks in the double-bond region). In the presence of a platinum catalyst, **A** absorbs one equivalent of hydrogen and is converted to **B** (C_6H_{12}). **B** is not optically active. Its ^1H nmr spectrum contains only these peaks: 0.9 ppm (doublet, 3H), 1.0 to 2.2 ppm (multiplet, 9H). Propose structures for **A** and **B**, and explain your reasoning.

Supplementary Reading

A. J. Gordon and R. A. Ford, *The Chemists' Companion,* John Wiley & Sons, New York, N.Y., 1972. This is a handbook of practical data, including extensive lists of ir, uv, and nmr data, as well as other information of value in organic chemistry, such as bond strengths, bond angles and lengths, acidities, and experimental techniques.

J. R. Dyer, *Applications of Absorption Spectroscopy of Organic Compounds,* Prentice-Hall, Inc., Englewood Cliffs, N.J., 1965.

J. R. Dyer, *Organic Spectral Problems,* Prentice-Hall, Inc., Englewood Cliffs, N.J., 1972.

A. Ault and G. O. Dudek, *NMR: An Introduction to Proton Nuclear Magnetic Resonance Spectroscopy,* Holden–Day, San Francisco, Calif., 1976.

E. Breitmayer and W. Voelter, *^{13}C NMR Spectroscopy,* Verlag Chemie, Weinheim/Bergstr., 2nd ed., 1978.

G. D. Levy, R. L. Lichter, and G. L. Nelson, *Carbon-13 Nuclear Magnetic Resonance Spectroscopy,* 2nd ed., Wiley-Interscience, New York, N.Y., 1980.

P. L. Fuchs and C. A. Bunnell, *Carbon-13 Nmr Based Organic Spectral Problems,* John Wiley & Sons, Inc., New York, N.Y., 1979.

H. C. Dorn, D. G. I. Kingston, and B. R. Simpers, Interpretation of a ^{13}C magnetic resonance spectrum, *J. Chem. Ed.,* **53,** 584 (1976).

Addition, Polymerization, and Biological Reactions of Alkenes 10

The σ bond and π bond which make up a carbon–carbon double bond differ in several important ways. One significant difference, noted in Chapter 7, is that the π electrons are, on the average, farther from the carbon nuclei than the σ electrons are. The result is that the π electrons are easily attacked by electron seeking reagents called **electrophiles**. This chapter introduces some of the most important reactions of alkenes with electrophilic reagents, all of which depend on the high reactivity of the carbon–carbon π bond. Examples are drawn from the laboratory, from the chemical industry, and from the living cell.

10:1 Introduction to Energy Changes and Mechanisms in π Bond Additions

Using the average bond energies given in Table 10–1, we can estimate a carbon–carbon π bond energy of 62 kcal/mol by subtracting the C—C σ bond energy (84 kcal/mol) from the total C=C ($\sigma + \pi$) bond energy (146 kcal/mol). This number can then be used in thermodynamic calculations such as those shown in equations 10.1 and 10.2. (The other bond energies

(10.1) $$\text{>C=C<} + \text{H—Br} \rightarrow \text{—}\overset{|}{\underset{\text{Br}}{\text{C}}}\text{—}\overset{\text{H}}{\underset{|}{\text{C}}}\text{—}$$

62 kcal (C=C π), 88 kcal (H—Br); $(62 + 88 = 150)$ — 68 kcal (C—Br), 98 kcal (C—H); $(68 + 98 = 166)$

$\Delta H = -16$ kcal/mol
$(150 - 166 = -16)$

(10.2) $$\text{>C=C<} + \text{H—H} \rightarrow \text{—}\overset{|}{\underset{\text{H}}{\text{C}}}\text{—}\overset{|}{\underset{\text{H}}{\text{C}}}\text{—}$$

62 kcal (C=C π), 104 kcal (H—H); $(62 + 104 = 166)$ — 98 kcal (C—H), 98 kcal (C—H); $(98 + 98 = 196)$

$\Delta H = -16$ kcal/mol
$(166 - 196 = -30)$

can be found in Table 5–1, p. 146). These equations are examples of *addition reactions,* in which all the atoms of the reactants are incorporated into a single molecule of product. The negative ΔH values indicate that the addition of molecules such as HBr and H_2 to the π bond is energetically favorable.

Table 10–1. Average Bond Energies of Some Single and Double Bonds (kcal/mol)

Single Bonds		Double Bonds	
C—C	84	C=C	146
N—N	60	N=N	100
C—N	79	C=N	142
C—O	86	C=O	176

One of the most characteristic types of reaction for compounds with multiple bonds is the addition reaction. The number of reagents which will add to double bonds is quite large, so addition reactions may be somewhat bewildering at first. Fortunately, the number of *mechanisms* which these reactions follow is quite small. We will therefore emphasize the mechanisms, since this approach provides a way of grouping the many reagents, thus simplifying the learning process.

One major mechanism is an ionic one in which the reagent AB transfers an electrophilic fragment, A^+, to the alkene. A carbocation intermediate is formed which captures the nucleophilic portion, B^-, of the reagent in a second step (eq. 10.3).

$$\text{(10.3)}\quad >C=C< \; + \; A^+B^- \rightarrow \underset{\text{Carbocation}}{>\underset{A}{\underset{|}{C}}-\overset{+}{C}<} \xrightarrow{B^-} >\underset{A}{\underset{|}{C}}-\underset{B}{\underset{|}{C}}< \qquad \text{Electrophilic addition}$$

There is also a two-step radical mechanism similar to the ionic process shown in Equation 10.3. This mechanism involves an initial attack by the radical R· on the alkene, generating a new radical intermediate. The intermediate then goes to product by abstracting a hydrogen atom from the addition reagent RH (eq. 10.4).

$$\text{(10.4)}\quad R\cdot + >C=C< \rightarrow >\underset{R}{\underset{|}{C}}-\dot{C}< \xrightarrow{RH} >\underset{R}{\underset{|}{C}}-\underset{H}{\underset{|}{C}}< + R\cdot \qquad \text{Radical addition}$$

Finally, there is a concerted mechanism in which both new σ bonds, C—A and C—B, are formed at the same time (eq. 10.5).

$$\text{(10.5)}\quad \begin{matrix} C=C \\ A-B \end{matrix} \rightarrow \begin{matrix} C-C \\ | \quad | \\ A \quad B \end{matrix} \qquad \text{Concerted addition}$$

Reactions that are initiated by attack of X^- or R^- on an isolated alkene double bond are rare.

The reaction of unsaturated compounds with H_2 was discovered well over 100 years ago and has become an important process for modification of alkenes. You've already seen that hydrogenation gives valuable thermochemical information about isomeric alkenes. Here we introduce the much wider use of hydrogenation as a synthetic tool.

Addition of hydrogen to a multiple bond is known as **reduction**. Although the addition is strongly exothermic (eq. 10.2), mixing hydrogen and an alkene at any reasonable temperature produces no change. The barrier to the reaction is the 104 kcal/mol needed to dissociate molecular hydrogen; hence, reductions with H_2 require a catalyst.

Numerous metals, such as platinum, can catalyze the reaction by adsorbing hydrogen molecules on their surfaces and causing them to dissociate into hydrogen atoms (thus providing adsorbed★ hydrogen atoms on the metal surface; Fig. 10–1). These metals adsorb organic molecules as well, particularly those with polarizable electrons such as alkenes. The metal surface thus acts as a molecular matchmaker, bringing together the reacting species. Most hydrogenation catalysts are insoluble in organic solvents, so the reaction is heterogeneous and takes place on the catalyst surface.

Platinum, palladium, and nickel are the metals most often used in hydrogenation. The catalyst surface is not uniformly active; instead there are regions called **active sites** that seem to be responsible for most of the catalytic activity. The means by which the catalyst is prepared affects the number and nature of these sites. The metals may be used by themselves (as pure platinum and nickel) or dispersed on the surface of a "support." Supporting materials have no catalytic properties of their own, but they modify properties of the dispersed metal. For example, the support may disperse the metal into fine particles and hence increase the surface area for adsorption. Certain supports can make the metal *less* active. Common supports include activated charcoal, various metal carbonates and sulfates, and alumina (Al_2O_3). Of the many catalysts, Pt, Pd/C (palladium on carbon), and Raney Ni (a catalyst formed from a Ni-Al alloy and NaOH) are the most common. Platinum is also often added in the form of PtO_2, platinum oxide, which is reduced to Pt in the presence of H_2.

★ *Adsorption* is a process in which gaseous or liquid molecules become bound to a solid surface.

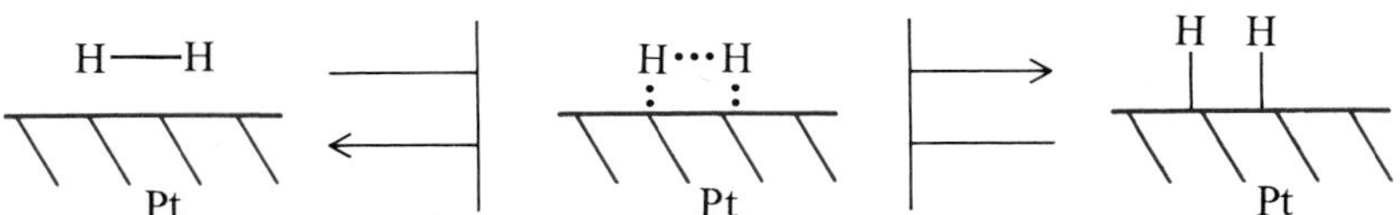

Figure 10–1. *Hydrogen molecule being adsorbed and dissociated by platinum (Pt) catalyst.*

Reductions usually give high yields of products (eqs. 10.6 to 10.8).

$$(10.6)\quad (CH_3)_2C{=}CHCH_3 + H_2\ (3\ atm) \xrightarrow[\text{ethanol, 25° (solvent)}]{Pt} \underset{(100\%)}{(CH_3)_2CHCH_2CH_3}$$

$$(10.7)\quad \text{Dihydropyran} + H_2\ (4\ atm) \xrightarrow[\text{25°, neat (no solvent)}]{\text{Raney Ni}} \text{Tetrahydropyran}\ (100\%)$$

$$(10.8)\quad \text{1,2-dimethylcyclohexene} + H_2 \xrightarrow[\text{acetic acid (solvent)}]{Pt} \text{1,2-dimethylcyclohexane}$$

(*cis*, 82 to 95%)
(*trans*, 5 to 18%)

Equation 10.8 shows a highly **stereoselective** reduction. A reaction is stereoselective if a single starting material can give rise to two or more stereoisomeric products and if one of these products is formed in much higher yield than the others (compare *stereospecific*, sec. 6:3). Stereoselective formation of *cis*-1,2-dimethylcyclohexane in preference to the *trans* isomer (eq. 10.8) is due to the transfer of two hydrogen atoms from the catalyst to the same face of the double bond. Thus, *syn* addition is favored by adsorption of the alkene to the catalyst surface.

Many vegetable oils are made up of molecules known as *fats* or *glycerides* (Chap. 17), and the alkyl side chains of these molecules may contain the C=C functional group. These *unsaturated fats,* which are oils, can be treated with H_2 and a catalyst in order to hydrogenate the alkene side chains. The products are *saturated fats* (eq. 10.9), which are waxy solids instead of oils.

$$(10.9)\quad \begin{array}{l} CH_2OC(=O)(CH_2)_7CH{=}CH(CH_2)_7CH_3 \\ | \\ CHOC(=O)(CH_2)_7CH{=}CH(CH_2)_7CH_3 \\ | \\ CH_2OC(=O)(CH_2)_7CH{=}CH(CH_2)_7CH_3 \end{array} \xrightarrow[\text{Ni catalyst}]{3H_2} \begin{array}{l} CH_2OC(=O)(CH_2)_{16}CH_3 \\ | \\ CHOC(=O)(CH_2)_{16}CH_3 \\ | \\ CH_2OC(=O)(CH_2)_{16}CH_3 \end{array}$$

Triolein (an unsaturated fat) — Tristearin (a saturated fat)

There is evidence that a balance between intake of saturated and unsaturated fats may be important in human nutrition, in the prevention of arteriosclerosis, and possibly carcinogenesis. This is an area of continuing research and controversy.

10–1 Write the structures of all the alkenes that can be reduced to each of the following alkanes, using H_2 or D_2:
a. methylcyclohexane
b. 3-methylhexane
c. butane-1,2-d_2($CH_3CH_2CHDCH_2D$)

10:3 Additions to Alkenes Under Acidic Conditions: The Markovnikov Rule

Since the π electrons of alkenes are easily polarized and are not sterically shielded, they are readily attacked by electrophiles. The simplest electrophile is the proton, which can be transferred from a hydronium ion (H_3O^+) obtained by dissolving an acid in water. Protonation of the alkene produces a carbocation intermediate, **1**, which is then captured by the nucleophile water (eq. 10.10 and Fig. 10–2). The net reaction is the addition of water to the alkene, with the acid acting as catalyst. This process is the reverse of acid-

$$\text{(10.10)}\quad CH_3CH{=}CH_2 + H{-}\overset{+}{\ddot{O}}H_2 \xrightarrow{\text{slow}} \underset{\substack{\mathbf{1}\\ \text{Carbocation}}}{CH_3\overset{+}{C}HCH_3} + H{-}\ddot{O}{-}H \xrightarrow{\text{fast}}$$

$$CH_3CH(\overset{+}{\ddot{O}}H_2)CH_3 \xrightarrow{\text{fast}} CH_3CH(\ddot{O}H)CH_3 + H^+$$

catalyzed dehydration of alcohols (sec. 7:5). The equilibrium is made to favor addition of water to the alkene (*hydration*) by the use of dilute acids near room temperature.

The acids usually chosen to catalyze a hydration are H_2SO_4, HNO_3, and $HClO_4$. These acids are favored because they contain weak nucleophilic anions (SO_4H^-, NO_3^-, ClO_4^-). This ensures that the carbocation will react

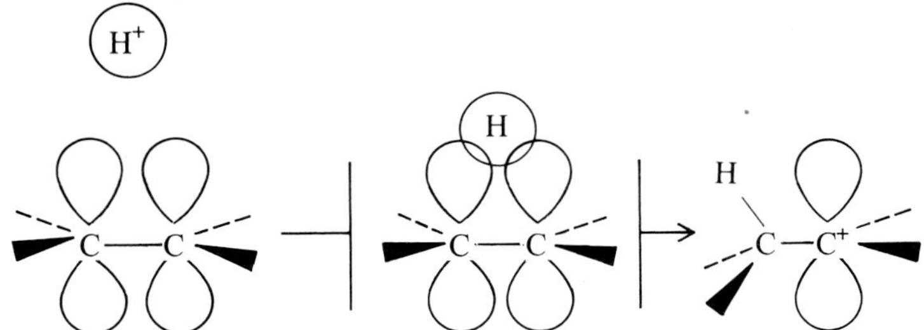

Figure 10–2. *Addition of a proton to the carbon–carbon π bond to produce a carbocation.*

preferentially with water (eq. 10.11) rather than with the anion.

$$(10.11)\qquad (CH_3)_2C{=}CH_2 + H_2O \underset{}{\overset{\text{dil } HClO_4}{\rightleftharpoons}} CH_3{-}\underset{OH}{\overset{CH_3}{C}}{-}CH_3 \qquad K^{50°} = 5.5$$

The practical limitation on this method is the poor solubility of most alkenes in water. The problem is often overcome by dissolving the alkene in strong, cold sulfuric acid, which converts it to an **alkyl hydrogen sulfate**. Hydrolysis of the sulfate then yields an alcohol and sulfuric acid (eq. 10.12). (The mechanism of hydrolysis is treated in Chapter 17.)

$$(10.12)\qquad (CH_3)_2C{=}C(CH_3)H \xrightarrow{80\%\ H_2SO_4} CH_3{-}\underset{OSO_3H}{\overset{CH_3}{C}}{-}CH_2CH_3 \xrightarrow{H_2O} CH_3\underset{OH}{\overset{CH_3}{C}}CH_2CH_3 + H_2SO_4$$

1,1-Dimethylpropyl hydrogen sulfate — 2-Methyl-2-butanol

Formation of the alkyl hydrogen sulfate, like hydration, is a two-step process proceeding by way of a carbocation (eq. 10.13).

$$(10.13)\qquad (CH_3)_2C{=}C(CH_3)H + HOSO_3H \rightarrow (CH_3)_2\overset{+}{C}{-}CH_2CH_3 + {}^{-}OSO_3H \rightarrow CH_3{-}\underset{OSO_3H}{\overset{CH_3}{C}}{-}CH_2CH_3$$

Equations 10.10 to 10.12 all involve addition of H—OH to unsymmetrical alkenes. In each case the reaction could, in principle, give two constitutional (structural) isomers as products. For example, hydration of propene (eq. 10.10) could yield either 1-propanol or 2-propanol; hydration of methylpropene (eq. 10.11) could give either isobutyl alcohol or *t*-butyl alcohol. What we observe in practice is that each alkene gives rise to only *one* product. Such reactions, which give only one of two possible constitutionally-isomeric products, are labelled *regiospecific*.

The two-step mechanism provides the clue to the regiospecificity of these additions. The rate-determining step is the formation of the carbocation. Remember that when two different carbocations can be formed from the same substrate, *the more stable carbocation is formed more rapidly* (sec. 6:7).

Starting from propene, 2-propyl cation is formed faster than 1-propyl cation, so we isolate only 2-propanol (top of eq. 10.14).

$$\text{(10.14)}\quad CH_3CH{=}CH_2 \xrightarrow{H^+} \begin{cases} \xrightarrow{\text{faster}} CH_3\overset{+}{C}H{-}CH_3 \text{ (2-Propyl cation)} \xrightarrow{H_2O} CH_3CH(OH)CH_3 \text{ (Only product isolated)} \\ \xrightarrow{\text{slower}} CH_3CH_2{-}CH_2^+ \text{ (1-Propyl cation)} \xrightarrow{H_2O} CH_3CH_2CH_2OH \end{cases}$$

The nineteenth century Russian chemist V. Markovnikov was the first to formulate a general rule describing the addition of H—Y reagents to unsymmetrical alkenes. In recognition of his work, reactions that give the more highly substituted product—as in the faster reaction of Equation 10.14—are classified as *Markovnikov* additions. Sections 10:5 and 10:6 introduce some contrasting *anti-Markovnikov additions*. We want to stress that the terms "Markovnikov" and "anti-Markovnikov" addition refer only to the product distribution and not to the mechanism of a reaction. It is often, though not always, true that Markovnikov and anti-Markovnikov additions follow different mechanisms.

Example

What product is formed when 1-hexene is treated with dilute aqueous sulfuric acid? What is the result for the same reaction with 2-hexene?

Answer

It is easier to remember products of reactions if the mechanism is written out until it is understood. 1-Hexene will first be protonated and two intermediate cations could be formed:

$$CH_2{=}CHCH_2CH_2CH_2CH_3 + H_2SO_4 \rightarrow CH_3\underset{+}{C}HCH_2CH_2CH_2CH_3 + {}^-HSO_4$$

2° Carbocation

$$(\text{or } \underset{+}{C}H_2CH_2CH_2CH_2CH_2CH_3)$$

1° Carbocation

The secondary carbocation is more stable than the primary and so is preferentially formed. It then reacts with the nucleophilic water to produce the major product, 2-hexanol.

$$CH_3\underset{+}{C}HCH_2CH_2CH_2CH_3 + H_2O{:} \rightarrow$$

$$CH_3\underset{\underset{+}{HOH}}{CH}CH_2CH_2CH_2CH_3 \rightarrow CH_3\underset{OH}{CH}CH_2CH_2CH_2CH_3$$

2-Hexanol

When 2-hexene is the reactant, the following two cations are formed:

$$CH_3CH{=}CHCH_2CH_2CH_3 + H_2SO_4 \rightarrow CH_3CH_2\overset{}{\underset{+}{C}}HCH_2CH_2CH_3$$

2° Carbocation

$$+\ CH_3\underset{+}{C}HCH_2CH_2CH_2CH_3$$

2° Carbocation

These two cations have different structures and hence would be of different energies, but since both are secondary cations the energy difference would be very small. One could not, in fact, predict which would be the more stable. The two alcohols, resulting from attack of water on the two cations, are formed in nearly equal yield:

$$\underset{\text{3-Hexanol}}{CH_3CH_2\underset{\large OH}{\underset{|}{C}}HCH_2CH_2CH_3} + \underset{\text{2-Hexanol}}{CH_3\underset{\large OH}{\underset{|}{C}}HCH_2CH_2CH_2CH_3}$$

Problem

10–2 Give the structures of the products formed in each reaction. Indicate whether any other products could, theoretically, be produced.

a. cyclopentene $\xrightarrow[HClO_4]{H_2O}$

b. 2-butene $\xrightarrow{\text{conc } H_2SO_4}$ **A** $\xrightarrow{H_2O}$

c. 2-ethyl-1-pentene $\xrightarrow[H_2SO_4]{H_2O}$

d. 1-methylcyclobutene $\xrightarrow[HNO_3]{H_2O}$

e. 2-pentene $\xrightarrow[HClO_4]{H_2O}$

Strong acids of the type HX, where X^- is a good nucleophile, will add to alkenes to give RX derivatives. Addition of HCl, HBr, and HI to alkenes is usually carried out neat (no solvent), or in nonnucleophilic solvents such as pentane, dichloromethane, and ether (to avoid forming products incorporating part of the solvent). The dry gas (HCl or HBr) is either passed into the solution or into the neat alkene (eq. 10.15). Stereoselective addition is common and the elements of H—X add either *anti* or *syn* to one another.

$$\underset{\text{(neat)}}{CH_3{-}\overset{\large CH_3}{\overset{|}{C}}{=}CH_2} + HCl \xrightarrow{-80^\circ} \underset{(88\text{ to }94\%)}{CH_3{-}\overset{\large CH_3}{\overset{|}{\underset{\large Cl}{\underset{|}{C}}}}{-}CH_3} \quad (10.15)$$

Equation 10.16 shows an example of predominant *anti* addition.

$$\text{1,2-dimethylcyclopentene} + HCl \xrightarrow[0^\circ]{\text{pentane}} \underset{(92\%)}{\text{1-chloro-1,2-dimethylcyclopentane (anti)}} + \underset{(8\%)}{\text{1-chloro-1,2-dimethylcyclopentane (syn)}} \quad (10.16)$$

The presence of carbocation intermediates during these proton-initiated additions is further indicated by the isolation of rearranged products. The product composition from some reactions indicates that a less stable carbocation has rearranged to a more stable one (eq. 10.17).

(10.17)
$$(CH_3)_2\overset{CH_3}{\underset{|}{C}}{-}CH{=}CH_2 \xrightarrow{HCl}$$
3,3-Dimethyl-1-butene (neat)

$$(CH_3)_2\overset{CH_3}{C}{-}\overset{+}{C}H{-}CH_3 \xrightarrow{CH_3^- \text{ shift}} (CH_3)_2\overset{+}{C}{-}\overset{CH_3}{C}H{-}CH_3$$
$$\downarrow Cl^- \qquad\qquad \downarrow Cl^-$$
$$(CH_3)_2\overset{CH_3}{C}{-}\overset{Cl}{C}H{-}CH_3 \quad + \quad (CH_3)_2\overset{Cl}{C}{-}\overset{CH_3}{C}H{-}CH_3$$
3-Chloro-2,2-dimethylbutane (50%) + 2-Chloro-2,3-dimethylbutane (50%)

Reactions of alkenes with HF often lead to extensive product degradation. However, a recent improvement involves the use of pyridine as solvent, which minimizes rearrangement and polymerization and gives the desired product in high yield (eq. 10.18).

(10.18) cyclohexene + HF $\xrightarrow{\text{pyridine}}$ fluorocyclohexane (80%)

Problems

10–3 Give the products of reaction of cyclohexene with each of the following reagents (pay attention to the mechanism):
a. HCl in CH_2Cl_2 solution
b. H_2SO_4, NaBr, H_2O
c. HI
d. HCl in ethanol solution

10–4 For each of the following reactions, show the structure of the final products.
a. 3,3-dimethyl-1-pentene $\xrightarrow[HClO_4]{H_2O}$
b. 2-methyl-2-butene $\xrightarrow{HCl}$
c. 1-methylcyclohexene $\xrightarrow[\text{pyridine}]{HF}$
d. 1-pentene $\xrightarrow[H_2SO_4]{H_2O,\ NaCl}$
e. (skeletal alkene) $\xrightarrow{HBr}$

10–5 When hydrogen chloride is added to neat 3-methyl-1-butene, two products are isolated, each having the formula $C_5H_{11}Cl$.
a. Write the structures of these products, and show how they are formed.
b. Assume that the two products were separated by distillation. How would you experimentally determine the structure of each product?

Table 10–2. Some Important Proton-Initiated Alkene Additions

$$\mathrm{H{-}Y} + \mathrm{C{=}C} \rightarrow \mathrm{H{-}\overset{|}{\underset{|}{C}}{-}\overset{|}{\underset{|}{C}}{-}Y}$$

Reagent (HY)	Catalysts	Product
H_2O	$HClO_4$, HNO_3, H_2SO_4	$\mathrm{H{-}\overset{|}{\underset{|}{C}}{-}\overset{|}{\underset{|}{C}}{-}OH}$ alcohol
ROH	$HClO_4$, HNO_3, H_2SO_4	$\mathrm{H{-}\overset{|}{\underset{|}{C}}{-}\overset{|}{\underset{|}{C}}{-}OR}$ dialkyl ether
H_2SO_4		$\mathrm{H{-}\overset{|}{\underset{|}{C}}{-}\overset{|}{\underset{|}{C}}{-}OSO_3H}$ alkyl hydrogen sulfate
HX(X = F, Cl, Br, I)		$\mathrm{H{-}\overset{|}{\underset{|}{C}}{-}\overset{|}{\underset{|}{C}}{-}X}$ haloalkane

Throughout this discussion of additions of strong acids to alkenes, we have assumed that the acid, HX, is capable of protonating the double bond. Weak acids such as HCN and H_2S do not dissociate sufficiently to protonate a double bond, and therefore will not take part in addition reactions under ordinary conditions. Table 10–2 summarizes the major proton-initiated additions to alkenes.

10:4 Preparation of Dihalides by Electrophilic Addition of Halogens to Alkenes

The last section showed that alkyl halides containing one halogen can be produced from the addition of HCl or HBr to alkenes. In a similar reaction, dihalides can be prepared by electrophilic addition of Cl_2 or Br_2 to alkenes. The addition of chlorine or bromine to carbon–carbon double bonds is exothermic, goes to completion, and is easily controlled (eqs. 10.19 and 10.20). Fluorine and iodine are rarely used in addition reactions. Direct fluorination is extremely exothermic and leads to extensive degradation of the molecule.

(10.19) cyclohexene + Br_2 $\xrightarrow{-5°,\ CCl_4}$ trans-1,2-dibromocyclohexane (Br, Br)

(95%)

(10.20) $$CH_3CH_2CH{=}CH_2 + Cl_2 \xrightarrow[-9°]{} CH_3CH_2CH(Cl)CH_2Cl \quad (97\%)$$

Iodination of alkenes is not useful because equilibrium does not favor the product.

You may note from Equation 10.19 the predominance of one addition product, the *trans* isomer. Addition of halogens (X_2) to alkenes is usually highly stereoselective, with only one of the two diastereomeric products being formed. The product is the result of net *anti* addition of the halogen atoms. We can explain this stereochemical result by proposing a mechanism in which a **halonium ion** is an intermediate. A halonium ion has two carbons bonded to a positively charged halogen. Equation 10.21 shows a bromonium ion (**2**) as an intermediate in the bromination of (Z)-2-butene.

(10.21)

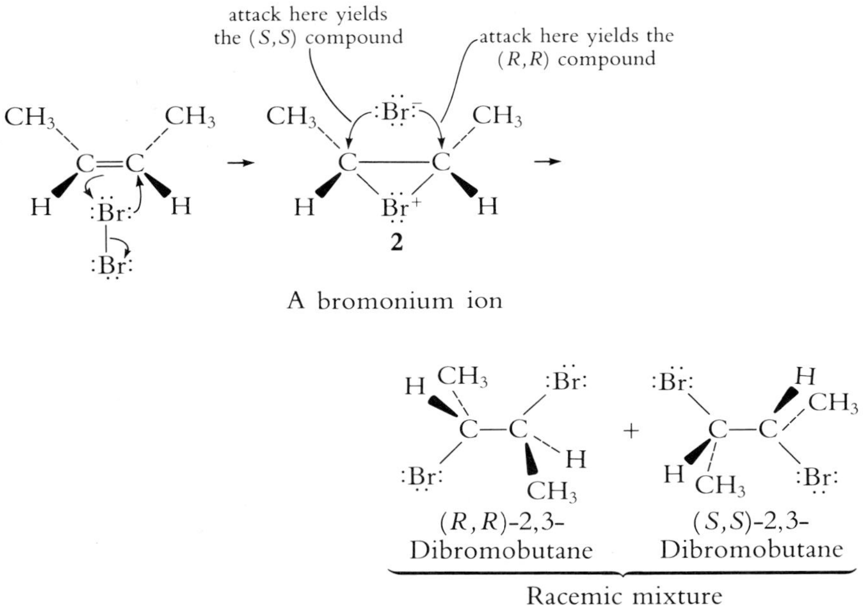

attack here yields the (S,S) compound

attack here yields the (R,R) compound

2

A bromonium ion

(R,R)-2,3-Dibromobutane + (S,S)-2,3-Dibromobutane

Racemic mixture

The positively charged intermediate, **2**, combines with the negatively charged bromide ion to form the final product. The reaction mechanism requires that the bromonium ion have the same stereochemistry as the starting alkene. The bromonium ion holds the two CH_3 groups in the original *cis* relationship and does not allow free rotation about the central C—C bond. The *incoming* Br^- must attack from the side *opposite* the bridging Br^+ (S_N2 reaction). This results in net *anti* addition of the two bromines to form two enantiomers. The last step resembles nucleophilic attack on epoxides (sec. 6:13).

According to the bromonium ion mechanism, we would expect the addition of bromine to (Z)-2-butene to yield only the racemic product, ($\pm$)-2,3-dibromobutane, and none of the *meso* compound, (R,S)-2,3-dibromobutane. The bromonium ion mechanism therefore satisfactorily explains our observed results (eq. 10.21).

The high stereoselectivity of halogen addition strongly favors the halonium ion mechanism, as opposed to one involving carbocations. Additional support comes from the lack of rearrangement in most halogen additions (eq. 10.22). (Compare Equation 10.22 with Equation 10.17, which shows a

$$(CH_3)_3CCH{=}CH_2 \xrightarrow[-10°,\ CCl_4]{Cl_2} (CH_3)_3C\underset{\displaystyle Cl}{\underset{|}{C}}HCH_2Cl \tag{10.22}$$

typical carbocation process.) Finally, halonium ions have been prepared and studied by nmr at low temperature, so we presume that they can also exist under ordinary reaction conditions.

Problem

10–6 One alternate mechanism of alkene halogenation involves the concerted *syn* addition of the halogen to the double bond (eq. 1).

H, H, C=C, H, H, Br—Br → H, H, H—C—C—H, Br Br (1)

If this mechanism were operating, what product would be formed by Br_2 addition to (Z)-2-butene? Do the experimental facts support the *syn* addition mechanism?

The capture of the halonium ion by the nucleophilic halide ion completes the addition process. If other nucleophiles are present, they too can react with the halonium ion. When halogen additions are carried out in nucleophilic solvents we do indeed isolate products which incorporate solvent molecules (eqs. 10.23 and 10.24).

$$Br_2 + CH_2{=}CH_2 \xrightarrow{CH_3OH} BrCH_2CH_2OCH_3 + BrCH_2CH_2Br + HBr \tag{10.23}$$

$$Br_2 + CH_2{=}CH_2 \xrightarrow[NaCl]{H_2O} BrCH_2CH_2Cl + \underset{\text{A bromohydrin}}{BrCH_2CH_2OH} + BrCH_2CH_2Br + HBr + NaBr \tag{10.24}$$

Example

What is the product of the reaction of Br_2 in CCl_4 with 1-butene?

Answer

In the first step of the reaction, Br_2 adds to the π bond.

$$\underset{\text{1-Butene}}{CH_2{=}CHCH_2CH_3} + Br_2 \rightarrow \underset{\underset{+}{Br}}{CH_2{-}CHCH_2CH_3} + Br^-$$

In the second step, $Br\colon^-$ attacks the bromonium ion intermediate.

$$\underset{\underset{+}{\diagdown Br \diagup}}{CH_2{-}CHCH_2CH_3} + Br\colon^- \longrightarrow \underset{\underset{Br}{|}}{BrCH_2CHCH_2CH_3}$$

1,2-Dibromobutane

In this case, the same product is obtained whether the Br^- attacks the bromonium ion at C-1 or C-2.

Problem

10–7 Write a mechanism that accounts for the formation of all the products shown in Equation 10.24.

It is instructive to compare the reactions of bromine with alkenes and alkanes. Bromine will not react with alkanes unless the mixture is either heated or irradiated with visible light. By contrast, bromine reacts with alkenes in the dark and at temperatures well below room temperature. One test for the presence of a double bond in a compound is adding a few drops of Br_2/CCl_4 solution to it. Instantaneous disappearance of the red-brown bromine color is evidence for the presence of a C=C group. These observations testify to the reactivity of the carbon–carbon π bond toward electrophiles.

Problems

10–8 The reaction of Cl_2 with (*E*)-2-butene under ionic conditions yields only *meso*-2,3-dichlorobutane. Does this result support the halonium ion hypothesis? Write a detailed reaction mechanism, showing the stereochemistry.

10–9 Give the products of each of the following reactions. If more than one diastereomer may be formed, then specify the stereochemistry of the product.

a. 2-methyl-1-butene + $Br_2 \xrightarrow[CCl_4]{}$

b. 1-methylcyclopentene + $Br_2 \xrightarrow[CH_2Cl_2]{}$

c. (*Z*)-2-butene + $Cl_2 \xrightarrow[CH_2Cl_2]{}$

d. propene + $Cl_2 \xrightarrow{H_2O}$

10–10 When cyclohexene is treated with Cl_2 in H_2O, the major product is 2-chlorocyclohexanol. What is its stereochemistry?

10:5 Free-Radical Additions to Alkenes

In Section 10:3 we discussed the regiospecificity of ionic (electrophilic) additions of H—X to the π bond of alkenes (eqs. 10.10 to 10.15). For example, when hydrogen bromide is added to propene, we would expect only one product, 2-bromopropane, to be formed.

Early investigators found that this was not always the case. They often obtained *both* 1- and 2-bromopropanes from this addition reaction and sometimes the 1-bromo compound was the major product. Subsequent studies of this puzzling behavior showed that when very pure propene and specially purified solvents were used, only 2-bromopropane—the regiospecific product of normal ionic addition—was formed. It was also necessary to run the reaction in an inert atmosphere—O_2 had to be excluded. Then, in the 1930s, the American chemists M. Kharasch and F. R. Mayo discovered that 1-bromopropane was formed by a *free-radical process* initiated by impurities, usually peroxides (RO—OR). Equations 10.25 to 10.28 give the steps for the free-radical addition. The energetics of the chain-carrying steps (eqs. 10.27 and 10.28) are favorable with HBr, but not with HCl or HI (see Prob. 10–35).

$$\text{R}\ddot{\underset{\cdot\cdot}{\text{O}}}\text{—}\ddot{\underset{\cdot\cdot}{\text{O}}}\text{R} \xrightarrow[\text{or heat}]{h\nu} 2\text{R}\ddot{\underset{\cdot\cdot}{\text{O}}}\cdot \tag{10.25}$$

$$\text{R}\ddot{\underset{\cdot\cdot}{\text{O}}}\cdot + \text{H}\ddot{\underset{\cdot\cdot}{\text{Br}}}\text{:} \rightarrow \text{R}\ddot{\underset{\cdot\cdot}{\text{O}}}\text{H} + \text{:}\ddot{\underset{\cdot\cdot}{\text{Br}}}\cdot \tag{10.26}$$

Initiation steps (10.25, 10.26)

$$\text{:}\ddot{\underset{\cdot\cdot}{\text{Br}}}\cdot + CH_2{=}CH{-}CH_3 \rightarrow BrCH_2\dot{C}HCH_3 \ (\textit{no}\ \dot{C}H_2{-}\underset{\text{:}\ddot{\underset{\cdot\cdot}{\text{Br}}}\text{:}}{\underset{|}{C}}HCH_3) \tag{10.27}$$

$$BrCH_2\dot{C}HCH_3 + HBr \rightarrow \underset{\text{1-Bromopropane}}{BrCH_2CH_2CH_3} + Br\cdot \tag{10.28}$$

Chain-carrying steps (10.27, 10.28)

The radical reaction of HBr can also be initiated by ultraviolet light alone (eq. 10.29).

$$CH_3(CH_2)_3CH{=}CH_2 + HBr \xrightarrow[4^\circ]{h\nu} \underset{(83\%)}{CH_3(CH_2)_3CH_2CH_2Br} + \underset{(10\%)}{CH_3(CH_2)_3\underset{\text{Br}}{\underset{|}{C}}HCH_3} \tag{10.29}$$

Note that in Equation 10.27, Br· adds to the double bond to give the secondary radical. This radical is more stable than the primary radical that would be formed if Br· added to C-2 of propene. Thus, the difference in regiospecificity between electrophilic addition and radical addition of HBr is simply the result of the manner in which the reaction is initiated. *In electrophilic additions the proton H^+ adds first (followed by Br^-), while in radical addition, the radical Br· adds first (followed by H·).* We call radical additions anti-Markovnikov because the product is not the one originally predicted by Markovnikov. It is important to recognize, however, that both Markovnikov and anti-Markovnikov additions are governed by the same principle—*the more stable of two possible intermediates is formed more rapidly.*

While free-radical addition of HBr to alkenes was originally just a nuisance, chemists later began to use free-radical additions for synthetic purposes. In these cases peroxides are deliberately added to the reaction mixture or the reaction is carried out under high light intensities (eq. 10.29). Many types of molecules can be synthesized by free-radical alkene addition. Thiols (RSH) react, for example, to form sulfides (eq. 10.30).

$$(CH_3)_2C{=}CH_2 + \underset{\text{Ethanethiol}}{CH_3CH_2SH} \xrightarrow[100^\circ]{\text{peroxides}} \underset{\substack{(94\%)\\ \text{A sulfide}}}{(CH_3)_2CHCH_2SCH_2CH_3} \tag{10.30}$$

10:6 Hydroboration and Anti-Markovnikov Synthesis of Alcohols

Acid-catalyzed hydration of alkenes produces only the alcohol from Markovnikov addition (eq. 10.31).

$$CH_3CH{=}CH_2 \xrightarrow[2.\ H_2O]{1.\ H_2SO_4} CH_3\underset{\displaystyle OH}{\underset{|}{C}}HCH_3 \qquad \text{Markovnikov alcohol synthesis} \tag{10.31}$$

A different reaction has been discovered that converts alkenes to the anti-Markovnikov alcohol (eq. 10.32). This reaction is both regiospecific and stereoselective (eq. 10.33).

$$CH_3CH{=}CH_2 \xrightarrow[2.\ H_2O_2,\ OH^-]{1.\ B_2H_6,\ THF} CH_3CH_2CH_2OH \qquad \text{Anti-Markovnikov alcohol synthesis} \tag{10.32}$$

(10.33) 1-methylcyclohexene $\xrightarrow[2.\ H_2O_2,\ OH^-]{1.\ B_2H_6,\ THF}$ trans-2-methylcyclohexanol (CH_3, OH) (no cis-2-methylcyclohexanol)

The first step of Equation 10.32 involves a reaction between the alkene and diborane, B_2H_6, a process known as **hydroboration** (eq. 10.34). In

(10.34) $CH_3CH{=}CH_2 + BH_3 \longrightarrow$ [Transition state: CH_3—$C^{\delta+}$(H)⋯C(H)—H, with H—$B^{\delta-}$(H)—H] $\longrightarrow CH_3CH_2CH_2BH_2 \xrightarrow{2CH_3CH{=}CH_2} (CH_3CH_2CH_2)_3B$

Transition state

Tripropylborane

solvents such as ether and THF (tetrahydrofuran; Table 6–5), B_2H_6 dissociates to BH_3, which is the active species. The boron in BH_3 has only six electrons in its valence shell and is therefore electrophilic. BH_3 attacks the π bond regiospecifically, so as to give the more stable transition state with partial positive charge on the most substituted carbon. A nucleophilic hydride ion then shifts from boron to the electron-deficient carbon, resulting in a net *syn* addition of B—H to the double bond. The process is repeated with two more alkene molecules, yielding a trialkylborane.

The trialkylborane is not isolated but oxidized directly with hydroperoxide ion, OOH^- (eq. 10.35). The products are three molecules of alcohol and

$$(CH_3CH_2CH_2)_3B + 3H_2O_2 + 3OH^- \rightarrow 3CH_3CH_2CH_2OH + BO_3^{3-} + 3H_2O \tag{10.35}$$

a borate ion. We will not be concerned with the mechanistic details of this second step except to note that —OH replaces —BR_2 stereospecifically, with retention of configuration at carbon.

Both steps in the hydroboration-oxidation sequence are highly specific with respect to regio- and stereochemistry, thus allowing reliable predictions of the product structure from a given starting material. Note from Equation 10.33 that the OH group ends up *trans* (not *cis*) to the methyl. Equation 10.36 shows how borane-d_3 can be used to synthesize a specifically deuterated alcohol.

$$3\ \begin{matrix} CH_3 & & CH_3 \\ & C{=}C & \\ H & & H \end{matrix} + BD_3 \xrightarrow[OH^-]{H_2O_2} 3\ \text{[product: H, D, CH}_3\text{ / H, CH}_3\text{, OH]} \quad (10.36)$$

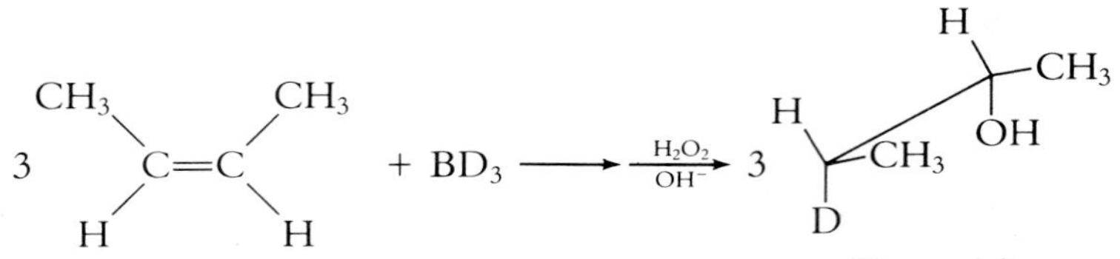

(Racemic)

Problem

10–11 a. What product(s) are formed when 1-methylcyclohexene (eq. 10.33) is hydrated in the standard way with $HClO_4/H_2O$?

b. What product(s) are formed when (*Z*)-2-butene (eq. 10.36) is treated with D_2O containing $DClO_4$? (Pay attention to stereochemistry.)

Example

Show how to prepare $CH_3CH_2CH_2CH_2CH_2OH$ from an alkene.

Answer

Alcohols are prepared from alkenes by hydration (sec. 10:3) or by hydroboration-oxidation (sec. 10:6). The starting alkene would be

$$CH_3CH_2CH_2CH{=}CH_2$$

Hydration (with aq H_2SO_4, for example) yields the Markovnikov product:

$$CH_3CH_2CH_2\underset{\displaystyle OH}{\underset{|}{C}}HCH_3$$

2-Pentanol

Hydroboration-oxidation would yield the desired anti-Markovnikov product:

$$CH_3CH_2CH_2CH_2CH_2OH$$

1-Pentanol

Problems

10–12 Show how the following alkenes may be converted to the specified alcohol.

a. 2-methylpropene → *t*-butyl alcohol
b. 2-methylpropene → isobutyl alcohol
c. methylenecyclopentane → cyclopentylmethanol
d. 2-methyl-1-hexene → 2-methyl-2-hexanol
e. 2-methyl-2-hexene → 2-methyl-3-hexanol

10–13 a. The conversion 1-pentene → 1-pentanol can be carried out with diborane *or* by a series of reactions starting with HBr addition. Write equations for both these processes, being careful to specify the reaction conditions.

b. Is there any way of converting 2-pentene regiospecifically to 2-pentanol? Explain your answer.

10–14 We want to convert methylpropene to 3-methylbutanenitrile (eq. 2). Suggest an efficient route for this conversion.

$$(CH_3)_2C{=}CH_2 \rightarrow (CH_3)_2CHCH_2C{\equiv}N \qquad (2)$$

10:7 Addition Polymerization of Alkenes

The preceding sections have highlighted the reactivity of alkenes toward electrophilic and free-radical reagents. In this section we extend the discussion to cover cases in which the attacking free radical or electrophile is itself derived from the alkene. Under these conditions the final product is a **polymer**, and the reaction is known as **polymerization** (from the Greek *polymeres,* meaning "of many parts"). Polymerization involves the joining of many small molecules, called **monomers** (from the Greek *monomeres,* meaning "of a single part"), to form one large molecule, a polymer or macromolecule.

There are several ways to polymerize alkenes, but one of the more common ways is through free-radical reactions. Recall from Section 5:4 that the radical chain mechanism involves initiation, propagation, and termination steps. Let us, then, consider the free-radical reactions in the polymerization of ethylene to form polyethylene.

One common way to initiate the polymerization is to add 0.01% oxygen to ethylene at 100° under high pressure. The diradical character of oxygen (eq. 5.21, p. 145) allows it to attack ethylene to produce a carbon radical (eq. 10.37). This step initiates the polymerization process. The free radical (**3**) which is formed can then propagate the reaction by adding to a second molecule of ethylene (eq. 10.38).

$$\cdot O{-}O\cdot + H_2C{=}CH_2 \rightarrow \underset{\mathbf{3}}{\cdot O{-}O{-}CH_2CH_2\cdot} \qquad (10.37)$$

$$\underset{\mathbf{3}}{\cdot O{-}O{-}CH_2CH_2\cdot} + H_2C{=}CH_2 \rightarrow \qquad (10.38)$$

$$\cdot O{-}O{-}CH_2CH_2CH_2CH_2\cdot \xrightarrow[\text{(excess)}]{H_2C=CH_2} \cdot OO{+}CH_2CH_2{+}_nCH_2CH_2\cdot$$

The chain mechanism continues until all the ethylene is used or until termination steps occur by radical couplings, reactions with impurities, or hydrogen abstractions (sec. 5:4). The letters Y and Z, symbols for the many possible end groups, indicate that both ends of the chain are terminated (eq. 10.39).

$$\cdot OO{-}(CH_2CH_2)_nCH_2CH_2\cdot \xrightarrow{\text{terminations}} YOO{-}(CH_2CH_2)_n{-}CH_2CH_2Z \qquad (10.39)$$

A typical value for n in the polyethylene product of Equation 10.39 is on the order of 40,000.

Termination is a statistical event, the result being that a sample of the polymer consists of molecules which do not all have a single molecular weight, but rather a range of molecular weights. We speak therefore of the *average molecular weight* of a polymer. The average molecular weights of most commercially important polymers lie in the range 10^4 to 10^6.

If the only reactions taking place during free-radical polymerization were those shown in Equations 10.38 and 10.39, the polyethylene produced would contain only *linear* molecules. In practice, however, we find that many of the polyethylene molecules are *branched.* This branching is understandable when we consider the behavior of free radicals. The terminal carbon atom with the unpaired electron may abstract a hydrogen atom from another carbon atom in the *same chain* (eq. 10.40). A new radical site is created, which serves as the starting point for a new polymer chain. *Inter*molecular hydrogen abstraction can also result in the formation of polymer branches.

(10.40)

$$\sim CH_2CH(H)-CH_2-CH_2-CH_2-CH_2\cdot \xrightarrow[1^\circ \rightarrow 2^\circ \text{ radical}]{\text{H transfer}} \sim CH_2\dot{C}H(CH_2)_3CH_3 \xrightarrow{CH_2=CH_2} \sim CH_2CH(CH_2CH_2\cdot)(CH_2)_3CH_3 \xrightarrow[\text{polymerization}]{\text{further}} \text{branched chain polymer}$$

The creation of branches is important because it affects the properties of the bulk polymer. In a perfectly linear polymer there would be maximum contact between individual polymer chains (Fig. 10–3), producing maximum van der Waals attraction of one molecule for another. The result is a polymer that is mechanically stiff and has a high tensile strength. Branching of the polymer chain interferes with intermolecular attraction, so the branched polymer is more deformable and less strong than the unbranched one. Free-radical polymerization of ethylene under high pressure yields *low-density polyethylene,* which contains a branch at approximately one carbon atom in 50. *High-density* polyethylene, which is far less branched, is produced by a non-free-radical process, discussed later in this section.

Derivatives of ethylene such as vinyl chloride, $CH_2{=}CHCl$, may also polymerize under free-radical conditions. Polyvinyl chloride is widely used in phonograph records, as a furniture covering, and in rainwear. As Equation 10.41 demonstrates, the polymerization of unsymmetrical alkenes al-

Figure 10–3. *Intertwining of linear polymer chains.*

most always produces *head-to-tail polymers*. A head-to-tail polymer is one

$$(10.41)\qquad 2nCH_2{=}CHCl \xrightarrow[\text{source}]{\text{free-radical}} \text{+}CH_2{-}\underset{\underset{Cl}{|}}{CH}{-}CH_2{-}\underset{\underset{Cl}{|}}{CH}\text{+}_n$$

Vinyl chloride — Polyvinyl chloride

formed by joining the nonidentical ends of the monomer molecules to one another. The mechanism of polymer growth is one in which the more highly substituted radical is the favored intermediate (eq. 10.42).

$$(10.42)\qquad R\cdot + \overset{\text{head}}{C}H_2{=}\overset{\text{tail}}{C}HCl \rightarrow RCH_2\underset{\underset{Cl}{|}}{\dot{C}H} \xrightarrow{CH_2{=}CHCl} RCH_2\underset{\underset{Cl}{|}}{CH}CH_2\underset{\underset{Cl}{|}}{\dot{C}H} \xrightarrow{\text{etc.}}$$

Teflon, another important commercial polymer, is made by radical polymerization of tetrafluoroethylene (eq. 10.43). The C—F bond is more inert

$$(10.43)\qquad nF_2C{=}CF_2 \xrightarrow{R\cdot} Y\text{+}CF_2CF_2\text{+}_nZ$$

Teflon

than is the C—H bond to many reagents, and hence Teflon is a more stable polymer than is polyethylene. Teflon is mechanically durable enough to be machined on a lathe, for instance, and we are all familiar with its use as a nonstick surface for kitchen utensils.

Not all vinyl polymers are made by free-radical processes. Sometimes ions, instead of free radicals, can propagate the polymerization process. This is called **ionic polymerization**. If the monomer can be converted to a relatively stable carbocation, then it can be polymerized under acidic conditions, as in the formation of polyisobutylene (eq. 10.44). Polyisobutylenes

$$(10.44)\qquad H^+ + CH_2{=}C(CH_3)_2 \rightarrow CH_3{-}\overset{\overset{CH_3}{|}}{\underset{\underset{CH_3}{|}}{C^+}} \xrightarrow{CH_2{=}C(CH_3)_2} \xrightarrow[\text{polymerization}]{\text{continued}}$$

Methylpropene
(*isobutylene*)

$$(CH_3)_3C\text{+}CH_2{-}\overset{\overset{CH_3}{|}}{\underset{\underset{CH_3}{|}}{C}}\text{+}_nCH{=}C(CH_3)_2$$

Poly(methylpropene)
(*polyisobutylene*)

are used as lubricants, while the copolymer of isobutylene and isoprene (Chap. 13) is butyl rubber.

Ethylene and its derivatives can also be polymerized by methods other than free radical and ionic processes. The polymers so produced have very desirable mechanical properties, but special catalysts are needed to effect this

Box 10–1. Polymers

More chemists work in the field of polymer chemistry than in any other area of chemistry. A quick rundown of the familiar names of some typical synthetic polymers will show how closely these products touch the lives of everyone: Cellophane, Bakelite, Nylon, PVC, Teflon, polyethylene, Styrofoam, Dacron, Mylar polyesters, polyurethanes. When we also consider some natural polymers such as proteins, carbohydrates, rubber, and lignin, the magnitude of polymer chemistry becomes apparent. So it is a surprising fact that the existence of high molecular weight polymers was not universally accepted by chemists until around 1930.

Classical nineteenth-century theories of molecular structure were completely compatible with the notion of very large molecules held together by normal valences. Nonetheless, at the end of that century, the prevalent idea was that high molecular weight compounds were merely colloidal aggregates of low molecular weight ones, joined by noncovalent forces such as dispersion forces and hydrogen bonds. In 1920 the German chemist H. Staudinger (Nobel Prize, 1953) published a very influential paper in which he argued that high molecular weight materials consisted of normal molecules held together by normal bonds. (It was Staudinger who, in 1924, coined the term *macromolecule*.)

It had been shown previously that natural rubber consisted of repeating *isoprene* units.

$$-CH_2-CH{=}\overset{\overset{\large CH_3}{|}}{C}-CH_2- \quad \text{An isoprene unit}$$

This conclusion was derived from the fact that ozonolysis of natural rubber yields only levulinaldehyde (ozonolysis involves oxidative cleavage of double bonds to give carbonyl groups; see sec. 10:10).

$$\text{natural rubber} \xrightarrow{O_3} \underset{\text{Levulinaldehyde}}{CH_3\overset{\overset{\large O}{\|}}{C}CH_2CH_2\overset{\overset{\large O}{\|}}{C}H}$$

The German chemist C. Harries (discoverer of ozonization) proposed that natural rubber consisted of small unsaturated rings held together by aggregation forces associated specifically with the double bonds. Staudinger showed that hydrogenation of natural rubber, which destroyed the double bonds, did not result in any decrease in its molecular weight. This result weighed heavily against the aggregation theory.

$$\underset{\substack{\text{Natural rubber}\\ \text{(high molecular}\\ \text{weight polymer)}}}{-\!\!\!+\!CH_2-CH{=}\overset{\overset{\large CH_3}{|}}{C}-CH_2\!+\!\!\!-_n} \xrightarrow[\text{catalyst}]{H_2} \underset{\substack{\text{High molecular}\\ \text{weight polymer}}}{-\!\!\!+\!CH_2-CH_2-\overset{\overset{\large CH_3}{|}}{C}H-CH_2\!+\!\!\!-_n}$$

These chemical results, combined with the crystallographic researches of the Austrian chemist H. Mark on natural polymers like cellulose, pretty well convinced chemists of the viability of macromolecules.

One of the first major American polymer chemists was W. H. Carothers of the DuPont Company, who in 1929 clearly drew the distinction between addition and condensation polymers. Carothers was the creator of nylon, the first totally synthetic fiber of commercial importance.

$$\underset{\substack{\text{Nylon 66}\\ \text{(a condensation polymer)}}}{-\!\!\!+\!NH(CH_2)_6NH\overset{\overset{\large O}{\|}}{C}(CH_2)_4\overset{\overset{\large O}{\|}}{C}\!+\!\!\!-_n}$$

One of the important discoveries about nylon was that it could be drawn out into a fiber when cold but not hardened. This gave it a silky appearance. The rapid drawing out of the soft material also oriented the polymer molecules in such a way that the strength of the fiber was substantially increased.

Another DuPont chemist is responsible for the accidental discovery that tetrafluoroethylene could be polymerized into Teflon. He was working with tetrafluoroethylene, a gaseous monomer which was being maintained under pressure in gas cylinders. One day the chemist turned the valve to obtain some of the gaseous monomer, but found none coming out even though the tank should have been nearly full. He sawed open the steel tank and found that the monomer had polymerized inside the tank to a hard, inert material. With the knowledge that tetrafluoroethylene could indeed be polymerized, chemists at DuPont were soon able to find useful ways of initiating the process in the laboratory.

$$\underset{\text{Tetrafluoroethylene}}{nCF_2{=}CF_2} \rightarrow \underset{\textit{Teflon}}{(CF_2-CF_2)_n}$$

kind of polymerization. The German and Italian chemists K. Ziegler and G. Natta were jointly awarded the Nobel Prize in 1963 for their development of such catalysts. A typical catalyst preparation involves the reaction between a trialkylaluminum and a transition metal salt in a hydrocarbon (eq. 10.45). The precipitate, which has a complex structure, is the active species.

(10.45) $$Et_3Al + TiCl_3 \xrightarrow[\text{heptane}]{} \underset{\text{catalyst}}{\text{Ziegler-Natta}} \downarrow$$

The growing polymer chain apparently becomes bonded to a titanium atom of the catalyst. There is a concerted addition of the Ti—C bond across the double bond of the alkene (analogous to B—H addition), so no organic radicals or ions are produced (eq. 10.46). The absence of ionic or radical intermediates greatly reduces the chance of branching. When ethylene is the monomer, the product is high-density polyethylene, which is stronger and stiffer than the low-density polymer.

(10.46) $$\sim CH_2CH_2 - Ti \text{ (Catalyst)} + CH_2{=}CH_2 \rightarrow$$

$$\sim CH_2CH_2CH_2CH_2 - Ti \xrightarrow[\text{continued addition}]{CH_2{=}CH_2} \text{high-density polyethylene}$$

Problems

10–15 a. Show how propene can be polymerized under acidic conditions, and illustrate the incorporation of three molecules of monomer into the chain. Why is a head-to-tail polymer formed?

b. Indicate how chain branching might arise in a polymer of this type.

10–16 Epoxides, or oxiranes (sec. 6:13), are useful monomers for polymerization. They are fundamental constituents of epoxy resins and cements, for example. With these monomers polymerization is usually conducted under basic conditions (eq. 3). Write a mechanism for the basic polymerization of ethylene oxide.

$$\underset{\textit{Ethylene oxide}}{CH_2{-}CH_2 \text{ (bridged by O)}} \xrightarrow{OH^-} \underset{\textit{Polyethylene oxide}}{HOCH_2CH_2{+}OCH_2CH_2{+}_nOCH_2CH_2OH} \quad (3)$$

10:8 Oxidative Additions to Alkenes: Preparation of Epoxides and Diols

We have described the synthesis of alcohols by both acid-catalyzed hydration of alkenes (sec. 10:3) and hydroboration-oxidation of alkenes (sec. 10:6). This section discusses the direct addition of *two* OH groups to alkenes to yield **diols**, alcohols with two OH groups. This *hydroxylation* reaction is carried out in the presence of metal oxidants in which the metal is in a high valence state: $KMnO_4[Mn(VII)]$ and $OsO_4[Os(VIII)]$. Much of the driving force for these reactions is the tendency for the metal ion to reach a lower valence state (be reduced). For example, treating an alkene with a dilute, basic solution of potassium permanganate ($KMnO_4$) in water or aqueous organic solvents at room temperature, results in the addition of one hydroxyl group at each end of the double bond and reduction of manganese to Mn(IV) (eqs. 10.47 and 10.48). The addition is stereospecifically *syn* and

(10.47) $3(CH_3)_2CHCH{=}CH_2 + 2KMnO_4 + 4H_2O \xrightarrow{25°}$ $3(CH_3)_2CHCH(OH)CH_2OH + 2MnO_2 + 2KOH$

3-Methyl-1-butene → 3-Methylbutane-1,2-diol (50%)

(10.48) $CH_3(CH_2)_7CH{=}CH(CH_2)_7COO^-$ (*cis*) $\xrightarrow[\text{dil aq soln}]{KMnO_4,\ OH^-}$ $CH_3(CH_2)_7CH(OH)CH(OH)(CH_2)_7COO^-$

Diol (*syn* addition) (60 to 90%)

involves a cyclic manganese derivative (a manganate ester) which accounts for the stereochemistry (eq. 10.49). The hydroxide ion catalyzes the hydrolysis of this manganate ester to the diol.

(10.49) cis-$CH_3CH{=}CHCH_3$ + MnO_4^- → Manganate ester $\xrightarrow[H_2O]{OH^-}$ $CH_3CH(OH)CH(OH)CH_3$ (*meso*)

Permanganate is a powerful oxidizing agent that can further oxidize the diol (see sec. 10:9). For this reason other reagents have also been used for the hydroxylation of alkenes. Osmium tetroxide, OsO_4, accomplishes the same type of stereospecific *syn* hydroxylation that potassium permanganate does, and with little danger of further oxidation (eq. 10.50). For this reason OsO_4 is often used for this purpose even though it is toxic and expensive.

(10.50) cyclohexene $\xrightarrow[\text{ether}]{OsO_4}$ $\xrightarrow{Na_2SO_3}$ *cis*-cyclohexane-1,2-diol

The first step of Equation 10.50 yields a cyclic ester, analogous to a manganate ester (10.49). The osmate ester is then decomposed by Na_2SO_3.

Hydroxylation of alkenes with either MnO_4^- or OsO_4 results in a diol in which the OH groups are *cis* to one another. In Section 6:13 we showed that hydration of an epoxide gives a 1,2-diol in which the OH groups are *trans* to one another (eq. 10.51, step 2). If we could oxidize alkenes to epoxides, without changing the stereochemistry (eq. 10.51, step 1), then by combining the two reactions we could effect a net *anti* addition of two OH groups to the double bond (eq. 10.51) to give a *trans* diol. This sequence offers a stereochemical alternative to the MnO_4^- and OsO_4 hydroxylations.

(10.51) Alkene $\xrightarrow{①}$ Epoxide $\xrightarrow[②]{H_2O,\ H^+}$ Diol (*anti* addition)

Alkenes may in fact be oxidized in one step to epoxides by *peroxyacids*

$$RC(=O)—OOH$$

which are derivatives of hydrogen peroxide and are mild oxidizing agents. The stereochemistry of the alkene is preserved in the epoxide (eqs. 10.52 and 10.53). The epoxide is then converted to the *trans* diol by acid-catalyzed hydration (eq. 10.51).

(10.52) $CH_3(CH_2)_7(H)C{=}C(H)(CH_2)_7CH_2OH$ (*cis*) + $CH_3C(=O)—O—OH$ (Peroxyacetic acid) $\xrightarrow[\text{(solvent), 25°}]{\text{ethyl acetate}}$

$CH_3(CH_2)_7$, $(CH_2)_7CH_2OH$ epoxide, *cis* (80%) + $CH_3C(=O)—OH$ (Acetic acid)

(10.53) $CH_3(CH_2)_7CH{=}CH(CH_2)_7CH_2OH$ (trans) $\xrightarrow{CH_3CO_3H}$ trans-9,10-epoxide

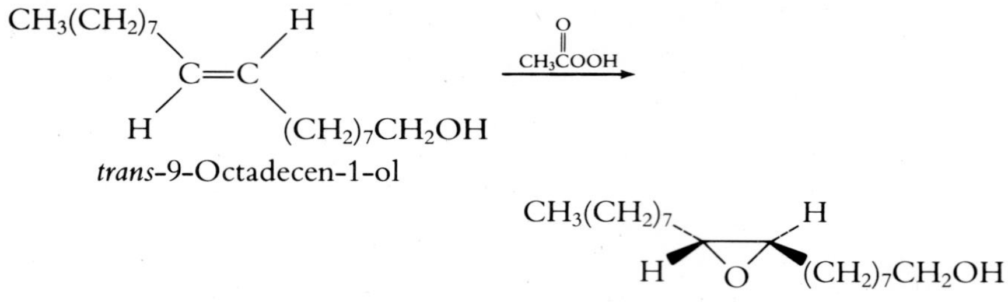

trans-9-Octadecen-1-ol

trans-9,10-Epoxy-1-octadecanol

The synthesis of *trans*-cyclopentane-1,2-diol from cyclopentene is an example of the overall epoxidation–hydration process (eq. 10.54).

(10.54) cyclopentene $\xrightarrow{CH_3CO_3H}$ cyclopentene oxide $\xrightarrow[H^+]{H_2O}$ *trans*-cyclopentane-1,2-diol

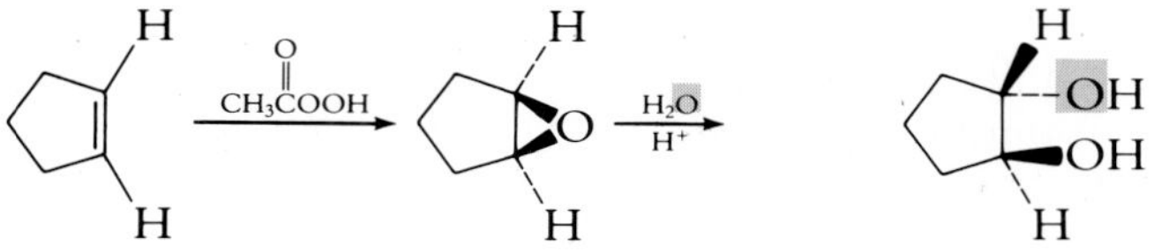

trans-Cyclopentane-1,2-diol

1,2-Diols, sometimes called *vicinal diols,* or *glycols,* are important compounds in many areas of chemistry. Ethane-1,2-diol (*ethylene glycol*) is manufactured from ethylene in enormous quantities by means of the epoxidation–hydration sequence. It is used extensively as an antifreeze and in the manufacture of polyesters. Sugars are naturally occurring compounds that contain numerous adjacent hydroxyl groups (Chap. 20). Many of them and their derivatives have been synthesized in the laboratory with the aid of the hydroxylation reactions described in this section.

10:9 Oxidative Cleavage of Diols and Alkenes to Produce Carbonyl Compounds and Acids

When $KMnO_4$ is used to prepare diols from alkenes, we need to pay close attention to the experimental conditions because the product diol can react further with $KMnO_4$. The C—C bond between the OH groups is easily broken and ketones or carboxylic acids, both of which contain the C=O group, are the resultant products (ketones and carboxylic acids are discussed in Chapters 11 and 16, respectively). Sometimes this *is* the desired reaction and then an alkene can be completely oxidized without isolation of the intermediate diol. Diols, but not alkenes, can also be cleaved by the mild oxidizing agent sodium periodate ($NaIO_4$). With either $KMnO_4$ or $NaIO_4$, the diol and the oxidant react to form a cyclic derivative, which then decomposes with electron rearrangement. Equation 10.55 is a schematic depiction of the process, in which M represents I(VII) or Mn(VII).

(10.55) $\text{C(OH)—C(OH)} + M^{n+} \rightarrow$ cyclic C—O—M—O—C ester $\rightarrow \text{C=O} + \text{C=O} + M^{(n-2)+}$

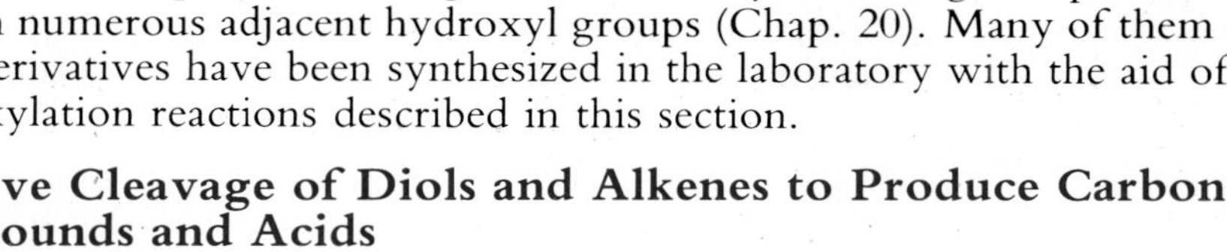

Two factors favor the oxidative cleavage: the reduction in the valence state of M, and the strength of the C=O π bonds being formed.

The products of the periodate cleavage of diols are ketones and/or aldehydes (another class of carbonyl compound; see Chap. 11), as shown in Equations 10.56 and 10.57.

(10.56) $CH_3CH(OH){-}C(CH_3)(OH){-}CH_3 \xrightarrow[pH\ 5]{NaIO_4}$ cyclic periodate ester (O—I(O)(O)(O$^-$)—O bridging the two carbons; cleavage between CH and C) $\rightarrow$

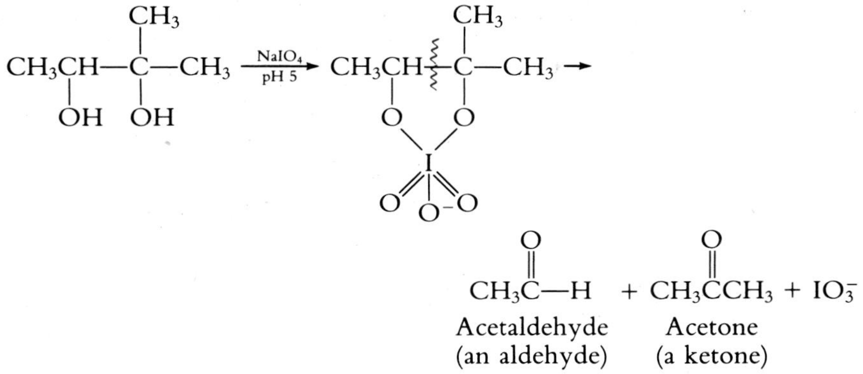

$CH_3\overset{O}{\overset{\|}{C}}{-}H + CH_3\overset{O}{\overset{\|}{C}}CH_3 + IO_3^-$

Acetaldehyde (an aldehyde) — Acetone (a ketone)

(10.57) cyclohexane-1,2-diol $\xrightarrow[H_2O]{NaIO_4}$ OHC–(CH$_2$)$_4$–CHO

A dialdehyde

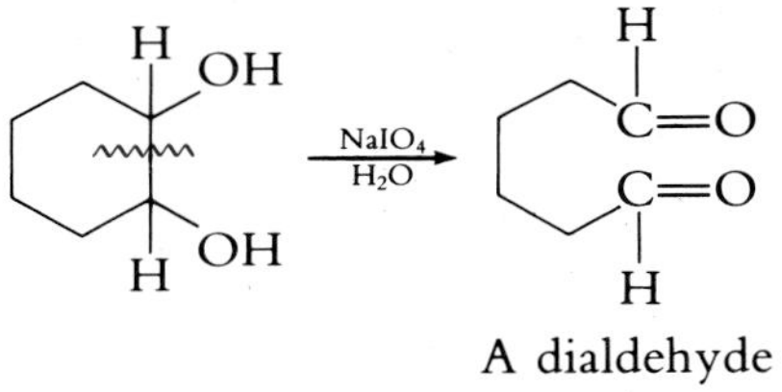

These reactions go in sufficiently high yield so that periodate can be used as a reagent for the quantitative determination of the number of adjacent diol groups in a molecule (Chap. 20).

Potassium permanganate can bring about either hydroxylation of the double bond (eqs. 10.47 and 10.48) or its complete cleavage (eq. 10.58). By controlling the reaction conditions we can promote one or the other of these reactions. A very important factor is the pH of the solution. The hydroxide ion helps to decompose the cyclic manganese intermediate (eq. 10.49) and prevent further oxidation. High pH (basic conditions) therefore favors hydroxylation over cleavage. High temperatures and low pH (acid conditions), on the other hand, favor the cleavage reaction.

(10.58) 1-methylcyclohexene $\xrightarrow{KMnO_4}$ [$CH_3C(=O)$–(CH$_2$)$_4$–C(=O)–H] $\xrightarrow{\text{further oxidation}}$ $CH_3C(=O)$–(CH$_2$)$_4$–C(=O)–OH

A ketocarboxylic acid

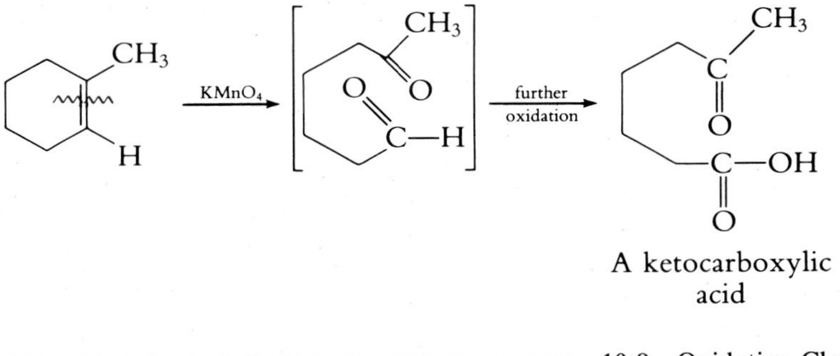

If one of the carbon atoms of the double bond carries a hydrogen atom, then the product will be a carboxylic acid (eq. 10.58). The strong oxidizing agent H_2CrO_4 can also be used to cleave alkenes in similar fashion.

Problem

10–17 Compound **A** (C_6H_{10}) absorbs one equivalent of H_2 when treated with H_2/Pt. **A** is transformed into **B** ($C_6H_{10}O_2$), as shown in Equation 4. The ^1H nmr spectrum of **B** contains only two singlets—one at 2.2 and one at 2.7 ppm, area ratio 3 : 2. The ir spectrum of **B** shows a very strong band at 1720 cm^{-1}.

$$\mathbf{A}(C_6H_{10}) \xrightarrow{OsO_4} \xrightarrow[pH\ 5]{NaIO_4} \mathbf{B}(C_6H_{10}O_2) \qquad (4)$$

a. Propose structures for **A** and **B**. Show all the chemistry involved in the conversion of **A** to **B**.

b. Show the structures of the products that would result if **A** were treated with peroxyacetic acid and the product then reacted with aqueous perchloric acid.

10:10 Structure Determination by Ozonization of Alkenes

Ozone, O_3, is also capable of cleaving the carbon–carbon double bond. In fact, ozone reacts with few functional groups other than double bonds, so it is more discriminating than $KMnO_4$. We obtain ozone in the laboratory by passing oxygen gas into an ozonizer, where a continuous electrical discharge produces a mixture of 3 to 8 weight percent ozone in oxygen. The gaseous mixture is then bubbled through a solution of the alkene in an inert solvent, such as $CHCl_3$, at low temperatures.

Ozonization proceeds through at least two intermediates. The first results from the attack of ozone on the π bond, giving a very unstable **primary ozonide** (eq. 10.59). The primary ozonide undergoes a rearrangement involving σ-bond cleavage, leading to the far more stable **ozonide** (second step of eq. 10.59).

(10.59)

Ozonides of high molecular weight compounds, such as steroids, are relatively stable, but those of low molecular weight compounds are very unstable and explosive; therefore, ozonides are not usually isolated. Instead, the solution containing them is treated with various reagents to decompose the ozonide and complete the cleavage of the original C=C bond. Equation 10.60 shows that oxidative decomposition (H_2O_2 and water) yields ketone or acid groups, while reductive decomposition (Zn and acetic acid, or dimethyl sulfide) gives aldehyde or ketone groups.

$$\text{(10.60)}\quad \text{1-methylcyclohexene} \xrightarrow{O_3} \text{ozonide} \begin{cases} \xrightarrow[H_2O]{H_2O_2} CH_3C(=O)CH_2CH_2CH_2CH_2C(=O)OH & \text{A ketocarboxylic acid} \\ \xrightarrow[\text{Zn, HOAc}]{(CH_3)_2S \text{ or}} CH_3C(=O)CH_2CH_2CH_2CH_2C(=O)H & \text{A ketoaldehyde} \end{cases}$$

The ozonization process has been very useful in the structure determination of unknown alkenes, such as the sex attractants isolated from the females of many insect species. These compounds show great promise in nonpesticide insect control programs. Ozonization of the unknown attractant, followed by cleavage, yields low molecular weight aldehydes or ketones that are easily identified by spectral comparison with standard (known) samples. For example, suppose that the unknown is found to have the molecular formula C_7H_{14}, and it also decolorizes Br_2 solution. This would mean that it has a double bond. Proof of structure could then be accomplished by ozonization-cleavage. For example, if only straight-chain structures are considered, Equation 10.61 shows the products expected for each of the three possible heptenes.

$$\text{(10.61)}\quad \underset{\text{3-Heptene}}{CH_3(CH_2)_2CH{=}CHCH_2CH_3} \xrightarrow[\substack{\text{2. Zn, HOAc} \\ \text{or } (CH_3)_2S}]{\text{1. } O_3} CH_3(CH_2)_2\overset{O}{\overset{\|}{C}}H + H\overset{O}{\overset{\|}{C}}CH_2CH_3$$

$$\underset{\text{2-Heptene}}{CH_3(CH_2)_2CH_2CH{=}CHCH_3} \xrightarrow[\substack{\text{2. Zn, HOAc} \\ \text{or } (CH_3)_2S}]{\text{1. } O_3} CH_3(CH_2)_2CH_2\overset{O}{\overset{\|}{C}}H + H\overset{O}{\overset{\|}{C}}CH_3$$

$$\underset{\text{1-Heptene}}{CH_3(CH_2)_2CH_2CH_2CH{=}CH_2} \xrightarrow[\substack{\text{2. Zn, HOAc} \\ \text{or } (CH_3)_2S}]{\text{1. } O_3} CH_3(CH_2)_2CH_2CH_2\overset{O}{\overset{\|}{C}}H + H\overset{O}{\overset{\|}{C}}H$$

Note that none of these degradative reactions gives any information about the *stereochemistry* of the original alkene—we must still rely on spectroscopic or other data for final proof of structure.

Table 10–3 summarizes the major oxidation reactions of alkenes.

Example

Give the major product(s) of the following reactions:

a. cyclohexene + $KMnO_4$, OH^- at 25°
b. 1-pentene + O_3, followed by Zn, HOAc

Answer

a. Under these conditions, a cyclic intermediate is formed when the permanganate ion attacks the C=C bond (eq. 10.49), and this is hydrolyzed by the OH^- to yield a *cis* glycol:

$$\text{cyclohexene} \xrightarrow[25^\circ]{KMnO_4,\ OH^-} \textit{cis}\text{-1,2-cyclohexanediol}$$

b. Ozonolysis followed by Zn, HOAc cleaves the alkene double bond to produce two carbonyl compounds:

$$H_2C{=}CHCH_2CH_2CH_3 \xrightarrow[2.\ Zn,\ HOAc]{1.\ O_3} H_2C{=}O + O{=}C(H)CH_2CH_2CH_3$$

Table 10–3. Major Oxidation Reactions of Alkenes

Oxidative Additions

	Reaction	Conditions	Product	Stereochemistry
1.	RR′C=CRR′	$KMnO_4$, 25°, basic solution	RR′C(OH)–C(OH)RR′	*syn* addition (sec. 10:8)
2.	RR′C=CRR′	OsO_4, 25°	RR′C(OH)–C(OH)RR′	*syn* addition (sec. 10:8)
3.	RR′C=CRR′	RCOOH (peracid: RC(=O)OOH) → epoxide; then H_3O^+	RR′C(OH)–C(OH)RR′	*anti* addition (sec. 10:8)

Oxidative Cleavages

	Reaction	Conditions	Products	
1.	RR′C(OH)–C(OH)RH	$NaIO_4$, pH 5	R—C(=O)—R + R—C(=O)—H	(sec. 10:9)
2.	$R_2C{=}CHR$	$KMnO_4$, heat, acidic or neutral solution	R—C(=O)—R + R—C(=O)—OH	(sec. 10:9)
3.	$R_2C{=}CHR$	O_3, −78°; then H_2O_2	R—C(=O)—R + R—C(=O)—OH	(sec. 10:10)
		O_3, −78°; then $(CH_3)_2S$ or Zn, HOAc	R—C(=O)—R + R—C(=O)—H	

10–18 Give the major product(s) from each of the following reactions:

a. 1-butene $\xrightarrow[25°]{KMnO_4,\ OH^-}$

b. (*E*)-2-butene $\xrightarrow[25°]{KMnO_4,\ OH^-}$

c. 1-methylcyclopentene $\xrightarrow[2.\ Na_2SO_3]{1.\ OsO_4}$

d. (*E*)-cyclooctene $\xrightarrow[2.\ Na_2SO_3]{1.\ OsO_4}$

e. cyclobutene $\xrightarrow[2.\ H_2O,\ H^+]{1.\ CH_3C(=O)OOH}$

f. propane-1,2-diol $\xrightarrow{NaIO_4}$

g. (*E*)-2-butene $\xrightarrow[2.\ Zn,\ HOAc\ or\ (CH_3)_2S]{1.\ O_3}$

h. (*Z*)-2-butene $\xrightarrow[2.\ Zn,\ HOAc\ or\ (CH_3)_2S]{1.\ O_3}$

10–19 a. Branch-chain structures were not considered for the unknown C_7H_{14} sex attractant discussed in Section 10:10. What would the unknown have been if the products isolated after ozonization and reductive cleavage were

$$CH_3\underset{\underset{O}{\|}}{C}(CH_2)_2CH_3 \text{ and } CH_3\underset{\underset{O}{\|}}{C}H?$$

What if the products were

$$CH_3\underset{\underset{O}{\|}}{C}(CH_2)_3CH_3 \text{ and } H\underset{\underset{O}{\|}}{C}H?$$

b. 2-Cyclopentyl-2-propanol (**I**) is heated with concentrated HBr. It yields several products, including two alkenes, **A** and **B** (eq. 5). These alkenes are C_8H_{14} isomers.

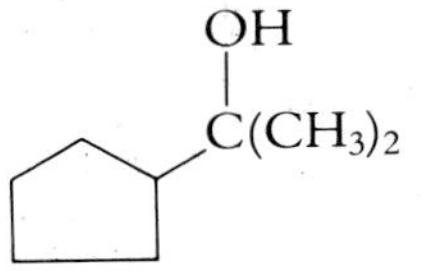

$$\text{cyclopentyl–}\overset{\overset{OH}{|}}{C}(CH_3)_2 \ (\mathbf{I}) \xrightarrow[heat]{HBr} \mathbf{A} + \mathbf{B} \qquad (5)$$

The ^1H nmr spectrum of **A** shows no peaks in the alkene region (5 to 7 ppm). When **A** is ozonized and worked up with H_2O_2 it gives two carbonyl compounds, C_5H_8O and C_3H_6O. The nmr spectrum of the C_3H_6O compound consists of a single peak.

Compound **B** has the following ^1H nmr spectrum: 1.00 ppm (6H, doublet); 1.48 to 2.50 ppm (7H, multiplet); 5.25 ppm (1H). Ozonolysis and oxidative cleavage of **B** gives a single compound, a ketocarboxylic acid. Propose structures for **A** and **B** and write a mechanism that explains their formation.

10:11 Enzyme Catalysis: Introduction

There is no doubt that some of the most challenging organic chemistry is found in the complex systems of living organisms. Yet the *individual* chemical events are relatively simple and quite analogous to those in the laboratory.

The most striking difference between similar reactions taking place *in vitro* and *in vivo* is that the latter usually proceed much faster, under much milder conditions (moderate temperature, pH near neutral), and with a high degree of stereospecificity. These differences can be attributed to the intervention of **enzymes**.

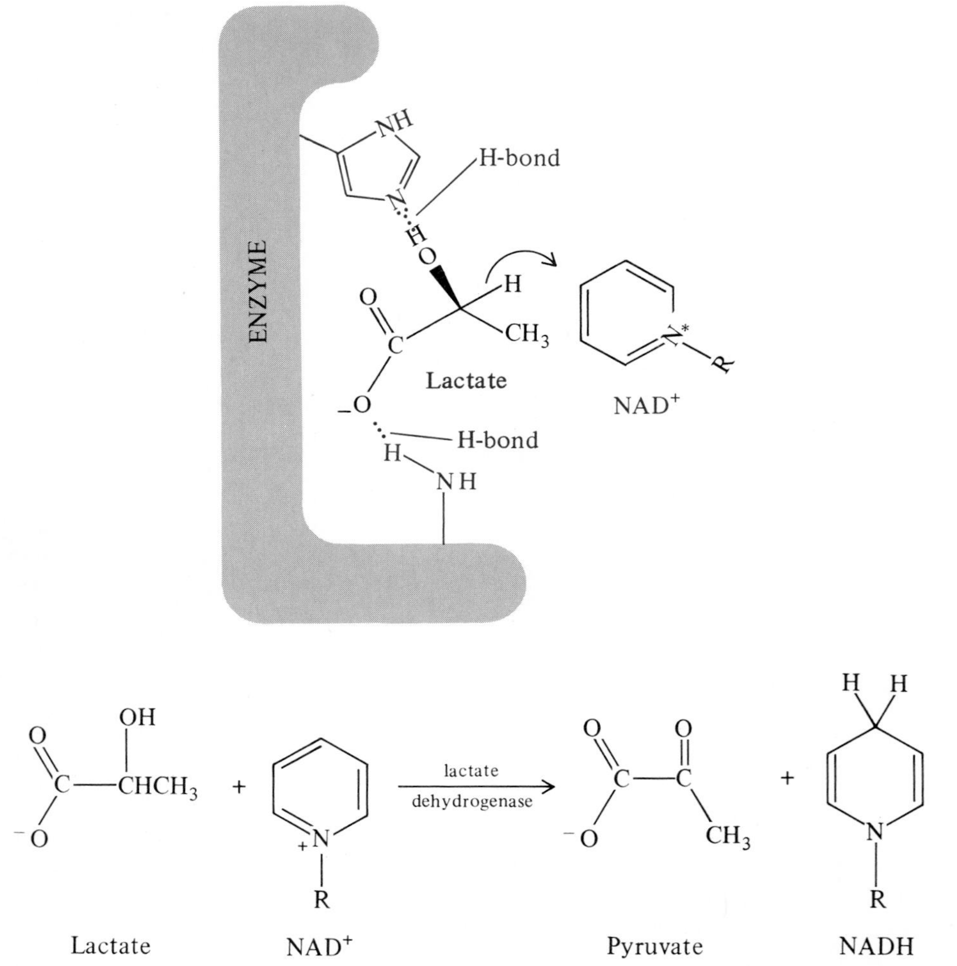

ENZYME

NH

H-bond

N

H

O

H

O

C

CH_3

Lactate

NAD$^+$

N

R

$^-$O

H-bond

H

NH

Lactate + NAD$^+$ $\xrightarrow[\text{dehydrogenase}]{\text{lactate}}$ Pyruvate + NADH

Lactate NAD$^+$ Pyruvate NADH

Figure 10–4. *Schematic drawing of a portion of the enzyme lactate dehydrogenase. The substrate lactate is held by hydrogen bonds to the enzyme forming the enzyme–substrate complex shown. The hydrogen of the OH in lactate is transferred to the enzyme and the hydrogen on carbon is transferred to NAD$^+$, thus completing the dehydrogenation reaction shown.*

Enzymes are biological catalysts. They all belong to a class of compounds known as **proteins** (biological polymers; see Chap. 22). Enzymes have high molecular weights, many chiral centers, and usually contain a number of acidic and basic groups. The reactant (or substrate) and the enzyme form a complex (intermediate) that subsequently decomposes, giving the product and regenerating the enzyme. The enzyme can bind the substrate covalently or by means of hydrogen bonds and electrostatic attractions, as is the case with the enzyme–substrate complex shown in Figure 10–4.

Enzymes often assume particular conformations due to intramolecular interactions, and the substrate must be able to fit itself to the enzyme. One or more substrates attach themselves to the enzyme in the conformation most favorable for reaction. The enzyme thereby sharply lowers the free energy of activation for the reaction, resulting in acceleration of the rate. In the next section we will consider an example of an enzyme-catalyzed hydration of a double bond. In this reaction the enzyme is bound to the alkene, the water molecule, and the proton that initiates the reaction. Since one of the most striking features of the reaction is its stereochemical specificity, let's look first at those stereochemical elements that come into play.

10:12 Prochirality and Its Application to Enzymatic Reactions

In Section 4:8 we mentioned that enzymes are able to discriminate between the enantiomeric forms of the same substrate with a high degree of specificity. In this section we'll show that enzymes are capable of making even more subtle distinctions between *apparently* equivalent groups within the same substrate molecule.

In order to appreciate these capabilities, you must be able to identify a **prochiral** carbon. This is a tetravalent carbon atom that is bonded to two substituents of the same kind, and to two other, different substituents. The carbon atom in bromochloromethane (CH_2BrCl) is prochiral, for example. It is called *pro*chiral because changing only one of the two identical substituents to an entirely new substituent makes the carbon atom chiral. Thus, if we substitute a D atom for one of the H atoms in bromochloromethane, we create a chiral molecule. Consider a molecule that contains only one prochiral center, such as ethanol (CH_3CH_2OH). The hydrogens at prochiral C-1 of ethanol are **enantiotopic** because selective replacement of each of them generates a pair of enantiomers (Fig. 10–5).★ Enantiotopic groups are chemically identical, have the same nmr chemical shifts in standard solvents, and are equally reactive under common reaction conditions.

Next we look at a molecule that has both a chiral and a prochiral center, such as 2-bromobutane (Fig. 10–6, **4,** R = CH_3). The hydrogens at the prochiral center, C-3, are classified as **diastereotopic** because replacement of each one selectively by deuterium creates a pair of diastereomers. Replacing H_a by D produces the *R* configuration at C-3, so H_a is called the *pro-R* hydrogen; analogously, H_b is called the *pro-S* hydrogen.

★ Note that C-2 of ethanol is **not** prochiral, because replacement of one of the CH_3 protons by D does **not** lead to stereoisomerism.

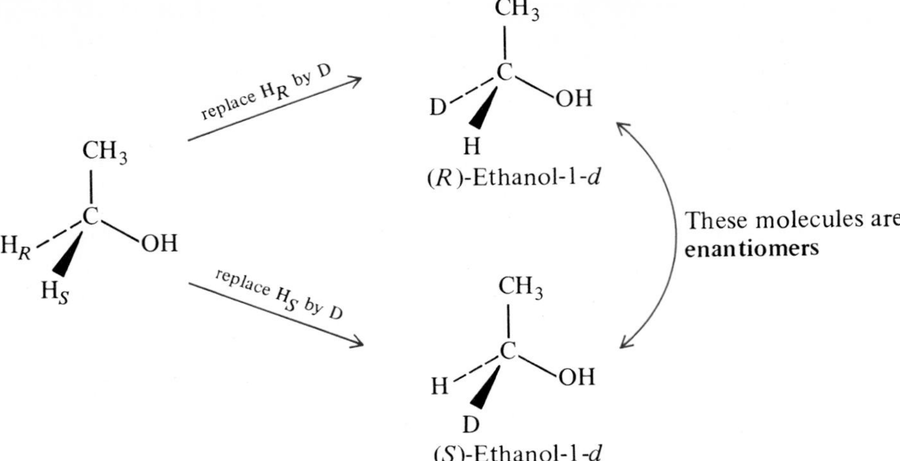

Figure 10–5. *Selective replacement of enantiotopic protons (H_S and H_R) on ethanol produces a pair of enantiomers.*

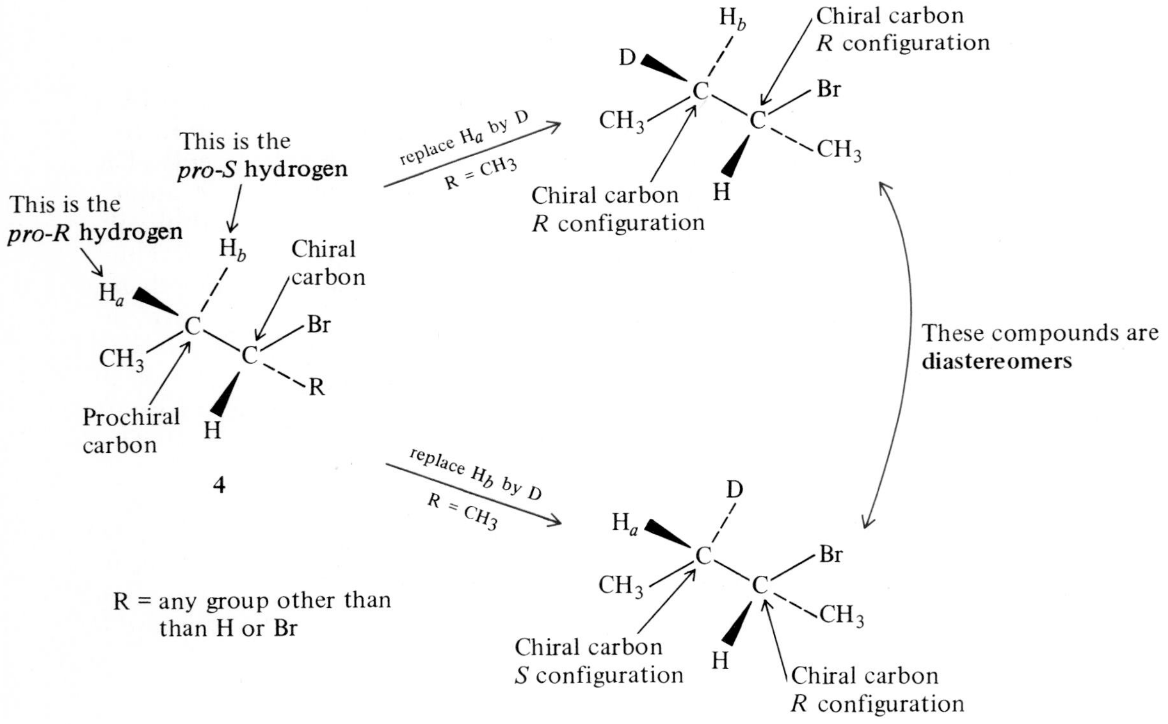

Figure 10–6. *How to classify the diastereotopic protons in 2-bromobutane and related compounds.*

Diastereotopic groups are chemically distinct and so should have different nmr chemical shifts. Figure 10–7 shows the 100 MHz ^1H nmr spectrum of a molecule of the type shown in Figure 10–6 (**4,** R = COOEt). The CH_2 protons give rise to a quintet because they are coupled equally to the four neighboring protons on CH_3 and CHBrR (sec. 9:7). The spectrum shows *two* closely overlapping quintets at 2.1 ppm, confirming that the diastereotopic protons do indeed have slightly different chemical shifts.

Problem

10–20 Which of the following compounds have prochiral carbons? What enantiotopic groups are present? Which compounds have diastereotopic groups?

a. $CH_3CH_2CH_3$
b. $CH_3CBr_2CH_2Br$
c. $CH_3CHBrCH_3$
d. $CH_3CHBrCH_2CH_2CH_3$
e. $BrCH_2CHClCH_3$

Since diastereotopic groups are chemically distinct, they should react at different rates even under non-enzymatic reaction conditions. In fact, diastereotopic hydrogens do sometimes show modest differences in reactivity in the course of standard laboratory reactions. Enzymes, however, make dramatic distinctions between reactive diastereotopic groups. We will illustrate this with a case in which one diastereotopic hydrogen is completely unreactive compared to the other.

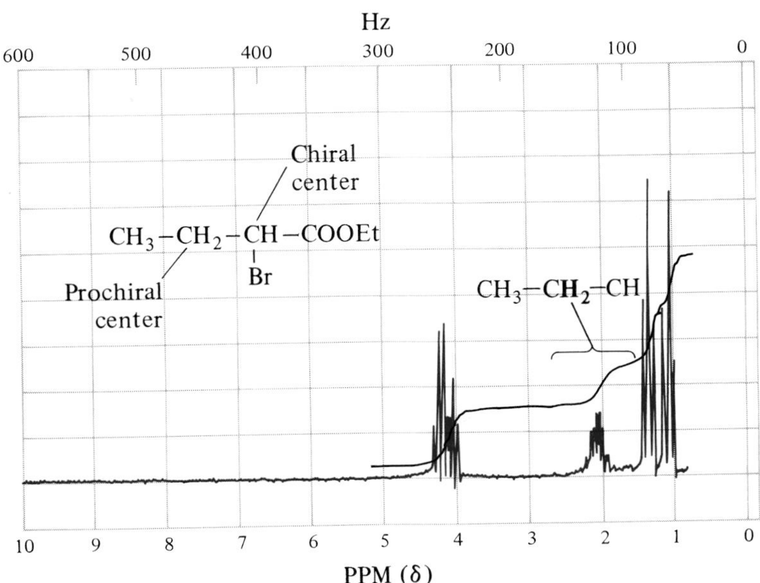

Figure 10–7. *A 100 MHz ^1H nmr spectrum of ethyl 2-bromobutanoate, showing a pair of quintets for the diastereotopic protons at the prochiral center.*

Fatty acids consist of hydrocarbon chains connected to a carboxyl group (—COOH). Biochemists have isolated a bacterial enzyme that catalyzes the addition of water to the double bond of the fatty acid, oleic acid (eq. 10.62).

$$\text{(10.62)}\quad \underset{\text{Oleic acid}}{\overset{CH_3(CH_2)_7\qquad (CH_2)_7\overset{O}{\overset{\|}{C}}OH}{\underset{H\qquad\quad H}{C{=}C}}} + H_2O \xrightarrow{\text{enzyme}} \underset{(R)\text{-10-Hydroxystearic acid}}{CH_3(CH_2)_7\text{—}\overset{H\quad OH}{C}\text{—}(CH_2)_8\overset{O}{\overset{\|}{C}}OH}$$

The addition is regiospecific and stereospecific and would be impossible to achieve in the absence of an enzyme or other chiral reagents. The acid produced is one pure enantiomer only, not a mixture of enantiomers. The chirality of the enzyme itself accounts for the stereospecificity of the reaction, and its binding properties (such as those illustrated in Fig. 10–4) account for the regiospecificity.

Figure 10–8 is a schematic representation of how the hydration of oleic acid may take place. B: and B—H$^+$ represent basic and acidic groups, respectively, on the enzyme. What this figure does not show, and what is crucial about enzyme chemistry, is the way in which the enzyme holds the oleic acid and the water in the optimum positions for the reaction to occur.

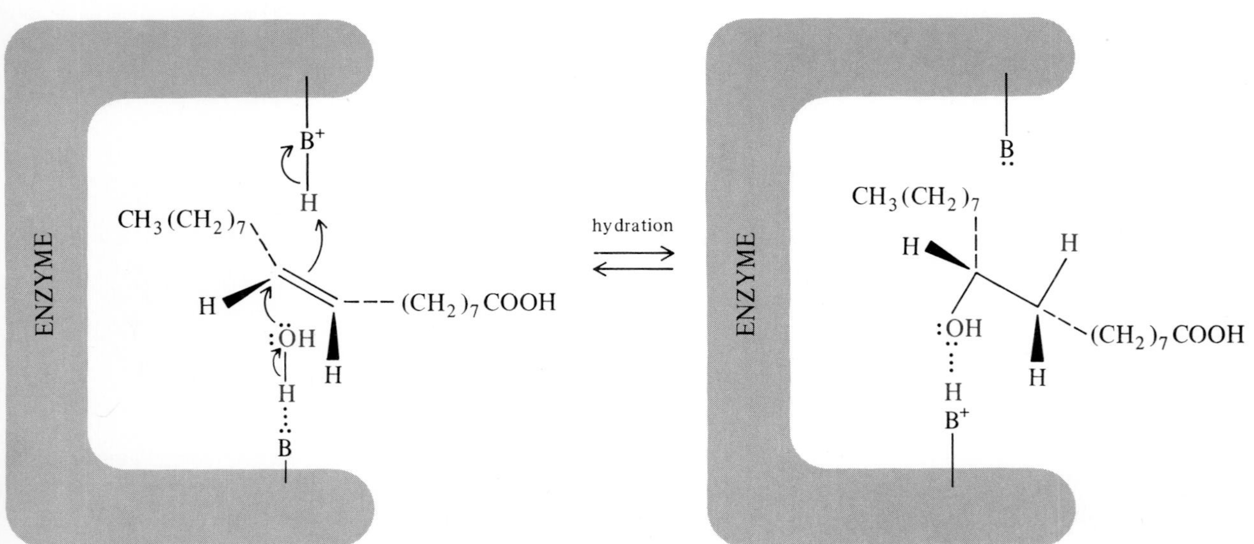

Figure 10–8. *Schematic mechanism for the enzymatic hydration of oleic acid to* (R)-*10-hydroxystearic acid.*

Most probably, the proton and the water molecule are held on opposite sides of the double bond, close to the carbon atoms to which they will be bound. The details of how substrates are positioned by enzymes is known for only a few reactions, such as the one we saw in Figure 10–4.

The hydration of oleic acid, like most biochemical processes, is reversible. The reverse reaction, dehydration of (*R*)-10-hydroxystearic acid, shows the same high degree of specificity as the forward reaction. Protons are removed only from C-9, and only the H_R proton of the diastereotopic pair at this carbon is involved (eq. 10.62).

You have seen (sec. 4:8) that chiral discrimination is not just the property of an enzyme, but can be true of chemical reagents as well, *if they are chiral*. Chemists have taken this clue from nature and are developing chiral reagents which give an excess of one enantiomer over another in certain reactions.

Summary

1 The C=C double bond consists of a σ bond and a π bond. The π bond is weaker and more polarizable than the σ bond. The characteristic reaction of π bonds is addition, usually initiated by the attack of an electrophile or a radical. Among the most important additions are those involving HX, X_2, ROH, RSH, and H_2. They are all thermodynamically favorable (sec. 10:1).

Hydrogen addition can be accomplished with metal catalysis. The catalysts—Pd, Pt, Ni—adsorb both alkene and H_2, and also bring about dissociation of the hydrogen molecule. *Syn* addition of H_2 to the double bond is usually the preferred stereochemical outcome (sec. 10:2).

2 When alkenes are treated with strong acids the π bond is protonated, giving a carbocation. The carbocation then captures any nucleophile(s) present, such as water, alcohols, halide ions, and bisulfate ion. If two energetically unequal carbocations can be obtained from a single alkene, the product isolated is the one coming from the more stable carbocation (modern version of Markovnikov's rule). Hydrogen halides are added to alkenes in the absence of solvent, or in nonnucleophilic solvents (sec. 10:3, Table 10–2).

3 Molecular halogens such as Br_2 and Cl_2 transfer X^+ to the alkene with formation of a cyclic halonium ion. This halonium ion is then attacked by a nucleophile on the side away from the positive halogen so that the overall addition is *anti*. The nucleophile may be a halide ion or the solvent (sec. 10:4).

4 In the presence of peroxides, or under uv illumination, HBr gives the anti-Markovnikov product from alkene addition. A free-radical chain is involved, in which the most stable radical intermediate predominates (sec. 10:5).

5 Hydroboration takes place with the addition of BH_3 to the double bond. The boron atom attaches itself to the less substituted carbon atom. The BH_3 reacts with three equivalents of alkene. Oxidation of the resulting trialkylborane with H_2O_2 and OH^- produces the anti-Markovnikov alcohol, in which H and OH have been added *syn* to the original double bond (sec. 10:6).

6 Radical initiators, when added to an alkene in small amounts, bring about polymerization of the alkene by way of a radical chain mechanism. Unsymmetrical alkenes are converted to head-to-tail polymers. The growing polymer chain may undergo intramolecular hydrogen transfer, resulting in polymer branching. Polymerization may also be initiated by acids. The use of Ziegler–Natta catalysts provides polymers that have desirable physical properties because their chains are almost completely linear (sec. 10:7).

7 Osmium tetroxide, OsO_4, and dilute potassium permanganate, $KMnO_4$, add an OH group to each end of a double bond. This hydroxylation reaction involves *syn* stereochemistry. Epoxidation of an alkene with a peroxyacid, followed by acid-catalyzed hydrolysis of the epoxide, also results in addition of two OH groups to the double bond. In this case the stereochemistry of addition is *anti* (sec. 10:8). The 1,2-diols produced by hydroxylation can be cleaved to carbonyl compounds by HIO_4 (sec. 10:9).

Alkenes can be directly cleaved to carbonyl compounds by hot $KMnO_4$ (sec. 10:9), or ozone (sec. 10:10). The products of the ozone reaction are ketones and/or aldehydes (reductive decomposition), or ketones and/or carboxylic acids (oxidative decomposition; see Table 10–3). The structure of the original alkene is inferred from the structures of the products (sec. 10:10).

8 Enzymes are biological polymers that serve as catalysts for cellular reactions. An enzyme binds all the reactants for the reaction it is catalyzing and places them in an optimal relationship to each other (sec. 10:11). Enzymatic reactions are notable for their high specificity in distinguishing between diastereotopic or enantiotopic groups at prochiral centers (sec. 10:12). Enzyme-catalyzed hydration of alkenes gives optically active alcohols. The chirality of the enzyme results in a regiospecific addition reaction. In the reverse reaction, one of two diastereotopic protons is removed exclusively (sec. 10:12).

Problems

10–21 Each of the following compounds is treated with Pt and excess H_2 at 25°. Write out each reaction, showing all products.

a. (*Z*)-2-butene
b. 1,3-butadiene
c. methylenecyclobutane
d. 1,3-dimethylcyclohexene
e. 2-ethyl-3-methyl-1-pentene
f. [cyclopentane ring bearing two adjacent =CH_2 groups]
CH_2
CH_2
g. mixed (*E*)- and (*Z*)-2-hexenes

10–22 Show how the following conversions can be made:

a. cyclopentene → cyclopentane
b. cyclohexene → cyclohexane-1,2-d_2
c. 2-bromopentane → pentane
d. 1-iodoheptane → heptane-1,2-d_2
e. 3-methyl-3-hexanol → 3-methylhexane
f. 2,2-dimethylcyclopentanol → 1,1-dimethylcyclopentane (free of other isomers)

10–23 Indicate all reagents and conditions necessary for carrying out the following conversions:

a. ethylene → ethanol

b. propene → $DCH_2\underset{\underset{\displaystyle OD}{|}}{C}HCH_3$

c. 1-pentene → 2-pentanol

d. 1-bromobutane → 2-butene

e. 2-bromopropene → 1,2-dibromopropane

f. chlorocyclopentane → *trans*-1,2-dichlorocyclopentane

g. 2-methyl-2-pentanol → 2,3-dibromo-2-methylpentane

h. (*E*)-3-hexene → 4-bromo-3-hexanol (show stereochemistry)

i. 2-methylbutane → 2-methylbutane-2,3-diol

10–24 The bromination of 3,3-dimethyl-1-butene in methanol is shown in Equation 6. Propose a mechanism that accounts for these results, and use the mechanism to explain why *no* 1-bromo-2-methoxy-3,3-dimethylbutane is found.

$$(CH_3)_3CCH{=}CH_2 \xrightarrow[25^\circ]{Br_2,\ CH_3OH}$$

$$\underset{(44\%)}{(CH_3)_3C\underset{\underset{\displaystyle Br}{|}}{C}HCH_2Br} + \underset{(45\%)}{(CH_3)_3\underset{\underset{\displaystyle Br}{|}}{C}HCH_2OCH_3} + \underset{(0\%)}{(CH_3)_3C\underset{\underset{\displaystyle OCH_3}{|}}{C}HCH_2Br} \quad (6)$$

10–25 The addition of HCl to 3,3-dimethyl-1-butene (eq. 10.17) produces two alkenes in addition to the chloroalkanes. The alkenes are isomers of the starting material. Give the structures of the alkenes and show how they were formed.

10–26 The *ionic* addition of HCl to 3,3,3-trifluoropropene gives only 3-chloro-1,1,1-trifluoropropane (eq. 7). Explain the regiospecificity of this anti-Markovnikov addition reaction. (*Hint:* Which is the more stable of the two intermediate cations that could be formed?)

$$CF_3CH{=}CH_2 \rightarrow CF_3CH_2CH_2Cl \quad (7)$$

10–27 Show the major products for each reaction. Indicate which reaction of each pair will be the faster, and explain your reasoning.

a. $CH_3CH_2CH{=}CH_2 \xrightarrow[CH_2Cl_2]{HBr}$ (8)

$CH_3CH_2\underset{\underset{\displaystyle CH_3}{|}}{C}{=}CH_2 \xrightarrow[CH_2Cl_2]{HBr}$ (9)

b. $-CH_3 \xrightarrow{H_2O,\ H_2SO_4}$ (10)

$\underset{|}{CH_3} \xrightarrow{H_2O,\ H_2SO_4}$ (11)

c. $FCH_2CH{=}CH_2 \xrightarrow{HI}$ (12)

$CH_3CH{=}CH_2 \xrightarrow{HI}$ (13)

10–28 The simple chemical 3-chloro-1,2-propanediol has been cleared for use against rodents. It acts by interfering with sperm production in male rodents. It is also being evaluated as a possible male contraceptive in humans. Suggest a short, high yield synthesis of this compound from inexpensive starting materials, which would be suitable for industrial use.

10–29 The addition of HI to 3,3-dimethyl-1-butene gives mostly unrearranged product (eq. 14). Contrast this result with that obtained in HCl addition,

$$(CH_3)_3CCH{=}CH_2 \xrightarrow{HI} \underset{(90\%)}{(CH_3)_3C\underset{\displaystyle I}{C}HCH_3} + \underset{(10\%)}{(CH_3)_2\underset{\displaystyle I}{\overset{\displaystyle CH_3}{C}}CHCH_3} \quad (14)$$

where much more rearrangement is observed (eq. 10.17, p. 353). Why do the two hydrogen halides produce different amounts of rearrangement?

10–30 The double bond of 1,2-dimethylcyclohexene is hydrated in the presence of HNO_3 (eq. 15). Both (Z)- and (E)-1,2-dimethylcyclohexanol are produced. Show how these results support the carbocation mechanism of alkene hydration.

$$\text{1,2-dimethylcyclohexene} \xrightarrow[HNO_3]{H_2O} \underset{Z}{\text{(CH}_3\text{, H / OH, CH}_3)} + \underset{E}{\text{(H, CH}_3\text{ / OH, CH}_3)} \quad (15)$$

10–31 When 5-methoxy-1-pentene is treated with Cl_2 in CCl_4, the expected product, 1,2-dichloro-5-methoxypentane, is obtained, along with substantial quantities of two unusual products: 1,5-dichloro-2-methoxypentane, and 2-chloromethyltetrahydrofuran (eq. 16). Write a mechanism that accounts for the formation of all three products. (Remember that the methoxy group can be an effective nucleophile.)

$$CH_2{=}CHCH_2CH_2CH_2OCH_3 \xrightarrow[CCl_4]{Cl_2}$$

$$ClCH_2\underset{\displaystyle Cl}{C}H(CH_2)_3OCH_3 + ClCH_2\underset{\displaystyle CH_3O}{C}H(CH_2)_3Cl + \text{2-(ClCH}_2\text{)-tetrahydrofuran} \quad (16)$$

10–32 The polymer nylon 10 is made by heating 11-aminoundecanoic acid, which is in turn prepared from 10-undecenoic acid, obtainable from castor oil (eq. 17).

$$\underset{\text{10-Undecenoic acid}}{CH_2{=}CH(CH_2)_8\overset{\displaystyle O}{\overset{\|}{C}}OH} \xrightarrow[\text{steps}]{\text{two}} \underset{\text{11-Aminoundecanoic acid}}{H_2N(CH_2)_{10}\overset{\displaystyle O}{\overset{\|}{C}}OH} \xrightarrow{\text{heat}} \underset{\text{Nylon 10}}{\left[HN(CH_2)_{10}\overset{\displaystyle O}{\overset{\|}{C}}\right]_n} \quad (17)$$

What are the two steps used to convert 10-undecenoic acid to 11-aminoundecanoic acid? (*Hint:* One of the steps is a substitution reaction.)

10–33 Vinyl chloride (chloroethylene) is the monomer from which polyvinyl chloride (PVC, sec. 10:7) is made. It is made in two steps from ethylene: chlorine is added to ethylene to produce 1,2-dichloroethane, which is then pyrolyzed to vinyl chloride and HCl (eq. 18).

$$CH_2{=}CH_2 \xrightarrow{Cl_2} ClCH_2CH_2Cl \xrightarrow[30\ \text{atm}]{500\text{–}600^\circ} CH_2{=}CHCl + HCl \qquad (18)$$

Propose a mechanism for the pyrolysis step. [*Hint:* This is a free-radical chain process; it will be helpful to review the mechanism for free-radical *addition* (sec. 10:5).] At 500° equilibrium favors $CH_2{=}CHCl$ and HCl; at room temperature, however, $ClCH_2CH_2Cl$ is favored. Explain this difference.

10–34 Three isomeric alkenes, **A**, **B**, and **C**, have the formula C_5H_8. They each absorb only one equivalent of hydrogen in the presence of a catalyst, and they all give the same saturated product.

When treated with HCl, isomers **A** and **B** give only one product, **D** (C_5H_9Cl). Compound **C** yields four products when treated with HCl, none of which are identical with **D**. Propose structures for **A**, **B**, and **C** and show their chemistry.

10–35 a. Use the bond energies of Tables 5–1 (p. 146) and 10–1 (p. 346) to calculate ΔH for each of the chain-carrying steps in the free-radical addition of HBr to propene (eqs. 10.27 and 10.28).

b. Carry out the same calculations for both HCl and HI as reactants.

c. Of the three hydrogen halides, only HBr readily undergoes anti-Markovnikov addition. Can you explain this observation in terms of your calculations?

10–36 Write the structure of the product(s) obtained by treating each of the following alkenes with diborane, followed by basic hydrogen peroxide solution.

a. cyclopentene
b. 1-hexene
c. 2-methyl-1-butene
d. (*Z*)-4-methyl-2-hexene
e. *trans*-3,4-diethylcyclobutene
f. 4-methycyclohexene

10–37 Write the structures of the organic products from the following reactions:

a. 1-pentene $\xrightarrow{OsO_4} \xrightarrow{Na_2SO_3}$

b. (*Z*)-2-butene $\xrightarrow[25^\circ]{\text{dil } KMnO_4}$

c. (*R*,*R*)-hexane-2,3-diol $\xrightarrow[\text{pH } 5]{NaIO_4}$

d. (*E*)-2-hexene $\xrightarrow[100^\circ]{KMnO_4}$

e. (*E*)-2-heptene $\xrightarrow{CH_3\overset{O}{\overset{\|}{C}}OOH}$

f. cyclobutene $\xrightarrow{CH_3\overset{O}{\overset{\|}{C}}OOH} \xrightarrow[H^+]{H_2O}$

g. (*Z*)-2-pentene $\xrightarrow{CH_3\overset{O}{\overset{\|}{C}}OOH}\xrightarrow{HBr}$

h. (*Z*)-1-bromo-4-chloro-2-butene $\xrightarrow{O_3}\xrightarrow{(CH_3)_2S}$

i. (*E*)-1-bromo-4-chloro-2-butene $\xrightarrow{O_3}\xrightarrow[H_2O]{H_2O_2}$

j. 1-methyl-2-ethylcycloheptene $\xrightarrow{O_3}\xrightarrow{Zn,\ HOAc}$

10–38 Each of the following alkenes was treated with ozone followed by $(CH_3)_2S$. Write out each reaction, showing structures of reactants and products.

a. 3,3-dimethylcyclopentene
b. 2-methyl-1,3-butadiene
c. 2,3-dimethyl-2-butene
d. 1,5-hexadiene

10–39 In each case, give the structures of reactants and products, including stereochemistry where appropriate.

a. (*Z*)-3,4-dimethyl-3-hexene + peroxyacetic acid →

b. 1-methylcyclohexene $\xrightarrow{Pt,\ D_2}$

c. (*R*)-3-methyl-1-pentene $\xrightarrow{OsO_4}\xrightarrow{Na_2SO_3}$

d. (*E*)-2-heptene $\xrightarrow[0°]{HBr\ (g)}$

10–40 A useful and mild method of cleaving alkenes involves treating them with an excess of $NaIO_4$ and a catalytic amount of $KMnO_4$ (eq. 19). This technique depends on the ability of IO_4^- to oxidize low-valent Mn species back to MnO_4^-.

$$CH_3CH_2\overset{CH_3}{\overset{|}{C}}{=}CHCH_2CH_3 \xrightarrow[KMnO_4\ (cat)]{NaIO_4} CH_3CH_2\overset{CH_3}{\overset{|}{C}}{=}O + CH_3CH_2\overset{O}{\overset{\|}{C}}OH \qquad (19)$$

Write all the individual oxidation steps involving organic species for the reaction shown in Equation 19.

10–41 Write the structure corresponding to each of the following names:

a. ethylene oxide
b. propylene oxide
c. *trans*-2-methylcyclohexanol
d. isobutyl ethyl sulfide
e. butyl hydrogen sulfate
f. *meso*-3,4-dichlorohexane
g. 1-methoxy-2-bromobutane

10–42 Alkene **A** has the molecular formula C_7H_{14}; its infrared spectrum contains a band at 970 cm^{-1}. When **A** is treated successively with ozone and dimethyl sulfide, it is transformed into a mixture of **B** and **C**. The ^1H nmr spectrum of **B**, C_3H_6O, consists of a triplet, a multiplet, and a triplet, the areas of which are in the ratio 3:2:1. The ^1H nmr spectrum of **C**, C_4H_8O, contains a doublet, a multiplet, and a doublet; area ratios are 6:1:1. Write structures for **A**, **B**, and **C**.

10–43 The sex attractant of the common housefly is a hydrocarbon called *muscalure*. Its mass spectrum shows a molecular ion at $m/z = 322$. When treated with H_2 and Pt, muscalure absorbs one equivalent of H_2. There is a band at 690 cm^{-1} in the ir spectrum of muscalure. When it is treated with ozone and then reduced, it yields two straight-chain aldehydes. The aldehydes have molecular ions at $m/z = 142$ and 212. Write the complete structure of muscalure.

10–44 Show three repeating units of the polymer formed from each of the following monomers:

a. propene

b. methylpropene

c. 1,1-dichloroethylene

d. $CH_2{=}CHOC(=O)CH_3$ (vinyl acetate)

10–45 Give the structure of a compound that fits the data in each case.

a. An optically active compound, **A** (C_8H_{16}), upon catalytic hydrogenation, gives an optically inactive product, **B** (C_8H_{18}).

b. An optically active alkene, **B** (C_7H_{14}), on ozonization followed by reduction, gives two products, one of which has the formula C_2H_4O.

c. Compound **C** (C_7H_{10}) undergoes catalytic hydrogenation to give a product, **D** (C_7H_{14}), which has one methyl group in it.

d. Compound **E** (C_6H_{10}) is found to give the same diol, **F**, whether treated with $KMnO_4$ solution or with peroxyacetic acid followed by hydrolysis. After catalytic hydrogenation **E** gives a saturated product, **G** (C_6H_{12}).

10–46 a. Under free-radical conditions, ethylene is much more difficult to polymerize than is propene. Write the first few steps for the free-radical polymerization of each monomer, and explain why the propene is more reactive.

b. The average length of a polymer chain can be affected by the presence of certain additives. For instance, the presence of a thiol, RSH, can result in a marked shortening of the average chain length. Show how this additive could interact with a growing polyethylene chain.

c. Would you expect the same relative reactivity (propene > ethylene) that is observed for radical polymerization to be true for cationic polymerization? Explain your answer.

10–47 Consider the starred atoms in each of the following structures. Which ones are prochiral? Of those that are prochiral, which have enantiotopic groups attached to them and which have diastereotopic groups?

a. ★CH_2CH_3 on C bearing F, F, F

b. ★CH_2CH_3 on C★ bearing H, Cl, Br

c. ★CH_2CH_3 on C★ bearing Br, Br, CH_3CH_2

d. O=P★ bearing F, OCH_3, CH_2CH_3

e. O=P★ bearing CH_3, CH_3, CH_2CH_3

f. Cyclohexane ring with ★ carbon bearing ★$CHBr_2$ and CH_3; ★ ring carbon

10–48 Both dichloromethane and bromochloromethane are achiral, but only the latter is prochiral.

a. What symmetry elements are present in each molecule? Does the presence or absence of a particular symmetry element seem to be related to prochirality?

b. Divide each molecule in half by means of a plane that cuts through the halogen atoms. What is the stereoisomeric relationship between the two halves of the dichloromethane molecule? What is the stereoisomeric relationship between the two halves of the bromochloromethane molecule?

10–49 The addition of water to the double bond of fumaric acid to yield (S)-malic acid (eq. 20) is an important step in the metabolic *tricarboxylic acid cycle* (see sec. 19:12).

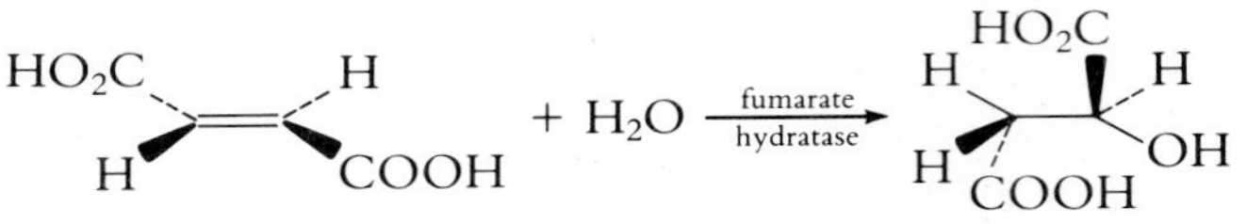

$$HO_2C(H)C{=}C(H)COOH \;(\text{Fumaric acid}) + H_2O \xrightarrow[\text{hydratase}]{\text{fumarate}} (S)\text{-Malic acid} \qquad (20)$$

Fumaric acid (S)-Malic acid

a. The addition of water can be either *syn* or *anti*. Give an example of each that would lead to the observed stereochemistry.
b. When D_2O is used, the only product is (2S,3R)-malic-3-d acid. Draw its structure, and suggest a mechanism for its formation.
c. Are the two ends of the double bond equivalent in *enzyme-bound* fumaric acid? Explain your answer.

10–50 The microorganism *Mycobacterium,* which is responsible for the disease tuberculosis, converts oleic acid to tuberculostearic acid (eq. 21) by addition of a methyl group, followed by reduction. The methyl donor is a sulfonium salt, (S)-adenosylmethionine (abbreviated $R_2\overset{+}{S}—CH_3$).

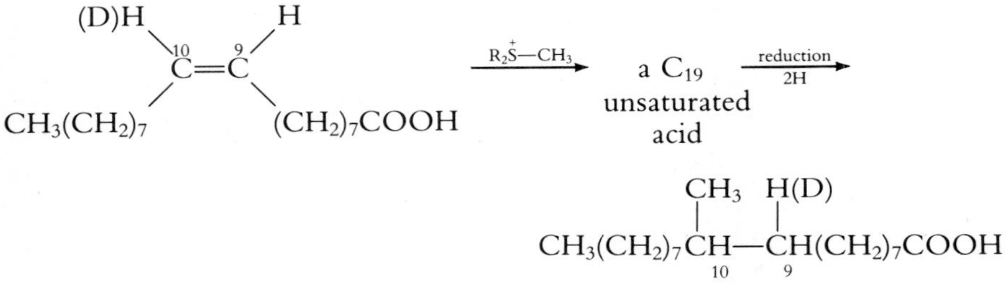

$$CH_3(CH_2)_7(D)H\overset{10}{C}{=}\overset{9}{C}H(CH_2)_7COOH \xrightarrow{R_2\overset{+}{S}—CH_3} \text{a } C_{19} \text{ unsaturated acid} \xrightarrow[2H]{\text{reduction}} CH_3(CH_2)_7\underset{10}{CH}(CH_3)—\underset{9}{CH}(H(D))(CH_2)_7COOH \qquad (21)$$

Tuberculostearic acid

When oleic-10-d acid is used, the deuterium eventually winds up at C-9 of tuberculostearic acid. It was shown that the methyl group at C-10 comes from (S)-adenosylmethionine. There is an intermediate unsaturated compound that is reduced to tuberculostearic acid. Propose a structure for this C_{19} unsaturated acid, and a mechanism for its formation.

Supplementary Reading

W. R. Dolbier, Jr., Electrophilic additions to olefins, *J. Chem. Ed.,* **46,** 342 (1969).

E. L. Eliel, Stereochemical non-equivalence of ligands and faces (heterotopicity), *J. Chem. Ed.,* **57,** 52 (1980).

N. Isenberg and M. Grdinic, A modern look at Markovnikov's rule and the peroxide effect, *J. Chem. Ed.,* **46,** 601 (1969).

F. W. Harris, Introduction to polymer chemistry, *J. Chem. Ed.,* **58,** 837 (1981).

D. C. Phillips, The three-dimensional structure of an enzyme molecule, *Scientific American* (*Sci. Am.*), **215,** 78 (1966).

R. P. Quirk, Stereochemistry and macromolecules, *J. Chem. Ed.,* **58,** 540 (1981).

C. Walsh, *Enzymatic Reaction Mechanisms,* W. H. Freeman and Co., San Francisco, Ca. (1979).

M. A. Wilson, Classification of the electrophilic addition reactions of olefins and acetylenes, *J. Chem. Ed.,* **52,** 495 (1975).

The Synthesis and Structure of Aldehydes and Ketones: Nucleophilic Additions to the Carbonyl Group 11

In this chapter we will examine the structure, synthesis, and reactions of two types of carbonyl compounds: aldehydes and ketones.

Carbonyl compounds contain the carbonyl group

$$>C=O$$

They exhibit many chemical properties similar to those of alkenes, including their tendency to undergo addition reactions. This is characteristic of functional groups containing π bonds. There are significant differences in reactivity, however, between alkenes and carbonyl compounds. Alkenes are attacked by electrophiles, and the reaction intermediates are often carbocations or species related to them. Carbonyl compounds, in contrast, are attacked rapidly by many *nucleophiles*. This is due to the fact that a *stable anionic intermediate, an alkoxide ion, is formed* (eq. 11.1). In this ion the negative charge is localized on the electronegative oxygen atom. Nucleophilic additions are often reversible because the C=O bond is a particularly strong one.

$$\text{Nu:}^- + \;>\overset{\delta+}{C}=\overset{\delta-}{\ddot{\underset{..}{O}}} \rightleftharpoons \text{Nu}-\overset{|}{\underset{|}{C}}-\ddot{\underset{..}{O}}\text{:}^- \tag{11.1}$$

Alkoxide ion

It will be helpful for you to remember this simple equilibrium. It represents the basic reaction initiation step for most processes discussed in this chapter, as well as for those in subsequent chapters dealing with other carbonyl compounds.

Problem

11–1 Use the data in Table 10–1 (p. 346) to calculate the bond strength of the C=O π bond and compare it to the C=C π bond strength. Why are C=O additions more likely to be reversible than C=C addition reactions?

11:1 Nomenclature of Aldehydes and Ketones

Before we can begin to examine the reactivity of the carbonyl group, we must first establish a basic working knowledge of nomenclature and structure.

A ketone is a carbonyl compound in which the carbonyl group is bonded to two other carbon atoms. In aldehydes the carbonyl group is bonded to at least one hydrogen atom. In the unique case of formaldehyde, the simplest aldehyde, the C=O is bonded to two hydrogen atoms. If the carbonyl carbon atom is attached to an atom other than carbon or hydrogen, the compound is not an aldehyde or a ketone.

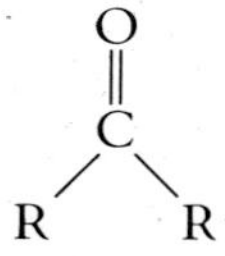

$$\underset{\text{Ketone}}{R\overset{O}{\overset{\|}{C}}R} \qquad \underset{\text{Aldehyde}}{R\overset{O}{\overset{\|}{C}}H} \qquad \underset{\text{Formaldehyde}}{H\overset{O}{\overset{\|}{C}}H} \qquad R = \text{alkyl or aryl group}$$

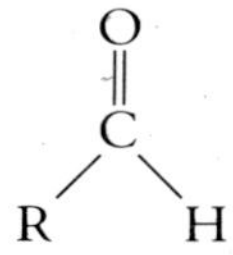

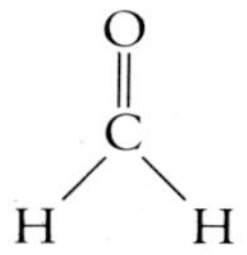

In the IUPAC system the name of an aldehyde or ketone is based upon the longest chain that includes the carbonyl group. The final "-e" is dropped from the hydrocarbon name denoting that chain, and the suffix "-al" is added to denote an aldehyde or "-one" to signify a ketone. The carbonyl carbon atom is understood to be at C-1 in aldehydes, but its position in a ketone must usually be specified with a number. The chain is numbered so as to give the lowest possible number to the carbonyl carbon atom. If the compound contains *both* types of carbonyl groups, then it is named as an aldehyde and the keto oxygen atom is treated as a substituent on the chain, denoted by the prefix "oxo-".

$$\underset{\text{Methanal}}{H_2C{=}O} \qquad \underset{\text{Propanal}}{CH_3CH_2\overset{O}{\overset{\|}{C}}H} \qquad \underset{\substack{\text{Propanone}\\(\textit{acetone})}}{CH_3\overset{O}{\overset{\|}{C}}CH_3}$$

$$\underset{\text{2-Methylbutanal}}{CH_3CH_2\underset{\substack{|\\CH_3}}{C}H\overset{O}{\overset{\|}{C}}H} \qquad \underset{\text{2-Pentanone}}{CH_3CH_2CH_2\overset{O}{\overset{\|}{C}}CH_3} \qquad \underset{\text{2,2-Dimethyl-3-pentanone}}{(CH_3)_3C\overset{O}{\overset{\|}{C}}CH_2CH_3}$$

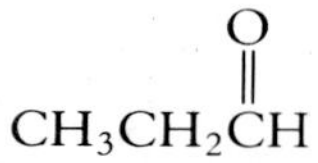

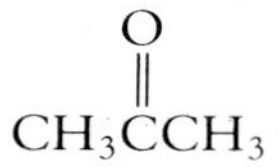

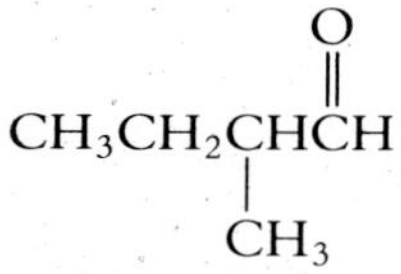

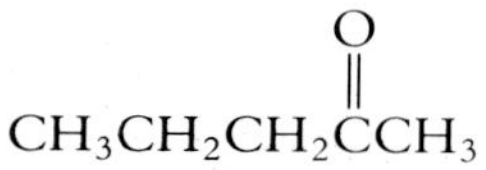

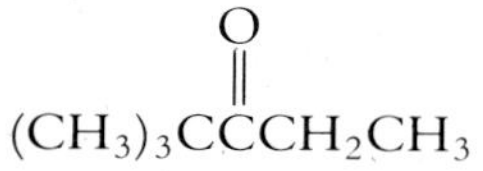

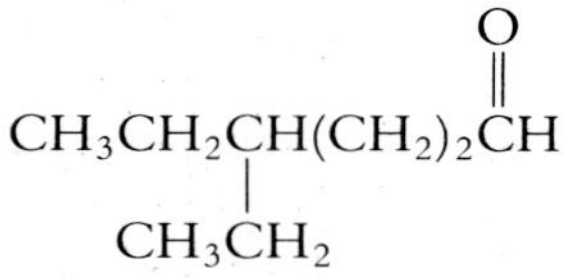

$$\underset{\text{4-Ethylhexanal}}{CH_3CH_2\underset{\substack{|\\CH_3CH_2}}{C}H(CH_2)_2\overset{O}{\overset{\|}{C}}H} \qquad \underset{\text{5-Methyl-8-oxodecanal}}{CH_3CH_2\overset{O}{\overset{\|}{C}}CH_2CH_2\underset{\substack{|\\CH_3}}{C}H(CH_2)_3\overset{O}{\overset{\|}{C}}H}$$

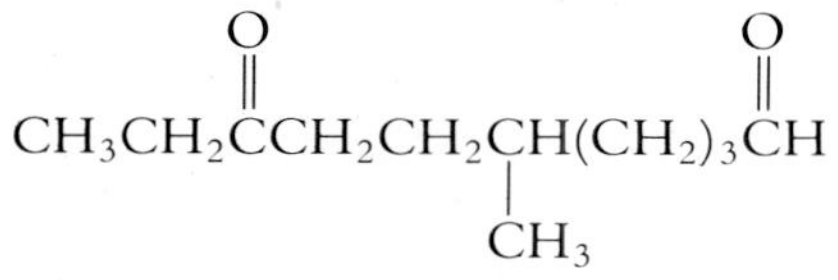

If the carbonyl group is part of a ring, then the compound must be a ketone and the root of its name is derived from the cyclic hydrocarbon. Numbering begins at the carbonyl carbon atom, which is understood to be number "1" in simple cyclic ketones; thus, its position need not be explicitly stated. When more than one carbonyl is present, the positions must be numbered so as to give the carbonyl groups the lowest possible numbers.

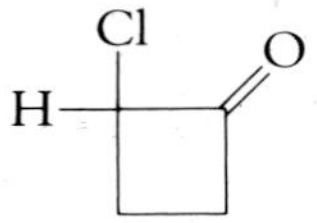

2-Chlorocyclobutanone

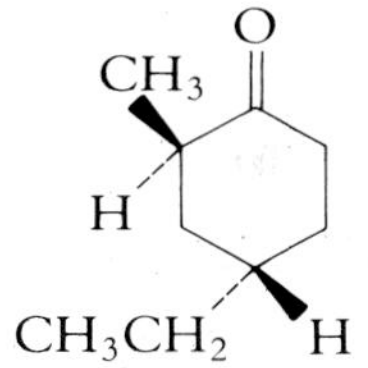

trans-4-Ethyl-2-methylcyclohexanone

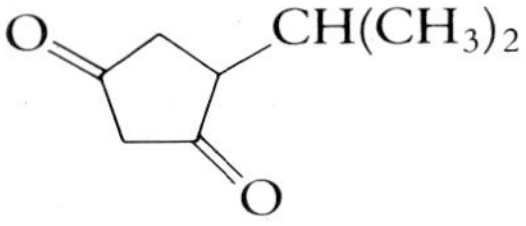

4-Isopropyl-1,3-cyclopentanedione

A special case occurs when an aldehyde group is directly attached to a saturated ring. In this case, the suffix "-carboxaldehyde" is used. Notice that the carbon atom *bonded to* the aldehyde group is understood to be number 1.

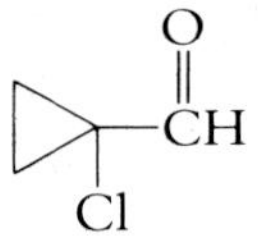

1-Chlorocyclopropanecarboxaldehyde

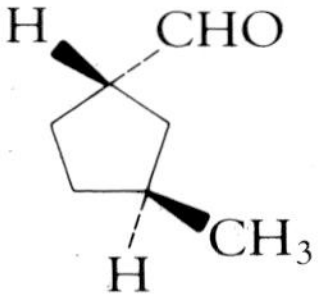

trans-3-Methylcyclopentanecarboxaldehyde

Certain non-IUPAC names for aldehydes and ketones are still in use today. For example, the C_1 and C_2 aldehydes are referred to in the chemical literature as *formaldehyde* and *acetaldehyde,* respectively. These names derive from formic acid and acetic acid, products of the easy oxidation of the aldehydes. Table 11–1 gives the structures and older common names of some low molecular weight aldehydes that are all derived from the names of the corresponding acids.

Table 11–1. Structures and Common Names of Some Simple Aldehydes

Structure	IUPAC Name	Common Name	Boiling Point (°C)
CH_2O	methanal	*formaldehyde*	−21
CH_3CHO	ethanal	*acetaldehyde*	20
CH_3CH_2CHO	propanal	*propionaldehyde*	50
$CH_3CH_2CH_2CHO$	butanal	*butyraldehyde*	76
$(CH_3)_2CHCHO$	2-methylpropanal	*isobutyraldehyde*	64
$CH_3(CH_2)_3CHO$	pentanal	*valeraldehyde*	103
$(CH_3)_2CHCH_2CHO$	3-methylbutanal	*isovaleraldehyde*	93
$(CH_3)_3CCHO$	2,2-dimethylpropanal	*pivaldehyde*	75

Table 11–2. Structures and Radicofunctional Names of Some Common Ketones

Structure	IUPAC Name	Radicofunctional Name	Boiling Point (°C)
CH_3COCH_3	propanone	—	56
$CH_3COCH_2CH_3$	butanone	ethyl methyl ketone	80
$CH_3COCH_2CH_2CH_3$	2-pentanone	methyl propyl ketone	102
$CH_3CH_2COCH_2CH_3$	3-pentanone	diethyl ketone	102
$(CH_3)_2CHCOCH_3$	3-methyl-2-butanone	isopropyl methyl ketone	95

There is also a radicofunctional system of nomenclature for ketones. In this system we name the two alkyl groups attached to the carbonyl group, and follow this by the word "ketone," to indicate the presence of the carbonyl group. Some examples are given in Table 11–2.

Problem

11–2 Give IUPAC names for the following structures:

a. 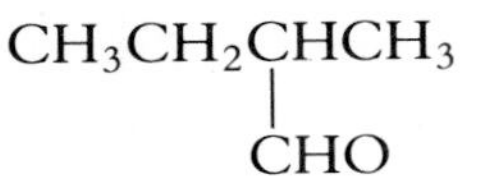 $CH_3CH_2CHCH_3$ with CHO on C-3 ($CH_3CH_2CH(CHO)CH_3$)

b. $CH_3CH_2CHCH_3$ with $CH_3C{=}O$ on C-3 ($CH_3CH_2CH(COCH_3)CH_3$)

c. Cyclohexane ring bearing H and CHO on one carbon, and H and CH_3 on the opposite carbon (wedge bonds to CHO and CH_3)

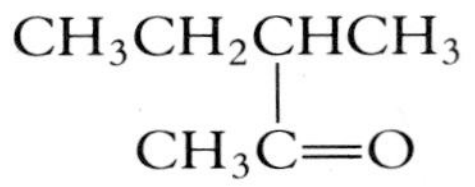

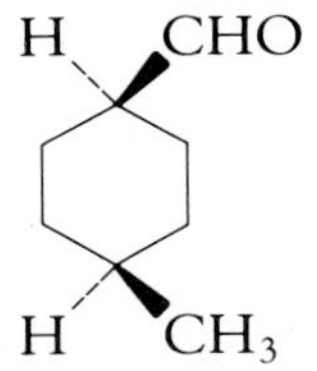

d. 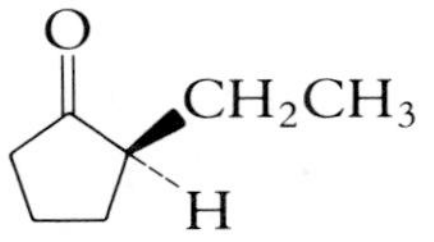 Cyclopentanone (C=O) with CH_2CH_3 (wedge) and H (dash) on the adjacent carbon

e. 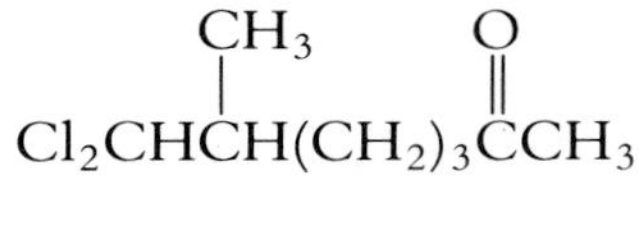 $Cl_2CHCH(CH_3)(CH_2)_3CCH_3$ with C=O ($Cl_2CHCH(CH_3)(CH_2)_3COCH_3$)

11:2 Physical Properties and Structures of Aldehydes and Ketones Revealed by IR and NMR Spectroscopy

In order to get some feel for the properties of the carbonyl group, it is useful to compare the physical properties of aldehydes and ketones (Tables 11–1 and 11–2) with those of related oxygen-containing molecules, such as ethers (Table 6–5, p. 199) and alcohols (Table 6–4, p. 198). Aldehydes and ketones have higher boiling points than do ethers of similar molecular weight because the carbonyl group, **1**, is more polar than the ether group. On the

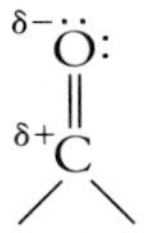

$\delta-$:Ö: ‖ $\delta+$ C (with two bonds)

Oxygen is more electronegative than carbon

1

other hand, aldehydes and ketones are not hydrogen bond donors, so they boil at lower temperatures than the corresponding alcohols. Because car-

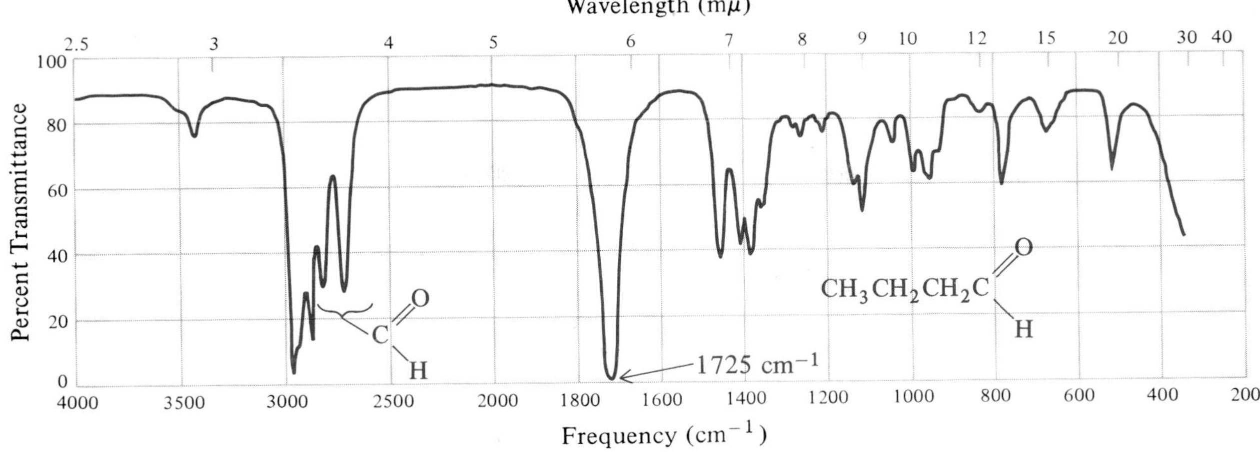

Figure 11–1. *Ir spectrum of butanal.*

bonyl compounds are hydrogen bond acceptors, the low molecular weight ones (four or fewer carbons) are reasonably water soluble.

The carbonyl group is usually easily identified by ir spectroscopy. All compounds containing carbonyl groups (including carboxylic acids and esters) exhibit a peak in their infrared spectra in the region 1660 to 1900 cm^{-1}. This peak is characteristic of the C=O stretching vibration, and for most aldehydes and ketones it lies in the 1690 to 1745 cm^{-1} region. It will usually be the most intense band in the spectra of simple molecules. In Chapter 8 you saw that ir band intensity is proportional to the change in dipole moment as the absorbing bond is stretched or bent. In order for there to be a substantial change in the dipole moment, the bond must be fairly polar. The high intensity of the C=O stretch is, therefore, another indication that the C=O bond is highly polarized.

The exact position of the C=O absorption varies somewhat depending upon whether the compound is an aldehyde or ketone, or has certain other structural features. The best way to ascertain the position of these peaks is to first consider a standard "normal" absorption and then predict how changes in the environment near the carbonyl will change the frequency. For example, Figure 11–1 gives the ir spectrum of butanal, which shows a carbonyl peak at 1725 cm^{-1}. Figure 11–2 gives the ir spectrum of 2-butanone, with a carbonyl absorption at 1715 cm^{-1}. These can be considered the normal absorptions for aldehydes and ketones, respectively. Thus, acetaldehyde would be expected to absorb near 1725 cm^{-1} and cyclohexanone to absorb near 1715 cm^{-1}. These expectations are met when we examine the ir spectra of these compounds.

Deviations of carbonyl frequencies from the normal values alert us to unusual structural features. For instance, with cyclic ketones smaller than cyclohexanone, the carbonyl frequencies increase steadily as the ring size decreases. We can correlate this trend with increasing angle strain in these compounds. Angle strain leads to a change in hybridization of the carbonyl carbon atom, which in turn affects the C=O stretching frequency. These

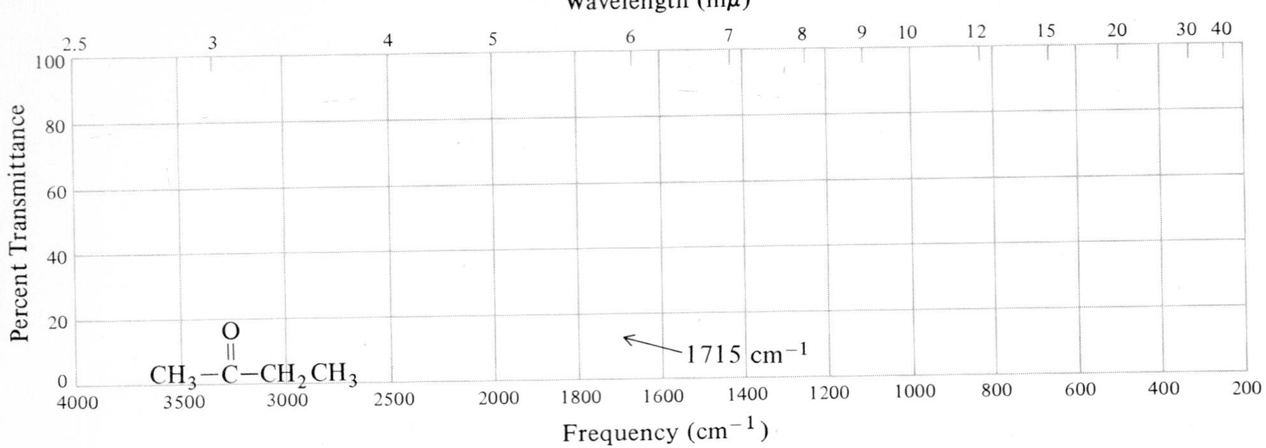

Figure 11–2. *Ir spectrum of 2-butanone.*

examples point again to the value of spectra as sources of structural, as well as analytical, information.

Cyclopentanone 1745 cm^{-1}

Cyclobutanone 1780 cm^{-1}

Cyclopropanone 1815 cm^{-1}

If a carbonyl group is next to a carbon–carbon double bond (—C=C—C=O), a shift of the carbonyl frequency to lower values (1670 to 1700 cm^{-1}) often occurs. This is the result of electron delocalization effects (Chap. 13).

Problem

11–3 The cyclohexanone ir carbonyl absorption is at 1715 cm^{-1}. What can you conclude about angle strain in this molecule?

Apart from their characteristic C=O ir bands, aldehydes can often be identified from ir absorptions due to the hydrogen attached to the carbonyl carbon atom. The aldehyde hydrogen gives rise to a double absorption whose peaks are near 2720 and 2820 cm^{-1} (Fig. 11–1). The lower of these two frequencies is valuable diagnostically because it occurs at an unusually low frequency for a C—H stretching motion.

The aldehyde hydrogen atom is also readily detected by ^{1}H nmr spectroscopy. It gives rise to a peak at very low field, usually in the range 9.6 to 10.0 ppm (Fig. 11–3). This hydrogen is strongly deshielded by the carbonyl group as the result of two factors:

1. the partial positive charge on the carbonyl group (**1**) pulls electrons away from neighboring atoms by way of inductive effect; and

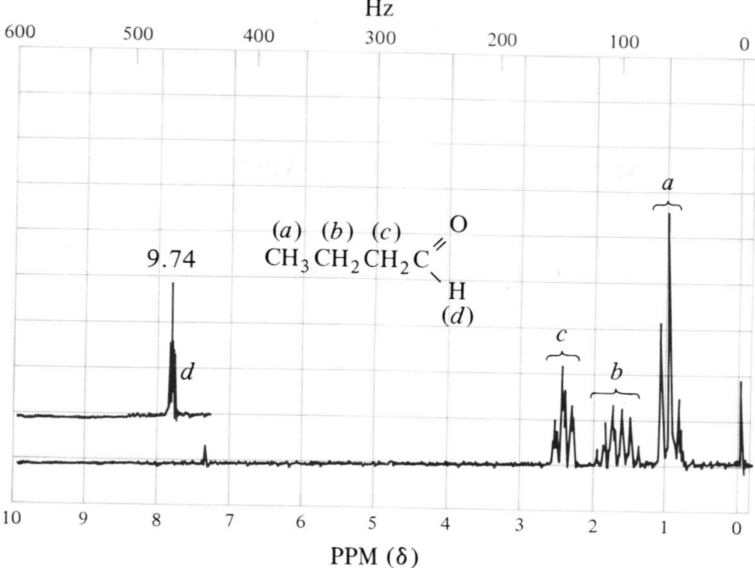

Figure 11–3. *¹H nmr spectrum of butanal.*

2. the π electrons of the carbonyl group create a magnetic field which deshields neighboring atoms.

These effects are also felt by methylene and methyl groups directly attached to the carbonyl group in both aldehydes and ketones. Their hydrogens give rise to peaks in the region 2.0 to 2.5 ppm, which is about 1 to 1.5 ppm lower than their positions in the spectra of alkanes (Figs. 11–3 and 11–4).

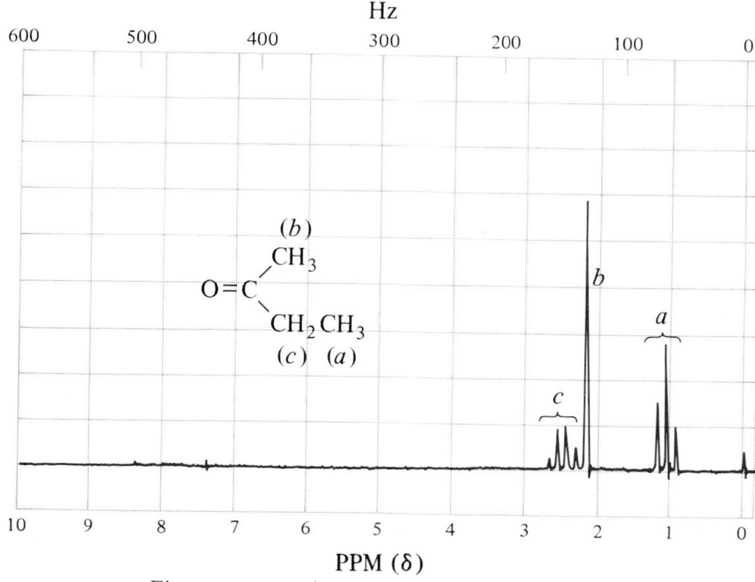

Figure 11–4. *¹H nmr spectrum of 2-butanone.*

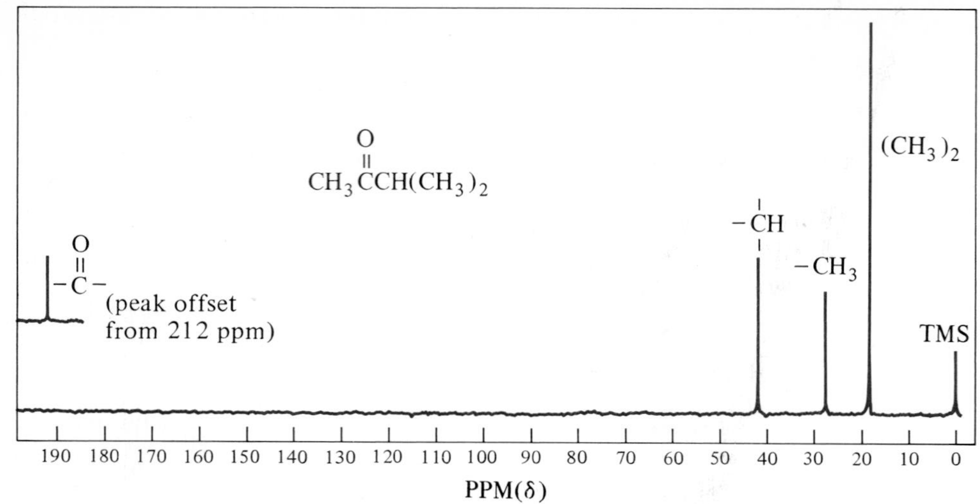

Figure 11–5. *Proton-decoupled ^{13}C nmr spectrum of 3-methyl-2-butanone.*

Carbonyl group deshielding affects ^{13}C nmr as well. Figure 11–5 gives the decoupled spectrum of 3-methyl-2-butanone. Note that the $^{13}C{=}O$ peak is at 212 ppm. This can be considered additional proof for the partial positively-charged character of the carbon in a carbonyl group.

Problem

11–4 Given below are ir and ^1H nmr spectra for an unknown compound, $C_6H_{10}O$. Interpret the spectral data and suggest a structure.

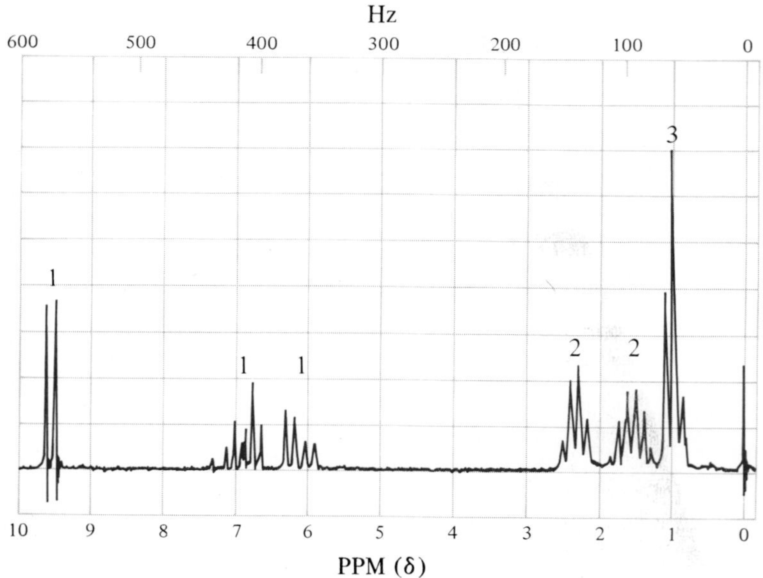

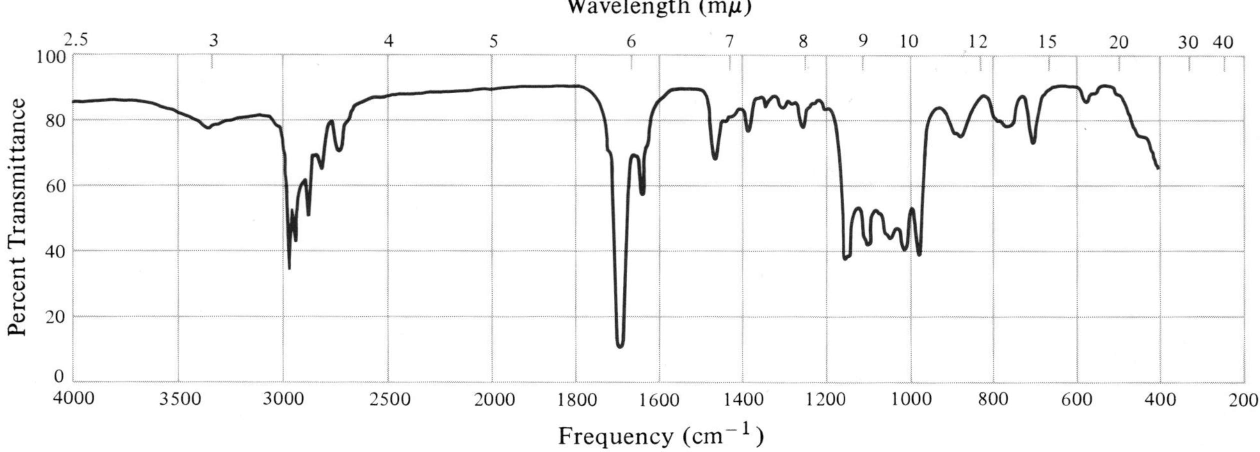

11:3 Synthesis of Ketones and Aldehydes by Oxidation of Alcohols

Recall from Section 10:2 that the addition of two hydrogen atoms to a π bond results in reduction of the alkene. When two hydrogens are removed to form a π bond, we have an oxidation reaction. Equation 11.2 shows a schematic representation of this process, where [O] represents any of a num-

$$\text{H—}\overset{|}{\underset{|}{\text{C}}}\text{—OH} \xrightarrow{[\text{O}]} \text{>C=O} + \text{H}_2\text{O} \qquad (11.2)$$

ber of chemical oxidants. The most widely used oxidants are inorganic chromium compounds containing Cr(VI), which is reduced to Cr(III) in the course of the reaction. Chromium trioxide, CrO_3, and sodium dichromate, $Na_2Cr_2O_7$, are the most common oxidizing agents. *Secondary alcohols can be oxidized to ketones, while primary alcohols can be oxidized to aldehydes* (eqs. 11.3 and 11.4). Oxidations of these two types of alcohols are carried out under slightly different conditions, so we will discuss their reactions separately.

$$\underset{\text{2° Alcohol}}{\text{R}\overset{\text{OH}}{\underset{\text{H}}{\text{C}}}\text{R}'} \xrightarrow{\text{Cr(VI)}} \underset{\text{Ketone}}{\text{R}\overset{\text{O}}{\overset{\|}{\text{C}}}\text{R}'} \qquad (11.3)$$

$$\underset{\text{1° Alcohol}}{\text{R}\overset{\text{OH}}{\underset{\text{H}}{\text{C}}}\text{H}} \xrightarrow{\text{Cr(VI)}} \underset{\text{Aldehyde}}{\text{R}\overset{\text{O}}{\overset{\|}{\text{C}}}\text{H}} \xrightarrow[\text{depending on conditions}]{\text{oxidation may proceed further}} \qquad (11.4)$$

The first step in the oxidation of secondary alcohols to ketones is the conversion of the oxidant to chromic acid. This is done in aqueous acid solution (eqs. 11.5 and 11.6). Most organic compounds are not very soluble

$$\underset{\text{Dichromate ion}}{Cr_2O_7^{2-}} + 2H^+ + H_2O \xrightarrow{\text{acetone}} \underset{\text{Chromic acid}}{2H_2CrO_4} \quad (11.5)$$

$$\underset{\text{Chromium trioxide}}{CrO_3} + H_2O \xrightarrow{\text{acetone}} H_2CrO_4 \quad (11.6)$$

in water; hence, an inert organic cosolvent such as acetone is added to the solution. As you can see in Equations 11.7 and 11.8, the oxidation of secondary alcohols usually produces good yields of ketones.

$$\underset{\text{2-Octanol}}{3CH_3\underset{\substack{|\\OH}}{C}H(CH_2)_5CH_3} + 2H_2CrO_4 + 6H^+ \xrightarrow[H_2O]{\text{reflux}} \underset{\substack{\text{2-Octanone}\\(96\%)}}{3CH_3\underset{\substack{\|\\O}}{C}(CH_2)_5CH_3} + 2Cr^{3+} + 8H_2O \quad (11.7)$$

(11.8) 3 (cyclohexanol: cyclohexane ring bearing H and OH) $+ 2H_2CrO_4 + 6H^+ \xrightarrow[H_2O,\ 60°]{}$ 3 (cyclohexanone, C=O) $+ 2Cr^{3+} + 8H_2O$

(85%)

The chromic acid oxidation of alcohols involves a complex multistep mechanism, the outline of which is shown in Equations 11.9 and 11.10 for the conversion of 2-propanol to acetone.

$$(CH_3)_2CHOH + HO\overset{\overset{O}{\|}}{\underset{\underset{O}{\|}}{Cr}}OH \xrightarrow[\text{steps}]{\text{several}} \underset{\text{2-Propyl hydrogen chromate}}{(CH_3)_2CHO\overset{\overset{O}{\|}}{\underset{\underset{O}{\|}}{Cr}}OH} + H_2O \quad (11.9)$$

(11.10) $(CH_3)_2C(H)$—Ö—Cr(=O)(=O)—OH (cyclic electron flow, arrows) $\rightarrow (CH_3)_2C{=}\ddot{O}: + H_2CrO_3$

In Equation 11.9 the alcohol reacts with the acid to form the *ester,* 2-propyl hydrogen chromate. (A detailed discussion of the mechanism of such reactions is given in Chapter 17.) The next step, shown in Equation 11.10, is the elimination of $HCrO_3^-$ from the ester.★ This is not the last step in the process because H_2CrO_3 (which contains Cr(IV)) must undergo a series of additional changes before ending up as Cr(III).

★ This is shown as a concerted process, but stepwise pathways are also possible.

Chromic acid can also be used to oxidize primary alcohols. The final product, however, is not the expected aldehyde but a carboxylic acid (eq. 11.11). An aldehyde *is* the first product of this reaction, but it reacts reversibly with water to produce a hydrate. This hydrate resembles a secondary alcohol and is subjected to further oxidation, yielding a carboxylic acid (eq. 11.12).

$$\text{(11.11)} \quad RCH_2OH \xrightarrow[H_3O^+]{CrO_3} \left[\begin{matrix} O \\ \| \\ RCH \end{matrix}\right] \xrightarrow[\text{oxidation}]{\text{further}} \begin{matrix} O \\ \| \\ RCOH \end{matrix}$$

Aldehyde — Carboxylic acid

$$\text{(11.12)} \quad \begin{matrix} O \\ \| \\ RCH \end{matrix} + H_2O \rightleftharpoons R{-}\begin{matrix} OH \\ | \\ C \\ | \\ H \end{matrix}{-}OH \xrightarrow{\text{oxidation}} R{-}\begin{matrix} O \\ \| \\ C \end{matrix}{-}OH$$

Aldehyde hydrate

Equation 11.13 gives an example of the H_2CrO_4 oxidation of a 1° alcohol to a carboxylic acid. Potassium permanganate, $KMnO_4$, produces the same result as H_2CrO_4.

$$\text{(11.13)} \quad 3CH_3CH_2CH_2OH + 4H_2CrO_4 + 12H^+ \xrightarrow{\text{heat}} 3CH_3CH_2\overset{\overset{\displaystyle O}{\|}}{C}OH + 4Cr^{3+} + 13H_2O$$

1-Propanol — Propanoic acid (65%)

Since it is the aldehyde hydrate (eq. 11.12) that gets oxidized to a carboxylic acid, it follows that if water is not present we should be able to stop the oxidation at the aldehyde stage. We therefore need a polar, *nonaqueous* solvent that is capable of dissolving both the CrO_3 and the organic alcohol substrate. It has been found that the organic base *pyridine* will complex with the CrO_3 and make it soluble in CH_2Cl_2.

Pyridine

Examples are given in Equations 11.14 and 11.15.

$$\text{(11.14)} \quad CH_3(CH_2)_6CH_2OH \xrightarrow[CH_2Cl_2,\ 25^\circ]{CrO_3,\ \text{pyridine}} CH_3(CH_2)_6\overset{\overset{\displaystyle O}{\|}}{C}H$$

1-Octanol — Octanal (90%)

$$\text{(11.15)} \quad \underset{\text{Hexane-1,6-diol}}{HO(CH_2)_6OH} \xrightarrow[CH_2Cl_2,\ 25°]{CrO_3,\ pyridine} \underset{\text{Hexanedial (68\%)}}{HC(=O)(CH_2)_4C(=O)H}$$

Example

What organic products are formed when the following two compounds are treated separately with CrO_3, H_3O^+, and with CrO_3 in pyridine/CH_2Cl_2?

$CH_3CH_2CH_2CH_2OH$ (1-Butanol) $CH_3CH(OH)CH_2CH_3$ (2-Butanol)

Answer

1-Butanol is a primary alcohol. It yields a carboxylic acid when oxidized with chromic acid and an aldehyde when the oxidation occurs with pyridine in CH_2Cl_2 (a nonaqueous solution). The reactions proceed as follows:

$$CH_3CH_2CH_2CH_2OH + CrO_3,\ H_3O^+ \rightarrow CH_3CH_2CH_2C(=O)OH$$

$$CH_3CH_2CH_2CH_2OH + CrO_3,\ \text{pyridine}/CH_2Cl_2 \rightarrow CH_3CH_2CH_2C(=O)H$$

2-Butanol is a secondary alcohol. In both cases, the product is the same:

$$CH_3CH(OH)CH_2CH_3 + CrO_3,\ H_3O^+ \rightarrow CH_3C(=O)CH_2CH_3$$

$$CH_3CH(OH)CH_2CH_3 + CrO_3,\ \text{pyridine}/CH_2Cl_2 \rightarrow CH_3C(=O)CH_2CH_3$$

Problem

11–5 Show the organic products formed when the following compounds are treated with CrO_3, H_3O^+, and with pyridine and CrO_3 in CH_2Cl_2.

a. $(CH_3)_2CHCH(OH)CH_3$ b. 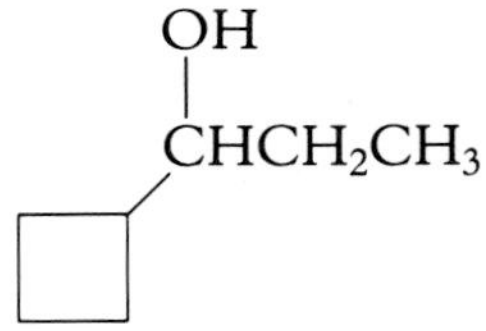 (cyclobutyl)$CH(OH)CH_2CH_3$ c. $CH_3CH_2CH(CH_2OH)CH_3$

d. 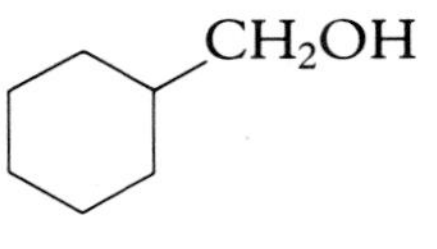 (cyclohexyl)CH_2OH e. $CH_3-C(CH_3)_2-OH$ f. 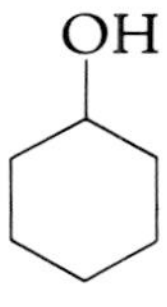 cyclohexanol (cyclohexane ring with OH)

Due to the fact that the chemical oxidants described in this section (and used in the laboratory) are too expensive for commercial use, industry has had to come up with alternative approaches to the synthesis of aldehydes and ketones. One of these is the direct removal of a molecule of hydrogen from the alcohol by passing the alcohol vapor over a catalyst at high temperature (eq. 11.16). This is the reverse of a carbonyl reduction described in Section 11:10.

$$CH_3CH_2OH \xrightarrow[300°]{Cu} CH_3CHO + H_2 \qquad (11.16)$$

Additional laboratory oxidations of alcohols using copper and silver oxidants are important in sugar chemistry and are discussed in Chapter 20.

11:4 Alcohol Synthesis by the Irreversible Addition of Grignard Reagents to Carbonyl Compounds

Let's reconsider the characteristic nucleophilic addition equilibrium of Equation 11.1 as we begin to examine some reactions of aldehydes and ketones in detail. If the nucleophile is very powerful, the equilibrium lies very far to the right and the addition is irreversible. Among the most powerful nucleophiles are carbanions, $R_3C\colon^-$, which react with carbonyl compounds to give alcohols. However, free carbanions are often too reactive to be stable, so we substitute organometallic compounds for them (sec. 6:14). In these compounds the carbon atom bonded to the metal carries a substantial negative charge because carbon is more electronegative than the metal ($R^{\delta-}$—$M^{\delta+}$).

The organometallic compound most often used to react with carbonyl groups is the Grignard reagent. The reaction of aldehydes and ketones with Grignard reagents (RMgX) is an extremely versatile method of synthesizing alcohols (eqs. 11.17 to 11.19). Depending on whether the carbonyl compound is formaldehyde (eq. 11.17), another aldehyde (eq. 11.18), or a ketone (eq. 11.19), the product will be, respectively, a primary, secondary, or tertiary alcohol.

$$C_6H_{11}MgCl + CH_2{=}O \xrightarrow{ether} \xrightarrow{H_3O^+} C_6H_{11}CH_2OH \qquad (11.17)$$

Cyclohexylmethanol
1° alcohol (65%)

$$(CH_3)_2CHMgBr + CH_3CH{=}O \xrightarrow{ether} \xrightarrow{H_3O^+} (CH_3)_2CHCH(OH)CH_3 \qquad (11.18)$$

3-Methyl-2-butanol
2° alcohol (54%)

$$(11.19)\quad CH_3CH_2CH_2CH_2MgBr + (CH_3)_2C{=}O \xrightarrow{\text{ether}} \xrightarrow{H_3O^+} CH_3CH_2CH_2CH_2\underset{\underset{\displaystyle CH_3}{|}}{\overset{\overset{\displaystyle CH_3}{|}}{C}}OH$$

2-Methyl-2-hexanol
3° alcohol (92%)

In a great many cases there will be two or more routes to the same compound. For example, we would have obtained the same product in Equation 11.18 if we had used 2-methylpropanal and methylmagnesium bromide as reactants. However, *it is general practice to choose the carbonyl compound with the least hindered carbonyl group*. Hindered carbonyls are much less susceptible to nucleophilic attack than are unhindered ones. Hence, the reaction shown in Equation 11.18 is the better choice.

We can write a simplified mechanism for this Grignard reaction by keeping in mind the high polarity of organometallic reagents such as RMgBr (eq. 11.20). The electron rich C-1 carbon in the R group of the Grignard reagent is attracted to the electron deficient carbonyl carbon, while the oppositely charged Mg and O atoms are similarly attracted to one another.

$$(11.20)\quad \overset{\delta+}{C}{=}\ddot{O}{:}^{\delta-} \;\; \overset{\delta-}{R}{-}\overset{\delta+}{MgBr} \longrightarrow -\underset{\underset{\displaystyle R}{|}}{\overset{|}{C}}-\ddot{\underset{\cdot\cdot}{O}}{:}MgBr \xrightarrow[H_2O]{H^+} \underset{\underset{\displaystyle R}{|}}{C}-\ddot{\underset{\cdot\cdot}{O}}H + Mg^{2+}\,(aq) + Br^-\,(aq)$$

Some detailed mechanism studies have shown that 2 mol of Grignard reagent and 1 mol of carbonyl compound are required to reach the transition state and that radical intermediates are sometimes generated. Hence, the first step of Equation 11.20 may involve a more complex mechanism than the one shown. Nonetheless, for the purpose of remembering the course and products of the reaction, the pathway shown in Equation 11.20 is satisfactory.

Also note that in order to get the alcohol product, it is necessary to decompose the intermediate alkoxymagnesium halide

$$R_2\underset{\underset{\displaystyle R}{|}}{C}OMgX$$

with aqueous acid. This is done in a separate step at the end of the reaction. Water must be excluded from the original reaction mixture containing the Grignard reagent, since the reagent is a powerful base and will react rapidly with an OH group (eq. 11.21).

$$(11.21)\quad CH_3MgBr + ROH \xrightarrow[\text{fast}]{\text{very}} CH_4 + Mg(OR)Br$$

Example

Choose the proper Grignard reagent and carbonyl compound to prepare

$$\underset{\displaystyle\text{OH}}{\text{CH}_3\underset{|}{\text{C}}\text{HCH}_2\text{CH}_2\text{CH}_3}$$

Answer

In a Grignard synthesis the new C—C bond is formed next to the carbon bearing the OH group. Therefore, we would rewrite the target structure as follows:

$$\text{CH}_3\!\not\!\!-\underset{\text{OH}}{\underset{|}{\text{C}}}\text{HCH}_2\text{CH}_2\text{CH}_3 \quad \text{or} \quad \text{CH}_3\underset{\text{OH}}{\underset{|}{\text{C}}}\text{H}\!\not\!\!-\text{CH}_2\text{CH}_2\text{CH}_3$$

Two syntheses would be available:

$$\text{CH}_3\text{MgBr} + \underset{\text{O}}{\underset{\|}{\text{C}}}\text{HCH}_2\text{CH}_2\text{CH}_3 \quad \text{or} \quad \text{CH}_3\underset{\text{O}}{\underset{\|}{\text{C}}}\text{H} + \text{BrMgCH}_2\text{CH}_2\text{CH}_3$$

Both procedures would work equally well because both carbonyl components are simple aldehydes. If one of the carbonyl components had been sterically hindered around the carbonyl group, that method would *not* have been chosen.

Problems

11–6 Give the product of the following Grignard reactions (Grignard reagent followed by acid hydrolysis):

a. propanal + $CH_3MgBr \rightarrow$

b. cyclopentanone + $CH_3CH_2CH_2MgBr \rightarrow$

c. cyclohexylmagnesium chloride + acetaldehyde $\rightarrow$

11–7 Choose the proper Grignard reagent and carbonyl compound to prepare each of the following compounds:

a. 3-heptanol

b. 2-methyl-1-butanol

c. 1-ethylcyclohexanol

d. 2,3-dimethyl-2-butanol

e. 2,4-dimethyl-1-pentanol

Alkyllithium compounds are polarized similarly to Grignard reagents and can add to carbonyl groups. They are more reactive and, hence, can be used when the substrate is a hindered ketone (eq. 11.22).

$$(11.22)\quad (\text{CH}_3)_2\text{CH}\overset{\text{O}}{\overset{\|}{\text{C}}}\text{CH}(\text{CH}_3)_2 + \overset{\delta-}{\text{CH}_3\text{CH}_2}\overset{\delta+}{\text{Li}} \xrightarrow[\text{ether}]{}$$

$$(\text{CH}_3)_2\text{CH}\overset{\text{OLi}}{\overset{|}{\underset{\text{CH}_3\text{CH}_2}{\underset{|}{\text{C}}}}}\text{CH}(\text{CH}_3)_2 \xrightarrow{\text{H}_3\text{O}^+} (\text{CH}_3)_2\text{CH}\overset{\text{OH}}{\overset{|}{\underset{\text{CH}_3\text{CH}_2}{\underset{|}{\text{C}}}}}\text{CH}(\text{CH}_3)_2$$

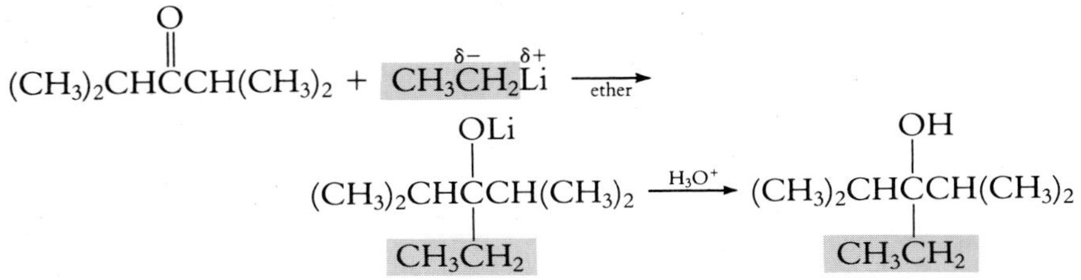

The Grignard reaction can be combined sequentially with the oxidation reactions discussed in Section 11:3. Their combination allows us to

Table 11–3. Grignard Synthesis of Alcohols

	Carbonyl Compound		Product
R-MgX +	CH_2O formaldehyde	$\xrightarrow{H_3O^+}$	**R**CH_2OH 1° alcohol
R-MgX +	RCHO any other aldehyde	$\xrightarrow{H_3O^+}$	**R**—CH(OH)R 2° alcohol
R-MgX +	RC(=O)R′ ketone	$\xrightarrow{H_3O^+}$	**R**C(OH)(R′)R 3° alcohol
R-MgX +	oxirane	$\xrightarrow{H_3O^+}$	**R**CH_2CH_2OH 1° alcohol

synthesize a large number of compounds exhibiting a variety of structures. For example, Equation 11.23 shows the preparation of a complex alcohol by a seven-step series of reactions starting from cyclopentanol.

(11.23) cyclopentyl–OH $\xrightarrow{HBr}$ cyclopentyl–Br $\xrightarrow[ether]{Mg}$ $\xrightarrow{CH_2{=}O}$ $\xrightarrow{H_3O^+}$ cyclopentyl–CH_2OH $\xrightarrow{CrO_3,\ pyridine}$ cyclopentyl–CHO $\xrightarrow[ether]{(CH_3)_3CMgCl}$ $\xrightarrow{H_3O^+}$ cyclopentyl–CH(OH)$C(CH_3)_3$

1-Cyclopentyl-2,2-dimethyl-1-propanol

A good review would be to follow each step closely and write the structures of the intermediate compounds that are not given. You may find it helpful to use Table 11–3, which summarizes the reactions of Grignard reagents with aldehydes and ketones. (Remember that 1° alcohols of the type RCH_2CH_2OH can be prepared from Grignard reagents and oxirane (sec. 6:14).)

11:5 Reversible Carbonyl Addition Reactions

The Grignard reaction discussed in the previous section gave us a glimpse of carbonyl group reactivity. We need to look now at a wider variety of carbonyl additions which differ from the Grignard reaction in that they are reversible. In this section we will illustrate the important mechanisms that govern the addition of protic reagents, $H^{\delta+}$—$Y^{\delta-}$, to aldehydes and ketones. These mechanisms will also be used to draw some generalizations about carbonyl addition reactions.

If either HY or Y^- is a strong nucleophile, the first step in the addition will be nucleophilic attack on the carbonyl group. An alkoxide intermediate is produced (**2**, eq. 11.24), which is rapidly protonated, giving the final product.

$$\text{(11.24)}\qquad Y\!:^- + \;>C{=}\ddot{O}: \rightleftarrows Y{-}\overset{|}{\underset{|}{C}}{-}\ddot{O}\!:^- \xrightleftharpoons{H^+ \text{ or } HB} Y{-}\overset{|}{\underset{|}{C}}{-}\ddot{O}H$$

2 Addition product

If neither HY nor Y^- is a good nucleophile, the reaction must be catalyzed by initial protonation of the carbonyl group. The intermediate in this case is a cation (**3**, eq. 11.25), which reacts rapidly even with weak nucleophiles.

$$\text{(11.25)}\qquad >C{=}\ddot{O}: + H^+ \rightleftarrows >C{=}\overset{+}{\ddot{O}}H \xrightleftharpoons{:Y{-}H} Y{-}\overset{|}{\underset{|}{C}}{-}\ddot{O}H + H^+$$

3 Addition product

The proton may come from HY or from an added acid catalyst. If the solution is acidic, the addition product may react further by loss of water.

The ionic intermediates **2** and **3** are fairly stable and are formed rapidly from either the carbonyl compound or the addition product. As a result, the reactants and products usually come quickly to equilibrium.

Some general conclusions can be offered about the equilibria themselves. In Section 10:1 we showed that C=C bonds undergo exothermic addition with a variety of reagents. Many of these reagents—HCl, HBr, Cl_2, Br_2—do *not* undergo measurable addition to the C=O bond. This lack of reactivity is due mainly to the high C=O π bond strength, which we can estimate at 90 kcal/mol with data from Table 10–1 (p. 346; see Prob. 11–1). Even those carbonyl additions that give good yields of product are only moderately exothermic (eq. 11.26; compare sec. 10:1). In addition, ΔS for this reaction is slightly negative, meaning that the entropy change is moderately unfavorable to addition. Thus, the entropy and enthalpy factors balance one

$$\text{(11.26)}\qquad >C{=}O + H{-}CN \rightarrow {-}\overset{CN}{\underset{|}{\overset{|}{C}}}{-}O{-}H \qquad \Delta H \simeq -6 \text{ kcal/mol}$$

another. The outcome is that many carbonyl addition reactions are rapidly reversible under the usual reaction conditions. In the following sections some of the more important reactions of this type are introduced.

Problem

11–8 Carbonyl addition reactions can be catalyzed by Lewis acids such as BF_3. Write the structure of the intermediate formed by the reaction of BF_3 with acetone. Show all the valence electrons around oxygen and boron, and any charges produced.

11:5 Reversible Carbonyl Addition Reactions

11:6 Addition of HCN, $NaHSO_3$, and H_2O to Carbonyl Compounds to Produce Important Addition Products

In mildly basic solutions the cyanide ion will add to the carbonyl groups of most aldehydes to produce stable **cyanohydrins** (eq. 11.27). This is also

$$(11.27)\quad CH_3\overset{\overset{\ddot{O}:}{\|}}{C}H + :CN^- \rightleftharpoons CH_3\overset{\overset{:\ddot{O}:^-}{|}}{\underset{\underset{H}{|}}{C}}{-}CN \overset{H_2\ddot{O}:}{\rightleftharpoons} CH_3\overset{\overset{:\ddot{O}H}{|}}{\underset{\underset{H}{|}}{C}}{-}CN + :\ddot{O}H^- \qquad K^{25°} = 1.4 \times 10^4$$

2-Hydroxypropanenitrile (*acetaldehyde cyanohydrin*)

true of some simple ketones, *with the reaction becoming less favorable as the ketone becomes more hindered.* Cyanohydrins of both aldehydes and ketones occur naturally as toxic components of several poisonous plants and are also part of the defense mechanism of some insects and animals. Cyanohydrins can also be used as intermediates for the synthesis of more complex compounds (see Chaps. 17 and 21).

When sodium bisulfite, $NaHSO_3$, is added to aldehydes and unhindered ketones (such as methyl ketones and cyclohexanone) the products are **bisulfite addition compounds** (eq. 11.28). These are crystalline solids that can be readily purified by recrystallization. If the reaction is then reversed, by treating the addition compound with a reagent that destroys the sodium bisulfite, we can regenerate the pure aldehyde or ketone (eq. 11.29). This addition→recrystallization→regeneration sequence is a purification method for liquid aldehydes and simple ketones.

$$(11.28)\quad Na^+\ H\ddot{S}O_3^- + CH_3\overset{\overset{O}{\|}}{C}H \rightarrow CH_3\overset{\overset{O^-}{|}}{\underset{\underset{H}{|}}{C}}{-}SO_3H + Na^+ \rightarrow CH_3\overset{\overset{OH}{|}}{\underset{\underset{H}{|}}{C}}{-}SO_3^-Na^+$$

Acetaldehyde bisulfite addition product

$$(11.29)\quad \underset{\text{Recrystallized}}{CH_3\overset{\overset{OH}{|}}{C}HSO_3^-Na^+} + HCl \rightarrow \underset{\text{Pure}}{CH_3CHO} + NaCl + SO_2 + H_2O$$

Aldehydes and ketones add water reversibly to give **hydrates.** Equation 11.30 shows that the equilibrium usually favors reactants, so that the hy-

$$(11.30)\quad (CH_3)_2CH\overset{\overset{O}{\|}}{C}H + H_2O \rightleftharpoons (CH_3)_2CH\overset{\overset{OH}{|}}{C}H{-}OH \qquad K = 0.6$$

2-Methylpropanal (*isobutyraldehyde*) → 2-Methylpropane-1,1-diol (*isobutyraldehyde hydrate*)

drates cannot normally be isolated. We have seen, however, that they are important intermediates in the oxidation of primary alcohols to carboxylic acids (sec. 11:3). The hydration reaction may be base-catalyzed (eq. 11.24, $Y^- = OH^-$) or acid-catalyzed (eq. 11.25, $HY = H_2O$).

If strong electron withdrawing groups are present in the aldehyde, the hydration reaction may actually be highly favored. Equation 11.31 shows the formation of chloral hydrate, a potent narcotic known as "knockout drops." The reversion of the hydrate to the aldehyde probably proceeds by

$$Cl_3C(=O)H + H_2O \rightleftarrows Cl_3C(OH)_2H \qquad K = 10{,}000 \tag{11.31}$$

Chloral hydrate

way of the carbocation (eq. 11.32), which is highly destabilized by the electron-withdrawing Cl groups.

$$Cl_3C(OH)_2H + H_3O^+ \rightarrow Cl_3C(OH)(\overset{+}{O}H_2)H \rightarrow Cl_3\overset{+}{C}(OH)H \rightarrow Cl_3C(=O)H \tag{11.32}$$

Unstable

In this section we have examined several carbonyl reactions which stop after addition has taken place. The next two sections deal with carbonyl reactions in which the addition product reacts further.

Problems

11–9 The equilibrium constant for reaction of HCN with acetaldehyde (eq. 11.27) is quite large. The equilibrium constant for formation of acetone cyanohydrin is much smaller (eq. 1). Explain this trend, taking note of the difference in steric crowding between the carbonyl compound and the product for each reaction.

$$CH_3C(=O)CH_3 + HCN \rightleftarrows CH_3C(OH)(CH_3)CN \qquad K^{25^\circ} = 33 \tag{1}$$

11–10 Write the structures of the products of each reaction.

a. pentanal + HCN →
b. 2-hexanone + HCN →
c. cyclopentyl methyl ketone + HCN →
d. cyclopentanone + $NaHSO_3$ →
e. $CF_3{-}C(=O){-}CF_3 + H_2O \rightarrow$

11:7 Reaction of Carbonyl Compounds with Amines: Addition Followed by Elimination to Produce Imines

Amines are organic derivatives of ammonia. Remember from Table 1–2 (p. 16) that they are classified as *primary* (**4**), *secondary* (**5**), or *tertiary* (**6**) (R represents an alkyl or aryl group). Because of the nonbonding electron pair on nitrogen, amines are good nucleophiles, and they readily attack carbonyl compounds. In this section we will focus on the reaction of primary amines with aldehydes and ketones. (Many more amine reactions are discussed in Chapter 21.)

$R\ddot{N}H_2$	$R_2\ddot{N}H$	$R_3\ddot{N}$
4	**5**	**6**
1° Amine	2° Amine	3° Amine

The first step is a reversible nucleophilic addition reaction (eq. 11.33), similar to the ones discussed in Section 11:6. The addition product (**7**) is not stable, however, and readily eliminates water, particularly when in the presence of acid (eq. 11.34).

$$\text{(11.33)}\quad R\overset{\overset{+}{\ddot{O}}:H}{\overset{\|}{C}}R' + H_2\ddot{N}\mathbf{R} \rightleftarrows RC(\ddot{O}:H)(R')\text{—}\overset{+}{N}H_2\mathbf{R} \rightleftarrows RC(\ddot{O}H)(R')\text{—}\ddot{N}H\mathbf{R}$$

Amine   **7**

$$\text{(11.34)}\quad RC(\ddot{O}H)(R')\text{—}\ddot{N}H\mathbf{R} \underset{}{\overset{H^+}{\rightleftarrows}} RC(H\ddot{O}H^+)(R')\text{—}\ddot{N}H\mathbf{R} \rightleftarrows$$

7

$$\begin{matrix}R \\ R'\end{matrix}\!\!>C{=}\overset{+}{N}\mathbf{R}(H) + H_2\ddot{O}: \rightleftarrows \begin{matrix}R \\ R'\end{matrix}\!\!>C{=}\ddot{N}\text{—}\mathbf{R} + H_3O^+$$

An imine

The loss of water is aided by the nonbonding electron pair on nitrogen. The products are **imines,** compounds containing the C=N double bond. When aldehydes are used as starting materials the resulting imines are also called *Schiff bases*.

There are several reasons for believing that the mechanism shown in Equations 11.33 and 11.34 is correct. For instance, when the reaction is monitored by uv spectroscopy, a time lag is found between the disappearance of the carbonyl compound and the appearance of the imine. This finding is consistent with the occurrence of an intermediate (such as **7**) between starting material and product. In addition, efforts to correlate rate changes and pH changes reveal that the reaction goes fastest at pH 5, and slows down at higher or lower pH (Fig. 11–6). The mechanism indicates that acid is beneficial because it catalyzes the loss of water (eq. 11.34). Excess acid can slow the rate, however, by protonating the amine and removing it as a

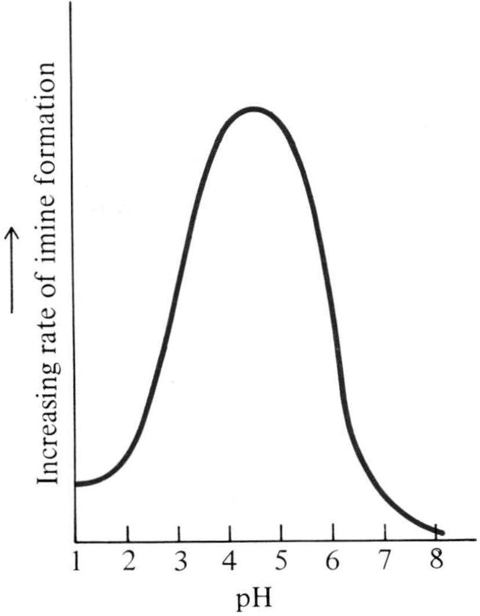

Figure 11–6. *How the rate of imine formation changes with changing pH.*

nucleophile (eq. 11.35). There are no nonbonded electron pairs left on RNH_3^+, so it is not nucleophilic.

(11.35)
$$\mathbf{R}NH_2 + H^+ \rightleftarrows \mathbf{R}NH_3^+$$
A nucleophile — Not a nucleophile

When all of the substituents on the C=N bond (R, R′, and **R**) are alkyl or hydrogen, the equilibrium constant for imine formation is small. In order to get a useful yield of product it is necessary to remove the water that is formed. This is done by adding Na_2SO_4 (eq. 11.36) or distilling off the H_2O as it is formed.

(11.36)
$$CH_3NH_2 + CH_3CHO \xrightarrow[\text{dry ether}]{Na_2SO_4\text{ (absorbs water)}} CH_3N{=}CHCH_3 + H_2O$$
(~40%)

Imines that contain an aryl substituent are substantially more stable, and more readily synthesized. Imines are not usually prepared for their own value, but are important reaction intermediates in many chemical and biochemical reactions, as we will show in Chapters 18 and 21.

Unusually stable imines are isolated when certain special derivatives of ammonia are allowed to react with carbonyl compounds. These reagents are distinguished by having an oxygen or nitrogen atom directly bonded to the NH_2 group. The most important are hydroxylamine (eq. 11.37), phenylhydrazine (eq. 11.38; C_6H_5 = —⟨phenyl ring⟩), and semicarbazide (eq. 11.39).

(11.37)
$$\text{(cyclopentanone)}{=}O + \ddot{N}H_2\ddot{O}H \xrightarrow[\text{EtOH, } H_2O]{\text{acetic acid}} \text{(cyclopentyl)}{=}\ddot{N}{-}\ddot{O}H + H_2O$$

stabilizing interaction

Cyclopentanone — *Hydroxylamine* — Cyclopentanone *oxime*

$$\text{(11.38)}\quad \underset{\text{Acetone}}{CH_3\overset{O}{\overset{\|}{C}}CH_3} + \underset{\textit{Phenylhydrazine}}{\ddot{N}H_2\ddot{N}HC_6H_5} \xrightarrow[\text{EtOH, } H_2O]{\text{acetic acid}} \underset{\substack{\text{Acetone}\\ \textit{phenylhydrazone}}}{CH_3\overset{\ddot{N}-\ddot{N}HC_6H_5}{\overset{\|}{C}}CH_3}$$

stabilizing interaction

$$\text{(11.39)}\quad \underset{\text{Butanal}}{CH_3CH_2CH_2CH{=}O} + \underset{\textit{Semicarbazide}}{\ddot{N}H_2\ddot{N}H\overset{O}{\overset{\|}{C}}\ddot{N}H_2} \xrightarrow[\text{EtOH, } H_2O]{\text{acetic acid}}$$

$$\underset{\text{Butanal } \textit{semicarbazone}}{CH_3CH_2CH_2CH{=}\ddot{N}-\ddot{N}H\overset{O}{\overset{\|}{C}}\ddot{N}H_2}$$

stabilizing interaction

The stability of the products—**oximes, phenylhydrazones,** and **semicarbazones**—is due mainly to a stablizing interaction between the C=N π bond and the lone electron pair on the neighboring O or N. In Chapter 13 we will explore the nature of this electron delocalizing interaction.

Oximes, phenylhydrazones, and semicarbazones are almost always solids, even when the aldehydes and ketones themselves are liquids. These derivatives are readily purified by recrystallization and have characteristic, sharp melting points. For these reasons, they are very useful for identifying unknown aldehydes and ketones. Oximes and phenylhydrazones can also be used in the synthesis of amines (Chaps. 21 and 25).

Example

What is the product of the reaction between acetone and hydroxylamine?

$$CH_3\underset{O}{\underset{\|}{C}}CH_3 + NH_2OH \rightarrow ?$$

Answer

Although it may seem a long way around at the beginning, it is best to arrive at the product by going through at least part of the mechanism (see eqs. 11.33 and 11.34 for the complete mechanism).

$$CH_3\underset{O}{\underset{\|}{C}}CH_3 + :NH_2OH \rightarrow CH_3\overset{^+NH_2OH}{\underset{O^-}{C}}CH_3 \rightarrow CH_3\overset{NHOH}{\underset{OH}{C}}CH_3 \rightarrow \underset{\text{Product}}{CH_3\overset{NOH}{\overset{\|}{C}}CH_3} + HOH$$

Problems

11–11 Give the products of the following reactions:

a. $CH_3CH_2CH{=}O + NH_2OH \rightarrow$

b. cyclohexanone + $C_6H_5NHNH_2 \rightarrow$

c. $(C_6H_5)_2C{=}O + CH_3CH_2NH_2 \rightarrow$

11–12 Many aldehyde and ketone derivatives exist in two isomeric forms. For example, acetaldehyde gives two phenylhydrazones, and 2-pentanone gives rise to two semicarbazones. In contrast, acetone and cyclohexanone give only one phenylhydrazone, one semicarbazone, and one oxime. Draw the structures of the derivatives named, paying particular attention to the molecular geometry. Use these structures to explain the origin of the isomerism.

11–13 Secondary amines react with aldehydes and ketones by way of an addition–elimination mechanism. The product is an *enamine,* such as **A** in Equation 2. The name enamine derives from the fact that **A** is both an alk*ene* and an *amine*. Suggest a mechanism for its formation.

$$CH_3\overset{\overset{\displaystyle O}{\|}}{C}CH_3 + CH_3\underset{\displaystyle H}{\underset{|}{\ddot{N}}}CH_3 \overset{H^+}{\rightleftharpoons} (CH_3)(CH_2{=})C{-}N(CH_3)_2 + H_2O \qquad (2)$$

An enamine
A

11:8 The Reaction of Carbonyl Compounds with Alcohols to Yield Hemiacetals, Acetals, Hemiketals, and Ketals

When an aldehyde is dissolved in an alcohol, a reversible addition reaction takes place. The product is a **hemiacetal** (eq. 11.40). In the presence of an anhydrous acid catalyst, equilibrium is rapidly established. The equilibrium favors the aldehyde (reactant).

$$(11.40) \quad CH_3CH_2\overset{\overset{\displaystyle O}{\|}}{C}H + CH_3OH \overset{H+}{\rightleftharpoons} CH_3CH_2\underset{\underset{\displaystyle H}{|}}{\overset{\overset{\displaystyle OH}{|}}{C}}{-}OCH_3 \qquad K < 1$$

Propanal (*propionaldehyde*)

1-Methoxy-1-propanol (*propionaldehyde methyl hemiacetal*)

The acid then catalyzes the reaction of the hemiacetal with additional alcohol, in which OR is substituted for OH. The product of this second step is an **acetal** (eq. 11.41).

$$(11.41) \quad CH_3CH_2\underset{\underset{\displaystyle H}{|}}{\overset{\overset{\displaystyle OH}{|}}{C}}{-}OCH_3 + CH_3OH \overset{H^+}{\rightleftharpoons} CH_3CH_2\underset{\underset{\displaystyle H}{|}}{\overset{\overset{\displaystyle OCH_3}{|}}{C}}{-}OCH_3 + H_2O$$

1,1-Dimethoxypropane (*propionaldehyde dimethyl acetal*)

Hemiacetals and acetals are important to us because they are incorporated into the cyclic structures of sugars and other carbohydrates (Chap. 20). In this section only basic mechanistic principles underlying the formation of these derivatives are considered.

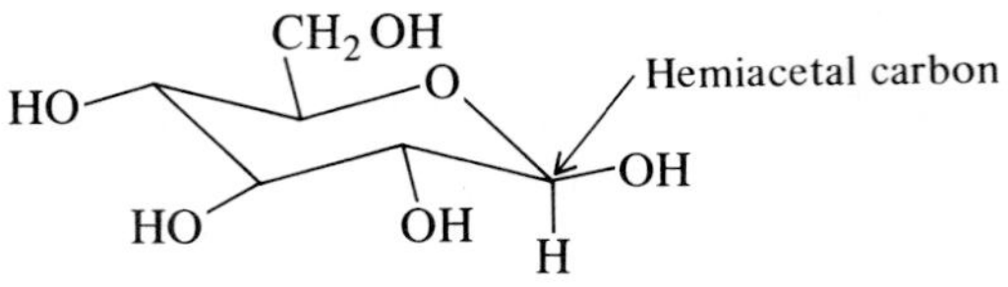

CH_2OH HO O Hemiacetal carbon OH HO OH H

β-Glucose
(a hemiacetal sugar)

You may have noticed that hemiacetals contain one OH group and an ether function, while acetals contain two ether groups. They can be represented by the following general structures:

$$\underset{\text{Hemiacetal}}{R-\overset{\overset{\displaystyle OH}{|}}{\underset{\underset{\displaystyle H}{|}}{C}}-OR} \qquad \underset{\text{Acetal}}{R-\overset{\overset{\displaystyle OR}{|}}{\underset{\underset{\displaystyle H}{|}}{C}}-OR}$$

When a substantial excess of alcohol is present, the aldehyde can be converted completely to the acetal. Once the reaction is completed, the acid is destroyed by addition of OH^-, allowing isolation of the acetal (eq. 11.42).

$$(11.42)\quad CH_3\overset{\overset{\displaystyle O}{\|}}{C}H + 2CH_3CH_2CH_2OH \xrightarrow[\text{excess alcohol}]{\text{HCl (cat)}} \xrightarrow[\text{2. distill}]{\text{1. OH}^-} CH_3CH(OC_3H_7)_2$$

Acetaldehyde dipropyl acetal (100%)

Similar reactions take place when a ketone is dissolved in an alcohol; the first product is a **hemiketal** (eq. 11.43).

$$(11.43)\quad CH_3\overset{\overset{\displaystyle O}{\|}}{C}CH_3 + CH_3CH_2OH \overset{H^+}{\rightleftharpoons} CH_3\overset{\overset{\displaystyle OH}{|}}{\underset{\underset{\displaystyle CH_3}{|}}{C}}-OCH_2CH_3 \qquad K << 1$$

Acetone ethyl hemiketal

Notice that the equilibrium constant for hemiketal formation is *very* small. There are two reasons for this—the entropy change is unfavorable (negative ΔS), and the hemiketal is much more crowded than the ketone. This very unfavorable equilibrium makes it impractical to synthesize **ketals** by allowing ketones to react with simple alcohols.

One way of overcoming these problems is to use a *diol* in place of a simple alcohol (eq. 11.44).

$$(11.44)\quad CH_3\overset{\overset{\displaystyle O}{\|}}{C}CH_3 + \underset{\displaystyle OH\quad OH}{\underset{|\qquad |}{CH_2{-}CH_2}} \xrightleftharpoons{H^+}$$

1,2-Ethanediol
(*ethylene glycol*)

$$\left[\begin{array}{c} OH\quad CH_2OH \\ |\qquad\quad | \\ CH_3C{-}OCH_2 \\ | \\ CH_3 \end{array}\right] \xrightleftharpoons{H^+} \begin{array}{c} CH_3\quad O{-}CH_2 \\ \diagdown\ \diagup\qquad | \\ C\qquad\ \ | \\ \diagup\ \diagdown\qquad | \\ CH_3\quad O{-}CH_2 \end{array} + H_2O$$

Hemiketal

Acetone ethylene ketal

The equilibrium can be driven toward products by removing the water. For this purpose benzene is used as the solvent, since water codistills with benzene (eq. 11.45). In any equilibrium, removing one of the products will push the equilibrium to that side (a statement of Le Chatelier's principle).

$$(11.45)\quad \text{cyclohexanone} + \underset{\displaystyle OH\ OH}{\underset{|\quad\ |}{CH_2CH_2}} \xrightarrow[\text{benzene, distill out water}]{\text{TsOH (cat)}} \text{cyclohexanone ethylene ketal} + H_2O$$

Cyclohexanone ethylene ketal
(80%)

Acetals and ketals have two alkoxy groups bonded to the same carbon atom. These aldehyde and ketone derivatives are formed from hemiacetals and hemiketals by way of a common set of mechanisms (eqs. 11.46 and 11.47). There is rapid equilibration between the carbonyl compound and the

$$(11.46)\quad R\overset{\overset{\displaystyle \ddot{O}:}{\|}}{C}R' \xrightleftharpoons{H^+} R\overset{\overset{\displaystyle +\ddot{O}H}{\|}}{C}R' \xrightleftharpoons{H\ddot{O}R} \underset{\displaystyle R'}{\underset{|}{R\overset{\overset{\displaystyle :\ddot{O}H}{|}}{C}}}{-}\underset{\displaystyle H}{\underset{|}{\overset{+}{\ddot{O}}\mathbf{R}}} \rightleftharpoons \underset{\displaystyle R'}{\underset{|}{R\overset{\overset{\displaystyle :\ddot{O}H}{|}}{C}}}{-}\ddot{\underset{\cdot\cdot}{O}}\mathbf{R} + H^+$$

Aldehyde or ketone

Hemiacetal or hemiketal

$$\text{(11.47)}\quad \underset{\substack{\text{Hemiacetal or}\\ \text{hemiketal}}}{\mathrm{R\overset{\overset{\ddot{O}H}{|}}{\underset{\underset{R'}{|}}{C}}-\ddot{O}R}} \overset{H^+}{\rightleftharpoons} \mathrm{R\overset{\overset{\overset{+}{H\ddot{O}H}}{|}}{\underset{\underset{R'}{|}}{C}}-\ddot{O}R} \rightleftharpoons$$

$$\underset{\substack{\text{Carbocation } \mathbf{(8)}\\ +\ H_2O}}{\mathrm{R\overset{+}{\underset{\underset{R'}{|}}{C}}-\ddot{O}R}} \overset{H\ddot{O}R}{\rightleftharpoons} \mathrm{R\overset{\overset{\overset{+}{H\ddot{O}R}}{|}}{\underset{\underset{R'}{|}}{C}}-\ddot{O}R} \rightleftharpoons \underset{\substack{\text{Acetal or}\\ \text{ketal}}}{\mathrm{R\overset{\overset{:\ddot{O}R}{|}}{\underset{\underset{R'}{|}}{C}}-\ddot{O}R}} + H^+$$

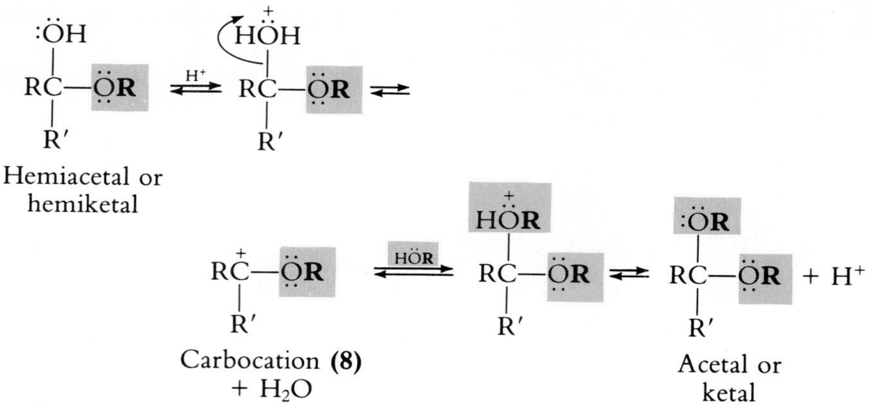

acetal or ketal, due to the unusual stability of the carbocation (**8**). The nonbonded electrons on oxygen help to satisfy the electron demand of the positive carbocation (this is explained more fully in Chap. 13).

Equation 11.47 shows that acid catalysis is vital for the conversion of the hemiketal (or hemiacetal) to the ketal (or acetal). The catalyst also makes possible the reverse reaction, hydrolysis of the acetal or ketal. If an acetal or ketal is isolated and then treated with aqueous acid, it is rapidly converted back to the carbonyl compound (eq. 11.48).

$$\text{(11.48)}\quad CH_3CH(OCH_2CH_2CH_3)_2 + H_2O \xrightarrow{H_3O^+} CH_3CHO + 2CH_3CH_2CH_2OH$$

You might ask why anyone would want to hydrolyze an acetal or ketal after having gone to the trouble of making one. We answer this question in the next section. In the meantime, Table 11–4 summarizes the important reversible addition reactions of aldehydes and ketones.

Example

What are the structures of the hemiacetal and acetal which are formed by the reaction of acetaldehyde with methanol?

Answer

Acetaldehyde reacts with one molecule of methanol to form the hemiacetal:

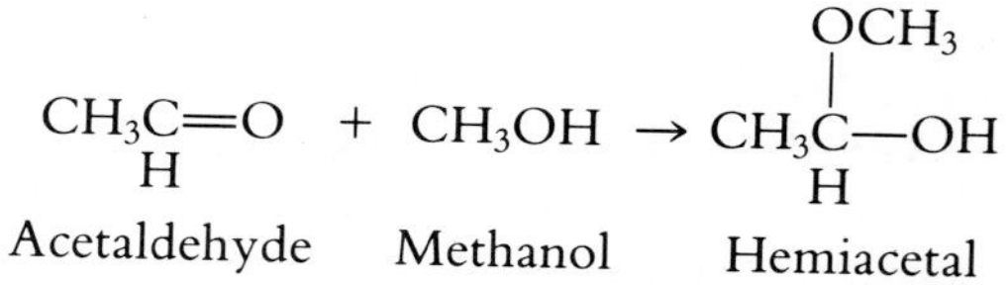

$$\underset{\text{Acetaldehyde}}{\mathrm{CH_3\underset{H}{C}{=}O}} + \underset{\text{Methanol}}{\mathrm{CH_3OH}} \rightarrow \underset{\text{Hemiacetal}}{\mathrm{CH_3\overset{\overset{OCH_3}{|}}{\underset{H}{C}}-OH}}$$

The hemiacetal then reacts with an additional molecule of methanol, in the presence of acid, to form the acetal:

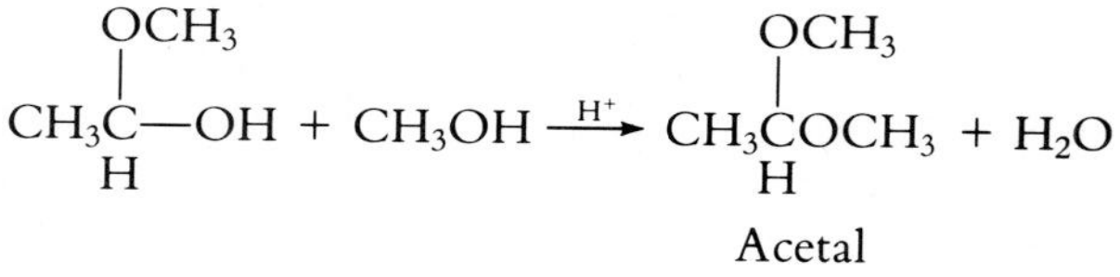

$$\mathrm{CH_3\overset{\overset{OCH_3}{|}}{\underset{H}{C}}-OH} + \mathrm{CH_3OH} \xrightarrow{H^+} \underset{\text{Acetal}}{\mathrm{CH_3\overset{\overset{OCH_3}{|}}{\underset{H}{C}}OCH_3}} + H_2O$$

Table 11–4. Reversible Additions to Aldehyde and Ketone Carbonyl Groups

Cyanohydrin formation (sec. 11:6)

$$\underset{\text{Aldehyde or ketone}}{R(R')C{=}O} + H^+ + CN^- \rightleftharpoons \underset{\text{Cyanohydrin}}{R{-}C(OH)(R'){-}CN}$$

Bisulfite addition (sec. 11:6)

$$\underset{\text{Aldehyde or ketone}}{R(R')C{=}O} + Na^+\ HSO_3^- \rightleftharpoons \underset{\text{Bisulfite compound}}{R{-}C(OH)(R'){-}SO_3^-\ Na^+}$$

Imine formation (sec. 11:7)

$$\underset{\text{Aldehyde or ketone}}{R(R')C{=}O} + \underset{1^\circ\ \text{Amine}}{\mathbf{RNH_2}} \rightleftharpoons R{-}C(OH)(R'){-}NH\mathbf{R} \overset{H^+}{\rightleftharpoons} \underset{\text{Imine}}{R(R')C{=}N\mathbf{R}} + H_2O$$

Acetal and ketal formation (sec. 11:8)

$$\underset{\text{Aldehyde or ketone}}{R(R')C{=}O} \underset{H^+}{\overset{ROH}{\rightleftharpoons}} \underset{\text{Hemicetal or hemiketal}}{R{-}C(OH)(R'){-}O\mathbf{R}} \underset{H^+}{\overset{ROH}{\rightleftharpoons}} \underset{\text{Acetal or ketal}}{R{-}C(O\mathbf{R})(R'){-}O\mathbf{R}}$$

Problems

11–14 Write a mechanism for the hydrolysis of acetaldehyde dipropyl acetal (eq. 11.48).

11–15 The equilibrium constant is usually small for hemiacetal formation (eq. 11.40), and the hemiacetal is too unstable to be isolated. An exception is provided by certain *intramolecular* hemiacetals, in which the alcohol and aldehyde groups are parts of the same molecule (eq. 3). Compare Equation 11.40

with Equation 3, and explain why hemiacetal formation is so much more favorable in the second case.

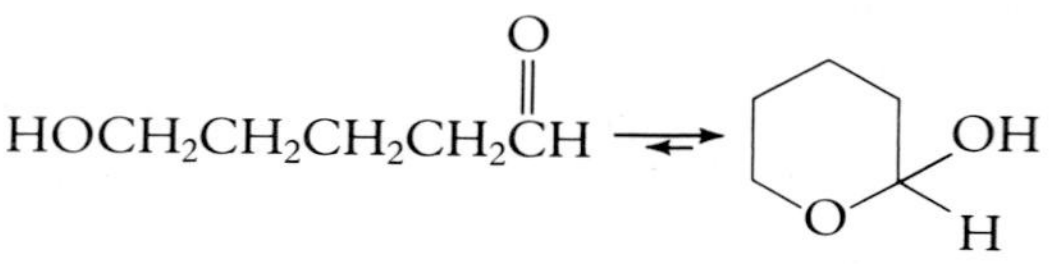

$$\mathrm{HOCH_2CH_2CH_2CH_2\overset{O}{\overset{\|}{C}}H} \rightleftharpoons \text{(cyclic hemiacetal: tetrahydropyran ring with OH and H on C-2)} \qquad K = 20 \qquad (3)$$

11–16 Write the structures of the products of each of the following reactions:

a. 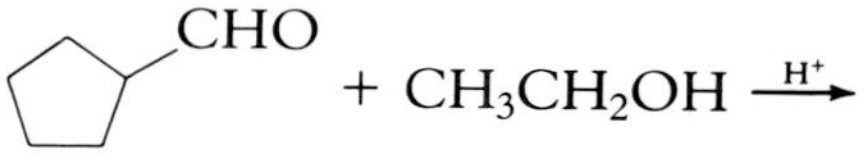 cyclopentyl–CHO $+ \ CH_3CH_2OH \xrightarrow{H^+}$ (excess)

b. 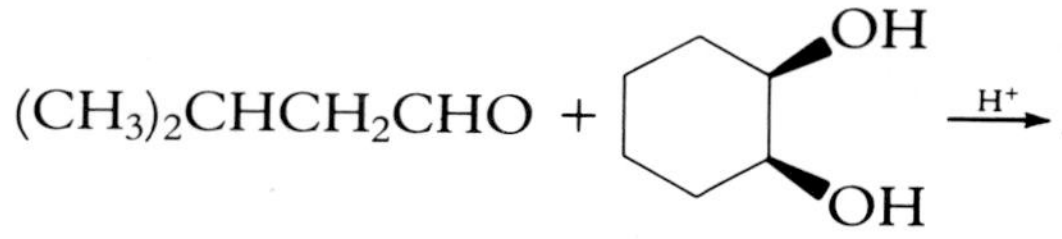 $(CH_3)_2CHCH_2CHO +$ cyclohexane-1,2-diol (OH, OH) $\xrightarrow{H^+}$

c. $CH_3\overset{O}{\overset{\|}{C}}CH_2CH_3 + HOCH_2CH_2OH \xrightarrow[\text{benzene}]{H^+}$

d. $CH_3\overset{O}{\overset{\|}{C}}CH_2CH_2CH_3 + HSCH_2CH_2SH \xrightarrow[\text{benzene}]{H^+}$

e. $HOCH_2\overset{O}{\overset{\|}{C}}H \rightarrow$ **A** $(C_4H_8O_4)$

11:9 Protecting Carbonyl and Hydroxyl Groups by Conversion to Acetals and Ketals

A common problem encountered in organic chemistry is to perform reactions at one functional group in a molecule while leaving the others unchanged. As an example, consider the conversion of 4-hydroxycyclohexanone to a 4-cyanocyclohexanone. A reasonable approach would involve converting the hydroxyl group to a good leaving group, such as tosylate, and displacing that with cyanide ion (eq. 11.49). We might anticipate problems with the second step of this sequence, since CN^- can attack the car-

$$\text{(11.49)} \quad \text{4-hydroxycyclohexanone (OH)} \xrightarrow[\text{OH}^-]{\text{TsCl}} \text{4-(OTs)cyclohexanone} \xrightarrow[\text{EtOH}]{\text{KCN}} \text{4-cyanocyclohexanone (CN)}$$

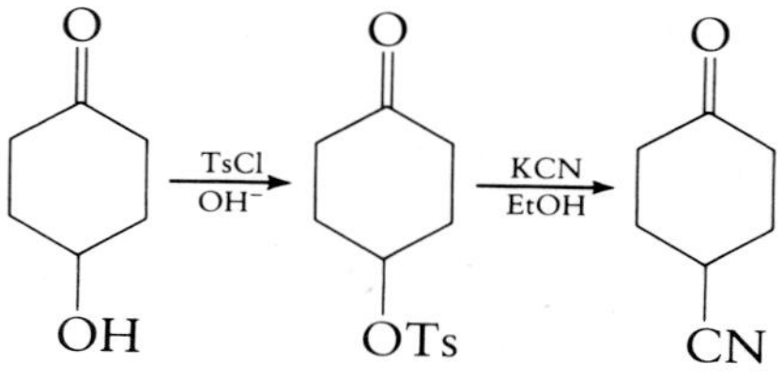

bonyl group as well as displacing tosylate. The problem can be avoided by reversibly *protecting* the carbonyl group. In this case we will protect it by converting it to a ketal. After the OTs^- group is displaced by CN^- and free CN^- is no longer present, the carbonyl group will be *deprotected* (freed) by hydrolysis (eq. 11.50).

(11.50) 4-hydroxycyclohexanone + $HOCH_2CH_2OH$ $\xrightarrow[\text{benzene (protecting step)}]{H^+}$ Ketal (OH) $\xrightarrow[OH^-]{TsCl}$ ketal (OTs) $\xrightarrow[EtOH]{KCN}$ ketal (CN) $\xrightarrow[\text{(deprotecting step)}]{H_3O^+}$ 4-cyanocyclohexanone (CN)

Acetal and ketal formation can also be used to protect the hdyroxyl group of alcohols. When alcohols are treated with the unsaturated ether *dihydropyran* and a drop of acid, they are converted to tetrahydropyranyl (THP) ethers which are, in fact, acetals (eq. 11.51).

(11.51) $CH_3CH_2\ddot{O}H$ + Dihydropyran $\xrightarrow{H^+}$

$CH_3CH_2\ddot{O}H$ + Stable carbocation $\rightarrow$ Ethyl tetrahydropyranyl ether (a mixed acetal) + H^+

These THP ethers are stable in base but hydrolyze easily in aqueous acid, releasing the alcohol. This property can be put to good use in synthesis. Equation 11.52 shows how the protection–deprotection sequence allows us to carry out a Grignard reaction at the carbonyl group of 4-hydroxycyclohexanone. The hydroxyl group, had it not been protected, would have destroyed the Grignard reagent (eq. 11.21).

(11.52) [dihydropyran] + HO–[cyclohexanone] $\xrightarrow[\text{(protection)}]{H^+}$

[THP ether of 4-hydroxycyclohexanone] $\xrightarrow[\text{ether}]{CH_3MgBr}$ [THP–O–cyclohexyl(OMgBr)(CH_3)] $\xrightarrow[\text{(hydrolysis and deprotection)}]{H_3O^+}$

THP ether

HO–[cyclohexane](OH)(CH_3)

1-Methyl-1,4-cyclohexanediol

Problem

11–17 Potassium permanganate readily oxidizes primary alcohols (RCH_2OH) to carboxylic acids (RCOOH) in neutral solution. Unfortunately, it also oxidizes aldehydes to carboxylic acids very easily (sec. 11:3). With these factors in mind, suggest a way of carrying out the transformation shown in Equation 4. (Protecting groups may be helpful.)

$$\overset{O}{\overset{\|}{H C}}(CH_2)_6CH_2OH \rightarrow \overset{O}{\overset{\|}{H C}}(CH_2)_6COOH \qquad (4)$$

11:10 Reduction of Aldehydes and Ketones to Alcohols

The carbonyl group of an aldehyde or ketone can be reduced to a hydroxyl group by using hydrogen and a catalyst. This reaction yields primary alcohols from aldehydes (eq. 11.53) and secondary alcohols from ketones (eq. 11.54). It is industrially useful when the carbonyl compounds are readily available but the alcohols are not.

(11.53) $$CH_3(CH_2)_5CHO \xrightarrow[C_2H_5OH]{H_2,\ Pt} CH_3(CH_2)_5CH_2OH$$

Hepatanal → 1-Heptanol (90 to 95%)

(11.54) [4-Methylcyclohexanone: CH_3, H on C-4; C=O] $\xrightarrow[\text{acetic acid}]{H_2,\ Pt}$

4-Methylcyclohexanone

[CH_3, H / OH, H cyclohexane] + [H, CH_3 / OH, H cyclohexane]

cis-4-Methylcyclohexanol (47%) *trans*-4-Methylcyclohexanol (53%)

These carbonyl reduction reactions are also very important in the laboratory. In this case the usual reductants are complex metal hydrides, *lithium aluminum hydride*, $LiAlH_4$, and *sodium borohydride*, $NaBH_4$. They are more convenient to handle than H_2 and are also more specific for the C=O group. Both $LiAlH_4$ and $NaBH_4$ act in a similar way, transferring the strongly nucleophilic hydride ion ($H:^-$) to the carbonyl carbon atom. Equations 11.55 to 11.58 give representative examples of the reduction of aldehydes and ketones to primary and secondary alcohols, respectively. Equation 11.56 shows the specificity of $NaBH_4$—it will usually *not* reduce isolated C=C bonds; the same is true of $LiAlH_4$.

$$(11.55)\quad 4CH_3\overset{\overset{\displaystyle O}{\|}}{C}CH_2CH_2CH_3 + NaBH_4 + 4CH_3OH \rightarrow$$

2-Pentanone

$$4CH_3\overset{\overset{\displaystyle OH}{|}}{C}HCH_2CH_2CH_3 + Na^+B(OCH_3)_4^-$$

2-Pentanol (100%)

$$(11.56)\quad CH_2{=}CHCH_2CH_2\overset{\overset{\displaystyle O}{\|}}{C}H + NaBH_4 \xrightarrow{C_2H_5OH} CH_2{=}CHCH_2CH_2CH_2OH$$

4-Pentenal → 4-Penten-1-ol

$$(11.57)\quad 4CH_3(CH_2)_5CHO + LiAlH_4 \xrightarrow{ether} \xrightarrow[H_2O]{4H_3O^+} CH_3(CH_2)_5CH_2OH$$

Heptanal → 1-Heptanol (86%)

$$(11.58)\quad CH_3\text{—}\langle C_6H_{10}\rangle{=}O + LiAlH_4 \xrightarrow{ether} \xrightarrow{H_3O^+}$$

4-Methylcyclohexanone

$$CH_3\text{—}\langle C_6H_{10}\rangle\text{—}OH + CH_3\text{—}\langle C_6H_{10}\rangle\text{---}OH$$

(25%) (75%)

The two reducing agents, $LiAlH_4$ and $NaBH_4$, have complementary uses. Lithium aluminum hydride is very reactive and will decompose exothermically in water or alcohols (often sufficiently to start a fire). It is slightly soluble in ethers, and is often used in these solvents. Sodium borohydride is a less vigorous reducing agent and reduces fewer functional groups than lithium aluminum hydride does. Unlike $LiAlH_4$, $NaBH_4$ can be used in water and alcohol solvents; it is practically insoluble in most ethers. A hydrolysis step is necessary to complete lithium aluminum hydride reductions, but this is not necessary with borohydride reductions that have been run in water or alcohols.

The mechanism for carbonyl reduction by metal hydrides is similar to that of other carbonyl reactions. The first step of Equation 11.59 (reduction of a ketone) shows that the reduction begins with transfer of a hydride ion ($H:^-$) acting as a nucleophile. In each subsequent step, another hydride is

$$\text{(11.59)}\quad H_3Al{-}H^- + R_2C{=}\ddot{O}: \longrightarrow R_2CH{-}\ddot{O}:^- + AlH_3$$

$$R_2CH{-}\ddot{O}{-}AlH_3^- \xrightarrow{R_2C=O} (R_2CH{-}O)_2AlH_2^- \xrightarrow{R_2C=O} (R_2CH{-}O)_3AlH^- \xrightarrow{R_2C=O} \underset{\mathbf{9}}{(R_2CH{-}O)_4Al^-} \xrightarrow{4H_3O^+} 4R_2CHOH + Al^{3+} + 4H_2O$$

transferred until all four of the original hydrogens on the AlH_4^- anion have added to the carbon of the carbonyl group. After all the hydrides are transferred, the alkoxyaluminum complex (**9,** eq. 11.59) is decomposed with acid. In the alcohol product, therefore, the C-H hydrogen comes from the $LiAlH_4$, but the O-H hydrogen comes from the H_3O^+. Thus, 1 mol of $LiAlH_4$ can reduce 4 mol of aldehyde or ketone (see balanced eq. 11.57). Borohydride reductions follow a similar mechanism.

Box 11–1. Accident and Creativity: H. C. Brown and Boron Hydride Chemistry

Discovery of the ability of $NaBH_4$ to reduce aldehydes and ketones and the development of the hydroboration reaction were major achievements which led to the awarding of the Nobel Prize in Chemistry to Herbert C. Brown in 1979. Both discoveries were results of accidents.

Brown, who was born in England, came with his parents to Chicago in 1914. When he was 14, his father died and Brown was forced to abandon his education in order to help support his family. But he eventually finished high school, took night courses, and studied at junior colleges. Finally, he entered the University of Chicago where he obtained a B.S. degree in chemistry in 1936 and his Ph.D. in 1938. While on the faculty at the University of Chicago Brown worked on the synthesis of $NaBH_4$, searching for a method of separating it from the $NaOCH_3$ formed in the following reaction:

$$4NaH + B(OCH_3)_3 \xrightarrow{250^\circ} NaBH_4 + 3NaOCH_3$$

Ether dissolved neither of the products and methanol dissolved both. When Brown tried acetone, he discovered that it reacted with the $NaBH_4$ to form 2-propanol. He did not get tenure at the University of Chicago and eventually went to Purdue where his major work was done (1947–1978). At Purdue he studied the ability of diborane (B_2H_6) to reduce esters to alcohols. When he tried an unsaturated ester, he found that the diborane reacted with the C=C bond and thus discovered the hydroboration reaction.

Less gifted scientists might have missed these discoveries, not because they wouldn't observe them, but because they might not have grasped their significance nor been able to develop them usefully. Indeed, as Louis Pasteur once commented: "In the field of observation, chance only favors the prepared mind."

Brown also did extensive studies on the role of steric effects in mechanisms and on the structure and reactivity of carbocations (carbonium ions). When Brown was awarded the Priestley medal of the American Chemical Society, a colleague, D. A. H. Taylor, commented:

". . . He invented the hydroboration reaction, and perhaps even more significantly the borohydrides and their many derivatives. This is of course true and very important. But many people have invented new methods and reagents, even if they can't all be the 'most important of the last three decades'. . . . Brown's unique quality lies not in this but in the fact that he has never been afraid to challenge established beliefs, whether about carbonium ions or the mechanism of the Hofmann elimination. . . . Whether he has been right or wrong in these matters, and I think he has usually been at least largely right, is not the point. It is in the readiness to challenge accepted ideas that Brown's real significance and greatness lie."

The conversion of carbonyl compounds to alcohols is a very common procedure. Less common, although occasionally very useful, is the conversion of a carbonyl group to a methylene (CH_2) group. While this conversion could be carried out in several steps, it is most efficient if accomplished in one step. Fortunately, we have available two separate methods for the one step carbonyl-to-methylene conversion.

One method uses strong base and the other strong acid. Which procedure we choose depends upon the stability of the other functional groups in the carbonyl compounds. The *Wolff-Kishner reduction* is used with compounds sensitive to acid. The *Clemmensen reduction* is used with compounds sensitive to base.

The Wolff-Kishner reduction involves the reaction between a carbonyl compound and hydrazine under strongly basic conditions at high temperature (eq. 11.60). Triethylene glycol serves as a high-boiling solvent.

(11.60) Cyclohexanone + NH_2NH_2 $\xrightarrow[\text{HO(CH}_2\text{CH}_2\text{O)}_3\text{H, 200°}]{\text{KOH}}$ Cyclohexane (80%) + N_2 + H_2O

A **hydrazone** is an intermediate in the reaction (step 1, eq. 11.61). The reaction depends on a base-catalyzed proton rearrangement (eq. 11.61, steps 2 and 3). The rearrangement is aided by the stability of the carbanion intermediate, **10.** In step 4, the driving force is the loss of a good leaving group, the stable N_2 molecule.

(11.61) $R_2C{=}\ddot{O}\!: \underset{①}{\overset{\ddot{N}H_2\ddot{N}H_2}{\rightleftharpoons}} R_2C{=}\ddot{N}{-}\ddot{N}H_2 \underset{②}{\overset{:\ddot{O}H^-}{\rightleftharpoons}}$

A hydrazone

$R_2\bar{\ddot{C}}{-}\ddot{N}{=}\ddot{N}H + H_2\ddot{O}\!: \underset{③}{\longrightarrow} R_2C(H){-}\ddot{N}{=}\ddot{N}{-}H + :\ddot{O}H^- \underset{④}{\longrightarrow}$

10

$R_2\ddot{C}H^- + N_2 + H_2\ddot{O}\!: \longrightarrow R_2CH_2 + :\ddot{O}H^- + N_2$

The Clemmensen reduction (eq. 11.62) is the second conversion process available. The reaction mixture is heterogenous, involving an organic layer

(11.62) $CH_3\overset{\overset{O}{\|}}{C}(CH_2)_5CH_3 + 2Zn(Hg) + 4HCl \xrightarrow{\text{heat}}$

2-Octanone

$CH_3CH_2(CH_2)_5CH_3 + 2ZnCl_2 + H_2O + Hg$

Octane

(the carbonyl compound), an aqueous layer containing concentrated hydrochloric acid, and a solid layer of amalgamated zinc. The mechanism for the reaction is a complex one which is not yet fully understood.

Problem

11–18 Give the proper reduction product for each of the following reactions:

a. $CH_2{=}CHCH_2CHO + H_2 \xrightarrow{Pt}$

b. $CH_2{=}CHCH_2CHO + LiAlH_4 \rightarrow$

c. $CH_2{=}CHCH_2\overset{\overset{\displaystyle O}{\|}}{C}CH_3 + NaBH_4 \rightarrow$

d. $CH_2{=}CH{-}\underset{\underset{\displaystyle CH_3}{|}}{\overset{\overset{\displaystyle CH_3}{|}}{C}}{-}\overset{\overset{\displaystyle O}{\|}}{C}CH_3 + NH_2NH_2 \xrightarrow[heat]{KOH}$

11:12 Oxidation and Reduction under Biological Conditions: The Reversible Interconversion of Alcohols with Aldehydes and Ketones

In Section 7:12 we mentioned that living cells get much of their energy from the controlled multi-step oxidation of organic compounds. In many cases one or more of those steps involves the oxidation of a hydroxyl group to a carbonyl group. Furthermore, the reverse reaction, reduction of a carbonyl group to a hydroxyl group, also plays an important role in cell metabolism. Many biological oxidations and reductions have in fact been shown to follow the *same* mechanism; only the position of equilibrium is different. The central feature of this mechanism is the transfer of a hydride ion ($H\colon^-$) to or from the carbon atom bound to the oxygen. In this respect there is a good deal of similarity between many biochemical reductions and those laboratory reductions which employ $LiAlH_4$ or $NaBH_4$. This section discusses one of the most common pathways for cellular oxidation and reduction.

Let's start by looking at the oxidation of ethanol to acetaldehyde—a normal task for human liver cells. An enzyme called *alcohol dehydrogenase* (or *oxidoreductase*) is required, though it alone is not capable of bringing about the oxidation. The reaction also requires a **coenzyme** called *nicotinamide adenine dinucleotide* (NAD). The business end of this coenzyme is nicotinamide, once known as vitamin B_3. The terms coenzyme and vitamin are almost synonymous. In the coenzyme, the pyridine nitrogen of nicotinamide is attached to a complex organic substituent called adenine dinucleotide.

O ‖ CNH₂ — N

Nicotinamide (vitamin B_3)

O ‖ CNH₂ — N+ — R

NAD^+

For now, we need only focus on the nicotinamide portion of the coenzyme.

The oxidized form of the coenzyme, NAD^+, removes a hydride ion ($H:^-$) from C-1 of ethanol, and is converted to the reduced form, NADH (eq. 11.63).

$$CH_3CH_2OH + NAD^+ \xrightleftharpoons[\text{dehydrogenase}]{\text{alcohol}} CH_3CHO + H^+ + NADH \tag{11.63}$$

NAD$^+$
(R = a complex organic substituent)

NADH

Although NAD^+ is a complicated molecule, the oxidation–reduction chemistry all takes place within the six-membered pyridine ring (sec. 11:3). In this section we are using an abbreviated representation of NAD^+ and NADH in which only this ring is drawn out; the full structure of NAD^+ was presented in Figure 10–4 (p. 374). In order to understand the stereochemical material that follows, you should keep in mind the fact that the R group contains many chiral centers.

Biochemical reactions are highly stereospecific, in contrast to many laboratory reactions (sec. 10:12). By using deuterium-labelled reactants it is possible to observe this stereospecificity in NAD-dependent reactions. For instance, deuterium-labelled ethanol converts NAD^+ to the deuterated version of the reduced coenzyme, NADH (eq. 11.64).

$$CH_3CD_2OH + NAD^+ \xrightarrow[\text{dehydrogenase}]{\text{alcohol}} CH_3CDO + H^+ + \text{NADH-4-}d \tag{11.64}$$

Ethanol-1,1-d_2 NAD$^+$

Acetaldehyde-1-d NADH-4-d
(stereospecifically deuterated)

The fourth carbon of the deuterated NADH is a chiral center, and the reduction shown in Equation 11.64 produces only one diastereomer.

The reverse of this stereochemical effect is observed when the (*R*)-NADH-4-*d* is used to reduce acetaldehyde (eq. 11.65). (The position of equilibrium in all these reactions is controlled by the reactant and product concentrations.) Only the deuterium is transferred, giving rise to ethanol-1-*d*. This alcohol is chiral, and only one of its enantiomers is obtained. The

$$\text{(11.65)}\quad H^+ + CH_3\overset{O}{\overset{\|}{C}}H + (R)\text{-NADH-4-}d \xrightarrow[\text{dehydrogenase}]{\text{alcohol}} \text{Ethanol-1-}d + NAD^+$$

labelling experiments show that the hydride is transferred to and from only one face of the pyridine ring. Furthermore, in the oxidation process (eq. 11.64), the enzyme–coenzyme complex attacks only *one* of the two hydrogens at C-1 of ethanol. While C-1 is not chiral, it is *prochiral* (sec. 10:12). The two identical groups at a prochiral center can be distinguished by a chiral reagent, and the enzyme–coenzyme complex is able to do just that with a very high degree of selectivity. With alcohol dehydrogenase, only the *pro-R* hydrogen (or deuterium) is removed.

The mechanistic aspects of NAD-dependent redox reactions are summarized in the proposed mechanism shown in Figure 11–7. It is believed that the

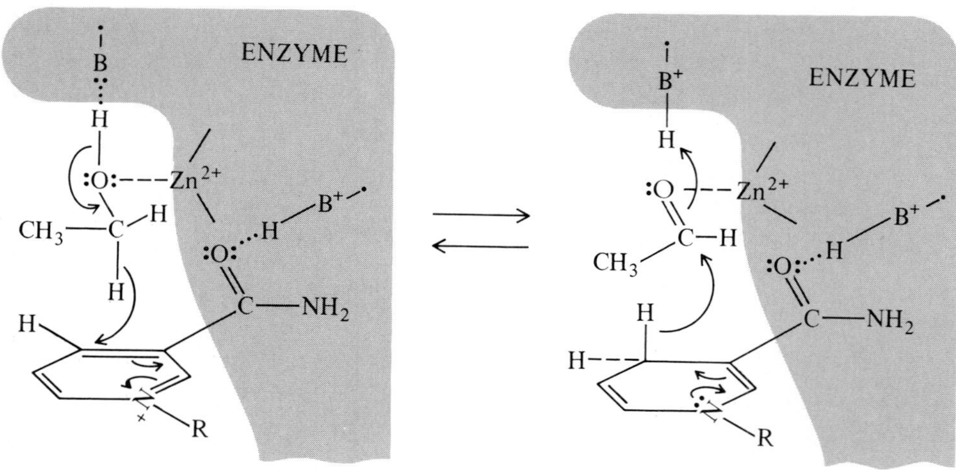

Figure 11–7. *A schematic mechanism for the enzymatic equilibration of ethanol/NAD⁺ and acetaldehyde/NADH.*

Zn^{2+} ion, which is part of the enzyme, helps to bind the carbonyl oxygen and polarize the carbonyl group, as Mg^{2+} does in the Grignard reaction (eq. 11.20).

Problem

11–19 Show the products of each of the following reactions, including their absolute configurations. All reactions are catalyzed by yeast alcohol dehydrogenase.

a. NAD^+ + (*S*)-ethanol-1-*d* →

b. NADH + acetaldehyde-1-*d* →

c. NAD^+-4-*d*

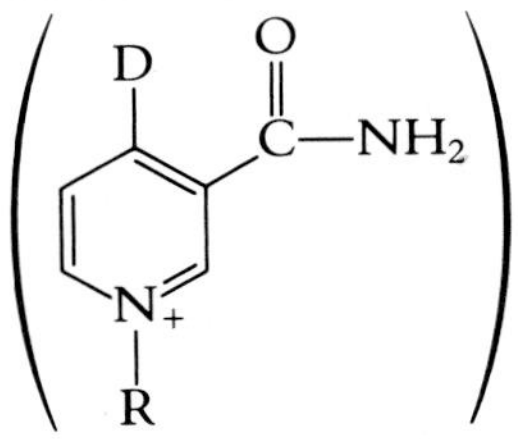

D, O, C—NH_2, N^+, R

is first reduced with ordinary ethanol, then oxidized by acetaldehyde-1-*d*

Summary

1 Aldehydes and ketones contain the carbonyl group (C═O) bonded to carbon or hydrogen atoms. The suffix "-al" denotes an aldehyde in IUPAC nomenclature (Table 11–1) and "-one" denotes a ketone (Table 11–2; sec. 11:1). Most common aldehydes and ketones have a characteristic, intense ir band in the 1690 to 1745 cm^{-1} region due to the carbonyl group. The carbonyl carbon is at very low field in the ^{13}C spectrum (often in the 200 ppm region). The aldehydic proton appears in the 9.6 to 10.0 ppm region of the ^1H nmr spectrum (sec. 11:2).

2 Oxidation of secondary alcohols with CrO_3 in acid gives ketones, while oxidation of primary alcohols with CrO_3 in nonaqueous pyridine solution gives aldehydes. Tertiary alcohols do not react with these reagents. Aldehydes are susceptible to further oxidation in aqueous CrO_3 solution, to yield carboxylic acids. Ketones are generally stable to further oxidation (sec. 11:3).

3 Grignard reagents add rapidly and irreversibly to carbonyl groups. A halomagnesium alkoxide is formed. This intermediate can be protonated in a second step to produce an alcohol. Formaldehyde gives primary alcohols, other aldehydes secondary alcohols, and ketones give rise to tertiary alcohols (Table 11–3; sec. 11:4).

4 Strong nucleophiles attack the carbonyl group directly to give an alkoxide ion. The ion is then protonated to give an addition product. If a weak nucleophile is used, protonation of the carbonyl oxygen must precede nucleophilic attack. Carbonyl additions are usually rapid, and the equilibrium constants not too large, so many reactions are readily reversible (sec. 11:5).

5 Addition of cyanide ion to aldehydes and ketones leads to cyanohydrin formation. Similarly, attack of bisulfite ion produces bisulfite addition products. Water can add to carbonyl compounds to give hydrates, although the equilibrium usually favors reactants (sec. 11:6).

6 Amines add to carbonyl compounds to give addition products that are not stable; they rapidly lose water in the presence of acid to yield imines. Most imines are easily hydrolyzed back to aldehydes and ketones. Very stable imines are obtained when the starting amines are hydroxylamine, phenylhydrazine, and semicarbazide. The solid products formed—oximes, phenylhydrazones, and semicarbazones—are useful for characterizing aldehydes and ketones (sec. 11:7).

7 Alcohols add to carbonyl compounds to give hemiacetals and hemiketals. Under anhydrous acidic conditions the hydroxyl group of the addition product can be replaced by another molecule of alcohol, producing an acetal or ketal. Aldehydes can be converted to acetals in good yield by reaction with simple alcohols; however, in order to get good yields of ketals from ketones, it is necessary to use diols (sec. 11:8). Acetal and ketal formation can be used as a way of protecting carbonyl compounds. Formation of the mixed acetals known as THP ethers is a common method for protecting alcohols (sec. 11:9).

8 Aldehydes can be reduced to primary alcohols, while ketones can be reduced to secondary alcohols. Hydrogen and a metal catalyst will accomplish these reductions of the C=O bond, just as they will reduce the C=C bond. The C=O bond is also reducible with the hydride reagents, lithium aluminum hydride ($LiAlH_4$) and sodium borohydride ($NaBH_4$), neither of which will reduce an isolated C=C bond (sec. 11:10).

9 The C=O group of aldehydes and ketones can be reduced directly to the methylene group (CH_2) by hydrazine in basic solution (Wolff-Kishner reduction) or by zinc amalgam in an acid mixture (Clemmensen reduction; sec. 11:11).

10 The enzymatic oxidation of alcohols to aldehydes and ketones, or the reverse reduction of aldehydes and ketones to alcohols, is mediated by a coenzyme (NAD) containing a pyridine ring. A hydride ion ($H{:}^-$) is selectively added to the pyridine ring or removed from it as the alcohol is oxidized or the carbonyl group is reduced. The process is stereospecific with respect to both C-4 of the pyridine ring and C-1 of the alcohol (sec. 11:12).

Problems

11–20 Give the IUPAC and common name (if there is one) of each compound.

a. $CH_3CH_2\overset{\overset{O}{\|}}{C}H$

b. $(CH_3)_2CH\overset{\overset{O}{\|}}{C}H$

c. $Cl_2CH\overset{\overset{O}{\|}}{C}H$

d. $(CH_3)_2CH\overset{\overset{O}{\|}}{C}CH(CH_3)_2$

e. 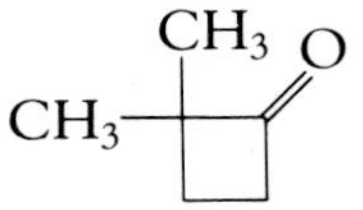

f. $CH_3\overset{\overset{O}{\|}}{C}CCl_3$

g. $\triangleright\!\!-\overset{\overset{\displaystyle O}{\|}}{C}CH_2CH_3$

11–21 Draw the structures that correspond to the following names:

a. bromoacetaldehyde
b. 2-methyl-3-hexanone
c. 2-ethyl-4,4-dimethylpentanal
d. 3,5-dioxooctanal
e. 1-cyclopentyl-2,2-dimethyl-3-pentanone

11–22 The following alcohols are treated with chromic acid in aqueous acetone at room temperature. Show the major organic product(s) in each case.

a. 2-pentanol
b. 3-pentanol
c. *trans*-3-chlorocyclopentanol
d. 1,4-hexanediol
e. 1-methyl-1,4-cyclohexanediol
f. 4,7-dibromo-5-isopropyl-3-octanol

11–23 Show the products that are formed when each of the following compounds is oxidized with CrO_3 in pyridine.

a. 2-butanol
b. 3-methylcyclopentanol
c. 4-bromo-1-butanol
d. acetaldehyde
e. 5-hexen-1-ol

11–24 Compound **A** has the formula $C_5H_{12}O$, and its ir spectrum shows that it contains an -OH group. It reacts very readily with a H_2CrO_4 solution at room temperature, giving rise to an acidic product. **A** reacts very slowly in aqueous HBr, however, and the solution must be heated to drive the reaction. Two isomeric products with the molecular formula $C_5H_{11}Br$ are obtained from the HBr reaction.

Compound **B** is an isomer of **A** and also contains an -OH group. It does *not* react with aqueous H_2CrO_4 at room temperature, as shown by the absence of a color change. **B** does react rapidly with aqueous HBr at room temperature to yield one product of formula $C_5H_{11}Br$.

Write structures for **A** and **B** that are in accord with the observations, and give equations for all the chemistry.

11–25 Specify the reagents and conditions needed to bring about the following transformations:

a. 1-hexanol → hexanal
b. 1-hexanol → hexanoic acid
c. cyclobutanol → cyclobutanone
d. cyclopropylmethanol → cyclopropanecarboxaldehyde
e. 6-methylheptane-1,5-diol → 5-oxo-6-methylheptanal

11–26 One of the most common methods for obtaining carbonyl compounds is by the oxidation of alcohols. Other groups may be converted to carbonyls if they can first be changed into hydroxyl groups. With this in mind, show how the following conversions can be carried out.

a. 1-bromopentane → pentanoic acid
b. pentyl tosylate → pentanal

c. chlorocyclohexane → cyclohexanone
d. 2-methoxy-4-methylheptane → 4-methyl-2-heptanone

11–27 When 3,3-dimethyl-2-pentanol is oxidized with chromic acid in an acetic acid/water mixture, the major product is 3,3-dimethyl-2-pentanone. Also isolated, however, are various quantities of acetaldehyde and 2-methyl-2-butanol. Suggest a mechanism for the formation of these "abnormal" oxidation products. Why are they found in this particular reaction? (Consider the decomposition of the chromate ester, as in Equation 11.8.)

11–28 Each of the following alcohols may be prepared by way of a Grignard reaction. Write a synthesis for each, using an alkyl halide and an aldehyde or ketone as starting materials. For some of these compounds more than one Grignard route may be available. In those cases, suggest a second pair of starting materials.

a. 2-pentanol
b. 3-pentanol
c. 1-pentanol
d. 2-methyl-2-propanol
e. 3-methyl-3-pentanol
f. 3-methyl-3-hexanol
g. 1-methylcyclopentanol
h. 1-cyclobutyl-1-propanol
i. 1-cyclobutyl-2-propanol

11–29 The following transformations may be made by combining Grignard reactions with other reactions. Propose a synthetic pathway for each one.

a. 2-pentanol → 4-methyl-3-heptanone (a compound that triggers alarm reactions in several insect species)
b. 1-butene → 2-methylbutanal
c. 1-butene → 2-methyl-3-hexanol
d. cyclopentanone → (cyclopentane bearing CN and CH_3 on adjacent carbons)

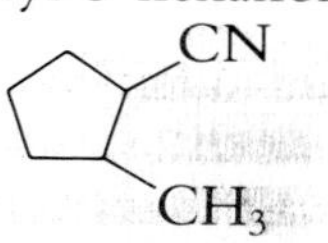

cis and *trans*

11–30 Show all the equilibrium products when the following pairs of compounds are allowed to react.

a. acetone and H_2O
b. formaldehyde and HCN
c. cyclohexanone and D_2O
d. 2-methylpropanal and HCN
e. dicyclopropyl ketone and H_2O

11–31 Thermodynamic calculations show that the enthalpy change for addition of a hydrogen halide (HX) to a carbonyl group is positive (endothermic). This reaction also involves an unfavorable entropy change, so the overall equilibrium strongly favors reactants (eq. 5).

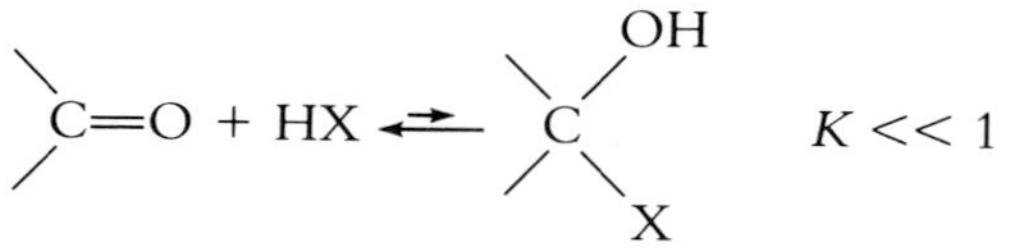

$$\geq C{=}O + HX \rightleftharpoons \geq C(OH)(X) \qquad K << 1 \tag{5}$$

An exception to this generalization is cyclopropanone. It reacts with HCl to yield the isolable compound 1-chlorocyclopropanol. Similarly, it reacts with water to give cyclopropane-1,1-diol, even though most other ketones do not form stable hydrates. Explain why cyclopropanone exhibits this unusual behavior.

11–32 When acetone bisulfite compound and propanal are mixed in water, propanal bisulfite compound and acetone are eventually formed (eq. 6). What is the driving force for this reaction? Propose a mechanism for it.

$$CH_3{-}\underset{CH_3}{\overset{OH}{\overset{|}{\underset{|}{C}}}}{-}SO_3^- \, Na^+ + CH_3CH_2CHO \rightarrow CH_3\overset{O}{\overset{\|}{C}}CH_3 + CH_3CH_2{-}\underset{H}{\overset{OH}{\overset{|}{\underset{|}{C}}}}{-}SO_3^- Na^+ \quad (6)$$

11–33 Acrylonitrile (**I**) and α-methylacrylonitrile (**II**) are important monomers for the manufacture of plastics, resins, and synthetic fibers.

$$\underset{\textbf{I}}{CH_2{=}CHCN} \qquad \underset{\textbf{II}}{CH_2{=}\overset{CH_3}{\overset{|}{C}}CN}$$

These compounds may be synthesized from acetaldehyde and acetone, respectively. Suggest how these syntheses might be carried out in the laboratory.

11–34 Write the structures of the products of the following reactions:

a. 2-methylpropanal $\xrightarrow{\text{KCN, } H_2O}$
b. 2-hexanone $\xrightarrow{NaHSO_3}$
c. 3-methylcyclopentanone $\xrightarrow{\text{hydroxylamine}}$
d. (*E*)-5-methyl-2-hexenal $\xrightarrow{\text{semicarbazide}}$
e. pentanal $\xrightarrow{\text{hydrazine}}$
f. cyclooctanecarboxaldehyde $\xrightarrow{\text{phenylhydrazine}}$
g. 5-ethyl-2-methyl-3-heptanone $\xrightarrow[\text{glycol, } H^+]{\text{ethylene}}$
h. (*cis*-2,*trans*-3-dimethylcyclobutanecarboxaldehyde $\xrightarrow{CH_3OH, \ H^+}$
i. cyclohexanone $\xrightarrow{\text{1,3-propanediol, } H^+}$

11–35 Show the products of each of the following reactions, including stereoisomers where possible. (*Hint:* Consider the possible *E*,*Z*-isomerism about the C=N bond.)

a. acetone + semicarbazide →
b. 3-methylcyclobutanone + hydroxylamine →
c. 2-butanone + phenylhydrazine →
d. 4-heptanone + HCN →
e. 2-methylcyclohexanone + phenylhydrazine →

11–36 2-[^{18}O]-Butanone may be prepared by treating the ordinary ketone with $H_2{}^{18}O$ containing a small quantity of acid. Suggest a mechanism for this reaction.

11–37 Reaction between alcohols and carbonyl compounds to produce hemiacetals and hemiketals is slow. It usually requires acid catalysis (eq. 7). The reaction between amines and carbonyl compounds, by contrast, proceeds rapidly in the absence of catalysts (eq. 8).

$$R_2C{=}O \overset{H^+}{\rightleftharpoons} R_2C{=}OH^+ \overset{R'OH}{\rightleftharpoons} R\overset{\overset{OH}{|}}{\underset{\underset{R}{|}}{C}}{-}\overset{+}{O}R'(H) \rightleftharpoons R\overset{\overset{OH}{|}}{\underset{\underset{R}{|}}{C}}{-}OR' + H^+ \qquad (7)$$

$$R_2C{=}O \overset{R'NH_2}{\rightleftharpoons} R\overset{\overset{O^-}{|}}{\underset{\underset{R}{|}}{C}}{-}\overset{+}{N}H_2R' \rightleftharpoons R\overset{\overset{OH}{|}}{\underset{\underset{R}{|}}{C}}{-}NHR' \qquad (8)$$

Explain the difference in reactivity between alcohols and amines.

11–38 Thiols (RSH) behave chemically like alcohols in many reactions. With this in mind, predict the product of each of the following reactions:

a. $CH_3CH_2\overset{\overset{O}{\|}}{C}H \xrightarrow{C_2H_5SH,\ H^+}$

b. $CH_3CH_2\overset{\overset{O}{\|}}{C}CH_2CH_3 \xrightarrow[H^+]{HSCH_2CH_2SH}$

c. cyclopentyl–CHO $\xrightarrow{CH_3(CH_2)_3SH,\ H^+}$

d. $(CH_3)_2CHCH_2SH \xrightarrow{\text{dihydropyran},\ H^+}$

11–39 a. When the ketoaldehyde 2-oxopropanal (*pyruvaldehyde*, **III**) is heated with an excess of methanol in the presence of acid, the only product isolated is the acetal **IV** (eq. 9). No ketal or acetalketal is found. Explain why reaction takes place only at the aldehyde carbonyl group.

$$\underset{\textbf{III}}{CH_3\overset{\overset{O}{\|}}{C}\overset{\overset{O}{\|}}{C}H} \xrightarrow[H^+]{CH_3OH} \underset{\textbf{IV}}{CH_3\overset{\overset{O}{\|}}{C}CH(OCH_3)_2} \qquad (9)$$

b. If ethylene glycol is used instead of methanol, reaction takes place at both carbonyl groups, the acetalketal **V** being the major product (eq. 10). Why does the use of ethylene glycol allow reaction to take place at both carbonyl groups?

$$\underset{\textbf{III}}{CH_3\overset{\overset{O}{\|}}{C}\overset{\overset{O}{\|}}{C}H} \xrightarrow[H^+]{HOCH_2CH_2OH} \textbf{V} \qquad (10)$$

(**V**: CH_3–C (in a dioxolane ring, O–CH_2CH_2–O) bonded to CH (in a second dioxolane ring, O–CH_2CH_2–O))

c. The reaction in Equation 10 also produced a compound that proved to be an isomer of **V**. It had the following ^1H nmr spectrum: 1.43 ppm (singlet, 3H), 4.26 ppm (multiplet, 8H), and 4.5 ppm (singlet, 1H). Propose a structure for this isomer, and propose a mechanism for its formation.

11–40 Show the structure of the tetrahydropyranyl ether of each of the following alcohols:

a. isopropyl alcohol
b. *s*-butyl alcohol
c. 4-penten-1-ol
d. cyclohexylmethanol
e. 2,2,2-trichloroethanol

11–41 a. A chemist wishes to make a sample of 1,2,5-pentanetriol. She plans to make it from 4-penten-1-ol (eq. 11) by hydroxylating the double bond with $KMnO_4$.

$$CH_2{=}CHCH_2CH_2CH_2OH \rightarrow HOCH_2\underset{\substack{|\\OH}}{C}HCH_2CH_2CH_2OH \qquad (11)$$

She knows, however, that $KMnO_4$ also oxidizes 1° alcohols to carboxylic acids (sec. 11:3). How can she solve the problem with the aid of a protecting group?

b. A synthesis of 6-hydroxy-2-hexanone is planned, using 5-chloro-2-pentanone as the starting material (eq. 12). The plan is to convert the chloro compound to a Grignard reagent, which can then be allowed to react with formaldehyde.

$$CH_3\overset{\overset{O}{\|}}{C}(CH_2)_3Cl \rightarrow CH_3\overset{\overset{O}{\|}}{C}(CH_2)_3CH_2OH \qquad (12)$$

The chemist in charge realized, however, that he can't form his Grignard reagent directly from the chloroketone. Explain why this is not feasible. Show how the transformation can be accomplished with the aid of a protecting group.

11–42 Write the structures of the products of the following reactions:

a. acetone $\xrightarrow[H_2O]{NaBH_4}$

b. (*E*)-4-hexen-2-one $\xrightarrow[CH_3OH]{NaBH_4}$

c. 4-methylpentanal $\xrightarrow[ether]{LiAlH_4} \xrightarrow{H_3O^+}$

d. 1,4-cyclohexanedione $\xrightarrow[ether]{LiAlH_4} \xrightarrow{H_3O^+}$

e. cyclopentanone $\xrightarrow[THF]{LiAlD_4} \xrightarrow{H_3O^+}$

f. cyclopentanone $\xrightarrow[THF]{LiAlH_4} \xrightarrow{D_3O^+}$

g. 2-cyclohexyl-3-pentanone $\xrightarrow[ether]{LiAlH_4} \xrightarrow{H_3O^+}$

h. 2-cyclohexenone $\xrightarrow[C_2H_5OH]{Pt,\ H_2}$

11–43 The rate of reduction of cyclobutanone by sodium borohydride in isopropanol is much faster than reduction of cyclohexanone.

a. Offer an explanation of this.
b. What would be the relative stabilities of the cyanohydrins of these two ketones?

11–44 Write a chemical reaction which is an example of each process.

a. The lithium aluminum hydride reduction of an aldehyde
b. A Clemmensen reduction of a ketone
c. The formation of a semicarbazone
d. The conversion of a diol to a cyclic ketal
e. The Wolff-Kishner reduction of a ketone
f. The formation of a hemiacetal from an aldehyde

11–45 (+)-Citronellal (**VI,** $C_{10}H_{18}O$) may be isolated from a number of natural sources, such as lemon oil and citronella oil.

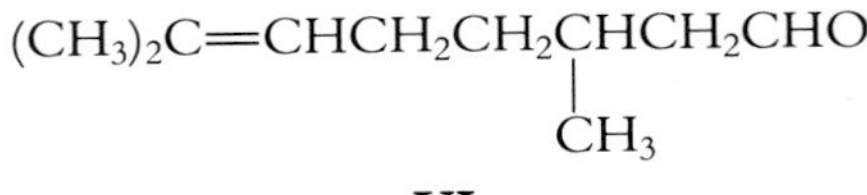

VI

The natural material has the (*S*)-configuration. Citronellal is heated with hydrazine and potassium hydroxide to give **A** ($C_{10}H_{20}$). Compound **A** is then treated successively with O_3 and Zn + HOAc to give **B** ($C_7H_{14}O$). Finally, **B** is heated with hydrazine and potassium hydroxide to give the optically active alkane **C.** Write all the chemistry for these reactions, showing absolute configurations.

11–46 Indicate how the following conversions can be carried out.

a. cyclopentanone → cyclopentyl tosylate
b. 3-methyl-1-pentanol → 3-methylpentanal diethyl acetal
c. pentanal → 3-cyanoheptane
d. 2-pentanol → pentane
e. 3-hexanone → 2-ethylpentanal

11–47 a. A compound **A,** C_8H_{16}, is oxidized with hot acidic potassium permanganate. Gas chromatographic analysis of the product reveals the presence of two compounds, **B** and **C,** which are separable by distillation. They each give a precipitate when treated with semicarbazide solution. Analytical results for **B** are compatible with the formula $C_5H_{10}O$. The nmr spectrum of **B** is as follows (ppm): 0.92 (3H, triplet), 1.6 (2H, multiplet), 2.18 (3H, singlet), 2.45 (2H, triplet). **C** is a volatile liquid, the nmr spectrum of which consists of a single peak. Write structures for **A, B,** and **C.**

b. Compound **D,** $C_5H_{12}O_2$, exhibits strong absorption in the 3400 to 3600 cm^{-1} region in its ir spectrum. When **D** is oxidized with chromium trioxide in pyridine, one of the isolated products is **E,** $C_5H_8O_2$. The ir spectrum of **E** contains an intense band at 1720 cm^{-1}. Its nmr spectrum is as follows (ppm): 2.20 (3H, singlet), 2.70 (4H, multiplet), 9.8 (1H, triplet). **E** gives a positive test with phenylhydrazine (precipitate).

E may be further oxidized with H_2CrO_4 to produce **F,** $C_5H_8O_3$, which also gives a precipitate with phenylhydrazine. If **D** is treated with sulfuric acid, it is transformed into **G,** $C_5H_{10}O$, which gives no reaction with phenylhydrazine, shows no ir band in the 3400 to 3600 cm^{-1} region, and does not decolorize solutions of bromine or chromic acid. Write structures for **D, E, F,** and **G.**

c. Compound **H,** $C_7H_{14}O$, gives a crystalline precipitate when treated with a solution of hydroxylamine. **H** is allowed to react with sodium borohydride in methanol and gives rise to two separable compounds. Both of these substances (**I** and **J**) have nearly identical mass spectra in which the parent peak is found at m/z = 116. Both **I** and **J** form crystalline tosylates when treated with tosyl chloride and base. The tosylates react with potassium *t*-butoxide to yield similar mixtures of isomers (MW = 98). Ozonization of these mixtures followed by treatment with $(CH_3)_2S$ yields two major products—acetaldehyde and **K.** Write structures for **H, I, J,** and **K.**

11–48 An unknown compound, $C_8H_{14}O$, has the following spectral characteristics: ir, 1720, 1640 cm^{-1}; ^1H nmr, 1.62 (3H, singlet), 1.70 (3H, singlet), 2.15 (3H, singlet), 2.1 to 2.6 (4H, multiplets), 5.16 (1H, triplet); ^{13}C nmr, 17.5 (quartet), 22.7 (triplet), 25.0 (quartet), 29.8 (quartet), 42.8 (triplet), 121.5 (doublet), 132.4 (singlet), 208.5 (singlet); mass spectrum, m/z = 126, 111, 43. Suggest a structure and assign the spectral data.

Supplementary Reading

S. Warren, *Chemistry of the Carbonyl Group: A Programmed Approach to Organic Reaction Mechanisms,* John Wiley and Sons, New York, N.Y., 1974.

C. D. Gutsche, *The Chemistry of Carbonyl Compounds,* Prentice–Hall, Englewood Cliffs, N.J., 1967.

F. A. Carey and R. J. Sundberg, *Advanced Organic Chemistry, Part A,* Chap. 8, Reactions of Carbonyl Compounds, Plenum Press, New York, N.Y., 1977.

J. S. Lowrey, T. A. Evans, and S. S. Wong, Stereospecificity of NAD^+/NADH reactions, *J. Chem. Ed.,* **58,** 816 (1981).

G. R. Meyer, Conjugate and nonconjugate reduction with $LiAlH_4$ and $NaBH_4$, *J. Chem. Ed.,* **58,** 628 (1981).

12 Chemistry of the Alkynes: Introduction to the Logic of Synthesis

Compounds containing carbon–carbon triple bonds are called **alkynes,** and they fit the general empirical formula C_nH_{2n-2} if only one triple bond is present. These compounds are also known as *acetylenes,* named after the first member of the class, HC≡CH (IUPAC name, ethyne).

In the IUPAC naming system, the suffix "-yne" is appended to a root which gives the number of carbon atoms in the longest chain containing the triple bond. A number denotes the position of the triple bond. An older system treated acyclic alkynes as derivatives of acetylene (names shown in italics).

$HC{\equiv}C(CH_2)_5CH_3$
1-Octyne
(*hexyl acetylene*)

$CH_3C{\equiv}CCH_2CH_3$
2-Pentyne
(*methyl ethyl acetylene*)

$CH_2{=}CHC{\equiv}CCH_3$
Pent-1-en-3-yne
(*methyl vinyl acetylene*)

$HC{\equiv}CCH(OH)C{\equiv}CH$
1,4-Pentadiyn-3-ol

Cyclooctyne

We will use the nonsystematic name acetylene, rather than ethyne, for HC≡CH, but all other alkynes will be named according to the IUPAC system.

Cycloalkynes are stable only if the ring is large. Because of the hybridization of the triply-bonded carbons, the atoms C—C≡C—C *must* lie in a straight line to avoid angle strain. Cyclooctyne is the smallest cyclic alkyne that is stable at 25°. Triple bonds can be generated in smaller rings, but the resultant compounds undergo instantaneous reactions unless the temperature is very low.

Acetylene is one of the few organic compounds that can be prepared directly and cheaply from what might be considered "inorganic" starting materials. Calcium carbide, which is prepared commercially from coke (coal) and limestone (CaO), can be treated with water to yield acetylene (eq. 12.1). During World War II, when coal was plentiful but supplies of petroleum were limited, German chemists developed an entire technology of hydrocarbon chemistry based upon acetylene.

$$(12.1) \qquad \underset{\text{Lime}}{CaO} + \underset{\text{Coke}}{3C} \xrightarrow{2000^\circ} CO + \underset{\substack{\text{Calcium}\\\text{carbide}}}{CaC_2}$$

$$CaC_2 + 2H_2O \rightarrow \underset{\text{Acetylene}}{HC{\equiv}CH} + Ca(OH)_2$$

The modern industrial synthesis of acetylene is through thermal cracking of methane.

The first part of this chapter shows the versatility of alkyne chemistry by describing the conversion of alkynes to alkenes, carbonyl compounds, and many other useful substances. With this information and the multitude of other reactions you have already learned, you will have adequate tools to plan the synthesis of relatively complex organic molecules. Unfortunately, simply memorizing reactions is not sufficient when planning a complex synthesis. Each target compound must be analyzed carefully and a logical plan established. We will discuss this type of thought process later in the chapter.

12:1 Orbital Description and Spectroscopic Properties of Alkynes

The carbon atoms in acetylene are *sp*-hybridized [Figs. 12–1 and 2–24 (p. 51)]. One *sp* orbital on each carbon atom is involved in the carbon–carbon σ bond, and the other one is part of the carbon–hydrogen bond. That leaves two perpendicular *p* orbitals on each carbon atom, which overlap with their opposite numbers to form two perpendicular π orbitals. The resulting π electron density around the carbon–carbon bond is cylindrically symmetrical (Fig. 12–1b). Due to the *sp* hybridization of the carbon atom, the entire molecule is linear.

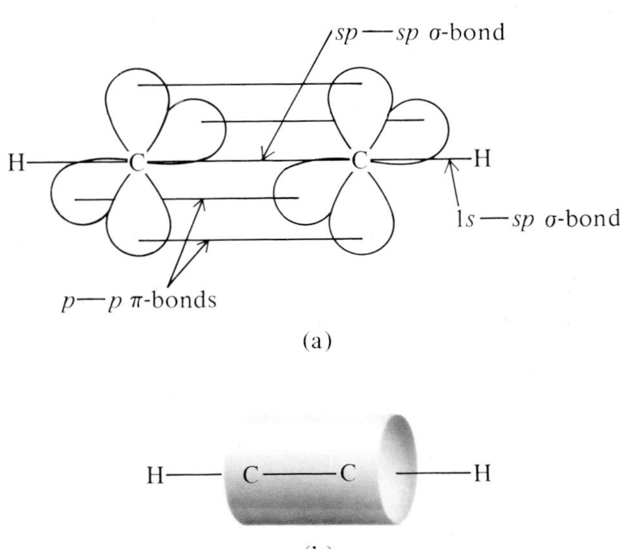

Figure 12–1. (a) *Overlap of* p *orbitals on adjacent* sp*-hybridized carbon atoms of produce the two* π *bonds of the carbon–carbon triple bond.* (b) *Cylindrical distribution of* π *electrons around the carbon–carbon triple bond.*

The —C≡C— stretching vibration occurs at a much higher frequency (2100 to 2250 cm^{-1}) than does the C=C vibration. Only one other functional group absorbs in this region (nitriles, R—C≡N, absorb at about 2250 cm^{-1}) and hence, this absorption band is very diagnostic for alkynes. *Internal* alkynes, such as 2-pentyne, are those in which the triple bond is not at the end of a chain. The stretching of their triple bonds causes almost no change in dipole moment, and the corresponding ir band is therefore quite weak. *Terminal* alkynes (those containing a ≡C—H group) are recognizable because of the characteristic sharp C—H band at 3300 cm^{-1} (Fig. 12–2). The only other functional group that gives a band at precisely this position is OH. This does not present a problem because the OH band in this region is always broad, while the alkyne C—H band is relatively sharp.

The ^1H nmr spectra of alkynes are also distinctive. This is due to the effect of the magnetic field on the cylindrically symmetrical π orbitals. When the axis of the triple bond is parallel to an applied magnetic field, the π electrons may circulate around the carbon–carbon bond (Fig. 12–3). The induced magnetic field caused by this electronic motion opposes the applied field. Other molecular orientations are possible and some of these give rise to the opposite magnetic effect, but the most powerful electron circulation occurs when the triple bond is parallel to the applied field. The result is that hydrogen atoms attached to the triply-bonded carbon atoms experience additional *shielding*. This effect even overcomes the deshielding due to the electronegative *sp*-hybridized carbon atoms. The characteristic alkyne proton resonance occurs at 2.0 to 3.0 ppm, well upfield from the alkene proton peaks.

The ^1H nmr spectrum of (*Z*)-1-methoxy-1-buten-3-yne (Fig. 12–4) is an instructive one. Note first of all that the signal from the alkynyl proton *a* is further upfield than the signals from the vinylic protons *c* and *d*. The two vinylic protons are readily differentiated because proton *c* is strongly coupled

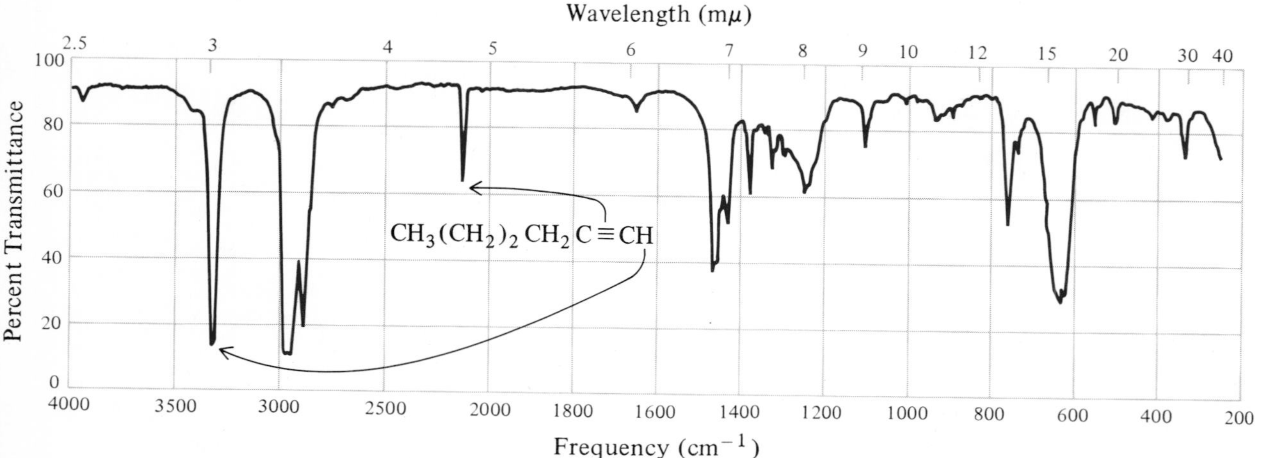

Figure 12–2. *Ir spectrum of the terminal alkyne 1-hexyne.*

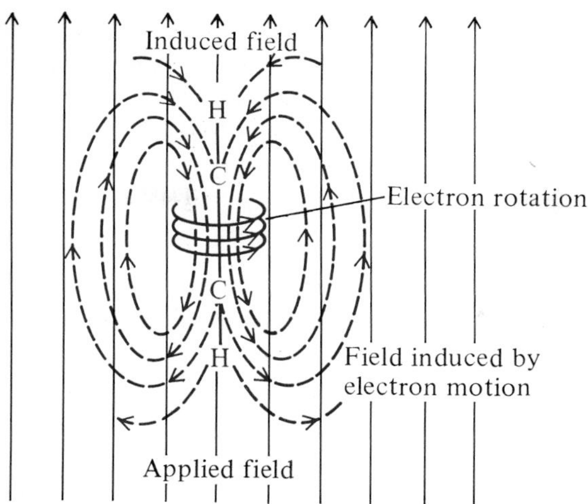

Figure 12–3. *Motion of π electrons in an acetylene molecule aligned parallel to an applied field.*

to both proton *a* (through the triple bond), and to proton *d*, while proton *d* is only weakly coupled to proton *a*. This is an example of *long-range coupling* between protons. The *a* hydrogen is coupled to the *c* proton that is three carbons away, and to the *d* proton that is four carbons away. Long-range coupling can be significant when there are π bonds in between the coupled protons.

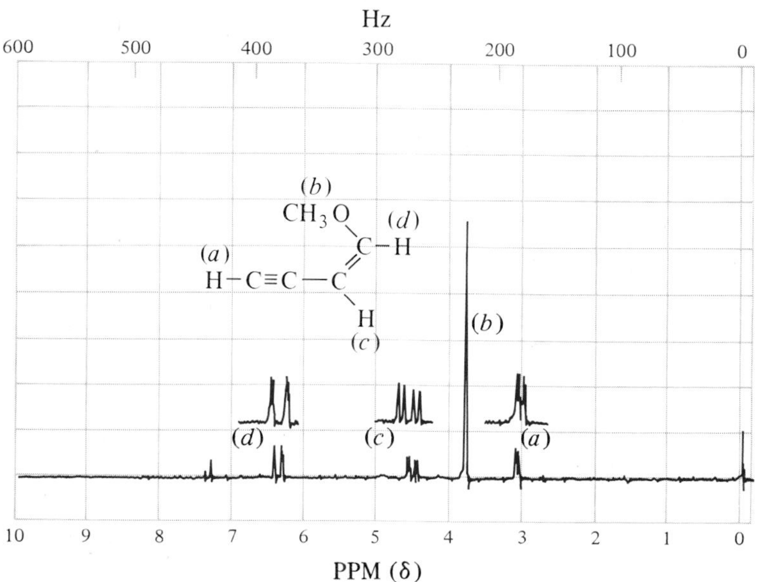

Figure 12–4. *[1]H nmr spectrum of* (Z)-*1-methoxy-1-buten-3-yne.*

Table 12–1. Acidities of Some Common Hydrocarbons

Hydrocarbon Acid	Conjugate Base	Approximate pK_a	Hybridization of Lone-pair Orbital
$RC{\equiv}C{-}H \rightleftarrows H^+ +$ alkyne	$RC{\equiv}C{:}^-$	26	*sp*
$R_2C{=}C(H)(R) \rightleftarrows H^+ +$ alkene	$R_2C{=}\ddot{C}^-{-}R$	36	sp^2
$R_3C{-}H \rightleftarrows H^+ +$ alkane	$R_3C{:}^-$	60	sp^3

12:2 The Acidity of Alkynes: A Comparison with Other Types of Hydrocarbons

Acetylene and other terminal alkynes are weak acids, capable of reacting with Na. Table 12–1 compares the acidity of alkynes with alkenes and alkanes. It is clear that terminal alkynes are the most acidic of these compounds. Generally, similar compounds differ in acidity because their conjugate bases differ in stability (sec. 2:8). Alkynes form a more stable anion than do alkanes and alkenes, hence their acid dissociation equilibrium lies farther to the right. We can conclude, then, that because the conjugate base of the alkyne (acetylide anion, $RC{\equiv}C{:}^-$) is more stable than the conjugate bases of both alkanes and alkenes, alkynes are the most acidic of the three types of hydrocarbons.

The trend in anion stability can be correlated with the differing hybridization of the orbital carrying the negative charge. As you can see from Table 12–1, acetylide ions have the negative charge localized in an *sp* orbital (50% *s*-character). With alkene conjugate bases the charge is in an sp^2 orbital (33% *s*-character), while the negative charge of an alkane conjugate base is contained in an sp^3 orbital (25% *s*-character). Thus, the greater the amount of *s*-character in the nonbonding orbital, the more stable the anion. This relationship makes sense because *s* electrons are closer to the nucleus than *p* electrons in the same shell, so *s* electrons are lower in energy. Overall, then, the most acidic hydrocarbon is the one giving the conjugate base with the largest percent *s* character in its nonbonding orbital. In the next section we will see how their acidity makes alkynes valuable in synthesis.

Problem

12–1 Indicate whether you expect the following equilibria to have constants much less than, much greater than, or approximately equal to one.

a. $CH_3C{\equiv}CH + CH_3^- \rightleftarrows CH_3C{\equiv}C^- + CH_4$
b. $CH_2{=}CH_2 + CH_3CH_2^- \rightleftarrows CH_2{=}CH^- + CH_3CH_3$
c. $CH_3^- + CH_3CH_3 \rightleftarrows CH_4 + CH_3CH_2^-$

d. $CH_4 + NH_2^- \rightleftarrows CH_3^- + NH_3$
e. $CH_3OH + HC{\equiv}C^- \rightleftarrows CH_3O^- + HC{\equiv}CH$

12:3 Formation and Reactions of Alkyne Salts

Alkynes ($pK_a = 26$) are intermediate in acidity between water ($pK_a = 15.7$) and ammonia ($pK_a = 35$). A consequence of this is that OH^- in water is not a sufficiently strong base to remove the proton from a terminal alkyne and create an acetylide anion (eq. 12.2).

$$RC{\equiv}CH + Na^+OH^- \rightleftarrows RC{\equiv}C\colon^- Na^+ + H_2O \tag{12.2}$$

On the other hand, NH_2^- in liquid NH_3 (boiling point −33°) *is* a strong enough base to remove the proton from an alkyne, and hence favors formation of the acetylide ion (eq. 12.3). The acetylide carbanion is a good nucleophile and a strong base, two properties we can use in synthetic reactions.

$$\underset{}{RC{\equiv}CH} + \underset{\text{Sodium amide}}{Na^+NH_2^-} \rightleftarrows \underset{\text{A sodium acetylide}}{RC{\equiv}C\colon^- Na^+} + NH_3 \tag{12.3}$$

A major use of acetylide ions, which we will emphasize, is in the synthesis of larger alkynes. Thus, the acetylide ion can act as a nucleophile in an S_N2 reaction and displace halide ion from an alkyl halide to form a new alkyne. Equations 12.4 and 12.5 show two examples of this. The first, where only

$$HC{\equiv}C\colon^- Na^+ + CH_3(CH_2)_2CH_2{-}Br \xrightarrow{NH_3,\ -33^\circ} \underset{\text{1-Hexyne (75\%)}}{CH_3(CH_2)_2CH_2C{\equiv}CH} + Na^+Br^- \tag{12.4}$$

$$HC{\equiv}CH + 2Na^+NH_2^- + 2CH_3CH_2Br \xrightarrow{NH_3,\ -33^\circ} \underset{\text{3-Hexyne (75\%)}}{CH_3CH_2C{\equiv}CCH_2CH_3} \tag{12.5}$$

one equivalent of sodium amide and alkyl halide is used, yields a monosubstituted alkyne, while the second procedure uses two equivalents of each to prepare a disubstituted alkyne. Lithium acetylides, prepared from $LiNH_2$, are used similarly as nucleophiles.

This reaction can only be used with primary halides. If we attempt to carry out the same type of reaction with a secondary or tertiary halide, the strongly basic acetylide anion will bring about an elimination instead (eq. 12.6).

$$HC{\equiv}C\colon^- + CH_3{-}\overset{\displaystyle CH_3}{\underset{\displaystyle CH_3}{C}}{-}Br \rightarrow HC{\equiv}CH + (CH_3)_2C{=}CH_2 + Br^- \tag{12.6}$$

The acidity of alkyne hydrogens also allows us to prepare alkynyl Grignard reagents (eq. 12.7). These Grignard reagents can be used exactly as

$$(12.7) \qquad HC{\equiv}CH + C_2H_5MgBr \xrightarrow[THF]{} C_2H_6 + \underset{\text{Ethynylmagnesium bromide}}{HC{\equiv}CMgBr}$$

were alkyl Grignard reagents (sec. 11:4) in the synthesis of alcohols from carbonyl compounds. Equation 12.8 gives an example.

$$(12.8) \quad (CH_3)_3CC{\equiv}CH + CH_3CH_2MgBr \xrightarrow[ether]{} (CH_3)_3CC{\equiv}CMgBr \xrightarrow[ether]{(CH_3)_2C{=}O} \xrightarrow{H_3O^+} (CH_3)_3CC{\equiv}C{-}C(CH_3)_2{-}OH$$

2,5,5-Trimethyl-3-hexyn-2-ol (81%)

Alcohol syntheses can also be carried out directly by reacting the carbonyl compounds with the alkali metal acetylides themselves (eq. 12.9).

(12.9) 3,3-dimethylcyclohexanone + $HC{\equiv}CLi$ (Lithium acetylide) $\xrightarrow[\text{benzene-THF}]{}$ $\xrightarrow{H_2O}$ 1-ethynyl-3,3-dimethylcyclohexanol (ring bearing OH and $C{\equiv}CH$; CH_3, CH_3)

1-Ethynyl-3,3-dimethylcyclohexanol (96%)

This reaction has been used to prepare norethynodrel (**1**), a component of the birth control pill, Enovid.

[Structure **1**: steroid skeleton with OH and $C{\equiv}CH$ at C-17, ketone O at C-3]

1

Many of the synthetic hormones used in oral contraceptives are modeled after progesterone, **2,** a natural hormone involved in the female ovulation cycle.

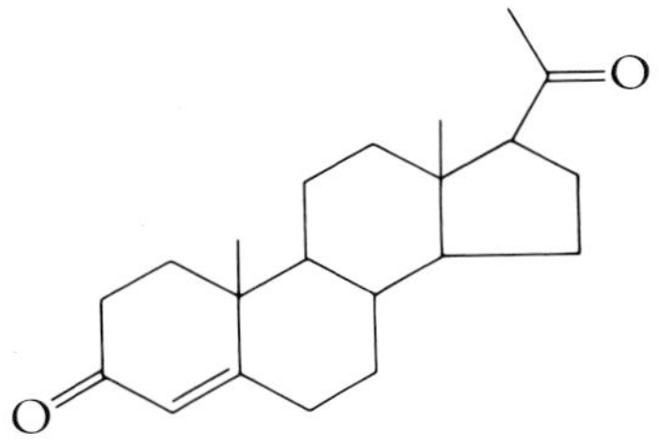

2

This type of synthesis, in which small changes are made in a known natural, physiologically-active compound, is called *molecular modification.* The synthetic compound can sometimes mimic certain actions of the natural substance or react at biological sites faster than the natural substance and hence block its action.

Example

Sodium amide ($NaNH_2$) is dissolved in liquid ammonia at $-33°$, to which propyne and then bromomethane are added.

What products remain after the liquid ammonia is evaporated?

Answer

The first reaction is that between NH_2^- and the propyne:

$$CH_3C{\equiv}CH + :NH_2^- \rightarrow CH_3C{\equiv}C:^- + NH_3$$

The anion then displaces the bromide in an S_N2 reaction:

$$CH_3C{\equiv}C:^- + CH_3{-}Br \rightarrow CH_3C{\equiv}CCH_3 + Br^-$$

After the liquid ammonia is allowed to evaporate, we are left with NaBr and a new alkyne, $CH_3C{\equiv}CCH_3$.

Problems

12–2 Give the organic products of the following reactions:

a. $HC{\equiv}CCH_2CH_3 + NH_2^- + CH_3Cl \xrightarrow[NH_3,\ -33°]{}$

b. $HC{\equiv}CH + 2NH_2^- + 2CH_3CH_2Br \rightarrow$

c. $CH_3C{\equiv}CH + CH_3MgBr \rightarrow$

d. $CH_3C{\equiv}CMgBr + CH_3\overset{\overset{O}{\|}}{C}H \longrightarrow \xrightarrow{H_3O^+}$

12–3 2-Methyl-1-buten-3-yne is an inexpensive commercial chemical. Suggest a simple synthesis for this compound based on inexpensive starting materials.

12:4 Synthesis of Alkynes Via Elimination Reactions

So far we have focused on the relatively high acidity of alkynes, which sets them apart from other hydrocarbons. The next several sections show the similarities between alkynes and alkenes, similarities that are due to the π bonds in both classes of compounds.

The π bonds of alkynes, like those of alkenes, can be readily formed by way of elimination reactions. One common way to synthesize alkynes is to use dihaloalkanes as substrates. Since alkynes contain two π bonds, the elimination of two equivalents of hydrogen halide is necessary. 1,2-Dihalides are the most commonly used since they are easily available from alkenes (sec. 10:4); 1,1-dihalides can also be used.

Under standard *E*2 elimination conditions (sec. 7:7), the first equivalent of hydrogen halide is eliminated easily, giving a mixture of haloalkenes (eq. 12.10).

(12.10) $CH_3(CH_2)_7CH(Br){-}CH_2Br \xrightarrow[\text{EtOH, 80°}]{\text{NaOH}}$

1,2-Dibromodecane

$CH_3(CH_2)_7C(Br){=}CH_2 + CH_3(CH_2)_7CH{=}CHBr$

2-Bromo-1-decene (*Z*)- and (*E*)-1-Bromo-1-decene

However, the haloalkenes are fairly unreactive toward further elimination. This barrier can be overcome by using a very strong base, such as $Na^+NH_2^-$ (eq. 12.11), or by using a conventional hydroxide in a high-boiling alcoholic solvent (eq. 12.12.).

(12.11) $CH_3(CH_2)_7C(Br){=}CH_2 + CH_3(CH_2)_7CH{=}CHBr \xrightarrow[\text{NH}_3,\ -33°]{\text{NaNH}_2} CH_3(CH_2)_7C{\equiv}CH$

1-Decyne (95%)

(12.12) $CH_3C(Br){=}CH_2 \xrightarrow[\text{BuOH, 118° (solvent)}]{\text{KOH}} CH_3C{\equiv}CH$

1-Propyne (85%)

Because the usual source of 1,2-dihalides is alkenes, this double elimination sequence permits us to convert alkenes to alkynes. Furthermore, it provides a route to alkynes with 2° or 3° alkyl groups attached to the triple bond (eq. 12.13), which are not available by reaction of acetylide ions with

(12.13) $(CH_3)_3CCH{=}CH_2 \xrightarrow{\text{Br}_2,\ \text{CCl}_4} (CH_3)_3CCH(Br){-}CH_2Br \xrightarrow[\text{NH}_3,\ -33°]{\text{NaNH}_2}$

3,3-Dimethyl-1-butene

$\left[(CH_3)_3CC(Br){=}CH_2\right] \xrightarrow[\text{NH}_3,\ -33°]{\text{NaNH}_2} (CH_3)_3CC{\equiv}CH$

Major product (not isolated) 3,3-Dimethyl-1-butyne

alkyl halides (sec. 12:3). Equation 12.13 shows that by using two equivalents of $NaNH_2$ we can avoid having to isolate the intermediate haloalkene.

Problems

12–4 Show how 3-methyl-1-butene can be converted to 2-methyl-3-heptyne.

12–5 a. The reactions that produce alkynes from bases and haloalkenes have been shown to be first order in each component. Using these results, suggest a mechanism for the reaction between $NaNH_2$ and 2-bromo-1-decene (eq. 12.11).

b. Both 2-bromo-1-decene and (*Z*)-1-bromo-1-decene react faster with $NaNH_2$ than does (*E*)-1-bromo-1-decene. What can you conclude about the preferred stereochemistry of elimination for haloalkenes? Compare your conclusion with what you know about the stereochemistry of elimination with haloalkanes (sec. 7:9).

Alkynes resemble alkenes not only in the way they are synthesized but also in their reactions with electrophilic reagents. These reactions are the main subjects of the next section.

12:5 Addition Reactions of the Triple Bond

The general principles that you learned for electrophilic alkene additions (sec. 10:3) apply to alkyne reactions as well. Acid-initiated addition reactions are most easily understood if we assume that a proton first attacks the π bond to give a **vinyl cation** (eq. 12.14).

$$CH_3-C\equiv C-H \xrightarrow[\text{slow}]{H^+} CH_3-\overset{+}{C}=CH_2 \xrightarrow{Y^-} CH_3-\underset{}{\overset{Y}{\overset{|}{C}}}=CH_2 \quad (12.14)$$

2-Propenyl cation (a vinyl cation)

$$CH_3-C\equiv C-H \xrightarrow{H^+} \text{(not formed)} \; CH_3-\overset{H}{\overset{|}{C}}=\overset{+}{C}H$$

This mechanism applies to the addition of water, alcohols, and hydrogen halides (HY = HOH, ROH, and HX) to alkynes. It also accounts for the regiospecificity of this reaction (eq. 12.15), in which only the more stable (highly substituted) of the two possible vinyl cations is formed.

$$CH_3(CH_2)_3C\equiv CH + HCl \rightarrow CH_3(CH_2)_3\underset{Cl}{\underset{|}{C}}=CH_2 \quad (12.15)$$

Since the product of HY addition is an alkene, it could react with a second mole of HY to give a disubstituted alkane. The addition of hydrogen halides to the triple bond may be controlled to give either the haloalkene or the dihaloalkane (eq. 12.16). The reaction in Equation 12.16 was carried out in the presence of free-radical inhibitors to prevent anti-Markovnikov addition (sec. 10:5).

$$(12.16)\quad CH_3(CH_2)_3C{\equiv}CH + HBr \xrightarrow{15^\circ} CH_3(CH_2)_3CBr{=}CH_2 \xrightarrow{HBr} CH_3(CH_2)_3CBr_2CH_3$$

Addition of HX to an internal alkyne should give a mixture of *cis* and *trans* alkenes if a vinyl cation is an intermediate (eq. 12.17).

$$(12.17)\quad R{-}C{\equiv}C{-}R \xrightarrow{H^+} R{-}C(H){=}\overset{+}{C}{-}R \xrightarrow{X^-} \underset{cis}{(H)(R)C{=}C(X)(R)} + \underset{trans}{(H)(R)C{=}C(R)(X)}$$

Vinyl cations are of comparatively high energy, particularly in solvents of low polarity. It is possible, therefore, to bypass free vinyl cations by loading the reaction medium with an excess of the nucleophile X^-. This leads to a *concerted* addition in which both the proton and the nucleophile become bonded to carbon at the same time. The major advantage of the concerted reaction is that it displays *anti* stereochemistry, leading to the *trans* isomer almost exclusively (eq. 12.18; the salt $(CH_3)_4N^+Cl^-$ is a source of chloride ions).

$$(12.18)\quad CH_3CH_2C{\equiv}CCH_2CH_3 + HCl \xrightarrow[25^\circ,\ \text{acetic acid}]{(CH_3)_4N^+Cl^-} \left[(CH_3CH_2)(\overset{\delta-}{Cl})C{=}C(H{-}{-}{-}\overset{\delta-}{Cl})(CH_2CH_3) \right] \longrightarrow \underset{(97\%)}{(CH_3CH_2)(Cl)C{=}C(H)(CH_2CH_3)} + Cl^-$$

Although the addition of hydrogen halides to alkynes is relatively rapid, the addition of water or alcohols to alkynes is a slow process. This is because alkynes are not very soluble in polar solvents, and because they are very weak bases toward hydronium (H_3O^+) and alkyloxonium (ROH_2^+) ions. The addition of a small amount of mercuric salt, HgX_2, to the reaction mixture vastly accelerates the reaction. The mercury cation (either Hg^{2+} or XHg^+) forms a π complex with the alkyne by overlap of an empty metal ion orbital with a filled π orbital (Fig. 12–5). The π complex is quite soluble in the reaction medium, and because it is positively charged, it is also susceptible to nucleophilic attack.

In the laboratory this reaction is usually performed with a terminal alkyne to produce a 2-alkanone because Markovnikov's rule is followed.

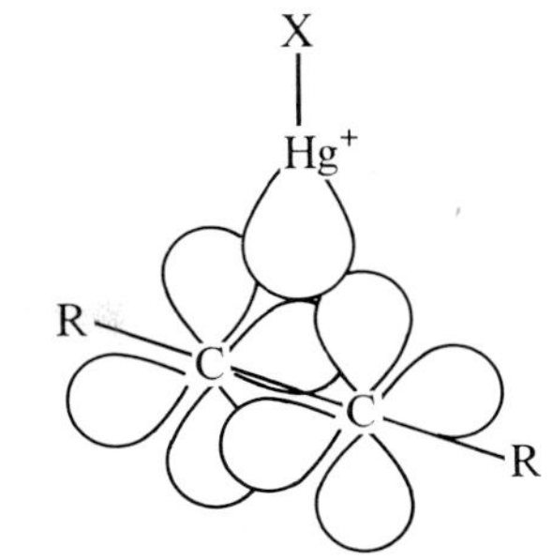

π-complex

Figure 12–5. *A π complex between Hg^{2+} and an alkyne. An empty orbital on Hg^{2+} overlaps with one of the π orbitals of the alkyne.*

Equation 12.19 shows the addition of water to 1-hexyne in the presence of mercuric sulfate.

$$\text{(12.19)}\quad \underset{\text{1-Hexyne}}{CH_3(CH_2)_3C\equiv CH} \xrightarrow[H_2SO_4]{HgSO_4,\ H_2O} \underset{\text{Enol}}{\left[CH_3(CH_2)_3\overset{\overset{\large OH}{|}}{C}=CH_2\right]} \longrightarrow \underset{\substack{\text{2-Hexanone}\\(80\%)}}{CH_3(CH_2)_3\overset{\overset{\large O}{\|}}{C}CH_3}$$

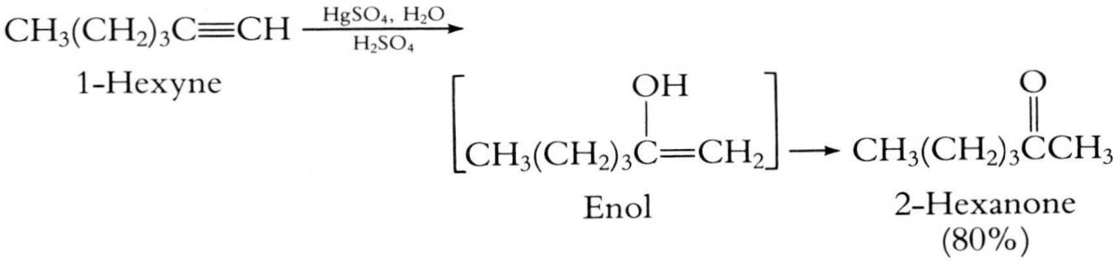

When acetylene is the reactant, ethanal (acetaldehyde) is the product. This route has been used to produce this important aldehyde commercially. In all cases an enol (eq. 12.20) is an intermediate. Enols rapidly isomerize to aldehydes or ketones, which are the more stable isomers (Chap. 18).

$$\text{(12.20)}\quad HC\equiv CH \xrightarrow[H_2SO_4]{H_2O,\ HgSO_4} \underset{\text{Enol}}{\left[H_2C=\overset{\overset{\large OH}{|}}{C}H\right]} \rightleftharpoons CH_3-\overset{\overset{\large O}{\|}}{C}H$$

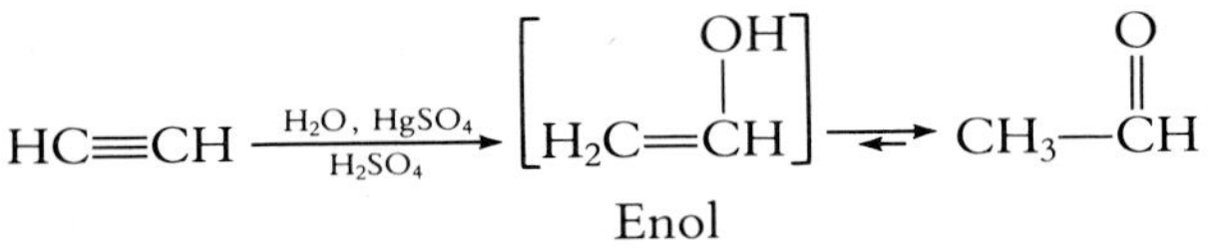

Equations 12.21 and 12.22 give a mechanism for the hydration of acetylene. When H_2SO_4 is a catalyst, X is probably HSO_4^-.

$$\text{(12.21)}\quad HC\equiv CH + HgX^+ \rightleftharpoons \underset{\pi\text{ complex}}{HC\overset{+HgX}{\equiv}CH} \xrightarrow[\text{slow}]{H_2O} \underset{OH}{HC}=C\!\begin{array}{l}HgX\\H\end{array} + H^+$$

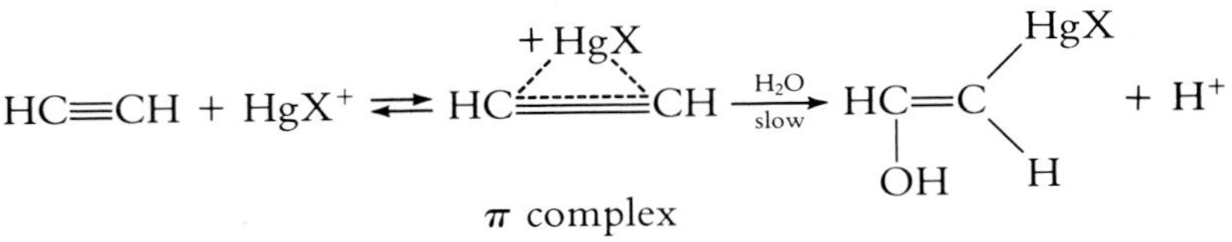

$$\text{(12.22)}\quad \underset{\underset{\large OH}{|}}{HC}=\overset{\overset{\large HgX}{|}}{C}H \xrightarrow{H^+} \underset{\text{Enol}}{\underset{\underset{\large OH}{|}}{HC}=CH_2} \rightleftharpoons H\overset{\overset{\large O}{\|}}{C}-CH_3$$

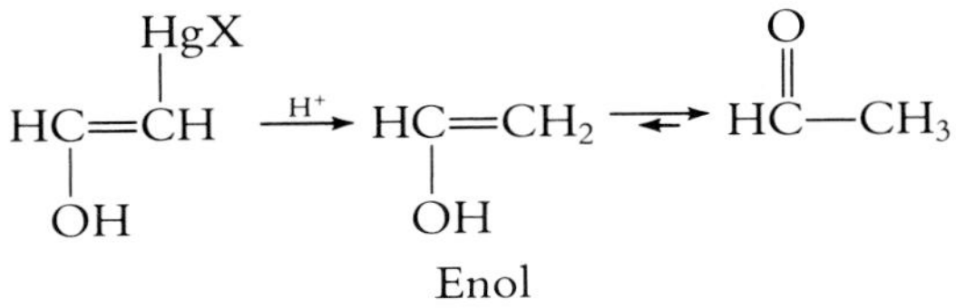

Because of their toxicity, mercury compounds must be carefully controlled. In the 1960s, a Japanese factory that manufactured acetaldehyde (eq. 12.20) discharged mercury compounds into the sea. This led to the deaths of over 50 people who ate fish contaminated with mercury.

Alkynes will add 1 mol of chlorine or bromine under ionic conditions, just as alkenes do. If the triple bond is connected only to alkyl groups or hydrogen, the addition is stereoselectively *anti* (eq. 12.23).

$$CH_3CH_2C{\equiv}CCH_2CH_2CH_3 \xrightarrow[\text{acetic acid}]{Br_2} (CH_3CH_2)(Br)C{=}C(Br)(CH_2CH_2CH_3) \quad (80\%) \tag{12.23}$$

The stereochemical results suggest that halogenation proceeds by way of a halonium ion (**3**), similar to the halonium ion that is an intermediate in the halogenation of alkenes (**4**) (sec. 10:4).

$$R{-}C{=}C{-}R \text{ bridged by } X^+ \quad \textbf{3}$$

$$H(R)C{-}C(H)R \text{ bridged by } X^+ \quad \textbf{4}$$

This mechanism explains the observation that *triple bonds are much less reactive toward chlorine and bromine than are double bonds,* as Equation 12.24 shows. According to this mechanism, halonium ion **3** suffers from much more angle strain than ion **4** and is therefore formed more slowly. Table 12–2 summa-

$$HC{\equiv}CCH_2CH{=}CH_2 \xrightarrow[-20^\circ,\ CCl_4]{Br_2} HC{\equiv}CCH_2CH(Br)CH_2Br \quad (90\%) \tag{12.24}$$

4

rizes the major electrophilic addition reactions of alkynes.

One property characteristic of alkynes, but not alkenes, is their ability to undergo *nucleophilic* additions with strong nucleophiles (eq. 12.25). These

$$CH_3C{\equiv}CH + EtOH \xrightarrow[150^\circ]{OH^-\ (cat)} \left[(CH_3)(EtO)C{=}\ddot{C}^-(H)\right] \xrightarrow{EtOH} CH_3C(OEt){=}CH_2 \tag{12.25}$$

A vinyl anion   Ethyl isopropenyl ether

reactions are made possible by the relative stability of the vinyl anion intermediate. Since nucleophilic addition results in only 1 mol of ROH being added, it is often preferable to the less easily controlled electrophilic addition.

Table 12–2. Major Electrophilic Additions to Alkynes

Reaction	Intermediate		Product
(1) $RC{\equiv}CR' \xrightarrow[(X = Cl, Br, I)]{HX, X^-}$	$(H)(R)C{=}C(R')(X)$ can be isolated	$\xrightarrow{HX}$	$RCH_2{-}CX_2R'$ can be isolated
(2) $RC{\equiv}CR' \xrightarrow[Hg^{2+}, H^+]{H_2O}$	$[RCH{=}C(R')(OH)]$ cannot be isolated	$\rightleftarrows$	$RCH_2C(=O)R'$ can be isolated
(3) $RC{\equiv}CR' \xrightarrow[(X = Cl, Br)]{X_2}$	$(R)(X)C{=}C(X)(R')$ can be isolated	$\xrightarrow{X_2}$	RCX_2CX_2R' can be isolated

Example

What are the products of the following reactions?

a. $CH_3C{\equiv}CH + HCl$

b. $CH_3C{\equiv}CH + H_2O + HgSO_4/H_2SO_4$

Answer

a. First, H^+ adds to the triple bond to give the more stable of two possible vinyl cations:

$$CH_3C{\equiv}CH + HCl \rightarrow CH_3\overset{+}{C}{=}CH_2 + Cl^-$$

Then Cl^- (the nucleophile) adds to the cation to give the product:

$$CH_3\overset{+}{C}{=}CH_2 + Cl^- \rightarrow CH_3C(Cl){=}CH_2$$

b. These are conditions for hydration of an alkyne. The Markovnikov addition of a molecule of water produces an enol intermediate.

$$CH_3C{\equiv}CH + H{-}OH \xrightarrow[H_2SO_4]{HgSO_4} CH_3C(OH){=}CH_2$$

The vinyl alcohol (or enol) is never seen because it rapidly isomerizes to the more stable ketone.

$$CH_3C(OH){=}CH_2 \rightarrow CH_3C(=O)CH_3$$

12–6 Give the products of the reaction of 2-butyne with the following reagents. Show the stereochemistry of the product if the reaction is stereoselective.
a. HCl ($Me_4N^+Cl^-$, acetic acid)
b. 2 mol of HBr (with radical inhibitors)
c. $HgSO_4$, H_2SO_4, H_2O
d. Br_2, acetic acid

12–7 a. Write mechanisms for both steps of Equation 12.25, including the role of the catalyst.
b. As Table 12–1 shows, a vinyl anion is the conjugate base of an alkene. Use the data of this table to explain why nucleophilic addition is much more rapid with alkynes than with alkenes.

12–8 Show *all* the possible products and intermediates expected when 1 mol of 2-pentyne is treated with 2 mol of HBr.

12:6 Selective Reduction of the Triple Bond

Chapter 10 showed us that alkenes are very useful because of the variety of reactions they undergo. For this reason much effort has gone into the synthesis of isomerically pure alkenes. The reduction of alkynes is a useful reaction because alkenes can be isolated before they are further reduced to alkanes. This means that alkynes are reduced to alkenes much faster than the alkenes are further reduced to alkanes. The rate difference can be explained by the fact that alkyne reduction (eq. 12.26) is more exothermic than alkene reduction (eq. 12.27).

$$CH_3(CH_2)_3C{\equiv}CH + H_2 \xrightarrow[\text{phase}]{\text{gas}} CH_3(CH_2)_3CH{=}CH_2 \qquad \Delta H = -39.5 \text{ kcal/mol} \tag{12.26}$$

$$CH_3(CH_2)_3CH{=}CH_2 + H_2 \xrightarrow[\text{phase}]{\text{gas}} CH_3(CH_2)_3CH_2CH_3 \qquad \Delta H = -30.0 \text{ kcal/mol} \tag{12.27}$$

Most reduction catalysts (sec. 7:3) are too active to discriminate between alkynes and alkenes. The problem has been solved by suspending a standard catalyst, palladium, on calcium carbonate and partially "poisoning" it with lead acetate [$Pb(OAc)_2$] and/or quinoline (an organic base). The poisoned catalyst (*Lindlar's catalyst*) is highly disciminatory and can reduce triple bonds without reducing double bonds (eq. 12.28). In addition, it is very stereoselective, producing almost pure *cis* alkenes from disubstituted alkynes (eq. 12.29).

$$HC{\equiv}C{-}\overset{\overset{\large CH_3}{|}}{C}{=}CHCH_2CH_2OH + H_2 \xrightarrow[\text{quinoline}]{Pd/CaCO_3} CH_2{=}CH{-}\overset{\overset{\large CH_3}{|}}{C}{=}CHCH_2CH_2OH \ (80\%) \tag{12.28}$$

(12.29) $CH_3(CH_2)_7C{\equiv}C(CH_2)_7COOH \xrightarrow[Pb(OAc)_2]{H_2,\ Pd/CaCO_3}$

$$\begin{array}{ccc} CH_3(CH_2)_7 & & (CH_2)_7COOH \\ & \diagdown \quad \diagup & \\ & C{=}C & \\ & \diagup \quad \diagdown & \\ H & & H \end{array}$$

(Z)-9-Octadecenoic acid
(*oleic acid,* 95%)

If *trans* alkenes are the desired product, we can react internal alkynes with sodium or lithium metal in liquid ammonia (eq. 12.30; compare with eq. 12.29). Carbon–carbon double bonds are not reduced by this reagent.

(12.30) $CH_3(CH_2)_7C{\equiv}C(CH_2)_7COOH \xrightarrow[NH_3]{Li} \xrightarrow{H^+}$

$$\begin{array}{ccc} CH_3(CH_2)_7 & & H \\ & \diagdown \quad \diagup & \\ & C{=}C & \\ & \diagup \quad \diagdown & \\ H & & (CH_2)_7COOH \end{array}$$

(E)-9-Octadecenoic acid
(*elaidic acid,* 97%)

Selective *cis* reduction can also be achieved by way of hydroboration (sec. 10:6). In order for this reaction to be useful, it has to be controlled so that the hydroboration reagent only attacks one of the two π bonds. This is done by using the modified borane $HBBr_2$ (dibromoborane), which only adds B—H *syn* across one π bond of a typical alkyne, such a 4-octyne (step 1, eq. 12.31). The addition product is an alkenylborane, which is usually not

(12.31) $CH_3CH_2CH_2C{\equiv}CCH_2CH_2CH_3 \xrightarrow[C_2H_4Cl_2\ (solvent)]{①\ HBBr_2}$

4-Octyne

$$\begin{array}{ccc} CH_3CH_2CH_2 & & CH_2CH_2CH_3 \\ & \diagdown \quad \diagup & \\ & C{=}C & \\ & \diagup \quad \diagdown & \\ H & & BBr_2 \end{array}$$

$$\Big\downarrow ②\ CH_3COOH,\ heat$$

$$\begin{array}{ccc} CH_3CH_2CH_2 & & CH_2CH_2CH_3 \\ & \diagdown \quad \diagup & \\ & C{=}C & \\ & \diagup \quad \diagdown & \\ H & & H \end{array}$$

(Z)-4-Octene
(88%)

isolated but is treated with a carboxylic acid to give the *cis* alkene (step 2, eq. 12.31). Notice that this process gives the same result as does hydrogenation with Lindlar's catalyst (eq. 12.29). Which method is chosen depends upon what other functional groups are present.

12:7 General Principles of Synthesis

In the introduction to this chapter we said that rote memorization of many single reactions is not sufficient for proper application of these reactions to a complex synthesis. The field of synthesis is one of the most challenging and intellectually stimulating areas of organic chemistry. Synthesis might involve the construction of previously unknown molecules or the discovery of more efficient ways of making known compounds. Today we have so many organic reactions available to us that a variety of pathways can usually lead to the desired compound. We think of organic synthesis as an art that consists of choosing those reactions that will be regio- and stereospecific, will give a high yield, and will take the fewest number of steps. This involves imagination and experience on the part of the chemist. Devising the best synthetic schemes is usually difficult for the beginning student, but there are some helpful general principles that apply to the synthesis of virtually all molecules, simple or complex.

To begin with, chemical reactions that have been discussed so far will be arbitrarily divided into two groups:

1. reactions that interchange functional groups; and
2. reactions that alter the molecular framework, or carbon skeleton, particularly carbon–carbon bond-forming reactions.

There are many reactions that do not fit neatly into either of these categories, but the classification will be useful nonetheless because it focuses attention on the two major tasks of the synthetic chemist.

Examples of functional group interchanges are the reduction of carbonyl groups to hydroxyls, or the replacement of hydroxyls by halides. Reactions that alter the molecular skeleton almost always involve the formation of one or more carbon–carbon bonds. The reaction of sodium acetylide with 1-bromobutane (eq. 12.4) is a characteristic example: a compound with a two-carbon skeleton and one with a four-carbon skeleton have been combined to yield a six-carbon product. The Grignard reaction (such as the one shown in eq. 12.9) is an example of a reaction in which both carbon–carbon bond formation and functional group transformation take place in one step.

12:8 Possible Syntheses of 3- and 4-Octanols

When dealing with practical syntheses (as opposed to textbook problems), you must be concerned with the economics of your synthesis. Are the starting materials too expensive? Are the yields sufficiently high? Are toxic, flammable, or explosive chemicals involved? These are questions that must be answered before choosing a specific synthetic pathway for commerical production.

Suppose we want 100 g each of racemic 3-octanol (**5**) and 4-octanol (**6**). (Line-angle structures will be used in order to focus on specific transformations, and so that you can practice using this common shorthand method of depicting structures.)

$$\begin{array}{c} OH \\ | \\ CH_3CH_2CHCH_2CH_2CH_2CH_2CH_3 \end{array} \qquad \begin{array}{c} OH \\ | \\ CH_3CH_2CH_2CHCH_2CH_2CH_2CH_3 \end{array}$$

OH OH

5 **6**

It is almost axiomatic in organic chemistry laboratories that it is cheaper (in terms of labor, material, and other costs) to purchase compounds rather than synthesize them if relatively small amounts are needed. Thus, we first consult chemical supply catalogs for 3- and 4-octanol. If neither compound is available, then syntheses are required. The first step is to examine the structure and see what simple functional group transformations might lead to the desired product, without altering the carbon skeleton. The reactions in Equations 12.32 to 12.34 are three possible choices for the synthesis of **6.**

(12.32) Br **7** $\xrightarrow{OH^-}$ **6**

(12.33) O **8** $\xrightarrow{LiAlH_4}$ **6**

(12.34) **9** $\xrightarrow[H_2O]{H_2SO_4}$ **6**

In order to choose among these three pathways, we consider the commercial availability of starting materials **7, 8,** and **9.** The 4-bromooctane (**7**) is not listed in chemical catalogs, while 4-octanone (**8**) can be purchased for about \$300 per mole, and *trans*-4-octene (**9**) for \$100 per mole. If the yields of the reactions in Equations 12.33 and 12.34 are similar, the cheaper alkene, **9,** would be chosen as the desired starting material. Thus, alkene hydration (eq. 12.34) appears to be the best route to **6.**

Now, if we want to prepare **5,** the same basic functional group transformations can be considered (eqs. 12.35 to 12.38). Again, we find that the bromide **10** (or the corresponding chloride) is not commercially available, while 3-octanone (**11**) can be purchased for about \$30 per mole. In the case of the alkene hydration reactions (eqs. 12.37 and 12.38), compound **5** would be available from either **12** or **13.**

(12.35) Br **10** $\xrightarrow{OH^-}$ **5**

(12.36) O **11** $\xrightarrow{LiAlH_4}$ **5**

(12.37) 12 $\xrightarrow[H_2O]{H_2SO_4}$ **5** + **6**

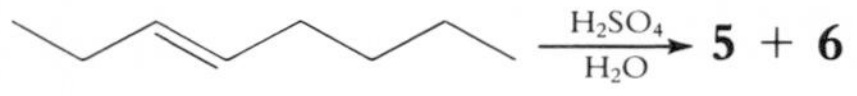

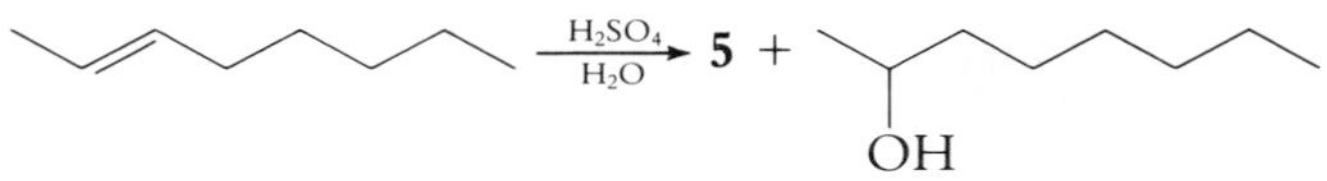

(12.38) 13 $\xrightarrow[H_2O]{H_2SO_4}$ **5** + OH

However, you may recall from the mechanism of hydration that two alcohols will result from hydration of either **12** or **13,** and these isomeric alcohols are very difficult to separate. Thus, for the preparation of **5,** ketone reduction (eq. 12.36) is probably the best reaction. In an industrial, rather than a laboratory setting, the reduction would be done with H_2 and a metal catalyst. This is cheaper than reduction with $LiAlH_4$.

Problem

12–9 a. Explain why the reaction of Equation 12.34 does not give two different alcohols.

b. Octane is an extremely cheap chemical ($10 per mole). Could it be used as a starting material in the preparation of **7, 8,** or **9**?

Until now we have considered only functional group transformations as synthetic possibilities. The octanols **5** and **6** could also be prepared by joining smaller molecules through carbon–carbon bond-forming reactions. In particular, the Grignard reaction is useful for the synthesis of secondary alcohols. Equations 12.39 and 12.40 give two Grignard methods for the preparation of **5,** either of which is a practical method for synthesizing the desired 3-

(12.39) Br ($8 per mole) $\xrightarrow[Et_2O]{Mg}$ MgBr $\xrightarrow[(\$1\ per\ mole)]{H,\ O}$ OH **5**

(12.40) Br ($2 per mole) $\xrightarrow[Et_2O]{Mg}$ MgBr $\xrightarrow[(\$10\ per\ mole)]{O,\ H}$ OH **5**

octanol. Naturally, the yield of product is also an important factor. The benefit of inexpensive starting materials is lost if reaction yields are poor.

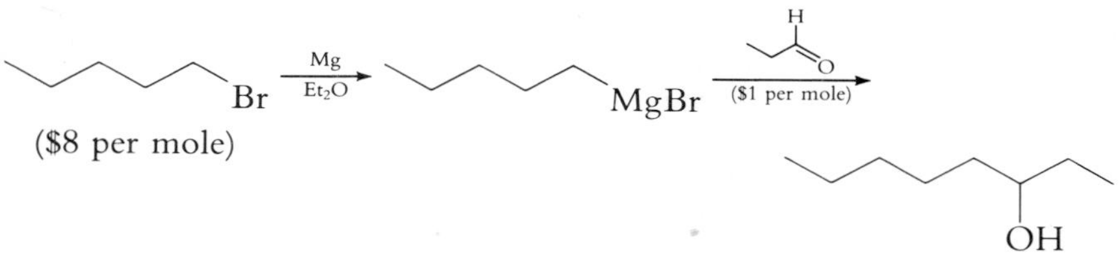

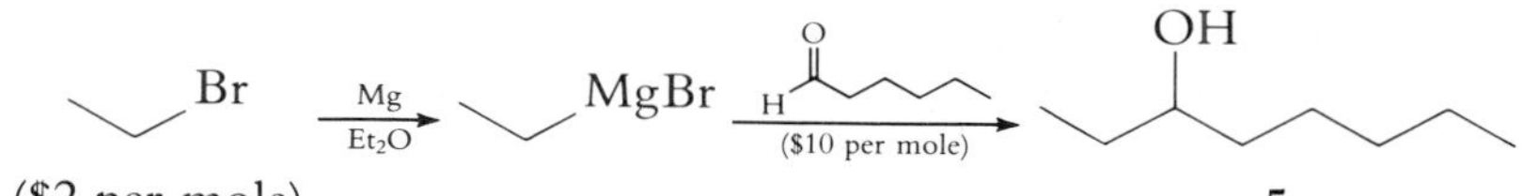

Problem

12–10 The reactions of acetylides with alkyl halides (eqs. 12.4 and 12.5) are useful for building carbon chains. Devise syntheses of **8, 9,** and **11** (if possible) using these reactions. If chemical catalogs are available, try to estimate if your products will be economically competitive with those available commercially. Assume 80% yields for each step.

Box 12–1. R. B. Woodward and the Art of Organic Synthesis

Professor Robert Burns Woodward (1917–1979) of Harvard University may well have been the most eminent organic chemist of the modern era. He was a child prodigy who conceived of the total synthesis of quinine at the age of 14, entered the Massachusetts Institute of Technology at the age of 16, and left with a Ph.D. at the age of 20. He went on to receive the Nobel Prize in Chemistry in 1965 for developments in "the art of organic synthesis."

Woodward was a hard taskmaster who would not allow for failure in his work. At one particularly discouraging point in his eventually successful total synthesis of strychnine, Woodward was quoted as saying, "If we can't make strychnine, we'll take strychnine!"

"R.B." (as he was known to some) was also a renowned lecturer whose eloquence and artistry at the blackboard were legendary. His students often felt dismayed at the end of a lecture because his multicolored sketches of reactions and mechanisms were works of art which should have been preserved, not erased. Though Woodward usually wore a dark blue suit, he was able to lecture and sketch vigorously without ever sullying his impeccable outfit with so much as a speck of chalk. In fact, it was this dark blue suit which probably brought an end to Woodward's own work at the laboratory bench. Woodward was working with his students in the laboratory at 3 A.M. when a separatory funnel he was shaking broke and spilled its contents on his pants. Thereafter, he was rarely seen performing his own experiments.

The elegance of the Woodward lecturing style, as well as his love for chemistry, shines through his final words when describing the total synthesis of the complex alkaloid, cholchicine:

"One aspect of work of this kind hardly shows through a dry account of the sort I have presented here and deserves explicit mention. Each of the intermediates along our progression to the colchicine molecule is a beautifully crystalline substance, an entirely new form of matter, persuaded into being in response to the challenge of an often remote objective. It is delightful to work with such things, and the delight which the experimenter experiences in his manipulation contributes in no small measure to the skill required to create them."

12:9 Synthesis of Disparlure

To illustrate how more complex syntheses are accomplished and to emphasize the concepts we have covered so far, let's consider a practical problem. An insect known as the gypsy moth was introduced into Massachusetts from Europe in 1869 to start a silk industry, but the insect soon escaped control. Since no natural predators were present in the new environment, an intense population growth occurred, and many orchards became completely defoliated. Partial control was not achieved until pesticides, such as DDT, were introduced in the late 1940s. Because of its slow rate of degradation and toxicity to birds, DDT use was discontinued in 1961 and the gypsy moth population began to increase again. By 1970 as many as 800,000 acres had been defoliated, and a search began for methods of biological control. One idea was to lure male moths into a trap by using a natural sex attractant, called a **pheromone,** obtained from female moths.* Consequently, the female sex attractant, named *disparlure,* was isolated and its structure determined by spectroscopic methods (Fig. 12–6).

Field tests showed that disparlure was highly active in attracting male moths to traps and might provide a better control method than random insecticide spraying. However, since only a few micrograms of pure

* A pheromone (from the Greek *pherein,* meaning "to carry," and *hormaein,* meaning "to excite") is a chemical secreted by an organism of one species to evoke a response in another member of the same species.

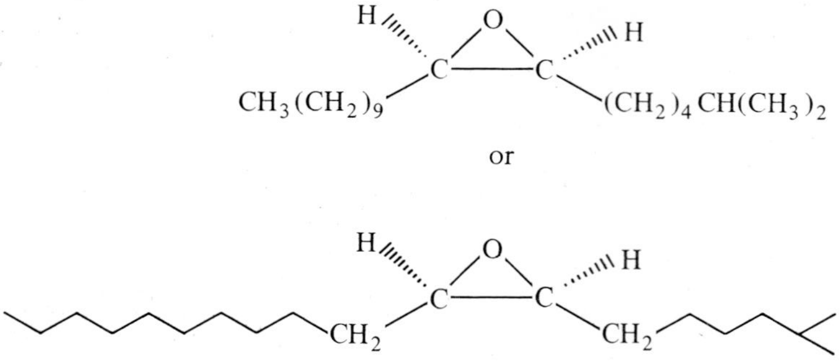

$$\text{CH}_3(\text{CH}_2)_9\text{—C(H)—O—C(H)—(CH}_2)_4\text{CH(CH}_3)_2$$

or

Figure 12–6. *Disparlure:* cis-*7,8-epoxy-2-methyloctadecane.*

disparlure could be obtained from many thousands of female moths, it was necessary to synthesize it.

As you saw in the last section, the best approach in devising a total synthesis is to work *backward* from the target molecule. This method is called **retrosynthesis.** The first step is to analyze the target molecule with respect to functional groups, stereochemistry, and the presence of *strategic bonds.* Strategic bonds are carbon–carbon bonds that, upon formation, allow the construction of a large molecule from smaller portions or substructures. They are usually found adjacent to a functional group or to one or two bonds removed from that group. In disparlure, the only functional group is the epoxide, and the two alkyl chains on the epoxide ring are *cis* to one another. The strategic bonds might be those connecting the alkyl chains to the epoxide ring.

Using the shorthand drawing in Figure 12–6 as a guide, we then begin to work backward from the target molecule (the epoxide) until we find an intermediate that can be constructed from commercially available starting materials. It is known that epoxidation of *cis* alkenes with peroxyacids will stereospecifically produce an epoxide with substituents *cis* to each other (sec. 10:8). Hence, the *last* step of the synthesis could be epoxidation (eq. 12.41).

(12.41) **14** (cis alkene, CH_2—CH=CH—CH_2) $\xrightarrow{\text{peroxybenzoic acid}}$ $CH_3(CH_2)_9$—C(H)—O—C(H)—$(CH_2)_4CH(CH_3)_2$

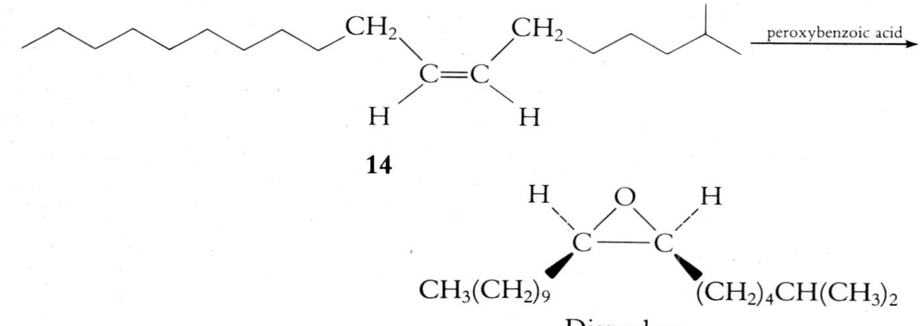

Disparlure

The total synthesis is now assured if the intermediate (**14**) can be prepared. Intermediate **14** can be analyzed in the same way as the original target molecule was. When alkynes are hydrogenated in the presence of Lindlar's catalyst, they can be converted to *cis* alkenes. A useful synthesis of **14** would therefore be from the alkyne **15** (eq. 12.42).

(12.42) $CH_2—C{\equiv}C—CH_2$ $\xrightarrow[Pd/CaCO_3]{H_2}$ **14**

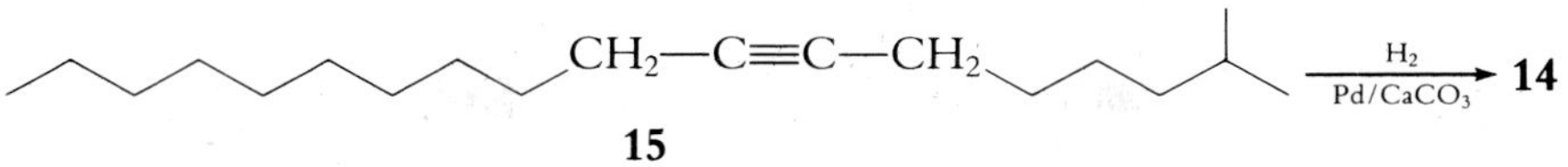

15

Now let's examine the intermediate **15** in terms of strategic bonds (this must be done at every stage of the synthesis).

The alkyne **15** has two strategic bonds on either side of the triple bond,

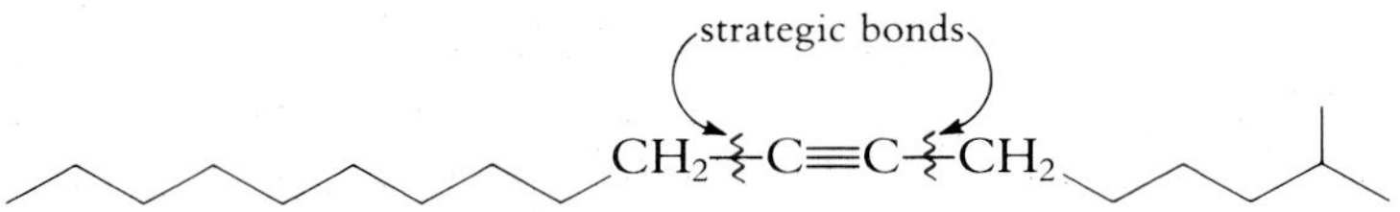

strategic bonds

$CH_2 \wr C{\equiv}C \wr CH_2$

because a standard method of making internal alkynes is by alkylating the anions of terminal alkynes. We have two possibilities for the synthesis of **15**, depending upon which side of the alkyne is alkylated last (eqs. 12.43 and 12.44).

(12.43) $CH_2—C{\equiv}C^-Na^+ + BrCH_2$

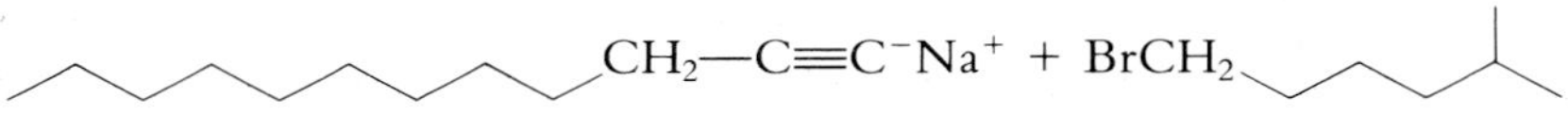

16 **17**

or $CH_3(CH_2)_9C{\equiv}C(CH_2)_4CH(CH_3)_2$

15

(12.44) $CH_2Br + Na^{+-}C{\equiv}C—CH_2$

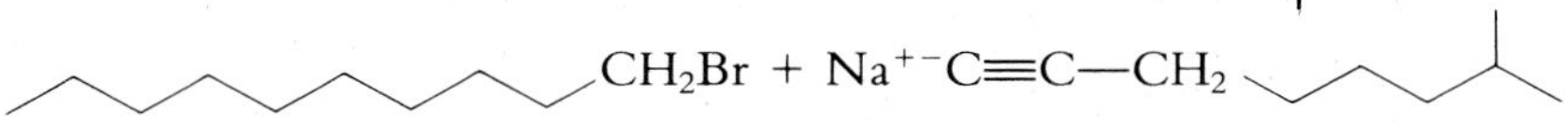

18 **19**

We have now (conceptually) broken the complex target molecule into smaller substructures whose availability from chemical suppliers can be checked. 1-Dodecyne (the precursor of **16**) is available at $400 per mole, as is 1-bromodecane (**18**) at $5 per mole. However, neither **17** nor the alkyne precursor of **19** are commercially available, so they would have to be prepared. The retrosynthetic analysis must be continued for synthesis of **17** if it is to be combined with 1-dodecyne (**16**). Compound **17** could be easily prepared from 5-methyl-1-hexene (**20**) by radical addition of HBr (sec. 10:5).

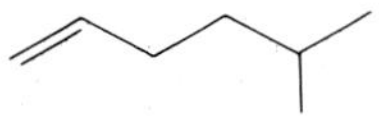

20

We find that **20** is commercially available at $450 per mole. A possible synthetic scheme for disparlure would then be: **20** → **17**; **17** + **16** → **15** → **14** → disparlure. In 1972 a disparlure synthesis very much like this one was reported; however, the route to compound **17** was different. Another approach to disparlure involving reactions we have not yet studied is given in Section 21:14.*

Note that disparlure is a chiral compound, so the outlined synthesis would yield a mixture of enantiomers. Natural products usually exist as one or the other pure enantiomer and usually only one enantiomer will be biologically active (Chap. 4). If we could prepare and use a pure enantiomer, its relative activity per gram would be twice that of the racemic mixture. In order to do this, a chiral reagent would have to be introduced at the point where enantiomers are generated. In the case of disparlure this would be the last step—epoxidation. For example, we might try to devise a chiral epoxidizing agent such as compound **21** which we would expect to produce an excess of *one* enantiomer of disparlure, thus achieving an **asymmetric synthesis.**

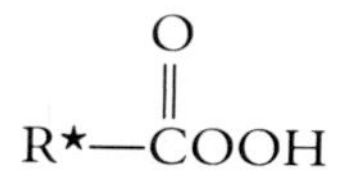

$$R^{\star}\text{—}\overset{\overset{\displaystyle O}{\|}}{C}OOH$$

(peroxy acid containing a chiral R group)
21

Example

Prepare the following compound from simple starting materials:

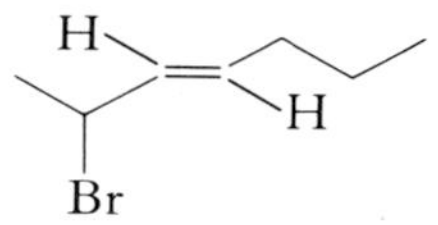

H, H, Br

Answer

A synthesis such as this can only be worked by going backward until you reach a molecule that can be easily prepared from smaller fragments. Sometimes false retrosynthetic analyses are made and one must simply start again.

We know that bromides can be prepared by reaction of PBr_3 with alcohols, so the *last* step of the synthesis might be

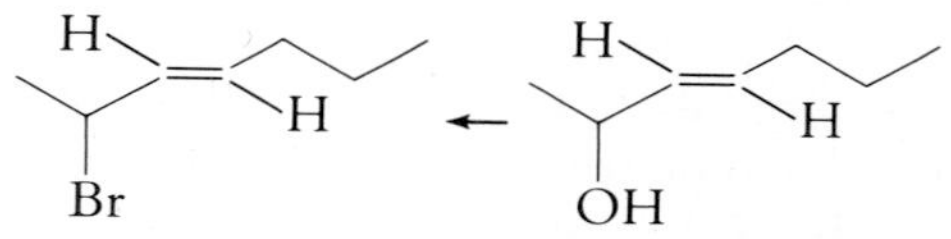

H, H, Br ← H, H, OH

Alcohols can be obtained from reduction of ketones, so the next to last step could be

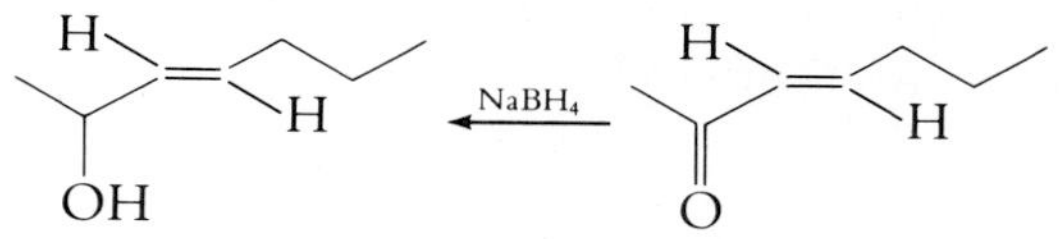

H, H, OH $\xleftarrow{NaBH_4}$ H, H, O

If you can't think of a way to prepare this new ketone target, try a different next to last step:

---

* Disparlure did not cure the gypsy moth problem. Control methods of this type were only partially successful, and the moth has spread to new areas.

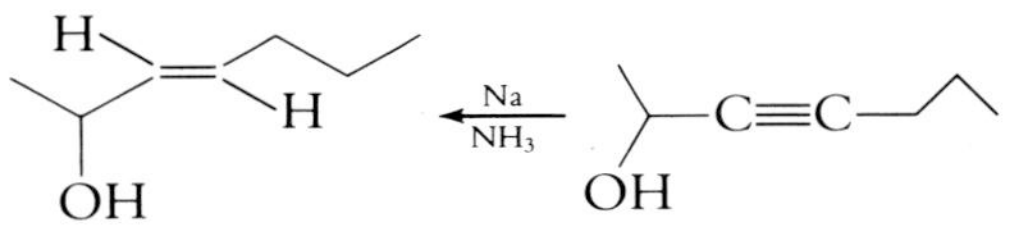

$$\text{(H, OH-substituted alkenol)} \xleftarrow[\text{NH}_3]{\text{Na}} \text{(alkynol)}-\text{C}\equiv\text{C}-$$

H, H, OH   OH

You may remember from Equations 12.8 and 12.9 that alkynols can be made by reacting a lithium alkyne (or Grignard reagent) with an aldehyde:

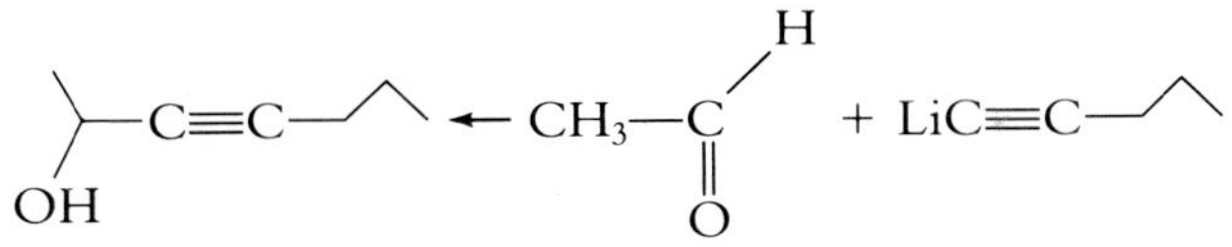

$$-\text{C}\equiv\text{C}- \leftarrow \text{CH}_3-\overset{\text{H}}{\text{C}}{=}\text{O} + \text{LiC}\equiv\text{C}-$$

OH

The needed 1-pentyne could be easily prepared from a reaction like that in Equation 12.4:

$$\text{HC}\equiv\text{C}- \leftarrow \text{HC}\equiv\text{C}^-\text{Na}^+ + \text{BrCH}_2\text{CH}_2\text{CH}_3$$

Problem

12–11 In the 1972 disparlure synthesis, compound **17** (1-bromo-5-methylhexane) was synthesized from propargyl alcohol, $HC{\equiv}CCH_2OH$. Suggest a reaction sequence for this synthesis. (Make use of the acidic alkynyl proton while remembering that the OH proton is *also* acidic.)

In a complex synthesis involving many steps, one must often begin with large amounts of starting materials and finish with milligrams or less of product after months of work. This was a common occurrence in the 1940s and 50s when many drug companies were attempting to prepare physiologically-active adrenal hormones. Some of the strain of this research is evident in the following quotation from the American chemist E. C. Kendall:

> "Only those who have worked in a chemical laboratory similar to that of Merck and Company can appreciate the contribution made by van de Kamp and Miller. For 11-dehydrocorticosterone, the starting material, deoxycholic acid, weighed 65 pounds. There was not much change in weight in the first few steps, but after many months of work the entire product could be contained in one small flask. If the flask were broken at that stage and the material lost, the project would have to be started over from the beginning. To continue under this psychic strain for more than a year requires more than technical ability and more than unusual patience and persistence; it requires strength of character. Had these two men failed in their assignment it is questionable whether anyone else would have ventured to make the attempt."

Summary

1 The triple bond in alkynes (acetylenes) consists of one *sp-sp* σ bond, and two *p-p* π bonds at 90° to one another. The π electron distribution is cylindrically

symmetrical about the carbon–carbon bond. Terminal alkynes can be distinguished from internal alkynes by their ir spectra and characteristic nmr peaks, due to the position of the C≡C—H proton signal (sec. 12:1).

2 Terminal alkynes contain a comparatively acidic proton, since their conjugate bases have the negative charge localized in an *sp* orbital (sec. 12:2). This makes the anion more stable than if the negative charge were localized in sp^2 or sp^3 orbitals (Table 12–1). Terminal alkynes can be converted to metallic derivatives (acetylides) by bases. The most important of these are the sodium, lithium, and halomagnesium derivatives. The sodium and lithium acetylides enter into S_N2 reactions with primary alkyl halides to yield alkynes of increased chain length. These salts also add to carbonyl groups, as do alkynyl Grignard reagents (sec. 12:3).

3 Alkynes may be synthesized by double elimination from 1,2-dihalides. Elimination of the first mole of hydrogen halide requires relatively mild conditions, while more stringent conditions are necessary to bring about loss of the second mole (sec. 12:4). The second elimination requires a strong base such as $NaNH_2$, or an ordinary base (NaOH) at high temperatures.

4 The triple bond undergoes addition with common electrophiles such as HX, X_2, and ROH with H^+. Electrophilic attack at triple bonds is usually slower than it is at double bonds, and Markovnikov addition prevails. It is often possible to stop the reaction after only 1 mol of the electrophile has been added (Table 12–2). The enol that is the initial product of addition of water to alkynes isomerizes completely to the isomeric ketone or aldehyde. Mercuric ions catalyze this addition (sec. 12:5).

Reduction of the triple bond may be halted after the alkene stage is reached. The use of a poisoned palladium catalyst (Lindlar's catalyst) and hydrogen results in the formation of *cis* alkenes almost exclusively. Alkynes react with $HBBr_2$ at only one π bond to give alkenylboranes, which then react with carboxylic acids to provide another route to *cis* alkenes. Reduction of alkynes with sodium in liquid ammonia gives mainly the *trans* alkenes (sec. 12:6).

5 A synthetic scheme for the preparation of a complex molecule is best devised by working backward, step-by-step, from the target molecule until simple, commercially available starting materials are reached. Most syntheses involve both functional group changes and new carbon–carbon bond formation. Grignard reactions and acetylide S_N2 displacements on halides are both important methods for constructing new carbon–carbon bonds (secs. 12:7 to 12:9). A practical, rather than textbook, synthesis will take into account prices of starting materials, yields in each step, safety considerations, and many other factors.

Problems

12–12 Give an IUPAC name for the following compounds:

a. $CH_3C{\equiv}CH$

b. $HOCH_2C{\equiv}CCH_2OH$

c. $HC{\equiv}CC{\equiv}CH$

d. $HC{\equiv}CCHCH_3$ (with OH on the CH carbon)

12–13 Write structures for the following compounds:

a. 2-butyne
b. dimethyl acetylene
c. vinyl acetylene
d. 4-bromo-1-pentyne
e. (*E*)-2-bromopent-2-en-4-yne

12–14 Name the following compounds by the IUPAC method:

a. $CH_3CH(Cl)CH_2C{\equiv}CH$
b. $CH_3CH_2C{\equiv}CCH(CH_2CH_3)CH_3$
c. $HC{\equiv}CCH_2CH_2C{\equiv}CH$
d. $HOCH_2CH_2CH_2C{\equiv}CH$
e. 1-ethynylcyclopentanol structure: cyclopentane ring bearing OH and $C{\equiv}CH$ on the same carbon

12–15 Show how 1-butyne would react with each of the following. Where the reaction involves an intermediate or intermediates, give their structures as well.

a. $LiNH_2$ in NH_3, $-33°$
b. CH_3MgBr
c. NaOEt in EtOH, heat
d. HBr
e. $HgSO_4$, H_2SO_4, H_2O
f. Cl_2, acetic acid
g. H_2, $Pd/CaCO_3$, quinoline
h. Na metal in NH_3, $-33°$

12–16 Show how acetylene might be converted to each of the following compounds. Indicate all reagents and conditions. Simple organic compounds may be used as reagents.

a. 1-hexyne
b. 3-decyne
c. (*E*)-3-hexene
d. (*Z*)-3-hexene
e. (*Z*)-1,4-octadiene
f. (*Z*)-4-bromo-4-octene
g. 3-butyn-2-ol
h. 2-butanone
i. 2-methyl-3-butyn-2-ol

12–17 Propose structures for compounds **A** to **D**, and **E** to **H**.

a. $HC{\equiv}CH \xrightarrow{EtMgBr} \mathbf{A} \xrightarrow{acetone} \xrightarrow{H_3O^+} \mathbf{B} \xrightarrow{PBr_3} \mathbf{C}\ (C_5H_7Br) + \mathbf{D}\ (C_5H_6)$

b. $F(CH_2)_8Br + HC{\equiv}CNa\ (1\ eq) \rightarrow \mathbf{E} \xrightarrow{LiNH_2} \mathbf{F} \xrightarrow{I(CH_2)_7Cl} \mathbf{G} \xrightarrow[DMSO]{NaCN} \mathbf{H}\ (C_{18}H_{30}FN)$

12–18 Show which carbonyl compound(s) will be formed when each acetylene is hydrated in the presence of Hg^{2+}.

a. 1-butyne
b. 1,4-pentadiyne
c. 2-methyl-3-hexyne
d. acetylene-d_2

12–19 Write out the reactions in each of the following cases, and indicate how many organic products would be formed.

a. 2-Methylcyclopentanone is treated with lithium acetylide and the product is hydrolyzed with aqueous acid.
b. The product from (a) is treated with $HgSO_4$ in aqueous H_2SO_4.
c. The product from (b) is treated with excess ethylmagnesium bromide, followed by cold dilute HCl.

12–20 The HO^--catalyzed addition of EtOH to propyne gives only the Markovnikov addition product (eq. 12.25). This result indicates that one of the two possible vinyl anion intermediates is more stable than the other. Draw the structures of these two intermediates and indicate which is more stable and why.

12–21 Show the products when alkynes (a) through (d) react with $HBBr_2$ followed by CH_3COOH.

a. acetylene
b. 4-methyl-1-pentyne
c. 2-heptyne
d. cyclooctyne

12–22 3-Hexyne is treated with hydrogen and poisoned $Pd/CaCO_3$ catalyst. The product of the reaction is brominated ($Br_2/CHCl_3$) and then reacted with potassium *t*-butoxide in *t*-butyl alcohol at room temperature. The final product from this series of reactions has the molecular formula $C_6H_{11}Br$. Write the structures of the product and intermediates, taking care to indicate stereochemistry without ambiguity.

12–23 Write the structure and IUPAC name of the alkyne you obtain by adding bromine to the following alkenes and then treating the product of the addition with excess $NaNH_2$.

a. 1-butene
b. (*Z*)-2,2,5,5-tetramethyl-3-hexene
c. cyclodecene

12–24 1,2-Dibromodecane is treated with sodium hydroxide in boiling ethanol (eq. 12.10). It yields three isomeric products: **A, B,** and **C** ($C_{10}H_{19}Br$). Their ir spectra are markedly different below 1000 cm^{-1} in the vinyl hydrogen out-of-plane bending region. Compound **A** gives rise to a band at 940 cm^{-1}, compound **B** at 890 cm^{-1}, and compound **C** at 690 cm^{-1}. Assign structures to each of the three products based on the spectral data. (A review of Section 8:9 may be helpful.)

12–25 Show all the steps involved in the following conversions:

a. 1-bromo-1-pentene to 1-pentyne
b. 2,3-dibromobutane to 2-butyne
c. ethane to acetylene

12–26 Provide efficient syntheses from available starting materials for the following compounds:

a. (*E*)-11-tetradecenal: $CH_3CH_2CH{=}CH(CH_2)_9CHO$ (the sex attractant of the spruce bud worm)
b. 1,2,9,10-decanetetrol: $HOCH_2\underset{\displaystyle OH}{\underset{|}{C}}H(CH_2)_6\underset{\displaystyle OH}{\underset{|}{C}}HCH_2OH$
c. $CH_3(CH_2)_7C{\equiv}C\underset{\displaystyle OH}{\underset{|}{C}}HCH{=}CH_2$

(an intermediate in the synthesis of megatomic acid, the sex attractant of the black carpet beetle)

12–27 An unknown compound (C_3H_4O) gives the ^1H nmr spectrum shown here. When a drop of D_2O is placed in the nmr tube, the mixture shaken, and the nmr spectrum rerun, the peak at 3.07 ppm disappears. Give a structure for the unknown and account for the behavior of the 3.07 ppm peak.

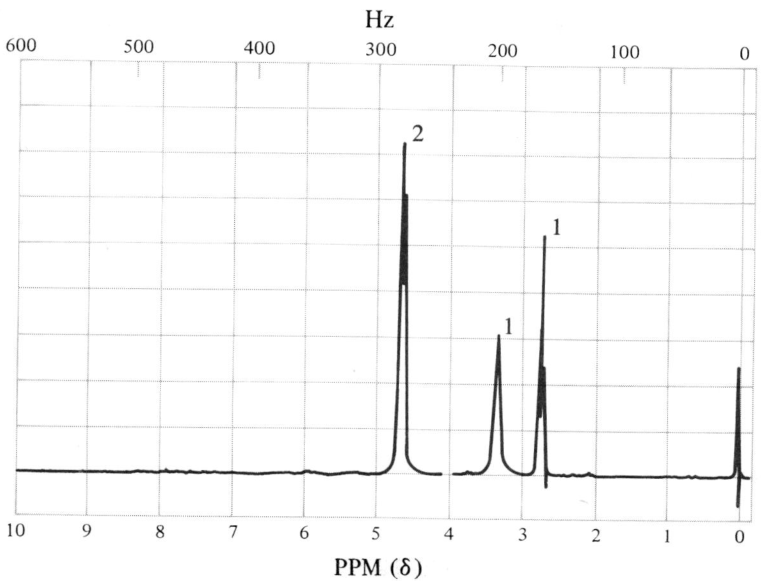

12–28 Which one of the compounds of Problem 12–17a would have the ^{13}C nmr spectrum given here? Assign the peaks.

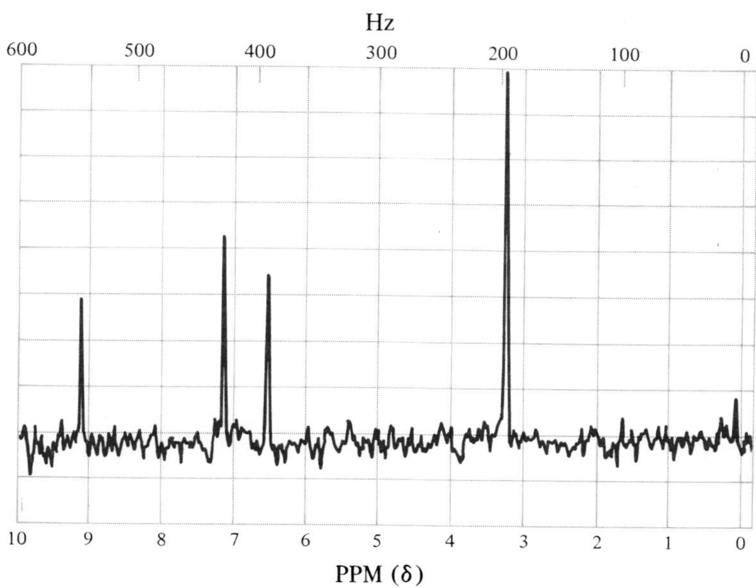

12–29 Write out the chemistry of each sequence, going forward from the starting material.

a. Propane is monobrominated under light catalysis, the product is converted to a Grignard reagent and is then allowed to react with 3-methylcyclopentanone. The reaction mixture is hydrolyzed to give a mixture of two compounds.

b. *cis*-2-Ethyl-3-butyloxirane is formed from an alkene by the action of peroxybenzoic acid; the alkene was made by hydrogenation (using Lindlar's catalyst) of an alkyne; the alkyne was made by two successive alkylations of acetylene, using $NaNH_2$.

c. 1-Pentanol is made from a Grignard reagent and formaldehyde, after which it is treated with hot aqueous HBr to give a product that with hot alcoholic KOH gives an alkene. Bromine is added to the alkene, and the product of this step is treated with excess $NaNH_2$ to give an alkyne.

12–30 a. If an 80% yield is assumed for each reaction in Section 12:9, what would be the cost of enough starting materials (compounds **20** and **16**) to prepare 30 g of disparlure? (Disparlure is so active in attracting male gypsy moths that 30 g has been suggested to be sufficient to bait 3,000,000 traps each year for 1000 years.)

b. The starting material 5-methyl-1-hexene (**20**) for the synthesis of disparlure is relatively expensive ($46 for 10 g). Suggest other methods for preparing compound **17** that would avoid this expense. (You will need a chemical catalog to compare costs.)

c. 1-Bromo-5-methylhexane (**17**) can be converted to the expensive compound **19.** Is this route economically competitive with the direct combination of **16** and **17** to yield **15**?

Supplementary Reading

T. F. Rutledge, *Acetylenic Compounds,* Reinhold, New York, N.Y., 1968.

T. F. Rutledge, *Acetylenes and Allenes,* Reinhold, New York, N.Y., 1969.

H. W. Orf, Computer-assisted instruction in organic synthesis, *J. Chem. Ed.* **52,** 464 (1975).

S. G. Smith and R. Chabay, Computer games in chemistry, *J. Chem. Ed.* **54,** 688 (1977). This article describes the use of the computer program PLATO in organic synthesis.

M. Jacobson, *Insect Sex Pheromones,* Academic Press, New York, N.Y., 1972. Describes the history and chemistry of disparlure and many other sex attractants.

B. A. Leonhardt and M. Beroza, *Insect Pheromone Technology: Chemistry and Applications,* American Chemical Society, Washington, D.C., 1982.

S. Warren, *Organic Synthesis: The Disconnection Approach,* John Wiley and Sons, Inc., Somerset, N.J., 1982.

Electron Delocalization in Aliphatic Molecules 13

Thus far, we have depicted molecules as being held together by electron pairs that are localized between pairs of nuclei. This localized bond model is successful in predicting molecular behavior or structure in a great many cases. There are, however, particular molecules and reactive intermediates whose properties cannot be explained with localized bond theory. You may already know from general chemistry that the properties of sulfur dioxide (SO_2) and nitrate anion (NO_3^-) cannot be accounted for solely on the basis of a single Lewis structure. We need to make use of multiple Lewis structures to account for the chemistry of these species. Similar cases arise in organic chemistry, where one structure alone does not provide an adequate basis for understanding the behavior of a molecule or ion. In this chapter we take up electron delocalization in aliphatic molecules. Chapter 14 extends the discussion to aromatic compounds, such as benzene.

13:1 Molecular Orbital and Resonance Descriptions of Electron Delocalization

In the discussion of covalent bond formation (sec. 2:3) we indicated that bonds are formed because bonding electrons, instead of being confined to atomic orbitals, could now move within the greater volume of molecular orbitals. That is, the electrons are more *delocalized* (spread out in space) in a molecular orbital than in an atomic orbital, and are therefore more stable. The additional stability arises because the electrons interact with an increasing number of positively charged nuclei as they spread out. If this explanation is accurate, then delocalization of electrons should always make for greater stability, and molecules and ions containing highly delocalized molecular orbitals should be unusually stable. Experimental evidence supports this hypothesis. As an illustration, let's reconsider an example from Chapter 1.

In Section 1:7 we pointed out that carboxylic acids (R—COOH) dissociate into ions in water (eq. 13.1).

$$CH_3{-}C(=O){-}\ddot{O}{-}H + H_2O \underset{}{\overset{K_a}{\rightleftharpoons}} CH_3{-}C(=O){-}\ddot{O}:^- + H_3O^+ \qquad (13.1)$$

Acetic acid (a carboxylic acid)

Acetate anion (a carboxylate anion)

$K_a = 1.76 \times 10^{-5}$
$pK_a = 4.75$

Carboxylic acids are similar to alcohols (R—OH) since both contain hydroxyl groups, yet alcohols are much weaker acids in water than carboxylic acids (eq. 13.2). This comparison leads to the conclusion that carboxylate

$$(13.2)\quad CH_3CH_2\ddot{\underset{\cdot\cdot}{O}}H + H_2O \overset{K_a}{\rightleftarrows} CH_3CH_2\ddot{\underset{\cdot\cdot}{O}}{:}^- + H_3O^+ \qquad K_a = 1 \times 10^{-16}$$

$$pK_a = 16$$

anions, R—COO$^-$, are more stable than alkoxide anions, R—O$^-$. (The relationship between acidity and conjugate base stability is explained in Section 12:2.) We now need to understand the source of the unusual stability of the carboxylate ion.

The carboxylate ion contains a C=O double bond and an oxygen atom with three lone pairs. One of these lone pairs occupies a *p* orbital on oxygen that is aligned with the C=O π orbital. Overlap of these *p* and π orbitals produces a *new* π orbital that is spread over the three-atom O—C—O unit (Fig. 13–1). Delocalization of electrons in this extended π orbital stabilizes the actual carboxylate ion relative to a hypothetical ion, in which the negative charge would be localized on a single oxygen atom. Molecules and ions in which electrons are delocalized in a π system extending over three or more contiguous atoms are said to be **conjugated.** The negative charge on oxygen in acetate anion (eq. 13.1) is conjugated with the π bond and hence can be delocalized. On the other hand, the negative charge on ethoxide anion (eq. 13.2) must remain localized on the oxygen. Acetate anion is therefore much more stable than ethoxide anion.

The molecular orbital description of delocalized systems is often inconvenient for the organic chemist's needs. A more convenient description for many purposes is provided by the resonance method, which uses two or more classical Lewis structures to depict a single molecule or ion.*

One characteristic of delocalized systems is that *they cannot be adequately represented by a single Lewis structure.* Consider, once again, the acetate ion. One Lewis structure for this ion is **1a,** while an equally good structure for

* Note that the molecular orbital and resonance descriptions are *alternate* ways of describing the same species. Their terminologies should not be mixed together.

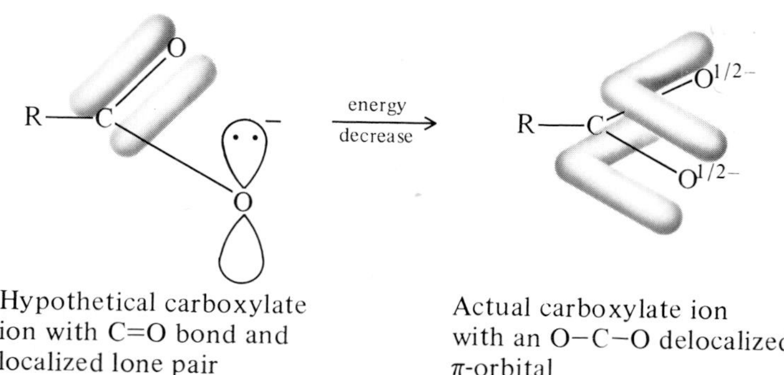

Figure 13–1. *Electron-localized and electron-delocalized structures for the carboxylate ion.*

the same ion is shown in **1b.** The double-headed arrow indicates that neither of these structures *by itself* can give an accurate picture of the bonding in the

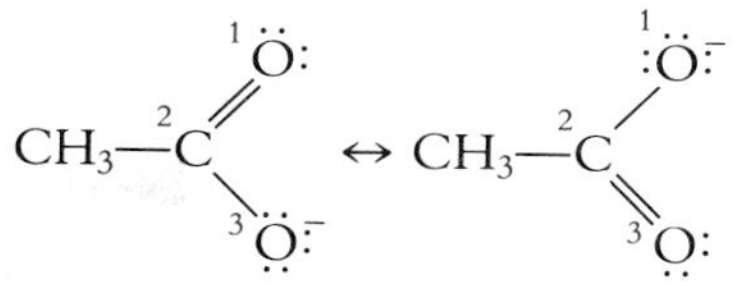

$$\mathrm{CH_3{-}C({=}\ddot{O}{:})(\ddot{O}{:}^-)} \leftrightarrow \mathrm{CH_3{-}C(\ddot{O}{:}^-)({=}\ddot{O}{:})}$$

1a **1b**

acetate ion. Rather, the actual acetate ion is a compromise, or *hybrid,* between these two hypothetical structures. The hybrid is called a **resonance hybrid,** and structures **1a** and **1b** are called **resonance structures.** (There is no connection between the use of the word "resonance" here and its use in nmr spectroscopy.)

Resonance structures provide a great deal of insight into the chemical properties of delocalized ions and molecules, and we will make extensive use of them throughout the remainder of this book. For this reason it is very important for you to understand the relationship between resonance structures, and to be able to draw them correctly. The most important fact to note about **1a** and **1b** is that *they differ* ***only*** *in the arrangement of their electrons;* their atomic connectivity and stereochemistry are identical. This criterion applies to all resonance structures of a given species—they can differ only in their electron distribution.*

Suppose we have one resonance structure for a given species. Can we derive the other resonance structure(s) from it? Figure 13–2 shows how this is done in the case of acetate ion. By shifting one pair of nonbonding electrons and one pair of π electrons we can derive **1b** from **1a,** and vice versa. It is generally true that the resonance structures of species with delocalized π orbitals will differ only in the arrangement of their nonbonding and π electrons.

Once we've derived **1b** from **1a,** then we have to ask if we have identified all the resonance structures for acetate ion. Although a large number of resonance structures can be written for most molecules and ions, only a few of them are informative and useful. How do we recognize a useful resonance structure? First, the octet rule should be satisfied for as many atoms in the structure as possible. In addition, negative charges should be placed on more electronegative atoms, and positive charges on less electronegative atoms,

* Two resonance structures *must* have the same number of paired electrons.

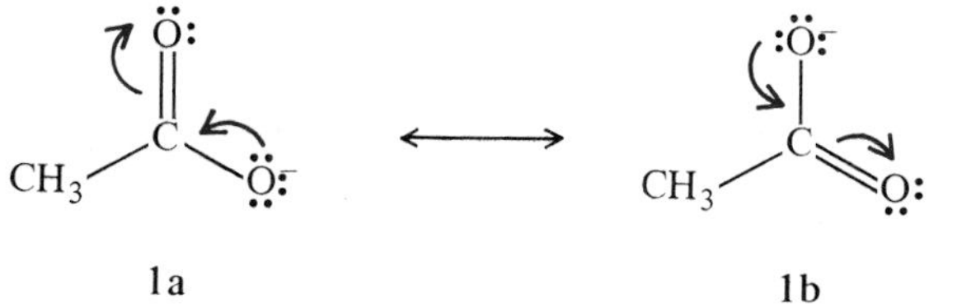

$$\mathrm{CH_3{-}C({=}\ddot{O}{:})(\ddot{O}{:}^-)} \longleftrightarrow \mathrm{CH_3{-}C(\ddot{O}{:}^-)({=}\ddot{O}{:})}$$

1a 1b

Figure 13–2. *How to derive one resonance structure of acetate ion from the other resonance structure by shifting one pair of π electrons and one pair of nonbonding electrons.*

provided that the octet rule is not violated. The operation of these "common sense" rules will be further illustrated in our continuing discussion of resonance. If, at this point, you examine the structure of acetate ion with these rules in mind, you will see that **1a** and **1b** are the only useful resonance structures.

Let's see where our line of reasoning has led us. We ascribe the acidity of acetic acid to the stability of the acetate ion, which is due to its π-electron delocalization. We have learned how to depict this delocalization, but we have not yet seen any direct, experimental proof for it. To do that we have to imagine what acetate ion would look like if it were *not* delocalized and if it could be adequately represented by a single Lewis structure such as **1a.** In that case we would expect to find two different carbon–oxygen bond distances: a longer C—O single bond length and a shorter C=O double bond length. However, X ray diffraction has shown that the two carbon–oxygen bonds in the acetate anion are *identical* in length, and that this length is intermediate between the usual C—O and C=O bond lengths. In addition, the infrared spectrum of the acetate anion shows a C=O stretching peak at about 1600 cm^{-1}, considerably different from the C=O peak of acetic acid (about 1715 cm^{-1}). The lower frequency absorption means that the C=O of acetate ion is much easier to stretch than is the C=O of acetic acid. These data are incompatible with a single, localized bond structure for acetate ion.

One way out of this dilemma would be to postulate a very rapid equilibrium between **1a** and **1b**:

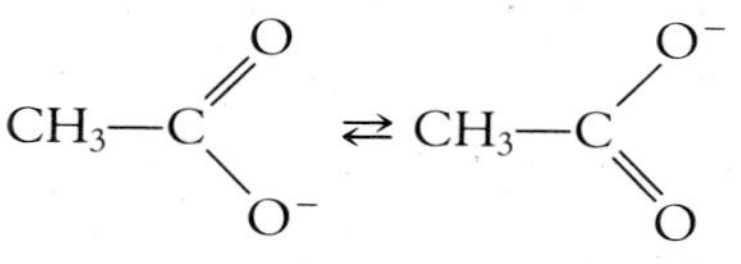

$$CH_3\text{—}C(\text{=O})\text{—}O^- \rightleftharpoons CH_3\text{—}C(\text{—}O^-)\text{=O}$$

1a **1b**

Equilibria can be slowed down at low temperature, so we might be able to see (spectroscopically) two kinds of C—O bonds. However, the spectroscopic data for acetate ion do *not* change as the temperature drops. We must therefore discard the equilibrium hypothesis, and accept the idea that acetate ion has a single structure in which the π electrons are delocalized. Because of this delocalization, acetate ion cannot be represented by a single Lewis structure. Chemists have borrowed the term hybrid from biology to explain this situation; we say that the acetate ion has a single structure that is a hybrid of **1a** and **1b.** To distinguish this from an equilibrium, a double-headed arrow is used:

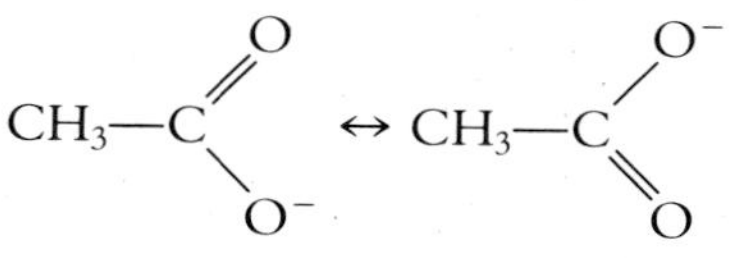

$$CH_3\text{—}C(\text{=O})\text{—}O^- \leftrightarrow CH_3\text{—}C(\text{—}O^-)\text{=O}$$

1a **1b**

Is there any way of depicting a delocalized species that avoids the somewhat cumbersome procedure of writing out all the resonance structures? This is possible for acetate ion, for which the single structure **1** is sometimes used. The dashed lines indicate partial bonds, which in this case would be

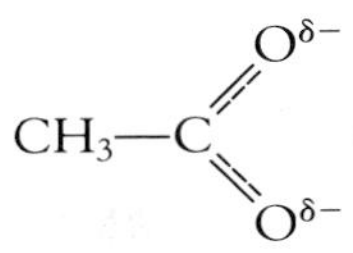

$$CH_3{-}C\overset{\displaystyle O^{\delta-}}{\underset{\displaystyle O^{\delta-}}{\lessgtr}} \equiv \mathbf{1a} \leftrightarrow \mathbf{1b}$$

1

half bonds. The symbol δ^- indicates that there is a *partial* negative charge on each oxygen atom, which for acetate equals half of an electronic charge. The electron arithmetic for structure **1** coincides with that obtained with more usual Lewis structures. Thus, the central carbon in **1** has three full bonds and two half bonds, for a total of four bonds and eight electrons.

The advantages of representation **1,** as opposed to **1a** ↔ **1b,** are that it is compact, and it conveys some aspects of the acetate ion structure more graphically: the equivalence of the two C⩵O bonds, and the equal distribution of the negative charge. On the other hand, electron counting is less straightforward with **1,** and it is not so easy to see the connection between **1** and acetic acid, from which it is derived (eq. 13.1). It is important to understand that neither **1** nor **1a** ↔ **1b** is the more correct. They represent *different* ways of depicting the same species, acetate ion. This elaborate approach is necessary because our conventional method of drawing structures is less successful for some species, like acetate ion and benzene, than it is for others, like ethoxide ion and methane.

Because the word "hybrid" is borrowed from biology, numerous biological examples have been cited as analogies to help explain the concept of resonance. For example, we can consider a resonance hybrid molecule as an analog of a mule. A mule is a hybrid betweeen a horse and a donkey. It is not a horse part of the time and a donkey part of the time, but is an independent entity that has characteristics of both the horse and the donkey. This is a somewhat faulty analogy since the horse and the donkey are also real entities, while neither of the resonance structures of an organic molecule actually exist. Another analogy, which avoids this problem but is a little fanciful, is to consider a rhinoceros as a hybrid of a unicorn and a dragon. In this case, only the hybrid truly exists.

In the resonance hybrid **1a** ↔ **1b,** the two individual resonance structures are exactly equivalent. We could tell them apart only if there were some way of labeling the individual carbons or oxygens (as we have done with numbers). Sometimes, we can write resonance structures that are not equivalent. Consider the carbanion formed by treating acetone with a strong base (eq. 13.3).

$$(13.3)\quad CH_3{-}\overset{\displaystyle \ddot{O}:}{\overset{\|}{C}}{-}CH_3 + :\ddot{N}H_2^- \rightarrow \underset{\mathbf{2a}}{CH_3{-}\overset{\displaystyle \ddot{O}:}{\overset{\|}{C}}{-}\ddot{C}H_2^-} \leftrightarrow \underset{\mathbf{2b}}{CH_3{-}\overset{\displaystyle :\ddot{O}:^-}{\overset{|}{C}}{=}CH_2} + \ddot{N}H_3$$

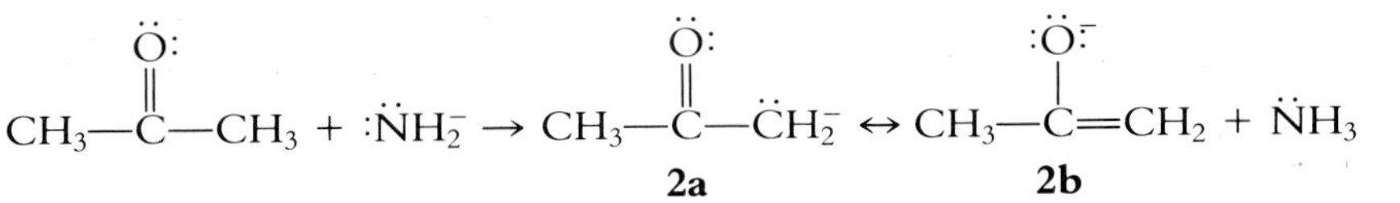

As indicated, the anion is a resonance hybrid depicted by the structures **2a** and **2b.** These two structures are not equivalent because **2b** has the negative charge on an electronegative oxygen rather than on carbon, as in structure

2a. Thus the hybrid, which can be represented by the single structure **2,** will

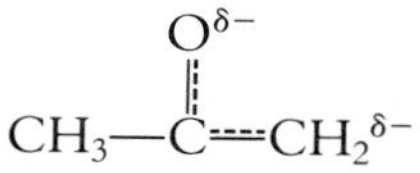

$$\begin{array}{c} O^{\delta-} \\ \| \\ CH_3—C \overset{---}{=} CH_2^{\ \delta-} \end{array}$$

2

be more like **2b** than it is like **2a;** there will be more negative charge on oxygen than on carbon. In this situation, it is said that the resonance structure **2b** contributes more to the hybrid than does **2a.**

If two (or more) nonequivalent resonance structures can be written for one species, each structure has to be examined separately. The structure that would be most stable, if it had an independent existence, will contribute most to the properties of the resonance hybrid.

Let's look at another example that uses some chemistry from Chapter 11. Protonation of a carbonyl group in equation 11.25 (p. 403) was depicted as follows:

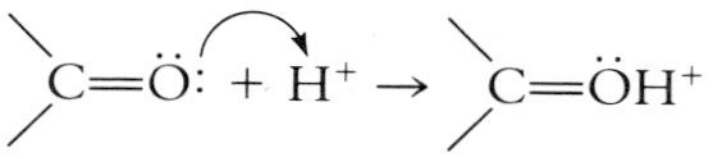

$$>C=\ddot{O}: + H^+ \rightarrow >C=\ddot{O}H^+$$

We could equally well have depicted the protonation reaction this way:

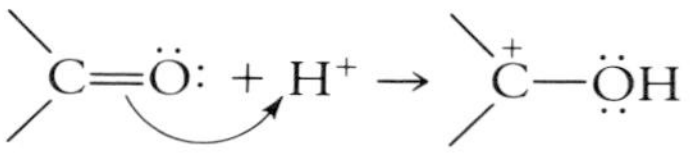

$$>C=\ddot{O}: + H^+ \rightarrow >\overset{+}{C}—\ddot{O}H$$

In reality these are two alternative ways of showing the same process, because the two product ion structures are nothing more than resonance structures:

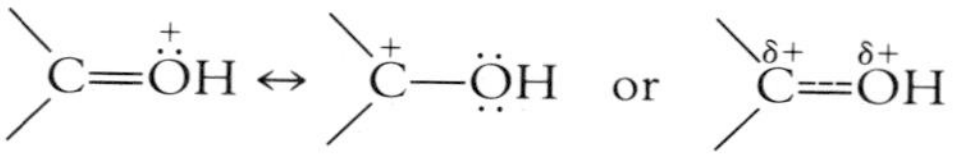

$$>C=\overset{+}{\ddot{O}}H \leftrightarrow >\overset{+}{C}—\ddot{O}H \quad \text{or} \quad >\overset{\delta+}{C}\overset{---}{=}\overset{\delta+}{\ddot{O}}H$$

Note that positive charge is placed on oxygen in one structure. Even though oxygen is more electronegative than carbon, the resonance structure with the positive charge on oxygen contributes more to the hybrid because all atoms have electron octets.

Example

Draw the resonance structure that would result from the electron movement indicated by the curved arrows:

a. 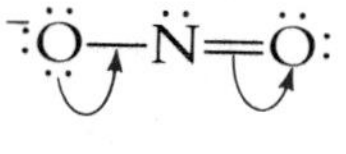 $^{-}:\ddot{O}—\ddot{N}=\ddot{O}:$

b. 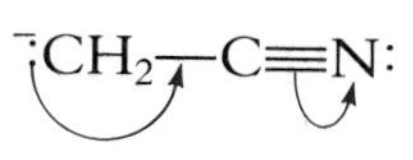 $^{-}:CH_2—C\equiv N:$

Answer

Each arrow means that an electron pair is moved, either to form a bond or to break a bond.

a. $:\ddot{O}=\ddot{N}—\ddot{O}:^{-}$

b. $CH_2=C=\ddot{N}:^{-}$

13–1 For each species, write all the resonance structures that are good Lewis structures, as well as a single hybrid structure. Which resonance structure has the greater weight in (f)?

a. $CH_2{=}CH{-}\ddot{C}H_2^-$

b. 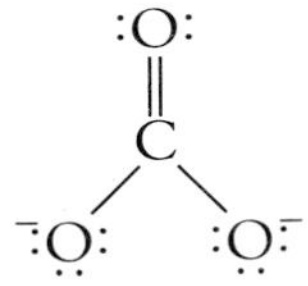 carbonate ion: $:O:$ double-bonded to C, with $^-:\ddot{O}:$ and $:\ddot{O}:^-$ single-bonded to C

c. $^-\ddot{C}H_2{-}C{\equiv}CH$

d. 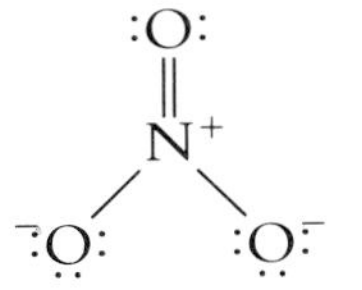 $:O:$ double-bonded to N^+, with $^-:\ddot{O}:$ and $:\ddot{O}:^-$ single-bonded to N

e. 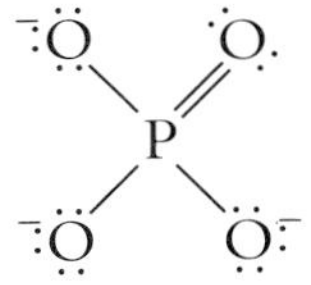 P bonded to $^-:\ddot{O}$, $\ddot{O}.$ (double bond), $^-:\ddot{O}$ and $\ddot{O}:^-$

f. 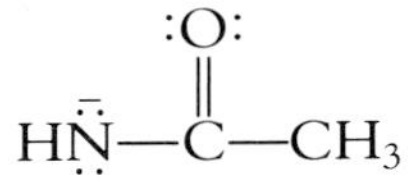 $H\ddot{N}^-{-}C({=}\ddot{O}){-}CH_3$

13–2 Propane is a very weak acid, with a pK_a of about 60 (eq. 1). Propene is also a very weak acid, but it is considerably stronger than propane (eq. 2). Explain the difference in acidity between these two hydrocarbons.

$$CH_3CH_2CH_3 \xrightleftharpoons{pK_a \simeq 60} CH_3CH_2\ddot{C}H_2^- + H^+ \quad (1)$$

$$CH_2{=}CH{-}CH_3 \xrightleftharpoons{pK_a \simeq 45} CH_2{=}CH{-}\ddot{C}H_2^- + H^+ \quad (2)$$

13:2 Resonance-Stabilized Ions, Radicals, and Molecules

Anions are not the only species which can exhibit electron delocalization and require the use of resonance structures for their representation. Delocalization is often an important stabilizing factor for all types of reactive intermediates, including carbocations and radicals. The following examples deal with carbocations, so we will review a bit of their chemistry (Chap. 6).

In polar hydroxylic solvents, such as water and ethanol, alkyl halides often undergo nucleophilic substitution by way of carbocations (S_N1 mechanism). When these reactions involve 1° halides their rate is low, because 1° carbocations are very high in energy and therefore slow to form (sec. 6:7, Fig. 6–12 (p. 193)). The reaction in Equation 13.4 is a typical example; **3** is a 1° carbocation. The reactions in Equations 13.5 and 13.6, on the other hand, do not fit our expectations. Although their substrates are 1° halides and they

(13.4) $$CH_3CH_2CH_2Br \xrightarrow[\text{slow}]{\text{aq EtOH}} \underset{\mathbf{3}}{CH_3CH_2CH_2^+} + Br^-$$

(13.5) $$\underset{\text{Allyl bromide}}{CH_2{=}CH{-}CH_2Br} \xrightarrow[\text{fast}]{\text{aq EtOH}} \underset{\mathbf{4}}{CH_2{=}CH{-}CH_2^+} + Br^-$$

(13.6) $$\underset{\text{Bromomethyl methyl ether}}{CH_3OCH_2Br} \xrightarrow[\text{very fast}]{\text{aq EtOH}} \underset{\mathbf{5}}{CH_3OCH_2^+} + Br^-$$

proceed by way of 1° carbocations, their reaction rates are thousands of times faster than the rate for the reaction in Equation 13.4. This unexpected behavior is observed because carbocations **4** and **5** can delocalize their charges, while **3** cannot. Delocalization stabilizes **4** and **5,** thus speeding up their rate of formation (sec. 6:7).

The **allyl cation, 4,** can be represented by two equivalent resonance structures. Each of these resonance structures can be derived from the other by movement of one pair of π electrons (Fig. 13–3). As in the case of acetate ion, the charge is equally distributed over the two terminal atoms. Contrast this with the methoxymethyl cation, **5,** the resonance structures for which are not equivalent. Note that one carbon atom in **5a** has only six valence

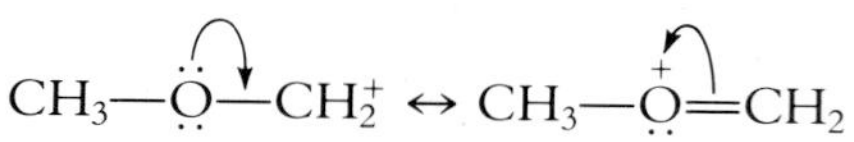

$$CH_3{-}\ddot{\underset{\cdot\cdot}{O}}{-}CH_2^+ \leftrightarrow CH_3{-}\overset{+}{\underset{\cdot\cdot}{O}}{=}CH_2$$

5a **5b**

electrons, while in **5b** all the atoms have complete octets. The most important resonance structures for any species are those that maximize the number of bonds and filled octets. Thus, **5b** will contribute more than **5a** to the hybrid methoxymethyl cation, even though the positive charge is on carbon in **5a,** which is preferable to having it on oxygen (**5b**).

Chemical species need not be charged in order to experience electron delocalization. The *allyl radical,* **6,** has an unpaired electron equally distributed over the two terminal carbon atoms (**6a** $\leftrightarrow$ **6b**).

$$CH_2{=}CH{-}\dot{C}H_2 \leftrightarrow \dot{C}H_2{-}CH{=}CH_2$$

6a **6b**

The chemistry of the allyl radical and its derivatives is completely in accord with the idea that the unpaired electron is delocalized, as we'll see in the next section. Furthermore, the chemistry of these radicals is of more than theoretical interest. The allyl radical is an intermediate in the *ammoxidation* of propene (eq. 13.7), a reaction used for the industrial production of acrolein (about 4.5 million tons a year worldwide).

$$(13.7)\quad CH_2{=}CH{-}CH_3 + NH_3 + \tfrac{3}{2}O_2 \xrightarrow[450^\circ]{Bi_2O_3,\ MoO_3} CH_2{=}CH{-}C{\equiv}N + 3H_2O$$

Acrylonitrile

Acrolein undergoes addition polymerization (sec. 10:7) to form important synthetic fibers and synthetic rubbers.

Finally, physical measurements such as spectroscopic data reveal electron delocalization even in stable molecules. Below are C=O stretching frequencies for two ketones.

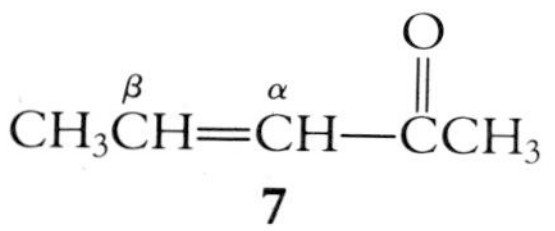

$$\overset{\beta}{CH_3CH}{=}\overset{\alpha}{CH}{-}\overset{O}{\overset{\|}{C}}CH_3 \qquad CH_3CH_2CH_2\overset{O}{\overset{\|}{C}}CH_3$$

7 — 1680 cm^{-1} **8** — 1715 cm^{-1}

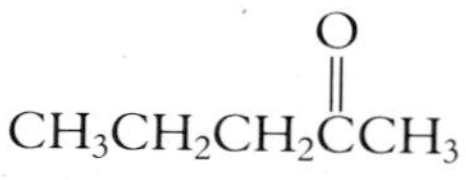

Compound **7** is called an **α,β-unsaturated** ketone because it has a C=C— next to the carbonyl group. It is a conjugated ketone (sec. 13:1) because there are π orbitals extending over four neighboring atoms. Compound **7,** like all α,β-unsaturated carbonyl compounds (aldehydes, acids, esters, ketones),

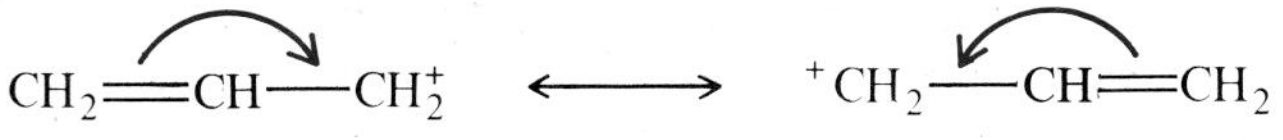

$$CH_2{=}CH{-}CH_2^+ \longleftrightarrow {}^+CH_2{-}CH{=}CH_2$$

4a **4b**

Figure 13–3. *How to derive the two resonance structures of allyl cation from one another by shifting a pair of π electrons.*

shows its C=O stretching absorption at a lower frequency than its corresponding saturated analog, **8.** This can be explained if compounds such as **7** are really resonance hybrids.

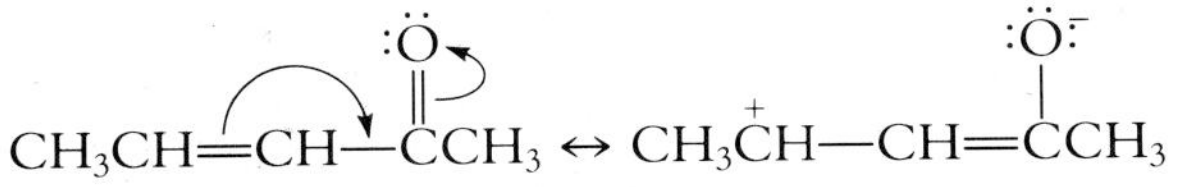

$$CH_3CH{=}CH{-}\overset{\overset{:\ddot{O}}{\|}}{C}CH_3 \leftrightarrow CH_3\overset{+}{C}H{-}CH{=}\overset{\overset{:\ddot{O}:^-}{|}}{C}CH_3$$

7 **7a**

In resonance structure **7a** the C—O bond is a single bond, rather than a double bond as in **7.** Therefore the C=O in the resonance hybrid is not a complete double bond and less energy is needed for stretching it. This means a lower stretching frequency for **7** than for **8.** (This same effect was noted in the acetate anion.)

Since all atoms contain octets in structure **7,** it contributes much more to the resonance hybrid than **7a** does. *Resonance structures without formal charges will always contribute more to the hybrid than will structures with formal charges.* Resonance structure **7a** was "created" from **7** by the conceptual electron movement shown. Suppose a different resonance structure is created by an opposite electron movement, **7** ↔ **7b.**

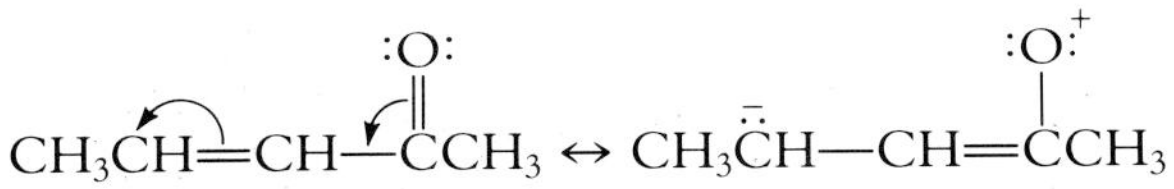

$$CH_3CH{=}CH{-}\overset{\overset{:O:}{\|}}{C}CH_3 \leftrightarrow CH_3\overset{-}{\ddot{C}}H{-}CH{=}\overset{\overset{:O:^+}{|}}{C}CH_3$$

7 **7b**

This is also an electron delocalization and should, in theory, contribute to stabilization. Resonance structures **7a** and **7b** have the same number of bonds and filled octets, but reversed charge distributions. In **7a,** the electronegative oxygen atom carries a negative charge, which is favorable; in **7b** oxygen has a positive charge and an electron sextet, which is unfavorable. Thus, **7b** is much less important than **7a,** and contributes little to the resonance hybrid.

We conclude this section by restating the rules that guide us in drawing useful resonance structures.

1. Structures with completely filled valence shells for all atoms are more important than those in which some atoms have unfilled shells.
2. Uncharged resonance structures are usually more important than charged resonance structures.
3. If all atoms have octets, charged structures that have negative charges on electronegative atoms are more important than those with negatively charged electropositive atoms. The reverse is true for positive charges.

In the following sections we will see how the concept of resonance stabilization can explain the rates and product distribution of a wide variety of chemical reactions.

Example

Draw resonance structures of each of the following compounds:

a. 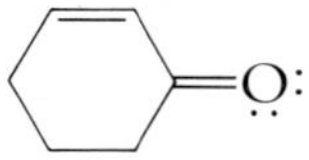 (cyclohexenone structure with $=\ddot{O}:$)

b. $CH_2{=}CH{-}C{\equiv}N:$

Answer

a. Use arrows to show how an electron pair from the double bond could be delocalized to the electronegative oxygen:

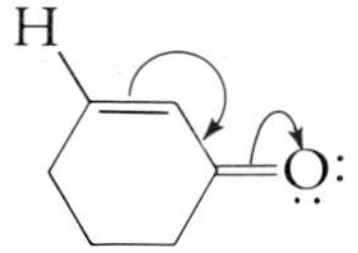

(structure with curved arrows)

Now draw the result:

(charged resonance structure with C^+ and $-\ddot{O}:^-$)

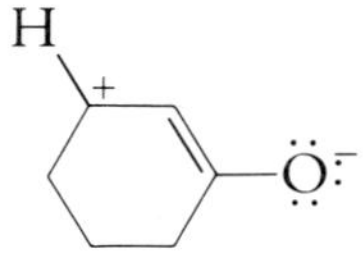

The uncharged resonance structure contributes more to the resonance hybrid.

b. Again, electron delocalization to the more electronegative nitrogen would give a favorable resonance structure:

$$CH_2{=}CH{-}C{\equiv}N: \leftrightarrow \overset{+}{C}H_2{-}CH{=}C{=}\ddot{N}:^-$$

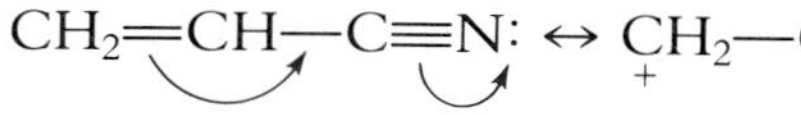

Of course, the uncharged original structure with eight electrons around each atom would be far more important than the charged structure.

Example

The two most reasonable resonance structures for carbon monoxide are shown below. Indicate which of the two you believe to be the more important, and why. Experimentally, we find a band at 2150 cm^{-1} in the ir spectrum of CO. How does this fact help you identify the major resonance structure?

$$:C{=}\ddot{O}: \leftrightarrow :\overset{-}{C}{\equiv}\overset{+}{O}:$$

A **B**

Answer

Resonance structure **A** lacks an octet around carbon, while resonance structure **B** suffers from the presence of charges. The rules regarding resonance structure indicate that the filling of octets has even higher priority than the avoidance of charges, so on that basis **B** should be chosen as the more important structure.

The 2150 cm^{-1} ir band falls in a region where triple bonds usually absorb. The ir data therefore reinforce our choice of **B** as the major resonance structure.

13–3 Cyclopentanol (eq. 3) and 2-cyclopentenol (eq. 4) each react with aqueous HCl to give the corresponding chlorides. Explain why the reaction in Equation 4 is much faster than the reaction in Equation 3. (You may want to review the mechanism of these reactions in Section 6:11.)

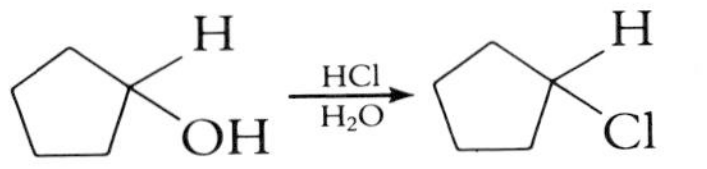

$$\text{Cyclopentanol (H, OH)} \xrightarrow[H_2O]{HCl} \text{Chlorocyclopentane (H, Cl)} \quad (3)$$

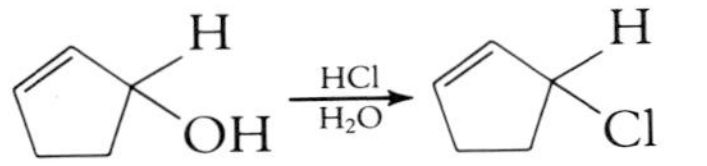

$$\text{2-Cyclopentenol (H, OH)} \xrightarrow[H_2O]{HCl} \text{3-Chlorocyclopentene (H, Cl)} \quad (4)$$

13–4 Write useful resonance structures for each of the following species, and predict their relative contributions to the hybrid.

a. 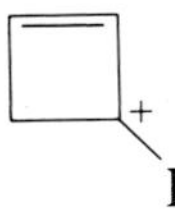 cyclobutenyl cation (ring C^+–H)

d. 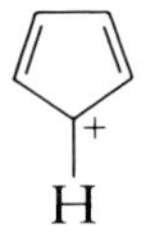 cyclopentadienyl cation (ring C^+–H) (5 resonance structures)

b. $CH_2{=}CH{-}C({=}\ddot{O}{:}){-}CH{=}CH_2$

e. $CH_3{-}\ddot{O}{-}CH{=}CH{-}\overset{+}{C}H_2$

c. $\overset{..}{\overset{-}{C}}H_2{-}CH{=}CH{-}CH{=}CH_2$

f. $CH_2{=}CH{-}\ddot{N}HCH_3$

13:3 Allylic Halogenation

A characteristic reaction of alkenes is the addition of a molecular halogen across the double bond, a reaction normally carried out in solution at room temperature or below. When alkenes and molecular halogens react at elevated temperatures in the gas phase, or under strong illumination, a different course is often followed—*substitution* of a halogen for a hydrogen is observed. For instance, the high-temperature chlorination of propene yields 3-chloropropene (*allyl chloride*) as the major product (eq. 13.8).

$$CH_2{=}CH{-}CH_3 + Cl_2 \xrightarrow[400^\circ]{} CH_2{=}CH{-}CH_2Cl + HCl \quad (13.8)$$

Propene → 3-Chloropropene (*allyl chloride*)

Halogenation is very specific for the hydrogen on the carbon atom *next to* the double bond—the allylic position (eq. 13.9). The mechanism of this

(13.9) Cyclohexene (Allylic positions) $\xrightarrow[\text{heat or light}]{Br_2 \text{ (dilute)}}$ 3-Bromocyclohexene (Br) ≡ (Br) (*no* 4-bromocyclohexene (Br) is formed)

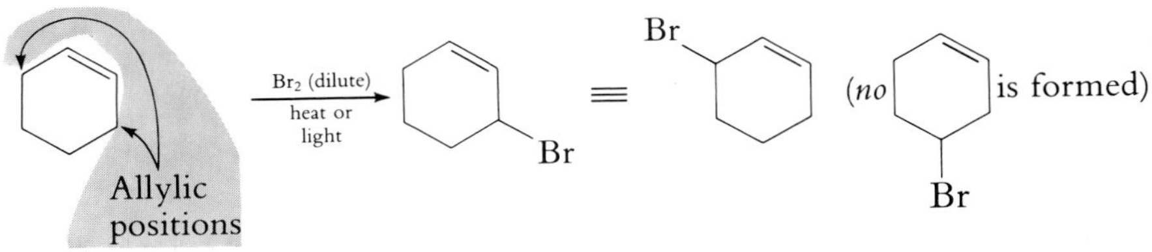

substitution is analogous to the free-radical chain mechanism for the halogenation of alkanes (sec. 5:4). Remember that the initial chain-carrying step in halogenation is formation of a carbon radical. Structures **9** and **10** are two possible intermediates for the halogenation of cyclohexene. The intermediate for allylic halogenation (**9**) is resonance stabilized and hence is formed much more rapidly.

Br· → [cyclohexenyl radical ↔ cyclohexenyl radical] + HBr **9**

Br· → [cyclohexenyl radical] + HBr **10**

You may wonder why the major products in Equations 13.8 and 13.9 are not the addition products (1,2-dichloropropane and 1,2-dibromocyclohexane, respectively). The reaction conditions are critical in determining the product distribution. Heat or light favor free-radical attack, and low concentrations of halogens favor substitution over free-radical addition.

Allylic halogenation is a valuable synthetic tool for functionalization of alkenes. Because of this, several reagents have been developed that are simpler to use and more specific than Cl_2 or Br_2. For allylic bromination the reagent of choice is *N*-bromosuccinimide, known universally as NBS (eq. 13.10). Under the reaction conditions NBS releases a steady, low concentration of Br_2, which minimizes double bond addition and promotes substitution.

(13.10) cyclohexene + N-Bromosuccinimide (NBS) $\xrightarrow{CCl_4,\ heat}$ 3-bromocyclohexene (85%) + succinimide (NH)

For allylic chlorination *t*-butyl hypochlorite is the most widely used reagent (eq. 13.11). *t*-Butyl hypochlorite dissociates into chlorine atoms and *t*-butoxy radicals; the latter initiate the chain by hydrogen abstraction.

(13.11) $(CH_3)_3C(H)C{=}C(H)CH_3$ + $CH_3—C(CH_3)_2—O—Cl$ (*t*-Butyl hypochlorite) $\xrightarrow{light,\ -78°}$ $(CH_3)_3C(H)C{=}C(H)CH_2Cl$ + $(CH_3)_3COH$

13:4 Allyl Cations and Radicals Lead to Multiple Products in Allylic Substitution Reactions

You've seen that the allyl cation is formed much more rapidly than other 1° carbocations (eq. 13.5). If we generate an unsymmetrically substituted allyl cation, an additional aspect of its behavior appears—it reacts with nucleophiles to give two isomeric products. Consider the solvolysis of 1-chloro-2-butene in water with Na_2CO_3 present. Ionization of this halide is the slow step of the reaction, yielding a methyl substituted allyl cation (eq. 13.12). The cation then reacts with water to give a mixture of alcohols **11** and **12** (eq. 13.13; we are not distinguishing between *cis* and *trans* isomer of **12**).

$$\text{(13.12)}\quad CH_3CH{=}CHCH_2Cl \xrightarrow{\text{slow}} Cl^- + [CH_3CH{=}CH{-}CH_2^+ \longleftrightarrow CH_3\overset{+}{C}H{-}CH{=}CH_2] \equiv CH_3\overset{\delta+}{C}H{\cdots}CH{\cdots}\overset{\delta+}{C}H_2$$

$$\text{(13.13)}\quad \overset{4}{C}H_3\overset{3\,\delta+}{C}H{\cdots}\overset{2}{C}H{\cdots}\overset{1\,\delta+}{C}H_2 \xrightarrow[CO_3^{2-}]{H_2O} \underset{\mathbf{11}}{\overset{4}{C}H_3\overset{3}{C}H(OH)\overset{2}{C}H{=}\overset{1}{C}H_2} + \underset{\mathbf{12}}{\overset{4}{C}H_3\overset{3}{C}H{=}\overset{2}{C}H\overset{1}{C}H_2OH} + HCO_3^-$$

Under the given conditions, with Na_2CO_3 present to neutralize H^+, more **11** is formed than **12.** This result is worth noting because the internal alkene **12** is more stable than the terminal alkene **11.** These results must mean that the less stable product (**11**) is formed faster than the more stable one (**12**). We may obtain results of this kind when the products are formed in an *exothermic* reaction, such as that between a carbocation and a nucleophile. When we deal with endothermic reactions, we usually observe the more stable product being formed more rapidly (Hammond's postulate, sec. 6:7). When we analyze a product mixture and find that the less stable product predominates, then we know that the less stable product is being formed faster than the more stable one, and we say that the reaction is under **kinetic control**.

In this particular case, **11** is produced faster than **12** because of the unequal charge distribution in the allylic cation. The secondary carbon carries a greater fraction of the positive charge than the primary carbon does, so the secondary carbon is attacked preferentially (Fig. 13–4). When we translate these relationships into an energy diagram, it shows a smaller E_a for formation of **11** than for formation of **12** (Fig. 13–5).

If the sodium carbonate is omitted from the reaction mixture, HCl accumulates as the reaction proceeds. The acid is a catalyst for reversing the final step (eq. 13.13) and regenerating the cation from alcohols **11** and **12** (eq.

$$\text{(13.14)}\quad \underset{\mathbf{11}}{CH_3CH(OH)CH{=}CH_2} + \underset{\mathbf{12}}{CH_3CH{=}CHCH_2OH} \xrightarrow{H^+} CH_3\overset{\delta+}{C}H{\cdots}CH{\cdots}\overset{\delta+}{C}H_2 + H_2O$$

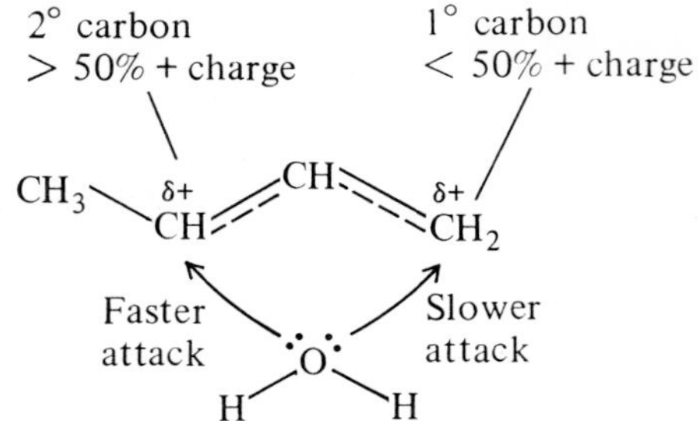

Figure 13–4. *Unequal charge distribution in the 2-butenyl cation. This leads to faster nucleophilic attack at the 2° carbon than at the 1° carbon.*

13.14). If the acidic mixture stands for a sufficiently long period, then the reactions in Equations 13.13 and 13.14 will bring it to equilibrium. At equilibrium, **12** will be the major product since it is the more stable. Such a reaction, in which the major product is the most stable one, is said to be under **thermodynamic control.**

Like unsymmetrical allyl cations, unsymmetrical allyl radicals also give rise to products that are constitutional isomers. Equations 13.15 and 13.16 show how this happens in the free-radical bromination (sec. 13:3) of 2-butene.

$$(13.15)\quad CH_3CH{=}CHCH_3 + Br\cdot \rightarrow HBr + [CH_3CH{=}CH\dot{C}H_2 \leftrightarrow CH_3\dot{C}HCH{=}CH_2] \equiv CH_3\overset{\delta\cdot}{C}H{\cdots}CH{\cdots}\overset{\delta\cdot}{C}H_2$$

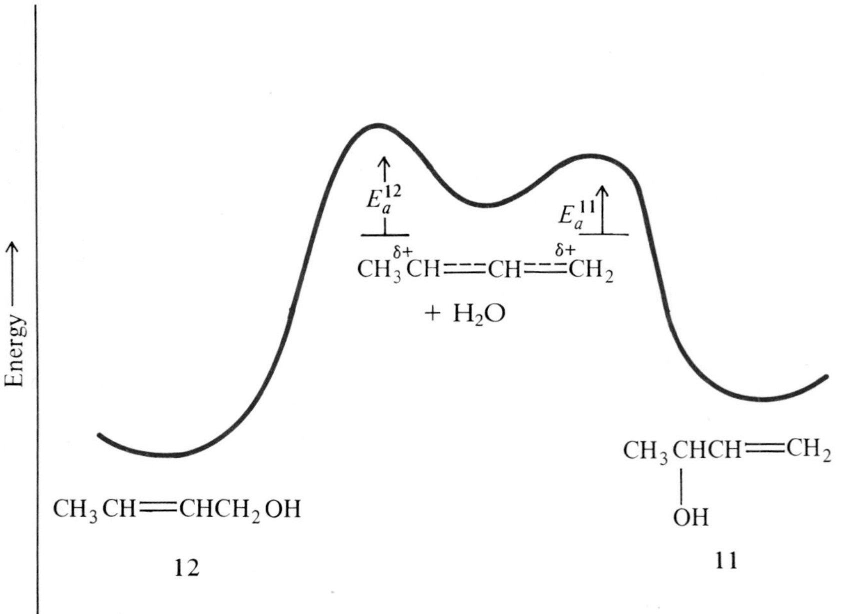

Figure 13–5. *Energy diagram for the reaction of 2-butenyl cation with water. The* E_a *is lower for formation of the less stable product than for formation of the more stable product.*

$$(13.16)\quad CH_3\overset{\delta\cdot}{C}H\text{---}CH\text{---}\overset{\delta\cdot}{C}H_2 + Br_2 \rightarrow$$

$$\underbrace{CH_3\underset{\displaystyle Br}{\underset{|}{C}}HCH{=}CH_2 + CH_3CH{=}CHCH_2Br}_{\text{Constitutional isomers}} + Br\cdot$$

Example

The following two compounds react together in the biosynthesis of steroids and terpenes (sec. 13:6). The reaction is started when *one* of them forms a carbocation by loss of the pyrophosphate (OPP) group. Which would more readily form a carbocation?

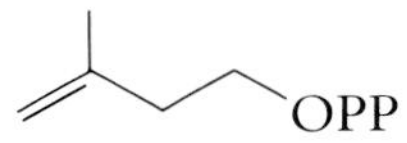

OPP
3-Isopentyl pyrophosphate

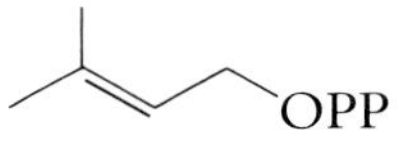

OPP
2-Isopentyl pyrophosphate
(γ,γ-dimethallylpyrophosphate)

Answer

First write the carbocation each would yield.

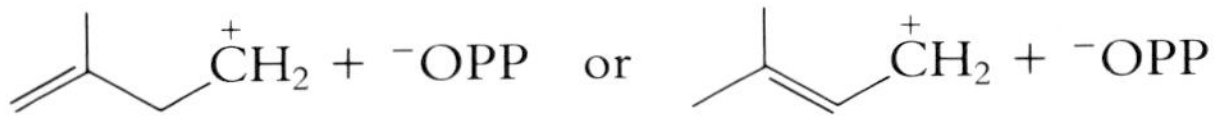

$\overset{+}{C}H_2 + {}^{-}OPP$ or $\overset{+}{C}H_2 + {}^{-}OPP$

Now decide which cation would be formed faster. Since the cation from γ,γ-dimethallylpyrophosphate is resonance stabilized, there would be a lower energy of activation to form that cation. Hence loss of pyrophosphate from the 2-isomer would be faster than from the 3-isomer.

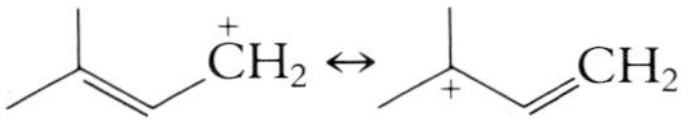

$\overset{+}{C}H_2 \leftrightarrow$ + CH_2

Problems

13–5 When *cis*-2-buten-1-ol is warmed with dilute sulfuric acid, analysis indicates the presence of two isomeric alcohols, **A** and **B.** The nmr spectrum of **A** contains two peaks in the aliphatic proton region, a doublet (3H) and a quartet (1H) that show signs of further splitting. There is a close resemblance between the nmr spectrum of **B** and that of *cis*-2-buten-1-ol, although there are clearly differences in several of the coupling constants. Suggest possible structures for **A** and **B** and explain how they are formed.

13–6 When 1-methyl-2-cyclopentenol (**I**) reacts with aqueous HBr it is converted to a mixture of two bromocyclopentenes. Draw the structures of these products and show the mechanism of their formation. Which product would you expect to be the major one under conditions in which the reaction is not reversible?

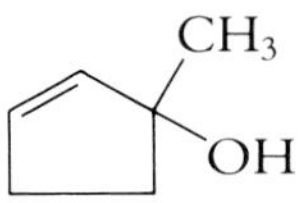

CH_3
OH

I

13–7 When 1-ethynylcyclohexanol (**II**) is treated with a catalytic amount of acid, it rearranges to the α,β-unsaturated aldehyde **III** (eq. 5). Show the mechanism for this process, paying particular attention to the intermediate and its resonance forms. [A review of Equations 12.19 and 12.20 (p. 443) may be helpful.]

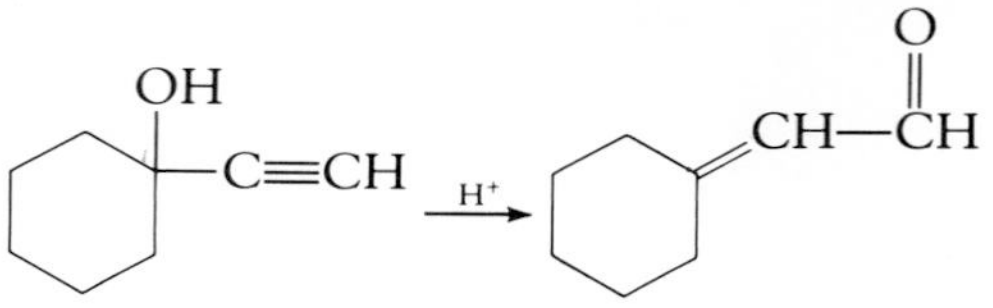

$$\text{1-ethynylcyclohexan-1-ol (C}_6\text{H}_{10}\text{(OH)C}{\equiv}\text{CH)} \xrightarrow{H^+} \text{C}_6\text{H}_{10}{=}\text{CH—CHO} \qquad (5)$$

II **III**

13–8 The free-radical chlorination of 4,4-dimethyl-2-pentene produces a minor product **IV** (7% yield), in addition to the major one shown in Equation 13.11. Give the structure of the allylic radical intermediate in this reaction, and show how it gives rise to **IV.**

$$(CH_3)_3C\underset{\displaystyle Cl}{\underset{|}{C}}HCH{=}CH_2$$

IV

13–9 What other alkene, in addition to 2-butene, would give the same product mixture as that in the reaction in Equation 13.16, upon free-radical (NBS) bromination? Explain your answer.

13:5 A Comparison of Allyl and Vinyl Halide Reactivity

It is useful to compare the reactivities of allyl halides with those of vinyl halides, $CH_2{=}CHX$, in which the halogen atom is directly bound to one of the sp^2-hybridized carbon atoms. While the *allyl* halides are *more* reactive toward nucleophiles than the corresponding alkyl halides, the *vinyl* halides are much less so.

$$XCH_2CH{=}CH_2 > XCH_2CH_2CH_3 \gg XCH{=}CHCH_3$$

Relative reactivity

The carbon–halogen bond is unusually strong in vinyl halides, partly because it uses a carbon sp^2 orbital, and partly because of the conjugation between the π bond and the nonbonding p orbitals on the halogen. Vinyl halides are resonance stabilized. If the reactant vinyl halide is stabilized, it is

$$CH_2{=}CH{-}\ddot{Cl} \leftrightarrow \overset{-}{C}H_2{-}CH{=}\overset{+}{Cl}$$

less likely to undergo nucleophilic substitution. With allyl halides, conjugation lowers the energy of the *transition state,* thereby promoting reaction.

13:6 Terpenes, Steroids, and Natural Rubber: The Role of Allylic Cations in Biosynthesis

From the earliest days of organic chemistry, there has been an intense interest in the chemical components of plants. Initial investigations showed that the volatile components of widely differing plants possessed certain features in common. One common feature was the fact that their carbon content was a multiple of five—the plants yielded C_{10}, C_{15}, and C_{20} components, and so on. These components were collectively called **terpenes,** because some of the first members known were isolated from oil of turpentine.

The terpenes appeared to have all been constructed from the same five-carbon unit. We can see that unit in the acyclic alcohol geraniol, and in the cyclic hydrocarbon limonene. It is characterized by its branched isopentane

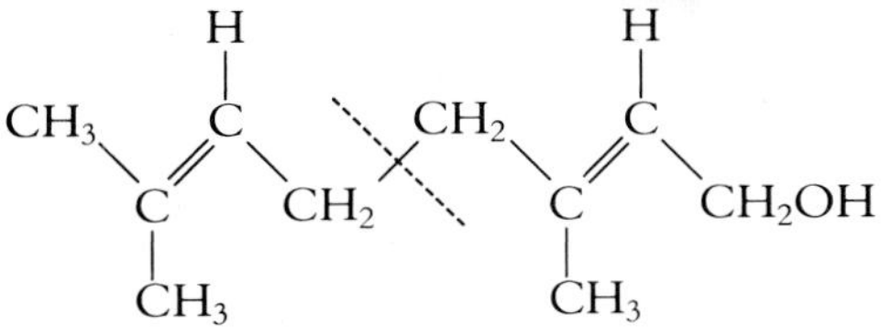

C_5 unit C_5 unit

Geraniol
(from geranium oil)

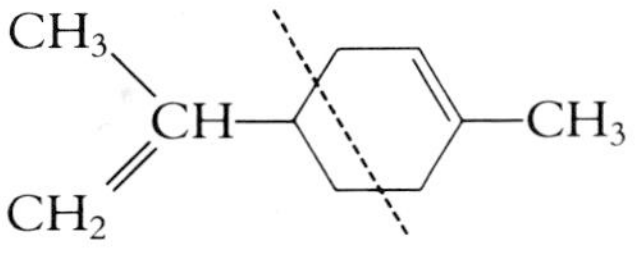

C_5 unit C_5 unit

Limonene
(from lemon, caraway, and dill oils)

chain skeleton (**13**), which is identical to that in *isoprene* (2-methyl-1,3-butadiene). Isoprene is obtained by pyrolysis of natural rubber and was thought for some time to be the five-carbon precursor of the terpenes.

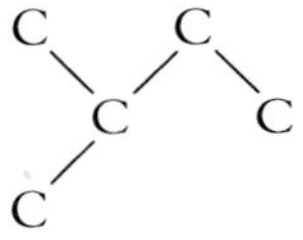

13

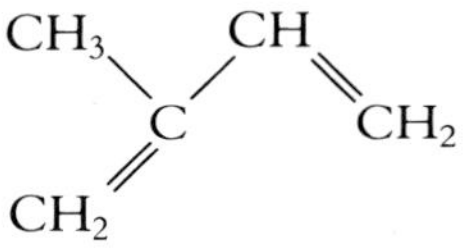

Isoprene

We know now that terpenes and steroids are not constructed from isoprene itself but rather from two isomeric five-carbon phosphate esters: 2-isopentenyl pyrophosphate (**14**) and 3-isopentenyl pyrophosphate (**15**). We will have more to say about phosphate esters in Chapter 17; for now it is sufficient to know that the pyrophosphate anion, abbreviated OPP^-, is a good leaving group, just like Br^- and OTs^-.

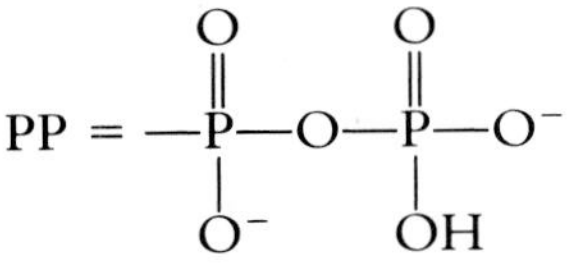

Pyrophosphoryl group

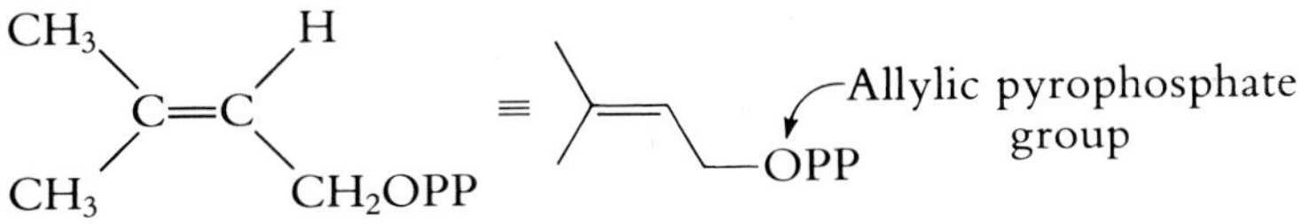

Allylic pyrophosphate group

(14)
2-Isopentenyl pyrophosphate

(15)
3-Isopentenyl pyrophosphate

Note that compound **14** is an allylic pyrophosphate. Loss of the pyrophosphate anion yields a stable allylic cation (eq. 13.17).

$$(13.17)\quad (CH_3)_2C{=}CH{-}CH_2OPP \; (\mathbf{14}) \rightarrow OPP^- + \left[(CH_3)_2C{=}CH{-}CH_2^+ \leftrightarrow (CH_3)_2\overset{+}{C}{-}CH{=}CH_2 \right]$$

Stable allylic cation

In the presence of the enzyme *prenyltransferase,* this cation will attack the π bond of a molecule of **15** (eq. 13.18, step 1) to give a carbocation intermediate. Loss of a proton from the intermediate gives a new double bond (eq. 13.18, step 2), and generates a new allylic pyrophosphate group. The C_{10} compound that arises from reaction between **14** and **15** is called geranyl pyrophosphate (**16**). One possible fate for **16** is that it can undergo hydrolysis, so that OPP is replaced by OH (sec. 13:4).

$$(13.18)\quad (CH_3)_2C{=}CH{-}CH_2^+ + CH_2{=}C(CH_3){-}CH_2{-}CH_2OPP \xrightarrow[\text{①}]{\text{prenyltransferase}}$$

$$(CH_3)C(CH_2){=}CH{-}CH_2{-}CH_2{-}\overset{+}{C}(CH_3){-}CH_2{-}CH_2OPP \xrightarrow{\text{②}}$$

$$(CH_3)_2C{=}CH{-}CH_2{-}CH_2{-}C(CH_3){=}CH{-}CH_2OPP + H^+$$

16
Geranyl pyrophosphate

In the geranium plant, compound **16** is hydrolyzed to geraniol. In the lavender plant, however, hydrolysis of **16** gives the isomeric alcohol linalool (eq. 13.19).

(13.19) 16 OPP $\xrightarrow[\text{geranium}]{H_2O}$ Geraniol (OH)

16 $\xrightarrow[\text{lavender}]{H_2O}$ Linalool (OH)

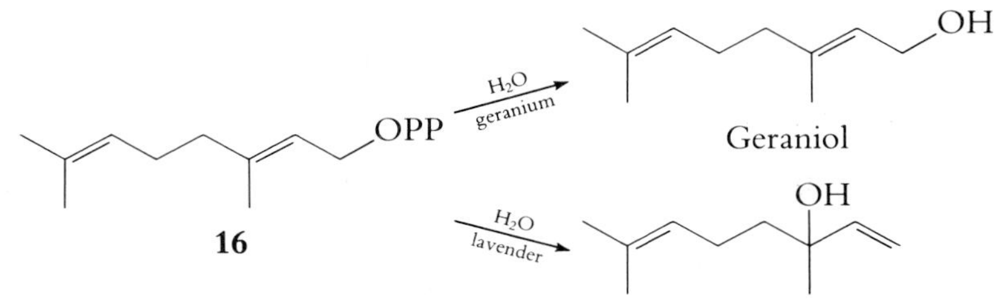

Geraniol and linalool are monoterpenes: C_{10} compounds that have been constructed from isopentene units. The biosynthesis of geraniol and linalool exemplifies once again the remarkable specificity of enzymatic reactions.

Box 13–1. Homoallylic and Nonclassical Carbocations

The halide whose structure is given below has a very fast rate of solvolysis even though it yields a cation that is *not* allylic.

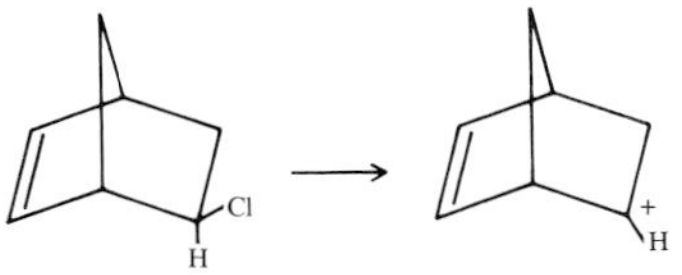

The molecule is rigid and bent so that the π electrons can still be delocalized to the cationic center even though it is removed one more carbon away than in an allyl cation.

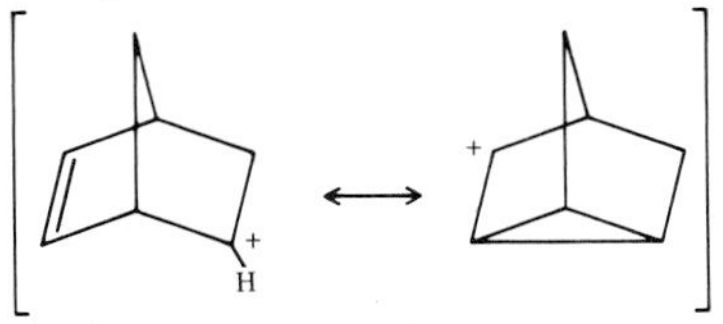

This is called a **homoallylic cation** (*homo*, meaning "one more carbon"). Another example occurs in steroid chemistry:

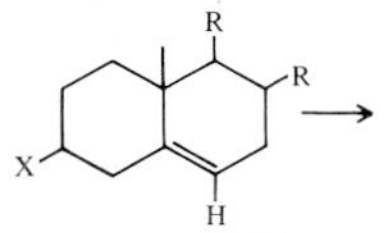

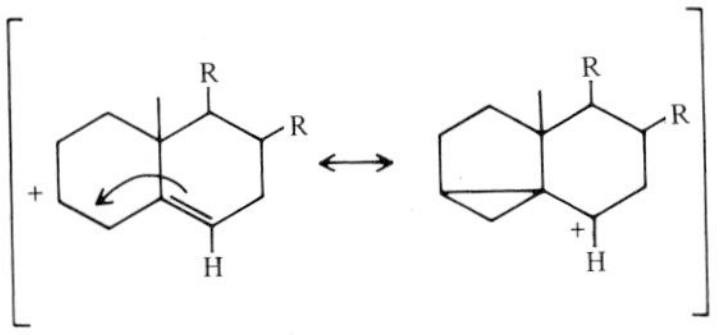

A great controversy arose among organic chemists when it was proposed that a σ bond could also undergo electron delocalization:

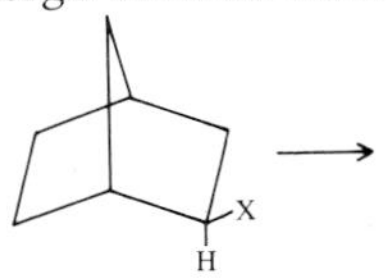

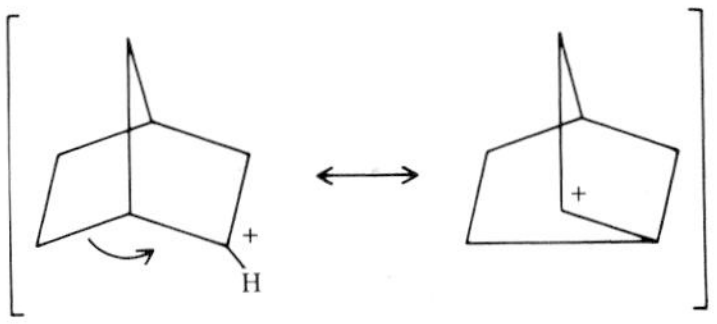

Although both resonance structures have normal or "classical" formulations, the resonance hybrid cannot be drawn without including partial σ bonds; hence this was termed a *nonclassical carbocation*.

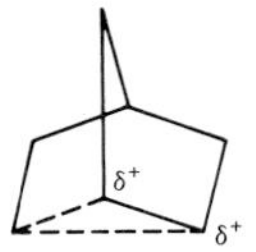

A nonclassical carbocation

One of the most vocal proponents of the nonclassical carbocation was Professor Saul Winstein of the University of California at Los Angeles. One of its opponents was Professor Herbert C. Brown (see Box 11–1, p. 418) who maintained that the chemistry of such systems was just as easily explained if there was a rapid equilibrium between classical carbocations:

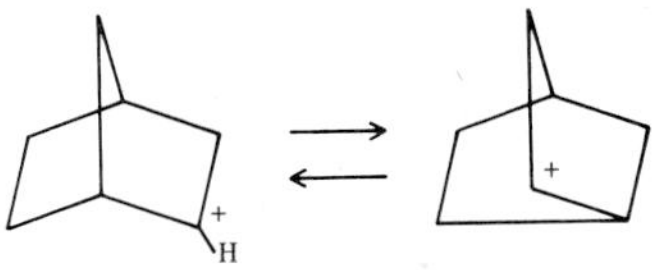

The controversy has raged for many years in the organic chemistry literature and probably few people will agree that it has been resolved yet. It has, however, helped teach a generation of organic chemists how to draw resonance structures and to understand the difference between resonance structures (A ↔ B) and equilibria (A ⇄ B).

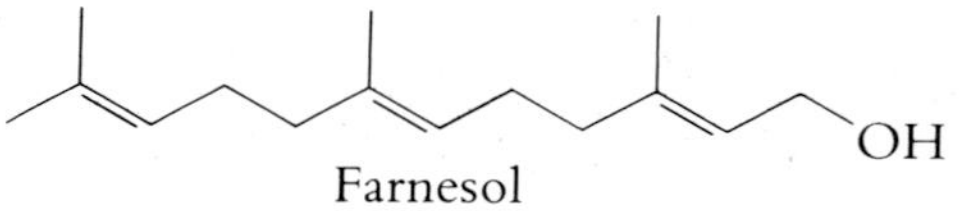

When an allyl halide hydrolyzes under nonenzymatic conditions, it yields a mixture of product alcohols (eq. 13.13). By contrast, the enzymes in geranium and lavender cells control the reaction so that only one alcohol is produced in each case.

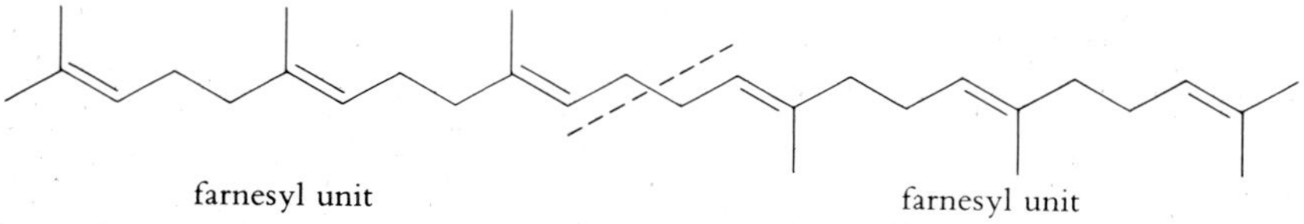

Geranyl pyrophosphate can also undergo further condensation with **15.** One product is the sesquiterpene (C_{15}) farnesol, the pyrophosphate of which is a precursor to many larger molecules. Tail-to-tail linkage of two molecules of farnesyl pyrophosphate produces the triterpene (C_{30}) squalene. Squalene is present in a wide variety of organisms, including man. Its epoxide serves as the starting material for the biosynthesis of many steroids by way of a remarkable cyclization reaction (Fig. 13–6).

Continued condensations of the terpenes result in the formation of an unsaturated polymer with all *cis* double bonds, natural rubber (Fig. 13–7).

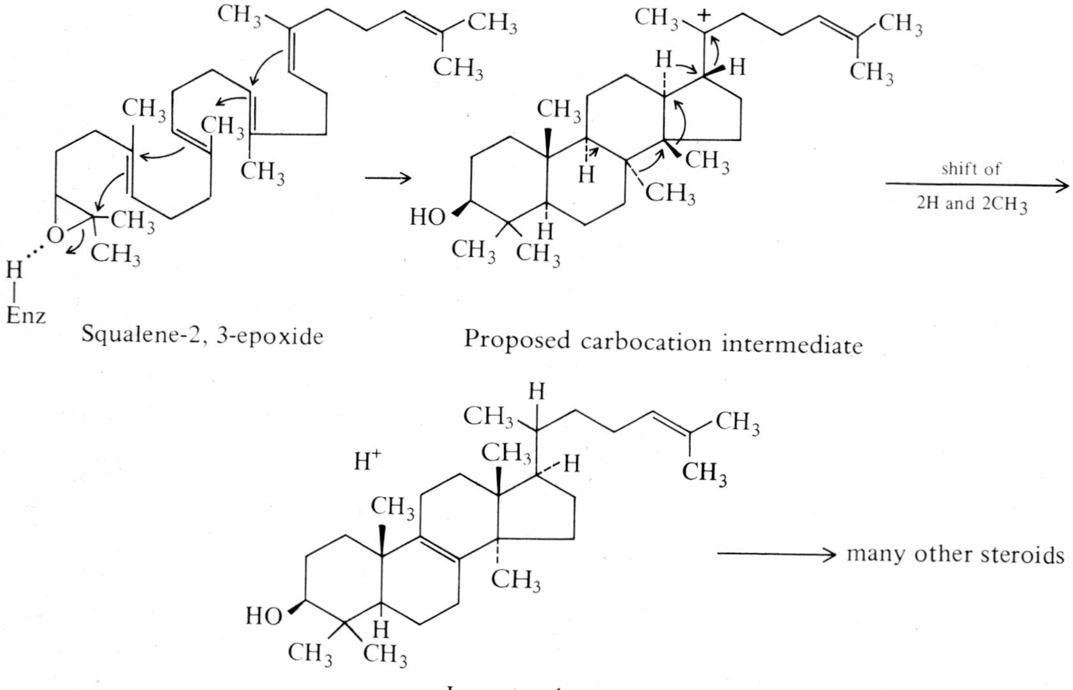

Figure 13–6. *The epoxide of squalene is the biosynthetic precursor of many steroids.*

Figure 13–7. *Natural rubber,* cis-*polyisoprene.*

Problem

13–10 a. Identify the isoprene units in the following structures. (*Hint:* Look for the basic isopentane (**13**) arrangement without worrying about double bond changes, shifts, or cyclizations.)

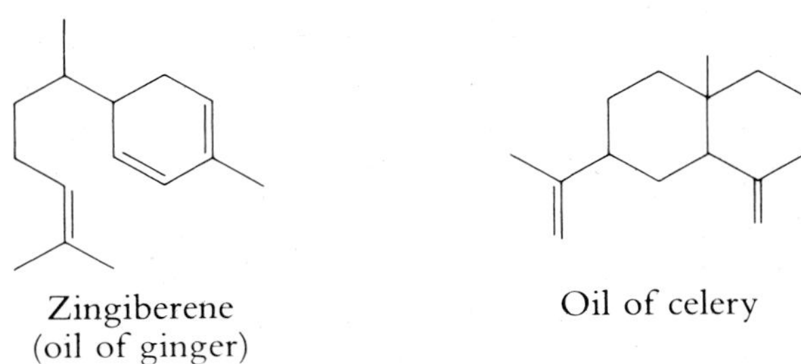

Zingiberene (oil of ginger) | Oil of celery

b. Show how the following terpenes could be synthesized in the living plant starting with compound **16**.

1) (myrcene: oil of bay leaves)

2) CHO (citronellal: citronella oil)

3) OH (α-terpineol: oil of turpentine)

(Consider *intramolecular* cyclization of an intermediate. Note also that allyl cations can change their double-bond geometry.)

13:7 Cumulated, Conjugated, and Nonconjugated Dienes

Three straight-chain alkenes having the empirical formula C_5H_8 (1,2-pentadiene, 1,3-pentadiene, and 1,4-pentadiene) have distinct chemical and physical properties. (A fourth, 2,3-pentadiene, has properties similar to those of 1,2-pentadiene.)

$CH_2{=}C{=}CHCH_2CH_3$ — 1,2-Pentadiene

$CH_2{=}CH{-}CH{=}CHCH_3$ — 1,3-Pentadiene

$CH_2{=}CHCH_2CH{=}CH_2$ — 1,4-Pentadiene

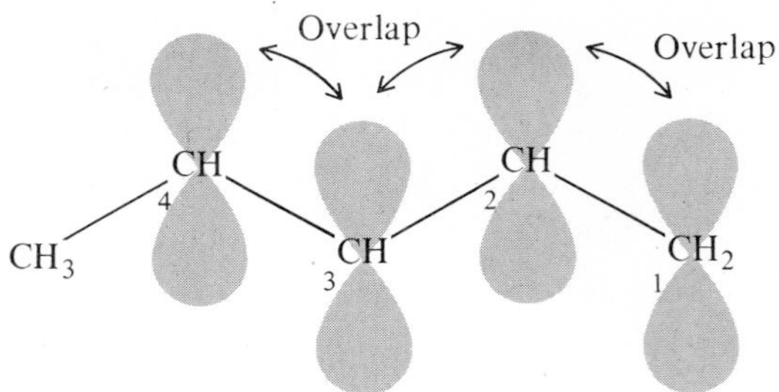

Figure 13–8. *The four neighboring* p *orbitals of 1,3-pentadiene overlap to form a* π *orbital that is delocalized over all four atoms. Maximization of overlap requires that the axes of all the* p *orbitals be parallel.*

In 1,4-pentadiene, the two double bonds are effectively isolated from each other by the sp^3-hybridized methylene carbon between them. Hence they do not interact chemically and are termed *nonconjugated* or *isolated* double bonds. Such nonconjugated dienes have chemical properties similar to those of simple alkenes. On the other hand, in 1,3-pentadiene there is an alternation of double, single, and double bonds that results in four neighboring sp^2-hybridized carbons carrying four neighboring *p* orbitals. These dienes (or extended systems such as 1,3,5-hexatriene) are *conjugated* and have very special chemical properties and reactivities.

When we say that 1,3-pentadiene is conjugated, we imply that there is a delocalized π orbital that extends from C-1 to C-4. Such an orbital is in fact present, formed by the four overlapping *p* orbitals on these carbons. For maximum overlap of the four neighboring orbitals, their axes must be parallel (Fig. 13–8). By maintaining the planar conformation shown in Figure 13–8, 1,3-pentadiene achieves maximum stability (see sec. 13:8).

The double bonds in 1,2-pentadiene and 2,3-pentadiene are said to be **cumulated**—that is, one *sp*-hybridized carbon atom is a part of two different double bonds, one on each side. Compounds with cumulated double bonds are called **allenes.** Allenes are not conjugated because the geometry of the *p* orbitals on the central *sp*-hybridized carbon causes the neighboring π systems to be at right angles to one another (Fig. 13–9; compare with Fig. 13–8).

Since the three C_5H_8 pentadienes just discussed are constitutional isomers of one another, spectroscopic methods are useful for identifying individual isomers. As is the case with many such isomers, nmr spectroscopy would

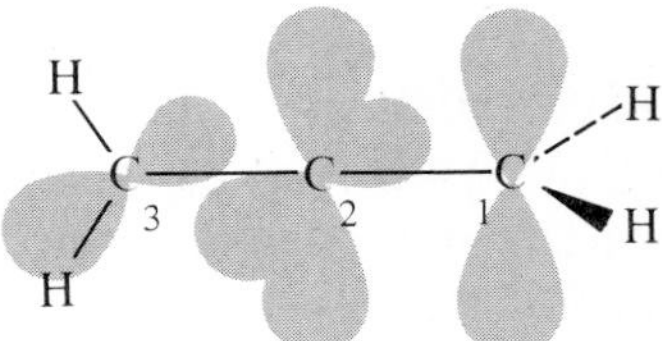

Figure 13–9. *The cumulated double-bond system of 1,2-propadiene, where* sp *hybridization of C-2 causes the methylene (CH_2) groups at C-1 and C-3 to lie at right angles to each other. (Compare this to the conjugated system shown in Figure 13–8.)*

distinguish among them because each has different kinds of hydrogens. In addition, allenes exhibit an infrared band at 1900 cm^{-1}, an anomalous position for a C=C stretch, which is normally in the 1640 to 1680 cm^{-1} region. Isolated and conjugated dienes are most easily distinguished by their ultraviolet-visible spectra (see sec. 13:13).

Problem

13–11 In each of the following compounds, identify those sets of double bonds which are conjugated, if any, and those which are cumulated, if any.

a. [cyclohexa-1,3-diene structure]

b. [cyclopentadiene structure]

c. $CH_2{=}CH{-}CH{=}CH{-}CH{=}CH_2$

d. $CH_3CH{=}C{=}CH{-}CH{=}CHCH_3$

e. $CH_3CH{=}CH{-}\overset{\overset{O}{\|}}{C}CH_3$

f. [cyclohex-2-enone structure]

g. $CH_2{=}CH{-}\underset{\underset{CH=CHCH_3}{|}}{C}{=}CH_2$

13:8 Hydrogenation of Conjugated Dienes, A Measure of Their Stability

Catalytic hydrogenation (sec. 10:2) of a nonconjugated diene such as 1,4-pentadiene releases about twice the amount of heat released when a simple alkene is hydrogenated (compare eqs. 13.20 and 13.21). These data indicate that there is no interaction between the two double bonds in 1,4-pentadiene. If there were a stabilizing interaction between the two double bonds, less heat would be liberated because stabilization of the alkene results in a decrease in the heat of hydrogenation. Reduction of the conjugated diene 1,3-butadiene reveals a different situation. The heat liberated is appreciably less

(13.20) $$\underset{\text{1,4-Pentadiene}}{CH_2{=}CHCH_2CH{=}CH_2} + 2H_2 \xrightarrow[\text{gas phase, 80°}]{Pt} CH_3CH_2CH_2CH_2CH_3$$

$$\Delta H = -60.8 \text{ kcal/mol}$$

(13.21) $$CH_3CH_2CH{=}CH_2 + H_2 \xrightarrow[\text{gas phase, 80°}]{Pt} CH_3CH_2CH_2CH_3$$

$$\Delta H = -30.3 \text{ kcal/mol}$$
$$(2 \times \Delta H = -60.6 \text{ kcal/mol})$$

(13.22) $$\underset{\text{1,3-Butadiene}}{CH_2{=}CHCH{=}CH_2} + 2H_2 \xrightarrow[\text{gas phase, 80°}]{Pt} CH_3CH_2CH_2CH_3$$

$$\Delta H = -57.1 \text{ kcal/mol}$$

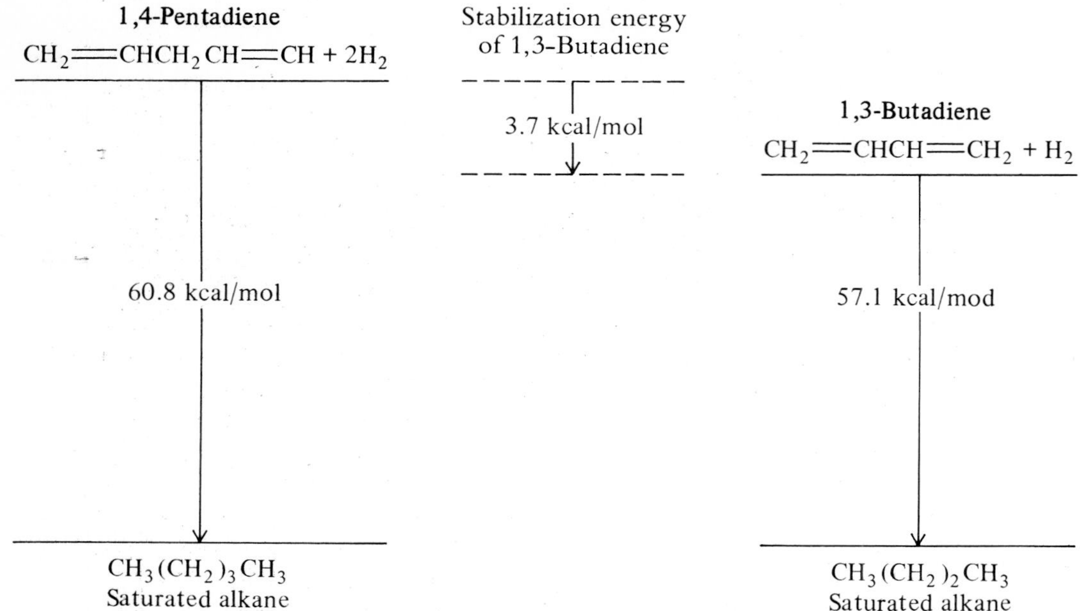

Figure 13–10. *The different heats of hydrogenation of 1,4-pentadiene and 1,3-butadiene, showing the stabilization of the conjugated butadiene.*

than twice the heat of hydrogenation of an isolated double bond (compare eqs. 13.22 and 13.21). The conjugated butadiene is about 3.7 kcal/mol more stable (Fig. 13–10). Several factors contribute to the extra stability of 1,3-butadiene; one of them is delocalization of the π electrons over all four carbon atoms.

Because of the delocalization in conjugated systems, they are more stable than their unconjugated isomers. A C=C bond will be stabilized by conjugation with almost any functional group containing π electrons, including aromatic rings (Chap. 14) and carbonyl groups. For instance, a double bond conjugated with a C=O group is reduced less exothermically than an unconjugated double bond (compare eqs. 13.23 and 13.24).

$$\underset{trans}{CH_3CH{=}CH\overset{\overset{\displaystyle O}{\|}}{C}H} + H_2 \xrightarrow[\text{gas phase}]{} CH_3CH_2CH_2\overset{\overset{\displaystyle O}{\|}}{C}H \qquad \Delta H = -25.2 \text{ kcal/mol} \tag{13.23}$$

$$\underset{trans}{CH_3CH{=}CHCH_3} + H_2 \xrightarrow[\text{gas phase}]{} CH_3CH_2CH_2CH_3 \qquad \Delta H = -27.6 \text{ kcal/mol} \tag{13.24}$$

Problems

13–12 The following equations describe the conversion of butyraldehyde and crotonaldehyde (2-butenal) to their respective methyl acetals.

$$CH_3CH_2CH_2CHO + 2CH_3OH \rightarrow CH_3CH_2CH_2CH(OCH_3)_2 + H_2O \qquad \Delta H = -8.8 \text{ kcal/mol}$$

$$CH_3CH{=}CHCHO + 2CH_3OH \rightarrow CH_3CH{=}CHCH(OCH_3)_2 + H_2O \qquad \Delta H = -3.7 \text{ kcal/mol}$$

What is the cause of the energy difference between these two related processes?

13–13 Explain the driving force for the reactions shown in Equations 6 and 7. For each of these reactions propose a mechanism (note that H^+ and OH^- are catalysts). (*Hint:* In Equation 6 the first step is a protonation. In Equation 7 the first step is abstraction of a proton by OH^-.)

1,4-cyclohexadiene $\xrightarrow[\text{heat}]{H^+}$ 1,3-cyclohexadiene (6)

$$CH_2{=}CH{-}CH_2{-}\overset{O}{\overset{\|}{C}}H \xrightarrow{OH^-} CH_3CH{=}CH{-}\overset{O}{\overset{\|}{C}}H \qquad (7)$$

13:9 Ionic Additions to Conjugated Dienes

The same reagents that add to simple alkenes (Chap. 10) also add to conjugated dienes. In both cases, these additions are initiated by electrophilic attack. However, the dienes (unlike alkenes) do not always undergo straightforward 1,2-addition reactions. When the electrophile attacks the conjugated π system, it bonds to one of the *terminal* atoms in order to produce the more stable intermediate carbocation—the allylic one (eq. 13.25).

(13.25) $X^+ + CH_2{=}CHCH{=}CH_2 \rightarrow XCH_2\overset{\delta+}{C}H{\cdots}CH{\cdots}\overset{\delta+}{C}H_2$ (More stable)

$X^+ + CH_2{=}CHCH{=}CH_2 \not\rightarrow CH_2{=}CH\overset{X}{\overset{|}{C}}H{-}CH_2^+$ (Less stable)

Reaction of this delocalized ion with a nucleophile can lead to two different products (neglecting *cis–trans* isomerism), deriving from overall 1,2- or 1,4-addition (eq. 13.26).

(13.26) $XCH_2\overset{\delta+}{C}H{\cdots}CH{\cdots}\overset{\delta+}{C}H_2 + Y^- \rightarrow$

$$\underset{\text{1,2-Addition product}}{XCH_2\underset{Y}{\underset{|}{C}}HCH{=}CH_2} + \underset{\text{1,4-Addition product}}{XCH_2CH{=}CHCH_2Y}$$

For example, the slow addition of 1 mol of bromine to 1 mol of 1,3-butadiene, under conditions where the products do not isomerize, leads to roughly equal amounts of 3,4-dibromo-1-butene and (*E*)-1,4-dibromo-1-butene (eq. 13.27). (No *Z* isomer from 1,4-addition is observed.)

(13.27) $CH_2{=}CH{-}CH{=}CH_2 + Br_2 \xrightarrow{\text{hexane}}$

$$BrCH_2\underset{\displaystyle Br}{\underset{|}{C}}HCH{=}CH_2 \;(54\%) \quad + \quad \begin{matrix} BrCH_2 & & H \\ & C{=}C & \\ H & & CH_2Br \end{matrix} \;(46\%)$$

Additions are much faster with conjugated dienes than with simple alkenes, so it is possible to isolate the dibromides before they undergo further addition to give tetrabromides. The relative proportion of the products certainly does not reflect their relative thermodynamic stabilities, as shown when they are allowed to equilibrate (eq. 13.28).

(13.28)
$$\begin{matrix} BrCH_2 & & H \\ & C{=}C & \\ H & & CH_2Br \end{matrix} \;(90.5\%) \quad \underset{61^\circ}{\rightleftharpoons} \quad BrCH_2\underset{\displaystyle Br}{\underset{|}{C}}HCH{=}CH_2 \;(9.5\%)$$

The addition shown in Equation 13.27 is therefore another example of a kinetically controlled reaction. The major product is the one that is formed fastest, not the one that is most stable.

It may seem that a contradiction has been introduced here. The last section showed that conjugated dienes are more stable than nonconjugated dienes, but this section shows that conjugated dienes are more reactive. This can be explained if you remember that rates of reactions depend upon the energy of activation, that is, the *difference* in energy between the reactant and the transition state. As we just showed, the addition reactions of conjugated dienes yield a very stable allylic cation intermediate and hence only a small activation energy is needed to reach that intermediate.

Problems

13–14 The intermediate in bromine addition to 1,3-butadiene (eq. 13.27) is a carbocation rather than the bromonium ion characteristic of Br_2 addition to alkenes (sec. 10:4). Explain the difference in behavior of alkenes and conjugated dienes.

13–15 Isoprene adds one equivalent of HOCl under mild conditions to give **A** (eq. 8). **A** has the following nmr spectrum: δ 1.3 (singlet, 3H), δ 2.8 (singlet, 1H), δ 3.6 (singlet, 2H), δ 5 to 6.3 (multiplet, 3H). The peak at δ 2.8 disappears when the sample is shaken with D_2O.

$$CH_2{=}\overset{\displaystyle CH_3}{\overset{|}{C}}{-}CH{=}CH_2 + HOCl \xrightarrow[0^\circ]{\text{pH } 6} \mathbf{A}(C_5H_9ClO) \qquad (8)$$

a. Propose a structure for **A,** and assign the nmr peaks.
b. What is the electrophile that initially adds to the diene? Why did it attack where it did?

13:10 The Diels–Alder Reaction

A characteristic reaction of conjugated dienes is their addition to alkenes to produce cyclohexene derivatives. The simplest example is the formation of cyclohexene from 1,3-butadiene and ethylene (eq. 13.29). This type of reaction was discovered by the German chemists Otto Diels and Kurt Alder, who in 1950 shared a Nobel Prize for their work. It is generally known as the **Diels–Alder reaction.** The product is sometimes called a Diels–Alder *adduct*. The Diels–Alder reaction provides a prime route to substituted six-membered cyclic compounds.

(13.29)

New σ bond

heat

New σ bond

Equation 13.29 does not depict a particularly efficient reaction. The rate of the Diels–Alder reaction is increased considerably if the alkene (known as the *dienophile*) is substituted with electron-withdrawing groups. For example, Equation 13.30 shows the reaction of 1,3-butadiene with a good dienophile, maleic anhydride. (Carbonyl groups are electron-withdrawing.)

(13.30)

benzene, 100°

1,3-Butadiene (diene) + Maleic anhydride (dienophile) → (100%)

The reaction is made even more favorable if the diene has electron-donating groups. Thus, the reaction of maleic anhydride with 2,3-dimethyl-1,3-butadiene is five times faster than the reaction in Equation 13.30. Other good dienophiles are propenal (*acrolein*), again because of the electron-withdrawing carbonyl group (eq. 13.31), and tetracyanoethylene, $(NC)_2C{=}C(CN)_2$. The latter reacts instantaneously at 0° with many dienes.

(13.31)

Me, Me, CHO

dioxane, 30°

2,3-dimethyl-1,3-butadiene (diene) + *Acrolein* (dienophile) →

Me, Me, CHO

3,4-Dimethyl-3-cyclohexenecarboxaldehyde

Alkynes that are substituted with electron-withdrawing groups are also excellent dienophiles (eqs. 13.32 and 13.33).

(13.32) [butadiene + NC–C≡C–CN —(THF, 100°)→ 1,2-dicyanocyclohexa-1,4-diene] (94%)

(13.33) [cyclopentadiene + MeOOC–C≡C–COOMe → bicyclic adduct with two COOMe] (84%)

Ethylene, the simplest dienophile (eq. 13.29), is a poor one because it has no electron-withdrawing groups. It requires elevated temperatures to react and the yields are usually low.

Most evidence indicates that the reaction of a diene and a dienophile is *concerted,* with both new σ bonds being formed at the same time. The observation leading to this conclusion is that *steric relationships present in the dienophile are retained in the product* (Fig. 3–11). Thus, a *trans* disubstituted dienophile produces a *trans* disubstituted cyclohexene (eq. 13.34), while a *cis* disubstituted dienophile yields a *cis* disubstituted cyclohexene (eq. 13.35).

(13.34) [butadiene + *trans* MeOOC–CH=CH–COOMe —(toluene, 150°)→ *trans* cyclohexene-4,5-dicarboxylate]

trans Dienophile → *trans* Product

(13.35) [isoprene (Me) + *cis* NC–CH=CH–CN → *cis* 4-methylcyclohex-4-ene-1,2-dicarbonitrile]

cis Dienophile → *cis* Product

The stereochemistry of the diene also plays an important role in Diels–Alder reactions. An open-chain diene such as 1,3-butadiene undergoes fairly free rotation around the C-2, C-3 bond, producing an infinite number of conformers. Nonetheless, only two of these conformers are particularly stable because they are planar and therefore maximize the conjugation between the double bonds (sec. 13:7). They are referred to as the *s-cis* and *s-trans,*★ and are in rapid equilibrium with each other at room temperature (eq. 13.36).

★ The terms *s-cis* and *s-trans* stand for *pseudo-cis* and *pseudo-trans*. The word *pseudo* indicates that neither configuration is permanently locked in.

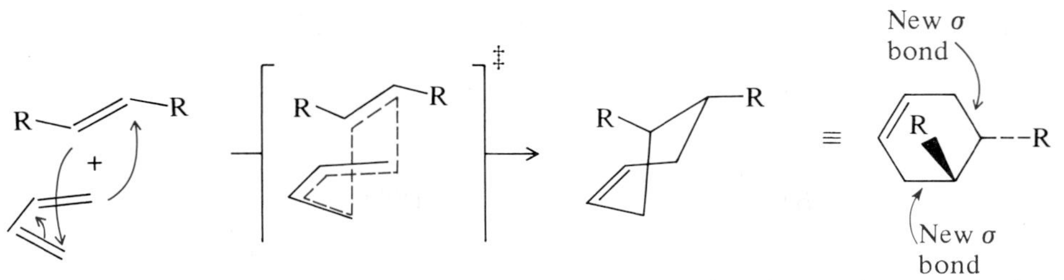

Figure 13–11. *Concerted Diels–Alder reaction between 1,3-butadiene and a* trans *disubstituted dienophile, showing the two reactants lying in parallel planes. At the transition state both new C—C σ bonds are formed simultaneously, so the two dienophile substituents remain* trans *in the product. (The arrows at the beginning are only used to keep track of the electrons.)*

(13.36)

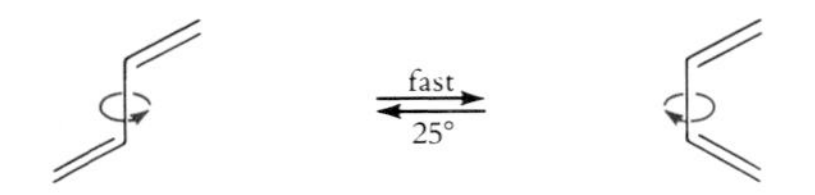

s-trans-1,3-Butadiene   *s-cis*-1,3-Butadiene

Based on our picture of the Diels–Alder reaction (Fig. 13–11), we would expect that only the *s-cis* conformation of the diene would be reactive. Here again the experimental evidence accords with this picture. Thus, a diene such as 3-methylenecyclohexene (**19**), which is *locked* in the *trans* configuration, does not undergo Diels–Alder reactions. By contrast, cyclopentadiene,

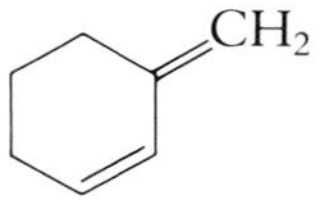

19

which is locked in a *cis* configuration, is a very reactive diene. It reacts with a number of dienophiles and is frequently used to synthesize bicyclo-[2.2.1]heptane (*norbornane*) derivatives, as shown in Equation 13.37. (Chlorinated norbornane derivatives are powerful, but environmentally persistent, pesticides. Two of these derivatives are named Dieldrin and Aldrin, after Diels and Alder.)

(13.37)

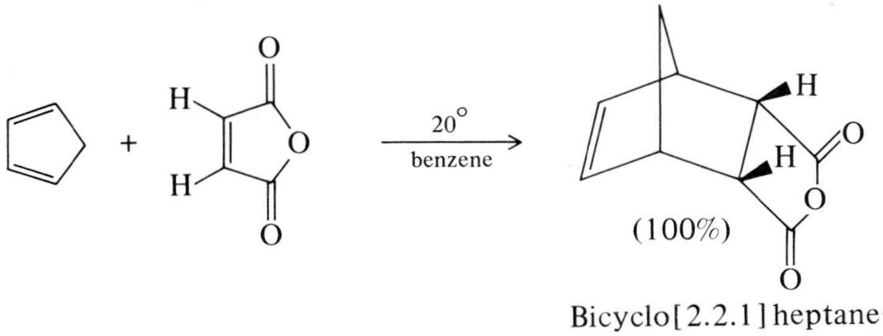

Bicyclo[2.2.1]heptane-2,3-dicarboxylic acid anhydride
(a norbornane derivative)

Cyclopentadiene is so reactive that it is not stable as a monomer at room temperature but exists as a Diels–Alder dimer (*dicyclopentadiene*). Cyclopentadiene is obtained by heating the dimer, which then undergoes a **retro Diels–Alder reaction** (eq. 13.38).

(13.38) Dicyclopentadiene $\xrightarrow{160^\circ}$ 2 cyclopentadiene

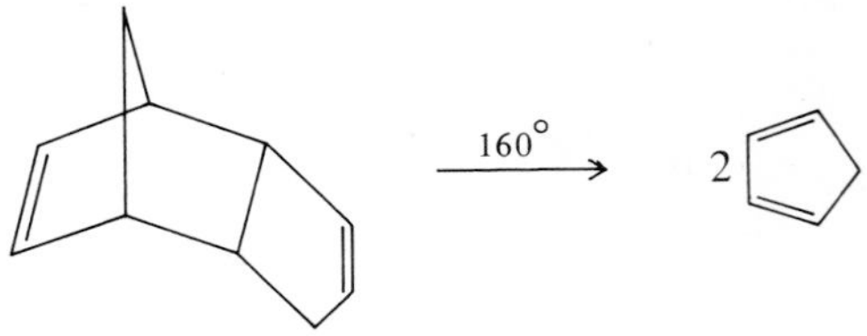

Dicyclopentadiene Collected at -78°

The Diels–Alder reaction is one of the most versatile routes to six-membered rings. Rings of other sizes can be prepared by analogous reactions involving unsaturated reactants. The success of these ring syntheses depends critically on reaction conditions—some are thermal processes, others require ultraviolet or visible light. To understand why the reaction conditions are so important, we need to examine the molecular orbital picture of alkenes.

Example

a. Give the structure of the product from the Diels–Alder reaction of 1,3-butadiene with vinyl chloride (chloroethene).
b. How would you prepare the following compound using a Diels–Alder reaction?

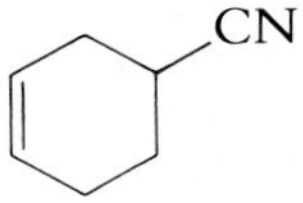

CN

Answer

a. 1,3-Butadiene is the diene reactant and vinyl chloride the dienophile.

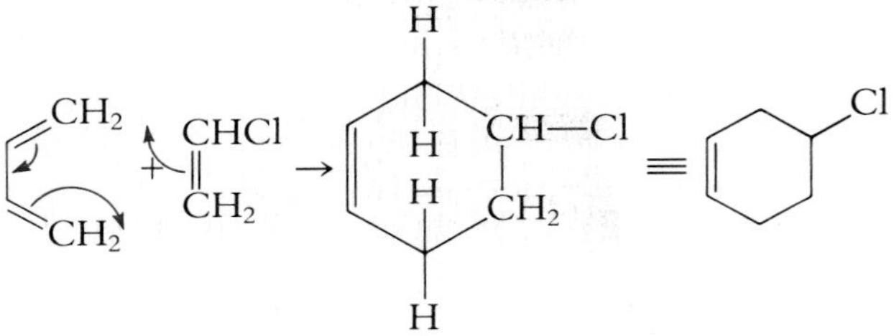

CH_2 CH_2 + CHCl CH_2 → H H H H CH—Cl CH_2 ≡ Cl

b. In the Diels–Alder adduct, carbons 1, 2, 3, and 6 came from the diene, carbons 4 and 5 from the dienophile. With this in mind, divide the molecule to show which bonds are formed in the Diels–Alder reaction.

6 CN 1 5 2 4 3

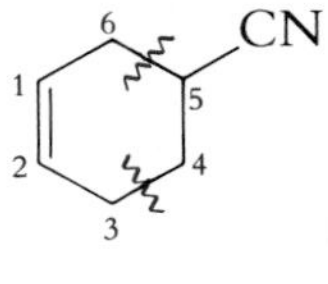

Now separate the diene and the dienophile.

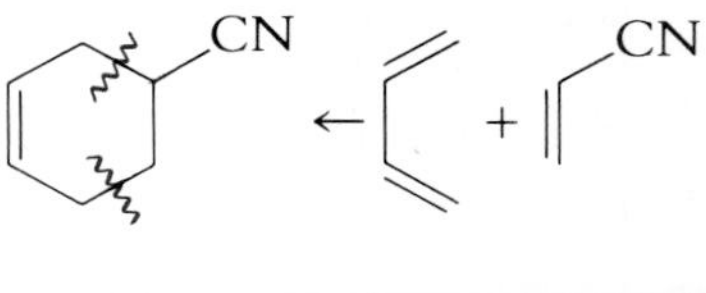

CN ← + CN

13–16 Give the structure of the product(s) from each of the following Diels–Alder reactions, and show the stereochemistry where appropriate.

a. 1,3-butadiene + (*Z*)-1,2-dichloroethene →

b. cyclopentadiene + propenal →

c. 2,3-dimethyl-1,3-butadiene + (*E*)-2-butenal →

d. 1,3-butadiene + EtOOC—C≡C—COOEt →

13–17 Show what diene and dienophile would be used to prepare the following compounds:

a. H Br

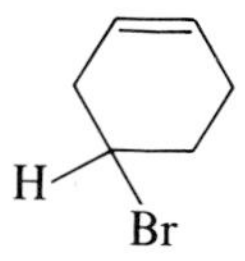

b. CH_3 C=O CH_3 CH_3

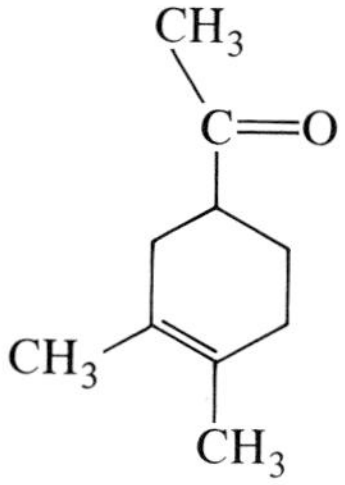

Box 13–2. Isoprene and Diene Polymers

Isoprene was the subject of intense investigations early in the history of organic chemistry. It was first prepared in 1826 by Michael Faraday who isolated a hydrocarbon, C_5H_{10}, from the dry distillation of rubber. Isoprene was also studied by A. Dumas and J. Dalton, so it had received attention from three "fathers" of chemistry who were better known for their contributions outside of organic chemistry. One of F. Wöhler's co-workers actually named the substance "faradin" but it was renamed isoprene in 1860.

Isoprene was manufactured in better yield in 1868 by passing turpentine over red-hot iron. In 1884 an even better procedure was discovered which involved taking only the turpentine fraction that contained C_{10} compounds and passing it over red-hot iron. The reverse reaction was also discovered: heating isoprene carefully allowed isolation of a C_{10} compound. Many years passed before it was determined that these represented Diels–Alder and retro Diels–Alder reactions:

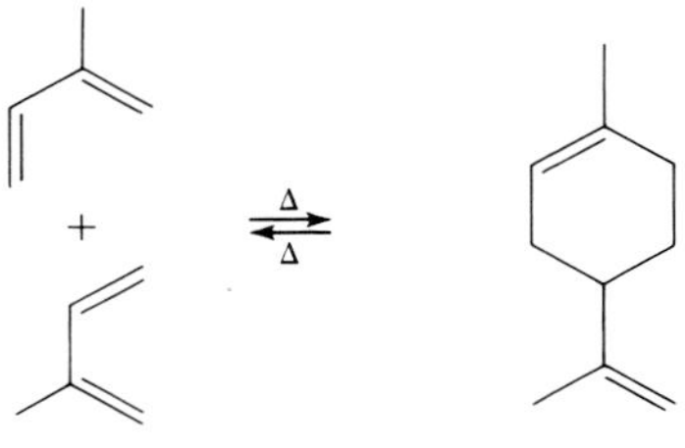

Two isoprenes ⇌ (Δ) (±)-Limonene (a turpentine constituent)

By the end of the nineteenth century, many chemists had turned their attention to natural rubber, which is simply *cis*-polyisoprene (Fig. 13–7). The first synthetic rubber material, called neoprene, was a polymer of 2-chloro-1,3-butadiene, whose structure replaces the CH_3 of isoprene with Cl. In spite of all efforts, and the synthesis of rubber materials from many different types of dienes, the polymerization of isoprene to give the natural rubber polymer remained elusive. Finally, in the mid-1950s, *cis*-polyisoprene was prepared in the laboratory. The key to its preparation was a complex aluminum and titanium metal catalyst developed by German chemist Karl Ziegler and Italian chemist Giulio Natta. Ziegler and Natta also developed catalysts for the polymerization of ethylene and propylene to give polyethylene and polypropylene; for this work they received the Nobel Prize in chemistry in 1963.

c. 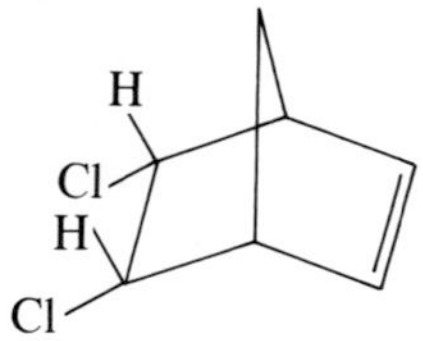 H, Cl, H, Cl

d. 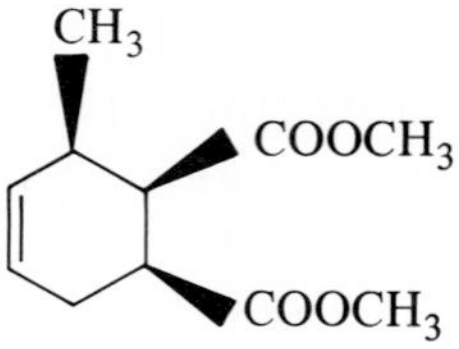 CH_3, $COOCH_3$, $COOCH_3$

13:11 Molecular π Orbitals: Their Nodes, Symmetries, and Energies

Many of the characteristic reactions of molecules containing π bonds are regiospecific and stereospecific. The specificity of the reactions makes them valuable because it allows us to predict and control the structure and configuration of the products. This section will explain why these reactions are so specific.

The explanation lies in the properties of the molecular orbitals most heavily involved in the reactions, the π orbitals. There are two orbital properties that are particularly important—the *number of nodes in an orbital,* and its *symmetry*. They are readily illustrated by reference to the molecular π orbitals of ethylene.

The two carbon 2*p* orbitals that give rise to the π orbitals may be in phase or out of phase when they combine (sec. 2:3), giving a bonding (π) orbital and an antibonding (π*) orbital, respectively (Fig. 13–12a). Both these orbitals have a planar node that coincides with the plane of the molecule. (The node is a region where the electron density goes to zero; Section 2:2). In addition, the π* orbital alone has another node, perpendicular to the molecular plane that bisects the π bond (Fig. 13–12a). The π orbital lacks any nodes perpendicular to the molecular plane. We see then that an increase in the number of nodes is associated with increasing orbital energy, since the π* orbital is higher in energy than the π orbital.

Next we have to consider the symmetry of these orbitals, which is defined relative to a plane that bisects the π bond (Fig. 13–12b). The π orbital is *symmetric*—the left half of the orbital is the mirror image of the right half. By contrast, the π* orbital is *antisymmetric*—every point in the left half of the orbital is equal to, but *opposite in sign,* to every point in the right half.

The ethylene example can be generalized as follows for any group of molecular π orbitals that are derived from a single set of equivalent atomic orbitals:

1. the number of nodes increases regularly with increasing energy of the molecular orbitals; and
2. all the molecular orbitals can be classified as symmetric or antisymmetric with relation to the same symmetry element.

These generalizations can be applied in a straightforward way to the π orbitals of conjugated systems. One of the simplest is the allyl radical, $CH_2{=}CHCH_2\cdot$ (sec. 13:2). The molecular orbitals are derived from three *p* orbitals, one from each sp^2-hybridized carbon atom. There must then be three π orbitals, since the number of molecular orbitals equals the number of atomic orbitals (sec. 2:3).

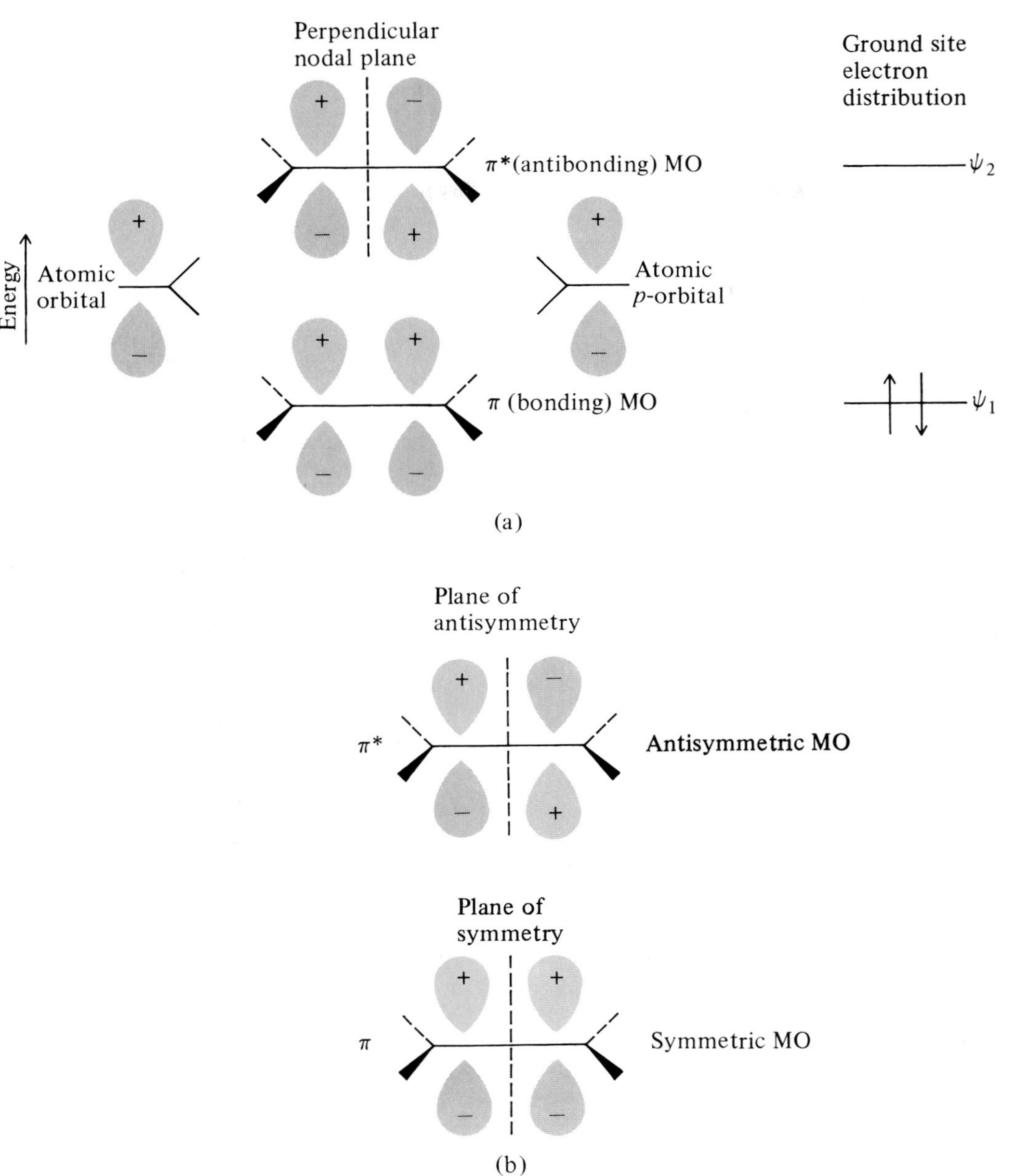

Figure 13–12. *(a) Formation of the bonding (π) and antibonding (π*) MOs of ethylene by overlap of two carbon 2*p *orbitals. In the ground state of ethylene, both π electrons are in the π orbital. (b) The symmetry of the π and π* orbitals with respect to a plane bisecting the π bond.*

Figure 13–13 shows the π orbitals of allyl radical, and shows how many electrons are in each orbital. The lowest energy π orbital, ψ_1, has no nodes perpendicular to the molecular plane, so the electrons in it are delocalized over all three carbon atoms. It is symmetric with respect to a plane through C-2 that cuts the molecule in half. The next higher orbital is antisymmetric,

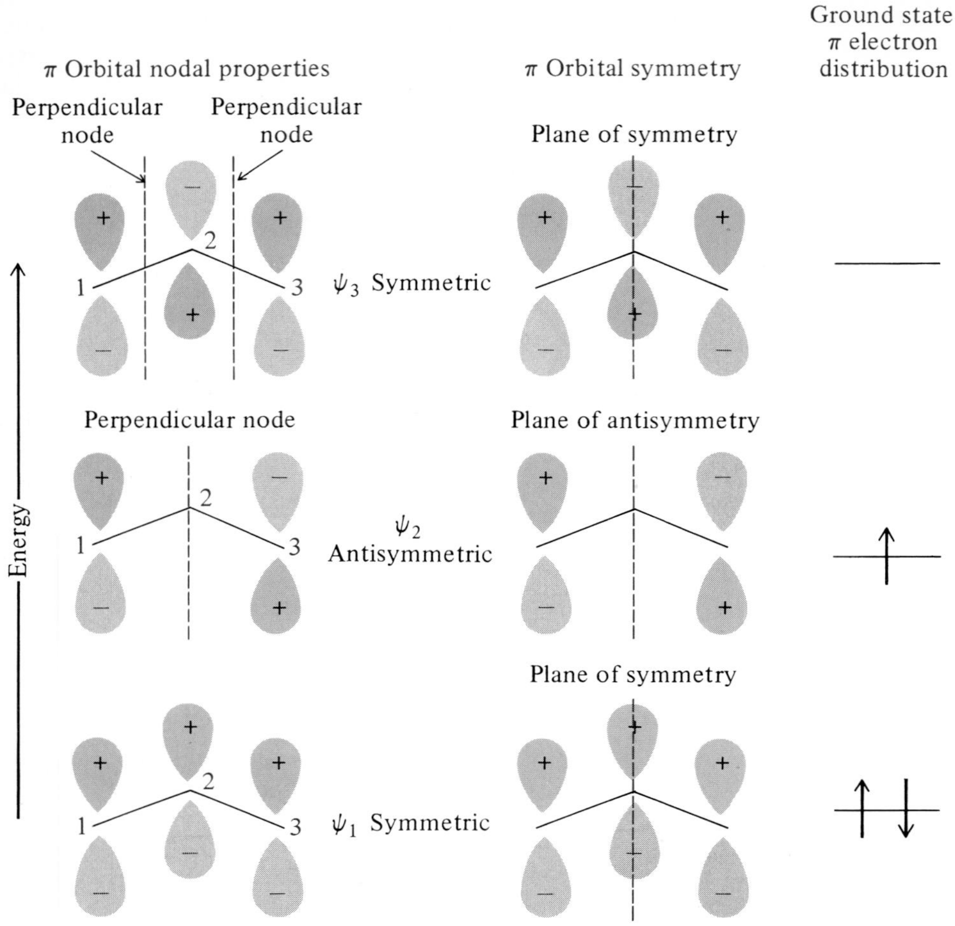

Figure 13–13. *π-orbital energies, nodal properties and symmetries, and ground state electron distribution for allyl radical.*

and contains one node perpendicular to the molecular plane. Note that the node passes through C-2, so that there is zero electron density in ψ_2 at this carbon atom. While this situation may at first seem strange, upon reflection you can see that the molecular orbital picture coincides with that given by resonance structures (sec. 13:2). They both show that the unpaired electron in ψ_2 divides its time between C-1 and C-3. Finally, Figure 13–13 shows that the highest energy π orbital, ψ_3, has two nodes through the molecular plane and is symmetric.

This pattern of a regular increase in the number of nodes with a corresponding increase in energy is further evident in the four π orbitals of 1,3-butadiene (Fig. 13–14). Each orbital beyond the lowest has one more node than the one below it, and the orbitals are alternately symmetric and antisymmetric. Figure 13–14 introduces some new terminology that will be very useful to us. Note that for ground state butadiene ψ_2 is the highest energy orbital in which there are electrons. Accordingly, ψ_2 is referred to as

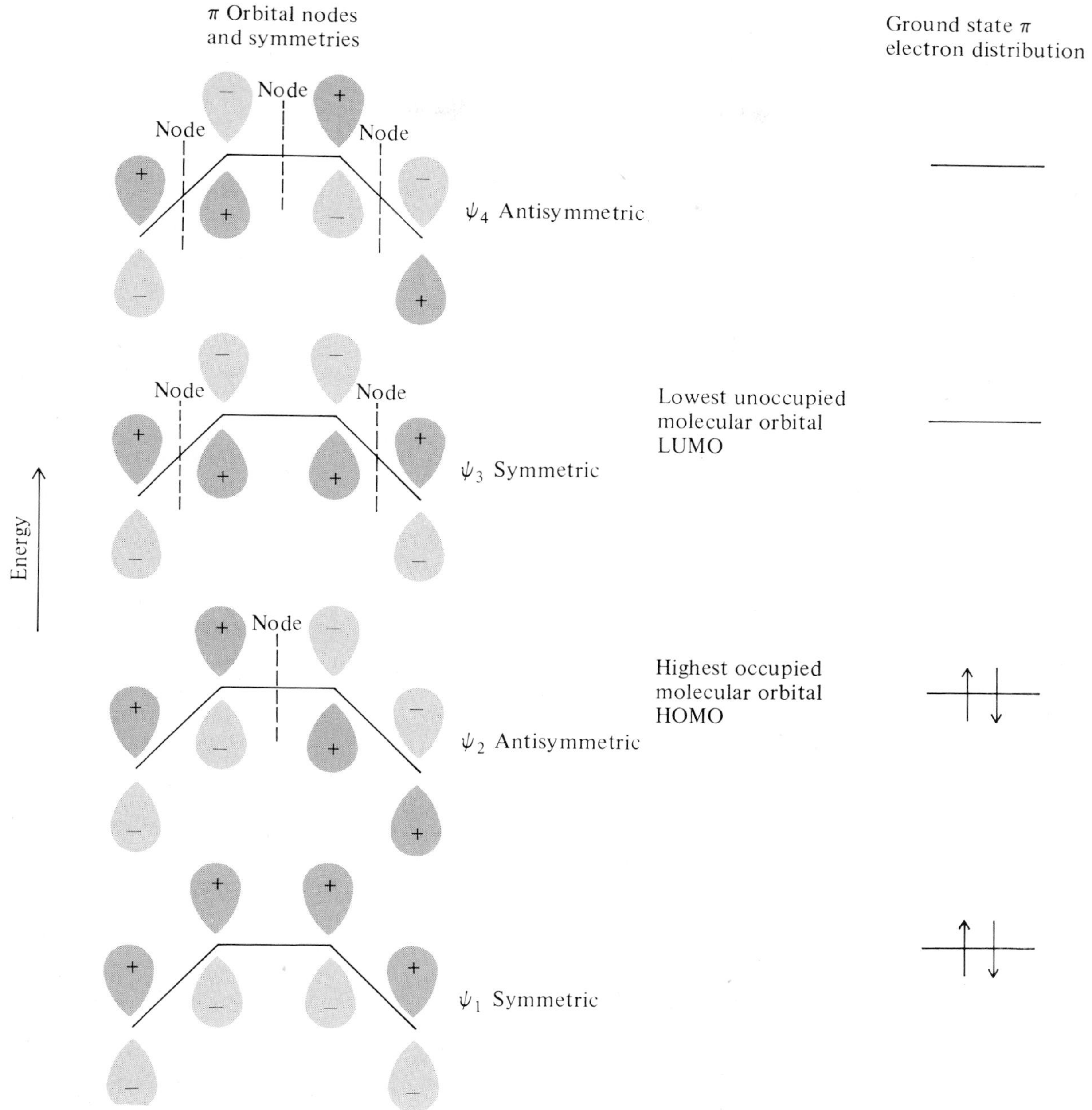

π Orbital nodes and symmetries

Ground state π electron distribution

Node

Node

Node

ψ_4 Antisymmetric

Node

Node

Lowest unoccupied molecular orbital LUMO

ψ_3 Symmetric

Energy

Node

Highest occupied molecular orbital HOMO

ψ_2 Antisymmetric

ψ_1 Symmetric

Figure 13–14. *π-orbital energies, nodal properties and symmetries, and ground state π electron distribution for 1,3-butadiene. In the ground state ψ_2 is the HOMO and ψ_3 the LUMO.*

the *highest occupied molecular orbital* (HOMO). Above it is ψ_3, the lowest energy orbital in which there are no electrons. It is labeled the *lowest unoccupied molecular orbital* (LUMO).

The regio- and stereospecific reactions of unsaturated compounds can be explained by the symmetry properties of their HOMO's and LUMO's. It is because of this important rule that we have taken the trouble to look in detail at the structures of π orbitals. The next section illustrates the operation of the rule.

Problem

13–18 Using the pattern we've found in the π orbitals of ethylene and butadiene, draw the π orbitals of 1,3,5-hexatriene. Arrange them in order of increasing energy, and label them symmetric or antisymmetric with respect to an appropriate symmetry element.

13:12 Concerted Thermal and Photochemical Reactions of Alkenes: The Woodward–Hoffmann Rules

Now that the necessary background on alkene MOs has been presented, we can consider the Diels–Alder reaction in more detail. The Diels–Alder reaction is an example of a **cycloaddition reaction.** In concerted cycloaddition reactions, an unsaturated molecule adds simultaneously to the ends of a π system, generating a new ring. In the Diels–Alder reaction, the four π electrons of the diene (butadiene) combine with the two π electrons of the dienophile (ethylene) (eq. 13.39), so it is termed a [4 + 2] cycloaddition. Since the reaction rate is increased by heat, it is a *thermal cycloaddition*.

(13.39) A thermal [4 + 2] cycloaddition

4π electrons 2π electrons

By contrast, two ethylenes *cannot* undergo concerted thermal cycloaddition. They can react, however, when irradiated with ultraviolet light. The result is formation of cyclobutane (eq. 13.40) in what is termed a [2 + 2] *photocycloaddition*. Incidentally, the [4 + 2] Diels–Alder reaction is *not* promoted by light, only heat.

(13.40) A [2 + 2] photocycloaddition

Cyclobutane

To help you understand these observations, let's consider the fact that *six* π electrons are involved in the reaction in Equation 13.39 (four from butadiene and two from ethylene), while *four* π electrons are involved in the reaction in Equation 13.40 (two from each ethylene). For these simple cycloaddition reactions, heat will catalyze concerted reactions with six π electrons, but not those involving four π electrons. The opposite is true for light-catalyzed reactions.* This remarkable specificity of cycloaddition reac-

* Recall from Chapter 6 that in a concerted reaction all reacting bonds are made and broken simultaneously, and no intermediate is formed. Concerted reactions tend to have low $\Delta H^{\ddagger}$s because the energetic cost of bond-breaking is considerably lowered by the simultaneous energetic gain from bond-making.

tions was explained by the American chemists R. B. Woodward and R. Hoffmann and the Japanese chemist K. Fukui, who established what are now known as the *Woodward–Hoffmann rules.*★

This approach involves looking at how the phases of the molecular orbitals of the reactants overlap as they are coming together to react. The theory is based upon the principle of **conservation of orbital symmetry.** If two compounds are to react together, the orbitals that overlap as the reaction proceeds must be in phase (have the same sign). If two reactant orbitals come together and both are occupied by electrons, they cannot overlap to form a new product orbital, since it would contain four electrons. If both orbitals are empty, no bonding can occur. Therefore, *one orbital must be occupied and the other must be empty for a reaction to take place.* The filled orbital is always the HOMO (highest occupied MO) of one reactant, and the empty orbital is always the LUMO (lowest unoccupied MO) of the other reactant. Now let's examine the orbitals of butadiene and ethylene and see how these molecules might react together.

Figure 13–15 shows that the HOMO of butadiene (ψ_2) and the LUMO of ethylene ($\pi^\star$) are both antisymmetric. Thus, these two orbitals are in phase and can overlap *at both ends,* resulting in concerted bond formation. Cyclohexene is the resultant product. This is called a **symmetry-allowed reaction.**

Now let's see what would happen if two ethylenes come together in the same manner. The HOMO (π) of one ethylene molecule would need to

★ Hoffmann shared the 1981 Nobel Prize in chemistry with Fukui for these and other theoretical studies.

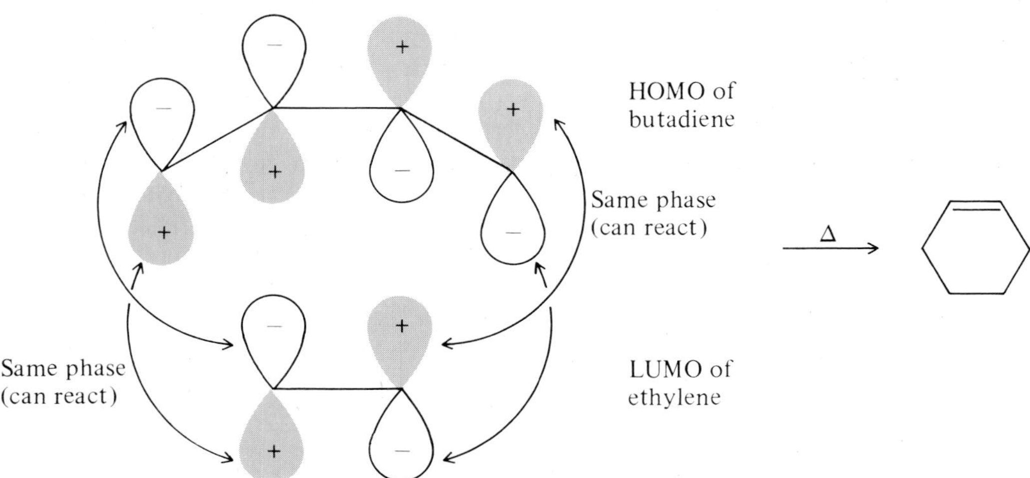

Figure 13–15. *The thermal reaction of butadiene with ethylene is symmetry allowed, since the HOMO of the first and the LUMO of the second have the same symmetry.*

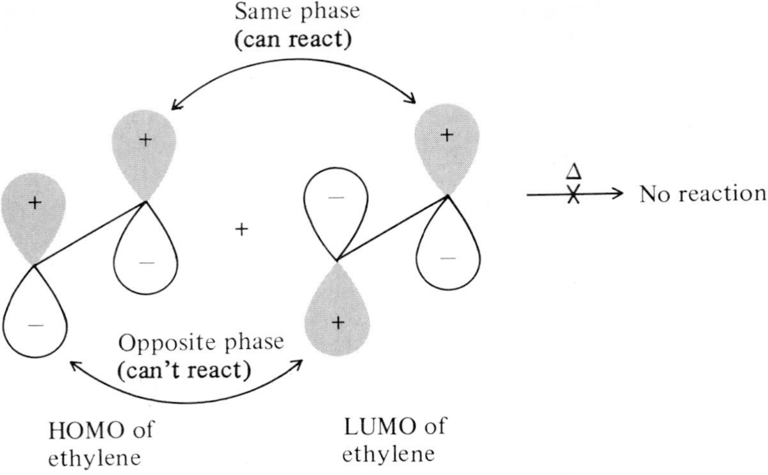

Figure 13–16. *The thermal reaction of two molecules of ethylene is forbidden by orbital symmetry, since HOMO and LUMO have different symmetries.*

react with the LUMO ($\pi^\star$) of another (Fig. 13–16). Since these orbitals are of different symmetries, they are in phase at only one end. Therefore, concerted bond formation cannot take place. This reaction is **symmetry forbidden.**

In order to complete the picture, we need to see why two ethylenes *do* undergo a cycloaddition when irradiated with ultraviolet light. Absorption of light by a molecule of ethylene will raise it to an excited state in which one electron is promoted to a $\pi^\star$ orbital (Fig. 13–17). Because there is now an electron in the $\pi^\star$ orbital, that orbital becomes the HOMO. If an excited state ethylene molecule now approaches a ground state ethylene, the HOMO of the excited state has the same symmetry as the LUMO of the ground state (Fig. 13–18). Concerted bond formation can now take place.

Another way of summarizing these reactions is to say that [4 + 2] cycloadditions are thermally allowed, but photochemically forbidden, while [2 + 2] cycloadditions are thermally forbidden, but photochemically allowed.

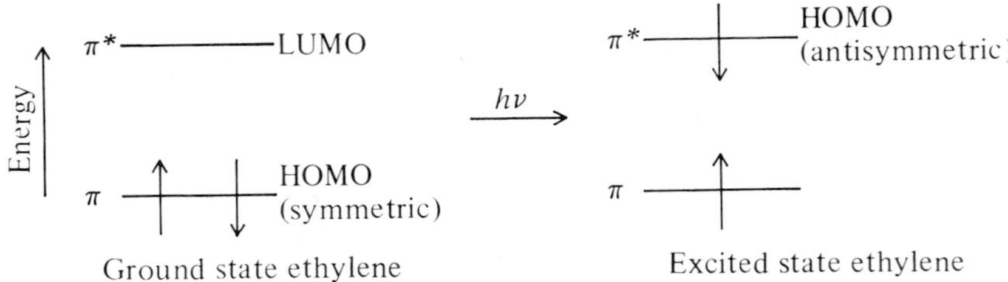

Figure 13–17. *Ultraviolet irradiation converts ethylene to an excited state, by promoting an electron from the π-orbital to the $\pi^\star$-orbital.*

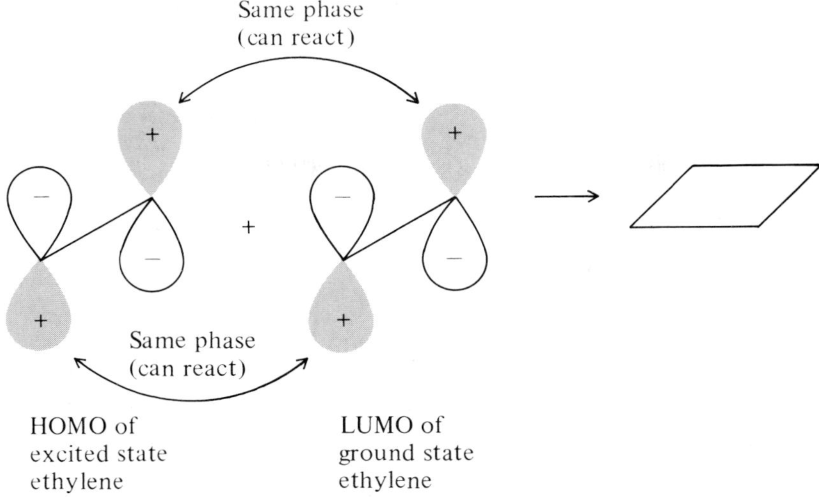

Figure 13–18. *Excited state ethylene (see Fig. 13–17) has the proper symmetry to react with ground state ethylene.*

Example

Why is the dimerization of butadiene to yield 1,5-cyclooctadiene thermally forbidden?

Answer

The HOMO of one butadiene is antisymmetric, while the LUMO of the other butadiene is symmetric. Figure 13–14 shows these orbitals, which would interact as follows:

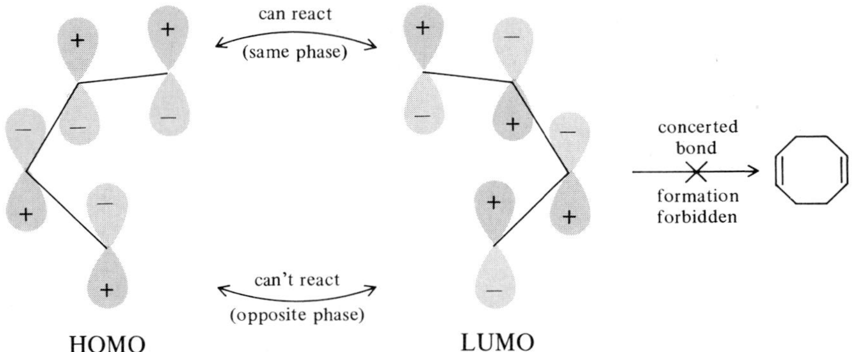

Problem

13–19 In Figure 13–15 we showed that the HOMO of butadiene has the proper symmetry to react thermally with the LUMO of ethylene. What is the case when the LUMO of butadiene reacts with the HOMO of ethylene?

Cycloadditions are biologically, as well as chemically, important. Ultraviolet light-induced mutations can be caused by a [2 + 2] photochemically allowed cycloaddition reaction between two thymine portions of the DNA in irradiated bacteria (eq. 13.41).

(13.41)

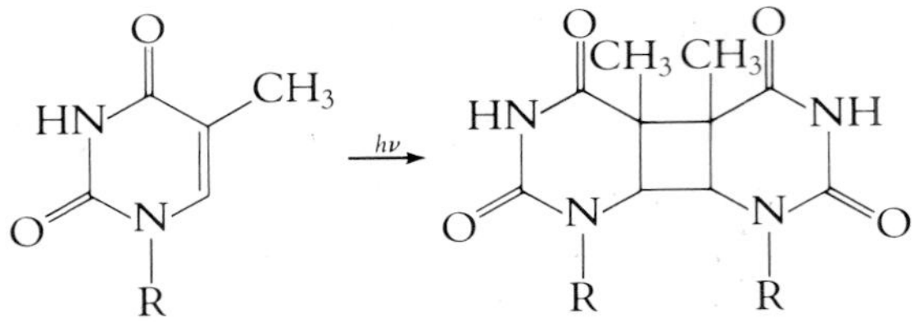

Thymine $\xrightarrow{h\nu}$ Thymine dimer

Other types of concerted reactions of alkenes can also be explained by the Woodward–Hoffmann rules. For example, Equation 13.42 shows an **electrocyclic reaction,** while Equation 13.43 gives an example of a **sigmatropic rearrangement.** An electrocyclic reaction is an *intramolecular* version of a cycloaddition. In a sigmatropic rearrangement, one σ bond is broken and another is formed *in addition to* π bond changes.

(13.42)

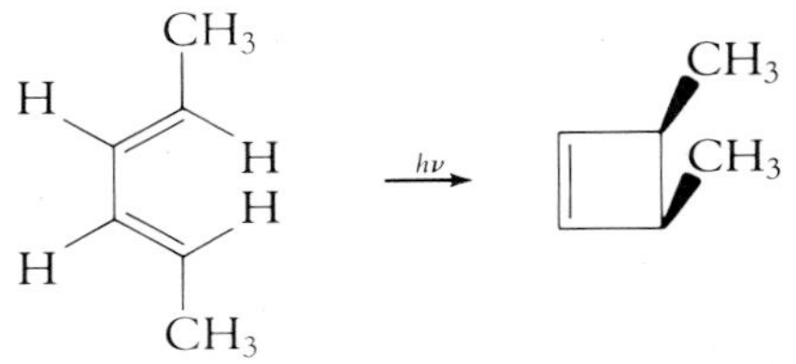

(*E,E*)-2,4-Hexadiene $\xrightarrow{h\nu}$ *cis*-3,4-Dimethylcyclobutene

An electrocyclic reaction

(13.43)

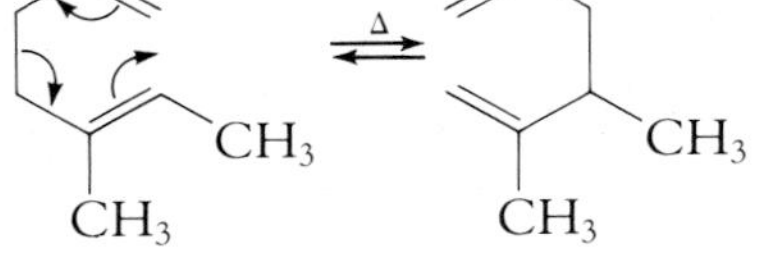

A sigmatropic rearrangement

The human body uses ultraviolet light to convert endogenous steroids to vitamin D compounds by way of both an electrocyclic ring opening and a sigmatropic rearrangement (eq. 13.44).

(13.44)

13:13 Ultraviolet-Visible Spectroscopy

All organic molecules can absorb electromagnetic radiation in the range of ultraviolet or visible light (Table 8–3, p. 276). The absorbed energy can promote an electron in a σ- or π-bonding orbital, or in a nonbonding orbital, to an empty antibonding orbital. This process is depicted in Figure 13–19 for ethylene and 1,3-butadiene. Electrons in σ bonds of these molecules could also absorb light energy, but the amount of energy needed is much greater than that needed for promotion of π electrons. Excitation of σ-bond electrons does not usually lead to useful spectroscopy.

The bonding and antibonding orbitals of ethylene are farther apart than the highest occupied bonding orbital (HOMO, ψ_2) and lowest antibonding orbital (LUMO, ψ_3) of butadiene. This difference in energy absorption can be detected with an *ultraviolet-visible* (uv-vis) *spectrometer*. The wavelength of light that contains excactly enough energy to promote the ethylene π electron is 165 nm, while that for 1,3-butadiene is 217 nm. (Remember that energy is inversely proportional to wavelength.) Most uv-visible spectrometers do not operate at wavelengths much lower than 200 nm, so uv-visible spectroscopy has its greatest importance for the study of molecules with conjugated π bonds such as 1,3-butadiene.

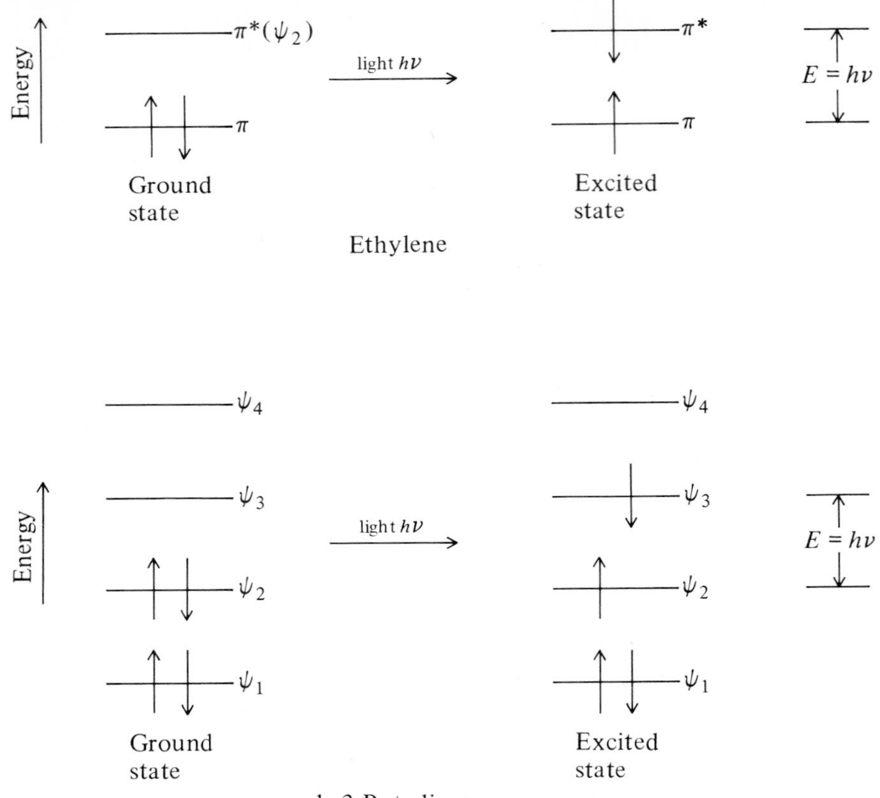

1, 3-Butadiene

Figure 13–19. *Orbital depiction of light absorption by ethylene and 1,3-butadiene.*

Now let's examine some actual spectra. Figure 13–20 shows the uv spectra of 1,3-butadiene and 2-butenal (crotonaldehyde, $CH_3CH{=}CH{-}CH{=}O$), and Figure 13–21 shows the uv-visible spectrum of a highly conjugated polyene, β-carotene. Examine the bottom scale of these graphs. This gives the wavelength of light, in nanometers, passing through the cell of the spectrophotometer. (The simple instrument diagram of Figure 8–9, p. 281, can serve for uv as well as ir spectroscopy. In this case, the light source emits uv-visible light, rather than ir light.) The peaks in Figure 13–20 tell us that butadiene and crotonaldehyde absorb light mainly at 217 and 218 nm, respectively, while β-carotene (Fig. 13–21) has a series of peaks, with the most intense at about 450 nm. Both butadiene and crotonaldehyde appear colorless to the eye, since they absorb light only in the uv region. β-Carotene, on the other hand, absorbs light in the visible region and appears

β-Carotene

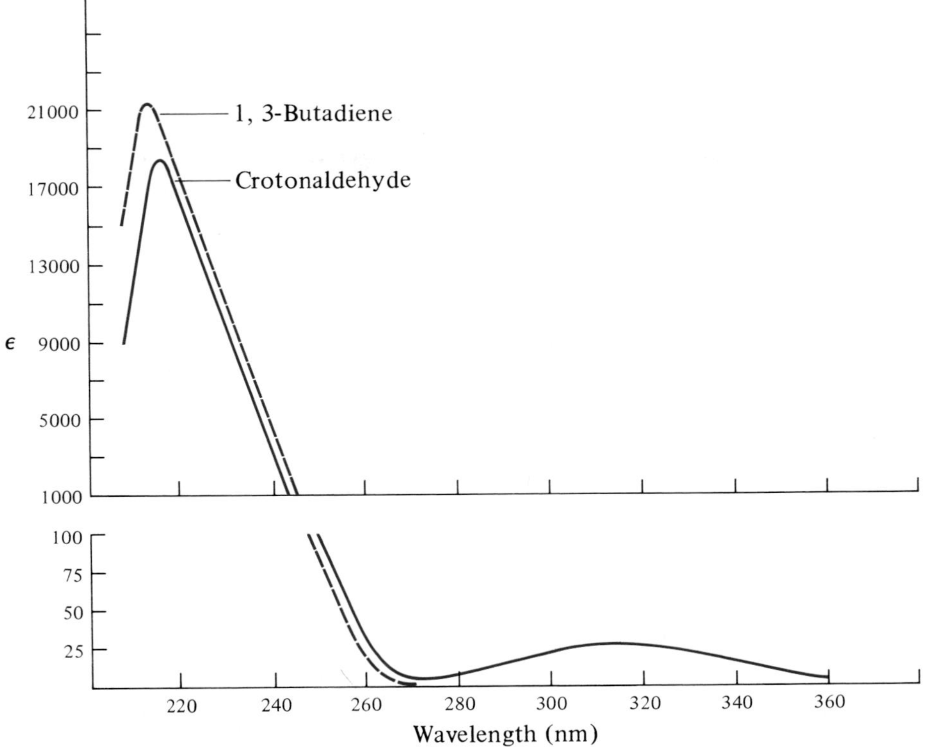

Figure 13–20. *Uv spectra of 1,3-butadiene (in hexane) and crotonaldehyde (in ethanol).*

orange. (It is the main pigment in carrots and is a precursor to vitamin A.) The wavelength(s) of the major peak (or peaks) in a spectrum is called λ_{max} (lambda max). For butadiene, λ_{max} is at 217 nm, while for β-carotene, the peaks occur at 420, 450, and 480 nm. Absorptions are also characterized by their intensity. The unit used is the **molar absorptivity,** ε, or a (sometimes

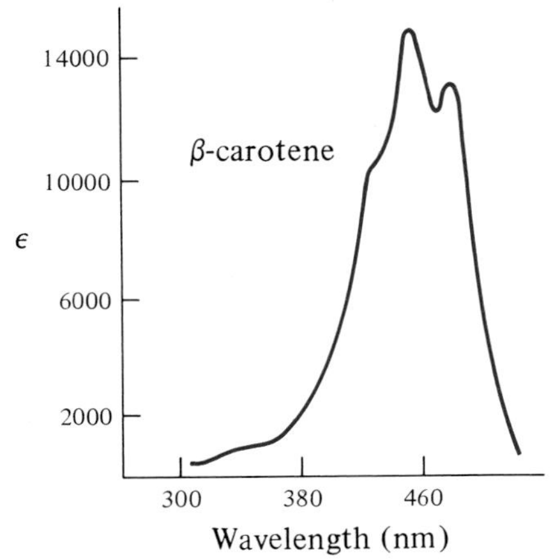

Figure 13–21. *The uv-visible spectrum of β-carotene.*

called the **molar extinction coefficient**). Section 13:14 will show how ε is calculated.

Table 13–1 gives the main absorptions for some conjugated compounds (along with the nonconjugated 1,4-pentadiene and acetone for comparison). Note that in general, as conjugation increases, so does λ_{max} and ε_{max} (ε_{max} is the value of ε at a wavelength that corresponds to λ_{max}). One exception is benzene, which shows an absorption peak similar to hexatriene, even though the intensity of the peak is much lower (low ε_{max}). The reasons for this are somewhat complex, but are partly due to poor overlap between the bonding π orbital and the antibonding orbital to which the promoted electron must go. Such absorptions are called forbidden transitions and always show weak intensities.

Another forbidden transition is the 320 nm absorption for crotonaldehyde (Fig. 13–20), which has an ε_{max} of only 30. This is typical of many compounds with carbonyl groups. This peak in the spectrum is not due to the excitation of a π electron, but rather to the excitation of a nonbonded electron on the oxygen to an antibonding π orbital (note the lack of overlap between these orbitals). Nonbonded electrons on carbonyl groups require less excitation energy than π electrons, and hence their absorptions appear at longer wavelengths than absorptions due to π electrons. The process of exciting π electrons is often called a $\pi \rightarrow \pi^\star$ transition, and the process of exciting nonbonded electrons is called an $n \rightarrow \pi^\star$ transition. The particular

Table 13–1. UV-Visible Spectral Absorptions of Some Compounds with π Bonds (nonpolar solvents)

Structure	λ_{max}	ε_{max}
$CH_2{=}CH{-}CH_2{-}CH{=}CH_2$ (1,4-pentadiene)	170	15,000
$CH_2{=}CH{-}CH{=}CH_2$ (1,3-butadiene)	217	21,000
$(CH_3)_2C{=}O$ (acetone)	188; 279	900; 13
$CH_3CH{=}CH{-}CH{=}O$ (crotonaldehyde)	214; 329	16,000; 25
$CH_2{=}CH{-}CH{=}CH{-}CH{=}CH_2$ (1,3,5-hexatriene)	263	52,500
 (benzene)	254	215
β-carotene (Provitamin A)	420; 450; 480	100,000; 150,000; 122,000

portion of a molecule that gives rise to a uv or visible absorption is called a **chromophore.** Thus, crotonaldehyde has two chromophores: the conjugated π system and the nonbonded electrons on the carbonyl group.

Structural changes other than conjugation can have effects on both λ_{max} and ε_{max}. Placing an alkyl substituent on a conjugated diene increases the wavelength of absorption about 5 nm. Two alkyl groups will increase λ_{max} about 10 nm, and so on.

$CH_2{=}CH{-}CH{=}CH_2$
1,3-Butadiene
(λ_{max} = 217 nm)

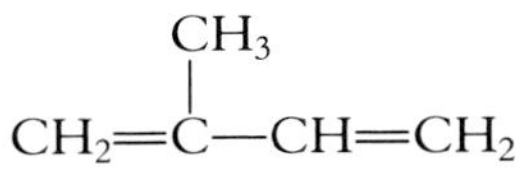

$CH_2{=}C(CH_3){-}CH{=}CH_2$
2-Methyl-1,3-butadiene
(λ_{max} = 222 nm)

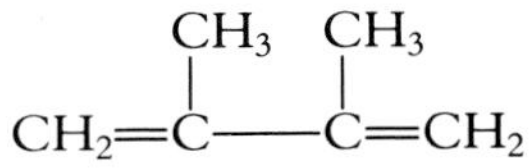

$CH_2{=}C(CH_3){-}C(CH_3){=}CH_2$
2,3-Dimethyl-1,3-butadiene
(λ_{max} = 227 nm)

When a double bond is in a ring (**20**), the same sort of increase in λ_{max} is observed.

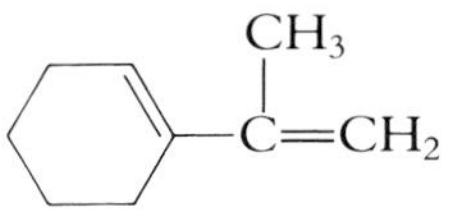

(cyclohexenyl)$-C(CH_3){=}CH_2$

20
(λ_{max} = 235 nm)

Problem

13–20 How many substituents are there on the diene system in compound **20**? By extrapolating the data from the other alkyl substituted butadienes, what λ_{max} would you predict for **20**?

Based on an extended set of correlations by R. B. Woodward, another American chemist, L. F. Fieser, proposed a series of rules for predicting λ_{max} for conjugated dienes and unsaturated carbonyl compounds. For example, the prediction that each alkyl substituent added to 1,3-butadiene will raise λ_{max} by 5 nm is an application of the Woodward–Fieser rules. These rules predict a 10 nm increase in λ_{max} for each alkyl substituent on an α,β-unsaturated carbonyl compound, such as crotonaldehyde. The rules can be very useful in deciding among different possible structures for an unknown compound.

Problems

13–21 Which of the compounds of Problem 13–11 would exhibit a uv absorption peak above 210 nm?

13–22 Show which orbitals of 1,3,5-hexatriene (Prob. 13–18) are involved in this compound's longest wavelength uv absorption at 258 nm. Give the π electron distribution before and after uv absorption.

13–23 The uv spectrum of an unknown compound is shown on the next page. It has the molecular formula C_6H_{10}. Draw the structures of at least two compounds that you would expect to produce a spectrum like this. Draw the

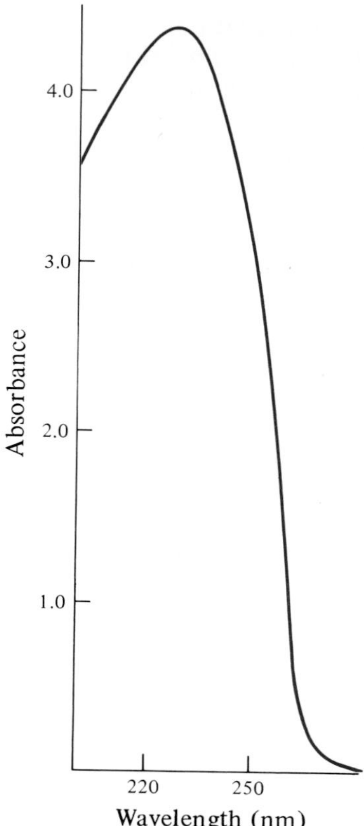

structures of several compounds not likely to give rise to such a uv spectrum. Are there any possible cyclic structures that could be responsible for this spectrum?

13:14 Quantitative Measurement of Absorption Intensities

We can calculate the intensity of a uv-visible absorption peak (ε_{max}) by taking the spectrum of a sample of known concentration. The number of photons absorbed by the sample will depend on the number of molecules in the light path. Photon absorption therefore depends on the sample concentration, on the thickness of the sample, and on the inherent absorbing power of the sample. Most spectrometers measure the amount of light absorbed by the sample and give a reading for absorbance (A) between 0 and 2. As mentioned in Section 13:13, we use a quantity called the molar absorptivity, ε, for making comparisons between different molecules. The molar absorptivity is calculated from Equation 13.45.

$$\varepsilon = \frac{A}{c \times l} \tag{13.45}$$

In this equation, c is the molar concentration of the compound being studied and l is the sample thickness (i.e., the cell length) in centimeters. Most uv-visible spectra of organic compounds are taken of a 10^{-2} to 10^{-6} molar solution in a 1 cm cell. Ethanol, methanol, and water are good solvents since they have no chromophores that absorb above 210 nm. For nonpolar solutes, saturated hydrocarbons are good solvents.

As an example, suppose we wanted to know the ε_{max} of the 2-cyclohexenone absorption at 225 nm. A trial spectrum would show that a few milligrams of it in 100 mL of ethanol gives a peak falling somewhere in the 0.5 to 2.0 range absorbance. We would then begin again by dissolving an exact amount, let's say 5.8 mg, in 1000 mL of ethanol. This is equivalent to 0.0058 grams/liter or 6.0×10^{-5} moles/liter, since the molecular weight of 2-cyclohexenone is 96. This solution would be put in a 1 cm quartz cell and the final spectrum taken. (Quartz is necessary since ordinary glass absorbs light below about 300 nm.) The absorbance on the spectrum would read 0.60 and the maximum would occur at 225 nm. The ε_{max} would then be calculated as follows (eq. 13.46):

$$\varepsilon_{225} = \frac{0.60}{(6.0 \times 10^{-5})(1)} = 10{,}000 \qquad (13.46)$$

Problem

13–24 Two compounds, **A** and **B,** are isomers of acetone and each has the molecular formula C_3H_6O. Sixteen milligrams of **A** is dissolved in 10 mL of ethanol and the solution is placed in a 1 cm cell. The uv spectrum is measured and shows a broad peak, $\lambda_{max} = 290$ nm, with absorbance 0.50. The spectrum of **B** shows no absorption in the uv-visible region. Calculate the molar absorptivity for **A.** Suggest structures for both compounds.

Summary

1 Carboxylate anions, which bear a negative charge on an oxygen next to a carbonyl group, are much more stable than alkoxide anions, where the oxygen-bearing negative charge is next to a saturated carbon. Allylic cations, anions, and radicals are notably more stable than their saturated counterparts. In each of these species, the charge (or unpaired electron) is delocalized by conjugation. Delocalized molecules and ions can be recognized by the fact that they require more than one Lewis structure to adequately describe their structural and chemical properties. Species that are represented this way are referred to as resonance hybrids (secs. 13:1 and 13:2).

2 Resonance structures in which all atoms have octets and in which the fewest charges are present contribute more to the resonance hybrid than structures in which an atom is lacking an octet or in which multiple charges occur. If charge does occur in a resonance structure, that structure which puts a negative charge on an electronegative atom contributes more to the hybrid (sec. 13:2).

3 Radical and cationic reactions take place readily at allylic positions. Allylic halides are unusually reactive because they form intermediate carbocations which are stabilized by resonance (secs. 13:3 and 13:4). Rearranged products can occur in both radical and ionic allylic substitution reactions, because the allyl intermediate has electron density or charge at more than one carbon.

The product obtained may be that which is formed fastest (kinetic control) or that which is more stable (thermodynamic control), depending upon reaction conditions (sec. 13:4).

Many natural products, including terpenes and steroids, contain a repeating branched-chain five-carbon unit. These compounds are derived from 2- and 3-isopentenyl pyrophosphate. Their biosynthesis involves intermediate allyl carbocations whose formation and subsequent reactions are under enzymatic control (sec. 13:6).

4 Conjugated dienes differ somewhat in their reactivity from nonconjugated dienes (alkenes with isolated double bonds). The heats of hydrogenation of conjugated dienes are lower than would be expected, meaning that they are more stable than expected (Fig. 13–10, sec. 13:8). This is due to the fact that they contain *p* orbitals on four or more adjacent carbons and hence have a delocalized molecular orbital of low energy (sec. 13:11). Similar stability is found in α,β-unsaturated carbonyl compounds. Ionic addition to conjugated dienes can give 1,2- or 1,4-addition products, depending upon whether the reaction is under kinetic or thermodynamic control. The intermediate is a stable allylic carbocation (sec. 13:9).

An important synthetic reaction of dienes with alkenes, used to prepare cyclohexenes, is the Diels–Alder reaction. Diels–Alder reactions are favored by electron-withdrawing substituents on the alkene (dienophile), and by electron-donating ones on the diene. Alkynes may serve as dienophiles. The two new σ bonds generated during the reaction are formed simultaneously, allowing steric relationships in the reactants to be retained in the product (Fig. 13–11, sec. 13:10).

5 In any molecule or ion containing π bonds, the number of π orbitals equals the number of atomic *p* orbitals used for π bonding. The π orbitals exhibit a regular increase in the number of nodes with increasing energy. All the π orbitals of a system may be classified as symmetric or antisymmetric with respect to a single symmetry element, usually a plane. The symmetry of two π orbitals in particular has a controlling effect on the reactivity of π systems—these are the highest occupied molecular orbital (HOMO) and the lowest unoccupied molecular orbital (LUMO; sec. 13:11).

In cycloaddition reactions, the LUMO of one component overlaps with the HOMO of the other. In order for the reaction to be concerted, these two orbitals must have the same symmetry so new σ bonds can be formed simultaneously by HOMO–LUMO overlap at both ends of the π system. By contrast, two ground state ethylenes cannot undergo concerted cycloaddition because the LUMO of one is antisymmetric, while the HOMO of the other is symmetric. However, when one ethylene is photoexcited, its new HOMO matches the other's LUMO in symmetry, and cyclobutane formation can occur concertedly. These cycloadditions are examples of reactions that obey the Woodward–Hoffmann principle of conservation of orbital symmetry (sec. 13:12).

6 Ultraviolet-visible spectroscopy provides a method for distinguishing between conjugated and nonconjugated π systems. The peaks in a uv-visible spectrum (λ_{max}) tell how much energy is needed to promote an electron from a bonding or nonbonding orbital to an unoccupied antibonding orbital. The more extensive the conjugation in a molecule, the greater is λ_{max} (sec. 13:13). The molar absorptivity (or molar extinction coefficient), ε, is a quantitative measure of the amount of light being absorbed at a given wavelength (sec. 13:14).

Problems

13–25 Pyrolysis of butane involves its initial dissociation to either two ethyl radicals or a propyl radical plus a methyl radical. What would you expect the predominant dissociation process(es) to be when 1-butene is pyrolyzed? Explain your answer.

13–26 Draw one or more additional resonance structures for each of the following ions:

a. $CH_3CH{=}CH{-}CH_2^+$

b. 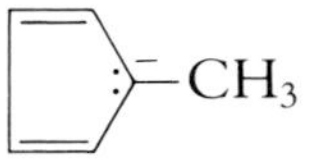 (cyclopentadienyl anion)$-CH_3$

c. 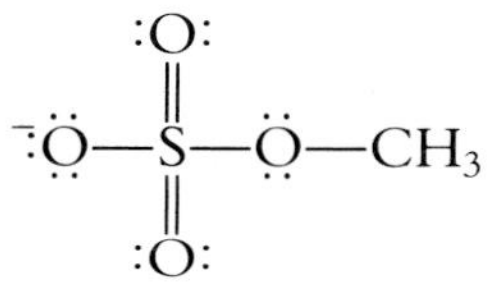 $^-\ddot{O}{-}S(=O)(=O){-}\ddot{O}{-}CH_3$ (with :O: above and below S)

d. $CH_2{=}CH{-}CH{=}CH{-}\ddot{O}:^-$

e. 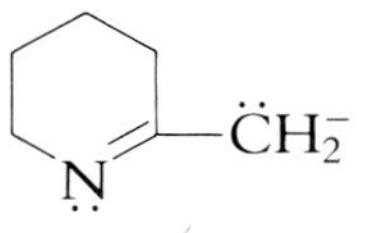 (tetrahydropyridine ring, N:)$-\ddot{C}H_2^-$

f. 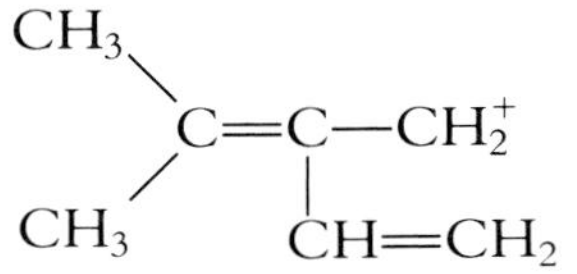 $(CH_3)_2C{=}C(CH{=}CH_2){-}CH_2^+$

13–27 Show the alcohol products that will be formed when each of the following cations reacts with water:

a. $^{14}CH_2{=}CH{-}CH_2^+$

b. 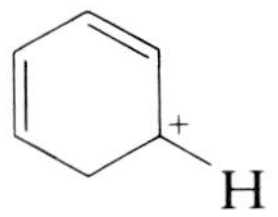 (cyclohexenyl cation, with + and H)

c. 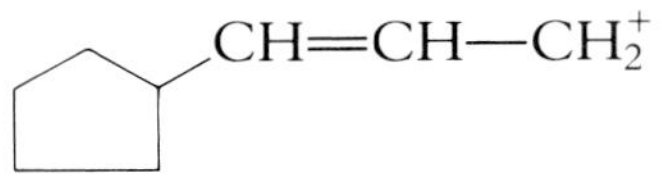 (cyclopentyl)$-CH{=}CH{-}CH_2^+$

d. $CH_2{=}CH{-}CH_2{-}\overset{+}{C}(CH_3)_2$

13–28 a. Addition of hydrogen chloride to vinyl chloride yields 1,1-dichloroethane as the predominant product. Account for this result with the aid of resonance structures.

b. Use resonance structures to predict the direction of addition of HCl to acrolein, $CH_2{=}CH{-}CHO$.

13–29 Explain each of the following observations, using appropriate resonance structures to illustrate your argument.

a. When 1,4-pentadiene is exposed to strong bases, it isomerizes to 1,3-pentadiene.

b. 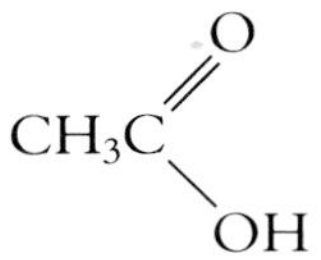 $CH_3C(=O)OH$ is a stronger acid than $CH_3C(=O)NH_2$

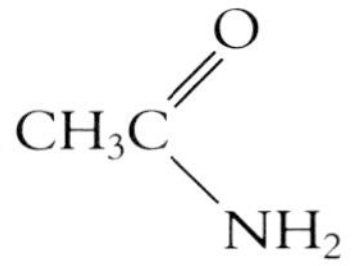

c. $CH_3OCH{=}CHCH_2Cl$ hydrolyzes much faster to the corresponding alcohols than $CH_3CH{=}CHCH_2Cl$ does.

13–30 Each of the following alkenes is allowed to react with one equivalent of *N*-bromosuccinimide to yield one or more bromoalkenes. Write the structures of all the products anticipated in each case.

a. cyclohexene
b. 3,3-dimethylcyclopentene
c. 2,3-dimethyl-2-butene
d. 3-methylcyclohexene

13–31 Resonance-stabilized intermediates play an important part in many of the carbonyl group reactions discussed in Chapter 11. We give some representative examples here.

a. Show the mechanism by which hemiacetals and hemiketals are rapidly converted to acetals and ketals (eq. 9; see sec. 11:8). Show the resonance structures for the intermediate.

$$R_2C(OH)(OR') \xrightarrow{R'OH,\ H^+} R_2C(OR')(OR') + H_2O \qquad (9)$$

b. In the Wolff–Kishner reduction of ketones (sec. 11:11), the hydrazone of the ketone is converted to a methylene group with loss of N_2. Part of this process is the rearrangement shown in Equation 10. Propose a mechanism for it, including the structure of the intermediate.

$$R_2C{=}NNH_2 \xrightarrow{OH^-} R_2CH{-}N{=}NH \qquad (10)$$

c. The formation of imines (sec. 11:7) requires the loss of water from the intermediate α-aminoalcohols (eq. 11). Give the mechanism for this reaction, and explain why it proceeds very rapidly.

$$R_2C(OH)(NHR') \xrightarrow{H^+} R_2C{=}NR' + H_2O \qquad (11)$$

13–32 Internal alkynes may be isomerized to terminal alkynes by strong bases such as $NaNH_2$ (eq. 12, step 1). The terminal alkynes are then deprotonated by one equivalent of the base (eq. 12, step 2), yielding the stable acetylide ion (sec. 12:3). Give a mechanism for the first step of this sequence.

$$R{-}C{\equiv}C{-}CH_3 \underset{①}{\overset{NH_2^-}{\rightleftharpoons}} RCH_2{-}C{\equiv}C{-}H \underset{②}{\overset{NH_2^-}{\longrightarrow}} RCH_2C{\equiv}C^- + NH_3 \qquad (12)$$

13–33 Write reaction sequences that would make possible the conversions shown below. Include all important reagents and conditions.

a. cyclohexanol → 3-chlorocyclohexene
b. acetone → 3-chloro-2-methylpropene
c. methyl isopropyl ketone → 1-bromo-2,3-dimethyl-2-butene
d. cyclopentylmethyl bromide → 1-(bromomethyl)cyclopentene

13–34 *N*-Bromosuccinimide reacts very slowly with 3,3-dimethyl-1-butene. How do you account for this observation?

13–35 a. 2-Methyl-2-buten-1-ol and 3-methyl-2-buten-1-ol are each converted to chlorides under ionizing conditions (concentrated HCl, heat). Which one should react more rapidly? Explain. Show the structures of the expected products.

b. 3-Chloro-1-butene and 3-chloro-2-methylpropene are each allowed to react with sodium iodide in acetone (S_N2 conditions). Which compound would be expected to be converted most rapidly to iodide? Explain.

13–36 Give structures for the products from the reaction of 1,3-butadiene and 1,3-cyclohexadiene with the following reagents:

a. 1 mol of Br_2
b. 2 mol of Br_2
c. 1 mol of HCl
d. 2 mol of HCl
e. 1 mol of H_2 with Pd catalyst
f. 1 mol of HBr with peroxides

13–37 a. When 2,3-dimethyl-1,3-butadiene is treated with a solution of chlorine in CCl_4, three products are formed: *trans*-1,4-dichloro-2,3-dimethyl-2-butene (**V**), 1-chloro-2,3-dimethyl-1,3-butadiene (**VI**), and 2-chloromethyl-3-methyl-1,3-butadiene (**VII**). Write a mechanism that accounts for the formation of all three compounds.

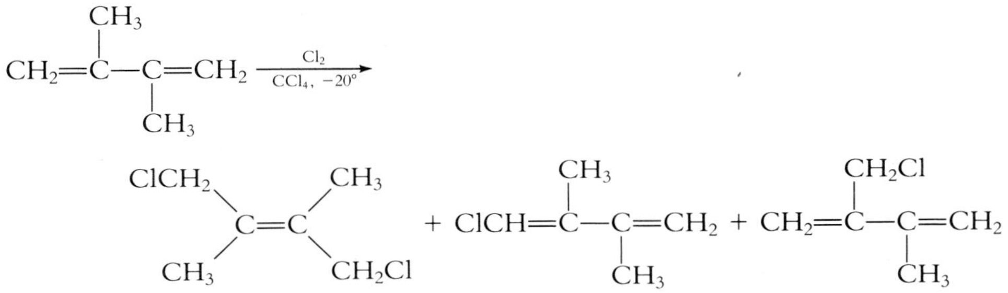

$$CH_2{=}C(CH_3){-}C(CH_3){=}CH_2 \xrightarrow[CCl_4,\ -20°]{Cl_2}$$

$$\underset{\mathbf{V}}{(ClCH_2)(CH_3)C{=}C(CH_3)(CH_2Cl)} + \underset{\mathbf{VI}}{ClCH{=}C(CH_3){-}C(CH_3){=}CH_2} + \underset{\mathbf{VII}}{CH_2{=}C(CH_2Cl){-}C(CH_3){=}CH_2}$$

b. The hydrogen chloride liberated during the formation of **VI** and **VII** reacts with the starting material. What products might be expected from this reaction?

13–38 Show clearly the isoprene units that make up each of the following compounds:

a. 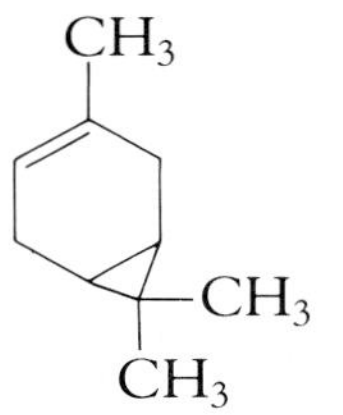 [structure: bicyclic compound with CH_3 on ring double bond carbon and a cyclopropane bearing two CH_3 groups]

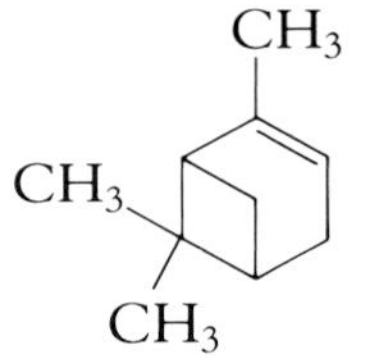

b. 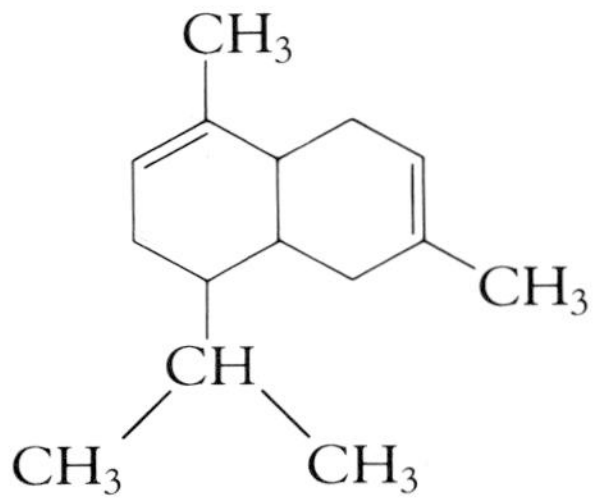 [structure: bicyclic compound with CH_3 groups and $CH(CH_3)_2$ substituent]

c. [structure: bicyclic compound with CH_3 on double bond carbon and bridge carbon bearing two CH_3 groups]

d. 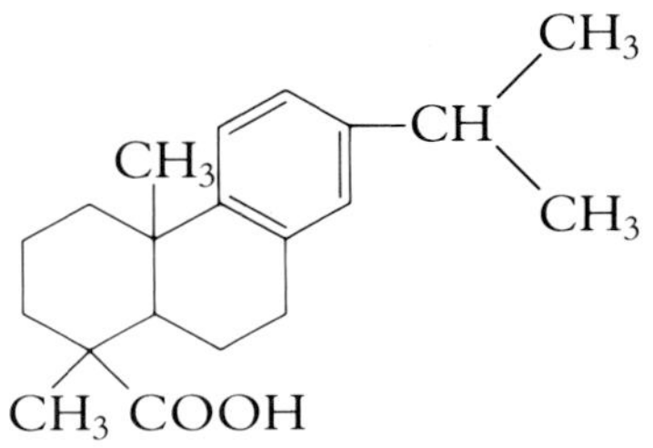 [structure: tricyclic compound with CH_3, $CH(CH_3)_2$ on aromatic ring, and CH_3, COOH substituents]

13–39 What isoprene units may have been used in the biosynthesis of eudesmol and guaiol? Why are these molecules called sesquiterpenes?

HO

OH

Eudesmol

Guaiol

13–40 1,3-Butadiene can dimerize to yield either 4-vinylcyclohexene, **VIII,** or 1,5-cyclooctadiene, **IX.** Under what conditions will each of these dimers be the predominant product?

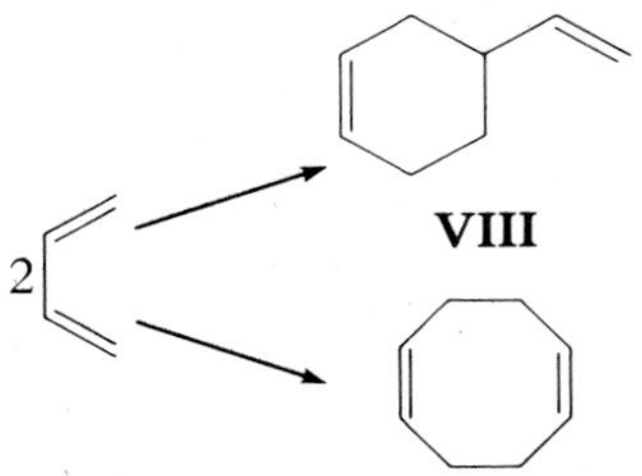

VIII

2

IX

13–41 Write the structures of the products from each of the following reactions. Be careful to show the stereochemistry, where appropriate.

a. 1,3-butadiene + 3-buten-2-one →
b. (*E,E*)-2,4-hexadiene + tetracyanoethylene →
c. 1,3-cyclohexadiene + maleic anhydride →
d. 2-methylbutadiene + propenal →
e. 5,5-dimethylcyclopentadiene + *trans*-1,2-dicyanoethylene →

13–42 The following cyclohexane derivatives may be synthesized by means of Diels–Alder reactions, followed by reduction of the double bond in the cyclohexene ring. Write all the possible Diels–Alder reactions that may lead to each product, and comment on their relative merits.

a. cyclohexanecarboxaldehyde
b. 3,4-dimethylcyclohexanecarboxaldehyde
c. *trans*-2-methylcyclohexyl cyanide

13–43 1,2-Dimethylenecycylobutane (**X**) reacts with one molecule of maleic anhydride at room temperature to yield a product, **A.** When **A** is heated to 150°, it isomerizes to **B,** which is capable of reacting with a second molecule of maleic anhydride to give **C.** Write structures for **A, B,** and **C.**

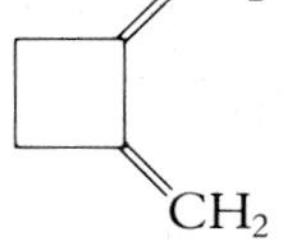

CH_2

CH_2

X

13–44 Two isomeric vinylcyclopentadienes, **A** and **B,** were obtained. One of them has structure **XI,** the other structure **XII. A** and **B** were separated and each allowed to react with the dienophile *N*-phenylmaleimide (**XIII**). Compound **A** reacted with only one equivalent of **XIII,** while **B** reacted successively with 2 mol of **XIII.** Assign structures to **A** and **B,** and show their reactions with **XIII.**

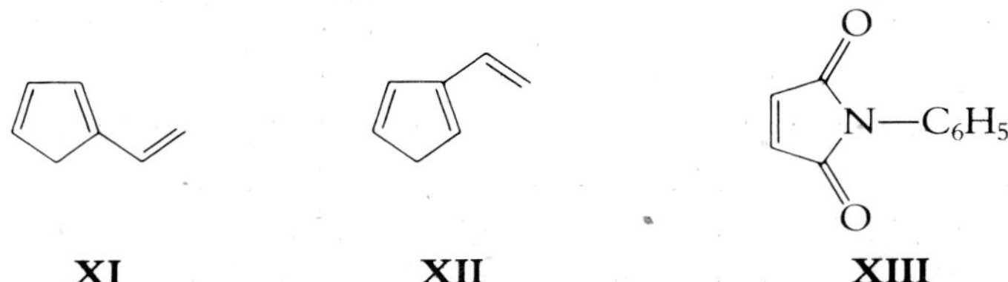

13–45 What is the difference in energy (in kilocalories per mole) needed for a $\pi \rightarrow \pi^\star$ electronic transition in $CH_2{=}CH{-}CH_2{-}CH{=}CH_2$ (1,4-pentadiene), as compared to $CH_2{=}CH{-}CH{=}CH_2$ (1,3-butadiene)? What difference in structural features between the two dienes accounts for the energy difference?

13–46 An unknown compound ($C_6H_{10}O$) shows λ_{max} at 237 and 315 nm. A solution containing 10.8 mg in 100 mL of ethanol gives an absorbance of 1.26 for the 237 nm peak and an absorbance of 0.078 for the 315 nm peak.

a. Calculate the molar absorptivity (molar extinction coefficient), ε_{max}, for each peak.

b. Give a structure for the unknown that fits the data (more than one may be possible). What structural features give rise to the two uv peaks?

13–47 An unknown compound, C_7H_{10}, shows a complex ^1H nmr absorption pattern in the 1.4 to 1.8 ppm region that integrates for six protons, and shows multiplets in the 5 to 6 ppm region integrating for four protons. The mass spectrum shows no fragmentation peaks corresponding to loss of 15 mass units from the molecular ion. In the ultraviolet absorption spectrum λ_{max} = 231 nm (ε_{max} = 21,000). Suggest a structure for the unknown, and interpret the spectral data.

Supplementary Reading

J. March, "Delocalized Chemical Bonding," pp. 32–37 in *Advanced Organic Chemistry,* 2nd ed., McGraw-Hill, New York, N.Y., 1977. This contains a detailed list of the so-called "Rules of Resonance."

F. A. Carey and R. J. Sundberg, "Kinetic Versus Thermodynamic Control," pp. 163–164 in *Advanced Organic Chemistry,* Part A, Plenum Press, New York, N.Y., 1977.

J. B. Hendrickson, *The Molecules of Nature,* W. A. Benjamin, Reading, Mass., 1973.

J. J. Vollmer and K. L. Servis, Woodward–Hoffmann rules: electrocyclic reactions, *J. Chem. Ed.,* **45,** 214 (1968).

R. B. Woodward and R. Hoffmann, *The Conservation of Orbital Symmetry,* Academic Press, New York, N.Y., 1970.

M. Orchin and H. H. Jaffe, *The Importance of Antibonding Orbitals,* Chap. 5, Houghton Mifflin, Boston, Mass., 1967.

R. M. Silverstein, G. C. Bassler, and T. C. Morrill, *Spectrometric Identification of Organic Compounds,* 4th ed., Chap. 6 (Ultraviolet Spectroscopy), J. Wiley and Sons, New York, N.Y., 1981.

14 Benzene and the Concept of Aromaticity

> Let us learn to dream, gentlemen, then perhaps we shall find the truth ... but let us beware of publishing our dreams before they have been put to the proof by the waking understanding.
>
> Friedrich August Kekulé

The dictionary defines aromatic as "sweet smelling." Since a number of the sweet smelling, naturally occurring compounds known in the nineteenth century were found to contain benzene rings, benzene and its derivatives were called "aromatic compounds." They also shared certain chemical properties that set them apart from aliphatic compounds, so the term aromatic eventually came to mean "resembling benzene in structure and in reactivity."

Although you will see that there are a variety of "nonbenzenoid" (not like benzene) aromatic compounds, the standard against which other aromatic compounds are judged is still benzene. Thus, it is appropriate for us to start this chapter by describing the discovery and structure determination of benzene.

14:1 History of the Benzene Problem

In 1823, Michael Faraday analyzed a sample of lighting fuel produced by the thermal cracking of whale oil. After repeated fractional distillation and fractional crystallization of this material, he obtained, among other products, a pure hydrocarbon that melted at 5.5°; he named it "bicarburet of hydrogen." Obtaining pure material was quite a feat for the time, since recent gas chromatographic studies show that the mixture from which the isolation was made contained literally hundreds of components. Undoubtedly, Faraday's isolation of "bicarburet of hydrogen" was made simpler by the fact that the melting point of the material was 5.5°, remarkably high for a hydrocarbon of low molecular weight.

Faraday reported the empirical formula of "bicarburet of hydrogen" as C_2H, based upon the then current belief that the atomic weight of carbon was six. Over the next four decades the concepts of atomic and molecular weights were more clearly defined, and the molecular formula of Faraday's compound was established as C_6H_6. The compound was renamed benzene by E. Mitscherlich, who showed that it could be obtained by pyrolyzing benzoic acid with calcium hydroxide. As chemical data about benzene and its derivatives began to accumulate, the stage was set for the appearance of Friedrich August Kekulé, one of the most creative thinkers in organic chemistry. Kekulé, who started out as a student of architecture, immersed himself in the study of some of the most important chemical questions of the day. He played a major part in establishing the tetravalency of carbon, and he was the first to hypothesize that carbon atoms could combine with one another to form chains. In Kekulé's time, most chemists viewed chemical formulas as symbols that depicted only atomic groupings that underwent transformation in the course of a reaction. These symbols were not thought to have any structural significance, that is, they did not represent the physical arrangement of the atoms in space.

Kekulé did not adhere to the traditional view. He believed that chemists *could* deduce, and symbolically represent, the structural features of molecules. Kekulé was not completely alone in his views but his papers were probably more influential than anyone else's in advancing these hypotheses.

During Kekulé's time, the concept of unsaturation was being formulated and the chemical properties of unsaturated compounds were first being identified. The molecular formula of benzene indicated that it should be highly unsaturated, yet its relative lack of reactivity contradicted this expectation. In the absence of light or catalysts, benzene reacts very slowly with bromine or chlorine and it is impervious to the usual oxidizing agents. Halogenation usually leads to *substitution* of a halogen atom for a hydrogen atom rather than the *addition* of a halogen molecule, which is so characteristic of unsaturated compounds such as alkenes and alkynes. In addition, these substitution reactions led to isomer patterns that were quite confusing.

Kekulé's solution was to postulate (in 1865) that the carbon atoms of benzene formed a ring, lying at the corners of a regular hexagon. He first proposed a regular plane hexagon as a symbol for benzene (**1**), but later added a system of conjugated double bonds to this formula (**2**) in order to account for the tetravalency of carbon.

H H H H H H

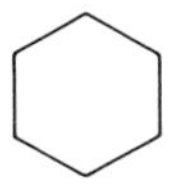

1 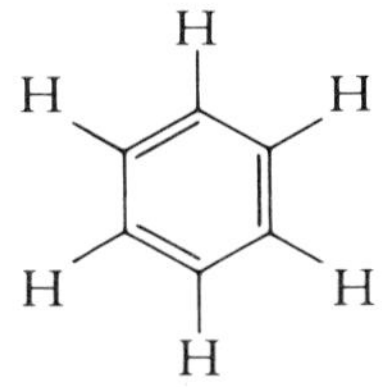 **2**

In an often repeated story, Kekulé related how the hexagon formula came to him one evening while he was dozing: chains of carbon atoms disported themselves like snakes before his eyes, until one snake seized its own tail with its mouth.

Structure **2** nicely accounts for the fact that for each monosubstituted benzene, C_6H_5X, only one isomer exists. It cannot, however, explain the number of known isomeric disubstituted benzenes. If **2** were really the formula for benzene, we would predict four possible disubstituted benzenes: two 1,2-disubstituted compounds (called **ortho** isomers, **3** and **4**); one 1,3-disubstituted compound (called the **meta** isomer, **5**); and one 1,4-disubstituted compound (called the **para** isomer, **6**). (The two postulated *ortho* isomers differ in the type of bond connecting the two substituted carbon atoms—in **3** it is a double bond, while in **4** it is a single bond.)

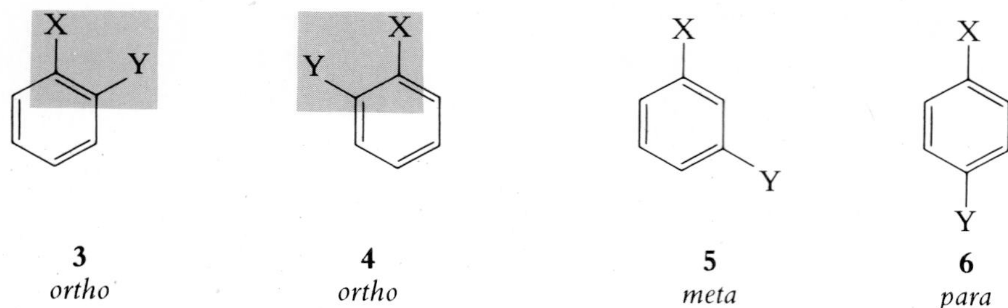

3 *ortho* **4** *ortho* **5** *meta* **6** *para*

However, no more than *three* isomers of a given disubstituted benzene had ever been found—one *ortho* (*o*), one *meta* (*m*), and one *para* (*p*). Kekulé resolved this problem by postulating that the benzene molecule constantly oscillates between two different but indistinguishable conjugated bond arrangements (eq. 14.1). These forms are now known as *Kekulé structures*.

(14.1)

Kekulé further stated that the interconversion between forms was so rapid that all the carbon–carbon bonds were effectively equivalent. Benzene therefore behaved *as if* it had the symmetry of a regular hexagon, and so only *one ortho* isomer could exist (eq. 14.2). In Section 14:3 we will discuss the modern theory that the Kekulé forms of benzene are really resonance structures, and hence there is *no equilibrium* between forms.

(14.2) Br Br **3a** ⇄ Br Br **4a** KeKulé's *incorrect* postulation of an equilibrium

Kekulé's proposals, effective as they were in rationalizing most of the available data, were not universally accepted. For one thing, it was by no means apparent from his formula why benzene did not behave like an alkene. Various alternatives were proposed, as chemists tried to utilize the fourth valence of each carbon atom in some way other than by incorporating it into a double bond. For example, one suggestion was a model with a long *para* bond (called Dewar benzene, **7**). In 1869, the German chemist Alfred Ladenburg offered a different model (now associated with his name) that has a prism formula (**8**). The Dewar and Ladenburg structures are still of interest; they have been recently synthesized and each shown to be different from benzene.

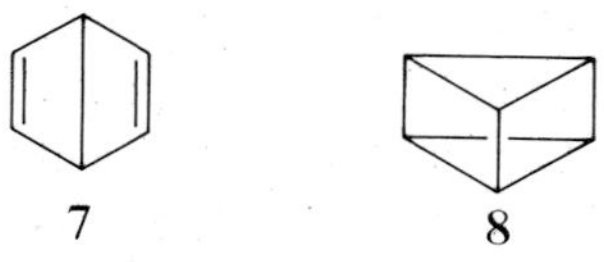

7 8

14–1 Kekulé's cyclic model for benzene came in part from the necessity of finding a structure of high symmetry that would account for the small number of disubstituted isomers. Draw some linear structures that correspond to C_6H_6. Do any of them fit the experimental data as far as the number of monosubstituted and disubstituted isomers is concerned?

14:2 Chemical Criteria for Aromaticity

Kekulé bestowed the generic name aromatic compounds on benzene and a number of its derivatives because of their aroma. After a time it was noted that most of these compounds had another common characteristic: resistance to chemical attack. Benzene does not react with halogens, acids, or oxidizing agents as do alkenes under the same conditions. When an aromatic compound *does* react, the product is usually one of substitution rather than addition, thereby maintaining the integrity of the aromatic ring. Aromaticity thus implied an unusual chemical stability of benzene, and other compounds were judged to be aromatic to the extent that they emulated benzene in chemical behavior. This criterion was often difficult to apply, since a given compound might appear to be aromatic toward one set of reagents but distinctly nonaromatic toward another set.

Near the end of the nineteenth century it was generally accepted that the aromaticity of benzene could be ascribed to its system of cyclic, conjugated double bonds. Other cyclic conjugated compounds were likewise expected to exhibit aromatic behavior. This theory was viable until cyclooctatetraene was synthesized in the early 1900s. This compound, though it is cyclic and possesses conjugated double bonds, demonstrates typical alkene behavior; it reacts readily with halogens, oxidizing agents, and so on.

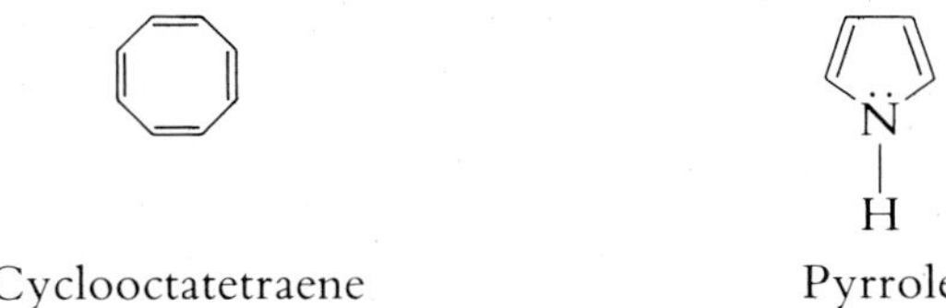

Cyclooctatetraene Pyrrole

This discovery led Robert Robinson to propose, in 1925, that aromaticity was characteristic of compounds that contained a set of *six* cyclic conjugated electrons. He noted that pyrrole behaves like benzene in many ways, even though they are structurally dissimilar. Like benzene, however, pyrrole does have six π electrons in a completely conjugated ring, if one includes the lone pair on nitrogen. Robinson was correct, and his six-electron rule was really a special case of a more general definition of aromaticity, which we will study further in Section 14:7.

14:3 Molecular Orbital Theory of Benzene

In order to explain the unusual properties of benzene we can update Kekulé's analysis. Instead of assuming that the two Kekulé structures represent independent equilibrating molecules (eq. 14.1), we view them instead as two equivalent resonance forms of a single molecule (**2a** $\leftrightarrow$ **2b**). Using this approach we can ascribe the high stability and low reactivity of benzene to

$\leftrightarrow$

2a **2b**

the fact that *the π electrons are symmetrically delocalized over the entire ring.* Thus, if **2a** and **2b** are resonance hybrid forms, then benzene itself should have six equivalent carbon–carbon bonds that are neither single nor double bonds, but have an intermediate value. The physical data fully support this interpretation (sec. 14:4).

One of the major reasons, then, for the stability of benzene is that its structure allows *complete delocalization of the π electrons.* Furthermore, a comparison of the chemical and physical characteristics of benzene and cyclooctatetraene leads to the conclusion that the π electrons in cyclooctatetraene are not delocalized. Recognizing this disparity forces us to confront some limitations of the resonance method. Two conjugated resonance structures each can be written for benzene and cyclooctatetraene, yet only benzene experiences π electron delocalization and the stability that comes from it. Thus, the fact that two or more equivalent resonance structures can be written for a cyclic conjugated molecule means that its π electrons may be delocalized, but does not *require* that they be delocalized. A dramatic instance of the second possibility is provided by cyclobutadiene, for which two equivalent resonance structures may be written. This molecule is only isola-

Cyclobutadiene

ble at $-196°$ or below; at $-100°$ it undergoes rapid Diels–Alder dimerization, hardly the behavior of a π-delocalized, aromatic molecule.

While resonance structures can indicate the possibility of aromaticity, we have to turn to molecular orbital theory to confirm or deny the actuality. We'll look next at the π-orbital structure of benzene and see what insights it gives us into benzene's unusual properties. (Although the resonance approach has limitations with respect to the prediction of aromaticity, it is valid and valuable in many other contexts, and we will continue to use it.)

Each carbon atom of benzene must be sp^2 hybridized since it takes part in three σ bonds and one π bond. This makes the σ framework a planar structure with six C—H bonds in the plane of the molecule. For the π-electron system there are six *p* orbitals, one at each carbon, perpendicular to the plane of the molecule and available for π-orbital formation. These six carbon atomic orbitals yield six molecular orbitals: three bonding and three antibonding. Since there are six π electrons available, ground state benzene has the three bonding orbitals exactly filled with two electrons each. The electron distribution is shown in Figure 14–1, along with the structures of the orbitals as determined from wave mechanical calculations. The lowest energy orbital is ψ_1 because its two electrons are delocalized over the entire ring; it has no nodal plane perpendicular to the molecular plane (sec. 13:11).

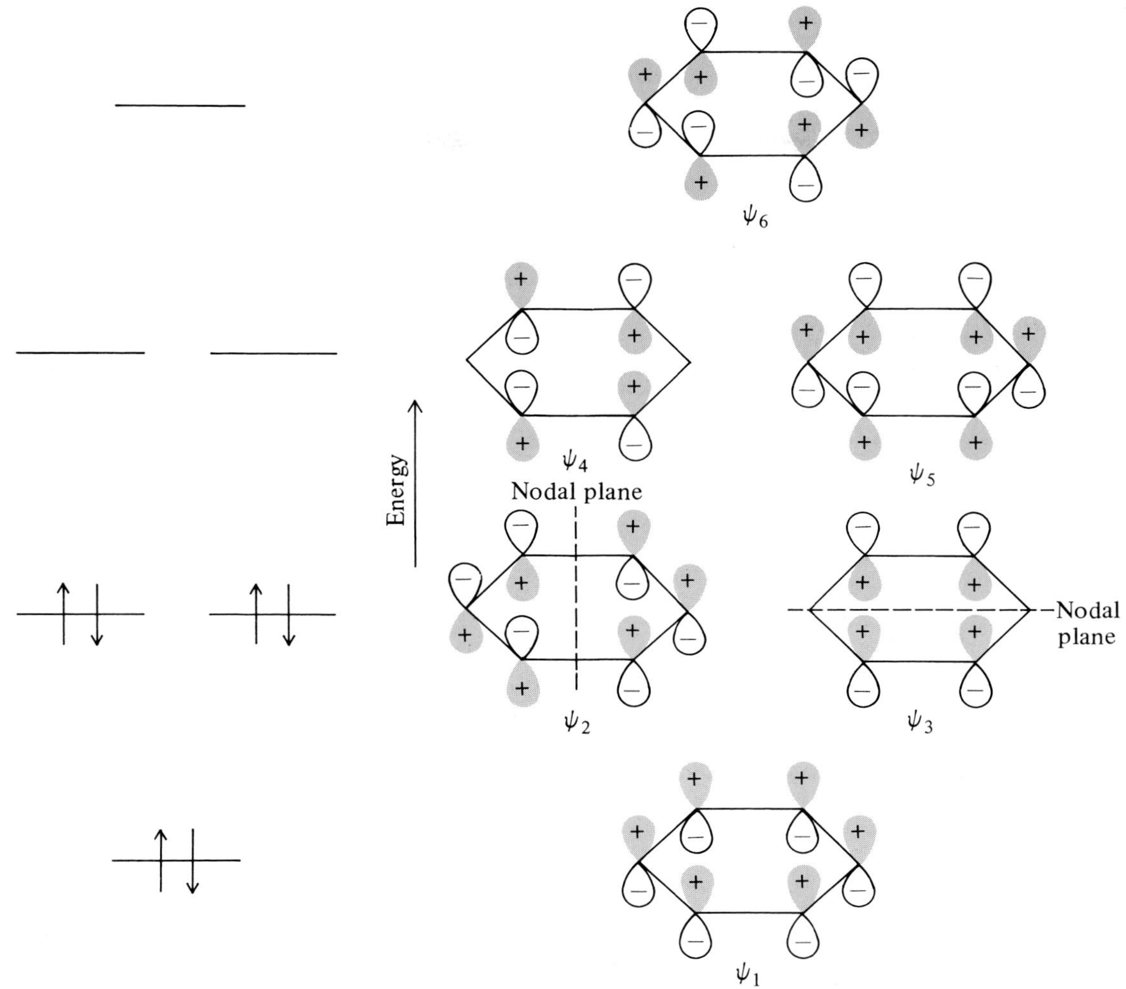

Figure 14–1. *Molecular π orbitals of benzene.*

Both ψ_2 and ψ_3 have one perpendicular node, so their electrons are correspondingly more restricted and higher in energy than those in ψ_1.

Figure 14–1 is a simplified diagram in that it ignores the wave character of electrons. Waves have crests and troughs. In each molecular orbital, therefore, the wavelike electron density is greater at some points and lesser at others. Thus, the carbons that have the greatest electron density from ψ_2 are precisely those that lie on the node in ψ_3. The result of this is that in the combination of ψ_1, ψ_2, and ψ_3 the *total* electron distribution is symmetrical and equal at each carbon.

In order to grasp the significance of Figure 14–1, it is instructive to compare the relative π-orbital energies of benzene with those of another conjugated, six π-electron system, 1,3,5-hexatriene (Fig. 14–2). Some very simple MO calculations show that the two lowest π orbitals of 1,3,5-hexatriene lie fairly close in energy to the π orbitals of benzene. The highest occupied π orbital (HOMO) of hexatriene, however, is substantially more energetic than the HOMO of benzene. This energy difference accounts in good part for the fact that benzene is aromatic, while 1,3,5-hexatriene behaves like a typical conjugated alkene.

This comparison suggests a new way of defining aromaticity:

If a cyclic conjugated molecules has less π energy than its linear counterpart, then the cyclic molecule is aromatic.

This approach to defining aromaticity is widely applicable. For instance, we have already seen chemical evidence that cyclobutadiene is not aromatic. In fact, molecular orbital calculations show that cyclobutadiene has *greater* π energy than its counterpart 1,3-butadiene, confirming cyclobutadiene's lack of aromaticity. In Section 14:7 we will expand on this approach.

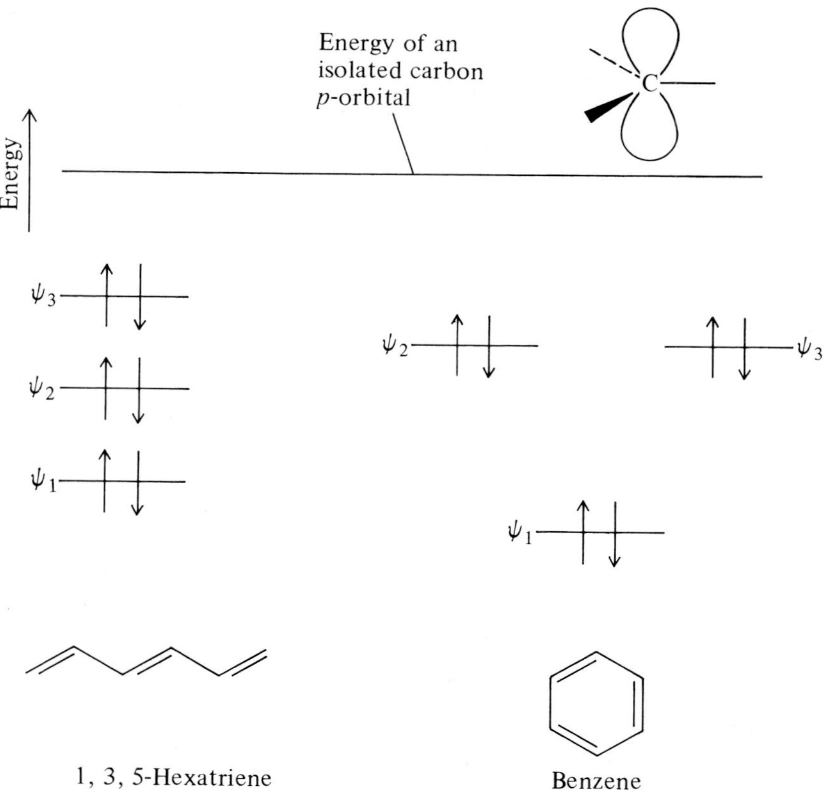

Figure 14–2. *Relative energies of the bonding π orbitals of 1,3,5-hexatriene and benzene. The total energy of the benzene π electrons is lower than the total energy of the hexatriene π electrons. This is due mainly to the fact that ψ_3 of benzene is lower than ψ_3 of hexatriene.*

14–2 Benzene reacts very slowly with bromine, and a catalyst is required in order for the reaction to be observed (eq. 1).

Br_2, $FeBr_3$ (cat) (1)

How would you expect 1,3,5-hexatriene to react with bromine? Consider both speed of reaction and the product(s) formed.

14:4 Structural Characteristics of Benzene and Different Ways of Depicting Them

X ray diffraction analysis shows that benzene is a regular planar hexagon in which all carbon–carbon bond lengths are identical and equal to 1.40 Å. The equilibrium bond length for a pure double bond is 1.34 Å, while that for an sp^2–sp^2 single bond is 1.49 Å (Table 2–3, p. 54). The benzene bond length is, therefore, intermediate between these two values, as we would expect for a delocalized system.

One criterion for designating a compound as aromatic is that the molecule in question should have equal, or nearly equal, carbon–carbon bond lengths.

This equality of bond lengths is an indication of complete delocalization of the π electrons. A conjugated system that is alkenelike, such as 1,3,5-hexatriene, has alternating single and double bonds whose lengths are close to the expected 1.49 and 1.34 Å, respectively.

Although the "problem" of the structure of benzene has been solved, there remains the question of how best to represent this structure on paper. Writing both resonance structures each time benzene is shown gets to be cumbersome, and so a general convention has been to use either **9** or **10** as a representation of benzene with the understanding that the position of the double bonds is not fixed. As in all structures drawn this way, we assume

9 **10**

that there is one hydrogen atom on each carbon atom unless another substituent is shown. This convention is used exclusively by some chemists and textbooks because, as you will see in later chapters, there are mechanistic parallels between the reactions of benzene and the reactions of conjugated alkenes.

A second convention for depicting benzene and its simple derivatives is

Box 14–1. Elemental Carbon

Pure carbon occurs on earth as graphite and as diamond. Graphite is made up of sheets of fused benzene rings. It has many conjugated double bonds whose π electrons absorb all visible wavelengths, giving graphite its black color. Its secondary structure is composed of

A section of graphite

large sheets that are weakly bound together so they can slide over each other, making graphite a useful lubricant. These sheets can be separated simply by exerting pressure against the graphite surface.

Diamond can be thought of as a network of cyclohexane rings joined in a large macromolecule. Unlike graphite, there are no weak points in its structure, which accounts for the hardness of diamond. In addition, all the carbon bonds are sp^3–sp^3 so there are no light-absorbing π bonds. Diamonds are therefore transparent.

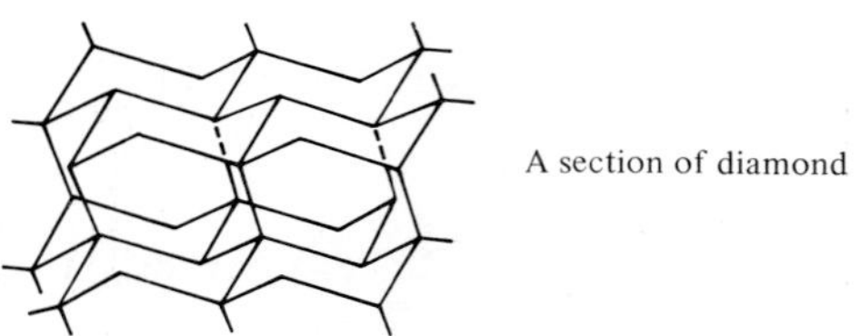

A section of diamond

structure **11.** It is *understood* that the circle represents six delocalized π electrons:

11

It takes less time to draw this structure, and it also gives a good representation of the uniform electron distribution in benzene. However, it is *not* useful when writing reaction mechanisms for benzene. Consequently, we recommend that either **9** or **10** (it does not matter which) be used when writing reactions involving the π electrons of benzene. Structures such as **11** can be used when the benzene ring itself is not undergoing reaction.

14:5 The Resonance Energy of Benzene: A Matter of Definition

The *resonance energy,* or *delocalization energy,* of benzene (its "extra" stabilization) can be defined as the difference in the heat of formation between benzene and cyclohexatriene, its hypothetical, bond-localized isomer (eq. 14.3).

all C–C bonds = 1.40 Å

$\leftrightarrow$

Benzene
(real molecule)

1.49 Å 1.34 Å

Cyclohexatriene
(hypothetical molecule)

$$\text{Resonance energy of benzene} = \Delta H_f(\text{cyclohexatriene}) - \Delta H_f(\text{benzene}) \tag{14.3}$$

Since cyclohexatriene does not exist, its heat of formation must be estimated and, hence, any value for the resonance (delocalization) energy of benzene will depend on the assumptions used to make this estimate.

In order to avoid these problems, an experimental approach to calculating the resonance energy of benzene has been used. The heat of hydrogenation of cyclohexene is -28.6 kcal/mol (eq. 14.4). We expect the hypothetical cyclohexatriene, then, to have a heat of hydrogenation of $3 \times -28.6 = -85.8$ kcal/mol.

(14.4) $+ H_2 \rightarrow$ $\Delta H = -28.6$ kcal/mol

When benzene is reduced to cyclohexane (eq. 14.5), the heat released amounts to only -49.8 kcal/mol.

(14.5) $+ 3H_2 \rightarrow$ $\Delta H = -49.8$ kcal/mol

Using the hydrogenation enthalpies of Equations 14.4 and 14.5, we estimate the resonance energy of benzene as (85.8 − 49.8) 36.0 = kcal/mol (Fig. 14–3). This value represents significant stabilization of benzene compared to typical alkenes. In fact, reduction of the first double bond in benzene is actually endothermic (eq. 14.6)—a very unusual result.

(14.6) $+ H_2 \rightarrow$ $\Delta H = 5.6$ kcal/mol

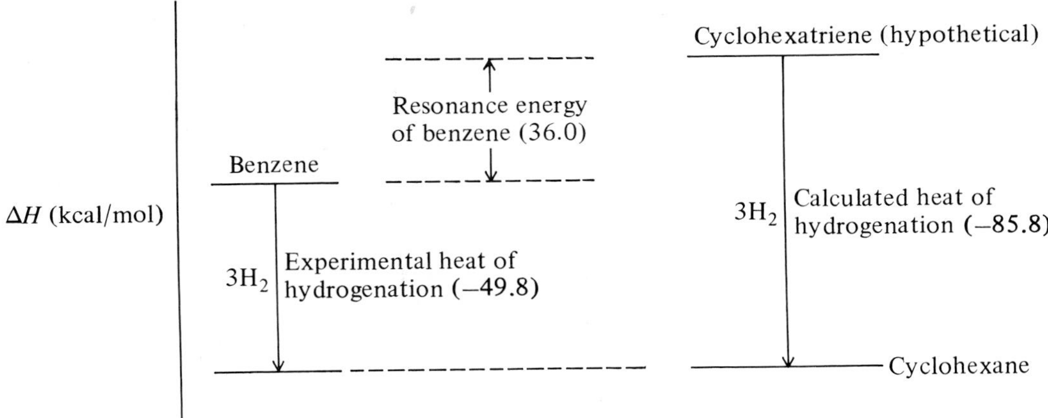

Figure 14–3. *Estimating the resonance energy of benzene by comparing its heat of hydrogenation to that calculated for the hypothetical cyclohexatriene. The calculated value is three times the heat of hydrogenation of cyclohexene.*

14–3 Based on the heat of hydrogenation data (eqs. 14.4 to 14.6), calculate the following quantities:

a. The enthalpy change when 1,3-cyclohexadiene is reduced to cyclohexene.

b. The resonance energy of cyclohexadiene.

c. The enthalpy change when 2 mol of 1,3-cyclohexadiene react to give 1 mol of benzene and 1 mol of cyclohexene.

14–4 a. The heat of hydrogenation of cyclooctatetraene to cyclooctane is −101 kcal/mol (eq. 2), while the heat of hydrogenation of cyclooctene to cyclooctane is −23 kcal/mol (eq. 3). Is there extensive electron delocalization in cyclooctatetraene? Explain your answer.

$$\text{cyclooctatetraene} + 4H_2 \rightarrow \text{cyclooctane} \qquad \Delta H = -101 \text{ kcal/mol} \tag{2}$$

$$\text{cyclooctene} + H_2 \rightarrow \text{cyclooctane} \qquad \Delta H = -23 \text{ kcal/mol} \tag{3}$$

b. X ray diffraction studies have shown that cyclooctatetraene is *not* planar. What effect would this have on overlap between neighboring π bonds?

14:6 Characteristic IR, NMR, and UV Absorptions of Benzene and Its Derivatives

Because benzene is highly symmetrical its infrared spectrum is comparatively simple. The C—H stretching motions give peaks above 3000 cm^{-1}, as expected for hydrogen bound to sp^2-hybridized carbon. Peaks of weak to medium intensity, assigned to stretching vibrations of the aromatic ring, lie in the 1450 to 1600 cm^{-1} region. The most intense band in the spectrum, found near 700 cm^{-1}, is due to the out-of-plane bending motions of the hydrogen atoms. Bands in this region are very characteristic of the aromatic ring, and their number and frequency are dependent on the number and location of the substituents attached to the ring. They have been used for deciphering the substitution pattern in substituted benzenes. The positions of substituents on a benzene ring are now more readily determined by ^1H nmr analysis.

The benzene protons are all equivalent and the proton nmr spectrum of benzene consists of a single peak at 7.27 ppm. It is significant that this peak is about 2 ppm below that usually found for hydrogen atoms connected to doubly-bonded carbon atoms. This additional deshielding tells us something about the electronic structure of benzene. The carbon ring can be viewed as a conducting wire, or solenoid. When this ring is placed in a magnetic field, the highly mobile, delocalized π electrons circulate around the ring, much as electrons do in a wire, giving rise to a ring current. This **ring current** induces a local magnetic field around the ring (Fig. 14–4).

The induced magnetic field is opposed to the applied field inside the ring,

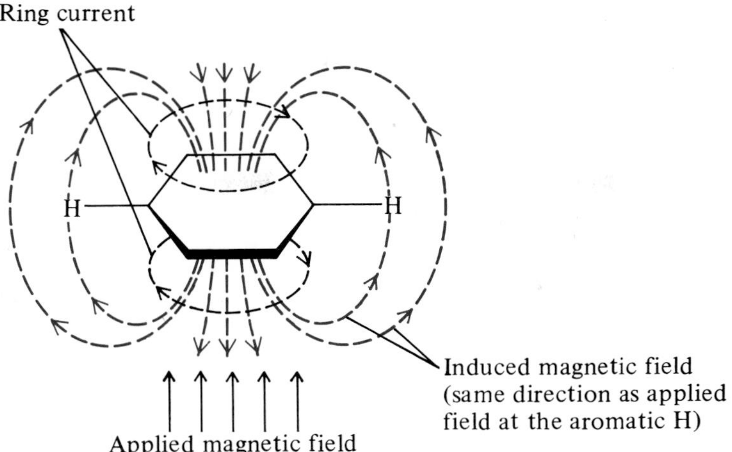

Figure 14–4. *Ring current in benzene due to the applied field. The ring current generates an induced field that deshields the protons on the ring.*

but it reinforces the applied field outside the ring. The benzene protons, which are on the outer periphery of the ring, are therefore *deshielded* by the induced magnetic field. If there were any protons inside the ring, they would be abnormally shielded by the induced magnetic field. *The ability of a compound to sustain a ring current is a useful criterion of aromaticity* and is easily determined because of the effect the ring current has on the chemical shift of the aromatic protons. An application of this criterion is given in Section 14:7.

The ^{13}C resonance for the carbons in benzene is at 128 ppm, and this position moves upfield or downfield with various substituents. Proton and carbon nmr spectroscopy, which are both very valuable for structural work on substituted benzenes, are discussed in more detail in Chapter 15.

The ultraviolet absorption spectrum of benzene, as determined in a vacuum spectrophotometer, is given in Figure 14–5. It shows the three main absorption bands, which differ in appearance and intensity. The intense band at 180 nm is beyond the range of most uv spectrometers, so it is the absorption bands at 203 and 254 nm that are the most useful. Note that the band at 254 nm consists of a series of sharp peaks. Each of these peaks represents a transition from the ground state to an excited state with the same electron distribution but with a slight difference in vibrational energy. That is why the peaks have slightly different wavelengths.

The 203 and 254 nm bands are useful in structure determination because their peak positions shift as different substituents are placed on the benzene ring. Here are a few examples to show what happens to the 203 nm band when a ring hydrogen is replaced by some other group. Introducing a methyl group (toluene) causes a small shift to longer wavelengths. A much larger shift to longer wavelengths is found with anisole, because the oxygen lone pairs are conjugated with the ring. Substituents that have π bonds

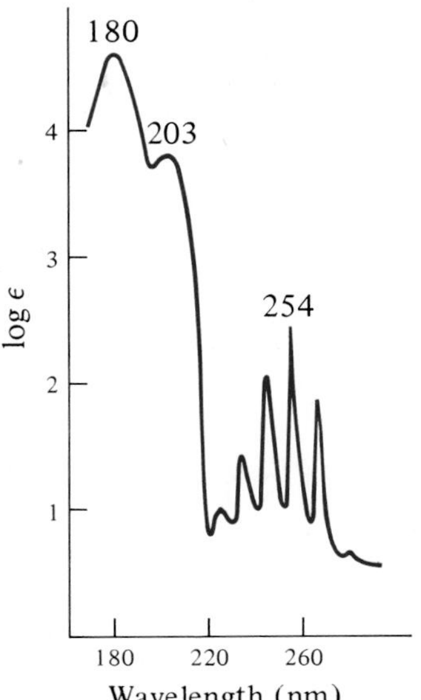

Figure 14–5. *Ultraviolet spectrum of benzene.*

conjugated with the ring produce the largest shifts of all, as you can see with styrene and acetophenone.

CH_3 OCH_3

Benzene $\lambda_{max} = 203$ nm

Toluene $\lambda_{max} = 206$ nm

Anisole $\lambda_{max} = 217$ nm

$CH{=}CH_2$ $\overset{O}{\overset{\|}{C}}CH_3$

Styrene $\lambda_{max} = 247$ nm

Acetophenone $\lambda_{max} = 245$ nm

The 254 nm band undergoes similar shifts upon ring substitution.

Problem

14–5 Preliminary investigation of an unknown compound, $C_{10}H_{12}$, shows that it is either 1-phenyl-1-butene (**I**) or 1-phenyl-2-butene (**II**). The uv spectrum of the unknown contains peaks at 251 and 284 nm.

$-CH{=}CHCH_2CH_3$ $-CH_2CH{=}CHCH_3$

I **II**

a. Indicate which structure is correct for the unknown, and explain how you arrived at your conclusion.
b. Suppose you had pure samples of both **I** and **II,** and hydrogenated each to 1-phenylbutane. Show which one would have the smaller heat of hydrogenation, and why.

14:7 Using Hückel's Rule to Identify Aromatic Molecules and Ions

You've seen that resonance structures alone cannot tell us whether a particular cyclic conjugated polyene is aromatic (sec. 14:3). One way to evaluate aromaticity is to compare the π orbital energies of the cyclic polyene with those of the related open-chain polyene. This approach, while very enlightening, is not very convenient. What would be ideal would be a method for telling from the structural formula alone whether or not a molecule is aromatic. Around 1930, the German theoretician E. Hückel proposed such a general rule for deciding whether or not appropriate molecules would be aromatic—that is, if they would be like benzene in stability and reduced reactivity. The rule applies to any *monocyclic* planar hydrocarbon composed **only** of sp^2-hybridized carbon atoms, and it states that the molecule will be aromatic *if the number of* π *electrons equals* ($4n + 2$)*, where* n *is a positive integer or zero.*

To determine what number of π electrons would give rise to aromaticity in a conjugated cyclic compound, we simply set n equal to zero, one, or two (or any other integer) and calculate the value of $(4n + 2)$. For example, if $n = 1$,

$$(4n + 2) = (4 \bullet 1 + 2) = 6$$

Therefore, Hückel's rule predicts that any planar, conjugated compound with six π electrons, such as benzene, is aromatic. If $n = 0$,

$$(4n + 2) = (4 \bullet 0 + 2) = 2$$

If $n = 2$,

$$(4n + 2) = (4 \bullet 2 + 2) = 10$$

Therefore, Hückel's rule predicts that compounds with 2 or 10 π electrons in cyclic conjugated systems are aromatic. Hückel's rule also predicts that neither cyclobutadiene (four π electrons) nor cyclooctatetraene (eight π electrons) should be aromatic, in accord with chemical experience. In fact, cyclooctatetraene is not even planar, and so fails to meet one of the necessary criteria for application of the rule (Prob. 14–4b).

Problem

14–6 Would Hückel's rule predict a compound with 12 π electrons in a cyclic conjugated system to be aromatic? A compound with 14 π electrons?

An aromatic ring need not contain only carbon atoms, even though Hückel's rule specifically refers only to hydrocarbons. Robinson's theory of the electron sextet (sec. 14:2) was prompted in part by the known benzene-like properties of thiophene and pyrrole.

Thiophene Pyrrole

The S and N atoms in these molecules each contribute one *p* orbital *and two electrons* to the π-electron system; the resultant number of π electrons is in accord with Hückel's rule, and pyrrole and thiophene are aromatic. The requirements of aromaticity impose sp^2 hybridization on the S and N atoms in these molecules.

Problem

14–7 Organic amines ($R\ddot{N}H_2$, $R_2\ddot{N}H$, and $R_3N{:}$) are basic and can react with proton sources ($R_3N{:} + H_3O^+ \rightarrow R_3\overset{+}{N}{:}H + H_2O$). Pyrrole, however, does not exhibit the normal basic properties of amines. Why not?

Hückel's rule says nothing about the number of atoms present in the aromatic ring—it requires only that there be $(4n + 2)$ π electrons. This leaves open the possibility of creating aromatic systems in which the number of π electrons would differ from the number of atoms—in other words, the aromatic rings would be anions or cations. The first aromatic anion discovered was **cyclopentadienide** ion. Cyclopentadiene exhibits remarkable acidity for a hydrocarbon. Its pK_a is 15, equal to that of water. The allylic hydrogens of propene, $CH_3CH{=}CH_2$, with which cyclopentadiene might be compared, have a pK_a of only about 45. Thus, cyclopentadiene is 10^{30} times more acidic than propene. Removing a proton from cyclopentadiene yields a negatively charged aromatic π-electron system (eq. 14.7), in which the charge is equally distributed over the five carbon atoms.

$$\text{(14.7)}\quad \underset{\text{Cyclopentadiene}}{C_5H_6} + Na \xrightarrow{\text{ether}} \tfrac{1}{2}H_2 + Na^+\left[C_5H_5^- \equiv \bigcirc^{-} \right]$$

(H H on C-5 of cyclopentadiene; H on the anion carbon)

(n = 1)
Cyclopentadienide ion

Note that C-5 of cyclopentadiene, which is sp^3 hybridized, changes to sp^2 hybridization upon losing the proton. (In localized carbanions, the carbon bearing the negative charge remains sp^3 hybridized.)

Aromatic cations that satisfy Hückel's rule for various values of n have also been synthesized. The first to be made, and the most widely studied, is the seven carbon, six π-electron cation known as **cycloheptatrienyl ion** (*tropylium ion*). Its fluoroborate (eq. 14.8) can be isolated and stored for indefinite periods. The fact that even three-membered aromatic cations can be isolated (eq. 14.9) must be due to the stabilization that comes with aromaticity, because the ring strain in this cation is considerable.

(14.8) [cycloheptatriene with H and OCH_2CH_3] + HBF_4 $\xrightarrow{\text{ethanol}}$ [cycloheptatrienyl cation ≡ (+)] BF_4^- + CH_3CH_2OH

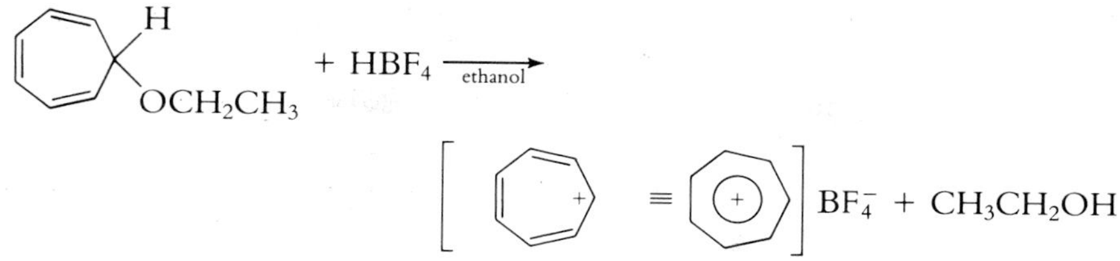

Tropylium fluoroborate ($n = 1$)

(14.9) [trimethylcyclopropene with CH_3, CH_3, CH_3, OCH_3] + HBF_4 → CH_3OH + [trimethylcyclopropenium] CH_3 BF_4^-

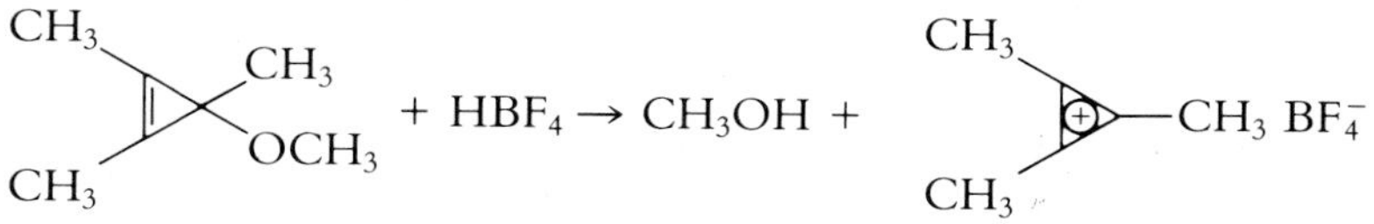

Trimethylcyclopropenium fluoroborate ($n = 0$)

The aromatic ions are much less reactive than the organic cations and carbanions discussed in previous chapters. Furthermore, they exhibit many physical properties associated with aromaticity—identical C—C bond lengths, equal distribution of the charge over all the ring carbon atoms, and the ability to sustain a ring current when placed in a magnetic field.

Benzene, cyclopentadienide ion (eq. 14.7), and cycloheptatrienyl cation (eq. 14.8) differ widely in reactivity even though they are all aromatic six π-electron systems. However, they share the property of each being less reactive than their open-chain counterparts. We've already noted that benzene is lower in π-electron energy and reactivity than its open-chain counterpart 1,3,5-hexatriene (sec. 14:3). Analogously, cyclopentadienide ion is less reactive and less energetic than pentadienide ion, $CH_2{=}CHCH{=}CH\ddot{C}H_2^-$. In assessing the aromaticity of a cyclic molecule, our standard is always the linear conjugated species with the same charge and number of π electrons.

Problem

14–8 Cyclic conjugated molecules that do not show electron delocalization are often more reactive than their linear counterparts. Such molecules have been classified as *antiaromatic*. Which compounds already discussed in this chapter qualify for this label? Suggest a general mathematical formula describing the number of π electrons in an antiaromatic compound.

Example

Are the following structures aromatic according to Hückel's rule?

a. 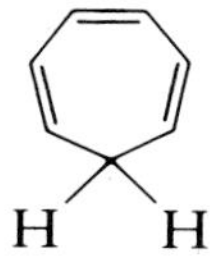 [cycloheptatriene, CH_2 with H H]   b. 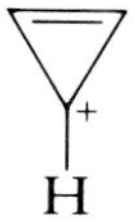 [cyclopropenyl cation with H, +]

Answer

a. This compound is cyclic and has six π electrons. There is, however, an sp^3 carbon between two of the double bonds. Hückel's rule applies only to cyclic arrays of sp^2-hybridized atoms, so this molecule is not aromatic.

b. In this structure there are two π electrons. If $n = 0$ in Hückel's rule $(4n + 2)$, then $(4 \cdot 0 + 2) = 2$, so the ion is aromatic.

Problems

14–9 Indicate whether or not Hückel's rule can be applied to each of the following structures. Explain your answer. If it can, then classify the structure as aromatic or not.

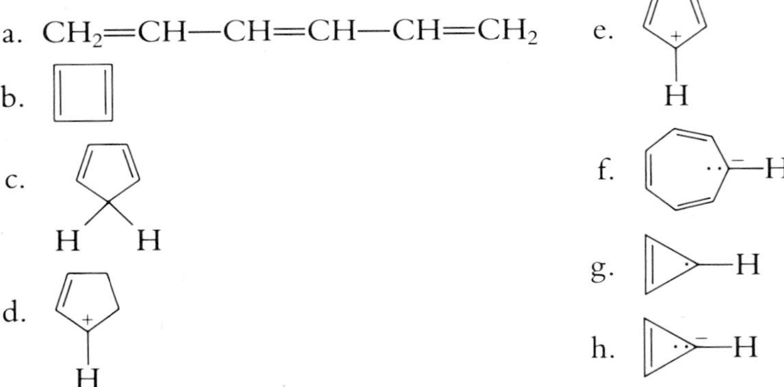

14–10 Pyridine is a much stronger base than pyrrole (see Prob. 14–7). Why?

14:8 The Benzyl System: How a Benzene Ring Affects the Reactivity of a Neighboring Carbon Atom

The benzene ring alters the reactivity of atoms and groups attached to it. This effect is most pronounced for reactive intermediates such as radicals and ions. Consider a CH_2 group attached to a benzene ring. This $C_6H_5{-}CH_2$ unit is the **benzyl** system. Depending on whether the CH_2 group holds zero, one, or two electrons, the structure is a benzyl cation, benzyl radical, or benzyl anion, respectively. These species are all considerably more stable than their simple alkyl counterparts, such as methyl cation, methyl radical, and methyl anion. Let's see why.

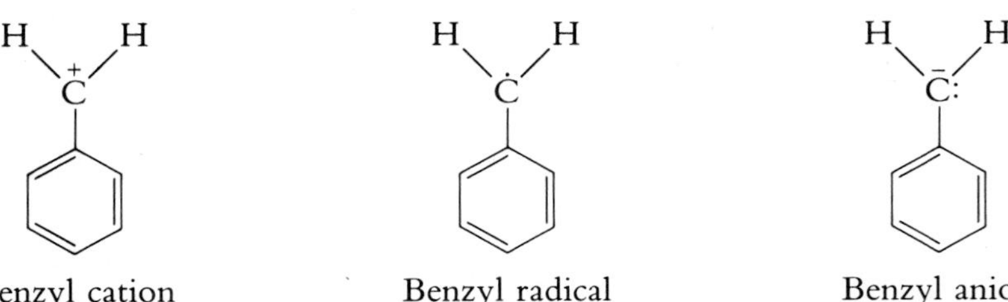

Benzyl cation Benzyl radical Benzyl anion

In a simple alkyl radical or ion, the unpaired electron or charge is mostly localized on a single carbon atom. In the case of a benzyl radical or anion, the unpaired electron is delocalized over the exocyclic (out of the ring) carbon atom and the *ortho* and *para* positions of the ring. This delocalization is made apparent by the resonance structures below.

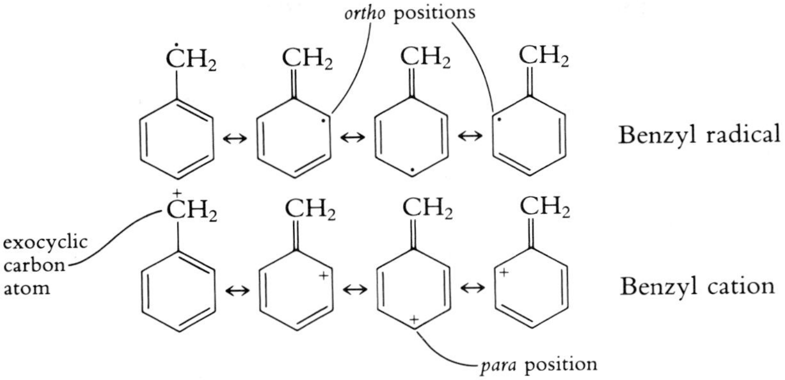

As we would expect, the benzyl radical and ions are stabilized by this additional electron delocalization. The stability is reflected in the value of the dissociation energy for the methyl C—H bonds of toluene (eq. 14.10) compared to that for ethane (eq. 14.11). It takes 10 kcal/mol less energy to dissociate the C—H bond in toluene.

(14.10) Toluene ($C_6H_5CH_3$) → $C_6H_5CH_2\cdot$ + H· $\Delta H = 88$ kcal/mol

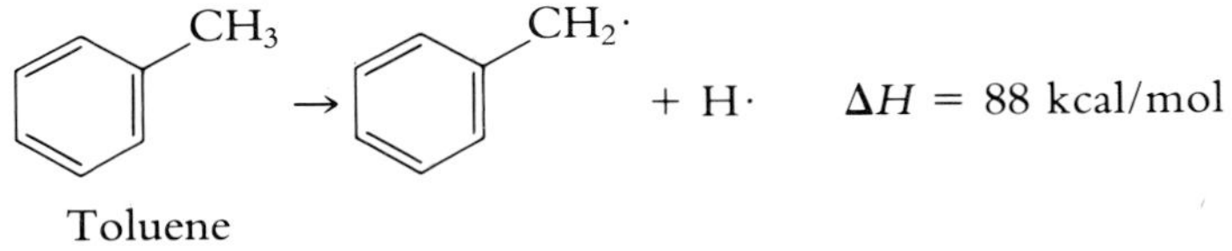

(14.11) $CH_3—CH_3 \rightarrow CH_3—CH_2\cdot + H\cdot \qquad \Delta H = 98$ kcal/mol

Since the benzyl radical is relatively stable, benzylic hydrogens are readily abstracted by other radicals, as are allylic hydrogens (sec. 13:3). Toluene, for example, is easily halogenated under free-radical conditions (eq. 14.12). If alkylbenzenes containing longer chains are halogenated, the benzylic hydrogens are replaced first (eq. 14.13).

(14.12) $C_6H_5CH_3 \xrightarrow[h\nu]{Cl_2} C_6H_5CH_2Cl \xrightarrow[h\nu]{Cl_2} C_6H_5CHCl_2 \xrightarrow[h\nu]{Cl_2} C_6H_5CCl_3$

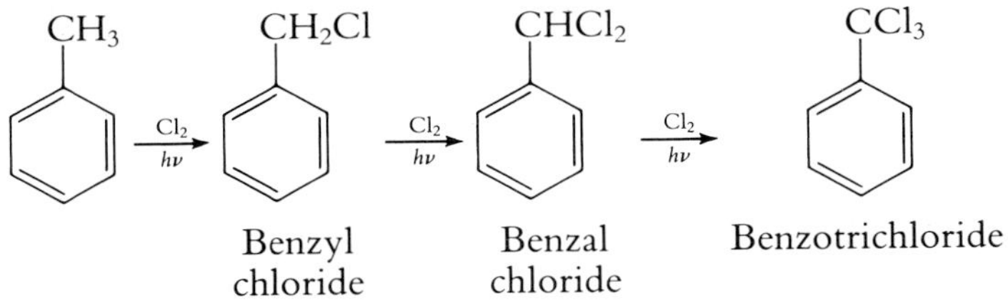

Benzyl chloride; Benzal chloride; Benzotrichloride

(14.13) $C_6H_5CH_2CH_3 + Br_2 \xrightarrow[0^\circ]{h\nu} C_6H_5CHBrCH_3$

Main product

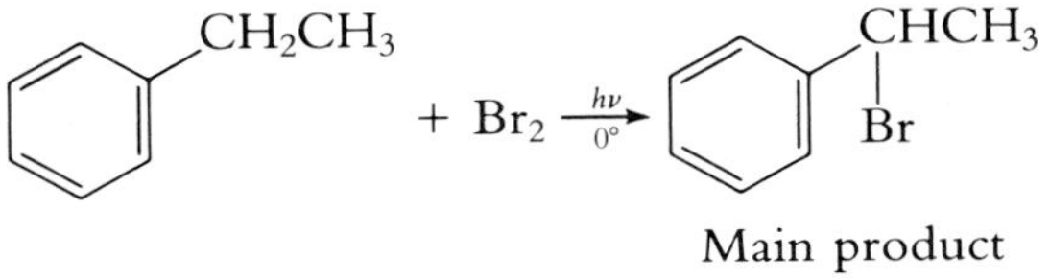

The presence of a **phenyl group** (C_6H_5) similarly enhances the ease of cation formation at the carbon atom to which it is attached.★ A phenyl group is often abbreviated as C_6H_5— and occasionally as Ph—. A comparison of Equations 14.14 and 14.15 indicates that phenyl is much more effective than methyl in promoting carbocation formation.

(14.14) $$(CH_3)_2CHBr + H_2O \xrightarrow[\text{aq acetone}]{} (CH_3)_2CHOH + HBr$$

S_N1/S_N2 borderline

(14.15) $$\underset{\text{Bromodiphenylmethane}}{Ph_2CHBr} + H_2O \xrightarrow[\text{aq acetone}]{} Ph_2CHOH + HBr$$

S_N1

Chlorotriphenylmethane, $(C_6H_5)_3CCl$ (known as *trityl chloride*), is so reactive that it is hydrolyzed at an appreciable rate even upon exposure to moist air.

Problems

14–11 Write resonance structures for the diphenylmethyl cation $(C_6H_5)_2CH^+$, showing its electron delocalization.

14–12 Toluene is a weak acid (pK_a ~42), although considerably stronger than saturated hydrocarbons (pK_as ~55 to 60). Show the reaction of toluene with a strong base Y^-, and indicate with the aid of resonance structures why the reaction equilibrium is more favorable with toluene than with saturated hydrocarbons.

14:9 Benzenoid (Benzenelike) Compounds from Coal Tar

The conversion of iron ore to metallic iron consumes large quantities of coke, the material that remains after the distillation of coal in the absence of air. The distillate consists of **coal gas** (which includes light hydrocarbons, ammonia, and hydrogen sulfide) and **coal tar.** Further refinement of the coal tar by extraction and distillation yields a number of commercially valuable benzene derivatives: benzene, toluene, the three xylenes (dimethylbenzenes), and phenol (hydroxybenzene) and its methylated derivatives. The aromatic compound pyridine and its methylated derivatives are also found in coal tar.

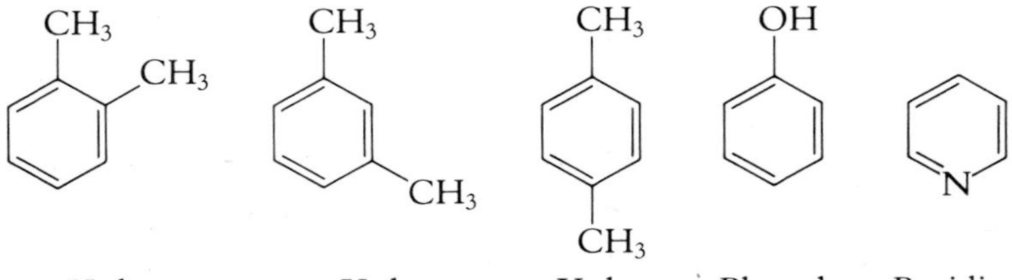

CH3 CH3 | CH3 CH3 | CH3 CH3 | OH | N

o-Xylene *m*-Xylene *p*-Xylene Phenol Pyridine

The chemistry and commercial importance of these substances is discussed in later chapters.

★ When the benzene ring itself is treated as a substituent on another atom, it is called a "phenyl" group, from the word *phène,* the name given to benzene by the French chemist A. Laurent.

Coal tar also yields a number of polycyclic hydrocarbons with **fused rings,** which are rings that share one or more common edges. These compounds are formed in part during the distillation of the coal tar. They are also present in tobacco smoke, as a product of burning tobacco. These polycyclic systems are able to withstand high temperatures because of the stability of their highly conjugated systems, characterized by large delocalization energies.

Several of the most common and important fused polycyclic benzenoid hydrocarbons are shown here, along with their characteristic numbering systems. Note that atoms common to two or more rings are *not* given new numbers. This is because these positions do not normally undergo substitution in the course of chemical reactions.

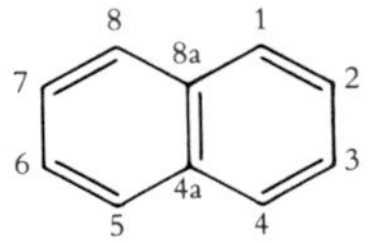

Naphthalene

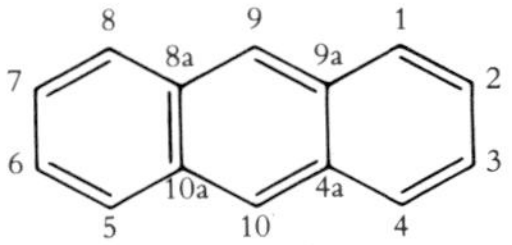

Anthracene

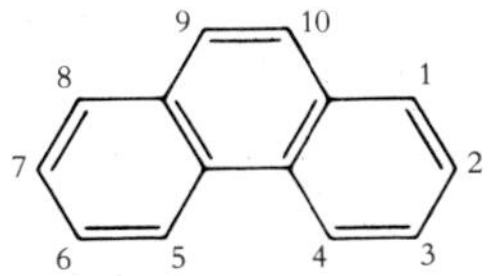

Phenanthrene

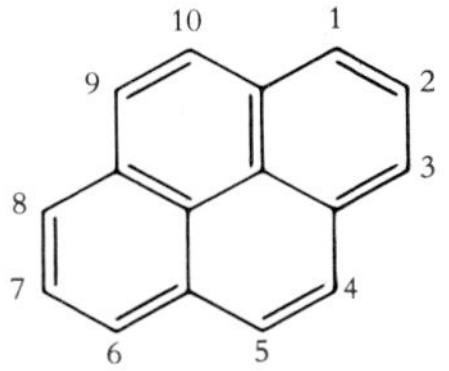

Pyrene

Problems

14–13 a. There are three resonance structures for naphthalene, four for anthracene, and five for phenanthrene. Draw all of them.

b. The shortest bond in phenanthrene is the C-9, C-10 bond. Use the resonance structures you have drawn to explain why this is so. What type of classical bond does the phenanthrene C-9, C-10 bond most resemble?

14–14 Hückel's rule was originally developed for monocyclic systems only. If it could be extended to polycyclic fused systems, would it predict aromaticity for naphthalene, phenanthrene, anthracene, and pyrene?

14:10 Chemical Behavior of Polycyclic Aromatic Hydrocarbons

Although each of the rings in the polycyclic hydrocarbons is benzenoid, there is an important difference between these rings and benzene. The benzene ring has an entire sextet of electrons to itself, while the fused rings must *share* their electron sextets with other rings. As a result, the resonance energy *per ring* in polycyclic compounds is less than the resonance energy of benzene.

The diminished resonance energy (increased reactivity) in the polycyclic systems results in their having a more varied chemistry than benzene. The characteristic reaction of benzene with electrophiles is substitution (Prob. 14–2), whereas the fused ring compounds readily undergo both substitution *and* addition. For instance, benzene will not take part in thermal Diels–Alder reactions. Naphthalene will do so only under extreme conditions, but anthracene reacts readily with dienophiles at the 9 and 10 positions (eq. 14.16).

(14.16)

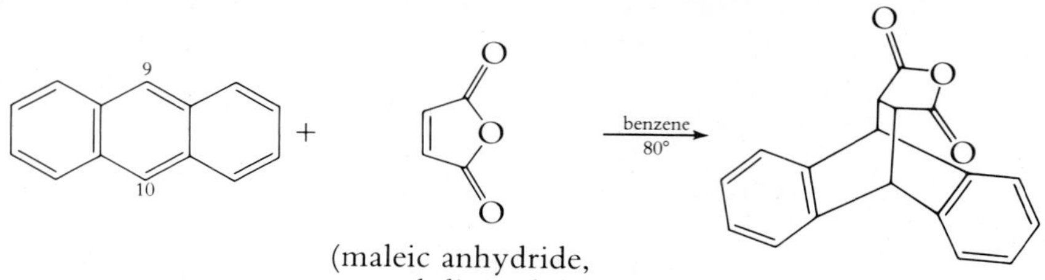

benzene
80°

(maleic anhydride,
a good dienophile)

Analogously, phenanthrene readily undergoes addition across its C-9, C-10 double bond (eq. 14.17).

(14.17) + Br_2 $\xrightarrow{CCl_4}$

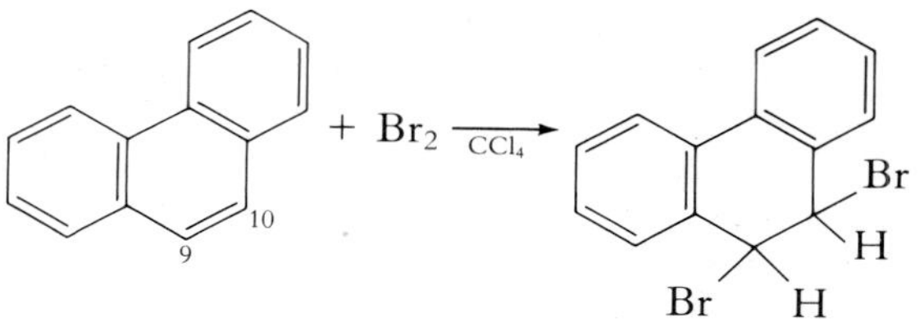

Both of these addition reactions result in some loss of total resonance stabilization present in the reacting hydrocarbon. This loss is partly compensated for by the fact that each product contains two benzene rings, each with its own electron sextet. Thus, the resonance stabilization of each of these *individual rings* has increased in the course of the reaction.

Problem

14–15 a. The addition of Br_2 to phenanthrene (eq. 14.17) yields a mixture of products, those coming from *syn* and from *anti* addition. Draw the structure of these products and write a mechanism for their formation. Your mechanism should make clear why the addition does not proceed solely by an *anti* mechanism, as in the case of simple alkenes (sec. 10:4).

b. When heated, the addition products readily lose HBr to yield 9-bromophenanthrene. Write a mechanism for this process.

The polycyclic aromatic hydrocarbons combine high thermal stability, which leads to their being products of many high-temperature reactions, with substantial chemical reactivity toward electrophiles. This combination of properties has given them an important place in human medical pathology, the subject of the next section.

14:11 Metabolism and Carcinogenicity of Polycyclic Aromatic Hydrocarbons

In 1775 a British physician, describing the high incidence of cancer of the scrotum among chimney sweeps, attributed this effect to the soot in the chimneys. At the beginning of the twentieth century this theory was revised to implicate the coal tars in soot as the carcinogenic agent, and other instances of a correlation between the development of cancers and exposure to coal tar were recorded. Finally, in the 1930s, it was shown that various polycyclic hydrocarbons in the coal tar, singly or as a mixture, could induce the growth of skin cancers in laboratory animals.

The current theory of carcinogenesis holds that chemical changes in DNA, the genetic material in the cellular nucleus, is responsible for initiating cancerous growth. It was initially difficult to see how the aromatic hydrocarbons, which are relatively unreactive, could bring about the required chemical changes in DNA. Beginning in the 1950s, various scientists began to hypothesize that the true carcinogens were not the hydrocarbons themselves but were one or more of their **metabolites.** (A metabolite is the product of one or more chemical transformations a living cell performs on an ingested compound.) Most of the evidence accumulated since then tends to support this hypothesis.

To illustrate the hypothesis we will follow part of the metabolic process for benzo[a]pyrene, a powerful carcinogen. The first step in the metabolism

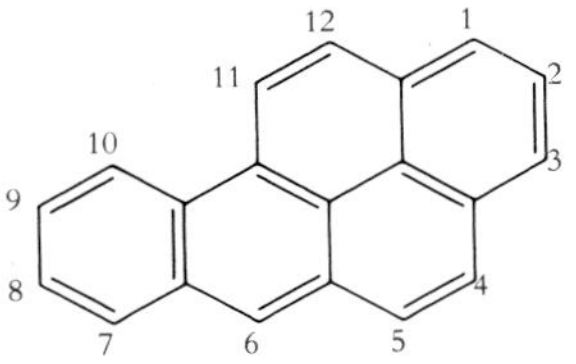

12 1 11 2 10 3 9 8 4 7 6 5

Benzo[a]pyrene

of a large number of aromatic polycycles is their conversion to epoxides. Equation 14.18 shows the oxidation of benzo[a]pyrene to its 7,8-epoxide.

(14.18) 10 9 8 7 $\xrightarrow[\text{enzyme}]{O_2,\ \text{NADPH}}$ O $\xrightarrow[\text{enzyme}]{H_2O}$

Benzo[a]pyrene 7,8-Epoxide

HO OH $\xrightarrow[\text{enzyme}]{O_2,\ \text{NADPH}}$ O HO OH

trans-7,8-Diol *trans*-7,8-Diol-9,10-epoxide

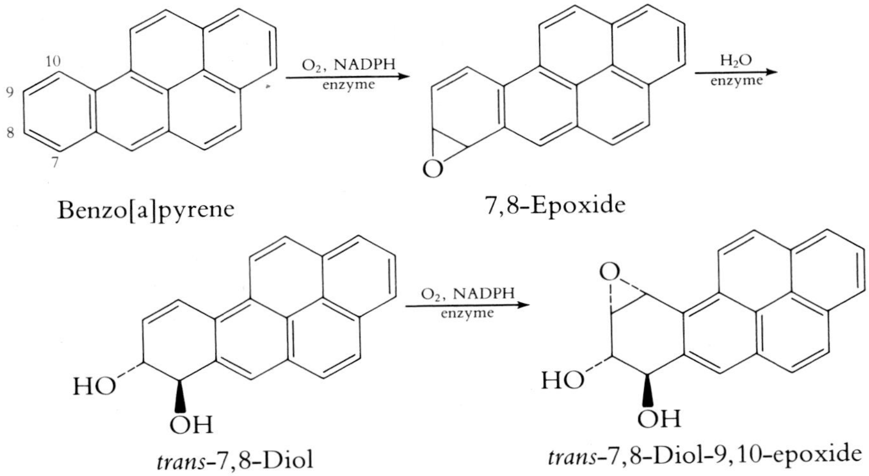

Oxidations such as this have a very important purpose in the normal organism—to make foreign compounds more water soluble and therefore excretable in urine.

You've already seen that in the presence of a catalyst, epoxides react with water to give 1,2-diols (sec. 6:13). The same reaction takes place in a cell, where the catalyst is an appropriate enzyme. There the reaction proceeds stereospecifically, the 7,8-epoxide giving only one *trans*-7,8-diol (eq. 14.18). Since the ring in which oxidation takes place is no longer aromatic, its remaining double bond undergoes further rapid oxidation. The product of this second oxidation is the *trans*-7,8-diol-9,10-epoxide. Most evidence points to this *diolepoxide* as the ultimate carcinogen.

The carcinogenicity of the diolepoxide rests on two properties:

1. The organic bases that are fundamental components of DNA are planar, aromatic molecules. As a result, the polycyclic aromatic hydrocarbons fit very well in between adjacent bases of the DNA helix.
2. The epoxide group, because of its inherent ring strain, is a powerful alkylating agent (sec. 6:13), capable of reacting with nucleophilic sites in DNA.

The final step in this proposed carcinogenesis mechanism involves the alkylation of a nucleophilic site, such as an NH_2 group in the DNA, by the epoxide function of the hydrocarbon (eq. 14.19). The diolepoxide is a relatively reactive alkylating agent, because any positive charge generated at C-10 from the epoxide ring opening will be stabilized by the adjacent aromatic ring. At the same time, its structure is such that it does not fit well on the enzyme which could hydrate its 9,10-epoxide, and thereby destroy its alkylating ability.

(14.19)

DNA alkylation is irreversible and disrupts a number of the normal functions of the DNA molecule. We do not yet know which of these disruptions is responsible for triggering carcinogenesis.

One of the ironic conclusions that came out of the polycyclic hydrocarbon investigation is that carcinogenesis is an *accidental* by-product of a very important cellular function, namely, oxidation and subsequent excretion of a foreign substance. A leading researcher in this area, R. G. Harvey, expressed the matter this way:

> Carcinogenesis appears to be the consequence of a rare event in which the biochemical machinery of enzymatic detoxification affords a reactive metabolite sufficiently stable to survive long enough to alkylate a critical cellular receptor, presumably DNA, in an appropriate molecular region to induce a nonlethal mutation ultimately resulting in tumor formation.

1 Benzene is a hydrocarbon with the molecular formula C_6H_6. The formula suggests extensive unsaturation, but benzene resists halogenation and oxidation under conditions that lead to rapid reaction with alkenes. Benzene also differs from alkenes in reacting by substitution rather than addition. Only one monosubstitution product of benzene with a given functional group is known, and only three disubstitution products with any particular pair of substituents are known. In order to account for this apparent high symmetry, Kekulé proposed a hexagonal structure for benzene with alternating single and double bonds that continually change positions with one another (secs. 14:1 and 14:2).

2 We now view the two Kekulé structures as equivalent resonance structures, and account for the aromatic stability of benzene in terms of the π-electron delocalization which these structures symbolize. Not all cyclic conjugated molecules exhibit π-electron delocalization, however. Molecular orbital calculations provide a better guide to aromaticity than do resonance structures. They show that the π electrons in benzene have considerably less total energy than the π electrons in 1,3,5-hexatriene, the open-chain analog of benzene (sec. 14:3).

X-ray analysis of benzene indicates that the carbon atoms lie at the vertices of a regular plane hexagon and that each one bears a hydrogen atom. The carbon–carbon distance is intermediate between that found in single bonds and in double bonds (sec. 14:4).

3 The resonance energy of benzene is defined as its stabilization energy relative to its hypothetical bond-localized isomer, cyclohexatriene. Heat of hydrogenation data allow us to estimate a value of 36 kcal/mol for this stabilization, which is due to delocalization of the benzene π electrons (sec. 14:5).

This delocalization becomes apparent when the molecule is placed in a magnetic field. A ring current is induced that leads, in turn, to the appearance of a local magnetic field around the benzene ring. The local field deshields the hydrogen nuclei, causing a low-field proton resonance for benzene. The uv spectrum of benzene is also very characteristic of its π system. The easily observable bands at 203 and 254 nm are diagnostically valuable, because their peaks shift when different substituents are attached to the ring (sec. 14:6).

4 "Aromaticity" (enhanced stability and diminished reactivity) is not a phenomenon unique to benzene. Hückel's rule predicts that monocyclic, planar conjugated hydrocarbons having $(4n + 2)$ π electrons will be aromatic. This prediction has been verified for a number of compounds, including ones in which a CH=CH unit is replaced by N (pyrrole) or by S (thiophene) (sec. 14:7).

Ions as well as neutral molecules may be aromatic. The best known aromatic ions are the CH homologs of benzene, the cyclopentadienide ion and cycloheptatrienyl cation (tropylium ion). Although these species differ in reactivity, they share the property of being more stable and less reactive than their open-chain analogs (sec. 14:7).

5 As a substituent, the benzene ring has a substantial stabilizing effect on attached radical or charge centers. Delocalization of the unpaired electron or charge delocalization in benzyl cations ($C_6H_5CH_2^+$), radicals ($C_6H_5CH_2\cdot$), or anions ($C_6H_5CH_2^-$) makes the formation of these reaction intermediates very favorable. The odd electron or charge is delocalized to the *ortho* and *para* ring carbons (sec. 14:8).

6 Coal tar, after appropriate work-up, yields substituted benzenes as well as a variety of aromatic systems with fused rings (sec. 14:9). These polycyclic aromatic hydrocarbons are generally more reactive than benzene, and as the number of rings increases, parts of the molecule become rather alkenelike in behavior (sec. 14:10). The carcinogenicity of polycyclic aromatic hydrocarbons is most likely due to their metabolic oxidation products, diolepoxides that can alkylate DNA (sec. 14:11).

Problems

14–16 a. Draw the structures of the three dimethylbenzenes [$C_6H_4(CH_3)_2$].
b. Each dimethylbenzene, upon reaction with molecular bromine, may be converted into one or more bromodimethylbenzene(s) [$C_6H_3Br(CH_3)_2$]. Write the structures of all the bromodimethylbenzenes that may be obtained from each dimethylbenzene. (You need to pay attention to the *symmetries* of the dimethylbenzenes.)

14–17 Körner's method (published in 1880) was a technique for deciding which isomer of a symmetrically disubstituted benzene is the *ortho* isomer, which is the *meta,* and which is the *para*. It depends on a study of the number of reaction products of each of the three isomers, as illustrated in Problem 14–16. Examine those products and see if you can determine how Körner's method would have worked. What experimental problems might interfere with the application of this method?

14–18 Write the structures of all the monochloro derivatives of each of the following aromatic hydrocarbons. Remember that each one can be represented by more than one Kekulé structure.

a. naphthalene
b. biphenyl (C_6H_5—C_6H_5)
c. anthracene
d. diphenylmethane
e. phenanthrene
f. terphenyl (1,4-diphenylbenzene)

14–19 Draw in the nodal planes for the antibonding orbitals of benzene (Fig. 14–1). Do the benzene π orbitals obey the general rule that increasing energy is associated with increasing number of nodes (sec. 13:11)? Choose an appropriate symmetry plane and classify each of the benzene π orbitals as symmetric or antisymmetric.

14–20 When the compound biphenyl, C_6H_5—C_6H_5, is reduced to cyclohexylcyclohexane, about 100 kcal/mol of heat are evolved. What does this tell us about the extent of conjugation between the two rings? What conclusions might you draw about the geometry of biphenyl?

14–21 The ^1H nmr spectrum of a benzene derivative with the formula $C_8H_{10}O$ is shown here (the numbers denote the relative number of protons). Propose a structure for this compound, including a specific isomer assignment if appropriate. Assign the peaks to various hydrogens in the molecule and explain your reasoning.

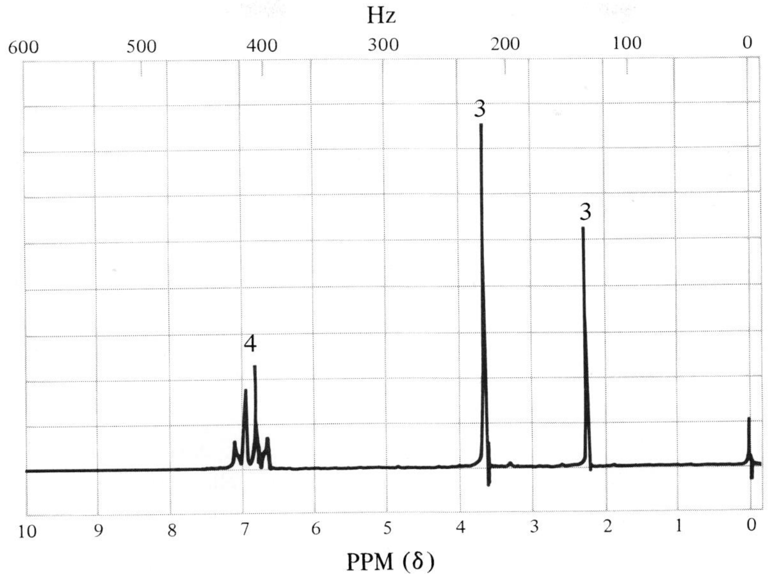

14–22 Indicate which of the following molecules you would expect to be aromatic and which you would not. Give your reasons.

a. 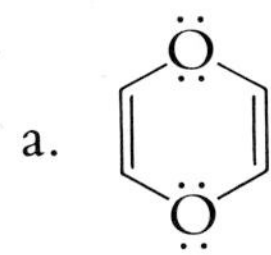

d. e.

b. 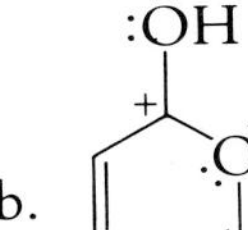

c.

14–23 Indicate whether or not the following molecules are aromatic (consider Hückel's rule):

a. 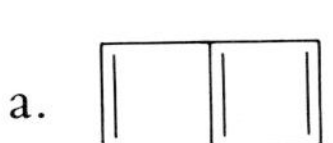

b. 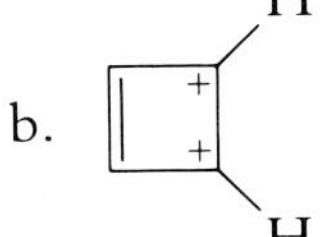

c. 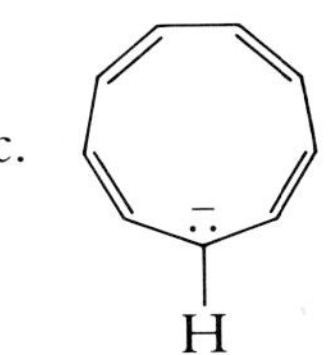

14–24 Would you expect the following carbonyl compounds to form enols readily (see eqs. 12.19 and 12.20, p. 443)? Explain your answer.

a. cyclobutenone   b. 2,4-cyclohexadienone

14–25 When dihydropentalene is treated with 2 mol of the strong base butyllithium, it is readily converted to a dianion, $C_8H_6^{2-}$ (eq. 4).

$$\text{(dihydropentalene)} + 2\text{BuLi} \xrightarrow{\text{THF}} C_8H_6^{2-} + 2\text{Li}^+ + 2C_4H_{10} \qquad (4)$$

Draw several resonance structures for the dianion $C_8H_6^{2-}$ and explain why it is formed with such relative ease.

14–26 a. 3-Iodo-1,5-pentadiene (**III**) solvolyzes *very* rapidly in ethanol to yield a mixture of ethoxypentadienes. By what mechanism does this reaction proceed? Draw the structure of the intermediate in this reaction.

b. 5-Iodo-1,3-cyclopentadiene (**IV**) is completely unreactive under the same conditions. Explain the difference in behavior between **III** and **IV**.

$CH_2{=}CH{-}CH(I){-}CH{=}CH_2$

III

IV

14–27 Addition of HCl separately to 1-phenyl-1-propene and 3-phenyl-1-propene gives different isomeric products. Draw the structures of these products and explain why each is formed from its starting material.

14–28 a. Is either the benzyl cation or benzyl anion aromatic? Explain your answer.

b. There is evidence to indicate that in the gas phase the benzyl cation rearranges to its isomer, the tropylium ion. What is the driving force for this rearrangement?

14–29 In the presence of free radicals, styrene is rapidly polymerized in a head-to-tail manner to yield polystyrene. This is a process of enormous commercial importance—in 1979 U.S. production of polystyrene and styrene copolymers came to about 4×10^6 tons!

$$2n\, C_6H_5CH{=}CH_2 \xrightarrow{R\cdot} {+}CH_2{-}CH(C_6H_5){-}CH_2{-}CH(C_6H_5){+}_n$$

Write a mechanism for the polymerization of styrene. Also explain why styrene is very easy to polymerize by comparison with other monomers, such as ethylene and vinyl chloride (sec. 10:7).

14–30 a. Draw the structures of all the isomers of phenylpropene, C_9H_{10}.
b. Each of these isomers was brominated with *N*-bromosuccinimide (NBS, sec. 13:3), leading to substitution of Br for H in the alkene portion of the molecule. Show all the possible products that could be formed from each starting material.
c. The rate-determining step in the NBS bromination is formation of an intermediate radical. Show this radical-forming step for each starting material and indicate their order of reactivity. Explain your answer, using appropriate resonance structures.

14–31 Trityl chloride (p. 532) undergoes S_N1 substitution far more easily than benzyl chloride. With the aid of resonance structures, explain this reactivity difference.

14–32 The addition of bromine to *cis*-2-butene gives only racemic 2,3-dibromobutane and none of the *meso*-diastereomer. This stereospecificity has been explained by the formation of a bromonium ion intermediate (sec. 10:4). By contrast, the addition of bromine to *cis*-1-phenyl-1-propene gives a mixture of the two diastereomeric 1,2-dibromo-1-phenylpropanes. Draw unambiguous structures for these two diastereomers, and also explain why bromine addition to phenylpropene is *not* stereospecific.

14–33 The reaction of benzyl bromide ($C_6H_5CH_2Br$) with water in acetone is rapid. Under the same conditions, 1-bromomethylnaphthalene (**V**) reacts even more rapidly. Show the products of each reaction and explain the relative reactivities. What is the structure of the intermediate in each reaction?

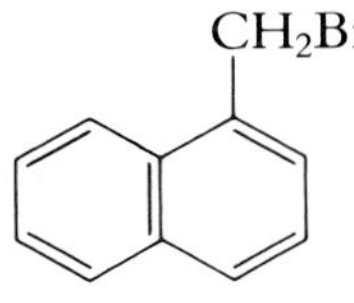

CH_2Br

V

14–34 The benzene ring current deshields not only protons directly attached to it, but also those in the vicinity of the ring. For instance, the methyl protons in toluene, $C_6H_5CH_3$, absorb at δ 2.3, about 1.3 ppm downfield from the position in aliphatic hydrocarbons. This deshielding effect provides an important clue to aid in solving the following problems; the use of a model may also help.

Addition of Br_2 to phenylacetylene gives a mixture of stereoisomeric α,β-dibromostyrenes (eq. 5).

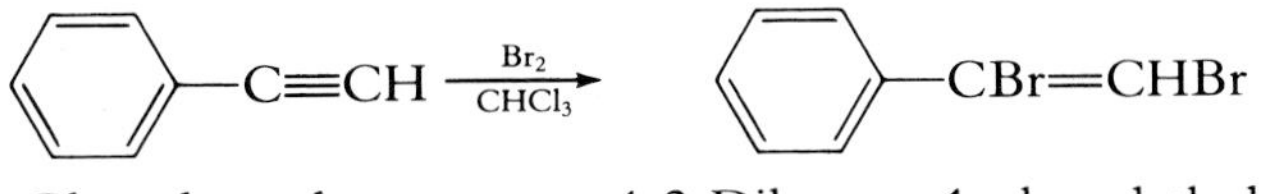

$$C_6H_5\text{—C}\equiv\text{CH} \xrightarrow[CHCl_3]{Br_2} C_6H_5\text{—CBr=CHBr} \quad (5)$$

Phenylacetylene — 1,2-Dibromo-1-phenylethylene (*α,β-dibromostyrene*)

a. The H on the double bond gives an nmr peak at δ 6.75 for one isomer and

δ 6.98 for the other isomer. Write the structures for each isomer and indicate which peak arises from which isomer. Explain your reasoning.

b. There is a constitutional isomer of the two α,β-dibromostyrenes that has a

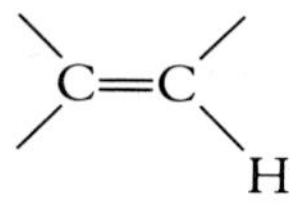

$$\underset{\diagup}{\diagdown}C{=}C\underset{\diagdown H}{\diagup}$$

resonance at δ 7.40. Write the structure of this isomer and explain why its proton is so deshielded.

14–35 1-Phenyl-2-propen-1-ol (**VI**) gives rise to two medium intensity uv peaks at 225 and 259 nm. When **VI** is treated with a catalytic amount of H_2SO_4 it rearranges to an isomer with an intense absorption at 251 nm. Assign a structure to the isomer, and explain the driving force for the rearrangement.

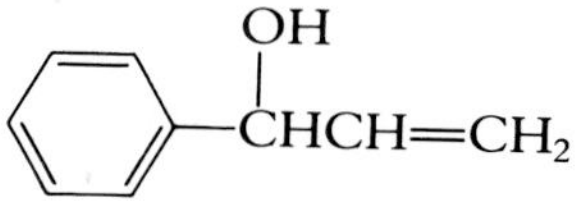

$$C_6H_5\text{-}\underset{}{\overset{OH}{\overset{|}{C}}}HCH{=}CH_2$$

VI

14–36 Large scale commercial production of isopropylbenzene, commonly known as *cumene,* is accomplished by the reaction of benzene with propene (see Prob. 15–9, sec. 15:6). Its principal use involves air oxidation to cumene hydroperoxide (eq. 6), which is used to make phenol (Chap. 24) and acetone.

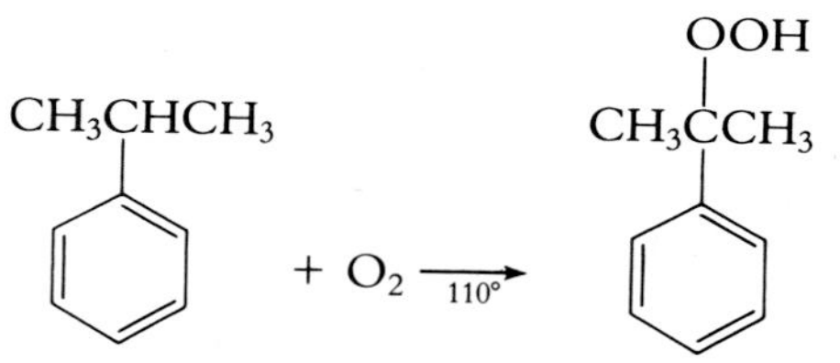

$$C_6H_5CH(CH_3)_2 + O_2 \xrightarrow{110^\circ} C_6H_5C(OOH)(CH_3)_2 \quad (6)$$

Cumene — Cumene hydroperoxide

Oxygen, you may remember, is a ground state diradical (sec. 5:13). Use this information to propose a mechanism for the reaction in Equation 6, and explain why the reaction takes place under relatively mild conditions.

14–37 Provide structures for compounds **A** to **D**.

$$\underset{\lambda_{max} = 246,281\text{ nm}}{\mathbf{A}(C_7H_6O)} \xrightarrow[\text{ether}]{EtMgBr} \xrightarrow{H_3O^+} \underset{\lambda_{max} = 222,258\text{ nm}}{\mathbf{B}(C_9H_{12}O)} \xrightarrow{H_2CrO_4}$$

$$\underset{\lambda_{max} = 241,279\text{ nm}}{\mathbf{C}(C_9H_{10}O)} \xrightarrow[\text{heat}]{N_2H_4,\ OH^-} \underset{\lambda_{max} = 206,260\text{ nm}}{\mathbf{D}(C_9H_{12})}$$

14–38 The ^{13}C nmr spectrum of ethylbenzene is given here. Draw the structure of ethylbenzene and assign the peaks to the various carbon atoms.

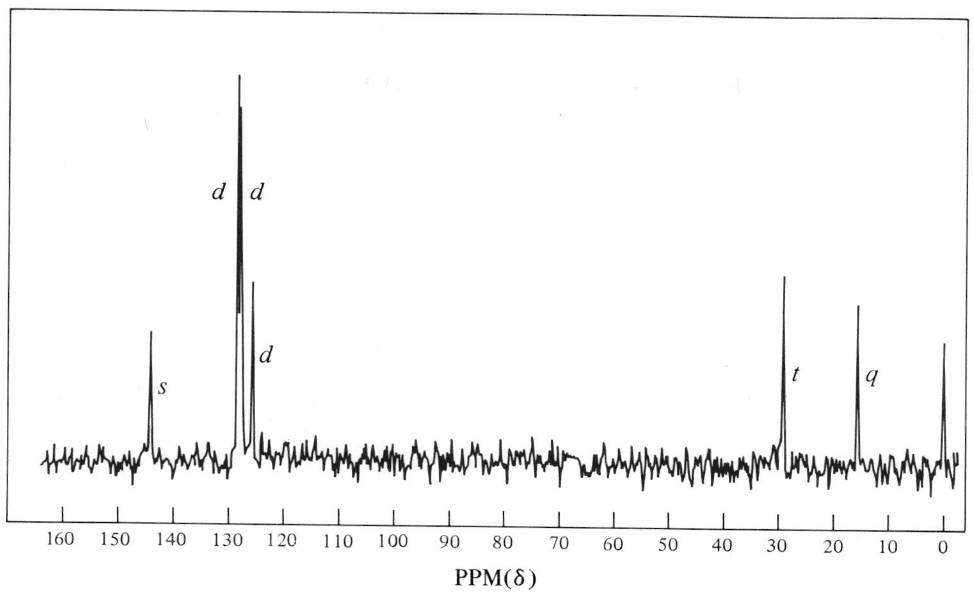

Supplementary Reading

Kekulé Symposium, *Advances in Chemistry Series,* No. 61, American Chemical Society, Washington, D.C., 1966.

R. Kaiser, The discovery of benzene, *Angewandte Chemie International Edition in English,* **7,** 345 (1968).

M. D. Saltzman, Benzene and the triumph of the octet theory, *J. Chem. Ed.,* **51,** 498 (1974).

K. Hafner, August Kekulé—The architect of chemistry, *Angew. Chem. Int. Ed. Eng.,* **18,** 641 (1979).

G. M. Badger, *Aromatic Character and Aromaticity,* London, Cambridge Univ. Press, 1969.

D. J. Sardella, Where does resonance energy come from, *J. Chem. Ed.,* **54,** 217 (1977).

R. G. Harvey, Polycyclic hydrocarbons and cancer, *American Scientist,* **70,** 386 (1982); Activated metabolites of carcinogenic hydrocarbons, *Accounts of Chemical Research,* **14,** 218 (1981).

D. A. Labianca, The chimney sweepers' cancer: An interdisciplinary approach to chemical carcinogenesis, *J. Chem. Ed.,* **59,** 843 (1982).

15 Electrophilic Aromatic Substitution

Benzene has some features in common with other unsaturated hydrocarbons, including alkenes (Chap. 10) and alkynes (Chap. 12). Like them, benzene reacts with typical electrophilic reagents such as molecular halogens, carbocations, and strong acids. Furthermore, all these hydrocarbons react by way of a two-step mechanism in which the intermediate is a cation. In spite of the similarities, there are some striking and characteristic differences between the reactions of benzene and those of the other unsaturated substrates. We've already alluded to the fact that benzene reacts very slowly with typical electrophiles (sec. 14:1). In practice it is usually necessary to coax benzene into reacting by adding a catalyst. Furthermore, benzene undergoes a substitution reaction in which a portion of the attacking reagent replaces one of the hydrogen atoms, leaving the π system intact. Alkenes and alkynes, by contrast, usually add the entire reagent and lose a π bond in the process.

Equation 15.1 shows the overall reaction mechanism for the $FeBr_3$ catalyzed bromination of benzene, with one resonance structure of the intermediate cation given.

(15.1) C_6H_5–H $+ Br_2 \xrightarrow{FeBr_3}$ [$C_6H_6Br^+$ (H, Br, +)] $+ FeBr_4^-$ $\longrightarrow$ C_6H_5–Br $+ HBr + FeBr_3$

If this reaction is defined in terms of the difference between reactants and products, it is a substitution reaction; Br has been substituted for H. If it is defined in terms of the mechanism of the reaction, it is an addition–elimination reaction. These typical reactions of benzene rings are usually called **electrophilic aromatic substitution reactions,** a broad sample of which are introduced in the next section.

15:1 Common Electrophilic Substitution Reactions of Benzene

Equations 15.2 to 15.8 are examples of the most widely used aromatic substitution reactions.

(15.2) C_6H_5–H $+ DF \xrightarrow{BF_3}$ C_6H_5–D $+ HF$ deuteration (sec. 15:2)

Benzene-*d*
(deuteriobenzene)

(15.3) Benzene (H) + HNO_3 $\xrightarrow{H_2SO_4}$ Nitrobenzene ($C_6H_5\overset{+}{N}(=O)O^-$) + H_2O nitration (sec. 15:3)

Nitrobenzene

(15.4) Benzene (H) + Cl_2 $\xrightarrow{AlCl_3}$ Chlorobenzene (Cl) + HCl chlorination (sec. 15:4)

Chlorobenzene

(15.5) Benzene (H) + Br_2 $\xrightarrow{FeBr_3}$ Bromobenzene (Br) + HBr bromination (sec. 15:4)

Bromobenzene

(15.6) Benzene (H) + SO_3 $\xrightarrow{H_2SO_4}$ $C_6H_5SO_3H$ sulfonation (sec. 15:5)

Benzenesulfonic acid

(15.7) Benzene (H) + CH_3CH_2Cl $\xrightarrow{AlCl_3}$ $C_6H_5CH_2CH_3$ + HCl alkylation (sec. 15:6)

Ethylbenzene

(15.8) Benzene (H) + $CH_3CH_2CH_2\overset{O}{\overset{\|}{C}}Cl$ $\xrightarrow{AlCl_3}$ $C_6H_5\overset{O}{\overset{\|}{C}}CH_2CH_2CH_3$ + HCl acylation (sec. 15:7)

1-Phenyl-1-butanone
(*butyrophenone*)

All of these reactions require an acid catalyst. Two of them (15.3 and 15.6) use a proton acid, while the remainder use Lewis acids. The Lewis and proton acids are all electron-deficient and are capable of accepting an electron pair. The acid reacts with the reagent to generate the electrophilic species that actually attacks benzene.

The next few sections present detailed mechanisms for each of these substitution reactions. We will begin with the hydrogen exchange reaction (eq. 15.2) because, from it, we can easily determine the structure of the cation intermediate common to all the reactions listed above.

15:2 Hydrogen Exchange in Benzene: Evidence for the Benzenium Ion

Hydrogen fluoride (HF) dissolves to some extent in bezene at low temperatures. Since benzene is a nonpolar solvent, the HF does not dissociate and the solution is nonionic. Subsequent addition of the Lewis acid BF_3 to this solution results in a marked change of solution behavior. In particular, electrical measurements show the formation of ions.

Boron trifluoride (BF_3) and hydrogen fluoride are known to form fluoroboric acid (eq. 15.9), a powerful electrophile. Nuclear magnetic resonance studies of benzene derivatives indicate that this acid protonates the benzene ring to form a carbocation called a **benzenium ion** (**1,** eq. 15.10).

(15.9) $$BF_3 + HF \rightleftarrows \underset{\text{Fluoroboric acid}}{H^+BF_4^-}$$

Unlike benzene, ion **1** is not aromatic, although it *is* resonance stabilized.

(15.10)

1
(benzenium ion)

The carbon atom bonded to the two hydrogens is sp^3 hybridized, resulting in an interruption of the cyclic conjugation. Intermediate **1** is sometimes referred to as the **σ-complex** because one pair of π electrons has been taken from the conjugated system and used to form a new σ bond to the electrophile (in this case, H^+).

The conversion of benzene to benzenium ion involves a loss of aromaticity and is therefore endothermic. Conversely, release of a proton from the sp^3 carbon atom of the ion results in aromaticity being regained, so this step is exothermic. We can generalize from these considerations and conclude that in electrophilic aromatic substitutions, the first of the two reaction steps (as in eq. 15.1) has the larger energy barrier and is therefore rate-determining.

By using deuterium fluoride (DF) instead of HF in the reaction in Equation 15.10 we can deuterate benzene (eq. 15.2). This reaction provides further evidence for a benzenium ion intermediate (**1,** eq. 15.11).

(15.11) Benzene (H) $\xrightleftharpoons{D^+BF_4^-}$ [benzenium ion resonance structures (H, D) $\longleftrightarrow$ (H, D) $\longleftrightarrow$ (H, D)] BF_4^- $\rightleftharpoons$ (D) $+ H^+BF_4^-$

1

Benzene-*d*

The energy diagram for deuteration conforms to the generalization just discussed—a large energy barrier for benzenium ion formation and a small energy barrier for its conversion back to benzene (Fig. 15–1).

As far as is known, all electrophilic substitution reactions of benzene and its derivatives involve benzenium ions as intermediates.

Let's examine benzenium ions in a little more detail. Although the benzenium ion is not aromatic, it is stabilized by delocalization of its four remaining π electrons over the five sp^2 carbons. As the resonance structures show (eqs. 15.10 and 15.11), the positive charge is concentrated mainly on three carbon atoms—two *ortho* and one *para* to the carbon that was protonated. In order to avoid writing all three resonance structures for each benzenium ion,

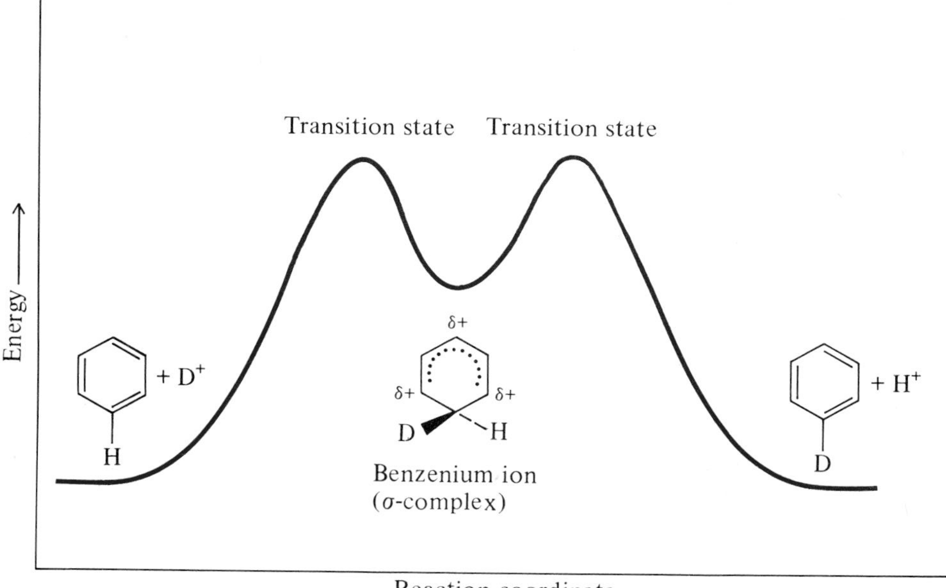

Figure 15–1. *Energy diagram for the deuteration of benzene with D^+. The diagram shows the benzenium ion intermediate and the energy of the transition states for its formation and decomposition.*

the abbreviated notation **2** can be used. This is the same type of notation used previously in depicting the allyl cation (Chap. 13).

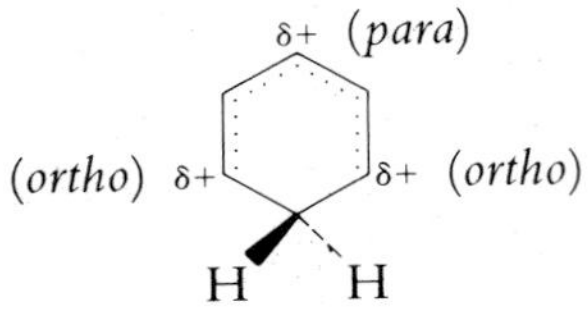

δ+ (*para*)

(*ortho*) δ+ δ+ (*ortho*)

H H

2

It is important to remember that **2** is identical to **1,** and to be able to draw both structures correctly.

Deuteration of benzene is not a widely used reaction. We took it up because it is a clear and convenient example for introducing the benzenium ion. By using electrophiles other than D^+ we can convert benzene to a variety of useful products. We'll begin by discussing the nitration of benzene.

Example

Draw all the resonance structures for the benzenium ion produced by protonation of methoxybenzene (anisole) at C-3.

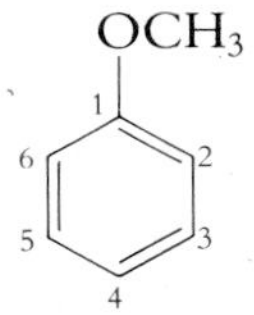

OCH_3

1 2 3 4 5 6

Answer

The first step is simply to add a proton to C-3 by using the π electrons in the C-3,C-4 double bond; this produces an electron deficiency at C-4.

OCH_3 H^+ → OCH_3 H H +

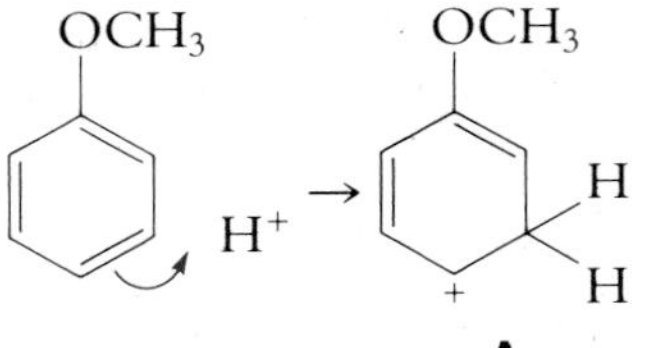

A

Now move π electrons from the C-5,C-6 double bond to relieve the electron deficiency at C-4. That, in turn, creates an electron deficiency at C-6.

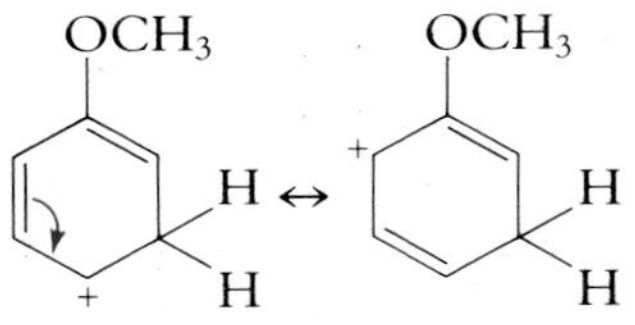

OCH_3 H ↔ OCH_3 H + H + H

A **B**

Repeat the process with the next double bond:

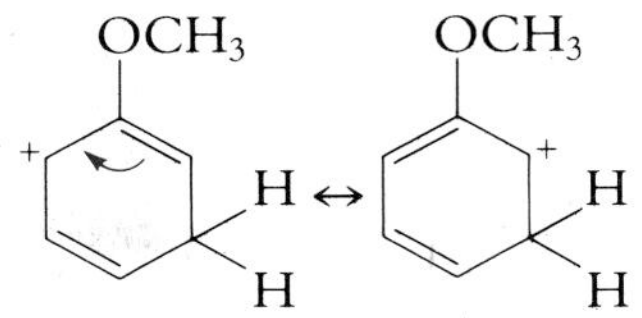

OCH$_3$ OCH$_3$

H ↔ H

H H

B **C**

The three resonance structures are **A, B,** and **C.**

Problem

15–1 Write all the resonance structures for the benzenium ions produced by protonation of the following hydrocarbons at the specified ring position. Draw a single structure for the resonance hybrid (refer to structure **2**).

a. toluene (methylbenzene) at C-4
b. 1,4-dimethylbenzene at C-2
c. naphthalene at C-1
d. anthracene at C-9

15:3 Nitration of Benzene

Benzene and its derivatives are most commonly nitrated with a solution of concentrated nitric acid in concentrated sulfuric acid. The H_2SO_4 is the acid catalyst required by this reaction, and the $HONO_2$ acts as a base. It has been shown that sulfuric acid dehydrates nitric acid to the **nitronium ion,** $O{=}\overset{+}{N}{=}O$ (eq. 15.12).

$$HONO_2 + H_2SO_4 \rightarrow H\overset{+}{O}(H){-}NO_2 + HSO_4^- \quad (15.12)$$

$$H\overset{+}{O}(H){-}NO_2 \rightarrow H_2O\colon + \; {}^+NO_2$$

Nitronium ion

The nitronium ion is a powerful electrophile and is the species that actually attacks the benzene ring. Salts containing this cation have been isolated and are capable of nitrating benzene in the absence of any additional reagents.

The mechanism of nitration involves attack of NO_2^+ on the π electrons of benzene, and the formation of a benzenium ion (eq. 15.13). Subsequent loss of the proton from the nitro-bearing carbon yields nitrobenzene (eq. 15.14). In Equation 15.14 we use one of the resonance structures to illustrate how the proton is lost. Of course, the ion that actually loses a proton to water is a hybrid of all three resonance structures.

(15.13) H O=N$^+$=O $\xrightarrow{\text{slow}}$

[H, O, N$^+$, O$^-$ ↔ H, NO$_2$ ↔ H, NO$_2$] ≡ δ+, δ+, δ+, H, O, N$^+$, O$^-$

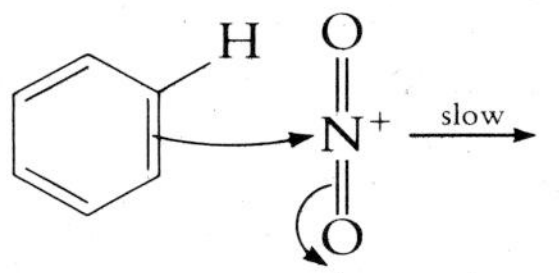

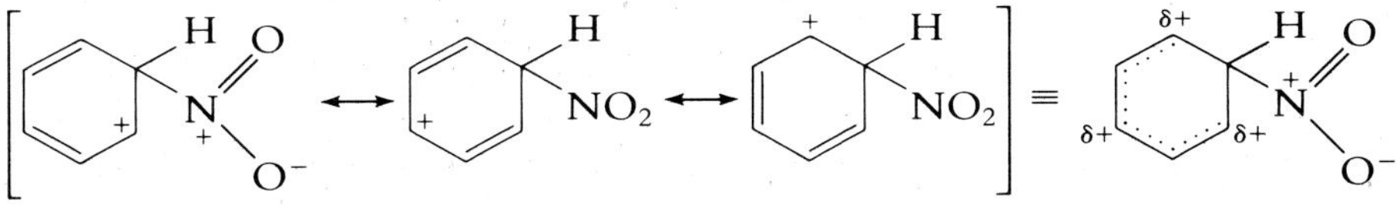

(15.14) [Chemical structure: benzenium ion with H and NO_2; $:OH_2$ removes H; $\xrightarrow{\text{fast}}$ nitrobenzene $+ H_3O^+$]

Nitrobenzene

The **nitro** group is attached to the benzene ring by way of the positively charged nitrogen atom, with the negative charge delocalized over both oxygen atoms.

[Chemical structures: two resonance forms of nitrobenzene connected by ↔]

Resonance structures for nitrobenzene

An energy diagram for the nitration reaction (Fig. 15–2) shows that it takes relatively little energy to get to the intermediate from the reactants. However, the energy barrier is high for returning to the intermediate from products. As a result, nitration is irreversible and acid-catalyzed denitration of nitrobenzenes is never observed.

Nitration of benzene is an important reaction because the nitro group can be converted into a large variety of other functional groups (Chap. 23). Halogenation of benzene is similarly important because halobenzenes serve as intermediates for the synthesis of many other substituted benzenes. The next section describes how halobenzenes are prepared.

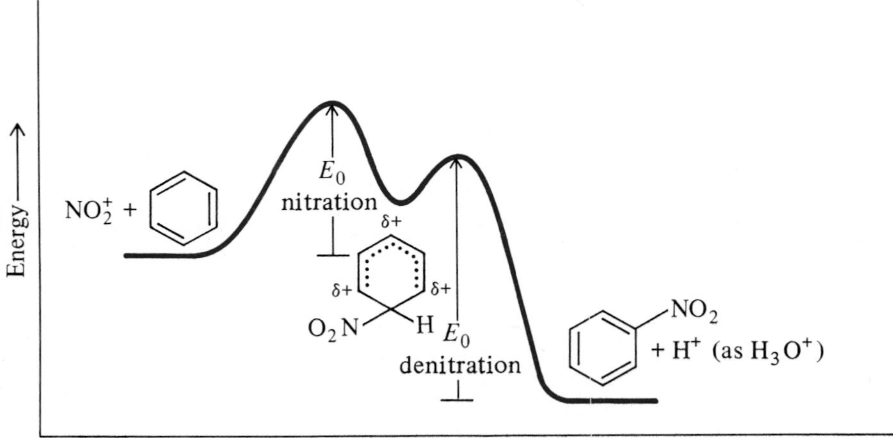

Figure 15–2. *Energy diagram for nitration of benzene. There is a large energy barrier between products and the benzenium ion, so denitration is very slow.*

Bromination and chlorination of aromatic compounds require the presence of a Lewis acid as catalyst—$AlCl_3$, $AlBr_3$, $FeCl_3$, and $FeBr_3$ are the most common. The function of the catalyst is to cause the halogen molecule to cleave *heterolytically,* by apparently pulling away Cl^- or Br^- and leaving behind the electrophile Cl^+ or Br^+ (eq. 15.15).

(15.15)
$$FeCl_3 + :\ddot{Cl}—\ddot{Cl}: \rightarrow [Cl—FeCl_3]^- + :\ddot{Cl}^+$$

It is probably more accurate to consider cleavage of the chlorine molecule, and attachment of one of the chlorines to the benzene, as a concerted process (eq. 15.16, slow step). As the catalyst begins to abstract Cl^- (or Br^-), the Cl^+ (or Br^+) is simultaneously transferred to the benzene ring, converting it to a benzenium ion. The benzenium ion then transfers a proton to the anion and yields the halobenzene.

(15.16)
$$C_6H_6 + Cl—Cl + FeCl_3 \xrightarrow{\text{slow}} [C_6H_6Cl]^{\delta+}\ \ H \cdots Cl—FeCl_3^- \xrightarrow{\text{fast}} C_6H_5Cl + H—Cl + FeCl_3$$

Chlorobenzene

An energy diagram of the chlorination reaction (Fig. 15–3) shows that the benzenium ion loses H^+ much more readily than Cl^+. This means that

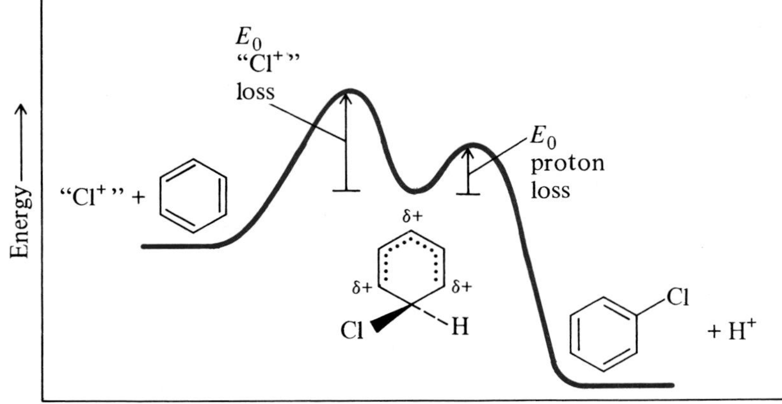

Figure 15–3. *Energy diagram for chlorination of benzene. The benzenium ion loses H^+ much faster than it loses Cl^+, making the reaction irreversible.*

chlorination, like nitration, is irreversible—once a Cl atom attaches itself to the ring it cannot be converted back to Cl_2.

In the absence of a Lewis acid catalyst, Cl_2 and Br_2 cleave homolytically, initiating free-radical reactions (Chap. 5). However, the electrophilic reactions are faster than the free-radical ones, so when a catalyst is present only the products of electrophilic substitution are found.

Problems

15–2 Write a complete mechanism for the bromination of benzene with aluminum bromide as the catalyst. Show all the resonance structures for the intermediate bromobenzenium ion. In this reaction, as in most electrophilic substitutions, formation of the benzenium ion is the rate-determining step. Draw an energy–reaction coordinate diagram for the reaction which reflects this fact.

15–3 Benzene can be chlorinated by hypochlorous acid, HOCl. The catalyst for this reaction is the hydrogen ion, H^+, and the other product is water. Write a mechanism for chlorination of benzene with hypochlorous acid, making sure you specify the role of the catalyst.

Direct iodination of benzene is thermodynamically unfavorable (eq. 15.17), in contrast to bromination and chlorination.

(15.17) $C_6H_5\text{–H} + I_2 \rightleftharpoons C_6H_5\text{–I} + HI \quad \Delta H = 12 \text{ kcal/mol}$

The reaction can nonetheless be carried out if I_2 is mixed with an oxidizing agent such as HNO_3 (eq. 15.18). The HNO_3 converts the I_2 to a powerful

(15.18) $C_6H_5\text{–H} \xrightarrow{I_2,\ HNO_3} C_6H_5\text{–I}$

Iodobenzene (86%)

electrophile, the structure of which is still not known with certainty. Since the reaction conditions shown in Equation 15.18 are fairly vigorous, they can bring about undesirable side reactions. For this reason, iodine is often introduced indirectly into organic molecules (Chap. 23).

Direct fluorination is very difficult to control because fluorine is so reactive; thus, like iodine, fluorine is usually introduced indirectly into benzene rings.

15:5 Sulfonation of Benzene and Naphthalene

Treatment of benzene with fuming sulfuric acid (concentrated sulfuric acid containing excess SO_3) leads to the formation of **benzenesulfonic acid.** The mechanism for its formation (eqs. 15.19 and 15.20) is analogous to the mechanisms presented for other electrophilic substitutions.

(15.19) $SO_3 + H_2SO_4 \rightleftarrows \left[O_2\overset{+}{S}{=}OH \leftrightarrow O_2S^{+}{-}OH \right] + HSO_4^-$

(15.20) [Benzene attacking $O_2S{=}\overset{+}{O}H$ (H, O, S, O, OH) $\rightleftarrows$ benzenium ion (δ+, H, $SO_2{-}OH$) ⟶ $C_6H_5SO_2{-}OH$ + H^+]

Benzenesulfonic acid

Very reactive compounds can be sulfonated by concentrated sulfuric acid alone because it contains a small amount of electrophilic SO_3H^+ from self-dehydration (eq. 15.21).

(15.21) $2HOSO_2OH \rightleftarrows H_2O + O_2\overset{+}{S}{=}OH\ HSO_4^-$

Benzenesulfonic acid is a strong acid with a pK_a similar to the first pK_a for sulfuric acid (eq. 15.22). The acid is therefore very soluble in water where

(15.22) $C_6H_5SO_3H + H_2O \rightleftarrows C_6H_5SO_3^- + H_3O^+ \quad pK_a = 0.3$

it dissociates completely to its anion and a hydronium ion. The closely related compound, *p*-toluenesulfonic acid, is very acidic and is also soluble in nonpolar solvents such as benzene. For these reasons it is widely used as an acid catalyst.

Unlike nitration and chlorination, sulfonation is reversible. Sulfonic acids convert back to benzenes and sulfuric acid when heated in hot aqueous acid (eq. 15.23). For both sulfonation and desulfonation, the formation of

(15.23) $C_6H_5SO_3H \xrightarrow{H_3O^+,\ heat}$ [benzenium ion, δ+, H, SO_3H] **3** $\rightleftarrows C_6H_5H + SO_3 + H^+$; $SO_3 \xrightarrow{H_2O} H_2SO_4$

the benzenium ion, **3,** is rate-determining. The energy of activation for

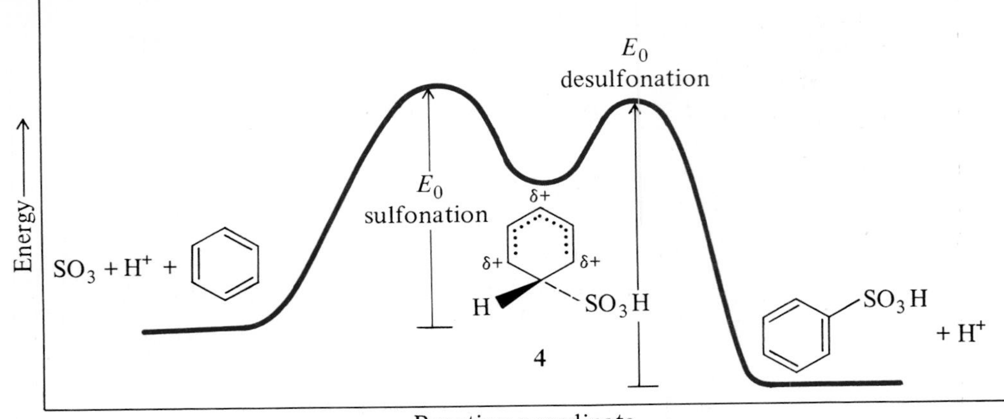

Figure 15–4. *Energy diagram for the sulfonation and desulfonation of benzene. Since the energy barrier to reach the benzenium ion is nearly the same for reactants and products, sulfonation is reversible.*

forming **3** from either reactants or products is not too different (Fig. 15–4), so the rate of desulfonation is competitive with the rate of sulfonation. Contrast this with the nitration and denitration of benzene (Fig. 15–2).

The reversibility of sulfonation explains why naphthalene gives different product distributions when sulfonated at different temperatures (eqs. 15.24 and 15.25). At 40° the two sulfonic acids are stable and do not revert back to

(15.24) $\xrightarrow[40^\circ]{H_2SO_4,\ SO_3}$

SO_3H (96%) + SO_3H (4%) kinetic control

(15.25) $\xrightarrow[160^\circ]{H_2SO_4,\ SO_3}$

SO_3H (15%) + SO_3H (85%) thermodynamic control

naphthalene and sulfuric acid. The product distribution reflects the kinetic advantage of the 1-isomer, which is formed faster than the 2-isomer and thus predominates (eq. 15.24). At 160°, however, the isomeric acids form rapidly *and* they also revert to starting materials rapidly. Since the 1-isomer is the less stable of the two, it reverts much more quickly to naphthalene and sulfuric acid, which then recombine to form the more stable 2-isomer. The product distribution at 160° is an equilibrium distribution, reflecting the greater thermodynamic stability of the 2-isomer (eq. 15.25). By starting with *either* of the pure isomeric acids and heating it to 160°, the same product distribution is obtained (eq. 15.26), identical to that formed by direct sulfonation of naphthalene at 160°.

(15.26) [1-naphthalenesulfonic acid, SO_3H] (100%) or [2-naphthalenesulfonic acid, SO_3H] (100%) $\xrightarrow[160°]{H_2SO_4}$ [1-naphthalenesulfonic acid, SO_3H] (15%) + [2-naphthalenesulfonic acid, SO_3H] (85%)

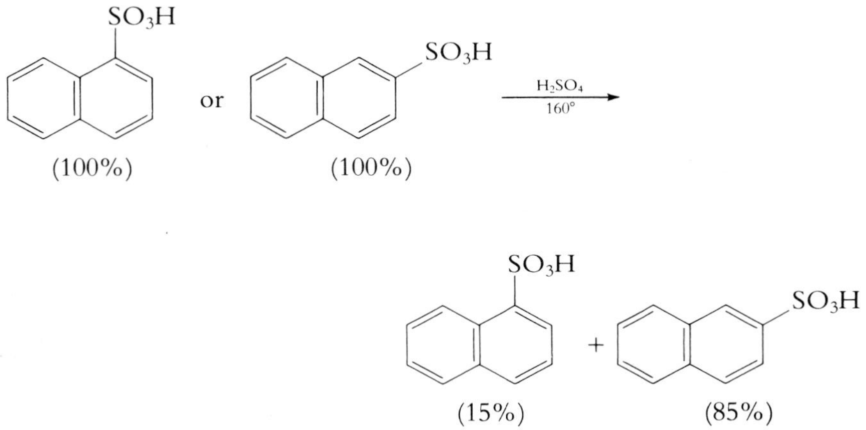

We can sum up the preceding discussion by saying that at 40° the sulfonation reaction is under kinetic control, while at 160° the reaction is under thermodynamic control. In other words, the kinetically favored product (the one that is formed faster) is the major one at low temperature, while the thermodynamically favored product (the one that is more stable) is predominant at high temperature. (You saw other examples of kinetic versus thermodynamic control in Sections 13:4 and 13:9.) Figure 15–5 summarizes the naphthalene sulfonation results.

All the electrophiles we've discussed so far have been inorganic. A very important class of organic electrophiles includes the carbocations. Their reactions with benzene lead to important products, as the next two sections will show.

Example

What are all the possible monosubstituted products from nitration of anisole?

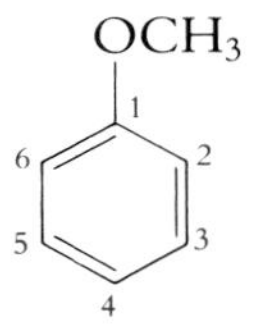

[benzene ring with OCH_3 at position 1; ring positions numbered 1–6]

Anisole

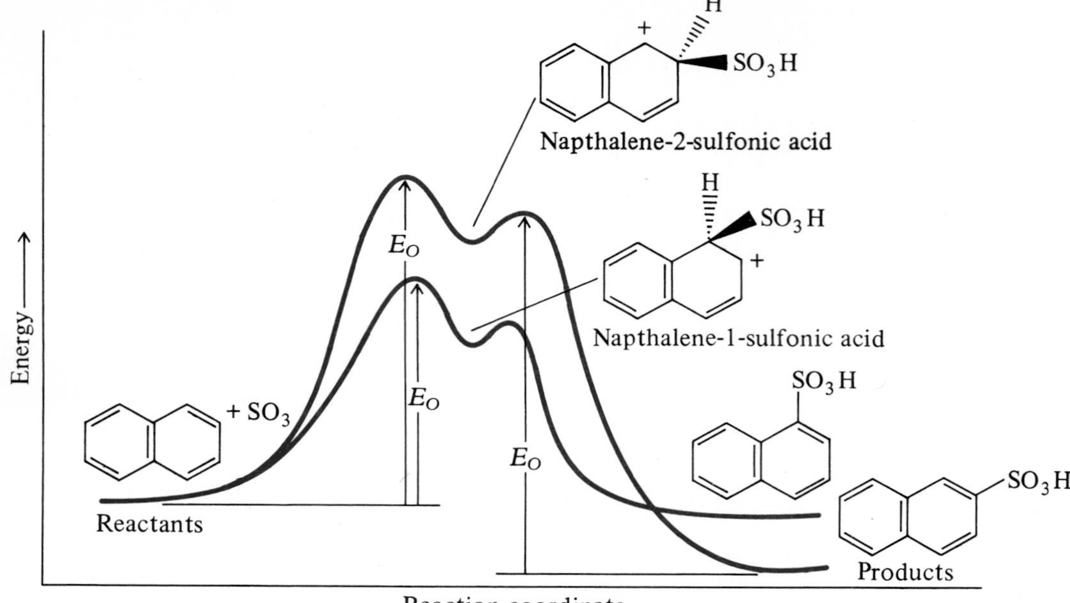

Figure 15–5. *Energy diagram for sulfonation of naphthalene. Naphthalene-1-sulfonic acid is formed faster since the energy barrier is lower than for formation of naphthalene-2-sulfonic acid. However, naphthalene-2-sulfonic acid is the more stable of the two isomers, so it becomes the major product when equilibrium is reached.*

Answer

Only three products are possible, those from nitration at carbons 2, 3, and 4:

OCH_3 / NO_2 — **A**　　OCH_3 / NO_2 — **B**　　OCH_3 / NO_2 — **C**

Carbon 5 is chemically equivalent to carbon 3, and carbon 6 is chemically equivalent to carbon 2.

OCH_3 / O_2N — same as **A**　　OCH_3 / O_2N — same as **B**

Box 15–1. Polychlorinated Biphenyls (PCBs)

For many years, a mixture of biphenyl and diphenyl ether was used as a heat exchange medium in electrical transformers. The mixture is extremely stable and so can absorb the

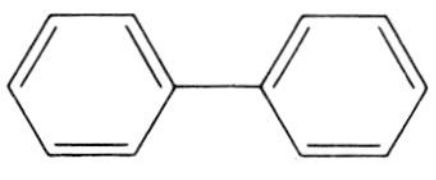

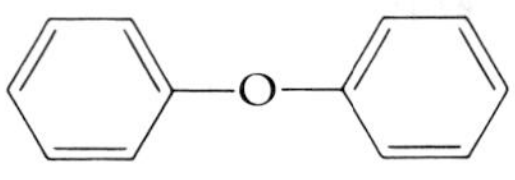

Biphenyl Diphenyl ether

heat (up to 400° C) generated in the transformer. It was then discovered that a complex mixture of products resulting from extensive chlorination of biphenyl, the so-called polychlorinated biphenyls, or PCBs, was an even better heat exchanger.

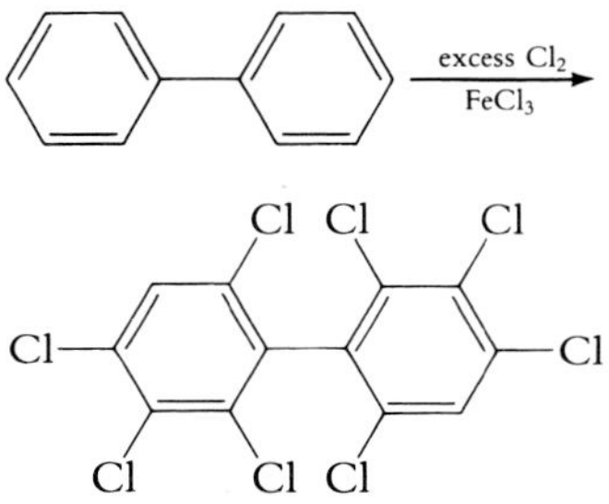

excess Cl_2, $FeCl_3$

+ many other chlorinated products

a PCB

Careless disposal of transformers and other implements contaminated with PCBs has caused extensive environmental contamination because PCBs are degraded extremely slowly in the environment. The metabolic oxidation products of PCBs are toxic to fish, birds, and people. They have been implicated in damage to embryos and in growth retardation of offspring. Since PCBs have been recognized as toxic, their use and disposal have been closely regulated; however, we are still faced with the problem of getting rid of large quantities of waste material.

The current method of PCB degradation, incineration, suffers from two disadvantages: it uses a good deal of energy, and it produces corrosive effluents such as HCl. Attention has therefore been focused on methods of chemically dechlorinating PCBs, since their toxicity is directly related to the chlorine content. One promising method that is economical and results in complete dechlorination involves the reaction of PCBs with sodium metal and oxygen from the air. The solvent is a mixture of polyethylene glycols, $HO\!\left(CH_2CH_2O\right)_nH$:

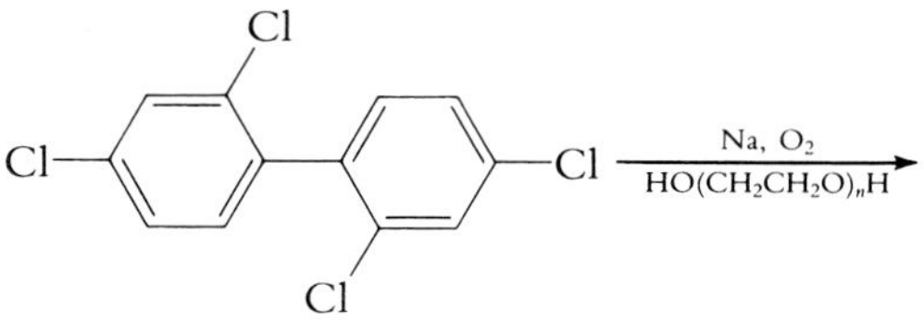

Na, O_2, $HO(CH_2CH_2O)_nH$

A PCB

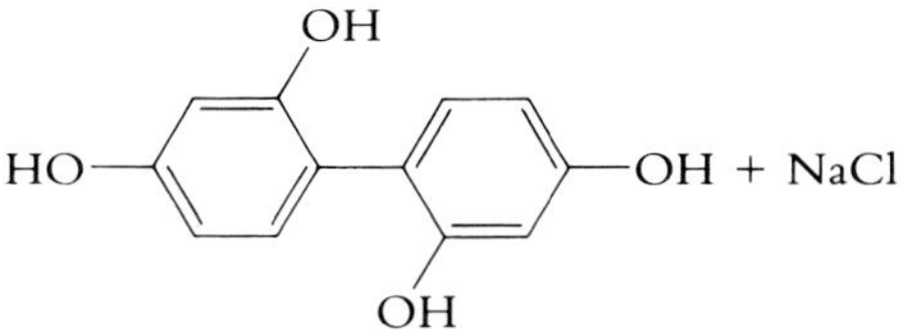

+ NaCl

A polyhydroxybiphenyl

The reaction mechanism is quite complex, involving both radical and ionic intermediates. The reaction products are environmentally benign and, in some cases, may have commercial value as antioxidants in foods and as monomers for polymerization.

Problems

15–4 Show all possible monosubstituted products formed when each of the following compounds is chlorinated or sulfonated:

a. toluene
b. 1,2-dimethylbenzene (*o*-xylene)
c. 1-bromo-4-methylbenzene
d. 1,3-dimethylbenzene (*m*-xylene)
e. fluorobenzene
f. anthracene

15–5 Benzenesulfonic acid and *p*-xylene (1,4-dimethylbenzene) are mixed together, along with a small quantity of H_2SO_4. The mixture is heated to 180°, cooled, and analyzed. In addition to the starting materials, two new products are found, benzene and *p*-xylenesulfonic acid. Write a mechanism that explains the formation of these products.

15–6 Draw resonance structures for the sulfonate group in $CH_3SO_3^-$, and explain why sulfonic acids are strong acids.

15:6 Friedel–Crafts Reactions: Carbocations as Electrophiles for Benzene

In 1877, Charles Friedel (a professor at the Sorbonne in France) and James Crafts (his American co-worker, on leave from MIT) discovered that numerous organic halides decompose in the presence of aluminum chloride (Lewis acid catalyst). When they carried out these decompositions in benzene solution, they obtained alkyl substituted benzenes in good yield.

Equation 15.27 is an example of this reaction, now known as a **Friedel–Crafts alkylation.**

(15.27) $C_6H_5\text{–}H + (CH_3)_2CHCl \xrightarrow{AlCl_3} C_6H_5\text{–}CH(CH_3)_2 + HCl$

Isopropylbenzene
(*cumene*)

The first step of this process is the reaction between the alkyl halide and aluminum chloride to form a carbocation (eq. 15.28). The second step involves the attack of the carbocation on benzene (eq. 15.29).

(15.28) $(CH_3)_2CH\text{—}Cl + AlCl_3 \rightleftarrows (CH_3)_2CH^+\ AlCl_4^-$

(15.29) $ClAlCl_3^-$; $C_6H_5\text{–}H + \overset{+}{C}H(CH_3)_2 \rightleftarrows$

$Cl\text{—}AlCl_3^-$; $\delta+$, $\delta+$, $\delta+$; H, $CH(CH_3)_2$ $\rightleftarrows C_6H_5\text{–}CH(CH_3)_2 + HCl + AlCl_3$

This mechanism is supported by the discovery that other sources of carbocations, such as alkenes in the presence of acid, are equally effective in alkylating benzene. For instance, the reaction in Equation 15.30 is a commercial process for the manufacture of ethylbenzene. Most of the ethylbenzene is, in turn, converted to styrene, the precursor to the polymer polystyrene, a familiar plastic material used in styrofoam cups and packaging materials (Prob. 14–29, p. 540).

$$(15.30)\quad CH_2{=}CH_2 \xrightleftharpoons{HCl,\ AlCl_3} CH_3{-}\overset{+}{C}H_2\ AlCl_4^- \xrightleftharpoons{C_6H_6} C_6H_5CH_2CH_3 + AlCl_3 + HCl$$

Ethylbenzene

United States production of ethylbenzene in 1979 amounted to a 4.2 × 10^6 tons, about 40% of which was made by alkylation of benzene.

As indicated in Equations 15.29 and 15.30, alkylations are reversible, particularly when the alkyl groups are secondary or tertiary. This reversibility sometimes produces undesirable mixtures of polyalkylated benzenes.

Primary haloalkanes do not react with benzenes by way of carbocations, presumably because of the high energy of the primary cations. Instead, the haloalkane and the catalyst ($AlCl_3$) form an electrophilic complex in which the positive charge rests on the halogen (eq. 15.31). The benzene then performs a nucleophilic displacement on this complex (eq. 15.32). As with other electrophilic substitutions, the Lewis acid ($AlCl_3$) is regenerated so only about 0.1 mol of it is required per mole of alkyl halide.

$$(15.31)\quad CH_3CH_2\ddot{\underset{..}{Cl}}: + AlCl_3 \rightleftarrows CH_3CH_2{-}\overset{+}{\ddot{\underset{..}{Cl}}}{-}\overset{-}{A}lCl_3$$

$$(15.32)\quad C_6H_6 + CH_3CH_2{-}\overset{+}{\ddot{\underset{..}{Cl}}}{-}\overset{-}{A}lCl_3 \rightleftarrows [C_6H_6(CH_2CH_3)]^{\delta+,\delta+,\delta+}\ AlCl_4^- \rightleftarrows C_6H_5CH_2CH_3 + HCl + AlCl_3$$

Ethylbenzene

One of the most serious problems encountered in Friedel–Crafts alkylation is the isolation of products with rearranged carbon skeletons (eq. 15.33).

$$(15.33)\quad C_6H_6 + CH_3(CH_2)_3Br \xrightarrow[70^\circ]{AlCl_3} C_6H_5(CH_2)_3CH_3 + C_6H_5CH(CH_3)CH_2CH_3$$

1-Bromobutane; 1-Phenylbutane (13%); 2-Phenylbutane (39%)

The carbocation mechanism fully accounts for this complication, since less stable carbocations are known to rearrange to more stable ones (sec. 6:8).

Even the electrophilic complex between a Lewis acid and a primary halide behaves like a carbocation, since it rearranges about as rapidly as it reacts with benzene (eq. 15.34).

(15.34) $CH_3CH_2CH(H)CH_2—\overset{+}{Br}—\overset{-}{Al}Cl_3 \xrightarrow{C_6H_6} C_6H_5(CH_2)_3CH_3$

↓ rearrangement

$CH_3CH_2\overset{+}{C}HCH_3 \; AlCl_3Br^- \xrightarrow{C_6H_6} C_6H_5CH(CH_3)CH_2CH_3$

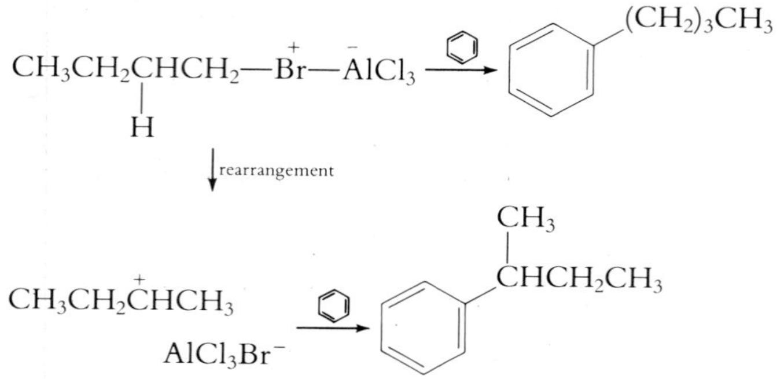

Example

Show how the following compound could be synthesized from benzene:

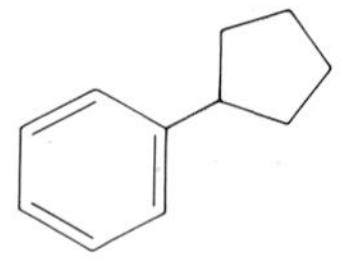

Phenylcyclopentane

Answer

To prepare phenylcyclopentane from benzene, you would need some way to form the C—C bond between the two rings:

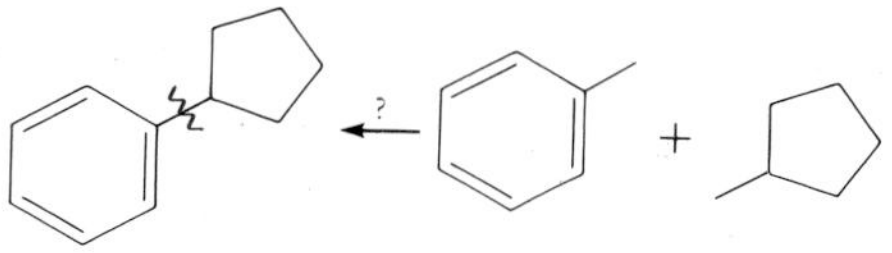

In this section, you have seen that Friedel–Crafts alkylations are useful for such transformations:

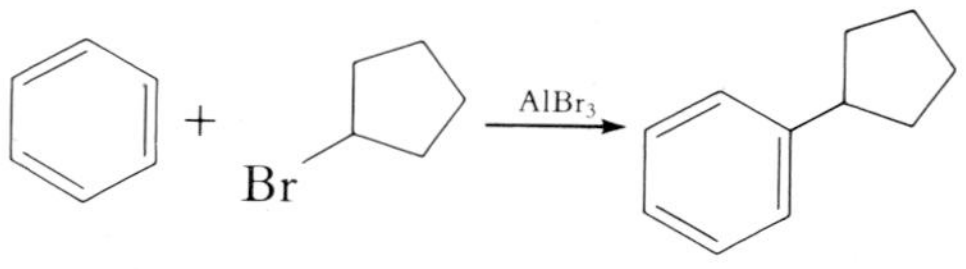

Benzene + bromocyclopentane $\xrightarrow{AlBr_3}$ phenylcyclopentane

Br

(You could have used chlorocyclopentane and $AlCl_3$ as well.)

Problems

15–7 Show how each of the following compounds can be synthesized from benzene or naphthalene. If you expect a mixture of products to result from the synthesis, show the structures of all products.

a. toluene
b. 2-phenylpentane
c. phenylcyclohexane
d. 1-isopropylnaphthalene
e. 2-propylnaphthalene

15–8 The powerful, but environmentally harmful insecticide, DDT, is manufactured by allowing chlorobenzene to react with 2,2,2-trichloroethane-1,1-diol (*chloral hydrate*), with sulfuric acid as a catalyst. Write a mechanism for this double Friedel–Crafts alkylation.

$$CCl_3CH(OH)_2 + 2\ C_6H_4\text{—}Cl \xrightarrow{H_2SO_4} CCl_3CH(\text{—}C_6H_4\text{—}Cl)_2 + 2H_2O$$

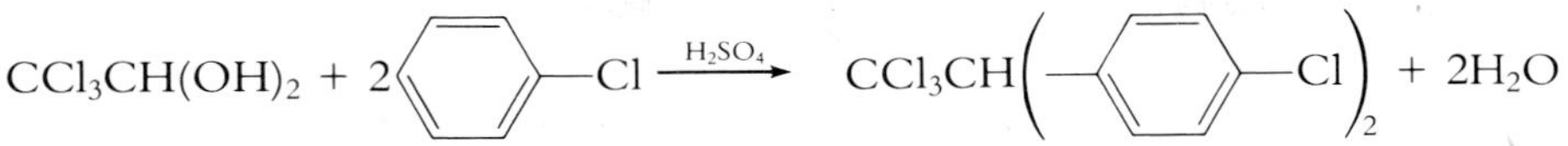

dichloro**d**iphenyl**t**richloroethane (DDT)

15–9 a. The commercially important compound isopropylbenzene (cumene) is manufactured by allowing benzene to react with propene. Propose a likely catalyst for this reaction, and show its mechanism.

b. When toluene is treated with 1-chloro-2-methylpropane and aluminum chloride, only the rearranged product 4-*t*-butyltoluene is obtained. Write a mechanism for this reaction.

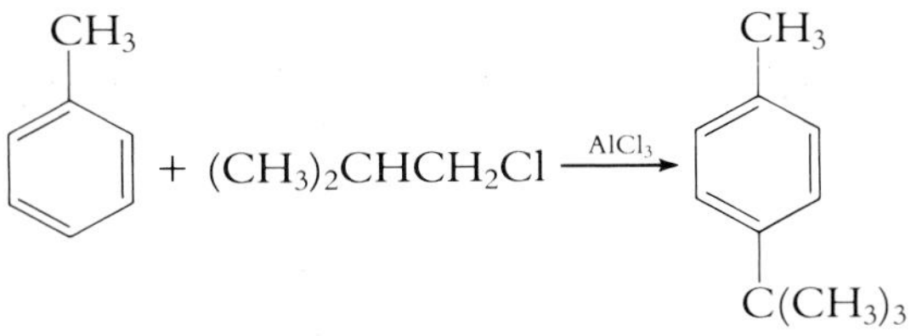

$$C_6H_5CH_3 + (CH_3)_2CHCH_2Cl \xrightarrow{AlCl_3} 4\text{-}CH_3C_6H_4C(CH_3)_3$$

In addition to rearrangement, Friedel–Crafts alkylations suffer from a tendency toward multiple substitution of the ring. This occurs because the monoalkylated benzene first formed is more reactive toward alkylation than benzene itself. Equation 15.35 shows a typical Friedel–Crafts alkylation in which both rearrangement and multiple substitution are found. Here, an alcohol is used to generate the carbocation.

$$C_6H_6 + CH_3CH_2CH_2OH \xrightarrow[60^\circ]{BF_3} C_6H_5CH(CH_3)_2 + CH_3CHCH_3\text{—}C_6H_4\text{—}CH_3CHCH_3 \quad (15.35)$$

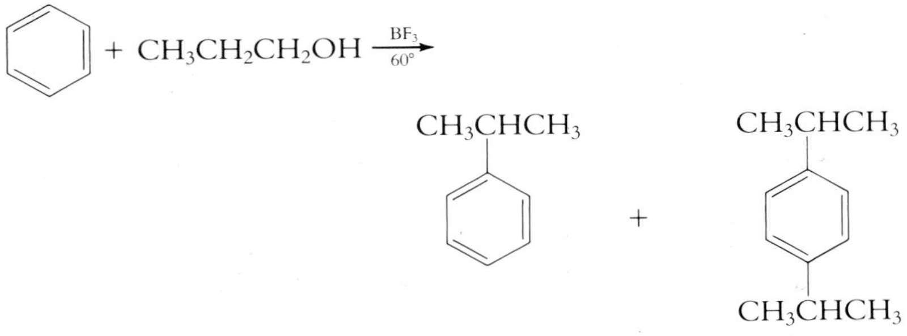

Isopropylbenzene (24%) 1,4-Diisopropylbenzene (13%)

15:7 Friedel–Crafts Reactions: Acylations with Acylium Ions

We can overcome the deficiencies of the alkylation reaction (rearrangement, multiple substitution) by using another reaction discovered by Friedel and Crafts. In this case the halide is an **acyl chloride** (or **acid chloride**),

$$R\overset{O}{\overset{\|}{C}}Cl$$

and the reaction is called a **Friedel–Crafts acylation** (eqs. 15.36 and 15.37). The name "acylation" derives from the fact that the group

$$R{-}\overset{\overset{\displaystyle O}{\|}}{C}{-}$$

is called an acyl group. Nomenclature of acid chlorides is discussed in Chapter 16.

(15.36) $$C_6H_6 + CH_3\overset{\overset{\displaystyle O}{\|}}{C}Cl \xrightarrow{AlCl_3} C_6H_5\overset{\overset{\displaystyle O}{\|}}{C}CH_3 + HCl$$

Acetyl chloride — 1-Phenylethanone (*acetophenone*, 97%)

(15.37) $$C_6H_6 + CH_3CH_2\overset{\overset{\displaystyle O}{\|}}{C}Cl \xrightarrow{AlCl_3} C_6H_5\overset{\overset{\displaystyle O}{\|}}{C}CH_2CH_3 + HCl$$

Propanoyl chloride — 1-Phenyl-1-propanone (*propiophenone*)

Friedel–Crafts acylation, unlike alkylation, involves no rearrangement. Furthermore, it yields only monosubstituted products because the aromatic ketones formed are *less* reactive to further substitution than is benzene. Alkylated benzenes are therefore often best obtained by a two-step route. The aromatic ring is first acylated, and the ketone group is then reduced to a methylene group, usually by the Clemmensen method (eq. 15.38; see sec. 11:11). Note that the reaction in Equation 15.38 yields 1-phenylbutane, free of 2-phenylbutane; compare this with the reaction in Equation 15.33.

(15.38) $$C_6H_6 + CH_3(CH_2)_2\overset{\overset{\displaystyle O}{\|}}{C}Cl \xrightarrow{AlCl_3}$$

$$C_6H_5\overset{\overset{\displaystyle O}{\|}}{C}(CH_2)_2CH_3 \xrightarrow[H_2O,\ toluene]{Zn/Hg,\ HCl} C_6H_5(CH_2)_3CH_3$$

(Clemmensen reduction)

Friedel–Crafts acylations, like the alkylations, involve carbocations. The catalyst ($AlCl_3$) removes the chloride ion from the acid chloride to yield an **acylium ion** (eq. 15.39). These carbocations are unusually stable because

$$\text{(15.39)}\quad CH_3CH_2\overset{\ddot{O}:}{\overset{\|}{C}}—Cl + AlCl_3 \rightarrow [CH_3CH_2\overset{+}{C}{=}\ddot{O}: \leftrightarrow CH_3CH_2C{\equiv}\overset{+}{O}:]AlCl_4^-$$

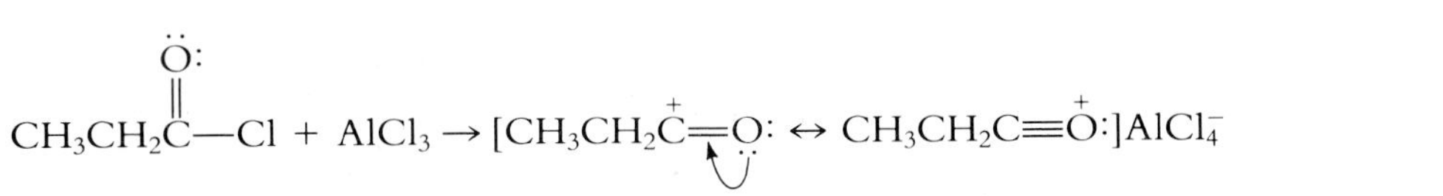

Propanoyl chloride — Propanoyl cation (an acylium ion)

the empty *p* orbital on carbon can overlap with one of the nonbonding orbitals of oxygen. Because of their stability the acylium ions do not rearrange. Friedel–Crafts acylations, like other electrophilic substitutions, go by way of benzenium ions (eq. 15.40).

$$\text{(15.40)}\quad CH_3CH_2CO^+ + C_6H_5H \longrightarrow$$

$AlCl_4^-$

$$[C_6H_6COCH_2CH_3]^{+}\ (\delta+,\ \delta+,\ \delta+)\ AlCl_4^- \longrightarrow C_6H_5COCH_2CH_3 + HCl + AlCl_3$$

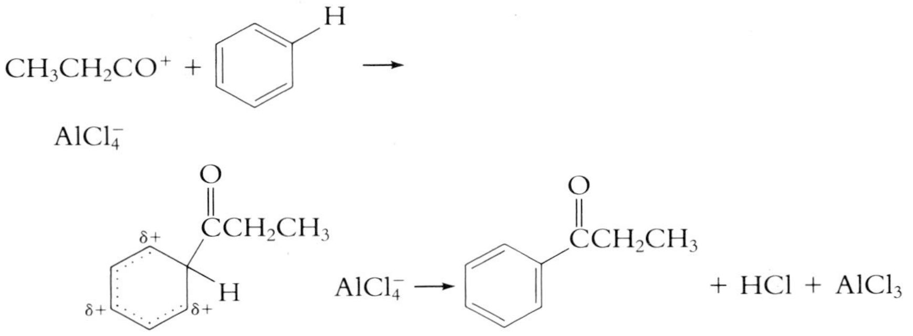

The $AlCl_3$ that is set free in the last step of Equation 15.40 is not available for further catalysis because, being a strong Lewis acid, it complexes with the carbonyl oxygen atom of the product. Thus, 1 mol of catalyst is required per mole of acyl chloride. At the conclusion of the reaction, water must be added to hydrolyze the $AlCl_3$ and free the ketone (eq. 15.41).

$$\text{(15.41)}\quad C_6H_5C(=\!\ddot{O}:)CH_2CH_3 \xrightarrow{AlCl_3} C_6H_5C(=\!\overset{+}{O}{:}\overset{-}{Al}Cl_3)CH_2CH_3 \xrightarrow{H_2O} C_6H_5C(=\!\ddot{O}:)CH_2CH_3 + Al^{3+} + HCl$$

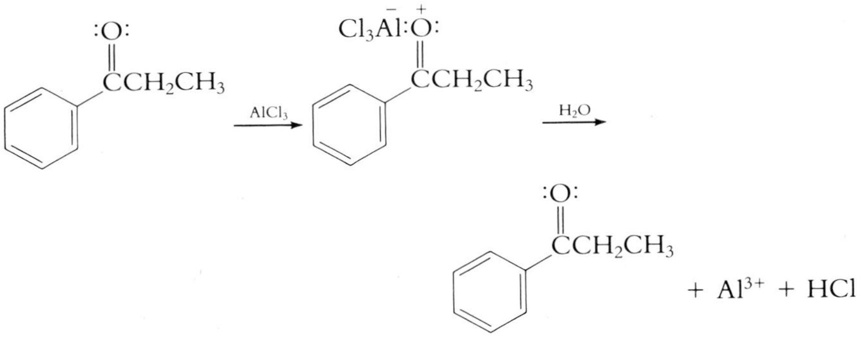

Example

Show how the following compounds can be made from benzene. Indicate any complications that might arise in either synthesis.

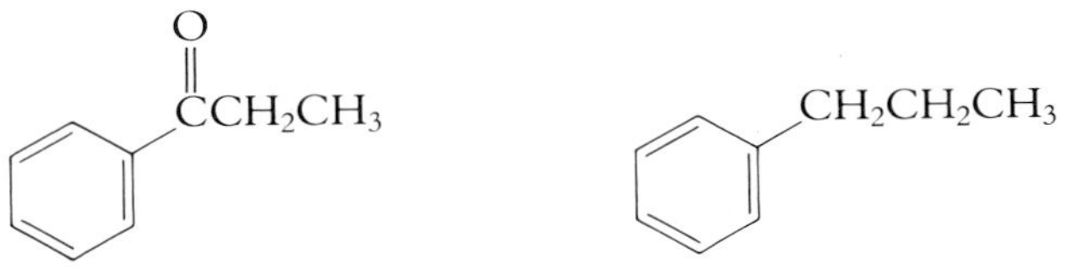

$$C_6H_5CCH_2CH_3\ (\text{with } C{=}O) \qquad C_6H_5CH_2CH_2CH_3$$

Answer

In each case we perform a retrosynthesis by disconnecting the bond between the aromatic ring and the side chain. The immediate precursors are then benzene and a carbocation:

disconnect here

$$C_6H_5\text{—}\overset{O}{\overset{\|}{C}}CH_2CH_3 \Longleftarrow C_6H_5\text{—}H + O{=}\overset{+}{C}CH_2CH_3$$

disconnect here

$$C_6H_5\text{—}CH_2CH_2CH_3 \Longleftarrow C_6H_5\text{—}H + \overset{+}{C}H_2CH_2CH_3$$

The carbocations, or complexes that act like them, are prepared by treating the corresponding organic halides with a Lewis acid such as $AlCl_3$:

$$AlCl_3 + Cl\text{—}\overset{O}{\overset{\|}{C}}CH_2CH_3 \rightarrow AlCl_4^- \quad O{=}\overset{+}{C}CH_2CH_3$$

$$AlCl_3 + Cl\text{—}CH_2CH_2CH_3 \rightarrow Cl_3\overset{-}{Al}\text{—}\overset{+}{Cl}\text{—}CH_2CH_2CH_3$$

The carbocation obtained from propanoyl chloride is stable and doesn't rearrange, so the synthesis is straightforward:

$$C_6H_6 + CH_3CH_2\overset{O}{\overset{\|}{C}}\text{—}Cl \xrightarrow{AlCl_3} \xrightarrow{H_2O} C_6H_5\text{—}\overset{O}{\overset{\|}{C}}CH_2CH_3$$

However, the complex obtained from propyl chloride and $AlCl_3$ can rearrange to a more substituted carbocation:

$$Cl_3\overset{-}{Al}\text{—}\overset{+}{Cl}\text{—}CH_2\overset{H}{\overset{|}{C}}HCH_3 \rightarrow AlCl_4^- \quad CH_3\overset{+}{C}HCH_3$$

Because of this rearrangement we can anticipate getting two products from the propyl chloride reaction:

$$CH_3CH_2CH_2{-}Cl + AlCl_3 \rightleftharpoons CH_3CH_2CH_2{-}\overset{+}{Cl}{-}\overset{-}{Al}Cl_3 \xrightarrow{C_6H_6} C_6H_5CH_2CH_2CH_3$$

$$CH_3CH_2CH_2{-}\overset{+}{Cl}{-}\overset{-}{Al}Cl_3 \longrightarrow CH_3\overset{+}{C}HCH_3\ AlCl_4^- \xrightarrow{C_6H_6} C_6H_5CH(CH_3)CH_3$$

To avoid this complication, we can take a slightly longer route that provides us with a single product isomer. Benzene is first acylated to produce propiophenone, as shown in the first part of this problem. The propiophenone is then reduced to propylbenzene, free of isopropylbenzene:

$$C_6H_5\overset{O}{\overset{\|}{C}}CH_2CH_3 \xrightarrow[\text{or } NH_2NH_2,\ KOH]{Zn/Hg,\ HCl} C_6H_5CH_2CH_2CH_3$$

Problems

15–10 Write the structures of the organic products of the following reactions:

a. $(CH_3)_2CHCH_2\overset{O}{\overset{\|}{C}}Cl + C_6H_6 \xrightarrow{AlCl_3}$

b. $CH_3{-}C_6H_4{-}\overset{O}{\overset{\|}{C}}Cl + C_6H_6 \xrightarrow{AlCl_3}$

c. $C_6H_5{-}CH_2\overset{O}{\overset{\|}{C}}Cl + C_6H_6 \xrightarrow{AlCl_3}$

d. $CH_3\overset{O}{\overset{\|}{C}}Cl + C_{10}H_8 \text{ (naphthalene)} \xrightarrow{AlCl_3}$

15–11 Show how each of the following compounds may be prepared from benzene, free of other products. You may use any other reagents you need.

a. 1-phenylpropane
b. 1-phenyl-3-methylbutane
c. 1,3-diphenylpropane
d. 4-chlorodiphenylmethane $C_6H_5{-}CH_2{-}C_6H_4{-}Cl$

Table 15–1. Important Electrophiles in Aromatic Substitution and Their Sources

Effective Electrophile	Source(s)
Cl^+	$Cl_2 + AlCl_3$ or $FeCl_3$
	$HOCl + H^+$
Br^+	$Br_2 + AlBr_3$ or $FeBr_3$
I^+	$I_2 + HNO_3$
SO_3H^+	$SO_3 + H_2SO_4$
NO_2^+	$HNO_3 + H_2SO_4$
R^+	$RX + AlX_3$ ($X = Cl$ or Br)
	$ROH + H^+$ or BF_3
	an alkene $+ H^+$
$R{-}\overset{+}{C}{=}O$	$RCOCl + AlCl_3$

Table 15–1 contains a summary of the important electrophiles used in aromatic substitution.

15:8 Nomenclature of Disubstituted Benzenes

Up to this point we have considered only reactions of benzene, and no attention has been given to reactions of substituted benzenes. Polysubstituted benzenes are important synthetic intermediates and their preparation and reactions need to be examined in detail. Before doing this, it is necessary to learn the nomenclature of such compounds.

The IUPAC system uses numbers to indicate the position of substituents on the ring. As usual, the ring is numbered in such a way as to use the smallest possible numbers. With disubstituted benzenes, the substituent names are listed in alphabetical order and the first substituent is given the lowest number. When naming sulfonic acids, the sulfonic acid group is always placed at C-1. Substituted methylbenzenes are treated as toluenes when the substituents are halogen, alkyl, and nitro groups; the methyl group is always placed at C-1 of the ring.

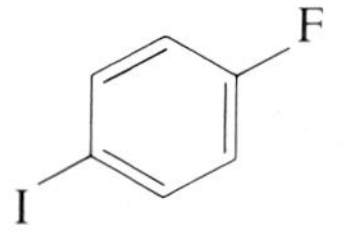

1-Fluoro-4-iodobenzene
(*p*-fluoroiodobenzene)

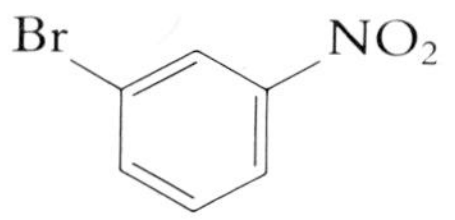

1-Bromo-3-nitrobenzene
(*m*-bromonitrobenzene)

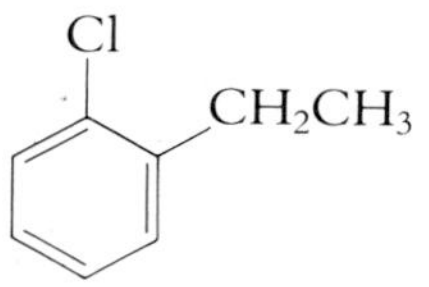

1-Chloro-2-ethylbenzene
(*o*-chloroethylbenzene)

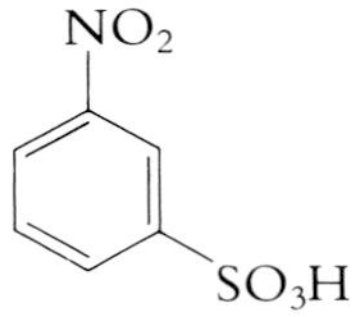

NO_2 / SO_3H CH_3 / NO_2 Br / CH_3

3-Nitrobenzenesulfonic acid
(*m*-nitrobenzenesulfonic acid)

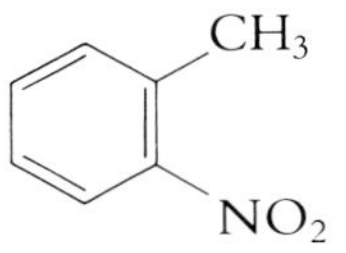

2-Nitrotoluene
(*o*-nitrotoluene)

4-Bromotoluene
(*p*-bromotoluene)

An older system designates 1,2-disubstituted benzenes as *ortho* (*o*) isomers; 1,3-disubstituted benzenes as *meta* (*m*) isomers; and 1,4-disubstitued benzenes as *para* (*p*) isomers. Names based on the *o, m, p* system are given above in parentheses.

Benzenes carrying more than two substituents should be named using the IUPAC system. The numbering is determined by the requirement that the numbers give the smallest possible sum (substituents are listed in alphabetical order).

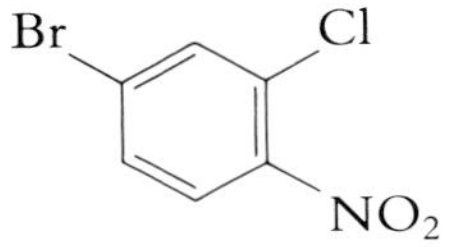

Br, Cl, NO_2 OCH_3, $CH_3(CH_2)_3$, $(CH_2)_2CH_3$

4-Bromo-2-chloro-1-nitrobenzene

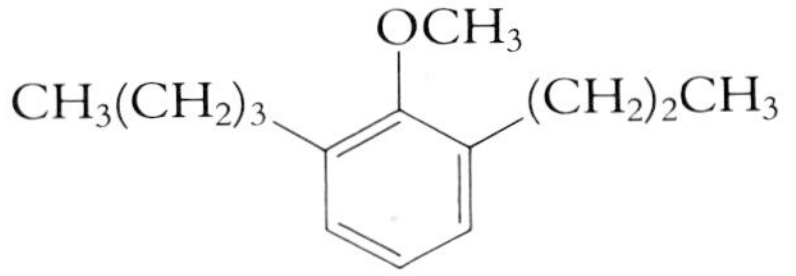

1-Butyl-2-methoxy-3-propylbenzene

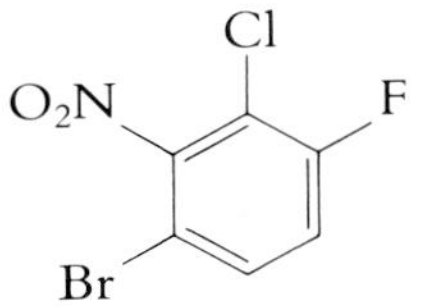

Cl, O_2N, F, Br

1-Bromo-3-chloro-4-fluoro-2-nitrobenzene

Benzene rings attached to alkyl chains can be named in one of two ways—as substituted benzenes or as substituted alkanes. A useful guideline is to choose the name that contains the maximum number of separate substituents.

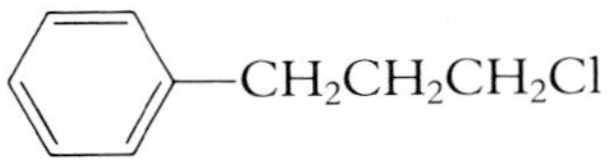

$-CH_2CH_2CH_2Cl$ $CH_2CH_2CH_3$, Cl

1-Chloro-3-phenylpropane

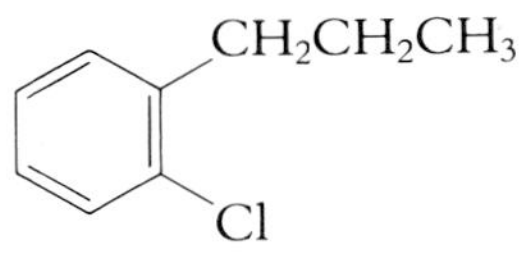

1-Chloro-2-propylbenzene

Problems

15–12 Provide names for each of the following compounds. More than one name may be acceptable.

a. F, NO_2

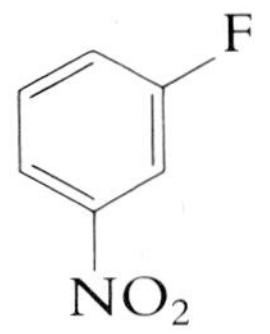

b. $CH(CH_3)_2$, Br

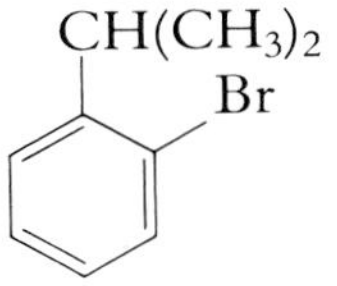

c. 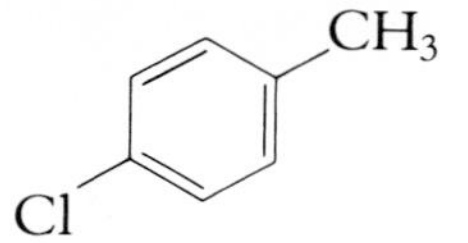 [structure: CH_3, Cl]

e. 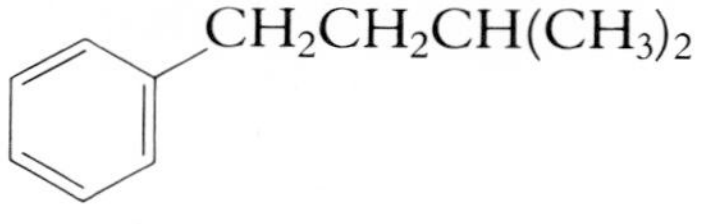 [structure: $CH_2CH_2CH(CH_3)_2$]

d. 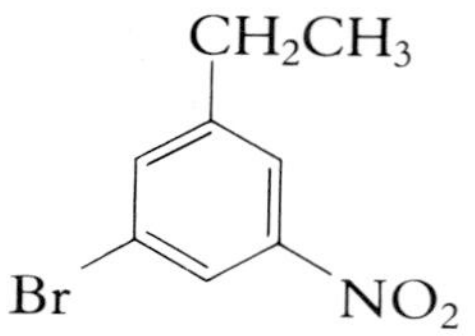 [structure: CH_2CH_3, Br, NO_2]

f. 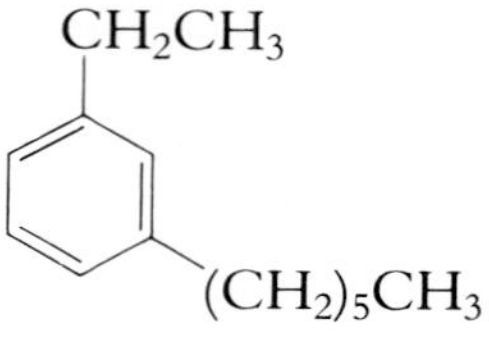 [structure: CH_2CH_3, $(CH_2)_5CH_3$]

15–13 Draw structures corresponding to the following names:

a. 2-chloro-6-nitrotoluene
b. *o*-chlorofluorobenzene
c. *m*-bromoacetophenone
d. 2-iodo-1,4-dimethylbenzene
e. 1,3-dibromo-2-phenylbutane

15:9 Substituent Effects in Aromatic Substitution: Introduction

When we carry out substitution reactions on already substituted benzenes, we have to account for certain factors which don't affect substitution reactions of benzene itself. We will study these differences here and in the following sections to gain a better understanding of the factors that control chemical reactivity, and to put ourselves in a better position to plan efficient syntheses. An analysis of the mechanisms will come first, followed by an examination of their practical consequences.

A substituent on a benzene ring can strongly affect the reactivity of the ring. For instance, a comparison of Equations 15.42 and 15.43 tells us that

(15.42) [benzene] + Br_2 $\xrightarrow[70°]{FeBr_3}$ [bromobenzene: Br] + HBr

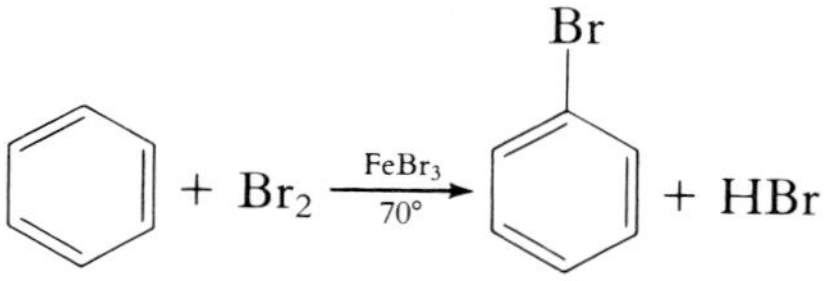

(15.43) [OCH$_3$ benzene] + Br_2 $\xrightarrow[25°]{\text{no catalyst}}$ [OCH$_3$, Br (ortho)] + [OCH$_3$, Br (para)] + HBr

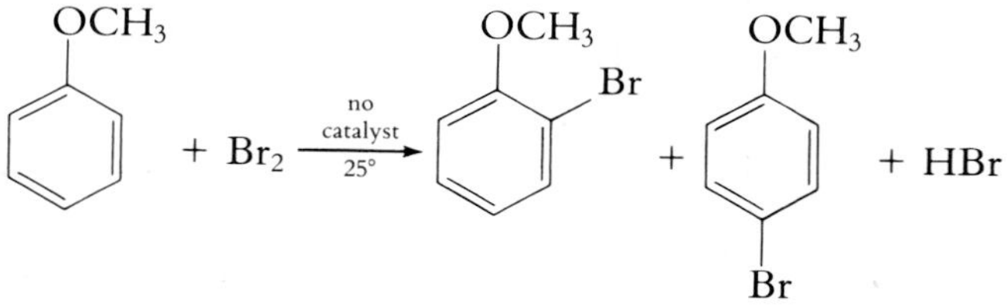

Anisole

(15.44) [NO$_2$ benzene] + Br_2 $\xrightarrow[140°]{FeBr_3}$ [NO$_2$, Br (meta)] + HBr

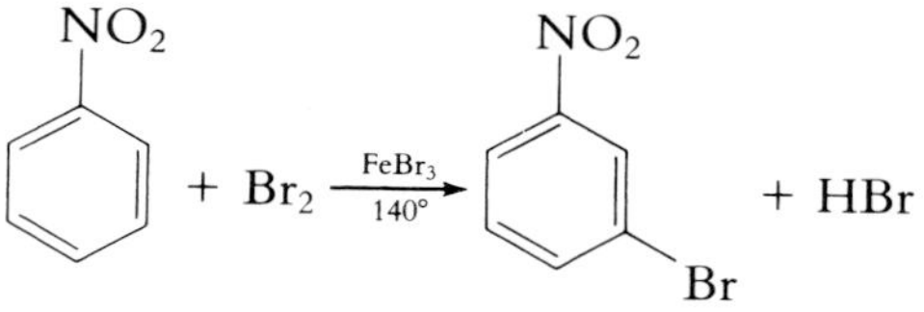

methoxybenzene (*anisole*) is much more reactive than benzene toward bromination. On the other hand, comparing Equations 15.42 and 15.44 tells us that nitrobenzene is much less reactive than benzene toward bromination. These facts can be summed up as follows:

1. When H on a benzene ring is replaced by OCH_3, the ring becomes **activated** toward electrophilic attack. Methoxyl is an **activating** substituent in electrophilic substitution.
2. When H on a benzene ring is replaced by NO_2, the ring becomes **deactivated** toward electrophilic attack. Nitro is a **deactivating** substituent in electrophilic substitution.

In addition to its activating or deactivating effect, the substituent also directs an incoming electrophile preferentially toward an ortho, meta, *or* para *position.*

Equation 15.43 shows that when a methoxyl group is on the ring, the incoming electrophile prefers to go to an *o* or *p* position. By contrast, a nitro group on the ring directs an incoming electrophile to the *m* position (eq. 15.44). In other words,

1. methoxyl is an ***o,p*-directing** substituent, and
2. nitro is an ***m*-directing** substituent.

Table 15–2 gives a list of activating and deactivating groups, along with their directive effects.

Table 15–2. Common Aromatic Substituents and Their Activating and Directing Effects

Activating Groups		Deactivating Groups	
$-R$	*ortho–para* directors	$-F$	*ortho–para* directors
$-Ar$		$-Cl$	
$-OH$		$-Br$	
$-OR$		$-I$	
$-OC(=O)R$			
$-NH_2$		$-NO_2$	*meta* directors
$-NR_2$		$-\overset{+}{N}R_3$	
$-NHC(=O)CH_3$		$-C(=O)R$	
$-SH$		$-C(=O)OR$	
$-SR$		$-CN$	
		$-SO_3H$	

In Section 15:10 we will show that the directing effects of a substituent and its activating or deactivating effects are connected. To do this we must first explain how the substituent affects the stability of the benzenium ion intermediate.

15:10 Relative Rates of Aromatic Substitution at Different Positions

The rates of electrophilic attack at different positions of a substituted benzene have been measured, with various electrophiles and various substituents. The results of these rate studies can help us understand both activating and directing effects in substituted benzenes.

Figure 15–6 shows energy diagrams for the rate-determining step in the bromination of benzene, nitrobenzene at the *meta* position, and nitrobenzene at the *para* position. Notice that the bromination of nitrobenzene at either position is slower than bromination of benzene. Furthermore, *m*-bromonitrobenzene is formed faster than *p*-bromonitrobenzene. (*o*-Bromonitrobenzene is formed even more slowly than the *p* isomer.)

The best way to make sense of these observations is to examine the structures of the benzenium ion intermediates. Structure **4** shows the three resonance structures of the benzenium ion formed by *meta*-bromination of nitrobenzene. The intermediate has a positively-charged ring and is electron deficient. The nitro group, with its positively charged nitrogen atom, tends to inductively pull electrons out of the ring. Since the ring is already electron

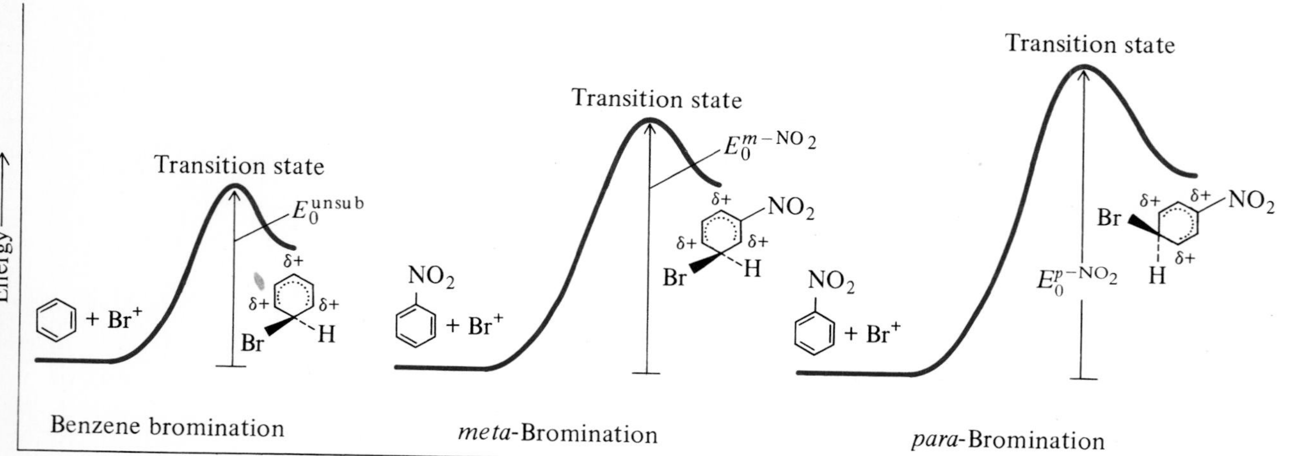

Figure 15–6. *Energy diagrams for bromobenzenium ion formation from benzene and from* meta *and* para *attack on nitrobenzene, showing increasing energy of the ion and of the transition state.*

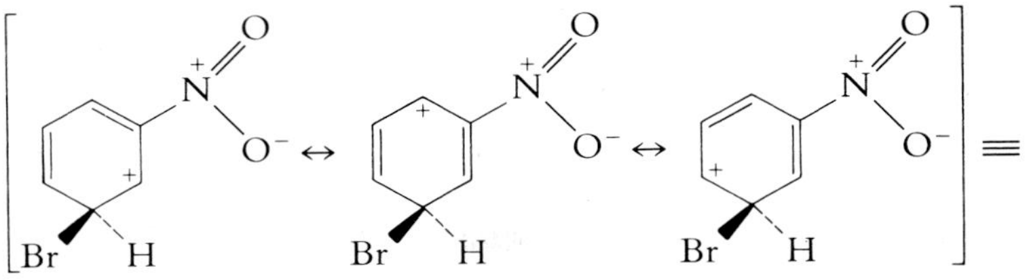

4

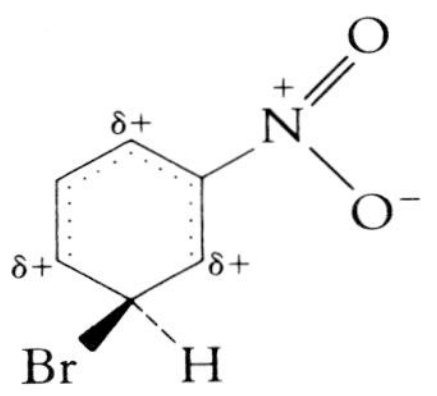

deficient, this electron withdrawal by the nitro substituent tends to destabilize it. We can say then that the nitro group destabilizes benzenium ions. That's why the energy barrier is larger for formation of **4** than for formation of an unsubstituted benzenium ion (Fig. 15–6).

Structure **5** shows the resonance forms for the benzenium ion formed by *para*-bromination of nitrobenzene. We can make the same comments about **5** that we made about **4**. However, there is one interaction present in **5** that was absent in **4**. Note that in **5** the electron withdrawing nitro group is directly attached to a ring carbon onto which the positive charge is delocalized (this is not so in **4**). This makes **5** even higher in energy than **4** because the positive charge on nitrogen is next to a positively-charged carbon.

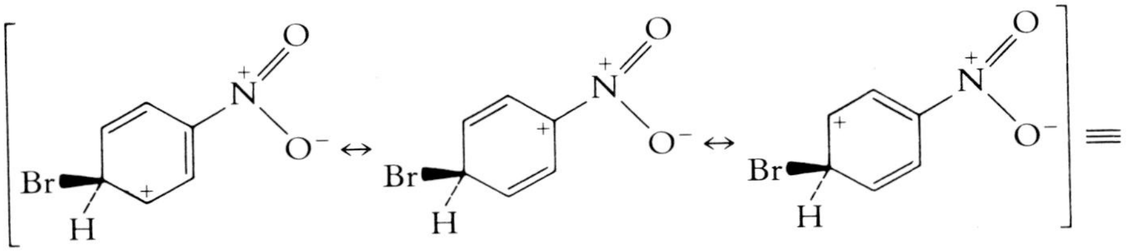

High-energy structure

5

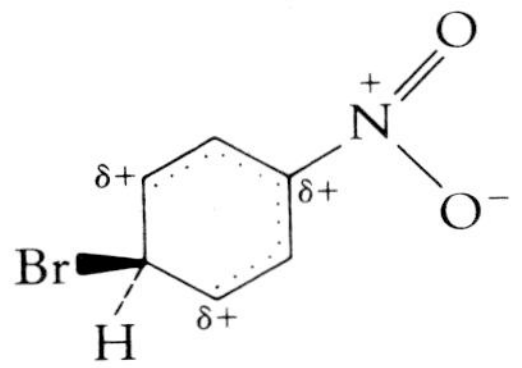

The benzenium ions we have discussed are arranged below in order of *decreasing* stability. You can see that *the relative rates of reaction parallel the relative stability of the intermediate benzenium ions.*

15:10 Relative Rates of Aromatic Substitution at Different Positions

Relative stability: **6** > **4** > **5**

δ+ δ+ δ+ H Br **6** > δ+ δ+ δ+ $\overset{+}{N}$ O O^- H Br **4** >

δ+ δ+ δ+ H Br $\overset{+}{N}$ O O^- or δ+ δ+ δ+ $\overset{+}{N}$ O O^- Br H

5

Problems

15–14 One mole of benzene is mixed with 1 mol of nitrobenzene and $\frac{1}{2}$ mol of bromine. Some $FeBr_3$ is added and the mixture is heated to reflux. What is the major reaction product? Explain your answer.

15–15 Draw the resonance structures for the benzenium ion formed by *ortho*-bromination of nitrobenzene. Use them to explain why virtually no *o*-bromonitrobenzene is formed upon bromination of nitrobenzene. Are there interactions present in this benzenium ion that are not present in **4** or **5**?

Figure 15–7 shows energy diagrams for the formation of bromobenzenium ions from benzene and from anisole by *meta* and *para* attack. The figure shows that the energy barrier for *meta*-bromination of anisole is close to that for bromination of benzene. However, the energy barrier for *para*- (and

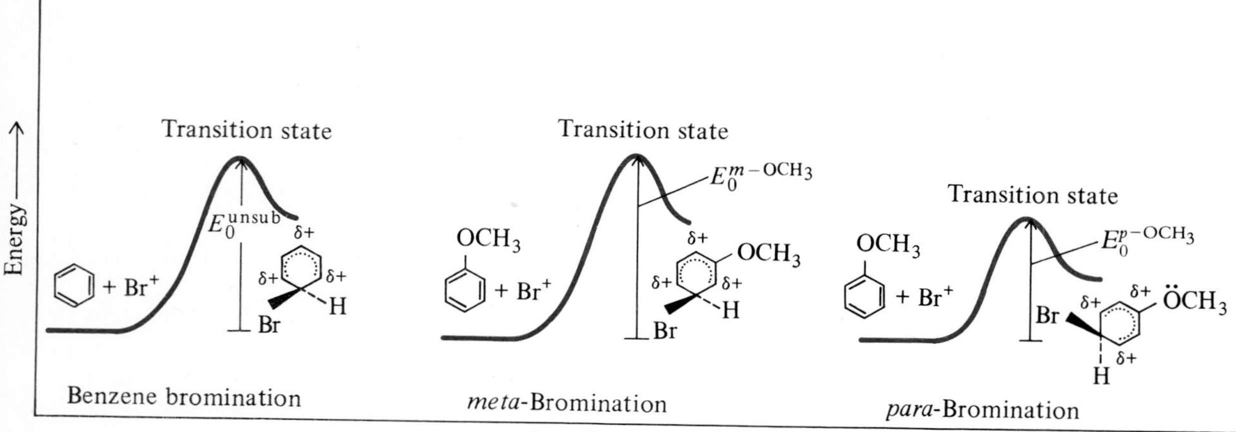

Figure 15–7. *Energy diagrams for formation of bromobenzenium ions from benzene and from* meta *and* para *attack on anisole. The* para-*substituted ion is lowest in energy due to electron donation by* OCH_3.

ortho-) bromination of anisole is considerably smaller than that for benzene bromination.

Once again, the structures of the benzenium ion intermediates provide a clear explanation of the rate differences for reactions at different positions. Structure **7** represents the benzenium ion produced by *meta*-bromination of anisole. In **7** there are no strong interactions between the ring and the substi-

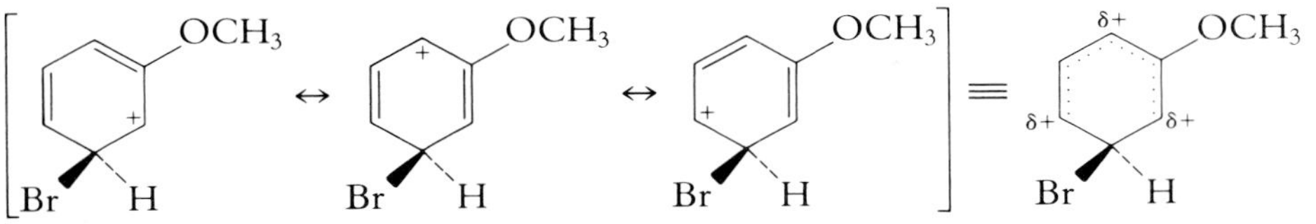

7

tutent because no partial positive charge is on the same carbon as the substituent. Since O is more electronegative than H, the OCH_3 group should be slightly electron-withdrawing, and **7** should be *slightly* less stable than the unsubstituted bromobenzenium ion, **6**. This conclusion agrees with the rate data.

The benzenium ion from *para*-bromination of anisole (**8**) exhibits a resonance interaction not found in the *meta* isomer (**7**). A nonbonding electron

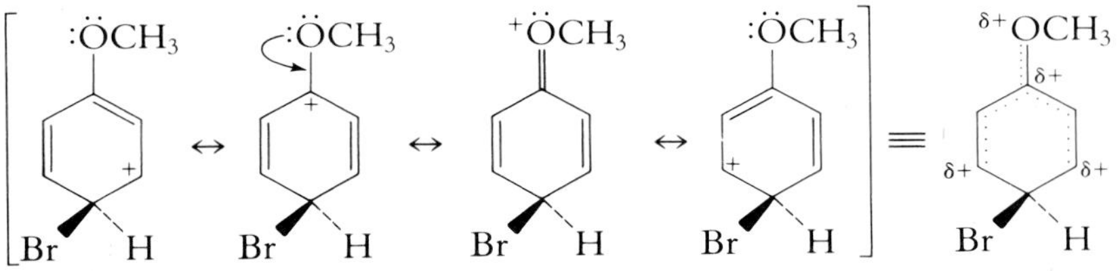

Low-energy structure

8

pair on oxygen can form a π bond to an electron-deficient ring carbon. This allows charge to be delocalized onto the oxygen atom, as structure **8** shows, which greatly stabilizes the benzenium ion. The electron-donating resonance effect of OCH_3 is sufficiently powerful to overcome the electron-withdrawing inductive effect of the OCH_3 group (**9a** and **9b**). This is confirmed by the fact that electrophilic attack is strongly favored *para* to OCH_3.

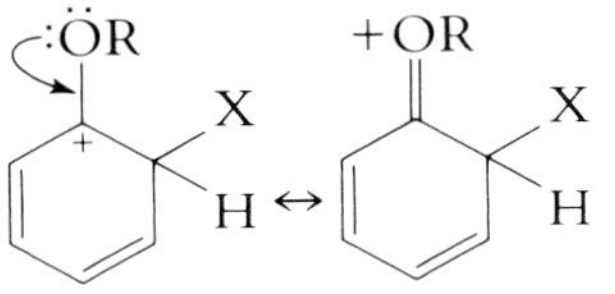

9a
A stabilizing resonance effect

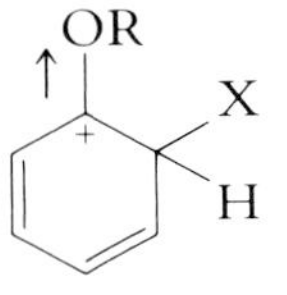

9b
A destabilizing inductive effect

The stability of the various benzenium ions lies in the order shown below.

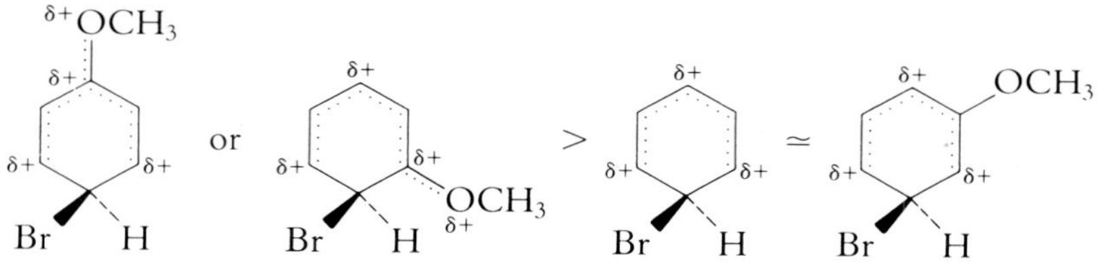

$\delta+$ OCH_3 ... **8** or ortho (OCH_3, $\delta+$) > **6** $\simeq$ **7** (OCH_3)

Br H   Br H   Br H   Br H

8   *ortho*   **6**   **7**

Problem

15–16 Draw the resonance structures for the benzenium ion produced by *ortho*-bromination of anisole. Use them to explain why OCH_3 is an *ortho*-director.

The following sections will show that this discussion of activating and directing effects is generally applicable to other substituents. Based on kinetic studies of a wide variety of substituted benzenes, the following overview of electrophilic aromatic substitution can be given:

1. Substituents exert their strongest activating or deactivating effect toward attack at the *ortho* or *para* positions.
2. If a substituent is deactivating, it deactivates the *ortho* and *para* positions most strongly, causing *meta* substitution. (The halogens are an important exception; see Section 15:13.)
3. If a substituent is activating, it activates the *ortho* and *para* positions most strongly, leading to *ortho,para* substitution.

15:11 *meta*-Directing Substituents

Equations 15.45 to 15.47 show the results from electrophilic substitution of several benzenes having electron-withdrawing groups. Note that all of the substituents are *m*-directors.

(15.45) NO_2-benzene $+ Cl_2 \xrightarrow{FeCl_3}$

NO_2, Cl + NO_2, Cl + NO_2, Cl

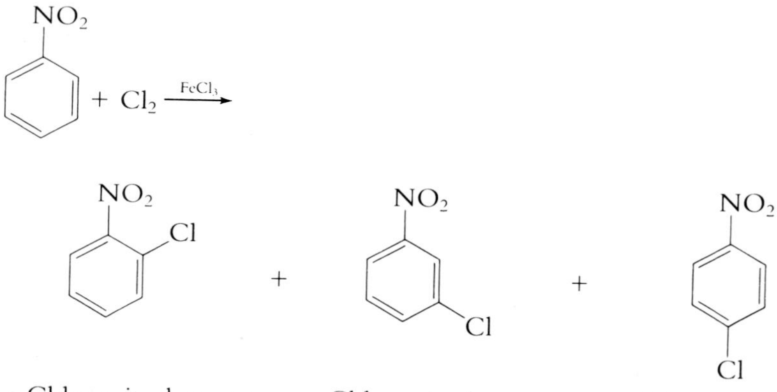

o-Chloronitrobenzene (1%)   *m*-Chloronitrobenzene (95%)   *p*-Chloronitrobenzene (4%)

(15.46)  Benzonitrile + HNO_3 $\xrightarrow[\text{heat}]{H_2SO_4}$ *o*-Nitrobenzonitrile (17%) + *m*-Nitrobenzonitrile (81%) + *p*-Nitrobenzonitrile (2%)

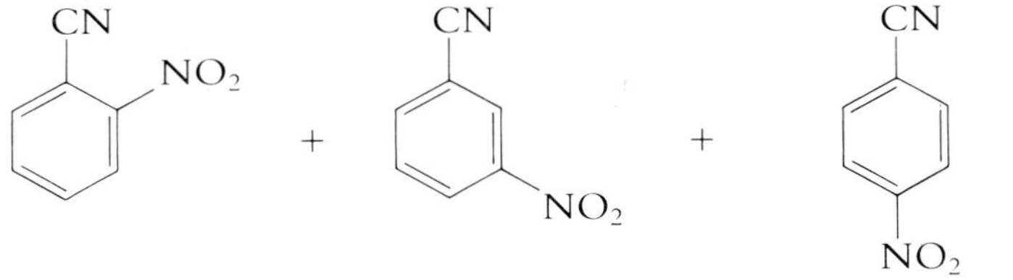

(15.47) 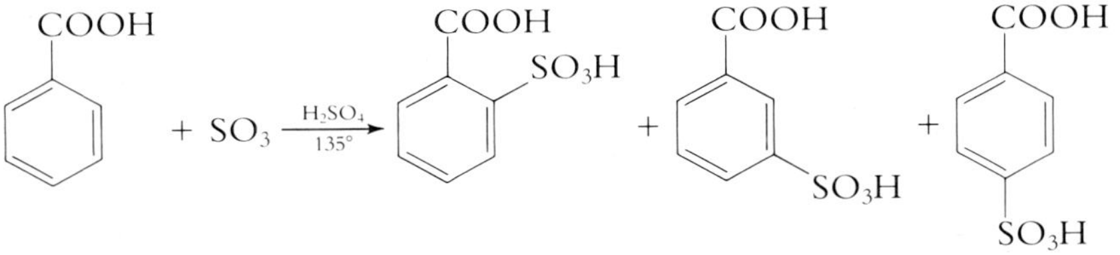 Benzoic acid + SO_3 $\xrightarrow[135°]{H_2SO_4}$ (1%) + (84%) + (15%)

As was shown before, m-*directors destabilize benzenium ions by electron withdrawal.*

A substituent will attract electrons away from the intermediate benzenium ion if the substituent bears a positively charged atom that is directly attached to the ring. This is the case with

$$-\overset{+}{N}(=O)-O^-$$

The atom attached to the ring may have a partial positive charge, and therefore be electron withdrawing, if it is connected to a more electronegative atom. This is the case for $\overset{\delta+}{C}{=}\overset{\delta-}{O}$, $\overset{\delta+}{S}{=}\overset{\delta-}{O}$, and $\overset{\delta+}{C}{\equiv}\overset{\delta-}{N}$, as shown in structures **10** and **11.** Thus, m-*directing substituents are characterized by a positively-charged*

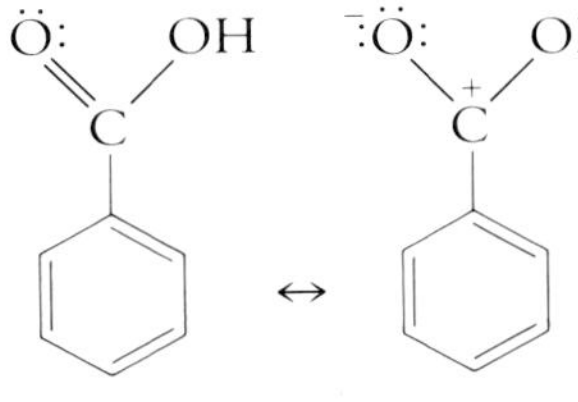

10 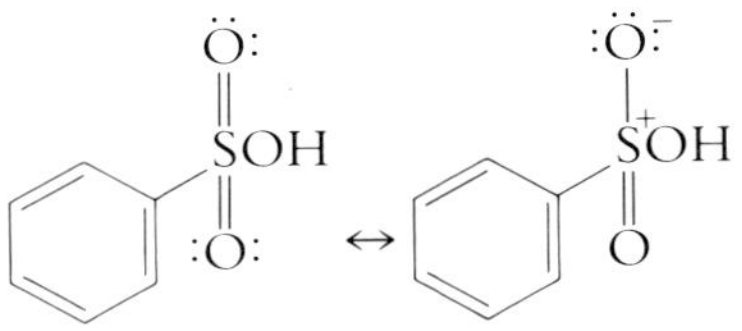   **11**

atom next to the ring, and/or a multiple bond to an electronegative atom conjugated with the ring.

Box 15–2. Nitrobenzene and Its Derivatives

Very pure nitrobenzene is colorless, but nearly all samples appear yellow from traces of decomposition products. It has an odor somewhat similar to that of bitter almonds and was once used as a perfume for soaps, a constituent of shoe polish, and a flavoring adulterant. Nitrobenzene is, however, quite toxic, and is readily absorbed through the skin. Its direct use in consumer products is now prohibited.

Nitration of toluene eventually leads to trinitrotoluene, or TNT, the well-known explosive. TNT can be purified by careful distil-

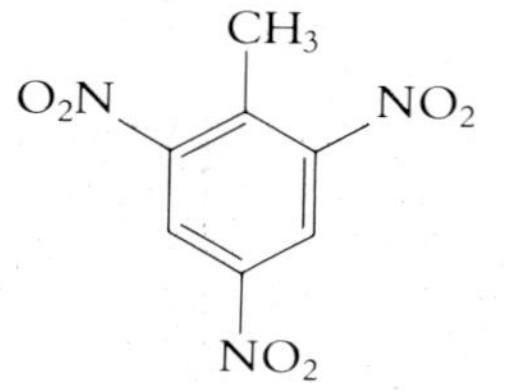

CH_3 O_2N NO_2 NO_2

TNT

lation, but it will explode when subjected to shock. Trinitrobenzene is actually a more powerful explosive than trinitrotoluene, but it cannot be made commercially in satisfactory yield. As each nitro group is introduced into toluene or benzene, it becomes more difficult to introduce another because the nitro group is so strongly deactivating. The third nitro group can be substituted onto toluene because the methyl group is activating and helps to counteract the deactivating effect of the nitro groups.

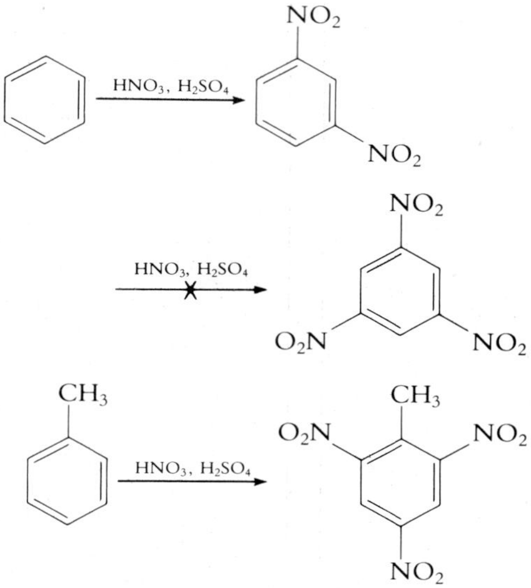

HNO_3, H_2SO_4 → (NO_2, NO_2)

HNO_3, H_2SO_4 ↛ (NO_2, O_2N, NO_2)

CH_3 HNO_3, H_2SO_4 → (CH_3, O_2N, NO_2, NO_2)

Problem

15–17 Using resonance structures, show why benzonitrile (C_6H_5CN) is chlorinated preferentially at the *meta* position.

15:12 *ortho,para*-Directing, Activating Substituents

Equations 15.48 to 15.50 show the electrophilic substitution of several benzene derivatives bearing electron-donating groups. The relative isomer yields show that these substituents activate *ortho* and *para* attack much more than they do *meta* attack.

(15.48) CH_3 Toluene + Cl_2 $\xrightarrow{FeCl_3}$

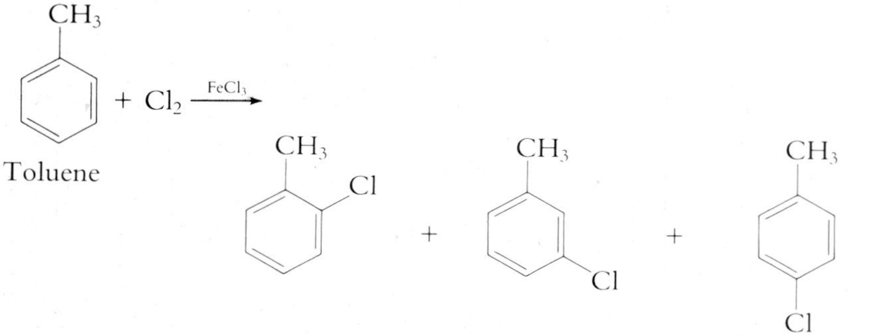

CH_3, Cl — *o*-Chlorotoluene (75%) + CH_3, Cl — *m*-Chlorotoluene (2%) + CH_3, Cl — *p*-Chlorotoluene (23%)

$$(15.49)\quad C_6H_5\ddot{N}HCOCH_3 + HNO_3 \xrightarrow[20^\circ]{H_2SO_4} o\text{-}O_2NC_6H_4\ddot{N}HCOCH_3 + m\text{-}O_2NC_6H_4\ddot{N}HCOCH_3 + p\text{-}O_2NC_6H_4\ddot{N}HCOCH_3$$

N-Phenylacetamide (*acetanilide*)

(19%) (2%) (79%)

$$(15.50)\quad C_6H_5OCH_3 + CH_3CCl\ (\text{C=O}) \xrightarrow{AlCl_3} p\text{-}CH_3OC_6H_4CCH_3\ (\text{C=O})$$

($\simeq$100%)

The *o,p*-directors can affect the intermediate benzenium ion in one of two ways. Those substituents that are capable of sharing an electron pair, such as $R\ddot{O}$— and $R_2\ddot{N}$—, stabilize the *ortho* or *para* substituted intermediate by delocalizing the charge, as in structure **9a.** Such groups are electron-donating by resonance. However, the oxygen (or nitrogen) atom of these substituents is electronegative and hence has an inductive effect that is in the opposite direction as in **9b.** This would destabilize the intermediate. Since alkoxy and amino groups are both strong *o,p*-directors (eqs. 15.49 and 15.50), the stabilizing interaction (**9a**) must be more powerful than the destabilizing one (**9b**).

Alkyl groups, such as methyl (eq. 15.48) or *t*-butyl (eq. 15.51), cannot have a resonance interaction with the benzenium ion intermediate. If, however, they are on a carbon that bears a positive charge, as in **12,** they have a

(15.51) $C_6H_5C(CH_3)_3 \xrightarrow{HOCl,\ H^+}$

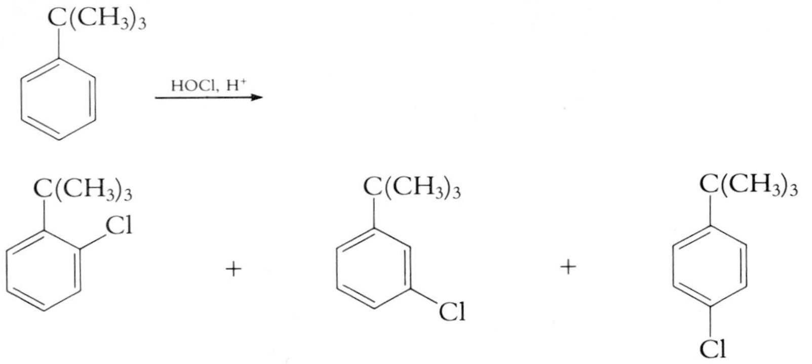

o-Chloro-t-butylbenzene (36%) + m-Chloro-t-butylbenzene (4%) + p-Chloro-t-butylbenzene (60%)

$C(CH_3)_3$, Cl (ortho)		$C(CH_3)_3$, Cl (meta)		$C(CH_3)_3$, Cl (para)
o-Chloro-*t*-butylbenzene (36%)	+	*m*-Chloro-*t*-butylbenzene (4%)	+	*p*-Chloro-*t*-butylbenzene (60%)

weak electron-donating inductive effect. This stabilizes the *ortho* (or *para*) substituted intermediate as it would any carbocation.

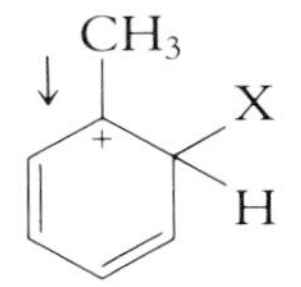

CH_3 ↓, +, X, H

12
A stabilizing inductive effect

Thus, o,p-*directing substituents are characterized by having either an atom with a lone electron pair connected to the ring, or electron-donating inductive ability.*

The above reasoning explains why electron-donating substituents direct *ortho* and *para,* rather than *meta.* The ratio of *ortho* to *para* products is, however, difficult to explain because it is often the result of a complex combination of factors. Steric hindrance often causes low *ortho* substitution percentages. For example, a comparison of Equation 15.51 with 15.48 shows that the presence of the bulky *t*-butyl group decreases the amount of *ortho* product. Reactions at carbonyl groups are known to be highly influenced by steric effects, and hence most Friedel–Crafts acylations (eq. 15.50) give an excellent yield of the *para* product but very little *ortho* product. It is worthwhile to note that Friedel–Crafts alkylations and acylations are extremely poor if *deactivators* are on the ring. Nitrobenzene, for instance, is completely unreactive.

15:13 The Halogens: *o,p*-Directing, Deactivating Substituents

The halogens are a special class of substituent. Halobenzenes are less reactive toward electrophilic substitution than benzene itself. They do, however, undergo primarily *o,p*-substitution (eqs. 15.52 and 15.53). In Equation 15.53 the catalyst is gallium tribromide, $GaBr_3$; it is a Lewis acid like $AlBr_3$, but less reactive.

(15.52) Fluorobenzene $+ HNO_3 \rightarrow$ o-fluoronitrobenzene (12%) + m-fluoronitrobenzene (1%) + p-fluoronitrobenzene (87%)

F, F, NO_2, F, NO_2, F, NO_2

(12%) (1%) (87%)

(15.53) Chlorobenzene $+ CH_3CH_2Br \xrightarrow{GaBr_3}$

Cl, CH_2CH_3, Cl, CH_2CH_3, Cl, CH_2CH_3

(42%) (16%) (42%)

Halogen behavior can be explained as a combination of inductive and resonance effects. Since the halogens are much more electronegative than carbon, they strongly withdraw electrons inductively. Structure **13** shows that chlorine withdraws electrons from the electron-deficient *meta* substituted benzenium ion, thus destabilizing it.

Cl, H, CH_2CH_3 — **13**

[:Cl: ↑, H, CH_2CH_3 ↔ $:Cl^+$, H, CH_2CH_3] — **14a**, **14b**

Structure **14a** shows the same destabilizing effect in the benzenium ion formed by *para* attack. Notice that the destabilizing inductive effect in this case can be partially offset by an electron-donating resonance effect (structure **14b**). Resonance electron donation is only possible when the electrophile attacks *ortho* or *para* to the halogen atom.

Because of their electron-withdrawing inductive effects, halogens destabilize intermediates from *ortho, meta,* or *para* attack, but due to conjugative electron donation, they destabilize intermediates from *ortho* and *para* attack *less* than they destabilize intermediates from *meta* attack (Fig. 15–8). This analysis once again emphasizes the importance of conjugation in stabilizing transition states leading to carbocations.

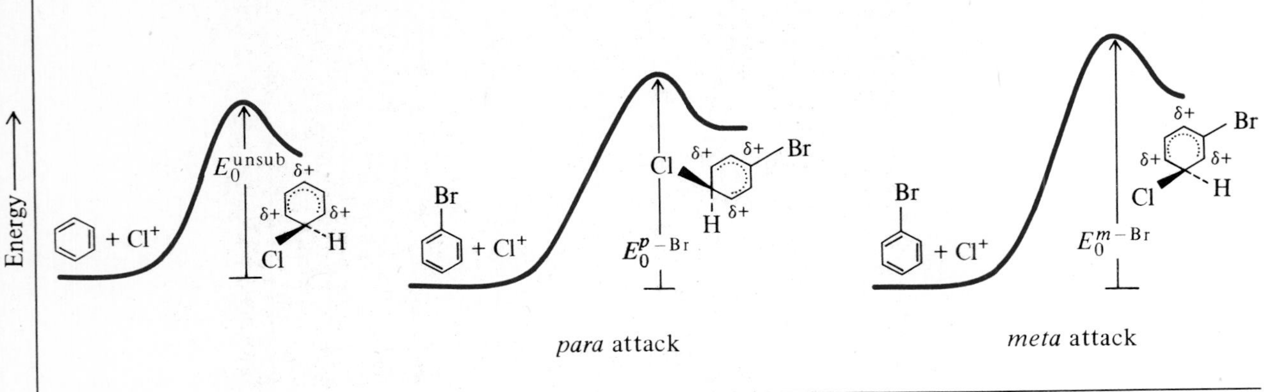

Figure 15–8. *Comparison of benzenium ion energies from chlorination of benzene, and of bromobenzene at the* meta *and* para *positions. Both ions from bromobenzene are higher in energy than the ion from benzene, but the* para*-substituted ion is the more stable of the two.*

Example

What is the major product from nitration of anisole and from nitration of benzoic acid?

OCH_3 $COOH$

Anisole   Benzoic acid

Answer

The methoxy group of anisole is an *ortho,para*-director. Therefore, a mixture of *o*- and *p*-nitroanisole will be obtained:

OCH_3 $\xrightarrow{HNO_3,\ H_2SO_4}$ OCH_3 (NO_2) + OCH_3 (NO_2)

The carboxyl group directs *meta* and hence the main product will be *m*-nitrobenzoic acid:

$COOH$ $\xrightarrow{HNO_3,\ H_2SO_4}$ $COOH$ (NO_2)

15–18 Indicate which product isomers will predominate when each of the following compounds is brominated with Br_2 and $FeBr_3$.

a. $C_6H_5SCH_3$ b. C_6H_5COOH c. $C_6H_5OCCH_3$ (with C=O) d. $C_6H_5SO_2CH_3$

15–19 Chloroethylene (*vinyl chloride*) reacts with hydrogen chloride to give 1,1-dichloroethane (eq. 1). Explain the following characteristics of this reaction: (1) chloroethylene reacts more *slowly* than ethylene; (2) the reaction is completely regiospecific, giving *no* 1,2-dichloroethane.

$$ClCH{=}CH_2 + HCl \rightarrow Cl_2CHCH_3 \qquad (1)$$

15:14 Some Practical Consequences of Substituent Directing Effects

A substituent already on a benzene ring will affect the placement of all subsequently introduced substituents, necessitating careful planning of syntheses of polysubstituted benzenes. The sequence of substitution reactions, in particular, is very important. Suppose that you are asked to synthesize 1-bromo-3-nitrobenzene from benzene. Since the substituents are to be *meta* to one another, the *first* one attached to the ring must be an *m*-director. The correct reaction sequence is given in Equation 15.54.

$$(15.54)\quad C_6H_6 \xrightarrow[H_2SO_4]{HNO_3} C_6H_5NO_2 \xrightarrow[AlBr_3]{Br_2} m\text{-}BrC_6H_4NO_2$$

Major product

Using the reverse sequence would yield the wrong isomers as major products (eq. 15.55).

$$(15.55)\quad C_6H_6 \xrightarrow[AlBr_3]{Br_2} C_6H_5Br \xrightarrow[H_2SO_4]{HNO_3} o\text{-}BrC_6H_4NO_2 + p\text{-}BrC_6H_4NO_2$$

Major products

You'll know you have a good understanding of this section if you can explain why the synthesis of *p*-ethylacetophenone requires an alkylation followed by an acylation, rather than the reverse.

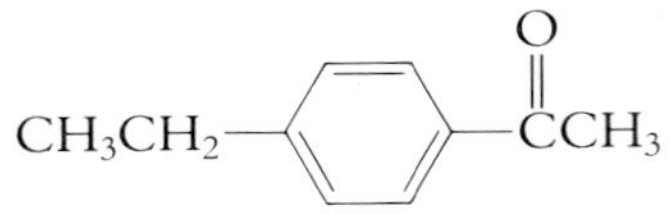

CH_3CH_2—C₆H₄—$\overset{O}{\overset{\|}{C}}CH_3$

p-Ethylacetophenone

Finally, with the information given thus far, would it be possible to plan a synthesis of *m*-dibromobenzene? The answer is that there is no direct reaction sequence that gives *m*-dibromobenzene as its major product. In general, we encounter difficulties trying to place two *o,p*-directors *meta* to one another or two *m*-directors *ortho* or *para* to one another. There are, however, some unique aromatic displacement reactions that permit us to overcome these problems, which we'll describe in Chapter 23.

Problems

15–20 Is the reaction sequence of importance in the synthesis of *m*-nitrobenzenesulfonic acid from benzene? Explain your answer.

15–21 Show how you would prepare each of the following compounds with benzene as the starting material.

a. 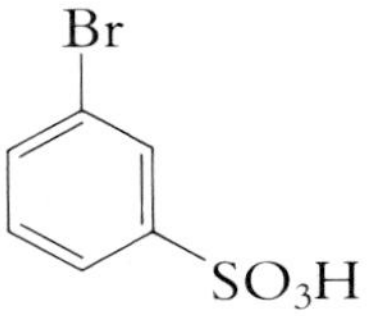 Br, SO_3H (meta) b. 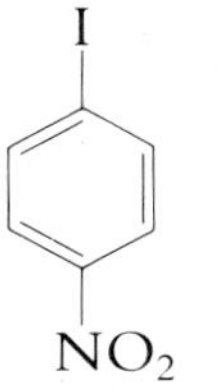 I, NO_2 (para) c. 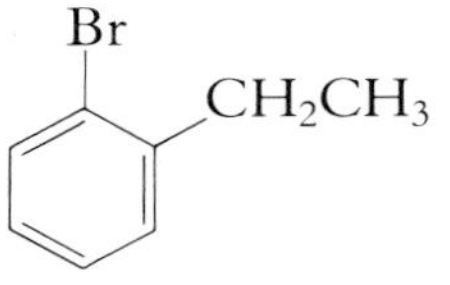 Br, CH_2CH_3 (ortho) d. 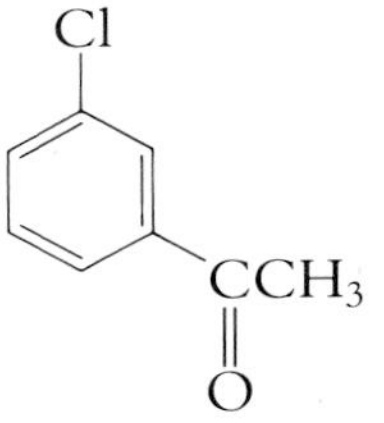 Cl, $\overset{O}{\overset{\|}{C}}CH_3$ (meta)

15:15 Orientation of Attack on Disubstituted Benzenes

There are some simple qualitative rules that allow us to predict the major product(s) from electrophilic attack on disubstituted benzenes. The easiest cases to treat are those in which the two substituents are so placed that their orienting effects reinforce one another. One such case is when one *o,p*-director and one *m*-director are either *ortho* or *para* to each other (**15**). If two *o,p*-directors (**16**) or two *m*-directors (**17**) are *meta* to each other, their orienting powers also reinforce each other.

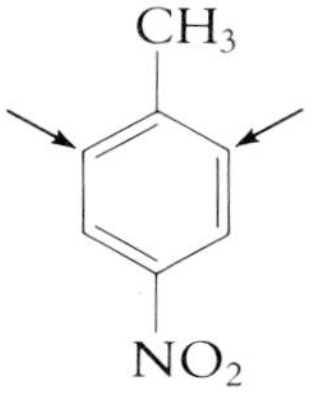

CH_3 / NO_2 **15** OMe / F **16** CO_2H / NO_2 **17**

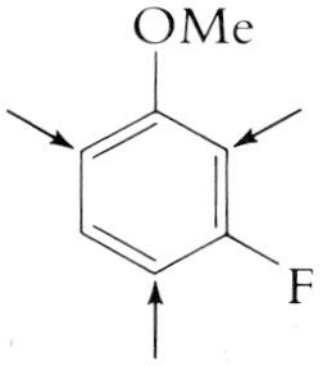

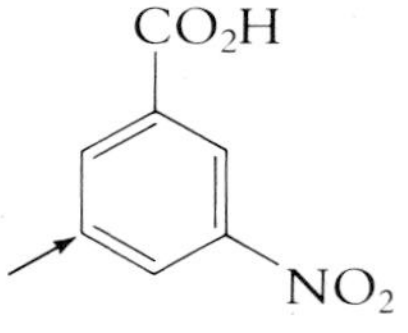

A different situation arises when the orienting effects of the two substituents are in opposition—for example, when two *o,p*-directors are *ortho* or *para* to each other (eqs. 15.56 and 15.57). If one of the substituents is substantially more activating than the other, it would determine the substituent pattern (eqs. 15.56 and 15.57).

(15.56) 4-Methylacetanilide ($NHCOCH_3$, CH_3) + Cl_2 $\xrightarrow{\text{acetic acid}}$ 2-chloro-4-methylacetanilide (Cl, $NHCOCH_3$, CH_3)

Two *o,p*-directors *para* to each other. $NHCOCH_3$ is a stronger director than CH_3.

(15.57) 2-Fluoroanisole (F, OCH_3) + HNO_3 → 2-fluoro-6-nitroanisole (F, OCH_3, NO_2) (66%) + 4-fluoro-3-methoxynitrobenzene (F, OCH_3, O_2N) (31%)

Two *o,p*-directors *ortho* to each other. OCH_3 is a stronger director than F.

The strongest activators are those that are the best resonance electron donors. Thus, the order of decreasing activating ability among activating substituents is $R_2N > HO > RO > CH_3CONH \gg$ alkyl. The halogens are deactivating and therefore will have the least influence, among *o,p*-directors, on the substitution patterns of disubstituted benzenes.

15:16 Biosynthesis of Thyroid Hormones: An *In Vivo* Electrophilic Aromatic Substitution

Reactions such as aromatic nitration, chlorination, or acylation are not usually found in biochemical systems. Aromatic iodination, however, is vital in human metabolism. The natural amino acid tyrosine undergoes iodination *ortho* to the powerfully activating hydroxyl group (eq. 15.58). The iodinat-

(15.58) Tyrosine $\xrightarrow{\text{I-Y}}$ 3-Iodotyrosine $\xrightarrow{\text{I-Y}}$ 3,5-Diiodotyrosine

ing agent *in vivo* is apparently not I_2, but some compound in which iodine is bonded to an electronegative atom such as O or N, represented in Equation 15.58 as I—Y. Although most iodine compounds are not very powerful electrophiles, phenol derivatives such as tyrosine are sufficiently activated to undergo rapid iodination.

Two molecules of diiodotyrosine are coupled together in a complex process to give a molecule of *thyroxine* (eq. 15.59). The iodination of tyrosine, and the subsequent biosynthesis of thyroxine, take place in the thyroid

(15.59) 3,5-Diiodotyrosine + 3,5-Diiodotyrosine → Thyroxine

gland. The thyroid gland releases the thyroxine into the bloodstream for circulation throughout the body. The major function of thyroxine is the control of metabolism and growth. Substances such as thyroxine that regulate bodily processes far from their point of origin, are known as **hormones** (from the Greek *horman,* meaning "to stimulate").

The biological iodinating agent for electrophilic substitution of tyrosine is derived from oxidation of the iodide ion. In many countries it has become common practice to add a little sodium iodide to ordinary table salt ("iodized salt") in order to prevent thyroid disorders such as goiterism.

15:17 NMR Spectroscopy of Substituted Aromatic Compounds

Both ^{1}H and ^{13}C nmr spectroscopy are used to determine the structure of unknown aromatic compounds. ^{1}H nmr also can provide spectroscopic evidence for some of the electron-withdrawing and donating effects that are critical to our understanding of electrophilic aromatic substitution. Figure 15–9 shows ^{1}H nmr spectra of *ortho-*, *meta-*, and *para*-disubstituted compounds. Let's examine the spectrum of *p*-nitroethylbenzene (Fig. 15–9a) first. The aromatic proton region is similar to an AB spectrum (sec. 9:7) in which the H_a protons *ortho* to the nitro group are equivalent and are split by the protons *ortho* to the ethyl group. Since the two components of each doublet are not the same height, it is apparent that some second-order effects are occurring. Nearly all *para*-disubstituted compounds with different substituents give splitting patterns similar to that observed in Figure 15–9a and this is their distinguishing feature (review Prob. 14–21, p. 539). Note that the downfield doublet at 8.06 ppm has been assigned to the protons *ortho* to the nitro group. Since there is a formal positive charge on the nitrogen of a nitro group, this group withdraws electrons from the vicinity of the *ortho* protons, causing them to be less shielded than in benzene. Electron-donating groups would shift *ortho* protons to higher field. ^{1}H nmr spectra for *meta-* and *ortho*-disubstituted benzenes are much more complex (Figs. 15–9b and 15–9c, respectively). Occasionally they can be analyzed if the spectra are expanded, as in the case of *o*-nitrophenol (Fig. 15–9c). We have analyzed the splitting of H_a, the proton *ortho* to the nitro, directly on the spectrum. Coupling constants between *ortho* protons generally have values of 6 to 9 Hz; between *meta* protons, 1 to 3 Hz; and between *para* protons, 0 to 1 Hz.

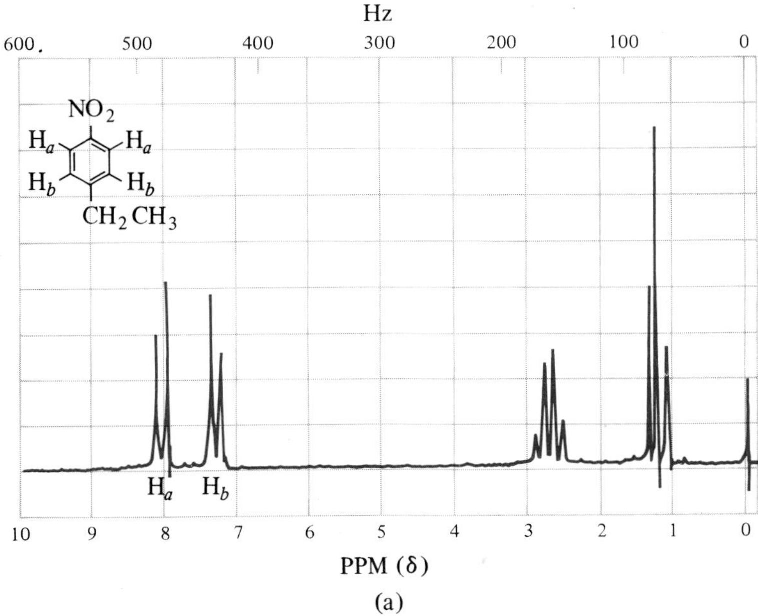

(a)

Figure 15–9. *(a)* ^{1}*H nmr spectrum of* p-*nitroethylbenzene.*

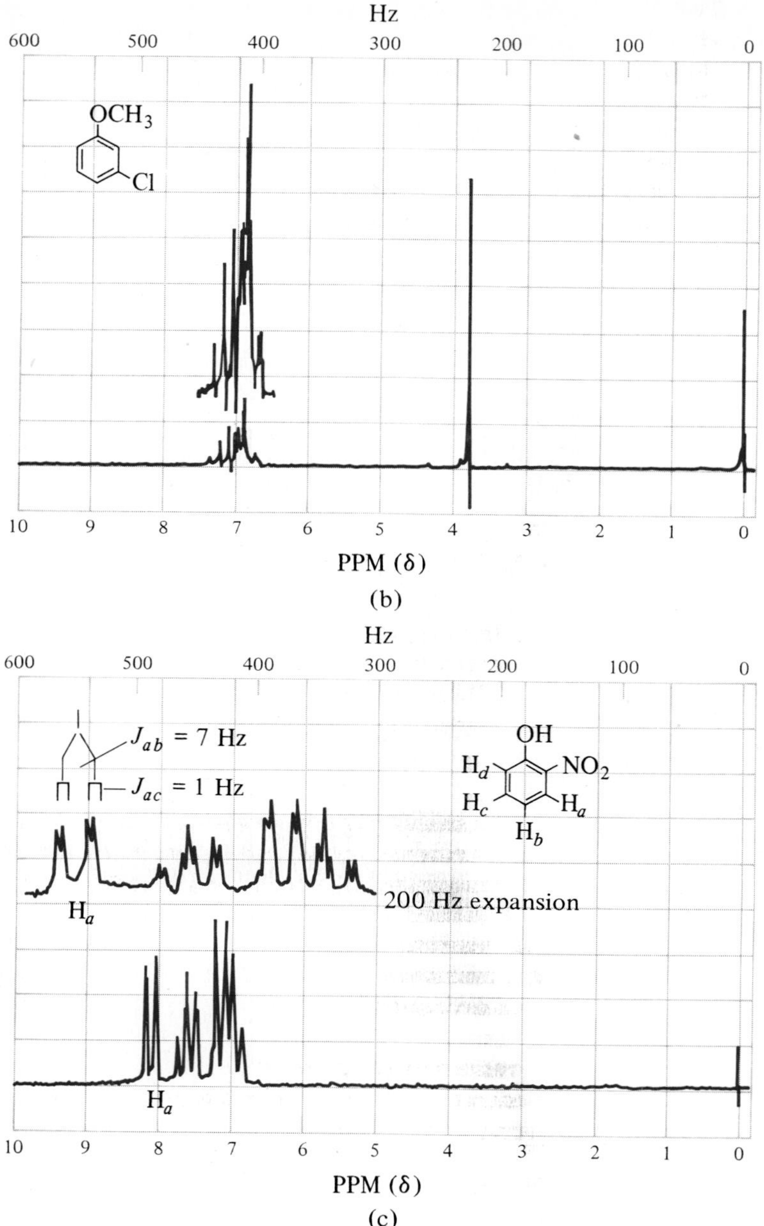

Figure 15–9. *¹H nmr spectra of (b)* m-*chloroanisole and (c)* o-*nitrophenol.*

Figure 15–10 gives the ^{13}C noise-decoupled nmr spectrum of 2-chloro-4-methoxyanisole. There are eight peaks and eight carbon atoms. Remember that the intensity of the peaks in ^{13}C noise-decoupled spectra cannot always be used as a measure of the number of each kind of carbon atom. An important factor in the intensity of peaks in ^{13}C spectra is whether or not the carbon has one or more hydrogens bonded to it. The peaks of low intensity

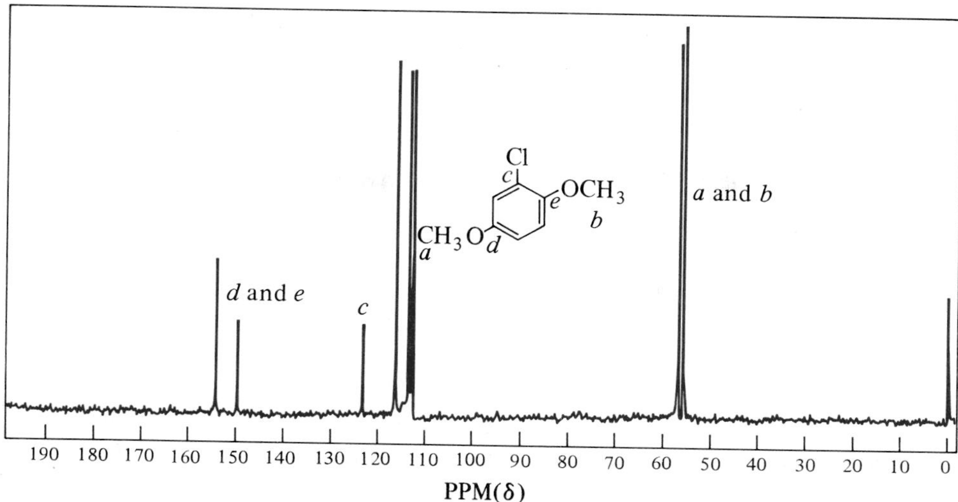

Figure 15–10. *^{13}C nmr spectrum of 2-chloro-4-methoxyanisole.*

(*c, d,* and *e*) do not bear hydrogen atoms, while those of high intensity (*a* and *b*, and the unlettered peaks of the remaining three aromatic carbons in the δ 113 to 116 region) do have hydrogens attached. The exact position of the ^{13}C absorptions is a sum of rather complex shielding and deshielding interactions which are as yet, not completely understood. Nevertheless, we can rely on the reproducibility of ^{13}C resonances when comparing very similar structures. For example, all methoxy methyls absorb very close to δ 58.

Summary

1 Aromatic compounds react with electrophiles generated by reaction of the attacking reagent with an acid catalyst. The overall reaction is a substitution. The electrophile uses two benzene π electrons to form a σ bond to one of the ring atoms. The resulting intermediate is a benzenium ion. In this ion the remaining four benzene π electrons are distributed over the five remaining sp^2 carbon atoms. The electron deficiency (positive charge) in the ion shows up at the atoms *ortho* and *para* to the carbon that was attacked by the electrophile E^+. The benzenium ion then loses H^+ and is converted back to an aromatic system. The proton is lost from the sp^3 carbon, which is the one bonded to the electrophile (sec. 15:2).

$$+ \; H^+$$

2 In nitration the electrophile is NO_2^+, produced by the reaction of HNO_3 with H_2SO_4 (sec. 15:3). In halogenation the electrophile is X^+ (X = Cl or Br), which is transferred from X_2 to the aromatic ring in the presence of a Lewis acid, usually AlX_3 or FeX_3. Iodination requires the use of I_2 in HNO_3 (sec. 15:4). In sulfonation the attacking electrophile is SO_3H^+, which is produced by a solution of SO_3 in H_2SO_4. Unlike nitration and halogenation, sulfonation is reversible (sec. 15:5).

Carbocations (R^+) and acylium ions (RCO^+) are generated by treating alkyl halides (RX) and acyl halides (RCOX), respectively, with Lewis acids such as AlX_3. These react with benzene in Friedel–Crafts reactions (secs. 15:6 and 15:7). Friedel–Crafts alkylations can also be carried out with alkenes or alcohols treated with Lewis or proton acids. Alkylations are complicated by rearrangement of the alkyl group and multiple substitutions of the aromatic ring (sec. 15:6). These complications are absent from Friedel–Crafts acylations. The acylation products can be reduced to isomerically pure alkylbenzenes (sec. 15:7).

3 When naming polysubstituted benzenes, list the substituent names in alphabetical order. The ring carbons are numbered so that the substituent numbers have the lowest possible sum (sec. 15:8).

4 Substituents on a benzenium ion affect its stability and therefore its rate of formation from the substituted benzene (secs. 15:9 and 15:10). Electron-withdrawing substituents destabilize benzenium ions and slow down their rate of formation. This deactivating effect is strongest when the electrophile attacks the ring *ortho* or *para* to the substituent. The deactivating effect is less severe when the electrophile attacks *meta* to the substituent. These substituents are classified as deactivating and *meta*-directing: NO_2, SO_3H, COR, CN, $\overset{+}{N}R_3$ (sec. 15:11).

Electron-donating substituents stabilize benzenium ions and speed up their rate of formation. This activating effect is strongest when the electrophile attacks the ring *ortho* or *para* to the substituent. Substituents that behave this way (R, Ar, OR, SR, NR_2) are classified as activating and *ortho,para*-directing (sec. 15:12). The groups OR and NR_2 are particularly powerful because of their conjugative effects.

Because of a delicate balance between their inductive electron-withdrawing and resonance electron-donating effects, halogen substituents are deactivating but *ortho,para*-directing (sec. 15:13).

5 The sequence of substitution reactions in the synthesis of a polysubstituted benzene is very important in determining which product isomers will be produced. For instance, if a *para* disubstituted benzene is the desired product, then the first substituent introduced into the ring must be an *o,p*-director. Similarly, the initial substituent should be an *m*-director if the desired product is the *m* isomer (sec. 15:14).

6 When two substituents are present on the same benzene ring, their directing effects may reinforce each other or oppose each other. If they oppose each other, the substituent that is most activating (or least deactivating) will exert the strongest directing effect.

7 Both ^1H and ^{13}C nmr spectroscopy are valuable for determining the structure of benzene derivatives because the position of several substituents on a ring can be located. Electron-withdrawing groups shift benzene proton absorptions to lower field and electron-donating groups cause shifts to higher field (sec. 15:17).

Problems

15–22 Give structures and names of the principal products expected from the ring nitration of each of the following compounds:

a. chlorobenzene
b. isopropyl phenyl ether
c. 2-phenylbutane
d. 1-chloro-3-phenylbutane
e. 1,8-dichloronaphthalene

15–23 Name each compound.

a. 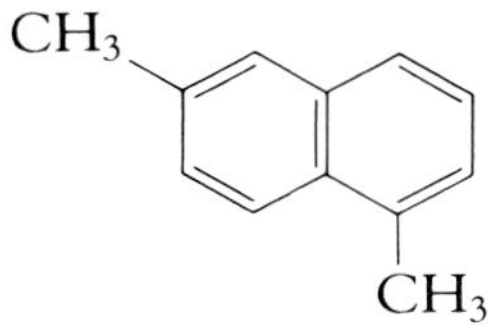 (naphthalene with CH_3 and CH_3 substituents)

b. 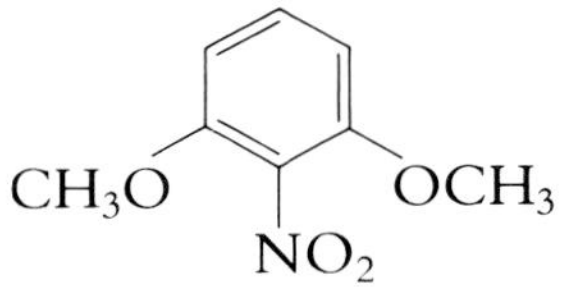 (benzene with CH_3O, NO_2, OCH_3 substituents)

c. 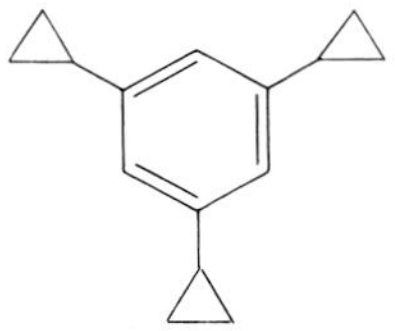 (benzene with three cyclopropyl substituents)

d. 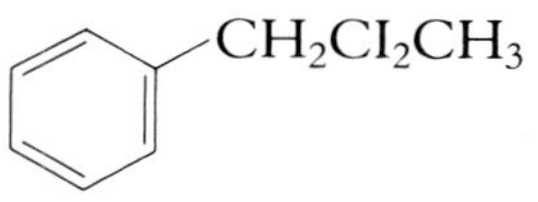 (benzene)–$CH_2CI_2CH_3$

e. 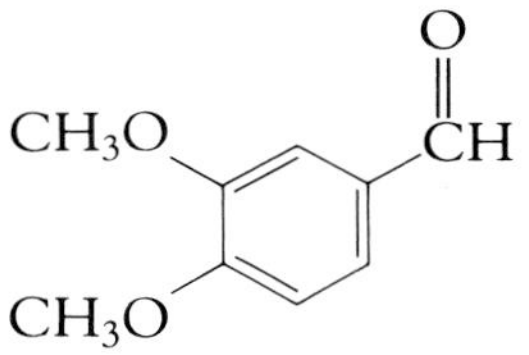 (benzene with CH_3O, CH_3O, and $\overset{O}{\overset{\|}{C}}H$ substituents)

15–24 Show all of the major monosubstitution products formed in each of the following reactions:

a. 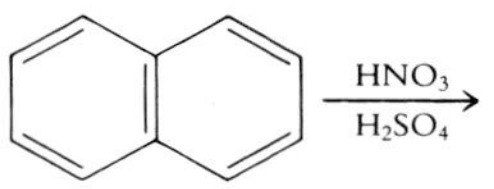 naphthalene $\xrightarrow[H_2SO_4]{HNO_3}$

b. 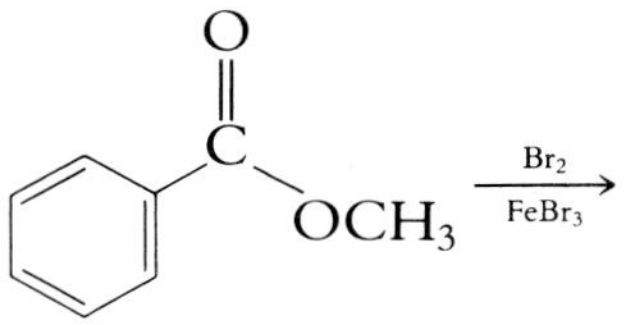 (benzene)–$\overset{O}{\overset{\|}{C}}$–$OCH_3$ $\xrightarrow[FeBr_3]{Br_2}$

c. (benzene)–CH_3 $\xrightarrow[HNO_3]{I_2}$

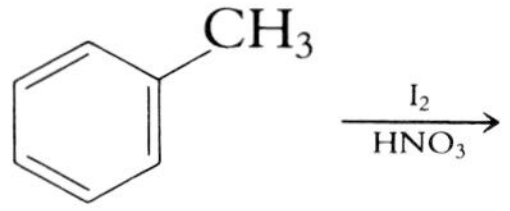

d. 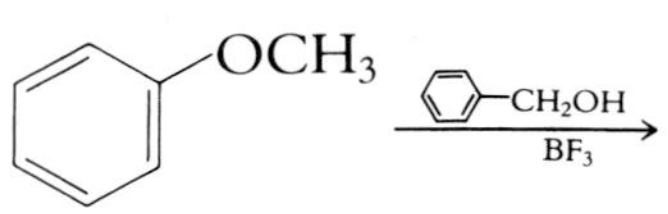 (benzene)–OCH_3 $\xrightarrow[BF_3]{C_6H_5CH_2OH}$

e. 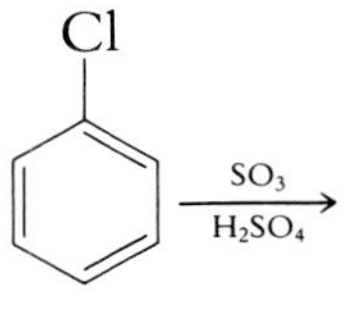 (benzene)–Cl $\xrightarrow[H_2SO_4]{SO_3}$

f. 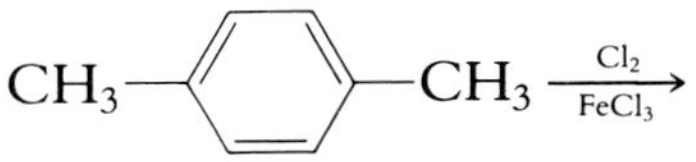 CH_3–(benzene)–CH_3 $\xrightarrow[FeCl_3]{Cl_2}$

g. 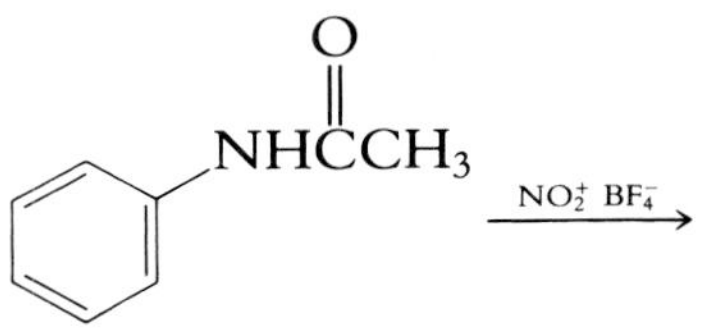 (benzene)–$NH\overset{O}{\overset{\|}{C}}CH_3$ $\xrightarrow{NO_2^+ BF_4^-}$

15–25 We showed that naphthalene is sulfonated more rapidly at C-1 than at C-2 (sec. 15:5). The same reactivity pattern is found with other electrophiles.

a. Draw an energy diagram for bromination of naphthalene. On this diagram, show the energy of the starting material and of the transition states and intermediates formed by Br^+ transfer to C-1 and C-2.

b. It has also been noted in this chapter that 2-substituted naphthalenes are more stable than their 1 isomers. Assume that this is true for the bromonaphthalenes. Complete your energy diagram by showing the relative energies of the two bromonaphthalenes and of the transition states immediately preceding them. (Remember that formation of the naphthalenium ions is the rate-determining step.)

c. Draw all the resonance structures for each bromonaphthalenium ion and use them to explain why bromination is faster at C-1 than at C-2. (Remember the importance of maintaining an intact aromatic sextet, as shown in Section 14:10.)

15–26 Molecular iodine is a weak electrophile toward benzene and its derivatives. It reacts slowly and the equilibrium constant for the reaction is unfavorable. Iodine chloride (ICl) has been effectively used to iodinate several benzene derivatives. Explain why ICl is a better iodinating agent, both kinetically and thermodynamically, than I_2.

15–27 Two isomers of formula C_6H_4BrCl are shown to be aromatic by means of ir and nmr spectroscopy. Isomer **A** is mononitrated, and it gives four different products (some in low yield). Mononitration of **B** yields two different products. Do these data allow you to assign a unique structure to either **A** or **B**? Explain your answer.

15–28 When benzene is treated with a mixture of formaldehyde and concentrated hydrocholoric acid, benzyl chloride ($C_6H_5CH_2Cl$) is produced in good yield (the *chloromethylation* reaction). The first reaction product is benzyl alcohol, which is rapidly converted to benzyl chloride by HCl. This synthesis can also be applied to other aromatics, as long as their reactivity equals or exceeds that of benzene. Write a detailed mechanism for the entire reaction, showing all important intermediates.

15–29 Substituted benzaldehydes may be synthesized by a procedure that is related to the Friedel–Crafts acylation. Carbon monoxide and hydrogen chloride are bubbled into a mixture of a substituted benzene (not strongly deactivated) and a Lewis acid (eq. 2). Write a mechanism for this reaction, showing explicitly the structure of the acylating agent.

$$CH_3-C_6H_5 + CO \xrightarrow[CuCl]{HCl} CH_3-C_6H_4-CHO \qquad (2)$$

p-Methylbenzaldehyde
(85%)

15–30 Show how the following ketones can be prepared from benzene and appropriate acid chlorides.

a. 1-phenyl-1-pentanone
b. 2-methyl-1-phenyl-1-propanone
c. cyclopentyl phenyl ketone
d. 1,8-diphenyl-1,8-octanedione
e. benzophenone (diphenyl ketone)

15–31 a. Draw structures for the ions formed in each of the following reactions:
1) $CH_3(CH_2)_3COCl + AlCl_3 \rightarrow$
2) $(CH_3)_2CHCOCl + AlCl_3 \rightarrow$
3) [cyclohexane ring bearing COCl and H] + $SnCl_4 \rightarrow$

b. None of the ions formed in these reactions show any tendency to rearrange. Explain this observation with the aid of resonance structures.

c. When benzoic acid (C_6H_5COOH) is dissolved in a strong acid, such as HF/BF_3, it gives rise to the same organic ion that can be obtained from benzoyl chloride (C_6H_5COCl) and $AlCl_3$. Write the structure of this ion, and show how it can be formed from benzoic acid.

15–32 Show how each of the following hydrocarbons can be prepared fairly pure, starting with benzene:

a. ethylbenzene
b. butylbenzene
c. *t*-butylbenzene
d. 4-propyltoluene

15–33 Arrange the following sets of compounds in order of decreasing reactivity toward acetyl chloride and $AlCl_3$.

a. benzene, methoxybenzene, *p*-dibromobenzene, toluene, chlorobenzene
b. bromobenzene, benzene, isopropylbenzene, *o*-xylene, benzenesulfonic acid

15–34 In each case indicate which ring you expect to be most rapidly nitrated, and what the principal product(s) will be.

a. $C_6H_5-CH_2C(=O)-O-C_6H_5$

b. $Cl-C_6H_4-C_6H_5$

c. $C_6H_5-CH_2CH_2-C(=O)-C_6H_5$

d. $HO_3S-C_6H_4-CH_2-C_6H_4-Br$

15–35 Devise efficient, high yield syntheses of the following compounds. Benzene, toluene, or benzoic acid (**I**) may be used as starting materials.

a. *m*-chloronitrobenzene
b. *p*-chloronitrobenzene
c. *p*-iodobenzenesulfonic acid
d. *o*-bromochlorobenzene
e. *m*-nitrobenzoic acid
f. 2-bromo-4-nitrotoluene
g. 4-chloro-3-nitrobenzenesulfonic acid
h. 3-bromo-5-nitrobenzoic acid
i. 1-chloro-2,4-dinitrobenzene

$C_6H_5-C(=O)OH$

I

15–36 Would you expect the following compounds and ions to undergo predominant *m* or *o,p* substitution? Briefly explain each case, using resonance structures.

a. C_6H_5–$\ddot{N}{=}\ddot{O}$: (nitrosobenzene)

b. C_6H_5–$\overset{+}{S}(CH_3)_2$

c. C_6H_5–C_6H_5

15–37 In the presence of a base, phenol undergoes electrophilic substitution even faster than it does in neutral or acid solution. What reaction might occur between phenol and a base? Why would this enhance the rate of electrophilic substitution?

15–38 Propose structures for the unknowns in the equations listed below. In all cases the unknowns represent the major product(s) of the indicated reaction. The starting materials are all monosubstituted benzenes.

a. $C_6H_5OCH_3$ (Anisole) $\xrightarrow{Br_2}$ **A** + **B** (both C_7H_7BrO)

A $\xrightarrow{HNO_3}$ **C** ($C_7H_6BrNO_3$)

B $\xrightarrow{HNO_3}$ **D** + **E**

b. $C_7H_6O \xrightarrow[SO_3]{H_2SO_4}$ **F** ($C_7H_6O_4S$) $\xrightarrow[H_2SO_4]{HNO_3}$ **G** ($C_7H_5NO_6S$)

c. $C_6H_5NHCOCH_3$ (Acetanilide) $\xrightarrow{Cl_2}$ **N** + **O**

N $\xrightarrow{H_2SO_4} \xrightarrow{H_3O^+}$ **P** + **Q** (both $C_6H_6ClNO_3S$)

O $\xrightarrow{H_2SO_4} \xrightarrow{H_3O^+}$ **R**

15–39 Propose syntheses for the following compounds using benzene, toluene, or one of the xylenes (dimethylbenzenes) as the starting material, and any other organic and inorganic reagents.

a. ethylbenzene
b. *p*-isopropyltoluene
c. *p*-*t*-butylethylbenzene
d. *p*-bromoacetophenone
e. *m*-bromoacetophenone
f. hexylbenzene

g. 1,2-dimethyl-4-propylbenzene
h. *p*-chloropropylbenzene
i. 2,4-dimethyl-1-nitrobenzene

15–40 Polymerization of styrene, $C_6H_5CH{=}CH_2$, gives the commercially useful polymer, polystyrene (Prob. 14–29). When the polymerization mixture includes 1,4-bis(chloromethyl)benzene and $SnCl_4$, a Lewis acid catalyst, the product is crosslinked polystyrene. This is a product in which the individual polymer chains of polystyrene have become chemically linked to one another. Show the structure of the crosslink, and how it is formed.

$ClCH_2-C_6H_4-CH_2Cl$

1,4-Bis(chloromethyl)benzene

15–41 *Motrin* [2-(*p*-isobutylphenyl)propanoic acid] is a compound that has been widely used as an antirheumatic and antiarthritic drug. Several syntheses of motrin proceed by way of *p*-isobutylacetophenone (eq. 3). Show how *p*-isobutylacetophenone can be prepared from benzene.

$$C_6H_6 \xrightarrow[\text{steps}]{\text{several}} (CH_3)_2CHCH_2-C_6H_4-\overset{O}{\overset{\|}{C}}CH_3 \xrightarrow[\text{steps}]{\text{several}} (CH_3)_2CHCH_2-C_6H_4-\overset{CH_3}{\overset{|}{C}}HCOOH \quad (3)$$

p-Isobutylacetophenone

Motrin

15–42 When hexamethylbenzene is treated with fluoroboric acid at −80°, it is protonated to yield a stable ion which gives the following ^1H nmr spectrum: δ 1.61 (doublet, 3H); δ 2.32 (singlet, 6H); δ 2.60 (singlet, 6H); δ 2.76 (singlet, 3H); and δ 4.01 (quartet, 1H). Draw a structure for the ion and assign all the nmr peaks to their various hydrogens. Explain why the three singlets have the relative positions that they do. What inferences can you make about the charge distribution in this ion from its nmr spectrum?

15–43 Bromination of benzenes is usually irreversible, but if the ring carries a strongly activating substituent, debromination and isomerization may be observed. When 2-bromo-*N,N*-dimethylaniline (**II**) is heated with HBr in chloroform, it is partially converted to *N,N*-dimethylaniline (**III**) and 4-bromo-*N,N*-dimethylaniline (**IV**) (eq. 4). No *m* isomer is found.

$$\underset{\textbf{II}}{o\text{-}BrC_6H_4NMe_2} \xrightarrow[CHCl_3]{HBr} \underset{\textbf{III}}{C_6H_5NMe_2} + \underset{\textbf{IV}}{p\text{-}BrC_6H_4NMe_2} \quad (4)$$

The evidence indicates that **III** is an intermediate in the formation of **IV**. Write a mechanism that accounts for the formation of **III** and **IV** and that is in accord with all the data.

15–44 The ease with which a benzenesulfonic acid may be desulfonated can be used advantageously. The sulfonic acid group has been used to *block* a position on the benzene ring, and has then been removed after its blocking action is no longer needed.

Bromination of anisole (eq. 15.43) gives only a 4% yield of *o*-bromoanisole. By using the sulfonation–desulfonation sequence, it is possible to obtain a high yield of pure *o*-bromoanisole. Outline the reactions involved in this preparation, starting with anisole.

15–45 The nitration of *N,N*-dimethylaniline [$C_6H_5N(CH_3)_2$] in 85% H_2SO_4 gives 45% *m*-nitro product and 38% *p*-nitro product. (The remainder is polynitrated material.) What is the predominant species actually undergoing nitration?

15–46 Nitration of toluene gives a mixture of 58% *o*-isomer, 5% *m*-isomer, and 37% *p*-isomer. Under the same conditions, the isomer proportions from benzyl chloride are 32, 16, and 52%; from benzal chloride (dichloromethylbenzene), 23, 34, and 43%; and from benzotrichloride (trichloromethylbenzene), 7, 64, and 29%.

a. How would you rationalize the trends in *o* and *m* substitution as you progressively replace methyl hydrogens by chlorine?
b. As hydrogen atoms are replaced by chlorine atoms, at what point does the methyl group change from being an *o,p*-director to a *m*-director? (What would be the distribution of products if the substituent exerted *no* directive power?)
c. Would the change in directive character of the methyl group come more or less quickly if the hydrogen atoms were being replaced by fluorine rather than chlorine? Explain your answer.

15–47 An unknown compound, $C_9H_{10}O$, has the ^1H nmr spectrum shown here. Treatment of the unknown with HCl gives a second compound, $C_9H_{11}OCl$. Give the structure of both compounds and a mechanism for the formation of the second compound.

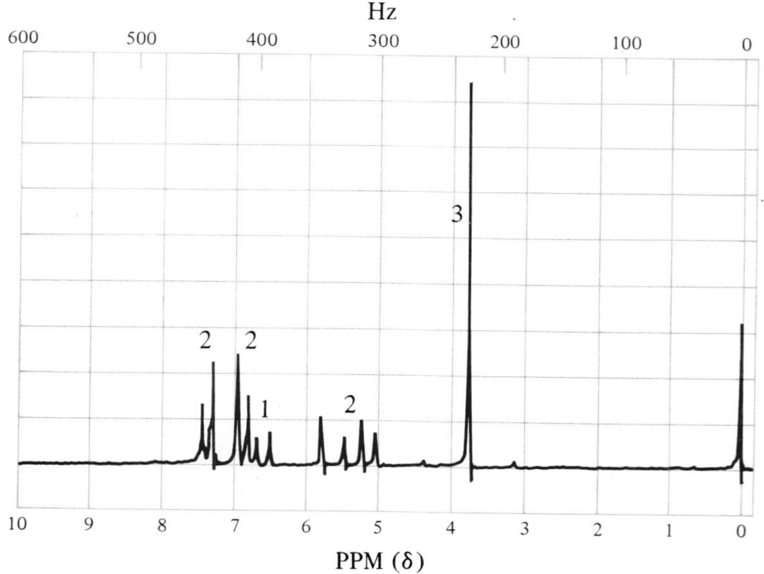

15–48 Propose syntheses of the following compounds using benzene and toluene as starting materials. It will be necessary to use reactions such as the Grignard reaction, in addition to electrophilic aromatic substituents, to reach the final product.

a. 1-phenylethanol
b. 2-(*p*-methylphenyl)-2-butanol
c. 1-phenylcyclopentene
d. 2-phenylethanol
e. 1-phenyl-1,2-dibromopropane

15–49 The ^{13}C nmr spectrum of 2-methylnaphthalene is shown here. To which carbon can peak *a* be assigned? To what carbons can the three peaks labeled *b* be assigned?

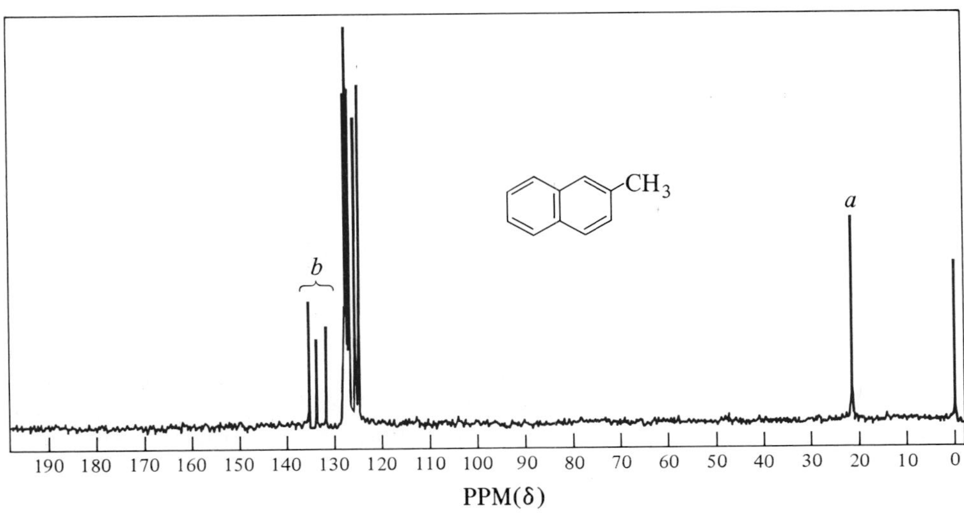

Supplementary Reading

P. Wiseman, *An Introduction to Industrial Organic Chemistry,* Chap. 5, Applied Science, London, 1979.

J. M. Tedder, A. Nechtval, and A. H. Jubb, *Basic Organic Chemistry. Part 5: Industrial Processes,* Chap. 5, John Wiley, New York, N.Y., 1975.

L. M. Stock, *Aromatic Substitution Reactions,* Prentice–Hall, Englewood Cliffs, N.J., 1968.

H. Duewell, Aromatic substitution, *J. Chem. Ed.,* **43,** 139 (1966).

16 Carboxylic Acids, Acid Halides, and Anhydrides

Among the first organic compounds to be isolated in relatively pure form were carboxylic acids,

$$\mathrm{R{-}\overset{\overset{\large O}{\|}}{C}{-}OH}$$

There are several reasons for this. Because of their acidity, carboxylic acids can be readily separated from neutral and basic materials. (Their salts have been used for thousands of years in the form of soap.) They are also widely distributed in natural sources. Acetic acid (CH_3COOH), for instance, may be obtained by distilling vinegar, which is simply sour wine.

The name "acetic acid" comes from the Latin *acetum,* meaning "vinegar." The nonsystematic names of most simple straight-chain acids reflect their origin. Formic acid (HCOOH) was obtained by pyrolysis of red ants (from the Latin *formica,* meaning "ant"). Butyric acid ($CH_3CH_2CH_2COOH$) is released when butter turns rancid (from the Latin *butyrum,* meaning "butter"). Caproic ($C_5H_{11}COOH$), caprylic ($C_7H_{15}COOH$), and capric ($C_9H_{19}COOH$) acids owe their names to their particularly penetrating odors (from the Latin *caper,* meaning "goat"). Many straight-chain monocarboxylic acids are present in animal fats and these compounds are often known as **fatty acids.** Carboxylic acids containing an even number of carbon atoms are much more abundant in nature than those with an odd number of carbon atoms. This is because successive steps of fatty acid biosynthesis result in the incorporation of two carbons at a time.

There are also many *derivatives* of carboxylic acids,

$$\mathrm{R{-}\overset{\overset{\large O}{\|}}{C}{-}Y}$$

where the Y group is neither a carbon nor a hydrogen. Two of these derivatives, acid halides and acid anhydrides, are useful synthetic intermediates and are discussed later in this chapter.

$$\mathrm{R{-}\overset{\overset{\large O}{\|}}{C}{-}Cl} \quad \text{or} \quad \mathrm{R{-}\overset{\overset{\large O}{\|}}{C}{-}Br} \qquad\qquad \mathrm{R{-}\overset{\overset{\large O}{\|}}{C}{-}O{-}\overset{\overset{\large O}{\|}}{C}{-}R}$$

Acid halides — Acid anhydrides

16:1 Nomenclature of Carboxylic Acids

In the IUPAC system, straight-chain acids are named by dropping the final "e" from the name of the hydrocarbon corresponding to the longest chain, and replacing it with the suffix -oic and the word "acid." The carbon atom of the carboxyl group is counted as part of the longest chain and is given the number one. Some important straight-chain acids, along with their systematic and common names, are listed in Table 16–1.

$$CH_3\overset{O}{\overset{\|}{C}}OH \qquad (CH_3)_2CHCH_2CH_2\overset{O}{\overset{\|}{C}}OH \qquad CH_3\underset{F}{\underset{|}{C}}HCH_2\overset{CH_3CH_2}{\overset{|}{C}}HCH_2\underset{F}{\underset{|}{C}}H\overset{O}{\overset{\|}{C}}OH$$

Ethanoic acid — 4-Methylpentanoic acid — 2,6-Difluoro-4-ethylheptanoic acid

Substituents on aliphatic carboxylic acids are usually designated by number. An older system, which uses the Greek letters α, β, γ, δ, and so forth, is still sometimes used when common names are employed, especially in biochemistry. The carbon of the carboxylic group is not given a letter.

$$\overset{\varepsilon}{C}H_3\overset{\delta}{C}H_2\overset{\gamma}{C}H_2\overset{\beta}{C}H_2\overset{\alpha}{C}H_2COOH \qquad CH_3\underset{Br}{\underset{|}{\overset{\beta}{C}}}H\overset{\alpha}{C}H_2COOH \qquad R\underset{NH_2}{\underset{|}{\overset{\alpha}{C}}}HCOOH$$

β-Bromobutyric acid — An α-amino acid

There is a series of important, naturally occurring straight-chain **dicarboxylic acids** (Table 16–2). The common names of the simple dicarboxylic acids are used almost exclusively. (An aid to remembering them is the mnemonic "*o*h *m*y *s*uch *g*ood *a*pple *p*ie.") In the IUPAC system, these

Table 16–1. IUPAC and Common Names of Some Straight-Chain Acids

	IUPAC Name	Common Name	
$HCOOH$	methanoic	formic	
CH_3COOH	ethanoic	acetic	
CH_3CH_2COOH	propanoic	propionic	
$CH_3(CH_2)_2COOH$	butanoic	butyric	
$CH_3(CH_2)_3COOH$	pentanoic	valeric	
$CH_3(CH_2)_4COOH$	hexanoic	caproic	
$CH_3(CH_2)_6COOH$	octanoic	caprylic	
$CH_3(CH_2)_8COOH$	decanoic	capric	
$CH_3(CH_2)_{10}COOH$	dodecanoic	lauric	
$CH_3(CH_2)_{12}COOH$	tetradecanoic	myristic	Fatty acids
$CH_3(CH_2)_{14}COOH$	hexadecanoic	palmitic	Fatty acids
$CH_3(CH_2)_{16}COOH$	octadecanoic	stearic	Fatty acids
$CH_3(CH_2)_7CH \overset{c}{=} CH(CH_2)_7COOH$	(*Z*)-9-octadecenoic	oleic	Fatty acids

Table 16–2. IUPAC and Common Names of Some Aliphatic Dicarboxylic Acids

	IUPAC Name	Common Name
HOOCCOOH	ethanedioic	oxalic
$HOOCCH_2COOH$	propanedioic	malonic
$HOOCCH_2CH_2COOH$	butanedioic	succinic
$HOOC(CH_2)_3COOH$	pentanedioic	glutaric
$HOOC(CH_2)_4COOH$	hexanedioic	adipic
$HOOC(CH_2)_5COOH$	heptanedioic	pimelic

compounds are named by adding the suffix -dioic and the word "acid" to the alkane name denoting the longest carbon chain *containing both carboxyl groups*.

When more than two carboxyl groups are present in the molecule, it is often convenient to treat the groups as substituents on the principal chain. In that case, the carboxyl carbon atoms are not counted as part of the chain. The same situation applies when carboxyl groups are attached directly to an alicyclic or aromatic ring.

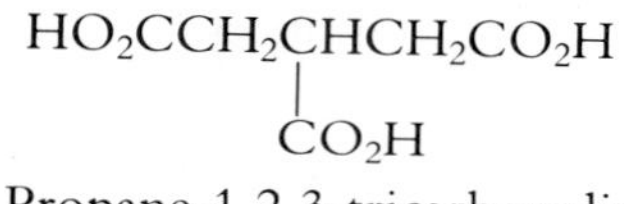

$HO_2CCH_2CHCH_2CO_2H$ (with CO_2H on the central CH)

Propane-1,2,3-tricarboxylic acid

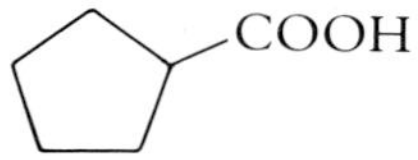

Cyclopentanecarboxylic acid

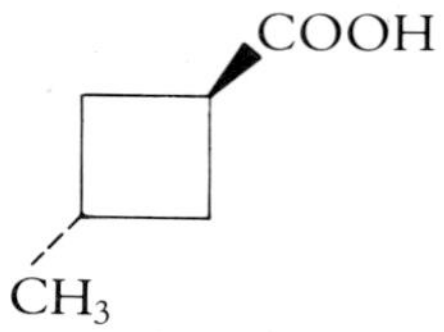

trans-3-Methylcyclobutane-carboxylic acid

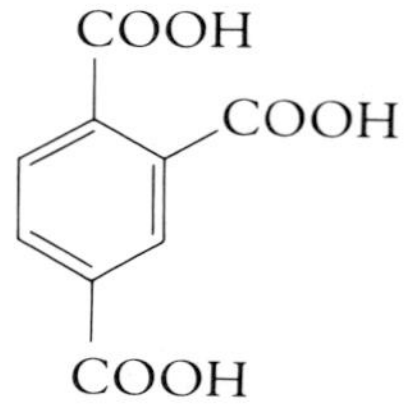

Benzene-1,2,4-tricarboxylic acid

Many aromatic carboxylic acids are known by their common names:

Benzoic acid

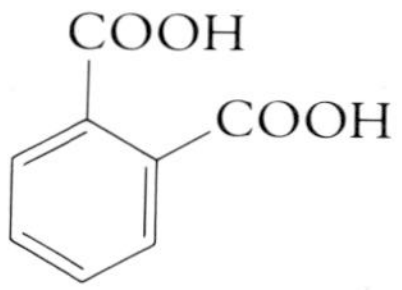

Phthalic acid

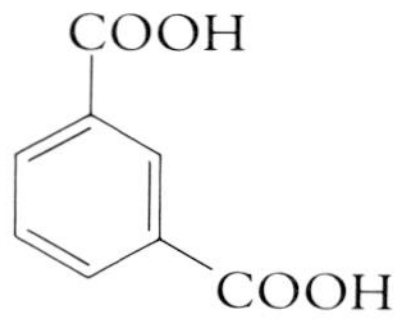

Isophthalic acid

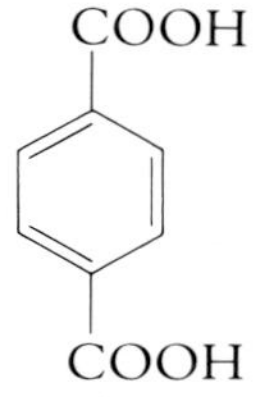

Terephthalic acid

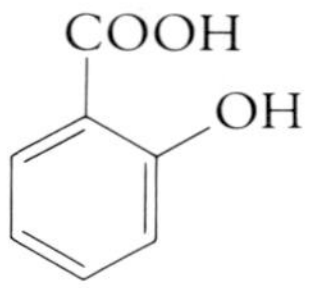

Salicylic acid

Finally, here are the structures of some biologically important carboxylic acids and their common names. All of these compounds are intermediates in the synthesis and breakdown of fats and sugars.

Maleic acid: $(H)(HOOC)C{=}C(H)(COOH)$ (cis)

Fumaric acid: $(H)(HOOC)C{=}C(COOH)(H)$ (trans)

Citric acid: $HOOCCH_2C(OH)(COOH)CH_2COOH$

Malic acid: $HOOCCH_2CH(OH)COOH$

Lactic acid: $CH_3CH(OH)COOH$

Problem

16–1 Draw structures corresponding to the names (a to e), and name the compounds whose structures are given (f to i).

a. 4-methyl-3-hexenoic acid
b. β-hydroxyvaleric acid
c. 2-bromo-3-methylheptanedioic acid
d. 2-methylcyclopropanecarboxylic acid
e. 3-bromophthalic acid
f. $CH_3CH(COOH)CH_2CH_2CH_3$
g. cyclohexane–COOH
h. benzene ring bearing NO_2, Br (adjacent to NO_2), and COOH
i. $HOOCCH_2CH(OCH_3)COOH$

Box 16–1. Prostaglandins and Essential Fatty Acids

It has been known for some time that mammals raised on fat-free diets suffer numerous developmental abnormalities, such as skin lesions, and that on-going fat deprivation can ultimately lead to death. However, the symptoms of fat deprivation can be completely reversed by the consumption of certain polyenoic (polyunsaturated) fatty acids classified as *essential fatty acids*.

The metabolic role of the essential fatty acids was long unknown, but the mystery began to unravel in Sweden in the 1930s. There it was found that certain substances, present in seminal fluid and prostate gland tissue, had the power to stimulate contraction of smooth muscle such as that found in the uterus. These compounds, called **prostaglandins** (PGs) because they were originally obtained from prostate tissue, were isolated and characterized in the 1960s. A large number of them have since been discovered in diverse tissues, exerting effects as varied as raising or lowering blood pressure, inhibiting stomach acid production, and causing pain.

Prostaglandins are synthesized from the C_{20} tetraenoic acid, arachidonic acid, which in turn is derived from the C_{18} dienoic acid, linoleic acid. Unlike plants, mammals are capable of biosynthesizing only saturated and monoenoic acids, such as oleic acid (Table 16–1). Thus, they must take in linoleic acid from plant sources.

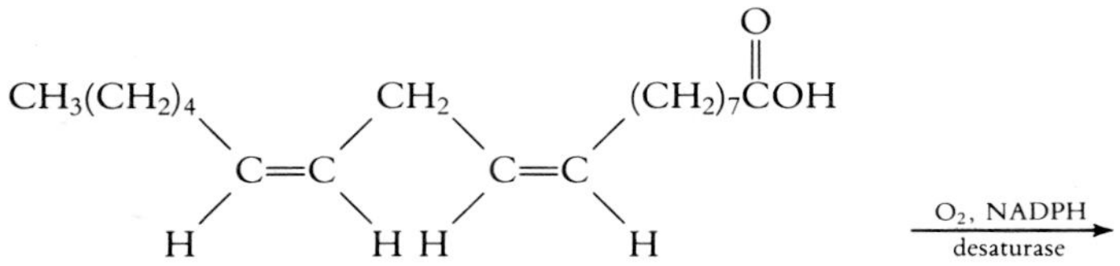

$CH_3(CH_2)_4CH{=}CHCH_2CH{=}CH(CH_2)_7COOH$

Linoleic acid

$\xrightarrow[\text{desaturase}]{O_2,\ NADPH}$

$CH_3(CH_2)_4CH{=}CHCH_2CH{=}CHCH_2CH{=}CH(CH_2)_4COOH$

$\xrightarrow[\text{chain extension (See sec. 19:13)}]{\text{two-carbon}}$ $\xrightarrow[\text{desaturase}]{O_2,\ NADPH}$

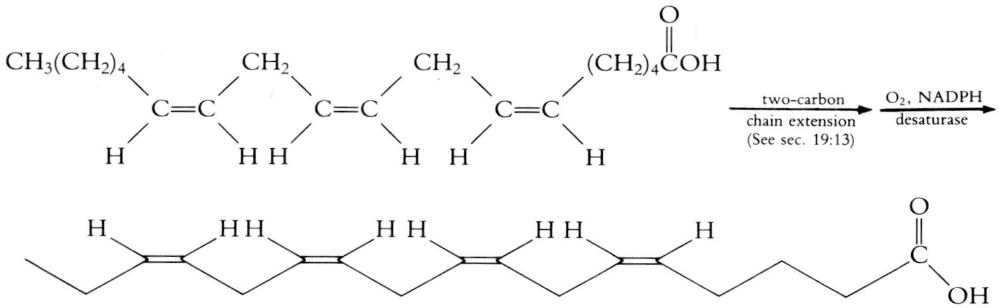

Arachidonic acid

In the presence of the enzyme fatty acid cyclooxygenase, arachidonic acid undergoes cycloaddition (sec. 13:12) with O_2, ultimately producing various prostaglandins. There are a series of other prostaglandins with some variations in functional groups. Further reactions convert prostaglandins to thromboxanes and prostacyclins.

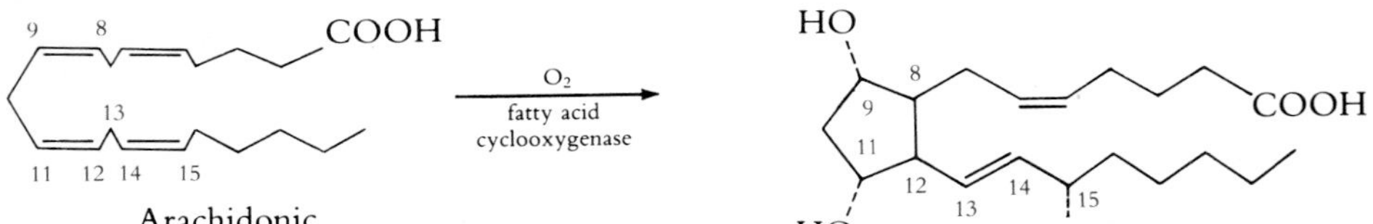

Arachidonic acid $\xrightarrow[\text{fatty acid cyclooxygenase}]{O_2}$ $PGF_{2\alpha}$

The 1982 Nobel Prize in Physiology and Medicine was awarded jointly to S. K. Bergstrom and B. I. Samuelsson of Sweden and J. R. Vane of England, for their work on prostaglandins.

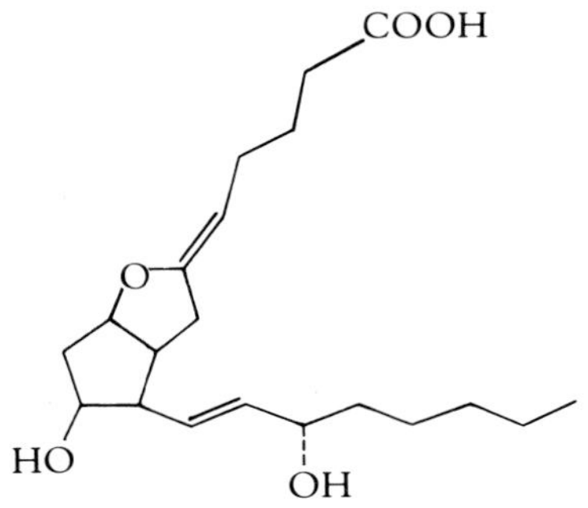

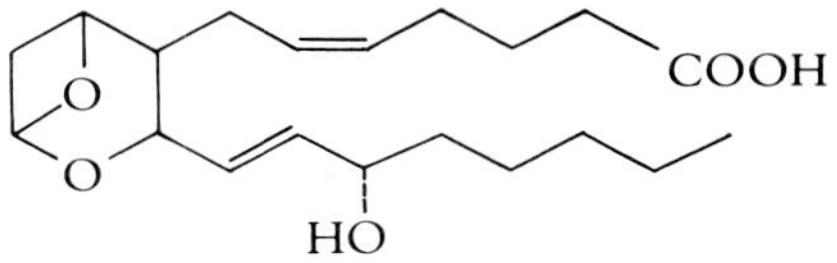

16:2 Physical Properties and Spectra of Carboxylic Acids

The straight-chain monocarboxylic acids up through nonanoic acid (C_9) are liquids at room temperature, and those through C_5 are water soluble. Benzoic acid and almost all substituted benzoic acids are solids and are only slightly soluble in water. Most carboxylic acids become water soluble above pH 5 because they are converted to anions (eq. 16.1). Pure liquid acids (other than formic acid) exist almost entirely in the form of molecular dimers (Fig. 16–1), in which two molecules are held together by a pair of hydrogen bonds.

(16.1) $$C_6H_5COOH + NaOH \rightarrow C_6H_5COO^-Na^+ + HOH$$

Intermolecular hydrogen-bonding, as shown in Figure 16–1, always produces characteristic effects in the ir spectrum. A typical carboxylic acid spectrum contains a broad peak between 2500 and 3500 cm^{-1} (Fig. 16–2). This absorption is characteristic of a strongly hydrogen-bonded OH, as in

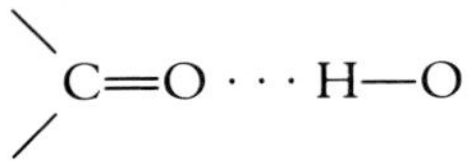

$$\gt C{=}O \cdots H{-}O$$

and is direct evidence for intramolecular hydrogen-bonding in carboxylic acid dimers. The characteristic carbonyl ir stretching absorption for carboxylic acid dimers is found at 1710 to 1720 cm^{-1}. This frequency can be modified by conjugation.

The ^1H nmr spectra of carboxylic acids are distinctive in that the

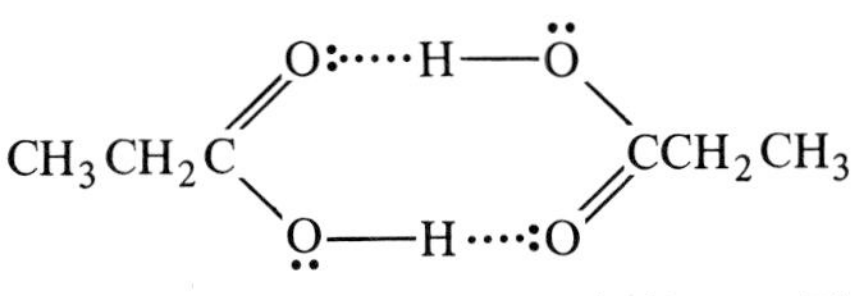

Figure 16–1. *Propanoic acid dimer. The two molecules are held together by hydrogen bonds between their carboxyl and hydroxyl groups.*

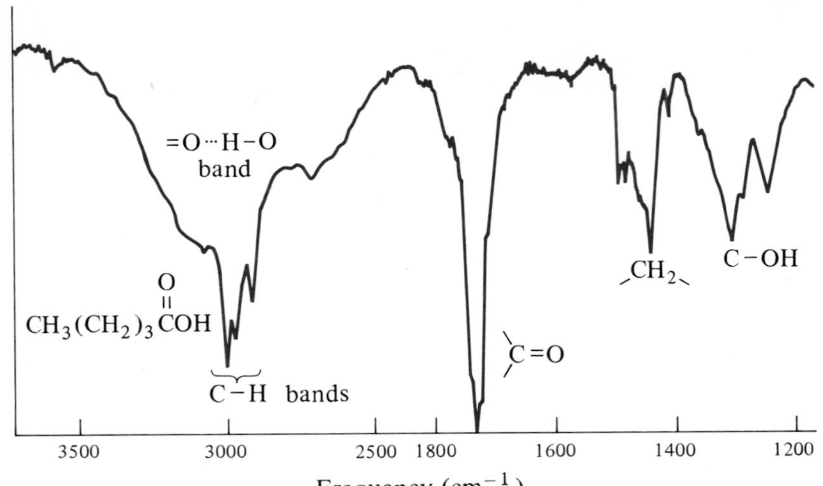

Figure 16–2. *Partial ir spectrum of pentanoic acid in* CCl_4*. Note the characteristic broad OH band between 2500 and 3400* cm^{-1}.

—COOH proton absorbs far downfield, giving rise to a singlet in the 10 to 13 ppm region. Only very acidic protons (e.g., $RSO_3\mathbf{H}$) appear in the same general area. Figure 16–3 shows the ^1H nmr spectrum of 2,2-dimethylbutanoic acid, where the acidic proton appears at 12.1 ppm.

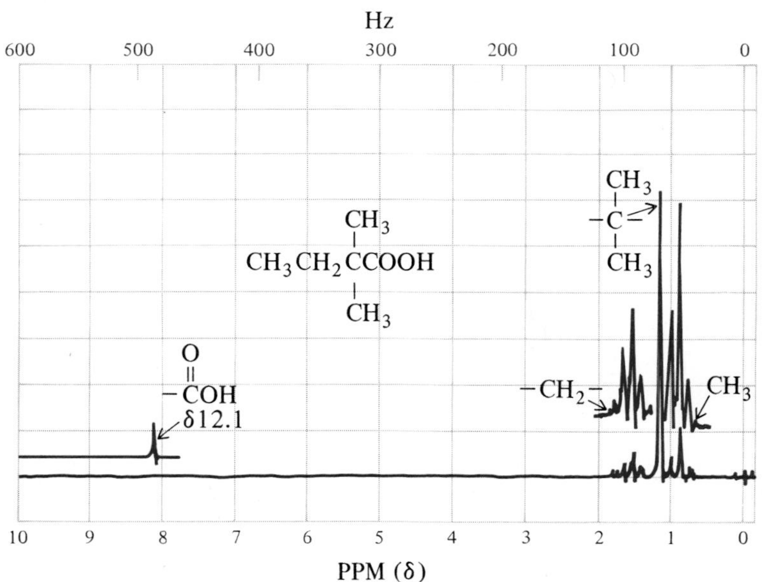

Figure 16–3. *^1H nmr spectrum of 2,2-dimethylbutanoic acid (COOH proton absorbance offset for 12.1 ppm).*

Problem

16–2 An unknown compound has the molecular formula $C_{12}H_{14}O_5$. Its ^1H nmr and ir spectra are given here. The one proton singlet raised on the right side is "offset," since it would have appeared off-scale on the left. Suggest a structure for the unknown based upon the spectral data.

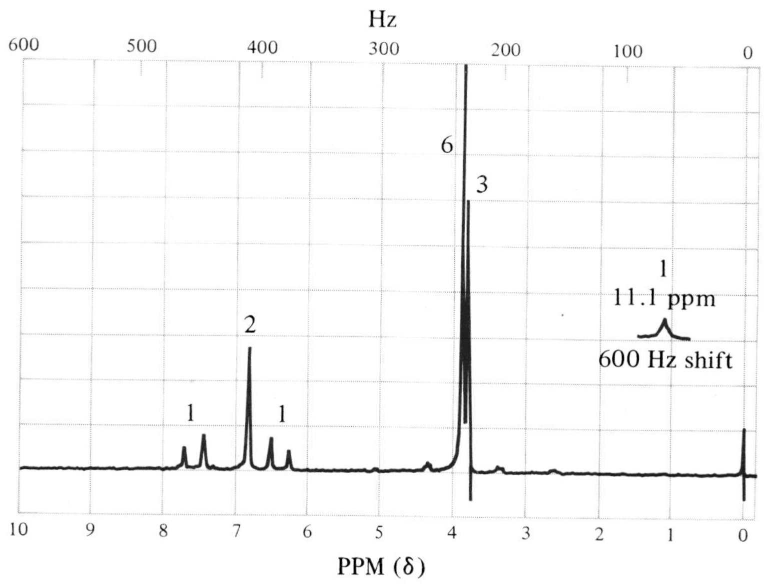

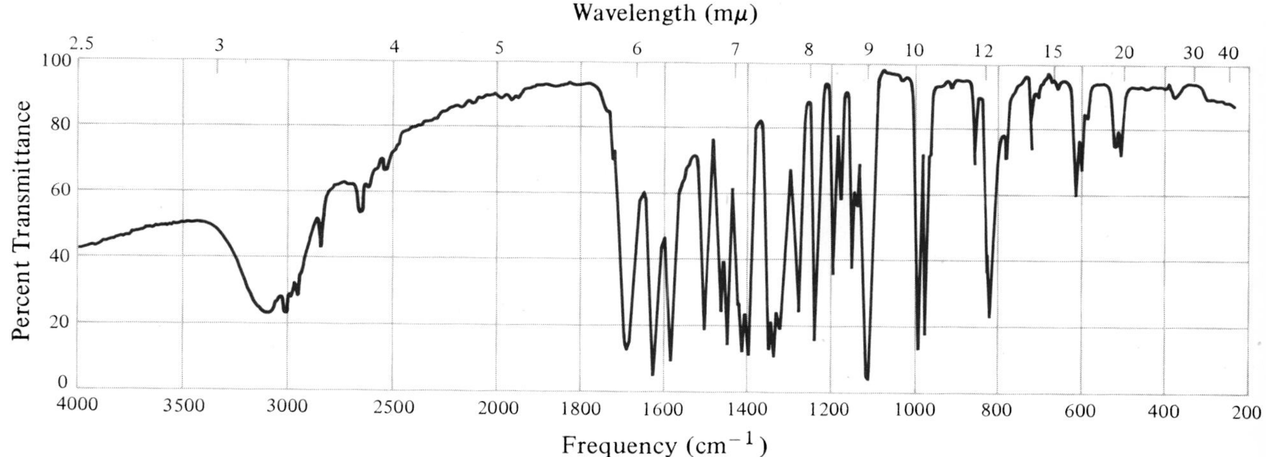

16:3 Ionization of Aliphatic Carboxylic Acids

Carboxylic acids rapidly give up a proton to a base and form a **carboxylate ion** (eq. 16.2; secs. 1:7 and 13:1).

$$CH_3\overset{\overset{\displaystyle O}{\|}}{C}OH + H_2O \rightleftarrows CH_3\overset{\overset{\displaystyle O}{\|}}{C}O^- + H_3O^+ \qquad (16.2)$$

Acetic acid — Acetate ion

$K_a^{25°} = 1.8 \times 10^{-5}$; $pK_a = 4.7$

Although most carboxylic acids are weak acids with pK_as in the range of 3.5 to 5.0, they are much more acidic than the corresponding alcohols containing the same number of carbon atoms (sec. 13:1). (Recall that pK_a = −log K_a. Therefore, the smaller the pK_a, the more dissociated is the acid at equilibrium. Very strong acids have negative pK_as.) An aqueous solution of a carboxylic acid turns litmus red; it can be neutralized with weak bases such as ammonia and its derivatives; it liberates carbon dioxide from sodium bicarbonate; and it reacts with a number of active metals with the release of hydrogen.

As we emphasized in Section 13:1, the most decisive factor affecting the acidity of carboxylic acids is the stability of the carboxylate ion, due to the delocalization of its π electrons (Fig. 16–4; structure **1**). The charge is shared equally between the two electronegative oxygen atoms, giving increased stability to the ion.

1

In the un-ionized acid (**2**), there is also some conjugative interaction between the carboxyl group and the hydroxyl oxygen atom. This interaction tends to stabilize the acid.

2

Nonetheless, resonance interactions are much less important in the acids, where the two resonance structures are of unequal energy (as in **2**), than they

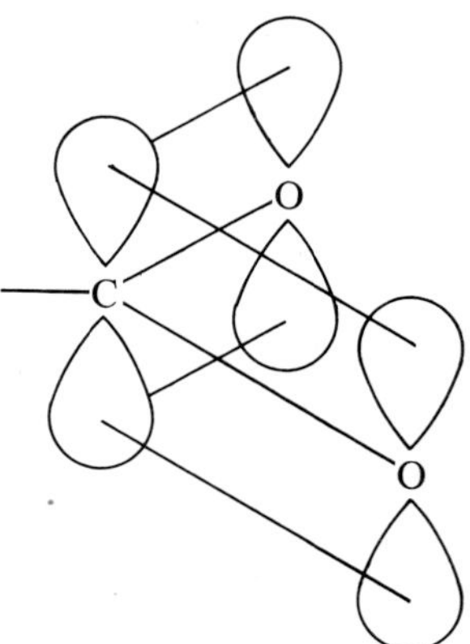

Figure 16–4. *The delocalized bonding π orbital in a carboxylate ion.*

are in the anions, where the two resonance structures are identical in energy (as in **1**). As a consequence, *there is a net* gain *in resonance energy upon ionization of a carboxylic acid.*

Substituents can exert a marked effect on the acidity of a carboxylic acid by affecting the stability of its anion. The most sizable changes occur when the substituent is close to the carboxyl group. Inductive electron withdrawal by a substituent helps to stabilize the carboxylate ion because it helps to disperse negative charge. This increases the acidity of the carboxylic acid. For example, Cl is electronegative and so the more Cl substituents there are, the more the negative charge will be drawn away from the carboxylate group and the more stable the anion will be. The trend in pK_as of the chloroacetic acids shows this effect.

Increasing acid strength →

	CH_3COOH	$ClCH_2COOH$	$Cl_2CHCOOH$	Cl_3CCOOH
pK_a	4.75	2.85	1.48	0.70

Increasing anion stability →

CH_3COO^- $ClCH_2COO^-$ Cl_2CHCOO^- Cl_3CCOO^-

$$\overleftarrow{Cl—C}—C(=O)O^-$$

A hydroxyl group also increases the acidity of an alkanoic acid by inductive electron withdrawal, which stabilizes the carboxylate ion.

	CH_3CH_2COOH	$\overleftarrow{HO—CH_2}COOH$
pK_a	4.87	3.83

Electron-donating groups such as alkyl have the opposite effect. They add electrons to the already negative anion, thereby *reducing* its stability and decreasing the acidity of the carboxylic acid. For instance, compare the acidities of propanoic and acetic acids. The additional CH_2 group in propanoic acid, which is electron-donating, lowers its acidity.

	$\overrightarrow{CH_3—CH_2}COOH$	$H—CH_2COOH$
pK_a	4.87	4.75
	Propanoic acid	Acetic acid

Table 16–3 presents the pK_as of some typical alkyl, halo, and hydroxy substituted acids.

In all of these (and many more) cases, the correct order of acidity can be predicted based upon what effect the substituent will have on the stability of the anion (eq. 16.2): Since acidity really deals with an equilibrium (eq. 16.2), why don't we worry about the stability of the acid itself? This is because *substituents have much less of an effect on an uncharged molecule than on a charged one.* Dispersal of charge in an ion by inductive or resonance effects will have a greater effect on stability because the ion is at a higher energy level than the

Table 16–3. pK_a Values for Some Aliphatic Carboxylic Acids Substituted with Alkyl, Halogen, and Hydroxyl Groups

Structure	pK_a (H_2O, 25°)
CH_3COOH	4.75
CH_3CH_2COOH	4.87
$(CH_3)_2CHCOOH$	4.84
$(CH_3)_3CCOOH$	5.03
$CH_3(CH_2)_{16}COOH$ (*stearic acid*)	4.89
$ClCH_2COOH$	2.85
$BrCH_2COOH$	2.96
ICH_2COOH	3.12
$CH_3CH(OH)COOH$ (*lactic acid*)	3.08

un-ionized acid. Neutral molecules are reasonably "happy" as they are and, being quite stable, are less affected by structural changes. There are, however, some cases where substituent effects on the acid itself are important (sec. 16:4).

Thus far we have painted a relatively simple picture of carboxylic acid ionization by ignoring the solvation differences on both sides of the equilibrium. Since the anion is charged, it will be more susceptible to solvation and other *environmental* effects than will the uncharged acid. In fact, the dissociation of many carboxylic acids has been shown to involve substantial entropy changes, which constitutes good evidence that solvation plays an important role in dissociation equilibria. Furthermore, the relative acidities of carboxylic acids are somewhat different in the gas phase from what they are in solution. There is an important lesson in this example. It often happens that a measurable physical property, such as pK_a, correlates nicely with a single molecular parameter, such as inductive effect. We often go on to infer that that parameter is the *only* important one that affects the measured property. As we have just seen, inferences of that kind are not always justified and often collapse when applied too broadly.

Example

The pK_a of $ClCH_2COOH$ is 2.85, while that for ICH_2COOH is 3.12. Which is the stronger acid? Why?

Answer

Chloroacetic acid has the lower pK_a. This means that its ionization equilibrium constant (eq. 16.2) is larger and that chloroacetic acid is a stronger acid than iodoacetic acid.

The strength of a carboxylic acid depends mainly on the stability of the carboxylate anion formed when a proton is given up (eq. 16.2). One must therefore decide which is the more stable anion,

$$ClCH_2COO^- \quad \text{or} \quad ICH_2COO^-$$

Chlorine is more strongly electron-withdrawing than iodine because it is more electronegative. Thus Cl can help stabilize a negative charge better. This makes the equilibrium of Equation 16.2 go further to the right; therefore, $ClCH_2COOH$ is a stronger acid than ICH_2COOH.

Problems

16–3 Trifluoroacetic acid, CF_3COOH, has a pK_a of 0.2, which puts it in the range of mineral acids such as HNO_3 (pK_a −1.4). Explain why CF_3COOH is such a strong acid.

16–4 Butanoic acid and 4-chlorobutanoic acid have nearly the same pK_a. What does this suggest about the inductive effect on acidity as substituents get further away from the α carbon?

16–5 Rank the following carboxylic acids in order of increasing acidity and give an approximate pK_a value for each.

a. FCH_2COOH

b. $ClCH_2CH_2COOH$

c. $CH_3CHCOOH$ (with CH_3CH_2 attached to the CH)

d. CH_3OCH_2COOH

16:4 Acidities of Substituted Benzoic Acids

The acidity of benzoic acid is subject to the same sort of substituent effects as is the acidity of acetic acid. *If substituents can donate electrons the acidity will be decreased, while electron-withdrawing substituents will increase acidity.* The main difference in the case of aromatic acids is that substituents can donate or withdraw electrons by resonance as well as by an inductive effect.

Table 16–4 lists the pK_a values for groups of substituted benzoic acids in which a single substituent is placed at the *ortho, meta,* and *para* positions. For some of these groups, the variation of acidity is small among the three isomers. In these cases it is difficult to decide whether acidity differences are due to electronic effects and/or to entropy and solvation effects. For some substituents the acidity differences among the three isomers are substantial, amounting to as much as 1.3 pK_a units. These compounds show some significant trends worth pursuing.

Table 16–4 shows that substituents cause both increases and decreases in the acidity of benzoic acid. The effect of the substituents can be generalized as follows:

1. *Meta* substituents cannot, by resonance, add or remove charge from the ring carbon holding the carboxyl group. Therefore *meta* substituents act mainly by way of an inductive effect.
2. *Para* substituents exert weak inductive effects, but have strong resonance interactions because they can add or remove charge from the ring carbon holding the carboxyl group.

Table 16–4. Comparative Acidities (pK_as) of Substituted Benzoic Acids

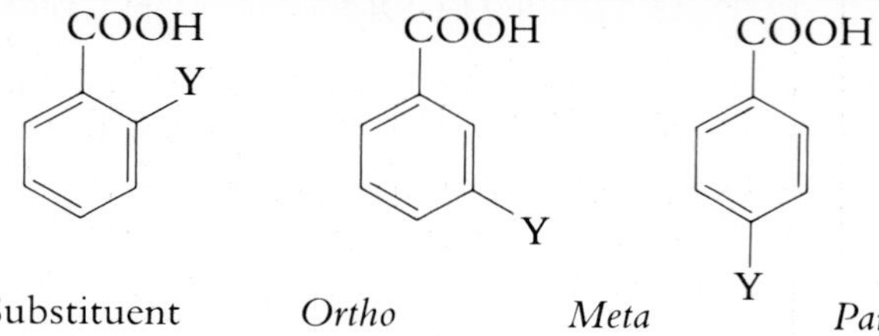

Y Substituent	*Ortho*	*Meta*	*Para*
H	4.20	4.20	4.20
OH	2.97	4.08	4.48
OCH_3	4.09	4.09	4.47
NO_2	2.17	3.49	3.42
F	3.27	3.86	4.14
Cl	2.92	3.82	3.98
Br	2.85	3.81	3.97
CH_3	3.91	4.27	4.37

3. *Ortho* substituents can exert strong inductive and resonance effects on the carboxyl group. In addition, there may be steric or hydrogen-bonding interactions between these substituents and the COOH group.

These substituent effects on acidity become clearer when we examine specific cases. For example, substitution of the strongly electron-withdrawing nitro group on the benzene ring increases the acidity (lowers the pK_a). Considering only the nitro group's electron-withdrawing effect, we would expect *o*-nitrobenzoic acid to be considerably stronger than *m*-nitrobenzoic acid. This is because the nitro group is closer to the carboxylate group in the *ortho* position than it is in the *meta* position. That is exactly what is observed. By the same reasoning, we would expect the *para* isomer to be the weakest acid, but in reality it is slightly *more* acidic than the *meta* compound (Table 16–4). This can be explained by looking at the different resonance structures for the *p*-nitrobenzoate (**3**) and *m*-nitrobenzoate (**4**) ions. Note that the resonance hybrid for the *para* substituted ion (**3**) has positive charge on the

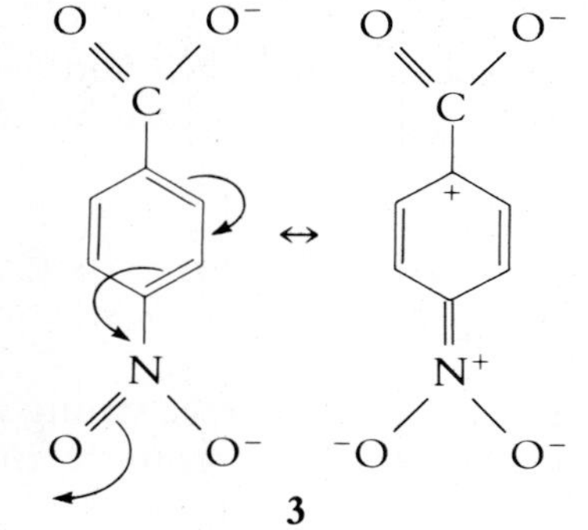

3

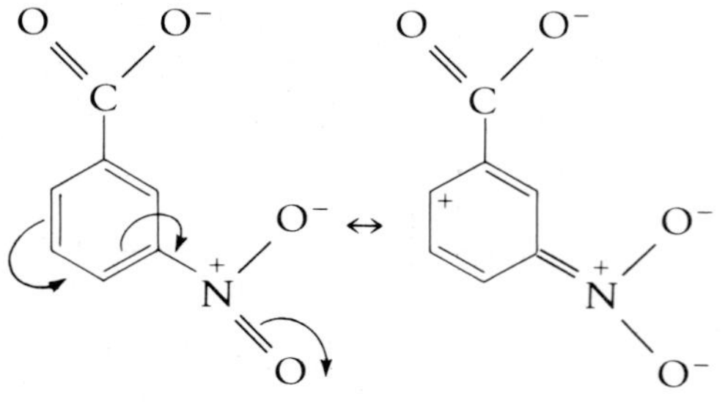

4

carbon that bears the carboxylate anion, while *no* positive charge is found on that carbon for the *meta* substituted ion (**4**). Electrons from the carboxylate group are therefore withdrawn to a greater extent in **3** than in **4.** Hence, *p*-nitrobenzoic acid is more acidic than we would expect if only the inductive effect were operating.

Another instructive example is provided by the hydroxybenzoic acids. *m*-Hydroxybenzoic acid (pK_a 4.08) is slightly stronger than benzoic acid (pK_a 4.20) because of the electron-withdrawing inductive effect of the OH group which stabilizes the carboxylate ion (**5**). *p*-Hydroxybenzoic acid (pK_a

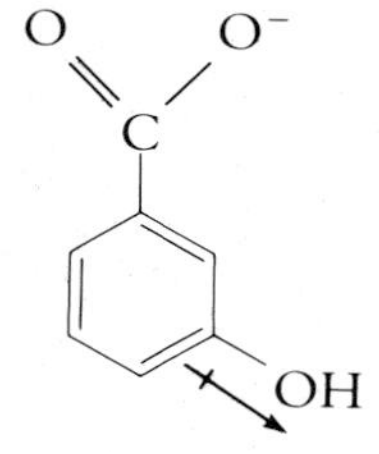

5

4.48), by contrast, is slightly *weaker* than benzoic acid. This is because the OH group has a resonance effect that *opposes* its inductive effect, and that resonance effect can only be important from the *ortho* and *para* positions. Resonance hybrid **6** shows that electrons can be donated to the carboxyl group of the un-ionized acid. Such electron donation stabilizes the acid so that it has less tendency to ionize than benzoic acid.

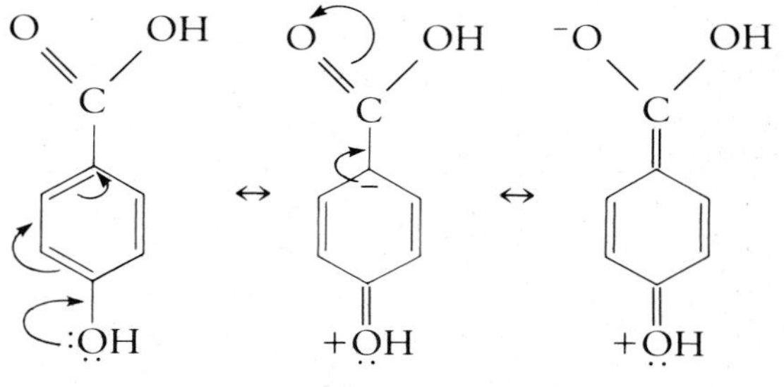

6

If the OH is in the *ortho* position, the inductive effect would make for a stronger acid than benzoic, but the resonance electron donation (as in **6**) would make for a weaker acid. In fact, neither of these effects is probably the most important factor in the *ortho* isomer. Examination of the *o*-hydroxybenzoate ion, **7**, shows that the OH group is able to effectively hydrogen

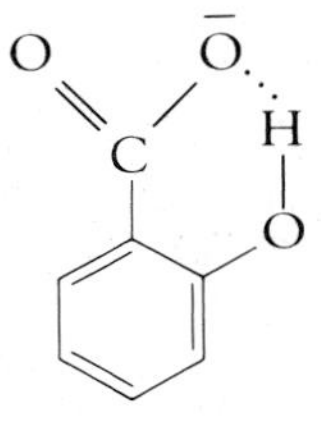

7

bond to the negative oxygen, stabilizing the anion and making *o*-hydroxybenzoic acid the strongest of the three isomers. Confirmation of this idea comes from the fact that *o*-methoxybenzoic acid (pK_a 4.09), which cannot hydrogen bond in a similar manner, is a much weaker acid than *o*-hydroxybenzoic acid.

The hydroxybenzoic acid system underscores the important point that we must pay attention to the *position* of a ring substituent as well as to its structure. A given substituent may even exert net *opposite* effects at different positions, as demonstrated with the OH group. A similar phenomenon was found with electrophilic aromatic substitution (sec. 15:10).

We can summarize the substituent effects on benzoic acid acidity as follows. Groups that are electron-donating, *by virtue of their structure and position on the ring,* tend to stabilize the *carboxylic acid* and decrease its acidity. Groups that are electron-withdrawing, *by virtue of their structure and position on the ring,* tend to stabilize the *carboxylate anion* and thereby increase the acidity of the acid (eqs. 16.3 and 16.4).

(16.3) *p*-HOC$_6$H$_4$COOH $\underset{}{\overset{K^p}{\rightleftharpoons}}$ *p*-HOC$_6$H$_4$COO$^-$ + H$^+$ $K^p < K_a$ for benzoic acid

OH (acid): Stabilizes the acid strongly
OH (anion): Stabilizes the anion weakly

(16.4) *m*-HOC$_6$H$_4$COOH $\overset{K^m}{\rightleftharpoons}$ *m*-HOC$_6$H$_4$COO$^-$ + H$^+$ $K^m > K_a$ for benzoic acid

OH (acid): Stabilizes the acid weakly
OH (anion): Stabilizes the anion strongly

Problems

16–6 With the aid of resonance structures, explain the following observations:

a. *p*-Cyanobenzoic acid is a stronger acid than benzoic acid.

b. *p*-Dimethylaminobenzoic acid [the dimethylamino group has the structure $(CH_3)_2N$—] is a weaker acid than benzoic acid.

$(CH_3)_2N$–C$_6$H$_4$–COOH

c. The dimethylamino group lowers the acidity of benzoic acid much more effectively from the *para* than from the *meta* position.

16–7 Biphenyl-3-carboxylic acid (**I,** pK_a 4.14) is a slightly stronger acid than benzoic acid. Its isomer, biphenyl-4-carboxylic acid (**II,** pK_a 4.21) is a tiny bit weaker than benzoic acid. Explain these data. (Note that **I** and **II** can be viewed as benzoic acids carrying a phenyl substituent.)

COOH

I

COOH

II

Note from Table 16–4 that *all ortho* substituents cause an increase in acidity of benzoic acid. Bulky *ortho* groups will force the C=O out of the ring plane, so that the π electrons of the C=O no longer overlap with the π electrons of the benzene ring. Loss of conjugation destabilizes the acid relative to the anion and causes increased acidity. This effect is sometimes called **steric inhibition of resonance.**

Although many of the substituent effects described thus far are the logical consequences of a set of relatively simple interactions, a closer examination shows that some of the data would need to be explained in other ways. For example, if the *meta* position effects are mainly inductive, why do all the *meta*-halogenated benzoic acids have almost identical pK_as? There is certainly a difference in the inductive ability of F, Cl, and Br, as evidenced by differences in the acidity of halogenated acetic acids (Table 16–3). Good, simple reasons for many of these acidity orders cannot be given because most of them involve a complex interplay of solvation, steric, resonance, and inductive effects. In spite of these complexities, chemists have found that so-called substituent effects are indeed characteristic of the various groups. That is, once we know the effect a given substituent has on the ionization of benzoic acid, we can make predictions about how that substituent will affect other reactions. The next section shows how this is done.

16:5 Substituent Effects and Reaction Mechanisms

We have considered the effects of substituents on the acidities of benzoic acids in detail because these effects have many valuable applications. For example, we can use them to study reaction mechanisms. Let's consider the reaction in Equation 16.5. The mechanism indicated is an S_N2 direct displacement.

$$\text{(16.5)}\quad Y\text{-}C_6H_4\text{-}\ddot{N}Me_2 + CH_3\text{—}I \rightarrow Y\text{-}C_6H_4\text{-}\overset{+}{N}(CH_3)Me_2 + I^-$$

ment. If this mechanism is correct, the availability of the nonbonded electrons on the nitrogen will be a major factor influencing the rate of the reaction. To test this mechanism, we measure the relative rates of several reactions: one reaction involving a substrate with no substituents (eq. 16.6), and the others with electron-releasing (eq. 16.7) or electron-withdrawing (eq. 16.8) substituents.

Relative rate

$$\text{(16.6)}\quad \text{H}-\text{C}_6\text{H}_4-\ddot{\text{N}}\text{Me}_2 + \text{CH}_3\text{I} \xrightarrow{k_H} \text{H}-\text{C}_6\text{H}_4-\overset{+}{\text{N}}\text{Me}_3 + \text{I}^- \qquad 1$$

$$\text{(16.7)}\quad \text{CH}_3-\text{C}_6\text{H}_4-\ddot{\text{N}}\text{Me}_2 + \text{CH}_3\text{I} \xrightarrow{k_{CH_3}} \text{CH}_3-\text{C}_6\text{H}_4-\overset{+}{\text{N}}\text{Me}_3 + \text{I}^- \qquad >1$$

$$\text{(16.8)}\quad \text{O}_2\text{N}-\text{C}_6\text{H}_4-\ddot{\text{N}}\text{Me}_2 + \text{CH}_3\text{I} \xrightarrow{k_{NO_2}} \text{O}_2\text{N}-\text{C}_6\text{H}_4-\overset{+}{\text{N}}\text{Me}_3 + \text{I}^- \qquad <1$$

A comparison of the relative rates of these reactions shows that electron-releasing groups do indeed enhance the rate, while electron-withdrawing groups slow down the rate. This is consistent with the mechanism of Equation 16.5, where electrons on the nitrogen are displacing the iodide in the transition state of the rate-determining step. In the above reactions, the methyl group will act to increase the electron density on nitrogen, while the nitro group will tend to draw electrons away from nitrogen.

Because substituent effects have great promise for providing mechanistic information, much effort has gone into quantifying these effects. The American chemist L. P. Hammett was the first to show that the change in an equilibrium or rate constant caused by a substituent was a product of two factors: one factor measured the effect of the substituent itself, while the other factor measured the sensitivity of the *reaction* to a change in substituent (the **Hammett correlation**). Once the dual nature of substituent effects had been demonstrated, a large body of data about many different substituents and reactions was collected. This led, in turn, to a wide application of substituent effect studies, as you'll see next.

Problems

16–8 The reaction in Equation 1 proceeds by way of a carbocation intermediate whose formation is the rate-determining step.

$$\text{Y}-\text{C}_6\text{H}_4-\text{C(CH}_3)_2-\text{Cl} + \text{H}_2\text{O} \rightarrow \text{Y}-\text{C}_6\text{H}_4-\text{C(CH}_3)_2-\text{OH} + \text{HCl} \qquad (1)$$

a. Write a mechanism for this reaction.

b. A p-OCH_3 group greatly increases the rate of the reaction. Use resonance structures to show why.

16–9 Use resonance structures to show why the reaction in Equation 16.8 is much slower than the reaction in Equation 16.6

16:6 How Substituent Effects Aid Drug Design

Substituent effects have been applied to the design of new medicines with improved biological activity. Many of our useful drugs are aromatic compounds, and the addition or removal of certain substituents often causes profound changes in their effectiveness. One way to synthesize a new drug with improved biological activity is to modify it with every possible substituent. This empirical (or random) approach, which requires numerous syntheses, has been used in the past and is still fruitful. Another approach is to synthesize only a limited number of compounds with different substituents and then, based upon biological testing data, predict what the best possible drug would be. This is done by using our knowledge of relative substituent effects.

As an example, consider structure **8,** a typical *sulfanilamide,* or *sulfa,* drug. Such compounds were among the first "wonder drugs" and were extensively used as antibacterial agents in the 1940s and 1950s. Their use in humans has generally been replaced by antibacterial penicillins and other

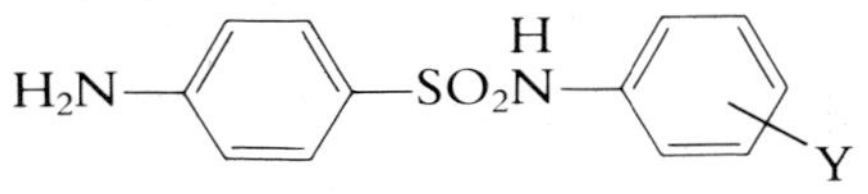

H_2N–C_6H_4–SO_2NH–C_6H_4–Y

8

antibiotics, but sulfa drugs are still widely used in veterinary medicine. The activity of many sulfas is strongly dependent upon the substituent Y. If Y is a *meta*-chloro substituent, the sulfa has a minimum inhibitory concentration (MIC) of 8 micromoles per liter against the common bacterium *Escherichia coli* (*E. coli*). If Y is a *meta*-methyl substituent, the sulfa has an MIC of 22 micromoles per liter. Thus, the *meta*-chloro compound is a more active drug because it inhibits bacterial growth at a lower concentration. Other MIC's are as follows: p-OCH_3 = 35; p-CH_3 = 27; m-OCH_3 = 11; m-NO_2 = 2 micromoles per liter. Given these data we can generalize and say that electron-withdrawing substituents will be effective in promoting antibacterial activity, while electron-donating substituents will have the opposite effect.

Generalizations of this type allow us to correctly predict, in many instances, the effect of structural changes on biological potency. However, there are also cases where such predictions are only partly successful. This means that there are factors we must account for in addition to the electron-donating or withdrawing effects of a substituent. One very critical factor in most drug action is the ability of the molecule to reach its reaction site. Cell walls, for example, are constructed of relatively nonpolar materials known as lipids. Drug molecules must have certain solubility in these lipids in order to pass through cell walls or membranes. However, the drug must also be reasonably soluble in more polar media in order to dissolve in cellular fluids.

Box 16–2. Some Carboxylic Acids That Affect Plant Growth

The bodily functions of humans and other animals are governed or strongly influenced by hormones such as steroids and prostaglandins, which are usually synthesized by the body in very minute amounts. Plants also synthesize regulating hormones, and those that control plant growth are particularly important. Of the five common types—auxins, gibberellins, ethylene, cytokinins, and abscisic acid—three are carboxylic acids:

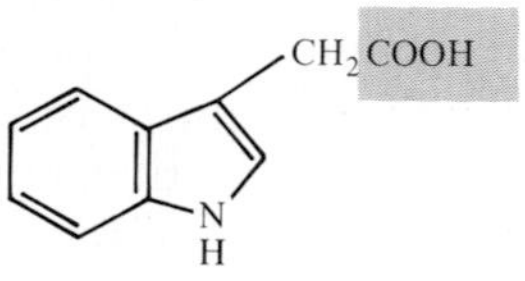

CH_2COOH

Indole-3-acetic acid (auxin)

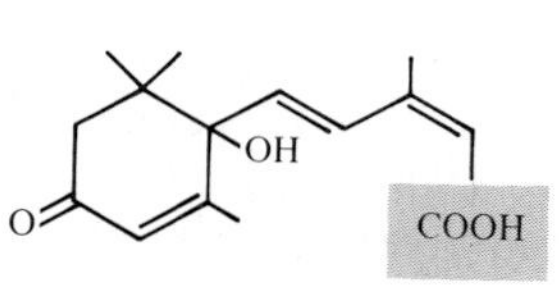

Abscisic acid (ABA)

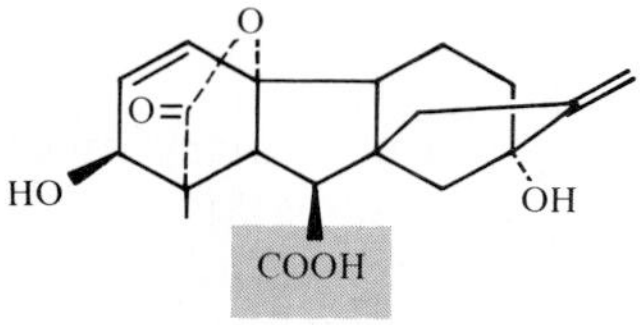

Gibberellic acid

Indole-3-acetic acid probably occurs minutely, but universally, in plant tissue, promoting cell enlargement of shoots but not roots. This is called auxin activity. In 1945 it was discovered that certain synthetic compounds, such as 2,4-dichlorophenoxyacetic acid, have a similar activity. When applied to broad-leaved plants these compounds cause uncontrolled growth, and, eventually, the death of the plant. Hence, they have been important herbicides for the control of weeds in food crops.

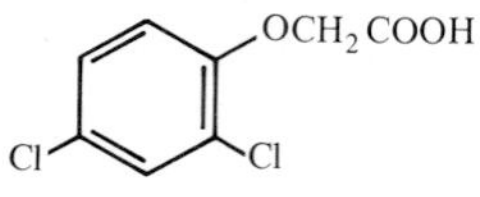

OCH_2COOH

2,4-Dichlorophenoxyacetic acid (2,4-D)

Substituent effect studies have shown that electron-withdrawing groups, such as halogens, on the benzene ring promote auxin activity, while electron-releasing groups inhibit activity. Note that the benzene ring in 2,4-D is two carbons and an oxygen away from the carboxyl group. This means that the substituents on the ring are probably too far removed to affect the acidity of the compound. The substituent effect is more likely creating an electron-deficient benzene ring which may be important either in binding the compound to an enzyme, or in transport of the molecule to the site of action.

Another synthetic material, naphthaleneacetic acid, has a different mode of action. It actually promotes *root* growth and hence is an ingredient in some so-called "rooting hormone" preparations.

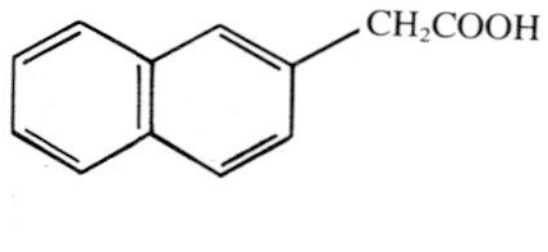

CH_2COOH

Naphthaleneacetic acid (NAA)

Even though a new synthetic drug might be expected to have improved activity, based on chemical criteria, introduction of the required substituent can alter the solubility properties in an undesirable manner. The American chemist C. Hansch has modified the equations that describe substituent effects so that they can take solubilities into account. The **Hansch correlation** is widely used to predict activity of synthetic drugs or to understand the mechanisms of drug action.

Although physical organic measurements such as the determination of quantitative substituent effects have added to our understanding of drug activity, this type of study is only in its infancy. The design of medicinals is still complex and difficult. Hammett and/or Hansch correlations cannot reveal what basic molecular structures will be physiologically active in the first place, but they can be used to improve activity by systematic molecular modification.

16:7 Preparation of Carboxylic Acids

Only a few carboxylic acids are available from natural sources, so numerous methods are available for synthesizing them. We will first look at some methods specific for aromatic acids.

Many substituted benzoic acids can be prepared by electrophilic substitution of benzoic acid. Due to the deactivating effect of the carboxyl group (sec. 15:11), this approach leads only to *meta*-substituted benzoic acids. The easiest way to prepare other substituted benzoic acids is to oxidize the alkyl side chain of a substituted alkylbenzene. Alkanes are generally not easily oxidized, but alkyl substituted benzenes are susceptible to attack at a benzylic hydrogen. Equations 16.9 and 16.10 show how they are oxidized with aqueous potassium permanganate.

benzylic hydrogen

(16.9) *o*-Chlorotoluene (CH_3, Cl) $+ KMnO_4 \xrightarrow[heat]{H_2O} \xrightarrow{HCl}$ *o*-Chlorobenzoic acid (77%) (COOH, Cl)

(16.10) (o-xylene: CH_3, CH_3) $\xrightarrow[H_2O,\ heat]{KMnO_4} \xrightarrow{HCl}$ Phthalic acid (COOH, COOH)

Because alkyl groups are *o,p*-directing, the side-chain oxidation method allows the preparation of *ortho*- and *para*-substituted benzoic acids. If the alkyl side chain is longer than methyl, oxidation will still give the benzoic acid as long as there is at least one benzylic hydrogen. Equation 16.11 shows an example involving oxidation of an isopropyl group, and also indicates that nitric acid can be used as an oxidizing agent of intermediate strength.

(16.11) (CH_3–CH–CH_3 on ring, para CH_3; benzylic hydrogen) $+ HNO_3(conc) \xrightarrow{heat}$ *p*-Methylbenzoic acid (55%) (COOH, CH_3)

After oxidation of one side chain, formation of the electron-withdrawing carboxyl group inhibits oxidation of the second alkyl group.

The oxidation of alkylbenzenes is an enormously important industrial process. It is used to convert toluene to benzoic acid and *p*-xylene to terephthalic acid, which is an important monomer for polyester synthesis (sec. 22:15). The oxidant is oxygen from the air, and the catalyst is Co^{3+}, which functions by generating a benzyl radical (eq. 16.12). All side-chain oxidations, regardless of oxidant, probably involve stable benzylic radical and/or cation intermediates (sec. 14:8).

$$C_6H_5CH_3 \xrightarrow[CH_3COOH]{Co^{3+}} C_6H_5\dot{C}H_2 + Co^{2+} + H^+ \xrightarrow[\text{many steps}]{O_2,\ Co^{3+}} C_6H_5COOH \tag{16.12}$$

The susceptibility of benzylic hydrogens toward oxidation is also seen in biological systems. Benzene, because it is so stable, cannot easily be metabolized by the human body (sec. 14:11). In addition, it is only slightly water soluble and so is excreted very slowly and tends to accumulate in the liver. Because of this, and the permeability of the skin and lungs to benzene, it is quite poisonous. Toluene, on the other hand, is transformed into benzoic acid by human metabolism, albeit slowly. Its carboxyl group allows benzoic acid to be converted into the highly water soluble hippuric acid ($C_6H_5CONHCH_2COOH$), which is excreted without difficulty. Toluene is therefore preferable to benzene in those instances when an aromatic solvent is required, even though toluene is by no means completely nontoxic.

Whereas oxidation of alkylbenzenes is a valuable method for synthesizing benzoic acids (eqs. 16.9 to 16.11), the oxidation of simple alkanes is usually not a useful technique for obtaining aliphatic carboxylic acids. In order to synthesize an aliphatic carboxylic acid by oxidation, it is necessary that one carbon atom already be oxygenated (1° alcohol or aldehyde; sec. 11:3), or that a double bond be present (alkenes; sec. 10:9). Ketones resist oxidation except under vigorous conditions.

Oxidation of primary alcohols and aldehydes leads to carboxylic acids with the same number of carbon atoms (eqs. 16.13 and 16.14). The most widely used oxidants are potassium permanganate and chromic acid.

$$\underset{\text{3-Methyl-1-butanol}}{(CH_3)_2CHCH_2CH_2OH} \xrightarrow[\text{acetone, 20° (solvent)}]{H_2CrO_4} \underset{\text{3-Methylbutanoic acid}}{(CH_3)_2CHCH_2COOH} \tag{16.13}$$

$$\underset{\text{Heptanal}}{CH_3(CH_2)_5CHO} \xrightarrow[20°]{KMnO_4} \xrightarrow{H_3O^+} \underset{\text{Heptanoic acid (77\%)}}{CH_3(CH_2)_5COOH} \tag{16.14}$$

One large scale commercial process for making acetic acid involves catalytic oxidation of acetaldehyde with air, which is much cheaper than oxidizing with potassium permanganate or chromic acid. This method is used frequently because of the ease with which aldehydic hydrogen atoms are abstracted by free radicals. Even when aldehydes are stored in sealed bottles they slowly accumulate acid impurities due to their reaction with air.

Another approach to carboxylic acids is by way of *nitriles,* which are organic cyanides. They are prepared from 1° or 2° haloalkanes by reaction with cyanide ions (S_N2 reactions; sec. 6:3). Base-promoted hydrolysis of the nitrile, followed by acidification, yields an acid *with one more carbon than the starting haloalkane* (eq. 16.15). (The mechanism for the hydrolysis is discussed

$$CH_3CH_2CH_2{-}Br \xrightarrow[DMSO]{NaCN} CH_3CH_2CH_2{-}CN \xrightarrow[heat]{OH^-,\ H_2O} \xrightarrow{H_3O^+} CH_3CH_2CH_2{-}\overset{O}{\overset{\|}{C}}OH \tag{16.15}$$

1-Bromopropane — Butanenitrile — Butanoic acid

in Section 17:12.) Aromatic carboxylic acids can also be made this way, in which case the starting material is often an aromatic amine (Chap. 23).

A widely used alternative to the nitrile route is the Grignard reaction. It is advantageous because it can be used with organic halides not suitable for the nitrile route, such as 3° halides and aryl halides (see Prob. 16–11). Carbon dioxide is the carbonyl compound that reacts with the Grignard reagent (eq. 16.16; see sec. 11:4). The product is a halomagnesium carboxylate, which is hydrolyzed to the carboxylic acid.

$$C_6H_5{-}MgBr + O{=}C{=}O \rightarrow C_6H_5{-}C({=}O){-}OMgBr \xrightarrow{H_3O^+} C_6H_5{-}COOH \quad (85\%) \tag{16.16}$$

We've already seen that carboxylic acids are produced by oxidation of alkenes carrying a hydrogen on the double bond (sec. 10:9). This reaction will give a carboxylic acid *with fewer carbon atoms than the starting alkene,* unless the alkene is cyclic.

Example

a. Show how to prepare *p*-chlorobenzoic acid from toluene.
b. How would you convert 1-bromobutane to butanoic acid?

Answer

a. Oxidation of toluene with $KMnO_4$ would give benzoic acid. If benzoic acid is treated with Cl_2 and $FeCl_3$, chlorination would give *m*-chlorobenzoic acid because the carboxyl group is *meta*-directing. Reversing the order of steps would allow for the synthesis of the desired isomer:

$$C_6H_5CH_3 \xrightarrow[FeCl_3]{Cl_2} p\text{-}ClC_6H_4CH_3 \xrightarrow[\Delta]{KMnO_4} p\text{-}ClC_6H_4COOH$$

Toluene — (separation from *ortho*-isomer) — *p*-Chlorobenzoic acid

b. There is no known reaction that will convert an alkyl halide directly to an acid with the same number of carbon atoms. We will first have to convert the bromobutane to an oxidizable functional group:

$$\underset{\text{1-Bromobutane}}{CH_3CH_2CH_2CH_2Br} \xrightarrow[\text{DMSO}]{OH^-} CH_3CH_2CH_2CH_2OH \qquad (S_N2\text{ reaction})$$

$$CH_3CH_2CH_2CH_2OH \xrightarrow[\text{acetone}]{H_2CrO_4} \underset{\text{Butanoic acid}}{CH_3CH_2CH_2COOH}$$

Table 16–5 summarizes the common methods used to synthesize carboxylic acids. The methods are grouped according to whether they lead to an increase, decrease, or no change in the number of carbon atoms.

Table 16–5. Methods of Synthesizing Carboxylic Acids

Methods that Leave the Number of Carbon Atoms Unchanged

1. Oxidation of toluenes

$$Y\text{-}C_6H_4\text{-}CH_3 \xrightarrow{KMnO_4} \xrightarrow{H_3O^+} Y\text{-}C_6H_4\text{-}COOH$$

2. Oxidation of aldehydes and primary alcohols

$$RCHO \text{ or } RCH_2OH \xrightarrow[\text{or } KMnO_4]{H_2CrO_4} RCOOH$$

Methods that Increase the Number of Carbon Atoms

3. Carbonation of Grignard Reagents

$$RX \xrightarrow[\text{ether}]{Mg} RMgX \xrightarrow{CO_2} \xrightarrow{H_3O^+} RCOOH$$

4. Hydrolysis of nitriles obtained from alkyl halides

$$RX \xrightarrow{CN^-} RCN \xrightarrow{OH^-} \xrightarrow{H_3O^+} RCOOH$$

Methods that Decrease the Number of Carbon Atoms

5. Oxidation of phenylalkanes (except toluenes)

$$Y\text{-}C_6H_4\text{-}CH(R)(R') \xrightarrow{KMnO_4} \xrightarrow{H_3O^+} Y\text{-}C_6H_4\text{-}COOH$$

6. Oxidation of alkenes (except cycloalkenes)

$$RCH{=}CHR' \xrightarrow{KMnO_4} RCOOH + R'COOH$$

16–10 Show how the following conversions could be carried out by way of nitrile intermediates.

a. 1-bromobutane → pentanoic acid
b. chlorocyclohexane → cyclohexanecarboxylic acid
c. 2-pentanol → 2-methylpentanoic acid
d. 3-heptanone → 2-ethylhexanoic acid

16–11 a. *t*-Butyl bromide may be converted to 2,2-dimethylpropanoic acid (*pivalic acid*) by way of a Grignard reaction (eq. 2).

$$(CH_3)_3CBr \xrightarrow{Mg} \xrightarrow{CO_2} \xrightarrow{H_3O^+} (CH_3)_3CCOOH \qquad (2)$$

This transformation cannot be carried out by way of the nitrile. Explain why the nitrile method would be unsuccessful.

b. What method would you use to convert 4-methoxybromobenzene (*p*-bromoanisole) to 4-methoxybenzoic acid?

16:8 Acid Halides

In the introduction to this chapter we mentioned two carboxylic acid derivatives: acid halides and acid anhydrides. In this and the following two sections we introduce some of the chemistry and properties of acid halides and acid anhydrides, and examine their use as synthetic intermediates.

If the hydroxyl group of a carboxylic acid is replaced by a halogen atom, the resultant compound is called an **acid halide** or **acyl halide.** Individual compounds are named by taking the name of the acids from which they are derived and replacing -ic acid by -yl halide.

$CH_3\overset{O}{\overset{\|}{C}}Cl$	$CH_3(CH_2)_4\overset{O}{\overset{\|}{C}}Br$	$Cl\overset{O}{\overset{\|}{C}}CH_2\overset{O}{\overset{\|}{C}}Cl$
Acetyl chloride	Hexanoyl bromide	Malonyl chloride

Acid chlorides are used most often because they can be prepared cheaply and are easier to store than the bromides and iodides. Acid chlorides are prepared from the reaction of carboxylic acids with thionyl chloride (eq. 16.17), phosphorus pentachloride (eq. 16.18), or phosphorus trichloride. Acid bromides may be prepared in the same way, from the corresponding inorganic bromides. Their use is much more restricted than that of the chlorides because the bromides are more expensive and less stable.

$$(16.17)\quad \underset{\text{Butanoic acid}}{CH_3CH_2CH_2\overset{O}{\overset{\|}{C}}OH} + \underset{\text{Thionyl chloride}}{SOCl_2} \xrightarrow{80^\circ} \underset{\text{Butanoyl chloride (85\%)}}{CH_3CH_2CH_2\overset{O}{\overset{\|}{C}}Cl} + SO_2 + HCl$$

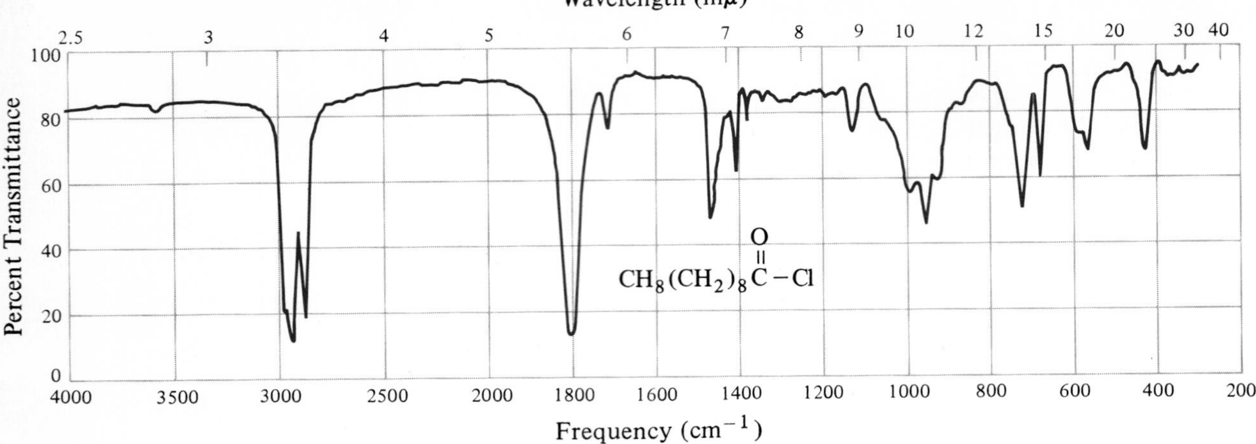

Figure 16–5. *Infrared spectrum of decanoyl chloride.*

$$(16.18) \quad C_6H_5\overset{O}{\overset{\|}{C}}OH + PCl_5 \rightarrow C_6H_5\overset{O}{\overset{\|}{C}}Cl + POCl_3 + HCl$$

Benzoic acid — Benzoyl chloride (90%)

The infrared C═O stretching frequency of acid chlorides lies in a characteristically high region, 1790 to 1810 cm^{-1} (Fig. 16–5). The reason for such a high frequency absorption is discussed in Chapter 17.

16:9 Acid Anhydrides

Acid anhydrides undergo many of the same reactions that acid halides do, and the two are often used interchangeably. Anhydrides are so named because they are *formally* derivable from two molecules of a monocarboxylic acid (eq. 16.19).

$$CH_3C(=O)\text{–}O\text{–}C(=O)CH_3 \qquad CH_3CH_2C(=O)\text{–}O\text{–}C(=O)CH_2CH_3 \qquad \text{cyclic } CH_2C(=O)\text{–}O\text{–}C(=O)CH_2$$

Acetic anhydride (Ac_2O) — Propanoic anhydride — Succinic anhydride

$$\text{(16.19)}\quad 2\,CH_3COOH \;\rightarrow\; (CH_3CO)_2O + H_2O$$

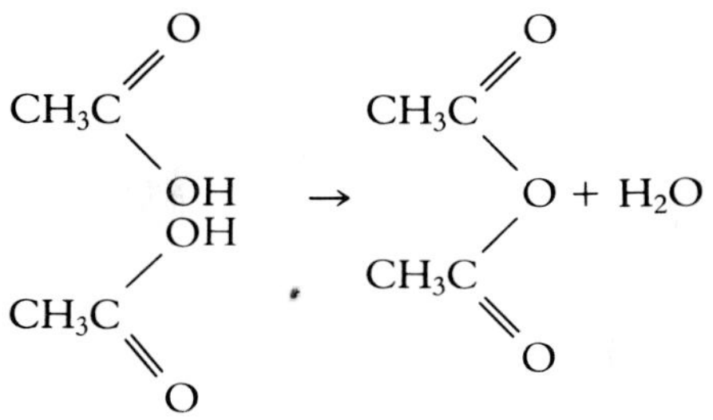

Two molecules of acetic acid → Acetic anhydride

Five- and six-membered cyclic acid anhydrides, such as succinic anhydride, can be prepared by heating the dicarboxylic acid so that a molecule of water is driven off (eq. 16.20).

$$\text{(16.20)}\quad C_6H_4(COOH)_2 \xrightarrow{\text{heat}} C_6H_4(CO)_2O + H_2O$$

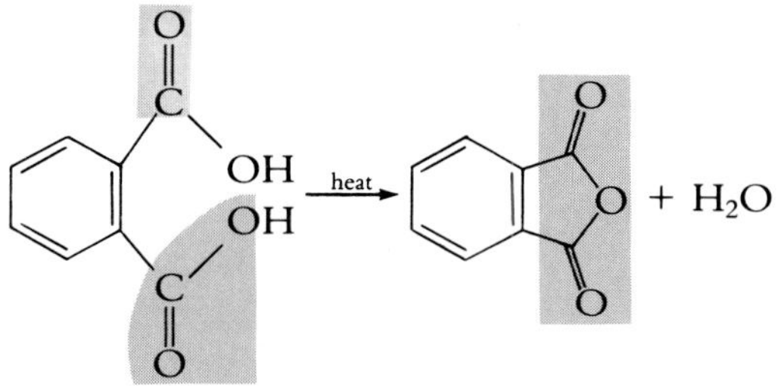

Phthalic acid → Phthalic anhydride

Open-chain anhydrides cannot be formed by simply heating the acid. A common method of forming them is to heat a carboxylic acid with acetic anhydride (eq. 16.21). An equilibrium is established between the desired anhydride (benzoic anhydride in this case) and acetic anhydride. By distilling out acetic acid, the most volatile component of the mixture, the reaction can be driven to completion.

$$\text{(16.21)}\quad 2C_6H_5COOH + (CH_3CO)_2O \xrightarrow[80^\circ]{\text{benzene}} \xrightarrow{\text{distill}} (C_6H_5CO)_2O + 2CH_3COOH\uparrow$$

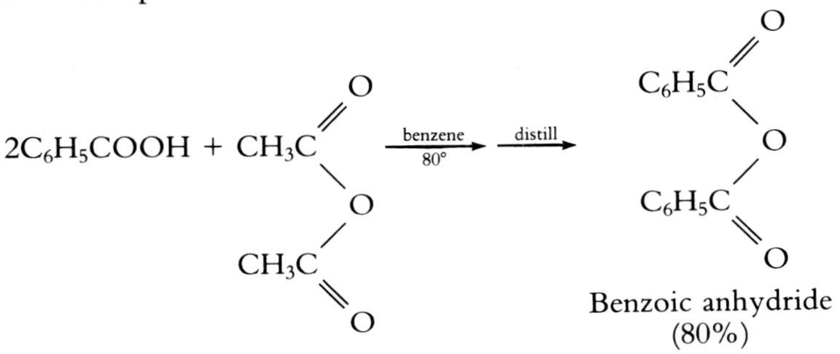

Benzoic anhydride (80%)

Problems

16–12 Show how each of the following compounds can be prepared from a carboxylic acid.

a. 2-methylbutanoyl chloride
b. *p*-nitrobenzoyl chloride
c. cyclobutanecarbonyl bromide
d. pentanoic anhydride
e. succinic anhydride
f. 3,4-dimethylbenzoic anhydride

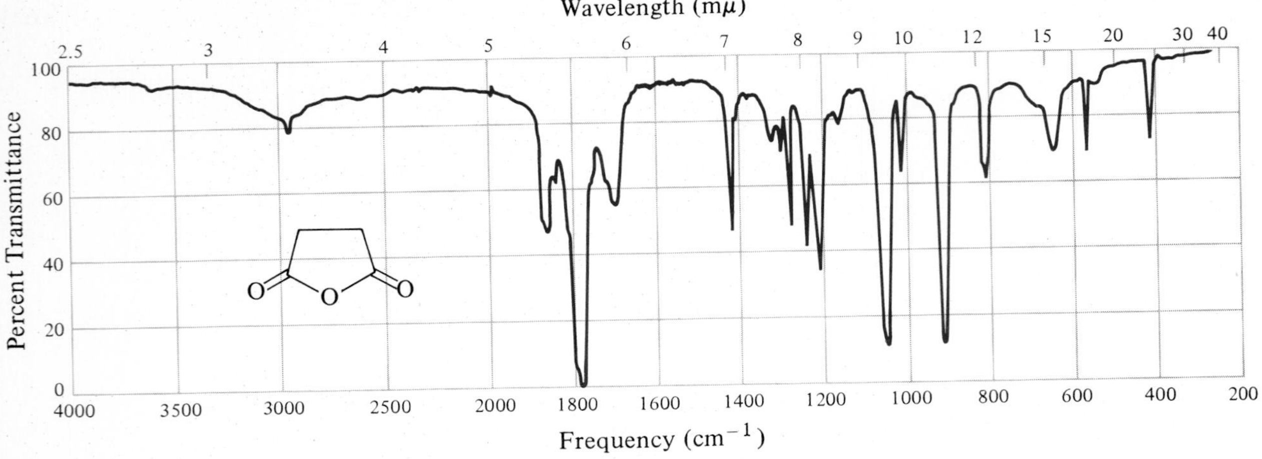

Figure 16–6. *Infrared spectrum of succinic anhydride.*

16–13 There are two isomers of butenedoic acid ($C_4H_4O_4$). One of them, maleic acid, easily loses water upon heating to form an anhydride (MW = 98). The other isomer, fumaric acid, loses water only at high temperatures to yield a high molecular weight, nonvolatile product. Assign structures to maleic and fumaric acids, and explain their chemistry.

Anhydrides are characteristically identified by their infrared spectra. They typically show *two* C=O absorptions in the 1835 to 1810 cm^{-1} and 1770 to 1750 cm^{-1} regions (Fig. 16–6).

16:10 Synthesis of Phenyl Ketones by Friedel–Crafts Acylation

We've already seen that acid chlorides are useful acylating agents under Friedel–Crafts conditions (sec. 15:7). Acid anhydrides can be used as well because, in the presence of Lewis acids, they also give rise to acylium ions (eq. 16.22), which are the actual electrophiles that attack the aromatic ring.

$$\text{(16.22)}\quad \mathrm{R\overset{\overset{\displaystyle O}{\|}}{C}Cl} \text{ or } \mathrm{(RC(=O))_2O} + \mathrm{AlCl_3} \rightarrow \left[\mathrm{R\overset{+}{C}{=}\ddot{O}\!: \leftrightarrow RC{\equiv}\overset{+}{O}\!:}\right] + \mathrm{X{-}AlCl_3^-}$$

An acylium ion (X = Cl or RCOO)

Two specific examples of Friedel–Crafts acylation are shown in Equations 16.23 and 16.24.

$$(16.23)\quad C_6H_6 + (CH_3)_2CH\overset{O}{\overset{\|}{C}}Cl \xrightarrow{AlCl_3} C_6H_5\overset{O}{\overset{\|}{C}}\underset{CH_3}{\underset{|}{C}H}CH_3 + HCl$$

2-Methylpropanoyl chloride — 1-Phenyl-2-methyl-1-propanone

$$(16.24)\quad C_6H_6 + CH_3\overset{O}{\overset{\|}{C}}O\overset{O}{\overset{\|}{C}}CH_3 \xrightarrow[(2\ mol)]{AlCl_3} \xrightarrow{H_3O^+} C_6H_5\overset{O}{\overset{\|}{C}}CH_3 + CH_3COOH$$

Acetic anhydride — Acetophenone (85%)

Some of the phenyl ketones produced in these reactions are so well known that their semisystematic names, rather than their IUPAC names, are commonly used (Table 16–6).

One of the more important Friedel–Crafts acylations is an *intramolecular* reaction that results in cyclization to form a fused ring. Examples are given in Equations 16.25 and 16.26. Note that acid chlorides and acids, but not anhydrides, are substrates in these reactions, because one of the two acyl groups in an anhydride is wasted in the Friedel–Crafts reaction (eq. 16.24).

Table 16–6. Names and Structures of Some Phenyl Ketones

Structure	IUPAC Name	Semisystematic Name
$C_6H_5\overset{O}{\overset{\|}{C}}CH_3$	1-phenylethanone	acetophenone
$C_6H_5\overset{O}{\overset{\|}{C}}CH_2CH_2CH_3$	1-phenyl-1-butanone	butyrophenone
$C_6H_5\overset{O}{\overset{\|}{C}}C_6H_5$	diphenylmethanone	benzophenone
(1-indanone structure)	1-indanone	1-indanone
(1-tetralone structure)	1,2,3,4-tetrahydronaphthalen-1-one	1-tetralone

(16.25)

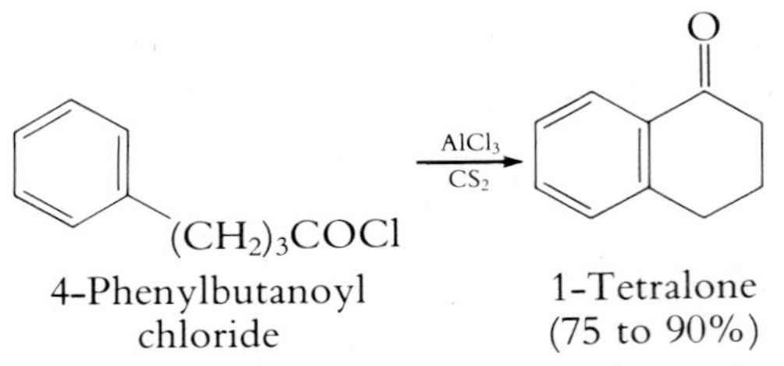

$(CH_2)_3COCl$ $\xrightarrow[CS_2]{AlCl_3}$

4-Phenylbutanoyl chloride — 1-Tetralone (75 to 90%)

(16.26)

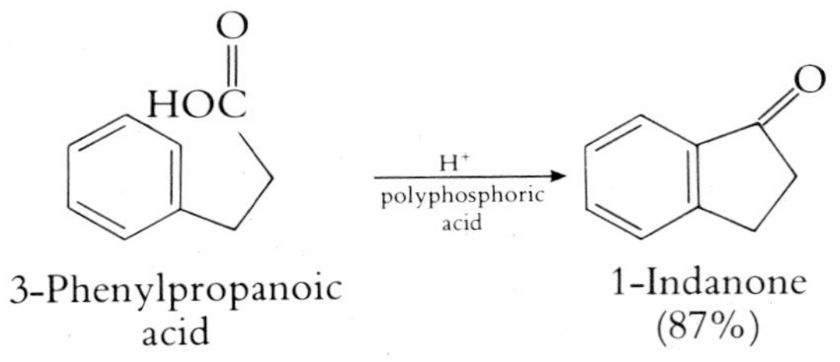

HOC(=O) $\xrightarrow[\text{polyphosphoric acid}]{H^+}$

3-Phenylpropanoic acid — 1-Indanone (87%)

Example

Show how the following compounds can be synthesized by way of Friedel–Crafts reactions.

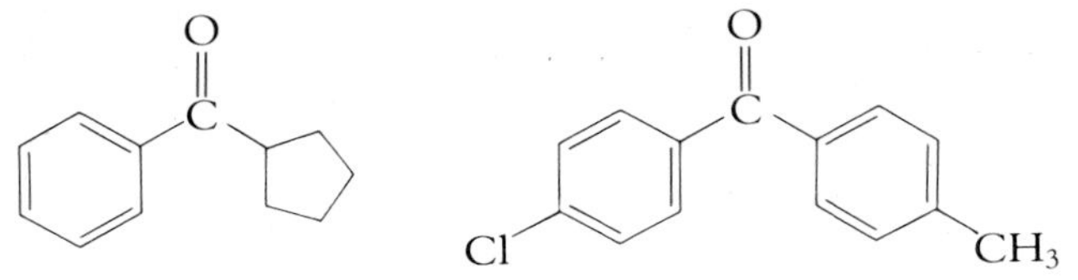

Answer

In the first compound only one of the rings is aromatic. Since the Friedel–Crafts reaction involves attack on an aromatic ring by an acylium ion, it must be the Ar—CO bond that is being formed:

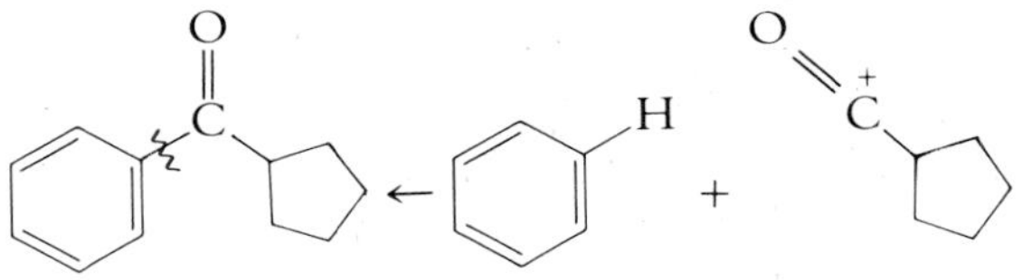

Acylium ions are generated from acid chlorides or anhydrides and Lewis acids, and so the full synthetic sequence is the following:

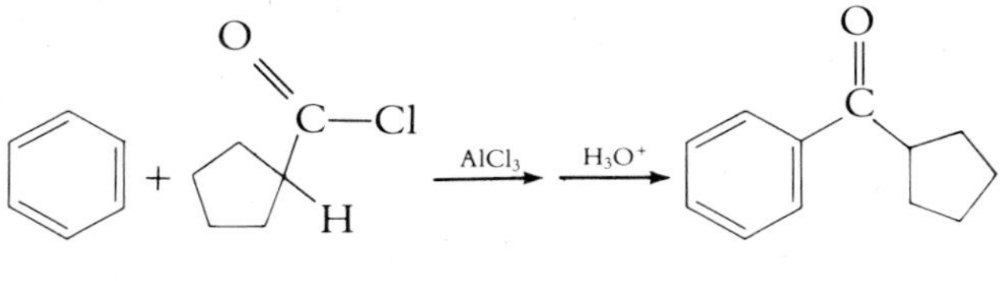

$\xrightarrow{AlCl_3} \xrightarrow{H_3O^+}$

In the second compound, both substituents on the carbonyl group are aromatic, so there are two possible synthetic paths:

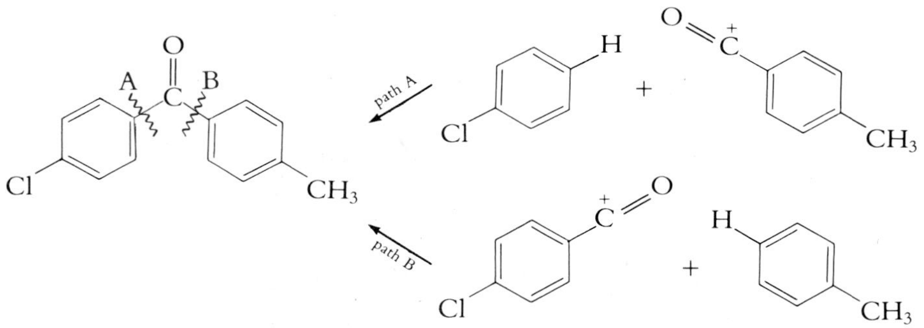

The Friedel–Crafts acylation works best when the ring undergoing substitution is as activated as possible. Thus a synthesis based on path B is the more desirable because CH_3 is more activating than Cl. The *para* product will predominate:

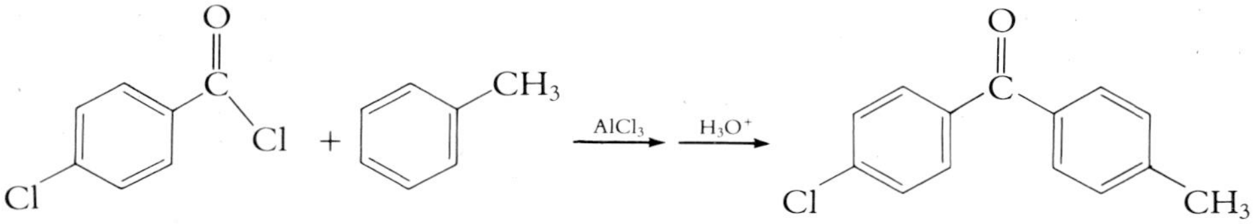

$$\text{Cl-C}_6\text{H}_4\text{-COCl} + \text{C}_6\text{H}_5\text{CH}_3 \xrightarrow{AlCl_3} \xrightarrow{H_3O^+} \text{Cl-C}_6\text{H}_4\text{-CO-C}_6\text{H}_4\text{-CH}_3$$

Problems

16–14 The cyclization step of the reaction in Equation 16.26 starts from the carboxylic acid and uses an inorganic acid as the catalyst. What is the structure of the probable intermediate in this reaction? Write a mechanism for its formation.

16–15 Show how the first four compounds in Table 16–6 can be synthesized, using acid halides or anhydrides as starting materials.

Friedel–Crafts acylations are strongly influenced by the presence of substituents on the ring. The reactions become very sluggish if electron-withdrawing groups are present. Acylations of halobenzenes often give poor yields, and nitrobenzene is completely inert. As a matter of fact, nitrobenzene is sometimes used as a *solvent* for Friedel–Crafts reactions.

Problem

16–16 Explain why the reactions in Equations 3 and 4 lead to different types of products.

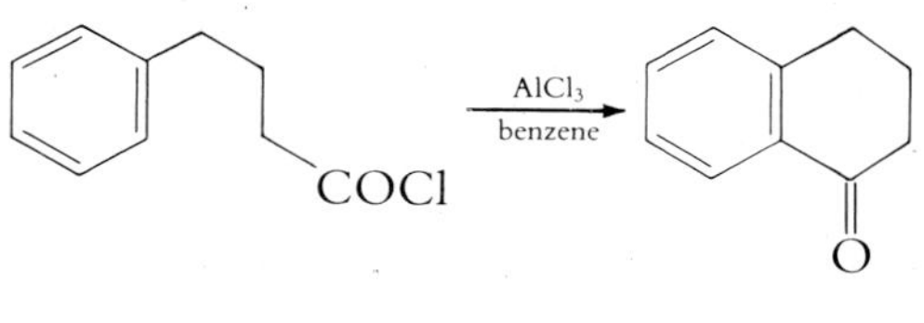

$$\text{C}_6\text{H}_5\text{CH}_2\text{CH}_2\text{CH}_2\text{COCl} \xrightarrow[\text{benzene}]{AlCl_3} \text{1-tetralone} \qquad (3)$$

$$C_6H_5COCH_2CH_2COCl \xrightarrow[\text{benzene}]{AlCl_3} C_6H_5COCH_2CH_2COC_6H_5 \qquad (4)$$

In the reactions shown in Equations 16.25 and 16.26, cyclization has taken place at the position *ortho* to the side chain. Even though *para* substitution is ordinarily the exclusive pathway in open-chain Friedel–Crafts acylations (eq. 16.27), cyclization occurs *ortho* because the side chain is not long enough to reach the *para* position.

$$(16.27) \qquad C_6H_5OCH_3 + R\overset{O}{\overset{\|}{C}}Cl \xrightarrow{AlCl_3} p\text{-}CH_3OC_6H_4COR \qquad \text{(no } ortho \text{ product)}$$

Summary

1 Carboxylic acids contain the carboxyl group, RCOOH, which ionizes in polar protic solvents such as water to yield carboxylate ions, $RCOO^-$. The broad ir absorption of carboxylic acids, between 2500 and 3500 cm^{-1}, shows that they exist as hydrogen-bonded dimers. The acidic proton gives rise to a

$$R\overset{O}{\overset{\|}{C}}OH + H_2O \rightleftarrows RC(O^{1/2-})_2 + H_3O^+$$

^1H nmr peak in the 10 to 13 ppm region (sec. 16:2). Acidity is expressed in terms of ionization constants, K_a, or their negative logs, pK_a. For acetic acid, $K_a = 1.8 \times 10^{-5}$ and p$K_a = 4.7$ (sec. 16:3).

2 Relative acidities of alkylcarboxylic acids are mainly dependent upon the inductive effects of substituents, which stabilize or destabilize the carboxylate anion. Thus, chloroacetic acid (pK_a 2.85) is a stronger acid than acetic acid because the electron-withdrawing chlorine substituent stabilizes the chloroacetate anion. Propanoic acid (pK_a 4.87) is weaker than acetic acid because the methyl substituent is weakly electron-donating and hence destabilizes the propanoate anion (sec. 16:3).

3 The relative acidities of substituted benzoic acids depend on an interplay among inductive, resonance, and steric effects. Substituents *meta* to a carboxyl group cannot interact with it by resonance, and hence affect acidities mainly by way of inductive effects. Substituents in the *ortho* and *para* positions can interact with the carboxyl group by way of resonance effects. Substituents *ortho* to a carboxyl group can affect acidity by a combination of all three (resonance, inductive, and steric) effects as well as by hydrogen-bonding to the anion (sec. 16:4).

A knowledge of substituent effects on benzoic acid acidities can be used to study reaction mechanisms (sec. 16:5) and to design improved medicinals (sec. 16:6).

4 Aromatic acids can be synthesized by oxidation of alkylbenzenes having an α-hydrogen. Aliphatic carboxylic acids are prepared by oxidation of alcohols and aldehydes. Reaction of 1° and 2° alkyl halides with the cyanide ion, followed by hydrolysis of the resulting nitrile, provides a good route to aliphatic acids. Addition of CO_2 to Grignard reagents can also be used to make aliphatic and aromatic acids from organic halides. These methods are summarized in Table 16–5 (sec. 16:7).

5 Acid halides are synthesized from carboxylic acids by replacing the OH group with a halogen (usually Cl; sec. 16:8). The major reagents for synthesizing these derivatives are phosphorus and thionyl halides. Acid anhydrides can be thought of as derivatives of acids formed by condensation of two molecules of acid with loss of a molecule of water (sec. 16:9). An important use of acid halides and acid anhydrides is in the preparation of phenyl ketones by means of Friedel–Crafts reactions. Both intermolecular and intramolecular Friedel–Crafts acylations can be used (sec. 16:10).

Problems

16–17 Show the structure of each of the following compounds:

a. acetyl bromide
b. propanoic anhydride
c. *p*-chlorophenylacetyl chloride
d. *m*-nitrobenzoyl chloride
e. succinic anhydride
f. bromoacetic anhydride

16–18 Show the structure of each of the following acids:

a. 3-phenylpropanoic acid
b. *p*-isopropylbenzoic acid
c. (*Z*)-3-phenylpropenoic acid
d. *meso*-2,3-dihydroxybutanedioic acid
e. cyclopropane-1,1-dicarboxylic acid
f. (*R*)-2-hydroxypropanoic acid
g. 2,4-dichlorophenoxyacetic acid

16–19 Fill in the values of K_a for each of the following acids:

	pK_a	K_a
a. octanoic acid	4.89	
b. chloroacetic acid	2.85	
c. *p*-nitrobenzoic acid	3.41	
d. *o*-toluic acid	3.91	

16–20 Which compound of each of the following pairs is the more acidic? Explain your answers.

a. HNO_2 or HNO_3
b. $HClO_3$ or $HClO_4$

c. C_6H_5—SO_3H or C_6H_5—SO_2H (benzenesulfinic acid)

d. FCH_2CO_2H or $ClCH_2CO_2H$
e. FCH_2CO_2H or $FCH_2CH_2CO_2H$
f. $CH_3OCH_2CO_2H$ or CH_3CO_2H

g. CH_3CO_2H or $CH_3\overset{O}{\overset{\|}{C}}OOH$ (peroxyacetic acid)

h. CH_3O—C_6H_4—CO_2H or C_6H_5—CO_2H

16–21 a. The first dissociation constants, K_1 (eq. 5) for oxalic, malonic, and succinic acids are 5.9×10^{-2}, 1.5×10^{-3}, and 6.9×10^{-5}, respectively. How do you account for this trend? Why is oxalic acid so much stronger than acetic acid?

$$(CH_2)_n(COOH)_2 \underset{}{\overset{K_1}{\rightleftharpoons}} (CH_2)_n(COO^-)(COOH) + H^+ \underset{}{\overset{K_2}{\rightleftharpoons}} (CH_2)_n(COO^-)_2 + H^+ \quad (5)$$

$n = 0, 1, 2$

b. The ratio of first and second dissociation constants (K_1/K_2) is 920 for oxalic acid and 740 for malonic acid. Why do the two dissociation constants for each acid differ so markedly? As a qualitative estimate, what ratio would you expect between the first and second dissociation constants of adipic acid ($n = 4$)?

16–22 3,5-Dimethylbenzoic acid (pK_a 4.41) is a slightly weaker acid than benzoic acid. By contrast, 2,6-dimethylbenzoic acid (pK_a 3.25) is a stronger acid than benzoic acid. Explain both of these observations.

16–23 The substitution of CH_3O for H in benzoic acid and in *p*-methylbenzoic acid produces the results shown. Explain why the methoxy group exerts opposite effects in these two systems.

p-H–C_6H_4–COOH (Stronger acid) > p-OCH_3–C_6H_4–COOH (Weaker acid)

p-CH_3–C_6H_4–COOH (Weaker acid) < p-CH_2—OCH_3–C_6H_4–COOH (Stronger acid)

16–24 Two unknown compounds, **A** and **B**, are isomers of molecular formula $C_7H_5O_2Cl$. Compound **A** has pK_a 2.92 and compound **B** has pK_a 3.82. Their ^1H nmr spectra are shown here. Give structures for the unknowns and detail your reasoning.

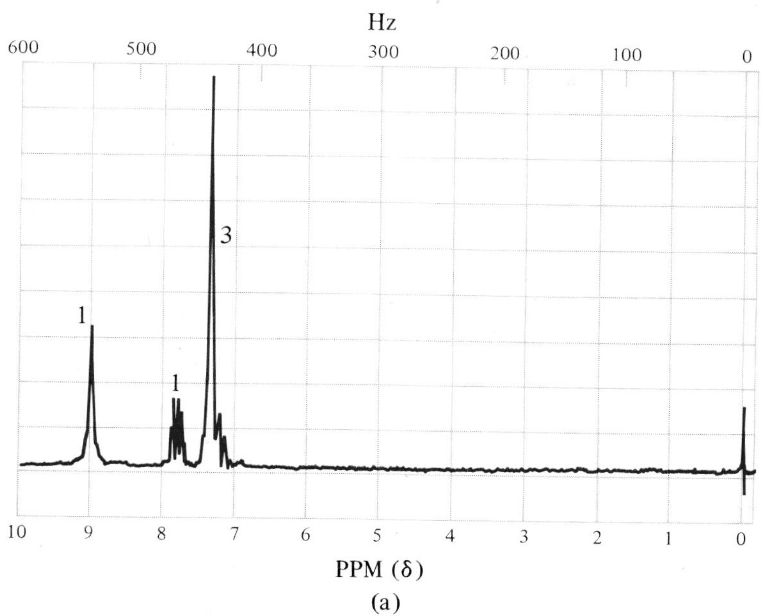

(a)

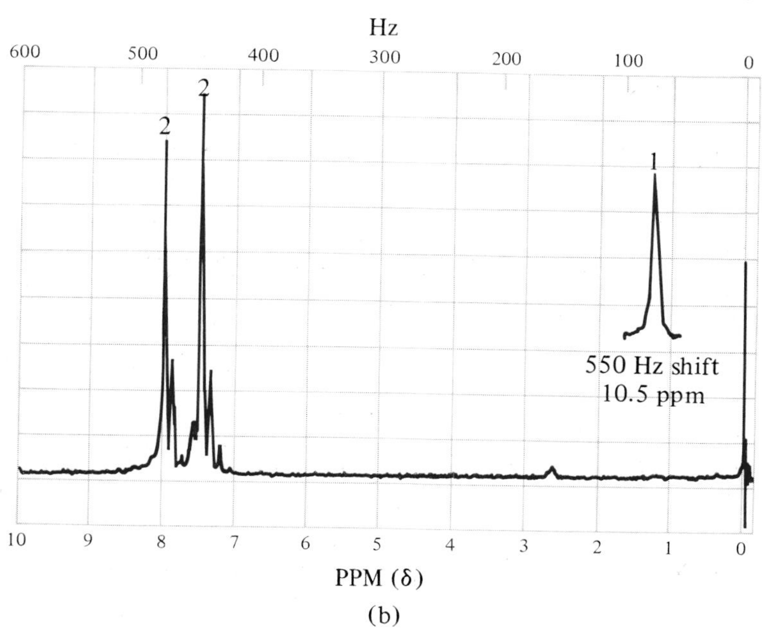

(b)

16–25 Carbon acids are extremely weak acids, but their acidity can still be measured (eq. 6). Use resonance and inductive effects to predict which of each pair is the stronger acid.

$$Y\text{-}C_6H_4\text{—}CH_3 + B:^- \rightleftarrows Y\text{-}C_6H_4\text{—}CH_2:^- + BH \qquad (6)$$

a. *m*-CH_3 or *p*-CH_3
b. *m*-OCH_3 or *p*-OCH_3
c. *m*-NO_2 or *p*-NO_2

16–26 Base-catalyzed exchange of hydrogen for deuterium takes place readily in phenylacetaldehyde. It is highly likely that the mechanism is a two-step one, with carbanion **III** as an intermediate (eqs. 7 and 8).

$$C_6H_5\text{—}CH_2\text{—}\overset{O}{\overset{\|}{C}}H + OD^- \rightarrow C_6H_5\text{—}\ddot{C}H^-\text{—}\overset{O}{\overset{\|}{C}}H + HOD \qquad (7)$$

III

$$C_6H_5\text{—}\ddot{C}H^-\text{—}\overset{O}{\overset{\|}{C}}H + D_2O \rightarrow C_6H_5\text{—}CHD\text{—}\overset{O}{\overset{\|}{C}}H + OD^- \qquad (8)$$

III

a. Use resonance structures to explain why the base preferentially attacks the methylene group to give **III.**
b. The exchange proceeds more rapidly with a *p*-nitro group on the benzene ring, and more slowly with a *p*-methyl group. Use these data to decide whether the reaction in Equation 7 or 8 is the rate-determining step of the exchange process. Explain your reasoning.

16–27 In each case, show how you would obtain the given carboxylic acid by means of an oxidation.

a. 4-nitrobenzoic acid
b. 3-methylbutanoic acid
c. 3-ethylbenzoic acid
d. 2,2-dimethylpentanoic acid

16–28 a. An optically active hydrocarbon, $C_{11}H_{16}$, upon vigorous oxidation gives terephthalic acid. Assign a structure to the hydrocarbon.
b. Is this the only possible structure? Explain your answer.

16–29 Show how each of the following acids could be prepared starting with either toluene or benzoic acid.

a. *o*-nitrobenzoic acid
b. *p*-methylbenzoic acid
c. *m*-bromobenzoic acid
d. *p*-*t*-butylbenzoic acid
e. 2-chloro-4-nitrobenzoic acid
f. 4-iodo-3-nitrobenzoic acid

16–30 Suggest efficient syntheses of each of the following compounds, based on alcohols or halides of five or fewer carbons.

a. pentanoic acid
b. 2-ethylbutanoic acid
c. heptanoic acid
d. 3,3-dimethylbutanoic acid
e. 3-cyclopentenecarboxylic acid
f. 1-methylcyclobutanecarboxylic acid

16–31 Show what acid(s) will be formed when each alkene is oxidized by excess hot $KMnO_4$ solution, followed by acidification.

a. 2-pentene
b. 2-methyl-2-pentene
c. cyclohexene
d. (α-pinene) 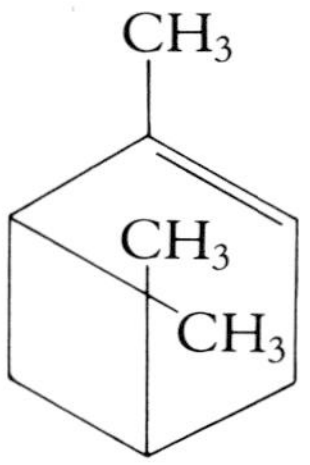
e. 1-phenyl-1-butene
f. 4-phenyl-1-butene

16–32 Optically active 4-methylhexanoic acid is found in the preen gland of Peking ducks. Suggest a synthesis of this acid from optically active 2-methyl-1-butanol, which is readily available from fusel oil (a collection of higher boiling alcohols formed by fermentation during ethanol production). (*Hint:* In solving this problem, think about a synthetic method that extends the chain by two carbon atoms in a single step.)

16–33 In the oxidation of *p*-isopropyltoluene to 4-methylbenzoic acid (*p*-toluic acid; eq. 16.11), the isopropyl group is attacked in preference to the methyl group. It is thought that HNO_3 oxidations involve initial attack by $\cdot NO_2$. Use this assumption to explain the preferential attack at the isopropyl group.

16–34 Motrin is an antiarthritic drug (Prob. 15–41). Several syntheses of this compound go by way of *p*-isobutylacetophenone, which is then converted to motrin (eq. 9). Show how this conversion can be carried out.

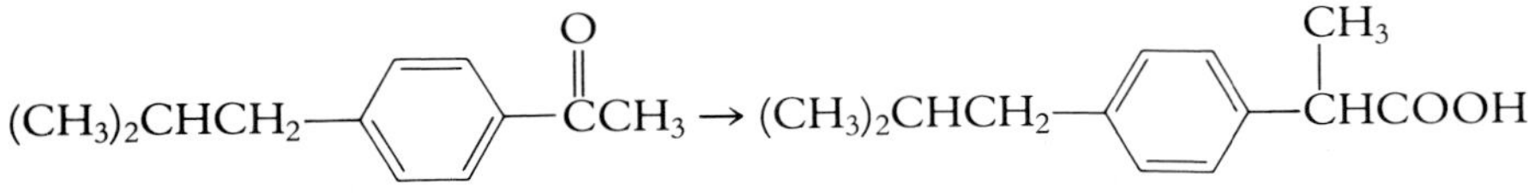

$$(CH_3)_2CHCH_2-C_6H_4-\overset{O}{\overset{\|}{C}}CH_3 \rightarrow (CH_3)_2CHCH_2-C_6H_4-\overset{CH_3}{\overset{|}{C}}HCOOH \quad (9)$$

Motrin

16–35 When octanedioic acid is heated with acetic anhydride, a small quantity of cyclic octanedioic anhydride is isolated from the reaction mixture. The major product is a high molecular weight material. Suggest a structure for this product. Why is the cyclic anhydride not a major product?

16–36 Human brain tissue contains numerous fatty acids, among them a compound with the formula $C_{24}H_{40}O_2$ (**A**). Hydrogenation of **A** yields the saturated straight-chain acid of formula $C_{24}H_{48}O_2$. Ozonization of **A,** followed by reduction with $(CH_3)_2S$, yields three compounds in the following proportions: hexanal (one eq.), 1,3-propanedial (*malondialdehyde,* two eqs.), and an aldehydic acid of formula $C_{12}H_{22}O_3$.

Propose a structure for **A,** consistent with the data. What structural features of **A** are not determined by the ozonization data? Would you expect **A** to show uv absorption? Explain your answer.

16–37 Indicate all steps in the synthesis of the following compounds, using benzene, toluene, or benzaldehyde as starting materials.

a. propiophenone (two methods)
b. α-methylbutyrophenone (two methods)

C_6H_5–$COCHCH_2CH_3$ (with CH_3 on the CH)

c. *p*-methylvalerophenone [1-(*p*-methylphenyl)-1-pentanone]
d. cyclohexyl phenyl ketone
e. benzophenone (two methods)
f. 4-nitrobenzophenone
g. γ-benzoylbutyric acid

C_6H_5–$\overset{O}{\overset{\|}{C}}(CH_2)_3\overset{O}{\overset{\|}{C}}OH$

h. 4-phenylbutanoic acid
i. 1-phenyl-3,4-dihydronaphthalene
j. 1-methyl-1,2,3,4-tetrahydronaphthalene

16–38 An unknown compound ($C_{10}H_{12}O$) has the ^{13}C and ^{1}H nmr spectra shown here. Give a structure for the compound and assign the nmr peaks in both spectra.

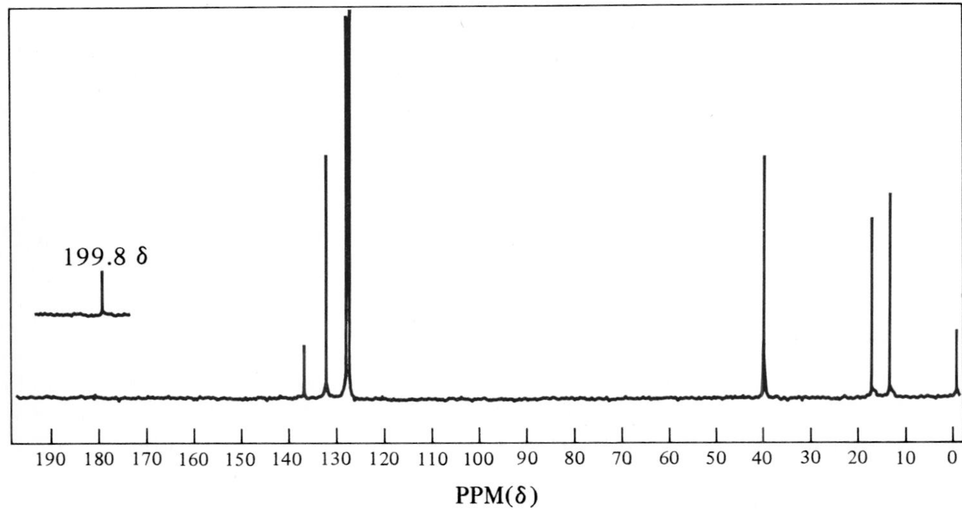

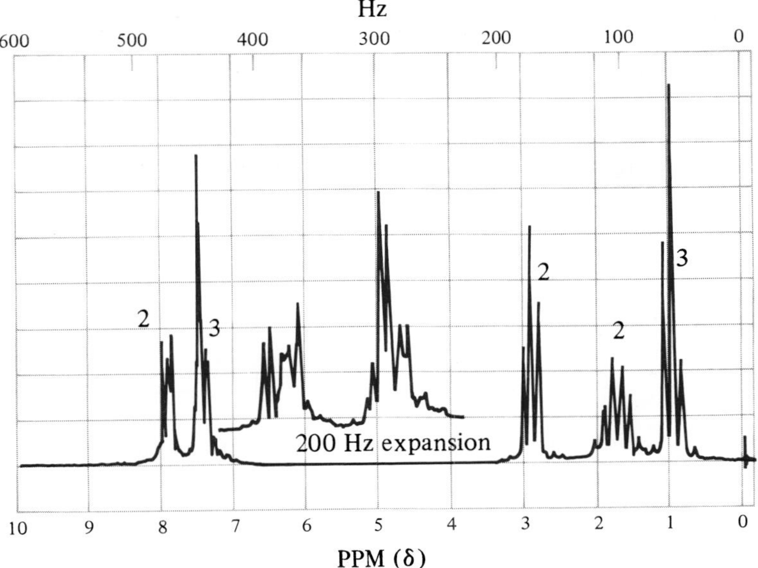

Supplementary Reading

G. V. Calder and T. J. Barton, Actual effects controlling the acidity of carboxylic acids, *J. Chem. Ed.* **48,** 338 (1971).

C. Hansch, Drug research or the luck of the draw, *J. Chem. Ed.* **51,** 360 (1974).

D. D. Perrin, B. Dempsey, and E. P. Serjeant, *pK$_a$ Prediction for Organic Acids and Bases,* Methuen, Inc., New York, N.Y., 1981.

J. Shorter, Linear free energy relationships, *Chemistry in Britain* **5,** 269 (1969).

R. W. Bragg *et al.,* Sweet organic chemistry, *J. Chem. Ed.* **55,** 281 (1978). An interesting example of structure–activity correlation.

17 Esters, Amides, and Nitriles: An Overview of Carboxylic Acid Derivatives and Their Chemistry

Acid halides and anhydrides, which we introduced in Chapter 16, are members of a family of compounds known as carboxylic acid derivatives. In this chapter we'll study the properties of the two most important members of this family, esters and amides. In addition, we'll take an overview of the entire family, whose members resemble one another in reactivity as well as in structure. Because all carboxylic acid derivatives (except nitriles) contain carbonyl groups, they can be easily interconverted, and we can use them to learn more about the relationship between structure and reactivity.

Table 17–1 lists the major derivatives. Note that although nitriles do not contain a carbonyl group, they are included in this list because they can be converted to carboxylic acids by hydrolylsis, and because their chemistry resembles that of the other compounds shown in Table 17–1. This chapter will first take up the carbonyl-containing compounds, then turn to nitriles, and conclude by briefly comparing carboxylic acid derivatives with derivatives of other types of acids.

17:1 Nomenclature of Esters and Amides

Esters can be prepared by reaction of an alcohol molecule with an acid molecule (usually in the presence of a catalyst; eq. 17.1). They are named after the alcohol and acid from which they are derived.

$$CH_3CH_2C(=O)OH + CH_3OH \xrightarrow{\text{catalyst}} CH_3CH_2C(=O)OCH_3 + H_2O \qquad (17.1)$$

Propanoic acid — Methyl alcohol — Methyl propanoate

The ester name consists of two words: the first word denotes the alcohol and the second denotes the carboxylic acid, with the -ic ending replaced by -ate. Substituents in the alcohol portion are noted in the alcohol name and are separate from the substituents in the acid portion.

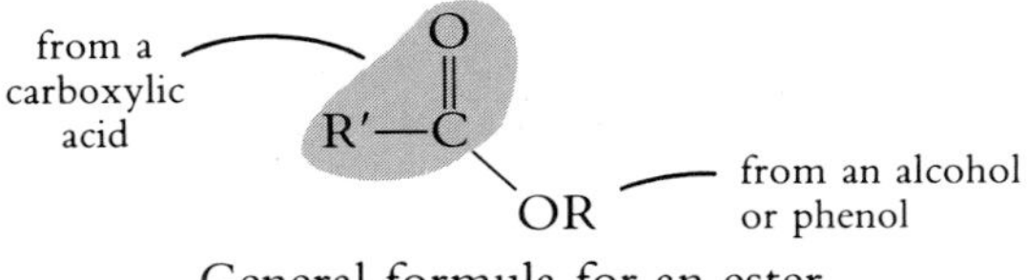

General formula for an ester

Table 17–1. Important Carboxylic Acid Derivatives

Functional Group	Name	Specific Examples
$-\overset{O}{\overset{\Vert}{C}}-X$ (X = F, Cl, Br, I)	acid (or acyl) halides	$CH_3(CH_2)_3\overset{O}{\overset{\Vert}{C}}F$ pentanoyl fluoride (*valeryl fluoride*)
$-\overset{O}{\overset{\Vert}{C}}-O-\overset{O}{\overset{\Vert}{C}}R$ (R = H, alkyl, aryl)	acid anhydrides	$CH_3CH_2\overset{O}{\overset{\Vert}{C}}O\overset{O}{\overset{\Vert}{C}}CH_2CH_3$ propanoic anhydride (*propionic anhydride*)
$-\overset{O}{\overset{\Vert}{C}}-OR$ (R = alkyl, aryl)	esters	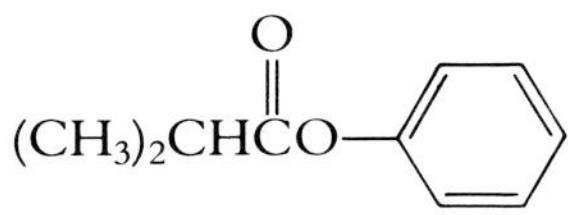 $(CH_3)_2CH\overset{O}{\overset{\Vert}{C}}O-C_6H_5$ phenyl 2-methylpropanoate (*phenyl isobutyrate*)
$-\overset{O}{\overset{\Vert}{C}}-NRR'$ (R,R′ = H, alkyl, aryl)	amides	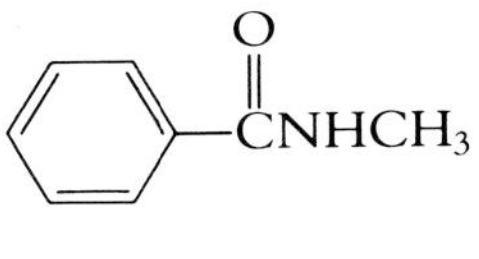 $C_6H_5-\overset{O}{\overset{\Vert}{C}}NHCH_3$ *N*-methylbenzamide
$-\overset{O}{\overset{\Vert}{C}}-O^-$	carboxylate ions	$(CH_3)_2CHCH_2\overset{O}{\overset{\Vert}{C}}-O^-Na^+$ sodium 3-methylbutanoate
$-C\equiv N$	nitriles	$CH_3CH_2CH_2CH_2C\equiv N$ pentanenitrile (*valeronitrile*)

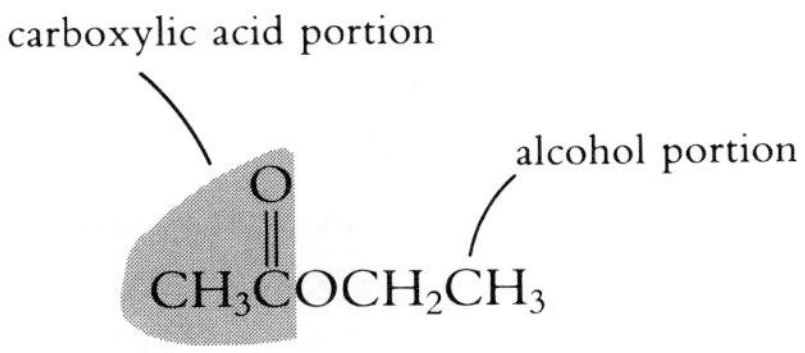

carboxylic acid portion — alcohol portion

$$CH_3\overset{O}{\overset{\Vert}{C}}OCH_2CH_3$$

Ethyl ethanoate
(*ethyl acetate*)

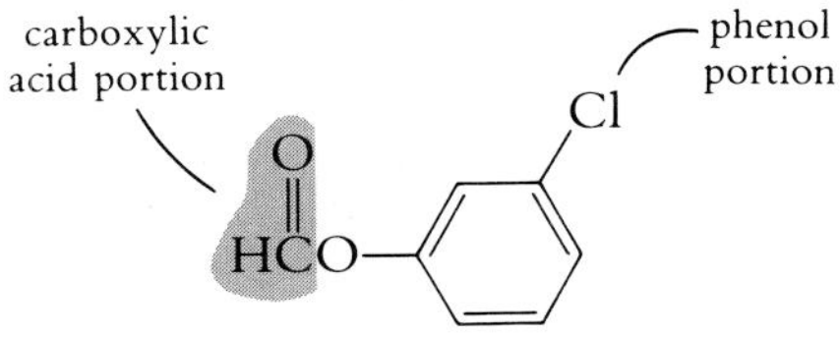

carboxylic acid portion — phenol portion

$$H\overset{O}{\overset{\Vert}{C}}O-C_6H_4Cl$$

m-Chlorophenyl formate

$$ClCH_2CH_2\overset{O}{\overset{\Vert}{C}}OCH_3$$

Methyl 3-chloropropanoate

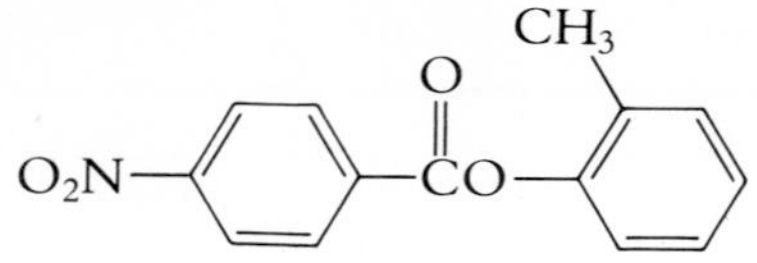

$$O_2N-C_6H_4-\overset{O}{\overset{\|}{C}}O-C_6H_4-CH_3$$

2-Methylphenyl 4-nitrobenzoate
(o-*cresyl* p-*nitrobenzoate*)

$$CH_3O\overset{O}{\overset{\|}{C}}(CH_2)_2\overset{O}{\overset{\|}{C}}OCH_3$$

Dimethyl butanedioate
(*dimethyl succinate*)

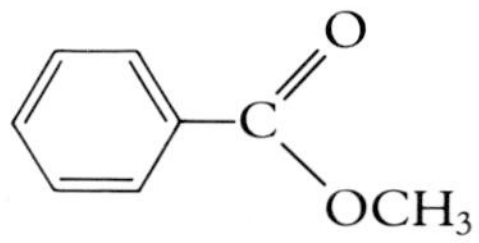

$$C_6H_5-C(=O)-OCH_3$$

Methyl benzoate

Esters are often written in condensed structures in which the C=O is understood, rather than written out. Thus, methyl benzoate may be written as $C_6H_5COOCH_3$, $C_6H_5CO_2CH_3$, $CH_3OOCC_6H_5$, or $CH_3O_2CC_6H_5$. All of these are equivalent to the line-bond structure given above.

Amides can often be prepared by heating together an acid and an amine or ammonia (eq. 17.2). In an amide, the OH group of the acid is replaced by an NH_2, NHR, or NR_2 group.

$$CH_3\overset{O}{\overset{\|}{C}}OH + NH_3 \rightarrow CH_3\overset{O}{\overset{\|}{C}}O^-NH_4^+ \xrightarrow{\text{heat}} CH_3\overset{O}{\overset{\|}{C}}NH_2 + H_2O \qquad (17.2)$$

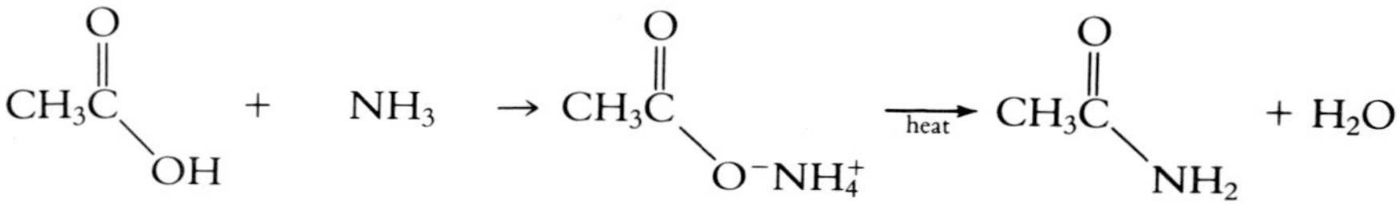

Acetic acid; Ammonia; Ammonium acetate; Ethanamide (*acetamide*)

$$R\overset{O}{\overset{\|}{C}}NH_2 \qquad R\overset{O}{\overset{\|}{C}}NHR \qquad R\overset{O}{\overset{\|}{C}}NR_2$$

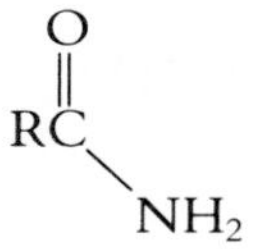

Amide (a 1° amide); *N*-Alkylamide (a 2° amide); *N*,*N*-Dialkylamide (a 3° amide)

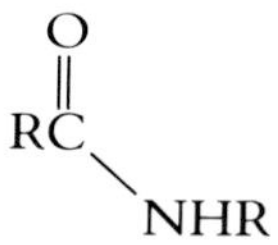

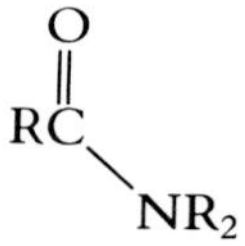

General formulas for amides

Names of amides are derived from the corresponding carboxylic acid names. The -ic or -oic acid ending is replaced by -amide. If the suffix -carboxylic acid appears in the name, it is replaced by -carboxamide. With amides it is necessary to distinguish between substituents on the carbon chain and substituents attached to the nitrogen atom. A group attached to the nitrogen atom has its location indicated by the prefix *N*, rather than by a number. For simple amides, the common names are still extensively used; for example, acetamide and formamide.

$$H\overset{O}{\overset{\|}{C}}NH_2 \qquad CH_3\overset{O}{\overset{\|}{C}}NH_2$$

Methanamide (*formamide*); Ethanamide (*acetamide*)

$(CH_3)_2CHCONH_2$ (C=O)

2-Methylpropanamide
(*isobutyramide*)

carboxylic acid portion — $m\text{-}O_2NC_6H_4CONH\text{-}C_6H_5$ — amine portion

N-Phenyl-*m*-nitrobenzamide
(m-*nitrobenzanilide*)

$HCON(CH_3)_2$

N,N-Dimethylmethanamide
(N,N-*dimethylformamide, DMF*)

$C_5H_9CONH_2$

Cyclopentanecarboxamide

$CH_3CH_2CH(CH_3)(CH_2)_4CONHCH_2CH_3$

N-Ethyl-6-methyloctanamide

Problems

17–1 Provide IUPAC names for each of the following compounds. Draw the structures of the alcohol and acid from which the esters are derived, and the acid from which the amides are derived.

a. $C_6H_5COOCH_2CH_2CH_3$

b. $Cl(CH_2)_4COOCH_2CH(CH_3)CH_2CH_3$

c. $m\text{-}CH_3C_6H_4CONH_2$

d. $BrCH_2CON(CH_2CH_3)_2$

e. $m\text{-}BrC_6H_4COOCH_2CH_2Br$

f. (cyclopent-2-enyl)$COOCH_2CH_2CH_3$

17–2 Write the structures that correspond to the following names:

a. methyl propanoate
b. 1-methylbutyl 3-chlorohexanoate
c. 3,4-dibromophenyl *cis*-2-methylcyclohexanecarboxylate
d. *N,N*-dimethylacetamide
e. *N*-(4-hydroxyphenyl)octanamide
f. *N*-chloro-*N*-propyl-2-naphthalenecarboxamide

17:2 Spectra of Carboxylic Acid Derivatives

As with all other compounds containing a carbonyl group, carboxylic acid derivatives give rise to strong C=O infrared bands. Table 17–2 lists the ranges for the carbonyl stretching frequency in the common carboxylic acid derivatives. You can see that the frequency varies markedly with structure, making these bands very useful for structure determination. There is a correlation between the reactivity of the carboxylic acid derivatives of Table 17–2 and their carbonyl stretching frequencies. The higher the frequency, the more reactive is the derivative (we will consider this further in sec. 17:3).

Table 17–2 shows that there are two bands for the C=O and C$\overset{\cdots}{=}$O stretches of anhydrides (sec. 16:9) and carboxylate ions, respectively. This is due to the interaction between the two groups as they are stretched. The lower frequency in each case is due to the *symmetrical* motion, in which both groups stretch and contract together. The higher frequency comes from the *asymmetrical* motion, in which one group stretches while the other contracts.

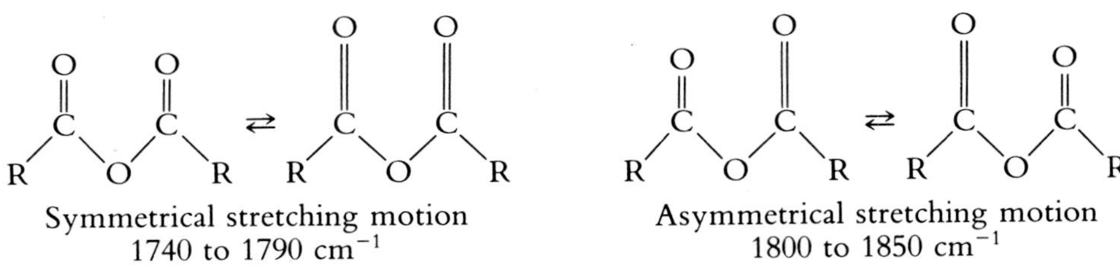

Symmetrical stretching motion 1740 to 1790 cm^{-1}

Asymmetrical stretching motion 1800 to 1850 cm^{-1}

Table 17–2. Carbonyl Stretching Frequencies of Some Carboxylic Acid Derivatives in Nonpolar Solvents

Structure	Frequency Range (cm^{-1})
$\mathrm{R\overset{O}{\overset{\|}{C}}Cl}$	1780 to 1815
$\mathrm{R\overset{O}{\overset{\|}{C}}O\overset{O}{\overset{\|}{C}}R}$	1800 to 1850 and 1740 to 1790
$\mathrm{R\overset{O}{\overset{\|}{C}}OR'}$	1735 to 1750
$\mathrm{R\overset{O}{\overset{\|}{C}}OH}$	1750 to 1770 (monomer) and 1710 to 1720 (dimer)
$\mathrm{R\overset{O}{\overset{\|}{C}}NR'_2}$	1630 to 1690
$\mathrm{R\overset{O}{\overset{\|}{C}}O^-}$	1550 to 1620 and 1390 to 1450

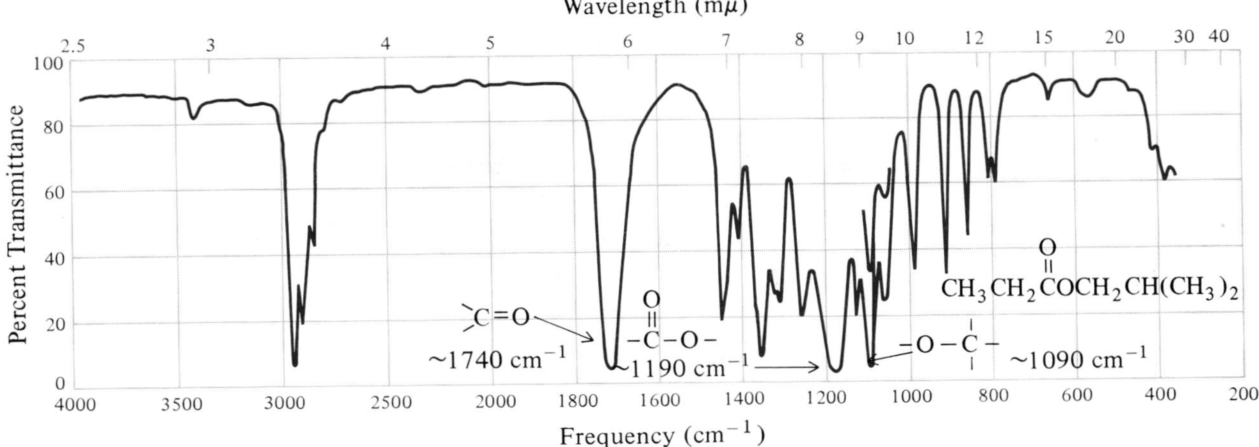

Figure 17–1. *Infrared spectrum of 2-methylpropyl propanoate.*

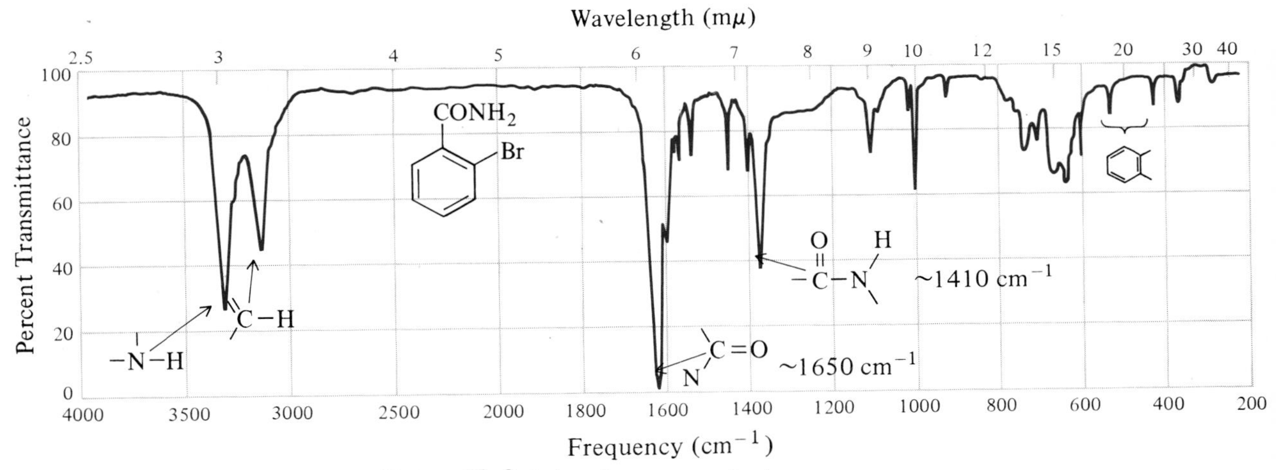

Figure 17–2. *Infrared spectrum of* o-*bromobenzamide.*

Table 17–2 also indicates that there are single carbonyl bands for esters, as shown in Figure 17–1, and amides, as shown in Figure 17–2. Note the characteristic intensity of these bands and their differing frequencies, which serve to identify the functional group in each case.

The mass spectra of carboxylic acid derivatives show many features similar to the mass spectra of ketones (Chap. 11). To explain the fragmentation pattern, we can assume that ionization occurs by loss of a nonbonding electron (eq. 17.3). The molecular ion can then decompose by loss of the organic group R·, or of the carbonyl substituent Y·, yielding, in either case, stable acylium ions.

$$\text{(17.3)}\quad R-\overset{\overset{\textstyle :O:}{\|}}{C}-Y \xrightarrow{\text{ionization}} R\!\!\not{}\,\overset{\overset{\textstyle :O\dot{}^{+}}{\|}}{C}\,\!\not{}\,Y \;\begin{cases} \xrightarrow{a} R-C{\equiv}O\dot{:}^{+} + Y\cdot \\ \xrightarrow{b} {}^{+}\!\!:O{\equiv}C-Y + R\cdot \end{cases}$$

b **a**

Chlorides, anhydrides, and esters lose Y· preferentially. Amides lose R· preferentially to form an ion strongly stabilized by the nitrogen lone pair (eq. 17.4).

$$\text{(17.4)}\quad R-\overset{\overset{\textstyle :O\dot{}^{+}}{\|}}{C}-NH_2 \rightarrow R\cdot + [\overset{+}{:O}{\equiv}C-\ddot{N}H_2 \leftrightarrow \ddot{O}{=}C{=}\overset{+}{N}H_2 \leftrightarrow O{=}\overset{+}{C}-\ddot{N}H_2]$$

$m/e = 44$

Problem

17–3 A compound of unknown structure has the molecular formula C_4H_9NO. Its ir spectrum contains a strong band at 1675 cm^{-1}, suggesting that it is an amide. Further investigation indicates that it is either *N*-methylpropanamide (**I**) or *N*-ethylacetamide (**II**). Show how mass spectrometry could allow you to deduce the correct structure.

$$CH_3CH_2\overset{\overset{\textstyle O}{\|}}{C}NHCH_3 \qquad\qquad CH_3\overset{\overset{\textstyle O}{\|}}{C}NHCH_2CH_3$$

I **II**

17:3 Reactivity of Carboxylic Acid Derivatives

Equations 17.5 to 17.7 show some characteristic reactions of carboxylic acid derivatives. In each of these reactions the acid derivatives are acylating (sec. 15:7) different nucleophiles.

$$\text{(17.5)}\quad \underset{\text{Acylating agent}}{CH_3\overset{\overset{\textstyle O}{\|}}{C}Cl} + \underset{\text{Nucleophile}}{CH_3CH_2\ddot{O}H} \rightarrow \underset{\text{Ester}}{CH_3\overset{\overset{\textstyle O}{\|}}{C}OCH_2CH_3} + HCl$$

$$\text{(17.6)}\quad \underset{\text{Acylating agent}}{C_6H_5-\overset{\overset{\textstyle O}{\|}}{C}OCH_3} + \underset{\text{Nucleophile}}{CH_3CH_2\ddot{N}H_2} \rightarrow \underset{\text{Amide}}{C_6H_5-\overset{\overset{\textstyle O}{\|}}{C}NHCH_2CH_3} + CH_3OH$$

$$(17.7)\quad \underset{\text{Acylating agent}}{(CH_3)_2CHCH_2\overset{\overset{\displaystyle O}{\|}}{C}OCH(CH_3)_2} + \underset{\text{Nucleophile}}{OH^-} \rightarrow$$

$$[(CH_3)_2CHCH_2\overset{\overset{\displaystyle O}{\|}}{C}OH + (CH_3)_2CHO^-] \rightarrow \underset{\text{Carboxylate ion}}{(CH_3)_2CHCH_2\overset{\overset{\displaystyle O}{\|}}{C}O^-} + (CH_3)_2CHOH$$

Since there are quite a few carboxylic acid derivatives (Table 17–1 gives only a sample) and a large number of nucleophiles, the number of possible reactions for this series of compounds is very great indeed. The topic can be made manageable by studying some general trends in the reactivity of these derivatives.

Problem

17–4 Write the structures of the products from each of the following reactions.

a. 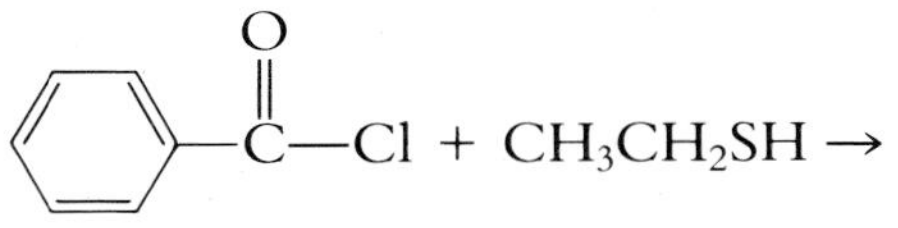 $C_6H_5\overset{\overset{\displaystyle O}{\|}}{C}\text{—}Cl + CH_3CH_2SH \rightarrow$

b. $CH_3CH_2\overset{\overset{\displaystyle O}{\|}}{C}\text{—}Cl + (CH_3)_2NH \rightarrow$

c. $CH_3CH_2\overset{\overset{\displaystyle O}{\|}}{C}O\overset{\overset{\displaystyle O}{\|}}{C}CH_2CH_3 + CH_3NH_2 \rightarrow$

d. $CH_3\overset{\overset{\displaystyle O}{\|}}{C}NHCH_3 + OH^- \rightarrow$

e. 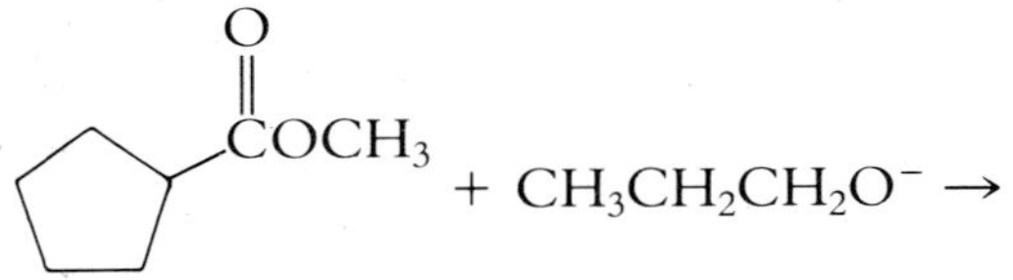 $C_5H_9\overset{\overset{\displaystyle O}{\|}}{C}OCH_3 + CH_3CH_2CH_2O^- \rightarrow$ (cyclopentyl ester)

The carboxylic acid derivatives of major interest fall into a well-defined pattern of relative reactivity toward nucleophiles:

$$\underset{\text{Acid halides}}{R\overset{\overset{\displaystyle O}{\|}}{C}X} > \underset{\text{Anhydrides}}{R\overset{\overset{\displaystyle O}{\|}}{C}O\overset{\overset{\displaystyle O}{\|}}{C}R} > \underset{\text{Esters}}{R\overset{\overset{\displaystyle O}{\|}}{C}OR'} > \underset{\text{Amides}}{R\overset{\overset{\displaystyle O}{\|}}{C}NR'_2} >> \underset{\text{Carboxylate ions}}{R\overset{\overset{\displaystyle O}{\|}}{C}O^-}$$

relative reactivity toward nucleophiles

The causes of this reactivity order will be discussed shortly, but for now, an important consequence of the relative reactivity will be examined.

Less reactive carboxylic acid derivatives (esters, amides) may be readily synthesized from more reactive ones (acid chlorides, anhydrides); however, the reverse substitution cannot be accomplished directly because it is thermodynamically unfavorable.

It follows from this that acid halides are versatile precursors to all the other carboxylic acid derivatives. Recall from Section 16:8 that preparation of the acid halides requires the use of highly reactive sulfur and phosphorus compounds (SOX_2, PX_3, PX_5; X = Cl or Br), which are themselves acid halides of inorganic acids (H_2SO_3, H_3PO_3, and H_3PO_4).

All carboxylic acid derivatives react with nucleophiles by a common two-step mechanism. This mechanism is represented schematically in Equation 17.8. In this mechanism the nucleophile first attacks the sp^2 carbonyl carbon to form a tetrahedral intermediate with an sp^3 carbon. In a second step, an sp^2 carbonyl carbon is again formed when the leaving group is expelled.

$$\text{(17.8)} \quad \mathrm{R\overset{O}{\overset{\|}{C}}{-}Y + {:}Nu^- \rightarrow R{-}\underset{Nu}{\underset{|}{\overset{^-O}{\overset{|}{C}}}}{-}Y \rightarrow R\overset{O}{\overset{\|}{C}}{-}Nu + Y{:}^-}$$

Tetrahedral intermediate

Suppose we try to explain reactivity by comparing the leaving groups ($Y{:}^-$) generated by nucleophilic attack at the carbonyl carbon atom (eq. 17.8). There is a clear correlation between decreasing reactivity and progressively poorer leaving groups (eqs. 17.9 to 17.13). This effect would, by itself, account for the observed reactivity order.

$$\text{(17.9)} \quad \mathrm{R\overset{O}{\overset{\|}{C}}Cl + Nu{:}^- \rightarrow R\overset{O}{\overset{\|}{C}}Nu + Cl^-}$$

$$\text{(17.10)} \quad \mathrm{R\overset{O}{\overset{\|}{C}}O\overset{O}{\overset{\|}{C}}R + Nu{:}^- \rightarrow R\overset{O}{\overset{\|}{C}}Nu + RCOO^-}$$

$$\text{(17.11)} \quad \mathrm{R\overset{O}{\overset{\|}{C}}OR' + Nu{:}^- \rightarrow R\overset{O}{\overset{\|}{C}}Nu + R'O^-}$$

$$\text{(17.12)} \quad \mathrm{R\overset{O}{\overset{\|}{C}}NR'_2 + Nu{:}^- \rightarrow R\overset{O}{\overset{\|}{C}}Nu + R'_2N^-}$$

$$\text{(17.13)} \quad \mathrm{R\overset{O}{\overset{\|}{C}}O^- + Nu{:}^- \rightarrow R\overset{O}{\overset{\|}{C}}Nu + O^{2-}}$$

↓ Poorer leaving group ↓ Increasing delocalization energy of starting carboxyl derivative ↓ Decreasing reactivity

There is an additional factor which reinforces the leaving group effect. All the carboxylic acid derivatives are stabilized to some extent by resonance interaction between their carbonyl π bonds and the lone pair on the attached heteroatom. When the carboxylic acid derivative reacts with a nucleophile to form a tetrahedral intermediate, this resonance stabilization is lost (eq. 17.14).

(17.14) [R–C(=O)–Y: ⟷ R–C(–O⁻)=Y⁺] + Nu:⁻ → R–C(O⁻)(Nu)–Y

Carboxylic acid derivative
stabilized by resonance

Tetrahedral intermediate
resonance stabilization lost

The energy barrier to nucleophilic attack is directly related to the amount of resonance stabilization. Acid chlorides have relatively little resonance stabilization so there is a small energy barrier to their reaction with nucleophiles. Amides, on the other hand, are strongly stabilized by resonance and their reaction with nucleophiles is subject to a large energy barrier (Fig. 17–3). Formation of the tetrahedral intermediate is usually the rate-determining step in these reactions, so there is a direct correlation between resonance stabilization and reactivity.

You have seen that resonance stabilization has a pronounced effect on reactivity. Let's now examine why different carboxylic acid derivatives differ in their degree of resonance stabilization.

In an acid chloride the nonbonded electrons on the chlorine are in $3p$ orbitals and would have to overlap with the $2p$ orbital of the carbonyl carbon. Since $2p$ and $3p$ orbitals have different sizes and extension in space, they can overlap only poorly and hence the resonance interaction is weak.

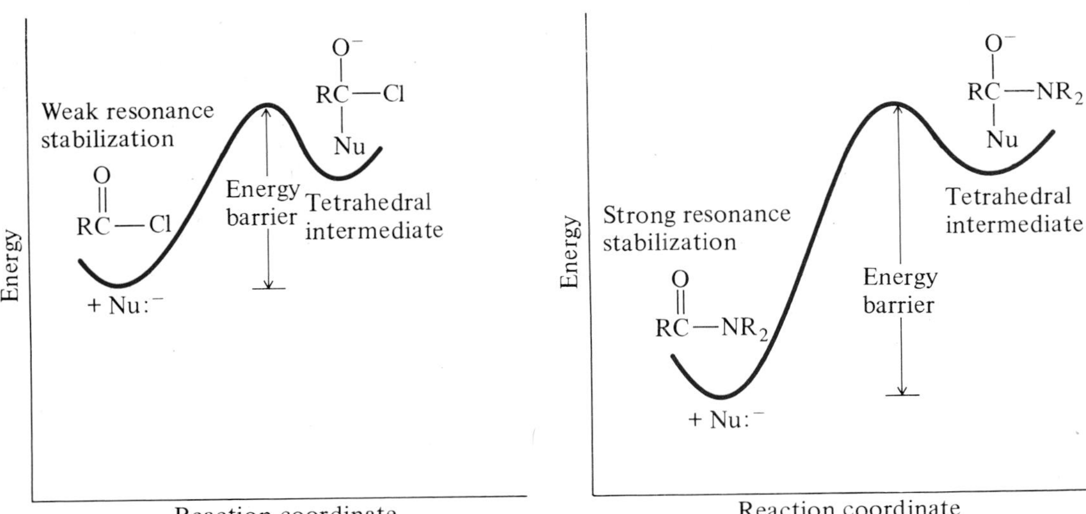

Figure 17–3. *Energy diagrams for the reaction of an acid chloride and an amide with a nucleophile. The resonance stabilization of the acid chloride is weak, so the barrier to reaction is small. The resonance stabilization of the amide is strong, leading to a large barrier to reaction.*

2p orbitals

$$\mathrm{R{-}C(=\ddot{O}:){-}\ddot{Cl}:} \longleftrightarrow \mathrm{R{-}C(-\ddot{O}:^{-})(=\overset{+}{Cl}:)}$$ Weak resonance interaction

3p orbitals

Acid chlorides (and bromides and iodides) are therefore highly reactive. Oxygen derivatives, on the other hand, have the nonbonded electrons in a 2*p* orbital. These electrons overlap easily with the carbonyl electrons in 2*p* orbitals, making for a strong resonance interaction.

2p orbitals

$$\mathrm{RC(=\ddot{O}:){-}\ddot{O}R'} \longleftrightarrow \mathrm{RC(-\ddot{O}:^{-})(=\overset{+}{O}R')}$$ Strong resonance interaction

2p electrons

Thus, oxygen derivatives have increased stability compared to acid halides. Esters and anhydrides are therefore less reactive than acid halides. In anhydrides the stabilization is less than in esters because in anhydrides the electron-withdrawing acyl group (shown below in color) makes the central oxygen less able to bear a positive charge.

destabilizing

$$\mathrm{RC(=O){-}\ddot{O}{-}C(=O)R} \leftrightarrow \mathrm{RC(-O^{-}){=}\overset{+}{O}{-}C(=O)R} \leftrightarrow \mathrm{RC(=O){-}\overset{+}{O}{=}C(-O^{-})R}$$

Resonance interactions should be stronger yet in amides, since nitrogen is less electronegative than oxygen and can more readily accept a positive charge. One of the most important consequences of this unusually strong

$$\mathrm{RC(=O){-}\ddot{N}R'_2} \longleftrightarrow \mathrm{RC(-O^{-})(=\overset{+}{N}R'_2)}$$ Very strong resonance interaction

1

resonance interaction is that the nitrogen atom, which is usually a basic site in organic compounds, is not basic in amides. For example, amines (R_3N:) are basic species because their nonbonded electron pair is readily protonated (Chap. 21). In amides, however, the contribution of resonance structure **1** makes the nonbonded electron pair on nitrogen unavailable for protonation. Hence, amides are not basic.

17–5 The reaction in Equation 1 is *not* a practical method of synthesizing acid chlorides. Explain why the left-to-right reaction is not favorable. (You should contrast this situation with that found in alcohol chemistry, where aliphatic chlorides can be synthesized from alcohols and HCl.)

$$RCOH(=O) + HCl \nrightarrow RCCl(=O) + H_2O \quad (1)$$

17–6 a. Amides can be protonated in strongly acidic solution. Nuclear magnetic resonance studies of this reaction suggest that the amide is protonated on *oxygen* rather than on nitrogen. Use resonance structures to explain why amides undergo preferential *O*-protonation.

b. It is also known that amides can function as weak *acids*—the pK_a of acetamide (CH_3CONH_2) in water is 15.08, for example. Write an equation showing the ionization of acetamide as an acid and explain the driving force for this ionization.

17:4 Introduction to the Synthesis of Esters and Amides

In the previous section we stated an important general rule: less reactive carboxylic acid derivatives are readily synthesized from more reactive ones. For example, esters are prepared from the reaction of acid chlorides or anhydrides with alcohols. The reaction of an acid chloride with an alcohol results in the release of 1 mol of HCl for each mole of ester formed (eq. 17.15). Pyridine solvent is used to neutralize the HCl. Equations 17.15 and 17.16 are representative examples.

$$C_6H_5C(=O)Cl + CH_3OH \xrightarrow{\text{pyridine}} C_6H_5C(=O)OCH_3 + HCl \quad (17.15)$$

Benzoyl chloride, Methyl alcohol, Methyl benzoate

$$CH_2(CCl(=O))_2 + 2(CH_3)_3COH \xrightarrow{\text{pyridine}} CH_2(COC(CH_3)_3(=O))_2 \quad (17.16)$$

Malonyl chloride, *t*-Butyl alcohol, Di-*t*-butyl malonate (83%)

When an anhydride is used to produce an ester, only one of its acyl groups is esterified; the other is converted to a carboxyl group (eq. 17.17).

$$\text{(17.17)}\quad (CH_3CH_2\overset{O}{\overset{\|}{C}})_2O + CH_3OH \rightarrow CH_3CH_2\overset{O}{\overset{\|}{C}}OCH_3 + CH_3CH_2COOH$$

Propanoic anhydride — Methyl propanoate — Propanoic acid

Amides can be prepared analogously to esters by allowing amines to react with acid chlorides and anhydrides (eqs. 17.18 to 17.20). In preparing amides from acid chlorides, the HCl produced can be neutralized by adding a base (NaOH in eq. 17.18) or by using an extra equivalent of amine (eq. 17.19).

$$\text{(17.18)}\quad C_6H_5\overset{O}{\overset{\|}{C}}{-}Cl + Et_2NH \xrightarrow[H_2O,\ 25^\circ]{NaOH} C_6H_5\overset{O}{\overset{\|}{C}}{-}NEt_2$$

Benzoyl chloride — Diethyl amine — *N,N*-diethylbenzamide (3° amide)

$$\text{(17.19)}\quad CH_3\overset{O}{\overset{\|}{C}}Cl \quad 2NH_3 \xrightarrow[H_2O,\ 25^\circ]{} CH_3\overset{O}{\overset{\|}{C}}NH_2 + NH_4^+Cl^-$$

Acetyl chloride — Ammonia — Acetamide (1° amide)

The reaction of an anhydride and an amine yields an amide and one equivalent of carboxylic acid, which is neutralized by excess amine (eq. 17.20).

$$\text{(17.20)}\quad 2\,C_6H_{11}{-}NH_2 + (CH_3CH_2CH_2\overset{O}{\overset{\|}{C}})_2O \xrightarrow[H_2O,\ 0^\circ]{}$$

Cyclohexylamine — Butanoic anhydride

$$C_6H_{11}{-}NH\overset{O}{\overset{\|}{C}}CH_2CH_2CH_3 + CH_3CH_2CH_2CO_2^-\ H_3\overset{+}{N}{-}C_6H_{11}$$

N-Cyclohexylbutanamide (2° amide)

The products of Equations 17.18, 17.19, and 17.20 are representative of the three different classes of amides (sec. 17:1). Acetamide, which has only the acyl group attached to the nitrogen atom, is a *primary amide* (also notice that 1° amides have *two* H atoms bonded to N). *N*-Cyclohexylbutanamide is a *secondary amide* because it has the acyl group and one alkyl (or aryl) group attached to nitrogen (it has *one* proton bound to N). *N,N*-Diethylbenzamide is a *tertiary amide* (it has *no* nitrogen-bound protons). An amine must possess at least one proton bound to nitrogen in order for it to be convertible to an amide. For example, 3° amines such as trimethylamine, $(CH_3)_3N$, do not react with acylating agents to give amide products under normal conditions.

Example

Synthesize isopropyl acetate.

$$CH_3\overset{\overset{\displaystyle O}{\|}}{C}OCH(CH_3)_2$$

Answer

The acyl portion must come from an acetic acid derivative, while 2-

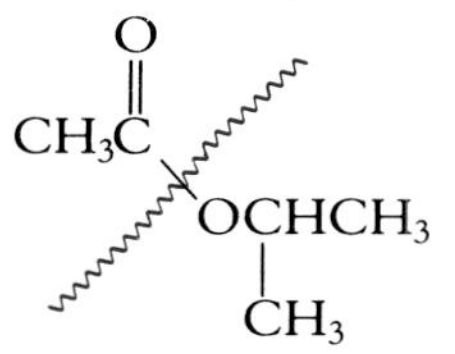

$$CH_3\overset{\overset{\displaystyle O}{\|}}{C}\text{—}\!\!\!\!\!\wr\;\; \underset{\underset{\displaystyle CH_3}{|}}{OCH}CH_3$$

propanol furnishes the alcohol portion. Therefore, one useful synthesis would involve the acid chloride and 2-propanol:

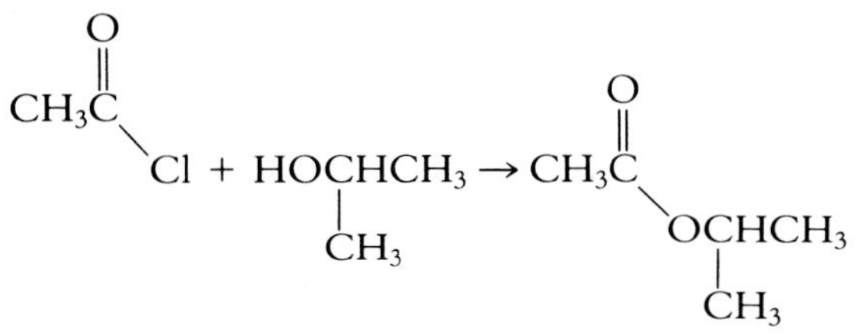

$$\underset{\text{Acetyl chloride}}{CH_3\overset{\overset{\displaystyle O}{\|}}{C}Cl} + \underset{\text{2-Propanol}}{HO\underset{\underset{\displaystyle CH_3}{|}}{C}HCH_3} \rightarrow \underset{\text{Isopropyl acetate}}{CH_3\overset{\overset{\displaystyle O}{\|}}{C}O\underset{\underset{\displaystyle CH_3}{|}}{C}HCH_3}$$

Problem

17–7 Suggest appropriate starting materials for the synthesis of each of the following compounds:
a. phenyl pentanoate
b. *N*-isopropylpentanamide
c. cyclopentyl 4-nitrobenzoate
d. *N,N*-dimethylheptanamide
e. 2-piperidone (**III**, *valerolactam*)

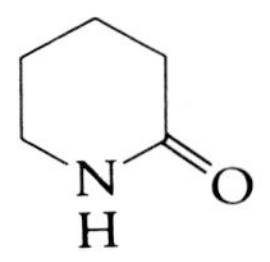

N O
H

III

17:5 The Tetrahedral Intermediate in Acid Halide and Anhydride Reactions

Equation 17.21 gives the mechanism for the formation of methyl acetate from methanol and acetyl chloride in an inert solvent. Note particularly the two intermediates, **2** and **3**, in which the carbonyl carbon atom is sp^3 hybridized.

$$\text{(17.21)}\quad CH_3C(=O)Cl + CH_3\ddot{O}H \rightleftarrows CH_3{-}\underset{\underset{H}{|}}{\overset{O^-}{\underset{+:\ddot{O}CH_3}{\overset{|}{C}}}}{-}Cl \rightleftarrows CH_3{-}\overset{O^-}{\underset{:\ddot{O}CH_3}{\overset{|}{\underset{|}{C}}}}{-}Cl + H^+ \rightarrow$$

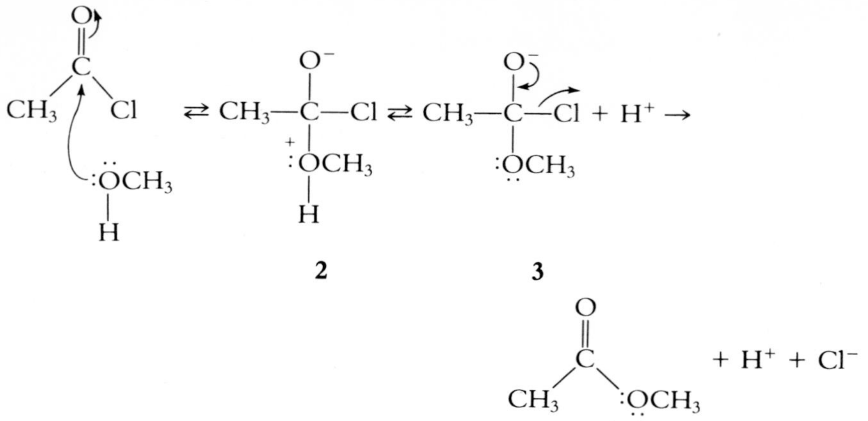

2 3

$$CH_3C(=O){:}\ddot{O}CH_3 + H^+ + Cl^-$$

Methyl acetate

Structures **2** and **3** are known as **tetrahedral intermediates**, and they are found in the reaction pathways of all carboxylic acid derivatives. The formation or breakdown of a tetrahedral intermediate is critical in virtually every reaction of these derivatives. Figure 17–4 shows the formation of intermediate **2**. The nucleophile approaches perpendicular to the planar carbonyl system; this is the most favorable approach, both electronically and sterically.

Because chloride ion is a much better leaving group than CH_3O^-, the breakdown of **3** leads irreversibly to the ester. The breakdown of the analogous tetrahedral intermediate (**4**; eq. 17.22) formed in the reaction between

$$\text{(17.22)}\quad CH_3OH + CH_3\overset{O}{\overset{\|}{C}}O\overset{O}{\overset{\|}{C}}CH_3 \rightleftarrows CH_3{-}\overset{^-O}{\underset{OCH_3}{\overset{|}{\underset{|}{C}}}}{-}O\overset{O}{\overset{\|}{C}}CH_3 + H^+ \rightarrow$$

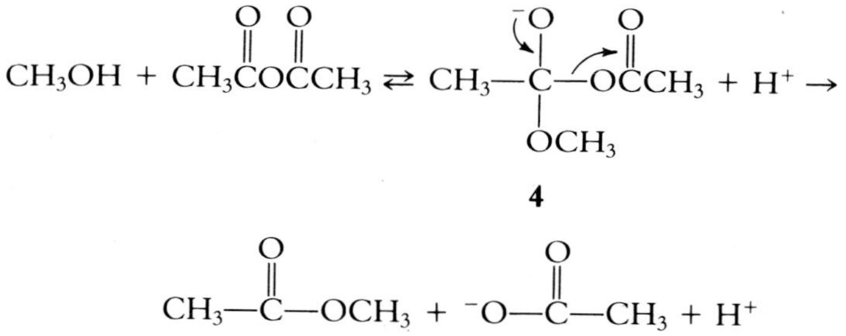

4

$$CH_3{-}\overset{O}{\overset{\|}{C}}{-}OCH_3 + {}^-O{-}\overset{O}{\overset{\|}{C}}{-}CH_3 + H^+$$

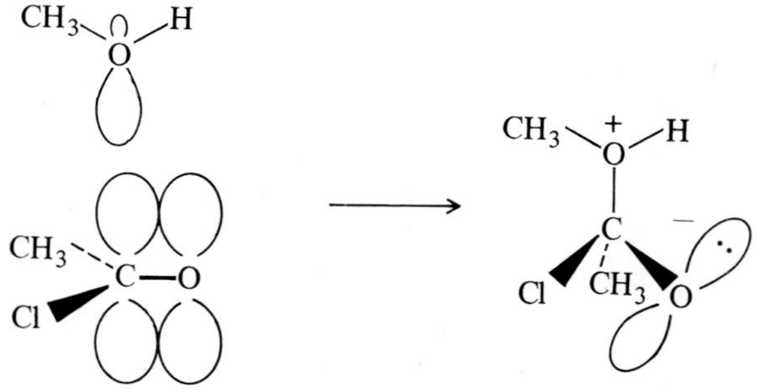

Figure 17–4. *Attack of methanol on the π bond of the carbonyl group of acetyl chloride.*

Box 17–1. Aspirin-like Pain Relievers

Aspirin, acetylsalicylic acid, was first introduced as an antifever (antipyretic) medicine in 1875 and soon thereafter as a pain reliever (analgesic). Aspirin is also recognized as an effective anti-inflammatory drug. The reason for aspirin's effectiveness was not discovered until the 1970s when it was found to inhibit prostaglandin synthesis. Aspirin acetylates and thereby deactivates the enzyme cyclooxygenase, which is involved in the biosynthesis of prostaglandins (sec. 16:1). While essential to our well-being, prostaglandins are involved in various painful physiological processes. For example, they are known to induce uterine muscle contractions and have also been implicated in the inflammatory disease, rheumatoid arthritis.

COOH, $OCCH_3$ (C=O)

Aspirin

Other compounds that react like aspirin and that are also available in over-the-counter pain relievers are acetaminophen (the ingredient in Tylenol—see Prob. 17–9) and acetophenetidin, or phenacetin. Because the mechanism

$HNCCH_3$ (C=O), OH

Acetaminophen
(*Tylenol*)

$NHCCH_3$ (C=O), OEt

Acetophenetidin
(phenacetin)

of action of compounds like aspirin is now known, a search is underway for other compounds that can block the enzyme cyclooxygenase.

methanol and acetic anhydride is also irreversible. These results mean that acid halides and anhydrides react completely with alcohols and amines, without requiring an excess of either the acylating agent or the nucleophile. For this reason, we make esters from acid halides or anhydrides when either the acylating agent or the alcohol is expensive or scarce. When they are not, we can synthesize esters from the less expensive, less reactive carboxylic acids, as the next section shows.

Problems

17–8 a. Write a mechanism for the reaction of methylamine (CH_3NH_2) with acetic anhydride to give *N*-methylacetamide. (Pay attention to the proton loss from the tetrahedral intermediate.)

b. Show how trimethylamine, $(CH_3)_3N$, would react with acetyl chloride, and why that reaction does not lead to an amide product. What happens to the initially formed tetrahedral intermediate?

17–9 When *p*-aminophenol is treated with acetic anhydride, the only product obtained is that resulting from nitrogen acylation, *N*-(4-hydroxyphenyl) acetamide, or *acetaminophen,* the aspirin-like pain reliever (**IV**, eq. 2). No ester product from oxygen acylation is found. Explain why the reaction is so selective. (It may help to write the mechanisms for each possible acylation.)

$$p\text{-}HOC_6H_4NH_2 + (CH_3\overset{O}{\overset{\|}{C}})_2O \xrightarrow{HCO_3^-} p\text{-}HOC_6H_4NH\overset{O}{\overset{\|}{C}}CH_3\ (\textbf{IV}) + CH_3COO^- \quad (2)$$

17:6 Esterification of Acids under Equilibrium Conditions

One of the first organic reaction equilibria to be studied quantitatively was that between ethanol and acetic acid to give ethyl acetate and water (eq. 17.23). This system reaches equilibrium only very slowly at room tempera-

$$(17.23)\quad CH_3\overset{O}{\overset{\|}{C}}OH + CH_3CH_2OH \underset{}{\overset{155°}{\rightleftharpoons}} CH_3\overset{O}{\overset{\|}{C}}OCH_2CH_3 + H_2O \qquad K = 4$$

ture, and therefore must be heated to fairly high temperatures in order to be studied. The equilibrium constant, which is not very large, indicates that there is not a great deal more stability in the products than in the reactants. This is reasonable, considering that carboxylic acids and esters are structurally very similar, particularly with regard to resonance interactions at the carbonyl groups.

$$\left.\begin{array}{l} R\overset{O}{\overset{\|}{C}}-\ddot{O}H \leftrightarrow R\overset{O^-}{\overset{|}{C}}=\overset{+}{O}H \\ R\overset{O}{\overset{\|}{C}}-\ddot{O}R' \leftrightarrow R\overset{O^-}{\overset{|}{C}}=\overset{+}{O}R \end{array}\right\} \text{Comparable resonance interactions}$$

There are several ways in which the esterification reaction can be driven to completion. The removal of water from the reaction mixture, either chemically or by means of azeotropic distillation with benzene or toluene, will drive the reaction to the right. Another common technique, as we mentioned before, is to use a large excess of the alcohol if it is an inexpensive one, as are methyl, ethyl, propyl, and butyl alcohol.

Notice that the reaction shown in Equation 17.23 is carried out at 155°. This is inconvenient because sealed vessels are needed to attain this temperature. The German chemist Emil Fischer discovered that addition of a small quantity of a strong inorganic acid, such as gaseous HCl or concentrated H_2SO_4, catalyzes the reaction so that it comes to equilibrium rapidly at the boiling point of the alcohol. Equations 17.24 and 17.25 show examples of what is now known as *Fischer esterification*.

$$(17.24)\quad \mathrm{NCCH_2\overset{\overset{\displaystyle O}{\|}}{C}OH} + \mathrm{CH_3CH_2OH} \xrightarrow[\text{3 hr, 80°}]{\mathrm{H_2SO_4(cat)}} \mathrm{NCCH_2\overset{\overset{\displaystyle O}{\|}}{C}OCH_2CH_3} + \mathrm{H_2O}$$

Cyanoacetic acid; (2.6-fold excess); Ethyl cyanoacetate (70%)

$$(17.25)\quad \mathrm{HO\overset{\overset{\displaystyle O}{\|}}{C}(CH_2)_4\overset{\overset{\displaystyle O}{\|}}{C}OH} + 2\mathrm{EtOH} \xrightarrow[\text{remove water with toluene}]{\mathrm{H_2SO_4},\ 75°} \mathrm{EtO\overset{\overset{\displaystyle O}{\|}}{C}(CH_2)_4\overset{\overset{\displaystyle O}{\|}}{C}OEt}$$

Hexanedioic acid (*adipic acid*); Diethyl hexanedioate (*diethyl adipate*, 95%)

Fischer esterification is normally used only with primary and secondary alcohols. The esterification equilibrium is sensitive to steric hindrance (eq. 17.26), allowing elimination reactions to compete with esterification. Hence, the reaction of a 3° alcohol with an acid is not practical. Tertiary alkyl esters are usually synthesized by way of acid chlorides (eq. 17.15).

$$(17.26)\quad \mathrm{CH_3COOH} + \mathrm{(CH_3)_3COH} \underset{}{\overset{155°}{\rightleftharpoons}} \mathrm{CH_3COOC(CH_3)_3} + \mathrm{H_2O} \qquad K = 4.9 \times 10^{-3}$$

When ^{18}O-labeled alcohols are esterified under equilibrium conditions, all of the label is found in the ester and none in the H_2O formed (eq. 17.27).

$$(17.27)\quad \mathrm{R\overset{\overset{\displaystyle O}{\|}}{C}OH} + \mathrm{H^{18}OR} \rightleftharpoons \mathrm{R\overset{\overset{\displaystyle O}{\|}}{C}{-}^{18}OR} + \mathrm{H_2O}$$

This result shows which bonds are broken and formed, and hence, allows the formulation of a mechanism for acid-catalyzed esterification of primary and secondary alcohols (eqs. 17.28 and 17.29). Initial protonation of the carbonyl oxygen promotes addition of the weakly nucleophilic alcohol to the carbonyl group.

$$(17.28)\quad \mathrm{RC(=\ddot{O}{:})\ddot{O}H} \underset{}{\overset{\mathrm{H^+}}{\rightleftharpoons}} \mathrm{RC(={}^{+}\ddot{O}H)\ddot{O}H} \xrightarrow[\text{slow}]{\mathrm{R\ddot{O}H}} \underset{\mathbf{5}}{\mathrm{RC(:\ddot{O}H)(:\ddot{O}H)}{-}\overset{+}{\ddot{O}}\mathrm{R'}(\mathrm{H})} \rightleftharpoons \underset{\mathbf{6}}{\mathrm{RC(:\ddot{O}H)(:\ddot{O}H)}{-}\ddot{O}\mathrm{R'}} + \mathrm{H^+}$$

$$(17.29)\quad \underset{\mathbf{6}}{\mathrm{RC(:\ddot{O}H)(:\ddot{O}H)}{-}\ddot{O}\mathrm{R'}} \underset{}{\overset{\mathrm{H^+}}{\rightleftharpoons}} \underset{\mathbf{7}}{\mathrm{RC(:\ddot{O}H)({}^{+}\ddot{O}H_2)}{-}\ddot{O}\mathrm{R'}} \rightleftharpoons \mathrm{RC(={}^{+}\ddot{O}H)\ddot{O}R'} + \mathrm{H_2\ddot{O}{:}} \rightleftharpoons \mathrm{RC(=\ddot{O}{:})\ddot{O}R'} + \mathrm{H_3O{:}^+}$$

This attack of the alcohol molecule on the protonated acid to yield the first tetrahedral intermediate, **5**, is the slow step of the reaction. The proton

Table 17–3. Common Methods for Synthesizing Esters

1. Reactions of alcohols with acid chlorides (sec. 17:4)

$$\text{RCOCl} + \text{R'OH} \xrightarrow{\text{base}} \text{RCOOR'}$$

2. Reactions of alcohols with acid anhydrides (sec. 17:4)

$$(\text{RCO})_2\text{O} + \text{R'OH} \rightarrow \text{RCOOR'} + \text{RCOOH}$$

3. Fischer esterification (for 1° and 2° alcohols only; sec. 17:6)

$$\text{RCOOH} + \text{R'OH} \xrightarrow[\text{H}_2\text{SO}_4]{\text{HCl or}} \text{RCOOR'} + \text{H}_2\text{O}$$

transfer steps to yield intermediates **6** and **7** are all very fast. Table 17–3 summarizes the common methods for synthesizing esters.

Equation 17.23 shows that an esterification equilibrium does not lie heavily on one side or the other. That being so, it should be possible to *hydrolyze* esters by simply heating them with an excess of water containing a small quantity of a strong acid catalyst (eq. 17.30). The mechanism is the reverse of that depicted in Equations 17.28 and 17.29.

$$\text{ClCH}_2\text{CH}_2\overset{\overset{\displaystyle\text{O}}{\|}}{\text{C}}\text{OCH}_2\text{CH(CH}_3)_2 + \text{H}_2\text{O} \xrightarrow{\text{H}^+} \qquad (17.30)$$

Isobutyl 3-chloropropanoate

$$\text{ClCH}_2\text{CH}_2\text{COOH} + (\text{CH}_3)_2\text{CHCH}_2\text{OH}$$

3-Chloropropanoic acid; Isobutyl alcohol

Problems

17–10 a. Write an equation showing the products of hydrolysis of ethyl propanoate with $H_2{}^{18}O$ in the presence of an acid catalyst.

b. Propose a mechanism for this reaction.

17–11 Studies of the acid-catalyzed hydrolysis of various alkyl acetates (CH_3COOR) have shown that the rate is fastest when the alkyl group (R) is small: $CH_3 > CH_3CH_2 > (CH_3)_2CH$. *t*-Butyl acetate (R = CMe_3), however, reacts more rapidly than any of these esters. Furthermore, when this ester is hydrolyzed in $H_2{}^{18}O$ the products are CH_3COOH and $(CH_3)_3C^{18}OH$. Propose a mechanism for hydrolysis of the *t*-butyl ester consistent with these facts. What property of the *t*-butyl group makes this mechanism possible?

One direct esterification reaction that takes place particularly rapidly is *intramolecular* esterification of a hydroxy acid. This reaction is possible because a hydroxy acid has both an alcohol (OH) portion and a carboxylic acid (COOH) portion. The products are known as **lactones**. Lactone formation is very favorable *if* the ring formed is five- or six-membered (eqs. 17.31 and

17.32). A carboxylic acid with a hydroxyl in the 4-position will form a γ-lactone (five-membered ring), while an acid with a hydroxyl in the 5-position will form a δ-lactone (six-membered ring).

(17.31) $HO\overset{\gamma}{C}H_2\overset{\beta}{C}H_2\overset{\alpha}{C}H_2\overset{O}{\overset{\|}{C}}OH \xrightarrow[\text{no catalyst added}]{\text{heat}}$ [γ-lactone ring with α, β, γ labels] $+ H_2O$

4-Hydroxybutanoic acid

4-Butanolide (*γ-butyrolactone*)

(17.32) HO–CH₂(δ)–CH₂(γ)–C(β)(HO)(CH₃)–CH₂(α)–C(=O)OH $\xrightarrow{\text{heat}}$ [δ-lactone ring with HO, CH₃ at β; α, β, γ, δ labels] $+ H_2O$

(*R*)-3,5-Dihydroxy-3-methylpentanoic acid (*mevalonic acid*)

(*R*)-3-Hydroxy-3-methylpentanolide (*mevalonolactone*, a δ-lactone)

If the hydroxy chain is too long to form a five- or six-membered lactone, then it will react *intermolecularly* to give polymers called **polyesters** (eq. 17.33). Polyesters are also formed by esterification of diacids with diols. These commercially important compounds are discussed in Chapter 22.

(17.33) $2n HO(CH_2)_5\overset{O}{\overset{\|}{C}}OH \xrightarrow[\text{heat}]{H^+} H{-}[O(CH_2)_5\overset{O}{\overset{\|}{C}}O(CH_2)_5\overset{O}{\overset{\|}{C}}]_n{-}OH$

(n = any integer)

A polyester

Problem

17–12 In Equation 17.32, the lactone was formed from reaction of the carboxyl group with the 5-hydroxyl group of mevalonic acid. Draw the structure of the lactone that would be formed by reaction with the 3-hydroxyl group, and explain why it is not formed.

17:7 Amide Formation and Hydrolysis

The direct reaction between a carboxylic acid and an amine in solution does not yield an amide. There is instead a very rapid proton transfer resulting in formation of a salt (eq. 17.34). However, if the amine and acid are mixed

(17.34) $RCOOH + R'_2NH \xrightarrow[\text{very fast}]{H_2O,\ 25^\circ} RCOO^- R'_2NH_2^+$

neat (no solvent) and the salt then heated to 200°, the amide can be obtained in a good yield (eq. 17.35). Since these conditions are rather vigorous, amides are usually made from acid halides or anhydrides (sec. 17:4).

(17.35) $C_6H_5COOH + C_6H_5NH_2 \xrightarrow{\text{neat}} [C_6H_5NH_3^+ \; ^-OOCC_6H_5] \xrightarrow[-H_2O]{190° \text{ to } 225°} C_6H_5NHCOC_6H_5$

N-Phenylbenzamide (80%)

The thermal method of amide synthesis is useful when both the amino and carboxyl groups are part of the same molecule, as in *amino acids*. Gamma and delta amino acids are the right length for rapid intramolecular reaction to produce stable five- or six-membered **lactams**, which is the name given to cyclic amides (eq. 17.36). We will learn more about lactams in Section 17:8.

(17.36) $CH_3\overset{\gamma}{C}H(NHCH_3)CH_2CH_2COOH \xrightarrow{140°} \rightarrow$ (cyclic γ-lactam: NCH₃ ring, C=O, γ-CH₃)

N-Methyl-4-pentanelactam
(a γ-lactam)

Table 17–4 summarizes the method for synthesis of amides. (Method 5 is discussed in Section 17:12.)

The hydrolysis of amides in aqueous solution is not favored thermodynamically (eq. 17.37). One way to overcome this barrier is by adding a

Table 17–4. Common Methods for Synthesizing Amides

1. Reactions of amines with acid chlorides (sec. 17:4)

$$RCOCl + 2R_2'NH \rightarrow RCONR_2' + R_2'NH_2^+Cl^-$$

2. Reactions of amines with acid anhydrides (sec. 17:4)

$$(RCO)_2O + 2R_2'NH \rightarrow RCONR_2' + R_2'NH_2^+RCOO^-$$

3. Reactions of amines with esters (sec. 17:3)

$$RCOOR'' + R_2'NH \rightarrow RCONR_2' + R''OH$$

4. Thermal reaction of acids and amines (sec. 17:7)

$$RCOOH + R_2'NH \xrightarrow{\text{neat}} R_2'NH_2^+RCOO^- \xrightarrow{\text{heat}} RCONR_2' + H_2O$$

5. Base-catalyzed hydrolysis of nitriles (sec. 17:12)

$$RC{\equiv}N + H_2O \xrightarrow[\text{heat}]{OH^-} R{-}C(=O)NH_2$$

$$\text{(17.37)}\quad CH_3\overset{\overset{\large O}{\|}}{C}NH_2 + H_2O \xrightarrow[H_2O]{} CH_3\overset{\overset{\large O}{\|}}{C}OH + NH_3 \qquad \Delta G = 6.4\ \text{kcal/mol}$$

strong acid to the solution, thereby removing the amine as an *ammonium ion* and driving the reaction to the right (eq. 17.38).

$$\text{(17.38)}\quad C_6H_5\overset{\overset{\large O}{\|}}{C}NH_2 + H_3O^+ \xrightarrow[\text{reflux}]{} C_6H_5COOH + \underset{\text{Ammonium ion}}{NH_4^+}$$

The acid also increases the *rate* of hydrolysis. It protonates the amide carbonyl group and promotes attack by water (eq. 17.39). The mechanism is very similar to that for acid-catalyzed ester hydrolysis.

$$\text{(17.39)}\quad R\overset{\overset{\large \ddot{O}:}{\|}}{C}{-}\ddot{N}H_2 \xrightarrow{H^+} RC(=\overset{+}{\ddot{O}}H){-}\ddot{N}H_2 \xrightarrow[\text{slow}]{:\ddot{O}H_2} RC(:\ddot{O}H)(^+\ddot{O}H_2){-}\ddot{N}H_2 \rightleftarrows RC(:\ddot{O}H)(:\ddot{O}H){-}\overset{+}{N}H_3 \rightleftarrows RC(={}^+\ddot{O}H)(\ddot{O}H) + \ddot{N}H_3 \rightarrow R\overset{\overset{\large \ddot{O}:}{\|}}{C}\ddot{O}H + NH_4^+$$

17:8 β-Lactam Antibiotics

There are two important classes of antibiotics containing a reactive four-membered lactam ring (β-lactam). These are the **cephalosporins** (**8**) and the **penicillins** (**9**). Both types of compounds are powerful bactericides, and were originally isolated from fungi. They are now available as semisynthetic derivatives that differ in structure by having differing R groups.

R—N(H)— [bicyclic β-lactam/dihydrothiazine ring with S, N, C=O, CH_2OOCCH_3, COOH] — β-lactam portion

8
A cephalosporin

These antibiotics work by interfering with the growing bacterium's synthesis of cell walls. The imperfectly constructed cell walls permit leakage of the cell's cytoplasm, resulting in the death of the organism. Penicillin acts by acylating one of the enzymes involved in cell wall biosynthesis (eq. 17.40). Amides are normally very poor acylating agents, but the bond angle strain in

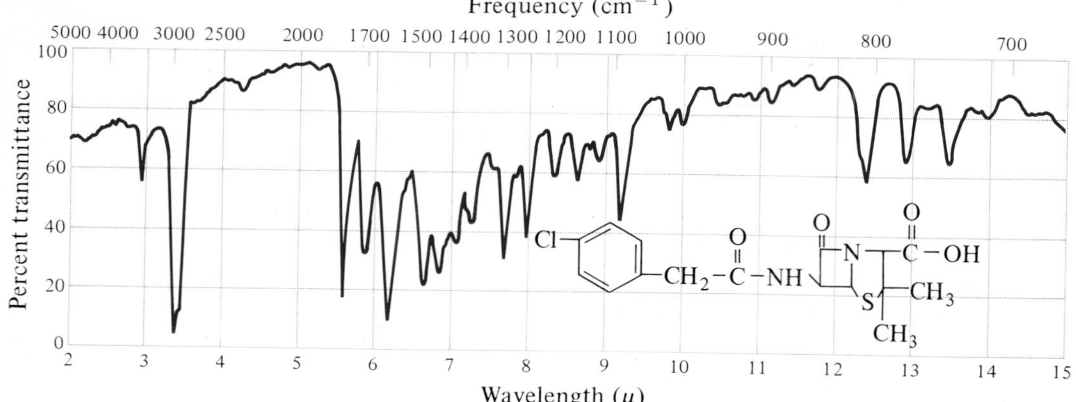

Figure 17–5. *Infrared spectrum of* p-*chloro-benzylpenicillin.*

the four-membered β-lactam ring makes it unusually reactive toward nucleophiles. Thus, the angle strain in penicillin is important for its biological potency.

(17.40)

O ‖ RCNH — S — CH_3 — CH_3 — COOH — N — O — H — Enz—N̈H_2 ⟶ ⟶ O ‖ RCNH — S — CH_3 — NH — CH_3 — COOH — O — HN H — Enz

9
A penicillin

Figure 17–5 shows the infrared spectrum of *p*-chlorobenzylpenicillin (**9**, R = $ClC_6H_4CH_2$). The intense band at 1780 cm^{-1} (5.61 μ) is due to the carbonyl group in the β-lactam portion of the molecule. This is an abnormally high frequency for an amide, and it is evidence of strain in the β-lactam ring.

Problem

17–13 There are two additional carbonyl stretching peaks in the ir spectrum of *p*-chlorobenzylpenicillin (Fig. 17–5). One is at 1700 cm^{-1} and the other at 1620 cm^{-1}. What functional groups give rise to these peaks?

17:9 Base-Promoted Hydrolysis of Esters and Amides

Equation 17.30 shows that esters can be hydrolyzed in the presence of a strong acid catalyst and an excess of water. An equally good method is to heat an ester in an aqueous solution of a strong base such as sodium or potassium hydroxide (eq. 17.41). This reaction is thermodynamically favor-

$$\text{(17.41)}\quad C_6H_5\overset{\overset{\displaystyle O}{\|}}{C}OCH_2CH_3 + OH^- \xrightarrow{H_2O,\ heat} C_6H_5C(O^{1/2-})(O^{1/2-}) + CH_3CH_2OH$$

Ester — Hydroxide — Carboxylate anion — Alcohol

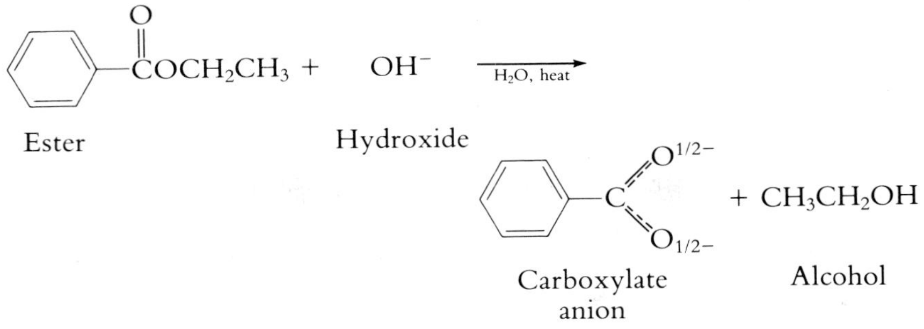

able because of the formation of the highly stable carboxylate ion. The process is often referred to as **saponification**, because the salts of long-chain carboxylic acids (C_{12} to C_{18}) are soaps (see Prob. 17–14). Note that saponification is not base-*catalyzed,* because the hydroxide ion is consumed.

In the mechanism for ester hydrolysis (saponification), the hydroxide ion attacks the unprotonated ester (eq. 17.42). Since the hydroxide ion is a much

$$\text{(17.42)}\quad R\overset{\overset{\displaystyle \ddot{O}:}{\|}}{C}-\ddot{O}R' \;(+\; :\ddot{O}H^-) \rightleftarrows \underset{\mathbf{10}}{R\underset{\underset{\displaystyle :\ddot{O}H}{|}}{\overset{\overset{\displaystyle :\ddot{O}:^-}{|}}{C}}-\ddot{O}R'} \rightleftarrows RC(=\ddot{O}:)\ddot{O}H + R'\ddot{O}:^- \rightarrow R\overset{\overset{\displaystyle O}{\|}}{C}O^- + R'OH$$

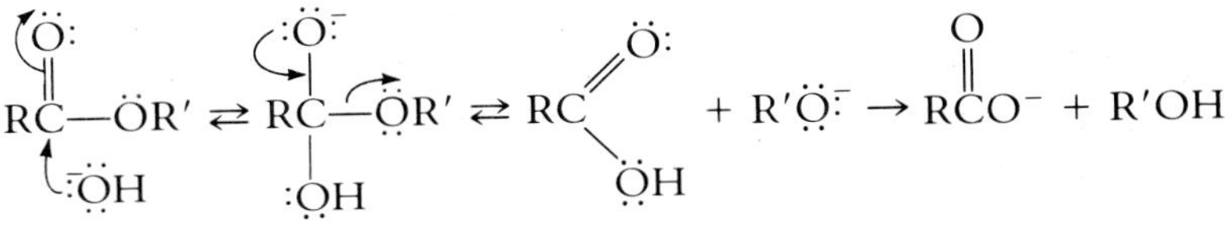

stronger nucleophile than water, the ester does not require activation by protonation. Depending on the structure of R and R′, either the formation *or* the breakdown of the tetrahedral intermediate **10** is rate-determining.

Problem

17–14 Fats and oils are triesters of *glycerin*

$$HOCH_2\underset{\underset{\displaystyle OH}{|}}{C}HCH_2OH$$

and are sometimes called glycerides (see also eq. 10.9, sec. 10:2). When these esters are heated with alkali metal hydroxides, soaps are formed. Write the structures of the products formed when the glyceride **V** is allowed to react with three equivalents of sodium hydroxide.

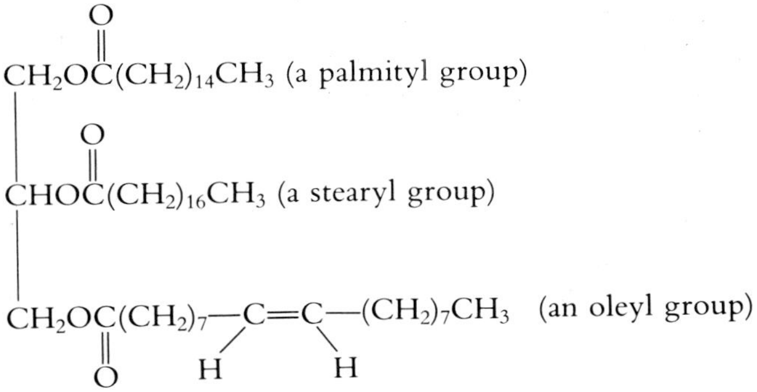

$$CH_2O\overset{\overset{\displaystyle O}{\|}}{C}(CH_2)_{14}CH_3 \text{ (a palmityl group)}$$
$$|$$
$$CHO\overset{\overset{\displaystyle O}{\|}}{C}(CH_2)_{16}CH_3 \text{ (a stearyl group)}$$
$$|$$
$$CH_2O\underset{\underset{\displaystyle O}{\|}}{C}(CH_2)_7-\underset{\underset{\displaystyle H}{/}}{C}=\underset{\underset{\displaystyle H}{\backslash}}{C}-(CH_2)_7CH_3 \text{ (an oleyl group)}$$

V

Amides also undergo hydrolysis in aqueous base (eq. 17.43) but the reaction is quite slow, so amides are usually hydrolyzed under acidic conditions (eq. 17.38).

$$(17.43)\quad C_6H_5\overset{O}{\overset{\|}{C}}NHMe + OH^- \xrightarrow[H_2O]{slow} C_6H_5COO^- + MeNH_2$$

We'll close this section with a brief description of a reaction not often used for preparative purposes, but which can be an annoying side reaction. This is *transesterification,* in which the ester of one alcohol is converted to the ester of another alcohol. For instance, when methyl benzoate is dissolved in ethanol it is partly converted to ethyl benzoate (eq. 17.44). Transesterification is catalyzed by either acids or alkoxide ions. The conclusion to be drawn from this is that when an ester is being reacted in an alcohol solvent, the solvent should be the same as the alkoxyl portion of the ester.

$$(17.44)\quad C_6H_5\overset{O}{\overset{\|}{C}}OMe + EtOH \underset{}{\overset{H^+ \text{ or } RO^-}{\rightleftharpoons}} C_6H_5\overset{O}{\overset{\|}{C}}OEt + MeOH$$

Problem

17–15 Show the mechanism by which methyl benzoate is converted to ethyl benzoate in the presence of sodium ethoxide (eq. 17.44).

Example

Show the products of the following reaction:

$$CH_3\overset{O}{\overset{\|}{C}}O{-}C_6H_5 + MeO^- \rightarrow ?$$

Answer

When the mechanism of a reaction is known, it is better to follow the mechanism through rather than try to guess the products.

$$CH_3\overset{O}{\overset{\|}{C}}O{-}C_6H_5 + MeO^- \rightarrow CH_3{-}\overset{O^-}{\underset{OMe}{C}}{-}O{-}C_6H_5 \rightarrow$$

$$CH_3\overset{O}{\overset{\|}{C}}OMe + {}^-O{-}C_6H_5$$

Box 17–2. Soaps and Detergents

Salts of long-chain fatty acids are soaps because their hydrophobic (water-hating) alkyl chains dissolve in oil and grease droplets, while their hydrophilic (water-loving) carboxylate ends stick out into the water, allowing the droplets to be dispersed in it.

Since the Ca^{2+} and Mg^{2+} salts of sulfonic acids are much more soluble than the corresponding salts of carboxylic acids, the synthetic detergents are preferable to soaps for use in hard water.

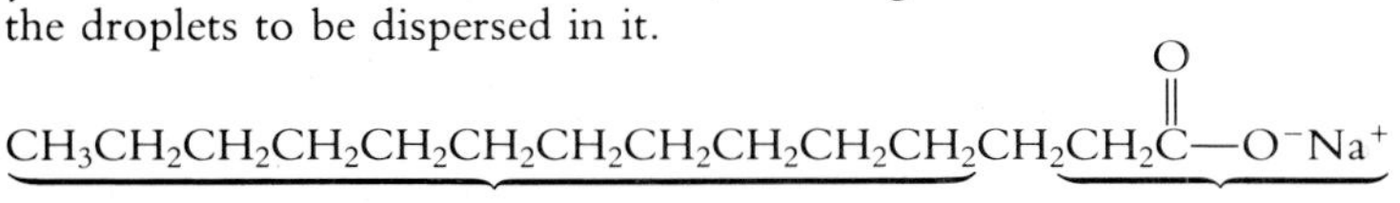

$$\underbrace{CH_3CH_2CH_2CH_2CH_2CH_2CH_2CH_2CH_2CH_2CH_2}_{\text{Hydrophobic portion (grease and fat soluble)}}\underbrace{CH_2CH_2\overset{O}{\overset{\|}{C}}—O^-Na^+}_{\text{Hydrophilic portion (water soluble)}}$$

If the wash water is "hard" (contains inorganic salts of Ca^{2+} and Mg^{2+}) the Na^+ ion is replaced by Ca^{2+} and Mg^{2+}, making the fatty acid salts water insoluble. This causes precipitates to form:

$$2CH_3(CH_2)_{14}COO^- + Ca^{2+} \rightarrow [CH_3(CH_2)_{14}COO]_2Ca$$

Calcium palmitate
(very limited water solubility; forms a scummy precipitate)

To help prevent these precipitates from forming, chemical research has developed synthetic detergents in which the carboxylate group is replaced by a sulfonate group. These detergents are made by Friedel–Crafts alkylation of benzene, followed by sulfonation of the ring:

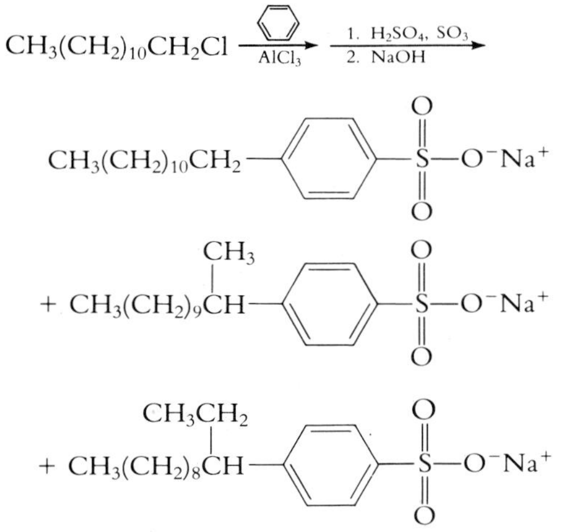

$$CH_3(CH_2)_{10}CH_2Cl \xrightarrow[AlCl_3]{C_6H_6} \xrightarrow[2.\ NaOH]{1.\ H_2SO_4,\ SO_3}$$

$$CH_3(CH_2)_{10}CH_2—C_6H_4—SO_2—O^-Na^+$$

$$+\ CH_3(CH_2)_9CH(CH_3)—C_6H_4—SO_2—O^-Na^+$$

$$+\ CH_3(CH_2)_8CH(CH_2CH_3)—C_6H_4—SO_2—O^-Na^+$$

Detergent use is not without its own problems, however. One of the highly desirable properties of soaps is their rapid biodegradability. Detergents tend to be less readily biodegraded, particularly when their alkyl chains are branched. Some branching is inevitable when the detergents are synthesized by way of Friedel–Crafts alkylations, because of the rearrangements inherent in such reactions (see above equation). The very first detergents had highly branched alkyl chains and were extremely resistant to microbial attack. Their lack of biodegradability resulted in foaming problems in streams and lakes, and they were eventually banned and replaced by the less highly branched alkylbenzenesulfonates shown in the equation above.

Even the newest alkylbenzenesulfonate detergents leave something to be desired with respect to biodegradability. A superior class of compound in this respect is the sulfuric acid monester salts, such as sodium lauryl sulfate. The detergents of this type are more expensive than the alkylbenzenesulfonates, and so are found largely in products oriented toward personal use, such as shampoos.

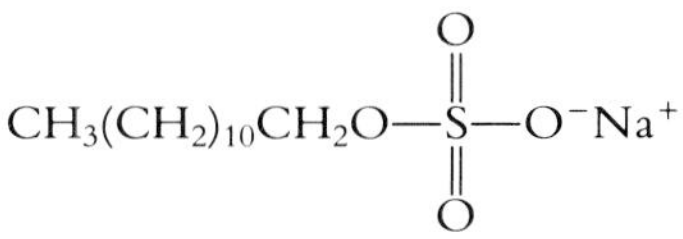

$$CH_3(CH_2)_{10}CH_2O—SO_2—O^-Na^+$$

Sodium lauryl sulfate

Problem

17–16 Show the products of each of the following reactions:

a. $EtOOCCH_2COOEt + OH^- \rightarrow$
(1 eq)

b. 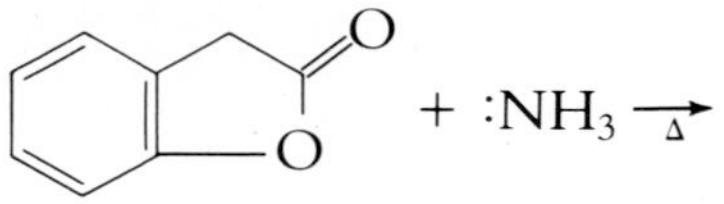 (lactone structure) + $:NH_3 \xrightarrow{\Delta}$

c. 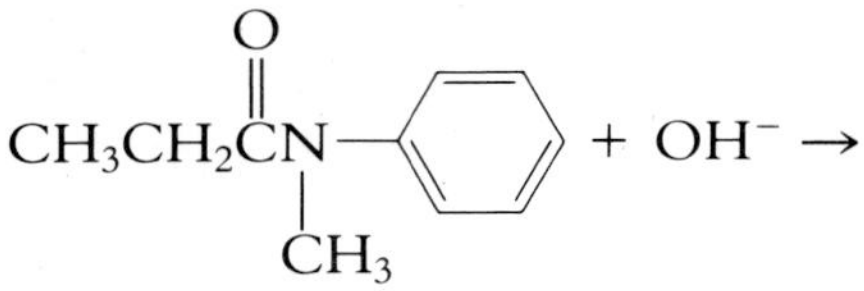 $CH_3CH_2\overset{O}{\overset{\|}{C}}N(CH_3)-C_6H_5 + OH^- \rightarrow$

d. 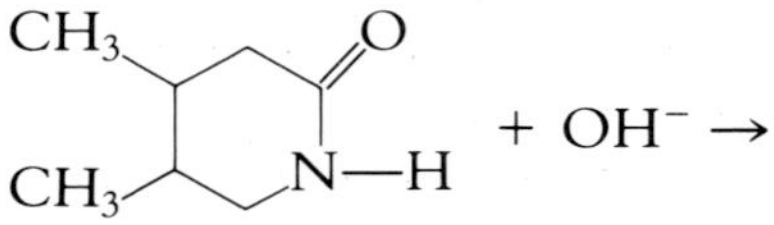 (4,5-dimethyl-2-piperidinone, N—H) + $OH^- \rightarrow$

17:10 Synthesis of Alcohols and Ketones by Reaction of Esters and Acid Chlorides with Organometallic Compounds

Like other nucleophiles, Grignard reagents add to the carbonyl groups of carboxylic acid derivatives. The tetrahedral intermediate thus formed breaks down to an aldehyde or ketone (eq. 17.45, step 2), which is rapidly attacked

(17.45)

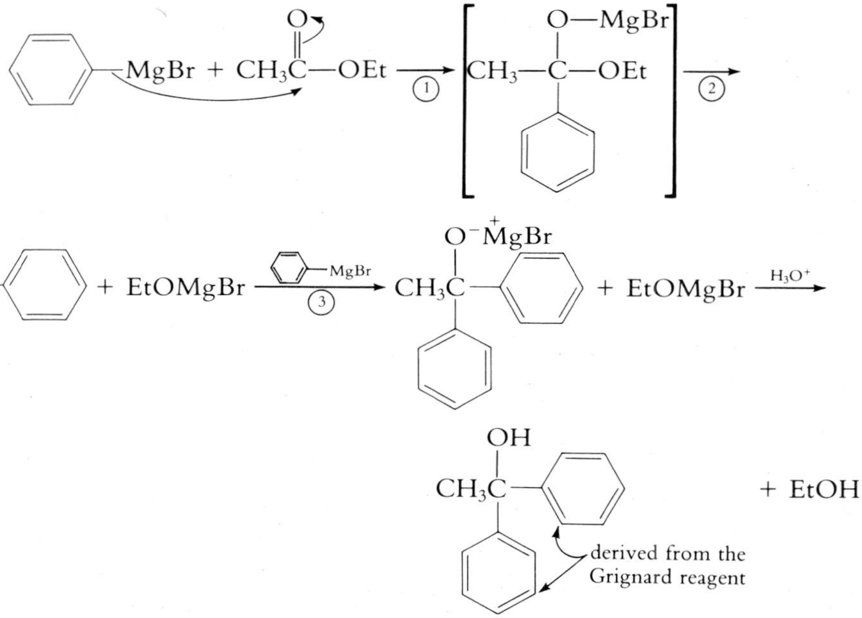

$$C_6H_5MgBr + CH_3\overset{O}{\overset{\|}{C}}-OEt \xrightarrow{(1)} \left[CH_3-\overset{O-MgBr}{\overset{|}{C}}(C_6H_5)-OEt \right] \xrightarrow{(2)}$$

$$CH_3\overset{O}{\overset{\|}{C}}-C_6H_5 + EtOMgBr \xrightarrow[(3)]{C_6H_5MgBr} CH_3\overset{O^-\overset{+}{M}gBr}{\overset{|}{C}}(C_6H_5)-C_6H_5 + EtOMgBr \xrightarrow{H_3O^+}$$

$$CH_3\overset{OH}{\overset{|}{C}}(C_6H_5)-C_6H_5 + EtOH$$

derived from the Grignard reagent

1,1-Diphenylethanol
(80%)

by additional Grignard reagent (step 3). The second Grignard attack takes place because aldehydes and ketones are more reactive than esters toward this reagent. Hydrolysis of the reaction mixture gives a product alcohol in which two of the groups are derived from the Grignard reagent. The alcohol portion of the starting ester is also isolated.

Although all carboxylic acid derivatives react with Grignard reagents, best results are obtained with esters. They are converted to 3° alcohols in good yield through reaction with at least 2 mol of a Grignard reagent. The structure of the alcohol can be controlled by appropriate choices of ester and Grignard reagent (compare eqs. 17.45 and 17.46).

$$\text{(17.46)} \quad C_6H_5\overset{O}{\overset{\|}{C}}OEt \xrightarrow{2CH_3MgI} \xrightarrow{H_3O^+} C_6H_5\underset{CH_3}{\underset{|}{\overset{OH}{\overset{|}{C}}}}-CH_3$$

derived from the Grignard reagent

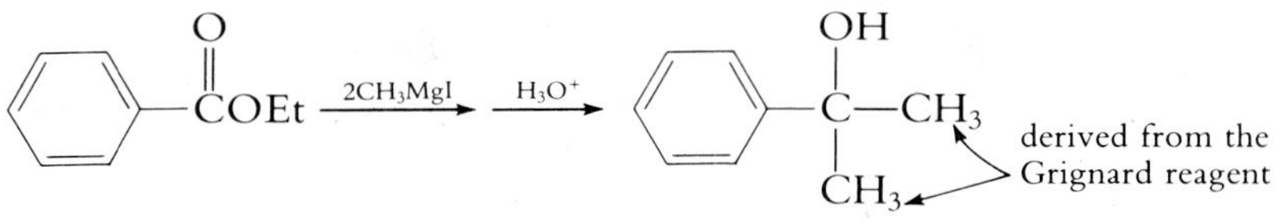

Ethyl benzoate — 2-Phenyl-2-propanol

Example

Give the structures of the ester and Grignard reagent that could be used to prepare 2-methyl-2-pentanol.

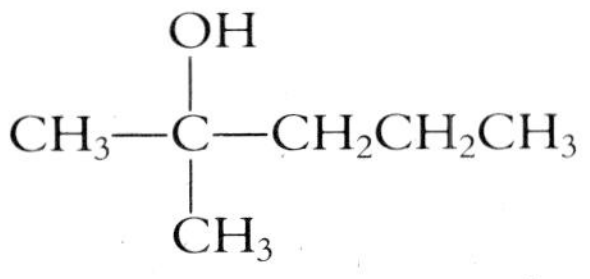

$$CH_3-\underset{CH_3}{\underset{|}{\overset{OH}{\overset{|}{C}}}}-CH_2CH_2CH_3$$

2-Methyl-2-pentanol

Answer

When analyzing this structure in terms of its precursor ester and Grignard reagent, the key is to remember that the oxygen of the alcohol comes from the carbonyl of the ester and that the two equivalent alkyl groups come from the Grignard reagent:

$$CH_3-\underset{CH_3}{\underset{|}{\overset{OH}{\overset{|}{C}}}}-CH_2CH_2CH_3$$

from ester

from Grignard reagent

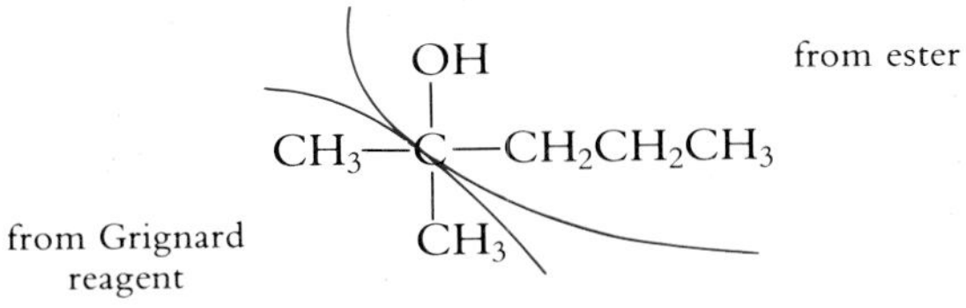

Therefore

$$RO\overset{O}{\overset{\|}{C}}CH_2CH_2CH_3 + 2CH_3MgBr \rightarrow \text{2-methyl-2-pentanol}$$

Problems

17–17 Write the structures of the ester and the Grignard reagent that could be used to prepare each of the following alcohols:

a. 4-ethyl-4-heptanol

b. 3-(p-chlorophenyl)-3-pentanol

c. cyclopentyldiphenylmethanol

d. 1,1-dicyclohexylethanol

e. 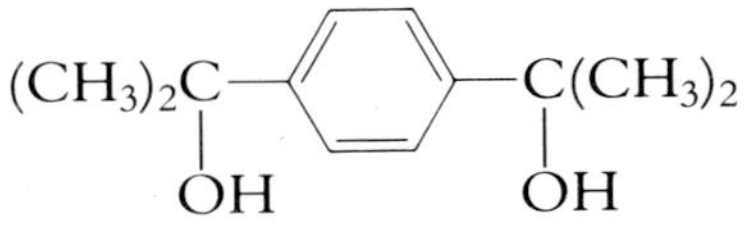 $(CH_3)_2\underset{OH}{\underset{|}{C}}-C_6H_4-\underset{OH}{\underset{|}{C}}(CH_3)_2$

17–18 When either propanoic acid or propanamide is treated with methyl magnesium bromide, there is immediate evolution of gas. After acidification of the reaction mixture, no alcohols are obtained; only the starting material (acid or amide) is recovered. Write equations that explain this chemical behavior.

In order to obtain ketones from carboxylic acid derivatives, it is necessary to use an organometallic reagent that is fairly unreactive toward aldehydes and ketones. The organic derivatives of copper(I) (sec. 6:14) fill the bill perfectly. Under the proper conditions they will react only with acid chlorides, and the reaction will proceed only to the ketone stage (eq. 17.47). (Note that in the reaction in Equation 17.47 only the acid chloride portion of the substrate reacts with the copper reagent; the ester portion does not.)

$$\text{(17.47)}\quad \mathrm{MeO\overset{O}{\overset{\|}{C}}CH_2CH_2\overset{O}{\overset{\|}{C}}Cl} \xrightarrow[\text{ether, }-78^\circ]{[CH_3(CH_2)_3]_2CuLi} \mathrm{MeO\overset{O}{\overset{\|}{C}}CH_2CH_2\overset{O}{\overset{\|}{C}}(CH_2)_3CH_3}$$

Acid chloride of monomethyl succinate — Methyl 4-oxooctanoate (a keto ester)

Problem

17–19 Show how to prepare the following compounds from simple acid chlorides. For (a) and (b), propose two routes to each compound.

a. $\mathrm{CH_3CH_2CH_2\overset{O}{\overset{\|}{C}}CH_2CH_3}$

b. $\mathrm{CH_3(CH_2)_4\overset{O}{\overset{\|}{C}}}-C_6H_5$

c. $\mathrm{CH_3\overset{O}{\overset{\|}{C}}CH_2CH_2CH_2C(=O)OH}$

17:11 Reduction of Carboxylic Acid Derivatives

The carbonyl group of acids, esters, and amides is so stable that catalytic hydrogenation of the compounds requires unusually high temperatures and pressures. In the laboratory, therefore, we employ chemical reduction of carboxylic acids and their derivatives. The reducing agents most commonly used are metal hydrides such as $LiAlH_4$ and B_2H_6. The catalytic reduction of esters to alcohols is still an important *commercial* process, however, because it is very inexpensive compared to chemical reduction.

An ester reacts with lithium aluminum hydride to yield two alcohols. One is from the alcohol portion of the ester group, and the other is an alcohol derived by reduction of the carboxyl portion of the ester (eq. 17.48). Equation 17.49 shows a practical example which also recalls the fact that lithium aluminum hydride does not reduce isolated C=C groups. Experi-

$$\text{(17.48)}\qquad \underset{\text{Ester}}{\mathrm{RC(=O)OR'}} + LiAlH_4 \rightarrow \underset{\text{Alcohol}}{RCH_2OH} + \underset{\text{Alcohol}}{R'OH}$$

$$\text{(17.49)}\quad 2CH_3CH{=}CHCH_2\overset{\overset{\displaystyle O}{\|}}{C}OCH_3 + LiAlH_4 \xrightarrow{\text{ether}} \xrightarrow{H_3O^+} 2CH_3CH{=}CHCH_2CH_2OH + 2CH_3OH$$

Methyl 3-pentenoate → 3-Pentene-1-ol (75%)

ments with an isotopically labeled reducing agent ($LiAlD_4$) show that the hydrogen added to the carbonyl carbon comes from the reducing agent (eq. 17.50). Hydrolysis with D_3O^+ shows that the hydrogen of the hydroxyl groups comes from the water (eq. 17.51).

$$\text{(17.50)}\quad C_6H_5{-}\overset{\overset{\displaystyle O}{\|}}{C}{-}OCH_3 \xrightarrow[\text{ether}]{LiAlD_4} \xrightarrow{H_3O^+} C_6H_5{-}CD_2OH + CH_3OH$$

$$\text{(17.51)}\quad C_6H_5{-}\overset{\overset{\displaystyle O}{\|}}{C}{-}OCH_3 \xrightarrow[\text{ether}]{LiAlH_4} \xrightarrow{D_3O^+} C_6H_5{-}CH_2OD + CH_3OD$$

A mechanism consistent with all the data is presented in Equations 17.52 and 17.53. It is exactly analogous to that given earlier (eq. 11.59) for other carbonyl metal hydride reductions. In all these cases, AlH_4^- acts as a source of the strongly nucleophilic hydride ion, H^-. An aldehyde is an intermediate in the ester reduction (eq. 17.52), but it is rapidly reduced further to a primary alcohol (eq. 17.53).

$$\text{(17.52)}\quad R\overset{\overset{\displaystyle O}{\|}}{C}{-}OR' + H{-}\overset{-}{A}lH_3 \rightarrow R{-}\underset{\underset{\displaystyle H}{|}}{\overset{\overset{\displaystyle O{-}AlH_3}{|}}{C}}{-}OR' \rightarrow R{-}\overset{\overset{\displaystyle O}{\|}}{C}{-}H + R'OAlH_3^-$$

Ester → Aldehyde

$$\text{(17.53)}\quad R\overset{\overset{\displaystyle O}{\|}}{C}H + H{-}\bar{A}lH_2OR' \rightarrow RCH_2O\bar{A}lH_2OR' \xrightarrow{RCOOR'} (RCH_2O)_2\bar{A}l(OR')_2 \xrightarrow{H_2O} 2RCH_2OH + 2R'OH$$

Two alcohols

Carboxylic acids are also reduced to primary alcohols by $LiAlH_4$ (eq. 17.54), although somewhat more slowly than the corresponding esters.

$$\text{(17.54)}\quad 3{-}CH_3O{-}C_6H_4{-}CO_2H \xrightarrow[\text{ether}]{LiAlH_4} \xrightarrow{H_3O^+} 3{-}CH_3O{-}C_6H_4{-}CH_2OH + H_2O$$

3-Methoxybenzoic acid → 3-Methoxybenzyl alcohol (85%)

One particularly effective reducing agent for acids is diborane. It reduces carboxyl groups faster than it does any other carbonyl groups, and so can be used for very selective reductions (eq. 17.55).

(17.55) Monoethyl terephthalate (p-$HOOC{-}C_6H_4{-}COOEt$) $\xrightarrow[\text{diglyme}]{B_2H_6}$ Ethyl 4-(hydroxymethyl)-benzoate (p-$HOCH_2{-}C_6H_4{-}COOEt$)

Diborane and $LiAlH_4$ are also used to reduce amides, in which case the products are *amines* (eq. 17.56).

$$(CH_3)_3C\overset{O}{\overset{\|}{C}}NHCH_3 + B_2H_6 \xrightarrow[65^\circ]{THF} \xrightarrow{H_2O} (CH_3)_3CCH_2NHCH_3 \quad (17.56)$$

(83%)

If the reduction of carboxylic acid derivatives could be halted halfway through, then a convenient route to aldehydes would be available (see eq. 17.52). This can be accomplished by using a very unreactive, and therefore highly selective, reducing agent such as lithium tri(*t*-butoxy)aluminohydride (eq. 17.57). Addition of this reagent to an acyl chloride at low temperatures results in formation of the aldehyde without further reduction. Many additional substituted $LiAlR_3H$ hydrides have been developed to perform specific reductions.

(17.57) Benzoyl chloride ($C_6H_5\overset{O}{\overset{\|}{C}}Cl$) + $Li[(CH_3)_3CO]_3AlH$ (Lithium tri (*t*-butoxy) aluminohydride) $\xrightarrow[\text{diglyme, } -75^\circ]{}$ Benzaldehyde ($C_6H_5\overset{O}{\overset{\|}{C}}H$) (85%)

17:12 Nitriles

Reactions of the cyano, or nitrile, group ($C{\equiv}N$) are intimately related to those of acids and amides:

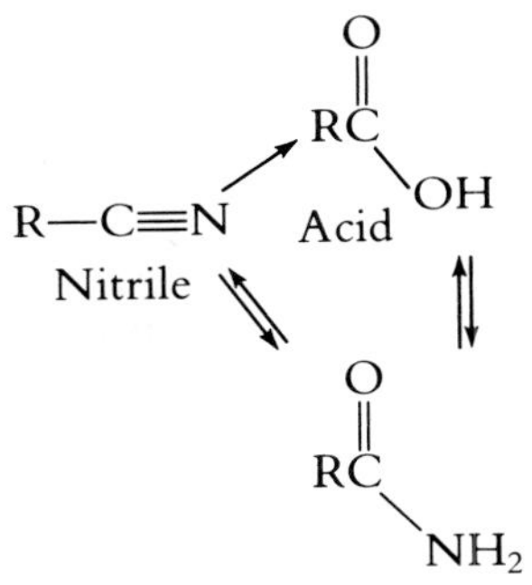

$$R{-}C{\equiv}N \text{ (Nitrile)} \rightarrow RC(=O)OH \text{ (Acid)}; \quad R{-}C{\equiv}N \rightleftharpoons RC(=O)NH_2 \text{ (Amide)} \rightleftharpoons RC(=O)OH$$

Because the C≡N is easily converted to acid and amide groups, it is useful to discuss some aspects of these relationships.

In the IUPAC system for naming open-chain compounds, the suffix -nitrile is added to the name of the longest carbon chain containing the C≡N group. The nitrile carbon is counted as part of this chain. When the CN group is attached to a ring, the suffix -carbonitrile is added to the name of the ring. An older nomenclature system is still used for nitriles related to common carboxylic acids. In this system the suffix -ic acid is replaced by nitrile, as in acetonitrile (from acetic acid) and benzonitrile (from benzoic acid). Finally, note that the CN group is sometimes treated as a substituent on a ring or chain, in which case the prefix cyano- is used, as in cyanoacetic acid, $NCCH_2COOH$.

CH_3CN	$CH_3(CH_2)_4CN$	$C_6H_5{-}CN$	$C_5H_9(H){-}CN$
Ethanenitrile (*acetonitrile*)	Hexanenitrile	Benzonitrile	Cyclopentane-carbonitrile

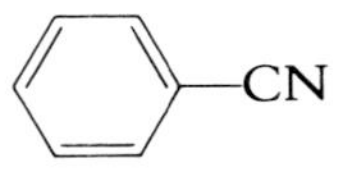

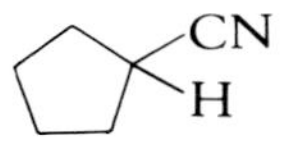

The carbon–nitrogen triple bond behaves very much like the carbon–oxygen double bond; it is susceptible to attack by a variety of nucleophiles. Nitriles can be hydrolyzed to amides (Table 17–4) and then to carboxylic acids. The reverse procedure may also be carried out, in which amides are dehydrated to nitriles.

Equation 17.58 shows a mechanism for the base-catalyzed hydrolysis of a nitrile to an amide. An isomer of the amide, called an imidic acid, is an

$$\text{(17.58)}\quad \underset{\text{Nitrile}}{RC{\equiv}N\colon} + {}^{-}OH \rightleftarrows RC(OH){=}\ddot{N}\colon^{-} \xrightleftharpoons{H_2O} \underset{\text{An imidic acid}}{RC(OH){=}\ddot{N}H} \xrightleftharpoons{OH^-} \underset{\mathbf{11}}{RC(\cdots\ddot{N}H^{\delta-})(\cdots O^{\delta-})} \xrightleftharpoons{H_2O} \underset{\text{An amide}}{RC(=O)\ddot{N}H_2}$$

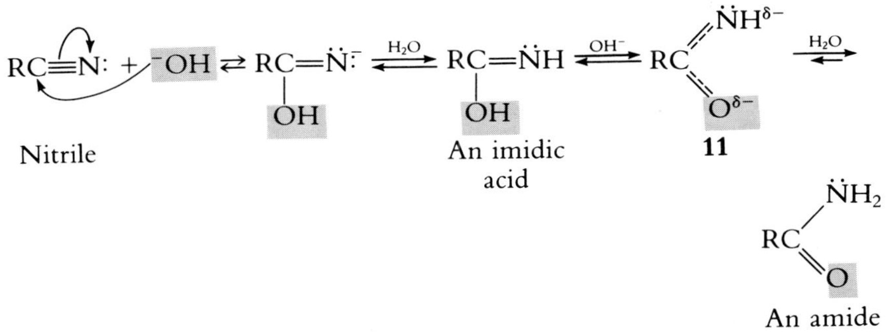

intermediate in the hydrolysis reaction. Such isomers are called **tautomers** (from the Greek *to auto,* meaning "the same," and *mer,* meaning "unit"). The term tautomers refers to isomers differing only in their single and double bond arrangements and/or proton positions (see eq. 12.20, sec. 12:5).

Problem

17–20 The imidic acid and amide tautomers are in equilibrium with one another under basic conditions, by way of the delocalized anion, **11**. In the great majority of cases, this equilibrium overwhelmingly favors the amide over the imidic acid. Using resonance theory, explain why the amide is the favored tautomer.

In the mechanism for acid-catalyzed nitrile hydrolysis, the first step is the reversible protonation of the nitrogen atom. The protonated nitrile is then attacked by the nucleophile H_2O:. *Under both acidic and basic conditions the amide may be further hydrolyzed to the carboxylic acid* (sec. 17:9). It is usually easier to stop the hydrolysis at the amide stage under basic conditions.

The behavior of nitriles toward reducing agents and Grignard reagents also is in accord with the similarity between the C≡N and C=O bonds. Nitriles are efficiently reduced to primary amines (eq. 17.59), while their

(17.59) $C_6H_5CH_2CN \xrightarrow[\text{diglyme, 25°}]{B_2H_6} \xrightarrow{OH^-,\ H_2O} C_6H_5CH_2CH_2NH_2$

Phenylacetonitrile → 2-Phenylethylamine (84%)

reaction with Grignard reagents yields ketones (eq. 17.60). The mechanism

(17.60) 9-Phenanthrenecarbonitrile (CN) $+ CH_3MgI \xrightarrow{\text{ether}} \xrightarrow{H_3O^+}$ 9-Acetylphenanthrene (CCH_3 with C=O) (55%)

of this ketone synthesis involves addition of the Grignard reagent across one of the nitrile π bonds. The intermediate is protonated to an imine, which hydrolyzes to the ketone during workup of the reaction (eq. 17.61).

$$(17.61)\quad \overset{\delta+}{R-C}\equiv\overset{\delta-}{N}: + \overset{\delta-}{R_1}-\overset{\delta+}{MgBr} \xrightarrow{\text{ether}} R(R_1)C=\overset{\delta-}{N}-\overset{\delta+}{MgBr} \xrightarrow{H_3O^+} \left[R(R_1)C=NH\right] \xrightarrow{H_3O^+} R(R_1)C=O + NH_4^+$$

Imine

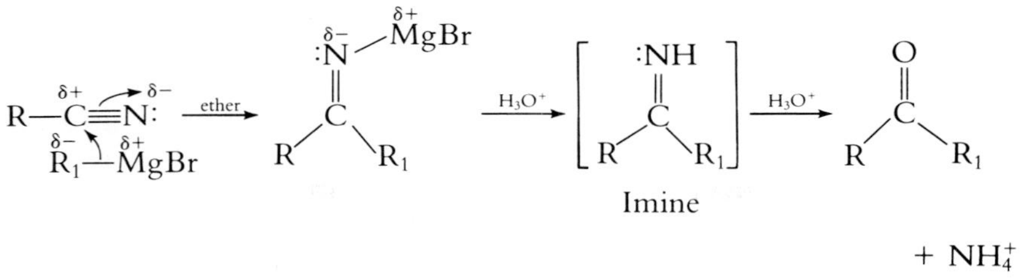

17:13 Thiolesters and Their Biological Significance

Thiols (RSH), which resemble alcohols in many respects (Chap. 6), react with acid chlorides and anhydrides to form *S*-substituted thioesters

$$(RC(=O)-SR)$$

also known as *thiolesters.* The thiolesters are somewhat more reactive than the corresponding oxygen esters, as Equation 17.62 shows (compare this

$$(17.62)\quad CH_3C(=O)SCH_2CH_3 + CH_3CH_2OH \overset{25^\circ}{\rightleftharpoons} CH_3C(=O)OCH_2CH_3 + CH_3CH_2SH \qquad K \sim 50$$

Thiolester (*S*-ethyl thioacetate); Ester; Ethanethiol

equilibrium constant with that of eq. 17.23). The difference in reactivity may be attributed to the weaker resonance interaction in thiolesters as compared to oxygen esters. The weak electron delocalization in thiolesters is due to the relatively poor overlap of the carbon 2*p* and sulfur 3*p* orbitals, the same

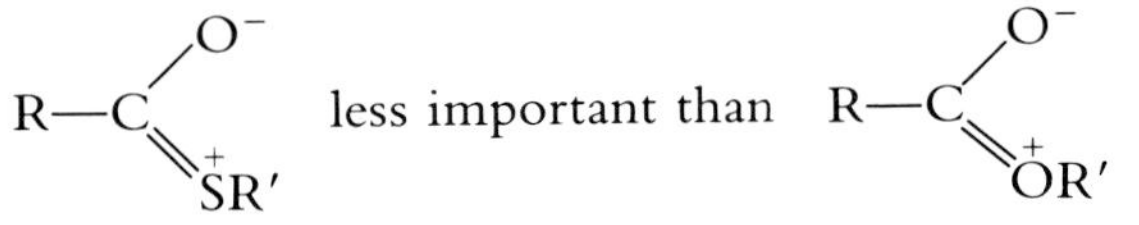

$$R-C(-O^-)=\overset{+}{S}R' \quad \text{less important than} \quad R-C(-O^-)=\overset{+}{O}R'$$

factor that accounts for the high reactivity of acid chlorides (sec. 17:3). Two thiolesters and their names are given here.

$$(CH_3)_2CHC(=O)S-C_6H_5 \qquad O_2N-C_6H_4-C(=O)SCH_2CH_3$$

S-Phenyl 2-methylthiopropanoate (*phenyl thiolisobutyrate*)

S-Ethyl 4-nitrothiobenzoate (*ethyl* p-*nitrothiolbenzoate*)

Thiolesters are not very important as laboratory chemicals, but they are biologically significant. One biological thiolester is involved in the key step of fatty acid biosynthesis in microbes, plants, and animals (including humans). Note in Table 16–1 (p. 597) that fatty acids only contain even numbers of carbon atoms. This is because such acids are biosynthesized by stringing together two-carbon units (the mechanism is discussed in detail in sec. 19:13). A complex organic molecule, known as acetyl coenzyme A (acetyl-CoA), contains the two-carbon unit, which is simply an acetyl group (CH_3CO). Figure 17–6 shows the structure of acetyl-CoA. Of this very large thiolester, the part that is involved in the biosynthesis reaction is only the shaded acetyl group portion. The rest of the molecule is simply a complex leaving group (eq. 17.63). Note that RS^- is a good leaving group from acyl carbon, but not from alkyl carbon (sec. 6:11). The difference arises because resonance effects are important in acyl derivatives, but not in alkyl derivatives.

$$CH_3C(=O)SCoA + Nu^- \rightarrow CH_3C(O^-)(Nu)SCoA \rightarrow CH_3C(=O)Nu + {}^-SCoA \quad (17.63)$$

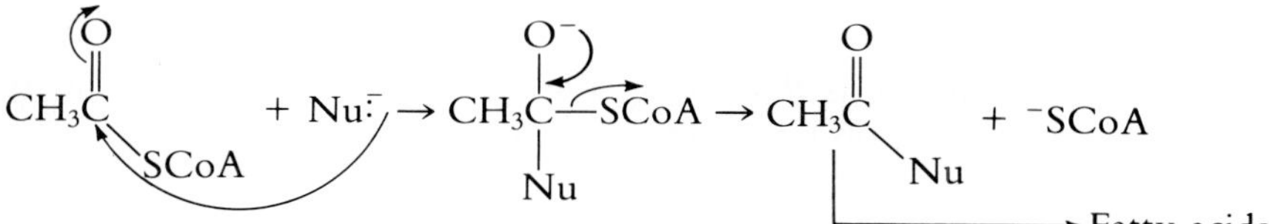

Acetyl-CoA

$CH_3C(=O)Nu \rightarrow$ Fatty acids

Because thiolesters are less resonance stabilized than oxygen esters (eq. 17.62), they are better acylating agents (eq. 17.63). As we'll see in Chapter 19, their acylating properties are vital to the biological role of thiolesters.

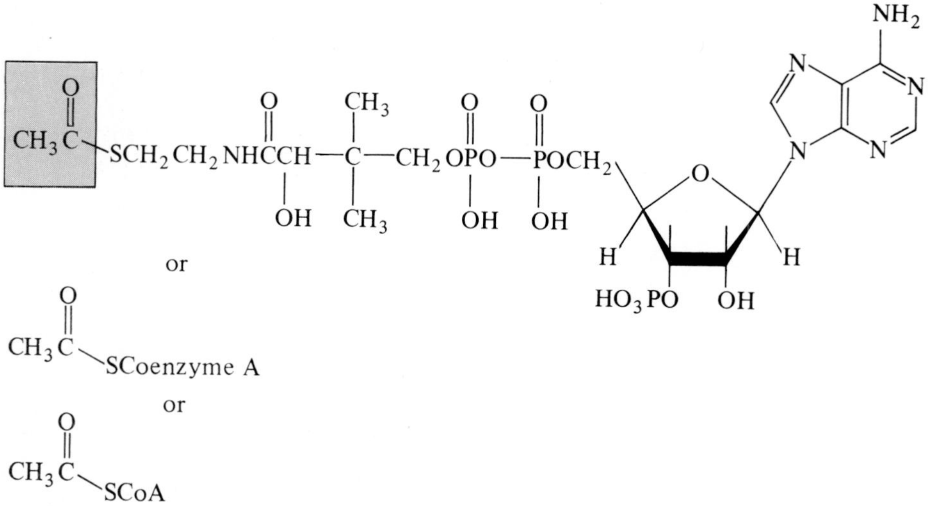

$CH_3C(=O)$–$SCH_2CH_2NHC(=O)CH(OH)$–$C(CH_3)_2$–$CH_2OP(=O)(OH)$–$OP(=O)(OH)OCH_2$–(ribose, 3'-OPO_3H, 2'-OH)–adenine (NH_2)

or

$CH_3C(=O)$SCoenzyme A

or

$CH_3C(=O)$SCoA

Figure 17–6. *Structure of acetyl-CoA.*

17:14 Comparative Chemistry of Carboxylate, Phosphate, and Sulfate Esters

The organic esters of sulfuric and phosphoric acids are of interest for both their chemical and biochemical properties. Phosphate esters, in particular, play a critical role in energy storage and release in all cells. The classes of esters discussed are shown below; R stands for any alkyl or aryl group.

$$\underset{\text{A carboxylic acid}}{RC(=O)OH} \qquad \underset{\text{A carboxylate ester}}{RC(=O)OR'}$$

$$\underset{\text{Phosphoric acid}}{HOP(=O)(OH)OH} \qquad \underset{\text{A phosphate ester}}{ROP(=O)(OR')OR''} \qquad \underset{\text{A phosphonic acid}}{RP(=O)(OH)OH} \qquad \underset{\text{A phosphonate ester}}{RP(=O)(OR'')OR'}$$

$$\underset{\text{Sulfuric acid}}{HOS(=O)_2OH} \qquad \underset{\text{A sulfate ester}}{ROS(=O)_2OR'} \qquad \underset{\text{A sulfonic acid}}{RS(=O)_2OH} \qquad \underset{\text{A sulfonate ester}}{RS(=O)_2OR'}$$

The sulfur and phosphorus esters resemble carboxylate esters in that they have S=O and P=O double bonds that are chemically similar to C=O groups. We anticipate that the chemistry of the S and P esters might be similar to, but not a carbon copy of, carboxylate ester chemistry.

In Chapter 6 you saw that hydrolysis of sulfonate esters takes place with nucleophilic attack at carbon. Thus, hydrolysis of optically active 2-octyl *p*-toluenesulfonate (tosylate) results in complete inversion of configuration at C-2 (eq. 17.64). Dialkyl sulfate esters also show a strong preference for nucleophilic attack at carbon (eq. 17.65).

(17.64)
$$H_2\ddot{O}: + \underset{(R)\text{-2-Octyl tosylate}}{CH_3(CH_2)_5CH(CH_3)-OSO_2-C_6H_4-CH_3} \longrightarrow \underset{(S)\text{-2-Octanol}}{HO-CH(CH_3)(CH_2)_5CH_3} + {}^-O_3S-C_6H_4-CH_3 + H^+$$

$$(17.65)\quad {}^{18}OH^- + CH_3\!-\!O\!-\!\overset{O}{\underset{O}{\overset{\|}{\underset{\|}{S}}}}\!-\!OCH_3 \longrightarrow H^{18}OCH_3 + {}^-OSO_3CH_3$$

Dimethyl sulfate — Methylsulfate ion

The reactivity of phosphate esters is intermediate between sulfate and carboxylate esters. Nucleophilic attack at carbon is competitive with nucleophilic attack at phosphorus, and the pathway followed is very sensitive to the nature of the nucleophile. Water and amines cleave the C—O bond (eq. 17.66).

$$(17.66)\quad C_6H_5\ddot{N}H_2 + CH_3\!-\!O\!-\!\overset{O}{\overset{\|}{\underset{OCH_3}{\underset{|}{P}}}}\!-\!OCH_3 \rightarrow C_6H_5\overset{CH_3}{\overset{|}{\underset{+}{N}}}H_2 + {}^-O\overset{O}{\overset{\|}{P}}(OCH_3)_2$$

Trimethyl phosphate — Dimethyl phosphate ion

By contrast, the hydroxide ion attacks at phosphorus, in a reaction that resembles the basic hydrolysis of carboxylate esters (eq. 17.67).

$$(17.67)\quad CH_3O\overset{O}{\overset{\|}{\underset{OCH_3}{\underset{|}{P}}}}OCH_3 \overset{OH^-}{\rightleftharpoons} CH_3O\overset{O}{\overset{\|}{\underset{OCH_3}{\underset{|}{P}}}}OH + CH_3O^- \rightarrow CH_3O\overset{O}{\overset{\|}{\underset{OCH_3}{\underset{|}{P}}}}O^- + CH_3OH$$

The compounds responsible for intracellular energy regulation are mostly esters of pyrophosphoric (**12**) and triphosphoric (**13**) acids. The ubiq-

$$HO\overset{O}{\overset{\|}{\underset{OH}{\underset{|}{P}}}}\!-\!O\!-\!\overset{O}{\overset{\|}{\underset{OH}{\underset{|}{P}}}}OH$$

12

$$HO\overset{O}{\overset{\|}{\underset{OH}{\underset{|}{P}}}}\!-\!O\!-\!\overset{O}{\overset{\|}{\underset{OH}{\underset{|}{P}}}}\!-\!O\!-\!\overset{O}{\overset{\|}{\underset{OH}{\underset{|}{P}}}}OH$$

13

uitous "high energy" compound in cells is **adenosine triphosphate**, ATP (Fig. 17–7). ATP is called a high energy compound because of its large negative free energy of hydrolysis (eq. 17.68).

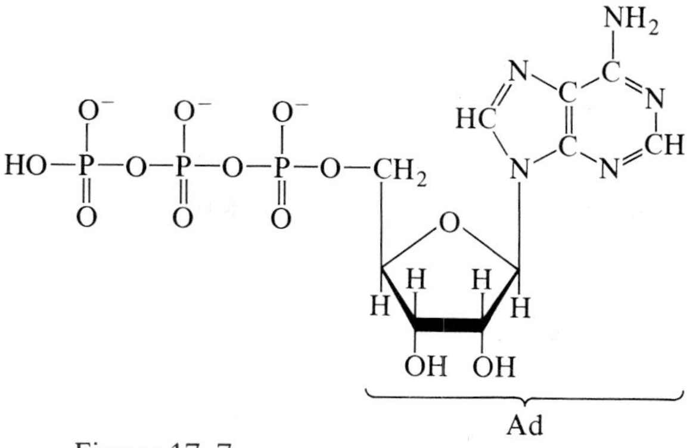

Figure 17–7. *Structure of adenosine triphosphate (ATP).*

$$\text{(17.68)}\quad \text{Ad—O—}\overset{\text{O}}{\underset{\text{O}^-}{\overset{\|}{\underset{|}{\text{P}}}}}\text{—O—}\overset{\text{O}}{\underset{\text{O}^-}{\overset{\|}{\underset{|}{\text{P}}}}}\text{—O—}\overset{\text{O}}{\underset{\text{O}^-}{\overset{\|}{\underset{|}{\text{P}}}}}\text{—OH} + H_2O \xrightarrow{\text{pH = 7}}$$

ATP

$$\text{Ad—O—}\overset{\text{O}}{\underset{\text{O}^-}{\overset{\|}{\underset{|}{\text{P}}}}}\text{—O—}\overset{\text{O}}{\underset{\text{O}^-}{\overset{\|}{\underset{|}{\text{P}}}}}\text{—OH} + {}^-\text{O—}\overset{\text{O}}{\underset{\text{O}^-}{\overset{\|}{\underset{|}{\text{P}}}}}\text{—OH} + H^+$$

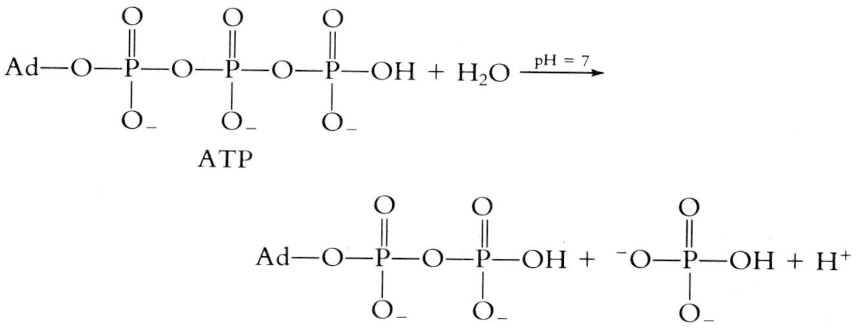

ADP $\qquad \Delta G° = -7.3$ kcal/mol

ATP is an anhydride of phosphoric acid, resembling carboxylic anhydrides, and is a good *phosphorylating* agent. Since it contains a number of basic P—O$^-$ groups, its degree of protonation and therefore its chemistry will be regulated by the pH of the medium.

Continual synthesis of ATP is one of the basic functions of virtually every cell. The energy for this synthesis is provided by the normal metabolic breakdown of substances such as lipids and carbohydrates. ATP is a form of stored cellular chemical energy, and large amounts of energy are released when ATP transfers a phosphoryl group (eq. 17.68). Because ATP is involved in almost all cellular processes that use or yield energy, it is sometimes thought of as having a "unique" high energy bond. In fact, the O—P bond *is* a high energy bond but it is not at all unusual. ATP is a phosphoric anhydride and as such is a naturally reactive phosphorylating agent, just as carboxylic anhydrides are reactive acylating agents.

Problem

17–21 Many phosphate esters are extremely active biologically and are used as insecticides and as nerve gases. In both applications they operate by reacting with a nucleophilic hydroxyl group in the enzyme *cholinesterase,* which is vital to normal nerve functioning. What are the structures of the products expected from the reaction of cholinesterase with tetraethylpyrophosphate (TEPP, eq. 3)?

$$\boxed{\text{enzyme}}\text{—CH}_2\text{OH} + \text{EtO—}\overset{\overset{\text{O}}{\|}}{\underset{\underset{\text{EtO}}{|}}{\text{P}}}\text{—O—}\overset{\overset{\text{O}}{\|}}{\underset{\underset{\text{OEt}}{|}}{\text{P}}}\text{—OEt} \rightarrow \qquad (3)$$

TEPP

Summary

1 There is a wide variety of C=O ir stretching frequencies among the different carboxylic acid derivatives (Table 17–2). This variation is useful for structure determination, and it often correlates with chemical reactivity of the carbonyl group in the derivative. The higher the C=O stretching frequency, the more reactive the acid derivative (sec. 17:2).

2 Interconversion of carboxylic acid derivatives takes place by way of tetrahedral intermediates. Reactive carboxylic acid derivatives are not strongly resonance stabilized, and they form tetrahedral intermediates easily. Unreactive derivatives are highly resonance stabilized, and they form tetrahedral intermediates relatively slowly. Reactive carboxylic acid derivatives contain good leaving groups, which permit the tetrahedral intermediate to decompose readily to products. Unreactive derivatives have poor leaving groups (sec. 17:3).

3 Alcohols react irreversibly with acid chlorides and anhydrides to yield esters. Analogously, 1° and 2° amines react with acid chlorides and anhydrides to yield amides. It is customary to conduct these reactions under basic conditions in order to neutralize the acid released (sec. 17:4). In all of these reactions, tetrahedral intermediates are formed from nucleophilic addition to the carbonyl group (sec. 17:5).

Carboxylic acids can be esterified with alcohols under reversible conditions, with HCl and H_2SO_4 as catalysts (Fischer esterification). The reaction is often driven to completion by removal of the water formed (Table 17–3). Amines react with carboxylic acids to yield salts, which are converted to amides at high temperatures (Table 17–4). Both esters and amides can be hydrolyzed back to carboxylic acids under acidic conditions (secs. 17:6 and 17:7).

Esters and amides are converted by aqueous base to carboxylate ions, and thus release alcohols and amines, respectively. Due to the stability of amides, their base-promoted hydrolysis is quite slow (sec. 17:9).

4 Esters and acid halides react with Grignard reagents and/or alkylcopper derivatives. Ester reactions with Grignard reagents produce tertiary alcohols. Alkylcopper reactions with acid chlorides are useful for the preparation of ketones (sec. 17:10).

$$\text{R}\overset{\overset{\text{O}}{\|}}{\text{C}}\text{OR}' + 2\textbf{R}\text{MgX} \xrightarrow[\text{ether}]{} \xrightarrow{\text{H}_3\text{O}^+} \textbf{R}\text{—}\overset{\overset{\text{OH}}{|}}{\underset{\underset{\text{R}}{|}}{\text{C}}}\text{—}\textbf{R}$$

Acids, esters, and amides can be reduced with $LiAlH_4$. Acids and esters are reduced to alcohols, and amides are reduced to amines. B_2H_6 is very effective at reducing carboxylic acids and amides, but not esters. Li(*t*-BuO)$_3$AlH is a useful reagent for converting acid chlorides to aldehydes (sec. 17:11).

5 Nitriles (prepared from alkyl halides and CN^-) can be hydrolyzed to yield amides or acids. They can also be reduced to amines. The nitrile group resembles the carbonyl group in much of its chemical behavior, as shown by its reactions with Grignard reagents to yield ketones after hydrolysis.

Amides can exist in two tautomeric forms: the imidic acid or the amide form. The amide form usually predominates. These tautomers are intermediates in nitrile hydrolysis (sec. 17:12).

6 Thiolesters resemble oxygen esters in their structure and chemistry. They are not as strongly resonance stabilized as are oxygen esters and hence are more reactive toward many nucleophiles. Thiolesters are very important biochemically in the form of acyl coenzyme A derivatives, such as acetyl-CoA (sec. 17:13).

Phosphate and sulfate esters, and the related phosphonate and sulfonate esters, behave differently from carboxylate esters. They often react with nucleophiles by way of mechanisms that involve alkyl–oxygen cleavage. Esters of di- and triphosphoric acids hydrolyze by cleavage of a P—O bond. This hydrolysis results in a large release of energy, and compounds like ATP are therefore important sources of energy for biological reactions (sec. 17:14).

Problems

17–22 Amines react with esters, although slowly, to yield amides (eq. 4). This reaction is sometimes useful if the ester is the only carboxylic acid derivative

$$\underset{\text{Ethyl formate}}{\overset{\overset{\displaystyle O}{\|}}{HC}OEt} + \underset{\text{Butylamine}}{CH_3(CH_2)_3NH_2} \rightarrow \underset{N\text{-Butylformamide}}{\overset{\overset{\displaystyle O}{\|}}{HC}NH(CH_2)_3CH_3} + EtOH \qquad (4)$$

conveniently available. Do you believe that a reaction such as that shown in Equation 4 is reversible or not? Explain your answer.

17–23 Write structures for all the anticipated products, if any, from the reaction of ethyl benzoate with the following reagents:

a. H_3O^+, reflux
b. OH^-, heat
c. $CH_3CH_2CH_2OH$, H^+
d. NH_3, H_2O
e. $(CH_3)_2NH$
f. NH_2NH_2
g. CH_3O^-, CH_3OH
h. $CH_3(CH_2)_3MgBr$, then H_3O^+
i. $LiAlH_4$, then H_3O^+

17–24 Describe efficient routes for carrying out the following conversions:

a. methyl benzoate → 4-phenyl-4-heptanol
b. butanoic acid → 3-ethyl-3-hexanol
c. butanoic acid → 2-methyl-4-heptanone
d. 3-methyl-1-pentanol → 3-methyl-5-nonanone

e. 3-methyl-1-pentanol → 7-methyl-4-nonanone
f. cyclopentylmethanol → cyclopentylmethanol-d_2

17–25 Write the structures of the organic products from the following reactions:

a. propyl benzoate $\xrightarrow{LiAlH_4}$ $\xrightarrow{H_3O^+}$

b. ethyl cyclobutanecarboxylate $\xrightarrow{LiAlH_4}$ $\xrightarrow{D_3O^+}$

c. methyl isobutyrate $\xrightarrow{LiAlD_4}$ $\xrightarrow{H_3O^+}$

d. diphenyl malonate $\xrightarrow{LiAlD_4}$ $\xrightarrow{D_3O^+}$

e. hexanoyl chloride $\xrightarrow{Li[(CH_3)_3CO]_3AlD}$

17–26 Esters can be synthesized by the reaction of carboxylate salts with alkyl halides (eq. 5; X = Br, I). These reactions are usually

$$RCOO^-Na^+ + R'X \rightarrow RCOOR' + NaX \quad (5)$$

carried out in dipolar aprotic solvents such as DMF, DMSO, and acetonitrile (sec. 6:10). This reaction involves some loss of resonance energy. What factors favor the reaction? What are its limitations?

17–27 The nmr spectrum of *N*,*N*-dimethylacetamide (**VI**), taken at room temperature, contains *three* methyl singlets of equal area.

$$CH_3\overset{\overset{\displaystyle O}{\|}}{C}-N\begin{matrix} \diagup CH_3 \\ \diagdown CH_3 \end{matrix}$$

VI

This finding indicates that the two *N*-methyl groups are *not* equivalent on the nmr time scale under these conditions. Suggest a reason for this phenomenon. What spectral change might occur if the sample were heated?

17–28 a. It is often the case that the stronger the electron-donating ability of Y, the lower the carbonyl frequency of

$$R\overset{\overset{\displaystyle O}{\|}}{C}-Y$$

(Table 17–1). Give a possible reason for this correlation, using resonance arguments.

b. An α,β-unsaturated carboxylic acid derivative

$$(RCH{=}CH\overset{\overset{\displaystyle O}{\|}}{C}Y)$$

always has a lower carbonyl stretching frequency than the corresponding saturated derivatives. Thus, for esters the C=O stretch is found at 1735 to 1750 cm^{-1}, while for α,β-unsaturated and benzoate esters the C=O stretch occurs at 1715 to 1720 cm^{-1}. Use resonance arguments to explain this difference.

17–29 The pyridine-promoted reaction of acid chlorides with alcohols is carried out as follows. The acid chloride is first added to cold, dry pyridine. There is immediate evolution of heat and a solid often precipitates (its precipitation suggests that it is a highly polar compound). The alcohol is added next, leading to rapid disappearance of the solid and formation of the product. This reaction is substantially faster than the reaction of the same alcohol and acid chloride in an inert solvent such as benzene.

Propose a structure for the precipitate. Explain why the acylation reaction is accelerated in pyridine. (Any other 3° amine will catalyze this reaction.)

17–30 Give the product and a detailed, step-by-step mechanism for each of the following reactions:

a. 5-hydroxypentanoic acid $\xrightarrow[\text{heat}]{}$

b. γ-butyrolactone $\xrightarrow[\text{heat}]{CH_3OH,\ H^+}$

17–31 Show how the following esters can be synthesized from carboxylic acids.

a. methyl 4-chlorobenzoate
b. cyclopentylmethyl hexanoate
c. diisopropyl succinate
d. ethyl 2-hydroxybutanoate
e. 3,4-dimethylphenyl bromoacetate
f. bis(2-hydroxyethyl) phthalate

17–32 Mesitoic acid (2,4,6-trimethylbenzoic acid) is esterified only very slowly under the usual conditions (excess alcohol and a trace of mineral acid). If mesitoic acid is dissolved in concentrated sulfuric acid and this solution poured into an excess of cold methanol, methyl mesitoate may be recovered in high yield. Why is mesitoic acid so unreactive under the usual conditions? What is the mechanism of the sulfuric acid-catalyzed esterification (see Probs. 15–31 and 16–14, pp. 591 and 625, respectively)?

17–33 *S*-Adenosylmethionine, **VII**, is synthesized by reaction of methionine with ATP (Prob. 6–64). In the presence of the proper enzyme, **VII** decomposes to compound **A** (eq. 6). Compound **A** undergoes enzymatic hydrolysis to homoserine. Propose a structure for **A**, and a mechanism for its formation.

$$AdCH_2\overset{+}{S}(CH_3)—CH_2CH_2CH(NH_3^+)COO^- \xrightarrow{\text{enzyme}}$$

S-Adenosylmethionine
VII

$$\mathbf{A}\ (C_4H_7NO_2) \xrightarrow[\text{enzyme}]{H_2O} HOCH_2CH_2CH(NH_3^+)COO^- \qquad (6)$$

Homoserine

17–34 Draw the structure of the tetrahedral intermediate formed by attack of OH^- on *N,N*-dimethylacetamide. Explain why this intermediate, in common with that from other amides, decomposes only occasionally to products, preferring instead to revert back to starting materials.

17–35 Ethyl acetate was hydrolyzed in aqueous acid solution containing $H_2{}^{18}O$. The reaction was halted before it had gone to completion, and the *unreacted* ethyl acetate recovered. Analysis of the recovered ester showed that some of it contained ^{18}O in the carbonyl group. Show how the unreacted ester could have incorporated ^{18}O in terms of the mechanism for reversible esterification-hydrolysis (eqs. 17.28 and 17.29). Explain how this experimental result supports the proposed mechanism.

17–36 Most lactams, like acyclic amides, are relatively resistant to hydrolysis. Compound **VIII**, however, undergoes hydrolysis at a surprisingly rapid rate. Offer an explanation for this finding. An important clue is that the C=O stretching frequency for **VIII** is much higher than is usual for amides and lactams.

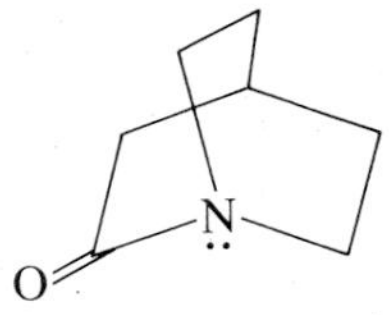

VIII

17–37 A natural glyceride (Prob. 17–14) was obtained optically active. The glyceride absorbed two equivalents of H_2 in the presence of a Pd catalyst. The reduced glyceride, which was no longer optically active, was saponified with aqueous NaOH. After saponification the reaction mixture was neutralized, after which it yielded one equivalent of glycerin and three equivalents of stearic acid, $CH_3(CH_2)_{16}COOH$. Provide a possible structure for the original, optically active glyceride.

17–38 Write the structures of the products of each of the following reactions:

a. 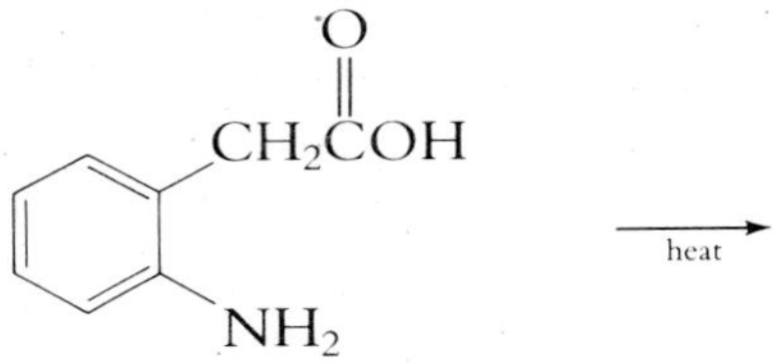 (2-aminophenyl)acetic acid, CH_2COOH / NH_2 $\xrightarrow{\text{heat}}$

b. 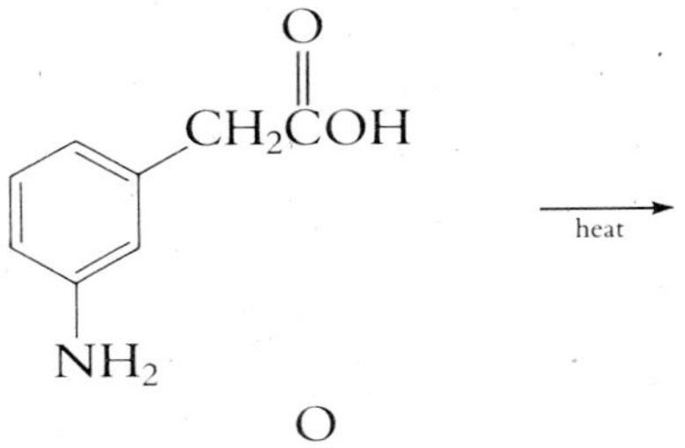 (3-aminophenyl)acetic acid, CH_2COOH / NH_2 $\xrightarrow{\text{heat}}$

c. 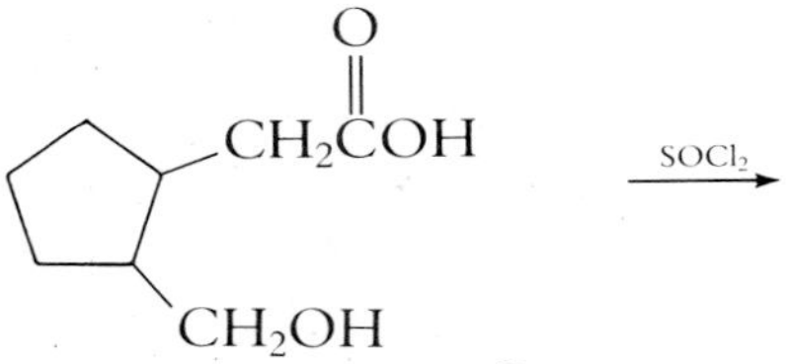 cyclopentane with CH_2COOH and CH_2OH $\xrightarrow{SOCl_2}$

d. 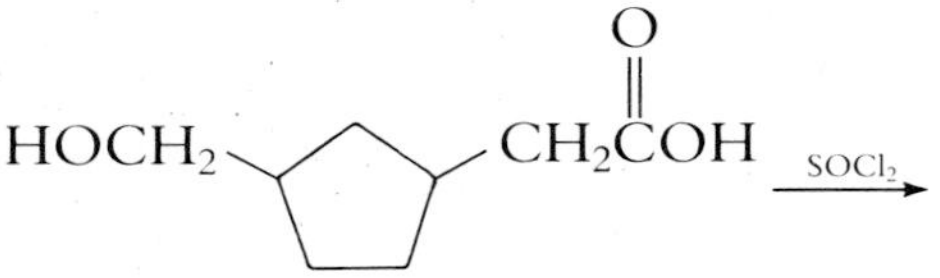 $HOCH_2$–cyclopentane–CH_2COOH $\xrightarrow{SOCl_2}$

17–39 a. What product would you expect from the reaction of hydrazine (NH_2NH_2) with:
1) one equivalent of methyl acetate?
2) two equivalents of methyl acetate?

b. The reaction of anhydrides and esters with hydroxylamine (NH_2OH) gives mainly hydroxamic acids (**IX**) and little or no *O*-acylhydroxylamines (**X**). Explain why the formation of **IX** is so favored over formation of **X**.

$$\underset{\textbf{IX}}{\mathrm{R\overset{\overset{\displaystyle O}{\|}}{C}NHOH}} \qquad \underset{\textbf{X}}{\mathrm{R\overset{\overset{\displaystyle O}{\|}}{C}ONH_2}}$$

17–40 Propose structures for compounds **A** to **D**.

o-Toluenesulfonyl chloride

B ($C_7H_7NO_4S$) $\xrightarrow{\text{heat}}$ **C** ($C_7H_5NO_3S$) $\xrightarrow{\text{NaOH}}$

D ($C_7H_4NO_3S^-$ Na^+; *sodium saccharin*)

17–41 Propose structures for compounds **A** to **I**.

a. Succinic acid $\xrightarrow{\text{heat}}$ **A**($C_4H_4O_3$) $\xrightarrow{CH_3NH_2}$

B($C_5H_9NO_3$) $\xrightarrow{\text{heat}}$ **C**($C_5H_7NO_2$; *N*-methylsuccinimide)

b. Phthalic acid $\xrightarrow[\text{anhydride}]{\text{acetic}}$ **D**($C_8H_4O_3$) $\xrightarrow{CH_3OH}$

E($C_9H_8O_4$) $\xrightarrow{SOCl_2}$ **F**($C_9H_7ClO_3$) $\xrightarrow{NH_3}$ **G**($C_9H_9NO_3$)

c. Phosgene ($COCl_2$) $\xrightarrow{CH_3OH}$ **H**($C_2H_3ClO_2$) $\xrightarrow{(CH_3CH_2)_2NH}$ **I**($C_6H_{13}NO_2$)

17–42 *S*-Ethyl-*O*-methylthiocarbonate (**XI**) reacts with one equivalent of ammonia to give the amide **XII** and ethanethiol as the only products (eq. 7). What other products might have been expected from this reaction? Why weren't they formed?

$$\underset{\textbf{XI}}{\mathrm{MeO\overset{\overset{\displaystyle O}{\|}}{C}SEt}} + NH_3 \rightarrow \underset{\textbf{XII}}{\mathrm{MeO\overset{\overset{\displaystyle O}{\|}}{C}NH_2}} + EtSH \qquad (7)$$

17–43 a. Compound **A** ($C_5H_{10}O_2$) is heated with an aqueous base and the solution is then extracted with ether. 1-Propanol is isolated from the ether extract. After acidification the aqueous layer yields compound **B**. Suggest structures for **A** and **B**.

b. When **C** ($C_6H_{13}NO$) is heated with an aqueous base, a gas is evolved. The solution is then acidified and extracted with ether, from which **D** may be isolated. **D** is heated with acidic methanol to give **E**, which can be reduced with lithium aluminum hydride. The reduction products are methanol and 4-methyl-1-pentanol. Write structures for **C**, **D**, and **E**.

c. Compound **F** ($C_7H_4Cl_2O$) is treated with isobutyl alcohol and **G** ($C_{11}H_{13}ClO_2$) is isolated. When **G** is heated with *n*-butylamine **H** is produced, which may be transformed by aqueous acid to a solid, **I** ($C_7H_5ClO_2$). Compound **I** may be heated with a mixture of nitric and sulfuric acids, giving rise to but a single product, **J** ($C_7H_4ClNO_4$). Propose structures for **F** to **J**.

17–44 Write structures for compounds **A** to **K**, showing their absolute configurations. Label the configurations for **B**, **E**, **G**, **I**, and **K**.

a. (*S*)-(+)-2-Butanol $\xrightarrow{PBr_3}$ **A** $\xrightarrow[DMF]{KCN}$ **B** $\xrightarrow[H_2O,\ \Delta]{H_2SO_4}$ (+)-**C**($C_5H_{10}O_2$)

(*S*)-(+)-2-Butanol $\xrightarrow[pyridine]{TsCl}$ **D** $\xrightarrow[DMSO]{NaCN}$ **E** $\xrightarrow[H_2O,\ \Delta]{H_2SO_4}$ (−)-**C**

b. (+)-**C** $\xrightarrow[H^+]{CH_3OH}$ **F** $\xrightarrow[2.\ H_3O^+]{1.\ LiAlH_4}$ (−)-**G**($C_5H_{12}O$) $\xrightarrow{PBr_3}$ **H** $\xrightarrow{(CH_3)_2CuLi}$ **I**(C_6H_{14})

c. **H** $\xrightarrow[DMF]{KCN}$ **J** $\xrightarrow[H_2O,\Delta]{H_2SO_4}$ (+)-**K**

17–45 Most esters hydrolyze extremely slowly at neutral pH. This is the case for phenyl hydrogen fumarate (**XIII**), which is in its anionic form at pH 7. By contrast, the saturated analog of **XIII**, the monophenyl succinate ion (**XIV**), does undergo hydrolysis to succinic acid at neutral pH. Furthermore, the *cis* isomer of **XIII**, the monophenyl maleate ion (**XV**), hydrolyzes 50 times faster than **XIV** under comparable conditions.

XIII: $H(C_6H_5OC(=O))C{=}C(H)CO_2^-$ (trans)

XIV: $C_6H_5OC(=O)-CH_2-CH_2-CO_2^-$

XV: $H(C_6H_5OC(=O))C{=}C(H)CO_2^-$ (cis)

It has been observed that during the hydrolysis of **XIV** and **XV** the phenoxide ion ($C_6H_5O^-$) is released faster than the final products (succinic and maleic acids) are formed. Write a mechanism for the hydrolysis of **XIV** and **XV** that is consistent with the data, and use it to explain why **XV** is more reactive than **XIV**.

17–46 Phenyl acetate reacts readily with amines to give amides (eq. 8). By contrast, triphenyl phosphate (**XVI**) is almost inert toward amines (eq. 9). Using the data already presented about the mechanisms of reaction of carboxylate and phosphate esters, suggest a reason for the differing behavior of **XVI** and phenyl acetate.

$$CH_3\overset{O}{\overset{\|}{C}}O\text{—}C_6H_5 + RNH_2 \rightarrow CH_3\overset{O}{\overset{\|}{C}}NHR + C_6H_5\text{—}OH \qquad (8)$$

$$(C_6H_5O)_3P{=}O + RNH_2 \rightarrow \text{no reaction} \qquad (9)$$

XVI

17–47 An unknown compound, $C_{10}H_{12}O_3$, has the ^1H nmr spectrum shown here, and an ir spectrum showing a strong peak at 1724 cm^{-1}. When the nmr spectrum is obtained with D_2O added, the absorption at 9.53 ppm disappears. Give a structure for the compound and assign all the nmr peaks.

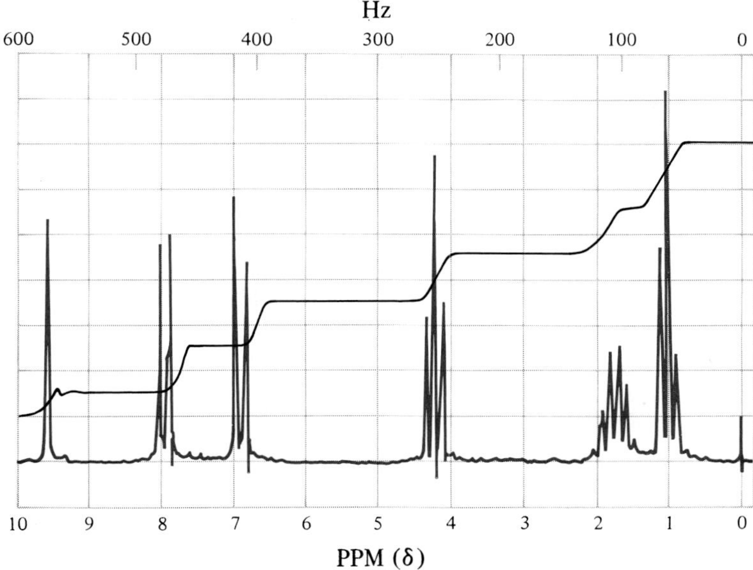

Supplementary Reading

J. March, *Advanced Organic Chemistry*, pp. 307–311, 347–357, McGraw-Hill, New York, N.Y., 1977. A general review of carboxylic acid derivative chemistry, with many references.

J. N. Lowe and L. L. Ingraham, *An Introduction to Biochemical Reaction Mechanisms*, Chaps. 2 and 4, Prentice–Hall, Englewood Cliffs, N.J., 1974.

S. Warren, *Chemistry of the Carbonyl Group*, Section 2, Wiley, New York, N.Y., 1974. A self-learning text allowing you to review and test your knowledge of carbonyl chemistry.

R. Singleton, Jr., The bioorganic chemistry of phosphorus, *J. Chem. Ed.* **50,** 538 (1973).

18 Aldol Condensation Reactions: Enols and Enolates

Chapters 11 and 17 stressed an important characteristic of carbonyl compounds—the reaction of the carbonyl functional group with nucleophiles. In this chapter we'll focus on another equally important feature of carbonyl compounds—the unusual acidity of the hydrogens on the carbons *next to* the carbonyl group. These **α hydrogens** are much more acidic than hydrogens on carbon atoms that are not next to carbonyl groups.

$$\underset{\substack{\uparrow \\ \text{Acidic} \\ \text{hydrogens}}}{-\overset{\alpha}{C}H_2\overset{O}{\overset{\|}{C}}-} + Y^{:-} \rightleftharpoons \underset{\text{Carbanion}}{-\overset{-}{\ddot{C}}H\overset{O}{\overset{\|}{C}}-} + YH$$

The carbanion formed by a proton loss is a nucleophile that attacks the carbonyl group of another molecule. Thus we can use this reaction to synthesize larger carbonyl compounds from smaller ones, as in Equation 18.1 where a new carbon–carbon bond is formed between the α carbon of one molecule and the carbonyl carbon of another.

$$2R\overset{\alpha}{C}H(\underset{\text{Acidic hydrogen}}{H})-\overset{O}{\overset{\|}{C}}R \rightarrow \underset{\text{Addition product}}{R-C(RCH_2)(OH)-\overset{\alpha}{C}(H)(R)-\overset{O}{\overset{\|}{C}}R} \rightarrow \underset{\text{Condensation product}}{(R)(RCH_2)\overset{\beta}{C}{=}\overset{\alpha}{C}(R)-\overset{O}{\overset{\|}{C}}R} + H_2O \qquad (18.1)$$

Note that two products can be obtained. An alcohol is formed first and can be isolated under certain conditions. The alcohol can also react further by elimination of a molecule of water to form an unsaturated carbonyl compound. This type of reaction has been given the generic name **aldol condensation**. The origin of the term *aldol* will be apparent shortly. The word *condensation* usually means the combination of two molecules, with the splitting out of a smaller one such as water or a simple alcohol. In this case, the reaction is usually called a condensation no matter which of the two products is isolated. In order to distinguish between the two products, we have labeled the alcohol an addition product and the unsaturated carbonyl compound a condensation product. This condensation product is often referred to as an α,β-unsaturated carbonyl compound.

By proper choice of catalyst or reaction conditions, we can isolate either of the products of Equation 18.1. Aldol condensations can be carried out with either base or acid catalysis, the details of which will be discussed subsequently. Since anions are not formed in the presence of acid, the acid catalysis mechanism will differ slightly from that shown in Equation 18.1. The structure of the carbonyl compound also plays an important role in determining which of the two products is obtained.

The aldol condensation finds wide use because of its versatility. Equation 18.2 shows how an addition product is an intermediate in the synthesis of an insect repellent. Equation 18.3 shows that a condensation reaction is the final

(18.2) $2CH_3CH_2\overset{\alpha}{C}H_2\overset{O}{\overset{\|}{C}}H \xrightarrow{OH^-}$

Butanal

$CH_3CH_2\overset{\alpha}{C}H\overset{O}{\overset{\|}{C}}H$ with $CH_3CH_2CH_2CHOH$ on the α carbon $\xrightarrow{H_2,\ Ni}$ $CH_3CH_2CHCH_2OH$ with $CH_3CH_2CH_2CHOH$

2-Ethyl-3-hydroxyhexanal (aldol addition product) — 2-Ethyl-1,3-hexanediol (insect repellent)

(18.3) [diketone with α-CH_2, O, CH_3] $\xrightarrow{OH^-}$ [cyclopentenone with α, CH_3]

cis-Jasmone (jasmine scent)
aldol condensation product

step in the synthesis of a compound responsible for the scent of jasmine which is used in perfumery. Nature uses an aldol condensation to synthesize the sugar fructose (eq. 18.4).

(18.4) $^-HO_3POCH_2CH(OH)CH{=}O$ + $\overset{\alpha}{C}H_2(OH)C({=}O)CH_2OPO_3H^-$ $\xrightarrow{enzyme}$

Glyceraldehyde-3-phosphate — Dihydroxyacetone phosphate

$^-HO_3POCH_2CH(OH)CH(OH)\text{—}\overset{\alpha}{C}H(OH)C({=}O)CH_2OPO_3H^-$

Fructose-1,6-diphosphate

In this chapter we'll be concerned mainly with the chemistry of the α positions of aldehydes and ketones. The corresponding chemistry of carboxylic acids and their derivatives will be discussed briefly, and treated more extensively in Chapter 19.

18:1 Base-Catalyzed Condensation of Aldehydes by Way of Enolate Ion Intermediates

It has been known for a long time that, in the presence of OH^-, two molecules of acetaldehyde will form one molecule of 3-hydroxybutanal (eq. 18.5).

$$\overset{O}{\overset{\|}{CH_3CH}} + \overset{\alpha}{} \overset{O}{\overset{\|}{CH_3CH}} \xrightarrow[H_2O,\ 5^\circ]{NaOH} \overset{OH\quad O}{CH_3\overset{|}{C}HCH_2\overset{\|}{C}H} \tag{18.5}$$

3-Hydroxybutanal
(*aldol*, 50%)

Since the product is both an aldehyde and an alcohol it was given the trivial name *aldol*, and this name was then applied to the entire class of related reactions (eq. 18.1).

In order to deduce its mechanism, the reaction in Equation 18.5 was studied kinetically. It was found to be second-order overall—first order in hydroxide ion and first order in acetaldehyde (eq. 18.6). This result is signifi-

$$\text{rate} = k[CH_3CHO][OH^-] \tag{18.6}$$

cant because the kinetics are only first order in aldehyde, even though two molecules of aldehyde are consumed overall. We must conclude that one molecule of aldehyde is reacting with the hydroxide ion in the first step, which is the rate-determining step. The product of this step is a reactive intermediate which then reacts rapidly with the second molecule of aldehyde. That reactive intermediate is an **enolate ion** (eq. 18.7). The enolate

$$H\ddot{O}:^- + H{-}CH_2{-}\overset{O}{\overset{\|}{C}H} \xrightleftharpoons{\text{slow}} H_2\ddot{O}: + \left[\bar{C}H_2{-}\overset{:\ddot{O}}{\overset{\|}{C}H} \leftrightarrow CH_2{=}\overset{:\ddot{O}:^-}{\overset{|}{C}H}\right] \equiv \overset{\delta-}{C}H_2{=\!=}\overset{O^{\delta-}}{\overset{\|}{C}H} \tag{18.7}$$

Acetaldehyde enolate ion

ion is stabilized by the delocalization of its negative charge (sec. 13:1), which is partially on the electronegative oxygen.

Since enolate ions have a negative charge, they are quite nucleophilic. They rapidly attack the carbonyl group of the second aldehyde molecule (eq. 18.8), and the resulting alkoxide ion is protonated (by the water or hydroxylic solvent) to give the final product (eq. 18.9). The overall reaction is a

$$CH_3\overset{O}{\overset{\|}{C}}{-}H + \overset{\delta-}{C}H_2{=\!=}\overset{\delta-O}{\overset{\|}{C}H} \xrightleftharpoons{\text{fast}} CH_3\overset{O^-}{\overset{|}{\underset{H}{\underset{|}{C}}}}{-}CH_2{-}\overset{O}{\overset{\|}{C}H} \tag{18.8}$$

Aldehyde — Enolate ion — Alkoxide ion

$$\text{(18.9)}\quad \mathrm{CH_3\overset{\overset{\large O^-}{|}}{C}H\!-\!CH_2\overset{\overset{\large O}{\|}}{C}H} + \mathrm{H_2O} \xrightleftharpoons{\text{fast}} \underset{\text{Aldol}}{\mathrm{CH_3\overset{\overset{\large OH}{|}}{C}HCH_2\overset{\overset{\large O}{\|}}{C}H}} + \mathrm{OH^-}$$

nucleophilic carbonyl addition reaction (sec. 11:5). The concentration of enolate ions is always very low because aldehydes are weak acids—K for the reaction in Equation 18.7 is only about 10^{-6}. Nonetheless, because of the high reactivity of the enolate ions, the system comes to equilibrium very quickly.

Other aliphatic aldehydes that contain α hydrogens undergo aldol condensation to form the addition product (eqs. 18.10 and 18.11). When basic catalysts are used,

$$\text{(18.10)}\quad \underset{\text{Propanal}}{\mathrm{CH_3CH_2\overset{\overset{\large O}{\|}}{C}H}} + \underset{\text{Propanal}}{\mathrm{CH_3\overset{\alpha}{C}H_2\overset{\overset{\large O}{\|}}{C}H}} \xrightarrow[\mathrm{H_2O,\ 5^\circ}]{\mathrm{NaOH}} \underset{\text{3-Hydroxy-2-methylpentanal}}{\mathrm{CH_3CH_2\overset{\overset{\large OH}{|}}{C}H\underset{\underset{\large CH_3}{|}}{C}H\overset{\overset{\large O}{\|}}{C}H}}$$

$$\text{(18.11)}\quad \mathrm{(CH_3)_2CH\overset{\overset{\large O}{\|}}{C}H} + \mathrm{(CH_3)_2\overset{\alpha}{C}H\overset{\overset{\large O}{\|}}{C}H} \xrightarrow[\mathrm{H_2O}]{\mathrm{NaOH}} \underset{(85\%)}{\mathrm{(CH_3)_2CH\overset{\overset{\large OH}{|}}{C}H\!-\!\underset{\underset{\large CH_3}{|}}{\overset{\overset{\large CH_3}{|}}{C}}\!-\!\overset{\overset{\large O}{\|}}{C}H}}$$

enolate ions are the reaction intermediates. *Aldehydes that lack α hydrogens cannot form enolate ions and therefore do not undergo self-condensation.* Benzaldehyde (eq. 18.12) is a typical example.

$$\text{(18.12)}\quad \mathrm{C_6H_5\overset{\overset{\large O}{\|}}{C}H} + \mathrm{C_6H_5\overset{\overset{\large O}{\|}}{C}H} \xrightarrow[\mathrm{H_2O,\ 5^\circ}]{\mathrm{OH^-}} \text{no reaction}$$

The relatively low temperatures (0 to 25°) used for the reactions in Equations 18.5, 18.10, and 18.11 allow the aldol addition product to be formed at a convenient rate, while keeping the rate of dehydration (formation of condensation product) small. Thus, *a combination of base catalysis and low temperature is the standard method for converting aliphatic aldehydes to aldol addition products.*

Example

What product is formed when 3-phenylpropanal is treated with dilute base at 5°?

$$\underset{\text{3-Phenylpropanal}}{\mathrm{C_6H_5CH_2CH_2\overset{\overset{\large O}{\|}}{C}H}}$$

Answer

Work with the problem mechanistically. You know that the base must produce a carbanion with its charge on the carbon next to the carbonyl group, by removing a proton from that carbon:

$$C_6H_5CH_2\underset{\displaystyle H}{\underset{|}{C}}H\overset{\displaystyle O}{\overset{\|}{C}}H + :\ddot{O}H^- \rightarrow C_6H_5CH_2\bar{\ddot{C}}H\overset{\displaystyle O}{\overset{\|}{C}}H + H_2O$$

The carbanion, called an enolate ion, is a powerful nucleophile and attacks the carbonyl group of an un-ionized aldehyde molecule:

$$C_6H_5CH_2CH_2\overset{\displaystyle O}{\overset{\|}{C}}H + \underset{\displaystyle CH_2C_6H_5}{\underset{|}{:\bar{C}}}H\overset{\displaystyle O}{\overset{\|}{C}}H \rightarrow C_6H_5CH_2CH_2\overset{\displaystyle O^-}{\overset{|}{C}}H\text{—}\underset{\displaystyle C_6H_5CH_2}{\underset{|}{C}}H\overset{\displaystyle O}{\overset{\|}{C}}H$$

The alkoxide ion produced by nucleophilic attack is rapidly protonated on oxygen by water:

$$C_6H_5CH_2CH_2\overset{\displaystyle O^-}{\overset{|}{C}}H\underset{\displaystyle C_6H_5CH_2}{\underset{|}{C}}H\overset{\displaystyle O}{\overset{\|}{C}}H + HOH \rightarrow C_6H_5CH_2CH_2\overset{\displaystyle OH}{\overset{|}{C}}H\underset{\displaystyle C_6H_5CH_2}{\underset{|}{C}}H\overset{\displaystyle O}{\overset{\|}{C}}H + OH^-$$

This product has two chiral centers, and is therefore a mixture of two racemic diastereomers.

Problems

18–1 a. Show the products formed when each of the following aldehydes is treated with dilute aqueous base at 5°.

1) 2-ethylbutanal
2) 3-methylbutanal
3) cyclopentanecarboxaldehyde
4) pentanal

b. In each of the condensations above, the reaction leads to the formation of new chiral centers in the product. Where more than one chiral center is produced, the product is actually a mixture of diastereomers (sec. 4:3). Identify the chiral centers in each product of (a) and decide which products are diastereomeric mixtures.

18–2 Write a mechanism for the addition reaction (eq. 18.10) in the aldol condensation of propanal.

18–3 Formaldehyde, CH_2O, lacks α hydrogens and cannot undergo self-condensation under the usual aldol conditions. However, it can react with other aldehydes under these conditions. With 2-methylpropanal and a base it produces compound **I**, for instance. Write a mechanism for this reaction.

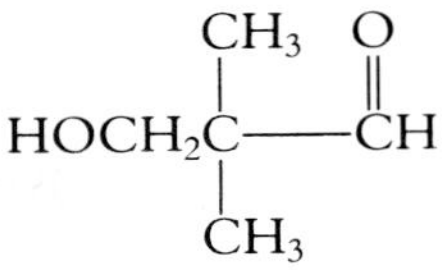

$$\mathrm{HOCH_2C(CH_3)_2{-}\overset{\displaystyle O}{\overset{\|}{C}}H}$$

I

18–4 Because the enolate negative charge is delocalized over the carbonyl oxygen and the α carbon, two constitutionally isomeric products could be expected from an aldol condensation (**II** and **III**). Use the average bond energies of Table 10–1 (p. 346) to explain why **II** is the only product isolated.

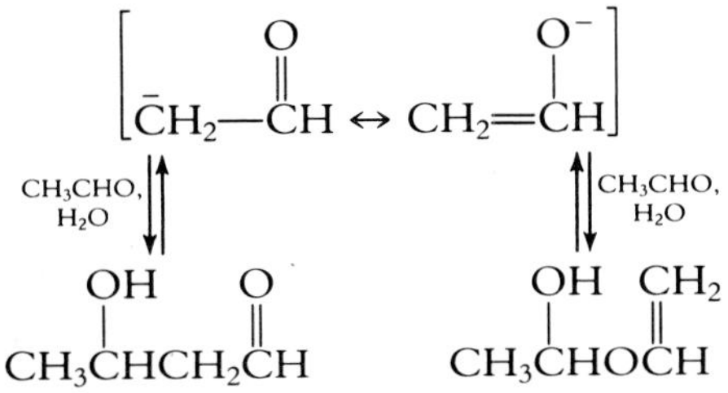

$$\left[\bar{C}H_2{-}\overset{\displaystyle O}{\overset{\|}{C}}H \leftrightarrow CH_2{=}\overset{\displaystyle O^-}{\overset{|}{C}}H\right]$$

$$\xrightarrow{CH_3CHO,\ H_2O} \quad \mathrm{CH_3\overset{\displaystyle OH}{\overset{|}{C}}HCH_2\overset{\displaystyle O}{\overset{\|}{C}}H}\ (\textbf{II}) \qquad \mathrm{CH_3\overset{\displaystyle OH}{\overset{|}{C}}HO\overset{\displaystyle CH_2}{\overset{\|}{C}}H}\ (\textbf{III})$$

The self-condensation of aliphatic aldehydes is a mildly exothermic reaction (eq. 18.13). This favorable energy change is counterbalanced, however,

$$2\mathrm{RCH_2\overset{\displaystyle O}{\overset{\|}{C}}H} \rightleftarrows \mathrm{RCH_2\overset{\displaystyle OH}{\overset{|}{C}}H\underset{\displaystyle R}{\underset{|}{C}}H\overset{\displaystyle O}{\overset{\|}{C}}H} \qquad \Delta H \simeq -4\ \text{kcal/mol} \tag{18.13}$$

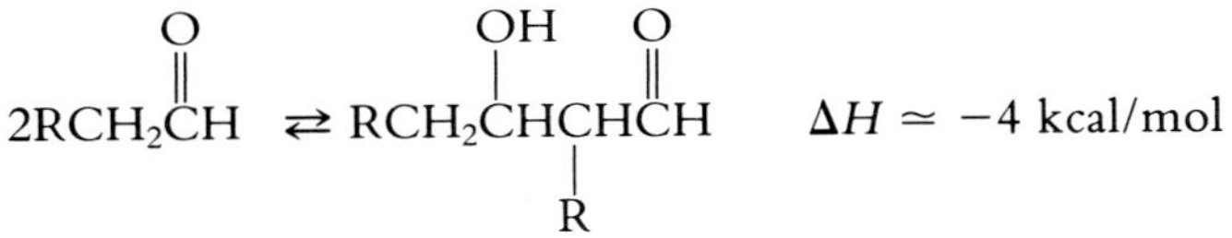

R = H, alkyl

by an unfavorable entropy change, since two molecules of reactant yield only one of product. As a result, the addition reaction is delicately balanced between opposing driving forces, and usually does not have a large equilibrium constant (eq. 18.5).

If you look again at the reaction in Equation 18.2 you will see that entropy favors the condensation product more than the addition product. This is because the condensation reaction yields two molecules of product from two molecules of reactant. By raising the temperature at which the aldol condensation is carried out we can speed up the rate of condensation product formation, as well as shifting the equilibrium in its favor. Thus, by adding acetaldehyde to aqueous base at 90° instead of at 5°, we produce 2-butanal (*crotonaldehyde*, eq. 18.14).

$$\text{(18.14)}\quad 2CH_3\overset{\overset{\displaystyle O}{\|}}{C}H \xrightarrow[H_2O,\ 90^\circ]{OH^-} \left[CH_3\overset{\overset{\displaystyle OH}{|}}{C}HCH_2\overset{\overset{\displaystyle O}{\|}}{C}H\right] \xrightarrow{OH^-} CH_3CH{=}CH\overset{\overset{\displaystyle O}{\|}}{C}H + H_2O$$

Aldol (not isolated) — *Crotonaldehyde*

The dehydration of aldol to crotonaldehyde can take place under basic conditions because of the acidity of the α hydrogens of aldol (eq. 18.15). Although

$$\text{(18.15)}\quad CH_3\overset{\overset{\displaystyle OH}{|}}{C}H{-}\underset{\underset{\displaystyle H}{|}}{\overset{\alpha}{C}}H\overset{\overset{\displaystyle O}{\|}}{C}H + :\ddot{O}H^- \rightleftarrows$$

$$CH_3\overset{\overset{\displaystyle OH}{|}}{C}H{-}\ddot{C}H\overset{\overset{\displaystyle O}{\|}}{C}H \rightarrow CH_3CH{=}CH\overset{\overset{\displaystyle O}{\|}}{C}H + OH^-$$

$$+ H_2\ddot{O}:$$

OH^- was not a useful leaving group in concerted elimination reactions (Chap. 7), here there is a strong driving force for the elimination of OH^-. The crotonaldehyde formed has the double bond in conjugation with the carbonyl group, making for a stable product. This fact, and the accumulation of negative charge on the α carbon, overcomes the hydroxide ion's poor leaving group ability.

Problems

18–5 Show the structure of the final product obtained after each of the following aldehydes is treated with aqueous NaOH at 90°.

a. propanal
b. 3-methylbutanal
c. 4-pentenal
d. cyclopentylacetaldehyde

18–6 The aldol condensation is useful synthetically because its products, β-hydroxycarbonyl compounds and α,β-unsaturated carbonyl compounds, can be further transformed into other materials. Show how each of the following compounds can be obtained from an aliphatic aldehyde by way of an aldol condensation.

a. 2-buten-1-ol (*crotyl alcohol*)
b. 2-butenoic acid (*crotonic acid*)
c. 1-butanol
d. 2-ethyl-1-hexanol
e. 2-methyl-1,3-pentanediol

If our intention is to obtain the condensation product (an α,β-unsaturated carbonyl compound), then, as we've just seen, we can accomplish this by carrying out the base-catalyzed aldol condensation at relatively high temperatures (90° and above). Another approach is to switch to acid catalysis, as the next section shows.

18:2 Acid-Catalyzed Condensation of Aldehydes by Way of Enol Intermediates

When aliphatic aldehydes are exposed to catalytic amounts of a strong acid, they undergo aldol condensation to α,β-unsaturated aldehydes. The addition product is usually dehydrated rapidly under acidic conditions, so only the condensation product is isolated (eq. 18.16).

$$CH_3\overset{O}{\overset{\|}{C}}H + CH_3\overset{O}{\overset{\|}{C}}H \xrightarrow{H_2SO_4} \left[CH_3\overset{OH}{\overset{|}{C}}HCH_2\overset{O}{\overset{\|}{C}}H\right] \xrightarrow{H_2SO_4} CH_3CH{=}CH\overset{O}{\overset{\|}{C}}H + H_2O \tag{18.16}$$

Aldol — Crotonaldehyde (60%)

As is the case with the base-catalyzed reaction, the acid-catalyzed aldol condensation depends on removal of an α hydrogen. Mechanistic studies show that the aldehyde is first protonated on the carbonyl oxygen. The protonated aldehyde then loses an α proton to water and is converted to an **enol** (eq. 18.17). Formation of this reactive intermediate is the slow step on

$$CH_3{-}\overset{:O:}{\overset{\|}{C}}H + H_3\ddot{O}^+ \xrightleftharpoons{\text{fast}} H_2\ddot{O}: + \underset{H}{\underset{|}{C}}H_2{-}\overset{\overset{+}{:O}H}{\overset{\|}{C}}H \xrightleftharpoons{\text{slow}} H_3\ddot{O}^+ + CH_2{=}\overset{:\ddot{O}H}{\overset{|}{C}}H \tag{18.17}$$

Protonated aldehyde — Acetaldehyde enol

the way to the addition product. The enol then reacts with another molecule of protonated aldehyde to give the addition product (eq. 18.18).

$$CH_3\overset{\overset{+}{:O}H}{\overset{\|}{C}}{-}H + CH_2{=}\overset{:\ddot{O}H}{\overset{|}{C}}H \rightleftarrows CH_3\underset{H}{\underset{|}{\overset{:\ddot{O}H}{\overset{|}{C}}}}{-}CH_2{-}\overset{\overset{+}{:O}H}{\overset{\|}{C}}H \rightleftarrows CH_3\overset{OH}{\overset{|}{C}}HCH_2\overset{O}{\overset{\|}{C}}H + H^+ \tag{18.18}$$

Protonated aldehyde — Acetaldehyde enol — Aldol

Dehydration of the addition product resembles the acid-catalyzed dehydration of alcohols (sec. 7:6). It is even faster because it involves the loss of an

α proton, which is easily removed (eq. 18.19). Thus, we see that acid cataly-

$$\text{(18.19)}\quad \mathrm{CH_3\overset{OH}{\overset{|}{C}}HCH_2\overset{O}{\overset{\|}{C}}H} \xrightleftharpoons{H_3O^+}$$

Aldol

$$\mathrm{CH_3\overset{\overset{+}{O}H_2}{\overset{|}{C}}H{-}\underset{\underset{H}{|}}{\overset{\alpha}{C}}H\overset{O}{\overset{\|}{C}}H} + \mathrm{H_2\ddot{O}:} \rightarrow \mathrm{CH_3CH{=}CH\overset{O}{\overset{\|}{C}}H} + \mathrm{H_2O} + \mathrm{H_3O^+}$$

sis of an aldol condensation is particularly useful when the desired product is the condensation product. The condensation product from acetaldehyde—crotonaldehyde—is a valuable commercial intermediate (see Prob. 18–6).

Problem

18–7 a. Draw the isomeric enols of propanal.
b. Show how one of the enols reacts with another molecule of propanal under acidic conditions to give the aldol addition product.

18:3 Why Base Catalysis Is Usually Not Effective for Condensation of Simple Ketones

Equation 18.20 shows that ketones, like aldehydes, undergo aldol condensations in basic solution at low temperatures. The addition product (called a *ketol*) may be isolated, but only in poor yield. The aldol condensation of ketones has a much less favorable equilibrium constant than that for aldehydes (compare eq. 18.20 with eqs. 18.5 and 18.11).

$$\text{(18.20)}\quad \mathrm{CH_3\overset{O}{\overset{\|}{C}}CH_3} + \mathrm{\overset{\alpha}{C}H_3\overset{O}{\overset{\|}{C}}CH_3} \xrightleftharpoons[H_2O,\ 25^\circ]{OH^-} \mathrm{CH_3\underset{\underset{CH_3}{|}}{\overset{OH}{\overset{|}{C}}}{-}CH_2\overset{O}{\overset{\|}{C}}CH_3}$$

4-Hydroxy-4-methyl-2-pentanone (*diacetone alcohol*, 2%)

We've already found in other cases that ketones are less reactive toward addition than are aldehydes (secs. 11:4, 11:6, and 11:8). The two major factors responsible for the difference are:

1. The ketone carbonyl group is more stable and therefore less reactive than the aldehyde carbonyl group.
2. The ketone addition product is more crowded, and therefore less stable, than the aldehyde addition product.

Fortunately, it is not the addition product, but the condensation product we desire, since it is more synthetically useful. So, we avoid the problem of isolating the addition product and drive the reaction to the condensation stage. Acid is most effective for this purpose, as we'll now see.

18–8 Write a mechanism for Equation 18.20, showing the structure of all intermediates.

18:4 Acid-Catalyzed Condensation of Ketones

Ketones, like aldehydes, are in equilibrium with their enols in the presence of acids (eq. 18.17). The concentration of enol is very small, but it is highly reactive and attacks a molecule of protonated ketone. The aldol addition product (ketol) forms and then rapidly dehydrates to the condensation product (eqs. 18.21 and 18.22). Even at low temperature the addition product is dehydrated too rapidly to be isolated.

$$CH_3CH_2C(=O)CH_2CH_3 + CH_3\overset{\alpha}{C}H_2C(=O)CH_2CH_3 \xrightarrow[25°]{HCl} \tag{18.21}$$

3-Pentanone

$$\left[CH_3CH_2C(OH)(CH_2CH_3)\!-\!\overset{\alpha}{C}H(CH_3)C(=O)CH_2CH_3 \right] \xrightarrow{HCl} (CH_3CH_2)_2C{=}C(CH_3)C(=O)CH_2CH_3 + H_2O$$

Ketol

5-Ethyl-4-methyl-4-hepten-3-one

(18.22) cyclohexanone + cyclohexanone (H–C$_\alpha$–H) $\xrightarrow{H_2SO_4}$ [Ketol: 2-(1-hydroxycyclohexyl)cyclohexanone (HO, H)] $\xrightarrow{H_2SO_4}$ 2-Cyclohexylidenecyclohexanone + H_2O

Ketol

2-Cyclohexylidene-cyclohexanone (85%)

The Lewis acid-catalyzed condensation of acetone is used in the commercial synthesis of methyl isobutyl ketone (MIBK), a solvent for paints. The acetone condensation product, mesityl oxide, is prepared in the presence of hydrogen and a palladium catalyst, and is immediately reduced to MIBK (eq. 18.23).

$$(18.23)\quad 2CH_3\overset{\overset{\displaystyle O}{\|}}{C}CH_3 \xrightarrow{\text{Lewis acid}} \left[\underset{\text{Diacetone alcohol}}{(CH_3)_2\underset{\underset{\displaystyle OH}{|}}{C}CH_2\overset{\overset{\displaystyle O}{\|}}{C}CH_3} \right] \longrightarrow$$

$$\underset{\text{Mesityl oxide}}{(CH_3)_2C{=}CH\overset{\overset{\displaystyle O}{\|}}{C}CH_3} \xrightarrow{H_2,\ Pd} \underset{\text{Methyl isobutyl ketone (MIBK)}}{(CH_3)_2CHCH_2\overset{\overset{\displaystyle O}{\|}}{C}CH_3}$$

Although both aldol addition and condensation products are synthetically useful, the condensation products are particularly desirable. Their conjugated C=C bonds have many useful chemical properties which unconjugated bonds do not (Chap. 19 shows how these properties are exploited). In the next two sections we explore a little further the formation of aldol condensation products—α,β-unsaturated aldehydes and ketones.

18:5 Intramolecular Condensation of Dicarbonyl Compounds to Produce Five- and Six-Membered Rings

In Sections 18:1 and 18:3 we noted that the aldol condensation equilibrium constant is not large, and is sensitive to many reaction parameters. One of these parameters is the entropy change. Previous examples in this book have shown us that intramolecular reactions always have more favorable entropy changes than do the corresponding intermolecular reactions. Based on these examples we should expect intramolecular aldol condensations to give high yields of product. That is what we find when the product is a five- or six-membered ring.

Equations 18.3, 18.24, and 18.25 are examples of intramolecular aldol condensations using dialdehydes and diketones as starting materials. Cyclopentenyl and cyclohexenyl carbonyl compounds are the isolated products. This is the case regardless of whether acid or base catalysis is used, and whether the starting materials are aldehydes or ketones.

(18.24) [Structure: HN: bearing CH₂–CH₂(α)–CH=O and CH₂–CH₂–HC=O chains] $\xrightarrow{HCl}$ [Structure: tetrahydropyridinium ring, H_2N^+, with α-carbon bearing CH=O] $Cl^- + H_2O$

(75%)

(18.25) 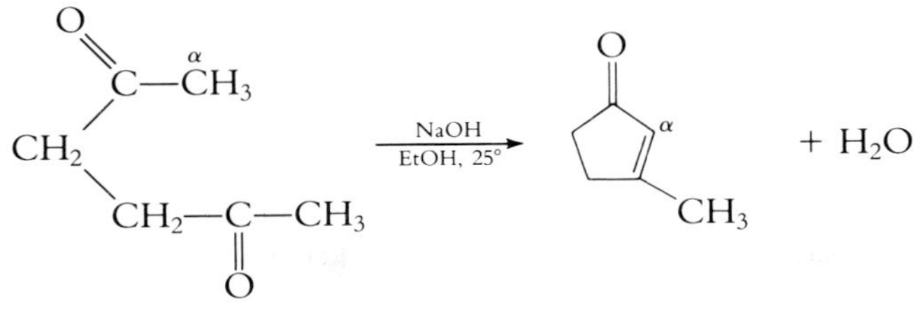 2,5-Hexanedione $\xrightarrow[\text{EtOH, 25°}]{\text{NaOH}}$ 3-Methyl-2-cyclopentenone (42%) + H_2O

Sections 18:1 through 18:5 showed that the product yields in aldol condensations depend on the type of catalysis, the structure of the reactants, and the temperature. To help you keep track of all these variables, here is an overview of the aldol condensation as applied to aliphatic aldehydes and ketones.

1. *Aliphatic aldehyde, base catalysis* At low temperatures the major product is a hydroxyaldehyde (addition product). At high temperatures the major product is an unsaturated aldehyde (condensation product).
2. *Aliphatic aldehyde, acid catalysis* The major product is an unsaturated aldehyde (condensation product).
3. *Aliphatic ketone, base catalysis* These conditions do not usually lead to useful yields of products.
4. *Aliphatic ketone, acid catalysis* The major product is an unsaturated ketone (condensation product).
5. *Dicarbonyl aldehydes or ketones, acid or base catalysis* The major products are five- or six-membered cyclic unsaturated aldehydes or ketones (intramolecular condensation products). Other ring sizes are not produced.

Example

Show how the following compound can be prepared by way of an aldol condensation.

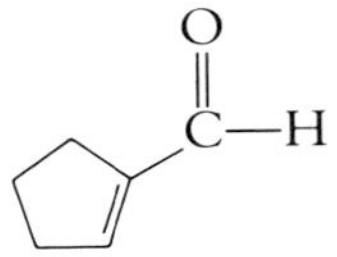

1-Cyclopentenecarboxaldehyde

Answer

We know that the double bond in such an α,β-unsaturated carbonyl compound was formed by dehydration of an alcohol. Hence, the *last* step is

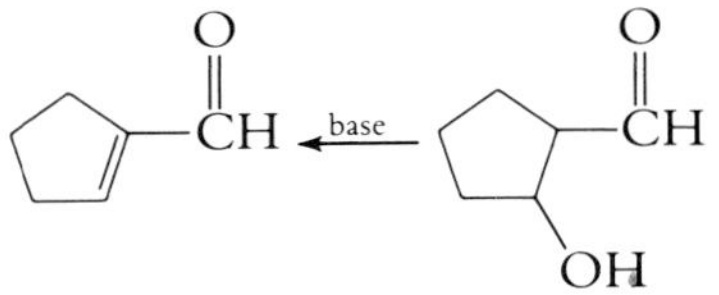

1-Cyclopentenecarboxaldehyde $\xleftarrow{\text{base}}$ 2-hydroxycyclopentanecarboxaldehyde (CHO, OH)

Since the aldol condensation involves forming a new carbon–carbon bond between an α carbon next to a carbonyl group and the carbonyl carbon of another carbonyl group, identification of the bond which was formed is the next step:

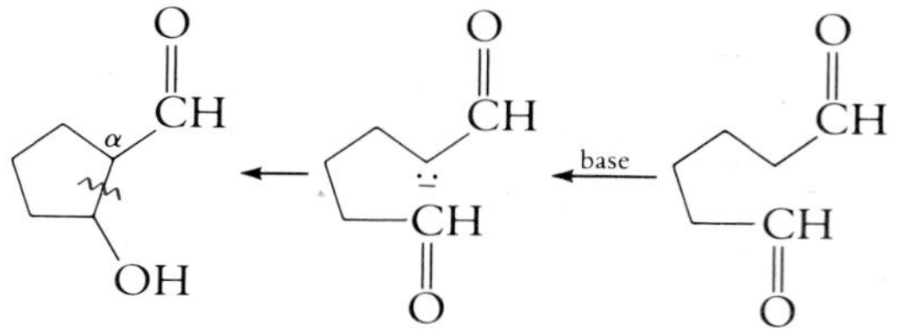

Problems

18–9 Compound **IV** is heated with *p*-toluenesulfonic acid in benzene, and water is removed by azeotropic distillation. Product **A** is formed in high yield. Suggest a structure for **A**.

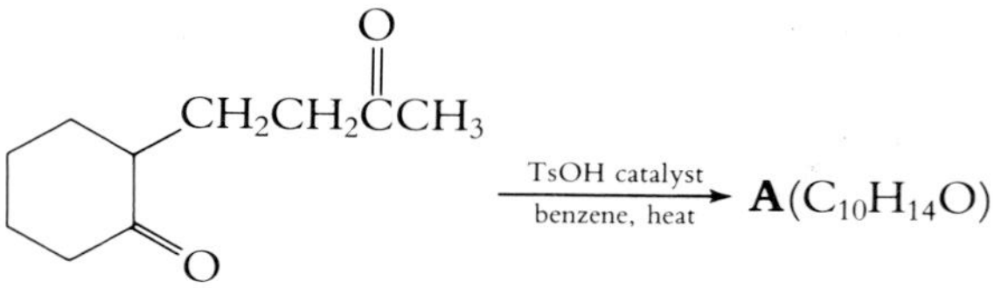

$$\textbf{IV} \xrightarrow[\text{benzene, heat}]{\text{TsOH catalyst}} \textbf{A}(C_{10}H_{14}O) \quad (90\%)$$

18–10 Show how the following compounds can be prepared by way of aldol condensations.

a. 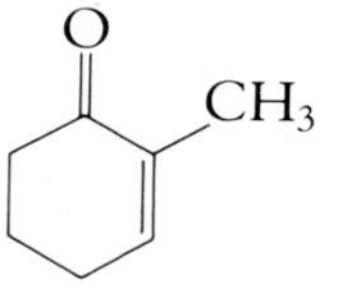 b. 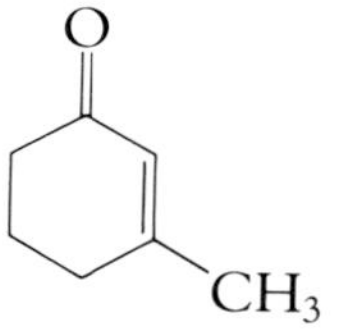 c.

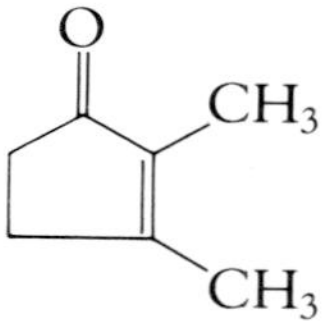

18:6 Why Mixed Aldol Condensations of Aliphatic Aldehydes Are Not Useful

As a general rule, mixed aldol condensations that use two *different* aldehydes are not preparatively useful because they give a mixture of too many products. For instance, given the reaction of acetaldehyde and propanal, four products come immediately to mind: one from self-condensation of propanal; one from self-condensation of acetaldehyde; and two mixed aldol condensation products (eqs. 18.26 and 18.27). (Actually, the product mixture is even more complex because of diastereomers of the products of eq. 18.27.)

(18.26) $\mathrm{CH_3\overset{O}{\overset{\|}{C}}H} \longrightarrow$
Acetaldehyde

$\mathrm{:\bar{C}H_2\overset{O}{\overset{\|}{C}}H}$ (Acetaldehyde enolate ion) $\xrightarrow{\mathrm{CH_3\overset{O}{\overset{\|}{C}}H}}$ $\mathrm{CH_3\overset{OH}{\overset{|}{C}}HCH_2\overset{O}{\overset{\|}{C}}H}$ self-condensation product

$\mathrm{:\bar{C}H_2\overset{O}{\overset{\|}{C}}H}$ $\xrightarrow{\mathrm{CH_3CH_2\overset{O}{\overset{\|}{C}}H}}$ $\mathrm{CH_3CH_2\overset{OH}{\overset{|}{C}}HCH_2\overset{O}{\overset{\|}{C}}H}$ mixed condensation product

(18.27) $\mathrm{CH_3CH_2\overset{O}{\overset{\|}{C}}H} \longrightarrow$
Propanal

$\mathrm{CH_3\ddot{\bar{C}}H\overset{O}{\overset{\|}{C}}H}$ (Propanal enolate ion) $\xrightarrow{\mathrm{CH_3CH_2\overset{O}{\overset{\|}{C}}H}}$ $\mathrm{CH_3CH_2\overset{OH}{\overset{|}{C}}H\underset{CH_3}{\underset{|}{C}}H\overset{O}{\overset{\|}{C}}H}$ self-condensation product

$\mathrm{CH_3\ddot{\bar{C}}H\overset{O}{\overset{\|}{C}}H}$ $\xrightarrow{\mathrm{CH_3\overset{O}{\overset{\|}{C}}H}}$ $\mathrm{CH_3\overset{OH}{\overset{|}{C}}H\underset{CH_3}{\underset{|}{C}}H\overset{O}{\overset{\|}{C}}H}$ mixed condensation product

The complexity of a mixed aldol condensation can be reduced considerably if only *one* of the reactants is **enolizable**. *An enolizable carbonyl compound is one that has α hydrogens and can give rise to an enol or enolate ion.* If only one enolate ion can be formed, for instance, then the number of possible products in a mixed aldol condensation is reduced. In fact, mixed aldol condensations utilizing aromatic aldehydes which are not enolizable, such as benzaldehyde (see eq. 18.12), are very common. They are described in the next section.

18:7 Mixed Aldol Condensations Using Aromatic Aldehydes

Among the most important and useful aldol condensations are those in which an aromatic aldehyde reacts with an aliphatic aldehyde or ketone. Since only the aliphatic carbonyl compound can enolize, only two products are possible—the *mixed* product from attack of the enolate ion on the aromatic aldehyde, and the self-condensation product of the aliphatic compound. When the experiments are actually carried out only the mixed product is isolated, usually in high yield (eqs. 18.28 and 18.29). Note particularly that this is the condensation product, even though the reaction is carried out under mild conditions with basic catalysts (compare eqs. 18.28 and 18.29 with eqs. 18.5 and 18.20).

(18.28) C_6H_5—CHO + $CH_3\overset{\overset{O}{\|}}{C}CH_3$ $\xrightarrow[H_2O,\ EtOH]{NaOH,\ 25°}$

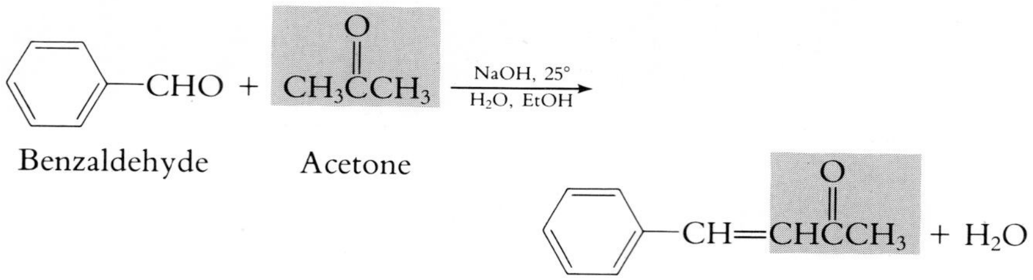

Benzaldehyde Acetone

C_6H_5—CH=CH$\overset{\overset{O}{\|}}{C}CH_3$ + H_2O

4-Phenyl-3-buten-2-one
(*benzalacetone,* 70%)

(18.29) 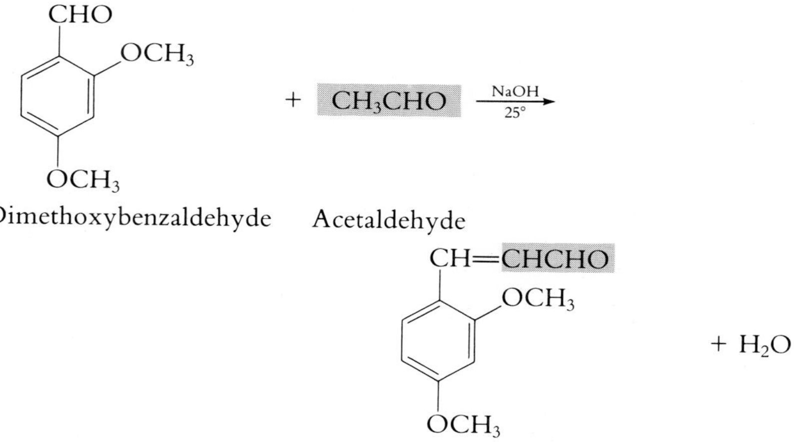 2,4-$(OCH_3)_2C_6H_3$—CHO + CH_3CHO $\xrightarrow[25°]{NaOH}$

2,4-Dimethoxybenzaldehyde Acetaldehyde

2,4-$(OCH_3)_2C_6H_3$—CH=CHCHO + H_2O

3-(2,4-dimethoxyphenyl)propenal
(*2,4-dimethoxycinnamaldehyde*)

These results can be explained by the fact that the mixed condensation product is the most stable possible product because of the extensive conjugation involving the C=O, C=C, and aryl groups. The intermediate aldol or ketol is rapidly dehydrated to yield such a stable product (eq. 18.30).

(18.30) C_6H_5—CHO + $CH_3\overset{\overset{O}{\|}}{C}R$ $\rightleftharpoons$ C_6H_5—$\overset{\overset{OH}{|}}{C}HCH_2\overset{\overset{O}{\|}}{C}R$ $\xrightarrow{fast}$

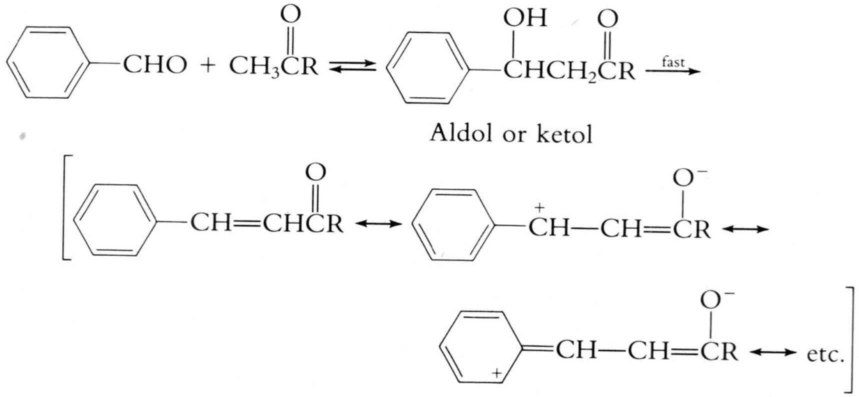

Aldol or ketol

[C_6H_5—CH=CH$\overset{\overset{O}{\|}}{C}R$ ⟷ C_6H_5—$\overset{+}{C}H$—CH=$\overset{\overset{O^-}{|}}{C}R$ ⟷

(+)C_6H_5=CH—CH=$\overset{\overset{O^-}{|}}{C}R$ ⟷ etc.]

Problem

18–11 Give the expected product(s) for the following reactions and show a detailed mechanism for their formation.

a. benzaldehyde + propanal $\xrightarrow[H_2O,\ EtOH]{NaOH}$

b. 2 benzaldehyde + acetone $\xrightarrow[H_2O,\ EtOH]{NaOH}$

If benzaldehyde reacts with an *unsymmetrical* ketone, condensation can occur at either α carbon. When the reaction is *base catalyzed,* condensation usually takes place at the methyl group (eq. 18.31). *Acid-promoted* condensations are more difficult to predict.

$$C_6H_5CHO + \overset{\alpha}{C}H_3\overset{O}{\overset{\|}{C}}\overset{\alpha}{C}H_2CH_3 \xrightarrow{NaOH} C_6H_5CH{=}CH\overset{O}{\overset{\|}{C}}CH_2CH_3 \quad (99\%) \tag{18.31}$$

Problem

18–12 Show the major product(s) of each of the following reactions:

a. $CH_3CH_2CH_2\overset{O}{\overset{\|}{C}}H \xrightarrow[25°]{NaOH}$

b. $CH_3(CH_2)_3\overset{O}{\overset{\|}{C}}H \xrightarrow[80°]{NaOH}$

c. $(CH_3)_2CH\overset{O}{\overset{\|}{C}}CH_3 \xrightarrow{H_2SO_4}$

d. $C_6H_5\overset{O}{\overset{\|}{C}}CH_3 \xrightarrow{H_2SO_4}$

e. cyclopentanone ($C_5H_8{=}O$) + $C_6H_5CHO \xrightarrow{KOH}$

f. $O_2N{-}C_6H_4{-}CHO + CH_3CH_2\overset{O}{\overset{\|}{C}}CH_3 \xrightarrow{NaOH}$

g. $C_6H_5\overset{O}{\overset{\|}{C}}CH_2CH_2\overset{O}{\overset{\|}{C}}CH_3 \xrightarrow{NaOH}$

18:8 Retrosynthetic Analysis of Aldol Products

It is unusual for a chemist to be presented with a set of reactions and be asked to predict their outcome, as we have asked you to do in Problem 18–12. (We use this approach as an aid in learning the course of the reactions.) More commonly, the chemist is presented with the structure of a *product,* and is asked to arrange a series of known reactions in order to achieve an efficient synthesis of that product. To do that, he or she must be able to perform a *retrosynthetic analysis* of the product structure (sec. 12:9). In this chapter and in Chapter 19, we will concentrate on solving synthetic problems built around enolate reactions. This should help to summarize some of the many reactions already shown.

The first step in retrosynthetic analysis is to identify a strategic carbon–carbon bond (sec. 12:9), whose formation would allow preparation of a complex *target molecule* from two simpler ones. In the case of aldol-type products, the strategic bond is the one that was originally formed between the α carbon of one molecule and the carbonyl carbon of the second molecule (see Example, p. 691). Equations 18.32 to 18.34 show retrosynthetic analyses of some simple products derived from aldol reactions.

(18.32) $\text{C}_5\text{H}_9\text{-CH(CHO)}\text{-CH(OH)-CH}_2\text{-C}_5\text{H}_9 \leftarrow \text{C}_5\text{H}_9\text{-CH}_2\text{-CHO} + \text{OHC-CH}_2\text{-C}_5\text{H}_9$

Target molecule

(18.33) bicyclic enone (hexahydroindenone) $\leftarrow$ 2-($\text{CH}_2\text{C(=O)CH}_3$)cyclohexanone

(18.34) $\text{C}_6\text{H}_5\text{-CH=C(CH}_3\text{)-CHO} \leftarrow \text{C}_6\text{H}_5\text{-CHO} + \text{CH}_3\text{CH}_2\text{-CHO}$

Target molecule

The target molecule can also be one in which a functional group change is necessary after the aldol reaction. Equations 18.35 and 18.36 give two examples.

(18.35) $\text{Cl-C}_6\text{H}_4\text{-CH=C(CH}_3\text{)COOH} \xleftarrow{\text{H}_2\text{CrO}_4}$

$\text{Cl-C}_6\text{H}_4\text{-CH=C(CH}_3\text{)CHO} \xleftarrow{\text{NaOH}} \text{Cl-C}_6\text{H}_4\text{-CHO} + \text{CH}_3\text{CH}_2\text{CHO}$

(18.36) 1-cyclopentyl-2-methoxycyclopentane (OCH_3) $\xleftarrow{\text{1. Na; 2. CH}_3\text{I}}$ 2-cyclopentylcyclopentanol (OH) $\xleftarrow[\text{cat}]{\text{H}_2}$

2-cyclopentylidenecyclopentanone $\xleftarrow{\text{H}_2\text{SO}_4}$ 2 cyclopentanone ($=\text{O}$)

What we look for in the intermediates or products are certain structural groupings, which could have resulted from an aldol condensation. One such grouping is two oxygen-bearing carbons in a 1,3-relationship. Examples are β-hydroxyaldehydes and ketones, and 1,3-diols. Another telltale group is an oxygen-bearing carbon next to a double bond, as in α,β-unsaturated carbonyl compounds, carboxylic acids, and alcohols.

It should be evident from these examples and the preceding problems that aldol reactions can be an important tool in the conversion of simple starting materials to complex organic compounds.

Problem

18–13 Use retrosynthetic analysis to suggest ways of preparing the following compounds through use of aldol-type reactions.

a. 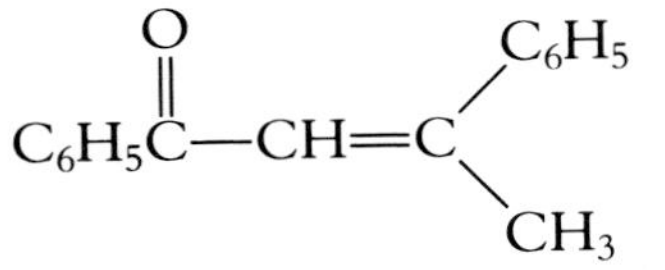 $C_6H_5C(=O)—CH=C(C_6H_5)(CH_3)$

b. CHO

c. HO OH

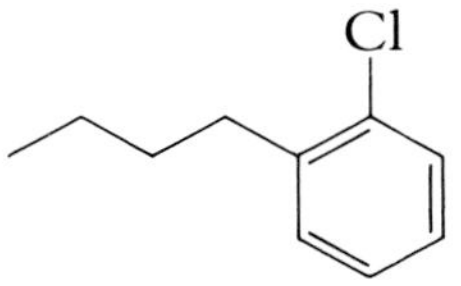

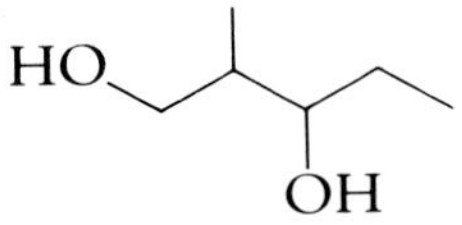

d. 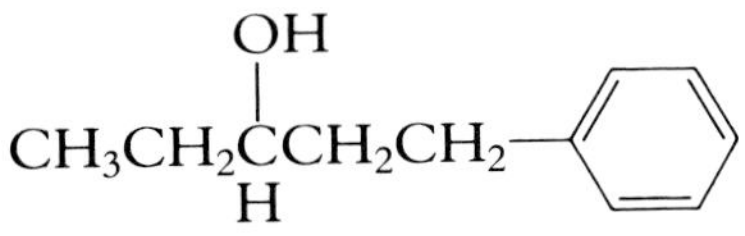 $CH_3CH_2CH(OH)CH_2CH_2—C_6H_5$

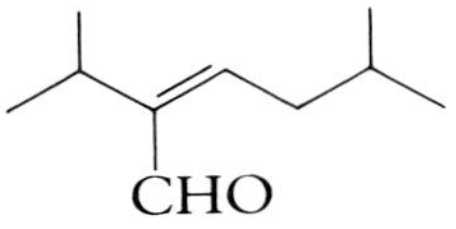

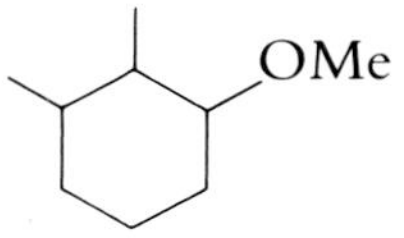

e. Cl

f. OMe

18:9 Reaction of Formaldehyde and Amines with Carbonyl Compounds: The Mannich Reaction

Amines react readily with carbonyl compounds (sec. 11:7) to give several types of products, some of which can take part in aldol-type condensations and related reactions. We'll explore one of these reactions—the Mannich reaction—in this section and introduce some others in Chapter 19.

The **Mannich reaction** involves mixing together one equivalent each of a secondary amine, formaldehyde, and a carbonyl compound. They react in the presence of an acid catalyst to yield a dialkylaminocarbonyl compound (eq. 18.37). Since formaldehyde is the more reactive of the two carbonyl

$$Et_2NH + CH_2O + CH_3\overset{O}{\overset{\|}{C}}CH_3 \xrightarrow[MeOH]{HCl} Et_2NCH_2CH_2\overset{O}{\overset{\|}{C}}CH_3 + H_2O \quad (18.37)$$

Diethylamine — 4-Diethylamino-2-butanone (70%)

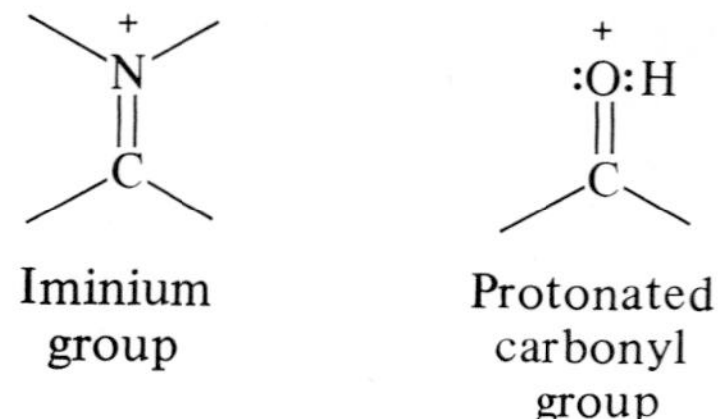

Figure 18–1. *Resemblance between iminium groups and protonated carbonyl groups.*

compounds, it reacts with the amine in the first step of the process. The product is an *iminium ion* (eq. 18.38), which strongly resembles a protonated carbonyl compound (Fig. 18–1). In the next step, the iminium ion reacts with the enol of the carbonyl compound (eq. 18.39). This reaction is very similar to an acid-catalyzed aldol condensation (eq. 18.18). Formaldehyde is used because it has no enolizable hydrogens and cannot undergo self-condensation.

$$\text{(18.38)}\quad Et_2\ddot{N}H + CH_2{=}O \rightleftharpoons Et_2\ddot{N}{-}CH_2OH \overset{H^+}{\rightleftharpoons} Et_2\overset{+}{N}{=}CH_2 + H_2O$$

Iminium ion

$$\text{(18.39)}\quad Et_2\overset{+}{N}{=}CH_2 + CH_2{=}\overset{\overset{:\ddot{O}-H}{|}}{C}CH_3 \longrightarrow Et_2\ddot{N}CH_2CH_2\overset{\overset{:O:}{\|}}{C}CH_3 + H^+$$

Iminium ion — Enol

The net effect of the Mannich reaction is to replace an α hydrogen of an aldehyde or ketone with the group R_2NCH_2. This change is shown in Equations 18.40 and 18.41.

$$\text{(18.40)}\quad (CH_3)_2NH + CH_2O + \text{cyclohexanone} \xrightarrow{H^+} \text{2-}[(CH_3)_2NCH_2]\text{cyclohexanone}$$

$$\text{(18.41)}\quad \text{pyrrolidine (N–H)} + CH_2O + H{-}\overset{\overset{CH_3}{|}}{\underset{\underset{CH_3}{|}}{C}}{-}\overset{\overset{O}{\|}}{C}H \xrightarrow{H^+} \text{(pyrrolidinyl)}NCH_2{-}\overset{\overset{CH_3}{|}}{\underset{\underset{CH_3}{|}}{C}}{-}\overset{\overset{O}{\|}}{C}H$$

The Mannich reaction is one of the key reactions in plant alkaloid biosynthesis. When the reaction occurs in the plant, rather than in the laboratory, the formation of the iminium ion (**7**) is often *intramolecular* because the aldehyde group and the amine are in the same molecule (step 1, eq. 18.42).

(18.42) [reaction scheme: N-methyl-4-aminobutanal cyclizes (H^+, step 1) to N-methylpyrrolinium ion **7**, which reacts (step 2) with the enol $CH{=}C(OH){-}CH_3$ bearing COOH, giving the pyrrolidinyl $CH(COOH){-}C(=O)CH_3$ product, which decarboxylates to the $CH_2C(=O)CH_3$ derivative, Hygrine]

The iminium ion can then react with an enol (step 2, eq. 18.42). Decarboxylation of the first product occurs in the plant to produce the alkaloid hygrine, which is a constituent of *Erythroxylon coca*. Cocaine, another alkaloid in the same plant, is also formed by a Mannich-type reaction.

Like hygrine, all alkaloids are naturally occurring nitrogen-containing compounds. They are usually physiologically active, and are found mainly in plants. Alkaloids are classified as **secondary metabolites** because they occur only sporadically, unlike primary metabolites such as amino acids and peptides, which are fundamental to all living matter. Secondary metabolites may serve as defense mechanisms since many are toxic and prevent the plants from being destroyed by herbivores.

Example

Show how to synthesize the following compound by a Mannich reaction.

$$CH_3CH_2\overset{O}{\overset{\|}{C}}CH_2CH_2NMe_2$$

Answer

The Mannich reaction involves combination of a carbonyl compound, formaldehyde, and an amine. The formaldehyde portion represents the carbon between the amine and the carbonyl compound. The synthetic target can thus be divided (and put together) as follows:

$$\underbrace{CH_3CH_2\overset{O}{\overset{\|}{C}}}_{\text{from the carbonyl compound}}\underbrace{CH_2}_{\text{from the formaldehyde}}\underbrace{CH_2NMe_2}_{\text{from the amine}}$$

The synthesis would therefore involve 2-butanone, formaldehyde, and dimethylamine:

$$CH_3CH_2\overset{O}{\overset{\|}{C}}CH_3 + H_2C{=}O + HNMe_2 \xrightarrow{H^+} CH_3CH_2\overset{O}{\overset{\|}{C}}CH_2CH_2NMe_2$$

Problem

18–14 Show how each of the following Mannich products can be synthesized from a carbonyl compound, an amine, and formaldehyde. (A retrosynthetic analysis will help.)

a. 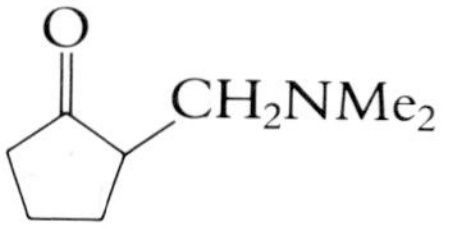 2-(CH_2NMe_2)cyclopentanone

b. 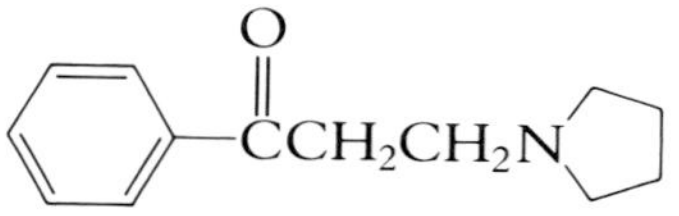 $C_6H_5\overset{O}{\overset{\|}{C}}CH_2CH_2N$(pyrrolidinyl)

c. $Cl-C_6H_4-\overset{O}{\overset{\|}{C}}\underset{CH_3}{\underset{|}{C}}HCH_2NEt_2$

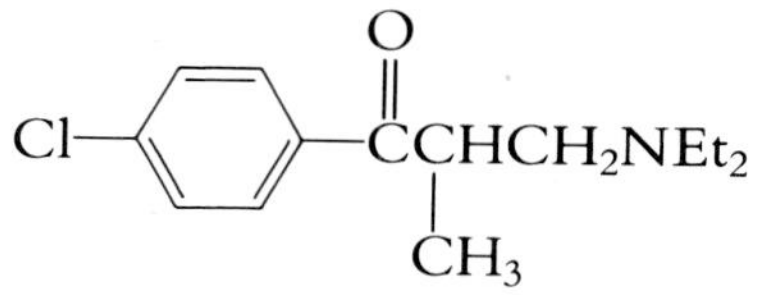

d. 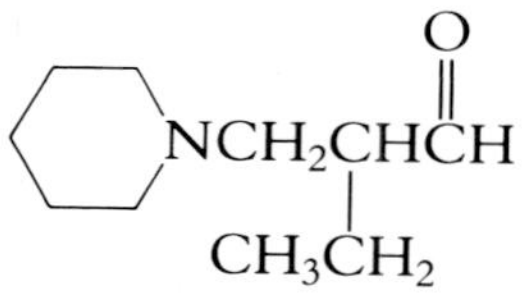 (piperidinyl)$N CH_2\underset{CH_3CH_2}{\underset{|}{C}}H\overset{O}{\overset{\|}{C}}H$

18:10 1,3-Dicarbonyl Compounds: Precursors to Stable Enols and Enolate Ions

Enols and enolates of monocarbonyl compounds are reactive and unstable intermediates, present in only low concentrations under equilibrium conditions. The small equilibrium concentration of acetaldehyde enol is typical (eq. 18.43).

$$CH_3\overset{O}{\overset{\|}{C}}H \rightleftarrows CH_2{=}\overset{OH}{\overset{|}{C}}H \qquad K = 5 \times 10^{-6} \tag{18.43}$$

Similarly, the relatively low acidity of monocarbonyl compounds means that their enolate ions are present to only a minor extent in protic solvents (eq. 18.44).

$$CH_3\overset{O}{\overset{\|}{C}}CH_3 + OH^- \overset{H_2O}{\rightleftarrows} [CH_3\overset{O^-}{\overset{|}{C}}{=}CH_2 \longleftrightarrow CH_3\overset{O}{\overset{\|}{C}}{-}\bar{C}H_2] + H_2O \qquad K \sim 10^{-5} \tag{18.44}$$

The situation is substantially different for 1,3-dicarbonyl compounds. The α hydrogens located *between* the two carbonyl groups are highly acidic and they enolize very rapidly. For instance, the nmr and ir spectra of 2,4-pentanedione indicate the presence of considerable amounts of the enol. In fact, in nonpolar solvents the enol tautomer predominates over the diketone (eq. 18.45).

$$CH_3\overset{O}{\overset{\|}{C}}CH_2\overset{O}{\overset{\|}{C}}CH_3 \underset{\text{hexane}}{\overset{0.1\ M\text{ in}}{\rightleftarrows}} CH_3{-}\overset{O-H\cdots O}{C{=}CH{-}C}{-}CH_3 \qquad K^{25°} = 11 \tag{18.45}$$

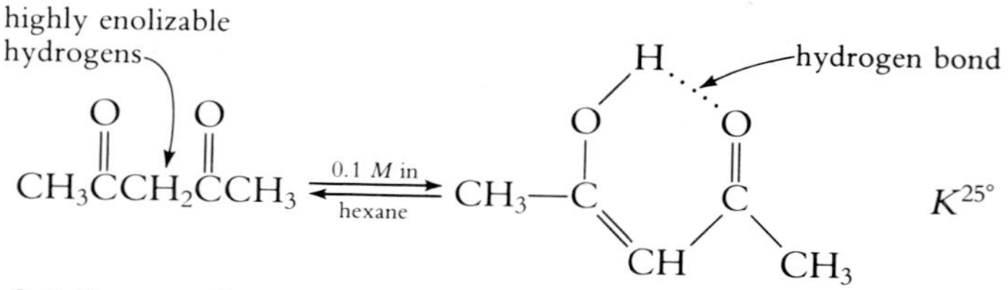

highly enolizable hydrogens (arrow to CH_2); hydrogen bond (arrow to $H\cdots O$)

2,4-Pentanedione (keto tautomer) — 2,4-Pentanedione (enol tautomer)

Part of the driving force for enol formation in 1,3-dicarbonyl compounds must be the creation of a conjugated unsaturated system. The enol C=C bond is conjugated with the remaining C=O group, and with the nonbonding $\ddot{O}H$ electrons. Furthermore, the enol forms of ethyl acetoacetate and of 2,4-pentanedione are also stabilized by intramolecular hydrogen-bonding through a six-membered ring (eq. 18.45). For ethyl 3-oxobutanoate, more commonly known as *ethyl acetoacetate,* there is an equal mixture of keto and enol tautomers in hexane solvent (eq. 18.46).

$$CH_3\overset{\overset{\displaystyle O}{\|}}{C}CH_2\overset{\overset{\displaystyle O}{\|}}{C}OEt \underset{\text{hexane}}{\overset{0.1\ M\ \text{in}}{\rightleftharpoons}} CH_3C(OH){=}CH{-}C({=}O)OEt \qquad K^{25°} = 1 \tag{18.46}$$

highly enolizable hydrogens (the CH_2 hydrogens of the keto tautomer); hydrogen bond (between the enol O–H and the C=O oxygen)

Ethyl acetoacetate (keto tautomer) — Ethyl acetoacetate (enol tautomer)

Problems

18–15 The nmr spectrum of ethyl acetoacetate (20% in CCl_4) is given below. Which peaks are due to the enol? Assign them to specific protons in the enol structure. What is the enol percentage in this solvent?

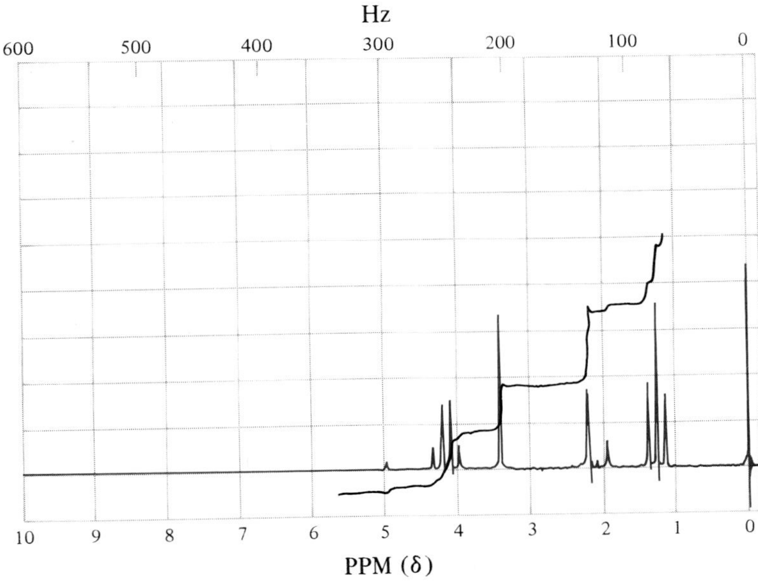

18–16 Simple carbonyl compounds have weak uv absorptions at 280 to 290 nm, and esters at 204 nm. However, a solution of ethyl acetoacetate in hexane shows strong absorption at 245 nm. Explain this finding. (A review of Section 13:13 may be helpful.)

Table 18–1. Acidities of Some Carbon Acids

Compound	Anion	pK_a
toluene, $PhCH_3$	$PhCH_2\colon^-$	41
acetylene, $HC{\equiv}CH$	$HC{\equiv}C\colon^-$	26
acetone, $CH_3\overset{O}{\overset{\parallel}{C}}CH_3$	$CH_3\overset{O}{\overset{\parallel}{C}}CH_2\colon^-$	20
diethyl malonate, $CH_2(COOEt)_2$	$^-\colon CH(COOEt)_2$	13
malononitrile, $CH_2(CN)_2$	$^-\colon CH(CN)_2$	11
ethyl acetoacetate, $CH_3\overset{O}{\overset{\parallel}{C}}CH_2COOEt$	$CH_3\overset{O}{\overset{\parallel}{C}}{-}\overset{..}{C}HCOOEt$ (anion)	11
nitromethane, CH_3NO_2	$^-\colon CH_2NO_2$	10
2,4-pentanedione, $CH_3\overset{O}{\overset{\parallel}{C}}CH_2\overset{O}{\overset{\parallel}{C}}CH_3$	$CH_3\overset{O}{\overset{\parallel}{C}}{-}\overset{..}{C}H\overset{O}{\overset{\parallel}{C}}CH_3$ (anion)	9
nitroacetone, $CH_3\overset{O}{\overset{\parallel}{C}}CH_2NO_2$	$CH_3\overset{O}{\overset{\parallel}{C}}{-}\overset{..}{C}HNO_2$ (anion)	5
dinitromethane, $CH_2(NO_2)_2$	$^-\colon CH(NO_2)_2$	4

1,3-Dicarbonyl compounds generate highly resonance-stabilized enolate ions when treated with bases. The stability of the enolate ions accounts for the fact that the 1,3-dicarbonyl compounds have pK_as considerably lower than 15 (that is, they are stronger acids than water; eqs. 18.47 and 18.48). The pK_as of some carbon acids are given in Table 18–1. Note the substantial difference between those for monocarbonyl compounds (such as acetone), and those for dicarbonyl compounds (diethyl malonate, ethyl acetoacetate).

$$H_2O + CH_2(\overset{O}{\overset{\parallel}{C}}OEt)(\underset{\underset{O}{\parallel}}{C}CH_3) \rightleftarrows H_3O^+ + \left[{}^-CH(\overset{:O:}{\overset{\parallel}{C}}OEt)(\underset{\underset{:O:}{\parallel}}{C}CH_3) \longleftrightarrow CH(\overset{:O:}{\overset{\parallel}{C}}OEt)({=}\underset{\underset{:\ddot{O}:^-}{|}}{C}CH_3) \longleftrightarrow CH({=}\overset{:\ddot{O}:^-}{\overset{|}{C}}OEt)(\underset{\underset{:O:}{\parallel}}{C}CH_3) \right] \tag{18.47}$$

Ethyl acetoacetate
$pK_a = 11$

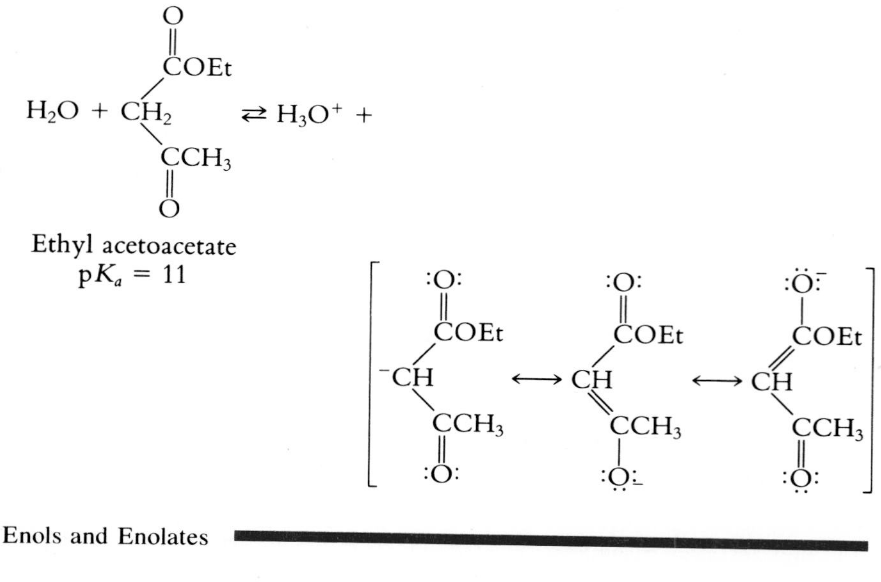

$$\text{(18.48)}\quad H_2O + \begin{array}{l}\overset{\displaystyle O}{\overset{\|}{C}}CH_3 \\ / \\ CH_2 \\ \backslash \\ \underset{\underset{\displaystyle O}{\|}}{C}CH_3\end{array} \rightleftarrows H_3O^+ + \left[{}^-CH(COCH_3)_2 \longleftrightarrow CH(=C(O^-)CH_3)(COCH_3) \longleftrightarrow CH(=C(O^-)CH_3)(COCH_3) \right]$$

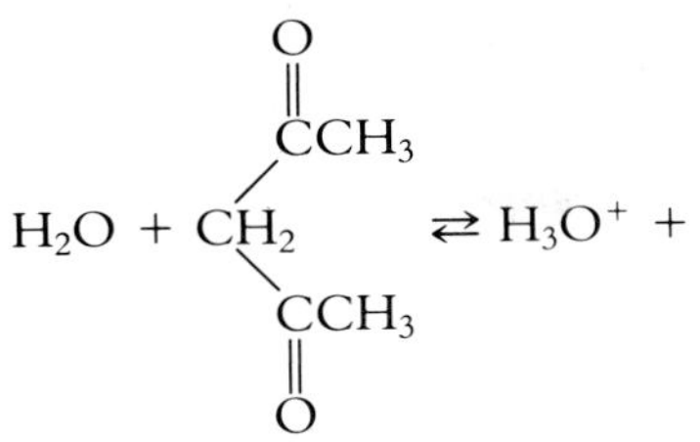

2,4-Pentanedione
$pK_a = 9$

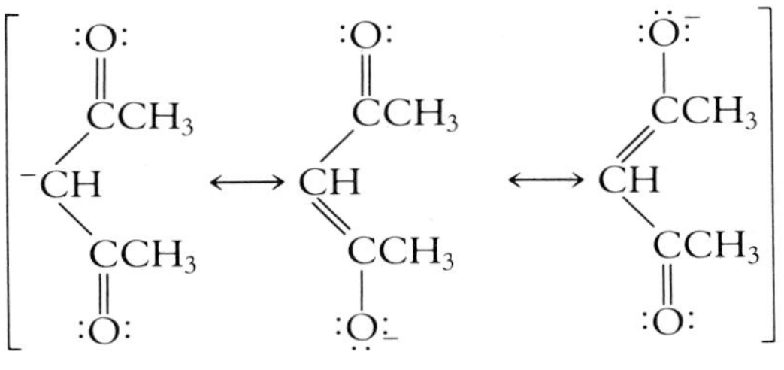

If you recall that the nitrile group behaves very much like a carbonyl group (sec. 17:12), then you will see that diethyl malonate and malononitrile are closely related. Both compounds generate relatively stable enolate ions (Table 18–1) that are important synthetic intermediates, and will be discussed again in Chapter 19.

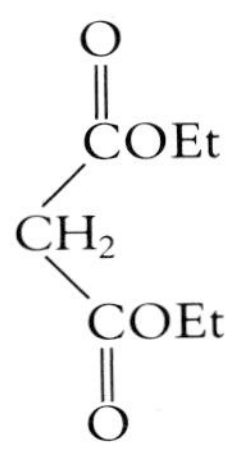

$$\begin{array}{l}\overset{\displaystyle O}{\overset{\|}{C}}OEt \\ / \\ CH_2 \\ \backslash \\ \underset{\underset{\displaystyle O}{\|}}{C}OEt\end{array} \qquad\qquad \begin{array}{l}C{\equiv}N \\ / \\ CH_2 \\ \backslash \\ C{\equiv}N\end{array}$$

Diethyl malonate 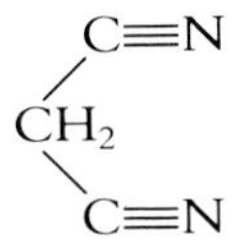 — Malononitrile

Problems

18–17 a. Draw all the resonance structures for the enolate ions that result when a methylene proton is removed from diethyl malonate and from malononitrile.

b. In the course of an attempted Grignard synthesis, a student added malononitrile to an ether solution of CH_3MgBr. There was immediate evolution of gas, and an organic salt was produced in solution. Adding HCl to this salt gave back malononitrile. Write the chemical reactions for the formation of the salt, and its conversion back to malononitrile. Why was the salt so readily formed?

18–18 Why does nitroacetone have a pK_a of 5 (Table 18–1)? This is about the same as the pK_a of acetic acid.

18:11 Halogenation of Ketones and Carboxylic Acids

In the presence of small quantities of acid, ketones can be halogenated at the α carbon atom with Br_2 or Cl_2 (eq. 18.49). The reaction proceeds readily in the dark (indicating that it is an ionic, rather than free-radical reaction), and attack occurs next to the carbonyl group exclusively.

(18.49) $$CH_3\overset{\overset{O}{\|}}{C}CH_3 + Br_2 \xrightarrow[H_2O,\ 65°]{CH_3COOH} CH_3\overset{\overset{O}{\|}}{C}CH_2Br + \text{dibromoacetones} + HBr$$
(50%)

Acetic acid is a common solvent for this reaction. The acidic solvent promotes enolization of the ketone, and the halogen reacts with the enol (eq. 18.50). In this reaction the bromine is electrophilic, while the enol is nucleophilic, just as it is in acid-catalyzed aldol condensations (sec. 18:2).

(18.50) $$CH_3\overset{\overset{O}{\|}}{C}CH_3 \overset{H^+}{\rightleftharpoons} \underset{\text{Enol}}{CH_3\overset{\overset{:OH}{|}}{C}{=}CH_2} \xrightarrow{Br{-}Br}$$

$$CH_3\overset{\overset{^+OH}{\|}}{C}CH_2{-}Br + Br^- \rightleftharpoons CH_3\overset{\overset{O}{\|}}{C}CH_2Br + HBr$$

The reaction of bromine with enols is extremely rapid—so rapid, in fact, that the enol content of ketones can be determined by rapid titration using very small concentrations of bromine. Reaction of the enol with bromine is complete long before there is any appreciable reestablishment of the keto–enol equilibrium.

In the case of acid-catalyzed halogenation of unsymmetrical ketones, the regioselectivity depends very much on the solvent used. Halogenation occurs at the more highly substituted position in *nonpolar solvents* (eq. 18.51; SO_2Cl_2 is a source of electrophilic Cl). The more stable of the two possible enols reacts with Cl_2 to give the product of Equation 18.51.

(18.51) 2-methylcyclohexanone (O, CH_3) $+ SO_2Cl_2 \xrightarrow{CCl_4,\ 25°}$ 2-chloro-2-methylcyclohexanone (O, CH_3, Cl) $+ HCl + SO_2$
(85%)

A halogen atom already in the molecule discourages further acid-catalyzed enolization, so it is possible to selectively introduce one halogen atom per molecule.

In methanol, a *polar* solvent, it is possible to brominate methyl alkyl ketones primarily at the methyl group (eq. 18.52).

(18.52) $$(CH_3)_2CHCOCH_3 \xrightarrow[CH_3OH]{Br_2} \underset{(95\%)}{(CH_3)_2CHCOCH_2Br} + \underset{(5\%)}{(CH_3)_2\overset{\overset{Br}{|}}{C}COCH_3}$$

Ketones may be brominated under basic conditions, but this always leads to polybromination, even with a deficiency of bromine. Base-promoted bromination is therefore not a useful method for preparing haloketones.

Carboxylic acids can be halogenated by a procedure involving Br_2 or Cl_2 and phosphorus (eq. 18.53). The phosphorus is converted to PBr_3, and the PBr_3, in turn, converts the acid to an acid bromide (eq. 18.54). It is the acid

$$(18.53)\quad CH_3(CH_2)_{13}CH_2COOH + Br_2 \xrightarrow[70°]{P} CH_3(CH_2)_{13}\underset{\displaystyle Br}{\underset{|}{C}}HCOOH + HBr$$

Hexadecanoic acid — 2-Bromohexadecanoic acid

$$(18.54)\quad 3RCOOH + PBr_3 \rightarrow 3R\overset{\displaystyle O}{\overset{\|}{C}}Br + H_3PO_3$$

halide that is actually brominated because it enolizes to a much greater extent than the acid. We can, in fact, substitute PCl_3 or PBr_3 for phosphorus, or discard them altogether by using an acid chloride or bromide directly as the reaction substrate. α-Halocarboxylic acids are useful precursors to amino acids (eq. 18.55).

$$(18.55)\quad CH_3\underset{\displaystyle Br}{\underset{|}{C}}HCOOH + NH_3\ (conc) \xrightarrow[25°]{H_2O} CH_3\underset{\displaystyle NH_3^+}{\underset{|}{C}}HCOO^- \quad (70\%)$$

Problem

18–19 Show how each of the following carboxylic acids can be converted into the desired product.

a. $(CH_3)_2CHCOOH \rightarrow (CH_3)_2\underset{\displaystyle Br}{\underset{|}{C}}COOH$

b. $CH_3CH_2CH_2COOH \rightarrow CH_3CH_2\underset{\displaystyle CN}{\underset{|}{C}}HCOOH$

c. $CH_3COOH \rightarrow CH_3NHCH_2COOH$

d. cyclopentanecarboxylic acid (cyclopentane–COOH) → cyclopent-1-enecarboxylic acid (cyclopentene–COOH)

Equation 18.53 suggests that the α hydrogens of carboxylic acids and their derivatives behave like the α hydrogens of aldehydes and ketones. This is indeed the case, and in Chapter 19 we will study the chemistry of enols and enolates of carboxylic acid derivatives.

Summary

1 The α hydrogens of aldehydes are unusually acidic. Bases remove these protons to yield nucleophilic enolate ions. The ions attack the carbonyl groups of other aldehyde molecules, yielding aldol addition products. These products are stable at room temperature and below, and may be isolated. When the reaction is carried out at higher temperatures the addition product loses water, giving an aldol condensation product (sec. 18:1).

$$RCH_2CHO + RCH_2CHO \xrightarrow{OH^-} RCH_2\overset{OH}{\overset{|}{C}}H\underset{\underset{R}{|}}{C}H\overset{O}{\overset{\|}{C}}H \xrightarrow[heat]{OH^-} RCH_2CH{=}\underset{\underset{R}{|}}{C}\overset{O}{\overset{\|}{C}}H$$

Addition product — Condensation product

2 Under acidic conditions aldehydes lose an α proton to form enols. Enols react with protonated aldehydes to give addition products that are quickly dehydrated to condensation products (sec. 18:2). Base-catalyzed condensation of simple ketones is not often used because ketol formation is not favored under equilibrium conditions (sec. 18:3). Acid-catalyzed ketone condensation, on the other hand, is an effective method for obtaining the condensation product (sec. 18:4). Intramolecular aldol condensation is very rapid when a five- or six-membered ring is formed. The condensation product is almost always obtained when either acidic or basic catalysts are used (sec. 18:5).

3 Mixed alcohol condensations are usually of limited preparative value when the reactants are all aliphatic (sec. 18:6). If one reactant is an aromatic aldehyde, it can be readily condensed with a variety of aliphatic aldehydes and ketones. In these cases the intermediate aldol or ketol is dehydrated in the course of the condensation, the product being a conjugated α,β-unsaturated aldehyde or ketone. The aromatic aldehyde condenses with the methyl group of methyl alkyl ketones if a basic catalyst is employed. (sec. 18:7).

4 Since the aldol condensation is very versatile, it is widely used for the synthesis of complex molecules. The initial aldol product may be subjected to additional functional group transformations before being converted to the final product. Retrosynthetic analysis shows that if the final product contains two oxygen-bearing carbons in a 1,3-relationship or an oxygen-bearing carbon connected to a double bond, then its synthesis could likely have incorporated an aldol condensation (sec. 18:8).

5 Amines react with formaldehyde in the presence of acid to give highly reactive iminium ions. These ions react with the enols of carbonyl compounds to give β-dialkylaminocarbonyl compounds. These so-called Mannich reactions are important not only in the laboratory, but also in the biosynthesis of alkaloids in plants (sec. 18:9).

6 The equilibrium enol content of 1,3-dicarbonyl compounds is much greater than the enol content of simple ketones and aldehydes. The 1,3-dicarbonyl enols are stabilized by conjugation and by intramolecular hydrogen-bond-

ing. 1,3-Dicarbonyl compounds are considerably more acidic than typical aldehydes and ketones (Table 18–1), because they form quite stable enolate ions (sec. 18:10).

7 Ketones and aldehydes are halogenated α to the carbonyl group, most commonly with bromine. These reactions are acid catalyzed and go by way of the enol. In nonpolar solvents the most highly substituted carbon is halogenated. In polar solvents the reverse is true (sec. 18:11).

Carboxylic acids can also be halogenated α to the carbonyl group. Phosphorus halides promote the reaction by converting the acids to acid halides. It is the acid halides that enolize and react to form α-halocarboxylic acids. These halo acids are useful synthetic intermediates in the formation of many α-substituted carboxylic acids (sec. 18:11).

Problems

18–20 Show all the products of each of the following reactions.

a. $CH_3CH_2CH_2\overset{O}{\overset{\|}{C}}H \xrightarrow[H_2O,\ 5^\circ]{NaOH}$

b. $CH_3CH_2\overset{O}{\overset{\|}{C}}H \xrightarrow[25^\circ]{HCl}$

c. $CH_3CH_2\overset{O}{\overset{\|}{C}}H \xrightarrow[H_2O,\ 85^\circ]{NaOH}$

d. $(CH_3)_2CHCH_2CH_2\overset{O}{\overset{\|}{C}}H \xrightarrow[CH_3COOH]{H_2SO_4}$

e. $H\overset{O}{\overset{\|}{C}}(CH_2)_4\overset{O}{\overset{\|}{C}}H \xrightarrow[H_2O,\ 50^\circ]{KOH}$

18–21 a. When acetaldehyde is heated with sodium ethoxide in ethanol, it condenses to crotonaldehyde, $CH_3CH{=}CHCHO$. Write a mechanism for this reaction, including the dehydration step.

b. The isolated yield of crotonaldehyde from this reaction is low, because the product condenses further with acetaldehyde and eventually yields a polymer. Write a mechanism for the reaction of crotonaldehyde with acetaldehyde. Which of the two molecules will be converted to its enolate ion? Explain your reasoning.

18–22 Write the structures of the products from each of the following reactions:

a. cyclopentanone $\xrightarrow{\text{dry HCl}}$

b. 3-methyl-2-butanone $\xrightarrow{H_2SO_4}$

c. *p*-methylbenzaldehyde + cyclohexanone $\xrightarrow[H_2O,\ C_2H_5OH,\ \Delta]{NaOH}$

d. *m*-nitrobenzaldehyde + 3-methylbutanal $\xrightarrow[H_2O,\ C_2H_5OH,\ \Delta]{NaOH}$

e. 3,4-dichlorobenzaldehyde + 2-pentanone $\xrightarrow[CH_3COOH,\ \Delta]{H_2SO_4}$

18–23 When acetone is dissolved in D_2O containing some NaOD, it is converted in low yield to diacetone alcohol (eq. 18.20). Most of the acetone is recovered. The recovered acetone is found to have exchanged all its hydrogen for deuterium (CD_3COCD_3). Suggest a mechanism for this exchange process, for which OD^- is a catalyst. This technique is an important one for introducing deuterium into organic molecules.

18–24 The ketone **V** has been obtained optically pure. When treated with aqueous sodium hydroxide, it undergoes racemization. The rate of racemization at a given pH is equivalent to the rate of hydrogen–deuterium exchange at the same pH (see Prob. 18–23). Write a mechanism accounting for both the racemization and the hydrogen exchange. What reactive intermediate is common to both reactions?

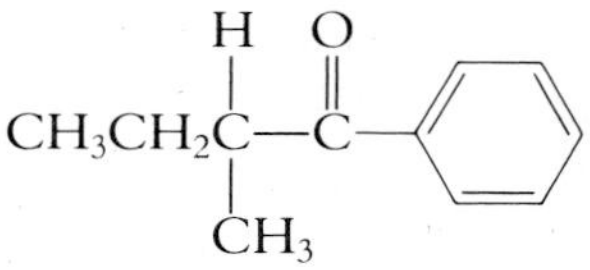

$$CH_3CH_2C(H)(CH_3)—C(=O)—C_6H_5$$

V

18–25 a. The intramolecular aldol condensation of 2,5-hexanedione (eq. 18.25) leads only to 3-methyl-2-cyclopentenone. What other intramolecular product could have reasonably been formed? Why wasn't it? What ring sizes would you expect to find from intramolecular aldol condensations?

b. Base-catalyzed aldol cyclization of 2,6-octanedione can lead to two six-membered cyclic products. Write their structures. The major product actually isolated is 2,3-dimethyl-2-cyclohexenone. Explain why this product is favored.

18–26 a. Sulfuric acid-catalyzed condensation of acetone leads to formation of 4-methyl-3-penten-2-one (*mesityl oxide*). Write a complete mechanism for this process.

b. A by-product of this reaction is *mesitylene* (1,3,5-trimethylbenzene). It has been shown that this compound arises from further acid-catalyzed condensation of mesityl oxide with acetone. Show how this reaction could take place. (A careful retrosynthetic analysis of the product will be helpful.)

18–27 The synthesis of each of the following compounds involves an aldol condensation as the final step. Show the structure of the reactants in each of those final steps.

a. 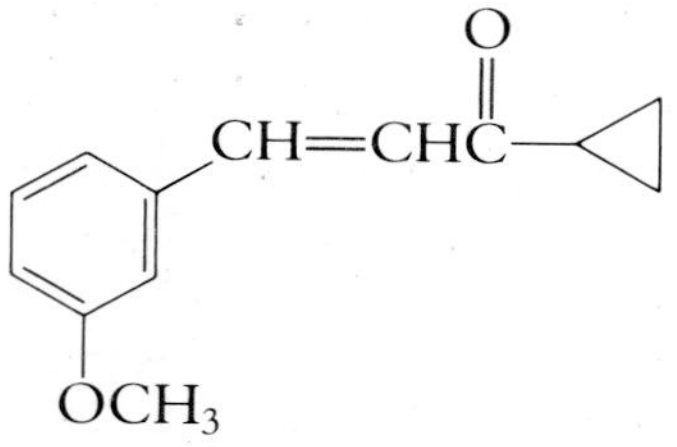 $3\text{-}CH_3O\text{-}C_6H_4—CH{=}CHC(=O)—\text{cyclopropyl}$ (CH=CHC(=O)–cyclopropyl; OCH_3)

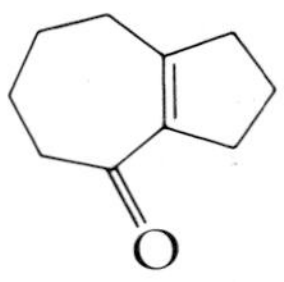

b. 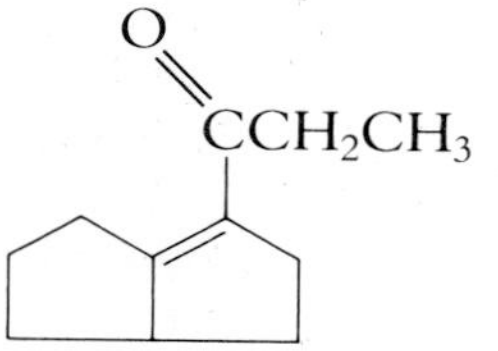 bicyclic (fused cyclopentene rings) bearing $C(=O)CH_2CH_3$

c. fused cycloheptanone–cyclopentene (C=O)

d. $PhCH_2CH_2CH{=}C(COOEt)_2$

18–28 The compounds [6]gingerol and [6]paradol have been found in certain pepper seeds. Show how each might be synthesized from vanillin.

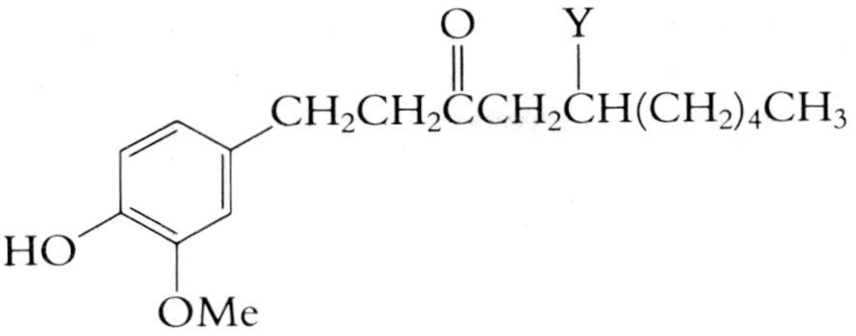

$$\text{Ar}-CH_2CH_2\overset{O}{\overset{\|}{C}}CH_2\overset{Y}{\overset{|}{C}}H(CH_2)_4CH_3 \quad (\text{Ar} = \text{4-HO-3-MeO-C}_6H_3)$$

Y = OH, [6]gingerol
Y = H, [6]paradol

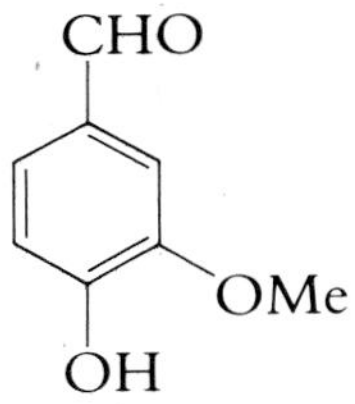

CHO, OMe, OH
Vanillin

18–29 The major aldehydes used in mixed aldol condensations are aromatic ones (sec. 18:7). Formaldehyde, which like the aromatic aldehydes lacks α hydrogen atoms, is also used in this way. Show how the following compounds can be synthesized by routes involving aldol condensations with formaldehyde.

a. 2,2-dimethyl-1,3-propanediol (*neopentyl glycol;* used in polyesters, paints, and lubricants)
b. $(ClCH_2)_2C(COOEt)_2$
c. 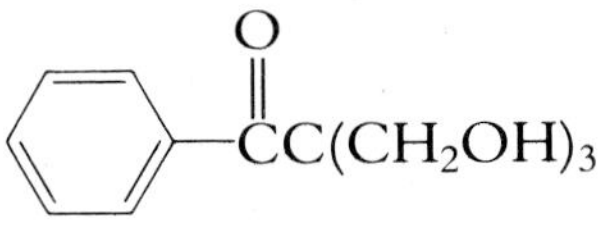 $C_6H_5\overset{O}{\overset{\|}{C}}C(CH_2OH)_3$

18–30 When a primary amine is used in a Mannich reaction, several products are often obtained. A study of the reaction in Equation 1 shows that **VII** could be obtained from **VI** by treating it with more formaldehyde and acetophenone. Show how **VI** and **VII** are probably formed.

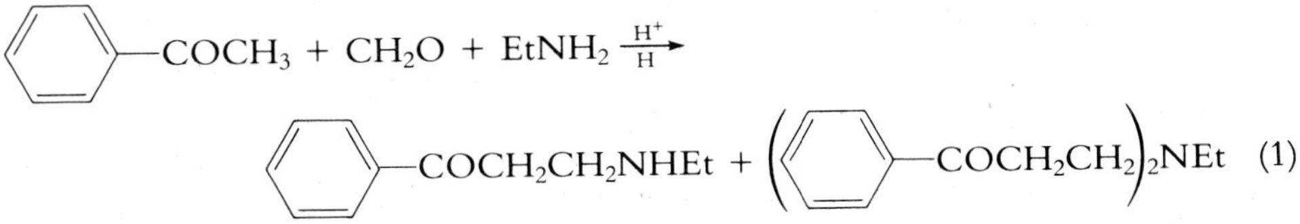

$$C_6H_5COCH_3 + CH_2O + EtNH_2 \xrightarrow[H]{H^+} \underset{\textbf{VI}}{C_6H_5COCH_2CH_2NHEt} + \underset{\textbf{VII}}{(C_6H_5COCH_2CH_2)_2NEt} \quad (1)$$

18–31 Ethyl acetoacetate is treated with sodium ethoxide, and a solution of the resulting sodium salt is cooled to −78°. The solution is acidified at that temperature, causing two products to precipitate. One is an oil that reacts instantaneously with bromine, while the second is crystalline and unreactive toward bromine. The compounds are isomers. Explain these observations.

18–32 Show by means of resonance structures why anions of the following compounds are relatively stable.

a. cyclopentane-1,3-dione
b. ethyl cyanoacetate
c. 3-oxobutanal (CH_3COCH_2CHO)
d. nitromethane

e. tribenzoylmethane

f. ethyl cyclopentanone-2-carboxylate

g. ethyl 2-cyclohexenone-4-carboxylate (**VIII**)

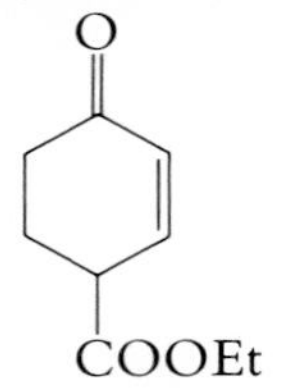

VIII

18–33 *Meldrum's acid* (2,2-dimethyl-1,3-dioxane-4,6-dione) is a relatively strong carbon acid with a pK_a of 5. Since Meldrum's acid is nothing more than a cyclic ester of malonic acid, its acidity is somewhat surprising (the common open-chain malonate esters have pK_as of 13; Table 18–1). Draw the structure of the enolate ion of this acid and explain why the acid is so strong. (Consider the geometry of the orbitals involved in the delocalized enolate ion.)

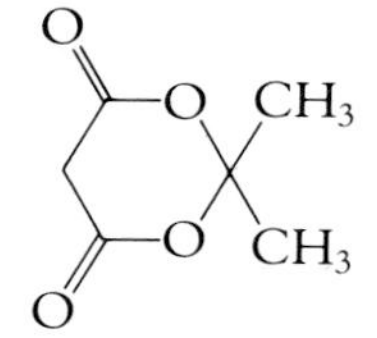

Meldrum's acid

18–34 2,4-Pentanedione (eq. 18.45) has a much greater tendency to enolize than does ethyl acetoacetate (eq. 18.46). A major reason is that conjugation of the C=C bond with the ketone group is stronger than its conjugation with the ester group. Explain the difference in conjugating ability of the two different carbonyl groups.

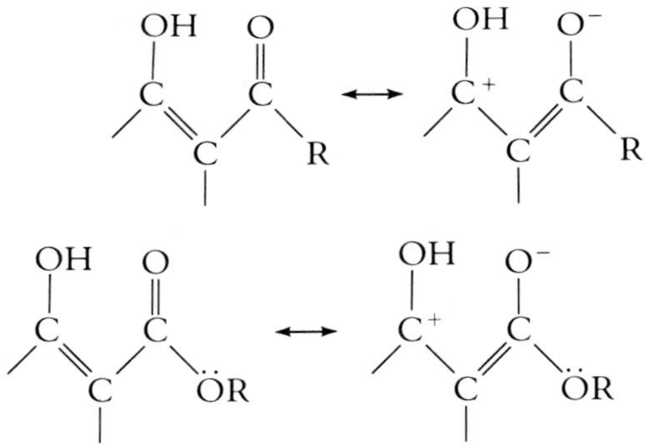

Stronger

Weaker

18–35 Use as starting materials carbonyl compounds containing five carbon atoms or less, and propose syntheses of the following compounds:

a. $CH_3CH_2CH_2CHCOOH$ (with Cl on the CH carbon)

$\quad\quad\quad\quad\quad\quad |$

$\quad\quad\quad\quad\quad\quad Cl$

b. $\mathrm{HOOC\underset{\displaystyle NH_2}{\underset{|}{C}}H\overset{\displaystyle CH_3}{\overset{|}{C}}HCH_3}$

c. $\mathrm{CH_3CH_2\overset{\displaystyle CH_3}{\overset{|}{C}}H\underset{\displaystyle Br}{\underset{|}{C}}HCOOH}$

18–36 a. The diester **IX** forms a fairly stable anion when treated with NaOEt. Show the structure of this anion, and account for its stability.

b. When 3-phenylpropanal is added to a mixture of **IX** and NaOEt, the reaction in Equation 2 takes place.

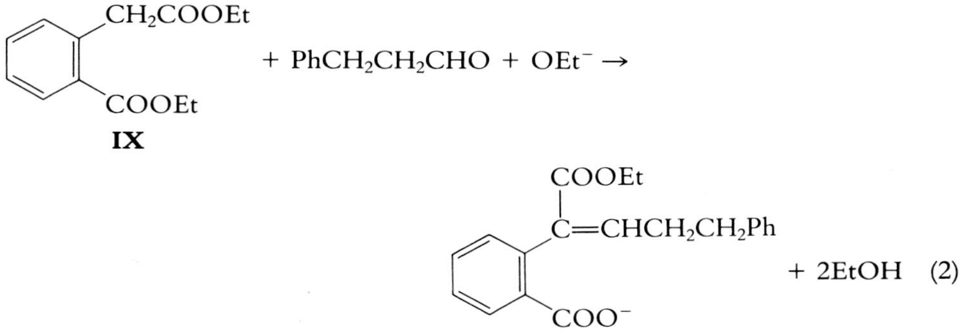

$$\text{IX (benzene ring with } CH_2COOEt \text{ and } COOEt \text{ ortho)} + PhCH_2CH_2CHO + OEt^- \rightarrow \text{(benzene ring with } \underset{\displaystyle |}{\overset{\displaystyle COOEt}{}}C{=}CHCH_2CH_2Ph \text{ and } COO^- \text{ ortho)} + 2EtOH \quad (2)$$

Write a mechanism for this reaction. Note particularly that one of the ester groups of **IX** has been hydrolyzed, although no water is present.

18–37 The following carbonyl compounds were treated with 1 mol of bromine in CCl_4. Show the monobromination products; for (d) and (e) indicate which, if any, of the isomeric products if favored.

a. 3-pentanone
b. cycloheptanone
c. phenylacetaldehyde
d. 2-hexanone
e. 3-methylcyclohexanone

18–38 Show how the following conversions could be carried out:

a. acetone → cyanoacetone
b. cyclopentanone → 2-methylthiocyclopentanone
c. hexanoic acid → 2-ethoxyhexanoic acid
d. octanoic acid → 2,2-dibromooctanoic acid
e. octanoic acid → 2-oxooctanoic acid $[CH_3(CH_2)_5\overset{\displaystyle O}{\overset{\|}{C}}CO_2H]$

18–39 Kinetic studies of the halogenation of ketones have shown that, except at very low concentrations of halogen, the reaction kinetics are first order in ketone concentration and *zero order* in halogen concentration. Furthermore, under identical conditions, halogenation and isotope exchange (Prob. 18–24) have been found to proceed at the same rate. How do these data fit the mechanism proposed in Equation 18.50? What is the rate-determining step in the halogenation reaction under normal conditions?

Supplementary Reading

S. Warren, *Chemistry of the Carbonyl Group,* Section IV, John Wiley and Sons, New York, N.Y., 1974.

F. A. Carey and R. J. Sundberg, *Advanced Organic Chemistry, Part B,* Sections 2.1 to 2.3, Plenum Press, New York, N.Y., 1983.

H. O. House, *Modern Synthetic Reactions,* Chaps. 9 and 10, W. A. Benjamin, New York, N.Y., 1972.

D. C. Ayres, *Carbanions in Synthesis,* Chaps. 4 and 5, American Elsevier, New York, N.Y., 1966.

H. Hart and M. Sasaoka, Simple enols: How rare are they? *J. Chem. Ed.,* **57,** 685 (1980).

Alkylation and Acylation of Enolates 19

In Chapter 18 you learned that enolate ions are powerful nucleophiles that readily attack the carbonyl group of aldehydes and ketones. In this chapter we'll show that enolates also react with alkylating agents, such as alkyl halides and toluenesulfonates, and with acylating agents such as esters. In addition to undergoing alkylation and acylation, enolates will add to a carbon–carbon double bond if it is conjugated with a strong electron-withdrawing group. These reactions have assumed great importance in the laboratory as a way of building large molecules from small ones. You'll also see that similar reactions take place in cells, for similar purposes.

19:1 Alkylation of Enolate Ions to Produce Alkylated Esters, Nitriles, and Ketones

The easiest enolate ions to work with are those that are relatively stable, which means that they are weaker bases than hydroxide or alkoxide ions. Dicarbonyl compounds give rise to stable enolate ions in which the negative charge is delocalized by two electron-withdrawing groups (sec. 18:10). Four typical substrates used to generate stable enolate anions are acetoacetic esters, malonic esters, malononitriles, and β-diketones.

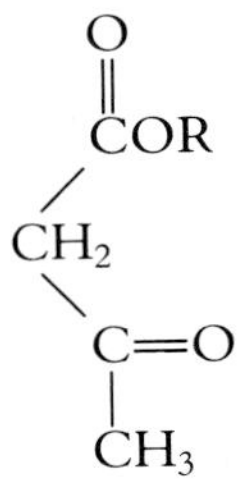

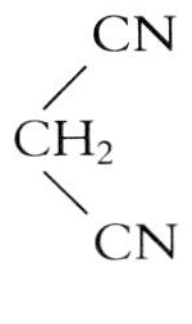

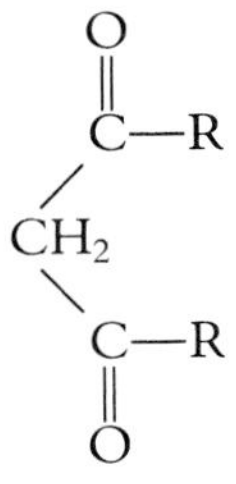

$$\underset{\text{Acetoacetic esters}}{\overset{O}{\overset{\|}{C}}OR-CH_2-C(=O)-CH_3} \qquad \underset{\text{Malonic esters}}{ROOC-CH_2-COOR} \qquad \underset{\text{Malononitrile}}{NC-CH_2-CN} \qquad \underset{\text{A }\beta\text{-diketone}}{R-CO-CH_2-CO-R}$$

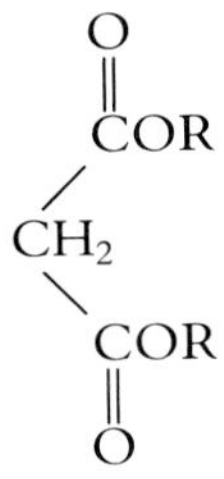

The enolate ions are made by allowing these acidic substrates to react with moderately strong bases such as alkoxide ions (eq. 19.1) and carbonate ions. These stable enolate ions then react easily with primary and secondary alkyl halides and sulfonates to give alkylated dicarbonyl compounds (eqs. 19.2 to 19.4).

$$\text{(19.1)}\quad \underset{\text{Acidic substrate}}{CH_2(COOEt)_2} + \underset{\text{Alkoxide ion}}{CH_3CH_2\ddot{O}:^-} \underset{}{\overset{CH_3CH_2OH}{\rightleftharpoons}} \underset{\text{Enolate ion}}{^-:CH(COOEt)_2} + CH_3CH_2OH$$

$$\text{(19.2)}\quad \underset{\text{1° Alkyl halide}}{CH_3(CH_2)_2CH_2{-}Br} + \underset{\text{Enolate ion}}{:\bar{C}H(COOEt)_2} \xrightarrow[C_2H_5OH]{} \underset{\text{Diethyl butylmalonate (80 to 90\%)}}{CH_3(CH_2)_2CH{-}CH(COOEt)_2} + Br^-$$

$$\text{(19.3)}\quad C_4H_9CH(CN)COOEt + (CH_3)_2CHOTs \xrightarrow[C_2H_5OH]{C_2H_5O} \underset{(87\%)}{C_4H_9C(CN)(CH(CH_3)_2)COOEt}$$

$$\text{(19.4)}\quad CH_3COCH_2COCH_3 + CH_3I \xrightarrow[\text{acetone (solvent)}]{K_2CO_3} \underset{\text{3-Methyl-2,4-pentanedione (75\%)}}{CH_3COCH(CH_3)COCH_3}$$

Enolate alkylation reactions are S_N2 displacements and only work well with methyl, 1°, and 2° halides. When tertiary halides are used as alkylating agents, the result is elimination rather than substitution (eq. 19.5).

$$\text{(19.5)}\quad CH_3{-}C(CH_2CH_3)(Cl){-}CH(H){-}CH_3 + {}^-:CH(COOEt)(COCH_3) \longrightarrow (CH_3)(CH_3CH_2)C{=}CH{-}CH_3 + CH_2(COOEt)(COCH_3) + Cl^-$$

In Section 19:7 you will see how alkylated malonic and acetoacetic esters are further transformed into carboxylic acids and ketones.

Example

Show the enolate ion and final product from the following reaction:

$$C_6H_5\overset{\overset{\displaystyle O}{\|}}{C}CH_2\overset{\overset{\displaystyle O}{\|}}{C}CH_3 + C_2H_5O^- \xrightarrow{CH_3CH_2Br} \quad ?$$

Answer

The base $C_2H_5O^-$ will remove an α hydrogen to give an enolate ion. The proton could be lost from either a CH_2 or CH_3 group, but only CH_2 deprotonation gives an enolate ion stabilized by *both* carbonyl groups.

$$C_2H_5O^- + CH_3\overset{\overset{\displaystyle O}{\|}}{C}\underset{\underset{\displaystyle H}{|}}{C}H\overset{\overset{\displaystyle O}{\|}}{C}C_6H_5 \rightarrow C_2H_5OH + [CH_3\overset{\overset{\displaystyle O}{\|}}{C}-\overset{-}{C}H-\overset{\overset{\displaystyle O}{\|}}{C}C_6H_5 \leftrightarrow$$

$$CH_3\overset{\overset{\displaystyle O^-}{|}}{C}=CH-\overset{\overset{\displaystyle O}{\|}}{C}C_6H_5 \leftrightarrow CH_3\overset{\overset{\displaystyle O}{\|}}{C}-CH=\overset{\overset{\displaystyle O^-}{|}}{C}C_6H_5]$$

The anionic carbon acts as a nucleophile and displaces the Br^- from bromoethane in an S_N2 reaction:

$$C_6H_5\overset{\overset{\displaystyle O}{\|}}{C}\overset{-}{C}H\overset{\overset{\displaystyle O}{\|}}{C}CH_3 + CH_3-CH_2-Br \rightarrow C_6H_5\overset{\overset{\displaystyle O}{\|}}{C}\underset{\underset{\displaystyle CH_2CH_3}{|}}{C}H\overset{\overset{\displaystyle O}{\|}}{C}CH_3 + Br^-$$

Problems

19–1 Show the enolate ions, and final products, from the following reactions:

a. $CH_3\overset{\overset{\displaystyle O}{\|}}{C}CH_2COOEt + CH_3(CH_2)_3I \xrightarrow[DMF]{K_2CO_3}$

b. $\underset{\underset{\displaystyle CN}{|}}{CH_2}COOMe + CH_3CH_2\underset{\underset{\displaystyle Cl}{|}}{C}HCH_3 \xrightarrow[CH_3OH]{CH_3ONa}$

c. $CH_2(COOEt)_2 + Br(CH_2)_5Br \xrightarrow[C_2H_5OH]{2C_2H_5ONa}$

d. (cyclohexane-1,3-dione) $+ CH_3(CH_2)_5Br \xrightarrow[DMF]{K_2CO_3}$

e. 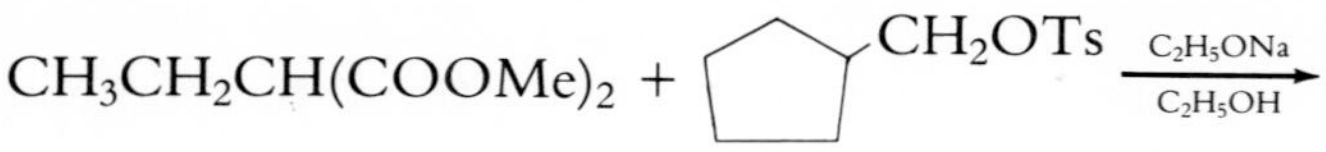 $CH_3CH_2CH(COOMe)_2$ + (cyclopentyl)CH_2OTs $\xrightarrow[C_2H_5OH]{C_2H_5ONa}$

f. malononitrile + *t*-butyl bromide $\xrightarrow[DMF]{K_2CO_3}$

19–2 Barbituric acid has structure **I**. If this ring is disubstituted, as in **II**, the resulting compounds (called *barbiturates*) are active as anesthetics, hypnotics, and sedatives. For instance, secobarbital (*Seconal,* **II**) has been widely used as a sedative.

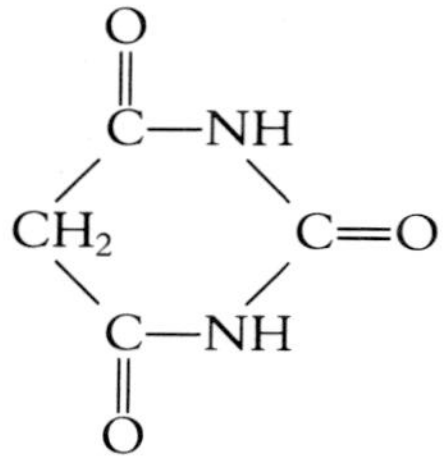

I

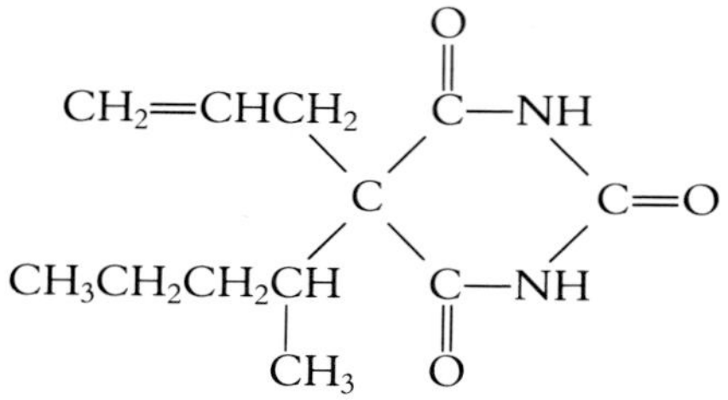

II

a. Compound **I** is made by allowing urea (H_2NCONH_2) to react with diethyl malonate in the presence of ethoxide catalyst. With this in mind, suggest a multistep synthesis of compound **II**. The reactants would be diethyl malonate, urea and whatever alkylating agents and other reagents you find necessary.
b. Barbiturates are often administered as their sodium salts. Show the product from the reaction of **II** with one equivalent of sodium ethoxide, and indicate with the aid of resonance structures why these compounds are called acids.

Compounds containing only *one* electron-withdrawing group (simple aldehydes, ketones, esters, or nitriles) are much weaker acids than those with two such groups (sec. 18:10). Thus, they require stronger bases than alkoxides for enolate formation, but use of such strong bases brings certain complications. Consider, for example, the reaction of cyclohexanone with the base sodium amide ($NaNH_2$). If the ketone is not completely converted to the anion, then the enolate ions will encounter un-ionized ketone molecules and we will get the aldol condensation product in addition to the alkylated product (eq. 19.6).

(19.6) $\xrightarrow{NaNH_2}$ O^-Na^+ $\longrightarrow$ O^-Na^+

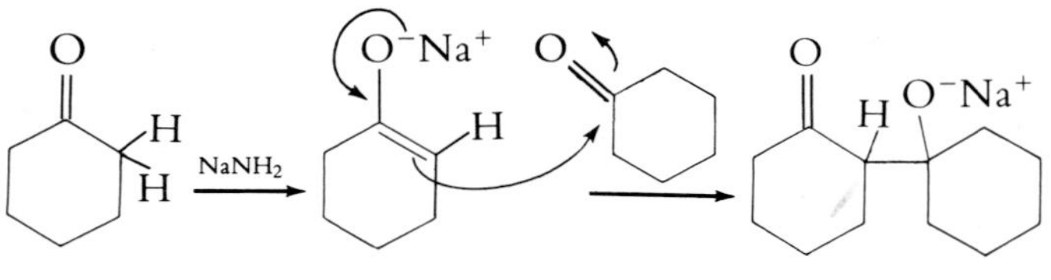

Under favorable conditions, however, it *is* possible to alkylate simple ketones, esters, and nitriles. The base most commonly used is lithium diisopropylamide, $[(CH_3)_2CH]_2NLi$ or (i-Pr_2NLi)], commonly known as LDA.

In order to minimize side reactions (eq. 19.6), it is necessary that LDA be soluble in the solvent, so that the enolate can be formed very rapidly. Ether solvents, such as 1,2-dimethoxyethane ($CH_3OCH_2CH_2OCH_3$, DME) are excellent for this purpose (eq. 19.7).

(19.7) $(CH_3)_3C$–C(=O)–CH_2–H + $\overset{\delta+}{Li}$–$\overset{\delta-}{N}Pr_2$-*i* $\xrightarrow{\text{DME}}$ $(CH_3)_3C$–C(O^-Li^+)=CH_2 + *i*-Pr_2NH $\xrightarrow{\text{I—}CH_2CH_3}$ $(CH_3)_3C$–C(=O)–$CH_2CH_2CH_3$

2,3-Dimethyl-2-butanone → 2,2-Dimethyl-3-hexanone

When ketones that can give two different enolates are used, mixtures result; however, the major product is often that from alkylation at the less substituted, and therefore less hindered, α position (eq. 19.8).

(19.8) 2-Methylcyclohexanone (CH_3, H on one α carbon; H, H on the other — less hindered) $\xrightarrow[\text{DME}]{i\text{-}Pr_2NLi}$ $\xrightarrow{C_6H_5CH_2Br}$ 2-methyl-6-benzylcyclohexanone (CH_3, H; $CH_2C_6H_5$, H) (*cis* and *trans*, 55%) + 2-benzyl-2-methylcyclohexanone ($C_6H_5CH_2$, CH_3) (8%) + dialkylated products

Equation 19.8 shows that alkylation of unsymmetrical ketones not only leads to isomeric products, but results in dialkylation as well. In Sections 19:10 and 19:11 we will discuss techniques used to avoid formation of these unwanted products.

Problems

19–3 Give the structures of all the products of each of the following reactions. If more than one may be formed, indicate which one should be the major product.

a. C_6H_5—C(=O)—CH_3 $\xrightarrow[\text{DME}]{i\text{-}Pr_2NLi(LDA)}$ $\xrightarrow{CH_3(CH_2)_4Br}$

b.  2-ethylcyclopentanone (cyclopentanone with CH_2CH_3 on the α-carbon) $\xrightarrow[\text{DME}]{\text{LDA}}$ $\xrightarrow{(CH_3)_2CHCH_2OTs}$

c. 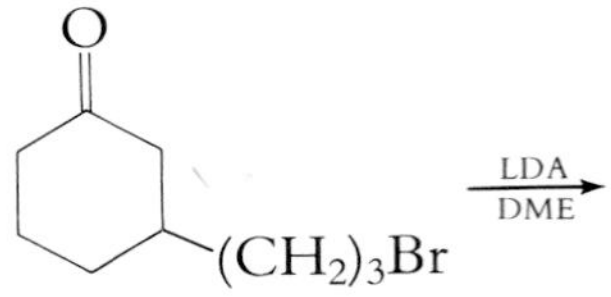 3-(3-bromopropyl)cyclohexanone (cyclohexanone with $(CH_2)_3Br$ on C-3) $\xrightarrow[\text{DME}]{\text{LDA}}$

19–4 All α-amino acids can be viewed as α-alkylated derivatives of glycine ($H_2N\overset{\alpha}{C}H_2COOH$). A simple synthesis of α-amino acids,

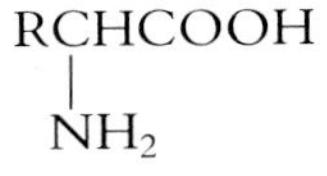

$$\underset{\displaystyle NH_2}{RCHCOOH}$$

based on glycine ethyl ester, is given in Equation 1.

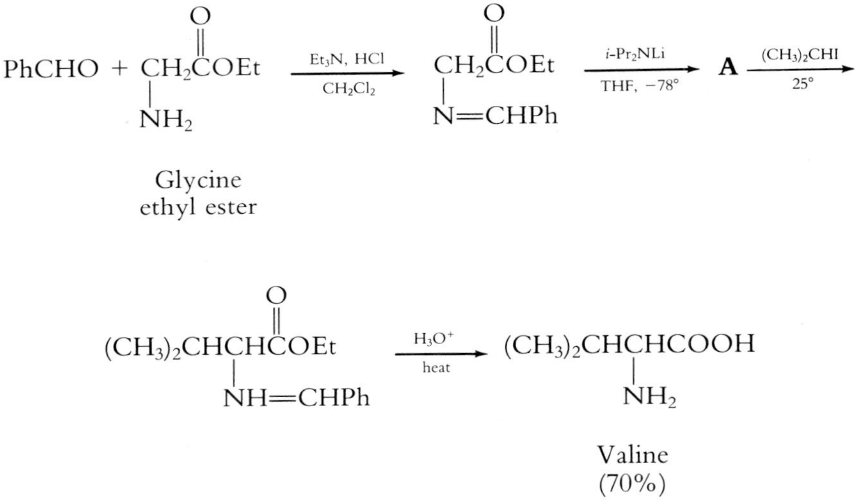

$$PhCHO + \underset{\displaystyle NH_2}{CH_2\overset{\displaystyle O}{\overset{\|}{C}}OEt} \xrightarrow[CH_2Cl_2]{Et_3N,\ HCl} \underset{\displaystyle N{=}CHPh}{CH_2\overset{\displaystyle O}{\overset{\|}{C}}OEt} \xrightarrow[THF,\ -78^\circ]{i\text{-}Pr_2NLi} \mathbf{A} \xrightarrow[25^\circ]{(CH_3)_2CHI}$$

Glycine
ethyl ester

$$\underset{\displaystyle NH{=}CHPh}{(CH_3)_2CHCH\overset{\displaystyle O}{\overset{\|}{C}}OEt} \xrightarrow[heat]{H_3O^+} \underset{\displaystyle NH_2}{(CH_3)_2CHCHCOOH} \qquad (1)$$

Valine
(70%)

a. Write a mechanism for the alkylation step in this equation, showing the structure of the intermediate **A**.
b. Why was the glycine ethyl ester condensed with benzaldehyde prior to alkylation?

19:2 1,4-Additions to Conjugated Carbonyl Systems

Section 19:1 showed that carbonyl compounds can give rise to nucleophilic enolate ions, which are then alkylated. In this section and the next you'll see that carbonyl compounds can also act as electrophilic alkylating agents *when the carbonyl group is conjugated with a double bond.* We'll first consider the reactions of these conjugated carbonyl compounds with simple nucleophiles such as amines and cyanide ion. For instance, 3-buten-2-one can alkylate amines (eq. 19.9). This reaction requires some comment, since simple al-

$$\text{(19.9)}\quad (CH_3)_2NH + CH_2{=}CH{-}\overset{\displaystyle O}{\overset{\|}{C}}CH_3 \xrightarrow[\text{MeOH}]{} (CH_3)_2NCH_2CH_2{-}\overset{\displaystyle O}{\overset{\|}{C}}CH_3$$

kenes are normally unreactive toward nucleophiles (Chap. 10). Indeed, alkenes show no tendency to react with amines at all (eq. 19.10).

$$\text{(19.10)}\quad (CH_3)_2NH + CH_2{=}CHCH_2CH_3 \xrightarrow[\text{MeOH}]{} \text{no reaction}$$

A large body of evidence indicates that the C=C bond in 3-buten-2-one (eq. 19.9) is reactive toward nucleophiles *only* because it is conjugated with the carbonyl group. Thus, when the amine attacks the C=C bond it produces not an unstable carbanion, but an enolate ion instead (eq. 19.11). The enolate ion is converted to the enol and then, by tautomeric equilibration, to the carbonyl product (eq. 19.11).

$$\text{(19.11)}\quad (CH_3)_2\overset{\displaystyle H}{\overset{|}{N}}{:} \curvearrowright CH_2{=}CH{-}\overset{\displaystyle O}{\overset{\|}{C}}CH_3 \rightleftarrows (CH_3)_2\overset{\displaystyle H}{\overset{|}{N}}{}^{+}{-}CH_2{-}CH{=}\underset{\text{Enolate}}{\overset{\displaystyle {}^{-}O}{\overset{|}{C}}CH_3} \rightleftarrows$$

$$\underset{\text{Enol}}{(CH_3)_2\ddot{N}{-}CH_2{-}CH{=}\overset{\displaystyle HO}{\overset{|}{C}}CH_3} \rightleftarrows (CH_3)_2\ddot{N}CH_2CH_2{-}\overset{\displaystyle O}{\overset{\|}{C}}CH_3$$

This mechanism nicely explains why dimethylamine adds readily to the C=C bond of 3-buten-2-one, but not to 1-butene. What it does not explain is why we get no products from attack of the amine at the carbonyl group. We know that such reactions take place easily (sec. 18:9). In fact, both processes do compete, but C=C addition produces a more stable product than C=O addition (eq. 19.12). So, if the system comes to equilibrium, the

$$\text{(19.12)}\quad (CH_3)_2NH + {}^{4}CH_2{=}{}^{3}CH{-}{}^{2}\overset{\displaystyle {}^{1}O}{\overset{\|}{C}}CH_3$$

$$\rightleftarrows \underset{\text{1,2-Addition product}}{CH_2{=}CH{-}\overset{\displaystyle OH}{\overset{|}{\underset{\displaystyle CH_3}{\underset{|}{C}}}}{-}N(CH_3)_2}\quad \text{less stable}$$

$$\rightleftarrows \underset{\text{1,4-Addition product}}{(CH_3)_2NCH_2{-}CH{=}\overset{\displaystyle HO}{\overset{|}{C}}CH_3} \overset{\text{fast}}{\rightleftarrows} (CH_3)_2NCH_2CH_2\overset{\displaystyle O}{\overset{\|}{C}}CH_3 \quad \text{more stable}$$

carbonyl product derived from **1,4-addition** (*conjugate addition*) completely outweighs the amino alcohol, or products derived from it, which arise by **1,2-addition**. The numbers 1 to 4 are applied only to the conjugated C=C—C=O unit, the oxygen being given the number one.

The greater stability of 1,4-addition products compared to 1,2-addition products is attributed to the higher bond energy of the C=O bond compared to that of the C=C bond (Table 10–1, p. 346). This difference is independent of the nucleophile involved in the reaction.

Amines are not the only nucleophiles that undergo 1,4-addition to conjugated carbonyl compounds. Other nucleophiles commonly used for this reaction are cyanide ion (eq. 19.13) and various sulfide ions (eq. 19.14).

$$\text{(19.13)}\quad \text{PhCH=CH}\overset{\text{O}}{\overset{\|}{\text{C}}}\text{Ph} + \text{CN}^- + \text{CH}_3\text{COOH} \xrightarrow{\text{ethanol}} \text{PhCH}(\text{CN})\text{—CH}_2\overset{\text{O}}{\overset{\|}{\text{C}}}\text{Ph} + \text{CH}_3\text{COO}^- \quad (95\%)$$

$$\text{(19.14)}\quad :\ddot{\text{S}}\text{H}^- + \text{CH}_2\text{=CH}\overset{\text{O}}{\overset{\|}{\text{C}}}\text{OMe} \xrightarrow[\text{CH}_3\text{OH}]{} \text{H}\ddot{\text{S}}\text{CH}_2\text{CH}_2\overset{\text{O}}{\overset{\|}{\text{C}}}\text{OMe} \xrightarrow[\text{(cat)}]{\text{OH}^-} {}^-:\ddot{\text{S}}\text{CH}_2\text{CH}_2\overset{\text{O}}{\overset{\|}{\text{C}}}\text{OMe} \xrightarrow[\text{CH}_3\text{OH}]{\text{CH}_2\text{=CH}\overset{\text{O}}{\overset{\|}{\text{C}}}\text{OMe}} :\ddot{\text{S}}(\text{CH}_2\text{CH}_2\text{COOMe})_2$$

Dimethyl 3,3-thiodipropanoate (75%)

Equation 19.15 shows clearly that an unconjugated double bond does not react with common nucleophiles and that 1,2-addition occurs normally.

$$\text{(19.15)}\quad \text{CH}_2\text{=CHCH}_2\text{CH}_2\overset{\text{O}}{\overset{\|}{\text{C}}}\text{CH}_3 + \text{CN}^- \xrightarrow{\text{H}_3\text{O}^+} \text{CH}_2\text{=CHCH}_2\text{CH}_2\text{C}(\text{OH})(\text{CN})\text{CH}_3$$

Problem

19–5 a. Write a mechanism for the reaction shown in Equation 19.13.

b. Write a mechanism for the reaction in Equation 2, paying careful attention to the order of the steps.

$$\text{CH}_2\text{=CH}\overset{\text{O}}{\overset{\|}{\text{C}}}\text{H} + \text{CN}^- \text{(excess)} \xrightarrow[\text{H}_2\text{O}]{\text{H}^+} \text{NCCH}_2\text{CH}_2\text{CH}(\text{OH})(\text{CN}) \qquad (2)$$

Amines and cyanide ion are nucleophilic enough to attack the conjugated carbonyl system at a convenient rate. Alcohols, by contrast, are weaker nucleophiles and react more slowly. The 1,4-addition reactions of alcohols can occur when they are catalyzed by acids (eq. 19.16).

(19.16) [2-cyclohexenone] + CH_3OH $\xrightarrow[\text{heat}]{\text{HCl (cat)}}$ [3-methoxycyclohexanone]

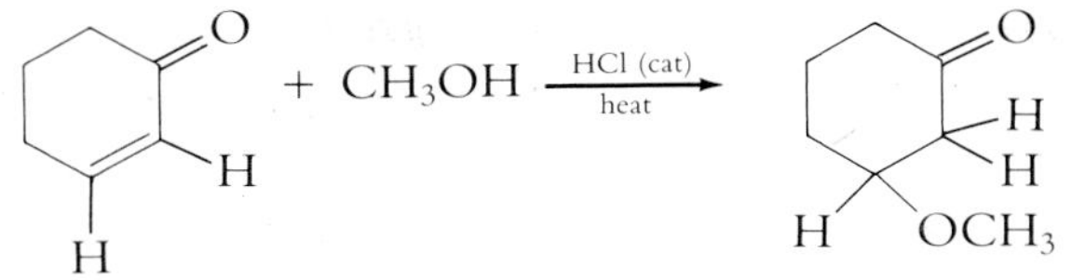

In this case, the proton attaches itself to the oxygen atom *before* the nucleophile adds to the 4-position (eq. 19.17). This provides an electrophilic carbon (C-4) to which the nucleophile can bind.

(19.17) $C^4{=}C^3{-}C^2{=}O^1$ $\overset{H^+}{\rightleftarrows}$ $\left[C{=}C{-}C{=}\overset{+}{O}H \leftrightarrow \overset{+}{C}{-}C{=}C{-}OH \right]$ $\xrightarrow{\text{Nu:}}$ $Nu{-}C{-}C{=}C{-}OH$ $\rightarrow$ $Nu{-}C{-}C(H){-}C{=}O$

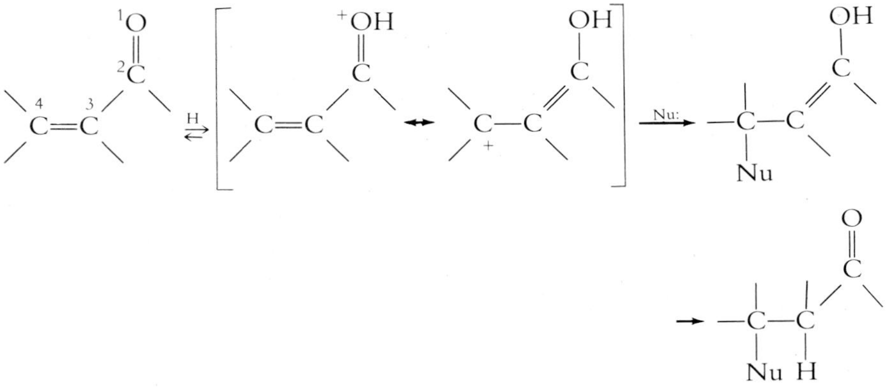

Derivatives of propenoic acid (also known as acrylic acid, $CH_2{=}CH{-}COOH$), are widely used substrates in conjugate addition reactions. Because these compounds also polymerize easily, their addition reactions are carried out at low temperatures, with low concentrations of acidic or basic catalysts.

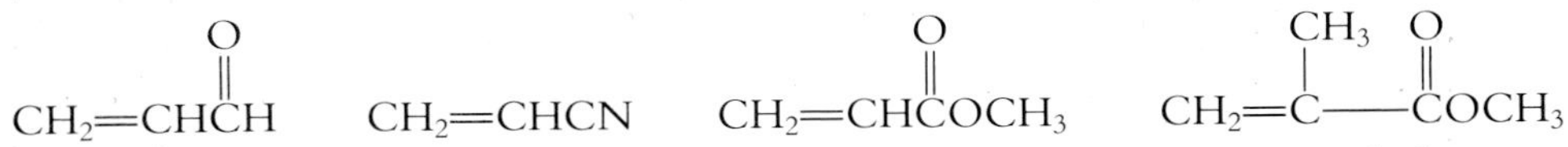

$CH_2{=}CH\overset{O}{\overset{\Vert}{C}}H$	$CH_2{=}CHCN$	$CH_2{=}CH\overset{O}{\overset{\Vert}{C}}OCH_3$	$CH_2{=}\overset{CH_3}{\overset{\vert}{C}}{-}\overset{O}{\overset{\Vert}{C}}OCH_3$
Acrolein	Acrylonitrile	Methyl acrylate	Methyl α-methacrylate

Section 19:1 demonstrated that enolate ions are reactive nucleophiles, while in this section we've seen that α,β-unsaturated carbonyl compounds are reactive alkylating agents. The next section describes what happens when the two are combined.

19:3 Michael Additions: How Enolate Ions React with α,β-Unsaturated Carbonyl Compounds

Stable enolate anions are sufficiently nucleophilic to add to α,β-unsaturated carbonyl compounds. This particular conjugate addition is known as the *Michael addition* (after the American chemist A. Michael). The catalysts are usually hydroxide or alkoxide ions (eqs. 19.18 and 19.19), which serve to generate the enolate ions.

(19.18) Diethyl fumarate (EtOOC(H)C=C(H)COOEt) + $CH_2(COOEt)_2$ (Enolate source) $\xrightarrow[\text{EtOH}]{\text{EtO}^-\text{Na}^+}$ $EtOOCCH_2CHCOOEt$ with $CH(COOEt)_2$ on the CH

(19.19) $CH_2{=}CHCN$ (Acrylonitrile) + C_6H_5—$\overset{\alpha}{C}H(CN)COOEt$ (Enolate source) $\xrightarrow[(CH_3)_3COH]{\text{KOH (cat)}}$ C_6H_5—$C(COOEt)(CN)CH_2CH_2CN$ (69 to 83%)

Equation 19.20 presents a typical mechanism for the addition of a β-dicarbonyl compound to a conjugated carbonyl compound.

(19.20) 1,3-cyclopentanedione $\xrightarrow[\text{MeOH}]{\text{MeO}^-}$ Stable enolate $\xrightarrow{CH_2{=}CH{-}COCH_3}$ 2-(CH_2—CH=$C(O^-)CH_3$)-1,3-cyclopentanedione $\xrightarrow{\text{MeOH}}$ 2-(CH_2—CH_2—$COCH_3$)-1,3-cyclopentanedione + MeO^-

Monocarbonyl compounds, such as simple aldehydes and ketones, do not give rise to enolates that are nearly as stable as those from β-dicarbonyl compounds. Thus very strong bases are required with monocarbonyl compounds, and these in turn bring about undesirable side reactions. The next section demonstrates how this problem can be avoided.

Problem

19–6 Write a mechanism for the reaction given in Equation 19.18, showing the structure of all intermediates.

When secondary amines react with aldehydes in the presence of acid, they produce iminium ions (sec. 18:9). If the aldehyde contains α hydrogens, then the iminium ion loses an α proton to produce an enamine (eq. 19.21; see Prob. 11–13, p. 409). Because enamines hydrolyze very readily back to carbonyl compounds which are more stable, the equilibrium in the reaction of Equation 19.21 is driven to the right by removal of water.

(19.21) $CH_3CH_2CH{=}O$ + pyrrolidine (N—H) $\xrightleftharpoons{H^+}$ $CH_3CH(H)_{\alpha}$—C(=N$^+$ pyrrolidinium)—H + H_2O ⇄ $CH_3CH{=}C(N\text{-pyrrolidinyl})—H + H_3O^+$

Propanal Pyrrolidine Iminium ion Propanal pyrrolidine enamine

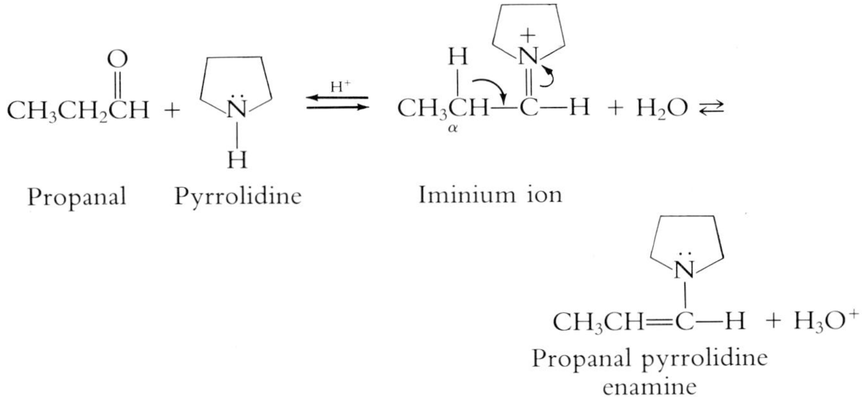

Enamines are very useful because their C=C bonds are quite nucleophilic, like the C=C bonds of enols and enolate ions (Fig. 19–1). Furthermore, they have some advantages over enols and enolates. Enamines usually do not undergo aldol condensations, which are a troublesome side reaction when enolate ions are being generated (eq. 19.6). In addition, carbonyl compounds can be completely converted to enamines, while the equilibrium concentration of enols and enolates is usually very small (sec. 18:1). The amines most frequently used for enamine formation are cyclic ones such as pyrrolidine and morpholine.

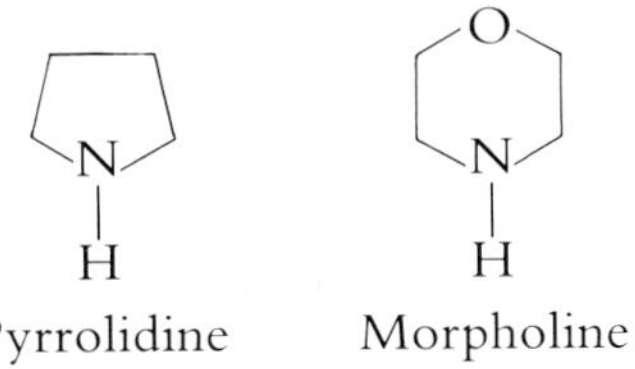

Pyrrolidine Morpholine

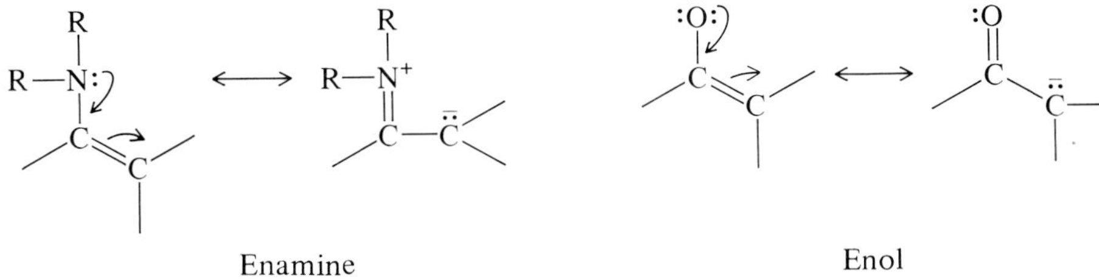

$R_2N{:}$—C=C $\longleftrightarrow$ R_2N^+=C—$\ddot{C}^-$ $\qquad$ ${:}\ddot{O}{:}^-$—C=C $\longleftrightarrow$ ${:}\ddot{O}$=C—$\ddot{C}^-$

Enamine Enol

Figure 19–1. *The similarity between enamines and enols, both of which contain nucleophilic double bonds.*

Because enamines behave like enolate ions, they can be used in Michael addition reactions. The aldehyde or ketone is converted to its enamine, and the enamine is reacted with the conjugated carbonyl compound in an aprotic solvent. Subsequent addition of aqueous acid leads to formation of the Michael addition product (eqs. 19.22 and 19.23).

$$(19.22)\quad (CH_3)_2\overset{\alpha}{C}HCH{=}O \xrightarrow[\text{benzene}]{\text{morpholine (N–H)}} (CH_3)_2C{=}CH{-}N(\text{morpholine}) \xrightarrow{CH_2{=}CHCN} \xrightarrow{H_3O^+} (CH_3)_2C(CH_2CH_2CN)CH{=}O + \text{morpholinium } (H_2N^+)$$

Isobutyraldehyde; Isobutyraldehyde morpholine enamine

$$(19.23)\quad \text{Cyclohexanone pyrrolidine enamine} + CH_2{=}CH\overset{O}{\overset{\|}{C}}CH_2CH_3 \xrightarrow{\text{dioxane}} \text{(iminium)}{-}CH_2CH{=}\overset{O^-}{C}CH_2CH_3 \xrightarrow{H_3O^+} \text{2-}(CH_2CH_2\overset{O}{\overset{\|}{C}}CH_2CH_3)\text{cyclohexanone } (65\%) + \text{pyrrolidinium } (\overset{+}{N}H_2)$$

Cyclohexanone pyrrolidine enamine

Table 19–1 summarizes the alkylation reactions of monocarbonyl and dicarbonyl compounds.

Problem

19–7 Show the structure of compounds **A**, **B**, and **C** in the following sequence:

$$\text{cyclopentanone} + \text{morpholine (N–H)} \xrightarrow[\text{benzene}]{H^+} \mathbf{A} \xrightarrow{CH_2{=}CHCOOCH_3} \mathbf{B} \xrightarrow{H_3O^+} \mathbf{C}$$

Alkylation of β-Dicarbonyl Compounds and Related Nitriles

1. Alkylation with alkyl halides and tosylates

$$\mathrm{Y\overset{O}{\overset{\|}{C}}{-}CH_2{-}\overset{O}{\overset{\|}{C}}Y + R'{-}X \xrightarrow{RO^-(1\ eq)}}$$

Y = R, OR   R′ = 1°, 2° alkyl
X = Cl, Br, I, OTs

$$\mathrm{Y\overset{O}{\overset{\|}{C}}{-}\underset{R'}{\underset{|}{C}H}{-}\overset{O}{\overset{\|}{C}}Y}$$ (Note that —C≡N can replace $-\overset{O}{\overset{\|}{C}}Y$)

2. Alkylation with α,β-unsaturated carbonyl compounds

$$\mathrm{Y\overset{O}{\overset{\|}{C}}{-}CH_2{-}\overset{O}{\overset{\|}{C}}Y + R_2C{=}C(R)\overset{O}{\overset{\|}{C}}Y \xrightarrow{RO^-\ (cat)}}$$

Y = R, OR

$$\mathrm{Y\overset{O}{\overset{\|}{C}}{-}\underset{CR_2\underset{R}{\underset{|}{C}H}\overset{O}{\overset{\|}{C}}Y}{\underset{|}{C}H}{-}\overset{O}{\overset{\|}{C}}Y}$$ (Note that —C≡N can replace $-\overset{O}{\overset{\|}{C}}Y$)

Alkylation of Monocarbonyl Compounds

1. Alkylation with alkyl halides and tosylates

$$\mathrm{R\overset{O}{\overset{\|}{C}}{-}CHR_2 \xrightarrow{\textit{i}\text{-}Pr_2NLi\ (LDA)} R\overset{O^-}{\overset{|}{C}}{=}CR_2 \xrightarrow{R'X} R\overset{O}{\overset{\|}{C}}{-}\overset{R'}{\overset{|}{C}}R_2}$$

2. Alkylation with α,β-unsaturated carbonyl compounds by way of enamines

$$\mathrm{R\overset{O}{\overset{\|}{C}}{-}CHR_2 \xrightarrow[H^+]{\text{pyrrolidine (N–H)}} R\overset{N(pyrrolidinyl)}{\overset{|}{C}}{=}CR_2 \xrightarrow[2.\ H_3O^+]{1.\ R_2C{=}C(R)\overset{O}{\overset{\|}{C}}Y} R\overset{O}{\overset{\|}{C}}{-}\underset{CR_2\underset{R}{\underset{|}{C}H}\overset{O}{\overset{\|}{C}}Y}{\underset{|}{C}R_2}}$$

Example

How would you prepare the following compound using a Michael reaction?

$$\mathrm{CH_3\overset{\overset{O}{\|}}{C}CH_2CH_2\overset{\overset{CN}{|}}{C}HCN}$$

Answer

The essence of a Michael addition is that a nucleophile attacks the carbon of a double bond that is conjugated with a carbonyl group. This means you must go two carbons away from the carbonyl carbon to find the new bond that was formed between the nucleophile and the unsaturated carbonyl compound:

$$\mathrm{CH_3\overset{\overset{O}{\|}}{C}{-}\underset{1}{CH_2}{-}\underset{2}{CH_2}\,\wr\,\overset{\overset{CN}{|}}{C}H{-}CN}$$

Thus, the nucleophile desired is derived from malononitrile and the reactants needed are

$$\mathrm{CH_3\overset{\overset{O}{\|}}{C}CH{=}CH_2 + H_2\overset{\overset{CN}{|}}{C}CN} + \text{base}$$

Problem

19–8 Each of the following conversions can be brought about by using a conjugate addition reaction in one of the steps. Write out complete schemes for each conversion.

a. $\mathrm{(CH_3)_2C{=}CH\overset{\overset{O}{\|}}{C}CH_3 \rightarrow (CH_3)_2\underset{\underset{NHCH_3}{|}}{C}CH_2\overset{\overset{O}{\|}}{C}CH_3}$

b. 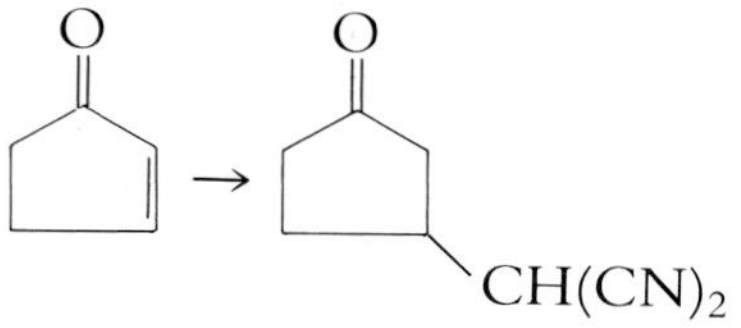 $\rightarrow$ CH(CN)$_2$

c. $\mathrm{CH_2{=}CH\overset{\overset{O}{\|}}{C}H \rightarrow EtOOC\underset{\underset{CN}{|}}{C}HCH_2CH_2CH(OEt)_2}$

d. $\mathrm{CH_3CH{=}C(COOEt)_2 \rightarrow C_6H_5\overset{\overset{O}{\|}}{C}CH_2\overset{\overset{CH_3}{|}}{C}HCH(COOEt)_2}$

e. $CH_3CH{=}CH\overset{O}{\overset{\|}{C}}H \rightarrow EtS\underset{CH_3}{\underset{|}{C}}HCH_2CH_2OH$

19:5 Selective Reactions of Organometallic Reagents with α,β-Unsaturated Carbonyl Compounds

Many of the reactions shown in Sections 19:2 to 19:4 are reversible, and so the major products isolated are those from the thermodynamically favored 1,4-addition. An important type of *irreversible* addition to α,β-unsaturated carbonyl systems is that involving the organometallic compounds RLi, RMgX, and "RCu."* In these cases the product distribution reflects the *rate* difference between 1,2- and 1,4-addition (kinetic control).

Equation 19.24 shows the reaction of a Grignard reagent with an unsaturated ketone. The major product arises from 1,2-addition. This result is

$$(19.24)\quad (CH_3)_2C{=}CH\overset{O}{\overset{\|}{C}}CH_3 + CH_3MgBr \xrightarrow[\text{ether}]{} \xrightarrow{H_3O^+} (CH_3)_2C{=}CH\underset{CH_3}{\underset{|}{\overset{OH}{\overset{|}{C}}}}{-}CH_3$$

1,2-Addition product (90%)

probably due at least in part to the steric hindrance of the two methyl groups at the β-carbon atom, which slows down 1,4-addition. Adding a catalytic quantity of CuCl to the reaction mixture leads to a marked enhancement of the rate of 1,4-addition (eq. 19.25). The second reaction goes by way of an organocopper species.

$$(19.25)\quad (CH_3)_2C{=}CH\overset{O}{\overset{\|}{C}}CH_3 + CH_3MgBr \xrightarrow[\text{ether}]{CuCl\ (1\%)} \xrightarrow{H_3O^+}$$

$$(CH_3)_2C{=}CH\underset{CH_3}{\underset{|}{\overset{OH}{\overset{|}{C}}}}{-}CH_3 + CH_3{-}\underset{CH_3}{\underset{|}{\overset{CH_3}{\overset{|}{C}}}}CH_2\overset{O}{\overset{\|}{C}}CH_3$$

1,2-Addition product (7%) | 1,4-Addition product (82%)

Based on a large number of experiments carried out under varying conditions, the following generalizations can be made about additions of organometallic reagents to α,β-unsaturated systems.

---

* The composition of the organocopper series can vary with the reactant stoichiometries. One of the most widely used has the composition R_2CuLi. Here we use RCu as a general symbol for an organocopper reagent.

1. The nature of the metal atom strongly influences the relative reaction rates. For a given R group, 1,2-addition is favored in the order RLi > RMgX > RCu, while 1,4-addition is favored in the order RCu > RMgX > RLi.
2. 1,4-Addition is greatly favored when there are *two* electron-withdrawing groups on one end of the double bond.
3. Steric hindrance at the carbonyl carbon atom or at the β-carbon atom discourages nucleophilic attack at those positions.

19:6 Conjugate Additions in Multistep Ring Syntheses

We have given much attention to the 1,4-addition reaction because of its usefulness in synthesis. The addition product can often undergo further *intramolecular* reactions which lead to cyclic molecules. Thus, the sequence of Michael addition followed by intramolecular aldol condensation is an important route to complex six-membered rings, such as those found in steroids. Equations 19.26 and 19.27 are representative examples of such reactions, commonly called Robinson **annulation** (or **annelation**) **reactions** (Latin *annulus,* meaning "ring") after Robert Robinson (sec. 14:2).

(19.26) $CH_3\overset{O}{\overset{\|}{C}}CH_2COOEt + CH_2{=}CHCHO \xrightarrow[\text{(Michael addition)}]{EtO^-}$ [$CH_3C(=O)CH(COOEt)CH_2CH_2CH{=}O$] $\xrightarrow[\text{(aldol condensation)}]{EtO^-}$ (cyclohexenone product bearing COOEt)

Acrolein

(19.27) $CH_3\overset{O}{\overset{\|}{C}}CH{=}CH_2$ + (2-methylcyclohexane-1,3-dione, CH$_3$, H) $\xrightarrow[\text{CH}_3\text{OH (Michael addition)}]{\text{KOH (cat)}}$

Methyl vinyl ketone

(2-methyl-2-($CH_2CH_2COCH_3$)cyclohexane-1,3-dione) $\xrightarrow[\text{remove water (aldol condensation)}]{\text{pyrrolidine, } C_6H_6}$ (bicyclic enedione with CH_3)

(65%)

The Michael addition reaction can be used in conjunction with standard alkylation methods to produce a wide variety of alkylated malonate and acetoacetate esters. We have already noted in Section 19:1 that these compounds serve as precursors to ketones and carboxylic acids. The following section shows how this is done.

Problem

19–9 Show how 3-penten-2-one can be converted to each of the following compounds:

a. 4-methyl-2-heptanone

b. 2-phenyl-3-penten-2-ol

c. (cyclohexenone bearing CH_3, $CH_3C(=O)$ and CH_3 substituents)

d. 4-methyl-2,6-heptanedione

19:7 Decarboxylation of β-Carbonylcarboxylic Acids

Ordinary carboxylic acids are stable to quite high temperatures. However, if β-dicarboxylic acids, such as malonic acid, are heated to 100°, they rapidly lose CO_2 and are converted to monocarboxylic acids (eq. 19.28). Similarly, the acid-catalyzed hydrolysis of ethyl acetoacetate gives acetoacetic acid (a **β**-keto acid), which undergoes rapid decarboxylation to yield CO_2 and acetone as final products (eq. 19.29).

$$\underset{\text{Malonic acid}}{\mathrm{HO}\overset{\mathrm{O}}{\overset{\|}{\underset{\beta}{\mathrm{C}}}}\!-\!\underset{\alpha}{\mathrm{CH_2}}\overset{\mathrm{O}}{\overset{\|}{\mathrm{C}}}\mathrm{OH}} \xrightarrow{\text{heat}} \underset{\text{Acetic acid}}{\mathrm{CH_3}\overset{\mathrm{O}}{\overset{\|}{\mathrm{C}}}\mathrm{OH}} + CO_2 \tag{19.28}$$

$$\mathrm{CH_3}\overset{\mathrm{O}}{\overset{\|}{\mathrm{C}}}\mathrm{CH_2}\!-\!\overset{\mathrm{O}}{\overset{\|}{\mathrm{C}}}\mathrm{OEt} \xrightarrow[\text{heat}]{H_3O^+} \underset{\text{Acetoacetic acid}}{[\mathrm{CH_3}\overset{\mathrm{O}}{\overset{\|}{\mathrm{C}}}\mathrm{CH_2}\!-\!\overset{\mathrm{O}}{\overset{\|}{\mathrm{C}}}\mathrm{OH}]} + \mathrm{EtOH} \longrightarrow \underset{\text{Acetone}}{\mathrm{CH_3}\overset{\mathrm{O}}{\overset{\|}{\mathrm{C}}}\mathrm{CH_3}} + CO_2 \tag{19.29}$$

Decarboxylation proceeds by a concerted mechanism in which there is a cyclic, six-membered transition state (eq. 19.30). The first product formed is the *enol*, which then tautomerizes to the carbonyl compound. This facile decarboxylation of β-carbonyl acids can be advantageous, as we'll now see.

$$\text{(19.30)}\quad \underset{CH_3\;\;CH_2}{\overset{O}{\overset{\|}{C}}}\text{—}\overset{H-O}{\overset{|}{C}}{=}O \xrightarrow{-CO_2} \underset{CH_3\quad CH_2}{\overset{H-O}{\overset{|}{C}}}\ \text{(Enol)} \rightleftharpoons \underset{CH_3\quad CH_3}{\overset{O}{\overset{\|}{C}}}$$

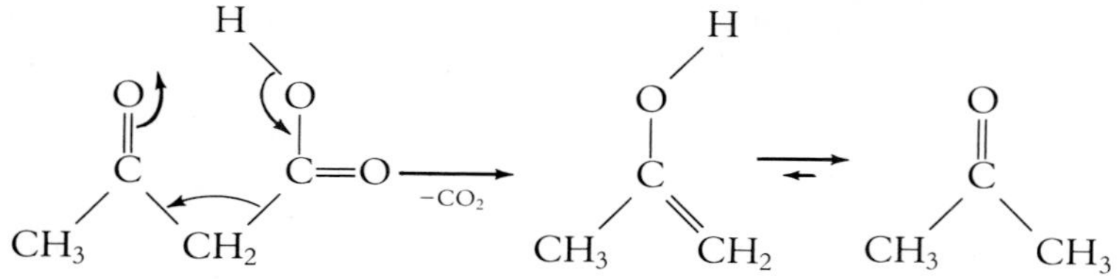

The β-ketoester enolate, **1**, is easier to generate and handle than the monocarbonyl enolate, **2**. Because any products derived from **1** can eventu-

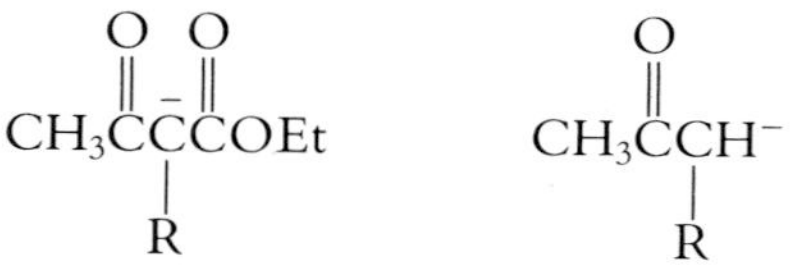

$$\underset{\textbf{1}}{CH_3\overset{O}{\overset{\|}{C}}\underset{R}{\underset{|}{\bar{C}}}\overset{O}{\overset{\|}{C}}OEt} \qquad \underset{\textbf{2}}{CH_3\overset{O}{\overset{\|}{C}}\underset{R}{\underset{|}{C}}H^-}$$

ally be decarboxylated, **1** *is functionally equivalent to* **2**. Thus, in all syntheses where one might use **2**, **1** can be used equally well and then decarboxylated later.

As an example, consider the synthesis of 2,6-heptanedione. Conceptually, this compound can be obtained by Michael addition of **2** to 3-buten-2-one. In practice, it is the more stable **1** that is allowed to react with the unsaturated ketone, and the product is then subjected to hydrolysis and decarboxylation (eq. 19.31). This approach is also a convenient alternative to using the enamine of acetone (sec. 19:4).

$$\text{(19.31)}\quad CH_3\overset{O}{\overset{\|}{C}}CH{=}CH_2 + CH_3\overset{O}{\overset{\|}{C}}\bar{C}HCOOEt \xrightarrow{EtOH}$$

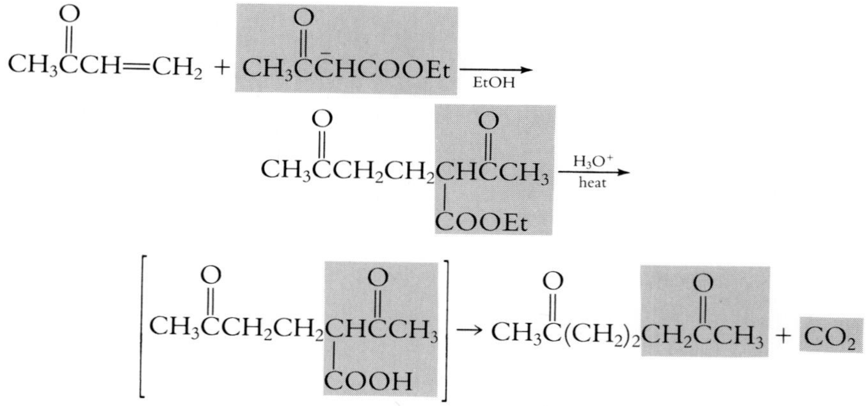

$$CH_3\overset{O}{\overset{\|}{C}}CH_2CH_2\underset{COOEt}{\underset{|}{C}H}\overset{O}{\overset{\|}{C}}CH_3 \xrightarrow[\text{heat}]{H_3O^+}$$

$$\left[CH_3\overset{O}{\overset{\|}{C}}CH_2CH_2\underset{COOH}{\underset{|}{C}H}\overset{O}{\overset{\|}{C}}CH_3\right] \rightarrow \underset{\text{2,6-Heptanedione}}{CH_3\overset{O}{\overset{\|}{C}}(CH_2)_2CH_2\overset{O}{\overset{\|}{C}}CH_3} + CO_2$$

We can also use decarboxylation reactions to synthesize carboxylic acids. One common method is to alkylate diethyl malonate and then hydrolyze and decarboxylate the substituted malonic ester (eq. 19.32).

$$\text{(19.32)}\quad (CH_3)_2CHCH_2Br + CH_2(COOEt)_2 \xrightarrow{C_2H_5O^-}$$

$$(CH_3)_2CHCH_2CH(COOEt)_2 \xrightarrow[\text{heat}]{H_3O^+}$$

$$[(CH_3)_2CHCH_2CH(COOH)_2] \rightarrow (CH_3)_2CHCH_2CH_2COOH + CO_2$$

In these reactions we see that the carboxyl group acts as an activator and is then removed in the last step when its work is accomplished. The activation–alkylation–decarboxylation sequence allows ethyl acetoacetate to be a good precursor for methyl ketones, and diethyl malonate a good precursor for carboxylic acids.

Nature has used an approach similar to that used in laboratory syntheses. In sections 19:12 and 19:13 some biological reactions are discussed in which decarboxylations of β-carbonylcarboxylic acids are important steps in metabolic processes. The seemingly wasteful sequence of adding a carboxyl group and then removing it a few steps later becomes understandable when the reaction mechanisms are known.

Example

Show how 2-hexanone could be synthesized using ethyl acetoacetate as one starting material.

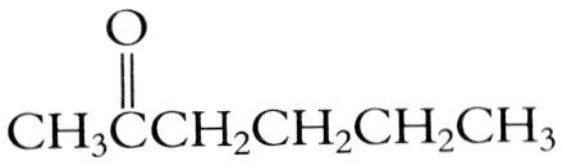

$$CH_3\overset{\overset{\large O}{\|}}{C}CH_2CH_2CH_2CH_3$$

2-Hexanone

Answer

In this section you saw that the carboxyl end of the ethyl acetoacetate can be lost in the last step of the synthesis. From the structure of ethyl acetoacetate, we can identify the position of the carboxyl group before the last synthetic step:

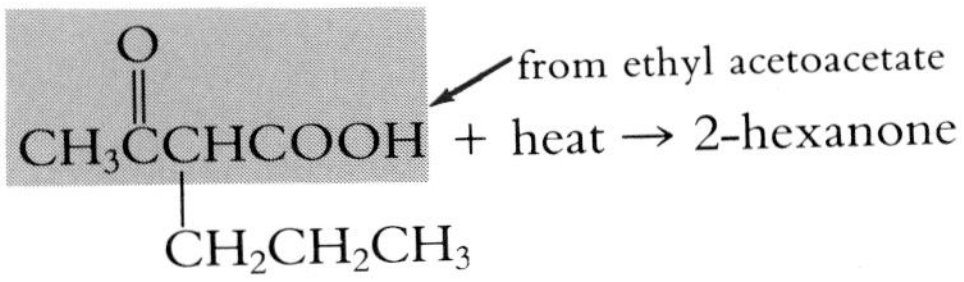

$$\underset{\textbf{A}}{CH_3\overset{\overset{\large O}{\|}}{C}\underset{\underset{\large CH_2CH_2CH_3}{|}}{C}HCOOH} + \text{heat} \rightarrow \text{2-hexanone}$$

(shaded portion: from ethyl acetoacetate)

This means that the first step will involve alkylation with 1-bromopropane:

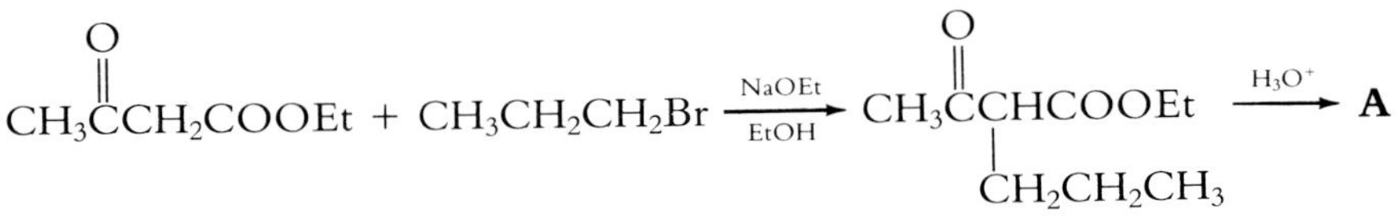

$$CH_3\overset{\overset{\large O}{\|}}{C}CH_2COOEt + CH_3CH_2CH_2Br \xrightarrow[\text{EtOH}]{\text{NaOEt}} CH_3\overset{\overset{\large O}{\|}}{C}\underset{\underset{\large CH_2CH_2CH_3}{|}}{C}HCOOEt \xrightarrow{H_3O^+} \textbf{A}$$

Problem

19–10 Show how each of the following compounds can be synthesized, using ethyl acetoacetate or diethyl malonate as starting materials.

a. 4-phenyl-2-butanone

b. 4-hexenoic acid

c. 5-oxohexanoic acid ($CH_3\overset{\overset{\large O}{\|}}{C}CH_2CH_2CH_2COOH$)

d. 2,6-dimethylheptanoic acid e. O= (cyclopentanone) CH_2CO_2H

19:8 Carboxylation of Enolates and Enols

Under normal conditions the decarboxylation of β-ketoacids proceeds irreversibly, because the carbon dioxide escapes. The reverse reaction ought to be feasible, however, if we generate an enol or enolate in an atmosphere of carbon dioxide.

A commercially important process of this type is the carboxylation of phenol and substituted phenols. *Salicylic acid* (*o*-hydroxybenzoic acid), which is the starting material for the manufacture of the mountains of aspirin consumed each year, is made in this way (eq. 19.33). Sodium phenoxide is, after all, just the very stable enolate of cyclohexa-2,4-dienone.

(19.33) O^-Na^+ (Sodium phenoxide) + CO_2 $\xrightarrow[150°]{5\ atm}$ $\xrightarrow{H^+}$ OH, COH (=O) (Salicylic acid (75%))

Sodium phenoxide

Salicylic acid (75%)

The mechanism involves nucleophilic attack by the phenoxide ion on carbon dioxide, followed by tautomerization to regenerate the aromatic ring (eq. 19.34).

(19.34) O^-Na^+ + O=C=O → **3** (Na⁺, O, O^-, C, H, O) → **4** (Na⁺, HO, O^-, C, O) $\xrightarrow{H^+}$ OH, COOH

3

4

Interestingly, when the *potassium* salt is used, a mixture of *o*- and *p*-hydroxybenzoic acids is obtained (eq. 19.35). The high yield of *ortho* product

(19.35) O^-K^+ $\xrightarrow[140°]{CO_2,\ 5\ atm}$ $\xrightarrow{H^+}$ OH, COOH (60%) + OH, COOH (40%)

(60%) (40%)

obtained from the sodium salt is due to the greater coordinating power of the sodium ion. The sodium ion is attracted to the oxygen atoms of both the phenoxide and the carboxylate groups, and thus stabilizes the *ortho* isomer.

The reactions in Equations 19.34 and 19.35 are unlike most of the reactions we have seen thus far. In previous cases it usually did not matter whether a metal cation was sodium, potassium, lithium, or others. With enol and enolate chemistry, however, there are often results that are cation dependent, and these can usually be attributed to different coordinating abilities of the cations.

Coordinating ability is particularly important when metal ions are associated with **chelating agents. Chelation** is a general phenomenon wherein a substrate (the chelating agent) forms *more than one* bond to a metal ion. The complex formed is called a **chelate** (from the Greek *chele,* meaning "crab's claw"). Structures **3** and **4** in Equation 19.34 are chelates. Chelates are important in nature, and include hemoglobin (a chelate containing Fe) and chlorophyll (a chelate containing Mg).

Equations 19.34 and 19.35 depict the carboxylation of enolate ions. Under biological conditions it is possible to carboxylate enols and their derivatives. This reaction plays an important role in some fundamental biochemical processes. One is photosynthesis, in which many plants and bacteria can use energy captured from sunlight to convert atmospheric CO_2 to carbohydrates (Chap. 20). This large scale utilization of solar energy is a major source of food for animals, including man. The carbohydrates so produced are metabolically burned to provide cellular energy. The metabolic process, called the *tricarboxylic acid cycle,* is a series of biochemical reactions (also common to humans) which we will describe in Section 19:12. For now we will focus on a reaction used by many plants and bacteria to synthesize oxaloacetic acid—a key substrate in the cycle. This reaction is a good biological example of an enol carboxylation reaction.

The substrate for this carboxylation process is phosphoenol pyruvate (PEP, sec. 7:12), a derivative of the enol of pyruvic acid. In PEP, the PO_3H^- group replaces the H of the enol. The carboxylation of PEP to produce oxaloacetic acid is shown in Equation 19.36.

(19.36) $CH_2{=}C(OH)COOH$ ⟶ [HO–C(=O)–O⁻ attacking P of $CH_2{=}C(COOH)–O–P(=O)(OH)O^-$] $\xrightarrow[Mn^{2+}]{enzyme}$

Enol of pyruvic acid

Phosphoenol pyruvate (PEP)

$HO–C(O^-)(O–PO_3H^-)–CH_2–C(=O)COOH$ ⟶ $HO\overset{O}{\overset{\|}{C}}CH_2\overset{O}{\overset{\|}{C}}COOH + HPO_4^{2-}$

Oxaloacetic acid

Because the carboxylation reaction takes place within a cell, there can't be much free CO_2. It is introduced in the form of bicarbonate ion, HCO_3^- ($CO_2 + OH^-$). In order to function, the enzyme requires Mn^{2+} which is thought to be chelated in much the same way as Na^+ is chelated in salicylic acid synthesis (eq. 19.34).

This is a good point from which to summarize much of what has been discussed in this chapter and in Chapter 18. We've seen that nucleophilic enols and enolate ions take part in addition reactions with aldehydes and ketones (secs. 18:1 to 18:4), and with carbon dioxide (sec. 19:8). They bring about substitution with alkyl halides and toluenesulfonates (sec. 19:1), and addition with conjugated alkenes (sec. 19:3). We will now look at one more related reaction of these versatile nucleophiles—their substitution reactions at the carbonyl groups of esters. The products of these reactions—β-carbonylesters—are exceedingly useful. Ethyl acetoacetate, for example, is extensively used as a precursor of other compounds (eq. 19.31); the same applies to diethyl malonate and its derivatives (eq. 19.32). Where do these compounds come from? The answer is found in the next few sections.

19:9 Acylation of Esters: The Claisen Condensation

When carboxylate esters are heated in the presence of bases such as alkoxides, they undergo a type of condensation in which one molecule of ester acylates another molecule at the α carbon (eq. 19.37).

$$CH_3\overset{O}{\overset{\|}{C}}OEt + \overset{\alpha}{C}H_3\overset{O}{\overset{\|}{C}}OEt \xrightarrow{EtO^-} \xrightarrow{H^+} CH_3\overset{O}{\overset{\|}{C}}CH_2\overset{O}{\overset{\|}{C}}OEt + EtOH \tag{19.37}$$

Ethyl acetoacetate

This reaction was discovered by the German chemist L. Claisen, who first synthesized ethyl acetoacetate by this route. It is called the **Claisen condensation**.

The entire mechanism for the Claisen condensation is shown in Equations 19.38 to 19.40. The first two steps are reversible, but the equilibrium for the last step lies strongly to the right, providing the driving force for the entire sequence.

$$EtO^- + CH_3\overset{O}{\overset{\|}{C}}OEt \rightleftarrows EtOH + {}^-CH_2\overset{O}{\overset{\|}{C}}OEt \tag{19.38}$$

Enolate ion

$$CH_3\overset{O}{\overset{\|}{C}}OEt + {}^-CH_2\overset{O}{\overset{\|}{C}}OEt \rightleftarrows CH_3\overset{O^-}{\overset{|}{C}}(CH_2COOEt)\text{—}OEt \rightleftarrows CH_3\overset{O}{\overset{\|}{C}}CH_2COOEt + EtO^- \tag{19.39}$$

Enolate ion

$$\text{(19.40)}\quad CH_3\overset{O}{\overset{\|}{C}}CH_2\overset{O}{\overset{\|}{C}}OEt + EtO^- \rightleftharpoons EtOH + \left[CH_3\overset{O}{\overset{\|}{C}}\bar{C}H\overset{O}{\overset{\|}{C}}OEt \leftrightarrow CH_3\overset{O^-}{\overset{|}{C}}{=}CH\overset{O}{\overset{\|}{C}}OEt \leftrightarrow CH_3\overset{O}{\overset{\|}{C}}CH{=}\overset{O^-}{\overset{|}{C}}OEt \right]$$

The last neutralization step (eq. 19.40) is an essential one in all alkoxide promoted acylations. Note that the product of this step is a highly stabilized enolate ion. This means that the α-carbon atom being acylated must have at least *two* hydrogen atoms. Thus, ethyl propanoate can also undergo Claisen condensation in the presence of sodium ethoxide (eq. 19.41), but esters with two α substituents cannot.

$$\text{(19.41)}\quad 2CH_3\overset{\alpha}{C}H_2\overset{O}{\overset{\|}{C}}OEt + EtO^- \xrightarrow[EtOH]{} CH_3CH_2\overset{O^-}{\overset{|}{C}}{=}\underset{CH_3}{\underset{|}{C}}\overset{O}{\overset{\|}{C}}OEt \xrightarrow{CH_3COOH} CH_3CH_2\overset{O}{\overset{\|}{C}}\underset{CH_3}{\underset{|}{C}H}\overset{O}{\overset{\|}{C}}OEt$$

(45%)

Example

What products will be formed when the following ester is treated with sodium methoxide, followed by an acid work-up?

$$C_6H_5CH_2CH_2\overset{O}{\overset{\|}{C}}OCH_3$$

Answer

As usual, the problem is best worked mechanistically. Sodium methoxide (MeO^-) forms the anion, **A**, by removing an α proton:

$$C_6H_5CH_2CH_2\overset{O}{\overset{\|}{C}}OCH_3 + MeO^- \rightarrow \underset{\mathbf{A}}{C_6H_5CH_2\bar{C}H\overset{O}{\overset{\|}{C}}OCH_3} + MeOH$$

Anion **A** then attacks the carbonyl group of another ester molecule:

$$C_6H_5CH_2CH_2\overset{O}{\overset{\|}{C}}OCH_3 + CH_3O\overset{O}{\overset{\|}{C}}\overset{-}{C}HCH_2C_6H_5 \rightarrow C_6H_5CH_2CH_2\overset{O^-}{\overset{|}{C}}(OCH_3)\text{—}CH(COCH_3{=}O)CH_2C_6H_5$$

A

$$\downarrow$$

$$C_6H_5CH_2CH_2\overset{O}{\overset{\|}{C}}\text{—}\overset{-}{C}(\overset{O}{\overset{\|}{C}}OCH_3)CH_2C_6H_5 + CH_3OH \leftarrow CH_3O^- + C_6H_5CH_2CH_2\overset{O}{\overset{\|}{C}}\text{—}CH(\overset{O}{\overset{\|}{C}}OCH_3)CH_2C_6H_5$$

Stable enolate

$$\downarrow H_3O^+$$

$$C_6H_5CH_2CH_2\overset{O}{\overset{\|}{C}}\text{—}CH(\overset{O}{\overset{\|}{C}}OCH_3)CH_2C_6H_5$$

Problems

19–11 What organic products will be formed when each of the following esters is treated with sodium ethoxide, followed by an acid work-up?

a. ethyl butanoate
b. ethyl 3-methylbutanoate
c. ethyl 2-methylbutanoate
d. ethyl phenylacetate

19–12 Ethyl thiolacetate will undergo Claisen condensation when treated with sodium ethoxide (eq. 3). This reaction is faster than the self-condensation of ethyl acetate (eq. 19.37), and the equilibrium favors the product. Explain these differences in rate and equilibrium. (A review of Section 17:13 may be helpful.)

$$2CH_3\overset{O}{\overset{\|}{C}}SEt \xrightarrow{EtO^-} \xrightarrow{H^+} CH_3\overset{O}{\overset{\|}{C}}CH_2\overset{O}{\overset{\|}{C}}SEt + EtSH \quad (3)$$

When two *different* esters, such as ethyl acetate and ethyl propanoate, are put into the reaction mixture, four products are formed, two from each enolate ion. This reaction, called a **crossed Claisen condensation**, is therefore of little use. It suffers the same defects as the mixed aldol condensation (sec. 18:6). However, the reaction *can* be useful if one of the two esters lacks α hydrogens and therefore cannot generate an enolate ion. Two types of nonenolizable esters most commonly used for these mixed condensations are ethyl formate (eq. 19.42), and aromatic esters such as ethyl benzoate (eq. 19.43).

$$\text{(19.42)}\quad \underset{\text{Ethyl formate}}{\text{H}\overset{\text{O}}{\overset{\|}{\text{C}}}\text{OEt}} + \underset{\text{Diethyl succinate}}{\begin{matrix}\text{CH}_2\text{COOEt}\\ |\\ \text{CH}_2\text{COOEt}\end{matrix}} \xrightarrow[\text{Na metal, ether}]{\text{EtO}^-\text{Na}^+} \xrightarrow{\text{H}^+} \underset{\text{Diethyl 2-formylsuccinate (60 to 70\%)}}{\begin{matrix}\text{H}\overset{\text{O}}{\overset{\|}{\text{C}}}\text{CHCOOEt}\\ |\\ \text{CH}_2\text{COOEt}\end{matrix}}$$

$$\text{(19.43)}\quad \underset{\text{Ethyl benzoate}}{\text{C}_6\text{H}_5\overset{\text{O}}{\overset{\|}{\text{C}}}\text{OEt}} + \underset{\text{Ethyl acetate}}{\text{CH}_3\overset{\text{O}}{\overset{\|}{\text{C}}}\text{OEt}} \xrightarrow[\text{Na metal}]{\text{EtO}^-\text{Na}^+} \xrightarrow{\text{H}^+} \underset{\text{Ethyl benzoylacetate (60\%)}}{\text{C}_6\text{H}_5\overset{\text{O}}{\overset{\|}{\text{C}}}\text{CH}_2\overset{\text{O}}{\overset{\|}{\text{C}}}\text{OEt}}$$

In Equations 19.42 and 19.43, sodium metal is used to react with the ethanol produced by converting it to sodium ethoxide, which helps drive the equilibrium toward products.

Problems

19–13 a. Show how methyl phenylformylacetate can be synthesized.

$$\text{H}\overset{\text{O}}{\overset{\|}{\text{C}}}\underset{\text{C}_6\text{H}_5}{\underset{|}{\text{C}}}\text{H}\overset{\text{O}}{\overset{\|}{\text{C}}}\text{OMe}$$

Methyl phenylformylacetate

b. Explain why the compound *cannot* be synthesized by way of a substitution reaction using

$$\underset{\text{O}}{\underset{\|}{\text{HC}}}\text{CH}_2\text{COOMe.}$$

19–14 Propose syntheses of the following compounds using crossed Claisen condensations, based on readily available starting materials.

a. $C_6H_5\overset{\overset{O}{\|}}{C}\underset{\underset{CH_3}{|}}{C}HCOOEt$

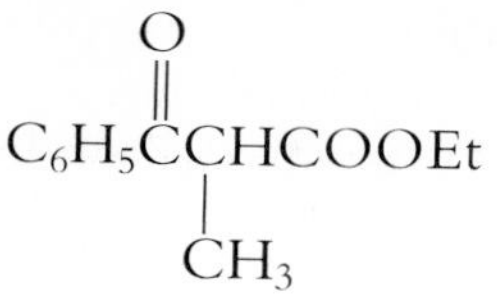

c. (six-membered lactone ring with CHO substituent)

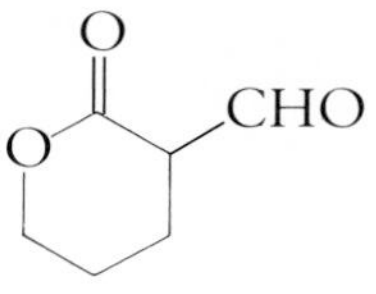

b. 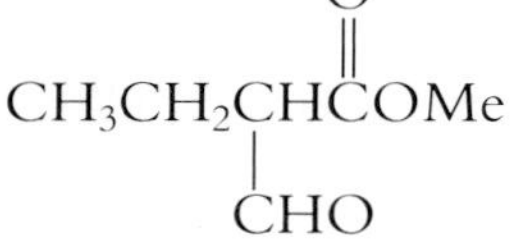 $CH_3CH_2\underset{\underset{CHO}{|}}{C}H\overset{\overset{O}{\|}}{C}OMe$

d. 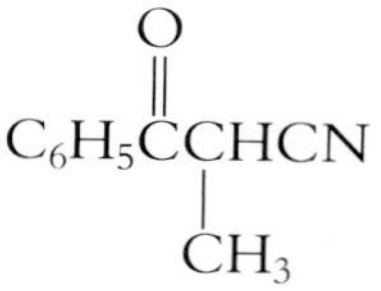 $C_6H_5\overset{\overset{O}{\|}}{C}\underset{\underset{CH_3}{|}}{C}HCN$

19:10 The Dieckmann Cyclization: Producing Five- and Six-Membered Rings by Intramolecular Acylation

The intramolecular acylation of diesters to produce cyclic ketoesters, which is a special case of the Claisen condensation, is called the **Dieckmann** cyclization (eq. 19.44).

(19.44) $(CH_2)_n$ bearing $CH_2\overset{O}{\overset{\|}{C}}R$ and $\underset{\underset{O}{\|}}{C}OR$ $\xrightarrow{\text{base}}$ $(CH_2)_n$ bearing $CH{=}\overset{O^-}{\overset{|}{C}}OR$ and $\underset{\underset{O}{\|}}{C}{-}OR$ $\rightleftarrows$ $(CH_2)_n$ ring with $\overset{H}{\overset{|}{C}}{-}\overset{O}{\overset{\|}{C}}OR$ and $\underset{\underset{^-O}{|}}{C}{-}OR$ $\rightleftarrows$

$(CH_2)_n$ ring with $\overset{H}{\overset{|}{C}}{-}\overset{O}{\overset{\|}{C}}OR$ and $\underset{\underset{O}{\|}}{C}$ $+ OR^- \rightarrow$ $(CH_2)_n$ ring with $\bar{C}{-}\overset{O}{\overset{\|}{C}}OR$ and $C{=}O$ $+ ROH \xrightarrow{H^+}$

$(CH_2)_n$ ring with $CH{-}\overset{O}{\overset{\|}{C}}OR$ and $C{=}O$ $+ ROH$

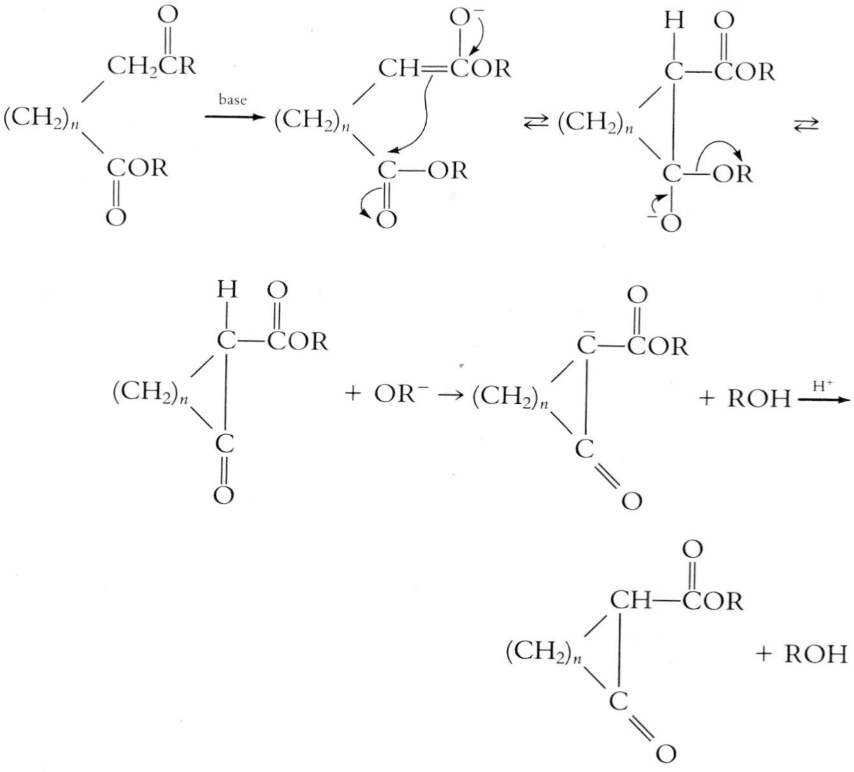

Five- and six-membered rings are formed readily by this approach (eq. 19.45), while attempts to form smaller rings usually fail. Larger rings can sometimes be prepared at high dilution where intramolecular reaction is favored over intermolecular reaction.

$$\text{(19.45)}\quad \begin{matrix} \overset{O}{\overset{\|}{C}}OEt \\ / \\ (CH_2)_4 \\ \backslash \\ \underset{O}{\underset{\|}{C}}OEt \end{matrix} \xrightarrow[\text{toluene}]{\text{NaH}} \begin{matrix} \overset{O}{\overset{\|}{C}}OEt \\ (CH_2)_3 \\ \underset{H}{\underset{|}{\bar{C}}}-\underset{O}{\underset{\|}{C}}OEt + H_2 \end{matrix} \rightarrow \text{2-oxocyclopentane}(H)(\underset{O}{\underset{\|}{C}}OEt) + EtO^- \rightarrow$$

Diethyl adipate

$$\text{2-oxocyclopentyl carbanion}-\overset{O}{\overset{\|}{C}}OEt + EtOH \xrightarrow{H_3O^+} \text{2-oxocyclopentane}(H)(\underset{O}{\underset{\|}{C}}OEt)$$

(65 to 80%)

Hydride ion is a powerful base that removes an α hydrogen to generate the ester enolate ion.

The Claisen and Dieckmann condensations may be used in conjunction with other enolate reactions (aldol condensation, alkylation, conjugate addition) to build up substituted carbon skeletons (eqs. 19.46 and 19.47).

$$\text{(19.46)}\quad CH_3CH_2CH_2\overset{O}{\overset{\|}{C}}OMe \xrightarrow[\substack{MeO^- \\ \text{(Claisen} \\ \text{condensation)}}]{HCOOMe} CH_3CH_2CH\begin{matrix} COOMe \\ CHO \end{matrix} \xrightarrow[\text{(alkylation)}]{CH_3CH_2CH_2I}$$

$$CH_3CH_2\underset{CHO}{\underset{|}{\overset{COOMe}{\overset{|}{C}}}}CH_2CH_2CH_3 \xrightarrow{H_3O^+} \left[CH_3CH_2-\underset{CHO}{\underset{|}{\overset{COOH}{\overset{|}{C}}}}-CH_2CH_2CH_3\right] \xrightarrow{\Delta} CH_3CH_2\underset{CHO}{\underset{|}{C}}HCH_2CH_2CH_3$$

$$\text{(19.47)}\quad CH_2(COOEt)_2 + 2CH_2{=}CHCOOEt \xrightarrow[\substack{\text{conjugate} \\ \text{addition}}]{EtO^-}$$

$$\left[\begin{matrix} EtOOC & & CH_2CH_2COOEt \\ & C & \\ EtOOC & & CH_2CH_2COOEt \end{matrix}\right] \xrightarrow[\substack{\text{(Dieckmann} \\ \text{cyclization)}}]{EtO^-}$$

$$\text{4,4-(EtOOC)}_2\text{-cyclohexanone, 2-COOET} \quad (\text{EtOOC}, \text{EtOOC}, {=}O, \text{COOET}) + EtOH$$

Problem

19–15 When diethyl 2-methylheptanedioate (**III**) is heated with sodium ethoxide, and the reaction mixture is then treated with acid, the only product isolated is ethyl 6-methylcyclohexanone-2-carboxylate (**IV**, eq. 4). Write a mechanism for the reaction shown in Equation 4, and explain why **IV** is the only isomer found.

$$(CH_2)_3(CH_2COOEt)(CHCOOEt\text{—}CH_3) \xrightarrow{NaOEt} \xrightarrow{H^+} \text{IV} \quad (4)$$

(**III**) (**IV**)

As you've seen, the Claisen condensation is subject to many variations. The reactants can be two molecules of the same ester, or two molecules of different esters, or a single molecule of a diester. All of these variants, however, share the same mechanism—an enolate ion is formed by removal of an α hydrogen, and this ion then attacks an ester group. In all the reactions we've just seen, the enolate ion has in fact been derived from an ester molecule, but this is not a necessary condition. Ketone enolates are perfectly good candidates for acylation, as the next section demonstrates.

19:11 Acylation of Ketones and Aldehydes

Esters containing no α hydrogens can be used to acylate ketones, in a variant of the crossed Claisen condensation (eqs. 19.48 and 19.49).

$$(19.48)\quad CH_3\overset{O}{\overset{\|}{C}}CH_3 + C_6H_5\overset{O}{\overset{\|}{C}}OEt \xrightarrow{EtO^-} \xrightarrow{H^+} C_6H_5\overset{O}{\overset{\|}{C}}CH_2\overset{O}{\overset{\|}{C}}CH_3$$

$$(19.49)\quad \text{cyclooctanone} + EtO\overset{O}{\overset{\|}{C}}OEt \xrightarrow[C_6H_6]{NaH} \xrightarrow{H_3O^+} \text{2-COOEt-cyclooctanone}$$

Diethyl carbonate (95%)

Once again, enolate ions are the nucleophiles in these reactions; in this case, they are derived from the ketone. Equation 19.50 gives a representative mechanism.

$$\text{(19.50)}\quad CH_3\overset{O}{\overset{\|}{C}}CH_3 \xrightarrow{EtO^-} CH_3\overset{O^-}{\overset{|}{C}}{=}CH_2 \xrightarrow{EtO-\overset{O}{\overset{\|}{C}}C_6H_5} CH_3\overset{O}{\overset{\|}{C}}-CH_2\overset{O}{\overset{\|}{C}}C_6H_5 + EtO^-$$

$$\rightarrow CH_3\overset{O}{\overset{\|}{C}}\overset{-}{\underset{CH}{\cdots}}\overset{O}{\overset{\|}{C}}C_6H_5 \xrightarrow{H^+} CH_3\overset{O}{\overset{\|}{C}}-CH_2-\overset{O}{\overset{\|}{C}}C_6H_5$$

Stable enolate
+ EtOH

Table 19–2 summarizes the major reactions involving acylation of enolates.

Table 19–2. Acylation Reactions of Enolates

1. **Claisen Condensation** Reaction between two molecules of a single ester, which has at least two α hydrogens.

$$RCH_2\overset{O}{\overset{\|}{C}}OR' + RCH_2\overset{O}{\overset{\|}{C}}OR' \xrightarrow[2.\ H^+]{1.\ R'O^-} RCH_2\overset{O}{\overset{\|}{C}}\underset{R}{\underset{|}{C}H}\overset{O}{\overset{\|}{C}}OR'$$

2. **Dieckmann Cyclization** Intramolecular acylation of a diester with a six- or seven-membered main chain.

$$ROOC(CH_2)_nCH_2COOR \xrightarrow[2.\ H^+]{1.\ RO^-} (CH_2)_n\begin{matrix} \diagup C{=}O \\ | \\ \diagdown CHCOOR \end{matrix} \qquad n = 3, 4$$

3. **Crossed Claisen Condensation** Two different esters react, only one of which has α hydrogens.

$$Ar\overset{O}{\overset{\|}{C}}OR' + RCH_2\overset{O}{\overset{\|}{C}}OR' \xrightarrow[2.\ H^+]{1.\ base} Ar\overset{O}{\overset{\|}{C}}\underset{R}{\underset{|}{C}H}\overset{O}{\overset{\|}{C}}OR'$$

$$H\overset{O}{\overset{\|}{C}}OR' + RCH_2\overset{O}{\overset{\|}{C}}OR' \xrightarrow[2.\ H^+]{1.\ base} H\overset{O}{\overset{\|}{C}}\underset{R}{\underset{|}{C}H}\overset{O}{\overset{\|}{C}}OR'$$

4. **Ketone Acylation** Reaction takes place between a ketone and an ester with no α hydrogens.

$$Y\overset{O}{\overset{\|}{C}}OR' + RCH_2\overset{O}{\overset{\|}{C}}R'' \xrightarrow[2.\ H^+]{1.\ base} Y\overset{O}{\overset{\|}{C}}\underset{R}{\underset{|}{C}H}\overset{O}{\overset{\|}{C}}R''$$

Y = H, Ar, OR′

19:11 Acylation of Ketones and Aldehydes

Box 19–1. Isolation of Tautomers

Although enol and keto tautomers of a single carbonyl compound are usually in rapid equilibrium, this does not mean that one of them cannot be isolated in pure form. The German chemist Ludwig Knorr isolated both tautomers of ethyl acetoacetate in 1911. He cooled a solution of ethyl acetoacetate in alcohol to −78°. A precipitate formed that did not react immediately with $FeCl_3$ (a test for an enolic hydroxyl group) and did not decolorize Br_2 (a test for C═C). (Remember that spectroscopic methods of analysis did not exist in 1911.) Knorr, therefore, assigned this carbonyl structure to the precipitate:

$$CH_3\overset{\overset{\displaystyle O}{\|}}{C}CH_2COOEt$$

When Knorr treated this compound with one equivalent of base and then neutralized the solution by passing in dry HCl gas, he obtained a solid that reacted immediately with both $FeCl_3$ and Br_2. To this compound he gave the enol structure

$$CH_3\overset{\overset{\displaystyle OH}{|}}{C}{=}CHCOOEt$$

Knorr had thus isolated both keto and enol forms. Each material was converted rapidly to an equilibrium mixture when the solution of either was warmed to room temperature. The interconversion has to be catalyzed by traces of acid or base, which are usually present in all glassware. In 1920, K. H. Meyer separated the tautomers of ethyl acetoacetate by distillation in a specially treated quartz apparatus.

Even the very reactive enol of acetaldehyde, $CH_2{=}CH{-}OH$ (*vinyl alcohol*) has been prepared by a special reaction—dehydration of $HOCH_2CH_2OH$ at 900° and low pressure. Vinyl alcohol was found to have a half-life of 30 minutes at 25° in ordinary glassware, before equilibrium was reached between it and acetaldehyde. These examples point out the difference between stability and reactivity. Both enols mentioned above are certainly "stable" molecules and would exist more or less forever if they were held completely isolated in some manner. They are very reactive, however, and are rapidly transformed in the presence of minute amounts of acids or bases.

19:12 The Tricarboxylic Acid Cycle

Many animals, including man, store the sugar glucose as a polymer called glycogen. This is used as a source of energy during periods of exercise. Glycogen is depolymerized in the body to give individual glucose molecules which are then degraded to pyruvic acid

$$CH_3\overset{\overset{\displaystyle O}{\|}}{C}COOH$$

in a series of reactions called glycolysis. Pyruvic acid then goes through a complex series of reactions designed to convert its carbon atoms to CO_2 and its hydrogen atoms ultimately to water. This series of reactions is known as the **tricarboxylic acid cycle**, or **Krebs cycle** (after its discoverer, the German Nobel laureate H. Krebs).

Figure 19–2 shows the structure of the various intermediates in the tricarboxylic acid cycle. While the cycle is very complex, close examination of the individual chemical reactions shows that all of the steps involve familiar types of organic reactions, such as conjugate additions, decarboxylations, and aldol condensations. These reactions are enzyme catalyzed and are extremely rapid. Nevertheless, their mechanisms are very similar to those of

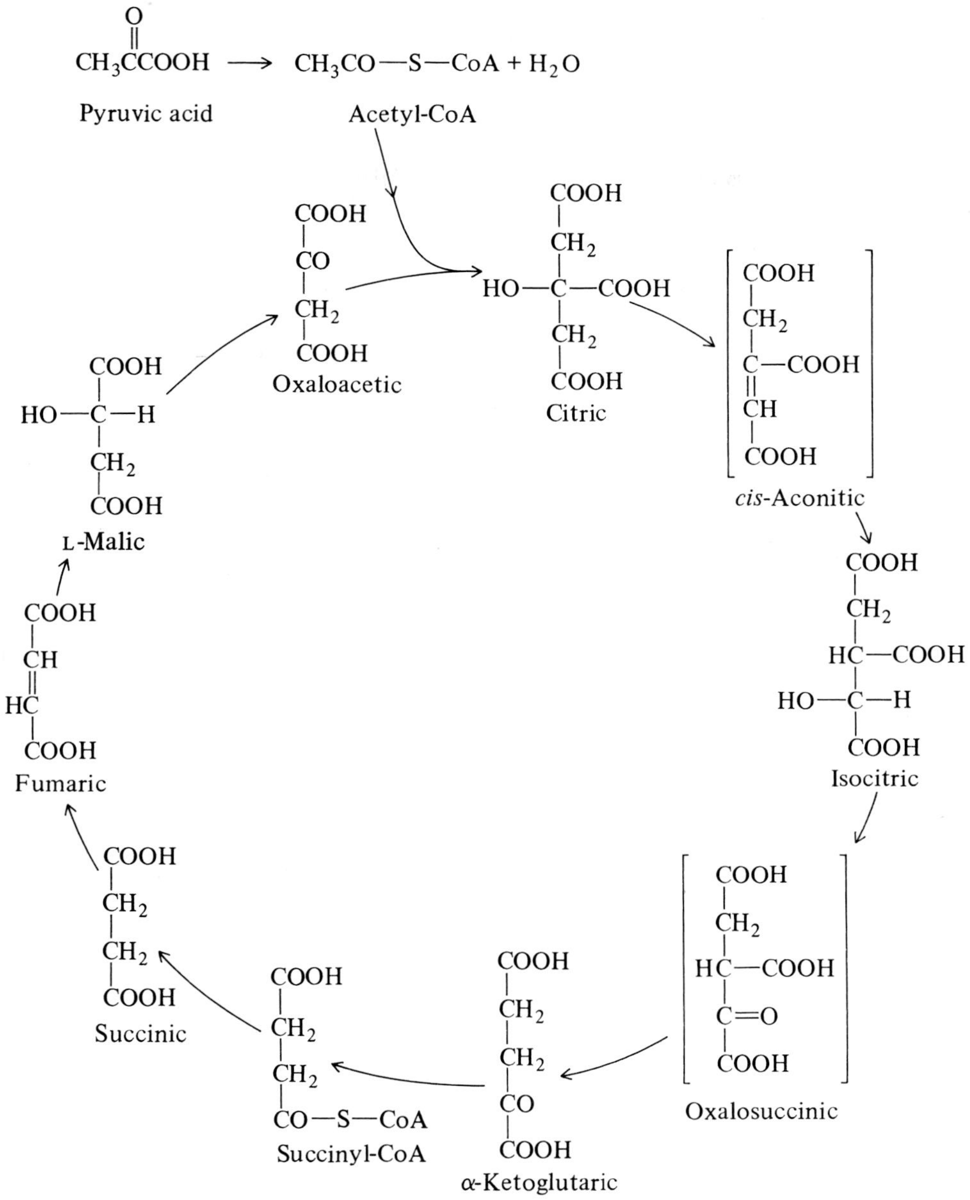

$$\mathrm{CH_3\overset{O}{\overset{\|}{C}}COOH \longrightarrow CH_3CO{-}S{-}CoA + H_2O}$$

Pyruvic acid　　Acetyl-CoA

COOH–CO–CH_2–COOH
Oxaloacetic

COOH–CH_2–C(OH)(COOH)–CH_2–COOH
Citric

[COOH–CH_2–C(COOH)=CH–COOH]
cis-Aconitic

COOH–CH_2–HC(COOH)–HC(OH)–COOH
Isocitric

[COOH–CH_2–HC(COOH)–C(=O)–COOH]
Oxalosuccinic

COOH–CH_2–CH_2–CO–COOH
α-Ketoglutaric

COOH–CH_2–CH_2–CO–S–CoA
Succinyl-CoA

COOH–CH_2–CH_2–COOH
Succinic

COOH–CH=HC–COOH
Fumaric

COOH–HO–C–H–CH_2–COOH
L-Malic

Figure 19–2. *The tricarboxylic acid cycle.*

their laboratory analogs. In the following paragraphs, the details of a few of the individual reactions in the tricarboxylic acid cycle will be examined.

The cycle begins (top of Fig. 19–2) with a reaction involving acetyl-CoA, which is synthesized from pyruvic acid (sec. 17:13). The acetyl-CoA undergoes aldol addition with the keto carbonyl group of oxaloacetic acid (2-oxosuccinic acid) to produce *citric acid* (eq. 19.51). Citric acid, which is widely distributed in nature, is particularly abundant in citrus fruits such as lemons and limes.

$$\text{(19.51)}\quad \text{CoA—S—}\underset{\text{CH}_3}{\overset{\text{O}}{\text{C}}} + \text{HOOC}\overset{\text{O}}{\overset{\|}{\text{C}}}\text{CH}_2\text{COOH} \xrightarrow[\text{citrate synthase}]{\text{H}_2\text{O}}$$

Oxaloacetic acid

$$\text{HOOC}\underset{\text{CH}_2\text{COOH}}{\overset{\text{OH}}{\text{C}}}\text{CH}_2\text{COOH} + \text{CoA—SH}$$

Citric acid

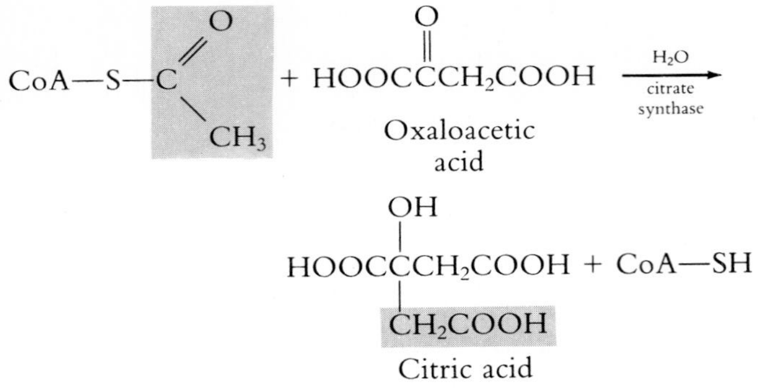

Aldol condensation

The reaction in Equation 19.51 can be initiated by formation of an acetyl-CoA enolate ion, **5**, or the related enol.

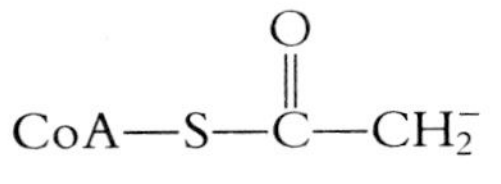

$$\text{CoA—S—}\overset{\text{O}}{\overset{\|}{\text{C}}}\text{—CH}_2^-$$

5

Citric acid is then dehydrated stereospecifically to *cis*-aconitic acid (eq. 19.52).

$$\text{(19.52)}\quad \text{HOOC}\underset{\text{HOOCCH}_2}{\overset{\text{OH}}{\text{C}}}\text{CH}_2\text{COOH} \xrightarrow{\text{aconitase}} \begin{array}{c}\text{HOOC} \quad\quad \text{CH}_2\text{COOH} \\ \text{C} \\ \| \\ \text{C} \\ \text{HOOC} \quad\quad \text{H}\end{array} + \text{H}_2\text{O}$$

cis-Aconitic acid

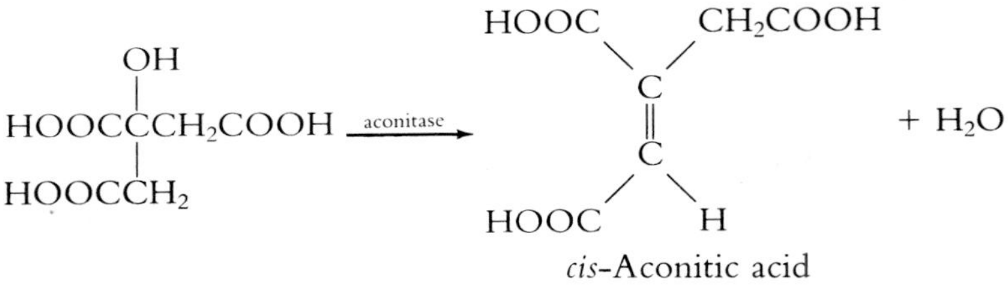

cis-Aconitic acid remains bound to the enzyme aconitase, where it undergoes 1,4-conjugate addition of water to the C=C double bond (eq. 19.53).

$$\text{(19.53)}\quad \begin{array}{c}\text{HOOCCH}_2 \quad\quad \text{H} \\ \text{C}=\text{C} \\ \text{HO—C} \quad\quad \text{COOH} \\ \| \\ \text{O} \\ \overset{+}{\text{B}}\text{—H}\end{array} + \begin{array}{c}\text{:OH} \\ | \\ \text{H} \\ \text{:B}\end{array} \xrightarrow{\text{aconitase}} \begin{array}{c}\text{HOOCCH}_2 \\ \text{C—CHOH} \\ \text{HO—C} \quad\quad \text{COOH} \\ | \\ \text{OH}\end{array}$$

$$\xrightarrow{\text{fast}} \begin{array}{c}\text{HOOCH}_2 \\ \text{CH—CHOH} \\ \text{HOOC} \quad\quad \text{COOH}\end{array}$$

Isocitric acid

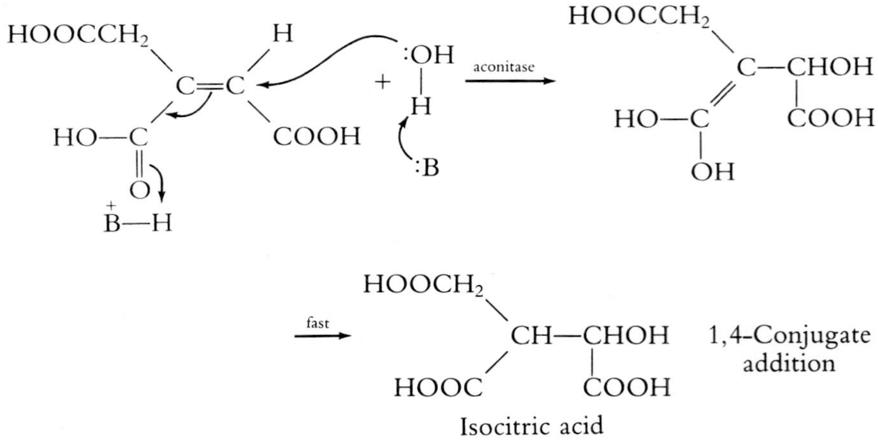

1,4-Conjugate addition

An examination of Figure 19–2 shows that the 2° hydroxyl group of isocitric acid is oxidized in the next step of the cycle. The product, *oxalosuccinic acid,* contains a carboxyl group β to the carbonyl, and it undergoes very rapid decarboxylation to *α-ketoglutaric acid* (eq. 19.54).

$$(19.54)\quad \underset{\text{Oxalosuccinic acid}}{HO_2C\underset{\underset{CH_2COOH}{|}}{C}H\overset{\overset{O}{\|}}{C}COOH} \rightarrow \underset{\alpha\text{-Ketoglutaric acid}}{\underset{\underset{CH_2COOH}{|}}{C}H_2\overset{\overset{O}{\|}}{C}COOH} + CO_2 \qquad \text{Decarboxylation}$$

Thus, of the first five steps in the tricarboxylic acid cycle, three involve reactions we have discussed in the present chapter: an aldol condensation, a 1,4-conjugate addition, and a decarboxylation. The next section illustrates the biological importance of another major topic of this chapter, acylation.

Problem

19–16 Write mechanisms for the following reactions, which are part of the tricarboxylic acid cycle (Fig. 19–2).
a. conversion of oxalosuccinic acid to α-ketoglutaric acid
b. conversion of fumaric acid to malic acid

19:13 Fatty Acid Metabolism

One of the major classes of metabolically important compounds is the fatty acids ($CH_3(CH_2)_nCOOH$) which are abundant in animals, plants, and microorganisms. During resting periods the organism builds up these long-chain acids by a process known as **anabolism**. When the organism requires energy, the long chains are broken down to smaller units by a process called **catabolism**. In both the construction and destruction of the long-chain acids, acylation reactions play a key role.

The biosynthesis is built around acetyl thioesters. First, a molecule of acetyl-CoA is carboxylated to malonyl-CoA (eq. 19.55). The malonyl group is then transferred to an *acyl carrier protein* (ACP-SH), which has a thiol group available for thioester formation (second step of eq. 19.55).

$$(19.55)\quad \underset{\text{Acetyl-CoA}}{CH_3\overset{\overset{O}{\|}}{C}S\text{—CoA}} + CO_2 \xrightarrow[\text{carboxylase}]{\text{acetyl-CoA}} \underset{\text{Malonyl-CoA}}{\underset{\underset{COOH}{|}}{C}H_2\overset{\overset{O}{\|}}{C}S\text{—CoA}} \xrightarrow{\text{ACP-SH}}$$

$$\underset{\substack{\text{Malonyl-ACP}\\ \text{(ACP = acyl carrier}\\ \text{protein)}}}{\underset{\underset{COOH}{|}}{C}H_2\overset{\overset{O}{\|}}{C}S\text{—ACP}} + \text{CoA—SH}$$

The malonyl-ACP is now acylated by an acetyl group that is bound to the enzyme by way of a thiol group (eq. 19.56). The loss of carbon dioxide during this reaction provides a thermodynamic driving force for the process.

$$\text{(19.56)}\quad CH_3\overset{O}{\overset{\|}{C}}S\text{—Enz} + \underset{\underset{\text{Malonyl-ACP}}{COOH}}{\underset{|}{CH_2}}\overset{O}{\overset{\|}{C}}S\text{—ACP} \xrightarrow[\text{synthase}]{\beta\text{-ketoacyl}}$$

$$CH_3\overset{O}{\overset{\|}{C}}CH_2\overset{O}{\overset{\|}{C}}S\text{—ACP} + \text{Enz—SH} + CO_2$$

Acetoacetyl-ACP

The product is a derivative of the very familiar acetoacetic acid.

The addition of a carboxyl group to acetyl-CoA (eq. 19.55) clearly has the effect of activating it toward acylation. One possibility is that the malonyl enolate ion is the actual species being acylated in the reaction of Equation 19.56. A plausible mechanism is given in Equation 19.57.

$$\text{(19.57)}\quad CH_3\overset{O}{\overset{\|}{C}}S\text{—Enz} + \overset{\overset{HO}{|}}{\overset{C=O}{|}}{}^{-}CHCOS\text{—ACP} \rightarrow \left[CH_3\overset{O}{\overset{\|}{C}}\text{—}\overset{\overset{H—O}{|}}{\overset{C=O}{|}}CHCOS\text{—ACP}\right] \rightarrow$$

Malonyl enolate ion

$$CH_3\overset{OH}{\overset{|}{C}}=CH\overset{O}{\overset{\|}{C}}S\text{—ACP} + CO_2$$

The intermediate ketocarboxylic acid would decarboxylate very rapidly to yield the conjugated enol of acetoacetyl-ACP.

Acetoacetyl-ACP is then reduced to a saturated fatty acid ester in three stages. Stereoselective reduction of the carbonyl group gives (*R*)-3-hydroxybutanoyl-ACP (eq. 19.58).

$$\text{(19.58)}\quad CH_3\overset{O}{\overset{\|}{C}}CH_2\overset{O}{\overset{\|}{C}}S\text{—ACP} + \text{NADPH} + H^+ \xrightarrow[\text{reductase}]{\beta\text{-ketoacyl-ACP}}$$

$$CH_3\overset{OH}{\overset{|}{\underset{H}{\underset{|}{C}}}}CH_2\overset{O}{\overset{\|}{C}}S\text{—ACP} + NADP^+$$

(*R*)-3-Hydroxybutanoyl-ACP

The hydroxy ester is next dehydrated to yield (*E*)-2-butenoyl-ACP (eq. 19.59).

$$\text{(19.59)}\quad CH_3\overset{OH}{\underset{H}{C}}CH_2\overset{O}{\overset{\|}{C}}S{-}ACP \xrightarrow[\text{hydratase}]{\text{enoyl-ACP}} \underset{(E)\text{-2-Butenoyl-ACP}}{(CH_3)HC{=}CH\overset{O}{\overset{\|}{C}}S{-}ACP} + H_2O$$

Finally, the unsaturated ACP ester is converted to a saturated ester by NADPH. This reaction may be viewed as a conjugate addition of H^- and H^+ to the α,β-unsaturated carbonyl system (eq. 19.60).

$$\text{(19.60)}\quad (CH_3)HC{=}CH\overset{O}{\overset{\|}{C}}S{-}ACP \xrightarrow[H^+]{\text{NADPH}} CH_3{-}\overset{H}{\underset{H}{C}}{-}CH{=}C(OH){-}S{-}ACP \rightleftharpoons CH_3CH_2CH_2\overset{O}{\overset{\|}{C}}S{-}ACP + NADP^+$$

NADPH

The butanoyl-ACP is then transferred to the synthase enzyme and used to acylate another molecule of malonyl-ACP to give a new β-ketoester (eq. 19.61).

$$\text{(19.61)}\quad CH_3CH_2CH_2\overset{O}{\overset{\|}{C}}S{-}Enz + \underset{COOH}{CH_2}\overset{O}{\overset{\|}{C}}{-}S{-}ACP \xrightarrow[\text{synthase}]{\beta\text{-ketoacyl-ACP}} CH_3CH_2CH_2\overset{O}{\overset{\|}{C}}CH_2\overset{O}{\overset{\|}{C}}{-}S{-}ACP$$

This β-ketoester will be subject, in turn, to reduction, dehydration, and another reduction. Note that in each step the chain is extended by two carbon units, which accounts for the observation that almost all naturally abundant fatty acids contain an *even* number of carbon atoms.

Fatty acids are not found free in animals but are stored as esters of glycerol (fats or glycerides; sec. 17:9). When the animal requires energy, these glycerides are hydrolyzed and the free fatty acids are then rapidly esterified with coenzyme A. They undergo catabolic breakdown as the CoA esters. The pathway of biodegradation is almost, but not quite, the exact reverse of the pathway for biosynthesis.

Problem

19–17 In 1904, the chemist F. Knoop did a series of experiments concerning the metabolism of fatty acids. He fed rabbits straight-chain fatty acids containing a phenyl group attached to the last carbon atom. If the chain consisted of an even number of carbon atoms, the rabbits excreted large quantities of phenylacetic acid. When the chain was odd-numbered, the major metabolic product was benzoic acid (eq. 5). What conclusions could reasonably be drawn from this experiment?

$$C_6H_5-(CH_2)_{n-1}COOH \xrightarrow{n\ =\ \text{even integer}} C_6H_5-CH_2COOH$$
$$C_6H_5-(CH_2)_{n-1}COOH \xrightarrow{n\ =\ \text{odd integer}} C_6H_5-COOH \qquad (5)$$

Summary

1 Stable enolate anions can be generated from acetoacetic esters, malonates, nitriles, and β-diketones by treating these substrates with alkoxide bases. These enolate ions can be selectively alkylated by reaction with alkyl halides and toluenesulfonates. Monocarbonyl compounds can also be alkylated, but only if strong bases such as LDA are used. This process is less selective than alkylation of the dicarbonyl enolates (sec. 19:1).

2 α,β-Unsaturated carbonyl compounds can undergo 1,4-conjugate addition with a variety of simple nucleophiles: $:CN^-$, $R_2\ddot{N}H$, $R\ddot{\underset{..}{S}}H$, and $R\ddot{\underset{..}{O}}H$ (sec. 19:2). When the nucleophile is an enolate anion [e.g., $^-\!:CH(COOEt)_2$], the reaction is called a Michael addition (sec. 19:3). It works well only if the enolate ion is a relatively stable one. The less stable enolates of simple aldehydes and ketones can be replaced in the Michael reaction by the corresponding enamines (sec. 19:4). The enamines are generated from secondary amines and carbonyl compounds.

Organometallic reagents (RLi, RMgBr, RCu) can add either 1,2 or 1,4 to unsaturated carbonyl compounds. Use of Cu(I) derivatives favors 1,4-addition, while Li and Mg reagents give predominantly 1,2-addition (sec. 19:5). Many of these conjugate additions, in conjunction with aldol reactions, provide methods for the synthesis of five- and six-membered rings (sec. 19:6).

3 Compounds that contain a carboxyl group β to a carbonyl group undergo decarboxylation upon heating. This reaction is put to use in the synthesis of substituted ketones and acids. An ester group is used to stabilize an enolate, and is later removed by hydrolysis and decarboxylation (sec. 19:7). Enolate nucleophiles, such as phenoxide ions, can add CO_2 in a carboxylation reaction (sec. 19:8). This is the reverse of decarboxylation. The carboxylation of enols is of fundamental biochemical importance.

4 Esters can undergo a base-promoted self-condensation known as the Claisen condensation. The products are β-ketoesters. If two different esters are used,

the reaction is known as a crossed Claisen condensation (sec. 19:9). Crossed condensations are most useful when only one of the esters can enolize.

Intramolecular acylation is valuable for the synthesis of five- and six-membered ring ketones, and is known as the Dieckmann cyclization (sec. 19:10). Esters that lack α hydrogens can be used to acylate ketones. The products are β-diketones (sec. 19:11).

5 Aldol condensations, conjugate additions, and decarboxylation reactions are important to living organisms. All of these reactions occur in the tricarboxylic acid cycle, or Krebs cycle (sec. 19:12).

Acylation reactions involving acetyl-CoA and malonyl-CoA are fundamental to the biosynthesis of long-chain fatty acids, $CH_3(CH_2)_nCOOH$ (sec. 19:13).

Problems

19–18 Propose a synthesis of the following compounds from readily available starting materials. (The syntheses will involve alkylations and conjugate additions, and may require multiple steps.)

a. cyclopentyl–$C(COOEt)_2$ with CH_3 on the same carbon (cyclopentyl–$C(CH_3)(COOEt)_2$)

b. cyclopentane bearing CN and COOEt on the same ring carbon

c. C_6H_5–$CH(CN)CH(COOEt)COCH_3$

d. $(CH_3)_3CCH(COOEt)_2$

e. cyclohexanone with $CH(COOEt)_2$ at C-3

f. $CH_3\overset{O}{\overset{\|}{C}}CH_2CH_2CH_2CN$

g. $CH_3CH(CH(CN)_2)CH_2CN$

h. $EtOOC–CH_2CH_2–C(CN)(COOEt)–CH_2CH_2–COOEt$

19–19 When 1,3-cyclohexanedione is treated successively with sodium methoxide and methyl iodide, two organic products are formed. One of these is the desired 2-methyl-1,3-cyclohexanedione. The other product, isomeric with the first, exhibits the following nmr spectrum: a multiplet in the methylene (CH_2) region (6H), a vinyl proton (1H), and a singlet (3H). Propose a structure for this compound, and rationalize its formation.

19–20 Indicate which compound of each of the following pairs will be more reactive toward a nucleophile in a 1,4-addition reaction.

a. $CH_2{=}CHCOOMe$ vs. $CH_2{=}C(COOMe)(CHO)$

b. 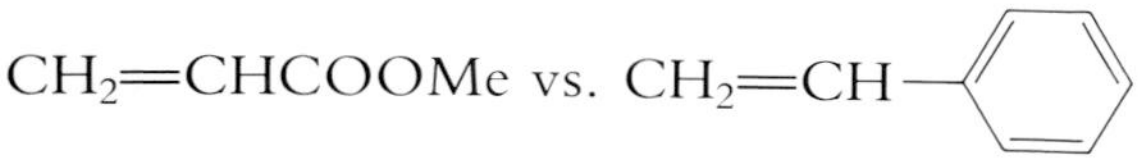 $CH_2{=}CHCOOMe$ vs. $CH_2{=}CH{-}C_6H_5$

c. $CH_2{=}CHCOOMe$ vs. $CH_3CH{=}CHCOOMe$

d. $CH_2{=}CHCOOMe$ vs. $CH_2{=}C(CH_3)COOMe$

e. 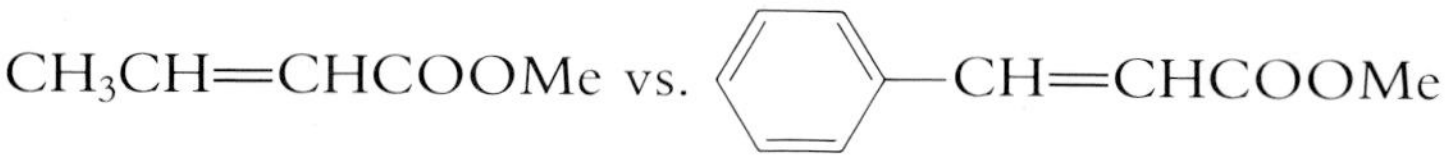 $CH_3CH{=}CHCOOMe$ vs. $C_6H_5{-}CH{=}CHCOOMe$

f. $CH_2{=}CHCOOMe$ vs. $CH_2{=}CHCOCH_3$

19–21 The reaction in Equation 6 is undetectably slow, while the reaction in Equation 7 is rapid even at 0°. Suggest a mechanism for this reaction, and explain why it is so much faster than that in Equation 6.

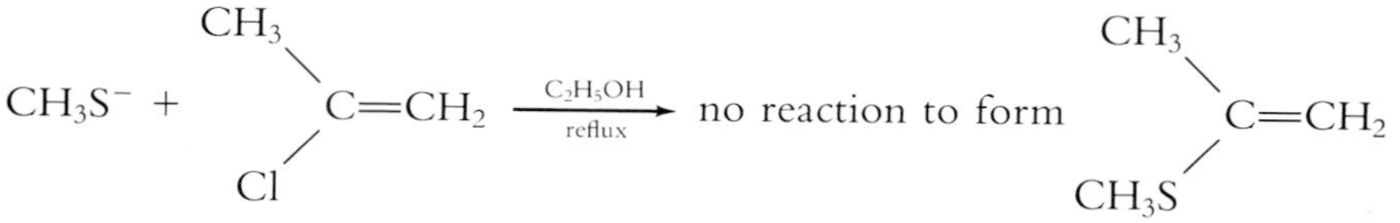

$$CH_3S^- + (CH_3)(Cl)C{=}CH_2 \xrightarrow[\text{reflux}]{C_2H_5OH} \text{no reaction to form } (CH_3)(CH_3S)C{=}CH_2 \qquad (6)$$

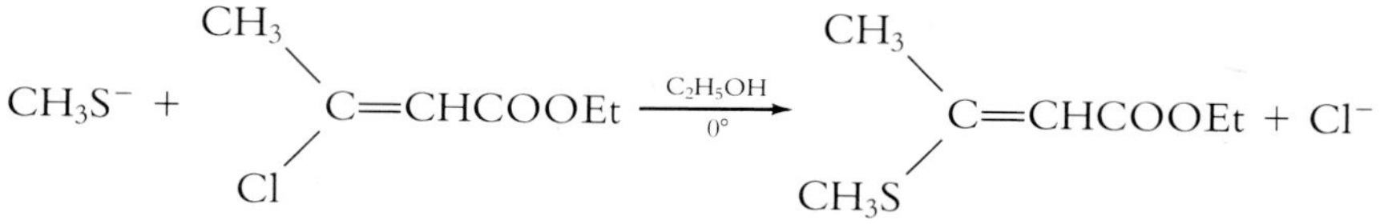

$$CH_3S^- + (CH_3)(Cl)C{=}CHCOOEt \xrightarrow[0^\circ]{C_2H_5OH} (CH_3)(CH_3S)C{=}CHCOOEt + Cl^- \qquad (7)$$

19–22 The compound 5-oxohexanenitrile is made commercially by the reaction between acetone and acrylonitrile, catalyzed by isopropylamine (eq. 8, first step). This compound is an intermediate in the synthesis of 2-methylpyridine, which is in turn converted to 2-vinylpyridine, a component of artificial rubber used in tires.

$$CH_3\overset{\overset{\displaystyle O}{\|}}{C}CH_3 + CH_2{=}CHCN \xrightarrow{(CH_3)CHNH_2} CH_3\overset{\overset{\displaystyle O}{\|}}{C}CH_2CH_2CH_2CN \rightarrow$$

5-Oxohexanenitrile

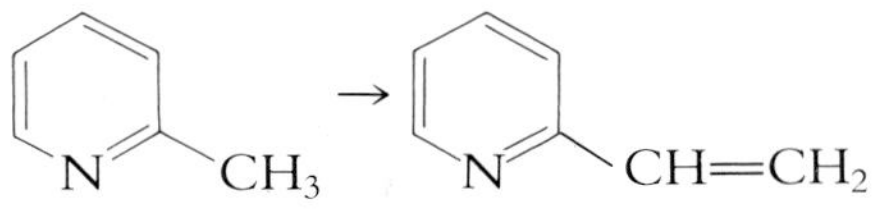

2-Methylpyridine → 2-Vinylpyridine (8)

Write a mechanism for the formation of 5-oxohexanenitrile.

19–23 Propose syntheses for the following acids and acid derivatives, based on common starting materials.
a. 2-ethylpentanoic acid
b. isobutyl 2,3-dimethylhexanoate
c. cyclopentanecarboxylic acid
d. 2-benzyl-3-phenylpropanamide
e. 2-methyl-4-oxopentanoic acid

19–24 The enzymatic conversion of fumaric acid to malic acid is catalyzed by the enzyme *fumarase*. It is known to take place by way of an *anti* addition of water to the double bond. The product has the *S* configuration (eq. 9).

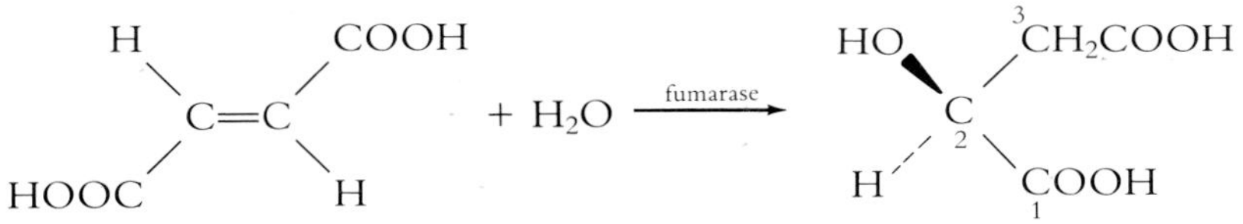

$$\text{Fumaric acid (H(COOH)C=C(H)COOH, trans)} + H_2O \xrightarrow{\text{fumarase}} \text{(S)-Malic acid: HO, H, CH}_2\text{COOH (C-3), COOH (C-1) on C-2} \qquad (9)$$

Fumaric acid

(*S*)-Malic acid

a. This hydration was carried out with D_2O to yield 3-deuteriomalic acid. What was the absolute configuration at C-3?
b. The 3-deuteriomalic acid was enzymatically oxidized to 3-deuterioox-aloacetic acid, which then entered the tricarboxylic acid cycle, along with some acetyl-CoA (Fig. 19–2). It was converted to citric acid and then to *cis*-aconitic acid. The *cis*-aconitic acid was shown to contain *no* deuterium. Based on this finding, and other data already presented, describe all the stereochemistry in the conversion of oxaloacetic acid to *cis*-aconitic acid.

19–25 Propose syntheses of the following compounds, based on readily available esters.
a. methyl *p*-chlorobenzoylacetate
b. ethyl 2-ethyl-3-oxohexanoate
c. 2-oxocyclopentanecarboxyaldehyde
d. $EtO_2C(CH_2)_3CH(COOEt)_2$

19–26 The compound iridodial is a defensive secretion of some ants. It has been proposed that it is biosynthesized from citronellol (eq. 10), which is found in numerous plants. Suggest reasonable intermediates in the biosynthesis. Show a reasonable mechanism for the ring-forming reaction. (There are enzymatic systems in tissue that can bring about the oxidation of CH_2 and CH_3 groups.)

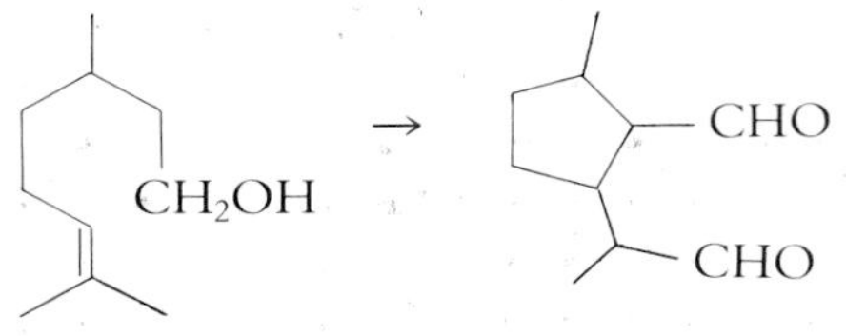

CH_2OH → CHO, CHO (10)

Citronellol Iridodial

19–27 When compound **V** is stirred with a catalytic amount of sodium ethoxide in ethanol, it slowly converts to the diester, **VI**. When **V** is heated with one equivalent of sodium ethoxide, and the solution is then neutralized, it is transformed into its isomer **VII**. Propose mechanisms for both these reactions, and explain why they lead to different products.

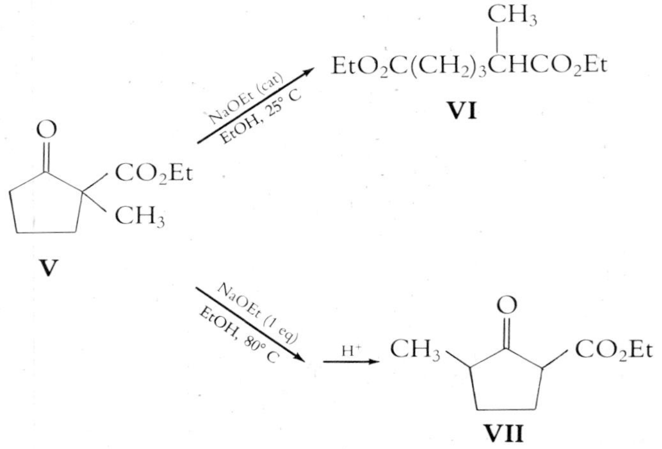

V (O, CO_2Et, CH_3) —NaOEt (cat), EtOH, 25° C→ $EtO_2C(CH_2)_3CH(CH_3)CO_2Et$ **VI**

V —NaOEt (1 eq), EtOH, 80° C→ —H^+→ CH_3, O, CO_2Et **VII**

19–28 The compound 2-methyl-3-buten-2-ol is one of three components of the sex pheromone of a destructive Scandinavian bark beetle. Draw the structure of this compound, and propose a synthesis from inexpensive starting materials.

19–29 One ecologically sound method of insect control is to expose young insects to naturally occurring substances that disrupt their sexual maturation. The compound precocene II, which is isolated from a common plant, possesses such activity. Write a mechanism for the first step of the laboratory precocene II synthesis shown in Equation 11.

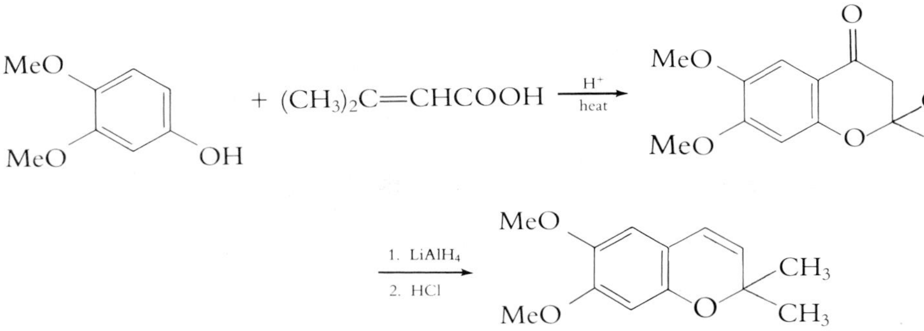

MeO, MeO, OH + $(CH_3)_2C{=}CHCOOH$ $\xrightarrow[\text{heat}]{H^+}$ (MeO, MeO, O, CH_3, CH_3, O)

$\xrightarrow[\text{2. HCl}]{\text{1. LiAlH}_4}$ (MeO, MeO, O, CH_3, CH_3) (11)

Precocene II

19–30 The compound nootkatone occurs in grapefruit and other citrus species, and its isomer isonootkatone is important in the manufacture of perfume. The construction of the carbon skeletons of these compounds illustrates the com-

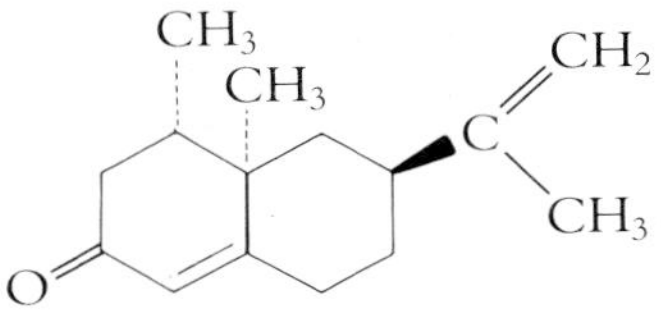

(CH_3, CH_3, CH_2, C, CH_3, O)

Nootkatone

(CH_3, CH_3, $C(CH_3)_2$, O)

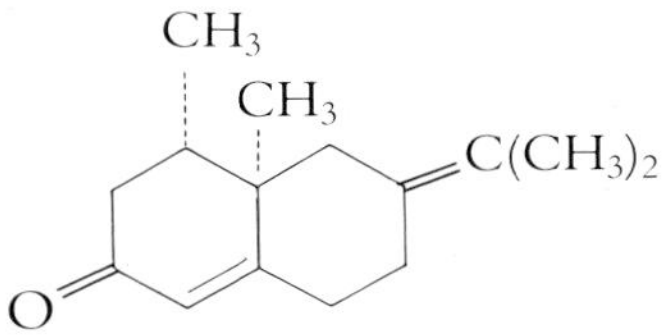

Isonootkatone

bined use of aldol condensations, acylations, and conjugate additions in organic synthesis. We shall concentrate on the carbon skeleton and ignore the stereochemical problems, although stereoselective syntheses are of prime importance in actual chemical research.

In the following reactions, fill in the structures of the compounds not shown, or the necessary reagents and conditions over the appropriate arrows.

a. $CH_2(COOEt)_2 \rightarrow (CH_3)_2C{=}C(COOEt)_2 \xrightarrow[\text{then } H_3O^+]{\text{LiAlH}_4,} \xrightarrow{\text{PBr}_3}$

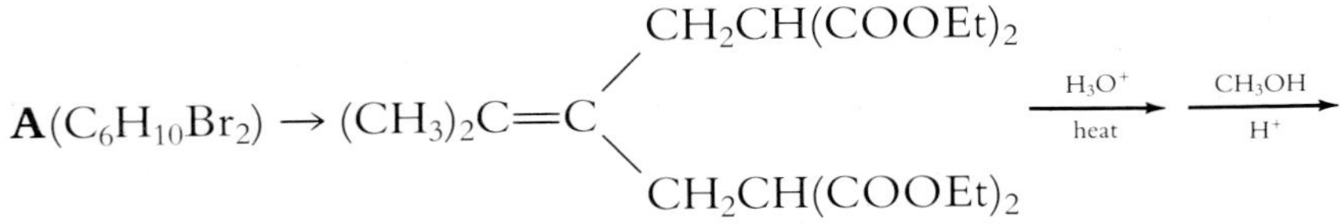

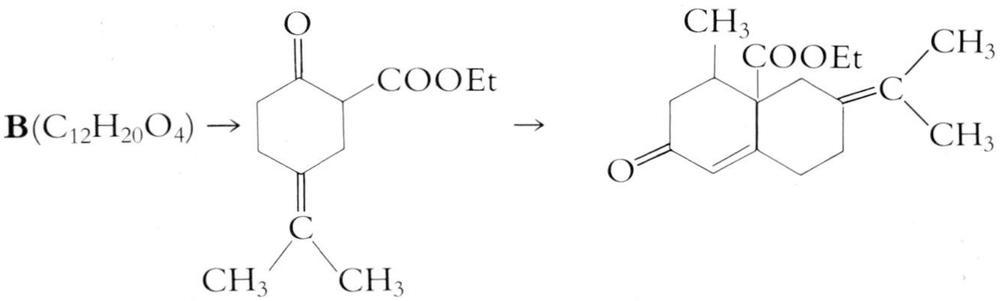

$\mathbf{A}(C_6H_{10}Br_2) \rightarrow (CH_3)_2C{=}C(CH_2CH(COOEt)_2)(CH_2CH(COOEt)_2) \xrightarrow[\text{heat}]{H_3O^+} \xrightarrow[H^+]{CH_3OH}$

$\mathbf{B}(C_{12}H_{20}O_4) \rightarrow$ (O, COOEt, C, CH_3, CH_3) $\rightarrow$ (CH_3, COOEt, CH_3, C, CH_3, O)

(an intermediate in the synthesis of isonootkatone)

b. [structure: 4-isopropenylcyclohexanone] → [structure: 2-formyl-4-isopropenylcyclohexanone] $\xrightarrow[K_2CO_3]{CH_3I}$ **C**($C_{11}H_{16}O_2$) →

[structure: dienone with CH_3 and $C(=CH_2)CH_3$ substituents] ⟶ [structure: enone with CH_3, CH_3 and $C(=CH_2)CH_3$ substituents]

(Comment on the selectivity of the last step.)

c. $CH_2(COOMe)_2 \rightarrow (NCCH_2CH_2)_2C(COOMe)_2 \xrightarrow[\text{heat vigorously}]{H_3O^+} \xrightarrow[H^+]{CH_3OH}$ **D**($C_{11}H_{18}O_6$)

D($C_{11}H_{18}O_6$) → [structure: 2-oxocyclohexane with $COOCH_3$ at C-1 and $COOCH_3$ at C-4] → [structure: 2-oxocyclohexane with CH_3 and $COOCH_3$ at C-1 and $COOCH_3$ at C-4]

→ [structure: 2-methylcyclohexanone with $COOCH_3$ at C-4]
(an intermediate in the formation of nootkatone)

19–31 The following sequence of reactions was employed in the total synthesis of one of the antibiotics of the tetracycline series. Supply the missing structures and reagents.

[structure: methyl 3-methoxybenzoate ($COOMe$, OMe)] + $CH_3COOMe \xrightarrow[DMF]{2NaH}$ **A**($C_{11}H_{11}O_4Na$) →

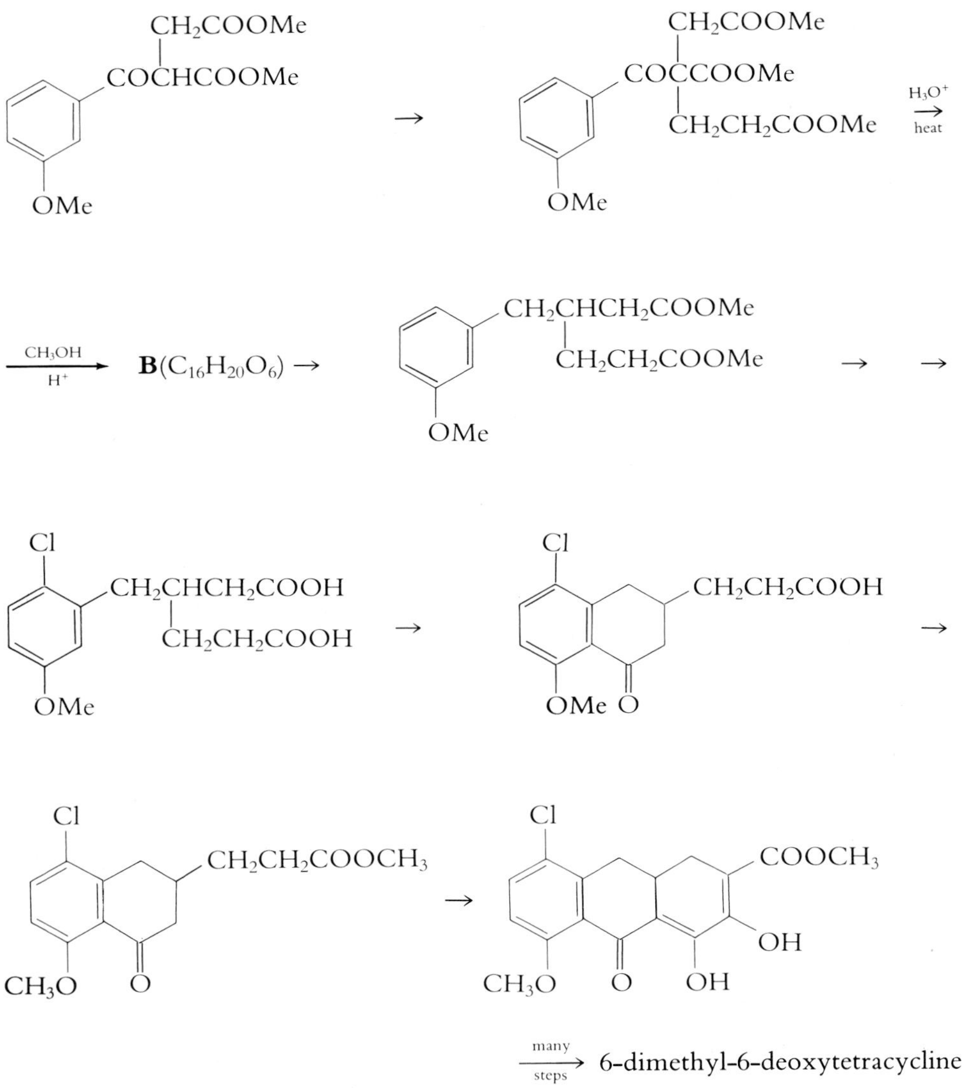

$\xrightarrow{\text{H}_3\text{O}^+,\ \text{heat}}$ $\xrightarrow[\text{H}^+]{\text{CH}_3\text{OH}}$ **B**($C_{16}H_{20}O_6$) →

→ →

→ 6-dimethyl-6-deoxytetracycline (many steps)

19–32 We have seen that $NaBH_4$ reduces C=O groups, but not isolated C=C groups (sec. 11:10). However, when α,β-unsaturated carbonyl compounds are reduced by $NaBH_4$ in alcoholic solvents, we often observe reduction of both functional groups (eq. 12).

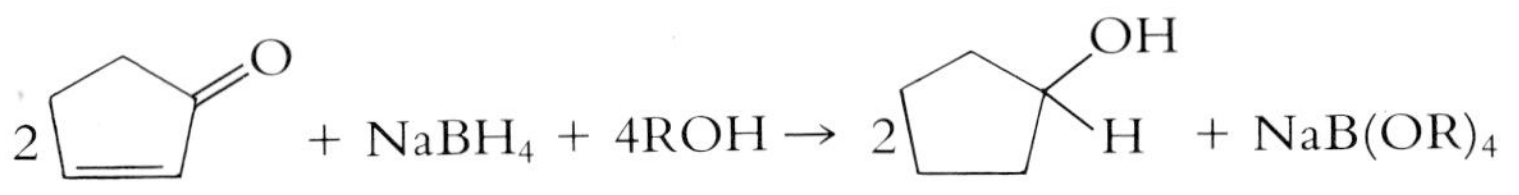

$$2\ \text{(cyclopentenone)} + NaBH_4 + 4ROH \rightarrow 2\ \text{(cyclopentanol)} + NaB(OR)_4 \qquad (12)$$

Suggest a mechanism for this reaction. Pay particular attention to the order of reduction of the two functional groups.

Supplementary Reading

S. Warren, *Designing Organic Syntheses,* pp. 45–75, John Wiley and Sons, New York, N.Y., 1978.

H. O. House, *Modern Synthetic Reactions,* Chaps. 9 and 11, W. A. Benjamin, Menlo Park, Calif., 1972.

J. Larner, *Intermediary Metabolism and Its Regulation,* Chap. 4, Prentice–Hall Inc., Englewood Cliffs, N.J., 1971.

J. Kirschbaum, Biological oxidations and energy conservation, *J. Chem. Ed.,* **45,** 28 (1968).

19 Alkylation and Acylation of Enolates

Sugars, Nucleosides, and Nucleic Acids 20

Sugars are the primary units of that class of compounds known as **carbohydrates**. **Monosaccharides**, such as glucose, are carbohydrates that contain one sugar unit and fit the molecular formula $C_n(H_2O)_n$. The formula for glucose is $C_6H_{12}O_6$. **Disaccharides**, such as sucrose, are carbohydrates that contain two sugar units and have the formula $C_n(H_2O)_{n-1}$. For example, sucrose has the molecular formula $C_{12}H_{22}O_{11}$. Carbohydrates consisting of between two and eight monosaccharide units are known collectively as **oligosaccharides**. Carbohydrates consisting of more than eight monosaccharide units are classified as **polysaccharides**. Sugar molecules are among the most important structural building blocks utilized by nature. Their enzymatic degradation provides life-sustaining energy for most organisms. Furthermore, as a by-product of sugar metabolism, green plants release into the atmosphere the oxygen upon which all higher forms of life depend. From a biochemical viewpoint, it would be difficult to find a more important chapter in this text than this one.

20:1 Structure of Monosaccharides: Straight-Chain Formulas

Formaldehyde, CH_2O, is the simplest compound that fits the $C_n(H_2O)_n$ empirical formula. We can consider all carbohydrates as polymers of formaldehyde, which can indeed be polymerized under certain conditions to yield a mixture of sugars.

The next members of this homologous series are glycolaldehyde, $C_2H_4O_2$ ($n = 2$), and glyceraldehyde and dihydroxyacetone, both $C_3H_6O_3$ ($n = 3$).

CHO | CH_2OH	CHO | CHOH | CH_2OH	CH_2OH | C=O | CH_2OH
Glycolaldehyde	Glyceraldehyde	Dihydroxyacetone

These low molecular weight compounds show chemical behavior and structural features that are also characteristic of their higher homologs.

As an example, consider the fact that in the solid state, all the compounds mentioned above exist preferably as their dimers. These dimers are formed by intermolecular hemiacetal formation, as shown for glycolaldehyde (eq. 20.1) and dihydroxyacetone (eq. 20.2). Thus, neither these sugars

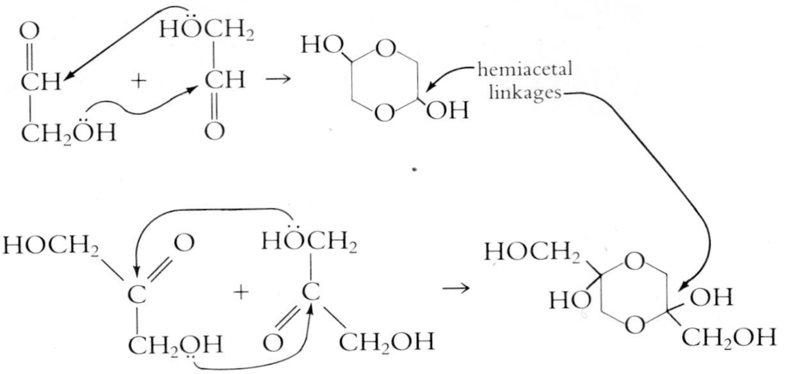

(20.1)

(20.2)

nor their higher homologs exist primarily in the carbonyl form. Rather, the carbonyl groups are masked as hemiacetals. Glycolaldehyde, glyceraldehyde, and dihydroxyacetone revert to their monomers when dissolved in water, but even in solution they probably exist as hydrates (sec. 11:6) rather than as free carbonyl compounds.

Stereochemistry is always an important aspect of structural studies. Emil Fischer, a renowned sugar chemist (see sec. 17:6), proposed a stereochemical convention that has been used for many years in sugar chemistry. He based the convention on the structure of (+)-glyceraldehyde, which can be represented as **1** or **2**. In this convention, structure **2** is flattened against the page to give **3**, which is known as a **Fischer projection**. In a Fischer

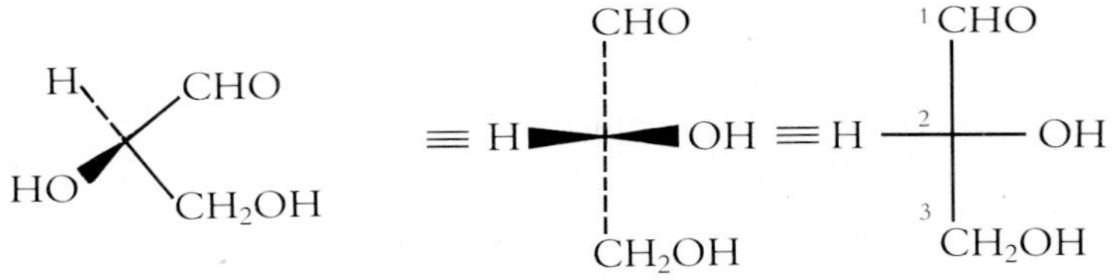

1 **2** **3**

(+)-Glyceraldehyde Fischer projection

projection, the horizontal bonds from the chiral center (those to H and OH) are *understood* to be projecting above the plane of the page (as in **2**). The vertical bonds to the chiral center are understood to lie behind the plane of the page.

In the Fischer convention, carbon chains are written vertically, with the carbon at the top (the aldehyde functional group) being given the lowest number (as in **3**). In most common sugars the chiral carbons (such as C-2 in structure **3**) are each connected to H and OH. If the *highest numbered* chiral carbon has the OH group on the right side, then the sugar is classified as a D sugar. Structure **3** is therefore a Fischer projection of D-glyceraldehyde. Its mirror image is labeled L-glyceraldehyde. (In *R,S* nomenclature, D-glyceraldehyde is *R*.) We will consider more examples shortly.

Beginning with sugars of four carbons and continuing through polymeric carbohydrates such as cellulose, the ending *-ose* denotes a sugar molecule. Trivial names are commonly used in sugar chemistry with each individual sugar having a specific name, often indicating its origin. For example, fructose is a sugar commonly found in fruit, and the sugar lactose is a component of milk.

If enantiomers are excluded, there are three possible four-carbon sugars: two aldehydes (called *aldotetroses*) and one ketone (called a *ketotetrose*). The aldotetroses each contain two chiral centers, so four stereoisomers exist (there are no *meso* structures). The ketotetrose has one chiral center, so two enantiomers exist. Their Fischer structures are written with the carbonyl group as near the top of the page as possible. The D or L designation is then conferred on the entire molecule on the basis of the configuration of the bottommost chiral carbon atom, since that will have the highest number. An L sugar is the enantiomer of the D sugar of the same name.

D-Erythrose (an aldose)	L-Erythrose (an aldose)	D-Threose (an aldose)	D-Erythrulose (a ketose)
CHO	CHO	CHO	CH_2OH
H—OH	HO—H	HO—H	C=O
H—OH	HO—H	H—OH	H—OH
CH_2OH	CH_2OH	CH_2OH	CH_2OH

D-Erythrose (an aldose) L-Erythrose (an aldose) D-Threose (an aldose) D-Erythrulose (a ketose)

Problem

20–1 Write Fischer projection structures for L-threose and L-erythrulose, showing the absolute configurations of all chiral atoms.

As the number of carbon atoms in the sugar increases, so does the possible number of isomers. For the pentoses (five-carbon sugars) there are eight aldopentoses, counting all enantiomeric possibilities. (Ketoses will be considered shortly.) At least one representative of each enantiomeric pair may be isolated from natural sources. Three of the most important aldopentoses are D-ribose (a constituent of ribonucleic acids), D-xylose (obtainable from the hydrolysis of straw, corn cobs, or wood), and L-arabinose (pectin sugar).

D-Ribose	D-Xylose	L-Arabinose
CHO	CHO	CHO
H—OH	H—OH	H—OH
H—OH	HO—H	HO—H
H—OH	H—OH	HO—H
CH_2OH	CH_2OH	CH_2OH

D-Ribose D-Xylose L-Arabinose

For the aldohexoses, there are eight different open-chain configurations, each of which gives rise to a pair of enantiomers. Of all the hexoses, by far the most important is glucose. It is, in the form of its polymers, the most abundant organic compound in nature. Two other important aldohexoses are D-galactose (cerebrose or brain sugar), and D-mannose, a common sugar of certain plants.

D-Glucose	D-Galactose	D-Mannose
CHO	CHO	CHO
H—C—OH	H—C—OH	HO—C—H
HO—C—H	HO—C—H	HO—C—H
H—C—OH	HO—C—H	H—C—OH
H—C—OH	H—C—OH	H—C—OH
CH_2OH	CH_2OH	CH_2OH

The number of keto sugars which *could* exist is quite large, since the carbonyl group could theoretically be at different positions in the chain of pentose and hexose sugars. Naturally occurring ketoses, however, only have the carbonyl group on the second carbon from one end, and thus they are much less abundant in nature than are aldoses. Nonetheless, there are several keto sugars that we should consider. Perhaps the most notable of these is D-fructose, which is an intermediate in the biosynthesis of all sugars. L-Sorbose, which is synthesized by bacteria, is used in the manufacture of 1000 tons of vitamin C (ascorbic acid) annually.

D-Fructose	L-Sorbose
CH_2OH	CH_2OH
C=O	C=O
HO—C—H	HO—C—H
H—C—OH	H—C—OH
H—C—OH	HO—C—H
CH_2OH	CH_2OH

Several keto sugars (as phosphate derivatives) are important intermediates in photosynthesis—the conversion of CO_2 into sugars and carbohydrates by plants and a few bacteria. Fructose occurs as the 6-phosphate and 1,6-diphosphate, ribulose as the 5-phosphate and 1,5-diphosphate, and sedoheptulose (a seven-carbon sugar) as the 7-phosphate and 1,7-diphosphate.

$$\begin{array}{c} CH_2OH \\ | \\ C{=}O \\ | \\ H{-}C{-}OH \\ | \\ H{-}C{-}OH \\ | \\ CH_2OH \end{array}$$

D-Ribulose

$$\begin{array}{c} CH_2OH \\ | \\ C{=}O \\ | \\ HO{-}C{-}H \\ | \\ H{-}C{-}OH \\ | \\ H{-}C{-}OH \\ | \\ H{-}C{-}OH \\ | \\ CH_2OH \end{array}$$

D-Sedoheptulose

The straight-chain formulas that have been given in this section for various sugars all have a C=O functional group. In spite of this, the infrared spectra of these compounds do *not* show a C=O stretch, and hence the structural formulas given must, in a sense, be incorrect. Section 20:2 shows that these sugars exist predominantly as cyclic hemiacetal structures. The straight-chain formulas are often used, however, because small amounts of the straight-chain isomers are present at equilibrium in solution, and hence they can take part in chemical reactions. Thus, sugars show typical carbonyl group reactivity. In addition, some of the stereochemical correlations between various sugars are often much easier to visualize with the open-chain formulas.

Problems

20–2 a. Write open-chain structures for L-ribose, D-arabinose, and L-xylose.
b. Write the structure of D-lyxose. (Lyxose is the one aldopentose *not* shown in this text.)

20–3 If D-glyceraldehyde is treated with HCN, two cyanohydrins are formed (sec. 11:6). Each cyanohydrin can have its CN group hydrolysed to COOH, and its CH_2OH group oxidized to COOH. In this way each cyanohydrin is converted to a tartaric acid (sec. 4:6). Give the structures of each of the tartaric acids, showing their absolute configurations. Refer to Section 4:6 and tell whether the optical rotation of each will be 0, +, or −. This set of transformations played an important role in confirming the structures of erythrose and threose.

20–4 Mild oxidation of aldoses with HNO_3 oxidizes the aldehyde function and the primary CH_2OH to carboxyl groups. Which tartaric acid is obtained by oxidation of D-erythrose? By oxidation of D-threose?

20:2 Cyclic Structures of Sugars

If a sample of D-glucose is crystallized from pure ethanol, the crystals so obtained have mp = 146° and $[\alpha]_D$ = +112.2°. This form of glucose is known as **α-D-glucose**. If another sample of glucose is recrystallized from a mixture of ethanol and water, crystals of mp = 148 to 155° and $[\alpha]_D$ =

+18.7° are obtained. This form of glucose is known as **β-D-glucose**. The optical rotations given for these two forms will be obtained *only* if the rotation is measured immediately after they are dissolved in water. The rotation of each solution changes with time and eventually reaches a stable value of +52.7°. This phenomenon of changing optical rotation with time is called **mutarotation**.

Thus, we have two problems with the straight-chain formulas of the last section. They do not account for the fact that no C=O absorption shows up in the ir spectrum, nor do they account for the fact that glucose (and other sugars as well) can exist in two different forms. Both problems were solved by Emil Fischer. We will present his results and also show how nmr spectroscopy can confirm Fischer's studies.

As we mentioned in the previous section, the open-chain form of glucose usually exists only in trace amounts. This is because *the OH group on C-5 reacts rapidly with the C=O group to form a cyclic hemiacetal* (sec. 11:8, Prob. 11–15, p. 413). This is shown in Figure 20–1. The equilibrium constant for conversion of the straight chain to the cyclic form is about 50,000. There is free rotation about the carbon bond joining the CHO to the remainder of the molecule, and therefore either of two cyclic hemiacetals can be formed, depending upon the conformation at the moment of reaction. The carbon atom that was the aldehydic carbon is called the **anomeric carbon** and the two cyclic forms are called the α-anomer and the β-anomer.* The two forms are also called **epimers**. Epimers are two stereoisomers that contain several chiral centers and are identical except for the configuration at one and only one of those centers. The cyclic structures of glucose are six-membered rings containing one oxygen. Such a heterocyclic ring is called a tetrahydropyran (Table 6–5, p. 199). Hence, the complete names for each of the glucose forms are **α-D-glucopyranose** and **β-D-glucopyranose** (Fig. 20–1).

Note that the β-anomer predominates at equilibrium. This is because the ring is in a chair conformation and all of the substituents are equatorial. (Recall that there is less steric strain when substituents are in the equatorial position.) The pyran rings exist in conformations essentially like those of cyclohexane and thus can be treated in the same manner. The other chair

* Remember from Chapter 3 (Fig. 3–34, p. 99) that an α isomer has the substituent "down" in the stereostructure, while the substituent is "up" in the β isomer.

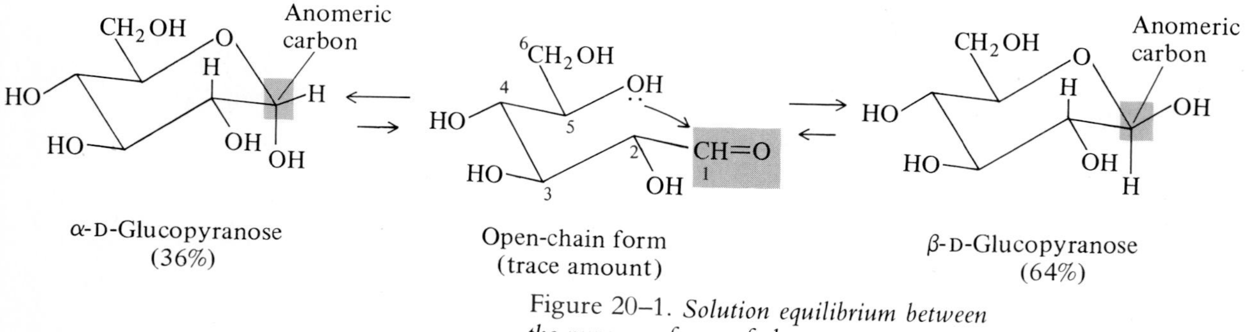

Figure 20–1. *Solution equilibrium between the pyranose forms of glucose.*

conformation of β-D-glucopyranose would have all substituents axial and hence need not be considered. In α-D-glucopyranose, the OH group at the anomeric carbon is axial, while the rest of the substituents are equatorial. The amount of each isomer in solution can be calculated from the known optical rotations of the pure epimers as compared to the final equilibrium value. A more direct method is through ^{1}H or ^{13}C nmr spectroscopy.

Figure 20–2 gives the ^{1}H nmr spectrum of the equilibrium solution in D_2O. D_2O is the chosen solvent for several reasons. First of all, glucose is more soluble in water than in any other medium because of the polar OH groups. In addition, by using D_2O we can bring about a considerable simplification of the nmr spectrum. All the OH groups of glucose are converted to OD, and thus do not give rise to any ^{1}H nmr absorptions (sec. 9:10). The exchange process that converts OH to OD also produces HDO, which generates the intense singlet at 4.21 ppm in Figure 20–2 (we have shaded this peak to indicate that it does not arise from glucose).

With the spectrum thus simplified, we can concentrate on the CH protons, in particular the one on the anomeric carbon atom. Since the anomeric carbon is bonded to two oxygen atoms, its proton will be more deshielded than all the other CH protons, and thus give rise to low field absorption. The α-anomeric proton will have a different chemical shift than the β-anomeric proton. Thus, we expect to find two peaks in the ^{1}H nmr spectrum at lower field than all the rest, one for the α-anomer and the other for the β-anomer. These peaks will be split into doublets, since the anomeric proton in each anomer has one neighboring proton. You can spot these low field doublets quite easily in Figure 20–2.

The axial proton on the anomeric carbon of β-D-glucopyranose gives rise to a doublet (J = 7.2 Hz) appearing at 4.6 ppm. The corresponding

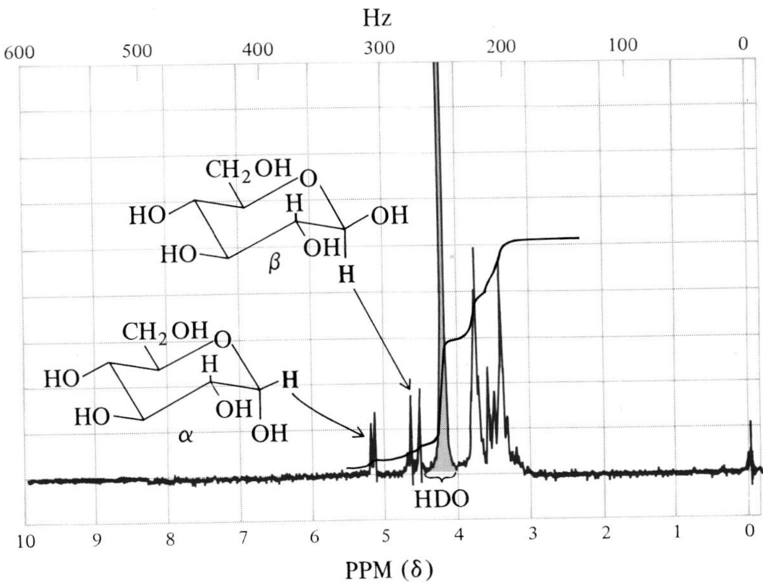

Figure 20–2. ^{1}H *nmr spectrum of* D*-glucose in* D_2O.

equatorial proton of the α-anomer appears at 5.2 ppm (J = 3.1 Hz). The coupling constant between two vicinal (neighboring) protons is usually closely related to the dihedral angle between those protons. The coupling is relatively large at 180° (an axial–axial 7.2 Hz coupling in the β form) and relatively small at 60° (an axial–equatorial 3.1 Hz coupling in the α form).

The relative size of each anomeric proton doublet shows that they are in the ratio of 60 parts β-D-glucose to 40 parts α-D-glucose. Thus, a direct instrumental proof is available for the existence of both conformers in solution. One can use ^{13}C nmr for the same analysis.

The cyclic forms of sugars are often represented in planar structures known as **Haworth projections**. The rings are certainly *not* planar, but this representation is still used because it is simple to draw and visualize isomers this way. The Haworth projections for α-D-glucopyranose and β-D-glucopyranose are given in Figure 20–3. These projections are to be visualized with the bottom part toward the viewer. Note that the hydroxyl at the anomeric carbon is depicted above the plane in the β form and below the plane in the α form. With this as a starting point, one can convert chair conformation drawings to Haworth projections by then noting whether hydroxyls on neighboring carbons are *cis* or *trans* to the hydroxyl on the anomeric carbon.

It is worth studying the relationship between the open chain and Haworth structures for D-glucose. For instance, the hydroxyl groups on C-2 and C-4, which lie to the right in the open-chain form, point down in the Haworth projection. The hydroxyl at C-3 is on the left in the acyclic structure, and up in the cyclic one. These correlations should allow easy interconversion of cyclic and acyclic structures.

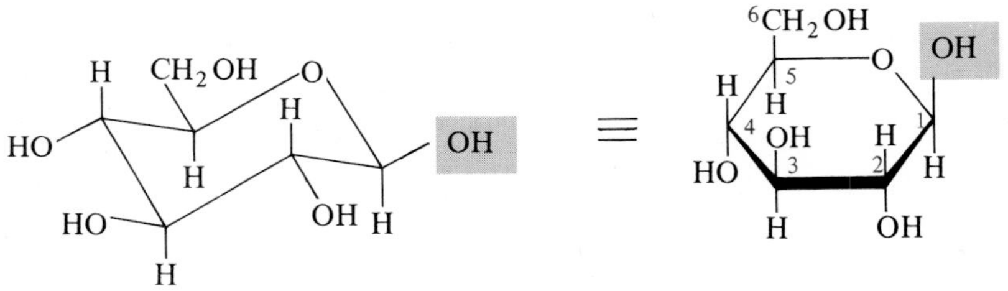

β-D-Glucopyranose

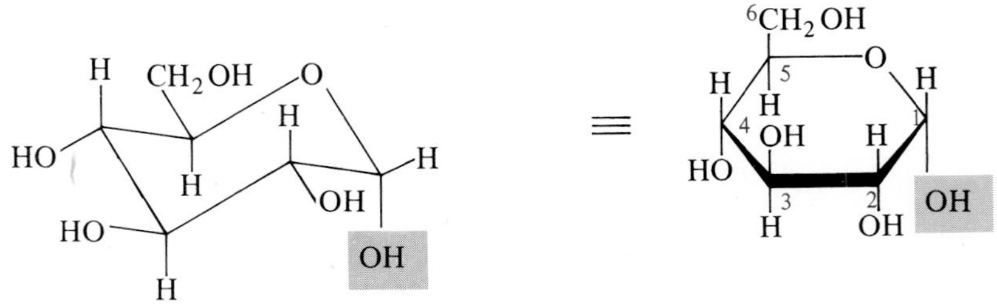

α-D-Glucopyranose

Figure 20–3. *Chair conformations and Haworth projections for α-D-glucopyranose and β-D-glucopyranose.*

Example

Draw chair conformations and Haworth projections for the α- and β-pyranose forms of D-allose.

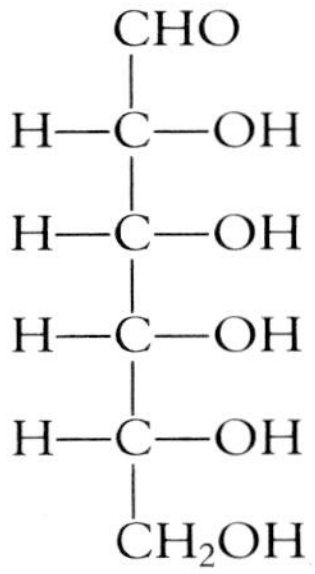

CHO
H—C—OH
H—C—OH
H—C—OH
H—C—OH
CH_2OH

D-Allose

Answer

The hydroxyl groups in the Fischer projection are to the right and they will therefore be *down* in the Haworth projection:

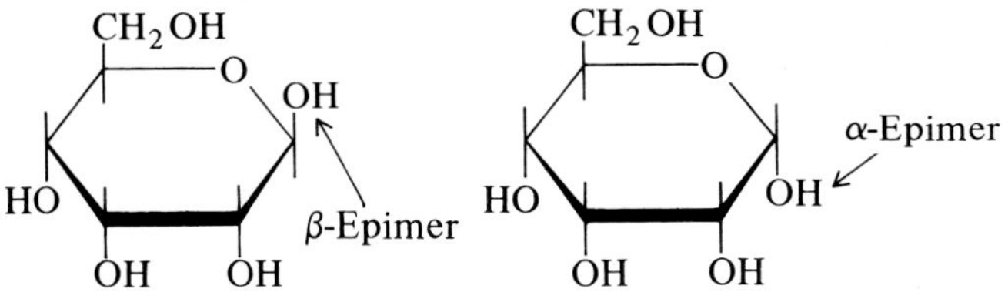

β-Epimer

α-Epimer

In the Haworth projection, the β-OH is up and the α-OH is down.

To convert to the chair conformation of the β-anomer, for example, remember that β is up and therefore should be equatorial. The OH at C-2 is *trans* to the OH on the anomeric carbon and therefore must be equatorial, too; the OH on C-3 is *cis* to that on C-2 and therefore must be axial; the OH on C-4 is *cis* to that on C-3 and therefore must be equatorial. Since this is a D-sugar, the CH_2OH group must be the same as in D-glucose. For the α-anomer, just change the OH on the anomeric carbon from equatorial to axial.

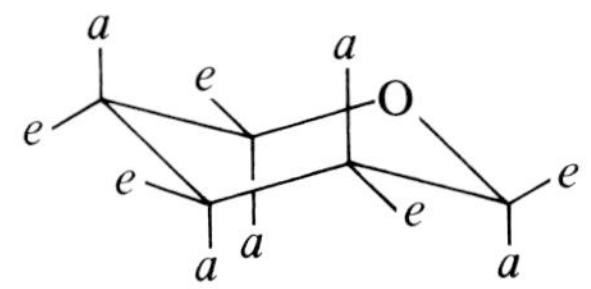

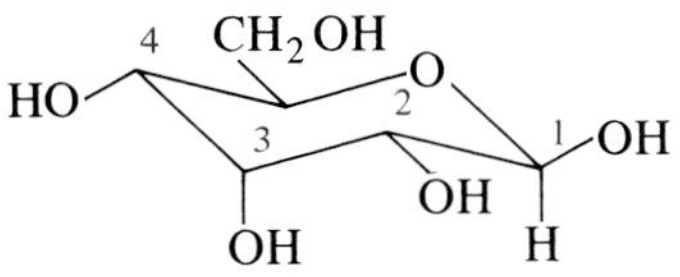

β-D-Allopyranose

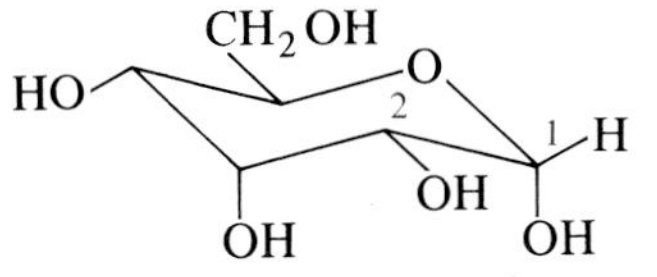

α-D-Allopyranose

Problem

20–5 Compare the open-chain structures for D-galactose and Dmannose (sec. 20:1) with that of D-glucose, and note carefully which carbon atoms have epimeric configurations in each case.

a. Draw cyclic chair conformations for the α- and β-pyranoses of D-galactose and D-mannose.

b. Draw Haworth projections of these four pyranoses.

With certain sugars, for example, fructose and ribose, different cyclic structures occur. Cyclic hemiacetal formation can yield a five-membered oxygen-containing ring (called the **furanose form**) as well as the six-membered pyranose ring. A ^{13}C nmr study of ribose in water showed that both furanose and pyranose forms were present at equilibrium, each isomer being a mixture of anomers (eq. 20.3). The ribofuranose epimers result from

(20.3)

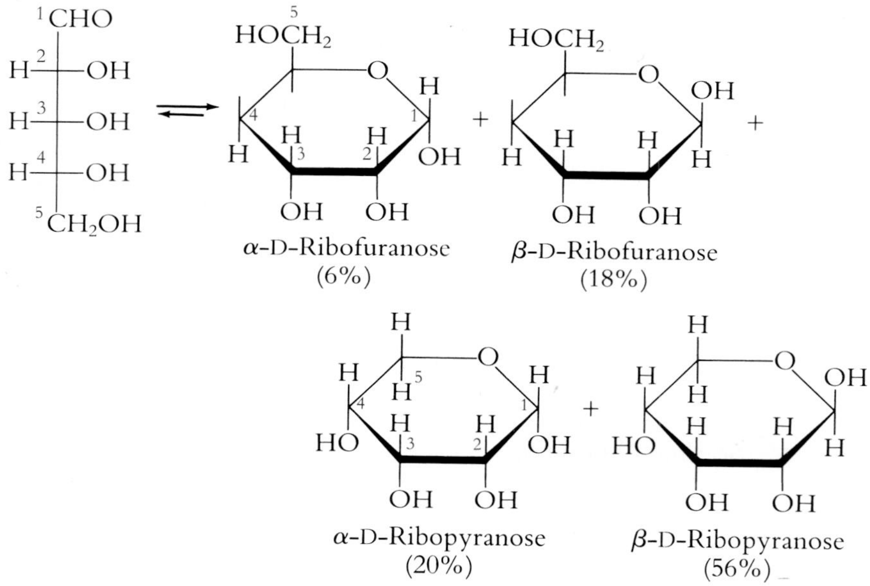

α-D-Ribofuranose (6%)

β-D-Ribofuranose (18%)

α-D-Ribopyranose (20%)

β-D-Ribopyranose (56%)

hemiacetal formation between the aldehyde carbonyl and the OH on C-4. The pyranose epimers result from reaction between the aldehyde group and the OH on C-5. The straight-chain isomer was present in too low a concentration to be detected by nmr spectroscopy.

Problems

20–6 Examine the stereochemistry of each of the four ribose isomers in Equation 20.3 and account for the order of relative abundance.

20–7 Write the structures of the furanose forms of D-erythrose and D-threose (sec. 20:1). Can these sugars exist in the pyranose form?

20:3 Glycoside Formation

The anomeric hydroxyl group, being part of a hemiacetal or hemiketal, is readily replaced when a sugar reacts with alcohols or phenols in the presence of acid. The products are acetals or ketals. (See Equations 11.46 and 11.47, pp. 411 and 412, respectively, for a complete mechanism.) For example, treatment of either α-D-glucopyranose or β-D-glucopyranose with methanol and a trace of acid yields a mixture of two acetals known as **glucosides** (eq. 20.4). Such products can be formed from any sugar, and the general term for such an acetal is **glycoside**. *Glycosides are formed readily because the loss of the*

(20.4)

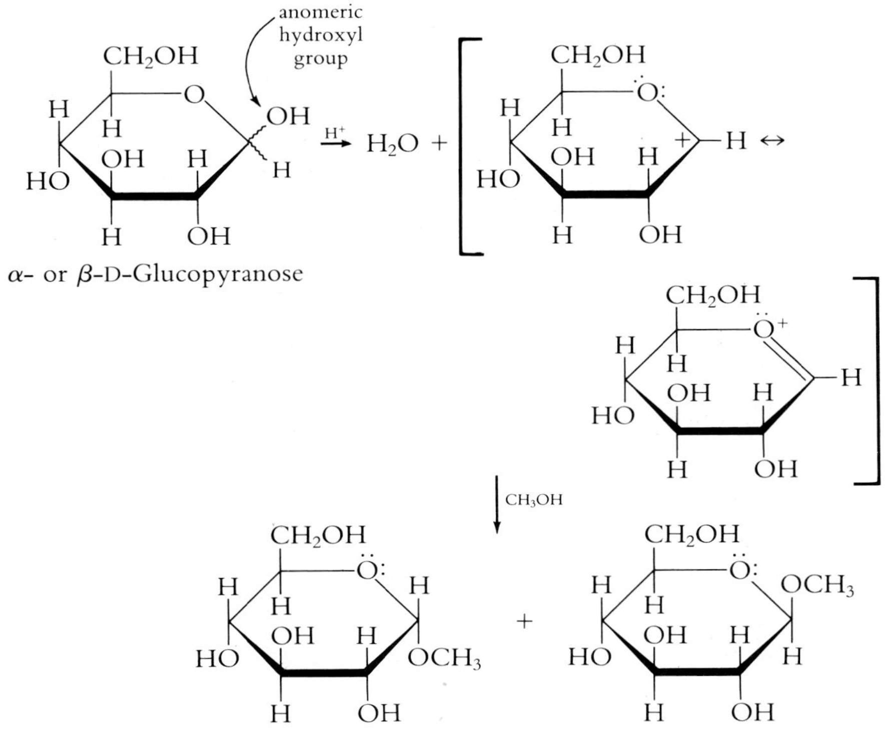

anomeric hydroxyl group

α- or β-D-Glucopyranose $\xrightarrow{H^+}$ H_2O + [oxygen-stabilized carbocation ↔ oxonium ion] $\xrightarrow{CH_3OH}$

Methyl α-D-glucopyranoside (a glucoside) + Methyl β-D-glucopyranoside (a glucoside)

protonated **anomeric** *OH group as H_2O results in formation of an oxygen-stabilized carbocation* (eq. 20.4). Since sugars contain OH groups, one sugar may form a **glycoside bond** to another sugar. This gives rise to dimeric sugars called **disaccharides**.

Problem

20–8 a. Draw the two anomers of methyl D-erythroside.
b. Which of the two anomers of methyl D-erythroside would you expect to be the most stable? Explain your answer.
c. Draw the anomers of methyl D-threoside. Which of these should be the most stable, and why?

20:4 Sugar Reactions

Sugars undergo numerous characteristic reactions which are used either to characterize them structurally, or to transform them for synthetic purposes. In this section we'll survey some of the most important of these reactions.

Osazone Formation

Although sugars exist in hemiacetal or hemiketal cyclic structures, they can undergo some of the typical reactions of carbonyl compounds. This is because their solutions at equilibrium contain at least trace amounts of the open-chain form with a free carbonyl. Reaction at the free carbonyl then shifts the equilibrium continually to the open-chain form, until all the sugar has reacted. For example, Emil Fischer found that sugars would react with phenylhydrazine to form phenylhydrazones (eq. 20.5).

(20.5)

$$\begin{array}{c} CHO \\ H-\!\!\!-\!\!\!-OH \\ H-\!\!\!-\!\!\!-OH \\ CH_2OH \end{array} + NH_2NH-C_6H_5 \rightarrow \begin{array}{c} CH{=}NNH-C_6H_5 \\ H-\!\!\!-\!\!\!-OH \\ H-\!\!\!-\!\!\!-OH \\ CH_2OH \end{array} + H_2O$$

D-Erythrose D-Erythrose phenylhydrazone

In the presence of excess phenylhydrazine, sugars react further to yield **osazones** (eq. 20.6).

(20.6)

$$\begin{array}{c} CHO \\ H-\!\!\!-\!\!\!-OH \\ H-\!\!\!-\!\!\!-OH \\ CH_2OH \end{array} + \underset{\text{(excess)}}{NH_2NHC_6H_5} \xrightarrow{\text{heat}} \begin{array}{c} CH{=}NNH-C_6H_5 \\ C{=}NNH-C_6H_5 \\ H-\!\!\!-\!\!\!-OH \\ CH_2OH \end{array}$$

D-Erythrosazone

The mechanism for osazone formation is complex and is not known in detail, but it begins with the tautomeric equilibrium of Equation 20.7. Osazones contain only two phenylhydrazone groups, since the reaction stops at that point. Intramolecular hydrogen-bonding between the two adjacent phenylhydrazone groups stabilizes them and does not allow for continued tautomeric changes down the chain.

(20.7)

$$\begin{array}{l} CH{=}NNH-C_6H_5 \\ | \\ CHOH \\ | \\ R \end{array} \rightleftarrows \begin{array}{l} CHNHNH-C_6H_5 \\ \| \\ C-OH \\ | \\ R \end{array} \rightleftarrows \begin{array}{l} CH_2NHNH-C_6H_5 \\ | \\ C{=}O \\ | \\ R \end{array}$$

Osazone formation results in the loss of chirality at C-2 of aldoses. D-Erythrose and D-threose thus give rise to the identical osazone. This confirms their configurational relationship (eq. 20.8). D-Glucose and D-fructose also give rise to the same osazone, since 2-ketohexoses react similarly to aldohexoses. The value of the osazone reaction lies precisely in its ability to tell us that two sugars are stereochemically identical at all carbons *except* C-1 and C-2.

(20.8)

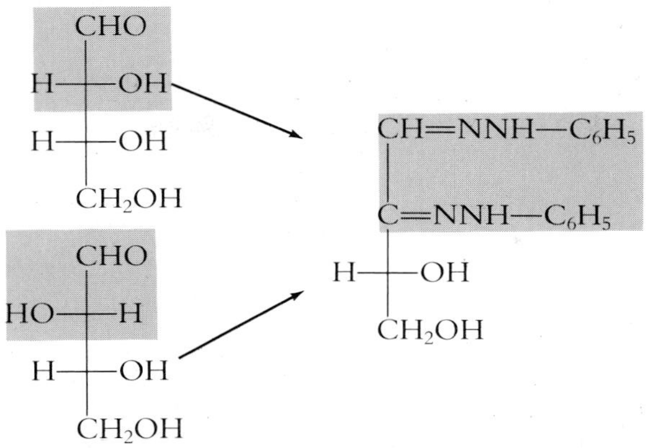

$$\begin{array}{c} CHO \\ H-\!\!\!+\!\!\!-OH \\ H-\!\!\!+\!\!\!-OH \\ CH_2OH \end{array} \quad \begin{array}{c} CHO \\ HO-\!\!\!+\!\!\!-H \\ H-\!\!\!+\!\!\!-OH \\ CH_2OH \end{array} \longrightarrow \begin{array}{c} CH{=}NNH{-}C_6H_5 \\ | \\ C{=}NNH{-}C_6H_5 \\ H-\!\!\!+\!\!\!-OH \\ CH_2OH \end{array}$$

Problems

20–9 Draw the structure of D-glucosazone. Three different sugars give this osazone: fructose, glucose, and another aldohexose. What are the structure and the name of this other aldohexose?

20–10 Considering only one of each enantiomeric pair, there are eight possible aldohexoses, but these give rise to only four osazones. Use structures to show why this is so.

Trimethylsilylation, Methylation, and Acetylation

Because sugars contain so many hydroxyl groups, they are very water soluble and are often high melting compounds. They are also sometimes difficult to obtain in good crystalline form and the physical properties of impure sugar are sometimes more like honey or caramel rather than table sugar. Therefore, several different sugar derivatives have been prepared and used in the laboratory. These derivatives are easier to study because they are more volatile and they are soluble in organic solvents. Sugar derivatives can be analyzed by mass spectrometry and gas chromatography, two techniques not easily applied to the sugars themselves.

One important derivative (which can also be formed from other alcohols) is the **trimethylsilyl** (TMS) ether. This is prepared according to Equation 20.9 by treating the sugar (or any alcohol) with trimethylsilyl chloride [$(CH_3)_3SiCl$, TMSCl] and an amine base (to scavenge released acid). The

(20.9)

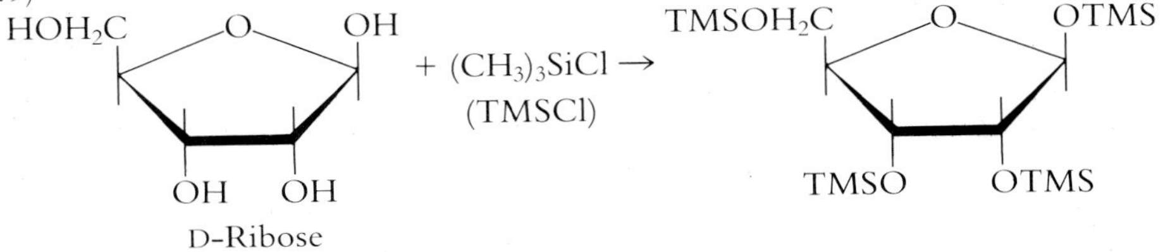

D-Ribose (furanose ring: HOH_2C, O, OH; OH, OH) $+ (CH_3)_3SiCl \rightarrow$ (TMSCl) → furanose ring: $TMSOH_2C$, O, OTMS; TMSO, OTMS

sugar trimethylsilyl ether is soluble in organic solvents and volatile. The nmr spectrum of the TMS ether can be taken in $CDCl_3$, and the ether can be analyzed by mass spectrometry. The silyl ether is stable in neutral and basic solvents, but mild acid treatment regenerates the sugar (or alcohol; eq. 20.10). Because of this easy reverse reaction, the trimethylsilyl ether is often used as an alcohol protecting group.

(20.10) [TMS-protected glucopyranose: $TMSOCH_2$, OTMS, TMSO, OTMS] $\xrightarrow{H_3O^+}$ [glucopyranose: $HOCH_2$, OH, HO, OH] $+ 4Me_3SiOH$

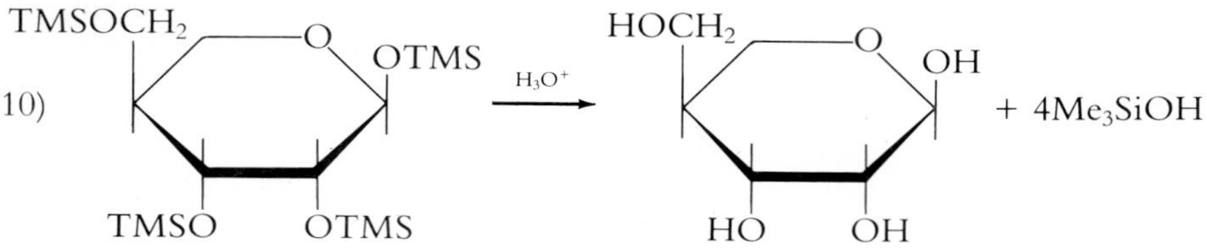

In a similar manner, one can make methyl ethers of sugars (eq. 20.11). These are somewhat more stable to hydrolysis than are trimethylsilyl ethers and are often used in the structure determination of disaccharides (sec. 20:6).

(20.11) [$HOCH_2$, OH, HO, OH, ~OH] $\xrightarrow[\text{2. } H_3O^+\text{, mild}]{\text{1. } Me_2SO_4\text{, } OH^-}$ [$MeOCH_2$, OMe, MeO, OMe, ~OH]

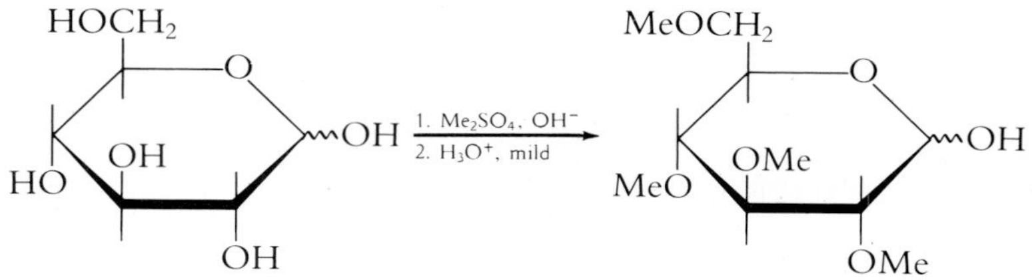

Note here the special character of the anomeric group. If some of the anomeric OH had been methylated in the first step, it would have been converted back to the OH group by the second step acid treatment. This is the result of the greater reactivity of *acetal* ethers over "normal" ones.

The O-acylation of sugars, particularly with acetic anhydride, is another useful reaction, illustrated in Equation 20.12 with D-glucose. The reaction is carried out at elevated temperature in order to obtain the more stable β-anomer of glucose pentaacetate.

(20.12) D-glucose $\xrightarrow[\text{NaOAc}]{Ac_2O}$ [CH_2OAc, OAc, OAc, AcO, OAc]

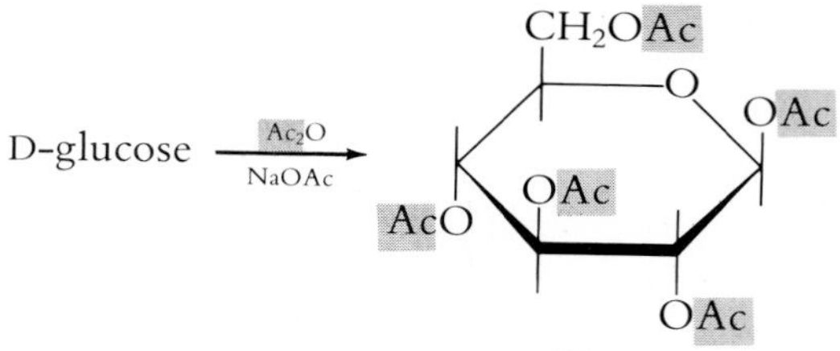

Glucose pentaacetate

Acetate derivatives of sugars have been used as starting materials for the synthesis of glucosides of many types. For this purpose the acetate is first treated with HBr, which results in replacement of the anomeric acetate (eq. 20.13) to produce acetobromoglucose.

(20.13)

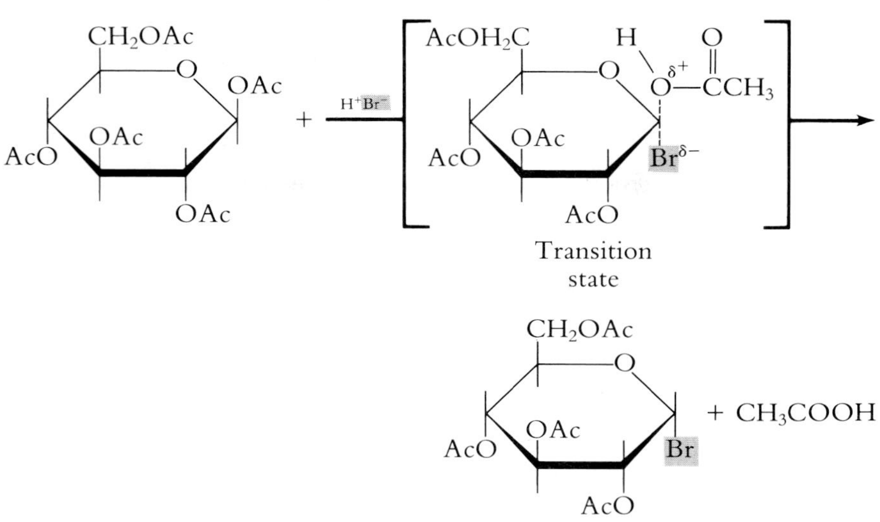

$$\text{Penta-O-acetyl-glucopyranose} \xrightarrow{H^+Br^-} \left[\text{Transition state}\right] \rightarrow \text{Tetra-O-acetyl-}\alpha\text{-D-glucopyranosyl bromide} + CH_3COOH$$

Transition state

Tetra-O-acetyl-α-D-glucopyranosyl bromide (*acetobromoglucose*)

Again, note the special reactivity of the anomeric group. This reaction is an S_N2 process, and proceeds with inversion of configuration. There is substantial positive charge on the anomeric carbon in the transition state (eq. 20.13) and as we've already seen, the oxygen lone pairs can stabilize this positive charge (sec. 20:3). This explains why only the anomeric acetate is replaced by Br.

Acetobromoglucose can then react with an alcohol to form the glucoside (eq. 20.14), a reaction which again proceeds with inversion of configuration.

(20.14)

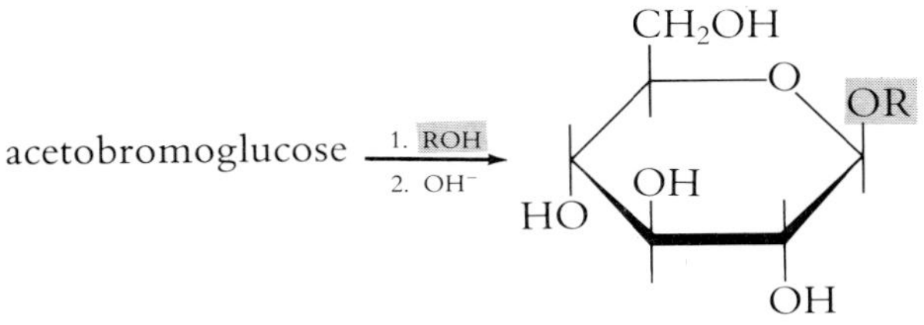

$$\text{acetobromoglucose} \xrightarrow[\text{2. } OH^-]{\text{1. ROH}} \text{A } \beta\text{-glucoside}$$

A β-glucoside

This roundabout method of glycoside synthesis (compare with sec. 20:3) is often used because it produces the β-glucoside, free of the α-anomer. Acetobromoglucose can also react with certain amines to form nucleosides. This reaction is a critical step in nucleic acid synthesis and will be discussed in Section 20:9.

Problem

20–11 In Section 11:8 the reaction of diols with a ketone to form cyclic ketals is described. This same reaction provides a rather selective way to derivatize certain sugars. Review Section 11:8 and then draw the structure of the cyclic diketal that is formed when D-galactopyranose reacts with two molecules of acetone.

Reduction

The carbonyl groups of aldoses and ketoses can be reduced in a similar manner to that used with other C=O groups. Both reduction and oxidation reactions have been used principally to determine sugar structures, although a few are important industrially.

Reduction of the aldehyde group of an aldose results in formation of a polyol, called an **alditol**. The reducing agent generally used in the laboratory is sodium borohydride (eq. 20.15). Catalytic hydrogenation is used industrially.

(20.15)

CHO / H—C—OH / HO—C—H / CH_2OH $\xrightarrow[2.\ H_3O^+]{1.\ NaBH_4}$ CH_2OH / H—C—OH / HO—C—H / CH_2OH

L-Threose → L-Threitol

For instance, in one preparation of ascorbic acid (vitamin C) the intermediate D-sorbitol is prepared by reduction of D-glucose (eq. 20.16).

(20.16)

D-glucose $\xrightarrow[\text{pressure}]{H_2/\text{cat}}$ CH_2OH / H—C—OH / HO—C—H / H—C—OH / H—C—OH / CH_2OH $\xrightarrow[\text{steps}]{\text{several}}$ Ascorbic acid

CH_2OH / H—C—OH / (ring O, C=O) / HO OH

D-Sorbitol

Ascorbic acid (vitamin C)

Oxidation

The oxidation of sugars was originally carried out mainly as an aid in structure determination. This function has been partly, but not completely, superseded by spectroscopic techniques. We will first deal with those oxidation methods that are still used to provide structural information, and then we'll touch upon oxidation as a means of producing synthetically useful sugar derivatives at the end of this section.

Of the oxidation reactions that are structurally informative, several are specific for carbonyl groups and adjoining hydroxyls. There are two types of carbonyl groups in sugars—aldehydes and α-hydroxyketones ($RCOCH_2OH$)—both of which are attacked by mild oxidizing agents. A silver-containing reagent known as Tollens reagent will oxidize compounds with either of these groups (eq. 20.17, for example). Silver ion is reduced to metallic silver during the reaction and the silver is deposited as a film on a clean glass surface. This is called the "silver mirror test."

$$\text{(20.17)}\quad RCHO + 2Ag(NH_3)_2^+ + 3OH^- \rightarrow RC(=O)O^- + 2Ag^0\downarrow + 2H_2O$$

A second oxidizing method involves the use of Fehling solution (Cu^{2+} and OH^-, eq. 20.18). Here, blue Cu^{2+} is reduced to Cu^{+1} and a red precipi-

$$\text{(20.18)}\quad RC(=O)CH_2OH \xrightarrow[\text{(blue)}]{Cu^{2+},\ OH^-} \underset{\text{(red)}}{Cu_2O\downarrow} + \left[RC(=O)C(=O)H\right] \xrightarrow{Cu^{2+}} \text{further oxidation products}$$

tate of Cu_2O is formed during the reaction. Sugars that are oxidized by these reagents (aldoses and 2-ketoses) are called **reducing sugars**, and those which are not oxidized are called **nonreducing sugars**. *If cyclic forms are considered, all sugars with an OH on the anomeric carbon are reducing sugars.* Glycosides will not be reducing sugars because an OR group is present on the anomeric carbon. These tests are commonly used to identify structural units in more complex oligosaccharides and polysaccharides (sec. 20:6).

Another reaction of value in structure determination is oxidation with periodate. Remember from Section 10:9 (eq. 10.56, p. 369) that $NaIO_4$ will oxidize vicinal diols, causing carbon–carbon bond cleavage between the oxygens. If a C=O group is next to an OH, or if two carbonyls are adjacent, cleavage will also occur. Equation 20.19 gives an example of the reaction sequence for glyceraldehyde. The —CH_2OH on the end of the chain is converted to formaldehyde, while the other carbons are eventually converted to formic acid. Equation 20.20 shows what happens when dihydroxyacetone, an isomer of glyceraldehyde, is subjected to periodate oxidation. The ketone carbonyl is converted to carbon dioxide and the other two carbons are oxidized to formaldehyde. A comparison of Equations 20.19 and 20.20 shows us that isomeric reactants can sometimes be distinguished by the differing product mixtures they generate.

$$\text{(20.19)}\quad CHO{-}CHOH{-}{}^{\star}CH_2OH \xrightarrow{NaIO_4,\ H^+} \underset{\text{Formaldehyde}}{{}^{\star}CH_2O} + \left[CHO{-}CHO\right] \xrightarrow{NaIO_4} \underset{\text{Formic acid}}{2HCOOH}$$

$$\text{(20.20)}\quad CH_2OH{-}C(=O){-}CH_2OH \xrightarrow{NaIO_4,\ H^+} 2CH_2O + CO_2$$

Periodate oxidations can be carried out quantitatively so that the number of moles of periodate consumed per mole of sugar oxidized can be calculated. This information can be quite helpful in structure determination, as the following example shows. Acetaldehyde reacts with the hydroxyls at C-4 and C-6 of glucose, giving the cyclic six-membered acetal **4** (eq. 20.21). Oxidation of **4** consumes two equivalents of periodate and yields two equivalents of formic acid and the erythrose derivative **5**.

(20.21)

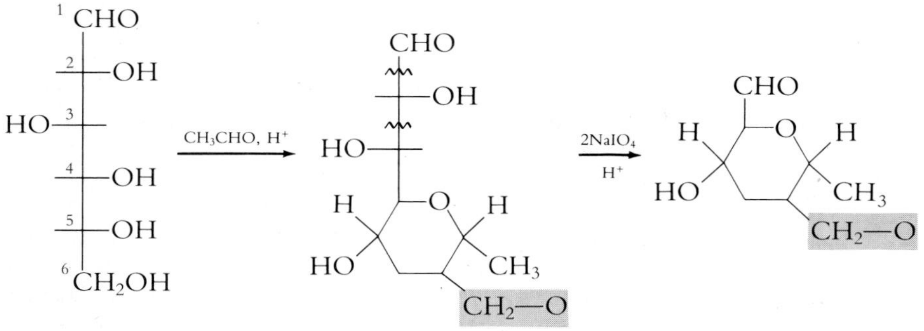

$\xrightarrow{CH_3CHO,\ H^+}$ **4** $\xrightarrow[H^+]{2NaIO_4}$ **5** + 2HCOOH

Compound **4** is not the only possible product of the glucose/acetaldehyde reaction. It could as well have been the five-membered acetal **6** (eq. 20.22). One method of determining which product, **4** or **6**, has been formed is to oxidize the product with periodate. If it is **4**, two equivalents of periodate will be consumed; if it is **6**, three equivalents will be consumed.

(20.22)

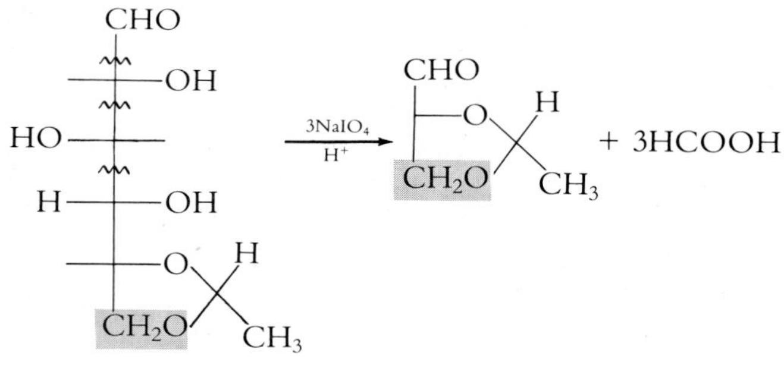

6 $\xrightarrow[H^+]{3NaIO_4}$ + 3HCOOH

Example

How many moles of periodate would be consumed by oxidation of D-allose (see Example in sec. 20:2)?

Answer

Cleavage occurs through each C—C bond that bears an oxygen function on each side:

```
       CHO                       HCOOH
        ~                          +
   H—C—OH                        HCOOH
        ~                          +
   H—C—OH          IO4−          HCOOH
        ~         ———→             +
   H—C—OH                        HCOOH
        ~                          +
   H—C—OH                        HCOOH
        ~                          +
      CH2OH                      HC=O
                                  |
                                  H
```

Therefore, 5 mol of periodate are used (one per cleavage), and the products are 1 mol of formaldehyde and 5 mol of formic acid.

Problems

20–12 When Emil Fischer began working on glucose in 1884, he knew that it was an aldohexose, but its stereochemical configuration was unknown. One experiment he performed to shed light on this mystery was to reduce the glucose to a hexol (eq. 20.16). He found that the hexol was optically active. Which possible aldohexose structures for glucose was Fischer able to *eliminate* as a result of this experiment? Explain your reasoning.

20–13 D-Galactose and L-galactose give the same alditol upon reduction with $NaBH_4$. Why?

20–14 a. How many moles of periodate will be consumed in the oxidation of L-arabinose (sec. 20:1)? What are the products?

b. An unknown compound consumed 5 mmol of periodate and yielded 4 mmol of formic acid and 2 mmol of formaldehyde. Propose a structure that fits these findings. Is your structure the only one that satisfies the data? Explain your answer.

20–15 How many moles of periodate will be consumed by 1 mol of each of the following compounds? Show the products of each oxidation.

a. glycerol (1,2,3-propanetriol)
b. 1,2,3,5-cyclohexanetetrol
c. glucose
d. methyl β-D-glucopyranoside

20–16 Erythrose derivative **5** was used in the synthesis of 2-deoxy-D-ribose (sec. 20:9). In the course of this synthesis, compound **5** was exposed to acid, which the investigators feared might have caused it to rearrange to the isomeric five-membered acetal **5a** (eq. 1).

```
        CHO                          CHO
  H      |         H                  |
   \     |—O      /                   |—OH
    \   /    \   /       ?            |
     ><       ><        ——→           |—O   H
    /   \    /   \                    |   \/
  HO     ——O      CH3              CH2O   /\
                                           CH3
```
(1)

5 → **5a**

a. Write a reasonable mechanism for the conversion of **5** to **5a**. (A review of Section 11:8 may be helpful.)
b. A straightforward chemical test showed that **5** had *not* rearranged to **5a**. What test was used?

As we noted earlier, sugars are oxidized for synthetic as well as analytical purposes. This is particularly true of glucose because it is so abundant and inexpensive. Since glucose is chiral and contains several functional groups, it serves as a versatile starting material for the synthesis of other chiral compounds, such as sorbitol (eq. 20.16).

Another useful modification of glucose is the selective oxidation of its aldehyde group to a carboxyl group. This is accomplished by the mild oxidizing reagent, bromine water (eq. 20.23).

$$\begin{array}{c} CHO \\ | \\ (CHOH)_4 \\ | \\ CH_2OH \end{array} \xrightarrow{Br_2,\ H_2O} \begin{array}{c} COOH \\ | \\ (CHOH)_4 \\ | \\ CH_2OH \end{array} \quad (20.23)$$

D-Gluconic acid

The product, D-gluconic acid, is valuable commercially because it forms strong complexes with many metal ions and prevents them from precipitating as insoluble salts.

Problem

20–17 An aldopentose is isolated, optically active, from a natural source. The alditol obtained from it by $NaBH_4$ reduction is also optically active. The aldopentose is degraded to an aldetetrose by oxidative removal of the —CHO group, and conversion of the hydroxyl at C-2 to an aldehyde. This aldotetrose is optically active, and its specific rotation is equal in value but *opposite* in sign to that of D-erythrose. Propose a structure for the aldopentose and name it. Show all the chemistry.

20:5 Sugar Biosynthesis: Fructose Formation

The construction of five- to seven-carbon sugars is a major metabolic process in both plants and animals. Green plants use solar energy to power the synthesis of sugar molecules from CO_2 and H_2O. As we mentioned in Chapter 19, this process of **photosynthesis** has been said to be earth's most important chemical reaction. For terrestrial organisms, photosynthesis is the ultimate source of energy.

The metabolic pathway for sugar synthesis is not identical in plants and animals, but in both types of organisms aldol condensations and retroaldol cleavages occur extensively. Let's examine the latter steps (common to both plants and animals) of the reaction sequence, starting with the dihydroxyaldehyde derivative (R)-3-phosphoglyceraldehyde (**7**). In the cell this compound is in equilibrium with its isomer dihydroxyacetone phosphate, **8** (eq. 20.24). This reversible interconversion takes place by way of the enediol **9**, which can be protonated at C-1 to give **8** or at C-2 to give **7**.

$$\text{(20.24)}\quad \underset{\substack{\mathbf{7}\\ \text{3-Phosphoglyceraldehyde}}}{\begin{array}{l} \text{H—C=O} \\ \text{H—C—OH} \\ \quad \text{CH}_2\text{OPO}_3\text{H}^- \end{array}} \rightleftharpoons \underset{\substack{\mathbf{9}\\ \text{Enediol}}}{\left[\begin{array}{l} \text{H, OH} \\ {}^1\text{C} \\ {}^2\text{C—OH} \\ {}^3\text{CH}_2\text{OPO}_3\text{H}^- \end{array}\right]} \rightleftharpoons \underset{\substack{\mathbf{8}\\ \text{Dihydroxyacetone phosphate}}}{\begin{array}{l} \text{CH}_2\text{—OH} \\ \text{C=O} \\ \text{CH}_2\text{OPO}_3\text{H}^- \end{array}}$$

The next step in the metabolic process is an aldol condensation between C-1 of **7** and C-1 of **8**, yielding fructose-1,6-diphosphate (eq. 20.25). The reaction is catalyzed by the *aldolase* enzymes.

(20.25)

$$\underset{\mathbf{7}}{{}^-\text{HO}_3\text{PO—}{}^3\text{CH}_2\text{—}{}^2\text{CH(OH)—}{}^1\text{CHO}} + \underset{\mathbf{8}}{\text{H—}{}^1\text{CH(OH)—}{}^2\text{C(=O)—}{}^3\text{CH}_2\text{—OPO}_3\text{H}^-} \xrightarrow{\text{aldolase}}$$

$${}^-\text{HO}_3\text{PO—CH}_2\text{—CH(OH)—}{}^4\text{CH(OH)—}{}^3\text{CH(OH)—C(=O)—CH}_2\text{—OPO}_3\text{H}^-$$

Fructose-1,6-diphosphate

It proceeds with the complete stereoselectivity associated with enzymatic reactions involving prochiral centers. Two new asymmetric centers are created, C-3 and C-4 of fructose (eq. 20.25); C-3 always has the *S* configuration and C-4 the *R* configuration. D-Fructose is then obtained by hydrolysis of fructose-1,6-diphosphate. This dephosphorylation reaction is catalyzed by the enzyme *phosphatase*.

It was discovered that during the condensation reaction, the dihydroxyacetone phosphate (**8**) becomes covalently bound to a nitrogen atom on the enzyme. The data support the idea that the ketone is converted to its enamine, using an amino group from the enzyme (eq. 20.26). The enamine then reacts with 3-phosphoglyceraldehyde, **7**, which is probably held to the enzyme by hydrogen bonds (Fig. 20–4).

$$\text{(20.26)}\quad \underset{\mathbf{8}}{\begin{array}{l} \text{CH}_2\text{OPO}_3\text{H}^- \\ \text{C=O} \\ \text{CH}_2\text{OH} \end{array}} + \text{H}_2\text{N—Enz} \xrightarrow{\text{H}^+} \begin{array}{l} \text{CH}_2\text{OPO}_3\text{H}^- \\ \text{C=}\overset{+}{\text{N}}\text{H—Enz} \\ \text{CH}_2\text{OH} \end{array} \xrightarrow{-\text{H}^+} \begin{array}{l} \text{CH}_2\text{OPO}_3\text{H}^- \\ \text{C—NH—Enz} \\ \text{C(HO)(H)} \end{array}$$

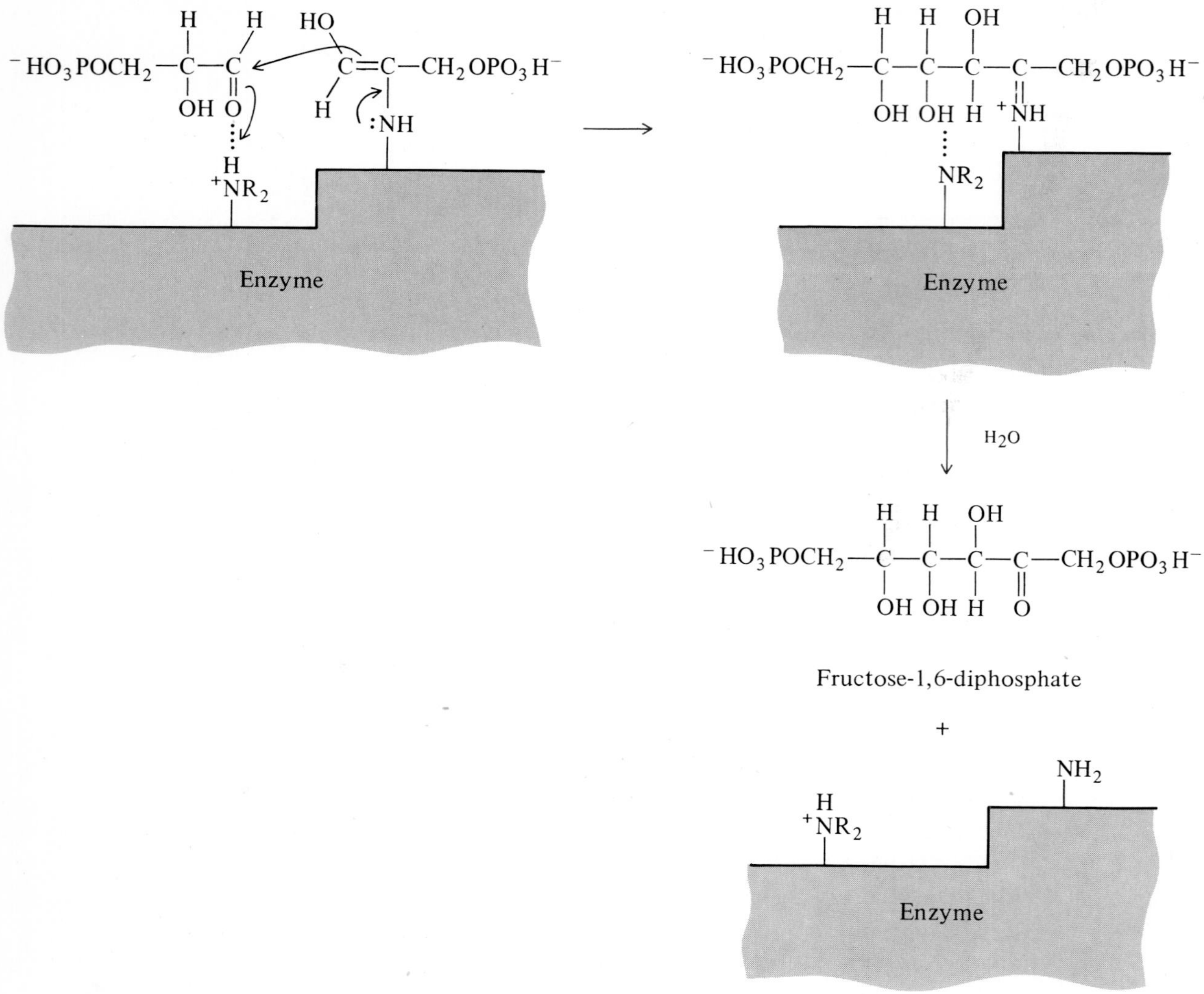

$$^{-}HO_3POCH_2-\underset{OH}{\overset{H}{C}}-\overset{H}{C}(=O\cdots H\overset{+}{N}R_2) \quad HO(H)C{=}C(-\ddot{N}H)-CH_2OPO_3H^{-}$$

Enzyme

$$\longrightarrow$$

$$^{-}HO_3POCH_2-\underset{OH}{\overset{H}{C}}-\underset{OH}{\overset{H}{C}}-\underset{H}{\overset{OH}{C}}-\underset{\overset{+}{N}H}{\overset{}{C}}-CH_2OPO_3H^{-}$$

NR_2

Enzyme

$$\big\downarrow H_2O$$

$$^{-}HO_3POCH_2-\underset{OH}{\overset{H}{C}}-\underset{OH}{\overset{H}{C}}-\underset{H}{\overset{OH}{C}}-\underset{O}{\overset{}{\overset{\|}{C}}}-CH_2OPO_3H^{-}$$

Fructose-1,6-diphosphate

+

NH_2

H
$^{+}NR_2$

Enzyme

Figure 20–4. *Enzymatic catalysis of fructose biosynthesis.*

It is interesting to note that the reverse of Equation 20.25 is part of what is known as the **glycolysis pathway** of metabolism. This pathway provides humans and many other organisms with most of their energy as glucose is broken down to pyruvic acid and eventually CO_2 and water (sec. 7:12).

Problem

20–18 In the normal metabolic cycle, fructose-1,6-diphosphate is hydrolyzed to fructose-6-phosphate, which is enzymatically isomerized to glucose-6-phosphate (eq. 2).

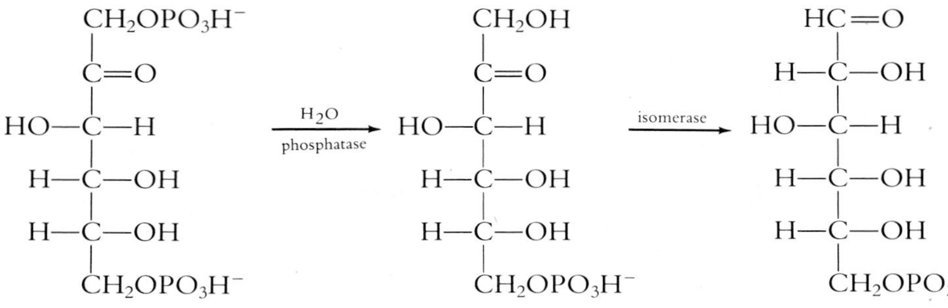

$$\begin{array}{ccc}
\begin{array}{c} CH_2OPO_3H^- \\ | \\ C{=}O \\ | \\ HO{-}C{-}H \\ | \\ H{-}C{-}OH \\ | \\ H{-}C{-}OH \\ | \\ CH_2OPO_3H^- \end{array}
\xrightarrow[\text{phosphatase}]{H_2O}
\begin{array}{c} CH_2OH \\ | \\ C{=}O \\ | \\ HO{-}C{-}H \\ | \\ H{-}C{-}OH \\ | \\ H{-}C{-}OH \\ | \\ CH_2OPO_3H^- \end{array}
\xrightarrow{\text{isomerase}}
\begin{array}{c} HC{=}O \\ | \\ H{-}C{-}OH \\ | \\ HO{-}C{-}H \\ | \\ H{-}C{-}OH \\ | \\ H{-}C{-}OH \\ | \\ CH_2OPO_3H^- \end{array}
\end{array} \quad (2)$$

Fructose-1,6-diphosphate   Fructose-6-phosphate   Glucose-6-phosphate

a. Write a mechanism for the isomerization, showing the structure of the intermediate.

b. The isomerization fructose → glucose can be carried out chemically (dilute OH^-) as well as enzymatically. In the chemical isomerization a third isomer, mannose, is also formed in the reaction. Like glucose, it is an aldose. What is the relationship of mannose to glucose? Why is it formed in the chemical isomerization?

20:6 Oligosaccharides and Polysaccharides

In nature, sugars are often found combined with one another in short chains called oligosaccharides or in high molecular weight polymers called polysaccharides.

The most widely known oligosaccharide is the disaccharide sucrose, which is obtained from sugar cane and sugar beets and is used as ordinary table sugar. Upon aqueous hydrolysis, sucrose yields one molecule of D-fructose and one of D-glucose. Sucrose itself is nonreducing (it does not react

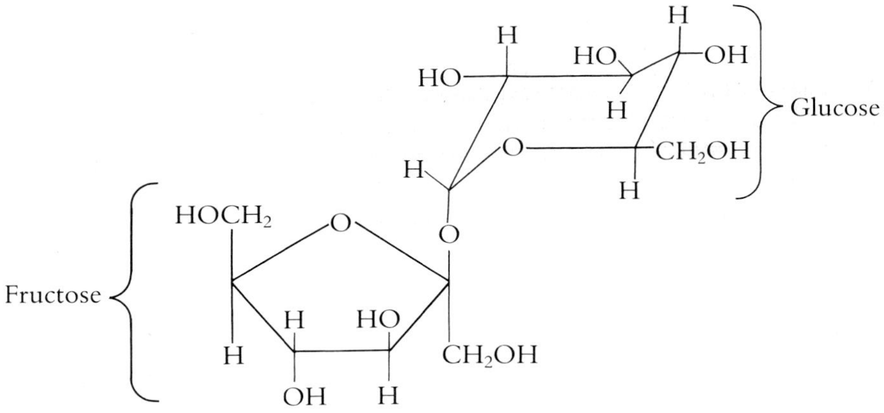

H
H
HO
OH
HO
H
Glucose
O
CH₂OH
H
H
HOCH₂
O
O
Fructose
H
HO
H
CH₂OH
OH
H

Sucrose

with either Tollens or Fehling reagent), which indicates that neither of the component sugars is in equilibrium with its straight-chain carbonyl form. Thus, *there cannot be a free anomeric hydroxyl group in sucrose.* The two monosaccharides, therefore, must be joined by a glycoside bond through loss of their anomeric hydroxyl groups (sec. 20:3).

For the chemist seeking to unravel the structures of oligosaccharides, determining the *stereochemistry* of the glycoside bond is perhaps the most difficult task. Nuclear magnetic resonance analysis can be used for simple disaccharides, but not for polysaccharides. The most reliable, but time-consuming, method is X-ray diffraction. It has been applied, for example, to sucrose. Carbohydrate chemists, however, still rely most heavily on enzyme-catalyzed oligosaccharide hydrolysis. This method allows them to take advantage of the well-documented stereochemical specificity of enzymatic catalysis.

There is an enzyme present in the human digestive tract that catalyzes the hydrolysis of sucrose to glucose and fructose. This enzyme has been found to be specific for the hydrolysis of β-fructofuranosides and is hence called β-fructofuranosidase. If it is able to hydrolyze an oligosaccharide such as sucrose, then we know that the saccharide must have a β-fructofuranoside linkage.

One method of following the course of such a hydrolysis is to monitor the change in optical rotation. Sucrose is dextrorotory, as is D-glucose (which is, for this reason, also known as dextrose). When sucrose is hydrolyzed to a mixture of D-glucose and D-fructose in their various cyclic forms, the net resulting optical rotation of the solution is levorotatory because one of the fructose forms is strongly levorotatory (see Prob. 20–19). This sugar solution is often called **invert sugar** because of the inversion in sign of rotation, and β-fructofuranosidase is often called **invertase**. Honey is mostly invert sugar. Equation 20.27 shows the action of invertase in the specific hydrolysis of a fructofuranoside in which the anomeric OR group is in the β configuration.

(20.27)

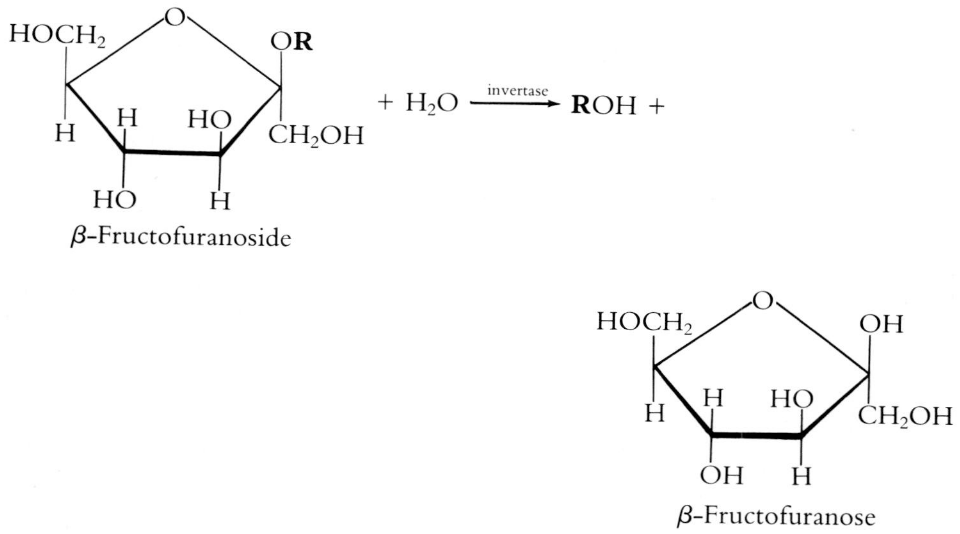

HOCH$_2$ O **OR** H H HO CH$_2$OH HO H

β-Fructofuranoside

$+ H_2O \xrightarrow{\text{invertase}}$ **ROH** +

HOCH$_2$ O OH H H HO CH$_2$OH OH H

β-Fructofuranose

Sucrose will also undergo hydrolysis in the presence of an α-glucosidase isolated from yeast. Sucrose is therefore an *α-glucopyranoside* as well as a β-

fructofuranoside, and hence is named α-D-glucopyranosyl-β-D-fructofuranoside. The -yl ending of the name of the first component indicates that it is connected to the second through its anomeric oxygen atom.

The anomeric carbon of an aldose is C-1 in the official numbering system, while the anomeric carbon of a ketose is C-2. Sucrose is therefore said to have a 1,2′-linkage.

Disaccharides are known with other types of linkages between the monosaccharides: 1,4′ and 1,6′ are two of the more common linkages. Aldose sugars with 1,1′ linkages are always nonreducing since they are joined at the anomeric carbons, while those with 1,4′ and 1,6′ linkages have a free anomeric carbon and hence are reducing. Several common reducing disaccharides are depicted below, along with their systematic names.

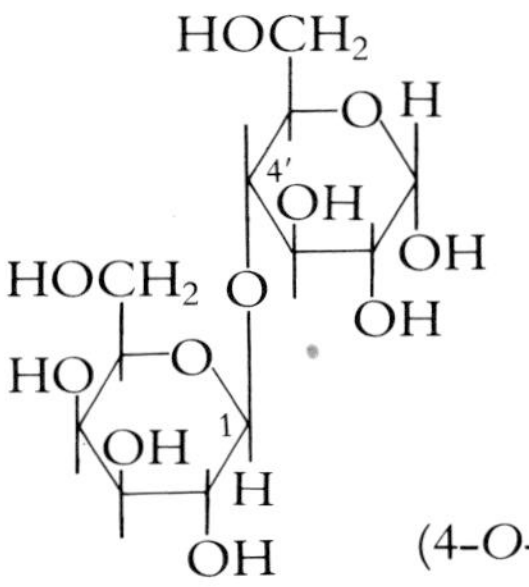

α-Lactose
(4-*O*-β-D-galactopyranosyl-α-D-glucopyranose)

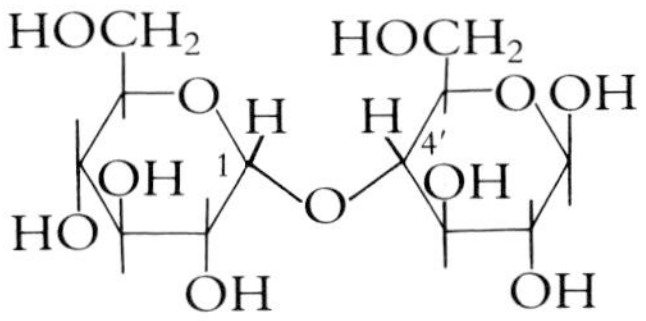

β-Maltose
(4-*O*-α-D-Glucopyranosyl-β-D-glucopyranose)

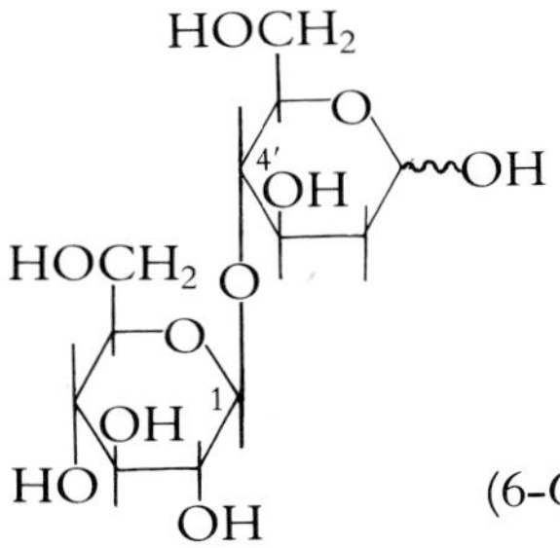

Cellobiose
(4-*O*-β-D-glucopyranosyl-D-glucopyranose)

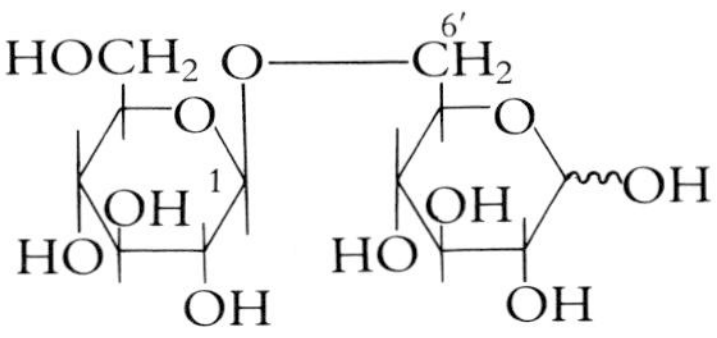

Gentiobiose
(6-*O*-β-D-glucopyranosyl-D-glucopyranose)

Maltose, cellobiose, and gentiobiose yield only glucose upon acidic hydrolysis. Cellobiose and gentiobiose are also hydrolyzed in the presence of an enzyme, isolated from plants, which hydrolyzes β-glucosides. Maltose is unaffected by this enzyme, but it will hydrolyze in the presence of the α-glucosidase *maltase.* Lactose is affected by neither of these enzymes, since it is a galactoside rather than a glucoside. These enzymes are specific for glucosides. All four disaccharides are reducing sugars, since they contain free anomeric hydroxyl groups.

Hydrolysis does not give any information on the type of linkage originally present in the disaccharide. For example, when maltose is hydrolyzed, it yields two molecules of glucose, but we don't know whether the molecules were joined 1,4′, 1,6′, or with some other linkage. The linkage position can be determined by first methylating the disaccharide and then performing the hydrolysis. If the sugars are joined at the 1,4′ positions, methylation will put methyls on all OHs except at the 1-position of one glucose molecule and the 4-position of the other glucose molecule. Hydrolysis will yield 2,3,4,6-tetra-*O*-methyl-D-glucose and 2,3,6-tri-*O*-methyl-D-glucose. Since the latter compound does not have a methyl on C-4, it must have been bound to the other glucose molecule at the oxygen of C-4.

Problems

20–19 a. The initial product of fructofuranoside hydrolysis is the β-fructofuranose (eq. 20.27), with almost none of the α-anomer formed. Even after the solution has reached equilibrium the concentration of α-fructofuranose is quite small. Explain why the β-anomer is so heavily favored.

b. β-D-Fructofuranose and α- and β-D-glucopyranose are all dextrorotatory. Thus, the solution obtained by rapid hydrolysis of sucrose is initially dextrorotatory. With time this solution eventually becomes levorotatory. The rotational change arises from the conversion of β-D-fructofuranose to an isomer that is strongly levorotatory. Draw the structure of this isomer and show how it is formed. What will be its predominant anomeric configuration?

20–20 a. Gentianose is a nonreducing trisaccharide composed of glucose and fructose. When treated with a β-glucosidase, gentianose is hydrolyzed to glucose and sucrose, while invertase-catalyzed hydrolysis results in the release of gentiobiose and fructose. Write the structure of gentianose.

b. Raffinose is a nonreducing trisaccharide, which upon acid hydrolysis yields 1 mol each of fructose, galactose, and glucose. Upon treatment with invertase, raffinose yields fructose and the disaccharide melibiose. After acid hydrolysis, melibiose gives glucose and galactose but it is resistant to an enzyme that hydrolyzes lactose. Suggest possible structures for melibiose and raffinose. What additional information would be needed to definitively assign structures to these compounds?

c. When gentiobiose is treated with dimethyl sulfate and sodium hydroxide it yields a permethylated derivative. If this methylated disaccharide is then hydrolyzed with acid, what methylated monosaccharides will be formed?

Polysaccharides are ubiquitous in nature. They serve as structural materials and as reserve foodstuffs. Starch and glycogen are storage forms of glucose in plants and animals, respectively. Starch contains two glucose polymers, *amylose* and *amylopectin,* each of which is a branched-chain consisting *mostly* of α-1,4-linkages. In amylopectin, there is extensive cross-linking of different saccharide chains by way of α-1,6-linkages (Fig. 20–5). Glycogen, the principal carbohydrate reserve of animals, is a heavily cross-linked poly-α-1,4-glucoside.

Cellulose is the cell wall material of higher plants. It too yields only glucose upon acid hydrolysis but, in contrast to starch and glycogen, it consists of very long 1,4-linked β-glucoside chains (Fig. 20–6). The great

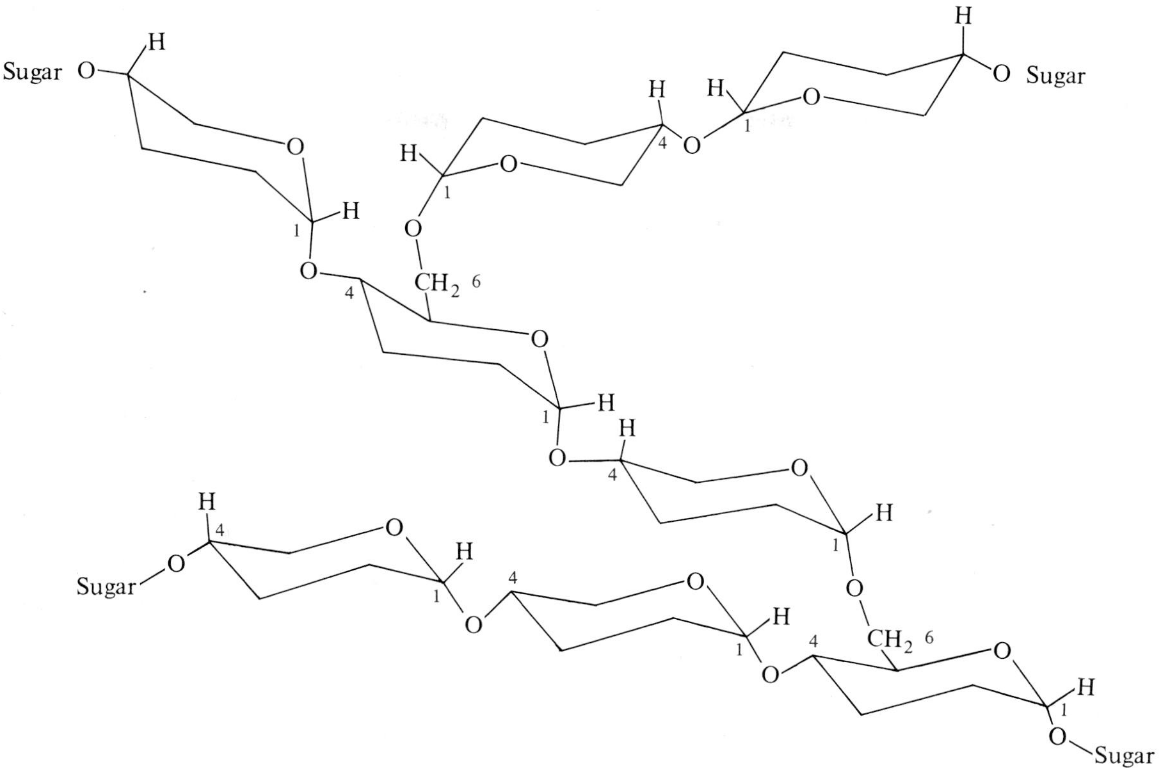

Figure 20–5. *Section of an amylopectin chain showing 1,6 cross-linking (OH groups on other carbons are removed for clarity.)*

strength and water insolubility of cellulose are due to the fact that the individual chains are held to one another by an extensive series of hydrogen bonds.

Figure 20–6. *Section of a cellulose molecule.*

Box 20–1. Genetic Variability in Sugars and Sugar Utilization

A whole new branch of carbohydrate chemistry has been opened up with the recent discovery that various human populations differ in their sugar chemistry.

One of the earliest (1953) findings in this area was that blood types are determined by sugars. In blood type A, for example, a monosaccharide known as acetylgalactosamine occurs as part of a polymer on the outside surface of the red blood cell, while in blood type B, that sugar is galactose. If these blood types are mixed in transfusions, there can be a life-threatening immunological reaction.

A number of people have only limited amounts of the enzyme lactase, which hydrolyzes milk sugar (lactose) into glucose and galactose. Only monosaccharides can pass into the blood stream and be utilized, and the unhydrolyzed disaccharide lactose passes into the large intestine undigested where it causes abdominal pain and diarrhea.

A much rarer variant of lactose intolerance is the familial defect of galactose metabolism called galactosemia. The galactose that is derived from lactose must be converted into glucose before it can be used. In galactosemia, an enzyme which is involved in this conversion is missing and hence galactose phosphate accumulates in the blood stream. In infants with galactosemia, the formation of the poisonous galactose phosphate can cause mental retardation and death.

Cellulose is the most abundant carbohydrate on earth, since it is a major constituent of plants. Unfortunately, we cannot utilize cellulose directly as a food source because we do not have enzymes that can hydrolyze β-1,4-linkages. Animals, such as cattle and sheep, that do utilize cellulose don't possess the proper enzymes either, but they do have a digestive organ called the rumen where bacteria break down cellulose. These bacteria have the enzyme β-1,4-glucosidase necessary for cellulose hydrolysis.

20:7 Natural and Artificial Sweeteners

For a molecule to taste "sweet," it must fit into a taste bud site where a nerve impulse can carry the message of sweetness from the tongue to the brain. Natural sugars have varying abilities to cause such a message to be transmitted, and individuals vary in their sensory perception of sweetness, as well as other taste sensations. One "sweetness" scale, developed by a panel of tasters, rates the relative sweetness of the common sugars as 1.00 for glucose, 1.45 for sucrose, and 1.65 for fructose, making fructose the sweetest of the natural sugars.

The most common sweetener, sucrose, is a pure compound which we use as table sugar. Honey, which is also used extensively, is a mixture of sugars and traces of other compounds transferred from flowers by the bees which manufacture the honey. The content of the various sugars in honey varies somewhat, but a typical analysis is 25% sucrose, 34% glucose, and 42% fructose. Some analyses show slightly more fructose.

The first artificial sweetener to be used extensively was saccharin, or its more soluble sodium salt. Saccharin was discovered by the American chemist Ira Remsen in the late 1800s and has been used since that time. Based on the typical substitution of a 1/2 grain (0.03 g) tablet of saccharin for a heaping teaspoon of sugar, saccharin is about 300 times sweeter than sucrose.

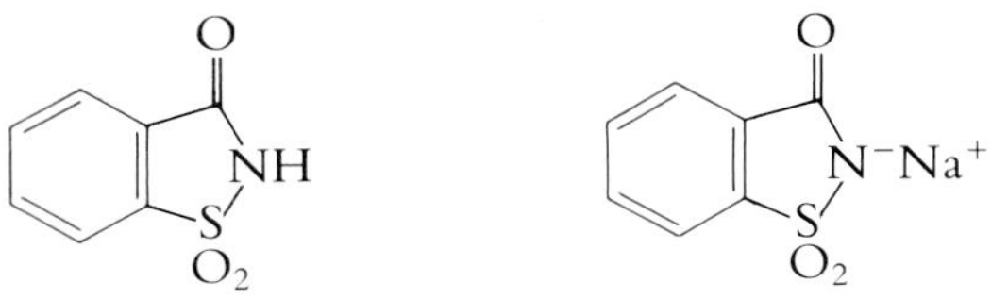

Saccharin   Sodium saccharin

A new artificial sweetener, aspartame, has a peptide structure and is discussed in Section 22:7. One conclusion that follows from a comparison of sucrose, saccharin, and aspartamine is that the sensation of sweetness is *not* induced solely by a particular molecular shape.

20:8 Sugars in Metabolism and Medicine

In mammalian tissue, metabolic by-products and toxic substances of external origin must be excreted. Excretion is facilitated if the substance is made more water soluble so that it can be eliminated rapidly in the urine. One way the body accomplishes this is to bind a sugar molecule with its many hydroxyl groups to the organic by-product or toxic substance. For example, the synthetic compound diethylstilbestrol (DES), **10**, has been used as an abortifacient and as a feed supplement for livestock. Compounds such as DES that contains a phenolic or alcoholic group are bound in the body to one or more molecules of **glucuronic acid**, **11**, and excreted as the sodium glucuronide, **12**.

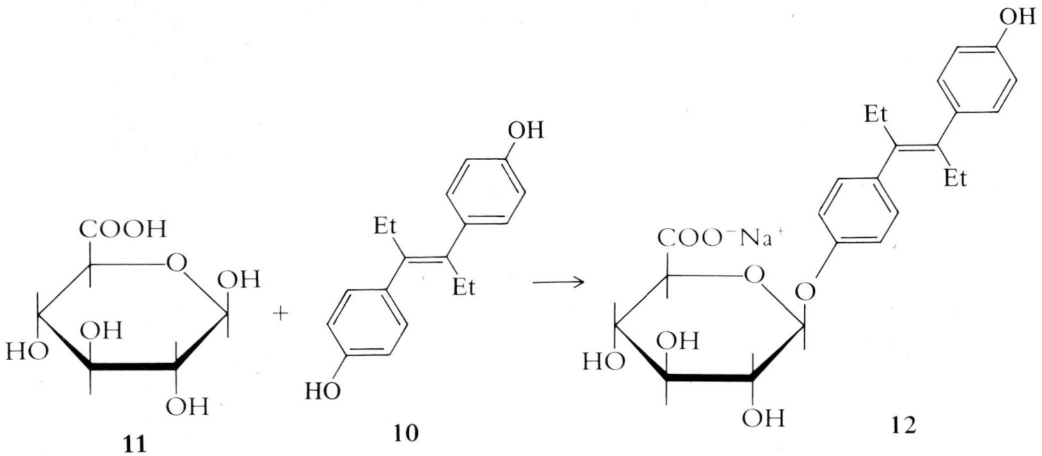

11 Glucuronic acid + **10** Diethylstilbestrol (DES) → **12** DES monoglucuronide

As another example of sugars binding to organic substrates, consider the fact that some of the most valuable antibiotics contain **aminosugar** portions. For example, the antibiotic adriamycin, **13**, also known as doxorubicin, is composed of two parts: the aminated sugar portion and the **aglycone** portion, which is made of four linear fused rings and resembles an important class of antibiotics called *tetracyclines*. Many antibiotics function because they attack cell membranes of the pathogen. Perhaps the presence of aminosugars in both antibiotics and cell membranes is related to this activity.

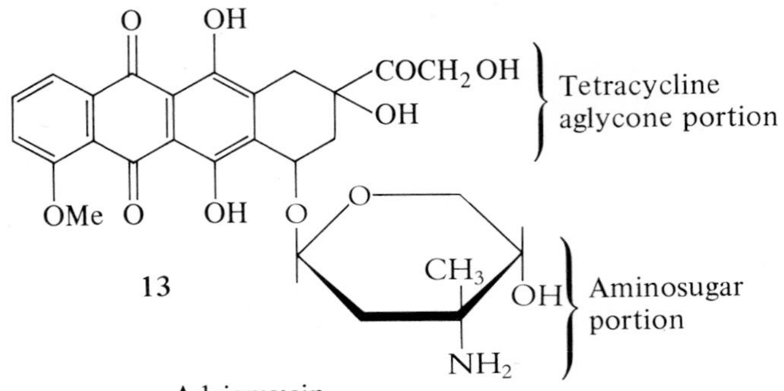

O OH
$COCH_2OH$
OH
Tetracycline aglycone portion
OMe O OH O
O
13
CH_3
OH
Aminosugar portion
NH_2

Adriamycin
(*doxorubicin*)

Aminosugars also occur in **glycoproteins** which, as their name implies, are proteins containing a number of carbohydrate groups. Glycoproteins occur in blood plasma, mucus secretions, and cell membranes. A widespread aminosugar is 2-deoxy-2-amino-D-glucose (D-glucosamine), which is incorporated into glycoproteins and polysaccharides as its *N*-acetyl derivative. The 1,4′-polymeric *N*-acetyl-*O*-glucosamine is called **chitin**. It makes up the hard outer coating of insects and crustaceans such as beetles and lobsters.

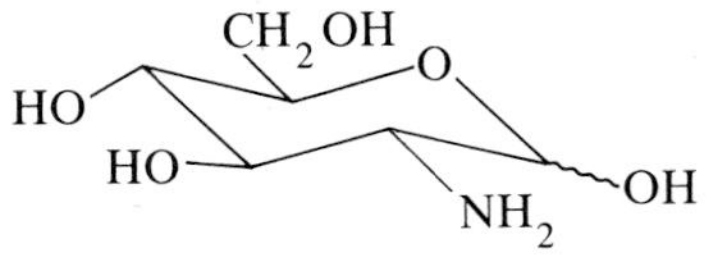

CH_2OH
O
HO
HO
OH
NH_2

2-Deoxy-2-amino-D-glucose
(D-Glucosamine)

20:9 Nucleic Acids

It is very common to find sugar structures incorporated into larger molecules that are not carbohydrates. You saw two examples of this in the previous section. Another example is the presence of ribose and deoxyribose in nucleic acids.

Nucleic acids are the substances most intimately linked with the transmission and utilization of genetic information in all living things. Complete hydrolysis of these high molecular weight (10^4 to 10^9) polymers reveals that they are composed of three different types of components: a purine or pyrimidine base (Chap. 25), an aldopentose, and phosphoric acid. Only two pentoses have been isolated from nucleic acids. One of these sugars is D-ribose (**14**), which is found in **ribonucleic acids (RNA)**. The second is 2-deoxy-D-ribose (**15**), found in **deoxyribonucleic acids (DNA)**.

In aqueous solution, both **14** and **15** consist of mixtures of furanose and pyranose isomers, in which the pyranose forms predominate. However, the biologically interesting derivatives of **14** and **15** are all derived from the furanose form. In these furanosides the anomeric hydroxyl at C-1 is replaced by some other group.

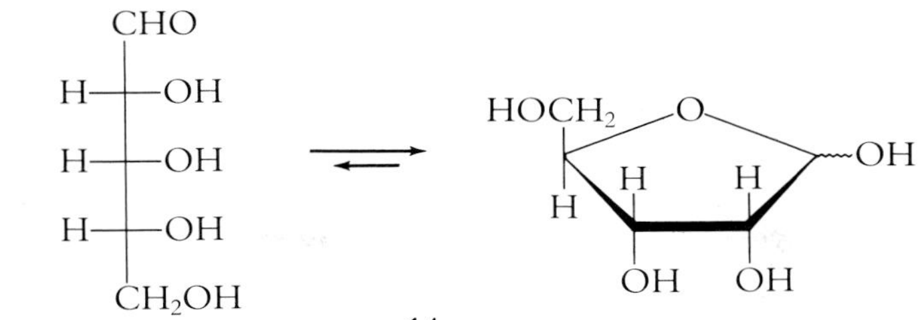

CHO
H—C—OH
H—C—OH
H—C—OH
CH_2OH

14

D-Ribose
(open-chain form)

D-Ribose
(furanose form)

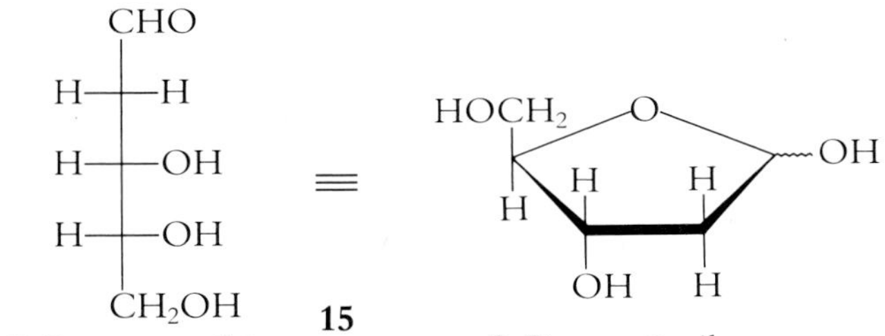

CHO
H—C—H
H—C—OH
H—C—OH
CH_2OH

15

2-Deoxy-D-ribose
(open-chain form)

2-Deoxy-D-ribose
(furanose form)

A heterocyclic base covalently bound to C-1 of these sugars constitutes a **nucleoside**. Depending on the sugar, these are either **ribonucleosides (16)** or **deoxyribonucleosides (17)**.

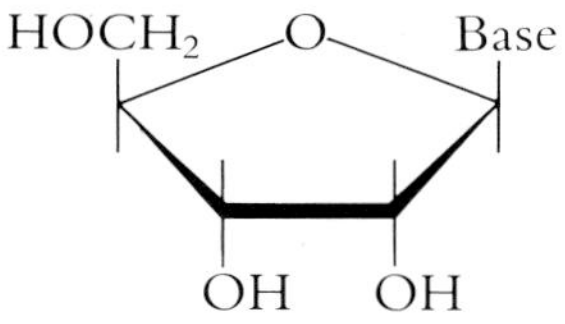

16

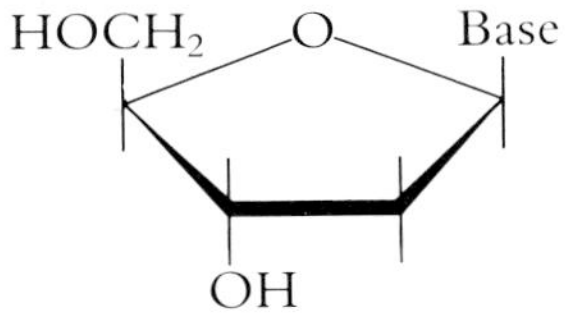

17

Nucleosides

If one of the sugar hydroxyls is esterified by phosphoric acid, **nucleotides** are produced. The schematic representations for **ribonucleotides (18)** and **deoxyribonucleotides (19)** are given below.

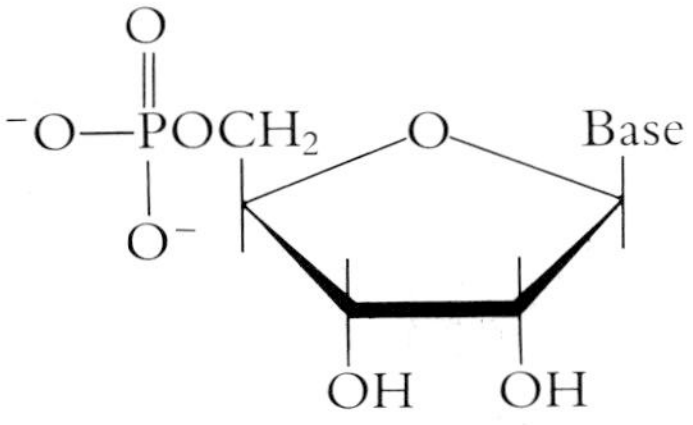

18

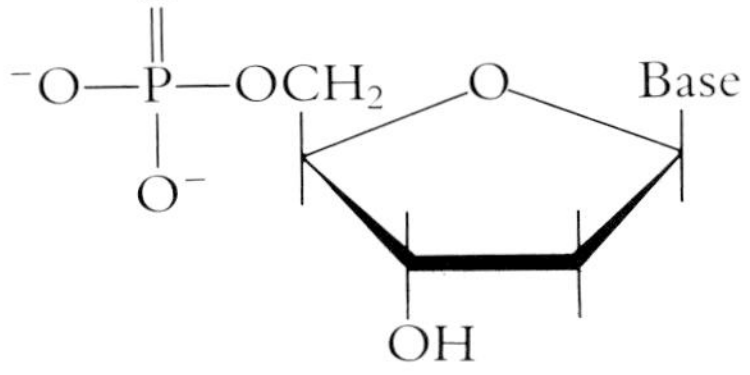

19

Nucleotides

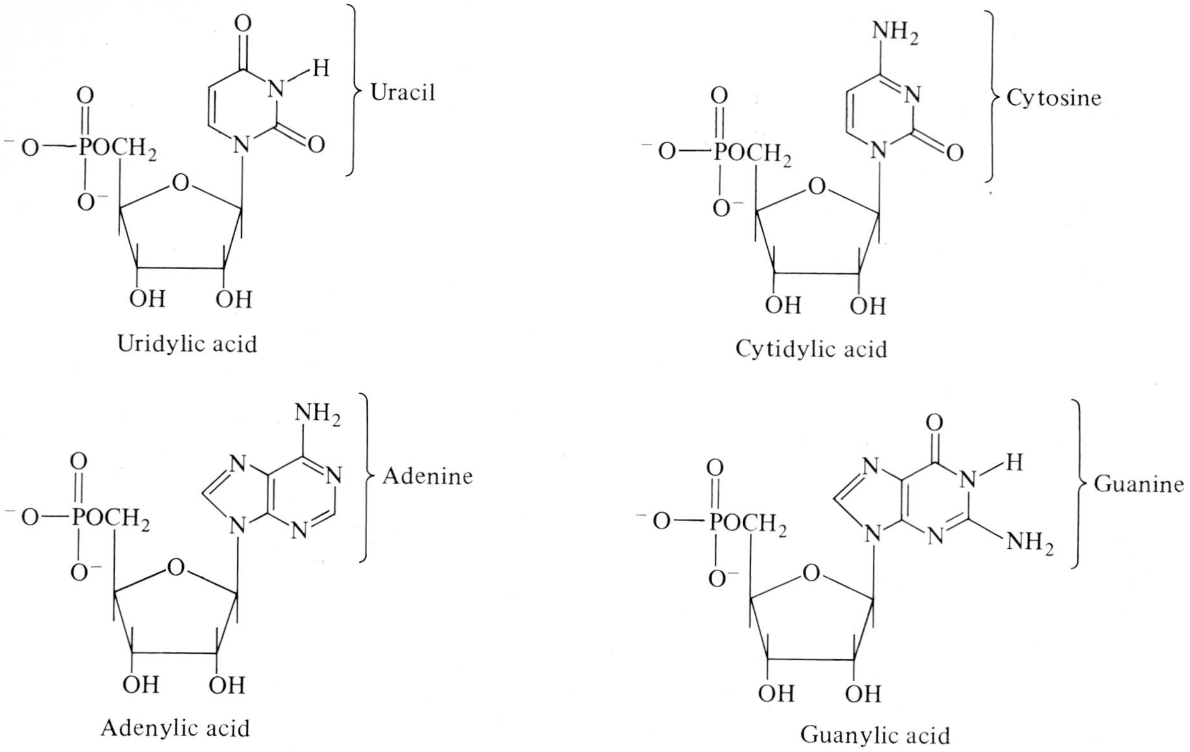

Uridylic acid

Cytidylic acid

Adenylic acid

Guanylic acid

Figure 20–7. *Nucleotides found in RNA.*

Nucleic acids are polymers of nucleotides. RNA is a polymer of **18** and DNA is a polymer of **19**. The nucleotides most commonly found in RNA are shown in Figure 20–7, while the common DNA nucleotides are shown in Figure 20–8. Note that thymidylic acid is absent from RNA, and uridylic acid from DNA. The bases adenine, guanine, and cytosine occur in both DNA and RNA, whereas the base uracil is specific to RNA and the base thymine is specific to DNA.

A portion of a DNA molecule is shown in Figure 20–9. The phosphate group at C-5′ of each nucleotide is esterified to the hydroxyl group at C-3′ of the neighboring nucleotide.★ This polymeric molecule can be cleaved by hydrolysis in aqueous acid. If the pH is low enough the DNA will be completely degraded to phosphoric acid, deoxyribose, and the organic bases. DNA molecules resist basic hydrolysis because the anionic phosphate group is attacked only very slowly by hydroxide ion. RNA, however, will undergo base-promoted hydrolysis. Figure 20–10 illustrates a mechanism that can account for the behavior of RNA. The hydroxyl at C-2′ (absent in DNA) provides a pathway for intramolecular hydrolysis. From this reaction various nucleotides are obtained, phosphorylated at C-2′ and C-3′.

★ Note that 5′ refers to the sugar moiety; the unprimed numbers refer to the base.

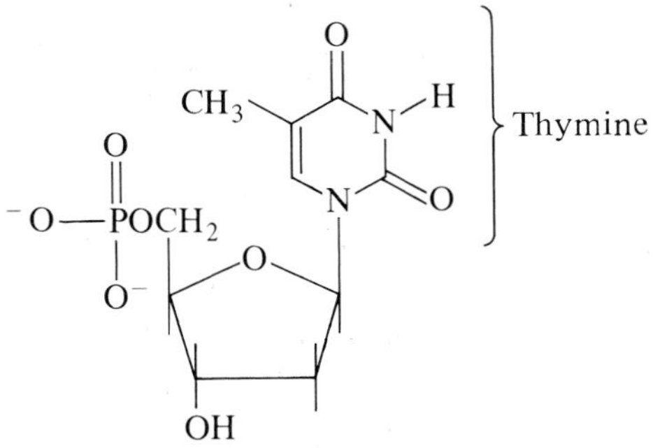

Thymine

Thymidilic acid

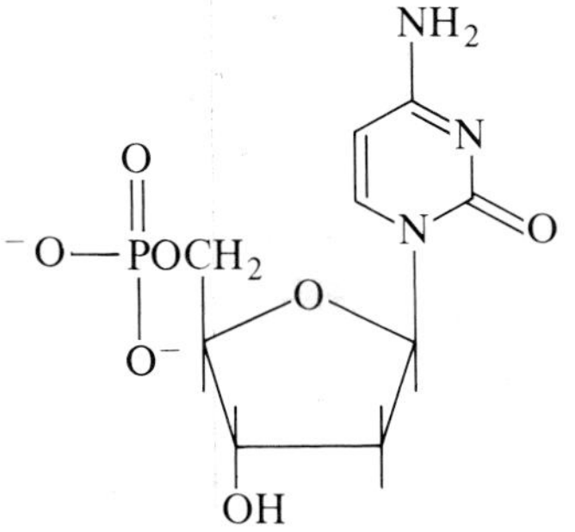

Deoxycytidylic acid

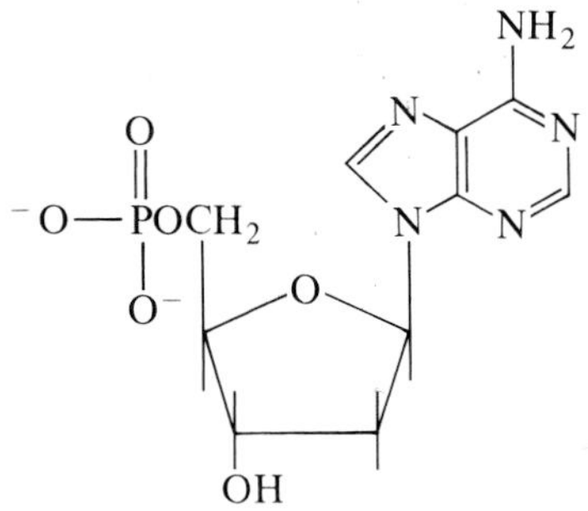

Deoxyadenylic acid

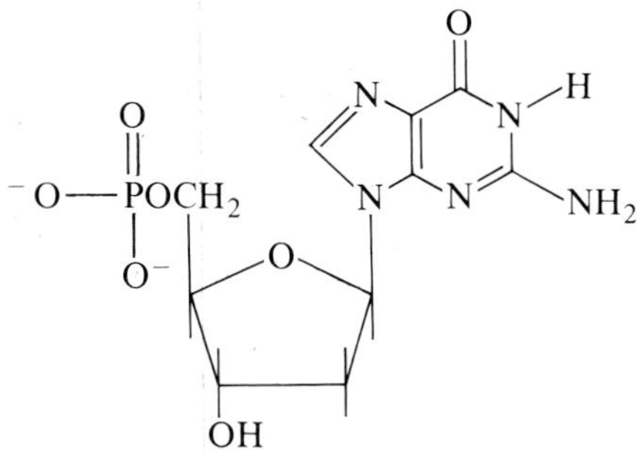

Deoxyguanylic acid

Figure 20–8. *Nucleotides found in DNA.*

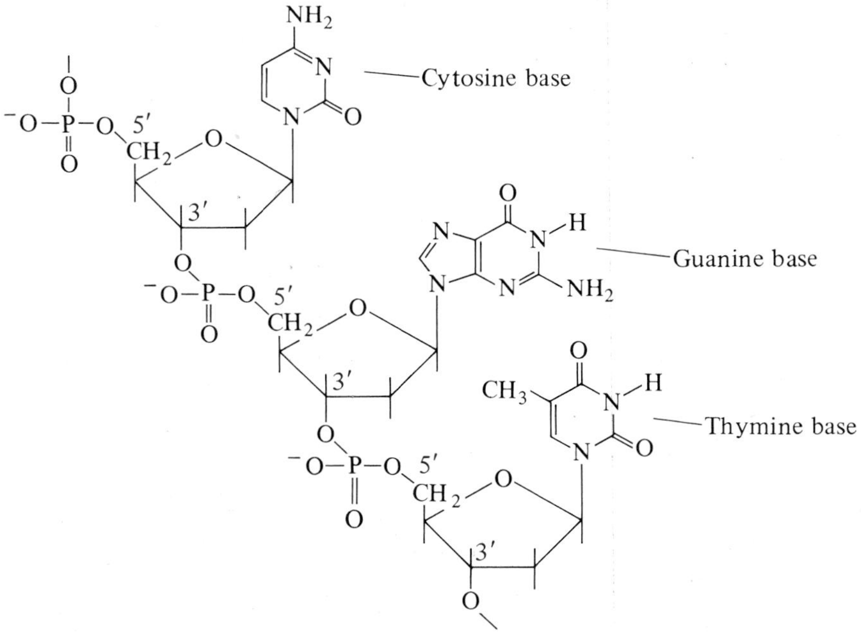

Cytosine base

Guanine base

Thymine base

Figure 20–9. *Portion of a DNA molecule showing linking of the nucleotides.*

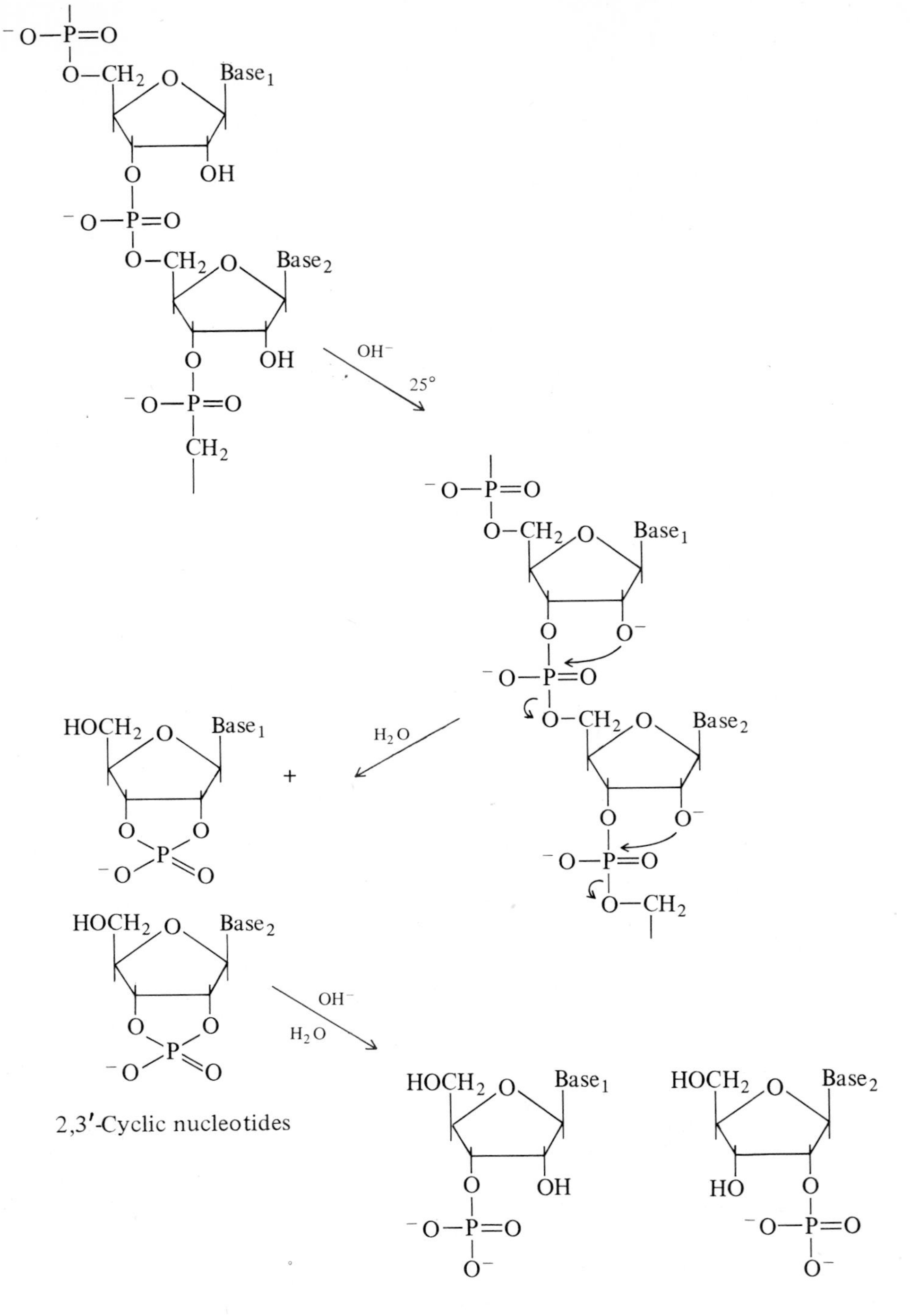

Figure 20–10. *Base-promoted cleavage of a RNA molecule.*

Nucleotides and nucleosides are found in monomeric form in cells, as well as in the polymeric nucleic acids. Some important coenzymes are built around them, such as NAD (nicotine adenine dinucleotide, sec. 11:12) and coenzyme A (sec. 17:13). (*S*)-Adenosylmethionine is a vital methylating agent (sec. 6:15; Prob. 6–64, p. 225), and adenosine is present in the most important cellular energy storing compound, ATP (adenosine triphosphate, sec. 17:14).

In Chapter 25 we will return to the fascinating subject of genetic transmission. We will focus specifically on the heterocyclic bases whose chemistry is directly involved with the transfer of genetic information.

Summary

1 Sugars (saccharides) are the primary units of the carbohydrate class of compounds. Monosaccharides, such as glucose and fructose, have the molecular formula $C_n(H_2O)_n$. Aldoses have an aldehyde group at one end of a chain, and ketoses have a keto carbonyl at the second carbon from one end (sec. 20:1). From the C_3 compounds on, most sugars have one or more chiral carbons.

In the straight-chain formula, sugars are often depicted in a Fischer projection (sec. 20:1). D-Sugars have the asymmetric carbon furthest from the carbonyl group in the same absolute configuration as D-(+)-glyceraldehyde.

2 Five- and six-membered sugar units exist mainly as cyclic hemiacetal structures. The five-membered rings are called furanoses and the six-membered rings pyranoses. The carbon of the C=O group in the straight-chain formula is called the anomeric carbon in the cyclic structure. The cyclic structure explains the phenomenon of mutarotation and the isolation of anomers. For example, glucose exists in solution as a mixture of β-D-glucopyranose and α-D-glucopyranose. These structures are epimeric at the anomeric carbon. We can distinguish between these epimers by using nmr analysis (sec. 20:2).

Acid-catalyzed reactions of sugars with alcohols can occur at the anomeric carbon to form glycosides. These reactions proceed by way of stabilized carbocations. Glycosides of glucose are called glucosides (sec. 20:3).

3 Sugars can react with phenylhydrazine to form crystalline phenylhydrazones. With excess reagent they yield osazones. Identical osazones arise from sugars which differ only at C-1 and C-2. Acetylation and trimethylsilylation of sugar OH groups provide volatile sugar derivatives that are easily analyzed. The aldehyde or ketone group can be reduced catalytically or with $NaBH_4$. Aldoses and 2-ketoses can be oxidized with Tollens reagent or Fehling solution. Sugars can be cleaved by oxidation with periodate. The formic acid and formaldehyde released by periodate can be measured quantitatively as an aid to structure determination (sec. 20:4).

4 Fructose-1,6-diphosphate is synthesized from glyceraldehyde phosphate and dihydroxyacetone phosphate. The major reaction is an aldol condensation in which iminium ions and enamines play a vital role (sec. 20:5). The reverse mechanism operates in glycolysis, in which sugars are broken down for energy production.

5 Two sugars can join to form disaccharides (e.g., sucrose). These can be reducing sugars (if one of the anomeric carbons still is attached to an OH group) or nonreducing sugars. The latter are joined through the anomeric carbons of each monosaccharide. The stereochemistry at the glycosidic bonds can be determined by enzymatic hydrolysis, since enzymes are specific for a particular anomeric configuration. Polysaccharides, such as cellulose, are common in nature. The stereochemistry of polysaccharide linkages is also established by specific enzymatic reactions (sec. 20:6).

6 Metabolites and toxic substances have to be made water soluble in order to be excreted. If they contain hydroxyl groups that can be combined with glucuronic acid, the resulting glycoside is eliminated in the urine. Amino sugars are found in antibiotics and combined with some proteins (sec. 20:8).

7 Nucleic acids carry the genetic information in living cells. One nucleic acid (RNA) contains ribose, the other (DNA) contains deoxyribose. The nucleic acids contain repeating units consisting of one sugar, one phosphoric acid molecule, and one heterocyclic base. These units are termed nucleotides. The ribose and deoxyribose components are present as furanoses. Nucleosides are the same units with the phosphoric acid removed.

RNA and DNA have three nucleotides in common. They each have a fourth that is unique to that nucleic acid. Both nucleic acids can be hydrolyzed by acid. RNA is also hydrolyzed in aqueous base by way of an intramolecular mechanism (sec. 20:9).

Problems

20–21 Give structural examples of the following terms:

a. a D sugar
b. an L sugar
c. an aldotetrose
d. a ketohexose
e. an anomeric carbon
f. two epimers that are also anomers
g. two epimers that are not anomers
h. a Haworth projection for α-D-xylopyranose
i. a Haworth projection for β-D-xylofuranose
j. a glycosidic bond
k. a reducing disaccharide

20–22 What is the structural relationship between D-glucose and L-glucose? Are these two compounds: enantiomers; diastereomers; epimers; mirror image forms; *E,Z* isomers; optical isomers? Draw the open-chain structure of L-glucose.

20–23 a. In Chapter 3 it was shown that any substituted cyclohexane consists of a mixture of *two* chair conformers in rapid conformational equilibrium with one another. Figure 20–1 shows only *one* chair conformer each for α- and β-D-glucopyranose. Draw the other chair conformer for these isomers. Would you expect appreciable amounts of these other conformers to be present at equilibrium? Explain your answer.

b. Draw the two chair conformations of α-D-ribopyranose and the two chair conformations of β-D-ribopyranose, and predict which conformation(s) will predominate.
c. One, and only one, naturally occurring aldohexose epimer has all equatorial substituents. What is the name of this compound?

20–24 What products would be expected from the reaction of D-lyxose (**I**) with a solution of HCl in methanol?

$$\begin{array}{c} \text{CHO} \\ | \\ \text{HO—C—H} \\ | \\ \text{HO—C—H} \\ | \\ \text{H—C—OH} \\ | \\ \text{CH}_2\text{OH} \end{array}$$

I
D-Lyxose

20–25 Show the products expected from periodate oxidation of the following compounds:
a. ethylene glycol
b. 1,2,3-butanetriol
c. 1,2,4-butanetriol
d. tartaric acid
e. erythrose
f. threose
g. methyl-α-D-lyxofuranoside (Prob. 20–24)
h. methyl-β-D-lyxopyranoside (Prob. 20–24)

20–26 Two sets of products are obtainable from erythrulose upon periodate oxidation, depending on the path taken by the oxidation. What are these two paths, and what are the products derived from each?

20–27 Erythritol is oxidized by sorbose bacteria to an optically active erythrulose. When the erythrulose is reduced by means of sodium borohydride, two polyols are obtained, one of which proves to be L-threitol.
a. Write the reactions described above, showing the structures of all reactants and products, including the second polyol.
b. What was the absolute configuration of the erythrulose?
c. What conclusions can be drawn about the stereospecificity of the bacterial oxidation?

20–28 Propose structures for compounds **A** to **E**.

$$\text{D-erythrose} \xrightarrow[H_2O]{Br_2} (+)\text{-}\mathbf{A} \xrightarrow{HNO_3} meso\text{-}\mathbf{B}(C_4H_6O_6) \xrightarrow{\text{heat}}$$

$$\mathbf{C}(C_4H_4O_5) \xrightarrow{CH_3OH} \mathbf{D}(C_5H_8O_6) \xrightarrow[THF]{BH_3} \mathbf{E}(C_5H_{10}O_5) \xrightarrow{H_3O^+} (\pm)\text{-}\mathbf{A}$$

20–29 a. Draw the pyranose forms of D-lyxose (see Prob. 20–24).
b. Pure, crystalline α- and β-D-lyxopyranoses may be obtained by selective

crystallization of the sugar from aqueous methanol. The α-anomer exhibits $[\alpha]_D^{20} = 5.6°$, while its β-epimer has $[\alpha]_D^{20} = -72.6°$. When *either* anomer is dissolved in water, the rotation changes until the equilibrium value of $-13.8°$ is reached. Show, with chair conformation structures, what is happening and calculate the percentage of each anomer at equilibrium.

20–30 a. When D-galactose is acetylated with acetic anhydride in the presence of sodium acetate, two galactose pentaacetates are isolated. Each of these may be isomerized to a different pentacetate with $ZnCl_2$, giving a total of four galactose pentaacetates in all. Draw their structures. What conclusions might you reach about the structures of D-galactose in solution?

b. How would you account for the difference in behavior of glucose (sec. 20:5) and galactose? (Compare the structures of their pyranose ring forms.)

20–31 Salicin is a compound found in the bark of several species of tree. Upon mild acid-catalyzed hydrolysis, salicin gives glucose and (*o*-hydroxyphenyl) methanol (eq. 3). If salicin is subjected to mild oxidation and then hydrolyzed, one can obtain glucuronic acid (**II**) and salicylic acid (eq. 4). Suggest a structure for salicin. What feature of the structure cannot be determined from the data given? How would you arrive at a definitive structure for salicin?

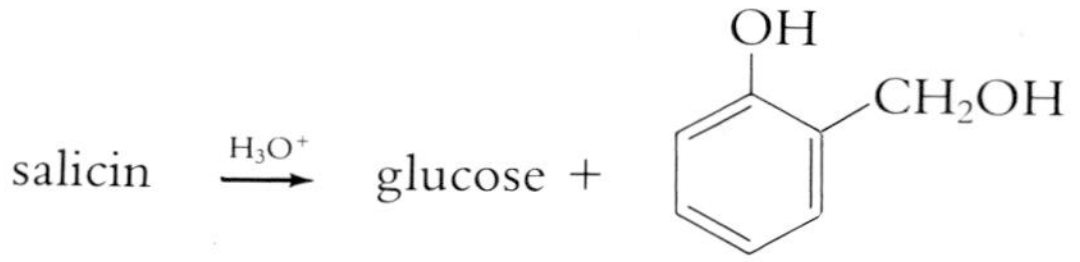

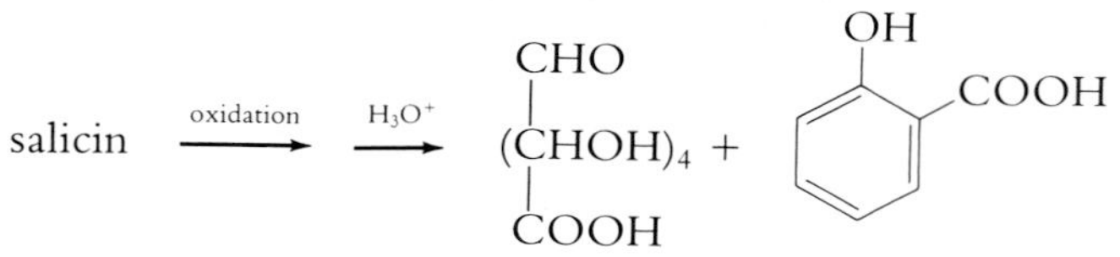

$$\text{salicin} \xrightarrow{H_3O^+} \text{glucose} + \text{(o-}HOC_6H_4CH_2OH) \quad (3)$$

$$\text{salicin} \xrightarrow{\text{oxidation}} \xrightarrow{H_3O^+} \underset{\textbf{II}}{CHO-(CHOH)_4-COOH} + \text{(o-}HOC_6H_4COOH) \quad (4)$$

20–32 a. In some plants, an important step in photosynthesis is the carboxylation of ribulose-1,5-disphosphate (eq. 5). Write a reasonable mechanism for this reaction, including the structure of the probable intermediate.

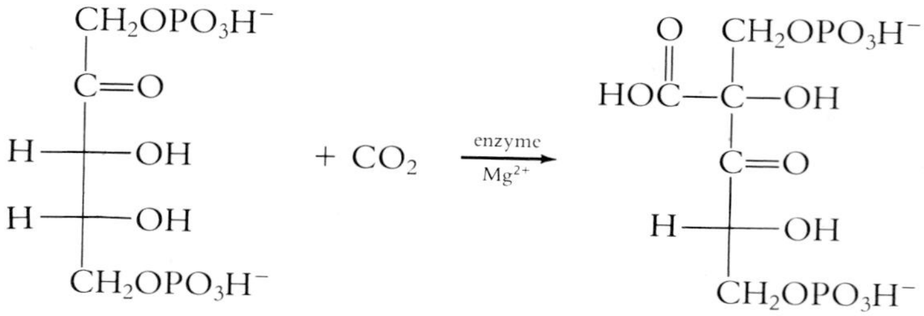

$$\underset{\text{Ribulose-1,5-diphosphate}}{CH_2OPO_3H^- - C(=O) - CH(OH) - CH(OH) - CH_2OPO_3H^-} + CO_2 \xrightarrow[Mg^{2+}]{\text{enzyme}} \underset{\textbf{III}}{HOC(=O) - C(OH)(CH_2OPO_3H^-) - C(=O) - CH(OH) - CH_2OPO_3H^-} \quad (5)$$

b. In the next step of this sequence, the carboxylic acid **III** reacts with water to produce two molecules of 3-phosphoglyceric acid (eq. 6). Write a mechanism for this reaction (note that it is a reverse acylation; sec. 19:9).

O CH$_2$OPO$_3$H$^-$
HOC—C—OH
C=O
H—C—OH
CH$_2$OPO$_3$H$^-$
+ H$_2$O $\xrightarrow{\text{enzyme}}$ 2 H—C(COOH)—OH, CH$_2$OPO$_3$H$^-$ (6)

3-Phosphoglyceric acid

20–33 The disaccharide *amygdalin* is found in the pits of many fruits, including apricot and bitter almond. It is the principal component of the controversial anticancer preparation, Laetrile. Those who decry the use of Laetrile argue that it is not only ineffective, but has actually proven fatal when ingested in large quantities. Show what would happen to amygdalin when it encountered a β-glucosidase in the digestive tract, and how that might account for the fatal properties of Laetrile.

CH$_2$OH, HO, HO, OH, O, CH$_2$OH, HO, OH, O—C(H)(CN)—Ph

Amygdalin

20–34 The ^{1}H nmr spectrum of sucrose in D$_2$O is shown here. How can proton nmr be used to identify the type of glycosidic bond in disaccharides?

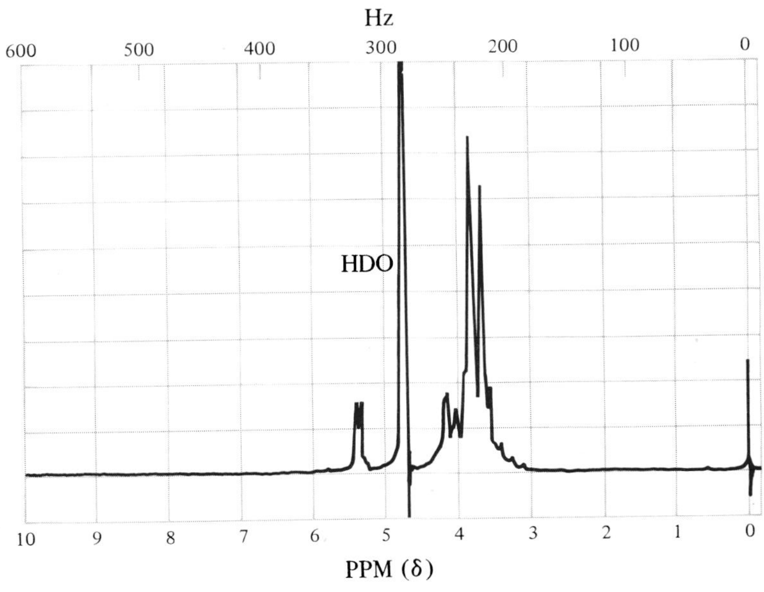

20–35 The ^{13}C nmr spectrum of D-galactose in D_2O is shown here. Why are so many peaks visible? If there were no overlapping of peaks, how many should be seen? Assign the peaks at 61.9, 62.1 (these two appear nearly overlapping), 93.2 and 97.3 ppm.

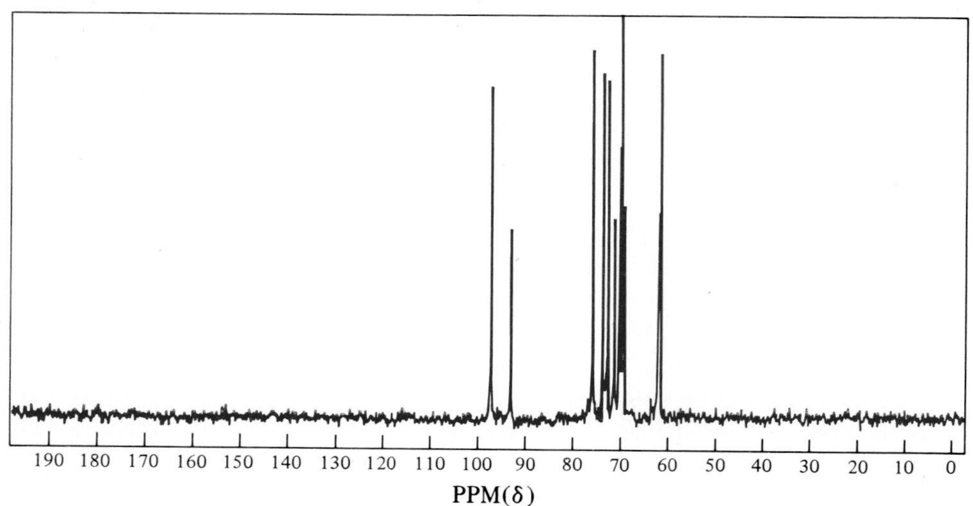

20–36 There is an inherited disease in humans which involves the failure to metabolize D-glucuronic acid (**11**, sec. 20:8). Those having this disease must avoid foods with large amounts of glucuronic acid. The first step in the metabolism of **11** is its conversion to L-gulonic acid (eq. 7). What has happened chemically in this reaction? What enzyme is probably involved in the reaction?

COOH, O, OH, OH, HO, OH

11

⟶

CH_2OH
HCOH
HOCH
HCOH
HCOH
COOH

(7)

D-Glucuronic acid　　　　L-Gulonic acid

Supplementary Reading

C. S. Hudson, Emil Fischer's proof of the configuration of glucose, *J. Chem. Ed.*, **18,** 353 (1941).

J. F. Stoddart, *Stereochemistry of Carbohydrates,* Wiley-Interscience, New York, N.Y., 1971.

N. Sharon, Carbohydrates, *Sci. Am.*, **243,** 90 (November 1980).

R. B. Barker, *Organic Chemistry of Biological Compounds,* Chaps. 5 and 8, Prentice–Hall, Englewood Cliffs, N.J., 1971.

B. Capon, Mechanism in carbohydrate chemistry, *Chem. Rev.*, **69,** 407 (1969).

A. Rich and S. H. Kim, The three-dimensional structure of transfer RNA, *Sci. Am.*, **238,** 52 (January 1978).

L. N. Ferguson, Bioorganic mechanisms: Chemoreception, *J. Chem. Ed.*, **58,** 456 (1981).

21 Amines, Phosphines, and Sulfides: Basicity, Nucleophilicity, and Valence Shell Expansion

Amines are derivatives of ammonia, NH_3, in which alkyl or aryl groups replace one or more hydrogens bonded to nitrogen—$R\ddot{N}H_2$, $R_2\ddot{N}H$, and $R_3\ddot{N}$. We have already encountered the amines in several of the preceding chapters, in their roles as either bases or nucleophiles. Both roles are made possible by the nonbonding electron pair on nitrogen.

Phosphorus, which lies directly below nitrogen in the Periodic Table, is found in a series of organic derivatives called **phosphines**, $R_3\ddot{P}$. The presence of the nonbonding electron pair in phosphines leads us to expect that they will behave like amines. In many cases they do, but there are also crucial differences in their chemical behavior. This is due to the difference in the electronic valence shells of nitrogen and phosphorus. In this chapter we will show how these differences have been exploited for synthetic purposes.

This chapter will also introduce the organic chemistry of sulfur, a neighbor of phosphorus. The organic derivatives of sulfur—thiols ($R\ddot{S}H$) and sulfides ($R\ddot{S}R$)—resemble alcohols and ethers in much of their chemistry. There are important differences, however, which, again, are due to differences in the properties of the valence shells. Phosphorus and sulfur, which have the same valence shell, also share many chemical characteristics.

Since nitrogen, phosphorus, and sulfur occur widely in the living cell, we will look at biochemical as well as chemical examples.

21:1 Nomenclature of Amines and Ammonium Compounds

Amines that contain only one organic group ($R\ddot{N}H_2$) are referred to as **primary amines**; those with two groups are **secondary amines** ($R_2\ddot{N}H$); and those with three groups are **tertiary amines** ($R_3\ddot{N}$). Since an additional organic group can be attached to the nonbonded pair, amine derivatives with four such groups can be prepared, R_4N^+. These quaternary ammonium cations must have a negative counterion (X^-); typical ones are Cl^-, Br^-, I^-, HSO_4^-, and OH^-. Together, the cation and anion constitute a **quaternary ammonium salt**. Quaternary ammonium salts are organic derivatives of ammonium salts, where the hydrogens of the latter are replaced by alkyl or aryl groups:

$NH_4^+Cl^-$	$NR_4^+Cl^-$
Ammonium chloride	A quaternary ammonium chloride

The nomenclature of amines is somewhat confusing, since simple and complex structures are named differently. In addition, the currently accepted IUPAC naming methods have not come into widespread use.

Alkyl amines carrying structurally simple substituents are most often named by a substitutive system, in which each organic group is considered to replace a hydrogen atom of ammonia.

$CH_3CH_2NH_2$	$CH_3CH_2NHCH_3$	$(CH_3)_3N$
Ethylamine (a 1° amine)	Ethylmethylamine (a 2° amine)	Trimethylamine (a 3° amine)

In the generally accepted system for naming primary amines containing a single nitrogen, the principal carbon chain, along with its substituents, is treated as a substituent on the NH_2 group, called "amine." The principal carbon chain is the one that (1) carries the amine substituent, and (2) has the greatest number of carbon atoms. The chain is numbered so that the amino group has the lowest possible locator number. If more than one NH_2 group is present, the name of the principal alkane has attached to it the suffixes *-diamine, -triamine,* and so on.

$(CH_3)_2CHCH_2NH_2$

2-Methyl-1-propylamine (*isobutylamine,* a 1° amine)

$CH_3C(CH_3)_2CH_2CH(NH_2)CH_3$

4,4-Dimethyl-2-pentylamine (a 1° amine)

$H_2N(CH_2)_4CH(NH_2)CH(NH_2)CH_3$

1,5,6-Heptanetriamine (a 1° amine)

In the IUPAC system, secondary and tertiary amines are named as derivatives of primary amines. The longest carbon chain is chosen as the primary one, and all other alkyl groups are treated as substituents. Alkyl groups that are substituted on the nitrogen atom are designated by the prefix *N*.

$CH_3CH_2CH_2N(H)CH_2CH_3$

N-Ethyl-1-propylamine (a 2° amine)

$C_5H_9N(CH_3)CH_2CH_3$

N-Ethyl-*N*-methylcyclopentylamine (a 3° amine)

If a molecule contains an —OH, —SH, or —COOH group in addition to $—NR_2$, then the amino group is named as a substituent.

$H_2NCH_2CH_2OH$

2-Aminoethanol

CH_3CH_2NH–C_6H_{10}–$COOH$

cis-4-(Ethylamino)cyclohexanecarboxylic acid

Benzenamine, $C_6H_5NH_2$, is always referred to by its common name, *aniline* (except in *Chemical Abstracts;* see Chap. 27). Other structurally uncomplicated aromatic amines are usually referred to by their systematic names.

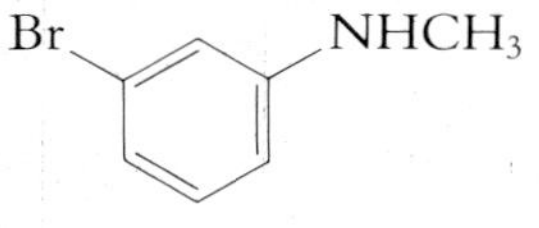

Br, $NHCH_3$

3-Bromo-*N*-methylaniline

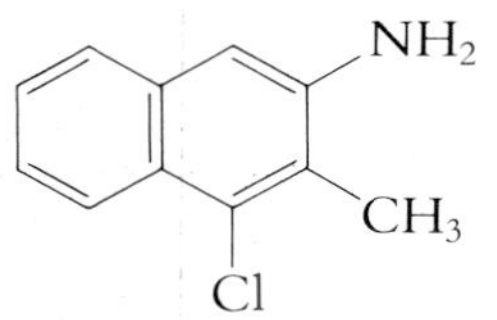

NH_2, CH_3, Cl

4-Chloro-3-methyl-2-naphthalenamine
(*4-chloro-3-methyl-2-naphthylamine*)

Many cyclic amines are universally known by their common (nonsystematic) names.

N H	N H	N H	N
Pyrrolidine	Pyrrole	Piperidine	Pyridine

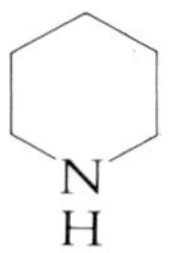

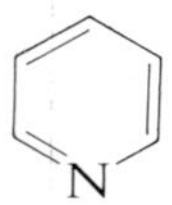

In cases where a cyclic amine does not have a nonsystematic name, the prefix *aza-* indicates that —NH— has been substituted for $—CH_2—$ in a hydrocarbon ring.

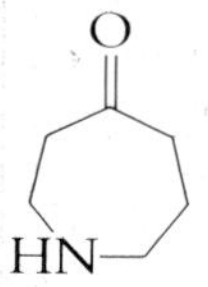

O, HN

4-Azacycloheptanone

CH_3, CH_2CH_3, $N—CH_3$

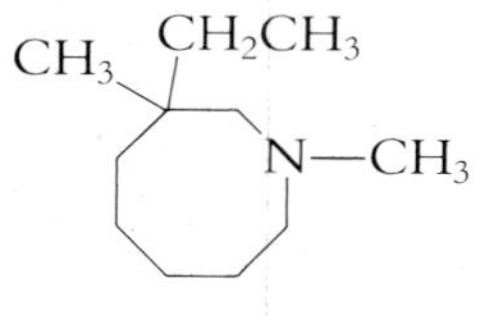

1,3-Dimethyl-3-ethylazacyclooctane

Finally, some amines or their derivatives that were found long ago in natural sources are known by special names derived from their chemical or biological source or function. For example, putrescine and cadaverine are odiferous products of bacterial decomposition of proteins.

$H_2N(CH_2)_4NH_2$	$H_2NCH_2CH_2CH_2CH_2CH_2NH_2$
Putrescine	*Cadaverine*
(1,4-butanediamine)	(1,5-pentanediamine)

A series of derivatives of 2-phenylethylamine are important in nerve metabolism and neurotransmission. They are linked by a series of biological hydroxylation and methylation steps. All are given common names:

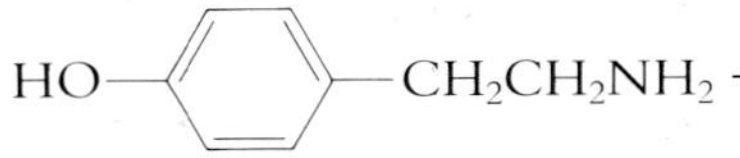

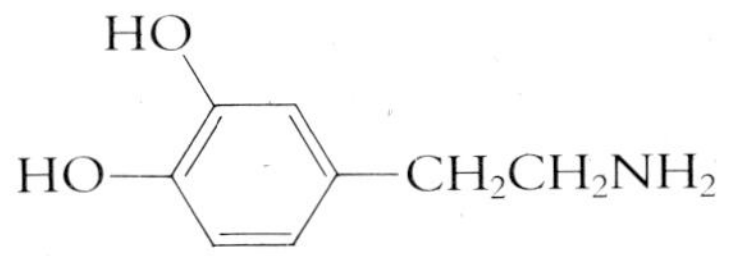

HO—C₆H₄—$CH_2CH_2NH_2$ → HO(HO)C₆H₃—$CH_2CH_2NH_2$

Tyramine
(from decarboxylation of the amino acid tyrosine)

Dopamine
(3,4-dihydroxyphenylethylamine)

↓

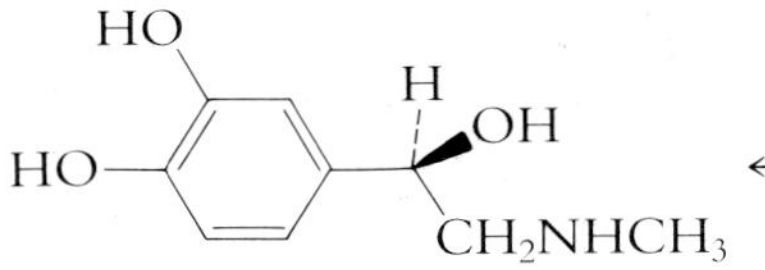

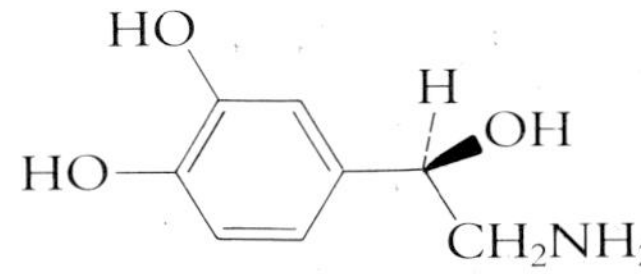

Epinephrine
(adrenaline)

←

Norepinephrine
(noradrenaline; the prefix *nor-* means "one carbon less than")

Closely related are two synthetic amines which are also active in the nervous system: amphetamine and methamphetamine. Section 21:12 discusses these substances in more detail.

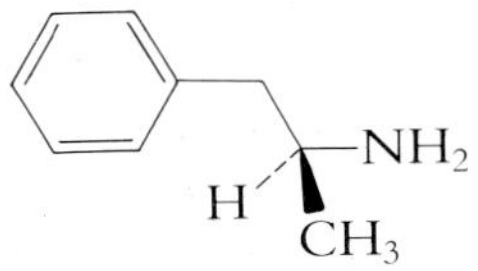

Dexedrine
(the ± form is known as amphetamine or benzedrine)

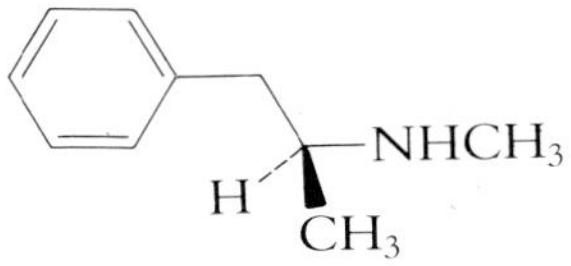

Methedrine
(the ± form is known as methamphetamine and colloquially as "speed")

Problem

21–1 Draw structures for the named compounds and write the names for the structures given. In each case, tell whether the amine is 1°, 2°, or 3°.

a. 2-ethylcyclohexylamine

b. *cis*-2-phenylcyclobutylamine

c. 2-ethoxy-*N*-methylethylamine

d. 4-bromo-*N,N*-dimethylaniline

e. $CH_3CH_2CH(NH_2)(CH_2)_2CH_3$

f. $CH_3CH(NH_2)CH_2CH_2NH_2$

g. $C_6H_5CH_2N(CH_3)_2$

h. CH_3, $NHCH_3$, CH_3CH_2, OCH_3 substituted C

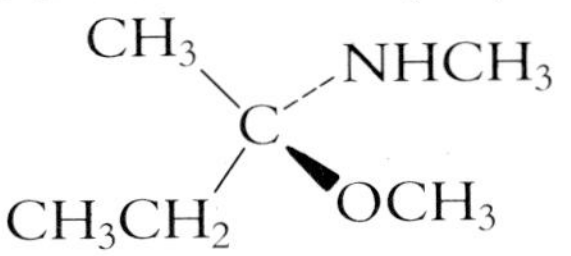

i. (naphthalene with NH_2 and OMe)

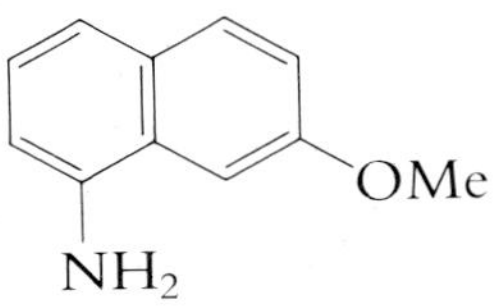

21:2 Spectroscopic Identification of Amines

Primary amines are readily identified by means of ir spectroscopy. The N—H stretching absorption is in the 3200 to 3500 cm^{-1} region and appears as a doublet because there are both symmetric and asymmetric stretching modes for the two N—H bonds. Secondary amines produce an NH band in the same region but its ir absorption is very weak. Figure 21–1 illustrates these characteristics with the ir spectra for butylamine (1°) and dibutylamine (2°).

In ^1H nmr spectra, hydrogens on carbons bonded to nitrogen (CH—N) are deshielded and appear at approximately 2.3 ppm for CH_3—N, 2.6 ppm for —CH_2—N, and 2.9 ppm for —CH—N. Many naturally occurring amines have the CH_3—N grouping. Structural variations in the rest of the molecule can have an effect that may result in a shielding or deshielding of

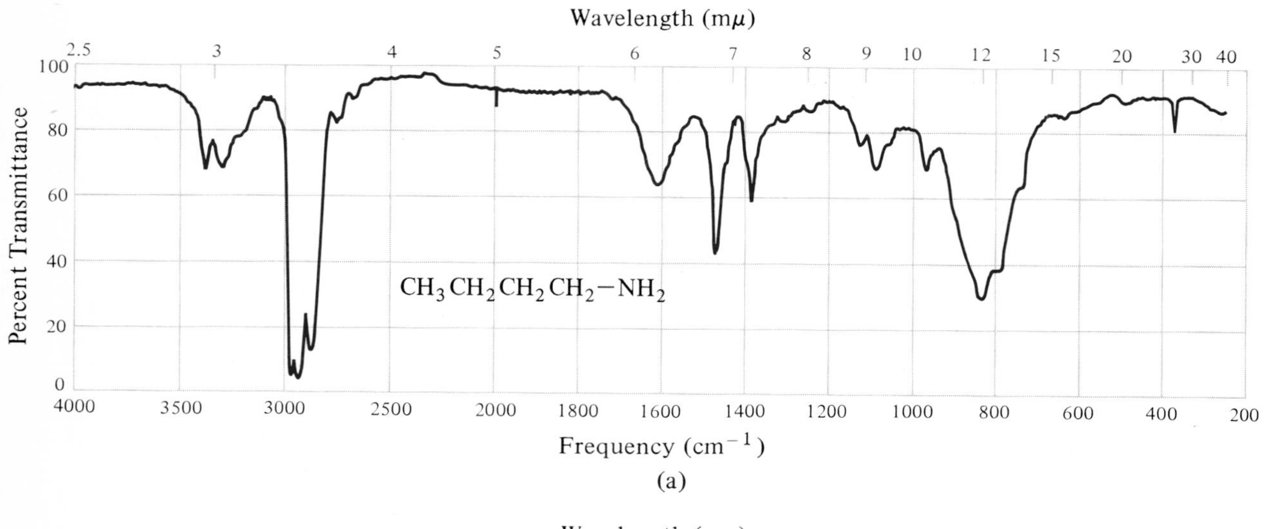

(a)

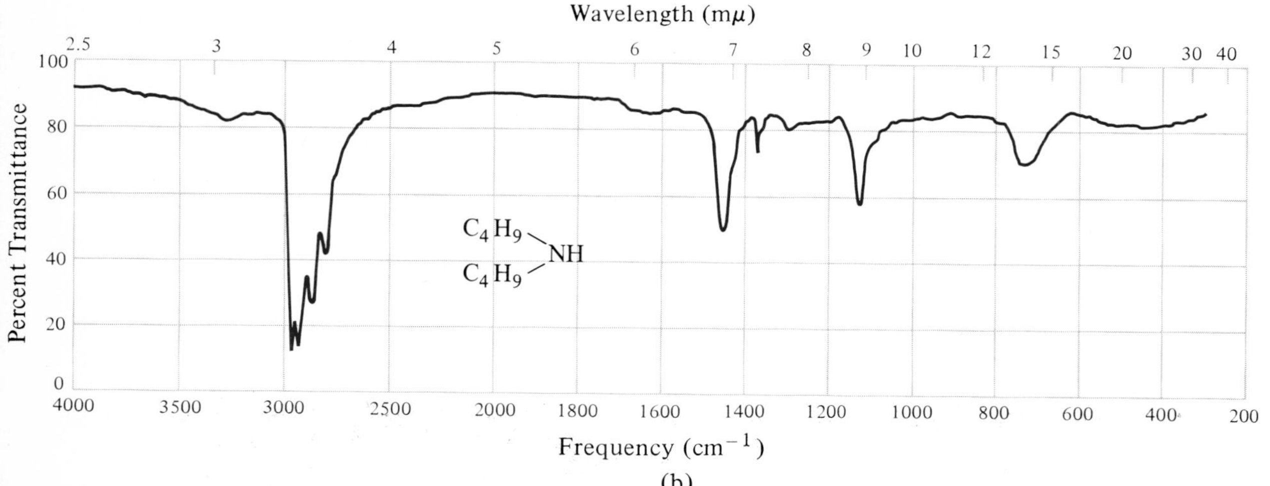

(b)

Figure 21–1. *Infrared spectra of (a) butylamine and (b) dibutylamine.*

the CH_3 absorption, causing its resonance to appear somewhat above or below 2.3 ppm. For instance, in methylammonium groups ($CH_3\overset{+}{N}R_3$), the methyl resonance is observed in the 3.9 to 4.2 ppm region. The positively-charged nitrogen is strongly electron-withdrawing and causes a large deshielding effect.

Fragmentation of amines in the mass spectrometer arises from a molecular ion that is formed when one of the nonbonded electrons on nitrogen is removed. Equation 21.1 shows the fragmentation mechanism for diethylamine.

$$(21.1)\quad CH_3CH_2\ddot{N}(H)\text{—}CH_2CH_3 \xrightarrow{-e} CH_3CH_2\dot{N}^{+}(H)\text{—}CH_2\text{—}CH_3 \rightarrow$$

m/z = 73 (30%)
Molecular ion

$$CH_3CH_2\overset{+}{N}(H){=}CH_2 + CH_3\cdot$$

m/z = 58 (100%)
Base peak

It is important to note that molecules containing an odd number of nitrogen atoms (1, 3, . . .) will have molecular ions with odd mass numbers. This assists in identifying nitrogen-containing unknowns, since compounds containing only C, H, and O have molecular ions with even mass numbers.

Problem

21–2 An unknown compound, $C_5H_{13}N$, gives the ir spectrum shown here. The mass spectrum shows a base peak at m/z = 44. Suggest a structure for the unknown, and give a mechanism for the mass spectral fragmentation.

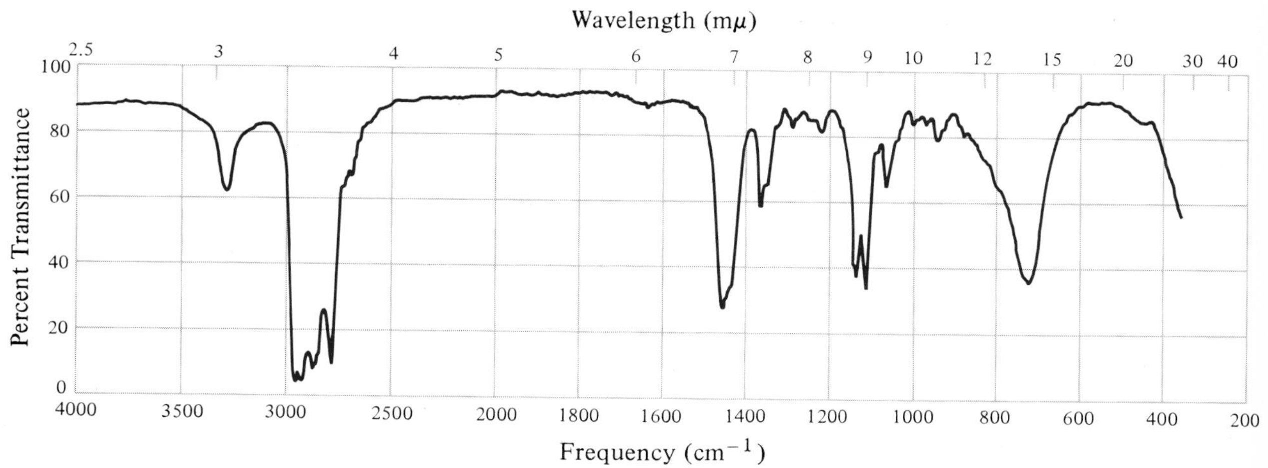

21:3 Structure of Amines: Fast and Slow Nitrogen Inversion

Ammonia is a pyramidal molecule (sec. 2:1) that undergoes very rapid inversion. A tertiary amine carrying three different alkyl groups, such as methylethylpropylamine, has a chiral nitrogen atom and can exist in two enantiomeric forms. Inversion of the nitrogen lone pair, however, converts one enantiomer into the other (eq. 21.2). Since the energy barrier for inversion is quite low, it is normally not possible to isolate the individual enantiomers.

(21.2) (Me)(Pr)(Et)N: (S) ⇌ [planar N with Pr, Me, Et] ⇌ :N(Me)(Pr)(Et) (R)

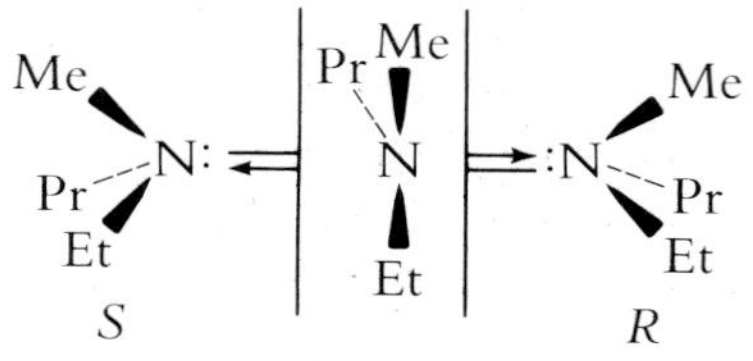

Inversion is impossible when the nitrogen lone pair is used for σ-bond formation because tetravalent nitrogen cannot invert. For instance, when methylethylpropylamine is treated with hydrochloric acid, we get a racemic mixture of two enantiomeric salts (eq. 21.3). Separation of the enantiomers

(21.3) Pr—N̈(Me)—Et (rapidly inverting) + HCl ⇌ H–N⁺(Pr)(Me)(Et) Cl^- (noninverting) + H–N⁺(Et)(Me)(Pr) Cl^- (noninverting)

Racemic mixture

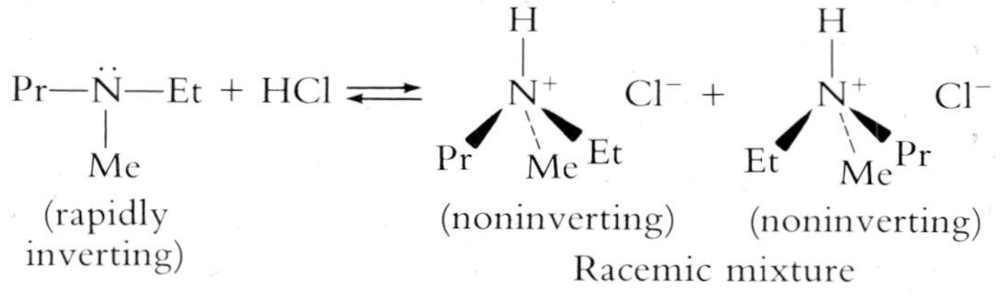

would be very difficult, however, because if the salt were dissolved in a medium in which it is able to dissociate to the tertiary amine, inversion would take place and the ion would be racemized.

The best chance for isolating pure enantiomers of a compound containing a chiral nitrogen atom is to have four nondissociating groups attached to the nitrogen atom. In fact, both the R and S forms of allylmethylphenylbenzylammonium iodide (**1**) were isolated at the turn of the century.

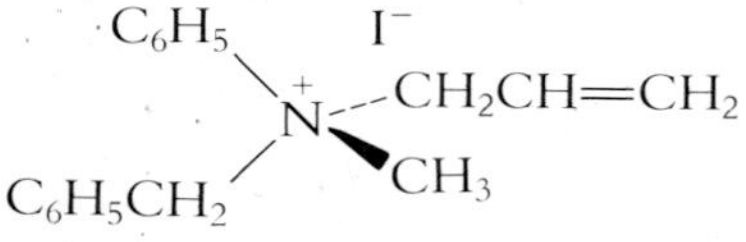

C_6H_5, $C_6H_5CH_2$, $CH_2CH{=}CH_2$, CH_3 on N^+ I^- — (R)-**1**

C_6H_5, $C_6H_5CH_2$, CH_3, $CH_2CH{=}CH_2$ on N^+ I^- — (S)-**1**

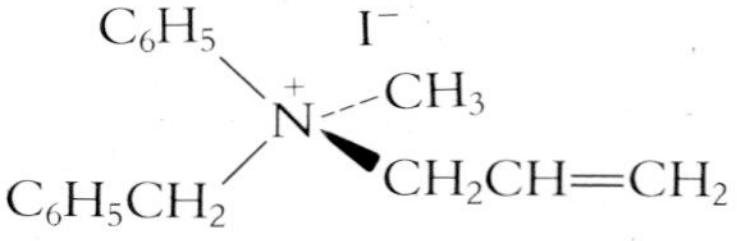

Amines are weak bases, similar to ammonia. One way to compare the basicity of different amines is to define an equilibrium constant, K_b, according to Equation 21.4. In this equation, B stands for the amine. In modern

$$\text{(21.4)} \qquad \underset{\text{Base}}{\text{B:}} + H_2\ddot{O}: \rightleftarrows \underset{\text{Conjugate acid}}{BH^+} + :\ddot{O}H^- \qquad K_b = \frac{[BH^+][OH^-]}{[B]}$$

practice, amine basicities are compared by looking at the *acidity* of the conjugate acid (BH^+) of the base. As with other acids, the pK_a notation is most often used.

Examine the equilibrium of Equation 21.5. If a base is very strong, the equilibrium will be shifted to the left and the K_a will be small. A strong base

$$\text{(21.5)} \qquad BH^+ + H_2\ddot{O}: \rightleftarrows B: + H_3O^+ \qquad K_a = \frac{[B][H_3O^+]}{[BH^+]}$$

will therefore have a small K_a and a large pK_a (since $pK_a = -\log K_a$); the opposite will be true of a weak base. By examining the pK_a's of a series of bases, we can discover how structure affects basicity (Table 21–1).

Ammonia is our reference base, and its conjugate acid has a pK_a of 9.25. *Replacing the hydrogens of ammonium ion by alkyl groups reduces its acidity.* That means that common alkylamines, such as the methyl- and ethylamines, are all stronger bases than ammonia. This is partly explained by the fact that the electron-donating alkyl groups help to stabilize the positively-charged conjugate acid (eq. 21.6).

$$\text{(21.6)} \qquad \underset{\text{Ammonium ion}}{NH_4^+} + (CH_3)_3N: \rightleftarrows \ddot{N}H_3 + \underset{\text{Trimethylammonium ion (stabilized by alkyl groups)}}{(CH_3)_3NH^+}$$

The inductive effect of the alkyl groups is important in determining basicity, but it is not the sole effect. Table 21–1 shows the trimethylamine is *less* basic than dimethylamine, and that triethylamine is *less* basic than diethylamine. These trends are the opposite of what would be expected if only the inductive effect were operating. They can be explained if we take into account the solvation changes that take place during the reaction. The alkyl groups on the nitrogen are much larger than protons, and hence they hinder the solvation of the ammonium cation. Thus the inductive and solvation effects of the alkyl groups have opposing effects on basicity, producing the reversal of amine basicities just noted.

Aryl substituents, unlike alkyl ones, strongly *depress* the basicity of amines (aniline derivatives). The major effect in these cases is the resonance interaction between the nonbonding electron pair and the aromatic ring. This interaction strongly stabilizes aniline (**2**), but it is lost when aniline is

Table 21–1. Acidities of the Conjugate Acids of Some Representative Amines

Amine	Structure	pK_a of the Conjugate Acid
piperidine	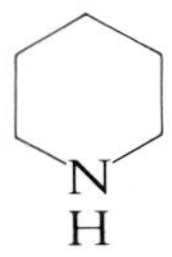 (piperidine ring, N–H)	11.1
diethylamine	$(CH_3CH_2)_2NH$	11.1
triethylamine	$(CH_3CH_2)_3N$	11.0
ethylamine	$CH_3CH_2NH_2$	10.8
dimethylamine	$(CH_3)_2NH$	10.7
methylamine	CH_3NH_2	10.7
trimethylamine	$(CH_3)_3N$	9.8
ammonia	NH_3	9.2
pyridine	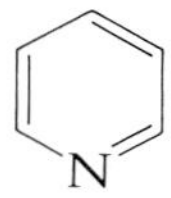 (pyridine ring, N)	5.2
N,N-dimethylaniline	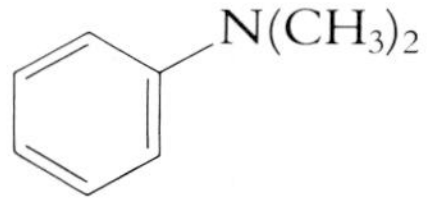 (benzene ring)–$N(CH_3)_2$	5.1
4-methylaniline	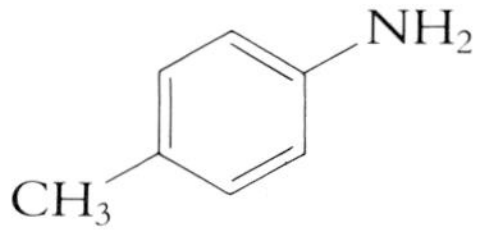 CH_3–(benzene ring)–NH_2	5.1
aniline	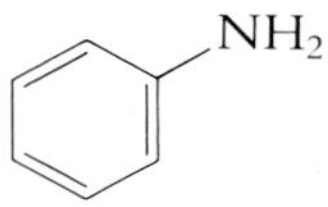 (benzene ring)–NH_2	4.6
3-nitroaniline	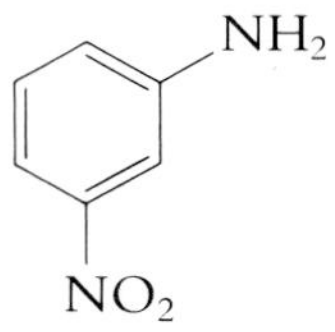 (benzene ring)–NH_2, NO_2	2.4
4-nitroaniline	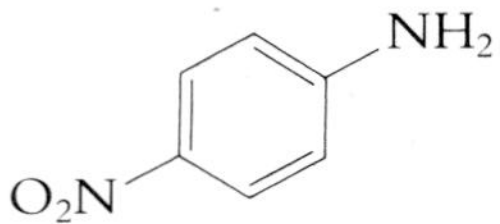 O_2N–(benzene ring)–NH_2	1.1
pyrrole	 (pyrrole ring, N–H)	0.4

Increasing Basicity ↑

protonated to give anilinium ion (**3**). The loss of resonance energy discour-

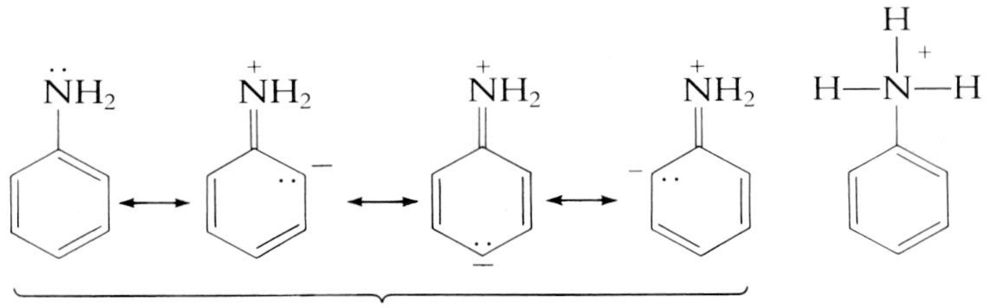

$\ddot{N}H_2$ ⟷ $\overset{+}{N}H_2$ ⟷ $\overset{+}{N}H_2$ ⟷ $\overset{+}{N}H_2$ $\qquad$ $H-\overset{+}{N}H-H$

No resonance possible involving nitrogen

2

Resonance delocalization of nitrogen lone pair

3

ages protonation of aniline, reducing its basicity. Substituted anilines behave predictably—electron-donating substituents increase basicity, while electron-withdrawing ones decrease it (4-methyl- and 4-nitroaniline, Table 21–1).

The low basicity of pyridines, compared to alkylamines, indicates that sp^2 nitrogen atoms are less basic than sp^3-hybridized ones. The reason for this is that sp^2 electrons are closer to the nucleus than sp^3 electrons and are therefore less available for bonding (sec. 12:2).

The same property that makes amines basic also makes them good acceptors for hydrogen bonds (sec. 6:9 and 6:10). When amines dissolve in water they are solvated by hydrogen-bonding water molecules (**4**). Amines

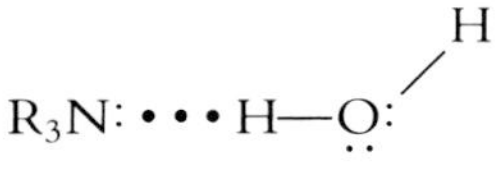

$$R_3N{:}\cdots H{-}\ddot{O}{:}{-}H$$

4

with small R groups are quite water soluble. If the amine contains N—H protons, it can also be a hydrogen-bond donor. The acceptor could be a water molecule (**5**).

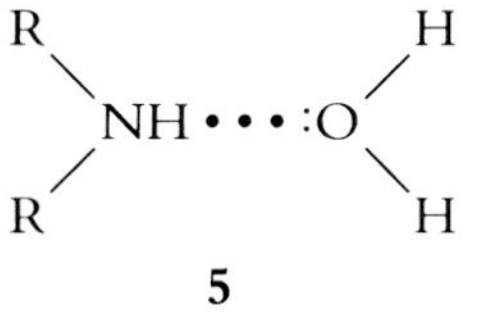

$$R_2NH\cdots{:}OH_2$$

5

As the R groups become large, these hydrogen-bonding effects are not strong enough to overcome the water insolubility of the alkyl chains. Hence, while hexylamine is water soluble, heptylamine is only slightly so.

Example

3-Nitroaniline is a base whose K_b in water is 4×10^{-12}. What is the pK_b? What is the pK_a of the conjugate acid? Aniline has a pK_a of 4.6. Which is the stronger base, aniline or 3-nitroaniline? Why?

Answer

The first thing to do in all acid–base questions is to write the equilibrium:

$$\text{3-nitroaniline } (\ddot{N}H_2) + H_2O \underset{}{\overset{K_b}{\rightleftharpoons}} \text{3-nitroanilinium } (\overset{+}{N}H_3) + {}^{-}:OH$$

If K_b is 4×10^{-12}, the pK_b equals $-\log(4 \times 10^{-12})$, or 11.6. To find the pK_a of the conjugate acid, subtract the pK_b from 14.0: $14.0 - 11.6 = 2.4$.

The stronger the base, the higher the pK_a; therefore aniline is a stronger base than 3-nitroaniline. In order to decide why, the equilibrium for aniline protonation must be compared with that for 3-nitroaniline.

$$\text{aniline } (\ddot{N}H_2) + H_2O \rightleftharpoons \text{anilinium } (\overset{+}{N}H_3) + {}^{-}:OH$$

The nitro group is electron-withdrawing. In the first equilibrium for 3-nitroaniline, a positive charge is present on nitrogen on the right side of the equilibrium. An electron-withdrawing group on a positively charged species is destabilizing. Therefore the equilibrium for 3-nitroaniline will not go as far to the right as the equilibrium for aniline, making aniline the stronger base.

Problems

21–3 The difference in acidity between acetic and benzoic acids is less than a factor of 4, while aniline is less basic than methylamine by a factor of 10^6. Why does the presence of an aromatic ring have such a different impact in the two systems? Explain with the aid of resonance structures.

21–4 2-Methoxyethylamine is less basic than ethylamine. Similarly, *m*-methoxyaniline is less basic than aniline. By contrast, *p*-methoxyaniline is more basic than aniline. Explain each of these observations by examining the effect of the substituent on both the amine and the ammonium ion derived from it by protonation.

21–5 Write structures for the products of the following reactions:

a. dimethylamine and hydrochloric acid
b. diethylisobutylamine and hydrobromic acid
c. *N*,3-diethyl-*N*-hexylaniline and dilute nitric acid
d. pyridine and sulfuric acid (1 mol)
e. *N*-isopropylpyrrolidine and sulfuric acid (0.5 mol)

f. *o*-methoxyaniline and *p*-nitrobenzoic acid
g. *N*-methyl-2-butylamine and propanoic acid

21–6 Explain why dimethylamine has a higher boiling point (6.9°) than trimethylamine (2.9°), even though trimethylamine has the higher molecular weight.

You have already seen how amines can be used in the resolution of racemic compounds. We showed schematically in Section 4:7 how a naturally occurring, enantiomerically pure amine, quinine, can be used to resolve racemic tartaric acid. Here we demonstrate the reverse—how an optically pure tartaric acid enantiomer, which is available from nature, serves as a resolving agent for a racemic amine. The amine in question, 1-phenylethylamine, has a chiral *carbon* atom, so its chirality is not affected by nitrogen inversion.

When equimolar quantities of the amine and the acid are allowed to react, they produce a mixture of diastereomeric salts. The salts are diastereomers because they are stereoisomers that are *not* mirror images of one another (eq. 21.7). The diastereomers differ in physical properties, such as

(21.7)

COOH
H—C—OH
HO—C—H
COOH
(+)-Tartaric acid

+

$PhCH\ddot{N}H_2$ (with CH_3 on the CH carbon)
Racemic 1-phenylethylamine

$\xrightarrow{\text{1. MeOH, heat; 2. cool}}$

CH_3, H, Ph, NH_3^+ · COO^- / H—C—OH / HO—C—H / COOH
(−)-1-Phenylethylammonium (+)-tartrate
(75%; precipitates from solution)

+

Ph, H, CH_3, NH_3^+ · COO^- / H—C—OH / HO—C—H / COOH
(+)-1-Phenylethylammonium (+)-tartrate
(remains in solution)

their solubility in methanol. The salt that is less soluble is isolated by filtration, and then treated with hydroxide ion to liberate the free *S* amine (eq. 21.8). The *R* enantiomer is obtained by further work-up of the original methanol solution.

$$\text{(21.8)}\quad \underset{\text{(−)-1-Phenylethylammonium (+)-tartrate}}{\text{CH}_3(\text{Ph})\text{CH–NH}_3^+ \;\; ^-\text{OOC–CH(OH)–CH(OH)–COOH}} + 2\text{NaOH} \xrightarrow{H_2O} \underset{\text{(S)-(−)-1-Phenylethylamine (55\%; extracted with ether)}}{\text{CH}_3(\text{Ph})\text{CH–}\ddot{\text{N}}\text{H}_2} + \text{Na}^+\,^-\text{OOC–CH(OH)–CH(OH)–COO}^-\text{Na}^+$$

(−)-1-Phenylethylammonium (+)-tartrate

(S)-(−)-1-Phenylethylamine (55%; extracted with ether)

21:5 Nucleophilicity and Basicity of Amines

The nonbonding electron pair on the nitrogen of an amine makes it a nucleophile as well as a base. For example, a tertiary amine can combine with a 1° or 2° alkyl halide to give a quaternary ammonium salt in high yield (eq. 21.9). This is a typical S_N2-type displacement (sec. 6:5) which is essentially

$$\text{(21.9)}\quad \underset{\text{Trimethylamine}}{\text{Me}_3\text{N:}} + \text{CH}_3\text{CH}_2\text{—Br} \rightarrow \underset{\text{Ethyltrimethylammonium bromide}}{\text{Me}_3\overset{+}{\text{N}}\text{—CH}_2\text{CH}_3\ \text{Br}^-}$$

irreversible and depends on the ability of the nitrogen to bind to a carbon atom. Trimethylamine is a good base as well (eq. 21.10), but the reaction

$$\text{(21.10)}\quad \text{Me}_3\text{N: + HA} \rightleftarrows \text{Me}_3\text{N:H}^+\ \text{A}^-$$

here (neutralization) is reversible (an equilibrium under thermodynamic control) and depends on the ability of the nitrogen to accept a proton. Thus, *although both nucleophilicity and basicity depend upon the availability of an electron pair, they are fundamentally different properties.*

A couple of examples may help to explain these ideas. Equation 21.11 shows that pyridine can be alkylated by iodomethane in a reaction that depends upon pyridine acting as a nucleophile. Equation 21.12 shows that this reaction does not occur with 2,6-dimethylpyridine. This is because the

$$\text{(21.11)}\quad \underset{\text{Pyridine}}{\text{C}_5\text{H}_5\ddot{\text{N}}} + \text{CH}_3\text{—I} \rightarrow \underset{N\text{-Methylpyridinium iodide}}{\text{C}_5\text{H}_5\overset{+}{\text{N}}\text{—CH}_3\ \text{I}^-}$$

(21.12) 2,6-Dimethylpyridine $+ CH_3I \rightarrow$ no reaction

bulky methyl groups next to the nitrogen cause steric interference in the crowded S_N2 transition state. But which is the stronger base, pyridine or 2,6-dimethylpyridine? Pyridine has a pK_a of 5.2, while the pK_a of 2,6-dimethylpyridine is 6.8. Thus, the methyl groups (electron donors) make 2,6-dimethylpyridine the stronger base even though pyridine is the better nucleophile.

You may remember from Section 7:7 that the differences between ethoxide (EtO^-) and *t*-butoxide (Me_3CO^-) were similarly explained. Ethoxide is both a base and nucleophile, while the bulky *t*-butoxide is a poor nucleophile in S_N2 reactions even though it is a stronger base than ethoxide.

Another nonnucleophilic, but very basic amine is 1,8-bis(dimethylamino)naphthalene, also known as Proton Sponge. There is steric interfer-

$Me_2N{:}$ ✕ $:NMe_2$

Proton Sponge (pK_a 12.3)

ence from the *peri*-dimethylamino groups which causes the nonbonded nitrogen pairs to no longer be conjugated with the aromatic ring (as they are in aniline). Proton Sponge is therefore a strong base that can accept a proton (which is very small), but is much too crowded around the nitrogen to allow for S_N2 attack at carbon. Thus, Proton Sponge is very effective at "soaking up" protons as they are formed in certain reactions, and helping to maintain a neutral pH.

What we have shown so far is that a nitrogen containing a lone pair, when not sterically blocked, is nucleophilic. If the nitrogen belongs to a tertiary amine, then alkylation gives a quaternary ammonium salt and no further reaction takes place (eqs. 21.9 and 21.11). Alkylation of ammonia and 1° and 2° amines, by contrast, is more complex because multiple alkylation occurs and this leads to mixtures of products. Multiple alkylation occurs because initial alkylation products are themselves nucleophilic and are subject to further alkylation. If we methylate ammonia, for instance, the methylamine produced (eqs. 21.13 and 21.14) is nucleophilic and is further alkylated to dimethylamine (eq. 21.15). The process eventually leads to the formation of the tetramethylammonium salt (eq. 21.16).

(21.13) $$\ddot{N}H_3 + CH_3I \rightarrow CH_3NH_3^+I^-$$
(1 mol) (1 mol)

(21.14) $$CH_3NH_3^+ + NH_3 \rightleftarrows CH_3\ddot{N}H_2 + NH_4^+$$
Methylamine

(21.15) $$CH_3\ddot{N}H_2 + CH_3I \rightarrow CH_3\underset{\displaystyle CH_3}{\underset{|}{N}}H_2^+I^- \xrightleftharpoons{NH_3} CH_3\ddot{N}HCH_3 + NH_4^+I^-$$
Dimethylamine

$$\text{(21.16)}\quad CH_3\ddot{N}HCH_3 + CH_3I \xrightarrow{NH_3} \underset{\text{Trimethylamine}}{CH_3\ddot{N}(CH_3)CH_3} \xrightarrow{CH_3I} \underset{\text{Tetramethylammonium iodide}}{(CH_3)_4N^+I^-}$$

The only useful reactions are therefore those in Equations 21.17 and 21.18. In Equation 21.17, if NH_3 is in excess it will have a statistically better

Box 21–1. Amines in the Nervous System

Acetylcholine, a quaternary amine synthesized and stored in nerve endings, is an essential transmitter of nerve impulses.

$$\underset{\text{Choline}}{(CH_3)_3\overset{+}{N}CH_2CH_2OH} + CoA{-}S{-}\overset{O}{\overset{\|}{C}}CH_3 \xrightarrow[\text{transferase}]{\text{choline acetyl}} \underset{\text{Acetylcholine}}{(CH_3)_3\overset{+}{N}CH_2CH_2O\overset{O}{\overset{\|}{C}}CH_3}$$

When a nerve cell receives an impulse the nerve endings release acetylcholine, which acts by altering the electrical potential across a nerve synapse (the region of contact between two nerve cells), allowing transmission of the impulse. The acetylcholine must then be quickly removed so that another impulse can be received, and hence the released acetylcholine is rapidly hydrolyzed enzymatically:

$$(CH_3)_3\overset{+}{N}CH_2CH_2O\overset{O}{\overset{\|}{C}}CH_3 \xrightarrow[H_2O]{\text{acetylcholinesterase}} (CH_3)_3\overset{+}{N}CH_2CH_2OH + CH_3\overset{O}{\overset{\|}{C}}OH$$

So-called "nerve gases," which are organophosphate esters (sec. 17:14), are acetylcholinesterase inhibitors that interfere with acetylcholine hydrolysis. Nerve impulses are thus continuous, leading eventually to paralysis and death. Several effective and relatively nontoxic insecticides, such as malathion, have been designed to inhibit nerve transmission in invertebrates (insects) but to have minimal or no effect on vertebrates (including man).

Cholinergic nerve pathways (those that use acetylcholine as their neurotransmitter) are known to be involved in Alzheimer's disease. This is a condition involving progressive, severe memory loss which can affect even relatively young people. It is associated with depressed levels of the enzymes responsible for acetylcholine synthesis and hydrolysis in a particular area of the brain.

Some forms of mental illness have been thought to involve disorders in the metabolism of neurotransmitter amines other than acetylcholine. Dopamine, norepinephrine, and their derivatives have at various times been implicated, but most hypotheses remain unproven. It *is* known, however, that in patients with Parkinson's disease, a degenerative nerve disease characterized by tremors, there is a deficiency of dopamine in the brain. The use of dopamine as a medicine for Parkinsonism was unsuccessful; it was discovered that dopamine could not cross what is known as the "blood-brain barrier." The derivative L-dopa was found to be an excellent medicine because it can cross from the bloodstream into the brain where it is decarboxylated enzymatically to dopamine.

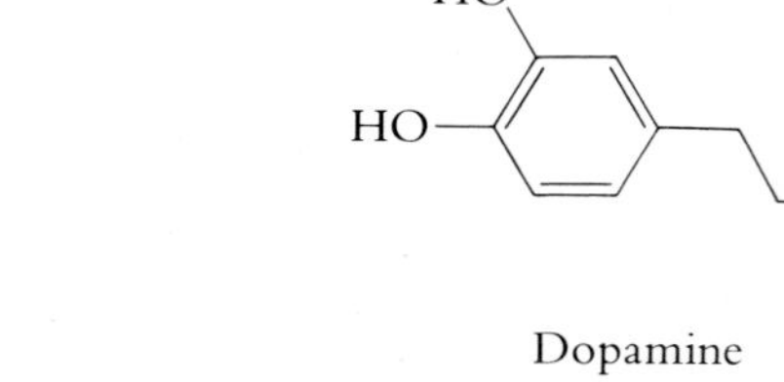

L-Dopa

Dopamine

(21.17) NH_3 (excess) + $CH_3I \rightarrow CH_3NH_2 + NH_4^+I^-$

(21.18) $NH_3 + CH_3I$ (excess) $\rightarrow (CH_3)_4N^+I^- + HI$

chance of reacting with CH_3I than the CH_3NH_2 product will. Even this is not a very good reaction with more complex alkyl halides. If there is a hydrogen on a carbon neighboring the halogen atom, then 1,2-elimination can take place, since ammonia (an amines) are bases and can cause *E*2 reactions. Alkenes are then additional by-products.

Because of these difficulties, somewhat more complex reactions have been developed to prepare amines in high yield. These are discussed in the next few sections.

Problem

21–7 Show the step-by-step mechanism for the formation of $CH_3CH_2N(CH_3)_2$ and $CH_3CH_2\overset{+}{N}(CH_3)_3I^-$ from the reaction of $CH_3CH_2NH_2$ with CH_3I.

21:6 Synthesis of Aliphatic Amines from Halides and Tosylates

As you saw in the previous section, direct alkylation of ammonia or of an amine leads to a mixture of products. This is because the amine formed (whether 1°, 2°, or 3°) can react further with the starting alkyl halide. This problem has been solved by first reacting the alkyl halide with some reagent which can, in a second step, be converted to the amine.

A typical example of this approach is to displace the halide ion from a 1° or 2° alkyl halide by azide ion, N_3^- (Table 6–2, p. 181), as shown in the first step of Equation 21.19. Azide ion is a good nucleophile, but the product alkyl azide is not nucleophilic and therefore cannot react further with the

(21.19) $CH_3CH(Br)CO_2H \xrightarrow[H_2O]{Na^+N_3^-} CH_3CH(N_3)CO_2H \xrightarrow[EtOH]{H_2,\ Pt} CH_3CH(NH_2)CO_2H$

(*R*)-2-Bromopropanoic acid → (*S*)-2-Azidopropanoic acid → (*S*)-Alanine (an amino acid)

alkyl halide. The alkyl azide can then be reduced to the amine (second step of Equation 21.19). Note that the displacement reaction is an S_N2 process and proceeds with inversion. Section 6:11 shows that the tosylate group is a good leaving group, so tosylates can be used in place of alkyl halides (eq. 21.20).

(21.20) *cis*-2-Methylcyclohexyl tosylate + NaN_3 $\xrightarrow{DMF\text{-}H_2O,\ 90°}$ (2-methylcyclohexyl azide, N_3) $\xrightarrow[2.\ H_2O]{1.\ LiAlH_4,\ Et_2O}$ *trans*-2-Methylcyclohexylamine

Reduction of the azide to the desired amine can be done with hydrogen and a catalyst (eq. 21.19) or with lithium aluminum hydride (eq. 21.20).

Remember from Chapter 6 that the OH group cannot be displaced from an alcohol directly. By first making the tosylate, however, and then going through the azide procedure, an alcohol can be converted to an amine (eq. 21.21).

(21.21) $$ROH + TsCl \rightarrow ROTs \xrightarrow{NaN_3} RN_3 \xrightarrow{\text{reduction}} RNH_2$$

Problem

21–8 Neosamine C is a diaminosugar that is a component of the antibiotic neomycin C (eq. 1). It has been synthesized from methyl *N*-acetyl-D-glucosamine, which in turn is obtained from the readily available 2-amino-2-deoxy-D-glucose (sec. 20:8). Show how the methyl glucoside can be converted to neosamine C.

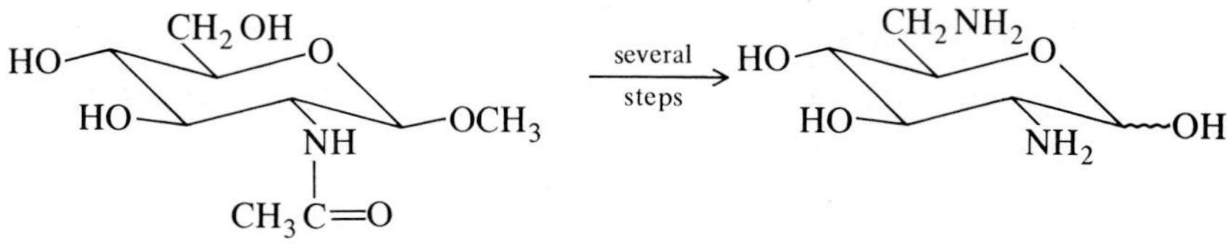

CH_2OH HO O HO OCH_3 NH $CH_3C{=}O$ $\xrightarrow[\text{steps}]{\text{several}}$ CH_2NH_2 HO O HO NH_2 OH

Methyl *N*-acetyl-D-glucosamine — Neosamine C

Related mechanistically to this azide path is the cyanide path, in which a nitrile is formed from an alkyl halide (eq. 21.22) and is then reduced to an amine. Only primary alkyl amines of the general type RCH_2NH_2 are available by this route (eqs. 21.22 and 21.23). (This reaction was previously mentioned in Chapters 16 and 17.)

(21.22) $$RCH_2Br + CN^- \rightarrow RCH_2CN \xrightarrow{\text{reduction}} RCH_2CH_2NH_2$$

(21.23) $$C_6H_5{-}CH_2CN \xrightarrow[130^\circ]{H_2,\ \text{Raney Ni}} C_6H_5{-}CH_2CH_2NH_2$$

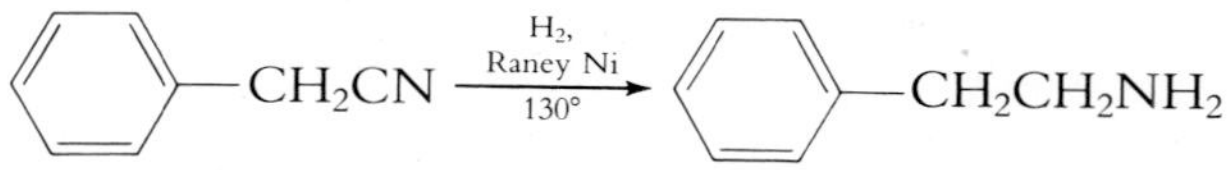

Benzyl cyanide — 2-Phenylethylamine (85%)

In this sequence the product amine has one more CH_2 group than the reactant alkyl halide.

Another way to produce primary amines is the **Gabriel synthesis**. It is similar to the azide and cyanide reactions just discussed in that it also involves a nucleophilic displacement—in this case, the nucleophile is the phthalimide anion. Phthalimide (**6**) is a weak acid ($pK_a = 8.3$) and is easily converted to its alkali metal salts (eq. 21.24). The phthalimide anion (**7**) can then be alkylated by primary and secondary alkyl halides to give *N*-alkylphthalimides (eqs. 21.25 and 21.26).

(21.24) [phthalimide] NH + KOH → [phthalimide] N^{-} K^+ + H_2O

6 7

(21.25) [phthalimide] N^{-} K^+ + $CH_3(CH_2)_3Br$ $\xrightarrow{DMF}$

7

[phthalimide] $N(CH_2)_3CH_3$ + K^+Br^-

N-Butylphthalimide

(21.26) [phthalimide] N^{-} K^+ + [benzene]$-CH_2Cl$ $\xrightarrow{DMF}$ [phthalimide] NCH_2—[benzene]

7

N-Benzylphthalimide

+ K^+Cl^-

The *N*-alkylphthalimides are nothing more than double amides, and it is necessary only to break the CO—N amide bond to recover the desired amine. Both acidic and basic hydrolyses have been used, although the conditions are often severe because the cyclic imide is quite stable. The mildest method available is to react the imide with hydrazine (eq. 21.27).

(21.27) [phthalimide] N—$CH_2CH(CH_3)_2$ + NH_2NH_2 $\xrightarrow[80°]{ethanol}$

[[benzene with C(=O)—$NHNH_2$ and C(=O)—$NHCH_2CH(CH_3)_2$]] → [phthalhydrazide, NH—NH] + $(CH_3)_2CHCH_2NH_2$

Isobutylamine

Problem

21–9 a. The synthesis of the antimalarial drug primaquine begins with the reaction of potassium phthalimide with 1,4-dibromopentane (eq. 2). Explain the selective displacement of the bromine at C-1.

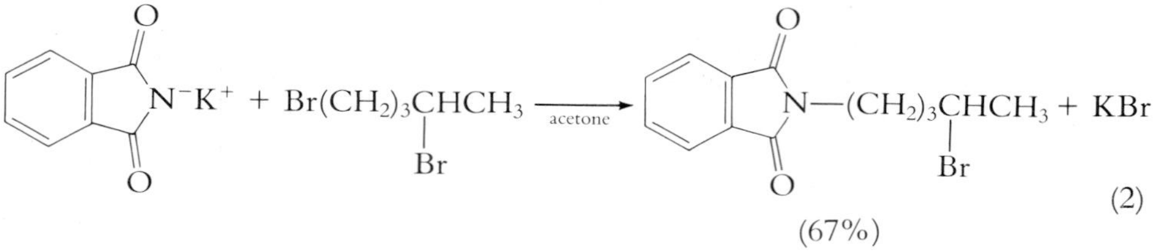

$$\text{(phthalimide)N}^-\text{K}^+ + \text{Br(CH}_2)_3\text{CH(Br)CH}_3 \xrightarrow{\text{acetone}} \text{(phthalimide)N—(CH}_2)_3\text{CH(Br)CH}_3 + \text{KBr} \quad (67\%) \qquad (2)$$

b. Show how the product of Equation 2 may be converted to primaquine (eq. 3). Comment on the order of steps in your synthesis.

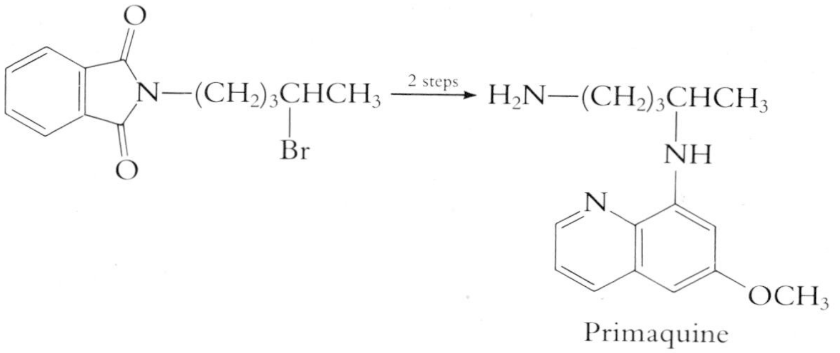

$$\text{(phthalimide)N—(CH}_2)_3\text{CH(Br)CH}_3 \xrightarrow{\text{2 steps}} \text{H}_2\text{N—(CH}_2)_3\text{CH(CH}_3\text{)—NH—(6-methoxyquinolin-8-yl)} \qquad (3)$$

Primaquine

The synthetic methods shown in this section are only useful for synthesizing primary amines that contain primary or secondary alkyl groups. In the following sections we will show you how to prepare secondary and tertiary amines and how tertiary alkyl and aryl groups may be attached to the nitrogen atom.

Problem

21–10 Show how the following conversions can be made, using the azide, nitrile, or Gabriel routes where appropriate.

a. 1-bromopentane → 1-pentylamine
b. 1-bromopentane → 1-hexylamine
c. (*R*)-2-chlorobutane → (*S*)-2-butylamine
d. *trans*-3-methylcyclopentanol → *cis*-3-methylcyclopentylamine

21:7 Aliphatic Amines from Carbonyl Compounds by Reductive Amination

Ketones and aldehydes are readily available starting materials, and two important methods are available for making the transformation from $R_2C{=}O$ to primary (R_2CHNH_2) or secondary (R_2CHNHR') amines. The carbonyl compound can be converted by hydroxylamine to an oxime in high yield (sec. 11:7). The oximes are readily reduced to primary amines by various reducing agents, including $LiAlH_4$ and catalytic hydrogenation (eq. 21.28).

(21.28)
$$\underset{\text{3-Heptanone}}{CH_3CH_2CH_2CH_2\overset{\overset{O}{\|}}{C}CH_2CH_3} + \underset{\text{Hydroxylamine}}{\ddot{N}H_2OH} \xrightarrow{H^+} \underset{\text{3-Heptanone oxime}}{CH_3(CH_2)_3\overset{\overset{\ddot{N}OH}{\|}}{C}CH_2CH_3}$$

$$\xrightarrow[\text{or } LiAlH_4]{H_2/Pt} \underset{\text{3-Heptylamine}}{CH_3(CH_2)_3\overset{\overset{\ddot{N}H_2}{|}}{C}HCH_2CH_3}$$

The second method for converting ketones to amines is through the imine. This gives secondary amines, which are not available by the methods so far discussed. The imine is obtained from the condensation of an aldehyde or ketone with a primary amine (sec. 11:7). It is then subjected to catalytic or hydride reduction (eq. 21.29).

(21.29)
$$\underset{\text{Imine}}{RR'C{=}NR''} \xrightarrow[\text{metal hydrides}]{H_2/\text{cat or}} \underset{2^\circ \text{ Amine}}{RR'CH{-}NHR''}$$

When *aliphatic* imines are the intermediates for dialkylamines, the reaction must be carried out in a special way. Aliphatic imines are unstable, difficult to isolate, and not the major species present at equilibrium. Thus, when the amine and carbonyl compound are mixed, a hydrogenation catalyst is added as well, and the reaction is then carried out in a hydrogen atmosphere. The small amount of imine initially formed is rapidly hydrogenated to the secondary amine. This causes more imine to form, and eventually a high yield of product is obtained (eq. 21.30).

(21.30)
$$\underset{\text{Carbonyl compound}}{(CH_3)_2C{=}O} + \underset{1^\circ \text{ Amine}}{H_2NCH_2CH_2OH} \overset{\text{fast}}{\rightleftharpoons}$$

$$H_2O + \underset{\text{Imine}}{(CH_3)_2C{=}NCH_2CH_2OH}$$

$$\xrightarrow[\text{EtOH}]{H_2,\ Pt} \underset{2^\circ \text{ Amine}}{(CH_3)_2CHNHCH_2CH_2OH}$$

When imines contain *aromatic* groups they are more stable and can often be isolated. Equation 21.31 shows an imine being isolated and then reduced with $NaBH_4$ to yield a 2° amine with an aryl group attached to N.

(21.31)

$$C_6H_5CHO + C_6H_5NH_2 \rightarrow C_6H_5CH{=}NC_6H_5 \xrightarrow[\text{MeOH}]{\text{NaBH}_4} C_6H_5CH_2NHC_6H_5$$

Benzaldehyde Aniline

N-Benzylaniline (97%)

In Section 17:11 you saw that amines, particularly secondary and tertiary amines, are readily synthesized when amides are reduced with lithium aluminum hydride or diborane (eq. 21.32).

(21.32)

$$C_6H_5NHCH_3 \xrightarrow[\text{OH}^-]{\text{CH}_3\text{CCl (O)}} C_6H_5N(CH_3)CCH_3(=O) \xrightarrow[\text{Et}_2\text{O}]{\text{LiAlH}_4} \xrightarrow{\text{H}_2\text{O}} C_6H_5N(CH_3)CH_2CH_3$$

N-Methyl-*N*-phenylacetamide

N-Ethyl-*N*-methylaniline (91%)

The amides are easily obtained by treating primary or secondary amines with acid chlorides or anhydrides, so amide reduction is an excellent route to secondary and tertiary amines.

Example

Show a good method for performing each of the following conversions:

a. $CH_3CH_2CH_2CH_2OH \rightarrow CH_3CH_2CH_2CH_2NH_2$
b. $CH_3CH_2CH_2CH_2OH \rightarrow CH_3CH_2CH_2CH_2NHCH_2CH_3$
c. $CH_3CH_2CH_2CH_2OH \rightarrow CH_3CH_2CH_2CH_2CH_2NH_2$

Answer

a. There is no one-step method for converting an OH group directly into an NH_2 group. The OH must first be converted to a good leaving group:

$$CH_3CH_2CH_2CH_2OH + PBr_3 \rightarrow CH_3CH_2CH_2CH_2Br$$

or

$$CH_3CH_2CH_2CH_2OH + TsCl \rightarrow CH_3CH_2CH_2CH_2OTs$$

Now either the bromide or the tosylate can be converted to the desired amine by your choice of (1) NaN_3 followed by H_2/catalyst or (2) phthalimide in base followed by heating with NH_2NH_2 (the Gabriel synthesis).

b. The product is a secondary amine which can be synthesized by first converting the OH group to the NH_2 group as you did in part (a) and then converting the amine to the *amide* or *imine*, either of which can be reduced to the desired product:

$$\underset{1^\circ \text{ Amine}}{CH_3CH_2CH_2CH_2NH_2} + CH_3COCl \rightarrow \underset{\text{Amide}}{CH_3CH_2CH_2CH_2NH\overset{\overset{O}{\|}}{C}CH_3} \xrightarrow{LiAlH_4}$$

$$\underset{2^\circ \text{ Amine}}{CH_3CH_2CH_2CH_2NHCH_2CH_3}$$

or

$$CH_3CH_2CH_2CH_2NH_2 + CH_3CHO \rightarrow \underset{\text{Imine}}{CH_3CH_2CH_2CH_2N{=}CHCH_3}$$

$$\xrightarrow{H_2/cat} CH_3CH_2CH_2CH_2NHCH_2CH_3$$

c. In this reaction note that we are starting with a *four*-carbon alcohol and are going to a *five*-carbon amine. We therefore need to add one more carbon atom to the reactant. This is typical of CN reactions. Again, the alcohol functional group must first be converted to a good leaving group, as in part (a).

$$CH_3CH_2CH_2CH_2OH \xrightarrow{PBr_3} CH_3CH_2CH_2CH_2Br \xrightarrow{CN^-} CH_3CH_2CH_2CH_2CN$$

$$\downarrow LiAlH_4$$

$$CH_3CH_2CH_2CH_2CH_2NH_2$$

Problem

21–11 a. A common industrial reaction involves mixing acetone and ammonia in a hydrogen atmosphere, in the presence of a reduction catalyst such as nickel. The major reaction products are isopropylamine and diisopropylamine. Write reactions showing how these products are formed.

b. How might the cheap amine diethylamine be synthesized commercially?

Table 21–2 summarizes the most common methods of synthesizing alkylamines. The methods are classified by whether they increase, decrease, or do not change the number of carbon atoms in the substrate. None of these methods is applicable to the preparation of aromatic amines, which we discuss in the next section.

21:7 Aliphatic Amines from Carbonyl Compounds

Table 21–2. Common Methods for Synthesizing Alkylamines

Methods that Do not Change the Number of Carbon Atoms

A. Primary Amine Syntheses

1. From 1° and 2° halides and tosylates

a. $R{-}X \xrightarrow{NH_3\ (excess)} R{-}NH_2$ (sec. 21:5)

b. $R{-}X \xrightarrow{NaN_3} R{-}N_3 \xrightarrow{H_2,\ Pt} R{-}NH_2$ (sec. 21:6)

c. $R{-}X \xrightarrow{\text{potassium phthalimide } (N^-K^+)} R{-}N(\text{phthalimide}) \xrightarrow{NH_2NH_2} RNH_2$ (sec. 21:6)

2. From carbonyl compounds

$$RR'C{=}O \xrightarrow{NH_2OH} RR'C{=}NOH \xrightarrow{\text{reducing agent}} RR'CH{-}NH_2 \quad \text{(sec. 21:7)}$$

B. Secondary and Tertiary Amine Syntheses

1. From amides

$$RC(=O){-}NR'R'' \xrightarrow{LiAlH_4} RCH_2{-}NR'R'' \quad \text{(sec. 21:7)}$$

Methods that Increase the Number of Carbon Atoms

A. Primary Amine Syntheses

1. From 1° and 2° halides and tosylates

$$R{-}X \xrightarrow{NaCN} R{-}CN \xrightarrow[\text{agent}]{\text{reducing}} R{-}CH_2{-}NH_2 \quad \text{(sec. 21:6)}$$

B. Secondary and Tertiary Amine Syntheses

1. From carbonyl compounds

$$RR'C{=}O + R''{-}NH_2 \xrightarrow{H_2,\ Ni} RR'CH{-}NH{-}R'' \quad \text{(sec. 21:7)}$$

Methods that Decrease the Number of Carbon Atoms

Primary, Secondary, and Tertiary Amine Syntheses

From carboxamides

$$R{-}CONH_2 \xrightarrow{Br_2,\ OH^-} R{-}NH_2 \quad \text{(sec. 21:9).}$$

21:8 Synthesizing Aromatic Amines from Nitroaromatic Precursors

The nitrogen-containing nucleophiles discussed thus far will not react directly with aromatic halides, except in certain restricted cases (Chap. 23). It is instead easier to attach a nitrogen atom to an aromatic ring by way of an *electrophilic* substitution. The nitration of aromatics, usually carried out with a nitric–sulfuric acid mixture, is the most common method. This approach is effective because the nitro group is readily reduced to an amino group. The typical synthetic sequence is aromatic → *nitro*aromatic → *amino*aromatic. You should remember that substituents initially present in the aromatic ring will have a directive effect on the incoming nitro group. The nitration step was discussed in Chapter 15; here we concentrate on nitro group reduction.

Catalytic reduction is often used because it entails mild reaction conditions, gives high yields, and does not disturb most other functional groups (eq. 21.33). In the presence of acid, metals such as zinc, iron, and tin may also be used to reduce the nitro group (eq. 21.34). Although the conditions

$$\text{(21.33)}\quad \text{(COOCH}_2\text{CH}_3\text{, NO}_2\text{)-benzene} + 3H_2 \xrightarrow[CH_3CH_2OH,\ 25^\circ]{Pt} \text{(COOCH}_2\text{CH}_3\text{, NH}_2\text{)-benzene}\ (95\%) + 2H_2O$$

$$\text{(21.34)}\quad \text{(CH}_3\text{, NO}_2\text{, NO}_2\text{)-benzene} + Fe \xrightarrow[\text{reflux}]{C_2H_5OH,\ HCl} \text{(CH}_3\text{, NH}_2\text{, NH}_2\text{)-benzene}\ (74\%)$$

of the reaction in Equation 21.34 are more severe than those in Equation 21.33, the latter has the advantage of not requiring special equipment. Both catalytic and chemical reduction have been used for the commercial production of aniline from nitrobenzene, which takes place on a scale approaching 1 billion pounds a year. For environmental reasons catalytic reduction is the preferred method. It is clean and the Pt catalyst can be reused. The chemical reduction method requires disposal of used acid and Fe residues.

The reduction of nitro groups by metals may be stopped at intermediate valence states of nitrogen, depending on the reagent and the pH of the medium (eq. 21.35). The most important of these reactions is the complete reduction to aniline.

$$\text{(21.35)}\quad C_6H_5NO_2 \rightarrow C_6H_5N{=}O \rightarrow C_6H_5NHOH \rightarrow C_6H_5NH_2$$

Nitrosobenzene N-Phenylhydroxylamine

Once present in an aromatic ring, the amino group can be used as a powerful activator and *o,p*-director for subsequent electrophilic substitutions. There are, however, a couple of problems. First, the acidic conditions necessary for these reactions convert the *o,p*-directing $—\ddot{N}H_2$ group to the *m*-directing $—\overset{+}{N}H_3$ group (eq. 21.36). Secnd, the high activating power of the amino group leads to polysubstitution, and sometimes oxidation, of the aromatic ring.

$$\text{(21.36)}\quad C_6H_5\ddot{N}H_2 + H^+ \rightleftarrows C_6H_5NH_3^+$$

o,p-Attack activated *m*-Attack deactivated

To avoid these problems the amino group is first acetylated with acetic anhydride (eq. 21.37), which reduces its basicity and activating ability. The resulting acetamido group is still *o,p*-directing, and the acetyl group is easily removed after the substitution has been carried out (eq. 21.38).

$$\text{(21.37)}\quad C_6H_5NH_2 \xrightarrow{(CH_3CO)_2O} C_6H_5NHCOCH_3$$

N-Phenylacetamide (*acetanilide*)

$$\text{(21.38)}\quad C_6H_5NHCOCH_3 \xrightarrow{Br_2} p\text{-}BrC_6H_4NHCOCH_3 \xrightarrow[2.\ OH^-]{1.\ H_3O^+} p\text{-}BrC_6H_4NH_2$$

Major product

p-Bromoaniline

If nitro group reduction is delayed until after the substitution reaction, a different isomer of the final product is obtained (eq. 21.39).

$$\text{(21.39)}\quad C_6H_5NO_2 \xrightarrow[FeBr_3]{Br_2} m\text{-}BrC_6H_4NO_2 \xrightarrow[Pt]{H_2} m\text{-}BrC_6H_4NH_2$$

m-Bromoaniline

Example

Start with benzene and prepare 1,3-diaminobenzene.

Answer

We have no reaction that will introduce an amino group directly into the benzene ring and hence most aminobenzenes are produced by reduction of the nitro compound:

$$\text{1,3-}C_6H_4(NO_2)_2 \xrightarrow{H_2/Pt} \text{1,3-}C_6H_4(NH_2)_2$$

We therefore need to prepare 1,3-dinitrobenzene. Luckily, the nitro group is a *meta*-director and hence nitration of benzene will work:

$$C_6H_6 \xrightarrow[\text{100°, 15 min}]{HNO_3,\ H_2SO_4} [C_6H_5NO_2] \rightarrow \text{1,3-}C_6H_4(NO_2)_2\ (75\%)$$

Problems

21–12 Show how each of the following compounds can be synthesized starting with a hydrocarbon.

a. *o*-chloroaniline
b. *m*-aminobenzenesulfonic acid
c. 4-nitro-1-naphthylamine
d. *p*-aminobenzoic acid
e. 3-bromo-*N*-isobutylaniline

21–13 Use resonance structures to help explain why acetanilide (eq. 21.35) is less basic and less activated than aniline.

21–14 *N*-(3,4-Dichlorophenyl)acetamide (*propanil*) is a widely used weed killer (herbicide). Work out a synthesis of propanil, starting from benzene.

$$\text{3,4-}Cl_2C_6H_3\text{–}NHCCH_2CH_3 \ (C{=}O)$$

Propanil

21:9 Hofmann Rearrangement of Amides to Amines

You've already seen that amines can be synthesized from amides by a reduction reaction that does not change the number of carbon atoms (sec. 21:7). It is also possible to chop one carbon atom out of the chain if one starts with an amide. This technique, in which the amide is treated with an alkaline sodium hypobromite (or hypochlorite) solution, is known as the **Hofmann rearrangement** (eqs. 21.40 and 21.41).

(21.40) $CH_3(CH_2)_4CONH_2 + OBr^- + OH^- \rightarrow$
hexanamide

$CH_3(CH_2)_4NH_2 + Br^- + HCO_3^-$
Pentylamine
(88%)

(21.41) $o\text{-}C_6H_4(CO_2^-Na^+)(CONH_2) \xrightarrow[\text{heat}]{OBr^-,\ OH^-} o\text{-}C_6H_4(CO_2^-Na^+)(NH_2)$
(85%)

The reaction is synthetically useful for preparing both alkyl and aryl amines. It has also been used for structure determination, in conjunction with other degradative reactions.

The steps in the Hofmann rearrangement include bromination of the amide on nitrogen, and then expulsion of the bromide ion, accompanied by a migration of the organic group attached to the carbonyl carbon atom (eqs. 21.42 and 21.43). The product of the first reaction stage is an organic *isocyanate* (**8**).

(21.42) $RC(=O)\ddot{N}H_2 + :\ddot{O}Br^- \rightleftarrows RC(=O)\ddot{N}H^- + Br\ddot{O}H \rightleftarrows RC(=O)\ddot{N}HBr + :\ddot{O}H^-$

(21.43) $RC(=O)\ddot{N}HBr + :\ddot{O}H^- \rightleftarrows \left[RC(=\ddot{O}:)\text{—}\bar{\ddot{N}}\text{—}Br \leftrightarrow (:\ddot{O}:^-)(R)C\text{=}\ddot{N}\text{—}Br \right] + H_2\ddot{O}: \xrightarrow{\text{slow}}$

$R\text{—}\ddot{N}\text{=}C\text{=}O + Br^-$
8
An isocyanate

In aqueous base, the isocyanate is very quickly attacked by water at its carbonyl group. The intermediate produced is the transitory and unstable carbamic acid, **9**, which decomposes and liberates the amine (eq. 21.44).

(21.44) $\underset{\mathbf{8}}{R\text{—}N\text{=}C\text{=}O} \xrightarrow{H_2O} \left[\underset{\mathbf{9}}{RNH\text{—}C(=O)OH} \right] \rightarrow RNH_2 + CO_2$

A carbamic acid

One virtue of the Hofmann rearrangement is that it does not depend on the nature of the R group. Therefore, it is used to prepare aryl amines (eq. 21.41) not easily available by the nitration route (sec. 21:8). The Hofmann rearrangement is also used to synthesize *t*-alkyl amines, which cannot be made by nucleophilic substitution (secs. 21:5 and 21:6) or by reductive amination (sec. 21:7). When the R group is tertiary, the intermediate isocyanate hydrolyzes very slowly in basic solution, so it is isolated and hydrolyzed in acid (eq. 21.45).

$$(CH_3CH_2CH_2)_3C\overset{O}{\overset{\|}{C}}NH_2 \xrightarrow{Br_2,\ OH^-} \quad (21.45)$$

2,2-Dipropylpentanamide

$$(CH_3CH_2CH_2)_3C{-}N{=}C{=}O \xrightarrow[heat]{H_3O^+} \xrightarrow{OH^-} (CH_3CH_2CH_2)_3C{-}NH_2$$

(90%) 4-Propyl-4-heptylamine (70%)

Problems

21–15 When the mixed acid amide *cis*-**I** is subjected to Hofmann rearrangement, it gives an amino acid **II** of unknown configuration. When **II** is heated it readily loses water to give **A** (eq. 4), which is not acidic.

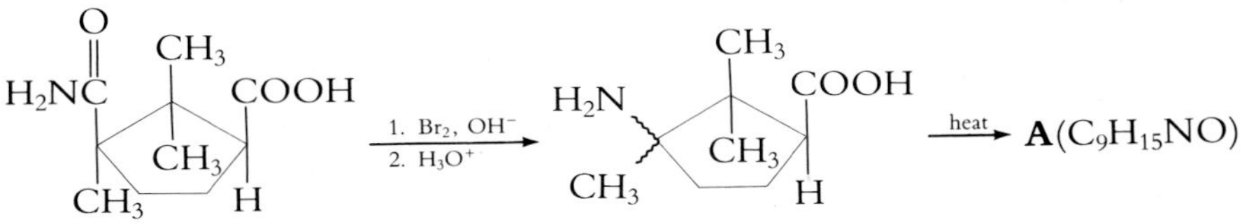

$$cis\text{-}\mathbf{I} \xrightarrow[2.\ H_3O^+]{1.\ Br_2,\ OH^-} \mathbf{II} \xrightarrow{heat} \mathbf{A}(C_9H_{15}NO) \quad (4)$$

(*cis*-I: cyclopentane ring bearing $H_2NC(=O)$ and CH_3 on one carbon, CH_3 on the adjacent carbon, CH_3 and COOH on the next carbon, H on a ring carbon; II: H_2N and CH_3 on one carbon, CH_3, COOH and H correspondingly)

Deduce a structure for **A** and from it, the stereochemistry of **II**. What does this sequence tell us about the stereochemistry of the Hofmann rearrangement? Is that stereochemistry compatible with the mechanism given in Equations 21.42 to 21.45?

21–16 1-Bromo-1-methylcyclopentane cannot be converted to 1-methylcyclopentylamine in one step. Why not? Suggest a multistep route that can accomplish this transformation.

21–17 Show how racemic 2-methylbutanoic acid could be converted to (*R*)-2-butylamine, if you had a sample of optically pure (−)-phenylethylamine (eq. 21.8) available. A review of Problem 21–15 and Section 21:4 may be helpful.

21:10 Tetraalkylammonium Ions as Phase-Transfer Catalysts

Tetraalkylammonium salts, $R_4N^+X^-$, possess many of the properties associated with inorganic salts, but there are some important and useful differences. Ammonium salts have much lower melting points than the corresponding alkali metal salts (Table 21–3). Some alkylammonium salts are even liquids at room temperature. The melting point data indicate that the

Table 21–3. Melting Points of Some Alkali Halides and Ammonium Halides

Salt	Melting Point (°C)	Cation Radius (Å)
Na^+Cl^-	801	1.17
K^+Br^-	730	1.33
Cs^+I^-	621	1.82
$NH_4^+Cl^-$	340 (subl)	1.63
$(CH_3)_4N^+Cl^-$	320 (decomp)	
$(C_3H_7)_4N^+Br^-$	252	4.35
$(C_4H_9)N^+I^-$	148	

ammonium salts have relatively low lattice energies compared to other salts. For example, it takes much less energy to separate $(CH_3)_4N^+$ and Cl^- ions than it does to separate Na^+ and Cl^- ions. The large radius of the ammonium cation results in a large lattice separation between the positively-charged nitrogen atom and the halide ion (Table 21–3). This leads to a reduced attraction between the ammonium cation and the negative counterion.

Another important consequence of the small lattice energy of tetraalkylammonium salts is their high solubility, compared to inorganic salts, in nonpolar organic solvents. The solubility in nonpolar solvents increases as the size of the organic cation increases, while the aqueous solubility decreases with increasing size. A number of the alkylammonium salts of intermediate size are both water soluble (hydrophilic) and soluble in nonpolar media (lipophilic). Such salts are called **amphiphilic,** and their property of solubility in both aqueous and organic phases allows them to be employed in a useful synthetic method known as **phase-transfer catalysis.**

As an example, consider the reaction shown in Equation 21.46. It consists of an aqueous phase containing NaOH, and an organic phase containing

$$CH_3(CH_2)_3Cl + CH_3(CH_2)_7OH \xrightarrow[\substack{Bu_4N^+HSO_4^- \\ 65°,\ 4\ hr}]{50\%\ aq\ NaOH} CH_3(CH_2)_3O(CH_2)_7CH_3 \quad (21.46)$$

(immiscible with water) (partly dissolved in water) (90%)

1-chlorobutane and 1-octanol. Since 1-octanol is slightly water soluble, it crosses into the aqueous phase and an equilibrium is set up with octylate ion (eq. 21.47). We want this alkoxide ion to react with the chlorobutane to give

$$CH_3(CH_2)_7\ddot{O}H + :\ddot{O}H^- \underset{phase}{\overset{aq}{\rightleftharpoons}} \underset{\text{Octylate ion}}{CH_3(CH_2)_7\ddot{O}:^-} + H_2\ddot{O}: \quad (21.47)$$

the product octyl butyl ether. The problem is that the octylate ion is very soluble in water, where it is well solvated (by hydrogen-bonding) but not very soluble in the organic phase. Addition of a quaternary ammonium salt such as $Bu_4N^+HSO_4^-$ solves the problem. Its cation, Bu_4N^+, is partly soluble in both phases, and thus pulls the octylate ion into the organic phase where it rapidly attacks the chlorobutane (eq. 21.48). The octylate ion is es-

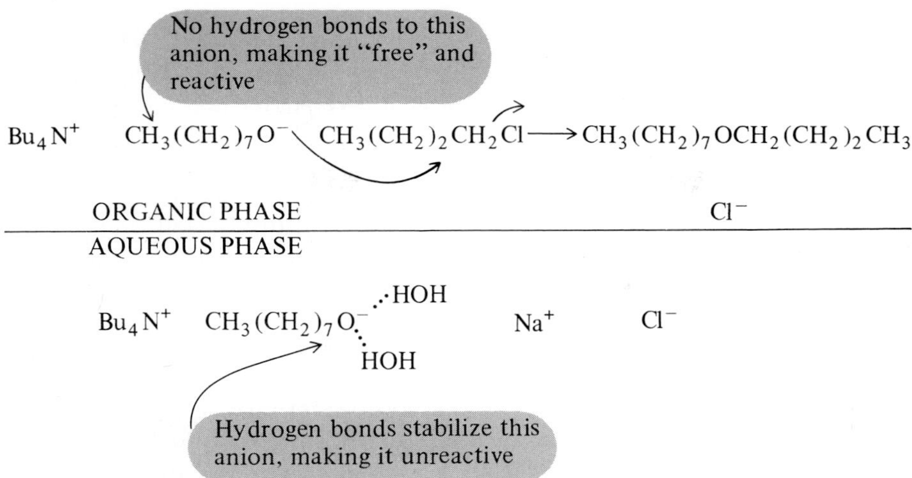

Figure 21–2. *Phase-transfer catalysis of a nucleophilic displacement reaction.*

pecially reactive in the organic phase because it is only poorly solvated there (Fig. 21–2).

$$\text{(21.48)}\quad CH_3(CH_2)_7\ddot{\underset{..}{O}}\!:^-\ Bu_4N^+ + \underset{\substack{\text{(solvent and}\\ \text{substrate)}}}{CH_3(CH_2)_2CH_2{-}\ddot{\underset{..}{Cl}}\!:} \xrightarrow[\text{phase}]{\text{org}} \underset{\text{Octyl butyl ether}}{CH_3(CH_2)_7\ddot{\underset{..}{O}}(CH_2)_3CH_3} + Bu_4N^+{:}\ddot{\underset{..}{Cl}}\!:^-$$

Another example of phase-transfer catalysis is shown in Equation 21.49. Here, the catalyst is a phosphonium salt, $R_4P^+X^-$, which is very similar to the corresponding ammonium salt.

$$\text{(21.49)}\quad CH_3(CH_2)_7Cl + NaCN\ \text{(aq sol)} \xrightarrow[105°,\ 2\ \text{hr}]{(C_{16}H_{33})\overset{+}{P}(C_4H_9)_3Br^-} \underset{(99\%)}{CH_3(CH_2)_7CN}$$

The reactions shown in Equations 21.46 and 21.49 proceed to only a negligible extent in the absence of the organic salts. This is testimony to the great power of phase-transfer catalysis.

Phase-transfer catalysis is of value both in the laboratory and on an industrial scale. The general phenomenon is of interest beyond the bounds of pure chemistry, however. Cells contain a number of amphiphilic substances, and a substantial fraction of them function as carriers for other molecules. Organic nutrients, drugs, and other biologically important molecules must cross cell walls to enter or leave these cells. The cell cytoplasm and most extracellular fluids are aqueous solutions, while the external and internal cell membranes contain a high proportion of lipids. Therefore, a good biological carrier substance must be both hydrophilic and lipophilic, as are phase-transfer catalysts.

21:11 Drug Transport and Physiologically Active Amines

In order for a medicine to be effective, it must be reactive at some site in the body and it must also be able to reach that site. Most drugs are at least partially amphiphilic—they contain both a lipophilic portion and a hydrophilic portion. The hydrophilic portion allows for some water solubility, and hence the drug can be carried by the blood or other body liquids. The molecule must also be able to pass through a cell wall to reach a site of action. For this reason a lipophilic (nonpolar) portion must be built into the drug to carry it through the nonpolar cell wall. A benzene or other aromatic ring is often used for the nonpolar part of a drug molecule.

These same sorts of ideas also apply to the chemical reaction that occurs at the site where the medicine is effective. Here it is important to correlate chemical structure with physiological activity. One of the most intriguing findings has been that compounds in which an aromatic ring is separated from an amine group by two or three carbon atoms often show potent biological activity. Figure 21–3 shows this relationship for several active compounds.

It is thought that these, and many other compounds, must fit into what is known as a **receptor site** in a biological macromolecule in order to exert their physiological activity. This is shown in Figure 21–4. One portion of

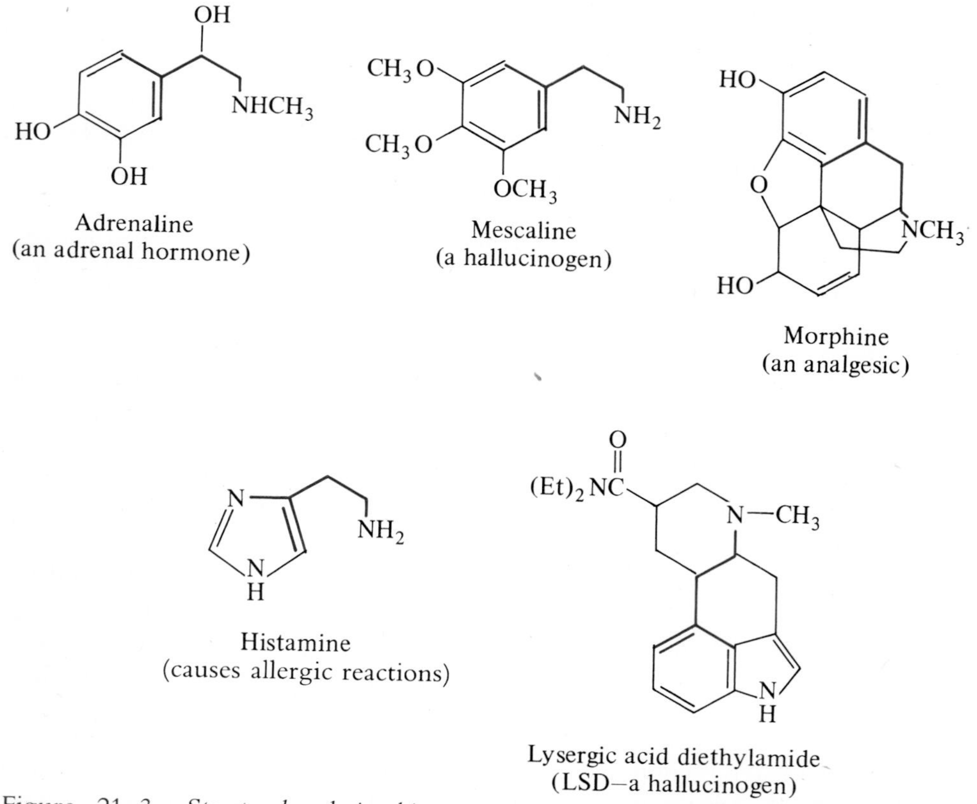

OH HO OH $NHCH_3$

Adrenaline
(an adrenal hormone)

CH_3O CH_3O OCH_3 NH_2

Mescaline
(a hallucinogen)

HO O HO NCH_3

Morphine
(an analgesic)

N N H NH_2

Histamine
(causes allergic reactions)

O $(Et)_2NC$ $N{-}CH_3$ N H

Lysergic acid diethylamide
(LSD—a hallucinogen)

Figure 21–3. *Structural relationships among some physiologically active amines. (In each, a nitrogen atom is two carbon atoms away from an aromatic ring.)*

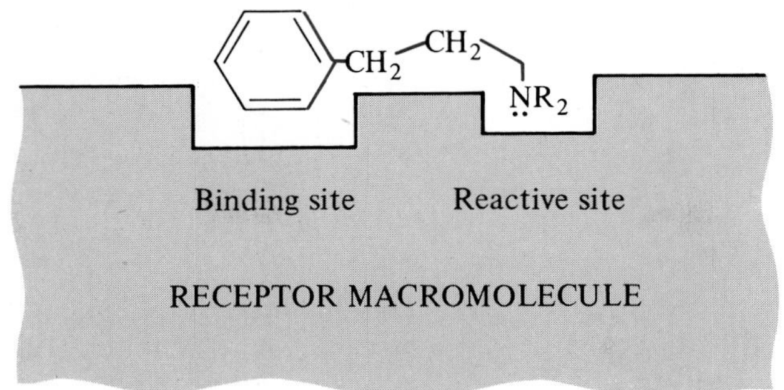

Figure 21–4. *Schematic diagram of a structural analogy for a biological receptor.*

the molecule (presumably the aromatic ring) is involved with simple binding to the macromolecule, while the second portion (the amine) participates in the chemical reaction that is the basis of its physiological activity. One can think of the receptor as a keyhole and the drug (molecule) as being the key. If the lock and key fit properly, then a chemical process occurs that gives the drug its activity.

21:12 Bonding in Sulfur and Phosphorus Compounds: Exceeding the Octet Rule

We turn now from nitrogen chemistry to the chemistry of two of its neighbors, phosphorus and sulfur. Many similarities will be seen in the chemical behavior of these three elements, but there are significant differences as well. In order to appreciate those differences, we have to understand the bonding in sulfur and phosphorus compounds.

Recall from Section 2:2 that second-period elements (Li to Ne) can accommodate a maximum of four electron pairs in their valence shells. This is because these shells have only four orbitals available for bonding (one *s* and three *p* orbitals). Third-period elements (Na to Ar, including P and S) have nine atomic orbitals in their valence shells (one *s*, three *p*, and five *d* orbitals). Although many sulfur and phosphorus compounds do obey the octet rule, they need not because they can theoretically accommodate up to 18 electrons in their valence shells. However, a completely full valence shell (18 e^-) is rarely observed because of repulsions between electrons and between substituent groups.

Hydrogen sulfide, phosphine, and phosphorus trichloride are examples of compounds that obey the octet rule.

H—S̈—H (with two lone pairs on S) — Hydrogen sulfide

PH_3 (P with one lone pair, bonded to three H) — Phosphine

PCl_3 (P with one lone pair, bonded to three Cl) — Phosphorus trichloride

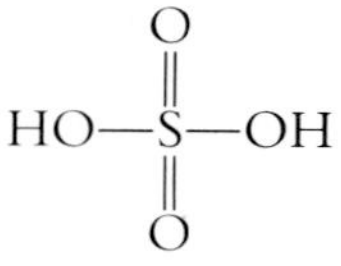

$$HO{-}\overset{\overset{O}{\|}}{\underset{\underset{O}{\|}}{S}}{-}OH$$

Sulfuric acid

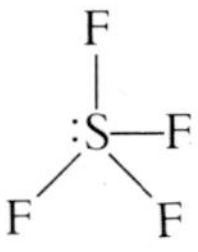

SF_4

Sulfur tetrafluoride

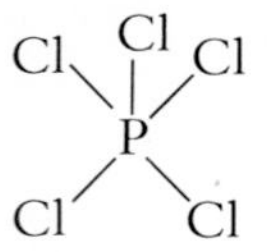

PCl_5

Phosphorus pentachloride

Some representative examples of sulfur and phosphorus compounds that exceed the octet rule are sulfuric acid, sulfur tetrafluoride, and phosphorus pentachloride. The ability of sulfur and phosphorus to increase and decrease the number of their valence shell electrons has a very important influence on the chemistry of their compounds.

Sulfur and phosphorus atoms are larger than their second-row analogs, oxygen and nitrogen. This factor has two major effects: (1) it enables S and P to tolerate charge better, because the charge is spread out within a larger volume; and (2) the valence electrons of S and P are relatively far from the nucleus, making them highly polarizable (sec. 2:7) and therefore easily attacked by electrophiles. Both of these features also bear on their chemistry.

21:13 Phosphines and Phosphonium Salts

The phosphorus analog of ammonia is phosphine, PH_3. Unlike ammonia, phosphine is extremely flammable (spontaneously in air) and not at all basic. The lack of basicity is due to the poor overlap between the larger valence orbitals of phosphorus and the small 1*s* valence orbital of the hydrogen ion, so PH_4^+ is less stable than NH_4^+.

Because phosphine is toxic and flammable, and a gas at room temperature, it is not a common laboratory chemical. Instead, tertiary phosphines, in which the hydrogens have been replaced by organic groups, are widely used. Like phosphine itself, tertiary phosphines are not bases, but they are very reactive nucleophiles (eqs. 21.50 and 21.51). The phosphonium salts produced by alkylation of tertiary phosphines resemble quaternary ammonium salts in their physical properties (sec. 21:10).

(21.50) $$(CH_3)_3P\colon + \colon\ddot{I}{-}CH_2CH_3 \xrightarrow{Et_2O} (CH_3)_3\overset{+}{P}CH_2CH_3 \; \colon\ddot{I}\colon^-$$

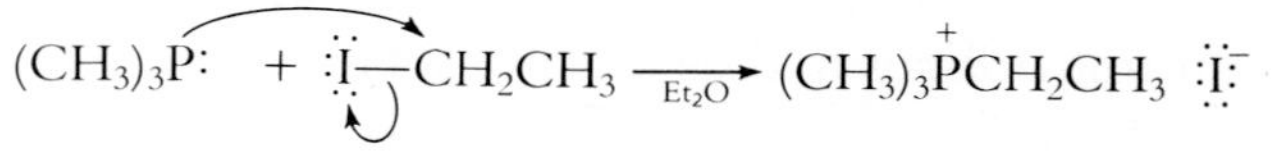

Trimethylphosphine — Ethyltrimethylphosphonium iodide

(21.51) $$(C_6H)_3P\colon + \colon\ddot{Cl}{-}CH_2CN \xrightarrow{C_6H_6} (C_6H_5)_3\overset{+}{P}CH_2CN \; \colon\ddot{Cl}\colon^-$$

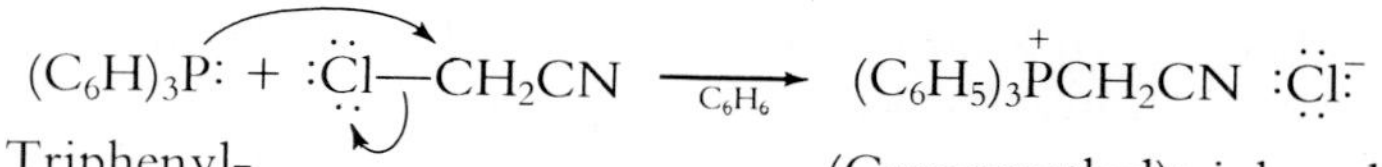

Triphenylphosphine — (Cyanomethyl)triphenylphosphonium chloride

We noted above that phosphine is spontaneously flammable in air. The same is true of the low molecular weight trialkylphosphines. They burn readily with the release of a good deal of heat (eq. 21.52). The exothermicity of this reaction is due to the high P=O bond strength, 130 to 150 kcal/mol.

$$(21.52) \quad (CH_3)_3P\!: + \tfrac{1}{2}O_2 \rightarrow [(CH_3)_3\overset{+}{P}—\ddot{\underset{..}{O}}\!:^- \leftrightarrow (CH_3)_3P{=}\ddot{O}\!:]$$

Trimethylphosphine oxide

$$\Delta H = -80 \text{ kcal/mol}$$

As indicated by the resonance structures (eq. 21.52), the partial double bond character of the PO bond is made possible because the phosphorus valence shell can accommodate 10 electrons. This tendency toward valence shell expansion, and the high affinity of phosphorus for oxygen, form the basis of the chemistry described in the next section. The starting phosphines for this chemistry are all triaryl derivatives, since they are stable in air.

21:14 The Wittig Reaction: Alkenes from Phosphorus Ylids

A remarkably useful reaction of phosphonium salts was discovered by the German chemist Georg Wittig (Nobel Prize winner, 1979) in 1953. He found that a phosphonium salt containing an alkyl group would lose a proton to a strong base (RLi, $NaNH_2$) to give a highly colored, oxygen- and water-sensitive product (eq. 21.53). This substance belongs to a class called **ylids**, which are neutral compounds containing positively- and negatively-

$$(21.53) \quad (C_6H_5)_3P\!: + CH_3Br \xrightarrow[C_6H_6]{} (C_6H_5)_3\overset{+}{P}—CH_3\ Br^- \xrightarrow[Et_2O,\ N_2]{C_6H_5Li}$$

$$[(C_6H_5)_3\overset{+}{P}—\overset{-}{\ddot{C}}H_2 \leftrightarrow (C_6H_5)_3P{=}CH_2] + LiBr + C_6H_6$$

Methylenetriphenylphosphorane
(an *ylid*)

charged atoms directly bonded to one another. The phosphorus ylids are named as derivatives of the pentavalent phosphorus hydride phosphorane, PH_5 (which has never been isolated). The stability of the phosphorus ylids is partly due to the ability of phosphorus to accommodate 10 electrons in its valence shell (eq. 21.53).

In an attempt to prepare a derivative of the ylid, Wittig allowed it to react with benzophenone. To his surprise, he obtained 1,1-diphenylethylene and triphenylphosphine oxide (eq. 21.54).

$$(21.54) \quad (C_6H_5)_2C{=}O + (C_6H_5)_3P{=}CH_2 \xrightarrow[ether]{} (C_6H_5)_2C{=}CH_2 + (C_6H_5)_3P{=}O$$

Benzophene — (84%)

Although this result was unexpected, Wittig was quick to see its significance for alkene synthesis. Thus, when the ylid, or *Wittig reagent,* was treated with cyclohexanone, it yielded methylenecyclohexane as the *sole* alkene (eq. 21.55).

(21.55) cyclohexanone (O) $+ (C_6H_5)_3P{=}CH_2 \xrightarrow[ether]{}$ methylenecyclohexane (CH_2) $+ (C_6H_5)_3PO$

(the only alkene, 40%)

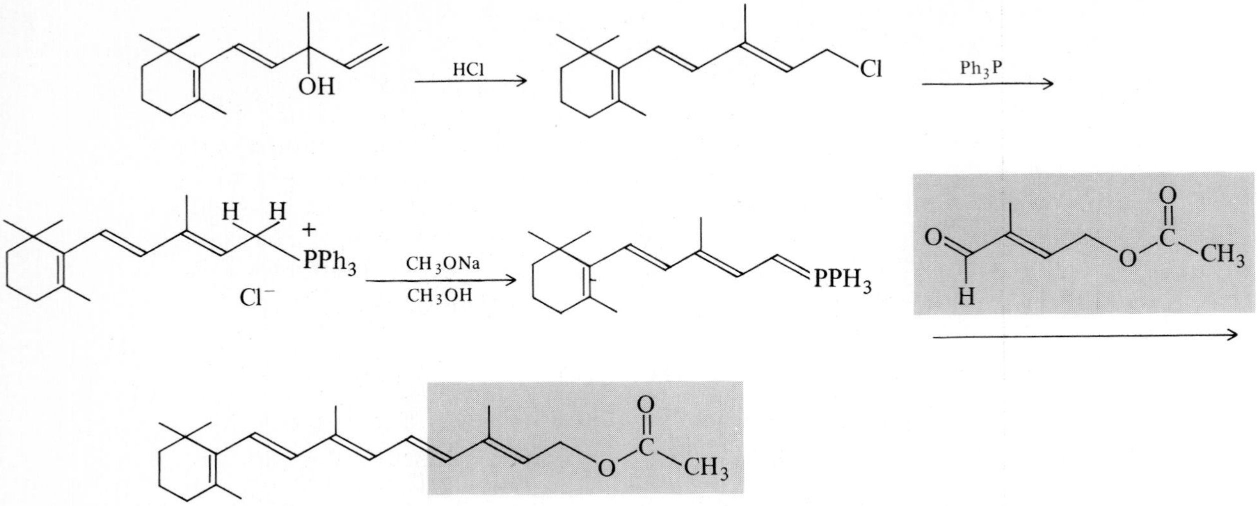

OH HCl Cl Ph_3P

H H + PPh_3 Cl^- CH_3ONa / CH_3OH PPH_3 O H O C CH_3 O

O C CH_3 O

Vitamin A acetate

Figure 21–5. *An industrial synthesis of vitamin A, based on the Wittig reaction.*

The value of the Wittig reaction, therefore, lies in its ability to produce pure alkenes *with known double bond positions*. Both the *E* and *Z* isomers can form, but under proper conditions the *E*/*Z* ratio can be controlled. Aldehydes as well as ketones can be used as starting materials. Figure 21–5 shows an industrial synthesis of vitamin A built around a Wittig reaction with an aldehyde.

Wittig assumed that the mechanism of his olefin synthesis began with an attack by the nucleophilic carbon of the ylid on the carbonyl group (eq. 21.56).

$$(21.56)\quad (C_6H_5)_3\overset{+}{P}-\ddot{C}H_2 + \begin{matrix}R\\ \diagdown \\ C=\ddot{O}: \\ \diagup \\ R'\end{matrix} \rightarrow \begin{matrix} & R & \\ CH_2- & C & -R' \\ | & | & \\ (C_6H_5)_3P^+ \leftarrow & :\ddot{O}:^- & \end{matrix} \rightarrow$$

$$\begin{matrix} & R & \\ CH_2- & C & -R' \\ | & | & \\ (C_6H_5)_3P- & \ddot{O}: & \end{matrix} \rightarrow CH_2=C\begin{matrix}R\\ \\ R'\end{matrix} + (C_6H_5)_3P=\ddot{O}:$$

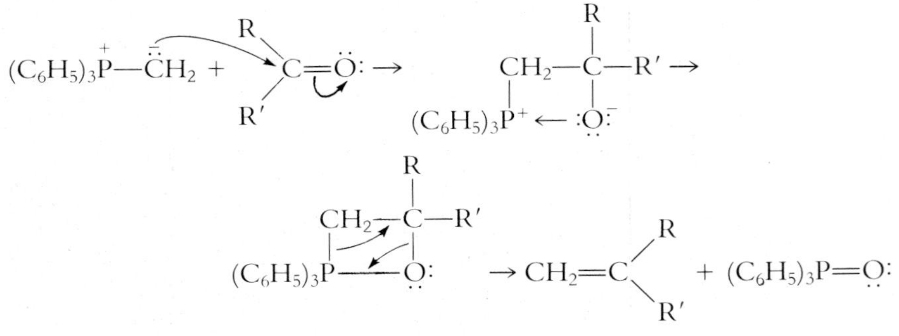

A four-membered ring is then formed which subsequently breaks down to give the products. This is now the generally accepted mechanism.

The Wittig reaction is quite general, and not confined to methylenephosphoranes. Thus, treating the $(C_6H_5)_3\overset{+}{P}CH_2CH_3$ cation with base yields the ylid $(C_6H_5)_3P{=}CHCH_3$; analogously, the $(C_6H_5)_3\overset{+}{P}CH(CH_3)_2$ cation gives rise to the ylid $(C_6H_5)_3P{=}C(CH_3)_2$.

In the following sections we will introduce organosulfur chemistry and show some reactions of sulfur ylids. Like phosphorus, sulfur is nucleophilic and is also capable of accommodating more than 8 electrons in its valence shell. Sulfur is more electronegative than phosphorus, however, so its chemistry does not completely mimic that of phosphorus.

Problems

21–18 a. Show the mechanism of the first step of the process depicted in Figure 21–5.

b. Why is CH_3ONa an adequate base for the fourth step of this process, while alkylphosphonium salts require strong bases such as RLi?

21–19 In Section 12:9 we described several syntheses of the insect sex attractant disparlure. The first laboratory synthesis of this compound utilized a Wittig reaction. Suggest a multistep route for the conversion of 6-methyl-1-heptanol to disparlure (eq. 5).

$(CH_3)_2CH(CH_2)_4CH_2OH \rightarrow \rightarrow$ [epoxide: $(CH_3)_2CH(CH_2)_4$ and $(CH_2)_9CH_3$ substituents, H, O, H] (5)

Disparlure

21–20 Show how the following compounds could be synthesized from a carbonyl compound and an appropriate ylid.

a. (cyclohexylidene)$=CHC_6H_5$

b. $CH_3(CH_2)_3CH{=}CHCH_3$

c. C_6H_5—$C(CH_3){=}CHCH_2CH_3$

d. (epoxide with CH_3, CH_3 on one carbon and C_6H_5, H on the other; O)

21:15 Nomenclature of Organosulfur Compounds

Sulfur is directly below oxygen in the Periodic Table, and there are several classes of organic compounds that are the sulfur analogs of common oxygen-containing organic compounds, such as alcohols, ethers, and esters. The sulfur analogs of alcohols are known as thiols (IUPAC), or mercaptans (semisystematic). The mercaptan terminology is used only for some of the simpler alkylsulfur compounds. In biochemical terminology, the —SH group is often called a **sulfhydryl** group.

CH_3SH	$CH_3CH_2CH_2CH_2SH$
Methanethiol	1-Butanethiol
(*methyl mercaptan*)	(*butyl mercaptan*)

Sulfur compounds with two organic substituents are generally known as *sulfides,* and a sulfide semisystematic nomenclature similar to that for ethers is common. The two organic substituents are named, followed by the word "sulfide." In the IUPAC nomenclature, the system is exactly analogous to the naming of ethers. The smaller alkyl group and the sulfur atom are combined as an alkylthio substituent on the larger organic group. *Disulfides* are compounds containing S—S bonds. The disulfide bonds in proteins

$CH_3CH_2SCH_2CH_3$
Ethylthioethane
(*diethyl sulfide*)

$$\begin{array}{c} \quad SCH_3 \\ \quad | \\ CH_3CHCH_2CH_3 \end{array}$$
2-Methylthiobutane
(s-*butyl methyl sulfide*)

(Chap. 22) play a major role in determining their three-dimensional shapes.

$C_6H_5SSC_6H_5$
Phenyldithiobenzene
(*diphenyl disulfide*)

One of the two nonbonded electron pairs of a sulfide can be used to form an additional alkyl or aryl bond. Trisubstituted sulfur compounds bear a

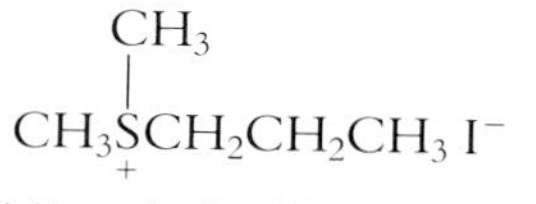

$$\begin{array}{l} \quad\; CH_3 \\ \quad\;\; | \\ CH_3\overset{}{\underset{+}{S}}CH_2CH_2CH_3\ I^- \end{array}$$
Propyldimethylsulfonium iodide

positive charge on sulfur and exist with a negative counterion as sulfonium salts. They are named similarly to ammonium compounds.

Low molecular weight thiols and sulfides have disagreeable skunklike odors and, indeed, several are constituents of the defensive secretions of the skunk.

Sulfides can be oxidized at the sulfur atom to yield *sulfoxides* and *sulfones.*

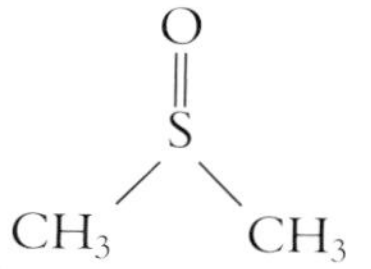

$$\begin{array}{c} O \\ \| \\ S \\ / \quad \backslash \\ CH_3 \qquad CH_3 \end{array}$$
Dimethyl sulfoxide

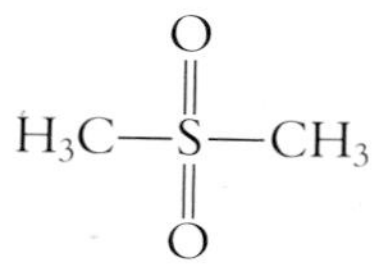

$$\begin{array}{c} O \\ \| \\ H_3C—S—CH_3 \\ \| \\ O \end{array}$$
Dimethyl sulfone

We begin our discussion of sulfur chemistry with hydrogen sulfide, H_2S, and progress to more and more substituted derivatives of it.

21:16 Thiols, Sulfides, and Sulfonium Salts: Their Synthesis and Nucleophilic Properties

Hydrogen sulfide is considerably more acidic than water; it has $pK_a^1 = 7.04$ and $pK_a^2 = 11.96$. Addition of hydrogen sulfide to an aqueous solution of sodium hydroxide results in consumption of the hydroxide ion and formation of the hydrogen sulfide ion (eq. 21.57).

$$(21.57)\quad :\ddot{O}H^- + H_2\ddot{S}: \rightleftharpoons H_2\ddot{O}: + :\ddot{S}H^-$$

Hydrogen-sulfide ion

The hydrogen sulfide ion is a very good nucleophile compared with the hydroxide ion, due to its more polarizable electrons and its weak hydrogen bonds (secs. 6:9 and 6:10). It reacts with primary or secondary alkyl halides to give thiols (eqs. 21.58 and 21.59).

$$(21.58)\quad CH_3(CH_2)_2CH_2—\ddot{Br}: + K^+:\ddot{S}H^- \xrightarrow{EtOH} CH_3(CH_2)_3\ddot{S}H + K^+:\ddot{Br}:^-$$

1-Butanethiol (54%)
(n-*butyl mercaptan*)

$$(21.59)\quad CH_3(CH_2)_5\underset{\displaystyle Br}{CH}CH_3 + K^+:\ddot{S}H^- \xrightarrow{EtOH} CH_3(CH_2)_5\underset{\displaystyle :\ddot{S}H}{CH}CH_3$$

2-Octanethiol (66%)
(2-*octyl mercaptan*)

In the presence of a base, thiols are converted to their sodium salts, which can then be alkylated to yield sulfides (eqs. 21.60 and 21.61).

$$(21.60)\quad CH_3CH_2\ddot{S}H \xrightarrow[C_2H_5OH]{C_2H_5\ddot{O}:^- Na^+} CH_3CH_2\ddot{S}:^- Na^+ \xrightarrow[C_2H_5OH]{CH_3(CH_2)_{15}—\ddot{Cl}:} CH_3CH_2\ddot{S}(CH_2)_{15}CH_3$$

(*ethyl hexadecyl sulfide*)

$$(21.61)\quad 2CH_3S^-Na^+ + BrCH_2CH_2Br \xrightarrow[C_2H_5OH]{} CH_3SCH_2CH_2SCH_3$$

1,2-Bis(methylthio)ethane (85%)

Dialkyl sulfides themselves are sufficiently nucleophilic to react with alkyl halides and tosylates to give sulfonium salts (eqs. 21.62 and 21.63).

$$(21.62)\quad (CH_3)_2\ddot{S}: + CH_3—\ddot{I}: \xrightarrow[C_2H_5OH,\ 25^\circ]{} (CH_3)_3\ddot{S}^+:\ddot{I}:^-$$

Trimethylsulfonium iodide (99%)

$$(21.63)\quad (CH_3)_2\ddot{S}: + BrCH_2COOC_2H_5 \xrightarrow[acetone]{} (CH_3)_2\overset{+}{\underset{..}{S}}CH_2COOC_2H_5\ Br^-$$

(90%)

Sulfides are much more nucleophilic than oxides (ethers), which is understandable since sulfur is less electronegative than oxygen. Sulfides are comparable in nucleophilicity to amines, and sulfonium salts resemble ammonium salts in many ways. For example, they undergo Hofmann elimination (eq. 21.64).

$$(21.64)\quad \underset{\displaystyle (CH_3)_2\ddot{S}^+\ Br^-}{CH_3CH_2CHCH_3} \xrightarrow[C_2H_5OH]{C_2H_5O^-Na^+} \underset{(26\%)}{CH_3CH{=}CHCH_3} + \underset{(74\%)}{CH_3CH_2CH{=}CH_2} + (CH_3)_2\ddot{S}:$$

Problem

21–21 Name each of the following compounds, and show how it can be synthesized starting with H_2S.

a. $CH_3CH_2CH_2SH$

b. $CH_3S(CH_2)_2CH(CH_3)_2$

c. $C_6H_5CH_2SH$ (benzene ring–CH_2SH)

d. $(CH_3CH_2)_2\overset{+}{S}CH_2CH{=}CH_2\ Cl^-$

e. cyclobutane bearing NC and $\overset{+}{S}(CH_3)_2\ I^-$ on opposite (1,3) carbons

21:17 Sulfur Ylids and Their Use

A positively-charged sulfur atom, as found in sulfonium ions, is a good electron acceptor. Treatment of sulfonium salts with strong bases leads to the formation of sulfur ylids (eq. 21.65). In the IUPAC system these ylids are named as derivatives of the hypothetical sulfur hydride, sulfurane, SH_4. The methylide nomenclature is more common, however.

$$(21.65)\quad (CH_3)_2\overset{+}{\underset{..}{S}}{-}CH_3\ Br^- \xrightarrow[DMSO]{NaH} [(CH_3)_2\overset{+}{\underset{..}{S}}{-}\overset{-}{\ddot{C}}H_2 \leftrightarrow (CH_3)_2\underset{..}{S}{=}CH_2]$$

Dimethyl(methylene)sulfurane
(*dimethylsulfonium methylide*)

The resonance structures shown in Equation 21.65 imply that there is π-bonding between carbon and sulfur. This is so, but the π bond is weak because $2p$ and $3p$ orbitals overlap poorly due to their size difference (sec. 17:13). A consequence of the weak π-bonding is that most of the negative charge remains on carbon, so that these sulfur-stabilized carbanions are quite nucleophilic. *Sulfur ylids react rapidly with carbonyl compounds* (eq. 21.66), *but give different products than do phosphorus ylids.* Dimethylsulfonium methylide

$$(21.66)\quad (CH_3)_2\overset{+}{\underset{..}{S}}{-}\overset{-}{\ddot{C}}H_2 + \underset{H}{\overset{:\ddot{O}}{\overset{\|}{C}}}C_6H_5 \xrightarrow{DMSO} (CH_3)_2\overset{+}{\underset{..}{S}}{-}CH_2{-}\underset{H}{\overset{:\ddot{O}:^-}{\overset{|}{C}}}C_6H_5 \rightarrow (CH_3)_2\ddot{S}: + \underset{\text{Phenyloxirane (65\%)}}{\overset{\ddot{O}}{CH_2{-}CHC_6H_5}}$$

transfers a methylene group to aldehydes and ketones, converting them to epoxides. The sulfur ylid behaves differently from the phosphorus ylid for two reasons:

1. The electronegativity of sulfur makes R_2S a good leaving group (second step of eq. 21.66); and
2. the S=O bond strength in sulfoxides, $R_2S{=}O$, is only 85 to 90 kcal/mol. Therefore formation of $R_2S{=}O$ is not as favored as formation of $R_3P{=}O$.

Problem

21–22 Sulfonium salts, such as $(CH_3)_3S^+I^-$, exchange H for D when they are dissolved in D_2O containing NaOD (eq. 6). Propose a mechanism for the exchange process.

$$(CH_3)_3S^+I^- \xrightarrow[D_2O]{OD^-} (CD_3)_3S^+I^- \qquad (6)$$

21:18 Thiols and Radical Inhibition

An important property of thiols, which is *not* exhibited by alcohols, is their superior reactivity with radicals (eq. 21.67). This property is a consequence

$$(21.67) \qquad R\cdot + \underset{\text{Butanethiol}}{CH_3CH_2CH_2CH_2SH} \rightarrow RH + \underset{\text{Butylthiyl radical}}{CH_3CH_2CH_2CH_2S\cdot}$$

of the low S—H bond strength, which averages 88 kcal/mol, lower than that of almost all C—H and O—H bonds. The reaction of thiols with radicals is so rapid that it is said to be **diffusion controlled**. That is, the slow step of reaction 21.67 is the rate at which the R· and butanethiol can diffuse together to meet in solution. Once the two components meet, they almost always react. Thiols are therefore **radical scavengers**. They are used to destroy, or "scavenge," highly reactive radicals and thereby interrupt radical-chain processes.

Thiols have also been investigated as possible **antiradiation agents**, one of the most studied being *2-mercaptoethylamine* (MEA, 2-aminoethanethiol). One type of radiation damage involves the production of alkyl radicals (eq. 21.68), which then affect living systems by further reaction with biochemi-

$$(21.68) \qquad RH \xrightarrow[\text{radiation}]{\text{high energy}} R\cdot + H\cdot$$

cally important molecules (sec. 5:13). If MEA is present in the system the reaction in Equation 21.69 can take place, effectively repairing the damage,

$$(21.69) \qquad R\cdot + \underset{\text{MEA}}{H_2NCH_2CH_2SH} \rightarrow RH + H_2NCH_2CH_2S\cdot$$

since the R· is rapidly converted back to unreactive RH. The $H_2NCH_2CH_2S\cdot$ species is much more selective in its reaction, and apparently causes little damage. It is also thought that the scavenging reaction can take place with

—SH-containing protein molecules to form **mixed disulfides**, **9**, which themselves protect against further cell damage.

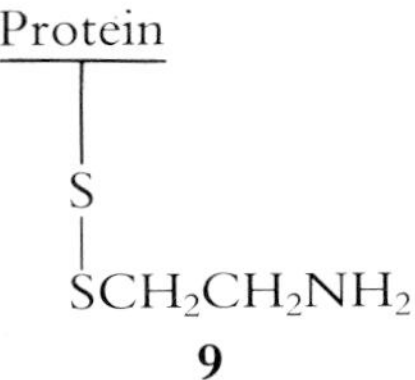

Summary

1 Primary amines are readily identifiable by ir spectroscopy, giving a double peak in the 3200 to 3500 cm^{-1} region. Secondary amines give rise to a single, weaker peak in this region. The mass spectra of amines are dominated by ions arising from cleavage of single bonds between atoms α and β to nitrogen (sec. 21:2). Nitrogen atoms bonded to three different groups are chiral, but the process of nitrogen inversion is too rapid to allow isolation of the separate amine enantiomers. Ammonium ions in which four organic groups are bonded to nitrogen form enantiomeric salts that are resolvable (sec. 21.3).

A comparison of the basicity of various amines can be made by comparing the pK_a's of the amine conjugate acids ($R_3\overset{+}{N}H \rightleftarrows R_3\ddot{N}: + H^+$). A strongly basic amine will have a large pK_a ($CH_3CH_2\ddot{N}H_2$ = 11.65). A weakly basic amine will have a small pK_a ($C_6H_5\ddot{N}H_2$ = 4.60). Basicity depends on the availability of the nonbonded pair to accept a proton. Electron-donating groups increase basicity; electron-withdrawing groups decrease it. The basicity of aromatic amines is affected by resonance as well as inductive effects (sec. 21:4).

2 The nucleophilic character of the nonbonded electron pair on nitrogen is seen in S_N2 reactions of amines (sec. 21:5). Nucleophilic reactions of amines are subject to steric hindrance, while their protonation reactions are not. Unhindered amines are easily alkylated by alkyl halides. When 1° and 2° amines are substrates, multiple alkylation leads to mixtures of products.

3 Aliphatic primary amines (alkylamines) are synthesized mainly by reduction of azides, cyanides (sec. 21:6), and oximes (sec. 21:7). Secondary and tertiary amines are obtained from imines and amides by reduction (sec. 21:7). The Gabriel synthesis (sec. 21:6) is a nonreductive route to primary aliphatic amines. A review of these syntheses is found in Table 21–2. Aromatic amines are prepared from the reduction of nitro compounds either chemically or catalytically (sec. 21:8).

Treatment of carboxamides with alkaline hypobromite solutions results in their conversion to amines, with loss of the carbonyl carbon (the Hofmann rearrangement). This is a useful synthesis for tertiary alkyl and aromatic amines (sec. 21:9).

4 Phase-transfer catalysis employs quaternary ammonium (or quaternary phosphonium) salts as catalysts to improve yields of reactions that proceed through anion intermediates (sec. 21:10). The quaternary cations carry the anions into the organic phase, where their nucleophilicity is increased.

5 Phosphorus and sulfur share two important characteristics: their valence shell electrons are highly polarizable, and their valence shells can accommodate more than eight electrons. These characteristics allow phosphorus and sulfur to be highly nucleophilic and to stabilize charges on themselves or on neighboring atoms (sec. 21:12).

Tertiary phosphines react rapidly with alkylating agents to give phosphonium salts. Strong bases convert these salts to ylids (Wittig reagents). The reaction between the ylids and aldehydes or ketones leads to formation of alkenes (secs. 21:13 and 21:14).

Under basic conditions, one or both of the hydrogens of H_2S can be replaced by alkyl groups. The reactions are nucleophilic displacements, and the products are thiols (RSH) and sulfides (RSR′). The sulfides may be further alkylated to produce sulfonium salts (sec. 21:16).

Sulfur can stabilize neighboring electron pairs by a combination of valence shell expansion and polarizability. Base treatment of sulfonium salts produce sulfur ylids, which convert aldehydes and ketones to epoxides (sec. 21:17).

6 The S—H bond is comparatively weak, so thiols are good scavengers for alkyl radicals. For this reason thiols have been tested as antiradiation agents (sec. 21:18).

Problems

21–23 Arrange the compounds in each group in decreasing order of basicity. Explain your reasoning.

a. aniline, cyclohexylamine, *p*-nitroaniline
b. *p*-chloroaniline, *p*-methoxyaniline, *p*-toluidine(*p*-methylaniline), *p*-cyanoaniline
c. aniline, *N*-methylaniline, *m*-methylaniline
d. aniline, benzylamine, 2-phenylethylamine
e. ethylamine, 2-aminoethanol, 2-fluoroethylamine

21–24 The addition of two methyl groups to the nitrogen atom of aniline causes a modest increase in basicity.

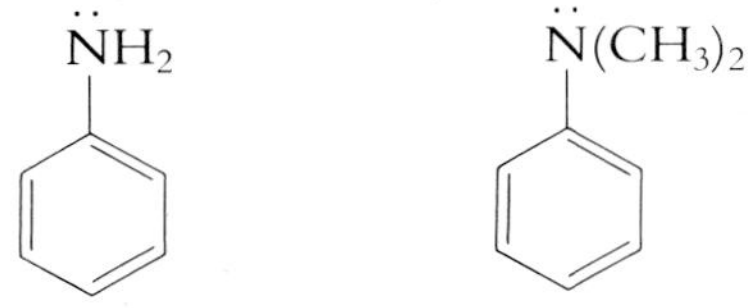

$\ddot{N}H_2$ (aniline): $pK_a = 4.60$ $\qquad$ $\ddot{N}(CH_3)_2$ (N,N-dimethylaniline): $pK_a = 5.04$

When the same operation is performed on 2,4,6-trinitroaniline, the increase in basicity exceeds 10^4. Explain the difference in behavior. (Consider the resonance interaction between nitrogen and the ring.)

$$\text{2,4,6-}(O_2N)_3C_6H_2\ddot{N}H_2 \qquad \text{2,4,6-}(O_2N)_3C_6H_2\ddot{N}(CH_3)_2$$

$$pK_a = -9.3 \qquad\qquad pK_a = -4.7$$

21–25 When an alcohol reacts with a cyclic anhydride, such as phthalic anhydride, the product is an acid ester (eq. 7).

$$\text{phthalic anhydride} + ROH \rightarrow C_6H_4(COOR)(COOH) \tag{7}$$

Suppose the alcohol is 2-butanol, which is racemic. Show how, with the aid of a chiral amine such as (−)-1-phenylethylamine (eq. 21.8), this reaction could be used to resolve the alcohol. Describe all steps explicitly.

21–26 Amines are more basic than imines, which are in turn more basic than nitriles. Offer an explanation for this trend. (You may find a review of Section 12:2 helpful.)

21–27 a. Amines form strong complexes with Lewis acids such as boron, aluminum, and ferric halides. Show the electronic structure of the boron trifluoride–trimethylamine complex.

b. When aniline derivatives such as *N,N*-dimethylaniline are subjected to Friedel–Crafts alkylation with an alkyl halide and aluminum chloride, the following observations are made:

1. More than 1 mol of catalyst per mole of amine is required to bring about reaction (other aromatic compounds usually require about 0.1 mol of catalyst).
2. A high proportion of *m*-alkylated aniline is found.

Explain these observations.

21–28 Complete the following reactions; show all products, including stereoisomers.

a. $CH_3CH_2NH_2(\text{excess}) + (CH_3)_2CHCH_2OTs \rightarrow$

b. $C_6H_5{-}NHCH_3 + CH_3CH_2I\ (\text{excess}) \rightarrow$

c. $(CH_3)_3N + (R)\text{-}CH_3\underset{\displaystyle Cl}{\underset{|}{C}}HCH_2CH_3 \rightarrow$

d. $CH_3CH_2\underset{\displaystyle CH_3}{\underset{|}{N}}CH_2CH_2CH_3 + (R)\text{-}CH_3CH_2\underset{\displaystyle CH_3}{\underset{|}{C}}HCH_2Br \rightarrow$

e. $CH_3\ddot{S}CH_3 + (S)\text{-}CH_3\underset{\displaystyle Br}{\underset{|}{C}}HCH_2CH_2CH_3 \rightarrow$

f. $CH_3CH(OH)CH_2CH_2Br \xrightarrow[\text{2. } H_2/\text{cat}]{\text{1. KCN}}$

21–29 a. Compounds in which there is a good leaving group α to nitrogen, such as α-chloroamines, decompose very rapidly in neutral or basic solution (eq. 8). This reaction is very fast compared with that of the carbon analog,

$$(CH_3)_2CH\underset{\displaystyle Cl}{\underset{|}{C}}HCH_3.$$

Furthermore, the reaction of the α-chloroamine is zero order in OH^-, while the reaction of the carbon compound is first order in OH^-.

$$(CH_3)_2N\underset{\displaystyle Cl}{\underset{|}{C}}HCH_3 + OH^- \rightarrow (CH_3)_2NH + CH_3\overset{\displaystyle O}{\overset{\|}{C}}H + Cl^- \qquad (8)$$

Propose a mechanism for the reaction in Equation 8.

b. β-Chloroamines are also unusually reactive in neutral or basic solution. The β-chloroamine shown in Equation 9 is many times more reactive than $(CH_3)_2CHCH_2CH_2Cl$, and its conversion to an alcohol is once again zero order in OH^-.

$$(CH_3)_2NCH_2CH_2Cl + OH^- \rightarrow (CH_3)_2NCH_2CH_2OH + Cl^- \qquad (9)$$

Propose a mechanism for the reaction in Equation 9. (See Box 6–1, p. 196.)

21–30 Describe efficient syntheses of the following amines, using appropriate alkyl halides, alcohols, aldehydes, ketones, and aromatic hydrocarbons as starting materials, plus other needed reagents.

a. 3-methyl-1-pentylamine
b. dipropylamine
c. *o*-toluidine (*o*-methylaniline)
d. *p*-bromoaniline
e. ethyl *m*-aminobenzoate
f. benzylamine
g. *p*-iodobenzylamine
h. 2-butylamine
i. ethyl(*t*-butyl)amine
j. methylcyclopentylbenzylamine

21–31 A supply of *cis*-4-methylcyclohexylamine is needed for a research project. The following related compounds are available:

a. *trans*-4-chloromethylcyclohexane

b. *cis*-4-methylcyclohexanol
c. *cis*-4-methylcyclohexanecarboxylic acid
Show how each compound could be converted to the desired amine.

21–32 The reagent *O*-2,4-dinitrophenylhydroxylamine, **III**, can convert acidic carbonyl compounds, such as diethyl malonate (sec. 18:10), to their amino derivatives in the presence of base (eq. 10). Write a mechanism for this reaction. Explain why it is necessary to have nitro groups on the aromatic ring.

$$CH_2(COOEt)_2 \xrightarrow[\text{NaOEt, EtOH}]{NH_2O-C_6H_3(NO_2)_2\ (\textbf{III})} H_2N{-}CH(COOEt)_2 \qquad (10)$$

21–33 The **Beckmann rearrangement** involves the acid-catalyzed conversion of oximes to amides; an example is given in Equation 11. The function of the catalyst is to initially remove the OH group from nitrogen.

$$CH_3C(=N{-}OH)CH_2CH_2CH_3 \xrightarrow[\text{Lewis acid}]{H^+ \text{ or}} CH_3C(=O)NHCH_2CH_2CH_3 \qquad (11)$$

Propose a mechanism for the rearrangement. (Notice the resemblance between this reaction and the Hofmann rearrangement, Section 21:9.)

21–34 The chlorination of anilines appears to proceed by way of *N*-chloroanilines, which rearrange to give the ring-substituted products that are normally isolated (eq. 12). The rearrangement is facilitated by electron-donating groups on either the nitrogen atom or the ring. Write mechanisms for the formation and the rearrangement of the *N*-chloroanilines.

$$C_6H_5NHCH_3 \xrightarrow{Cl_2} C_6H_5N(Cl)CH_3 + HCl \rightarrow o\text{-}ClC_6H_4NHCH_3 + p\text{-}ClC_6H_4NHCH_3 \qquad (12)$$

21–35 Trimethyl phosphite rearranges exothermically to dimethyl methylphosphonate, as shown in Equation 13. A small quantity of methyl iodide catalyzes this reaction. Propose a mechanism for it, and indicate what you think is the driving force.

$$\underset{\text{Trimethyl phosphite}}{(CH_3O)_3P:} \xrightarrow[\text{heat}]{CH_3I\ (cat)} \underset{\text{Dimethyl methylphosphonate}}{(CH_3O)_2P(=O)CH_3} \qquad (13)$$

21–36 Phosphorous acid has the molecular formula H_3PO_3. It exists as an equilibrium mixture of two isomers; in one of them the octet rule is met, in the other the octet rule is exceeded. Suggest structures for the two isomers, showing all the valence electrons. Remember that phosphorus forms strong double bonds to oxygen.

21–37 Each of the following compounds has been synthesized by way of a Wittig reaction. Suggest two possible routes to each product, starting with triphenylphosphine in each case and using any other needed reagents.

a. $CH_3(CH_2)_2CH{=}C(CH_3)CH_2CH_3$

b. $CH_2{=}CHCH{=}C(CH_3)COOCH_3$

c. (2-cyanophenyl)–CH=CH–(1,4-phenylene)–CH=CH–(2-cyanophenyl) (a commercial optical brightener in detergents)

21–38 Devise syntheses for the following compounds, using easily available organic starting materials and any necessary inorganic reagents.

a. $(CH_3)_2CHCH_2SH$

b. $CH_3CH_2CH(CH_3)S$–cyclopentyl

c. thiane (six-membered ring containing S)

d. $HOCH_2CH_2SH$

e. cyclohexane with CH_3 and SCH_3 on adjacent carbons (both wedged)

f. $CH_3CH_2\overset{+}{S}(CH_3)CH(CH_3)_2\ I^-$

21–39 The reaction shown in Equation 14 leads almost exclusively to the rearranged product shown. Propose a mechanism that accounts for the course of this reaction (review $SOCl_2$ reactions; sec. 6:11).

$$C_6H_5SCH(CH_3)CH_2OH + SOCl_2 \rightarrow CH_3CH(Cl)CH_2S{-}C_6H_5 + SO_2 + HCl \quad (14)$$

21–40 a. Dimethyl sulfoxide can be described as a resonance hybrid of the two structures **IV** and **V**. Dimethyl sulfoxide is found to be appreciably more polar than acetone. Explain.

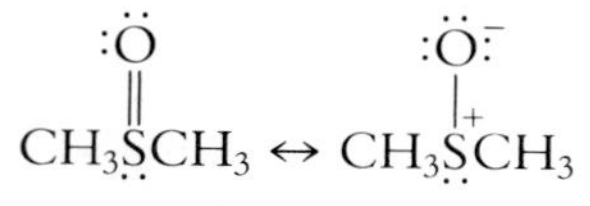

$$\underset{\textbf{IV}}{CH_3\overset{\overset{:\ddot{O}}{\|}}{\underset{..}{S}}CH_3} \leftrightarrow \underset{\textbf{V}}{CH_3\overset{\overset{:\ddot{O}:^-}{|+}}{\underset{..}{S}}CH_3}$$

b. Arrange dimethyl sulfoxide, dimethyl sulfide, and dimethyl sulfone (sec. 21:15) in increasing order of ability to exchange protons with D_2O containing OD^- (see Prob. 21–22). Indicate the mechanism of exchange, and justify your answer.

c. Dimethyl sulfoxide reacts with methyl iodide to produce two isomeric compounds, C_3H_9OSI. One or the other may be preferentially produced by the appropriate choice of conditions. Both are soluble in water and insoluble in benzene. Write their structures.

21–41 Cyclohexanone and the sulfonium salt **VI** are dissolved in a heterogenous mixture of CH_2Cl_2 and 15 *M* NaOH solution, which is then stirred for 10 hours at 25°. Isolation and distillation of the CH_2Cl_2 layer gives a high yield of methylenecyclohexane oxide (eq. 15). Propose a mechanism for this reaction, paying attention to the two-phase solvent. What is the purpose of having the long-chain alkyl group in the sulfonium ion?

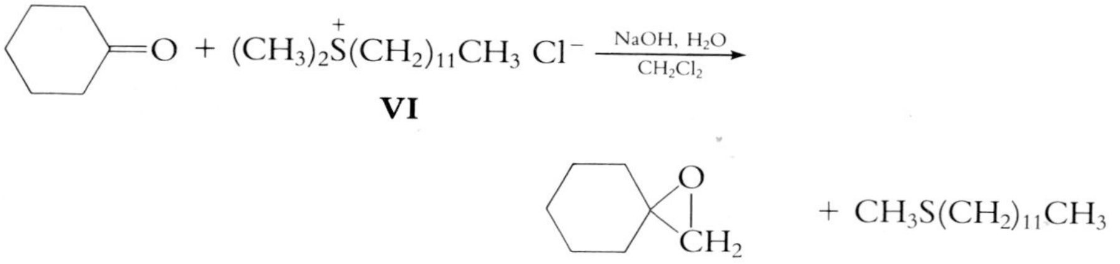

$$\text{C}_6\text{H}_{10}{=}O + \underset{\textbf{VI}}{(CH_3)_2\overset{+}{S}(CH_2)_{11}CH_3\ Cl^-} \xrightarrow[CH_2Cl_2]{NaOH,\ H_2O} \underset{\text{methylenecyclohexane oxide}}{\text{C}_6\text{H}_{10}\overset{O}{\diagdown}CH_2} + CH_3S(CH_2)_{11}CH_3 \quad (15)$$

21–42 Both anisole (Y = O) and thioanisole (Y = S) undergo Friedel–Crafts acetylation when treated with acetyl chloride and $AlCl_3$ (eq. 16). Anisole reacts about 40 times faster than thioanisole. Explain this difference in reactivity with the aid of resonance structures (it may help to review the mechanism of this reaction in sec. 15:7).

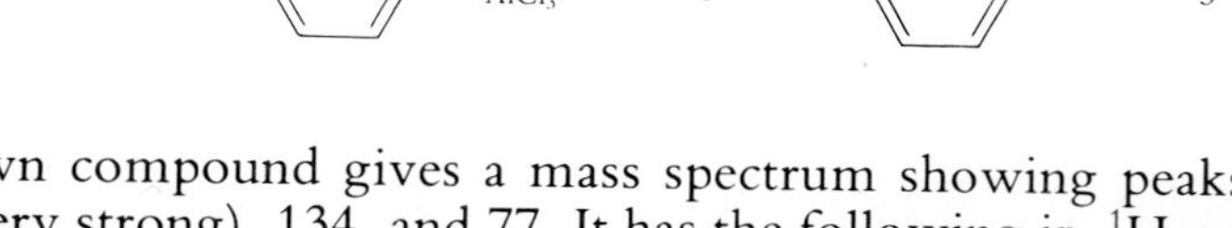

$$CH_3{-}Y{-}C_6H_5 \xrightarrow[AlCl_3]{CH_3\overset{O}{\overset{\|}{C}}Cl} CH_3{-}Y{-}C_6H_4{-}\overset{O}{\overset{\|}{C}}CH_3 \quad (16)$$

21–43 An unknown compound gives a mass spectrum showing peaks at m/z = 149, 148 (very strong), 134, and 77. It has the following ir, ^1H nmr, and ^{13}C spectra. Suggest a structure and identify the nmr peaks.

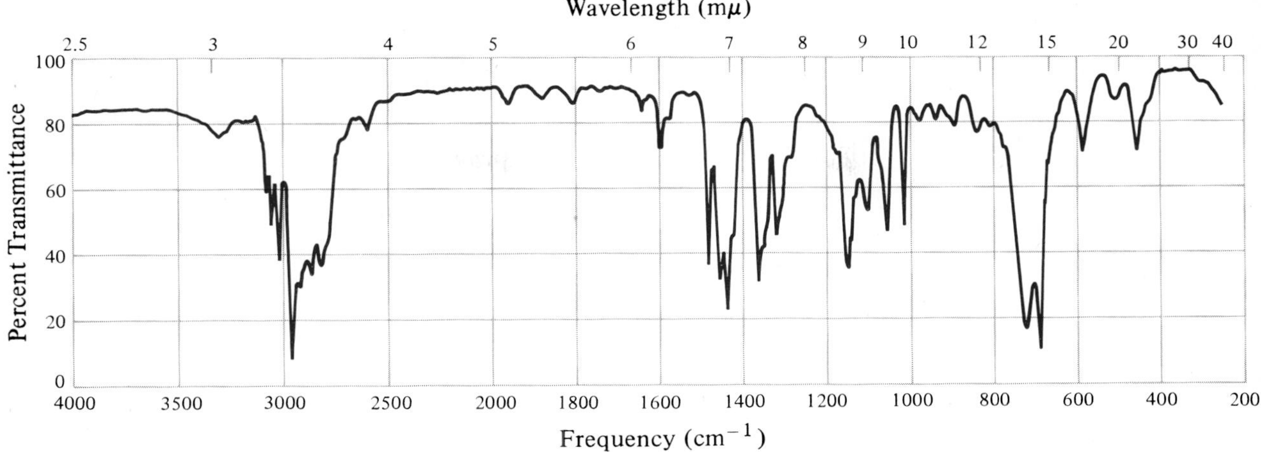

Wavelength (mμ)
2.5
3
4
5
6
7
8
9
10
12
15
20
30
40
100
80
60
40
20
0
Percent Transmittance
4000
3500
3000
2500
2000
1800
1600
1400
1200
1000
800
600
400
200
Frequency (cm⁻¹)

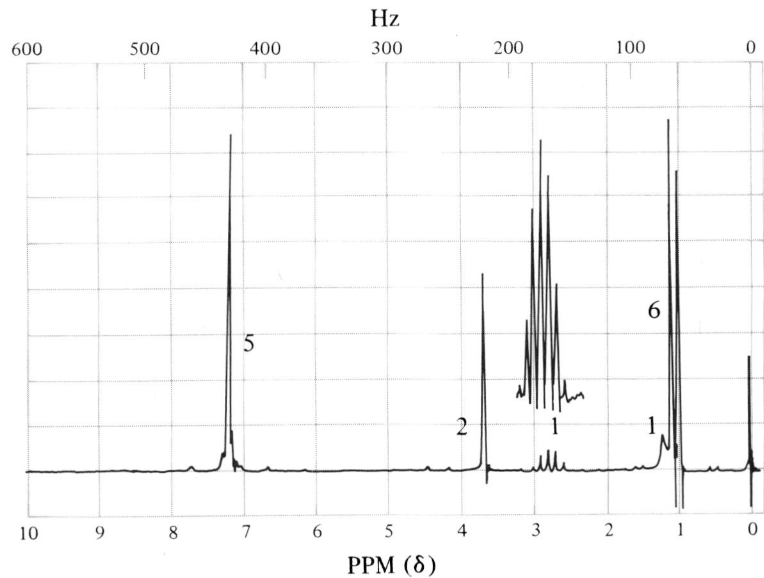

Hz
600
500
400
300
200
100
0
5
2
1
6
1
10
9
8
7
6
5
4
3
2
1
0
PPM (δ)

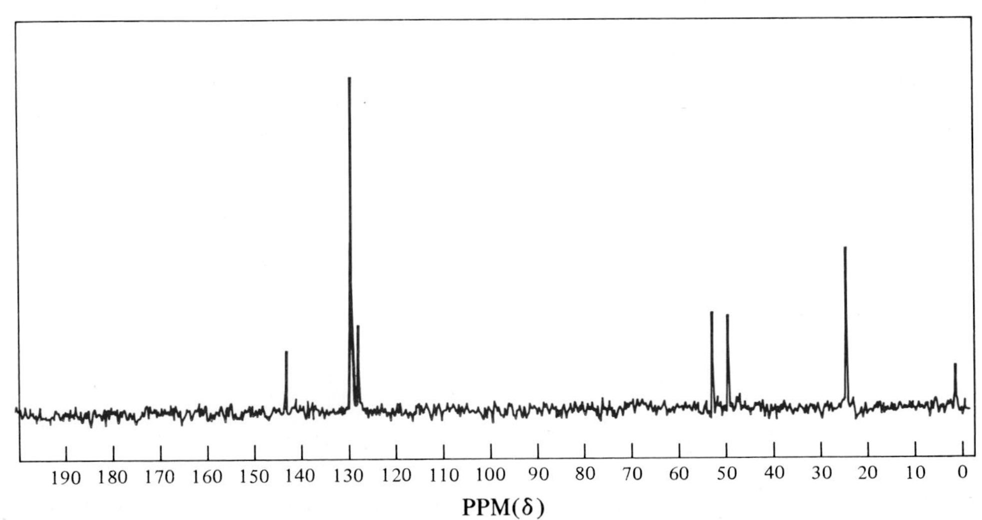

190
180
170
160
150
140
130
120
110
100
90
80
70
60
50
40
30
20
10
0
PPM(δ)

P. Varughese, Quaternary ammonium salts: Some recent applications in organic synthesis, *J. Chem. Ed.,* **54,** 666 (1977).

G. W. Gokel and W. P. Weber, Phase-transfer catalysis, *J. Chem. Ed.,* **55,** 350, 429 (1978); J. M. McIntosh, Phase-transfer catalysis using quaternary 'onium salts, *J. Chem. Ed.,* **55,** 235 (1978).

H. Pommer, The Wittig reaction in industrial practice, *Angewandte Chemie International Edition in English,* **16,** 423 (1977).

F. Leh and K. M. Chan, Sulfur compounds: Pollution, health effects, and biological function, *J. Chem. Ed.,* **50,** 246 (1973).

P. Grund, J. D. Andose, J. B. Rhodes, and G. M. Smith, Three-dimensional molecular modeling and drug design, *Science,* **208,** 1425 (1980).

L. N. Ferguson, Bioactivity in organic chemistry courses, *J. Chem. Ed.,* **49,** 721 (1972).

22 Amino Acids, Peptides, and Proteins: Natural and Synthetic Condensation Polymers

In Chapters 16 and 21 we discussed the chemistry of the carboxyl and amino functional groups. When both groups are present in the same molecule, the resultant bifunctional **amino acids** have some of the properties of both amines and acids. These amino acids also show unique chemistry which makes them worthy of special attention, but more importantly, they are the building blocks of natural polymers known as peptides and proteins. As such, amino acids are components of all living systems.

Sections 10:7 and 13:6 introduced the preparation and properties of some macromolecules known as addition polymers. Amino acids form peptides and proteins, which belong to a second class of macromolecule—the condensation polymers. In addition to the naturally occurring members of this class, which includes polysaccharides as well as proteins, there are synthetic condensation polymers that go into the make-up of fabrics, plastics, automobile tires, and many other useful materials. Mechanistically, the biological and synthetic polymers are both formed in a similar fashion and hence there are many analogies between their chemistry. The importance of both these areas can perhaps be judged by the fact that over 50% of all chemists employed in industry are involved in some way with polymers.

After a brief background section, this chapter will discuss the structure and properties of individual amino acids, their synthesis, and the ways in which they combine to form peptides and proteins. The chapter will finish with a survey of the industrial preparation of synthetic polymers designed to substitute for, or improve upon, natural polymers.

22:1 Definitions: Amino Acid, Peptide, Protein, and Condensation Polymer

The amino acids that are found as components of living systems are all α-amino acids (**1**). They are so designated because the amino group is attached to the α-carbon. There are 20 α-amino acids that are widely found in nature, each differing in its R substituent. (A cyclic amino acid, proline, is the only one *not* fitting the general structure, **1**.) Any two amino acids can be com-

$$H_2N-\overset{\alpha}{C}H(R)-COOH$$

An α-amino acid
1

bined through formation of an amide bond between the NH_2 of one amino acid and the carboxyl group of the second amino acid, with the loss of a molecule of water (eq. 22.1). (The mechanism of amide formation was

(22.1) $H_2N—CH(R)—COOH + H_2N—CH(R')—COOH \rightarrow$

Amino acid   Amino acid

$H_2N—CH(R)—C(=O)—NH—CH(R')—COOH + H_2O$

peptide bond

A dipeptide

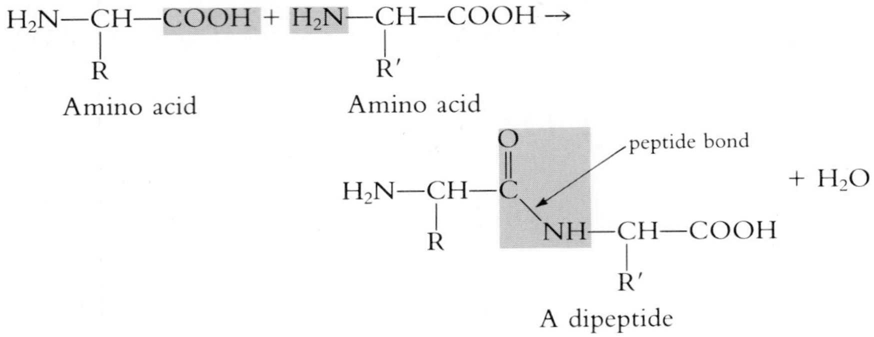

discussed in Sections 17:4 and 17:7.) The product amide is known as a **dipeptide** and the amide bond is often called a *peptide bond.* Additional amino acids can be added to either end of the dipeptide to form tripeptides, tetrapeptides, and so on. These are given the general name **polypeptides** until approximately 50 amino acid residues have been reached. After that, these large macromolecules are called **proteins**. The difference between polypeptides and proteins is only one of definition and does not involve an important or sharp distinction. Alternatively, a molecular weight of 5000 has

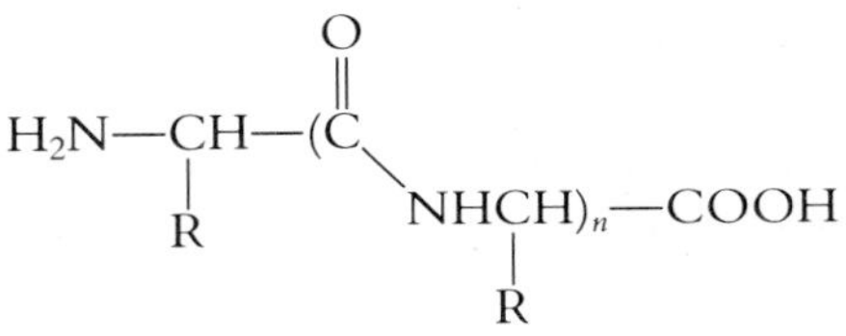

$H_2N—CH(R)—(C(=O)—NHCH(R))_n—COOH$

$n < 50$ (a polypeptide)
$n > 50$ (a protein)

been suggested as the dividing line between polypeptides and proteins. Proteins are the most abundant organic molecules found in cells and make up about 50% of the cellular dry weight. Proteins are thus available in quantity from natural sources and can be hydrolyzed to yield individual amino acids.

Polymers containing amide bonds are not limited to natural peptides and proteins. One of the most common and important groups of synthetic fibers is the nylons, which are **polyamides**. One of them, nylon 66, is synthesized from a diacid (adipic acid, a 6-carbon acid) and a diamine (hexamethylenediamine, a 6-carbon amine), rather than from an amino acid (eq. 22.2).

Equations 22.1 and 22.2 are examples of **condensation polymerizations**. In these polymerizations, two difunctional monomers are combined with loss of a small molecule such as water, to form a new molecule which still is difunctional (see the dipeptide of eq. 22.1). This new molecule can continue to grow at both ends to form a long chain.

$$\text{(22.2)}\quad \underset{\text{Adipic acid}}{HOOC(CH_2)_4COOH} + \underset{\text{Hexamethylenediamine}}{H_2N(CH_2)_6NH_2} \xrightarrow{250^\circ} HOOC(CH_2)_4\overset{\overset{\displaystyle O}{\|}}{C}{-}NH(CH_2)_6NH_2 + H_2O$$

$$\downarrow \text{continued heating}$$

$$\leftarrow H_2N(CH_2)_6NH\overset{\overset{\displaystyle O}{\|}}{C}(CH_2)_4\overset{\overset{\displaystyle O}{\|}}{C}{-}NH(CH_2)_6NH\overset{\overset{\displaystyle O}{\|}}{C}(CH_2)_4COOH$$

$$(2n - 1)H_2O + \underset{\substack{\text{Nylon 66} \\ (n \approx 44,\ MW \simeq 10{,}000)}}{H[HN(CH_2)_6NH\overset{\overset{\displaystyle O}{\|}}{C}(CH_2)_4\overset{\overset{\displaystyle O}{\|}}{C}]_nOH}$$

22:2 Structures and Properties of Amino Acids

The 20 amino acids that are the major components of peptides and proteins are often called *standard amino acids*. Their structures, names, and common abbreviations are given in Table 22–1. The abbreviations are used when we are representing the structures of polypeptides and proteins.

In Table 22–1, the amino acids have been arranged in groups based on the properties of their side chains **R** (in **R**—$CH(NH_2)COOH$). These side chains not only determine the properties of the individual amino acids to a great degree, they also exert a decisive influence on the reactivity and conformation of the protein chain.

Some of the amino acids are simple derivatives of other acids. Glutamine and asparagine are monoamides of glutamic and aspartic acids, respectively. Tyrosine is the *p*-hydroxy derivative of phenylalanine. Cysteine (pronounced sí-stĕ-ēn) is usually found in proteins in the form of its oxidative dimer cystine (pronounced sí-stēn; eq. 22.3). The oxygen present in air is sufficient to effect the oxidative coupling of cysteine to cystine. The easy removal of its —SH hydrogen atom enables cysteine to play a special role in the structure of some proteins and in radiation-induced cellular processes (sec. 21:18).

$$\text{(22.3)}\quad \underset{\text{Cysteine}}{\begin{array}{c} H_2NCHCO_2H \\ | \\ CH_2 \\ | \\ SH \end{array}} \underset{\text{reduction}}{\overset{\text{oxidation}}{\rightleftharpoons}} \underset{\text{Cystine}}{\begin{array}{c} H_2NCHCO_2H \\ | \\ CH_2 \\ | \\ S \\ | \\ S \\ | \\ CH_2 \\ | \\ H_2NCHCO_2H \end{array}}$$

Table 22–1. Structures and Names of Standard Amino Acids

Name	Structure	Abbreviation
	Nonpolar Side Chains	
glycine	$\mathbf{H}$—CHCOOH NH_2	Gly
alanine	$\mathbf{CH_3}$—CHCOOH NH_2	Ala
leucine	$\mathbf{(CH_3)_2CHCH_2}$—CHCOOH NH_2	Leu
isoleucine	$\mathbf{CH_3CH_2CH}$—CHCOOH $\mathbf{CH_3}$ NH_2	Ile
valine	$\mathbf{(CH_3)_2CH}$—CHCOOH NH_2	Val
proline	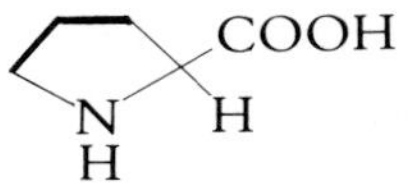 COOH N H H	Pro
phenylalanine	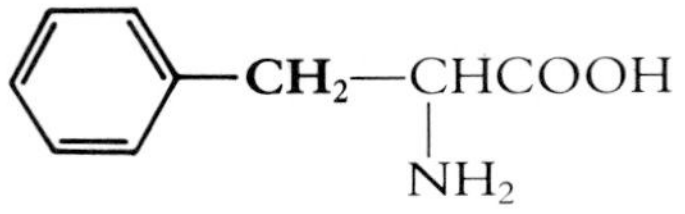 —$\mathbf{CH_2}$—CHCOOH NH_2	Phe
tryptophan	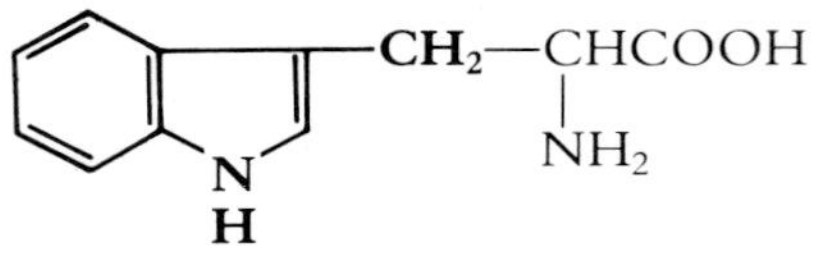 —$\mathbf{CH_2}$—CHCOOH NH_2 **N** **H**	Trp
methionine	$\mathbf{CH_3SCH_2CH_2}$—CHCOOH NH_2	Met
	Polar, Nonionizing Side Chains	
asparagine	**O** ‖ $\mathbf{H_2NCCH_2}$—CHCOOH NH_2	Asn
glutamine	**O** ‖ $\mathbf{H_2NCCH_2CH_2}$—CHCOOH NH_2	Gln

Table 22–1. cont.

Name	Structure	Abbreviation
cysteine	$\mathbf{HSCH_2}{-}\underset{NH_2}{\underset{\vert}{C}}HCOOH$	Cys
serine	$\mathbf{HOCH_2}{-}\underset{NH_2}{\underset{\vert}{C}}HCOOH$	Ser
threonine	$\mathbf{CH_3}\underset{\mathbf{OH}}{\underset{\vert}{\mathbf{C}}}\mathbf{H}{-}\underset{NH_2}{\underset{\vert}{C}}HCOOH$	Thr
tyrosine	$\mathbf{HO{-}C_6H_4{-}CH_2}{-}\underset{NH_2}{\underset{\vert}{C}}HCOOH$	Tyr
	Ionizable Side Chains	
aspartic acid (acidic)	$\mathbf{HOOCCH_2}{-}\underset{NH_2}{\underset{\vert}{C}}HCOOH$	Asp
glutamic acid (acidic)	$\mathbf{HOOCCH_2CH_2}{-}\underset{NH_2}{\underset{\vert}{C}}HCOOH$	Glu
arginine (basic)	$\mathbf{HN{=}\underset{NH_2}{\underset{\vert}{C}}NHCH_2CH_2CH_2}{-}\underset{NH_2}{\underset{\vert}{C}}HCOOH$	Arg
histidine (basic)	$\mathbf{C_3H_2N_2H{-}CH_2}{-}\underset{NH_2}{\underset{\vert}{C}}HCOOH$	His
lysine (basic)	$\mathbf{H_2NCH_2CH_2CH_2CH_2}{-}\underset{NH_2}{\underset{\vert}{C}}HCOOH$	Lys

A common physical characteristic of amino acids is their high melting point. Alanine, for example, melts at 297° with decomposition. This is difficult to reconcile with its low molecular weight if alanine has a structure corresponding to the general formula **1**. Such a structure is also inconsistent

$$H_2N{-}\underset{CH_3}{\underset{|}{C}}H{-}COOH$$

Alanine
(from general formula **1**)

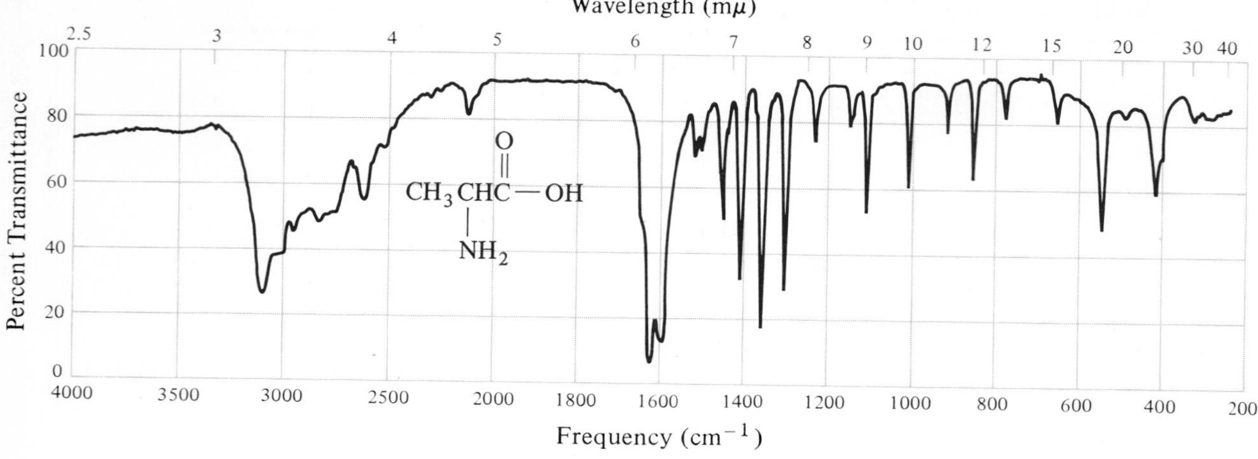

Figure 22–1. *Infrared spectrum of alanine.*

with the infrared spectrum of alanine (Fig. 22–1). The spectrum doesn't show the expected N—H or O—H stretching bands for the $—NH_2$ or —COOH groups, and there is no C=O stretch in the 1700 cm^{-1} region. Its high melting point, ir spectrum, and the fact that alanine is soluble in water (a *polar* solvent) can be explained if alanine (and other amino acids) exist as dipolar **zwitterions** (German *zwitter,* meaning "hybrid").

$$H_3\overset{+}{N}—CH(CH_3)—C(=O)—O^-$$

Alanine (mp 297°)
(zwitterion structure)

Each amino acid contains both an acidic (COOH) and basic (NH_2) group which are able to undergo an *internal acid–base reaction to yield the dipolar zwitterion.* Amino acids in either aqueous solution or the crystalline state are predominantly in the zwitterionic form, which contains a cationic ammonium group and an anionic carboxylate group (identifiable by the ir absorption at about 1600 cm^{-1}; Fig. 22–1). It has been estimated that the zwitterionic tautomer of glycine, $\overset{+}{N}H_3CH_2COO^-$, is more stable than the uncharged tautomer, NH_2CH_2COOH, by about 8 kcal/mol in aqueous solution. For the sake of convenience, however, we will sometimes draw amino acids in the form shown in structure **1**, because they do undergo reactions characteristic of free amino and carboxyl groups.

22:3 The Effect of pH on Amino Acid Structure: The Isoelectric Point and Electrophoresis

The fact that the amino and carboxyl groups of amino acids exist primarily in their ionized form has important biological and chemical consequences.

Chemically, we take advantage of the zwitterionic nature of amino acids to separate them; the same principle is also applicable to peptides and proteins. In order to understand how this separation method works, we must see what happens to amino acids in solution as the pH changes.

Let's start by considering a solution of alanine in aqueous HCl. The acid protonates the basic —COO$^-$ group, converting alanine to a cation (**2**, eq. 22.4). If NaOH is added to this solution, it neutralizes the —COOH group

$$(22.4)\quad \underset{\substack{\mathbf{2}\\ \text{Alanine cation}\\ \text{(predominant form in acid)}}}{\overset{+}{H_3N}\text{—}\underset{|\atop CH_3}{C}HCOOH} \underset{OH^-}{\overset{H^+}{\rightleftharpoons}} \underset{\substack{\mathbf{3}\\ \text{Alanine zwitterion}\\ \text{(predominant form at the isoelectric point, pH 6)}}}{\overset{+}{H_3N}\text{—}\underset{|\atop CH_3}{C}HCOO^-} \underset{H^+}{\overset{OH^-}{\rightleftharpoons}} \underset{\substack{\mathbf{4}\\ \text{Alanine anion}\\ \text{(predominant form in base)}}}{H_2N\text{—}\underset{|\atop CH_3}{C}HCOO^-}$$

of **2**, which is more acidic than the —NH_3^+ group. Addition of the correct quantity of NaOH will convert all of the alanine to the zwitterionic form **3**; the pH of the solution at this point is called the **isoelectric point**, or **isoelectric pH**. Addition of more base will make the solution basic, and result in conversion of alanine to an anion (**4**, eq. 22.4) by removal of an —NH_3^+ proton. Generalizing from this example we can say that *any amino acid will exist partly as a cation at pHs below its isoelectric point, and partly as an anion at pHs above the isoelectric point.*

All amino acids containing one basic and one acidic group have isoelectric points at about pH 6. However, the exact isoelectric pH will differ slightly for each amino acid because the nature of the R group (structure **1**) has some effect on the isoelectric point. Thus, alanine (pH 6.12) and phenylalanine (pH 5.91) have measurably different isoelectric points. It is these differences which allow separations to be made.

Suppose we introduce a mixture of alanine and phenylalanine into a solution buffered at exactly pH 6.12. Since this is the isoelectric pH for alanine it will be present as a zwitterion, and the *net* charge on its molecules will be zero. However, pH 6.12 is above the isoelectric point of phenylalanine, so it will be present as a mixture of zwitterion and anion (eq. 22.5).

$$(22.5)\quad \underset{\substack{\text{Phenylalanine}\\ \text{zwitterion}}}{\overset{+}{H_3N}\text{—}\underset{|\atop CH_2Ph}{C}HCOO^-} + H_2O \overset{\text{pH } 6.12}{\rightleftharpoons} \underset{\substack{\text{Phenylalanine}\\ \text{anion}}}{H_2N\text{—}\underset{|\atop CH_2Ph}{C}HCOO^-} + H_3O^+$$

Thus, there will be a net partial negative charge on the phenylalanine molecules. When the amino acid mixture is placed in an electric field between two charged plates, the negatively-charged phenylalanine molecules will migrate toward the positive plate, while the alanine molecules will not move. This is the basis for separating mixtures of amino acids and peptides by means of **electrophoresis** (Fig. 22–2). The solution of amino acids is usually placed on a buffered gel or piece of paper.

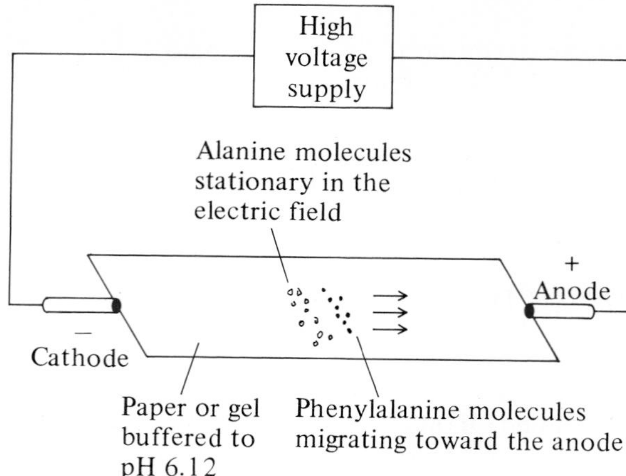

Figure 22–2. *Separation of alanine and phenylalanine by electrophoresis at the isoelectric pH of alanine.*

Electrophoretic separations become much easier if one of the amino acids in the mixture carries more than one acidic or basic group. Consider the case of aspartic acid. At pH 6, carboxyl group 1 and the α-amino group have neutralized one another, just as in alanine. However, carboxyl group 2 can partly dissociate (eq. 22.6), so that at pH 6, aspartic acid molecules carry a

$$\text{(22.6)} \quad \underset{\text{Aspartic acid zwitterion}}{\underset{\displaystyle CH_2COOH}{H_3\overset{+}{N}\text{—}\overset{|}{C}HCOO^-}} + H_2O \xrightleftharpoons{\text{pH 6}} \underset{\text{Aspartic acid monoanion}}{\underset{\displaystyle CH_2COO^-}{H_3\overset{+}{N}\text{—}\overset{|}{C}HCOO^-}} + H_3O^+$$

net negative charge. In order to convert aspartic acid completely to the zwitterionic form, it is necessary to go down to pH 2.98 (aspartic acid isoelectric point).

Analogously, amino acids carrying a second basic group, such as lysine, will have a net positive charge at pH 6 (eq. 22.7). Its isoelectric point is at pH

$$\text{(22.7)} \quad \underset{\text{Lysine zwitterion}}{\underset{\displaystyle CH_2(CH_2)_3NH_2}{H_3\overset{+}{N}\text{—}\overset{|}{C}HCOO^-}} + H_2O \xrightleftharpoons{\text{pH 6}} \underset{\text{Lysine monocation}}{\underset{\displaystyle CH_2(CH_2)_3\overset{+}{N}H_3}{H_3\overset{+}{N}\text{—}\overset{|}{C}HCOO^-}} + OH^-$$

9.47 (since lysine is a *basic* amino acid). Thus, lysine and aspartic acid will move in opposite directions during an electrophoretic separation carried out at pH 6.

Example

What is the predominant species present in solution when methionine is dissolved in water at (a) pH 1; (b) pH 6; (c) pH 12?

Answer

At pH 1, the solution is strongly acidic and therefore the carboxyl group and the amine are protonated:

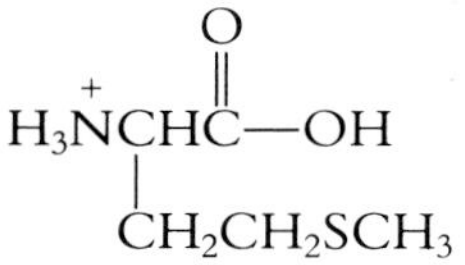

$$\overset{+}{H_3N}CH(CH_2CH_2SCH_3)\overset{O}{\overset{\|}{C}}—OH$$

Methionine is neither a basic nor acidic amino acid and hence its isoelectric pH is near 6. At this pH, the zwitterion is always at maximum concentration:

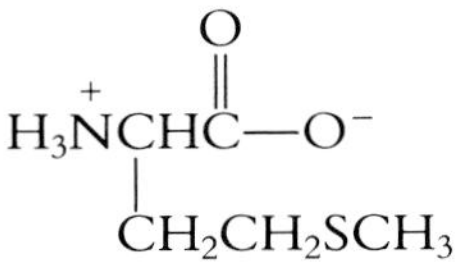

$$\overset{+}{H_3N}CH(CH_2CH_2SCH_3)\overset{O}{\overset{\|}{C}}—O^-$$

At pH 12, the solution is basic enough to remove a proton from the $—NH_3^+$ group:

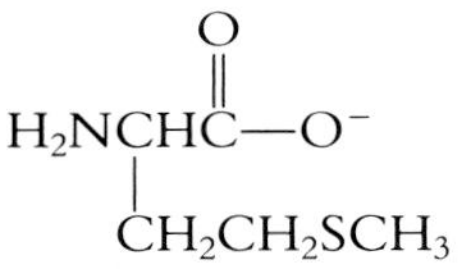

$$H_2NCH(CH_2CH_2SCH_3)\overset{O}{\overset{\|}{C}}—O^-$$

Problems

22–1 What is the predominant species present in solution when the following amino acids or peptides are dissolved in water at (a) pH 1; (b) pH 6; (c) pH 12?

a. alanine
b. threonine
c. lysine
d. arginine
e. glutamic acid
f. asparagine
g. tyrosine
h. histidine
i.  $H_2NCH(CH_3)CONHCH_2COOH$
j.  $H_2NCH(CH_3)CONHCH(COOH)(CH_2)_4NH_2$

22–2 When aspartic acid cation is treated with one equivalent of base, the carboxyl group on the carbon next to the nitrogen transfers its proton (eq. 1). Explain why *this* carboxyl group, rather than the more distant carboxyl group, gives up its proton.

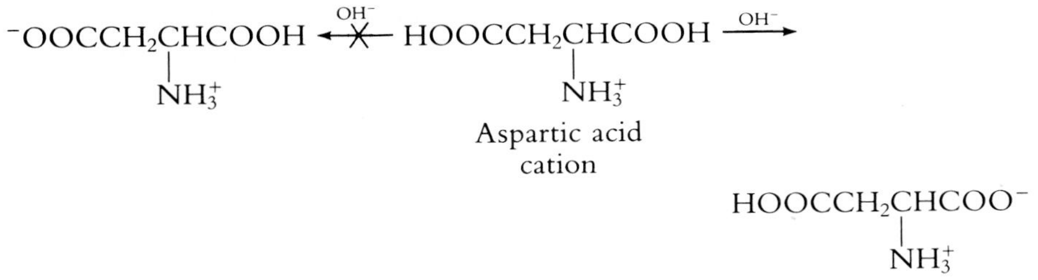

$$^-OOCCH_2CH(NH_3^+)COOH \xleftarrow[\times]{OH^-} HOOCCH_2CH(NH_3^+)COOH \xrightarrow{OH^-} HOOCCH_2CH(NH_3^+)COO^- \quad (1)$$

Aspartic acid cation

22:4 Absolute Configuration of Amino Acids: Application of the Fischer Convention to Amino Acids

Except for glycine, all the amino acids listed in Table 22–1 contain a chiral α-carbon atom. Several of them possess an additional chiral center as well. In general, only one enantiomer of each amino acid is prevalent in nature. Furthermore, with few exceptions, natural amino acids have the *same* absolute configuration at the α-carbon atom. Several amino acid enantiomers with the "abnormal" configuration at the α-carbon atom have been isolated from microorganisms. Interestingly, these abnormal amino acids have been found in antibiotics of fungal and bacterial origin. Humans and other higher animals cannot utilize amino acids of the "wrong" configuration as food.

All absolute configuration assignments are originally based upon an X-ray diffraction analysis. In the case of the standard amino acids, such analysis has shown that they all can be represented by structure **5**.

H R

H_2N COOH

5

Absolute configuration at
α-carbon of standard amino acids

In the *R,S* nomenclature system, the H will always be of lowest priority and the NH_2 group of highest priority. In all amino acids except cysteine, the COOH group has second priority and the chiral center will hence be designated *S*. In cysteine, however, the R group contains a carbon bonded to sulfur which takes priority over the COOH group (carbon bonded to oxygen). Cysteine is therefore designated *R*. Remember that the *R,S* system is merely a method of nomenclature. All standard amino acids have the *same absolute configuration*.

The *R,S* system of nomenclature is, in practice, rarely used with natural amino acids. Instead, the D,L system which was developed for sugars (sec. 20:1) is still generally employed. This was based on the structure of D-glyceraldehyde. Chemical transformations have shown that the natural standard amino acid serine is related in absolute configuration to L-glyceraldehyde, **6**, rather than to D-glyceraldehyde, and hence it was assigned the

H CH_2OH

HO CHO

6

L-Glyceraldehyde

structure **7**. Since all the standard amino acids have the same absolute configuration, they are called L-amino acids. Remember that the observed direc-

H CH_2OH

H_2N COOH

7

L-Serine

tion of optical rotation does not necessarily correlate with the absolute configuration. This is apparent from the specific rotations (sodium D line) of the following amino acids: L-alanine, +1.8°; L-leucine, −11.0°; L-phenylalanine, −34.5°; L-lysine, +13.5°. Furthermore, optical rotation depends upon solvent, temperature, and concentration.

22:5 Natural Sources of Amino Acids

The synthesis of peptides and proteins depends upon the availability of enantiomerically pure amino acids. We might imagine that naturally available proteins themselves would be a good source of these compounds, but the need is not so simply met. Most proteins contain a wide variety of amino acids, and large scale separation and purification of each of the components is often too laborious to be worthwhile.

There are, however, some proteins that are extremely rich in only a few amino acids, and are therefore useful sources. For instance, the protein fiber of silk consists almost entirely of glycine, L-alanine, L-serine, and L-tyrosine, and it serves as a source for the last three. Tyrosine is easily separated because of its low solubility in water, and serine and alanine are selectively precipitated as their aromatic sulfonic acid salts. Similarly, L-asparagine is widely distributed in plants and is conveniently obtained from certain seedlings; L-cystine is available from hair.

One modern method for obtaining enantiomerically pure amino acids is through fermentation. There are certain microorganisms which, when fed

Box 22–1. Dating of Organic Materials in Archaeological Research

For many years, measurement of the amount of radioactive ^{14}C present in organic compounds has been used to date ancient artifacts or plant remains. This is based on the assumption that all living plants and animals have the same ratio of the abundant ^{12}C isotope to the minor radioactive ^{14}C isotope. When the plant or animal dies, it no longer exchanges carbon with its surroundings and the amount of ^{14}C begins to decrease due to radioactive decay (^{14}C has a half-life of 5760 years).

It was more recently discovered that after an organism dies, the optical rotations of its amino acids change due to racemization at their chiral carbons. The percentage of amino acids that have changed from L to D is a measure of how long the organism has been dead. Attempts to use amino acid racemization for archaeological or geological dating must be applied with caution, however. While nuclear processes, such as radioactive decay, are not dependent upon temperature, pH, or other properties of the environment, racemization rates are. Racemization is a chemical process involving a bond-breaking and bond-making sequence that is surely dependent upon temperature, pH, moisture content, and other factors. As an example, consider what effect trace amounts of base might have on the rate of the following reaction:

D-Amino acid residue

Racemization is apparently complete in fossils from the Miocene Age (12 million years ago) and hence the dating method cannot be applied to samples originating before that time. It is possible that racemization processes also could have occurred in the carbonaceous meteorites; therefore, the existence of racemic amino acids in the meteorites is not conclusive proof of their abiogenic (nonbiochemical) origin.

the proper diet, produce pure L-amino acids as metabolic by-products. This approach provides several amino acids that are used on a commercial scale as supplements in human food and in animal feed.

22:6 Laboratory Syntheses of Amino Acids

Because only a few amino acids can be obtained from natural protein sources, much effort has been spent on amino acid synthesis. There are a number of instances in which the use of synthetic amino acids may be necessary or desirable: (1) the natural amino acid is not readily available by protein hydrolysis; (2) an isotopically labeled amino acid is required; (3) a totally new amino acid is desired.

The simplest route for the synthesis of α-amino acids involves the reaction of an α-halo acid with ammonia (eq. 22.8; sec. 18:11).

$$\text{(22.8)} \qquad \underset{\displaystyle X}{RCHCOOH} + 2NH_3 \rightarrow \underset{\displaystyle NH_2}{RCHCOOH} + NH_4^+X^-$$

Glycine is made commercially from chloroacetic acid and ammonia, and virtually all of the other aliphatic amino acids have been synthesized *in their racemic form* from the appropriate α-bromoacid and ammonia. Optically active α-bromo acids are occasionally used and, under most conditions, this reaction proceeds with inversion of configuration.

The α-haloacids can also be reacted with nitrogen nucleophiles other than ammonia. For example, the Gabriel synthesis (sec. 21:6) uses potassium phthalimide as a starting material (eq. 22.9).

$$\text{(22.9)} \qquad \text{(phthalimide)}N^-K^+ + ClCH_2COOC_2H_5 \rightarrow$$

$$\text{(phthalimide)}N\text{—}CH_2COOC_2H_5 \xrightarrow[\text{2. HCl}]{\text{1. }N_2H_4} \underset{\displaystyle NH_3^+}{CH_2COO^-} + \text{(phthalhydrazide, NH—NH)}$$

Glycine (80%)

Several synthetic methods are known that involve extension of the carbon chain. One of these is the **Strecker synthesis** (eq. 22.10) in which an aldehyde or ketone is reacted with ammonia and hydrogen cyanide (formed in the flask from a mixture of ammonium chloride and potassium cyanide) to yield an α-aminonitrile as an intermediate. The aminonitrile is then hydrolyzed in acid to yield the amino acid.

$$\text{(22.10)}\quad RCHO + NH_3 + HCN \xrightarrow[H_2O]{} H_2O + \underset{\displaystyle \alpha\text{-Aminonitrile}}{\underset{\displaystyle NH_2}{\overset{|}{RCHCN}}} \xrightarrow[heat]{H_3O^+} \underset{\displaystyle NH_2}{\overset{|}{RCH{-}COOH}} \quad \text{Strecker synthesis}$$

Formation of the aminonitrile proceeds through the imine, which undergoes nucleophilic attack by CN^- (eq. 22.11).

$$\text{(22.11)}\quad RCHO + NH_3 \rightarrow RCH{=}\ddot{N}H + H_2O \xrightarrow{HCN} RCH{=}\overset{+}{N}H_2 + CN^- \rightarrow \underset{\displaystyle CN}{\overset{|}{RCH{-}\ddot{N}H_2}}$$

Esters are also the starting materials for amino acid syntheses which rely on the acidity of the α hydrogens of the ester (one example was given in Prob. 19–4, p. 718). Equation 22.12 shows a process which is used industrially to produce L-phenylalanine and its derivatives. The first step of the

$$\text{(22.12)}\quad PhCHO + \underset{\displaystyle NHCOCH_3}{\overset{|}{CH_2COOH}} \xrightarrow[(CH_3CO)_2O]{CH_3COO^-} \underset{\displaystyle NHCOCH_3}{\overset{|}{PhCH{=}CCOOH}} \xrightarrow[\text{2. } H_3O^+]{\text{1. } H_2\text{, chiral Rh catalyst}} \underset{\displaystyle \text{L-Phenylalanine}}{PhCH_2\text{–}C(H)(NH_2)\text{–}COOH}$$

synthesis is a variant of the mixed aldol condensation (sec. 18:7) in which the acid is the source of the enolate ion. The next step, catalytic reduction, is carried out as usual in the presence of a transition metal catalyst, except that the catalyst is bound to a chiral arylphosphorus compound. Because of this, the reduction gives predominantly one product enantiomer. This is an excellent example of a modern trend toward using chiral reagents for synthesis so that a chiral product, rather than an enantiomeric mixture, is obtained. This process is sometimes called **asymmetric induction**.

Other laboratory preparations of amino acids yield racemic mixtures. However, enantiomerically pure amino acids would be needed to synthesize biologically active proteins. For this reason, a number of racemic α-amino acids have been resolved in the traditional way. The amino group is acylated and the *N*-acylamino acid is then allowed to react with an optically active base such as brucine or strychnine. The diastereomeric salts are separated and the free acid then regenerated (sec. 21:4).

Another method of resolution depends on the preference shown by enzymes for catalyzing reactions of L-amino acid derivatives only. The first step is to acylate the racemic amino acid with acetic anhydride. The acylamino acid is then exposed to a deacylase, an enzyme that catalyzes the hydrolysis of *N*-acylated amino acids (eq. 22.13).

$$\text{(22.13)}\quad \underset{\text{Racemic alanine}}{\text{DL-}CH_3\underset{NH_2}{\underset{|}{C}}HCOOH} \xrightarrow{(CH_3CO)_2O} \underset{\text{Racemic } N\text{-acetylalanine}}{\text{DL-}CH_3\underset{CH_3CONH}{\underset{|}{C}}HCOOH} \xrightarrow[\text{deacylase, } H_2O]{\text{hog kidney}}$$

$$CH_3COOH + \text{L-}CH_3\underset{NH_2}{\underset{|}{C}}HCOOH + \text{D-}CH_3\underset{CH_3CONH}{\underset{|}{C}}HCOOH \xrightarrow{\text{separate}} \underset{\text{Pure L-alanine}}{\text{L-}CH_3\underset{NH_2}{\underset{|}{C}}HCOOH}$$

Only the L enantiomer is affected by the deacylase, and a mixture of L-amino acid and D-acylamino acid is obtained. The two components are then easily separated by crystallization.

Problems

22–3 Use the indicated synthetic method to synthesize the following amino acids. Show all intermediates that occur during the process.

a. alanine (Gabriel)
b. valine (Strecker)
c. phenylalanine (Br displacement by NH_3)
d. L-isoleucine (chiral synthesis)
e. methionine (Strecker)
f. glutamic acid (Br displacement by N_3^-

22–4 For the synthesis of labeled amino acids, the cheapest method is usually that which allows one to purchase the labeled starting material in the simplest form (e.g., $^{14}CN^-$, $^{15}NH_3$, $^{13}CN^-$). What method(s) might you choose to prepare the following labeled amino acids?

a. $CH_3\underset{NH_2}{\underset{|}{C}}H^{13}COOH$

b. $CH_3\underset{^{15}NH_2}{\underset{|}{C}}HCOOH$

c. $CH_3—^{14}\underset{NH_2}{\underset{|}{C}}HCOOH$

22:7 Small Peptides: An Introduction to Their Sources and Bioactivity

Individual amino acids occur to some extent in cells in the "free" state, that is, as unbound, separate molecules. In this form they have little, if any, physiological activity. Of much more importance is their occurrence in the polymeric form as peptides and proteins (sec. 22:1).

Even low molecular weight peptides often have physiological activity. Thus, L-aspartyl-L-phenylalanine methyl ester (Aspartame) is a synthetic sweetener many times sweeter (per unit weight) than ordinary table sugar (sucrose, sec. 20:7). Some slightly more complex peptides, such as the penta-

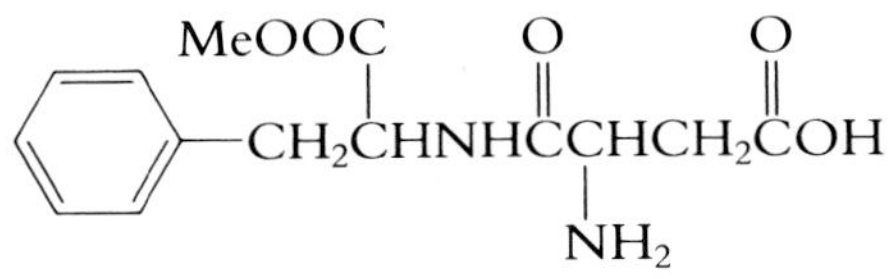

MeOOC O O

C_6H_5—$CH_2CHNHCCHCH_2COH$

NH_2

Aspartame
(L-aspartyl-L-phenylalanine methyl ester)

peptide known as methionine-enkephalin, have been found to have analgesic and other opiate effects similar to those of morphine (sec. 21:11).

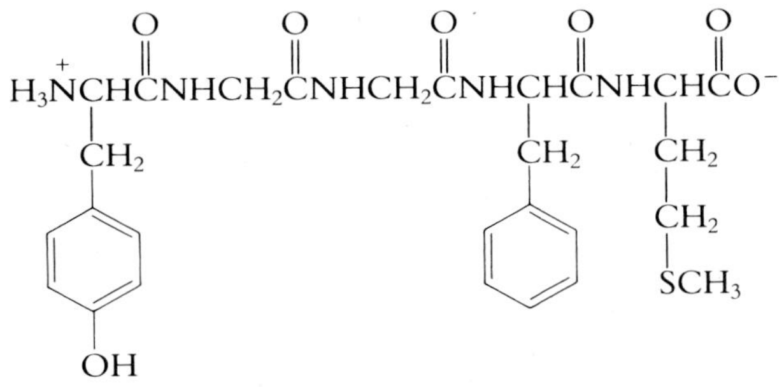

O O O O O

$H_3\overset{+}{N}CHCNHCH_2CNHCH_2CNHCHCNHCHCO^-$

CH_2 CH_2 CH_2

CH_2

SCH_3

OH

Methionine-enkephalin

Small peptides such as methionine-enkephalin form a portion of longer polypeptides known as *endorphins*. Endorphins are active in minute traces and have been implicated in analgesia, narcotic addiction, and some forms of mental illness.

Small peptides occur in the brain and have been found to play an important role in the control of endocrine glands such as the thyroid, the gonads, and the adrenal cortex. For example, one molecule that regulates the thyroid gland is called the thyrotropin-releasing factor (TRF) and was found to be a simple tripeptide. R. Guillemin, R. S. Yalow, and A. Schally shared the Nobel Prize in 1977 for work which, in part, culminated in the identification of TRF and other brain peptides. It took 300,000 sheep hypothalamus glands

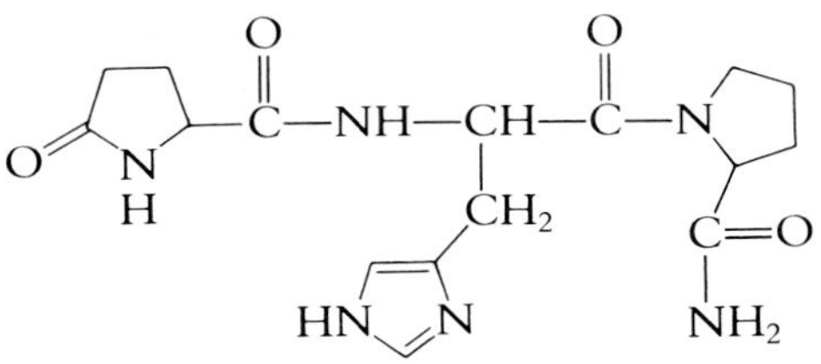

O O

O N H —C—NH—CH—C—N

CH_2 C=O

HN N NH_2

TRF

to yield 1.0 mg of pure, biologically-active TRF. Its structure was determined by infrared and mass spectrometry. This experiment illustrates the remarkably minute natural concentration of such hormones and the ability of modern spectroscopy to deduce structure from small quantities of material.

Bradykinin is a natural nonapeptide with hypotensive (blood pressure lowering) activity. It is called the "pain peptide" because of its potent pain-producing action. The nonapeptide oxytocin is a uterine muscle contractant. Phalloidine is a cyclic hexapeptide found to be the toxic substance of the deadly poisonous mushroom, *Amanita phalloides*.

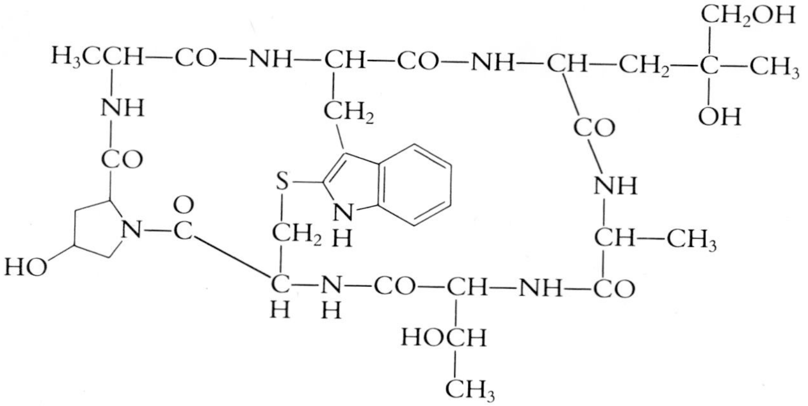

```
                                                            CH2OH
                                                              |
H3CCH—CO—NH—CH—CO—NH—CH—CH2—C—CH3
   |              |           \          |
   NH            CH2           CO        OH
   |               \             \
   CO          S—(indole)         NH
   |           |    N             |
  (ring)N—C    CH2  H            CH—CH3
 HO        O \  |                 |
               C—N—CO—CH—NH—CO
               H  H     |
                       HOCH
                        |
                        CH3
```

Phalloidine
(from *Amanita phalloides* mushroom)

Our study of peptides and proteins commences with an investigation of methods for their structure determination.

22:8 Amino Acid Analysis and Structure

Suppose that a new peptide or protein has been discovered in nature. How can we determine its structure? Because of the size and complexity of some peptides and proteins, ordinary methods of infrared, nuclear magnetic resonance, and mass spectrometry are often not directly applicable, although they can be used for low molecular weight peptides.

The first step after isolation and purification of the protein is to determine its molecular weight. This can be done by subjecting the protein solution to very powerful accelerations in an **ultracentrifuge** and studying the sedimentation rate. Larger proteins will have a higher sedimentation constant than smaller proteins. With a set of standard proteins of known molecular weight, one can compare the unknown with the knowns and arrive at a molecular weight for the new protein.

The second step is to determine the number and kinds of amino acids that make up the protein; for example, how many serines, tyrosines, and so forth are present. In order to determine the amino acid content of the protein, all the peptide bonds in the molecule must be hydrolytically cleaved. This can be done by heating a solution of the peptide in 6 *M* hydrochloric acid for a 24-hour period. A separation of the mixture of individual amino acids thus obtained can be accomplished by **paper** or **thin-layer chromatography**. A small portion of the solution containing the mixture is evaporated near the bottom of a strip of paper or a small glass slide coated with cellulose or silica gel. The paper or slide is then placed in a small trough of solvent and the solvent is allowed to rise by capillary action. Individual amino acids migrate different distances as the solvent moves. The ratio of migratory distance to solvent movement is called the R_f value. The migration distance for each amino acid depends on its size and polarity, and upon the composition and pH of the solvent.

The presence of the amino acid on the chromatogram can then be detected by spraying with a visualization agent, a reagent that reacts with the amino acid to form a colored substance. The most common reagent is a solution of **ninhydrin**, a compound that reacts with amino groups to form a purple dye (Ruhemann's purple; eq. 22.14). Proline, which contains a secondary rather than a primary amino group, reacts to give a yellow color instead of the purple dye.

$$\text{Ninhydrin} + H_2N\underset{\displaystyle R}{\underset{|}{C}}HCO_2H \xrightarrow[\text{steps}]{\text{several}} \text{Ruhemann's purple } (\lambda_{max} \text{ at } 570 \text{ nm}) \qquad (22.14)$$

The chromatogram that we are left with at the end of this process is a strip of paper, or a slide, containing a number of colored spots at varying distances from the bottom. We can identify the unknown amino acids by comparing the positions of the spots to those in a chromatogram carried out under identical conditions with known amino acids.

Quantitative analysis of an amino acid mixture can be accomplished by elution chromatography on **ion exchange resin** columns in an instrument known as an **amino acid analyzer**. An ion exchange resin is an organic polymer that contains ionic groups. The most frequently used resin in amino acid analysis is a polymer of sodium *p*-styrenesulfonate. If an amino acid mixture in a buffer solution at an acidic pH is passed through such a resin, the cationic $—\overset{+}{N}H_3$ groups of the amino acids will be attracted to the negatively-charged sulfonate groups, thereby slowing their passage through the column. The styrene portion of the resin is nonpolar (lipophilic) and amino acids with nonpolar side chains will also pass through the column slowly. Because of a balance between these two factors, each individual amino acid has a characteristic elution time as buffered solution is passed through the column. This elution time can be compared with that of the known standard amino acid. As each acid comes off the column it enters a cell where it reacts with a solution of ninhydrin, and the strong absorption at 570 nm is then monitored.

Figure 22–3 shows the amino acid analysis of a mixture obtained by

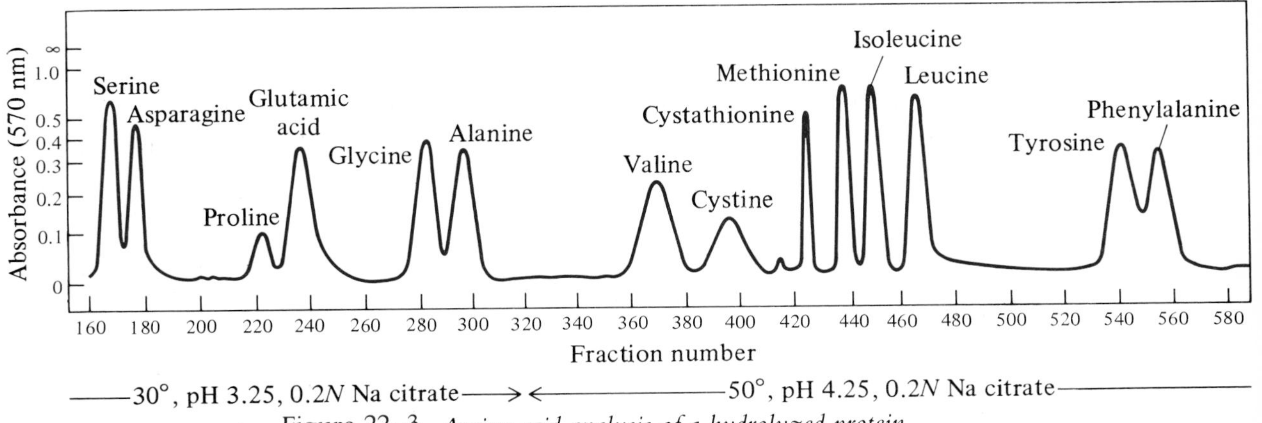

Figure 22–3. *Amino acid analysis of a hydrolyzed protein.*

hydrolyzing an unknown protein. By use of the amino acid analyzer, a complete quantitative analysis of a complex protein can be accomplished in about 24 hours. This analysis tells us, according to the area under each peak, how many residues of each type are in the protein molecule, but nothing about their arrangement or *sequence*. We take up that topic next.

22:9 Amino Acid Sequence Determination in Peptides and Proteins

Once the total amino acid composition of a protein or peptide has been established, the structural problem has just begun; the infinitely more challenging task of determining the sequence of amino acids in the chain lies ahead. A tetrapeptide, for instance, which contains only four different amino acids, can exist in 24 different sequential arrangements!

Before going into the details of amino acid sequence determination, it is worthwhile to look at the structure of a typical peptide. Figure 22–4 gives the structure of a tetrapeptide, alanylglycyltyrosylglutamine, which can be abbreviated as Ala-Gly-Tyr-Gln. In all noncyclic peptides, one of the end (terminal) amino acids will have a free α-amino group not bound in an amide bond. This residue is known as the N-*terminal amino acid.* For the peptide in Figure 22–4, the *N*-terminal amino acid is alanine. The amino acid at the other end of the peptide chain has a free carboxyl group and is known as the C-*terminal amino acid* (glutamine in Fig. 22–4). Peptides are named as derivatives of the *C*-terminal amino acid, the names of the other amino acids being given in sequence, starting with the *N*-terminal acid. *It is customary to write the formula of the peptide with the* N-*terminal acid (numbered one) on the left and the* C-*terminal acid on the right.* When the abbreviated name is given, a notation is sometimes used where H— marks the position of the free α-amino group and —OH locates the free α-carboxyl group, as in Figure 22–4.

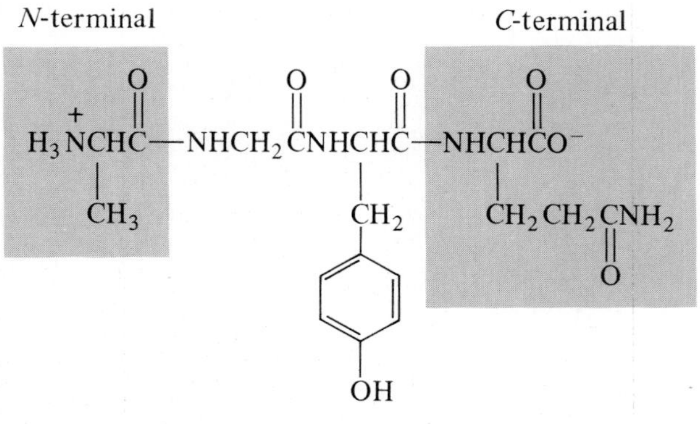

N-terminal   C-terminal

$$H_3\overset{+}{N}\underset{\displaystyle CH_3}{\underset{|}{CH}}\overset{O}{\overset{\|}{C}}—NHCH_2\overset{O}{\overset{\|}{C}}NH\underset{\displaystyle CH_2C_6H_4OH}{\underset{|}{CH}}\overset{O}{\overset{\|}{C}}—NH\underset{\displaystyle CH_2CH_2\underset{\underset{\displaystyle O}{\|}}{C}NH_2}{\underset{|}{CH}}\overset{O}{\overset{\|}{C}}O^-$$

Full name: alanylglycyltyrosylglutamine

Abbreviated name: H—Ala—Gly—Tyr—Gln—OH

Figure 22–4. *The structure of the tetrapeptide alanylglycyltyrosylglutamine.*

Chemically, the easiest amino acids to identify and place in sequence are the terminal ones, particularly the *N*-terminal residues. Most of the methods currently used depend upon the reactivity of the free amino group of the *N*-terminal acid. One example involves reacting the peptide with a reagent known as *dansyl chloride* (5-*d*imethyl*a*mino*n*aphthalene-1-*s*ulfon*yl chloride*). Equation 22.15 shows that a strong sulfonamide bond is made to the *N*-terminal acid, and that the dansyl peptide can then be hydrolyzed to give the free amino acids along with a dansyl amino acid from the *N*-terminal residue. The dansyl amino acid is insoluble in water and can be extracted into an

(22.15) Dansyl chloride (naphthalene ring with NMe_2 and SO_2Cl) + Ala-Gly-Tyr-Gln $\rightarrow$

Dansyl peptide (naphthalene ring with NMe_2 and SO_2–NH–$CH_3CHC(=O)$–Gly-Tyr-Gln) $\xrightarrow{H^+}$ Dansyl *N*-terminal amino acid (naphthalene ring with NMe_2 and SO_2–NH–$CH_3CHCOOH$) + glycine + tyrosine + glutamine

organic solvent. It is thus separated from the water-soluble amino acids that arise from the remainder of the peptide chain. The dansyl derivative is used because it is highly fluorescent and thus can be detected in trace amounts. The end result is that we have detected and separated the *N*-terminal amino acid, and it is now ready to be analyzed.

Sensitive detection methods using reagents such as dansyl chloride are important when analyzing a protein for its *N*-terminal amino acid, since there is only one *N*-terminal acid among the many other amino acid residues. However, this type of sequence determination method suffers from the fact that the entire peptide is hydrolyzed to obtain knowledge of just the *N*-terminal amino acid. To avoid this problem the most widely used method for *N*-terminal analysis is probably the **Edman degradation**. In the Edman procedure, phenyl isothiocyanate is reacted with the *N*-terminal amino acid (eq. 22.16).

$$(22.16)\quad C_6H_5\text{—}N\text{=}C\text{=}S + CH_3CH(NH_2)C(\text{=}O)\text{—Gly-Tyr-Gln} \xrightarrow{\text{base}}$$

Phenyl isothiocyanate

$$C_6H_5\text{—}NHC(\text{=}S)\text{—}NHCH(CH_3)C(\text{=}O)\text{—Gly-Tyr-Gln} \xrightarrow{H_3O^+ \text{(mild)}}$$

$$\text{Gly-Tyr-Gln} + C_6H_5N\text{(ring: C=S, NH, CH}(CH_3)\text{, C=O)}$$

this part derived from the *N*-terminal amino acid

Alanine phenylthiohydantoin (PHT-alanine)

The *N*-terminal acid is isolated as a phenylthiohydantoin (PHT) derivative; the hydrolysis is mild enough so that the remainder of the peptide chain is recovered intact. This is the great advantage of the Edman degradation, since the residual peptide fragment can be recycled through further Edman degradations. The entire process has been automated with an instrument that permits rapid automatic determination of the sequence of a polypeptide up to approximately 60 amino acid residues.

In recent years, mass spectral analysis has been applied to the determination of amino acid sequences. Since a peptide such as Ala-Gly-Ser is a different molecule from Ala-Ser-Gly, the two peptides will have different fragmentation patterns. Computer analysis of the spectrum is usually necessary, and peptides of up to 20 amino acid residues have been sequenced in this way.

Example

Draw the structures of Ala-Gly (alanylglycine) and Gly-Ala (glycylalanine).

Answer

For Ala-Gly, the *N*-terminal amino acid will be alanine and the *C*-terminal acid will be glycine:

N-terminal — *C*-terminal

$$H_2NCH(CH_3)C(\text{=}O)\text{—}NHCH_2COOH$$

Ala-Gly

For Gly-Ala, the *N*-terminal amino acid will be glycine and the *C*-terminal acid will be alanine:

Box 22–2. Enkephalins and Endorphins

Interest in amino acid polymers smaller than proteins has been stimulated by the recent discovery of biological activity in endorphins, which are high molecular weight peptides, and enkephalins, which are small peptides containing only four or five amino acid residues. Some endorphins and enkephalins apparently play important roles in epilepsy, memory, pain perception, and a number of other biological processes.

The opiate morphine has been used to relieve pain for over 400 years. It probably exerts its pain-relieving action by binding to a receptor site involved in the perception of pain. [This is fundamentally different from the action of aspirin (sec. 19:8).] Now it has been discovered that enkephalins (Met enkephalin, for example; sec. 22:7) have a similar receptor site and also influence pain perception. A problem with the search for new pain relievers similar to morphine has always been the fact that some people find the pain relief from a sugar pill (placebo) equal to that of morphine if they are first told that the medicine they are receiving will stop their pain. So it seems possible that the human body can manufacture its own pain killers (enkephalins) in some circumstances. For example, it is a common occurrence that soldiers who are wounded in battle or athletes who suffer an injury during competition continue fighting or playing and only feel pain *after* the stress of the moment has passed. In such instances, the stress of injury causes the secretion of enkephalins that relieve pain. Once the stress has passed, the compounds are no longer synthesized and the ones secreted are rapidly hydrolyzed back to the component amino acids.

Enkephalins are also part of the "alarm" system which mammals have evolved in order to escape from predators. The initial fright causes the body to manufacture enkephalins which then trigger adrenaline release and cause an immediate response of flight from danger. The cliché about "adrenaline flowing" in competitive athletes is a result of peptide chemistry in the body.

N-terminal → $H_2NCH_2C(=O)$—$NHCH(CH_3)COOH$ ← *C*-terminal

Gly-Ala

Problems

22–5 Give structures for peptides (a–c) and name peptides (d–e).

a. His-Ser

b. methionylcysteine

c. Pro-Tyr

d. $H_2NCH_2C(=O)NHCH_2COOH$

e. $H_2NCH(CH_2COOH)C(=O)NHCH(CH_2COOH)C(=O)NHCH_2COOH$

22–6 Identify the amino acid residues present in methionine-enkephalin and phalloidine (sec. 22:7). There are some nonstandard amino acid residues in phalloidine. What are their structures?

22–7 Insulin was the first protein to have its complete amino acid sequence determined (MW about 6000). This was reported in 1953 by the English chemist F. Sanger, who received the Nobel Prize in Chemistry for this work and for his development of sequencing techniques. For *N*-terminal analysis, Sanger used the reagent 2,4-dinitrofluorobenzene (DNFB), which reacts as shown in Equation 2.

$$\text{DNFB (1-F-2,4-}(NO_2)_2C_6H_3) + H_2N\text{-peptide} \rightarrow \text{2,4-}(NO_2)_2C_6H_3\text{-HN-peptide} + F^- \quad (2)$$

Show the products that would result if each of the following peptides were first treated with DNFB, and then hydrolyzed.

a. Phe-Gly-Ser b. Leu-Ala-Asp c. Ile-Ser-Lys

22:10 Partial Hydrolysis of Proteins

The methods of sequence determination discussed so far are impractical for a total sequence analysis of a protein or polypeptide with more than 50 amino acid residues. The key to successful sequence analysis of a large molecule is to hydrolyze it partially to small fragments whose structure can be determined by using one of the methods discussed in the previous section. It is vital that the fragments overlap in amino acid content. Otherwise there would be no clue as to how the fragment peptides fit together to form the complete chain. Partial hydrolysis of a protein can be achieved by decreasing the acid strength, the temperature, or the time of hydrolysis.

In the structure determination for insulin (Prob. 22–7), the protein was partially hydrolyzed to obtain a number of small peptides which were then individually analyzed. For example, one fragment obtained was an octapeptide whose amino acid analysis showed the following composition: Ala, Lys, Phe_2, Pro_2, Thr, Tyr. This is not a sequence, simply an alphabetical listing of the contained amino acids, with the subscripts indicating the number of each kind of amino acid residue. End-group analysis was used to identify phenylalanine as the *N*-terminal amino acid of the octapeptide. The octapeptide was then hydrolyzed further and five small peptides obtained, the sequence of each of which was readily determined by *N*-terminal and/or *C*-terminal analysis. These peptides were Pro-Ala-Lys, Phe-Phe-Tyr, Thr-Pro, Lys-Pro, and Tyr-Thr. Because phenylalanine was known to be *N*-terminal in the original peptide, and because the other amino acids occurred in more than one fragment peptide, the sequence of the octapeptide was defined as shown in Figure 22–5. [Try it for yourself, starting with a fragment containing the *N*-terminal residue (Phe-Phe-Tyr) and then looking for residues with overlapping amino acids)]. This type of fragment analysis is continued until the amino acid analysis and sequence of the entire insulin chain is complete.

Acid-catalyzed hydrolysis of peptides is rather nonspecific and produces a large number of fragments. This makes it difficult to obtain enough fragments in sufficient quantity to carry out a complete analysis. A much more subtle and economical technique is to use various proteolytic enzymes, the naturally occuring catalysts whose function it is to hydrolyze peptides and proteins. They split the undecoded peptide into small fragments, but they do

Phe–Phe–Tyr
⋮
Tyr–Thr
⋮
Thr–Pro
⋮
Pro–Ala–Lys
⋮
Lys–Pro

Phe–Phe–Tyr–Thr–Pro–Ala–Lys–Pro

Figure 22–5. *Sequence determination of an octapeptide from insulin by fitting together the hydrolysis fragments of the octapeptide.*

so in specific, known ways, in contrast to the random acid-catalyzed hydrolyses.

For investigating the *C*-terminal amino acid residues, *carboxypeptidases* (enzymes secreted by the pancreas) are used. They will only attack the amide bonds of terminal amino acids containing free α-carboxyl groups. A carboxypeptidase will relentlessly eat its way down a peptide chain, continually releasing the *C*-terminal amino acid. In order to get information about the sequence, it is necessary to study the concentration of released amino acids as a function of time. While it is usually not possible to unravel an entire peptide chain sequence using carboxypeptidase alone, a time study of the enzymatic hydrolysis will often suffice to completely unravel the sequence in a pentapeptide or hexapeptide fragment.

Two more proteolytic enzymes, *trypsin* and *chymotrypsin,* also originate in the pancreas, though they exert their digestive action in the small intestine. Trypsin shows a preference for hydrolyzing amides in which the carbonyl group is part of an arginine or lysine residue. Chymotrypsin tends to catalyze hydrolysis at the carbonyl groups of aromatic amino acid residues (phenylalanine, tryptophan, tyrosine), although it will also attack amide bonds from a few other amino acids.

Subtilisin and *papain* are enzymes of very low specificity and are useful for reducing the peptide chain to small fragments. The use of papain as a meat tenderizer is based upon this peptide-hydrolyzing ability.

Problems

22–8 When a peptide whose amino acid analysis shows His, Phe, Ser, Tyr, and Val is partially hydrolyzed, it yields the dipeptides Val-Ser, Tyr-His, Phe-Val, and Ser-Tyr. What is the structure of the original peptide?

22–9 What smaller peptides would be formed if the following peptide is enzymatically hydrolyzed with chymotrypsin: Ser-Ala-Ala-Glu-Glu-Phe-Val-Met-Tyr-Met-Glu-Trp-Gly-Gly-Phe?

Sections 22:8 and 22:10 were devoted to one of the major activities of protein chemists—analysis of the content and structure of peptides and proteins. Due to the pervasive and indispensable role of these molecules in the living cell, it is not surprising that so much effort is expended on their analysis. But analysis is not the only concern of protein chemists. They are also concerned with the synthesis of peptides and proteins, for several reasons. Synthesis is an indisputable method for verifying a structure deduced from analytical studies. Furthermore, synthesis of unnatural peptides can tell us a great deal about the biological activity of the natural ones. For instance, in a peptide containing 50 amino acid residues, it may be that the biological activity is due primarily to only a half dozen. One can discover this fact by synthesizing analogs of the natural peptide in which the amino acid sequence is altered in a systematic way. For these reasons, we'll devote the next three sections to peptide synthesis.

22:11 Peptide Synthesis: Protecting and Activating Groups

Because peptides and proteins are precisely-ordered heteropolymers, their synthesis utilizes different techniques than the syntheses of low molecular weight molecules, and of simple homopolymers. In peptide synthesis, the trick is to cause the amino acids to condense together under mild conditions *and in the correct sequence.* Amides can be prepared by heating amino acids at high temperature to drive off water formed in the condensation. If we wanted to prepare Gly-Ala this way, the reaction would unfortunately yield four dipeptides (eq. 22.17). This mixture of products results because the

$$\text{(22.17)}\quad \underset{\text{Glycine}}{H_2NCH_2COOH} + \underset{\text{Alanine}}{H_2N\underset{\displaystyle CH_3}{\underset{|}{C}}HCOOH} \xrightarrow{\text{heat}}$$

$$\underset{\text{Gly-Ala}}{H_2NCH_2\overset{\displaystyle O}{\overset{\|}{C}}\text{—}NH\underset{\displaystyle CH_3}{\underset{|}{C}}HCOOH} + \text{Ala-Gly} + \text{Ala-Ala} + \text{Gly-Gly}$$

nucleophilic NH_2 group of each acid can react equally well with the carboxyl of the other acid or with the carboxyl of another molecule of the same acid.

Selectivity in peptide synthesis is therefore a necessity in order to avoid indiscriminate reaction. It is achieved mainly by *blocking* the reactive amino group of one acid with a protective acyl group, and also *activating* the carboxyl group of this same acid, usually by converting it to a reactive ester. More specifically, in order to prepare Gly-Ala, we want the amine group of alanine to attack the carboxyl group of glycine. We therefore activate glycine's carboxyl group and deactivate its amino group. Alanine remains unchanged. Equation 22.18 gives a general outline of the method. Section 17:3 presented a detailed discussion of the relative reactivities of the carboxyl, ester, and amide carbonyl groups. Peptide synthesis, as exemplified in Equation 22.18, makes practical use of these relative reactivities.

$$\text{(22.18)}\quad H_2NCH_2COOH \xrightarrow{RCOCl} \overset{O}{\overset{\|}{RC}}NHCH_2COOH \xrightarrow{R'OH} \overset{O}{\overset{\|}{RC}}NHCH_2COOR'$$

Protecting group (RC=O) — Activating group (R')

$$\overset{O}{\overset{\|}{RC}}NHCH_2\overset{O}{\overset{\|}{C}}OR' + H_2N\underset{CH_3}{\underset{|}{C}}HCOOH \rightarrow RCONHCH_2\overset{O}{\overset{\|}{C}}NH\underset{CH_3}{\underset{|}{C}}HCOOH \xrightarrow{H^+} \text{Gly-Ala}$$

Acylated dipeptide — Dipeptide

Note that in order to obtain the desired Gly-Ala, the acylated dipeptide must be hydrolyzed to remove the protecting acyl group. Since the dipeptide amide bond can also be hydrolyzed, it is important to choose protective groups that can be easily removed without affecting the peptide bond. This is done by using a *carbamate ester* protecting group (eq. 22.19). Carbamate esters can be easily removed under mild conditions to give back the original amine (eq. 22.19). The critical requirement then revolves around the structure of the R group in the ester, which also must be removable under mild conditions.

$$\text{(22.19)}\quad R'O\overset{O}{\overset{\|}{C}}Cl + H_2N\underset{R''}{\underset{|}{C}}HCOOR \rightarrow R'O\overset{O}{\overset{\|}{C}}NH\underset{R''}{\underset{|}{C}}HCOOR \xrightarrow[\text{conditions}]{\text{mild}} H_2N\underset{R''}{\underset{|}{C}}HCOOR + R'H + CO_2$$

Carbamate ester

One protective group developed for amines is the *benzyloxycarbonyl* group (also called the carbobenzoxy group, and abbreviated Z in the chemical literature). The benzyloxycarbonyl group is introduced by way of the acid chloride, **8**. *N*-Acylation of the amino acid proceeds easily and in good yield in alkaline solution, and subsequent removal of the acyl group is equally easy (eq. 22.20). The second step of Equation 22.20 is a hydrogenol-

$$(22.20)\quad C_6H_5\text{-}CH_2O\overset{O}{\overset{\|}{C}}Cl + H_2N\underset{CH_3}{\underset{|}{C}}HCOOH \xrightarrow[\text{(blocking)}]{\text{NaOH, 25°}}$$

8
Benzyloxycarbonyl chloride

H-Ala-OH

$$C_6H_5\text{-}CH_2O\overset{O}{\overset{\|}{C}}NH\overset{CH_3}{\overset{|}{C}}HCOOH \xrightarrow[\text{(deblocking)}]{H_2,\ Pd/C}$$

A carbamate (90%)
Z-Ala-OH

$$H_2N\underset{CH_3}{\underset{|}{C}}HCO_2H + CO_2 + C_6H_5\text{-}CH_3$$

ysis which forms toluene and the unstable carbamic acid (HOOCNHR), which decarboxylates. In a real synthesis, the carbamate would be coupled with the second amino acid before hydrogenolysis. The hydrogenolysis would not affect the peptide bond.

A protective group related to benzyloxycarbonyl is *t-butoxycarbonyl* [$(CH_3)_3COCO$—, abbreviated Boc]. This group is attached to an amino group by way of the anhydride $(Boc)_2O$ (eq. 22.21).

$$(CH_3)_3CO\overset{O}{\overset{\|}{C}}O\overset{O}{\overset{\|}{C}}OC(CH_3)_3$$

$(Boc)_2O$

$$(22.21)\quad [(CH_3)_3CO\overset{O}{\overset{\|}{C}}]_2O + H_2N\underset{HOCH_2}{\underset{|}{C}}HCOOH \xrightarrow{\text{dioxane}}$$

$(Boc)_2O$ H-Ser-OH

$$(CH_3)_3CO\overset{O}{\overset{\|}{C}}\text{—}NH\underset{CH_2OH}{\underset{|}{C}}HCOOH + (CH_3)_3COH + CO_2$$

Boc-Ser-OH
(99%)

The Boc group has properties complementary to those of the benzyloxycarbonyl (Z) group. Unlike Z, Boc cannot be removed by hydrogenolysis. However, Boc is very acid-sensitive, and can be removed under mildly acidic conditions (eq. 22.22) that do not affect the peptide linkage (acidic hydrolysis is easily accomplished because of the stability of the *t*-butyl cation; see Prob. 17–11, p. 652).

$$(CH_3)_3CO\overset{\overset{\displaystyle O}{\|}}{C}\text{—Phe-Lys-OH} \xrightarrow[\text{acetic acid, 25°}]{1\ M\ \text{HCl in}} \text{H-Phe-Lys-OH} + (CH_3)_2C{=}CH_2 + CO_2 \tag{22.22}$$

(Boc-Phe-Lys-OH)

As we mentioned before, amide formation between a carboxyl group of one amino acid and a free (unprotected) amine (see eq. 22.18) can be achieved by transforming the carboxyl group to a more reactive functionality. Esters are useful carboxyl derivatives for coupling purposes and, in order to maximize the rate and yield of amide formation, esters containing good leaving groups are used. The *p*-nitrophenyl (—ONp) ester has an excellent leaving group (eq. 22.23), because the nitro group *para* to the oxygen stabilizes the phenoxide anion.

$$\text{Z—NH}\underset{\underset{\displaystyle CH(CH_3)_2}{|}}{CH}\overset{\overset{\displaystyle O}{\|}}{C}\text{—O—}C_6H_4\text{—}NO_2 + H_2N\overset{\overset{\displaystyle CH_3}{|}}{C}HCOOH \rightarrow \tag{22.23}$$

Z-Val-ONp (blocked and activated) H-Ala-OH

$$\text{Z—NH}\underset{\underset{\displaystyle CH(CH_3)_2}{|}}{CH}\overset{\overset{\displaystyle O}{\|}}{C}\text{—NH}\overset{\overset{\displaystyle CH_3}{|}}{C}HCOOH + HO\text{—}C_6H_4\text{—}NO_2$$

Z-Val-Ala-OH

One of the best methods of coupling amino acids is with the reagent *N,N'*-dicyclohexylcarbodiimide (DCCD, eq. 22.24). This reagent activates the carboxyl group *in situ,* that is, directly in the reaction flask, and a prior activating step is not necessary. Note that the amino acid with the free amino group, glycine, is present as its ethyl ester. This is necessary to prevent the DCCD from reacting with the carboxyl group of glycine.

(22.24) $Z\text{—NHCH}_2\overset{O}{\overset{\|}{C}}\text{NH}\underset{\text{PhCH}_2}{\underset{|}{\text{CH}}}\text{COOH} +$

(blocked but not activated)

H_2NCH_2COOEt + $C_6H_{11}\text{—N=C=N—}C_6H_{11}$ $\xrightarrow{\text{THF, 25°}}$

H-Gly-OEt DCCD

$Z\text{—NHCH}_2\overset{O}{\overset{\|}{C}}\text{NH}\underset{\text{PhCH}_2}{\underset{|}{\text{CH}}}\overset{O}{\overset{\|}{C}}\text{NHCH}_2\text{COOEt} + C_6H_{11}\text{—NH}\overset{O}{\overset{\|}{C}}\text{NH—}C_6H_{11}\downarrow$

(87%) *N,N′*-Dicyclohexylurea

The initial reaction takes place between the free carboxyl group and the carbodiimide to yield the reactive intermediate, **10** (eq. 22.25). The amine then attacks at the carbonyl group of **10**, generating the coupled product.

(22.25) RCOOH + $C_6H_{11}\text{—}\ddot{N}\text{=C=}\ddot{N}\text{—}C_6H_{11}$ ⇄

$C_6H_{11}\text{—N=C=}\overset{+}{N}(H)\text{—}C_6H_{11}$ + $RCOO^-$ → $R\overset{O}{\overset{\|}{C}}\text{—O}\overset{N\text{—}C_6H_{11}}{\overset{\|}{C}}\text{NH—}C_6H_{11}$ (**10**) $\xrightarrow{R'\text{—NH}_2}$

$R\overset{O}{\overset{\|}{C}}\text{NHR}' + C_6H_{11}\text{—NH}\overset{O}{\overset{\|}{C}}\text{NH—}C_6H_{11}$

Finally, it should be noted that amino groups are not the only nucleophilic centers requiring protection in amino acids during peptide synthesis. The thiol group of cysteine, the heterocyclic nitrogen atom of histidine, and the phenolic hydroxyl group of tyrosine are examples of other functional groups that must be protected.

$\underset{\underset{\uparrow}{SH}}{\underset{|}{CH_2}}\text{—}\underset{NH_2}{\underset{|}{CH}}\text{COOH}$ Cysteine

→ HN (imidazole ring) $\text{—CH}_2\underset{NH_2}{\underset{|}{CH}}\text{COOH}$ Histidine

$\underset{\uparrow}{HO}\text{—}C_6H_4\text{—CH}_2\underset{NH_2}{\underset{|}{CH}}\text{COOH}$ Tyrosine

Benzyl sulfides, amines, and ethers are usually the protected forms of these functional groups. Equation 22.26 shows how tyrosine can be protected with a benzyl group that is subsequently removed by hydrogenolysis.

$$(22.26)\quad HO-C_6H_4-CH_2CH(NH_2)COOH \xrightarrow[^-OH]{C_6H_5CH_2Br} C_6H_5CH_2O-C_6H_4-CH_2CH(NH_2)-COOH \xrightarrow[\text{(deblocking)}]{H_2,\ Pd/C} HO-C_6H_4-CH_2CH(NH_2)COOH + C_6H_5CH_3$$

22:12 Summary of Methods for Peptide Synthesis

The previous section showed that there are several different protecting and activating groups which can be used in peptide syntheses. Therefore, a given peptide may be prepared in various ways. To summarize, we will follow two syntheses of Gly-Ala, using abbreviations for protective groups and reactants and omitting intermediates and mechanisms. The sequence of steps for one method of synthesis is given in Equations 22.27 to 22.29.

$$(22.27)\quad H_2NCH_2COOH + Z{-}Cl \rightarrow ZNHCH_2COOH \quad \textbf{blocking}\ (Z = C_6H_5CH_2OC(=O){-})$$

$$(22.28)\quad ZNHCH_2COOH + HONp \rightarrow ZNHCH_2COONp \quad \textbf{activating}\ (Np = O_2N{-}C_6H_4{-})$$

$$(22.29)\quad ZNHCH_2COONp + H_2NCH(CH_3)COOH \xrightarrow{\text{coupling}} ZNHCH_2C(=O)NHCH(CH_3)COOH \xrightarrow[\textbf{deblocking}]{H_2,\ Pd/C} H_2NCH_2C(=O)NHCH(CH_3)COOH$$

Gly-Ala

An alternate method of preparation of Gly-Ala is given in Equations 22.30 and 22.31. There is no *a priori* way of telling what particular route may be best in any given peptide synthesis. Experimentally, it is often found that one method or sequence of steps may give a better yield of a certain peptide while a different method may be preferable for an alternate peptide. Thus, one approach is often chosen over another based only on experience, not on factors predetermined from theory. Many of the special reactions shown have been devised to involve very mild conditions, since a strong acid or base could cause racemization of the amino acids.

(22.30) $\underset{NH_2}{CH_2}COOH + (Boc)_2O \rightarrow \underset{BocNH}{CH_2}COOH$ **blocking** (Boc = $(CH_3)_3CO\overset{O}{\overset{\|}{C}}^-$)

(22.31) $\underset{BocNH}{CH_2}COOH + CH_3\underset{NH_2}{CH}COOEt \xrightarrow[\text{coupling}]{\text{DCCD}} \underset{BocNH}{CH_2}C(=O)NH\underset{CH_3}{CH}COOEt$

deblocking ↓ H_3O^+

$\underset{NH_2}{CH_2}C(=O)NH\underset{CH_3}{CH}COOH$

Gly-Ala

Example

Prepare Phe-Gly.

Answer

Phenylalanine will be the *N*-terminal amino acid, so we must begin by blocking its NH_2 group so that it will not react when the peptide bond is formed. The benzyloxycarbonyl (Z) or *t*-butoxycarbonyl (Boc) group can be used equally well.

$$H_2N\underset{CH_2Ph}{CH}COOH + ZCl \rightarrow ZNH\underset{CH_2Ph}{CH}COOH$$

or

$$H_2N\underset{CH_2Ph}{CH}COOH + (Boc)_2O \rightarrow BocNH\underset{CH_2Ph}{CH}COOH$$

In the next step, you could use DCCD to couple blocked phenylalanine with the ethyl ester of glycine. DCCD forms the peptide bond *only* between a carboxyl group and an amine:

$$\underset{\displaystyle \mathrm{CH_2Ph}}{\mathrm{BocNH\overset{|}{C}HCOOH}} + \mathrm{H_2NCH_2COOEt} \xrightarrow{\mathrm{DCCD}} \mathrm{BocNH\underset{\displaystyle CH_2Ph}{\underset{|}{C}}H\overset{\displaystyle O}{\overset{\|}{C}}NHCH_2COOEt}$$

The last step would be removal of the blocking Boc group and conversion of the ethyl ester to the free carboxyl. Acid hydrolysis accomplishes both changes:

$$\mathrm{BocNH\underset{\displaystyle CH_2Ph}{\underset{|}{C}}H\overset{\displaystyle O}{\overset{\|}{C}}NHCH_2COOEt} + \mathrm{H_3O^+} \rightarrow \underset{\text{Phe-Gly}}{\mathrm{H_2N\underset{\displaystyle CH_2Ph}{\underset{|}{C}}H\overset{\displaystyle O}{\overset{\|}{C}}NHCH_2COOH}}$$

Another alternative would be to convert the blocked phenylalanine to the *p*-nitrophenyl ester and then react that activated amino acid with glycine. Hydrolysis of the coupled product would yield Phe-Gly.

Problem

22–10 a. Show how you would synthesize Ala-Gly using benzyloxycarbonyl as an amino protecting group and *p*-nitrophenyl as a carboxyl activator. Write out all steps with complete structures and then with the use of standard abbreviations.

b. Repeat the synthesis, using *t*-butoxycarbonyl as the amino protecting group and the dicyclohexylcarbodiimide (DCCD) condensation reagent.

22:13 Solid Phase Peptide Synthesis

One of the great problems in the synthesis of complex peptides is the necessity of isolating and purifying the growing chain after each step, before a new amino acid residue can be added. The reaction mixture will contain coupling reagents, remnants of protecting groups, and other by-products. The desired peptides themselves are often not crystalline and are therefore difficult to handle. In purification steps, some of the product is always lost. Any lowering of overall yields can be disastrous in a synthetic scheme needed to prepare, for example, a peptide of 25 amino acid residues. Such a synthesis may (with the need for preparing blocked and activated residues) require four steps for each residue added, and thus 100 separate steps could be necessary for a 25 amino acid peptide. If every step went in 90% yield, 100 grams of starting material in the first step would give about 0.003 grams of the final polypeptide.

In 1962, R. B. Merrifield and J. M. Stewart reported an ingenious technique for avoiding these obstacles—**solid phase peptide synthesis**. The *C*-terminal amino acid residue was bound to an insoluble polymeric resin. After each coupling step, the reaction by-products could simply be washed away from the polymer that contained the bound, growing peptide chain. When the synthesis was finished, the polypeptide could be removed from the polymer resin by hydrolysis. The first polymer prepared by this method was the nonapeptide bradykinin. The synthesis was accomplished in 32 hours and gave an 85% yield. The classical method of preparing the same peptide took eight days and a yield of only 68% was obtained. One of the greatest successes of the method was the preparation of the enzyme ribonuclease, which contains 124 amino acid residues. Synthetic ribonuclease was prepared in the remarkable overall yield of 18% and it exhibited the expected enzymatic activity.

The polymeric resin most often used in solid phase synthesis is a polystyrene. The polystyrene structure is modified (eq. 22.32) by a Friedel–Crafts

(22.32) ~CHCH₂~ (phenyl) + $CH_3OCH_2Cl \xrightarrow{SnCl_4}$ ~CHCH₂~ (p-CH_2Cl phenyl) + CH_3OH

Polystyrene — Chloromethylated polystyrene

chloromethylation reaction that introduces an active benzylic halide group. There are many such active sites on the polymer surface. The protected *C*-terminal amino acid is then attached by an S_N2 displacement of the chloride from the reactive benzyl group (first step of eq. 22.33). The amino group is then deblocked (second step of eq. 22.33). This is followed by a reaction

(22.33) ~CHCH₂~ (p-CH_2—Cl phenyl) + $^-O(O=)C$—CHR(NHBoc) → ~CHCH₂~ (p-$CH_2OC(=O)CHR(NHBoc)$ phenyl) $\xrightarrow[CH_2Cl_2]{CF_3COOH}$ ~CHCH₂~ (p-$CH_2OC(=O)CHR(NH_2)$ phenyl)

with the second amino acid residue, whose amino group is blocked with (usually) a *t*-butoxycarbonyl group (eq. 22.34). The reaction is promoted by

(22.34) $\sim CHCH_2 \sim$ (polystyrene resin, p-phenylene) $CH_2OC(=O)CHR(NH_2)$ + $HO-C(=O)-CH(NHBoc)-R'$ $\xrightarrow{DCCD}$ $\sim CHCH_2 \sim$ (p-phenylene) $CH_2OC(=O)CHR-NHC(=O)-CHR'-NH-Boc$

DCCD. The blocking group is then hydrolyzed and washed away (eq. 22.35), leaving a dipeptide on the resin, ready for the next amino acid residue. Finally, the completed polypeptide is cleaved from the polystyrene by catalytic hydrogenation. All the steps are repetitive, allowing peptide synthesis by the solid phase method to be automated and computer controlled.

(22.35) $\sim CHCH_2 \sim$ (p-phenylene) $CH_2OC(=O)CHR-NHC(=O)-CHR'-NH-Boc$ $\xrightarrow[H_3O^+]{mild}$ $\sim CHCH_2 \sim$ (p-phenylene) $CH_2OC(=O)CHR-NHC(=O)-CHR'-NH_2$

22:14 Proteins: Secondary, Tertiary, and Quaternary Structures

Section 22:1 indicated that the term *protein* is used to describe natural polypeptides that contain approximately 50 or more amino acid residues. Proteins are set apart from smaller peptides because of their often different biological function, but also because the structures of such large macromolecules display certain unique features. All molecules can be described as having a **primary structure**. This is the arrangement of atoms that arises from the various covalent bonds between those atoms, including stereochemical configurations, and is the structure depicted in the usual structural formulas. For peptides and proteins, the primary structure is determined by the sequence of amino acid residues (sec. 22:9). Thus far we have depicted the primary structure of these molecules as linear chains. However, long-chain proteins (and various other macromolecules) do not exist as random chains, but instead adopt helical or other ordered arrangements, which arise from the preferred conformations around the various single bonds. Such arrangements are called **secondary structures** and are due mainly to *intramolecular* hydrogen bonds and to restricted rotation about the amide (peptide) bonds. Interactions with the solvent (e.g., water) cause the helical chains of proteins to bend back on themselves to form folds and twists, giving the entire protein an overall three-dimensional shape (globular, elongated, etc.). This is known as the **tertiary structure**. In many proteins an important determinant of tertiary structure is the presence of disulfide bridges, which we discuss further in Section 22:16.

Finally, some proteins form aggregates in which several separate molecules unite to give a single biologically active unit. This is called **quaternary structure** and is due to a variety of interactions: *intermolecular* hydrogen bonds, coordination with metal ions, or mutual aggregation of lipophilic side chains repelled by the water solution. Let's take a detailed look at each of these structural features.

Peptide chains are somewhat more rigid than analogous hydrocarbon chains. This is due to the steric requirements for maximal overlap in the peptide bond between the nitrogen atom lone pair and the carbonyl group π bond (Fig. 22–6). This resonance interaction gives double bond character to the amide bond and promotes rigidity. All the atoms immediately attached to the peptide linkage prefer coplanarity, with adjacent R groups *trans* to one another.

In small peptides, the principal conformational restraint on the molecule is that due to the hindered rotation about the N—CO bond. As the length of the peptide chain increases, various intramolecular interactions begin to influence the conformational behavior as well. *Chief among these is hydrogen-bonding.* The NH groups of amides are good hydrogen-bond donors, while the carbonyl oxygen atoms are very effective hydrogen-bond acceptors. Hydrogen bonding between *adjacent* NH and CO groups is not favored, however, because the three atoms involved in a hydrogen bond should ideally be colinear. The criterion for colinearity can readily be satisfied if an NH group in a given peptide linkage interacts with a carbonyl group separated from it by several amino acid residues (Fig. 22–7). Use of molecular models shows that *an NH group can form a particularly strong hydrogen bond to a carbonyl group three peptide linkages distant.* Furthermore, this type of bonding can be established along the entire length of the chain if it assumes a spiral, or helical, conformation called an α-helix. Experimental measurements of optical rotation, ultraviolet and infrared absorption, and viscosity changes show that a large number of peptides and proteins possess at least some sections

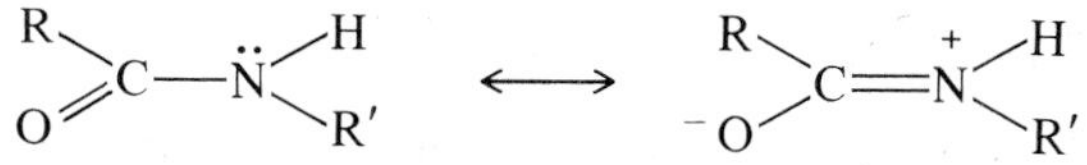

$$\mathrm{R}\!-\!\mathrm{C}(=\mathrm{O})\!-\!\ddot{\mathrm{N}}(\mathrm{H})\mathrm{R'} \longleftrightarrow \mathrm{R}\!-\!\mathrm{C}(\mathrm{O}^-)\!=\!\overset{+}{\mathrm{N}}(\mathrm{H})\mathrm{R'}$$

Figure 22–6. *Preferred planar stereochemistry about the peptide bond. The partial double bond character of the C—N bond leads to restricted rotation.*

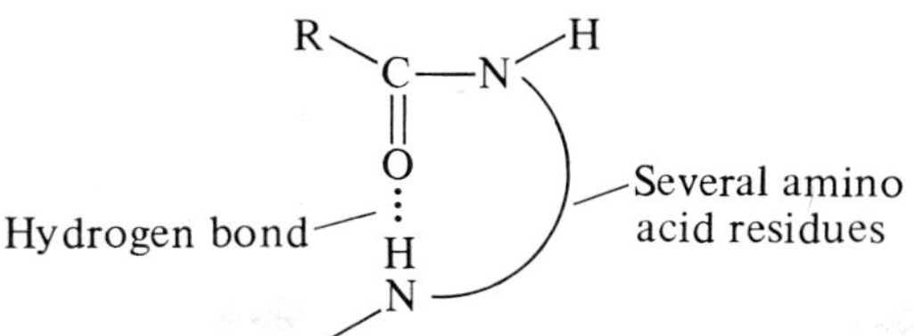

Figure 22–7. *An intramolecular C=O··· HN hydrogen bond in a protein chain.*

that are in the form of α-helices. The American chemist L. Pauling (Nobel Prize winner in 1954 for his studies of chemical bonding) was one of the first to emphasize the helical structures of proteins.

An α-helix is not superimposable on its mirror image and hence can confer chirality on a protein. The presence of helical sections thus gives the molecule high optical rotatory power (in addition to that caused by the individual amino acids). This phenomenon provides a useful device for studying helical conformations in protein chains. The helix may be right- or left-handed (Fig. 22–8). When the constituents are L-amino acids, the right-handed helix is sterically favored.

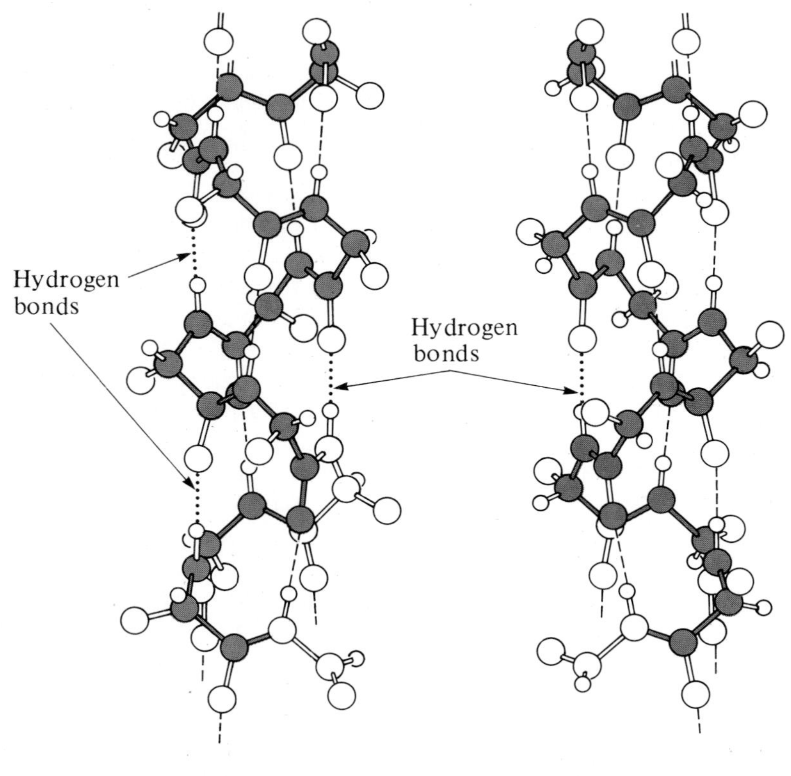

Figure 22–8. *Models of (a) left- and (b) right-handed peptide α-helices. These are examples of secondary protein structure.*

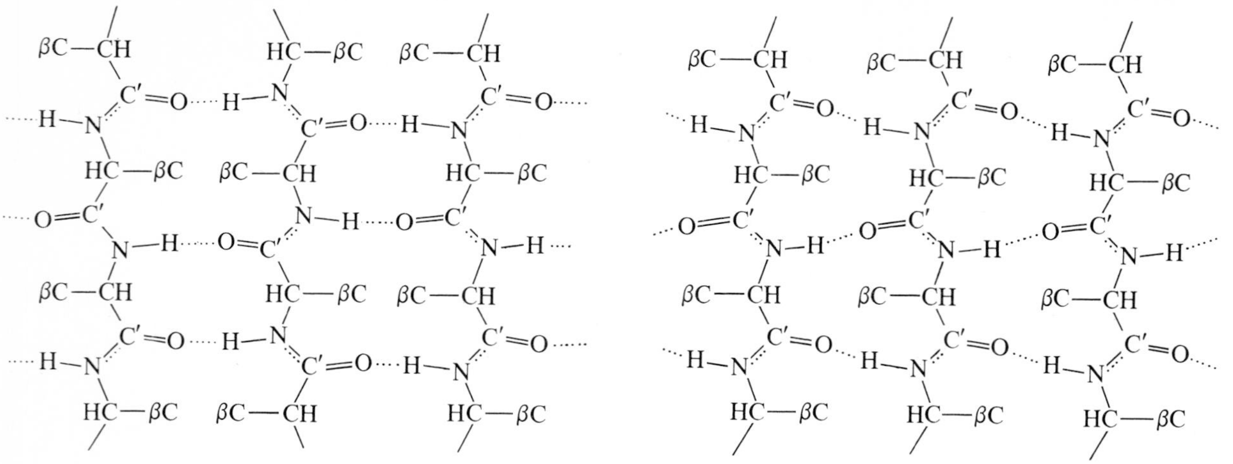

Figure 22–9. *Parallel (a) and antiparallel (b) pleated sheet structures in peptide chains.*

The adoption of the helical conformation in proteins is usually energetically favorable because it promotes intramolecular hydrogen-bonding. On the other hand, the helical molecule is more restricted than a random coil and entropy effects work against helix formation. For a number of molecules, these two tendencies balance one another and the helix rapidly forms and unwinds in solution.

In any given protein not all sections of the molecule will be equally amenable to α-helix formation. For example, the presence of a proline residue, which contains the rigid pyrrolidine ring, interferes with helix formation. Valine tends to inhibit α-helix formation because of the steric effect of its isopropyl group. Protein sections that do not coil into helices may still engage in intramolecular hydrogen-bonding. The chain may double back on itself and the two strands zip together by way of NH···O═C bonds to give a secondary structural feature known as a pleated sheet, or β-structure (Fig. 22–9).

To illustrate the effect of secondary structure on physical properties, let's consider α-keratin, the protein of hair. Hair can be stretched to almost double its length by the application of moist heat. These conditions cause breakage of the intramolecular hydrogen bonds and conversion of the normal helical structure into an extended zigzag chain. Upon cooling, hydrogen bonds are reestablished, the helix forms once again, and the hair is returned to its normal length.

For its tertiary structure, the entire protein molecule will assume an overall three-dimensional shape that is intimately connected with its biological function. The overall shape of the enzyme chymotrypsin is shown in Figure 22–10. One of the important determinants of tertiary structure is the tendency of hydrophilic functional groups, such as amino and carboxyl groups, to be on the outside of the molecule and in contact with water. Hydrophobic side chains such as alkyl and phenyl, on the other hand, prefer to interact with one another in the interior of the molecule, away from the

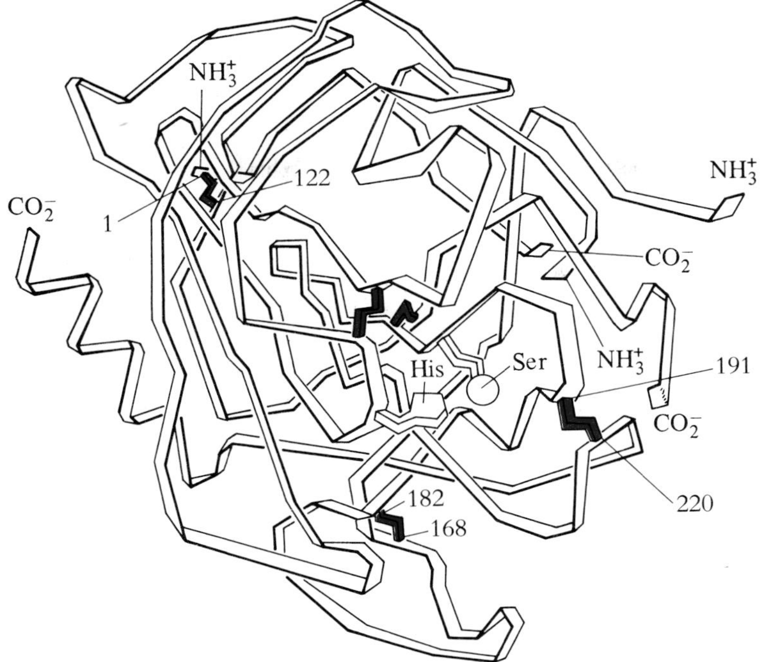

Figure 22–10. *The conformation of chymotrypsin, showing its tertiary structure. It has three chains held together by disulfide bridges and electrostatic interactions. (Disulfide bridges are shown in color.)*

influence of water molecules. If tertiary structure is disrupted by heating, or by introducing a solvent that interferes with intramolecular hydrogen-bonding, then the protein generally loses all its biological potency. This process is known as **denaturation**. A common example of denaturation is the irreversible change accompanying the cooking of egg white.

Some proteins consist of several individual folded or helical chains combined by weak interactions into an aggregate, or oligomer (quaternary structure). One of the most important oligomeric proteins is hemoglobin, the oxygen carrier of blood erythrocytes. Hemoglobin consists of four interwoven polypeptide chains and four **heme** groups. The heme groups are aromatic, nitrogen-containing rings within which iron atoms are bound. The entire hemoglobin protein forms a relatively spherical mass and hence is called a **globular** protein (Fig. 22–11). Its subunits by themselves are not biologically active.

The only method available for directly determining the complete structure of proteins is X-ray crystallography. It is possible to delineate the secondary, tertiary, and quaternary structure of proteins in this way but it is difficult, time consuming, and expensive. The English researchers J. C. Kendrew and M. Perutz in 1962, and D. C. Hodgkin in 1964, received Nobel Prizes for their work in X-ray crystallography of proteins. Information on the structure of proteins in solution, as contrasted to the crystal, can be obtained from nmr experiments. Solution and crystal conformations may differ. Since proteins react in solution, knowing the most stable solution

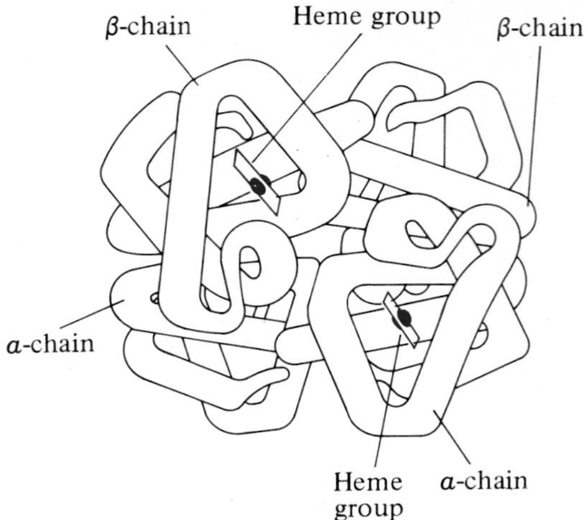

Figure 22–11. *The quaternary structure of hemoglobin, showing the four heme-containing peptide chains.*

conformation is important for understanding structure–activity relationships.

Box 22–3. Quotations from "The Physics and Chemistry of Lemon Meringue Pie"*

"A meringue dish raises several questions about the protein chemistry of eggs and the physics of heat transfer. . . . Recently, after yet another embarrassing failure with a lemon meringue pie, I began to wonder what really happens to the egg whites when I beat them. To explain what is happening to the whites, I shall first describe the protein in the egg white. What a good cook is doing is properly preparing that protein for the oven. What I am doing in my lack of experience is ruining the protein."

"The whites of chicken eggs are composed of several types of protein, principally ovalbumin, globulin, and ovomucin. . . . When a cook forces a whisk through egg whites, shearing the fluid, some of the weaker bonds are ruptured and parts of the tertiary structure of the proteins are destroyed. . . . As a result of the whisking, a protein is gradually unraveled from its initial spaghetti-like sphere (ovalbumin and globulin). Any such altering of the structure of protein is called denaturing. . . . Once the protein molecules are partially unraveled they begin to attach themselves to one another to form a three-dimensional gel. This attraction is unlikely before denaturation because the proteins are globular and relatively few of their sites for possible bonds are exposed. As the whites are beaten, air bubbles are formed and trapped in the gel mesh, along with a layer of water. . . . Excessive beating ruins the meringue because it unravels the proteins too much. As a result, the mesh becomes too firm and water separates from the proteins."

"Yolk must be rigorously excluded from the meringue. . . . When yolk is present, the lipid (fat) molecules in it attach themselves to the hydrophobic sites on the protein by means of van der Waals forces. Hence those sites are no longer available for bonding with other proteins to create a gel mesh. It is then harder to trap air, and the volume that can be obtained by beating the whites is smaller. . . . The presence of 10% of yolk of one egg in the whites of three eggs will decrease the beaten volume to about one-half."

* Jearl Walker, *Scientific American,* **244,** 194 (1981). Used with permission.

22:15 Synthetic Condensation Polymers

As you've seen, biological polyamides are synthesized by condensation of amino acids. Commercial polyamides, by contrast, are made by two different routes. One imitates nature by relying on the condensation of amino acids, often in the form of their lactams (eq. 22.36). The other method utilizes the condensation of diamines with dicarboxylic acids (eq. 22.37).

(22.36)
$$n\ \text{(2-pyrrolidinone)} \xrightarrow[\text{heat}]{OH^-} \text{+}NHCH_2CH_2CH_2\overset{\overset{O}{\|}}{C}\text{+}_n$$

Nylon 4

(22.37)
$$n\text{HOOC}(CH_2)_4\text{COOH} + n\text{H}_2\text{N}(CH_2)_6\text{NH}_2 \xrightarrow{200\text{ to }300^\circ} \text{+}\overset{\overset{O}{\|}}{C}(CH_2)_4\overset{\overset{O}{\|}}{C}NH(CH_2)_6NH\text{+}_n + 2nH_2O$$

Nylon 66

Both methods produce synthetic polyamides known as **nylons**. Nylons are stable to dilute acids or bases and often have high melting points, due in part to hydrogen-bonding between N—H and O=C groups on different chains. Nylon 66 has a molecular weight of about 10,000 and has a melting point of 260°. The numbers following the name denote the number of carbon atoms in the monomer(s) from which the polymer is generated. If the monomer is an amino acid, then only one number is required (nylon 4); when a pair of monomers is employed, then the number of carbons in each must be specified (nylon 66).

Hydrogen-bonding between NH and O=C groups also gives nylon its solidity and strength so that it can be molded into hard blocks for use in machine gear manufacture, for example. The nylon used as fiber is spun from the melted material and then drawn into long filaments.

Another important type of man-made fiber is **polyester**, which is usually synthesized by reacting a diol with an ester of a dicarboxylic acid (eq. 22.38). Polyesters such as Dacron have properties somewhat different from

(22.38)
$$n\ \begin{matrix} CH_2OH \\ | \\ CH_2OH \end{matrix} + n\ EtOOC\text{-}C_6H_4\text{-}COOEt \xrightarrow{\text{heat}} \text{+}\overset{\overset{O}{\|}}{C}\text{-}C_6H_4\text{-}COOCH_2CH_2O\text{+}_n + 2nEtOH$$

Ethylene glycol — Diethyl terephthalate — Dacron

those of polyamides such as nylon. Polyesters are not nearly as high-melting and thus cannot usually be used in molded form. This is undoubtedly due to their lack of —NH— groups for formation of hydrogen bonds, and to freer rotation in ester groups than in amide groups.

Not all important polyesters are man-made. Polyesters are vital constituents of the protective layers of certain plant cells. For example, the biopolyester *cutin* is the main component of the plant cuticle, the exterior of the epidermal cell wall.

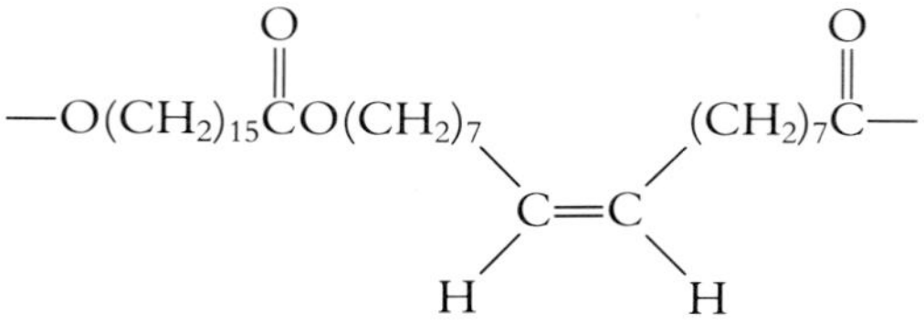

$$-O(CH_2)_{15}\overset{O}{\overset{\|}{C}}O(CH_2)_7-CH=CH-(CH_2)_7\overset{O}{\overset{\|}{C}}- \quad (\text{cis: both H on the same side})$$

A portion of a cutin molecule

Problems

22–11 Nylon 610 is made from hexamethylenediamine and sebacoyl chloride (eq. 3). Give the structure of nylon 610 and show at least two repeating units.

$$\underset{\text{Hexamethylenediamine}}{H_2N(CH_2)_6NH_2} + \underset{\text{Sebacoyl chloride}}{Cl\overset{O}{\overset{\|}{C}}(CH_2)_8\overset{O}{\overset{\|}{C}}Cl} \rightarrow \text{nylon 610} \qquad (3)$$

22–12 Maleic anhydride forms linear polymers when heated with diols such as ethylene glycol. Write a structure for a section of the maleic anhydride–ethylene glycol copolymer, showing two of the repeating units.

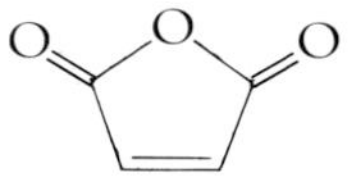

Maleic anhydride

22:16 Cross-linking in Natural and Synthetic Polymers

Any two polymers composed of different monomers will naturally have different properties. For example, the difference in physical properties among polyethylene (eq. 10.38, p. 361), Teflon (eq. 10.43, p. 363), and nylon (eq. 22.37) are probably familiar to you. Proteins that make up muscle are of different monomer amino acid content than are those that make up hemoglobin. Polymer properties can also be changed by taking a given polymer and modifying it through reaction in some way. One of the most important modifying reactions is cross-linking.

In proteins, the disulfide bond linkage (see eq. 22.39) often serves to cross-link separate polymer chains, or widely separated parts of the same polymeric chain. Both results are seen in chymotrypsin (Fig. 22–10), and depend on the oxidative coupling of two cysteine residues to form a disulfide bridge. The coupling of cysteines 1 and 122 (Fig. 22–10, upper left) serves to link the A and B chains of chymotrypsin. The coupling of cysteines 191 and 220 (lower right) causes the C chain to curve back on itself.

(22.39) 2∿C(=O)NHCH(CH₂SH)C(=O)∿ → ∿C(=O)NHCH(C(=O)∿)CH₂—S—S—CH₂CH(C(=O)∿)NHC(=O)∿

$$2\sim\mathrm{C(=O)NHCH(CH_2SH)C(=O)}\sim \;\rightarrow\; \sim\mathrm{C(=O)NHCH(C(=O)\sim)CH_2{-}S{-}S{-}CH_2CH(C(=O)\sim)NHC(=O)}\sim \quad (22.39)$$

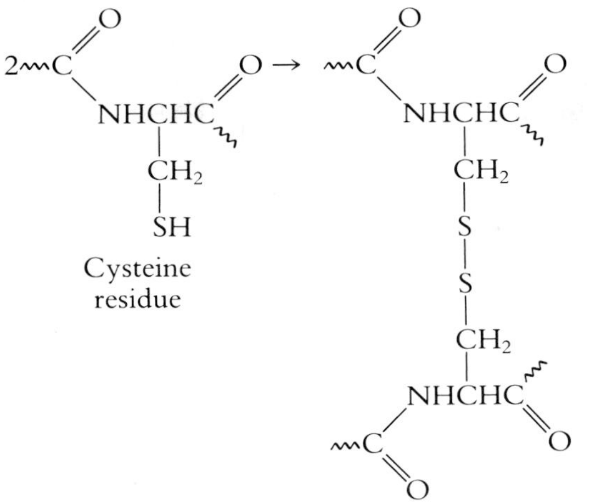

Cysteine residue

A disulfide bond cross-link in a protein

These disulfide cross-links impart rigidity not present in the individual chains. The keratin protein of hair is very highly cross-linked in this way. In the "permanent wave" process, hair is treated with a thiol that reductively cleaves the S—S linkages and allows the hair to become straightened. The hair is then reset and air oxidation forms new S—S cross-links, thus fixing strands in the new "curled" position.

Natural rubber (polyisopropene, Fig. 13–7, p. 481) has few cross-links and hence becomes soft when heated slightly, and brittle when cooled. Because of these undesirable properties, early use of rubber was only for waterproofing of fabrics. In 1839, Charles Goodyear was experimenting with the addition of sulfur to rubber to improve its properties when he accidentally spilled one of his experiments on a hot stove. The heating of natural rubber with the sulfur incorporated S—S cross-links into the rubber molecules, and greatly improved heat and cold properties of the rubber. This process, later termed *vulcanization,* was the foundation of commercial rubber technology. Although Goodyear held all the original patents on vulcanization, he remained a poor man because he was constantly in court defending his patents against infringement.

Consider as another example the commercially important urethane polymers. These are formed by co-polymerization of two monomers: an aryldiisocyanate and a polyfunctional alcohol (eq. 22.40). If condensation is

$$\mathrm{O{=}C{=}NArN{=}C{=}O} + \mathrm{HOCH_2CH(OH)CH_2OH} \rightarrow \quad (22.40)$$

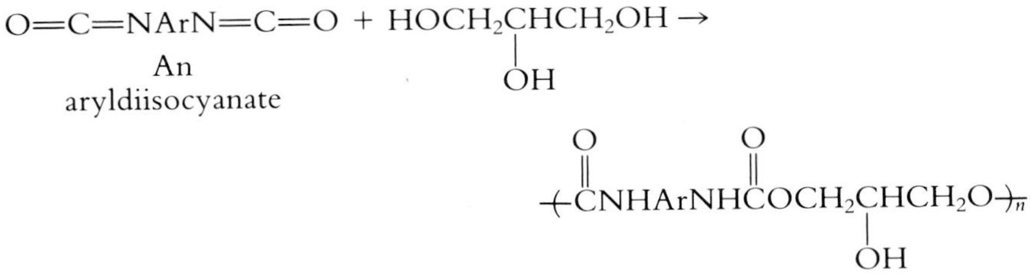

An aryldiisocyanate

$$\mathrm{\left[C(=O)NHArNHC(=O)OCH_2CH(OH)CH_2O\right]_n}$$

Spandex

controlled so that highly linear polymerization occurs, the product is an elastic fiber known as *Spandex*. If a rigid polymer is desired, then sufficient diisocyanate is included in the reaction mixture so that cross-linking between strands containing the free secondary OH groups can occur. If a low-boiling liquid such as CH_2Cl_2 is placed in the reaction mixture, it will volatilize during the polymer hardening process and cause expansion of the rigid polymer to form polyurethane foams.

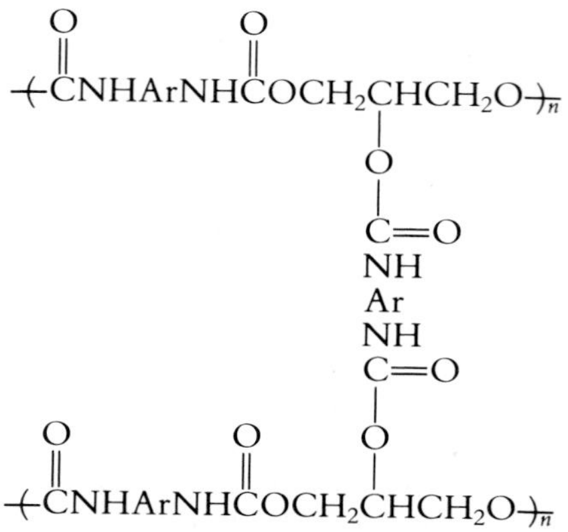

$$\left(\overset{\displaystyle O}{\overset{\|}{C}}NHArNH\overset{\displaystyle O}{\overset{\|}{C}}OCH_2\underset{\displaystyle |}{C}HCH_2O \right)_n$$
$$\begin{array}{c} O \\ | \\ C{=}O \\ NH \\ Ar \\ NH \\ C{=}O \\ | \\ O \\ | \end{array}$$
$$\left(\overset{\displaystyle O}{\overset{\|}{C}}NHArNH\overset{\displaystyle O}{\overset{\|}{C}}OCH_2CHCH_2O \right)_n$$

A section of rigid,
cross-linked polyurethan

Problem

22–13 Give a mechanism for the reaction in Equation 22.40.

Summary

1 α-Amino acids are the building blocks of peptides and proteins. Two amino acids combine with loss of water to form a peptide containing an amide (CO—NH) bond between the amino acid residues. This process may be continually repeated. If more than 50 amino acid units are combined, the resulting macromolecule is called a protein.

Synthetic polymers can be made by similarly combining difunctional acids and amines to yield polymers known as polyamides (e.g., nylon). These are termed condensation polymers (sec. 22:1).

2 There are 20 standard amino acids of major biochemical importance. Since they contain both acidic (COOH) and basic (NH_2) groups in the same molecule, amino acids exist as zwitterions (sec. 22:2). The pH of an aqueous amino acid solution determines the predominant form in which the amino acid is present. The isoelectric point is the pH at which the zwitterionic form is present in highest concentration and at which there is thus no migration during electrophoresis. At other pH values, positively- or negatively-charged species may predominate (sec. 22:3). Since different amino acids have different isoelectric pHs, they can be separated by electrophoresis.

Except for glycine, all α-amino acids contain a chiral α-carbon atom. When isolated from proteins, these acids are each enantiomerically pure and have the same absolute configuration. This configuration has been related to that of L-glyceraldehyde and the amino acids have therefore also been given the L designation (sec. 22:4).

3 Many amino acids can be obtained in commercial quantities by hydrolysis of proteins, or as fermentation by-products (sec. 22:5). Synthetic procedures are also available for amino acid synthesis in the laboratory. The Gabriel method gives a racemic product with the same number of carbons as the starting haloacid. The Strecker synthesis also yields a racemic product, but with one more carbon than the starting aldehyde. Racemic amino acids can be resolved enzymatically. A crossed aldol condensation with an ester, followed by reduction over a chiral catalyst, yields the chiral product directly (sec. 22:6).

4 Peptides and proteins can be hydrolyzed and their amino acid content determined by chromatography. This is accomplished by an amino acid analyzer that employs a ninhydrin color test to determine the quantity of each amino acid isolated (sec. 22:8).

The sequence of amino acids in a peptide or protein can be determined through partial hydrolysis and end-group analysis. The *N*-terminal amino acid is usually determined by forming the dansyl derivative or by the Edman degradation (sec. 22:9). Enzymatic hydrolysis is a powerful technique for sequence determination. The *C*-terminal amino acids are often determined by the use of the enzyme carboxypeptidase. Other enzymes cleave peptides specifically between certain amino acid residues. Analysis of the fragments allows the sequence of the original peptide to be calculated (sec. 22:10).

5 Peptide syntheses involve the combination of two amino acids to form an amide. A fundamental problem is to coax the correct amino and carboxyl groups into reacting with each other while, simultaneously, preventing undesired amide formation between the wrong amino and carboxyl groups. Prevention of the latter depends largely on the use of protecting groups that severely depress the nucleophilicity of an amino group. One is the benzyloxycarbonyl (Z) group, which not only fulfills its protective function but may also be removed reductively under conditions such that the amide bonds of the peptide are not disturbed. The *t*-butyloxycarbonyl (Boc) group is of even greater value and may be easily removed by mild acid hydrolysis (secs. 22:11 and 22:12).

Free carboxyl groups will not readily condense to form amides so it is necessary to activate them. The activating group must promote rapid reaction with the nucleophile without also promoting racemization at the α-carbon atom. A very useful method is to condense the free carboxylic acid and the amine in the presence of dicyclohexylcarbodiimide (DCCD; secs. 22:11 and 22:12).

One of the most significant techniques in peptide synthesis is the solid state method. The *C*-terminal acid, whose amino group is protected with Boc, is attached to a chloromethylated polystyrene resin. After removal of the protecting group, the polymer-bound amino acid is condensed with another protected amino acid with the aid of dicyclohexylcarbodiimide. The insoluble product is thoroughly washed and the cycle is then repeated; isolation and purification of the growing polypeptide chain is not necessary. One of the greatest virtues of this technique is that it lends itself to automation (sec. 22:13).

6 Overall protein structure depends on several components: primary (amino acid sequence), secondary (conformational effects), tertiary (three-dimensional chain folding), and quaternary (aggregation). Two types of hydrogen-bonded secondary structures are particularly prevalent, the α-helix and the pleated sheet. The hydrogen-bond donors are NH groups and the acceptors are amide CO groups. Tertiary structure involves the overall shape of the molecule and is affected by disulfide bridges and by interaction of hydrophobic and hydrophilic side chains with the solvent. The quaternary structure involves molecular aggregates (more than one protein chain). Loss of tertiary or quaternary structure by denaturation leads to loss of biological function (sec. 22:14).

7 Some common synthetic fibers are prepared from dicarboxylic acids and diols (polyesters, such as Dacron), or diacids and diamines (nylon). The physical properties of polymers can be modified by cross-linking individual polymer strands. Examples of cross-linking are through disulfide bonds, or through the use of trifunctional monomers, as in polyurethans (sec. 22:15).

Problems

22–14 a. Suppose you have a mixture of histidine, tyrosine, glutamic acid, and glycine. Upon electrophoresis at pH 6, which amino acid(s) would you expect to remain at the origin, which would move to the cathode (negative plate), and which would move to the anode (positive plate)?

b. At pH 2 and pH 10, would the following peptides move toward the anode, cathode, or remain at the origin? Explain your answer.
1) Lys-Ala-Gly　　2) Glu-Ala-Glu

22–15 a. Arginine is the most basic of the amino acids (isoelectric pH 10.76). Use resonance structures to explain this observation. Draw structures of the arginine monocation and dication.

b. Aspartic acid (isoelectric pH 2.98) is the most acidic of the amino acids, and is slightly more acidic than glutamic acid (isoelectric pH 3.08). Explain these data.

c. Tyrosine (isoelectric pH 5.63) is somewhat more acidic than phenylalanine (isoelectric pH 5.91), although the two amino acids are very similar in structure. Account for the additional acidity of tyrosine.

22–16 a. Draw a projection formula of D-alanine.

b. Draw the structure of a tetrapeptide composed of four D-alanine molecules linked together as in Equation 22.1. Be sure to show the stereochemical configuration.

c. Now draw the formula of a tetrapeptide composed of alternating D- and L-alanine molecules. How does this tetrapeptide differ from the one you drew in (b)? How might this difference affect the biochemical properties of these two tetrapeptides?

22–17 a. Oxidation of (+)-glyceraldehyde gives (−)-glyceric acid

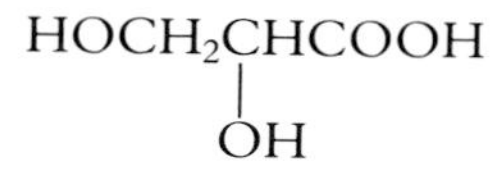

(−)-Glyceric acid has been shown to have the same absolute configuration as (−)-lactic acid

$$\underset{\displaystyle\mid\atop \displaystyle OH}{CH_3CHCOOH}$$

On the basis of the absolute configuration of (+)-glyceraldehyde, assign absolute configurations to (−)-glyceric acid and (−)-lactic acid.

b. Upon treatment with aqueous hydroxide ion (+)-2-bromopropanoic acid gives (+)-lactic acid. When allowed to react with aqueous ammonia, (+)-2-bromopropanoic acid yields (+)-alanine. Assign absolute configurations to (+)-2-bromopropanoic acid and to (+)-alanine.

c. Explain how, even without knowing the true absolute configurations, this sequence of reactions establishes the relative configurations of (+)-glyceraldehyde and (+)-alanine.

22–18 Show the structure of an arginine molecule that is obtained by hydrolysis of a peptide molecule after treatment with dansyl chloride when (a) arginine is the *N*-terminal residue, and when (b) arginine is in the center of the molecule.

22–19 Sickle-cell anemia is the result of a genetic error which slightly alters the blood protein hemoglobin. In some cases, isocyanic acid (HN=C=O) has been found to return these altered hemoglobin proteins to normal. It is believed that isocyanic acid reacts with an *N*-terminal valine of the protein. Show how the *N*-terminal valine might react with isocyanic acid.

22–20 Several amino acids are synthesized on a commercial scale from inexpensive starting materials. Show how the following amino acids can be prepared from the indicated starting material. In the first two cases, the Strecker reaction is used.

a. methionine from acrolein (propenal)
b. lysine from acrylonitrile (propenenitrile)
c. lysine from 6-aminohexanoic acid (the lactam of this amino acid is the monomer for production of nylon 6)

22–21 Numerous laboratory syntheses of amino acids use diethyl *N*-acetylaminomalonate (**I**) as their starting point. Advantage is taken of the high acidity of the proton located between the two ester groups (sec. 19:1). Given here are several amino acid syntheses based on **I**. In each case, provide the structures of the intermediates and final product.

a. $\underset{\textbf{I}}{CH_3\overset{O}{\overset{\|}{C}}NHCH(COOEt)_2} \xrightarrow{NaOEt} \xrightarrow{(CH_3)_2CHCH_2Br}$

$$\textbf{A} \xrightarrow[heat]{OH^-} \xrightarrow[heat]{H_3O^+} \textbf{B}(C_6H_{13}NO_2)$$

b. $\textbf{I} \xrightarrow[NaOEt\ (cat)]{CH_2{=}CHCN} \textbf{C} \xrightarrow[2.\ H_3O^+,\ heat]{1.\ OH^-,\ heat} \textbf{D}(C_5H_9NO_4)$

22–22 Show how you would synthesize the following amino acids from common, nonamino acid starting materials.

a. (+)-isoleucine b. L-cysteine c. D-tyrosine

22–23 An unknown tetrapeptide is hydrolyzed to yield two tripeptides, **A** and **B**. Tripeptide **A** contains Gly, Ile, Ala. Peptide **B** contains Ile, Val, Gly. The *N*-terminal residue of **B** is found to be glycine and the *C*-terminal residue is valine. What is the structure of the original tetrapeptide?

22–24 Give a step-by-step synthesis of the following peptides:
a. Met-Leu (a "classical" synthesis)
b. Try-Gly-Ala (a solid phase synthesis)
c. Lys-Glu

22–25 In an attempt to obtain the methyl ester of alanylalanine, the methyl ester of L-alanine was allowed to undergo self-condensation (eq. 4). A different product, **A**, was obtained instead. Propose a structure for **A**, and a mechanism for its formation.

$$CH_3\underset{\displaystyle NH_2}{\underset{|}{C}}HCOOCH_3 \rightarrow \mathbf{A}(C_6H_{10}N_2O_2) \qquad (4)$$

22–26 In solid phase peptide syntheses it is sometimes found that the unfinished peptide is cleaved from the polystyrene resin when acid is added to deprotect an amino group. This can be prevented by nitrating the chloromethylated polystyrene. Explain why the presence of the nitro group inhibits acid-catalyzed cleavage of the peptide from the resin.

In order to remove the peptide from the nitrated resin at the conclusion of the synthesis, it is necessary to treat the resin with ammonia or hydrazine. In what form does the polypeptide come off after treatment with these nucleophiles?

22–27 A chemist wants to perform an acid-catalyzed dehydration of 10-hydroxydecanoic acid (eq. 5), but obtains a very poor yield. Part of the failure was due to a high yield of polymeric material. Give a structure for the polymer by-product.

$$HOCH_2CH_2(CH_2)_7COOH \xrightarrow{H^+} CH_2{=}CH{-}(CH_2)_7COOH + \text{polymer} \qquad (5)$$

22–28 It is possible to cross-link the polymer described in Problem 22–12 by adding an alkene such as styrene, along with a peroxide catalyst and heat. Write the structure of the cross-linked polymer and a mechanism for its formation.

22–29 The lactam of 6-aminohexanoic acid (*caprolactam*) is very inexpensive and available in large quantities, being made from cyclohexane. It is polymerized by being heated with a small quantity of water (eq. 6). Propose a mechanism for the polymerization to form nylon 6, showing the condensation of two monomer units.

$$\text{caprolactam (cyclic, C=O, NH)} \xrightarrow[\text{heat}]{H_2O\ (\text{cat})} \left[NH(CH_2)_5\overset{O}{\overset{\|}{C}}\right] \qquad (6)$$

Nylon 6

22–30 Acrylamide can be polymerized under different conditions and thereby give different products. In the presence of strong nucleophiles the product is nylon 3 (eq. 7a). However, the use of radical initiators leads to a polyethylene derivative (eq. 7b). Provide mechanisms for each polymerization process, and speculate as to why they take different courses.

$$\underset{\text{Acrylamide}}{CH_2{=}CH\overset{\overset{\displaystyle O}{\|}}{C}NH_2} \xrightarrow{RO^-} \underset{\text{Nylon 3}}{\left(CH_2CH_2\overset{\overset{\displaystyle O}{\|}}{C}NH\right)_n} \qquad (7a)$$

$$\underset{\text{Acrylamide}}{CH_2{=}CH\overset{\overset{\displaystyle O}{\|}}{C}NH_2} \xrightarrow{R\cdot} \left(CH_2\underset{\underset{\displaystyle CONH_2}{|}}{CH}CH_2{-}\underset{\underset{\displaystyle CONH_2}{|}}{CH}\right)_n \qquad (7b)$$

22–31 Show the structure of the repeating polymer unit formed when each of the following mixtures is heated.

a. toluene-2,4-diisocyanage and 1,4-butanediol
b. dimethyl terephthalate and benzene-1,4-diamine
c. adipic(hexanedioic) acid and ethylene glycol
d. phthalic anhydride and glycerol(propane-1,2,3-triol)
e. succinic anhydride and $C(CH_2OH)_4$(pentaerythritol)

Supplementary Reading

H. D. Law, *The Organic Chemistry of Peptides,* Wiley-Interscience, New York, N.Y., 1970.

M. L. Bender and L. J. Brubacher, *Catalysis and Enzyme Action,* McGraw-Hill, New York, N.Y., 1973.

R. E. Dickerson and I. Geis, *The Structure and Action of Proteins,* Harper and Row, New York, N.Y., 1969.

D. E. Metzler, *Biochemistry,* Chap. 2, pp. 47–48, Academic Press, New York, N.Y., 1975.

D. C. Neckers, Solid phase synthesis, *J. Chem. Ed.,* **52,** 695 (1975).

D. Blackman, Amino acid sequence diversity in proteins, *J. Chem. Ed.,* **54,** 170 (1977).

M. Morton, *et al.,* Learning from the macromolecule, *J. Chem. Ed.,* **50,** 731 (1973).

N. Sharon, Glycoproteins, *Sci. Am.,* **230,** 78 (May, 1974).

P. E. Kolattukudy, Biopolyester membranes of plants: Cutin and suberin, *Science,* **208,** 990 (1980).

J. Preston, High-strength/High-modulus fibers, *J. Chem. Ed.,* **58,** 935 (1981).

23 Nucleophilic Aromatic Substitution: The Chemistry of Aryl Halides and Aryl Diazonium Ions

In this chapter we return to the chemistry of aromatic compounds, particularly aryl halides and amines. Chapter 15 showed that attack by various electrophiles on the ring provides a major route for preparing substituted aromatics. We might also expect that *nucleophilic* attack on aryl halides would be a promising approach to aryl ethers, amines, and so forth. After all, nucleophilic substitution is probably the most widely used technique for preparing substituted aliphatic compounds (Chap. 6).

The supposition is not only reasonable but also correct—nucleophilic substitution is a useful way of preparing various aryl derivatives. However, it is not as versatile with aromatic substrates as with aliphatic ones. All *aliphatic* nucleophilic substitutions can be neatly classified as S_N1 or S_N2, even though they may differ substantially with respect to substrate structure, leaving group, and so on. *Aromatic* nucleophilic substitutions, on the other hand, are much more diverse mechanistically; in this chapter we'll discover five different mechanisms. In *none* of these reactions is there a concerted, direct displacement analogous to the S_N2 reaction; all aromatic substitutions involve intermediates. The operation of these different mechanisms depends strongly on the structure of the substrate, the nucleophile, and the leaving group.

Since nucleophilic aromatic substitution seems to be more complex than electrophilic substitution, why do we deal with it? For one thing, it makes available many aromatic compounds not easily synthesized by electrophilic reactions. In addition, nucleophilic substitution is a major reaction pathway for the synthesis of substituted *heterocyclic* aromatic compounds (Chap. 25). Finally, by adopting a systematic approach, we'll see that the diversity in this area can be mastered.

23:1 Nucleophilic Substitution Promoted by Electron-Withdrawing Groups: The S_NAr Mechanism

In Problem 22–7 (p. 867) we noted that 2,4-dinitrofluorobenzene has been used to label *N*-terminal amino acids in peptides. It does this by reacting with the free amino group in a nucleophilic aromatic substitution. The reaction can take place under mild conditions because of the presence of the nitro groups on the aromatic ring; unsubstituted fluorobenzene does not react at all with amino groups under mild conditions.

This reaction is one example of a general phenomenon—nucleophiles displace halide ions from aryl halides under relatively mild conditions when there are electron-withdrawing substituents *ortho* and *para* to the halogen. These substituents include nitro, cyano, and carbonyl groups. Equations 23.1 to 23.3 provide a representative sample of such halide reactions. In the following discussion, we will use the nitro group to exemplify the role of the substituent.

(23.1) 2-fluoronitrobenzene + $(CH_3)_3CO^-$ $\xrightarrow{(CH_3)_3COH,\ 50°}$ 2-$OC(CH_3)_3$-nitrobenzene + F^-

(23.2) methyl 4-bromo-3-nitrobenzoate (Br, NO_2, COOMe) + $C_6H_5S^-$ $\xrightarrow{CH_3OH,\ 0°}$ MeOOC–$C_6H_3(NO_2)$–S–C_6H_5 + Br^-

(23.3) 4-chloro-3-nitrobenzonitrile (Cl, NO_2, CN) + 2 piperidine (NH) $\xrightarrow{C_6H_6,\ 45°}$ NC–$C_6H_3(NO_2)$–N(piperidino) + piperidinium ($\overset{+}{N}H_2$) Cl^-

The effect of the nitro groups on the reaction rate is cumulative. Thus, chloride ion can be displaced from 2,4,6-trinitrochlorobenzene (*picryl chloride*) by the weak nucleophile, water, at 50° (eq. 23.4). When only one nitro group is present in the molecule, the stronger nucleophile, OH^-, is required to displace the chloride ion at a substantially higher temperature (eq. 23.5).

(23.4) Picryl chloride (Cl; O_2N, NO_2, NO_2) + H_2O $\xrightarrow{50°}$ Picric acid (OH; O_2N, NO_2, NO_2) + HCl

Picryl chloride

Picric acid

(23.5) p-Nitrochlorobenzene + OH^- $\xrightarrow{H_2O,\ 160°}$ (OH, NO_2) + Cl^- $\xrightarrow[\text{fast}]{OH^-}$ (O^-, NO_2) $\xrightarrow{H^+}$ p-Nitrophenol

Note that phenols, which are the products of the reactions in Equations 23.4 and 23.5, are weak acids (Chap. 24). In the presence of OH^- they are converted to their salts (eq. 23.5); addition of a strong acid regenerates the parent phenol.

The *position* of the electron-withdrawing group with respect to the leaving group also strongly influences the reaction rate. In all the examples so far shown, the electron-withdrawing groups were either *ortho* or *para* to the leaving group. Nitro groups in the *meta* position are very poor at promoting nucleophilic attack.

The mechanism for these reactions shows why an electron-withdrawing group in the *ortho* and/or *para* position is effective. The nucleophile first adds to the ring by attacking the carbon holding the leaving group (eq. 23.6). The intermediate **benzenide ion** (**1**) is stabilized by resonance delocalization of the negative charge *ortho* and *para* to the point of attack. The nitro groups at

(23.6) (Cl, NO_2, NO_2) + ^-OH $\xrightarrow{\text{slow}}$ [Cl OH, NO_2, NO_2 ↔ Cl OH, NO_2, NO_2 ↔ Cl OH, NO_2, NO_2] ≡ Cl OH, NO_2, δ−, δ−, δ−, NO_2

A benzenide ion
1

1

the *ortho* and *para* positions delocalize the charge even further (see resonance structures below). This increases the stability of the benzenide ion and therefore increases its rate of formation.

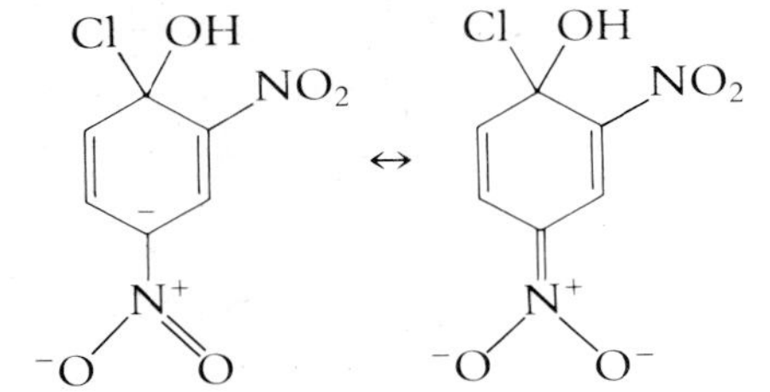

Charge stabilization by a *para* nitro group

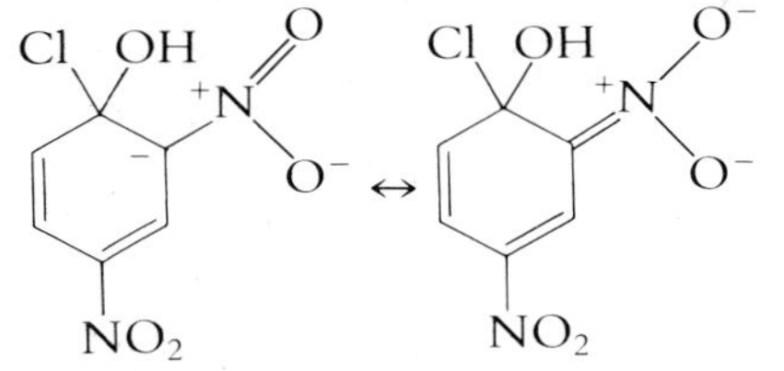

Charge stabilization by an *ortho* nitro group

In the second step of the mechanism the leaving group is expelled, and the ring becomes aromatic once again (eq. 23.7).

(23.7)

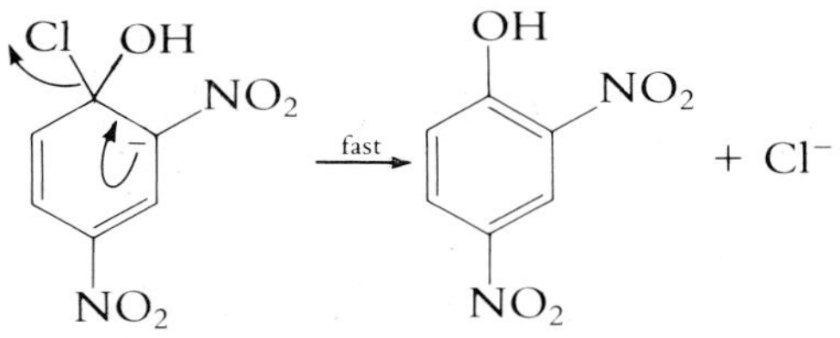

$\xrightarrow{\text{fast}}$ + Cl^-

The resonance forms for the benzenide ion produced from a *meta*-nitro-substituted aryl halide are shown below. You can see that the negative charge *cannot* be delocalized into the nitro groups. Thus the *meta*-substituted ion is much less stable than those produced from *ortho*- and *para*-nitrosubstituted reactants and the *meta*-substituted aryl halide is relatively unreactive.

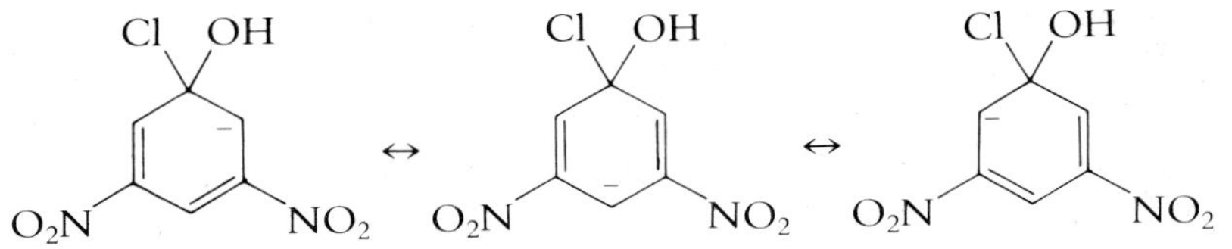

In most cases it is the first step of the reaction, addition of the nucleophile to the ring, that is rate-determining. This is shown by the reactivity order for halonitrobenzenes: F > Cl > Br > I. Since F^- is the poorest leaving group, this reactivity order would be difficult to explain if the second step of the reaction (eq. 23.7) were the rate-determining one. However, the powerful inductive effect of F does stabilize the benzenide ion, and thus speeds up its rate-determining formation (eq. 23.6). Figure 23–1 shows an energy diagram for the overall mechanism, which is referred to as addition–elimination, or S_NAr (*substitution, nucleophilic, aromatic*).

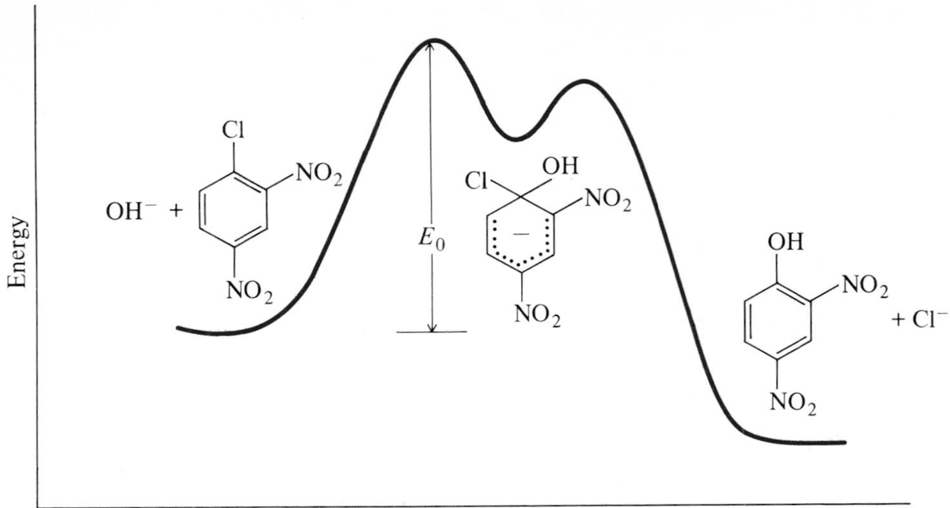

Reaction coordinate

Figure 23–1. *Energy diagram for the reaction of hydroxide ion with 2,4-dinitrochlorobenzene. The slow step is the formation of the benzenide ion intermediate.*

Example

Show how the following compound could be prepared from an aromatic halide.

OEt
COOEt
COOEt

Answer

The ester groups are strongly electron-withdrawing and will assist nucleophilic substitution at positions *ortho* and *para* to them. The ethoxy group is *ortho* to one COOEt and *para* to the other and can therefore be introduced by displacing a halogen at that position:

Br, COOEt, COOEt $\xrightarrow{Na^+\bar{O}Et}$ OEt, COOEt, COOEt

Problems

23–1 Many different nucleophiles will attack reactive aryl halides. Show the products of reaction of 1-chloro-2,4-dinitrobenzene with each of the following nucleophiles in ethanol solution.

a. sodium azide
b. sodium ethoxide
c. potassium cyanide
d. potassium hydrogen sulfide

23–2 Give the structures of the benzenide intermediates in the reactions in Equations 23.1 to 23.3. Use resonance structures to show why they can form under relatively mild conditions.

23–3 Show how the following compounds could be prepared from aromatic halides.

a. 2,4-dinitrobenzenethiol
b. *N*-ethyl-4-cyanoaniline
c. diethyl *o*-nitrophenylmalonate
d. MeOOC–C_6H_3(COOMe)–CN
e. 2-methoxy-1,3-dinitrobenzene

In Section 15:10 we discussed the *deactivating* effect of nitro groups on *electrophilic* substitution. With *nucleophilic* substitution, the opposite effect is observed. *What was stabilizing to the positively-charged benzenium ion is destabilizing to the negatively-charged benzenide ion,* and vice versa. Thus, we expect that all groups that are deactivating for electrophilic substitution (Table 15–2, p. 569) should promote nucleophilic attack. For example, halogens, being electronegative and therefore electron-withdrawing, were shown in Section 15:13 to be deactivating for electrophilic substitution. We therefore expect that halogens would be activating for nucleophilic substitution, but less effective than nitro or cyano groups, since halogens withdraw electrons by an inductive, rather than a resonance, effect. This is indeed what we find experimentally.

Equation 23.8 shows the industrially important conversion of 1,2,4,5-tetrachlorobenzene to 2,4,5-trichlorophenol. This reaction goes by way of an intermediate benzenide ion that is stabilized by the inductive effect of the

(23.8) 1,2,4,5-Tetrachlorobenzene $\xrightarrow[\text{ethylene glycol, 200°}]{\text{KOH}}$ $\xrightarrow{H^+}$ 2,4,5-Trichlorophenol

2,4,5-Trichlorophenol $\xrightarrow[OH^-]{ClCH_2CO_2H}$ $\xrightarrow{H^+}$ 2,4,5-Trichlorophenoxyacetic acid (2,4,5-T, a powerful herbicide)

2,4,5-Trichlorophenol $\xrightarrow{CH_2{=}O}$ Hexachlorophene (a bactericide)

chlorines. Note the rather stringent conditions required (compare with eq. 23.4). Trichlorophenol is used in the synthesis of some biologically potent compounds, such as 2,4,5-T and hexachlorophene (eq. 23.8).

Problem

23–4 A dangerous by-product in the manufacture of 2,4,5-trichlorophenol (eq. 23.8) is 2,3,7,8-tetrachlorodibenzo-*p*-dioxin (**I**). **I** is a powerful **teratogen** (a substance causing deformities in developing fetuses) which was released in large amounts when a chemical reactor used to manufacture the phenol exploded in Seveso, Italy in 1976. The effects on the surrounding area were disastrous. Show how **I** might have been formed in the course of the trichlorophenol synthesis.

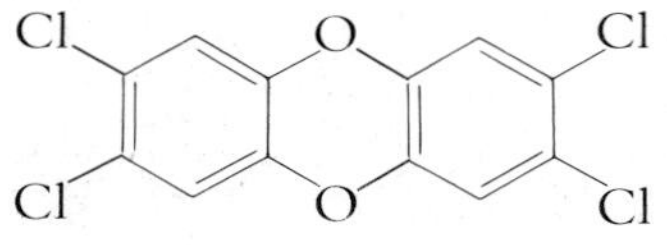

Cl, O, Cl, Cl, O, Cl

I

In this section we've stressed the powerful activating effect of electron-withdrawing groups in nucleophilic aromatic substitution. Does this mean that in the absence of such groups aryl halides will not undergo nucleophilic substitution? No. Unsubstituted aryl halides can be made to react with nucleophiles, but *severe* conditions are required. For instance, Equation 23.9 describes the *Dow process,* which was first used industrially in 1928 and was for many decades thereafter the major industrial route to phenol (it has now been superseded by a very different process; see Prob. 14–36, p. 542). Note the very high temperatures required.

(23.9) Chlorobenzene (Cl) + 2NaOH (2*M*) $\xrightarrow{370°}$ O^-Na^+ (sodium phenoxide) $\xrightarrow{H^+}$ Phenol (OH) + H_2O + NaCl

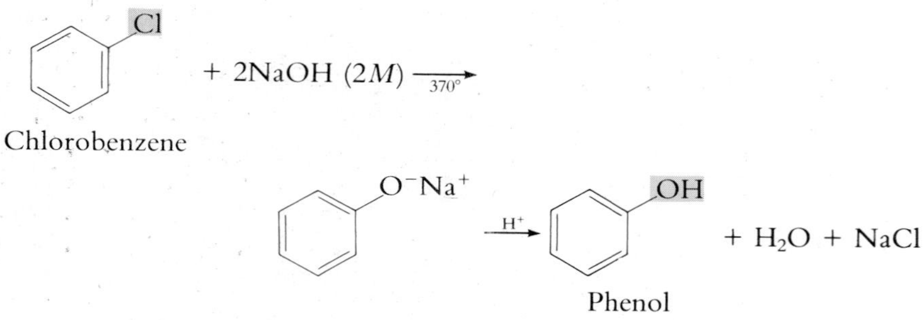

Chlorobenzene will undergo nucleophilic substitution at much lower temperatures if the nucleophile is also a powerful base. As you'll see in the next section, employing severe conditions—high temperatures or powerful bases—results in a new substitution mechanism.

Treating an unsubstituted aryl halide such as chloro-, bromo-, or iodobenzene with amide ion in liquid ammonia (−33°) results in its rapid conversion to aniline (eq. 23.10). Although NH_2^- is a better nucleophile than OH^-, that

(23.10) Br-benzene + $K^+NH_2^-$ $\xrightarrow[-33°]{NH_3\ (liq)}$ NH_2-benzene + KBr

difference cannot explain why the reaction in Equation 23.9 requires a temperature of 370°, while the reaction in Equation 23.10 proceeds at −33°. The critical difference between NH_2^- and OH^- is not in their nucleophilicity but in their *basicity*.

We can confirm the importance of basicity with a mechanistic study of the reaction in Equation 23.10 using chlorobenzene-1-^{14}C as the substrate (eq. 23.11; the asterisk denotes the position of ^{14}C). Analysis of the aniline product shows that the amino group is attached to the ^{14}C in just over half of the product. The remainder of the carbon label is in the two equivalent *ortho* positions (eq. 23.11).

(23.11) Cl ★ 100% $\xrightarrow[NH_3,\ -33°]{K^+NH_2^-}$ NH_2 ★ + NH_2 ★

(53%) (47%)

These data make sense if we assume that the amide ion *first reacts as a base* toward the aryl halide, eliminating H and Cl to form **benzyne** (eq. 23.12).

(23.12) Cl ★ H :NH_2^- $\xrightarrow{slow}$ ★ + NH_3 + Cl^-

Benzyne

The triple bond in benzyne is highly reactive, and it is attacked by a second amide ion that acts as a nucleophile (second step of eq. 23.13). Notice that amide attack can occur at either end of the triple bond. This gives a phenyl anion which is finally protonated (third step of eq. 23.13) to yield aniline.

(23.13) Cl ★ H $\xrightarrow[①]{NH_2^-}$ ★ $\xrightarrow[②]{NH_2^-}$ ★ NH_2 $\xrightarrow[③]{NH_3}$ H ★ NH_2

$\xrightarrow[②]{NH_2^-}$ NH_2 ★ $\xrightarrow[③]{NH_3}$ NH_2 ★ H

The label is scrambled in the final product because both ends of the triple bond are chemically equivalent, but only one end is labeled.

This mechanism is known as the benzyne, or *elimination–addition,* mechanism. It is most likely to occur when the aryl halide is treated with very strong bases such as salts of ammonia (KNH_2) or amines (R_2NLi), or organolithium compounds (RLi). When weaker bases are used, higher temperatures are required to initiate benzyne formation (compare eqs. 23.9 and 23.10). However, OH^- (H_2O, $pK_a = 15$) is so much less basic than NH_2^- (NH_3, $pK_a = 35$) that even at very high temperatures the reaction between OH^- and chlorobenzene only partly follows the benzyne mechanism. Under these extreme conditions the S_NAr mechanism manages to compete to a small extent, even in the absence of an electron-withdrawing group (eq. 23.14).★

(23.14) OH^- + chlorobenzene $\xrightarrow{340°}$ benzyne path → benzyne $\xrightarrow[2.\ H_2O]{1.\ OH^-}$ phenol (Major path); S_NAr path → [δ− intermediate with OH and Cl] → phenol (Minor path)

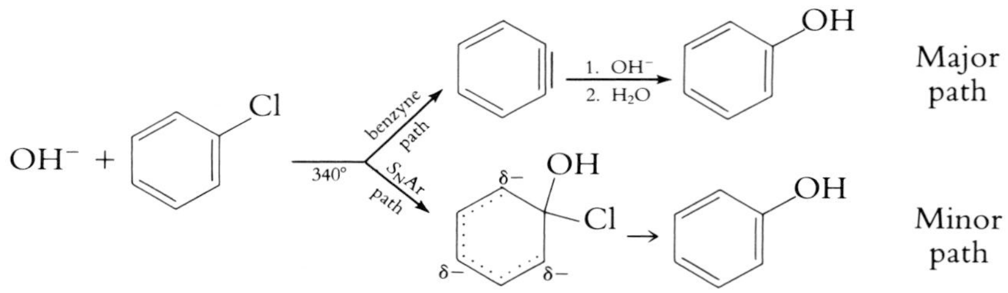

Problems

23–5 In S_NAr reactions fluorobenzenes are the most reactive halobenzenes (sec. 23:1). In benzyne reactions the reverse is true—fluorobenzenes are the least reactive. Explain this difference. What does this indicate about the rate-determining step in the benzyne reaction?

23–6 Provide a complete mechanism that can explain the product distribution of the reaction in Equation 1.

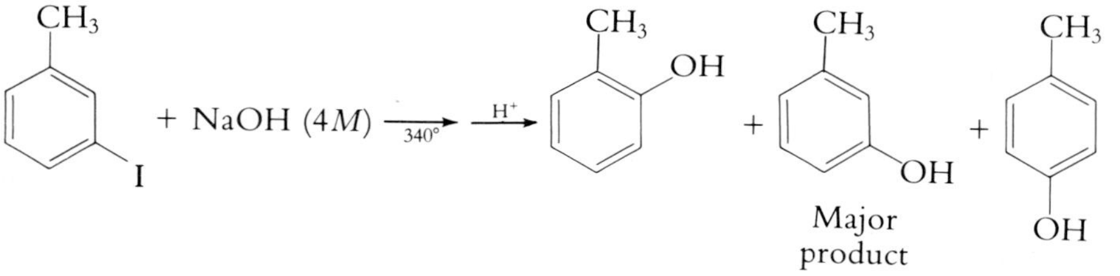

3-iodotoluene + NaOH (4*M*) $\xrightarrow{340°}$ $\xrightarrow{H^+}$ o-cresol + m-cresol (Major product) + p-cresol (1)

23–7 The bromine atom of 1-bromonaphthalene can be replaced by piperidine. The rates and product distributions vary widely depending on whether piperidine alone is used (eq. 2) or piperidine containing sodium amide (eq. 3). Propose reaction mechanisms that explain these results.

★ There must also be some S_NAr participation in Equation 23.11 as well. A strict benzyne mechanism would give a 50:50 ratio rather than a 53:47 ratio.

Br + N H → (230°, 48 hr) N (2)

Br + N H → (NaNH$_2$, 106°, 2 hr) N + N (3)

Since benzynes are so reactive, it is not possible to isolate and study them at room temperature. Benzyne itself has been generated and frozen at 8° K in an inert medium, where its ir spectrum has been studied. The aromatic π electron system seems to be substantially intact, because the carbon skeleton is not too badly distorted from a regular hexagonal geometry. The two electrons of the triple bond are probably paired, although there is only weak overlap between the two sp^2 orbitals (Fig. 23–2).

Figure 23–2. *π bond formed by weak overlap of two* sp^2 *orbitals in benzyne.*

Box 23–1. A Scientific Publication in Free Verse

The scientific literature is generally written in such a way that individual style is sacrificed for a more or less uniform "scientific-ese" which varies little from paper to paper or from author to author. This tradition was broken by J. F. Bunnett and F. J. Kearley, Jr. whose paper entitled "Comparative Mobility of Halogens in Reactions of Dihalobenzenes with Potassium Amide in Ammonia"* was written entirely in free verse. It begins as follows:

"Reactions of potassium amide
With halobenzenes in ammonia
Via benzyne intermediates occur.
Bergstrom and associates did report,
Based on two-component competition runs,
Bromobenzene the fastest to react,
By iodobenzene closely followed,
The chloro compound lagging far behind,
And fluorobenzene to be quite inert
at reflux (−33°). . . .
  The irregular mobility rank
Explanation finds in the mechanism
Whereby arynes are formed. There are
two steps:
Abstraction of the *ortho* proton
And then expulsion of the halogen
From the anion intermediate."

* From *Journal of Organic Chemistry*, **36**, 184 (1971). Reprinted with permission. Copyright 1982 American Chemical Society.

23:3 Copper-Assisted Aromatic Substitutions

In the previous sections you saw that severe conditions (strong bases, elevated temperatures) are needed for replacing halogens in aromatic rings that do not have electron-withdrawing groups. Such reactions can occur under less severe conditions in the presence of Cu(I) (cuprous ion), which markedly accelerates the reaction rate. This is advantageous in terms of energy consumption, and in terms of avoiding rearranged by-products (see Prob. 23–6).

The chemical industry makes use of copper's catalytic ability in various commercially important syntheses—for example, the synthesis of aniline from chlorobenzene. The aryl halide is treated with a concentrated aqueous ammonia solution in the presence of a cuprous salt (eq. 23.15). This route is economically competitive with the classical synthesis of aniline from nitrobenzene (sec. 21:8).

$$C_6H_5Cl + 2NH_3 \xrightarrow[200^\circ]{Cu^+,\ H_2O} C_6H_5NH_2 + NH_4^+Cl^- \quad (23.15)$$

A major laboratory application of copper-assisted aromatic substitutions is the synthesis of aromatic nitriles, starting with aryl bromides or iodides and cuprous cyanide (eq. 23.16). This process is valuable because the nitriles

$$p\text{-}BrC_6H_4NH_2 + CuCN \xrightarrow[\text{DMF, }160^\circ,\ 4\text{ hr}]{} p\text{-}NCC_6H_4NH_2 + CuBr \quad (23.16)$$

p-Bromoaniline → *p*-Aminobenzonitrile (83%)

can be converted to other classes of compounds such as carboxylic acids, primary amines, and ketones (sec. 17:12). Another valuable feature of this reaction is that we are converting an *o,p*-directing halogen to an *m*-directing cyano group. Therefore, the positional isomerism of a disubstituted product can be controlled by choosing different points in the reaction sequence at which to change the halide to CN (compare eqs. 23.17 and 23.18).

$$C_6H_5Br \xrightarrow[H_2SO_4]{HNO_3} p\text{-}BrC_6H_4NO_2 \xrightarrow{CuCN} p\text{-}NCC_6H_4NO_2 \quad (23.17)$$

p-Nitrobenzonitrile

(23.18) Bromobenzene $\xrightarrow{CuCN}$ Benzonitrile (CN) $\xrightarrow[H_2SO_4]{HNO_3}$ m-Nitrobenzonitrile (CN, NO_2)

m-Nitrobenzonitrile

We have already discussed the effectiveness of organic derivatives of Cu(I) (sec. 6:14) in coupling with alkyl or vinyl halides to give alkanes or alkenes with a longer carbon chain. The same approach can be used to make alkyl-substituted aromatic compounds (eq. 23.19).

(23.19) $C_6H_5I + (C_4H_9)_2CuLi \rightarrow C_6H_5(CH_2)_3CH_3 + C_4H_9Cu + LiI$

For all of the copper-assisted reactions discussed in this section, the reactivity of the aryl halides follows the order I > Br > Cl. This finding is compatible with mechanisms in which the rate-determining step involves breaking the carbon–halogen bond. Beyond that, not many mechanistic details are known with certainty. It is believed that a number of these reactions go by way of intermediates that contain derivatives of Cu(III). These highly oxidized intermediates then decompose to organic products and Cu(I) derivatives (eq. 23.20).

(23.20) $RCu + R'X \rightarrow R(R')CuX \rightarrow R{-}R' + CuX$

Cu(I) species	Cu(III) species	Organic product	Cu(I) species

23:4 Formation of Diazonium Ions from Primary Aromatic Amines

As we said in the introduction of this chapter, aromatic halides are never displaced by nucleophiles in a concerted, S_N2-type mechanism. Remember from Chapter 6 (Fig. 6–4, p. 183) that these reactions are initiated by nucleophilic attack on the backside of the carbon atom that bears the leaving group. In an aromatic halide, the backside of the carbon in question is inside the aromatic ring. Therefore S_N2 reactions do not take place. Can S_N1 mechanisms occur with aromatic halides? The rate-determining step in an S_N1 reaction is loss of the halide to form a cation. With aromatic halides, the carbon–halogen bond is too strong to be easily broken, probably due to resonance between the halogen and the aromatic ring.

$C_6H_5{-}\ddot{X}: \leftrightarrow {}^{-}C_6H_5{=}\ddot{X}^{+}$ — resonance strengthens C—X bond

An S_N1-type mechanism *could* occur if the benzene ring carried a better leaving group than halide ion. One class of compounds that does provide such a leaving group is **arenediazonium salts**. Diazonium salts, which are made from aromatic amines, can be converted to a variety of substituted aromatics, often through an S_N1-type mechanism. Thus, they are important intermediates in the conversion of other aromatic amines to a number of aromatic compounds, many of which are not readily available by other nucleophilic or electrophilic substitution reactions. This section discusses formation of diazonium salts, while Sections 23:5 to 23:8 discuss their reactions.

When aromatic primary amines such as aniline are treated with nitrous acid (HONO) and a mineral acid, they are converted into diazonium salts (eq. 23.21).

(23.21) $C_6H_5NH_2 + HONO + HCl \rightarrow C_6H_5\overset{+}{N}{\equiv}N\ \ Cl^- + 2H_2O$

Benzenediazonium chloride

The HONO is generated *in situ* from sodium nitrite ($NaNO_2$). Salts such as diazonium chlorides and bromides are rarely isolated, as they are explosive when dry. If, however, fluoroboric acid is used instead of hydrochloric or hydrobromic acid, then the final product is the slightly soluble diazonium fluoroborate (eq. 23.22), which is quite stable in the solid state.

(23.22) $p\text{-}NO_2C_6H_4NH_2 + HONO + HBF_4 \rightarrow O_2N{-}C_6H_4{-}\overset{+}{N}{\equiv}N\ \ BF_4^- + 2H_2O$

p-Nitrobenzenediazonium fluoroborate (100%)

Before the chemistry of diazonium ions is discussed, brief attention will be given to the mechanism of their formation. As a subsequent section shows, this mechanism is biologically, as well as chemically, important. The starting point for this discussion is the chemistry of nitrous acid. This weak acid cannot be isolated, but is formed as needed in the reaction solution by protonation of nitrite ion (eq. 23.23). The proton is supplied by a strong acid such as HCl and the nitrite ion is supplied by sodium nitrite.

(23.23) $$^-O{-}N{=}O + H^+ \rightleftarrows HO{-}N{=}O$$

In acidic solution, nitrous acid reacts with anions such as Cl^- or NO_2^- to give derivatives of the type NOX (eq. 23.24). These derivatives react readily

(23.24) $$HO{-}N{=}O + H^+ + X^- \rightarrow X{-}N{=}O + H_2O$$
(X = Cl, ONO, etc.)

with nucleophiles such as amines which displace X^-, thus transferring the *nitroso group*, —N=O, to the amine. The process is referred to as *nitrosation*.

Nitrosation of a primary amine such as aniline (eq. 23.25) first yields an *N*-nitrosamine, which tautomerizes readily to a diazohydroxide (**2**, eq. 23.26).

(23.25) $C_6H_5\ddot{N}H_2 + O{=}\ddot{N}{-}X \rightarrow C_6H_5\overset{+}{N}H_2{-}\ddot{N}{=}O + X^- \rightarrow C_6H_5\ddot{N}H{-}\ddot{N}{=}O + HX$

N-Nitrosoaniline
(unstable)

(23.26) $Ar\ddot{N}H{-}\ddot{N}{=}\ddot{O} \overset{H^+}{\rightleftharpoons} Ar{-}\ddot{N}H{-}\ddot{N}{=}\overset{+}{\ddot{O}}H \rightleftharpoons Ar{-}\ddot{N}{=}\ddot{N}{-}\ddot{O}H \overset{H^+}{\rightleftharpoons}$

2

$Ar\ddot{N}{=}\ddot{N}{-}\overset{+}{\ddot{O}}H_2 \xrightarrow{H^+} Ar{-}\overset{+}{N}{\equiv}N{:} + H_2O$

Diazonium
ion

The hydroxide, **2**, is protonated and loses water rapidly to yield a diazonium ion.

Problem

23–8 Write the structures of the products of the following reactions:

a. 4-methylaniline $\xrightarrow[HCl]{NaNO_2}$

b. 2,4-dichloroaniline $\xrightarrow[HBF_4]{NaNO_2}$

c. 5-nitro-2-naphthylamine $\xrightarrow[HBF_4]{NaNO_2}$

d. 2-aminobenzoic acid $\xrightarrow[HCl]{NaNO_2}$ **A** $\xrightarrow{Ag_2O}$ $C_7H_4N_2O_2$
an internal salt

23:5 Nucleophilic Substitution by Way of Phenyl Cations Generated by Diazonium Ion Decomposition

Diazonium ions contain the best leaving group in organic chemistry—*molecular nitrogen*. There is a strong thermodynamic driving force for the decomposition of benzenediazonium ion to phenyl cation and nitrogen because of the stability of N_2 (eq. 23.27; Fig. 23–3).

(23.27) $C_6H_5\overset{+}{N}{\equiv}N{:} \xrightarrow{heat} C_6H_5^+ + {:}N{\equiv}N{:}$

Phenyl
cation

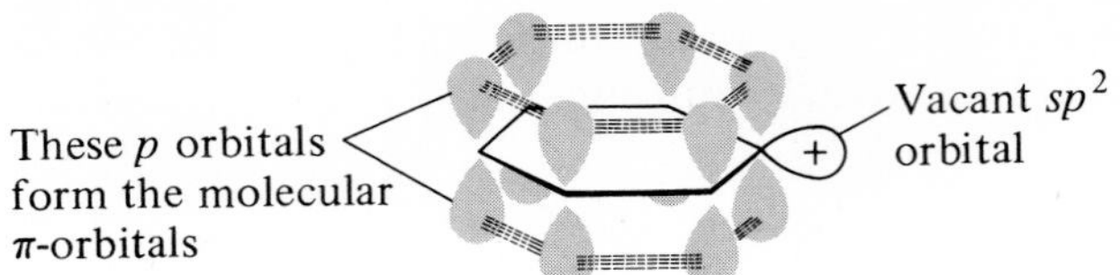

Figure 23–3. *Phenyl cation, showing the lack of overlap between the vacant* sp^2 *orbital and the molecular* π *orbitals.*

Note that the empty orbital carrying the positive charge of this ion is at right angles to the π system (Fig. 23–3), so the ion *cannot* be resonance stabilized. For this reason phenyl cation is highly reactive and not very discriminatory. We therefore try to generate it under conditions where only one nucleophile is present in high concentrations. Thus, arenediazonium ions are converted to phenols by heating in water. This is a standard method of synthesizing substituted phenols (eqs. 23.28 and 23.29) from amines. The

(23.28) m-Bromoaniline (NH_2, Br) $\xrightarrow[H_2O,\ H_2SO_4]{NaNO_2,\ 0°}$ $N_2^+HSO_4^-$ (Br) $\xrightarrow[H_2SO_4]{H_2O,\ 130°}$ OH (Br) + N_2

m-Bromobenzenediazonium bisulfate

m-Bromophenol (66%)

(23.29) NH_2 (F, OCH_3) $\xrightarrow[H_2O,\ H_2SO_4]{NaNO_2,\ 0°}$ $N_2^+HSO_4^-$ (F, OCH_3) $\xrightarrow[H_2SO_4]{H_2O,\ 130°}$ OH (F, OCH_3) + N_2

2-Fluoro-4-methoxyphenol (70%)

procedure consists of adding the solution of the diazonium bisulfate to hot aqueous sulfuric acid. The reaction mechanism involves capture of water by phenyl cation (eq. 23.30). This is an analog of a standard S_N1-type mechanism.

(23.30) [Reaction scheme: benzenediazonium ion $\xrightarrow{\text{heat}}$ phenyl cation $+ N_2$ $\xrightarrow{H_2O:}$ protonated phenol (H_2O^+–C$_6$H$_5$) $\rightarrow$ phenol (OH) $+ H^+$]

Another useful synthetic method that goes by way of a phenyl cation intermediate is the **Schiemann reaction**, which involves heating *dry* diazonium fluoroborates until they decompose. (The fluoroborates are kept dry to prevent reaction with water.) This route is one of the few available for preparation of aryl fluorides (eq. 23.31).

(23.31) [m-aminopropiophenone (NH_2, $O{=}CCH_2CH_3$) $\xrightarrow[HBF_4]{NaNO_2}$ $N_2^+BF_4^-$ diazonium salt ($O{=}CCH_2CH_3$) $\xrightarrow[95°]{\text{dry salt}}$ F-substituted product ($O{=}CCH_2CH_3$) $+ BF_3 + N_2$]

m-Fluoropropiophenone
(68%)

Example

Prepare 3-hydroxybenzoic acid from benzoic acid.

Answer

We have no good ways of introducing a hydroxyl group directly into a benzene ring, but in this section it was shown that an amine can be converted to a phenol through the diazonium salt:

[3-aminobenzoic acid (NH_2, COOH) $\xrightarrow[2.\ H_2O,\ \Delta]{1.\ NaNO_2,\ H_2SO_4}$ 3-hydroxybenzoic acid (OH, COOH)]

If we can prepare 3-aminobenzoic acid, the synthesis is complete. Amino groups cannot be directly introduced into a benzene ring either, but are readily available by reduction of the nitro compound. Therefore, the synthesis is completed as follows:

[benzoic acid (COOH) $\xrightarrow[\Delta]{HNO_3,\ H_2SO_4}$ 3-nitrobenzoic acid (NO_2, COOH) $\xrightarrow{H_2/\text{cat}}$ 3-aminobenzoic acid (NH_2, COOH)]

23–9 Write a mechanism for the conversion of benzenediazonium fluoroborate to fluorobenzene.

23–10 Show how the following compounds can be prepared from the appropriate amine by way of the diazonium ion.
a. 4-methyl-1-naphthol
b. 1-fluoro-3-nitrobenzene
c. ethyl 4-hydroxybenzoate
d. 4-chloro-3-pyridinol
e. 4,4′-difluorobiphenyl

23:6 Phenyl Radicals from Diazonium Ions: The Sandmeyer Reaction

There are a number of other reactions in which the diazonium group is replaced by a nucleophile, but these proceed by a mechanism different from the S_N1-type. This new mechanism can occur when arenediazonium ions react with easily oxidizable ions or molecules. Diazonium ions are highly electrophilic and are therefore prone to accept electrons and decompose to aryl *radicals*, which can react further to give substituted arenes. In this section we'll explore some synthetically important reactions that follow this path.

If sodium or potassium iodide is added to a solution of an arenediazonium chloride, a rapid reaction occurs. Nitrogen is given off and an aryl iodide separates from the reaction mixture (eq. 23.32). There are several

(23.32) o-(COOH)C$_6$H$_4$$N_2^+Cl^-$ $\xrightarrow[25°]{KI}$ o-(COOH)C$_6$H$_4$I + N_2 + KCl

o-Iodobenzoic acid

pieces of evidence which strongly suggest that these are not ionic reactions: the reaction takes place at room temperature, no chlorobenzenes are isolated, and biphenyls (Ar-Ar) are often obtained as by-products.

This evidence is indicative of a free-radical reaction. The diazonium and iodide ions combine to form a diazoiodide, which is very unstable and rapidly decomposes to radicals and N_2 (eq. 23.33). Simple recombination of the phenyl and iodine radicals can lead to iodobenzene (eq. 23.34).

(23.33) $C_6H_5-\overset{+}{N}\equiv N:$ + $:\ddot{I}:^-$ → [C_6H_5–N=N–I:] → $C_6H_5\cdot$ + N_2 + $:\ddot{I}\cdot$

Diazoiodide   Phenyl radical

(23.34) $C_6H_5\cdot$ + $:\ddot{I}\cdot$ → $C_6H_5-\ddot{I}:$

The simple radical combination of Equation 23.34 probably doesn't tell the full story. Recent evidence suggests that a phenyl radical can react with anions to give **radical anions**. For example, attack of the phenyl radical on iodide ion would yield iodobenzene radical anion (**3**, eq. 23.35). The radical anion is an excellent electron donor, and could reduce a molecule of diazonium ion to a phenyl radical (eq. 23.35, second step). The phenyl radical then continues the radical anion chain mechanism. The extra electron in **3** can be either mainly centered on the iodine, or delocalized in the ring.*

$$(23.35)\quad C_6H_5\cdot + I^- \xrightarrow{①} [C_6H_5I]^{\cdot -} \xrightarrow[②]{C_6H_5N_2^+} C_6H_5I + C_6H_5\cdot + N_2$$

3
A radical anion

This reaction works well with iodide ion because I^- is easily oxidized. Other nucleophiles of interest, such as Cl^-, Br^-, and CN^-, are more difficult to oxidize and react only very slowly with diazonium ions. However, addition of a catalytic amount of a cuprous salt to the reaction mixture leads to formation of aryl halides or cyanides in good yields (the **Sandmeyer reaction**, eqs. 23.36 and 23.37).

$$(23.36)\quad m\text{-}OHC\text{-}C_6H_4\text{-}NH_2 \xrightarrow[HCl]{NaNO_2} m\text{-}OHC\text{-}C_6H_4\text{-}N_2^+Cl^- \xrightarrow[HCl,\ 75°]{CuCl} m\text{-}OHC\text{-}C_6H_4\text{-}Cl + N_2$$

m-Chlorobenzaldehyde

$$(23.37)\quad o\text{-}CH_3\text{-}C_6H_4\text{-}NH_2 \xrightarrow[HCl]{NaNO_2} o\text{-}CH_3\text{-}C_6H_4\text{-}N_2^+Cl^- \xrightarrow[0\ \text{to}\ 5°]{CuCN,\ KCN} o\text{-}CH_3\text{-}C_6H_4\text{-}CN$$

o-Methylbenzonitrile
(*o-tolunitrile*, 70%)

The Cu^+ acts as a catalyst because it is easily oxidized by loss of an electron to give Cu^{2+}. The electron is used to reduce the benzenediazonium ion (eq. 23.38). In the last step of the reaction, Cu^{2+} is converted back to Cu^+

* Remember that according to MO theory, even if all atoms of a given molecule have eight electrons in their valence shells, empty antibonding orbitals can still accept additional electrons.

(eq. 23.39). The mechanism illustrated is for the synthesis of chlorobenzene, but is also applicable to bromide and cyanide reactions.

(23.38) $$CuCl + Cl^- \rightleftarrows CuCl_2^- \xrightarrow{C_6H_5N_2^+} C_6H_5\cdot + N_2 + CuCl_2$$

(23.39) $$C_6H_5\cdot + CuCl_2 \rightarrow C_6H_5Cl + CuCl$$

Since direct iodination of benzene (sec. 15:4) is difficult, and direct cyanation not possible, synthesis of iodo- and cyanobenzenes by way of diazonium salts is of major importance. On the other hand, Sandmeyer synthesis of chlorobenzenes and bromobenzenes is usually less efficient than one-step direct halogenation of benzene (sec. 15:4). Exceptions occur when the starting material is a substituted benzene and the desired product isomer is difficult to obtain because of the directing effects of the substituent(s). For instance, *m*-bromochlorobenzene is only a minor product from chlorination of bromobenzene or bromination of chlorobenzene. It can be made isomerically pure, however, starting with nitrobenzene, as shown in Equation 23.40. Thus, using this sequence, we are able to put two *ortho,para* directing groups *meta* to one another.

(23.40) $$C_6H_5NO_2 \xrightarrow[AlCl_3]{Cl_2} m\text{-}ClC_6H_4NO_2 \xrightarrow[Pt]{H_2} m\text{-}ClC_6H_4NH_2 \xrightarrow[HBr]{NaNO_2} \xrightarrow{CuBr} m\text{-}ClC_6H_4Br$$

Problem

23–11 Design efficient reaction sequences for making the conversions shown below.

a. nitrobenzene → *m*-bromocyanobenzene
b. nitrobenzene → *p*-chlorocyanobenzene
c. 1-methylnaphthalene → 1-iodo-4-methylnaphthalene
d. benzoic acid → 3-bromo-5-iodobenzoic acid

23:7 Replacement of the Diazonium Group by Hydrogen

Another way to place two *ortho,para* directing groups *meta* to one another makes use of the fact that the N_2^+ group can be replaced by H when a diazonium salt reacts with hypophosphorous acid, H_3PO_2 (eq. 23.41).

(23.41) $$ArN_2^+Cl^- + H_3PO_2 \xrightarrow[0^\circ]{H_2O} ArH + N_2 + H_3PO_3 + HCl$$

Suppose that we want to prepare 3,5-dibromotoluene. Bromination of tolu-

3,5-Dibromotoluene

ene would not be useful because the methyl group is *o,p*-directing. The desired product can, however, be reached by first nitrating toluene and then reducing the nitro group to an amino function (eq. 23.42). The amino group

(23.42) toluene $\xrightarrow[H_2SO_4]{HNO_3}$ 4-nitrotoluene (CH_3, NO_2) $\xrightarrow[Pt]{H_2}$ 4-methylaniline (CH_3, NH_2)

is a much stronger *o,p*-director than the methyl group and hence bromination would introduce the bromines next to the amine (first step of eq. 23.43). The amino group could then be removed by conversion to the diazonium salt, followed by treatment with hypophosphorous acid (second and third steps of eq. 23.43).

(23.43) 4-methylaniline (CH_3, NH_2) $\xrightarrow[CCl_4]{Br_2}$ 2,6-dibromo-4-methylaniline (Br, Br, NH_2, CH_3) $\xrightarrow[HCl]{HONO}$ diazonium salt (Br, Br, $N_2^+Cl^-$, CH_3) $\xrightarrow{H_3PO_2}$ 3,5-dibromotoluene (Br, Br, H, CH_3)

If 3-bromotoluene was the synthetic goal, the reaction sequence would be modified by first converting the strongly activating amino group into the less active $NHCOCH_3$ group (sec. 15:14). Now monobromination is possible; the bromine still enters *ortho* to the $NHCOCH_3$ group, since the latter is a better electron donor than CH_3. The diazotization-reduction sequence then gives the desired product (eq. 23.44). Hypophosphorous acid reduction of

(23.44) 4-methylaniline (CH_3, NH_2) $\xrightarrow[OH^-]{(CH_3CO)_2O}$ 4-methylacetanilide (CH_3, $NHCOCH_3$) $\xrightarrow{Br_2}$ $\xrightarrow{\text{hydrolysis}}$ 2-bromo-4-methylaniline (CH_3, Br, NH_2) (65%) $\xrightarrow[130°]{1.\ NaNO_2,\ HCl;\ 2.\ H_3PO_2,\ H_2O}$ 3-bromotoluene (CH_3, Br) (90%)

diazonium salts is probably a free-radical process somewhat similar to the Sandmeyer reaction.

Example

Show how to convert *p*-nitrotoluene to pure 2-bromobenzoic acid (no isomer separation in the synthesis).

CH₃ / NO₂ (p-nitrotoluene) $\xrightarrow{?}$ COOH / Br (2-bromobenzoic acid)

Answer

One can convert a methyl group on a benzene ring to the carboxyl group by oxidation. The nitro group can be removed by converting it to the diazonium salt and treating the salt with H_3PO_2. This leaves at least two possible schemes:

p-nitrotoluene $\xrightarrow{KMnO_4}$ p-nitrobenzoic acid $\xrightarrow[FeBr_3]{Br_2}$ 2-bromo-4-nitrobenzoic acid $\xrightarrow{H_2/eat}$ 4-amino-2-bromobenzoic acid $\xrightarrow[H_3PO_2]{NaNO_2,\ H^+}$ 2-bromobenzoic acid

or

p-nitrotoluene $\xrightarrow[FeBr_3]{Br_2}$ 2-bromo-4-nitrotoluene $\xrightarrow{H_2/cat}$ 4-amino-2-bromotoluene $\xrightarrow[H_3PO_2]{NaNO_2,\ H^+}$ 2-bromotoluene $\xrightarrow{KMnO_4}$ 2-bromobenzoic acid

In the first series of reactions, the second step is bromination with *meta* directors on the ring. A by-product would be the isomer with Br next to the nitro group. Bromination would be very slow, since both substituents are deactivating. These problems are avoided in the second sequence by brominating in the first step, since only one product is formed (CH_3 is *ortho*-

directing and NO_2 is *meta*-directing) and the activating CH_3 group speeds up bromination.

Problem

23–12 Show how *p*-nitrotoluene can be converted to each of the following isomerically pure compounds.

a. *o*-methylbenzenesulfonic acid
b. *m*-nitrotoluene
c. 3,4-dichlorotoluene
d. 3-methylbenzophenone

23:8 Diazonium Ions as Electrophiles: Azo Coupling Reactions as Sources of Dyes

There are a number of reactions in which the diazonium ion acts as an electrophile toward various aromatic molecules. This reaction provides compounds of the type R—N=N—R′, which are known as **azo compounds** (eq. 23.45). Azo compounds constitute the most common type of synthetic dyes. The discovery and exploitation of the azo coupling reaction fathered the rise of the enormous dye industry, particularly in Germany, toward the end of the nineteenth century.

$$ArN_2^+ + Ar'H \rightarrow \underset{\text{Aromatic azo compound}}{Ar{-}N{=}N{-}Ar'} + H^+ \tag{23.45}$$

What makes azo compounds so suitable as dyes? In the first place the azo group, —N=N—, serves as a link between the two aromatic systems, resulting in an extensive conjugated system. The more extensive the molecular conjugation, the smaller the energy gap between the highest filled π orbital and the lowest empty $\pi^\star$ orbital (sec. 13:13). If this gap is small enough, then even low energy light in the visible range may be sufficient to excite the molecule. Absorption of visible light results in the compound appearing colored. A substance such as a dye shows a color *complementary* to that of the wavelength of visible light actually absorbed. That is, if the dye absorbs blue wavelengths from white light, the reflected light will be relatively rich in red wavelengths, giving the dyed object a red color. Figure

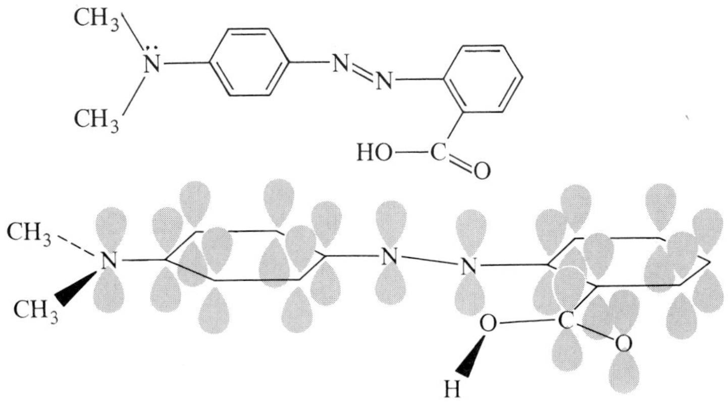

Figure 23–4. *Extended π system of the dye and indicator, methyl red.*

23–4 shows the extensively delocalized π system of the pH indicator methyl red.

Unlike all of the other reactions discussed in this chapter, azo coupling is an *electrophilic* aromatic substitution reaction. Because the diazonium ion is a weak electrophile, the rate of the process is greatly increased by the presence of electron-donating substituents in the ring being attacked. Thus, most of the aromatic rings used in coupling reactions contain either amino or hydroxy groups (eq. 23.46 and 23.47). The diazonium ion attacks preferentially *para* to the substituent but if the *para* position is blocked, it will go to the *ortho* position.

(23.46)

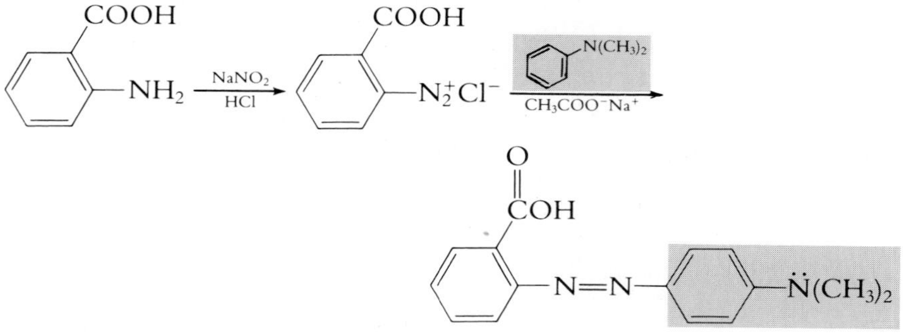

Methyl red (65%)

(23.47)

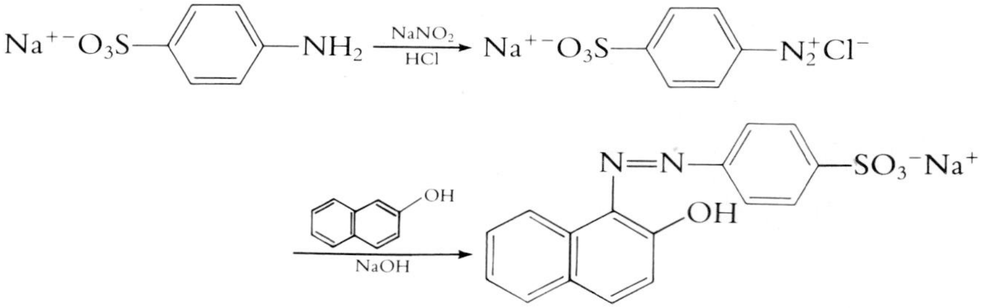

Orange H (80%)

The mechanism of azo coupling is given in Equation 23.48.

(23.48)

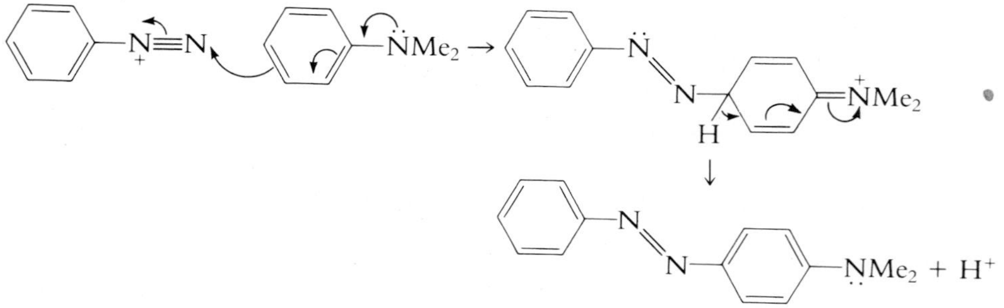

Box 23–2. The First Synthetic Commercial Dye

The first important synthetic commercial dye, mauve, was synthesized in 1856 by the English chemist W. H. Perkin. Perkin's achievement predated the discovery of the diazotization reaction by six years. The enormous commercial success of mauve, an *azine dye,* spurred an active search for other synthetic dyes and undoubtedly contributed in that way to the creation of the azo dyes.

At the age of 18, Perkin was working in the laboratory of August von Hofmann at the Royal College of Chemistry. Hofmann was interested in the synthesis of the antimalarial alkaloid, quinine, because of the high rate of malaria deaths in the British colonies. Hofmann suggested, solely on the basis of molecular formula, that quinine might be synthesized from a toluidine (methylaniline) derivative, so Perkin spent a vacation period in his home laboratory studying the oxidation of toluidines.* His failure to obtain quinine induced him to study the oxidation of aniline, which had been obtained from benzene by way of nitrobenzene. Since benzene from coal tar was contaminated with toluene, Perkin's aniline was contaminated with toluidines. Oxidation of this mixture with chromic acid yielded a black precipitate from which Perkin isolated a brilliant purple material which could dye silk. He called the dye aniline purple, but French dyers named it "mauve" after a flower (called "mallow" in English) that has blossoms of deep lavender.

$$C_6H_5NH_2 + CH_3C_6H_4NH_2 \xrightarrow[H_2SO_4]{Na_2Cr_2O_7}$$

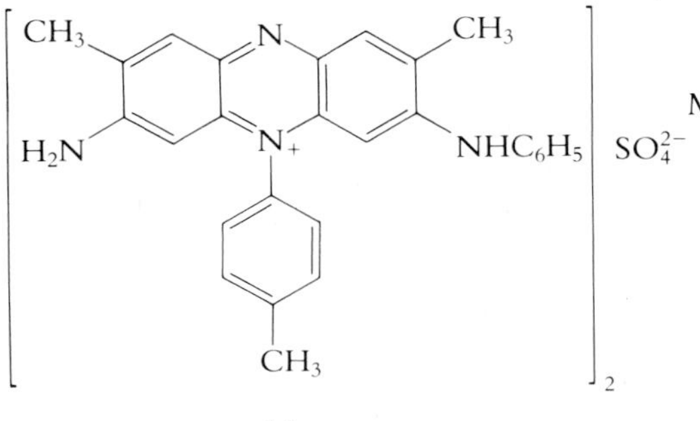

Mauve

Perkin's discovery of mauve has an element common to many other great chemical discoveries—*serendipity,* the faculty for making happy discoveries by accident. What was *not* serendipitous was Perkin's recognition of the significance of his discovery and his subsequent exploitation of it. Using borrowed family funds, he set up a factory for the manufacture of mauve, which made him a wealthy man and led eventually to his being knighted. The passion for mauve was so intense that the last decade of the nineteenth century was known as the Mauve Decade. In 1872 Perkin retired from business and devoted himself to scientific research.

Perkin's success inspired many other chemists to oxidize anilines and their derivatives in the hope of obtaining useful dyes. For instance, milder oxidating agents than chromic acid, acting on aniline/methylaniline mixtures, produce *rosaniline,* or *fuchsin,* dyes:

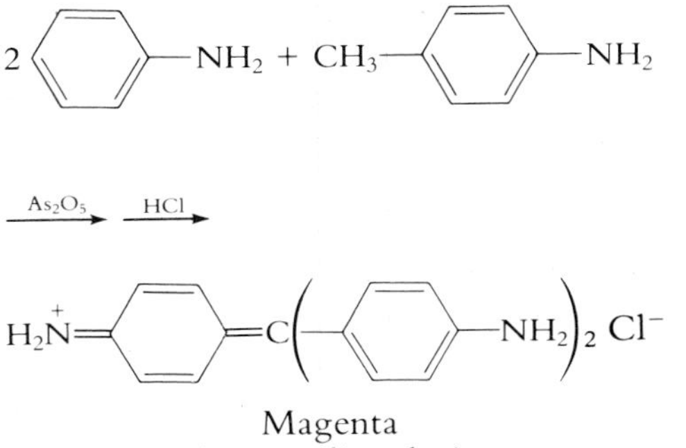

$$2\,C_6H_5NH_2 + CH_3C_6H_4NH_2 \xrightarrow{As_2O_5} \xrightarrow{HCl}$$

Magenta
(a rosaniline dye)

The rosaniline dyes have the useful property of being acid–base indicators because at high pH they are converted to nonionic forms which are no longer colored. This is because they no longer have extended conjugated systems.

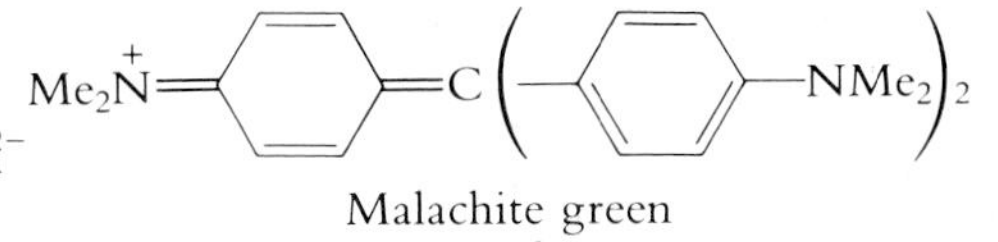

Malachite green
(green form)

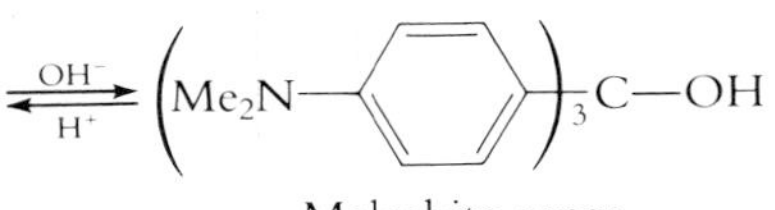

$$\underset{H^+}{\overset{OH^-}{\rightleftharpoons}}$$

Malachite green
(colorless (leuco) form)

* We now know that quinine, which was synthesized in 1944 by R. B. Woodward and W. Doering, is structurally unrelated to the methylanilines.

23:9 Reactions of Other Amines with Nitrous Acid

All amines (not just primary aromatic amines) react in one way or another with nitrous acid. *Aliphatic* primary amines are converted to diazonium ions, just like their aromatic counterparts. Aliphatic diazonium ions are much less stable than aromatic ones, however, and decompose to carbocations as soon as they are formed (eq. 23.49). The products isolated from these reactions are the ones expected from a carbocation intermediate (eq. 23.50).

$$\text{(23.49)} \quad RCH_2NH_2 \xrightarrow[H^+]{NaNO_2} RCH_2\overset{+}{N}{\equiv}N \xrightarrow[\text{fast}]{\text{very}} RCH_2^+ + N_2$$

$$\text{(23.50)} \quad CH_3CH_2CH_2NH_2 + NaNO_2 + HCl \xrightarrow{H_2O}$$

$$\underset{(7\%)}{CH_3CH_2CH_2OH} + \underset{(28\%)}{CH_3\underset{\displaystyle OH}{\underset{|}{C}}HCH_3} + \underset{(34\%)}{CH_3CH{=}CH_2} + N_2 + NaCl$$

Secondary amines, whether aliphatic or aromatic, are also nitrosated by nitrous acid (see eq. 23.25). These nitrosamines cannot rearrange to diazonium ions, and can therefore be isolated (eq. 23.51). The nitrosamines are not soluble in dilute acid, unlike the amines from which they are made, and they separate out as they are formed.

$$\text{(23.51)} \quad (CH_3)_2NH \xrightarrow[HCl,\ 75°]{NaNO_2} (CH_3)_2N{-}NO$$

N-Nitrosodimethylamine
(90%)

Aromatic tertiary amines are nitrosated on the ring preferentially *para* to the amino group (eq. 23.52). Under the usual nitrosation conditions (0 to 5°) aliphatic tertiary amines are not affected by nitrosating agents; at higher temperatures they decompose by C—N cleavage.

$$\text{(23.52)} \quad C_6H_5NMe_2 \xrightarrow[HCl]{NaNO_2} \xrightarrow{\text{neutralize}} p\text{-}O{=}N{-}C_6H_4{-}NMe_2$$

N,N-Dimethyl-*p*-nitrosoaniline
(85%)

Problems

23–13 Show the product(s) of coupling between benzenediazonium chloride and each of the following compounds. Explain your answers, making reference to the reaction intermediate (eq. 23.48).

a. 1,3-dimethoxybenzene
b. 1-naphthylamine
c. sodium *p*-hydroxybenzene-sulfonate
d. 9-anthracenol (secs. 14:9 and 14:10)

23–14 Indicate which member of the following pairs of coupling reactions will proceed faster, and why.

a. benzenediazonium chloride with phenol, or with sodium phenoxide
b. benzenediazonium chloride with aniline, or with *o*-chloroaniline

c. 1-naphthol with benzenediazonium chloride, or with *p*-nitrobenzenediazonium chloride

23–15 Because nitrosamines are not soluble in dilute acid, they are very weak bases. Show with the aid of resonance structures why nitrosamines are less basic than simple alkylamines.

23:10 Nitrosamines as Carcinogens

It has been known for some time that dialkylnitrosamines are acutely toxic. *N*-Nitrosodimethylamine has even been used as a poison in several murders. Nitrosamines are both carcinogenic and mutagenic and produce tumors of the liver, kidney, and lung when injected into rats. It is significant that they do not induce tumors in the brain or intestines even though the nitrosamines are absorbed by these organs. This finding suggests that nitrosamines are converted metabolically into some other compound within the affected organs, and that this metabolic product is the true carcinogen. (The same situation is found with the polycyclic aromatic hydrocarbons, Section 14:11.)

It was noted in Section 14:11 that the liver contains high concentrations of oxidation enzymes. These are called **mixed function oxidases**. The purpose of these enzymes is to hydroxylate all types of foreign substances, in order to make them more water soluble and therefore more easily excreted. Based on this knowledge and direct experimental evidence, the following mechanism has been proposed for the metabolic "activation" of *N*-nitrosodimethylamine (eq. 23.53). After one of the methyl groups is hydroxylated by the enzyme (first step of eq. 23.53) it is rapidly lost as formaldehyde (second step), leaving behind a diazohydroxide. The diazohydroxide very quickly decomposes to methanediazonium ion.

$$(CH_3)_2N{-}N{=}O \xrightarrow{\text{oxidase}} CH_3N(CH_2OH){-}N{=}O \rightarrow \quad (23.53)$$

Hydroxymethyl derivative

$$CH_3{-}N{=}N{-}OH + CH_2{=}O \xrightarrow{H^+} CH_3{-}\overset{+}{N}{\equiv}N$$

Diazohydroxide + $CH_2{=}O$ — Methanediazonium ion

The methanediazonium ion is a very powerful methylating agent and attacks any nucleophilic sites available to it. The nucleophilic groups we show in Equation 23.54 are typical of those found in enzymes and ribonucleic acids. Thus, it has been found that administration of *N*-nitrosodi-

$$\text{(23.54)}\qquad CH_3{-}\overset{+}{N}{\equiv}N \begin{cases} \xrightarrow{RSH} RSCH_3 + N_2 + H^+ \\ \xrightarrow{R_3N:} R_3\overset{+}{N}CH_3 + N_2 \\ \xrightarrow{R\ddot{O}H} ROCH_3 + N_2 + H^+ \end{cases}$$

methylamine leads to *O*- and *N*-methylation of a nitrogen base in ribonucleic acids. These methylated bases interrupt the proper coding patterns for protein synthesis, thus leading to mutant cells and, in some cases, cancer formation.

These findings are of more than peripheral interest. Sodium nitrite has been used over a long period of time as a food preservative. It is found in relatively large amounts in preserved meats, such as salami, bacon, and frankfurters. It has been suggested that ingestion of sodium nitrite could lead to the formation of nitrous acid, since the stomach operates at pH 1. The body contains numerous secondary amine and amide groupings, and it is conceivable that they could be nitrosated by nitrous acid generated in the digestive tract. At present there is no evidence that links sodium nitrite ingestion directly to an increased incidence of cancer, although research is continuing. The subject is complicated by the fact that a "normal" diet contains large amounts of *nitrate,* which is reduced by enzymes in the saliva to nitrite. The human body contains more nitrite from this process than it gains from ingestion of cured meats.

Summary

1 Halobenzenes that have strongly electron-withdrawing groups (NO_2, CN, carbonyl) in the *ortho* or *para* positions undergo substitution readily when treated with nucleophiles. This substitution occurs by way of an S_NAr mechanism involving a resonance-stabilized benzenide intermediate (sec. 23:1). This reaction is particularly useful for preparing phenol, thiophenol, and aniline derivatives.

If simple (unsubstituted) aryl halides are treated with a *very* basic nucleophile, such as amide ion (NH_2^-), an elimination reaction can take place to form a benzyne intermediate (sec. 23:2). Benzyne intermediates are attacked readily by nucleophiles to form substituted arenes. Weaker bases, such as OH^-, give partial benzyne formation at high temperatures.

Copper-assisted nucleophilic substitution reactions are valuable for replacing a halide substituent by a cyano (CN) group (sec. 23:3). Alkylbenzenes can also be prepared from organocopper intermediates by reaction with aromatic halides.

2 Arenediazonium ions are useful reactive intermediates because they release the powerful leaving group N_2. They are made by reaction of primary aromatic amines with nitrous acid ($NaNO_2$ + HX) (sec. 23:4). Diazonium salts which are formed in H_2SO_4 and then heated, give a phenyl cation that reacts with water to form a phenol (sec. 23:5). Heating dry diazonium fluoroborates converts them to fluorides (Schiemann reaction).

Diazonium ions can be converted to aryl iodides with KI and to other halides or cyanides with CuCl, CuBr, or CuCN (Sandmeyer reaction; sec. 23:6). These reactions go through phenyl radical intermediates.

The diazonium group can be replaced by a hydrogen with H_3PO_2 (sec. 23:7). This allows the synthesis of substituted benzenes which cannot be made by direct substitution reactions. Diazonium ions couple with phenols or anilines to give aromatic azo compounds (Ar—N=N—Ar; sec. 23:8). These substances constitute a major class of dyes.

3 Aliphatic primary amines are converted to diazonium ions, but the latter rapidly lose nitrogen to yield carbocations. Secondary aliphatic or aromatic amines give water-insoluble *N*-nitroso compounds. Aromatic tertiary amines are nitrosated in the ring, while aliphatic tertiary amines are unreactive at 0° (sec. 23:9). Some *N*-nitroso compounds (also known as nitrosamines) are potent carcinogens. They are metabolically converted to diazonium ions, which alkylate nucleophilic sites in enzymes and nucleic acids (sec. 23:10).

Problems

23–16 Start with any monosubstituted benzene and devise syntheses of the following compounds:

a. *N*-phenyl-4-nitroaniline
b. methyl 3-hydroxybenzoate
c. 1-ethylthio-2-nitrobenzene
d. 1,4-benzenedicarbonitrile (1,4-dicyanobenzene)
e. 3,5-dinitrophenol
f. *p*-ethoxyaniline

23–17 2,4-Dinitrophenylhydrazine is a valuable reagent for preparing crystalline hydrazones of aldehydes and ketones. Suggest a route for its synthesis.

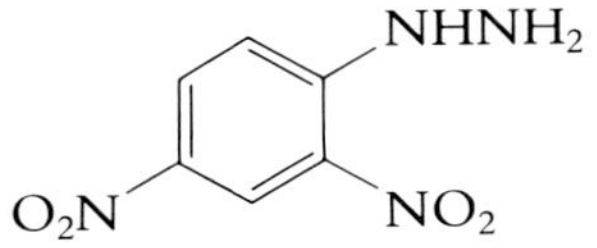

$NHNH_2$

O_2N NO_2

2,4-Dinitrophenylhydrazine

23–18 2,4-Dinitrofluorobenzene has been used to identify the free amino groups of peptides (Prob. 22–7, p. 867). Write a mechanism for the reaction of DNFB with the free amino group of glycylvalylalanine.

23–19 Phenol was produced commercially by heating chlorobenzene with NaOH solution (eq. 23.9). A by-product of this process is diphenyl ether, $C_6H_5—O—C_6H_5$.

a. Propose two different mechanisms that could account for the formation of diphenyl ether under the reaction conditions.
b. Phenoxide ion ($C_6H_5O^-$) is a weaker nucleophile than hydroxide ion. In light of this, which of your mechanisms seems more plausible? Explain your answer.

23–20 a. Equation 4 is an example of a reaction that has considerable potential for synthesis of indoles. What is the probable mechanism for the reaction?

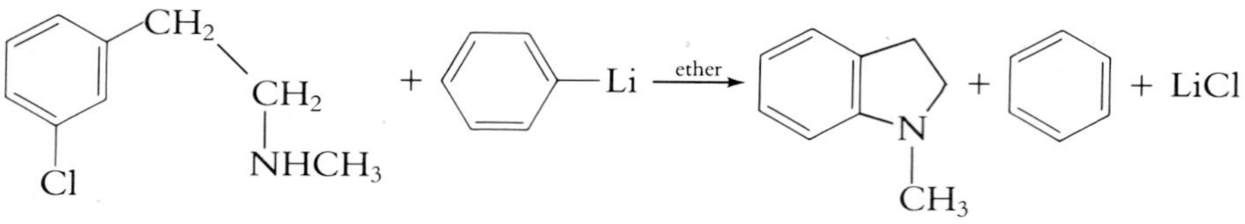

$$m\text{-}ClC_6H_4CH_2CH_2NHCH_3 + C_6H_5Li \xrightarrow{\text{ether}} \text{(N-methylindoline)} + C_6H_6 + LiCl \quad (4)$$

An indole

b. When compound **II** is treated with $NaNH_2$, two isomeric products of formula $C_{19}H_{31}N$ are formed. Write their structures.

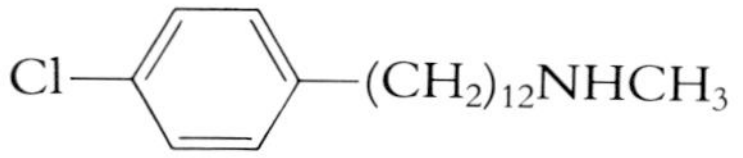

$$Cl-C_6H_4-(CH_2)_{12}NHCH_3$$

II

23–21 Show how the following compounds could be prepared from appropriately substituted aryl halides by way of Cu(I) catalyzed reactions.

a. *p*-methylbenzonitrile
b. *m*-fluoroaniline
c. methyl *o*-propylbenzoate
d. 1,5-dicyanonaphthalene
e. 4-methoxybenzylamine
f. *m*-acetylbenzoic acid

23–22 Propose efficient methods for bringing about the following conversions.

a. phenol → 4-butoxybenzonitrile
b. bromobenzene → benzene-1,3-dicarbonitrile
c. 1-nitronaphthalene → 6-cyano-1-naphthylamine
d. 2-iodonaphthalene → 1-chloro-2-naphthalenecarboxylic acid

23–23 Describe efficient methods for converting aniline to the following compounds:

a. bromobenzene
b. phenol
c. 1-phenylethanol
d. *p*-nitrobenzonitrile
e. ethyl *m*-bromobenzoate
f. *m*-nitroacetophenone
g. *m*-dichlorobenzene
h. 5-bromo-2-chloronitrobenzene

23–24 Show how the following compounds can be synthesized from any monosubstituted aromatic starting material.

a. *m*-hydroxybenzenesulfonic acid
b. *p*-fluorotoluene
c. 4-nitro-1-naphthol
d. 3-chloro-4-methylphenol
e. 1,2-dichloro-4-fluorobenzene

23–25 Two resonance structures for a diazonium ion are **IIIa** and **IIIb**. An ir study of a series of arenediazonium ions showed that the diazonio group absorbed at about 2250 cm^{-1} (Table 8–4, p. 279). Which of the two resonance structures is the best approximation to the true structure? Explain your reasoning.

$$R-\overset{+}{N}\equiv N: \leftrightarrow R-\ddot{N}=\overset{+}{N}:$$

IIIa **IIIb**

23–26 Sometimes it is very difficult to hydrolyze primary amides to carboxylic acids using the usual acids or bases. In these cases, nitrous acid is used (eq. 5). Propose a mechanism for this process.

$$\mathrm{R\overset{O}{\overset{\|}{C}}NH_2 + HNO_2 \xrightarrow{H^+} R\overset{O}{\overset{\|}{C}}OH + N_2 + H_2O} \tag{5}$$

23–27 Diazonium chlorides will react with alkenes in which the double bond is conjugated with an electron-withdrawing group. This reaction is catalyzed by cupric chloride, which contains a little cuprous chloride (eq. 6). Aryl chlorides are often side products in these reactions. This method has been used to synthesize ring-substituted phenylalanines. Propose a mechanism for the reaction shown in Equation 6.

$$\mathrm{O_2N{-}C_6H_4{-}\overset{+}{N}{\equiv}N\ Cl^- + CH_2{=}CHCOOCH_3 \xrightarrow[CuCl]{CuCl_2}}$$

$$\mathrm{O_2N{-}C_6H_4{-}CH_2\underset{\underset{Cl}{|}}{C}HCOOCH_3} \tag{6}$$

23–28 When benzenediazonium chloride is coupled with aniline in the presence of sodium acetate, it yields compound **A** ($C_{12}H_{11}N_3$). If hydrochloric acid is added to the reaction mixture, **A** is irreversibly isomerized to *p*-aminoazobenzene (**IV**), the "normal" coupling product (eq. 7). Propose a structure for **A**, and a mechanism for its conversion to **IV**. Show how the sodium acetate facilitates the formation of **A**.

$$\mathrm{C_6H_5{-}N_2^+Cl^- + C_6H_5{-}NH_2 \xrightarrow{Na^+OAc^-} A \xrightarrow{HCl}}$$

$$\mathrm{C_6H_5{-}N{=}N{-}C_6H_4{-}NH_2} \tag{7}$$

IV

23–29 Indicate what the likely products would be from treatment of the following amines with nitrous acid in the cold.

a. methylethylamine
b. *N*-methyl-2-butylamine
c. diethylisopropylamine
d. *N*,4-diethylaniline
e. *t*-butylamine
f. 2-butylamine

23–30 The structures of a number of common dyes are given below. With the aid of a chemical catalog, decide which readily available compounds can be used to synthesize these dyes. Write reactions for these syntheses.

a.

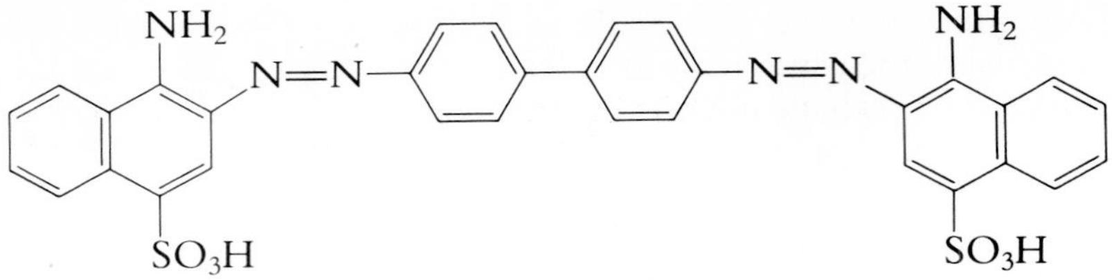

Congo red

b.

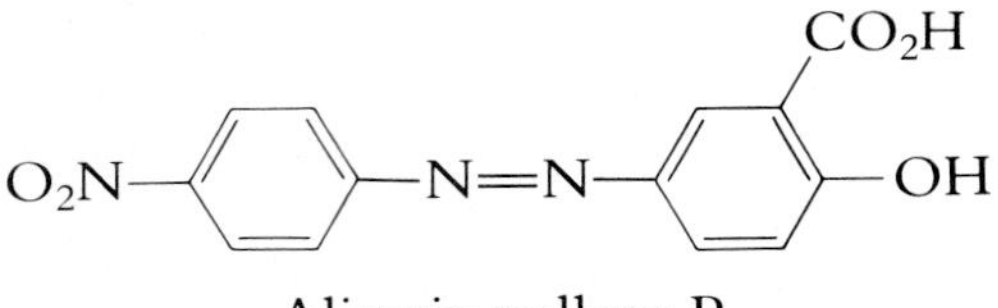

Alizarin yellow R

c.

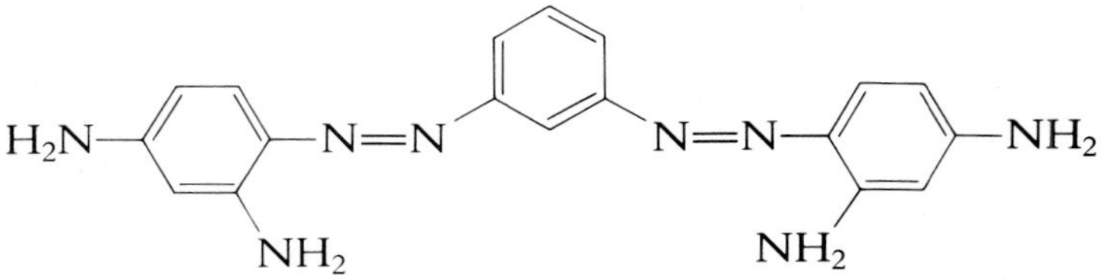

Bismarck brown B

23–31 Benzenediazonium ion can couple with 7-amino-2-naphthol (**V**) in either ring. At pH 5, coupling takes place *ortho* to the amino group; at pH 9, attack occurs *ortho* to the hydroxyl (eq. 8). Explain why the reaction takes different courses at different pHs.

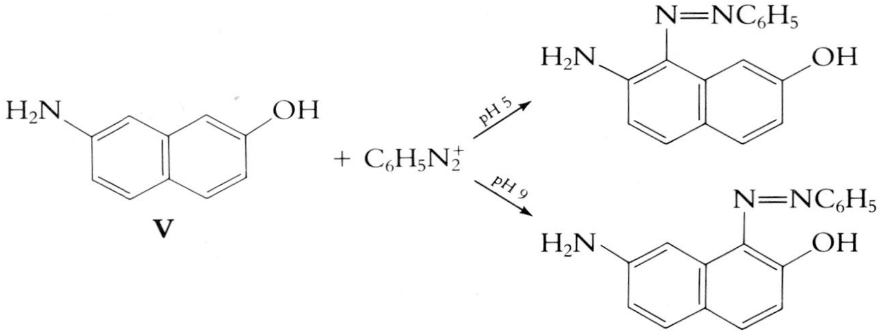

$$\mathbf{V} + C_6H_5N_2^+ \xrightarrow{\text{pH 5}} \text{(coupling ortho to } NH_2\text{, } N{=}NC_6H_5\text{)} \quad ; \quad \xrightarrow{\text{pH 9}} \text{(coupling ortho to OH, } N{=}NC_6H_5\text{)} \tag{8}$$

Supplementary Reading

J. F. Bunnett, The remarkable reactivity of aryl halides with nucleophiles, *J. Chem. Ed.,* **51,** 312 (1974).

R. A. Rossi and R. H. de Rossi, *Aromatic Substitution by the $S_{RN}1$ Mechanism,* American Chemical Society, Washington, D.C., 1983.

P. Rademacher, Chemical carcinogens, *J. Chem. Ed.,* **53,** 757 (1976).

J. S. Wishnok, Formation of nitrosamines in food and in the digestive system, *J. Chem. Ed.,* **54,** 440 (1977).

M. Sequin-Frey, The chemistry of plant and animal dyes, *J. Chem. Ed.,* **58,** 301 (1981).

Phenols and Quinones 24

We have just seen that phenols and their ethers are available from nucleophilic aromatic substitutions: via the S_NAr path (sec. 23:1), via benzyne intermediates (sec. 23:2), and via diazonium ions (sec. 23:5). In addition to being valuable synthetic intermediates, phenols occur naturally in such diverse materials as coloring pigments in flowers, lignin structures in wood, antibiotics, and alkaloids. Phenols are easily oxidized to carbonyl compounds known as quinones. A series of quinones which occur in green plants are involved in one of the first chemical steps of photosynthesis, the conversion of energy from the sun into chemical energy for the plant.

In this chapter we will study two fundamental properties of phenols, their acidity and their ease of oxidation, in order to gain insight into their reactivity. This is followed by a discussion of phenol biosynthesis. Let's first look at the structures of some common phenols.

Simple monohydroxybenzenes are available in large quantities from coal tar and are usually given common names. The cresols form the basis of a crude coal tar mixture called creosote, which is used as a wood preservative for telephone poles and fence posts. Many simple phenols, including phenol

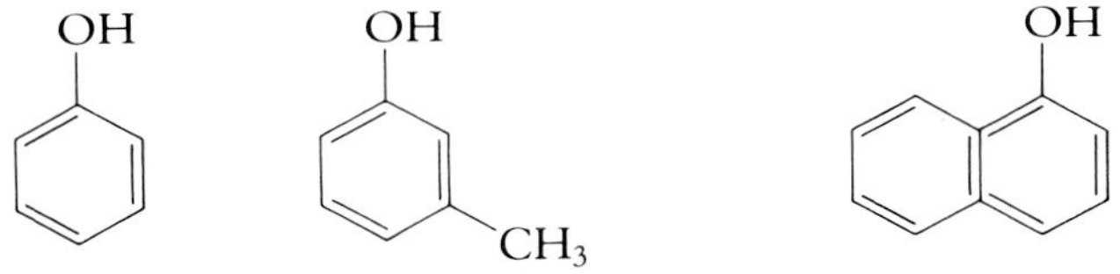

OH OH OH

CH_3

Phenol *m*-Cresol 1-Naphthol

itself, are active against fungi and microbes. In fact, phenol was the first generally used antiseptic, being introduced into medical practice in 1867 by Joseph Lister. Because of its high toxicity, especially toward skin cells, phenol has been replaced as an antiseptic by a number of its derivatives which are both less toxic and more effective. The antiseptic activity of these derivatives is measured by their *phenol coefficient,* which compares their effectiveness toward *Staphylococcus aureus* to that of phenol itself. A representative phenol derivative is hexylphenol, which has a phenol coefficient of about

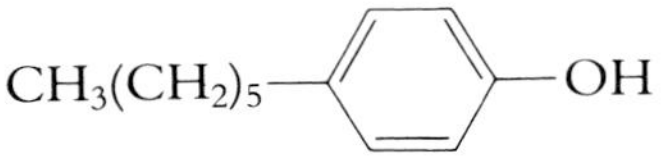

$CH_3(CH_2)_5$—⟨benzene ring⟩—OH

Hexylphenol

500. The length of the side chain affects the solubility of the phenol in both the aqueous and lipid portions of the cell, and is clearly important in determining its biological potency (see Box 17–2, p. 659).

All the dihydroxybenzenes are readily available materials. Catechol is used in photography, resorcinol in the manufacture of resins, and hydroquinone as a photographic reducer and developer, among other uses. (We will

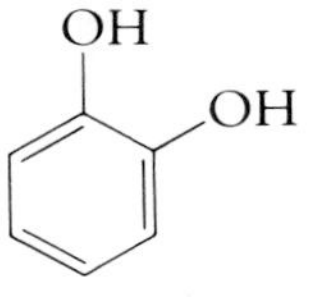

Catechol 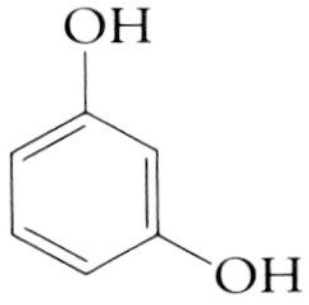 Resorcinol  Hydroquinone

discuss the chemistry of hydroquinone in more detail in Section 24:7.) The *o*-dihydroxybenzene group is common in nature; its presence in compounds such as dopamine and the epinephrines (sec. 21:1) gives those substances their generic name, *catecholamines.*

The *o*-dihydroxybenzene group is also found in the *urushiols,* a group of severe irritants present in such plants as poison ivy, oak, and sumac. The different members of the group differ only in the length and degree of unsaturation of their side chains.

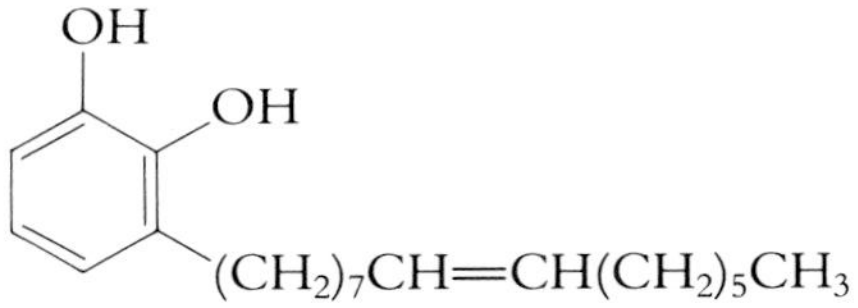

$(CH_2)_7CH{=}CH(CH_2)_5CH_3$ $CH_2(CH_2)_4CH_3$

A urushiol 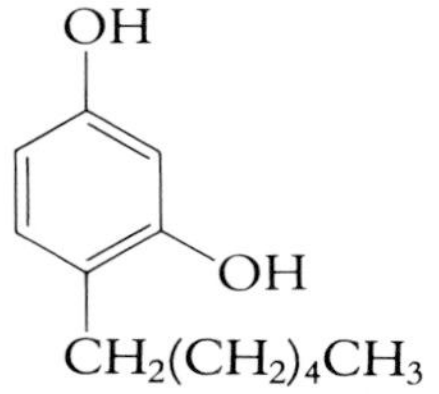 Hexylresorcinol

The relative placement of the hydroxyl groups in the dihydroxybenzene has a profound effect on its biological activity. While the urushiols are powerful *vesicants* (blistering agents), hexylresorcinol is an antiseptic mild enough to be used in throat lozenges.

24:1 Phenols as Acids

Most aliphatic alcohols are weak acids (eq. 24.1), even weaker than water (pK_a 16). By contrast, phenol is a stronger acid than water (eq. 24.2). Its solutions are acidic enough to turn litmus red, although phenol is a weak acid compared to carboxylic acids. (The K_a for acetic acid is larger by 10^5 than the K_a of phenol.)

(24.1) $$C_2H_5OH + H_2O \rightleftarrows C_2H_5O^- + H_3O^+ \qquad K_a = 10^{-17}, \quad pK_a = 17$$

(24.2) $$C_6H_5OH + H_2O \rightleftarrows C_6H_5O^- + H_3O^+ \qquad K_a = 1.28 \times 10^{-10}, \quad pK_a = 9.9$$

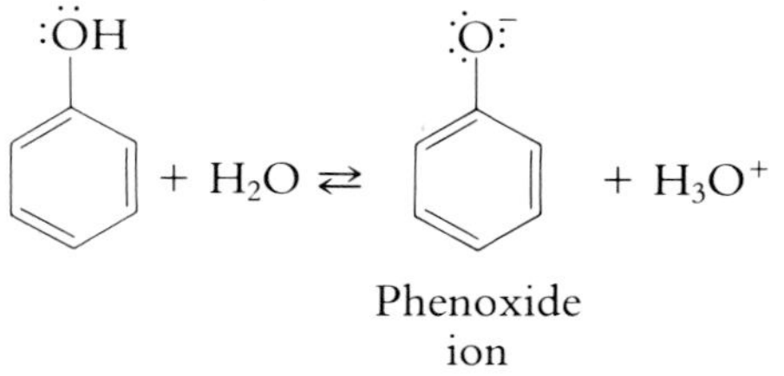

Phenoxide ion

Phenol acidity is mainly due to stabilization of the phenoxide ion, by delocalization of the negative charge into the benzene ring.

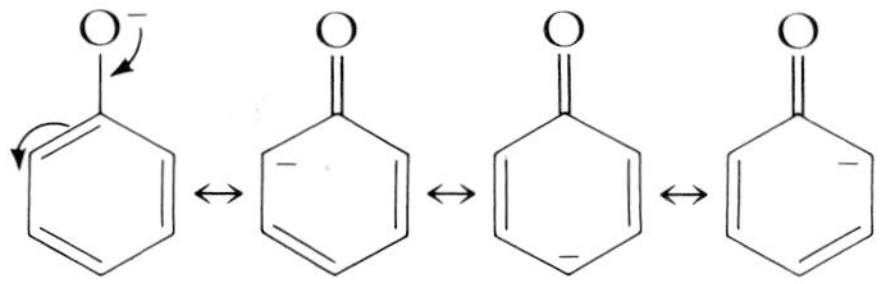

Phenoxide ion

Substituents in the aromatic ring will raise or lower the acidity of phenol in a predictable way. Electron-withdrawing substituents stabilize the phenoxide ion, making the phenol more acidic, thus shifting the equilibrium of Equation 24.2 to the right (eq. 24.3). Electron-donating substituents raise the energy of the phenoxide ion, and lower the acidity of the phenol (eq. 24.4). The strongest effects will be exerted by conjugating substituents at the

(24.3) *p*-Chlorophenol $+ H_2O \rightleftarrows$ $p\text{-}ClC_6H_4O^-$ $+ H_3O^+$ $K_a = 6.3 \times 10^{-10}$

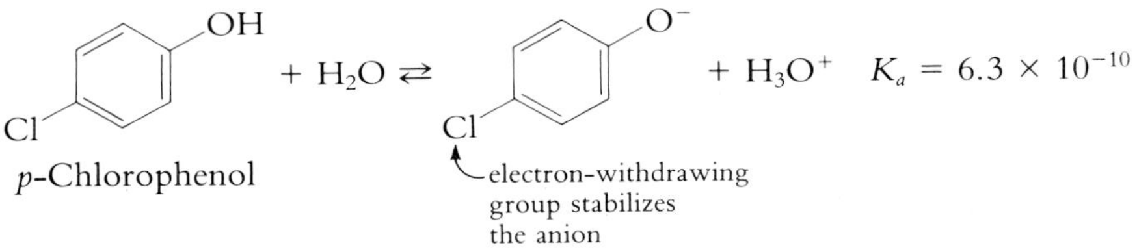

electron-withdrawing group stabilizes the anion

(24.4) *p*-Cresol $+ H_2O \rightleftarrows$ $p\text{-}CH_3C_6H_4O^-$ $+ H_3O^+$ $K_a = 0.67 \times 10^{-10}$

electron-donating group destabilizes the anion

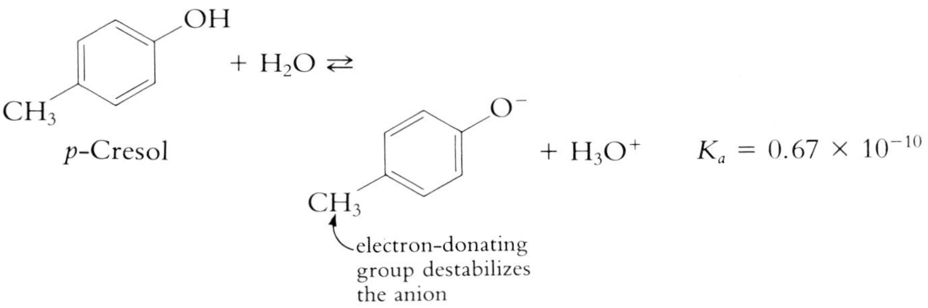

ortho or *para* positions. Resonance structures show how well *p*-nitrophenoxide ion is stabilized by conjugative electron withdrawal.

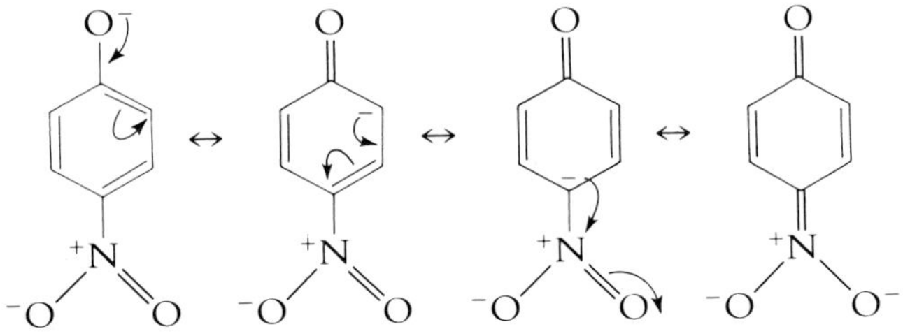

p-Nitrophenoxide ion

The number of electron-withdrawing substituents will also affect acidity. For example, 2,4,6-trinitrophenoxide ion is very stable (NO_2 groups in all *ortho* and *para* positions). In fact, 2,4,6-trinitrophenol (*picric acid,* pK_a 0.4) is as strong an acid as some mineral acids.

Ultraviolet-visible spectroscopy is the most convenient method for establishing the presence of a phenolic group. The unknown compound is dissolved in a solvent such as ethanol and its uv spectrum is taken in the usual way. Then one or two drops of concentrated aqueous NaOH are added. This converts any phenolic group present to a phenoxide ion. The phenoxide ion shows absorption at longer wavelengths and with an increased absorptivity compared to the phenol. This so-called **base shift** is strongly indicative of the presence of a phenolic group.

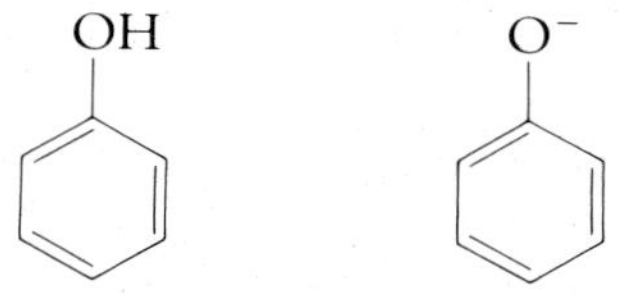

OH — λ_{max} 270 nm (ε = 1450)

O^- — λ_{max} 287 nm (ε = 2600)

Because most phenols are weaker acids than carboxylic acids, the two can be easily separated. An aqueous solution of the weak base sodium bicarbonate, $NaHCO_3$, will dissolve carboxylic acids as their sodium salts, but will not affect most phenols. Thus, given an organic solution containing a mixture of carboxylic acids and phenols, extraction with aqueous $NaHCO_3$ selectively removes the carboxylic acids as salts. Subsequent extraction with NaOH solution removes the phenols as phenoxides.

Example

Which is the stronger acid, *p*-hydroxybenzaldehyde or *p*-bromophenol? Why?

Answer

Each phenol will react with base to give its respective anion:

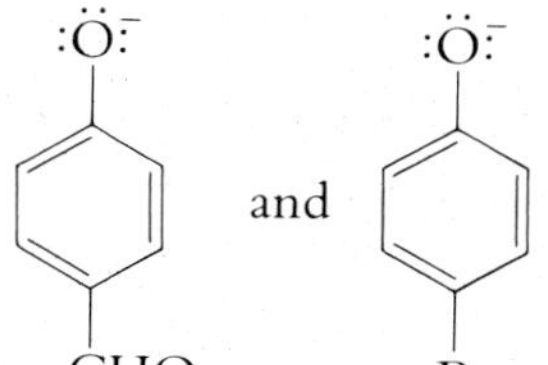

$:\ddot{O}:^-$ (CHO) and $:\ddot{O}:^-$ (Br)

In order to determine which is the stronger acid, one must determine which gives the more stable anion. This can be decided by looking at resonance structures:

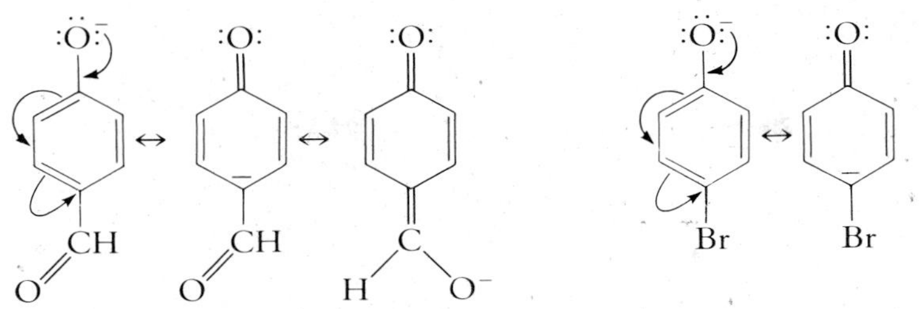

This anion is more stable because charge is delocalized by resonance.

This anion is less stable because charge is only delocalized inductively.

The CHO group can delocalize negative charge by resonance while the Br group cannot; thus, *p*-hydroxybenzaldehyde should be the stronger acid. This prediction is borne out by experiment (*p*-hydroxybenzaldehyde, pK_a 7.62; *p*-bromophenol, pK_a 9.34).

Problems

24–1 Arrange the following phenols in order of increasing acidity. Explain your reasoning.

a. *m*-cresol, *p*-cresol, *m*-bromophenol, phenol
b. *p*-methoxyphenol, 2,4-dinitrophenol, *p*-cyanophenol, *p*-fluorophenol
c. *p*-aminophenol, hydroquinone, catechol

24–2 Explain the following data:

a. The pK_a of 2,4-dinitrophenol is 3.96, while the pK_a of 3,5-dinitrophenol is 6.73.
b. The pK_a of 2,4,6-trichlorophenol is 6.00, while the pK_a of 2,4,6-trinitrophenol is 0.38.

24–3 Resonance plays a very important role in stabilizing phenoxide ion. Resonance interaction between the OH group and ring is also possible in an undissociated phenol. Illustrate this interaction with the aid of resonance structures, and explain why it is less significant than resonance in the phenoxide ion.

24:2 Electrophilic Substitution of Phenols

When phenols are subjected to reaction in polar media there are two reactive species present. One is the undissociated phenol, and the other is the phenoxide ion. The concentration of the ion may be small (0.001% in water at pH 5), but its reactivity is very high in electrophilic substitution reactions. Remember that these reactions involve electron withdrawal from the benzene ring by the electrophile. This withdrawal is compensated for by the powerful electron donation of the negatively-charged phenoxide oxygen. It is estimated that the phenoxide ion is 10^6 to 10^7 times more reactive than phenol toward electrophiles.

We can learn something about phenoxide ion reactivity by examining the bromination of phenol under different reaction conditions. Bromination of phenol in water leads to the isolation of 2,4,4,6-tetrabromocyclohexadienone, which precipitates from solution (eq. 24.5). This reaction, which

(24.5) phenol $+ 3Br_2 \xrightarrow{H_2O}$ [2,4,6-tribromophenol] $+ 3HBr \xrightarrow{Br_2}$ 2,4,4,6-tetrabromocyclohexadienone (Precipitates) $\xrightarrow{NaHSO_3}$ 2,4,6-tribromophenol

takes place even when phenol is present in excess, goes by way of 2,4,6-tribromophenol. The isolated tetrabromocyclohexadienone can be converted to 2,4,6-tribromophenol by treating with $NaHSO_3$ (eq. 24.5). When the reaction is carried out in a nonpolar solvent, such as CS_2 or CCl_4, only the monobrominated product is formed (eq. 24.6).

(24.6) phenol $+ Br_2 \xrightarrow{CS_2,\ 0°}$ 4-bromophenol (80%)

The reason for the observed difference is that in a polar solvent such as water, the species actually being brominated is the phenoxide ion (**1**, eq. 24.7). Equation 24.7 also shows that the monobromophenol produced is more acidic than phenol and hence there is a higher concentration of bromophenoxide ion, **2**, than there is of phenoxide ion, **1**. The same argument

(24.7)

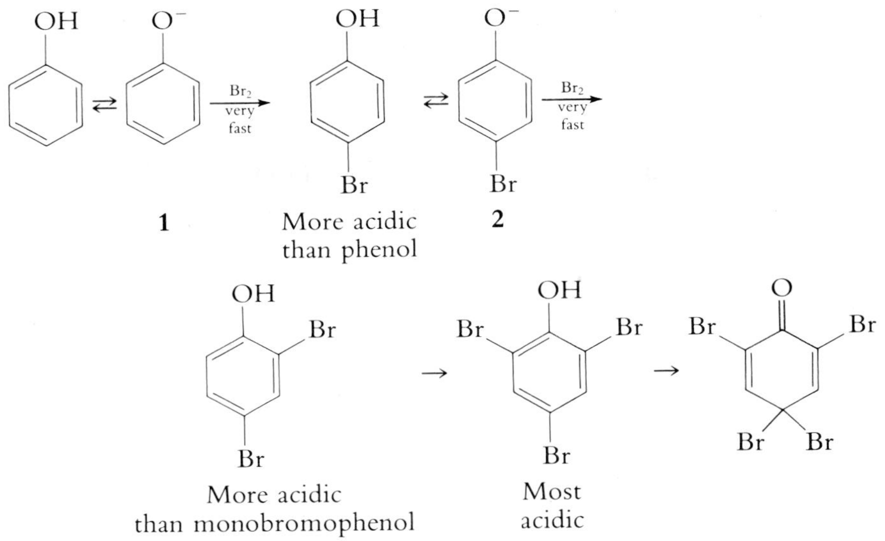

1

More acidic than phenol

2

More acidic than monobromophenol

Most acidic

holds after formation of 2,4-dibromophenol, which is brominated even more rapidly, and so on.

On the other hand, the nonpolar solvent (carbon disulfide) cannot act as a base. Hence, no phenoxide ions are present in CS_2 solution, and phenol itself is the reacting substance. The reaction follows the expected course—introduction of one halogen atom deactivates the ring toward further attack. The observation that anisole, which cannot dissociate into the phenoxide ion, yields only monobrominated product (eq. 24.8) is further evidence of the high reactivity of the phenoxide ion.

(24.8)

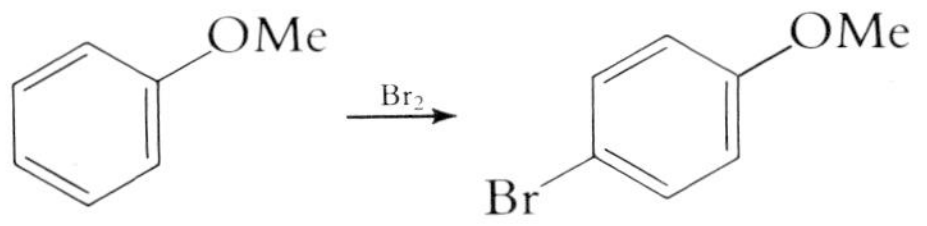

Anisole $\xrightarrow{Br_2}$ *p*-Bromoanisole (major product)

We can usefully generalize from the bromination reaction as follows: if we want to obtain a monosubstituted phenol by electrophilic substitution, we need to prevent formation of the phenoxide ion. For this reason phenol ethers and esters are often used as substrates (eq. 24.8). Their chemistry is described in the next section.

Problems

24–4 Describe the preparation of the following compounds starting, in each case, from an appropriate phenol.

a. 2,6-dichloro-4-methylphenol
b. 4-hydroxybenzenesulfonic acid
c. 4-iodo-1-naphthol

24:2 Electrophilic Substitution of Phenols

24–5 a. Show how 2,4,4,6-tetrabromocyclohexadienone is formed from 2,4,6-tribromophenol (eq. 24.5).

b. When 2,6-dimethylphenol is brominated in acetic acid, an intermediate builds up in the course of the reaction. This intermediate has λ_{max} = 252 nm ($\varepsilon \sim$ 13,000). The intermediate decomposes rapidly to give the final product, 4-bromo-2,6-dimethylphenol. Give a structure for this intermediate and show how it is formed and how it decomposes.

24–6 Compound **A** has the molecular formula $C_8H_{10}O$. **A** is soluble in aqueous NaOH but not in $NaHCO_3$ solution. The nmr spectrum of **A** is given here.

When **A** is treated with Br_2 in water, it is readily converted to a dibromo derivative, $C_8H_8Br_2O$, which is also soluble in NaOH. Write structures for **A** and its dibromo derivative.

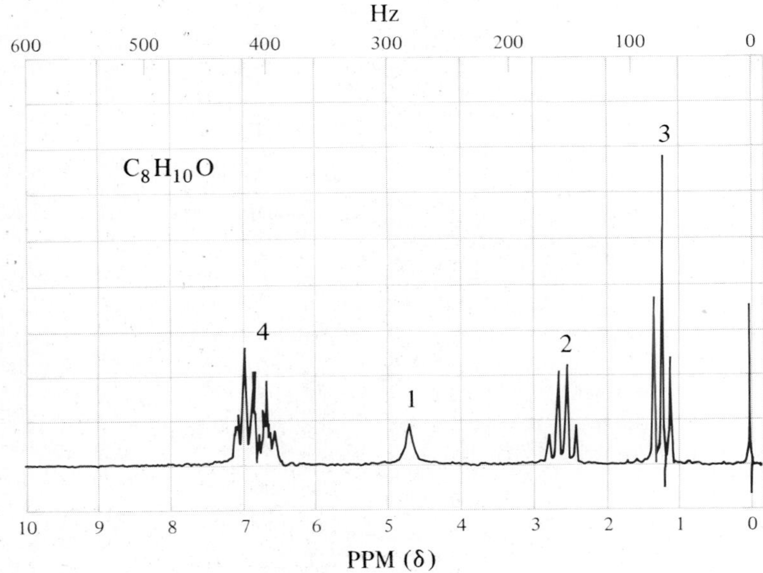

24:3 Reactions of the Phenolic Hydroxyl Group

Phenols undergo many reactions characteristic of aliphatic alcohols, but they are much less nucleophilic than alcohols. Thus, if a phenol and a carboxylic acid are refluxed together with a trace of mineral acid, only a small quantity of ester is formed because of the unfavorable equilibrium constant (sec. 17:6). If water is removed continuously, however, then the phenol may be satisfactorily esterified by this method (eq. 24.9).

(24.9) CH_3–C_6H_4–OH + $CH_3(CH_2)_4COOH$ $\xrightarrow[\text{remove water}]{\text{toluene, } H_2SO_4}$ CH_3–C_6H_4–$OC(=O)(CH_2)_4CH_3$

p-Cresol

p-Cresyl hexanoate (99%)

Phenyl esters are also prepared by *O*-acylation of the appropriate phenol or its salts with the desired acid chloride or anhydride (eqs. 24.10 and 24.11). These reactions are often conducted in the presence of a base (AcO^- or OH^-), which first converts the phenol to the much more reactive phenoxide ion.

(24.10) *m*-bromophenol + $(CH_3CO)_2O$ $\xrightarrow[\text{reflux}]{CH_3COO^-Na^+}$ *m*-Bromophenyl acetate ($OCCH_3$, C=O) + CH_3COOH

m-Bromophenyl acetate

(24.11) *p*-chlorophenol + *m*-O_2N–C_6H_4–COCl $\xrightarrow{\text{10\% NaOH}}$ Cl–C_6H_4–OC(=O)–C_6H_4–NO_2

p-Chlorophenyl *m*-nitrobenzoate

Phenols may also be converted to phenyl ethers. The reactants are usually a phenoxide salt and an alkyl halide (eq. 24.12). The other combination

$$\text{(24.12)}\quad Br\text{–}C_6H_4\text{–}O^-K^+ + CH_3I \xrightarrow{\text{EtOH}} Br\text{–}C_6H_4\text{–}OCH_3 + KBr$$

Potassium *p*-bromophenoxide; 1-Bromo-4-methoxybenzene

(alkoxide salt and aryl halide) is much less commonly used because halogen atoms bound to aromatic rings are quite difficult to displace (sec. 23:1).

Since the negative charge of phenoxide ion is delocalized in the ring, both *C*-alkylation and *O*-alkylation should be possible. *O*-Alkylation (eq. 24.12) is the common reaction because it avoids disruption of the aromatic ring. *C*-Alkylation may be observed (in *addition* to *O*-alkylation) when very reactive alkylating agents, such as allyl or benzyl halides, are used (eq. 24.13; recall from sec. 5:6 and Table 5–2, p. 151, that very reactive reagents tend to be unselective). The relative proportions of *C*- and *O*-alkylation also depend on the solvent.

(24.13) [phenoxide ion] + $CH_2{=}CHCH_2Br$ →(ether) [2-allyl-cyclohexadienone intermediate, $CH_2CH{=}CH_2$] →(fast) 2-Allylphenol (60%); and $OCH_2CH{=}CH_2$ Allylphenyl ether (40%)

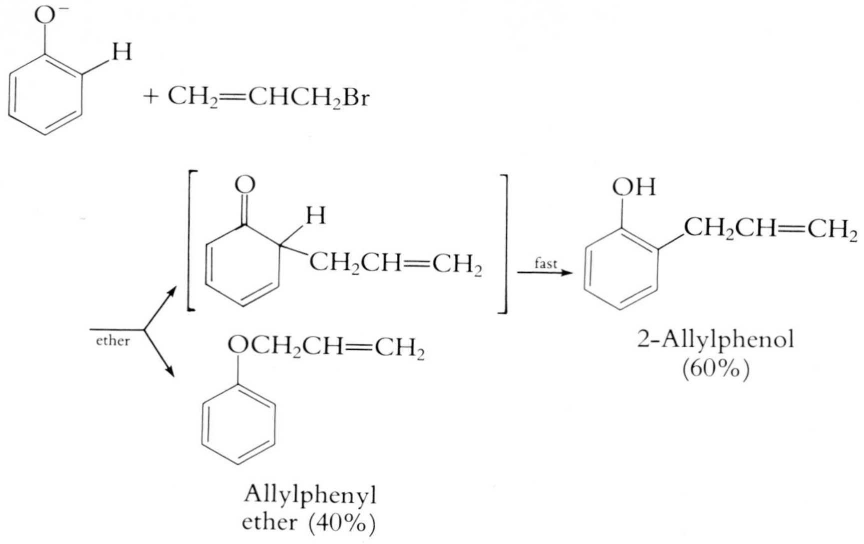

Alkyl phenyl ethers can be cleaved by the same reagents that attack dialkyl ethers (sec. 6:11). Due to the very strong phenyl–oxygen bond, only *alkyl*–oxygen fission is observed (eq. 24.14), leading to phenols and alkyl halides.

(24.14) [anisole, OCH_3] + HI →(reflux) [phenol, OH] + CH_3I (*not* [iodobenzene, I] + CH_3OH)

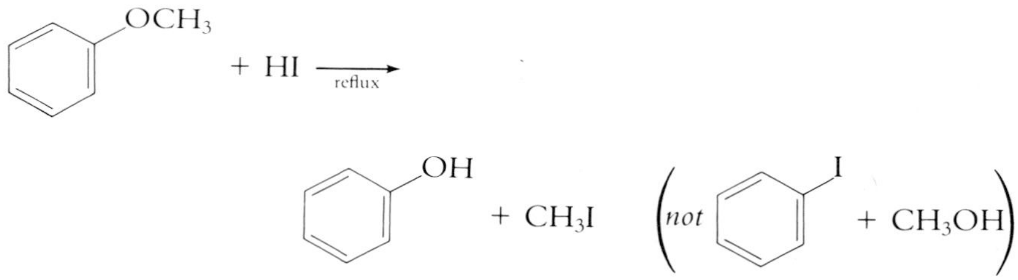

Problems

24–7 Show how phenol can be converted to each of the following compounds:

a. *p*-chlorophenol
b. 1-chloro-4-propoxybenzene
c. 2,4-dibromo-1-butoxybenzene
d. *p*-iodophenyl butanoate
e. *p*-isopropylanisole
f. methyl 4-ethoxybenzenesulfonate

24–8 How can the two isomeric reaction products of Equation 24.13 be separated? (*Hint:* Remember the acidic property of phenols.)

24–9 The tribenzyl ether (**II**) of phloroglucinol (**I**) is synthesized by treating the starting material with sodium hydride, and then with benzyl chloride (eq. 1). A by-product of this reaction is compound **A** ($C_{34}H_{30}O_3$). Propose a structure for **A** and show how both it and **II** are formed from **I**.

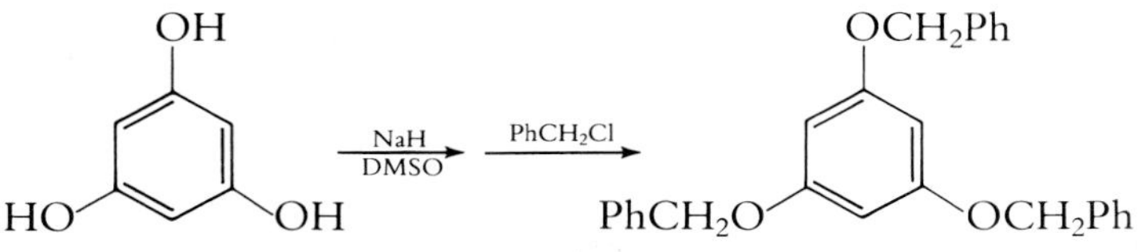

Phloroglucinol (I) $\xrightarrow{NaH/DMSO}$ $\xrightarrow{PhCH_2Cl}$ II (1,3,5-tris(benzyloxy)benzene) + $\mathbf{A}(C_{34}H_{30}O_3)$ (1)

Box 24–1. Bakelite and Its Offspring: Phenolic Polymers

The first commercially successful synthetic plastic (1909) was a substance called Bakelite, named for its inventor, the Belgian-born American chemist L. Baekland. It was produced from two very plentiful and inexpensive monomers, phenol and formaldehyde, which condense readily in the presence of a base.

The reaction of the phenoxide ion with formaldehyde is reminiscent of the Kolbe carboxylation of phenol to salicylic acid (sec. 19:8):

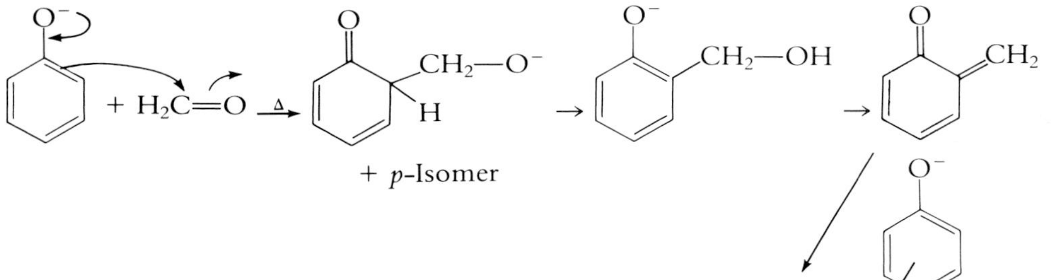

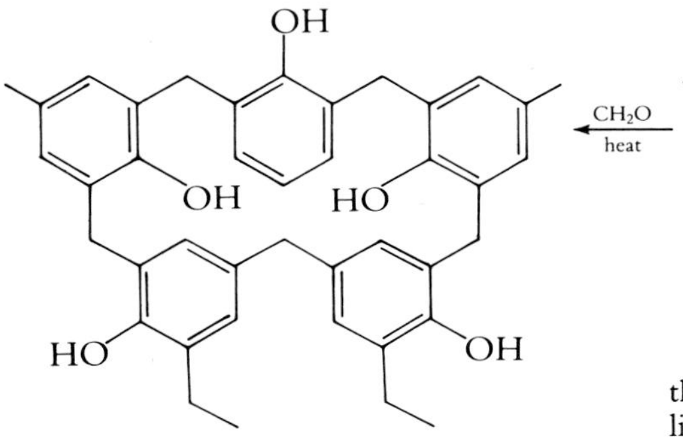

$$C_6H_5O^- + H_2C{=}O \xrightarrow{\Delta} \text{(2-(}CH_2{-}O^-\text{)cyclohexadienone)} + p\text{-Isomer} \rightarrow o\text{-}(CH_2{-}OH)C_6H_4O^- \rightarrow \text{o-quinone methide (}{=}CH_2\text{)}$$

o-quinone methide + $^-OC_6H_4CH_2OH$ → Linear polymer $\xrightarrow[heat]{CH_2O}$ Cross-linked polymer

Linear polymer

Cross-linked polymer
Bakelite

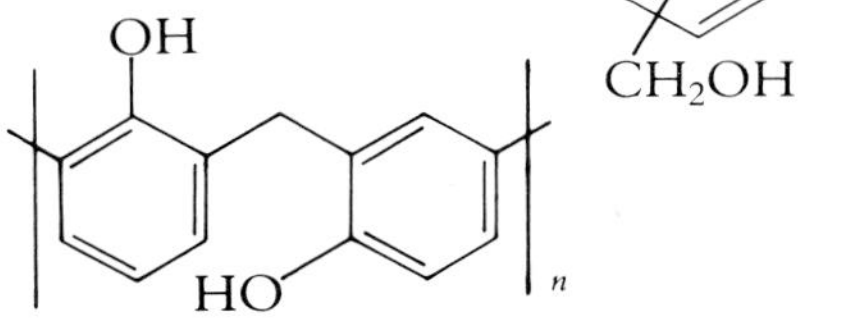

The polymerization was carried out in several stages, one intermediate being a *thermoplastic* resin—one that softened upon being heated. This material was molded into the desired shape and then taken to higher temperatures and pressures where it was converted to a *thermoset* resin—a rigid material that no longer softened thermally, due to extensive cross-linking.

Bakelite was first used as an electrical insulator and later to form plastic parts for the automotive and radio industries. It was the forerunner of a large number of phenol–formaldehyde plastics. In fact, Baekland himself took out more than 400 patents during research and development of Bakelite.

Baekland was an outstanding example of the scientist–entrepreneur, a person who deliberately uses a sound knowledge of contemporary science to invent a product that is socially useful and personally profitable. Indeed, Bakelite is just one example of Baekland's enterprising ability. Before Baekland even started work on Bakelite, he developed a photographic paper called Velox. Not being equipped to market it himself, he sold it to George Eastman of Kodak for a million dollars. That money financed his work on Bakelite.

24–10 Show how phenol can be converted to *p*-nitrophenol by way of a phenyl ester or ether. Direct nitration of phenol often gives low yields because of competing oxidation reactions (see sec. 24:4).

24:4 Oxidation of Phenols: Oxidative Coupling Reactions

Because of the strong electron-donating effect of the OH group, the aromatic ring of a phenol is electron rich. It easily releases electrons to oxidizing agents. For instance, the nitration of phenols gives low yields of nitrophenols because a competing oxidation reaction consumes much of the starting material (Prob. 24–10). While phenol oxidation may sometimes be a nuisance, as in nitration, it can also lead to useful products, as we'll discover in this section.

When phenols are treated with an oxidizing agent that accepts one electron, the products contain two (or more) molecules of the starting material (eq. 24.15). These so-called **oxidative coupling** reactions are carried out under neutral or basic conditions.

(24.15) [4-methylphenol] $\xrightarrow[OH^-]{K_3Fe(CN)_6}$ [dimer] + [trimer]

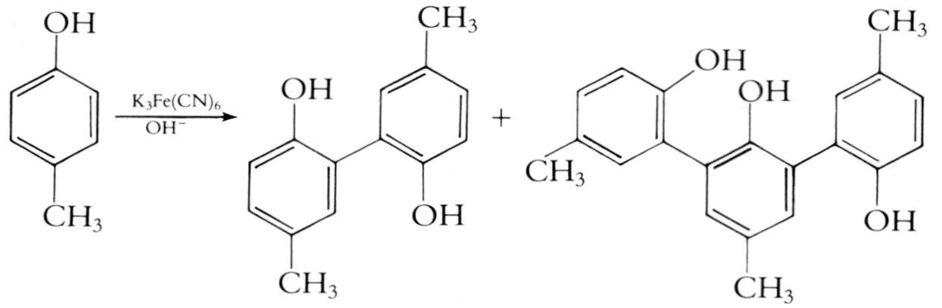

Oxidative coupling products

This reaction has been interpreted in the following way. The oxidizing agent removes one electron from the phenoxide ion to produce a phenoxy radical, **3**, which is stabilized by extensive electron delocalization (eq. 24.16).

(24.16) [4-methylphenoxide] $+ Fe(CN)_6^{3-} \rightarrow Fe(CN)_6^{4-} +$ [resonance structures ↔ ↔ ↔]

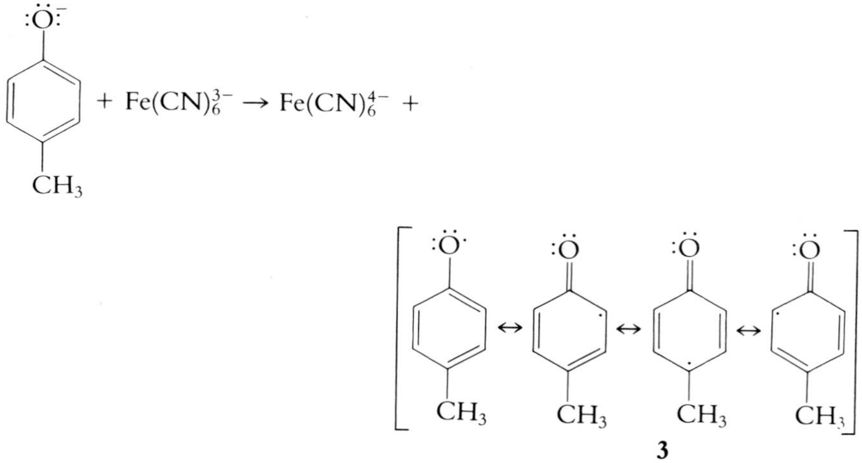

3

4-Methylphenoxy radical

Two phenoxy radicals then couple together to give a stable product with paired electrons (eq. 24.17).

(24.17)

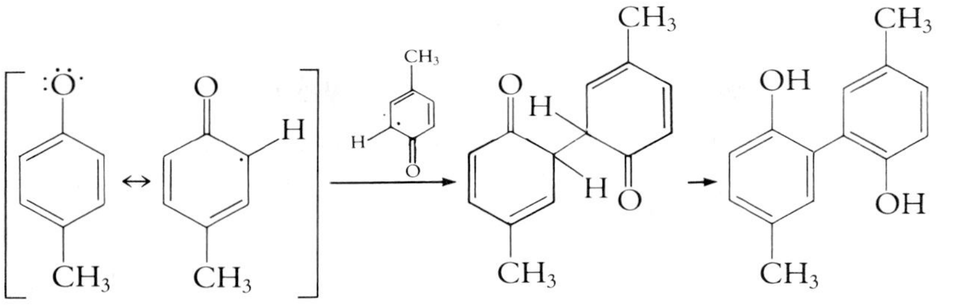

Since the phenoxy radical is a resonance hybrid, it has radical character on oxygen and hence one might expect coupling between two oxygens to form a peroxide. Products of O—O coupling are not found since the peroxide bond is weak and such products, if formed, would rapidly dissociate back to phenoxy radicals.

If the *ortho* position is blocked, coupling can occur at the *para* position.

If the *ortho* and *para* positions of the phenol are occupied by bulky groups, dimerization of the phenoxy radical is very slow, and the radical may even be isolated. 2,4,6-Tri-*t*-butylphenoxy radical (**4**, eq. 24.18), which is deep blue in color, is stable both in solution and in the crystal.

(24.18)

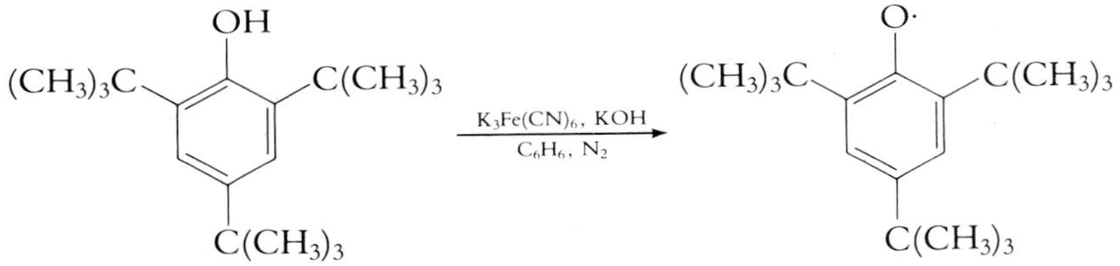

$\xrightarrow[C_6H_6,\ N_2]{K_3Fe(CN)_6,\ KOH}$

4

Formation of a relatively stable phenoxy radical is an important function of many food antioxidants (preservatives). The most common food antioxidant is 2,6-di-*t*-butyl-4-methylphenol, which is given the common name *butylated hydroxytoluene* (BHT).

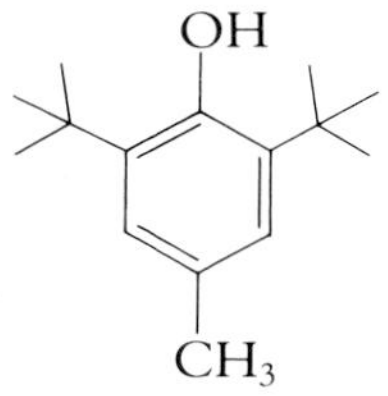

BHT

Radicals are formed when oxygen in the air reacts with compounds in food (eq. 24.19). BHT disrupts the autooxidation chain reaction (eq. 24.20) by forming a stable (unreactive) radical (eq. 24.21). This prevents the formation of the oxidized compounds that are the main causes of food spoilage.

(24.19) $$\underset{\text{(compound in food)}}{R{:}H} + O_2 \rightarrow R\cdot + HOO\cdot$$

(24.20) $$R\cdot + O_2 \rightarrow ROO\cdot \xrightarrow{RH} ROOH \rightarrow \text{aldehydes, acids, other spoilage products}$$

(24.21) R· (or ROO·) + [2,6-di-tert-butyl-4-methylphenol (OH)] → RH (or ROOH) + [2,6-di-tert-butyl-4-methylphenoxy radical (O·)] (less reactive with food compounds than R·)

Problems

24–11 Explain how the trimeric product of Equation 24.15 is formed. Write a mechanism that includes all intermediates.

24–12 Show the structures of the major dimeric products from oxidation of the following phenols under basic conditions.

a. phenol $\xrightarrow{FeCl_3}$ 3 products

b. 1-naphthol $\xrightarrow{FeCl_3}$ 3 products

24:5 Biosynthesis of Phenolic Compounds

We've already pointed out that many important, naturally occurring organic compounds (antibiotics, alkaloids such as morphine, and plant pigments) are phenols or phenol derivatives. There are two main pathways in nature for the synthesis of phenolic compounds.

One biosynthetic sequence is called the **polyketide** pathway because the immediate precursor to the phenolic compound is a polyketone. In Section 19:12 you saw that fatty acid biosynthesis starts from acetic acid and proceeds according to Equation 24.22. In fatty acid metabolism, the product

(24.22) $$2CH_3COOH \rightarrow \underset{\text{Acetoacetic acid}}{CH_3\overset{O}{\overset{\|}{C}}CH_2\overset{O}{\overset{\|}{C}}OH}$$

acetoacetic acid was reduced, then dehydrated, and again reduced to form butanoic acid. An alternative metabolic transformation of acetoacetic acid involves further condensation with acetate to produce a polyketide. This process is shown in Equation 24.23, where the dot indicates which carbon in

(24.23) $$nCH_3\dot{C}OOH + CH_3\overset{O}{\overset{\|}{\dot{C}}}CH_2\dot{C}OOH \rightarrow CH_3\overset{O}{\overset{\|}{\dot{C}}}\!-\!\!(CH_2\overset{O}{\overset{\|}{\dot{C}}})_n\!-\!CH_2\dot{C}OOH$$

the polyketide chain was originally a carboxyl carbon of acetate. Polyketides undergo aldol-type cyclizations to form phenols, as in Equation 24.24. This example shows the presumed biosynthesis of orsellinic acid, a constituent of certain fungi.

(24.24) $CH_3COOH \xrightarrow[\text{steps}]{\text{several}}$ $CH_3-CO-CH_2-CO-CH_2-CO-CH_2-COOH$ → [cyclic intermediate] → Orsellinic acid

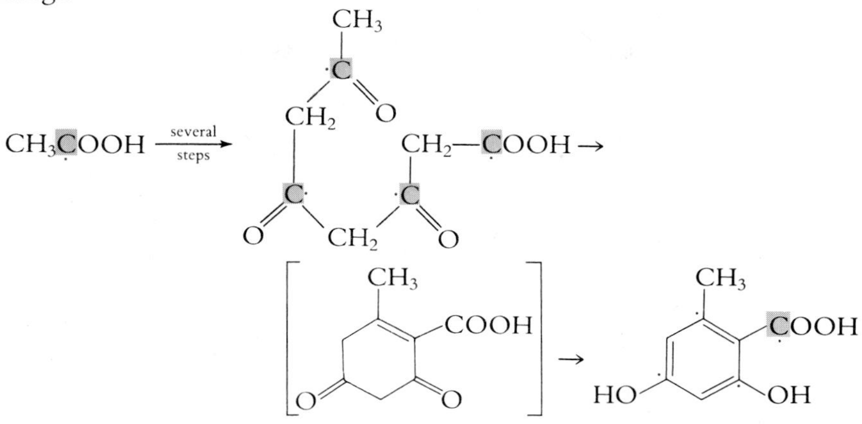

Such pathways have been verified by the use of acetate labeled with ^{14}C or ^{13}C in the carboxyl and/or methyl groups.

Sometimes the phenolic OHs are not completely retained in the product. For example, another constituent of certain fungi is 6-methylsalicylic acid, the precursor of which is undoubtedly orsellinic acid.

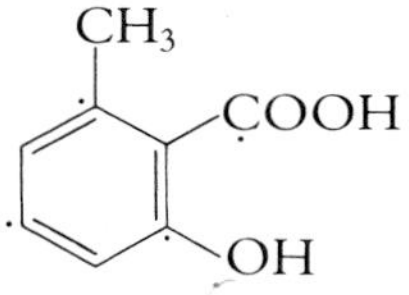

6-Methylsalicyclic acid

One consequence of the polyketide pathway for the biosynthesis of phenols is that, if the phenolic OHs are retained in the final product, they are *meta* to each other. This result is in clear contrast to the biosynthetic pathway we describe next.

Problem

24–13 Show what polyketide might have been the biosynthetic precursor of alternariol, a metabolite of *Alternaria tenuis* (a bacterium).

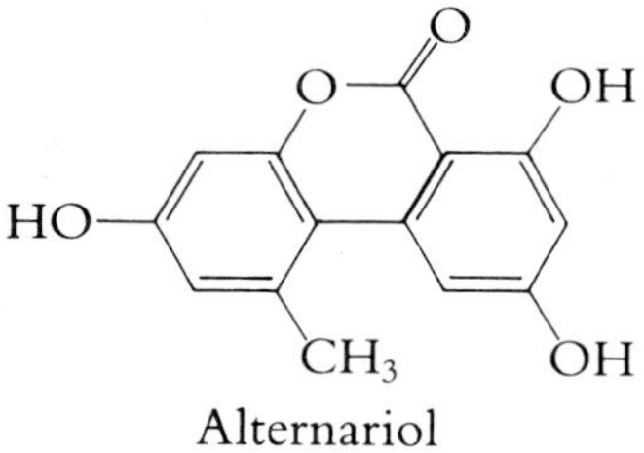

Alternariol

The second major biosynthetic pathway for natural phenols comes from a compound known as **shikimic acid**, which is synthesized from sugars (eq. 24.25), rather than from acetate.

(24.25) Erythrose phosphate ($OHC-CH(OH)-CH(OH)-CH_2OPO_3^{2-}$) + Phosphoenol pyruvate (PEP) ($CH_2=C(OPO_3^{2-})-COOH$) $\xrightarrow{\text{several steps}}$ Shikimic acid

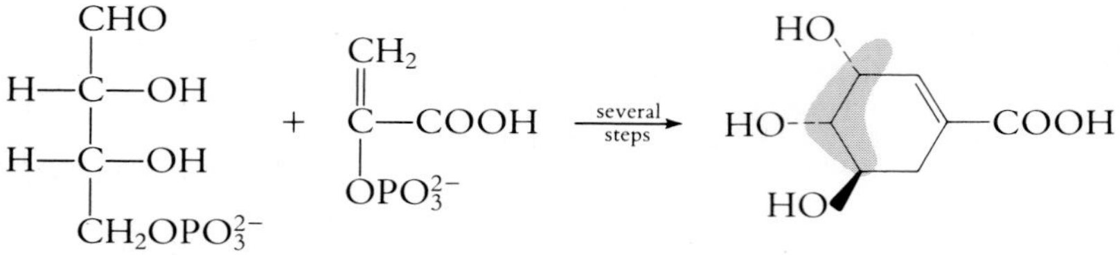

Erythrose phosphate

Phosphoenol pyruvate (PEP)

Shikimic acid

Shikimic acid is the precursor of substances such as gallic acid and coniferyl alcohol. In coniferyl alcohol, one phenolic OH has been lost and a second methylated.

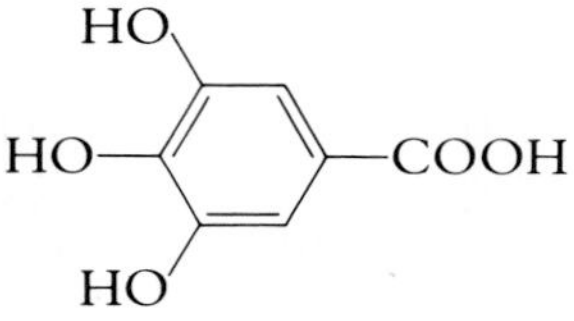

Gallic acid (widespread in plants)

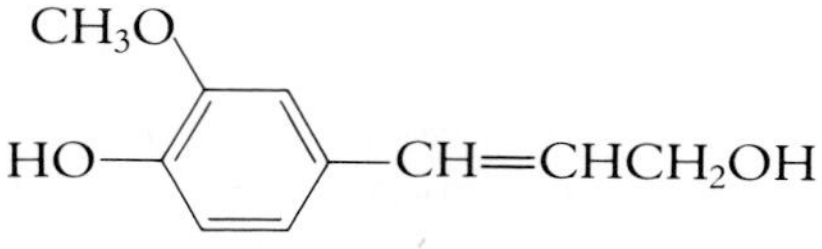

Coniferyl alcohol (found in conifers [pine trees])

Unlike the polyketide pathway, the shikimate pathway yields phenols with phenolic groups usually *ortho* to one another. Thus, the positions of the phenolic groups in the product phenols give a clue as to whether the biosynthesis is via a polyketide or shikimate pathway.

Problem

24–14 Flavonoids are common constituents of plants, where they and their derivatives make up many of the colored pigments. One group of flavonoids is called the flavonols, and a common flavonol is quercetin. By examining the relative placement of the phenolic groups, suggest how each of the aromatic A and B rings are formed.

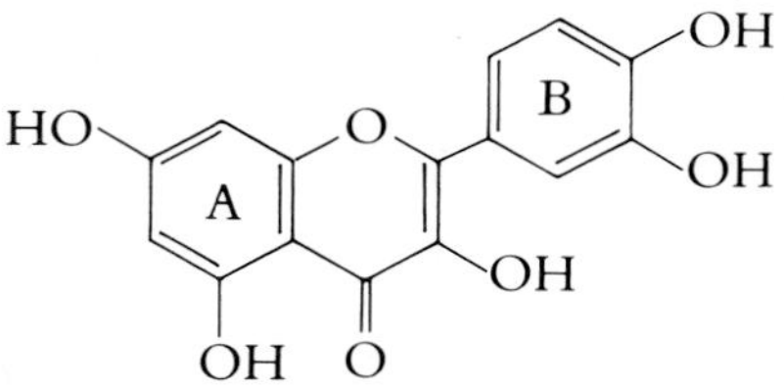

Quercetin

Box 24–2. Biological Hydroxylations and Phenylketonuria

The general formation of phenols *directly* from aromatic hydrocarbons is a reaction that the chemist has not yet been able to do cleanly and in high yield. However, hydroxylation of aromatic systems is a common *biological* reaction (sec. 14:11) that allows for the solubilization and easy secretion of foreign substances.

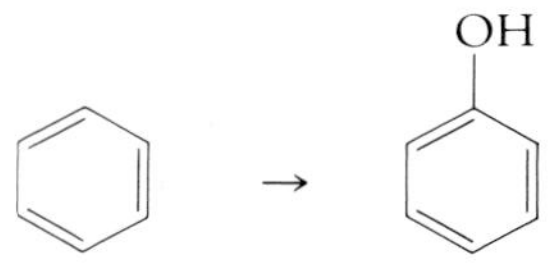

Benzene → Phenol (OH)

H_2O insoluble; poorly excreted — H_2O soluble; easily excreted

The hydroxylation of natural substances is also important biosynthetically; for example, some organisms are able to convert phenylalanine to tyrosine:

$$H_2N{-}CH(CH_2C_6H_5){-}COOH \rightarrow H_2N{-}CH(CH_2C_6H_4OH){-}COOH$$

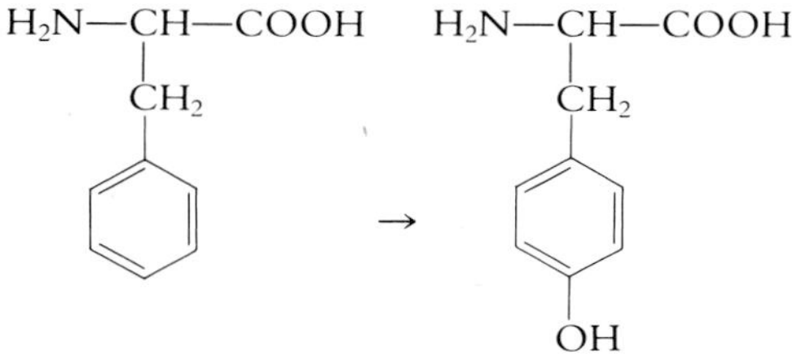

Phenylalanine — Tyrosine

This is the major pathway for metabolism of phenylalanine in humans. About one in 10,000 individuals is born with a genetic defect called *phenylketonuria* (PKU) which prevents the oxidation of phenylalanine to tyrosine. In the absence of the hydroxylation pathway, individuals suffering from PKU convert phenylalanine to phenylpyruvate by way of a *transamination* reaction (see sec. 25:13):

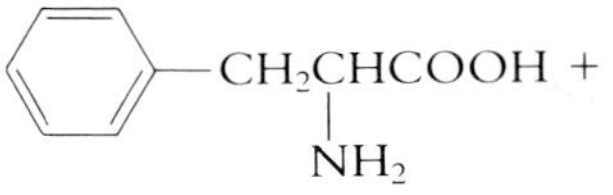

$$C_6H_5CH_2CH(NH_2)COOH + HOOCCH_2CH_2C(=O)COO^- \xrightarrow[\text{transaminase}]{\text{vitamin } B_6} C_6H_5CH_2C(=O)COO^- + HOOCCH_2CH_2CH(NH_2)COOH$$

Phenylalanine; 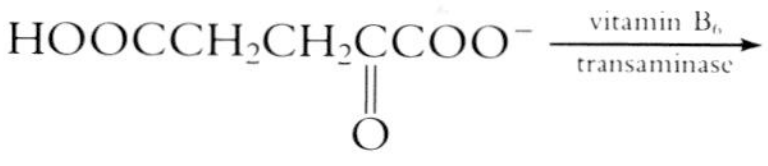 α-Ketoglutarate; 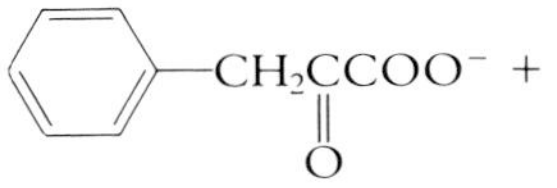 Phenylpyruvate;  Glutamic acid

It is thought that the mental retardation associated with PKU may be due to the buildup of phenylpyruvate.

Although PKU is not yet curable, its effects can be avoided by placing the infant immediately on a diet low in phenylalanine. This requires that the condition be detected very soon after birth, so newborns are now routinely tested for PKU. The test is designed to uncover the elevated levels of phenylalanine present in the blood. A blood sample is added to a culture of a bacterial strain that requires phenylalanine for growth; a high rate of bacterial reproduction indicates a high level of phenylalanine.

24:6 The Formation of Quinones

Oxidations of phenols carried out under acidic conditions take a different course than do those under neutral or basic conditions (sec. 24:4). It is necessary to use stronger oxidizing agents, and **quinones** are the products isolated (eq. 24.26).

$$\text{2-ethylphenol (OH, } CH_2CH_3\text{)} \xrightarrow[CH_3COOH]{K_2Cr_2O_7} \text{2-ethyl-}p\text{-benzoquinone} \qquad (24.26)$$

2-Ethyl-*p*-benzoquinone

Quinones are unsaturated cyclohexanediones in which the two carbonyl groups are *ortho* or *para* to one another. *o*-Quinones are much less stable than *p*-quinones and undergo further reaction. Thus, the *p*-quinone is the only one isolated in the reaction shown in Equation 24.26.

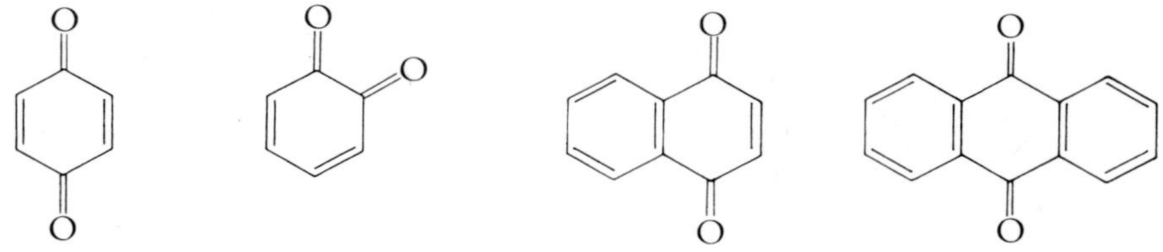

p-Benzoquinone *o*-Benzoquinone 1,4-Naphthoquinone 9,10-Anthraquinone

Under acidic conditions, the major reactive intermediate in the oxidation of a phenol is probably not a phenoxy radical but rather a *phenol radical cation* (**5**, eq. 24.27).

(24.27)

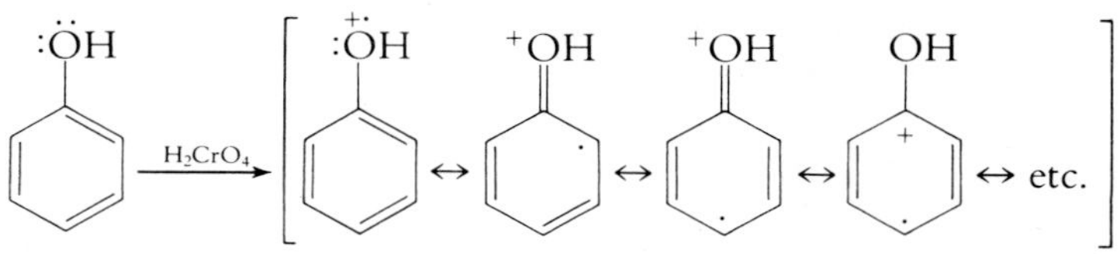

5

A phenol radical cation

These radical cations do not couple quickly because of mutual charge repulsion and hence survive long enough to react further with the oxidizing agent (eq. 24.28). The second oxygen that is added probably comes from H_2CrO_4.

(24.28) $+ H_2CrO_4 + 2H^+ \rightarrow$ $+ Cr^{3+} + 3H_2O$

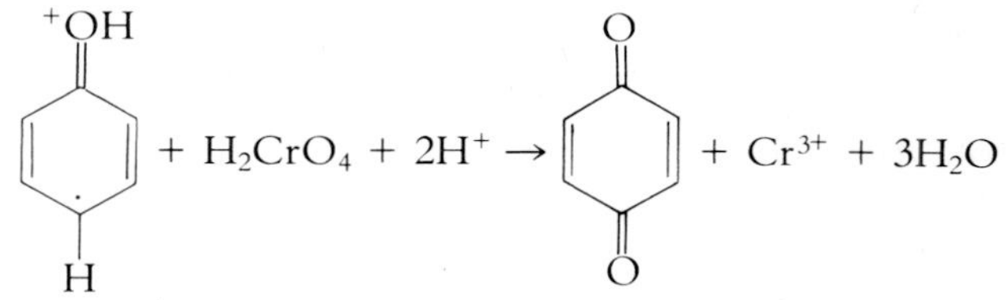

Oxidation of phenols does not provide a generally useful route to quinones. The yields are variable and the side reactions numerous. A more reliable method uses hydroquinones (eq. 24.29) as precursors of *p*-benzo-

(24.29)

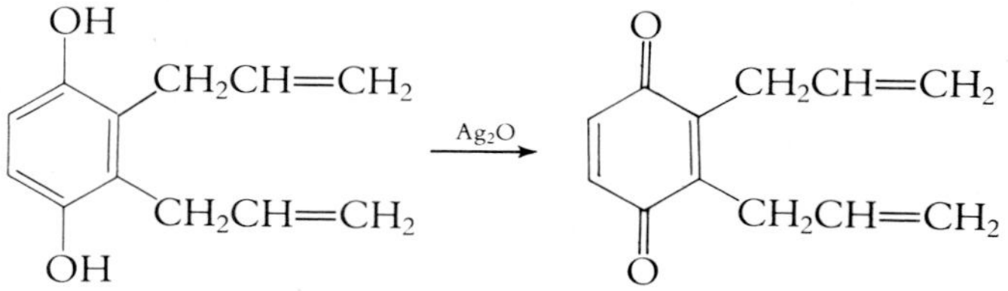

2,3-Diallylhydroquinone 2,3-Diallyl-*p*-benzoquinone (96%)

quinones. Hydroquinones are so easily oxidized that very mild, neutral conditions can be used. Catechols give *o*-benzoquinones (eq. 24.30), while *o*-

and p-disubstituted polycyclic aromatics yield the corresponding quinones (eqs. 24.31 and 24.32). Note that in these cases aminophenols are used as starting materials. They are as effective as hydroquinones and catechols as precursors of quinones.

(24.30) Tetrachlorocatechol $\xrightarrow{Ag_2O}$ Tetrachloro-o-benzoquinone

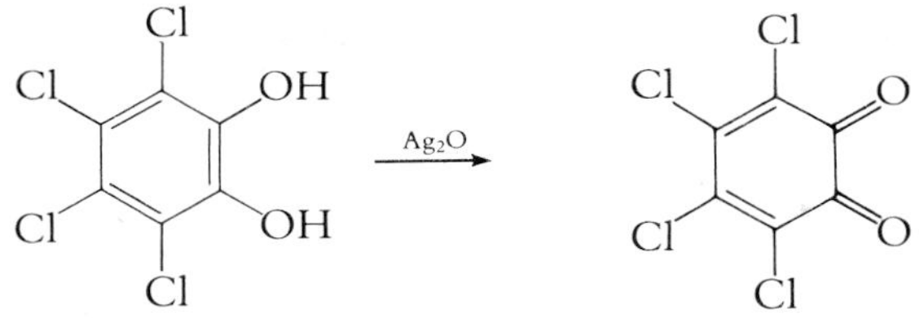

(24.31) 1-Amino-2-naphthol $\xrightarrow[\text{dil } H_3O^+]{FeCl_3}$ 1,2-Naphthoquinone

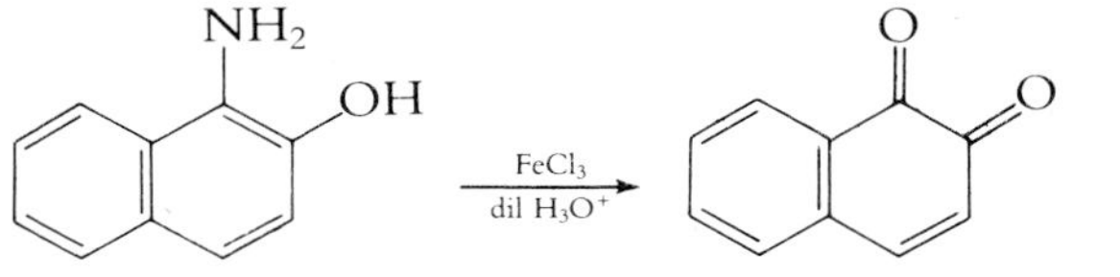

(24.32) 4-Amino-1-naphthol $\xrightarrow[\text{dil } H_2SO_4]{K_2Cr_2O_7}$ 1,4-Naphthoquinone

The aminophenols are well known to photographers. After exposure to light, a photographic film is developed by washing it with a solution of p-aminophenol. The phenol is oxidized by those Ag^+ ions that have been activated by light, and it is converted to a quinonimine. The Ag^+ ions are simultaneously reduced to silver metal (eq. 24.33). In the fixing process that

(24.33) p-aminophenol $+ 2Ag^+ \rightarrow$ A quinonimine $+ 2Ag^0\downarrow + 2H^+$

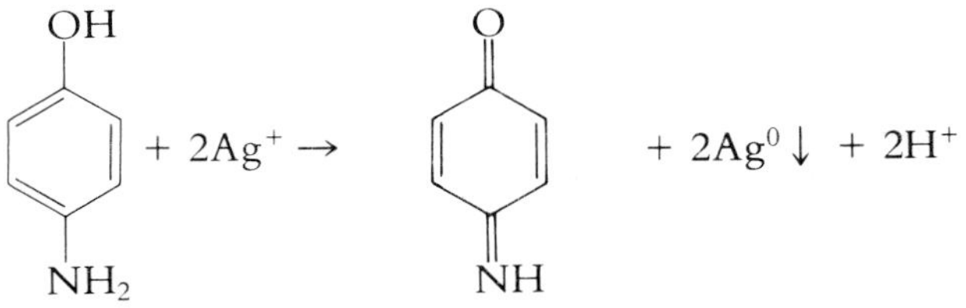

follows, sodium thiosulfate ($Na_2S_2O_3$) solution is used to wash out the unreacted silver halide, leaving only the silver metal to indicate exposed portions of the film. When color film is developed, complex but stable quinonimine dyes are formed and remain at the site of silver reduction.

24:7 Reversible Redox Processes Involving Quinones and Quinols

When hydroquinone is oxidized, a dark green solid precipitates early in the reaction. The precipitate is eventually consumed and the color changes to the characteristic yellow of benzoquinone. The green crystals, which may be isolated and stored, consist of a molecular complex containing one molecule of hydroquinone and one of benzoquinone, held together by hydrogen bonds. The complex is known as *quinhydrone* (eq. 24.34). The oxidation of

(24.34)

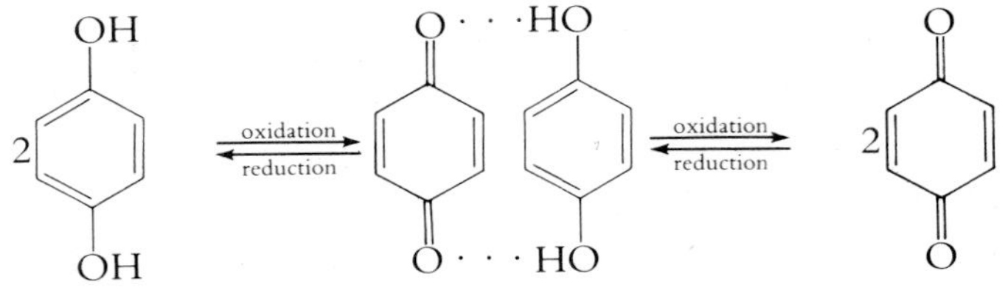

2 Hydroquinone ⇌ (oxidation / reduction) Quinhydrone ⇌ (oxidation / reduction) 2 Benzoquinone

hydroquinone to benzoquinone and the related reduction, both by way of quinhydrone, are completely reversible in acid. In the next section we will discuss the importance of this redox process in the function of biological cells.

The equilibrium of Equation 24.34 is affected by substituents on the ring. Electron-donating substituents stabilize the benzoquinones compared to hydroquinones. A methoxy group, for example, brings about this stabilization by a contribution from resonance forms such as **6**. Electron-withdrawing substituents stabilize hydroquinones compared to quinones, largely through contribution of resonance forms such as **7**.

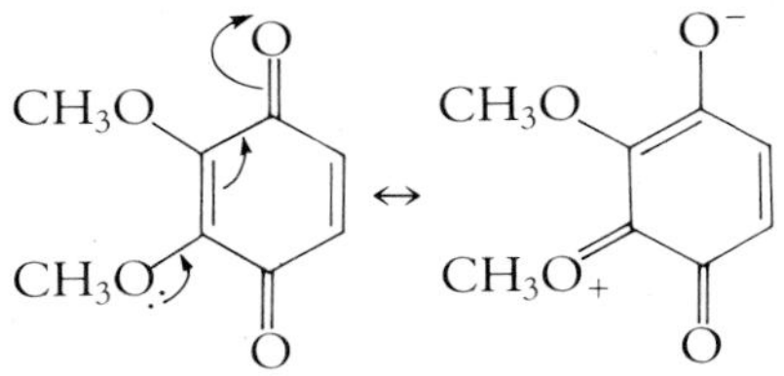

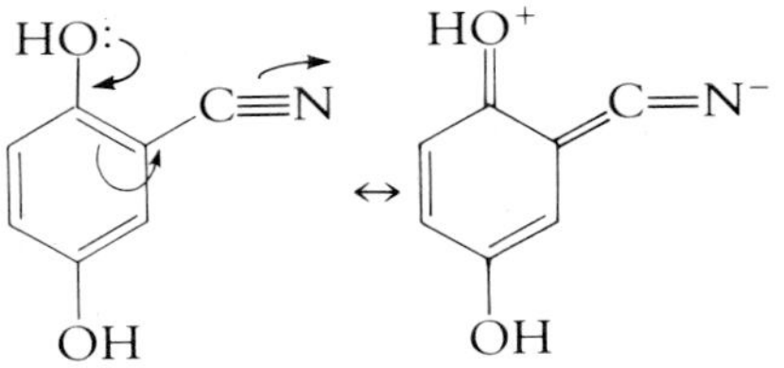

CH_3O, CH_3O ↔ CH_3O, CH_3O^+, O^-

6

Electron donation stabilizes quinones

HO:, C≡N, OH ↔ HO^+, $C{=}N^-$, OH

7

Electron withdrawal stabilizes phenols

The presence of several electron-withdrawing groups on a quinone make it readily convertible to the hydroquinone if hydrogens for the reduction are available. Because of this, substances such as 2,3-dichloro-5,6-dicyano-1,4-benzoquinone (DDQ) are very powerful oxidizing agents that are particularly useful for dehydrogenating certain types of organic compounds (eq. 24.35).

(24.35)

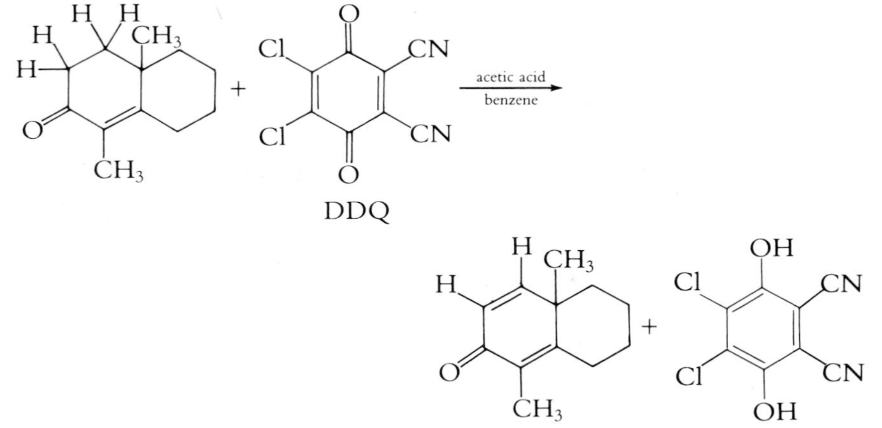

DDQ

Quinones and quinols are widely distributed in the plant kingdom, and some play an important role in animal metabolism as well. (*Quinol* is a generic name for the reduction product of a quinone.) One of the most unusual uses of quinones is as an insect defensive weapon. When attacked, the bombardier beetle secretes a mixture of hydroquinone and methylhydroquinone into a special compartment containing aqueous hydrogen peroxide. Under enzymic catalysis, the peroxide oxidizes the hydroquinones to benzoquinones with almost explosive rapidity. The boiling hot mixture, containing the highly irritating quinones, discharges onto the luckless aggressor.

Problem

24–15 a. The naphthoquinone *naphthazarin* (5,8-dihydroxy-1,4-naphthoquinone) displays some unusual properties. It undergoes a rapid rearrangement that makes all the protons on the aromatic carbons equivalent. Show this rearrangement and its mechanism. (Note that the dimethyl ether of naphthazarin does not undergo the rearrangement.)

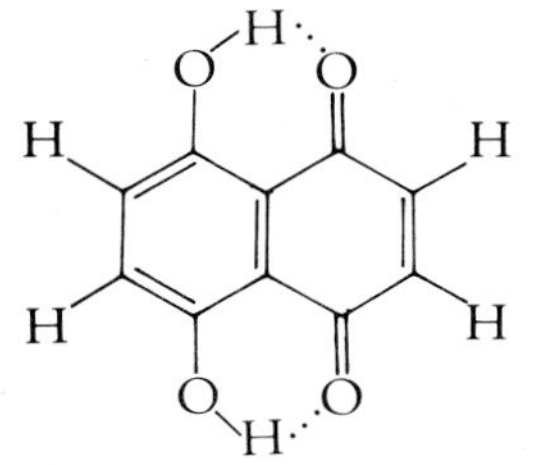

Naphthazarin

b. Ring substituents on naphthazarin slow the rearrangement considerably and favor one of the two possible isomeric structures. The favored structures for hydroxy- and acetylnaphthazarin are shown below. Explain why these structures are the more stable. (Naphthazarin-type pigments are found in some marine animals such as sea urchins.)

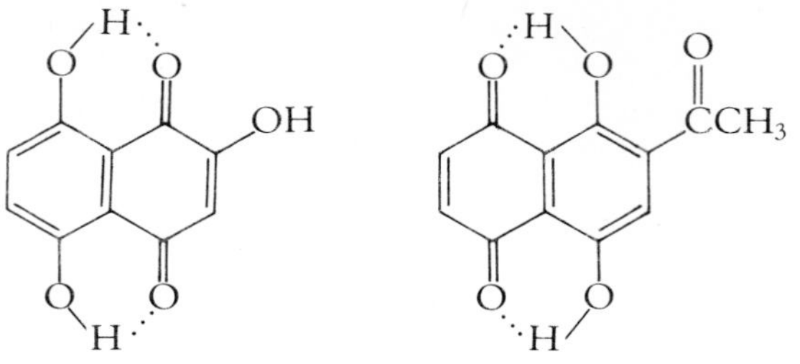

Hydroxynaphthazarin Acetylnaphthazarin

24:8 Some Biologically Significant Quinones

Biological functions for quinones have been discovered in cellular oxidation–reduction systems. As you will soon see, their function relies on their ability to reversibly interconvert from quinones to quinols, as was shown in the last section. The *menaquinones* ($n = 0$ to 12) take part in electron transport processes in bacteria. Menaquinones are members of the vitamin K family, and several of them ($n = 4$, 6, 7) play a vital role in maintaining blood clotting ability in higher animals. *Plastoquinone* is widely distributed in the

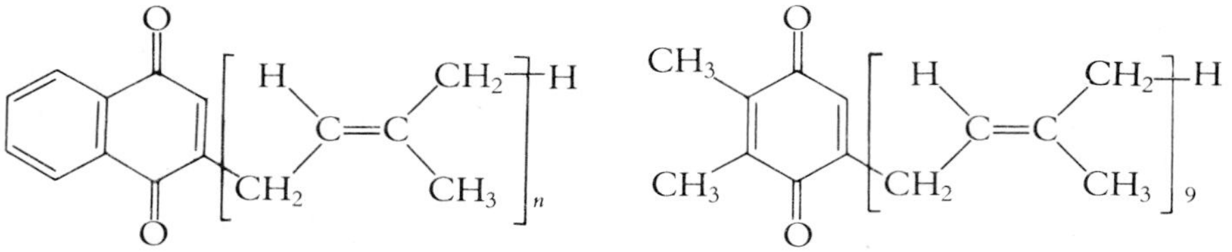

Menaquinones Plastoquinone

chloroplasts of photosynthetic plants. The chloroplasts use solar energy to oxidize water to molecular oxygen, in the course of which electrons are released. At some point in the photosynthetic cycle these electrons are used for the reduction of plastoquinone to plastoquinol. The quinol is subsequently oxidized back to the quinone. The chemical energy released by this oxidation–reduction cycle is used in the biosynthesis of glucose.

The *ubiquinones* (coenzymes Q_n, $n = 6$ to 10) were given that name because of their ubiquitous (widespread) distribution in nature. They are found in the mitochondria of microorganisms, algae, higher plants, invertebrates, and vertebrates, where they participate in the electron transfer system which leads ultimately to the generation of ATP (recall that ATP is the cellular energy storage medium; sec. 17:14).

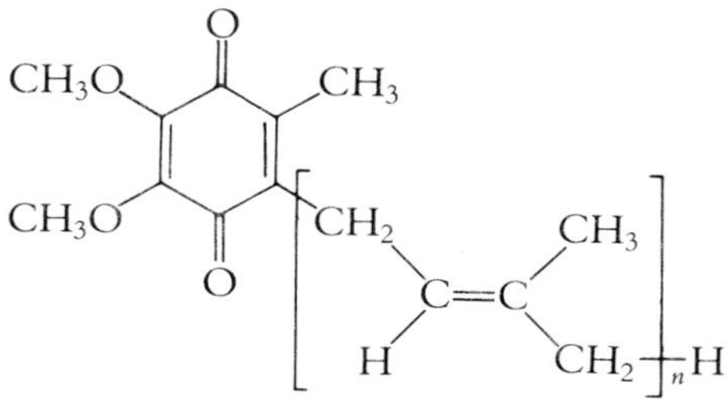

Ubiquinones

A ring-closed relative of plastoquinone is α-tocopherol, also known as vitamin E. Vitamin E is an antioxidant that may act similarly to BHT (sec. 24:4). For instance, vitamin E is an efficient inhibitor of lipid peroxidation.

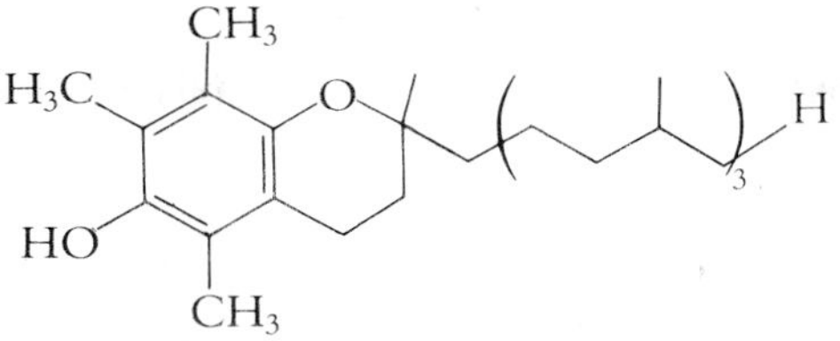

α-Tocopherol (vitamin E)

This may be the reason that vitamin E is effective in protecting cell membranes from deterioration, since the membranes contain numerous lipids susceptible to oxidation. Vitamin E has been shown to prolong the life span of cultured human cells and of lower animals such as the roundworm. It has been claimed that vitamin E inhibits carcinogenesis and promotes fertility in humans, but these claims are the subject of much controversy, and are by no means universally accepted.

Summary

1 Phenols are much more acidic than alcohols because the phenoxide ion, which is formed by loss of a proton, is resonance stabilized. The presence of electron-withdrawing substituents on the aryl ring increases the acidity of the phenol by stabilizing the anion. Electron-donating substituents have the opposite effect. Because of their conversion to phenoxides, phenols show a marked uv absorption shift when dissolved in base (sec. 24:1).

2 Phenols undergo electrophilic substitution in polar solvents so readily that it is difficult to isolate monosubstitution products under these conditions. This is because phenoxide ions are the species undergoing reaction. If the substitution is carried out in a nonpolar solvent, monosubstitution products can be obtained (sec. 24:2).

Phenols can be converted to esters and to ethers by acylation and alkylation of their anions. Alkylation may take place on the oxygen or on the *o*- and *p*-positions of the ring (sec. 24:3).

Phenols give rise to phenoxy radicals when they are treated with one-electron oxidizing agents under basic conditions. These radicals can couple with one another. This process, known as oxidative coupling, appears to be widespread in biological systems, as well as being synthetically useful (sec. 24:4).

3 In plants, phenols are produced mainly in two ways: the polyketide pathway (which proceeds from acetate), and the shikimic acid pathway (which is based on sugars). The former pathway yields phenols with OH groups *meta* to each other and the latter gives compounds with *ortho* OH groups (sec. 24:5).

4 If phenols are oxidized in acid, they yield quinones. High yields of quinones are best obtained by oxidation of *ortho*- and *para*-dihydroxybenzenes and aminophenols (sec. 24:6). Hydroquinone and benzoquinone undergo a mutual oxidation–reduction interconversion. The position of equilibrium is affected by ring substituents—electron-donating substituents favor the quinone while electron-withdrawing ones favor the quinol (sec. 24:7). The reversibility of this redox process accounts for the importance of many naturally occurring quinones to various biological processes (sec. 24:8).

Problems

24–16 With the aid of resonance structures, explain the following observations:

a. 1-Naphthol (pK_a 9.31) is a slightly stronger acid than 2-naphthol (pK_a 9.55).

b. Both naphthols are slightly stronger acids than phenol.

24–17 Write the structure of the major organic product(s) from each of the following reactions:

a. CH_3O–C_6H_4–CH_3 $\xrightarrow{Br_2}$

b. C_6H_5–O–C_6H_5 $\xrightarrow{HNO_3}$

c. C_6H_5–OC(=O)–C_6H_5 $\xrightarrow[AlCl_3]{CH_3CCl \text{ (O)}}$

d. C_6H_5–OH $\xrightarrow[DMSO]{NaH}$ $\xrightarrow{(CH_3)_2CHCH_2I}$

e. C_6H_5–OH $\xrightarrow[NaOH]{Ac_2O}$

24–18 a. Electrophilic substitution of phenol and phenyl ethers gives mainly *para* substitution, along with lesser amounts of *ortho* substitution. Products of *meta* substitution are present in only minute quantities. Explain all these observations, making use of resonance structures.

b. Certain reagents can be used to circumvent the preference for *para* substitution in phenols. For instance, use of *t*-butyl hypochlorite as a chlorinating agent leads to almost exclusive *ortho* chlorination (eq. 2). Propose a mechanism which explains the *ortho* selectivity of this reaction.

1-Naphthol + *t*-BuOCl $\xrightarrow{CCl_4,\ 77^\circ}$ 2-Chloro-1-naphthol (77%) + *t*-BuOH (2)

24–19 Polyhydroxyaromatic compounds undergo electrophilic substitution with remarkable ease. *Pyrogallol* (1,2,3-trihydroxybenzene) is acylated by carboxylic acids in the presence of the Lewis acid boron trifluoride (eq. 3). Only 4-substituted derivatives are obtained; no 5-substitution is observed. Explain this selectivity.

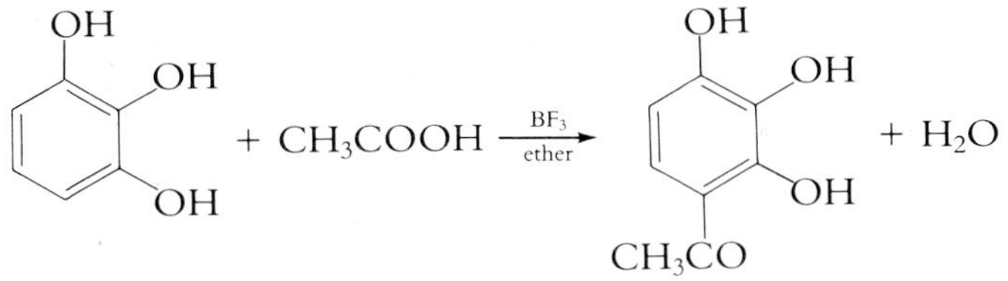

$$\text{Pyrogallol} + CH_3COOH \xrightarrow[\text{ether}]{BF_3} \text{4-}CH_3CO\text{-pyrogallol } (90\%) + H_2O \qquad (3)$$

24–20 a. Chlorination of pentachlorophenol (a common preservative) yields the two hexachlorocyclohexadienones **III** and **IV** (eq. 4). Show the mechanism for formation of each.

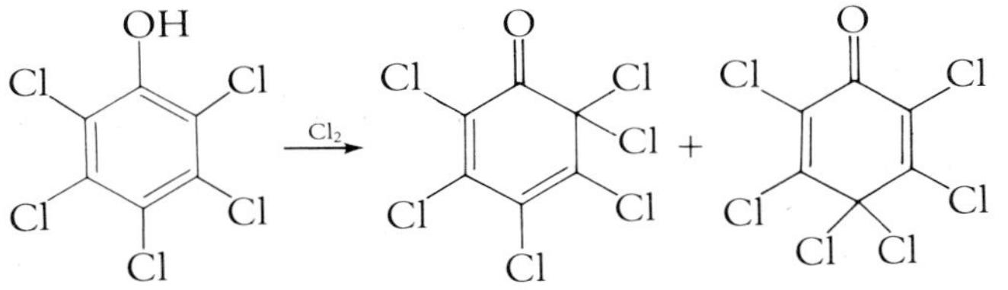

$$\text{pentachlorophenol} \xrightarrow{Cl_2} \textbf{III} + \textbf{IV} \qquad (4)$$

b. Both **III** and **IV** are useful as mild chlorinating agents. **III**, for example, reacts with phenol to produce mainly *o*-chlorophenol (eq. 5). Write a mechanism for the reaction in Equation 5 and use it to explain why the reaction proceeds to the right.

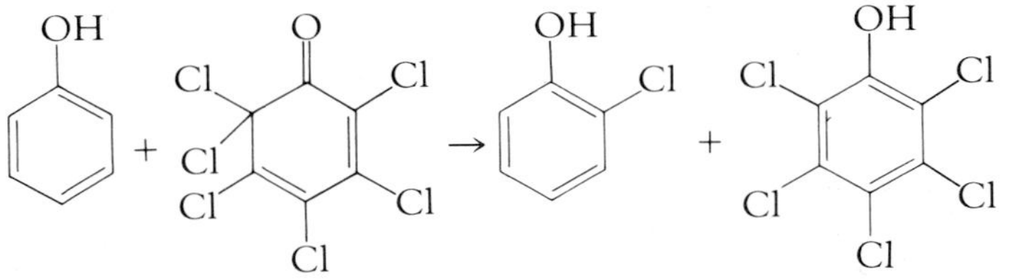

$$\text{phenol} + \textbf{III} \rightarrow o\text{-chlorophenol} + \text{pentachlorophenol} \qquad (5)$$

c. Suggest ways in which the two products of Equation 5 can be easily separated.

24–21 Phenols are enols. These enols are normally more stable than their tautomeric ketones because of aromatic stabilization. In certain cases, however, the equilibrium can be made to shift in favor of the keto form. For instance, *phloroglucinol* (1,3,5-trihydroxybenzene) shows a normal phenolic ring ^1H nmr absorption at δ 6.05. Its monoanion also behaves normally, giving a ^1H nmr absorption at δ 6.02 (eq. 6). However, the dianion (**A**, eq. 6) shows very different properties—it gives two nmr peaks, equal in intensity, at δ 5.08 and δ 3.05. Propose a structure for **A**, and account for its stability compared to the expected structure for the dianion.

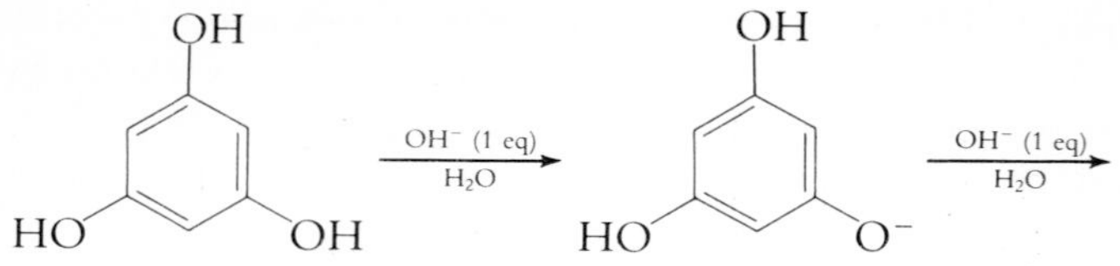

(nmr: δ 6.05) $\xrightarrow[H_2O]{OH^-\ (1\ eq)}$ (nmr: δ 6.02) $\xrightarrow[H_2O]{OH^-\ (1\ eq)}$

A($C_6H_4O_3$; nmr: δ 3.05, 5.08) (6)

24–22 The major source of commercial phenol is the readily available, inexpensive cumene hydroperoxide (Prob. 14–36). Acid-catalyzed rearrangement converts the hydroperoxide to a hemiketal, which rapidly decomposes to phenol and acetone (eq. 7). Over 2.5 billion pounds of phenol are manufactured this way each year. Its advantage over the Dow process (sec. 23:1) is due in part to the economic demand for acetone. Write a mechanism for the rearrangement.

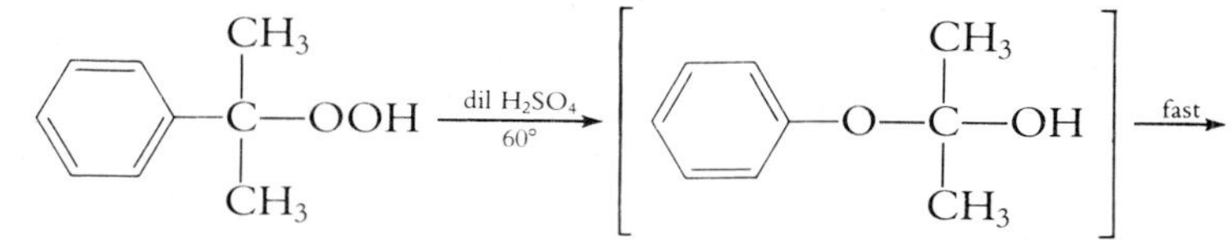

$$C_6H_5C(CH_3)_2OOH \xrightarrow[60^\circ]{\text{dil } H_2SO_4} [C_6H_5-O-C(CH_3)_2-OH] \xrightarrow{\text{fast}} C_6H_5OH + CH_3\overset{O}{\overset{\|}{C}}CH_3 \quad (7)$$

Cumene hydroperoxide — Hemiketal

24–23 When *p*-aminophenol is treated with slightly less than one equivalent of acetyl chloride at (a) low pH, (b) pH 7, and (c) high pH, what is the predominant product in each case? Explain the effect of pH on the site of reaction.

24–24 Primary and (some) secondary alkly phenyl ethers can be synthesized by allowing the appropriate alkyl halides to react with phenoxide ion. The reaction of phenoxide ion with tertiary halides leads to no ether product. Explain.

24–25 a. The saponification of phenyl acetate (R = C_6H_5) proceeds much more rapidly than the corresponding reaction of ethyl acetate (R = C_2H_5) (eq. 8). Use the mechanism for ester hydrolysis (sec. 17:9) to explain the observed relative reactivities.

$$CH_3\overset{O}{\overset{\|}{C}}OR + OH^- \rightarrow CH_3COO^- + ROH \quad (8)$$

b. Arrange the following compounds in order of increasing rate of saponification.

1. ethyl acetate, phenyl acetate, *m*-cresyl acetate (3-methylphenyl acetate)
2. *m*-chlorophenyl butyrate, phenyl butyrate, 4-*t*-butylphenyl butyrate, 3,4-dichlorphenyl butyrate, *p*-nitrophenyl butyrate
3. phenyl *p*-toluate, *o*-cresyl *p*-toluate, *p*-cresyl *p*-toluate

4. p-aminophenyl succinate, p-trimethylammoniophenyl succinate, phenyl succinate

24–26 In addition to the products shown in Equation 24.15, oxidation of p-cresol also produces Pummerer's ketone.

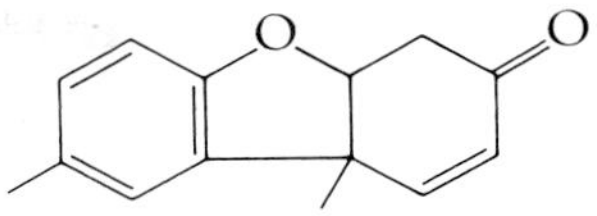

Pummerer's ketone

Show how this compound was formed. (Pay attention particularly to the way in which the two radicals couple, and subsequent reaction of the coupling product.)

24–27 The commercial polymer polyphenylene oxide is a much used electrical insulator. It is made by oxidation of 2,6-dimethylphenol with O_2 in pyridine (eq. 9). A Cu(II) salt is an essential catalyst in this reaction. Write a mechanism showing the formation of several units of the polymer chain, paying particular attention to the role of pyridine and Cu^{2+}.

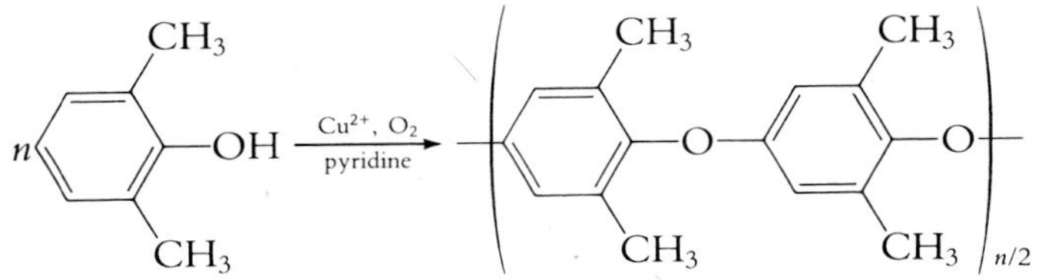

$$n\,\text{(2,6-dimethylphenol)} \xrightarrow[\text{pyridine}]{Cu^{2+},\ O_2} \left[\text{—(2,6-dimethylphenylene)—O—(2,6-dimethylphenylene)—O—}\right]_{n/2} \qquad (9)$$

Polyphenylene oxide
(MW ~ 2.8×10^4)

24–28 Show how the following compounds could be prepared by oxidation of substituted phenols.

a. (Et, HO, OH, Et substituted biphenyl)

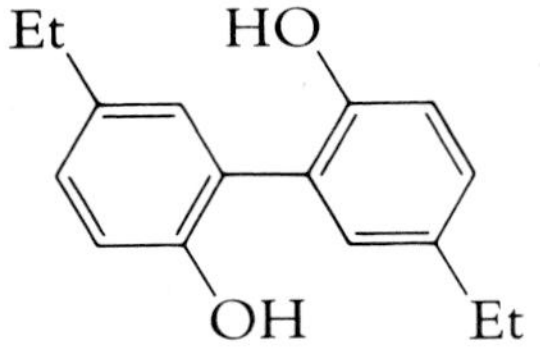

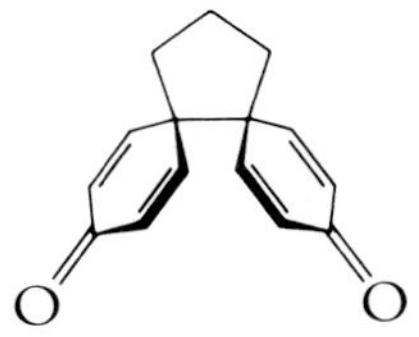

b. Me— (biphenyl with OH, Me, OH, HO substituents) —OH

c. (spiro cyclopentane bis-cyclohexadienone; O, O)

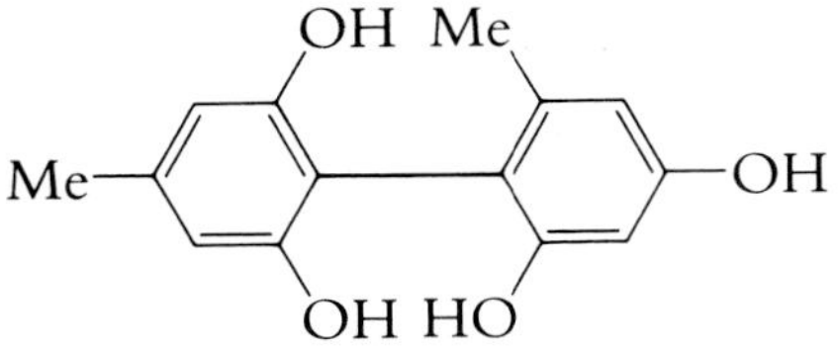

24–29 When aniline is oxidized with MnO_2 and aqueous sulfuric acid, it is converted to a polymer which, upon hydrolysis, yields equimolar amounts of p-benzoquinone and 1,4-benzenediamine. The 1,4-benzenediamine may be further oxidized to p-benzoquinone. Suggest a structure for the polymer, and propose a mechanism for its formation from aniline.

24–30 Indicate whether the equilibrium constants for each of the following reactions will be substantially greater than, substantially less than, or essentially equal to one. Justify your answers.

a. 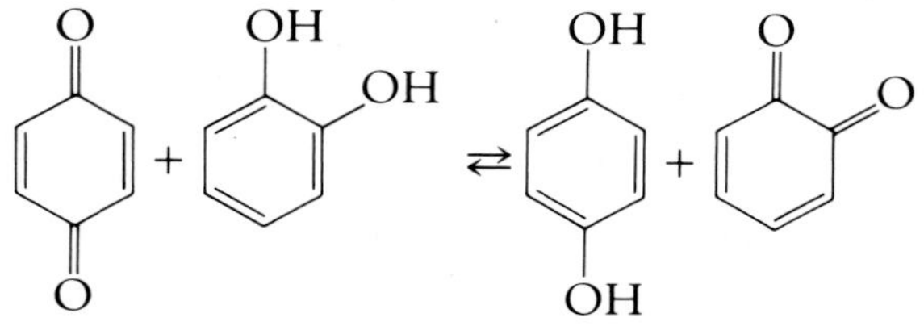 [structures: p-benzoquinone + catechol ⇄ hydroquinone + o-benzoquinone]

b. 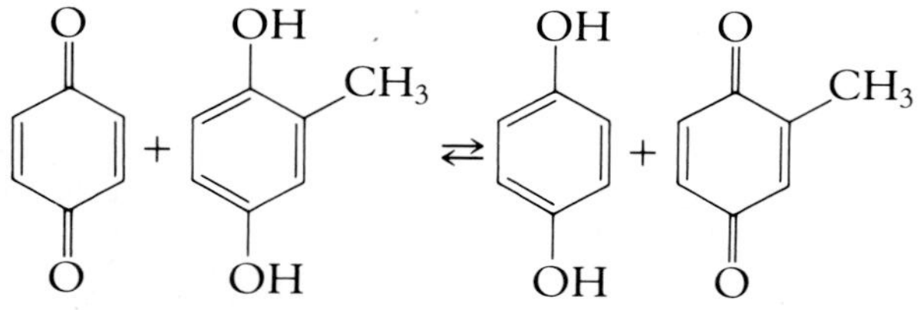 [structures: p-benzoquinone + 2-methylhydroquinone ⇄ hydroquinone + 2-methyl-p-benzoquinone]

c. 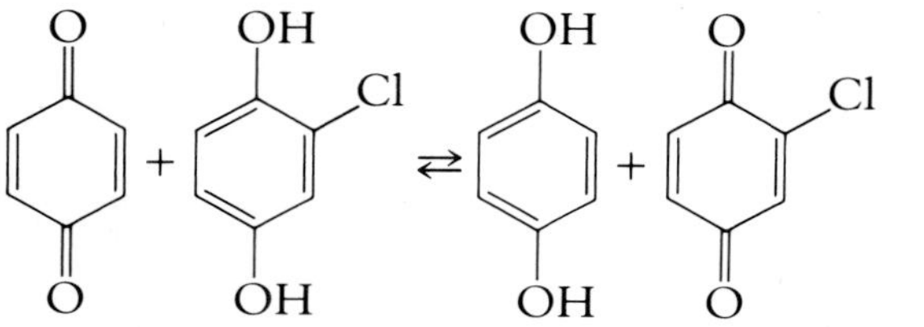 [structures: p-benzoquinone + 2-chlorohydroquinone ⇄ hydroquinone + 2-chloro-p-benzoquinone]

d. 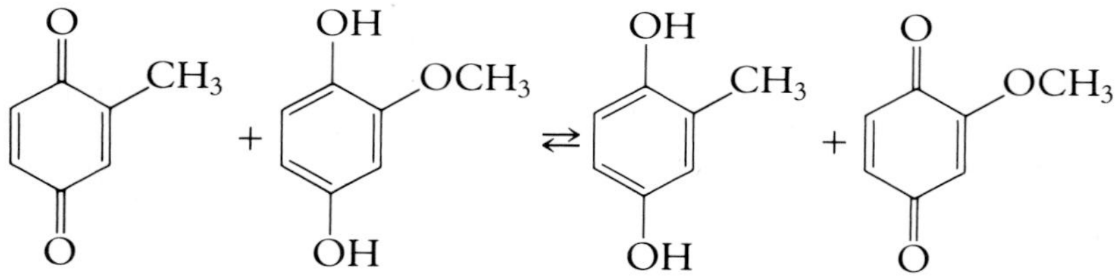 [structures: 2-methyl-p-benzoquinone + 2-methoxyhydroquinone ⇄ 2-methylhydroquinone + 2-methoxy-p-benzoquinone]

24–31 Like phenols, amines are readily oxidized by one-electron oxidizing agents. When *N,N,N′,N′*-tetramethyl-1,4-benzenediamine is mixed with tetrachloro-*p*-benzoquinone in a polar solvent, one electron is transferred from the amine to the quinone (eq. 10). The product is a salt consisting of an amine radical cation (**V**, a *Wurster cation*) and a semiquinone radical anion. With the aid of resonance structures, show why the cation and anion are stable.

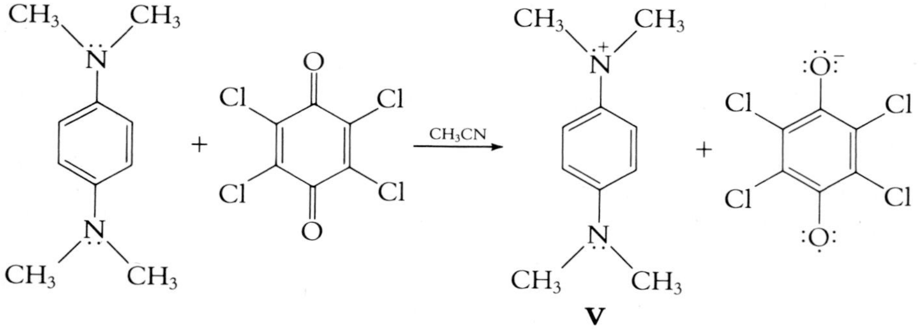

[structures: N,N,N′,N′-tetramethyl-1,4-benzenediamine + tetrachloro-p-benzoquinone —CH_3CN→ radical cation **V** + tetrachlorosemiquinone radical anion] (10)

24–32 2,3-Dicyanohydroquinone (**VI**) is synthesized from *p*-benzoquinone and HCN (eq. 11). **VI** is useful as a precursor of 2,3-dicyano-*p*-benzoquinone and of DDQ. The mechanism for this synthesis is believed to involve three separate steps, shown in Equations 12 to 14.

$$2\,(p\text{-benzoquinone}) + 2KCN + H_2SO_4 \xrightarrow{H_2O} \text{hydroquinone} + \text{2,3-dicyanohydroquinone (IV)} + K_2SO_4 \quad (11)$$

$$p\text{-benzoquinone} + HCN \rightleftarrows \text{2-cyanohydroquinone} \quad (12)$$

$$\text{2-cyanohydroquinone} + p\text{-benzoquinone} \rightleftarrows \text{2-cyano-}p\text{-benzoquinone} + \text{hydroquinone} \quad (13)$$

$$\text{2-cyano-}p\text{-benzoquinone} + HCN \rightleftarrows \text{2,3-dicyanohydroquinone} \quad (14)$$

a. Suggest a mechanism for the addition of HCN to the quinones (eqs. 12 and 14). It will be helpful to remember that the quinones are α,β-unsaturated ketones. Are the addition reactions thermodynamically favored? Justify your answer.

b. Does the equilibrium shown in Equation 13 favor reactants or products? Explain your answer.

c. What is the driving force for the overall reaction (eq. 11)?

24–33 The structure of tetrahydrocannabinol (THC) is

Me

OH

Me O

Me

$CH_2CH_2CH_2CH_2CH_3$

THC

a. Show which carbons would be labeled if the *Cannabis sativa* plant (from which THC can be extracted) were fed CH_3COO^- (acetate labeled with ^{14}C or ^{13}C in the carboxyl group).

b. A portion of the THC molecule is derived from a precursor containing two isoprene units (sec. 13:6). Show how two isoprene units could be incorporated into THC.

24–34 An unknown alkaloid ($C_{10}H_{15}NO$) is obtained from a plant by extraction of the plant with acid, followed by basification of the aqueous solution and extraction with $CHCl_3$. If the solution is made highly basic (pH 11), the alkaloid is not extractable from the aqueous solution. It can, however, be removed into the $CHCl_3$ layer if the solution is weakly basic (pH 9). The ultraviolet spectrum of the alkaloid in ethanol is given below (spectrum A). When several drops of a solution of NaOH in ethanol are added to the uv cell, spectrum B is obtained. The nmr spectrum of the unknown alkaloid is also given below.

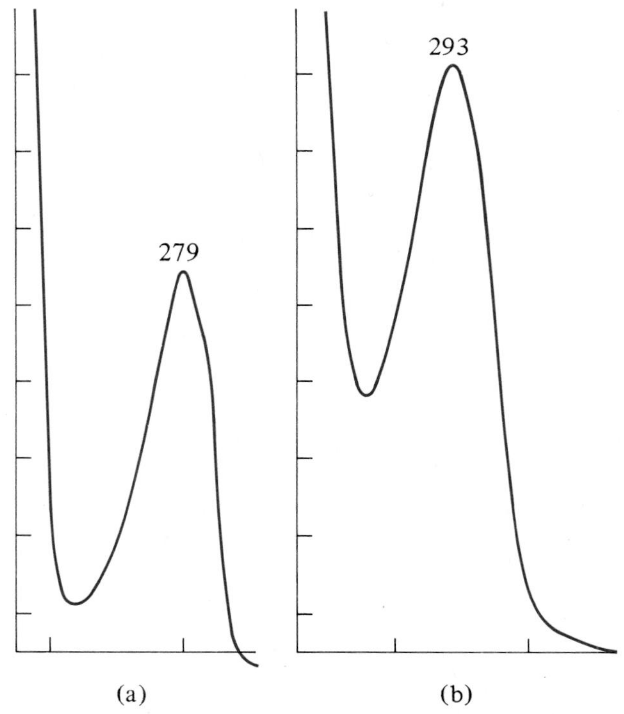

(a) (b)

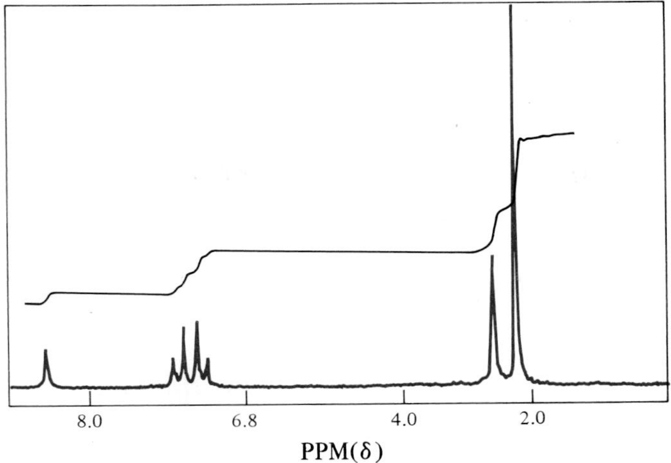

a. Suggest a structure for the alkaloid.
b. Why is the alkaloid extractable at pH 9 but not at pH 11?
c. Why is the uv spectrum of the alkaloid different in neutral and basic conditions?

24–35 Vitamin E is soluble in fats and oils. Vitamin C (ascorbic acid), on the other hand, is water soluble. Vitamin C serves as an antioxidant in the aqueous portions of cells. What stable radical might be formed from vitamin C by hydrogen abstraction? Write its resonance structures.

$HOCH_2CHOH$ O O H HO OH

Vitamin C

Supplementary Reading

J. B. Hendrickson, *The Molecules of Nature,* Chap. 2 and 3, W. A. Benjamin, Reading, Mass, 1973.

D. E. Metzler, *Biochemistry,* Academic Press, New York, N.Y., 1977, pp. 576–600.

E. C. M. Chen and S. L. Sjoberg, The kinetics and thermodynamics of the phenol from cumene process, *J. Chem. Ed.,* **57,** 458 (1980).

M. C. Nagel, Dr. Baekland's Bakelite, *J. Chem. Ed.,* **57,** 811 (1980).

H. G. Fravel, Jr., The manufacture of diphenyl oxide, *J. Chem. Ed.,* **57,** 873 (1980).

H. Hart and J. L. Reilly, Oxidative coupling of phenols, *J. Chem. Ed.,* **55,** 120 (1978).

25 The Chemistry of Aromatic Heterocycles

There are many cyclic molecules in which elements other than carbon are part of the ring. Such molecules are called **heterocycles**. The most common heterocycles are those in which one or more atoms in the ring skeleton are nitrogen, oxygen, or sulfur. You have already encountered a few examples in previous chapters. Chemical behavior of *saturated* heterocycles can usually be predicted on the basis of our knowledge of open-chain compounds containing a heteroatom. This chapter will therefore be devoted almost exclusively to *aromatic* heterocycles.

Chemical differences between heterocycles and their all-carbon analogs may usually be traced to one or both of the following facts:

1. The electronegativity difference between the heteroatom and the carbon atom results in the presence of polarized bonds in the molecule.
2. The heteroatoms possess nonbonding electron pairs.

Among aromatic heterocycles there are two fundamental classes, one made up of five-membered rings and the other made up of six-membered rings. Their chemical properties are substantially different. Five-membered heterocycles are usually electron-rich compared to benzene, and so undergo electrophilic substitution quite readily. Six-membered heterocycles, by contrast, are usually electron-poor compared to benzene, and thus tend to undergo nucleophilic substitution fairly easily. The reasons for this difference in behavior will become apparent when we study the structure of the heterocycles.

This chapter will first treat the structure and chemistry of the five-membered heterocycles, followed by a similar, systematic treatment of the six-membered compounds. We will then use the chemical characteristics of these ring systems to explore their biochemical roles.

25:1 Five-Membered Aromatic Monoheterocycles

Heterocycles are treated separately from carbocycles mainly because of the unique chemistry of aromatic heterocyclic compounds. This behavior is more easily understood if we begin by grouping the compounds according to ring size. The most important five-membered aromatic heterocycles with one heteroatom are furan, pyrrole, and thiophene. There are also numerous

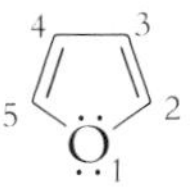

Furan

Pyrrole

Thiophene

five-membered aromatic rings that contain two or more heteroatoms. We shall concentrate on two of them, imidazole and thiazole.

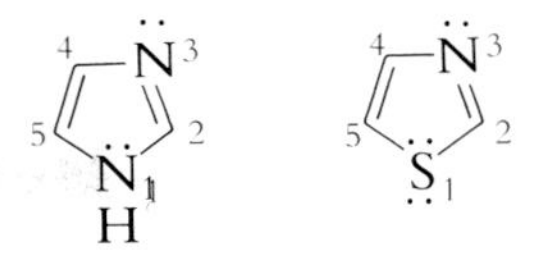

Imidazole Thiazole

In our discussions of heterocyclic compounds, it is often necessary to refer to a specific atom in the ring by using a simple numbering system. In monoheterocycles, the heteroatom is always given the number one. In the polyheterocyclic systems, the atom of highest atomic number is given the number one, and the rest of the molecule is numbered in such a way as to place the other heteroatoms at the lowest numbered positions.

The 5-membered monoheterocycles all contain six delocalized π electrons, two of which are contributed by the heteroatom. The presence of a π-electron sextet leads us to expect aromatic-type behavior, and this expectation is borne out. The heterocycles are not so highly aromatic as benzene, however. For one thing, the π-electron density is not distributed symmetrically. Examination of the resonance forms for these compounds reveals that there is a net shift of electron density away from the heteroatom and toward the carbon atoms. The π-resonance effect outweighs the inductive pull on the σ electrons exerted by the electronegative heteroatoms.

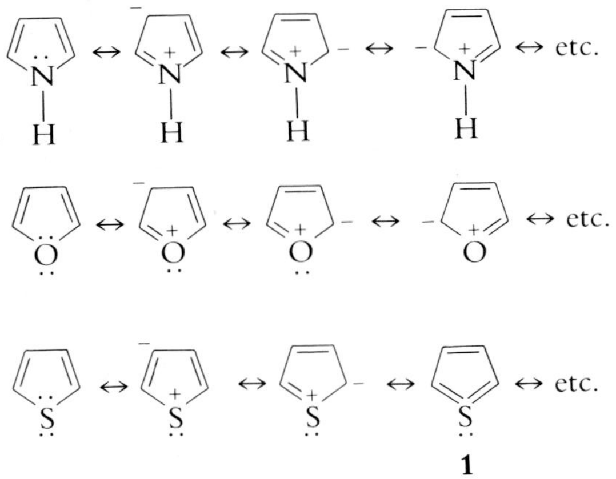

Resonance effects that shift π-electron density from heteroatoms to carbon

Because sulfur has *d* orbitals available, thiophene has a resonance structure (**1**) without charge separation, in which sulfur contains 10 electrons in its valence shell. Calculations indicate that thiophene has the greatest resonance stabilization of any of the monoheterocyclic aromatics, as much as 80% of the resonance stabilization of benzene. Furan is the least stable; it has less than 50% of the resonance stabilization of benzene. The relative instability of furan is understandable because electron delocalization in this molecule requires the removal of electron density from the highly electronegative oxygen atom. Here, the electronegativity effect is beginning to balance the resonance effect. These theoretical arguments are often, but not always matched by the experimental results. The relative "aromaticity" of thiophene, furan, and pyrrole is still a controversial matter.

Problem

25–1 Pyrrole and furan each have a total of five important resonance structures. Thiophene has six. Draw the resonance structures not given in the text for these compounds.

Electron delocalization in the aromatic heterocycles manifests itself in a variety of ways. Pyrrole, for example, shows none of the basicity of a typical secondary amine because the nonbonded electron pair participates in the aromatic sextet and hence is not available for bonding to a proton.

Imidazole, on the other hand, is a weak base with a pK_a of 7. The site of protonation of imidazole is *not* N-1, because N-1 donates its electron pair to the aromatic sextet. Rather, N-3 accepts a proton because it possesses an electron pair perpendicular to the π system (eq. 25.1 and Fig. 25–1).

(25.1) Imidazole + H^+ ⇄ Imidazolium ion (two resonance structures ↔)

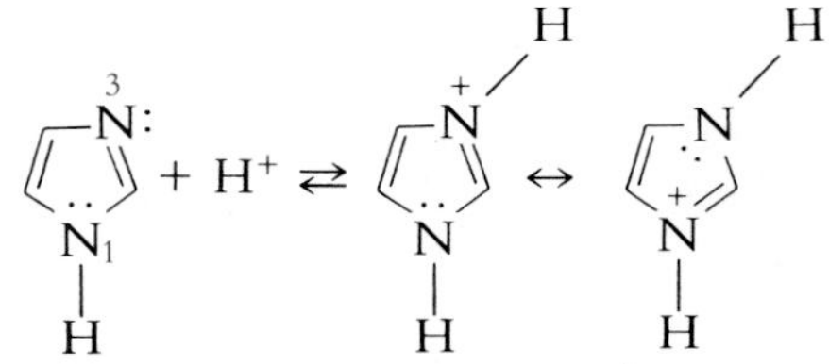

Imidazole   Imidazolium ion

Imidazole is less basic than ordinary alkyl amines (p$K_a \sim 11$) because the electron pair being protonated is in an sp^2 orbital rather than an sp^3 orbital. An electron pair in an sp^2 orbital is closer to the nitrogen nucleus and therefore more tightly held, than it would be in an sp^3 orbital. Hence, it is less available for bonding to a proton. For this same reason, pyridine is also much less basic (pK_a = 5.2) than saturated amines (sec. 21:4). The difference in pK_a between imidazole and pyridine indicates that imidazole is about 100 times more basic than pyridine. If pyridine is protonated, the positive charge is localized on one nitrogen atom. Note, however, from Equation 25.1 that the positive charge in protonated imidazole is delocalized over *two* nitrogen

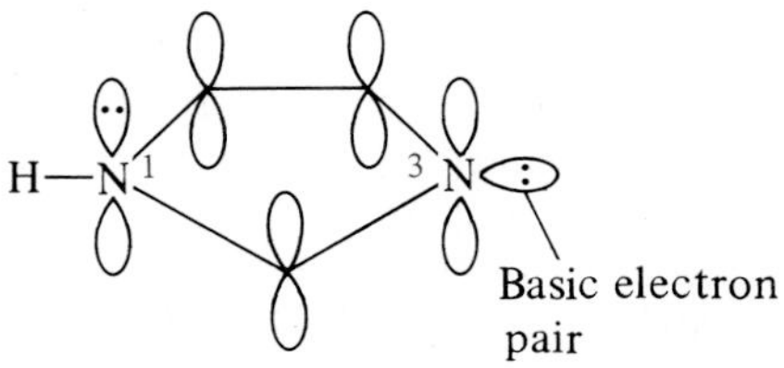

H—N 1   3 N   Basic electron pair

Figure 25–1. *Atomic* p *orbitals of imidazole. The N-3 atom can act as a basic site because it has an electron pair perpendicular to the* π *system.*

atoms. Thus, protonated imidazole is much more stable than is protonated pyridine and imidazole is therefore the stronger base. Although imidazole is only occasionally used in the laboratory as a basic catalyst, it is an extremely important base and nucleophile in numerous enzymes (see sec. 25:8).

Problems

25–2 Oxazole has the structure shown. Oxazole can be protonated to give a cation. Show the important resonance structures for this ion, and use them to decide whether oxazole is more or less basic than imidazole. Explain your reasoning.

Oxazole

25–3 When treated with one equivalent of CH_3I and a base, imidazole is converted to 1-methylimidazole. Addition of a second equivalent of CH_3I yields 1,3-dimethylimidazolium iodide. Write the structures of both products and show how they are formed.

25–4 The ^1H nmr spectrum of furan is shown here. Assign the peaks. Why are triplets observed? The singlet ^1H absorption for benzene is at 7.2 ppm. Why does one triplet for furan appear at such high field (6.38 ppm)?

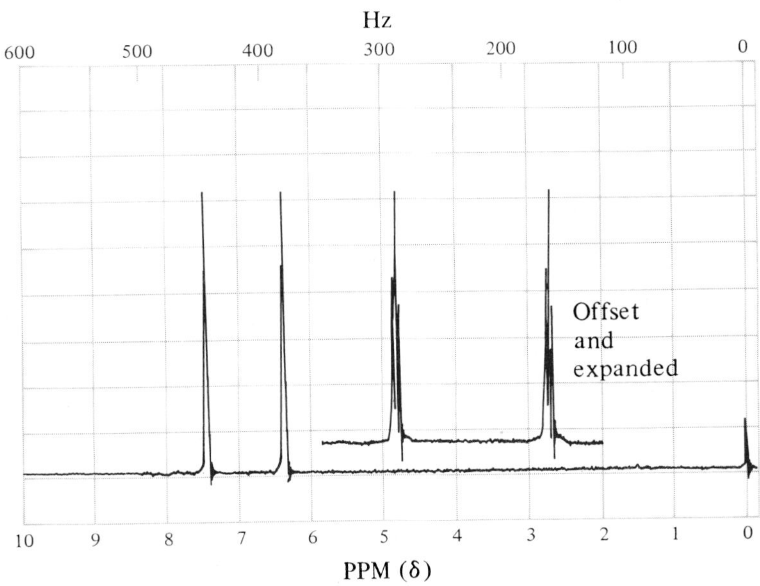

25:2 Electrophilic Substitution of Five-Membered Heterocycles

Like most aromatic compounds, *all five-membered unsaturated heterocycles undergo electrophilic substitution*. In the monoheterocycles there are only two distinguishable positions for monosubstitution, C-2 and C-3 (sometimes called the α- and β- positions). Substitution at C-2 is usually much faster than substitution at C-3. A good example of this is the acid-catalyzed deuterium exchange in thiophene (eq. 25.2). In this reaction, the electrophile is a deuteron, D^+.

(25.2) $\frac{k_1}{k_2} = 1045$

The preference for substitution at C-2 can be understood if we examine the intermediate cations that are produced in the course of the reaction.

The intermediate arising from attack at C-2 (**2**, eq. 25.3) is capable of greater charge delocalization than is the intermediate produced by attack at C-3 (eq. 25.4). Intermediate **2** has one more contributing resonance structure than intermediate **3**.

(25.3)

Z = N—H, O, S

(25.4)

Pyrrole and furan are attacked readily by electrophilic reagents and are sometimes compared to aniline and phenol in their reactivities. Thiophene is less reactive than pyrrole and furan, but slightly more reactive than benzene. Pyrrole yields tetrabromopyrrole upon treatment with Br_2 in methanol (eq. 25.5), and gives 2-acetylpyrrole upon heating with acetic anhydride, *without a Friedel–Crafts catalyst* (eq. 25.6).

(25.5) + Br_2 (excess) $\xrightarrow{MeOH}$

(25.6) + Ac_2O $\xrightarrow{\Delta}$

2-Acetylpyrrole

Both pyrrole and furan are too reactive to be sulfonated with sulfuric acid; only resins and polymers result. Thiophene, however, can be readily sulfonated (eq. 25.7). Thiophene and benzene are very similar in their physi-

(25.7) Thiophene + H_2SO_4 (95%) $\xrightarrow{30°}$ Thiophene-2-sulfonic acid (70%) (2-SO_3H)

cal properties, and technical-grade benzene from coal tar always contains traces of thiophene that are not separable by distillation. If such benzene is washed with sulfuric acid, the thiophene is sulfonated and becomes soluble in the sulfuric acid. In this way, thiophene-free benzene is obtained.

Special conditions (to be discussed shortly) are necessary for the nitration of furan and pyrrole, but thiophene can be nitrated directly (eq. 25.8).

(25.8) Thiophene + HNO_3 $\xrightarrow[(CH_3CO)_2O]{CH_3CO_2H}$ 2-NO_2 thiophene (70 to 85%) + 3-NO_2 thiophene (5 to 15%) + 2,4-di-NO_2 thiophene Minor

Notice that Equations 25.6 to 25.8 show the characteristic preference for electrophilic substitution at C-2 (α-substitution) in these heterocycles.

The effects of substitution on the orientation of incoming electrophiles are similar to those observed for electrophilic substitution in benzene and its derivatives (Chap. 15). It is important to keep in mind that, unlike benzene, the heteroatom of the ring exerts a preference of its own in addition to the influence of any attached substituents.

Because heteroatoms promote reaction at the α-position of the rings, the presence of one α-substituent on the ring will cause reaction to take place at the free α-position (C-5). This behavior is observed with substituents that vary widely in electrical character (eqs. 25.9 and 25.10).

(25.9) 2-CH_3 thiophene + HNO_3 $\rightarrow$ 5-O_2N-2-CH_3 thiophene (70%) + 3-NO_2-2-CH_3 thiophene (30%)

(25.10) Furan + $2SO_3$ $\xrightarrow{pyridine}$ HO_3S–furan–SO_3H Furan-2,5-disulfonic acid (80%)

Example

Predict the product(s) from acylation (acetic anhydride and $AlCl_3$) of 3-methylthiophene.

Answer

A comparison of Equations 25.3 and 25.4 shows that the major products would be from attack next to the sulfur at the 2 and 5 positions:

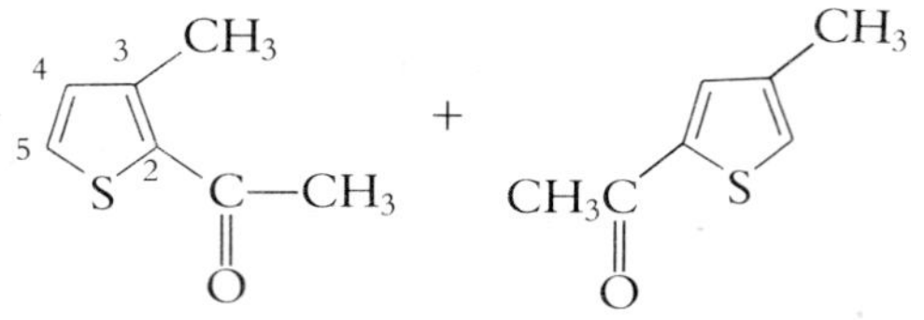

To predict which of these two products would be the major one, it is necessary to draw resonance structures of the intermediates.

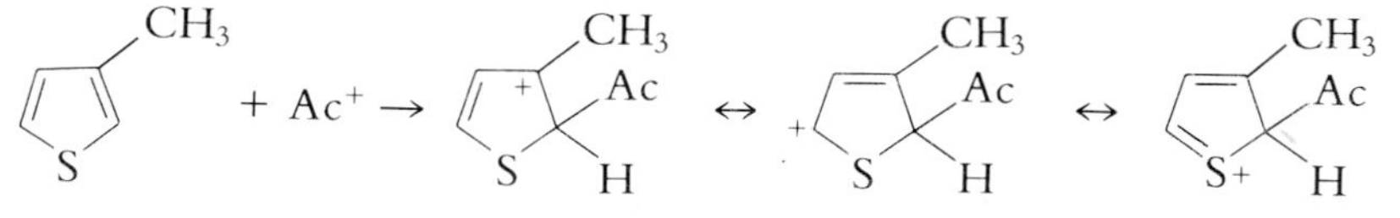

Structures **B** and **C** are essentially identical in energy to structures **B′** and **C′**. Structure **A**, however, is lower in energy than structure **A′** because **A** has a positive charge on the carbon that bears the electron-donating methyl group. Acetylation should therefore be favored at position 2 rather than position 5. Experimentally it is found that the reaction yields 80% 2-acetyl-3-methylthiophene and 20% 2-acetyl-4-methylthiophene.

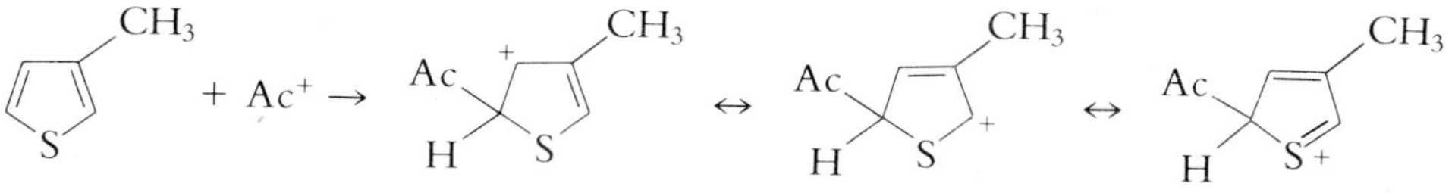

Problems

25–5 Write the structures of the predominant products of the following reactions:

a. 3-nitrothiophene $\xrightarrow[H_2SO_4]{HNO_3}$

b. thiophene $\xrightarrow{2Br_2}$

c. pyrrole-2-carboxylic acid $\xrightarrow[CH_3COOH]{HNO_3}$

d. furan-2-carboxylic acid $\xrightarrow{Br_2}$

e. thiophene + CH_2O + $(CH_3)_2NH$ $\xrightarrow{H^+}$

25–6 Nitration of furan, thiophene, and pyrrole takes place under very mild conditions (eq. 25.8, for instance). Imidazole, by contrast, requires rather strenuous conditions for nitration—a boiling mixture of nitric and sulfuric acids. What property of imidazole leads to its low reactivity under these conditions (review eq. 25.1)?

25:3 Addition Reactions of Furan: Chemical Evidence of Diminished Aromaticity

Electrophilic attack on furan must proceed initially by way of the intermediate ion **4** (eq. 25.11). If **4** loses a proton, as does a benzenium ion in similar circumstances, then the substitution product is obtained directly.

(25.11) furan + A^+B^- → [4] → A–furan + H^+B^- (substitution); [4] $\xrightarrow{B^-}$ **5** and/or **6** (addition)

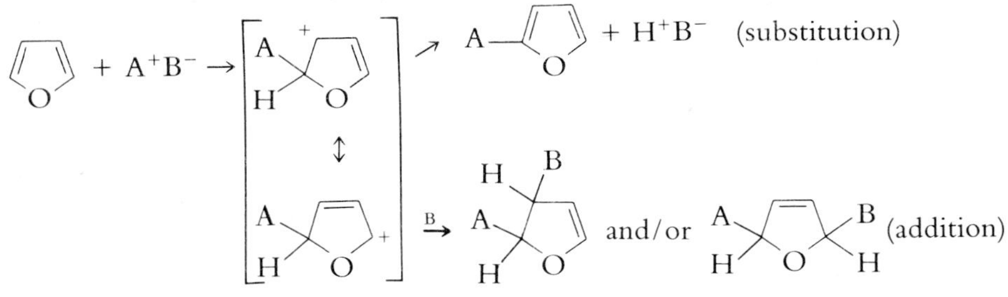

4 **5** **6**

The carbocation, **4**, may also react with nucleophiles present in solution to give one or both of the addition products **5** and **6**. A net substitution may also occur by this path, if the addition product readily undergoes elimination of HB (eq. 25.12).

(25.12) **6** → H^+ + A–furan–H + B^-

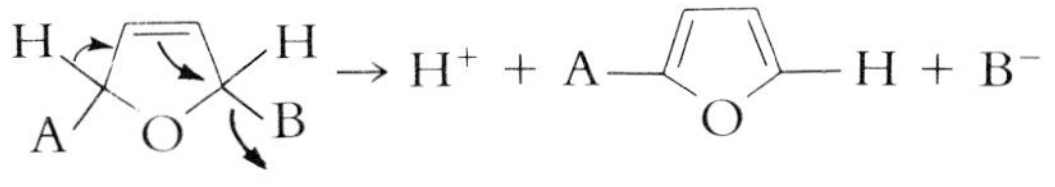

Furan has less resonance stabilization than benzene. Hence, in the case of furan, there is less driving force to lose a proton from **4** and go directly to the substituted furan (substitution path, eq. 25.11). This allows time for the addition products **5** and/or **6** to form. Eventually, the elimination reaction (eq. 25.12), or the corresponding reaction from **5**, will occur and the final substituted furan will be produced.

For example, the nitration of furan in acetic anhydride occurs by such an addition–elimination reaction. (Furan cannot be directly nitrated.) Nitric acid is first added to acetic anhydride to form acetyl nitrate (eq. 25.13).

$$(CH_3\overset{O}{\overset{\|}{C}})_2O + HONO_2 \rightleftarrows CH_3\overset{O}{\overset{\|}{C}}ONO_2 + CH_3COOH \qquad (25.13)$$

Acetyl nitrate

When a furan is added to this solution, the acetyl nitrate adds 1,4 to the furan to yield an unstable addition product (eq. 25.14). This intermediate, upon treatment with a weak base such as pyridine, is converted to 2-nitrofuran (eq. 25.15).

$$CH_3\overset{O}{\overset{\|}{C}}ONO_2 + \text{furan} \xrightarrow[-10^\circ,\ CH_3CO_2H]{} \text{2-}(CH_3\overset{O}{\overset{\|}{C}}O)\text{-5-}(NO_2)\text{-2,5-dihydrofuran} \qquad (25.14)$$

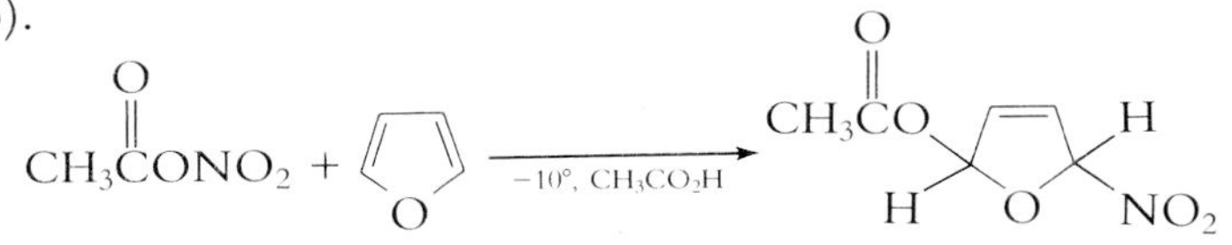

Addition product

(25.15) [Reaction scheme: addition product (CH_3CO_2-, H, H, NO_2 dihydrofuran) + pyridine → 2-nitrofuran + pyridinium (N^+–H) + CH_3COO^-]

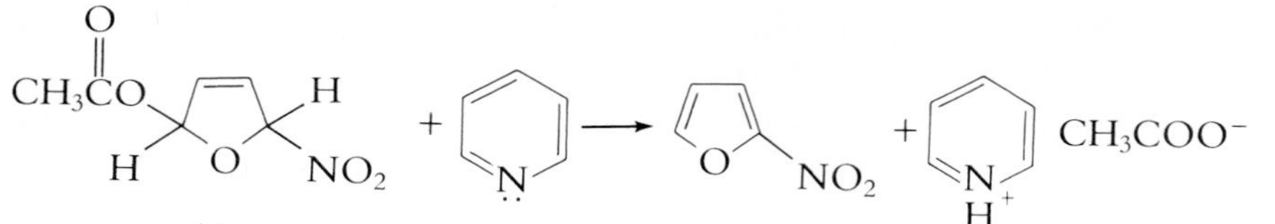

Addition product

Another indication of the high reactivity of furan is its ready participation as a diene in the Diels–Alder reaction (eq. 25.16). Neither thiophene nor pyrrole is active as a diene in the Diels–Alder reaction except under forcing conditions.

(25.16) CH_3O–(furan)–CH_3 + (maleic anhydride) $\xrightarrow{\text{ether, 25°}}$ [adduct: CH_3, O, H, H, OCH_3]

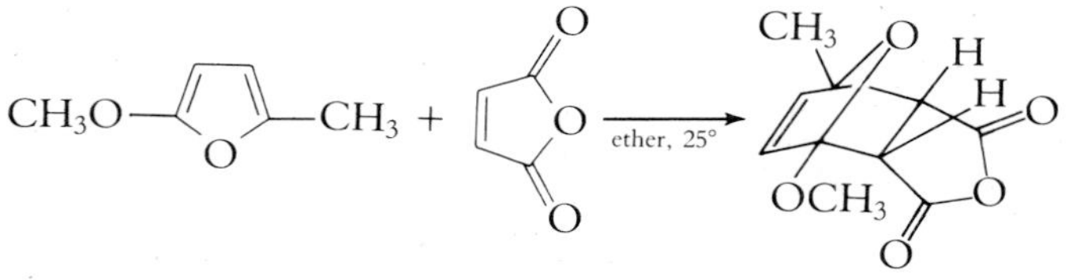

Problems

25–7 Predict the product(s) from each of the following reactions. Where more than one product is expected, tell which would be the major product and which would be the minor product.

a. pyrrole + $2H_2/Pt$
b. furan + Cl_2 (at $-40°$)
c. furan + acetic anhydride (0°)
d. pyrrole + HNO_3 in acetic anhydride (0°)

25–8 2-(Chloromethyl)furan (**I**) is a highly reactive compound. Its reaction with cyanide ion in water (eq. 1) has the following characteristics: the kinetics are first order, the rate depending only on the concentration of **I**, and the reaction yields two products, **II** and **III**. Suggest a mechanism for this reaction that accounts for both the rate and product data.

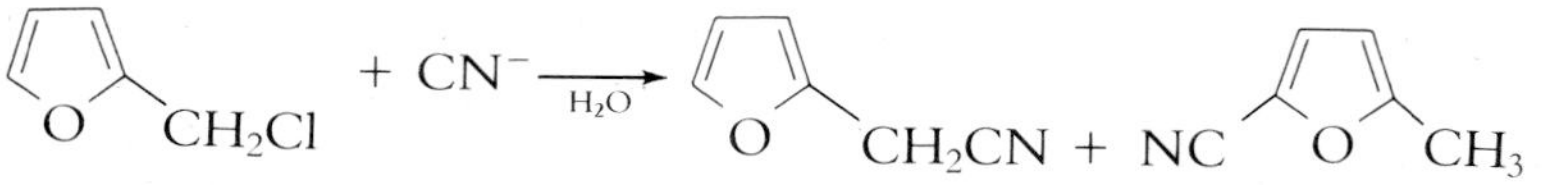

$$\text{(furan)–}CH_2Cl \; (\mathbf{I}) + CN^- \xrightarrow{H_2O} \text{(furan)–}CH_2CN \; (\text{Minor } \mathbf{II}) + NC\text{–(furan)–}CH_3 \; (\text{Major } \mathbf{III}) \quad (1)$$

25–9 *Furfural* (furan 2-carboxaldehyde) reacts with Br_2 to give a furfural monobromide. Which one is it? Why doesn't polysubstitution by Br_2 occur?

25–10 The compound 2,5-dihydroxyfuran (**IV**) does not exist as such. Why not? (What is its predominant tautomeric structure?) Explain why **IV** and hydroquinone (sec. 24:7) differ with respect to predominant tautomeric form.

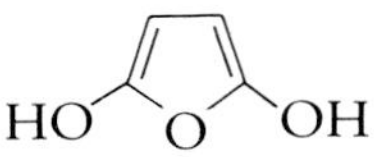

$$HO-C_4H_2O-OH$$

IV

25:4 Five-Membered Heterocycles Fused to Benzene Rings

Among the most common compounds consisting of five-membered heterocycles fused to benzene rings are indole, benzo[*b*]thiophene, and benzo-

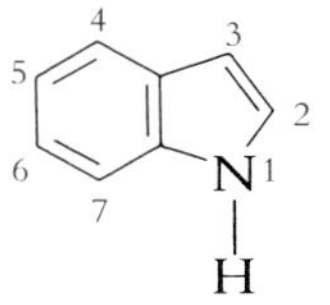

Indole 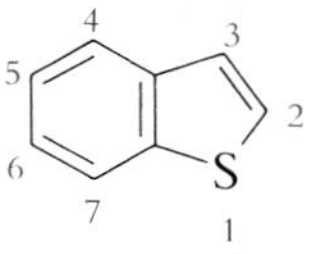   Benzothiophene 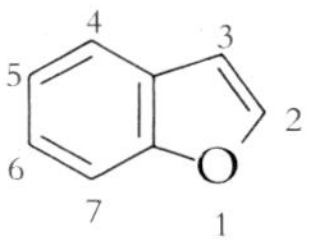   Benzofuran

[*b*]furan. Of all these ring systems, indole is far and away the most important because of the physiological activity of so many of its derivatives.

Electrophilic substitution of indole takes place almost exclusively in the pyrrole ring, at C-3. Equation 25.17 is an example of a Friedel–Crafts reaction that yields a 3-substituted derivative.

(25.17)

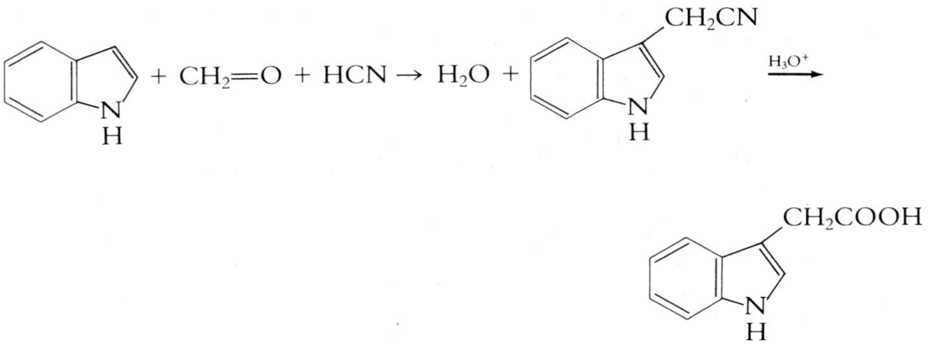

$$\text{indole} + CH_2{=}O + HCN \rightarrow H_2O + \text{3-}(CH_2CN)\text{indole} \xrightarrow{H_3O^+} \text{3-}(CH_2COOH)\text{indole}$$

3-Indoleacetic acid (a plant hormone)

We can understand why substitution occurs at C-3 rather than at C-2 by examining the reaction intermediates. The greater stability of intermediate **7** over intermediate **8** is believed due to the fact that charge delocalization may be accomplished in **7** without disturbing the aromaticity of the benzene ring. That is not possible in **8**.

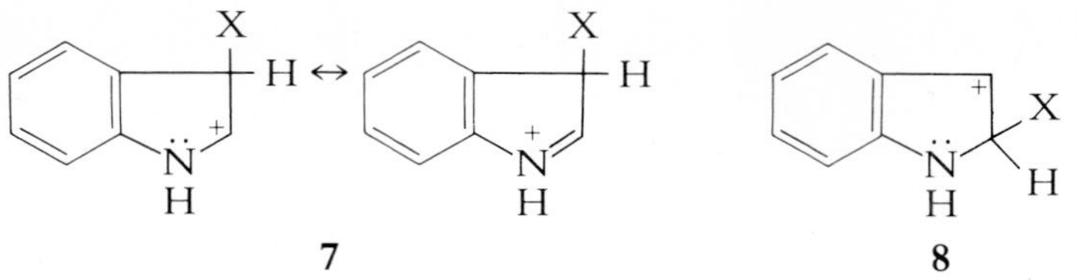

7 8

Physiologically important indole derivatives are widespread. Indole-3-acetic acid (*heteroauxin,* eq. 25.17) is an important growth factor in seeds. The amino acid tryptophan has an indole ring as does skatole, its physiological decomposition product which has a penetrating fecal odor. Lysergic acid is a well-known indole alkaloid (recall that an alkaloid is a nitrogen compound derived from plants). Its diethylamide is the hallucinogen LSD.

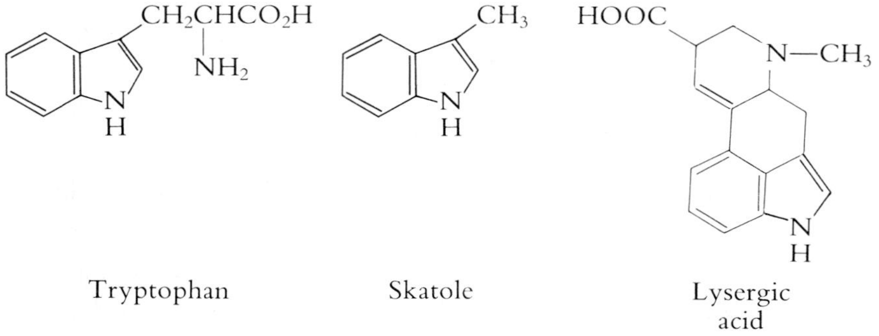

Tryptophan Skatole Lysergic acid

A versatile method of obtaining substituted indoles is the *Fischer indole synthesis,* in which the phenylhydrazones of aldehydes and ketones are used as starting materials. Either Lewis or proton acid catalysts are usually necessary to convert these to the indole (eqs. 25.18 and 25.19).

(25.18)

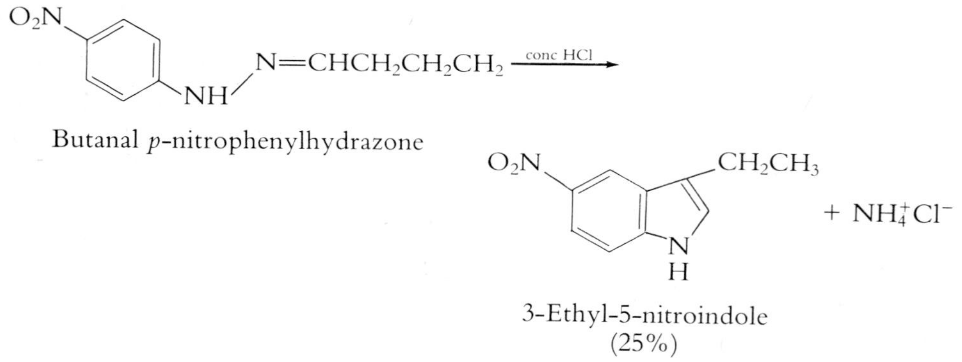

Butanal *p*-nitrophenylhydrazone $\xrightarrow{\text{conc HCl}}$ 3-Ethyl-5-nitroindole (25%) + $NH_4^+Cl^-$

(25.19)

$\xrightarrow[C_2H_5OH]{HCl}$ 1-Methyl-2-indolecarboxylic acid

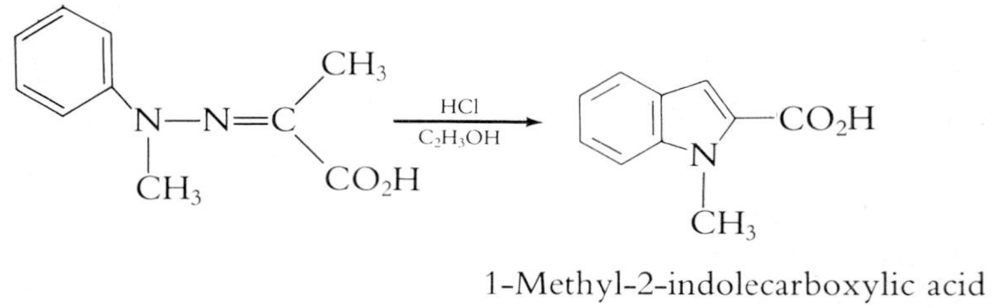

25:5 Natural and Synthetic Five-Membered Heterocycles

Pyrrole and thiophene are found in coal tar, and pyrrole may also be obtained from several plant sources. Furan is produced commercially by decarbonylation of furfural (furan-2-carboxaldehyde), which is in turn obtained from the acid degradation of oat hulls and corncobs. These substances contain sizable quantities of five-carbon sugars that are dehydrated to furfural (eq. 25.20).

(25.20) CHO–CHOH–CHOH–CHOH–CH_2OH $\xrightarrow[\text{heat}]{H^+}$ Furfural (furan-2-CHO) $\xrightarrow[400°]{\text{catalyst}}$ Furan + CO

Furfural Furan

Alkyl-substituted five-membered aromatic heterocycles are usually synthesized directly in the laboratory rather than by substitution of the parent ring. 1,4-Dicarbonyl compounds are invaluable starting materials for such syntheses (eq. 25.21). The dicarbonyl compounds are converted to furans by

(25.21) RC(=O)–CHR′–CHR″–C(=O)R‴ $\xrightarrow[\Delta]{P_4S_6 \text{ or } P_4S_{10}}$ thiophene (R, R′, R″, R‴ substituted, S)

RC(=O)–CHR′–CHR″–C(=O)R‴ $\xrightarrow[\Delta]{\text{acid}}$ furan (R, R′, R″, R‴ substituted, O) + H_3O

RC(=O)–CHR′–CHR″–C(=O)R‴ $\xrightarrow[\Delta]{GNH_2}$ pyrrole (R, R′, R″, R‴ substituted, N–G) + $2H_2O$

acid-catalyzed dehydration, to pyrroles by reaction with ammonia or primary amines, and to thiophenes by reaction with a phosphorus sulfide (eq. 25.21).

Problems

25–11 2,5-Hexanedione is heated to 100° with ammonium carbonate, $(NH_4)_2CO_3$. A liquid product is obtained which has the following nmr spectrum: δ 2.2(s, 6H), δ 6.0(s, 2H), δ 7.9(broad s, 1H). Give the structure of the compound and assign the nmr absorptions.

25–12 Show how each of the following heterocycles can be synthesized from a dicarbonyl compound.

a. 3,4-dimethylfuran

b. 2-ethyl-5-methylthiophene

c. 2,5-diphenylpyrrole

d. 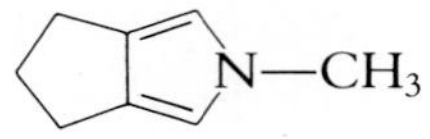 N—CH_3

25:6 Porphyrins

Without a doubt, the most important pyrrole derivatives are macrocyclic rings which incorporate four pyrrole molecules, known collectively as **porphyrins**. *Chlorophyll*★, which captures the sunlight that powers photosynthesis, contains a porphyrin ring. So does *hemoglobin,* the component of the red blood cells responsible for oxygen transport in the blood of vertebrates (Fig. 25–2). *Cytochrome P-450,* which takes part in the liver's oxidation of

★ The Greek word *chloros* means green, from which we derive the words *chlorine,* Cl_2, a green gas, and *chlorophyll,* a green pigment.

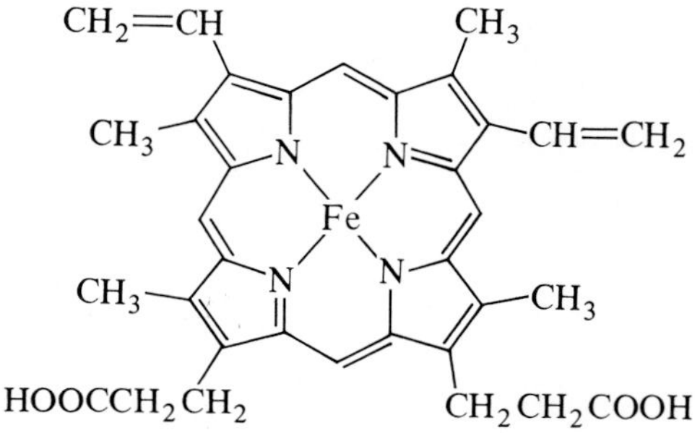

Heme

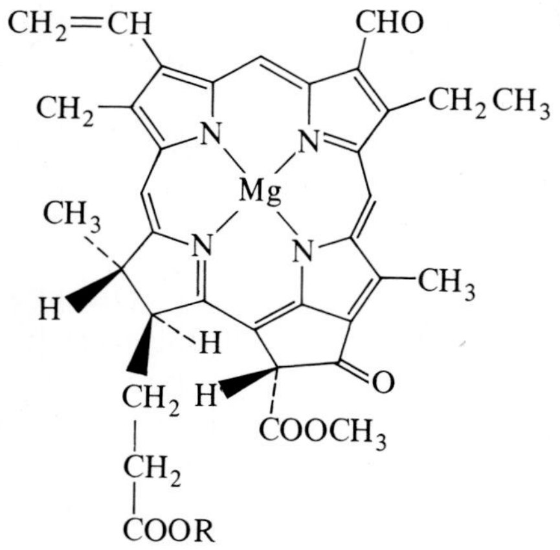

Chlorophyll *b*

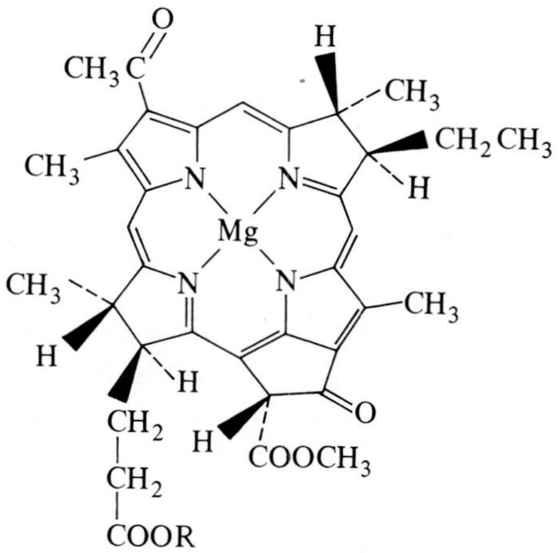

Bacteriochlorophyll *a*

$$R = H(CH_2\overset{\displaystyle CH_3}{\overset{|}{C}}HCH_2CH_2)_3CH_2\overset{\displaystyle CH_3}{\overset{|}{C}}{=}CHCH_2$$

Figure 25–2. *The structures of the natural metalloporphyrins heme, chlorophyll* b *and bacteriochlorophyll* a.

foreign substances to facilitate their excretion, is yet another porphyrin-containing compound. The list could be multiplied almost endlessly; furthermore, it is not confined to biological compounds. The common commercial pigment *cyan blue* contains a **phthalocyanine** ring in which four CH groups of porphyrin are replaced by N atoms and four benzene rings are fused to the pyrrole rings.

Here is the structure of the porphyrin ring:

Porphyrin

Positions 5, 10, 15, and 20 are known as the *meso* positions. In the phthalocyanines, the *meso* CH groups are replaced by N atoms. Note that the double bonds indicated in boldface form an aromatic 18 π-electron system (Hückel's $4n + 2$ rule, $n = 4$), which confers great stability on the porphyrin ring. For instance, porphyrins are often synthesized in the laboratory by heating together a pyrrole derivative with an aldehyde, the temperatures often reaching 300 to 400°. That the porphyrins survive these conditions testifies to their thermodynamic stability.

Porphyrin is always found in a substituted form in nature. There are numerous groups such as methyl, vinyl, acetic acid, and propionic acid attached to the peripheral carbon atoms. In addition, the ring is usually coordinated to a metal ion which replaces the N—H protons and is bonded to all four nitrogens. In mammals the most common metal is iron, as in *heme*. Heme is the metalloporphyrin component of hemoglobin, which actually binds O_2. Hemoglobin consists of four polypeptide chains, each of which contains a heme unit. Each heme unit is bound to a polypeptide by coordination of two imidazole rings from each polypeptide to the iron atom, above and below the plane of the porphyrin ring (see Fig. 22–11, p. 884).

In the oxygenated form of hemoglobin (oxyhemoglobin), the Fe(II) is six-coordinate. Four coordination sites are occupied by the nitrogen atoms of the porphyrin ring. A fifth coordinating nitrogen atom is provided by an imidazole ring from a histidine unit in the peptide side chain. A molecule of oxygen is bound at the sixth coordination site (Fig. 25–3). After delivering oxygen to a cell, the hemoglobin also helps to transport carbon dioxide away from it.

As the formula of porphyrin shows, the 7,8 and 17,18 double bonds lie outside the aromatic system. Many natural porphyrins have one (*chlorins*) or both (*bacteriochlorins*) double bonds missing (see Fig. 25–2). This is particularly true of the chlorophylls, which are partly reduced magnesium porphyrins.

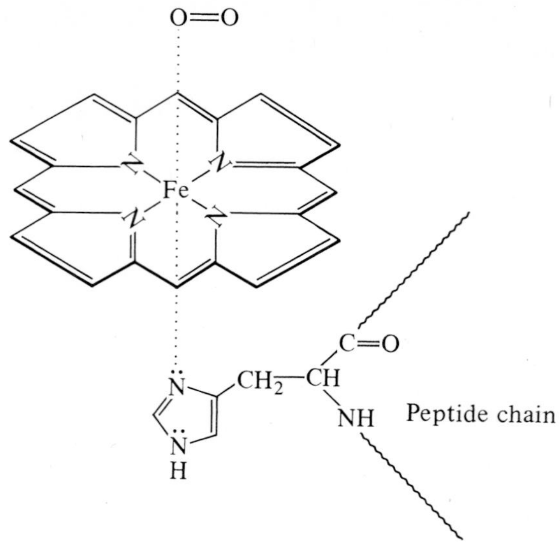

Figure 25–3. *Partial structure of oxyhemoglobin.*

25:7 Imidazole in Enzyme Catalysis

Enzymes bring about enormous increases in reaction rates compared to the rates of uncatalyzed reactions—rate enhancements of 10^{10} and 10^{12} times the uncatalyzed rate are not uncommon. Such rate accelerations are vital because the reactions in the human body must proceed rapidly at 37° at a pH near 7 in aqueous solution. In physiological systems one cannot employ the extremes of temperature, pressure, pH, and concentration that are possible in the laboratory.

The enzyme chymotrypsin (Fig. 22–10, p. 883), which hydrolyzes proteins, has been particularly well studied and is a good example of how a heterocyclic ring in an enzyme exerts its catalytic effects. Although enzymes are very large proteins, their reactivity is usually confined to a single, small portion of the molecule known as the *active site.* Chymotrypsin consists of three polypeptide chains held together by cystine disulfide bonds (shaded in Fig. 22–10). A combination of chemical and physical methods have shown that the active site is the relatively open region at the central part of the figure where there are histidine and serine amino acid residues in close proximity.

The kinetics of enzyme reactions indicate that the first step is usually binding of the substrate (in this case a polypeptide) to the enzyme in the vicinity of the active site. This so-called **enzyme–substrate complex** can be held together by dispersion forces and/or hydrogen bonds. Although the *exact* molecular mechanism of what occurs at the active site in chymotrypsin is difficult to determine, a highly probable pathway can be suggested based upon a knowledge of organic reaction mechanisms. The first step involves transfer of the OH proton of serine to the basic nitrogen of the imidazole ring of histidine (eq. 25.22). The active site histidine and serine amino acids are shown in Figure 22-10. This is followed (step 2, eq. 25.22) by binding of

(25.22)

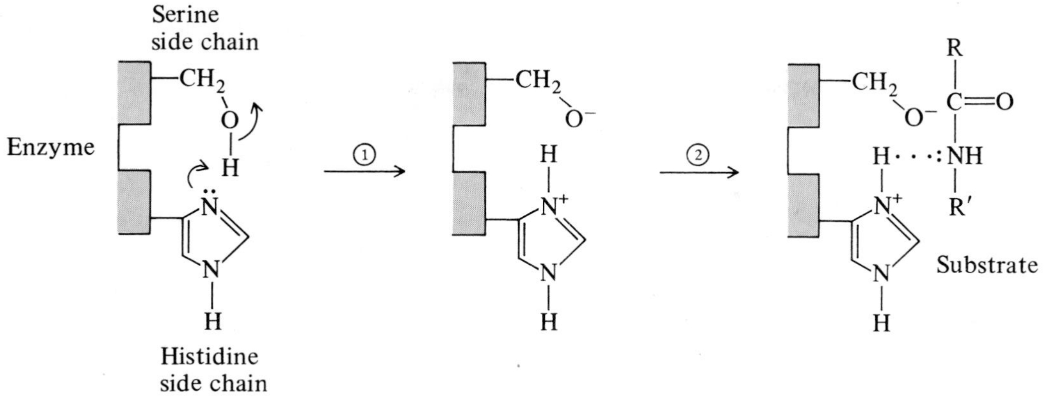

Serine side chain

Enzyme

Histidine side chain

Substrate

the peptide substrate to the resultant imidazolium ion. The peptide bond of the complexed substrate would then be in perfect position for attack by the alkoxide anion of the serine residue (step 3, eq. 25.23). This is followed by electron rearrangement to form an acyl enzyme (step 4), with simultaneous release of that portion of the peptide bound to the amide nitrogen.

(25.23)

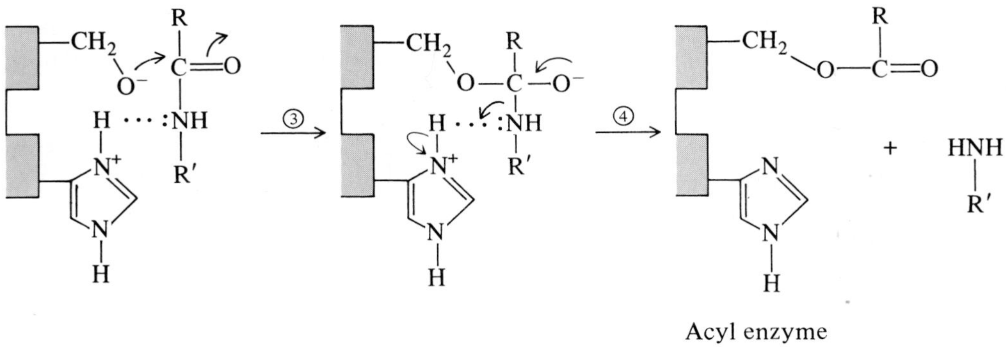

Acyl enzyme

The acyl enzyme then reacts with water to regenerate the free enzyme and liberate the remainder of the peptide originally bound to the amide carbonyl (eq. 25.24). This hydrolysis is also accelerated by intramolecular catalysis by the imidazole ring.

(25.24)

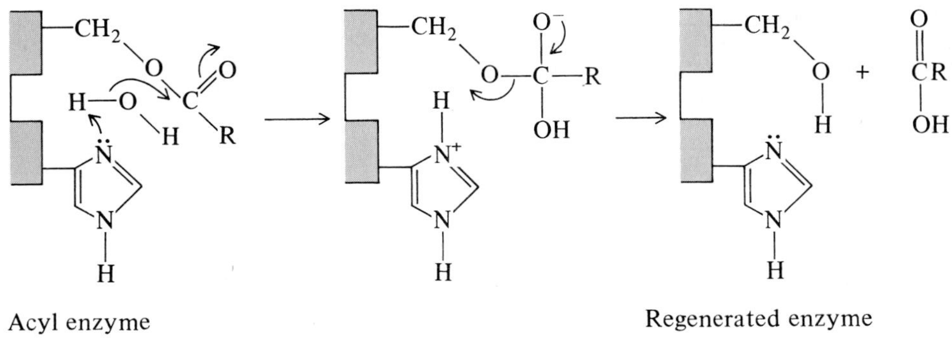

Acyl enzyme

Regenerated enzyme

25:8 Pyridine and Related Compounds: Structure and Basicity

In previous chapters we have seen several uses of pyridine as a basic catalyst, and we have also discussed its role in biological redox reactions. The next few sections amplify the chemistry of pyridine and also discuss chemistry of two fused ring derivatives of pyridine. These are quinoline and isoquinoline, whose ring structure is a component of several physiologically-active alkaloids and medicinals. Pyridine, quinoline, and their alkylated derivatives can be obtained from coal tar. This route is being challenged, however, by industrial syntheses based on ammonia and acetaldehyde.

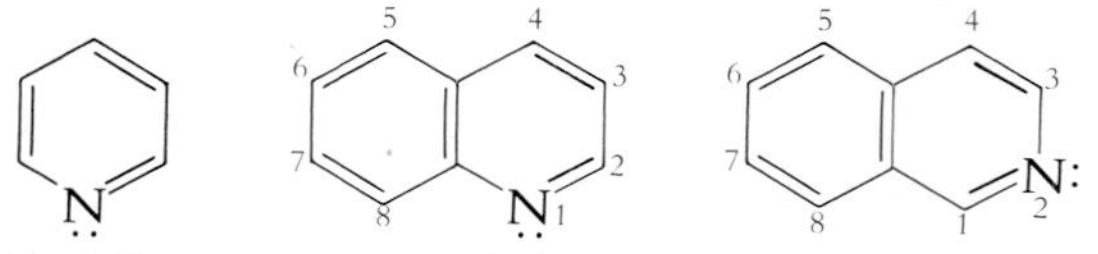

Pyridine   Quinoline   Isoquinoline

In these compounds, nitrogen possesses a nonbonded electron pair that is perpendicular to the π system, similar to the arrangement shown in Figure 25–1. These molecules are therefore bases. They are weaker bases than trialkylamines (sp^3 hybridized) because the nonbonding electrons occupy an sp^2 orbital (sec. 25:1).

The π-electron system of pyridine closely resembles that of benzene, and pyridine has a high degree of resonance stabilization. The resemblance is not complete, however. Due to its nonbonding electron pair, nitrogen may become attached to a proton or other electrophile, in the course of which it will acquire a positive charge (eq. 25.25). This pyridinium ion is similar to nitrobenzene in its unreactivity toward electrophilic substitution.★ On the other hand, the pyridinium ion is much more reactive than pyridine itself toward *nucleophiles,* again resembling nitrobenzene in its reactivity.

(25.25)

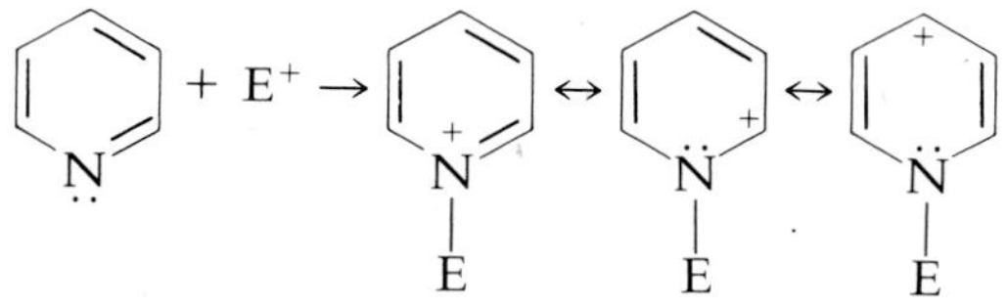

Pyridinium ion

Problem

25–13 a. The pK_a of pyridine is 5.36 while the pK_a of aniline is 4.60. Give a possible explanation for the fact that pyridine is a slightly stronger base than aniline.

b. The pK_a for 3-aminopyridine is 6.6 and that for 4-aminopyridine is 9.1. Which nitrogen is protonated in these aminopyridines? With the aid of resonance structures, explain why the 4-isomer is more basic than the 3-isomer.

★ *Two* positive charges would be present in the intermediate if pyridinium ion were attacked by an electrophile.

c. 3-Aminopyridine will react with two equivalents of HCl to form a salt containing a dication. Write its structure. 4-Aminopyridine reacts with only one equivalent of HCl. Explain the difference in behavior of these two isomers.

25:9 Electrophilic Substitution of Pyridine and Quinoline

Under the conditions necessary for electrophilic substitution of the deactivated pyridine ring (high concentration of a strong proton donor or Lewis acid catalyst), it is completely converted to a pyridinium ion (eq. 25.25). The pyridinium ion is one of the most deactivated aromatic systems known. Conditions for nitration, sulfonation, and halogenation are accordingly very severe, while Friedel–Crafts reactions cannot be carried out at all. Substitution takes place almost exclusively at C-3, *meta* to the nitrogen atom (eq. 25.26).

(25.26) pyridine $\xrightarrow[\text{KNO}_3,\ 330°]{\text{fuming } H_2SO_4}$ $\xrightarrow{OH^-}$ 3-nitropyridine (NO_2) (15%)

Problem

25–14 Show with the aid of resonance structures why nitronium ion, NO_2^+, attacks the pyridinium ion preferentially at C-3, rather than at C-2 or C-4.

The same factors that control the electrophilic substitution of pyridine also determine the electrophilic reactivity of its fused derivatives, such as quinoline. Quinolines can undergo electrophilic substitution on the nonheterocyclic ring because the reactivity of the heterocyclic ring is decreased by protonation of the nitrogen atom. For example, bromination of quinoline in sulfuric acid, with the aid of silver sulfate, results in bromination of the benzene ring rather than the pyridine ring (eq. 25.27). Under these condi-

(25.27) quinoline + Br_2 $\xrightarrow[H_2SO_4]{Ag_2SO_4}$ 5-bromoquinoline (28%) + 8-bromoquinoline (29%) + 5,8-dibromoquinoline (43%)

tions, the nitrogen of the quinoline ring is protonated so that reaction in the nitrogen-containing ring is slowed down, and substitution occurs in the other ring. However, if the bromination is carried out in carbon tetrachloride in the presence of pyridine, which reacts with the hydrogen bromide released, substitution takes place in the pyridine portion of the quinoline ring (eq. 25.28).★

(25.28) quinoline + Br_2 $\xrightarrow[CCl_4,\ 77^\circ]{\text{pyridine (base)}}$ 3-bromoquinoline (82%) + 6-bromoquinoline (2%)

25:10 Electrophilic Substitution of Pyridine *N*-Oxide

It is evident from Equation 25.26 that direct electrophilic substitution of pyridine is of limited synthetic use. Pyridine can, however, be converted to an *N*-oxide, which undergoes electrophilic substitution with ease. The *N*-oxide group can subsequently be removed. Equation 25.29 shows the formation of pyridine *N*-oxide with peroxyacetic acid, and Equation 25.30 shows a typical substitution reaction of this compound. Equation 25.31 shows the removal of the *N*-oxide.

(25.29) pyridine + $CH_3C(=O)OOH$ $\xrightarrow{100^\circ}$ pyridine *N*-oxide + CH_3COOH

Pyridine *N*-oxide (80%)

(25.30) pyridine *N*-oxide + HNO_3 $\xrightarrow[170^\circ]{H_2SO_4}$ $\xrightarrow{Na_2CO_3}$ 4-nitropyridine *N*-oxide

4-Nitropyridine *N*-oxide (72%)

(25.31) 4-nitropyridine *N*-oxide + PCl_3 $\xrightarrow[70\text{ to }80^\circ]{CHCl_3}$ 4-nitropyridine + $POCl_3$

4-Nitropyridine (79%)

★ The pyridine ring portion of quinoline is more reactive than is pyridine itself, and the added base is not brominated under the given conditions.

Substitution occurs at C-2 and C-4 of pyridine *N*-oxides. The intermediate for attack at the 4-position, for example, will have structure **9**, and it is the contribution from **9c** that stabilizes the intermediate. The *N*-oxide may be removed with simultaneous reduction of the nitro group (eq. 25.32).

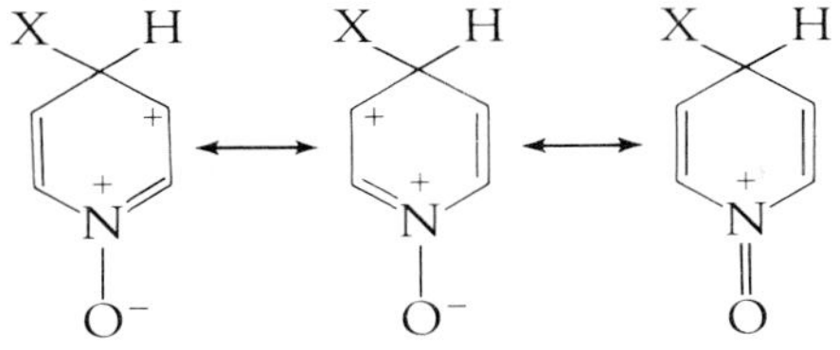

9a **9b** **9c**

(25.32) 4-nitropyridine N-oxide $\xrightarrow[CH_3COOH]{H_2,\ Pd/C}$ 4-aminopyridine

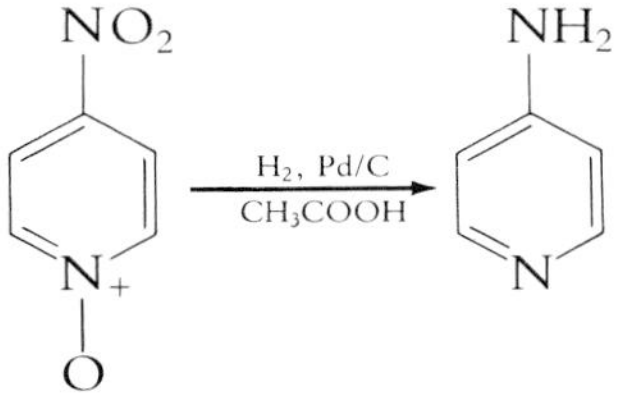

25:11 Nucleophilic Substitution of Pyridine Rings

Electron-deficient aromatics such as nitrobenzenes and pyridine, which are poor substrates for *electrophilic* attack because of destabilization of the positively-charged intermediate (eq. 25.25), are good substrates for *nucleophilic* attack (sec. 23:1). Equation 25.33 shows that the strong nucleophile NH_2^- can add to pyridine to form an intermediate (**10**) which is stabilized because the negative charge is on nitrogen. The final products are formed by loss of a hydride ion, $:H^-$ (eq. 25.34).

(25.33) pyridine + $Na^+\bar{N}H_2$ $\xrightarrow{110°}$ [**10**, Na^+] $\xrightarrow{H_2O}$ 2-Aminopyridine (70%) $+ H_2 + NaOH$

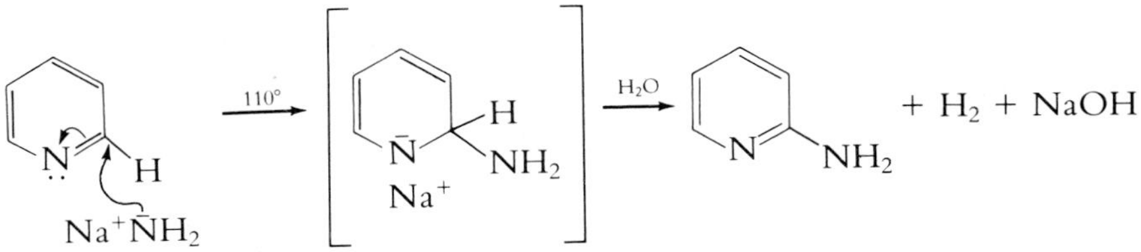

(25.34) **10** $\rightarrow$ 2-aminopyridine $+ H^-$

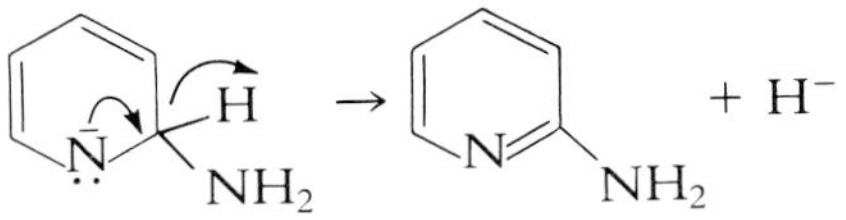

$$H^- + H_2O \rightarrow H_2 + OH^-$$

Nitrogen is more electronegative than carbon and hence nucleophilic substitution can take place easily on pyridine, but not on benzene. Substitution at C-2 is strongly preferred, probably because of complex formation between the Na^+ and the ring nitrogen in intermediate **10**. Substitution will take place at the 4-position if the 2-position is blocked (eq. 25.35).

(25.35) Quinoline-2-carboxylic acid $\xrightarrow[NH_3\text{ (liq)}, 25°]{2K^+NH_2^-}$ $\xrightarrow{H_3O^+}$ 4-Aminoquinoline-2-carboxylic acid (81%)

Organolithium reagents provide carbanion nucleophiles powerful enough to react with pyridine, as shown in Equation 25.36.

(25.36) 4-methylpyridine + phenyllithium $\xrightarrow[\text{reflux}]{\text{ether}}$ 4-methyl-2-phenylpyridine (39%) + LiH

Problem

25–15 a. When pyridine or substituted pyridines are treated with $NaNH_2$, the preferred position of attack is at C-2, followed by C-4. The amide ion does *not* attack at C-3. Explain this finding with the aid of resonance structures.

b. 3-Aminopyridine is commercially available. One synthesis of it starts with the readily available pyridine-3-carboxylic acid (nicotinic acid). Show how the compound can be converted to 3-aminopyridine (see sec. 21:9).

25:12 Synthesis of Pyridines and Isoquinolines

Pyridine, quinoline, and isoquinoline, and some of their methyl derivatives are all commercially available. Simple substituted derivatives can be made by the processes discussed in Sections 25:10 and 25:11. The most interesting derivatives (because of their physiological activity as medicinals) are, however, often substituted in such a manner that they cannot be directly prepared from the basic ring systems. In these cases, syntheses which involve formation of the heterocyclic ring are used. There are many methods for the formation of heterocyclic rings, but only a few examples will be given here.

Pyridines are most often prepared by the *Hantzsch synthesis,* based on β-dicarbonyl compounds such as ethyl acetoacetate (eq. 25.37). The synthesis involves the formation of a 1,4-dihydropyridine which is then oxidized in the same flask by adding nitric acid after the condensations have taken place.

(25.37) Reaction scheme: CH_3CHO + $EtOOCCH_2C(=O)CH_3$ + $CH_3C(=O)CH_2COOEt$ + NH_3 → $EtOOCCH{=}C(CH_3)NH_2$ + $CH_3CH{=}C(COOEt)C(=O)CH_3$ → A 1,4-dihydropyridine → (HNO_3) → diethyl 2,4,6-trimethylpyridine-3,5-dicarboxylate

A 1,4-dihydropyridine

A number of biologically active alkaloids, such as morphine (Fig. 21–3, p. 828), have structures based upon a 1-benzyltetrahydroisoquinoline skeleton. Plants cannot always provide sufficient amounts of these alkaloids for medicinal use, and furthermore the natural alkaloid may produce undesirable side effects. In order, therefore, to increase the availability of natural alkaloids and/or to modify their physiological effects through structural changes, many syntheses of benzyltetrahydroisoquinolines have been devised.

One route to 1-benzyltetrahydroisoquinolines is via the *Pictet–Spengler* method (eq. 25.38), in which a 2-phenylethylamine reacts with a phenylacetaldehyde.

(25.38) 2-(3,4-dimethoxyphenyl)ethylamine (NH_2) + (3,4-dimethoxyphenyl)acetaldehyde ($HC{=}O$) —H^+→ iminium ion ($\overset{+}{N}H{=}CH$) + H_2O → cyclized intermediate ($Me\overset{+}{O}{=}$, H) → H^+ + A 1,2,3,4-Tetrahydro-1-benzylisoquinoline

A 1,2,3,4-Tetrahydro-1-benzylisoquinoline

This reaction is interesting because it parallels the plant biosynthetic pathway. This was one of the first laboratory syntheses of a complex natural product to be carried out under "physiological conditions." Simply mixing the two starting materials in a buffered solution at pH 7 results in product formation, although acid catalysis (as shown in eq. 25.38) does improve the yield. This synthesis is another instance where a laboratory process has already been anticipated in nature.

Problems

25–16 The second step of Equation 25.38 (the ring closure) is a variant of an electrophilic substitution reaction. Draw a resonance structure of the iminium ion which makes this clear.

25–17 Equation 2 gives the overall stoichiometry for the commercial synthesis of 2-methylpyridine (*2-picoline*). Suggest the structures of some reasonable intermediates in this reaction, which probably involves aldol condensations (Chap. 18) and conjugate additions (Chap. 19). The 2-picoline is the starting material for the manufacture of the important monomer 2-vinylpyridine.

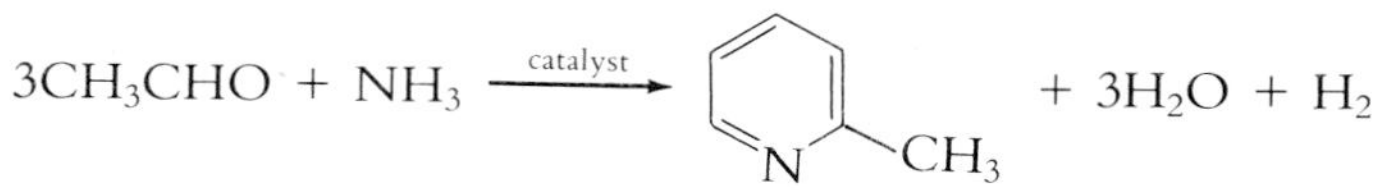

$$3CH_3CHO + NH_3 \xrightarrow{\text{catalyst}} \text{(2-methylpyridine, ring N with } CH_3\text{)} + 3H_2O + H_2 \quad (2)$$

25:13 The Pyridoxal Phosphate Coenzyme: Reaction Catalysis by a Vitamin

By now you are familiar with several examples of the pyridine ring's biological versatility; for instance the chemistry of the pyridine derivative NAD^+/NADH as a coenzyme in biological reductions and oxidations (eqs. 11.63 to 11.65, pp. 421–422). Another example is provided by pyridoxal phosphate (PLP), a coenzyme derived from pyridoxine (vitamin B_6). The terms coenzyme and vitamin are often used interchangeably.

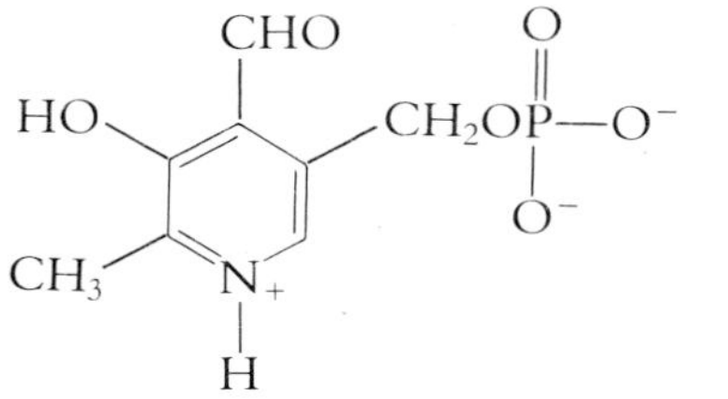

(Structures: pyridinium ring with CHO, HO, CH_3, $CH_2OP(=O)(O^-)$—O^-, N^+—H; pyridine ring with CH_2OH, HO, CH_2OH, CH_3, N)

Pyridoxal phosphate (PLP) 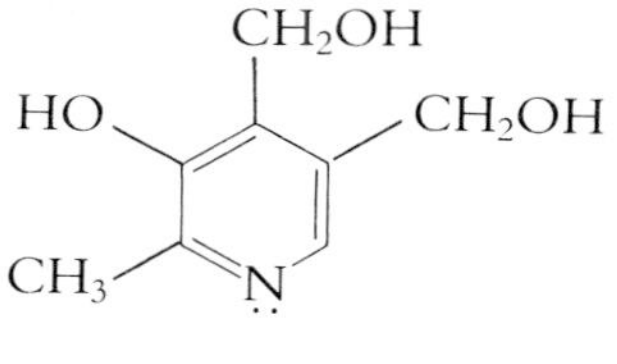   Pyridoxine

The vitamin B_6 group includes pyridoxal phosphate, pyridoxine, and pyridoxamine (where the CH_2OH is replaced by CH_2NH_2). These compounds or their derivatives are found in many foods and hence a vitamin B_6 deficiency is rarely found among people who have an adequate supply of food. Vitamin B_6 catalysts are among the most versatile in nature and their reactions are fundamental to nitrogen metabolism in the body. For example, pyridoxal phosphate is a key catalyst in amino acid biosynthesis. Equations

25.39 to 25.41 show schematically the main steps in the formation of an amino acid. Ammonia reacts with α-ketoglutaric acid (a product of the

$$\text{(25.39)}\quad N_2 \rightarrow NO_3^- \rightarrow NO_2^- \rightarrow NH_3 \qquad \text{Nitrogen fixation}$$

$$\text{(25.40)}\quad NH_3 + HOOCCH_2CH_2\underset{\underset{O}{\|}}{C}COOH \rightarrow$$

α-Ketoglutaric acid

$$HOOCCH_2CH_2\underset{\underset{NH_2}{|}}{C}HCOOH \qquad \text{Reductive amination (sec. 21:7)}$$

Glutamic acid

$$\text{(25.41)}\quad HOOCCH_2CH_2\underset{\underset{NH_2}{|}}{C}HCOOH + R\underset{\underset{O}{\|}}{C}COOH \xrightarrow{PLP}$$

A keto acid

$$HOOCCH_2CH_2\underset{\underset{O}{\|}}{C}COOH + R\underset{\underset{NH_2}{|}}{C}HCOOH \qquad \text{Transamination}$$

An amino acid

tricarboxylic acid cycle, sec. 19:12) to form glutamic acid (eq. 25.40). Glutamic acid can then transfer its amino group to any α-ketocarboxylic acid to produce a different amino acid (eq. 25.41). This last step is called **transamination** and is the step requiring pyridoxal phosphate. A detailed examination of the transamination reaction will show how nature uses pyridine heterocycle reactivity. In the reactions that follow, be certain to focus your attention on the reactive sites of the molecules. This will make these seemingly complex reactions much more readily understood.

The transamination reaction (eq. 25.41) begins with imine formation between the glutamic acid and the coenzyme (eq. 25.42). An imine is proposed because the enzymatic reaction is inhibited by $NaBH_4$, which reduces —C=N— bonds. The internal hydrogen-bonding in the product appar-

(25.42)

$$HOOC(CH_2)_2\underset{\underset{NH_2}{|}}{C}HCOOH + \text{PLP} \rightarrow \text{Imine intermediate} + H_2O$$

PLP: CHO, HO, $CH_2OPO_3^{2-}$, CH_3, N^+–H (pyridinium ring)

Imine intermediate: $HOOC(CH_2)_2CHCO_2H$, N, CH, H···O, $CH_2OPO_3^{2-}$, CH_3, N^+–H

ently exerts an important effect on the stability and reactivity of this imine. The imine may lose a proton from the α-position of the amino acid portion (step 1, eq. 25.43).

(25.43)

A dihydropyridine

The product, a dihydropyridine, is then reprotonated on the exocyclic carbon atom to yield a new imine (step 2, eq. 25.43).

Hydrolysis of this new imine (eq. 25.44) yields α-ketoglutaric acid, while pyridoxal phosphate has been reduced to pyridoxamine phosphate. If the pyridoxamine phosphate reacts with a different α-keto acid (as in eq. 25.41), then by a reversal of Equations 25.44 to 25.42 the keto acid will be

(25.44)

Pyridoxamine phosphate

α-Ketoglutaric acid

converted to a new α-amino acid and pyridoxal phosphate will be regenerated. This completes the transamination reaction and the biosynthesis of a new amino acid.

The catalytic properties of both pyridoxal and NAD^+/NADH are largely determined by the chemical properties of the pyridine. Both of these coenzymes can exist in an oxidized pyridinium ion form (**11**) and a reduced dihydropyridine form (**12**). The catalytic mechanisms require that neither form be markedly more stable than the other, which is exactly the case. Reduction of the pyridinium ion is favored by neutralization of the positive charge on nitrogen, so the pyridinium ion is an electron acceptor—an electron "sink" (**11**). The electron pair can come from a C—H bond (PLP reduction) or from an attacking hydride ion (NAD^+ reduction).

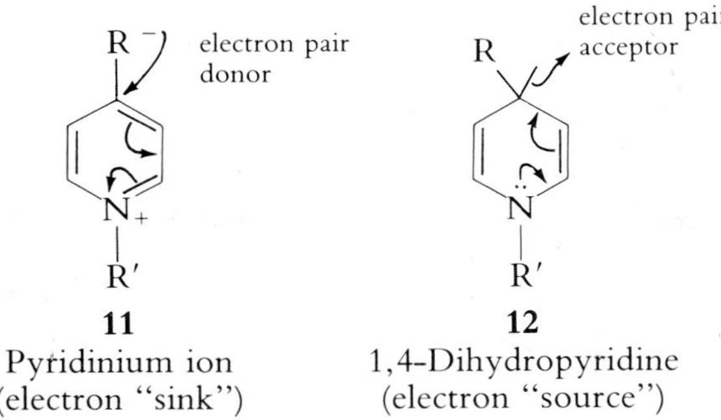

11 Pyridinium ion (electron "sink")

12 1,4-Dihydropyridine (electron "source")

On the other hand, oxidation of the dihydropyridine form is favored by the gain in resonance energy. Thus, the dihydropyridine easily donates an electron pair—it acts as an electron "source" (**12**). For this reason, the dihydropyridine intermediate in PLP catalysis is easily protonated on the exocyclic carbon atom (step 2, eq. 25.43). The balance between oxidizing and reducing forces allows PLP and NAD^+/NADH to convert rapidly between pyridinium and dihydropyridine forms.

Problem

25–18 In the presence of PLP and a transaminase enzyme, glycine and pyruvic acid will react to form alanine and glyoxylic acid (eq. 3). In the presence of $NaBH_4$, the initial product from reaction of glycine with PLP is converted to a compound that cannot continue the reaction sequence. What is the structure of this compound?

$$\underset{\text{Glycine}}{\underset{NH_2}{\underset{|}{CH_2}}COOH} + \underset{\text{Pyruvic acid}}{CH_3\underset{O}{\underset{\|}{C}}COOH} \xrightarrow[\text{enzyme}]{\text{PLP}} \underset{\text{Glyoxylic acid}}{OHCCOOH} + \underset{\text{Alanine}}{CH_3\underset{NH_2}{\underset{|}{C}}HCOOH} \qquad (3)$$

25:14 Purines and Pyrimidines: Their Roles in Genetic Information Storage and in Fighting Disease

There are a great many heterocyclic systems which contain more than one heteroatom. Some are remarkable for their special properties rather than being of general chemical interest. For example, pyrazines (1,4-diazabenzenes) in trace amounts account for many flavor properties of foods. The presence of 2-methoxy-3-ethylpyrazine at a level of 10 ppm in liquids gives them the flavor of raw potatoes. Even more potent is 2-methoxy-3-isobutylpyrazine, which can be detected at the 0.002 ppb (parts per billion) level and has the flavor of green bell peppers.

(Pyrazine ring: N, N; substituents CH_2CH_3, OCH_3)
2-Methoxy-3-ethylpyrazine
(potato flavor)

(Pyrazine ring: N, N; substituents $CH_2CH(CH_3)_2$, OCH_3)
2-Methoxy-3-isobutylpyrazine
(bell pepper flavor)

Of all the polyazaheterocycles (two or more N atoms in the ring), two of the most important are pyrimidine and purine. The latter consists of a pyrimidine ring and an imidazole ring fused together. (A common purine is the natural stimulant caffeine [see Box 25–1].) Of greater consequence are the two series of bases derived from pyrimidine and purine that comprise the

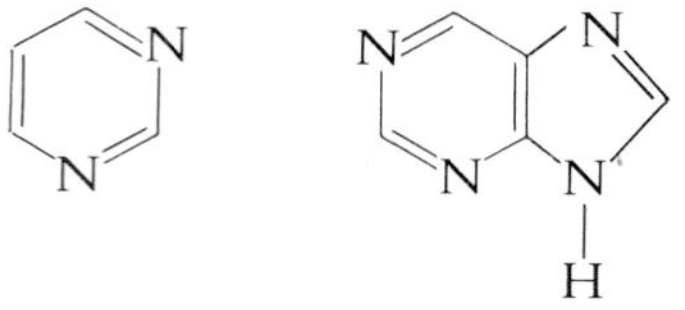

Pyrimidine   Purine

vital components of DNA and RNA. DNA contains the purines adenine and guanine and the pyrimidines thymine, cytosine, and 5-methylcytosine. In RNA, adenine, guanine, cytosine, and uracil are found.

Pyrimidines: Cytosine   Uracil   Thymine   5-Methylcytosine

Purines: Adenine   Guanine

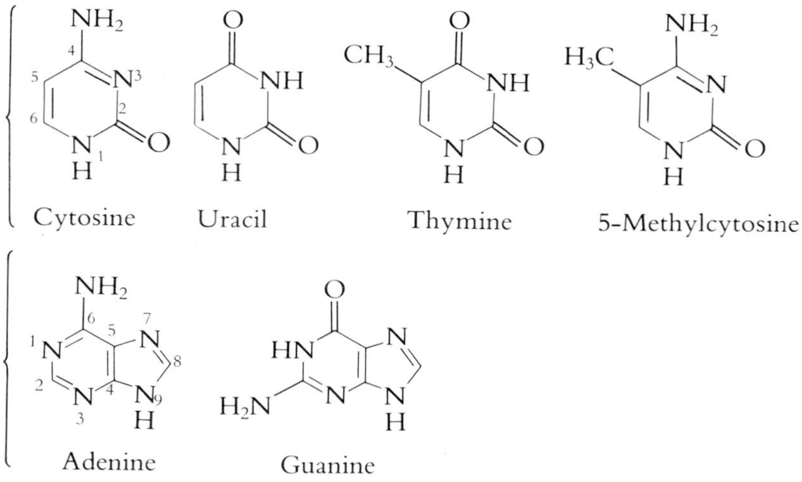

In Section 20:9 we showed how each of these heterocyclic bases is attached to ribose or deoxyribose to form nucleosides. Addition of phosphate groups converts nucleosides to nucleotides. These are joined together (as shown in Figure 20–9, p. 789) to form a long polynucleotide strand of RNA or DNA. There are many kinds of DNA and RNA polynucleotides that differ in their sequence or arrangement of heterocyclic base components.

Two DNA chains intertwine to form a double helix held together by hydrogen bonds between bases on each of the chains (Fig. 25–4a). The proper intermeshing of hydrogen bonds depends upon the specific structure of each heterocycle. For example, adenine binds *only* to thymine and guanidine binds *only* to cytosine, because this arrangement maximizes the number and strength of hydrogen bonds between each base pair (Fig. 25–4b).

Note further that only the correct tautomeric form of each base can effectively participate in hydrogen-bonding. The structure given above for thymine represents only one tautomer, the lactam form, which is the pre-

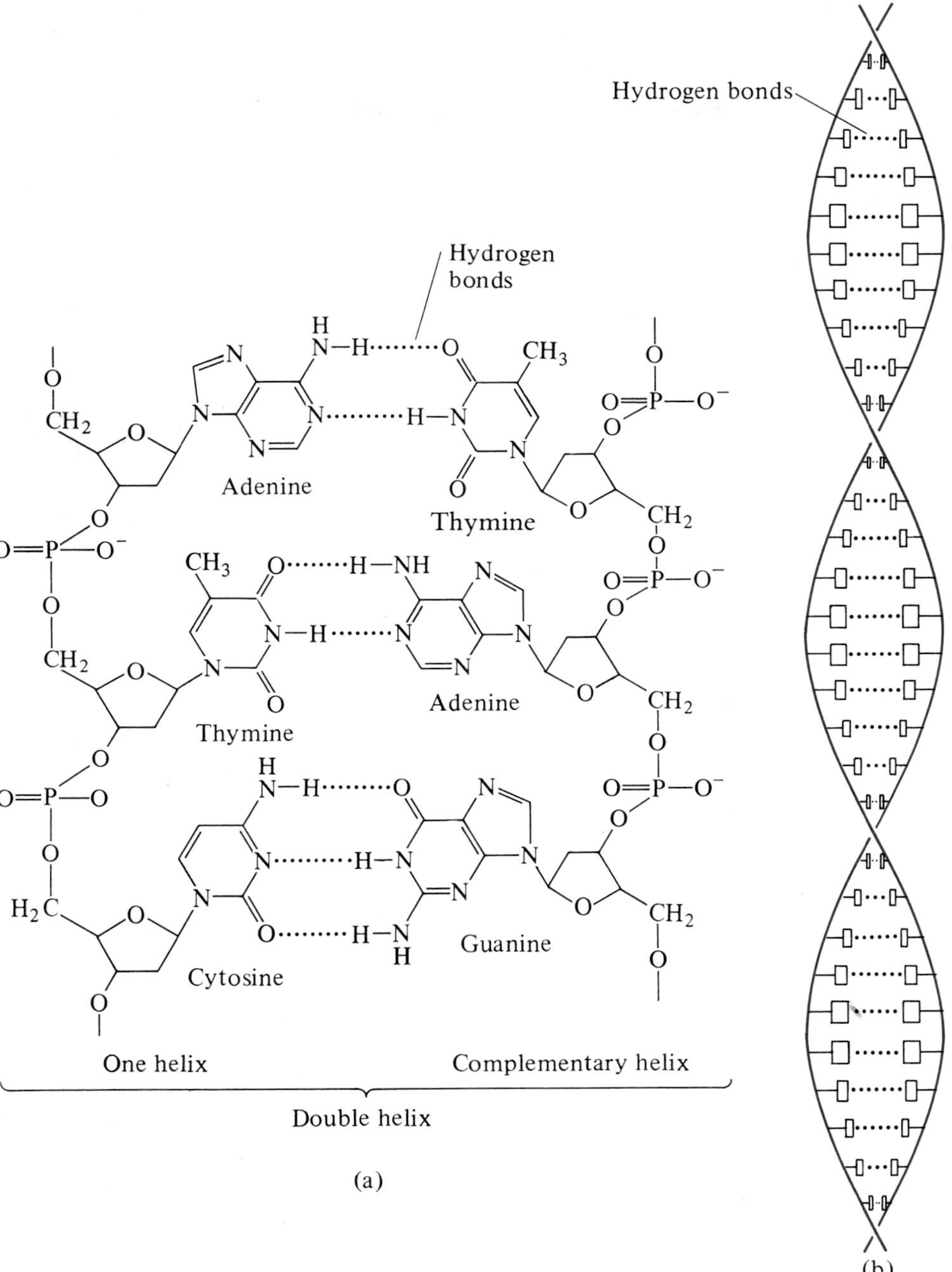

Figure 25–4. *Base pairing (a), and the structure of DNA (b).*

dominant form in DNA. The equilibria of Equation 25.45 show, however, that thymine could exist in two other tautomeric forms, the lactim and the double lactim. The lactam form is the predominant one in solution at pH 7 and in the solid state.

$$\text{(25.45)}\quad \underset{\text{Lactam}}{\text{HN}-\text{C}(=\text{O})-\text{C}(\text{CH}_3)=\text{CH}-\text{NH}-\text{C}(=\text{O})-} \rightleftarrows \underset{\text{Lactim}}{\text{N}=\text{C}(\text{OH})-\text{C}(\text{CH}_3)=\text{CH}-\text{NH}-\text{C}(=\text{O})-} \rightleftarrows \underset{\text{Double lactim}}{\text{N}=\text{C}(\text{OH})-\text{C}(\text{CH}_3)=\text{CH}-\text{NH}-\text{C}(\text{OH})=}$$

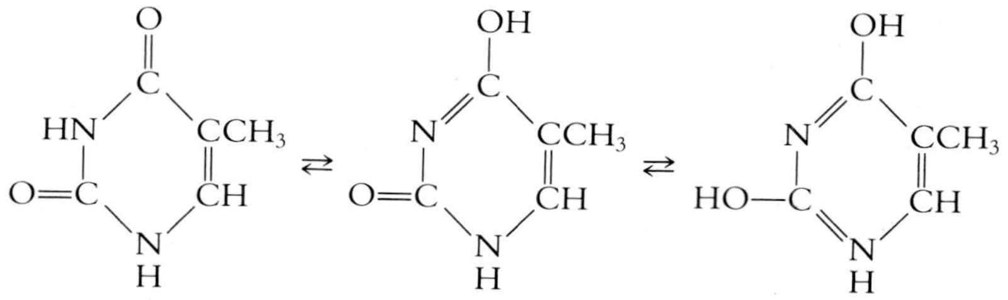

The selective hydrogen-bonding between base pairs—such as adenine with thymine, guanine with cytosine—is the chemical basis for the transmission and utilization of genetic information. From this simple physiochemical effect enormous consequences flow. Consider, for example, what happens when a cell divides into two new cells (**mitosis**). Each DNA helix *of the original cell* unravels into its two separate strands, one of which goes into each daughter cell. Thus, the new cells initially have only half their proper complement of DNA. They remedy this by synthesizing a new "partner" for each single strand of DNA, thus reconstituting the original double helix.

How does the cell insure that the new partner has exactly the right sequence of bases? It uses the original single strand of DNA as a blueprint, or template. Suppose, for example, that daughter cell A received the left hand DNA strand of Figure 25–4. To make a new partner for this strand a thymine base, in the form of thymidylic acid, is "fitted" to each adenine by way of hydrogen bonds. Similarly, to each cytosine on the original strand a molecule of guanidylic acid is fitted, and so on. Finally, the individual nucleotides of the new strand are all stitched together by the enzyme *DNA polymerase,* thus completing formation of the complementary DNA strand which, together with the original, exactly reproduces the original double helix of the parent cell.

A cell's DNA molecules are found in its nucleus, bound into giant macromolecules called **chromosomes**. Their function is to regulate the synthesis of new cell proteins needed as enzymes, membrane components, and so on. *It is the sequence of bases on the DNA strands that determines the sequence of amino acids in each protein.* A sequence of three bases on the DNA strand, called a **codon**, is the blueprint for a single amino acid. For example, the sequence uracil-uracil-guanine (UUG) is a codon for phenylalanine, while G-G-C codes for glycine. A portion of the DNA that reads -U-U-G-G-G-C- would therefore be responsible for directing the synthesis of a portion of a protein molecule with sequence -Phe-Gly-.

In order to transmit its sequential information, each DNA helix must temporarily unravel, and a single strand is used as a template for the synthesis of a molecule of **messenger RNA**. As before, the proper base sequence in the messenger RNA is assured by the preferential hydrogen-bonding between complementary bases. Finally, the messenger RNA interacts with **transfer RNA**, which carries the amino acid necessary to form a peptide. Once again there is an interaction through hydrogen bonds to bring about the ultimate goal, synthesis of a specific protein.

Since the nuclear DNA contains the blueprints for all the cell functions, any defects in its structure will interfere with cell growth and reproduction. Attention has therefore been directed toward attempts to alter the DNA molecules of tumor cells (see, for instance, Box 6–1). One aspect of this search for new antitumor agents is focused on the synthesis of **antimetabolites**. Antimetabolites are structurally similar to normal metabolites, but they incorporate a small structural change. When they are taken up by the cell, antimetabolites interfere with normal cell metabolism. In the antitumor field, two clinically useful antimetabolites are 6-mercaptopurine and 5-fluorouracil (also see sec. 25:15). These compounds replace adenine and uracil, respectively, in DNA and hence retard cell growth. The basis of the chemotherapeutic effect is that the antimetabolites will have a more important effect on the fast growing cancer cells than on normal cells, and will therefore destroy the cancer cells before appreciably damaging the normal cells.

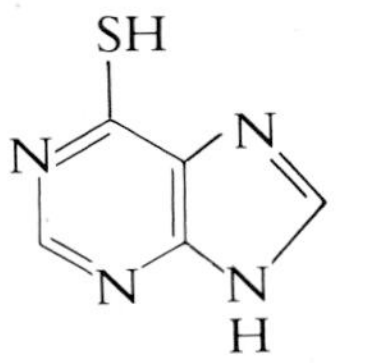

6-Mercaptopurine 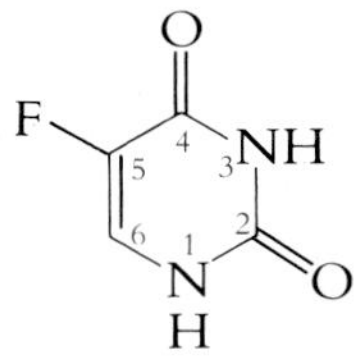   5-Fluorouracil

Problems

25–19 Barbituric acid (**V**, R = H) and 5,5-disubstituted barbiturates (**V**, R = alkyl or aryl groups), which are pyrimidine derivatives, are acids and will react with sodium ethoxide (Prob. 19–2, p. 716). The pK_a of barbituric acid is

Box 25–1. Caffeine

Caffeine, the stimulatory alkaloid in coffee, has been called "the most widely used psychoactive substance on earth," but its mode of action has not been known until recently. New research indicates that caffeine's structural relationship with the nucleoside adenosine may be the cause.

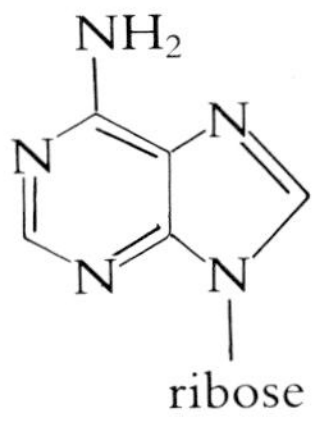

Adenosine has been found to depress nerve-cell firing in the brain. It does so by inhibiting the release of neurotransmitters (Box 21–1) which carry nerve impulses between neurons. In order to do this, adenosine must bind to receptors on neuronal membranes. Caffeine is able to bind to these same receptors and when it does, it prevents the attachment of adenosine. Caffeine does not, however, depress nerve-cell firing as the adenosine would have done. Hence, nerve action continues undiminished, leading to the stimulatory effect that most people experience after drinking beverages containing caffeine.

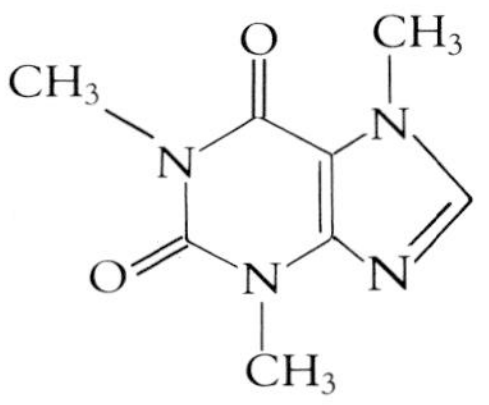

Adenosine   Caffeine

4.0, while the pK_as for the 5,5-disubstituted barbiturates lie in the range 7.4 to 8.4. Write reactions showing the ionization of barbituric acid and 5,5-diethylbarbituric acid. Note the large difference in pK_a. With the aid of resonance structures, suggest an explanation for this difference.

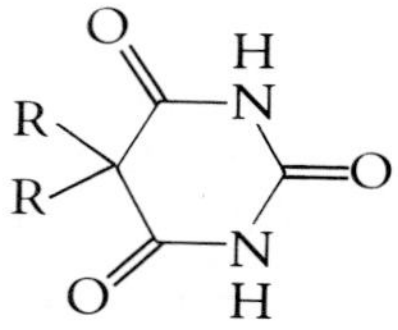

V

25–20 2-Pyridinol is in equilibrium with its tautomer 2-pyridone (eq. 4). The position of the equilibrium depends strongly on the medium. In the gas phase 2-pyridinol predominates, while in solution 2-pyridone is the major tautomer. The solution equilibrium constant is also affected by the solvent.

a. Indicate why 2-pyridinol is the favored tautomer in the gas phase.
b. In solution, molecules of 2-pyridone form hydrogen-bonded dimers, which tips the equilibrium in its favor. Suggest a structure for these dimers, which are present even in dilute solution.

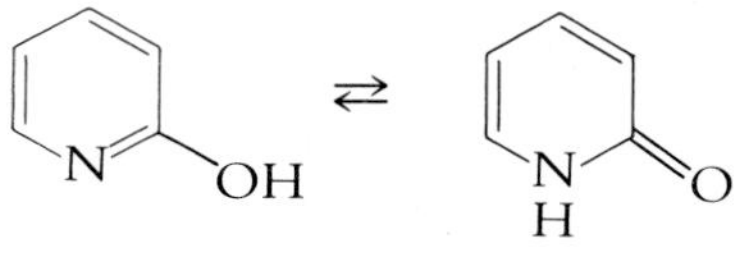

$$\text{2-Pyridinol} \rightleftarrows \text{2-Pyridone} \quad (4)$$

25:15 Nucleoside Synthesis

Because nucleosides and nucleotides are fundamental building blocks of DNA and RNA, their synthetic preparation has been an important area of research. In the search for antitumor agents, for example, nucleoside analogs, which have a chemical structure slightly different from natural nucleosides, have been prepared. When this "foreign" nucleoside is incorporated into the DNA of a cancer cell, it could interfere with metabolism and be fatal to the cell. Viral nucleic acids could be attacked in the same way, thus stimulating a similar search for antiviral nucleoside analogs.

As mentioned in the last section, one compound recognized as an effective chemotherapeutic agent is 5-fluorouracil, an analog of uracil. Our cells use uracil as a precursor to thymine, which is a component of DNA. The substitution of one atom—an F for an H at the 5-position—makes the uracil derivative an ineffective substrate for thymine synthesis. Thus the cell is derived of thymine and dies.

To prepare a nucleoside analog, one approach is to first modify the purine or pyrimidine base and then attach it to the sugar. For example, 5-fluorouracil (sec. 25:14) can be prepared from uracil, as shown in Equation 25.46. In this reaction, CF_3OF (fluoroxytrifluoromethane) acts as a source of

(25.46)

Uracil $\xrightarrow[CF_3COOH,\ H_2O]{CF_3OF}$ 5-Fluorouracil (85%)

electrophilic fluorine because of the high polarity of the RO—F bond: $\overset{\delta-}{F_3CO}$—$\overset{\delta+}{F}$. The second step is to react the fluoropyrimidine with a protected sugar. This reaction is facilitated by first converting the pyrimidine base to a protected enol (eq. 25.47). In the protected enol, the nitrogen nonbonded pair is much more nucleophilic in the subsequent reaction (eq. 25.48).

(25.47)

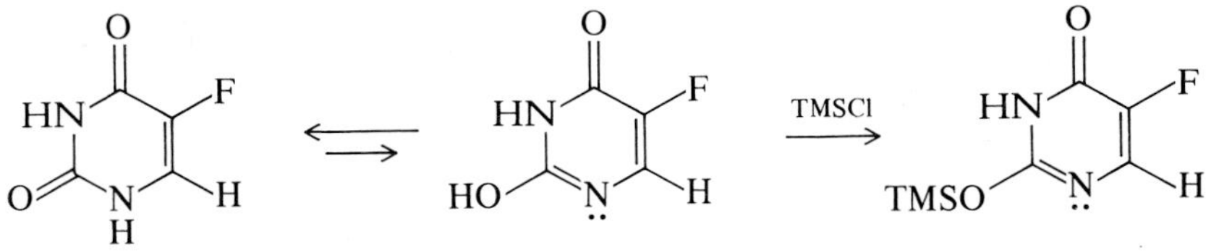

TMSCl

(25.48)

SnCl4

H_3O^+, Δ

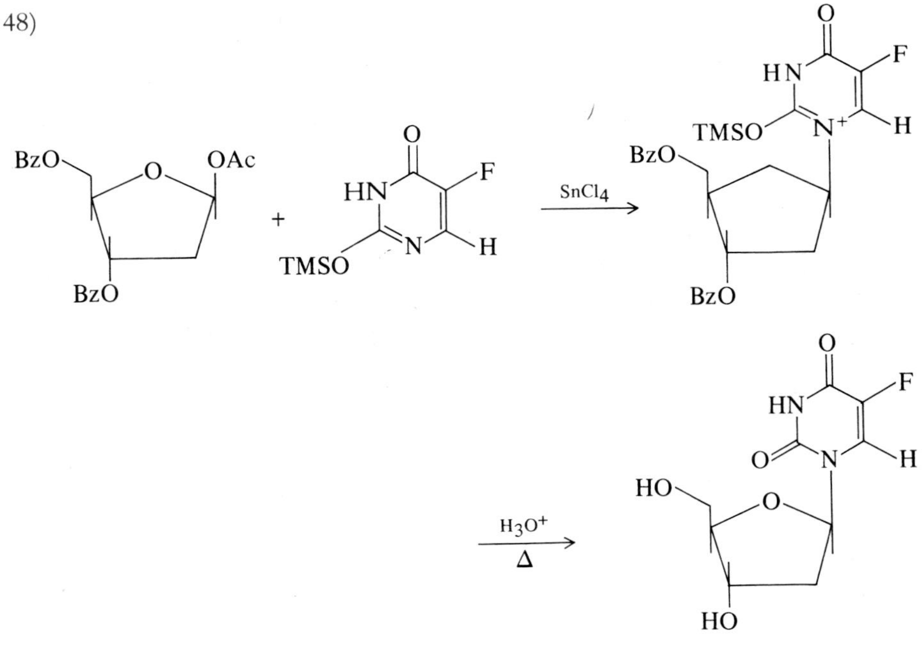

Lewis acid catalysts such as tin(IV) chloride are often used to effect coupling, since they provide the sugar carbocation (eq. 25.49) that reacts with the pyrimidine base.

(25.49)

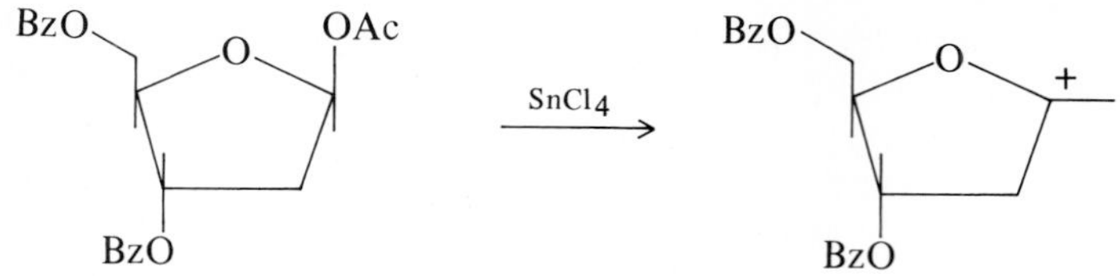

BzO OAc → ($SnCl_4$) BzO +

BzO BzO

Problem

25–21 Give a mechanism for the electrophilic substitution reaction of Equation 25.46. Show what intermediate would have been formed from attack at C-6, and explain why attack is favored at C-5 rather than C-6.

Nucleoside analogs of purine bases can be made by way of similar reactions. Equation 25.50 provides an example.

(25.50)

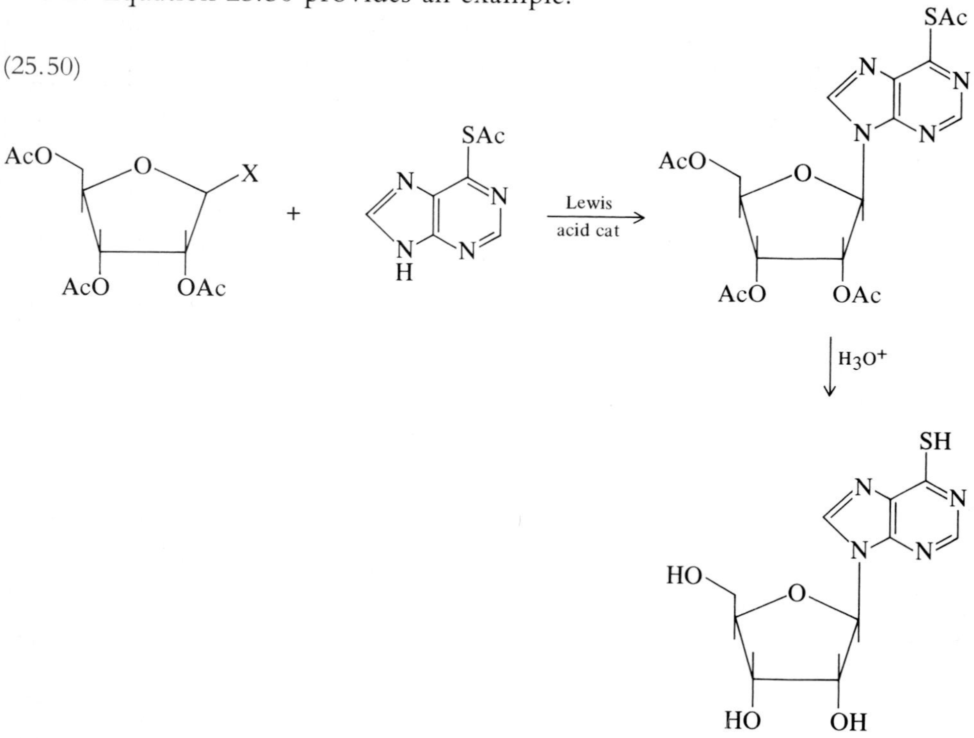

AcO X AcO OAc + SAc N N N N H → (Lewis acid cat) SAc N N N N AcO AcO OAc

↓ H_3O^+

SH N N N N HO HO OH

25:16 Genetic Engineering and the Sequencing and Synthesis of DNA

You have just seen several examples in which nucleosides are synthesized for use as agents in cancer chemotherapy. The rapidly growing field of **genetic engineering** depends on our ability to determine the nucleoside sequence of a desirable gene, synthesize that sequence, and then introduce the synthetic gene into an organism in which we wish to have those characteristics which derive from the gene. The field of genetic engineering has become of major importance because of three scientific advances of the 1970s. The first was biological and involved the development of methods for introducing non-bacterial genes into bacteria. The other two advances were chemical and involved (1) the use of enzymatic and chemical reactions to recognize and characterize unique DNA sequences, and (2) development of automatic methods for the synthesis of those DNA sequences (the gene). We will briefly outline some of the chemical methods used in genetic engineering.

The sequencing of DNA is conceptually similar to the sequencing of peptides and proteins (sec. 22:9). It is presently easier than peptide sequencing and of more potential benefit. Either chemical or enzymatic methods are used to partially cleave the DNA and hence generate a map of overlapping DNA segments (similar to the amino acid residue map shown in Fig. 22–5, p. 869). The chemical method uses a reaction which modifies only a specific base in the sequence and then allows cleavage of the DNA chain at those positions. It has been found, for example, that methylation of DNA with dimethyl sulfate can be made to occur specifically at guanine residues (step 1 of Fig. 25–5). The guanine residue is now susceptible to ring opening with hydroxide ion and cleavage with piperidine (steps 2 and 3 of Fig. 25–5). The final step (4) involves further reaction with OH^- to cleave the DNA into fragments terminating with phosphate groups from the 5′ and 3′ posi-

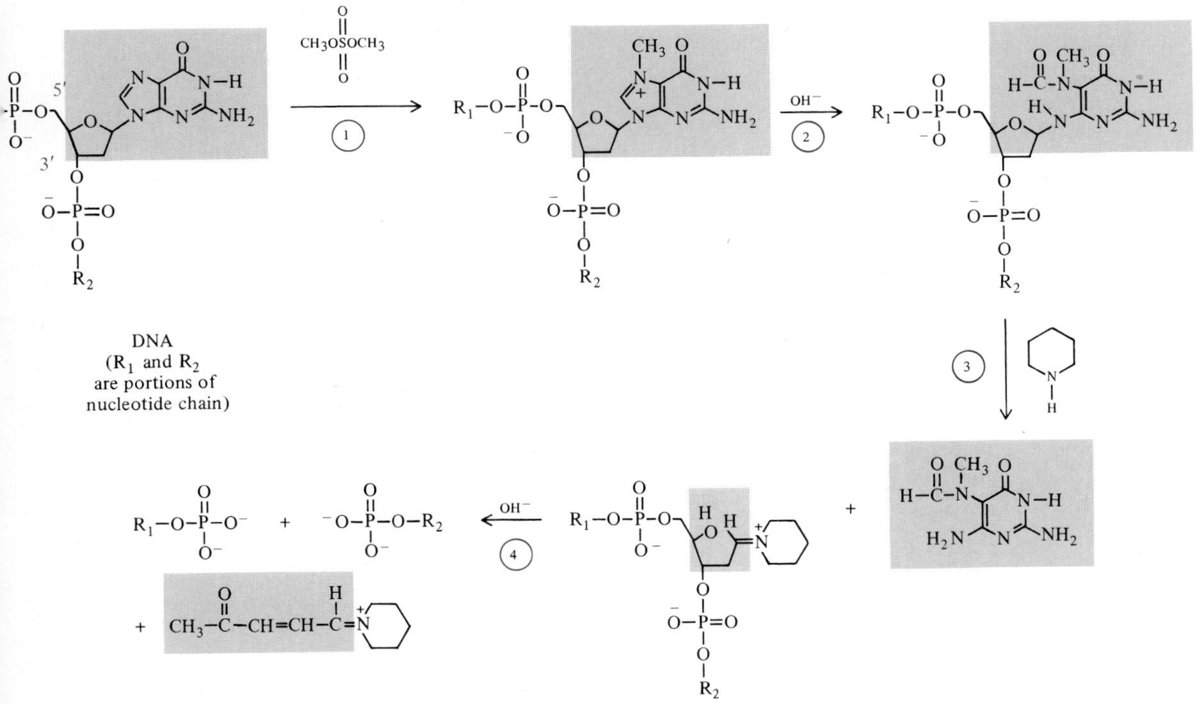

Figure 25–5. *Cleavage of a DNA chain by reaction of dimethyl sulfate and piperidine with a guanosine residue.*

tions. Reagents specific for other bases in the DNA chain are also known. Hydrazine, NH_2NH_2, reacts specifically with thymine and cytosine residues. After the various cleavage reactions have been run, fragments are separated and identified by electrophoresis.

Once the sequence of nucleosides in the DNA or DNA fragment are known, the gene can be synthesized. The synthetic material is then injected into the host bacterium, plant, or animal and the host takes over replication of the injected sequence. DNA synthesis uses ideas similar to those developed for the synthesis of proteins: protecting groups, coupling reactions, and solid phase methods (secs. 22:11 to 22:13).

In order to prepare a fragment of DNA, it is necessary to have the four nucleosides deoxyadenosine, deoxyguanosine, deoxycytidine, and deoxythymidine. Figure 25–5 shows the four bases and their attached deoxyribose sugars as they are present in sections of DNA. The synthesis of a DNA fragment involves the proper formation of the phosphate ester bonds that bind together individual nucleosides in the correct sequence.

The nucleosides themselves are readily available from the degradation of naturally occurring DNA. For example, DNA is obtained from sources such as salmon cells and is then hydrolyzed. The nucleosides are separated and offered for sale by biochemical supply firms.

One method of DNA fragment synthesis (eq. 25.51) starts with a nucleoside blocked at the 5′ position and, using DCCD (sec. 22:11), condenses that with a nucleotide to form a dinucleotide.

(25.51) Ar_3CO–(5′) [deoxyribose, O, Base; 3′-OH] + HO—P(=O)(OH)—O–(5′) [deoxyribose, O, Base; 3′-OAc] $\xrightarrow{\text{DCCD}}$ Ar_3CO– [deoxyribose, O, Base]–O—P(=O)(HO)—O– [deoxyribose, O, Base; OAc]

Dinucleotide

The acyl protecting group is removed with base and a new nucleotide is coupled (eq. 25.52). At the end of the synthesis, the blocking group at the 5′ position can be removed with dilute acetic acid. This synthetic method has been called the **diester approach**. A modification called the **triester approach** involves condensation of two nucleotides (phosphate esters at 5′ and 3′) with a different coupling reagent.

(25.52) Ar_3CO— O Base

HO—P=O

O— O Base

OH

+ HO—P=O (OH)

O— O Base

OAc

DCCD →

Ar_3C—O— 5′ O Base

HO—P=O

O— O Base

HO—P=O

O— O Base

3′

OAc

Trinucleotide

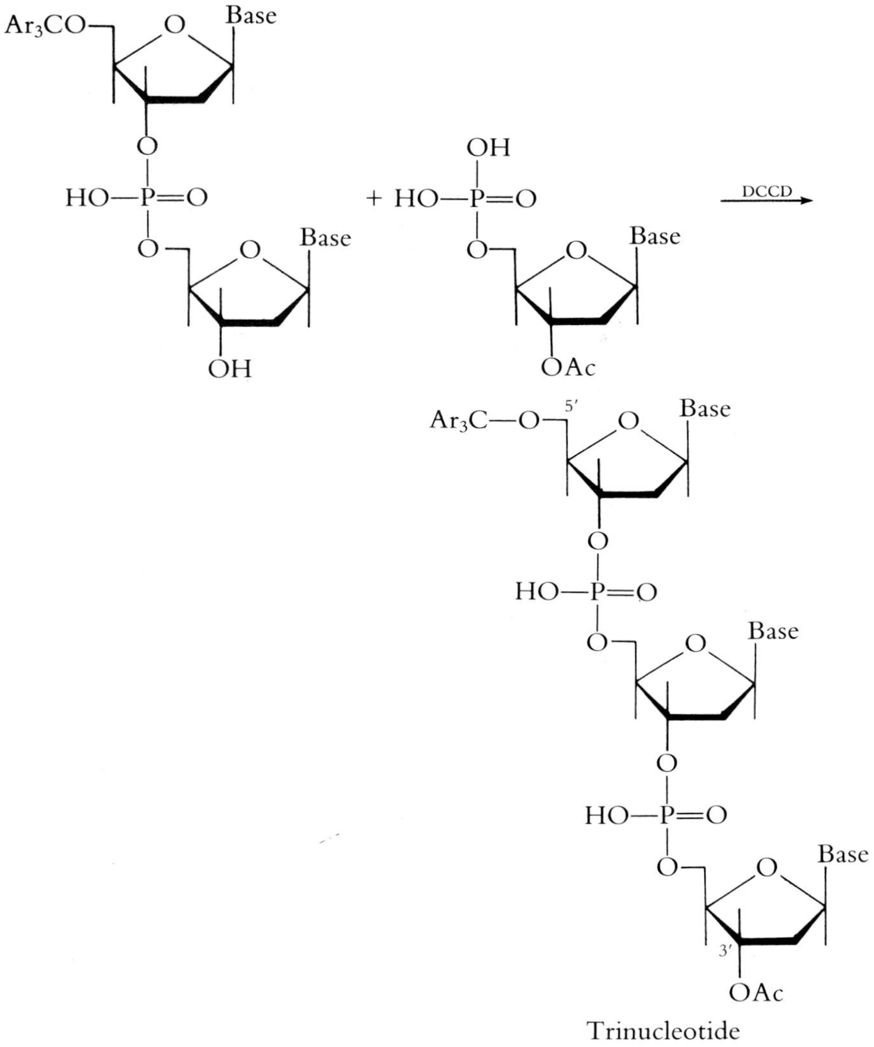

A third method of synthesis is called the **phosphite coupling approach**, since the blocked nucleoside reacts with a phosphite (eq. 25.53). This method starts with the 3′ OH attached to the carboxyl group of solid silica gel. Both the phosphite approach and a solid phase analog of the triester approach have been automated and developed for use in commercial DNA/RNA synthesizers.

(25.53)

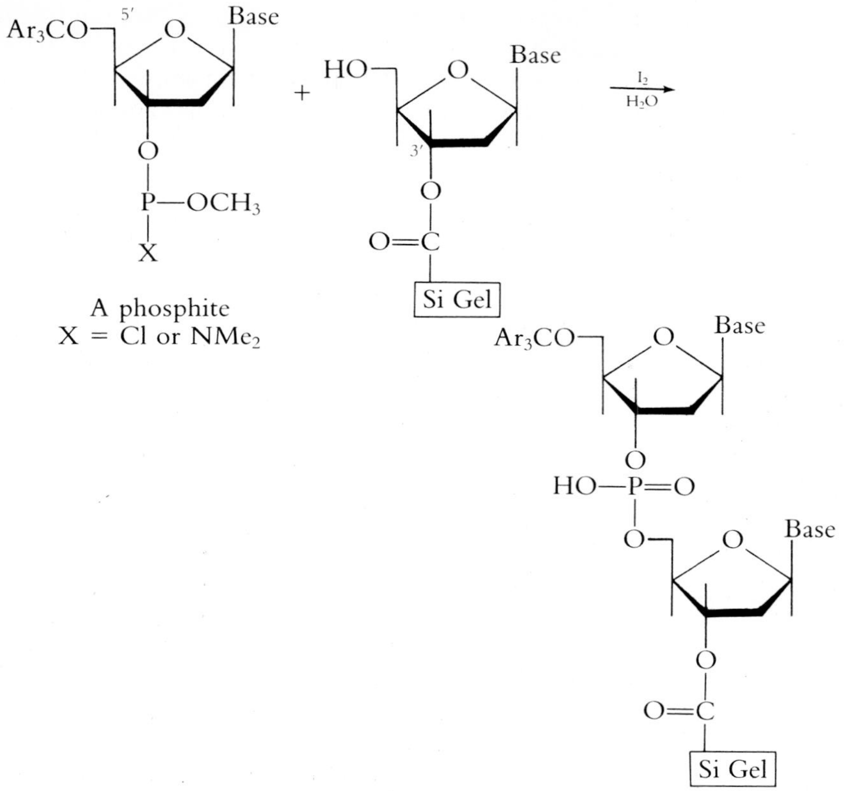

Ar_3CO— 5′ O Base; 3′-O—P(OCH_3)—X

A phosphite
X = Cl or NMe_2

+ HO— O Base; 3′-O—C(=O)—Si Gel

$\xrightarrow[H_2O]{I_2}$

Ar_3CO— O Base; O—P(=O)(OH)—O— O Base; O—C(=O)—Si Gel

Summary

1 Heterocycles are ring compounds in which at least one carbon is replaced by another element. The most common aromatic heterocycles contain one or more nitrogens, oxygens, or sulfurs. There are two fundamental classes of aromatic heterocycles, those containing five-membered and six-membered rings. The five-membered ring compounds are characterized by electron transfer from the ring heteroatom to the carbon atoms. The nitrogen lone pair in pyrrole is part of the aromatic sextet and hence pyrrole is not basic. Imidazole is basic, because one of its lone pairs is not part of the aromatic sextet (sec. 25:1).

2 Pyrrole and furan are very reactive (similar to aniline and phenol) in electrophilic substitution reactions; thiophene is less so. In all three, the α positions are more reactive than the β positions toward halogenation, nitration, and so on (sec. 25:2). Furan is so reactive that its derivatives are often formed by way of an addition–elimination sequence (sec. 25:3).

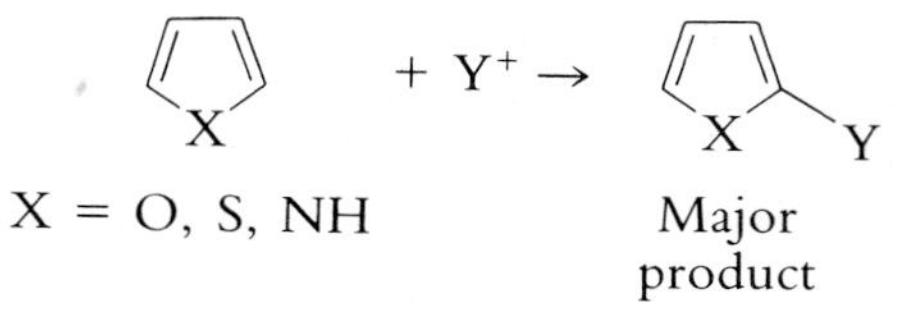

$$\text{(ring with X)} + Y^+ \rightarrow \text{(ring with X, 2-Y)}$$

X = O, S, NH

Major product

Fusing a five-membered heterocycle to a benzene ring alters its reactivity pattern, but not its high susceptibility to electrophilic substitution. Thus, indole undergoes electrophilic substitution only in the pyrrole ring, at C-3. Many 3-substituted indoles are physiologically active and are found in nature. Indoles are synthesized in the laboratory from phenylhydrazones (Fischer synthesis; sec. 25:4).

3 Some five-membered heterocycles are obtained from natural sources. Highly substituted ones are made by allowing appropriate 1,4-diones to react with acids, amines, or phosphorus sulfides (sec. 25:5).

4 Macrocycles in which four pyrrole rings are cyclically condensed, separated by one-carbon bridges, are called porphyrins. Porphyrins are found extensively in nature in metalloporphyrins, where they act as ligands for metal ions. Hemoglobin is an Fe(II) metalloporphyrin; chlorophyll is a Mg metalloporphyrin. Metalloporphyrins act by transporting small molecules and electrons (sec. 25:6).

5 Imidazole plays a vital role in the functioning of proteolytic enzymes such as chymotrypsin, which hydrolyze peptides and proteins. It serves to deprotonate and reprotonate reacting portions of the enzyme and substrate (sec. 25:7).

6 In pyridine and its benzo derivatives (quinoline, isoquinoline), the nitrogen lone pair lies outside the aromatic π-electron system. The nitrogen-containing ring is very sluggish toward electrophilic substitution (sec. 25:8). This is particularly true under acidic conditions when the pyridine is converted to a pyridinium ion. Thus, pyridine undergoes nitration and halogenation only at C-3 (sec. 25:9).

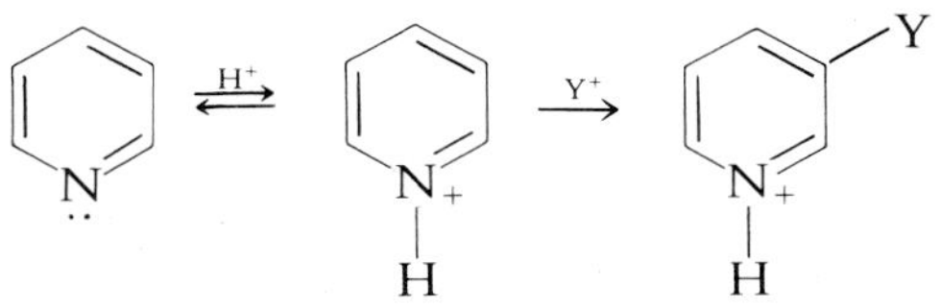

Pyridine *N*-oxide is much more reactive toward electrophiles than is pyridine itself. Furthermore, it reacts at C-2 and C-4. The oxide group may be removed after the substitution is completed (sec. 25:10). Pyridine can be attacked by strong nucleophiles such as $NaNH_2$ and RLi, leading to 2-substituted pyridines (sec. 25:11).

Pyridines can be synthesized from ammonia and dicarbonyl compounds (Hantzsch synthesis). Tetrahydrobenzylisoquinolines are obtained when 2-phenylethylamines react with aldehydes under acidic conditions (Pictet–Spengler synthesis; sec. 25:12).

7 A pyridine derivative, pyridoxal phosphate, is a coenzyme (vitamin) that catalyzes transamination and other reactions in living cells. The pyridoxal phosphate reacts with an amino acid and is converted to pyridoxamine, which then reacts with an α-ketoacid and converts it to a new amino acid. Pyridine-based coenzymes rely on the electron "sink" properties of pyridinium ions and the electron "source" properties of dihydropyridines (sec. 25:13).

Purine and pyrimidine bases are involved, by hydrogen-bond formation, in maintaining the double helix of DNA. These bases occur mainly in the lactam form, which provides molecular sites for the required hydrogen-bonding. The selective hydrogen-bonding between specific base pairs is responsible for the maintenance, transmission, and expression of the genetic code which regulates cellular life (sec. 25:14).

8 Nucleoside analogs have been synthesized as anticancer and antiviral chemotherapeutic agents (sec. 25:15). Genes that govern the creation of desirable cell characteristics can also be synthesized. The first step in this process is to determine the nucleoside sequence of the desired portion of DNA. Once the sequence is known, the gene can be synthesized by a series of reactions whose key step is the formation of the internucleotide phosphate ester bonds. The synthetic gene is then injected into the host species where it becomes part of the host's genetic material (sec. 25:16).

Problems

25–22 Name the following compounds:

a.

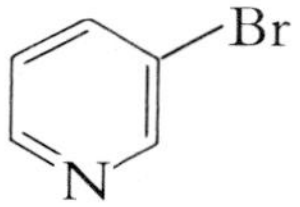

c.

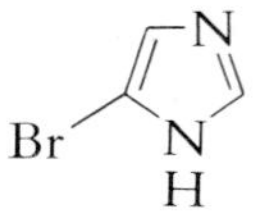

b.

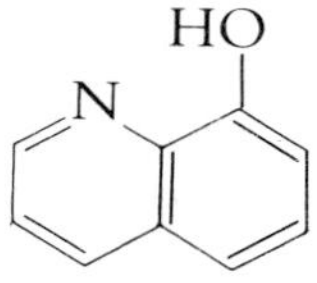

d.

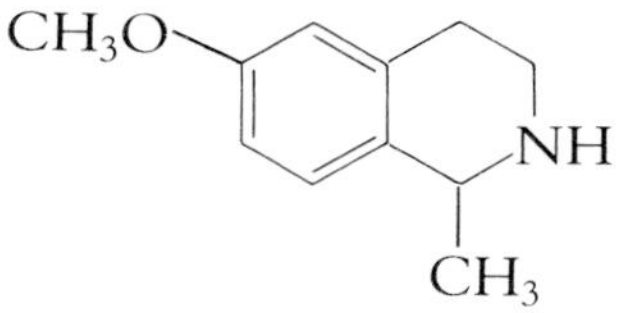

25–23 Give a structure for each of the following compounds:

a. 3-acetylthiophene
b. 1-methylindole
c. 2-ethylbenzofuran
d. 4-methoxythiazole
e. 5-mercaptopyrimidine
f. 2-methyladenine
g. 5-ethoxycytosine
h. 5,6-dihydrouracil
i. 1-methylguanine

25–24 a. There are two tautomeric isomers of pyrrole: 2*H*-pyrrole and 3*H*-pyrrole. Attempts to prepare either of them always result in the isolation of pyrrole instead. Explain.

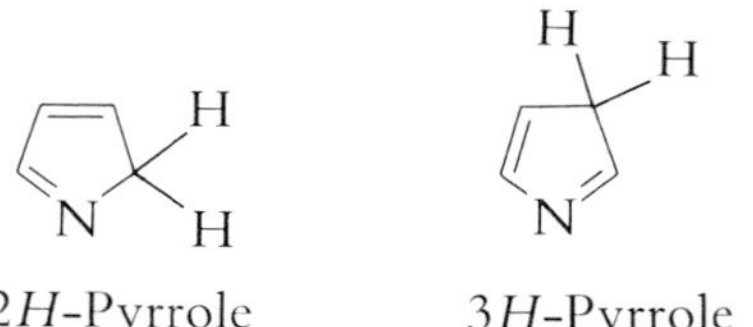

2*H*-Pyrrole 3*H*-Pyrrole

b. Pyrrole is a very weak base. When subjected to strong mineral acids, however, it may be protonated. Nuclear magnetic resonance studies reveal that protonation takes place at C-2 or C-3 rather than at nitrogen. Explain this observation.

25–25 Which compound would you expect to be a stronger base, aniline or 2-aminothiophene? Justify your answer.

25–26 A pyrrole-2-carboxyaldehyde will condense with another pyrrole molecule under acid conditions. The products are **dipyrrylmethenes**, isolated as their salts (**VI**, eq. 5). These dipyrrylmethenes are valuable synthetic intermediates because condensation of appropriately substituted ones can yield porphyrins (sec. 25:7). Suggest a mechanism for the reaction shown in Equation 5. Salts such as **VI** are highly stable. Account for this finding.

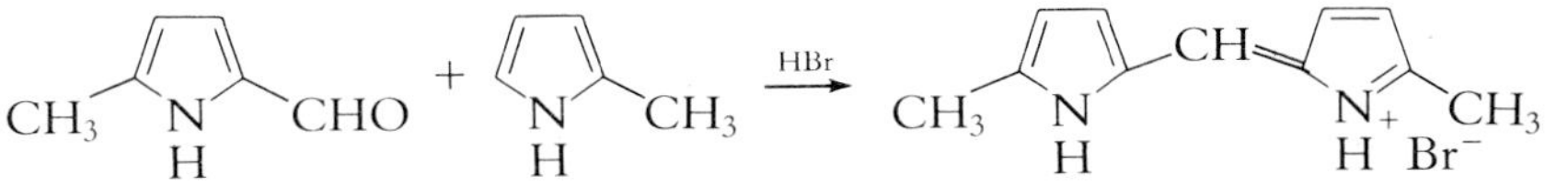

CH$_3$ N(H) CHO + N(H) CH$_3$ $\xrightarrow{HBr}$ CH$_3$ N(H) CH= N$^+$(H) CH$_3$ Br$^-$ (5)

VI

25–27 The thiazole derivative **VII** reacts with the pyrimidine derivative **VIII** to produce thiamine (vitamin B$_1$) as its bromide salt (eq. 6). Propose a structure for thiamine.

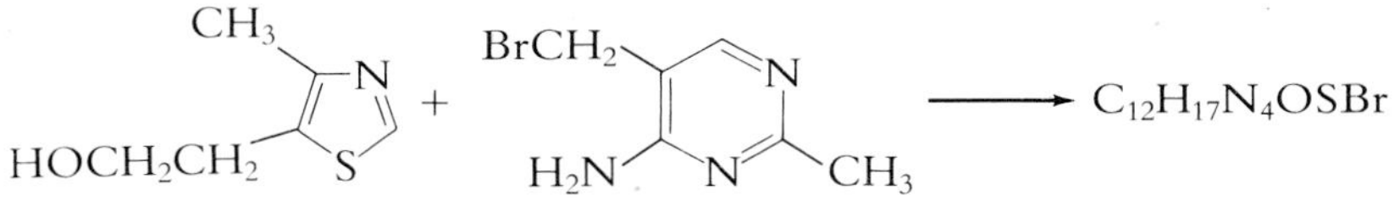

CH$_3$, HOCH$_2$CH$_2$ (thiazole N, S) + BrCH$_2$, H$_2$N (pyrimidine N, N) CH$_3$ $\longrightarrow$ $C_{12}H_{17}N_4OSBr$ (6)

VII **VIII** Thiamine bromide

25–28 Base-promoted hydrolysis of amides usually takes place very slowly at room temperature. By contrast, *N*-acetylpyrrole hydrolyzes rapidly in aqueous base at 25°. Write the structure of this compound, and explain its high reactivity.

25–29 Predict the major product(s) from mononitration of the following compounds:

a. 2-methylfuran
b. 3-chlorothiophene
c. 5-methoxy-2-methylpyrrole
d. 3-chloro-4-nitrothiophene
e. methyl 2-methylpyrrole-4-carboxylate
f. 3-methylindole
g. 3-nitroindole
h. dibenzothiophene

25–30 As a general rule, the reactivity of five-membered heterocycles toward electrophiles lies in the order pyrrole > furan > thiophene. How would you explain this order?

25–31 Bromination of benzothiophene leads to substitution at C-3 (eq. 7). If benzothiophene is treated with a strong alkylating agent, it can be converted to a sulfonium salt. This salt will *add* bromine to the double bond rather than undergo substitution (eq. 8). Write mechanisms for both the substitution and addition reactions. Why do the neutral compound and the sulfonium salt behave differently toward Br$_2$?

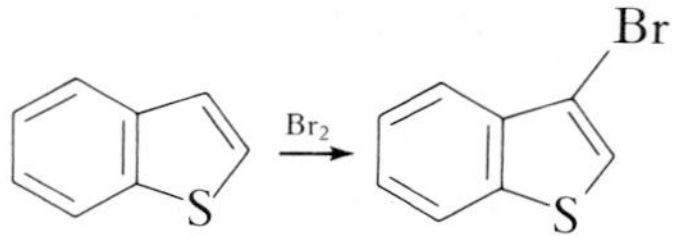

$$\text{benzothiophene} \xrightarrow{Br_2} \text{3-bromobenzothiophene} \tag{7}$$

$$\text{S-methylbenzothiophenium } BF_4^- \xrightarrow{Br_2} \text{2,3-dibromo-2,3-dihydro-S-methylbenzothiophenium } BF_4^- \tag{8}$$

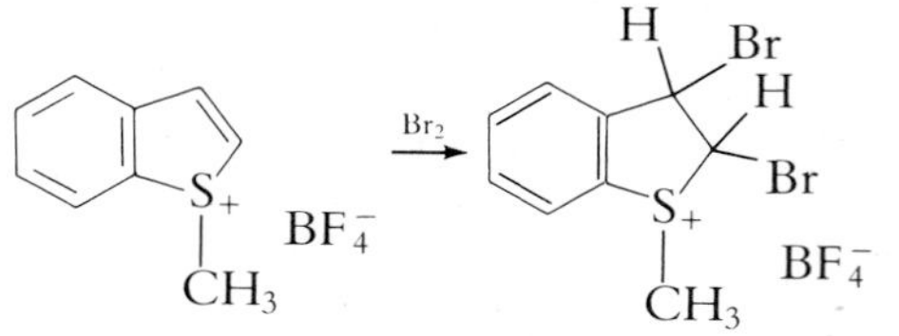

25–32 Using appropriately substituted phenylhydrazines and carbonyl compounds, propose syntheses of the following indoles:

a. 2,3-dimethylindole

b. 5-chloro-1,3-dimethylindole

c. indomethacine

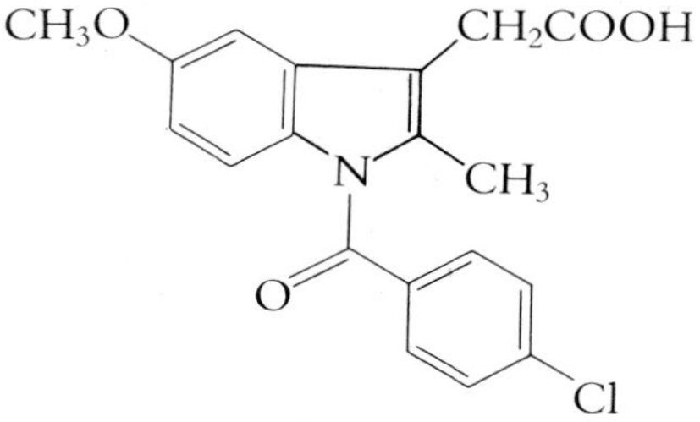

25–33 Propose syntheses of the following compounds from readily available starting materials.

a. 5-methoxy-2-methylindole (MeO, CH_3, NH)

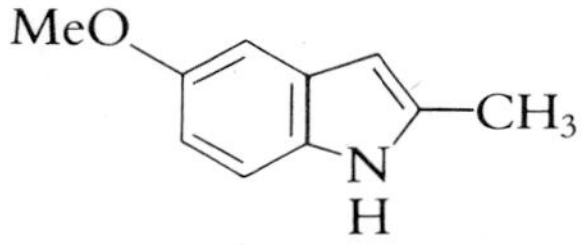

b. 1,3,4-trimethylpyrrole (CH_3, CH_3, N–CH_3)

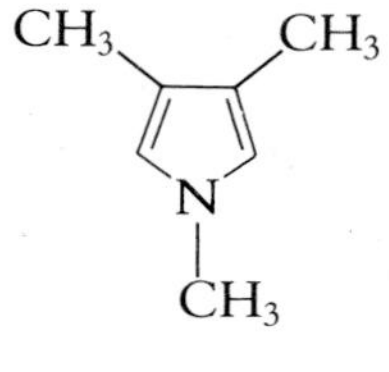

c. 2-acetyl-3,4-diethylthiophene (Et, Et, S, $\overset{O}{\overset{\|}{C}}CH_3$)

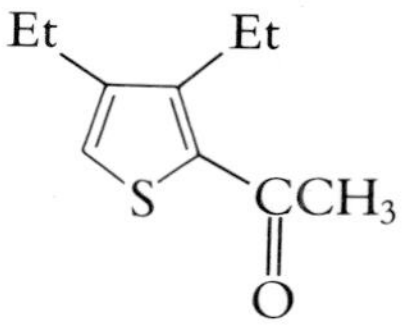

25–34 Suggest syntheses of the following compounds. Use pyridine or quinoline, or one of their methyl derivatives, as the starting materials.

a. pyridine-4-carboxylic acid

b. 4-methylpyridine-3-sulfonic acid

c. 1-ethylpyridinium iodide

d. 2-methylpyridine-*N*-oxide

e. 2-methyl-4-nitropyridine

f. 4-nitroquinoline

g. 4-amino-2-methylquinoline

h. 2-amino-5-bromopyridine

i. 2,6-diethylpyridine

25–35 a. Pyridine reacts rapidly with 1-bromobutane at 100° in dioxane to give 1-butylpyridinium bromide. This alkylation goes somewhat more rapidly with 3- or 4-methylpyridine (*3-* or *4-picoline*), but more slowly with 2-

methylpyridine (*2-picoline*). Explain the effects of the substituents on the reaction rate.

b. Pyridine reacts with Lewis acids, such as BF_3, as well as with protons. Write the structure of the product from the pyridine/BF_3 reaction. Then explain the following trends in the heats of reaction.

Reaction with H^+: pyridine < 2-methylpyridine < 2,6-dimethylpyridine (most exothermic)

Reaction with BF_3: pyridine > 2-methylpyridine > 2,6-dimethylpyridine (least exothermic)

25–36 Nicotine is a very toxic alkaloid that is found mainly in tobacco. When nicotine is treated with one equivalent of iodomethane in methanol, two isomeric quaternary ammonium iodides, **A** and **B**, are produced ($C_{11}H_{17}N_2I$). The ratio of **A** : **B** is about 2 : 5. When **A** and **B** are each treated with one equivalent of iodomethane, they both give rise to **C** ($C_{12}H_{20}N_2I_2$). Propose structures for **A**, **B**, and **C**. Explain why more **B** than **A** is produced during the first methylation step.

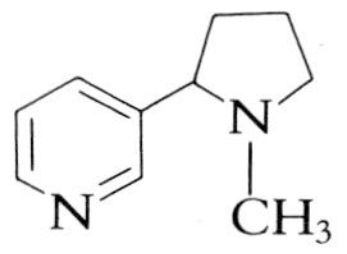

N N CH_3

Nicotine

25–37 a. Derivatives of nicotinamide (such as NAD^+) react rapidly under certain conditions to give 1,4-dihydropyridines. Give a mechanism for the reaction shown in Equation 9.

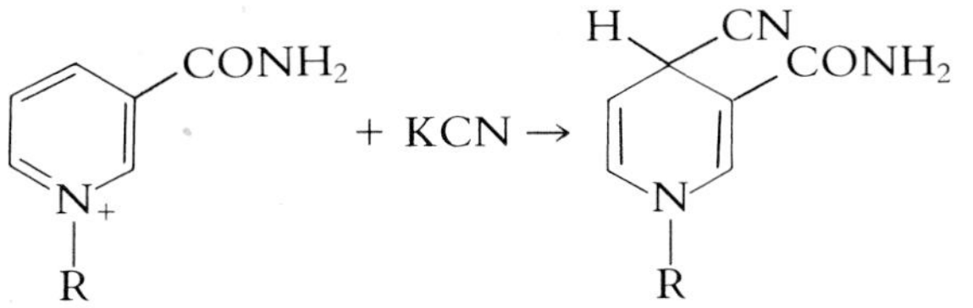

$CONH_2$ N^+ R + KCN → H CN $CONH_2$ N R (9)

b. The antitumor compound nitidine chloride has an ultraviolet-visible spectrum with λ_{max} = 272, 292, 328, 380 nm. If the solution is made basic with OH^-, the spectrum has λ_{max} = 228, 280 nm. Show what happens to cause the spectrum change. (A similar reaction with nucleophilic groups in proteins or enzymes may be responsible for the antitumor activity.)

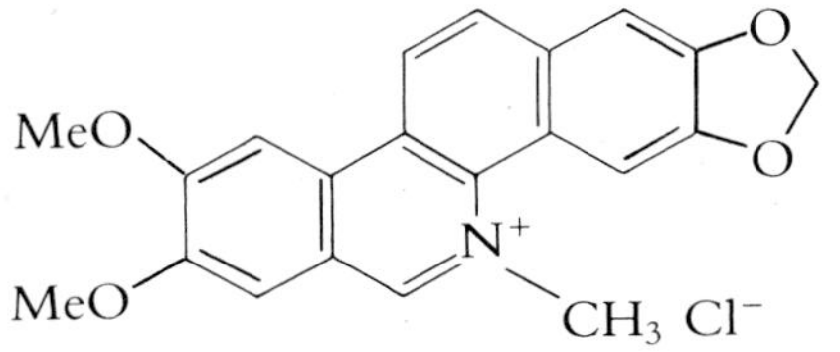

MeO MeO N^+ O O CH_3 Cl^-

Nitidine chloride

25–38 Nucleophilic substitutions can be carried out on pyridine-*N*-oxides by way of the sequence shown in Equation 10. Propose a plausible mechanism for these reactions, showing the structure of all intermediates.

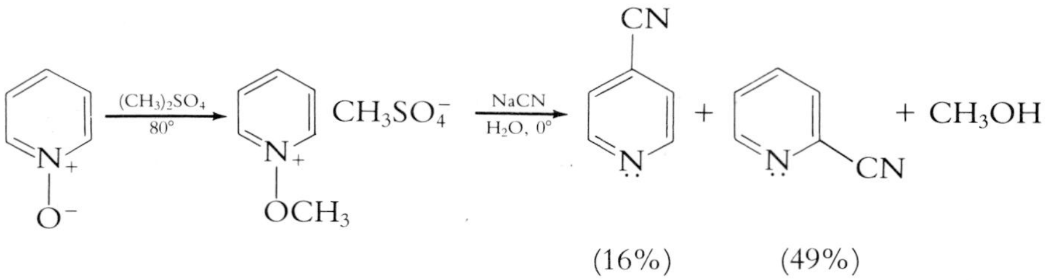

$$\text{pyridine-}N\text{-oxide} \xrightarrow[80^\circ]{(CH_3)_2SO_4} N\text{-methoxypyridinium } CH_3SO_4^- \xrightarrow[H_2O,\ 0^\circ]{NaCN} \text{4-cyanopyridine } (16\%) + \text{2-cyanopyridine } (49\%) + CH_3OH \quad (10)$$

25–39 4*H*-Pyran-4-one (**IX**) and 4*H*-thiopyran-4-one (**X**) are readily accessible. Their carbonyl groups exhibit low ir frequencies (1658 cm^{-1} and 1690 cm^{-1}, respectively). They react sluggishly with normal carbonyl reagents; for instance, their oximes cannot be prepared. In addition, compound (**X**) forms a stable hydrochloride salt in which the proton is attached to the carbonyl oxygen atom. 4*H*-Thiopyran-4-one 1,1-dioxide (**XI**), by contrast, forms normal carbonyl derivatives, and cannot form a stable hydrochloride salt. Explain these data.

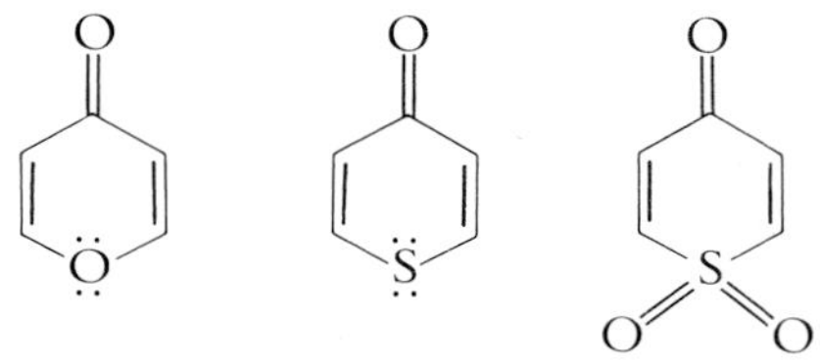

IX **X** **XI**

25–40 Methoxide ion displaces chloride from 4-chloropyrimidine (eq. 11). This reaction is about 10^6 times faster than the corresponding reaction of methoxide ion with 2- or 4-chloropyridine. Write a mechanism for the reaction shown in Equation 11. Show the structure of the intermediate and use it to explain why the reaction is so much faster than with the analogous chloropyridine.

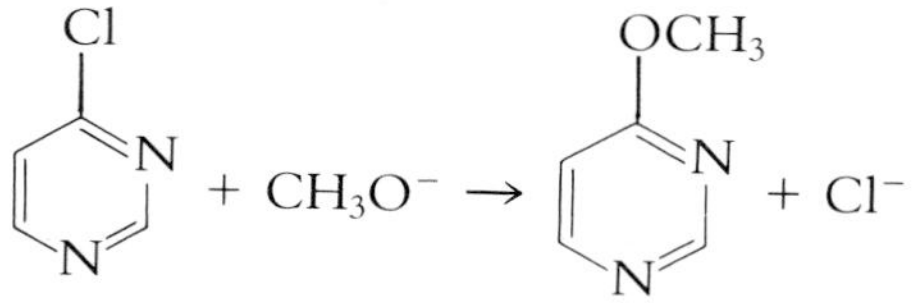

$$\text{4-chloropyrimidine} + CH_3O^- \rightarrow \text{4-methoxypyrimidine} + Cl^- \quad (11)$$

25–41 Pyridoxal phosphate (PLP, sec. 25:14) catalyzes the decarboxylation of α-amino acids (eq. 12). Using the reactions shown in Equations 25.42 to 25.44, and the electron sink–electron source concept, propose a mechanism for the reaction in Equation 12.

$$RCH(NH_2)COOH \xrightarrow{PLP} RCH_2NH_2 + CO_2 \quad (12)$$

25–42 An unknown heterocycle is found to have the molecular formula $C_5H_5N_2OCl$. Its ultraviolet spectrum shows that it is neither a pyrazine nor a pyrimidine. The ^1H and ^{13}C nmr spectra of the unknown are given below. Suggest a structure and assign the peaks in the spectra.

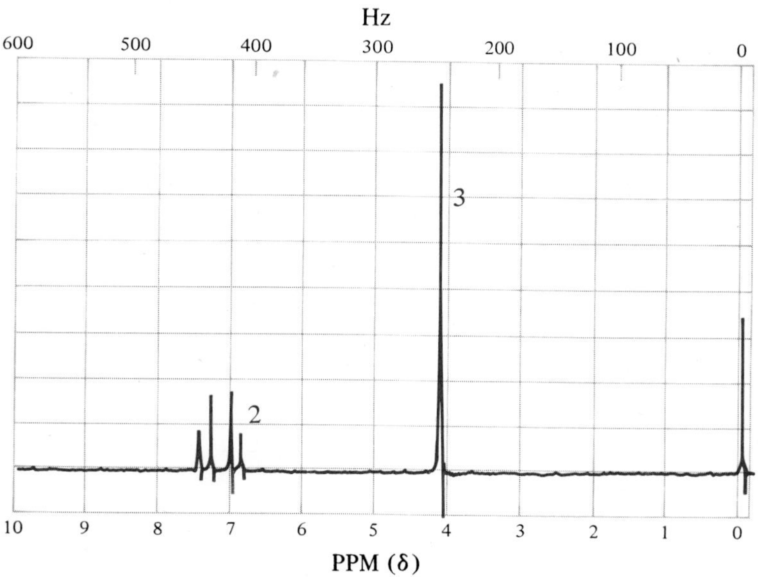

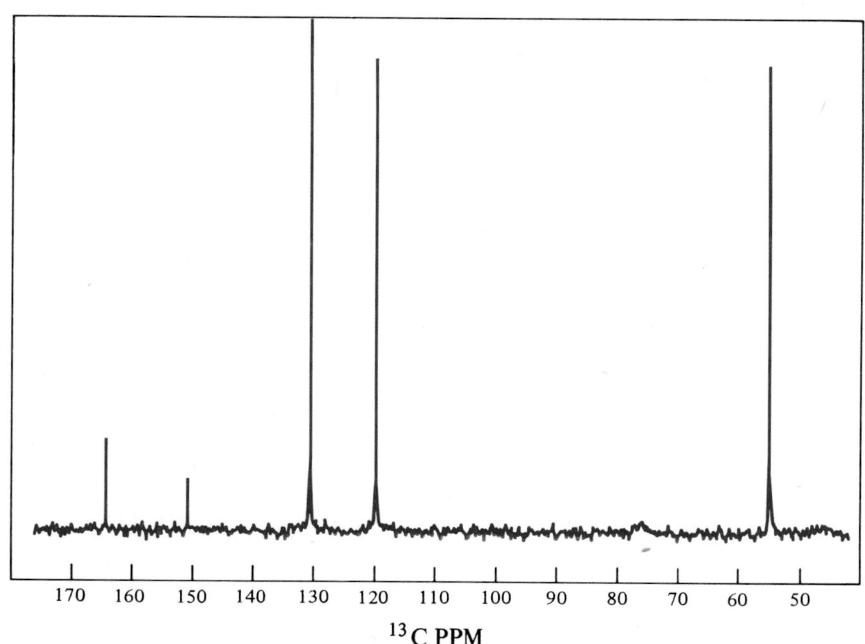

25–43 When thymine is in the double lactim form (eq. 25.45), it can still hydrogen-bond to adenine but in a different manner than the hydrogen-bonding in DNA (Fig. 25–4). Draw an alternate H-bonded dimer of adenine and the double lactim form of thymine.

Supplementary Reading

G. R. Newkome and W. W. Paudler, *Contemporary Heterocyclic Chemistry,* John Wiley, New York, N.Y., 1982.

R. M. Acheson, *An Introduction to the Chemistry of Heterocyclic Compounds,* John Wiley, New York, N.Y., 1976.

N. M. Semozand and R. L. Hunt, Hemoglobin: Its occurrence, structure, and adaptation, *J. Chem. Ed.,* **59,** 173 (1982).

L. S. Alexander and H. M. Goff, Chemicals, cancer, and cytochrome P-450, *J. Chem. Ed.,* **59,** 179 (1982).

J. N. Lowe and L. L. Ingraham, *An Introduction to Biochemical Reaction Mechanisms,* Chap. 3, "Pyridoxal and Schiff Bases", Prentice–Hall, Englewood Cliffs, N.J., 1974.

F. Sanger, Determination of nucleotide sequences in DNA, *Science,* **214,** 1205 (1981).

K. K. Ogilvie, *et al.,* Automated synthesis of gene fragments, *Science,* **214,** 270 (1981).

R. M. Stroud, A family of protein cutting enzymes, *Sci. Amer.,* **231,** 74 (July, 1974).

26 Photochemically Induced Excited States, Carbenes, and Organotransition Metal Intermediates

Very few organic molecules react when they are simply mixed together. At the end of a course in organic chemistry, this statement should come as no surprise. You have seen how acids or bases, radical initiators, or special solvents are necessary to produce reactive intermediates such as carbocations, carbanions, and free radicals. We find it necessary to apply heat to most reactions in order to provide reactant molecules with enough energy to overcome the activation barrier which leads to products.

Two methods of coaxing organic molecules to do our bidding have up to now not received the detailed treatment they deserve. The first of these is the use of visible or ultraviolet light to lift molecules to reactive excited states. This is the field of photochemistry. Sections 13:11 and 13:12 gave an introduction to this field; this chapter provides much more information on photochemical reactions.

Another useful way to activate organic molecules is to attach them to metal atoms. You're already familiar with the high reactivity and versatility of Grignard and organolithium reagents. These reagents contain magnesium and lithium, respectively, which are *main group metals.* Main group metals are those in which only *s* and *p* orbitals are available in the metal atom's valence shell for bonding. *Transition metals,* by contrast, also have valence shell *d* orbitals available for bonding. When organic groups become attached to transition metal atoms they exhibit a diverse range of chemical behavior, far greater than anything observed in main group organometallic compounds. Organocopper chemistry has provided us with a brief introduction to this area (secs. 6:14, 17:10, 19:5, and 23:3), but it is just the beginning. In this chapter we'll take a more comprehensive view of this fastest growing area of organic chemistry, which shows as much promise for industry as it does for the research laboratory.

Finally, in this chapter you will also encounter a new reactive intermediate, the **carbene**.

$$R{:}\ddot{C}{:}R$$

A carbene

This is a neutral species that has only six electrons around carbon and hence is highly reactive as an electrophile. Carbenes can be generated photochemically and stabilized by reaction with transition metals. Although their chemistry is not as widely developed and widely applicable as that of the other reactive intermediates, carbenes do provide us with some unique capabilities; for example, they are particularly valuable for the synthesis of cyclopropanes.

26:1 Light Absorption and Excited State Processes

The manner in which the absorption of light by a molecule gives rise to an ultraviolet-visible spectrum was discussed in Section 13:13. Before talking further about photochemical reactions, we need to look in more detail at exactly what happens to molecules after they absorb ultraviolet or visible light.

In Section 13:12 we showed that absorption of light energy by a molecule involves promotion of electrons from bonding or nonbonding orbitals to higher energy antibonding orbitals. Before absorbing light energy, most molecules have all their electrons paired (opposite spins). This is called a **singlet state**. (Very few molecules have two unpaired (same spin) electrons in their ground state; this is called a **triplet state**.)

When a singlet ground state molecule (symbolized as S_0) absorbs light energy, the spin of the electron that is promoted remains the same as it was in the ground state. Thus, the first reactive species produced is an *excited*

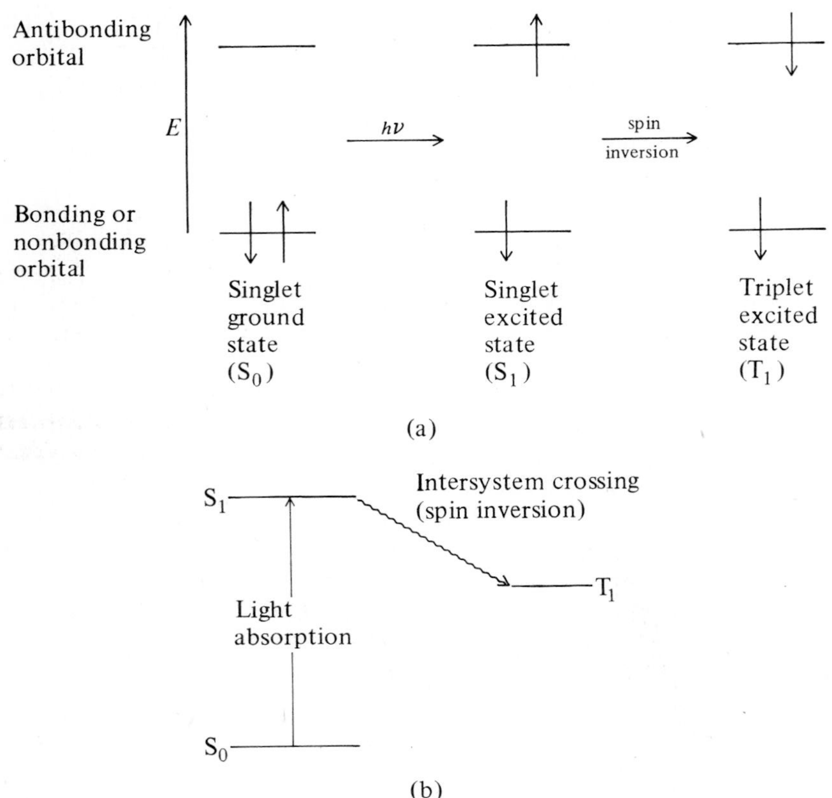

Figure 26–1. *(a) Formation of a singlet excited state by photoexcitation of a molecule in its singlet ground state. The excited singlet may undergo electronic spin inversion, converting it to a triplet excited state. (b) The energy relationships among S_0, S_1, and T_1.*

singlet state (S_1). After a time the spin of the promoted electron may invert, leading to an excited triplet state (T_1), which is usually lower in energy than S_1 (Fig. 26–1). (The name triplet arises from the fact that the two unpaired electrons have *three* possible energy states in a magnetic field; their spin vectors can be aligned with the field, against the field, and perpendicular to the field.) This process is called **intersystem crossing**; in some molecules it is very fast, in others very slow.

In general, a molecule has many excited singlet and triplet states; S_1 and T_1 are simply the lowest energy excited states having these spins. Most of the chemical processes which will interest us involve only S_1 and T_1, so we will concentrate on them.

What happens to a molecule in an excited state? It can lose its extra energy simply by colliding with other molecules, or by emitting it as a photon of light. A singlet excited state can return directly to the ground state (S_0) by releasing light energy, called **fluorescence** (Fig. 26–2). If the molecule returns to the ground state from T_1, the emitted radiation is called **phosphorescence**. If the extra energy is not lost in this way, *it can provide the driving force for either intramolecular or intermolecular reaction.*

Plant photosynthesis, the manufacture of vitamin D in the skin of animals, and the initiation of some types of cancers and mutations are all dependent upon excited state chemistry. These processes, as well as many laboratory photochemical reactions, proceed from the excited states of compounds containing carbon–carbon double bonds or carbonyl groups. We will therefore look a little more closely at the electronic structure of these two functional groups before giving examples of their reactivity.

Problems

26–1 Direct absorption of 268 nm radiation by benzene produces excited state 1. Excited state 1 then undergoes slow conversion to excited state 2. Benzene can return to its ground state (S_0) from excited state 2 by emitting radiation of wavelength 344 nm. Using the relationship between wavelength and energy (sec. 8:4), calculate the energies of excited states 1 and 2 relative to the ground state. Assign the correct spin designation to each excited state, and summarize all this information on an energy state diagram like Figure 26–1b.

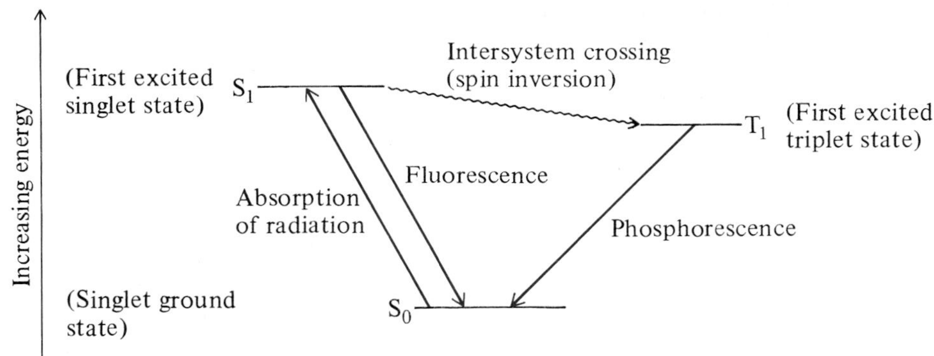

Figure 26–2. *Formation, interconversion, and energy loss of electronic excited states.*

26–2 Most stable molecules have singlet ground states. An important exception is oxygen, O_2. This molecule has a triplet ground state, which accounts for its free-radical-like chemical behavior (sec. 5:13). The first triplet excited state of O_2 is 103 kcal/mol above the ground state. Subsequently, by intersystem crossing, a singlet excited state is formed, 23 kcal/mol above the ground state. Draw an energy state diagram like Figure 26–1b for O_2. Show all the processes which convert one state to another.

26:2 Photoexcitation Processes in Alkenes and Carbonyl Compounds

In order to understand which of the many types of electrons in a complex molecule are candidates for photoexcitation, we need to look at the kinds of orbitals which they occupy and into which they can be promoted. We'll build on the discussion in Sections 13:11 and 13:12.

A localized molecular orbital diagram for the C═C bond of ethylene (CH_2═CH_2) is given in Figure 26–3. (Orbitals for the CH bonds are neglected.) Included in the figure are both the ground state and excited singlet state electronic configurations, along with terminology used to describe the excitation process and the excited state. As you can see, electrons in the C═C σ orbital have lower energies than those in the π orbital, so the π orbital is the HOMO (sec. 13:11). Conversely, the $\sigma^\star$ orbital lies higher than the $\pi^\star$ orbital, so the $\pi^\star$ orbital is the LUMO. Consequently, the excitation process requiring the least amount of energy is promotion of one electron from the π orbital to the $\pi^\star$ orbital. This is called a $\pi \rightarrow \pi^\star$ transition and the excited state reached is called a $\pi,\pi^\star$ state. $\sigma \rightarrow \pi^\star$ or $\sigma \rightarrow \sigma^\star$ transitions would require more energy than the $\pi \rightarrow \pi^\star$ transition. In a uv spectrum, the $\pi \rightarrow \pi^\star$ transition would appear at a *longer* wavelength than the other two transitions, since it involves less energy. The $\pi \rightarrow \pi^\star$ transition in ethylene occurs at 175 nm (163 kcal/mol). The singlet $\pi,\pi^\star$ state could be converted to a triplet $\pi,\pi^\star$ state of lower energy by electron spin inversion.

The same orbital analysis can be applied to carbon–oxygen double bonds, with one important difference. Oxygen carries two nonbonding

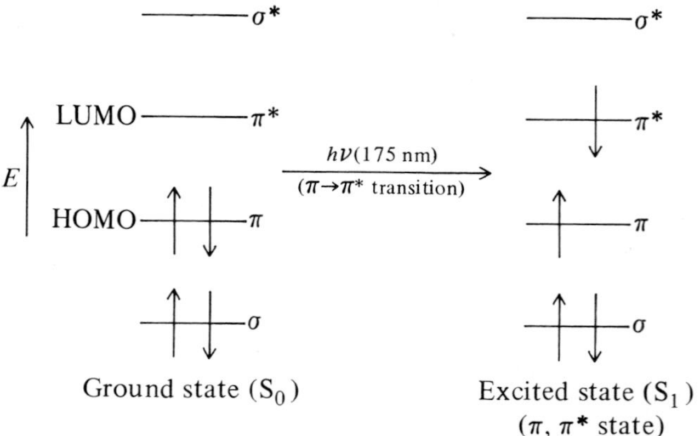

Figure 26–3. *Electron distribution in the C═C molecular orbitals for ground and excited states of ethylene.*

electron pairs (n electrons), which are higher in energy than the π electrons. In molecules containing this functional group, the $n \rightarrow \pi^\star$ transition usually requires the least amount of energy, about 100 kcal/mol, and gives rise to uv absorption in the range 275 to 290 nm.

Figure 26–4 shows the orbitals and electronic configuration for the ground state of acetone (again neglecting orbitals for bonds other than the C=O). The absorption spectrum of acetone shows peaks at 275 and 188 nm, and these are assigned to the $n \rightarrow \pi^\star$ and $n \rightarrow \sigma^\star$ transitions, respectively. The $\pi \rightarrow \pi^\star$ is at ~150 nm.

In order to promote a molecule to an excited state from which it can enter into chemical reactions, it must be irradiated with light of the proper wavelength (energy). Thus, to get ethylene to react, it must be irradiated with light of 175 nm wavelength. For acetone to react, it could absorb light of either 275 or 188 nm. In general, the advantage of using light rather than heat to perform reactions is that the energy is more or less concentrated in the light-absorbing portion of the molecule, called the **chromophore**. When a molecule is heated thermally, the energy goes into vibrations and

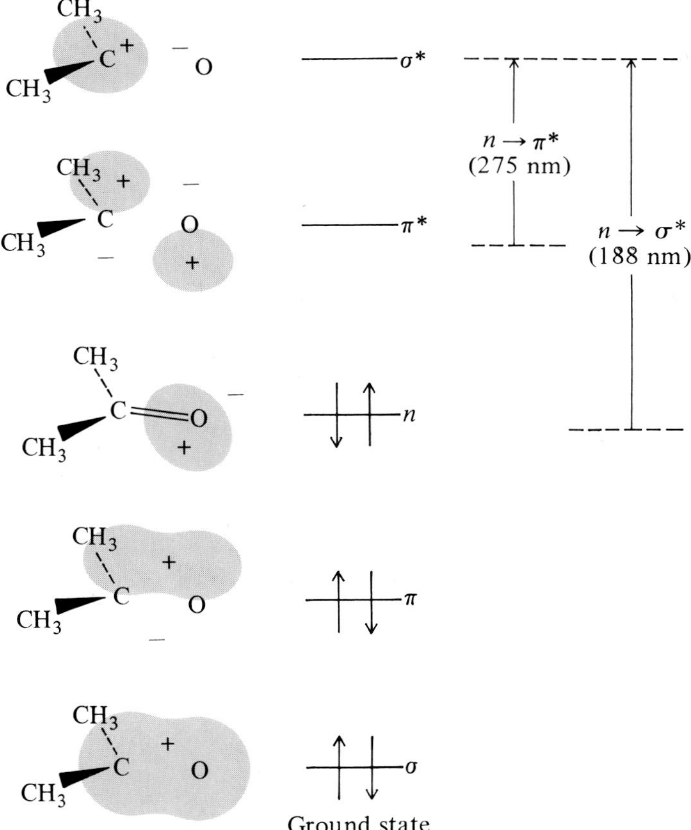

Figure 26–4. *Some localized molecular orbitals of the acetone C=O group and electronic transitions between them.*

other motions of the entire molecule, and the resulting chemistry is therefore less specific. For example, consider the molecule 1-phenyl-2-propanone. When this molecule is irradiated with light of 275 nm, that light will be

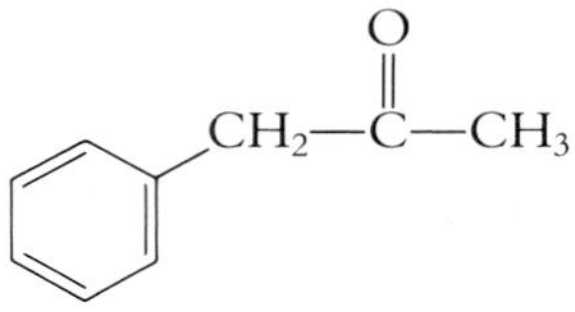

$$C_6H_5{-}CH_2{-}\overset{\overset{\displaystyle O}{\|}}{C}{-}CH_3$$

1-Phenyl-2-propanone

absorbed by the carbonyl chromophore ($n \rightarrow \pi^\star$ transition) and only that portion of the molecule will gain energy (~100 kcal/mol). Essentially, no energy will be absorbed by the aromatic ring or C—H bonds. If instead, 1-phenyl-2-propanone is *heated,* then the entire molecule will have greatly increased kinetic and potential energy. *Thus, through absorption of light, a molecule may be given a large amount of energy at a very specific site.*

Finally, it is important to recognize why photoexcited chromophores are chemically reactive. In most ground state molecules, all the bonding and nonbonding orbitals are completely filled and all the antibonding orbitals completely empty. Absorption of radiation changes this situation, as Figures 26–3 and 26–4 show. One of the previously filled molecular orbitals becomes half empty, and one of the previously empty orbitals becomes half filled. *The change in electron distribution in the HOMOs and LUMOs* (sec. 13:11) *is what accounts for photochemical reactivity.*

Problem

26–3 You've already seen that conjugated alkenes have smaller HOMO–LUMO energy gaps than unconjugated alkenes. 1,3-Butadiene shows uv absorption peaks at 230 nm and 193 nm. Assign the electronic transitions corresponding to these absorptions, using the energy diagram for 1,3-butadiene (Fig. 13–14, p. 495).

26:3 Some Photoreactions of the Carbon–Carbon Double Bond

Our first mention of photochemical reactions was the [2 + 2] cycloaddition of ethylenes to give cyclobutanes (eqs. 13.40 and 13.41, p. 500). Production of cyclobutane itself from two ethylenes is not a particularly high yield reaction, but some *substituted* cyclobutanes can be obtained easily and in high yield using this method. Equations 26.1 and 26.2 give two examples.

(26.1) 2 (H)(Ph)C=C(NO$_2$)(H) $\xrightarrow[H_2O,\ stir]{h\nu}$ cyclobutane with Ph, NO$_2$ (top) and NO$_2$, Ph (bottom) (70%)

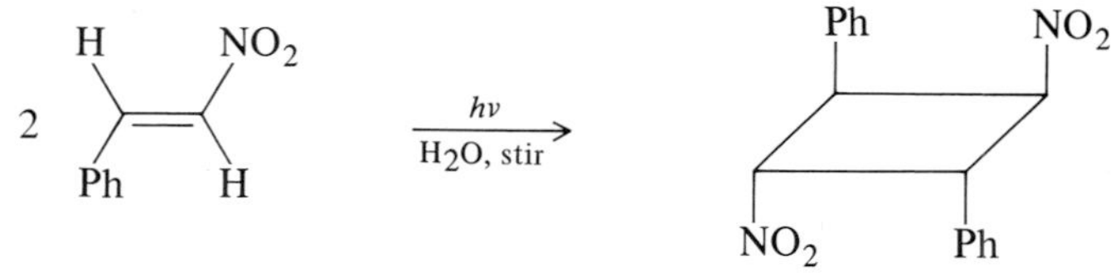

(26.2) $2\,\text{PhCH=CHCOOH} \xrightarrow[\text{H}_2\text{O, stir}]{h\nu}$ cyclobutane dimer (Ph, COOH, COOH, Ph) (56%)

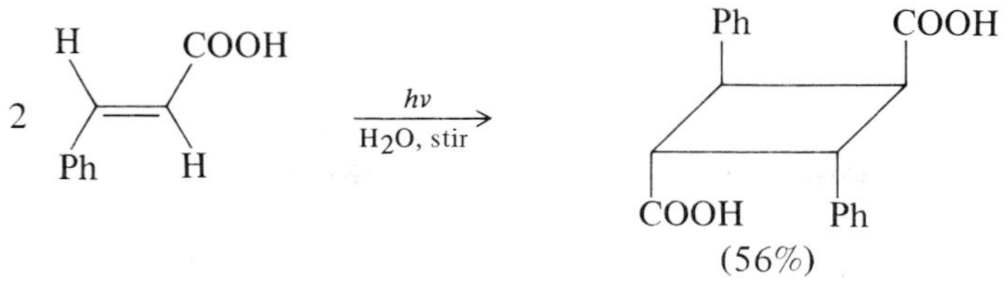

These reactions are stereospecific because they are actually taking place in the solid state and the stereochemistry observed is directly related to the way in which the molecules are packed in the crystal. The starting materials are not water soluble and hence are merely suspended in the medium as microcrystals.

As a biological analog, consider the irradiation of bacterial DNA. Subsequent chemical degradation results in isolation of the thymine dimer (eq. 13.41, p. 500, *cis* ring junctions) as the major product. We can consider this biological example to be like a solid state reaction because the thymines of the DNA are held in somewhat rigid positions relative to one another in the DNA chain, just as they would be in a crystal. (If the reactions in Equations 26.1 and 26.2 had been carried out in an organic solvent such as ether, the reaction would not have been stereospecific but would instead have yielded mixtures of several isomers of the dimers found in those equations.)

One dream of photochemists is to have an organic molecule absorb energy from the sun and then store that energy indefinitely for future use. The formation of cyclobutanes from alkenes has been explored for just that purpose (as have other photochemical reactions). Consider the reaction in Equation 26.3, which has been carried out with a variety of norbornadiene derivatives. If a derivative that absorbs solar energy strongly can be used,

(26.3) Norbornadiene $\xrightarrow[\text{ether}]{h\nu}$ Quadricyclane $\quad \Delta H = +21$ kcal/mol

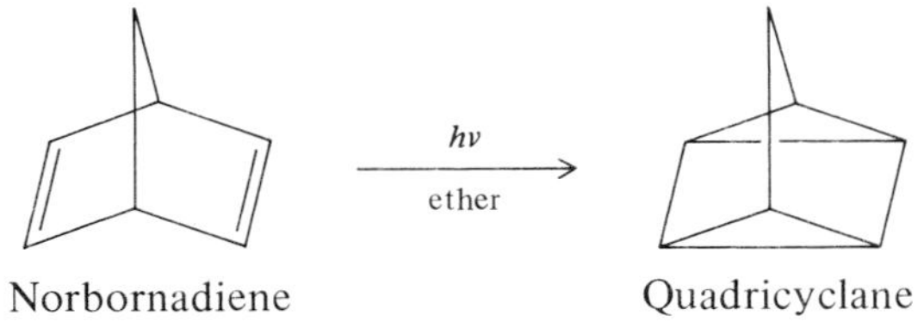

then quadricyclane formation could be useful for energy storage since the product contains a large amount of strain energy. Although strained molecules such as quadricyclanes are reactive to a number of reagents, they are stable indefinitely if stored where they do not contact such reagents. When Equation 26.3 is reversed by metal catalysts, the stored energy is indeed released, and hence a method for storing and recovering solar energy is possible. For such a process to be practical it must, however, be close to 100% efficient in both directions and this has not been achieved with this reaction (or any other reaction). Formation of by-products drains off the starting material before enough cycles can be accomplished for useful energy storage.

Box 26–1. Light Sources for Photochemical Reactions

The source of light, $h\nu$, for photochemical reactions in the laboratory most often comes from ultraviolet lamps. These lamps emit light of nearly all wavelengths, although some specific wavelengths (254, 313, and 355 nm, for example) are of much greater intensity than others. Special filters can be used between the lamp and the reactants in order to isolate one or more of the irradiating frequencies. Sunlight can also be used as a source of direct irradiation. The atmosphere filters out most wavelengths below about 280 nm, so compounds which absorb only below that wavelength will not be affected. (Pyrex glass has about the same filtering properties, while quartz allows much shorter wavelengths of light to pass.) There are numerous compounds whose photochemistry can be explored just with the use of sunlight. Indeed, in the early 1900s, photochemists such as the Italian G. Ciamician often covered the roofs of their laboratories with innumerable flasks containing reactants in solution. Ciamician, in addition to being one of the first organic photochemists, was also a visionary. He dreamed of replacing sooty and smelly chemical factories with acres of glass tubing in desert areas of the world. Here the clean, inexhaustible supply of sunlight would be used as an energy source for chemical industries based upon photochemistry.

26:4 Photoisomerizations and a Photocyclization

The products obtainable upon irradiation of a given molecule can vary quite markedly, depending upon reaction conditions. The photochemistry undergone by stilbene ($C_6H_5CH{=}CHC_6H_5$) provides a good example. When either *cis*- or *trans*-stilbene is irradiated as a solid or in concentrated solution, tetraphenylcyclobutanes are obtained, as the last section would lead us to expect.

In relatively dilute solutions, however, *cis–trans* isomerization is faster than dimerization (eq. 26.4). Notice that, at 313 nm, the photochemical

(26.4) *trans* (7%) $\underset{}{\xrightleftharpoons{h\nu,\ 313\ \text{nm}}}$ *cis* (93%)

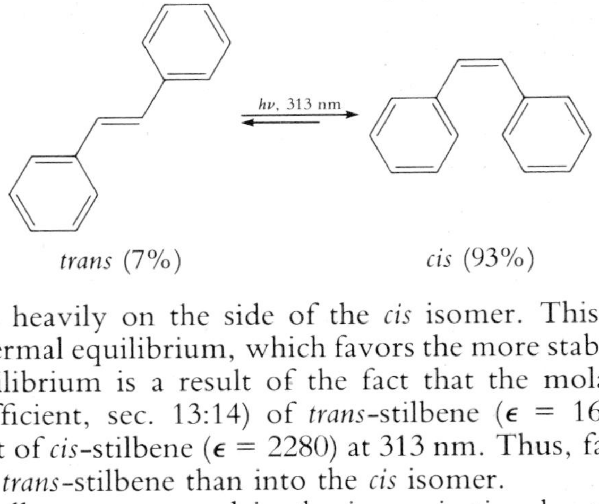

equilibrium lies heavily on the side of the *cis* isomer. This is exactly the opposite of a thermal equilibrium, which favors the more stable *trans* isomer. The photo equilibrium is a result of the fact that the molar absorptivity (extinction coefficient, sec. 13:14) of *trans*-stilbene (ϵ = 16,300) is much greater than that of *cis*-stilbene (ϵ = 2280) at 313 nm. Thus, far more energy is pumped into *trans*-stilbene than into the *cis* isomer.

Mechanistically, we can explain the isomerization by presuming that promotion of one π electron into an antibonding orbital effectively breaks the π bond and allows relatively free rotation. This situation is expressed by Equation 26.5, where the $\star$ symbol refers to an electron in an antibonding orbital of the excited stilbene. Rotation around the double bond takes place easily in the excited state because the electron in the antibonding orbital considerably weakens the double bond. Some time after rotation takes place, the promoted electron falls back into the bonding orbital, and the molecule returns to the ground state.

(26.5)

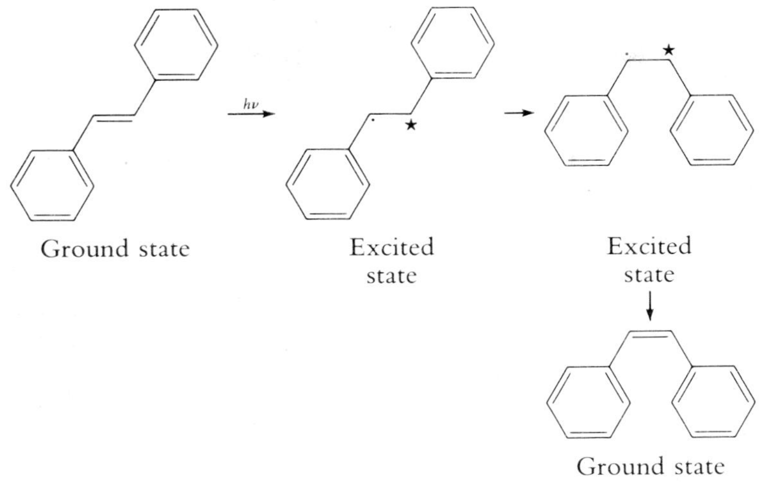

Ground state Excited state Excited state

Ground state

If the photoisomerization is carried out in the presence of oxygen or I_2, then a photocyclization takes place and phenanthrene can be isolated in high yield. Equation 26.6 shows how this takes place. Without I_2 or O_2 to remove

(26.6)

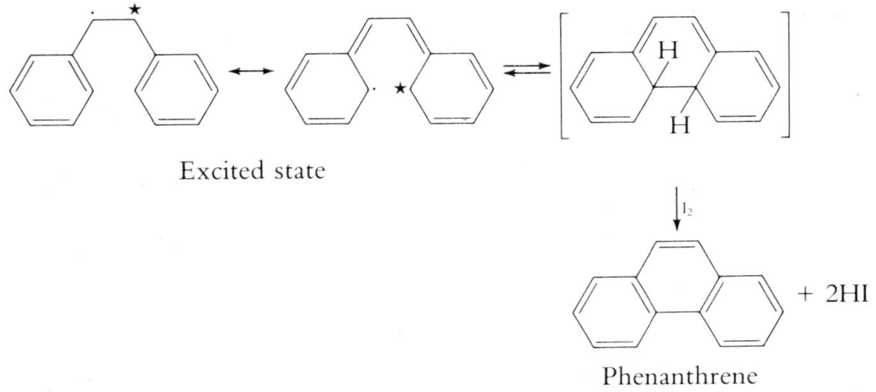

Excited state

$+ 2HI$

Phenanthrene

the hydrogens, the intermediate (shown in brackets) will simply revert to *cis*-stilbene.

Problems

26–4 What is the total number of cyclobutane isomers that could be formed in the reaction in Equation 26.1?

26–5 Psoralenes (**I**) are compounds that occur in plants such as parsnips. Animals that eat these plants become very sensitive to sunlight. Other psoralenes, when applied to the skin and irradiated with uv light, are used to alleviate psoriasis. What product(s) might be formed by irradiation of psoralene?

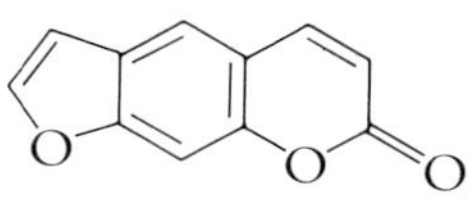

I
Psoralene

26–6 How might you try to prepare **II**, starting with aniline and benzaldehyde?

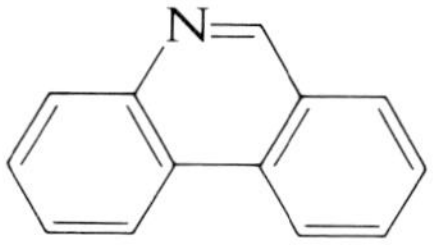

II

26:5 Some Photoreactions of the Carbonyl Group: Similarity Between Photochemical and Mass Spectral Reactions

The photochemical [2 + 2] cycloaddition reaction can also take place between the π bond of a *carbonyl group* and an alkene. Equation 26.7 shows that

$$Ph_2C{=}O + NC{-}CH{=}CH{-}CN \xrightarrow[355\ \text{nm}]{h\nu} \tag{26.7}$$

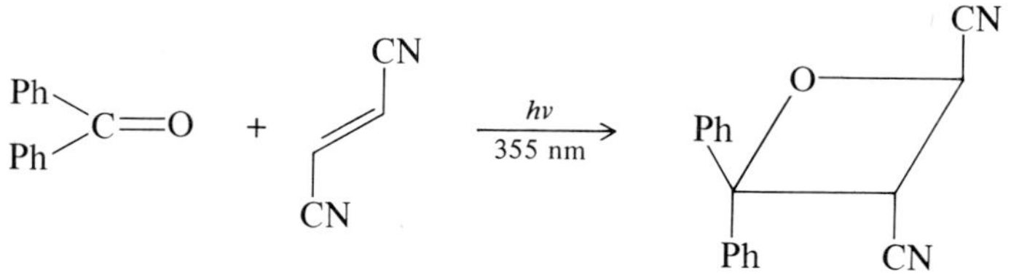

trans-2,3-Dicyano-4,4-diphenyloxetane

this can be accomplished by selectively irradiating the carbonyl group (the alkene does not absorb light at 355 nm). This process, which yields an oxetane, is called the *Paterno–Büchi reaction.*

An important characteristic of a photoexcited carbonyl group is that the oxygen end of the molecule acts as if it were a free radical, since one of its *n*-orbitals is only half occupied (Fig. 26–4). Thus it can abstract a hydrogen atom from a C—H bond nearby. Equation 26.8 shows photoexcited benzophenone abstracting a hydrogen atom from benzhydrol (diphenylmethanol), to give two identical, highly stable radicals. These radicals combine to form benzpinacol (tetraphenyl-1,2-ethanediol).

(26.8) PhCPh (O=) $\xrightarrow[355\text{ nm}]{h\nu}$ $^{\star}[\text{PhC(O}\cdot\text{)Ph}]$ $\xrightarrow[\text{benzhydrol}]{\text{H—C(Ph)}_2\text{—OH}}$ $\text{Ph—}\dot{\text{C}}\text{(OH)—Ph} + \text{Ph—}\dot{\text{C}}\text{(OH)—Ph}$ $\longrightarrow$ $\text{Ph}_2\text{C(OH)—C(OH)Ph}_2$

Benzophenone — Excited state — Benzpinacol

This reaction is mainly of value when only one product is formed.

An intramolecular variant of this reaction is also of great significance. It can lead to both cyclobutanol and to cleavage of the molecule (eq. 26.9). *This path requires a hydrogen atom γ to the carbonyl group.* There have been

(26.9) $\text{CH}_3\text{C(=O)CH}_2(\alpha)\text{CH}_2(\beta)\text{CH}_2(\gamma)\text{—H} \xrightarrow{h\nu}$ $\text{CH}_3\dot{\text{C}}\text{(OH)CH}_2\text{CH}_2\dot{\text{C}}\text{H}_2$ $\xrightarrow{\text{cleavage}}$ $\text{CH}_3\text{C(OH)=CH}_2 + \text{CH}_2\text{=CH}_2$; $\text{CH}_3\text{C(OH)=CH}_2 \longrightarrow \text{CH}_3\text{C(=O)CH}_3$

$\text{CH}_3\dot{\text{C}}\text{(OH)CH}_2\text{CH}_2\dot{\text{C}}\text{H}_2$ $\xrightarrow{\text{oxetane formation}}$ 1-methylcyclobutanol (HO, CH_3)

some attempts to use the cleavage reaction to prepare photodegradable plastics. Polyethylene, for example, is composed of long chains of unreactive CH_2 groups and hence is environmentally very stable. If light-absorbing C=O groups were incorporated into the long polymer chain, sunlight could be used to initiate cleavage reactions such as that shown in Equation 26.9. This would cause degradation of the polymer into smaller, biodegradable molecules.

The reaction in Equation 26.9 is known as a **Norrish Type II cleavage** of ketones after the English chemist R. Norrish (Nobel Prize, 1967). Carbonyl compounds, particularly in the gas phase, can also cleave *next* to the C=O group and this is termed a **Norrish Type I cleavage** (eq. 26.10).

$$\text{(26.10)}\quad CH_3\overset{\overset{\displaystyle O}{\|}}{C}CH_2CH_3 \xrightarrow[\text{vapor phase}]{h\nu} [CH_3\cdot + O{=}\dot{C}CH_2CH_3 + CH_3\dot{C}{=}O + \cdot CH_2CH_3]$$

$$\downarrow$$

$$CH_3\underset{\underset{\displaystyle O}{\|}}{C}CH_3 + CH_3\underset{\underset{\displaystyle O}{\|}}{C}\underset{\underset{\displaystyle O}{\|}}{C}CH_3 + CH_3CH_2\underset{\underset{\displaystyle O}{\|}}{C}CH_2CH_3 + CH_3CH_3 + \text{etc.}$$

Some interesting comparisons can be made between the photochemistry and mass spectra of carbonyl compounds. Remember from Sections 8:1 to 8:3 that mass spectrometry involves removing one electron completely from the molecule, rather than merely promoting it to a higher orbital. This results in formation of a radical cation called the molecular ion. Molecular ions of carbonyl compounds undergo subsequent decompositions similar to photochemical cleavages. Equation 26.11 gives an example of a mass spectral fragmentation similar to a Norrish Type I cleavage.

$$\text{(26.11)}\quad CH_3\overset{\overset{\displaystyle :O:}{\|}}{C}CH_2CH_3 + e^- \rightarrow$$

$$2e^- + \underset{\text{Molecular ion}}{CH_3\overset{\overset{\displaystyle :O^{\cdot +}}{\|}}{C}{:}CH_2CH_3} \rightarrow CH_3C{\equiv}\ddot{O}^+ + \cdot CH_2CH_3$$

The Norrish Type II cleavage also has a mass spectral analog called the **McLafferty rearrangement** (eq. 26.12). Molecules that have a hydrogen atom γ to the carbonyl group often undergo this type of fragmentation.

$$\text{(26.12)}\quad CH_3\overset{\overset{\displaystyle :O^{\cdot +}}{\|}}{C}\underset{\alpha}{CH_2}\underset{\beta}{CH_2}\underset{\gamma}{CH_2}H \rightarrow CH_3\overset{\overset{\displaystyle :\dot{O}^{+}:H}{|}}{C}{=}CH_2 + \underset{CH_2}{\overset{CH_2}{\|}}$$

Example

What product would you expect from irradiation of acetaldehyde in ethanol solution?

Answer

Acetaldehyde absorbs light, is promoted to an excited state, and can then abstract a hydrogen from ethanol. The C(1)—H bond is the one most readily broken (Table 5–1, p. 146):

$$CH_3\overset{\overset{\displaystyle H}{|}}{C}{=}O + h\nu \rightarrow \left[CH_3\overset{\overset{\displaystyle H}{|}}{C}{=}O\right]^\star \xrightarrow{CH_3CH_2OH} CH_3\overset{\overset{\displaystyle H}{|}}{\dot{C}}{-}OH + CH_3\dot{C}HOH$$

Now the free radicals which were formed can combine:

$$CH_3\dot{C}HOH + CH_3\dot{C}HOH \rightarrow \begin{array}{c} CH_3CHOH \\ | \\ CH_3CHOH \end{array}$$

Note that Norrish Type II cleavage isn't possible.

Problems

26–7 Give the products of each of the following reactions, and show the intermediates which lead to those products.

a. acetone + 2-propanol $\xrightarrow{h\nu}$

b. acetophenone + 1-phenylethanol $\xrightarrow{h\nu}$

c. $CH_3\overset{\overset{\displaystyle O}{\|}}{C}CH_2CH_2CD_2CH_3 \xrightarrow[\text{inert solvent}]{h\nu}$

26–8 The mass spectrum of 3-methylbutanamide shows a major peak at m/z 59. What fragment might account for this peak? How is it formed?

$$\begin{array}{c} CH_3\underset{\underset{\displaystyle CH_3}{|}}{C}HCH_2\underset{\underset{\displaystyle O}{\|}}{C}NH_2 \end{array}$$

3-Methylbutanamide

26:6 How Excited States Are Produced and Destroyed by Energy Exchange Between Molecules

In the past few sections we have shown that the reactions of molecules in excited states can be rationalized using the same types of intermediates and reaction paths known from ground state chemistry. To completely determine a photochemical mechanism, we must also find out whether *the reaction proceeds from the excited singlet or excited triplet state*. Remember from Figures 26.1 and 26.2 that light absorbed by a singlet ground state molecule first promotes the molecule to an excited singlet which *may* then be converted to the excited triplet state. Reactions could involve either excited state.

Excited state identification is often made by adding a **quencher** or a **sensitizer** to the reaction mixture. A quencher is a molecule that takes the energy away from an excited molecule before it can react. A sensitizer is a molecule that can itself absorb light energy and then transfer that energy to a ground state molecule. The fundamental idea behind both of these reactions is that there can be a transfer of excited state energy from one molecule to another. Some concrete examples will make these concepts clear.

If benzophenone ($C_6H_5COC_6H_5$) is irradiated at 355 nm, the ground state molecule (S_0) is converted to the first excited singlet (S_1) which is then rapidly converted to the triplet excited state (T_1). These processes are depicted in Equation 26.13.

$$C_6H_5\overset{\overset{\displaystyle O}{\|}}{C}C_6H_5\ (S_0) \xrightarrow{h\nu,\ 355\ \text{nm}} C_6H_5\overset{\overset{\displaystyle O}{\|}}{C}C_6H_5\ (S_1)^{\star} \xrightarrow{\text{fast}} C_6H_5\overset{\overset{\displaystyle O}{\|}}{C}C_6H_5\ (T_1)^{\star} \tag{26.13}$$

One of these excited states ($S_1^\star$ or $T_1^\star$) of benzophenone can initiate the reaction in Equation 26.8, for example. One way to determine which state *is* the reactive one is to add some 1,3-pentadiene (*piperylene*, $CH_3CH{=}CH{-}CH{=}CH_2$) to the reaction mixture of Equation 26.8. Piperylene is a *known* triplet quencher. The result of this experiment is that *no* benzpinacol is formed in the presence of piperylene. This indicates that the reactive state of benzophenone is the triplet state.

What happens in the quenching process is that the excited triplet energy of benzophenone is transferred to piperylene (eq. 26.14).

$$(26.14)\quad C_6H_5\overset{\overset{\displaystyle O}{\|}}{C}C_6H_5\ (T_1)^\star + CH_3CH{=}CH{-}CH{=}CH_2\ (S_0) \xrightarrow[\text{transfer}]{\text{energy}} C_6H_5\overset{\overset{\displaystyle O}{\|}}{C}C_6H_5\ (S_0) + CH_3CH{=}CH{-}CH{=}CH_2\ (T_1)^\star$$

Piperylene is then converted to its excited triplet state and the benzophenone returns to its ground state, where it is unreactive. Good quenchers must react rapidly and quantitatively with excited molecules. The triplet state of the quencher must be lower in energy than that of the excited molecule in order for energy transfer to take place.

Good quenchers should also be specific for certain excited states. If the reaction in Equation 26.8 involved benzophenone in its excited singlet state (S_1), piperylene would not have stopped (quenched) the reaction. Oxygen, O_2, is a good triplet state quencher as well and hence many photochemical reactions must be run in the absence of oxygen or air.

The process of sensitization is complementary to that of quenching. To explain this concept we will examine the reaction in Equation 26.3 more closely. If we attempted to promote this reaction using solar energy we would be sadly disappointed. The reaction does not proceed at all in sunlight because the atmosphere filters out all wavelengths below about 280 nm, and norbornadiene only absorbs light considerably below 280 nm. Thus we cannot conveniently excite norbornadiene by direct irradiation. Through use of a sensitizer that will absorb sunlight and transfer its energy to norbornadiene, we can make reaction 26.3 useful in sunlight. This is shown in Equation 26.15 where benzophenone is used as a sensitizer. Light is absorbed by

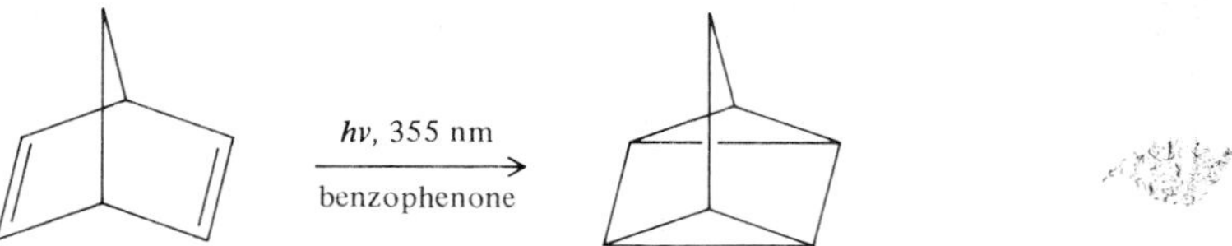

benzophenone, which is converted to its triplet state by the series of steps shown in Equation 26.13. Upon collision with a molecule of norbornadiene, excited benzophenone transfers its energy to the diene and is then converted to its ground state. Norbornadiene accepts the energy and is converted to its triplet excited state, from which it undergoes the cyclization.

Note that the processes of quenching and sensitization are just two ways of applying the same phenomenon: energy transfer from an excited molecule to a ground state molecule.

Problem

26–9 Explain the following observations.

a. Irradiation of

$$C_6H_5\underset{\underset{O}{\|}}{C}CH_2CH_2CH_3$$

gives Norrish Type II cleavage products, but not in the presence of piperylene. Irradiation of

$$CH_3\underset{\underset{O}{\|}}{C}CH_2CH_2CH_3$$

gives Norrish Type II products which are still formed (but in much lower yield) when piperylene is present.

b. Compound **A** gives products **B** and **C** when irradiated directly at 254 nm. When benzophenone is present and irradiation is at 355 nm, **A** yields products **D** and **E**.

26:7 Photochemical Production of Carbenes and Nitrenes

When a molecule absorbs a photon of ultraviolet or visible light it acquires a considerable quantity of energy—in the case of uv light (~280 nm), for instance, it amounts to about 96 kcal/mol. As the Norrish photochemical reactions show, this quantity of energy is enough to bring about bond cleavage and molecular fragmentation. Fragmentation can be a highly favorable reaction pathway for a photoexcited molecule if one of the fragments is particularly stable. This section is concerned with several photochemical processes that fit the pattern.

One class of compounds that meets this criterion is the **diazo** compounds, which contain the functional group

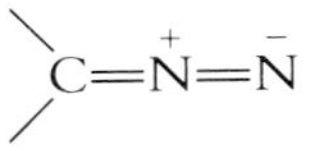

$$>C{=}\overset{+}{N}{=}\overset{-}{N}$$

Upon irradiation they decompose to a highly stable nitrogen molecule and a carbene (eq. 26.16).

$$\text{(26.16)}\quad C_6H_5\text{—}\underset{\underset{H}{|}}{C}{=}\overset{+}{N}{=}\ddot{N}{:}^{-} \longleftrightarrow C_6H_5\text{—}\underset{\underset{H}{|}}{\overset{-}{\ddot{C}}}\text{—}\overset{+}{N}{\equiv}N{:} \xrightarrow{h\nu} C_6H_5\text{—}\ddot{C}\text{—}H + N_2$$

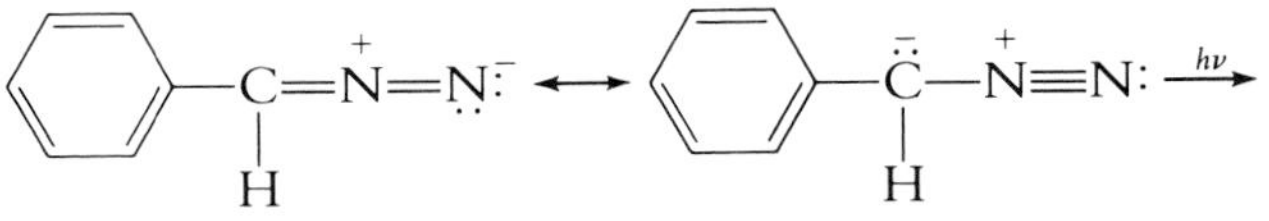

Phenyldiazomethane

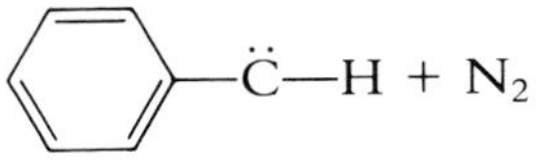

Phenylcarbene

Diazo compounds are available from a variety of precursors. One common route to simple diazo compounds is by oxidation of ketone hydrazones (eq. 26.17). When the diazo group is next to a carbonyl group, the diazotization of the amine can be used; the intermediate diazonium ion (sec. 23:9) rapidly loses a proton (eq. 26.18).

$$(26.17)\quad (C_6H_5)_2C{=}NNH_2 + HgO \rightarrow (C_6H_5)_2C{=}\overset{+}{N}{=}\overset{-}{N} + H_2O + Hg$$

Benzophenone hydrazone — Diphenyldiazomethane

$$(26.18)\quad H_2NCH_2COOEt \xrightarrow[HCl]{HNO_2} [N{\equiv}\overset{+}{N}{-}CH_2COOEt] \rightarrow \overset{-}{N}{=}\overset{+}{N}{=}CHCOOEt + H^+$$

Ethyl glycinate — Ethyl diazoacetate

Organic diazo compounds are so unstable that heating will also decompose them to carbenes and nitrogen. However, photolysis usually gives higher yields of carbene-derived products than does thermolysis.

In a similar manner, organic azides can be photochemically fragmented (eq. 26.19) to give nitrogen and a **nitrene**.

$$(26.19)\quad C_6H_5\ddot{N}{=}\overset{+}{N}{=}\ddot{\bar{N}}: \xrightarrow{h\nu} C_6H_5{-}\dot{N}: \quad + N_2$$

Phenyl azide — Phenyl nitrene

Alkyl azides can be synthesized from alkyl halides and azide ion (sec. 6:3); aromatic azides are made by treating arenediazonium ions (sec. 23:4) with azide ion (eq. 26.20).

$$(26.20)\quad Br{-}C_6H_4{-}N_2^+ + N_3^- \rightarrow Br{-}C_6H_4{-}N_3 + N_2$$

p-Bromophenyl azide

Problem

26–10 Calculate the formal charge (sec. 1:3) on carbon for a carbene and on nitrogen for a nitrene.

Because the carbon in carbenes and the nitrogen in nitrenes lack an octet of electrons, they are highly electrophilic. For instance, carbenes react rapidly with π bonds just as other electrophiles do (nitrenes do so much less effectively). The product (eq. 26.21) is a cyclopropane. Because of the mild,

$$(26.21)\quad CH_2N_2 \xrightarrow{h\nu} H_2C: \quad + N_2 \xrightarrow{H_2C=CH_2} \text{cyclo-}(CH_2)_3$$

Diazomethane — Methylene

neutral conditions in which carbenes are generated, they can be used to prepare some strained, unstable compounds (eqs. 26.22 and 26.23).

(26.22) $CH_2N_2 + CH_2{=}C{=}O \xrightarrow{h\nu}$ [cyclopropanone]$=O$ $+ N_2$

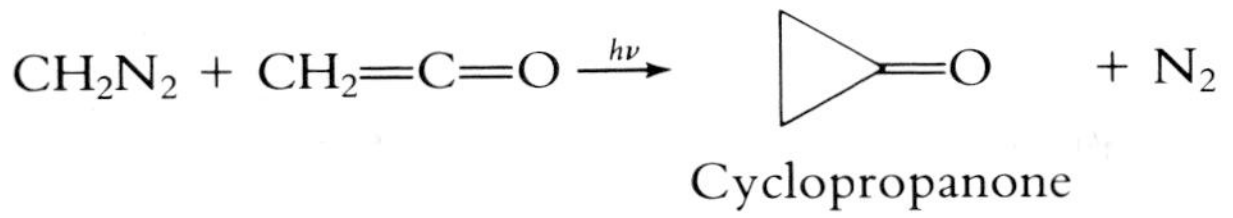

Cyclopropanone

(26.23) $CH_2{=}CH{-}CH_2{-}CHN_2 \xrightarrow{h\nu} CH_2{=}CH{-}CH_2{-}\ddot{C}H + N_2 \rightarrow$ [bicyclobutane]

1-Diazo-3-butene Bicyclobutane

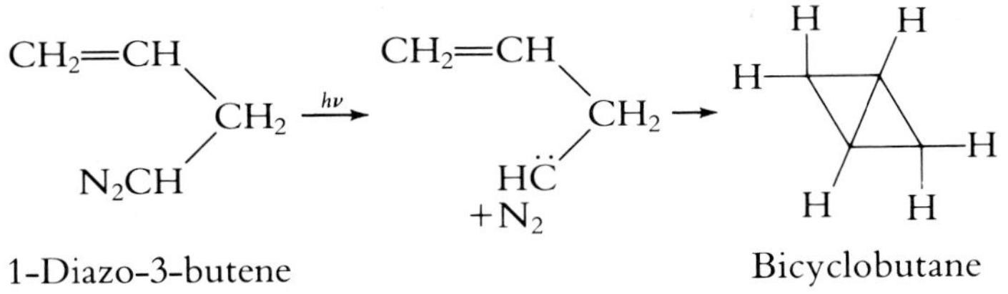

Carbenes and nitrenes are so in need of electrons that they will even attack a C—H bond to share its bonding electrons. This very unusual reaction is known as an insertion, and is depicted in Equation 26.24 for a carbene, and in Equation 26.25 for a nitrene. Equation 26.24 is an *inter*molecular insertion reaction, while Equation 26.25 is an *intra*molecular insertion reaction. In each case the reactive species is inserted between C and H atoms that were originally bonded to one another.

(26.24) [cyclobutane C—H] + $:CH_2$ $\xrightarrow{\text{gas phase}}$ [transition state] $\rightarrow$ [cyclobutyl-CH_3]

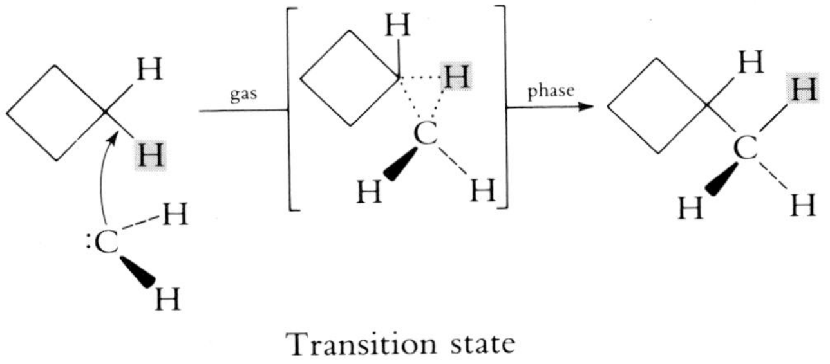

Transition state

(26.25) [o-(sec-butyl)phenyl azide, N_3] $\xrightarrow{h\nu}$ [nitrene, $\ddot{N}$] $\rightarrow$ [3-methylindoline, N—H]

(15%)

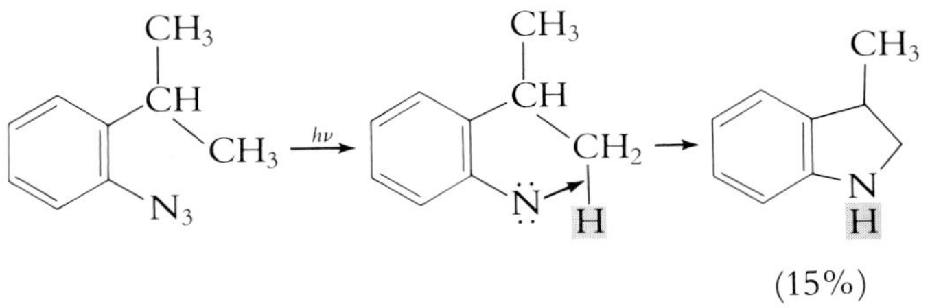

When a carbene undergoes insertion, a new C—C bond is formed and the reaction is exothermic by about 80 kcal/mol. When a reaction is this exothermic the reagent is not expected to be very selective (sec. 5:6), which is exactly what is found. Equation 26.26 shows that the product ratio is close

$$(26.26)\quad CH_3\text{—}CH_2\text{—}CH_3 + CH_2N_2 \xrightarrow{h\nu} CH_3\text{—}CH_2\text{—}CH_2\text{—}CH_3 + CH_3\text{—}CH(CH_3)\text{—}CH_3$$

(70%) (insertion at the 1° C—H bond) (30%) (insertion at the 2° C—H bond)

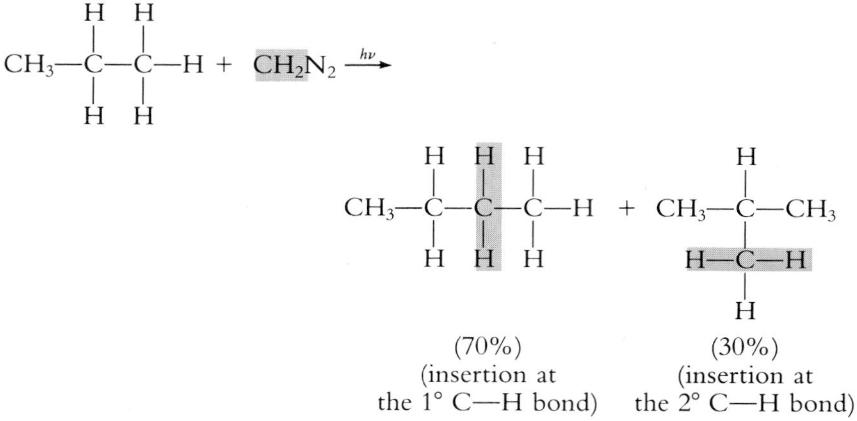

to the 75:25% expected for random attack. You can compare this with the results of Equation 5.35 (p. 151) where a very selective reagent (Br·) is attacking the same starting material. In fact, :CH_2 has about the same low selectivity as F· (Table 5–2, p. 151). Because of the mixture of products obtained, intermolecular insertion reactions such as that in Equation 26.26 are usually not useful synthetically.

Although carbenes are nonselective in the C—H insertion process, they do add to π bonds with high stereospecificity under some conditions (eq. 26.27).

$$(26.27)\quad CH_2N_2 + \underset{\text{(liq, excess)}}{cis\text{-}CH_3CH{=}CHCH_3} \xrightarrow{h\nu}$$

cis-1,2-dimethylcyclopropane (47.5%) + trans-1,2-dimethylcyclopropane (0.4%) + C—H insertion products

Addition products

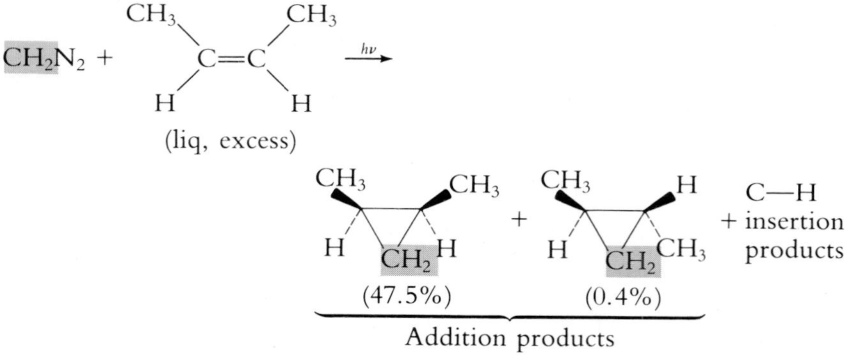

This π-bond addition reaction is the principal synthetic use made of carbenes, because it provides a route to cyclopropanes which are not easily obtained otherwise. α-Diazocarbonyl compounds are very often used for π-bond addition (eq. 26.28). The intramolecular version of this reaction provides access to many unusual and fascinating molecules (eq. 26.29).

$$(26.28)\quad \text{norbornene} + N_2CHCOEt \ (\text{C}{=}\text{O}) \xrightarrow{h\nu} \text{tricyclic cyclopropane}\text{—}\overset{O}{\overset{\|}{C}}OEt + N_2$$

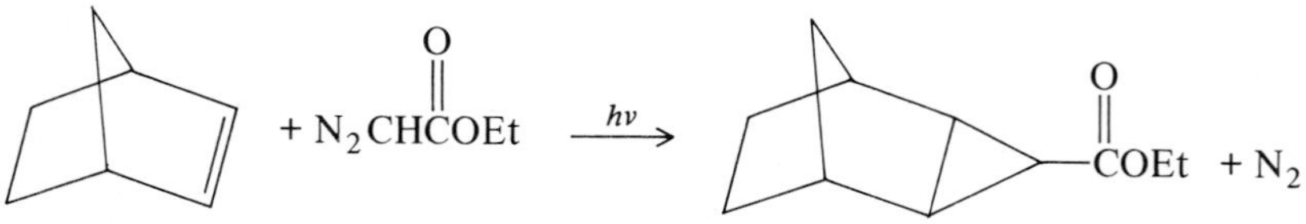

(26.29) [structure: 1,2-dimethylcyclopropene bearing $-\overset{O}{\overset{\|}{C}}CHN_2$ and H] $\xrightarrow{h\nu}$ [tricyclic ketone product bearing H, H and two CH_3 groups]

Problems

26–11 Show how each of the following compounds can be synthetized from an alkene and a diazo compound. More than one combination may be available.
a. ethyl *trans*-2-methylcyclopropanecarboxylate

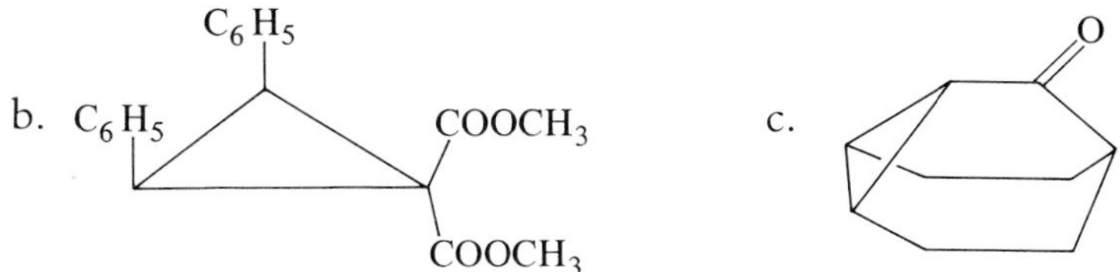

26–12 In the synthesis of bicyclobutane (eq. 26.23), a major by-product is 1,3-butadiene. Show how this compound is formed. The process that leads to 1,3-butadiene is one of the major reaction pathways of alkylcarbenes.

26–13 What two major products would you expect from the following reaction? (Pay attention to product stereoisomerism.)

$$C_6H_5CHN_2 + \underset{H}{\overset{CH_3}{}}\!\!\!\!\;C{=}C\!\!\!\!\;\underset{H}{\overset{CH_3}{}} \xrightarrow{h\nu}$$

26–14 The gas-phase photolysis of 1-diazo-2-methylpropane leads to the products shown in Equation 1. What is the likely intermediate in this reaction? Show how each of the products can be derived from this intermediate.

$$(CH_3)_2CHCHN_2 \xrightarrow{h\nu} (CH_3)_2C{=}CH_2 + \text{methylcyclopropane } (CH_3\text{–}\triangle) + CH_3CH{=}CHCH_3 \quad (1)$$

(*cis* and *trans*)

The formula used so far to depict methylene (CH_2) has left out some very important information about its *electron distribution*. This information is necessary to more fully understand the chemical behavior of carbenes.

In methylene, two of the four carbon orbitals are used in C—H bonds. That leaves two orbitals to accommodate the two nonbonding electrons. There are two important electronic states that account for the chemistry of methylene. The triplet state has one electron in each nonbonding orbital, with parallel spins. The lowest energy singlet state has both electrons in the lower energy nonbonding orbital, with spins paired (Fig. 26–5).

Triplet methylene Singlet methylene

Figure 26–5. *Geometry of triplet and lowest energy singlet methylene.*

Spectroscopic experiments have shown that the triplet state is the ground state of methylene. Nonetheless, photolysis of diazomethane yields singlet methylene which can react before it undergoes spin inversion to the triplet. Direct insertion into C—H bonds (eq. 26.26) and stereospecific addition to alkenes (eq. 26.27) are reactions characteristic of singlet carbenes. These reactions are orbital symmetry-allowed concerted reactions.

Triplet methylene cannot undergo concerted C—H insertion or stereospecific addition to alkenes. It reacts with alkenes by way of a hydrogen abstraction mechanism (eq. 26.30). The products obtained are typical of a free-radical process.

(26.30) $$R_3C{-}H + {}^3CH_2{:} \rightarrow R_3C\cdot + \cdot CH_3 \rightarrow R_3C{-}CH_3$$

Most of the nitrenes studied also appear to have triplet ground states. They often show a mixture of triplet (H abstraction) and singlet (C—H insertion) reactivity (eq. 26.31).

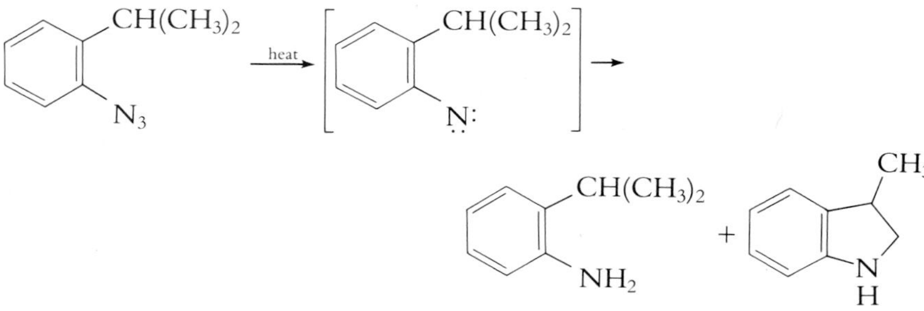

Abstraction product (37%)

Insertion product (15%)

Problem

26–15 Explain, in terms of electron repulsion (sec. 2:1), why the bond angle in ${}^3CH_2$ is greater than the ideal 120° for sp^2 hybridization while the bond angle for ${}^1CH_2$ is less than 120°.

Although diazo compound decomposition is one of the most general routes to carbenes, it is not the only route. Another important path is ***α*-elimination**, which is used to generate *halocarbenes,* $R\ddot{C}X$. In this reaction, a compound containing two or three halogens on the same carbon is treated with a base, so that it loses both H and X from that carbon atom (eq. 26.32).

(26.32)
$$RCHX_2 + Y^- \rightleftarrows R\ddot{C}^{-}(X)\text{—}X \rightarrow R\ddot{C}X + X^-$$
$$(R = H, Ar, X) \qquad + YH$$

In order to undergo α-elimination the halide must *not* be capable of undergoing β-elimination (Chap. 7), which is the favored reaction path when available.

Halocarbenes react by π addition much faster than they do by σ insertion. α-Elimination was first applied to haloforms to produce good yields of dihalocyclopropanes (eq. 26.33), and later to generate phenylhalocyclopropanes (eq. 26.34). Note the use of a phase-transfer catalyst (sec. 21:10) in Equation 26.33 to carry OH^- into the chloroform phase.

(26.33)
$$CHCl_3 \xrightarrow[R_4N^+X^-\ (cat)]{OH^-,\ H_2O} :CCl_2 \xrightarrow{C_6H_5CH=CH_2}$$ 1-phenyl-2,2-dichlorocyclopropane (H, C_6H_5 on one carbon; Cl, Cl on the other)

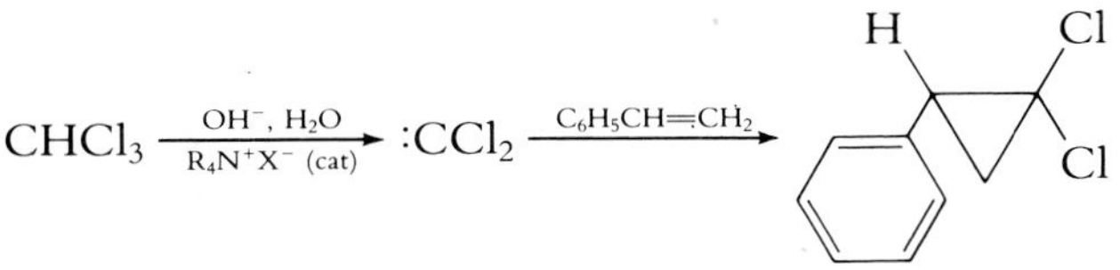

(26.34)
$$(CH_3)_2C=CH_2 + C_6H_5CHFBr + KOBu\text{-}t \rightarrow$$ 1-fluoro-1-phenyl-2,2-dimethylcyclopropane (F, C_6H_5 on one carbon; CH_3, CH_3 on the other) $+ t\text{-BuOH} + KBr$

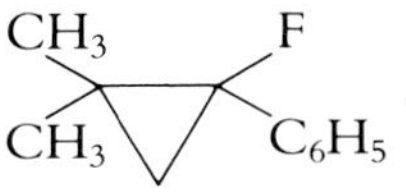

Example

a. What product would you expect from each of the following reactions?

a. $CH_3C(CH_3){=}N\text{—}C_6H_5 + CHCl_3 \xrightarrow{NaOMe} ?$

b. $CH_3CH_2CH_2CH_3 + CH_2N_2 \xrightarrow{h\nu} ?$

Answers

a. Treatment of $CHCl_3$ with strong base converts it to dichlorocarbene:

$$CHCl_3 + MeO^- \rightarrow :CCl_2 + MeOH + Cl^-$$

Dichlorocarbene adds rapidly to the π bond to form an aziridine:

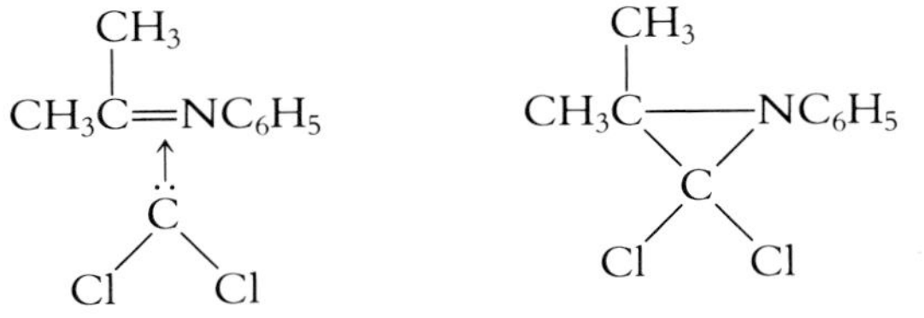

$$CH_3C(CH_3){=}NC_6H_5 \;\uparrow\; :CCl_2 \qquad\qquad CH_3C(CH_3)\text{—}NC_6H_5 \text{ (ring closed through } CCl_2)$$

b. Irradiation of CH_2N_2 produces methylene:

$$CH_2N_2 \xrightarrow{h\nu} :CH_2 + N_2$$

The very reactive methylene can insert into the C—H bonds of butane:

$$CH_3CH_2CH_2\overset{H}{\underset{H}{\overset{|}{\underset{|}{C}}}}—H + :CH_2 \rightarrow CH_3CH_2CH_2\overset{H}{\underset{H}{\overset{|}{\underset{|}{C}}}}—CH_2H \equiv CH_3CH_2CH_2CH_2CH_3$$

$$CH_3CH_2—\overset{H}{\underset{H}{\overset{|}{\underset{|}{C}}}}—CH_3 + :CH_2 \rightarrow CH_3CH_2—\overset{H}{\underset{CH_2H}{\overset{|}{\underset{|}{C}}}}—CH_3 \equiv CH_3CH_2\overset{H}{\underset{CH_3}{\overset{|}{\underset{|}{C}}}}CH_3$$

Problems

26–16 The generation and capture of carbenes from two mixed haloforms is shown in Equations 2 and 3. What carbene was generated in each of these reactions? What other carbenes *could* have been generated in each case, but were not? Use the proposed mechanism of halocarbene formation and explain why only one of the two possible carbenes was formed in each case.

$$CHCl_2Br + (CH_3)HC{=}CH(CH_3) \xrightarrow{OH^-} \text{1,1-dichloro-2,3-dimethylcyclopropane (CH}_3\text{, H; CH}_3\text{, H; Cl, Cl)} + H_2O + Br^- \quad (2)$$

$$CHCl_2F + (CH_3)HC{=}CH(CH_3) \xrightarrow{OH^-} \text{1-chloro-1-fluoro-2,3-dimethylcyclopropane (CH}_3\text{, H; CH}_3\text{, H; Cl, F)} + H_2O + Cl^- \quad (3)$$

26–17 Show how each of the following cyclopropanes can be synthesized. Indicate what complications might ensue.

a. 7,7-dibromobicyclo[4.1.0]heptane (cyclohexane fused to cyclopropane bearing Br, Br)

b. cyclopropane with Br, Br on one carbon and CH_3, H on another

c. cyclopropane with H, C_6H_5 on one carbon and Cl, C_6H_5 on another

26:8 Transition Metals: An Introduction

Most of the organometallic compounds encountered previously seem to react by furnishing carbanion-type intermediates. For example, the reactions of Grignard reagents and organolithium compounds can be explained in terms of the *strong* R^-—M^+ polarization in the carbon–metal σ bond. A different type of intermediate must be postulated in order to explain the organometallic chemistry that arises from the use of transition metals. Transition metals strongly affect the reactivity of π-bonded systems. As we will see shortly, the transition metals can accept electrons to fill *d* orbitals and thus will bind to many different **ligands** which have electrons to donate. A ligand is an ion or molecule attached directly to a metal atom or ion. The loss of electrons from the donor ligand makes it susceptible to nucleophilic attack. Equation 26.35 shows this general process, where M is a transition metal or ion and an alkene is the ligand. Transition metals can also be used to stabilize otherwise unstable ligands, as in the formation of transition metal carbene complexes.

$$\text{(26.35)}\quad M + H_2C{=}CH_2 \longrightarrow H_2C{\overset{}{=}}CH_2\ (\text{bound to } M) \xrightarrow{Nu:} H_2C(M)\text{—}CH_2(Nu) \xrightarrow{-MH} H_2C{=}CH\text{—}Nu$$

$$H_2C(M)\text{—}CH_2(Nu) \xrightarrow{+H^-} H_3C\text{—}CH_2\text{—}Nu$$

We've already seen some examples of the unique properties of transition metals in binding π ligands. For instance, the common catalysts for alkene (sec. 10:2) and alkyne (sec. 12:6) hydrogenation are Pd, Pt, and Ni—all transition metals. These and other transition metals (Ti, Cr, Co, Rh) are also used extensively in the commercial synthesis of petrochemicals on a billion pound per year scale. Furthermore, the use of transition metals in the laboratory has given the synthetic chemist new possibilities for bond formation and new degrees of control. Before dealing with a few specific examples from this vast literature, we need first to study the transition metals themselves in more detail.

The transition metals are defined as those elements that have partially filled *d* orbitals. Examples include Fe ($3d^64s^2$), Rh ($4d^85s$), and Ni ($3d^84s^2$). Table 26–1 contains an uncommon representation of the transition metals, since they do not all have the electronic configuration implied by their position in the Periodic Table. The *s* and *p* orbitals of higher valence shells are close in energy to the *d* orbitals of the next lower shell. As a result, variations in nuclear charge can cause the relative energies of these orbitals to change. That, in turn, will lead to a shift of electron distribution.

As an example, consider the elements in the eighth column of Table 26–1. The highest one, nickel, has the electron configuration $3d^84s^2$. We would therefore expect the next element in this column, palladium, to have

Table 26–1. The Transition Metals★

IIIB	IVB	VB	VIB	VIIB	VIIIB			IB
21 (Ar) $3d^1$ $4s^2$ **Sc**	22 (Ar) $3d^2$ $4s^2$ **Ti**	23 (Ar) $3d^3$ $4s^2$ **V**	24 (Ar) $3d^5$ $4s^1$ **Cr**	25 (Ar) $3d^5$ $4s^2$ **Mn**	26 (Ar) $3d^6$ $4s^2$ **Fe**	27 (Ar) $3d^7$ $4s^2$ **Co**	28 (Ar) $3d^8$ $4s^2$ **Ni**	29 (Ar) $3d^{10}$ $4s^1$ **Cu**
39 (Kr) $4d^1$ $5s^2$ **Y**	40 (Kr) $4d^2$ $5s^2$ **Zr**	41 (Kr) $4d^4$ $5s^1$ **Nb**	42 (Kr) $4d^5$ $5s^1$ **Mo**	43 (Kr) $4d^5$ $5s^2$ **Tc**	44 (Kr) $4d^7$ $5s^1$ **Ru**	45 (Kr) $4d^8$ $5s^1$ **Rh**	46 (Kr) $4d^{10}$ **Pd**	47 (Kr) $4d^{10}$ $5s^1$ **Ag**
57 (Xe) $5d^1$ $6s^2$ **La★**	72 (Xe) $4f^{14}$ $5d^2$ $6s^2$ **Hf**	73 (Xe) $4f^{14}$ $5d^3$ $6s^2$ **Ta**	74 (Xe) $4f^{14}$ $5d^4$ $6s^2$ **W**	75 (Xe) $4f^{14}$ $5d^5$ $6s^2$ **Re**	76 (Xe) $4f^{14}$ $5d^6$ $6s^2$ **Os**	77 (Xe) $4f^{14}$ $5d^7$ $6s^2$ **Ir**	78 (Xe) $4f^{14}$ $5d^9$ $6s^1$ **Pt**	79 (Xe) $4f^{14}$ $5d^{10}$ $6s^1$ **Au**

★ The *lanthanides,* elements 58 through 71, have been omitted.

the configuration $4d^8 5s^2$. However, experiment shows that Pd has the configuration $4d^{10} 5s^0$. Despite the difference in electron distribution, these two elements, (Ni and Pd) have much in common chemically. This finding points to the importance of the *total number* of electrons in the outer shells.

The transition metals can usually attain stable states if they fill their outermost *s* and *p* orbitals, and the *d* orbitals of the next outermost shell. To completely fill these *s*, *p*, and *d* orbitals, a total of 18 electrons is required. Much transition metal chemistry can be explained in terms of an 18-electron rule, which is analogous to the octet rule that governs the chemistry of main group elements.

The metals themselves do not have enough electrons to fill these *s*, *p*, and *d* orbitals, so electrons must be obtained from the ligands to which they are bound. A very effective ligand toward many zerovalent metals is carbon monoxide. For instance, nickel metal reacts with carbon monoxide at atmospheric pressure to yield the volatile, highly toxic tetracarbonylnickel (eq. 26.36). Similarly, at higher temperatures and pressures, iron can be induced

(26.36)
$$\text{Ni} + 4\text{:C}\equiv\text{O:} \xrightarrow[\text{1 atm}]{25^\circ} \text{Ni(:C}\equiv\text{O:)}_4$$

Tetracarbonylnickel
(or nickel carbonyl)

to form pentacarbonyliron (eq. 26.37). The reaction shown in Equation 26.36 is used in the refining of nickel (the *Mond process*), while thermal decomposition of $Fe(CO)_5$ yields iron powder of high purity.

(26.37)
$$\text{Fe} + 5\text{CO} \rightarrow \text{Fe(CO)}_5$$

Nickel ($3d^8 4s^2$) requires eight additional electrons to attain the stable $3d^{10} 4s^2 4p^6$ configuration, and iron ($3d^6 4s^2$) requires ten additional electrons (Table 26–1). Each carbon monoxide (:C≡O:) can share its carbon atom's nonbonding electrons with the metal atom, thus accounting for the stoichiometry of Equations 26.36 and 26.37.

If the metal atoms in these carbonyl complexes only accepted electrons from the ligands they would be very electron-rich, and the complex would not be stable. The complexes are stabilized by simultaneous electron donation from the metal to the ligand, called **back bonding**. The orbitals involved in metal–CO bonding are depicted in Figure 26–6. A filled nonbonding *d* orbital on the metal overlaps the empty $\pi^\star$ orbital of CO, forming a π bond which allows electron density to flow from the metal back to the ligand. A good ligand, then, is often one which has high-energy filled orbitals for electron donation, and low-lying empty orbitals for electron acceptance. For instance, a singlet carbene (**1**) contains the necessary empty and filled orbitals to be effective as a ligand, and several transition metals form carbene complexes.

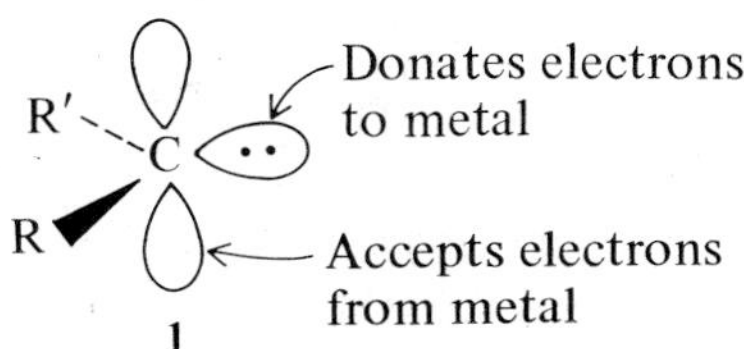

We've seen that molecules containing localized π bonds, such as ethylene (eq. 26.35) and carbon monoxide (eq. 26.36), are good ligands for transition metals. Organic compounds that contain delocalized π systems, such as butadiene and benzene, are also effective ligands. For example, chromium ($3d^5 4s^1$) requires 12 electrons from its ligands for optimal stabilization. In benzenetricarbonylchromium (**2**), six of the electrons are provided by the three carbonyl groups and six more are provided by the π electrons of the benzene ring. (The number of π bonds to benzene in **2** is really three, but the

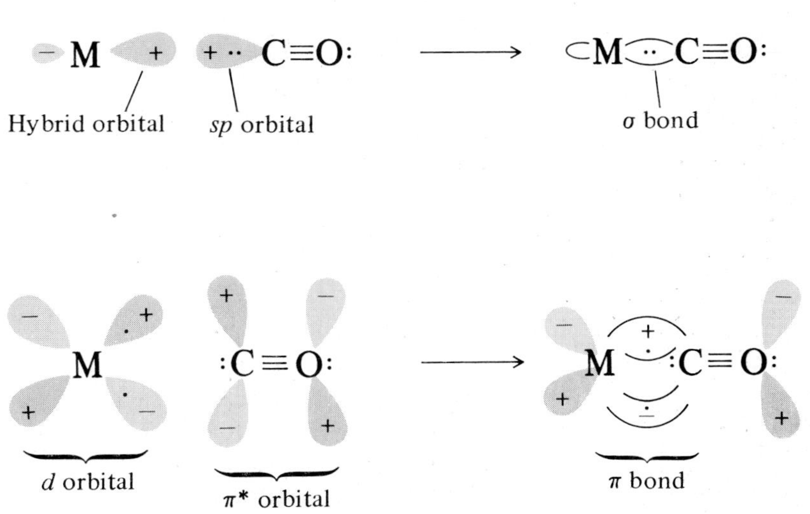

Figure 26–6. *σ and π bonds between a metal and carbon monoxide.*

convention of using a single line is widely used.) A comparable complex is cyclobutadienetricarbonyliron (**3**).

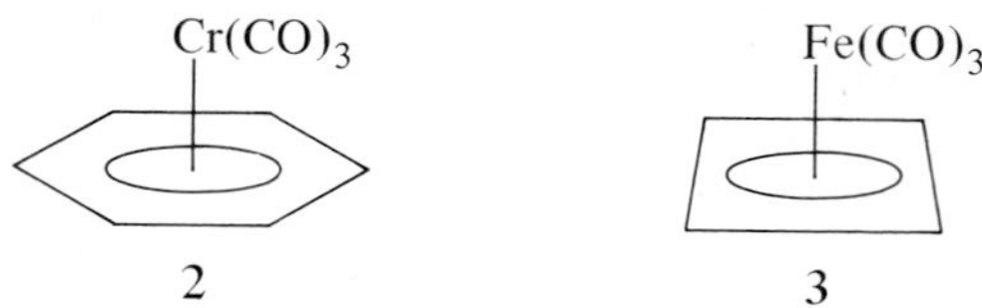

Metal ions, as well as zerovalent metals, may be used to form organometallic π complexes. Since Fe(0) requires 10 electrons from its ligand, Fe(II) will require 12. Ferrous ion reacts with two aromatic, six π electron cyclopentadienide ions (sec. 14:7) to yield *ferrocene* (eq. 26.38). Ferrocene is an example of a *sandwich compound.*

(26.38) $Fe^{2+} + 2\,(C_5H_5^-) \longrightarrow$ Fe (Ferrocene)

The π bond to the metal in these complexes is quite stable. The rings may be subjected to electrophilic (**3** and ferrocene) or nucleophilic (**2**) substitution without breaking the bond to the metal. Although free cyclobutadiene is extremely reactive, the complex **3** is stable and unreactive. Some organometallic complexes are stable enough to be used as gasoline additives for reducing engine knock.

Example

Iron(II) forms a stable complex in which both CO and $C_5H_5^-$ (cyclopentadienide) are ligands. What is the structure of this complex?

Answer

Iron(II) requires 12 electrons from its ligands. Cyclopentadienide supplies six, so the other six must come from CO. Since each CO provides two electrons, the formula of the complex is $C_5H_5Fe(CO)_3$.

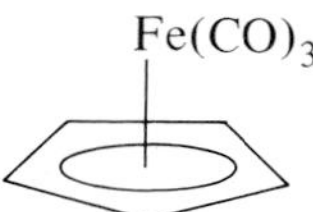

Example

Cobalt(III) forms a stable complex ion having only $C_5H_5^-$ ligands. What is the formula of and charge on this ion?

Answer

Cobalt(0) has the electron configuration ($3d^7 4s^2$), which means it has nine valence shell electrons. Therefore Co(III) has six valence shell electrons and requires 12 from its ligands. It can accommodate two cyclopentadienides.

Since each cyclopentadienide has a charge of −1, and Co(III) has a charge of +3, the net charge on the complex ion is +1.

$$2C_5H_5^- + Co^{3+} \rightarrow (C_5H_5)_2Co^+$$

Example

Pentacarbonyliron reacts with 1,3-butadiene to displace some CO ligands and give a butadiene–carbonyl complex. What is its formula?

Answer

1,3-Butadiene has four π electrons available for bonding to the metal. Thus, it can displace two CO molecules.

$$\text{(butadiene)} + Fe(CO)_5 \rightarrow C_4H_6Fe(CO)_3 + 2CO$$

Problems

26–18 Use Table 26–1 and the 18-electron rule to predict the formulas of the stable carbonyl derivatives of the following metals:
a. chromium
b. tungsten
c. ruthenium

26–19 a. All the simple carbonyl derivatives of cobalt contain more than one cobalt atom per molecule. The lowest in molecular weight is octacarbonyldicobalt, $Co_2(CO)_8$. Give a possible structural formula for this molecule. Why can a stable cobalt carbonyl with only one cobalt atom not be isolated?
b. Some metal carbonyls are stable as anions, such as $V(CO)_6^-$, $Mn(CO)_5^-$, and $Co(CO)_4^-$. Account for the stability of these anions.

26–20 Alkenes, dienes, and allyl groups may serve as π ligands for transition metals. In each of the following complexes, calculate how many electrons are being donated to the metal by the organic ligands. Assume that the metal is in the oxidation state shown in brackets. (Allyl is denoted by π-C_3H_5, and cyclopentadienyl by π-C_5H_5.)
a. $(C_6H_5C{\equiv}CC_6H_5)(\pi\text{-}C_5H_5)Mn(CO)_2$ [Mn(I)]
b. $(\pi\text{-}C_5H_5)Rh(CH_2 = CH_2)_2$ [Rh(I)]
c. $(\pi\text{-}C_5H_5)Ni(\pi\text{-}C_3H_5)$ [Ni(II)]
d. $(\pi\text{-}C_5H_5)(\pi\text{-}C_3H_5)Cr(CO)_2$ [Cr(II)]
e. $(\pi\text{-}C_3H_5)Fe(CO)_3Cl$ [Fe(II)]

26:9 Organometallic π Complexes in Organic Synthesis

Organometallic complexes are increasingly important as catalysts and as reactants in organic synthesis. There are a number of features that make them particularly effective for these purposes:

1. In some organometallic π complexes, one or more ligands dissociate readily from the metal. This makes it easy to replace one ligand by another.
2. Two ligands attached to the same metal atom are often in a very favorable position to react with each other.

3. The metal atom can change oxidation state, and in that way activate the ligands by either donating electrons to them or removing electrons from them.
4. For a given oxidation state, the metal will have a preferred hybridization state and a preferred number of ligands. The number of ligands may not be sufficient to supply the metal with needed number of electrons, so that the metal will be *coordinatively unsaturated.* It will then be reactive toward additional ligands.

It will not be possible to give even a representative sample of reactions in this vast area. Instead, we will describe a few characteristic reactions of palladium, cobalt, and nickel.

We'll start with a commercially important example of the general reaction found in Equation 26.35. Ethylene, which is available in large quantities from natural gas and petroleum refining, can be converted to acetaldehyde with palladium chloride and water (eq. 26.39). This reaction is known as the *Wacker process.*

$$H_2C{=}CH_2 + PdCl_2 + H_2O \rightarrow CH_3\underset{\displaystyle H}{\underset{|}{C}}{=}O + Pd^0 + 2HCl \quad (26.39)$$

In the most likely mechanism (eq. 26.40), water and ethylene become li-

(26.40) $PdCl_2 \xrightarrow[H_2O]{CH_2{=}CH_2}$ $Cl_2Pd(\eta^2\text{-}CH_2{=}CH_2)(OH_2)$ $\xrightarrow{:OH_2}$ $\left[Cl_2Pd(OH_2)(CH_2CH(H){-}\ddot{O}{-}H)\right]^- + H^+ \rightarrow CH_3CHO + Pd^0 + 2Cl^- + H_3O^+$

gands on the $PdCl_2$. Coordination to palladium makes the ethylene electron poor and susceptible to nucleophilic attack by another water molecule. The last step involves an intramolecular electron transfer where the organic ligand is oxidized and the palladium reduced. The high oxidation potential of Pd is a major driving force for this last step.

Note that 1 mol of $PdCl_2$ is used for every mole of acetaldehyde formed. Palladium is a rare metal and its cost (and that of its derivatives) rises and falls with the general price of metals such as gold and platinum. The price of $PdCl_2$ is generally over \$1000 per mole, while acetaldehyde generally sells for around 3¢ per mole. How can one use an expensive chemical to make a cheap one? This is only possible because the Pd(0) formed in the reaction is easily and cheaply reconverted to $PdCl_2$ with $CuCl_2$. The CuCl by-product is itself reoxidized to $CuCl_2$ with the cheapest oxidant, aerial O_2. In the overall reaction, then, ethylene is oxidized with O_2 from the air.

Another industrially important method of producing aldehydes is the *hydroformylation reaction* (also known as the *oxo process*). The name indicates that a hydrogen atom and a formyl group

$$\mathrm{HC{-}} \atop {\Vert \atop \mathrm{O}}$$

have been added to the double bond of an alkene. For example, a mixture of butanal and 2-methylpropanal is produced commercially from propene by this method (eq. 26.41), with butanal being the major product. The catalyst

$$\text{(26.41)}\quad CH_3CH{=}CH_2 + CO + H_2 \xrightarrow[150°,\ 200\ \text{atm}]{Co_2(CO)_8} CH_3CH_2CH_2CHO + CH_3\underset{\displaystyle CHO}{\underset{|}{C}}HCH_3$$

is octacarbonyldicobalt, $Co_2(CO)_8$ (Prob. 26–19). Under the reaction conditions, the catalyst is partially converted to $HCo(CO)_4$, which is the active species. Rhodium has recently been used in place of cobalt in this catalyst because is allows less stringent operating conditions.

The hydroformylation mechanism illustrates the general principles listed in the introduction to this section. Dissociation or intramolecular reaction of ligands can convert a species with 18 electrons to one with 16 electrons. The 16-electron species is coordinatively unsaturated and can bind a new ligand. A portion of the hydroformylation mechanism is shown in Equations 26.42 to 26.44; the number under each intermediate indicates the number of electrons in the outermost valence shell of the metal.

$$\text{(26.42)}\quad \underset{18}{HCo(CO)_4} \xrightarrow{-CO} \underset{16}{HCo(CO)_3} \xrightarrow{CH_2{=}CHCH_3} \underset{18}{HCo(CO)_3(C_3H_6)}$$

$$\text{(26.43)}\quad \underset{18}{(CO)_3\overset{H}{\overset{|}{C}}o\text{--}\Vert{\scriptstyle CH_2 \atop CHCH_3}} \longrightarrow \underset{16}{(CO)_3Co{-}\overset{CH_3}{\overset{|}{C}}HCH_3} \xrightarrow{CO} \underset{18}{(CO)_4Co{-}CH(CH_3)_2}$$

$$\text{(26.44)}\quad \underset{18}{(CO)_3\overset{\overset{O}{\Vert}}{\overset{C}{\Vert}}\!\!\!Co{-}CH(CH_3)_2} \longrightarrow \underset{16}{(CO)_3Co{-}\overset{O}{\overset{\Vert}{C}}CH(CH_3)_2}$$

$$\xrightarrow{HCo(CO)_4} H\overset{O}{\overset{\Vert}{C}}CH(CH_3)_2 + \underset{16}{Co_2(CO)_7}$$

Nickel π complexes are known for their ability to bring about the oligomerization (dimerization, trimerization, etc.) of alkenes and alkynes. Some examples are given in Equations 26.45 and 26.46.

(26.45) $$4HC{\equiv}CH \xrightarrow{Ni(CN)_2}$$

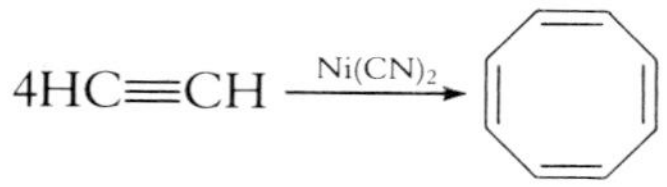

(26.46) $$3CH_2{=}CHCH{=}CH_2 \xrightarrow{(C_8H_{12})_2Ni}$$

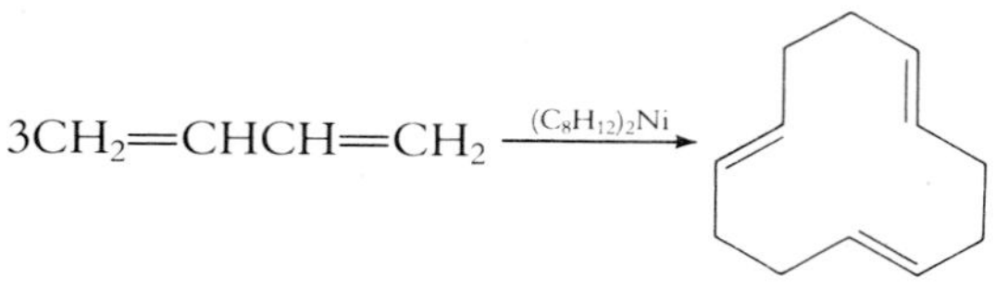

The reaction in Equation 26.45 was developed by the German chemist W. Reppe during World War II to provide a nonpetroleum source of gasoline. Germany had plentiful supplies of coal from which coke could be produced and, in turn, CaC_2. Treatment of CaC_2 with water produces acetylene (eq. 12.1, p. 433). Various transition metal catalysts were used by Reppe to make trimers and tetramers (as in eq. 26.45) for conversion to gasoline feedstocks. Although this work was done in the 1940s, the exact mechanism of this reaction is still not known.

The cyclododecatriene produced in Equation 26.46 is useful because it can be reduced to cyclododecene, which can then be converted to $HOOC(CH_2)_{10}COOH$ and $H_2N(CH_2)_{12}NH_2$. These are condensed to give nylon 12,12 (sec. 22:15), making an important commercial polymer available from the abundant 1,3-butadiene.

The mechanism of Equation 26.46 has many steps and is still under active investigation, so we cannot give a detailed account of it. Nonetheless, a partial description of the mechanism is useful because it illustrates some of the principles described earlier. The first step involves replacement of the two C_8H_{12} (1,5-cyclooctadiene) ligands on Ni by two butadienes (eq. 26.47). Because of their proximity and the effect of the metal on their π systems, the butadienes couple to form an organonickel intermediate (**4**).

(26.47)

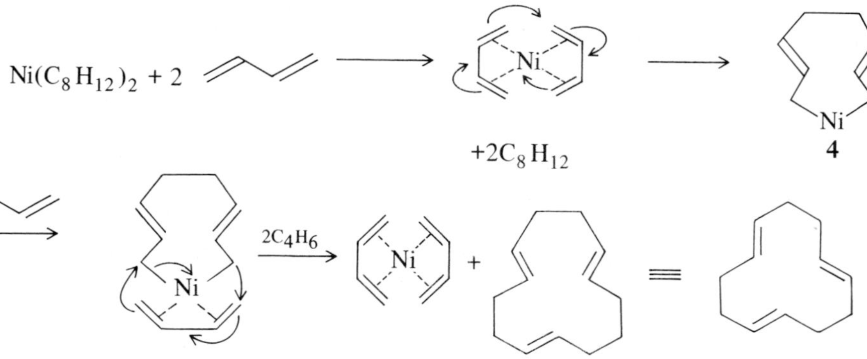

$Ni(C_8H_{12})_2$ + 2 ⟶ Ni ⟶ Ni **4**

$+2C_8H_{12}$

⟶ Ni $\xrightarrow{2C_4H_6}$ Ni + ≡

This intermediate binds a third butadiene molecule, which couples to the C_8 unit to give the final product and regenerate the nickel catalyst.

It should be clear from the reactions given in this section that organic compounds behave very differently when they are complexed to transition metals than when they are free. In fact, there is no way to carry out the reactions shown above in the absence of transition metals.

26 Photochemically Induced Intermediates

Example

A rhodium complex used for hydroformylation reactions is $HRh(CO)(PPh_3)_3$. What product(s) are formed when this catalyst is used as follows:

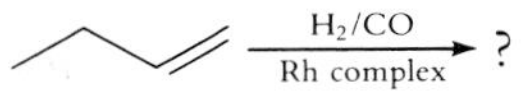

$$\xrightarrow[\text{Rh complex}]{H_2/CO} ?$$

Answer

The result of hydroformylation is to convert an alkene to an aldehyde:

$$\rightarrow \text{CHO}$$

Major product

Usually the straight-chain isomer is predominant, but some of the branched-chain isomer can also be formed:

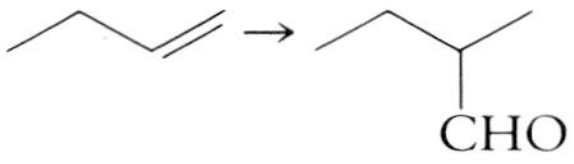

$$\rightarrow \text{CHO}$$

Problems

26–21 The cyano group coordinates with metals in a manner similar to carbon monoxide (eq. 5). Suggest a mechanism for Equation 6.

$$Ni + HCN \rightarrow H{-}Ni{-}CN \quad (5)$$

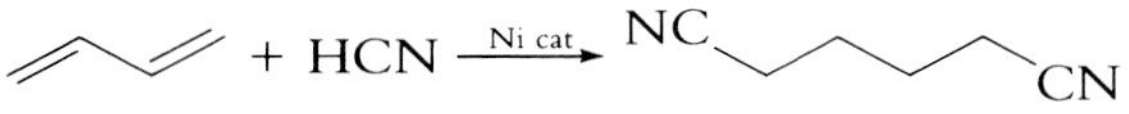

$$+ \; HCN \xrightarrow{\text{Ni cat}} NC\text{~~~~}CN \quad (6)$$

26–22 Give a mechanism for the following reaction. (Note the similarity to the Wacker process.)

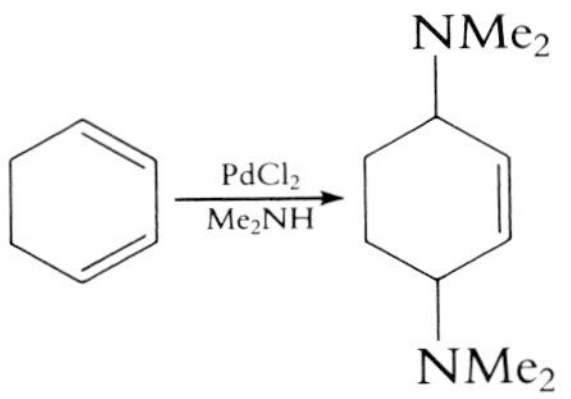

$$\xrightarrow[Me_2NH]{PdCl_2}$$

NMe_2

NMe_2

Summary

1 An organic molecule can absorb light and be promoted to an excited singlet state, which can then convert to an excited triplet state. Singlet states have all electrons paired (opposite spins), while triplet states have two unpaired electrons (same spin). An excited singlet state can lose its energy as emitted light called fluorescence, while triplet states can emit light called phosphorescence. The excited molecules can also undergo chemical reactions (sec. 26:1).

2 The photoexcitation of alkenes usually involves the promotion of a π electron to an antibonding orbital, giving a $\pi,\pi^\star$ excited state. The same transition can take place upon irradiation of a carbonyl group. In addition, carbonyl groups have a lower energy excitation pathway involving the n-electrons, leading to an $n,\pi^\star$ excited state. In both cases the light energy is concentrated in the light-absorbing chromophore. This makes possible specific reactions for photoexcited molecules (sec. 26:2).

Photoexcited alkenes can dimerize to yield cyclobutanes (sec. 26:3). They also undergo *cis–trans* isomerization and, in the case of stilbenes, the *cis* isomer can photocyclize to a phenanthrene when O_2 or I_2 is present (sec. 26:4).

Photoexcited carbonyls can react with alkenes to give oxetanes, and they can abstract hydrogens from alcohols to yield pinacols. An intramolecular variation of the latter reaction can yield cyclobutanols. This is accompanied by a fragmentation called Norrish Type II cleavage. An analogous electron impact fragmentation (mass spectrometric reaction) is called the McLafferty rearrangement (sec. 26:5).

3 Carbenes and nitrenes, which are neutral but highly reactive species with only six valence shell electrons, can be generated from diazo and azido compounds by light irradiation. These intermediates are highly electrophilic and add to π bonds to form three-membered rings. Carbenes and nitrenes also undergo insertion into C-H bonds. Halocarbenes can be generated from di- or trihalides by α-elimination; they undergo selective π-bond addition. Carbenes have either singlet or triplet electronic configuration. Singlets and triplets differ in geometry and in energy; for many carbenes, the lowest energy form is the triplet (sec. 26:7).

4 Transition metals such as Co, Ni, Fe, Pt, and Pd can accept electrons from donors such as alkenes, arenes, and carbon monoxide. The number of ligands which can donate electron pairs depends on the particular transition metal and its oxidation state. Most of the stable metal–ligand complexes obey the 18-electron rule (sec. 26:8). The complexes formed are very reactive; coordinated alkenes, for example, are often susceptible to nucleophilic attack. Ligands attached to the same metal atom are often in a favorable relationship to react with one another. Addition of new ligands can take place as coordinatively saturated complexes rearrange to form coordinatively unsaturated ones (sec. 26:9).

Problems

26–23 What products could be formed from ultraviolet irradiation of the following compounds?

a. fumaric acid

b. acetophenone dissolved in cyclohexene

c. F–(3-fluorophenyl)–CH=CH–(3-fluorophenyl)–F

d. $CH_3CH_2OCH_2CH_2CH_2C(=O)CH_2CH_3$

26–24 Irradiation of 2-butylquinoline yields 2-methylquinoline and propene (eq. 7). 4-Butylquinoline is stable to light irradiation. Propose a mechanism for this reaction. (Note its similarity to ketone photocleavage.)

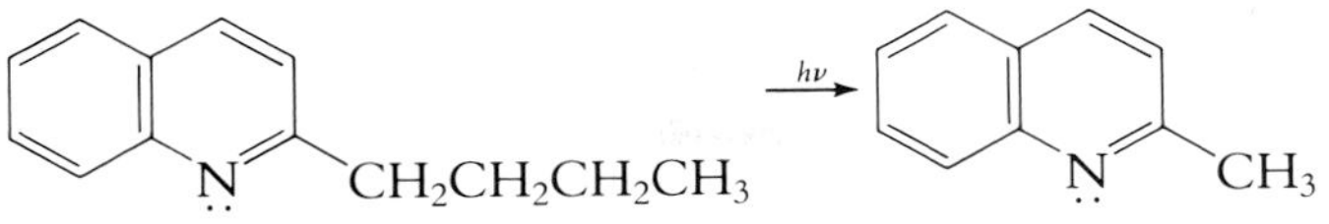

$$\text{2-butylquinoline } (CH_2CH_2CH_2CH_3) \xrightarrow{h\nu} \text{2-methylquinoline } (CH_3) + CH_2{=}CH{-}CH_3 \quad (7)$$

26–25 Show how each of the following compounds could be synthesized from an appropriate alkene.

a. cyclopropyl methyl ketone
b. ethyl 2,2-diphenylcyclopropanecarboxylate
c. 1,1-dichloro-*trans*-2,3-diphenylcyclopropane
d. 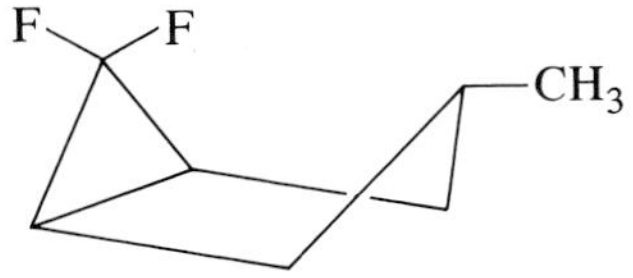 F F CH$_3$

26–26 The insertion of $^3CH_2$ into C—H is believed to involve hydrogen abstraction followed by combination of the two radicals so produced (eq. 26.30). If this mechanism is correct, two additional products should be present. Give their structures and show how they were formed. In several reactions of CH_2, such additional products have indeed been found, suggesting that the triplet rather than the singlet was the reactive species.

26–27 The *Reimer–Tiemann reaction* consists of treating phenols with aqueous base and chloroform. When phenol is treated this way, the products obtained after acidification are salicylaldehyde (*o*-hydroxybenzaldehyde) and *p*-hydroxybenzaldehyde (eq. 8). When *p*-cresol is subjected to the Reimer–Tiemann reaction, it yields a methylsalicylaldehyde as the major product, and the cyclohexadienone (**III**) as the minor product (eq. 9). Suggest a mechanism that can rationalize the formation of the products shown in Equations 8 and 9. It will be helpful to consider the carbene intermediate which could be produced from $CHCl_3$ and OH^-, and how it might yield product **III**.

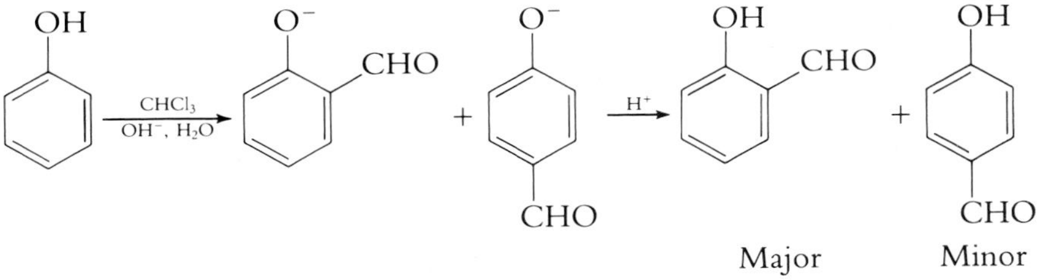

$$\text{phenol (OH)} \xrightarrow[OH^-,\ H_2O]{CHCl_3} \text{o-}O^-\text{, CHO} + \text{p-}O^-\text{, CHO} \xrightarrow{H^+} \text{o-OH, CHO (Major)} + \text{p-OH, CHO (Minor)} \quad (8)$$

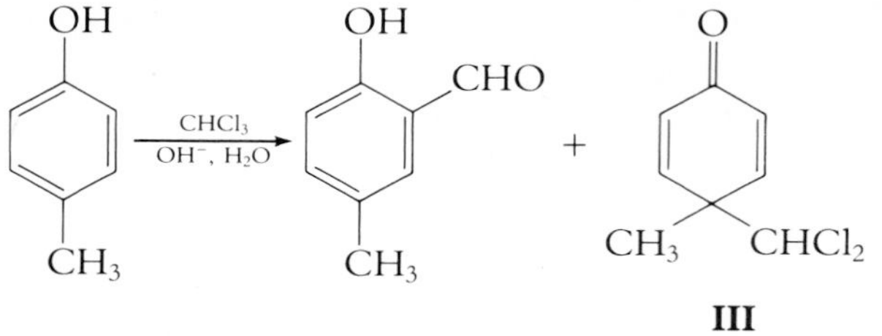

$$\text{p-cresol} \xrightarrow[\text{OH}^-,\ \text{H}_2\text{O}]{\text{CHCl}_3} \text{2-hydroxy-5-methylbenzaldehyde (CHO)} + \text{4-methyl-4-(dichloromethyl)cyclohexa-2,5-dienone } (\mathbf{III}) \tag{9}$$

26–28 Diphenylacetylene and *benzal chloride* (dichlorophenylmethane) are dissolved together in benzene and the solution is then allowed to react with potassium *t*-butoxide. Work-up of the benzene layer affords a white crystalline material, **A**, $C_{25}H_{24}O$. The nmr spectrum of this compound shows absorption due to aromatic protons, as well as a singlet in the methyl region. The ratio of aromatic to methyl peak areas is 5:3. When **A** is dissolved in ether and dry hydrogen bromide is then passed in, compound **B** precipitates ($C_{21}H_{15}Br$, mp 270°). **B** is insoluble in benzene, ether, and chloroform but soluble in aqueous ethanol. Write structures for **A** and **B**, and account for the properties of **B**. (The properties of **B** may be clarified by reference to sec. 14:7.)

26–29 Ethyl azidoformate may be decomposed either thermally or photochemically. When the decomposition is allowed to proceed in cyclohexene, the products **IV**, **V**, and **VI** are isolated (eq. 10). Suggest a mechanism for the formation of products **IV** to **VI**.

$$\underset{\text{Ethyl azidoformate}}{C_2H_5O\overset{O}{\overset{\|}{C}}N_3} \xrightarrow{\text{cyclohexene}}$$

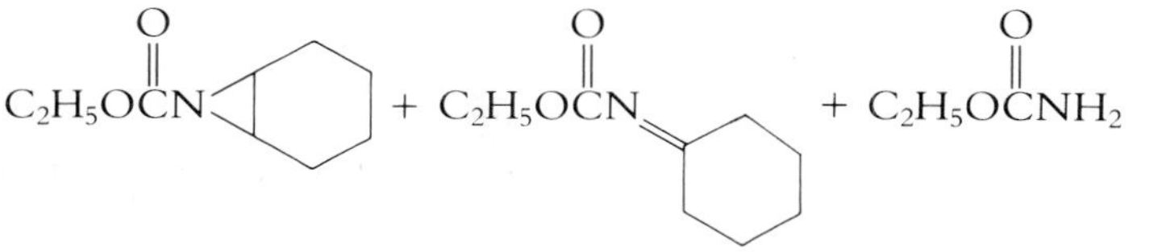

$$\underset{\mathbf{IV}}{C_2H_5O\overset{O}{\overset{\|}{C}}N\text{(7-azabicyclo[4.1.0]heptane)}} + \underset{\mathbf{V}}{C_2H_5O\overset{O}{\overset{\|}{C}}N{=}\text{cyclohexylidene}} + \underset{\mathbf{VI}}{C_2H_5O\overset{O}{\overset{\|}{C}}NH_2} \tag{10}$$

26–30 One convenient method of carrying out carbene reactions is to decompose the tosylhydrazones of aldehydes or ketones in the presence of base (the *Bamford–Stevens reaction*). The tosylhydrazones are made by allowing the carbonyl compound to react with *tosylhydrazine*. The decomposition of the

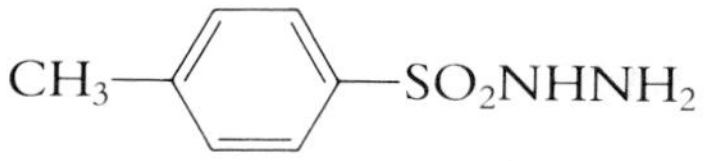

$$CH_3\text{—}C_6H_4\text{—}SO_2NHNH_2$$

Tosylhydrazine

cyclopentanecarboxaldehyde tosylhydrazone is shown in Equation 11. What are the likely intermediates in this reaction? Show how they are formed from the starting materials, and how they yield the observed products.

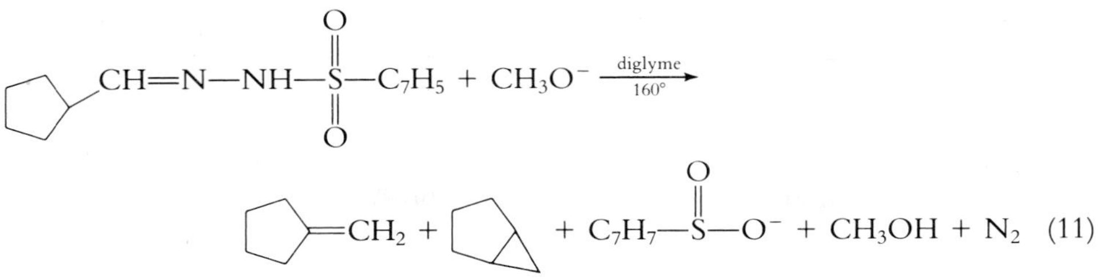

$$\text{C}_5\text{H}_9\text{—CH=N—NH—}\overset{\text{O}}{\underset{\text{O}}{\overset{\|}{\underset{\|}{\text{S}}}}}\text{—C}_7\text{H}_5 + \text{CH}_3\text{O}^- \xrightarrow[160°]{\text{diglyme}}$$

$$\text{(cyclopentylidene)=CH}_2 + \text{(bicyclo[3.1.0]hexane)} + \text{C}_7\text{H}_7\text{—}\overset{\text{O}}{\overset{\|}{\text{S}}}\text{—O}^- + \text{CH}_3\text{OH} + \text{N}_2 \quad (11)$$

(27%) (73%)

26–31 Suggest structures for the product of each rection.

a. $CH_3\text{—C}_6H_5 + Cr(CO)_6 \xrightarrow{\text{heat}}$

b. $CH_3I + Na^+\ Mn(CO)_5^- \xrightarrow[\text{THF}]{}$

c. $[(C_3H_5)PdCl]_2 + 2Na^+C_5H_5^- \rightarrow$

d. $(C_5H_5)_2Fe + CH_3\overset{\text{O}}{\overset{\|}{\text{C}}}Cl \xrightarrow{AlCl_3}$
Ferrocene

e. $CH_2\text{=}CHCOOCH_3 \xrightarrow[Co_2(CO)_8]{H_2,\ CO}$

f.  (cyclopentene) $\xrightarrow[\text{Ni}]{\text{HCN}}$

26–32 Calculate the oxidation state of the metal in each of the following complexes. To do so you need to keep in mind the charges on each of the ligands—some are neutral (CO, C_2H_4), some are mononegative (X^-, CN^-, $C_5H_5^-$), and so on.

a. $Fe(CO)_4Br_2$
b. $Co(CO)_2(\pi\text{-}C_5H_5)$
c. $[PtCl_3(C_2H_4)]^-$
d. $Mn(CO)_3(C_6H_6)$
e. $Ni(CO)I(\pi\text{-}C_5H_5)$
f. $Mo(CO)_2(\pi\text{-}C_3H_5)(\pi\text{-}C_5H_5)$

26–33 Rhodium is now partially supplanting cobalt as the catalytic metal in the commercial hydroformylation process (sec. 26:9). Would you expect the intermediates in the Rh-catalyzed reaction to have the same structures as those in the Co-catalyzed reaction (eqs. 26.42 to 26.44)? Explain your answer.

Supplementary Reading

N. J. Turro, *Modern Molecular Photochemistry,* Benjamin/Cummings, Menlo Park, Calif., 1978.

A. Padwa, *Organic Photochemistry*, Marcel Dekker, Inc., New York, N.Y., 1979.

C. A. Buehler, Carbenes in insertion and addition reactions, *J. Chem. Ed.*, **49**, 239 (1972).

T. L. Gilchrist and C. W. Rees, *Carbenes, Nitrenes and Arynes*, Appleton-Century-Crofts, New York, N.Y. 1969.

P. R. Mitchell and R. V. Parish, The 18-electron rule, *J. Chem. Ed.*, **46**, 811 (1969).

J. P. Collman and L. S. Hegedus, *Principles and Applications of Organotransition Metal Chemistry*, University Science Books, Mill Valley, Calif., 1980.

G. W. Parshall, *Homogenous Catalysis*, John Wiley, New York, N.Y., 1980.

27 Introduction to Chemical Literature: Books, Abstracts, and Journals

A textbook of organic chemistry can cover only the barest minimum of concepts and facts related to this continually growing science. Even with a good index, a textbook makes a relatively poor reference volume in the practical world of chemical research and development. If you become involved in undergraduate research or even in open-ended laboratory courses in organic chemistry, some introduction to the literature where you can find properties of compounds or reaction procedures will be of value. The first section of this chapter outlines how and where new facts or ideas in organic chemistry are published. A more detailed description is then given of how to search the literature for information on specific compounds or reactions.

3136 *J. Am. Chem. Soc.* **1983**, *105*, 3136–3147

Stereoselectivity of Diels–Alder Cycloadditions to Norbornyl- and Norbornenyl-Fused Spirocyclic Cyclopentadiene Systems[1]

Leo A. Paquette,*[2a] Pana Charumilind,[2a] Michael C. Böhm,[2b] Rolf Gleiter,*[2b] Lawrence S. Bass,[2c] and Jon Clardy[2c]

Contribution from the Evans Chemical Laboratories, The Ohio State University, Columbus, Ohio 43210, Institut für Organische Chemie der Universität Heidelberg, D-6900 Heidelberg, West Germany, and Department of Chemistry-Baker Laboratory, Cornell University, Ithaca, New York 14853. *Received July 15, 1982*

Abstract: The question of long-range stereoelectronic control in Diels–Alder reactions was studied by examining the stereoselectivity of dienophile addition to the four spirocyclic dienes **1–4**. The two spirocyclopropane derivatives (**1** and **2**) exhibit a strong predilection for below-plane attack with a wide range of dienophiles to give adducts having *syn*-sesquinorbornene geometry. In contrast, spirocyclopentane **3** enters into [4 + 2] cycloaddition totally by top-face bonding to generate *anti*-sesquinorbornene derivatives, except when dimethyl acetylenedicarboxylate is involved. The added double bond in **4** induces a loss in stereoselectivity. The PE spectra of **1** and **3** have been recorded and the ground-state electronic character of these dienes has been analyzed by detailed computational methods. The range of factors which are potentially able to account for these stereochemical phenomena is discussed. The best rationalization to this time deals with σ orbtial mixing with the π_s diene orbital, such interactions serving to tilt the diene orbitals with resultant minimization of antibonding influences on the bottom surface of **1** and top face of **3**.

This report deals with the stereochemical features of Diels–Alder reactions to four spirocyclopentadienes (**1–4**) in which the

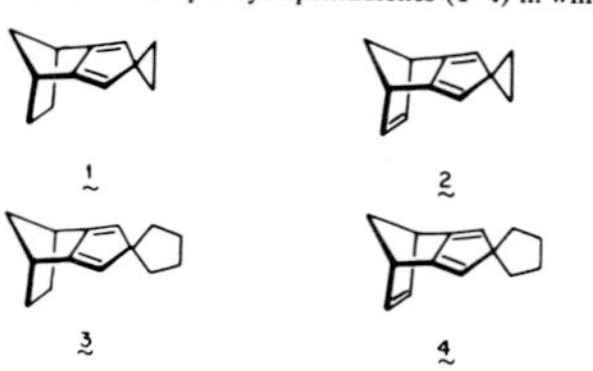

unsaturated five-membered rings are fused to norbornyl and norbornenyl moieties. These substrates, made available through spiroalkylation of the corresponding cyclopentadienide anions,[1]

Table I. Selected ^{13}C Chemical Shifts in the Cycloadducts Derived from 1 (ppm, $CDCl_3$ Solution)[a]

compd	C_2', C_3'	C_{12}'	C_{11}'	$C_{8}'a$, $C_{10}'a$
5	25.49	50.19	44.24	55.16
6	25.79	50.43	52.31	52.80
7	25.05	50.15	44.08	66.46
	26.07			30.15
8	26.22	46.75	53.65	66.22
	26.51			31.36
9	25.88[b]	49.91	45.44	25.59[b]
13	25.29	50.01	41.56	52.00
14	25.73	50.25	57.05	49.76
16	22.58	48.01	66.22	150.06

[a] See formula **6** for numbering scheme. [b] These values are possibly interchanged.

the gem-dimethyl analogues is reported in the ensuing paper.[6b]

Figure 27–1.

27:1 How New Discoveries Are Made Known

When new chemical reactions or theories are developed, or new compounds are synthesized or discovered in nature, the information is first made public in refereed journals or patents. The chemist usually submits the details of his or her discovery to one of many monthly or biweekly journals. The submitted manuscript is read by two or three other chemists knowledgeable in the field, known as referees, who accept or reject the paper for publication. If the process or compound discovered appears to have commercial value or possibilities for wide use, the information might first be disclosed in a patent application rather than in a journal. Disclosure in a patent often takes several years, while journal publication usually takes three to nine months after the manuscript has been accepted. Figure 27–1 shows part of the title page and part of the Experimental Section of an article from the *Journal of the American Chemical Society*, a biweekly publication. Figure 27–2 shows the title page

United States Patent [19] [11] **3,950,236**

Miller et al. [45] **Apr. 13, 1976**

[54] **PRODUCTION OF ANGULAR ALKYLATED POLYCYCLIDES BY ELECTROCHEMICAL ANNELATION**

[75] Inventors: **Larry L. Miller; Frank R. Stermitz,** both of Fort Collins, Colo.; **J. Russell Falck,** Albuquerque, N. Mex.

[73] Assignee: **The United States of America as represented by the Secretary of Health, Education and Welfare,** Washington, D.C.

[22] Filed: **July 24, 1974**

[21] Appl. No.: **491,610**

[52] **U.S. Cl.** **204/78**

[51] **Int. Cl.**^{2} **C25B 3/02**

[58] **Field of Search** 204/78

[56] **References Cited**

UNITED STATES PATENTS

2,960,452 11/1960 Slager et al. 204/78

Primary Examiner—R. L. Andrews
Attorney, Agent, or Firm—Holman & Stern

[57] **ABSTRACT**

A process of producing angular alicylated polycyclides by electrochemical annelation wherein a biphenyl compound containing an alkyl group is oxidized at the anode in an electrolyte to form the corresponding dieone.

6 Claims, No Drawings

3,950,236

1

PRODUCTION OF ANGULAR ALKYLATED POLYCYCLIDES BY ELECTROCHEMICAL ANNELATION

The subject invention is directed to a novel process for producing cyclic compounds having an angular alkyl group. In particular, the instant invention is directed to the anodic cyclization and rearrangement of alkoxy bibenzyl compounds to the corresponding phenanthrone compounds which are useful as precursors in the preparation of both steroids and terpenoids.

Annelelation reactions, notably those which result in an angular alkyl group, occupy a prominent position in numerous synthetic sequences. This is especially true in the total synthesis of complex molecules such as steroids and terpenoids wherein the presence of angular methyl groups impose stringent limitations on the number and scope of suitable methods of preparation.

The anodic coupling of dimethoxybenzene rings has recently been demonstrated to have utility and resulted in the production of a variety of morphinandienones.

2

prior work. However, the preparation of such cyclic compounds wherein an angular alkyl group is present has created substantial difficulty. Moreover, the preparation of such compounds by electrochemical means has previously not been disclosed.

Therefore, it is an object of the subject invention to disclose a process for preparing a polycyclic compound having an angular alkyl group.

A further object of the subject invention is to prepare such polycyclic compounds by electrochemical means.

A still further object of the subject invention is to prepare phenanthrone compounds by electrochemical means, which compounds are useful in the preparation of steroid and terpenoid compounds.

These and other objects of the subject invention will become more evident from the following more detailed disclosure thereof.

As noted, the subject invention is directed to a process of producing angular alkylated polycyclides by electrochemical annelation. The reaction involved is not yet fully understood, but it is hypothesized that a very clean coupling reaction is produced which unex-

Figure 27–2.

and part of the Introduction to a United States patent. Journals and patents are often referred to as the **primary literature**.

At least a small part of the information in a journal or patent article will appear at least once more in print. This will be when an abstract of the publication appears in *Chemical Abstracts. Chemical Abstracts* publishes short summaries of articles (Fig. 27–3) from the primary literature and lists (by structural formula and name) essentially all compounds whose chemistry is described in journal and patent articles. *Chemical Abstracts* starts a new volume every six months and numbers its abstracts consecutively. The number of chemical articles abstracted is very large. In 1982 for example, there were 457,789 articles abstracted! *Chemical Abstracts* summarizes articles from *all* fields of chemistry, but it is probably fair to estimate that about one-half of the nearly 460,000 articles in 1982 contained information at least partly related to organic chemistry.

Because of the huge amount of literature available in primary journals, **secondary sources** of information are of importance. *Chemical Abstracts* is one secondary source. There are also review journals (*Chemical Reviews, Accounts of Chemical Research*) where articles summarizing the results of numerous primary publications are found. In addition, there are annual compi-

78165 *Chemical Abstracts* Vol. 79, 1973

alkyl halide, and $AgBF_4$ gave mixts. of primary and secondary sulfonium salts. However, pure primary alkyl diphenylsulfonium salts can be prepd., although in low yield, by the reaction of diphenyl sulfide with *n*-alkyl trifluoromethanesulfonates.

78166j **Reaction of vinyl ethers with chlorocarbene.** Shostakovskii, S. M.; Aksenova, L. N.; Erofeev, V. I.; Aksenov, V. S. (USSR). *Dokl. Vses. Konf. Khim. Atsetilena, 4th* **1972**, 2, 354–9 (Russ). From *Ref. Zh., Khim.* **1973**, Abstr. No. 4Zh120.

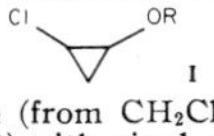

Reaction of chlorocarbene (from CH_2Cl_2 and BuLi in Et_2O or pentane at −60° to +10°) with vinyl ethers gave the *cis, trans*-cyclopropanes (**I**); the isomer ratio did not depend on solvent or temp. (cis isomers predominated). Parameters of the reaction were discussed. **I** isolated were (R and configurations given): Bu, *cis;* Bu, *trans;* Et, *cis;* vinyl, *cis*.

78167k **gem-Chloro(methylthio)cyclopropanes from gem-chloro(methylthio)carbene.** Moss, Robert A.; Pilkiewicz, Frank G. (Sch. Chem., Rutgers State Univ., New Brunswick, N.J.). *Synthesis* **1973**, (4), 209–11 (Eng). Inverse addn. of solid Me_3COK to solns. of $MeSCHCl_2$ in $Me_2C{:}CMe_2$, $Me_2C{:}CHMe$, $Me_2C{:}CH_2$, and *trans*-2-butene gave the corresponding cyclopropane derivs. (**I**; R, R^1, R^2, R^3 and % yield given): R = R^1 = R^2 = R^3 = Me, 28; R = R^1 = R^2 = Me, R^3 = H and R = R^1 = R^3 = Me, R^2 = H, 15; R = R^1 = Me, R^2 = R^3 = H, 17; R = R^3 = Me, R^1 = R^2 = H, 8 and a carbene dimer MeSClC:CClSMe in 23–72% yield in every case. The NMR spectra of **I** were discussed.

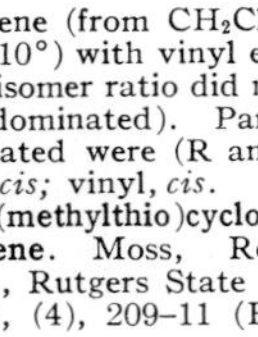

78168m **General method for the conversion to cyclobutanones of cyclopropanols and cyclopropyl silyl ethers carrying an acceptor carbon in position 1.** Barnier, J. P.; Garnier, B.; Girard, C.; Denis, J. M.; Salaun, J.; Conia, J. M. (Lab. Carbocycles, Univ. Paris-Sud, Orsay, Fr.). *Tetrahedron Lett.* **1973**, (20), 1747–50 (Fr). 1-Hydroxycyclopropanes and the corresponding Me or Me_3Si ethers, when also 1-substituted with a partially pos. group [CH_2Br, CH_2Cl, $CH_2O_3SC_6H_4Me$-*p*, CH_2OH, CMe_2OSiMe_3, $CMe(OMe)_2$, $CH{:}CH_2$, COMe, CHO] underwent ring enlargement to cyclobutanones under acidic, basic, or thermal conditions; e.g., 1-methylenecyclopropane reacted with *N*-bromosuccinimide in H_2O-Me_2SO to give 1-(bromomethyl)cyclopro-

Katal. Pererab. Uglevodorod. Syr'ya **1971**, No. 5, 146–51 (Russ). From *Ref. Zh., Khim.* **1973**, Abstr. No. 3N136. Optimum conditions for hydroisomerization of C_6H_6 to methylcyclopentane over the bifunctional catalyst D-31 (2CoO-MoO_3 15, partially chlorinated $Al(OH)_3$ 85, Fe oxide 0.5%) were 400°, 120 atm, and space velocity 0.25 hr^{-1}; the yield was 72%.

78173j **Production of carbon-14-labeled methylcyclopentane through isomerization of carbon-14-labeled cyclohexane.** Russow, R. (VEB Otto "Grotewhol" Boehlen, Boehlen, E. Ger.). *Isotopenpraxis* **1973**, 9(6), 223–4 (Ger). Refluxing cyclohexane-$^{14}C_6$ 6 hr over $AlCl_3$ gave 28% methylcyclopentane-$^{14}C_6$.

78174k **Hydroisomerization of benzenes over bifunctional catalyst D-30.** Khasanov, A. K.; Kadyrov, I. (USSR). *Katal. Pererab. Uglevodorod. Syr'ya* **1971**, No. 5, 133–9 (Russ). From *Ref. Zh., Khim.* **1973**, Abstr. No. 3N250. A catalyst, consisting of Ni and Mo oxides, supported on partially chlorinated $Al(OH)_3$, was prepd. and used for hydroisomerization of benzenes to cyclopentanes.

78175m **Stereoselective total synthesis of (±)-prostaglandin E_1.** Taub, D.; Hoffsommer, R. D.; Kuo, C. H.; Slates, H. L.; Zelawski, Z. S.; Wendler, N. L. (Merck Sharp and Dohme Res. Lab., Merck and Co., Inc., Rahway, N.J.). *Tetrahedron* **1973**, 29(11), 1447–56 (Eng). Prostaglandin E_1 was prepd. stereoselectively from 5-methoxy-1-indanone through the keto ester (**I**), which was converted to the *cis*-hydrindenone (**II**). Oxidative cleavage of the ethylene acetal of **II** formed an epimerizable Ac group to give the all-trans intermediate [**III**, R = $(CH_2)_2CO_2Me$, R^1 = Ac], which was converted to the aldehyde (**III**, R = CHO, R^1 = AcO) and thence to (±)-prostaglandin E_1.

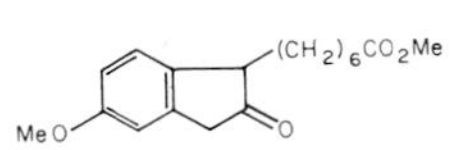

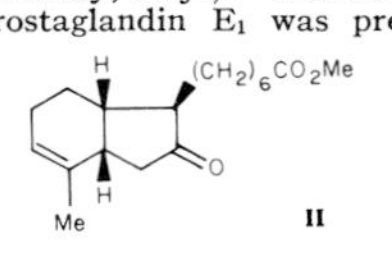

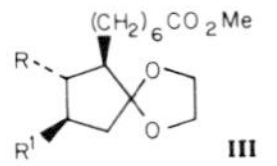

78176n **Approach to the prostaglandin skeleton through organoborane reactions.** Attanasi, Orazio; Baccolini, Graziano; Caglioti, Luciano; Rosini, Goffredo (Lab. Eteroatomi, Cons. Naz. Ric., Ozzano Emilia, Italy). *Gazz. Chim. Ital.* **1973**, 103-

Figure 27–3.

lations of the more important journal articles in a given field (e.g., *Advances in Photochemistry* or *Progress in Physical Organic Chemistry*). Certain reactions or syntheses of wide interest may also be described in special secondary sources (e.g., *Organic Reactions* or *Organic Syntheses*).

Physical properties of compounds taken from the primary literature appear again in handbooks and/or dictionaries. The most comprehensive of these is *Beilstein's Handbuch der Organischen Chemie* (a German publication commonly referred to simply as *Beilstein*). It contains data, such as physical constants, on *all* organic compounds ever described in the chemical literature, but is currently many years behind in its compilation. *Beilstein* also contains some information on synthesis and reactions of compounds listed. Other handbooks are far more restricted and may contain data for only a few compounds (e.g., *The Handbook of Chemistry and Physics* and *Lange's Handbook of Chemistry*) or for specialized ones (e.g., *The Merck Index*, which is a good source of data on compounds of medicinal or toxicological interest).

The advantage of secondary sources of information is that they are much more easily or rapidly searched than primary sources. Their disadvantages are that many of them contain summaries or only part of the data required, and that all are not *comprehensive* in their coverage.

27:2 How to Search for Physical Properties of Specific Compounds

In a way, searching the chemical literature for information is often the reverse procedure from that used to get information into the literature. The procedure followed is different depending upon whether one desires information on a specific compound or on a reaction or mechanism. In this section, methods of rapidly finding out specifics about a given compound (such as its physical properties) are examined.

As an example, suppose that you want to know the melting point of 2,4-dichlorophenol in order to verify the structure of an unknown isolated from the chlorination of phenol. A first approach would be to look for this infor-

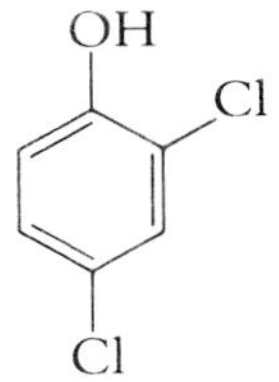

OH
Cl
Cl

2,4-Dichlorophenol

mation in a handbook. *Lange's Handbook of Chemistry* (with listings of 6500 compounds) gives the data on the page reproduced in Figure 27–4, which indicates a melting point of 45° C. The *Handbook of Chemistry and Physics*, published by the Chemical Rubber Company, lists the same compound (Fig. 27–5) but under the heading "Phenol, 2,4-dichloro-." Note that both handbooks provide data such as crystalline form, solubility, and a *Beilstein* reference.

An alternative method for finding the same data would have been to go directly to *Beilstein*. *Beilstein* is published in German and no English translation is available. German is often a required language for the B.S. degree in chemistry, but even those who know no German can readily learn abbrevia-

512 **PHYSICAL CONSTANTS OF ORGANIC COMPOUNDS** 513

No.	Name	Synonym	Formula	Beil. Ref.	Formula Weight	Crystalline Form and Color	Specific Gravity	Melting Point °C.	Boiling Point °C.	Solubility in 100 Parts: Water	Alcohol	Ether
	Dichloro											
2021	4-nitroaniline	(2,6;4,1)	$Cl_2C_6H_2(NO_2)NH_2$	XII-735	207.03	yel. nd./ac.	...	194-5	...	i. h. HCl	s.	...
2022	nitrobenzene (2,4)	NO_2-Cl-benzene	$Cl_2C_6H_3{\cdot}NO_2$	V-245	192.00	nd./al.	$1.439^{80°}$	33	258.5	i.	v. s. h.	∞
2023	nitrobenzene (2,5)	...	$Cl_2C_6H_3{\cdot}NO_2$	V-245	192.00	tri./al.	$1.669^{22°}$	54.6	266	i.	v. s. h.	v. s. bz.
2024	nitrobenzene (2,6)	...	$Cl_2C_6H_3{\cdot}NO_2$	V-246	192.00	mn. pr.	$1.603^{17°}$	72.5	130^{8mm}	i.	s.	s. CS_2
2025	nitrobenzene (3,4)	...	$Cl_2C_6H_3{\cdot}NO_2$	V-246	192.00	nd./al.	$1.456^{7.5°}_{4}$	42-3(α)*	255-6	...	...	...
2026	nitrobenzene (3,5)	...	$Cl_2C_6H_3{\cdot}NO_2$	V-246	192.00	yel. mn.	$1.692^{14°}$	65.4	...	i.	s.	s. ac.
2027	1-nitroethane (1,1)	...	$Cl_2C(NO_2){\cdot}CH_3$	...	143.96	lq.	$1.405^{20°}_{20}$	...	122-5	$<0.5^{20°}$	...	...
2028	nitrohydrin	diCl-Pr-nitrate	$ClCH_2{\cdot}CHCl{\cdot}CH_2{\cdot}NO_3$	I-356	173.98	lq.	$1.37°$	...	180	...	...	...
2029	nitrophenol	(4,6;2,1)	$Cl_2C_6H_2(NO_2)OH$	VI-241	208.00	yel. mn./al.	$1.822^{20°}_{4}$	122-3	subl. <100	sl. s.	sl. s.; s. bz.	s; s. chl.
2030	nitrophenol	(2,6;4,1)	$Cl_2H_6H_2(NO_2)OH$	VI-241	208.00	yel. pl./et.	...	125 d.	...	i.; sl. s. bz.	s. h.	v. s.
2031	nitropropane	(1,1,1)	$C_2H_5{\cdot}C(NO_2)Cl_2$	...	157.98	lq.	$1.314^{20°}_{20}$	...	141-4	$<0.5^{20°}$	...	...
2032	1-nitrosoethane (1,1)	...	$CH_3{\cdot}CCl_2{\cdot}NO$	I-99	127.96	b. oil	$1.252^{19°}$	...	68	i.	...	...
2033	pentane (1,4)	...	$CH_3CHCl(CH_2)_3Cl$	I-131	141.04	col. lq.	...	...	$58\text{-}60^{15mm}$	i.	∞	∞
2034	pentane (1,5)	amylene chloride	$ClCH_2(CH_2)_3CH_2Cl$	I-131	141.04	col. lq.	$1.094^{25°}_{4}$	...	180-1	i.; s. CS_2	s.; s. chl.	s.
2035	pentane (2,3)	...	$C_2H_5(CHCl)_2CH_3$	I-131	141.04	col. lq.	...	...	138-9	i.	∞	∞
2036	pentane (2,4)	...	$(CH_3CHCl)_2{:}CH_2$	*I-43	141.04	col. lq.	$1.063^{18°}$	...	147-50	i.	∞	∞
2037	pentane	1,4-dichloro-2-methyl-butane	$ClCH_2{\cdot}CH(CH_3){\cdot}CH_2{\cdot}CH_2Cl$	*I-47	141.04	col. lq.	$1.103^{21°}_{4}$	...	170-2	i.	∞	∞
2038	pentane	2,3-dichloro-2-methyl-butane	$(CH_3)_2CCl{\cdot}CHCl{\cdot}CH_3$	I-135	141.04	lq.	$1.068^{15°}_{4}$	...	130-5	i.	∞	∞
2039	pentane	2,4-dichloro-2-methyl-butane	$(CH_3)_2CCl{\cdot}CH_2{\cdot}CH_2Cl$	I-135	141.04	col. lq.	$1.065^{20°}_{4}$	...	152-4 sl. d.	i.	∞	∞
2040	pentane	3,4-dichloro-2-methyl-butane	$(CH_3)_2CH{\cdot}CHCl{\cdot}CH_2Cl$	I-135	141.04	col. lq.	$1.092^{17.5°}$	...	143-5	i.	∞	∞
2041	pentane	4,4-dichloro-2-methyl-butane	$(CH_3)_2CH{\cdot}CH_2{\cdot}CHCl_2$	I-135	141.04	col. lq.	$1.05^{24°}$	...	130	i.	∞	∞
2042	phenol (2,4)	...	$Cl_2C_6H_3{\cdot}OH$	VI-189	163.00	nd./bz.	$1.383^{60°}_{25}$	45	209-10	$0.45^{20°}$ v. s. chl.	v. s.; $150^{25°}$ CCl_4	v. s.; v. s. bz.
2043	phenyl-3,4-dichlorobenzene sulfonate (2,4)	...	$Cl_2C_6H_3SO_3{\cdot}C_6H_3Cl_2$	...	372.06	col. cr.	...	81-2	...	i.	s. h.	s.
2044	phenylhydrazine	(2,5)	$Cl_2C_6H_3{\cdot}NH{\cdot}NH_2$	XV-431	177.03	nd./aq.	...	105	...	sl. s. h.	s.	s.; s. ac.
2045	phenylhydrazine-4-sulfonic acid	(2,5)	$Cl_2C_6H_2(SO_3H){\cdot}NH{\cdot}NH_2$	XV-643	257.10	nd./aq. HCl	...	...	...	...	...	...
2046	phenyl-phenyl phosphate (*o,o'*)	phosphen 4	$C_6H_5{\cdot}O{\cdot}PO{:}(ClC_6H_4{\cdot}O)_2$	...	395.18	col. lq.	$1.34^{25°}_{25}$	<0	$255\text{-}7^{5mm}$	i.; ∞ bz.	v. s.	∞ CCl_4
2047	phenyl-4-toluene sulfonate	...	$CH_3{\cdot}C_6H_4{\cdot}SO_3{\cdot}C_6H_3Cl_2$	...	317.19	col. nd.	...	118-9	...	i.	s. h.	i.

Figure 27–4.

PHYSICAL CONSTANTS OF ORGANIC COMPOUNDS (Continued)

No.	Name	Synonyms and Formula	Mol. wt.	Color. crystalline form. specific rotation and λ_{max} (log ε)	m.p. °C	b.p. °C	Density	n_D	Solubility: w	al	eth	ace	bz	other solvents	Ref.
	Phenol														
Ω p623	—,—,dihydro-chloride*	Amidol. $C_6H_8N_2O.2HCl$. *See* p555	197.07	nd $\lambda^{w,\,pH=3}$ 235 sh (3.6), 280 (3.3)	230–40d	...	...	...	v	δ	δ	...	...	...	**B13**^{2}, 308
p624	—,**2,5-diamino-***	$C_6H_8N_2O$. *See* p555	124.15	nd	68	...	...	...	v					...	**B13**^{2}, 312
p625	—,**3,4-diamino-***	$C_6H_8N_2O$. *See* p555	124.15	nd	170–2	...	...	...	...	...	...	...	...	...	**B13**^{1}, 210
p626	—,**3,5-diamino-***	$C_6H_8N_2O$. *See* p555	124.15	nd or pr (chl)	168–70 (180)	...	...	...	v	...	δ	...	...	chl s^h	**B13**, 567
Ω p627	—,**2,4-dibromo-***	$C_6H_4Br_2O$. *See* p555	251.92	nd (peth)	40	238–9 177^{17}	...	...	δ s^h	v	v	...	v	CS_2 v	**B6**^{2}, 188
p628	—,**2,6-dibromo-***	$C_6H_4Br_2O$. *See* p555	251.92	nd (w) $\lambda^{0.1N\,NaOH}$ 367.5 (3.57)	56–7	162^{21} sub	...	...	s^h	v	v	...	...	...	**B6**^{2}, 189
Ω p629	—,**2,6-dibromo-4-nitro-***	$C_6H_3Br_2NO_3$. *See* p555	296.92	pa ye pr or lf (al)	145–6	d > 144	...	...	i	v^h	v^h	...	δ	CS_2 v^h aa δ lig i	**B6**^{2}, 234
Ω p630	—,**2,3-dichloro-***	$C_6H_4Cl_2O$. *See* p555	163.00	cr (lig, bz) $\lambda^{0.1N\,HCl}$ 277 (3.52), 280 (3.30), 283 (3.29)	57–9	...	...	...	...	s	s	...	s^h	lig s^h	**B6**^{3}, 699
Ω p631	—,**2,4-dichloro-***	$C_6H_4Cl_2O$. *See* p555	163.00	hex nd (bz) $\lambda^{0.1N\,NaOH}$ 304 (3.50)	45	210^{760} $145\text{–}7^{110}$	...	...	δ	s	s	...	s	chl s	**B6**^{3}, 699
Ω p632	—,**2,5-dichloro-***	$C_6H_4Cl_2O$. *See* p555	163.00	pr (bz, peth) $\lambda^{0.1N\,HCl}$ 280 (3.41)	59	211^{744}	...	...	δ	v	v	...	s	peth s^h	**B6**^{3}, 712
Ω p633	—,**2,6-dichloro-***	$C_6H_4Cl_2O$. *See* p555	163.00	nd (peth) $\lambda^{0.1N\,NaOH}$ 238 (3.85), 301 (3.71)	68–9	$219\text{–}20^{740}$ $80\text{–}5^{4}$	...	...	...	v	v	...	s	peth s^h	**B6**^{3}, 713
Ω p634	—,**3,4-dichloro-***	$C_6H_4Cl_2O$. *See* p555	163.00	nd (bz-peth) $\lambda^{0.1N\,NaOH}$ 244 (4.12), 302 (3.51)	68	253.5^{767}	...	...	δ	v	v	...	s	peth s^h	**B6**^{3}, 715
Ω p635	—,**3,5-dichloro-***	$C_6H_4Cl_2O$. *See* p555	163.00	pr (peth) λ^{al} 220 (3.86), 278 (3.23)	68	233^{757} $122\text{–}4^{8}$	...	...	δ	v	v	...	...	peth s^h	**B6**^{3}, 715

Figure 27–5.

tions for melting point (F.) and boiling point (Kp.) and find these data in the text. Figure 27–6 shows a portion of the general subject index (*Generalsachregister*) of *Beilstein* with the entry for dichlorophenol indicated. The boldface **6** indicates Volume VI. The following numbers (189, 190) refer to the pages in Volume VI on which information about dichlorophenols (and their simple derivatives) can be found. *Beilstein* has been published in a main series and first, second, third, and fourth supplements (*Ergänzungswerke*). In the entry in Figure 27–6 for dichlorophenol, the numbers "I 102, 103" and "II 178, 179" refer to the pages for Volume VI of the first supplement (I) and the second supplement (II). The third and fourth supplements are not as yet complete and hence do not have a cumulative index. Each individual volume is indexed, however, and one can check Volume VI of the third and fourth supplements for additional data on 2,4-dichlorophenol. Figure 27–7 gives a reproduction of the *Beilstein* main series entry for 2,4-dichlorophenol. Preparations and reactions of this compound (with references to the primary journal articles), as well as physical properties, are given. The main series through the third supplement covers organic compounds appearing in the literature through 1949 (approximately). The fourth supplement covers 1950 to 1959 and is nearing completion.

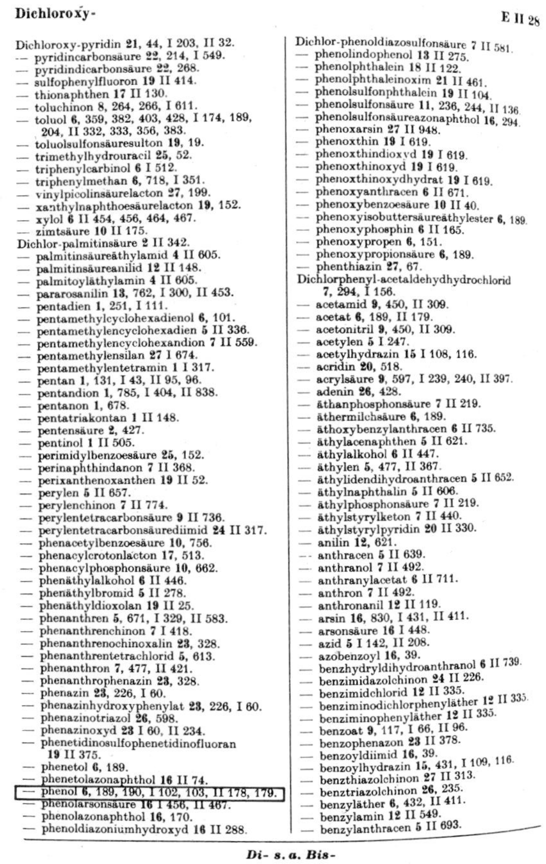
Dichloroxy-   E II 28

Dichloroxy-pyridin 21, 44, I 203, II 32.
— pyridincarbonsäure 22, 214, I 549.
— pyridindicarbonsäure 22, 268.
— sulfophenylfluoron 19 II 414.
— thionaphthen 17 II 130.
— toluchinon 8, 264, 266, I 611.
— toluol 6, 359, 382, 403, 428, I 174, 189, 204, II 332, 333, 356, 383.
— toluolsulfonsäuresulton 19, 19.
— trimethylhydrouracil 25, 52.
— triphenylcarbinol 6 I 512.
— triphenylmethan 6, 718, I 351.
— vinylpicolinsäurelacton 27, 199.
— xanthylnaphthoesäurelacton 19, 152.
— xylol 6 II 454, 456, 464, 467.
— zimtsäure 10 II 175.
Dichlor-palmitinsäure 2 II 342.
— palmitinsäureäthylamid 4 II 605.
— palmitinsäureanilid 12 II 148.
— palmitoyläthylamin 4 II 605.
— pararosanilin 13, 762, I 300, II 453.
— pentadien 1, 251, I 111.
— pentamethylcyclohexadienol 6, 101.
— pentamethylencyclohexadien 5 II 336.
— pentamethylencyclohexandion 7 II 559.
— pentamethylensilan 27 I 674.
— pentamethylentetramin 1 I 317.
— pentan 1, 131, I 43, II 95, 96.
— pentandion 1, 785, I 404, II 838.
— pentanon 1, 678.
— pentatriakontan 1 II 148.
— pentensäure 2, 427.
— pentinol 1 II 505.
— perimidylbenzoesäure 25, 152.
— perinaphthindanon 7 II 368.
— perixanthenoxanthen 19 II 52.
— perylen 5 II 657.
— perylenchinon 7 II 774.
— perylentetracarbonsäure 9 II 736.
— perylentetracarbonsäurediimid 24 II 317.
— phenacetylbenzoesäure 10, 756.
— phenacylcrotonlacton 17, 513.
— phenacylphosphonsäure 10, 662.
— phenäthylalkohol 6 II 446.
— phenäthylbromid 5 II 278.
— phenäthyldioxolan 19 II 25.
— phenanthren 5, 671, I 329, II 583.
— phenanthrenchinon 7 I 418.
— phenanthrenochinoxalin 23, 328.
— phenanthrentetrachlorid 5, 613.
— phenanthron 7, 477, II 421.
— phenanthrophenazin 23, 328.
— phenazin 23, 226, I 60.
— phenazinhydroxyphenylat 23, 226, I 60.
— phenazinotriazol 26, 598.
— phenazinoxyd 23 I 60, II 234.
— phenetidinosulfophenetidinofluoran 19 II 375.
— phenetol 6, 189.
— phenetolazonaphthol 16 II 74.
— phenol 6, 189, 190, I 102, 103, II 178, 179.
— phenolarsonsäure 16 I 456, II 467.
— phenolazonaphthol 16, 170.
— phenoldiazoniumhydroxyd 16 II 288.
Dichlor-phenoldiazosulfonsäure 7 II 581.
— phenolindophenol 13 II 275.
— phenolphthalein 18 II 122.
— phenolphthaleinoxim 21 II 461.
— phenolsulfonphthalein 19 II 104.
— phenolsulfonsäure 11, 236, 244, II 136.
— phenolsulfonsäureazonaphthol 16, 294.
— phenoxarsin 27 II 948.
— phenoxthin 19 I 619.
— phenoxthindioxyd 19 I 619.
— phenoxthinoxyd 19 I 619.
— phenoxthinoxydhydrat 19 I 619.
— phenoxyanthracen 6 II 671.
— phenoxybenzoesäure 10 II 40.
— phenoxyisobuttersäureäthylester 6, 189.
— phenoxyphosphin 6 II 165.
— phenoxypropen 6, 151.
— phenoxypropionsäure 6, 189.
— phenthiazin 27, 67.
Dichlorphenyl-acetaldehydhydrochlorid 7, 294, I 156.
— acetamid 9, 450, II 309.
— acetat 6, 189, II 179.
— acetonitril 9, 450, II 309.
— acetylen 5 I 247.
— acetylhydrazin 15 I 108, 116.
— acridin 20, 518.
— acrylsäure 9, 597, I 239, 240, II 397.
— adenin 26, 428.
— äthanphosphonsäure 7 II 219.
— äthermilchsäure 6, 189.
— äthoxybenzylanthracen 6 II 735.
— äthylacenaphthen 5 II 621.
— äthylalkohol 6 II 447.
— äthylen 5, 477, II 367.
— äthylidendihydroanthracen 5 II 652.
— äthylnaphthalin 5 II 606.
— äthylphosphonsäure 7 II 219.
— äthylstyrylketon 7 II 440.
— äthylstyrylpyridin 20 II 330.
— anilin 12, 621.
— anthracen 5 II 639.
— anthranol 7 II 492.
— anthranylacetat 6 II 711.
— anthron 7 II 492.
— anthronanil 12 II 119.
— arsin 16, 830, I 431, II 411.
— arsonsäure 16 I 448.
— azid 5 I 142, II 208.
— azobenzoyl 16, 39.
— benzhydryldihydroanthranol 6 II 739.
— benzimidazolchinon 24 II 226.
— benzimidchlorid 12 II 335.
— benziminodichlorphenyläther 12 II 335.
— benziminophenyläther 12 II 335.
— benzoat 9, 117, I 66, II 96.
— benzophenazon 23 II 378.
— benzoyldiimid 16, 39.
— benzoylhydrazin 15, 431, I 109, 116.
— benzthiazolchinon 27 II 313.
— benztriazolchinon 26, 235.
— benzyläther 6, 432, II 411.
— benzylamin 12 II 549.
— benzylanthracen 5 II 693.

Di- s. a. Bis-

Figure 27–6.

Syst. No. 522.] DICHLORPHENOLE. 189

2.4-Dichlor-phenol $C_6H_4OCl_2 = C_6H_3Cl_2 \cdot OH$. *B.* Entsteht neben geringeren Mengen 2.6-Dichlor-phenol bei 2-tägigem Einleiten von Chlor in 500 g geschmolzenes Phenol (F. FISCHER, *A. Spl.* **7**, 180; vgl. LAURENT, *A. ch.* [2] **63**, 27, 37; [3] **3**, 210; *A.* **23**, 61; **43**, 212), sowie beim Chlorieren von Phenol mittels alkal. Natriumhypochloritlösung (CHANDELON, *B.* **16**, 1751). Bei der Chlorierung von Salicylsäure in Kalilauge (in Gegenwart von mehr als 3 KOH) (TARUGI, *G.* **30** II, 489). Durch Erhitzen von 3.5-Dichlor-2-oxy-benzoesäure mit Kalk (HECHT, *Am.* **12**, 505; ZINCKE, *A.* **261**, 254; T., *G.* **30** II, 488). — Sechsseitige Nadeln (aus Benzol). Geruch unangenehm und anhaftend (F.). F: 43° (F.). Kp: 209—210° (F.). 100 Tle. Wasser lösen bei 20° ca. 0,45 Tle. (MOSSO, *J.* **1887**, 1300). Leicht löslich in Alkohol, Äther, Benzol und Chloroform (F.). Elektrolytische Dissoziationskonstante k bei 25°: $3,1 \times 10^{-8}$ (HANTZSCH, *B.* **32**, 3070). Die Lösung in schwachem Alkohol reagiert sauer (F.). Salzbildungsvermögen: THIEL, ROEMER, *Ph. Ch.* **63**, 735. Zersetzt beim Kochen mit Wasser kohlensaure Alkalien und Erden, wird aber bei gewöhnlicher Temperatur aus seinen Salzen durch CO_2 größtenteils wieder ausgeschieden (F.). Bestimmung des Hydrolysegrades von 2.4-Dichlor-phenol-Natrium durch Messung der Verseifungsgeschwindigkeit des Methylacetats: HANTZSCH, *B.* **32**, 3083. — Bei der Einw. von rauchender Salpetersäure erhält man 4.6-Dichlor-2-nitrophenol (F.). Das Kaliumsalz liefert mit CO_2 bei 140° 3.5-Dichlor-2-oxy-benzoesäure (T.). — $NH_4C_6H_3OCl_2$. Tafeln. Verliert an der Luft alles Ammoniak (F.). — $KC_6H_3OCl_2 + {}^1/_2H_2O$. Täfelchen. Gibt an Wasser freie Säure ab (F.). — $AgC_6H_3OCl_2$. Gelblicher amorpher Niederschlag. Sehr unbeständig (F.). — $HO \cdot PbC_6H_3OCl_2$. Weißer, amorpher, in Wasser unlöslicher Niederschlag (F.). — Diäthylaminsalz. *B.* Aus Diäthylamin und 2.4-Dichlor-phenol in Äther bei —70° (PETERS, *B.* **39**, 2783). Krystalle (aus Äther); löslich in warmem Äther; beständig bei Zimmertemperatur (P.).

Methyl-[2.4-dichlor-phenyl]-äther, 2.4-Dichlor-anisol $C_7H_6OCl_2 = C_6H_3Cl_2 \cdot O \cdot CH_3$. *B.* Durch Chlorieren von Anisol (HUGOUNENQ, *A. ch.* [6] **20**, 510). Aus 4-Chlor-2-amino-anisol nach SANDMEYER (REVERDIN, ECKHARD, *B.* **32**, 2624). — Rhombische (MOREL, *A. ch.* [6] **20**, 514) Prismen. F: 27—28°; $Kp_{743,5}$: 232—233° (korr.); Kp_{10}: 125° (H.).

Äthyl-[2.4-dichlor-phenyl]-äther, 2.4-Dichlor-phenetol $C_8H_8OCl_2 = C_6H_3Cl_2 \cdot O \cdot C_2H_5$. *B.* Durch Kochen von 2.4-Dichlor-phenol-Kalium mit Äthyljodid (F. FISCHER, *A. Spl.* **7**, 183). — Flüssig. Kp: 236—237°. Wird von wäßr. Kalilauge nicht verändert.

Figure 27–7.

Beilstein groups compounds in various volumes according to structural class and assigns classes various "System Numbers." Searching according to *Beilstein* class is complex and is not commonly done. If more than just physical properties and a few leading references to chemistry is needed, then a complete search of *Chemical Abstracts* must be done (see sec. 27:3).

Finally, a word needs to be said about nomenclature, subject indexes, and formula indexes. Various methods are employed in indexing names, as was seen from the simple case of 2,4-dichlorophenol, found under "Dichloro-" in one handbook and in *Beilstein,* and under "Phenol, dichloro-" in another handbook. There is no general agreement on indexing and it may be alphabetical according to the IUPAC name (Dichlorophenol), or based upon what might be considered the parent compound (Phenol, dichloro-). The latter method is used by *Chemical Abstracts* (although *Chemical Abstracts* has changed its indexing methods several times through the years). With other compounds, additional problems may arise when one consults an index. For example, 2,4-dichloroaniline is listed alphabetically in *Lange's Handbook,* under "Aniline, 2,4-dichloro-" in the *Handbook of Chemistry and Physics* and in *Beilstein,* but under "Benzeneamine, dichloro-" in *Chemical Abstracts* (since 1972★).

To avoid the problem of not knowing just how compounds are listed in a subject index, try using the formula index first. In formula indexing, compounds are listed in order of increasing number of carbon atoms. For 2,4-dichlorophenol, one would look for $C_6H_4Cl_2O$ (elements after carbon and hydrogen are listed alphabetically) and then scan the names given under

---

★ *Chemical Abstracts* used "Aniline, 2,4-dichloro-" up to that time.

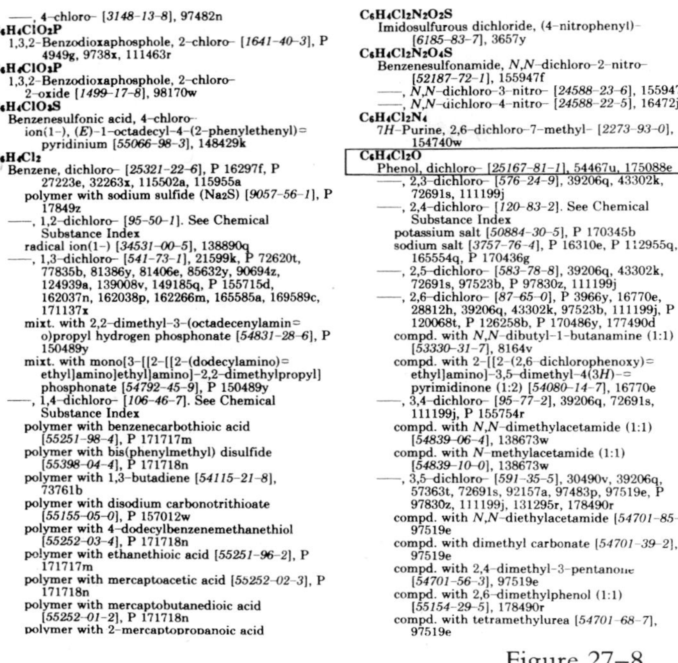

C_6H_4ClO — *CHEMICAL ABSTRACTS-VOL. 82* — 1043

—, 4-chloro- [*3148-13-8*], 97482n

$C_6H_4ClO_2P$
1,3,2-Benzodioxaphosphole, 2-chloro- [*1641-40-3*], P 4949g, 9738x, 111463r

$C_6H_4ClO_3P$
1,3,2-Benzodioxaphosphole, 2-chloro-, 2-oxide [*1499-17-8*], 98170w

$C_6H_4ClO_3S$
Benzenesulfonic acid, 4-chloro-
ion(1-), (*E*)-1-octadecyl-4-(2-phenylethenyl)= pyridinium [*55066-98-3*], 148429k

$C_6H_4Cl_2$
Benzene, dichloro- [*25321-22-6*], P 16297f, P 27223e, 32263x, 115502a, 115955a
polymer with sodium sulfide (Na_2S) [*9057-56-1*], P 17849z
—, 1,2-dichloro- [*95-50-1*]. See Chemical Substance Index
radical ion(1-) [*34531-00-5*], 138890q
—, 1,3-dichloro- [*541-73-1*], 21599k, P 72620t, 77835b, 81386y, 81406e, 85632y, 90694z, 124939a, 139008v, 149185q, P 155715d, 162037n, 162038p, 162266m, 165585a, 169589c, 171137x
mixt. with 2,2-dimethyl-3-(octadecenylamin= o)propyl hydrogen phosphonate [*54831-28-6*], P 150489y
mixt. with mono[3-[[2-[[2-(dodecylamino)= ethyl]amino]ethyl]amino]-2,2-dimethylpropyl] phosphonate [*54792-45-9*], P 150489y
—, 1,4-dichloro- [*106-46-7*]. See Chemical Substance Index
polymer with benzenecarbothioic acid [*55251-98-4*], P 171717m
polymer with bis(phenylmethyl) disulfide [*55398-04-4*], P 171718n
polymer with 1,3-butadiene [*54115-21-8*], 73761b
polymer with disodium carbonotrithioate [*55155-05-0*], P 157012w
polymer with 4-dodecylbenzenemethanethiol [*55252-03-4*], P 171718n
polymer with ethanethioic acid [*55251-96-2*], P 171717m
polymer with mercaptoacetic acid [*55252-02-3*], P 171718n
polymer with mercaptobutanedioic acid [*55252-01-2*], P 171718n
polymer with 2-mercaptopropanoic acid

$C_6H_4Cl_2N_2O_2S$
Imidosulfurous dichloride, (4-nitrophenyl)- [*6185-83-7*], 3657y

$C_6H_4Cl_2N_2O_4S$
Benzenesulfonamide, *N,N*-dichloro-2-nitro- [*52187-72-1*], 155947f
—, *N,N*-dichloro-3-nitro- [*24588-23-6*], 155947f
—, *N,N*-dichloro-4-nitro- [*24588-22-5*], 16472j

$C_6H_4Cl_2N_4$
7*H*-Purine, 2,6-dichloro-7-methyl- [*2273-93-0*], 154740w

$C_6H_4Cl_2O$
Phenol, dichloro- [*25167-81-1*], 54467u, 175088e
—, 2,3-dichloro- [*576-24-9*], 39206q, 43302k, 72691s, 111199j
—, 2,4-dichloro- [*120-83-2*]. See Chemical Substance Index
potassium salt [*50884-30-5*], P 170345b
sodium salt [*3757-76-4*], P 16310e, P 112955q, 165554q, P 170436g
—, 2,5-dichloro- [*583-78-8*], 39206q, 43302k, 72691s, 97523b, P 97830z, 111199j
—, 2,6-dichloro- [*87-65-0*], P 3966y, 16770e, 28812h, 39206q, 43302k, 97523b, 111199j, P 120068t, P 126258b, P 170486y, 177490d
compd. with *N,N*-dibutyl-1-butanamine (1:1) [*53330-31-7*], 8164v
compd. with 2-[[2-(2,6-dichlorophenoxy)= ethyl]amino]-3,5-dimethyl-4(3*H*)-= pyrimidinone (1:2) [*54080-14-7*], 16770e
—, 3,4-dichloro- [*95-77-2*], 39206q, 72691s, 111199j, P 155754r
compd. with *N,N*-dimethylacetamide (1:1) [*54839-06-4*], 138673w
compd. with *N*-methylacetamide (1:1) [*54839-10-0*], 138673w
—, 3,5-dichloro- [*591-35-5*], 30490v, 39206q, 57363t, 72691s, 92157a, 97483p, 97519e, P 97830z, 111199j, 131295r, 178490r
compd. with *N,N*-diethylacetamide [*54701-85-8*], 97519e
compd. with dimethyl carbonate [*54701-39-2*], 97519e
compd. with 2,4-dimethyl-3-pentanone [*54701-56-3*], 97519e
compd. with 2,6-dimethylphenol (1:1) [*55154-29-5*], 178490r
compd. with tetramethylurea [*54701-68-7*], 97519e

Benzenethiol, 2,5-dichloro- [*5858-18-4*], 105843y

$C_6H_4Cl_2S_2$
1,3-Benzenedisulfenyl dichloride [*53938-78-6*], P 59914x

$C_6H_4Cl_3FN_2$
2-Pyridinamine, 6-fluoro-4-(trichloromethyl)- [*52413-69-1*], P 4135p, P 170724z

$C_6H_4Cl_3F_2N_3OS$
Acetamide, 2-chloro-*N*-[5-(2,2-dichloro-1,1-= difluoroethyl)-1,3,4-thiadiazol-2-yl]- [*53644-35-2*], P 156326q

$C_6H_4Cl_3N$
Benzenamine, trichloro- [*54686-91-8*], P 113239w
—, 2,3,6-trichloro-
hydrochloride [*56059-53-1*], 155665n
—, 2,4,5-trichloro- [*636-30-6*], 86179t, P 87681n, P 99929n, P 139631t, P 139800x
8-quinolinecarboxylate [*55706-66-6*], P 156130w
—, 2,4,6-trichloro- [*634-93-5*], 12245q, 27198a, P 139800x, 162254f
compd. with 2-propanone (1:1) [*54524-46-8*], 85602p
compd. with 1,3,5-trinitrobenzene (1:1) [*54524-47-9*], 85602p, 97526e
—, 3,4,5-trichloro- [*634-91-3*], 162271j
Pyridine, 2,6-dichloro-3-(chloromethyl)- [*41789-37-1*], P 72799h, P 156110q, P 156113t
—, 2,3,5-trichloro-4-methyl- [*53939-29-0*], P 39563d

$C_6H_4Cl_3NO$
Pyridine, 2,3,5-trichloro-6-methoxy- [*31557-34-3*], P 170723y
3-Pyridinemethanol, 2,5,6-trichloro- [*55483-86-8*], P 156110q

$C_6H_4Cl_3NO_2S$
Benzenesulfonamide, *N,N*,4-trichloro- [*17260-65-0*], 16472j, 155686v, 155946e, 155947f

$C_6H_4Cl_3NO_2SSe$
Imidoselenious dichloride, [(4-chlorophenyl)= sulfonyl]- [*15855-28-4*], 15822m

$C_6H_4Cl_3N_2O_2P$
Phosphorimidic trichloride, (2-nitrophenyl)- [*54998-94-6*], 16894y

$C_6H_4Cl_3N_3O_2$
4-Pyridazinol, 3,5,6-trichloro-
methyl carbamate (ester) [*54699-98-8*], P 57718n

Figure 27–8.

that formula. One can usually pick out with relative ease the name of the desired compound from among the names of other compounds with the same molecular formula. A portion of the formula index for one volume of *Chemical Abstracts* is given in Figure 27–8.

Problem

27–1 a. What is the boiling point of 2-heptanone?
b. What is the melting point of 1-methoxy-2,4-dinitrobenzene?
c. Is 9-ethylanthracene soluble in ethanol?
d. What is the melting point of 2-methoxy-1-methylnaphthalene?

27:3 How to Search for General Information (Syntheses, Reactions, Mechanisms)

Searching the literature for more than the physical properties of common compounds is often a complex process. In many cases what may be desired is a description of an exact laboratory preparation or details about the mechanism of a reaction. For this sort of information, it is often necessary to read the original primary journal article. The most important key to finding such an article is *Chemical Abstracts*.

Suppose you were interested in methods for the synthesis of prostaglandins such as **1**. *Chemical Abstracts* contains a General Subject Index, a Chemical Substance Index, a Formula Index, and an Author Index. The first approach in this case would be to consult the General Subject Index for "prostaglandins." As an example, Figure 27–9 gives a reproduction of part of a General Subject Index page for Volume 79 (1973). Entries of interest ("total synthesis, 78175m" and "total synthesis of, R31524j") are marked with an arrow and should be consulted. Abstracts are numbered consecu-

1

Prosimulium mysticum
hibernation of, physiol. of, 76134y
Prostaglandins
R 454g, R 9785y, R 14466z, R 14482b, R 27573g, R 27575j, R 35049f, R 38711b, R 38713d, R 38715f, R 49375w, R 61804t, R 61805u, R 63706m, R 100777h, R 100778j, R 100779k, R 103093t, R 118265z, R 122003e, R 122007j, R 122009m, R 133085j
A and E, muscle response to, 143139u
acylation of, at 15-hydroxy position, P 78212w
9-*O*-acyl derivs. in $F_{2\alpha}$ series, P 91676v
adenylate kinase and ATPase of cell membranes in response to, 73887k
adenyl cyclase activation by, in heart, 111829c
adrenal gland and kidney in response to, R 49368w
of adrenergic nerve endings, function of, R 29680p
alkoxytetrahydropyranyl ethers of, prepn. of, P 42036c
allergy of skin in relation to, 61897a
analogs, P 5047m
anaphylactic slow-reacting substances in relation to, R 144723s
antibodies to, R 100775f
antihypertensive, neutral renomedullary lipids in relation to, 51480v
as antihypertensives, R 49367v
appetite response to, body wt. and hypothalamus in relation to, 100848g
B, blood vessel and heart in response to, 100858k
biochem. and pharmacol. of, R 111811r
blood coagulation and pressure in relation to, 133186t
blood platelet aggregation and thrombosis response to, R 49366u
in blood platelet aggregation by arachidonic acid, 112122d
blood pressure response to, in hypertension, 73972j
of blood serum
neoplasms in induction of, 16591k
sex in relation to, 16327d
capillary permeability response to propranolol in relation to, 13619q
carbohydrate and fat metab. in response to, R 49373u
of cerebrospinal fluid, pyrogen and antipyretics effect on, 142949w
chromatog. of, silver nitrate in ethylene glycol as stationary phase for high-speed liq., 121611q
circulation in response to, 630m
circulation of lung in response to, 87754b
conformation of, NMR in relation to, 14797h
cyclic AMP formation by bladder and kidney in response to, after vasopressin stimulation of, 27634c
cyclic AMP formation by oviduct and uterus in response to, 524e
cyclic AMP of uterus in response to, estradiol and adrenergic receptors in relation to, 133092j
cytochrome P 450 binding of, 122648u
2-decarboxy-2-(1*H*-tetrazol-5-yl)-, P 31581a
detn. of
gas chromatog.-mass spectrometric, 2177t
in kidney by high-speed chromatog., 123220k
difluoromethylene, 52837d, 92580w
8,12-diisoprostanoic acid derivs., P 78211v
double bond configuration of, europium induced PMR shifts in relation to, 77951z
E
in body temp. regulation, R 73882e
cyclic AMP formation by ovary in response to, 73933x
ductus arteriosus relaxation by, in oxygen deficiency, 13849q
formation of, by eye, indomethacin inhibition of, 122088m
of kidney, sodium chloride effect on, 62169b
noradrenaline secretion by nerve after elec. and potassium stimulation in relation to, 3273h
release from heart, ATP and ADP stimulation of, 14054p
gastrointestinal tract in relation to, R 50951n
gastrointestinal tract in response to, R 38714e
glycerides of, by esterification, P 136619x
gonadotropins and prolactin secretion in response to, after hypothalamic administration, 111872m
half-life prolongation of, in blood, P 115213v
heart contraction response to, 87686f
hormone regulation by, R 122004f
hydroxy and tetrahydropyranyloxy, P 42039f
in inflammation, R 73871a, R 87623h
inflammation and mucopolysaccharides of skin in response to, 74080d
inflammation inhibitors and, R 49374v
insulin secretion stimulation by, cyclic AMP in relation to, 143078y
intestine in response to, ADP effect on, 73925w
in kidney, in hypertension, R 144571r
of kidney
in cancer and hypertension, 51554x
hypertension and sodium resorption in relation to, R 49367v
hypertension in relation to, R 463j
of kidney medulla, doca effect on, 111907b
kidney renomedullary body contg., 63826a
mechanism of, R 133062z
of neoplasms, of polyoma virus-transformed kidney fibroblast, 64272d
noradrenaline inhibition by α-sympathomimetics in relation to, 61416t
noradrenaline secretion by vas deferens in response to indomethacin in relation to, 73502f
ovary response to gonadotropins in relation to, 38812k
ovulation and parturition response to, in domestic animals, R 143034f
in parturition induction
74035t
estradiol in relation to, 73891g
pharmaceutical, for stomach acid secretion inhibition, P 35155n
pharmacology of, R 100776g
potassium and sodium transport by eye lens in response to, 73926x
prepn. and activity of, R 91641e ←
progesterone formation by placenta in response to, 143070q
receptors for, of uterus, 144512x
release of
by kidney, angiotensin analogs effect on, 73927y
by spleen, 27672p
by uterus, in contraceptive device presence, 90130a
in reprodn., R 16142q, R 61806v, R 133081e
of reproductive tract of female, R 73872b
reproductive tract of female in response to, R 27574h
resolution and configurational assignment of, 18175b
resolution of aldehydes and ketones in prepn. of, 65874p
series E, from F_α prastaglandins, 78216a
sodium of diet effect on, of blood plasma, 30014f
sperm transport in oviduct in response to, fertilization in relation to, 122083f
spleen smooth muscle response to, R 121659m
sterically controlled synthesis of, P 115652n, P 115653p, P 115656s, P 115657t, P 126538f
tetrazolyldecarboxy, P 78816w
thin-layer chromatog. of, 29187h
total synthesis, 78175m
total synthesis of, R 31524j ←
transport of, by eye, R 143040e
uterus contraction in response to bradykinin and, 38756v
uterus in response to, polyphloretin phosphate inhibition of, 87700f
uterus response to, abortion in relation to, 143128q
in veterinary medicine, R 111805s
β, phosphorylation by mitochondria activation by, 13764h

Figure 27–9.

tively, so you would go to the 1973 volume containing abstract number 78175 (the "m" in 78175m is used for computer checking of the reference and is not needed for literature searching). Abstract 78175 is given in Figure 27–3. It presents a very brief description of an original article that appeared in the journal *Tetrahedron,* Volume 29, pages 1447 to 1456, in 1973. The abstract should provide sufficient information so that you could tell whether or not you want to consult the original article from which the abstract was taken. Next, abstracts R31524 and R91641 would be consulted. The "R" prefix stands for "Review" and therefore these abstracts would refer to journal articles containing reviews of various methods for the total synthesis or preparation of prostaglandins.

Chemical Abstracts is, of course, also important for obtaining information on specific compounds. These can be found either through the Chemical Substance Index (Fig. 27–10) or the Formula Index (Fig. 27–8). A strict IUPAC naming system is *not* used by *Chemical Abstracts,* although it is adhered to where possible. The methods of naming and indexing have also changed over the years and you may encounter compounds under one name in the 1976 index, under a different name in the 1956 index, and under still a different name in the 1936 index. At the present time *Chemical Abstracts* publishes an Index Guide, which contains cross-references to trivial or semi-systematic names. Time can often be saved if the Index Guide is consulted

2927CS — *JULY-DEC. 1975-CHEM. SUBSTANCE INDEX* — **Isoquinoline, 4-(chloromethyl)-1,2,3,4-...**

8-Isoquinolinamine [*23687-27-6*]
reaction of, with diethoxytetrahydrofuran, P 10034s
——, **5-chloro-** [*55766-82-0*]
reaction of, with diethoxytetrahydrofuran, P 10034s
——, **7-chloro-** [*55766-90-0*]
reaction of, with diethoxytetrahydrofuran, P 10034s
——, **7-methoxy-** [*55766-74-0*]
oxidn. of, Fremy's salt, 79198r
reaction of, with diethoxytetrahydrofuran, P 10034s
——, **7-nitro-** [*56623-92-8*]
prepn. and reductive cyclization of, with formic acid, 114292r
——, 1,2,3,4-**tetrahydro-2-methyl-4-phenyl-** (*nomifensine*) [*24526-64-5*]
behavior response to, dopaminergic and sertoninergic systems of brain in relation to, 53414h
brain elec. activity response to, 72108c
(*Z*)-2-butenedioate (1:1) (*nomifensine hydrogen maleate*) [*32795-47-4*]
dopaminergic agonist, 53455x
motor activity and dopamine metab. by brain response to, 188284m
4-Isoquinolinaminium
——, **1,2,3,4-tetrahydro-*N*,*N*,*N*,2-tetramethyl-=1-oxo-**
iodide [*54931-74-7*], 42464y
salt with 2,4,6-trinitrophenol (1:1) [*54931-76-9*], 42464y
Isoquinoline [*119-65-3*]

[structure: isoquinoline ring, positions 1, N 2, 4, 5, 8a]

27980v, 146716y
from absorption oils, by acidic extn., 118345h
addn. reactions with nitromethane and acetylacetylene, 9719n
addn. reaction with nitromethane and Me propiolate, 9718m
alkylation of, P 58667c
bond energy of, heat of formation from, 163433w
bromination of, 9731k
carbon-13 NMR of, calcns. for, 77959r
catalysts, for oxidative dehydrogenation of
nickel complex [*14552-74-0*], octahedral-tetrahedral equil. of, in chlorobenzene, 153485a
2-oxide [*1532-72-5*]
acylamination of, by *N*-phenylbenzimidoyl chloride, 43263u
photoelectron spectrum of, 192063f
photolysis of, on eosin-dyed alumina, green-light, electron transfer in, 155613q
reaction of, with *O*-benzoyl benzaldehyde cyanohydrins, 9726n
reaction of, with cyanogen bromide, 9734p
reaction of, with methacrylonitrile, 58627q
reaction of, with tricyclohexylborane, mechanism of, 146716y
2-oxide, compd. with 2,4,6-trinitrophenol (1:1) [*55417-68-0*], 9734p
palladium complex [*23807-51-4*], catalyst, contg. molybdenum oxide and chromium hexacarbonyl for reaction of dinitrotoluene with carbon monoxide, P 113938u
perchlorate [*57147-10-1*], cyclocondensation reaction with cyclic and open-chain ketones, 147382y
phosphate (1:1) [*22307-63-7*], soly. of, in absorption oils, 118345h
reaction product with epichlorohydrin polymers, quaternary ammonium salts, inhibitors, for acid systems, P 208762w
rhodium complex [*55220-47-8*], 21127a
rhodium complex [*55220-49-0*], 21127a
rhodium complex [*55220-51-4*], 21127a
sulfate (1:1) [*5984-79-2*], soly. of, in absorption oils, 118345h
tin complex [*33790-28-2*], 164320u
tin complex [*33790-40-8*], 164320u
tin complex [*57306-93-1*], 164320u
tin complex [*57306-94-2*], 164320u
tin complex [*57306-95-3*], 164320u
tin complex [*57306-96-4*], 164320u
tin complex [*57306-97-5*], 164320u
tin complex [*57306-98-6*], 164320u
tin complex [*57306-99-7*], 164320u
tin complex [*57307-00-3*], 164320u
tin complex [*57307-23-0*], 164320u
——, **2-acetyl-1,2,3,4,5,8-hexahydro-** [*57699-01-1*], 193046q
——, **7-(acetyloxy)-1-[[4-(acetyloxy)phenyl]=methyl]-8-chloro-1,2,3,4-tetrahydro-6-=methoxy-2-(trifluoroacetyl)-**
(±)- [*58093-41-7*], 206461v

Consult Vol. 76 Index Guide
And Most Recent Supplement
Before Using Index

reductive debromination of, by methoxide, parameters and mechanism of, 205464q
2-oxide [*3749-21-1*]
reaction of, with cyanogen bromide, 9734p
reaction with phosphorus oxybromide, 9732m
——, **1-[2-(6-bromo-1,3-benzodioxol-5-yl)ethy=l]-3,4-dihydro-6-methoxy-7-=(phenylmethoxy)-** [*56920-63-9*]
prepn. and debenzylation of, 131429h
——, **8-bromo-3,4-dihydro-6-methoxy-7-=(phenylmethoxy)-1-[[4-=(phenylmethoxy)phenyl]methyl]-**
compd. with 2,4,6-trinitrophenol (1:1) [*58093-45-1*], 206461y
phosphorodichloridate [*58093-44-0*], prepn. and methylation of, 206461y
——, **1-[2-(2-bromo-4,5-dimethoxyphenyl)ethy=l]-3,4-dihydro-6,7-dimethoxy-** [*56920-56-0*]
reaction of, with sodium methylsulfinylmethanide, 131429h
——, **1-[2-(2-bromo-4,5-dimethoxyphenyl)ethy=l]-3,4-dihydro-6-methoxy-7-=(phenylmethoxy)-** [*31735-15-6*]
debenzylation of, 131429h
——, **1-[(2-bromo-4,5-dimethoxyphenyl)methy=l]-3,4-dihydro-6,7-dimethoxy-** [*41823-67-0*], 178762b
compd. with 2,4,6-trinitrophenol (1:1) [*56982-35-5*], 178762b
——, **1-[(3-bromo-4,5-dimethoxyphenyl)methy=l]-3,4-dihydro-6,7-dimethoxy-** [*56982-06-0*], 178762b
——, **1-[(2-bromo-4,5-dimethoxyphenyl)methy=l]-3,4-dihydro-7-methoxy-6-=(phenylmethoxy)-**
hydrochloride [*32151-61-4*], prepn. and hydride redn. of, 59120f
——, **1-[(2-bromo-4,5-dimethoxyphenyl)methy=l]-1,2,3,4-tetrahydro-7-methoxy-6-=(phenylmethoxy)-**
(±)- [*56533-84-7*], 59120f
hydrochloride, (±)- [*56533-85-8*], prepn. and

Figure 27–10.

kinetics of——see also *Kinetics of van Alphen–Huettel rearrangement*
Vancide 40
See *Morpholine, 4-(2-nitrobutyl)-* [2224-44-4]
Vancide 89RE
See *1H-Isoindole-1,3(2H)-dione, 3a,4,7,7a-tetrahydro-2-[(trichloromethyl)=thio]-* [133-06-2]
Vancide BL
See *Phenol, 2,2'-thiobis[4,6-dichloro-* [97-18-7]
Vancide BN
See *Phenol, 2,2'-thiobis[4,6-dichloro-*, disodium salt [6385-58-6]
Vancide PA
See *Propane, 1,1'-[1,2-ethenediylbis(sulfonyl)]bis-, (E)-* [1113-14-0]
Vancide PB
See *Benzene, 2,3,4-trichloro-1,5-dinitro-* [6379-46-0]
Vancide TH
See *1,3,5-Triazine, 1,3,5-triethylhexahydro-* [7779-27-3]
Vancide Z
See *Zinc, bis(dimethylcarbamodithioato-S,S')-, (T-4)-*, mixt. with 2(3*H*)-benzothiazolone zinc salt [8003-42-7]
Vancosamine
See *L-lyxo-Hexose, 3-amino-2,3,6-trideoxy-=3-C-methyl-* [36480-36-1]
Vandem VAC
See *Zinc oxide* [1314-13-2]
Vandem VOC
See *Zinc oxide* [1314-13-2]
Vandem VPC
See *Zinc oxide* [1314-13-2]
Van den Bergh reaction
See also *Diazo reaction*
catalysts——see *Van den Bergh reaction catalysts*
enthalpy of——see *Heat of van den Bergh reaction*
kinetics of——see also *Kinetics of van den Bergh reaction*
Van der Waals equation
See *Equation of state*
Van der Waals force
See *Force*
Vandid
See *Benzamide, N,N-diethyl-4-hydroxy-3-=methoxy-* [304-84-7]
Vandride 4
See *1,3-Isobenzofurandione, 3a,4,7,7a-=tetrahydromethyl-* [26590-20-5]
Vandrikidine
See *Aspidospermidine-3-carboxylic acid, 2,3,6,7-tetradehydro-20-hydroxy-16-=methoxy-*, methyl ester, (5α,12β,19α)- [50656-92-3]
Vanilloylacetyl
See *1,2-Propanedione, 1-(4-hydroxy-3-=methoxyphenyl)-* [2034-60-8]
Vanillyl alcohol
See *Benzenemethanol, 4-hydroxy-3-methoxy-* [498-00-0]
o-Vanillyl alcohol
See *Benzenemethanol, 2-hydroxy-3-methoxy-* [4383-05-5]
Vanillylamine
See *Phenol, 4-(aminomethyl)-2-methoxy-* [1196-92-5]
Vanilmandelic acid
See *Benzeneacetic acid, α,4-dihydroxy-3-methoxy-* [55-10-7]
Vanilpyruvic acid
See *Benzenepropanoic acid, 4-hydroxy-3-=methoxy-α-oxo-* [1081-71-6]
Vanirex HW
See *Lignosulfonic acid*, sodium salt [8061-51-6]
Vanirex N
See *Lignosulfonic acid*, sodium salt [8061-51-6]
Vanisperse
See *Lignosulfonic acid*, sodium salt [8061-51-6]
Vanitiolide
See *Morpholine, 4-[(4-hydroxy-3-=methoxyphenyl)thioxomethyl]-* [17692-71-6]
Vanitrope
See *Phenol, 2-ethoxy-5-(1-propenyl)-* [94-86-0]
Vanlube AZ
See *Zinc, bis(dipentylcarbamodithioato-S,S')-, (T-4)-* [15337-18-5]
Vanoxy
See also *Epoxy resins*
Vanoxy 126
See also *Epoxy resins*
Vanquin
See *Quinolinium, 6-(dimethylamino)-2-[2-(2,=5-dimethyl-1-phenyl-1H-pyrrol-3-yl)=ethenyl]-1-methyl-*, salt with 4,4'-methylenebis[3-hydroxy-2-=naphthalenecarboxylic acid] (2:1) [3546-41-6]
Vanquish
See *Benzoic acid, 2-(acetyloxy)-*, mixt. with aluminum hydroxide, 3,7-dihydro-1,3,7-=trimethyl-1*H*-purine-2,6-dione, *N*-(4-hydroxyphenyl)acetamide and magnesium hydroxide [53908-22-8]
Van Slyke reaction
catalysts——see *Van Slyke reaction catalysts*
enthalpy of——see *Heat of van Slyke reaction*
kinetics of——see also *Kinetics of van Slyke reaction*
Vanstar 7
See *Vanadium alloy, base*, V,Cr,Fe,Zr (Vanstar 7) [12605-40-2]

Figure 27–11.

before seeking a specific compound in the General Subject or Chemical Substance Indexes. A portion of one Index Guide is given in Figure 27–11.

In order to assist in literature searches, *Chemical Abstracts* publishes a *Collective Index* in addition to the semiannual indexes that have been shown so far. Originally, decennial collective indexes were published, but since 1957 new collective indexes have appeared every five years.

Problems

27–2 The reproduction of the prostaglandin references (Fig. 27–9) is for 1973, not long after these important hormones were characterized chemically. Use several more recent *Chemical Abstracts* indexes and plot a growth curve (number of references versus time) for chemical research in the prostaglandin field.

27–3 Use *Chemical Abstracts* to find an abstract or abstracts dealing with the detection or isolation from human urine of the adrenalin metabolite, **I**, whose common name is vanilmandelic acid (VMA).

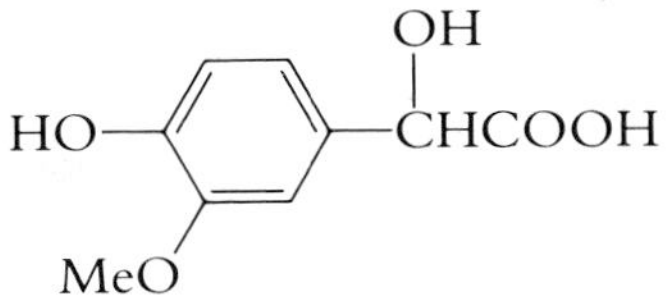

OH
HO—⟨benzene ring⟩—CHCOOH
MeO

I

27:4 Research Journals of Interest in Organic Chemistry

It has been pointed out that original work in chemistry is usually first published in journals. Some of these journals are devoted only to organic chemistry (e.g., *The Journal of Organic Chemistry* and *Tetrahedron*), while others may contain articles from all branches of chemistry. Table 27–1 lists some of the more important journals, the languages used in these publications, and their standard abbreviations as employed in *Chemical Abstracts*. In addition to these journals, most of which are relatively general in the field of organic chemistry, there are specialized organic journals of more restricted interest. Some of these are the *Journal of Organometallic Chemistry, Photochemistry and Photobiology, Organic Mass Spectroscopy, Organic Magnetic Resonance,* the *Chemical and Pharmaceutical Bulletin (Japan),* and *Phytochemistry*.

Journals from the Soviet Union have not been listed in Table 27–1 since they are in somewhat of a special category. Because most Western scientists cannot read Russian, English translations of Russian journals are available. The translations appear six months to one year after publication of the original journals in Russian, but it is the original articles that are abstracted in *Chemical Abstracts*. Table 27–2 lists some journals from the Soviet Union that are of interest in organic chemistry.

Publication in primary journals is in the form of full papers (which include exact experimental details) and communications. Communications are for matters of special interest where very rapid publication is of value. These articles do not usually contain experimental details, but the details are often included in a subsequent full paper.

27:5 Monographs, Advanced Textbooks, and Compilations of Spectral Data

There are innumerable monographs about specific areas of organic chemistry which may appear as single volumes or as a continuing series. Three of the more important series are mentioned here, along with several books of value.

Organic Reactions is a monograph series published regularly (Volume 1 appeared in 1942 and Volume 28 in 1982). Each volume treats several general reaction types (e.g., the acetoacetic ester condensation or the Diels–Alder reaction) as comprehensively as possible, and includes yields and conditions for many examples of the reaction under discussion. There are usually four to seven different reaction types discussed in each volume and the number of references to primary journal articles often reaches several hundred for each

Table 27–1. Some Major Research Journals Containing Organic Chemistry Articles

Journal	Language(s) of Articles	Abbreviation
Acta Chemica Scandinavica	English	*Acta Chem. Scand.*
Angewandte Chemie	German	*Angew. Chem.*
Angewandte Chemie International Edition in English	English	*Angew. Chem. Int. Ed. Eng.*
Australian Journal of Chemistry	English	*Aust. J. Chem.*
Bulletin of the Chemical Society of Japan	English	*Bull. Chem. Soc. Jpn.*
Bulletin de la société chimique de France	French	*Bull. Soc. Chim. Fr.*
Canadian Journal of Chemistry	English, French	*Can. J. Chem.*
Collection of Czechoslovak Chemical Communications	German, English	*Collect. Czech. Chem. Commun.*
Chemische Berichte	German	*Chem. Ber.*
Gazzeta Chimica Italiana	Italian, English	*Gazz. Chim. Ital.*
Helvetica Chimica Acta	German, English, French	*Helv. Chim. Acta*
Indian Journal of Chemistry	English	*Indian J. Chem.*
Journal of the American Chemical Society	English	*J. Amer. Chem. Soc.*
Journal of the Chemical Society, Chemical Communications	English	*J. Chem. Soc., Chem. Commun.*
Journal of the Chemical Society, Perkin Transactions I (or *II*)	English	*J. Chem. Soc., Perkin Trans. I* (or *II*)
Journal of Heterocyclic Chemistry	English	*J. Heterocycl. Chem.*
Journal of Medicinal Chemistry	English	*J. Med. Chem.*
Journal of Natural Products	English	*J. Nat. Prod.*
Journal of Organic Chemistry	English	*J. Org. Chem.*
Journal of Organometallic Chemistry	English	*J. Organomet. Chem.*
Justus Liebig's Annalen der Chemie	German	*Ann.*
Synthesis	English	*Synthesis*
Tetrahedron	English, German, French	*Tetrahedron*
Tetrahedron Letters	English, German, French	*Tetrahedron Lett.*

type of reaction. A sample page from the Chapter and Topic Index (cumulative for all volumes) is given in Figure 27–12.

Organic Syntheses is a monograph series that appears yearly (Volume 1

Table 27–2. Soviet Research Journals Containing Organic Chemistry Articles

Journal	Abbreviation
Journal of General Chemistry (*Zhurnal Obscheii Khimii*)	*Zh. Obshch. Khim.*
Bulletin of the Academy of Sciences of the U.S.S.R. (*Izvestii Akademii Nauk SSSR*)	*Izv. Adak. Nauk SSSR*
Reports of the Academy of Sciences of the U.S.S.R. (*Doklady Akademii Nauk*)	*Dokl. Akad. Nauk*
Chemistry of Heterocyclic Compounds (*Khimiya Geterotsikli Soedinenii*)	*Khim. Geterotsikl. Soedin.*
Journal of Organic Chemistry (*Zhurnal Organicheskii Khimii*)	*Zh. Org. Khim.*

was published in 1921 and Volume 60 in 1982). Each volume is devoted to 25 or 30 synthetic preparations that have been submitted as important new methods or modifications of methods. Exact experimental conditions are given. Figure 27–13 shows a typical Table of Contents, and Figure 27–14, a portion of one preparation.

CHAPTER AND TOPIC INDEX, VOLUMES 1–24

Many chapters contain brief discussions of reactions and comparisons of alternative synthetic methods which are related to the reaction that is the subject of the chapter. These related reactions and alternative methods are not usually listed in this index. In this index the volume number is in BOLDFACE, the chapter number in ordinary type.

Figure 27–12.

CONTENTS

Anodic Oxidation of Acids: Dimethyl Decanedioate 1
Atropaldehyde 6
Carboxylic Acids From the Oxidation of Terminal Alkenes by Permanganate: Nonadecanoic Acid 11
Conversion of Ketones to Cyanohydrins: Benzophenone Cyanohydrin 14
1,2-Cyclobutanedione 18
Cyclobutanone 20
Cyclopropanecarboxaldehyde 25
Deoxygenation of Epoxides With Lower Valent Tungsten Halides: *trans*-Cyclododecene 29
2,6-Di-*tert*-Butyl-4-methylpyridine 34
Diels-Alder Reaction of 1,2,4,5-Hexatetraene: Tetramethyl [2.2]Paracyclophane-4,5,12,13-tetracarboxylate 41
N-(2,4-Diformyl-5-hydroxyphenyl) acetamide 49
1,2-Diphenyl-3-methylcyclopropene: Preparation of Chlorophenyldiazirine and Thermal Generation of Chlorophenylcarbene 53
Electrohydrodimerization of an Activated Alkene: Tetraethyl 1,2,3,4-Butanetetracarboxylate 58
Epoxidation of Olefins by Hydrogen Peroxide–Acetonitrile: *cis*-Cyclooctene Oxide 63
Homoconjugate Addition of Nucleophiles to Cyclopropane-1,1-dicarboxylate Derivatives: Preparation of 2-Oxo-1-phenyl-3-pyrrolidinecarboxylic Acid 66
Imidazole-2-carboxaldehyde 72
Indirect Electrolysis: Tetramethyl 1,1,2,2-Ethanetetracarboxylate 78
Methyl 4-Hydroxy-2-butynoate 81
Methylene Ketones and Aldehydes by Simple, Direct Methylene Transfer: 2-Methylene-1-oxo-1,2,3,4-tetrahydronaphthalene ... 88
(*R,S*)-Mevalonolactone-2-^{13}C 92
2-Nitropropene 101
2-Phenyl-2-adamantanamine Hydrochloride 104
Reduction of Ketones to Hydrocarbons with Triethylsilane: *m*-Nitroethylbenzene 108

Figure 27–13.

EPOXIDATION OF OLEFINS BY HYDROGEN PEROXIDE–ACETONITRILE: *cis*-CYCLOOCTENE OXIDE

(*cis*-9-Oxabicyclo[6.1.0]nonane)

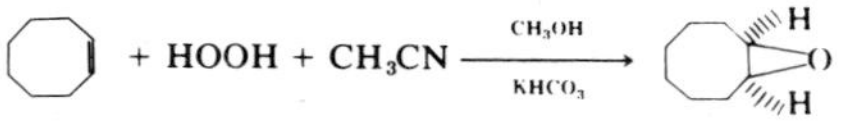
+ HOOH + CH3CN
CH3OH
KHCO3
H
O
H

Submitted by R. D. Bach and J. W. Knight[1]
Checked by K. W. Fowler and G. Büchi

1. Procedure

Caution! Organic-soluble peroxides may be explosive (Note 4)!

In a three-necked, 5-L, round-bottomed flask fitted with a mechanical overhead stirrer, addition funnel, and thermometer are placed 484 g (4.4 mol) of *cis*-cyclooctene, 3 L of reagent methanol (Note 1), 330 g (8.04 mol) of acetonitrile, and 77 g (0.77 mol) of potassium bicarbonate (Note 2). To the resulting heterogeneous mixture is added dropwise 522 g (4.6 mol) of 30% hydrogen peroxide with cooling at a rate that maintains the temperature of the reaction at 25–35°C (Note 3). Following the addition of hydrogen peroxide, the ice bath is removed and the reaction mixture is allowed to stir at room temperature overnight. The reaction mixture is divided in half, and each portion is diluted with 500 mL of a saturated sodium chloride solution. Each portion is then extracted with four 500-mL portions of methylene chloride (Note 4). The organic phases are combined, dried over magnesium sulfate, and concentrated at reduced pressure by rotary evaporation. Short-path distillation of the crude product (Note 5) under reduced pressure gives 333–337 g (60–61%) of *cis*-cyclooctene oxide, bp 85–87°C (20 mm), as a white solid, mp 53–56°C (Note 6).

Figure 27–14.

A Textbook of Practical Organic Chemistry (by A. I. Vogel, published by John Wiley and Sons), provides practical hints and methods of value in the laboratory, including directions for performing often-used reactions (e.g., esterification, formation of carbonyl derivatives, and simple functional group changes).

Quite frequently, reactions cannot be performed directly with solvents or reagents as they come from a purchased bottle. Such reagents and solvents often need to be purified or dried before use. *Purification of Laboratory Chemicals* (by D. D. Perrin, W. L. F. Armarego, and D. R. Perrin, published by Pergamon Press) is a good source for directions in this area. An invaluable guide to the preparation of various reagents for use in the laboratory is the series *Reagents for Organic Synthesis* (by L. F. Fieser and M. Fieser, published by John Wiley and Sons).

Primary literature journals do not usually reproduce uv-visible, ir, and nmr spectra, but simply describe the peaks. Exact comparison of spectra is, however, often necessary in organic chemistry and hence volumes devoted to the reproduction of various spectra of organic compounds are published. The most extensive of these are published by the Sadtler Research Laboratories and are known as *The Sadtler Standard Spectra.* Separate compendia (to which additions are continually being made) are available for uv-visible, ir, and ^{1}H and ^{13}C nmr spectra. The Aldrich Chemical Company (Milwaukee, Wisconsin) is one of the major suppliers of laboratory organic chemicals and has published *The Aldrich Library of NMR Spectra* and *The Aldrich Library of IR Spectra.* These are far less comprehensive than Sadtler, but considerably cheaper. The best source of uv-visible absorption spectra is probably *Organic Electronic Spectral Data* (Interscience, New York). These yearly volumes abstract most of the uv-visible spectral data published in primary journals. A good summary of ir data, arranged according to functional groups, is found in *Infrared Absorption Spectroscopy* (K. Nakanishi and P. H. Solomon, 2nd Edition, Holden–Day, Inc., 1977). The Supplementary Reading lists at the end of Chapters 8 and 9 give additional sources of information on spectroscopy.

Several advanced chemistry textbooks that are often used in graduate chemistry courses provide many more detailed mechanistic and synthetic schemes than can be provided in an intermediate text. Some of these are listed below.

Advanced Organic Chemistry (Part A—Structure and Mechanism; Part B—Reactions and Synthesis), 2nd ed., F. A. Carey and R. J. Sundberg, Plenum Press, New York, N.Y., 1983.

Modern Synthetic Reactions, 2nd ed., H. O. House, W. A. Benjamin, Inc., Menlo Park, Calif., 1972.

Advanced Organic Chemistry, 2nd ed., J. March, McGraw-Hill, New York, N.Y., 1977.

Mechanism and Theory in Organic Chemistry, T. H. Lowry and K. S. Richardson, 2nd ed., Harper & Row, New York, N.Y., 1981.

Concepts in Theoretical Organic Chemistry, J. A. Hirsch, Allyn and Bacon, Boston, Mass., 1974.

Introduction to Physical Organic Chemistry, R. D. Gilliom, Addison-Wesley, Reading, Mass., 1970.

27:6 Science Citation Index: An Alternate Literature Searching Method

It is probably evident from the preceding discussions that literature searching can be a time-consuming task. A special key to the literature, called *Science Citation Index* (*SCI*), has been developed to assist the researcher in rapidly finding the information desired. *SCI* is published by the Institute for Scientific Information, Inc., Philadelphia, Pennsylvania. This index is based upon the references cited by an author in any given publication. It is particularly designed to answer the question, "What has happened since?"

For example, S. Winstein and N. Holness were among the first to suggest that essentially all the molecules of *t*-butylcyclohexane would have the *t*-butyl group in the equatorial position at room temperature (sec. 3:10). Because of the size of the *t*-butyl group, there would be virtually no molecules in the conformation with that group axial. This suggestion was published in the *Journal of the American Chemical Society* in Volume 77, page 5562, in 1955. You can turn to the *Citation Index* section of *SCI* in any subsequent year and

WINSTEIN S

			VOL	PG	YR
53 J AM CHEMICAL SOC 75 155					
BURDON J	TETRAHEDRON		36	1423	80
53 J AM CHEMICAL SOC 75 2297					
BILLUPS WE	J ORG CHEM		45	4636	80
54 J AM CHEMICAL SOC 76 2597					
BELETSKA.IP	BK# 11436	R	1	119	79
BERONIUS P	ACT CHEM A		33	675	79
BUSHBY RJ	J CHEM R-S		1980	306	80
CACACE F	J AM CHEM S		102	3511	80
DANEY M	J ORGMET CH		188	277	80
EDLUND U	OMR-ORG MAG		12	661	79
HUYSKENS PL	B S CHIM BE		89	937	80
KARGER BL	BK# 14874	R	1	113	80
LEVY GC	OMR-ORG MAG		13	403	80
RAPPOPOR.Z	J ORG CHEM		45	4309	80
SONNEY JM	HELV CHIM A		63	1016	80
54 J AM CHEMICAL SOC 76 4368					
BRUSOVA GP	DAN SSSR		253	349	80
YOKOYAMA Y	MACROMOLEC		13	252	80
55 EXPERIENTIA S 2 137					
DONG DC	J ORG CHEM		45	2395	80
55 J AM CHEMICAL SOC 77 3747					
FUNG CW	J CHEM S P2		1980	267	80
55 J AM CHEMICAL SOC 77 4183					
GASSMAN PG	J AM CHEM S		102	4138	80
ILIC P	PUR A CHEM		52	1495	80
SAUNDERS M	BK# 14270	R	42	1	80
TAKASUKA M	J CHEM S P2		1980	486	80
55 J AM CHEMICAL SOC 77 5562					
ACCROMBE.GC	J ORG CHEM		45	4139	80
BANKS RM	J CHEM S P2		1980	427	80
BEGTRUP M	"		1980	544	80
CARLEER R	OMR-ORG MAG		12	673	79
COREY EJ	J ORG CHEM		45	765	80
HUFFMAN JW	"		45	3088	80
KIKUCHI H	CHEM LETT		1980	209	80
KONOPELS.JP	J AM CHEM S		102	2737	80
"	J ORG CHEM		45	2297	80
LUTZ ETG	SPECT ACT A		36	177	80
MILLER LS	J ORG CHEM		45	641	80
NACHTERG.WA	B S CHIM BE		89	525	80
PONS A	TETRAHEDRON		36	2219	80
SEEMAN JI	J AM CHEM S		102	7741	80
"	TETRAHEDRON		36	1173	80
TOROMANO.E	"		36	1971	80
56 CHEMISTRY IND-LONDON 56					
GARRATT DG	CAN J CHEM		58	1327	80
KATSUSHI.T	B CHEM S J		53	2031	80
56 EXPERIENTIA 12 138					
ATKINSON RS	J CHEM S P1		1979	3017	79
BECKWITH AL	BK# 14270	R	42	161	80
CHATTERJ.S	J CHEM S P1		1979	2914	79
GHOSAL PK	I J CHEM B		17	315	79
TIECCO M	ACC CHEM RE	R	13	51	80
56 J AM CHEMICAL SOC 78 328					
GOERING HL	J AM CHEM S		102	1628	80
JOHNSON JE	J ORG CHEM		45	4144	80
LEE CC	BK# 16160		5	1	80
"	CAN J CHEM		58	2369	80
MILLER LS	J ORG CHEM		45	641	80
RAPPOPOR.Z	J AM CHEM S		102	3837	80
"	J ORG CHEM		45	4309	80
VANGINKE.FI	J AM CHEM S		102	7514	80

WINSTEIN S

			VOL	PG	YR
ILIC P	PUR A CHEM		52	1495	80
STUHLER HO	Z NATURFO B		35	843	80
65 J AM CHEMICAL SOC 87 5247					
BAZBOUZ A	TETRAHEDRON		36	2745	80
BOWMAN WR	J CHEM S P2		1980	731	80
LENOIR D	ANGEW CHEM	N	19	318	80
MARTIN HD	OMR-ORG MAG		14	192	80
STEELE WV	J CHEM THER		12	187	80
65 J AM CHEMICAL SOC 87 5249					
STEELE WV	J CHEM THER		12	187	80
65 19 CHEM SOC SPEC PUB 109					
BENTLEY TW	J CHEM S P2		1980	1244	80
JOHNSON JE	J ORG CHEM		45	4144	80
JONES P	J CHEM S P2		1979	1679	79
RAPPOPOR.Z	J ORG CHEM		45	4309	80
SONNEY JM	HELV CHIM A		63	1016	80
VANGINKE.FI	J AM CHEM S		102	7514	80
WILLIAMS DL	BK# 16160	R	5	147	80
66 J AM CHEMICAL SOC 88 1319					
LAINE RM	TRANSIT MET		5	158	80
66 J AM CHEMICAL SOC 88 2047					
CHILDS RF	J AM CHEM S		102	4159	80
FERBER PH	AUST J CHEM		33	1569	80
ILIC P	PUR A CHEM		52	1495	80
66 TETRAHEDRON S8 621					
ILIC P	PUR A CHEM		52	1495	80
WALKOWIC.M	B POL CHIM		27	541	79
67 AROMATICITY 47					
ILIC P	ISR J CHEM		20	258	80
67 J AM CHEMICAL SOC 89 3656					
GILLISSE.HM	J ORG CHEM		45	319	80
JEFFORD CW	B S CHIM BE		88	969	79
PEOPLES PR	J AM CHEM S		102	4709	80
67 J AM CHEMICAL SOC 89 3956					
JOHNSON TH	SYN COMMUN		10	381	80
67 21 CHEM SOC SPEC PUB					
ILIC P	PUR A CHEM		52	1495	80
69 CHEM SOC SPEC PUBL 19 109					
LEE CC	CAN J CHEM		58	2369	80
69 QUART REV CHEM SOC 23 141					
CHILDS RF	J AM CHEM S		102	4159	80
GATI G	MAGY KEM FO		86	193	80
HAIGH CW	PROG NUCL	R	13	303	79
MURATA I	ISR J CHEM		20	244	80
OLAH GA	NOUV J CHIM		4	691	80
SCHWESIN.R	CHEM BER		113	3127	80
YAMAOKA H	B CHEM S J		53	469	80
72 CARBONIUM IONS 3 965					
CHEREMIS.AA	OMR-ORG MAG		14	327	80
ILIC P	ISR J CHEM		20	258	80
72 J ORGANIC CHEMISTRY 37 825					
ANDO T	B CHEM S J		53	2348	80
DANA G	CAN J CHEM		58	1451	80
HELLWINK.D	CHEM BER		113	3275	80
72 J AM CHEMICAL SOC 94 2286					
CANE DE	TETRAHEDRON	R	36	1109	80
OVERTON KH	CHEM SOC RE		8	447	79
73 J AM CHEMICAL SOC 95 2651					
ILIC P	PUR A CHEM		52	1495	80
76 J AM CHEMICAL SOC 98 6913					
KUROSAWA H	J AM CHEM S		102	6996	80
LAPPIN AG	BK# 10425	R	6	69	79
TROST BM	J AM CHEM S		102	3572	80

Figure 27–15.

find out which authors of other papers have quoted the work of Winstein and Holness. In this way, you can keep track of what research has been done in this area of conformational analysis since 1955. Figure 27–15 shows a portion of the Citation Index for 1980, with the pertinent section marked. This shows that 16 papers were published in the primary literature of 1980 wherein the work of Winstein and Holness was quoted. These papers could be consulted to see what modern developments had taken place in this area. Saul Winstein was a physical organic chemist at the University of California, Los Angeles, who made many contributions to the chemistry of carbocations (carbonium ions), as well as to conformational analysis. It is interesting to note from this section of *SCI* that *116 different papers* published by Winstein were referred to by other authors in 1980. Thus, *SCI* can be used to measure the impact which a given scientist has had on the field of chemistry; that of Winstein was certainly lasting.

SCI also contains a *Source Index*. This index lists all the publications of a given author for each year. A sample portion is shown in Figure 27–16. The *Source Index* is often used when one wishes to consult all the papers of an authority in a particular field. For example, A. I. Scott has been a leading researcher in the biosynthesis of porphyrins (sec. 25:7). The *Source Index* for 1976 shows that Scott published five papers (see box, Fig. 27–16) on porphyrin biosynthesis in 1976. If you examine these papers in the primary journals listed, many other references to porphyrin biosynthesis will be found. You can turn to the *Citation Index* for years after 1976 and find out

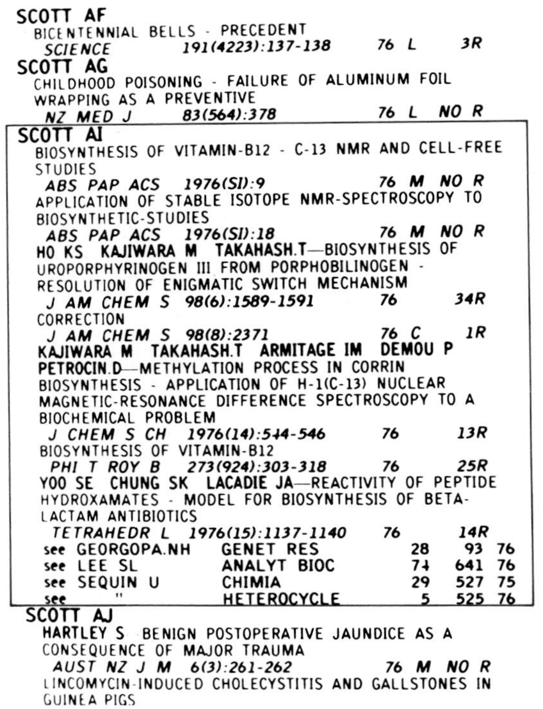

SCOTT AF
BICENTENNIAL BELLS - PRECEDENT
SCIENCE *191(4223):137-138* *76 L* *3R*
SCOTT AG
CHILDHOOD POISONING - FAILURE OF ALUMINUM FOIL WRAPPING AS A PREVENTIVE
NZ MED J *83(564):378* *76 L NO R*
SCOTT AI
BIOSYNTHESIS OF VITAMIN-B12 - C-13 NMR AND CELL-FREE STUDIES
ABS PAP ACS *1976(SI):9* *76 M NO R*
APPLICATION OF STABLE ISOTOPE NMR-SPECTROSCOPY TO BIOSYNTHETIC-STUDIES
ABS PAP ACS *1976(SI):18* *76 M NO R*
HO KS KAJIWARA M TAKAHASH.T—BIOSYNTHESIS OF UROPORPHYRINOGEN III FROM PORPHOBILINOGEN - RESOLUTION OF ENIGMATIC SWITCH MECHANISM
J AM CHEM S *98(6):1589-1591* *76* *34R*
CORRECTION
J AM CHEM S *98(8):2371* *76 C* *1R*
KAJIWARA M TAKAHASH.T ARMITAGE IM DEMOU P PETROCIN.D—METHYLATION PROCESS IN CORRIN BIOSYNTHESIS - APPLICATION OF H-1(C-13) NUCLEAR MAGNETIC-RESONANCE DIFFERENCE SPECTROSCOPY TO A BIOCHEMICAL PROBLEM
J CHEM S CH *1976(14):544-546* *76* *13R*
BIOSYNTHESIS OF VITAMIN-B12
PHI T ROY B *273(924):303-318* *76* *25R*
YOO SE CHUNG SK LACADIE JA—REACTIVITY OF PEPTIDE HYDROXAMATES - MODEL FOR BIOSYNTHESIS OF BETA-LACTAM ANTIBIOTICS
TETRAHEDR L *1976(15):1137-1140* *76* *14R*

see GEORGOPA.NH	GENET RES	28	93	76
see LEE SL	ANALYT BIOC	74	641	76
see SEQUIN U	CHIMIA	29	527	75
see "	HETEROCYCLE	5	525	76

SCOTT AJ
HARTLEY S BENIGN POSTOPERATIVE JAUNDICE AS A CONSEQUENCE OF MAJOR TRAUMA
AUST NZ J M *6(3):261-262* *76 M NO R*
LINCOMYCIN-INDUCED CHOLECYSTITIS AND GALLSTONES IN GUINEA PIGS

Figure 27–16.

what other chemists quoted any of these papers. Thus, immediate access to most of the important research papers in this area will have been achieved.

Problem

27–4 The first announcement of solid phase peptide synthesis (sec. 22:13) was by R. B. Merrifield, *J. Amer. Chem. Soc.*, **85**, 2149 (1963). Count the number of citations of this article in the 1965, 1970, and 1977 literature.

Supplementary Reading

H. M. Woodburn, *Using the Chemical Literature—A Practical Guide,* Marcel Dekker, New York, N.Y., 1974.

R. T. Bottle, *Use of the Chemical Literature,* Butterworths, Woburn, Mass., 1976.

G. Gorin, An approach to teaching chemical information retrieval, *J. Chem. Ed.*, **59,** 991 (1982).

Answers to Selected Problems

Answers to selected problems from the text are given below. Some are complete, while others contain only partial answers or hints on how to solve the problem. For complete answers to all problems, the Student Solutions Manual should be consulted. When comparing the answers below to those in the solutions manual remember that: (1) structures that are actually identical may be written differently; (2) there may be more than one "correct" answer to a given problem, especially in synthesis; and (3) mechanisms and exact *interpretations* of data may vary from instructor to instructor.

Chapter 1

1–1 a. :C̤l:C(:C̤l:)(:C̤l:):H **d.** H:B(H):H **e.** :N̤(H):Ö:H **1–2 a.** H:C(H)(H):C(H)(H):B̤r: H—C(H)(H)—C(H)(H)—Br CH_3CH_2Br

b. H:C(H)(H):C(H)(H):C(H)(H):H H—C(H)(H)—C(H)(H)—C(H)(H)—H $CH_3CH_2CH_3$

e. H:C(H)(H):C(H)(H):N̤(H): H—C(H)(H)—C(H)(H)—N(H): $CH_3CH_2NH_2$

1–3 a. H:C:::C:H H—C≡C—H **b.** H:C:::N: H—C≡N: **c.** H:C(H)::N(H): H—C(H)=N(H):

d. H:C(H)(H):N::Ö: H—C(H)(H)—N=Ö: **f.** H:C(H)::C::Ö: H—C(H)=C=Ö: **1–4** H—C(H)(H):⁻

1–5 a. +1 **b.** −1 **c.** 0 **d.** −1, +1 **e.** 0, −1 **1–6 a.** No charges on first structure; −1 on C, +1 on O of second structure. First structure is more stable. **c.** No charges on either structure. Comparable stability. **d.** First structure: −1 on C, 0 and +1 on O. No charges on second structure, which is more stable.

1–8 a. H—C(H)(H)—C(H)(H)—C(H)(H)—O—H H—C(H)(H)—C(H)(O—H)—C(H)(H)—H H—C(H)(H)—C(H)(H)—O—C(H)(H)—H

e. Br—C(H)(Br)—C(H)(H)—C(H)(H)—C(H)(H)—H H—C(H)(Br)—C(Br)(H)—C(H)(H)—C(H)(H)—H

1–8 e. (cont.)

H H H H   H H H H   H H H H
H—C—C—C—C—H   H—C—C—C—C—H   H—C—C—C—C—H
Br H Br H   H Br Br H   Br H H Br

H H Br H   H H H   H Br Br   H H Br
H—C—C—C—C—H   H—C—C—C—Br   H—C—C—C—H   H—C—C—C—Br
H H Br H   H | H   H | H   H | H
H—C—Br   H—C—H   H—C—H
H   H   H

1–9 a. $CH_3CH_2CH_2I$, /\/I, an alkyl iodide. **b.** $CH_3CHCH_2CH_2CH_3$ (with NH_2 on C-2), (skeletal, NH_2), a secondary amine.

d. $CH_3CHCH_2C{=}O$ (with Br on C-2 and OH on C=O), (skeletal, Br, O, OH), a carboxylic acid and an alkyl halide.

e. $CH_3OCH_2CH_2NHCH_3$, (skeletal, O, NH), a secondary amine and an ether.

1–10 a. $C_{12}H_{13}NO_2$ **b.** $C_{10}H_{14}O_3$ **1–11** $C_{13}H_{20}N_2$, $C_{20}H_{32}N_3$, $C_{13}H_{21}N_2$ **1–12 a.** 98 **b.** 98 **c.** 82

1–13 a. C with H, H, CH_3, OH **b.** N̈ with H, H, CH_3 **c.** C=O with CH_3, CH_3

1–14 a. CH_3CHF_2, FCH_2CH_2F

b. $CH_3CH_2CH_2CH_3$, CH_3CHCH_3 (with CH_3 on C-2)

c. CH_3OOCH_3, $HOCH_2OCH_3$, $HOCH_2CH_2OH$, $CH_3CH(OH)_2$, CH_3CH_2OOH

d. $CH_3C{\equiv}CH$, $H_2C{=}C{=}CH_2$, $CH{=}CH$ (ring with CH_2)

f. $CH_3CH_2CH_2NH_2$, $CH_3CH_2NHCH_3$, CH_3NCH_3 (with CH_3 on N), CH_3CHNH_2 (with CH_3 on C)

1–16 a. $CH_3CH_2\ddot{C}^{-}$ (with H, H on C), $CH_3\ddot{C}^{-}HCH_3$

1–17 Hydrobromic acid, Br^-; sulfuric acid, $^-OSO_3H$; benzoic acid, $C_6H_5COO^-$; phenol, $C_6H_5O^-$
1–18 CH_3COOH and NaF
1–19 HC—OH (with =O on C) $K_a = 1.71 \times 10^{-4}$, $pK_a = 3.77$

Chapter 2

2–1 90°. Repulsions between H atoms are greater because they are closer. **2–2** 109.5°. There is a greater repulsion between the two nonbonded electron pairs than between the two Hs which forces the Hs closer together. **2–3 a.** Cl—C—Cl 109.5° **b.** H—N—H 109.5° **c.** H—C—H less than 109.5°, H—C—Cl ~109.5°, Cl—C—Cl more than 109.5° **d.** H—B—Cl slightly less than 120°, Cl—B—Cl slightly greater than 120° **h.** H—C—H slightly less than 120°, H—C—C slightly more than 120° **j.** H—C—H 120° **2–4 a.** By loss of an electron Li reaches the stable helium configuration. Helium already is in the stable configuration and loss of an electron would destabilize it. **b.** Removal of an electron from the inner level: $1s^2 2s^1 \rightarrow 1s^1 2s^1 + e^-$. The result is an excited state. **2–6 a.** $3s + 3s$ **b.** $3p + 3p$ **c.** $1s + 4p$ **d.** $5p + 3p$ **2–7 a.** C sp^3, C—Cl sp^3-p **b.** C sp^3, C—H sp^3-s, C—Cl sp^3-p **c.** B sp^3, B—F sp^3-$2p$ **d.** C and N sp^3, C—H sp^3-s, C—N sp^3-sp^3, N—H sp^3-s **e.** C and O are sp^3, C—H sp^3-s, C—O sp^3-sp^3 **f.** C and S are sp^3, C—H sp^3-s, C—S sp^3-sp^3 **2–8 a.** B sp^2, B—Cl sp^2-p **b.** C, N sp^2, C=N sp^2-sp^2 and p-p **c.** C sp, O sp^2, C=O sp-sp^2 and p-p **f.** First O sp^3, N sp^2, second O sp^2; H—O s-sp^3, O—N sp^3-sp^2, N=O sp^2-sp^2 and p-p **2–9** Dimethyl ether 754 kcal/mol, ethanol 759 kcal/mol

2–10 a. $\overset{\delta-}{N}—\overset{\delta+}{H}$ **b.** $\overset{\delta+}{C}—\overset{\delta-}{Cl}$ **c.** $\overset{\delta+}{C}—\overset{\delta-}{O}$, $\overset{\delta-}{O}—\overset{\delta+}{H}$ **d.** $\overset{\delta-}{S}—\overset{\delta+}{H}$

2–11 HCl. F is more electronegative, but Cl is larger.

2–12 $CH_3NH_2 + H_2O \rightleftarrows CH_3\ddot{N}H^- + H_3O^+$.

Nitrogen donates the H because $CH_3\ddot{N}H^-$ is more stable than $^-:CH_2NH_2$. N is more electronegative than C. **2–13 a.** CH_4; electron pair compresses angle in $:CH_3^-$ **c.** $H_2C{=}O$; H—C—H is 120° **d.** CH_3^+; H—C—H is 120° **2–14 a.** Four; CH_4, sp^3 **b.** Three; BH_3, sp^2 **c.** Two; BeH_2, sp **e.** Five; PH_3, sp^3 **2–15 a.** sp^3 **b.** sp^3 **c.** sp^3 **d.** sp^2 **e.** sp^2 **f.** sp^3 **g.** sp^3 **h.** sp^2 **i.** sp^2 **j.** sp^3 **2–16 a.** $CH_3\cdot$ **b.** $CH_3OH_2^+$ **c.** $2CH_3O\cdot$ **d.** $CH_3CO\ddot{O}^-$ **e.** $CH_3\ddot{S}^-$ **2–18 a.** Two **b.** Three **2–19 a.** Electron pair sp^3, N sp^3 **b.** Electron pairs sp^3, C and O sp^3 **d.** Electron pair sp, C and N sp **f.** Electron pairs on O sp^2 and on Cl p, first C sp^3, second C sp^2, O sp^2, Cl not hybridized **2–20 a.** C—C **b.** C—H **c.** C—O **d.** O—O **e.** N—N **f.** C—N **g.** C—C **h.** C—H **i.** Si—Si **2–21 a.** −116 kcal/mol **b.** −212 kcal/mol **c.** −10 kcal/mol **d.** −6 kcal/mol **2–22 a.** Cyclopentane more stable by 22 kcal/mol **b.** Acetaldehyde more stable by 21 kcal/mol **c.** $CH_2{=}NNH_2$ more stable by 26 kcal/mol

2–24 a. $\overset{\delta+}{Li}—\overset{\delta-}{F}$ **b.** $\overset{\delta+}{C}—\overset{\delta-}{O}$ **d.** $\overset{\delta+}{C}—\overset{\delta-}{F}$ **g.** $\overset{\delta+}{Al}—\overset{\delta-}{C}$

2–25 Compounds **(a–d)**, **(f–h)**, and **(m)** are covalent. **2–26** $CH_3NH_3^+$ is more stable than $CH_3OH_2^+$ because N is less electronegative than O and better supports a positive charge.

2–27 $ROH + Na \rightarrow H_2 + R\ddot{O}^-$ (relatively stable); $RCH_3 + Na \not\rightarrow H_2 + RCH_2^-$ (very unstable)

Chapter 3

3–1 $CH_3CH_2CH_2CH_2CH_2CH_3$ $CH_3CHCH_2CH_2CH_3$ (with CH_3 on C-2)

$CH_3CH_2CHCH_2CH_3$ (with CH_3 on C-3) $CH_3CCH_2CH_3$ (with two CH_3 on C-2) $CH_3CHCHCH_3$ (with CH_3 on C-2 and C-3)

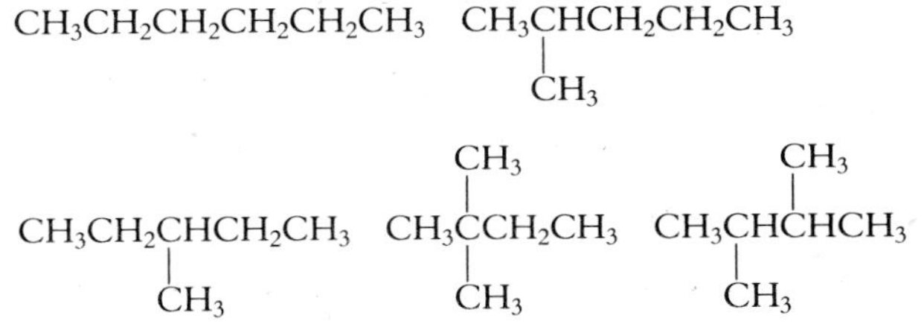

3–2 a. 1° **b.** 3° **c.** 2° **d.** 2° **e.** 1°

3–3 a. 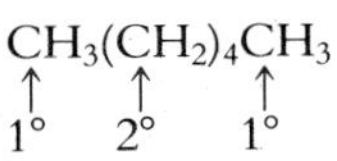 $CH_3(CH_2)_4CH_3$ (1°, 2°, 1°)

d. 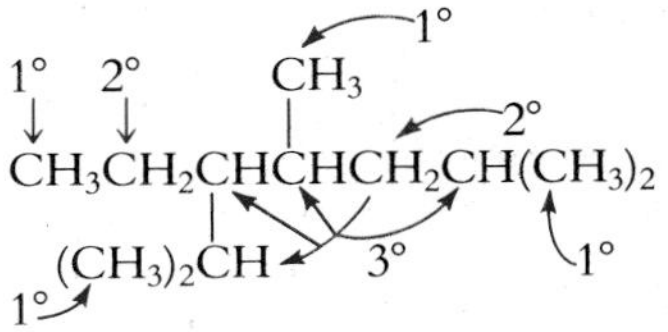 $CH_3CH_2CHCHCH_2CH(CH_3)_2$ (with CH_3 and $(CH_3)_2CH$ substituents; labels 1°, 2°, 1°, 2°, 3°, 1°, 1°)

3–4 a. 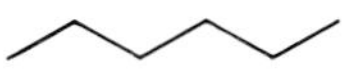 **b.** 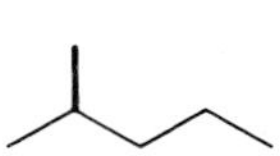 **c.** **d.** 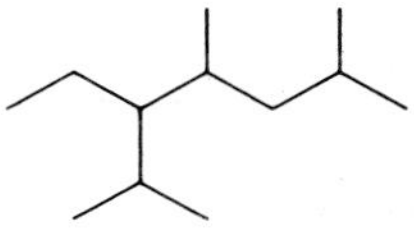 **e.** 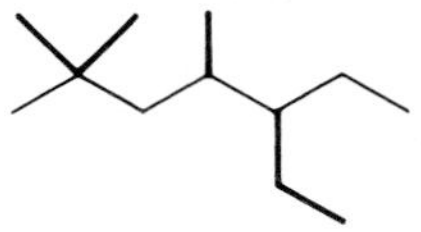 **f.**

3–5 a. 2-methylbutane **b.** 2,4-dimethylpentane **c.** 5-ethyl-2-methylheptane **d.** 5-methyldecane **3–6** Energy minima at 0°, 120° and maxima at 60°, 180° **3–7** Three; three; no

3–9

1 2 3

2 and **3** are equivalent in energy, and both are more stable than **1**.

3–10 a.

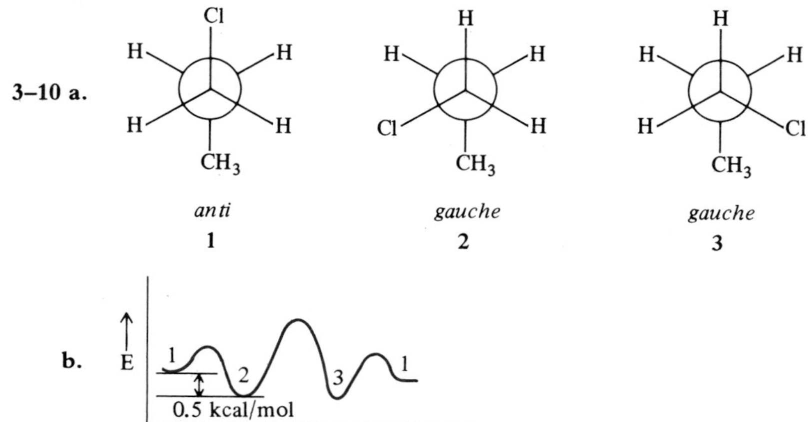

3–11 Conformations **2** and **3** will predominate at low *and* high temperature since they are favored both by enthalpy and entropy. **3–12 a.** One molecule going to two increases disorder and entropy. *b.* An ordered crystal (NaCl) dissociates to disordered ions. **c.** There is only one *anti* but *two gauche* conformations. **3–13** Cyclopropane, because it has the highest heat of combustion per CH_2 group.

3–14 a. **b.**

3–15. The twist boat has the greater entropy.

3–17 a. H H H H

b. (**NOTE ERRATUM:** Although it is possible to estimate the total strain energy from Figure 3–10, it would be more helpful to refer to Figure 3–11.) 24 kcal/mol

3–18 The *cis* compound has two neighboring methyls that are close enough to interact if cyclobutane were planar. In the folded conformation they would still be in a *gauche* relationship. **3–19** The *cis*- and *trans*-1,2-dimethylcyclopentanes are stereoisomers of one another. The *cis*- and *trans*-1,3-dimethylcyclopentanes are stereoisomers of one another.

3–20

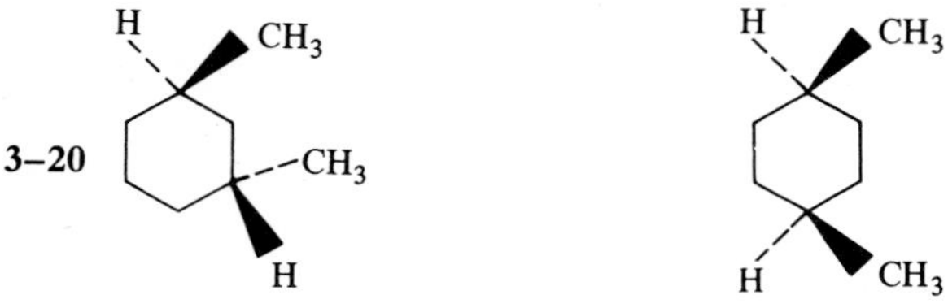

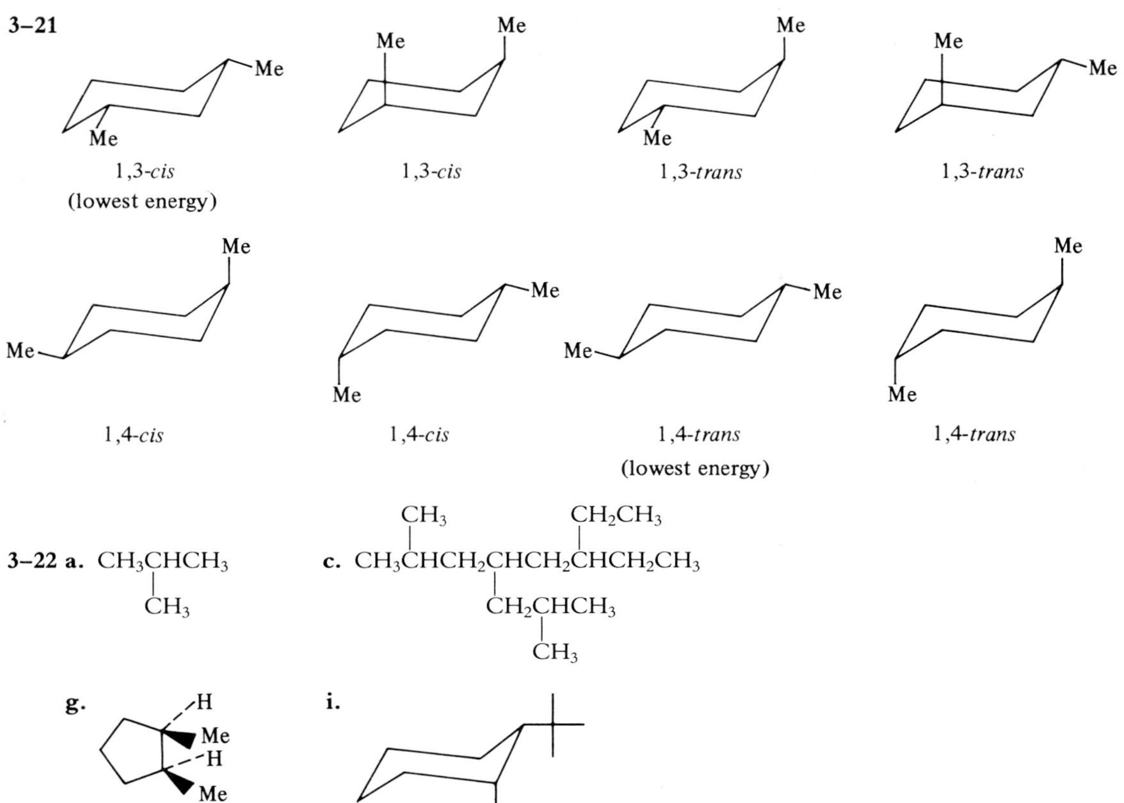

3–23 a. 3-methylheptane **b.** 3-isopropyl-2-methylhexane **c.** 7-ethyl-2,8-dimethyldecane **d.** (2-methylpropyl)cyclohexane (or isobutylcyclohexane)

3–25 a. $CH_3CH(CH_3)CH_2CH_2CH_3$ $CH_3CH_2CH(CH_3)CH_2CH_3$

b. $(CH_3)_2CHCH(CH_3)_2$

3–26 $CH_3CH_2CH(CH_3)CH(CH_3)CH(CH_3)CH_2CH_3$

3,4,5-trimethylheptane

3–30 Halogens are more polarizable than methyl groups are not. When halogens are close together one becomes slightly positive, the other slightly negative, and there is a resultant attractive force. **3–33** Cyclopropane 11.9, cyclobutane 11.7, cyclopentane 11.3 kcal/g; cyclopropane 499.8 kcal/mol, cyclobutane 655.2 kcal/mol, cyclopentane 791.0 kcal/mol; cyclopropane 166.6 kcal/mol/CH_2, cyclobutane 163.8 kcal/mol/CH_2, cyclopentane 158.2 kcal/mol/CH_2. Total strain energies: cyclopropane 27.6 kcal/mol, cyclobutane 25.6 kcal/mol, cyclopentane 4.0 kcal/mol. **3–35** *cis*-1,4-Dimethylcyclohexane will have the higher heat of combustion since one of the methyls is always in the axial position. This is of higher energy than the *trans* which can exist in a diequatorial conformation. **3–37** 1,2-*trans*, 1,3-*cis*, or 1,4-*trans*

3–38 (a–c) *trans* is the more stable isomer.

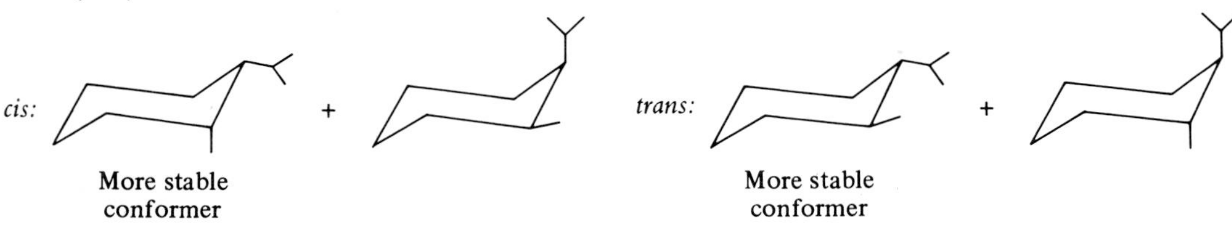

3–41

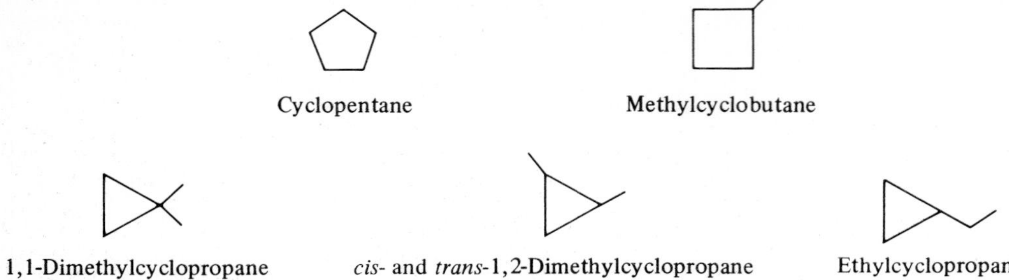

3–42 1,1-dimethylcyclobutane, ethylcyclobutane, *cis*- and *trans*-1,2-dimethylcyclobutane, *cis*- and *trans*-1,3-dimethylcyclobutane **3–45** *cis*-1,2-dimethylcyclopropane

Chapter 4

4–1 Neither is chiral. Stereoisomers (diastereomers). Constitutional isomers.

4–2

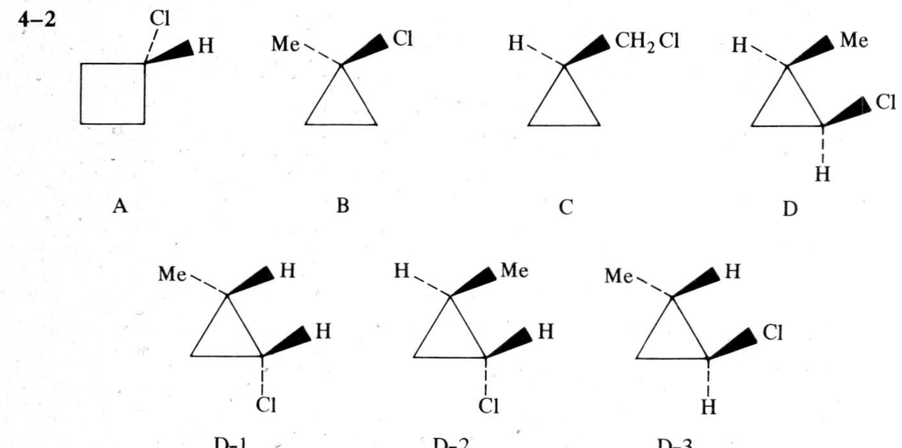

A, B, C, D: constitutional isomers. D and D-1 are diastereomers of D-2 and D-3. D and D-1 are enantiomers. D-2 and D-3 are enantiomers.

4–4a. A plane perpendicular to the ring plane and bisecting the 1,3-carbons. **b.** As in **(a)** plus a plane perpendicular to the ring plane and bisecting C-2 and C-4. **c.** No plane of symmetry. **d.** A plane perpendicular to the ring plane and bisecting the CH_2 between the two chlorine-containing carbons. **4–5 a.** Plane **b.** Plane and center **c.** None **4–6 (a–c)**, **(e)**, **(g)**, and **(i)** are achiral. **4–7 a.** No chiral carbons **b.** *meso* (*cis*-1,3-dimethylcyclopentane) **c.** Chiral centers but no plane of symmetry **d.** *meso* possible **e.** Same as **(c)** **f.** No chiral centers, no *meso* **4–8** Chiral. They are conformational enantiomers. **4–9** The two conformers are nonsuperimposable mirror images of one another and hence there would be a rapid interconversion of "enantiomers." **4–10** −16.0° **4–12 a.** *S* **b.** *R* **c.** *S*, *S*

4–13

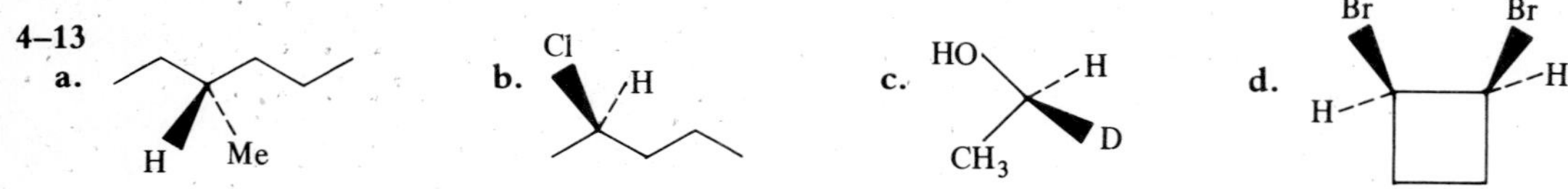

4–14 9a is 1*S*, 2*R*; **9b** is 1*R*, 2*S*; **10a** is 1*S*, 2*S*; **10b** is 1*R*, 2*R* **4–15** (*S*)-baseH^+(*R*)-lactate$^-$
4–17 a. $(CH_3)_2CHCH_2CH_2CH_2CH_3$ achiral **b.** $(CH_3)_2CClCH_2CH_2CH_2CH_3$ achiral **c.** $ClCH_2\overset{\star}{C}H(CH_3)CH_2CH_2CH_2CH_3$ chiral **d.** $CH_3\overset{\star}{C}H(Br)CH_2\overset{\star}{C}H(Br)CH_2CH_3$ chiral **f.** $CH_3CH_2\overset{\star}{C}H(I)\overset{\star}{C}H(CH_3)CH_2CH_2CH_3$ chiral **g.** Achiral **h.** Two chiral centers

4–19 $CH_3CH_2CH(CH_3)_2 + Br_2$ gives $CH_3\overset{\star}{C}H(Br)CH(CH_3)_2$ chiral + $CH_3CH_2\overset{\star}{C}H(CH_3)CH_2Br$ chiral **4–20 a.** Identical **b.** Enantiomers **c.** Identical **d.** Diastereomers **e.** Enantiomers **f.** Diastereomers **g.** Diastereomers **4–22** $CH_3CHOHCH_2CH_2CH_3$ or $(CH_3)_2CHCHOHCH_3$ or $CH_3CH_2CH(CH_3)CH_2OH$ **4–23 a.** +66.55° **b.** 0.24 moles per liter **4–24 a.** Dextrorotatory **b.** Levorotatory **c.** − **d.** + **e.** Clockwise order of descending priority **f.** Counterclockwise order of descending priority **g.** A compound with more than one chiral center which has a plane of symmetry **h.** Specific rotation at the sodium D line at 25°C **i.** Nonidentical mirror images **j.** Stereoisomers which are not enantiomers **k.** The actual arrangement in space of atoms in a molecule **l.** The separation of a racemic mixture into a pure enantiomer or enantiomers **4–25 a.** *R* **b.** *R* **c.** *S* **d.** *S* **e.** *R* **f.** *S* **g.** *S*, *S* **4–26 a.** *S* **b.** *R* **c.** *R* **d.** *S* **4–28 (d)**, **(e)**, and **(f)** can be *meso*. **4–29 a.** Two *meso*, four chiral **b.** Eight **4–30** 12° **4–31** 6.8 g **4–32 a.** No. The (*S*) enantiomer is present in larger amounts. **b.** 76% (*S*), 24% (*R*) **4–35 a.** 9-(*S*), 10-(*R*) in **V**; 9-(*S*), 10-(*S*) in **VI**. C-9 underwent retention, C-10 inversion.

Chapter 5

5–1 $Br_2 \rightarrow 2Br\cdot$, $Br\cdot + CH_4 \rightarrow HBr + CH_3\cdot$, $CH_3\cdot + Br_2 \rightarrow CH_3Br + Br\cdot$, etc.
5–2 $Cl\cdot + CH_3Cl \rightarrow HCl + \cdot CH_2Cl$, $\cdot CH_2Cl + Cl_2 \rightarrow ClCH_2Cl + Cl\cdot$, etc.
5–3 a. +13 kcal/mol **b.** The overall reaction is endothermic.
5–4 $CH_3{-}H\ (104) + Cl\cdot \rightarrow CH_3\cdot + H{-}Cl\ (103)$ $\Delta H° = +1$ kcal/mol
$CH_3\cdot + Cl{-}Cl\ (58) \rightarrow CH_3{-}Cl\ (83.5) + Cl\cdot$ $\Delta H° = -25.5$ kcal/mol
1 − 25.5 = −24.5, the same result as obtained in Equation 5.22. **5–5** +33 kcal/mol and −20 kcal/mol. An alternate route would be to use $CH_3N{=}NCH_3$ as a source of $CH_3\cdot$ to react with I_2. **5–6 a.** $ClCH_2CH_2CH_2CH_2CH_3$ (minor), $CH_3CH(Cl)CH_2CH_2CH_3$ (major), and $CH_3CH_2CH(Cl)CH_2CH_3$ (minor). The last two are favored by selectivity, the major product also favored by number of Hs replaced. **5–7** Br_2 because it is more selective than F_2 or Cl_2 and I_2 is unreactive. Selectivity is highest at 25°. **5–8 (a)**, **(d)**, and **(g)** 1°; **(c)** and **(h)** 2°; **(b)**, **(e)**, and **(f)** 3°. **5–11** 2-bromo-2-methylpropane, 2-iodopropane, 1-fluoro-2-methylpropane **5–12 a.** 1-bromo-1-chlorocyclobutane, 1-bromo-2-chlorocyclobutane (*cis* and *trans*), 1-bromo-3-chlorocyclobutane (*cis* and *trans*) **b.** 1-bromo-3-fluoropentane, 2-bromo-3-fluoropentane, 3-bromo-3-fluoropentane **c.** 1-bromo-2-methylpentane, 2-bromo-2-methylpentane, 3-bromo-2-methylpentane, 2-bromo-4-methylpentane, 1-bromo-4-methylpentane
5–13 Rate = $k[C_2H_5O^-]\,[C_6H_5COCH_3]^2$ **5–14** Rate = $k[(CH_3)_3COH]\,[H^+]$
5–15 $H_3\overset{\delta\cdot}{C}\cdots Cl \cdots \overset{\delta\cdot}{Cl}$
5–17 a. $\Delta H° = -15$ kcal/mol **b.** The selectivity of HO· would be greater than that of F· but less than that of Cl· or Br·. The more reactive a radical is, the less selective it will be. **5–18 a.** $CH_3CH_2O\cdot$, $CH_3\dot{C}HOH$, $\cdot CH_2CH_2OH$ **b.** $\Delta H°$ = −15, −25.5, and −21 kcal/mol. Favored product: $CH_3\dot{C}HOH$ **5–19 a.** $CH_3CH_2\cdot$ **b.** $(CH_3)_2CH\cdot$ **d.** $(CH_3)_3C\cdot$ **5–20 a.** 2-bromo-4-methylpentane **b.** 1,1-dichloro-4-methylpentane **c.** 2-bromo-1-fluoro-4-methylpentane **d.** 3-chloro-1,1-dimethylcyclohexane **e.** 1,1-dibromo-2-fluorocyclopentane
5–22 $Me_4Pb \rightarrow Me_3Pb + CH_3\cdot$, $CH_3\cdot + Cl_2 \rightarrow CH_3Cl + Cl\cdot$
Lower temperatures can be used because the $CH_3{-}Pb$ bond is weak.
5–23 $\Delta H°$ = +20.5 and −45 kcal/mol. The first step is much more endothermic than the first step of the accepted mechanism (eq. 5.26). **5–24 a.** 4 compounds **b.** 1 **c.** 4 **d.** 4 **e.** 4 **5–25 a.** 4 compounds **b.** 1 **c.** 8 **d.** 7 **e.** 3 **5–26 a.** $CH_3CH_2CH_2CH_3$ or $(CH_3)_3CH$ **b.** cyclopentane **c.** $(CH_3)_2CHCH(CH_3)_2$ **5–29** *t*-BuOO*t*-Bu → 2*t*-BuO· or *t*-BuO· + Br_2 → *t*-BuOBr + Br· or *t*-BuO· + C_6H_{12} → *t*-BuOH + $C_6H_{11}\cdot$ **5–30** Abstraction of the 3° H from (*R*)-3-hexane yields a radical that is sp^2 hybridized, planar, and hence achiral. **5–33 a.** C—H bonds are more reactive than C—D bonds towards radicals. **b.** Through mass spectrometry **5–36 a.** Five products including three enantiomeric pairs (eight total including enantiomers) **5–37 a.** 2-bromo-2-methylpropane **b.** 2-bromohexane and 3-bromohexane **e.** 1-bromo-1,2-dimethylcyclobutane (*cis* and *trans*)

Chapter 6

6–1 (a–c), **(f)**, and **(g)** are nucleophilic. **6–2** Replace Cl with I in each case. Chloromethane > 1-chlorobutane > 2-chlorobutane > chlorocyclopentane > 2-chloro-2-methylbutane **6–3 a.** *trans*-1-iodo-2-methylcyclohexane **b.** *cis*-1-iodo-4-ethylcyclohexane **c.** (*S*)-2-iodobutane **d.** (*S*)-1-iodo-2-methylbutane **6–5** The three methyls cause steric crowding with the entering I^- and leaving Br^- in the transition state.

6–6 $CH_3CH_2\overset{+}{C}(CH_3)CH_2CH_3 \rightarrow$

$CH_3CH_2C(N_3)(CH_3)CH_2CH_3 + CH_3CH_2C(OH)(CH_3)CH_2CH_3 + CH_3CH_2C(OEt)(CH_3)CH_2CH_3$

6–7 In an S_N1 reaction the *rate* is independent of azide ion, but the *amount of t*-butyl azide formed will increase with increasing azide concentration.

6–8 a. $CH_3CH_2C(Br)(CH_3)_2 \rightarrow CH_3CH_2\overset{+}{C}(CH_3)_2 \rightarrow CH_3CH_2C(OAc)(CH_3)_2$

b. Same cation intermediate $\rightarrow CH_3CH_2C(OMe)(CH_3)_2$.

The rate-determining step is still cation formation.

6–10 $CH_3C(CH_3)_2CH_2Br \rightarrow CH_3C(CH_3)_2CH_2^+ \rightarrow CH_3\overset{+}{C}(CH_3)CH_2CH_3 \rightarrow CH_3C(OR)(CH_3)CH_2CH_3$

The S_N2 reaction is slowed by the hindrance in the transition state caused by the methyls on C-2. **6–11** A carbocation is formed that reacts with acetic acid to give the first product. The cation also undergoes a hydride shift to form a new cation which leads to the second product. **6–12 a.** 5-methyl-3-hexanol **b.** 1-bromo-2-pentanol **c.** 1,3-butanediol **d.** $HOCH_2C(CH_3)_2CH_2CH_3$ **e.** $CH_3CH(OH)CF(CH_3)_2$ **f.** $HOCH_2CH(CH_3)CH_2CH_2OH$ **6–13 a.** 1-methoxypropane, methyl propyl ether **b.** 2-ethoxypropane, ethyl isopropyl ether **c.** 2-propoxycyclobutane, cyclobutyl isopropyl ether **e.** $HOCH_2CH(OH)CH_2CH_2OCH_2CH_3$ **6–14** RSH is weakly or not at all hydrogen-bonded, while ROH is strongly bonded. **6–15** $(CH_3)_2CHO{:}^-$ and $HO{:}^-$ are formed initially, then H_2O reacts with isopropoxide to produce more $HO{:}^-$. The equilibrium lies on the hydroxide side. **6–16** Methanol solvates the F^- making it less reactive, while DMF only solvates the K^+, leaving the F^- free to react.

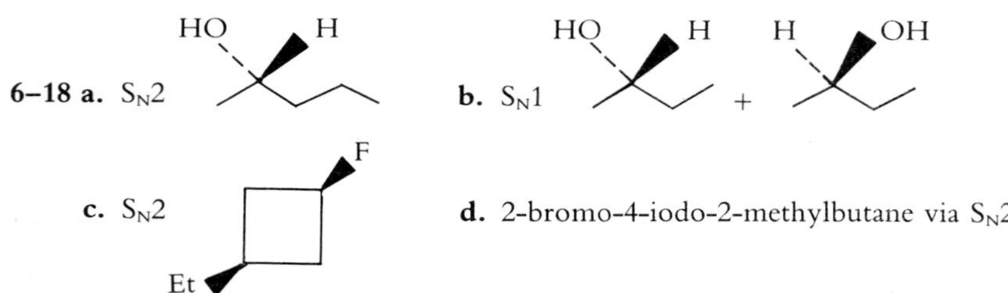

6–18 a. S_N2 **b.** S_N1 **c.** S_N2 **d.** 2-bromo-4-iodo-2-methylbutane via S_N2

6–19 a. bromocyclohexane **b.** 1,6-diiodohexane **c.** 2-chloro-2-methylpentane **d.** *cis-* and *trans*-1-chloro-2-methylcyclopentane and 1-chloro-1-methylcyclopentane **e.** propyl fluoride **f.** no reaction **g.** 2-bromo-4-methylpentane **6–20 a.** hexanol + HCl or $SOCl_2$ or PCl_3 **b.** 2-pentanol + PBr_3 or HBr **c.** cyclobutanol + TsCl followed by F^- **d.** (*R*)-2-methyl-1-butanol + HCl or $SOCl_2$ **e.** 3,3-dimethyl-2-pentanol + PI_3, but hindrance might cause slow reaction. **6–21** The protonated alcohol can undergo S_N2 displacement by Cl^- to give **A**, but can also lose H_2O to yield a carbocation. The carbocation can undergo methyl migration to yield a new cation that reacts with Cl^- to give **B**. The new cation can undergo hydride shift to form another cation which then reacts with Cl^- to form **C**. **6–22** Reaction 7 involves S_N2 displacement at a tertiary center, which is very unfavorable. Reaction 8 could give some product since it is S_N2 at a 1° center. The *t*-butoxide is, however, a very large and hence poor nucleophile even at 1° centers. Various *t*-butyl ethers are sometimes made by an S_N1 reaction yielding a *t*-butyl cation which can react with an alcohol. **6–23 a.** $EtOH \rightarrow Et\overset{+}{O}H_2$ A second EtOH molecule displaces H_2O to yield $Et\overset{+}{O}(H)Et$ which deprotonates to form EtOEt. **b.** No, since it would yield a mixture of EtOEt, EtOPr, and PrOPr. **6–24 a.** 2-propanol + Na and then 1-bromopentane **b.** *trans*-2-methylcyclohexanol + Na and then CH_3I

6–25 $CH_3CH(Cl)CH_2Cl + OH^- \rightarrow CH_3CH(Cl)CH_2OH \rightarrow CH_3CH(Cl)CH_2O^- \rightarrow$ **V**
The 1° Cl in the starting material is more easily displaced than the 2° in the first step.
6–26 As the (−)-isomer is formed, it can be converted back to the (+)-isomer, and a 50:50 mixture eventually results. **6–27** Formation of the tosylate doesn't involve the chiral center, but the second step to form the chloride is S_N2 and results in inversion. **6–28** Iodide displaces Cl^- and the iodopropane is converted to the alcohol with water. This releases I^-, which can restart the cycle. **6–29** Excess CN^- might form some dicyanohexane. **6–30 a.** 2° **b.** 3° **c.** 1° → 3° and 2° **d.** 1° → 2° **e.** 1° → 2° by ring expansion; 1° → 3° by hydride shift **6–31** Me_3COMe, Me_3COEt, Me_3COH, Me_3CI **6–33 a.** 1,3-difluoro-2,2,5-trimethylheptane **b.** dicyclopropylmethanol **c.** 1-bromo-2-cyclobutylpropane **d.** 1-chloro-3-ethoxy-1-methoxybutane **6–34 a.** Five alcohols **b.** Eight ethers **6–35** Seven tertiary alcohols **6–36** Eight alcohols; four 1°, three 2°, one 3°. **6–37** $EtOH + OH^- \rightleftarrows EtO^- + H_2O$ **6–40** In water, larger anions are less easily solvated and hence more available as nucleophiles. In acetone, the smaller nucleophile reacts more rapidly. **6–41** MeOH + HCl followed by Cl_2 and light or heat **6–43 a.** CH_3Br **c.** 2-bromopentane **d.** 4-bromo-1-butanol and 1,4-dibromobutane **6–44** **A** is (*R*)-1-bromo-2-methylbutane, formed by alcohol protonation and Br^- displacement of H_2O. **B**, 2-bromo-2-methylbutane, is the result of 1° carbocation formation from the alcohol, hydride shift, and reaction with Br^-. **6–46 a.** PBr_3 or HBr **c.** PBr_3, then KCN **e.** Na, then 1-bromopentane **g.** PBr_3, then cyclohexanol treated with Na **6–48 a.** cyclopentanol + Na, then EtBr **b.** 2-butanol + Na, then PrBr **6–49 b.** butyl methyl ether, 2-methoxybutane, isobutyl methyl ether, *t*-butyl methyl ether **6–51 a.** (*S*)-2-cyanopentane **d.** *cis*-1-bromo-2-methylcyclobutane **g.** (*R*)-2-cyano-1-propanol **6–54** The *t*-butyl carbocation is formed much more rapidly than one from EtOH, so it is the only cation in solution. The carbocation is too bulky to react with *t*-butyl alcohol, so it only reacts with EtOH to give *t*-butyl ethyl ether. **6–58** Start with tetrahydrofuran, protonate the ether oxygen, and attack the neighboring carbon with H_2O: as a nucleophile. **6–60 a.** The 3° OH is protonated, H_2O is lost to give a cation with a plane of symmetry, and this cation is attacked intramolecularly by the 1° OH. **b.** The tosylate is formed at the 1° OH and is then displaced intramolecularly by the 3° OH in an S_N2 fashion. **6–63 a.** The N displaces the Cl intramolecularly to give a three-membered ring with positively-charged N. OH^- then attacks the CH_2 of the ring to reopen and form $Me_2NCH_2CH_2OH$.

Chapter 7

7–1 a. $CH_2{=}CHC(CH_3)_2CH_3$ **b.** $CH_2{=}C(CH_2CH_3)_2$

c. 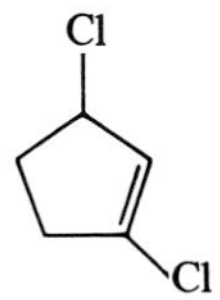 Cl Cl **d.** 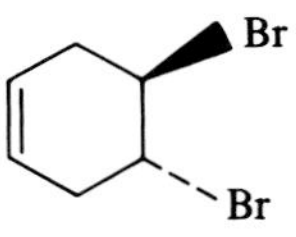 Br Br

e. $(CH_3)_2C(OH)CH(CH_3)CH{=}CH_2$
7–3 4-chloro-1-butene, 3-chloro-1-butene, 2-chloro-1-butene, (*Z*)- and (*E*)-1-chloro-1-butene, (*Z*)- and (*E*)-1-chloro-2-butene, (*Z*)- and (*E*)-2-chloro-2-butene, 1-chloro-2-methylpropene, 3-chloro-2-methylpropene **7–4 a.** 2-Methylpropene yields isobutane rather than butane. **b.** 4.2 kcal/mol **7–5** There is a large steric interaction between *cis* alkyl groups in the (*Z*)-isomer. **7–6** For Equation 7.15: $RCH_2C(CH_3)_2OH \rightarrow RCH_2C(CH_3)_2OH_2^+ \rightarrow RCH_2\overset{+}{C}(CH_3)_2 \rightarrow RCH{=}C(CH_3)_2 + RCH_2C(CH_3){=}CH_2$ **7–7 a.** $CH_2{=}C(CH_3)CH_2CH_3$, $(CH_3)_2C{=}CHCH_3$, $(CH_3)_2C(OMe)CH_2CH_3$

b. 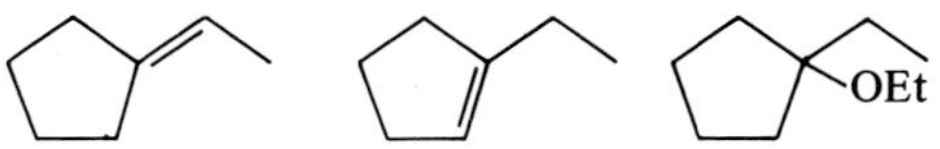 OEt

c. $CH_3CH_2C(CH_3){=}C(CH_3)_2$, $CH_2{=}C(CH_2CH_3){-}CH(CH_3)_2$, (*Z*)- and (*E*)-$CH_3CH{=}C(CH_3)CH(CH_3)_2$,

$CH_3CH_2C(OMe)(CH_3)CH(CH_3)_2$

7–8 a. $(CH_3)_2C{=}CHCH_3$ major, $CH_2{=}C(CH_3)CH_2CH_3$ minor **c.** $CH_3CH_2C(CH_3){=}C(CH_3)_2$ major **d.** $CH_3CH{=}CHCH_3$ major

7–9 a. HO, H_2O^+, +, II, +, +, III

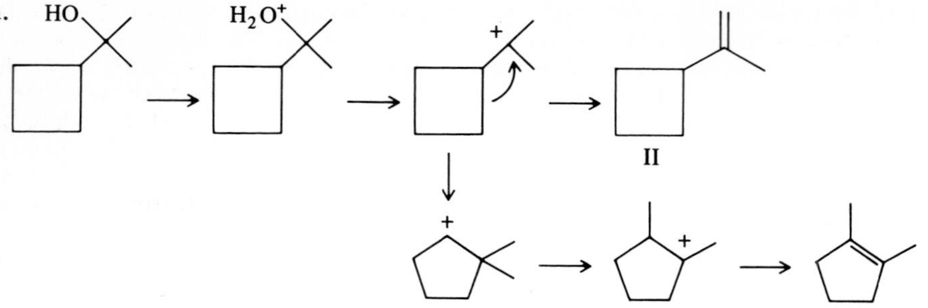

b. The cyclobutane ring is opened because it is strained and can form a more stable cyclopentane. **7–10 a.** cyclopentene and methoxycyclopentane **b.** $Et_2CHCH_2OH + Et_2C{=}CH_2$

c. $EtCH(OiPr)(CH_2)_4CH_3$ + (*Z*)- and (*E*)-$EtCH{=}CHBu$ + (*Z*)- and (*E*)-$MeCH{=}CH(CH_2)_4CH_3$
d. (*Z*)- and (*E*)-$Me_3CCH{=}CHEt$
7–11 First half: H_2SO_4 dehydration yields 3-methyl-2-pentene. Second half: PBr_3 gives the bromide which, with *t*-BuO^-, gives mainly 3-methyl-1-pentene.

7–12 a. $R\overset{+}{N}(Et)_2CH_2CH_3 + OH^- \rightarrow RNEt_2 + CH_2{=}CH_2 + HOH$ **b.** Two alkenes are formed: $CH_3CH{=}CH_2$ and $CH_2{=}CH_2$ (major). **7–13 a.** Draw Newman projections with H and Br *anti* on neighboring carbons.

7–14 H, Me, H, Me, D, Br → Me, D, Me, H; D, Me, H, H, Me, Br → H, Me, Me, H

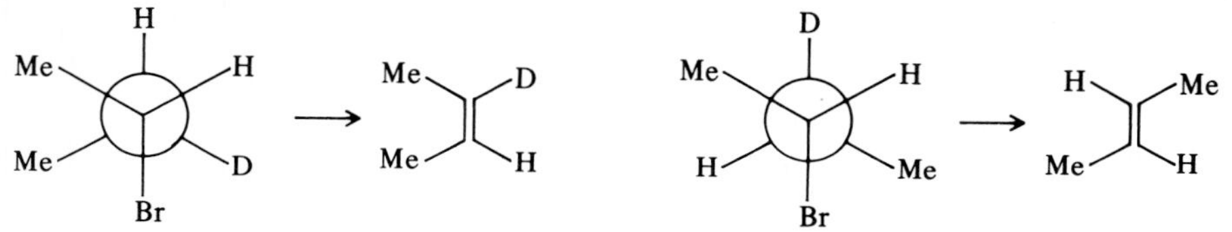

7–15 OTs, H, Me, Me, Et, H → H, Me, Me, Et

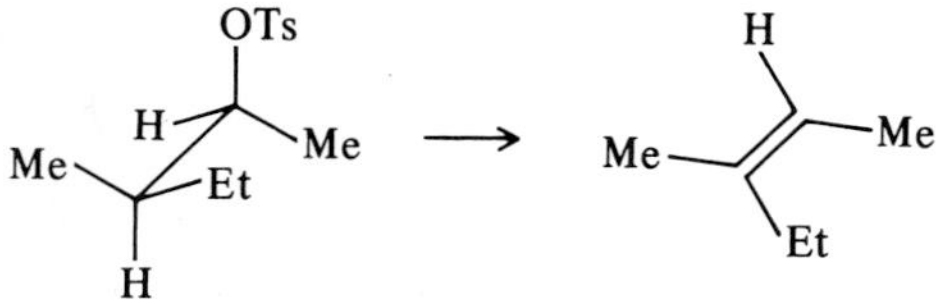

VI (*Z*)-alkene

7–16 The *cis* compound has one Br in an axial position and hence there is an axial H on the neighboring carbon for fast elimination. In the diaxial conformation of the *trans* dibromide, the H on the neighboring carbon (the one with the other Br) is equatorial. There is, however, an axial H on the other neighboring carbon (without the Br) and elimination occurs in that direction. **7–17 a.** Br_2, heat; RO^- **b.** Br_2, heat; I^-, acetone **c.** H_2SO_4, heat **d.** TsCl; RO^- **e.** H_3PO_4, heat **7–18 a.** 5,5-dimethyl-1-hexene **b.** (*Z*)-6-bromo-2-hexene **c.** (*Z*)-2,4-dichloro-5-methyl-4-decene **d.** *cis*-3-ethyl-4-methylcyclohexene **e.** (2*Z*,5*Z*)-2,5,7-trimethyl-2,5-octadiene

7–19 a. $CH_2{=}CH(CH_2)_3CH_3$ **b.** (*Z*)-2-hexene (*E*)-2-hexene

c. $(CH_3)_2C{=}CHCH_2CH_3$
7–20 a. $Et_2C{=}CHCH_2CH_3$ **b.** $Cl_2C{=}CHCH_2CHCl_2$ **e.** $HOCH_2(CH_2)_4CH{=}CH_2$ **f.** $F_3CCF{=}CF_2$ **7–22 a.** 3 alkenes **b.** 1 **c.** 4 **d.** 5 **e.** 12 **7–23 a.** 1-octene **b.** (*E*)-1-bromo-1-chloro-1-pentene **e.** (*E*)-5-bromo-3-ethyl-2-pentene **7–25 a.** Both enthalpies are negative because more highly substituted double bonds are formed. Equation 6 is less negative than Equation 7 because there is extra strain involved with a double bond in a four-membered ring. **b.** Probably positive because the amount of ring strain introduced would overcome the gain in having a more substituted double bond. **7–26 a.** 2 is more stable than 1. **c.** 2 and 3 are about equal, and both more stable than 1. **7–28.** $\Delta H = -1.37$ kcal/mol. The

isomerization is favorable because a more highly substituted double bond is formed. **7–31 a.** propene **c.** 1-pentene **d.** 1-pentene + (*Z*)- and (*E*)-2-pentene **7–32 a.** 2-methyl-2-propanol + H_2SO_4, heat **c.** 3-heptanol + H_2SO_4, heat (both *cis* and *trans* isomers will be formed) **e.** from 4-methylcyclohexanol **7–35** 1-pentene, (*Z*)- and (*E*)-2-pentene, 2-isopropoxypentane **7–36 a.** Br_2, heat, and RO^- **c.** PBr_3 and RO^- **7–39 A** 2,3-dimethylbutane; **B** 1-chloro-2,3-dimethylbutane; **C** 2-chloro-2,3-dimethylbutane; **D** 2,3-dimethyl-1-butene; **E** 2,3-dimethyl-2-butene **7–40 a.** $Me_2NCH_2CH_2CH{=}CH_2$ **b.** 1,3-butadiene **c.** A cyclic amine will still contain N while an acyclic amine will not. **7–42 a. A** (*Z*)-2-bromo-2-pentene; **B** (*Z*)-3-bromo-2-pentene, or vice versa **b. A** and **B** are the only products possible from *anti* elimination. **7–45 a.** Br_2, light; KI, acetone **b.** Br_2, light; NaSH

Chapter 8

8–1 Yes. $C_3H_7NO_3$ = 105.0424 **8–2** 49 is $^{35}ClCH_2^{+\cdot}$, 51 is $^{37}ClCH_2^{+\cdot}$ **8–3** $(CH_3)_2C{=}OH^+$ *m/z* = 59 from *t*-butyl alcohol. Major ion from 1-butanol: $CH_2{=}OH^+$ *m/z* = 31. Major ion from 2-butanol: $CH_3CH_2CH{=}OH^+$ and $CH_3CH{=}OH^+$ *m/z* = 45. **8–4 a.** 1-pentanol: *m/z* = 31; 2-pentanol: *m/z* = 45, 73; 3-pentanol: *m/z* = 59 **8–5** 1-methoxybutane: $CH_2{=}\overset{+}{O}CH_3$; 1-ethoxypropane: $CH_2{=}\overset{+}{O}CH_2CH_3$ and $CH_3CH_2CH_2\overset{+}{O}{=}CH_2$ **8–6 a.** Visible light ranges from 28.6 to 81.6 kcal/mol. This is not sufficient energy to break the C—C bond in ethane (84 kcal/mol). **b.** Yes, O—O bond energy is 45 kcal/mol. **c.** 490 nm **d.** No. **e.** 10^5 kcal/mol can break many bonds. **8–7** Spectrum A: tetrahydrofuran shows no O—H or C=O. Spectrum B: 2-butanone C=O at 1710 and no OH stretch. Spectrum C: 2-buten-1-ol OH at 3300 cm^{-1}, C=C at 1650. **8–9** CH_3CH_2CN from CN triple bond at 2225 cm^{-1} **8–10 a.** No **b.** 1650 cm^{-1} indicates a C=C: $CH_3CH_2OCH{=}CH_2$ **8–12** $CH_2{=}C{=}CHCH_3$, $HC{\equiv}CCH_2CH_3$, 1-methylcyclopropene, 3-methylcyclopropene

8–13 CH_3CN: 2200–2300 cm^{-1}; $CH_2{=}C{=}NH$: 1600–1690, 3400 cm^{-1};

$HC{\equiv}CNH_2$: 2100–2200 cm^{-1}; N H 1600–1680 cm^{-1}; N 1630–1690 cm^{-1};

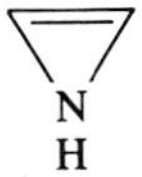

8–14 a. 1 **b.** 3000 cm^{-1} C—H; 1030–1130 cm^{-1} C—O

c. C=O, O—H, C=C **d.** 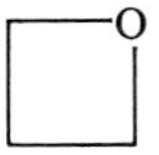 O , O

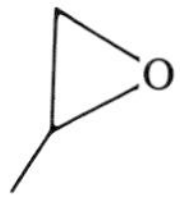

8–16 a. *m/z* = 66: $^{37}\overset{+}{Cl}CH_2CH_3$; *m/z* = 64: $^{35}\overset{+}{Cl}CH_2CH_3$; *m/z* = 51, 49: $ClCH_2^{+\cdot}$; *m/z* = 29: $CH_3CH_2^{+\cdot}$; *m/z* = 27: $CH_2{=}CH^+$ **8–17** Diethyl ether cleaves to $CH_3CH_2\overset{+}{O}{=}CH_2$. The same process with tetrahydrofuran opens the ring to give $\cdot CH_2CH_2CH_2\overset{+}{O}{=}CH_2$ which has the same *m/z* as the molecular ion. **8–18 a.** 86.073 **b.** 86.109 **8–19 a.** $CH_3CH_2C(CH_3){=}OH^+$, $CH_3CH_2CH_2C(CH_3){=}OH^+$, $CH_3CH_2CH_2C(CH_2CH_3){=}OH^+$ **8–21 a.** CH_3OCl, $ClCH_2OH$: only the latter has an O—H (3400 cm^{-1}). **8–22 a.** $:\overset{-}{C}{\equiv}\overset{+}{O}:$ **b. V** = $CH_3\overset{+}{C}{\equiv}O:$ **8–23** C_2H_6O = ethanol; $C_3H_6O_2$ = $CH_3C(OMe){=}O$

Chapter 9

9–1 a. 3; 2:3:3 **b.** 120 Hz, 2.17 ppm; 205 Hz, 3.45 ppm; 240 Hz, 4.00 ppm **c.** $CH_3OCH_2C(CH_3){=}O$ **9–2** Structure **I** is correct because there are two equivalent signals and six total H's, indicating the presence of two CH_3 groups. 3.75 ppm is $\mathbf{CH_3}$O singlet. **9–3 a.** Two kinds: two triplets **b.** Three kinds: singlet, quartet, triplet **c.** One kind: singlet **d.** Three kinds: singlet, triplet, triplet **9–4 a.** $BrCH_2CH_2Br$ and Br_2CHCH_3 **b.** $BrCH_2CH_2Br$ **c.** Two peaks: a doublet and a quartet **9–5** Doublet 1.15 ppm: CH_3 next to CH; apparent quartet 1.43 ppm: CH_2 next to CH_3; triplet 0.92 ppm: CH_3 next to CH_2; singlet 1.8 ppm: next to OH; sextet 3.7 ppm: next to CH bearing OH **9–6 a.** $CH_3CH_2CH{=}O$ **9–7** Hydrogens on C-1: triplet 3.57; on C-2: apparent quartet 2.25; on C-3: apparent quartet 4.32; on C-4: doublet 1.65 **9–8** $C_4H_9Br(CH_3)_2CHCH_2Br$; ^{13}C nmr: 21 ppm, 2 $\mathbf{CH_3}$ groups; 31 ppm, **C**H; 42 ppm, $\mathbf{CH_2}$Br **9–11** $C_2H_3Cl_3$: $ClCH_2CHCl_2$ $C_4H_{11}N$: $(CH_3)_3CNH_2$ C_3H_7ClO: $ClCH_2OCH_2CH_3$ C_4H_8O: $CH_3CH_2C(CH_3){=}O$ **9–12** $C_4H_{11}NO_2$: $(CH_3O)_2CHCH_2NH_2$ **9–14** $CH_3CH_2CH_2OH$ **9–15** $Cl_3CCH(CH_3)OH$ **9–16 b.** $BrCH_2CH_2CH_2Br$ **9–17** Pr_2O **9–19** C_4H_8O: $CH_2{=}CHOCH_2CH_3$; $C_5H_{10}O$: $CH_2{=}CHCH_2CH(OH)CH_3$

Chapter 10

10–1 a. 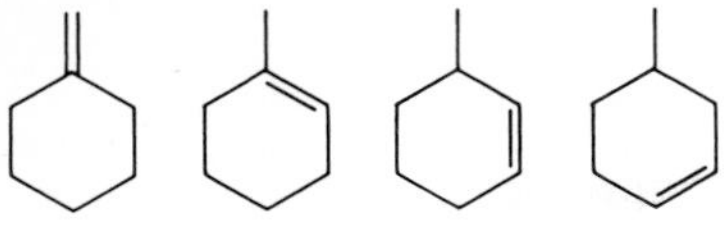 **b.**

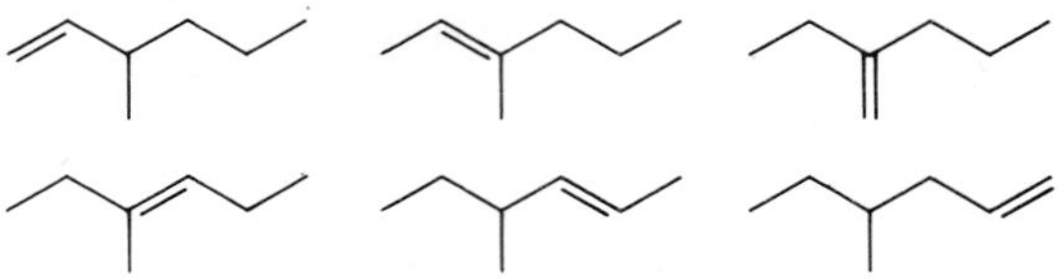

10–2 a. Cyclopentanol **b.** 2-butanol **c.** 3-methyl-3-hexanol **d.** 1-methylcyclobutanol **e.** 2-pentanol, 3-pentanol **10–3 a.** Chlorocyclohexane **b.** Bromocyclohexane **c.** Iodocyclohexane **d.** Chlorocyclohexane and ethoxycyclohexane **10–4 a.** 2,3-dimethyl-3-pentanol, 2,3-dimethyl-2-pentanol **b.** 2-chloro-2-methylbutane **c.** 1-fluoro-1-methylcyclohexane **d.** 2-chloropentane and 2-pentanol **e.** 2-bromopentane and 3-bromopentane **10–5 a.** 2-chloro-2-methylbutane, 2-chloro-3-methylbutane **b.** ^{1}H or ^{13}C nmr **10–6** *meso*-2,3-dibromobutane; no **10–7** $CH_2{=}CH_2 + Br_2 \rightarrow$ bromonium ion (3-membered ring) + Br^-. The bromonium ion reacts with Br^- to give $BrCH_2CH_2Br$; with Cl^- to give $BrCH_2CH_2Cl$; and with H_2O to give $BrCH_2CH_2OH$. **10–9 a.** 1,2-dibromo-2-methylbutane **b.** *trans*-1,2-dibromo-1-methylcyclopentane **c.** (*R*,*R*)- and (*S*,*S*)-2,3-dichlorobutane **d.** 1-chloro-2-propanol **10–10** *trans* **10–11 a.** 1-methylcyclohexanol **10–12 a.** H_2SO_4, H_2O **b.** B_2H_6; H_2O_2, OH^- **c.** B_2H_6; H_2O_2, OH^- **d.** H_2SO_4, H_2O **e.** B_2H_6; H_2O_2, OH^- **10–13 a.** Either B_2H_6 then H_2O_2, OH^- or HBr, peroxides, and OH^- **b.** No. Either of the above reactions would lead to a mixture of 2-pentanol and 3-pentanol. **10–14** HBr, peroxides, and CN^-

10–15 a. $CH_2{=}CHCH_3 \rightarrow CH_3\overset{+}{C}HCH_3 \rightarrow (CH_3)_2CHCH_2\overset{+}{C}HCH_3 \rightarrow (CH_3)_2CHCH_2CH(CH_3)CH_2\overset{+}{C}HCH_3$. Head-to-tail addition gives the more stable of the two possible cations.
b. *Hint:* Use an *intermolecular* hydride transfer.
10–16 ethylene oxide + $OH^- \rightarrow HOCH_2CH_2O^- \rightarrow HOCH_2CH_2OCH_2CH_2O^- \rightarrow$ etc.
10–17 a. 1,2-dimethylcyclobutene (**A**), $CH_3COCH_2CH_2COCH_3$ (**B**) **b.** *trans*-1,2-dimethyl-1,2-cyclobutanediol **10–18 a.** 1,2-butanediol **b.** 2,3-butanediol **c.** 1-methyl-1,2-cyclopentanediol **d.** *trans*-1,2-cyclooctanediol **e.** *trans*-1,2-cyclobutanediol **f.** $CH_3CH{=}O$ and $H_2C{=}O$ **g.** $2CH_3CH{=}O$ **h.** $2CH_3CH{=}O$ **10–19 a.** 3-methyl-2-hexene and 2-methyl-1-hexene

b. A **B**

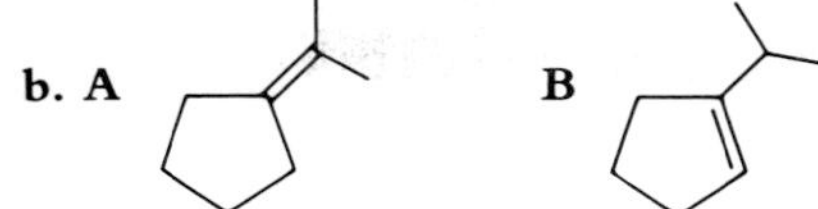

10–20 a. None **b.** A prochiral CH_2 carbon, a prochiral CBr_2 carbon, and enantiotopic Br's **c.** A prochiral CH carbon, enantiotopic CH_3 groups **d.** Two prochiral CH_2 carbons, two sets of diastereotopic hydrogens (on the CH_2 groups) **e.** A prochiral CH_2 carbon, diastereotopic hydrogens **10–21 a.** butane **b.** butane **c.** methylcyclobutane **d.** *cis*- and *trans*-1,3-dimethylcyclohexane **10–22 a.** H_2/cat **d.** *t*-BuO^- and D_2/cat **10–23 a.** H_2SO_4, H_2O **b.** D_2SO_4, D_2O **d.** *t*-BuO^- and HBr, dark, then *t*-BuO^- **10–24** Bromonium ion intermediate. Attack by CH_3OH at one carbon is hindered by the three methyl groups. **10–26** $CH_3\overset{+}{C}HCF_3$ is less stable than $\overset{+}{C}H_2CH_2CF_3$ because of the strong electron-withdrawing inductive effect of the CF_3 group. **10–28** $CH_2{=}CHCH_2OH + Cl_2$, OH^- **10–32** HBr, peroxides, and NH_3 **10–34 A:** 1-methylcyclobutene; **B:** methylenecyclobutane; **C:** 3-methylcyclobutene; **D:** 1-chloro-1-methylcyclobutane **10–35** $\Delta H = -8$ and -7 kcal/mol **b.** H = +8.5 and −21.5 kcal/mol for HCl; H = +6 and −23.5 kcal/mol for HI **c.** Only with HBr are both steps exothermic. **10–36 a.** cyclopentanol **c.** 2-methyl-1-pentanol **10–37 a.** 1,2-pentanediol **c.** $CH_3CH{=}O + CH_3CH_2CH_2CH{=}O$ **g.** 3-bromo-2-pentanol and 2-bromo-3-pentanol **10–38 b.** $2H_2C{=}O + CH_3COCH{=}O$

10–44 a. $-CH_2\overset{\overset{CH_3}{|}}{C}HCH_2CH(CH_3)CH_2\overset{|}{C}HCH_3$ **c.** $-CH_2CCl_2-CH_2CCl_2-CH_2CCl_2-$

10–45 A: 3-ethyl-1-hexene; **B:** 4-methyl-2-hexene **10–47 a.** Prochiral, enantiotopic Hs **c.** Both prochiral, enantiotopic Br's and Hs

Chapter 11

11–1 C=O 90 kcal/mol; C=C 62 kcal/mol. Carbonyl additions are reversible because of the high C=O bond strength. **11–2 a.** 2-methylbutanal **b.** 3-methyl-2-pentanone **c.** *cis*-4-methylcyclohexanecarboxaldehyde **d.** (*S*)-2-ethylcyclopentanone **e.** 7,7-dichloro-6-methyl-2-heptanone **11–3** There is no angle strain since the ir absorption is at the same position as in an acyclic ketone. **11–4** $CH_3CH_2CH_2CH{=}CHCHO$ **11–5 a.** $Me_2CHC(Me){=}O$ **b.** Cyclobutyl ethyl ketone **c.** $CH_3CH_2CH(CH_3)COOH$, $CH_3CH_2CH(CH_3)CH{=}O$ **d.** $C_6H_{12}COOH$, $C_6H_{12}CHO$ **e.** No reaction. **f.** Cyclohexanone **11–6 a.** 2-butanol **b.** 1-propylcyclopentanol **c.** $C_6H_{12}CH(CH_3)OH$ **11–7 a.** Propanol + BuMgBr **b.** Formaldehyde + *s*-BuMgBr **c.** Cyclohexanone + EtMgBr **e.** 3-Methylbutanone + methylmagnesium bromide **11–8** $Me_2C{=}\overset{+}{\underset{..}{O}}{-}\overset{-}{B}F_3$ **11–9** Bulky groups disfavor sp^3 (109°) geometry compared to sp^2 (120°) geometry.

11–10 a. BuCH(OH)CN **b.** $CH_3C(OH)Bu$ with CN on C **c.** 1-cyclopentyl-1-hydroxy(CN) structure: HO, CN **e.** $(CF_3)_2C(OH)_2$

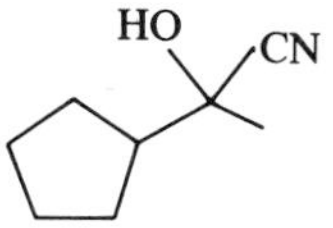

11–11 a. EtCH=NOH **b.** Cyclohexylidene=NNHPh **c.** $(C_6H_5)_2C{=}NCH_2CH_3$

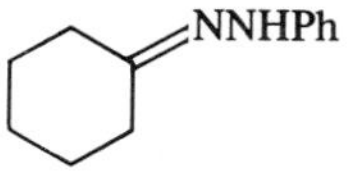

11–12 Acetaldehyde: $CH_3CH{=}N{-}NHPh$ (two isomers: NHPh *syn* and *anti* to CH_3)

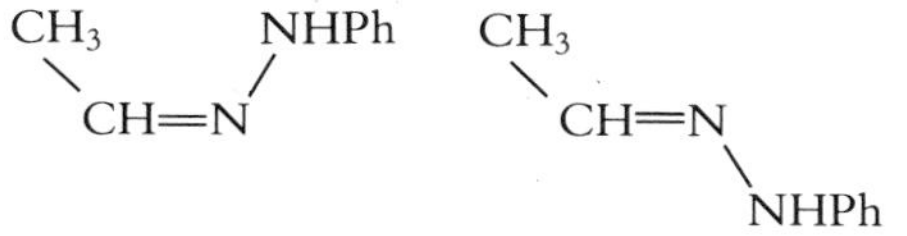

11–13

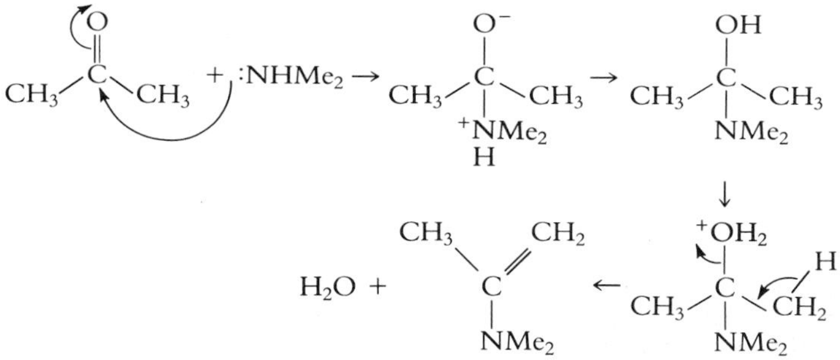

$$CH_3C(=O)CH_3 + {:}NHMe_2 \rightarrow CH_3C(O^-)(CH_3)\overset{+}{N}Me_2H \rightarrow CH_3C(OH)(CH_3)NMe_2$$

$$\downarrow$$

$$H_2O + CH_3C(NMe_2){=}CH_2 \leftarrow CH_3C(\overset{+}{O}H_2)(CH_2{-}H)NMe_2$$

11–14

$$CH_3CH(\ddot{O}R)_2 \xrightarrow{H^+} CH_3{-}CH(\ddot{O}R)(\overset{+}{O}R\,H) \rightarrow CH_3CH{=}\overset{+}{O}R + HOR$$

$$\downarrow H_2O$$

$$HOR + CH_3CH{=}\overset{+}{O}H \leftarrow CH_3{-}CH(\overset{+}{HOR})(\ddot{O}H) \leftarrow CH_3CH(OR)(\overset{+}{O}H_2)$$

$$\downarrow$$

$$CH_3CH{=}O + H^+$$

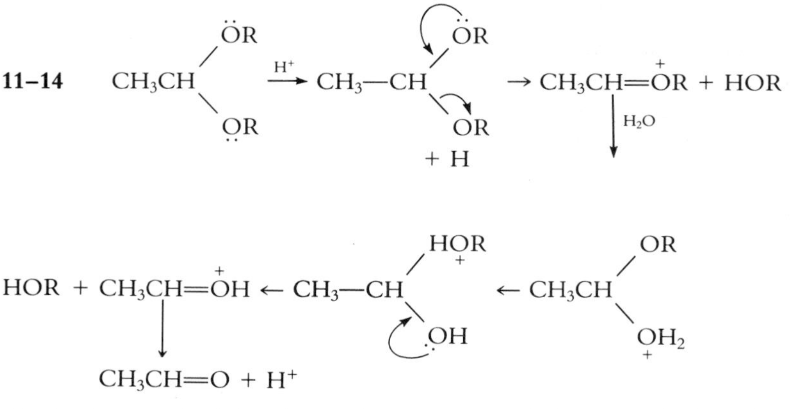

11–15 Equation 11.40 involves making one molecule from two (unfavorable entropy), while Equation 3 involves converting one molecule into another (more favorable entropy).

11–16 a. Cyclopentyl–$CH(OEt)_2$ **b.** Cyclohexane-fused 1,3-dioxolane with 2-CH_2CHMe_2

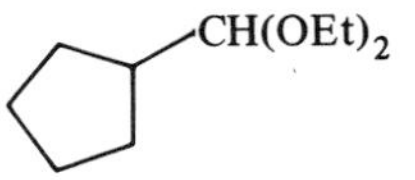

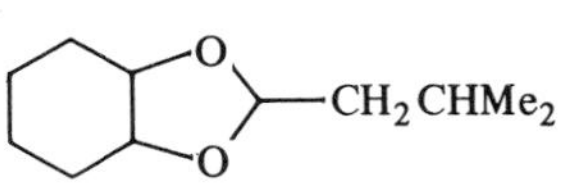

11–17 $O{=}CH(CH_2)_6CH_2OH \rightarrow (RO)_2CH(CH_2)_6CH_2OH \rightarrow (RO)_2CH(CH_2)_6COOH \rightarrow O{=}CH(CH_2)_6COOH$

11–18 a. BuOH **b.** $CH_2{=}CHCH_2CH_2OH$ **c.** $CH_2{=}CHCH_2CH(OH)CH_3$ **d.** $CH_2{=}CHC(CH_3)_2CH_2CH_3$
11–19 a. NADH + $CH_3CD{=}O$ **b.** (*S*)-ethanol-1-*d* **11–21 a.** $BrCH_2CHO$ **c.** $Me_3CCH_2CH(CHO)CH_2CH_3$ **d.** $CH_3CH_2CH_2COCH_2COCH_2CHO$ **11–22 a.** $CH_3COCH_2CH_2CH_3$ **b.** EtCOEt **d.** $EtCOCH_2CH_2COOH$ **11–24 A:** Me_3CCH_2OH; **B:** $Me_2C(OH)CH_2CH_3$ **11–25 a.** CrO_3, pyridine, CH_2Cl_2 **c.** H_2CrO_4, H_2O or CrO_3, pyridine, CH_2Cl_2 **d.** CrO_3, pyridine, CH_2Cl_2 **11–26 a.** OH^-; H_2CrO_4 **b.** OH^-; CrO_3, pyridine **d.** 48% HBr; H_2CrO_4 **11–28 a.** Butanal + MeMgBr or acetaldehyde + PrMgBr **c.** Formaldehyde + BuMgBr **e.** 3-pentanone + MeMgBr or 2-butanone + EtMgBr **g.** Cyclopentanone + MeMgBr **11–29 a.** PBr_3; Mg, Et_2O; propanal then H_3O^+; H_2CrO_4 **c.** B_2H_6 then H_2O_2, OH^-; CrO_3, pyridine, CH_2Cl_2; isopropyl magnesium bromide then H_3O^+ **11–33** Acetaldehyde + HCN then TsCl then OH^- → **I** **11–34 a.** $Me_2CHCH(OH)CN$ **d.** $Me_2CHCH_2CH{=}CHCH{=}NNHCONH_2$ **11–38 a.** $EtCH(SEt)_2$
11–41 a. $CH_2{=}CHCH_2CH_2CH_2OH$ + dihydropyran, H^+ → $CH_2{=}CHCH_2CH_2CH_2OR$ → $HOCH_2CH(OH)CH_2CH_2CH_2OR$ → $HOCH_2CH(OH)CH_2CH_2CH_2OH$
11–42 a. 2-propanol **b.** 4-hexen-2-ol **c.** 4-methyl-1-pentanol **g.** 2-cyclohexyl-3-pentanol **11–46 a.** $LiAlH_4$; TsCl, OH^- **c.** EtMgBr then H_3O^+; TsCl, OH^-; KCN, DMF **11–47 A:** 2,3-dimethyl-2-hexene; **B:** 2-pentanone; **C:** acetone **11–48** 6-methyl-5-hepten-2-one

Chapter 12

12–1 a. > 1; $CH_3C{\equiv}C^-$ more stable than CH_3^- **b.** > 1; $CH_2{=}CH^-$ more stable than CH_3^- **c.** very slightly < 1; CH_3^- more stable than $CH_3CH_2^-$ **d.** < 1; NH_2^- more stable than CH_3^- **e.** > 1; CH_3O^- more stable than $HC{\equiv}C^-$ **12–2 a.** $CH_3C{\equiv}CCH_2CH_3$ **b.** $EtC{\equiv}CEt$ **c.** $MeC{\equiv}CMgBr + CH_4$ **d.** $CH_3CH(OH)C{\equiv}CCH_3$ **12–3** Acetone + $HC{\equiv}CMgBr$, then dehydration **12–4** Br_2; $NaNH_2$; NH_2^- and PrBr **12–5 a.** Concerted *anti* elimination of H and Br across C=C **b.** The (*E*)-decene does not have H and Br *trans* to each other. **12–6 a.** (*Z*)-2-chloro-2-butene **b.** 2,2-dibromobutane **c.** 2-butanone **d.** (*E*)-2,3-dibromo-2-butene
12–7 a. $EtOH + OH^- \rightarrow EtO^-$
$MeC{\equiv}CH + EtO^- \rightarrow MeC(OEt){=}CH^- \xrightarrow{EtOH} MeC(OEt){=}CH_2 + EtO^-$
b. The $NuCH{=}CH^-$ formed is more stable than $NuCH_2CH_2^-$. **12–9 a.** Addition in both senses gives the same product. **b.** No. There would be too many halogenated octanes to separate. **12–11** Start with $HC{\equiv}CCH_2OR$, couple to Me_2CHCH_2Br, deprotect the OH, reduce the alkyne, and convert to the bromide. **12–12 a.** propyne **b.** 2-butyne-1,4-diol **c.** 1,3-butadiyne **d.** 3-butyne-2-ol **12–13 a.** $MeC{\equiv}CMe$ **b.** $MeC{\equiv}CMe$ **c.** $CH_2{=}CHC{\equiv}CH$ **12–15 a.** $RC{\equiv}C^-Li^+$ **b.** $RC{\equiv}CMgBr + CH_4$ **c.** $CH_3CH_2C(OEt){=}CH_2$ **d.** $CH_3CH_2C(Br){=}CH_2$ **e.** $CH_3CH_2COCH_3$ **12–16 a.** $LiNH_2$ (2) BuBr **c.** (1) $NaNH_2$ (2) EtBr (3) Na, NH_3 **g.** acetaldehyde + $HC{\equiv}CMgBr$ **12–17 A:** $HC{\equiv}CMgBr$; **B:** $HC{\equiv}CC(Me)_2OH$; **E:** $FCH_2(CH_2)_7C{\equiv}CH$ **12–24 A:** $CH_3(CH_2)_7CH{=}CHBr$, (*E*)-isomer; **C:** (*Z*)-isomer; **B:** $CH_3(CH_2)_7C(Br){=}CH_2$ **12–27** $HC{\equiv}CCH_2OH$; 3.07 ppm is **OH**.

12–26 b. Retrosynthetically, HO–[chain with OH, OH, OH] ← [diene chain]

← $HC{\equiv}C$–[chain]–$C{\equiv}CH$ ← $HC{\equiv}CH$ + Br–[chain]–Br

Chapter 13

13–1 a. $^-CH_2CH{=}CH_2$; $\overset{\delta-}{C}H_2\cdots CH\cdots \overset{\delta-}{C}H_2$ **b.** $O{=}C(O^-)O^-$ ↔ $^-O{-}C(O^-){=}O$; $C(O^{\delta-})_3$ (delocalized)

f. $HN{=}C(O^-)CH_3$; $\overset{\delta-}{HN}\cdots C(\overset{\delta-}{O})CH_3$

13–2 The allyl anion from propene is stabilized by electron delocalization: $CH_2{=}CHCH_2^- \leftrightarrow {}^-CH_2CH{=}CH_2$.

13–3 The cation that is formed is stabilized by charge delocalization:

13–4 a. [cyclobutenyl cation resonance structures] equal **b.** $^{+}CH_2CH{=}C(O^{-})CH{=}CH_2$

and $CH_2{=}CHC(O^{-}){=}CHCH_2^{+}$, both of equal weight

e. $CH_3O\overset{+}{C}HCH{=}CH_2 \leftrightarrow CH_3\overset{+}{O}{=}CHCH{=}CH_2$ (most)

13–5 A: $HOCH_2CH{=}CHCH_3 \rightarrow {}^{+}CH_2CH{=}CHCH_3 \leftrightarrow CH_2{=}CH\overset{+}{C}HCH_3 \rightarrow CH_2{=}CHCH(OH)CH_3$

B: (E)-$HOCH_2CH{=}CHCH_3$

13–7 [cyclohexyl cation]$-C{\equiv}CH \leftrightarrow$ [cyclohexylidene]$=C{=}{}^{+}CH \longrightarrow$ [cyclohexylidene]$=C{=}CHOH \longrightarrow$ **III**

13–8 $Me_3CCH{=}CHCH_3 \rightarrow Me_3CCH{=}CHCH_2\cdot \leftrightarrow Me_3C\dot{C}HCH{=}CH_2 \rightarrow$ **IV** **13–9** 1-butene

13–10 a. [structures] **b.** (1) [geranyl OPP → allylic cation → tertiary cation] → myrcene

b. (3) [OPP → cyclization + → cation +] → terpineol

13–11 (b), (c), (e), and **(g)** are conjugated; **(d)** is conjugated and cumulated.

13–12 $CH_3CH{=}CHCH{=}O \leftrightarrow CH_3\overset{+}{C}HCH{=}CH{-}O^{-}$ resonance stabilizes crotonaldehyde.

13–13 [cyclohexa-1,4-diene → protonated cation → allylic cation → cyclohexa-1,3-diene]

$CH_2{=}CHCH_2CH{=}O \rightarrow CH_2{=}CH\overset{-}{C}HCH{=}O \leftrightarrow {}^{-}CH_2CH{=}CHCH{=}O \rightarrow CH_3CH{=}CHCH{=}O$

13–14 $CH_2{=}CHCH{=}CH_2 \rightarrow BrCH_2\overset{+}{C}HCH{=}CH_2 \leftrightarrow BrCH{=}CHCH_2^{+}$

The resonance hybrid is more stable than the bromonium ion.

13–15 A: $ClCH_2C(OH)(CH_3)CH{=}CH_2$

13–16 a. [Cl, Cl] **b.** [H, CHO] **c.** [CHO, CH_3] **d.** [COOEt, COOEt]

13–17 a. Butadiene + vinyl bromide **b.** 2,3-dimethylbutadiene + 3-buten-2-one **c.** Cyclopentadiene + (*Z*)-1,2-dichloroethene **13–19** Same result **13–20** Three; 232 nm
13–21 (b–g) 13–22 One electron is promoted from the HOMO to the LUMO
13–23 Possible: $CH_3CH{=}CHCH{=}CHCH_3$, $CH_2{=}C(CH_3)CH{=}CHCH_3$. Not possible: $CH_2{=}CH_2CH(CH_3)CH{=}CH_2$ or any ring structures. **13–24 A:** propanal ($\varepsilon = 18$); **B:** cyclopropanol or oxetane
13–25 $CH_3\cdot$ + $\cdot CH_2CH{=}CH_2$ (resonance stabilized) **13–28 a.** $CH_2{=}CHCl \rightarrow CH_3\overset{+}{C}HCl \leftrightarrow CH_3CH{=}Cl^+$ **13–29 a.** $CH_2{=}CHCH_2CH{=}CH_2 \rightarrow CH_2{=}CH\overset{-}{C}HCH{=}CH_2 \leftrightarrow CH_2{=}CHCH{=}CHCH_2^- \rightarrow$ 1,3-pentadiene. **c.** $CH_3OCH{=}CHCH_2^+$ is formed, which is resonance stabilized: $CH_3\overset{+}{O}{=}CHCH{=}CH_2$ **13–31 a.** $R_2\overset{+}{C}OR' \leftrightarrow R_2C{=}\overset{+}{O}R'$ **13–33 a.** H_2SO_4; *N*-chlorosuccinimide or *t*-butylhypochlorite **b.** MeMgBr; H_2SO_4; *t*-butylhypochlorite
13–34 There are no allylic hydrogens. **13–35 a.** $HOCH_2CH(CH_3){=}CH_2 \rightarrow {}^+CH_2CH(CH_3){=}CH_2 \leftrightarrow CH_2{=}C(CH_3)\overset{+}{C}HCH_3 \rightarrow CH_2{=}C(CH_3)CH(Cl)CH_3$

$HOCH_2CH{=}C(CH_3)_2 \rightarrow {}^+CH_2CH{=}C(CH_3)_2 \leftrightarrow CH_2{=}CH\overset{+}{C}(CH_3)_2 \rightarrow CH_2{=}CHC(Cl)(CH_3)_2$

The second reaction should be more rapid because the resonance hybrid has a + charge on a 3° carbon.
13–40 VIII will be formed with heat, **IX** by light.
13–42 a. (1) 1,3-butadiene + propenal followed by H_2/cat, or (2) 2,4-pentadienal + ethylene followed by H_2/cat. (1) is better because an electron-withdrawing group (CHO) is on the dieneophile. **13–45** 36 kcal/mol. The HOMO and LUMO of conjugated dienes are closer together in energy than the corresponding orbitals for isolated double bonds. **13–47** 1,3-cycloheptadiene

Chapter 14

14–1 $H_2C{=}CHCH{=}C{=}C{=}CH_2$, $HC{\equiv}CCH{=}CHCH{=}CH_2$, etc. No. **14–2** $H_2C{=}CHCH{=}CHCH{=}CH_2 + Br_2 \rightarrow BrCH_2CH(Br)CH{=}CH{=}CH_2 + BrCH_2CH{=}CHCH(Br)CH{=}CH_2$ fast
14–3 a. −26.8 kcal/mol **b.** 1.8 kcal/mol **c.** −32.4 kcal/mol **14–4 a.** $-23 \times 4 = -92$ kcal/mol calculated. Cyclooctatetraene is therefore *less* stable than calculated, and there is no electron delocalization. **b.** No overlap of neighboring atomic *p* orbitals. **14–5 a.** Only **I** would have the 284 nm absorption; **II** would be similar to toluene. **b. I** is conjugated, would be more stable, and would have the smaller heat of hydrogenation. **14–6** 12 no; 14 yes **14–7** The nonbonded electron pair on N which would ordinarily be basic, is used in pyrrole to make up an aromatic sextet and is delocalized through the ring. **14–8** Cyclobutadiene and cyclooctatetraene; $4n$ **14–9 a.** Not applicable; not cyclic **b.** Applicable; no **c.** Not applicable; has an sp^3 carbon in the ring **(e–h)** Applicable, but none are aromatic **14–10** The nonbonded pair on N in pyridine is in an sp^2 orbital perpendicular to the π electrons and is not needed for the aromatic sextet; hence it is basic.

14–11 $C_6H_5\overset{+}{C}HPh \longleftrightarrow$ (cyclohexadienyl cation)$=CHPh \longleftrightarrow$ etc.

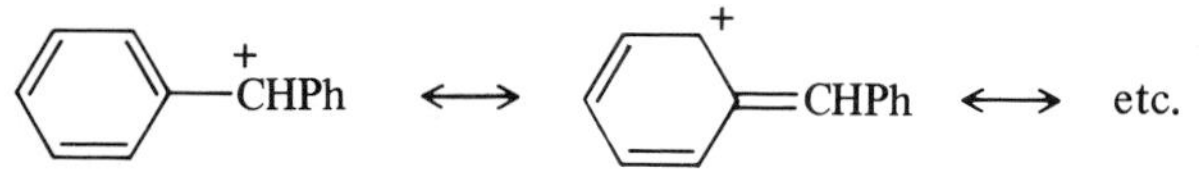

14–12 $C_6H_5CH_3 \longrightarrow C_6H_5\overset{-}{C}H_2 \longleftrightarrow$ (cyclohexadienyl anion)$=CH_2 \longleftrightarrow$ etc.

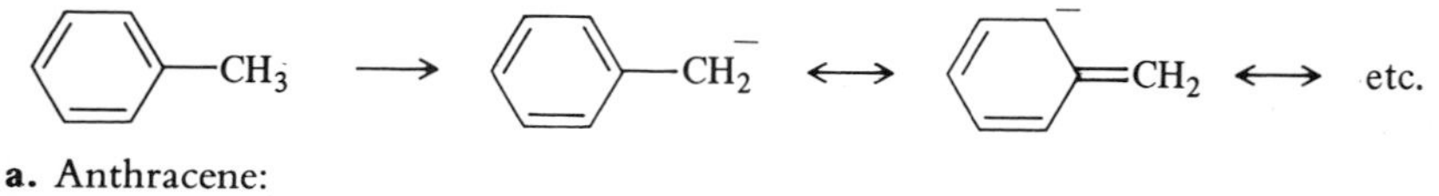

14–13 a. Anthracene:

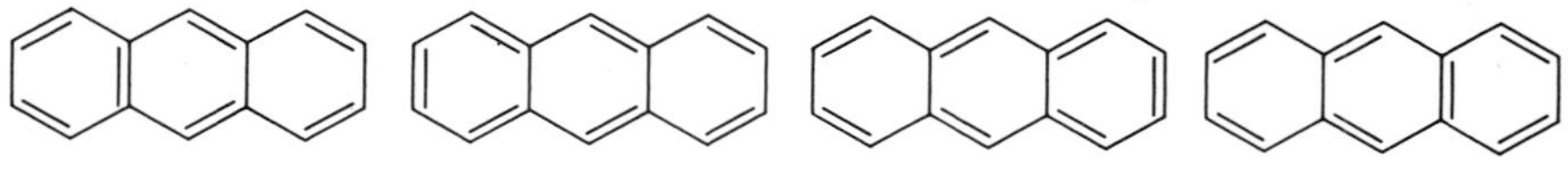

14–14 Yes; yes; yes; no (see Solutions Manual)

14–16 a. 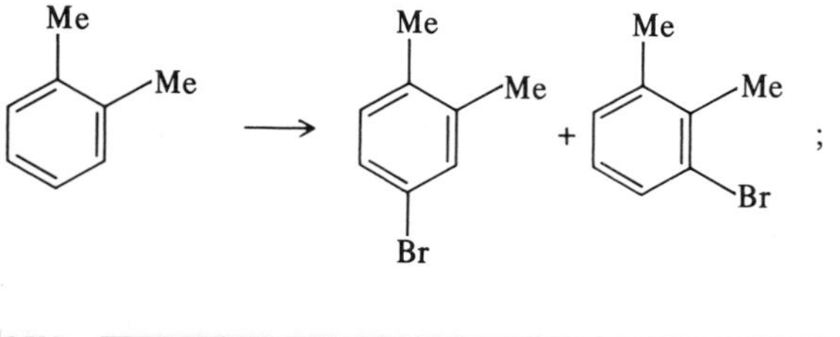 o-xylene (Me, Me) $\longrightarrow$ 4-bromo-1,2-dimethylbenzene (Me, Me, Br) + 3-bromo-1,2-dimethylbenzene (Me, Me, Br) ;

m-xylene (Me, Me) → 2-bromo-1,3-dimethylbenzene + 4-bromo-1,3-dimethylbenzene + 5-bromo-1,3-dimethylbenzene ; p-xylene → 2-bromo-1,4-dimethylbenzene

14–21 MeO–C_6H_4–Me (para)

14–24 a. No, because the enol is antiaromatic. **b.** Yes, because the enol is aromatic.

14–26 a. $CH_2{=}CHCH(I)CH{=}CH_2 \rightarrow CH_2{=}CH\overset{+}{C}HCH{=}CH_2$ (a resonance-stabilized cation) **b.** The cation formed from **IV** is antiaromatic.

14–27 $PhCH{=}CHCH_3 + HCl \rightarrow Ph\overset{+}{C}HCH_2CH_3 \rightarrow PhCH(Cl)CH_2CH_3$

The intermediate cation can be stabilized by charge delocalization through the ring.

14–35 $PhCH(OH)CH{=}CH_2 \rightarrow Ph\overset{+}{C}HCH{=}CH_2 \leftrightarrow PhCH{=}CHCH_2^+ \rightarrow PhCH{=}CHCH_2OH$
14–37 A: $PhCH{=}O$ **B:** $PhCH(OH)CH_2CH_3$ **C:** $PhCOCH_2CH_3$ **D:** $PhCH_2CH_2CH_3$
14–38 CH_3: 16 ppm; CH_2 30 ppm; benzene ring carbons with H: 125–130 ppm; benzene ring carbon with the Et group: 145 ppm

Chapter 15

15–1 a. Me / Me / Me (three resonance structures of the arenium ion, H H at the sp³ carbon)

c. H H / H H / H H / H H / H H (five resonance structures of the naphthalene arenium ion)

15–2 $AlBr_3 + Br_2 \rightarrow AlBr_4^-Br^+$ benzene $\xrightarrow{\text{“}Br^+\text{”}}$ (arenium ion: Br, H) $\longleftrightarrow$ (Br, H) $\longleftrightarrow$ (Br, H)

Energy diagram as in Figure 15–3.

15–3 $HOCl + H^+ \rightarrow H_2\overset{+}{O}Cl \rightarrow H_2O + Cl^+$
15–4 a. 2-chlorotoluene, 3-chlorotoluene, 4-chlorotoluene. Sulfonation similar. **b.** 3-chloro-1,2-dimethylbenzene, 4-chloro-1,2-dimethylbenzene **c.** 1-bromo-2-chloro-4-methylbenzene, 1-bromo-3-chloro-4-methylbenzene

15–6 $MeS(=O)_2{-}O^- \leftrightarrow MeS(=O)(O^-){=}O \leftrightarrow MeS(=O)(O^-){=}O$

15–7 a. CH_3Cl, $AlCl_3$ **b.** 2-chloropentane, $AlCl_3$ **c.** chlorocyclohexane, $AlCl_3$ **d.** 2-chloropropane, $AlCl_3$

15–8 *Hint:* $Cl_3CH(OH)_2 + H_2SO_4 \rightarrow Cl_3C\overset{+}{C}HOH$ starts the reaction. **15–9 a.** $CH_3CH{=}CH_2 + HF \rightarrow CH_3\overset{+}{C}HCH_3$, which adds to benzene.

15–10 a. $PhCOCH_2CHMe_2$ **c.** $PhCH_2COPh$ **15–11 a.** $CH_3CH_2C(Cl){=}O$, $AlCl_3$ + benzene; then H_2NNH_2, base or Zn(Hg), HCl **b.** same as **(a)**, starting with $Me_2CHCH_2C(Cl){=}O$ **c.** same as **(a)**, starting with $O{=}(Cl)CCH_2C(Cl){=}O$ **15–12 a.** 1-fluoro-3-nitrobenzene, *m*-fluoronitrobenzene **b.** 1-bromo-2-isopropylbenzene, *o*-bromoisopropylbenzene **c.** 4-chlorotoluene, *p*-chlorotoluene

15–13 a. 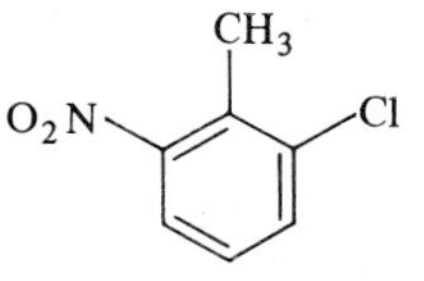 O_2N, CH_3, Cl **b.** 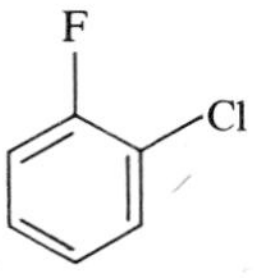 F, Cl **c.** 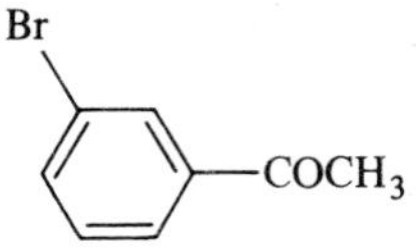 Br, $COCH_3$ **d.** 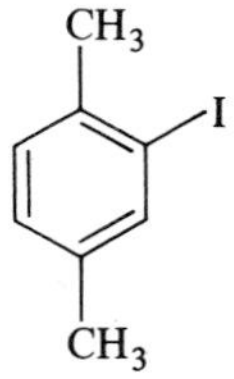 CH_3, I, CH_3

15–14 Bromobenzene, since benzene reacts faster than nitrobenzene. **15–15** There is steric hindrance between the large Br and the NO_2 group

15–16 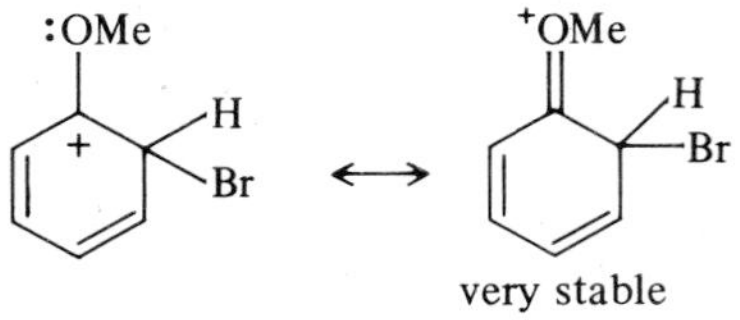 :OMe, +, H, Br ⟷ $^+$OMe, H, Br
very stable

15–17 Addition *ortho* or *para* is slow because of the unfavorable structure: $^{\delta+}C{\equiv}N^{\delta-}$, +, H, Cl

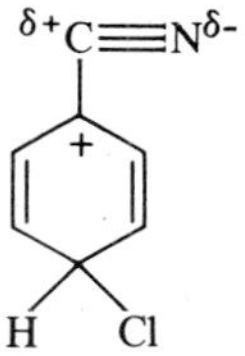

15–18 a. *o,p* **b.** *m* **c.** *o,p* **d.** *m* **15–19** $Cl\overset{+}{C}HCH_3$ intermediate is destabilized by the inductive effect of the Cl (slow rate) but stabilized by resonance: $\overset{+}{Cl}{=}CHCH_3$.
15–20 No. Both are *meta* directors. **15–21 a.** H_2SO_4, SO_3; Br_2, $FeBr_3$ **b.** I_2, HNO_3; HNO_3, H_2SO_4 **c.** EtCl, $AlCl_3$; Br_2, $FeBr_3$ **d.** CH_3COCl, $AlCl_3$; Cl_2, $FeCl_3$ **15–24 a.** 1- and 2-nitro **b.** 3-bromo **e.** 2- and 4-sulfonates **g.** 2- and 4-nitro **15–27** The *o*- and *m*-disubstituted compounds give four products, while *p*-chlorobromobenzene can give only two and is therefore **B.**

15–28 $H_2C{=}O + HCl \rightarrow H_2\overset{+}{C}{-}OH \rightarrow PhCH_2OH \rightarrow PhCH_2Cl$ **15–29** $:\overset{-}{C}{\equiv}\overset{+}{O}: + HCl \rightarrow H\overset{+}{C}{=}O$
15–30 a. BuCOCl, $AlCl_3$ **d.** $ClCO(CH_2)_6COCl$, $AlCl_3$ **e.** PhCOCl, $AlCl_3$ **15–35 a.** $C_6H_6 + HNO_3$, H_2SO_4; Cl_2, $FeCl_3$ **e.** PhCOOH + HNO_3, H_2SO_4 **h.** PhCOOH + HNO_3, H_2SO_4; Br_2, $FeBr_3$

15–36 a. *ortho, para:* +, $\ddot{N}{=}O$, X, H ⟷ $\overset{+}{N}{=}O$, X, H

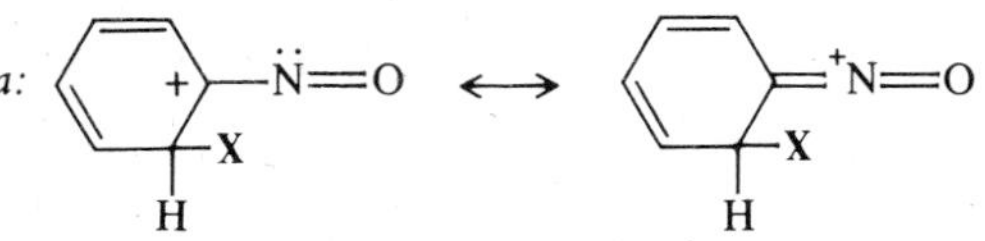

15–37 PhOH + base → PhO^-. Very strong *o,p*-director by resonance-stabilized intermediate **15–38 A:** *p*-bromoanisole; **B:** *o*-bromoanisole **15–39 b.** toluene + 2-chloropropane, $AlCl_3$ **d.** $C_6H_6 + Br_2$, $FeBr_3$; CH_3COCl, $AlCl_3$ **15–44** H_2SO_4, SO_3; Br_2, $FeBr_3$; H_3O^+

15–45 $Ph\overset{+}{N}HMe_2$ **15–47** $C_9H_{10}O$: *p*-methoxystyrene **15–48 a.** $C_6H_6 + CH_3COCl$, $AlCl_3$; $LiAlH_4$ **d.** Product from **(a)**; H_2SO_4, heat; B_2H_6, H_2O_2, OH^-

Chapter 16

16–1 a. $CH_3CH_2C(CH_3){=}CHCH_2COOH$ **b.** $CH_3CH_2CH(OH)CH_2COOH$ **c.** $HOOCCH(Br)CH(CH_3)(CH_2)_3COOH$

d.  COOH **e.** 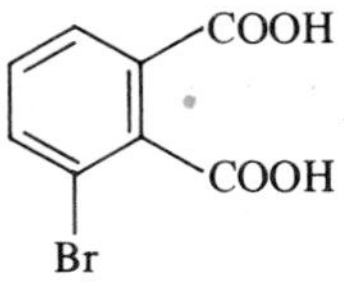 COOH, COOH, Br

f. 2-methylpentanoic acid **g.** cyclohexanecarboxylic acid **h.** 4-bromo-3-nitrobenzoic acid **i.** 2-methoxysuccinic acid or 2-methoxybutanedioic acid

16–2 3,4,5-trimethoxycinnamic acid or 3-(3,4,5-trimethoxyphenyl)propenoic acid **16–3** CF_3COO^- is stabilized by the inductive electron withdrawal of the CF_3 group. **16–4** A rapid fall-off in effect. **16–5 a.** 2.5 **b.** 4 **c.** 5 **d.** 3

16–6 a. $N^-{=}C{=}C_6H_4^{+}{-}COO^-$ **b.** $Me_2\overset{+}{N}{=}C_6H_4^{-}{-}COO^-$ **c.** No resonance interaction from *meta*

16–8 a. $Ph\overset{+}{C}Me_2$ intermediate **b.** $MeO{-}C_6H_4{-}\overset{+}{C}Me_2 \longleftrightarrow Me\overset{+}{O}{=}C_6H_4{=}CMe_2$

16–9 $O{=}\overset{+}{N}(O^-){-}C_6H_4{-}\ddot{N}Me_2 \longleftrightarrow ({}^-O)_2\overset{+}{N}{=}C_6H_4{=}\overset{+}{N}Me_2$

16–10 a. CN^-; H_2O, OH^- **b.** Same as **(a)** **c.** PBr_3; CN^-; H_2O, OH^- **d.** $LiAlH_4$; PBr_3; CN^-; H_2O, OH^-
16–11 a. S_N2 displacements don't work at 3° carbons. **b.** Mg; CO_2
16–12 a. $SOCl_2$ **b.** $SOCl_2$ **c.** $SOBr_2$ **d.** Ac_2O, heat **e.** heat **f.** Ac_2O, heat
16–14 $RCOOH + PPA \rightarrow RCO\overset{+}{O}H_2 \rightarrow R\overset{+}{C}{=}O$
16–15 C_6H_6 + CH_3COCl, $CH_3CH_2CH_2COCl$, $PhCOCl$ and $AlCl_3$ for compounds 1–3; $PhCH_2CH_2COCl$, $AlCl_3$ for compound 4
16–16 In Rxn (eq. 4), the C=O group deactivates the aromatic ring.
16–20 a. HNO_3 **b.** $HClO_4$ **c.** $PhSO_3H$ **d.** FCH_2COOH **e.** FCH_2COOH **f.** CH_3OCOOH **g.** CH_3COOH **h.** PhCOOH
All give the more stable anion due to resonance or inductive effects.
16–21 b. $^-OOCCOO^-$ is unstable (two negative charges)
16–24 A: *m*-chlorobenzoic acid; **B:** *p*-chloro
16–25 a. *m*-CH_3 **b.** *m*-OCH_3 **c.** *p*-NO_2
16–27 a. toluene + HNO_3, H_2SO_4; $KMnO_4$
16–30 b. 2-bromopentane + CN^-; OH^-, H_2O
16–37 a. CH_3CH_2COCl, $AlCl_3$, or benzaldehyde + EtMgBr; CrO_3 **g.** C_6H_6 + glutaric anhydride, HF

Chapter 17

17–1 a. propyl benzoate, PhCOOH, $CH_3CH_2CH_2OH$ **b.** 2-methylbutyl 5-chloropentanoate, $Cl(CH_2)_4COOH$, $HOCH_2CH(CH_3)CH_2CH_3$ **c.** 3-methylbenzamide, 3-methylbenzoic acid **d.** 2-bromo-*N*,*N*-diethylacetamide, $BrCH_2COOH$ **e.** 2-bromoethyl 3-bromobenzoate, 3-bromobenzoic acid, $HOCH_2CH_2Br$ **f.** propyl 3-cyclopentenecarboxylate
17–2 a. $CH_3CH_2COOCH_3$ **b.** $CH_3CH_2CH_2CH(Cl)CH_2COOCH(CH_3)CH_2CH_2CH_3$ **d.** CH_3CONMe_2 **e.** $CH_3(CH_2)_6CONHC_6H_4OH$
17–3 I forms $CH_3\overset{+}{N}HC{=}O$, **II** forms $CH_3CH_2\overset{+}{N}HC{=}O$
17–4 a. $PhCOSCH_2CH_3$ **b.** $CH_3CH_2CONMe_2$ **c.** EtCONHMe **d.** CH_3COO^- + CH_3NH_2 **e.** $C_5H_9COOCH_2CH_2CH_3$

17–6 a. $R\overset{O}{\overset{\|}{C}}{-}\ddot{N}H_2 \leftrightarrow R\overset{O^-}{\overset{|}{C}}{=}\overset{+}{N}H_2$ **b.** $R\overset{O}{\overset{\|}{C}}NH_2 \rightarrow R\overset{O}{\overset{\|}{C}}{-}NH^- \leftrightarrow R\overset{O^-}{\overset{|}{C}}{=}NH$

Resonance stabilization of the anion.
17–7 a. BuCOCl + PhOH **b.** BuCOCl + Me_2CHNH_2 **c.** 4-nitrobenzoyl chloride + cyclopentanol **d.** heptanoyl chloride + Me_2NH **e.** $H_2N(CH_2)_4COCl$

17–8 a. 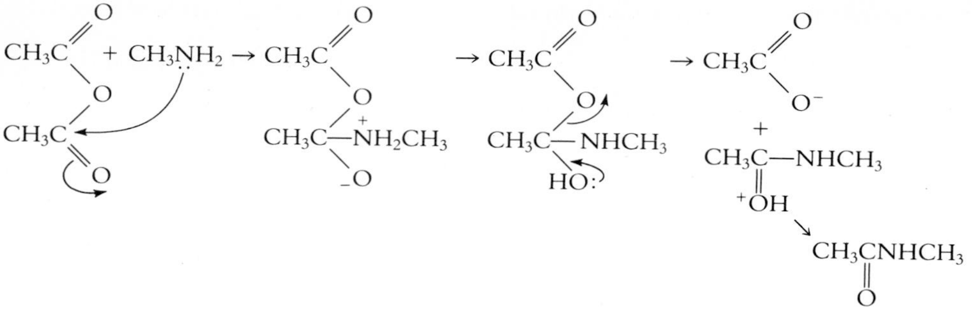 $CH_3C(=O)–O–C(=O)CH_3 + CH_3\ddot{N}H_2 \rightarrow CH_3C(=O)–O–C(–O^-)(CH_3)–\overset{+}{N}H_2CH_3 \rightarrow CH_3C(=O)–O–C(CH_3)(OH)–NHCH_3 \rightarrow CH_3C(=O)O^- + CH_3C(=\overset{+}{O}H)–NHCH_3 \rightarrow CH_3C(=O)NHCH_3$

17–10 b. 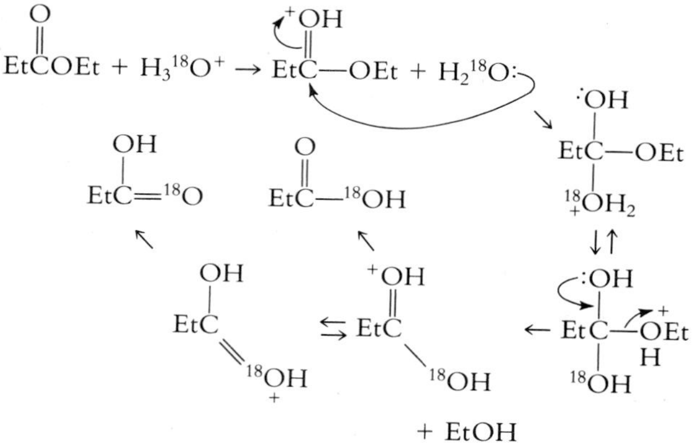 $EtC(=O)OEt + H_3{}^{18}O^+ \rightarrow EtC(=\overset{+}{O}H)–OEt + H_2{}^{18}O: \rightarrow EtC(OH)(OEt)({}^{18}\overset{+}{O}H_2) \rightleftarrows EtC(:OH)(\overset{+}{O}(H)Et)({}^{18}OH) \rightarrow EtC(=\overset{+}{O}H)({}^{18}OH) \rightleftarrows EtC(OH)(={}^{18}\overset{+}{O}H)$ + EtOH

$EtC(=\overset{+}{O}H)({}^{18}OH) \rightarrow EtC(=O)–{}^{18}OH$; $EtC(OH)(={}^{18}\overset{+}{O}H) \rightarrow EtC(OH)={}^{18}O$

17–11 $CH_3COOCM_3 \rightarrow CH_3COO^- + {}^+CMe_3 \quad Me_3C^+ + H_2{}^{18}O \rightarrow MeC^{18}OH$
17–12 It would be a strained four-membered lactone.
17–13 COOH 1700; CONH 1620 cm^{-1}
17–14 $HOCH_2CH(OH)CH_2OH + CH_3(CH_2)_{16}COOH + CH_3(CH_2)_{14}COOH + CH_3(CH_2)_7CH{=}CH(CH_2)_7COOH$

17–15 $PhC(=O)OMe + {}^-\!:OEt \rightarrow Ph–C(O^-)(OEt)–OMe \rightarrow PhC(=O)–OEt + {}^-\!:OMe$

17–16 a. $^-OOCCH_2COOEt$ **c.** $CH_3CH_2COO^- + CH_3NHPh$ **d.** $H_2NCH_2CH(CH_3)CH(CH_3)CH_2COO^-$
17–17 a. EtCOOEt + PrMgBr **b.** Ethyl *p*-chlorobenzoate + EtMgBr **c.** Ethyl cyclopentanecarboxylate + PhMgBr **d.** Ethyl acetate + cyclohexyl MgBr **e.** $EtCOOC_6H_4COOEt$ + MeMgBr **17–18** RCOOH + CH_3MgBr: MgBr → RCOO + CH_4 **17–19 a.** PrCOCl + Et_2CuLi or EtCOCl + Pr_2CuLi **b.** AmCOCl + Ph_2CuLi or PhCOCl + Am_2CuLi **17–23 a.** PhCOOH + EtOH **b.** $PhCOO^-$ + EtOH **c.** PhCOOPr **d.** $PhCONH_2$ **e.** $PhCONMe_2$ **f.** $PhCONHNH_2$ **g.** $PhCOOCH_3$ **h.** $PhC(OH)Bu_2$ **i.** $PhCH_2OH$ **17–24 b.** MeOH, H^+; EtMgBr **d.** CrO_3, H^+; $SOCl_2$; Bu_2CuLi **e.** PBr_3; Li; CuI; PrCOCl **17–25 a.** $PhCH_2OH$ + PrOH **d.** $DOCD_2CH_2CD_2OD$ + PhOD

17–27 $(^-O)(H_3C)C{=}\overset{+}{N}(CH_3)(CH_3)$ has three different methyls.

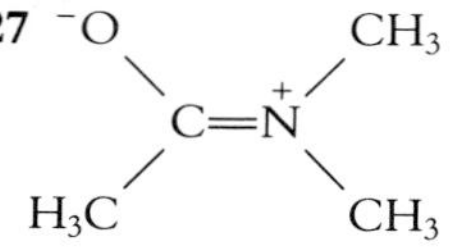

17–30 a. 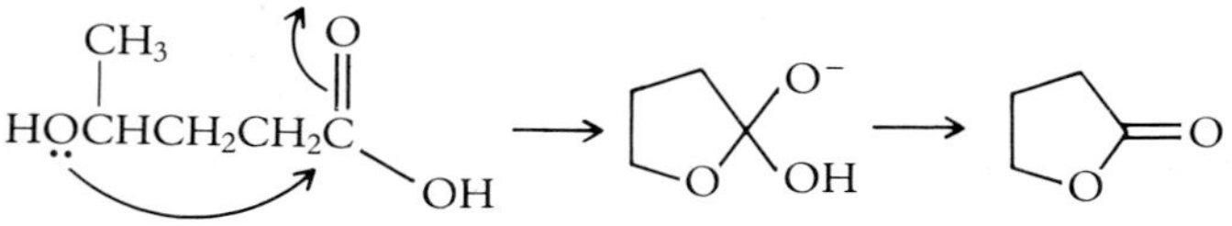 $H\ddot{O}CH(CH_3)CH_2CH_2C(=O)OH \rightarrow$ (cyclic tetrahedral intermediate with O^- and OH) $\rightarrow$ (five-membered lactone, C=O)

17–31 a. 4-chlorobenzoic acid + methanol **c.** succinic acid + $SOCl_2$; 2-propanol

17–34 $CH_3-\overset{O^-}{\underset{OH}{C}}-NMe_2$ OH^- is a better leaving group than Me_2N^-.

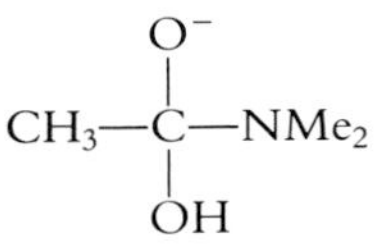

17–40 A CH_3 SO_2NH_2 **B** COOH SO_2NH_2 **17–41 C** O NMe O **F** COOMe COCl

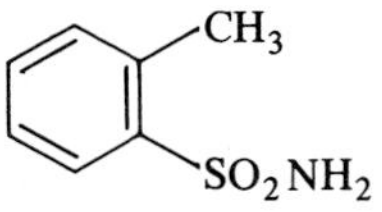

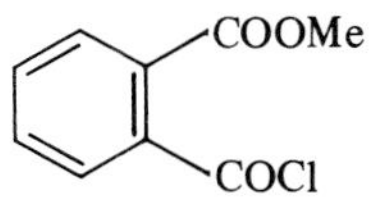

17–43 B: CH_3COOH **I:** COOH Cl

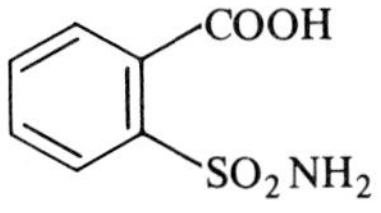

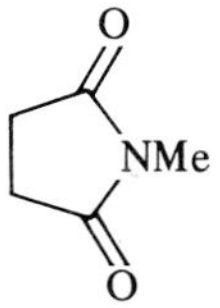

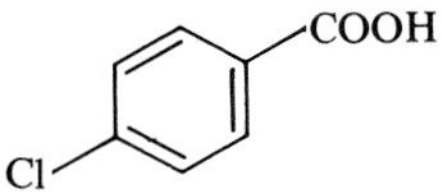

17–44 B: H CN **D:** H OTs (−)**C:** H COOH

17–47 HO—C_6H_4—$COOCH_2CH_2CH_3$

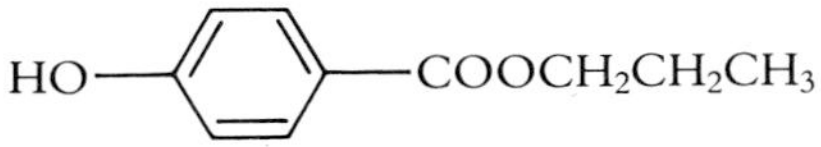

Chapter 18

18–1 a. (1) $CH_3CH_2CH(Et)CH(OH)C(CHO)(Et)CH_2CH_3$ (2) $(CH_3)_2CHCH_2CH(OH)CH(CHO)CH(CH_3)_2$ (4) BuCH(OH)CH(CHO)Bu

18–2 $CH_3\bar{C}HCHO + CH_3CH_2\underset{H}{C}=O \rightarrow CH_3CH_2\underset{CH(CHO)CH_3}{CH}-O^- \rightarrow CH_3CH_2\underset{CH(CHO)CH_3}{CHOH}$

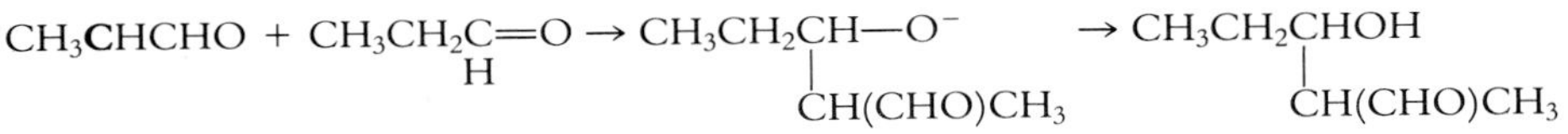

18–3 $Me_2\bar{C}CHO + H_2C=O \rightarrow \underset{C(CHO)Me_2}{H_2C}-O^- \rightarrow \underset{C(CHO)Me_2}{H_2COH}$

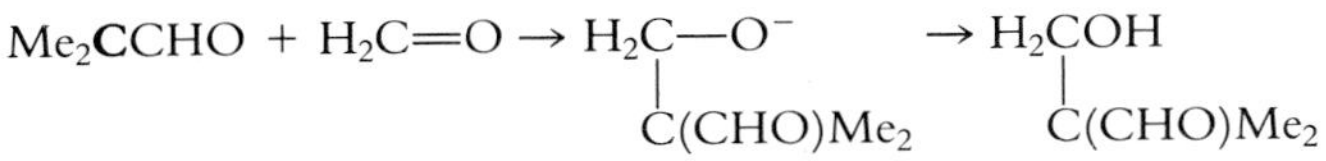

18–4 II: 2C—C + C═O = 344 kcal/mol; **III:** 2C—O + C═C = 318 kcal/mol

18–5 a. $CH_3CH_2CH=C(CH_3)CHO$ **b.** $Me_2CHCH_2CH=C(CHO)CHMe_2$ **c.** $CH_2=CHCH_2CH_2CH=C(CHO)CH_2CH=CH_2$

18–6 a. $CH_3CHO \rightarrow CH_3CH(OH)CH_2CHO \rightarrow CH_3CH=CHCHO \rightarrow CH_3CH=CHCH_2OH$ **b.** $CH_3CH=CHCHO \rightarrow CH_3CH=CHCOOH$ **c.** $CH_3CH=CHCHO \rightarrow CH_3CH_2CH_2CH_2OH$ **d.** Start with $CH_3CH_2CH_2CHO$ **e.** Start with CH_3CH_2CHO

18–7 b. 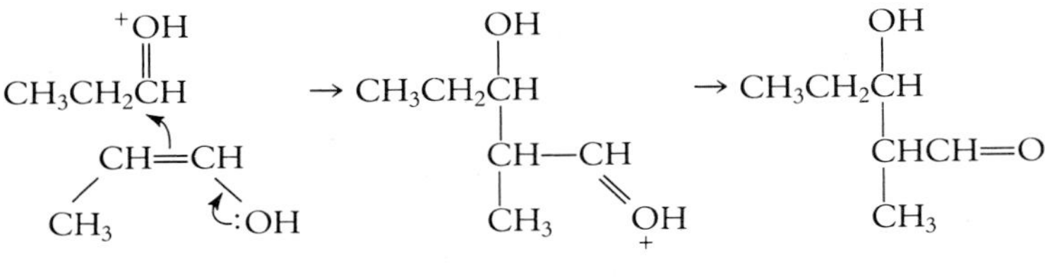 $CH_3CH_2\overset{^+OH}{\overset{\|}{C}H}$ (CH═CH, CH_3, :OH) $\rightarrow CH_3CH_2\overset{OH}{CH}$ (CH—CH═$\overset{+}{O}H$, CH_3) $\rightarrow CH_3CH_2\overset{OH}{CH}$ (CHCH═O, CH_3)

18–8 $CH_3\overset{O}{\overset{\|}{C}}CH_3 + HO^- \rightarrow CH_3\overset{O}{\overset{\|}{C}}\bar{C}H_2 \xrightarrow{CH_3\overset{O}{\overset{\|}{C}}CH_3} CH_3\overset{O^-}{C}CH_3$ ($CH_2-\overset{O}{\overset{\|}{C}}CH_3$) $\rightarrow CH_3\overset{OH}{C}CH_3$ ($CH_2\overset{O}{\overset{\|}{C}}CH_3$)

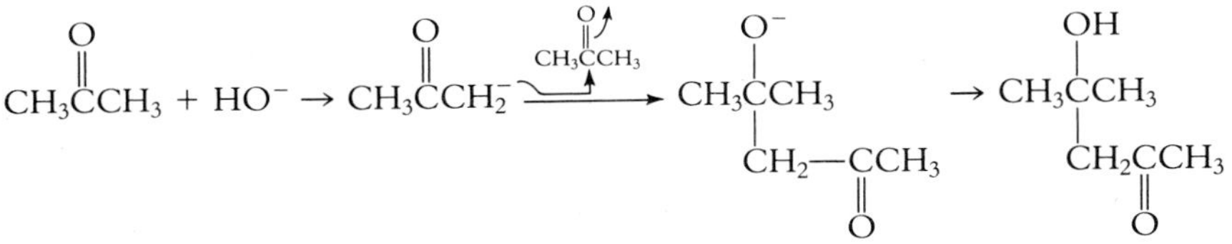

18–9

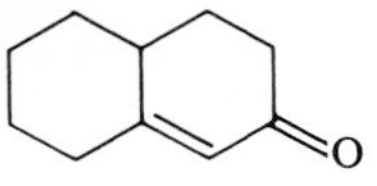

18–10 a.

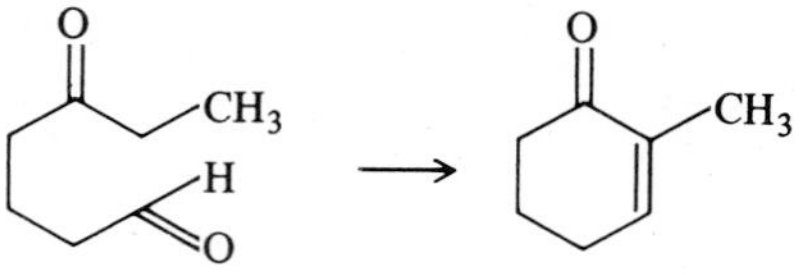

18–11 a. 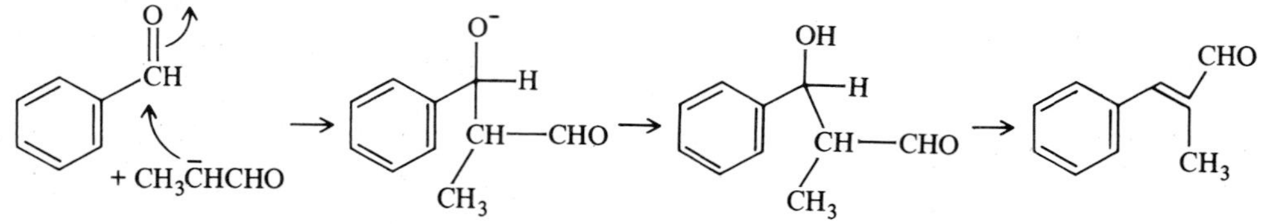 $+ CH_3\bar{C}HCHO$

18–12 a. $CH_3CH_2CH_2CH(OH)CH(CHO)CH_2CH_3$ **b.** $CH_3(CH_2)_3CH{=}C(CHO)CH_2CH_2CH_3$ **c.** $Me_2CHC(Me){=}CHCOCHMe_2$

d. $PhC{=}CHCOPh$ with CH_3 on the first C **e.** 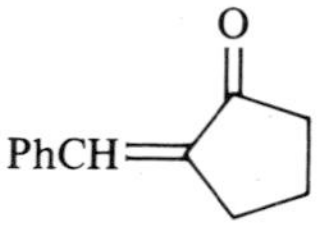 PhCH=(2-cyclopentanone)

18–13 a. $2PhCOCH_3$ **b.** $2Me_2CHCH_2CHO$ **c.** $2CH_3CH_2CHO$ **d.** PhCHO + EtCOMe

18–14 a. Cyclopentanone + CH_2O + $HNMe_2$ **b.** Acetophenone + CH_2O + HN(pyrrolidine)

18–15 $CH_3C(OH){=}CHCOOEt$: CH_3 at 1.9 ppm, vinyl H at 4.95 ppm. About 12% by comparing the CH_3 integration with the CH_3 for the ketone form at 2.14 ppm.
18–16 This is the conjugated carbonyl absorption for the enol form.

18–18 $CH_3C(=O)\bar{C}H{-}\overset{+}{N}(O^-){=}O \leftrightarrow CH_3C(=O)CH{=}\overset{+}{N}(O^-){-}O^-$

Stabilization of the anion by resonance to the nitro group and to the C=O
18–19 a. Br_2, P **b.** Br_2, P; KCN **c.** Br_2, P; CH_3NH_2 **d.** Br_2, P; *t*-BuO^-
18–20 a. $CH_3CH_2CH_2CH(OH)CH(CHO)CH_2CH_3$ **c.** (*Z*)- and (*E*)-$CH_3CH_2CH{=}C(CHO)CH_3$ **e.** 1-Cyclopentenecarboxyaldehyde

18–22 b. $Me_2CHC(Me){=}CHCOCHMe_2$ **c.**

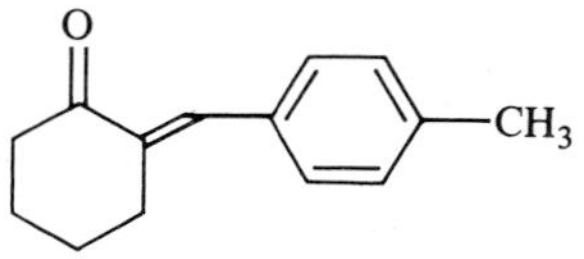

18–26 a. 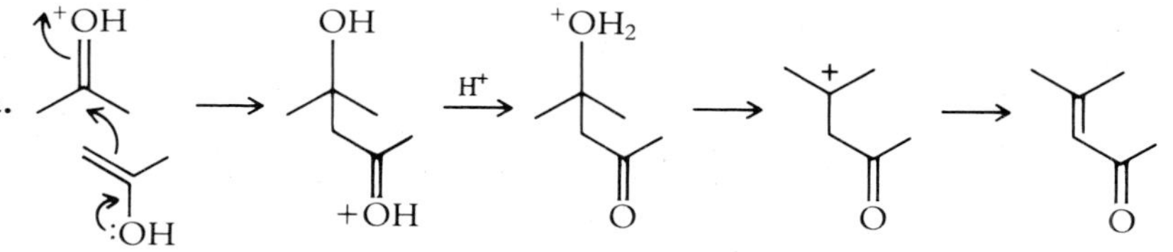 ^+OH … :OH → OH … +OH $\xrightarrow{H^+}$ $^+OH_2$ … O → + … O → O

18–27 b. 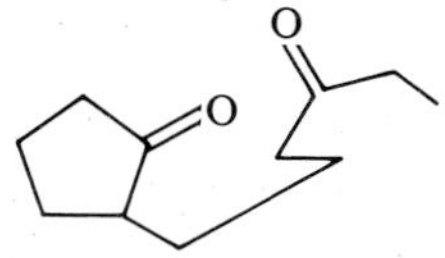 **d.** $PhCH_2CH_2CHO + H_2C(COOEt)_2$

18–29 a. $CH_2O + (CH_3)_2CHCHO$ **c.** $PhCOCH_3 + CH_2O$ **18–30** $PhCOCH_3 + CH_2{=}\overset{+}{N}HEt \rightarrow PhCOCH_2CH_2NHEt, + CH_2O \rightarrow PhCOCH_2CH_2\overset{+}{N}(Et){=}CH_2 \rightarrow$ **VII**
18–38 b. Br_2; MeSH, base **e.** Br_2, P; HO^-; H_2CrO_4

19–1 a. $CH_3CO\bar{C}HCOOEt$, $CH_3COCH(Bu)COOEt$ **b.** $NC\bar{C}HCOOMe$, $EtCH(Me)CH(CN)COOMe$

c. $^-CH(COOEt)_2$, [cyclohexane-1,1-dicarboxylate: COOEt, COOEt] **d.** 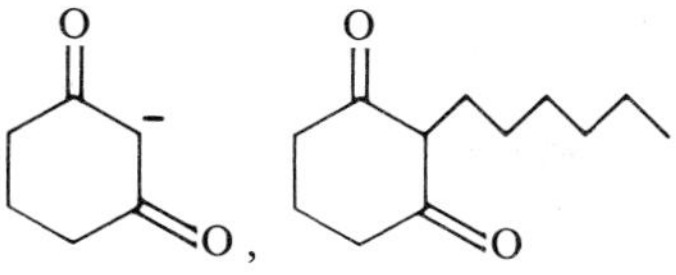 [cyclohexane-1,3-dione anion], [2-pentylcyclohexane-1,3-dione]

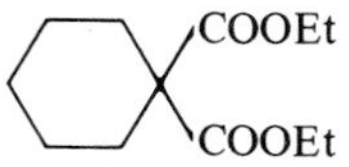

f. $^-CH(CN_2)$, $CH_2{=}CMe_2$

19–2 a. Alkylate ethyl malonate with the proper halides, then react with NH_2CONH_2.

19–3 a. $PhCOCH_2(CH_2)_4CH_3$ **b.** 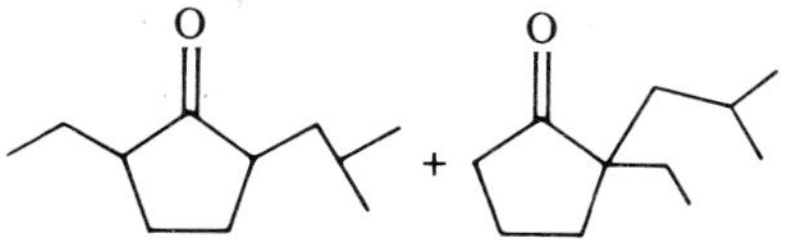 [structures] + [structures] Major Minor **c.** 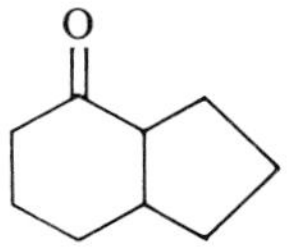 [structure]

19–4 a. A: $PhCH{=}N\bar{C}HCOOEt$ **b.** LDA is a strong base and would remove a proton from the NH_2.

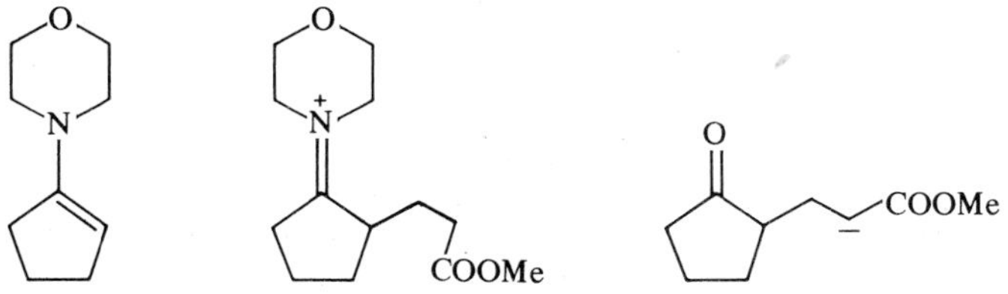

19–5 a. 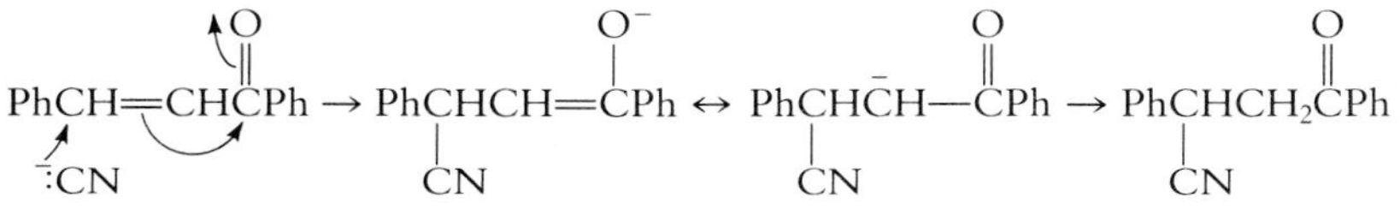 $PhCH{=}CHCPh$ (C=O; :CN⁻ attacks) → $PhCH(CN)CH{=}C(O^-)Ph$ ↔ $PhCH(CN)\bar{C}H{-}C(=O)Ph$ → $PhCH(CN)CH_2C(=O)Ph$

19–7 [structures] A B (COOMe) C (COOMe)

19–8 b. 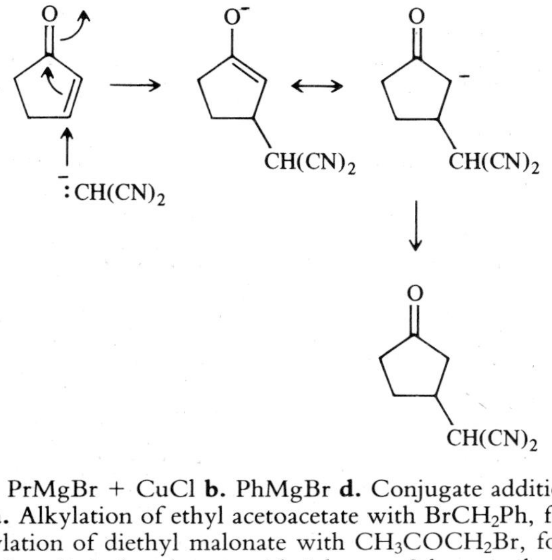 [structures] → [structure: CH(CN)₂] ↔ [structure: CH(CN)₂] ↓ [structure: CH(CN)₂]

$:\bar{C}H(CN)_2$

19–9 a. $PrMgBr + CuCl$ **b.** $PhMgBr$ **d.** Conjugate addition of acetone anion to $MeCOCH{=}CHMe$
19–10 a. Alkylation of ethyl acetoacetate with $BrCH_2Ph$, followed by hydrolysis and decarboxylation.
c. Alkylation of diethyl malonate with CH_3COCH_2Br, followed by hydrolysis and decarboxylation.
b. Alkylate diethyl malonate with 1-bromo-2-butene; decarboxylate. **(c)** and dehydrate with H_2SO_4.
19–11 a. $PrCOCH(COOEt)Et$ **b.** $Me_2CHCH_2COCH(COOEt)CHMe_2$ **d.** $PhCH_2COCH(COOEt)Ph$

19–12 $^{-}O{-}C(CH_3){=}\overset{+}{O}Et$ is a more important resonance structure in the hybrid than $^{-}O{-}C(CH_3){=}\overset{+}{S}Et$. Also, EtS^- is a better leaving group than EtO^-.
19–13 a. Ethyl formate + $PhCH_2COOMe$ **b.** The process would require nucleophilic displacement on bromobenzene.
19–14 a. Ethyl benzoate + ethyl propionate **b.** Methyl butyrate + methyl formate **c.** Ethyl formate + δ-valerolactone **d.** Ethyl benzoate + CH_3CH_2CN

19–15 (diester, COOEt / COOEt) → (anion, COOEt / COOEt) → (cyclohexanone with COOEt and =O)

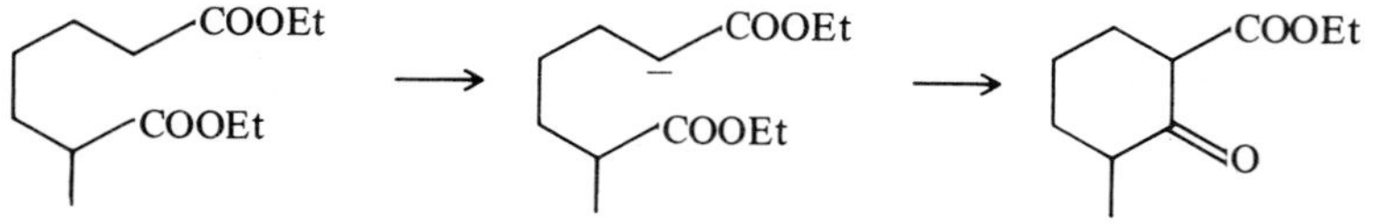

The other possible anion would be 3° rather than 2° and less stable. It would lead to a product without an α-H.

19–16 a. $HOOCCH_2CH(C(=O){-}O{-}H)(C{=}O\ COOH)$ → $HOOCCH_2CH{=}C(OH)COOH + CO_2$ → $HOOCCH_2CH_2C(=O)COOH$

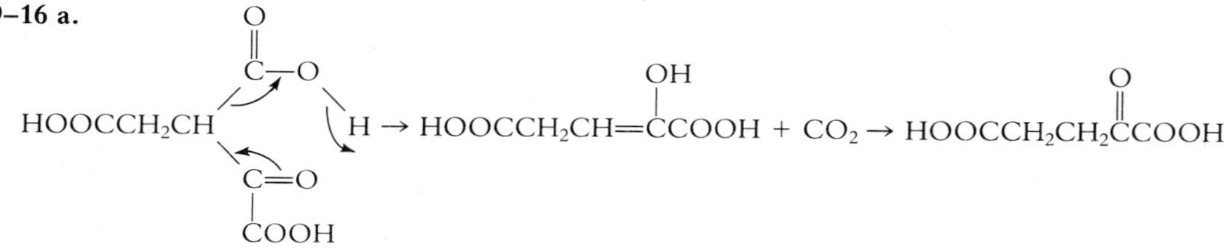

19–17 Degradations must be by successive loss of fragments containing two carbons each.
19–18 a. Alkylation of diethyl malonate with CH_3Br and bromocyclopentane **b.** 1,4-dibromobutane + $EtOOCCH_2CN$ **c.** $PhCH(Br)CN + CH_3COCH_2COOEt$ **e.** 2-Cyclohexenone + diethyl malonate (*Note:* 3-Halocyclohexanones are not available.)

19–20 a. $CH_2{=}C(COOMe)(CHO)$ **b.** $CH_2{=}CHCOOMe$ **c.** $CH_2{=}CHCOOMe$
e. $CH_3CH{=}CHCOOMe$ **f.** $CH_2{=}CHCOCH_3$

19–22 *Hint:* The mechanism involves 1,4-addition by anenamine. **19–23 a.** Diethyl malonate + EtBr, PrBr, and then hydrolysis and decarboxylation. **d.** Alkylation of diethyl malonate with $PhCH_2Br$ (twice), hydrolysis, decarboxylation, and amide formation. **19–24 a.** (*R*) **19–25 a.** Ethyl acetate + ethyl *p*-chlorobenzoate **c.** Cyclopentanone + ethyl formate **19–28** $CH_3COCH{=}CH_2$ + MeMgBr → $CH_2{=}CHC(OH)Me_2$ **19–30 A:** $Me_2C{=}C(CH_2Br)_2$ **D:** $MeOOCCH(CH_2CH_2COOMe)_2$

Chapter 20

20–1

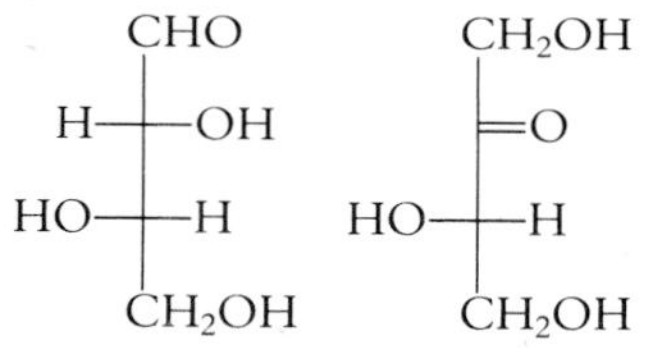

L-Threose	L-Erythrulose
CHO	CH_2OH
H—OH	=O
HO—H	HO—H
CH_2OH	CH_2OH

20–2 a.

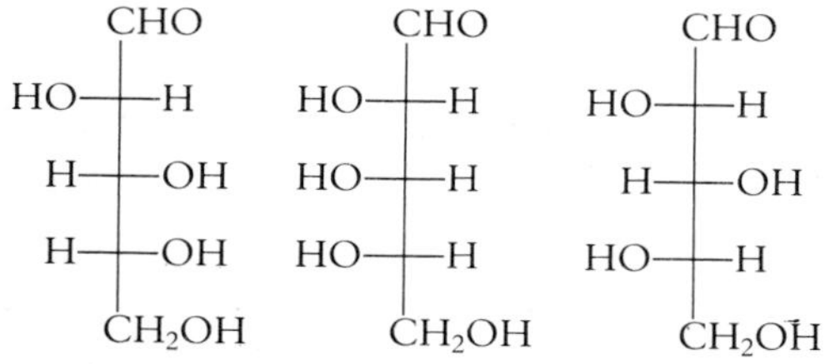

D-Arabinose	L-Ribose	L-Xylose
CHO	CHO	CHO
HO—H	HO—H	HO—H
H—OH	HO—H	H—OH
H—OH	HO—H	HO—H
CH_2OH	CH_2OH	CH_2OH

b.

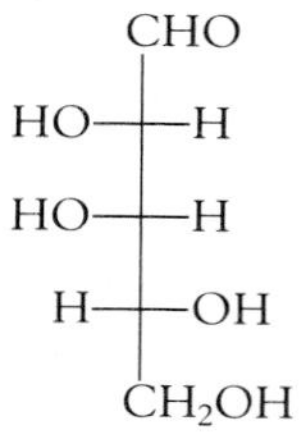

D-Lyxose
CHO
HO—H
HO—H
H—OH
CH_2OH

20–3 CHO; H—OH; CH_2OH → **A**: CN; HO—H; H—OH; CH_2OH + **B**: CN; H—OH; H—OH; CH_2OH

A → (−)-(*S*,*S*)-Tartaric acid

B → *meso*-Tartaric acid $[\alpha] = 0°$

20–4 D-erythrose → *meso*-tartaric acid

20–5

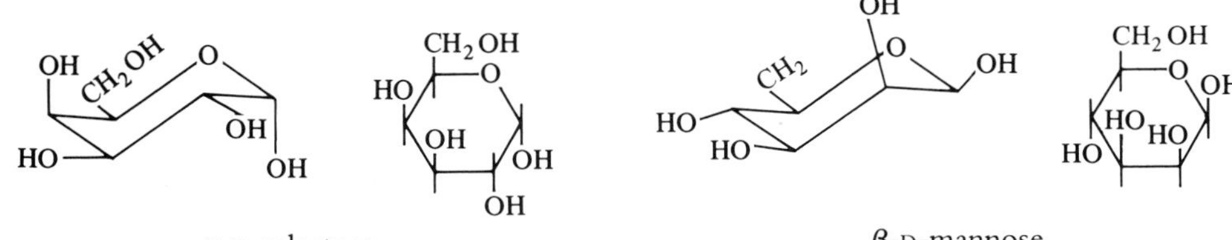

α-D-galactose β-D-mannose

20–6 The β-D-ribopyranose has one axial group, the α-anomer has two. Both are more stable than the furanoses, of which the α-anomer is less stable since it has three neighboring eclipsed OH groups.

20–7

O, OH, HO, OH — Erythrose

O, HO, OH, OH — Threose

20–8 a. O, OMe, HO, OH; O, OMe, HO, OH

20–9 CH=NNHPh; C=NNHPh; HO—H; H—OH; H—OH; CH_2OH

The osazone of mannose also has this structure.

20–11 CH_2OH, O, O, O, O, O

20–13 CH_2OH; H—OH; HO—H; HO—H; H—OH; CH_2OH ≡ CH_2OH; HO—H; H—OH; H—OH; HO—H; CH_2OH

20–14 a. 4 mol; 4 HCOOH, 1 CH_2O **20–15 a.** 2 mol; HCOOH + $2CH_2O$ **b.** 2 mol; HCOOH + $OHCCH_2CH(OH)CH_2CHO$ **c.** 5 mol; 5 HCOOH, 1 CH_2O

20–18 CH_2OH—C(=O)—R → ^-CHOH—C(=O)—R ↔ HC—OH=C—O^-—R → HC—O^-=C—OH—R ↔ HC=O—^-C—OH—R → HC=O—HC—OH—R

20–20 a. 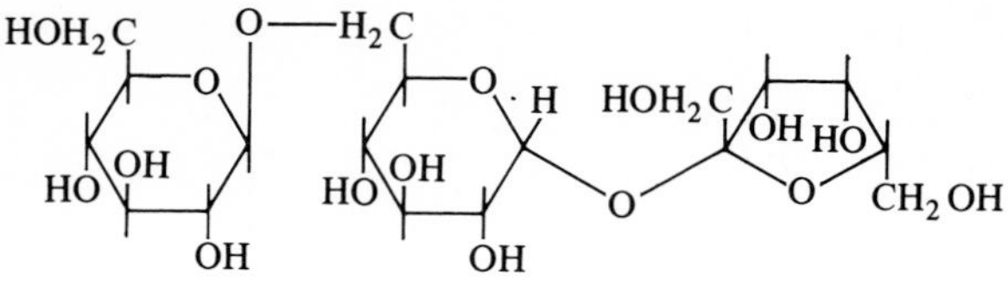 [structure]

c. 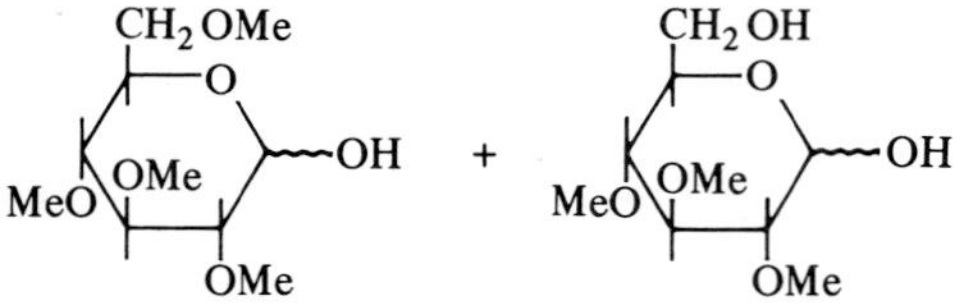 [structures] +

20–22 Enantiomers; mirror image isomers; optical isomers **20–23 b.** 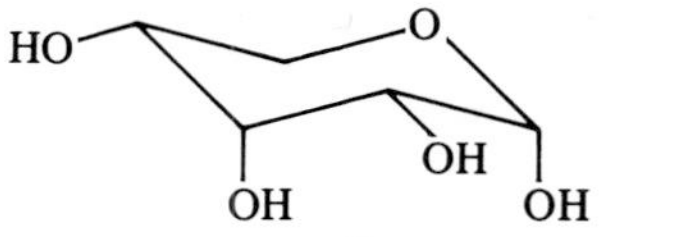 [structures: α, β]

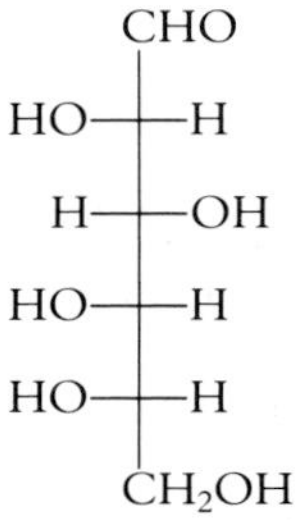

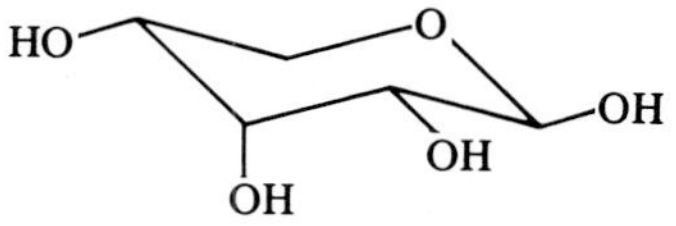

CHO
HO—H
H—OH
HO—H
HO—H
CH_2OH

20–25 a. $2CH_2O$ **d.** $2HC(=O)$—COOH **e.** 3HCOOH, 1 CH_2O

20–28 A: $HOCH_2CH(OH)CH(OH)COOH$ **C:** A cyclic anhydride

20–31 [structure]

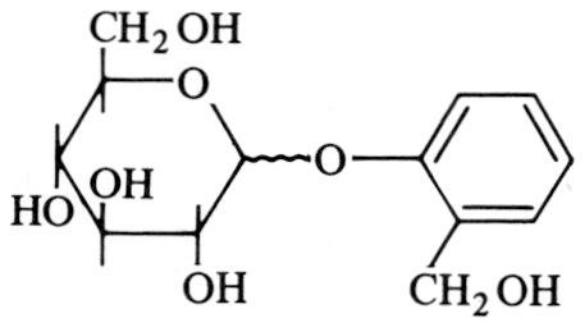

You would not know which anomer was correct for salicin. This is most easily determined by ^{1}H nmr.
20–33 Amygdalin → glucose + PhCH(CN)OH
This is the cyanohydrin of benzaldehyde, which could decompose to give CN^-.
20–35 There are two anomers in solution. Twelve peaks should be seen, one for each carbon. The peaks at 61.9 and 62.1 ppm are the $\mathbf{CH_2}$OH carbons; at 93.2 and 97.3 ppm are the anomeric carbons.

Chapter 21

21–1 a. 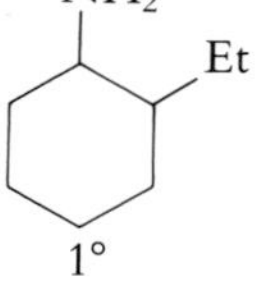 [structure] 1° **b.** 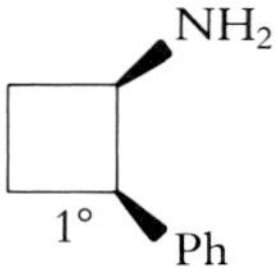 [structure] 1° **c.** $EtOCH_2CH_2NHCH_3$, 2° **d.** 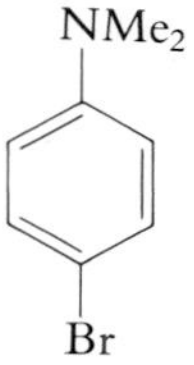 [structure] 3°

e. 3-hexylamine (3-hexanamine), 1° **g.** benzyldimethylamine, 3° **h.** (*R*)-2-methoxy-N-methyl-2-butylamine, 2° **i.** 7-methoxy-1-naphthylamine, 1°.

21–2 $CH_3CH_2CH_2$—CH_2—$\dot{N}^{+}HCH_3$ → $CH_3CH_2CH_2\cdot$ + CH_2=$\overset{+}{N}HCH_3$ (m/z = 44)

21–3 The nonbonded electron pair on N in aniline is next to the ring and can participate in resonance structures: $\ddot{N}{-}C{=}CH \leftrightarrow \overset{+}{N}{=}C{-}CH\colon^{-}$. The carboxyl anion cannot do so.

21–4 The methoxy group destabilizes the ammonium ion by an inductive electron withdrawal: $CH_3OCH_2CH_2\overset{+}{N}H_3$. In the *para* benzene derivative there is *resonance stabilization* of the ion:

$$MeO{-}C_6H_4{-}\overset{+}{N}H_3 \leftrightarrow Me\overset{+}{O}{=}C_6H_4^{-}{-}\overset{+}{N}H_3.$$

21–5 a. $Me_2NH_2^+Cl^-$
21–6 Dimethylamine can form hydrogen-bonded structures: $Me_2NH\cdots NHMe_2$.

$$\textbf{21–7}\ CH_3CH_2\ddot{N}H_2 + CH_3{-}I \rightarrow CH_3CH_2\overset{CH_3}{\overset{|}{N}}H_2^+\ I^- \rightleftarrows CH_3CH_2NHCH_3 + HI \rightarrow \text{etc.}$$

21–9 a. S_N2 displacement is favored at the 1° carbon over the 2°.
21–10 a. Gabriel or azide **b.** CN^- then $LiAlH_4$ **c.** Gabriel or azide **d.** TsCl then azide
21–11 a. $Me_2C{=}O + NH_3 \rightarrow Me_2C{=}NH \rightarrow Me_2CHNH_2 \rightarrow Me_2CHN{=}CMe_2 \rightarrow Me_2CHNHCHMe_2$ **b.** $CH_3CHO + NH_3$ **21–12 a.** benzene + (1) Cl_2, $FeCl_3$; (2) HNO_3, H_2SO_4; (3) H_2, Pt **b.** benzene + (1) H_2SO_4, SO_3; (2) HNO_3, H_2SO_4; (3) H_2, Pt **c.** naphthalene + (1) HNO_3, H_2SO_4; (2) H_2, Pt; (3) Ac_2O; (4) HNO_3, H_2SO_4; (5) H_3O^+ **d.** toluene + (1) HNO_3, H_2SO_4; (2) $KMnO_4$; (3) H_2, Pt **e.** benzene + (1) HNO_3, H_2SO_4; (2) Br_2, $FeBr_3$; (3) H_2, Pt; (4) $(CH_3)_2CHCHO$, H_2
21–14 (1) Cl_2, $FeCl_3$; (2) HNO_3, H_2SO_4; (3) Cl_2, $FeCl_3$; (4) H_2, cat; (5) EtCOCl **21–17** Resolve with an optically active base, then treat the pure enantiomer with Br_2, OH^- after converting it to the amide.

$$\textbf{21–18}\ R\underset{OH}{\underset{|}{C}}(CH_3)CH{=}CH_2 \rightarrow R\overset{+}{C}(CH_3)CH{=}CH_2 \leftrightarrow RC(CH_3){=}CHCH_2^+ \rightarrow RC(CH_3){=}CHCH_2Cl$$

b. Because the anion formed would be allylic and therefore stable. The allylic H is more acidic than in the alkyl phosphonium salt. **21–19** 6-methyl-1-heptanol + HBr, $\rightarrow Me_2CH(CH_2)_4CH_2Br$. Convert to the Wittig reagent and react with $CH_3(CH_2)_9CHO$. The resulting alkene can be converted to the epoxide with peracid.

21–20 a. cyclohexanone + $Ph\overset{-}{C}H\overset{+}{P}R_3$ **c.** acetophenone + $Et\overset{-}{C}H\overset{+}{P}R_3$ **21–21 a.** 1-propanethiol; PrBr + H_2S, OH^- **b.** 1-methylthio-2-methylbutane; $Me_2CHCH_2CH_2Br$ + MeSH, OH^- **c.** benzylthiol; $PhCH_2Br + H_2S$, OH^- **d.** allyldiethylsulfonium chloride **e.** *cis*-(3-cyanocylobutyl)dimethylsulfonium iodide **21–22** $Me_2\overset{+}{S}CH_3 + DO^- \rightarrow Me_2\overset{+}{S}CH_2^- + DOH \rightarrow Me_2\overset{+}{S}CH_2D$ **21–23 a.** cyclohexylamine, aniline, nitroaniline **c.** *N*-methylaniline, 3-methylaniline, aniline **21–24** There is a steric inhibition of resonance, and the nonbonded pair is not delocalized through the ring in *N,N*-dimethyl-2,4,6-trinitroaniline.

21–27 a. $Me_3\overset{+}{N}{:}\overset{-}{B}F_3$ **b.** $AlCl_3$ forms a complex as in (a) with the nitrogen nonbonded electron pair. This is now a *meta*-directing group.

21–28 a. Me_2CHCH_2NHEt **c.** $(S)\text{-}CH_3\underset{^+NMe_3}{\underset{|}{C}H}CH_2CH_3$ **f.** $CH_3CH(OH)CH_2CH_2CH_2NH_2$

21–30 a. $CH_3CH_2CH(CH_3)CH_2Br$ + (1) NaCN, DMF; (2) H_2, Pt
c. toluene + (1) HNO_3, H_2SO_4; (2) H_2, Pt

$$\textbf{21–33}\ Pr(Me)C{=}NOH \rightarrow Pr(Me)C{=}\ddot{N}\overset{+}{O}H_2 \rightarrow Pr(Me)C{=}\ddot{N}^+ \rightarrow$$

$$MeC{\equiv}\overset{+}{N}Pr \rightarrow Me\underset{OH}{\underset{|}{C}}{=}NPr \rightarrow Me\underset{O}{\underset{\|}{C}}NHPr$$

$$\textbf{21–36}\ HO{-}\underset{OH}{\underset{|}{\ddot{P}}}{-}OH \qquad HO{-}\underset{O}{\underset{\|}{P}}H{-}OH$$

21–37 a. $PrCHO + Et(Me)\overset{-}{C}\overset{+}{P}Ph_3$ or 2-butanone + $Pr\overset{-}{C}H\overset{+}{P}Ph_3$ **21–38 a.** $Me_2CHCH_2Br + H_2S$, OH^- **e.** *trans*-2-bromo-1-methylcyclohexane + MeSH, OH^-

21–39 The 1° chloride is first formed, but undergoes intramolecular formation of a 3-membered ring sulfonium salt which is opened by Cl^-. **21–40 b.** Look at the number of resonance structures that can be drawn from each anion.

21–41 First step is formation of the ylid from VI: $^-CH_2(CH_3)\overset{+}{S}R$. The long alkyl chain on R makes the ylid soluble in the organic CH_2Cl_2 phase where it is not encumbered with Na^+. It is therefore very free to react rapidly with the carbonyl (see eq. 21.66). **21–43** benzylisopropylamine

Chapter 22

22–1 a. (a) $CH_3CH(NH_3^+)COOH$; (b) $CH_3CH(NH_3^+)COO^-$; (c) $CH_3CH(NH_2)COO^-$ **c.** (a) $^+H_3N(CH_2)_4CH(NH_3^+)COOH$; (b) $^+H_3N(CH_2)_4CH(NH_3^+)COO^-$; (c) $H_2N(CH_2)_4CH(NH_2)COO^-$ **d.** (a) $^+H_2N{=}C(NH_2)NH(CH_2)_3CH(NH_3)^+COOH$; (b) $^+H_2N{=}C(NH_2)NH(CH_2)_3CH(NH_3^+)COO^-$; (c) $HN{=}C(NH_2)NH(CH_2)_3CH(NH_2)COO^-$ **e.** (a) $HOOCCH_2CH_2CH(NH_3^+)COOH$; (b) $^-OOCCH_2CH_2CH(NH_3^+)COO^-$; (c) $^-OOCCH_2CH_2CH(NH_2)COO^-$ **f.** *Hint:* The amide N is not basic. **g.** *Hint:* Phenol is a weak acid. **i.** (a) $^+H_3NCH(CH_3)CONHCH_2COOH$; (b) $^+H_3NCH(CH_3)CONHCH_2COO^-$; (c) $H_2NCH(CH_3)CONHCH_2COO^-$

22–2 In $HOOCCH_2CH(NH_3^+)COO^-$ the two opposite charges are closer together, thus it is more stable. **22–3 a.** $CH_3CH(Br)COOH$ + (1) phthalimide, base; (2) N_2H_4; (3) HCl **b.** Me_2CHCHO + (1) NH_3, CN^-; (2) H_3O^+ **c.** $PhCH_2CH_2COOH$ + (1) Br_2, P; (2) NH_3 **d.** 2-butanone + $AcNHCH_2COOH$ + (1) AcO^-; (2) H_2, chiral Rh catalyst

22–5 a. $H_2NCH(CH_2\text{-imidazolyl})CONHCH(CH_2OH)COOH$ **b.** $H_2NCH(CH_2CH_2SCH_3)CONHCH(CH_2SH)COOH$ **c.** 2-pyrrolidinyl–$CONHCH(CH_2C_6H_4OH)COOH$

d. glycylglycine **e.** Asp-Asp-Gly

22–6 Met-Enkephalin: Tyr-Gly-Gly-Phe-Met Phalloidine: Ala, hydroxyproline, Cys derivative, Thr, Ala, Try derivative, $H_2NCH(COOH)CH_2C(OH)(CH_3)CH_2OH$ **22–7 a.** $ArNHCH(COOH)CH_2Ph$, Gly, Ser

22–8 Phe-Val-Ser-Tyr-His **22–9** Ser-Ala-Ala-Glu-Glu-Phe, Val-Met-Tyr, Met-Glu-Trp, Gly-Gly-Phe

22–10 a. $H_2NCH(CH_3)COOH + PhCH_2OCOCl \rightarrow PhCH_2OCONHCH(CH_3)COOH \rightarrow PhCH_2OCONHCH(CH_3)COOC_6H_4NO_2 \rightarrow PhCH_2OCONHCH(CH_3)CONHCH_2COOH \rightarrow$ Ala-Gly by H_2/Pd

or

$H_2NCH(CH_3)COOH + Z{-}Cl \rightarrow ZNHCH(CH_3)COOH \rightarrow ZNHCH(CH_3)COONp \rightarrow ZNHCH(CH_3)CONHCH_2COOH \rightarrow H_2NCH(CH_3)CONHCH_2COOH$

22–11 $H_2N(CH_2)_6NHCO(CH_2)_8CONH(CH_2)_6NHCO(CH_2)_8COOH$

22–12 $HOCH_2CH_2OCOCH{=}CHCOOCH_2CH_2OCOCH{=}CHCOOCH_2CH_2OH$

22–13 *Hint:* The first step involves attack by HOR on the carbonyl of $RN{=}C{=}O$.

22–14 a. histidine: cathode, tyrosine: origin, glycine: origin, glutamic acid: anode **b.** (1) pH 2: cathode, pH 10: origin (2) pH 2: origin, pH 10: anode

22–15 a. $HN{=}C(R){-}NH_2 + H^+ \rightarrow H_2\overset{+}{N}{=}C(R){-}NH_2 \leftrightarrow H_2N{-}C(R){=}\overset{+}{N}H_2$

22–17 a. H, CH_2OH, HOOC, OH on C — (−)-Glyceric acid; H, CH_3, HOOC, OH on C — (−)-Lactic acid

22–19 $RNH_2 + HN{=}C{=}O \rightarrow R\overset{+}{N}H_2{-}C(O^-){=}NH \rightarrow RNH{-}C(OH){=}NH \rightarrow RNHC({=}O){-}NH_2$

22–20 a. *Hint:* First step involves 1,4-addition. **22–21 a. A:** $CH_3CONHC(COOEt)_2CH_2CHMe_2$
22–22 b. $HSCH_2CHO$ + (1) HCN, NH_3; (2) H_3O^+ **22–23** Ala-Gly-Ileu-Val

22–24 a. Met + BocCl → $MeSCH_2CH_2CH(COOH)NHBoc$ →

$MeSCH_2CH_2CHNHBoc$ with $O{=}CNHCH(COOH)CH_2Me_2$ attached at CH

by reaction with DCCD and $Me_2CHCH_2CH(NH_2)COOH$. Then use acid to cleave the Boc group. **22–25** *Hint:* **A** is the result of an intramolecular attack of the end NH_2 with the ester function. **22–26** The polypeptide comes off as an amide. **22–27** $HOCH_2CH_2(CH_2)_7COOCH_2CH_2(CH_2)_7COO$— etc.

22–30 (7a) $RO^- + CH_2{=}CHC({=}O)NH_2 \rightarrow CH_2{=}CHC({=}O)NH^-$,

which then adds 1,4 to another acrylamide. In (7b) there would be 1,2-addition since only anions remove H^+ from the $CONH_2$ group.
22–31 b. $H_2NC_6H_4NHCOC_6H_4CONHC_6H_4NHCO$— etc.

Chapter 23

23–1 b. 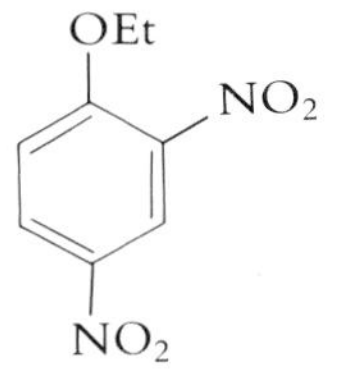 [structure: benzene ring with OEt, NO_2 (ortho), NO_2 (para)] **c.** 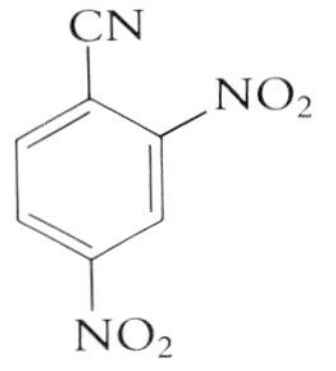 [structure: benzene ring with CN, NO_2 (ortho), NO_2 (para)] **23–2** [structure: cyclohexadienyl anion with H, SAr, $\overset{+}{N}(=O)O^-$, COOMe] ↔ [structure: with H, SAr, $=N(O^-)O^-$, COOMe]

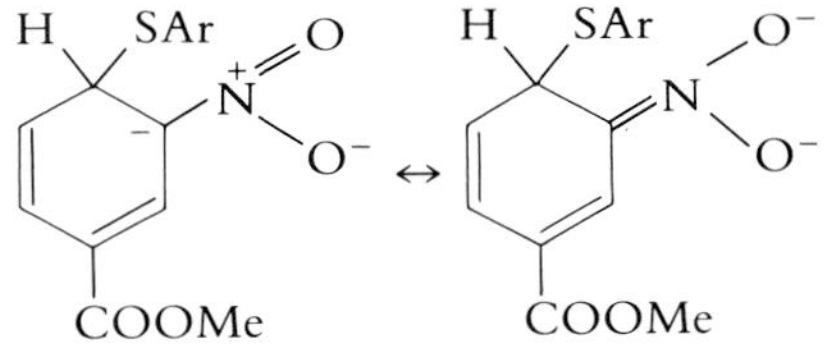

23–3 a. 2,4-dinitrochlorobenzene + H_2S, base **b.** 4-cyanochlorobenzene + $EtNH_2$ **c.** 2-chloronitrobenzene + diethylmalonate, LDA **e.** 2-chloro-1,3-dinitrobenzene + CH_3O^-

23–4 [structure: 2,4,5-trichlorophenol (Cl, OH, Cl, Cl)] + [structure: trichlorophenoxide ($^-$O, Cl, Cl, Cl)] → [structure: Cl, Cl, $^-$O, Cl diphenyl ether, O bridge, Cl, Cl] → etc.

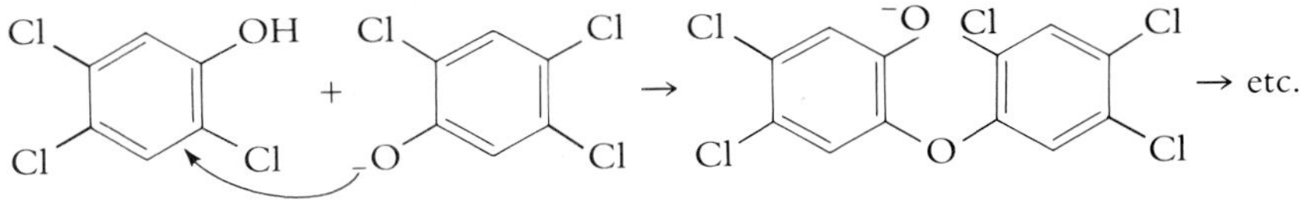

23–5 In the S_NAr reaction, nucleophilic addition is the slow step, while in the benzyne reaction, fluoride removal is the slow step. **23–6** The products come from two different benzynes, [structures: two methylbenzynes, CH_3 + CH_3] as well as by S_NAr.

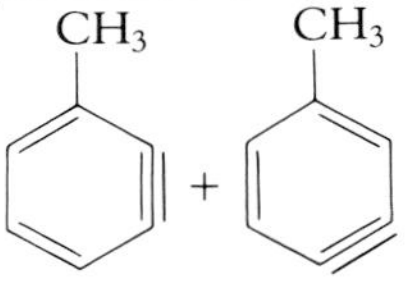

23–7 Reaction (2) is S_NAr, reaction (3) is benzyne (naphthalyne).
23–8 a. $CH_3C_6H_4N_2^+Cl^-$ **b.** $ArN_2^+BF_4^-$ **d.** $^-OOCC_6H_4N_2^+$ + AgCl ppt
23–10 a. 1-amino-4-methylnaphthalene + (1) $NaNO_2$, HCl; (2) H_3O^+, 130° **b.** 3-nitroaniline + (1) $NaNO_2$, HBF_4; (2) heat **c.** ethyl 4-aminobenzoate + (1) $NaNO_2$, HCl; (2) H_3O^+ **23–11 a.** (1) Br_2, $FeBr_3$; (2) H_2, Pt; (3) $NaNO_2$, HCl; (4) CuCN, KCN **b.** (1) H_2, Pt; (2) Ac_2O; (3) Cl_2, $FeCl_3$; (4) H_3O^+; (5) $NaNO_2$, HCl; (6) CuCN, KCN **c.** (1) HNO_3, H_2SO_4; (2) H_2, Pt; (3) $NaNO_2$, HCl; (4) KI **d.** (1) HNO_3, H_2SO_4; (2) Br_2, $FeBr_3$; (3) H_2, Pt; (4) $NaNO_2$, HCl (5) KI
23–12 a. (1) H_2SO_4, SO_3; (2) H_2, Pt; (3) $NaNO_2$, HCl; (4) H_3PO_2, H_2O, heat **b.** (1) H_2, Pt; (2) Ac_2O; (3) HNO_3, H_2SO_4; (4) H_3O^+, heat; (5) $NaNO_2$, HCl; (6) H_3PO_2

23–13 **a.** 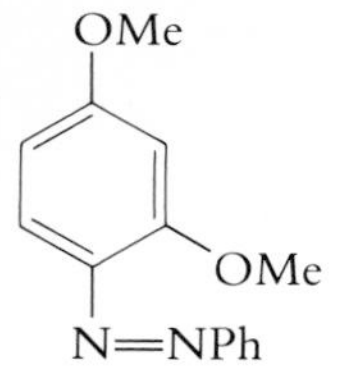 (structure: 2,4-dimethoxy azobenzene: OMe, OMe, N=NPh) **b.** 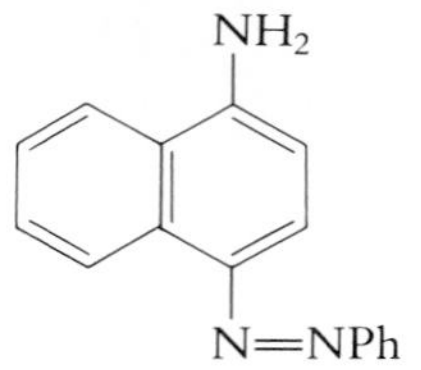 (NH$_2$, naphthalene, N=NPh) **c.** 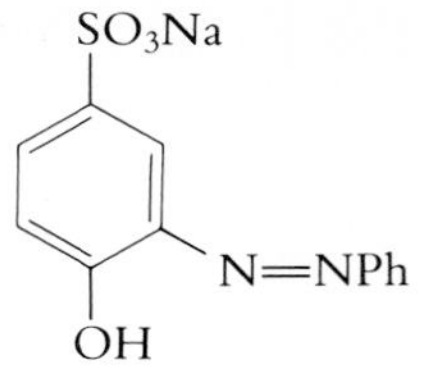 (SO$_3$Na, OH, N=NPh)

23–14 **a.** sodium phenoxide **b.** aniline **c.** *p*-nitrobenzenediazonium chloride

23–15 $Me_2\ddot{N}{-}N{=}O \leftrightarrow Me_2\overset{+}{N}{=}N{-}O^-$ **23–16 b.** PhCOOH + (1) HNO_3, H_2SO_4; (2) H_2, Pt; (3) $NaNO_2$, HCl; (4) H_3O^+, heat; (5) MeOH, H^+ **23–16 c.** chlorobenzene + (1) HNO_3, H_2SO_4; (2) EtSH, base **23–19 a.** S_NAr with OH^- yields phenol. This can form the phenoxide anion and react with chlorobenzene to give diphenyl ether. Chlorobenzene could also yield a benzyne that would react with phenol to give diphenyl ether. **b.** The benzyne mechanism. **23–20 a.** A benzyne that reacts intramolecularly with the nonbonded electron pair on the $RNHCH_3$.
23–21 a. 4-chlorotoluene + CuCN **f.** 3-bromoacetophenone + (1) CuCN; (2) H_3O^+
23–22 a. (1) base, BuBr; (2) Br_2, $FeBr_3$; (3) CuCN **b.** (1) CuCN; (2) Br_2, $FeBr_3$; (3) CuCN
23–23 a. (1) $NaNO_2$, HCl; (2) CuBr **d.** (1) Ac_2O; (2) HNO_3, H_2SO_4; (3) H_3O^+ (4) $NaNO_2$, HCl; (5) CuCN, KCN **23–24 a.** benzenesulfonic acid + (1) HNO_3, H_2SO_4; (2) H_2, Pt; (3) $NaNO_2$, HCl; (4) H_3O^+, heat **23–26** $RCONH_2 + NaNO_2, HCl \rightarrow RCON_2^+Cl^- \rightarrow RCOOH$ **23–29 a.** MeN(NO)Et **c.** none **23–31** At pH 5, the position next to the NH_2 is more reactive than the one next to the OH. At pH 9, the OH is converted to RO^-, which yields a more reactive position next to this functional group.

Chapter 24

24–1 **a.** *m*-cresol, *p*-cresol, phenol, *m*-bromophenol **b.** catechol, hydroquinone, *p*-aminophenol **c.** 2,4-dinitrophenol, *p*-cyanophenol, *p*-fluorophenol, *p*-methoxyphenol

24–2 **a.** 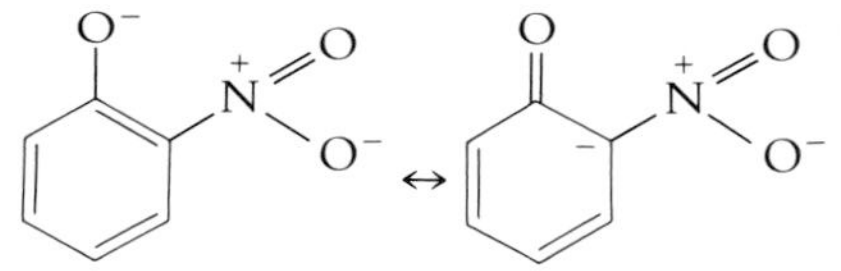 (resonance structures of *o*-nitrophenoxide: O^-, $\overset{+}{N}$, O, O^- $\leftrightarrow$ O, $\overset{+}{N}$, O, O^-)

This resonance hybrid has a negative charge on a carbon which bears a NO_2 group. This is not true of the 3,5-isomer.

b. 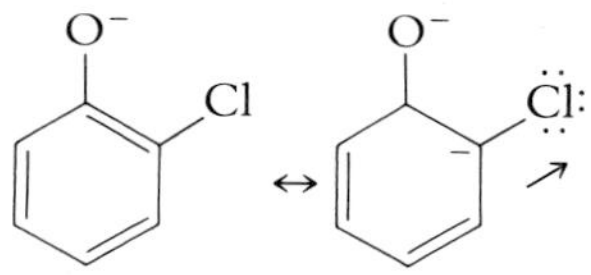 (resonance structures of *o*-chlorophenoxide: O^-, Cl $\leftrightarrow$ O^-, $\ddot{Cl}$:)

The Cl is not as electron-withdrawing as the NO_2 group, therefore, it doesn't stabilize the phenoxide anion as much.

24–3 In the phenoxide ion, resonance delocalizes charge. In the phenol, an electron pair is delocalized but charge is being created. **24–4 a.** f-cresol, Cl_2 **b.** H_2SO_4, SO_3 **c.** I_2, CS_2, **24–6 A:** 2-ethylphenol → 2,4-dibromo-6-ethylphenol
24–7 a. Cl_2, CCl_4 **b.** (1) base, PrI; (2) Cl_2, $FeCl_3$ **e.** (1) base, MeI; (2) 2-bromopropane, $AlCl_3$
24–8 Dissolve the mixture in $CHCl_3$ and extract with aqueous NaOH. The allylphenyl ether remains in the $CHCl_3$ and can be recovered by evaporating the $CHCl_3$. The aqueous NaOH is made acidic and extracted with $CHCl_3$. The $CHCl_3$ is evaporated to leave the 2-allylphenol. **24–9** *Hint:* **A** is the product of *C*-alkylation (on the ring). **24–10** (1) Ac_2O; (2) H_2SO_4, HNO_3; (3) H_3O^+, heat

24–11 (structures: O, CH_3 radical → dimer with O, H, H, O, CH_3, CH_3 → biphenyl with OH, OH, CH_3, CH_3)

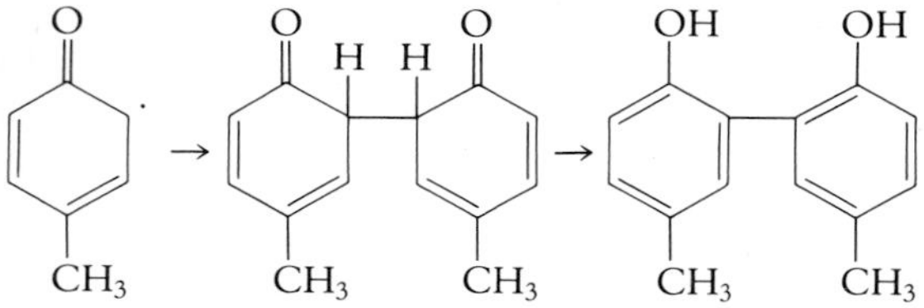

24–12 a. OH OH + OH OH + HO— —OH

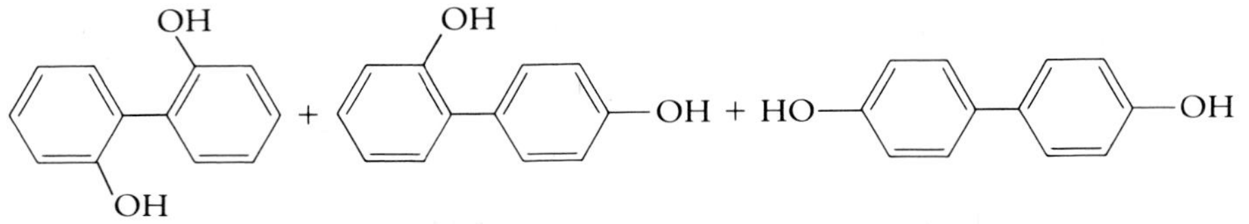

24–13 COOH O; H_2C—C; CH_3; CH_2—C=O; O; CH_2; O=C; CH_2—C; C=O; CH_2—C; CH_2; O

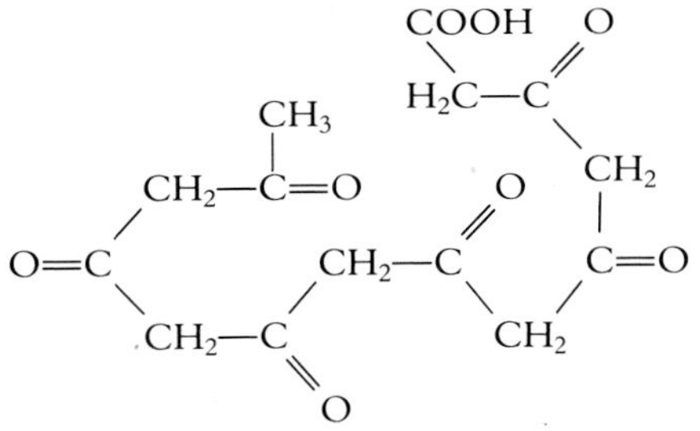

24–14 Ring A is from the polyketide pathway and ring B from shikimic acid.
24–16 a. 1-Naphthol anion has two resonance structures that do not interrupt the second benzene ring, while 2-naphthol has only one. **24–17 a.** 3-bromo-4-methoxytoluene **b.** 4-phenoxynitrobenzene **d.** isobutoxybenzene **e.** phenyl acetate

24–22 $PhC(Me)_2O\overset{+}{O}H_2 \rightarrow PhC(Me)_2O^+ \rightarrow {}^+C(Me)_2OPh \rightarrow$ hemiketal
24–23 a. $AcOC_6H_4NH_3^+$ **b.** HOC_6H_4NHAc **c.** $AcOC_6H_4NH_2$ **24–25 a.** Look for the best (most stable) leaving group. **b.** (1) ethyl, *m*-cresyl, phenyl acetates (3) *o*-cresyl = *p*-cresyl, phenyl **24–27** ArOH + pyridine → $ArO^- \rightarrow ArO\cdot + Cu^{+1} \rightarrow$ etc.

24–29 $Ph\ddot{N}H_2 \rightarrow Ph\overset{\cdot+}{N}H_2 \rightarrow$ coupling reactions **24–30 a.** <1 **b.** >1 **c.** <1 **d.** >1

24–33 a. O; C—CH_3; CH_2; O O O; C; C—$CH_2CCH_2CCH_3$; O CH_2 → OH; R; R; O; $CH_2CH_2CH_2CH_2CH_3$

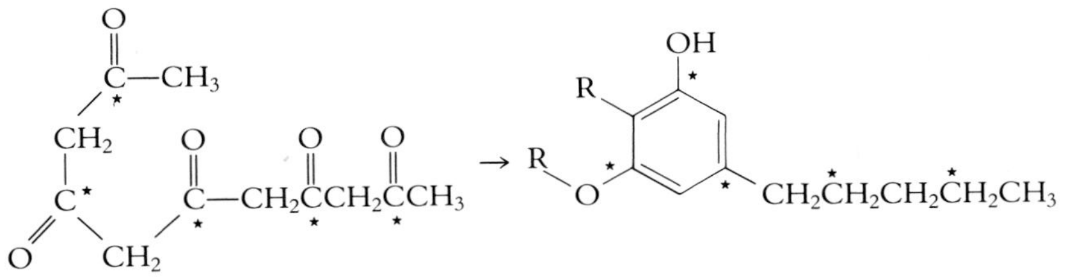

24–34 a. NMe_2; HO

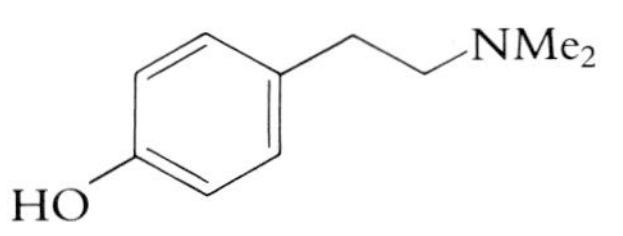

b. It's a phenol. **c.** The phenol (279 nm) is converted to the phenoxide (293 nm).

25–1

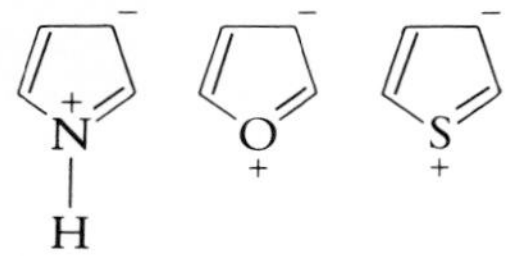

25–2 ↔ ↔

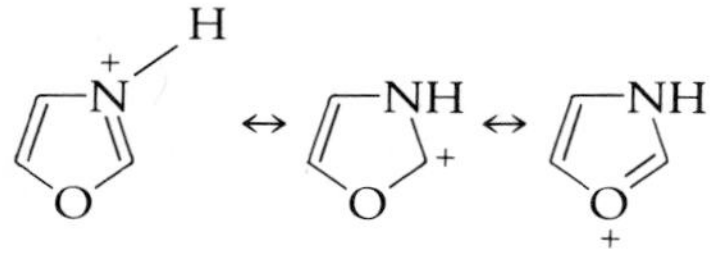

25–4 ↔ H 6.38

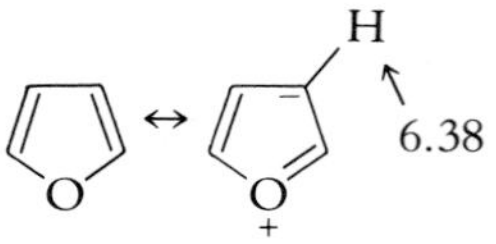

Imidazolium ion is more stable because it has charge delocalized on two Ns, rather than N and O.
25–5 a. 2,4-dinitrothiophene **b.** 2,5-dibromothiophene **c.** 5-nitropyrrole-2-carboxylic acid

e. CH_2NMe_2

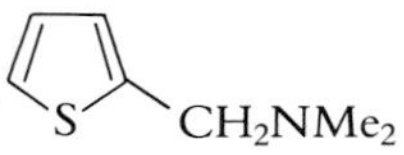

25–6 Imidazole is protonated by the acid mixture and the imidazolium ion undergoes electrophilic substitution with difficulty. The other heterocycles are not protonated.
25–7 a. pyrrolidine **b., c.** *Hint:* Addition products, not substitution.

25–8 CH_2^+ ↔ → CH_2CN + NC H → NC CH_3

25–9 5-Bromofuran-2-carboxaldehyde. Both groups are deactivators.
25–10 The tautomer of **IV** is maleic anhydride. It has two carbonyl groups whose high bond strength makes up for the loss of aromaticity in **IV**. **IV** is much less aromatic than benzene and hence the same situation does not hold for hydroquinone.

25–11 CH_3 N H CH_3

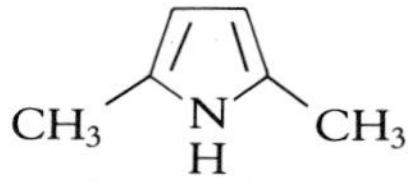

25–12 a. $OHCCH(CH_3)CH(CH_3)CHO$ **b.** 2,5-heptanedione + P_4S_6 **c.** $PhCOCH_2CH_2COPh + NH_3$ **d.** cyclopentane-1,2-dicarboxaldehyde + methylamine
25–13 a. The nonbonded electron pair on N in pyridine is free to be protonated (it is not part of the aromatic system), but the nonbonded pair on N in aniline can be delocalized throughout the benzene ring.

b. $:NH_2$ ↔ $^+NH_2$

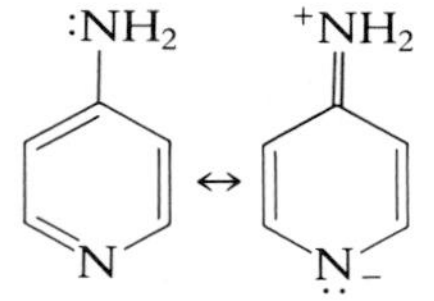

This gives a very basic ring nitrogen which is easily protonated.

25–14 NO_2 H 2+ N H — Less stable; NO_2 H + + N H — More stable

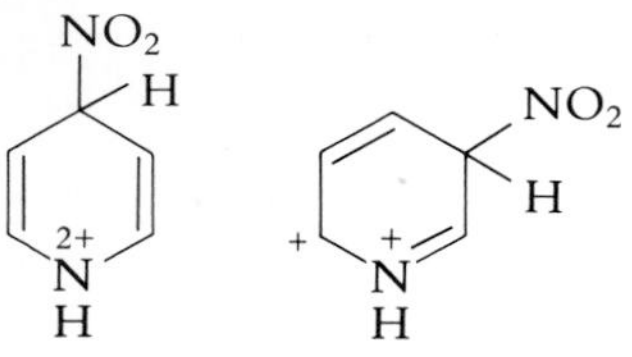

25–15 a. [structures: 2-amino-2H-pyridine anion (More stable); 4-amino-4H-pyridine anion (Less stable)] More stable 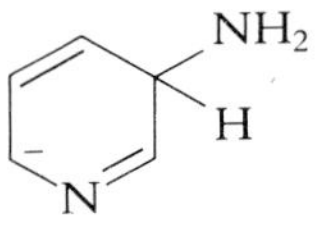   Less stable   **b.** Hofmann rearrangement

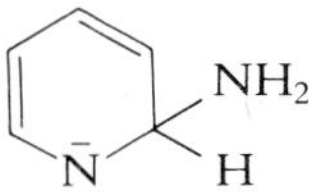

25–16 $RCH{=}NHR \leftrightarrow R\overset{+}{C}H{-}\ddot{N}HR$ **25–17** $CH_3CH(OH)CH_2CHO \rightarrow CH_3CH{=}CHCHO \rightarrow CH_3CH(NH_2)CH_2CHO \rightarrow CH_3CH(NH_2)CH_2CH(OH)CH_2CHO \rightarrow$ cyclization $\rightarrow$ etc.
25–18 $ArCH_2NHCH_2COOH$ **25–21 a.** In the gas phase, the aromatic tautomer is more stable. **b.** In solution:

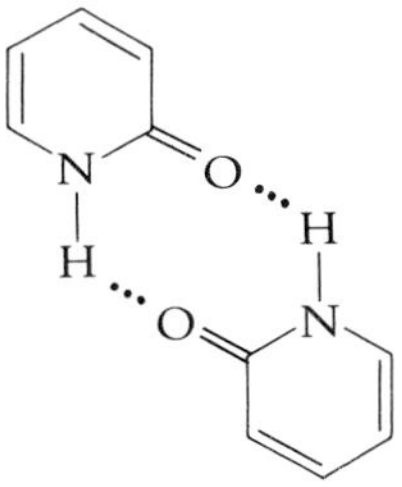

[structure: hydrogen-bonded dimer of 2-pyridone, N–H···O]

25–22 a. 3-bromopyridine **b.** 8-hydroxyquinoline **c.** 5-bromoimidazole **d.** 6-methoxy-1-methyl-1,2,3,4-tetrahydroisoquinoline

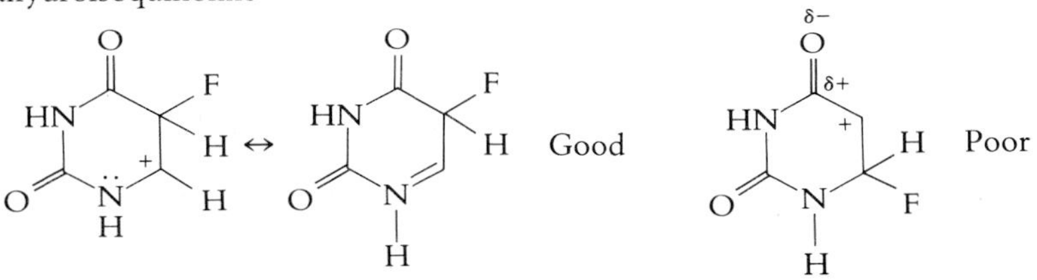

[structures: resonance forms of fluorinated pyrimidinedione cation (HN, F, H, O, $\overset{+}{\ddot{N}}$H) ↔ (Good); alternative with δ− on O, δ+ on C (Poor)] Good   Poor

25–24 a. They tautomerize to pyrrole. **b.** The electron pair on N is involved in the aromatic sextet and won't be protonated. **25–29 a.** 5-nitro **b.** 5-nitro **f.** 2-nitro **g.** 5-nitro **25–32 a.** $PhNHNH_2$ + 2-butanone
25–33 b. $CH_3NH_2 + OHCCH(CH_3)CH(CH_3)CHO$
25–34 a. 4-methylpyridine + $KMnO_4$ **d.** 2-methylpyridine + AcOOH **h.** pyridine + (1) NH_2^-; (2) Br_2, $FeBr_3$ **25–38** *Hint:* Nucleophilic additions. **25–41** This is an S_NAr reaction. Resonance structures of your intermediate should place negative charge on *both* nitrogens.

25–43 [structure: 3-chloro-6-(hydroxymethyl)pyridazine: Cl, $HOCH_2$, N, N]

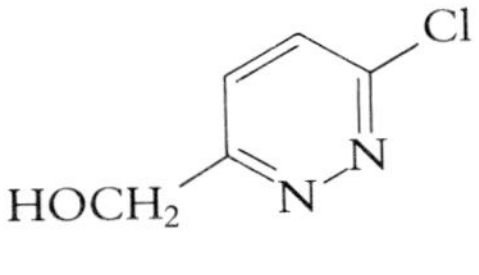

Chapter 26

26–1 $S_0 \rightarrow S_1$ 268 nm = 107 kcal/mol; $T_1 \rightarrow S_0$ 344 nm = 83 kcal/mol
26–3 230 nm = $\psi_2 \rightarrow \psi_3$; 193 nm = $\psi_1 \rightarrow \psi_3$ **26–4** Four, if dimerization does not also involve isomerization of the starting alkene.
26–5 Cyclobutane dimers from the C=C in the lactone or furan rings.
26–6 $PhNH_2 + PhCHO \rightarrow PhN{=}CHPh \rightarrow$ **II** by irradiation. **26–7 a.** $Me_2\dot{C}OH \rightarrow Me_2C(OH)C(OH)Me_2$ **b.** $Ph\dot{C}(OH)CH_3 \rightarrow PhC(CH_3)(OH)C(CH_3)(OH)Ph$

c. $CH_3COCH_2D + CH_2{=}CDCH_3$ + [structure: cyclobutane with OD, Me, D, H]

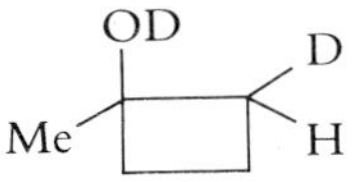

26–8 $H_2N-C(=O:)-CH_2-CH=CH_2(H)-CH_3 \rightarrow H_2N-C(^{+}OH)=CH_2 + CH_2=CH-CH_3$

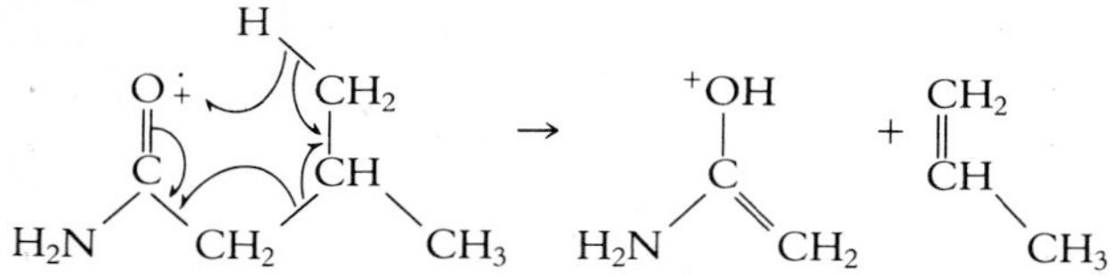

$m/z = 59$

26–9 a. $PhCOCH_2CH_2CH_3$ cleaves only from the triplet state, which is quenched by piperylene. $CH_3COCH_2CH_2CH_3$ cleaves from both excited singlet and triplet states, but only the latter is quenched by piperylene. **b.** The direct irradiation of **A** gives **B** and **C** through the excited singlet state. When irradiation is in the presence of benzophenone, the benzophenone, not **A**, absorbs light and then transfers energy to **A**. **A** is now in the triplet excited state and yields **D** and **E**. **26–10** Both are neutral, no formal charge. **26–11 a.** CH_2N_2 + (*E*)-$CH_3CH{=}CHCOOEt$ **b.** $C(COOMe)_2N_2$ + *cis*-stilbene **c.** An intramolecular reaction from (cycloheptenyl)$-C(=O)CHN_2$

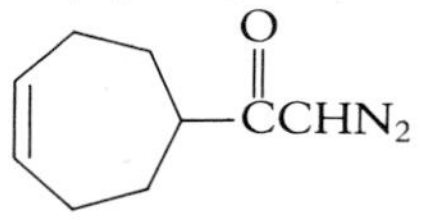

26–12 $CH_2{=}CH{-}CH(H){-}\ddot{C}H$

has an empty *p*-orbital in the singlet state and H migrates as in carbocation rearrangements.

26–13 (cyclopropane: CH_3, CH_3, Ph, H) + (cyclopropane: CH_3, CH_3, H, Ph)

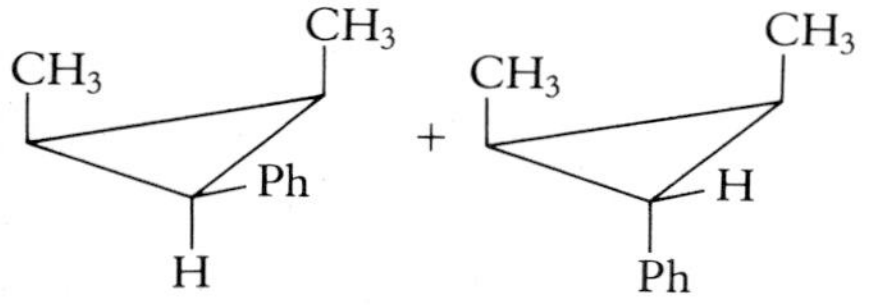

26–14 $(CH_3)_2C{=}CH_2$ as in Problem 26–12. $CH_3{-}CH(:CH){-}CH_3 \xrightarrow[\text{migration}]{\text{methyl}} CH_3CH{=}CHCH_3$; $\xrightarrow[\text{insertion}]{\text{C—H}}$ H_3C-cyclopropane

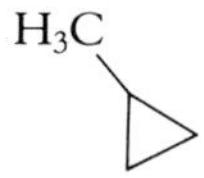

26–16 (2): Cl_2C: (Br^- is a better leaving group than Cl^-); (3): Cl(F)C: (Cl^- is a better leaving group than F^-).

26–17 a. cyclohexene + $HCBr_3$, OH^- **b.** $PhCH{=}CH_2$ + $PhCHCl_2$, OH^- or $PhCH{=}C(Cl)Ph$ + CH_2N_2, light **c.** propene + $CHBr_3$, OH^- or $CH_3CH{=}CBr_2$ + CH_2N_2, light

26–18 a. $Cr(CO)_6$ **b.** $W(CO)_6$ **c.** $Ru(CO)_5$

26–19 a. $(CO)_4Co{-}Co(CO)_4$. Because Co has an odd number of valence electrons (9). **b.** V ($3d^3\,4s^2$) + $e^- \rightarrow$ 6 valence electrons. $18 - 6 = 12$, or 6 CO groups. Mn ($4d^5\,4s^2$) + $e^- \rightarrow$ 8 electrons. $18 - 8 = 10$, or 5 CO groups. Co ($3d^7\,4s^2$) + $e^- \rightarrow$ 10 electrons. $18 - 10 = 8$, or 4 CO groups.

26–20 a. $Mn^7 - 1e^- \rightarrow$ Mn(I) with 6 valence electrons. $18 - 6 = 12$. The two CO groups furnish two electrons each, leaving eight to be furnished by the organic ligands. **b.** 10 **c.** 10 **d.** 14 **e.** 10

26–21 $CH_2(HNiCN){=}CHCH{=}CH_2 \rightarrow NC{-}CH_2CH_2CH{=}CH_2 \rightarrow$ etc.

26–22

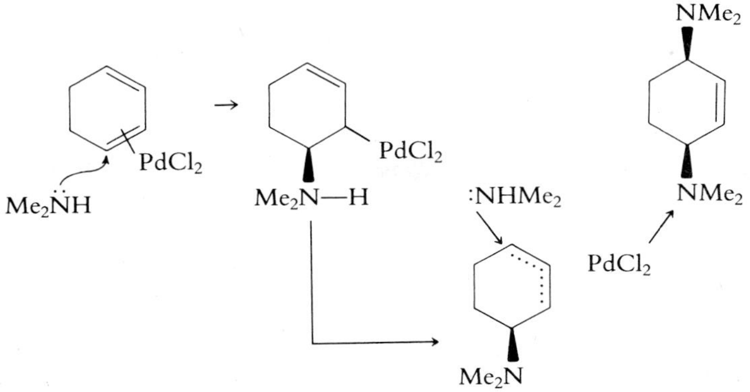

$Me_2\ddot{N}H$ + cyclohexadiene–$PdCl_2$ → Me_2N—H / $PdCl_2$ adduct; :$NHMe_2$; Me_2N; $PdCl_2$; NMe_2 / NMe_2

26–23 a. *Hint*: two cyclobutanes **b.** $CH_3C(OH)(Ph)$—(cyclohexenyl) **d.** *Hint:* Norrish Type II cleavage

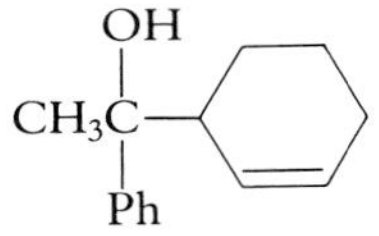

26–24 a. *Hint:* Norrish Type II cleavage **26–25 a.** $CH_2{=}CHCOCH_3$ + CH_2N_2, light **d.** 4-methycyclopentene + $CHClF_2$, OH^-

26–28 A: Ph, Ph, Ph, O*t*Bu (cyclopropene) **B:** Ph, Ph, Ph (cyclopropenium) + Br^-

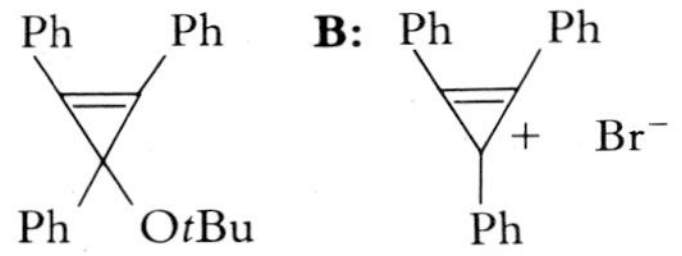

26–30 $RCH{=}NNHSO_2R \rightarrow RCH{=}N\underline{N}SO_2R \rightarrow RCH{:} + N_2 + {}^-O_2SR$
26–31 a. *Hint:* The toluene acts as a ligand. **b.** $H_3CMn(CO)_5$
26–33 Yes. They have the same electronic configuration: 9 valence electrons.

Appendix

Some Important Acids, Their Conjugate Bases and pK_as

Acid	Conjugate Base	pK_a
HI	I^-	~ −9
HCl	Cl^-	~ −7
H_2SO_4	HSO_4^-	−5
$CH_3OH_2^+$	CH_3OH	−2
H_3O^+	H_2O	−2
HNO_3	NO_3^-	−1
CF_3COOH	CF_3COO^-	0.2
$C_6H_5SO_3H$	$C_6H_5SO_3^-$	0.5
HF	F^-	3
$C_6H_5NH_3^+$	$C_6H_5NH_2$	5
CH_3COOH	CH_3COO^-	5
N^+ H	N:	5
HCN	CN^-	9
NH_4^+	NH_3	9
C_6H_5OH	$C_6H_5O^-$	10
$CH_3NH_3^+$	CH_3NH_2	11
CH_3COCH_2COOEt	$CH_3CO\bar{C}HCOOEt$	11
$CH_2(COOEt)$	$\bar{C}H(COOEt)_2$	13
H_2O	OH^-	16
C_2H_5OH	$C_2H_5O^-$	16
H H	H ⊖	16
CH_3COCH_3	$CH_3COCH_2^-$	20
HC≡CH	$HC≡C^-$	26
NH_3	NH_2^-	35
—CH_3	—CH_2^-	~42
CH_4	CH_3^-	~57

Index

A4
B5
C6
D7
E8
F9

Usual ^{13}C NMR Chemical Shift Values (ppm)

Nonaromatic Compounds		Benzene Derivatives		
CH_3C	10 to 30			
$CH_3C{=}O$	20 to 30			
		Y	α	β
CH_3N	30 to 50			
CH_3O	50 to 60	H (benzene)	128.5	128.5
CH_3F	75.4	CH_3	139.4	129.2
CH_3Cl	24.9	OH	155.4	115.8
CH_3Br	10.0	CN	113.1	132.1
CH_3I	−20.7	NH_2	146	115
C=C	110 to 140	F	163	115
C=O	155 to 220	Cl	134	128

Usual ^{1}H NMR Chemical Shift Values (ppm)

Functional Group	Approximate Range
$\mathbf{CH_3}$—C	0.8 to 1.9
$\mathbf{CH_3}$C=O	1.9 to 2.2
$\mathbf{CH_3}$—N	2.1 to 3.1
$\mathbf{CH_3}$—O	3.2 to 4.1
$\mathbf{CH_3}$—F	4.26
$\mathbf{CH_3}$—Cl	3.05
$\mathbf{CH_3}$—Br	2.69
—$\mathbf{CH_2}$—C	1.0 to 2.0
—$\mathbf{CH_2}$C=O	2.0 to 2.4
—$\mathbf{CH_2}$—N	2.3 to 3.6
—$\mathbf{CH_2}$—O	3.3 to 4.5
—C=**CH**— (alkene)	4.5 to 7.0
—C=**CH**— (aromatic)	6.5 to 8.5
—C≡**CH**	2.0 to 3.0
—**CH**=O	9.0 to 10.0
—COO**H**	10.4 to 12.1